BIOLOGY

THE JONES AND BARTLETT SERIES IN BIOLOGY

THE JONES AND BARTLETT/BOOKMARK SERIES IN BIOLOGY

Biology: Investigating Life on Earth, Second Edition
Vernon L. Avila

Jones and Bartlett Publishers, Inc., in a joint venture with Bookmark Publishers,
provides the second edition of *Biology* by Vernon L. Avila. This title is the
first in a series of new offerings for Introductory Biology curricula.

BIOLOGY

INVESTIGATING LIFE ON EARTH

Second Edition

Vernon L. Avila

San Diego State University

JONES AND BARTLETT PUBLISHERS

Boston and London

BOOKMARK PUBLISHERS

Jamul, California

The first edition was published under the title *Biology: A Human Endeavor*.

International Standard Book Number: 0-86720-942-9

Library of Congress Cataloging-in-Publication Data

Avila, Vernon L.
  Biology : investigating life on earth / Vernon L. Avila.—2nd ed.
    p.   cm.—(The Jones and Bartlett/Bookmark series in biology)
  Includes bibliographical references (p.   ) and index.
  ISBN 0-86720-942-9
  1. Biology. I. Title. II. Series.
QH308.2.A955   1995
574—dc20   94-39326
                                                        CIP

Communications and orders for additional textbooks may be addressed to:

Editorial, Sales, and Customer Service Offices

Jones and Bartlett Publishers	Jones and Bartlett Publishers International
One Exeter Plaza	7 Melrose Terrace
Boston, MA 02116	London W6 7RL
617-859-3900	England
800-832-0034	

V.P. and Publisher: Clayton E. Jones
V.P., Production and Manufacturing: Paula Carroll
Sponsoring Editor: David E. Phanco
Editorial Assistant: Deborah Haffner
Director of Marketing: David P. Geggis
Manufacturing Buyer: Dana L. Cerrito
Production Manager: Hal Lockwood
Production Assistant: Jane Brundage
Art Director: Hal Lockwood
Art Editor: Jane Brundage
Text Designers: Catherine Dorin, Renee Deprey, Paula Goldstein
Cover Designer: Marshall Henrichs
Copyeditor: Elizabeth Judd
Illustrators: Wayne Clark, Cecile Duray-Bito, Chris Hayden, Deborah Ivanoff, Tim Krasnansky,
  Paula MacKenzie, Sandy MacMahon, Linda McVay, Elizabeth Morales-Denney,
  Greg O'Leary, Carla Simmons, Lin Teishman, Mary Ann Tenario, Tom Webster
Photo Researchers: Stuart Kenter, Deborah Haffner
Photographers: Don Martin, Darin Norfleet
Indexers: Information Bank, Ruth Anne Lowe
Compositors: Dix Type Inc., Rick Gordon, Myrna Vladic
Color Separator: H & S Graphics
Cover Printer: Henry N. Sawyer Co.
Printer and Binder: R. R. Donnelley and Sons

The GeoSphere Images on the cover and Unit 1 are specially created cloudless images of the earth as seen from space. They are used with permission of the GeoSphere Project, which was assisted by Eyes On Earth, NOAA, NASA, and the National Geographic Society.

1 2 3 4 5 6 7 8 9 10—RRD—00 99 98 97 96 95 94

Printed in the United States of America

This book is dedicated to my family,
    Patty, Patrick, Raquel, Pam, and Paula,
and to the memory of my grandfather,
    Juan Manuel Olguin.

—VERNON AVILA

The GeoSphere Image is a spectacular view of the earth and marks a milestone in history. This color composite is the first visually accurate, three-dimensional map of the earth as seen from space. Thousands of 4-kilometer-resolution pixels (picture elements) taken by the NOAA TIROS satellites were chosen so that no cloud cover would obstruct the view of Earth.

The purpose of the GeoSphere Project is to make complex global systems understandable. Through the use of technology, we are creating visualizations that illustrate and enhance understanding of earth processes, thereby encouraging better global resource mangement. It is our hope that as people of all nations view such images, the fragile interdependence of all life will be better understood and each person might make informed choices and use his/her individual talents to help preserve the health and beauty of planet Earth.

— TOM VAN SANT
    CEO, The GeoSphere Project

Brief Contents

Contents

3

The Molecules of Life 51

4

The Cell: Where It All Happens 75

5

Membranes, Molecules, and Cell Function 101

UNIT 3
GENETIC CONTINUITY 211

10
Cell Reproduction 213

11
DNA Replication 241

12
DNA and Protein Synthesis 259

13
Recombinant DNA: Genes by Design 285

14
Genetics: Patterns of Inheritance 307

15
Human Genetics 335

UNIT 4
EVOLUTION AND DIVERSITY 363

16

Life: How Did It All Begin? 365

17

Evolution: Living Things Changing Over Time 381

18

Diversity and Taxonomy 409

UNIT 5
PLANTS   433

19
Plants: A Survey   435

20
Plant Structure and Growth   467

21
Transport and Regulation   491

UNIT 6
ANIMALS   513

22

An Introduction to Animal Form and Function: Protection, Support, and Motion 515

23

Digestion 543

24

Circulation 571

29

Cell-to-Cell Communication: Hormonal Signals 707

30

Cell-to-Cell Communication: Neural Signals 735

31

Animals: An Evolutionary Perspective — Invertebrates 767

32

Animals: An Evolutionary Perspective — The Chordates 791

UNIT 7
BEHAVIOR AND ECOLOGY   809

33

Behavior 811

34

Introduction to Ecology: Biomes 831

35

Populations 853

Profiles

Preface to the Second Edition

It is inevitable that over the course of 23 years of teaching general biology at the university level and 10 years as a high school biology teacher, one should form strong opinions about how best to teach the course. My own opinions have been shaped by my varied experiences as a biologist — as a researcher, as a consultant, and as an administrator. To me, the essence of being a biologist is that instant of insight when all the hours of preparation pay off with the revelation of something new, something I didn't know before. It is the excitement generated by such a moment that I think holds the key to successful teaching of biology. If we can make our students feel that excitement and follow the thoughts of scientists as they work, we can make biology more than an accumulation of facts they must memorize.

This book is the manifestation of that simple idea. The idea has led me to certain conclusions about how to reach students, conclusions that I hope I have put into action here. The first is that the best way to engage students in the thinking behind the biology is to ask questions and to question answers. Throughout the book I pose problems, asking readers to anticipate a discovery or to predict a mechanism. By inducing readers to think, I hope to get them to participate in what they are reading.

The second is that I believe it is important for students, especially nonmajors, to understand that scientists are not a breed apart. Each chapter of the book profiles a biologist, discussing not only his or her major contribution but also how he or she became interested in biology. The individuals profiled are a varied lot. In fact, they share only an irrepressible curiosity about living things, a curiosity I hope to awaken in students. A secondary purpose of these profiles is to provide role models from a broad scope for our varied student population.

In keeping with the investigative theme, Human Endeavors boxes are found in many of the chapters and are used to trace the history of experiments and observations that have enhanced our understanding of life.

No one needs to be reminded of the increasing impact of current biological research on our everyday lives. An understanding of basic biological concepts is now essential for the modern voter, consumer, and parent — in fact, for many of the roles our students will soon play. The second edition of *Biology: Investigating Life on Earth* (the first edition was titled *Biology: A Human Endeavor*) is intended to prepare them for those roles, and especially to help them become biologically literate.

SECOND EDITION HIGHLIGHTS

It is always rewarding to have one's efforts validated by colleagues and students. The very positive reception of the first edition of this text exceeded my highest expectations. It has been particularly gratifying to hear from instructors that the book has been used successfully with biology students as well as those not majoring in science.

Hundreds of colleagues have read the book, and thousands of students have used the text as they endeavored to become biologically literate. Their encouraging comments and letters helped inspire this second edition.

As in the first edition, the basic philosophy is that the context in which material is presented is more important than content; therefore, concepts are emphasized. Students "can see the forest through the trees."

When questioning colleagues and students about what changes they would make in the second edition of the text, the overwhelming comments were "leave well enough alone," they liked the text as is — the conceptual approach; the conversational writing; the outstanding art program; the Focus boxes and student learning aids; the Profiles; and the organization. One consistent suggestion was to add more information on the animals, both the invertebrates and the vertebrates. I have added that material in the second edition in Chapters 31 and 32. The emphasis has been conceptual; the important evolutionary relationships among various animal groups are emphasized in an attempt to avoid a descriptive survey of the phyla. The sense of enthusiasm of zoologists who work in these areas is apparent in the chapters.

Another request was to expand the ecology section. The first edition consisted of only two chapters devoted to ecology. This edition has an ecology section expanded to five chapters — Chapters 34, 35, 36, 37, and 38. These chapters help students understand the important ecological concepts. Chapter 38 is devoted to the human impact on the ecosystem, how we as humans are destroying the very planet that provides us life. Reviewers have concurred that the these chapters are accurate and correct. I am especially indebted to Dr. George Cox and Dr. Kathy Williams, ecologists in the Biology Department at San Diego State University, for their consultation in the development and writing of these chapters. This enhanced section of the text served as inspiration for the new cover design and subtitle for the text. I wish to thank Tom Van Sant of the GeoSphere Project for his beautiful image of Earth, which adds something special to our understanding of it all.

CHANGES IN THE SECOND EDITION

As mentioned, the general consensus of colleagues was to expand the ecology section and add two chapters on animals. In addition, new Profiles were written to introduce students to scientists and to stress that scientists are varied in personality and background. The individuals selected provide a diverse multicultural sample. Feedback from students and teachers who have used the first edition have liked the fact that the scientists selected are multicultural and provide role models for a wide range of students.

As requested by colleagues, Human Endeavors boxes have been continued and three new Human Endeavors boxes — on oncogenes, retroviruses, and the origin of life — have been added. Focus boxes have been changed and updated.

Every chapter was thoroughly reviewed and updated and, when necessary, entire sections were rewritten to assure clear understanding of a concept.

ORGANIZATION

The organization of topics was developed to enhance the student's ability to grasp the biological principles presented. It reflects my philosophy that an understanding of the molecular and cellular principles of biology is key to fully appreciating the many intricacies of life. These principles, then, are introduced early in the text and integrated throughout.

The 38 chapters have been divided into 7 units. Beginning with Unit 1, an introduction to the basic concepts, controversies, and wonders of biology, the text progresses to Unit 2, where the molecular machinery of the cell is explored. With this foundation laid, the concepts of genetics are discussed in Unit 3. Unit 4 ties the concept of evolution to the genetics of the previous unit and introduces the diversity of living organisms. The chapters in Unit 5 (Plants) and Unit 6 (Animals) emphasize current research and relevant examples. The final unit, Behavior and Ecology, brings the text to a close by integrating the ideas of energy transduction and evolution in a discussion of an organism's interaction with its environment.

SPECIAL FEATURES

More specifically, the features of this book that reflect my philosophy of teaching are as follows:

1. *Clear and direct writing style.* The first requirement of textbook writing is that it be clear. Concepts must be explained in a logical, step-by-step progression that allows students to follow the thinking. Writing that is clear need not be dry. I have tried in this book to speak directly to students, making connections for them between what they are learning and what they already know, without draining the life out of the science.

2. *Asking questions.* Part of this direct approach, as I've already mentioned, is using questions to lead the reader through the logical relationships of the concepts presented.

3. *Profiles.* Each chapter profiles one or more scientists, introducing students to the people who are biologists. Students see the human side of biology: the curiosity that led these men and women to study biology and the challenges they face.

4. *Human Endeavors boxes.* These essays or visuals provide students with the opportunity to follow historically the thinking and experiments that have led scientists to the development of particular concepts.

5. *Focus boxes.* Each chapter contains one or more Focus boxes that highlight current applications of biology, present recent research, or discuss current controversies — all relating biology to our lives. More and more, we are all asked to form opinions on issues requiring some understanding of biology. We cannot, of course, teach our students all the facts they will need to know to understand the issues they will face during their lifetimes, but we can teach them the underlying principles and the general approaches that will allow them to assimilate the new knowledge they acquire.

6. *Photosynthesis and respiration overview chapter.* As a result of classroom testing, it was determined that an overview chapter on photosynthesis and respiration followed by more in-depth chapters on these topics is the optimal approach. This gives instructors the flexibility to choose the level of detail they want to emphasize in their coverage of this material.

7. *Strong pedagogy.* Each chapter includes pedagogical aids that will help the student, especially the nonmajor, master the increasing complexity of today's biology:

The spot summaries accompanying headings guide the reader through the chapter and later serve as a review as the student prepares for exams.

End-of-chapter summaries, key terms, and page-referenced questions help the student focus on the important concepts.

The glossary at the end of the book provides clear definitions of all key terms.

8. *Illustration program.* The art in this book deserves special mention, because in the first edition it was the first major biology text to have most of its art rendered on the computer. The result is a full-color illustration program that has been carefully reviewed and imaginatively presented. The program provides accurate, clear, and useful figures that not only help students visualize the more difficult concepts but also convey a sense of the wonder of biology.

ANCILLARY PACKAGE

Biology: Investigating Life on Earth comes with a complete array of ancillary materials. A 325-page *Study Guide* contains chapter summaries, review questions, and such exercises as working with figures, tables, Greek and Latin roots, reactions — testing areas where nonmajors are likely to have difficulty.

The *Instructor's Manual* consists of two sections. The first section contains chapter outlines and teaching tips. The second section is the test bank, with fill-in-the-blank, true/false, matching, multiple choice, and essay questions. There are about 50 questions for each chapter, for a total of about 1900 questions. The test bank is also available in computerized form for both the PC and the Macintosh.

Two free educational resources will be made available to adopters of *Biology: Investigating Life on Earth*. First, instructors will receive more than 50, author-selected, four-color overhead transparencies of art taken directly from the text. These transparencies illustrate major concepts and are adapted for effective projection use. Second, qualified adopters may select from the Jones and Bartlett audio-visual library. Please call the Marketing Department for a listing of these selections.

ACKNOWLEDGMENTS

Writing a textbook can be an individual task: take one biologist with several years' experience in teaching and research, add paper and pencil, reference material, and self-discipline, and sooner or later a manuscript appears. Developing a textbook from that manuscript, on the other hand, is a team effort, which in today's textbook market has become ever more complex. It is the numerous team members that I want to thank here.

First, I thank the biologists who have reviewed this manuscript. Their comments, suggestions, and assistance played a major role in its development. During the writing of the first edition of the book, I was fortunate to work at the University of Puerto Rico, Rio Piedras campus for two years, and at the National Institutes of Health. I thank my colleagues at those institutions for their suggestions and encouragement. I also thank my colleagues at San Diego State University for allowing me to drop into their laboratories and ask for a photograph or to stop them in the hallway and ask them to check this paragraph or that to ensure accuracy.

Since the beginning of this endeavor — six years for the first edition and another three years for the second edition — numerous editors have helped me through the developmental stages.

The production of *Biology: Investigating Life on Earth* was handled by a crew of extraordinarily talented people at Bookman Productions and Jones and Bartlett Publishers. Hal Lockwood coordinated the enormously complex project with his usual award-winning style. His energies and wealth of textbook production experience are largely responsible for the excellent quality of the textbook. Marc Deprey and Illustrious, Inc., created the artwork for the first edition and Deborah Ivanoff of Media

Resources expanded the art program for the second edition. The whole art team have provided a new standard in the preparation of textbook illustrations using computer-rendered art.

At Jones and Bartlett I want to thank David E. Phanco, Acquisitions Editor, for his insight and experience in pursuing and completing this joint venture; to Clayton E. Jones, Publisher, for his recognition of the quality of our textbook and the experience to make it a leader in the twenty-first century; to Deborah L. Haffner, Editorial Assistant, for her coordination of this effort and for photo research; to Paula Carroll, Vice President, Production and Manufacturing, for working to see that the final textbook lives up to the high standards of Jones and Bartlett and Bookmark Publishers; to Dana L. Cerrito, Manufacturing Buyer; and to David P. Geggis, Marketing Director, for his excellent marketing and advertising campaigns. In addition, I would like to thank Marshall Henrichs for the cover design and Tom Van Sant for the cover image. For providing the excellent photographs that truly enhance the visual impact of this text, I would also like to thank Stuart Kenter, photo researcher, Don Martin and Darin Norfleet, photographers, and my colleagues around the world. And to all others whose efforts are often not acknowledged but absolutely essential to the process of producing a quality textbook, I express my gratitude.

To the typists who turned my handwritten drafts into legible manuscripts, I am ever thankful: Maria Hedrich, who transcribed almost the entire first edition draft, Diane Weeks, who did some typing and also assisted in the preliminary editing, and Kathy Aragon, Barbara Groeger, and Maureen Gibbins for the first edition, who helped type some chapters, and Joline Treanor and Danielle Hoelzer for the second edition.

I would also like to thank Patricia G. Avila, M.D., M.P.H., for her excellent review of the human anatomy and physiology portions of this text.

Finally, I would like to thank my parents, Isabelle and Eloy Avila, for their support — without them, there would not be me.

Reviewers for *Biology*

I wish to thank my colleagues at San Diego State University (especially those listed) and the numerous other reviewers for their careful reading of the manuscript, the benefit of their expertise, and many valuable suggestions.

First Edition

Mike Atkins
San Diego State University

Karl Aufderheide
Texas A & M University, College Station, TX

Frank Awbrey
San Diego State University

Wayne M. Becker
University of Wisconsin, Madison

Annalisa Berta
San Diego State University

Marlin Bolar
California State University, Sacramento

Jack L. Bond
Ricks College, Rexburg, ID

Richard T. Briggs
Smith College, Northampton, MA

L. Herbert Bruneau
Oklahoma State University, Stillwater

Jerry Button
Portland Community College, Portland, OR

Joseph Chinnic
Virginia Commonwealth University, Richmond

Jose Cintron
University of Puerto Rico, Rio Piedras

Galen E. Clothier
Sonoma State University, Rohnert Park, CA

David J. Cotter
Georgia College, Milledgeville, GA

D. G. Davis
University of Alabama, University, AL

Donald J. Defler
Portland Community College, Portland, OR

Jean DeSaix
University of North Carolina, Chapel Hill

Gary E. Dolph
Indiana University at Kokomo

Lee Ehrman
State University of New York, Purchase

George Fleck
Smith College, Northampton, MA

Gregory Florant
Swarthmore College, Swarthmore, PA

Terrence Frey
San Diego State University

Laurence Fulton
American River College, Sacramento, CA

Michael S. Gaines
University of Kansas, Lawrence

L. Lee Grismer
La Sierra University, Riverside, CA

George Hennings
(retired)

Arthur Jantz
West Oklahoma State College, Altus

David T. Jenkins
University of Alabama at Birmingham

Stephen H. Loomis
University of California, Davis

Philip M. Mathis
Middle Tennessee State University, Murfreesboro

Lee McClenaghan
San Diego State University

Darrel L. Murray
University of Illinois, Chicago

Lowell P. Orr
Kent State University, Kent, OH

Thaddeus Osmolski
University of Lowell, Lowell, MA

Kathryn Stanley Podwall
Nassau Community College, Garden City, NY

Samuel B. Rhodes
Franklin College of Indiana, Franklin

R. Harvard Riches
Pittsburg State University, Pittsburg, KS

Robert Rinehart
San Diego State University

David Sadava
The Claremont College, Claremont, CA

C. W. Schaefer
University of Connecticut, Storrs

Paul B. Shubeck
Montclair State University, Upper Montclair, NJ

Michael Simpson
San Diego State University

Ted L. Swensen
Portland Community College, Portland, OR

Thomas M. Terry
University of Connecticut, Storrs

Teresa Thomas
Southwestern Community College, Chula Vista, CA

Stephen Tilly
Smith College, Northampton, MA

Mala Wingard
San Diego State University

About the Author

VERNON L. AVILA is a biology professor at San Diego State University. He received his B.S. from the University of New Mexico, his M.A. from Northern Arizona University, and his Ph.D. from the University of Colorado, Boulder. Dr. Avila was a high school biology teacher for ten years and was nominated for the National Association of Biology Teachers' Outstanding Biology Teacher Award. He has also received special recognition from the American Institute of Biological Sciences as having one of the four most successful and effective general biology courses in the United States, and was selected by the Educational Testing Service in Princeton, New Jersey, to write, construct, and evaluate the the Graduate Record Examination Subject Area Test in Biology. Dr. Avila is also a reader for the Advanced Placement Biology Examination. He has served as an Expert Consultant for the National Cancer Institute and for the Division of Research Resources at the National Institutes of Health, has had several articles published in professional journals, and has over 23 years' experience in the teaching of introductory biology courses at the university level.

Note to the Student

This biology text is written for you. I have attempted to share my joy and enthusiasm for biology with you by guiding you in a logical manner to biological literacy. To me, a textbook should be the primary learning device: a student in complete isolation (imagine being stranded on a tropical island) should be able to gain a fundamental understanding of the subject after reading the text. I have assumed that you have a limited background in chemistry and biology and that many of you are not aware of the relevance of biology in your everyday lives. Therefore, in the early chapters, biological vocabulary is held to a minimum, and a logical, step-by-step approach leads you to basic biological and chemical principles required to understand life.

We have all heard about cancer genes, AIDS research, genetic engineering, and so on. Current developments in these areas — what you read about in the newspapers — are based on the molecular level of biology. Thus one of my aims is to help you understand the basic concepts underlying this approach.

I am also aware that the student population is changing: not everyone is from the same background or has the same interests. This is also true of scientists, as you will see from the Profiles of scientists in the chapters. As you read the Profiles you can see that scientists are human, subject to all the strengths and weaknesses of the human condition. To me there is nothing worse than taking the approach that all science is absolute, that biology is just a collection of facts to be recalled for an examination. Biology is alive and growing and thus subject to change and controversy. Biologists don't just answer questions, they question answers, and you should, too. I have tried to share the vitality of biology with you by discussing the excitement and controversy in the field.

Best Wishes,

Vernon L. Avila, Ph.D.

All right, enough of the rationale as to why I want to convince all of you to become biologically literate. What are some of the ways that I try to help you reach that goal? As you read the text, notice that I have included preview summaries that accompany the major headings within the chapter. These summaries tell you what the section will be about. Later, you can skim these summaries as you review the material for exams. Also, I have provided summaries at the end of chapters, page-referenced review questions, with answers in the appendix, and key terms (defined in the glossary at the end of the book) that reinforce your understanding of the material. After you have read the chapter, see if you can answer the discussion questions; they will help you determine which sections of the chapter you understand and which you might want to review. When a term is first introduced it is defined to ensure that you understand the material. In the early chapters, I have made the vocabulary light, but as your understanding and confidence with the subject increases, the chapters become more detailed. The Focus boxes provide interesting, current topics that will demonstrate the relevance and just plain fun of biology. The Human Endeavors boxes allow you to follow the logic, observation, and experimentation that have made some of the major biological advances possible.

This textbook is your friend. Use it, write notes in the margins, highlight important passages. After all, it should be by your side as a reference source throughout your education. If you get the chance, I would appreciate hearing from you. Let me know if my text did what I intended it to do — provide you with the opportunity to share in the joy I find in the study of biology.

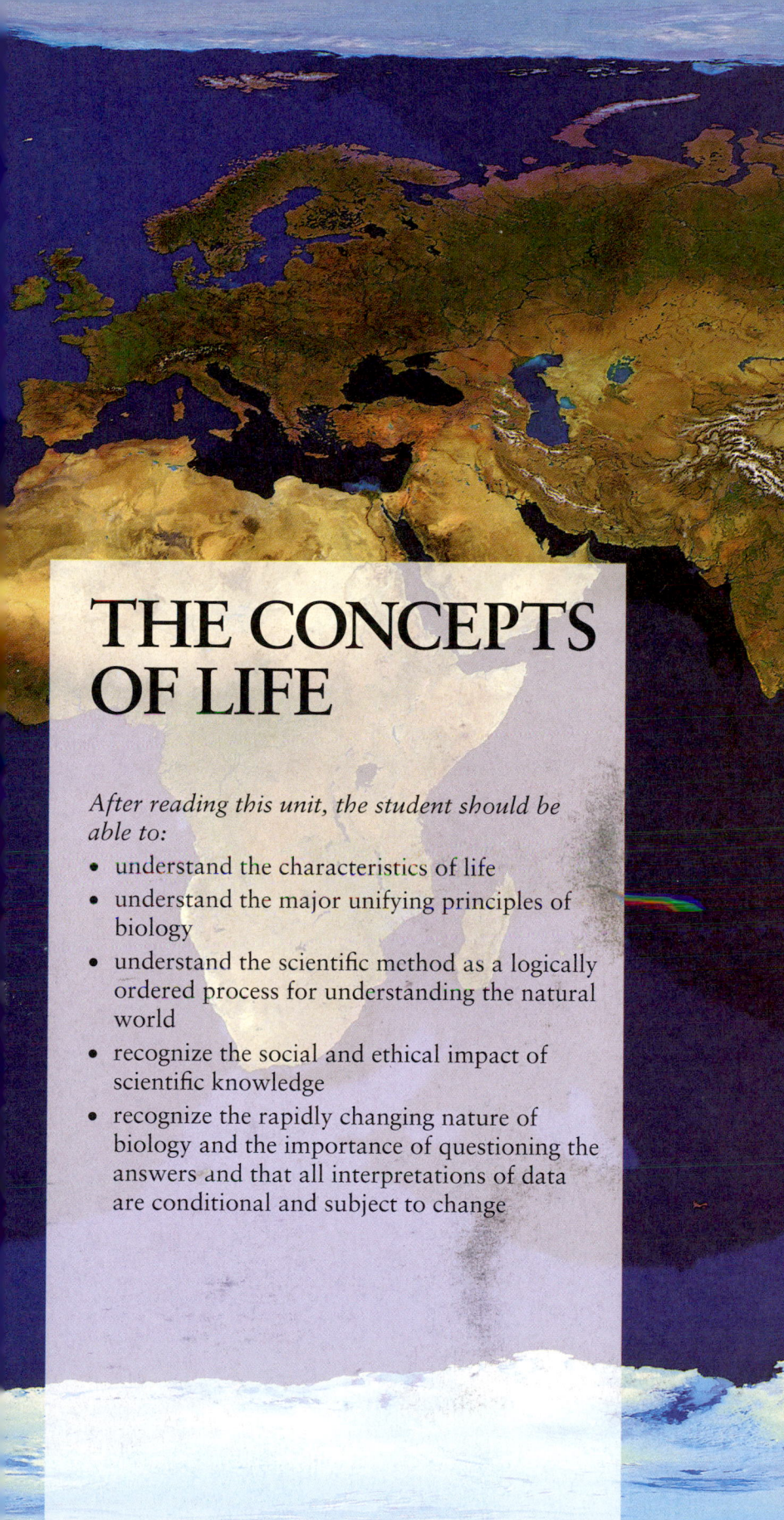

THE CONCEPTS OF LIFE

After reading this unit, the student should be able to:

- understand the characteristics of life
- understand the major unifying principles of biology
- understand the scientific method as a logically ordered process for understanding the natural world
- recognize the social and ethical impact of scientific knowledge
- recognize the rapidly changing nature of biology and the importance of questioning the answers and that all interpretations of data are conditional and subject to change

1

Biology: A Contemporary Subject

All of us wonder about the phenomenon of life. Our curiosity is part of the human condition. We watch the birth of a loved one with awe. The antics of the monkey in the zoo or children playing in a swimming pool's cool spray impress us with life's vitality (Figure 1.1). Helpless, we watch life slip away from a loved pet or fellow human from age, sickness, or accident. In the arid southwestern deserts we experience the silence of nature and realize our vulnerability to its harshness. In moments like this we wonder, what is this phenomenon we call life, shared by all life forms on our planet, from bacteria to whales?

All of us are life watchers. But when someone gives that activity a name, calls it biology (the science of life), we groan and say, almost in unison,

"Oh no, I can't stand science. I hate biology. It's too hard. It's a subject that only people with 'scientific minds' can understand." But consider the contradiction implicit in such an attitude. We all understand things about life, whatever our educational level or major interests. Studying biology only makes us look at life very closely.

The questions and choices that biology raises make it a very relevant subject indeed. Because we live at this time in human history — at the beginning of what many believe to be a biological age — we must understand the subject well enough to discuss our options intelligently and make rational decisions about the impact of our knowledge on society. The theme of biology's relevance to our lives is very important in this text.

In fact, none of us can go through a single day without experiencing the impact of the science of biology on our everyday life. Listening to the radio, we hear that biomedical researchers are embark-

Above: Scientists use various techniques to study life. These scientists are looking at x-ray patterns of chemicals to learn more about life.

(a)

(b)

FIGURE 1.1 Life
We are all life watchers, whether
(a) the antics of a gorilla playing in
the zoo or (b) children playing in the
water of a swimming pool on a hot
summer day.

ing on a project to map the human genome — the hereditary material that makes us what we are.

Later, when we read a newspaper over our morning coffee, we see articles about biology's impact on our everyday lives. One on "The Depletion of the Ozone Layer" raises the question of whether our chemical wastes are destroying the ozone layer that protects us from harmful ultraviolet light — a serious health hazard that can increase the risk of skin cancer. Another article details the latest attempts to treat acquired immune deficiency syndrome (AIDS). It is the same when we read magazines or watch television's evening news. We hear stories on diet, nutrition, health, exercise, drugs, sex, behavior, life, death, food, genetics, disease, and treatment. We are surrounded by scientific debates: pollution, overpopulation, test-tube babies, use of fetal tissue for brain implants, surrogate parents, wildlife habitat devastation, cancer, should we, shouldn't we, is it safe, how safe is safe? In short, we are bombarded with information, questions, and ethical dilemmas, all related to the science of biology (Figure 1.2).

As you read through this text, you will notice that every chapter contains a biographical profile of a scientist and many chapters have "Human Endeavors" boxes that discuss the development of important biological concepts. Those essays will tell you about some of the people who have contributed much to our understanding of life, and you will discover that each of them is not some "special sort of person" who, unlike you, understands science. All of them are as human as you or I, for biology is truly a human endeavor. Shall we begin?

WHAT IS LIFE?

When we walk across a street or through a forest and see something that we suspect is alive, we try to find out if it is. A child will poke at a rock with a toe or a stick to see if it is really a rock or is a turtle in disguise. When we try to define life, though, we discover that the task is more difficult than we thought it would be. Usually, we resort to a list of the characteristics: "I know it was alive, because it moved, or it grew — it reproduced — it responded to my touch."

Biologists also have a list of the characteristics of living things. Some of those characteristics are evident when we watch movies or television. When someone is brought into an emergency room, the person in charge looks for specific, fairly well accepted criteria to determine if the victim is alive or dead. Any list would vary slightly from individual to individual, based on experience, but it is most likely that each of us would include the following: *organization, maintenance, growth, response to stimuli, reproduction, variation,* and *adaptation.*

(a)

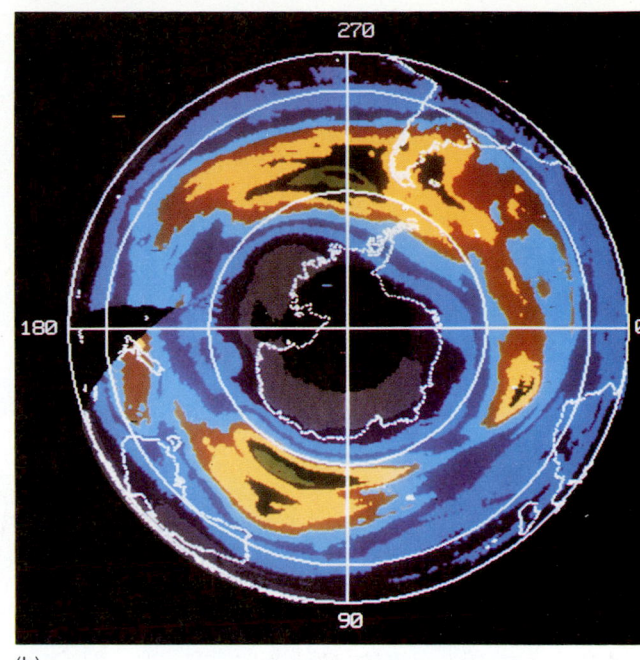

(b)

(c)

(d)

FIGURE 1.2   Biology: A Relevant Subject
(a) A scientist studies a cancer under an inverted
microscope. (b) A satellite photograph reveals the severe
depletion of the earth's ozone layer over Antarctica. (c) In
the name of progress, humans are depleting our rain
forests. (d) Overpopulation is one of the most serious
problems facing humanity today.

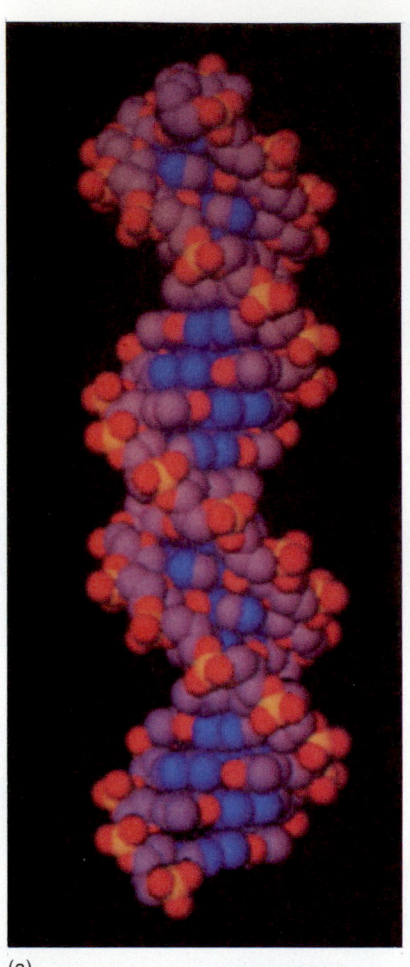

(a)

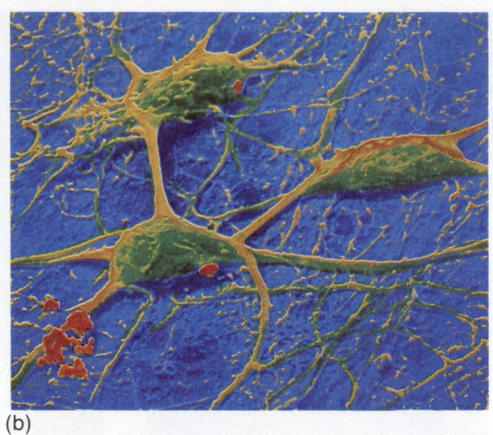

(b)

(c)

FIGURE 1.3 Organization Complex structures organize life from the molecular to the individual level. (a) Notice the high degree of molecular organization in this computer-generated model of DNA. (b) These nerve cells are organized into a complete network or neural system. (c) The beauty of these flamingos is the result of tissues, organs, and systems working together as a unified, living organism.

Organization

■ **All living things are organized and structured at the molecular, cellular, tissue, organ, system, and individual level. Organization also exists at levels beyond the individual, such as populations, communities, and ecosystems.**

When we look at a crowd of people or a dense forest, we see that living things are organized and structured. Small atoms are arranged into molecules, molecules into cells, cells into tissues, tissues into organs, and organs into systems, resulting in the living organism. In fact, in one sense the organization of life goes farther than the living organism itself. Living populations, communities, and ecosystems are also organized, as we will discuss in depth in Chapters 32 and 33 (Figure 1.3). Although nonliving things are also made of structured atoms and molecules, they never exhibit the degree of complexity found in living things.

Like a wrecking crew, the forces of nature are always trying to tear down living structures. Physicists like to explain these forces in terms of the second law of thermodynamics (a scientific law dealing with heat and energy levels), which states that in the universe there is a tendency toward increasing *entropy* — or, simply stated, a tendency toward disorder. When a living thing dies, disorder takes over, structures break down, and the remains no longer look like a living form. How do living things overcome entropy? An answer to that question leads to the second characteristic of life.

Maintenance/Metabolism

■ **To overcome entropy, living things use energy to maintain homeostasis. Metabolism is a collective term to describe the chemical and physical reactions that result in life.**

To overcome the wrecking-crew tendencies of nature, living things use energy to maintain an organized structure. The individual molecules in our bodies are constantly changing. In other words, you are not made up of the identical cells and molecules today that you were a year ago. You have the same types of molecules and cells but not the same ones.

For instance, your system requires stomach acid in order to digest your food. As your stomach digests that food, the amount of the acid molecules decreases. That means that your body must continually resupply new acid molecules in order to continue the digestive process.

In addition, to overcome the tendency toward disorder, a part of the maintenance plan of living things is to *repair*. Living things can repair themselves. Energy is needed constantly to rebuild molecules and cells.

Considered in this way, life is in a constant state of flux in order to stay the same, almost like running in place. By changing to stay the same, an organism maintains a constant, structured internal environment. Maintaining its sameness is referred to as **homeostasis** (Figure 1.4).

Closely related to maintenance is a characteristic of life known as **metabolism,** a collective term that means all the chemical and physical reactions that normally occur in a living organism. Because metabolism is necessary for the maintenance of life, some biologists like to group metabolism and maintenance under the same heading. In summary then, living things take in chemicals from the environment and, through a series of chemical reactions known collectively as metabolism, maintain the life process.

Growth

- ■ **Living things grow. The size and shape of an individual are determined by its genetic makeup and by the environment.**

Living things increase in cell size or cell number within limitations predetermined by their genetic inheritance. Although nature has produced a wide variety of growth plans, each individual takes on the characteristic form of its species. So, even though there is some variation within any group, people look like people, pine trees like pine trees, dogs like dogs, molds like molds. And, as we will discuss in the chapter on human genetics, even growth variations within a group are determined by genetic inheritance. For example, individual characteristics such as your height and weight are the result of the genes that your parents gave you, as well as of environmental factors like your diet, exercise, and so on.

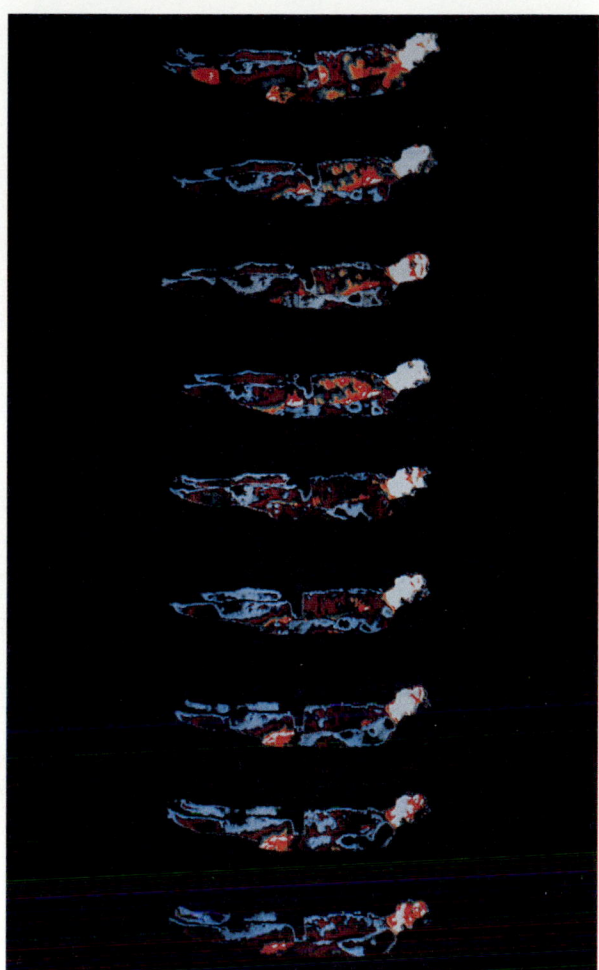

FIGURE 1.4   Homeostasis and Metabolism
This sequence of nine photographs of a man sleeping, taken at half hour intervals, shows changes in skin temperature. These skin temperature variations are due to the changing metabolism of the individual. As chemical reactions occur, some of the energy given off is released as heat.

Response to Stimuli

- ■ **Living things react to information that comes from outside or inside of themselves — they respond to stimuli.**

Simply put, a stimulus is any information coming from inside or outside a living system that elicits a reaction of some sort. The kinds of stimuli to which a living thing responds vary according to group, so every organism has certain stimuli that are crucial to its continued existence. When someone calls your name, you respond. Snakes track their prey by smell or by means of heat patterns. Male gypsy moths find their mates through pheromones, scents given off by the females. Dogs respond to smell. Plants

FIGURE 1.5   Response to Stimuli
Some plants respond to touch, such as this Venus flytrap, which has just trapped a young frog.

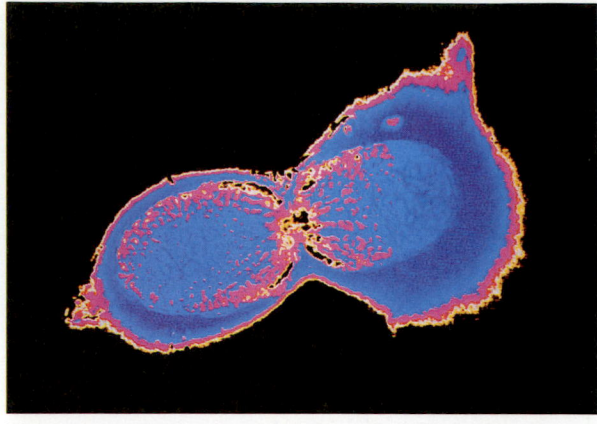

(a)

(b)

FIGURE 1.6   Reproduction
Living creatures produce like creatures. (a) Organisms made up of only one cell, such as this bacterium, reproduce asexually by undergoing simple cell division. (b) Lady bugs (multicellular organisms) mate and produce other lady bugs through sexual reproduction.

respond to sunlight, and some to touch (Figure 1.5). Indeed, the way an object responds to a stimulus is one way we determine if it is a living thing. When you see an unknown, buglike creature lurking under your bed, you touch it. If it scuttles out of reach, you respond by saying, "It's alive!" Living things have the ability to detect a stimulus and respond accordingly.

Reproduction

■ **Living things reproduce themselves. They also reproduce at the subcellular and cellular levels. In some instances, genetic information is altered. These alterations (mutations) and genetic recombinations give rise to variations in a species.**

We have all marveled at the birth of a puppy or a human infant or at the germination of a seed. **Reproduction** — the ability to produce offspring — is a characteristic of all living things. Humans produce humans, birds produce birds, and apple trees produce apple trees, through the process of reproduction and what is called **genetic recombination** (which simply means that new variations result). Reproduction ensures that the genetic information of the parents is replicated and passed on to the next generation (Figure 1.6).

However, replications are not always perfect. Occasionally, some genetic information is altered because of a chance error. Such an "error" may be helpful or harmful to the individual's chances for survival, depending on the nature of the change and the environment in which it lives. Changes in genetic makeup are called mutations, and in many cases such mutations are passed on to new cells or to offspring. Inherited mutations, along with genetic recombinations, give rise to *variation* within a species.

Variation

■ **Living things are varied because of mutations and genetic recombination. Variation may affect an individual's appearance or chemical makeup, and many genetic variations are passed from one generation to the next.**

Because of mutations and genetic recombination, a collection of living things usually exhibits variation, not just between different groups but also within groups (Figure 1.7). Not all beagles look exactly alike, nor do all Arabian horses or all fiddlehead ferns. Even if all the individuals of a species appear to be completely alike (like species of algae or amoeba), they are not, because variations can be chemical as well as physical. Because natural variations are genetic, they can be passed on from one

(a)

FIGURE 1.7 Variation
(a) Within this group of cats, several different varieties are apparent. (b) There is variation within a species, as illustrated by the students at this kindergarten Halloween party.

(b)

generation to the next. However, those variations due to environmental factors are not.

For example, if someone in your classroom released the contents of a culture flask containing the organisms that cause Legionnaires' disease, not all the students would become ill. Because of genetic variation, some would have inborn resistance to the disease, and they would not be affected. These individuals, in all probability, will pass on this resistance to their offspring.

Adaptation

■ **Living things adapt to changes in their environments.**

All living things must be able to **adapt,** to change in order to be better suited to their environment.

These adaptations assure survival. For example, when it gets hot, your body can adapt immediately through physical processes (such as perspiring) that regulate internal temperature. Behavioral adaptations are also a way to adjust to environmental changes. For example, an organism may move from an environment that has become too hot or too cold, too wet or too dry, to another place more suitable for survival. Some adaptations are of a more permanent nature, and we see examples of them when we consider how animals or plants have adapted to their changing environment (Figure 1.8). For example, as wetlands change to deserts over time, those animals and plants that are best adapted to survive in the desert environment live and reproduce, and their genetic adaptations are passed on to their offspring.

(a)

(b)

FIGURE 1.8   Adaptation
(a) These sac-winged bats have adapted behaviorally to escape predators by looking like thorns. (b) Because of their moisture-retaining characteristics, these cacti are well adapted to the desert environment.

As you can see, there is no single, simple definition of life. Instead, we use a list of related characteristics that allow us to distinguish the living from the nonliving. In its complexity and its simplicity, how wondrous is this phenomenon we call life!

VITALISM VERSUS MECHANISM: HOW CAN WE STUDY LIFE?

■ **Vitalism is a philosophy that explains life in terms of a life force. It is untestable. Mechanism is a testable philosophy that attributes life to natural laws.**

We all watch life, all wonder about it. But you may feel that scientists look at things differently. After all, it's one thing to sing Irish folk songs and another to be a folk song scholar. And you are right — there is a difference. So, before we look at the way biologists talk about the characteristics of life, we need to discuss the philosophical framework that allows us to study life.

At some time in your life you may have butchered an animal, cleaned a fish, or watched an organism die. And at that time you thought, "One moment this was alive and doing all the things that living things do, and the next moment it was a corpse. Life is gone forever. What happened in the moment life left the body?"

Throughout their history, humans have asked the question "What is life — and death?" and have developed various philosophies to explain the death of an organism. They say such things as "The soul left the body," "The forces of life are gone," or "The fire of life is out." This philosophy is referred to as **vitalism.** The vitalist states that things are alive because they possess a vital life force, and when it leaves the body, life ceases.

But then what about other life forms? Do flowers or grasses have souls? How can you study a life force? How can you discern cause-effect relationships? The concept of a soul or life force is untestable — it is a matter of faith, an article of belief, or a question that philosophers call *metaphysical* (meaning beyond or above physical reality). In contrast to vitalism is the philosophical position called **mechanism.** Mechanism states that since life is a natural phenomenon, life processes are subject to the natural laws of physics and chemistry.

Mechanism is not a new idea — one of the earliest major supporters of this philosophy was the French philosopher René Descartes (1596–1650). Even though mechanism received a lot of criticism at the time (as it still does today), mechanists began to develop experiments to test their philosophy. They reasoned that if a reaction thought to occur only in a living organism could be duplicated without an organism, a vital life force could no longer be said to be the cause of the reaction. Up until that time, scientists defined *organic* chemicals as substances produced only by living things (as opposed to *inorganic* compounds, which were not produced by living things). Then, Friedrich Wöhler (1800–1882) was able to convert an inorganic substance called ammonium cyanate into urea, an organic compound that is the major component of mammalian urine. By producing an organic compound without the intervention of a living organism, Wöhler's experiment provided major support for mechanism.

Several other successful experiments followed. In 1898, Eduard and Hans Buchner demonstrated that a substance extracted from yeast cells could ferment sugar outside a cell. This substance ultimately came to be called an *enzyme*. Enzymes enable reactions to occur in living things. During the past century, chemists have accumulated a wealth of evidence of the importance of enzymes in chemical reactions.

The majority of biologists today follow mechanistic philosophy in their work, since mechanism can generate the testable questions that vitalism cannot. Today's biologists study life by trying to understand more about the natural laws of chemistry and physics, and how they apply to the phenomenon we call life (see Profile page 14).

Understanding what happens to enable a butterfly to fly or a fish to respire does not deny the wondrousness of such events. The beauty of studying life is that no matter how much we learn, it remains almost a miracle.

UNIFYING PRINCIPLES IN BIOLOGY

■ **A set of unifying generalizations guide biology, like all areas of study. These concepts, based on the characteristics of life, guide investigations and organize information.**

All of us, whether we are vitalists, mechanists, or something in between (that gray area in which many people find themselves), accept the characteristics of life that we listed earlier. But because biology is a science, it has organized the characteristics of life into *unifying principles,* or concepts, that serve as its major generalizations. This process is not a strange quirk of biologists. Generalizations are found in every subject you study because they organize the details of a subject area into logical divisions. Without generalizations, people might become lost in the details — they "can't see the forest for the trees."

Many biologists state that biology consists of at least seven unifying concepts: *unity and diversity, genetic continuity, homeostasis and regulation, the*

biological basis of behavior, the complementarity of structure and function, the interplay of the organism and the environment, and *evolution.* As you read through the subsequent chapters, you will see that these concepts are woven throughout this text, wherever specifics are discussed. I hope that the short discussions here will help you gain a better understanding of these concepts, and I urge you to refer to this chapter again from time to time, to refresh your awareness of biological principles. They will help you understand how the specifics and the generalizations join to create the entire discipline of biology.

Unity and Diversity

■ **The concept of unity and diversity is based on the observation that all living things share many characteristics, while at the same time expressing those characteristics in a wide variety of life forms.**

All life forms share many characteristics, some of which we discussed earlier. Biologists refer to these common characteristics as the unity among living things. Even if it is difficult to see how a giraffe and a one-celled animal like a *Paramecium* are alike, nonetheless they are. Similarities between species exist at molecular, cellular, physiological, and anatomical levels. With very few exceptions, all living things are composed of *cells,* the fundamental units of life, and it is at the cellular level that the unity among living things is most notable. For example, the genetic material, deoxyribonucleic acid (DNA), with few exceptions, is universal among organisms, as is the role of adenosine triphosphate (ATP) as a mechanism for energy transfer in cells.

Living things are united because of their common chemical, structural, and physiological characteristics. On the other hand, literally thousands of different patterns of life forms have evolved on this planet. Biologists combine these two facts in the concept of unity and diversity (Figure 1.9).

Genetic Continuity

■ **The concept of the genetic continuity of life states that all living things reproduce offspring of their own kind and that reproductive variations are one way to assure continuation of the species.**

The life characteristics of reproduction and variation join in the concept of the **genetic continuity** of life. All species perpetuate their kind, resulting in the continuation of life on earth. In biology, however, reproduction includes more than just birth. Living things reproduce at the subcellular level

FIGURE 1.9 Unity and Diversity: Flowers
Not only is unity and diversity exhibited among living things that appear very different, such as trees and bacteria, but among very similar organisms. For example, flowers, despite their similarities, are also very diverse in form and color.

(when cell parts produce cell parts), at the cellular level (when cells produce cells for growth and repair), and at the level of the entire organism.

One way to look at life is to think of the individual of a species as nothing more than a disposable container of genetic material (DNA). In this scheme of things, the individual's sole purpose is to maintain and incubate this genetic material, which, through reproduction and genetic recombination, allows for variation and for continuation of the species. The container (the individual) is disposable and can be thrown away, but the contents (DNA) must be perpetuated through the individual's offspring. The individual body dies, but the species continues, and the genetic information lives on.

Homeostasis and Regulation

■ **The concept of homeostasis and regulation states that living things take in energy and change it into useful forms in order to maintain life. Enzymes mediate energy transformations, and genetic material regulates enzyme production. Secretions by nerves and hormone producers mediate cell-to-cell regulation.**

The principle of homeostasis and regulation combines the characteristics of maintenance and growth. Earlier, when we discussed maintenance, we said living things must overcome entropy to remain alive, to maintain homeostasis. The French physiologist Claude Bernard is usually credited as one of the first individuals to develop the concept of homeostasis. Because energy is needed to maintain homeostasis, living things must be *energy transducers,* or energy changers. Through the process of photosynthesis, certain cells, such as some found in green plants, are capable of taking light energy and changing it (transducing it) to chemical energy that is then stored in food. Other cells, such as plant or animal cells, take the energy stored in food and transduce it to other energy forms that can be used to overcome entropy.

Enzyme Mediation. The metabolic reactions maintaining homeostasis are mediated by enzymes. Enzymes are organic substances that help regulate chemical reactions by lowering the amount of energy required for the reactions to occur. Gustav Kirchhoff (1824–1887) was one of the first individuals to describe enzymes when he prepared a wheat

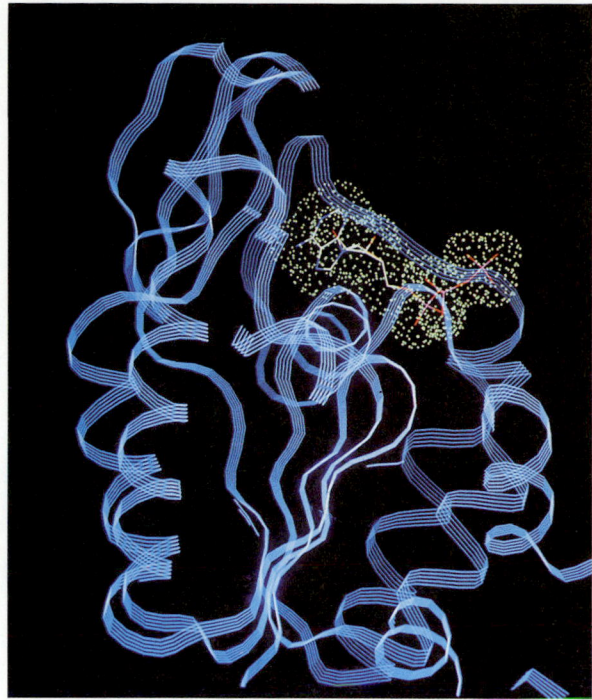

FIGURE 1.10   Enzymes: Metabolic Mediators
An enzyme is a special kind of protein that facilitates various steps in biochemical reactions. The enzyme shown in this computer-generated model facilitates the release of energy used in living cells.

grain extract that could convert starch to sugar. What he called *ferments* we now call *enzymes* (Figure 1.10).

Enzymes are forms of protein that regulate chemical reactions. But what regulates the synthesis (the production) of enzymes? Not only is the DNA molecule involved in the transmission of genetic information, but this genetic information also regulates protein synthesis. Some of those proteins are enzymes.

Remember that to maintain homeostasis, cells have to be energy transducers. Enzymes mediate metabolic processes, and the genetic code regulates protein synthesis.

Regulation. Regulation occurs not only at the cellular level. Organisms that are many-celled, such as humans, have evolved mechanisms that enable cells to communicate with each other. This communication is chemical in nature. In some cases, in animals, the chemical communication is the result of secretions produced by nerve cells. (The knee jerk reaction of your leg to a physician's hammer is an example.) In other cases, chemicals called **hormones** (chemicals produced by specialized glands or cells in animals or by special areas in plants) allow for the regulation and integration of life activities.

We have all heard of the mother who is in a car accident and suddenly has the strength to lift an automobile to free her trapped child. And we have all seen how houseplants grow in the direction of the bright light from a window. Try to imagine the extensive events within a living cell that allow it to regulate and integrate the chemical reactions that result in life. These are only two examples of the reactions of multicellular organisms to chemical communications sent by hormones.

The Biological Basis of Behavior

■ **Biologists are determining that behavior has a biological basis.**

The characteristic of stimulus-response falls under the principle of the biological basis of behavior. Living things behave — that is, they act in any number of ways. They respond to stimuli. Simple unlearned reflexes, such as the sucking reflex of a baby or the pecking of a gull chick at its father's beak for food, are based directly on the organism's heredity. Plants also have simple behavior in response to their environment. Roots grow toward gravity, and leaves grow toward the sun. More complex behavior, such as learned behavior, also has a biological basis, and so does the behavior of organisms in groups or societies (Figure 1.11).

FIGURE 1.11   Biological Basis of Behavior
The complex interaction of nerves and muscles provides the biological basis for the events that result in the learned behavior of this chimpanzee who is using a leaf to drink water.

Scientists Talk About Science

Because the study of biology is a human endeavor, each chapter of this text will include a profile, a biographical sketch of a scientist (or a group of scientists) and his or her contribution to biology. As an introduction to this distinguished group of people, listen to a few eminent biologists discussing their feelings about science.

Dr. Kary Mullis
Biochemist,
Nobel Prize Laureate

When asked about his development of the technique of polymerase chain reaction (PCR), considered by many the most important discovery in genetic research since the discovery of the structure of deoxyribonucleic acid (DNA), Mullis responded by saying, "I was playing. I think really good science doesn't come from hard work. The striking advances come from people on the fringes of being playful." It was that playful mind that was at work on a beautiful spring evening in 1983 when the idea for PCR came to Mullis. He was pondering how to find a particular section of the DNA molecule and isolate it. From that thought came the idea that one should be able to mark off a section of DNA and then, using enzymes, force it to copy itself over and over again, almost as if the DNA section

was part of a computer program in which the same process is repeated over and over again. If this could be done with a section of DNA, then the DNA would copy itself over and over again — two, four, eight, exponentially — so that in 20 cycles there would be 1,048,576 copies and so on. Over a billion copies in less than three hours.

Because of the PCR technique developed by Mullis, scientists now have the ability to create billions of copies of genes within the DNA molecule in a few hours. These strands of DNA can be utilized in many ways: to help convict or free accused criminals, to locate genes, and to copy genes from DNA samples of creatures that died millions of years ago, among other applications.

Mullis has engaged in the fun of science all of his life. As a young boy, he would constantly experiment. He studied biochemistry as a graduate student at the University of California, Berkeley.

Mullis has a lifestyle and personality that do not fit what most people think of when they imagine a Nobel Prize laureate. He is a man that many of his colleagues would call quirky and outlandish. He is more likely to be found in the hot tub at his ranch in the Anderson Valley region of the California wine country, or surfing, or sunning near

his home in La Jolla. He is a scientist whose personality elicits reactions from colleagues ranging from giggles to awe. Kary Mullis is truly, as stated by some, "the Quirky Genius who has changed the world."

He, like most who have made major contributions to science, did not continue to just accept one way of looking at a phenomenon, but rather, used curiosity and insight to view a process. A major characteristic of good scientists is their ability to question the answers and not just answer questions. Kary Mullis definitely is an example of an individual who constantly questions the answers in order to further scientific knowledge.

Source: Parade, October 10, 1993.

Behavior has a biological basis, even in highly complex life forms (including humans). For instance, we now know that some mental disorders, such as depression or schizophrenia, are mediated by an imbalance of chemicals in the brain, and a carefully monitored program of drug treatment may be an essential part of the treatment of these disorders. On the other hand, we all know (unfortunately) about drug abuse and the dramatic effect that chemicals such as cocaine, heroin, and alcohol have on behavior. The more we learn about the interaction of chemicals and behavior, the better off we will be.

The Complementarity of Structure and Function

■ **The shape of a biological structure often reflects its function.**

When you look at the workings of an electric mixer or a computer, you can infer something about the

Dr. Barbara McClintock
Geneticist, Nobel Prize Laureate

The late Dr. Barbara McClintock was the first person to discover and report that genes were not fixed in place like pearls on a string. Rather they could move from place to place on a chromosome. This discovery flew in the face of established dogma, and her work was ignored until the advent of recombinant DNA research. When asked about her discovery of jumping genes and its rejection by the established scientific community, the private, shy woman said, "They thought I was crazy, absolutely mad. I knew what I was doing [many years ago] wasn't acceptable. That was all right. . . . When you know you're right, you

don't care. You know sooner or later it will come out in the wash."

When asked about receiving the Nobel Prize, McClintock stated, "I was overwhelmed on receiving news of the Nobel Committee's decision this morning; the prize is such an extraordinary honor. It might seem unfair, however, to reward a person for having so much pleasure over the years, asking the maize plant [corn plant] to solve specific problems and then watching its responses. . . . I was having too good of a time doing the work."

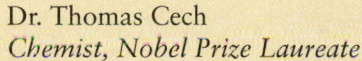

Source: Discover, December 1983.

Dr. Thomas Cech
Chemist, Nobel Prize Laureate

Dr. Thomas Cech of the University of Colorado at Boulder and Sidney Altman from Yale were awarded the Nobel Prize for Chemistry in 1989 for showing that ribonucleic acid (RNA) could have enzyme activity. This finding went against the established dogma that DNA regulated the synthesis of RNA, thought to provide the basis for the synthesis of proteins. Cech's finding is changing molecular biology.

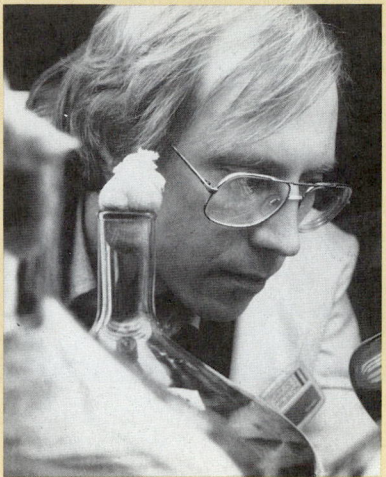

Cech is often asked, "Isn't it the dream of every scientist to win the Nobel Prize?" Cech's reply: "It feels real good. But I would never trade the day-to-day excitement of doing science. It's the process, not the end that I've always been after."

Source: Discover, November 1990.

function of any part by examining its structure and its relationship to other parts. This is because the function of a thing is often strongly tied to its form. Such a relationship, called complementarity of structure and function, is found in living things as well. For example, the structure of a leaf is related to its functions, as are the structures of muscles.

Function also follows form at the cellular and molecular level. The structure of a *mitochondrion* (a small structure within cells, involved in energy production) is related to its function. Its membrane

holds molecules in the proper relationship for the sequence of reactions that occurs within the structure. Specific molecular structures determine how a molecule is going to react with other molecules. The three-dimensional structure of an enzyme determines to a large extent how it is going to function in a reaction, since an enzyme's shape enables it to join only with a specific substance. (The fit is rather like a lock and key.) Much of what we know about biology today was learned by observing the shapes of things, inferring the function that such a shape

FIGURE 1.12 Chameleon Catching a Fly
The long tongue of this chameleon dramatically illustrates the concept of the complementarity of structure and function. The structure of the tongue (its length) directly relates to its function (getting food).

would allow, and testing those inferences until we found out what was going on (Figure 1.12).

The Interplay of the Organism and the Environment

■ **The concept of the interplay of organism and environment states that all living things are influenced by their environment. This relationship is reciprocal, so the environment is also affected by living things.**

Organisms interact with the environment at all biological levels. For example, at the cellular level, in the cellular environment, the parts of a cell interact with the cell as a whole. The individual cell is acted on by adjacent cells, and it acts on them in turn. At the organismic level, the individual exists in an environment and is acted on by both living and nonliving factors of that environment.

For instance, at this moment, as I write, it is raining outside, and this environmental factor affects me and other living and nonliving entities. The low barometric pressure may make my bones ache, the water may nourish my garden, and the runoff will carry some of the topsoil from my yard into the street.

However, there are no single-sided events in the complex relationships between life forms and their environment. As a result, the rain will also be affected. The heat of my cement driveway will cause some raindrops to evaporate. Some of the oil that dripped from my car's leaky oil pan onto the concrete will float away with the water into the street.

The concept of the reciprocity, or interplay, between the organism and the environment is known as **ecology.** Ecology is much more than cleaning up the environment and recycling beer cans. It is the scientific study of the interaction of the organism and its environment. Many of humankind's environmental problems are vivid examples of this interaction. Our unthinking abuse of the earth has resulted in many of the problems humans and their living neighbors now face — acid rain, pollution, the greenhouse effect, poisons of all types, resource or ozone depletion, the premature extinction of species. These and many other careless acts have taken place either with no understanding of how human beings and the environment interact or with a total disregard for the place of humans in nature (Figure 1.13).

Evolution

■ **The concept of evolution states that the genetic makeup of a population changes over time. The basis of the evolutionary process is genetic variation, and one mechanism of the evolutionary process is natural selection.**

All the characteristics of life unite in one of the major unifying principles in biology: **evolution,** the idea that living things change over time or, more precisely, that the genetic makeup of a population changes over time. Although it may seem that the idea of evolution is a fairly recent one, it is not. Philosophers discussed it from the time of the ancient Greeks, but not until Charles Darwin published his work entitled *On the Origin of Species by Means of Natural Selection* (1859) was there an explanation of evolution that was scientifically acceptable. How does evolution happen?

(a)

(b)

FIGURE 1.13   Acid Rain and Solid Waste Pollution
(a) We continuously pollute our environment with automobile exhaust and with smoke from our factories. The acid rain that results is slowly destroying our environment. (b) This unsightly garbage dump illustrates how our self-centered consumption and careless disposal of waste is affecting our environment.

The Role of Variation and Adaptation. When we look at the living world around us, we see the variation within any given species, and we know that these variations can be transmitted from parent to offspring (she has her mother's eyes, or he has his father's nose). Natural selection that may lead to evolution is based on the idea that some variations have *adaptive value.* In other words, they determine how well the individual will be able to compete in the environment — for food, shelter, or mates.

Long-term adaptations are the result of genetic variations that have proved to have what biologists call *selective advantage.* Remember that variations are the result of mutations (changes in the DNA molecule) and/or genetic recombinations (new combinations of genetic factors). Not all genetic variations are beneficial to an organism. However, a selectively beneficial mutation or genetic recombination in an individual enhances the degree of success of its offspring (that is, the number that survive and reproduce). Such variation provides selective advantages to that group of individuals (Figure 1.14). For example, during the evolution of humans, they acquired sweat glands that allowed them to perspire when they become overheated, opposable thumbs that have led some anthropologists to call us *the tool users,* and many other characteristics that have given us a distinctive selective advantage.

FIGURE 1.14   Pine Tree
Since it has grown in solid rock soil, a harsh environment, this pine tree possessed the genetic makeup that allowed for its survival. Pine seedlings that lacked a suitable makeup for this environment were selected against and died.

The Role of Natural Selection. The success of an adaptation is measured by the number of offspring that are produced and reproduce in contrast to members of a group who lack that adaptation. Over generations, environmental conditions are more favorable to some variations than to others. In evolutionary terms, that means that the individual will survive, will mate, and will transmit its genetic inheritance to its offspring. The process by which an individual is tested by the environment is called natural selection. The key here is that variations occur, and the environment *selects* those that will continue and ultimately become a characteristic of the group.

So variation serves as the raw material, or clay of life, and natural selection by the environment serves as the sculptor that determines the form into which life will be molded. Some individuals, because of adaptations that enhance their chance of survival, mature and reproduce, thereby leaving more offspring than others. The more offspring there are, the more chances for genetic variation through reproduction. Over time, the result has been a tremendous diversity among living things on earth.

So the *basis* of the evolutionary process is genetic variation, and one of the *mechanisms* of the evolutionary process is natural selection. Variation within a species is due to genetic differences among individuals. Successful individuals can compete better for their needs. When successful individuals reproduce, they pass on their traits to their offspring, whose chances for success are greater than the chances of those who do not possess those beneficial traits. Since evolution is a continuous process, over time, through the means of natural selection, variation and adaptation shape the diversity of life forms on our planet.

This list of generalizations and principles is not meant to be exhaustive or complete. Rather, it provides an introduction to basic concepts that guide the study of biology and is also a means for understanding how the characteristics of life interact. The principles have been formulated through careful investigations and observations of the natural world. What is more, no matter how much knowledge we gain, the quest for more knowledge continues. Whatever we learn is subject to revision when careful study dictates.

HOW DO WE STUDY LIFE?

Now that we have discussed the general characteristics of life, have a philosophical perspective, and are aware of the basic generalizations that guide biology, *how* do we study life? To explain that, we need to understand a few basic terms.

For instance, what is meant by the word *science?* Looked at in one way, as an activity, science is an attempt to understand ourselves and the natural world. Looked at in a slightly different way, science is the body of accumulated knowledge about the natural world recorded as men and women have tried to understand it over the centuries.

Since both of these definitions are true, science must change as new ideas and new techniques allow us to obtain more information and come to a better understanding of any phenomenon—and it does. For instance, with the invention of microscopes, our knowledge about cells and their structure changed, and as these tools become more and more powerful, our knowledge will continue to grow and change. So the science of biology is a changing body of information about life in the natural world. Biologists are willing to test an idea again and again. What they know may change, and their understanding will be clarified.

The Scientific Method

■ **The scientific method is a logically ordered procedure for solving problems and answering questions. Carefully structured questions arise from what is already known, and possible answers are investigated. For results to be considered valid, they must be predictable and repeatable.**

It is obvious that humans are curious animals. (Have you ever talked to a four-year-old?) Scientists just happen to be a little more curious about natural phenomena than most, and they are willing to test their observations and curiosities before they draw interpretations or conclusions. The method that scientists use to investigate the workings of the living world is the **scientific method.** Undoubtedly you have heard of it. Unfortunately, the scientific method is often presented as if it were something special, limited only to science. We all use a scientific method almost automatically, however, as we solve our everyday problems.

The Beginning — The Idea. If you were a scientific detective and wanted to learn more about a natural phenomenon, how would you go about it? In all probability, your task would include two phases. First, you would collect information, through either observation or experimentation, and second, you would subject these data to organization and interpretation. Basically, that's what scientists do (Figure 1.15).

(a)

(b)

FIGURE 1.15   Field Biology
Biologists investigate life in a variety of settings, from the laboratory to the field. These field biologists are conducting a fish population study by (a) netting fish and (b) gathering data.

A scientist begins by being curious about the world and recognizing the need to inquire about a natural event. He or she then goes to scientific journals, which report the results of other scientists' work, in order to find out what is already known about the event. With this information in hand, the scientist formulates a carefully worded statement of the problem and from that develops a **hypothesis,** or a generalization based on his or her research that will be proved or disproved through experimentation. Once the hypothesis is formulated, the next step is to develop a controlled experiment or plan an observational scheme in order to gather information or data.

As an overall plan for solving a problem, this is not much different from trying to figure out where you left your shoes. You begin with a general review: when you last saw them, where you were, and what you were doing. You review what is known about the phenomenon: where you usually leave them (under the bed, behind the chair in the corner, next to the bathtub). Having reviewed past history (they aren't in any of those places), you then formulate a guess, a hypothesis, as to their probable location. You last saw them after coming back from swimming at the pool. You know that you immediately went into the kitchen for a snack and then watched TV for an hour. Even if your original hypothesis is wrong, you still learn something (where they aren't). Logic will guide your research until you solve your problem.

Controls and Control Groups. What makes the scientific method special is the concept of a control.

By using controls, scientists try to eliminate any factor that might influence their results, other than the element that they want to investigate. For example, if you wanted to investigate the effect of temperature variation on the rate of growth of bean plants, you would arrange the plants into two groups, the experimental group and the control group. In the experimental group, you vary the temperature. In the control group, however, you maintain a constant temperature (Figure 1.16).

FIGURE 1.16   Experimental and Control Groups
This experiment investigated the effect of temperature on plant growth for a period of two weeks. The control group on the left was grown at a temperature of 70° F. The experimental group on the right was grown at a temperature of 40° F. The plants grown at the lower temperature grew smaller than the control group.

FOCUS 1.1
Which Came First, the Enzyme or the Nucleic Acid? Questioning the Answers

"Which came first, the enzymes responsible for nucleic acid replication, or the nucleic acids responsible for synthesizing the proteins that function as enzymes?" (See Figure A.) We may have an answer, though a controversial one. Dr. Sidney Altman and Dr. Norman Pace have observed that one kind of RNA can function as a enzyme. Think how significant this finding is: It goes against the dogma that we associate with replication and function of nucleic acid. Enzymes will have to be redefined. The RNA role in cellular activities will have to be reexplained.

The RNA that demonstrates enzyme activity was first detected in the RNA-processing enzyme RNAse P, found in the bacterium *Escherichia coli*. This enzyme is necessary for the maturation of a transfer RNA essential in protein synthesis. The enzyme contains not only the protein portion expected, but RNA as well; that is, the enzyme actually consists of a protein and a nucleic acid. Other investigators working with the same enzyme in other species of bacteria found that when

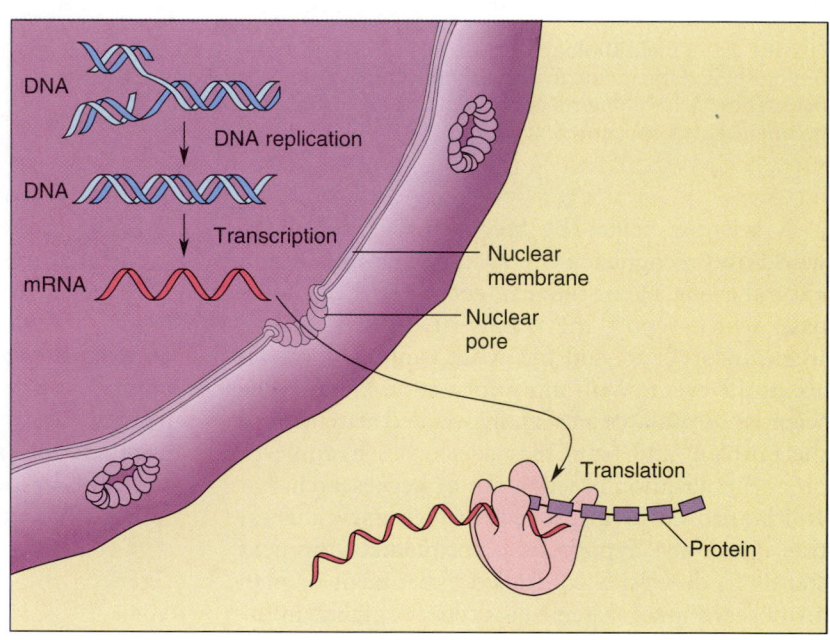

FIGURE A
A simple model of protein synthesis.

the protein and RNA components of RNAse P were separated , the RNA component acted as a catalyst for producing transfer RNA. The RNA was functioning as an enzyme.

Investigators have always thought that the splicing and ligase enzymes were necessary for RNA maturation so that RNA could perform its function in protein synthesis. Investigators at the University of Colorado at Boulder have, however, have found a type of RNA that can cut and splice itself without help from enzymes.

But knowing what you already do about plants, you know other factors can affect growth — light, water, soil, and nutrients. Any factor that can affect the results of an experiment is called a *variable*. To be sure that your results are reliable, you must control *all* variables that could influence the results in both the experimental and control groups. In other words, both groups must be grown under the same conditions. The only variable that can change is temperature, which is called the *experimental variable*.

Data. You then record measurements of the two groups of plants over a period of time. The measurements constitute your data. If the rate of growth is almost the same for both groups, you conclude that this slight variation is due not to temperature variation but to chance. However, if the results show that the experimental group has a statistically significant difference in the rate of growth, you conclude that this difference is due to the experimental variable, not to chance.

How can you be sure that your results are correct? The ultimate test of the scientific explanation is predictability, or repeatability. In other words, can a scientist predict that the same experiment, under the same conditions and with the same careful procedures, will produce the same results? If so, the results (the knowledge gained) are most likely accurate. If not, the original idea is most likely incorrect or the procedure is flawed. It is the quest for predictability that makes the Food and Drug Administration conduct research on new medications for several years before it will approve a new treatment for any disease, such as arthritis or cancer.

Scientists not only repeat, or *replicate,* their experiments or observations to verify that the results they obtained are valid. When appropriate, they subject their data to statistical analysis and submit their results to the judgment of other scientists (peers) through scientific journals. Before publication, a manuscript is given a peer review to determine the scientific merit of the work and to decide whether the results should be published. This process allows science to police itself and brings a high probability of reliability to reports on research. Peer review is one way of assuring the scientific community that the science appearing in such journals is valid.

Overview. These, then, are the steps a scientist usually follows when investigating a phenomenon. He or she:

1. Develops an inquiry about a phenomenon
2. Formulates a specific problem statement
3. Develops a hypothesis
4. Investigates the phenomenon
5. Gathers data
6. Interprets data
7. Draws conclusions about whether the hypothesis is correct or needs to be refined or reformulated
8. Publishes the results for at least three reasons: (a) to communicate scientific information to all of the scientific community, (b) for the scrutiny of his or her peers, and (c) to provide others the opportunity to expand on what has been learned

In all probability, no real scientist goes through all these steps. No specific steps or procedures can truly explain the trials, insights, errors, frustrations, imagination, and luck that enable a scientist to gather information, just as no one can truly outline the specific steps that enable a detective to solve a crime.

Keep in mind that the scientific method is one of our most powerful tools of inquiry, but since it is we humans who are obtaining this knowledge, whatever we learn is subject to the limits of our senses, our understanding, and the machines at our disposal. Therefore, scientific knowledge is subject to change, although this change can occur only if newly acquired data refute existing information and can stand the rigors of constant testing (Focus 1.1).

Scientists and Society

Obviously, scientists are human beings who "do science" — that is, they investigate the world in which they live. However, as you read the biographical sketches found in each chapter of this text, you may be surprised to find out that some scientists not only gather scientific information; they also help to establish political procedures for determining the direction of science research. Decisions of this kind are often difficult.

Many questions before the biological community today have moral and ethical dimensions that are very difficult to resolve. For instance, should taxpayers support research for test-tube babies? (What happens if something goes wrong: If you destroy the developing embryo, is that murder?) Who should receive an organ transplant? Should human genes be inserted in domestic animals? When academic institutions hire a scientist and provide a salary and research laboratory, who owns the results of the research? (Many of the funds for research are from taxpayers' dollars through federal agencies that provide grant money for research — no one wants the research that they pay for to profit the researcher primarily.)

Focus 1.2 discusses some of the issues we face in deciding how government money should be spent on scientific research. Concerns like these have convinced many scientists that they can no longer be just scientists. They must also be politicians and

FOCUS 1.2
Commentary on Scientific Funding

In these times of worries about balanced budgets and governmental agencies that waste our money, people in the United States hear almost every day about one more way to cut federal spending, and one way is to cut federal funds for scientific research. A senator gives his "Golden Fleece" award (an honorary award for wasteful research dedicated to causes he judges unworthy of our support) to an agency that funded research on the sex life of the gypsy moth. We hear that funding for basic research is decreasing. Scientists solemnly declare that this is dangerous, and politicians say that they must be the final judge on how the taxpayers' money is spent. What's the fuss all about?

Before we can decide how we want our tax money spent, we must become familiar with two terms, applied and basic research. *Applied research* is research conducted with a product or a goal in mind. For in-stance, a drug company will run tests to determine the effectiveness of a certain drug against the common cold. Or a researcher in a private laboratory will try to determine ways to combat lung cancer. *Basic research* has no such specific goal in mind. Instead, a scientist doing basic research may be trying to determine what exactly triggers cell replication or reaction to infection. The goal is to understand the mechanism of biological function more clearly. Applied research builds on the discoveries made during basic research.

The distinction between applied and basic research is an important one, and one that is all too easy to misunderstand. Unfortunately, some individuals feel that we should support only practical or applied research because the "payoff" is much more easily grasped. After all, why would anyone be interested in the sex life of a gypsy moth (besides, of course, another gypsy moth)? With funds as tight as they are becoming, it is imperative that scientists be able to explain fully and completely the significance and importance of basic research, for if basic research is discontinued, applied research will suffer as well. In a sense, basic research is the ground breaker; applied research only plants there once the land is tamed.

It is imperative that our lawmakers understand the need for support of basic as well as applied research. Our knowledge about the types of cancer, the variety of causes, and potential cures has come from basic research on how cells function. Without that research, we would still be no closer to answers than we were twenty years ago, no matter how much applied research had taken place. Research on the life cycles of insects has resulted in techniques of dealing with farm and forestry pests, like the gypsy moth,

must be responsible for the moral and ethical issues raised by some of their research. Often our answers give rise to more problems. Gone is the day when scientists could hide, cloistered in their laboratories, practically immune to the activities of the outside world. From some of our scientific knowledge have come moral and social dilemmas, and the people most involved must act responsibly.

SUMMARY

1. All living beings share a number of characteristics: organization, maintenance, growth, response to stimuli, reproduction, variation, adaptation.
2. The organization of living systems is constantly being countered by the force of nature called entropy (the tendency toward disorder). Living sys-

without the use of dangerous pesticides that have damaged our environment. Instead, researchers use *pheromones,* which are chemical "odors" given off by insects during mating, to attract the insects to a trap (Figure A). No other life forms are damaged, because pheromones are *species-specific,* which means only that a particular type of pest is attracted to the odor. Much of our information about genetics, the study of heredity and variation, has come from studies of the life cycles of sweet peas and of fruit flies.

In some of these examples, the "payoff" has been fairly obvious. However, much basic research has led us to new and useful ways of using biology only after a long period of time. For example, in 1946, Lederberg and Tatum received funding to conduct research on whether or not bacteria reproduced sexually. They wanted to find out simply because everyone was convinced that bacteria did not reproduce sexually. This rather fundamental research, which established that bacteria did indeed reproduce sexually, began a line of investigation that ultimately led to the development of genetic engineering, an area of science that will have a greater impact on our everyday lives than even the development of the birth control pill (Figure B).

When these men originated their research, no one knew that they had begun a project that could have such an impact, and the same has been true of many basic research projects. If Golden Fleece awards had been given in 1946, I wonder if Lederberg and Tatum would have received one? In fact, given the diminished funds and the attitudes toward basic research today, it is difficult to say whether they even would have received funding to conduct their research.

We have to be very careful as taxpayers to make sure that our research dollars are well spent, but we must also be sure that we do not stop funding the type of basic research from which all our knowledge and technological advances arise.

FIGURE A
The male gypsy moth can detect over large distances pheromones emitted by the female gypsy moths.

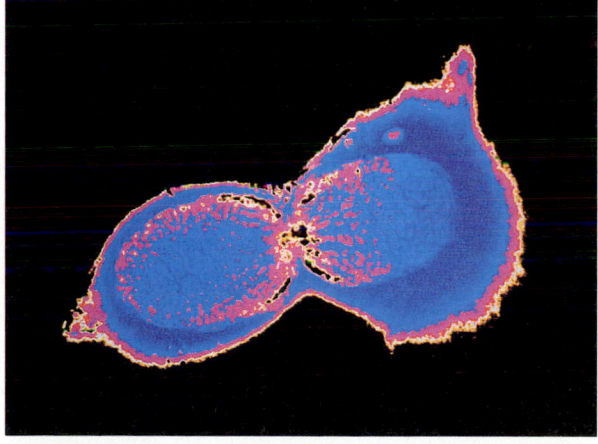

FIGURE B
Prior to 1946 most biologists thought that bacteria reproduced only asexually. Today we know that some bacteria do reproduce sexually.

tems use energy to overcome this disorder and maintain homeostasis (a constant internal environment).
3. Metabolism is the sum of the chemical and physical reactions in living systems.
4. Vitalism, the philosophy that a vital life force is responsible for life, is not testable. The philosophy of mechanism, which states that life is a natural phenomenon and subject to natural laws, is testable, and serves as the philosophical perspective for the study of life.
5. The principle of unity and diversity states that all living things share common processes, especially at the molecular and cellular level. However, living things also reveal great diversity within and among species.

6. Living things have the ability to reproduce their kind ("like begets like"), thus ensuring that their genetic heritage will continue in their offspring. This is called genetic continuity.

7. Homeostasis depends on thousands of chemical reactions, mediated by enzymes, which produce a dynamic sameness. Genetic material controls the synthesis of proteins, including enzymes. Organisms regulate themselves at the cellular and multicellular levels through chemical communication.

8. The behavior of both individuals and groups of organisms has a biological basis.

9. The relationship between structure and function is complementary.

10. A reciprocal association occurs between organisms and the environment.

11. Living things evolve as a result of the interaction of genetic variations and the selective forces of the environment. Genetic variations less suited to the environment are eliminated by natural selection.

12. Scientists generally follow certain steps (called the scientific method) when investigating a phenomenon: inquiry, formulation of a problem, development of a hypothesis, investigation, data collection, interpretation of data, development of a conclusion.

13. Directly and indirectly, scientific investigations have a political, social, moral, and ethical impact on us. Scientists must be aware of the effects of their knowledge.

KEY TERMS

homeostasis	genetic continuity
metabolism	hormone
reproduction	ecology
genetic recombination	evolution
adaptation	scientific method
vitalism	hypothesis
mechanism	

REVIEW QUESTIONS

True/False

1. The study of biology is based on several unifying principles. Among them are homeostasis and regulation, genetic continuity and evolution.
   true   false   (page 11)

2. The statement, "Living things are disposable containers of DNA," is a way of thinking of the concept of the genetic continuity of life.
   true   false   (page 12)

3. In order to remain alive, cells must be energy transducers.
   true   false   (page 12)

Fill in the Blank

1. _____ are proteins and lower the amount of energy required for a chemical reaction to occur. *(page 13)*

2. The molecule known as _____ is the chemical basis of the gene or source of genetic information. *(page 12)*

3. The collective term _____ is used to describe all or some of the chemical reactions occurring in a living organism. *(page 7)*

Multiple Choice

1. The statement, "Life cannot just follow natural and physical laws; there must be something more," would most likely have been made by an individual with the philosophical perspective of:
   a. A vitalist
   b. A mechanist
   c. An existentialist
   d. An atheist
   (page 10)

2. The following question relates to a biologist's study of human reproduction. The biologist reasons that glands produce the uterine cycle in females. This statement could best be defined as:
   a. A hypothesis
   b. A conclusion
   c. An observation
   d. An interpretation
   (page 19)

3. A biologist wants to investigate the role of male hormones in baby chicks. Two groups of chicks are established — one receives the injection of hormone in oil; the other group just receives an injection of oil. The group receiving only an injection of oil is called the:
   a. Experimental group
   b. Placebo
   c. Control group
   d. Observational group
   (page 19)

4. When looking at a slide of the cells lining the inside of your mouth, you notice that the cells are thin and flat. This observation illustrates the general biological concept of:
   a. Variations
   b. Interaction of form and function
   c. Interaction of the organism and environment
   d. Energy transduction
   (page 15)

Discussion

1. List and describe five major characteristics of all living things.

2. If you found an undescribable mass under your bed, how would you test to see if it were alive?

3. Discuss some examples from your experiences that demonstrate that biology is a relevant subject.

4. What is homeostatic control, and why is metabolic activity required to maintain this homeostasis?

5. Have you ever used the scientific method of logic to investigate an everyday problem? What was the problem, and how did you use this logic?

6. We hear a lot about scientists and what they do. Discuss how scientists do science. Does their work have any impact on you?

7. As an educated taxpayer, why must you understand some of the politics of federal scientific research funding?

8. Do you agree with the statement that we must continue to support basic scientific research? Why or why not?

9. Why is it useful to think of biology as a group of major biological generalizations?

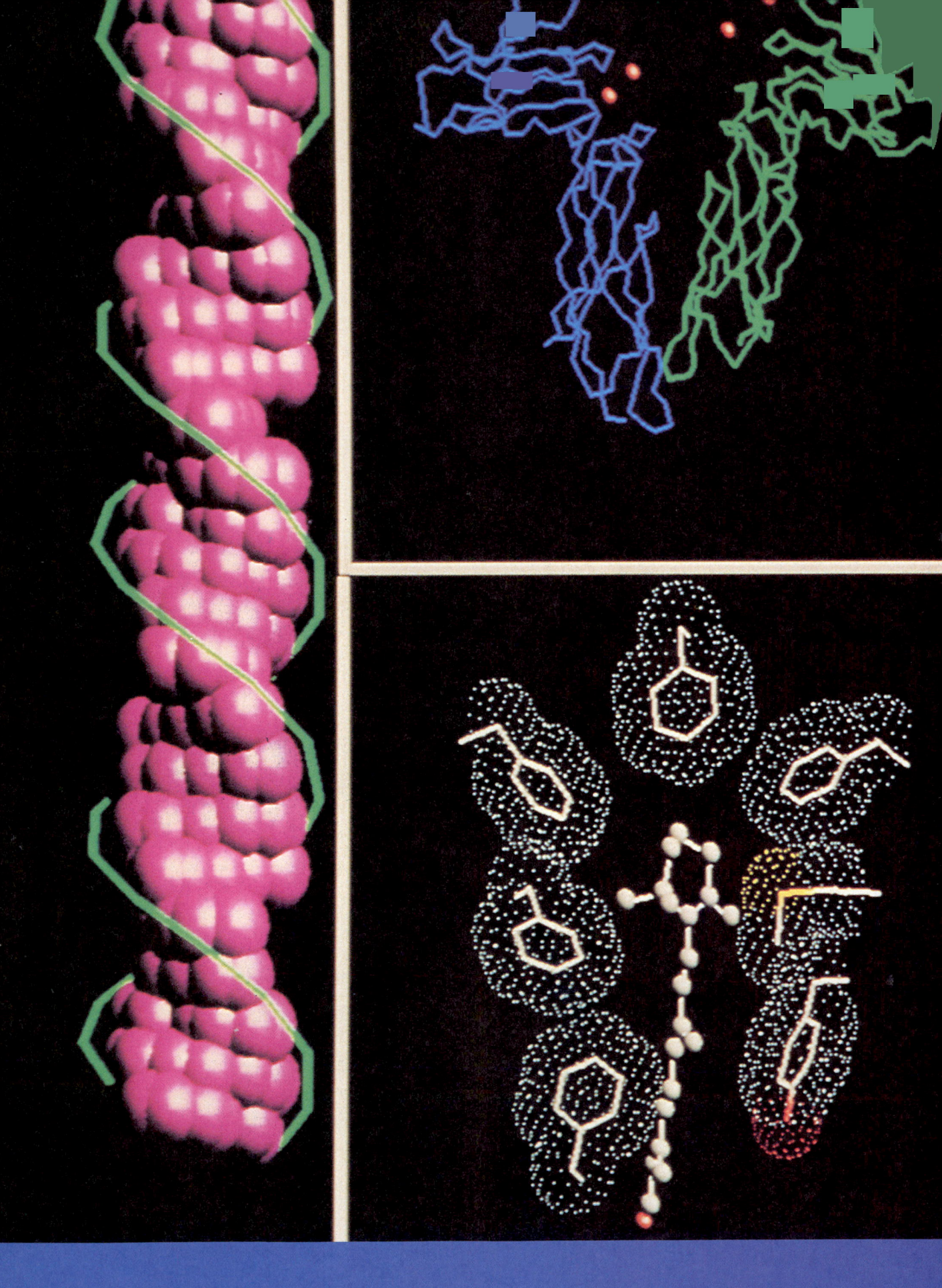

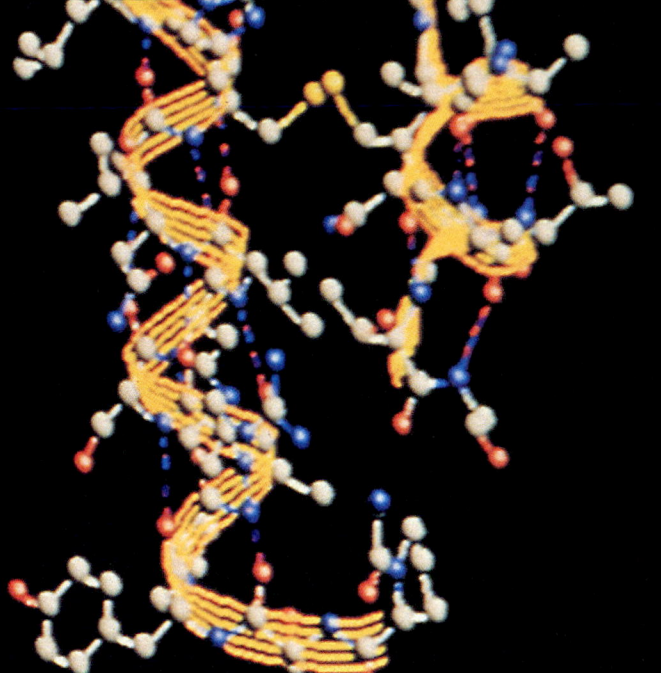

UNITY: A MOLECULAR/ CELLULAR PERSPECTIVE

After reading this unit, the student should be able to:

- understand that cells are the basic units of life and that cellular processes resulting in the phenomenon that we call life are based upon physical and chemical changes
- understand that all organisms are united by the very fact that they are alive
- realize that in order to understand life, we need to understand the relationship between matter, energy, and organization that results in life
- understand the relationship between structure and function at the atomic, molecular, and cellular levels
- understand the energy relationships in living things that result in energy capture and utilization

2

The Chemical Basis of Life

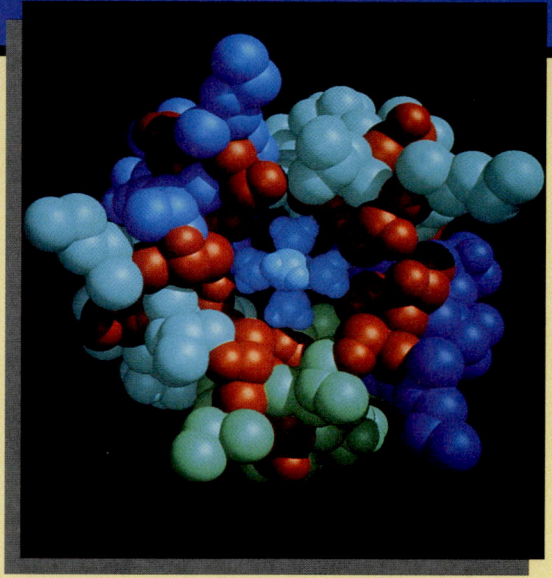

In our journey toward a more scientific understanding of life, we must first place ourselves in the world of the nonliving — the world of atoms, molecules, and the chemicals of life. How did an atom become "Adam" (Figure 2.1)? In order to answer that question, we have to understand and visualize the atomic world and the relationships among matter, energy, and life. That means we have to understand chemistry. Obviously, those biologists whose work involves observing animals as they interact with each other and with their environment can do their work without emphasizing the chemical reactions that produce the phenomenon we call life. However, it is biochemistry — the chemistry of life — that has brought us many of the major advances in medicine and a more thorough understanding of life processes. Life is truly a chemical phenomenon.

To understand the part that chemistry plays in biology, we must begin with the smallest parts. What are atoms and molecules, and how are they involved in producing the life within us? In a sense, we are all chemical machines designed to incubate the self-replicating molecule, DNA. We make our living by controlling, through DNA, the myriad of chemical reactions that result in life.

Although chemistry may seem difficult or beyond your abilities, it is an interesting human endeavor in its own right. As a biologist, I feel that chemistry is even more interesting when we discuss how it applies to life. So we will now look at chemistry, only to the degree that is necessary to understand biological processes. To learn more, you will need to explore the topic in advanced courses in biochemistry.

Above: Computers can be used to produce graphic representations of molecules such as this neural secretion receptor.

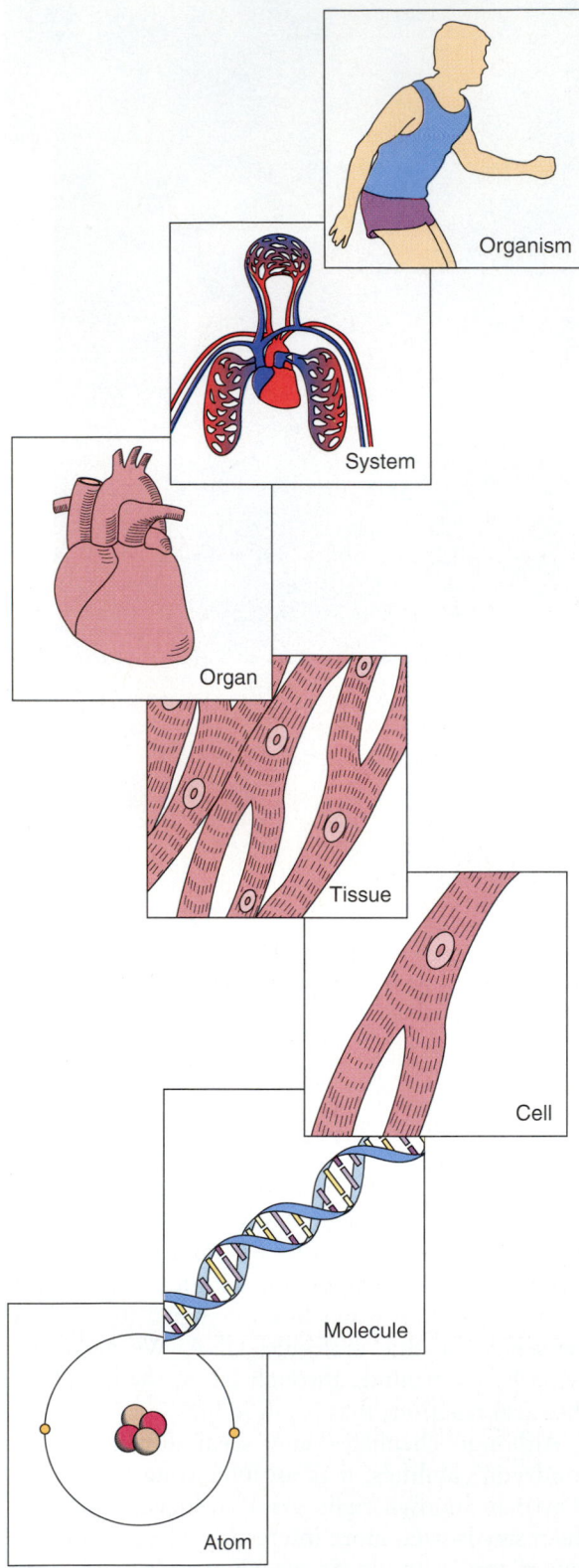

FIGURE 2.1   Levels of Organization
The most basic level is the atom. Atoms form molecules; molecules comprise cells. Cells of a similar type (like these heart muscle cells) form tissues, which form the organ — the heart. Various organs form a system, and various systems make up the living individual.

MATTER AND LIFE

■ **Matter, anything that has mass and occupies space, may exist in a solid, liquid, or gaseous state.**

If you have ever miscalculated as you left a room and banged your hip on a table edge or a door frame, you have had a concrete example of an encounter with matter. And although we rarely think about it, we walk through air all the time, and air is also a form of matter. Matter can be simply defined as anything that has mass and occupies space. Matter may exist in any of three states, as a *solid*, a *liquid*, or a *gas*, depending on the amount of energy in the system. For example, water is one of the most common substances on earth and one of the few commonly found in all three states of matter: solid (ice), liquid, and gas (water vapor).

As you look around the world we live in, almost everything you perceive is composed of matter. The exception is energy. **Energy** can be defined for our purposes as the capacity to do work. In biological terms, that work involves everything from throwing a telephone pole across a field in the Scottish game called "throwing the caber" to the slightest activity in a single cell. It is obvious that energy is important to life processes, and we will discuss energy in more detail in Chapter 6.

Atoms and Elements

■ **Atoms are the fundamental units of all matter. Elements are substances containing only one type of atom. There are 92 naturally occurring elements. The elements carbon, hydrogen, oxygen, nitrogen, sulfur, and phosphorus constitute the basis of 99 percent of all living matter.**

Of what is matter composed? Suppose you had a bar of pure gold (one should be so fortunate!) and divided it in half, and then the halves in half, and the halves of the half in half, and so on. Would you ever get to the point that if you made one more division you would no longer have gold? The answer is yes, and that point is the atom — that indivisible part of matter. An **atom** is the fundamental unit of matter — the smallest particle of an element that can enter into a chemical reaction. As early as 600 to 400 B.C., the Greek philosophers Leucippus, Democritus, and Epicurus were already formulating the concept of an atom, using philosophical logic rather than the experimental methods we now use to investigate the basic structure of matter. As a matter of fact, today we can actually see images of atoms (see Human Endeavors box in this chapter).

Why did I select a gold bar for my example? Gold is what is called an **element** — a substance made up of only one kind of atom. There are 92 elements that occur naturally on this earth, ranging from hydrogen, which is the smallest, to uranium, which is the largest. In addition to those 92, scientists have successfully produced at least 14 more elements in the laboratory. These 14, of which plutonium is one, do not occur in nature but must be produced under the artificial conditions of the laboratory. The elements that comprise 99 percent of the chemical makeup of living things are carbon, hydrogen, oxygen, nitrogen, sulfur, and phosphorus.

In order to make it easier to refer to elements, chemists have developed a shorthand of symbols to represent them. For example, in a chemist's shorthand the elements found most often in living things are represented by capital letters: C (carbon), H (hydrogen), O (oxygen), N (nitrogen), S (sulfur), and P (phosphorus). As you can see, the shorthand is usually based on the first letter (or first two letters) of the name of the element. This isn't always the case, however, since there are so many elements. Therefore, in some instances chemists use the letters of the Latin or Greek name. For example, *sodium* in Latin is *natrium*, so the symbol for sodium is Na. When the symbol has two letters, the second letter is always lowercase.

ATOMIC STRUCTURE

■ **Atoms are composed of a nucleus surrounded by one or more energy levels. The nucleus contains protons and neutrons, and the energy levels contain electrons, moving in orbitals at the speed of light.**

What are atoms composed of? If you were a chemist or an atomic physicist, your reply would be "hundreds of subatomic particles, from hadrons to quarks," and you would stress that atoms are extremely complex. However, for our purposes we need to consider only three subatomic units — neutrons, protons, and electrons.

We can describe the atom as consisting of two regions. First, there is a center core, or **nucleus,** in which we find the **protons** (positively charged particles) and **neutrons** (particles with no charge). The second is a region surrounding the nucleus where **electrons** (negatively charged particles) travel around the nucleus in paths called **energy levels.** Electrons travel at extremely high speed. In a sense, electrons form a cloud as they whiz around the nucleus (Figure 2.2). Each energy level is made up of

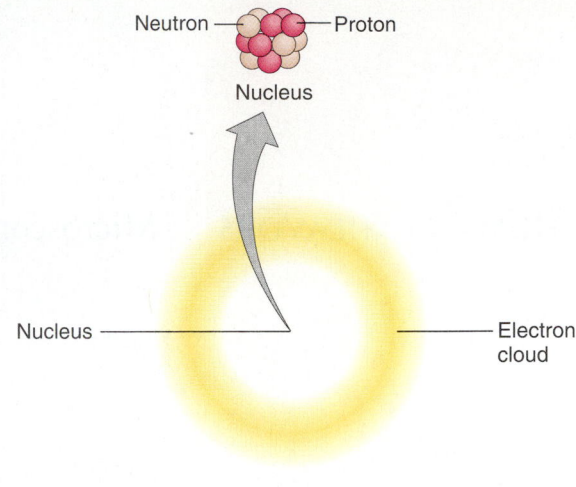

FIGURE 2.2   An Atom
Located in the nucleus of this atom of carbon are six protons and six neutrons. Its six electrons move rapidly about the nucleus and are represented by an electron cloud.

one or more **orbitals** — specific regions, at some distance from the nucleus, in which an electron is found at least 90 percent of the time. The shape of an orbital will vary, depending on how many electrons are whizzing around the nucleus. The atoms of different elements have different numbers of protons, neutrons, and electrons.

The nucleus of an atom is very dense, compared with the energy levels around it. Try to envision it in the following way. If you had an atom with a diameter as long as two football fields, the nucleus in the center would have a diameter only as thick as five dimes stacked together. All around the nucleus, electrons would be mere specks whizzing around to form a cloud. In reality, most of the structure of an atom is empty space.

Electrostatic Charges

■ **Electrostatic charges hold atoms together. Protons carry a positive charge, neutrons are neutral, and electrons are negative. Most atoms are electrically balanced and have an equal number of positive and negative particles.**

What holds atoms together? They are held together by electrical charges. The proton has a positive charge ($+$), hence its name. A neutron has no electrical charge, and the electron has a negative charge ($-$). As is the case with magnets, unlike charges attract — a positive charge attracts a negative one, and vice versa ($+$ to $-$). Again like magnets, like charges repel each other — a positive charge repels

Microscopes: From Animalcules to Atoms

From Leeuwenhoek's first report of microscopic animals (animalcules) in 1676 to the viewing of atoms in 1983, humans have endeavored to probe the invisible through the development of better microscopes and other technology.

In 1676 Antonie van Leeuwenhoek reported to the Royal Society in London that he had observed an invisible world made visible by the microscope, a world in which small animals, living creatures that he called animalcules, were alive in a drop of rainwater. Thus began the first step in a human endeavor that would lead to the cell theory, to germ theory, and eventually to our imaging of atoms and even to detecting the changes that occur in atoms as they undergo chemical reactions. Imagine, only a century ago scientists debated the existence of atoms, and until January 1983 no one had ever seen the image of an atom. Today, using modern scanning-probe microscopes, scientists are not only seeing atoms but can move them about. All this is due to our new microscopic technology.

Since the early seventeenth century, lens makers were viewing the microscopic world through their very primitive light microscopes, and the *light optical microscope* was the primary research tool of biologists until the 1930's. But visible light has limits as to how small an object you can see. This limit is due to a phenomenon known as *resolving power*. Resolving power is the ability to distinguish between two points and is due to the wavelength of light energy used. Light travels in waves, and the distance between the

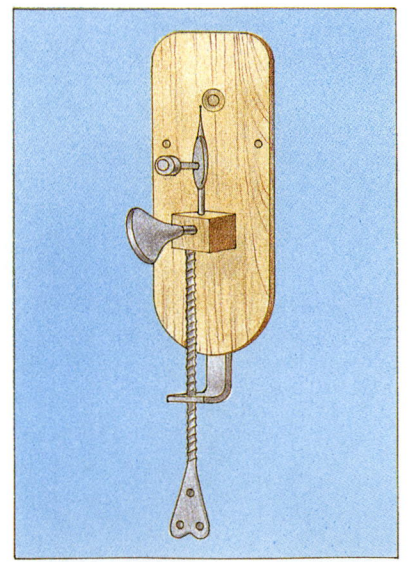

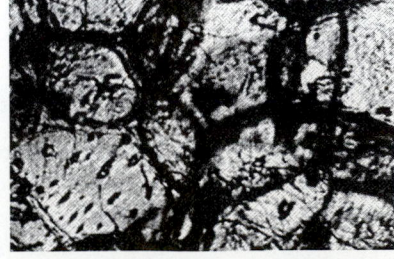

(a) Leeuwenhoek's microscope

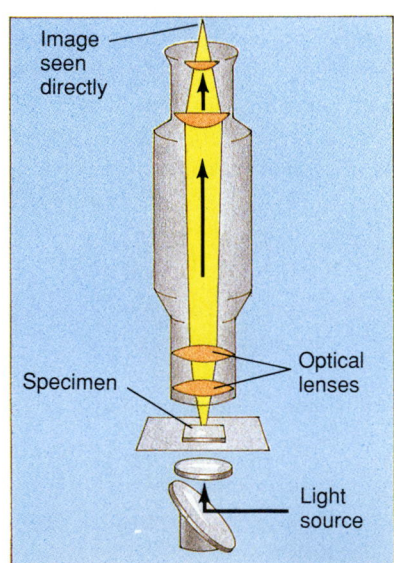

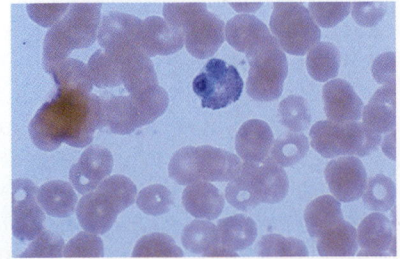

(b) Light optical microscope

Chapter 2   The Chemical Basis of Life

crests of the waves is known as its *wavelength*. The resolving power is equal to one-half of the wavelength of light used. For example, because visible light ranges from 380 to 720 nanometers, or billionths of a meter (a nanometer is equal to 1 billionth of a meter), visible light can only distinguish objects about 160 nanometers in diameter. Obviously, if you used a smaller wavelength, for example the wavelength of electrons (which is about 1 nanometer in length)—you can see objects 0.5 nanometers in size. In the 1930s, a new type of microscope was developed, the *electron microscope*, which could be used to see very small objects. Electron microscopes use electrons rather than light and magnetic lenses rather than the glass lens used in the optical microscope to view specimens, but the electron microscope also has its limitations. One limitation is that the electrons have to be beamed in a vacuum to avoid scattering. Because of this, the specimen has to be specially prepared and sectioned for viewing. For the most part, except when using the high-voltage electron microscope, no living specimens can be viewed. If you try to see atoms, because the electron beam is so concentrated, the energy of the beam would destroy the fragile atoms in organic molecules or pass right through them, showing nothing. Scientists also developed modifications of the electron microscopes, like the *scanning electron microscope*, which has the advantage of scanning the surface of the specimen with a narrow beam of electrons. This allows for the production of a three-dimensional image that can be viewed on a visual monitor (like a computer screen). Also, computer programs can enhance the image and even add color. But still, the limits of the wavelength prevent the viewing of atoms. But what if we imaged by feel (touch), not sight?

In 1981, Gerd Binning and Heinrich Rohrer in Zurich, Switzerland, developed the *scanning-probe microscope*. The scanning-probe mi-

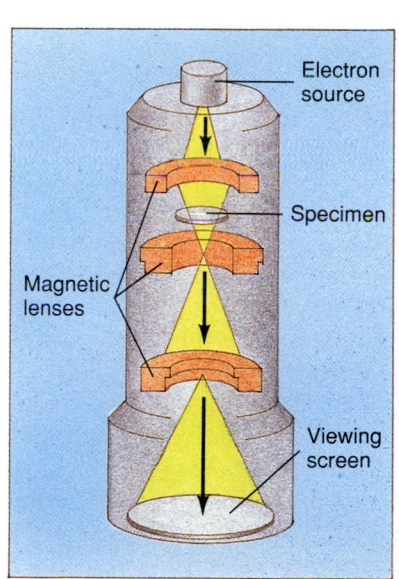

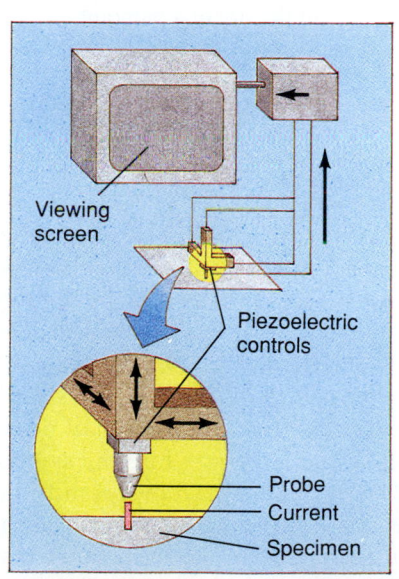

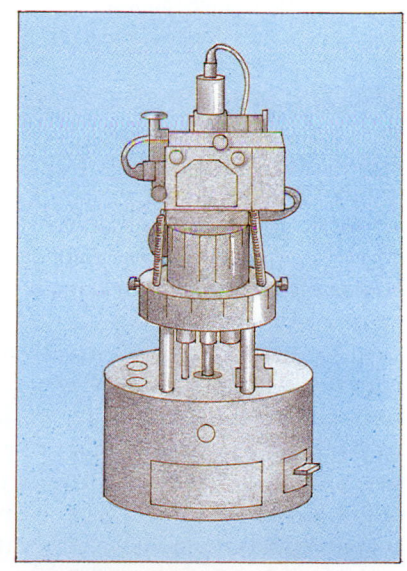

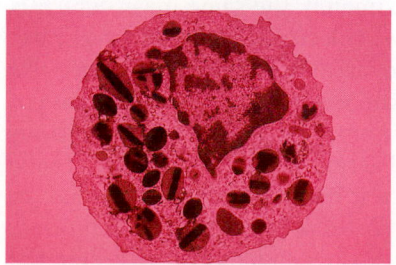

(c) Electron microscope

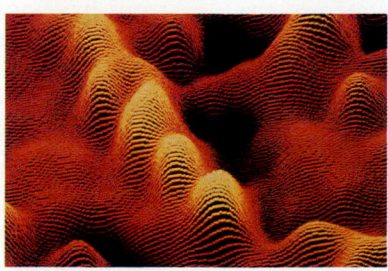

(d) Scanning-tunneling microscope

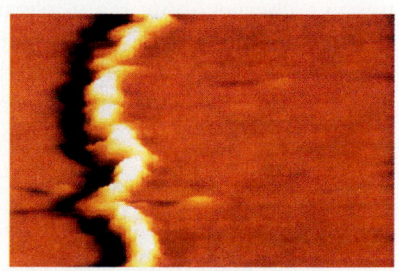

(e) Atomic-force microscope

croscope is based on the principle that if you place a probe near the surface of a sample and measure the distance between the probe and the sample, you can produce an accurate image of the sample. It is sort of like being in the dark and running (or scanning) your hands over the surface of eggs sticking out of an egg carton to detect the nature of the surface of the eggs, rather than using light to see them. The probes are like your hands, and they detect the peaks and valleys of the image; as you move your hands, you can scan the sample, but in order to have an accurate image of atoms, you have to be able to move the probe in controlled steps, each step less than the diameter of an atom. The technological breakthrough that allowed for that exact scanning was the development of piezoelectric crystals—material that can change shape as electric current is applied, which is translated into minute movement of the probe. Three sets of crystals are used to allow for a three-dimensional image. Today, using a fine-wire platinum probe and piezoelectric crystals, scientists are literally tunneling into atoms. The *scanning-tunneling microscope* charts the shape of the surface being scanned, just like you touching the eggs to detect their size, shape, number, and so on. And the computer message, the electrical current between the probe and the specimen, is used to produce an image of the atom or molecule. But what if the specimen does not conduct electrons? Today that limitation of the scanning-tunneling microscope has been overcome by the *atomic-force microscope*, developed in 1985. The atomic-force microscope uses a diamond point glued to a piece of silicon about the size of a pinpoint. This forms a silicon cantilever, sort of like a diving board. The repulsive forces produced by the electron cloud of the atom bend the silicon cantilever. A laser beam can then measure the amount of bending and produce the image of the specimen. Using this microscope, scientists can image very delicate molecules like DNA and other organic materials and even film the molecular events that lead to a biological process, like the formation of protein fibers in a blood clot.

As you can see, we literally can image and see atoms. But this is just the beginning. In Focus 3.1, we will discuss how this technology at the nanometer level is rapidly evolving into nanotechnology, where we can actually micromanipulate atoms and image the changes in atoms as chemical reactions occur.

Source: James Trefil, "Seeing Atoms," *Discover,* June 1990.

---

a positive (+ to +), and a negative charge repels a negative (− to −). These same forces are responsible for the energy relationships that exist within the structure of an atom.

It is the attraction between the positively charged protons and the negatively charged electrons that keeps the electrons from flying off into space. When an atom has the same number of protons and electrons, it is said to be *electrically balanced,* and is called a neutral atom. The number of positive charges equals the number of negative charges. For example, since a carbon atom has six protons, or positive charges, how many electrons, or negative charges, would it need to be electrically balanced? As we would expect, carbon, with its six protons, does have six electrons orbiting the nucleus.

Atomic Number and Atomic Mass

■ **The number of protons in an atom is called its atomic number. The atom's protons and neutrons added together determine its atomic mass. When the atoms of an element have differing atomic masses, each atom is called an isotope.**

The number of protons in an atom is called the **atomic number.** Every atom of a particular element has the same atomic number. For example, each atom of the element hydrogen has one proton, so it has an atomic number of 1, expressed as $_1H$. Carbon has six protons and is expressed as $_6C$. Note the small number that is set slightly low, right before the symbol for the element. Atomic number is always indicated by a subscript preceding the symbol of the element.

In most instances, the number of protons and neutrons in an atom is the same. Protons and neutrons have approximately the same mass, which is arbitrarily called 1. (Electrons also have mass, but it is so small compared with that of protons and neutrons that we can ignore it.) The **atomic mass** of an atom is the number of protons and neutrons added together. Therefore, carbon has an atomic number of 6 (since it has 6 protons), but it has an atomic mass of 12 (6 protons plus 6 neutrons), which we indicate with a superscript preceding the chemical symbol: ^{12}C. Table 2.1 lists the chemical symbol, atomic number, and atomic mass of some biologically important elements.

Chapter 2   The Chemical Basis of Life

TABLE 2.1
Some Essential Elements in Living Organisms

Element[1]	Symbol	Atomic number	Atomic mass[2]	Examples of biological roles
Carbon	C	6	12.0	Backbone of organic molecules
Hydrogen	H	1	1.0	Component of water and of most organic molecules; used in cellular respiration
Oxygen	O	8	16.0	Component of water and most organic molecules; used in cellular respiration
Phosphorus	P	15	31.0	Energy-rich bond of ATP
Potassium	K	19	39.1	Generation of nerve impulses
Iodine	I	53	126.9	Component of a hormone
Nitrogen	N	7	14.0	Component of proteins and nucleic acids
Sulfur	S	16	32.0	Component of many proteins
Calcium	Ca	20	40.0	Component of bone; functions in muscle contraction
Iron	Fe	26	55.8	Component of oxygen-carrying pigment of many animals
Magnesium	Mg	12	24.0	Component of photosynthetic pigment; essential to some enzyme mediation
Copper	Cu	29	63.5	Component of oxygen-carrying pigment of mollusk blood
Zinc	Zn	30	65.4	Essential to functioning of the alcohol-oxidizing enzymes
Molybdenum	Mo	42	95.9	Essential to some enzyme reactions
Chlorine	Cl	17	35.5	Essential in digestion in humans
Cobalt	Co	27	58.9	Components of vitamin B_{12}

[1] There are many ways to arrange elements. This table is arranged to help you remember these essential elements. Just remember the following slogan:

See	Hopkins	Cafe Management:	Cousin	Moe	Chlorine	Company
C	HOPKINS	CaFe Mg	CuZn	Mo	Cl	Co

[2] Rounded to the nearest tenth. It should be noted that atomic mass varies among the atoms of a specific element because not all atoms of a specific element contain the same number of neutrons (remember isotopes). Therefore, the atomic mass of some elements is expressed as a whole number plus a fraction.

Isotopes

In nature, each atom of an element does not always contain an equal number of protons and neutrons. In the case of carbon, for example, there are usually 6 neutrons, but sometimes carbon atoms will have as many as 7 or 8 neutrons. Carbon 14, for example, has 6 protons in its nucleus, but it has 8 neutrons. This atom, therefore, has an atomic mass of 14 (6 protons plus 8 neutrons). Carbon can have an atomic mass of either 12, 13, or 14. When the atoms of an element have differing atomic masses, each atom is called an **isotope.** Carbon, for instance, has three isotopes: ^{12}C, ^{13}C, and ^{14}C.

Carbon 14 is radioactive — that is, its nucleus is unstable and gives off energy. Hence carbon 14 is called a **radioisotope** — an isotope that emits energy. The fact that this substance is radioactive makes it very useful in scientific research, because the energy emitted by radioactive atoms can be detected from a distance, just like the sound emitted by a cow's bell. By putting a "bell" (actually it is called a "tag") on a molecule with a radioactive isotope, scientists can follow it around by detecting the energy produced by the isotope as it moves through the body. With the help of isotopes, scientists also can develop techniques to measure small amounts of chemicals in an organism (see Focus 2.1 and Profile).

FOCUS 2.1
Use of Radioisotopes in Biological Research

When most people think of radiation, they usually picture it in its most destructive form — that emitted from a nuclear bomb. However, controlled radiation has its positive uses. As we discuss in this chapter, all atoms of an element have the same number of protons and electrons. However, not all atoms of a given element have the same atomic mass (remember that *mass* equals the number of protons plus the number of neutrons). For example, the most common mass number of hydrogen is ^1H, because it has only one proton in its nucleus. Sometimes, however, hydrogen also has a neutron in addition to its proton; hence its mass is ^2H, and it is called deuterium. In other instances, hydrogen has two neutrons, so it has a total mass of three (^3H). This isotope is called tritium.

Multiple forms of the same element, which are due to a change in the number of neutrons in the nucleus, are called *isotopes*. Some isotopes are unstable — that is, they *decay*, or change to achieve a more stable form. To decay, isotopes spontaneously give off energy or subatomic particles. In other words, they emit radiation, which is why they are called *radioisotopes*. Tritium (^3H) or carbon 14 (^{14}C) are two radioisotopes that you may have heard of. The released energy or subatomic particles can be detected by various counting instruments, such as Geiger counters or devices like photographic film. Their detectability makes them extremely valuable in biological research.

One of the major uses of radioisotopes is to label, or "tag," a molecule. You can label any molecule with radioactive isotopes and then trace its pathway through a cell or organism. For example, let's say that you allow a bean plant to take up radioactive phosphorus (^{32}P) through its roots for a period of time. If you then press the plant onto a sheet of flat photographic film in the dark, the radiation emitted by the radioisotopes will develop the film, revealing the path that the liquid has taken through the plant. This process, called *autoradiography*, is a useful technique for locating the position of a radioactive substance (Figure A).

Radioisotopes can be used to study many molecular processes. For example, if researchers inject radioactive amino acids (the molecular subunits that make up proteins) into a growing organism or cell, they can then isolate proteins from various tissues and find the proportion that are radioactive. This will help them determine the rate of protein synthesis. Radioisotopes are also very useful in medicine, not only as diagnostic tools but also as a means

Overview

At this time a short summary may be useful. An atom is the fundamental unit of matter. At the center of an atom is a nucleus, which is made up of positively charged ($+$) protons and uncharged neutrons. Surrounding the nucleus are energy levels in which negatively charged ($-$) electrons whiz around the nucleus in a general path called an orbital. Usually, for any atom, the number of protons equals the number of electrons. The atomic number refers to the number of protons. The number of protons plus neutrons is referred to as the atomic mass. The chemical notation $^{12}_{6}C$ indicates a carbon atom having an atomic mass of 12 and an atomic number of 6. Differing numbers of neutrons, and hence differing atomic masses, produce different isotopes of the same element.

of destroying cancerous tissues and as an aid in other methods of cure (Figure B).

Because isotopes emit energy in order to achieve a more stable form, they change from one isotope to another at fixed rates. Since we know the rate of isotope decay, we can determine the age of fossils and rocks by measuring the relative amounts of carbon radioisotopes in a sample. For example, ^{14}C has a half-life of 5700 years. So by determining the ratio of ^{14}C to nonreactive ^{12}C in a sample, we can find out how old it is. Using this method, scientists were able to determine the age of the Dead Sea scrolls with an accuracy of $+5$ to -5 percent.

Radioisotopes are very useful for tracing and identifying molecules as they pass through organisms and for determining the age of substances. This makes radioisotopes fundamental to a number of biological techniques. Much of the spectacular progress made in cell biology, biochemistry, and molecular genetics is a direct result of the use of radioisotope techniques. Throughout this text you will find references to experiments using valuable controlled radioactive isotopes.

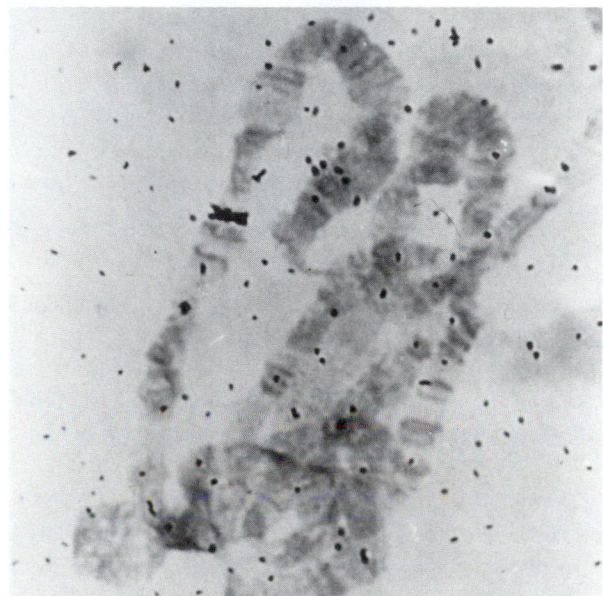

FIGURE A
The darkened areas on these chromosomes from a fruit fly indicate the presence of radioactive DNA. The DNA was tagged with radioactive hydrogen (^3H). When the DNA was placed on some unexposed film, the products of the radioactive decay of ^3H exposed the film leaving the dark areas—an autoradiograph.

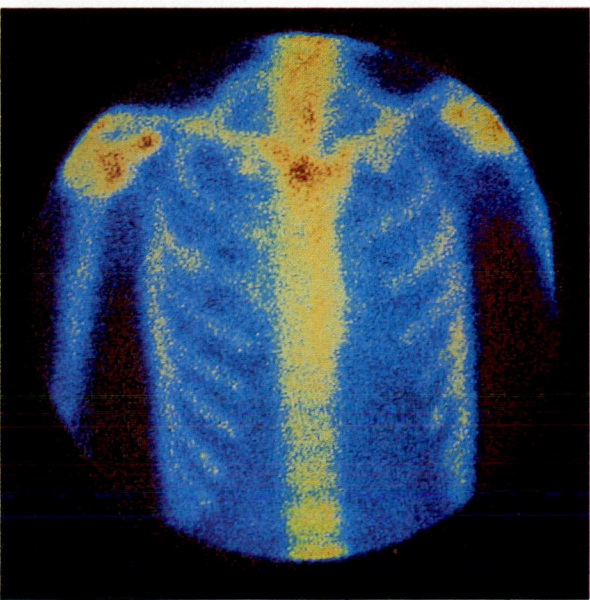

FIGURE B
Through the use of radioactive substances physicians can view structures within the body. This front-view, false-color bone scan of a normal chest illustrates the concentration of a radioactive substance in the bone.

ATOM-ATOM INTERACTIONS

■ The first energy level of an atom can hold up to two electrons, the second up to eight, and subsequent levels can hold more than eight. Above the first energy level, all levels having eight electrons act as if they were filled. Valence electrons determine the atom's reactivity.

Do atoms always exist alone in living or nonliving things? Obviously they do not. It is the ways that atoms interact that determine the nature of the substances that make up the living world.

Energy Levels

In order to understand how atoms join, we have to look again at their structure. Remember that we said that electrons whiz around the nucleus of the

Dr. Rosalyn Yalow
Nobel Laureate, Physicist, Biomedical Researcher

On a not-so-unusual Saturday morning, Rosalyn Yalow takes time from her very hectic schedule to talk to gifted elementary and high school students in Manhattan about her life in science. When these young students ask, "Why did you enter science?", she replies, "I wandered into it oddly, because of things I couldn't do. I couldn't draw. I'm tone-deaf, so that ruled out music. I've no athletic talents. My chemistry teacher taught me the joys of working with my hands and brains."

Truthful, candid words from a gifted woman. In 1977, Rosalyn Yalow was awarded the Nobel Prize for her development of and subsequent work with a technique called radioimmunoassay (RIA). In RIA, radioisotopes are used to measure very small amounts of hormones (chemical messengers of the body) and other chemicals within an organism. The method is so sensitive that it is similar to being able to measure the equivalent of a cup of sugar in Lake Michigan. Today, by using RIA, scientists can detect abnormalities in the levels of hormones and other chemicals in the body.

Rosalyn Yalow was born of immigrant parents in the Bronx. At nine-teen she graduated Phi Beta Kappa from Hunter College in Manhattan. As a chemistry and physics major, she should have had a bright future, but because it was the Great Depression and because she was a Jew and a woman, her opportunities were limited. Eventually in 1941 she was accepted into graduate school at the University of Illinois — the first woman accepted since 1917. There she met her husband-to-be, Aaron Yalow, a physics graduate student. His interest in medical physics helped plant the seeds that eventually led Rosalyn into biomedical research.

After graduation Yalow worked at a hospital using radioisotopes to treat cancer, and that led to her interest in using radioactive material to trace chemical reactions. She began to collaborate with Solomon Berson, and together they developed the RIA technique while investigating diabetes (a disease in which individuals cannot make or use the insulin that is required for the body to utilize sugar properly).

Yalow's work does not make up her entire life, however. She also has two children, a son Benjamin and daughter Elenna. In 1952, Yalow ignored a 1952 Veterans Administra-tion rule that pregnant women take a leave of absence after their fifth month of pregnancy. Instead, she worked until her son was born, took one week off, then was back on the job.

In a sense, Yalow has always been an outsider in the male-dominated world of biomedical research. She could never be one of the "good old boys." However, her devotion to the search for scientific truth has truly earned her the respect of all of her colleagues.

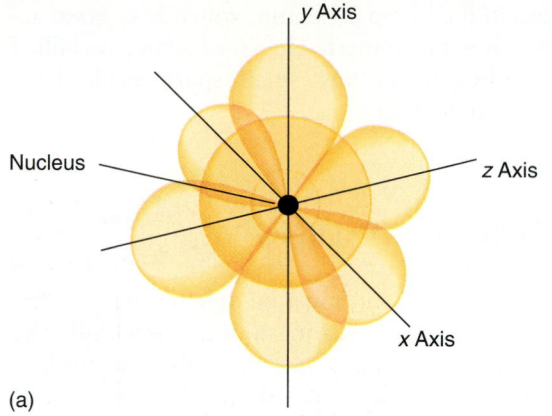

Nucleus

y Axis

z Axis

x Axis

(a)

FIGURE 2.3   Orbitals
(a) The energy levels of atoms are divided into orbitals.
(b) The first energy level consists of only a spherical orbital that can contain two electrons. The second energy level consists of four orbitals, each of which can contain two electrons, for a total of eight.

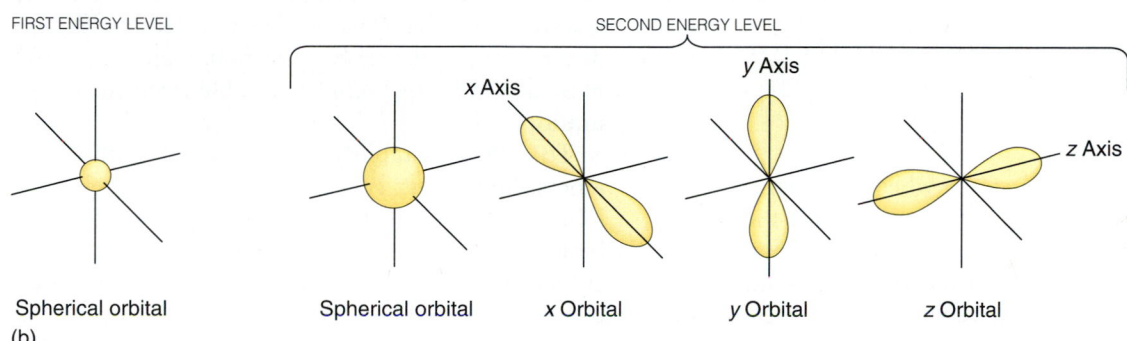

FIRST ENERGY LEVEL   SECOND ENERGY LEVEL

x Axis   y Axis   z Axis

Spherical orbital   Spherical orbital   x Orbital   y Orbital   z Orbital
(b)

atom in what are called orbitals. Orbitals follow a specific pattern, and in each orbital electrons are distributed in a specific pattern. What's more, there is a maximum number of orbitals that an energy level can contain. It is the interaction of these orbitals that determines the shape of the atom.

The lowest energy level (often called the *first energy level*) has a single spherical orbital that can hold a maximum of *two electrons*. The *second energy level* has four orbitals, and each orbital holds

a maximum of two electrons, for a total of eight at that level. One of these orbitals is spherical; the other three are dumbbell-shaped and designated as *x*, *y*, or *z* (Figure 2.3). Subsequent energy levels can contain more than eight electrons, but whenever the outermost energy level of an atom contains exactly eight electrons, this atom is usually considered to be stable. A stable atom is nonreactive — that is, it does not usually combine with other atoms to form molecules.

TABLE 2.2
Electron Distribution in the First Two Orbitals of Selected Elements

Elements	Atomic number	First energy level	Second energy level			
		Spherical orbital	Spherical orbital	x Orbital	y Orbital	z Orbital
Hydrogen	1	1				
Helium	2	2				
Carbon	6	2	2	1	1	
Nitrogen	7	2	2	1	1	1
Oxygen	8	2	2	2	1	1
Neon	10	2	2	2	2	2

Atom-Atom Interactions

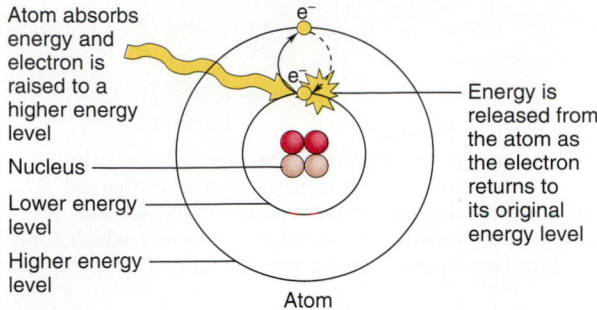

FIGURE 2.4 Energy and Energy Levels
When energy strikes an atom or molecule, the energy is absorbed, electrons are moved to higher energy levels, and the energy is stored. When the electrons return to their original energy level, energy is released. Photosynthesis can serve as an example of this energy relationship.

An atom is most stable when all of its electrons are at their lowest energy level. Hence the energy levels are filled in order. The spherical, first energy level is filled first, followed by the second energy level, and then each succeeding energy level in order (see Table 2.2).

As we said earlier, an atom is held together by the attraction between its positively charged nucleus and its negatively charged electrons. It takes energy to pull the electrons and the nucleus apart. Thus it takes energy to move an electron to a higher energy level — that is, to move it farther from the nucleus (Figure 2.4). As we will see in later chapters, this fact is crucial for the chemical processes of life. Energy is stored in atoms when electrons move to higher energy levels. Energy is released when electrons move to lower energy levels.

Valence Electrons

The outermost energy level is referred to as the *valence energy level,* and the electrons in this outermost orbital are called *valence electrons.* It is primarily the arrangement of the valence electrons that determines the reactivity or chemical properties of the atom. For example, the hydrogen atom in its common form contains only one proton and one electron. Having only one electron in its first energy level, hydrogen is quite reactive. On the other hand, helium has two protons, two neutrons, and two electrons. Because its outermost energy level is full, helium is not reactive under normal circumstances (Figure 2.5). The difference between these two elements explains why the Goodyear blimp will never explode as the German zeppelin *Hindenburg* did.

The gas in the blimp is helium, which is so nonreactive that it is called inert. The *Hindenburg* was filled with hydrogen, so any small spark could have caused it to explode.

CHEMICAL BONDING

You can see why you must be able to visualize the distribution of electrons in an atom, especially the valence electrons, since their number determines how one atom will react with another. What happens when an atom with only one energy level has less than two electrons, or when an atom with two or more energy levels has less than eight electrons? How does an atom achieve a stable electron configuration?

To achieve stability, atoms unite with other atoms. Sometimes two atoms of the same element join together to form a molecule of that element. In biology, one of the most important of such molecules is hydrogen. A hydrogen molecule is indicated by the notation H_2. The subscript following the symbol indicates the number of atoms in the molecule.

When the atoms of two or more elements combine chemically in a definite ratio, they form a **compound.** For instance, two hydrogen atoms and an oxygen atom produce the compound we know as water (H_2O). When atoms combine to form compounds, they do so in a specific way, according to their valence electrons. The compounds formed by the elements have different chemical properties than the elements have when functioning alone.

In order to form compounds, atoms are held together with **chemical bonds.** How do chemical bonds form, and what are they? Chemical bonds are not actually objects. Instead they represent an

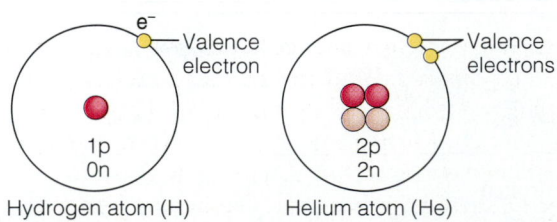

FIGURE 2.5 Valence Electrons
The outermost orbital of an atom determines its reactivity. An atom with less than a full complement of valence electrons (like H) is more reactive than one that has a full valence orbital (like He).

Chapter 2 The Chemical Basis of Life

energy relationship between the atoms involved in a compound. Bonds describe what is called the "attractive force" between two or more atoms.

The attractive force is the glue that holds the atoms together. At some time you may have done some playing around with iron filings and magnets. If you spread the filings over a piece of paper and then put the magnet under the paper, the filings respond to the attraction of the magnet by making a design on the paper. If you put two magnets beneath the paper, the patterns made by the magnetized filings allow you to see how the magnets are attracting each other. Those patterns are a concrete example of the attractive forces between the magnets at work.

The chemical bonds between atoms are similar to the patterns of those iron filings. They are electrostatic attractions between one or more atoms. There are two types of chemical bonds between atoms that are important in biology: the *ionic bond* and the *covalent bond*.

Ionic Bonds

■ **An ion is an atom or molecule that has gained or lost one or more valence electrons and therefore possesses a charge. Ionic bonds, which are relatively weak, are formed by the electromagnetic attraction of opposite-charged ions.**

Remember that we said earlier that most atoms are electrically neutral. That is, the number of positive charges equals the number of negative charges. As a result, the electrons of an atom stay with it because they are attracted by the positive charges of the protons in the nucleus.

However, sometimes this attraction can be overcome. One or more electrons can be stolen away from the valence energy level of an atom. Or an atom can gain one or more electrons. Atoms that have lost or gained electrons and are therefore positively or negatively charged are called **ions**.

Ions. When such a change takes place, the atom becomes *charged*. That means that now one of its protons is not balanced by a negative charge, or one of its electrons is not balanced by a positive charge. Whether the atom (or group of atoms) lost or gained electrons, it is now called an ion. For example, a hydrogen atom consists of one proton and one electron. If it loses its electron, it exists as a hydrogen ion. It now has a positive charge (because of the proton in the nucleus) but no negative electron to balance that charge. To indicate its status as

an ion, a plus sign is added to its chemical symbol, so the notation for this atom is now H^+. As a matter of convention, negatively charged ions are called *anions* and positively charged ions are called *cations*.

Ionization. When one atom loses or gains an electron, the process is called ionization. This process involves a complete transfer of electrons between atoms. A classic example of ionization is that of sodium (Na) and chlorine (Cl).

First, in order to visualize the changes that take place in a sodium atom, you must understand its structure. Sodium has an atomic number of 11 (in other words, it has 11 protons). It also has 11 electrons, so it has a net charge of 0. According to what we have learned so far, its first energy level will contain 2 electrons, and there will be 8 in the second. The third energy level, then, will contain only 1 electron. In order to have a stable structure, with an outermost energy level containing 8 electrons, would it be easier to lose 1 electron or to gain 7? Obviously, it is easier to lose 1, and that is what sodium does. As a result, it now becomes an ion having 11 positive charges and only 10 negative charges. The atom's net charge (protons minus electrons) is plus 1, so in chemical shorthand the notation for this charged atom is now Na^+, since it is an ion.

Chlorine, on the other hand, has 17 protons and 17 electrons. Like sodium, it has a net charge of 0. There are 2 electrons in its first energy level, 8 in its second, and 7 in its third. Chlorine too requires an outer energy level containing 8 electrons, and thus it must gain 1 electron or lose 7. It most often gains 1, so it also becomes an ion. When it gains that extra electron to fill its valence orbital, it has 17 (+) charges in its nucleus, but it has 18 electrons for 18 (−) charges. The atom, now an ion, has a net charge of minus 1, written as Cl^- and called a chloride ion.

Thus sodium's electron is transferred to chlorine. Because these two ions now have opposite charges, they are electrostatically attracted to one another. This attraction is called an **ionic bond** and the result is the compound NaCl, which is common table salt (Figure 2.6).

Ionic bonds are relatively weak. As a result they can be broken when placed in water, because water pushes the atoms apart. When the atoms are separated, they are said to *dissociate* to become *free ions*. The relative ease with which ionic bonds dissociate is important in living things, because an ionic **solution** (a uniform mixture of two or more

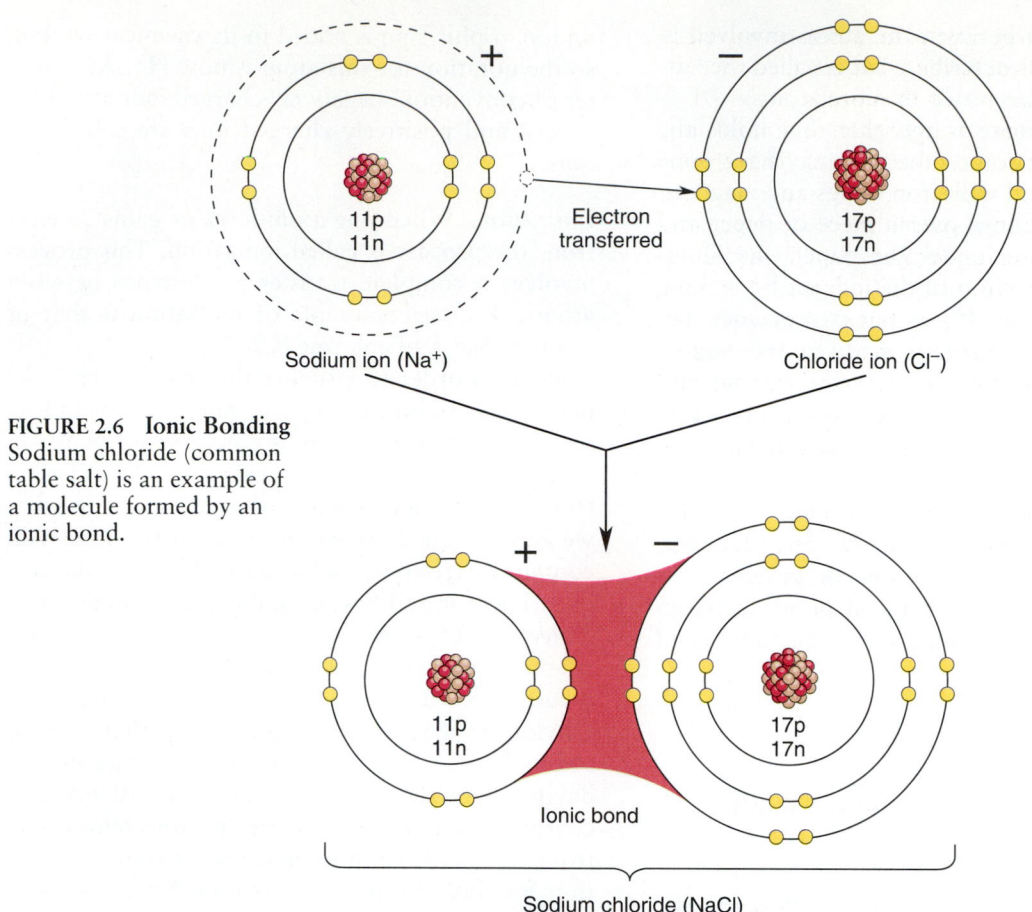

FIGURE 2.6 Ionic Bonding Sodium chloride (common table salt) is an example of a molecule formed by an ionic bond.

Sodium ion (Na⁺)

Chloride ion (Cl⁻)

Electron transferred

11p
11n

17p
17n

+ −

11p
11n

17p
17n

Ionic bond

Sodium chloride (NaCl)

types of molecules), often called an *electrolytic* solution, conducts electric current. What is more, as we will see in later chapters, the movement of ions into and out of cells is a very necessary process in sustaining life.

Covalent Bonds

■ **Covalent bonds are bonds that are created when atoms share one or more valence electrons. These bonds are stronger than ionic bonds. The number of dashes present in a structural formula represents the number of pairs of shared electrons.**

Keep in mind that an ionic bond involves a complete transfer of electrons between atoms. This characteristic is in contrast to another type of chemical bond, one that is probably the most common in living systems. (Ionic bonds, on the other hand, are more characteristic of inorganic compounds.) A **covalent bond** is formed by the cooperative *sharing* of valence electrons between two or more atoms, in order to satisfy the orbital configuration of the

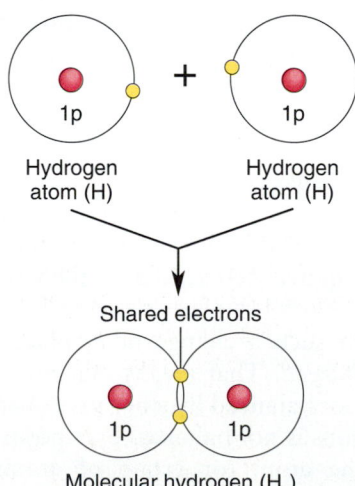

Hydrogen atom (H)   Hydrogen atom (H)

1p   1p

+

Shared electrons

1p   1p

Molecular hydrogen (H₂)

FIGURE 2.7 Covalent Bonding A molecule of hydrogen is formed when two hydrogen atoms share their valence electrons, forming a covalent bond.

atoms involved. Atoms held together by covalent bonds form **molecules.** Because the electrons are attracted to the nuclei of all the atoms involved, a covalent bond is much stronger than an ionic bond. For instance, hydrogen, with its single electron in the first energy level, rarely exists as just a single atom. Instead, hydrogen atoms usually form covalent bonds and exist in pairs, as H_2 (Figure 2.7). The two electrons travel around both nuclei, so that in effect each atom has a first energy level with two electrons in it — its outermost energy level is full.

Another example of a covalent bond is the methane molecule CH_4 (Figure 2.8). Each hydrogen atom has one electron in its outer shell, and carbon has four electrons in its outer shell. By sharing electrons, each hydrogen atom has two electrons in its outer shell (its first energy level), and the carbon has eight in its outer shell (its second energy level).

When chemists refer to the hydrogen molecule, they often write out the notation in what is called a structural formula, as H—H. The dash indicates that the two hydrogen atoms share a pair of electrons with each other. The notation H—H means the same as H_2, but, as you can see, a structural formula provides a more specific representation of the atom's position within the molecule.

Atoms sometimes share more than one pair of electrons. When two atoms share two pairs of electrons, a double bond is formed. When they share three pairs, a triple bond is formed. For example, oxygen has six electrons in its outer shell; it needs two to make a complete shell of eight. If two oxygen atoms share two pairs of electrons — that is, form a double bond — each will complete its outer shell. The structural formula for this molecule is O=O (Figure 2.9).

The Carbon Base of Life Forms

■ **A carbon atom contains four valence electrons, which can bond with up to four other atoms. Carbon atoms can join in chains or rings of various shapes. To a large extent, the shape of a molecule determines its chemical properties.**

Substances that contain carbon are called *organic,* because all carbon-based substances had their origin in living things. The life forms that have evolved on this planet are said to be carbon-based — in other words, the element carbon has been the primary building block of living things on earth. (We will discuss this in detail in Chapter 16.) But with 92 naturally occurring elements, why is life on this earth based on carbon? What is so unique about

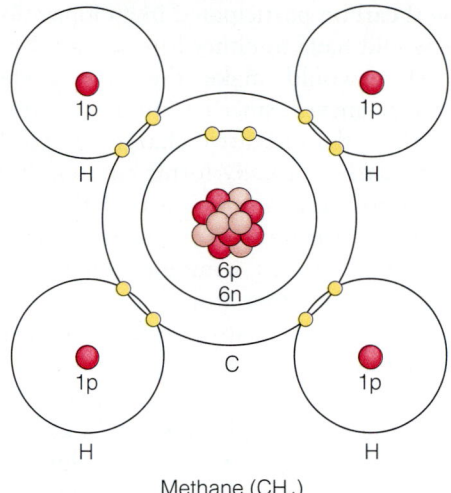

Methane (CH_4)

FIGURE 2.8 Methane
In methane, one carbon atom and four hydrogen atoms share their valence electrons, forming four covalent bonds.

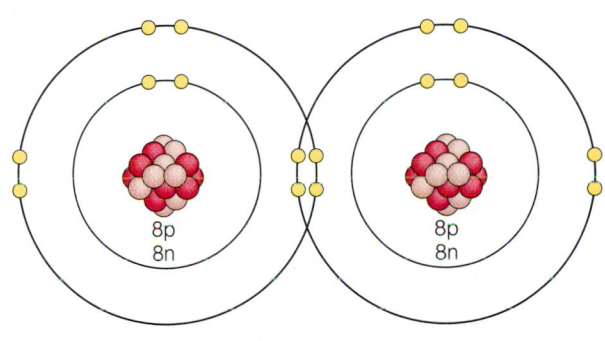

Molecular oxygen (O_2)

FIGURE 2.9 Double Bonding
Two oxygen atoms share two pairs of electrons between them to form a double bond.

this atom? To understand that you must look carefully at carbon's structure.

Carbon's Valence Electrons. Carbon's uniqueness and versatility are due to its electronic configuration — that is, the arrangement of electrons in its outermost energy level. Remember that we said earlier that carbon has six protons and six electrons. The first energy level contains two electrons, so how many electrons does the second energy level have? Correct, it has four. Because of this fact, carbon can either gain or lose four electrons in order to satisfy its electron requirements, and that fact makes carbon a very versatile atom.

Now if carbon participated in an ionization process, it would have to either lose or gain four electrons. That would make the atom's electrical balance very uneven, since it would have either four more positive than negative charges, or vice versa. Therefore, carbon usually forms covalent bonds — it *shares four electrons*. Because of this ability to share four electrons, carbon is the most versatile of all the atoms. It can combine convalently very readily with many other atoms.

For example, as we saw earlier, carbon can combine with four hydrogen atoms to form methane (swamp gas), a type of compound called a hydrocarbon, because it is made up of hydrogen and carbon. (There are numerous other hydrocarbon molecules besides methane.) Or carbon can join with two oxygen atoms to form carbon dioxide ($O=C=O$). Carbon bonds very readily with other elements as well, such as nitrogen, sulfur, chlorine, and phosphorus. It can also combine with one or more other carbon atoms, by forming a single ($C—C$), double ($C=C$), or triple ($C≡C$) bond. You will see part of the importance of this fact when we discuss carbon chains in saturated and unsaturated bonds in the next chapter.

Carbon also bonds with functional groups. A **functional group** is a group of atoms that provides a distinctive chemical property to any molecule of which the group is a part. For example, OH is what is known as a *hydroxyl group*, which is sometimes called an alcohol group. A hydroxyl group imparts its properties to all alcohols. When a hydrogen atom of methane (CH_4) is replaced by a hydroxyl, it becomes methanol (CH_3OH), or wood alcohol, which is poisonous. When ethane (C_2H_6) has a hydrogen atom replaced by a hydroxyl, it becomes ethyl alcohol (C_2H_5OH), the grain alcohol contained in vodka. We will take a closer look at functional groups in Chapter 3.

Thinking of Carbon-Based Molecules in Three Dimensions. To picture a carbon-based molecule accurately, you have to realize that the structural formula of molecules actually represents a three-dimensional object (Figure 2.10), even though, for convenience, they are usually drawn in only two dimensions. As we'll see in the next chapter not only can carbon combine with a variety of other atoms and with functional groups, but the combinations come in a variety of shapes. In some molecules, carbon atoms form a basic backbonelike arrangement for an array of atoms or functional groups that branch out to one side or another. At other times, carbon atoms are bonded to each other to form rings:

$$
\begin{array}{ccc}
 & C—C & \\
 & | & \\
C—C—C—C & \\
 & | & \\
 & C &
\end{array}
\qquad
\begin{array}{ccc}
 & C & \\
C & & C \\
| & & | \\
C & & C \\
 & C &
\end{array}
$$

The shape of a carbon-based molecule is important because, as you will learn in the next chapter, some molecules (called *isomers*) have the same number of atoms, but because their shapes are different, they exist as different molecules. As a result, they react with other molecules in different ways, and they serve different purposes.

So the importance of the chemistry of carbon is its ability to combine in several ways with itself, other atoms, and molecules. As a matter of fact, because of their significance in living things, carbon-containing molecules are called **organic** (life-form–related), as opposed to inorganic (noncarbon, nonlife molecules). Life is truly carbon-based.

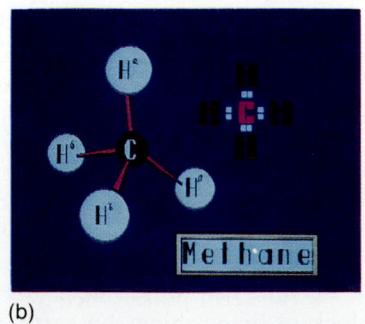

Structural formula
(a)

Ball-and-stick model

Space-filling model

(b)

FIGURE 2.10 Different Representations of Methane and Other Molecules
(a) Although the structural formula of methane shows the bonding of atoms in the molecule, the three-dimensional structure can be better shown by spatial models or computer modeling. (b) Computer model of a methane molecule.

Polar Covalent Bonds

■ **In some molecules, atoms share electrons unequally, developing areas that have slightly positive and negative charges, which form a polar covalent bond.**

When atoms form covalent bonds, they often share the electrons equally, so that the charges are equally distributed. However, in some special cases, the shared electrons are drawn to one section of the molecule because the electrostatic pull of that section is stronger. In these cases, the bond is called a **polar covalent bond.**

For instance, think about the structure of a water molecule. Most of us already know that water is H_2O — two hydrogen atoms and one oxygen molecule. And we know that hydrogen has an atomic number of 1. Oxygen has an atomic number of 8, which means that it has 2 electrons in its first energy level and 6 electrons in its second energy level. In a water molecule, oxygen and two hydrogen atoms form covalent bonds in order to fill oxygen's second energy level and hydrogen's first energy level. The structure that results is represented in this way:

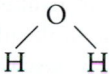

However, even though the three atoms share electrons in order to fill their energy levels, the oxygen nucleus, with its eight positively charged protons, has a stronger attraction for electrons than the hydrogen, with only one proton. As a result, the majority of the electrons are near the oxygen portion of the molecule, which gives that region a slight negative charge. (Remember, electrons have a negative charge.) The two hydrogens, on the other hand, have fewer electrons near them, so they have a slight positive charge. As you can see in Figure 2.11, the water molecule assumes a shape having four corners, with two positively charged regions and two negatively charged regions. In other words, there is an unequal distribution of charges in the molecule.

Because given portions of the molecule have differing electrostatic charges, the molecule is said to be **polar,** and the bonds are said to be *polar covalent bonds.* (The use of the word *polar* here is meant in the same sense as the term *poles* when you are dealing with magnets.) The polarity of the water molecule leads us to the discussion of the next type of bond that is important in biological systems, the hydrogen bond.

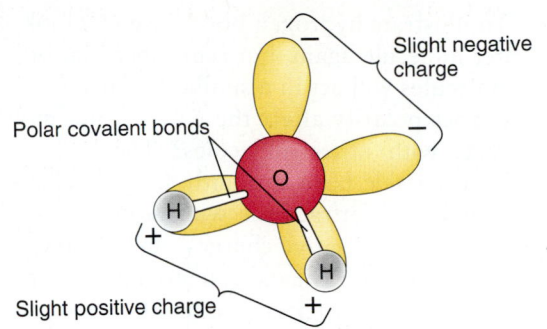

FIGURE 2.11 Polar Covalent Bonds
The polar nature of the water molecule is due to an unequal distribution of electrons. Oxygen with its eight protons exerts a greater pull on the shared electrons, giving the oxygen "end" of the molecule a slightly negative charge.

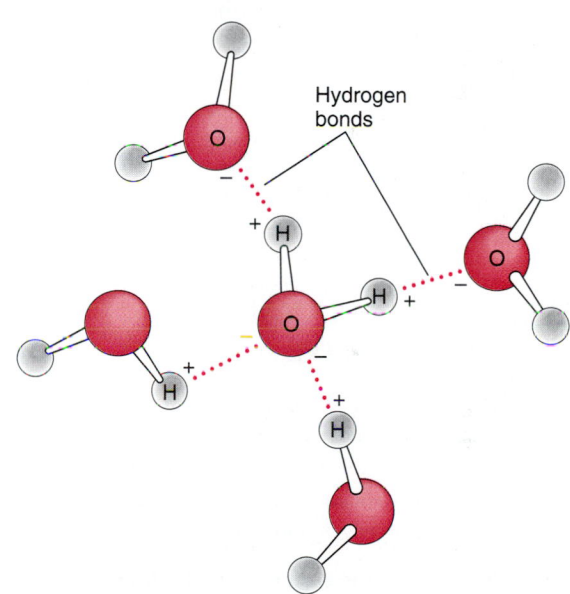

FIGURE 2.12 Hydrogen Bonding
Hydrogen bonds may occur when a hydrogen atom is shared between the negative regions of two molecules. Given their polar nature, water molecules readily form hydrogen bonds.

Hydrogen Bonds

■ **Hydrogen bonds are weak bonds between polar molecules. An especially important hydrogen bond occurs between the oxygen of one water molecule and the hydrogen of another.**

Ionic and covalent bonds form between the atoms of a molecule. Hydrogen bonds, however, form between molecules, not between atoms, and are made possible because of the polarity of certain molecules. To illustrate hydrogen bonds, we will look at the water molecule again, but remember that other polar molecules will act in a similar manner.

How does polarity affect the way a water molecule reacts with other molecules? The hydrogen atoms at each side have a slightly positive charge, so they will form attractions with atoms in other molecules having negative charges, such as oxygen, nitrogen, or fluorine. The oxygen atom in the center has a slightly negative charge, so it will form attractions with atoms or molecules having positive charges. This weak electrostatic attraction is called the **hydrogen bond.** For example, in water a hydrogen atom in one molecule (the relatively positive end) is attracted to the oxygen atom (the relatively negative portion) in another molecule, forming a hydrogen bond (Figure. 2.12).

Hydrogen bonds are very important in biology, in part because they are so easy to break. For example, hydrogen bonds are responsible for holding certain portions of the DNA molecule together, and their weak hold allows the molecule to separate very easily in order to replicate itself. Hydrogen bonds also play a role in the activities of enzymes in chemical reactions.

WATER, THE CRUCIAL MOLECULE OF LIFE

■ **Water is essential to life. Because of its polar structure, water has several important properties: it is a universal solvent and a temperature stabilizer, it can dissociate ionic compounds, and it has great tensile strength and cohesiveness.**

The hydrogen bonding that occurs in water has profound biological consequences. Some astronomers refer to our planet as "the water planet," for in many ways the existence of water makes earth unique in our solar system. In some ways, all living things on earth are "water beings," because about 50 to 90 percent of the bodies of living things consists of water. What is more, it is the special structure of water, with its polarity and its ability to form weak hydrogen bonds between one water molecule and another, that made it possible for life to arise. Because of hydrogen bonding, water is, among other things, a universal solvent and a remarkable temperature stabilizer, and it has great cohesiveness and tensile strength.

A Universal Solvent

A **solvent** is any substance in which another substance (called a **solute**) can dissolve. Water's polar

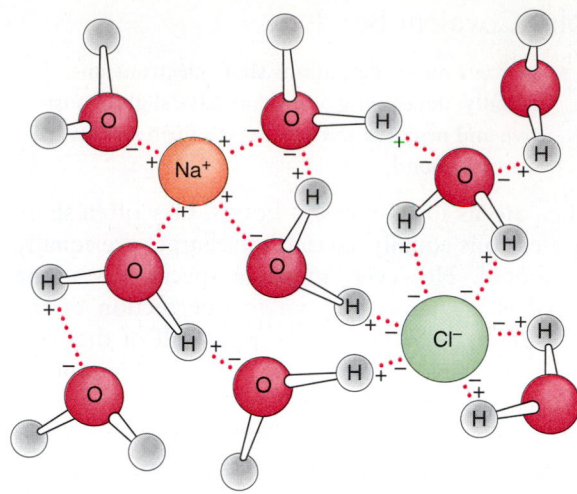

FIGURE 2.13 Water as a Solvent
The polar nature of water molecules tends to separate charged particles (either ions like the sodium and chloride ions in this illustration or other polar molecules) so that they become dissolved in water.

nature makes it an excellent solvent. As we discussed earlier, because there is an unequal distribution of electrons between the atoms of the molecule, the atom of oxygen behaves as if it had a negative charge (or as if it were *electronegative*). The hydrogen regions behave as if they had positive charges (or as if they were *electropositive*). When a polar molecule is placed with other polar substances, there is an electrical interaction and the molecules are dispersed — spread out. In other words, they dissolve (Figure 2.13).

Polar molecules that form weak hydrogen bonds with water molecules and thus dissolve are said to be **hydrophilic** (water-loving). Some molecules are not polar, however, so they do not dissolve in water. Instead they are usually insoluble in water, so they are called **hydrophobic,** or water-fearing. For example, if you mix oil and water, the oil (a liquid fat) is nonpolar, so it remains in the form of droplets in the water (Figure 2.14). The oil and water separate. That is why you need to remix your salad dressing every so often. As we will see when we discuss cell membranes, the hydrophobic and hydrophilic properties of molecules are very important in the movement of molecules through membranes.

Ionization

With a predictable frequency, water itself can ionize into a hydrogen ion (H^+) and hydroxide ion (OH^-),

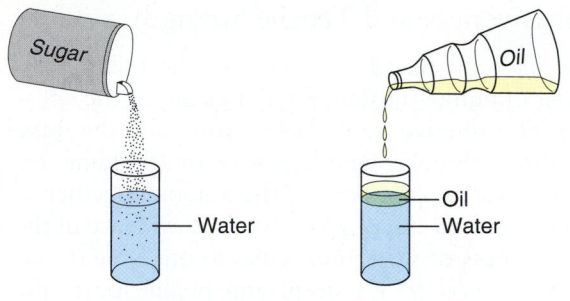

(a) Sugar is hydrophilic. It dissolves in water.

(b) Oil is hydrophobic. It does not dissolve in water.

FIGURE 2.14   Hydrophilic Versus Hydrophobic
(a) Because water is a polar solvent and sugar is also polar, sugar is hydrophilic (water loving) and will dissolve in water. Oil is nonpolar and since water is polar, oil will not dissolve in water. Oil is hydrophobic (water fearing).

thus providing the ions required in many fundamental reactions essential to the continuation of life. In addition, other molecules containing H^+ and OH^- groups dissociate in water. A substance that releases hydrogen ions (H^+) is called an **acid.** Substances that release hydroxide ions (OH^-) in water are called **bases.** (This is not a complete definition of an acid or base, but it will serve our purposes.) We use a scale called **pH**, which runs from 0 to 14, to indicate the relationship between these two ions in solutions. In pure water, the numbers of hydro-

gen ions (H^+) and hydroxide ions (OH^-) are equal; hence the water is said to have a neutral pH, or a pH of 7. When the number of H^+ exceeds the number of OH^-, the solution is said to be *acidic,* and has a pH of less than 7. Conversely, when the number of OH^- exceeds the number of H^+, the solution is said to be basic or *alkaline*; it has a pH between 7 and 14 (Figure 2.15). The scale is based on powers of 10, so that the difference of just one indicates a change ten times as great. For example, stomach acid has an acidity level around 2. Apples have a pH of about 5. That means that stomach acid is 1000 times as acidic as apples.

Chemical reactions of living organisms usually occur at a pH range of 6.9 to 7.5, a range that is called "neutral." (There are exceptions, such as the highly acidic environment inside your stomach.) However, many of the chemical reactions that occur in aqueous solutions either release or utilize hydrogen, which affects the pH. How does the cell prevent pH shifts away from neutrality? The maintenance of the internal pH of all cells is primarily due to buffers, chemical substances that play one of two roles. When there are too many hydrogen ions (when the solution is acidic), buffers combine with excess hydrogen ions to bring the solution to a neutral state. When there are too many hydroxide ions (when the solution is basic), buffers combine with hydroxide ions to bring the solution to a neutral state. Hence neutrality is maintained.

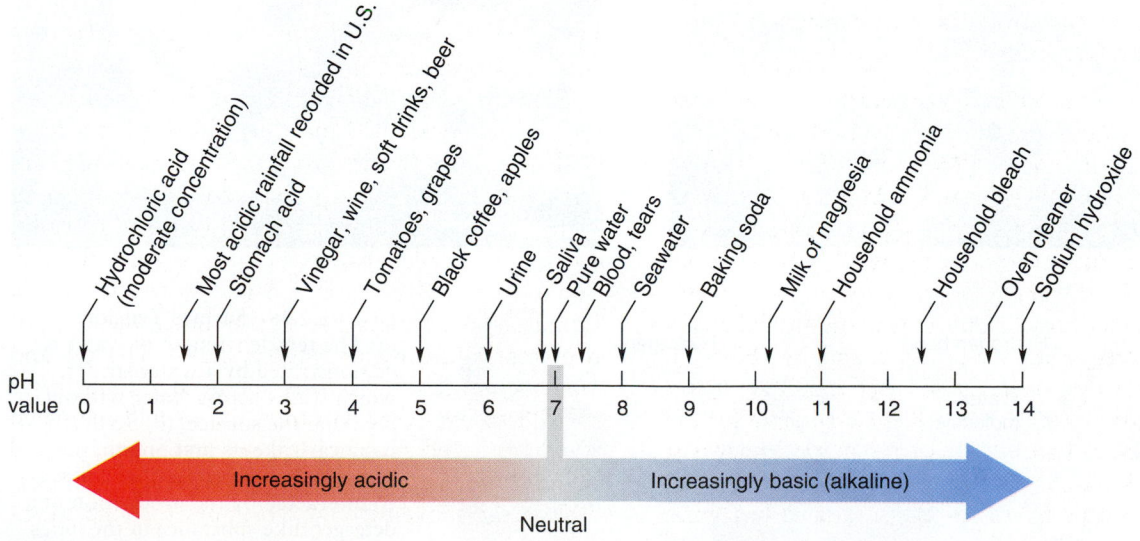

FIGURE 2.15   The pH Scale
The Ph scale is used to describe the relative concentration of hydrogen (H^+) and hydroxide (OH^-) ions in a solution. The scale is based on powers of 10, called a logarithmic scale. In other words, one change on the scale is equal to a tenfold change in ion concentration. For example, tomatoes (pH4) are ten times more acidic than black coffee (pH5).

Water, The Crucial Molecule of Life

Carbonic acid is one of the major buffering substances of blood. Carbonic acid when present in water dissociates into bicarbonate and hydrogen ions. When chemical reactions in the body cause a high concentration of hydrogen ions in the blood, they combine with the bicarbonate to form carbonic acid, thus removing the hydrogen ions from the blood. When there is an excess of hydroxide ions, they combine with the hydrogen ions to form water.

Temperature Stabilization

Because of the hydrogen bonds that form between water molecules, it has a high **specific heat** (the amount of heat required to raise the temperature of a substance by a specific amount). That means that for a given rate of heat input, the temperature of water will rise more slowly than that of almost any other molecule. Conversely, the temperature of water falls more slowly as the heat is removed. This is important in temperature regulation in organisms. The ability to maintain a constant temperature is essential to life, since life can exist only within certain temperature limits and within a certain pH limit.

Cohesiveness and Tensile Strength

You have all heard of *adhesion* (the holding together of unlike substances). For example, the adhesion of adhesive tape holds you and the tape together. There is another way of "holding together," called *cohesion,* or the holding together of like substances. All of you have seen evidence of the cohesiveness of water molecules at one time or another. As you try to sleep, and finally locate the dripping faucet that has been keeping you awake, you notice that for a few moments before it falls, the trickle of water clings to the faucet as it forms a drop. Others have seen water striders run across the surface of a pond. All these phenomena are due to the *surface tension* of water. The water moisture in our lungs also exerts a surface tension that we must counteract to avoid the collapse of the air sacs in our lungs. Surface tension is a result of the hydrogen bonds that have formed because of the electronegative and electropositive qualities of water molecules.

Because of these cohesive properties, water also has a remarkable tensile strength — in other words it resists being pulled apart. As a matter of fact, under certain conditions the tensile strength of

(a)

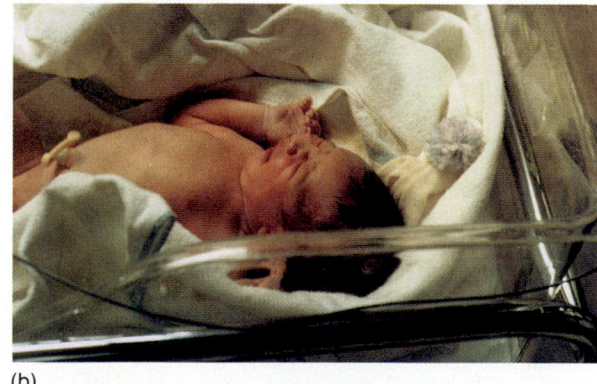

(b)

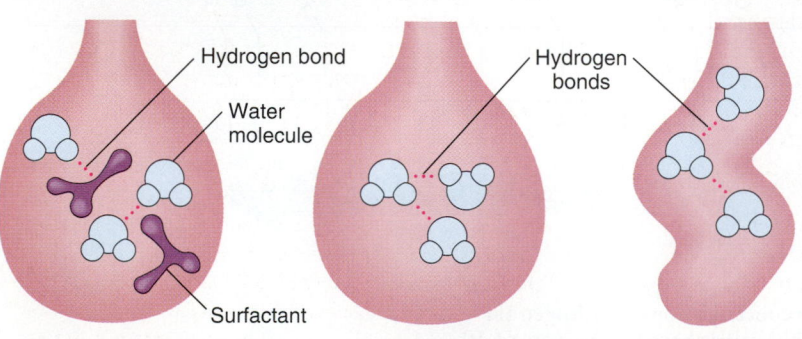

Hydrogen bond

Water molecule

Surfactant

Air sac of lung with surfactant

(c)

Hydrogen bonds

Air sac of lung without surfactant
(air sac collapses)

FIGURE 2.16 Surface Tension
(a) The tensile strength of water is demonstrated by a water strider, which walks across water without breaking the surface. (b) So that a baby can take its first breath, the surface tension property of water is counteracted by the production of a detergentlike substance in the lungs called surfactant. (c) Surfactant coats the air sacs in the lungs and prevents the hydrogen bonds formed by the water lining the sacs from pulling so tight as to collapse the air sacs.

Chapter 2   The Chemical Basis of Life

water exceeds that of steel wire. The cohesiveness and tensile strength of water help to explain how water can rise hundreds of feet from roots to the needles of a giant redwood tree.

So now you know the basic chemistry that governs the functioning of living things. Indeed, living systems like you and me are marvelous chemical factories producing life. But how do these chemicals work together? What are the biological compounds, the molecules of life? This is the topic of our next chapter.

SUMMARY

1. Matter is defined as anything that has mass and occupies space. Depending on the amount of energy in the system, matter exists in one of three physical states: as a solid, a liquid, or a gas.

2. The fundamental indivisible unit of matter is the atom. Elements are made up of only one type of atom; 92 elements occur in nature. Scientists have produced at least 14 more elements in laboratory settings.

3. The central region of the atom, called the nucleus, contains positively charged protons and neutral particles called neutrons. Negatively charged particles called electrons whiz around the nucleus and form an electron cloud.

4. Atoms are electrically balanced. In other words, the number of positive charges equals the number of negative charges.

5. Electrons orbit the nucleus in energy levels. Within these energy levels are distinct orbitals, in which the electrons are found 90 percent of the time. Some orbitals are spherical and some are dumbbell-shaped.

6. Generally speaking, the maximum number of electrons contained in the energy level nearest the nucleus is two. The maximum for the second energy level is eight. When an atom has a maximum number of electrons in its outermost energy level, it is stable, or usually nonreactive.

7. The atomic number of an atom is equal to the number of protons. The atomic mass is equal to the number of protons plus the number of neutrons.

8. Isotopes are atoms of the same element with different atomic masses, due to different numbers of neutrons. For instance, since carbon can have an atomic mass of either 12, 13, or 14, carbon has three isotopes: ^{12}C, ^{13}C, ^{14}C.

9. The number of electrons in an atom's outermost energy level, known as the valence energy level, determines the chemical properties of an atom.

10. An atom that lacks a full complement of electrons in its valence energy level can gain stability, either by gaining or losing electrons or by sharing electrons with one or more other atoms.

11. When whole atoms combine or bond chemically, they form molecules. The ways that atoms join depend on the number of electrons in the valence energy level. Atoms take on new characteristics when they form a molecule.

12. Chemical bonds join the atoms of a molecule. Chemical bonds are not physical objects, but rather an energy relationship between atoms or molecules.

13. When an atom loses or gains one or more electrons, it acquires an electrical charge and is called an ion. An atom that gains electrons is negatively charged, because it has more electrons than protons. An atom that loses electrons is positively charged, because it has more protons than electrons.

14. Water can push apart the ionically bonded atoms of ionic compounds. This process is called ionization.

15. When ions are attracted to each other by opposing electrostatic forces, they join by means of an ionic bond.

16. Ionic bonds are relatively weak, so that when ionic compounds are placed in water, the bonds are easily broken by a process called dissociation. Dissociation produces free ions, and the solution is called electrolytic, because it can conduct electric current.

17. Covalent bonds involve the sharing of electrons between two or more atoms.

18. Because carbon has four electrons in its outermost energy level, it can combine covalently very readily to produce uncountable combinations with other atoms.

19. Some covalent bonds are called polar covalent bonds because the electrons shared by the atoms are located near one atom more often than near the others. This configuration gives one section of the molecule a slightly positive charge and the other parts of the molecule a slightly negative charge. Molecules containing polar bonds are called polar molecules.

20. Polar molecules can form bonds with each other, through the attraction of the positively charged region of one molecule and the negatively charged region of another. Such a molecule-to-molecule bond, called a hydrogen bond, is seen most often between a hydrogen atom of one molecule and an oxygen, nitrogen, or fluorine atom covalently bonded in another molecule.

21. Water is essential to life, and its special properties are related to its polar structure.

22. Water will ionize with a predictable frequency into H^+ and OH^- ions. When the number of H^+ and OH^- ions is equal, the solution is said to be neutral. When the number of H^+ ions is greater than the number of OH^- ions, the solution is said to be acidic. If there are more OH^- ions than H^+ ions, the solution is said to be alkaline. The ratio of H^+ to OH^- is measured by means of a sliding scale called the pH scale. The pH scale goes from 0 to 14, with 0 (although rarely found) being the most acidic, 7 being neutral, and 14 being the most basic.

23. Water has a high specific heat, which means it can gain and lose large amounts of heat with very little change in its own temperature. As a result, water is very useful in temperature stabilization.
24. Under certain conditions, water has a high tensile strength (resistance to being pulled apart) because of the hydrogen bonds between its molecules.

KEY TERMS

energy	solution
atom	covalent bond
element	molecule
nucleus	functional group
proton	organic
neutron	polar covalent bond
electron	polar
energy level	hydrogen bond
orbital	solvent
atomic number	solute
atomic mass	hydrophilic
isotope	hydrophobic
radioisotope	acid
compound	base
chemical bond	pH
ion	specific heat
ionic bond	

REVIEW QUESTIONS

True/False

1. Matter can be defined as anything that has mass and occupies space.
   true   false   (page 30)

2. The nucleus of an atom consists of protons, neutrons, and electrons.
   true   false   (page 31)

3. The atomic mass of an atom is equal to the number of protons.
   true   false   (page 34)

4. Covalent bonds are formed when atoms lose or gain valence electrons.
   true   false   (page 42)

Fill in the Blank

1. An atom that has gained or lost an electron or electrons is called a(n) _____. *(page 41)*

2. Molecules that have an unequal distribution of electrical charges due to the shape of the molecules are said to be _____. *(page 45)*

3. When two molecules have the same number of atoms but the arrangement of the atoms within the molecules differs, they are called _____. *(page 45)*

4. Because of its _____ structure, water is an excellent solvent. *(page 46)*

Multiple Choice

1. The most common elements found in living things are:
   a. carbohydrates, fats, proteins
   b. carbohydrates, lipids, water
   c. carbon, hydrogen, oxygen, nitrogen
   d. carbon, zinc, sulfur, nitrogen
   (page 31)

2. The atomic number of the element is determined by the number of:
   a. nuclei
   b. atoms
   c. neutrons
   d. protons
   (page 34)

3. The type of bond most often found in carbon compounds is:
   a. a metallic bond
   b. a covalent bond
   c. an ionic bond
   d. a metallic and covalent bond
   (page 44)

4. Covalent bonds are formed from:
   a. differences in ionic charge
   b. fusions of nuclei from several atoms
   c. sharing of electrons among atoms
   d. the attraction of positive charges
   (page 42)

Discussion

1. Why is it necessary to understand chemistry in order to understand life?

2. Define an atom, element, compound, molecule, and ion.

3. Why is the distribution of electrons in the outermost orbital so important to the chemical properties of an atom?

4. Carbon has an atomic number of 6. How many electrons are in its first energy level? And in its second energy level?

5. How are the properties of water related to its polar structure?

6. What are ionic bonds? Give an example. Why are they important to living systems?

7. Explain covalent bonding. Use a simple line drawing to represent a covalently bonded molecule.

8. What are hydrogen bonds?

9. Why can we say that life is carbon-based?

10. What is the pH scale? Why is pH important in living systems?

3

The Molecules of Life

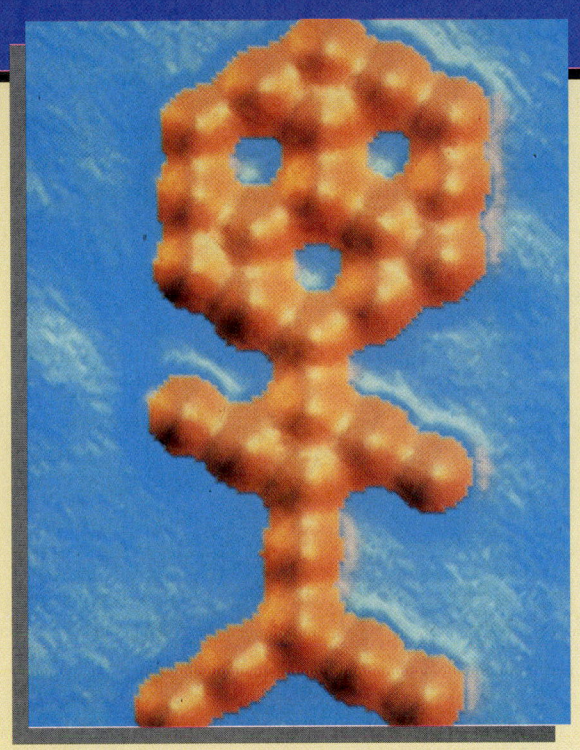

In the last chapter we discussed the idea that life on this earth can be thought of as a chemical phenomenon. Digestive enzymes break down complex food molecules that are carried by our blood to our cells for use as energy or building materials. Bacterial and viral invaders wage war on our bodies. We defend ourselves by our chemical arsenal of proteins known as antibodies. A thought is an electrochemical signal sent across an interconnected grid of brain cells. Plants take in light energy from the sun and change it into chemical energy in the form of food.

All these examples lead to the conclusion that life is the result of the organization, behavior, and interaction of atoms and molecules. Biologists know that six elements make up over 98 to 99 per-

cent of the total composition of living things — carbon, hydrogen, oxygen, nitrogen, sulfur, and phosphorus. The question is, how are these organized to form the molecules of life?

CHEMICAL REACTIONS AND THE MOLECULES OF LIFE

I just finished eating breakfast, which today consisted of a glass of low-fat milk and French toast covered with peanut butter and syrup. (I happen to like peanut butter on my pancakes and French toast — you should try it.) How do we as living things break this meal down into its constituent parts and reconstitute or assimilate the chemical components into ourselves?

Although my breakfast consisted of the things I just mentioned, to the biological system that is my

Above: This image is actually made up of carbon monoxide molecules arranged by human technology to form a stick figure.

body, my meal consisted of *carbohydrates, lipids,* and *proteins,* the large molecular constructions that are found in all foods. It also contained water, minerals, and *nucleic acids.* The cells of my body and of other living things do not distinguish between a meal of insects or of steak. Protein is protein, and it will be broken down into the *amino acids* that make up that protein. The amino acids will eventually enter my cells and be reunited into new amino acid sequences that form the protein in my cells. How does nature *degrade,* or break down, large molecules into their component subunits? And how does nature take these component subunits and reunite them to form other complex molecules?

To begin with, all these activities are chemical reactions; so in order to answer our questions, we must first understand how chemical reactions take place.

Collision Theory

■ **The collision theory states that chemical reactions occur when molecules collide. The energy required to start a reaction is called the minimum energy of activation. There are four general types of chemical reactions: rearrangement, synthesis, degradation, and displacement.**

Today is a very exciting time in chemistry, since scientists can actually move atoms and also image the progress of chemical reactions (Focus 3.1). But, what are chemical reactions?

Simply stated, **chemical reactions** involve breaking and reforming chemical bonds. In order to do this, energy is required, and the amount of energy available in the system determines whether or not a reaction will occur. For example, if you assume that all atoms, ions, and molecules are constantly moving, then a chemical reaction can occur when they collide with one another. This is what happens when two atoms of hydrogen unite with a molecule of oxygen to form water (H_2O). Chemists call this explanation the *collision theory.*

For example, when you played with your old chemistry set you learned that if you simply mixed two chemicals together, it usually took a long time for a chemical reaction to occur. However, if you heated the mixture (which is one way to put energy into a system), the reaction proceeded much more quickly. Basically, according to the collision theory, the addition of heat increased the speed at which the atoms and molecules were moving, and thus increased the likelihood that they would collide and react. (There are other ways to add energy to a material — by shaking or increasing pressure, for instance — but heat is the one that is most common, especially in biological processes.)

Adding just a little heat will not necessarily get results. Before a reaction can occur, a certain minimum amount of energy must be added, which chemists call the *minimum energy of activation.* What is more, the level of the minimum energy of activation depends on the substances involved. For instance, if you light a match and toss it into a small pan of gasoline, you provide enough energy (in the form of heat) to combine the gasoline and oxygen to form carbon dioxide, water, and a tremendous amount of liberated energy (in other words, the explosion will be something to see). However, if you fill the pan with another liquid, such as plain water, the lighted match will have no effect.

Types of Chemical Reactions

For convenience, chemists divide chemical reactions into four groups (Figure 3.1). As we talk about the chemistry of living systems, all reactions will fall into one of these categories. Very complex processes, such as photosynthesis, include several types of reactions.

Rearrangement Reactions. In rearrangement reactions, atoms are rearranged within a molecule. For example, in the process of releasing the energy in a glucose molecule, one step involves rearranging its atoms into the molecule fructose.

Synthesis Reactions. **Synthesis** reactions occur when small molecules are combined to make larger molecules: A + B → AB. A relatively simple synthesis reaction occurs when two hydrogen molecules combine with one oxygen molecule to make two water molecules: $2H_2 + O_2 → 2H_2O$. (Note that this reaction is *balanced* — there are an equal number of atoms on either side of the equation.) Making protein out of amino acids is another example of a synthesis reaction.

Degradation Reactions. Degradation reactions, which are the reverse of synthesis reactions, "tear down," or *degrade,* large molecules into their smaller parts. In other words, large molecules are broken down into smaller molecules: AB → A + B. If you break down two molecules of water, you divide them into two hydrogen molecules and one oxygen molecule: $2H_2O → 2H_2 + O_2$. (This reaction is also balanced.) Digestion is an example of a degradation reaction involving the breaking of the bonds in molecules.

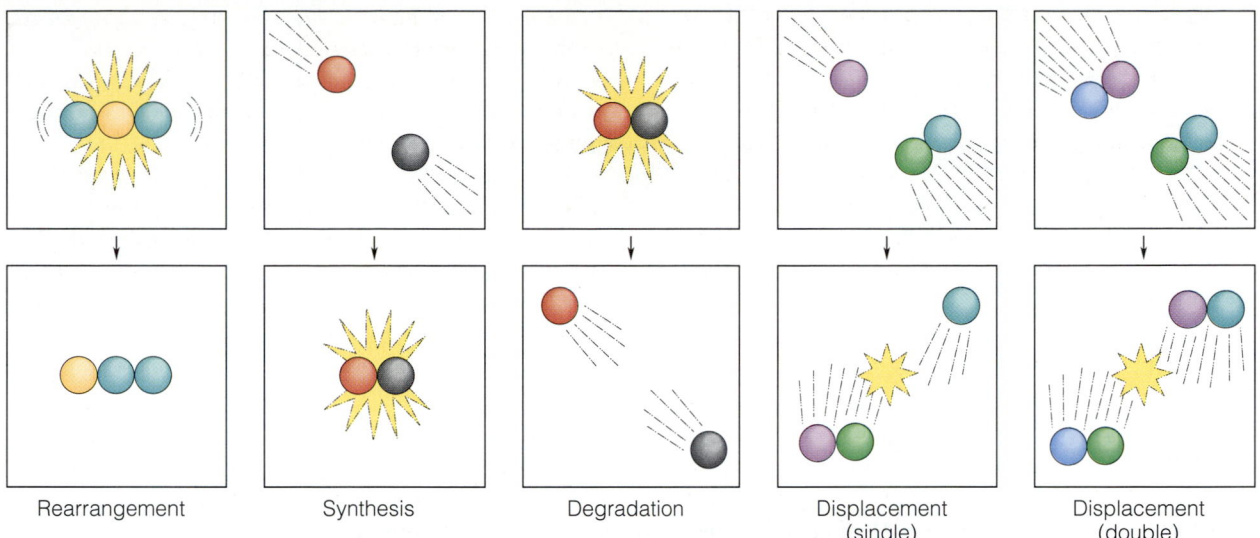

| Rearrangement | Synthesis | Degradation | Displacement (single) | Displacement (double) |

FIGURE 3.1   Four Types of Chemical Reactions

Displacement Reactions. In displacement reactions, there is an exchange of atoms (or groups of atoms) between molecules. There are two types of displacement reactions, single and double displacement. In a *single displacement* reaction, such as A + BC → AC + B, one element shifts position. In living systems, for example, hemoglobin (an iron-containing molecule) can combine with CO_2 from the cells to form carbamino hemoglobin. When this carbamino hemoglobin is transported by the blood to the lungs, the carbon dioxide is displaced by O_2. This single displacement reaction can be represented as follows:

$HbCO_2$	$+ O_2$	$\rightarrow HbO_2$	$+ CO_2$
Carbamino hemoglobin	Oxygen	Hemoglobin-oxygen complex	Carbon dioxide

In a *double displacement*, as in AB + CD → AC + BD, two elements shift position. A double displacement reaction occurs when silver nitrate reacts with hydrochloric acid to form silver chloride and nitric acid.

$AgNO_3$	$+ HCl$	$\rightarrow AgCl$	$+ HNO_3$
Silver nitrate	Hydrochloric acid	Silver chloride	Nitric acid

All four types of reactions take place during the various life processes in living things, such as the utilization of my breakfast. Often the reactions occur in pairs — something is taken apart (degraded) in order to be put together (synthesized) in another arrangement. Such an arrangement is called

a *paired reaction*. It is rather like cutting apart several types of board in order to build a bookcase. You have the same wood when you have finished that you had when you started, but the structure and arrangement are different. Paired reactions are an essential characteristic of chemical reactions in living things. In other words, if you put things together, you can usually break them apart. In fact, most chemical reactions are reversible. Chemists use a pair of arrows to indicate a reversible reaction:

A + B ⇌ AB

Chemical Reactions in Biological Terms

■ **During hydrolysis macromolecules are cleaved into their smaller subunits by adding water. Dehydration synthesis is the linking of smaller subunits by removing water to form larger molecules.**

Let us see how some of these types of reactions affect my meal. Earlier I said that my meal actually consisted of large molecules (often called **macromolecules** or **polymers**) of proteins, carbohydrates, and lipids. To change those macromolecules into the types of proteins, carbohydrates, and lipids that my body can use, a set of paired reactions occurs in my cells: *hydrolysis* and *dehydration synthesis* (Figure 3.2). Both of these reactions are mediated by enzymes.

The first type of reaction is a degradation reaction called hydrolysis, a term that means *water*

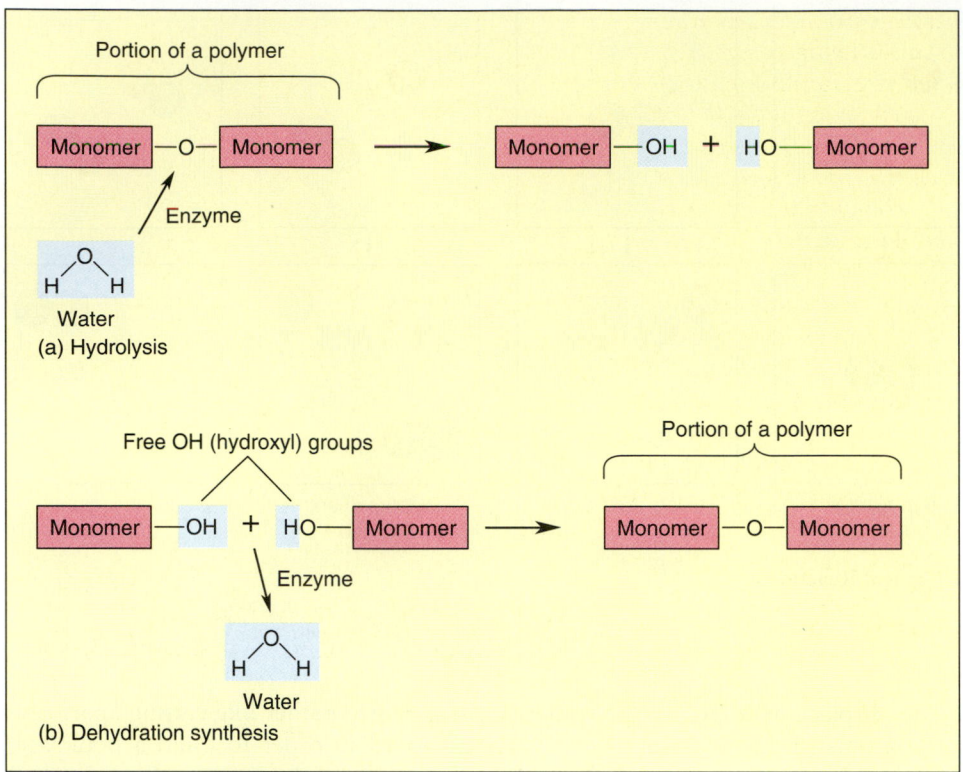

FIGURE 3.2 Hydrolysis and Dehydration Synthesis
(a) Hydrolysis is the degradation of larger molecules to smaller molecules by inserting a water molecule to split them apart. (b) Dehydration synthesis involves the removal of a water molecule between smaller molecules to form larger molecules, including the macromolecules found in organisms.

splitting. In **hydrolysis,** water molecules are enzymatically added to macromolecules, splitting them into their component subunits. Such small molecules are called *monomers.*

The second type of reaction is a synthesis reaction called **dehydration synthesis.** This step recycles the monomers by removing the water used in hydrolysis. The monomers are then joined to form *polymers,* such as the proteins, carbohydrates, and lipids that my body needs. Hence the name of the process, which means *synthesis* (or putting together) by *dehydration* (the removal of water).

So I benefit from my breakfast because of a pair of reactions that are the reverse of each other. One reaction adds water to a macromolecule to degrade it into monomers, and the other process removes water from monomers to form macromolecules. At present I am hydrolyzing, or breaking down, my meal into its component parts, its monomers. Once these simple molecules get into my cells, they will be utilized as sources of energy or as the raw materials used to synthesize the macromolecules that make up me.

So through these two paired reactions, monomers are synthesized into macromolecules that are of biological importance. What are the properties of these biological compounds, and what are their functions? Let us discuss them.

There are four groups of biological compounds that are important in living things: *carbohydrates, lipids, proteins,* and *nucleic acids.* Each group is divided into smaller subgroups that play an important role in the chemical activities of living things.

CARBOHYDRATES

■ **Carbohydrates function as the primary cellular energy source. They have an empirical formula of $[C(H_2O)]_n$. The main types of carbohydrates are monosaccharides, disaccharides, and polysaccharides.**

Carbohydrates are not just those things that you try to avoid when you are on a diet. All living things (plants, animals, fungi, and microorganisms) contain and utilize carbohydrates for their primary source of energy. In addition, carbohydrates also serve a structural function in all living things. For example, plants are supported structurally by a carbohydrate called *cellulose,* which exists in their cell walls. Both plants and animals store reserves of excess carbohydrates, as either *starch* (in plants) or *glycogen* (animal starch). Carbohydrates fall into three major categories: *monosaccharides, disaccharides,* and *polysaccharides.*

An empirical formula is one that indicates the type and number of atoms within a molecule. When we compare the empirical formulas of most carbohydrates, we find carbon, hydrogen, and oxygen. *Carbohydrates* are literally watered (or *hydrated*) carbons — $C(H_2O)$ — as the name carbohydrate suggests, and their generalized empirical formula is $[C(H_2O)]_n$. The n designates the number of carbohydrate units in the molecule. The category to which a specific carbohydrate belongs depends on the number of units it possesses.

Monosaccharides

■ The simplest carbohydrate unit is called a monosaccharide. In some cases, monosaccharides (as well as other molecules) have identical empirical formulas, but their structural formulas differ, giving each variety unique chemical properties. Such molecules are called isomers.

The simplest units of carbohydrates (which are the building blocks of all complex carbohydrates) are monomers known as simple sugars, or **monosaccharides.** (The term *mono-* means *one,* and the term *-saccharide* means *sugar unit.*) Typically, simple sugars consist of three to seven carbon atoms. Of special importance to biologists are two types of monosaccharides: the five-carbon (or pentose) molecules, such as those that are part of DNA and RNA; and the six-carbon (hexose) molecules, such as glucose and fructose. The common ending *-ose* indicates a sugar.

Glucose and fructose are examples of what are called *isomers* — molecules that have identical empirical formulas but have different properties. We touched briefly on this topic in Chapter 2. The empirical formula for both glucose and fructose is $C_6H_{12}O_6$, indicating that each consists of 6 atoms of carbon, 12 atoms of hydrogen, and 6 atoms of oxygen.

In an instance like this, in order to see the differences between the two kinds of molecules, a *structural formula* is more useful than a general formula. (A structural formula shows the position of the atoms within a molecule.) This type of formula il-

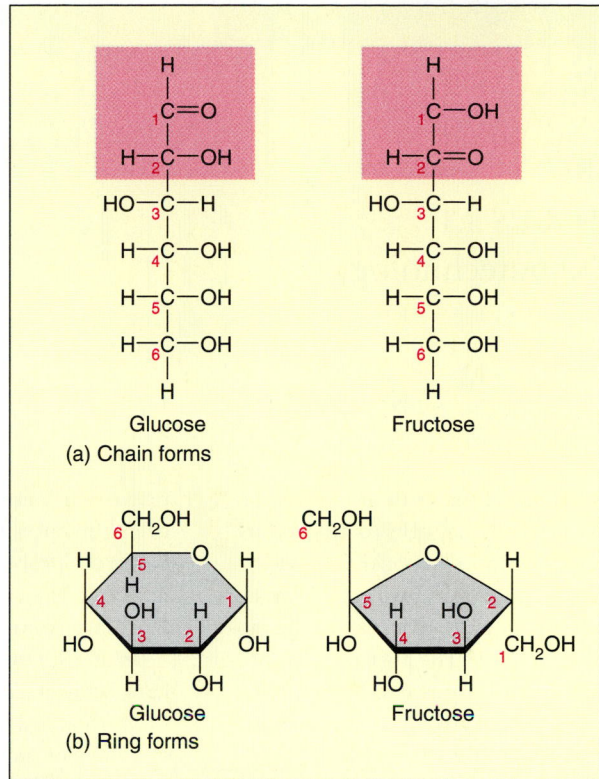

FIGURE 3.3 Glucose and Fructose
The structural formulas for these molecules can be drawn in a straight chain form (a) or ring form (b). These molecules are isomers of one another; they share the same empirical formula but differ in structure about the first and second carbons. This structural difference gives the molecules different chemical properties. Glucose has a terminal carbonyl group, so it is an aldehyde, and fructose has an internal carbonyl so it is a ketone.

lustrates the significant nature of each isomer by showing variations in molecular arrangement (Figure 3.3). The differences in the bonds between the carbons change the properties of the chemicals. For example, because of the molecular structure of fructose, it stimulates our taste buds more than glucose, so it tastes sweeter than glucose.

One comment about the notation you will see in a structural formula of carbon molecules. Some monosaccharides, especially those having five or six carbon atoms, form a ring when placed in water, as in the watery interior of living cells (Figure 3.3b). In a structural formula, as a matter of convention, the locations of the carbon atoms are numbered for reference, and the numbers serve as an indication of the placement of the carbons at the intersection of the links in the ring.

In Chapter 2 we mentioned functional groups in molecular compounds, but we now need to clarify

FOCUS 3.1
Nanotechnology

Can we see the birth of molecules? What really happens to atoms during chemical reactions? Can we move atoms? Can we actually visualize biological processes at the molecular level? The answer to all of these questions is yes. Today, because of the advent of new instrumentation like the family of scanning-tunneling microscopes, scientists can now pick up and move atoms, and, with the use of molecular cameras, they can take photographs as fast as one quadrillionth of a second (a femtosecond). Scientists can image the step-by-step changes that occur in atoms and molecules as chemical reactions progress.

As was discussed in the Human Endeavors box in Chapter 2, scanning-tunneling microscopes "feel" the surface of the molecular or atomic specimens to form the image. A tiny platinum probe is given a small electrical charge and the specimen is scanned, atomic diameter by atomic diameter, across its surface. As the probe moves, electrons respond to the voltage differential between the probe and the surface of the specimen by tunneling across the gap, which produces a very small current. The variations in the current—which are due to the miniscule differences in the distance between the probe and the object—are detected. The computer produces the image from these distances.

In 1989, Donald Eigler and Erhard Schweitzer of IBM in San Jose, California, lowered the temperature of a plate of nickel to about absolute zero ($-456°$ F), which minimized atomic vibrations. Then, by spraying atoms of xenon gas over the plate and by increasing the electrical charge on the probe of the scanning-tunneling microscope, they were able to drag the xenon atoms across the nickel to spell out "IBM" in letters only five atoms tall, the entire logo only about 660 billionths of an inch long (Figure A). This example of the manipulation of materials, be they inorganic or organic, atom by atom, is what is meant by *nanoengineering*. The new nanotechnology combines the technological capabilities of the scanning-tunneling microscope with laser cameras that can detect changes that occur in quadrillionths of a second, and you can image what happens in atoms and molecules as they react. The synchroton ray camera can do just that. When you pulse an atom with a specific frequency of energy, the atom will absorb that energy and the electrons are moved to a higher energy state. When the electrons return to their original ground state, energy of a specific frequency is emitted. Using this principle, scientists can pulse molecule A and molecule B and detect how their frequencies change to form molecule C. But how can you do that and how can you record the snapshots or "frames" of this continuous event, since it only occurs in quadrillionths of a second? Figure B illustrates how this camera works.

So today we can see the advent of a new technology even more exciting or as exciting as genetic engineering. Not only can we now manipulate genetic variation, but we can actually manipulate atoms and molecules. With this technology our advances in science, including biology, will truly expand. Just think, today not only can we image atoms, but we can manipulate them as well and detect the actual events in the birth of new molecules as atoms and molecules react.

Source: Zewail, Ahmed H., "The Birth of Molecules," *Scientific American,* 263:6, Dec. 1990, pp. 76–82.

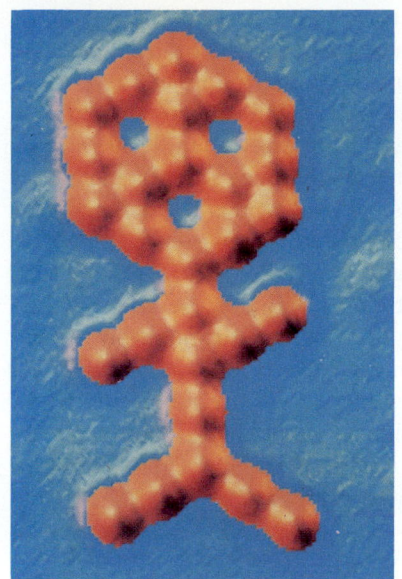

FIGURE A
The photograph is a scanning-tunneling microscope image of carbon monoxide molecules (on a platinum crystal) arranged to form a stick figure. This stick figure was prepared by scientists at IBM's Almaden Research Center.

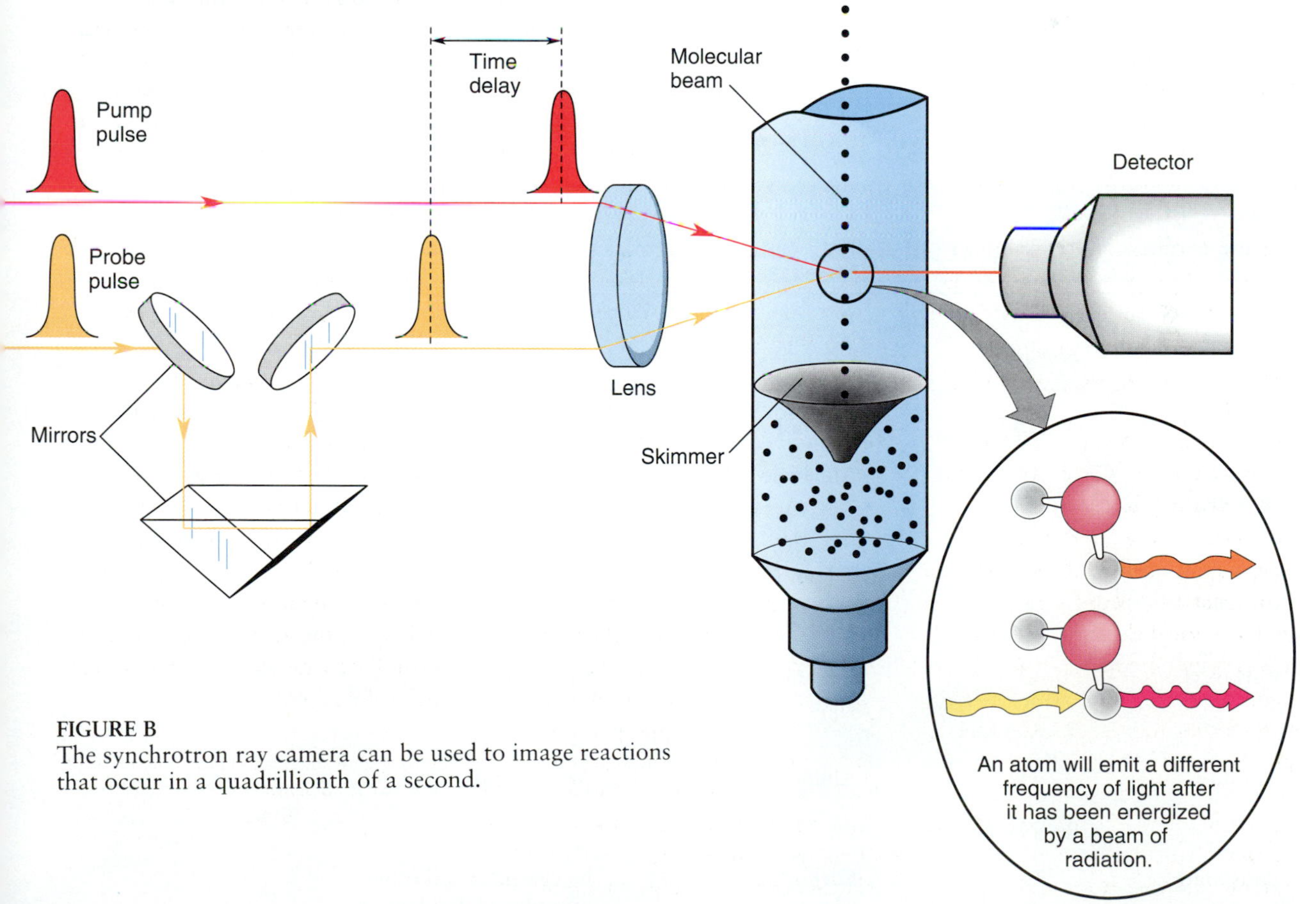

Time delay

Pump pulse

Probe pulse

Mirrors

Lens

Molecular beam

Skimmer

Detector

An atom will emit a different frequency of light after it has been energized by a beam of radiation.

FIGURE B
The synchrotron ray camera can be used to image reactions that occur in a quadrillionth of a second.

what functional groups are. A functional group is a set of atoms that, working together, impart certain characteristics to the molecules in which they are found. For instance, a hydroxyl functional group (OH) is polar, and when it appears in a molecule, it makes that molecule highly soluble in water. (Remember, water is also a polar molecule.) Sugars have hydroxyl groups, and that is why sugar dissolves so easily when you add it to your coffee.

There are six major functional groups that are important in biology: *hydroxyl, carbonyl (terminal), carbonyl (internal), carboxyl, amino,* and *phosphate.* Table 3.1 details the structure and characteristics of some of these six functional groups. Here, and in many structural formulas, the *R-group* represents the *radical,* or, for our purposes, the *remainder of the molecule.*

A functional group's position in a molecule is an important factor in determining that molecule's chemical properties. A terminal carbonyl functional group, which forms an aldehyde, appears at the end of a carbon chain. On the other hand, a ketone, which is an internal carbonyl, is flanked on either side by carbon atoms. The position of the functional group often affects how the molecule reacts, and is an important consideration in isomers.

For instance, in glucose the carbonyl functional group is found at the end of the molecule's long carbon chain; so it is an aldehyde and acts like other aldehydes. Fructose, however, has the carbonyl attached to the second carbon in the straight and chain form (see Figure 3.3a). So fructose is a ketone. As a result, even though both glucose and fructose are six-carbon sugars, and even though both contain a carbonyl functional group, the chemical and physical properties of glucose are more similar to other aldehydes, like formaldehyde, than to fructose. Fructose, on the other hand, has more similarities to other ketones, such as acetone.

Disaccharides

■ **Disaccharides are formed when two monosaccharide units covalently bond together through dehydration synthesis. A common example is sucrose, or table sugar.**

TABLE 3.1
Some Biologically Important Functional Groups

Group	Formula *	Class of compound	Occurs in
Hydroxyl	R—OH	Alcohol	Sugars, fats, alcohols
Carbonyl (terminal)	$$R-\overset{\displaystyle O}{\overset{\displaystyle \|}{C}}-H$$	Aldehyde	Sugars
Carbonyl (internal)	$$R-\overset{\displaystyle O}{\overset{\displaystyle \|}{C}}-R$$	Ketone	Sugars
Carboxyl	$$R-\overset{\displaystyle O}{\overset{\displaystyle \|}{C}}-OH$$	Organic acid	Fats, amino acids
Amino	$$R-\overset{\displaystyle H}{\overset{\displaystyle \|}{N}}-H$$	Amine	Amino acids
Phosphate	$$R-\overset{\displaystyle O}{\overset{\displaystyle \|}{\underset{\displaystyle O^-}{\underset{\displaystyle \|}{P}}}}-O^-$$	Orthophosphate	Nucleotides and other phosphorylated fats and sugars

* For our purposes here, R represents the rest of the molecule.

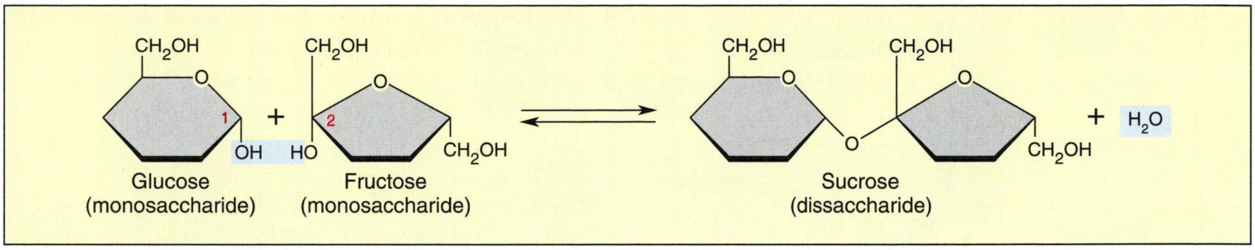

FIGURE 3.4   A Disaccharide
The formation of a molecule of the disaccharide sucrose from glucose and fructose by dehydration synthesis. The structural detail has been simplified.

When the process of dehydration synthesis covalently bonds two monosaccharides together, the result is a **disaccharide.** (Remember that saccharide means sugar. The prefix *di-* means two.) We are all aware of many disaccharides. Common table sugar, *sucrose,* is a disaccharide, composed of a molecule of glucose bonded to a molecule of fructose (Figure 3.4). *Lactose,* or milk sugar, is a disaccharide composed of glucose combined with the monosaccharide galactose. *Oligosaccharides* consist of two to ten monosaccharide units bonded together, and some types can cause very unique problems, such as the excessive gas and stomach distress you get when you eat beans. Beans contain large amounts of oligosaccharides.

Some adults lose the ability to digest the milk sugar lactose, because their bodies produce insufficient amounts of the enzyme lactase. Lactase is responsible for the hydrolysis of lactose to glucose and galactose. Individuals with this condition suffer from gas, diarrhea, and cramps after the ingestion of lactose. The condition is more prominent in blacks than in members of other races, compounding the difficulty of relief efforts in African countries suffering from famine due to years of drought. A major food supplement shipped to famine-stricken areas is dried milk, because it is light and easier to transport than other protein sources. However, because of enzyme deficiencies, the recipients of this aid do not always benefit as much as their benefactors intend.

Polysaccharides

■ **Polysaccharides are carbohydrates containing more than 10 monosaccharides; they function in energy storage and structural support in plants and animals. The way that monomer units are linked together determines the polysaccharide's properties.**

When dehydration synthesis covalently bonds more than 10 monosaccharides together, the result is a macromolecule, sometimes called a polymer. Macromolecules containing monosaccharides are called **polysaccharides,** meaning *many sugars.* A polysaccharide may have a branched shape, or it may be arranged like a long linear chain. The chemical properties of polysaccharides are due to the different types of linkages that can occur between the monomer units.

Storage Polysaccharides.   Both glycogen (animal starch stored in the liver and muscles of higher animals) and plant starch consist of several hundred glucose units. The two types of polysaccharides react differently because of the way their long chains are arranged. In glycogen, the glucose units are aligned in such a way that they form highly branched forms. In plant starch, the glucose units are aligned to form a twisted, coiled structure. The properties of polysaccharides are due to different types of linkage alignments that can occur between monomer units (Figure 3.5).

Polysaccharides serve as a means of storing carbohydrates. However, before their elements can be transported through living systems and used for energy, they must undergo hydrolysis and be broken down into monosaccharides or disaccharides. Many times large carbohydrates are formed of only one type of monosaccharide, but some are composed of two, three, or four different types.

Structural Polysaccharides.   Polysaccharides not only function in energy relationships in cells. They are also very important structural components of living things. For example, cellulose, which is made up of glucose monomers, is a fibrous, water-insoluble substance that functions as the main structural material in plant cell walls. In fact, cellulose may well be the most abundant organic chemical in the

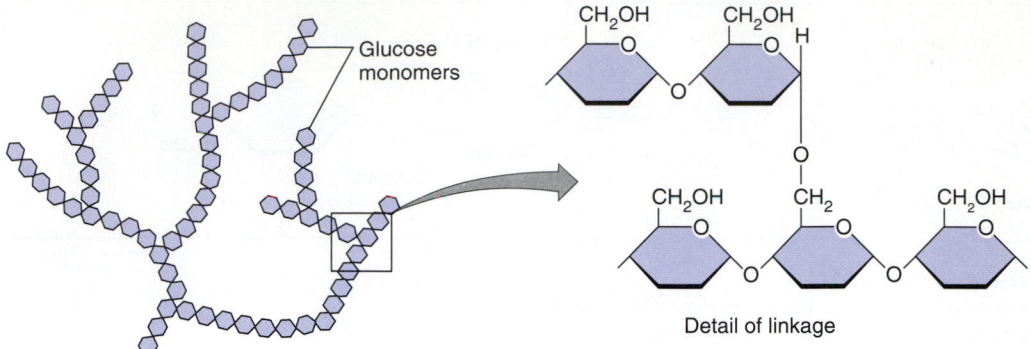

Glucose monomers

Detail of linkage

(a) Highly branched chain of glycogen molecule

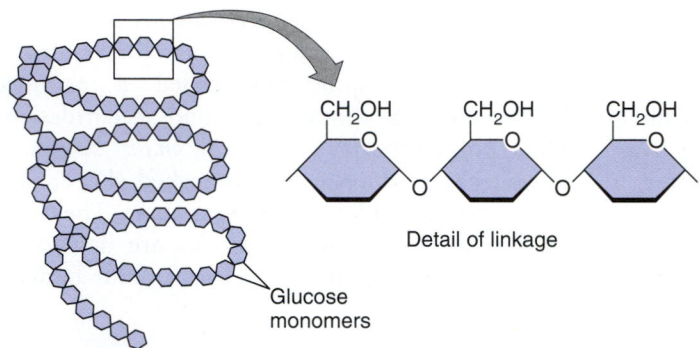

Detail of linkage

Glucose monomers

(b) Coiled, unbranched chain of starch molecule

FIGURE 3.5 Storage Polysaccharides
Although formed from chains of the same basic unit of glucose, the storage polysaccharides of (a) glycogen and (b) plant starch form very different polymers because of the way in which the glucose monomers form either branches or coils.

world. Humans benefit from this abundance, for we use this molecule for such things as lumber, rope fiber, and paper (Figure 3.6a).

Chitin is another polysaccharide, one that is called a *modified carbohydrate* because it contains nitrogen atoms in its monomer units (Figure 3.6b). Chitin, which is secreted by the outer tissue layer of the animal group that includes lobsters and insects, is the main structural component of the animals' outer skeletons (or *exoskeleton*). Until recently chitin was thought to have little food value for humans, since humans do not have enzymes to break down this polysaccharide to simpler units. However, biologists now know that some bacteria contain enzymes that can break down chitin into sugars so that the material could be used as a dietary energy source. As a result, some molecular biologists are trying to determine the nature of the gene that controls the production of the enzyme so that we can produce it in large amounts. With this enzyme we can transform the chitin previously believed to be of no use to humans and break it down into sugar. (This is only one possible application of molecular biology technology. For more examples of the potential of this field, see Chapter 13.)

LIPIDS

■ **Lipids are a diverse group of molecules that consist mainly of carbon, hydrogen, and oxygen and are nonpolar. Lipids function in energy storage and structural support.**

Lipids are a very diverse group of organic compounds. They consist of *triglycerides, phospholipids, steroids, terpenes,* and *glycolipids.* Because they are so diverse, you might wonder why they are all grouped together. They are grouped together because they are all nonpolar molecules (hydrophobic) and do not dissolve in water. Most lipids usually have an oily or waxy consistency, and they function primarily as a means of energy storage and as structural components for living cells. They also play a role in insulation, in padding (for protection), and as an insect attractant in flowers.

Approximately 15 to 25 percent of the human body is made of lipids, the majority of which are stored in fat cells. The two genders tend to distribute their fat cells unevenly. Women tend to store fat in the breasts, hips, and thighs, resulting in their characteristic shape. Men tend to store fat around their waist, hence their characteristic shape.

(a) Cellulose

Hydrogen bonds link cellulose chains

Modified monomer

(b) Chitin

FIGURE 3.6 Structural Polysaccharides
(a) The main structural support in plant cell walls is cellulose as seen in the cell walls of this dead leaf. (b) Chitin provides support in the exoskeletons of insects and crustaceans.

Triglycerides, or Fats

■ **Triglycerides consist of a glycerol (3-carbon alcohol) and three fatty acids. In saturated fats, all the carbon molecules in the carbon skeleton of the fatty acid are bonded with hydrogen atoms. In unsaturated fats, some carbon atoms in the carbon skeleton are double bonded to each other rather than to hydrogen. Triglycerides function in energy storage.**

Fats are composed only of atoms of carbon, hydrogen, and oxygen (C, H, and O), which are arranged into two subunits: a glycerol unit (which is a three-carbon alcohol) and a fatty acid unit (Figure 3.7). Glycerol is composed of three carbons, to which three alcohol (OH) groups are attached. Fatty acids consist of an unbranched carbon skeleton to which hydrogen and a carboxyl group (COOH), an organic acid functional group, is attached.

Fats are formed by the dehydration synthesis of fatty acid chains to a glycerol molecule. As many as three fatty acid chains can bond to glycerol. When three fatty acids bind with glycerol, the fat is called a *triglyceride,* since *tri-* means three. Since the type of fatty acids attached to the glycerol can differ from one triglyceride to another, there are many types of this polymer. In Figure 3.7, stearic, oleic, and palmitic acids form the fatty acid tails.

Fats contain fewer oxygen atoms than other macromolecules and have more hydrogen bonds. For this reason these compounds hold more energy than other molecules. (The reason for this will be clear when we discuss oxidation and reduction in Chapter 6.) More oxygen is required to break down fat molecules during energy-releasing activities in cells. Since fat molecules weigh less than carbohydrate molecules, fats are ideal for the efficient storage of energy.

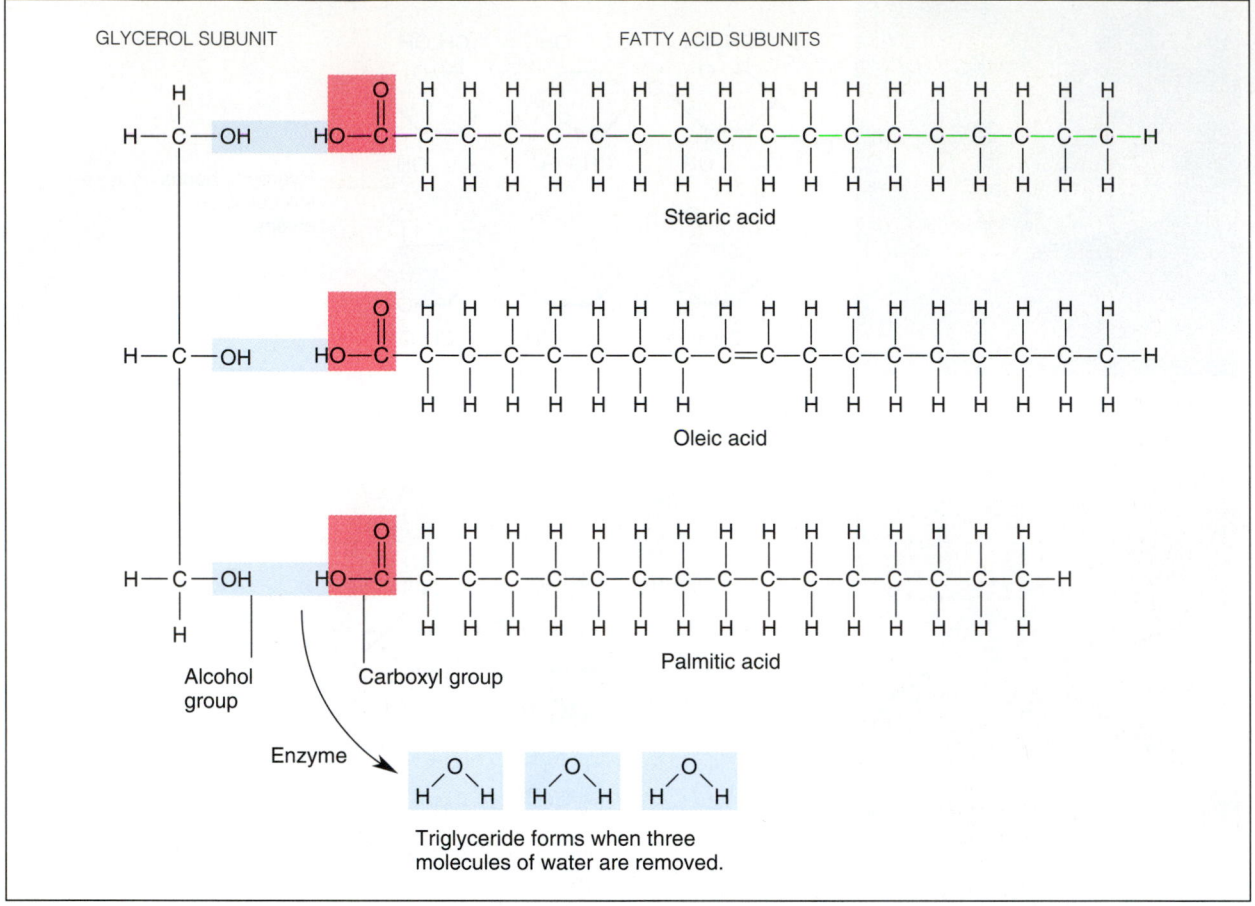

GLYCEROL SUBUNIT   FATTY ACID SUBUNITS

Stearic acid

Oleic acid

Alcohol
group

Carboxyl group   Palmitic acid

Enzyme

Triglyceride forms when three
molecules of water are removed.

FIGURE 3.7   Components of Triglycerides
A triglyceride is formed when three fatty acids covalently bond to glycerol.
Each chemical bond results from the removal of a water molecule between the
glycerol and the fatty acid.

Having fewer oxygen atoms, fats lack the electropositive regions that would attract the hydrogen atoms (electronegative regions) in water molecules. In other words, fat molecules are nonpolar and *hydrophobic* — they do not dissolve in water, only in other nonpolar substances like ether or methane. For instance, did you know that the wax you apply to a kitchen floor is a fat? That's why you cannot remove it by simply washing the floor with water — water molecules are polar. You have to use wax remover (which is nonpolar) to remove wax, since fats are hydrophobic. In general, we can say that like dissolves like. That is, polar substances dissolve polar substances, and nonpolar dissolve nonpolar. That's why wax, a fat and therefore nonpolar, will dissolve only in a nonpolar substance.

As we mentioned in Chapter 1, living things are constantly changing their chemical makeup in order to remain the same, to maintain homeostasis. Fats are constantly being turned over, or converted. They are hydrolyzed to fatty acids and glycerol, and when completely broken down they yield as much as 9 Calories per gram, almost twice as much energy as a gram of carbohydrate, which yields 3.5 to 4 Calories per gram. (A **Calorie** is the amount of heat energy required to raise the temperature of 1000 grams of water 1 degree Celsius; see page 125.) This high-energy yield is due to the fact that fats contain a higher proportion of carbon-hydrogen bonds, or C—H, than any other class of molecules in living things.

Triglycerides contain large amounts of chemical energy and are stored in seeds or in the fatty tissue of animals. Triglycerides are also used for thermal insulation and protection in animals such as the arctic seal.

The fatty acid units in a triglyceride can be *saturated* or *unsaturated*. In a **saturated fat**, the carbon

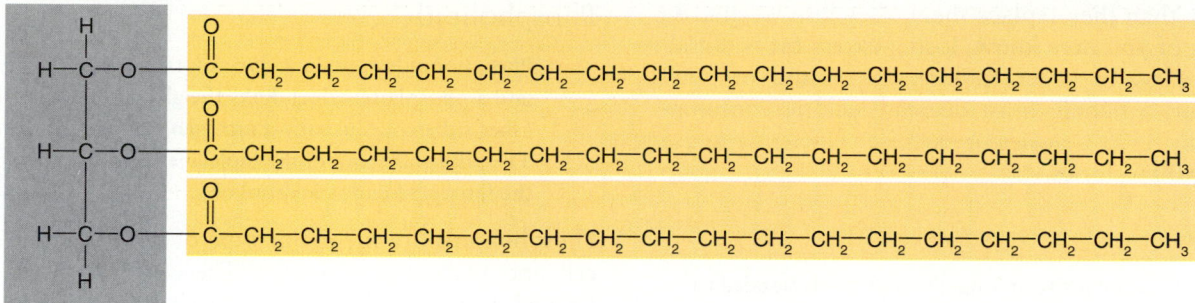

(a) Saturated fat

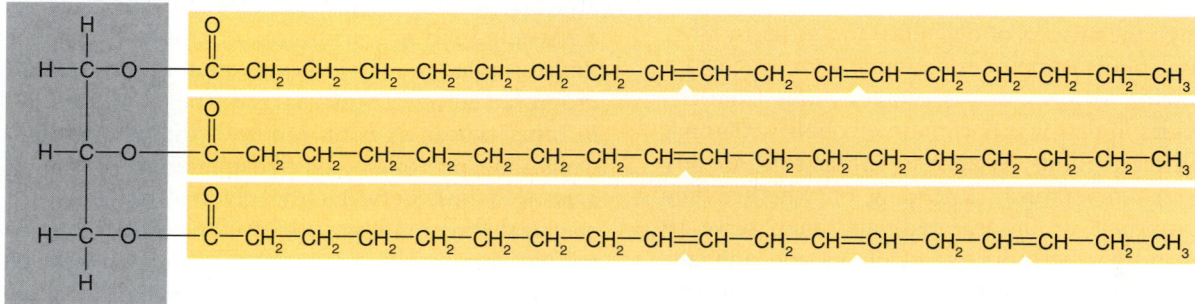

(b) Unsaturated fat

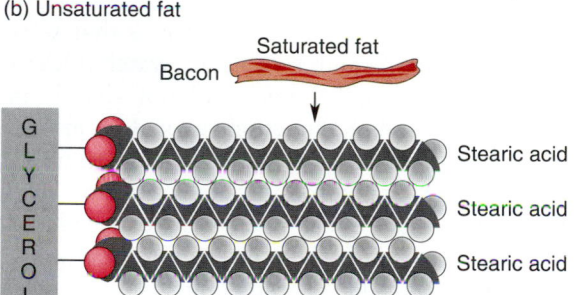

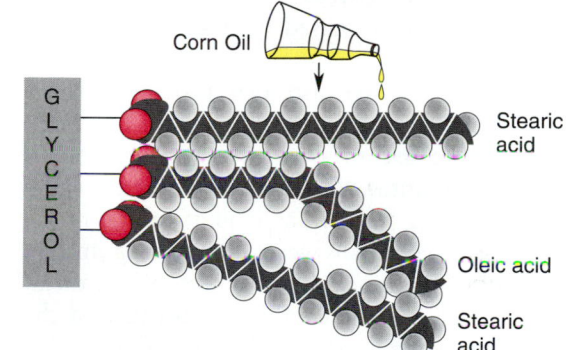

(c) Space-filling models

FIGURE 3.8   Saturated and Unsaturated Fats
(a) A saturated fat has closely fitted molecules that make it more difficult to break down than an unsaturated fat. (b) The double bonds between carbon atoms in the unsaturated fat mean that fewer hydrogen atoms can bond to the molecule. (c) Double bonds produce kinks in a chain, preventing a tight fit between fatty acid molecules.

skeleton of the fatty acid is saturated with hydrogens — in other words, there are hydrogen atoms covalently bonded to all possible locations on the carbon atoms (Figure 3.8a). Animal fats, such as butter, lard, and bacon fat, are saturated fats. Since saturated fats usually have fairly high melting temperatures, they are solid at room temperature.

In an **unsaturated fat**, on the other hand, some carbon atoms form double bonds with other carbon atoms (C=C), so that the molecule contains fewer hydrogen atoms than a saturated fat (Figure 3.8b). The formation of these double bonds produces kinks in the fatty acid chains, preventing a tight fit

between the fatty acid molecules (Figure 3.8c). The "looseness" of the molecule's structure affects its consistency. For example, plant fats (called *oils*) are usually unsaturated. This type of fat generally has a low melting point, which is why vegetable oils are usually liquid at room temperature.

When individuals, be they human or house pets, take in more energy in the form of food than they require, the excess energy is stored in fat, and excess fat can lead to obesity. Obesity is the most common chronic medical problem in the United States today. When individuals diet, they are living off their stored fat reserves. If they use up these fat reserves faster than they replace them, they lose weight. As some of you may know, losing excess fat is not all that easy. In order to lose one pound of body weight, a human must decrease caloric consumption by 3500 Calories or exercise sufficiently to use up 3500 Calories.

Waxes

■ **Waxes consist of long-chain alcohols bonded to long-chain fatty acids and are used for protective coverings in living things.**

Through the process of dehydration synthesis, long-chain alcohols sometimes combine with long-chain fatty acids to form waxes. Waxes differ from triglycerides in that waxes contain alcohols with more than three carbons. Waxes are either solid or oily at room temperature, depending on whether they are saturated or unsaturated, and they serve to cover and protect portions of plants and animals. For example, as you walk through a park, run your finger lightly over the surface of one of the green leaves growing on the trees there. The slightly waxy feel of the surface is caused by a substance called *cutin* (Figure 3.9).

Solid and oily waxes also form protective coverings for insects, skin, fur, and feathers. The mink oil that people use to waterproof their boots is made from a natural wax. If you have ever seen a duck preening its feathers as it floats along in the water, you have actually seen it spreading an oily wax over the feathers in order to waterproof them. Without lipids, a duck would sink because of the weight of the water soaked up by its feathers.

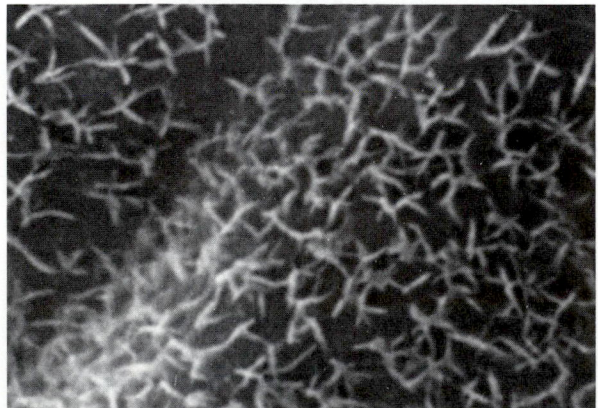

FIGURE 3.9 Waxes
This micrograph shows the undersurface of a soybean leaf with its heavy deposits of cutin. The cutin, which coats the leaves, keeps water loss to a minimum.

Phospholipids

■ **Phospholipids contain a glycerol, two fatty acids, and a phosphate, which links the glycerol to another substance (usually a nitrogen compound). The phosphate end of the molecule is hydrophilic; the fatty acid end is hydrophobic.**

Phospholipids are the major type of lipid found in cell and plasma membranes. These polymers are formed during dehydration synthesis from glycerol and fatty acids. Rather than having a fatty acid attached at the number three carbon of the glycerol, a phospholipid has a phosphate group (PO_4^{-3}). (In this formula, the $^{-3}$ indicates that the phosphate group has an electrical charge of negative three — in other words, it is an ion with three more electrons than protons.) The phosphate group in turn acts as a link between the glycerol and another chemical functional group, usually a nitrogen-containing compound. So, a phospholipid consists of two fatty acids and a phosphate bonded to a glycerol molecule (Figure 3.10a).

In Chapter 2 we discussed molecules that were polar, hydrophilic, and soluble in water. Others were nonpolar, hydrophobic, and not soluble in water. A phospholipid functions as it does in part because its phosphate end, with its attached nitrogen compound, is hydrophilic, but the fatty acid end is hydrophobic. When phospholipids are placed in water, they arrange themselves with their polar, water-soluble phosphate heads immersed in water, and the nonpolar, fatty acid ends pointing away from water (Figure 3.10b). The dual nature of phospholipids helps explain the structure of cell membranes (Chapter 5).

Glycolipids and Lipoproteins

■ **Glycolipids and lipoproteins play a part in the chemical communication between cells.**

Glycolipids (a combination of water-soluble carbohydrates and lipids) and lipoproteins (a combination of lipids and proteins) are both important components of cell membrane structure. These two molecules are responsible, in part, for the chemical communication between cells, by serving as what are called **receptor sites,** or receivers, for chemical messages on the membrane. We will be discussing glycolipids and lipoproteins in more detail in Chapter 5.

Steroids and Their Relatives

■ **Steroids, like all lipids, are insoluble in water and consist of carbon atoms arranged into four rings.**

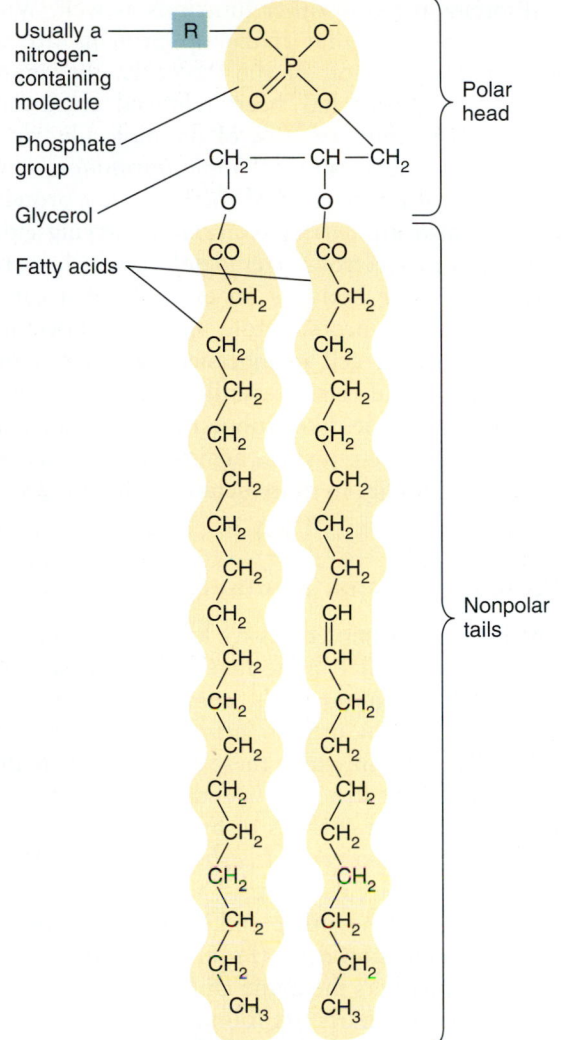

Usually a nitrogen-containing molecule

Phosphate group

Glycerol

Fatty acids

Polar head

Nonpolar tails

(a)

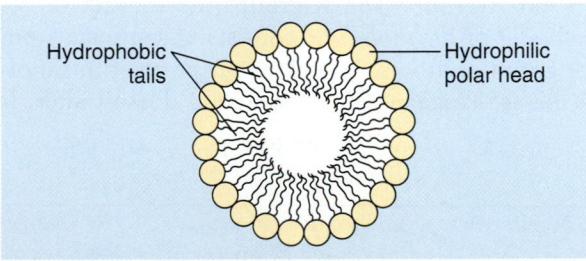

Hydrophobic tails

Hydrophilic polar head

(b)

FIGURE 3.10 Phospholipids
(a) Note the polar head and the nonpolar tails of the phospholipid. (b) Because phospholipids consist of a polar head and nonpolar tails, when placed in water they tend to arrange themselves with the head end pointing into the water and the nonpolar tail region pointing away from the water. The head is said to be hydrophilic ("water-loving"), and the tails are said to be hydrophobic ("water-fearing").

Sex hormones (such as estrogens and androgens), cholesterol, and vitamin D are examples of **steroids.** Steroids are made up of 17 carbon atoms, which form four rings. Even though the structure of steroids is not like that of most lipids, they are still classified as lipids because the two groups are insoluble in water.

Cholesterol is familiar because of its association with heart disease. However, cholesterol is important in humans for other reasons. For instance, other steroids, such as sex hormones, are derived and synthesized from cholesterol. Humans either take in cholesterol that is already formed, by eating foods containing the substance, or their livers synthesize the substance from fats. Because of the existing evidence that a high cholesterol level in the blood is responsible for arteriosclerosis (hardening of the arteries) and other cardiovascular disorders, physicians often put heart patients on a low-cholesterol diet, to remove one source of the problem (Figure 3.11).

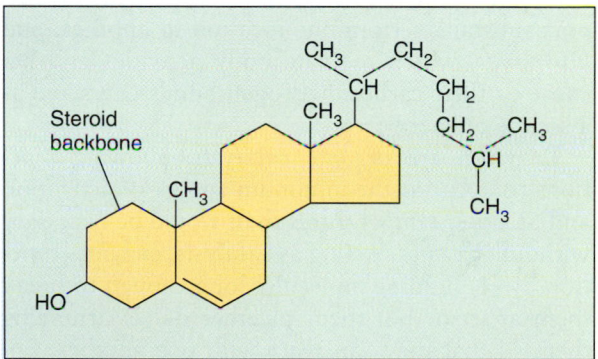

Steroid backbone

(a) Cholesterol

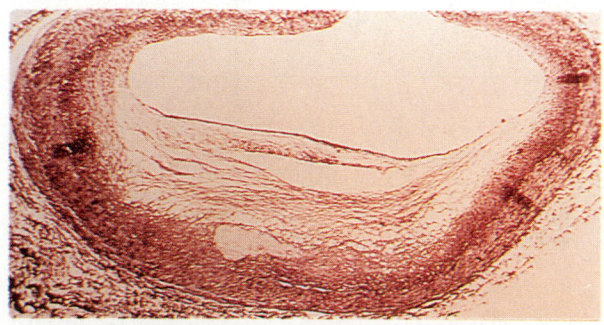

(b) Partially clogged artery

FIGURE 3.11 Cholesterol
(a) The structural formula of cholesterol shows the distinctive four-ring structure common to steroids. (b) This section taken from a coronary artery shows an inner ring of cholesterol deposits that have effectively reduced the diameter of this major blood vessel by half.

Terpenes are steroidlike lipids derived from five-carbon subunits. Usually these subunits form a long, water-soluble polymer. This polymer is a very important part of the *chlorophyll* molecule, which is the green pigment in plants responsible for trapping light energy during photosynthesis.

PROTEINS

■ **Proteins are composed of amino acids. There are 20 different types of amino acids. When amino acids bond together, the linkage is called a peptide linkage. Proteins provide structure, function as enzymes, carrier molecules, hormones, and provide energy.**

Proteins are involved in all aspects of life. These versatile molecules, which are made up of *amino acids,* are often called the building blocks of life because they serve as the major structural components of living things. Proteins also function in a variety of roles, including the following: *enzymes,* which, as we have already pointed out, mediate chemical reactions; *carrier molecules,* which transport substances from one location to another; and *hormones,* which regulate body activities and because of their carbon hydrogen bonds can serve as a source of energy.

Enzymes are *organic catalysts,* allowing reactions to occur at the minimum energy of activation and at lower temperatures than would be necessary without enzymes. Acting as catalysts, enzymes function by (1) holding molecules long enough to break them apart or put them together or (2) arranging them so that they can react with one another. Enzymes participate in almost all chemical reactions in living cells, from digesting food to synthesizing DNA.

Proteins perform other functions as well. When the amino acids that make up proteins are *deaminated,* a portion of the molecule, the amino-functional group, has been removed chemically. Once this is done, the rest of the molecule can be used as an energy source. Some *hormones,* which are chemical messengers of the body, are proteins. Proteins also are *carrier molecules,* carrying either molecules or electrons. Hemoglobin, which carries oxygen in your body, is an example of a carrier molecule. Other protein molecules are important in the structure of cell membranes. Obviously, proteins are a very diverse, versatile group of biological compounds. Some, like globular proteins called antibodies, even help protect against disease. We will discuss antibodies in more detail in Chapter 23.

Protein Subunits: Amino Acids

The large macromolecules we know as proteins are made up of smaller monomers, called **amino acids.** There are 20 amino acids, composed of atoms of carbon, hydrogen, oxygen, nitrogen, and sometimes sulfur. The way in which these atoms are structured is important, as structure is with all chemical units. Each amino acid consists of five parts. (1) Basic to the monomer is a *central carbon atom.* (2) At one end of the molecule there is an *amino functional group.* (3) At the other end, there is a *carboxyl group,* which is an acid. (From these two sections, the monomer gets the name *amino acid.*)

The last two parts are found between the amino end and the acid end: (4) a *hydrogen molecule,* and (5) an *R-group.* The R-group (the radical, or *remainder of the molecule*) consists of a specific atom or group of atoms. The first four parts of this molecule are constant. Their structure doesn't alter. It

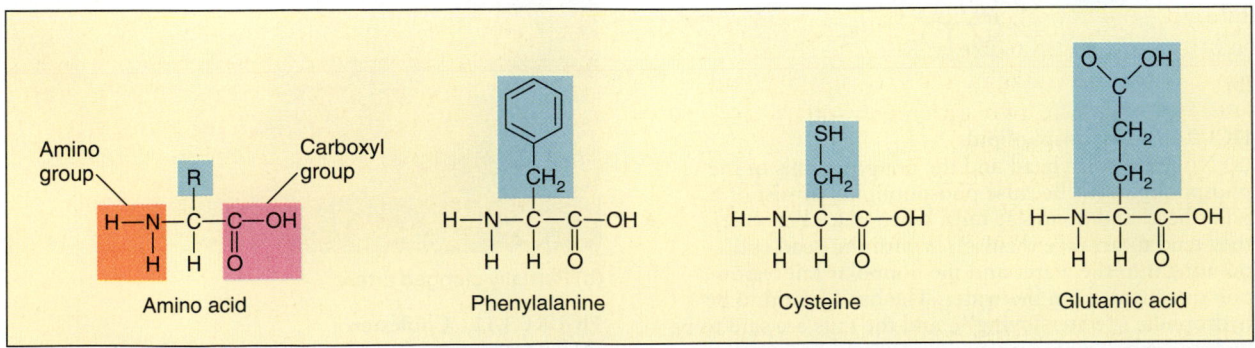

FIGURE 3.12 Amino Acids
All amino acids have an amino group, a carboxyl group, and a hydrogen atom grouped around a carbon atom. Amino acids differ only in their R-groups, highlighted here in a blue-green color.

Chapter 3   The Molecules of Life

is the **R-group** that varies in chemical structure and determines the identity of the amino acid (Figure 3.12).

Peptide Linkage. When a living organism ingests protein (as I did with my breakfast), which type of reaction occurs in order to break down that protein into its amino acid subunits? That's right, a hydrolysis (or water-splitting) reaction. Then, in order to form the proteins that my body needs, a dehydration synthesis takes place which, as you recall, means that water is removed from the protein. To form the H_2O, the needed atoms are taken from adjacent amino acids. A hydroxyl (OH) is taken from the carboxyl group at the end of one amino acid, which leaves a carbon atom capable of bonding with another atom. A hydrogen atom (H) is taken from the amino group of the adjacent amino acid, which makes a nitrogen atom (N) capable of bonding with another atom. The C and the N of the two adjacent amino acids then are covalently linked together in what is called a **peptide linkage.**

When two amino acids are linked together, the result is a dipeptide (Figure 3.13). But as you can see, at one end of the dipeptide there is still a free carboxyl group, and at the other end there is a free amino group. By repeating the linkage procedure, more amino acids can be added to create a longer and more complex chain of amino acids. Medium-sized polymers are called **polypeptides.** When the polymer becomes more than about 50 amino acids long, it is called a protein.

The Amino Acid Alphabet. Which amino acids can join together? Since the structure of all includes a carboxyl and an amino group, any one of the 20 amino acids can link up with any other, which results in the possibility of tremendous variety. The identity of a specific protein is the result of the number and sequence of the amino acid subunits. If you think of the amino acids as letters in an alphabet, and proteins as the words created by those letters, you can begin to imagine the diversity that can result. For example, two different proteins may be made up of the same 100 amino acids, but because the sequence of those amino acids differs, the proteins are different, just as the order of the letters *a*, *t*, and *p* determines whether the word they make is *tap* or *pat*.

To fully appreciate the significance of the flexibility of protein construction, compare it with the diversity you find in the English language. It contains 26 letters, and some letter combinations never occur! In fact, if you work out the calculations, you will find that in a protein of only 100 amino acids,

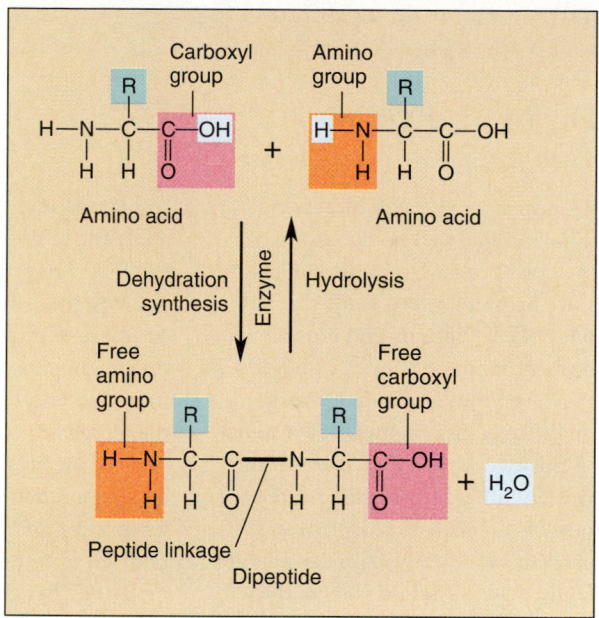

FIGURE 3.13   The Synthesis and Hydrolysis of a Dipeptide
Because of the free amino group at one end of the molecule and the carboxyl group at the other end, amino acids can bond together to form long chains called polypeptides.

there could be 20^{100} possible sequences. That number means 20 times 20 calculated 100 times, and the result would be millions of times larger than the number of grains of sand on the earth.

Protein: Levels of Organization

■ **The levels of organization of proteins can be described as primary, secondary, tertiary, and quaternary. The structure of the protein determines its biological function.**

The diversity we find in proteins is not only a result of the number of amino acids and their various sequences. Their shape, or configuration, also dictates how they will interact chemically. Biologists talk about four levels of protein structure: *primary, secondary, tertiary,* and *quaternary* structure.

The **primary structure** of a protein molecule refers simply to the linear sequence of its amino acids.

Secondary structure refers to the way a protein molecule folds or coils in space. As a protein is formed, there is repeated hydrogen bonding between every fourth amino acid in the chain. In some molecules this creates a circular staircase shape, called a *helical pattern.* This pattern,

Dr. Linus Pauling
Theoretical Chemist, Peace Activist

"A neat desk is a sign of a sick mind" might well be the motto of the late chemist Dr. Linus Pauling, since he preferred to wade through piles of calculations and notes heaped on his desk or elsewhere. However, though Pauling's desk might have been cluttered, the mind of this two-time Nobel Prize winner was anything but the same. Even into his nineties, this controversial chemist led as active a life as anyone could desire. Pauling still conducted research into the structure of molecules, advocated vitamin C in large amounts to fight off colds and even increase intelligence, and supported antiwar nuclear arms control. His life was full.

The son of a pharmacist, Pauling was born in 1901 in Portland, Oregon. A naturally curious individual, at an early age he took an active interest in rock collecting and chemistry, even conducting experiments in a laboratory in his home. In 1922, Pauling graduated from Oregon Agricultural College (now Oregon State University) with a degree in Chemical Engineering. From there he moved on to the California Institute of Technology, where he plunged into more chemistry and was stimulated by the enormous impact that atomic physics was having on his field. At the same time he began to develop his interests in the nature of the chemical bond. He completed his Ph.D. in 1925 and went to study physics in Munich, Germany.

In Germany, Pauling began to apply the laws of quantum mechanics (then a new area in physics, dealing with the nature of energy) to the structure of molecules and to the chemical bond. In 1927, Pauling returned to Cal Tech, and in 1930, in the *Journal of the American Chemical Society,* he published his work in a paper entitled "The Nature of the Chemical Bond." This paper revolutionized scientific thinking about molecular structure and chemical bonding. The value of Pauling's work was formally recognized in 1954 when he was awarded the Nobel Prize in Chemistry.

In 1935, Pauling turned his interests to biochemistry and began to study human hemoglobin, the oxygen-carrying molecule in the blood. Ultimately, he was able to describe its structure. He studied sickle-cell anemia, a genetic disorder prevalent among blacks. The major symptoms are the distinctive sickle shape of the red blood cells and their reduced oxygen-carrying capabilities. Pauling determined that the symptoms are caused by a change in the sequence of amino acids in the hemoglobin molecule. This was the first direct evidence relating a disease to a defect in a molecule.

Pauling continued to have an impact on the scientific community. He also came up with the idea that many reactions are a result of the shape of the molecules involved. For example, an antibody identifies an infectious agent in the bloodstream

by its molecular configuration.

Like most scientists, Pauling did not fit the stereotype of the scientist who hides in the laboratory, protected from and unaware of society. Rather he was a social activist. As a measure of his dedication to issues involving work for peace and humanitarian causes, he was awarded the Nobel Peace Prize in 1962 — his second Nobel Prize.

Linus Pauling was a respected chemist, a strong supporter of vitamin C megadoses to avoid colds (a position that drew heavy criticism from some other scientists), and an outspoken opponent of war and nuclear armament. He was a man whose scientific and social endeavors affected and will continue to affect all our lives.

Dr. Pauling died in the summer of 1994.

which was first described in 1948 by Linus Pauling, is called an alpha helix (see Profile). Usually fibrous proteins, such as those in hair, have only primary and secondary structures; there are other secondary structures such as the β pleated structure and random coil.

Tertiary structure of a protein describes the way some proteins fold back on themselves to create a three-dimensional structure. Such proteins are called *globular proteins.* In order to take on this structure, the R-groups in the protein molecule interact with each other, *or* the carbon backbone interacts with itself. The protein becomes permanently folded into a unique, specific three-dimensional shape. For instance, cysteine amino acids contain a sulfur atom. When two cysteine amino acids exist in a protein, they form a strong disulfide bridge (S—S). This chemical linkage twists the chain to form a permanent fold or kink in the protein's shape. The unique shape of a protein helps to determine how the molecule will interact with other molecules, especially in the case of enzymes.

A protein exhibits **quaternary structure** when two or more individual polypeptide globular chains interact and twine around themselves to form a single complex protein molecule. Hemoglobin, the oxygen-carrying pigment in the blood, is a good example of a protein with quaternary structure. It contains four polypeptide chains.

Figure 3.14 shows the levels of protein organization in the hemoglobin molecule.

Hemoglobin is a good example of how the four structural characteristics of proteins interact with each other. Two of the chains are called *alpha chains,* and each contains 141 amino acids. The other two chains are called *beta chains,* and they each contain 146 amino acids. Biologists now know that about 200 possible variations can occur in their order (their primary structure). Some variations do not affect the function of the molecule, but others can cause severe problems. One such abnormality is responsible for the disorder known as sickle-cell anemia, found predominantly among blacks. In sickle-cell hemoglobin, only one amino acid out of 146 in a beta chain differs from the normal hemoglobin amino acid sequence. In the sixth position, the amino acid valine is substituted for glutamic acid.

Because of this one substitution, each polypeptide chain forms an altered helical twist (its secondary structure), the twists create different bonds

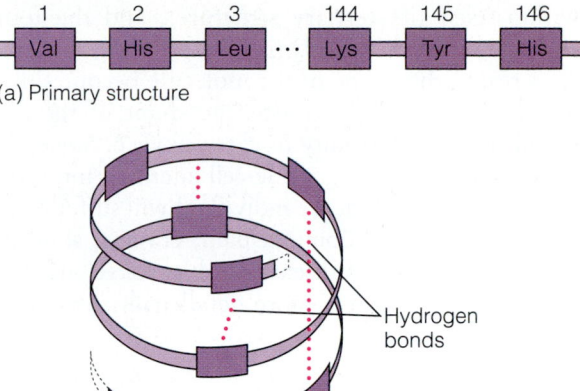

(a) Primary structure

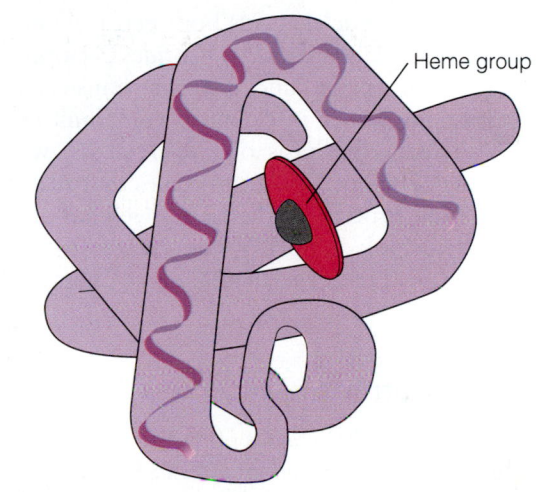

(b) Secondary structure

Hydrogen bonds

Heme group

(c) Tertiary structure

Polypeptide chain

Polypeptide chain

Polypeptide chain

Polypeptide chain

(d) Quaternary structure

FIGURE 3.14 Protein Structure—Hemoglobin
(a) The primary structure of a protein is its sequence of amino acids. (b) The secondary structure is the shape the protein assumes as hydrogen bonds form between its amino acids. (c) The tertiary structure is the three-dimensional folding of the polypeptide chain. (d) The quaternary structure is the ultimate shape of a protein consisting of more than one polypeptide chain.

within itself (its tertiary structure), and the four chains interact differently (its quaternary structure). As a result, the shape of the molecule becomes distorted, which in turn affects the shape of the red blood cell and its ability to carry oxygen. Since the cells of anyone having sickle-cell anemia cannot receive enough oxygen, the individual will suffer from severe joint and abdominal pain, cramps, skin ulcers, chronic kidney disease, and unconsciousness. (See Chapter 15 for a more detailed discussion of sickle-cell anemia.)

NUCLEOTIDES AND NUCLEIC ACIDS

Nucleotides and nucleic acids both perform unique functions in living systems. *Nucleotides* perform the basic tasks of energy and electron transfer, and work with hormones in chemical communication. They also form the basis of *nucleic acids,* which form either DNA or RNA, molecules that regulate reproduction, protein synthesis, and cellular activities. We will look at nucleotides and nucleic acids here briefly, and return to them later in Chapters 11 and 12.

Nucleotide Structure and Function

■ **Nucleotides, the subunits of nucleic acids, also perform specific functions in living things by holding or transferring energy.**

Nucleotides, which are the subunits (building blocks) of nucleic acids, are important in their own right, because they function in cells as energy carriers, chemical messengers, and electron transporters. There are three basic components in each nucleotide (Figure 3.15): (1) a *five-carbon sugar,* which is either ribose or deoxyribose; (2) a *nitrogen-containing base,* either a purine (which has two carbon-based ring structures, called a *double ring*), or a pyrimidine (which has one carbon-based ring structure, called a *single ring*); (3) and a *phosphate group.*

There are a number of nucleotides that you will read about again and again as you work through this text. One is called *adenosine triphosphate* (ATP), and it is often called the energy currency of the cell. ATP has gained this nickname because it creates bonds that hold a high level of chemical energy. Other nucleotides, such as *nicotinamide adenine dinucleotide* (NAD⁺) and *flavin adenine dinucleotide* (FAD), function as electron transporters, playing an essential role in the processes by which living things change energy from one form to another. Still another nucleotide, known as *cyclic adenosine monophosphate* (cAMP), functions as a chemical messenger, assisting many hormone molecules as they carry their chemical messages in cells.

Nucleic Acid Structure and Function

■ **Nucleic acids are single or double nucleotide strands. DNA is a double strand that regulates protein synthesis and reproduction. RNA is a single strand that is involved in protein synthesis.**

When nucleotides join together, they create **nucleic acids.** Nucleic acids are large polymers that can be

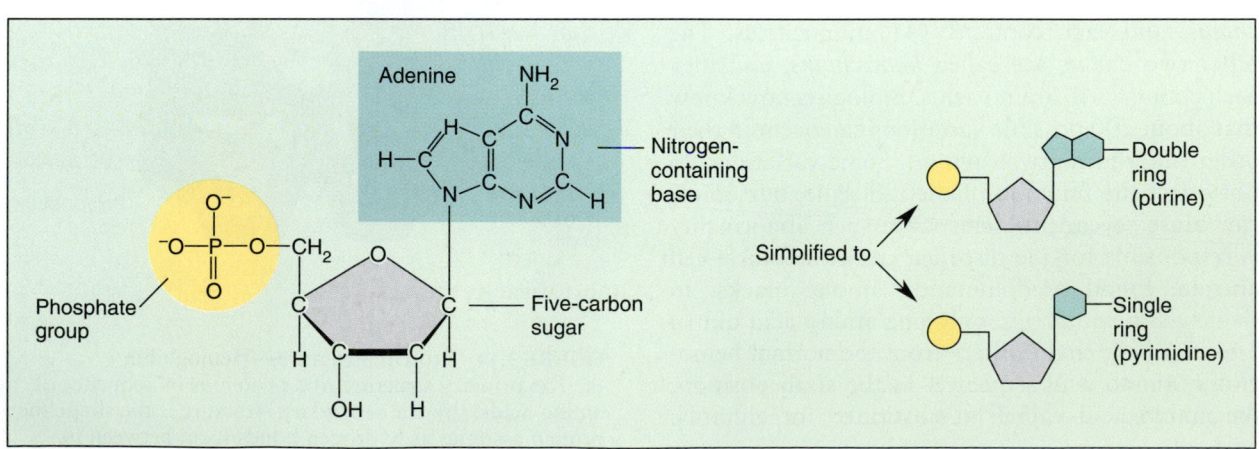

FIGURE 3.15 A Nucleotide
A nucleotide is composed of a five-carbon sugar, a phosphate group, and a nitrogen-containing base (in this case, adenine).

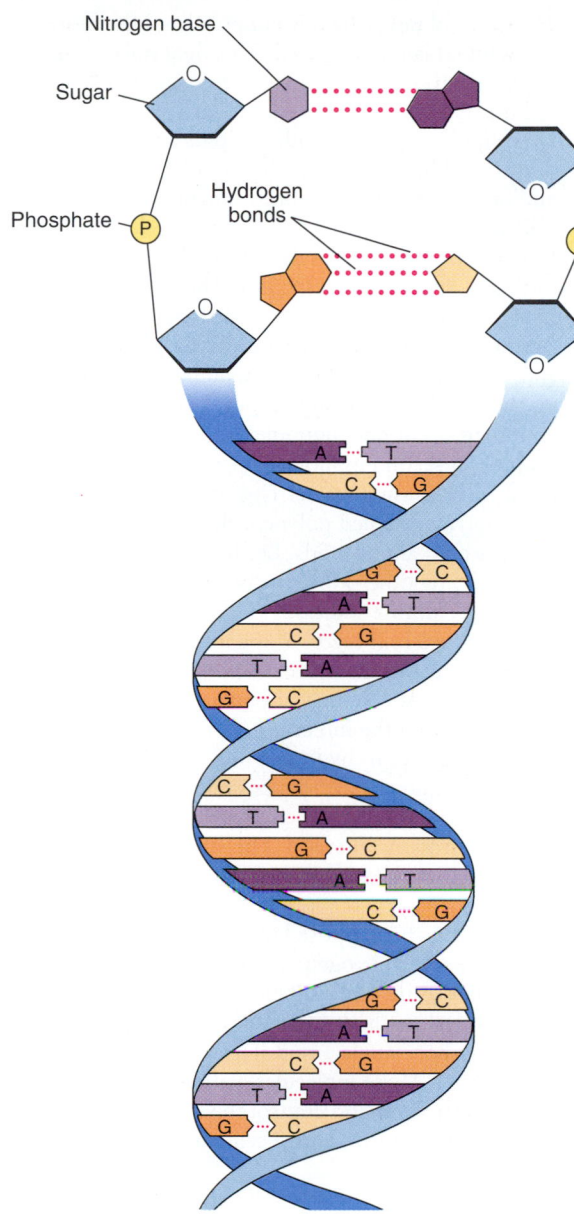

(a)

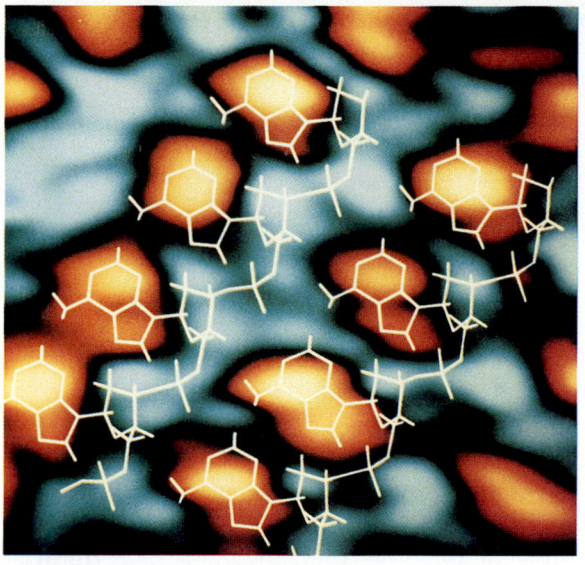

(b)

FIGURE 3.16   Deoxyribonucleic Acid
(a) Deoxyribonucleic acid (DNA) is a double-stranded string of nucleotides, held together by hydrogen bonds. (b) A scanning-tunneling micrograph shows single-stranded DNA consisting of four nucleotides of deoxyadenine (magnification 25,000,000×).

ample, an adenine nucleotide includes an adenine group. Since the DNA molecule consists of two long strands arranged in a twisted ladder shape, it is said to be a **double helix.** The strands are held together by hydrogen bonds between the purines and pyrimidines — adenine is bonded to thymine and guanine is bonded to cytosine (Figure 3.16).

Other important nucleic acids are known as **ribonucleic acids** (**RNA**), of which there are several types. All RNA molecules consist of nucleotides that have the sugar *ribose* instead of deoxyribose (hence the name ribonucleic acid). In addition, they are usually single-stranded. The nucleotides in RNA can contain three of the nucleotides found in DNA — adenine, cytosine, and guanine — but the pyrimidine uracil replaces the thymine found only in DNA.

Carbohydrates, lipids, proteins, and nucleic acids — as foreign as they may seem — are the essence of living things. When you look about you, at your reflection in the mirror or at other living things — dogs, cats, blossoms on a dogwood tree, organisms swimming in a drop of pond water — it is difficult to realize that the life you see is the result of chemical reactions. Living things are indeed made up of molecules, and it is the functioning of these molecules that eventually results in life.

either single- or double-stranded. Some are cloverleaflike with the leaves pointing in or out in a three-dimensional representation; others form long threads or circles.

The double-stranded nucleic acid **deoxyribonucleic acid** (**DNA**) is the molecule that functions as the basis of heredity. DNA consists of a string of nucleotides, and each nucleotide contains (1) the sugar *deoxyribose;* (2) a *phosphate;* and (3) a *nitrogen base,* either a purine (adenine or guanine) or a pyrimidine (thymine or cytosine). The nitrogen base gives its name to the specific nucleotide — for ex-

SUMMARY

1. In general, the important biological compounds found in all living things can be grouped into carbohydrates, lipids, proteins, and nucleotides.

2. It is widely accepted that chemical reactions occur because of the collision of atoms and molecules. The amount of energy needed to activate the reaction is called the minimum energy of activation.

3. There are many types of chemical reactions, but for convenience scientists divide them into four groups: rearrangement, synthesis, degradation, and displacement reactions.

4. During hydrolysis, polymer molecules are broken down to their monomers by adding water. During dehydration synthesis, the simple monomer units are strung together to form polymers by removing water.

5. Carbohydrates have the basic $[C(H_2O)]_n$ structure, and function as the primary cellular energy source. Starches and sugars are the chief examples. Some carbohydrates such as cellulose provide structure to cells.

6. The simplest unit of a carbohydrate is called a monosaccharide. Typically it consists of sugars having three, four, five, or six carbons.

7. When the molecular formula for two compounds is the same, but the structural formula is different, the compounds are called isomers. Isomers have different chemical properties.

8. When two monosaccharides are bonded together through dehydration synthesis, a disaccharide is formed.

9. When long chains of sugar molecules are bonded together through dehydration synthesis, a polysaccharide is formed. Polysaccharides function as structural components and important energy-storage molecules.

10. Lipids are made up of carbon (C), hydrogen (H), and oxygen (O). They do not have the $[C(H_2O)]_n$ ratio of carbohydrates. Sometimes they have phosphate, nitrogen, or other molecules associated with them. Lipids are insoluble in water but soluble in nonpolar solvents. They function in energy storage, hormones, and important cellular structures, such as membranes.

11. True fats, triglycerides, are made up of three fatty acids that are bonded to a three-carbon alcohol (usually glycerol) through dehydration synthesis. In unsaturated fatty acids, not all the carbons are bonded to hydrogen atoms. Instead the carbons form double bonds. In saturated fatty acids, all the carbon molecules in the carbon skeleton are bonded with hydrogen molecules.

12. Phospholipids are composed of glycerol and two fatty acids, with a phosphate bonded to the glycerol. When the molecules are placed in water, the phosphate end (which is polar and therefore water-soluble) is attracted to the polar water molecules.

The glycerol tail (which is nonpolar and therefore nonsoluble) is not attracted. This dual polarity is important to the functioning of phospholipids in cell membranes.

13. Glycolipids are compounds of lipids and carbohydrates, and lipoproteins are compounds of lipids and protein. Both compounds function in cell-to-cell communication.

14. Proteins are made up of carbon, hydrogen, oxygen, nitrogen, and sometimes sulfur. The monomers that make up their structures are called amino acids.

15. Amino acids are composed of an acid group on one end, a variable central portion, and an amino group on the other end. During dehydration synthesis, the amino group of one monomer links with the acid group of a second monomer, forming a peptide bond and releasing water. This linkage can form large polymers called polypeptides.

16. There are 20 amino acids. Because each kind can link with any other kind, the potential for variety in proteins is almost endless.

17. Proteins have four levels of organization: primary (which is the linear sequence of amino acids in the chain), secondary (a helical pattern due to hydrogen bonding within the molecule), tertiary (more complex folding due to hydrogen and disulfide bonding, forming a unique three-dimensional shape), and quaternary structure (in which two or more globular proteins interact). The structure of the protein molecule determines how the protein will function.

18. Nucleotides are composed of a sugar (ribose or deoxyribose), a nitrogen base (either a purine or pyrimidine), and a phosphate group.

19. Nucleotides are the subunits of nucleic acids, such as deoxyribose nucleic acid (DNA) or ribonucleic acid (RNA). These molecules contain the cells' genetic information and control the synthesis of proteins, which in turn control cellular functions.

KEY TERMS

chemical reaction	steroid
synthesis	protein
macromolecule	amino acid
polymer	R-group
hydrolysis	peptide linkage
dehydration synthesis	polypeptide
carbohydrate	primary structure
monosaccharide	secondary structure
disaccharide	tertiary structure
polysaccharide	quaternary structure
lipid	nucleotide
Calorie	nucleic acid
saturated fat	deoxyribonucleic acid (DNA)
unsaturated fat	double helix
phospholipid	ribonucleic acid (RNA)
receptor site	

REVIEW QUESTIONS

True/False

1. The removal of a molecule of water from two molecules to link the two molecules together is called dehydration synthesis.
   true   false   (page 54)

2. The collision theory is one explanation for how chemical reactions occur.
   true   false   (page 52)

3. The empirical formula of a molecule gives the scientist more information than a structural formula.
   true   false   (page 55)

4. Phospholipids are important structural components in cell membranes.
   true   false   (page 64)

Fill in the Blank

1. The following reaction is an example of a _____ reaction:
   $2H_2 + O_2 \rightarrow 2H_2O$ (page 52)

2. The general term for the simplest unit of a carbohydrate is a _____. (page 55)

3. The _____ _____ of activation is required for a chemical reaction to occur. (page 52)

4. The linear sequence of amino acids in a protein is called its _____ structure. (page 67)

Multiple Choice

1. Which of these combinations would be found in a nucleotide?
   a. base-acid-salt
   b. adenine-thymine-uracil
   c. base-sugar-phosphate group
   d. DNA-RNA-nucleus
   (page 70)

2. If two amino acids are joined together, the resultant molecule is:
   a. a protein   c. a fatty acid
   b. a disaccharide   d. a dipeptide
   (page 67)

3. A triglyceride is composed of:
   a. monosaccharides and glycerol
   b. fatty acids and enzymes
   c. fatty acids and glycerol
   d. peptides and fatty acids
   (page 61)

4. Organic catalysts such as enzymes are important because:
   a. they are structural components
   b. they lower the minimum energy of activation
   c. they raise the minimum energy of activation
   d. they are carbohydrates
   (page 66)

Discussion

1. List and describe the four important biological compounds in living things.

2. Many of the synthetic reactions in living things involve dehydration synthesis. What is dehydration synthesis?

3. Many of the degradation reactions in living things involve a type of reaction known as hydrolysis. What is hydrolysis?

4. What are phospholipids, and why is the polar nature of a phospholipid important?

5. Distinguish between saturated and unsaturated fats. How do these relate to human health conditions?

6. What are proteins? Discuss the two major functions of proteins in living systems.

7. Why is the structure of a protein so important to its function?

8. Nucleotides and nucleic acids are essential in living things. Why?

4

The Cell: Where It All Happens

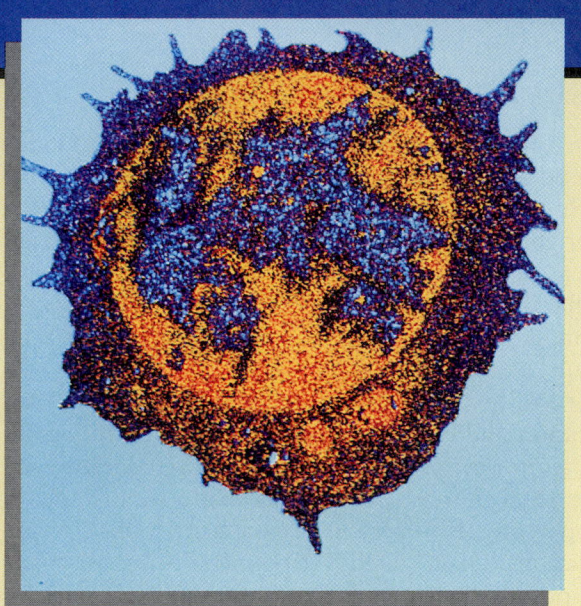

So far our journey through life has concentrated on the amazing chemical reactions that produce the phenomenon we call life. The next obvious question is "Where do all these reactions happen?" You probably know the answer. They happen in that fundamental unit of life known as the cell.

But what are cells? How can we study them? Why do cells exist? What are they made up of? What do they do? These are the types of questions biologists have asked innumerable times over the centuries and continue to ask today. However, the answers to many of them could not be found until the technology of the laboratory had advanced. Because the average cell is invisible to the naked eye (that is, it is microscopic), mechanical means had to assist biologists in observing this uncharted world.

Above: A transmission electron microphotograph of a human white blood cell has been given color by use of the computer.

Until scientists had access to today's powerful lenses, our journey into inner space could only barely begin.

MICROSCOPES AND THE CELL THEORY

■ **Cell theory, which originated in the early 1800s, states that almost all living things are made of cells and that all cells arise from preexisting cells.**

Human curiosity about cells and their invisible activities is not new. The ancient Aztecs experimented with lenses and in all probability viewed the microscopic world. Early in the seventeenth century, Galileo used lenses to view various biological specimens. For a time in that century, in both France and England, lenses were considered merely fascinating toys. Then individuals like Anton van Leeuwenhoek and Hans and Zacharias Janssen incorporated lenses into the first microscopes and began to examine the microscopic world in more detail.

Early Explorers

In 1665, an English scientist named Robert Hooke used a primitive microscope in a radically new way. Hooke sliced off a very thin piece of cork, shined a bright light through it, and examined it with his lens. There he observed tiny empty compartments that reminded him of empty rooms (or *cellulae*), similar to the ones in which monks lived. As a result, Hooke called them "cells."

Most early scientists limited their work to observation and did not attempt to interpret the significance of the cells Hooke had discovered. However, in 1805, Lorenz Oken, a German naturalist, developed the **cell theory** in response to Hooke's observations. Oken stated, "All organic beings *originate* from and *consist* of vesicles or cells." Note that this theory contains two important components: (1) all living things come from preexisting cells, and (2) all living things are composed of cells. Oken's theory is still accepted in biology today, even though biologists now know that there are some exceptions, such as viruses.

Even though Oken originated the cell theory and went on to develop it, credit for it, as sometimes happens in the world, has been given to someone else — usually to two German biologists, Matthias Jakob Schleiden, a botanist, and Theodor Schwann, a zoologist. Although Schleiden and Schwann were working apart, in 1839 they simultaneously published essentially the same statement that Oken had made almost 40 years earlier. Fifty years after Schleiden and Schwann, another German, Rudolf Virchow, condensed the cell theory even further when he said, "All cells from cells."

Seeing: A Matter of Size?

Cells vary in size (Figure 4.1). They can be as small as mycoplasmas, which are the smallest living cellular organisms — 0.2 to 1.5 micrometers — or as large as an ostrich egg, which is as large as a softball. (A micrometer, which scientists abbreviate μm, is one-millionth of a meter, or about 1/1000 of the thinness of a dime.) Nerve cells can be very long, even though they are too thin to be seen with the naked eye. For example, a nerve cell in your leg may be as much as a meter long, reaching from the base of your spinal cord down to your toes.

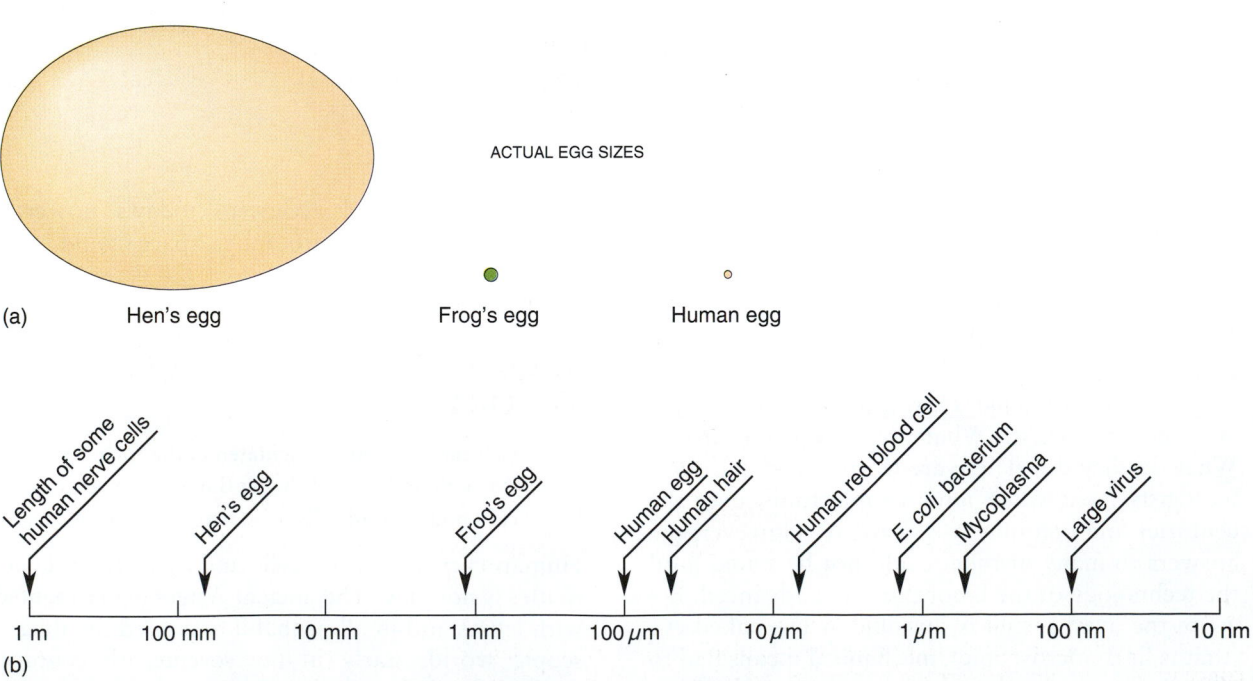

FIGURE 4.1 Range of Cell Sizes
All three eggs shown in (a) are specialized cells and dramatically illustrate the differences in cell sizes. (b) This scale portrays other interesting size relationships. The difference in size between a hen's egg and a human egg is about the same as the difference in size between a human egg and a large virus.

The majority of cells, however, range from 10 to 100 micrometers (10 to 100 μm), which means they are microscopic. Today, for the most part, biologists use three types of microscopes: the *light microscope*, the *transmission electron microscope*, and the *scanning electron microscope*. Each type serves special needs, and each has special strengths and weaknesses, which relate to how a microscope works and to the character of light.

We all know that a microscope magnifies, or enlarges, the apparent size of an object. Even more important than magnification is what is called a microscope's *resolving power*. So, before we discuss the types of microscopes, we need to discuss some of the characteristics of light.

Resolving Power and Microscopes

■ **Light travels in waves of various lengths. The shorter the wavelength, the greater the resolving power, or the ability to distinguish between two points. Light microscopes cannot clearly present structures that are less than 200 nanometers in size.**

Resolving power is a factor in all three types of microscopes and in unaided vision as well. Have you ever looked at a car's headlights off in the distance coming toward you on a summer's night? At first, the two headlights appear blurred and as one, but as the car comes closer, the image separates, and you can distinguish two distinct headlights.

This ability to distinguish between two points as separate entities is called **resolving power.**

Resolving power has to do with light and its characteristic wavelengths, which we will mention here only briefly (see Chapter 8 for more detail). Light travels in waves of various lengths. The range of the light waves makes up what is called light's *spectrum*. The colors that you see in a rainbow or in the pattern created when you pass light through a prism reveal a portion of the spectrum of light. You see a number of different colors because each wavelength results in its particular color. In other words, the red and blue of a rainbow are the result of two different wavelengths.

The unaided human eye uses only part of the light spectrum (or electromagnetic spectrum) in order to see. That part is called **visible light.** (Compare this with the ability of some snakes to detect heat, because of special sensors near their noses that can detect infrared light.) The wavelengths of visible light are fairly long, but some wavelengths are too long to be detected by the human eye. Others, such as ultraviolet — the type of light that is one of the prime causes of skin cancer — are too short to be visible to the human eye.

The **light microscope** works by passing light through the specimen and then through lenses that magnify and focus the image. The resolving power of a light microscope is limited by the wavelengths of light: The microscope cannot distinguish between objects that are closer together than the length of one wavelength (Figure 4.2a). The best light microscopes have a resolving power of about

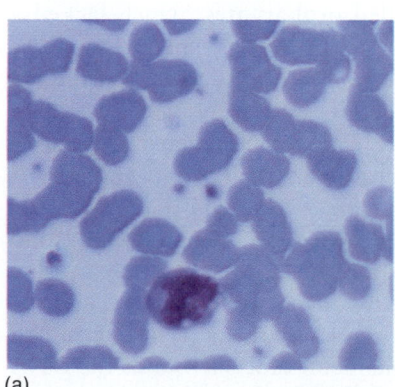
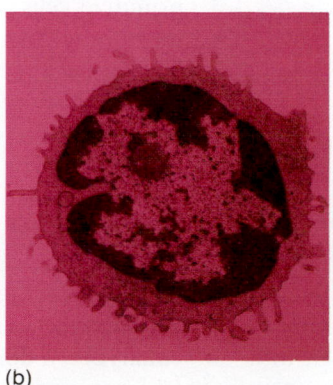
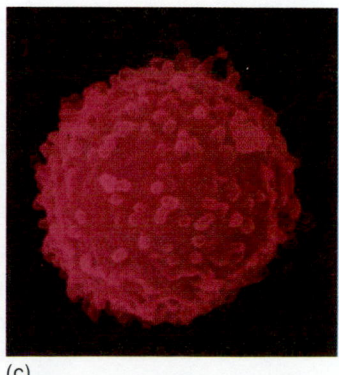

(a) (b) (c)

FIGURE 4.2 Viewing the Microscopic World
(a) A light micrograph of this monocyte (white blood cell) shows its relative concentration among the surrounding red blood cells. Enough detail of the nuclear material (stained purple) is visible to be able to identify the cell.
(b) With the greater resolution of this transmission electron micrograph, the internal details of the monocyte become more clearly defined. (c) A scanning electron micrograph of monocytes completes the picture by giving us a sense of its external, three-dimensional structure.

0.2 μm, or 200 nanometers (nm). (A nanometer is 0.001 μm.) This means that light microscopes cannot clearly show any object smaller than 200 nanometers.

Transmission Electron Microscopes

■ **Transmission electron microscopes direct electrons through a specimen and can present clear images up to 0.5 nanometer in size.**

If you use shorter wavelengths, like the wavelengths of vibrating electrons, you can greatly increase the resolving power of a microscope to about 0.5 nm, which is approximately 200,000 times greater than the unaided eye. The **transmission electron microscope** (TEM) applies this principle to cell observation by directing electrons through the cell (Figure 4.2b). Another type of electron microscope is the high-voltage electron microscope, which has enough power to assure that an electron beam can pass through an intact living cell. This enables scientists to view living cells or thick sections of cells or tissues.

The complex of equipment required for a transmission electron microscope is substantial. Believe it or not, some high-voltage transmission electron microscopes, like the one at the University of Colorado at Boulder, are so large that maintenance workers must use ladders to work on the top of the two-story structure. Because of the size (and the expense) you won't find many such microscopes in the typical college biology lab.

Scanning Electron Microscopes

■ **Scanning electron microscopes pass a thin beam of electrons over the surface of a specimen and present the illusion of a three-dimensional image.**

One type of microscope is important for reasons other than its resolving power. The **scanning electron microscope,** or SEM, uses an electron beam to scan, or pass over the surface of a specimen in a specific pattern. Passing the beam over the specimen produces a pattern of scattered electrons from the specimen (the specimen itself also emits some electrons). The electrons produce a three-dimensional image on a television screen (Figure 4.2c). So, even though the SEM has a resolving power of only about 10 nm, the three-dimensional aspects of the image make it a very valuable scientific tool (see Focus 4.1).

CELLS AND THEIR BASIC PARTS

■ **Prokaryotic cells lack a membrane-bound nucleus and membrane-bound organelles. Eukaryotic cells contain a cytoplasmic region, a membrane-bound nucleus, and membrane-bound organelles.**

Today we know that **cells** are the fundamental units of life, whose dynamic functioning results in what we call life. They are three-dimensional entities enclosed by membranes, and unlike Robert Hooke's empty cork cells, they contain a fluidlike substance. In 1820, Robert Brown, a botanist, gave the first description of the fluid content of the cell, which is called the *cytoplasmic region.* Brown paid special attention to a structure within the cell now known as the *nucleus.* Today we know that not all cells have nuclei. We also know that a large group of cells contain other membrane-bound structures called *organelles,* or "little organs." The difference between cells with and without internal structures is a significant one; so before we examine cells closely, we need to distinguish between the two types.

Prokaryotic and Eukaryotic Cells

In the world of living things there are two major types of cells, called *prokaryotic* and *eukaryotic* cells. Later in this chapter we will discuss the first primitive cells, called **prokaryotes,** which lacked a membrane around their DNA (genetic material). The modern representatives of those first, primitive prokaryotes are bacteria and cyanobacteria (which until recently were called blue-green algae). In a prokaryote, the area in which its DNA material is located is called the *nucleoid region* (Figure 4.3).

Members of the second fundamental cell group are called eukaryotes. All remaining life forms — that is, everything except viruses, bacteria, and cyanobacteria — are made up of eukaryotic cells. In a **eukaryote,** a double membrane surrounds the DNA to form a structure within the cell called the *nucleus.* This membrane separates the nuclear material from the cytoplasmic region.

Table 4.1 compares structure and function in prokaryotic and eukaryotic cells. With the distinction between the two cell types firmly in mind, let us discuss an imaginary, composite eukaryotic cell.

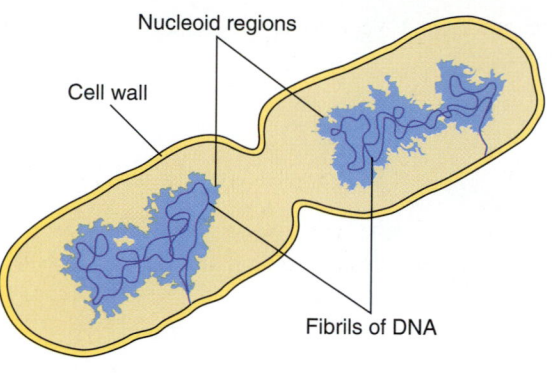

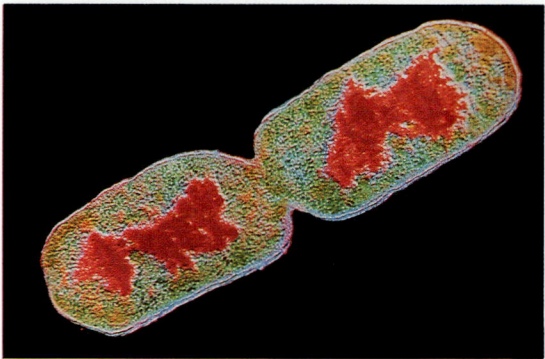

FIGURE 4.3   Prokaryotic Cells
Perhaps one of the most studied organisms of all is the prokaryote *Escherichia coli* —the *E. coli* bacterium. Prokaryotic cells consist of an outer cell boundary (either a cell wall or cell membrane) with the genetic material (DNA) limited to a certain region or regions of the cell. (Because this cell is about to divide, two nucleoid regions are visible.) The genetic material is not surrounded by a membrane.

TABLE 4.1
A Comparison of Prokaryotic and Eukaryotic Cells

Structure	Function	Prokaryotic	Eukaryotic
Cell covering	Support and protection	Capsule, wall	Cell walls in plants
Cell membrane	Regulation of transport of molecules into and out of cell; site of reactions, cell recognition	+, site of reactions	+
Nucleoid	Region of DNA location	+	−
Nucleus	Region of genetic material surrounded by a membrane	−	+
Nuclear membrane	Encloses the nucleus	−	+ with pores
Chromosomes (DNA and protein)	Genetic material	+, but not contained in a nucleus	+
Nucleoli	Ribosome synthesis	−	+
Endoplasmic reticulum	Transportation within the cell	−	+
Ribosome	Site of protein synthesis	+	+
Golgi apparatus	Processing and packaging of materials	−	usually +
Lysosome	Controlled hydrolysis of certain materials	−	+
Mitochondria	ATP production	−	+
Cytoskeletal structure	Structure, support, cytoplasmic organization	−	+
Microtubules	Movement and locomotion	+	+
Flagella/cilia	Movement and locomotion	+, but lacking microtubules	+ with microtubules
Centrioles	Basal bodies of cilia and flagella; cell division	−	+ in all but higher plants
Photosynthetic membranes	Photosynthesis	+ in some	+ in the thylakoid of the chloroplast
Chloroplasts	Photosynthesis	−	+ in plants
Plastids	Pigmented organelles	−	+ in plants

Key: + = present
    − = absent

FOCUS 4.1
How Are Cells Prepared for Viewing?

As you read through this text, you will find many photographs of cells. Some will be of slides under a light microscope, some will be from transmission electron microscopes, and still others will be from scanning electron microscopes. The cytologist must choose from a wide variety of preparation techniques before beginning to examine the tissue or cells. Each technique has a specific purpose, and some are specifically designed to meet the requirements of a given microscope. Let us consider some of these techniques.

If you have ever looked at a living cell under a light microscope, one of the first things you noticed was that living cells are translucent — in other words, most of the light from the scope passes through the cell. The translucence of subcellular structures presents a cytologist with a major problem. How can you make a living "cyte," or cell, less transparent? In other words, how do you increase the contrast between the various cell parts so you can study the cell's structure?

One answer is to stain the cells. Biologists use several pigmented chemicals for staining. And over the years, they have learned that often a given stain is "selective." That is, it will have more affinity for some cellular structures than for others (Figure A). For example, the outer covering of some bacterial cell walls will retain violet dye while that of others will not. This difference in staining properties is due to the affinities of the individual chemicals found in the cell wall. These differential staining techniques have given

rise to an entire branch of cytology, *histochemistry* (from the Greek word *histo,* meaning tissue).

Although the use of a stain can increase the contrast between subcellular structures, it usually kills the cell. In some instances, of course, this is no problem, but sometimes a researcher wants to observe a living cell. To meet this problem, scientists had to develop new techniques, most of which utilize different aspects of the properties of light. For instance, the desire to view living cells led to the development of two new microscopes, the *phase contrast* and *differential-interference* microscopes, both of which enhance the contrast in living cells (Figure B). New video technology also is being incorporated into the biology laboratory. For instance, one method combines a videocamera and a microscope. By using this combination, biologists can view and record activities of living cells immediately. Even

sound waves are being investigated and used to study cells (sort of an "acoustic microscope"). This type of microscope bounces sound waves off a cell in the same way that radar bounces sound waves off an object. Scientists can then observe the shape and structure of the cell they wish to study, as revealed by the sound image.

Biologists observe both cells and tissues, which are groups of cells having a similar structure and func-

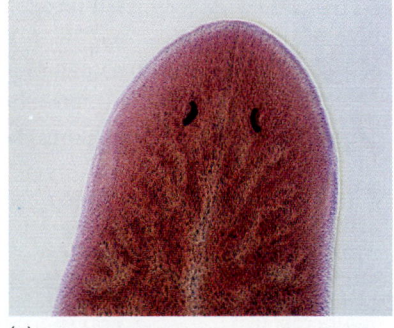

(a)

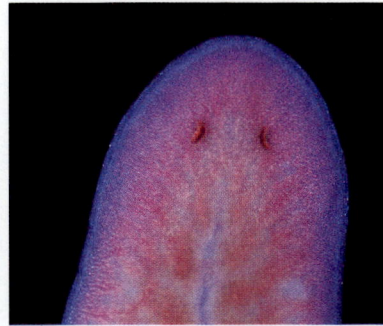

(b)

FIGURE B
The techniques of (a) phase contrast and (b) differential-interference allow biologists to view living specimens, like these planaria (flatworms).

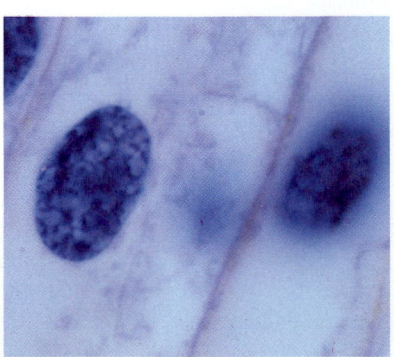

FIGURE A
Selective use of stains in preparing slides allows certain types of cells or parts of cells to be more clearly delineated.

tion. Both cells and tissues are fragile, and sometimes they are also thick. To prevent damaging any specimen, biologists have developed techniques for fixing and sectioning a specimen. (*Sectioning* involves making thin slices.) These two techniques are employed in order to prepare the best specimen possible for observation and study.

First the cells are killed and fixed — treated with chemicals that prevent them from becoming distorted and that also make them more receptive to the stains. Then the cells are embedded in a wax or plastic resin so that they can be sliced very thin. An instrument called a *microtome* is used to slice pieces of the specimen in sections, and then finally the specimen is stained.

Whether a biologist wants to use a light microscope or a transmission electron microscope, the techniques for fixing and sectioning are very similar. The major exception is that in electron microscopy, the specimen is embedded in a plastic resin rather than in wax, is usually fixed with a substance called *gluteraldehyde,* and

is usually supported on a copper grid instead of a glass slide.

However, preparing a specimen for viewing with the scanning electron microscope requires different techniques because, as we mentioned, the electrons in this type of microscope do not pass through the specimen but are deflected by it. When a specimen is prepared, its surface is coated with a thin metal layer. When the electrons from the scanning electron microscope scan over the specimen, secondary electrons from the metal are emitted. These emitted electrons produce a pattern that is detected by an instrument similar to a video camera, and the three-dimensional image is produced on a screen.

Cytologists have developed some techniques that are valuable no matter what kind of microscope is involved. Some of these techniques have been developed to separate and isolate specific cellular organelles. For instance, in any solution of cells, some organelles are denser than others. If you had a solution containing a number of cells. the denser organ-

elles would settle out (or separate themselves from the other cellular structures) faster and accumulate at the bottom of the container. The process is called *sedimentation*. The organelles that are less dense will float on top of the solution. To speed up the time required to separate organelles, biologists use a technique called *centrifugation*. In this technique, the solution is spun in a centrifuge so that the particles suspended in the solution settle out more rapidly than they would normally. Ultracentrifugation can be used to separate out cellular parts or organelles because of their different sedimentation rates. Using a technique such as this, biologists can have as many mitochondria or other organelles to study as they want (Figure C).

Biologists have devised many other techniques that will give us a more accurate, detailed representation of cells and tissues. As new techniques arise, not only can we better image cells but molecules and atoms as well (see the Human Endeavors box in Chapter 2 and Focus 3.1).

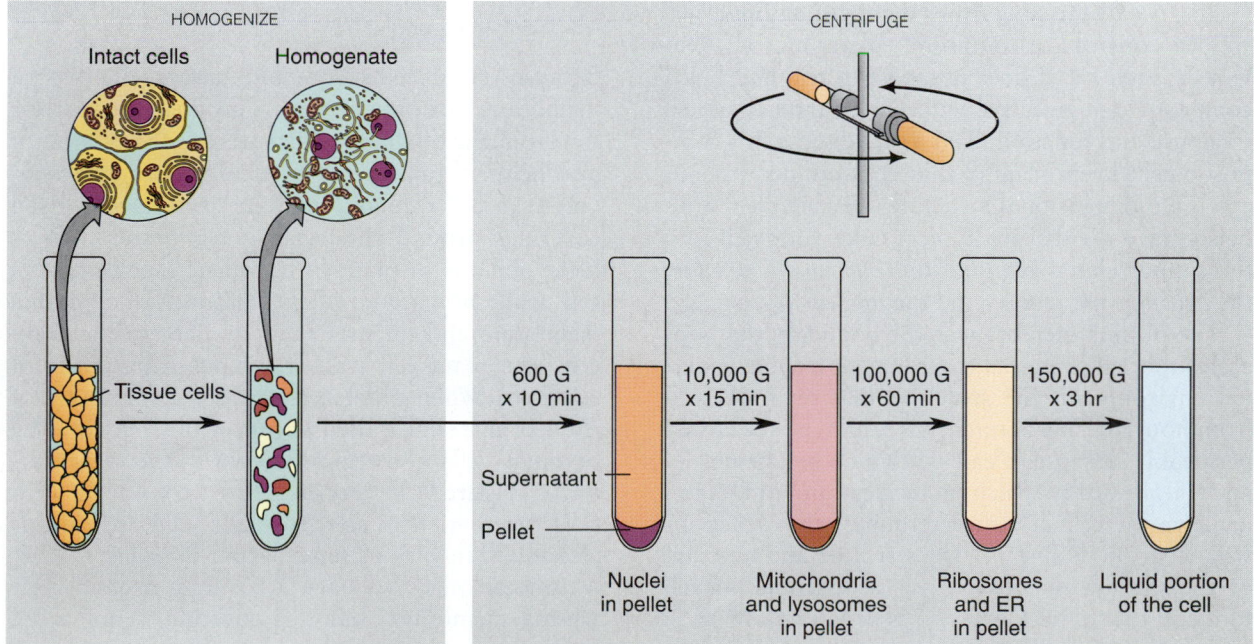

FIGURE C
Cells or tissues are broken into pieces by the use of grinding, ultrasound, or other techniques. Once homogenized, the homogenate (a semiliquid mixture) is spun at various speeds in a centrifuge. As the homogenate spins at increasing speeds, first larger, then smaller cell components are separated out into the pellet.

Cell Structures — A Composite View

■ Eukaryotic cells exhibit tremendous diversity, depending on the function that each serves. However, despite that diversity, they all are cells and contain similar structures, so we can imagine a composite cell that includes universal cellular structures.

When you begin to observe cells, you will quickly learn that there is a tremendous diversity among them. For example, of the more than 60 trillion cells in your body, there are over 200 different types. In addition, there are numerous other kinds of animal cells, plant cells, fungal cells, and bacterial cells. This diversity is due to the complementarity of form and function: Cells have a particular structure because they perform a particular task. Liver cells perform liver functions, chlorophyll-containing plant cells perform photosynthetic functions, and so on. In many respects, the liver cells of a human and a mouse are more similar than are a human's heart cells and liver cells. Function is reflected in form.

However, cells exhibit unity as well as diversity. Despite their differences, they are also similar and possess similar structures. Remember, every cell must carry out essentially the same life functions, and thus cells are similar at both the biochemical level and the subcellular structural level. This does not mean that they are just bags of liquids or that they function like bricks in a wall. Instead they are dynamic, integrated units.

Because of the similarities that exist among cells, we can construct a composite eukaryotic cell. Obviously such a cell does not exist in real life, but a composite will give us a better understanding of cell structure and function. And as we discuss this composite cell, keep in mind that cells are not flat but are three-dimensional structures. Remember, biologists have divided the typical eukaryotic cell into three interrelated regions: the *plasma membrane*, the *cytoplasmic region*, and the *nuclear region*.

The **plasma membrane** is the boundary that separates the cell's internal environment from its external environment, the structure that separates cell from noncell. This boundary is said to be **selectively permeable** because it can determine biochemically and biophysically which molecules can bind, enter, or leave the cell. It also determines what reactions will occur on its surface. The structure and function of the plasma membrane are so essential to cell function that all of Chapter 5 is devoted to them.

The **cytoplasmic region** is the area enclosed by the plasma membrane, with the exception of the nucleus (the region that encloses the genetic material). Within this semiliquid cellular region are embedded the subcellular structures known collectively as organelles. **Organelles** are distinct structures that perform different activities essential to the life of a cell. Organelles are not distributed at random but are organized within the cytoplasm according to the functions that they perform.

The third cellular component is the **nucleus,** which contains the cell's genetic material (DNA). The nucleus is surrounded by a double membrane that separates the nuclear region from the cytoplasmic region.

All eukaryotic cells include a plasma membrane, a cytoplasmic region, and a nuclear region. But animal and plant cells are not alike. As we create our composite cell, we will occasionally note that a structure is found usually in either animal or plant cells alone. Keep these distinctions in mind as they occur. Now, with composite drawings of an animal and a plant cell to guide us, let us discover the parts of a typical eukaryotic cell (Figure 4.4).

COVERINGS AND THE PLASMA MEMBRANE

■ Some cells have coverings that surround their plasma membrane. In plant cells, the covering constitutes a cell wall. In animal cells, the covering consists of glycoproteins, which are involved with cell recognition.

Because of the cell's liquidlike nature, the plasma membrane not only serves to produce the discrete three-dimensional entity that we call the cell, but it also helps to maintain a particular form and is involved in cell recognition. Some cells have evolved coverings around the plasma membrane. This is most obvious in plant cells, which usually have a **cell wall,** a material that surrounds the plasma membrane and imparts rigid strength to the plant cell. The plant cell wall is composed primarily of cellulose. When plant cells divide, a thin layer of jellylike material, called pectin, forms between the two new cells in a region known as the *middle lamella* (Figure 4.5). Pectin helps to hold adjacent cells together. It is used to make jellies gel and is also found in various remedies for diarrhea.

Animal cells also have a covering around their plasma membrane, but this covering is not a cell wall composed of cellulose. Instead the covering is composed of molecules of combined carbohydrates

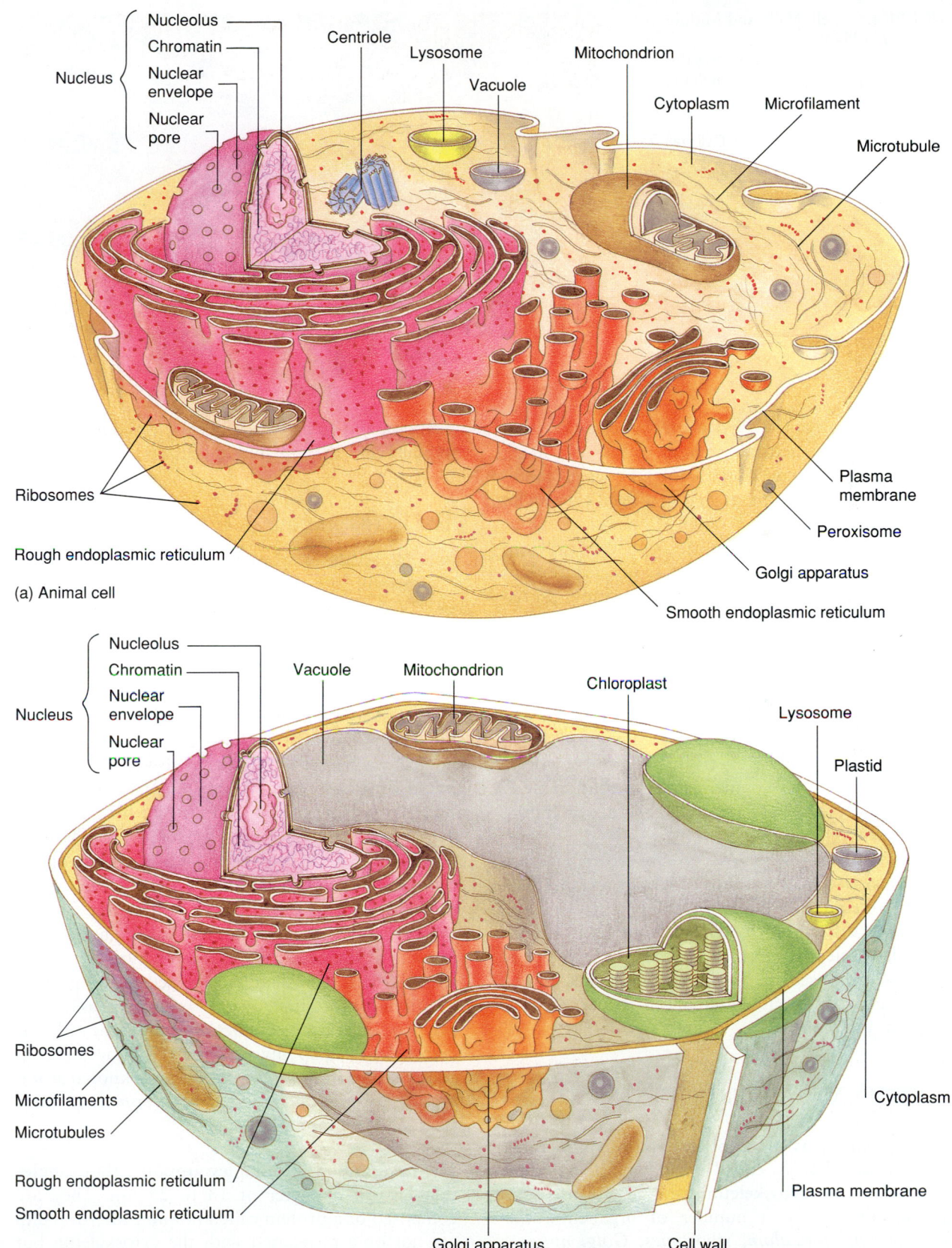

Nucleus
- Nucleolus
- Chromatin
- Nuclear envelope
- Nuclear pore

Centriole

Lysosome

Vacuole

Mitochondrion

Cytoplasm

Microfilament

Microtubule

Ribosomes

Rough endoplasmic reticulum

(a) Animal cell

Plasma membrane

Peroxisome

Golgi apparatus

Smooth endoplasmic reticulum

Nucleus
- Nucleolus
- Chromatin
- Nuclear envelope
- Nuclear pore

Vacuole

Mitochondrion

Chloroplast

Lysosome

Plastid

Ribosomes

Microfilaments

Microtubules

Rough endoplasmic reticulum

Smooth endoplasmic reticulum

Golgi apparatus

Cell wall

Cytoplasm

Plasma membrane

(b) Plant cell

FIGURE 4.4 Eukaryotic Cells
Composite, idealized drawings of (a) an animal cell and (b) a plant cell, illustrating some of the major cellular structures. Eukaryotic cells have a distinct nucleus.

FIGURE 4.5 Cell Walls and Middle Lamella in Plant Cells
This electron micrograph shows the relationship of the middle lamella to the cell walls.

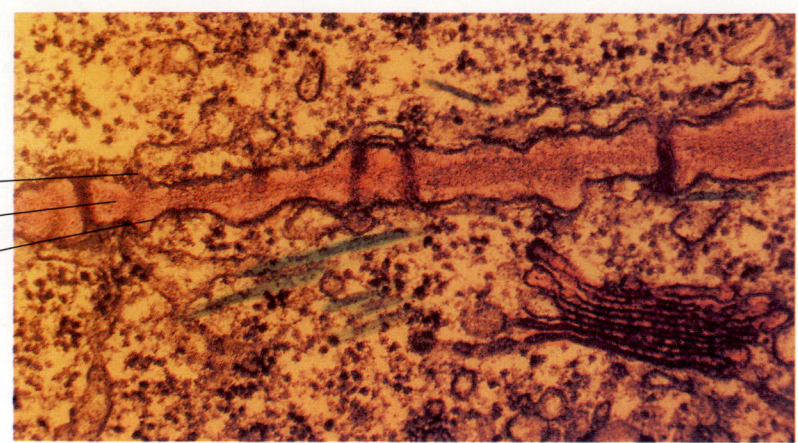

Cell wall ⎯
Middle lamella ⎯
Cell wall ⎯

and proteins (called *glycoprotein molecules*) that impart some structural strength to the cell. More importantly, the covering on animal cells promotes cell-to-cell recognition, as well as the cell's reactivity toward other cells (see Profile).

For example, biologists studying cell structure and function use what are called *tissue culture techniques* to grow the cells they wish to study. (In tissue culture, cells are grown in glass or plastic containers by providing them with nutrients, proper temperature, and other factors necessary for growth.) When scientists combine muscle cells from the heart with liver cells, each type of cell clumps only with its own kind to form heart muscle tissue and liver tissue. In other words, each type of cell recognizes its own kind. This is due, in part, to the glycoprotein covering of animal eukaryotic cells.

Bacterial cells (which, as we saw earlier, are prokaryotes) also may produce coverings around their plasma membrane. A cell envelope, or capsule, of this type differs biochemically from both animal and plant cell coverings. In bacteria, the envelope is usually composed of special organic acids, rather than the glycoprotein of animal cells or the cellulose of plant cell walls.

CYTOPLASMIC STRUCTURES

The cytoplasm is a complex of fibrous structures that form what is called the *cytoskeleton*. Suspended in the cytoskeleton, and in a sense supported by it, are a number of organelles: *the endoplasmic reticulum, ribosomes, Golgi apparatus, lysosomes, peroxisomes, mitochondria, vacuoles,* and *plastids.* Each structure has a special purpose, and some structures are found more often in plant cells than in animal cells and vice versa.

The Cytoskeleton

■ **Within the cytoplasm is a cytoskeleton that consists of microtubules, microfilaments, and intermediate fibers. These elements give shape to the cell, hold various organelles in place, arrange and rearrange cellular structures, and participate in cellular movement.**

Until just a few years ago, biologists thought that the cytoplasm was a fluid in which organelles floated freely. However, the high-voltage electron microscope has expanded our understanding of the cell. Recently, Dr. Keith Porter and his colleagues at the University of Colorado, and others as well, have used this microscope to view thick sections of cells. Those sections revealed a fibrous network in the cytoplasm of cells, sort of a three-dimensional **cytoskeleton** that is dynamic and can change or shift its form as the need arises. The function of the network is (1) to give shape to the cell, (2) to provide locations for the attachment of organelles, and (3) to arrange and rearrange structures within the cell. The cytoskeleton, also involved in cell movement, is composed of three separate structures known as *microtubules, microfilaments,* and *intermediate fibers* (Figure 4.6).

Microtubules. **Microtubules** are long, thin tubules that have a diameter of 18 to 25 nm. They are composed of a protein called *tubulin.* Microtubules are not only associated with the cytoskeleton but are also responsible for movement. In some cases, they form cilia and flagella, which are microscopic

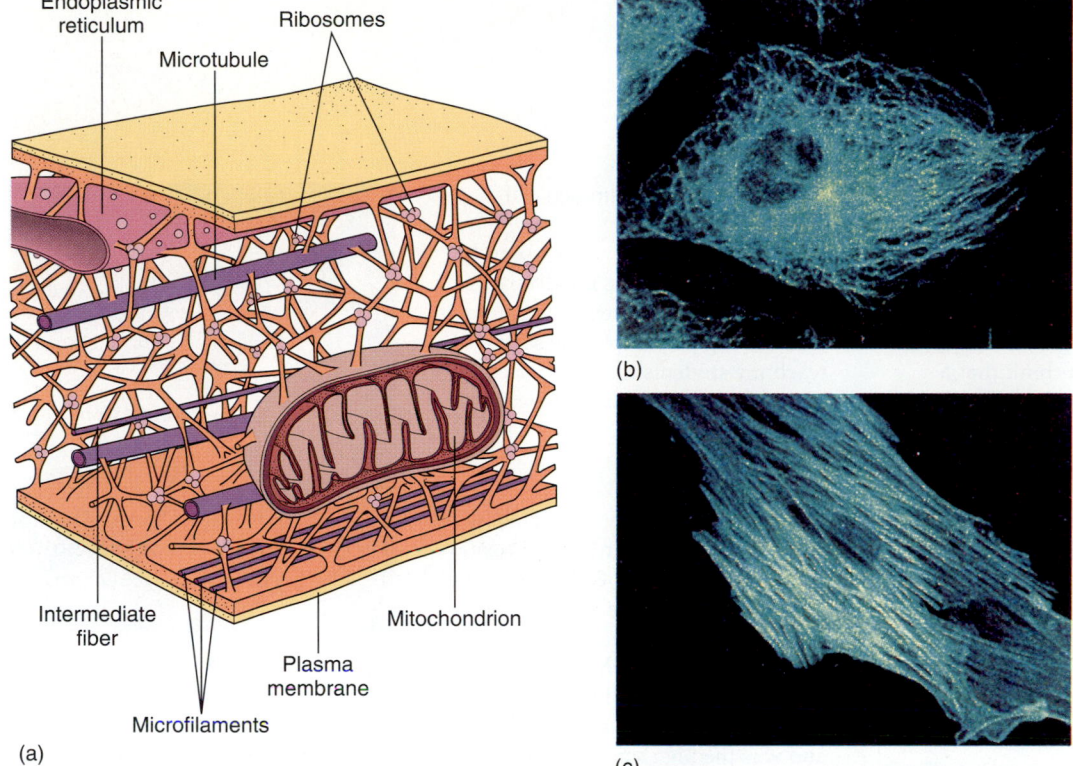

FIGURE 4.6   Cytoskeleton
(a) A simplified drawing shows how the cytoskeleton supports and positions organelles within the cell. It also plays a vital role in cell movement and reproduction. (b) A fluorescence micrograph provides a dramatic overview of microtubules. (c) This micrograph shows the microfilaments of the cytoskeleton fibers.

hairlike structures that function somewhat like oars. Microtubules are also associated with amoebalike movement. These structures enable the cytoplasm to flow, by sending out a *pseudopodium* (or false foot). The cytoplasm then flows into the foot, which results in movement of the cell from place to place. We will discuss cellular movement in more detail later in this chapter. Microtubules are also involved with the process of depositing cellulose as plant cell walls are formed and with chromosome movement in cell division (Chapter 10).

Microfilaments. **Microfilaments** are very thin protein threads with a diameter of 3 to 6 nm. Microfilaments differ from microtubules in that they are composed of a globular protein called *actin,* rather than tubulin. Microfilaments are also smaller in diameter.

Microfilaments function in cell contraction and cell movement. For instance, when we discussed cell recognition, we said that if heart and liver cells are mixed together in a tissue culture, individual cells will clump only with their own kind. In order to unite, the cells must move together, and microfilaments are responsible for that movement. When an animal cell divides, microfilaments form a ring around the middle of the cell, contract, and pinch the cell in two. In part, microfilaments are responsible for the movement of cytoplasm in cells.

Intermediate Fibers. **Intermediate fibers** are about 7 to 9 nm in diameter, somewhat intermediate between the diameter of microfilaments and microtubules. The fibers are composed of fibrous proteins similar to the protein collagen, which is found in bones and tendons, for instance. Because of their protein nature, intermediate fibers are strong and are found in cells that require or provide mechanical strength.

Although biologists have learned much about the cytoskeleton of the cell, there is still much left to discover. For instance, what mechanism signals

Cytoplasmic Structures

Dr. Jerry C. Guyden
Cell Molecular Biologist

On a cold, blustery, winter day in New York City, Dr. Jerry C. Guyden is in the basement of his home practicing his golf swing. This winter has been especially harsh, and he can't wait for spring to play golf. Most nonscientists often think that a respected cell biologist only thinks of the laboratory and research. While it is true that Dr. Guyden devotes most of his time and energy to his research, he is a multifaceted individual. Not only does he love his work, golf, and his family, he is also a published poet. One of his poems follows:

me and the sea

The ocean waves at my feet as
  I wonder . . .
Where it all started.
Did the H_2O that now bathes
  my soul
Begin together or parted . . . ?
Was it me or the sea that changed
  its mind
For us to meet today . . . ?
Whatever the case I'm glad
  to find
That it ended up this way.

In the laboratory at the City College of New York, Professor Guyden is investigating cells of the immune system — more specifically, thymic nurse cells. Thymic nurse cells are one of two cell types in the body that actually engulf another cell type as a function of development. Developing thymocytes must be internalized into thymic nurse cells during their maturation in the thymus. Professor Guyden has been able to capture thymic nurse cell internalization on video. Furthermore, he and his colleagues have shown that thymic nurse cells induce *apoptosis* (cell deletion by fragmentation) in a subset of the internalized thymocyte population, while rescuing the remaining fraction from the process of programmed cell death.

Not only does he value research, he also recognizes the importance of quality teaching. When asked about his teaching philosophy, he states, "I teach my students to strive for excellence. Excellence is the only way to participate in basic research." His quest for excellence is typical of his personality.

Dr. Guyden was born in Dallas, Texas, and grew up in Bryan, Texas, the home of Texas A & M University. His parents were both teachers. His mother taught mathematics and his father was a coach and physical education teacher. While in high school, Dr. Guyden excelled in track and was the high school state champion in the pole vault. In part because of his athletic ability, he was accepted at the University of North Texas. While at the University of North Texas, he continued to excel in track, but soon he realized that the real reason for attending the university was not athletics but education. In his junior year, as a mathematics major, he began to apply himself academically.

In 1974 he was accepted into medical school. That summer he was invited to participate in a summer research program at North Texas by his genetics professor, Dr. Bushee, on a cancer research project. The work done that summer with Dr. Bushee was successful and was published. This was his first taste of research, a taste he liked so much he never attended medical school. It was then that he decided to do research. He earned his masters degree at the University of North Texas and then went on to the University of California, Berkeley, for his doctorate. In addition to doing research and earning his Ph.D. at Berkeley, Dr. Guyden was very committed to the recruitment of minority students into

Ph.D. programs at Berkeley. Of the 20 students he recruited to Berkeley from 1976–1981, 19 earned the Ph.D., which is an outstanding record. Among those he recruited was Dr. Wilfred Denetclaw, profiled on page 119 of this text, and Dr. Estralita Martin, also profiled in this text on page 533. Others whom he recruited — such as Dr. Sheila McClure at Spellman University, Dr. Maria Nieto at California State University, Hayward, and Dr. Timothy Turner at the University of Alabama, Birmingham — are highly respected, productive scientists.

Now at the City College of New York, he continues his life-long passion to "do research and to get folks together." His laboratory is made up of people with many different ethnicities from all over the world. "The way I see it, we have two major responsibilities in life — to enjoy what we do and to try and make this a better world. I truly enjoy my research, with its combination of competition and discovery. And I try to bring people with different backgrounds together around science. Not only is competitive science done in such an atmosphere, but we also learn about each other and discover that we all are not so different."

a cell to divide? How do microtubules and micro-filaments function in cell movement, division, and cellular dynamics? An understanding of these structures will have deep biological and biomedical importance in the future, especially in cancer research and disorders in cell movement and contraction of muscle cells, such as muscular dystrophy.

Endoplasmic Reticulum

■ **The endoplasmic reticulum is an intracellular transport system. On the surface of rough endoplasmic reticulum there are ribosomes, which are involved in protein synthesis. Smooth endoplasmic reticulum contains no ribosomes and is involved in lipid synthesis.**

One of the most obvious structures in the cytoplasm appears to be a continuation of the plasma membrane itself. This structure is called the **endoplasmic reticulum** (ER), which is an interconnecting transport system, a double membrane that produces a series of channels, tubes, and flattened sacs spread throughout the cytoplasm. There are two forms of endoplasmic reticulum: *rough ER* and *smooth ER*. If we were exploring a prokaryotic cell, we would not find an endoplasmic reticulum.

Rough Endoplasmic Reticulum. ER having attached ribosomes appears rough or granular in an electron microphotograph, so it is called rough endoplasmic reticulum (RER) (Figure 4.7). Normally, proteins synthesized for export (or secretion) outside the cell, such as protein hormones like insulin, are produced by the ribosomes of the RER. These proteins are then transported through the channel of the RER, where they undergo biochemical modification. RER channels lead into the other membrane systems such as the smooth endoplasmic reticulum and Golgi apparatus, which we will discuss shortly. There they are incorporated into **vesicles,** which are small droplets of material bound by a single membrane. The vesicles then fuse with the plasma membrane, and the proteins leave the cell.

Smooth Endoplasmic Reticulum. The second type of ER lacks ribosomes. As a result, it appears smooth, or agranular, and is called smooth endoplasmic reticulum (SER) (Figure 4.7). Like RER, SER is also involved in transport. However, SER also contains the enzymes involved in the synthesis of lipids, such as steroids and triglycerides. Therefore, the cells of body structures like the adrenal gland, which produces a lot of steroid hormones, have a large amount of SER, as do intestinal cells, which produce large amounts of triglycerides.

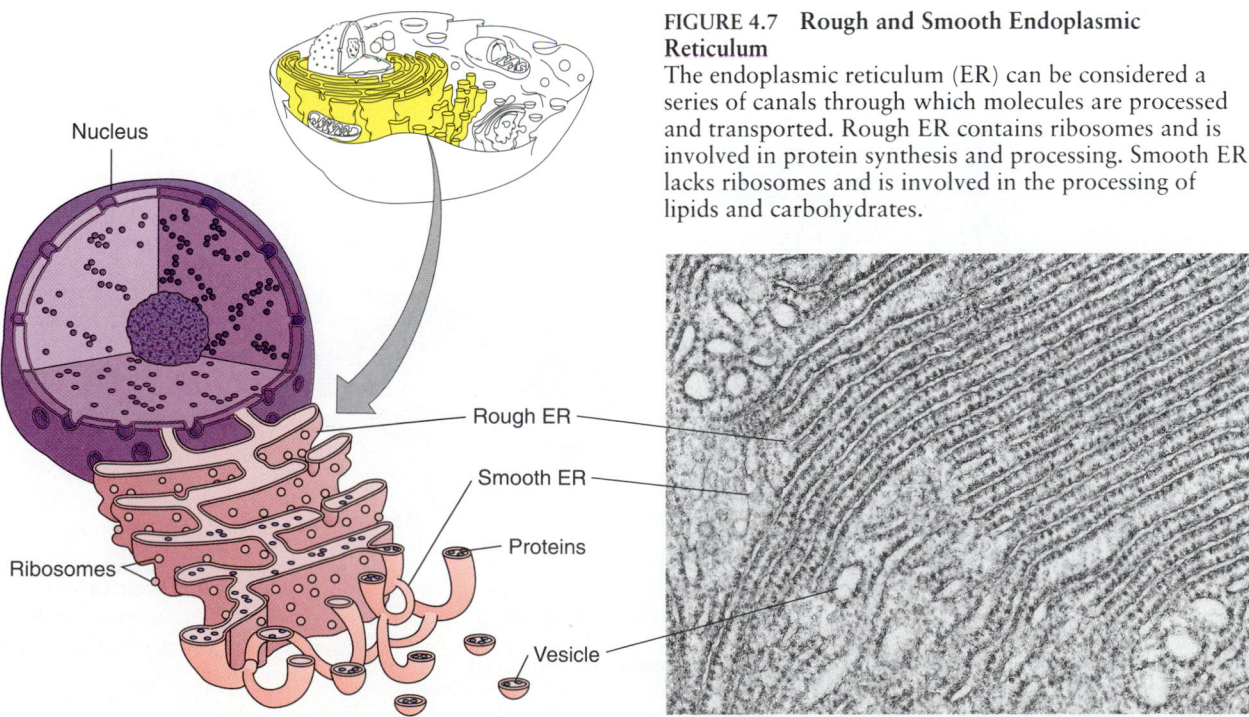

FIGURE 4.7   Rough and Smooth Endoplasmic Reticulum
The endoplasmic reticulum (ER) can be considered a series of canals through which molecules are processed and transported. Rough ER contains ribosomes and is involved in protein synthesis and processing. Smooth ER lacks ribosomes and is involved in the processing of lipids and carbohydrates.

Nucleus

Ribosomes

Rough ER

Smooth ER

Proteins

Vesicle

Ribosomes

- Ribosomes, which consist of two subunits, are not enclosed in a membrane and are the sites of protein synthesis in both eukaryotic and prokaryotic cells.

One of the most numerous structures that we see on our journey through the cell is the ribosome. **Ribosomes,** found in both prokaryotic and eukaryotic cells, are small particles consisting primarily of protein and the nucleic acid ribonucleic acid (RNA). This organelle has two main characteristics: (1) it is made up of two subunits, one large and one small, each made up of RNA and protein; (2) it is not surrounded by a membrane.

Ribosomes function as the site for protein synthesis, which is so important a process that we will devote a large portion of Chapter 12 to it. For now, we need only note that the ribosome functions in the assembly of amino acids to form proteins. During the assembly of proteins, several ribosomes may link together to form a *polyribosome*. The ribosomes then work together to build the protein, like workers on an assembly line.

The distribution of the ribosomes in the cytoplasmic region depends on how the protein they synthesize is going to be used. For example, in embryonic cells, which make protein primarily for their own use, ribosomes are distributed throughout the cytoplasm. In situations like this, recent evidence demonstrates that they are not floating free but are associated with and possibly attached to the cytoskeleton. In digestive cells and others involved in producing proteins to be exported (or secreted) elsewhere in the body, ribosomes are attached to the ER.

Golgi Apparatus

- The Golgi apparatus is formed from portions of ER that bud off, or break away. It receives protein macromolecules produced in ER and further processes them into complex macromolecules that the cell or organism requires for life functions.

The **Golgi apparatus,** which is named after its discoverer, Camillo Golgi, consists of a collection of flattened, disklike membranous sacs (cisternae). At present, biologists believe that the structures are formed by a process through which portions of ER or plasma membrane flatten and then bud off (Figure 4.8), but more research is required before biologists will fully understand the mechanism.

The Golgi apparatus functions as a processing and packaging center for the macromolecules produced by the ER. The vesicles and their contents, formed by the ER, are passed on to the Golgi apparatus, where the macromolecules undergo further processing, ultimately becoming polysaccharides, glycoproteins, and lipoproteins, to name just a few. The Golgi apparatus then forms secretory vesicles containing the processed macromolecules and releases the vesicles. The vesicles move to other locations within the cell or to the plasma membrane, where the contents are released to the outside of the

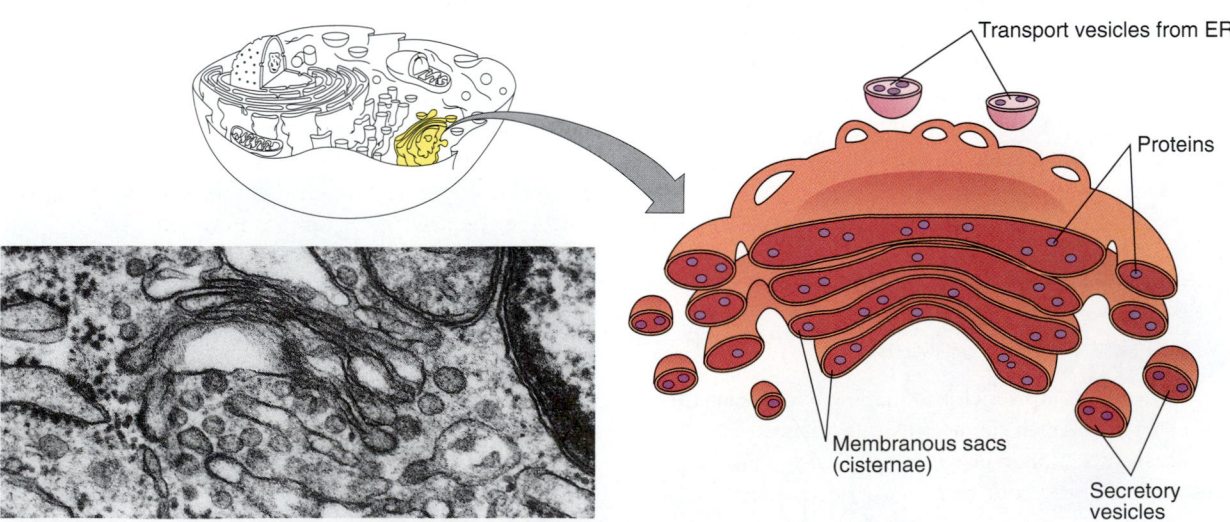

Transport vesicles from ER

Proteins

Membranous sacs (cisternae)

Secretory vesicles

FIGURE 4.8 Golgi Apparatus
The Golgi apparatus is formed by portions of the endoplasmic reticulum. The apparatus functions in the receipt and further processing of proteins into complex macromolecules that are packaged into vesicles and secreted to the outside of the cell.

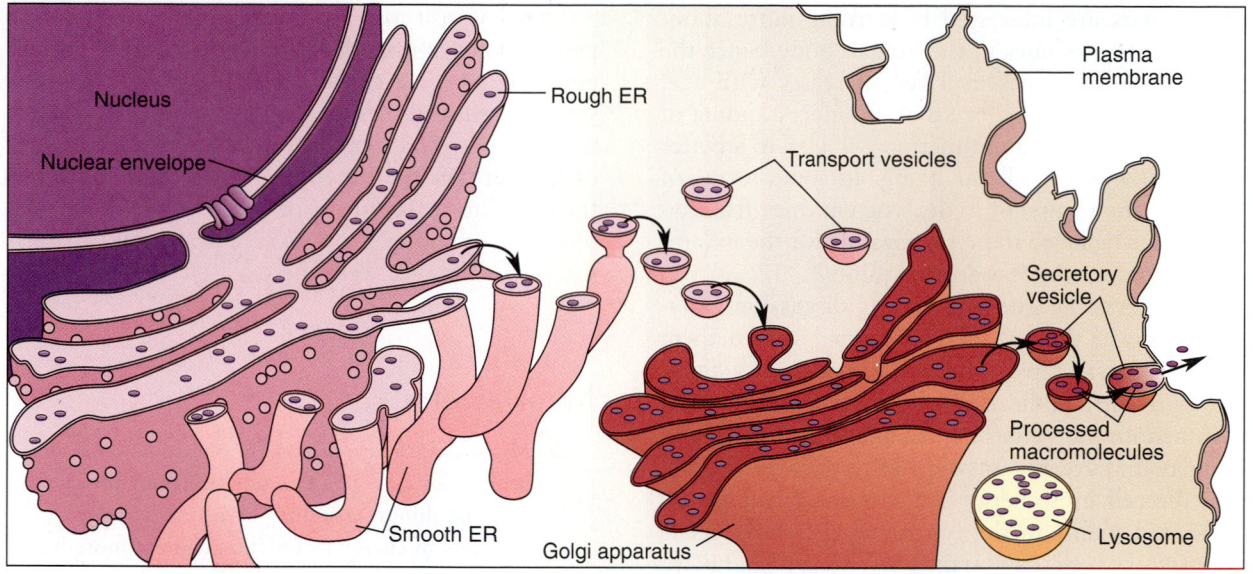

FIGURE 4.9   Movement and Processing of Macromolecules
Macromolecules produced by the endoplasmic reticulum are eventually transported to the Golgi apparatus where they are processed further. They are then packaged into either secretory vesicles that eventually expel the molecules from the cell, or they are packaged into lysosomes. Lysosomes retain the processed molecules for use within the cell.

cell. In short, the Golgi apparatus accepts the vesicles pinched off the ER, modifies the vesicles and their contents, and delivers and releases the processed macromolecules. Figure 4.9 illustrates the interrelationship between the SER and the Golgi apparatus as a cell secretes substances.

The Golgi apparatus is found in most eukaryotic cells, especially those that function in the secretion of chemicals, as you might expect. Therefore, the pancreas, the glands such as the thyroid, and nerve cells (which also secrete many substances) contain large numbers of Golgi apparatus. Plant cells also contain large numbers of Golgi apparatus, which are integral to the process of assembling components that make up the cell wall. They are not, however, in prokaryotic cells.

Lysosomes and Peroxisomes

■ **Lysosomes are vesicles containing hydrolytic enzymes, which can destroy cell contents. Peroxisomes contain enzymes that destroy the hydrogen peroxide formed during some cellular processes.**

If you have ever wondered about the mechanism responsible for the disappearance (or *reabsorption*) of the tail of a tadpole as it develops into the adult frog, the cellular explanation, at least in part, has

to do with the structure known as the lysosome. The **lysosome** is a small, single-membrane-bound vesicle usually formed by the Golgi apparatus. This small structure is essentially a sac of hydrolytic enzymes (remember that hydrolysis is the addition of water to a macromolecule in order to break it down into smaller components). The organelle's membrane separates its contents from the rest of the cell, but when the lysosome membrane breaks open, the hydrolytic enzymes it contains are released, and they digest the cell. In the case of the tadpole's transformation, the enzymes in the lysosomes of the tail's cells break down the cellular components. The frog then uses the resulting molecules for energy and for the raw materials for new cells.

The Nobel Prize laureate Christian de Duve is given credit for the discovery of lysosomes, and because they were elements in a cell that could destroy it, he called them "suicide bags." Although their role seems most obvious in the life cycle of a frog, lysosomes are important to other life forms as well, including humans. For instance, lysosomes play a role in the way that the human body protects itself from disease. A white cell in the blood can move like an amoeba and engulf an invading bacterium. Once the bacterium is engulfed, a *vacuole* (membrane-bound sac) surrounds the bacterium. Lysosomes from the white cell unite with the vacuole, release their enzymes, and digest the bacterium.

Scientists are interested in learning more about the role of lysosomes in the human body. Since the number of lysosomes in a cell increases with age, some investigators believe that an understanding of lysosome structure and function will lead to a better understanding of cellular aging. In some forms of arthritis, the escape of hydrolytic enzymes from lysosomes is believed to be responsible for the inflammation and tenderness around joints.

The inevitable question in any discussion of lysosomes is if the enzymes contained within them can be so destructive, why don't the enzymes digest the lysosome membrane? When they do digest their own membrane, what has changed to allow this to happen? In part, the answer has to do with the fact that the cell uses energy in order to constantly rebuild and maintain the lysosomal membrane. Ordinarily the cell can maintain enough energy to prevent the enzymes' escape into the cytoplasm. When it cannot, the organelle self-destructs and the lysosomal enzymes flow into the cytoplasm and destroy the cell.

Lysosomes are usually found in animal cells, but not in prokaryote cells. Structures that resemble lysosomes are found in fungi, plants, and single-celled animals.

Peroxisomes are small organelles that act much like lysosomes but under different circumstances. Recall that in Chapter 3 we noted the fact that amino acids can be deaminated — in other words, the amino group can be removed. The remainder of the molecule can then be used as an energy source in order to synthesize ATP, the "energy currency of the cell." One of the by-products of this process is hydrogen peroxide (H_2O_2), a substance that is poisonous to living cells. (The toxic nature of hydrogen peroxide is what makes it a good disinfectant.) At that point, peroxisomes release other enzymes that break down the by-product (H_2O_2) to water and oxygen gas ($2\ H_2O_2 \rightarrow 2\ H_2O + O_2$), thus preventing the destruction of the cell (Figure 4.10).

Some of you may have observed the activity of peroxisomes firsthand. If you ever placed hydrogen peroxide on a cut finger, you noticed that the substance forms bubbles as soon as it comes in contact with the cut. That reaction was due to the response of the peroxisome enzymes in the exposed cells of the cut. They react to the presence of hydrogen peroxide by breaking it down into water and oxygen gas — the bubbles were the released O_2 gas.

Mitochondria

■ **Mitochondria are organelles that function in the production of ATP, an energy-carrying molecule in living things. Structures that require large amounts of energy to function contain more mitochondria than structures that do not.**

Some ATP synthesis can occur in the cytoplasm in both prokaryotic and eukaryotic cells, but the bulk of ATP synthesis occurs within the membrane of the eukaryotic cell organelles that biologists call **mitochondria** (plural). These organelles are not found in prokaryotic cells.

A mitochondrion (singular) is fairly large — 1 to 3 μm in length, or about the size of a bacteria cell. The organelle is usually oval in shape and is surrounded by a double membrane that divides it into two compartments, an outer compartment and an inner compartment called the matrix. The outer membrane separates the mitochondria from the other parts of the cell, and the inner membrane folds and doubles on itself to form the *cristae* (pronounced like Corpus Christi, Texas). The infolding of the inner membrane increases the surface area available for reaction sites (Figure 4.11). In the cristae you will find particles that contain enzymes used in the production of ATP. As you might expect, the larger the energy requirements of the cell, the more mitochondria per unit volume of cell.

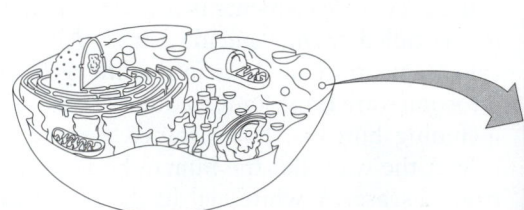

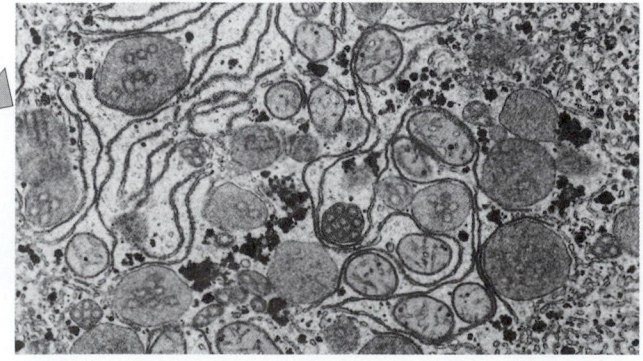

FIGURE 4.10 Peroxisomes
Amino acids and components of fats are substances broken down in peroxisomes. Hydrogen peroxide is a toxic by-product. Peroxisomes contain enzymes that help change hydrogen peroxide to water and oxygen.

Chapter 4   The Cell: Where It All Happens

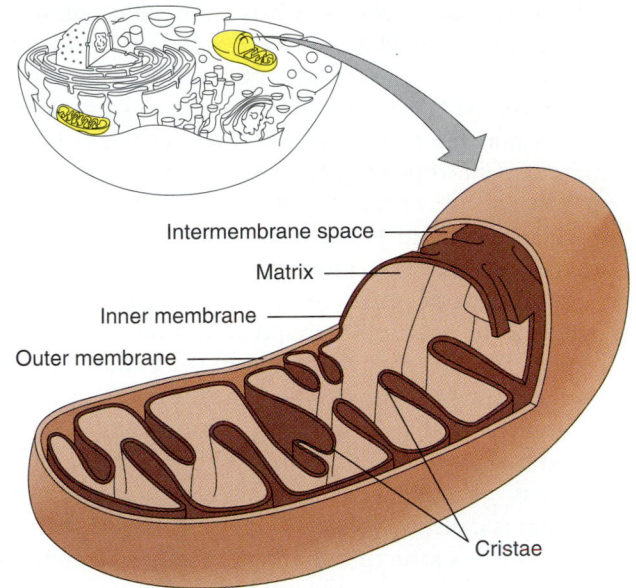

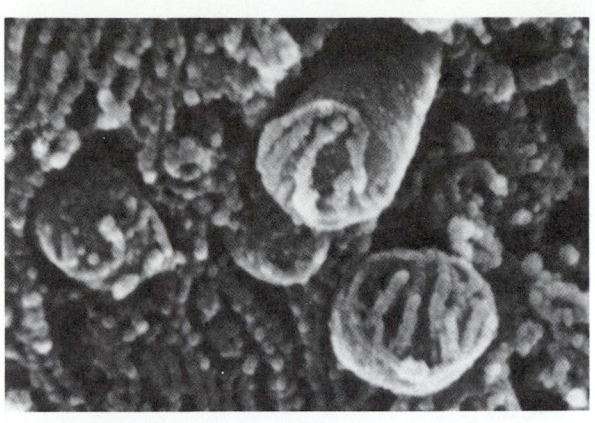

(a)

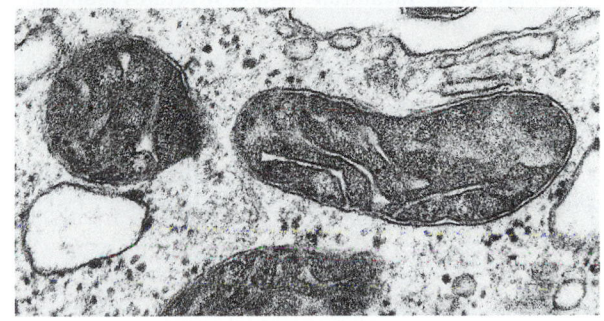

(b)

FIGURE 4.11 **Mitochondria**
The majority of the all-important energy compound, ATP, is formed in mitochondria. (a) Note the cristae within this SEM photograph of mitochondria viewed in cross-section. (b) In this TEM, the infolding of the inner membrane to form the cristae is apparent.

Labels in Figure 4.11:
- Intermembrane space
- Matrix
- Inner membrane
- Outer membrane
- Cristae

Vacuoles

■ **Vacuoles are fluid-filled structures found most often in plant cells. These structures store food produced by the organism, support the cell, and increase its surface area.**

Often large vesicle-like structures known as vacuoles occur in the cytoplasm of cells, typically in plant cells (Figure 4.12). A **vacuole** is a large space, usually filled with fluid, that contains dissolved substances. The space is surrounded by a single membrane.

Vacuoles seem to play a crucial role in food storage and in cell size and rigidity in plant cells. An experimenter who observes the development of plant cells will always see that immature plant cells consist of numerous small vacuoles. As the cell matures, these eventually come together to form a very large central vacuole. In some plant cells, the central vacuole comprises as much as 90 percent of the cell's volume. Both large and small vacuoles function as storage areas for metabolic products such as salts, sugars, and other substances associated with the particular plant cell.

Because of the change in the vacuoles in plant cells, botanists (scientists who study plant life) believe that an additional function of the vacuole is to aid in the elongation of young plant cells. In other

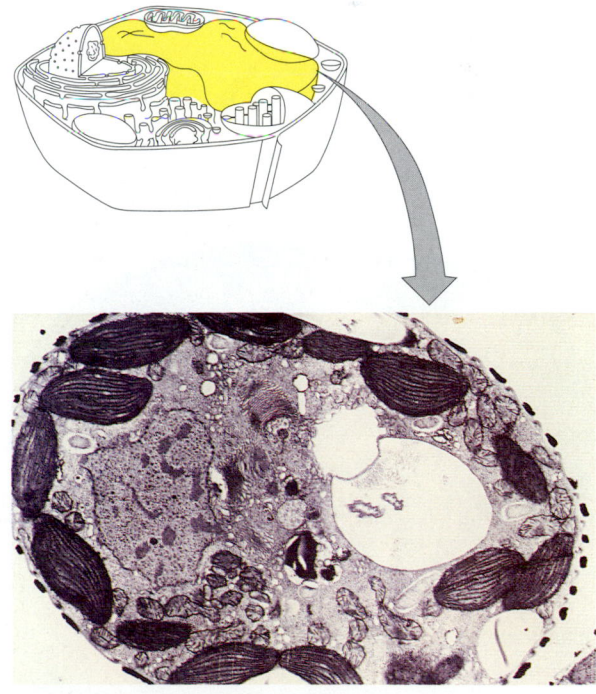

FIGURE 4.12 **Vacuoles**
Vacuoles store many substances such as water, ions, and glucose. Sometimes they contain the pigments that give flower petals their color. Still others function as containers for harmful wastes.

words, as the vacuole fills with fluid, the result is an increase in internal pressure. The pressure pushes against the young cell wall, increasing the size of the cell and providing it with additional rigidity.

In animal cells, membrane-bound fluids are usually called *vesicles* rather than vacuoles. They function primarily in the storage and transport of materials within the cells. Vesicles are usually smaller than plant vacuoles. While vacuoles are usually more than 100 nm in diameter, vesicles are usually less than 100 nm in diameter.

Plastids

- **Plastids are membrane-bound structures found in plant cells.**

Plant cells frequently contain **plastids,** which are organelles surrounded by a double membrane and are usually involved in food production and storage. (These organelles are not found in prokaryotic cells.) Botanists usually describe three types of plas-

tids: *leucoplasts, chromoplasts,* and *chloroplasts.* Leucoplasts are white plastids and are found in storage compartments and plant stems. Chromoplasts are the colored plastids that are so striking in yellow and red fruits, such as strawberries and tomatoes. Chloroplasts are green plastids. We will discuss the role of the chloroplast in photosynthesis in Chapters 7 and 8. Right now, let us concentrate on the physical characteristics of chloroplasts alone.

In eukaryotic plant cells, the light-energy-trapping **chloroplasts** are surrounded by a double membrane. Within the chloroplast is a fluid-filled portion known as the *stroma.* Located in the stroma are structures that look like stacks of coins. These are known as the *grana* (the singular form of this word is *granum*). Each coinlike structure in a granum is called a *thylakoid,* and it is on the membranes of this structure that light energy is trapped by chlorophyll and converted to ATP and other substances, which ultimately form energy-rich carbon compounds (Figure 4.13).

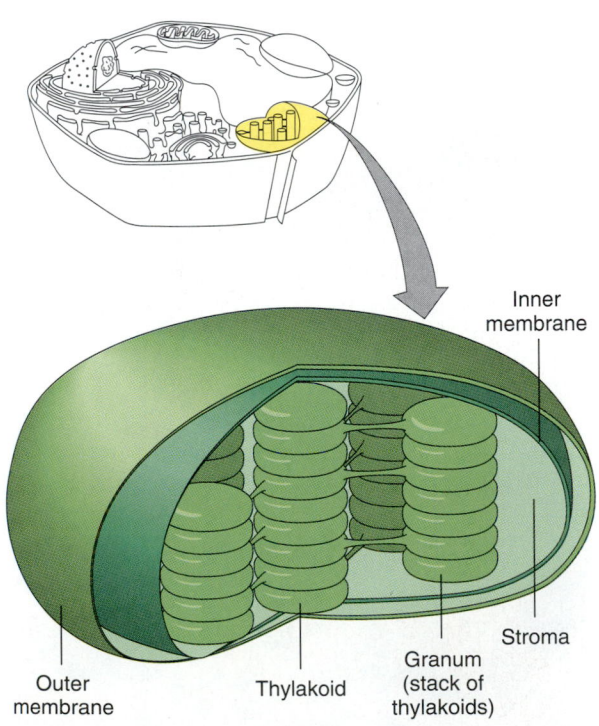

Inner membrane

Outer membrane

Thylakoid

Granum (stack of thylakoids)

Stroma

FIGURE 4.13 Chloroplasts
(a) The small green bodies within these plant cells are chloroplasts. (b) An electron microphotograph of a chloroplast illustrates its internal structure.

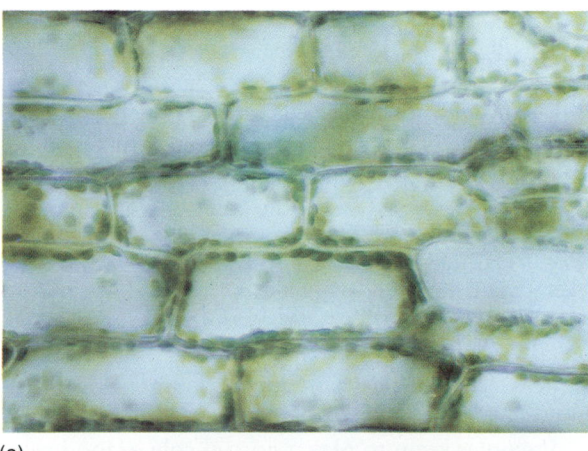

(a)

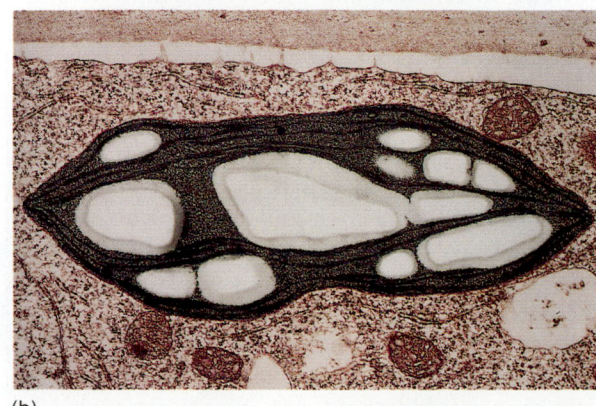

(b)

THE NUCLEAR REGION

■ **The nuclear region is bound by a double membrane called the nuclear envelope. Inside the envelope are found the nucleolus, nucleoplasm, and chromatin.**

As we continue our journey through the eukaryotic cell, we find the nucleus, a region that is isolated from the remainder of the cell by a double membrane and that contains a nucleolus, nucleoplasm, and chromatin, which is the loosely organized genetic material that controls cellular activities and also cellular reproduction (Figure 4.14). These two processes are so important to life that we will de-

vote Chapters 10 through 13 to a discussion of them. For now, a simple overview of nuclear structure and function will be sufficient.

The Nuclear Envelope

The nuclear region in a eukaryotic cell, unlike that in a prokaryotic cell, is enclosed in a double membrane called the **nuclear envelope.** It consists of two membranes separated by a space of about 30 to 40 nm. Many tiny porelike protein channels, called *nuclear pores,* perforate both layers of the entire envelope to allow the entrance and exit of material between the nucleus and the cytoplasm of the re-

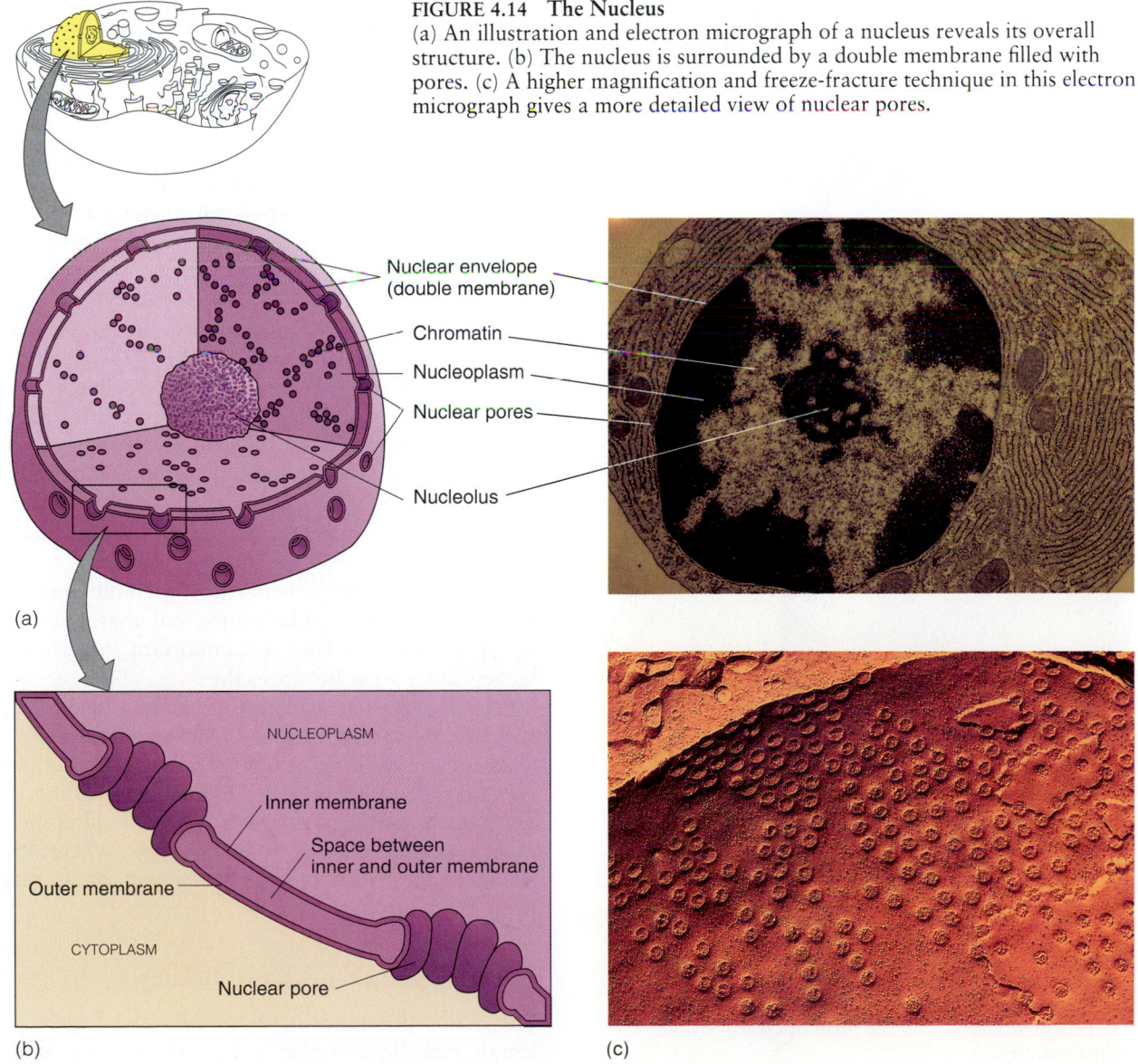

FIGURE 4.14 The Nucleus
(a) An illustration and electron micrograph of a nucleus reveals its overall structure. (b) The nucleus is surrounded by a double membrane filled with pores. (c) A higher magnification and freeze-fracture technique in this electron micrograph gives a more detailed view of nuclear pores.

Nuclear envelope (double membrane)

Chromatin

Nucleoplasm

Nuclear pores

Nucleolus

(a)

NUCLEOPLASM

Inner membrane

Space between inner and outer membrane

Outer membrane

CYTOPLASM

Nuclear pore

(b)

(c)

mainder of the cell. The outer nuclear membrane is connected to the endoplasmic reticulum, and there is some speculation that ER may originate from the envelope, although the research is not yet conclusive. During cell division the nuclear membrane breaks down, and when cell division is complete, the membrane is reformed — a process that researchers still do not completely understand.

Nucleoplasm and Chromatin

Within the nuclear envelope is the nucleoplasm. Nucleoplasm is a general term used to describe the contents of the nucleus. The nucleoplasm consists primarily of masses of DNA and protein, which are called **chromatin** (colored complex) because they absorb stain very easily. As a cell prepares to divide, chromatin condenses into long strands visible under the light microscope called *chromosomes* or *colored bodies* (see Figure 4.15). Do not let the two terms confuse you. The distinction is rather like the one between a hand and a fist — both terms refer to the same structure, except that one is relaxed (chromatin) and the other is not (chromosomes).

Chromatin threads are difficult to view in detail because they are so spread apart during normal cell functions. However, when cell division is occurring, the chromatin coils upon itself to form distinct chromosomes in the nucleus.

Within the nucleoplasm is a large, dark mass called the **nucleolus** (see Figure 4.14), which is composed of RNA and protein. The nucleolus functions in the synthesis of ribosomes, which as we mentioned earlier, are involved in protein synthesis.

CELL MOVEMENT

■ Cell movement is related to cell structure, such as the cytoskeleton, cilia, and flagella. The centrioles found in eukaryotic cells arise from basal bodies, found in all cilia and flagella.

A cell's structure and many of its organelles are involved in one basic cellular activity that we need to discuss a little further — cell movement. Some cell movements occur within a cell, and others serve to move the cell from place to place. We have already mentioned that, because of the cytoskeletal structure, cytoplasm can produce the movement typical of the amoeba and of other cells, such as human white blood cells. However, much cell movement is related to special cell structures such as cilia and flagella. Both are constructed mainly of microtubules and are found only in eukaryotic cells. (Prokaryotic cells have similar structures, but they are composed of a protein called *flagellin* and lack microtubules.)

Cilia and **flagella** (singular forms: cilium and flagellum) are thin, hairlike projections that give rise to cell movement. They are important for cellular locomotion — for instance, the long, taillike structure of a sperm cell is a flagellum. In modified form, cilia are associated with sensory structures in animals, as in the photoreceptor cells of the human eye. Modified cilia also are found in the lining of tubes such as the respiratory tract (to filter and cleanse inhaled air), or in the oviducts, where, for instance, they promote the movement of an egg from the ovary to the uterus.

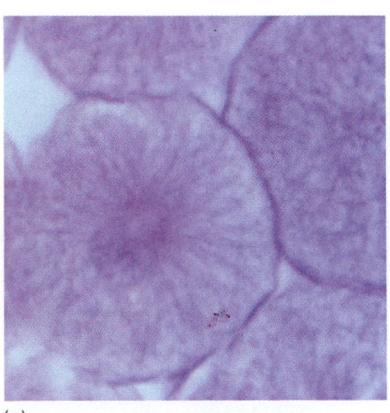

(a)

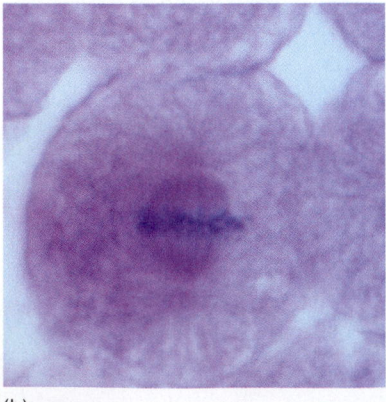

(b)

FIGURE 4.15 Chromatin
In cells that are not dividing (a) the chromatin is not formed into distinct chromosomes. In dividing cells (b) the genetic material is in the form of clearly visible chromosomes.

Cilia, Flagella, and Basal Bodies

The major difference between cilia and flagella is length. Flagella are larger and usually fewer in number, and cilia are shorter and more numerous. As we just mentioned, structurally, cilia and flagella are composed of microtubules. The microtubules are grouped in pairs and arranged in a circular pattern of nine pairs around the periphery and two solitary microtubules in the center (Figure 4.16). This arrangement is called a 9 + 2 pattern. These microtubules are responsible for the movement of the cilia and flagella, which in turn move the cell.

At the base of each flagellum and cilium are many mitochondria that supply the energy required to fuel movement, as well as a structure called the basal body. The **basal body** consists of nine microtubular triplets arranged around the periphery of the structure. The two single microtubules found in cilia and flagella do not enter the basal body. There-

fore, basal bodies do not have the 9 + 2 pattern found in the cilia and flagella.

A basal body is responsible for the formation of a cilium or flagellum. For instance, as a sperm develops, a basal body moves near the cell membrane, and the flagellum begins to develop from the membrane.

Centrioles

Structures called **centrioles,** which are similar to basal bodies, are found in the center of animal cells, near the nucleus. (These structures are rarely found in plant cells and never in prokaryotic cells.) Centrioles are structurally identical to the basal bodies

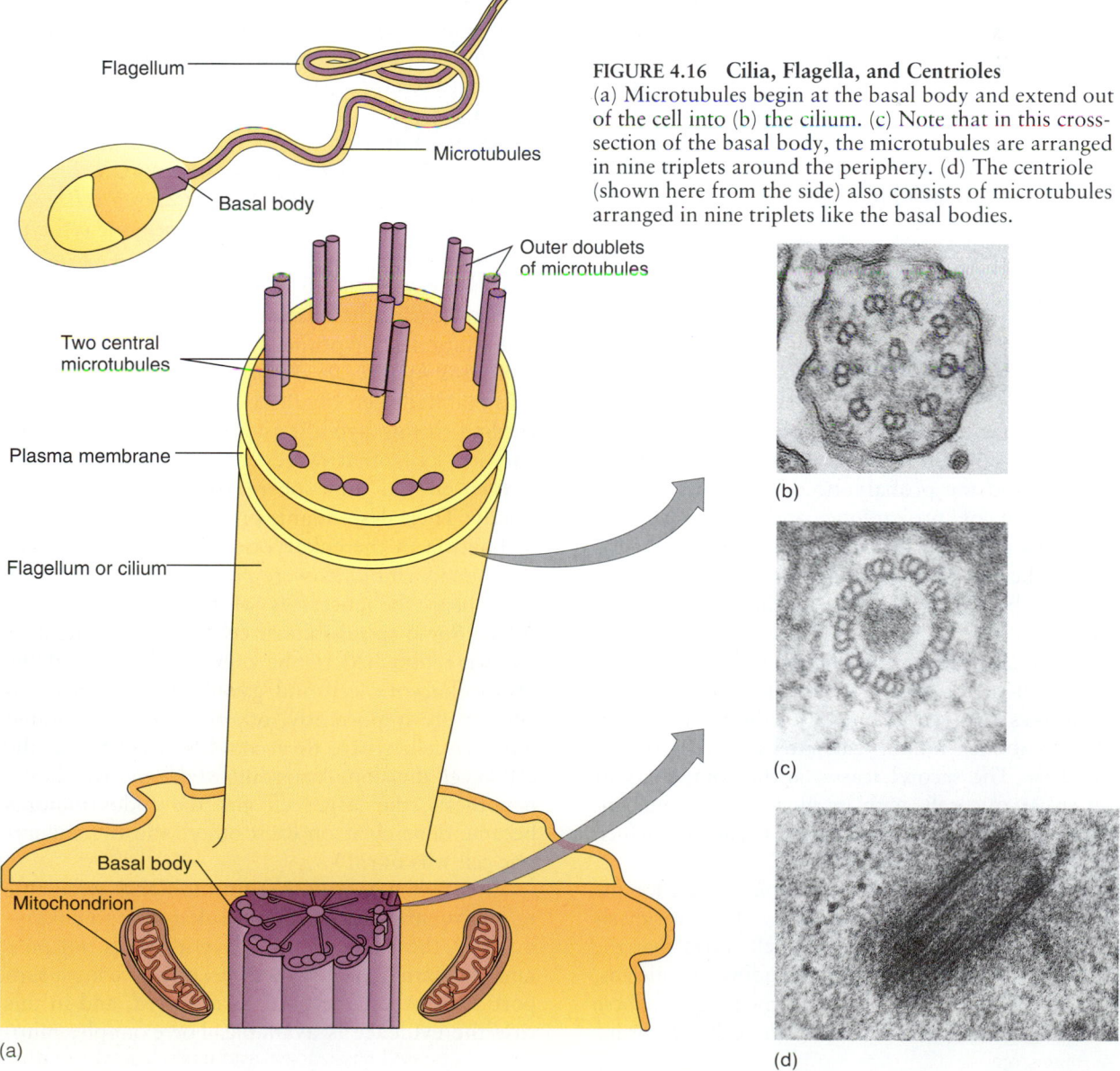

FIGURE 4.16   Cilia, Flagella, and Centrioles
(a) Microtubules begin at the basal body and extend out of the cell into (b) the cilium. (c) Note that in this cross-section of the basal body, the microtubules are arranged in nine triplets around the periphery. (d) The centriole (shown here from the side) also consists of microtubules arranged in nine triplets like the basal bodies.

— that is, they have nine microtubules in a triplet arrangement around their periphery (see Figure 4.16).

Usually found in pairs, centrioles establish the *plane of cell division*. When a cell prepares to divide, the centrioles play a role in developing the *spindle*, a structure that is involved in the proper movement and arrangement of cellular components during cell division. During the reproduction process, the spindle functions in organizing microtubules in the dividing cell.

The more biologists understand about the function of the spindle, the more they understand about the function of other cellular structures. Since the spindle arises from centrioles, and since centrioles are structurally similar to the basal bodies that give rise to flagella and cilia, the predominant theory at this time is that both centrioles and basal bodies are probably organizers of microtubules. Further research will enable us to know more.

THE ORIGIN OF THE EUKARYOTIC CELL

■ **Many biologists think that the eukaryotic cell evolved when a primitive prokaryotic cell engulfed several smaller prokaryotic cells but did not digest them. The large cell and its smaller elements became mutually dependent, creating a eukaryotic cell.**

As we have explored our composite eukaryotic cell, we have discovered differences again and again between it and the prokaryotic cell. You will find a comparison of prokaryotic and eukaryotic cell structures and their functions in Table 4.1. What is the significance of those differences?

When biologists consider the history of life on earth, they believe that prokaryotes existed on earth before eukaryotes, for two reasons. The first is that there is fossil evidence that establishes the fact that eukaryotes evolved about 1.5 billion years ago, which is about 2 billion years after the first prokaryotic cells. The second reason is the complexity of the eukaryotic cell, with its membrane-bound nucleus and other membrane-bound organelles, which are not found in prokaryotes.

There are several tentative explanations for the evolution of the eukaryotic cell. One hypothesis of the origin of the eukaryotic cell states that the membrane-bound organelles, including the nucleus, began as infoldings of the cell membrane. These infoldings then separated from the cell membrane to become separate organelles. Another hypothesis presumes that the nucleus was the primary structure and that the endoplasmic reticulum and all other membranes were added later as extensions of the nuclear membrane. These are mainly armchair hypotheses, because they are not testable. In other words, no experiments could be designed to support or refute these hypotheses.

In the late 1960s, many biologists were speculating about the origin of eukaryotic cells. In 1967, Dr. Lynn Margulis, at Boston University, developed a hypothesis based on the written information available from a number of descriptions of cells. Margulis hypothesized that the primitive eukaryotic cell evolved from large, primitive, prokaryotic cells that contained other, smaller, primitive prokaryotic cells. In other words, some of the small prokaryotic cells were engulfed by larger prokaryotes (much as a small cell is engulfed by an amoeba) but were not digested. Rather, a host-guest relationship developed between the small cells and the larger cell (Figure 4.17). The resulting cell was not yet a eukaryote, but it no longer would be a prokaryote either, so it is called a *protoeukaryotic cell*. This term means *first true nucleus-containing cell*, indicating that it was a transition between one cell type and another.

When two or more organisms live together for mutual benefit, the relationship is called *symbiosis*, and Margulis's hypothesis is known as the *endosymbiosis hypothesis* (from the term for the small cells that had been engulfed — *endosymbionts*). According to this hypothesis, some of the small prokaryotic cells that were engulfed could use the oxygen in the atmosphere to respire more efficiently. Their descendants probably became the mitochondria in the eukaryotic cells of today. Other small endosymbionts were photosynthetic and may have been the ancestors of modern-day chloroplasts. Some spiral-shaped cells were not engulfed but were attached to the outer membrane of the protoeukaryotic cell and gave rise to flagella and cilia. More importantly, others were incorporated into the cell, where they could be used to aid the cell in cell division. According to the endosymbiosis hypothesis, the large cell and the endosymbionts became dependent on each other, and the eukaryotic cell was born.

Although it is true that no biologists were able to observe the change from prokaryotic to eukaryotic cell, the endosymbiosis hypothesis is not pure fantasy. Rather, as in all science, evidence must support a hypothesis before it is accepted, and in this case the evidence is available. For example, mito-

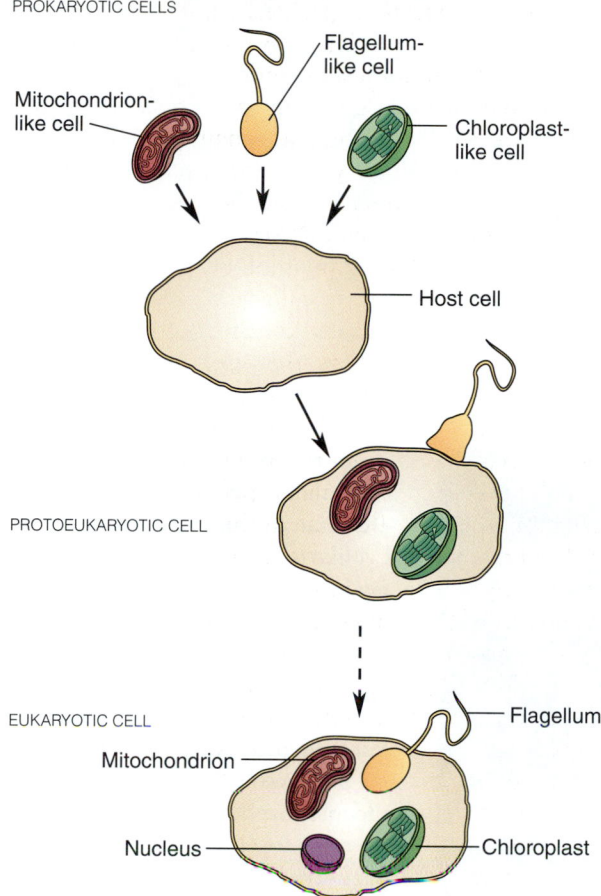

PROKARYOTIC CELLS

Flagellum-
like cell

Mitochondrion-
like cell

Chloroplast-
like cell

Host cell

PROTOEUKARYOTIC CELL

EUKARYOTIC CELL

Flagellum

Mitochondrion

Nucleus

Chloroplast

FIGURE 4.17 Endosymbiotic Origins of the Eukaryotic Cell
The theory of endosymbiosis suggests that the eukaryotic cell originated from a prokaryotic cell that had engulfed and incorporated smaller organelle-like prokaryotes, to become the protoeukaryotic cell that evolved into the eukaryotic cell.

chondria and chloroplasts have their own DNA and ribosomes, and the ribosomes found in them are like ribosomes found in bacteria, which are prokaryotic. The DNA composition of the mitochondria and chloroplasts differs from the nuclear DNA in the cell each inhabits. This evidence is consistent with the endosymbiosis hypothesis.

Some investigators do not accept this hypothesis. They state that the host cell also needs to provide DNA for the synthesis of mitochondrial enzymes and structural proteins. If the endosymbiosis hypothesis were correct, they believe that mitochondria and chloroplasts would be more independent of nuclear DNA. However, the majority of scientists accept the endosymbiosis hypothesis as the best current explanation for the origin of the eukaryotic cell.

As this chapter demonstrates, human beings, and all living things, are made of cells that consist of several different kinds of structures. This fact gives rise to the question of who a person really is. Is a person merely an accumulation of parts? Is he or she simply an accumulation of different types of cells? The course of the evolution of the eukaryotic cell raises another issue. Most humans do not consider that part of our bodies is made up of another life form, the bacteria. Yet, as the essay by Lewis Thomas points out (see Focus 4.2), human beings are not alone — they carry within their cells microscopic journeyers, companions in the passage through life. What are the implications of this fact as individuals attempt to define who and what they are? What is it that makes you you? Your individual cells? Your cells working together? Your cells and their tiny inhabitants? The answer is not as clear as we often think.

We have taken a journey through the cell and have discussed cellular structures and their functions in the phenomenon we call life. Truly, the basic processes that we associate with life happen within the cell.

SUMMARY

1. Cell theory states basically that all living things consist of cells and that all cells are formed from other cells.
2. The microscopes most commonly used to observe cells are the light microscope, transmission electron microscope (TEM), and scanning electron microscope (SEM).
3. The limiting factor in microscopy is resolving power, which is the property of light that allows one to distinguish between two points. Magnification is the property of light whereby objects can be enlarged.
4. There are two fundamental cell types — the prokaryotic cell (first cell) and eukaryotic cell (true cell). Prokaryotic cells lack a membrane-bound nucleus. Eukaryotic cells have a membrane-bound nucleus and other membrane-bound organelles.
5. The typical composite eukaryotic cell is said to have three regions — the plasma membrane, the cytoplasmic region, and the nucleus, which contains genetic material.
6. Some cells have structures that surround the plasma membrane. In plant cells, the cell wall imparts strength and form to the cell. The plasma membranes of prokaryotic cells are often surrounded by protective capsules, or envelopes, of organic acids.

FOCUS 4.2
The Ecosystems of
Our Cells

Lewis Thomas, physician, educator, writer, and former medical administrator, has long been an observer of the natural world. His interest reaches from the complex workings of the cell to the varied interrelationships between organisms. Thomas published The Lives of a Cell *in 1973, from which we take these comments on our relationship with our mitochondria.*

A good case can be made for our nonexistence as entities. We are not made up, as we had always supposed, of successively enriched packets of our own parts. We are shared, rented, occupied. At the interior of our cells, driving them, providing the oxidative energy that sends us out for the improvement of each shining day, are the mitochondria, and in a strict sense they are not ours. They turn out to be little separate creatures, the colonial posterity of migrant prokaryotes, probably primitive bacteria that swam into ancestral precursors of our eukaryotic cells and stayed there. Ever since, they have maintained themselves and their ways, replicating in their own fashion, privately, with their own DNA and RNA quite different from ours. They are as much symbionts as the rhizobial bacteria in the roots of beans. Without the mitochondria, we would not move a muscle, drum a finger, or think a thought. [See Figure A.]

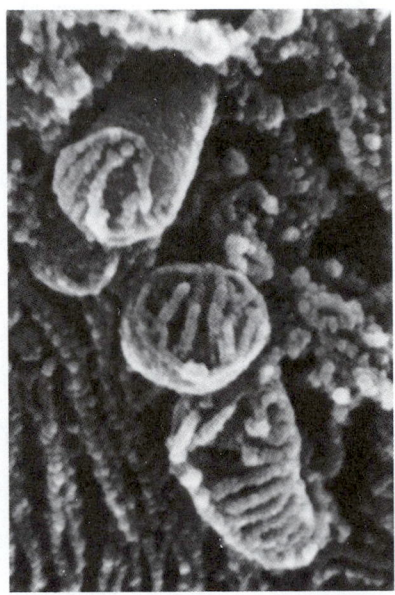

FIGURE A
In this SEM microphotograph of mitochondria, the existence of separate entities within the cell is apparent.

Mitochondria are stable and responsible lodgers, and I choose to trust them. But what of the other little animals, similarly established in my cells, sorting and balancing me, clustering me together? My centrioles, basal bodies, and probably a good many other more obscure tiny beings at work inside my cells, each with its own special genome, are as foreign, and as essential, as aphids in anthills. My cells are no longer the pure line entities I was raised with; they are ecosystems more complex than Jamaica Bay.

I like to think that they work in my interest, that each breath they draw they draw for me, but perhaps it is they who walk through the local park in the early morning, sensing my senses, listening to my music, thinking my thoughts.

I am consoled, somewhat, by the thought that the green plants are in the same fix. They could not be plants, or green, without their chloroplasts, which run the photosynthetic enterprise and generate oxygen for the rest of us. As it turns out, chloroplasts are also separate creatures with their own genomes, speaking their own language.

We carry stores of DNA in our nuclei that may have come in, at one time or another, from the fusion of ancestral cells and the linking of ancestral organisms in symbiosis. Our genomes are catalogues of instructions from all kinds of sources in nature, filed for all kinds of contingencies. As for me, I am grateful for differentiation and speciation, but I cannot feel as separate an entity as I did a few years ago, before I was told these things, nor, I should think, can anyone else.

Source: Lewis Thomas, *The Lives of a Cell.* Copyright © 1973 by the Massachusetts Medical Society. Reprinted by permission of Viking Penguin, Inc.

Chapter 4   The Cell: Where It All Happens

7. The cytoplasmic region consists of a fibrous cytoskeleton, which functions in support, movement, and organization within the cytoplasm. The cytoskeleton is composed of microtubules, microfilaments, and intermediate fibers.

8. The endoplasmic reticulum (ER) is a network of channels, tubes, and flat disks that functions as an internal transport and storage system. ER having attached ribosomes (sites of protein synthesis) is called rough ER (RER). ER lacking ribosomes is called smooth ER (SER). SER is the site of lipid synthesis.

9. Ribosomes, which function in protein synthesis, are comprised of two subunits, each consisting of protein and ribonucleic acid (RNA). Sometimes several ribosomes are linked together to form a polyribosome.

10. The Golgi apparatus is a collection of flattened, disklike membranes involved in packaging and secreting chemicals produced by the cell.

11. Lysosomes contain hydrolytic enzymes and are responsible for digestion of cell parts.

12. Peroxisomes are small vesicles containing an enzyme that can break down hydrogen peroxide produced by cellular activities.

13. Mitochondria are membrane-bound organelles that serve as the primary site for cellular processes that result in ATP synthesis.

14. Vacuoles, which are most often found in plant cells, are fluid-filled membranous sacs that function in storage and in increasing surface area. Similar structures found in animal cells are usually smaller and are often called vesicles.

15. Chloroplasts are membrane-bound, chlorophyll-containing organelles that function in photosynthesis.

16. The nucleus is the center for cellular control and cellular reproduction. It consists of a nuclear envelope and nucleoplasm, which contains strands of DNA and protein called chromatin. During cell division, chromatin coils upon itself to form chromosomes. The large dark mass in the nucleus, called the nucleolus, consists of RNA and functions in ribosome synthesis.

17. Cilia and flagella are hairlike projections used for cellular locomotion. Both types have a structure formed by microtubules arranged in a 9 + 2 pattern.

18. The basal body is located at the point where a cilium or flagellum meets the cell membrane. A basal body is composed of microtubules arranged in nine sets of three microtubules around the periphery.

19. Centrioles establish the plane of cell division by forming the spindle in eukaryotic cells. When centrioles are present, they are usually located in pairs just outside the nuclear envelope.

20. According to the endosymbiosis hypothesis, eukaryotic cells originated when large, primitive prokaryotic cells established an endosymbiotic relationship with other, smaller prokaryotic cells by engulfing them but not digesting them. The cells developed a host-guest relationship.

KEY TERMS

cell theory	microfilament
light microscope	intermediate fiber
resolving power	endoplasmic reticulum
visible light	vesicle
transmission electron microscope	ribosome
scanning electron microscope	Golgi apparatus
cell	lysosome
prokaryote	peroxisome
eukaryote	mitochondrion
plasma membrane	vacuole
selectively permeable	plastid
cytoplasmic region	chloroplast
organelle	nuclear envelope
nucleus	chromatin
cell wall	nucleolus
cytoskeleton	cilia
microtubule	flagella
	basal body
	centriole

REVIEW QUESTIONS

True/False

1. The first cells on earth were probably eukaryotic.
   true   *false*   *(page 78)*

2. The limiting factor in microscopy is magnification.
   true   *false*   *(page 77)*

3. In general, eukaryotic cells consist of a nucleus and a cytoplasmic region.
   true   *false*   *(page 82)*

4. The function of a cell is related to its structure.
   true   *false*   *(page 82)*

Fill in the Blank

1. The ability to distinguish two points as separate entities is called _____ _____.
   (page 77)

2. The double-membraned intercellular transport system with ribosomes is called the _____ _____ _____. *(page 87)*

3. Secretory cells would in all probability contain more _____ _____ than nonsecretory cells. *(page 87)*

4. It was once believed that cellular organelles floated in the cytoplasm of cells. Today, however, it has

been discovered that a _____ provides a site of attachment for many cellular organelles. *(page 84)*

Multiple Choice

1. Cells that lack membrane-bound organelles are called:
   a. Prokaryotic
   b. Eukaryotic
   c. Sperm
   d. Egg
   (page 78)

2. In the cell the major enzymes involved in the synthesis of ATP are found in the:
   a. Golgi apparatus
   b. Mitochondria
   c. Nuclei
   d. Centrioles
   (page 90)

3. The rough endoplasmic reticulum owes its rough surface to:
   a. Mitochondria
   b. Proteins
   c. Ribosomes
   d. DNA particles
   (page 87)

4. Many of the micrographs in this book have a three-dimensional quality. These must have been taken with a:
   a. Transmission electron microscope
   b. Light microscope
   c. Scanning electron microscope
   d. High-voltage transmission electron microscope
   (page 78)

Discussion

1. What is a cell, and what is meant by the cell theory?

2. Compare and contrast the light microscope, transmission electron microscope, and scanning electron microscope.

3. Define and give examples of a prokaryotic cell and a eukaryotic cell.

4. What is meant by the endosymbiosis hypothesis?

5. Compare the structure of a typical prokaryotic cell to that of a eukaryotic animal cell.

6. How are plant and animal cells alike? How do they differ?

7. Discuss the function of the endoplasmic reticulum, Golgi apparatus, ribosomes, lysosomes, and peroxisomes.

8. What is the function of a chloroplast? A mitochondrion?

9. What is meant by the cytoskeleton? What is it comprised of? What function does it perform?

10. What is the function of the nucleus, and what is it comprised of?

5

Membranes, Molecules, and Cell Function

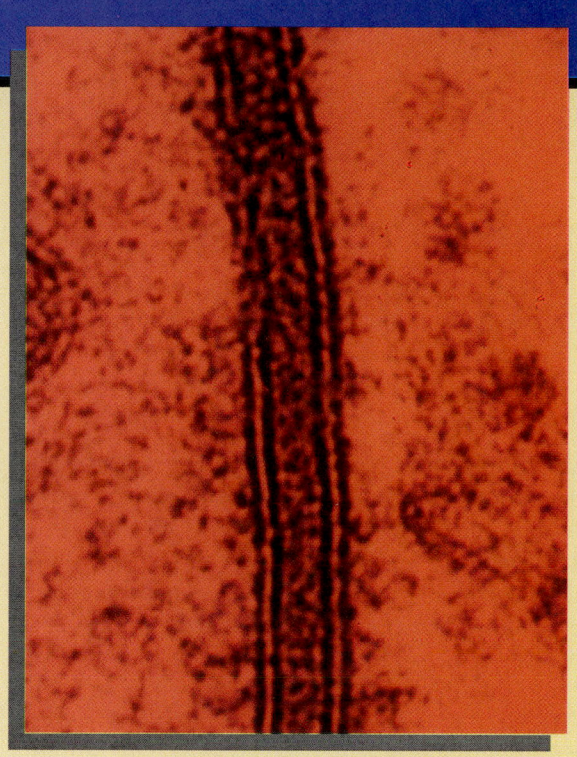

What causes cells to divide? How do cells recognize self from nonself? How do they control what molecules enter and leave their interiors? Early biologists thought that *cell membranes*, or *plasma membranes* as they are also often called, served primarily to separate the cell from its surrounding environment. Later, researchers learned that plasma membranes are *selective* — that is, membranes allow only certain substances to enter and leave the cell. This selectivity plays an important role in many of the functions occurring in living organisms.

The cell surface is the channel of communication between the cell and the exterior world. Cells are, by their very nature, islands separated by a thin

Above: A transmission electron microphotograph of two adjacent cell membranes. Note the dark region in the center, which is the space between the two cells.

membrane from the rest of the world. And what a membrane! It admits no large molecules across itself (with certain exceptions such as DNA, which is taken up during transformation) and only a relative handful of small charged molecules. Each of these requires the presence of a specific carrier molecule. In multicellular organisms, the cell membrane is studded with receptors and identifying "tags" of polysaccharides that allow the cell to be recognized by other cells and communication chemicals. The cell membrane can recognize specific hormones and specific cells. It can distinguish self from nonself. Indeed the cell membrane is a dynamic structure essential for the maintenance of life.

As we begin our study of plasma membranes, it is important to remember that the cell is not the only structure that is surrounded by a membrane. Membranes also enclose cellular organelles such as mitochondria, chloroplasts, and the nucleus. What is more, many cellular transport systems (such as the endoplasmic reticulum and the Golgi apparatus) are membranous structures. With this information in mind, let us begin with a discussion of membrane structure. Then we can move on to how molecules are recognized and transported through membranes.

MEMBRANE STRUCTURE

■ **Today the fluid mosaic model is the accepted explanation of the structure of the plasma membrane. This model proposes that the membrane consists of a phospholipid bilayer with embedded globular proteins and exterior glycoproteins.**

Even though the **plasma membrane** is a common structure in living organisms, it is difficult to observe, so biologists know very few specifics about its structure from direct observation. Part of the difficulty arises from the fact that the structure is very thin — too thin to be seen with a light microscope. The plasma membrane measures only about 65 to 100 angstroms (Å) thick (an angstrom is only 1/10,000,000,000, or one ten-billionth of a meter or, said another way, $1/10$ of a nanometer). Without visible evidence in hand, biologists have had to observe membrane behavior and form theories based on their observations. A review of those theories will tell us much about the structure and functions of the plasma membrane (see Human Endeavors).

Early Theories

Because of its thinness, and because early investigators lacked the electron microscope, anyone who wanted to study the membrane before 1950 was forced to rely on chemical experiments. For example, since lipids and lipid-soluble substances pass through the membrane with ease but ions do not, biologists concluded that the chemical composition of the membrane must include lipids (because lipids are soluble in lipids). Experiments about the membrane's surface tension and elasticity led to the conclusion that proteins also must be involved in the plasma membrane structure.

In 1940, H. Davson and J. Danielli developed a description (what scientists call a *model*) of the plasma membrane, based on their work and the synthesis of the available research information. Their model proposed that the plasma membrane consisted of two lipid layers sandwiched between two protein layers (sort of a fat sandwich, with layers of protein for bread).

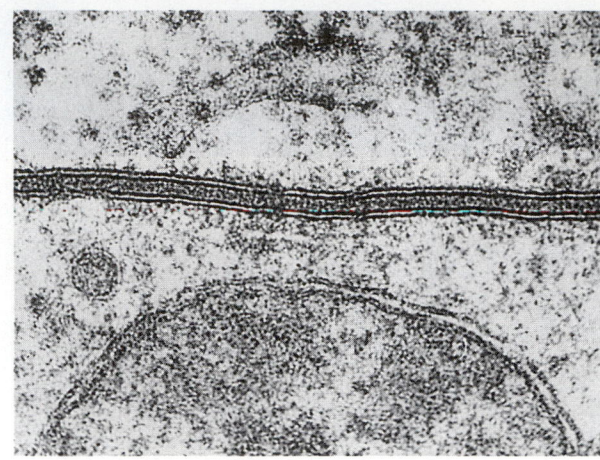

FIGURE 5.1 The Plasma Membrane
An electron microgram of a cross-section of a red blood cell shows the plasma membrane as a three-layer structure: a light band (lipid layer) sandwiched between two dark bands (protein).

Then, in the 1950s, as a result of the development of the electron microscope, J. D. Robertson was able to photograph the plasma membrane. He found that its structure appeared as a light line sandwiched between two dark lines (Figure 5.1). This observation supported the Davson/Danielli model.

Fluid Mosaic Model

Today, biologists know that the Davson/Danielli model was not completely valid. With the advent of better electron microscopes and new techniques of specimen preparation, especially two techniques called *freeze fracture* and *freeze etching*, biologists were able to describe a more accurate model. (These techniques are illustrated in Figure 5.2.) In 1972, S. Singer and G. Nicolson proposed the fluid mosaic model, which is accepted by most investigators today. The fluid mosaic model presents the membrane as a more dynamic, changing structure than that proposed by Davson and Danielli.

The Lipid Bilayer

■ **The lipid bilayer of the plasma membrane is composed of phospholipids. The polar, hydrophilic heads of the phospholipids face outward. The nonpolar, hydrophobic tails face inward, toward the center of the membrane.**

According to the **fluid mosaic model,** all membranes, including those that surround organelles,

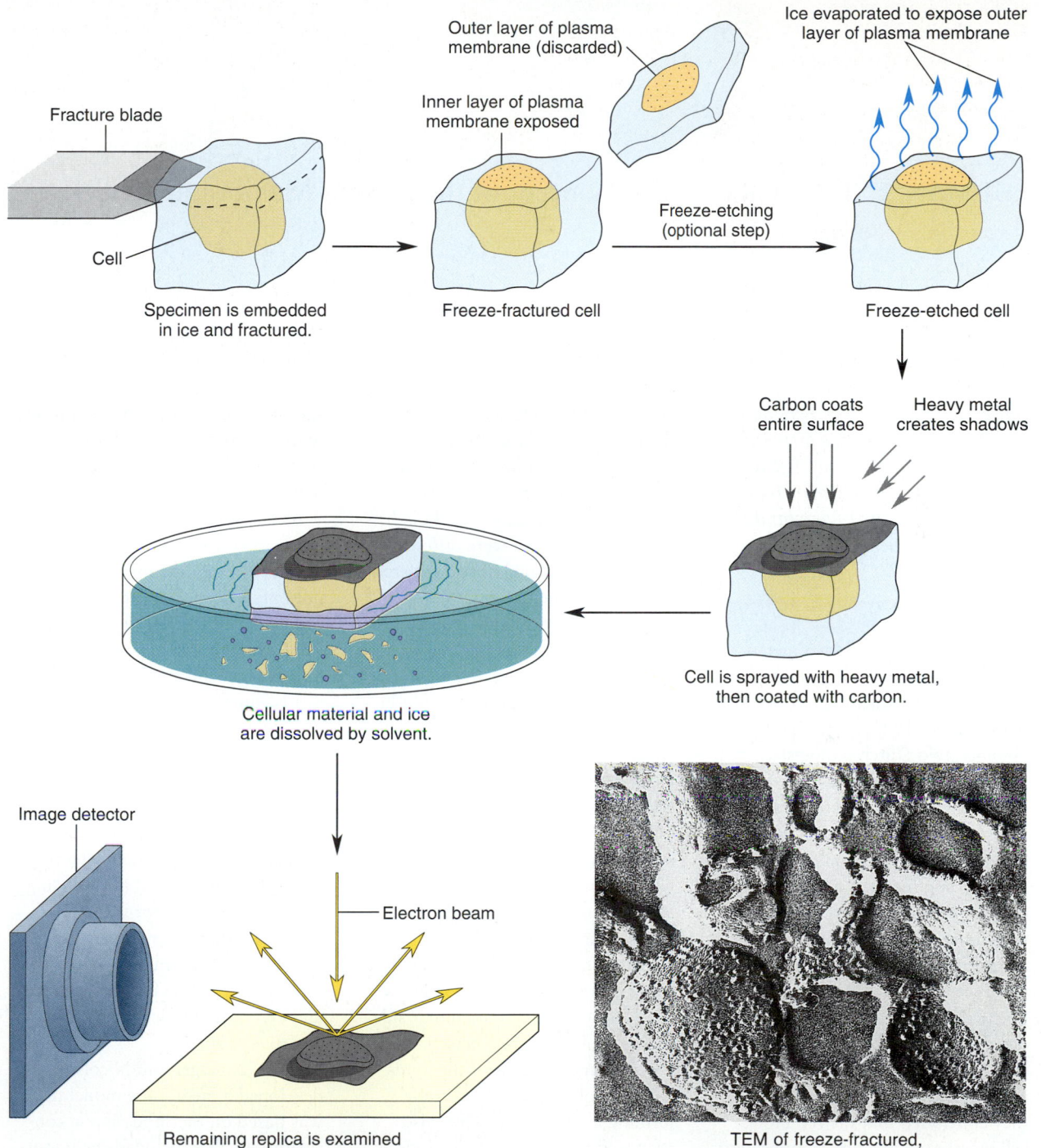

Fracture blade

Cell

Specimen is embedded
in ice and fractured.

Outer layer of plasma
membrane (discarded)

Inner layer of plasma
membrane exposed

Freeze-fractured cell

Freeze-etching
(optional step)

Ice evaporated to expose outer
layer of plasma membrane

Freeze-etched cell

Carbon coats
entire surface

Heavy metal
creates shadows

Cell is sprayed with heavy metal,
then coated with carbon.

Cellular material and ice
are dissolved by solvent.

Image detector

Electron beam

Remaining replica is examined
using electron microscope

TEM of freeze-fractured,
freeze-etched plasma membrane

FIGURE 5.2   Freeze Fracturing and Freeze Etching

Freeze fracturing and freeze etching are important techniques to cell
biologists. As a cold fracture blade cuts through the frozen material, cells are
literally fractured. In the freeze-fracturing process these cell fragments are
coated with carbon and a heavy metal such as platinum. When the frozen
material and cell fragments are removed, a replica remains. An additional step
is taken in freeze etching. Before the fragments are coated, part of the ice is
sublimated (removed by evaporation of a solid directly into vapor). This
exposes more surfaces, and the replicas then reveal additional features of the
cell.

The Concept of the Plasma Membrane Model

Since the 1920s, researchers have proposed models for the structure of the plasma membrane. Let us follow the human endeavor that has resulted in our understanding of the plasma membrane.

Is the cell surrounded by a membrane? And if so, what is its structure? Early investigators asked those questions. In 1925, Gorter and Grendel, two Dutch biologists, thought that the cell was surrounded by a membrane. Specifically, they investigated red blood cells, then called *chromocytes*, and observed that if you punctured a red blood cell, the contents leaked out. Hence, there must be a covering or a membrane enveloping the cell. From this observation and by measuring the lipid content of the membrane, they determined that the ratio of the surface area of the cell to the surface area of the lipid was 2:1. Therefore, they concluded that the red blood cell membrane must be composed of a bilayer (two layers) of lipids, since the lipid surface area was twice as much as the cell's surface area. It's sort of like having enough wrapping paper to cover a package two times. Because the chemists of the day knew that lipids would spontaneously form a layer in water,

Gorter and Grendel proposed that the red blood cell membrane must be composed of a lipid bilayer. They didn't generalize to other cells, because at that time it was thought that since cell types were different, each cell type would have a different membrane structure.

In 1934, Davson and Danielli proposed that the cell membrane must have proteins because if it were only composed of lipids, it would not be as strong as cell membranes actually were. So they proposed that the membrane bilayer surface must be coated by globular proteins. With further investigation, it was determined that the proteins could bind with the lipid bilayer more tightly. So they modified their model to indicate that the globular proteins' hydrophobic side chains (remember, water-fearing, nonpolar) somehow unrolled so that these nonpolar side chains could penetrate between the polar (charged) heads of the lipids and nonpolar lipid tails, thereby forming a tight union. They also proposed that there must be loosely attached globular proteins that would allow for the passage of nonpolar substances through the membrane (remember that, in general, nonpolar substances are solvent in nonpolar solvents and polar substances are soluble in polar solvents). Therefore, proteins must

somehow serve as the pores through which polar substances could pass through this highly nonpolar lipid membrane.

As Davson and Danielli thought more about the passage of substances through the membrane, it not only became apparent that the proteins were necessary for structural strength and permeability, but also that this permeability could not just be simple diffusion. The proteins must form a channel through which polar molecules could pass during facilitated transport. So, in 1952, they proposed the *unit membrane* model, in which they concluded that separate proteins on the surface functioned in providing structural integrity for the membrane and that other proteins might form channels to interact with other molecules to allow for the entrance and exit of molecules by facilitated, or active, transport.

In 1972, Singer and Nicolson developed a membrane model that was based on thermodynamic considerations that included the important energy relationships between the lipids and proteins. They proposed that there were no surface globular proteins lining the membrane with the lipids in between (sort of like a cheese sandwich, with the protein layers serving as the two pieces of bread

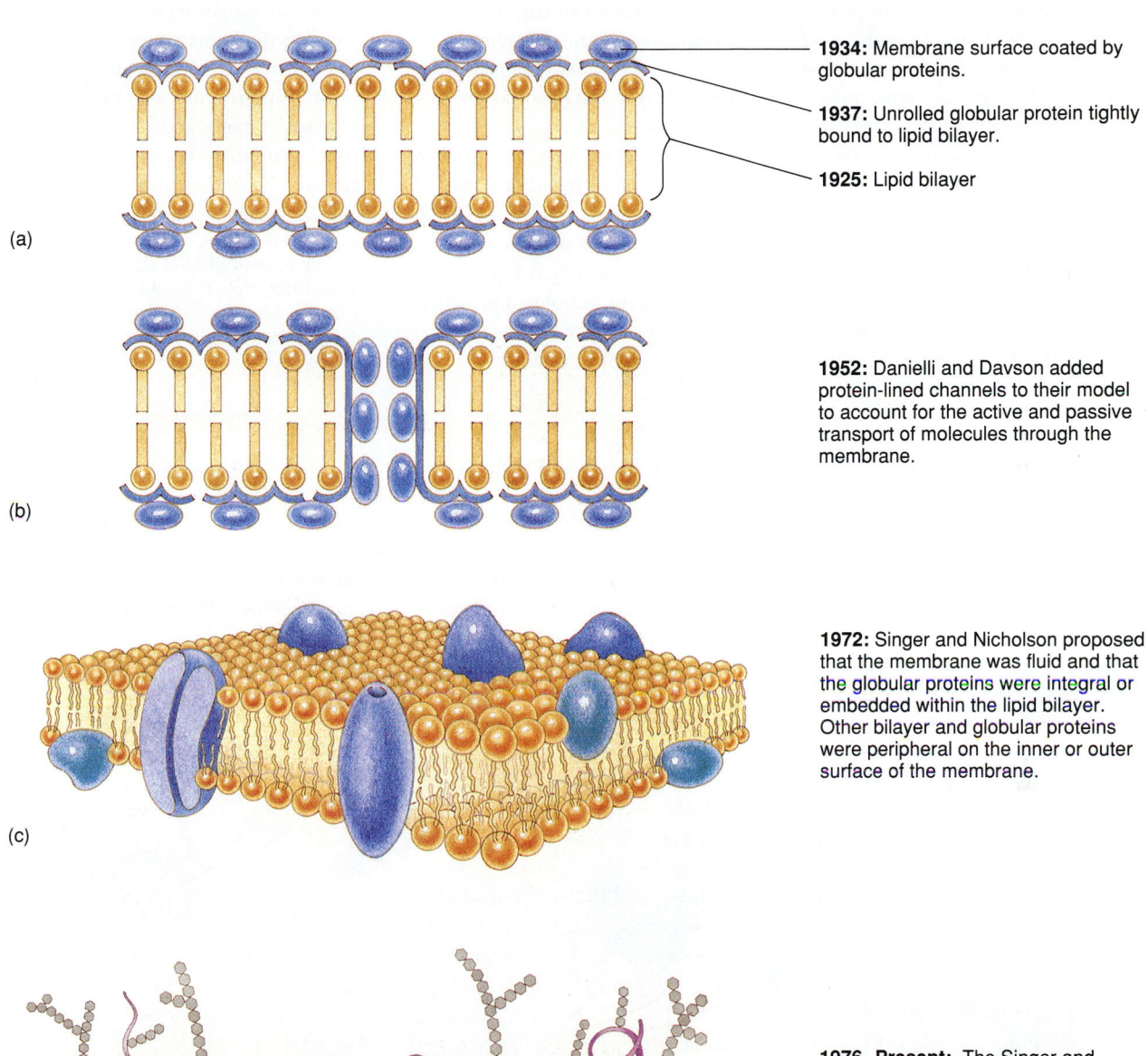

(a)

1934: Membrane surface coated by globular proteins.

1937: Unrolled globular protein tightly bound to lipid bilayer.

1925: Lipid bilayer

(b)

1952: Danielli and Davson added protein-lined channels to their model to account for the active and passive transport of molecules through the membrane.

(c)

1972: Singer and Nicholson proposed that the membrane was fluid and that the globular proteins were integral or embedded within the lipid bilayer. Other bilayer and globular proteins were peripheral on the inner or outer surface of the membrane.

(d)

1976–Present: The Singer and Nicholson model was modified to account for new findings about membranes. No partially embedded proteins were found in the bilayer; all spanned the bilayer and the membrane was fluid. Some globular proteins were anchored and some inner peripheral proteins were attached to cytoskeletal components. More is being learned about the role of the receptor proteins on the outer surface.

and the cheese the lipid bilayer), but rather the globular proteins were randomly distributed through the lipid bilayer. They said that the membrane was fluid and the proteins could float around in this fluid bilayer. Even though they proposed that proteins were different, they were all similar because of the arrangement of amino acids, the hydrophilic ends and hydrophobic centers that allowed them to cross the lipid bilayer. This model was called the *Singer-Nicolson Fluid Mosaic Model*. In 1976, biologists modified the Singer-Nicolson model to account for situations in which the membrane was less fluid. Researchers knew that many factors affected the fluidity of the membrane—temperature, cholesterol content, lipid-protein interaction, and others. Therefore, the membrane was not as fluid as Singer and Nicolson originally proposed.

Since 1976, we have learned more about the plasma membrane. In 1989, Singer described several kinds of transmembrane protein molecules. Other researchers have emphasized the peripheral proteins on the inner surface of the membrane that form or interact with the cell's cytoskeletal structure. This interaction helps us to understand cell mobility and changes in the cell's shape. So as you can see, several investigators have participated in the human endeavor that has allowed us to better understand membrane structure and function. Our understanding of this will continue to improve because of new research by investigators.

Source: Adapted from "Evolutions: The Plasma Membrane," originally published in *The Journal of NIH Research*, May–June 1989, Vol. 1, pp. 131–132. Illustration adapted by Elizabeth Morales-Denney from original artwork by Sally Bensusen.

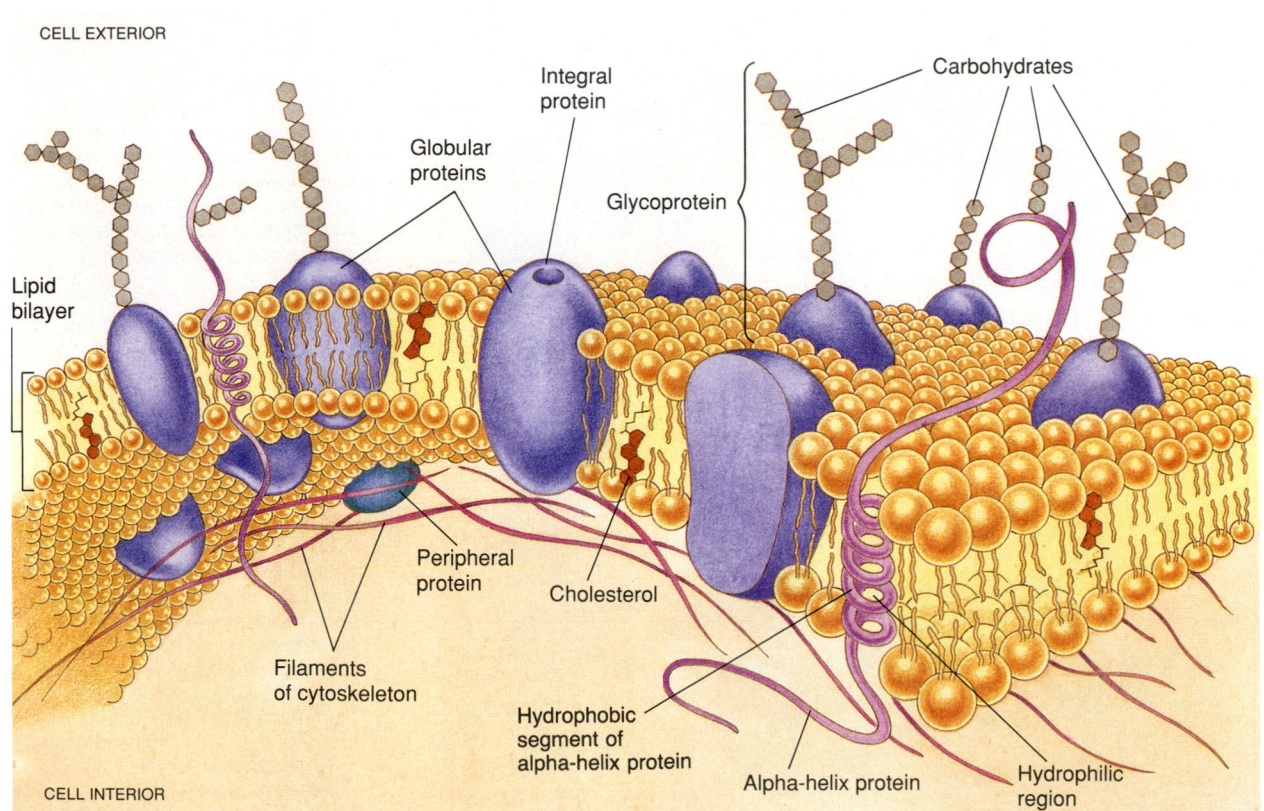

FIGURE 5.3   Fluid Mosaic Model
This model of the plasma membrane depicts the membrane as a dynamic structure composed of lipids, proteins, and carbohydrates. The lipids (largely phospholipids) are arranged in a double layer with the tails facing inward. Protein molecules lie on both the inner and outer surfaces, with many penetrating the entire lipid bilayer.

consist of a *double lipid layer,* or lipid bilayer. Embedded within the layer or peripheral to it are *large globular proteins.* Some of the protein molecules are situated only on the outer surface, others only on the inner surface, and still others extend through both lipid layers. For convenience, biologists like to refer to the proteins as either *integral proteins,* which are embedded in the lipid bilayer and span the entire width of the membrane, or *peripheral proteins,* which are not embedded in the lipid bilayer but rather are attached to integral proteins or other parts of the cell. For example, on the inner side of the membrane some peripheral proteins are attached to the cytoskeleton. The lipid layers themselves are fluid, having about the same consistency as light vegetable oil (Figure 5.3). Some globular proteins float within the layer or migrate from one location to another within this dynamic structure.

The lipid bilayer is composed primarily of *phospholipids.* In Chapter 3 we discussed the dual polarity of these macromolecules. The head portion contains the phosphate, is polar, and is hydrophilic (water-loving), or water-soluble. The tail portion contains two fatty acid chains, is nonpolar, and is hydrophobic (water-fearing), or not water-soluble (see Figure 3.10). Dispersed within the lipid bilayer are unsaturated fatty acid tails and steroids such as cholesterol. These substances help keep the layer in a fluid state.

Because of the dual polarity of phospholipid molecules, the head ends of the lipid bilayer point toward the outer and inner surfaces of the membrane (toward water), and the tails face inward, toward the center of the layer and away from water (Figure 5.4).

Globular Proteins

■ **Globular proteins in the plasma membrane organize cellular processes and transport molecules through the membrane. They also function in cytoskeleton attachment.**

In Chapter 3 we discussed the four structural levels of protein molecules. We mentioned there (and elsewhere) that the geometry of a protein determines to a large extent how it will react with other molecules. That is also the case with proteins of the plasma membrane: The protein composition of a membrane is directly related to the membrane's function.

For instance, biologists theorize that certain proteins are involved in molecular transportation. Serving as molecular carriers, the protein molecules

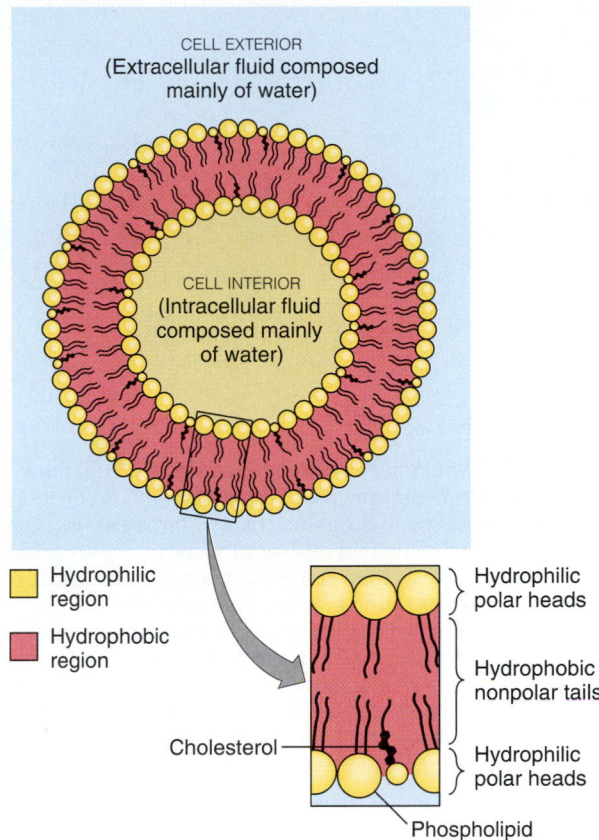

FIGURE 5.4 Polarity and the Plasma Membrane
The tightly packed molecules of phospholipids and cholesterol form a bilayer that effectively separates their hydrophobic (nonpolar) tails from a watery (polar) environment.

are just long enough to extend through the lipid bilayer and form hydrophilic pores, or channels (due to their alpha helical structure) through which substances can pass. We consider these topics in detail later in the chapter. In Chapters 6 through 9, we discuss other proteins that are enzymes regulating membrane-bound chemical reactions, such as the ordered passing of electrons in photosynthesis and respiration. Other globular proteins function as receptor sites to receive and bind with a specific chemical messenger like a hormone. Once the messenger is received, the receptor protein undergoes a change in shape that initiates a change in the shape of the protein inside the cell. This causes a series of chemical reactions that result in the changes the hormone initiates. Other membrane proteins hold cells together in a variety of intercellular junctions. Still other proteins are fixed in place and are thought to be part of the system that anchors the

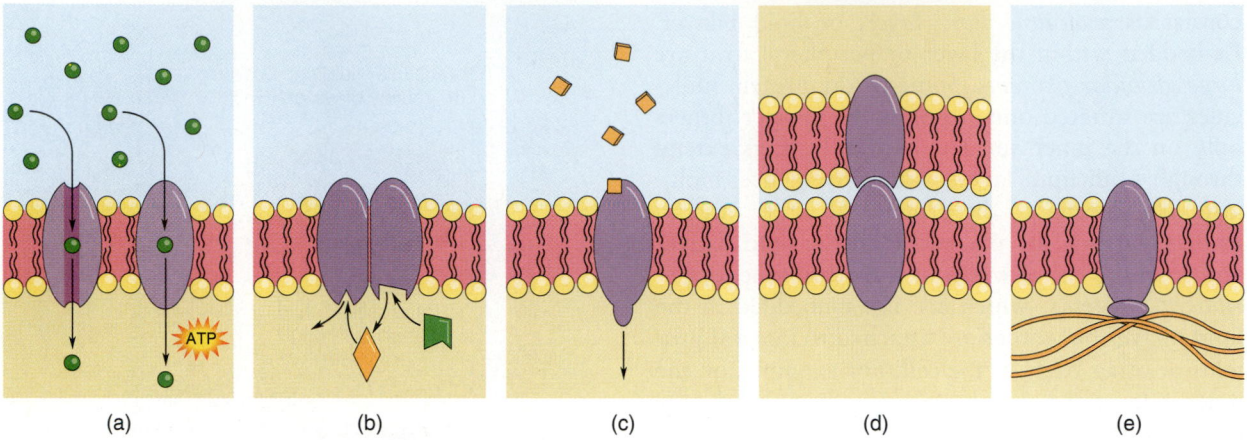

FIGURE 5.5 Summary of Membrane Protein Functions
The globular membrane proteins perform many functions: (a) Transport
proteins contain a hydrophilic channel through which polar molecules are
transported in facilitated diffusion. Some transport proteins use ATP as a
source of energy as they pump substances from a place of lower concentration
to higher concentration in active transport. (b) Membrane proteins may be
enzymes or electron carriers that function in sequential reactions. (c) The
outer protein of some membrane proteins serve as receptor sites for the
binding of hormones and other substances. (d) In cell adhesion, some
membrane proteins play a role in holding adjacent cells together. (e) Some of
the inner membrane proteins are attached to the cytoskeletal actin microfilaments.

cytoskeleton structure to the plasma membrane.
Figure 5.5 summarizes membrane protein function.

Carbohydrates on Membranes

■ Carbohydrates on the membrane surface are involved in cell-to-cell recognition, cell adhesion,
and cell receptors, but more research is needed before we know all their functions.

Some membranes, especially the plasma membrane,
contain a significant carbohydrate content on their
outer surfaces. For example, a red blood cell membrane is approximately 40 percent phospholipid, 52
percent protein, and 8 percent carbohydrate. When
carbohydrates are present, most (over 90 percent)
are bonded to proteins to form *glycoproteins* (sugar
proteins). The remainder are bound to lipids to
form *glycolipids*. More research is necessary before
we will fully understand the function of glycolipids
and glycoproteins, but we do know that they are
involved in cell-to-cell recognition and in the adhesion of cells and that they serve as receptors for
hormones, viruses, bacteria, and other chemicals.

SOME TERMINOLOGY

■ A membrane is selectively permeable because only
certain molecules or ions can pass through it.

In Chapter 1 we discussed the fact that *homeostasis*
is a crucial characteristic of living things. In order
to maintain a homeostatic condition, the flow of
materials into and out of cells cannot be random.
Molecular movement must be regulated, and that
regulation is one of the major functions of the cell
membrane.

Our discussion about membrane function centers
on both the environment outside the cell (extracellular fluid) and the environment inside the cell (intracellular fluid). The discussion also concerns how
molecules move from one location to the other. As
we discuss molecular movement, we use terminology that describes the environment in which an activity is taking place and the types of movement that
can occur. So before we examine membrane function, we need to discuss the terms that biologists
use. Many terms appear again in this text.

One of the major characteristics of membranes
is that they are selectively permeable. In other
words, some molecules and some ions pass through
the membranes with ease, but others do not.
Whether a molecule can pass through or not is regulated by two factors: (1) its size, in relation to the
protein channel in the membrane through which the
molecule must pass; and (2) its chemical nature
(Figure 5.6). The permeability of membranes to specific substances can change with different cellular
environments.

Chapter 5   Membranes, Molecules, and Cell Function

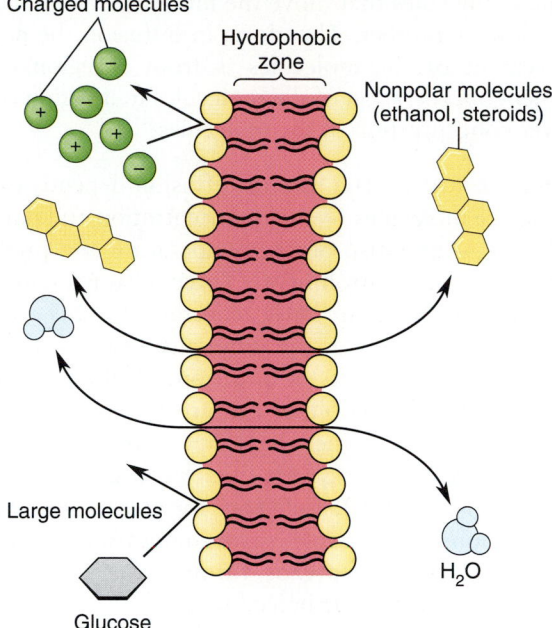

Charged molecules

Hydrophobic zone

Nonpolar molecules (ethanol, steroids)

Large molecules

Glucose

H_2O

FIGURE 5.6 Membrane Selectivity
Because of the chemical nature of the membrane, it is selectively permeable. Water and most nonpolar molecules can pass through the membrane. However, large molecules and charged ions cannot pass through the membrane unless associated with specific transport proteins.

In Chapter 2 we discussed the characteristics of solvents and solutes. There we said that a *solvent* is a chemical in which something is dissolved, and a *solute* is the substance that is dissolved in the solvent. A solution is a combination of a solvent and a solute. The relationship between the level of dissolved substances in a cell (its solutes) and that in the environment in which the cell is placed is called **tonicity.**

PASSIVE TRANSPORT

- **Passive transport is a method of moving molecules through a membrane without requiring any energy from the cell.**

The movement of molecules across membranes can be divided into two types. (1) In **passive transport,** molecules move through the membrane without any expenditure of cellular energy. (2) In **active transport,** molecules move through a membrane with an expenditure of cellular energy. We will begin by looking at three kinds of passive transport: *diffusion, osmosis,* and *facilitated transport.* (Table 5.1 provides an overview of the types of membrane transport — passive and active — discussed in this chapter.)

TABLE 5.1
Membrane Transport Mechanisms

Passive transport (requires no direct cellular energy)	*Active transport (requires direct expenditure of cellular energy)*
Diffusion Molecules move from a region of higher concentration to a region of lower concentration. In cells this movement involves both passage through a membrane and transport within the cell.	*Active Transport* Movement of molecules or ions against a diffusion gradient through the protein channels of the membrane. ATP serves as the energy source.
Osmosis Diffusion of a solvent (usually water) through a membrane from a region of higher concentration to a region of lower concentration.	*Endocytosis* Large particles are moved *into* the cell by being engulfed by the cell membrane. ATP is the energy source.
Facilitated Transport Carrier proteins bind with specific molecules and facilitate their passive transport through the protein-lined pore.	*Receptor-Mediated Endocytosis* Specific molecules are brought into the cell by binding with specific sites (receptors) on the cell membrane. Once inside the cell, coated vesicles surround the molecules, forming a transport vesicle.
	Exocytosis Expulsion of particles or substances enclosed in vesicles from the cell. The vesicles fuse with the cell membrane, then rupture to expel their contents.

Diffusion

■ **Diffusion is due to the random movement of molecules. Molecules move along a concentration gradient from a higher concentration to a lower concentration until a state of equilibrium is reached. The rate of diffusion depends on the size and concentration of the solutes, and on temperature and pressure.**

Diffusion is responsible for the movement of some simple molecules like carbon dioxide across membranes and for the movement of materials within the cell itself. Simply put, **diffusion** is the random movement of molecules that ultimately results in their even distribution. While that may sound complicated, it really is not.

We are all familiar with the fact that an odor coming from something in one corner of a room will eventually permeate the entire room. In a similar way, if you place a drop of food coloring in water, it will eventually spread (or *diffuse*) through all the water. (To use the terms introduced earlier, the food coloring is the *solute,* the water is the *solvent,* and the combination is the *solution.*) Diffusion can take place in a gas, liquid, or solid. For instance, if you place a drop of food coloring on a piece of clear gelatin (like uncolored Jell-O), the coloring will diffuse through the substance.

In all these examples, the source (the odor or the concentrated dye) is the place of *higher concentration,* and the areas that the source spreads to are areas of *lower concentration.* The movement of molecules from a place of higher concentration to a place of lower concentration is called diffusion.

Concentration Gradient. The difference between the place where most of the substance is and the place that it is moving toward is called a **concentration gradient.** Molecules move along the concentration gradient from a region of higher concentration to a region of lower concentration. Molecules continue to move until eventually they are equally distributed throughout the area that encloses them. When this state is reached, biologists call it an equilibrium. The molecules will continue to move randomly, but the movement in one direction is balanced by movement in other directions, so that the net effect is that the molecules remain evenly distributed.

To understand how the molecules are evenly distributed, we need to refer to the discussion of *collision theory* in Chapter 3. There we mentioned that molecules are in constant random motion and are continually colliding with one another. It is this motion and the impact of the collisions between individual molecules that move the molecules from one location to another. Therefore, in diffusion the net movement of the molecules is from a region of higher concentration of the molecules to a region of lower concentration of the molecules.

Diffusion Rates. The *rate* of diffusion depends on the solute molecules — their concentration and particle size — and also on the temperature and pressure of their environment. Three characteristics of diffusion remain fairly constant in any situation: (1) the greater the difference in concentration, the faster the rate of diffusion; (2) the smaller the particle of the solute, the greater the rate of diffusion; (3) the higher the temperature, the more energy there is in the system (which causes a greater number of collisions), and hence the greater the rate of diffusion. For example, sugar in hot tea dissolves more quickly and spreads through the tea at a greater rate than sugar in iced tea.

Because diffusion is responsible for moving materials within the cell, diffusion rates affect cell size. The process of diffusion is essentially quite slow over large distances and fairly rapid over small distances. As a result, one of the major factors that limit cell size is a cell's dependence on diffusion. Also, to have the mechanics of diffusion function efficiently, cells must maintain a high surface area–to–volume ratio. Only a small cell has the necessary ratio. The rationale behind this concept is explained in Chapter 10.

Osmosis

■ **Osmosis is the diffusion of water molecules from a region of higher concentration to a region of lower concentration through a membrane.**

As we've seen, membranes are selectively permeable: Some molecules and ions pass through the membranes but others do not. When diffusion involves the passage of water through a selectively permeable membrane from a region of higher concentration to a region of lower concentration, the process is called **osmosis.** The movement of water molecules into or out of cells is an example of osmosis (remember, the direction of movement is dependent on the concentration gradient).

Tonicity

■ **When a cell is in a solution, the surrounding solution is described as isotonic, hypertonic, or hypotonic when compared with the solution inside the membrane. Tonicity refers to concentration of solutes in solution.**

When you combine sugar and water to make Kool-Aid, you have a solution of sugar water, made up of the solute sugar and the solvent water. You can make that solution very sweet if you add a lot of sugar — in that case, you have a *high concentration* of solute in the solution. In scientific terms, the ratio of solute to solvent is high. To make the sugar-water solution only slightly sweet, you add a small amount of sugar — then you have a *low concentration* of solute in the solution. In other words, the ratio of solute to solvent is low.

The plasma membrane around a cell (or any membrane that surrounds a subcellular body) separates two solutions. One solution is within the cell itself, and the other is outside of the cell. In order to indicate the differences between those two solutions, chemists and biologists use the terms *isotonic*, *hypertonic*, and *hypotonic*. These terms are derived from Latin roots: *iso-* means equal, *hyper-* means above or high, and *hypo-* means below or low. The common suffix of the three terms is *-tonic*, which means strength. (This is why people who exercise refer to "muscle tone.") In discussions about cells and their surroundings, *strength* (or *tonicity*) always refers to the *concentration of solutes* in the solution.

For convenience's sake, let us assume that we have a cell containing a sugar-water solution, and we place this cell in a solution of sugar water. If we say that a solution is **isotonic**, what do we mean? That is correct — we are saying that the concentration of sugar and water molecules in the solution surrounding the cell is equal to the concentration of sugar and water inside the cell. In an isotonic solution like this, individual molecules may move from one side of the membrane to the other, but the sugar-water ratio always stays the same. There is no net molecular movement; so the size of the cell does not change (Figure 5.7a).

If the concentration of sugar in the solution outside the cell is greater than the concentration of sugar in the cell, the solution outside the cell is said to be **hypertonic** — *above strength* or *high strength*. (The solution surrounding the cell is sweeter than the solution inside the cell.) In scientific terms, the concentration of sugar in the sugar-water solution is greater than the concentration of sugar in the cell. The mechanism of osmosis will then operate in order to equalize the concentration on both sides of the membrane. On balance, more water molecules move from the cell and into the surrounding solution until the solution outside the cell and that inside the cell have equal concentrations of sugar. Because water molecules have left the cell, the cell shrinks — a process called *plasmolysis* (Figure 5.7b).

The opposite of hypertonic is **hypotonic** — *below strength* or *low strength*. When a solution is hypotonic, there is a lower concentration of solute and a higher concentration of water outside the cell. The concentration of sugar in the solution outside the cell is lower than the sugar concentration inside the cell. (The solution outside the cell is not as sweet as the solution inside the cell.) In this situation, the mechanism of osmosis moves more water molecules

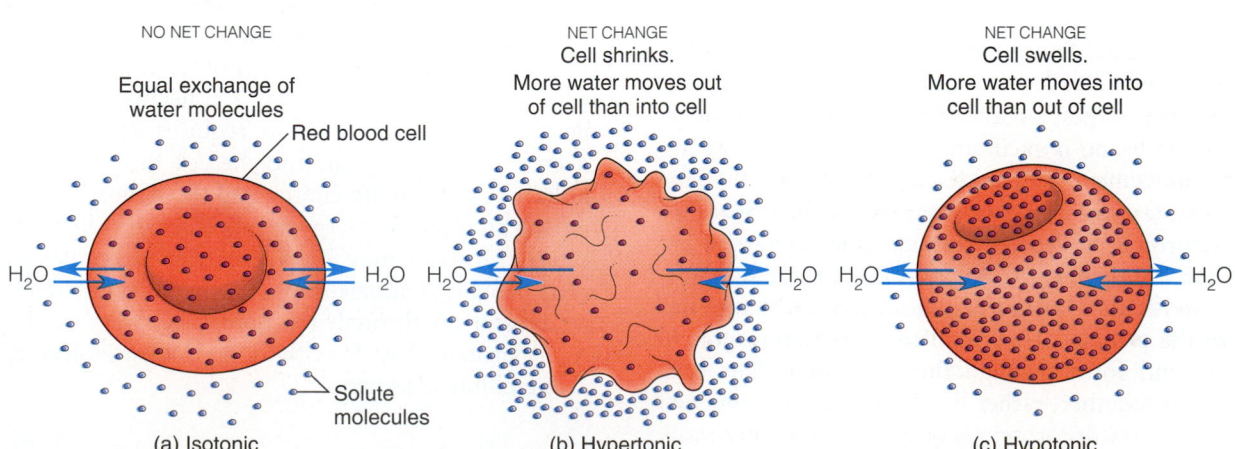

FIGURE 5.7 Effects of Tonicity on Human Red Blood Cells
(a) In an isotonic solution, the concentration of solute is the same inside and outside the cells. The amount of water leaving the cells equals the amount entering. (b) In a hypertonic solution, the concentration of solute is greater outside the cells, and although some water enters the cell, the amount that leaves is far greater. (c) In a hypotonic solution, there is a greater concentration of solute inside the cell and far more water enters.

out of the solution and into the cell to reach equilibrium. As a result, in a hypotonic solution, the cell increases in size (Figure 5.7c). The process can continue until the cell bursts.

In all cases, as you can see, osmosis results in equalizing the concentration in both solutions by moving molecules through a membrane. The mechanism for doing this is the movement of solvent molecules, in this case water, along a concentration gradient from the area of greater concentration to an area of lesser concentration. The movement of solvent molecules can affect cell size. By knowing the type of solution in which a cell is located, you automatically know the direction of molecular movement and the effect of that movement on the size of the cell.

We make use of our knowledge about osmosis often in daily life. For example, when we preserve fruit or pickle cucumbers, we place them in a solution that is hypertonic to the bacterial cells that cause spoilage. Water molecules flow out of the bacterial cell, causing the cell to shrink and die. As a result, the bacteria cannot affect the fruit or cucumbers.

Facilitated Transport

■ **Facilitated transport is passive transport in which transport proteins in the lipid membrane move molecules through it along a concentration gradient. Transport proteins are specific to certain substances.**

Now that we know about the movement of water through the membrane, what about the movement of the solutes? Specifically, how do the polar molecules get through the nonpolar lipid layer? The transport of these polar molecules and ions through the membrane is facilitated by transport proteins. This movement is called **facilitated transport.**

The tertiary and quaternary structures of transport protein molecules are important factors in facilitated transport. In general one can say that only if a molecule's spatial configuration (its shape) fits into the protein carrier will the carrier transport it back and forth through the membrane. The two must fit together, rather like pieces of a jigsaw puzzle. As a result, protein molecules like carrier molecules are said to be very *specific*. (As you will learn in Chapter 6, enzymes also are very specific protein molecules.)

Glucose is an example of a molecule that relies on facilitated transport to pass through a plasma membrane, from a region of higher concentration to a region of lower concentration. Remember that we said earlier that glucose is polar and lipids are nonpolar. That means that glucose is insoluble in the membrane's lipid bilayer. As a result, biologists know that the mechanism of movement cannot be simple diffusion. So, how does glucose get into the cell?

The accepted explanation of facilitated transport is that some of the membrane transport proteins in the membrane bind with the transport specific solutes. These proteins are linked with hydrophilic groups along their interior that attract other hydrophilic groups in molecules such as glucose. Because the hydrophilic groups in the proteins extend from one side of the membrane to the other, the proteins form pores, or channels, that function as gates through which the molecules squeeze as they move in or out of the cell. The movement of the molecule through the membrane is facilitated by changes in the shape of the protein channel as the molecule moves through it (Figure 5.8).

In facilitated transport, molecules move from an area of higher concentration to one of lower concentration. The energy for this movement comes solely from the concentration gradient that exists between the inside and the outside of the cell. The cell itself uses no energy to move the molecules.

ACTIVE TRANSPORT

Facilitated transport requires no energy input from the cell, since the molecules are moved from a region of higher concentration to one of a lower concentration. However, many times molecules are pumped from a region of *lower* concentration to one of *higher* concentration. In other words, molecules are transported *against* a concentration gradient. This type of transport requires active input of energy from the cell, and the cell draws on the molecule ATP for its energy supply. Transport requiring an input of energy is called *active transport.* Active transport may involve carrier molecules in some protein channels in packaging and transporting substances through the pore. The shape of the carrier protein must fit with the passenger molecule, as in facilitated transport.

Active Transport by Transport Proteins

■ **In active transport, energy is required as transport proteins pump substances from lower concentrations to higher concentrations.**

Active transport is not yet well understood. One of the most studied examples of it is the sodium-

Chapter 5   Membranes, Molecules, and Cell Function

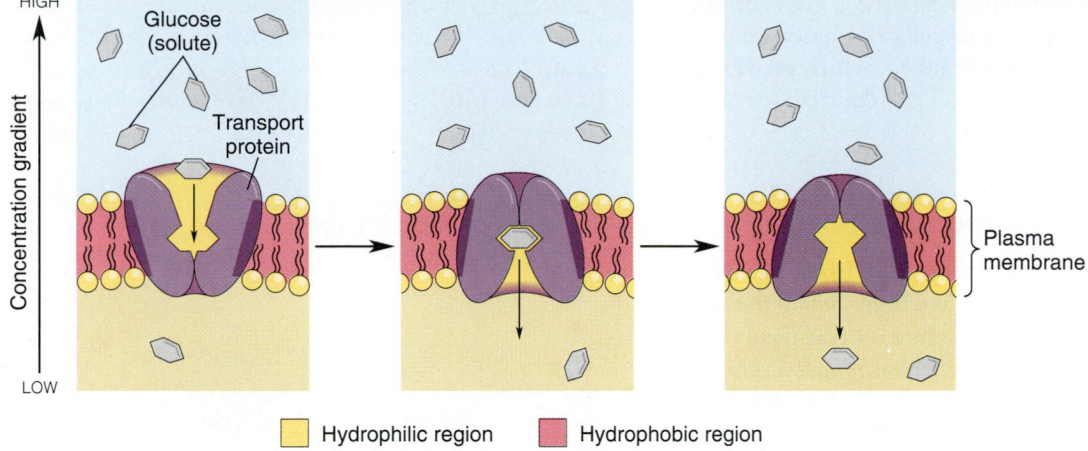

HIGH

Concentration gradient

LOW

Glucose (solute)

Transport protein

Plasma membrane

Hydrophilic region   Hydrophobic region

FIGURE 5.8   Facilitated Transport
Facilitated transport depends on transport proteins in the membrane. Hydrophilic solute molecules move through the proteins along a concentration gradient. Without the proteins facilitating their movement, the solute molecules would be unable to cross the membrane's lipid layers.

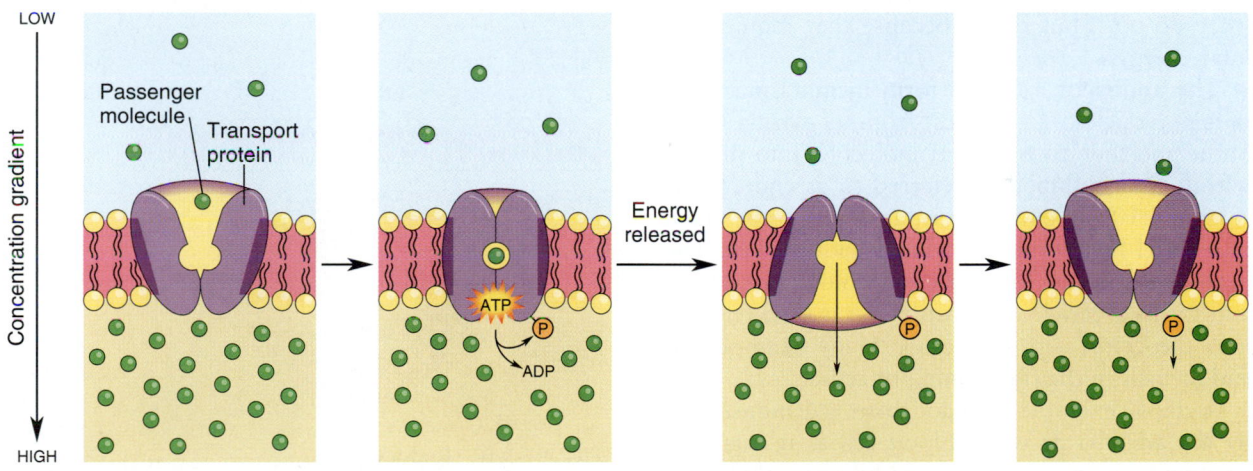

LOW

Concentration gradient

HIGH

Passenger molecule

Transport protein

ATP

ADP

Energy released

FIGURE 5.9   Active Transport
To move molecules and ions against a concentration gradient (from an area of low concentration to an area of high concentration) requires energy. Cells use the energy in ATP to pump those molecules and ions into areas of high concentration. Transport proteins are an integral part of the process.

potassium pump, which cells (especially nerve cells) use to maintain unequal concentrations of sodium (Na^+) and potassium (K^+) ions both inside and outside the nerve cell. This status, called a *differential concentration,* enables a nerve cell to transmit impulses. (We discuss this in more detail later, in Chapter 30.) During active transport, when the passenger molecule binds to the transport protein within the pore of the membrane, biologists theorize that the process causes the transport protein to

change in shape and squeeze the passenger molecule or ion through the pore and eject it on the other side. Figure 5.9 presents a simplified version of active transport.

Another example of active transport occurs in the liver. Again working against the concentration gradient, glucose is transported from the blood, in which there is a low glucose concentration, into liver cells, where the glucose concentration is higher. Kidney tubule cells also actively transport many ions and substances against a concentration gradient. As a result, the cells of these structures contain numerous mitochondria, which provide the adenosine triphosphate (ATP) required for active transport.

Endocytosis and Exocytosis

■ **Endocytosis is the process whereby the membrane surrounds a substance and brings it into the cell. Packaging a liquid substance is called pinocytosis. During exocytosis, a vesicle surrounds a substance and migrates to the cell surface, where the substance is secreted from the cell.**

As you might expect, cellular processes that involve forming compartments around particles or substances to be brought into or removed from the cell require an energy output. Two mechanisms of active transport engage in processing materials — *endocytosis* and *exocytosis*. These processes are classified as forms of active transport, not because they move materials against a diffusion gradient (they do not), but rather because they require cellular energy.

The ability of a cell to form membranous containers (vesicles or vacuoles) from its plasma membrane and then to transport molecules into the cell within the containers is referred to as **endocytosis** (Figure 5.10). In endocytosis, a particle to be taken into a cell first comes in contact with the cell. The plasma membrane folds inward around the particle and forms a vesicle or vacuole that is then released into the cytoplasm. There are many examples of endocytosis in the functioning of cells.

A classic example of endocytosis is demonstrated by the amoeba or white blood cell as it engulfs a single-celled organism, such as bacteria. Biologists call this special type of endocytosis **phagocytosis,** which means cell-feeding. Many single-celled organisms depend on this process for getting food and other substances.

If the substance engulfed is small and liquid, rather than a formed particle, biologists use the term **pinocytosis,** which means cell-drinking. Pinocytosis is found not only in single-celled organisms but also in multicellular organisms. For example, the human egg is surrounded by several nurse cells that provide the egg with nutrients. The egg takes in many of these nutrients through pinocytosis.

Pinocytosis is an important phenomenon from a medical perspective. Sometimes a drug molecule is so large or its chemical nature is such that it cannot easily pass through the plasma membrane. To

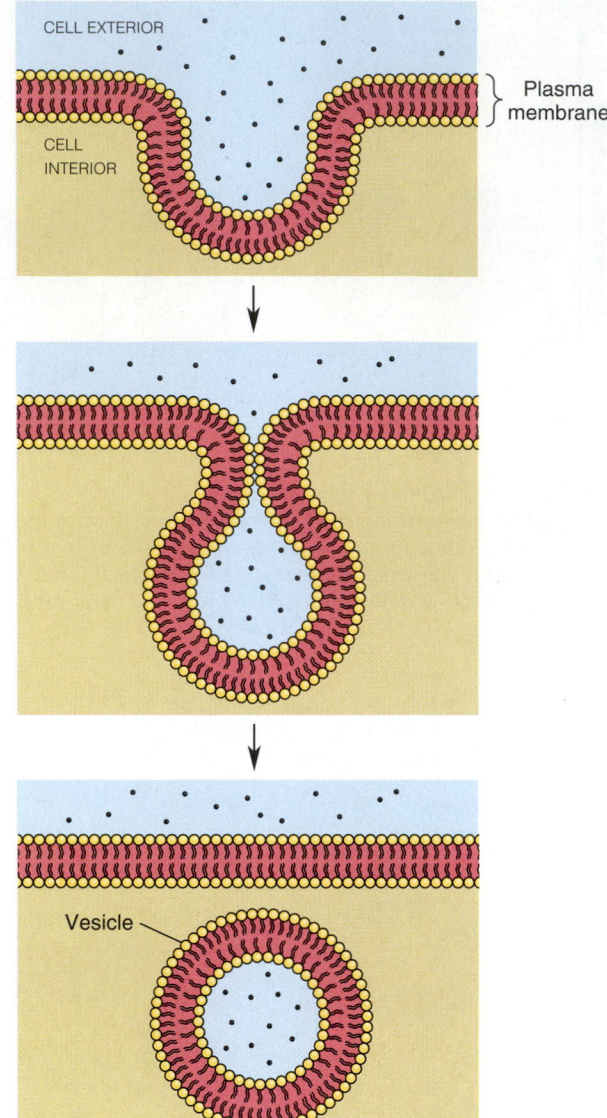

FIGURE 5.10 Endocytosis
Substances can be surrounded by membranes and brought into the cell. Because this requires an expenditure of energy by the cell, it is a form of active transport.

ease the penetration of the drug through the membrane, manufacturers mix a pinocytotic-stimulating substance such as albumin (egg protein) with the drug. The cell then pinocytotically engulfs the two together.

In receptor-mediated endocytosis specific sites on the membrane (receptors) bind to specific protein molecules (Figure 5.11). The membrane pit then folds around the protein molecule to bring it inside

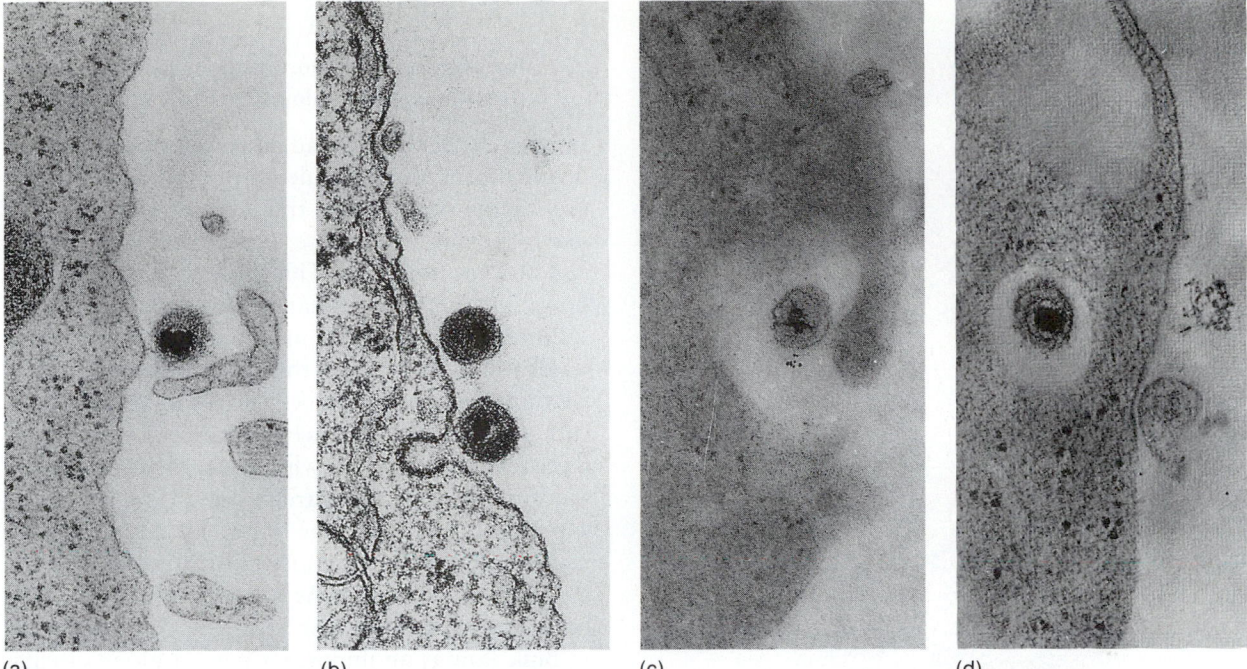

(a)   (b)   (c)   (d)

FIGURE 5.11   Receptor-Mediated Endocytosis
In this series of transmission electron microscope photographs, a virus binds to receptors on
the cell membrane. A vesicle forms as the virus is taken into the cell.

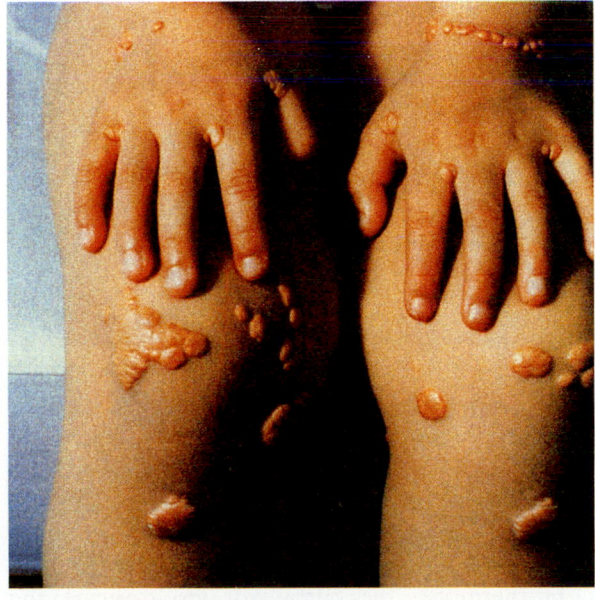

FIGURE 5.12   Familial Hypercholesterolemia
Because of an inherited disorder, the plasma membrane
receptors for low-density lipoproteins cannot take in
cholesterol through receptor-mediated endocytosis.
Cholesterol is carried in the blood by low-density
lipoproteins. The yellow deposits under the skin of this
individual are deposits of cholesterol and other lipids
due to this genetic defect.

the cell. When it is inside the cell, a coated vesicle
is formed from proteins of the cytoskeleton and the
contents remain inside the vesicle until the vesicle
lyses (breaks open) and the contents are freed into
the cell. An example of receptor-mediated endocytosis relative to humans has to do with the cell membrane receptors for low-density lipoproteins (LDL).
LDLs remove cholesterol from the blood by receptor-mediated endocytosis. The cholesterol can be
used for the synthesis of membranes and other steroids. Some people have too few LDL receptors,
so the level of cholesterol in their blood is very
high. This condition is called *hypercholesterolemia*
(Figure 5.12). These increased amounts of cholesterol can lead to early deposits of fat in the lining
of blood vessels, sometimes resulting in blood clots
that can lead to strokes and heart attacks.

Many cells, such as secretory or glandular cells,
do the reverse of endocytosis. This is called **exocytosis** (Figure 5.13). The Golgi apparatus forms
vacuoles or vesicles, which then are transported to
the plasma membrane. The vesicles fuse with the
plasma membrane and release their contents to the
outside of the cell. Endocrine glands (glands that
produce hormones) use this process constantly to
secrete substances into the bloodstream.

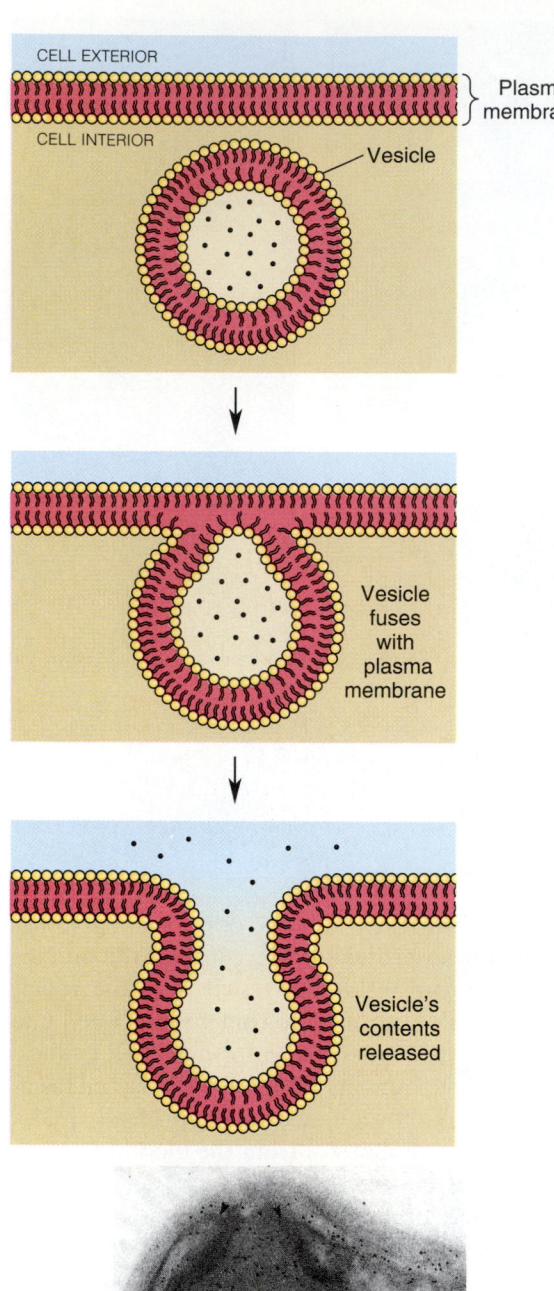

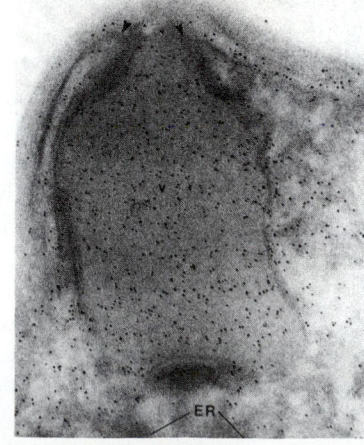

FIGURE 5.13 Exocytosis
The process by which substances are released from the cell is called exocytosis. The micrograph shows an early stage in which the vesicle has fused with the membrane and the contents are about to be released.

BULK FLOW

■ **Bulk flow is a transport method that moves fluids through organisms along a pressure gradient.**

In passive, facilitated, and active transport, only individual ions or molecules are transported. However, living systems also transport fluid mixtures of molecules and ions per se — in other words, solute and solvent together. This process is called **bulk flow.** Bulk flow is similar to diffusion in that the solutions move together in response to a gradient.

However, instead of the concentration gradient we discussed earlier, bulk flow responds to a pressure gradient (a difference in pressure between two fluid regions). Anyone who has turned on a water faucet connected to a garden hose has experienced a pressure gradient. The water in the pipes connected to the faucet is under more pressure than the air in the empty hose. As a result, water moves in bulk into the garden hose.

Bulk flow is an important part of blood circulation. The transport of molecules and ions from capillaries (microscopic blood vessels) to the tissues is due to the bulk flow of the blood that carries the molecules. The contraction of the heart serves as a pump for the bulk flow of blood from a region of higher pressure (the contracting heart) to a region of lower pressure (the tissue fluids). When you have your blood pressure taken, you are receiving a measure of bulk flow in your system.

Bulk flow is also involved in another familiar natural process: The spring rising of sap in trees is due to the bulk flow of sucrose in water from the roots to the leaves and other parts of the tree.

OTHER MEMBRANE FUNCTIONS

Just as important as the transport of substances into and out of cells is the content of the plasma membrane's outer surface. Here, specific proteins, lipids, and carbohydrates have the ability to recognize and serve as receptor sites for the attachment and binding of a variety of substances. For example, viruses attach to a host cell because of the carbohydrates on the surface of the host cell's membrane.

Furthermore, the carbohydrates covering the surface of a plasma membrane are crucial components in the immune response, the mechanism by which self recognizes nonself. Chapter 25 discusses the immune response, which is of great concern in disease prevention, allergic response, and heart and other transplant operations. Receptor molecules function in a variety of other roles as well, some of which we still do not fully understand.

FOCUS 5.1
Computer Graphic Model of Receptor Function

Cell membranes are not just gossamer sacs that separate the cell from its environment. Rather, as we have discussed, they are dynamic structures essential for many of the functions of life. One of these essential functions has to do with membrane receptors and their association with ion channels. Investigators at the University of California at San Francisco have used computer graphics to understand the relationship of the acetylcholine receptor, a membrane protein, to the ion channel opening. Acetylcholine is a chemical produced by nerve cells that conveys the neural message to adjacent nerve cells or muscle cells. How does this chemical messenger affect the membrane protein and allow its message to be passed on to the cell?

Dr. Robert Stroud at the University of California at San Francisco has made an important contribution to our understanding of how the receptor changes in the presence of acetylcholine. Dr. Stroud has used computer graphics to model the three-dimensional structure and changes that occur in the funnel-shaped protein subunits that make up the ion channel (see Figure A). The computer-generated model shows how the acetylcholine receptor changes shape as it opens and closes to allow an ion to pass through the membrane. Once inside, the ion induces changes resulting in the transmission of the nerve impulse.

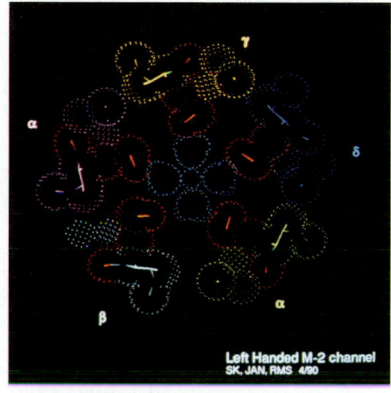
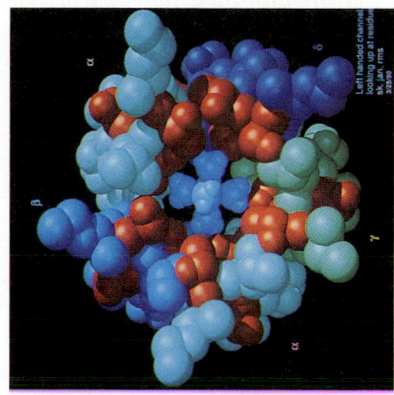
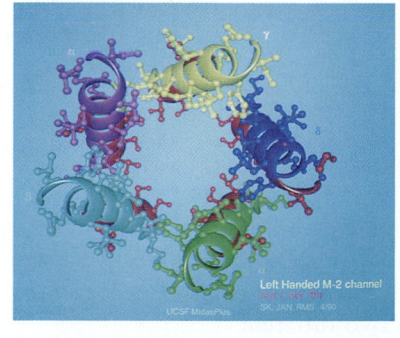
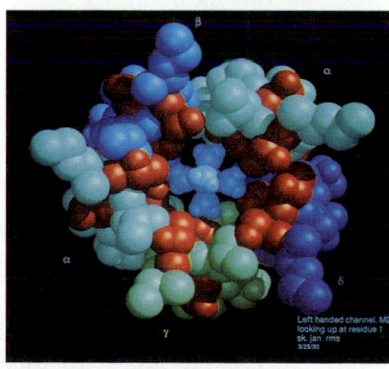

FIGURE A
The acetylcholine receptor is shown in these computer graphic representations. Knowledge of the receptor's structure has allowed us to understand how one particular region of the receptor can function as the channel through which ions pass.

Computer models are a definite asset to our understanding of large molecules. You can imagine how long it would take to build a ball-and-stick model of these large molecules and how large such a model would have to be. Computer graphic models are small, can be constructed rapidly, and can enhance certain regions like the active site. The model can easily be modified to study changes in configuration as the molecule interacts with other molecules. In addition, computer-generated models can provide a great deal of useful quantifiable data like bonding angles and bond energies.

Membrane-Bound Reactions

When we discussed globular proteins, we mentioned that they played a role in providing an ordered sequence of enzymes for membrane-bound reactions. Speaking in evolutionary terms, since many biochemical reactions must proceed in an ordered sequence (such as the passing of electrons), it is a selective advantage to have the molecules that transport these electrons be in a sequenced order. Physically locating them with the membrane can provide this order. The membrane serves as a base or location for the ordered attachment of enzymes and other molecules in some sequential reactions. The reaction of respiration within the mitochondrial membranes and photosynthesis within the membrane of the chloroplasts are good examples of the importance of membranes in sequential reactions (Chapters 7 through 9).

Receptor Site Reactions

Consider that hormones target in only on certain cells or that cells can recognize self from nonself. For example, certain sex hormones bind only to the cells lining the womb. The hormone recognizes the target cell because of receptor proteins on the cell membrane. Plant cell membranes also have receptor sites, but not as many as animal cells (see Focus 5.1).

Cell-to-Cell Recognition

How do cells recognize cells of their own kind? What is the basis of cellular recognition? It is the proteins and carbohydrates on the cell membrane that are responsible for the mediation of cell-to-cell interactions that result in cell recognition. For example, when sponge cells from two different sponge species are mixed together, the cells group with their own kind. Like recognizes like. Reactions occurring on the surface of the cells are responsible for this recognition of self (Figure 5.14).

Contact Inhibition

To repair a wound, cells divide. When the wound heals, *contact inhibition* occurs, signaling the cells to stop dividing. Cancerous cells, on the other hand, lose their inhibition and just keep on dividing. The mechanism of contact inhibition is believed to be mediated by cell-surface proteins and carbohydrates, resulting in membrane changes. Basic cell research in the area of contact inhibition may

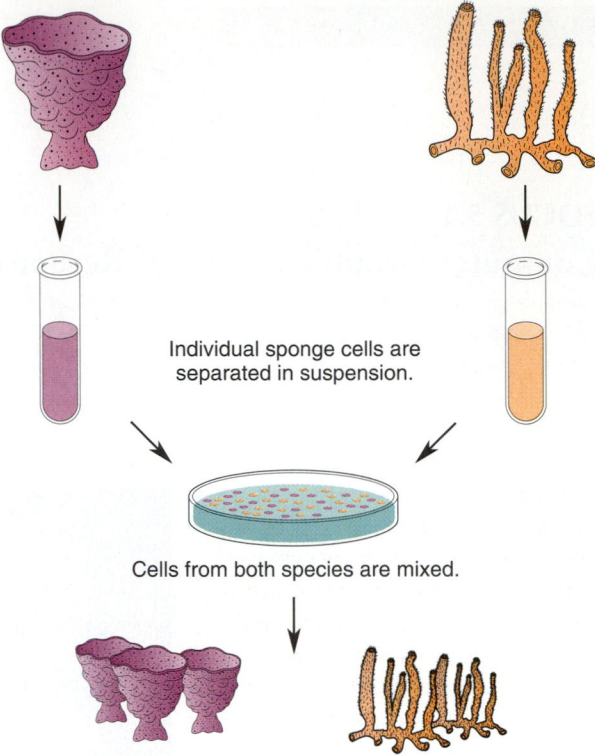

Individual sponge cells are separated in suspension.

Cells from both species are mixed.

Like cells clump together, forming individual tiny sponges that are species-specific.

FIGURE 5.14 Cell-to-Cell Recognition
The cells of two different species of sponge have different sets of receptor sites on their surfaces. When two cells of the same species come into contact, they can "recognize" each other because their receptor sites are the same.

well provide understanding of cell growth and the ability to regulate the uninhibited cell growth of cancer tumors (see Chapter 10).

Functioning of Surface Proteins

Usually a hormone or other substance binds only to the surface of the membrane — it does not enter the cell. How, then, does it produce its effect? Studies have demonstrated that the effect occurs because receptor proteins pass through the entire bilayer. When a substance such as a hormone binds with a receptor, the new chemical configuration changes the protein. (We discuss this phenomenon in more detail in Chapter 29.) The protein can now bind with an enzyme, thus causing a specific reaction to occur within the cell (Figure 5.15).

The plasma membrane is not the static, porous bag that surrounds cells, as biologists previously

Dr. Wilfred F. Denetclaw

Cell Biologist

As he rides horseback through the vastness of the high New Mexico desert, Wilfred Denetclaw's thoughts are on the activities associated with the spring birthing of lambs. All around him is nature: birth, death, the earth, and sky. Nature and its wonder are all a part of his culture, for Wilfred is a full-blooded Navajo of the Many Hogan and Salt Clan. Here on the Navajo reservation nature is abundant, but opportunities for the formal study of nature are scarce. Little did Wilfred suspect that one day his journey would take him to the Department of Molecular and Cell Biology at the University of California at Berkeley, where he would earn his Ph.D. in 1991 and then do postdoctoral work with some of the world's most respected biologists.

His long journey began with his education in the small town of Shiprock, New Mexico. He graduated from Shiprock High School in 1977 and was most noted for his football talents. He was honored as an All District defensive tackle, and for Wilfred, athletics seemed like a way to a successful life. None of his family had pursued a formal college education.

One of his aunts, however, told him of a biomedical research program at Navajo Community College at Shiprock in which the National Institutes of Health Minority Biomedical Research Support (MBRS) program was looking for talented students who might want to enter research. Wilfred applied and was selected to work with Dr. Lora M. Shields. His talent and interest in research grew under the tutelage of Dr. Shields. He investigated ß hemolytic streptococci infections. His desire to do research grew, and after

graduating from Navajo Community College, he transferred to Ft. Lewis College in Durango, Colorado, where he earned his B.S. in biology in 1983.

Dr. Lora Shields, his MBRS mentor, and Dr. Richard A. Steinhardt, his Ph.D. adviser, were extremely encouraging and assisted him on his research journey.

When he attended a national meeting in Albuquerque, New Mexico, in 1982, he was looking for a graduate program to attend. He talked to the recruiter from Berkeley, Dr. Jerry Guyden, who encouraged him to apply to U.C. Berkeley. To Wilfred, being able to work on his Ph.D. at U.C. Berkeley exceeded his dreams. But to his delight he was accepted in 1983. His research interests are in intracellular signaling (how cells communicate) and membrane transport.

Most recently, the research team of Drs. Fong, Turner, Denetclaw, and Steinhardt published a research paper, "Increased Activity of Calcium Leak Channels in Myotubes of Duchenne Human and Mouse Cells," in *Science*. They found that due to an abnormality in the Ca^{++} channels of dystrophic muscle cells, there was an increased concentration of Ca^{++} ions within these muscle cells. This increased concentration of Ca^{++} can cause increased protein degradation in muscle cells, suggesting that increased Ca^{++} entry through Ca^{++} leak channels may lead to the pathology of dystrophic muscle—an important finding in the understanding of Duchenne muscular dystrophy.

When asked about being a graduate student, and especially what it is like being one of the very few Native American students, Wilfred says that

to him science is "fun and exciting," he feels "like an explorer," and being a Native American in science is sometimes difficult, since there are very few people who understand his culture. He says that "the Navajo Way is important, but you must be true to yourself and don't compromise your integrity for another person's interpretation of what is right culturally."

He still misses the open vastness of the New Mexico desert, but he has the support of his wife and family. Some people would say that Wilfred is a solitary person. But he likes to think of himself as an individual who is comfortable alone or with a few friends.

It was difficult for Wilfred to say goodbye to his family back on the reservation, and the journey has been hard. But, he knows that his love for biology will result in his making a contribution to our knowledge. It is this journey that he will continue.

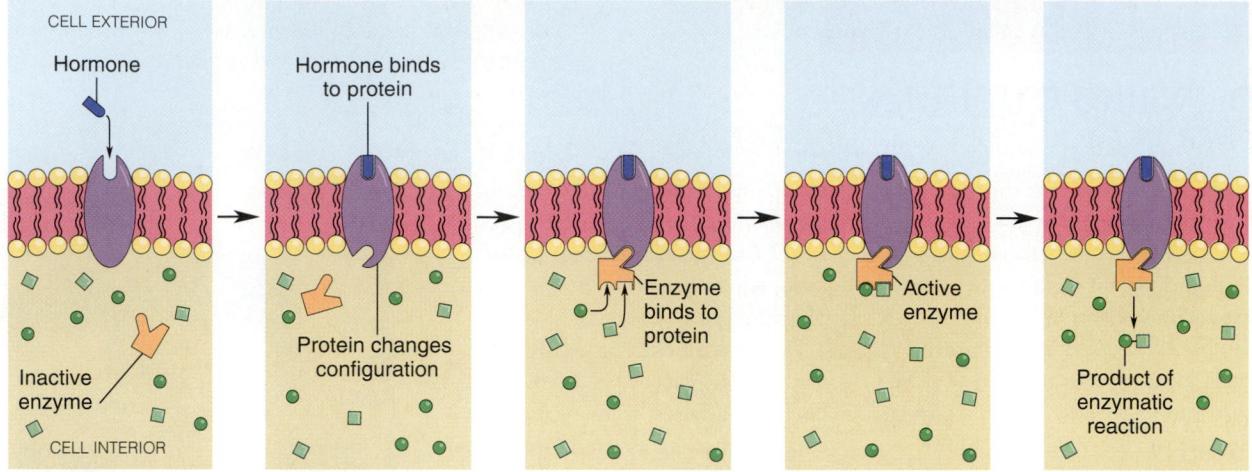

FIGURE 5.15 Protein Receptors
Some substances don't need to enter a cell to cause an effect within the cell. A hormone can bind to a protein in the plasma membrane, causing a change in the configuration of the protein. In turn this change allows an inactive enzyme inside the cell to bind to the protein. The enzyme then assumes its active form and catalyzes a reaction.

believed. Instead, membranes are living, dynamic, vital cellular structures constantly working to maintain life's functions. The study of membranes is one of the most dynamic areas in cellular biology today — an area that will provide us with the answers to vital questions currently confronting humankind (see Profile). Questions concerning cancer, cell development, and immunology all lie within the province of membrane study. Clearly, our futures are bound up with the endeavor to better understand the boundary of the cell.

SUMMARY

1. Membranes have three basic functions: (1) They regulate the movement of materials into and out of the cell, (2) they serve as a surface on which several step-by-step reactions occur, and (3) they function in cell surface and substance recognition.

2. The fluid mosaic model proposes that all membranes consist of a double lipid layer. Large globular proteins are embedded in the lipid bilayer. Some of these proteins have hydrophilic pores through which substances pass. Some membranes, especially cell membranes, have glycoproteins and glycolipids associated with their outer surfaces.

3. Materials enter and leave the cells through two major mechanisms. In passive transport, the cell does not expend cellular energy as molecules pass from one side of the membrane to the other. In ac-

tive transport, the cell expends energy transporting molecules.

4. Diffusion is a form of passive transport occurring in solids, liquids, and gases. Molecules move at random along a concentration gradient until the solution achieves a state of equilibrium. The rate of diffusion depends on the size and concentration of the solutes, and on their temperature and pressure.

5. Osmosis is a form of passive transport that involves the movement of solvent molecules (usually water) through the membrane. The solvent molecules move along the concentration gradient, ultimately reaching a state of equilibrium.

6. The relationship between the concentration of solutes in the cell and the environment can be described as isotonic, hypertonic, or hypotonic. Depending on the type of solution, cell size may remain the same, shrink, or expand.

7. Dialysis is a form of passive transport that moves solute molecules through a membrane.

8. Facilitated transport is a form of passive transport in which carrier protein molecules that span the lipid membrane move molecules through the membrane along a concentration gradient.

9. During active transport, molecules and ions are moved through the cell membrane against the concentration gradient, and ATP is used to supply the energy. Proteins that span the membrane move substances from lower to higher concentrations.

10. During endocytosis, the membrane surrounds a substance and brings it into the cell. If the substance is a solid, the process is called phagocytosis. If the substance is a liquid, the process is called pinocyto-

Chapter 5   Membranes, Molecules, and Cell Function

sis. If the substance binds with a specific receptor, the mechanism is called receptor-mediated endocytosis.

11. During exocytosis, the cellular membrane surrounds a substance and secretes it from the cell.

12. Specific proteins and carbohydrates on the surface of the cell membrane serve as the receptor sites for the attachment of several substances and for recognition of self from nonself.

KEY TERMS

plasma membrane	hypertonic
fluid mosaic model	hypotonic
tonicity	facilitated transport
passive transport	endocytosis
active transport	phagocytosis
diffusion	pinocytosis
concentration gradient	exocytosis
osmosis	bulk flow
isotonic	

REVIEW QUESTIONS

True/False

1. Diffusion is the movement of molecules from a place of higher concentration to a place of lower concentration in a liquid, solid, or gas.
   true   false   (page 109)

2. Osmosis is the passive transport of a solvent molecule, usually water, from a place of higher concentration through a membrane to a place of lower concentration.
   true   false   (page 109)

3. Facilitated transport requires the expenditure of energy.
   true   false   (page 109)

4. Plasma membranes are not the site of many cellular reactions; rather they are just static structures that separate cells from each other.
   true   false   (page 108)

Fill in the Blank

1. Active transport is the movement of molecules against a diffusion gradient from _____ concentration to _____ concentration. *(page 109)*

2. Active transport and facilitated transport are alike in that they require _____ proteins, but they differ in that in active transport _____ is required. *(page 109)*

3. The type of endocytosis known as "cell feeding" is called _____. *(page 114)*

4. As dividing cells contact one another this inhibits the cell from dividing further. This phenomenon is believed to be due to the cell surface _____ and _____. *(page 118)*

Multiple Choice

1. The energy necessary for active transport across cytoplasmic membranes is believed to come from:
   a. ATP
   b. Diffusion
   c. Osmosis
   d. Kinetic energy
   (page 112)

2. The cell membrane is composed primarily of:
   a. Cellulose
   b. Chitin
   c. Lipids
   d. Lipids and proteins
   (page 102)

3. Transport proteins are required for:
   a. Diffusion
   b. Osmosis
   c. Facilitated transport
   d. Facilitated transport and active transport
   (page 109)

4. What scientific evidence do the current models of membrane structure take into account?
   a. Membranes are dynamic systems.
   b. Proteins are globular units embedded in a lipid matrix.
   c. Many complex energy reactions occur on membranes.
   d. All of the above.
   (pages 102, 104)

Discussion

1. Identify and discuss the three basic functions of the plasma membrane.

2. Describe the Davson/Danielli model. What type of evidence did Robertson provide with respect to the Davson/Danielli model?

3. Draw, label, and describe the fluid mosaic model.

4. Discuss the differences between active and passive mechanisms of transport through cell membranes.

5. Define diffusion, osmosis, isotonic, hypertonic, and hypotonic.

6. Why don't cucumbers spoil when pickled in a salt solution?

7. Discuss facilitated and active transport.

8. What is meant by bulk flow of materials?

9. What is meant by endocytosis and exocytosis?

10. Discuss the importance of cell surface proteins and carbohydrates.

6

Energy Relationships in Living Things

Motion, whether the disjointed, boneless maneuvers of a break dancer or a gentle, silken gesture of a ballerina's arm, is a marvel to behold. You are amazed by the boundless energy of children at play or by ants scurrying down trails worn smooth by hundreds of tiny footsteps. You marvel as you peer into the microscopic world and see an amoeba extend part of its body to engulf a bacterium. What provides energy for all those activities?

Just imagine what you would see if you could shrink yourself small enough to be able to enter a cell and observe the origin of all that energy. There

would be enzymes catalyzing reactions, bursts of energy being transferred carefully from molecule to molecule, reactions being controlled as enzymes act together. All this is happening in just one cell. Multiply this activity by the hundreds of billions of cells that make up your body, and you have some idea of the complexity of reactions occurring in your body right now, as you study your biology. How do we and other living things trap, transfer, and utilize energy?

Living things are energy transformers, or energy *transducers*. In this chapter, we look at the general principles involved in energy relationships within cells. We discuss how cells transform, regulate, and utilize energy. With this overview of the energy relationships in cells, our journey in later chapters through the major metabolic pathways in living cells will be easier to understand.

Above: The movement of this rock climber is the result of energy being transformed, regulated, and used by body cells.

FIGURE 6.1 **Energy and Life**
Without a constant source of energy all life would cease. Plants transform energy from sunlight into chemical energy, a form that can be used by cells and organisms in meeting the demands of their existence. Without plants, animals would lose their link with energy from the sun and would be unable to survive.

ENERGY, THE ESSENCE OF LIFE

■ **Living things require a constant input of energy. Energy is defined as the capacity to do work. Potential energy is stored energy. Kinetic energy is the energy of motion.**

Energy is required for all life processes. Without it, no plants grow, no flowers blossom, no babies cry, no birds sing. In fact, without a continual input of energy, life ceases. What is this entity we call energy, which is so essential to life? In simplest terms, energy can be defined as the capacity to do work. In biological systems, that work can signify an entire range of activities, because energy is required to maintain an even temperature, clear wastes from a cell, breathe the morning air, call someone to dinner, see (or form) a flower, or run after (or away from) another organism (Figure 6.1).

States of Energy

For convenience, scientists say that energy exists in two *states* — potential energy and kinetic energy. **Potential energy** is stored energy. For example, a bullet in a gun has potential energy before it is fired. A candy bar, before it is eaten, has potential energy. In these examples, the potential energy exists in chemical form. As we will discover, there are other forms of potential energy.

A lead weight, before it is dropped on your toe, has potential energy, because of the pull of gravity on the lead's mass and the relationship between the position of the weight and your toe. When the lead weight is released and is falling toward your toe, the energy changes to another state — the energy of motion, **kinetic energy.** As the lead weight drops, the amount of potential energy decreases and the amount of kinetic energy increases. The same change occurs when someone fires a bullet. Potential energy is transformed to kinetic energy (Figure 6.2).

The process can also happen in reverse: kinetic energy can change into potential energy. For example, when sunlight, the initial form of energy on earth, strikes our planet, it arrives in waves (in other words, as kinetic energy, the energy of motion). The amount of energy that comes to the earth from the sun in just one year is equal to all the electricity that all our power plants could produce in 1.5 billion years. The green pigment molecules in plants (chlorophyll) trap the radiant energy of light waves and transform it into potential energy, which is then stored in the chemical bonds of food molecules. When chlorophyll molecules convert light energy to chemical energy, they complete the first step in the process known as *photosynthesis*, which is a set of reactions using light energy. (Photosynthesis will be discussed in more detail in Chapters 7 and 8.) So as you can see, potential energy can be transformed into kinetic energy and vice versa.

Forms of Energy

■ **Many forms of energy are important to living things — for example, chemical, mechanical, radiant, and electrical.**

Not only are there two major energy states, but there are several *forms* of energy important to living

Chapter 6   Energy Relationships in Living Things

FIGURE 6.2 Potential and Kinetic Energy
Potential energy is usable energy, but energy that is not causing anything to happen. When energy affects motion or causes change, it becomes kinetic energy. The wheelchair racer is converting biological potential energy into kinetic energy.

things: chemical, mechanical, radiant, and electrical. Chemical energy is the energy contained in the chemical bonds of molecules. For instance, a hiker who snacks on chocolate bars for quick energy is tapping the chemical energy they contain. Mechanical energy is the energy directly involved in moving matter. For instance, when little children jump on new furniture, their body systems are transforming the chemical energy of food to contract and expand their young muscles.

Radiant energy, sometimes called electromagnetic energy, is the energy that travels in waves. Radiant energy is the primary source of energy for all living things — from heat (which moves molecules, causing collisions that result in chemical reactions) to photosynthesis (in which radiant energy from the sun is converted to the chemical energy that drives living systems). Electrical energy results from the flow of electrons or ions. Nerve impulses are transmitted along a nerve cell by means of electrical energy.

Measuring Energy

■ **One unit of energy measurement is the kilocalorie (or Calorie), which is the amount of heat energy required to raise the temperature of 1000 grams (1 kilogram) of water by one degree Celsius.**

Since we cannot see energy, how do we measure this entity so essential for life? We all know one way, especially those of us who are diet-conscious. The unit we call the calorie is a unit of energy measurement. The *calorie* (from the Latin *calor,* which, as all Spanish-speaking people and Latin students will recognize, means heat) is the amount of energy required to raise the temperature of one gram of water one degree Celsius. A *kilocalorie,* or Calorie (with a capital C), is a thousand calories. Most diet information is expressed in kilocalories; so when a diet book tells you that 4 ounces of lean meat contains 220 "calories," it really means 220 kilocalories, or 220,000 calories. Since biologists almost always use the Calorie (or kilocalorie) when they talk about energy, we will follow that practice here.

Biologists also measure the amount of energy in a specific quantity of matter. For example, in chemistry, a *mole* is a unit of quantity, like a dozen eggs. A mole refers to a set number of particles, specifically 6.023×10^{23}, no matter whether those particles are atoms, ions, or molecules. One mole of glucose contains 6.023×10^{23} glucose molecules, one mole of ATP contains 6.023×10^{23} ATP molecules, and one mole of H^+ contains 6.023×10^{23} hydrogen ions.

Because the quantity of a mole is so specifically defined, it is possible to measure the amount of energy contained within a substance fairly accurately. To determine the energy content, a mole of a substance is burned in a device called a *calorimeter.* The change in temperature caused by the burning is measured; the figures are then converted to Calories. Scientists can then compare the energy contained in several substances. It is generally accepted that one mole of glucose possesses the energy content of 686 Calories when converted to CO_2 and H_2O, and a mole of ATP releases 7.3 Calories when broken down to ADP and an inorganic phosphate.

Physicists have even more ways to measure energy, but for our purposes as biologists, the Calorie will be sufficient. Biologists use the Calorie as the unit of energy when they measure the amount of energy required for a chemical reaction to occur or the amount of energy liberated by a chemical reaction.

THE LAWS OF THERMODYNAMICS

■ **The first law of thermodynamics states that energy can be neither created nor destroyed, only changed in form. The second law of thermodynamics states that during these changes in form some usable energy dissipates. Hence, the amount of usable (or free) energy decreases, and the amount of entropy increases.**

The laws of thermodynamics, which explain energy relationships, are what are called universal laws, and as such they apply to all things found in the universe, not just to living things. When scientists discuss the implications of these laws, they do not always discuss all of the universe, however. Instead, they often refer to a *system,* by which they mean matter in some defined region, and the *surroundings,* which are all things outside that system. Anything can be a system, whether it is a relatively simple bacterium, an ecosystem, or a planet. Since all living things are systems that require energy for their existence to continue, the laws of thermodynamics apply to biology as well as physics.

The First Law of Thermodynamics

The **first law of thermodynamics** deals with the *quantity of energy* in the universe, or within a system in that universe. This law simply states that, in ordinary chemical reactions, energy can be neither created nor destroyed, only changed in form. All living things follow this law as they change one form of energy to another. You change chemical energy to mechanical energy as you draw on the reserves in food cells to wash your car or run for a bus. As you read this printed page, your eyes transform light energy to chemical energy, and then chemical energy to electrochemical energy as the nerve cells in your eyes send an image to your brain.

The importance of the first law of thermodynamics lies in the fact that it tells us that, since energy can be neither created nor destroyed, we can only get the energy out of a system that we put into it. To put energy into a system, we must bring that energy into the system from its surroundings. In other words, you can't get something for nothing.

The Second Law of Thermodynamics

The **second law of thermodynamics** deals with the *quality of energy* in the universe or within a system in that universe. In any system, this law states, part of the total energy released in a reaction is in a form called *usable (or free) energy* and is capable of performing work, and some is dissipated into the surroundings. Dissipated energy is usually found in the form of low-level heat, and it is no longer available to do work.

There is the tendency in the universe for energy to become disordered (to become random and unavailable for useful work). This disorder is called **entropy.** For example, as we transform the potential chemical energy in candle wax into the kinetic energy of light, some of the wax's useful energy is dissipated as heat rather than transformed into light. The energy that manifests itself as heat has not been destroyed, but it no longer has the *quality* required to do useful work (that is, create light). That quality has been lost or diminished.

The universe is what we call a *closed system.* That simply means that it has no surroundings, no place outside of itself from which energy can be taken to add to the energy it already has. The amount of energy in the universe is constant; so the amount available to do work is decreasing and entropy is increasing. As a result, most scientists hypothesize that the universe will some day run out of useful energy. The twinkling stars will burn out, and life will come to an end.

How does the second law of thermodynamics affect living systems? Why doesn't an amoeba or a tulip or an antelope run out of energy? How do living things counteract the tendency toward disorder? They do so by putting useful energy into their systems. Remember, the universe has no surroundings, but living things do. They are *open systems.* In other words, living things can bring in usable energy from outside of themselves.

For instance, photosynthetic organisms, the energy trappers, capture light energy and store it in complex food molecules. Living cells convert the chemical energy in food molecules into the energy needed to overcome entropy, to maintain life. Consider what happens when useful energy input is unavailable — death and immediate disorder, decomposition, and decay. No organism is ever totally successful as it converts energy from one form to another — that is, all organisms always lose

     Chapter 6   Energy Relationships in Living Things

FIGURE 6.3 Heat
A mother penguin with her chick nestled next to her belly can transfer some of her body heat to her young chick.

some energy (Figure 6.3), usually as low-level heat, as they convert energy from one form to another. (When baby chicks nestle under their mother's feathers, they benefit from the low-level heat that radiates from her body.) But they are successful enough to counteract the tendency toward entropy.

Human beings are no exception. As living organisms, when we eat we take in energy. That energy supplies our bodies with useful energy to replace the energy that we have used to do work *and* the energy that we have lost because of entropy. The amount of energy flowing into any living system equals the amount of energy flowing out of that system. In this way living systems overcome the tendency toward disorder, and structured life continues.

COUPLED REACTIONS

■ Chemical reactions that liberate energy are called exergonic. Chemical reactions that use energy are called endergonic. Endergonic and exergonic reactions are often coupled.

Chemical reactions are due to the collision of molecules as they move about. During those reactions, bonds are broken and reformed and electrons can shift from one energy level to another. Some molecules join together to form new molecules, and others split apart. From an energy perspective, some reactions require extra energy in order to take place, and some reactions have energy left over once they are complete. Activities that *require* energy, such as the synthesis of a starch molecule from simple sugars, are **endergonic** (energy-in) reactions. Reactions that *liberate* energy, such as degradation of starches to simple sugars, are called **exergonic** (energy-out) reactions (Figure 6.4). Endergonic and

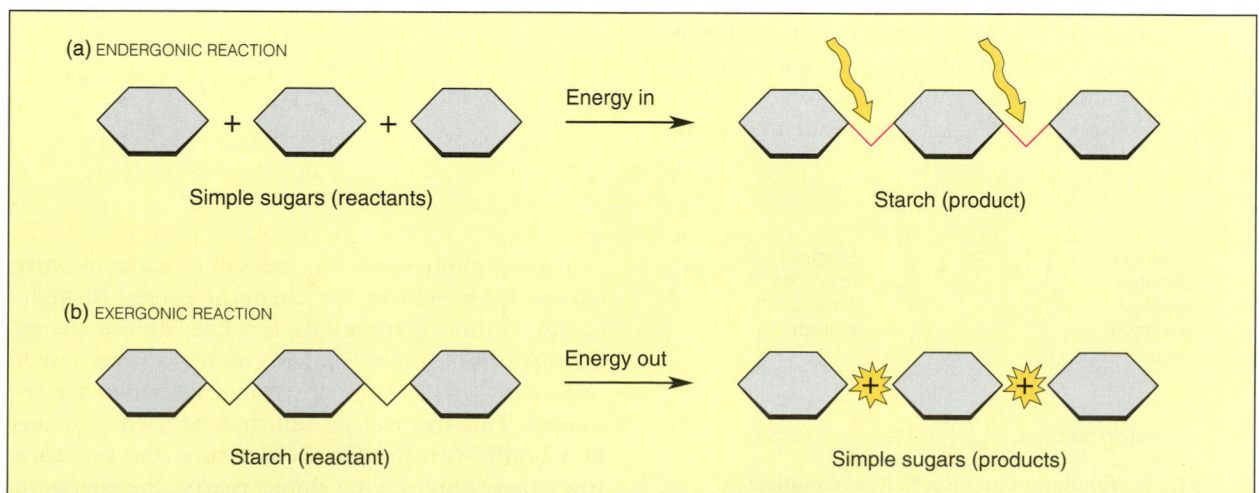

FIGURE 6.4 Endergonic and Exergonic Reactions
(a) When organisms make starches from simple sugars, they must use energy. These reactions are endergonic. (b) When starches are broken down to form simple sugars, energy is released in exergonic reactions.

exergonic reactions are coupled and reversible, because the energy liberated by one reaction can supply the energy used by the other.

Reduction and Oxidation

■ **Reduction and oxidation always occur together and are two biologically significant, coupled reactions. In reduction reactions, an atom or molecule *gains* one or more electrons and stores energy. In oxidation reactions, the reverse occurs: An atom or molecule *loses* one or more electrons and gives off energy. For every oxidation reaction there is a reduction reaction.**

The most important coupled reactions in the living world are those called reduction and oxidation reactions, or redox reactions. Redox reactions are important in that they release or consume large amounts of free energy needed by cells to maintain life. These reactions have to do with the transfer of one or more electrons. In some cases, oxygen plays a part in oxidation reactions, but for the time being we will concentrate only on electrons. Hydrogen ions play a role in many biological redox reactions, as we will see a little later in this chapter.

When an atom or molecule loses one or more electrons (in other words, the atom or molecule is an electron donor), energy is liberated, and the electron donor has undergone an **oxidation** reaction. (The name *oxidation* comes from the fact that oxygen is often the acceptor for these donated electrons.) When an atom or molecule gains one or more electrons (in other words, the atom or molecule becomes an electron acceptor), the energy is stored, and the electron acceptor has undergone a **reduction** reaction. If one molecule is donating electrons, a second must accept them. An important

fact about redox reactions is that they are coupled reactions (Figure 6.5). Many biological processes are based on coupled, reduction-oxidation reactions. The oxidation of one compound provides the energy to reduce another compound.

Photosynthesis and Respiration

■ **With respect to carbon compounds, photosynthesis can be thought of as a reduction reaction in which carbon dioxide is reduced, forming food, which is a form of stored (potential) energy. Respiration can be thought of as an oxidation reaction in which the chemical energy stored in food (carbon compounds) is released.**

The more you learn about energy relationships in the living world, the more you will see that they are merely oxidation and reduction reactions. In reality, photosynthesis and respiration include both oxidation and reduction reactions (since they are coupled), but in general, **photosynthesis** (energy storage) is a process in which CO_2 is reduced to produce carbohydrate, and H_2O is oxidized to produce O_2. Chlorophyll-containing plants take light energy, convert it to chemical energy, and then, by means of a series of reactions, ultimately store this chemical energy in a reduced form of carbon (high-energy carbon compounds), such as carbohydrates. Following is a simplified illustration of photosynthesis, which will be discussed in more detail in Chapters 7 and 8. In this equation and the one that follows, CH_2O represents any high-energy carbon component or unit. These units are, in a sense, food.

$$CO_2 + 2\,H_2O + \overset{\text{light}}{\text{energy}} \rightarrow CH_2O + O_2 + H_2O$$

Carbon dioxide   Water   High-energy carbon compound (food)   Oxygen

During photosynthesis, reduction reactions store energy from light in the chemical energy of high-energy carbon compounds. To liberate the energy in these molecules, plants, animals, and other organisms break them down and electrons are removed. This free energy can then be used to maintain bodily functions and overcome the tendency toward entropy. Living things release the energy in food molecules in an oxidation reaction known as **respiration.** Respiration is presented in a simplified illustration below; the various steps involved in this process will be discussed in Chapter 8. The released

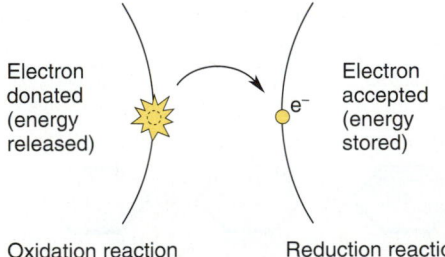

Electron donated (energy released)   Electron accepted (energy stored)

e^-

Oxidation reaction   Reduction reaction

FIGURE 6.5 Oxidation and Reduction; Coupled Reactions
When an atom or molecule loses an electron (e^-), energy is liberated and the molecule is said to be oxidized. When an atom or molecule gains an electron, energy is stored and the molecule is said to be reduced.

Chapter 6   Energy Relationships in Living Things

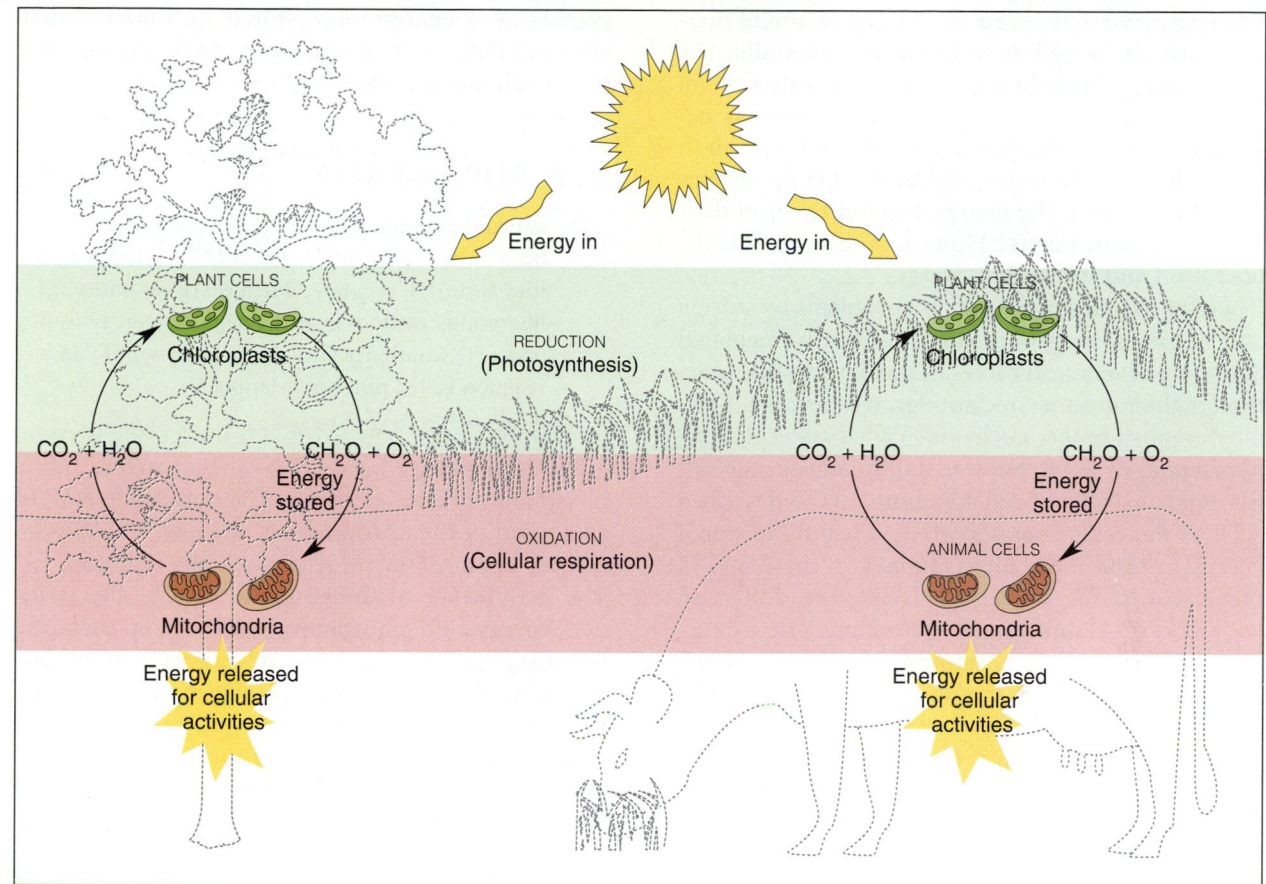

FIGURE 6.6 Biological Energy Relationships
Plants produce a reduced carbon compound through photosynthesis (simplified here to CH_2O). The energy required for this endergonic process comes from the sun. When the reduced carbon compound is oxidized in cellular respiration, the stored energy is released in exergonic reactions and used for cellular activities.

energy can be used to overcome the tendency toward entropy.

$$CH_2O + O_2 \rightarrow CO_2 + H_2O + energy$$

High-energy carbon compound (food)   Oxygen   Carbon dioxide   Water

In this reaction, as food is oxidized, electrons and hydrogen ions are removed from food and eventually combine with oxygen to form water. Energy and CO_2 remain.

The energy relationship in living systems includes many reduction-oxidation reactions found in the processes of photosynthesis (which stores energy) and respiration (which liberates energy). In essence the major energy-storage reaction for living things is photosynthesis, and the major energy-liberating reaction is respiration. Figure 6.6 illustrates the flow of energy available for biological purposes.

ATP SYNTHESIS

■ **Adenosine triphosphate (ATP) is the primary "energy-currency" molecule in living systems.**

How is the energy stored and released in redox reactions really utilized? Electrons that are exchanged do not just whiz around and bounce off one another. Such a process would be like a windmill that isn't connected to anything. As the blades spin, they would produce heat in the gear system and little more. But connect the shaft of a windmill to a pulley and gears, regulate the release of the harnessed energy, and you can pump water, circulate air, even generate electrical power, as well as perform many other useful tasks.

Energy in living systems also needs to be managed effectively. This becomes clear when you realize that unmanaged energy can destroy life rather than perpetuate it. Earlier we said that a mole of glucose has the energy potential of 686 Calories. If this energy were released all at once, it would literally burn up the cell. It would be like exploding the whole tank of gasoline in your car at one time in order to make it move. Your car's engine is designed to control the reactions that occur within it, by producing minute gas explosions. Living systems also must control the energy reactions within their cells in minute bursts. How do they regulate the liberation and transfer of energy?

Living systems manage these problems by storing and releasing energy in small, efficiently regulated amounts. Several molecules that can serve this function, including some nucleotides, have evolved. The most important is a nucleotide called *adenosine triphosphate,* or **ATP.** Several cell biologists (specialists who investigate cell functions) have described ATP as the cell's energy currency, while the stored energy in food molecules is money in the bank. In other words, the cell's ATP is like the dollars in your pocket — immediately spendable energy currency. The energy is stored in small enough amounts to serve even the simplest of purposes, like a dollar bill that is handy for all occasions. Should you need a large amount of money, you withdraw it from a bank — you don't need instant access to funds of that size. Living systems bank, or store, these large amounts of excess "cash" in energy-containing molecules like food and draw on them as they need them. You could, of course, carry around a $1000 bill with you, but if you tried to purchase a hamburger at your local fast food restaurant, you would have trouble. The same is true with living systems — if energy were stored in amounts that were too large, cells would have problems, too, and of a much more serious kind.

ATP, ADP, and AMP

■ **When the terminal energy-rich bond in ATP is broken, energy is made available for cellular activities and ADP remains. When energy from energy-liberating reactions is available, inorganic phosphate (P_i) combines with ADP, forming ATP in a reaction called phosphorylation.**

In Chapter 3 you learned that a nucleotide is composed of a nitrogen base, a 5-carbon sugar, and a phosphate functional group. The nucleotide ATP is composed of the nitrogen base adenine, ribose (a 5-carbon sugar), and three phosphates (Figure 6.7). The significance of the structure of ATP lies in its two **energy-rich phosphate bonds.** In a chemical formula, such a bond is illustrated as a wavy line ($\sim$). The energy in ATP's terminal bonds is due to their position and to the number of electrons in the ATP molecule. When hydrolysis breaks one of these energy-rich bonds, the released energy is made available to the cell in an accessible, convenient form.

A mole of ATP usually yields about 7 Calories. What remains is an adenosine and two phosphates, *adenosine diphosphate,* or **ADP,** and a free phosphate. (Although the symbol for phosphate is HPO_4^{-2}, chemists often use the symbol P_i to indicate

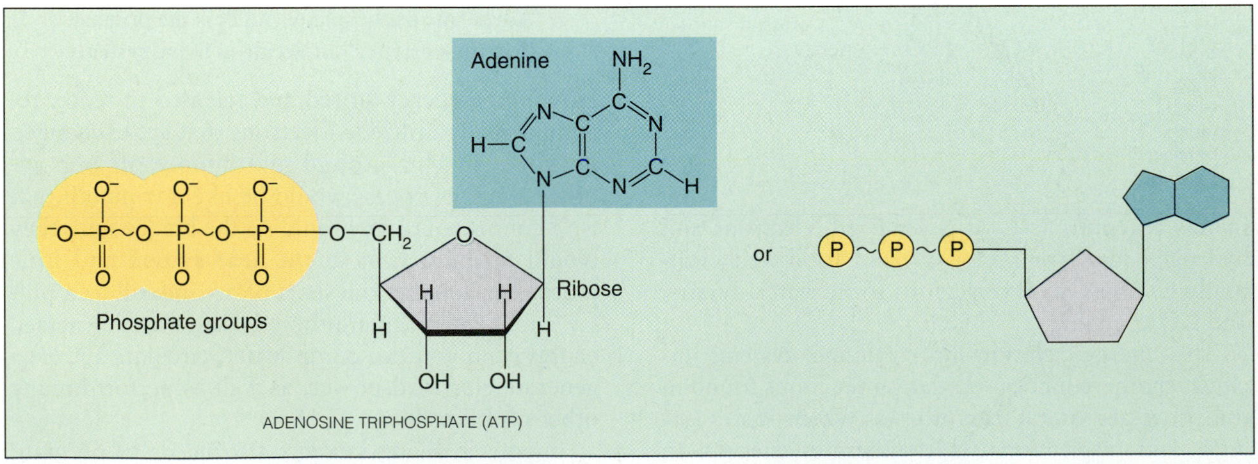

FIGURE 6.7   Adenosine Triphosphate (ATP)
ATP is a major energy-bearing biological molecule. As cells gradually oxidize food molecules, energy is released in small doses. The energy is stored in molecules of ATP and used as needed by the cell.

inorganic phosphate, a practice we will also follow.) The reaction is illustrated like this:

$$ATP + H_2O \xrightarrow{\text{Enzyme}} ADP + P_i + \text{energy for cellular activities}$$

With one phosphate removed, the molecule now has only two terminal phosphates, one of which has an energy-rich bond. In some cases, this terminal phosphate of ADP can be removed to release the energy it contains. The molecule then becomes *adenosine monophosphate,* or **AMP,** which indicates an adenosine and one phosphate. This rarely happens, however. In most cases, only one phosphate is removed; ATP becomes ADP.

Phosphorylation

We see that when a cell requires energy, a phosphate is removed from ATP and the energy in the bond is released. But how is ATP recreated — in other words, how does a cell add phosphates to adenosine diphosphate to form ATP? If 7.3 Calories of energy are released when a phosphate is removed, what is required to add a phosphate to ADP and make ATP? Energy is required. To synthesize ATP (that is, to add the phosphate to ADP) requires

7.3 Calories per mole. (Remember, when ATP lost its terminal bond and energy was released, some of that energy was lost as entropy. To compensate for that loss, more energy is required to add a phosphate than is gained by releasing one.) The process of adding a phosphate to ADP (or any other molecule) is called **phosphorylation.**

So, in order to keep cellular energy available in the form of ATP, ADP has to be constantly resupplied with energy and phosphate, or "recharged." How does phosphorylation come about? Light or oxidation of food supplies the energy that is needed to recharge the ADP and form ATP. The formula for this process is shown below.

$$ADP + P_i + \text{energy} \xrightarrow{\text{Enzyme}} ATP + H_2O$$

Because this is a reversible process, a single molecule of ATP can be broken down and rephosphorylated thousands of times a day. Figure 6.8 summarizes the relationship among energy, ADP, and ATP.

As you can see, in most instances life forms are powered by ATP molecules, whose ultimate energy source is the sun. The radiant energy was stored in food, then oxidized to synthesize ATP from ADP in order to power living things.

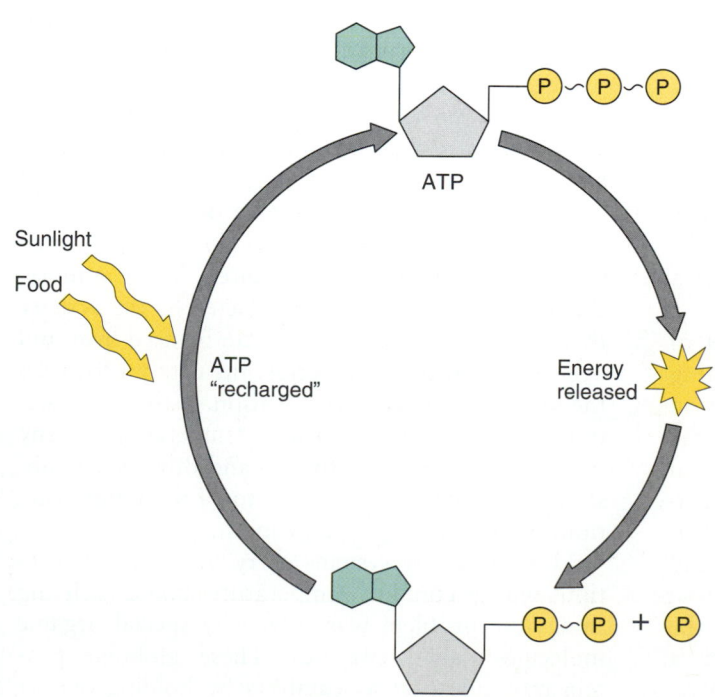

Sunlight
Food

ATP

ATP "recharged"

Energy released

ADP

FIGURE 6.8   Recycling of ATP
ATP is constantly recycled in the cell. When it loses a phosphate group, energy is released and ATP becomes ADP. Energy from the sun and that stored in food can be used to reattach a phosphate group and reestablish the terminal high-energy bond (phosphorylation). In this flashlight fish, some of the chemical energy from ATP is used to convert certain chemicals to a form that emits light energy (bioluminescence).

ENZYMES — PROTEINS AT WORK

In the early chapters of this book we mentioned that enzymes mediate chemical reactions. But what exactly does that mean? In living things, enzymes perform an essential role in many cell processes, including the creation of ATP. To understand why and how enzymes do this, we must first discuss how chemical reactions take place.

Minimum Energy of Activation

- **The mimimum energy of activation is the level of energy that must be present in order for a reaction to occur.**

In Chapter 3 we mentioned that chemical reactions occur when molecules collide and bonds break or are formed, and that a certain amount of energy is required to assure that reactions occur. Sometimes the system contains enough energy to set off the reaction. However, if the energy level is too low, there will be no reaction. The energy level required to achieve a molecular collision and get a chemical reaction started is called the **minimum energy of activation.**

The minimum energy of activation plays a role in processes that you take for granted every day. For instance, when you start your car, you use the energy stored in its battery to create a spark that ignites the gasoline. Once this minimum energy is provided, the gasoline (a hydrocarbon) is oxidized, which releases the energy stored in its chemical bonds. That energy helps the reaction to continue. Without the energy initially provided by the battery, not enough energy is present in the gasoline molecules themselves to produce the reaction. The same is true with the energy in the molecules of cells in living systems. The average energy in the molecules within the cell is not enough to produce a reaction.

In the laboratory, scientists usually use heat to increase the kinetic energy of molecules and bring about a reaction. But living things cannot rely on heat to produce reactions in their cells, because the temperature level required is so high that it would destroy a cell. For example, consider the temperature of burning sugar. Human beings burn (or oxidize) sugars in their cells all the time. To understand the temperatures involved, compare the oxidation of the sugars in your cells with the oxidation of the sugar in a marshmallow. Did you ever burn a marshmallow over a campfire? Did it take a lot of heat energy to burn the marshmallow? Did you

even burn your fingers as you took the smoldering marshmallow off the fork? Your fingers burned because the burning of the sugar of the marshmallow released a great deal of heat.

So the question then is, how can we burn (oxidize) sugar in our cells at temperatures that won't destroy our cells? Living things need a way to lower the energy of activation in molecules so that reactions can occur within a temperature range that will not destroy them. How does this occur? It occurs through the use of catalysts.

Catalysts

- **Enzymes are organic catalysts that make reactions possible within the temperature limits of living things by lowering the minimum energy of activation.**

As you might imagine, the amount of energy required to achieve a reaction varies with the kind of molecules involved and the conditions under which the reaction takes place. However, in many cases it is possible to lower the minimum energy of activation by using substances called **catalysts.**

At some time in your life, you may have taken chemistry or received a chemistry set and attempted to make anything that would stink or explode. When you took a vial of hydrogen peroxide (H_2O_2) and tried to convert it to H_2O and O_2, more than likely nothing happened until you added some iron filings. At room temperature, this combination produced visible O_2 gas bubbles and formed water.

In this experiment, the iron filings acted as a catalyst. They provided favorable conditions because the filings increased the surface area on which reactions could occur. As a result, the molecules could concentrate on the iron surface of the filings and collide at room temperature. No outside energy source, such as heat, was required. What is more, the filings themselves are unaffected by the activity they are promoting, and they can be used over and over again. In scientific terms, the catalyst lowered the minimum energy of activation so that the reaction could proceed at room temperature. Many industries use iron, platinum, and other such substances as catalysts, in order to bring about reactions without a large energy input.

How do living systems carry on chemical reactions within confined temperature limits? In living things the problem was solved by special organic molecules called **enzymes.** These globular protein molecules act as catalysts by holding one or more molecules long enough to break them apart,

FIGURE 6.9   Minimum Energy of Activation
Before chemical reactions can occur, a certain amount of energy must be present. Catalysts lower the amount of activation energy needed. In living cells enzymes act as catalysts, allowing many critical reactions to occur with much less energy of activation and in a much shorter period of time.

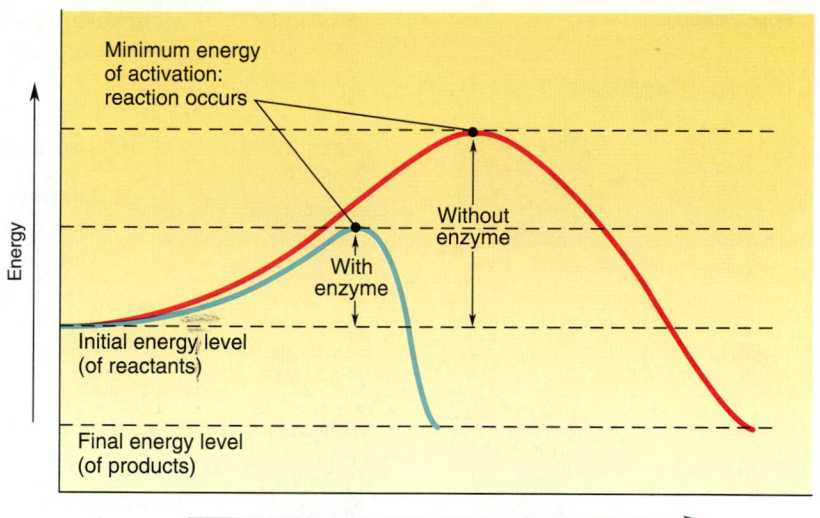

rearrange their structures, or put them together. In this way enzymes lower the minimum energy of activation (Figure 6.9). They are called *organic catalysts* because the enzyme molecules contain carbon.

Each kind of enzyme combines with very specific substances, so specific, in fact, that the common name of an enzyme is based on the name of the reaction it catalyzes. Then the suffix *-ase* is added to the term. For example, malt*ase* is the enzyme that catalyzes reactions involving maltose molecules. The molecule that is changed is called the **substrate**. Therefore, maltose is the substrate of maltase. What do you think the dehydrogenases do? Yes, they remove hydrogens from molecules in which hydrogens are found.

How Enzymes Function

How does an enzyme function, and does its shape make any difference? There is one basic principle to remember: The structure of an enzyme is the key to its functions. Here is a prime example of the concept of the complementarity of function and structure. The structure (shape) of an enzyme will determine to a large extent how it can "fit" (interact) with other molecules (see Profile, page 137).

There is a very definite correlation between the shape of an enzyme and the task it performs. Enzymes are proteins. Some are pure protein, and others contain specific nonprotein chemical groups. Each type of enzyme has a very specific three-dimensional shape, or folding, which produces a

specific region (or regions) that will interact only with its substrate. This region is called the **active site.**

One way to describe enzyme function is to say that the substrate molecule fits into the active site in the enzyme molecule somewhat like a key into a lock, thus enabling a reaction to occur. More precisely, the active site of the enzyme contains atoms that combine with certain atoms on the substrate and, because the fit is not perfect, it effectively weakens the substrate's chemical bonds. The weakening lowers the minimum energy of activation, and a reaction occurs.

Enzymes do not function just by weakening bonds, however. Some work by positioning molecules to just the right orientation that they need in order to react. For example, in *synthesis reactions*, very specific areas of two or more molecules must meet before any reaction can take place (Figure 6.10a). When heat is applied to a mixture, the number of collisions increases, thus increasing the odds that the molecules will collide in the correct orientation. In this type of reaction, the enzyme functions by bringing reactants close enough to each other and in the correct orientation for the reaction to occur. Because of the enzyme's function less energy is required for them to react.

How do enzymes function in other types of reactions? In a *rearrangement reaction* the active site of the enzyme can cause the internal arrangement of atoms within the molecules to change, producing a different product (Figure 6.10b). In a *degradation reaction* the enzyme can split the substrate into two

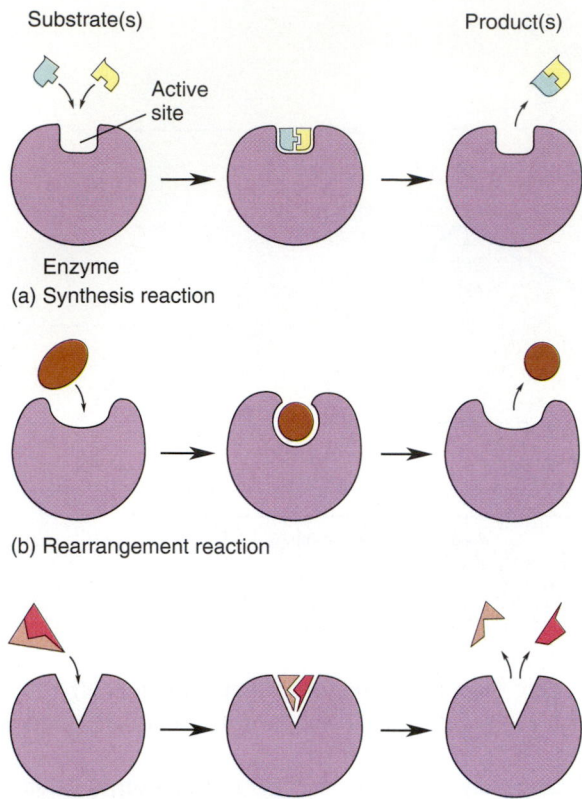

Substrate(s)   Product(s)

Active
site

Enzyme

(a) Synthesis reaction

(b) Rearrangement reaction

(c) Degradation reaction

FIGURE 6.10 Enzymes and Substrates
(a) Some enzymes play a role in synthesizing
compounds. An enzyme, for example, is involved in
synthesizing ATP from ADP and a phosphate. (b) Other
enzymes rearrange the atoms in particular molecules,
forming new compounds. This occurs in some of the
reactions in which food molecules are oxidized. (c) A
third category of enzymes degrades substances. You may
recall the powerful enzymes in lysosomes that break
down foreign or substances of no use to the cell.

or more products. Once the enzyme has catalyzed
the reaction, it releases the products and the enzyme
can then be reused (Figure 6.10c).

Biologists are still learning about how enzymes
function. One current theory, called the **induced-fit
theory,** suggests that the fit between an enzyme's
active site and its substrate is not perfect. As the
enzyme and substrate join, the imperfect fit strains
the enzyme and the substrate, thus altering their
shapes. Because of the strain, the enzyme grips the
substrate so tightly that the substrate's bonds are
distorted and weakened, and the reaction can occur
(Figure 6.11).

We do know one important fact about enzymes
— they are not altered or destroyed by the reactions
that they assist. Once its reaction is complete, an
enzyme releases its substrate and can repeat the process, again and again, thousands and thousands of
times. Because of the presence of enzymes, the

chemical reactions of life can occur within the temperature constraints of living systems (see Focus
6.1).

Cofactors and Coenzymes

■ **Enzymes often have cofactors associated with
their protein portions. Some are ions. Others are
vitamins. The symptoms of vitamin-deficiency diseases result from failure of some enzyme-mediated
reactions to occur because the vitamin coenzyme
is lacking.**

One interesting characteristic of some enzymes is
that they work with partners, separate molecules
that are essential in the function of a number of
enzymes. This partner is called the **cofactor.** There
are two basic types of cofactors: coenzymes and
metal ions.

Coenzymes are not protein, as are enzymes, but
are small organic molecules. Coenzymes function
by serving as carriers of atoms or functional groups
that adapt or assist enzymes in their reactions. They
attach to the enzymes at specific sites, just as substrates do, and the two together then form a complete active site (Figure 6.12). The substrate will
now fit correctly, and the reaction can occur. During the reaction the enzyme catalyzes a chemical
change in the makeup of the coenzyme, just as it
does in the substrate.

Vitamins usually are the precursor molecules
that are eventually converted by the cell to coenzymes. (The name *vitamin* is a misnomer that arose
because biochemists of the past thought these elements were "vital amines," hence vitamins.) As research has provided more knowledge about the
functioning of enzymes and coenzymes, we have
learned much about vitamins that has helped solve
medical problems that have plagued humankind for
most of its history.

Some of you may have heard of vitamin-deficiency diseases such as beriberi, scurvy, and pellagra. In 1914, pellagra was widespread in the United
States. The patient had diarrhea, roughening of the
skin, and mental confusion. Physicians thought that
pellagra was spread by germs, but Dr. Joseph Gold-
berger disagreed. He proved his point by having his
wife, himself, and 22 volunteers swallow capsules
containing blood, feces, and urine of pellagra patients (what one will not do for science!). None
developed the disease. Later it was learned that if
the diet contained high levels of the B vitamin
known as niacin, the symptoms of pellagra would
disappear.

We now know that patients with vitamin-
deficiency disorders exhibit the symptoms that

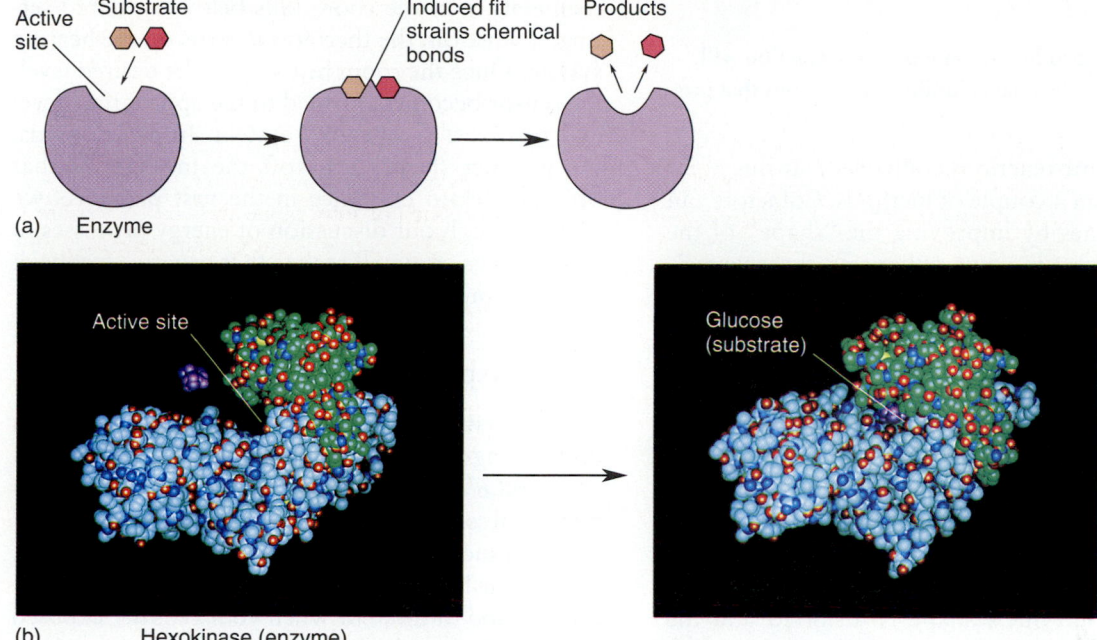

FIGURE 6.11 Induced-Fit Theory
(a) This theory of enzyme action proposes that a substrate molecule does not fit exactly in the active site of the enzyme, which induces a change in the enzyme's conformation to make a closer fit. This induced fit causes stress on particular bonds of the substrate, resulting in a catalyzed reaction. (b) In this computer-generated model of the enzyme hexokinase reacting with its substrate, glucose (shown in purple), the configuration of the enzyme changes when the substrate binds to the enzyme.

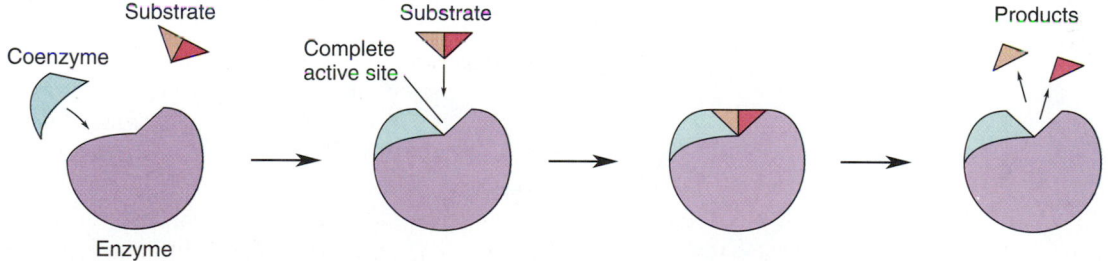

FIGURE 6.12 Coenzymes
Some enzymes cannot function unless a coenzyme is associated with their active site. Some coenzymes are bound to the enzyme tightly and permanently; others are not. In either case, the enzyme cannot catalyze the reaction without the coenzyme.

they do because there are insufficient amounts of the vitamin (that is, the coenzyme) necessary for certain enzyme-mediated reactions. As a result, the reaction that they regulate cannot occur. Today, because of the variety in the diets of most people in the United States, the major dietary problem is not vitamin deficiencies but obesity. In poorer and developing parts of the country and the world, however, vitamin deficiencies persist.

Metal Ions. Some enzymes have metal ions as a cofactor. These ions can assist the enzyme in adding or removing electrons from the substrate. For example, because of its positive charges, the *ferric ion* (symbolized by Fe^{+++}) can attract an electron away from a substrate, thereby oxidizing that substrate and releasing energy. In turn, the ferric ion (Fe^{+++}) becomes reduced to a *ferrous ion* (Fe^{++}), because it acquired the electron (which, remember, is negative) from the substrate.

Enzyme Control

■ **Often the product of an enzymatic reaction will, by negative feedback, inhibit the reactions that produced it.**

How are enzyme reactions controlled? At this point we will mention a couple of methods. Cofactors can *activate* enzymes by improving the "shape" of the enzyme's active site, thus enhancing the enzyme's ability to react. Other molecules *inhibit* enzyme reactions by changing the shape of the enzyme's active site or by interfering with the enzyme substrate complex.

One method of control involves a group of enzymes called **allosteric enzymes** (other shapes). In addition to its active site, an allosteric enzyme has another binding site, called its **allosteric site**. A molecule not meant to participate in the enzyme's reaction can bind at the allosteric site, and when that happens, the enzyme's shape is distorted and the enzyme cannot join with its normal substrate. The overall effect is like punching a pillow and changing its shape.

The molecule that binds to the allosteric site is usually the end product of a reaction. Enzyme inhibition of this type is a form of **negative feedback** in which the product turns around and inhibits the reaction that produced it (Figure 6.13). An analogy for negative feedback in a cell is the thermostat that regulates the temperature of your house. When the

temperature in the room falls below a desired setting, a sensor in the thermostat turns on the heating system. Once the room heats up to the desired level, the sensor becomes warmed to the appropriate level and turns off the heating system. In other words, the product (heat) turns off the mechanism that brought it into existence in the first place. As we work through our discussion of energy processes in living things, you will realize that negative feedback is a very important concept in biology.

Other Aspects of Enzymes

An understanding of enzymes helps to explain many things. For example, why is our body temperature 98.6° Fahrenheit, or 37° Celsius? As you might guess, 98.6°F or 37°C is the optimal temperature for most enzyme reactions in the human body. When you have a very high fever (say, over 103°F for a period of time) or when your cells are exposed to temperatures above the optimal level, the protein in the enzyme may become *denatured* — that is, lose its three-dimensional shape. For example, when you cook an egg, the heat you apply changes the physical characteristics of the egg white's protein. It is denatured. That means that the hydrogen bonds break and the structure or shape of the enzyme (protein) changes.

Enzymes are also very sensitive to pH levels, and many function best when the pH is around 7, that

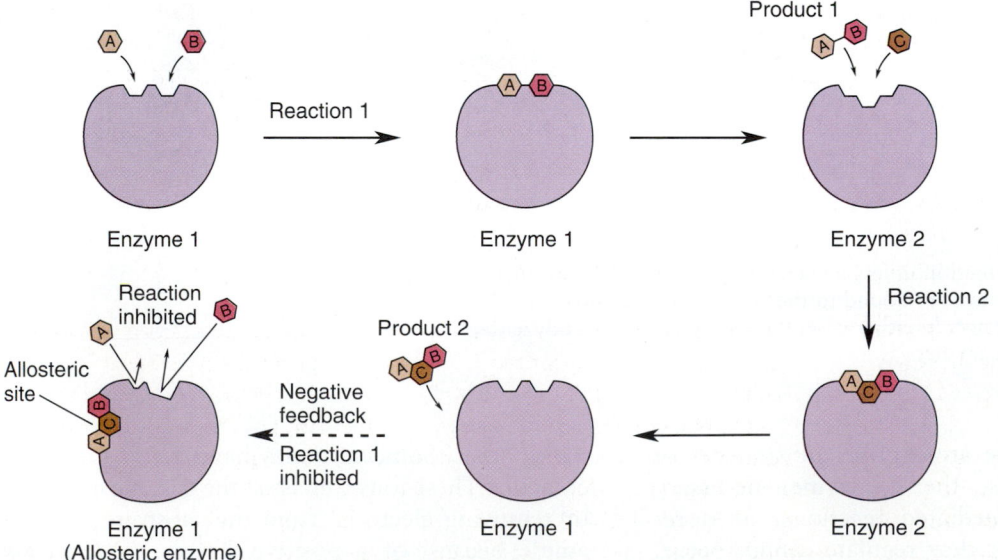

FIGURE 6.13 Allosteric Enzymes and Negative Feedback
Sometimes the end product of an enzymatic reaction can bind with the allosteric site of the enzyme, distorting the enzyme and preventing the reaction from continuing. This is an example of negative feedback; the end product is the means by which the reaction is turned off.

Chapter 6   Energy Relationships in Living Things

Dr. Christian B. Anfinsen
Protein Chemist, Nobel Prize Laureate

When Dr. Christian Anfinsen is not working in his laboratory or participating in international human rights activities, he loves to go sailing. However, finding time for it is somewhat of a problem. As a Nobel laureate, Anfinsen has been increasingly involved in promoting the development of science in Israel and other Middle Eastern countries. In 1981, he retired from his position as Chief of Chemical Biology for the Institute of Arthritis, Diabetes, Digestive and Kidney Disorders of the National Institutes of Health in Bethesda, Maryland. He then took a position as Chief Scientist of Taglit (which means *discovery* in Hebrew), a company that is a part of the Weismann Institute of Science in Rehovot, Israel. At present he is working at Johns Hopkins Medical School.

In 1972, Dr. Anfinsen was awarded the Nobel Prize for demonstrating that the linear sequence of amino acids in a protein is responsible for its three-dimensional configuration. This concept is one of the most unifying in protein chemistry. Anfinsen's work demonstrated that the enzyme *ribonuclease,* which catalyzes the breakdown of RNA, was composed of a single polypeptide chain. The sequence of the amino acids in the chain was responsible for its spontaneous folding, and that folding created the shape that allowed the enzyme to become biologically active. From this discovery came information about the structure of proteins and an understanding of how a protein's structure determines its biological function.

We now know, because of Anfinsen's work, that the biological function of a protein can be thought of as a consequence of protein geometry.

Dr. Anfinsen was born in Monessen, Pennsylvania, in 1916. His father was an engineer and wanted his son also to pursue a career in engineering, but Anfinsen elected to become a biological scientist. Anfinsen jokingly says, "My pop wanted me to be an engineer, but science, which was second best, was as close as I could come." In 1937, he received a B.A. degree from Swarthmore College, and in 1939, he received an M.S. in organic chemistry from the University of Pennsylvania. From 1939 to 1940 he was a visiting scientist at the Carlsberg Laboratories in Copenhagen, Denmark, before returning to the United States to complete work on a Ph.D. in biochemistry at Harvard Medical School. After completion of his Ph.D., he remained at Harvard as a professor for seven years. In 1963, because of what he felt were excessive pressures to write grant proposals for funding, which took time away from his work, he joined the staff of the National Institutes of Health Laboratories, a place rather sheltered from that problem.

When asked by a newspaper reporter about how he spent the $49,000 cash award then associated with the Nobel Prize, Anfinsen says, "I didn't do anything humanitarian with it. I just spent it stupidly like any other person would." In the world of finance he may not be a ge-

nius, but in the laboratory he truly is. He is also a true humanitarian. He has devoted a tremendous amount of time and energy to activities concerned with the advancement of human rights, especially those of scientists. In 1977, he went to Uruguay and Argentina to convince their governments to release some of the scientists imprisoned as political prisoners. Anfinsen's efforts were partially successful, since some were freed after his visit. When asked about human rights, he says, "I find myself using whatever clout I have to further those causes along."

From advancing our knowledge of protein chemistry to meeting with Pope John Paul II as a member of the Pontifical Academy of Science, Dr. Christian Anfinsen continues to pursue those interests that have made him an outstanding member of the scientific community.

is, neutral. (There are, of course, exceptions. Some enzymes, like the ones in your stomach, function best in an acid pH. Other enzymes function best in a basic pH.) When pH changes too far in one direction or another, again enzymes are denatured. Denaturing is one factor in the destructiveness of acid rain in lakes and forests. As the pH level becomes lower (in other words, as the water becomes more acid), many enzymes can no longer function, and life processes cease.

Many of the poisons we know, such as heavy metals like arsenic, lead, or mercury, damage living organisms because they distort an enzyme's shape and thereby inhibit enzymatic reactions. This phenomenon is the basis of many incidences of lead poisoning in urban areas — cases in which young children become seriously ill and sometimes die from eating chips of the lead-based paint that was used years ago to paint the walls. (Manufacturers are no longer allowed to utilize lead as a base for any paint to be used for a house.) Also, researchers have found that the lead content of the water in many schools is high, possibly due to the lead used in connecting the copper pipes in plumbing and in water coolers.

We make use of our knowledge about enzyme substrate specificity when we manufacture drugs. For example, some drugs and medications inhibit coenzyme activity. Other antibiotics inhibit bacterial growth, because the antibiotic distorts the active site, binds with the enzyme's allosteric site, or geometrically *mimics* the natural substrate. Such a mimic is called an *analog*. An analog combines with an enzyme in such a way that the enzyme cannot catalyze its normal reaction. For instance, the drug aminopterin is used in cancer treatment because it resembles the coenzyme folic acid. Normally, folic acid must be present in order for a cell to synthesize nuclear material. When the folic acid analog is present, the hungry cancer cells, which require more folic acid than noncancerous cells, cannot synthesize sufficient amounts of nuclear materials; so they die, and the patient's condition improves. Unfortunately folic acid analogs also affect some normal cells, which results in the side effects associated with chemotherapy.

TYING ENERGY LAWS AND PROCESSES TOGETHER

So now you know about energy laws and redox reactions and enzymes. But how are they inter-related in the processes of life? The answer is that they weave together in the *electron transfer process,* the *electron transport chain,* and *chemiosmosis.*

Electron Carriers

■ **Electron carrier molecules are enzymes that either (1) float free in the cytoplasm or (2) are embedded in membranes. These enzymes transfer electrons, protons, and/or energy from one molecule to another. Among the most significant carriers are NAD+ and NADP+, and those in the electron transport chain.**

The molecules that facilitate the electron transfers that are so basic to redox reactions are called **electron carrier** molecules. There are two types of electron carrier molecules: (1) those that float free in the cytoplasm of a cell and (2) those that are associated with membranes, such as the ones in the chloroplast and the mitochondria.

The *free-moving electron carriers* transfer electrons from one reaction site to another within the cell. The two most important of these carriers are the nucleotide coenzyme molecules known as nicotinamide adenine dinucleotide (NAD+ and nicotinamide adenine dinucleotide phosphate (NADP+) (Figure 6.14). Both are derivatives of the B vitamin known as niacin.

NAD+ accepts hydrogen ions and electrons from food molecules as they are metabolized, and is thus reduced to NADH — or, actually, NADH + H+ — but we will use NADH to represent the reduced form. NADH can then transfer a hydrogen ion (which is a single proton), electrons, and energy (when it is oxidized) to another molecule. When **NADP+** is reduced, it becomes NADPH — or, actually, NADPH + H+ — but we will use NADPH to represent the reduced form (Figure 6.15). NADPH functions in photosynthesis by transferring electrons and hydrogen ions from the sites where the energy of sunlight is harvested to the sites where that energy will be converted to the chemical energy found in food molecules.

Electron carriers that are associated with membranes are arranged in a very specific order with other molecules. *Membrane-bound carriers* undergo a series of sequential enzyme-mediated redox reactions. A pair of electrons is passed from one molecule to another, reducing the molecules as they receive the electrons and oxidizing them as they lose the electrons. This series of reactions, called an *electron transport chain,* is found in the processes of photosynthesis and cellular respiration.

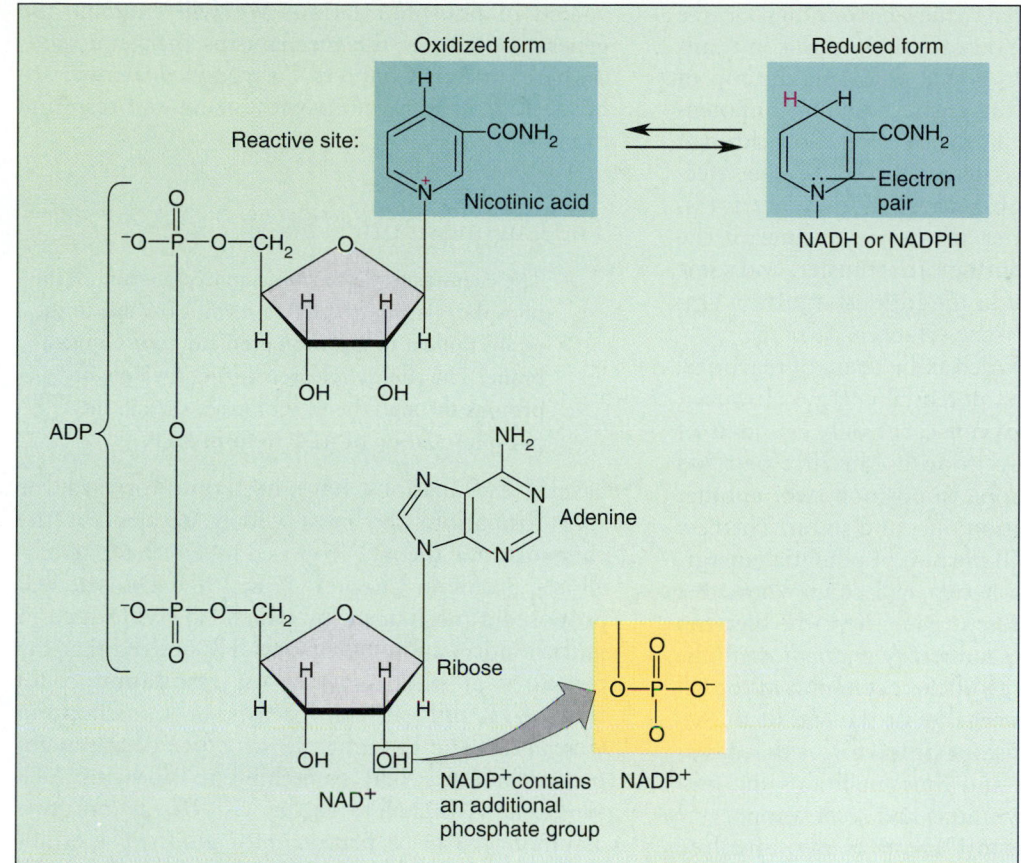

FIGURE 6.14 NAD+ and NADP+
Because oxidation/reduction reactions are the foundation of energy transfers in cells, mobile electron carriers like NAD+ and NADP+ are important. The two carriers have the same structure except that NADP+ has a third phosphate group.

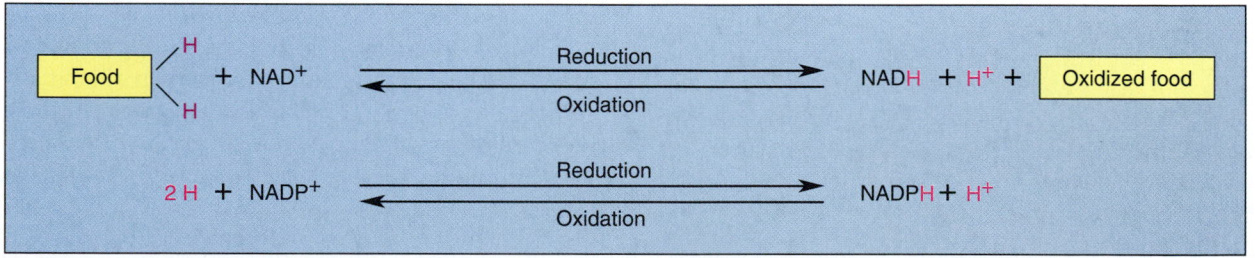

FIGURE 6.15 NADH and NADPH
When in reduced form, NADH and NADPH carry electrons and energy. Oxidation of these molecules to NAD+ and NADP+ is highly exergonic. NAD+ is reduced to NADH when food molecules are oxidized. NADP+ is reduced to NADPH in the process of photosynthesis.

The Electron Transport Chain

■ **In the electron transport chain, electrons move along an electrical gradient from a higher to a lower level of potential energy. The entropy of the system increases as the electrons are passed from molecule to molecule. Usable energy is lost.**

It is during the processes of the electron transport chain that two principles of the laws of thermodynamics play an important role. Remember that potential energy is stored energy. Remember also that the second law of thermodynamics discusses entropy — the loss of usable energy. When electrons enter the transport chain, they have a high degree of potential energy. (You can think of this in terms of a ball rolling down a stairway — at the top of the stairs the ball has the greatest amount of potential energy.) However, as the electrons flow through the transport chain, entropy increases. The electrons carry less energy because as they are transferred from one carrier to the next some of the energy is lost as heat during each transfer, and some of the energy is stored in the form of a proton gradient across the membrane (Figure 6.16).

As an electron flows down the chain of reactions, it contains less and less potential energy. However, an electron chain does not carry only one pair of electrons at a time. As soon as the first pair has passed from position one to position two, another pair can move to position one, and so on. The new pair will contain its full portion of potential energy, but the pair in position two will contain less. Because entropy increases at each step, the electron transfer chain becomes an *energy gradient*, with the highest potential energy at the beginning of the sequence and the lowest energy at the end of the sequence. The gradient is one of the forces that keeps the sequence moving and thus maintains the processes of both photosynthesis and respiration.

The electron transport chain is also involved with another process, the phosphorylation of ADP to form ATP, which occurs in both photosynthesis and cellular respiration. A theory called the chemiosmotic theory explains the mechanisms in-volved in phosphorylation. We will examine the general outline of the mechanisms of chemiosmosis here, and in Chapters 7 through 9 we will see how it operates in photosynthesizing and respiring organisms.

The Chemiosmotic Theory

■ **The chemiosmotic theory is an explanation of the phosphorylation of ADP to form ATP, due to the establishment of a proton gradient across a membrane. The energy released during the flow of protons through the membrane is used in the phosphorylation of ADP to form ATP.**

Several explanations have been put forward for ATP formation; the most widely accepted is the **chemiosmotic theory**, proposed by Dr. Peter Mitchell (see Profile in Chapter 7). In 1961, Dr. Mitchell proposed a mechanism by which ATP is formed in mitochondria and chloroplasts. He noticed that the processes of photosynthesis or respiration could form ATP only if chloroplasts or mitochondria were intact; he thought the structure of the membrane therefore had something to do with ATP synthesis. Mitchell's theory is based on two assumptions: (1) a *proton* (H^+) *gradient* is established across membranes found in both the mitochondrion and the chloroplast, and (2) the gradient stores *potential energy,* which is used to form ATP from ADP plus P_i. Since Mitchell published his findings, independent research has indicated that if there is damage to the inner membrane of the mitochondrion or to structures called the thylakoid disks found in chloroplasts, ATP production ceases. Other experiments have shown that if researchers artificially increase the H^+ concentration on one side of a membrane, ATP can be produced. The accumulated evidence supports Mitchell's theory.

Mitchell's chemiosmotic theory is built around the idea that the electrical gradient exists between one side of a membrane and the other because one side has a greater concentration of H^+ (with its positive charge) than the other. This proton gradient occurs because it is only electrons that are passed along the electron transport chain. As electrons are passed from molecule to molecule, some of the released energy is used to pump protons (H^+) across a membrane into a proton reservoir.

The existence of a proton reservoir establishes an electrical gradient, since there are more H^+ ions (with their positive charge) on one side of the membrane than on the other. The reservoir also creates

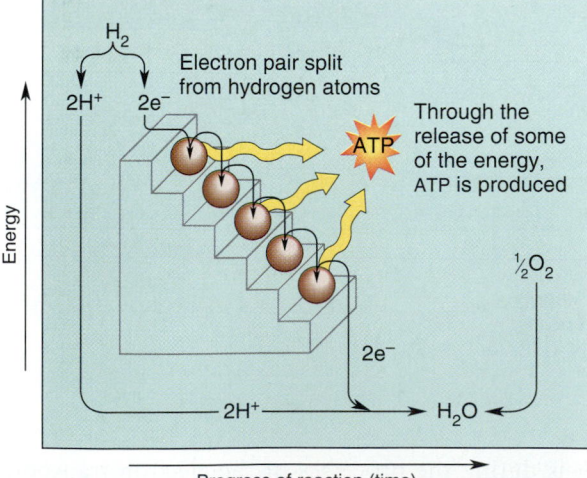

FIGURE 6.16 Electron Transport Chain
The series of electron transfers in this chain is a series of reductions and oxidations. The electrons gradually lose energy, and their stored energy is released in the chain step by step.

FOCUS 6.1
Enzymes and Other Tailor-Made Proteins

With the knowledge biochemists have gained about the special shape of protein molecules, today's molecular biologists can synthesize and redesign enzymes for many specific purposes. For instance, when someone is vaccinated against any disease, he or she is usually injected with a vaccine made from weakened or dead disease-causing organisms. However, scientists can now copy short sections of protein molecules from a variety of disease-causing organisms and inject the copies into the body. The copies stimulate the body's defense mechanisms without producing the sometimes dangerous side effects that can result from traditional vaccines. A synthetically produced vaccine against the hepatitis B virus is currently on the market. In addition, one of the avenues now being investigated in the search for an AIDS vaccine is the protein GP120 found in the HIV virus's outer shell. Computers are being used to outline the shape of specific proteins like enzymes. Once the shape is known, a search for molecules that would fit into the enzyme and thus disable it is conducted. For example, the AIDS virus has a critical enzyme known as *protease,* which "snips" out the proteins that make the new virus. When the enzyme is blocked, the virus cannot reproduce. Using the computer, scientists have found a molecule that will block the enzyme.

The possibilities presented by protein research are truly amazing and sometimes sound more like science fiction than science. Some researchers are looking for "magic bullets" — substances that can carry toxic chemicals attached to a specific type of protein but are so selective that they will bind to and destroy diseased cells only, leaving the healthy cells alone. Other theorists talk about proteins that will stimulate the body to regenerate nerve fibers, or about enzymes and proteins that will mimic embryonic regulators and effect the growth of new organs for transplantation. If the latter process is perfected, each of us could serve as our own organ bank.

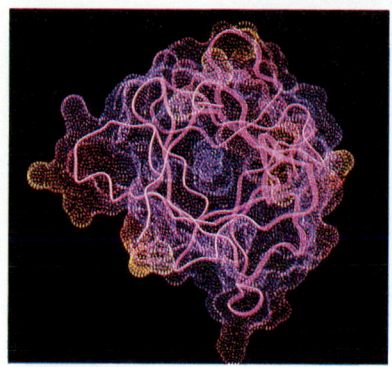

FIGURE A
The protein interleukin is tailor-made for use in immunotherapy.

Such applications of protein engineering lie far in the future, if in fact they are developed at all. More mundane uses that may be closer to becoming realities are a new protein in detergents that would resist being broken down by bleach, a surgical adhesive developed from proteins similar to those that allow mussels to cling to rocks under water, and a protein that would bind with toxins to remove them from wastewater.

Today's scientists and protein engineering corporations are investing time and money on research to explore the commercial applications of protein synthesis (Figure A). What will be the result of the work that bioorganic experimenters are now developing? As they work on reforming proteins, will they also reform our lifestyles?

Perhaps the most significant thing to realize is that every potential application is the result of previous basic research, in which scientists searched for knowledge about protein structure for knowledge's sake rather than for monetary reward or with practical, pragmatic goals in mind. The hundreds of potential applications of protein research are the results of our understanding of the basic three-dimensional structure of proteins and of the way that proteins and substrates fit and interact.

chemical gradients, since the accumulation of H^+ ions on one side of the membrane results in a lower pH level than on the other side of the membrane. The two together are called an *electrochemical gradient*. This gradient establishes a difference in potential energy. (Think of this in terms of the ball at the top of the stairs.) As protons are passed back through the membrane (or the ball rolls down the stairs) the potential energy of the gradient becomes available to form the energy-rich bonds in ATP.

The structures and mechanisms involved in chemiosmosis vary according to the process being discussed, whether it is photosynthesis or cellular respiration, but the general features are the same: (1) Some of the energy released as electrons pass along the electron transport chain is used to pump protons into a proton reservoir, resulting in a difference in H^+ concentration. (2) The difference in H^+ concentration creates an electrochemical gradient that establishes an increase in potential en-

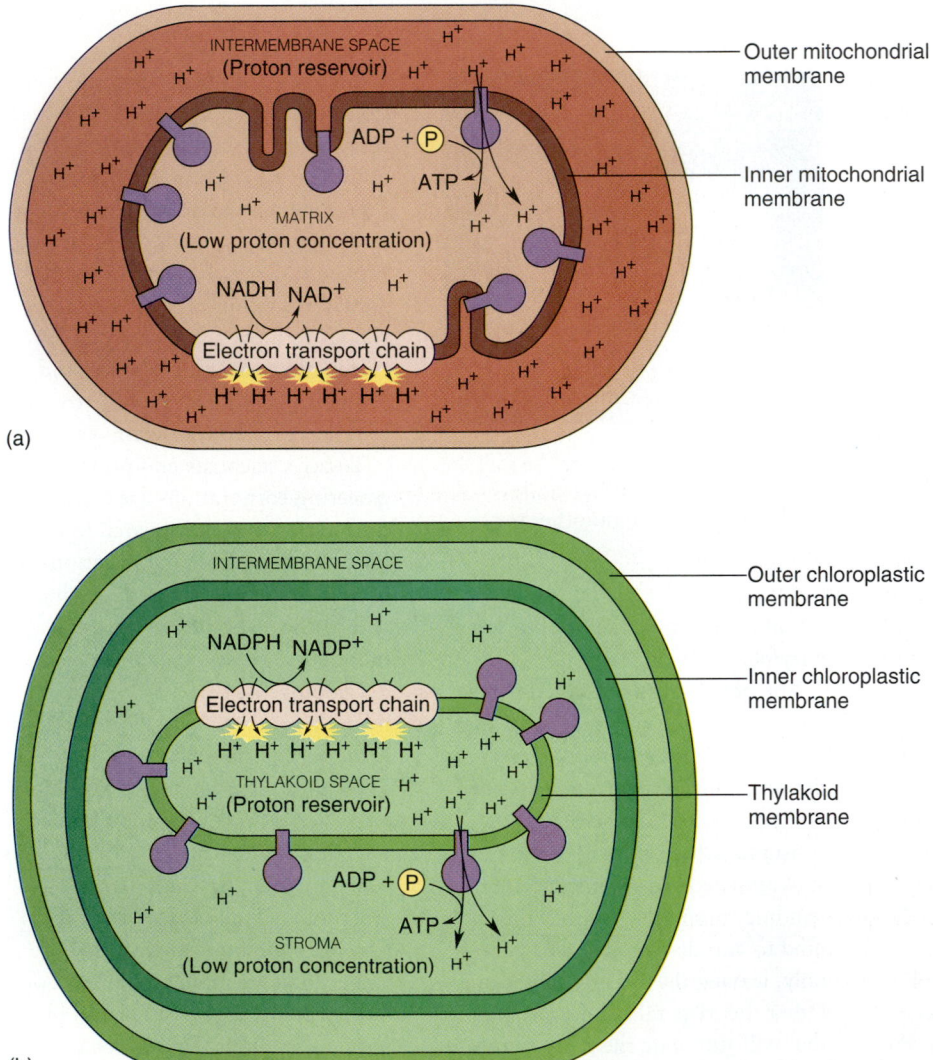

FIGURE 6.17 Chemiosmosis
(a) Energy released when electrons flow through the electron transport chain is used to pump protons (H^+) into the intermembrane space against a concentration gradient. That is, energy is used to pump protons from an area of low concentration to an area of high concentration. ATP is synthesized using energy released as protons flow back into the matrix through special channels. (b) The principle is the same in chloroplasts, where chemiosmotic phosphorylation also occurs.

ergy. (3) Protons flow back through the membrane through a large enzyme that contains a channel for the protons and an active site that captures the potential energy from the protons and uses it to synthesize ATP from ADP and P_i. Figure 6.17 illustrates the chemiosmotic mechanism within a mitochondrion and a chloroplast.

Dr. Mitchell's theory is an explanation of how phosphorylation takes place. In recognition of his work, he received the Nobel Prize in 1978. Although some of the details of the chemiosmotic process are still unknown, it provides a general explanation that is more satisfactory than any other theory put forward at this time. It successfully combines the insights that have come to biology from physics and chemistry about the nature of energy and the laws that govern its movement and its changes.

SUMMARY

1. Living things require a constant input of energy, which they transform to overcome the tendency toward entropy (increased disorder).

2. Energy, the capacity to do work, can exist in two forms: potential energy (stored energy) and kinetic energy (energy of motion). Among the forms of energy important to living things are chemical, mechanical, radiant, and electrical energy.

3. One unit of energy measurement used in biology is the kilocalorie, or Calorie, which is the amount of heat energy required to raise the temperature of 1000 grams (1 kilogram) of water by 1°C.

4. The first law of thermodynamics states that in ordinary chemical reactions energy can be neither created nor destroyed, only changed in form. The second law of thermodynamics states that during changes in form the amount of usable energy available is reduced (in other words, entropy increases).

5. In reduction reactions, an atom or molecule gains one or more electrons and energy is stored. In oxidation reactions an atom or molecule loses one or more electrons and releases energy. For every oxidation reaction there is a reduction reaction. These processes power both photosynthesis and respiration.

6. Adenosine triphosphate (ATP) is the primary energy-currency molecule in living systems. This is due to its two terminal energy-rich phosphate bonds. When one of these bonds is broken, energy is released, and ATP becomes ADP. When energy is available, inorganic phosphate (P_i) is added to an ADP molecule to form ATP. Converting ADP to ATP is called phosphorylation.

7. Enzymes are proteins that act as organic catalysts. By lowering the minimum energy of activation, en-

zymes enable chemical reactions to take place within the temperature limitations of living cells.

8. Each type of enzyme combines with a specific substrate. Following the reaction, the enzyme is released and can be used again.

9. Enzymes often work in conjunction with cofactors. Some cofactors are ions and some are coenzymes.

10. Enzymes are affected by several factors — temperature and pH, for example. They can be destroyed or inactivated by analogs and heavy metal ions.

11. Electron carrier molecules such as NAD^+ and $NADP^+$ are special coenzymes that transfer electrons, protons, and energy from one area to another within a cell. Electron carrier molecules either float in the cytoplasm or are embedded in membranes.

12. The electron carrier molecules of the electron transport chain pass electrons along an electrical gradient from higher levels to lower levels of potential energy. Some usable energy is lost as heat.

13. The chemiosmotic theory is based on establishing a proton gradient across a membrane. The energy provided by this gradient is used for the phosphorylation of ADP to ATP.

KEY TERMS

potential energy	phosphorylation
kinetic energy	minimum energy of
first law of	activation
thermodynamics	catalyst
second law of	enzyme
thermodynamics	substrate
entropy	active site
endergonic	induced-fit theory
exergonic	cofactor
oxidation	coenzyme
reduction	vitamin
photosynthesis	allosteric enzyme
respiration	allosteric site
ATP	negative feedback
energy-rich phosphate	electron carrier
bond	NAD^+
ADP	$NADP^+$
AMP	chemiosmotic theory

REVIEW QUESTIONS

True/False

1. Living things can be thought of as energy transducers.
   true   *false*   *(page 123)*

2. The second law of thermodynamics states that the amount of useful energy in the universe is decreasing.
   true   *false*   *(page 126)*

3. The gain of an electron or electrons by a molecule is termed reduction.
   true   false   (page 128)

4. Enzymes are inorganic catalysts that raise the energy of activation.
   true   false   (page 133)

Fill in the Blank

1. Living things require energy to overcome the natural tendency toward _____ in the universe. *(page 127)*

2. The molecule most often associated with the release of energy in small biologically useful units in the cell is the nucleotide called _____ . *(page 130)*

3. A reaction in which an electron is added to a molecule and energy is stored is known as a _____ reaction. *(page 128)*

4. The three-dimensional folding of an enzyme produces a specific shape. This forms a specific site within the molecule that reacts with the specific substrate. This site is called the _____ . *(page 133)*

Multiple Choice

1. A molecule is oxidized when:
   a. Electrons are removed from the molecule.
   b. Hydrogen is added to the molecule.
   c. Water is removed from the molecule.
   d. Any of the above.
      (page 128)

2. ATP acts as the energy currency of the cell because:
   a. It is a nucleotide.
   b. It is a polar molecule.
   c. It contains energy-rich phosphate bonds.
   d. It contains a sugar.
      (page 130)

3. Vitamins are important in living cells because:
   a. They are vital amines.
   b. They are required in order for some enzymes to function.
   c. They are water-soluble.
   d. They are fat-soluble.
      (page 134)

4. All of the following statements about enzymes are true except:
   a. A single enzyme molecule can be used over and over again.
   b. Each enzyme catalyzes one specific reaction.
   c. Some enzymes contain an essential nonprotein component.
   d. Enzymes raise the minimum energy of activation.
      (pages 133–38)

Discussion

1. Why do living things require energy? Relate your answer to the first and second laws of thermodynamics.

2. Distinguish between potential energy and kinetic energy. Give examples of each.

3. Why are oxidation and reduction reactions considered so important to the energy relationships in the living world?

4. Why is ATP important in living things? What is a phosphorylation reaction?

5. What are enzymes, and how are they believed to lower the energy of activation?

6. Why can it be stated that enzyme substrate specificity is an example of the concept of the complementarity of structure and function?

7. Based on what you have read in this chapter, what is the biological basis for vitamin-deficiency diseases?

Chapter 6   Energy Relationships in Living Things

7

Photosynthesis and Respiration: An Overview

The Flow of Energy in Living Systems
Photosynthesis
Cellular Respiration
Summary

Now we have arrived at one of the most interesting points in our discussion of biology: How do living things obtain and manage the energy necessary to drive the phenomenon we refer to as life? As we discussed in the previous chapter, living things are energy transducers; that is, living things convert energy into various forms and require energy to overcome the tendency toward disorder.

This chapter presents a general overview of the ongoing flow of energy that occurs through photosynthesis and respiration. Chapters 8 and 9 offer a more in-depth discussion of photosynthesis and respiration, respectively. Whether you read all three chapters or only this one depends on your objectives and the objectives of your instructor. Now, let's enter the exciting world of cellular energetics.

Above: The frog and the grass illustrate the ongoing flow of energy through photosynthesis and respiration.

THE FLOW OF ENERGY IN LIVING SYSTEMS

■ **Light energy is converted to food by the process of photosynthesis. Food is converted to cellular energy by the process of cellular respiration.**

How is light energy converted to food? How is the energy found in the chemical bonds of those food molecules made available in useful forms to sustain life? Those are the questions that we will answer in this chapter.

Light energy is converted to chemical energy (food) by the series of reactions known as **photosynthesis**. The energy stored in food is made available to drive living cells by a process known as *cellular respiration*. In summary, photosynthetic organisms capture light energy and convert it to food. Then, as cells respire, the energy stored in the food is converted to energy that can be used as the

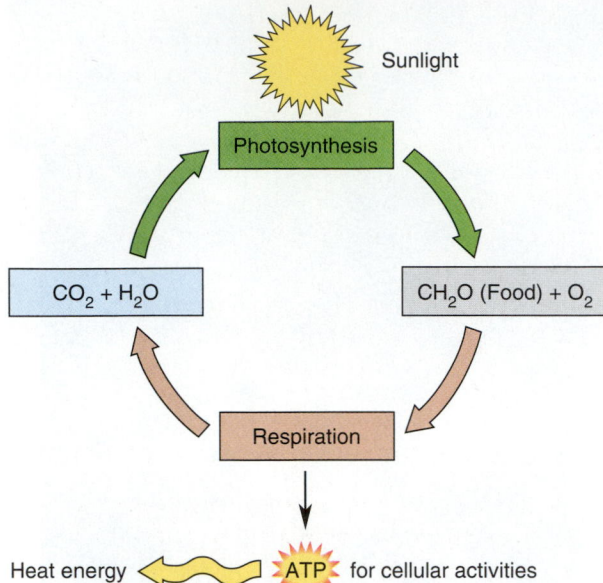

FIGURE 7.1 Energy Flow Through Living Systems
The sun is the ultimate form of energy on the earth. Plants absorb light energy and through the process of photosynthesis convert light energy to the chemical energy latent in food. In photosynthesis, carbon dioxide and water are used to create food, and oxygen is produced as a byproduct. Through the process of respiration, the chemical energy stored in the chemical bonds in the food molecule is broken down to carbon dioxide, water, and energy to form adenosine triphosphate (ATP). The carbon dioxide and water can then be recycled into photosynthesis, and the cycle of energy continues.

source of energy for life (Figure 7.1). Let's begin our discussion of this energy flow by first considering photosynthesis; then we'll turn to respiration.

PHOTOSYNTHESIS

■ **Photosynthesis requires photosynthetic pigments and light energy as well as the raw materials, CO_2 and H_2O. Photosynthesis can be thought of as consisting of two phases: the light-dependent phase (occurring in the membranes of the thylakoid), which produces ATP and NADPH, and the light-independent phase (occurring in the stroma of the chloroplast) in which CO_2 is converted to high-energy carbon compounds (food).**

A simplified equation for *photosynthesis* is:

$$CO_2 + 2\ H_2O \xrightarrow[\text{Chloroplasts}]{\text{Light energy}} (CH_2O)_n + H_2O + O_2$$

As you can see from this equation, carbon dioxide and water serve as the raw materials. The hydrogens from water are used to form the $(CH_2O)_n$ carbohydrate units. For example, if the carbohydrate synthesized is glucose, which has six carbons, then the CO_2 would be increased by six in order to have the raw material to synthesize glucose (see the following equation).

$$12\ H_2O + 6\ CO_2 \rightarrow C_6H_{12}O_6 + 6\ H_2O + 6\ O_2$$

Light energy serves as the energy source, and in eukaryotic cells, these events occur in the green chloroplasts. Photosynthesis takes place in photosynthetic membranes in prokaryotic cells. For convenience, biologists describe photosynthesis as consisting of two phases: a phase that is directly dependent on light, the **light-dependent reaction**, and a phase that is independent of light, the **light-independent reaction**. Through these two phases, ATP is synthesized, CO_2 is eventually reduced (hydrogens are added) to form those carbohydrate molecules CH_2O we call food, and O_2 is released from H_2O.

The Light-Dependent Reactions

■ **In the light-dependent reactions, specific wavelengths of light are absorbed by chlorophyll arranged in photosystems. The photosystems consist of numerous pigment and protein molecules. Light strikes a chlorophyll molecule within these photosystems and two electrons are energized and combine with an electron acceptor molecule.**

Light travels in waves and consists of packets of energy called **photons**. The distance between two crests of a light wave is called the *wavelength*, and the energy available is inversely proportional to the wavelength. In other words, the smaller the wavelength, the more energy; the larger the wavelength, the less energy. Figure 7.2 illustrates the wave and photon nature of light. When light of a specific wavelength strikes a green plant, special structures and pigments located within the inner membrane of the chloroplasts trap the light and convert it into chemical energy. Let's see how, but first we should discuss the structure of a chloroplast.

In Chapter 4, we described a **chloroplast** as a membranous organelle containing a fluidlike matrix called the **stroma**, which is crossed by a series of continuous membrane structures. When the membrane structures appear like flattened discs they are called **thylakoids**; when diffuse, they are called *lamellae*. When the thylakoids appear stacked one on the other like a stack of coins, the stack is

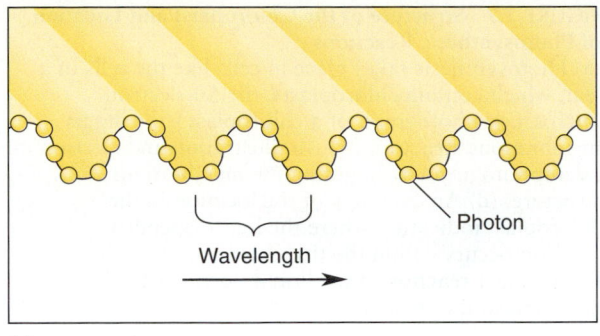

Photon

Wavelength

(a)

Amount of light absorbed

Chlorophyll *a*

Chlorophyll *b*

Prism

Gamma rays	X rays	Ultraviolet	Visible light	Infrared	Microwaves	Radio waves

Electromagnetic spectrum

High energy
(short wavelength)

Low energy
(long wavelength)

(b)

FIGURE 7.2 Light
(a) Light energy travels in waves and consists of packets of energy called photons. (b) Only a small range of wavelengths out of the entire electromagnetic spectrum is visible to the human eye. This is termed *visible light*. To determine which wavelengths are absorbed by photosynthetic pigments, visible light is passed through a prism, producing a rainbow of colors, then a specific wavelength is passed through chlorophyll or another pigment and an instrument called a spectrophotometer measures how much light energy is absorbed at that wavelength. The amount of light absorbed is plotted on a graph to determine the absorption spectrum of the pigment. As you can see from this graph, chlorophyll *a* absorbs most of the light in the violet and red-orange wavelengths, which chlorophyll *b* overlaps to some extent, but chlorophyll *b* is broader, with less absorption than chlorophyll *a*.

called the *granum* (plural, *grana*). Figure 7.3 illustrates the structure of the chloroplast and shows the location of the photosynthetic reactions. The light-dependent reaction takes place within the membranes of the thylakoid. Here, the light-capturing proteins and pigments, electron carriers, and enzymes and other proteins required for the conversion of light energy to chemical energy are located. The trapped light energy is used to form ATP and the reduced form of nicotinamide adenine dinucleotide phosphate (NADPH) (Chapter 6).

Light Absorption and Electron Transport

■ **In the light-dependent reaction, solar energy is used to strip electrons from water. In the non-cyclic electron pathway, the energy released is used to produce ATP and NADPH . At other times, water is not split, and the cyclic electron pathway is used to produce only ATP.**

The first step in photosynthesis is the absorption of light by the **chlorophyll** (the green pigment in plants) and other pigments in the membrane of the thylakoid. When a specific wavelength of light strikes a chlorophyll molecule (in less than a billionth of a second), two electrons are moved to a higher energy level. From this level the electrons can be captured by other molecules called *electron acceptors*. By accepting electrons, the electron acceptor molecules are reduced. A number of electron acceptor molecules are arranged in a series within the thylakoid membrane called an **electron transport chain**. Each electron acceptor is at a lower energy level than the one before it, much like a series of steps. As the electrons are passed down the "steps," electron acceptor molecules gain and then lose electrons. The acceptor molecules are reduced, then oxidized. The energy from oxidation is used to pump protons (hydrogen atoms that have lost their electrons, H^+) into the inner compartment of the thylakoid.

What happens to the electron-deficient chlorophyll molecules? There are two pathways that restore electrons to the chlorophyll molecule. One pathway is cyclic, and the electrons return to the same chlorophyll molecule from which they came. The other pathway is noncyclic, because the electrons that replace the lost electrons come from other sources (such as the water molecules that are present in the thylakoid compartment). When water molecules are broken down, oxygen gas is liberated. The hydrogen atoms lose their electrons and are converted to a free proton, H^+. The H^+ remains in

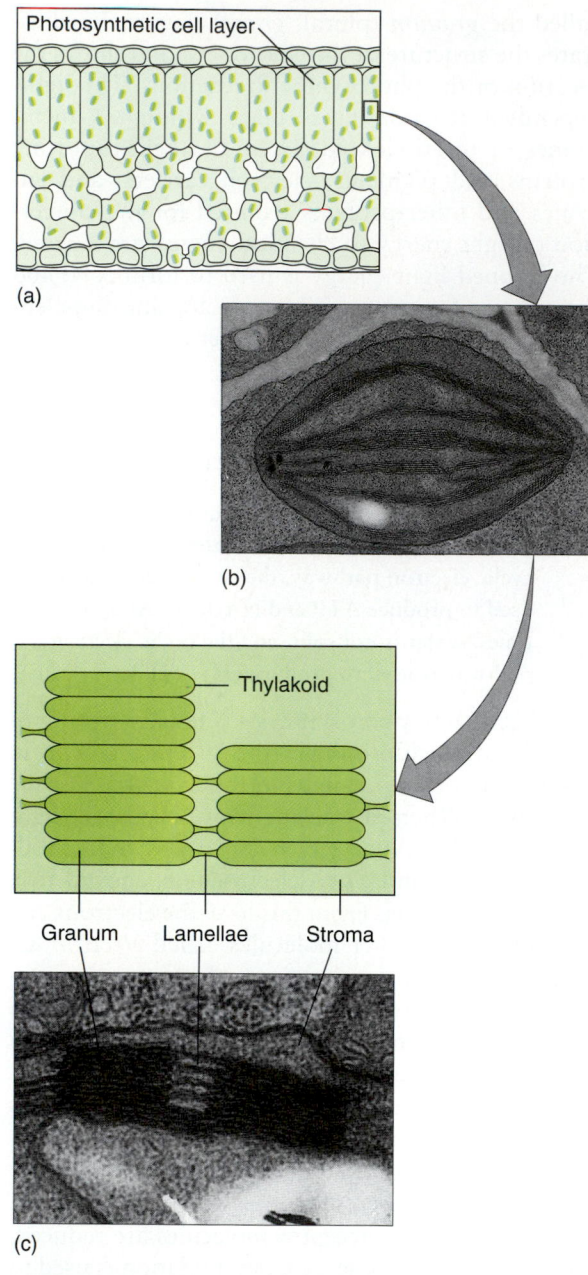

(a)

(b)

Thylakoid

Granum   Lamellae   Stroma

(c)

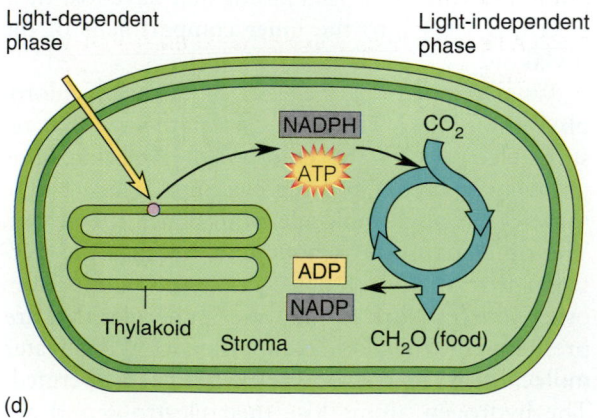

Light-dependent phase

Light-independent phase

NADPH

ATP

CO₂

ADP

NADP

CH₂O (food)

Thylakoid   Stroma

(d)

◀ **FIGURE 7.3   Structure of the Chloroplast and Location of Photosynthetic Reactions**
(a) Photosynthesis takes place in cells like the cells of a leaf, which contain chloroplasts. (b) An electron microscope photograph of a chloroplast showing its membranous structure. (c) An illustration and electron microphotograph highlighting the membranous structure. (d) An overview of the location in the chloroplast indicating where the light-dependent reaction occurs within the thylakoid and the light-independent reaction in the liquid portion of the chloroplast, the stroma.

the thylakoid compartment, and the electrons are used to replace the missing electrons in the electron-deficient chlorophyll molecule.

These two electron pathways eventually result in the phosphorylation of ATP. The pathway is called **cyclic photophosphorylation** when the electrons that cycle through the electron carriers return to the same type of chlorophyll from which they came. The pathway is called **noncyclic photophosphorylation** when the electrons used to replace the missing electrons in the electron-deficient chlorophyll molecules come from another source, such as the water present in the cell.

The light-capturing pigments are not randomly organized in the membrane. Instead, they are organized into clusters of 200 to 300 pigment and protein molecules called *photosystems*. **Photosystem I** (named photosystem I because scientists believed it evolved first) contains chlorophyll *a* molecules that absorb light having a wavelength of 700 nanometers (nm). Hence, the chlorophyll molecule is called pigment 700 or P700. It is within photosystem I that the cyclic pathway of electron transport occurs which results in ATP synthesis (Figure 7.4a).

The noncyclic pathway includes both photosystems I and II (Figure 7.4b). **Photosystem II** consists of a chlorophyll molecule that absorbs light having a wavelength of 680 nm (P680). When light is absorbed by the P680 chlorophyll molecule, P680 loses its electrons to a series of the electron acceptor molecules, which transfer them to a P700 chlorophyll molecule of photosystem I. During this process, ATP is synthesized. How are the electrons replaced in the electron-deficient P680? As we just mentioned, these electrons come from the breakdown of H_2O, to O_2, H^+, and electrons. The splitting of water using light energy is called **photolysis**. Photolysis is the major source of the oxygen that is present in our atmosphere. As electrons from the

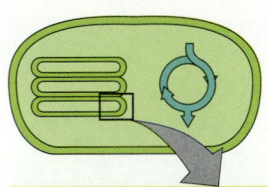

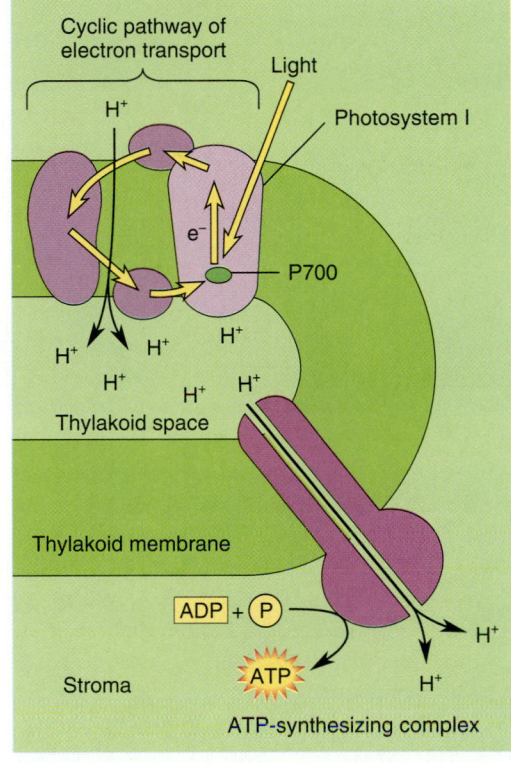

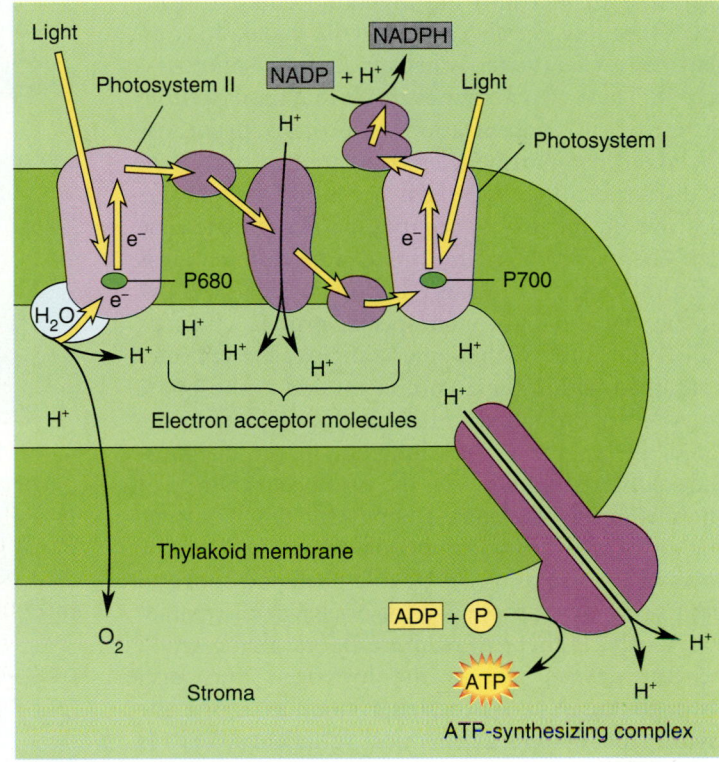

(a)

(b)

FIGURE 7.4 Cyclic and Noncyclic Photophosphorylation
(a) In cyclic photophosphorylation, the electrons flow in a sequential, circular manner and return to the same type of chlorophyll molecule from which they came. (b) In noncyclic photophosphorylation, the electrons from the chlorophyll molecules do not flow in a circular manner. Electrons produced from splitting water replace the missing electrons in chlorophyll molecules in the photosystem II complex, and the electrons from the chlorophyll in the photosystem II complex replace the missing electrons in the chlorophyll molecules in the photosytem I complex.

chlorophyll molecules (P700) of photosystem I are passed through a second series of electron acceptors, two electrons and a hydrogen ion combine with NADP⁺ to reduce it to NADPH:

$$NADP^+ + 2\,e^- + H^+ \rightarrow NADPH$$

Three important events occur in the noncyclic pathways: synthesis of ATP, the photolysis of water, and the reduction of $NADP^+$ to NADPH. The NADPH will serve as the source of hydrogen to reduce CO_2 in the light-independent reaction. ATP provides the energy to make the light-independent reaction run. Before studying this reaction, we need to discuss how ATP is formed in the thylakoid.

ATP Synthesis

■ ATP is synthesized as a result of the free energy available across the thylakoid membrane as a chemiosmotic gradient is produced. The inner compartment becomes more positive as protons are pumped into it, while the stroma outside the thylakoid becomes more negative. As ions flow from the inside of the thylakoid to the stroma through the ATP-synthesizing complex, enzymes phosphorylate ADP to ATP.

In Chapter 6, we mentioned that one method of putting an energy-rich phosphate onto a molecule is by *chemiosmotic phosphorylation*. This process can occur in either a mitochondrion or a chloroplast (see Profile). As the electrons that were emitted from the chlorophyll molecules are passed along a succession of electron carriers, the energy that is given off is used to pump hydrogen ions (protons) from the stroma into the thylakoid compartment. Hydrogen ions are also added to the thylakoid compartment when water is broken down in photosystem II. These two activities result in a thousandfold difference in concentration of protons inside the thylakoid compartment as compared to outside it. A difference in charge exists across the membrane, positive on the inside, negative on the outside. The free energy produced by this charge differential is used to form ATP. As the hydrogen ions flow out of the thylakoid through special protein channels called the ATP-synthesizing complex, the enzymes contained within these complexes add a phosphate to ADP to form ATP. Hence, ADP is phosphorylated due to the energy inherent in the chemiosmotic differential between the inside and outside of the membrane. A rough analogy for this would be the rush of a crowd of people at a rock concert toward a single door. As people push and shove to get to that door, there is a lot of energy dissipated. Obviously, if there were more doors, there would be less energy. So, as the hydrogen ions rush through the ATP-synthesizing complex along the hydrogen ion electrochemical gradient, ATP is formed (Figure 7.5). As you can see in the figure, ATP synthesis takes place due to an electrochemical gradient in which there is a separation of electrical charge across the membrane. The ATP and the NADPH are used to drive the light-independent reaction in which CO_2 is reduced to form carbohydrate. Now let's look at the light-independent reactions.

The Light-Independent Reactions

■ **In the light-independent phase, the ATP and NADPH produced in the light-dependent reactions are used to reduce CO_2. One set of reactions in which CO_2 is converted to high-energy compounds is called the Calvin-Benson cycle.**

Now that we have discussed the formation of ATP and NADPH, how are they used to form those high-energy carbon compounds we call food? As we mentioned, the actual synthesis of food occurs in the light-independent reactions in the stroma of the chloroplasts.

Biologists have designed a series of elaborate experiments to describe the events that occur in the

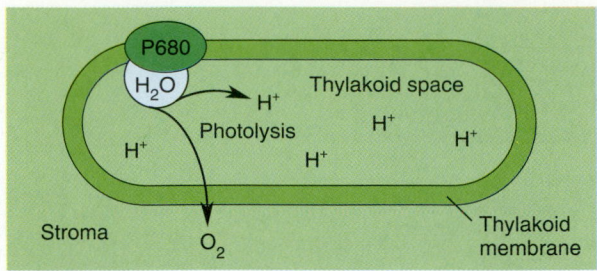

(a)

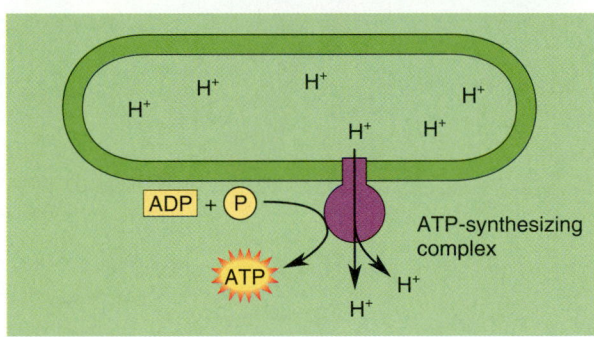

(b)

(c)

FIGURE 7.5 ATP Synthesis
ATP synthesis takes place in the thylakoid due to a chemiosmotic gradient, in which there is a higher concentration of H^+ in the thylakoid compartment relative to the outside of the thylakoid. This difference in H^+ concentration is due to: (a) the photolysis of water and (b) the pumping of protons into the thylakoid compartment as electrons are transported. (c) As the H^+ pass through the ATP-synthesizing complex, enzymes are used to phosphorylate ADP to form ATP.

light-independent reactions. Melvin Calvin and Andrew Benson were the first to describe the series of 15 or 16 steps involved in the cyclic events resulting in the conversion of CO_2 to food. Hence, this cycle is called the **Calvin-Benson cycle** (Figure 7.6).

For convenience, the Calvin-Benson cycle can be thought of as having three phases: a carbon-fixing phase, a reduction phase, and a regeneration phase. The cycle begins with the addition of CO_2 to a five-

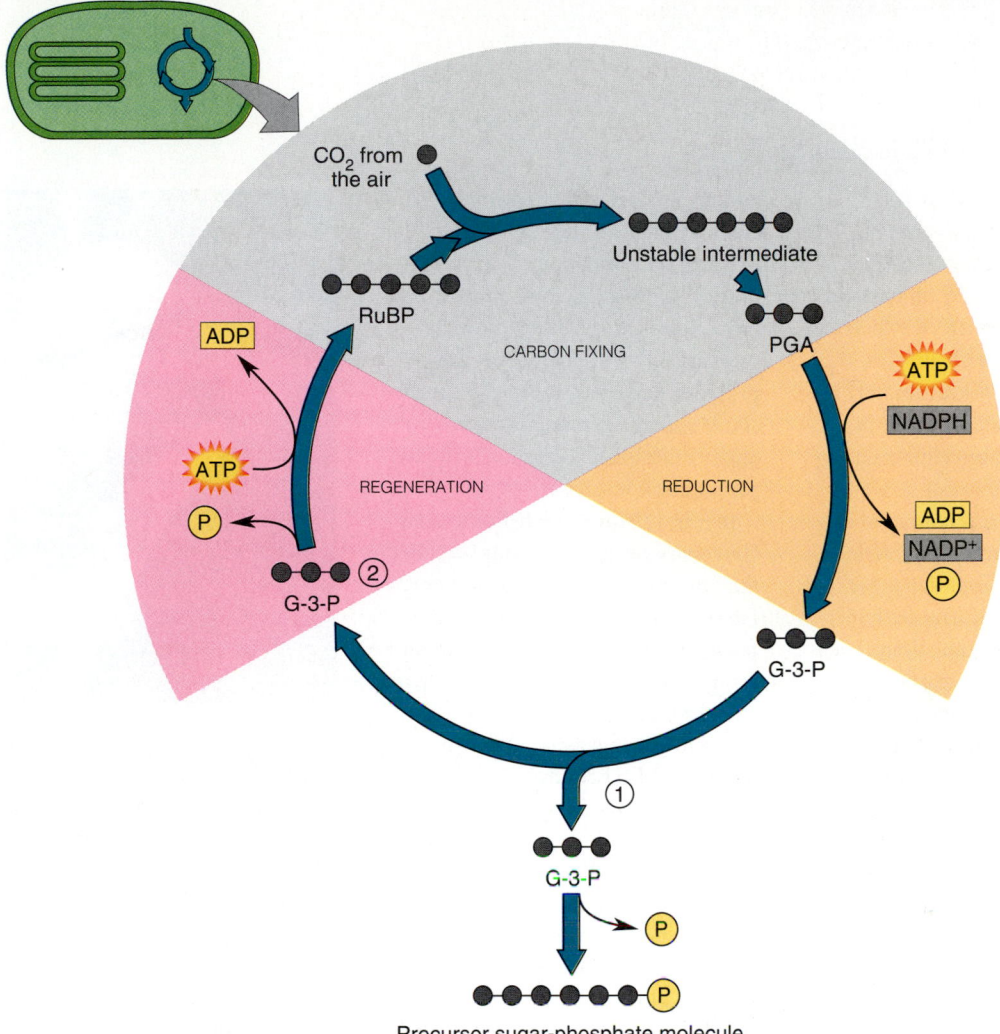

FIGURE 7.6 **The Calvin-Benson Cycle**
A summary of the light-independent reactions in which carbon is fixed (added) to a five-carbon sugar, converted to an intermediate three-carbon compound, then reduced by the addition of hydrogen from NADPH to form G-3-P, which can then be (1) converted to sugars and more complex molecules or (2) used to regenerate RuBP to keep the cycle going.

carbon compound called ribulose bisphosphate (RuBP) to form an unstable intermediate. This intermediate splits right away into two three-carbon compounds called phosphoglycerate (PGA). Therefore, in the carbon-fixing phase, gaseous CO_2 is now incorporated into an organic molecule. As the cycle continues, PGA is first phosphorylated (phosphate is added) and then reduced by the acceptance of a hydrogen from NADPH to form glyceraldehyde-3-phosphate (G-3-P). In the reduction phase,

NADPH provides a hydrogen to reduce the three-carbon intermediate. The next series of reactions results in either the G-3-P (1) being converted to glucose and other sugars or (2) undergoing a series of reactions that result in the regeneration of the RuBP to keep the cycle going.

The $NADP^+$ and the ADP are reused in the light-dependent reactions, where they are reconverted to ATP and NADPH. The glucose and other carbohydrates produced in the Calvin-Benson cycle are then

Peter Mitchell

Nobel Prize-winning Biochemist

Peter Mitchell is not what you would call a typical scientist, because of his unorthodox laboratory (he works at home) and unorthodox thought processes, both of which led him to formulate his chemiosmotic hypothesis. That work ultimately led to a Nobel Prize in Chemistry in 1978, for his work in cellular energy production.

Mitchell is truly a scientist who is a creative thinker and is not afraid to challenge scientific dogma. He formulated his chemiosmotic hypothesis in 1961. At that time, scientists knew that energy-transduction reactions, like photosynthesis and respiration, produced ATP to power the cell, but they were not sure how this ATP was formed. Nonetheless, most biochemists were satisfied with the textbook accounting system method of explaining ATP synthesis, which was based on a theory of coupled molecular isolation. For example, part of the explanation stated that a molecule NADH was reduced by a pair of electrons during the process. For every reduced molecule of NADH, exactly three ATPs were formed.

Mitchell, however, was not satisfied. In 1961, he proposed a rather radical idea. He suggested that the key to ATP synthesis was the formation of an electrical gradient (a gradient based on differing levels of voltage) on opposite sides of membranes in bacteria, chloroplasts, and mitochondria. It was the flow of protons across this gradient that provided the energy for ATP synthesis. Mitchell's explanation went against established dogma, and few people took it seriously.

In 1964, Mitchell left Edinburgh University because of what he called the "overstimulating environment." He purchased a dairy farm on the outskirts of Cornwall, England, and set up his own private laboratory in a regency-style mansion that resembles a medieval castle. Over the next 15 years, Mitchell literally milked the funds to conduct his research from his dairy herd earnings, along with a few small grants. During that time, he and his colleague Jennifer Moyle conducted experiments that demonstrated that their chemiosmotic explanation of ATP synthesis in chloroplasts and mitochondria was credible and that the traditional explanation was not. Ultimately his work was accepted by the rest of the scientific community.

Peter Mitchell is a unique scientific entrepreneur who enjoys working in small groups, away from the large bureaucratic research laboratory

ries typical of many research institutions. When he received the Nobel Prize, someone asked, "What are you going to do with the funds that you received?" Mitchell simply said that he would add two more people to his laboratory staff of six.

Not only has Peter Mitchell's human endeavor into cellular energetics provided us with a greater understanding of the mechanism of membrane ATP synthesis, but the chemiosmotic hypothesis may suggest any number of other applications, from fighting bacterial infections to more efficient utilization of solar energy.

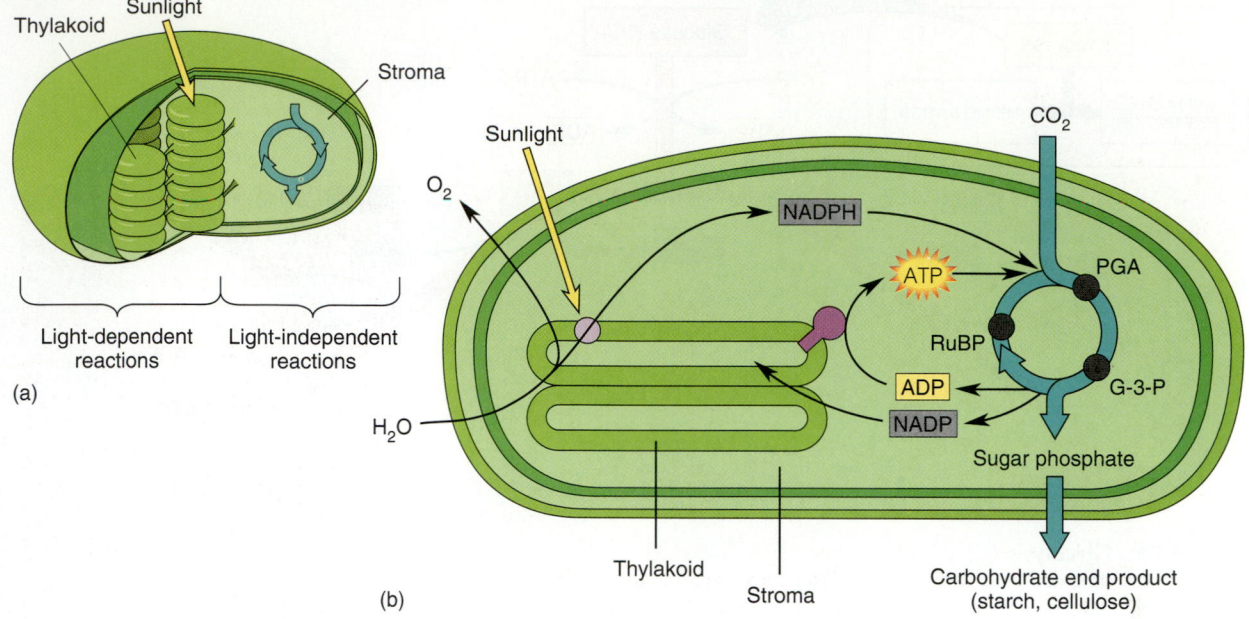

FIGURE 7.7 Summary of Photosynthesis
(a) The locations of the light-dependent and light-independent reactions within the chloroplast. (b) A summary of the major reactions leading to the production of food per the equation

$$12\ H_2O + 6\ CO_2 \xrightarrow{\text{Sunlight}} 6\ O_2 + C_6H_{12}O_6 + 6\ H_2O$$

further metabolized into the myriad of other molecules that make up the plant. Figure 7.7 provides an overall summary of photosynthesis and the light-independent reactions.

Now that we have discussed photosynthesis, let's turn our attention to cellular respiration — the process by which food is converted to usable energy.

CELLULAR RESPIRATION

■ **The series of reactions that occurs in cells and releases energy from food to form ATP is called cellular respiration.**

The process by which cells use the chemical energy trapped in carbon compounds to form ATP is called **cellular respiration**. As you know, we eat a variety of foods, but in reality those foods can be grouped into carbohydrates, fats, or proteins. Food can be used as a source of energy or as the raw material to build cellular structures. Of the foods we eat, carbohydrates and fats are usually thought of as our major sources of energy. Let's focus our discussion of cellular respiration on the carbohydrate known as glucose. Metabolism of glucose can occur in the absence of oxygen, a process known as fermenta-

tion, or in the presence of oxygen. In the presence of oxygen, the metabolism of glucose involves three major processes: glycolysis, the Krebs cycle, and electron transport resulting in chemiosmotic phosphorylation. These processes occur in a specific sequence starting with glycolysis.

Glycolysis

■ **The first step in cellular respiration, glycolysis, results in the breakdown of glucose to pyruvate . There is also a net production of two ATP.**

During the first stage of cellular respiration, whether oxygen is present or not, glucose, a six-carbon compound, is split into two molecules of pyruvate, a three-carbon compound. As you can see, because a glucose molecule is split, **glycolysis** (from *lysis,* "to split") is a very descriptive name. Two ATP molecules are used to initiate the reaction. Four ATP molecules are produced. Therefore, there is a net gain of two ATP molecules. Hydrogen ions and electrons are released from intermediate compounds during glycolysis, and they reduce the electron acceptor nicotinamide adenine dinucleotide (NAD$^+$) to NADH (see Chapter 6):

$$NAD^+ + 2\ H^+ + 2\ e^- \rightarrow NADH + H^+$$

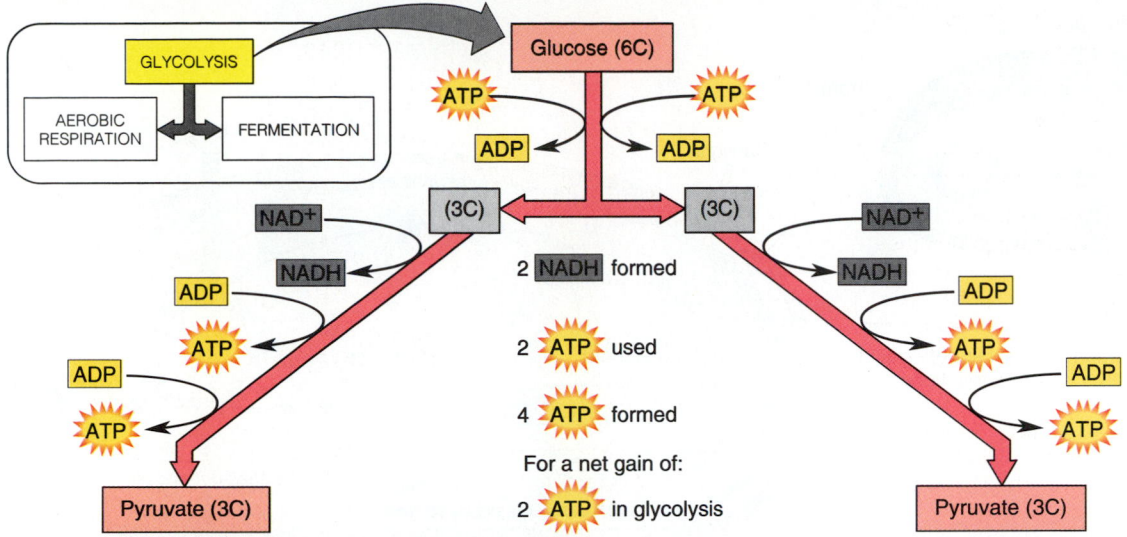

FIGURE 7.8 Glycolysis
An overview of glycolysis, in which glucose, a six-carbon compound, is metabolized into two three-carbon compounds called pyruvate. In this process, a net of two ATP molecules are synthesized.

Figure 7.8 provides an overview of glycolysis, which takes place in the cytoplasm of cells.

Up to this point, the energy gained has been very small. However, we still have NADH and pyruvate. What happens to them?

NADH and pyruvate can take either of two routes, depending on the type of cell and the amount of oxygen available. If the cell is a eukaryotic cell and oxygen is present, the pyruvate is converted to an acetyl group attached to coenzyme A. The substance, called acetyl coenzyme A (acetyl CoA), enters a mitochondrion (see Human Endeavors box) for a more complete oxidation. The NADH is used in electron transport. If there is no oxygen present, the pyruvate undergoes a process called **fermentation** within the cytoplasm of the cell.

Anaerobic Respiration and Fermentation

■ **If little or no oxygen is present, the pyruvate is metabolized by a process called fermentation. An intermediate such as ethyl alcohol or lactate serves as the hydrogen acceptor to oxidize NADH to NAD⁺. The NAD⁺ is then recycled to continue glycolysis.**

In some cells, such as muscle cells and yeast cells, the pyruvate must follow the pathway of **anaerobic respiration** when there is no oxygen present. *Anaerobic* means respiration without oxygen. In aerobic respiration, oxygen is the ultimate electron acceptor. When anaerobic respiration occurs, the cells must use another compound to replace oxygen as the ultimate electron acceptor. In other words, that molecule will remove the electrons from NADH so that the NAD^+ can be recycled back to glycolysis, allowing glycolysis to continue. This final electron acceptor in anaerobic respiration is either ethyl alcohol or lactate (Figure 7.9).

In alcoholic fermentation, the two pyruvate molecules produced by glycolysis are further metabolized to acetylaldehyde. The acetylaldehyde then accepts the electrons and protons from NADH and is reduced to ethyl alcohol. In alcoholic fermentation, CO_2 is released as a gas. CO_2 is what produces the bubbles in bread dough that cause it to rise as well as the bubbles in beer. The regenerated NAD is reused in glycolysis.

In lactate fermentation (which often occurs in muscle cells when oxygen is not available), the two pyruvate molecules are converted to a pair of three-carbon lactate molecules. You will notice that lactate is the final hydrogen acceptor. The NAD^+ that is regenerated is reused in glycolysis.

Very little usable energy in the form of ATP is produced by the combination of glycolysis and fermentation. The other reactions that take place in the presence of oxygen are where most of the cellular energy is produced. Let's now discuss the pathway of pyruvate when oxygen is available.

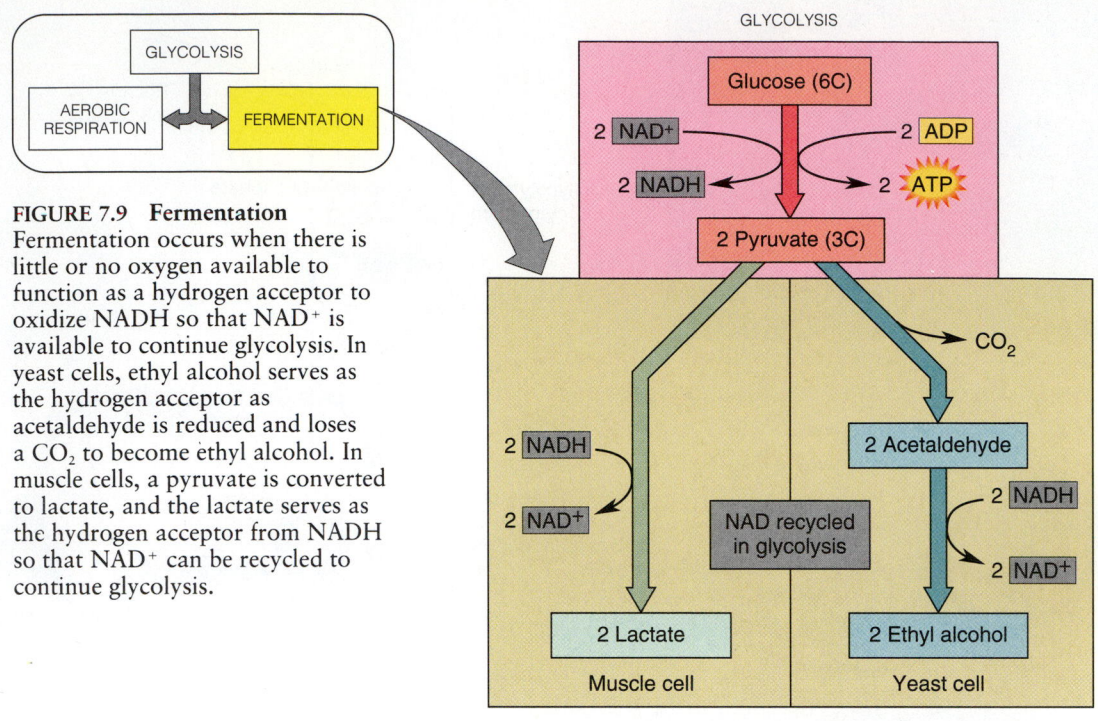

FIGURE 7.9 Fermentation
Fermentation occurs when there is little or no oxygen available to function as a hydrogen acceptor to oxidize NADH so that NAD⁺ is available to continue glycolysis. In yeast cells, ethyl alcohol serves as the hydrogen acceptor as acetaldehyde is reduced and loses a CO_2 to become ethyl alcohol. In muscle cells, a pyruvate is converted to lactate, and the lactate serves as the hydrogen acceptor from NADH so that NAD⁺ can be recycled to continue glycolysis.

Aerobic Respiration

■ **If oxygen is present, pyruvate is converted in the mitochondrion of the cell to acetyl CoA, which enters a series of reactions known as the Krebs cycle. In the Krebs cycle hydrogens are removed from carbon compounds. The hydrogen ions (protons) and electrons that are released are used in the electron transport chain and chemiosmotic phosphorylation. The electrons are transported through the electron transport chain, and the energy released is used to pump protons into the space between the inner and outer mitochondrial membrane. This concentration of the protons forms a chemiosmotic gradient. Because of this gradient, free energy is released and used to add a phosphate to ADP to form ATP.**

We refer to respiration as aerobic when it occurs in the presence of oxygen. The oxygen is used as the final electron acceptor during **aerobic respiration** (Figure 7.10). Aerobic respiration starts with a series of reactions known as the **Krebs cycle** (named after Hans Krebs). In the Krebs cycle, electrons and H⁺ are removed from pyruvate, forming three molecules of CO_2.

The Krebs Cycle. The Krebs cycle occurs in the matrix of the mitochondrion. Before the Krebs cycle

begins, enzymes remove a CO_2 from each pyruvate molecule. At the same time, other enzymes remove hydrogens (protons) and electrons, forming NADH. The remaining two-carbon molecule, called an acetyl group, binds with coenzyme A (CoA) to form acetyl CoA, the first molecule to enter the Krebs cycle. The acetyl group of acetyl CoA binds with oxaloacetate, a four-carbon compound, to form a six-carbon compound called citrate. The CoA is reused to bind with another acetyl group.

Then, through a series of enzyme-mediated reactions, three-carbon molecules are rearranged and broken down. Hydrogens (protons) and electrons are removed from molecules and are accepted by electron acceptors. NAD⁺ is used in these reactions; a second electron carrier — flavin adenine dinucleotide or FAD — is also used.

$$FAD + 2 H + 2 e^- \rightarrow FADH_2$$

CO_2 is removed. In some cases, energy is used directly to form ATP at the substrate level (Chapter 6). Finally, oxaloacetate is reformed and made available to combine with another acetyl group from acetyl CoA, starting the cycle again.

Only two additional ATP molecules are formed, one for each acetyl group that enters the Krebs cycle. What happens to all of the molecules of

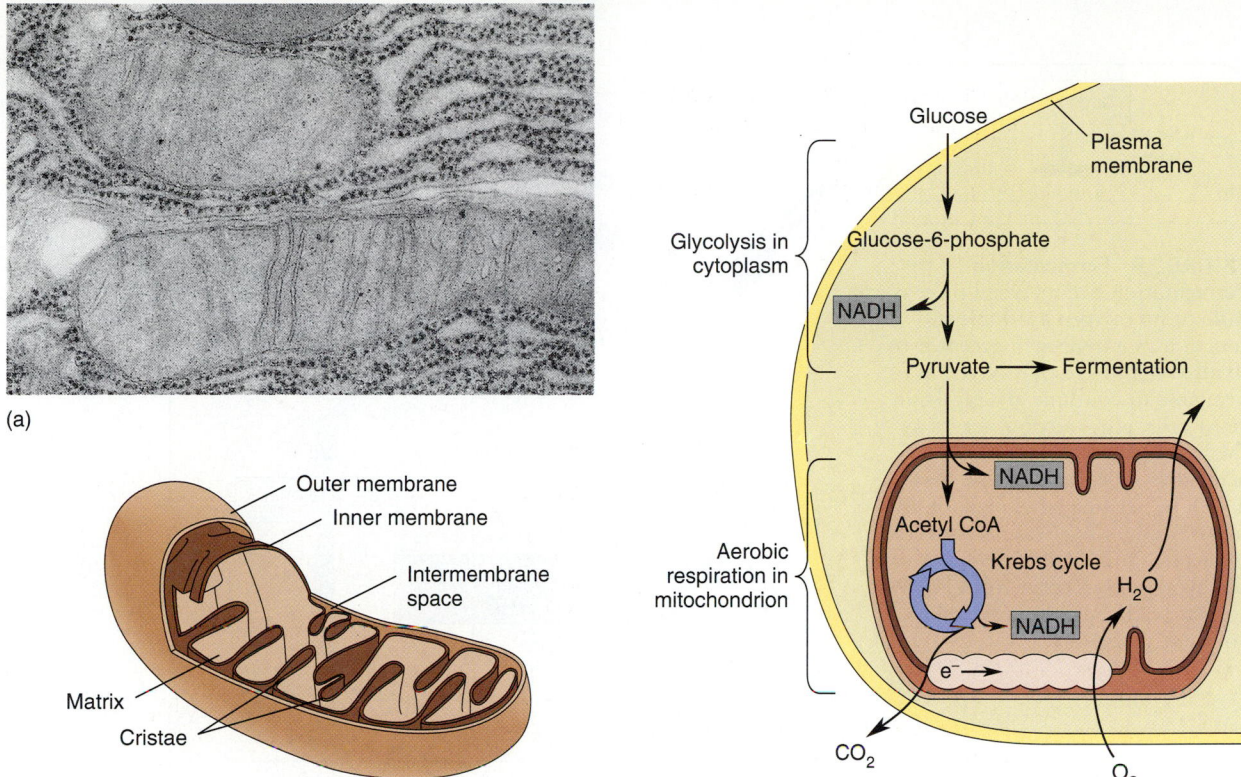

(a)

(b)   (c)

FIGURE 7.10   An Overview of Aerobic Respiration
Glycolysis takes place in the cytoplasm of the cell. In eukaryotic cells, the
majority of the energy-liberating reactions that result in ATP synthesis occurs
in the mitochondrion. (a) An electron microphotograph of mitochondria,
(b) a sectional diagram of the mitochondrion, and (c) the locations where
aerobic respiration reactions occur within the cell and in the mitochondrion.
It should be noted that even though fermentation is anaerobic, it is illustrated
here because glycolysis is often considered a precursor to aerobic respiration
in spite of the fact that glycosis does not require oxygen.

NADH and FADH$_2$? You have probably already
guessed it! The electrons are transferred to electron
acceptor molecules within the inner membrane of
the mitochondrion, to begin electron transport and
chemiosmotic phosphorylation.

*Electron Transport and Chemiosmotic Phosphory-
lation.* Electrons are accepted by electron acceptor
molecules in the inner membrane of the mitochondrion, just as they are in the thylakoid membrane
of the chloroplast. As the electrons are passed between electron acceptors, each acceptor is at a
lower energy level than the one before it. The energy released in these transfers is used to pump hydrogen ions (protons) from the matrix to the inner
membrane space, establishing a chemiosmotic gradient across the mitochondrial inner membrane.
This is very similar to what happens in the chloroplast. As the positively charged H$^+$ (protons) accu-

mulate, a large electrical gradient develops with
more positive charges outside the inner membrane
and more negative charges inside. The H$^+$ are allowed to return to the matrix by passing through
the ATP-synthesizing complex that contains enzymes which phosphorylate ADP to form ATP. In
cellular respiration, initial electron acceptors are
either nicotinamide adenine dinucleotide (NAD$^+$)
or flavin adenine dinucleotide (FAD), which accept
electrons and are reduced to NADH and FADH.

$$NAD^+ \xrightarrow{2\,H^+} NADH + H^+$$
$$FAD \xrightarrow{2\,H^+} FADH_2$$

NAD$^+$ and FAD — coenzymes associated with dehydrogenase — remove hydrogens from molecules.
The ultimate electron acceptor is oxygen, which accepts electrons and hydrogen ions to form water.

In summary, the Krebs cycle provides a source of
hydrogen ions and electrons. The electrons are used

by the electron transport system to provide energy to pump protons from the matrix side of the inner mitochondrial membrane to the inner membrane space of the mitochondrion, establishing a chemiosmotic gradient. The free energy established by this gradient is then used to phosphorylate ADP to form ATP. Okay, but how much energy is actually produced by one molecule of glucose? Let's see.

Aerobic Respiration Energy Yield

- The complete oxidation of one molecule of glucose produces 36 or 38 molecules of ATP. Other compounds can enter glycolysis or the Krebs cycle to be used as sources of energy or to be synthesized to other compounds.

For you accounting majors, and of course because we as biologists are interested in the total energy yield, let's tally up the number of ATPs synthesized from one molecule of glucose through aerobic respiration. During glycolysis, as you remember, there is the net synthesis of two ATPs, through substrate-level phosphorylation. During the Krebs cycle there are also two substrate-level phosphorylations. So far, four ATPs have been produced. But as you might suspect, most ATP synthesis occurs during chemiosmotic phosphorylation. In the average body cell, 32 ATPs are produced during chemiosmotic phosphorylation, for a grand total of 36 ATPs per glucose molecule. In some cells, like those of the heart and liver, 34 ATPs are produced by chemiosmotic phosphorylation, for a total of 38 ATPs. Figure 7.11 summarizes the energy yield for a molecule of glucose by aerobic respiration.

Obviously cells don't get all their energy from glucose, nor do they use only food molecules for energy. Food can be used to build cellular components. Other energy-containing molecules such as proteins, fats, nucleic acids, and other polysaccharides can enter glycolysis or the Krebs cycle at different points, and the energy contained in their chemical bonds can be used to synthesize ATP. Also, intermediates formed during glycolysis or the Krebs cycle can be used in biosynthetic reactions to form molecules needed by the cell such as amino acids, nucleic acids, fatty acids, and steroids.

The condition of the cell and the internal environment signal what pathway the molecule will take. For example, when a human is starving, the supply of glucose falls and the cell must metabolize animal starch reserves or use fats and proteins. As these molecules are broken down, some of their subunits are converted to pyruvate or intermediates

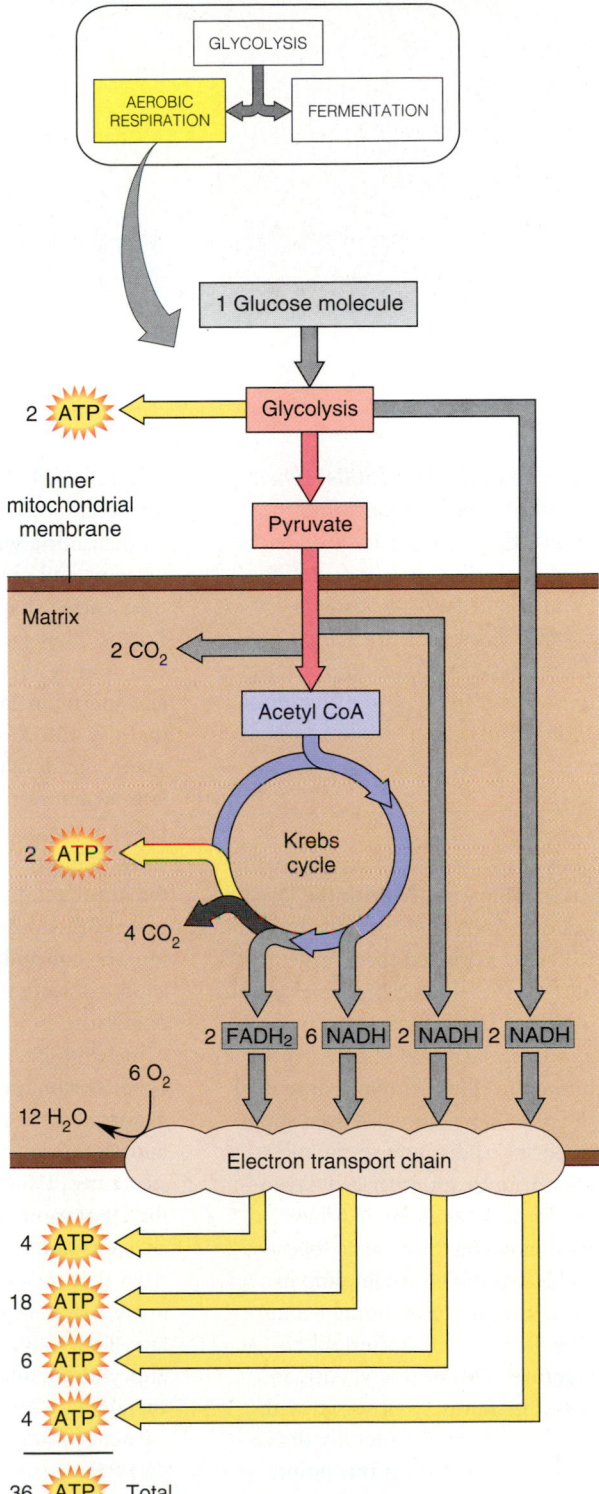

FIGURE 7.11 **Aerobic Energy Yield**
During the complete oxidation of one molecule of glucose, enough energy is made available to synthesize 36 to 38 ATP molecules, and 12 molecules of H_2O and 6 molecules of CO_2 are produced.

Evolution of Our Understanding of the Mitochondrion

Ever since the late 1800s, when the mitochondrion was first observed, scientists have endeavored to understand its functions. (You might be interested in reading Focus 4.2, in which Lewis Thomas shares his wonderful essay with us on his views of the mitochondrion.)

Between 1850 and 1900 microscopists were fascinated with the large granular inclusions in cells. In 1888, Köllicker removed these large granules from the muscles of insects and observed that they swelled in water and were surrounded by a membrane. In 1890, Altmann described the mitochondrion and used stains (fuchsin) to distinguish the mitochondrion from other cellular inclusions. Altmann called these mitochondria "bioblasts" because he had seen them swimming in the cytoplasm and thought that they were simple bacterialike creatures. Interestingly, Altmann was apparently so obsessed with bioblasts that they literally drove him to insanity. (On this point, see E. B. Cowdry, "Historical Background of Research on Mitochondria," *Journal of Histochemistry and Cytochemistry*, vol. 1, p. 183, 1953).

Around the turn of the century, microscopists were becoming very interested in utilizing stains that were selective for certain cellular structures. Michaelis used many different stains and determined that the mitochondria were composed of proteins and phospholipids despite their varied shapes and sizes. Because they were found only in cells with a nucleus, including egg and sperm, in 1918 Meves proposed that the mitochondrion changed shape and developed into the various structures in differentiated cells. Other investigators disagreed. In 1912, Kingsbury observed that the mitochondrion had the ability to oxidize various stains. Therefore, he concluded that the mitochondria were the sites of cellular oxidation.

In 1934, Bensley and Hoerr developed a technique that used centrifugation to separate and isolate mitochondria from other cellular structures. Using this technique for the separation of mitochondria, biochemists and cell biologists hypothesized that the enzyme systems for many reactions, like the Krebs cycle, fatty acid oxidation, oxidative phosphorylation, and others, must be correlated with the structure of the mitochondrion. Hogeboom, Schneider, Palade (see Profile in Chapter 9), and Lehninger showed that the respiratory apparatus was located in intact mitochondria. They also found RNA in the mitochondria, but at that time they didn't find DNA.

In 1953, Palade described the outer and inner membranes, the cristae, and matrix granules. But now the question was, what is the structure and function of the inner and outer membranes?

As different cell types were compared, electron microscopists determined that the location of the mitochondrion within the cell and the number of cristae per mitochondrion varied with the energy requirements of the cell. In the 1960s, researchers used biochemical and electron microscopy techniques to determine that the respiratory apparatus was located on the cristae. Racker made "submitochondrial particles" of the inner surface of the inner membrane and demonstrated that this surface is the site of oxidative phosphorylation and that the respiring electron transporting components are embedded within the inner membrane, not on the surface of the inner membrane.

In 1966, it was determined that the mitochondrion also contained DNA, and Borst and others determined that the DNA was circular, like bacterial DNA. This finding led to the idea that the mitochondrion originated as bacteria that were incorporated into the cells that engulfed them. From this, a symbiotic relationship developed (see the section titled "The Origin of the Eukaryotic Cell" on pages 96–97).

From 1970 until the present, our understanding of mitochondria has grown considerably. Griffiths isolated yeast cells that lacked respiratory enzymes. Thus, after several experiments using these mutants it was determined that the mitochondria have genes for their own proteins as well as their own ribosomal and transfer RNA, even though most of the proteins that function in the mitochondria are produced by the genes in the nucleus of the cell. The finding that mitochondrial DNA controlled for some of the mitochondrial proteins opened up the new field of mitochondrial genetics. It is known that because of abnormal mitochondria, several diseases can result. For example, in 1989 Doug Wallace and G. P. Singh argued that a disease known as Leber's hereditary optic neuropathy, which results in complete or partial blindness due to the degeneration of the optic nerve, is due to a single mutation in the DNA of the mitochondrion. Since you only get your mitochondria from your mother, this is inherited maternally.

In the 1980s researchers sequenced the entire genetic makeup of the mitochondrion from several species and found that the genetic makeup of mitochondria varies greatly between species.

Source: Adapted from "Evolutions: Mitochondria," originally published in *The Journal of NIH Research*, November–December 1989, vol. 1, p. 136. Illustration adapted by Elizabeth Morales-Denney from original artwork by Sally Bensusen.

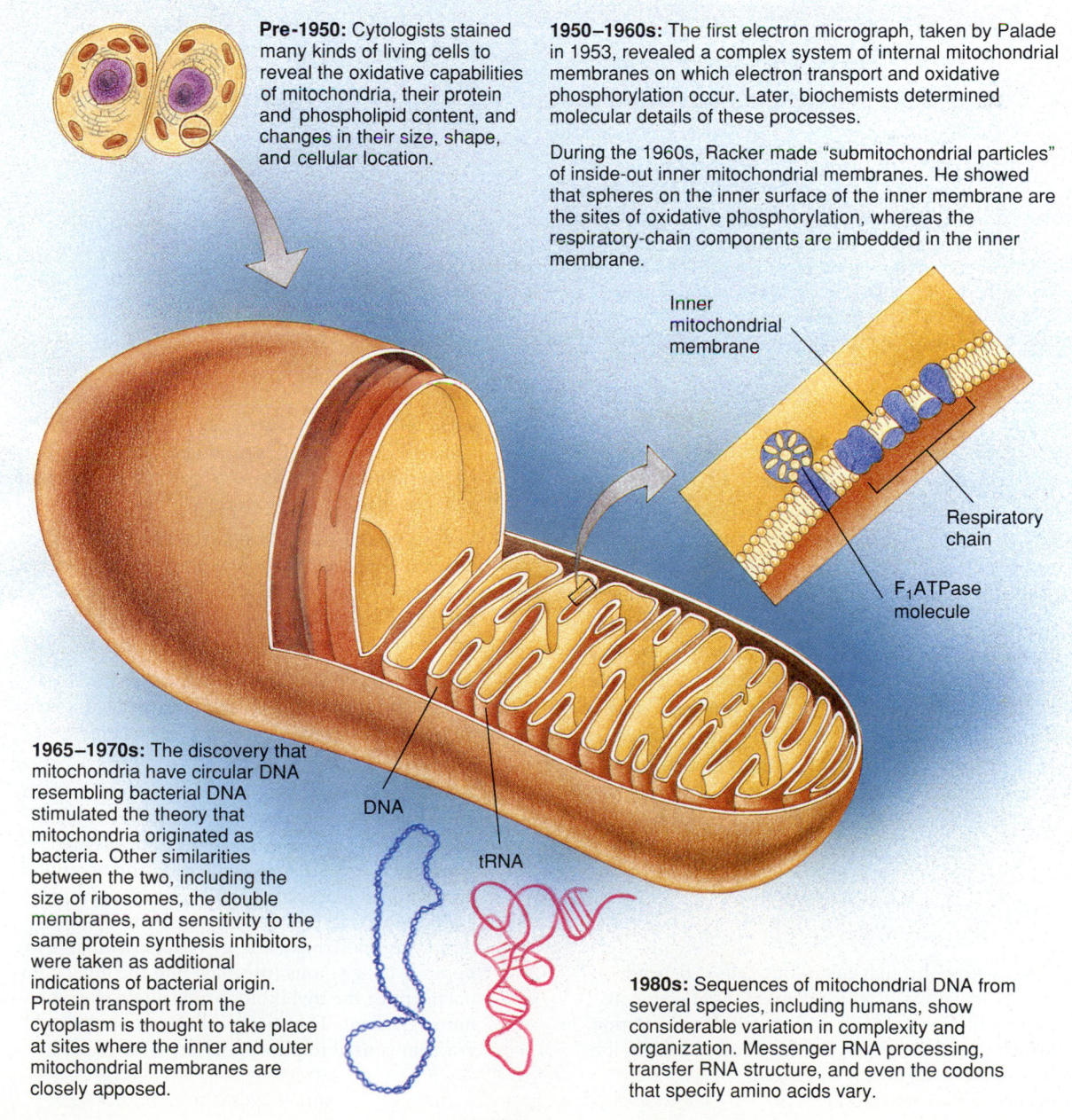

Pre-1950: Cytologists stained many kinds of living cells to reveal the oxidative capabilities of mitochondria, their protein and phospholipid content, and changes in their size, shape, and cellular location.

1950–1960s: The first electron micrograph, taken by Palade in 1953, revealed a complex system of internal mitochondrial membranes on which electron transport and oxidative phosphorylation occur. Later, biochemists determined molecular details of these processes.

During the 1960s, Racker made "submitochondrial particles" of inside-out inner mitochondrial membranes. He showed that spheres on the inner surface of the inner membrane are the sites of oxidative phosphorylation, whereas the respiratory-chain components are imbedded in the inner membrane.

Inner mitochondrial membrane

Respiratory chain

F_1ATPase molecule

1965–1970s: The discovery that mitochondria have circular DNA resembling bacterial DNA stimulated the theory that mitochondria originated as bacteria. Other similarities between the two, including the size of ribosomes, the double membranes, and sensitivity to the same protein synthesis inhibitors, were taken as additional indications of bacterial origin. Protein transport from the cytoplasm is thought to take place at sites where the inner and outer mitochondrial membranes are closely apposed.

DNA

tRNA

1980s: Sequences of mitochondrial DNA from several species, including humans, show considerable variation in complexity and organization. Messenger RNA processing, transfer RNA structure, and even the codons that specify amino acids vary.

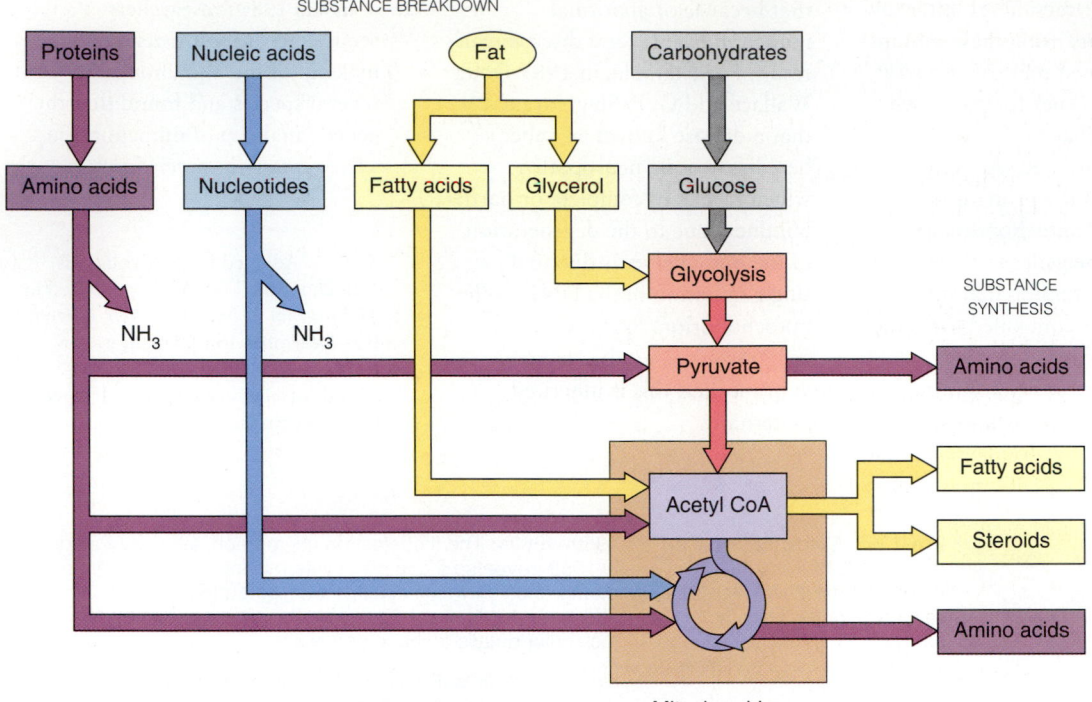

SUBSTANCE BREAKDOWN

SUBSTANCE SYNTHESIS

Mitochondrion

FIGURE 7.12   Breakdown and Synthesis of Various Substances
The Krebs cycle is central to a number of reactions. For example, fats, proteins, nucleic acids, and other substances can be metabolized into forms that enter the Krebs cycle and can be used as sources of energy to synthesize ATP. Also, many of the intermediates of the Krebs cycle can be used as building blocks to synthesize new material for the cell.

of the Krebs cycle. These molecules then can be used as sources of energy (Figure 7.12).

We have now discussed in a general sense how energy is made available to living organisms, through the marvelous reactions of photosynthesis and cellular respiration. If we reflect for a moment, we can see that life is sustained because, through a constant input of energy, living things can overcome entropy or the tendency toward disorder. Living things are truly energy transducers, and an understanding of energy flow through living cells helps us to understand that natural phenomenon we call life.

SUMMARY

1. Energy flow through eukaryotic cells is dependent on photosynthesis, which converts light energy to chemical energy (food), and on cellular respiration, which uses the energy stored in food to sustain life.

2. Chlorophyll and other photosynthetic pigments photochemically convert CO_2 to food with oxygen being released as a by-product.

3. Photosynthesis consists of two phases, a light-dependent phase in which light energy is converted to chemical energy stored in ATP and NADPH, and a light-independent phase in which CO_2 is converted to a high-energy compound (food).

4. The light-dependent reactions occur in the thylakoid membrane of the chloroplast, while the light-independent reactions occur in the stroma of the chloroplast.

5. When specific wavelengths of light strike a chlorophyll molecule, two electrons are sent to a higher energy level. These electrons are then accepted by an electron acceptor molecule. Electron acceptors are arranged in a series. As the electron acceptors are first reduced (accept electrons) and then oxidized (lose electrons), energy is made available to pump hydrogen ions (protons) into the inner compartment of the thylakoid, producing a chemiosmotic gradient. The energy produced by this gradient is used to phosphorylate ADP to ATP.

6. The electrons emitted by the chlorophyll molecules can take one of two electron transport pathways: (1) in cyclic photophosphorylation, the electrons are passed along an electron transport chain and return to the same chlorophyll molecule; (2) in noncyclic photophosphorylation, electrons do not return to the original chlorophyll molecule. The now electron-deficient chlorophyll obtains its electrons from other chlorophyll molecules or by splitting a water molecule.

7. Two light-absorbing systems have evolved in plants. The first one to evolve is called photosystem I. Photosystem I contains a chlorophyll molecule that absorbs light having a wavelength of 700 nm. Hence, it is called P700. Photosystem II, which contains a chlorophyll molecule that absorbs light having a wavelength of 680 nm, is called P680.

8. When light strikes photosystems I and II, high-energy electrons are released and captured by electron acceptors. Their increased energy is used to produce a chemiosmotic gradient across the thylakoid membrane. Ultimately, the electrons are accepted by NADP to form NADPH.

9. During noncyclic photophosphorylation, water is split, releasing 2 H$^+$ (protons), 2 e$^-$, and oxygen. The oxygen is eventually released into the atmosphere. The protons are used to help establish the chemiosmotic gradient within the thylakoid compartment. The electrons are captured by the electron-deficient P680 chlorophyll molecules.

10. The ATP and NADPH produced by the light-dependent reactions are used to reduce CO_2 to a three-carbon compound called glyceraldehyde-3-phosphate (G-3-P) by a series of reactions known as the Calvin-Benson cycle.

11. The reactions of the Calvin-Benson cycle can be grouped into three phases: (1) a carbon-fixing phase in which CO_2 is added (fixed) to a five-carbon compound, followed by a series of reactions that split the newly synthesized six-carbon compound component into two three-carbon compounds; (2) the three-carbon compound is phosphorylated by accepting a phosphate from ATP and then reduced by accepting a hydrogen from NADPH in the reduction phase; and (3) in the regeneration phase, the three-carbon compound formed, G-3-P, can then either follow a pathway that results in the formation of more complex molecules such as proteins and polysaccharides or it can undergo a series of reactions to be converted back to the original five-carbon compound to complete the cycle.

12. Cellular respiration results in the production of CO_2 and H_2O and in the release of energy from food to form ATP. In general, respiration consists of three sets of reactions: glycolysis, the Krebs cycle, and chemiosmotic phosphorylation.

13. The first stage in cellular respiration is called glycolysis. Glycolysis occurs in the cytoplasm of the cell. In glycolysis, glucose (a six-carbon molecule) is metabolized into two three-carbon molecules of pyruvate. In the process, two ATPs are used and four ATPs are synthesized by substrate-level phosphorylation, resulting in a net gain of two ATPs. In addition, NAD$^+$ is reduced to form NADH.

14. If oxygen is not available, the two pyruvate molecules undergo a process known as fermentation. There are two types of fermentation. In alcoholic fermentation, ethyl alcohol is the ultimate electron acceptor, NADH is oxidized to produce NAD, and CO_2 is produced. In lactate fermentation, lactate serves as the ultimate electron acceptor, oxidizing NADH to form NAD to be recycled to glycolysis.

15. If oxygen is available, pyruvate enters the mitochondria and is converted to acetyl CoA in the matrix. The acetyl CoA enters the Krebs cycle. In the Krebs cycle, there is a series of reactions in which the remains of the pyruvate molecules are rearranged and broken down. CO_2's are removed and electrons, with their associated H$^+$ (protons), are released and accepted by electron carrier molecules. Some energy is used to phosphorylate ADP to form ATP at the substrate level.

16. The electrons captured by NAD$^+$ and FAD during cellular respiration are passed on through the electron transport chain. The energy released is used to pump H$^+$ (protons) into the space between the inner and outer mitochondrial membranes, producing a chemiosmotic gradient. When the excess protons move through the ATP-synthesizing complex to return to the matrix, enzymes are used to add phosphates to ADP to form ATP. The protons that pass through these channels pick up electrons to form hydrogen and are combined with oxygen to form water.

17. From the complete aerobic respiration of one molecule of glucose, 36 to 38 molecules of ATP are formed. This ATP provides the chemical energy for life.

18. Many other foods (carbohydrates, fats, and proteins) can be metabolized to intermediates that can enter glycolysis or the Krebs cycle and be used as sources of energy or synthesized into other substances.

KEY TERMS

photosynthesis	chloroplast
light-dependent reaction	stroma
light-independent reaction	thylakoid
	chlorophyll
photon	electron transport chain

cyclic photophosphoryla-
tion
noncyclic photophosphor-
ylation
photosystems I and II
photolysis
Calvin-Benson cycle

cellular respiration
glycolysis
fermentation
anaerobic respiration
aerobic respiration
Krebs cycle

REVIEW QUESTIONS

True/False

1. The ongoing flow of energy through living systems is often described as occurring because of the reactions of photosynthesis and respiration.
   true   false   (page 145)

2. In photosynthesis, light energy is converted to chemical energy.
   true   false   (page 145)

3. Respiration is the process whereby the energy available in the chemical bonds of food is made available to sustain life.
   true   false   (page 145)

4. Respiration can only occur in the presence of oxygen.
   true   false   (page 153)

Fill in the Blank

1. Photosynthesis can be thought of as consisting of two phases, the _____ - _____ reactions and the _____ - _____ reactions.
   (page 146)

2. In the Calvin-Benson cycle, _____ is reduced to form food.
   (page 150)

3. Most of the ATP synthesized in respiration is due to _____ phosphorylation in the mitochondrion.
   (page 156)

4. Fermentation occurs in the _____ of oxygen.
   (page 154)

Multiple Choice

1. What would happen to life on earth if a mysterious disease evolved that destroyed chloroplast membranes?

   a. Photosynthesis in plants would cease.
   b. Living forms would begin to die.
   c. Life would continue as it is.
   d. a and b
   (page 150)

2. The major source of free oxygen on earth is due to:
   a. Glycolysis
   b. Krebs cycle
   c. Light-dependent reactions of photosynthesis
   d. Light-independent reactions of photosynthesis
   (page 146)

3. If an animal had abnormal mitochondrial membranes,
   a. It might be fatigued.
   b. It might be hyperactive.
   c. It might have vision problems.
   d. All of the above.
   (page 157)

4. The major event that occurs in the Krebs cycle is:
   a. Direct formation of ATP
   b. Removal of hydrogens to be used to establish a H^+ concentration gradient
   c. Formation of oxygen
   d. Regeneration of glucose
   (page 155)

Discussion

1. Discuss the interrelationship of photosynthesis and respiration.

2. Relate the structure of a chloroplast and mitochondrion to its role in ATP production.

3. Why are membranes important to energy reactions in living cells?

4. Discuss in a general sense the light-dependent and light-independent reactions.

5. What is respiration? Where does it occur?

6. Briefly discuss the importance of fermentation to organisms or cells.

7. Why do you think an understanding of photosynthesis and respiration is important to you?

Photosynthesis: Light for Life

Above: Within the green cells of this tree, light energy is converted to chemical energy through photosynthesis.

Every time we drink a glass of orange juice, eat an apple, or eat or drink anything else, we are consuming energy from sunlight. Unlike animals, plants (and some single-celled organisms) have the ability to trap the energy from light, convert it into chemical energy, and store this energy in food, through the process of photosynthesis. Photosynthesis is a group of reactions through which plants form the high-energy carbon compounds we call food.

All living things (with the exception of some microorganisms) are dependent on the products of photosynthesis for their energy, for the raw materials necessary for growth and repair, and for the oxygen necessary for life. **Autotrophs,** or self-feeders, can synthesize their own food from simple in-organic substances (Figure 8.1). Some autotrophs, like many of the bacteria, are called **chemosynthetic** because they use the energy from inorganic reactions rather than from light to power other reactions that result in the synthesis of needed food molecules. Photosynthetic organisms use the energy from light to synthesize food. To understand why some poets say all flesh is grass, let us discuss photosynthesis.

THE BASIC REQUIREMENTS OF PHOTOSYNTHESIS

■ Photosynthesis is the process by which plants convert the radiant energy of light into chemical energy, which is stored in food molecules. For photosynthesis to occur, there must be light, chlorophyll in a photosynthetic membrane, water, and carbon dioxide. The initial products are carbohydrate molecules and oxygen gas.

FIGURE 8.1 Autotrophs
Autotrophic organisms do not need to ingest food to live. Instead, they can make their own food. (a) Some, like many bacteria, can make their own food by using the energy from inorganic reactions, and they are called chemosynthetic. (b) Other autotrophs, such as the green plants, are called photosynthetic because they use the energy from light to make food.

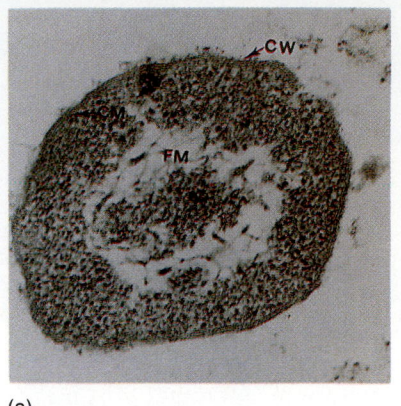

(a)

(b)

Biologists have known the basic formula for photosynthesis for some time. Even though continuing research reveals more of the complexities of the process, that basic formula has not changed. Here is the basic equation for photosynthesis:

$$CO_2 + H_2O \xrightarrow{\text{Light energy}} (CH_2O) + O_2 + H_2O$$

Carbon dioxide	Water	Chlorophyll and other pigments arranged in a photosynthetic membrane	Carbohydrate food unit	Oxygen Water gas

As noted in the chemical equation, the raw materials for photosynthesis are *carbon dioxide* (CO_2), *water* (H_2O), *light energy,* and *photosynthetic pigments.* (Pigments are the coloring substances in cells.)

The chlorophylls and other photosynthetic pigments are very specifically organized in membranes. In prokaryotic cells, photosynthetic pigments are found within the cell's photosynthetic membranes. In eukaryotic cells, photosynthetic pigments are organized within a membrane that is contained within a structure called the *chloroplast* (see Chapter 4). Numerous chloroplasts are found in the plant cells responsible for photosynthesis. Although the photosynthetic structures in prokaryotes and eukaryotes differ, the processes are similar. In this chapter, we focus on eukaryotes.

For photosynthesis to occur, all the basic materials must be present. Because of their importance in the reaction, let us first discuss light and the structure of the chloroplast.

Light

■ **The spectrum of light is made of light waves of various frequencies and discrete particles of energy called photons. The visible spectrum of light, especially the violet and red wavelengths, is used in photosynthesis.**

In Chapter 4 we discussed some of the characteristics of light. Now we need to look at this energy source in more detail. Scientists can describe light in terms of its wavelength or energy content, those discrete particles of light energy called photons (Figure 8.2). Let's first discuss wavelength, then photons.

The theory of light waves comes to us from physics. Physicists use models to explain light, and one model describes the **electromagnetic spectrum** as a continuous spectrum of radiation composed of varying wavelengths in which only a small portion is visible light. Wavelengths are the distance from the peak of one wave to the peak of the next wave.

Chapter 8 Photosynthesis: Light for Life

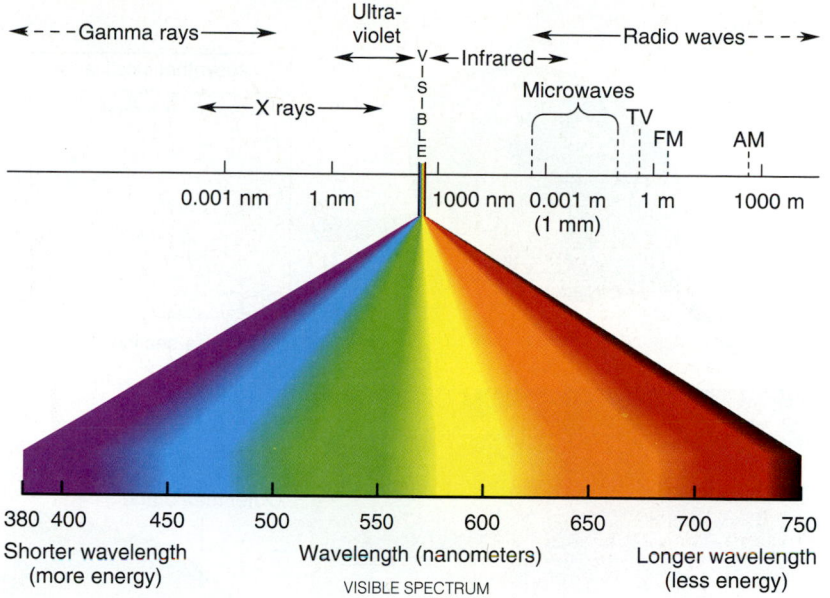

FIGURE 8.2   The Electromagnetic Spectrum
Visible light is a small part of the entire electromagnetic spectrum. The light used by plants in photosynthesis is, in turn, only a small part of the visible spectrum.

Wavelengths range from very short, high-frequency waves (such as gamma rays) to very long, low-frequency waves (radio waves). The amount of energy contained in any kind of light is inversely proportional to the wavelength and directly proportional to the frequency. This means that short waves, found at the violet end of the visible spectrum, have more energy than the long waves found at the red end of the spectrum. In fact, violet light has almost twice the energy of red light.

Albert Einstein helped support a second light model, called the *particle model*, which complements the wave model. According to the particle model, light is composed of particles called **photons.** The energy contained in a photon is inversely proportional to the wavelength — the shorter the wave, the more energy in the particles, and vice versa. Both the wavelength and photon models are used to explain the qualities of light.

Plants absorb only a portion of the electromagnetic spectrum for photosynthesis — those wavelengths from about 380 to 750 nm, which is almost the same visible spectrum that the human eye can detect. The primary photosynthetic pigments in chloroplasts are most responsive to the short light waves found at the violet-to-blue end of the spectrum, as well as to the longer wavelengths of orange to red.

However, photosynthesis depends on more than light energy levels. *Intensity,* or the amount of light shining on a given area for a specific length of time, also plays a part in photosynthesis. If the light intensity is decreased, photosynthesis slows, because the amount of ATP made is proportional to the number of photons that strike the chloroplasts. If the intensity is too high, photosynthesis ceases, because very high light energy destroys the photosynthetic abilities of the chloroplasts.

Photosynthetic Membranes

■ **The reactions in which light is trapped occur within membranes. In eukaryotic cells, the chloroplast, a double-membrane organelle, is the photosynthetic structure. Within the chloroplast are a dense liquid called the stroma and membranous disc-shaped structures called thylakoids, which contain protein and pigment molecules. Grana are stacklike arrangements of individual thylakoids.**

In photosynthetic prokaryotic cells like cyanobacteria, the photosynthetic apparatus is located on parts of the surface of the cell membrane or in subcellular membranous structures that are found free in the cell's cytoplasm. In eukaryotic plant cells the photosynthetic structure is the cellular organelle called the chloroplast.

The chloroplast is a double-membrane-bound organelle that contains green chlorophyll pigments and other pigments as well. Chloroplasts are abundant in green leaves — there can be as many as 500,000 chloroplasts per square millimeter of leaf surface (a millimeter, or mm, equals 1/1000 meter — about the thickness of a dime). As with other cellular structures, the structure of the chloroplast illustrates the complementarity of form and function. For photosynthesis to occur, the pigments are

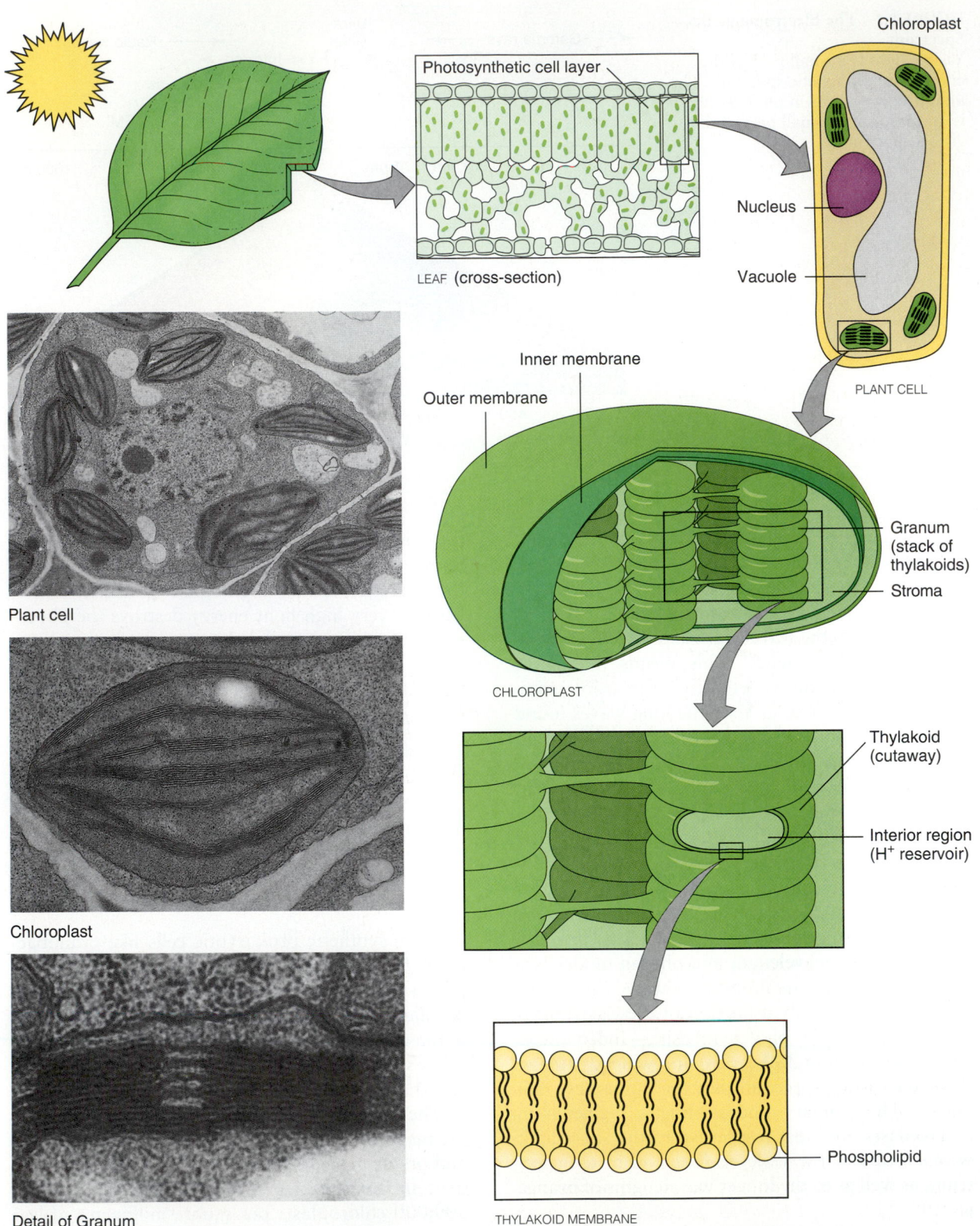

FIGURE 8.3 Chloroplasts
Within green plant cells are the chlorophyll-containing organelles called chloroplasts. The chloroplast is surrounded by a membrane and contains membranous structures in a liquid stroma. The membranes appear stacked like coins, and this stack is called a granum. Each disclike structure in the granum is called a thylakoid. A thylakoid is formed by a membrane that contains the pigments, proteins, and other structures required for photosynthesis. The inner portion of the thylakoid forms a compartment or region in which H⁺ accumulate for chemiosmotic phosphorylation of ADP to ATP.

Chapter 8   Photosynthesis: Light for Life

associated with membrane proteins that are involved in electron transfer reactions. All must be in the proper physical association. The internal structure of a chloroplast is illustrated in Figure 8.3.

The inside of the chloroplast contains a dense solution known as the **stroma,** and within the stroma are the organized membranous components so essential in photosynthesis. These membranes are continuous, but some of them form flattened disclike structures called **thylakoids.** Each thylakoid is a hollow disc whose membranes contain the chlorophylls and other pigments and structures necessary for photosynthesis, including electron acceptors. During photosysthesis, water is split and the hydrogen atoms are broken into their component parts, a proton (H^+) and an electron (e^-). The interior of the thylakoid forms a compartment called the *thylakoid region* or thylakoid space that provides a reservoir for hydrogen ions (H^+). The stacklike arrangement of the thylakoids (like a stack of coins) is called the **granum** (plural, *grana*). These structures are illustrated in Figure 8.3.

Chlorophyll

■ The primary photosynthesizing pigment in the chloroplast is chlorophyll, which has an absorption spectrum complemented by other pigments in the chloroplast. Light energy excites a pair of electrons in the chlorophyll, sending them to a higher energy level. In chloroplasts, as the electrons return to their normal level, the energy can be converted to useful chemical energy for photosynthesis.

Have you ever wondered why a strawberry is red or why plants are green? The pigments in a strawberry absorb every wavelength except red, which is reflected. Your eyes detect the reflected wavelength, so you perceive the strawberry as red.

The same principle applies to the appearance of leaves. Leaves look green because the major photosynthetic pigment in plants is **chlorophyll,** which, owing to its molecular structure, absorbs most other wavelengths except green and yellow, which are reflected.

There are two important chlorophyll molecules in photosynthesizing organisms: (1) chlorophyll *a,* which is found in all photosynthesizing organisms, and (2) chlorophyll *b,* which is found in the more advanced land plants (Figure 8.4). Both contain carbon, hydrogen, oxygen, nitrogen, and magnesium (Mg). Each type of chlorophyll absorbs different wavelengths of light.

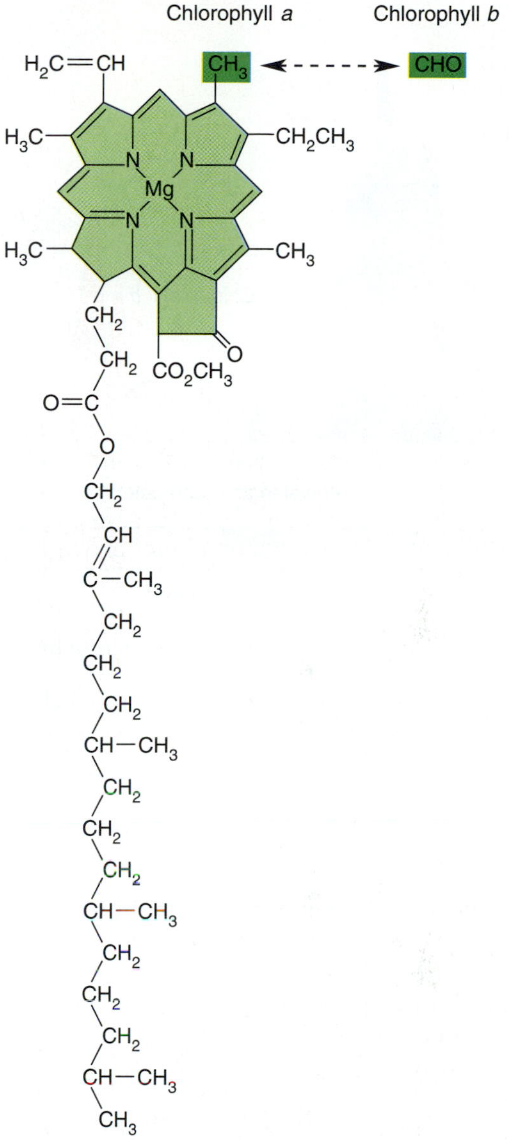

Chlorophyll pigments are not the only pigments to absorb light in photosynthesis. Other accessory

FIGURE 8.4   Chlorophyll *a* and *b*
Chlorophyll *a* and *b* are very similar in structure. Chlorophyll *a* has a CH_3 group and chlorophyll *b* has a CHO group, two very similar pigments that play a vital role in the conversion of sunlight to chemical energy. When light energy excites a pair of electrons in chlorophyll *a* or chlorophyll *b,* a sequence of events begins that eventually leads to the production of ATP.

pigment molecules, such as the *carotenoids* (red, orange, or yellow pigments), are also involved. In the fall, when leaves cease to produce chlorophyll, their color changes from green to yellow, orange, or red. That is because you are then seeing the color of the accessory pigments, which in the summer were masked by the color of chlorophyll.

Having a variety of pigments gives plants a distinct advantage. Each individual pigment absorbs light waves in a definite, measurable pattern, called

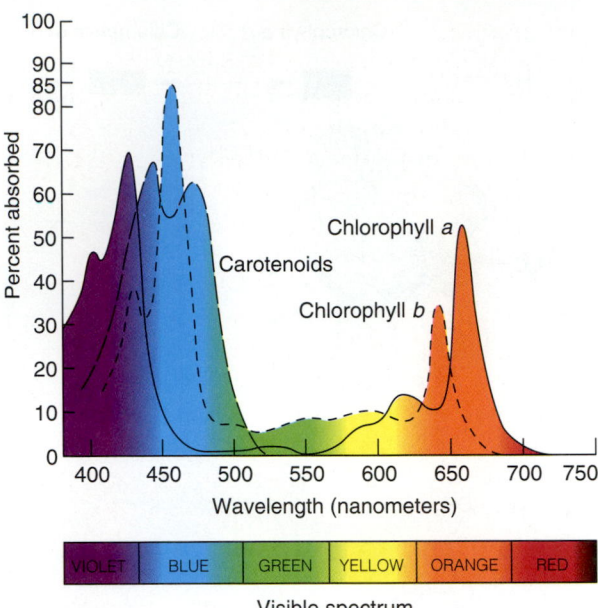

FIGURE 8.5 **Absorption Spectrum for Chlorophyll and Other Pigments**
Chlorophyll and other plant pigments absorb light primarily in the violet, blue, and orange-red portions of the visible spectrum. Green, which is usually the color we see reflected from leaves, is not absorbed.

the pigment's **absorption spectrum.** Chlorophyll absorbs light energy primarily in the red and the violet-blue wavelengths of the visible spectrum, producing two peaks of absorption. The other pigments supplement the chlorophyll by absorbing

those waves that chlorophyll does not. The result is a relatively efficient "harvesting" of light energy (Figure 8.5).

When a photon of visible light is absorbed by the special light-trapping chlorophyll molecules, the chlorophyll molecules become *excited*, or energized. The energy of the light excites one electron at a time, in sequence, until a pair of electrons in one orbital of the chlorophyll molecule is bumped to a higher energy level. In about one-billionth of a second, *if* there is no electron acceptor molecule to capture them, the pair of electrons fall back to their original energy level (Figure 8.6).

Test-tube experiments have shown that the energy that started the process may be reemitted in the form of heat and light. For example, if you shine a light on a test tube filled with chlorophyll, electrons in the chlorophyll molecules become excited. When the electrons fall back to their original energy level, the energy that excited them is reemitted as light and heat energy, not as chemical energy, such as that stored in ATP. As a result, the molecules *fluoresce*, emitting photons of light and heat.

In living plants, however, chlorophyll converts light energy to chemical energy. What makes the difference? Over a long period of time, the photosynthetic membrane evolved as the mechanism to process light energy. Chlorophyll can convert light energy to chemical energy only when it is associated with certain electron acceptors, proteins, enzymes, and other molecules organized within a membrane, which catalyze the transfer of the excited electrons along an electron transport chain.

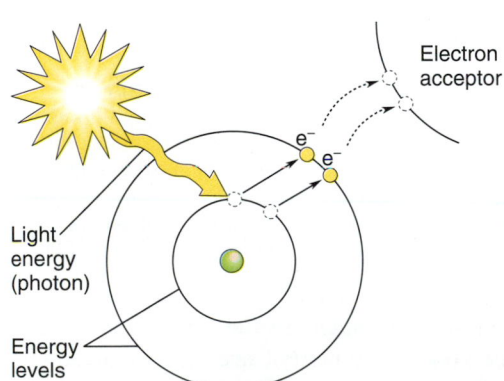

(a) Energy is "captured" when excited electrons are transferred to electron acceptors.

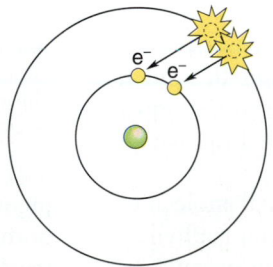

(b) Energy is emitted as light and heat when no acceptors are available—as in this fluorescent solution of chlorophyll.

FIGURE 8.6 **Chlorophyll and Electrons**
(a) If excited electrons in a chlorophyll molecule are not transferred to an electron acceptor, they release their energy as light and heat (b), as illustrated by the fluorescence of chlorophyll molecules contained in this flask.

Carbon Dioxide

■ **Carbon dioxide serves as the source of carbon for the carbon skeleton of carbohydrate molecules synthesized by photosynthesis. The atmosphere of earth today consists of about three to four molecules of CO_2 per 10,000 molecules of other gases. This concentration is sufficient for photosynthesis to take place.**

Carbon dioxide (CO_2) serves as the source of carbon in photosynthesis. Remember that in Chapter 6 we noted that reduction occurs when a hydrogen ion is added to a molecule. That is the case in photosynthesis. CO_2 is reduced by the addition of hydrogen ions (remember, a hydrogen ion is a proton). The carbon of the molecule becomes part of a carbon skeleton of a carbohydrate food unit. Energy is stored, and the hydrogens bond to the carbon skeleton.

Water

■ **Water is a source of hydrogen needed in photosynthesis to reduce carbon dioxide. Water also is the source of the oxygen by-product of photosynthesis.**

Water is essential for life's activities, but why is it so special in photosynthesis? Why do you have to water the plants in your dormitory room or at home? Well, combine those questions with this one: Where do the hydrogen ions come from to reduce carbon dioxide? There is only one source for hydrogen atoms — water. What is more, where do the oxygen atoms come from? Again, the source is water.

Now look back at the overall equation for photosynthesis. Oxygen (O_2) is one of the products of photosynthesis, but where does that oxygen come from — the carbon dioxide (CO_2) or the water (H_2O)? For a long time, many biologists thought that carbon dioxide was the source of O_2 gas. That assumption changed because of work done in the early 1930s by C. B. Van Niel, a graduate student at Stanford University who was studying photosynthesis in purple sulfur bacteria.

Purple sulfur bacteria use hydrogen sulfide (H_2S) instead of water for their hydrogen source. The photosynthesizing process for the bacteria is illustrated in this way:

$$CO_2 + 2\,H_2S \longrightarrow (CH_2O) + H_2O + 2\,S$$

| Carbon dioxide | Hydrogen sulfide | Light | Carbohy-drate | Water | Sulfur globules |

Van Niel's most important observation was that the sulfur bacteria produced sulfur globules rather than oxygen. There were two ways to explain the bacteria's strange reaction. Either (1) these organisms were examples of the evolution of an entirely different photosynthetic process, or (2) the existing explanation for photosynthesis was inaccurate. The second alternative seemed more likely, and other experimental evidence supported it. Further research determined that when water was prepared with ^{18}O, an oxygen isotope that is heavier than the more abundant ^{16}O, and that water was then used in photosynthesis experiments, the oxygen given off by plants was indeed the same ^{18}O. Therefore, the oxygen by-product had to originate in the water molecules, not in carbon dioxide molecules.

As a result of the photosynthetic process, plants give off oxygen into the atmosphere. So if you talk to your plants, as I do, the next time you water your philodendron, instead of saying, "Here is your drink," tell it, "Here are your hydrogen sources, so you can reduce your carbon dioxide to make your food and to produce oxygen for me."

PHOTOSYNTHETIC REACTIONS: AN OVERVIEW

■ **Photosynthesis is a highly organized process that is divided into two major stages: light-dependent reactions and light-independent reactions. Light-dependent reactions produce ATP and NADPH and liberate O_2. The NADPH and ATP produced, along with carbon dioxide, are used to synthesize carbohydrates and other high-energy carbon compounds in the light-independent reactions.**

Now that we know the structures and materials required for photosynthesis, let us discuss the reactions that take place. The chemical equation for photosynthesis that we looked at earlier is a generalized expression of the processes involved in photosynthesis. In actuality, photosynthesis is much more complicated. You cannot just combine some carbon dioxide and chlorophyll in a test tube, shine a light on it, and make food molecules. Photosynthesis must take place in an ordered series of steps and in an organized set of molecules in order to produce basic carbohydrate food units.

The need for order is apparent in many chemical reactions. For instance, a car battery is made up of a specific set of substances, organized in a specific manner. If you took those substances, ground them all up, and combined them, the mixture would not store electrical energy. The various components of the battery must be structured and intact before the battery can perform its function. In photosynthetic organisms, the same is true. It is the manner in which photosynthetic membranes are organized that provides the structure necessary for the conversion of light energy to chemical energy.

To expand our general equation (and to make the process easier to explain), it is best to divide photosynthesis into two interrelated stages: a *light-dependent stage* and a *light-independent stage* (Figure 8.7). In the **light-dependent reactions** (once called the *light reactions*), light energy strikes a chlorophyll molecule and other pigment molecules embedded in the photosynthetic membranes of the thylakoids of the chloroplast. There the light is converted to chemical energy, and during the conversion process the energy is used to split water into its component parts, to reduce the electron carrier molecule $NADP^+$ to NADPH, and to synthesize ATP.

Both ATP and NADPH then move to the **light-independent reactions** (once called by the less accurate name *dark reactions*). Here, the products of the light-dependent reactions are used to reduce CO_2 in order to form high-energy carbon compounds that we call food, such as glucose. In the course of the light-independent reactions, hydrogen is removed from NADPH to produce $NADP^+$, which then cycles back to the light-dependent reactions. This interchange is only one of the ways in which the light-dependent and light-independent reactions are interrelated.

With this introduction in mind let us now discuss the light-dependent and light-independent reactions in more detail.

THE LIGHT-DEPENDENT REACTIONS

■ **Light energy is absorbed by absorption antennas, which transfer the energy to one of two reaction centers. Reaction centers that contain chlorophyll *a* P700 molecules are called photosystem I centers. Reaction centers that contain chlorophyll *a* P680 molecules are called photosystem II centers.**

Light absorption takes place in a collection of pigment molecules embedded in the thylakoid membrane. These light-harvesting molecules are often referred to as **light-harvesting antennas,** since they act like little satellite television antennas. These antennas consist of clusters of about 200 to 300 pigment molecules: chlorophyll *a*, chlorophyll *b*, and carotenoids (which are orange and yellow pigments) bound to special protein molecules. Each type of pigment is responsible for the absorption of a specific set of light-wave frequencies.

Reaction Centers

Light-harvesting antennas are differentiated by the type of chlorophyll *a* that is found in what is called the antenna's reaction center (Figure 8.8). As light strikes the antenna molecules, the molecules transmit the light's energy to the reaction center, which consists of a chlorophyll *a* molecule and a protein.

In one reaction center, the chlorophyll *a* molecule absorbs wavelengths of light up to 700 nm, and as a result it is called **pigment 700** or **P700**. The

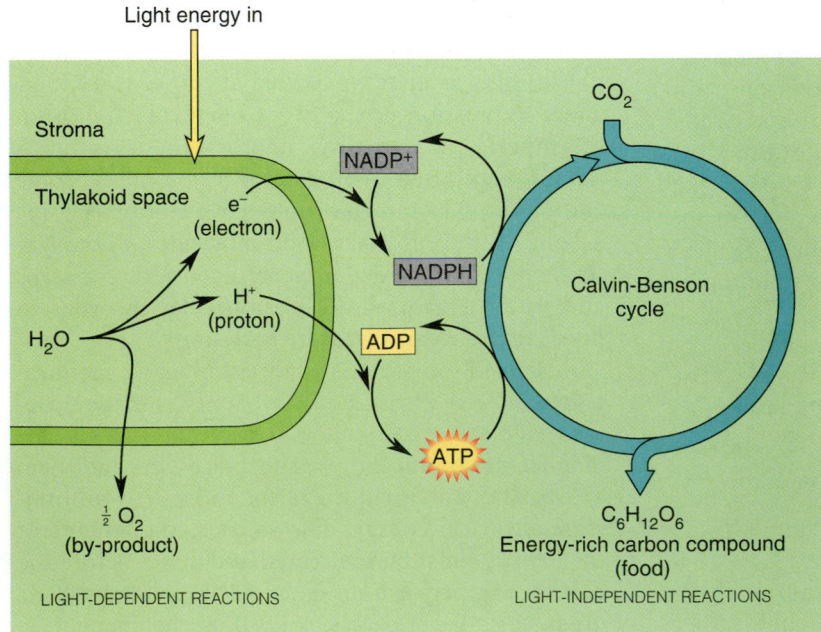

FIGURE 8.7 The Two Stages of Photosynthesis
In the light-dependent stage of photosynthesis, ATP and NADPH are produced, which are then used for energy and reducing power in the light-independent stage, in which molecules of carbon dioxide are reduced and become food molecules. When living cells subsequently oxidize these food molecules, the stored energy is released and available to provide the energy required for life. The light-dependent stage occurs within the thylakoid, and the light-independent phase occurs in the stroma of the chloroplast.

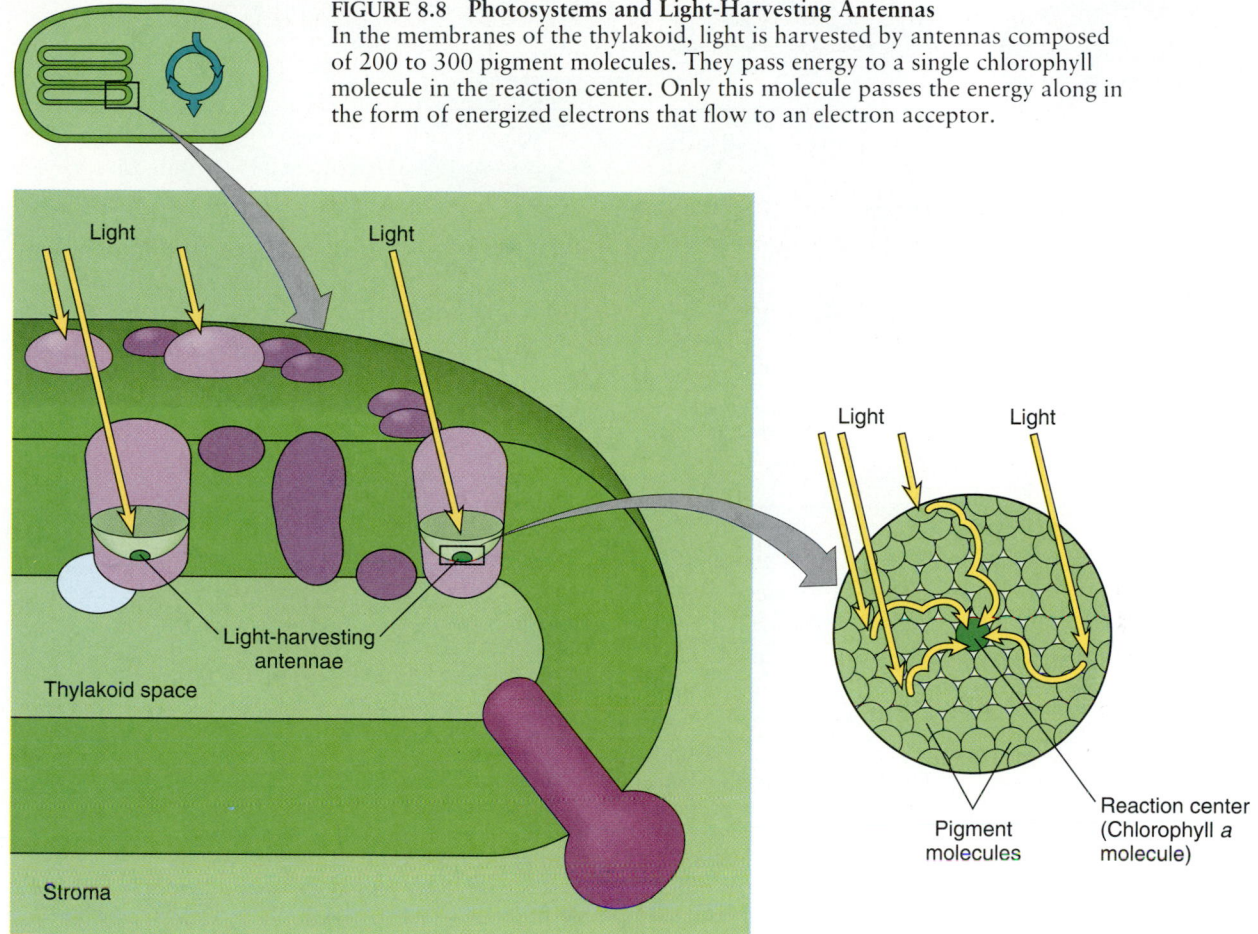

FIGURE 8.8 Photosystems and Light-Harvesting Antennas
In the membranes of the thylakoid, light is harvested by antennas composed of 200 to 300 pigment molecules. They pass energy to a single chlorophyll molecule in the reaction center. Only this molecule passes the energy along in the form of energized electrons that flow to an electron acceptor.

Light

Light

Light

Light

Light-harvesting antennae

Thylakoid space

Stroma

Pigment molecules

Reaction center (Chlorophyll *a* molecule)

(a) Cross-section of thylakoid membrane

(b) Light-harvesting antenna (top view)

entire complex of the light-receiving system is called **photosystem I.**

The second type of reaction center contains a chlorophyll *a* that absorbs a slightly shorter wavelength. The wavelength is 680 nm, and so this particular molecule is called **pigment 680** or **P680.** This light-receiving system is called **photosystem II.**

The Electron Transport Chain

■ **When the reaction center collects enough radiant energy, electrons become energized and are raised to a higher energy level. Electron carrier molecules pass the energized electrons along an electron transport chain, where the acquired energy from light is harvested in order to phosphorylate ADP to form ATP.**

The first step of the light-dependent reaction is the capture of light energy and the transfer of an activated electron to adjacent electron acceptor molecules in the thylakoid. Once the antenna system

absorbs light within its chlorophyll-containing reaction center, some chlorophylls become excited, and that state in turn sends one and then another of the molecules' electrons to a higher energy level. The excited electron can then be transferred to other electron acceptor molecules embedded in the thylakoid membrane.

In both photosystems I and II, electrons from the reaction center are accepted by the first electron acceptors, and then the electrons pass along an electron transport chain in the membrane of the thylakoid. But the acceptor does not pass on the proton. Instead, the proton is transported into the interior region of the thylakoid. This results in a higher concentration of H^+ ions in the inner compartment of the thylakoid than is found in the stroma, and the difference in the concentration of hydrogen ions is instrumental in ATP production (Figure 8.9). Protons are pumped through the channels and through protein complexes called the **CF_1 complex.** We discuss ATP production later, at the end of our discussion of the light-dependent reactions.

The Light-Dependent Reactions

171

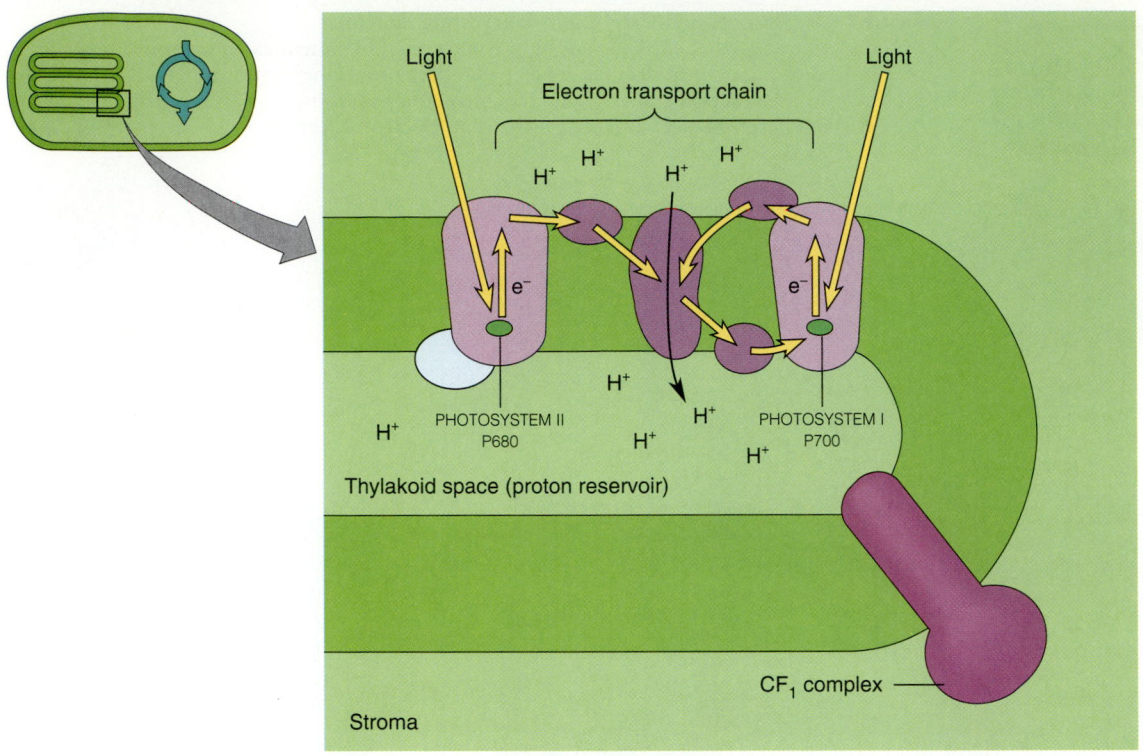

FIGURE 8.9   Electron Transport
As energized electrons pass through the electron transport chain, they become associated with protons. Energy from the electrons is used to pump these protons into the interior compartment of the thylakoid, where they are concentrated. Just as in mitochondria, the resulting proton gradient provides the energy to form ATP.

Ultimately the electrons have released almost all of the energy that the antennas harvested from sunlight and are at the end of the electron transport chain. What happens to these electrons once they leave their final electron acceptor molecule? They usually can move through one of two electron transfer pathways: *cyclic photophosphorylation* and *noncyclic photophosphorylation*.

Cyclic Photophosphorylation

■ **In cyclic photophosphorylation, which involves electrons from photosystem I reaction centers, excited electrons move down the electron transport chain, then recycle to the P700 reaction center. ATP is formed from the harvested energy.**

The simplest pathway, which involves only photosystem I, is called **cyclic photophosphorylation.** This pathway is called *cyclic* because the electron returns to the same type of chlorophyll molecule from which it came. Because today's primitive photosynthetic bacteria express only cyclic photophosphorylation, botanists think that this system is probably similar to the reactions of the first primitive photosynthetic organisms. In fact, during times of stress (for instance, when there is not enough rain), photosystem I is the primary photosynthetic process in seed plants. However, because the process usually produces only one ATP for every two activated electrons, it is not very efficient. In addition, photosystem I does not result in the formation of high-energy carbon compounds such as glucose.

The processes of cyclic photophosphorylation are illustrated in Figure 8.10, and you may find it helpful to glance at the figure as you read through the details of the discussion. (As you read along, keep in mind that a hydrogen atom only has one proton and one electron. As a result, when hydrogen becomes an ion, it loses an electron and only the proton remains.)

When light is absorbed by the P700 molecules in a photosystem I reaction center, the energized electron from the chlorophyll *a* reaction center is bumped to a higher energy level and is then passed along the electron carrier molecules of the electron transport chain within the thylakoid membrane.

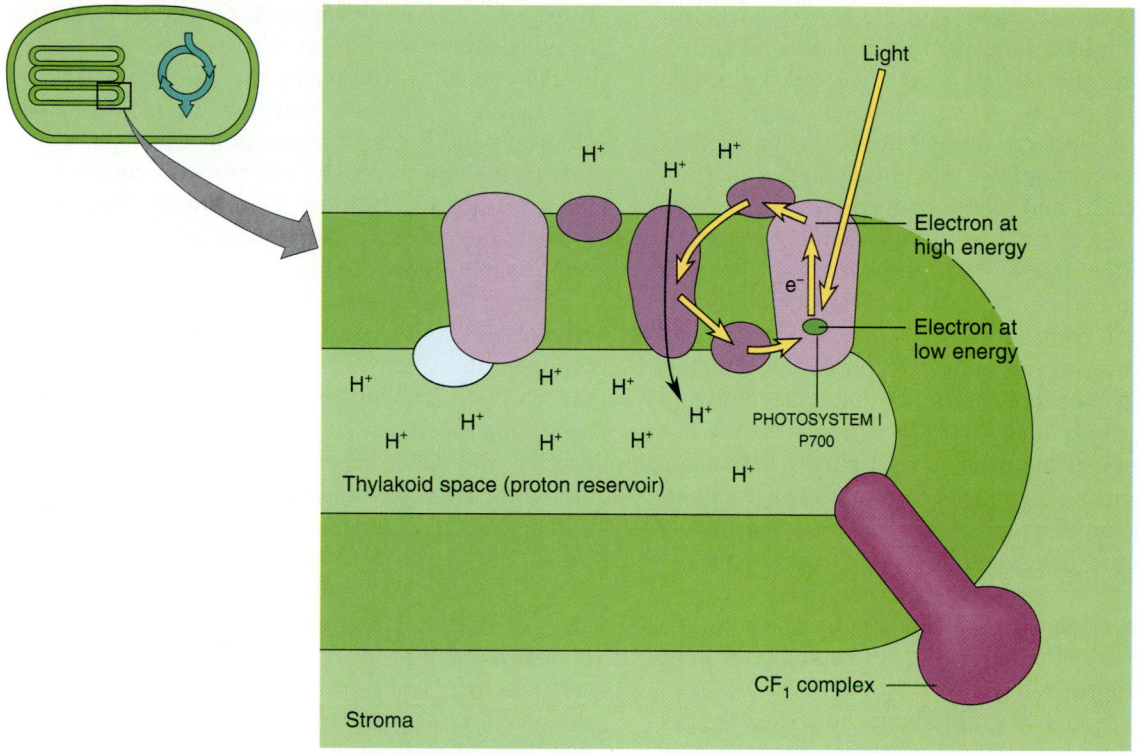

FIGURE 8.10   Cyclic Photophosphorylation
The energized electrons that have left the chlorophyll molecules in photosystem I cycle back to the same type of chlorophyll from which they left. In the process, protons are pumped into the interior compartment of the thylakoid to maintain the H⁺ gradient required for the chemiosmotic phosphorylation of ADP to ATP.

This causes the transport of protons from the stroma to the inner compartment of the thylakoid. Stripped of its excess energy, the electron then returns to fill the hole in the same P700 molecule from which it came. The cycle can then occur again. In most cases, during the cycling of photosystem I, one ATP is formed for every two electrons that cycle through the electron transport chain.

Noncyclic Photophosphorylation

■ Noncyclic photophosphorylation involves reactions in both photosystems I and II. In photosystem II, light excites the electrons of the P680 reaction center and splits water through photolysis. The electrons from the dissociated water fill the P680 molecule, the protons remain in the stroma, and the oxygen is eventually passed to the air. The energized electrons pass through the electron transport chain, causing transport of protons into the inner compartment of the thylakoid, and then fill the P700 molecule of photosystem I. The excited electrons from photosystem I pass through

a series of electron transport molecules and, with hydrogen ions, finally join NADP⁺ to form NADPH. NADPH is transferred to the light-independent reactions.

Over time, a second photosynthetic system evolved. This system is called **noncyclic photophosphorylation,** a combination of the mechanisms of photosystem I and photosystem II that proved more efficient than cyclic photophosphorylation. Noncyclic photophosphorylation (1) results in the formation of an H⁺ gradient that provides energy for ATP synthesis, (2) splits hydrogens from water to use to reduce NADP⁺ to NADPH, and (3) passes NADPH to the light-independent reactions to serve as the hydrogen source to reduce CO_2. Every two electrons entering this pathway yield energy that is usually sufficient to produce two ATPs, form one NADPH, and split water to oxygen gas, hydrogen ions, and electrons.

This system is called *noncyclic* because the energized electrons from the P680 chlorophyll molecule of the photosystem II reaction center do not return to the molecule from which they came. Instead, the electrons eventually feed into the electron-deficient P700 chlorophyll molecule in the photosystem I reaction center. For clarity's sake, we will begin with photosystem II.

Before we begin to discuss noncyclic photophosphorylation in detail, we need to set forth a caution. Photosystems I and II work together, much like an assembly line in an automobile factory. Materials enter the line at one end, and a finished product leaves at the other. Activity on the assembly line is constant and continuous.

The processes of noncyclic photophosphorylation exhibit the same continuity that you would find on the assembly line. All the reactions that we discuss as steps in the process are actually happening *simultaneously*. Light energy is activating a light-harvesting antenna at the same time that water molecules are being split. NADP$^+$ is receiving hydrogens. The energy from the energized electron is being used to generate an H$^+$ gradient, which will result in energy that is available to add a phosphate to ADP to form ATP.

So even though we talk about the first or the last step in photosystem II, the most significant characteristic of noncyclic photophosphorylation is its continuity. Presenting the process as a series of steps is a matter of convenience, a way of discerning the interrelationships in what is actually a highly interwoven set of reactions that began somewhere in the early moments of the life of an organism. Indeed, there are still details that biologists do not understand, partly because of the amazing complexity of the processes of photosynthesis.

The Activities of Photosystem II. Light is used for *two* purposes in photosystem II. In one, light activates an electron from the P680 chlorophyll molecule of the reaction center. In the second, a process called **photolysis,** the energy from light is indirectly used to split a water molecule into its component parts: two electrons, two protons, and one atom of oxygen. As a result there are four elements that are instrumental in photosystem II: (1) energized electrons from the P680 molecules, (2) electrons from photolysis, (3) protons from photolysis, and (4) oxygen from photolysis. What happens to all of them? The reactions in photosystem II are illustrated in Figure 8.11. Each element plays a particular role.

The energized electrons are carried through the electron transport chain by electron carrier molecules. The transport chain functions just like the one in photosystem I, passing the electrons along an electrical gradient until they have lost most of their acquired energy. The processes of the electron transport chain establish a proton gradient that will provide energy for ATP synthesis.

The electrons released by the photolysis of water are used to fill the hole that now exists in the P680 chlorophyll. Since energization and photolysis occurred simultaneously, the reaction center is capable of absorbing more light energy almost immediately, thus maintaining the continuity of the photosystem II processing.

The protons released by the photolysis of water are now free in the inner compartment. Some membrane-bound electron carriers accept protons from the stroma of the chloroplast, then pump those protons into the inner compartment, bolstering even further the electrochemical gradient that exists across the thylakoid membrane.

The oxygen molecule released by photolysis passes out of the thylakoid, into the cell, and eventually into the atmosphere. That means that oxygen, so essential to most living things, including plants, is a by-product of the photosynthetic process. To help you keep track of the paths taken by the products of photolysis, here is an equation you may find helpful:

$$H_2O \rightarrow 2\ e^- \qquad +\ 2\ H^+ \qquad +\ \tfrac{1}{2}\ O_2$$

| To P680 chlorophyll *a* | To inner compartment of thylakoid | To atmosphere |

Photosystem I Reactions. To understand what happens next, we first need to attend to the reactions that are occurring (again simultaneously) in photosystem I. There the P700 chlorophyll molecule in the reaction has been absorbing light. The excited electron passes to an electron acceptor molecule and moves through the electron transport chain, losing energy at each step in the chain.

However, the electron does not return to fill the hole in the P700 molecule. Instead, two electrons are combined with the electron transport molecule NADP$^+$ to reduce NADP$^+$ to NADPH. Then NADPH is cycled to the *light-independent reactions,* where ultimately it will participate in the formation of food molecules.

However, there is still a hole in P700 that must be filled. Are there any spare electrons floating around (so to speak) that can fulfill this function? Look back at the processes of photosystem II and you will find that the energized electrons that passed through the electron transport chain have so far not been assigned a role. These electrons move to fill the P700 chlorophyll molecule. The interaction between photosystems I and II in noncyclic photophosphorylation is illustrated in Figure 8.12.

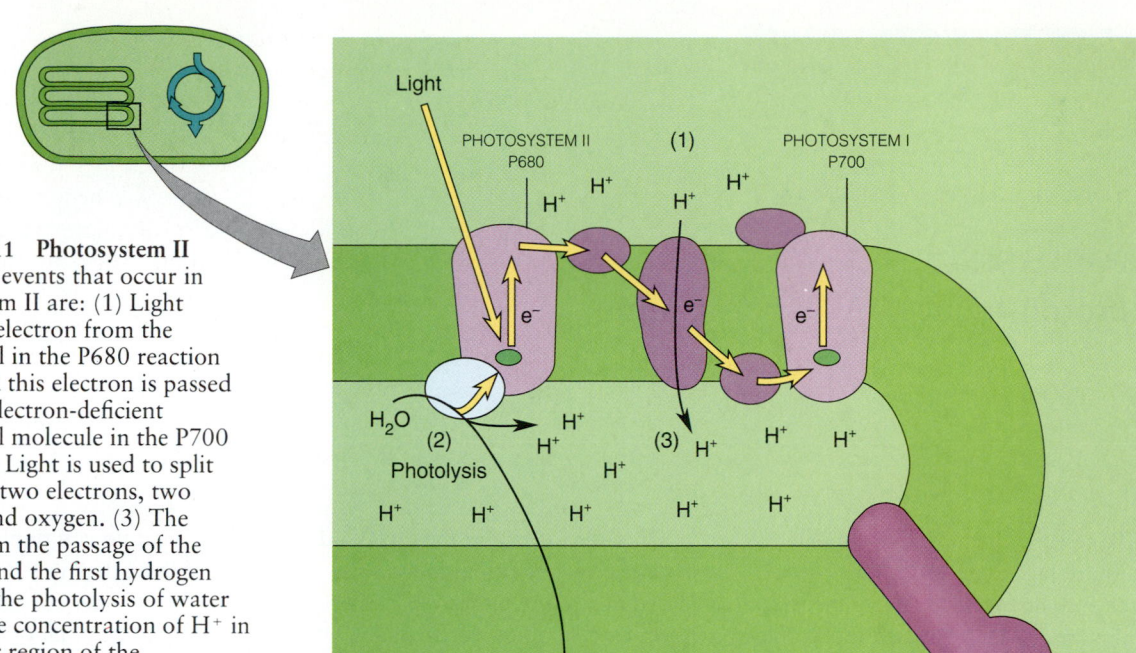

FIGURE 8.11   Photosystem II
The major events that occur in photosystem II are: (1) Light excites an electron from the chlorophyll in the P680 reaction center, and this electron is passed on to the electron-deficient chlorophyll molecule in the P700 system. (2) Light is used to split water into two electrons, two protons, and oxygen. (3) The energy from the passage of the electrons and the first hydrogen ions from the photolysis of water increase the concentration of H^+ in the interior region of the thylakoid, establishing the chemiosmotic gradient necessary for the phosphorylation of ADP to ATP.

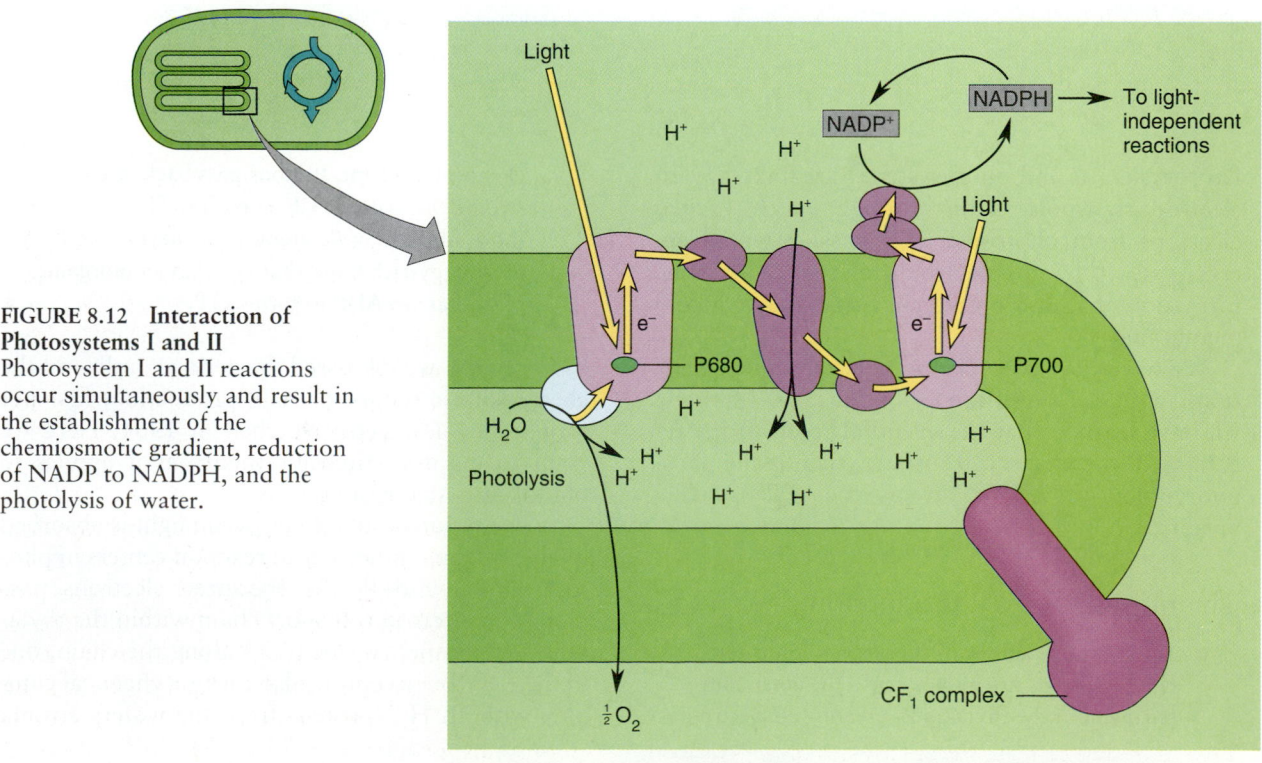

FIGURE 8.12   Interaction of Photosystems I and II
Photosystem I and II reactions occur simultaneously and result in the establishment of the chemiosmotic gradient, reduction of NADP to NADPH, and the photolysis of water.

Proton-Induced ATP Formation

To test the chemiosmotic model, André Jagendorf and Ernest Uribe devised an ingenious experiment (Figure A). They isolated intact chloroplasts from spinach and placed them in the dark so that there would be no ATP synthesis due to the light-dependent reactions. The chloroplasts were then placed in a solution of low pH (remember, a low pH indicates a high concentration of H^+). Through the process of diffusion, the stroma and then the inner compartment of the thylakoid became acidic.

Once the interior of the thylakoid became acidic, the chloroplast was placed in a solution that was more neutral, or, in other words, had a higher pH. Now the situation was such that the interior of the thylakoid was more acidic than the area outside of the thylakoid. This situation created a pH gradient that mimicked the condition set up during the light-dependent reactions, when actions of the electron transport chain pumped protons into the thylakoid compartment.

And what do you think was the result of the differences between the pH inside the inner compartment of the thylakoid and that outside of the thylakoid? As you might have guessed, even though light was unavailable, the pH gradient immediately resulted in ATP synthesis. Jagendorf and Uribe's experiment demonstrated that the chemiosmotic model did explain ATP synthesis.

Photosystem II and photosystem I are linked by an electron transport chain through which excited electrons from photosystem II pass. Potential energy is stored by this electron transport chain in the formation of a proton gradient across the thylakoid membrane.

We have now covered the light-dependent reactions, with the exception of one matter. All along we have learned that the eventual result of the reactions of the electron transport chain is the phosphorylation of ATP. How is the ATP actually produced?

Chemiosmotic ATP Production

■ **During the reactions of the electron transport chain, protons are pumped into the inner compartment of the thylakoids, creating an electro-chemical gradient. Protons pass back into the stroma through the CF$_1$ complex. The energy of the gradient is used, along with enzymes, to form the energy-rich bond that attaches an inorganic phosphate to ADP to form ATP.**

In Chapter 6 we discussed the general outline of the chemiosmotic theory of ATP production, first put forward by Dr. Peter Mitchell in 1961. Here we want to discuss Mitchell's theory as it applies to photosynthesizing organisms.

As we pointed out earlier, when light is absorbed by the antenna molecules in reaction centers in photosystems I and II, the energized electrons pass along the electron transport chain within the thylakoid membrane. At one point along the chain, one of the electron acceptors also forms a chemical complex with an H^+ (proton) from the watery stroma

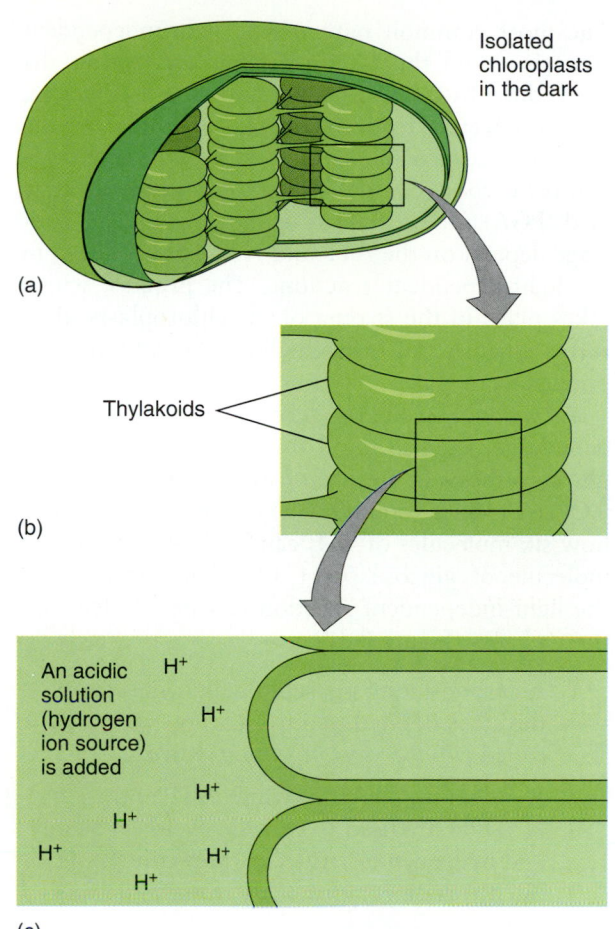

(a)

(b)

Thylakoids

(c)

An acidic solution (hydrogen ion source) is added

H⁺ H⁺ H⁺ H⁺ H⁺ H⁺ H⁺

Isolated chloroplasts in the dark

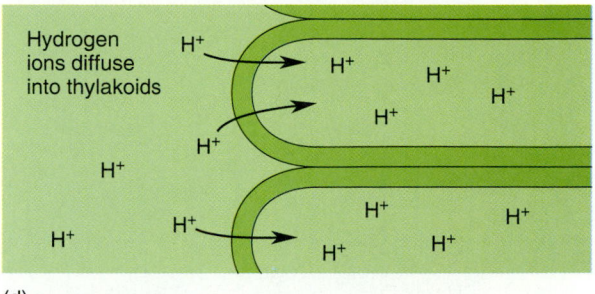

(d)

Hydrogen ions diffuse into thylakoids

H⁺ H⁺ H⁺ H⁺ H⁺ H⁺ H⁺ H⁺ H⁺ H⁺ H⁺ H⁺

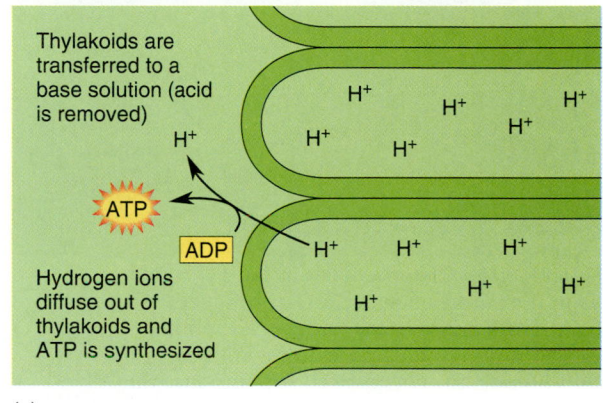

(e)

Thylakoids are transferred to a base solution (acid is removed)

ATP
ADP

Hydrogen ions diffuse out of thylakoids and ATP is synthesized

H⁺ H⁺ H⁺ H⁺ H⁺ H⁺ H⁺ H⁺ H⁺ H⁺ H⁺ H⁺ H⁺ H⁺

FIGURE A
André Jagendorf and Ernest Uribe devised an experiment that demonstrated that intact, isolated chloroplasts could produce ATP in the dark if protons were allowed to diffuse into the thylakoid inner compartment.

region, and the proton is transported into the inner compartment of the thylakoid (see Focus 8.1).

As these events are repeated thousands of times and H⁺ continues to be pumped into the inner compartment of the thylakoid, additional H⁺ are also generated in the thylakoids by photolysis. The growing concentration of H⁺ establishes a highly acidic region having a pH level of about 4. On the other hand, on the outside of the thylakoid membrane, in the stroma, there is a greater concentration of OH⁻ (hydroxide ion), which is basic rather than acidic. The stroma has a pH of about 7.5, which means there is about a 5000-fold difference in H⁺ concentration, a *chemical concentration gradient*, between the two regions.

At the same time, since the hydrogen ions have a positive charge and hydroxide ions have a negative charge, an *electrical gradient* also exists. The

stroma is electronegative in comparison with the inner compartment of the thylakoid. The two gradients together make an *electrochemical gradient*. The differences between the two regions are great enough that they constitute a substantial store of potential energy, which sets the stage for chemiosmosis.

Because of the electrochemical gradient, free energy is available for the synthesis of ATP, which takes place in the CF₁ complex, structures that stud the outside of the thylakoid (Figure 8.13). A CF₁ complex consists of two parts: a *protein channel*, which connects the inner compartment with the stroma, and a *spherical head*, which contains the enzymes that are involved in adding a phosphate to ADP.

As the protons move through the protein channel in the CF₁ complex, they combine with OH⁻ to

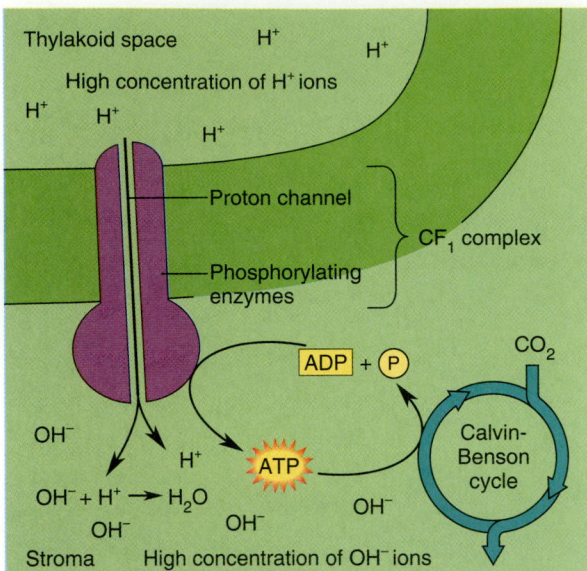

FIGURE 8.13 Chemiosmosis in the Thylakoid
As in the mitochondrion, ATP is produced as protons flow across a membrane from an area of higher concentration to an area of lower concentration. In chloroplasts, the energy needed to produce the concentration gradient comes from sunlight. In mitochondria, the energy comes from oxidation of food molecules.

form water. The energy released is used to attach an inorganic phosphate to ADP to form ATP, by means of the phosphorylating enzymes in the head of the particle.

THE LIGHT-INDEPENDENT REACTIONS

The light-independent reactions take the products of the light-dependent reactions and use them to form the high-energy carbon compounds that are the ultimate product of photosynthesis. The light-independent reactions can follow one of two paths. Most photosynthesizing organisms follow the Calvin-Benson or C₃ cycle; others, particularly those that exist in drought conditions, follow the C₄ cycle.

The Calvin-Benson Cycle

■ **One type of light-independent stage reaction is called the Calvin-Benson, or C₃ cycle. The three steps of the process are carbon fixing, reduction, and regeneration.**

The most common path of the light-independent stage is called the **Calvin-Benson cycle,** after the work done by Melvin Calvin and Andrew Benson, who discovered its mechanisms (see Profile), or the **C₃ cycle,** because an early product of the process is a three-carbon compound called phosphoglyceric acid (PGA). The reactions of the light-independent stage depend on the ATP and NADPH produced in the light-dependent reactions. The process, which takes place in the stroma of the chloroplasts, then forms glucose and other high-energy carbon compounds from CO_2.

As we analyze the reactions of the light-independent stage, keep in mind that the overall result of photosynthesis is the use of light energy to reduce CO_2 to produce food. Our objective now is to learn how six molecules of CO_2 can be used to form one molecule of glucose ($C_6H_{12}O_6$). For convenience, the light-independent reaction can be divided into three phases (Figure 8.14):

1. A carbon-fixing phase, in which CO_2 from the air is bonded to a five-carbon sugar, called *ribulose bisphosphate* (RuBP).

2. A reduction phase, in which ATP and NADPH are used to provide chemical energy and hydrogen to reduce carbon-containing intermediate compounds to food molecules.

3. A regeneration phase, in which RuBP is synthesized to cycle around again and bond with more CO_2.

Carbon dioxide fixation occurs when a molecule of CO_2 combines with the five-carbon sugar, ribulose bisphosphate (RuBP). The combination results in a very unstable six-carbon sugar, which reacts with water and immediately breaks into two molecules of the three-carbon compound called *phosphoglyceric acid* (PGA). For every 6 molecules of CO_2 and RuBP, 12 molecules of PGA are produced.

The reduction phase combines the 12 PGA molecules, 12 ATP molecules, and the hydrogen atoms from 12 NADPH (produced in the reactions of the light-dependent phase), to reduce the PGA molecules. These molecules are reduced and converted to molecules of *glyceraldehyde-3-phosphate (G-3-P),* a three-carbon aldehyde. G-3-P is then rearranged to form several intermediate compounds. Some go on to become fructose, glucose, starch, and other carbohydrates. Others are enzymatically converted to fats, or (with the addition of nitrogen) they can become amino acids. So, in a sense, G-3-P is the molecule that serves as the basic carbon structure to which atoms are added to form the substances

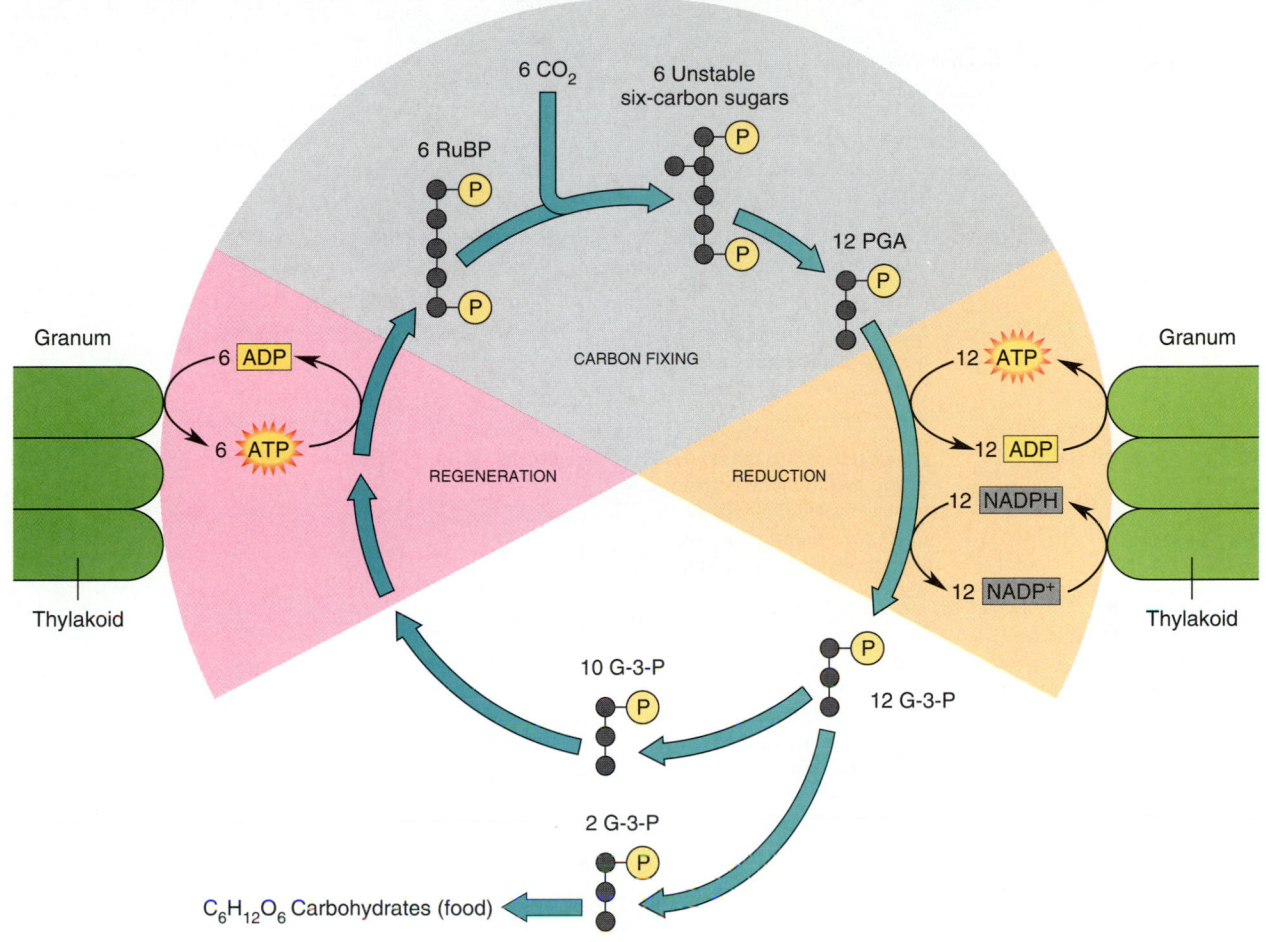

FIGURE 8.14 The Calvin-Benson Cycle
In this cycle, energy from the sun is incorporated into the chemical bonds of food, such as sugar (carbohydrate) molecules. Overall, the Calvin-Benson cycle consists of three phases: (1) a carbon-fixing phase, (2) a reduction phase, and (3) a regeneration phase.

we call food. G-3-P could be said to be the molecule that feeds the world.

If all the G-3-P were made into food, the photosynthetic process could not continue, for there would be no RuBP to pick up the next CO_2. Therefore, only two 3-G-P are used in the synthesis of glucose and 10 G-3-P are used to resynthesize six more RuBP. These reactions are completed through the use of several intermediate compounds and ATP. In addition, NADP+ and ADP are recycled from the stroma to the thylakoid, where the light-dependent reactions occur.

Table 8.1 is a summary of the light-dependent and light-independent reactions of photosynthesis. In our discussion we have emphasized the events that occur in the light-dependent reactions, since

they have been the subject of more research investigation than the light-independent reactions. However, some researchers are also investigating alternative carbon-fixing schemes. One that has received much attention lately is the C_4 system, a carbon-fixing scheme found in some plants.

The C_4 Cycle

■ The C_4 system is another light-independent scheme. C_4 plants take up carbon dioxide much more efficiently than C_3 plants, which is an advantage in dry environments.

Most of the original research on photosynthesis was conducted using either *chlorella* (a green alga) or

TABLE 8.1
A Summary of Photosynthesis

Reaction	Event	Components	Products
Light-dependent reactions (occur in the thylakoid)	Light energy is used to split H_2O, reduce $NADP^+$, and synthesize ATP	Specific wavelengths of light and specific chlorophyll molecules P680 and P700	
Light-harvesting reactions	Electrons in the chlorophyll acceptor molecules are excited and accepted by electron acceptors	Proteins, chlorophylls, and light	Electrons
Electron transport	As electrons are transported along a chain of electron acceptor molecules within the thylakoid membrane, some eventually reduce $NADP^+$. As water is split, some H^+ accumulates within the inner compartment of the thylakoid	Electrons, H_2O, $NADP^+$	$NADPH + O_2 + H^+ + e^-$
Chemiosmosis	As H^+ are pumped across the thylakoid membrane, a proton gradient is formed as the return across the membrane is used to synthesize ATP in the CF_1 complex	Membrane, H^+ gradient, ADP, and P_i	ATP
Light-independent reactions	Carbon dioxide is reduced by the use of an organic molecule to a high-energy carbon compound	Ribulose, bisphosphate, CO_2, ATP, NADPH	High-energy carbon compound, $ADP + P_i$, $NADP^+$

spinach. However, when investigators studied other plants, such as sugarcane, desert plants, and some grasses, they found that they do not exclusively use the Calvin-Benson scheme we just examined. In the C_4 photosynthetic process, CO_2 does not join RuBP immediately. Instead, CO_2 is bound to a three-carbon compound called *phosphoenolpyruvate* (PEP) to form a four-carbon acid, called *oxaloacetic acid* (Figure 8.15). This structure is called a C_4 *acid,* and the system that incorporates it is called the C_4 **carbon-fixing process.** The oxaloacetic acid then gives up CO_2, forming pyruvate. The released CO_2 can then enter the Calvin-Benson cycle. The process is also called the *Hatch-Slack* scheme, after its discoverers M. D. Hatch and C. R. Slack, who described it in the 1970s.

Although the first stage of carbon fixing in C_4 contains an extra step, that of forming oxaloacetic acid, from then on the process is the same as the Calvin-Benson cycle. Since there is only that one difference, why did the C_4 scheme evolve in some plants? You would think that they would go right into the C_3 scheme, rather than using energy to carry out an extra step. However, biologists have discovered the advantage of the C_4 scheme — it helps plants cope with difficult weather conditions.

In areas where rainfall is uncertain or sparse, plants like the desert plants, crabgrass, and others must conserve water. Most of the water lost by plants passes out through pores in the leaf called **stomata** (singular, *stoma*). By closing the stomata, water is conserved. However, carbon dioxide also enters the plant through these pores; so when they close to prevent water loss, the plant's ability to take up carbon dioxide is also reduced.

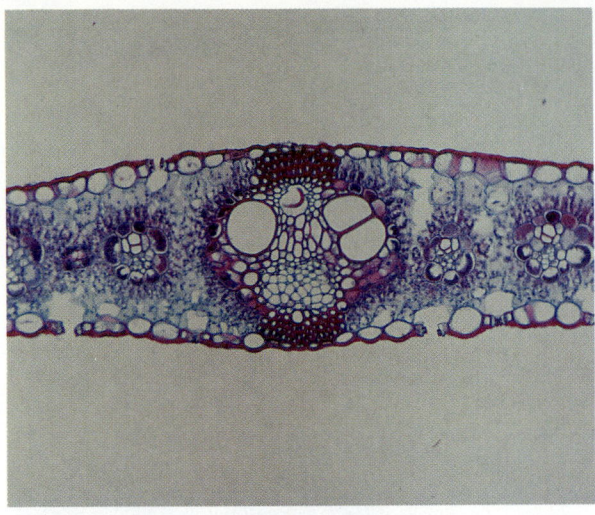

(a)

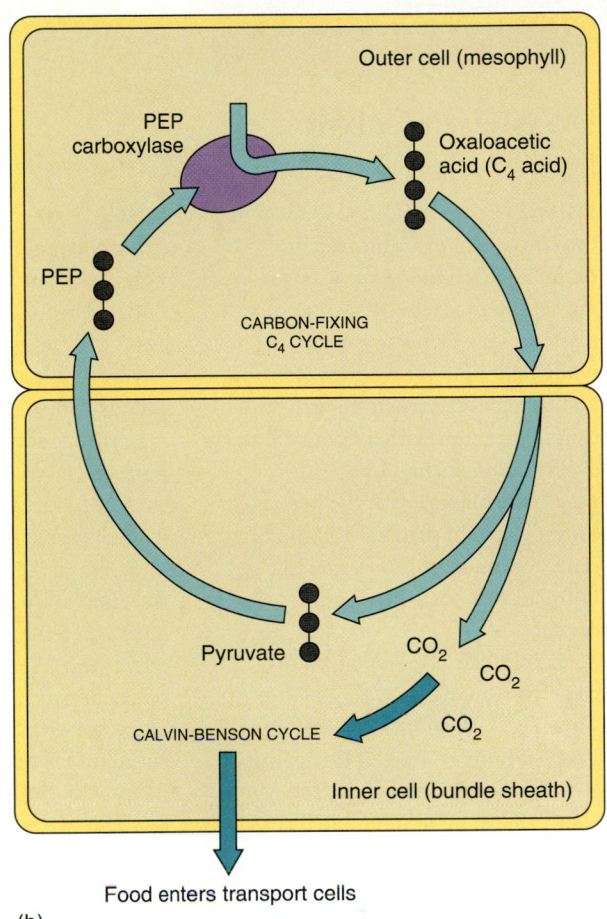

Outer cell (mesophyll)

PEP carboxylase

Oxaloacetic acid (C_4 acid)

PEP

CARBON-FIXING C_4 CYCLE

Pyruvate

CO_2

CO_2

CO_2

CALVIN-BENSON CYCLE

Inner cell (bundle sheath)

Food enters transport cells

(b)

FIGURE 8.15 C_4 Carbon Fixing
(a) A cross-section of a leaf from a corn plant, a C_4 plant. (b) The C_4 carbon-fixing cycle enables plants to take up and use carbon dioxide more efficiently than C_3 plants, and it occurs in different cells of the leaf than the Calvin-Benson (C_3) cycle does. C_4 cells are in general nearer the surface of the leaf than the C_3 cells. The enzyme (PEP carboxylase) that binds CO_2 to oxaloacetate in the C_4 system has a much greater affinity for CO_2 than the enzyme that binds CO_2 to RuBP in the C_3 cycle. So, when in short supply, CO_2 can be captured more efficiently. This CO_2 can then be passed along to the inner cells, where the C_3 cycle takes place. Sugarcane and desert plants are important C_4 plants.

Thus, in some plants, an alternative method of taking in carbon dioxide evolved. Because their cells contain special enzymes such as *PEP carboxylase*, C_4 plants can actively bind, or gulp in, more CO_2 than their C_3 counterparts from a lower concentration of CO_2. What is more, their carbon dioxide is more concentrated. Their stomata can remain almost closed to conserve water and still acquire enough CO_2 for photosynthesis to continue. That's why, on those hot, dry days of summer, Kentucky bluegrass, a C_3 plant, finds it tough going, but C_4 crabgrass does well. In times of water shortage, a C_4 plant has an advantage over a C_3 plant, because the oxaloacetic acid can serve as a CO_2 reservoir, holding extra CO_2 until it is needed by the plant.

As of this writing over 100 different kinds of plants are known to be C_4 plants. At least 15 of those plants can use both the C_3 and C_4 schemes, depending on environmental conditions, an ability that is another example of the environment's natural selection of organisms.

PHOTOSYNTHESIS AND RESPIRATION

The energy relationships of living things are fascinating in many ways. Even though you have just worked your way through a very complex process, one striking feature is the simplicity and conservation of nature. In photosynthesis, light energy is trapped and converted by chlorophyll and other pigments to chemical energy. This chemical energy is used to form ATP and other important energy carriers, such as NADPH. These energy-carrying molecules can use their energy and hydrogen from water as sources to reduce carbon dioxide and form food. Oxygen gas is a waste product of the system. Look again at the simplified overall equation:

$$CO_2 + H_2O \xrightarrow{\text{Light}} (CH_2O) + O_2 + H_2O$$

Carbon dioxide   Water   Chlorophyll, other pigments, enzymes   High-energy carbon compound   Oxygen

Dr. Melvin Calvin

Biochemist, Photosynthesis Researcher, Nobel Laureate

In 1952, on a small, indoor farm at the University of California at Berkeley, Melvin Calvin was working on the problem of carbon dioxide fixation by photosynthesizing plant cells. Calvin, Andrew Benson, and others were trying to answer a simple question with a complex answer. Biologists had known for years that plants, with the help of sunlight, could turn carbon dioxide (CO_2) and water into sugar. But how did this basic life process occur?

Calvin had been trying to discover the answer to this question since 1937. Born in 1911, the son of Russian immigrants, he earned his B.S. degree in chemistry at the Michigan College of Mining and Technology in 1931, and his Ph.D. degree in chemistry from the University of Minnesota in 1935. Following the receipt of his Ph.D. he worked at the University of Manchester, England. He returned to the United States and took a position at the University of California at Berkeley in 1937.

To find out the secrets of photosynthesis, Calvin and his coworkers had been using radioactive carbon 14 (^{14}C) to trace the metabolic pathways of CO_2 from the air to its incorporation into sugar. Slowly, step by step, Calvin and his team would expose living plant cells to ^{14}C for a very short time (measured in seconds). Then they would kill the plant cells and extract and identify any molecules that had incorporated the ^{14}C. The plan was to create a map of the movements of the carbon molecules.

After 15 years of following those tagged atoms, Calvin still could not understand the pathway — until one day, in a moment of brilliant insight, he grasped the significance of their movement. And he did not figure it out bent over a table in a sterilized laboratory. Instead, Dr. Calvin was sitting in his car, waiting for his wife to come back from an errand, when he suddenly realized that the pathway of the CO_2 fixation in plants was not a direct, straight route but was rather like his drive to work — full of hairpin turns and places where the path doubled back on itself.

The cycle that Calvin and his coworkers described is now called the Calvin-Benson cycle, or C_3 photosynthesis, and it does seem to proceed in an unusual way. The first step is the formation of the three-carbon acid (phosphoglyceric acid), which is followed by the formation of a six-carbon substance similar to sugar. It is at this point that Calvin's insight came into play. He realized that the six-carbon intermediate substance did not simply continue in a straight path. The path was not that simple and direct. Instead, the six-carbon molecule doubles back to yield more three-carbon substances. Only then is a six-carbon sugar produced.

In 1961, as a result of that afternoon's insight, Dr. Melvin Calvin was awarded the Nobel Prize for his work in photosynthesis. From that time on his colleagues gave him the title "Mr. Photosynthesis."

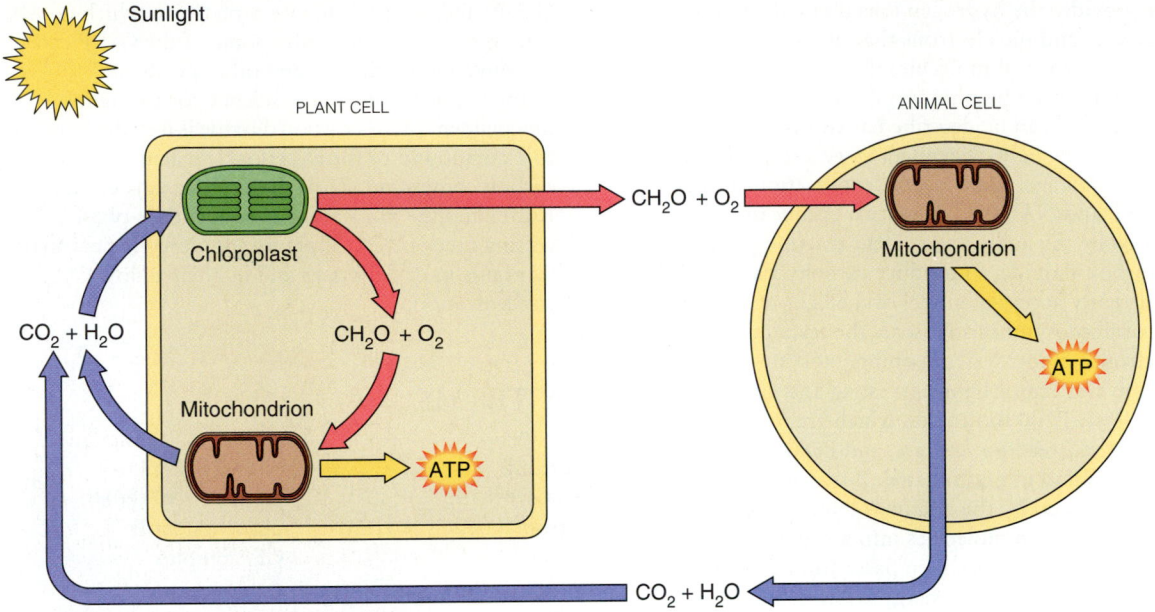

FIGURE 8.16 Photosynthesis and Cellular Respiration
Photosynthesis and cellular respiration are two reactions that complement each other. The by-products of one provide the fuel that drives the other. And in the process, ATP is constantly "recharged," providing the energy for the activities of life: self-maintenance, growth, and reproduction.

To appreciate the interdependence of the life processes, we need to look ahead for a moment. In the next chapter we examine *cellular respiration,* the process by which food molecules are broken down in cells and the stored energy in the chemical bonds of food is used to form ATP and other energy-carrying molecules. These energy-liberating molecules are used to provide the energy for cellular work. Water and carbon dioxide are the waste products of cellular respiration. The overall summary of cellular respiration is as follows:

$$CH_2O \ + \ O_2 \ \xrightarrow{\text{Respiring cell}} \ CO_2 \ + \ H_2O \ + \ \text{energy}$$

High-energy carbon compound (food)   Oxygen   Carbon dioxide   Water

If you examine the reactions of both photosynthesis and cellular respiration, you will see that they are the reverse of one another. Photosynthesis is a reducing process, one in which energy is stored and oxygen is a waste product. Cellular respiration is an oxidation process, one in which energy is released. It uses oxygen, and its waste products (carbon dioxide and water) are used as the basic materials in photosynthesis. Photosynthesis and res-

piration are linked together and provide the energy to drive life (Figure 8.16).

SUMMARY

1. Photosynthesis is the major energy-trapping reaction in nature. Green plants convert light energy to chemical energy to reduce CO_2 and form food.
2. All life depends on light energy. Animals obtain their energy from plants or from animals that feed on plants.
3. The raw materials for photosynthesis are carbon dioxide (CO_2), water (H_2O), light energy, chlorophylls, other pigments, and enzymes organized within the chloroplast.
4. Photosynthesis traps wavelengths of light in the visible spectrum. The absorption spectra of pigments in chloroplasts complement each other, with the predominant absorbing waves in the red, blue, and blue-violet range.
5. The chloroplast is the photosynthetic organelle in plants. It encloses a thick liquid called stroma and membranous, disclike structures called thylakoids. Thylakoids stacked on top of each other form grana.
6. Carbon dioxide provides the carbon for the carbon skeleton to which hydrogens are attached to form carbohydrates.

7. Water provides the hydrogen ions that reduce carbon dioxide and the electrons that fill electron-deficient chlorophyll molecules. It is also the source of oxygen released by photosynthesis.

8. Photosynthesis can be described as consisting of two phases: the light-dependent stage and the light-independent stage. Light-dependent reactions produce ATP and NADPH. These two chemicals are then used by the light-independent reactions, along with carbon dioxide, to produce carbohydrates.

9. The currently accepted model of the light-dependent phase of photosynthesis presents the reactions in two schemes, photosystem I and photosystem II.

10. There are two major light-harvesting antennas in chloroplasts: P700 molecules, which are found in photosystem I reaction centers, and P680 molecules, which are found in photosystem II reaction centers.

11. Energy harvested by the antennas bumps electrons in the photosystem molecules into a higher energy level; those electrons are then passed along an electron transport chain, the energy is harvested, and ATP is formed.

12. In cyclic photophosphorylation, which occurs only in photosystem I, the energized electrons that are stripped of their extra energy cycle back to the P700 molecule from which they came.

13. In noncyclic photophosphorylation, which involves both photosystems I and II, the electrons do not cycle back to the same chlorophyll molecule from which they originated. Light energy excites the P680 molecules and splits water molecules into two protons, two electrons, and one oxygen atom. The electrons fill the P680 molecule, the protons remain in the stroma, and the oxygen passes to the air.

14. The energized electrons from P680 move through the electron transport chain. At one point in the chain, H^+ are pumped from the stroma into the inner compartment of the thylakoid, where the ions accumulate. The electrons then fill the P700 molecule.

15. Photosystem II absorbs light energy and electrons through the electron transport chain. The electrons then pass to photosystem I, where they fill the P700 molecule.

16. The electrons released through the photolysis of water join with protons in the stroma and FAD on the surface of the thylakoid to form $FADH_2$. $FADH_2$ reduces $NADP^+$ to NADPH, which then cycles to the light-independent reactions.

17. In the C_3 scheme of the light-independent phase (called the Calvin-Benson cycle), CO_2 is reduced to form high-energy carbon compounds. There are three stages: (1) a carbon-fixing stage, in which carbon dioxide is bonded to five-carbon ribulose bisphosphate (RuBP) — the resulting unstable molecule immediately forms two molecules of the three-carbon glyceraldehyde-3-phosphate (G-3-P); (2) a reduction phase, in which ATP and NADPH reduce the intermediate carbon compounds to form

G-3-P; and (3) a regeneration phase, in which G-3-P is used to synthesize RuBP. Some of the G-3-P can be converted to glucose and other products.

18. In the C_4, or Hatch-Slack, scheme for the light-independent phase, carbon dioxide is initially bonded to a compound to form oxaloacetic acid, a four-carbon compound. Then the CO_2 is transferred to RuBP, and the normal light-independent-phase processing takes place. C_4 plants are more efficient than C_3 plants in dry climates and in areas of low CO_2 concentration.

KEY TERMS

autotroph	photosystem I
chemosynthetic	pigment 680 (P680)
electromagnetic spectrum	photosystem II
photon	CF_1 complex
stroma	cyclic
thylakoid	photophosphorylation
granum	noncyclic
chlorophyll	photophosphorylation
absorption spectrum	photolysis
light-dependent reaction	Calvin-Benson cycle
light-independent	C_3 cycle
reaction	C_4 carbon-fixing process
light-harvesting antennas	stomata
pigment 700 (P700)	

REVIEW QUESTIONS

True/False

1. Photosynthesis is the major energy-trapping reaction in living systems.
   true   *false*   *(page 163)*

2. Water is required in photosynthesis as the source of hydrogen for the reduction of CO_2.
   true   *false*   *(page 169)*

3. Leaves look green because the chlorophyll they contain reflects all wavelengths but green and yellow.
   true   *false*   *(page 167)*

4. The light-independent reactions occur only in the absence of light.
   true   *false*   *(page 170)*

5. Although the C_3 scheme involves an extra step, it is more efficient for desert plants than the C_4 scheme.
   true   *false*   *(page 180)*

Fill in the Blank

1. A major event in noncyclic photophosphorylation is the synthesis of _____ and _____.
   (page 173)

2. Another major event in the light-dependent reactions known as noncyclic photophosphorylation is the _____ of water, producing H^+ (protons), electrons, and oxygen. *(page 174)*

3. The actual synthesis of ATP in the light-dependent reactions is due to the differences in concentration of _____ between the inside and outside of the thylakoid. *(pages 176–177)*

4. C_4 carbon-fixing plants can bind _____ more efficiently than C_3 plants. *(page 180)*

Multiple Choice

1. Which of the following best describes the initial use of light in photosynthesis?
   a. It splits ribulose bisphosphate into G-3-P molecules.
   b. It excites the electrons that then leave the chlorophyll molecule.
   c. It is necessary to initiate the light-independent reactions.
   d. It produces $NADH_2$. *(page 170)*

2. What would happen if the chlorophyll were removed from a photosynthesizing cell?
   a. Continued photosynthesis would require more light.
   b. Other pigments would take over the function of chlorophyll and photosynthesis would continue.
   c. Photosynthesis activity would decrease slightly.
   d. Photosynthesis would stop. *(page 167)*

3. The formation of ATP during the light reactions is due to:
   a. Differences in hydrogen-ion concentration between the stroma and the thylakoid inner compartment

   b. Substrate-level phosphorylation
   c. Oxidative phosphorylation
   d. Electron-deficient chlorophyll molecules *(page 176)*

4. The major event that occurs in the light-independent reaction is:
   a. Synthesis of ATP
   b. Photolysis of water
   c. Formation of carbohydrates
   d. Production of hydrogen and oxygen gas *(page 178)*

Discussion

1. What are the raw materials required for photosynthesis, and how do they function in producing the overall products of the process?

2. Draw and label a chloroplast.

3. Why can it be stated that "all flesh is grass"?

4. Describe (in a general sense) the major events that occur in light-dependent and light-independent reactions of photosynthesis.

5. Discuss the advantages and disadvantages of C_4 photosynthesis.

6. Discuss how the chemiosmotic theory helps explain the synthesis of ATP during the light reactions.

7. Write the simplified general equations for photosynthesis and respiration. Discuss the interrelationships between these two reactions.

9

Respiration: Energy Liberation

In the last chapter we discussed the process of photosynthesis, the way that plants harvest the sun's energy and transform it into the chemical energy of the bonds of carbohydrate molecules. Photosynthetic organisms create their own food from water, sunlight, and carbon dioxide. Because of this ability, such organisms are called *autotrophs,* or self-feeders.

Animals, fungi, and some bacteria cannot manufacture their own food supply, so they are called **heterotrophs,** which means "other-feeders." Heterotrophs get their energy by eating other organisms — either plants or animals (Figure 9.1). Ultimately all heterotrophs depend on the energy that photosynthesizing organisms trap from the sun.

The lion on the Serengeti Plain may eat zebras and gazelles, but those zebras and gazelles graze on the grass of the plains. Thus the lion indirectly lives on the energy provided by the grass.

Thus both autotrophs and heterotrophs rely on the energy stored by photosynthesis. How do they release the energy in those high-energy carbon compounds that we discussed in Chapters 7 and 8, and use it to overcome entropy and shape the raw materials for the growth and maintenance of life?

The answer is that organisms release the energy stored in food by one of two processes: (1) aerobic cellular respiration, in which large amounts of energy are released from food molecules and in which oxygen is utilized as the final electron acceptor, or (2) anaerobic cellular respiration (sometimes called fermentation), in which energy is released from food molecules without utilizing oxygen as the final electron receptor and which results in smaller

---

Above: The energy produced by cellular respiration enables these sockeye salmon to swim up this waterfall.

FIGURE 9.1   Heterotrophs
Zebras get their energy from the sun indirectly, by feeding on plant life. They in turn
provide the link to the sun's energy for their predators, the lions.

(a)

(b)

(c)

(d)

FIGURE 9.2   Aerobic and Anaerobic Organisms
(a) The cells in this caribou, and in the vegetation that it feeds on, all
respire aerobically. (b) This is true of the cells in these fish and in the
coral reef as well. These organisms need oxygen to survive. (c) The
organism we know as brewers yeast respires both aerobically and
anaerobically. When denied oxygen, the product of their respiration
is alcohol. (d) *Clostridium botulinum,* the anaerobic bacterium
responsible for the often fatal disease botulism, cannot survive in the
presence of oxygen.

amounts of energy release than respiration (Figure 9.2).

CELLULAR RESPIRATION

■ **In cellular respiration, the products of photosynthesis are broken down in cells and the released energy is used to manufacture ATP and other energy-carrying molecules.**

The process of releasing the energy stored in carbohydrate molecules is called **cellular respiration.** This is the process by which food molecules are broken down in cells and the stored energy in their chemical bonds is used to form ATP and other energy-carrying molecules. These energy-liberating molecules are then used to provide the energy for cellular work (see Profile).

Respiration is not the same as breathing. Breathing is a muscular process in which there is an exchange of gases, usually carbon dioxide and oxygen. Respiration is a chemical process; cellular respiration involves the oxidation or release of energy from the chemical bonds contained in food molecules.

Here is a summary of the process of respiration in a simplified, balanced equation using glucose as the carbohydrate:

$$\underset{\substack{\text{Carbo-}\\\text{hydrate}}}{C_6H_{12}O_6} + \underset{\text{Oxygen}}{6\,O_2} \xrightarrow{\text{Enzymes}} \underset{\substack{\text{Carbon}\\\text{dioxide}}}{6\,CO_2} + \underset{\text{Water}}{6\,H_2O} + \underset{\substack{\text{cellular}\\\text{activity}}}{\text{energy for}}$$

This equation is a bare-bones presentation of respiration in the presence of oxygen. The actual process involves a series of reactions, mediated by enzymes, in which a high-energy carbon compound, such as glucose, is completely oxidized to CO_2 and H_2O. There is an orderly release of energy as the chemical bonds in the reduced organic compounds are broken. Much of that order results from the reactions of electron transporting molecules, and the released energy is used to phosphorylate ADP to ATP. As we mentioned in Chapter 6, ATP serves as the energy currency of the cell. The energy released when a phosphate is removed can then be used for the large number of energy-requiring reactions necessary to sustain life.

Some of the respiratory stages occur in the cytoplasm of the cell, and others occur within a specialized organelle in eukaryotic cells known as the *mitochondrion* (see Chapter 4). Unless otherwise noted, the discussion of cellular respiration will concentrate on the reactions taking place in the eukaryotic cells.

AN OVERVIEW OF RESPIRATION

■ **Glycolysis, which requires no oxygen, precedes most forms of cellular respiration. Once glycolysis is completed, its products can take either of two routes. One requires oxygen and is called aerobic respiration. The other, which requires no oxygen, is called anaerobic respiration.**

Respiration is a continuous process that usually begins with *glycolysis*, the breakdown of glucose. The products of glycolysis then take either of two routes: aerobic respiration, respiration that uses oxygen; or anaerobic respiration, respiration that does not use oxygen (Figure 9.3). We begin with a discussion of glycolysis and go on to an overview of the stages of aerobic respiration: the Krebs cycle and electron transport. We then consider three types of anaerobic respiration.

One note before we begin. Many high-energy carbon compounds, such as carbohydrates, fats,

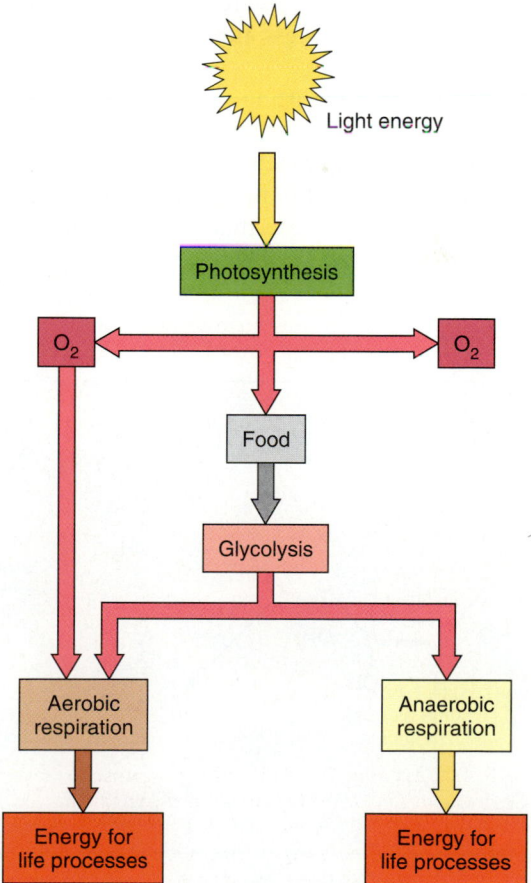

FIGURE 9.3 Glycolysis and Respiration
Respiration begins with the process of glycolysis. The products of glycolysis can then follow one of two pathways: aerobic respiration or anaerobic respiration.

and proteins, are a source of energy for living cells. However, because many of the components that are used by cells are changed into glucose (or other carbon compounds that are similar to glucose) during cellular respiration, our discussion will concentrate on glucose.

Glycolysis

■ **Glycolysis is a set of reactions occurring in the cell's cytoplasm, in which glucose is metabolized to pyruvate and a net of two ATP molecules are produced. The process requires no oxygen.**

Glycolysis, which means "glucose-splitting," occurs in the cytoplasm of the cell and does not require oxygen. During the process, a series of rearrangement and oxidation reactions results in the breakdown of the six-carbon molecule glucose into two molecules of a three-carbon compound known as **pyruvate** (Figure 9.4). (*Pyruvate* is a term used for the ionized form of pyruvic acid. Most acids exist in an ionized form in living systems. We will follow conventional usage here, which dictates that we use the suffix *-ate* to express the ionized form of an acid.)

As a result of the splitting of one mole of glucose, about 143 Calories are released. The two moles of pyruvate still contain much stored chemical energy, which can be removed in further respiratory stages. For each mole of glucose that passes through glycolysis, enough chemical energy results to produce four moles of ATP from ADP. But as you will see, because two ATP molecules are used, the net production is two molecules of ATP. (For a quick review of the quantity called a mole, see Chapter 6.) In addition, hydrogens and electrons are transferred to NAD^+, forming the reduced electron carrier NADH.

Stages of Aerobic Respiration

■ **Aerobic respiration consists of two basic stages: the Krebs cycle, in which the carbon skeleton of the original glucose molecule is dismembered, CO_2 is released, and hydrogen molecules are removed; and the electron transport chain, during which most of the ATP production takes place by chemiosmosis. The ultimate electron acceptor is oxygen.**

Aerobic respiration is the form of respiration that utilizes oxygen as the final electron acceptor. Although the Krebs cycle itself does not require oxygen, it does donate the electrons that are eventually passed to the electron transport chain, and the electron transport chain does require oxygen. Thus, we can say that there are two stages in aerobic respiration: the Krebs cycle and the electron transport chain.

Krebs Cycle. One mole of the pyruvate molecules produced by glycolysis contains approximately 680 Calories of energy within the molecule's chemical

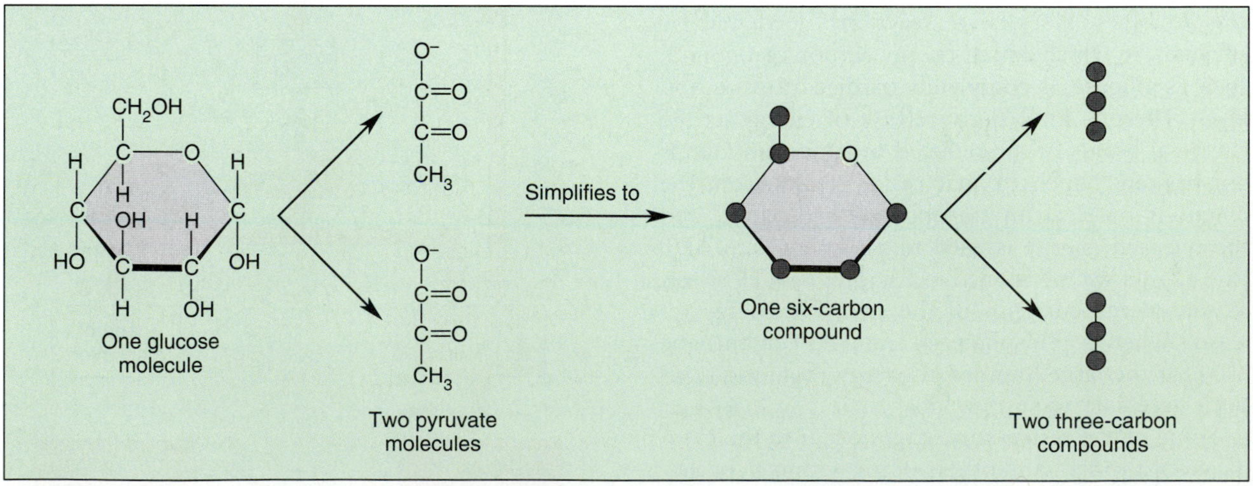

FIGURE 9.4 Glucose Splitting
During glycolysis a molecule of glucose, a six-carbon sugar, is broken down into molecules of pyruvate, a three-carbon compound. Here we illustrate how we will simplify the molecular structures.

bonds. That is a lot of stored energy. The products of glycolysis, the pyruvate molecules, move to the mitochondrion where the next process, the *Krebs cycle*, takes place. The enzymes responsible for the processing that takes place during the Krebs cycle are located in the matrix of the mitochondrion. The molecule's carbon skeleton is dismembered, carbons are removed as CO_2, and hydrogens and electrons are removed through oxidation.

Electron Transport. The electrons from the hydrogen atoms that were removed in the Krebs cycle and glycolysis are transferred, step by step, to a series of electron acceptor molecules in an electron transport chain located in the inner membrane of the mitochondria. Once the electrons have passed through the transport chain, they are passed on to oxygen, which serves as the final electron acceptor. The oxygen then combines with hydrogen ions plus electrons (which form hydrogen atoms) to form water: $\frac{1}{2}O_2 + 2e^- + 2H^+ \rightarrow H_2O$.

Figure 9.5 presents an overview of the reactions required for the complete oxidation of glucose. The total ATP presented there indicates the approximate number of ATP molecules that are synthesized as one molecule of glucose is oxidized. A more detailed accounting of ATP yield will be discussed later in the chapter.

Stages of Anaerobic Respiration

■ **Anaerobic respiration is conducted by specific organisms and in vertebrate muscle tissue.**

Anaerobic respiration, or anaerobic electron transport, is a type of respiration that takes place when oxygen is not available as the final electron acceptor once the electron reaches the end of the electron transport chain. This pathway is the one used by the bacteria found in the intestines of humans and other animals, in yeast, and in many prokaryotes.

Fermentation is a form of anaerobic respiration. These reactions take place within the cytoplasm. No membrane-bound enzymes are involved. Once glycolysis (a process that does not require oxygen) is complete, the pyruvate generated can ultimately be converted to carbon dioxide and ethyl alcohol in yeast cells. Remember, glycolysis is a part of fermentation. Fermentation begins with glycolysis and then, depending on the type of cell and the amount of oxygen available, pyruvate can be converted to ethyl alcohol or sometimes to lactate.

In human muscles and in some bacteria, anaerobic fermentation does not end in alcohol formation.

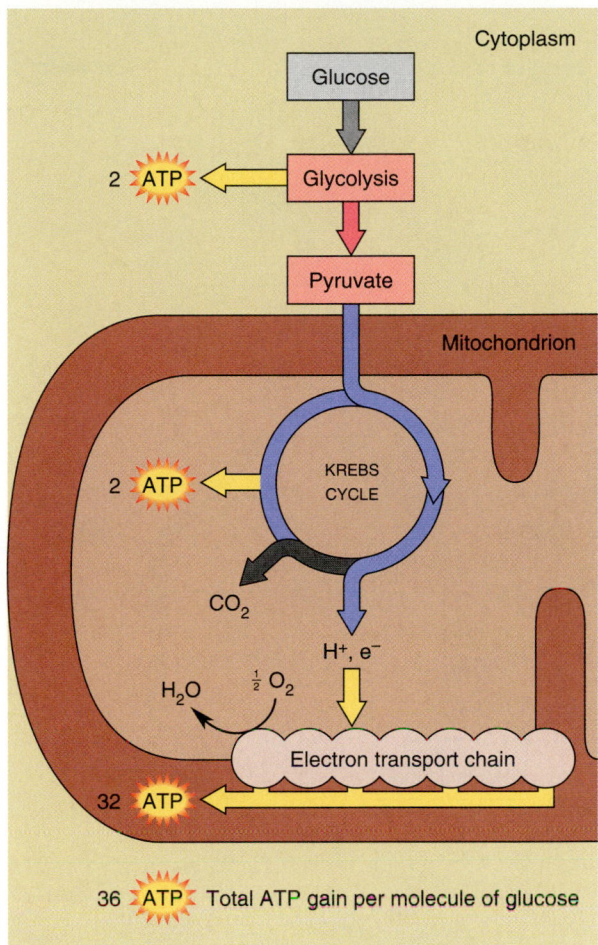

FIGURE 9.5   Glycolysis and Aerobic Respiration
In eukaryotic cells the aerobic stages of respiration take place in mitochondria. The complete oxidation of glucose yields a significant amount of the energy-producing molecule ATP.

Instead, in the final step of anaerobic respiration the pyruvate produced by glycolysis is converted to *lactate*.

Figure 9.6 shows two possible anaerobic pathways of pyruvate, depending on the organism and the environmental conditions. Fermentation begins with glycolysis, in which glucose is converted to pyruvate, a net of two ATPs are produced, and NAD^+ is reduced to NADPH. Note that during fermentation no more ATP has been synthesized. So if fermentation does not produce any ATP, what is its function? One of the major functions of fermentation is to strip off the hydrogen from NADH so that NAD^+ can be regenerated to be used again in the ATP-producing stages of glycolysis.

As you read on through the more detailed discussions of the various types of respiration, keep these

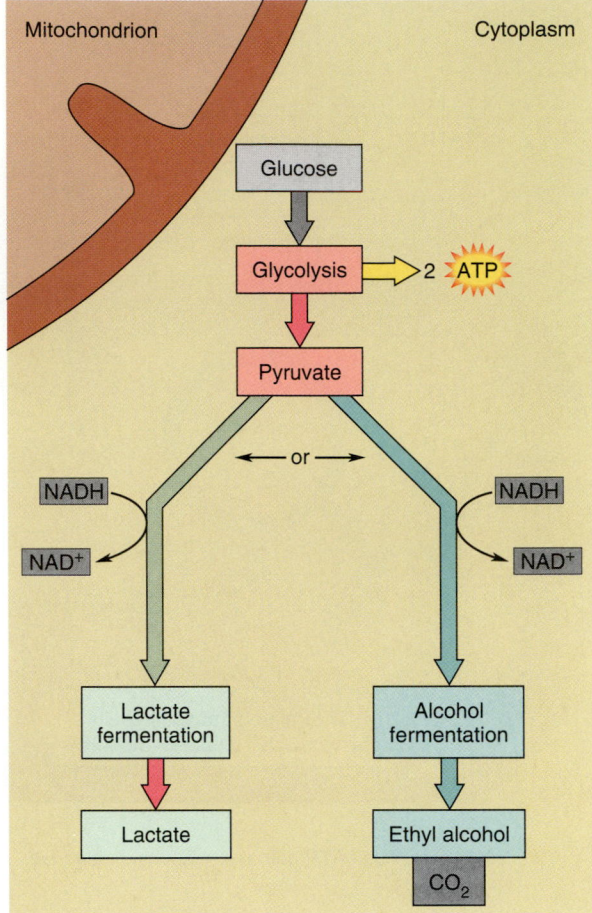

FIGURE 9.6 Glycolysis and Fermentation
Under anaerobic conditions, glycolysis may take one of
two pathways, each leading to a form of fermentation.
Fermentation regenerates the NAD⁺ needed for
glycolysis.

short discussions in mind. They will help you keep
the basic processes well in hand.

GLYCOLYSIS:
GLUCOSE SPLITTING

- **During glycolysis, glucose is degraded to pyruvate.
  During the process a net of two ATP molecules
  are synthesized and two NAD⁺ are reduced to
  two NADH.**

When a molecule of glucose enters the cytoplasm of
the cell, a series of enzyme-mediated reactions assure the orderly step-by-step degradation of glucose. During this process, ATPs are synthesized and
NAD⁺ is reduced to NADH.

To break down glucose, there must be an input
of energy that will activate the reaction. This energy

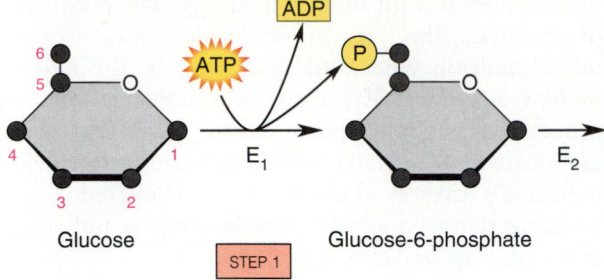

FIGURE 9.7 Glycolysis
Glycolysis serves as the primary respiratory route in the
cell. Oxygen is not required for glycolysis. In fact,
primordial prokaryotes are believed to have carried on
glycolysis millions of years ago before oxygen was even
present in the earth's atmosphere.

is made available from two molecules of ATP.
While it may seem strange to use ATP in order to
make ATP, the sense of it is not difficult to grasp.
Just think of it in the same terms as pushing a car
over a hill: It takes a little energy to get the car
moving, but once it is moving, the energy that is
liberated will more than make up for the energy
required for the initial push.

Because glycolysis is a good example of a biochemical process in which an overall reaction proceeds in several small steps, we will look at the
individual steps in glycolysis one by one, in order
to follow one small metabolic pathway that illustrates the complexity of the chemistry of living
processes. In particular, note the way in which the
six-carbon sugar is prepared to be cleaved into the
two three-carbon compounds and the points at
which ATP is synthesized and NAD⁺ is reduced. As
you read through the explanation, follow the presentation in Figure 9.7.

Step 1: Glucose Phosphorylation

The first step in glycolysis involves the phosphorylation of glucose — that is, the transfer of a phosphate molecule to glucose. The necessary phosphate
and the energy required for the activation of glucose
are provided by ATP. When the phosphate group
bonds to the glucose, the newly formed molecule
becomes glucose-6-phosphate (a six-carbon sugar
with a phosphate bonded to the number 6 carbon).

The process is actually a set of coupled reactions.
In one reaction, ATP is hydrolyzed to ADP + P_i.

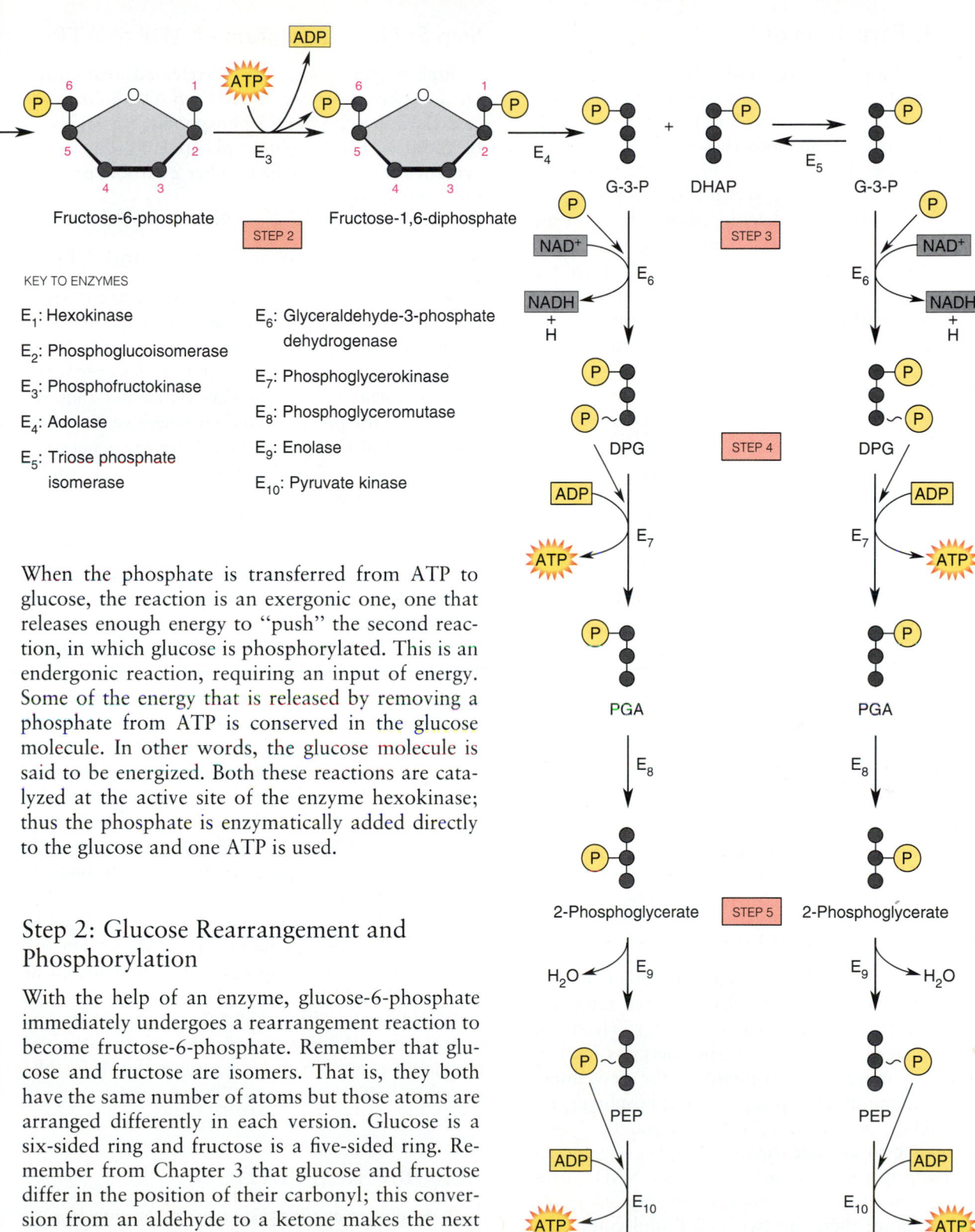

KEY TO ENZYMES

E_1: Hexokinase

E_2: Phosphoglucoisomerase

E_3: Phosphofructokinase

E_4: Adolase

E_5: Triose phosphate isomerase

E_6: Glyceraldehyde-3-phosphate dehydrogenase

E_7: Phosphoglycerokinase

E_8: Phosphoglyceromutase

E_9: Enolase

E_{10}: Pyruvate kinase

When the phosphate is transferred from ATP to glucose, the reaction is an exergonic one, one that releases enough energy to "push" the second reaction, in which glucose is phosphorylated. This is an endergonic reaction, requiring an input of energy. Some of the energy that is released by removing a phosphate from ATP is conserved in the glucose molecule. In other words, the glucose molecule is said to be energized. Both these reactions are catalyzed at the active site of the enzyme hexokinase; thus the phosphate is enzymatically added directly to the glucose and one ATP is used.

Step 2: Glucose Rearrangement and Phosphorylation

With the help of an enzyme, glucose-6-phosphate immediately undergoes a rearrangement reaction to become fructose-6-phosphate. Remember that glucose and fructose are isomers. That is, they both have the same number of atoms but those atoms are arranged differently in each version. Glucose is a six-sided ring and fructose is a five-sided ring. Remember from Chapter 3 that glucose and fructose differ in the position of their carbonyl; this conversion from an aldehyde to a ketone makes the next step possible. The fructose uses another molecule of ATP to form fructose-1,6-diphosphate, meaning that there are phosphates bonded to the number 1 and 6 carbons. (Note that by this point, the process has used two ATPs, but none have been synthesized.)

Step 3: Formation of G-3-P

Because of the rearrangement reaction in step 2, the single fructose molecule now can be enzymatically split into two molecules. Specifically, fructose-1, 6-diphosphate is split into two three-carbon compounds, dihydroxyacetone phosphate (DHAP), which is a ketone, and glyceraldehyde-3-phosphate (G-3-P), which is an aldehyde. (Remember that the terms *ketone* and *aldehyde* refer to the position of a functional group in a carbon molecule. Here the carbonyl functional groups (C=O) either are flanked by carbon molecules, in the ketone, or are at the end of the carbon chain, in the aldehyde.)

$$\underset{\substack{\text{Dihydroxyacetone} \\ \text{phosphate (DHAP)}}}{\begin{array}{l} 1\ CH_2OH \\ | \\ 2\ C{=}O \\ | \\ 3\ CH_2{-}O{-}\text{\textcircled{P}} \end{array}} \quad \overset{\substack{\text{Triose phosphate} \\ \text{isomerase}}}{\underset{}{\rightleftarrows}} \quad \underset{\substack{\text{Glyceraldehyde-3-phosphate} \\ \text{(G-3-P)}}}{\begin{array}{l} H \\ | \\ 1\ C{=}O \\ | \\ 2\ CHOH \\ | \\ 3\ CH_2{-}O{-}\text{\textcircled{P}} \end{array}}$$

It should be noted that DHAP and G-3-P are interconvertible — in other words, through the actions of an isomerase enzyme DHAP can be converted to G-3-P. DHAP can follow a separate pathway that ultimately leads to fat formation; however, in order to concentrate on the process of glycolysis, we concern ourselves here with the process that takes place when DHAP is converted to G-3-P. That means that for all practical purposes, two molecules of G-3-P have been produced from one glucose molecule. Therefore, beyond this point each step occurs twice, since two G-3-P molecules are produced from one glucose molecule.

Step 4: Oxidation of G-3-P

Each G-3-P is now oxidized — in other words, hydrogen ions with their electrons are removed. The oxidation process releases energy, electrons, and hydrogen ions. Some of the energy is used to attach an inorganic phosphate to the new three-carbon compound to produce 1,3-diphosphoglycerate (DPG). (Remember, by convention the suffix -*ate* is used to indicate the ionized form of an acid.) The hydrogen ions are used to reduce NAD^+, producing the hydrogen energy-carrying molecule NADH. Since there are two G-3-P molecules, two DPG and two NADH molecules are also formed. This is the first reaction in the cell in which energy is harvested. To indicate the presence of the stored energy, biologists use a ~ (wavy line) in the structural formula, where the phosphate attaches to the number 1 carbon.

Step 5: Phosphorylation of ADP to ATP

A high-energy phosphate is released from each DPG to phosphorylate one ADP to ATP. In the process DPG has become 3-phosphoglycerate (PGA). In addition, the remaining phosphate group is enzymatically moved to the number 2 carbon, producing 2-phosphoglycerate.

Step 6: Generation of Pyruvate and ATP

Through a dehydration and rearrangement reaction, each 2-phosphoglycerate molecule is rearranged to form phosphoenolpyruvic acid (PEP). Each PEP transfers its energy-rich phosphate to ADP to form ATP. (You may remember that we met PEP in the previous chapter when we discussed the C_4 carbon-fixing pathway.) With the generation of two more ATP molecules, two pyruvate molecules are formed, and glycolysis is completed.

In summary, through a series of reactions, one glucose molecule has been converted to two pyruvate molecules. Enough energy has been liberated to synthesize four ATP molecules for a net gain of two molecules of ATP (remember that two molecules of ATP were used to get the reaction started) and two NADH molecules.

Glycolysis is not a very efficient method of energy liberation, since the pyruvate molecule is still highly reduced and still contains 680 Calories of potential energy. Once glycolysis is complete, the pyruvate molecules can follow various metabolic pathways. For instance, they can be used to synthesize amino acids, which will become proteins. If pyruvate enters the pathway of anaerobic respiration, the molecule can be fermented to form ethyl alcohol or lactate, depending on the type of cell in which the reaction occurs. If pyruvate enters the aerobic respiration pathway, it will be oxidized to form CO_2 and H_2O.

We will now follow the pathway of aerobic respiration into the mitochondrion and see how the remaining energy in the pyruvate molecules is liberated to synthesize more ATP molecules.

AEROBIC RESPIRATION

■ **Aerobic respiration takes place in the mitochondria. The Krebs cycle occurs in the matrix; the electron transport chain occurs in the cristae.**

Once glycolysis is complete, pyruvate moves from the cytoplasm of the cell into the mitochondria (Chapter 4). Most important to respiration is the double membrane that surrounds each mitochondrion. The outer membrane separates the mitochondrion from the other parts of the cell. A space exists between the outer membrane and the inner membrane; this space is called the *inner membrane space (outer compartment)*. The inner membrane folds and doubles on itself to form what are called *cristae*. Because of the infolding of the inner membrane, it has a much greater surface area than that of the outside membrane (Figure 9.8a). You can see the effect if you take a gift box and wrap it with a piece of tissue paper. Both have the same surface area. Now crumple the tissue and put it inside the box. More than likely you will have enough space left over to add another piece of tissue equal in size to the first. Thus, the surface area of the inner sheets will be greater than that of

the outer one. The same phenomenon takes place in a mitochondrion.

The cristae of the mitochondrion's inner membrane increase the area that is available for chemical reactions that take place during respiration (Figure 9.8b). Molecules of the ATP synthetase, an enzyme that phosphorylates ADP to produce ATP, are found embedded in the inner membrane (the cristae). The remaining space inside the mitochondrion (the inner compartment) is filled with a thick fluid substance called the **matrix.** The first stage of reactions in aerobic respiration, the Krebs cycle, occurs in the matrix. Electrons are then moved from the matrix to the membrane-bound electron carrier molecules in the inner membrane during the electron transport chain.

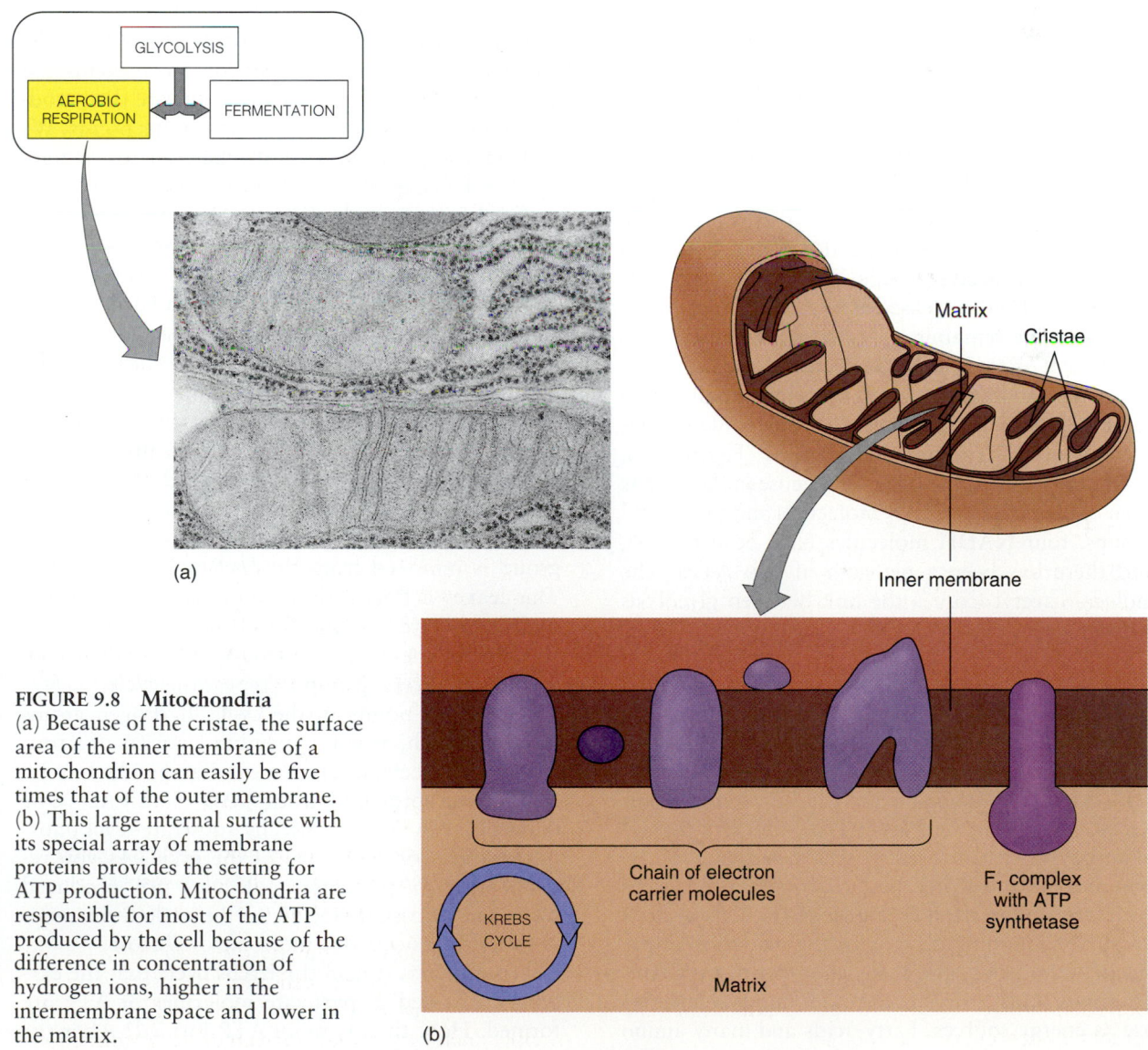

(a)

(b)

FIGURE 9.8 Mitochondria
(a) Because of the cristae, the surface area of the inner membrane of a mitochondrion can easily be five times that of the outer membrane. (b) This large internal surface with its special array of membrane proteins provides the setting for ATP production. Mitochondria are responsible for most of the ATP produced by the cell because of the difference in concentration of hydrogen ions, higher in the intermembrane space and lower in the matrix.

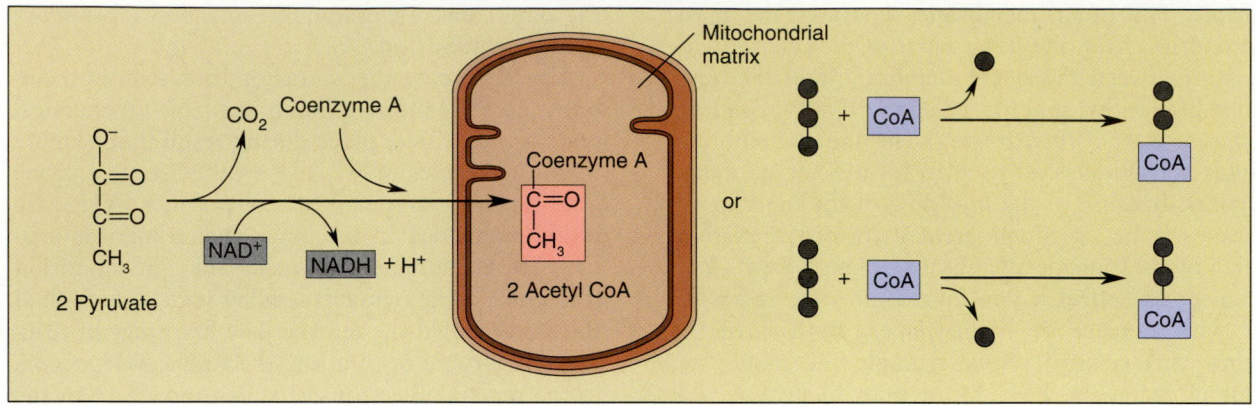

FIGURE 9.9 Entering the Krebs Cycle
Pyruvate is transported from the cytoplasm into the matrix of the mitochondrion, where it loses one carbon atom and becomes associated with coenzyme A. Acetyl coenzyme A can now enter the Krebs cycle within the mitochondrion. The circles to the right of the mitochondrion represent a simplified version of the conversion of pyruvate to CoA.

Preparation for Cycling

■ **In preparation for the Krebs cycle, pyruvate is oxidized to acetyl CoA. The released carbon is given off as carbon dioxide. The released energy is used to reduce NAD$^+$ to NADH.**

Once pyruvate passes from the cytoplasm into the matrix of the mitochondria, the molecule is converted to two acetyl groups. To conduct this process, one carbon is released as CO_2 and the carbonyl carbon of the remaining 2-carbon fragment is oxidized (that is, it loses two electrons and a hydrogen). A two-carbon acetyl group remains and combines with a coenzyme molecule called coenzyme A (CoA) to form acetyl CoA (Figure 9.9).

In other words, the original glucose molecule has been oxidized to two CO_2 molecules and two acetyl groups, four NADH molecules have been formed, and there has been a net gain of two ATPs. The molecule acetyl CoA is the link between glycolysis and the Krebs cycle.

Krebs Cycle

The **Krebs cycle,** sometimes referred to as the *citric acid cycle,* was named after the biochemist Hans Krebs, who won the Nobel Prize in 1953 for explaining the aerobic breakdown of pyruvate to carbon dioxide and water. The reactions of the Krebs cycle are summarized in Figure 9.10. This cycle is the final common pathway for the oxidation of pyruvate, fatty acids, and the carbon-hydrogen chains of amino acids, substances that living systems also use as energy sources. Fatty acids and many amino acids can be converted to acetyl CoA and enter the Krebs cycle at this point. In addition, proteins can be broken down to amino acids that are then modified and enter the Krebs cycle at various points. We will return to the topic of alternate energy sources later in the chapter.

The first step in the Krebs cycle is the transfer of the two-carbon acetyl group from CoA to the four-carbon compound called oxaloacetate. Once this is done, the CoA molecule can cycle back to the matrix of the mitochondrion and be used to pick up more acetyl groups. The remainder of the cycle consists of a series of *decarboxylations* (removal of CO_2), *dehydrogenations* (removal of hydrogens), and *make-ready* reactions (in which atoms within molecules are sometimes rearranged through the enzymatic addition of water in preparation for the next metabolic step). First one, then another acid group is removed from the fuel molecule as CO_2. This leaves a four-carbon compound that is eventually converted to the four-carbon oxaloacetate, completing the cycle. Oxaloacetate can then pick up the next acetyl group to repeat the cycle.

At certain points in the cycle, hydrogen atoms (each of which, as you remember, consists of a proton and an electron) are removed and picked up by electron acceptor molecules NAD$^+$ and FAD. One molecule of ATP is formed by substrate-level phosphorylation for each turn of the cycle. As you remember, two pyruvate molecules are produced from one glucose molecule, so the cycle turns twice in order to process one molecule of glucose.

In summary, it is in the Krebs cycle that the carbons contained in pyruvate molecules of ATP are formed. How then is more ATP formed? After the

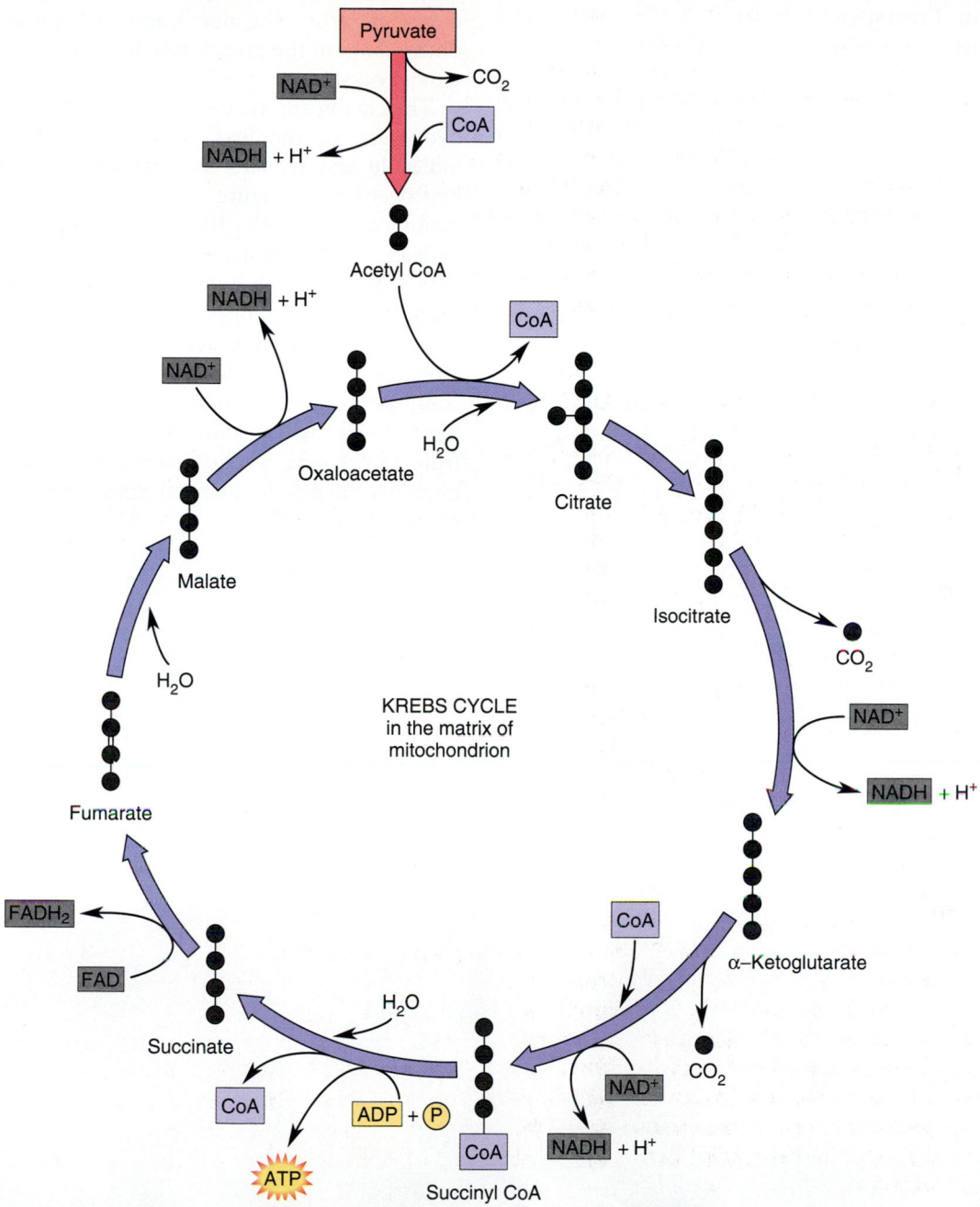

FIGURE 9.10 The Krebs Cycle
Two pyruvate molecules from glycolysis enter the Krebs cycle as acetyl groups attached to coenzyme A. Each molecule will yield only one molecule of ATP, for a net gain of two ATP. Once both molecules have cycled through, most of the ATP generated in aerobic respiration is derived from the NADH and FADH$_2$ as they pass electrons through the electron transport chain.

hydrogens are stripped off the carbon skeleton by dehydrogenation, their electrons are transported through the electron transport chain molecules in the inner membrane of the mitochondrion. The energy released during the electron transport chain will be used to establish a chemiosmotic gradient in the mitochondrion. If this sounds familiar, it should. The electrochemical gradient of chemiosmosis has figured prominently in the last three chapters.

Electron Transport Chain and Chemiosmotic Phosphorylation

■ **The electron transport chain functions in two processes: cycling the energized electrons down an electrical gradient, and moving protons into the inner membrane space of the mitochondrion. The energy due to the electrochemical gradient is used to phosphorylate ADP to ATP in the F_1 complex with ATP synthetase embedded in the cristae.**

We now need to look at the chemiosmotic theory again, this time in light of the processes and structures involved in cellular respiration. The reactions of the Krebs cycle take place in the matrix of the mitochondrion, and those of the electron transport chain take place in the cristae of the mitochondria. Protons and electrons are found in the matrix. The electrons flow down the energy gradient from a higher energy level to a lower energy level, through a series of electron acceptors called the **cytochromes,** which are proteins and enzyme molecules

containing iron. The membrane holds the molecules of the chain in the proper relationship so that transfer can occur.

The chemiosmotic cycle begins as NADH carries hydrogens to the first electron transport carrier molecule and reduces that carrier by passing two hydrogens to it (Figure 9.11). (The NAD^+ that remains can return the Krebs cycle to pick up more hydrogens.) The first carrier then releases two protons (one from each hydrogen) into the inner membrane space on the opposite side of the membrane. The two electrons are accepted by the next carrier molecule in a redox reaction. The redox reactions continue along the transport chain. As the electrons move down the transport chain, protons are moved from the matrix to the inner membrane space to establish the electrochemical gradient necessary for chemiosmosis.

Coenzyme Q. During the course of the transport chain, the electrons are passed to a carrier coen-

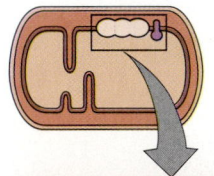

FIGURE 9.11 The Electron Transport Chain
Energy derived from electrons (e^-) as they move along the electron transport chain is used to pump protons (H^+) into the intermembrane space of the mitochondrion, against a strong concentration gradient. It is in this stage of aerobic respiration that oxygen is used—as the final acceptor of the electrons.

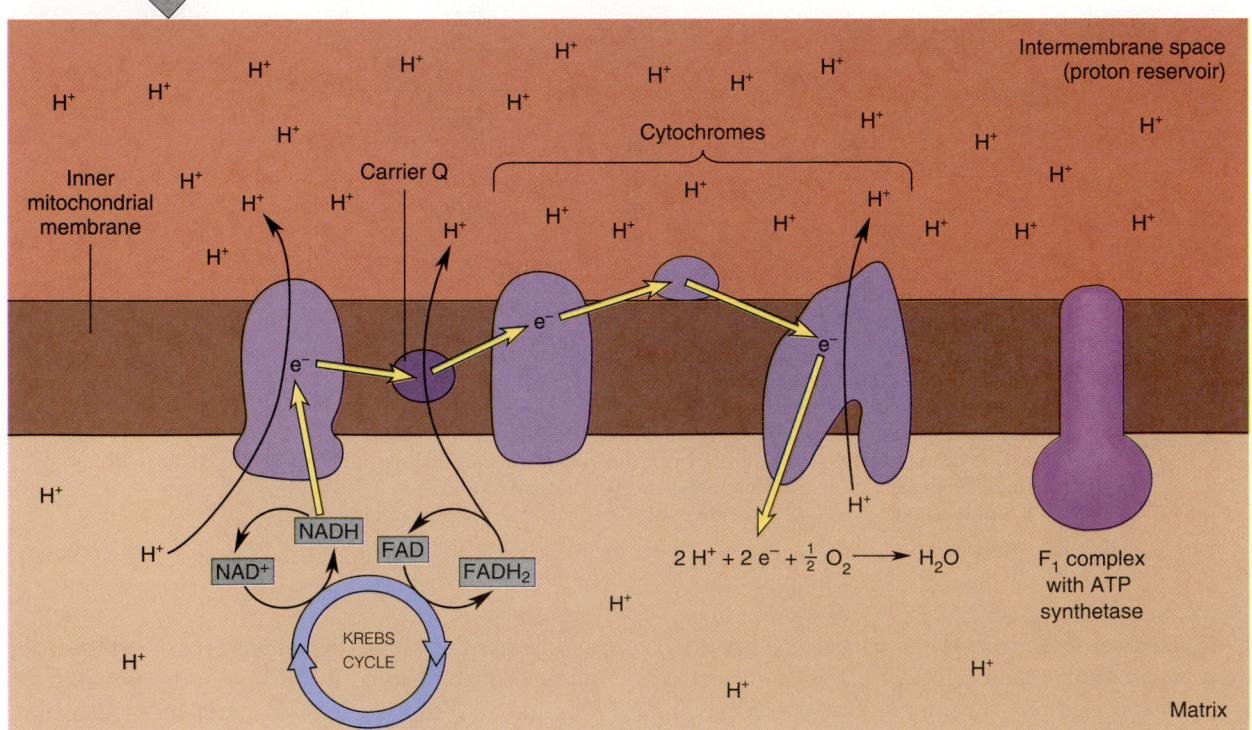

zyme called Q. Carrier (coenzyme) Q is important because it is instrumental in moving protons from the mitochondrial matrix to the intermembrane space, as well as functioning as a carrier of electrons. Carrier Q performs one other function. Because it has a lower level of potential energy than NADH, reduced flavin adenine dinucleotide ($FADH_2$) enters the electron transport chain at carrier Q. As $FADH_2$ enters the electron transport chain, it also contributes its protons to the intermembrane space. Because it has a lower level of energy potential, $FADH_2$ can only pump four protons into the intermembrane space. (Each NADH provides the energy to pump six protons across the membrane.)

With the protons shifted to the intermembrane space, the electrons continue to be carried through the various electron complexes until eventually their energy is almost depleted. The final carrier uses the remaining energy to pump two more protons into the intermembrane space. In aerobic respiration, oxygen is the ultimate electron acceptor. At the end of the transport chain the electrons are accepted by oxygen, and the oxygen in turn combines with hydrogen ions to form water. (That is the source of some of the water vapor in your breath — the water is carried to your lungs, which produce water vapor).

F_1 Complex with ATP Synthetase. Given the processes taking place, the intermembrane space has a much higher concentration of H^+ (protons) than the matrix. As happened in the thylakoid and stroma in photosynthesis, the result is a chemiosmotic gradient; therefore, the phosphorylation of ADP to form ATP can occur (Figure 9.12). Phosphorylation takes place within structures that connect the mitochondrion's outer and intermembrane regions, called the F_1 complex. The F_1 complex contains the enzyme ATP synthetase that is used to phosphorylate ADP to form ATP. Protons pass into the protein channels of the F_1 particle, which couples the movement of protons back into the matrix, and then the enzyme ATP synthetase is used to phosphorylate ADP to produce ATP.

At the same time, ADP and the needed inorganic phosphates are transported to the "head" of the complex. Hydroxyl ions (OH^-) are also transported to the complex. When the protons meet hydroxyl ions to form water, free energy is transferred in order to add an inorganic phosphate to ADP to form ATP. This process is catalyzed by the head or F_1 portion of the complex that has ATP synthetase.

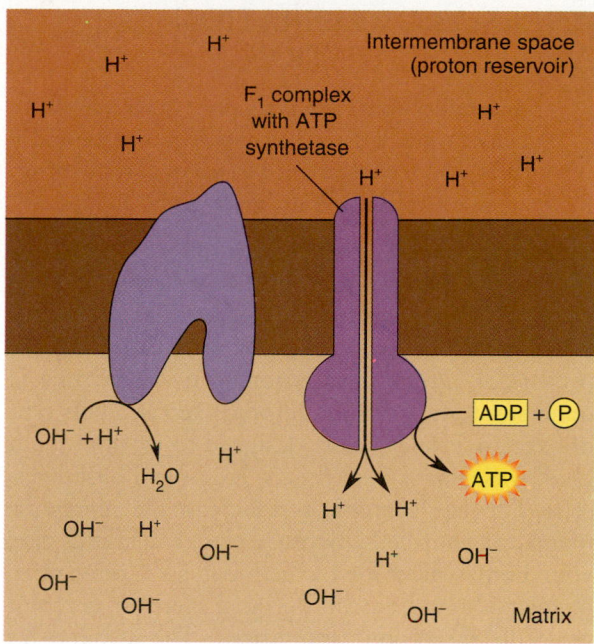

FIGURE 9.12 Generating ATP Through Chemiosmosis The concentration of protons in the intermembrane space of the mitochondrion is 10 to 100 times greater than that in the matrix. Energy freed as the protons flow through the F_1 ATP synthetase complex back into the matrix is used to generate the large amounts of ATP associated with aerobic respiration.

ATP Accounting. Many cell physiologists have stated almost emphatically that 38 molecules of ATP are formed for every molecule of glucose that is completely oxidized to CO_2 and water. But Peter Mitchell's chemiosmotic theory has made such an accounting much more difficult. Because of our understanding of chemiosmosis, many of the processes that occur in the oxidative (chemiosmotic) phosphorylation of glucose are unknown, and the results are far less predictable.

For example, if there are proton leaks in the mitochondrial membrane, the chemiosmotic differential will be reduced, which means that less ATP can be synthesized. Also, the energy stored in the proton gradient is used for other important processes such as transport of molecules into the mitochondria. As a result, it is impossible to pinpoint an exact accounting of ATP synthesis. At best we can only give a range. Most estimates state that through oxidative phosphorylation 25 to 36 ATP molecules can be synthesized. However, even if the smaller number is the correct one, the process is still very efficient.

ANAEROBIC RESPIRATION AND FERMENTATION

■ Anaerobic respiration takes place in several ways, two of which are ethyl alcohol fermentation and lactic acid fermentation. The major purpose of anaerobic respiration is to regenerate NAD$^+$ for use in glycolysis.

Not a day goes by that we don't hear about the value of aerobic exercise. We are told to jog, or swim, or dance. Now that you have examined the process of aerobic respiration, you will be able to understand the value of aerobic exercise. This type of exercise is specifically designed to work on your cardiovascular system (heart, lungs, and blood vessels). Through exercise, you increase the energy demands of your cells. Because energy demands rise, your body's need for oxygen increases (remember, oxygen is required for the final electron acceptor molecule in aerobic respiration). The demand for oxygen increases your cardiac output, since it is red blood cells that carry oxygen to cells. The demand for increased cardiac output results in an increase in the rate and intensity of the pumping of your heart.

However, during long, intense periods of exercise, something happens to the mechanics of aerobic respiration. You have seen an extreme example of the demands that aerobic exercise has on the human system if you have ever watched a marathon or ever run one yourself. At about the 18- to 20-mile point, marathoners experience what they call "hitting the wall"; they feel as if they simply can't go any further. Their reserves are gone. They lose all strength and experience leg pains of varying intensity; they may feel nauseous. Successful runners are the ones who continue, who push their bodies beyond the physiological response to aerobic exercise to finish the race (Figure 9.13).

Hitting the wall is something that training cannot prevent, because its existence has nothing to do with fitness. Instead it involves a basic biological reaction. To understand the change in the physiological response in such a case, we need to explore a second kind of respiration, called *anaerobic respiration*.

Putting it simply, when we discuss anaerobic respiration we are discussing respiratory reactions that occur without using oxygen as the final electron acceptor. Instead, another molecule — usually nitrogen — serves as the final electron acceptor. Using this logic, let us discuss the method of anaerobic electron transport and two types of anaerobic

FIGURE 9.13 Hitting the Wall
When the stress of marathon running overloads aerobic metabolic processes, anaerobic processes are activated causing discomfort and sometimes putting the runner at physiological risk.

respiration: ethyl alcohol fermentation and lactic acid fermentation.

Anaerobic Electron Transport

Anaerobic electron transport is quite similar to aerobic respiration; however, in the anaerobic electron transport chain oxygen does not function as the final electron acceptor. Instead, once glycolysis is complete, pyruvate cycles through until another inorganic molecule (such as nitrogen) is the final electron acceptor of electrons donated by NADH. One of the major functions of anaerobic respiration is to regenerate NAD$^+$ to recycle back to glycolysis to accept more electrons so that ATP can be synthesized in glycolysis.

As you will learn in Chapter 16, biologists hypothesize that the first organisms on the earth evolved in an environment that was relatively oxygen-free. In the past, the simpler prokaryotic cells and eukaryotic cells sometimes used anaerobic fer-

Dr. Roger A. Sabbadini
Muscle Membrane Physiologist

In the mid-1960s Dr. Roger Sabbadini was a student attending the University of California, Davis, and involved in protests against the Vietnam War. At that time, Sabbadini felt that his traditional college classes were irrelevant to the problems he saw in the real world. The real world was made up of the Vietnam War and racial injustice. Basically Sabbadini was more interested in fighting what he saw as atrocities perpetuated by the establishment, and that establishment included universities.

It should come as no surprise that Sabbadini's grades were low. He missed classes often, devoting most of his time to what he considered the most critical question of the day — the war.

But Sabbadini, like most students, needed money. He obtained a job in the work-study program, washing dishes in the laboratory of muscle physiologist Professor R. J. Baskin. And as he washed dishes he began to see another side of the university — one that was real and relevant. Dr. Baskin noticed Sabbadini's growing interest, so he had him do a few simple experiments. Over time Sabbadini became fascinated with science and with muscular dystrophy and the physiology of muscle contraction in particular.

With his growing interest in sci-

ence, he began to take school more seriously. He completed a B.S. in psychology at Davis and stayed on, working in Dr. Baskin's laboratory, this time concentrating his course work on physiology. In 1974, he completed work on a Ph.D. in physiology. When he graduated he was an honors student and was the recipient of UC Davis's Hertzendorf Memorial Award in physiology.

If Sabbadini's attitude toward science had changed, his attitude toward scientists had also changed. During his graduate studies, he attended his first major research conference in 1972, a meeting of world-renowned muscle physiologists at Cold Springs Harbor, Maine. He found that, for all their fame, these research leaders were normal people who had normal human interests — they enjoyed sailing and baseball, could laugh and have a good time. They were regular people involved in important work that could better people's lives. Sabbadini knew then that he too wanted to be a research scientist.

Today, Dr. Roger Sabbadini is an established muscle physiologist whose work is respected by his peers. Working at San Diego State University, he has devoted much of his research energy to the role of transverse tubules, the sarcoplasmic reticulum, and other membranes in

the contraction of skeletal and cardiac muscles, with an emphasis on the mechanisms of membrane activities.

The Roger Sabbadini of old might have trouble understanding the Dr. Sabbadini of the present, for this Dr. Sabbadini says that he loves his research, his teaching, and the feeling of purpose and the ability to help others that his work provides. He also says that science teaches him the limits of our knowledge. Science also teaches him that people are but one small component of life on earth, but, unfortunately, with a disproportionate impact on the environment.

mentation. How were these organisms able to respire without oxygen? These primitive organisms released energy by means of anaerobic respiration. During this process, pyruvate formed in glycolysis can be converted to CO_2 and ethyl alcohol (ethanol). The term *fermentation* is often used to mean anaerobic respiration, but in reality it is a type of respiration in which an organic molecule rather than oxygen is used as the final hydrogen acceptor.

Alcohol Fermentation

■ **Alcohol fermentation is anaerobic respiration that produces ethyl alcohol and carbon dioxide.**

Yeast cells, including one called *Saccharomyces cerevisiae*, undergo one type of anaerobic respiration, *alcohol fermentation*. (Spanish-speaking stu-

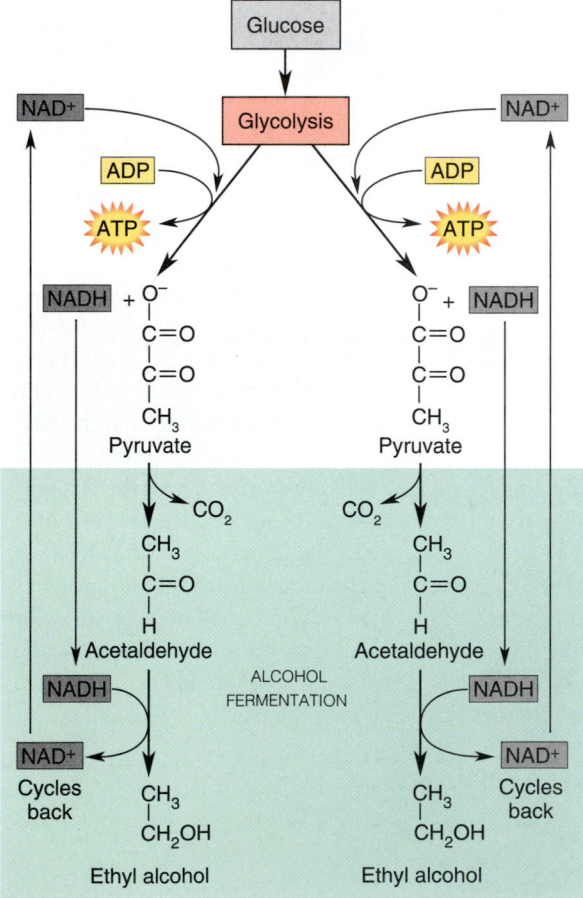

FIGURE 9.14 Alcohol Fermentation
The alcohol fermentation process allows for the oxidation of NADH; thus the NAD⁺ can be reused in glycolysis to accept hydrogens. Without this recycling of NAD⁺, glycolysis would stop.

dents will recognize that *cerevisiae* is related to *cervesa*, Spanish for "beer.") How do yeast cells respire, or ferment, and how do they produce alcohol?

In alcohol fermentation, once pyruvate is formed through glycolysis, the next step is that a carbon atom is removed from the pyruvate molecule to form acetaldehyde, a two-carbon compound, and CO_2, which leaves the cell. The NADH that was formed during glycolysis serves as a hydrogen source to reduce acetaldehyde to form ethyl alcohol. (The NAD⁺ returns to glycolysis to accept hydrogens.) It is from this end product that this cycle gets the name alcohol fermentation (Figure 9.14).

People have used this process to produce alcoholic beverages since the beginning of recorded history. Alcohol fermentation produces enough energy from one molecule of glucose to yield two molecules of ATP, not nearly as efficient as aerobic respiration but enough to sustain life in an environment free of oxygen. There is ample evidence of the calories that remain in alcohol in the "beer belly" that is often the result of the regular consumption of alcohol (Focus 9.1).

Lactic Acid Fermentation

■ **Lactic acid fermentation is a type of anaerobic respiration that results in lactic acid formation.**

Another possible route for pyruvate in anaerobic respiration is lactic acid fermentation, in which lactic acid is formed. (Because the acid form is the more common usage we will use it rather than the ionic term lactate.) This process is usually described in human muscle cells, although it can occur in the skeletal muscles of all vertebrates and in bacteria as well. Many of the lactate-fermenting bacteria are found in milk and are responsible for its souring. In this form of respiration, the pyruvate formed in glycolysis is reduced by the hydrogen from NADH to form lactic acid (lactate) (Figure 9.15).

Here, as with alcohol fermentation, enough energy is released to form two ATP molecules. NADH is oxidized and the NAD⁺ is recycled through glycolysis.

Aerobic and Anaerobic Respiration: Levels of Efficiency

■ **On the basis of caloric output and ATP production, aerobic respiration is more efficient than anaerobic respiration.**

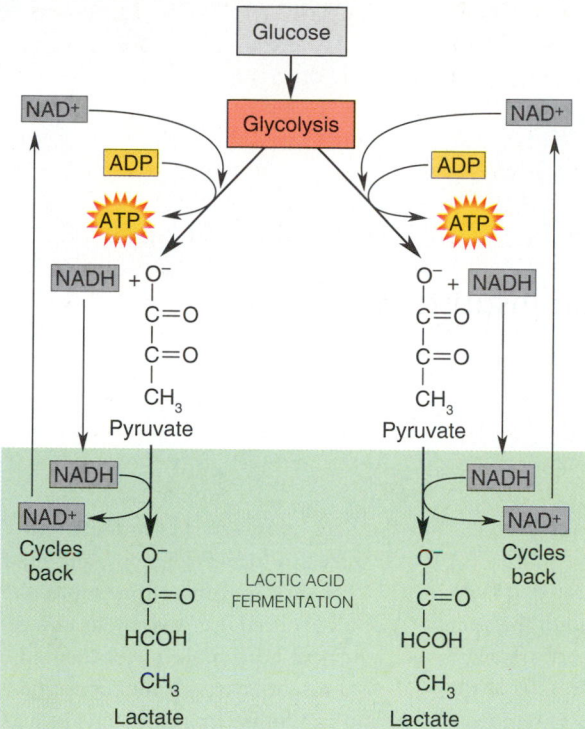

FIGURE 9.15   Lactic Acid Fermentation
As in the alcohol fermentation process, lactic acid fermentation allows for the oxidation of NADH to NAD^+, which is then reused in glycolysis. However, lactic acid fermentation usually occurs in muscle cells and some bacteria.

It should be noted that forms of anaerobic respiration produce only about 150 Calories of energy per mole of glucose, while aerobic respiration can produce about 680 Calories. If the two types of respiration are compared on the basis of the efficiency of ATP production, aerobic respiration is much more efficient than anaerobic respiration. This difference exists because a great deal of energy remains in the molecules that are the end product of anaerobic respiration. That explains why alcohol burns — enough calories remain in the molecules, even after they have gone through a fermentation process, so that energy can still be liberated.

Another Look at Aerobic Exercise

■ **Strenuous exercise depletes the system of oxygen supplies, which switches the form of respiration from aerobic to anaerobic. When carbohydrate reserves are also depleted, organisms draw on other energy sources.**

Let's now discuss strenuous exercise. Under normal conditions your body provides your muscle cells with enough oxygen for aerobic respiration, but many athletes undergo such strenuous exercise that their bodies can no longer keep up with the oxygen demands of aerobic respiration. The electron transport chain becomes loaded with electrons that it cannot donate to oxygen, because there is not enough oxygen to serve as the ultimate electron acceptor. Since the chain cannot get rid of the electrons through normal oxidative phosphorylation, NADH is no longer oxidized, so there is no NAD^+. With a shortage of NAD^+, the Krebs cycle cannot function. With no Krebs cycle, pyruvate builds up.

The buildup of pyruvate signals muscle cells to shift metabolic gear to the less efficient anaerobic respiration or lactic acid fermentation. But lactic acid fermentation produces a buildup of lactic acid in muscle tissue, and lactic acid is toxic to human systems, which is why the buildup can produce muscle pains and cramps during strenuous exercise.

When you stop exercising and rest, your circulatory system removes the excess lactic acid and increases the oxygen level in the cells. The increased oxygen supply allows your muscle cells to switch back to aerobic respiration and to convert some lactic acid into pyruvate. Some pyruvate is oxidized to CO_2 and H_2O to provide ATP, which is used to convert the remaining pyruvate into glucose and glycogen, which can be stored and used at a later time. The following equation illustrates what happens when excessive lactic acid builds up in a system.

$$\text{Excess lactic acid} \xrightarrow{NAD^+ \rightarrow NADH} \text{Pyruvate}$$

$$\text{Pyruvate} \rightarrow CO_2 + H_2O + ATP$$

$$\text{Pyruvate} \xrightarrow{ATP} \text{Glycogen} \rightarrow \text{Glucose}$$

Now we are ready to return to the problems that face our marathoner. The wall that plagues marathoners is not caused by lactic acid buildup alone. In this case, the cells run out of stored carbohydrates and glycogen (which is being converted to glucose), and they have to switch to metabolizing fat molecules. Some women athletes do better than men in ultramarathons, which are 50 to 100 miles long. Biologists hypothesize that female systems can switch over more easily to fat metabolism than can those of males.

FOCUS 9.1
Alcoholism and Cellular Respiration:
The Health Problems Arising from a Social Institution

As we all know, alcohol is the most abused drug in the United States. What is more, because of our social drinking pattern, alcohol is considered socially acceptable. As college students, you know that 90 percent of college students drink, and 1 in 12 will have a serious drinking problem as an adult. Approximately 100 million people in the United States drink alcohol, and of those drinkers about 9 million are addicted. When that 9 million is broken down according to ethnic groups, Native American youths have the highest proportion of heavy drinkers (16.5 percent), followed by Orientals (13.5 percent), Chicanos (10.9 percent), whites (10.7 percent), and blacks (5.7 percent). These are very sad statistics, indeed, especially when you consider the serious health problems and the mortality rate attributed to the disease of alcoholism.

What does alcohol do when it enters the body? Why are heavy drinkers prone to serious liver diseases? What is cirrhosis of the liver, and why is it the leading cause of death among New Yorkers between the ages of twenty-five and sixty-five? From a biological viewpoint, the answers to these questions are quite simple, involving an understanding of how the liver metabolizes alcohol.

When alcohol is ingested, it is absorbed through the wall of the stomach, enters the circulatory system, and is taken to the liver. One of the functions of the liver is to remove toxic substances from the body, such as the ethanol (ethyl alcohol) found in alcoholic beverages. This takes place in a series of steps.

1. One of the B vitamins, niacin, is used to form NAD^+, which is necessary for the breakdown of ethanol. Another B vitamin, thiamin, is also necessary for the breakdown of ethanol.

2. An enzyme in the liver, *alcohol dehydrogenase,* removes two hydrogen atoms from ethanol, thus converting it to acetaldehyde. The two hydrogen atoms are transferred to NAD^+. The reaction is as follows:

$$CH_3CH_2OH + 2\ NAD^+ \rightarrow$$
Ethanol

$$CH_3CHO + 2\ NADH + 2\ H^+$$
Acetaldehyde

3. The acetaldehyde, along with another molecule, of which thiamin is a part, is converted to acetyl CoA.

4. The acetyl CoA enters the Kreb cycle and is eventually oxidized to form ATP, CO_2, and H_2O.

The metabolizing of one molecule of alcohol requires two molecules of niacin and one molecule of thiamin. As a result, heavy alcohol consumption can utilize large amounts of niacin and thiamin, depleting the amount of these substances in the body. Since niacin and thiamin are necessary components of electron transfer molecules, they are necessary for proper glucose metabolism. Depletion of these two vitamins can affect every cell in the body. The effect is particularly profound in brain cells, since they can use only glucose (not fats or protein) for the production of their energy.

But how does the processing of alcohol affect the liver? As you remember, in the normal course of events NADH goes to the electron transport chain in the mitochondria, where it is oxidized to produce ATP and H_2O. Because NADH can be produced from the hydrogens in alcohol, the liver cell has high levels of NADH formed from alcohol metabolism. As a result, the liver does not need the NADH normally produced from glycolysis and the Krebs cycle. That in turn means that sugars, fatty acids, and proteins are not broken down completely but rather are converted to fats, which are stored in the liver cells (Figure A). Metabolism becomes disproportionate because the Krebs cycle is almost shut down and the electron transport chain works much harder, which causes the mitochondria to become enlarged and distended. Other cellular structures proliferate, such as the smooth endoplasmic reticulum, because of the accumulation of fats.

Eventually the liver cells fill with fat and can no longer function; so they die. The liver becomes inflamed and impaired, a condition known as *alcohol hepatitis*. If this continues long enough, the dead liver cells are replaced by scar tissue, which interferes with normal liver cell metabolism and the blood supply to the liver, a condition known as *cirrhosis*. Liver cells cannot function, nitrogen wastes are not broken down, toxic poisons accumulate, and the person can die. Should the individual continue to drink, cirrhosis can worsen into a condition known as *ascites*. In this condition, the passage of blood to the liver is impaired and liver fluid accumulates in the abdominal cavity, producing an extremely large distended abdomen.

The accumulation of fats and early alcoholic hepatitis, the earlier stages of liver damage, can be reversed through proper diet and reduced alcohol consumption. However, cirrhosis of the liver cannot be reversed and is the seventh leading cause of death in the United States.

Liver damage is a fairly well-known complication of heavy alcohol use, but it is not the only problem. Alcoholism can also weaken muscle tissue, impair brain activities, and adversely affect sperm production and sex drive. Expectant mothers who consume alcohol during pregnancy can bear children suffering from *fetal alcohol syndrome*.

This syndrome is the result of the fact that the nervous tissues of the developing fetus were deprived of necessary niacin and thiamin for proper glucose usage. Fetal alcohol syndrome can produce both mental and physical retardation. The brain damage is irreversible; children so affected are doomed to a life of mental retardation. The amount of alcohol needed to produce this syndrome is still under investigation, but some investigations say even small amounts may be harmful. Many doctors now recommend that their pregnant patients consume no alcohol.

The effects of high alcohol consumption are clear, and the costs, even if you consider only the health problems that can arise, are tremendous. All members of our society should be aware of the possible complications that can arise from the use and abuse of our "great social custom."

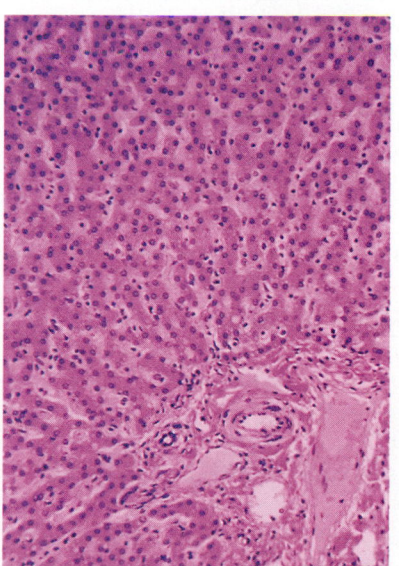 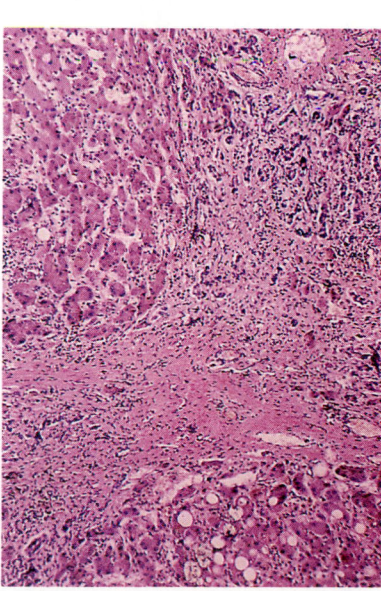

FIGURE A
Normal human liver cells (left) as compared with those taken from a human liver affected by cirrhosis (right). Fat globules are readily obvious in the affected cells. Both micrographs were taken at the same magnification.

Anaerobic Respiration and Fermentation

RESPIRATION OF FATS AND PROTEINS — OTHER ENERGY PATHWAYS

The way that a marathoner's system functions as it searches for new sources of energy once its carbohydrate reserve is gone shows us that cells can switch from one energy source, like starch or glucose, to other energy sources, such as fats or proteins. What happens when other foods are the energy source (Figure 9.16)?

To burn foods such as fats or proteins, the molecules must undergo a series of make-ready rearrangement reactions before enzymes can remove hydrogen. The make-ready rearrangements result in a molecule that can be oxidized to release the energy required to phosphorylate ADP to ATP.

For instance, proteins are hydrolyzed to amino acids. The amino acids enter the cell, where they may be used as building blocks to form other proteins, or as sources of energy. To be used as an energy source, the amino acid is *deaminated* (that is, the amino group is removed). The remaining carbon skeleton can be converted to an acetyl group or other molecules that function in the glycolytic pathway or the Krebs cycle. If the amino group is not reused, it is eventually removed from the body as urea or other nitrogen-containing wastes. That's why (as parents discover when they change their baby's diapers) urine sometimes smells like ammonia — certain bacteria can convert urea to ammonia.

Fats are digested to the three-carbon alcohol, glycerol, which enters the glycolytic pathway at the point when DHAP (step 3, page 194) is formed. DHAP can also serve as a fat precursor. The fatty acids are converted into two-carbon fragments (acetyl groups) and combined with CoA to form acetyl CoA and enter the Krebs cycle.

So, as you can see, living things do not live by bread alone but can utilize other carbon compounds, such as proteins and fats, as sources of energy, especially during dieting and during starvation.

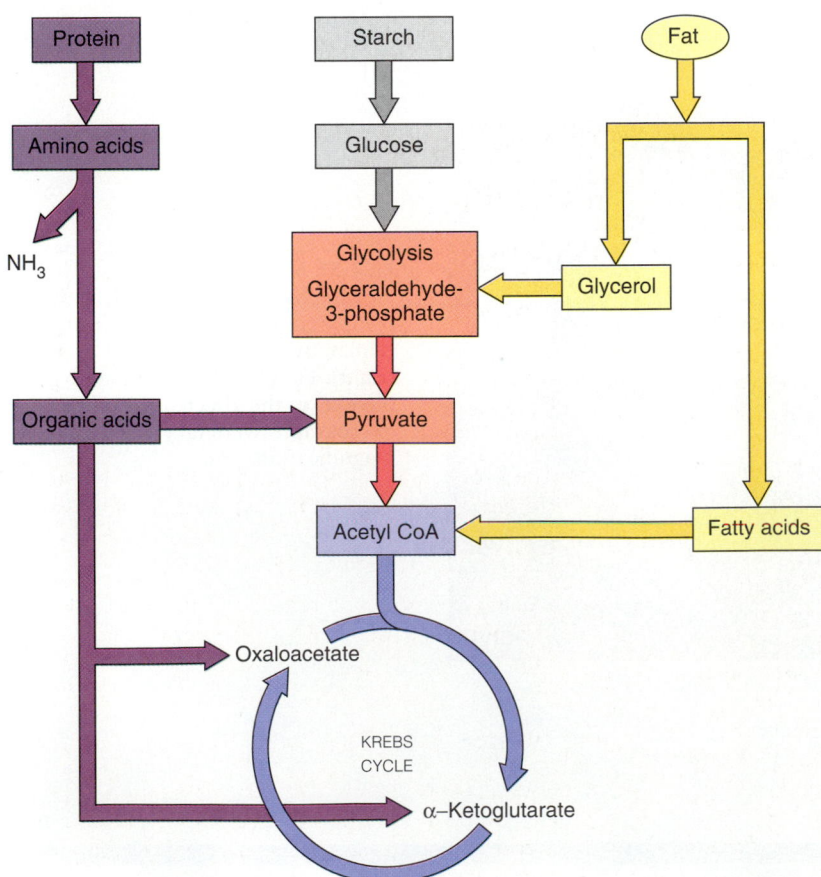

FIGURE 9.16 Other Energy Pathways
With relatively minor alterations, a variety of food molecules can enter glycolysis or the Krebs cycle.

The connections between photosynthesis and respiration are even closer than is apparent in this chapter. As we will discuss in Chapter 32, on our planet autotrophs and heterotrophs are interdependent. In other words, heterotrophs are as essential to the life of autotrophs as autotrophs are to heterotrophs, in that the decay of the bodies of heterotrophs supplies minerals to the soil. Without those materials, plants could not grow and develop to produce the food that heterotrophs need for the energy and raw materials for their growth.

So, in order to maintain life, two interrelated systems have evolved, photosynthesis and respiration. In fact, evidence for evolution includes the similarities in the ways each system manages energy and transfers it from one molecule to another. One such similarity involves the chemiosmotic reactions of the electron transport chains in both photosynthesis and respiration. Another is the universal presence of the energy molecule ATP, the coenzymes that function as electron carriers like NAD^+ and $NADP^+$, and the important electron carriers found within membranes. For all the differences between the living things on our planet, how similar we are.

SUMMARY

1. Cellular respiration is a series of step-by-step, enzymatically mediated oxidation reactions that liberate energy for the synthesis of ATP.
2. Respiration is a continuous process. It is divided into two types: aerobic respiration (respiration that uses oxygen) and anaerobic respiration (respiration that does not use oxygen). Both types begin with glycolysis.
3. In glycolysis, which occurs in the cytoplasm of the cell, one molecule of glucose is converted through a series of rearrangement reactions into two molecules of pyruvate. There is a net gain of two ATPs.
4. Aerobic respiration has two stages: the Krebs cycle, which removes hydrogen and carbon atoms, and the electron transport chain.
5. The Krebs cycle, which occurs within the matrix of the mitochondria, consists of a series of nine reactions in which dehydrogenation (hydrogen removal) and decarboxylation (removal of CO_2) occur.
6. The hydrogen atoms that are removed during the Krebs cycle enter the inner mitochondrial membrane and are picked up by the molecules of the electron transport chain. As the electrons are transferred from one electron acceptor to another, their excess energy is removed.
7. The chemiosmotic theory explains the actual formation of ATP, which takes place in the F_1 complex

with ATP synthetase that connects the matrix and inner membrane space in the mitochondrion. Protons (H^+) are transferred into the intermembrane space, establishing an electrochemical gradient across the inner mitochondrial membrane. ATP formation occurs in the F_1 ATP synthetase complex.
8. Anaerobic respiration occurs in three forms: the anaerobic electron transport chain, which occurs in prokaryotes, ethyl alcohol fermentation, which occurs in yeast, and lactic acid fermentation, which occurs in vertebrate muscle cells and certain bacteria.
9. In ethyl alcohol fermentation, pyruvate is broken down to carbon dioxide and ethyl alcohol. In lactic acid fermentation, pyruvate is broken down to lactic acid. In other cells, when oxygen is available, the pyruvate is further oxidized in the Krebs cycle.
10. Other food substances, such as fats, carbohydrates other than glucose, and proteins, can be used as sources of energy to form ATP.

KEY TERMS

heterotroph	anaerobic respiration
cellular respiration	fermentation
glycolysis	matrix
pyruvate	Krebs cycle
aerobic respiration	cytochromes

REVIEW QUESTIONS

True/False

1. Respiration and breathing mean the same in terms of the physiological processes that release energy in living organisms.
   true   false   (page 189)

2. All organisms require oxygen to release the energy stored in food molecules.
   true   false   (page 189)

3. The Krebs cycle can be defined as an orderly series of reactions in which glucose is degraded to three-carbon compounds.
   true   false   (pages 196–197)

4. The biochemical events that occur in the electron transport chain occur only in the cytoplasm.
   true   false   (page 198)

Fill in the Blank

1. Some microorganisms and even human muscle cells can respire without oxygen. This type of respiration is called _____ _____ fermentation.
   (page 202)

2. The cytochrome molecules required for the orderly passage of electrons in the mitochondrion are embedded in the _____ of the mitochondrion. *(page 198)*

3. The major events in the Krebs cycle are the removal of _____ to the atmosphere and hydrogens that are carried to the electron transport chain. *(page 196)*

4. According to the chemiosmotic theory the energy that is used to synthesize ATP from ADP is due to the greater concentration of _____ between the inner and outer portions of the inner membrane of the mitochondrion. *(pages 198–199)*

Multiple Choice

1. Aerobic cellular respiration is a process by which cells of an organism:
   a. Use oxygen
   b. Produce carbon dioxide
   c. Produce energy
   d. Form ATP
   e. All of these
      (page 190)

2. Before lipids can yield energy in respiration they must be converted to:
   a. Glucose
   b. Protein
   c. Glycogen
   d. Intermediates of glycolysis and acetyl CoA
      (page 206)

3. Cell M contains more mitochondria than cell N. All the following are probable characteristics of cell M *except*:
   a. Potential ATP production is greater than in cell N.
   b. Energy requirements are greater than those of cell N.

c. Volume of O_2 produced per unit of time is greater than that produced by cell N.
   d. Mechanisms for oxygen diffusion are different from that of cell N. *(pages 198–199)*

4. Which of the following stages in respiration occurs in the cytoplasm of the cell?
   a. Glycolysis
   b. Krebs cycle
   c. Cyclic photophosphorylation
   d. Electron transport chain
      (page 190)

Discussion

1. Why is respiration important in living things?

2. Briefly describe glycolysis, the Krebs cycle, and the electron transport system.

3. Why is aerobic respiration more efficient than anaerobic respiration?

4. How does the chemiosmotic theory help to increase our understanding of ATP synthesis?

5. From a cellular respiration perspective, explain why strenuous aerobic exercise may produce muscle cramps.

6. From a cellular respiration perspective, do living things live by bread alone? In other words, can other foods be utilized as sources of energy? Why or why not?

7. What would happen if all the mitochondria were removed from an animal cell? Explain your answer relative to your understanding of the chemiosmotic theory.

8. Why is oxygen important in the oxidative phosphorylation of ADP to form ATP?

GENETIC CONTINUITY

After reading this unit, the student should be able to:

- understand the molecular and cellular events that result in the passage of structural and functional information from one generation to the next
- understand the processes that result in the packaging of genetic information into chromosomes and the events that result in genetic continuity at the cellular level
- understand the structure and function of RNA, DNA, and proteins and their role in genetic continuity and gene regulation
- understand the bioethical issues of genetic recombinant technology and human genetics
- understand patterns of inheritance as they apply to organisms, including humans

10

Cell Reproduction

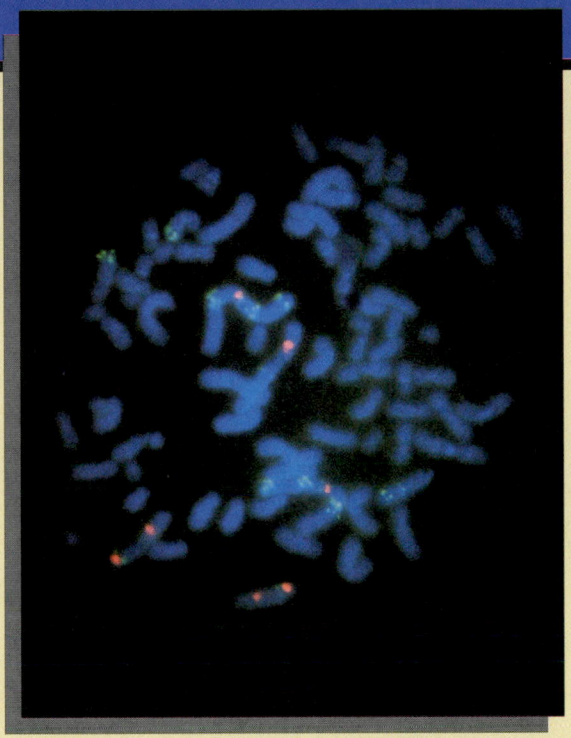

No one knows exactly why, but one or more of the cells in Henrietta Lacks's uterus had lost the ability to regulate that fine balance between maturation and reproduction. Somehow, an alteration had taken place in the DNA, and it had lost its ability to regulate cell division. Now these altered cells were selfishly responding to a chemical urgency to reproduce their altered state. As their ability to communicate through cell surface contact inhibition was lost, the cells returned to an almost undifferentiated, unspecialized state, absorbing nutrients and producing no good for the individual. In short, Henrietta Lacks had uterine cancer.

What began slowly accelerated quickly, for the number of cells doubled every time, increasing their mass at an incredible pace. First there were two

cells, then four, then eight, then sixteen, then thirty-two, until after twenty-seven divisions there were thousands of rogue cells. The cancer cells began to migrate to other parts of Henrietta Lacks's body — the growth was spreading — and in 1952 she died. The woman died, that is. The cells from her cancerous tumor (now known as HeLa cells) live on to this very day in tissue cultures that began in 1952. Why? They have provided material for cancer research. In fact, not only have Henrietta Lacks's cells outlived her, when added altogether there are literally trillions of HeLa cells growing in culture flasks in laboratories scattered all over the world.

We all know that cancer cells do not behave like normal cells (Figure 10.1). Why don't cultured cancerous cells die after a genetically predetermined number of generations as cultured noncancerous cells do? What is this unregulated cell division known as cancer? How do cells divide, and why? We begin life as a single cell — how do we become

---

Above: A special type of fluorescence microscopy was used to photograph these human chromosomes.

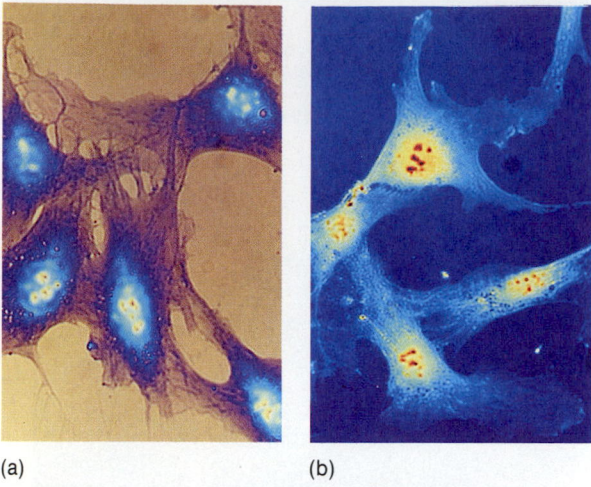

(a)　　　　　　(b)

FIGURE 10.1 Normal and Cancerous Cells
(a) Normal cells of the retina and (b) the cells of a malignant tumor. Cancerous cells divide uncontrollably, piling on top of one another. Growth in normal cells is carefully regulated.

such a complicated multicellular organism, consisting of several trillion (10^{12}) specialized cells? Let's begin the search for answers to these intriguing questions (see Profile).

All living organisms, whether composed of prokaryotic or eukaryotic cells, grow and reproduce by processes occurring at the cellular level. Our cells take in the raw materials necessary for life and use these nutrients for growth and repair, until the cell reaches a specific size. It then reproduces itself through cell division. No one knows what mechanism initiates cell division, but in all probability it involves a series of genes that either initiate or inhibit DNA replication and cytoplasmic division. Once the cell divides, the result is two daughter cells that, in most instances, are exact duplicates of the original. How does the dividing cell bridge the gap between generations? To understand that, we need to begin our discussion with an overview of the major concepts involved in cell reproduction.

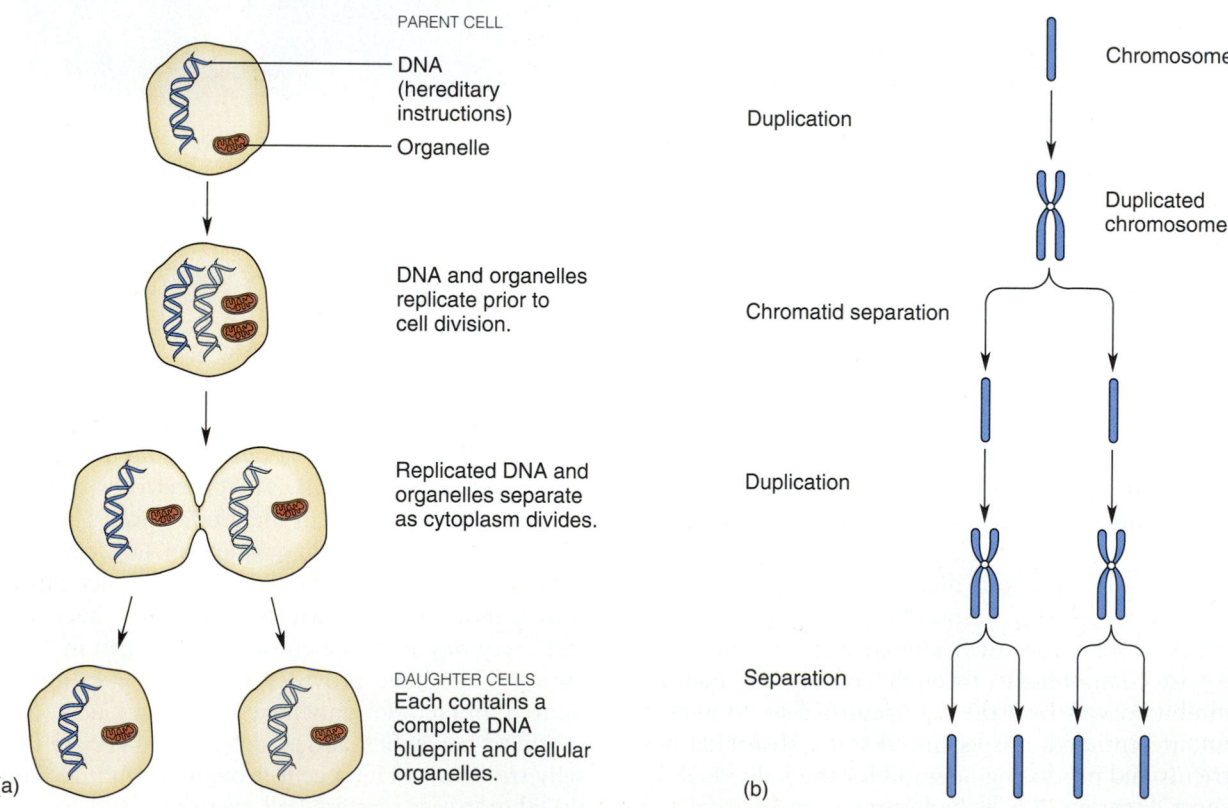

FIGURE 10.2 Cell Division and Replication
(a) During cell division, a complete set of genetic information (DNA) is passed on from the parent cell to the daughter cells, as are some organelles like the mitochondrion. (b)Notice that the replicated DNA is packaged into chromosomes. During cell division the duplicated chromosomes halve (the two chromatids separate) resulting in daughter cells each with the same chromosome number as the parent cell. In the next generation, and subsequent generations of cells, the process is repeated.

OVERVIEW OF CELLULAR REPRODUCTION

When cells reproduce, the daughter cells must receive the genetic information from the parent cell. Therefore, cellular reproduction includes a series of processes in which genetic information is packaged to be, shall we say, shipped to the next cell generation. This is just like when you move; to avoid losing things, you condense them into boxes for moving. Cells do a similar thing prior to cell reproduction. Genetic information is duplicated and then packaged into "boxes"—chromosomes. When information is packaged, the possibility of losing important information is decreased. But what is the genetic information that is being packaged? As we discussed in Chapter 3, DNA is the genetic material; therefore, each new daughter cell must receive a duplicate of the parental DNA. Within the DNA is the genetic code that provides the instructions for producing the RNA used in making the proteins, which either functions as enzymes, other regulators, or structural components of the cell. But the daughter cell not only receives genetic information from the parent cell, it also inherits cytoplasmic material like enzymes, and in eukaryotic cells, cellular organelles—enough to sustain its life until it becomes fully developed and functioning. Then it reproduces and the cycle continues (Figure 10.2). With that general overview in mind, let us first look at cellular reproduction in prokaryotic cells, and then let us look at the structures involved in cellular reproduction in eukaryotic cells.

CELLULAR REPRODUCTION: EUKARYOTIC EMPHASIS

- **Cell reproduction in prokaryotic cells involves DNA replication. In eukaryotic cells, cell reproduction is more involved, partly because of the nuclear membrane.**

Prokaryotic cells reproduce themselves by a process known as *cellular fission,* or *binary fission,* as it is sometimes called. That means that the process of cell division results in reproduction of that organism, since a one-celled organism recreates itself when it divides. Fission in a prokaryote is a relatively simple matter. The DNA replicates, but since it is not surrounded by a nuclear membrane, the genetic material does not have to be repackaged in new nuclear membranes. Instead, the two sets of DNA are attached to two different sites along the cell membrane, and the cell membrane then pinches

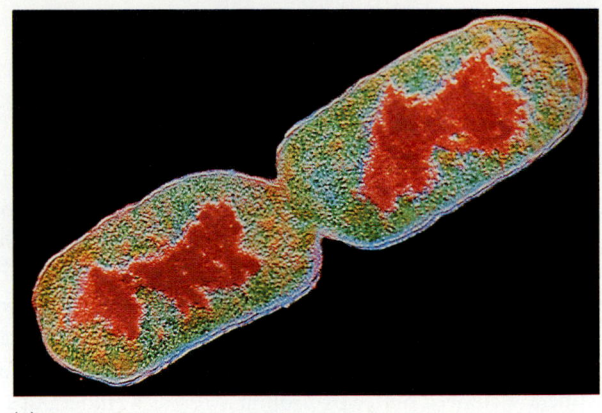

(a)

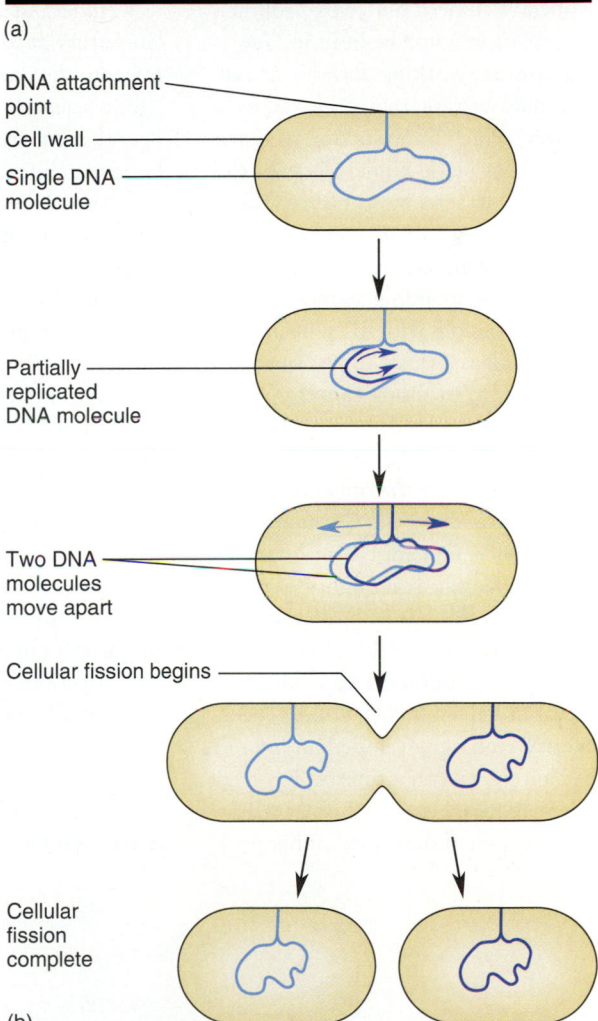

DNA attachment point
Cell wall
Single DNA molecule

Partially replicated DNA molecule

Two DNA molecules move apart

Cellular fission begins

Cellular fission complete

(b)

FIGURE 10.3 Cellular Division in a Prokaryotic Cell
(a) A light micrograph showing a thin section through a dividing bacterial cell (prokaryote). (b) Basic steps involved in DNA replication and cell division in prokaryotic cells.

in two to divide the cell. Each of the new cells has a complete set of hereditary DNA material (Figure 10.3). In Chapter 13 we look at prokaryotic division in more detail.

Dr. Robert Weinberg
Experimental Biologist

It is not easy to find free time, given the demands at the cancer research laboratory at the Massachusetts Institute of Technology, but with a little effort one can manage. For instance, you can sometimes find Dr. Robert Weinberg puttering around the vacation home he built in New Hampshire, working with his prized begonias, or researching his family's history, which interests him so much that he has made numerous trips to Germany to search out old records. And in spite of the demands on his time, Weinberg is very much a family man, spending as much time as possible with his wife and children. It is a busy life for one of the world's leading cancer researchers.

Robert Weinberg, born in Pittsburgh in 1942, is the only child of parents who fled Nazi Germany. Given the significance of his work in cancer research, you might be surprised to learn that he had no interest in science as a child, although he did decide to pursue a premedical major at MIT. Once in college, however, Weinberg developed an interest in biology, and that interest has never abated. He stayed on at MIT and completed the work on his

Ph.D. He was soon identified as an aspiring researcher and was accepted as a postdoctoral fellow at the Weismann Institute in Israel, which was followed by a fellowship at the Salk Institute in La Jolla, California.

Weinberg usually arrives in his laboratory at MIT around 10 A.M. to begin the pursuit of what he calls "good science" and the training of "good scientists." He often refers to his research team as "friendly rivals," since they work together toward the common goal of obtaining a better understanding of the genetic basis of cancer.

Even though Weinberg was one of the scientists responsible for identifying the specific proto-oncogenes involved in certain forms of cancer, this man is not the type to rest on his past accomplishments. (Proto-oncogenes are genes that regulate cell growth but sometimes are changed so that they convert normal cells to cancer cells.) Instead, Weinberg prefers to discuss some of the new directions of oncogene research. For example, he and his staff of about 12 highly talented scientists are currently pursuing the answers to questions such as these: How do cancer-producing agents convert the

proto-oncogene into a cancer gene? How does the abnormal protein produced by the oncogene function in the regulation of cell division? What are the other events that occur in the development of cancer?

Robert Weinberg is one of the pioneers in oncogene research. His work, along with that of numerous others, is helping to solve the mysteries of cell divison regulation and has obvious implications for understanding the molecular basis of the causes of cancer.

Parents and Daughters: An Overview

■ The original cell is called the parent cell, and the two cells that result from division are called the daughter cells. Each contains the necessary organelles, nucleus, and DNA in order to function.

In the process of cell reproduction, the original cell is called the **parent cell.** The two cells that result when the parent cell divides are called **daughter cells.** When a eukaryotic parent cell divides, it does not merely pinch in two. In order to function, the new cells must have the necessary assortment of cytoplasmic structures, such as mitochondria, chloroplasts, and centrioles. A daughter cell must also receive a complete set of hereditary instructions, or DNA.

As a result of evolution, each new daughter cell receives a complete set of cytoplasmic structures and hereditary instructions, ensuring continuity in the genetic information from generation to generation. Before cell division, the cell replicates its genetic instructions (see Figure 10.2) and cellular organelles for the cell reproductive process.

DNA: The Special Molecule of Heredity

■ DNA is the substance that carries genetic information, important for heredity and for protein synthesis. The information is encoded in units called genes.

Genetic information, the blueprint for life, is contained in the molecule known as deoxyribonucleic acid, or DNA. The linear sequence of nucleotides within the DNA molecule serves as the genetic code. The alphabet of life is spelled by the varied arrangements of four nucleotides (adenine, guanine, thymine, and cytosine) within the DNA molecule (Chapters 11 and 12). Every sequence of nucleotides that functions as a unit serves as a **gene** (see Chapter 12).

DNA serves two main purposes in the functioning of an organism. The first is to regulate the cell's activities. It does this by determining which proteins will be synthesized. Some proteins will function as enzymes or other regulatory molecules, moderating which reactions will occur within the cell and which will not. The second major role of DNA is its role in heredity, the transmission of qualities from parents to offspring. The ability of the DNA molecule to make copies of itself allows it to carry information for regulating the cell's activities from one generation to the next.

Chromosomes: Packaged DNA and Proteins

■ Relaxed or dispersed DNA is in the form of chromatin. Condensed DNA is in the form of chromosomes. Chromosomes occur in homologous pairs. The left and right branch of each chromosome is called a chromatid. Chromatids are joined at the centromere region.

In eukaryotes, the genetic material, DNA, is packaged with proteins into structures called **chromosomes,** which usually occur in pairs. The members of the pair are called **homologues,** or *homologous chromosomes,* one of which is maternal and the other paternal. Each eukaryotic species usually has a particular number of pairs of chromosomes. When biologists discuss pairs of chromosomes, they refer to the **diploid (2n) number.** Each parent contributes one member of each chromosome pair, so biologists refer to the maternal and the paternal chromosome. Most eukaryotic cells have two or more pairs of chromosomes, depending on the species being discussed. Humans, for example, have 46 chromosomes, arranged in 23 pairs, so their diploid number is 46. Table 10.1 compares the diploid number of chromosomes of several species.

In a sexually reproducing organism, body cells are called *somatic cells;* the cells that are involved in reproduction are called *germ cells.* As we noted, in humans each somatic cell contains 23 pairs of chromosomes, one from each parent. As you can see in Figure 10.4, each chromosome has a distinctive shape and size, and it is relatively simple to

TABLE 10.1 Diploid Chromosome Number of Some Organisms			
Alligator	32	Grasshopper	24
Amoeba	50	Housefly	12
Cat	32	Human	46
Chicken	78	Lettuce	18
Chimpanzee	48	Marijuana	20
Corn	20	Penicillium	2
Dog	78	Pigeon	80
Earthworm	36	Rhesus monkey	42
Fruit fly	8	Tobacco	48
Garden pea	14	Yucca	60

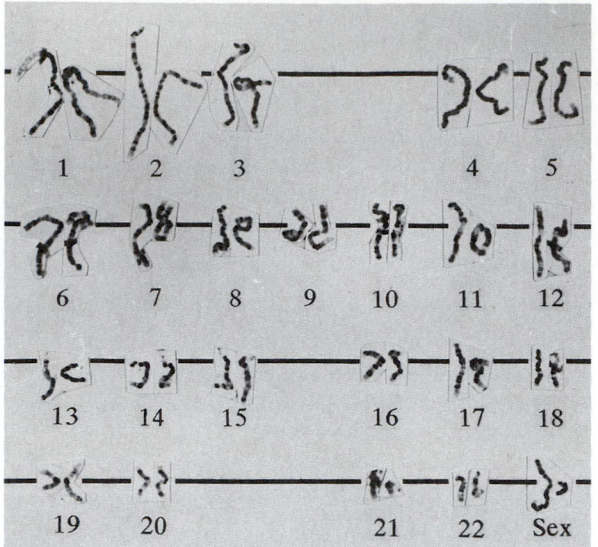

FIGURE 10.4 Human Karyotype
This prepared karyotype (male) shows the 23 pairs of human chromosomes arranged in their homologous pairs. Each parent contributes one member to each pair.

group them into matching pairs according to their appearance. The technique in which chromosomes are photographed and matched up for study is called *karyotyping*. In this procedure, one simply takes a photograph of the chromosomes, cuts out each one, and matches it up with its partner. The chromosomes can be counted, the chromosome numbers obtained and identified, and abnormalities noted.

But what is the structure of a chromosome? A chromosome in reality is packaged DNA. Prior to cell reproduction, DNA is replicated, condensed, and packaged into units. This condensed DNA, along with numerous proteins, is called *chromatin*, "colored material," due to its staining properties when viewed by a light microscope in an undividing cell. Chromatin is composed of both DNA and two general categories of proteins: **histone** and non-histone. The nonhistone proteins are acidic because of their negatively charged R groups (remember functional groups, Chapter 3) and are found in loose association with the DNA. The function of nonhistone proteins is not known; however, it is speculated that they may be involved in gene regulation. Histone proteins are positively charged proteins and bind to the DNA, which is negatively charged. There are five types of histone proteins: H1, H2A, H2B, H3, and H4. As the chromosome forms, the negatively charged DNA (due to its phos-

phates) is wound around eight (positively charged) histone molecules consisting of two copies of H2A, H2B, H3, and H4. This forms a unit called a **nucleosome**. The nucleosome and DNA look like pearls on a string, each pearl being the nucleosome and the DNA between the pearls—called *linker DNA*—linking the nucleosomes together. The nucleosome is the basic unit of DNA packaging. The chromosome continues to condense, forming tighter coils. For example, H1 helps to hold nucleosomes closer together to form a fiber that has six nucleosomes, together called the *30-nm fiber*, and so on. Figure 10.5 and the Human Endeavors box illustrate chromatin and chromosome formation. As the condensation continues, the chromosome is formed.

When you examine the chromosomes shown in Figure 10.5, you see that they appear like two strands of twine (two arms) held together at some point by a structure that looks like a narrow band. Since there are two sides to each chromosome, the chromosomes are said to be bivalent. The region where they are connected is called the **centromere** region. The arms of the chromosome (each side), called **chromatids**, consist of coils of protein surrounded by DNA. The two together are called *sister chromatids* (Figure 10.5). The centromere is the chromosome's "waist," which seems to tie the sister chromatids together.

It is important to remember that by the time the chromosome is visible with a light microscope the cell is ready to undergo cell division. Part of the preparation for division is duplication of the cell's organelles and replication of the DNA. (We discuss the replication of DNA in Chapter 11.) When the chromatin condenses, DNA has already replicated. The result is the chromosome, having two arms that are identical. In other words the sister chromatids of a chromosome contain identical genetic information.

Also remember that chromosomes occur in homologous pairs, each chromosome being called a homologue. The maternal homologue is inherited from the female, and the paternal one is inherited from the male. By observing the location of homologous chromosomes and of chromatids within the cell, we can determine the stage of cell division.

For the present, we focus our discussion on cell duplication in eukaryotic animal cells. (When differences between plant and animal cells occur, we will note them, but we will not concentrate on them here.) The process of cellular division is more complex in eukaryotic cells than in prokaryotic cells,

FIGURE 10.5 Basic Chromosome Structure
(a) An SEM of a chromosome showing the two sister chromatids held together at the centromere region. (b) As the chromosomes are formed, there are several levels of organization and packaging. 1–6 illustrate these various levels, which eventually result in the chromosome strand that forms the arms of the condensed chromosome.

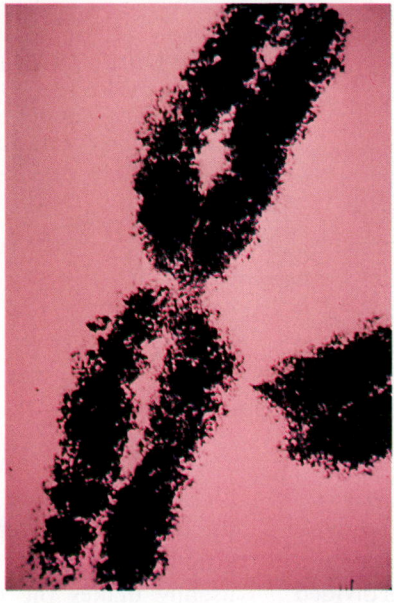

(a)

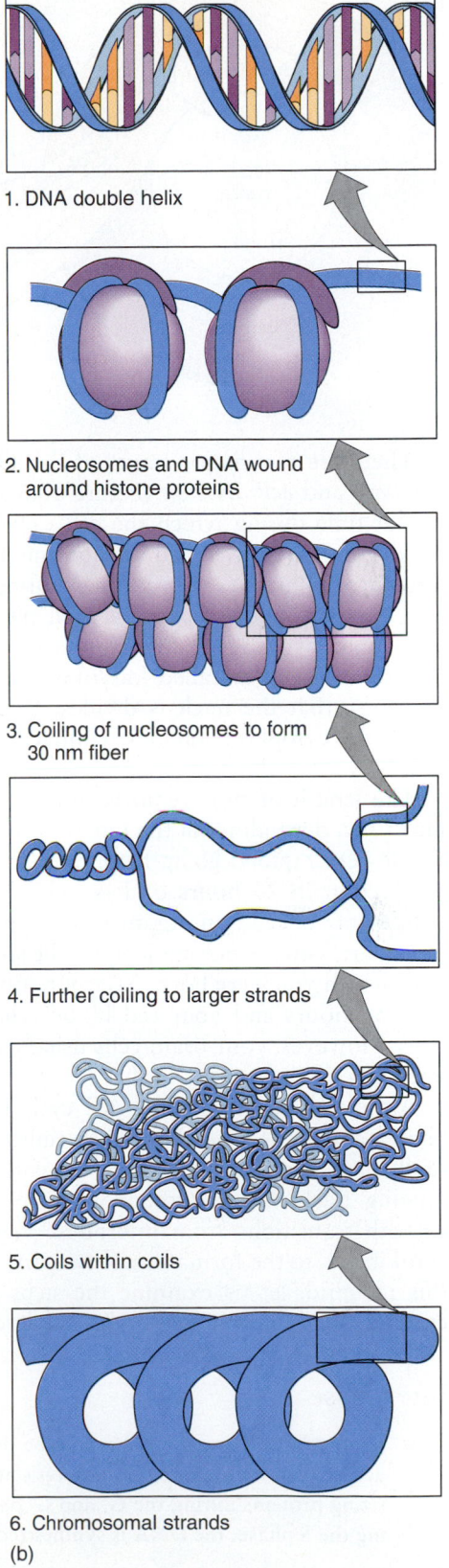

1. DNA double helix

2. Nucleosomes and DNA wound around histone proteins

3. Coiling of nucleosomes to form 30 nm fiber

4. Further coiling to larger strands

5. Coils within coils

6. Chromosomal strands

(b)

partly because the DNA of a eukaryotic cell is surrounded by a membrane, forming the nucleus. In order for the cell to reproduce, the DNA, its membrane, and the cell itself must be replicated. Then too, eukaryotic DNA is usually complexed with proteins and other substances, and its nucleotide sequences usually are longer than those of prokaryotic cells.

For all these reasons, reproduction of the eukaryotic cell has to be a highly organized process. Understanding the mechanisms that regulate this process is of fundamental biological importance, so it is not surprising that a major portion of biological and biomedical research is directed toward the mechanisms of cell division.

Cell Cycle

- **The cell cycle of most cells consists of two stages: interphase and cell division. Cell division is further broken down into mitosis and cytokinesis.**

In a unicellular organism, cellular reproduction results in two brand-new individuals of the species. In a multicellular organism, cellular reproduction results in the formation of two new cells. Almost all cells — whether in a unicellular or a multicellular organism — repeat the cycle of growth, maintenance of life, and cell reproduction. In so doing, the cell passes through several stages that comprise the **cell cycle.**

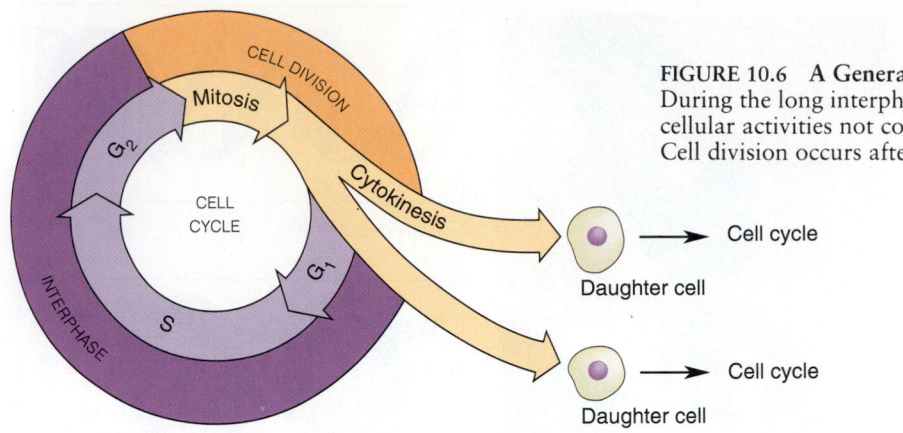

FIGURE 10.6 A Generalized Cell Cycle
During the long interphase, the cell carries on all normal cellular activities not concerned with cellular division. Cell division occurs after the events of interphase.

There are two major stages of the cell cycle: *interphase* and *cell division* (Figure 10.6). Interphase is that time during which the cell's main activities are devoted to growth and metabolism. It is divided into three phases: G_1, S, G_2. It is during this stage that the chromatin replicates. Cell division is divided into two phases. It is during the first phase, *mitosis* (sometimes called karyokinesis, or nuclear division), that the nucleus divides. During *cytokinesis,* the cytoplasm divides, completing cell division.

The length of time required for a certain stage may vary, depending on the kind of cell. Some cells divide very rapidly, going from their "birth" to reproduction in 20 hours or less, while others may divide only once in an organism's life span. For the most part, you are not made up of the identical cells with which you were born. Your intestinal cells live only 36 hours and your red blood cells only 120 days. However, your brain cells usually divide only once in a lifetime.

Although we discuss the cell cycle in terms of phases and steps, you should remember that it actually is a series of continuous events, with one flowing naturally into the next. The cell makes smooth transitions from interphase to mitosis, to cytokinesis, to the formation of two new cells. With this in mind, let us examine the steps of the cell cycle, divided into phases for ease in description.

Interphase

■ During interphase, the cell carries on its normal activities of maintaining homeostasis and synthesizing proteins during the G_1 and G_2 phases. During the S phase, the DNA is synthesized.

For most of its life the cell is in the nondividing phase known as **interphase.** Strictly speaking, inter-phase is not a phase of cell division at all but rather the nondividing preparatory stage between cell divisions. (It is the length of the interphase stage that usually makes the difference in the life spans of various types of cells.) During interphase the cell is at work maintaining homeostasis, synthesizing important molecules, and regulating the myriad reactions necessary for life, as well as preparing for division. If you examine a cell in interphase, you cannot see the chromosomes in the nucleus. The DNA is still in the form of fine threads of chromatin. The nuclear membrane is intact, and the nucleolus is visible. In animal cells, adjacent to the nucleus you can see the small centrioles.

Interphase is composed of three phases often designated G_1, S, and G_2. The G stands for gap of time before and after DNA synthesis, and the S represents DNA synthesis. During the G_1 **phase** of interphase, the cell conducts normal homeostatic metabolic events that maintain it as it grows and prepares for DNA replication. At the end of the G_1 phase, the S **phase** begins, during which DNA is synthesized. The genetic material is still in chromatin form; individual chromosomes are not visible. Once DNA replication has occurred, the G_2 **phase** begins and the cell is further readied for division through the replication of the organelles necessary for conducting life processes.

Mitosis

■ Mitosis (karyokinesis) ensures that the daughter cells have the same number of chromosomes as the parent cell. It is divided into four stages: prophase, metaphase, anaphase, and telophase.

The next phase in the life cycle of the cell is known as **mitosis,** which results in a separation of sister chromatids. During mitosis the chromatin (which

Chromatin Structure

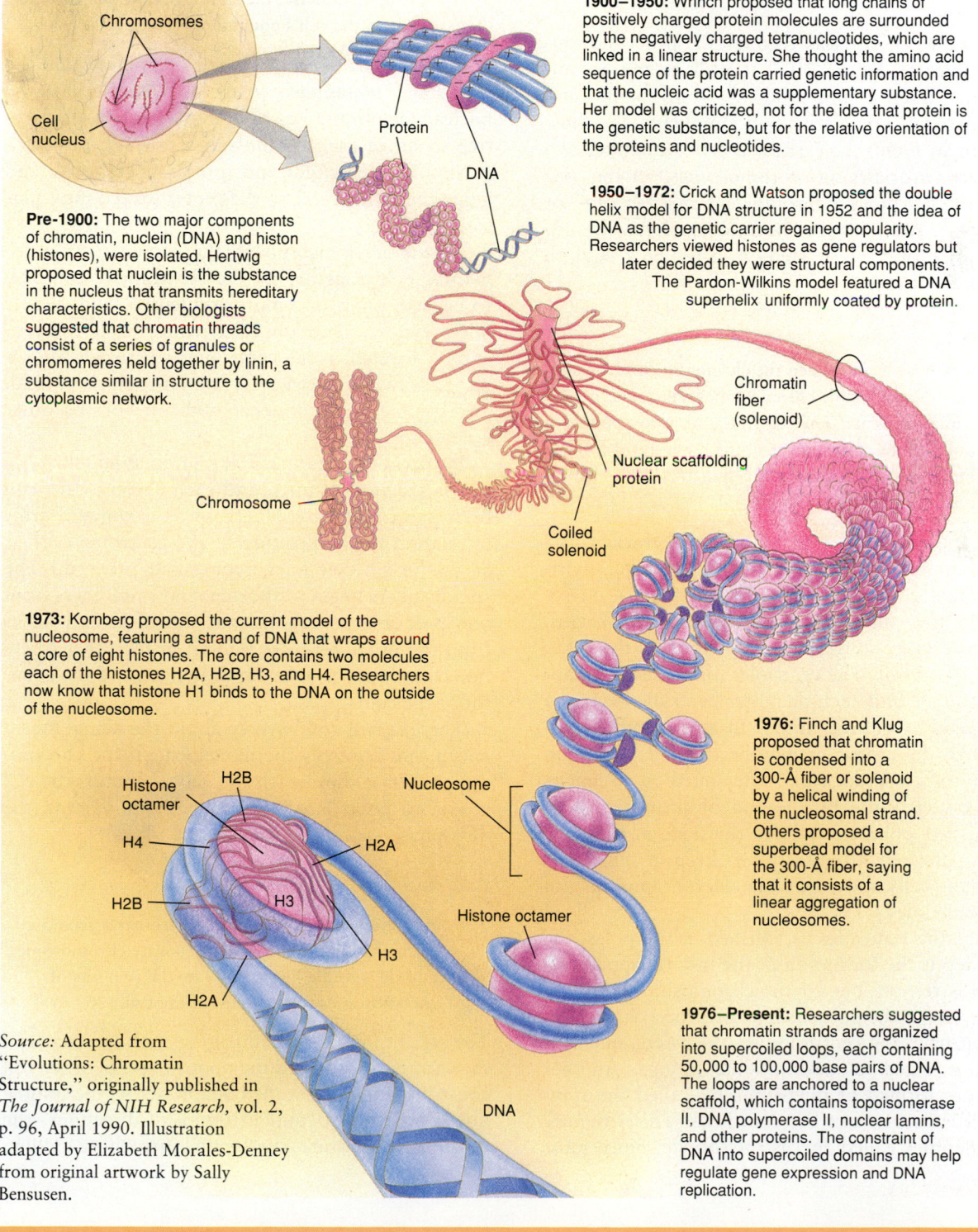

Pre-1900: The two major components of chromatin, nuclein (DNA) and histon (histones), were isolated. Hertwig proposed that nuclein is the substance in the nucleus that transmits hereditary characteristics. Other biologists suggested that chromatin threads consist of a series of granules or chromomeres held together by linin, a substance similar in structure to the cytoplasmic network.

1900–1950: Wrinch proposed that long chains of positively charged protein molecules are surrounded by the negatively charged tetranucleotides, which are linked in a linear structure. She thought the amino acid sequence of the protein carried genetic information and that the nucleic acid was a supplementary substance. Her model was criticized, not for the idea that protein is the genetic substance, but for the relative orientation of the proteins and nucleotides.

1950–1972: Crick and Watson proposed the double helix model for DNA structure in 1952 and the idea of DNA as the genetic carrier regained popularity. Researchers viewed histones as gene regulators but later decided they were structural components. The Pardon-Wilkins model featured a DNA superhelix uniformly coated by protein.

1973: Kornberg proposed the current model of the nucleosome, featuring a strand of DNA that wraps around a core of eight histones. The core contains two molecules each of the histones H2A, H2B, H3, and H4. Researchers now know that histone H1 binds to the DNA on the outside of the nucleosome.

1976: Finch and Klug proposed that chromatin is condensed into a 300-Å fiber or solenoid by a helical winding of the nucleosomal strand. Others proposed a superbead model for the 300-Å fiber, saying that it consists of a linear aggregation of nucleosomes.

1976–Present: Researchers suggested that chromatin strands are organized into supercoiled loops, each containing 50,000 to 100,000 base pairs of DNA. The loops are anchored to a nuclear scaffold, which contains topoisomerase II, DNA polymerase II, nuclear lamins, and other proteins. The constraint of DNA into supercoiled domains may help regulate gene expression and DNA replication.

Source: Adapted from "Evolutions: Chromatin Structure," originally published in *The Journal of NIH Research*, vol. 2, p. 96, April 1990. Illustration adapted by Elizabeth Morales-Denney from original artwork by Sally Bensusen.

contains the replicated DNA) contracts to form chromosomes, and these are parceled out to the new daughter cell in a series of activities that, for convenience, are divided into four distinct phases: *prophase, metaphase, anaphase,* and *telophase.* Mitosis ensures that each daughter cell has a complete copy of the hereditary instructions contained in the original parent cell, and the process's precision is responsible for the fact that the chromosome number remains constant from cell generation to generation.

As we work our way through the process of mitosis, refer to Figure 10.7 as well as the accompanying micrographs (taken from plant and animal cells). For clarity and simplicity, the illustrations in Figure 10.7 show a cell containing two homologous pairs of chromosomes, giving our sample cell a diploid chromosome number of 4, or two pair). As the discussion moves along, keep in mind that our set of steps is actually a smooth continuous process.

Prophase

■ **Prophase begins when the chromatin condenses to become the visible chromosomes, and the centrioles replicate. The centrioles move to opposite poles of the cell, and the mitotic apparatus forms. Finally the nuclear membrane breaks down, as does the nucleolus.**

When the chromatin begins to shorten, thicken, and coil upon itself to form distinct chromosomes (visible when stained and observed under a microscope), the stage known as **prophase** begins. Prophase takes longer to complete than the other stages of mitosis. The centrioles (discussed in Chapter 4) divide and each daughter centriole begins to move away from the other. (In animal cells, separation of the centrioles may be the first indication that mitosis is under way.) They move in a semicircular pattern and, by the end of prophase, have established poles of cellular division, at opposite ends of the cell. (Remember to keep in mind that the plant cells shown in the micrographs do not have centrioles.)

With the centrioles established at the poles, the next step is the formation of the mitotic apparatus, which is responsible for the chromosomes' orderly movement to the poles as mitosis proceeds. Radiating out from the centrioles are the **asters,** microtubules arranged in starburst-like groups. Another structure formed of microtubules, called the **spindle,** extends between the centrioles. This structure is sometimes called the *astral spindle.* Since most plant cells lack centrioles and asters but do have spindle fibers, the mitotic structure in plants is called the *anastral spindle.* The three elements — the centriole, the asters, and the spindle — form the mitotic apparatus. Near the end of the prophase, the nuclear membrane disintegrates and the nucleolus disappears.

Metaphase

■ **During metaphase, the spindle guides the chromosomes to the cell's equator.**

Prophase takes the most time to complete, and the next stage, **metaphase,** takes the least. Because the nuclear membrane has broken down, the spindle is free to move the chromosomes through the cytoplasm. The centrioles and spindle fibers move and guide the centromere to the center (called the equator) of the cell.

During early metaphase, the pair of sister chromatids, held together by the centromere, continue to become more and more condensed, and spindle fibers attach to specialized structures called *kinetochores,* located in the centromeres. In each centromere are two kinetochores, one for each chromatid, and spindle fibers connect with one or the other kinetochore.

In middle metaphase the individual chromosomes are very distinct, and each chromosome can be recognized by the length of its chromatid arms in relation to the position of the centromere. (It is at this point that karyotyping can proceed.) The spindle fibers begin to shorten, and since fibers from one pole are pulling in that direction and fibers from the other pole are pulling in the other direction, the chromosomes are drawn toward the equator of the cell.

There is still controversy about the actual mechanism that controls the movement of the chromosomes. When they reach the cell's equator, the cell is said to be in late metaphase. At this point, the next phase, *anaphase,* begins.

Anaphase

■ **Anaphase begins when the centromeres separate. Each chromatid becomes a separate chromosome, and the spindle fibers separate the two sets of chromosomes to the poles of the cell.**

The separation of the centromere and movement of the chromatids toward the opposite poles of the cell constitute the phase called **anaphase.** As the spindle fibers continue to pull in opposite directions, the centromere separates, with one kinetochore moving

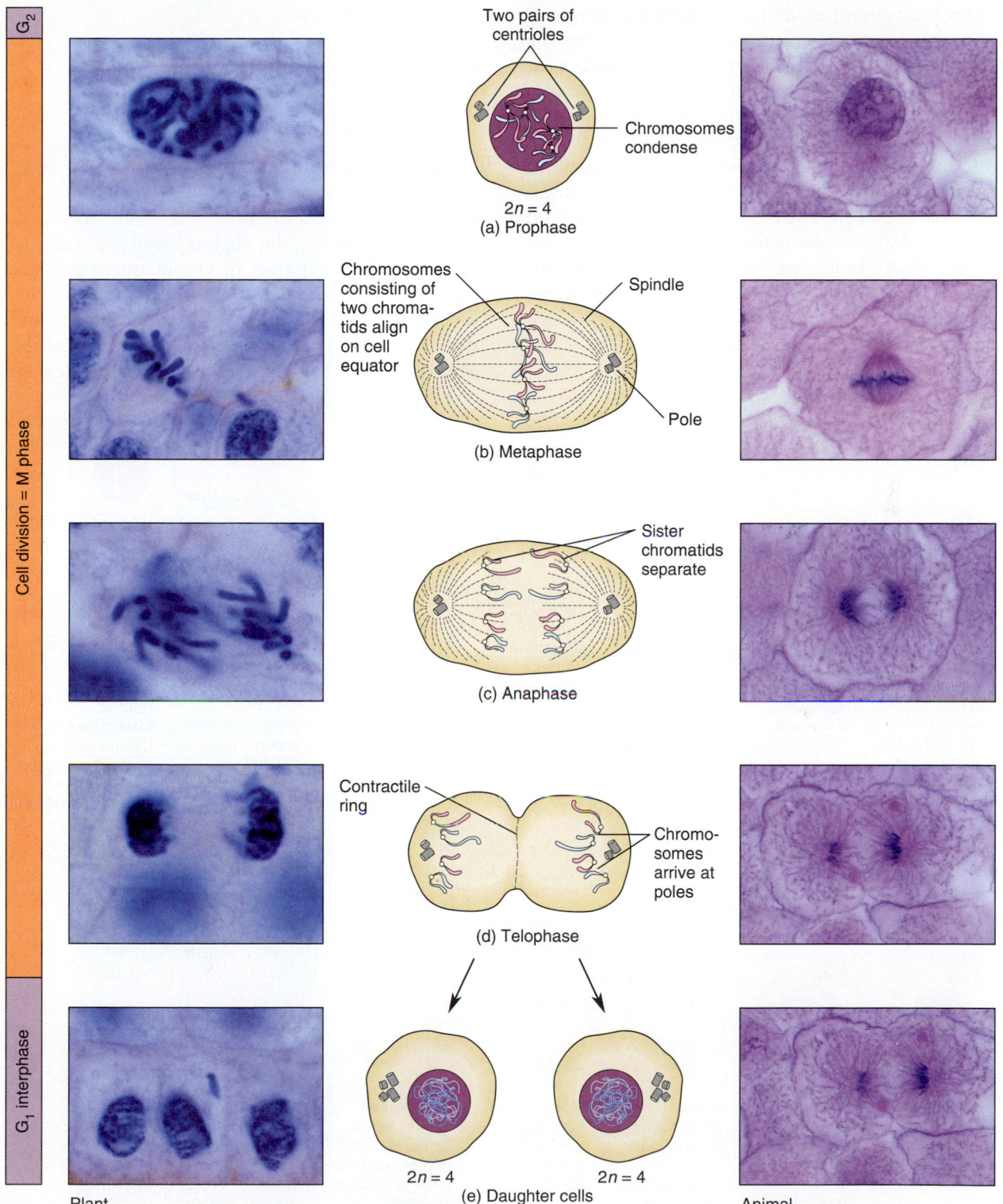

G₂

Cell division = M phase

G₁ interphase

Two pairs of centrioles

Chromosomes condense

2n = 4
(a) Prophase

Chromosomes consisting of two chromatids align on cell equator

Spindle

Pole

(b) Metaphase

Sister chromatids separate

(c) Anaphase

Contractile ring

Chromosomes arrive at poles

(d) Telophase

2n = 4

2n = 4

(e) Daughter cells

Plant

Animal

FIGURE 10.7   Overview of Mitosis and Cytokinesis
This simplified diagram shows an animal cell with a diploid number of 4 producing daughter cells with the same diploid number.

in each direction, so that one chromatid moves to one pole, and the other chromatid moves to the other. Once the centromere has separated, each chromatid becomes a chromosome in its own right, because each contains a centromere and has a complete complement of DNA information. The spindle microtubules guide the chromosomes as they move apart in an orderly manner. The exact procedures involved are not known, but the microtubules (and in all probability the cytoskeleton as well) are involved. Some biologists believe that one set of microtubules is involved in pulling the chromosomes to the poles, while another set is involved with elongating the cells at the poles, readying the cytoplasm for division.

Telophase

■ **During telophase, a new nuclear membrane forms around each set of chromosomes, the nucleolus reappears, and the chromosomes uncoil into their chromatin form.**

Anaphase ends once the chromosomes arrive at the opposite poles of the cell, and **telophase** begins. The chromosomes begin to uncoil and assume their condition as chromatin, the nuclear membrane reforms around each new parcel of chromatin to produce two new nuclei, and the nucleolus reforms. One daughter nucleus possesses the same number and type of chromosomes as its sister nucleus, and the contents of both daughter nuclei are identical to

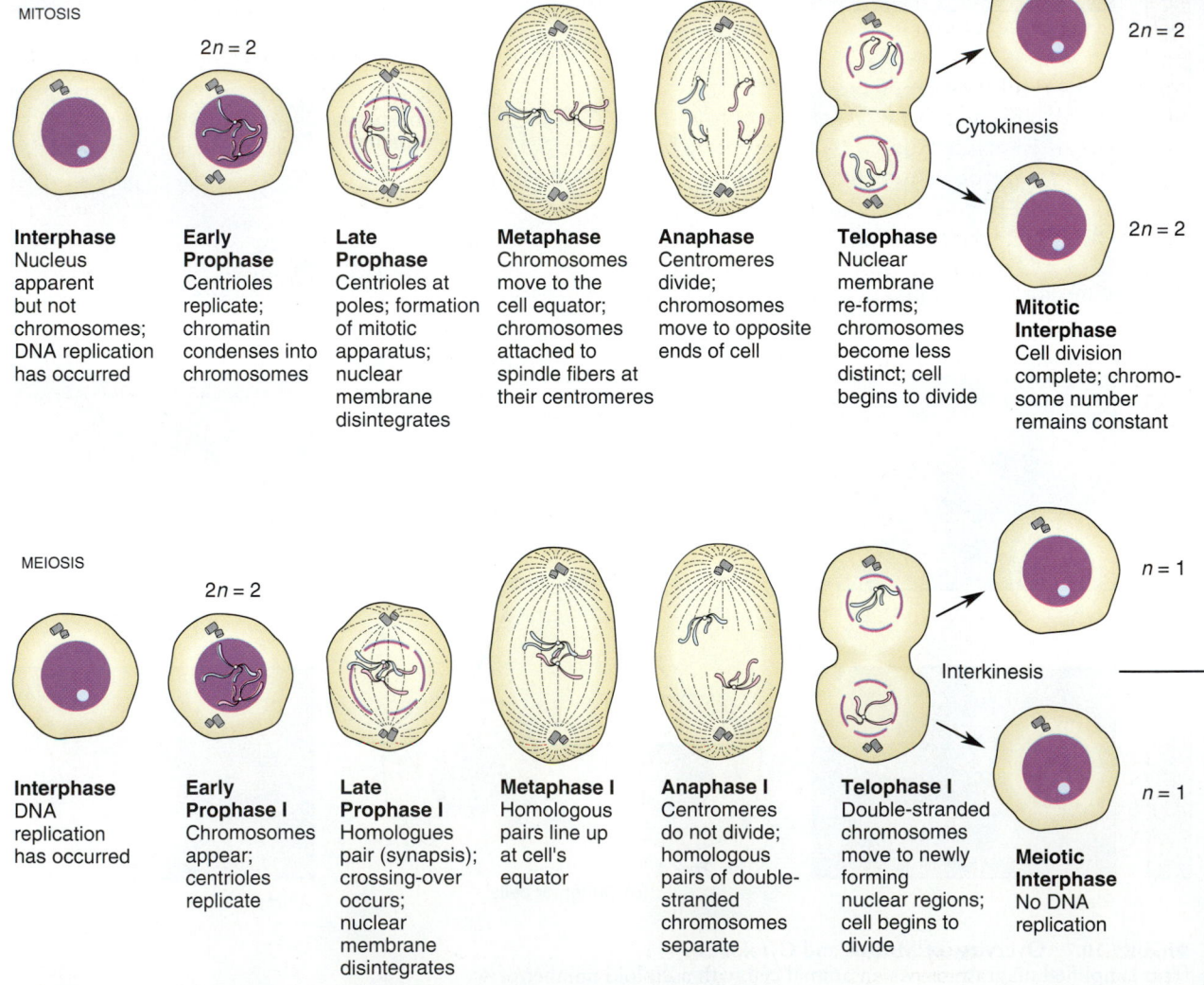

FIGURE 10.8   A Comparison of Mitosis and Meiosis
The illustrations compare mitosis and meiosis in an animal cell with a $2n = 2$ chromosome number. Note that in mitosis the chromosome number remains the same ($2n$-to-$2n$). In meiosis the chromosome number is reduced from a $2n$ of 2 to an n of 1.

Chapter 10   Cell Reproduction

that present in the original parent nucleus. At this point, mitosis, which involves only nuclear division, is complete. But we also have a cell that has two nuclei. What is the final stage? In most instances, telophase also marks the beginning of cytokinesis, the division of the cytoplasm that results in the division of the cell itself.

Cytokinesis

■ **During cytokinesis, the contents of the cell — cytoplasm and organelles — are divided into each section of the cell. A cell furrow forms in animals; in plants a cell plate develops. The cell divides in two.**

The nucleus has now divided, but the cell has not. In some cells, such as those found in your skeletal muscles and in some insects, plants, and fungi, the cellular unit does not divide after mitosis. The result is a multinucleate cell. Usually, however, mitosis is followed by **cytokinesis,** a division of the cytoplasm into two parts. During this phase the cytoplasm and organelles are divided into two approximately equal separate components. In most animal cells a ring of microtubules forms at about the center of the cell. This region begins to constrict, producing a cleavage furrow, which eventually pinches the cell into two. In plant cells, the cytoplasmic division is due to the formation of a partition, or wall, called the *cell plate,* that divides the cell into two.

In summary, mitosis is the orderly duplication of cells in which there is an equal distribution of chromosomal (genetic) material from parent to daughter cell, resulting in the production of two daughter cells having the same genetic content as that of the original parent cell. In addition, some cytoplasmic organelles (like the mitochondria) also replicate themselves and are distributed to the new daughter cells during cytokinesis.

MITOSIS VERSUS MEIOSIS (REDUCTION DIVISION)

■ **Mitosis produces cells that have the diploid ($2n$) number of chromosomes. Meiosis produces cells having only one homologue of each chromosome pair, or the haploid (n) number.**

Mitosis results in an increase in the number of cells, all having the same DNA complement. We humans, like other multicellular organisms, begin life as a single cell, but through the process of mitosis the number of cells increases until that large multicellular mass of specialized cells known as the individual organism is produced.

During some types of sexual reproduction (reproduction involving an exchange of genetic material), **gametes,** or sex cells, are formed. The male sex cell is called the **sperm** and the female sex cell is called the **egg.** When these two sex cells unite,

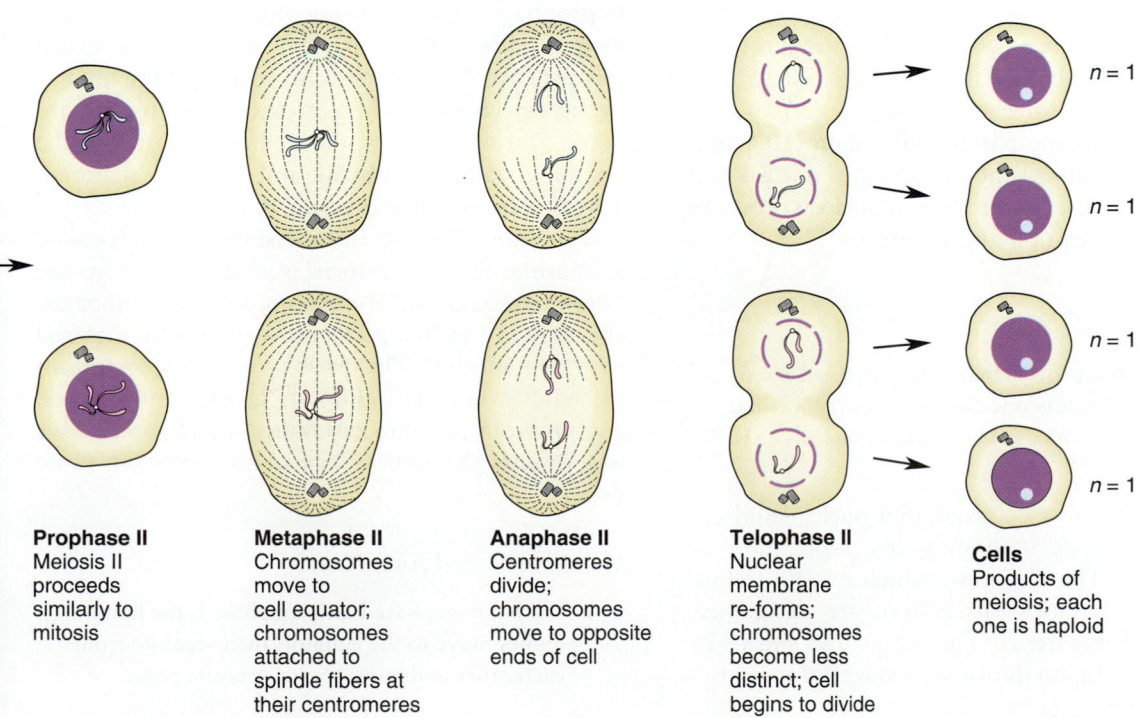

Prophase II
Meiosis II proceeds similarly to mitosis

Metaphase II
Chromosomes move to cell equator; chromosomes attached to spindle fibers at their centromeres

Anaphase II
Centromeres divide; chromosomes move to opposite ends of cell

Telophase II
Nuclear membrane re-forms; chromosomes become less distinct; cell begins to divide

Cells
Products of meiosis; each one is haploid

$n = 1$

$n = 1$

$n = 1$

$n = 1$

the resulting cell possesses both a maternal and a paternal set of genetic instructions. But how does this occur? If each parent contributed a full set of chromosomes, the chromosome number in the offspring should be double that of the parents, but it is not.

For example, you, like most humans, have 46 chromosomes. (In rare cases the diploid number is not 46 — we will discuss these cases in Chapter 15.) If you contributed 46 chromosomes to the next generation and your partner also contributed 46 chromosomes, your offspring would have 92 chromosomes. The following generation would have 184, and so on. The chromosome number of each generation would continue to double until the cell literally becomes filled with chromosomes. That would cause the cell to enlarge, but after a few generations, the surface-to-volume relationship would have to be violated to allow for the large mass of DNA, and the cell would die.

Obviously, chromosomal doubling does not happen, and each individual organism of the species generally has the diploid (2n) number of chromosomes characteristic of its species. How does the chromosome number remain constant, and how is chromosomal overload prevented? To maintain the diploid number, a method of cellular reproduction called **meiosis** evolved, in which only one member of a chromosome pair of the parent is passed on to the offspring. Cells produced by this process contain one chromosome from each parental chromosome pair — what is called the **haploid (n) number of chromosomes.** In essence *mitosis* preserves the chromosome number, while *meiosis* reduces the chromosome number of the daughter cell to one-half (one member of each chromosome pair) the number found in the parent cell. Figure 10.8 presents a comparison of the processes of mitosis and meiosis. As we discuss the steps of meiosis, you can refer to this illustration and Figure 10.9.

MEIOSIS

■ **Meiosis is a process that entails two cell divisions, preceded by only one chromosome replication. Meiosis is divided into two steps: meiosis I and meiosis II.**

Meiosis consists of two sequential nuclear and cytoplasmic divisions with only one replication of chromosomes. The divisions, which usually follow one another in quick succession, are designated *meiosis I* and *meiosis II*. The schematics shown in Figure 10.9 illustrate the various stages of meiosis.

Meiosis I

■ **Meiosis I has four stages: prophase I, metaphase I, anaphase I, and telophase I. During these stages, the centromeres do not divide and homologous chromosome pairs are separated into different cells.**

A cell that will go through meiosis passes through interphase, just as it would if it were going on to mitosis. During interphase, the DNA replicates, some cellular organelles replicate, and the cytoplasm is prepared for division. The cell then enters meiosis I, involving the separation of homologous chromosomes, which are, as you should remember, the two members of each pair of chromosomes in a diploid cell. The centromere of each chromosome still appears to bind the two sides of the bivalent chromosomes together, so the kinetochores and therefore the sister chromatids remain together. (Mitosis, on the other hand, involves separation of sister chromatids.) The four steps of meiosis I are named *prophase I*, *metaphase I*, *anaphase I*, and *telophase I*.

Prophase I

■ **Synapsis is the process by which homologous chromosomes join. Following synapsis, branches of the chromatids cross over each other at points called chiasmata, break, and exchange genetic information. This process, called crossing over, increases genetic variability.**

In prophase I, the first stage of meiosis I, each homologue pairs with its partner in a process called **synapsis.** At this time, the chromatids of the homologous chromosomes lie across or close to one another. During synapsis, adjacent chromatids of the homologous pair can break and exchange material. This process is called **crossing over.**

The point at which the crossover occurs is called a **chiasma** (the plural form is *chiasmata*) from the Greek letter chi (χ). The chromosomes break at the chiasma and exchange sections of genetic material with one another. This process is a very important mechanism for reassembling genetic information and providing genetic variability. In Chapter 14 we will discuss the importance of this event in more detail.

Metaphase I and Anaphase I

■ **During metaphase I and anaphase I, the homologues move to the equator, then separate from each other and move to the opposite poles.**

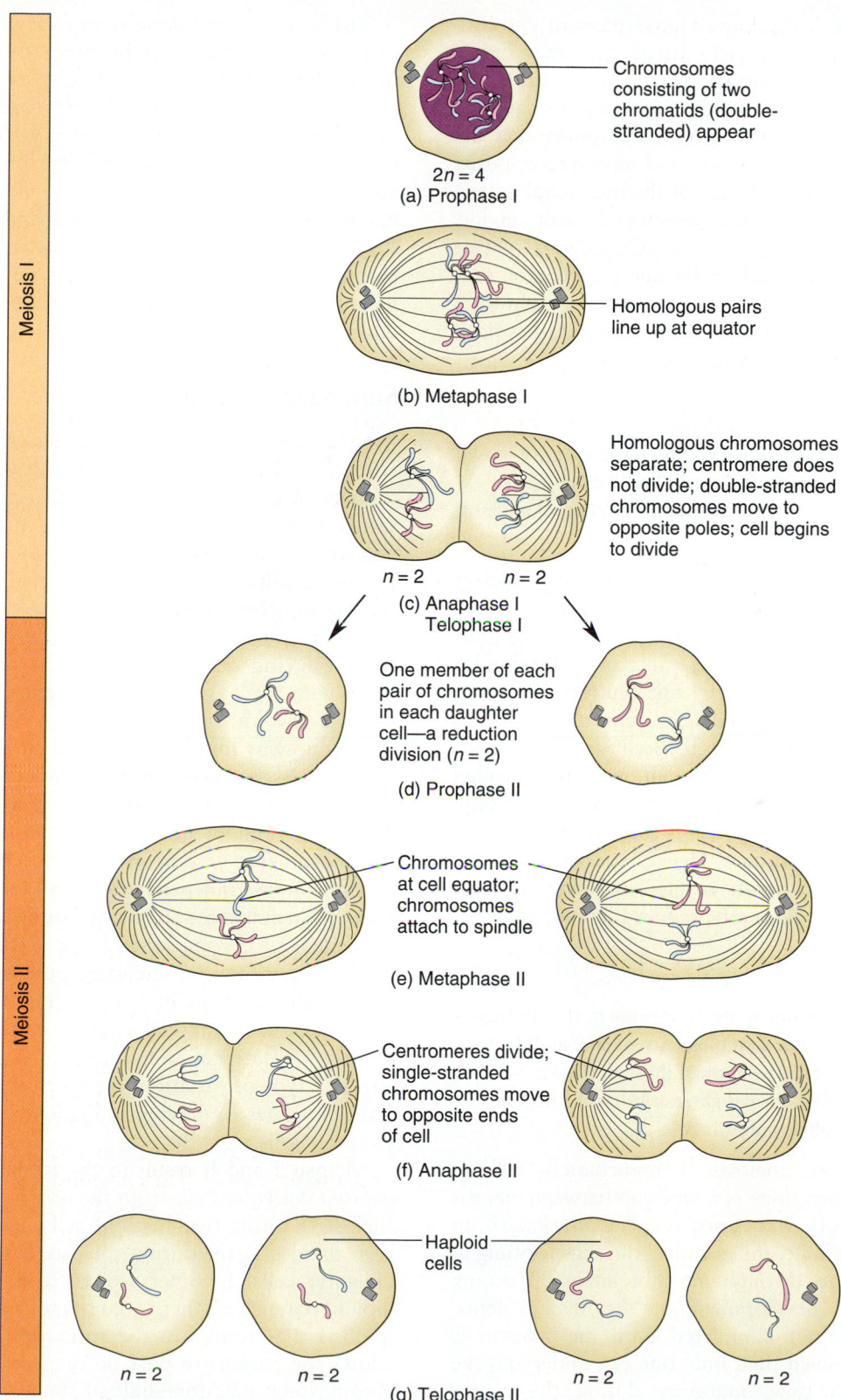

Meiosis I

Meiosis II

Chromosomes consisting of two chromatids (double-stranded) appear

$2n = 4$
(a) Prophase I

Homologous pairs line up at equator

(b) Metaphase I

Homologous chromosomes separate; centromere does not divide; double-stranded chromosomes move to opposite poles; cell begins to divide

$n = 2$ $n = 2$
(c) Anaphase I
Telophase I

One member of each pair of chromosomes in each daughter cell—a reduction division ($n = 2$)

(d) Prophase II

Chromosomes at cell equator; chromosomes attach to spindle

(e) Metaphase II

Centromeres divide; single-stranded chromosomes move to opposite ends of cell

(f) Anaphase II

Haploid cells

$n = 2$ $n = 2$ $n = 2$ $n = 2$
(g) Telophase II

FIGURE 10.9   Meiosis
The two sequential divisions of meiosis in an animal cell with a diploid number of 4 are illustrated. Note that the daughter cells produced have only one member of each chromosome pair and the number is haploid (2).

In metaphase I, the homologous pairs of chromosomes line up at the cell's equator, as they did in the metaphase of mitosis. During anaphase I, the differences between meiosis and mitosis become clear. The homologous pairs of chromosomes are separated from one another and moved to opposite poles, but the centromeres of the individual maternal and paternal chromosomes do not divide. Therefore, each chromosome still consists of two chromatids held together by one centromere. It is important to note that when you count chromosomes, you count the number of centromeres present, not the number of chromatids.

Telophase I

■ **During telophase I, the chromosomes move to two regions forming at the cells' poles. Each newly forming nuclear region has the haploid number of chromosomes.**

The chromosomes reach the newly forming nuclear region in telophase I, and shortly thereafter the cell divides (cytokinesis). The two daughter cells thus formed contain only one member of each homologous pair, held together by one centromere, which means each daughter cell's chromosome number is half that of the parent cell. Because the chromosome number is reduced by half, meiosis I is called the *reduction division*. Again, remember that only the number of centromeres are counted when counting chromosomes. Hence, because the centromeres do not divide in meiosis I, the chromosome number is said to be reduced by one-half.

Meiosis II

■ **During the four stages of meiosis II, the chromatids of the lone homologues separate, and the four daughter cells are formed through cytokinesis. Each of the four daughter cells has the haploid number of chromosomes.**

In most cases, meiosis II immediately follows meiosis I. When there is a time gap between meiosis I and meiosis II, that period is called *interkinesis*. In the human female, for instance, the eggs forming in the ovaries go through most of meiosis I during growth and development of the female fetus. Meiosis I is not completed until the woman is mature, and even then only one egg undergoes the process at a time, maturing during the course of the menstrual cycle. Biologists hypothesize that the length of interkinesis may be one factor that gives rise to increased danger of genetic disorders in children born to women over the age of 35.

The length of interkinesis can vary, but in most cases meiosis II begins right away. In the now familiar terminology, meiosis II consists of the steps of *prophase II, metaphase II, anaphase II,* and *telophase II,* as you can see in Figure 10.9. Meiosis II is similar to mitosis in that the centromere divides, and the two sister chromatids are pulled apart. The major differences between mitosis and meiosis are that in meiosis each cell (1) is genetically different because of crossing over and the random assortment of chromosomes and (2) is haploid rather than diploid.

Summary of the Results of Meiosis

Two basic activities occur in meiosis. The first is the reduction of chromosome number. If you count the number of chromosomes (remembering to concentrate on the number of centromeres) in a cell before and after completion of meiosis, you will see that meiosis results in a reduction from the diploid to the haploid condition.

Second, meiosis also provides two mechanisms for producing genetic variation in offspring. To begin with, because homologous pairs function independently during division — in other words, there is no way for all the paternal chromosomes to be drawn to one pole of the cell and all the maternal chromosomes to be drawn to the other pole — there is a random mix of maternal and paternal chromosomes in each newly formed cell. Each chromosome moves independently of all the others. The fate of chromosome 1 is not bound up with the movement of chromosome 2, for instance; so maternal 1 may move to the same pole as paternal 2. Thus, meiosis results in a reshuffling of the genetic plan. Each resulting cell ends up with a different mixture of maternal and paternal chromosomes. In a second type of variation, crossing over can further reshuffle genes by having individual chromosomes swap sections.

Meiosis I and II result in the formation of *four haploid daughter cells* from the original parent cell. Meiosis I forms two haploid cells, each of which then divides again during meiosis II to form four daughter cells. In other words, from the original diploid parent cell, there are now four haploid cells. Meiosis that results in the formation of sex cells allows the parents to pass on one member of each chromosome pair (one-half of their DNA) to succeeding generations. The individual of a species is eventually disposed of, but the DNA lives on through reproduction. Somatic cells die, but the germ cells — egg and sperm — provide for the immortality of the species.

MEIOSIS AND THE LIFE CYCLE

■ Meiosis is the process by which haploid cells are produced for sexual reproduction in plants and animals. By the joining of two haploid cells, the diploid number remains constant generation after generation.

You are no doubt wondering where meiosis occurs in relation to the life cycle of organisms, including humans. The answer is fairly obvious: For ourselves, meiosis occurs in the formation of the gametes — the sex cells, which are also sometimes called *germ cells*. In an adult diploid animal, sex cells are produced in the sex organs, or **gonads.** Typically, specialized cells produced in the gonads undergo meiosis to become haploid gametes as the animal becomes sexually mature. This process is called **gametogenesis.**

Stages of Gametogenesis

■ Meiosis, which results in the production of a specific sex cell (gamete), is called gametogenesis. If the gamete is an egg, the process is called oogenesis; if a sperm, spermatogenesis.

In the male, the specific term for the process that forms the male sex cell (sperm) is called **spermatogenesis** (Figure 10.10a). The meiotic process gives rise to four spermatids for each dividing cell; each spermatide develops a flagellum and becomes a mature haploid sperm. In the mature female, the female gamete, or egg, is produced through a process called **oogenesis** (Figure 10.10b). This process is similar to spermatogenesis, except that the cytoplasm of the original parent cell is divided unequally during cytokinesis. As a result, oogenesis produces only *one* fully developed haploid egg cell,

FIGURE 10.10   Spermatogenesis and Oogenesis
Mature sex cells (gametes) are produced from diploid parent cells during the meiotic processes of (a) spermatogenesis and (b) oogenesis. The sperm and ovum that result are haploid (*n*) and are able to fuse to form a new diploid (*2n*) individual during fertilization.

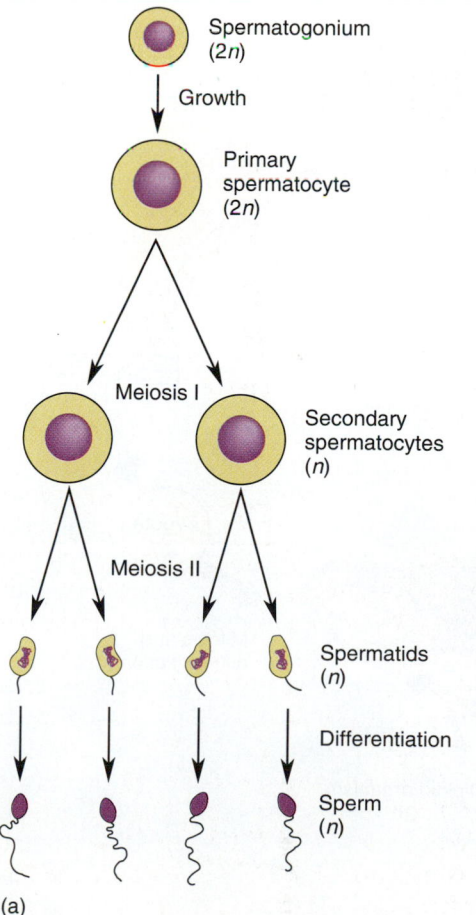

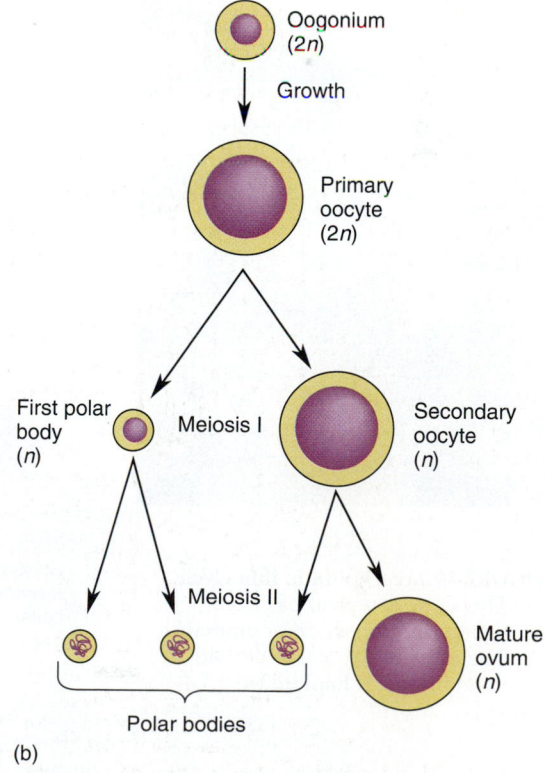

which includes an accumulated source of nutrients in an enlarged cytoplasm. The other three cells, called **polar bodies,** produced through oogenesis are much smaller than the egg cell and do not function as gametes. Instead they eventually disintegrate. In this way the egg cell is provided ample nutrients for the potential new organism.

Earlier we said that during cell reproduction a new set of cellular organelles must be reproduced. In the higher animals, only the egg donates mitochondria to the new organism. Although sperm do contain mitochondria in their flagellum, these organelles are not passed on to the new offspring. Instead, when a sperm penetrates an egg, the flagellum, along with its mitochondria, falls off. Only the genetic material combines with the egg to produce the next generation. Therefore, all of the mitochondria present in humans originated from the egg, truly an example of "woman power." As a matter of fact many biologists are seeking our mitochondrial Eve (see the Focus in Chapter 17).

During fertilization, the haploid gametes — sperm and egg — fuse, and the new diploid cell of the new generation, the **zygote,** forms. The zygote divides through mitosis, cells become specialized, and the organism develops and becomes the mature adult, able to repeat the life cycle again. This life cycle is sometimes called a **diplontic life cycle,** since the adult is diploid. Diplontic life cycles occur in humans as well as most other multicellular animals (Figure 10.11).

Haplontic Life Cycle. Some organisms, such as the lower unicellular plants, protista, and many fungi, have a **haplontic life cycle** in which the adult is haploid rather than diploid. *Chlamydomonas,* a microscopic, unicellular alga, is a good example of this. Here the adult is haploid, and it produces gametes that are also haploid. Under favorable conditions, fertilization produces a diploid zygote that immediately undergoes meiosis; these haploid cells

(a)

FIGURE 10.11 Diplontic Life Cycle
(a) This cycle is typical of orangutans and most other animals.
(b) The adult individual is diploid (2*n*) and produces haploid (*n*) gametes.

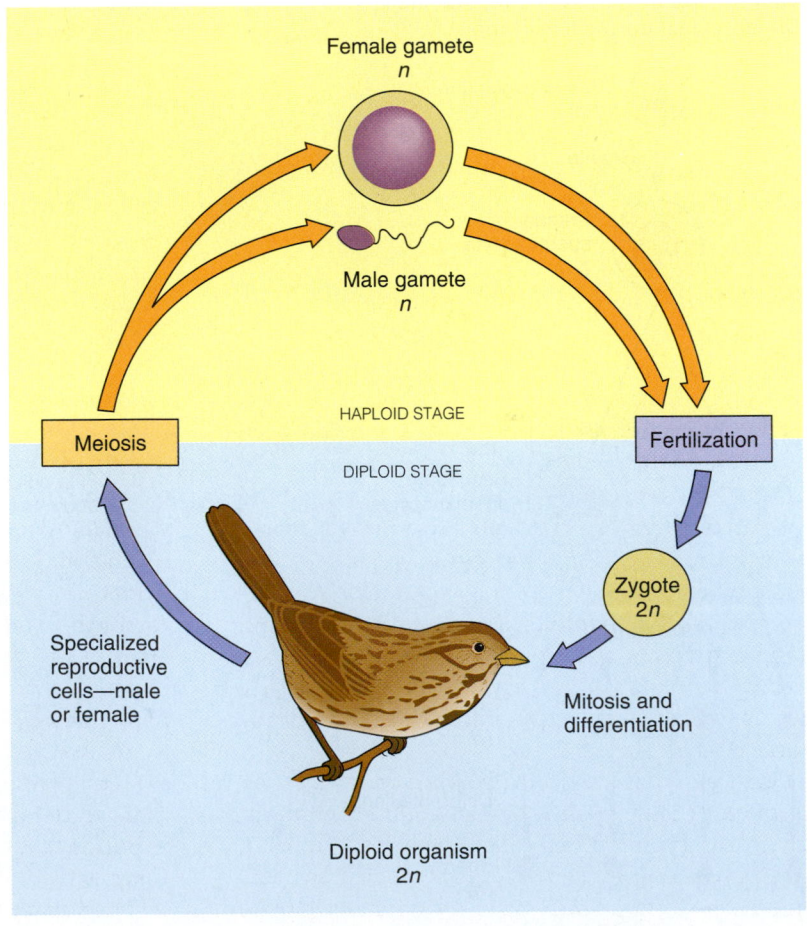

(b)

Chapter 10   Cell Reproduction

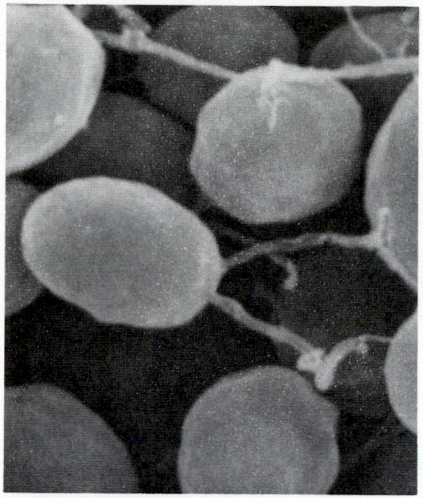
(a)

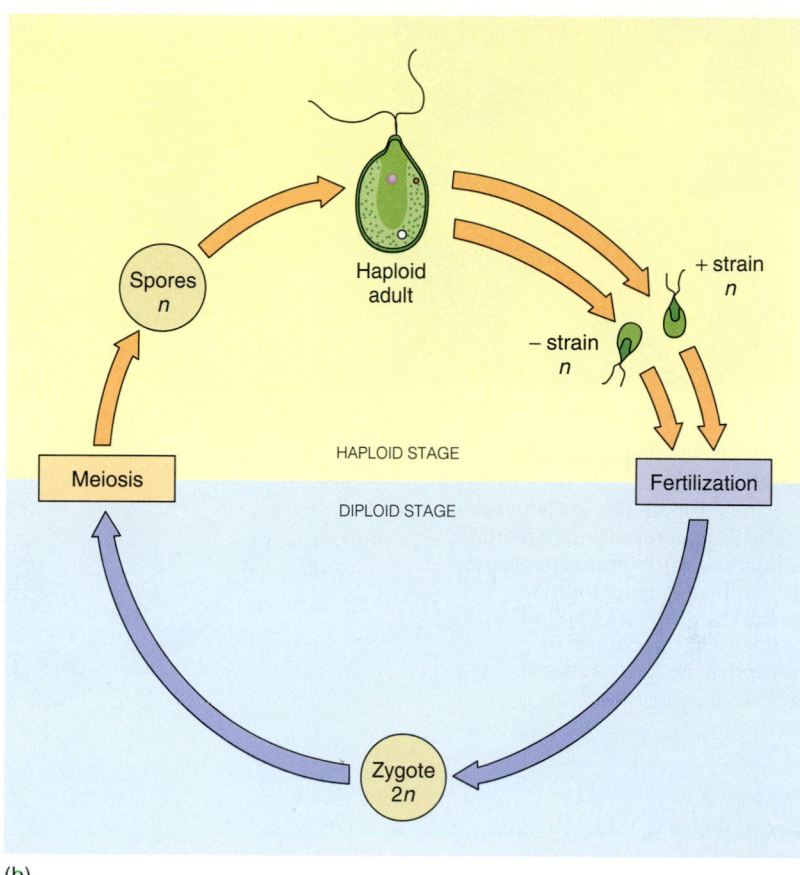
(b)

FIGURE 10.12 Haplontic Life Cycle
(a) Lower unicellular plants, like *chlamydomonas*, protista, and many fungi typify this cycle. (b) The adult is haploid (*n*) and produces haploid (*n*) gametes.

then undergo mitosis to become the haploid adult (Figure 10.12).

Diplohaplontic Life Cycle. Plants combine a haploid stage and a diploid stage into one complex life cycle. The haploid multicellular body is called the *gametophyte,* or a gamete-producing plant, because it produces the cells that function as sex cells. When these gametes fuse, a diploid zygote is formed. The zygote then undergoes several mitotic divisions and becomes the diploid, multicellular adult called the *sporophyte,* which is a spore-producing plant. The sporophyte produces spores through *meiosis;* therefore, the spore is haploid. For the present, we can define a spore as a reproductive cell that develops directly into a haploid body. In a plant life cycle such as this, meiosis occurs in the formation of the spores (Figure 10.13).

Don't get confused. I realize that we usually do not think of a life cycle with two adult generations, but in almost all plants this is, indeed, the case. They have a diploid generation in which some cells undergo meiosis to become spores. These haploid spores develop through mitosis (remember, in mitosis the chromosome number remains the same) into a gamete-producing generation. When mature, the gametophyte gives rise to sex cells, sperm or egg, which are haploid. The sperm and egg fuse, fertilization occurs, and a diploid zygote is produced that develops through mitosis into the sporophyte generation. In a life cycle such as this, the gametophyte generation alternates with the sporophyte generation. The process is called **alternation of generations** and the life cycle itself is called a **diplohaplontic life cycle,** because there is both a diploid and a haploid adult.

You may be more familiar with diplohaplontic plants than you know. For example, consider the rosebush. The rosebush is the form taken by this plant in its sporophyte generation. Within the rose blossom two microscopic gametophyte generations (male and female) are formed. A pollen grain is, in reality, the male gametophyte, and within the female part of the flower the female gametophyte is located (Figure 10.13). We discuss the importance of this phenomenon in more detail in Chapter 19.

The variety that exists in the life cycles of various living organisms is a perfect example of the unity

(a)

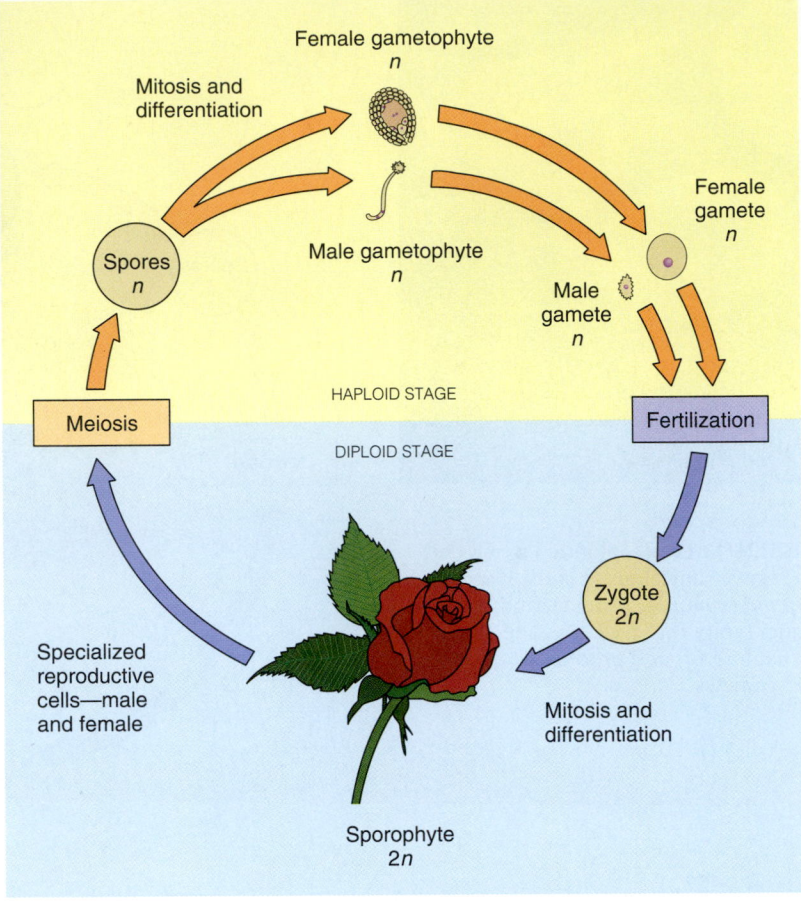

FIGURE 10.13 Diplohaplontic Life Cycle
(a) Plants display this cycle, which results (b) in alternating haploid (*n*) and diploid (2*n*) adult generations. The haploid or gametophyte generation produces haploid (*n*) gametes and the diploid or sporophyte generation produces haploid (*n*) spores.

(b)

and diversity in living things. The basic processes of meiosis are the unifying characteristic. Diversity is exhibited in the fact that the position of meiosis in the life cycle varies in different types of organisms.

THE BIOLOGICAL BASIS OF CANCER

■ **Cancerous cells exhibit unrestrained and unregulated mitosis and cytokinesis, which results in a tumor. These cells also can metastasize, or spread to distant sites within the body.**

Henrietta Lacks had cancer and died. And, as we all know, her case is not an isolated instance. In humans, we know of over 100 clinically distinct forms of cancer, some of which are noted in Figure 10.14. In the United States today, cancer accounts for more than 450,000 deaths per year, or about 20 percent of the total death toll every year. What is cancer? How can it be detected? How can it be treated?

As we have just seen, living cells have the capacity to grow and divide. In multicellular organisms,

this increase is not simply an increase in the number of cells; the cells also differentiate and specialize. Some cells become muscle cells, others become nerve cells, and so on. Multicellular plants also exhibit cellular differentiation.

Usually the process of cell growth and division is orderly, but sometimes it becomes disorderly. Unrestrained proliferation of undifferentiated cells — unregulated mitosis and cytokinesis — results in a mass of abnormal undifferentiated cells known as a tumor, which can form in both animal and plant life. Most tumors are benign (noncancerous). Some tumors are malignant (cancerous), because of many changes, including changes that occur in the cell's surface membrane. Specifically, the membrane of cancerous cells lacks certain properties that control cell growth. As a result, the cells can go on dividing indefinitely. Cancer cells, in addition to their production of an undifferentiated mass, also have the ability to break off from the original mass and be spread to other parts of the body through the circulatory system. This process is called **metastasis.**

In normal cells, cell division is regulated by genes

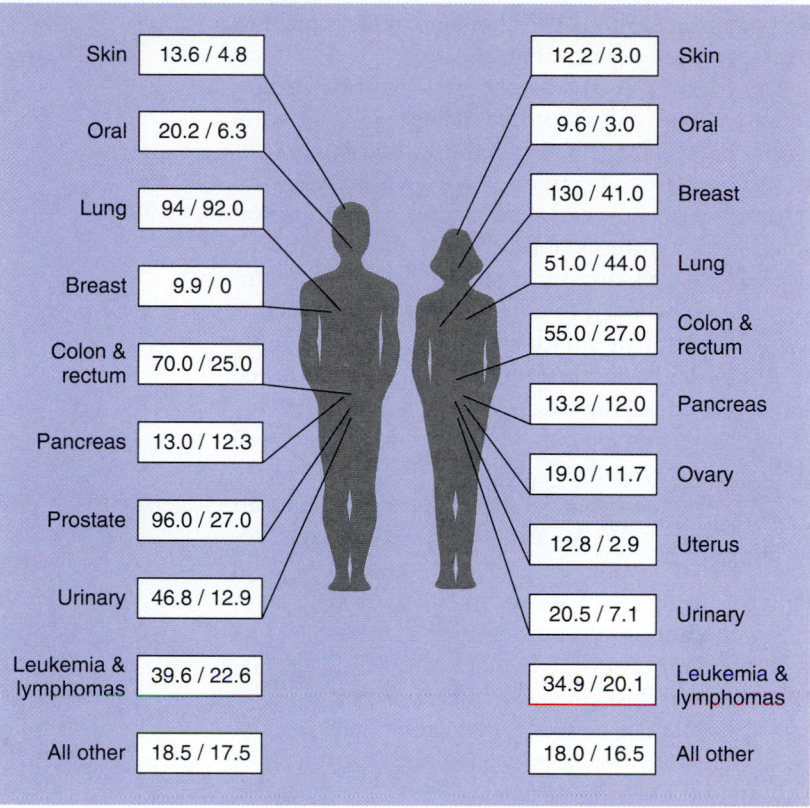

FIGURE 10.14 Cancer Incidence Numbers listed are in thousands as per 1989 estimates by incidence and deaths.

Skin	13.6 / 4.8
Oral	20.2 / 6.3
Lung	94 / 92.0
Breast	9.9 / 0
Colon & rectum	70.0 / 25.0
Pancreas	13.0 / 12.3
Prostate	96.0 / 27.0
Urinary	46.8 / 12.9
Leukemia & lymphomas	39.6 / 22.6
All other	18.5 / 17.5

12.2 / 3.0	Skin
9.6 / 3.0	Oral
130 / 41.0	Breast
51.0 / 44.0	Lung
55.0 / 27.0	Colon & rectum
13.2 / 12.0	Pancreas
19.0 / 11.7	Ovary
12.8 / 2.9	Uterus
20.5 / 7.1	Urinary
34.9 / 20.1	Leukemia & lymphomas
18.0 / 16.5	All other

in the nucleus. These genes also regulate chemical reactions at the cell surface, one of which is contact inhibition, which we discussed in Chapter 5. Under normal circumstances contact inhibition signals that the cell is adjacent to another. For example, a wound in your skin produces a gap or space between cells, so that cells on either side of the cut are not touching. Because cells have lost contact, the chemical signals of contact inhibition are deactivated, signaling cell division to begin. The cells promptly divide and repair the damage. What is the signal that maintains the normal specialized cellular condition and regulated cell division? As of this date, we do not know for sure, but we have some good ideas.

Oncogenes and Proto-Oncogenes

■ **It is thought that all or most cancers are caused by a number of genetic events that result in cells growing out of control. Experiments have demonstrated that cancer may be caused by oncogenes contained in normal DNA.**

Scientists have formulated several hypotheses about the causes of cancer; mutations, chemicals known as **carcinogens** (cancer-causing agents), viruses, and many others are all under consideration. Looking carefully at the possibilities, researchers form a hypothesis that appears to underlie all theories about the causes of cancer: At the fundamental molecular level, all or most cancers are caused by a number of genetic events that result in cells growing out of control. That means that if we can understand the molecular mechanisms of cell activity, we will be able to understand the changes that result in unhindered cell division in cancer (see Human Endeavors).

Cancer gene experiments have demonstrated that it is not foreign matter that causes some cancers but rather a few of the genes contained in all normal DNA that have the potential to become cancer genes called **oncogenes.** For example, if the genes from the cancer cells are first extracted, then the DNA extracted, crystallized, and added to normal cells, some cells will take up the cancerous DNA and become cancerous.

At present, over 36 oncogenes have been found (and the number is growing), in the DNA of normal animal cells, and 11 have been found in the DNA of normal human cells. Robert Weinberg at MIT (see the Profile at the beginning of this chapter), Geoffrey Cooper at the Sidney Farber Cancer Institute in Boston, and Michael Wigler of Cold Spring Harbor have identified the actual genes responsible for human cancers of the bladder, lung, colon, and breast, and for a form of leukemia.

These genes are normally switched off; in that state they are called **proto-oncogenes,** which are genes that regulate normal cell division. When cancer occurs, proto-oncogenes are switched on and become oncogenes, and these oncogenes are responsible for the synthesis of proteins responsible for the unregulated cell division that we call cancer. Oncogenes are believed to consist of two components: (1) the actual gene that dictates the linear sequence of amino acids in the protein (it should be noted that DNA codes the mRNA, which determines the sequence of amino acids in a protein, as discussed in Chapter 12) and (2) the control element — the DNA sequence that turns the oncogenes on and off (Figure 10.15). What turns on these proto-oncogenes? What is the chemical difference between a benign (harmless) proto-oncogene and one that is transformed into an oncogene?

In September 1982, Douglas Lowe of the National Cancer Institute, working with the gene involved in bladder cancer, came up with an answer. He and his collaborators determined that the DNA code in the proto-oncogene was G-G-C (guanine, guanine, and cytosine). However, in the oncogene it was G-T-C. In other words, the middle guanine had been replaced by thymine (see Chapter 11). That difference resulted in the production of large amounts of an abnormal protein that somehow disrupts cellular regulation. Is cancer of the bladder spelled GTC? That is probably a gross oversimplification of a complex question, but the answer is getting closer, and researchers appear to be on the right path.

Oncogene Products

Investigators at the Salk Institute in San Diego have provided evidence as to the nature of the protein coded for by oncogenes and also how the protein functions in normal and cancerous cells. A significant part of their work involved the use of viruses, which are organisms that must live within another cell (see Chapter 18). When researchers infected normal cells with a virus containing an oncogene, the whole cell produced a group of enzymes known as *tyrosine kinases.* (Some researchers are suggesting that the production of chemicals like tyrosine kinases could be the basis of an early-detection screening process.) This virally dictated protein was produced continuously by the whole cell and initiated events leading to cell division and to changes in the cell's shape that resulted in a tumor. When normal cells were exposed to growth factors (chem-

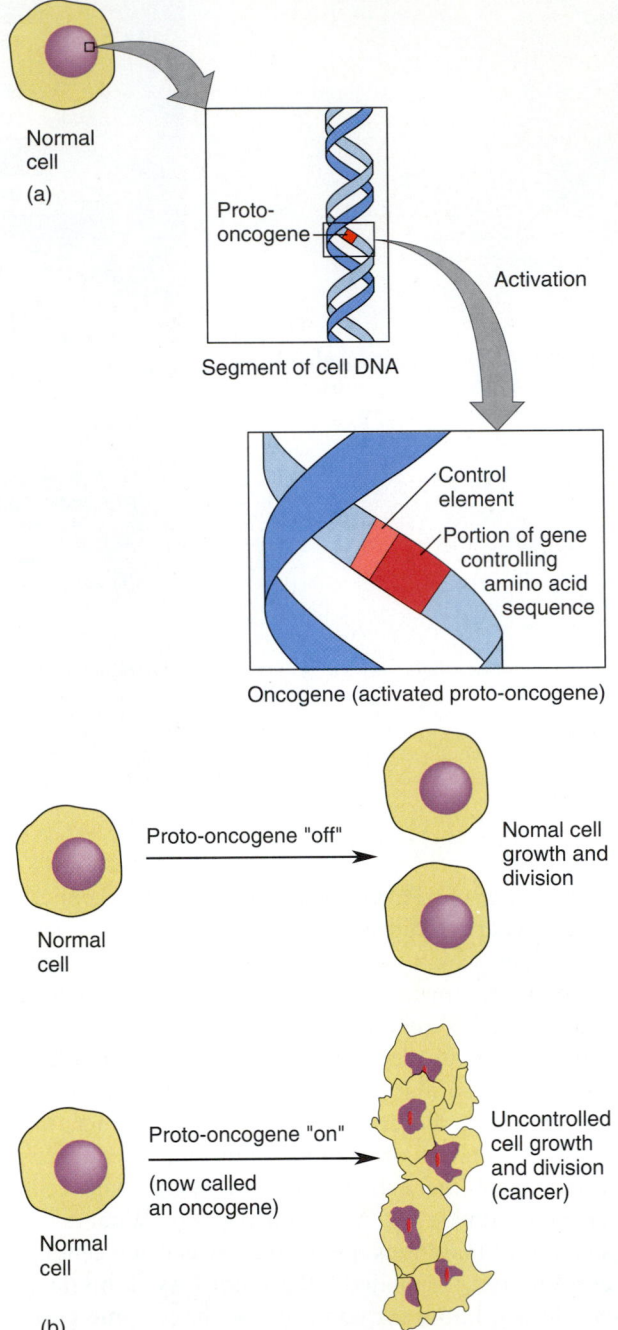

FIGURE 10.15 Oncogenes and Cancer
(a) Proto-oncogenes regulate cell division. (b) When a proto-oncogene is activated it is known as an oncogene and can direct unregulated cell growth and division known as cancer.

icals known to increase growth), the normal cells produced tyrosine kinase but *without* tumor formation. What was the difference that would cause the oncogene to form or to turn on in the normal cell?

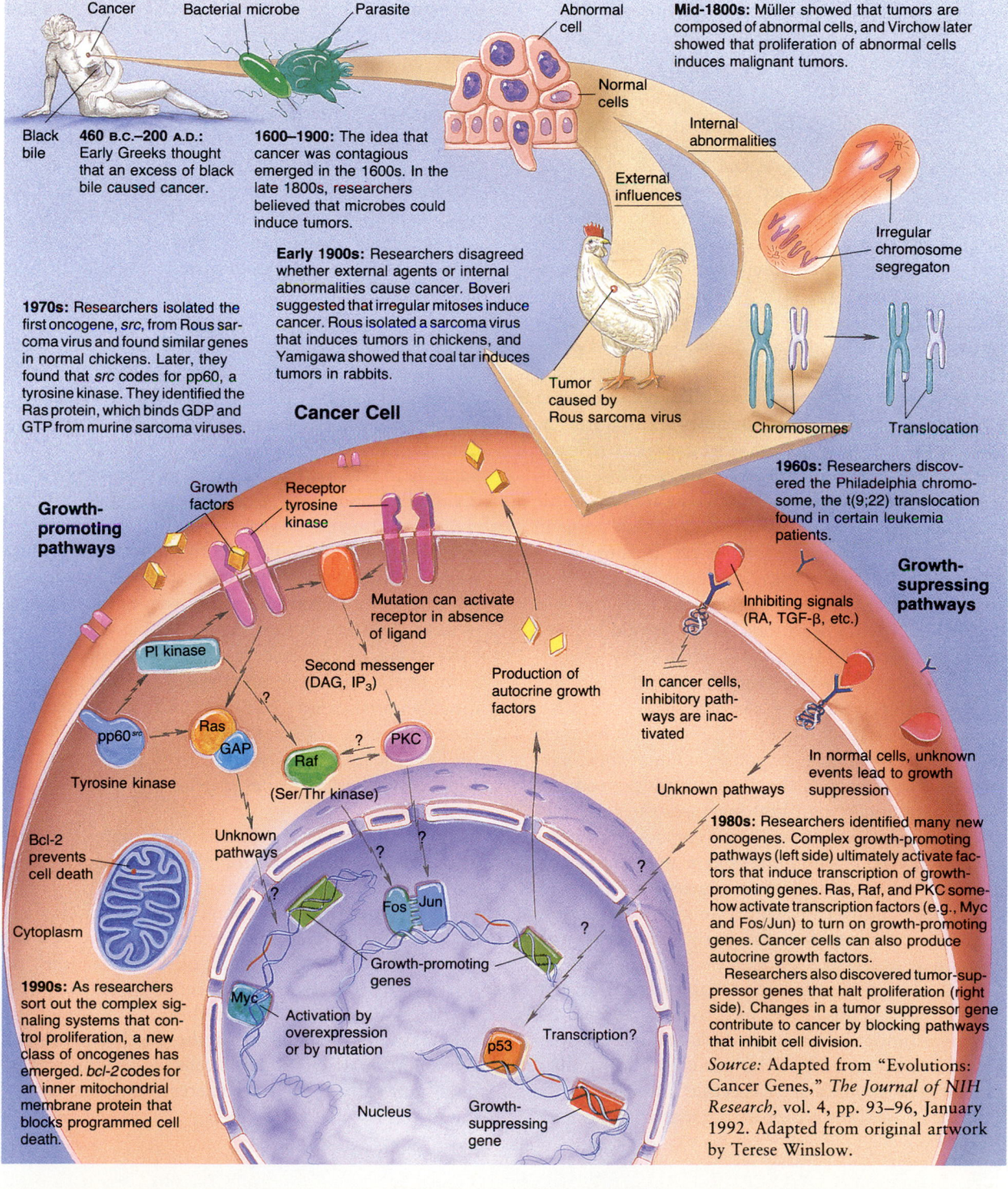

Cancer

Bacterial microbe

Parasite

Abnormal cell

Normal cells

Mid-1800s: Müller showed that tumors are composed of abnormal cells, and Virchow later showed that proliferation of abnormal cells induces malignant tumors.

Black bile

460 B.C.–200 A.D.: Early Greeks thought that an excess of black bile caused cancer.

1600–1900: The idea that cancer was contagious emerged in the 1600s. In the late 1800s, researchers believed that microbes could induce tumors.

Internal abnormalities

External influences

Irregular chromosome segregaton

Early 1900s: Researchers disagreed whether external agents or internal abnormalities cause cancer. Boveri suggested that irregular mitoses induce cancer. Rous isolated a sarcoma virus that induces tumors in chickens, and Yamigawa showed that coal tar induces tumors in rabbits.

1970s: Researchers isolated the first oncogene, *src*, from Rous sarcoma virus and found similar genes in normal chickens. Later, they found that *src* codes for pp60, a tyrosine kinase. They identified the Ras protein, which binds GDP and GTP from murine sarcoma viruses.

Cancer Cell

Tumor caused by Rous sarcoma virus

Chromosomes

Translocation

1960s: Researchers discovered the Philadelphia chromosome, the t(9;22) translocation found in certain leukemia patients.

Growth-promoting pathways

Growth factors

Receptor tyrosine kinase

Mutation can activate receptor in absence of ligand

Second messenger (DAG, IP$_3$)

Production of autocrine growth factors

Inhibiting signals (RA, TGF-β, etc.)

In cancer cells, inhibitory pathways are inactivated

Growth-supressing pathways

PI kinase

pp60^{src}

Ras

GAP

Raf

(Ser/Thr kinase)

PKC

Tyrosine kinase

Unknown pathways

Unknown pathways

In normal cells, unknown events lead to growth suppression

Bcl-2 prevents cell death

Cytoplasm

Fos   Jun

Growth-promoting genes

Myc

Activation by overexpression or by mutation

p53

Transcription?

Nucleus

Growth-suppressing gene

1990s: As researchers sort out the complex signaling systems that control proliferation, a new class of oncogenes has emerged. *bcl-2* codes for an inner mitochondrial membrane protein that blocks programmed cell death.

1980s: Researchers identified many new oncogenes. Complex growth-promoting pathways (left side) ultimately activate factors that induce transcription of growth-promoting genes. Ras, Raf, and PKC somehow activate transcription factors (e.g., Myc and Fos/Jun) to turn on growth-promoting genes. Cancer cells can also produce autocrine growth factors.

Researchers also discovered tumor-suppressor genes that halt proliferation (right side). Changes in a tumor suppressor gene contribute to cancer by blocking pathways that inhibit cell division.

Source: Adapted from "Evolutions: Cancer Genes," *The Journal of NIH Research,* vol. 4, pp. 93–96, January 1992. Adapted from original artwork by Terese Winslow.

The research done at the Salk Institute demonstrated that in normal cells and in those infected with viral oncogenes, tyrosine kinase initiates cell division. In normal cells the growth factor is received at the membrane, which initiates tyrosine kinase production. The growth factor receptor complex is then destroyed, thereby regulating cell division. In cells infected with the virus, the membrane receptor complex was bypassed. As a result, because the viral tyrosine kinase is not linked to the membrane-bound growth factors, controls are absent. The cell-division signal instructs the cell to keep on dividing (see Figure 10.16).

Progress in Our Battle Against Cancer

The implications of oncogene research are clear. Researchers have provided evidence that there may be a common central mechanism for cancer — the small number of genes known as oncogenes. An understanding of the protein synthesized by oncogenes could pave the way for the development of diagnostic tools capable of detecting the presence of cancer-producing protein at the earliest stages of the disease. With this evidence in hand, investigators can then devise methods to block the deadly cancer message before the cancerous mass does its irreversible damage. Early detection of cancer is still one of the major keys to survival.

There are reasons for optimism in our battle against cancer. For our purposes, at this point in the text, monoclonal antibodies can be simply defined as specially designed proteins that can bind with cancer cells and not with normal cells and thus can serve as what are often called "magic bullets" (see Chapter 25). These antibodies carry chemotherapeutic medicines to target in on and kill cancer cells without destroying normal body cells. Research also is teaching us more about substances that modify biological responses. One of these substances is **interferon,** a chemical substance produced by the immune system that increases the resistance of cells to viral infections. Such response modifiers enhance the body's immune system and may be able to shrink and control tumorous growths.

Another approach to controlling cancer is to prevent its spread within the body. Scientists now know that in order for a cancer cell to metastasize it must develop pseudopodia ("false feet" similar to those of an amoeba) to allow it to migrate from the primary tumor site through the circulatory system to the secondary sites. The pseudopodia of these cancer cells contain a twentyfold increase in the amount of laminin receptors compared to noncancerous cells. Laminin is a protein found in the basement membrane that forms the tissue barriers separating tissue compartments. For metastasis to occur, the cancer cell must bind to the laminin to pass into and out of the circulatory system. Scientists have cloned the gene for the laminin receptor and are developing ways to inhibit the cancer cell from binding with laminin, thus preventing metastasis. One way is by developing a synthetic laminin so that the cancer cell laminin receptors bind with the synthetic laminin rather than the laminin of the basement membrane.

Recently scientists have also discovered that there are genes called *tumor-suppressor genes* that prevent a cell from dividing. But for some reason, in certain cells these tumor-suppressor genes can be turned off, and the cell loses its ability to regulate its reproduction. Therefore, cancer can be thought of as a result of the conversion of proto-oncogenes to oncogenes, promoting cell growth, and the turning off of tumor-suppressor genes, again promoting cell growth.

Obviously, these are exciting times in cancer research, and many scientists are confident that the answer to the question of cancer is near. It should be noted, however, that not all cancer researchers are enthusiastic about finding a cure for cancer — in fact some feel that a cure is many years away. To back their arguments, they cite the fact that an individual's chances of dying from cancer have increased in the last 40 years, and the major progress that has occurred resulted from the medical profession's increased ability to determine the presence of the disease much earlier than in years past. Because of advances in detection, treatment has begun at an early stage of the disease, when success is much more likely. But some researchers maintain that although statistics say that more individuals are surviving longer with cancer, in reality they are surviving longer only because their disease was diagnosed earlier. In other words, the statistics are misleading.

No matter whether specific individuals are optimistic or pessimistic about finding a cure for cancer, all are amazed at the astonishing progress that is being made in our understanding of the fundamental mechanisms of normal and unregulated cell division. There is no doubt that our understanding of the basic mechanisms of cancer is increasing, and that fact should allow us to think more optimistically. Maybe the reason for the deadly, persistent growth of Henrietta Lacks's cells — and the cells of numerous other cancer sufferers — will soon be found.

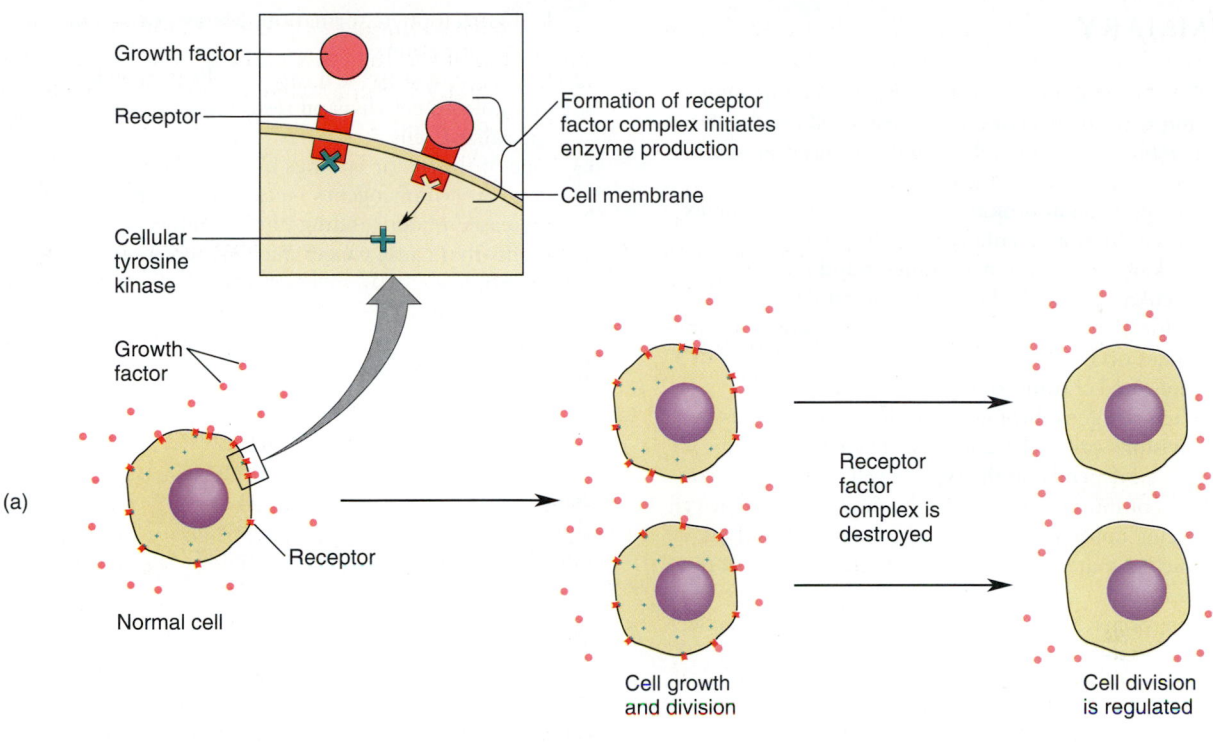

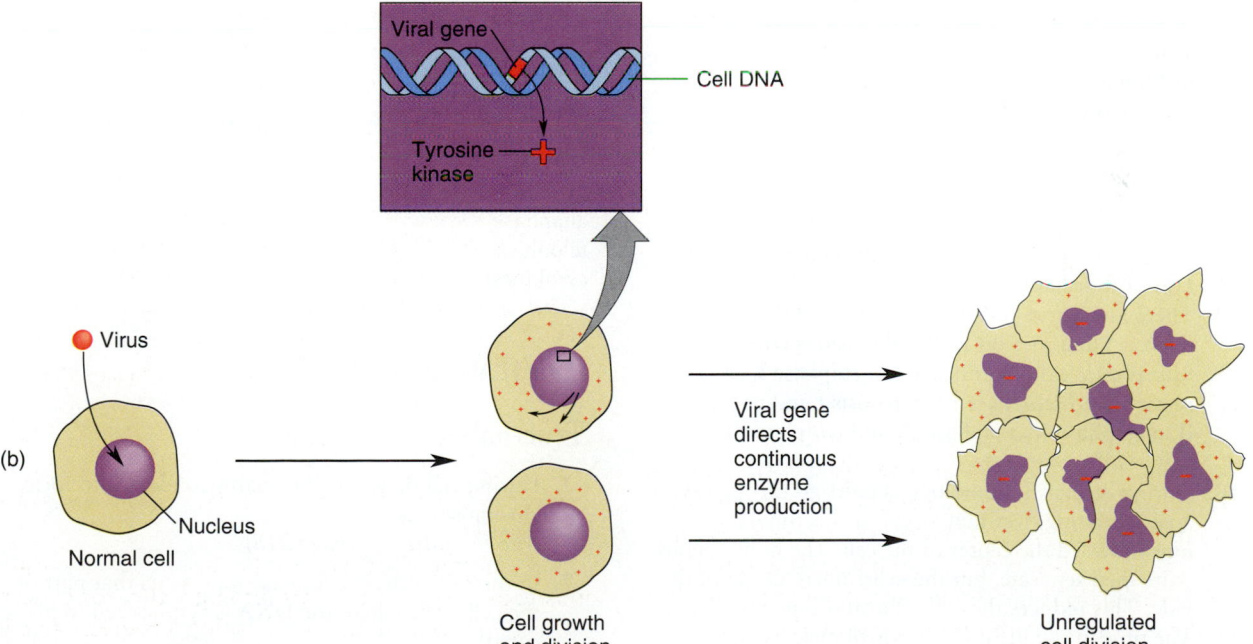

FIGURE 10.16 Chemical Control of Cell Division
The presence or absence of certain chemicals may play a role in regulating cellular division and growth. (a) When a particular cellular enzyme is produced in the presence of another chemical (growth factor), cell division occurs. When enzyme production ceases, cell division stops. (b) However, if the controlling enzyme is continuously produced (in this case, under the direction of a viral gene), cell division continues and unregulated cell growth results.

SUMMARY

1. DNA is responsible for the maintenance, growth, and reproduction of cells. When a cell divides, DNA is replicated and the daughter cells have the same amount of DNA as the original cell.

2. Cell division in prokaryotic cells is not as complex as cell division in eukaryotic cells, in part because prokaryotes lack a membrane-bound nucleus.

3. In eukaryotic cells the DNA plus proteins condense to form chromosomes during cell division. Chromosomes occur in homologous pairs. Each "arm" of a replicated chromosome is called a chromatid.

4. Each cell goes through a number of stages between divisions called the cell cycle. The length of time involved depends on the type of cell. A cell cycle consists of interphase (the nondividing phase of the cell cycle), mitosis (division of chromosomes and the nuclear materials), and cytokinesis (division of the cell).

5. Interphase, the nondividing phase of the cell cycle, consists of three stages: G_1, S, and G_2. In the G_1 stage, the cell carries on normal functions and prepares for cell division; in the S stage, DNA is replicated; and in the G_2 phase, the cell is readied for mitosis (nuclear division).

6. Interphase is followed by mitosis, or nuclear division, in which the chromosome number remains the same. Mitosis consists of four phases: prophase, metaphase, anaphase, and telophase.

7. Mitosis is usually followed by a cellular division known as cytokinesis, in which the cytoplasm and organelles are divided equally between the two new daughter cells.

8. Meiosis is a type of cell division in which the chromosome number is halved, resulting in the formation of four haploid daughter cells.

9. There are two sequential divisions in meiosis: meiosis I and meiosis II. Both divisions consist of four phases called prophase I, metaphase I, anaphase I, and telophase I for meiosis I and prophase II, metaphase II, anaphase II, and telophase II for meiosis II.

10. During meiosis I, homologous pairs go through synapsis (crossing over may occur at this time) and move toward the center of the cell. The homologous pairs then separate, but the centromere does not divide. This reduces the cell's chromosome number.

11. The time in the life cycle at which meiosis occurs depends on the organism. In a diplontic life cycle, meiosis occurs in the formation of the sex cells. In haplontic organisms (such as many of the unicellular plants and some simple multicellular plants), meiosis occurs after the formation of the zygote, producing a haploid adult. Diplohaplontic organisms, such as multicellular plants, have alternating diploid and haploid adult cycles. A diploid adult generation, the sporophyte, alternates with the hap-loid gametophyte formation. Meiosis occurs in spore formation.

12. Cancer, in a sense, is a case of cells gone wild — an unregulated cell division resulting in a mass of unspecialized cells.

13. One of the major findings in cancer research is the discovery of oncogenes — cancer-producing genes. Increased understanding of the number of genetic events that cause cancer may enable researchers to control it.

KEY TERMS

parent cell	gamete
daughter cell	sperm
gene	egg
chromosome	meiosis
homologue	haploid (n) number
diploid (2n) number	synapsis
histone	crossing over
nucleosome	chiasma
centromere	gonad
chromatid	gametogenesis
cell cycle	spermatogenesis
interphase	oogenesis
G_1 phase	polar body
S phase	zygote
G_2 phase	diplontic life cycle
mitosis	haplontic life cycle
prophase	alternation of generations
aster	diplohaplontic life cycle
spindle	metastasis
metaphase	carcinogen
anaphase	oncogene
telophase	proto-oncogene
cytokinesis	interferon

REVIEW QUESTIONS

True/False

1. During cell division, chromatin condenses to form chromosomes.
   true   false   (page 218)

2. In the cell cycle, the S phase represents that part of the cell cycle where the DNA is synthesized (replicated).
   true   false   (page 220)

3. During mitosis, the resulting daughter cells have the same chromosome number as the parent cell.
   true   false   (page 220)

4. Meiosis can be defined as a reduction in which the daughter cells produced contain one member of each chromosome pair.
   true   false   (page 226)

Fill in the Blank

1. Cell reproduction is not just the result of an arbitrary pinching in two of a cell. Rather, it is important that a complete set of _____ material be transferred from one cell generation to the next. *(page 217)*

2. Cell reproduction usually consists of two sequential processes, _____, or nuclear division, and _____, or cell division. *(pages 222–225)*

3. During meiosis, the diploid chromosome number is reduced to the _____ number. *(page 226)*

4. _____ cells can, in a general sense, divide indefinitely. This is because they have lost their cell surface properties that inhibit growth. *(page 232)*

Multiple Choice

1. In an animal in which the haploid chromosome number is 4, each gamete or sex cell will contain how many chromosome(s)?
   a. 2
   b. 4
   c. 8
   d. 16
   (page 229)

2. Reduction division occurs in:
   a. Gamete formation in a diploid life cycle
   b. Spore formation in a diplohaplontic life cycle
   c. Meiosis
   d. All of the above
   (pages 230–231)

3. Crossing over is an important mechanism for:
   a. Increasing genetic recombination
   b. Preventing mutation
   c. Producing diploid gametes
   d. Increasing chromosome number
   (page 226)

4. At the fundamental biological level, cancer is believed to be caused by:
   a. A number of genetic events causing the cell to divide out of control
   b. The activities of the lysosome
   c. Lack of chloroplasts
   d. Lack of antibodies
   (page 233)

Discussion

1. Discuss the structure of a eukaryotic chromosome.

2. Briefly summarize the stages in a typical cell cycle.

3. Discuss mitosis. Be sure to include its biological purpose, where it occurs in the cell cycle, and the four phases involved.

4. How does cytokinesis differ from mitosis?

5. What is meiosis? From a biological perspective, why did meiosis evolve?

6. Briefly discuss the events in meiois I and II.

7. What are spermatogenesis and oogenesis?

8. Does meiosis always occur at the same time in the life cycle of all organisms? Why or why not?

9. What are oncogenes? Speculate on the impact our understanding of oncogenes may have in the future.

11

DNA Replication

*The Search for the Genetic Material —
A Human Endeavor*
The Race for Discovering DNA Structure
The Process of Replication in DNA
Enzyme Activity in DNA Replication
Summary

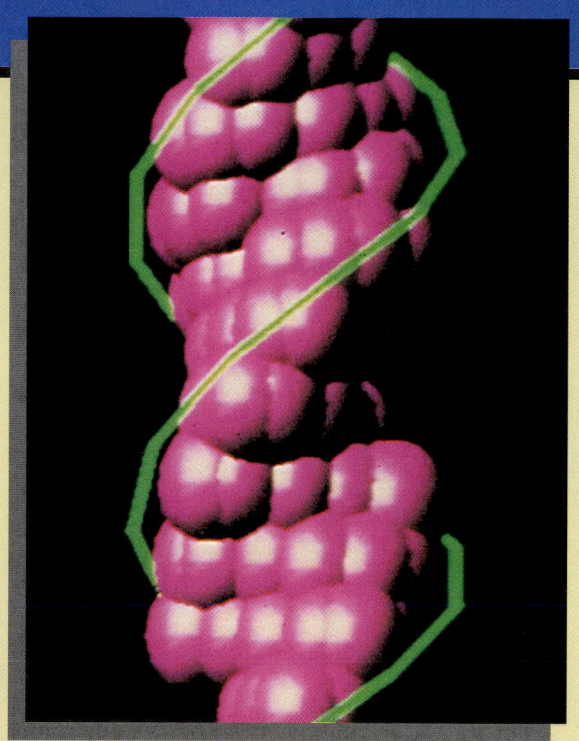

When you look at your family portrait, you are seeing a picture painted by genes. In the faces, the stature, the color of hair and eyes, you probably see certain similarities between the members pictured there. The family ties, the common heritage, are evident. Nonetheless, each individual in the portrait also exhibits distinguishing characteristics. In fact, each individual in that portrait (and everywhere else) is unique — a plan never to be repeated.

The mixture of similarities and uniqueness is due to the fact that we are each following the genetic design that we inherited from our ancestors. We share this journey with all living creatures, for at

the molecular level genes provide the necessary information to produce all living things. Genes are, indeed, the blueprints of life; they specify the architecture and program the activities of every being (Figure 11.1). What is the mysterious molecule capable of providing the information to turn an egg into a chicken, to produce a cell from a cell, and to dictate the molecular instructions that result in life?

You may already know some of the answers, for we have discussed this molecule before: **deoxyribonucleic acid (DNA)** — the code of life. However, biologists have known the significance of the DNA molecule only for about forty years. What makes DNA so special, and what tests did it have to meet in order to become recognized? To learn the answers, let us follow the work of the men and women who have enabled us to understand a part of the wondrous mystery contained in the gene.

Above: A computer graphic image of a molecule of DNA, the code of life.

FIGURE 11.1 Genes: The Blueprints of Life
Caterpillars, seals, and flowers all develop from a single egg. The genetic diversity represented here all begins at the molecular level — with the gene.

THE SEARCH FOR THE GENETIC MATERIAL — A HUMAN ENDEAVOR

Scientists had searched for the heredity molecule for a very long time. They knew it must fulfill some very specific requirements. The genetic material of life had to be capable of performing two basic functions:

1. It must be able to replicate itself (that is, produce exact copies of itself but yet allow for changes to be passed on from cell to cell and generation to generation).

2. It must be able to control life's activities.

In the late 1800s, biologists first observed that when the nucleus of a cell was destroyed or removed, the cell lost its ability to reproduce itself and to control its chemical activity. Scientists also learned that when the nucleus was replaced or when one was transplanted into a cell without a nucleus, many cellular activities resumed. Therefore, most biologists concluded that the logical place to look for genetic material was in the nucleus of the cell.

Learning About the Nucleus

■ **In 1869, the chemical composition of nuclein was first described as being rich with deoxyribonucleic acid.**

One of the earliest experiments occurred in 1869. Friedrich Miescher, a Swiss chemist, was investigating the chemical composition of the material contained in the nucleus of the cell. He was working with two sources: white blood cells that had stuck to the bandages of wounds, and the sperm obtained from salmon (sperm consist primarily of nuclear material). Miescher isolated the nuclear material from these two sources and found that the material (which he named *nuclein*) consisted of an acid rich in phosphorus. Later, this material was identified as deoxyribonucleic acid. Few scientists appreciated the influence that the material Miescher extracted from pus and sperm would have on our understanding of life. In fact, it would not be until approximately 75 years later that the genetic code contained in the nucleic acid Miescher had named would be deciphered.

DNA Staining Techniques

■ **In 1924, the discovery of a dye especially suited for staining DNA led to another discovery: that the amount of DNA was constant in body cells and was halved in sex cells.**

In 1924, another step was taken in our understanding of the structure of DNA when Robert Feulgen developed a simple staining technique to identify the presence of DNA. Feulgen found that the dye *basic fuchsin* turned bright purple in the presence of DNA. Then, after much work, Alfred Mirsky and his colleagues discovered that the body cells of any given organism contained identical amounts of DNA but that the sex cells contained only half that amount of DNA.

To understand the significance of this discovery, remember that the genetic material must be capable of transferring its information from generation to generation. Therefore, it was significant when Mirsky measured the amount of DNA after a sperm cell had fertilized an egg and discovered that the fertilized egg contained the same amount of DNA as the adult body cells. This finding suggested that DNA might be the molecule that transferred information from one generation to the next.

Transformation

■ **The next significant finding came from experiments on two variations of pneumonia bacteria. Somehow a substance or substances within a bacterial cell could transform rough (R) bacteria into smooth (S) bacteria.**

Who would believe that a clue pointing to DNA as the genetic material would come from an experiment to produce a pneumonia vaccine? In 1928, Fred Griffith was working with the bacterium *Diplococcus pneumoniae* (this is a pair of round bacterial cells that cause a type of pneumonia). There are two varieties (or strains) of pneumococcal bacteria: one called *smooth* (S), because the bacteria have a smooth protective capsule and produce smooth shiny colonies when grown in the laboratory, the other called *rough* (R), because the bacteria lack this capsule and produce rough colonies.

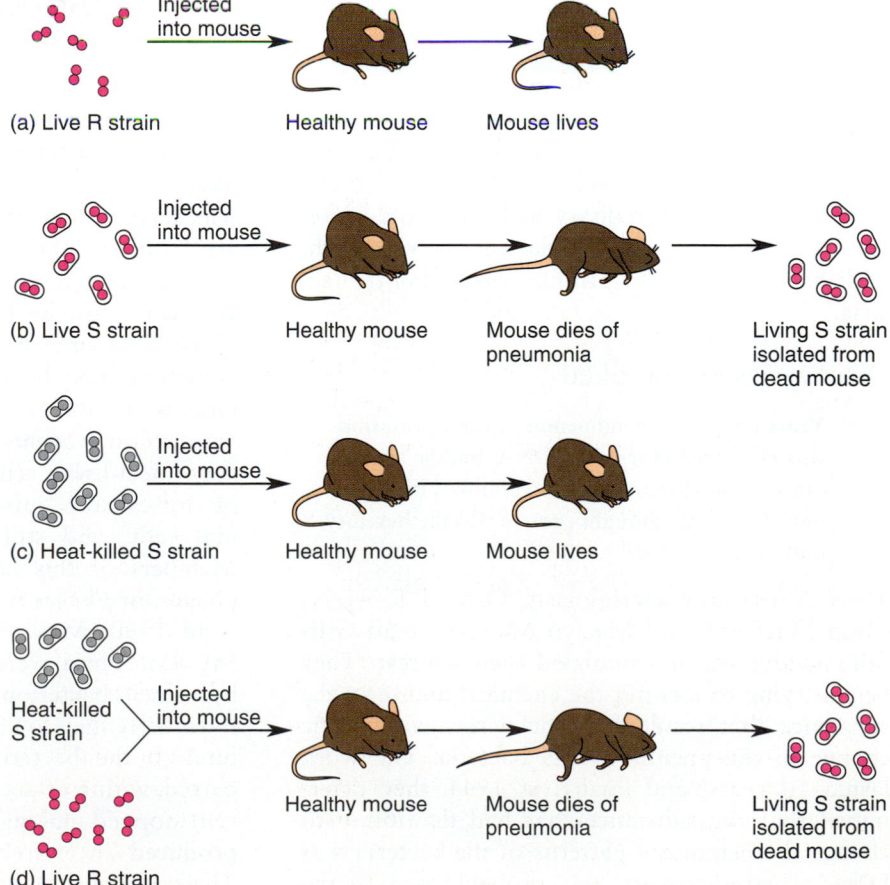

FIGURE 11.2 The Griffith Transformation Experiments While experimenting with mice to develop a vaccine against pneumonia (a–c), Frederick Griffith found that a certain "substance," which was apparently passed on from dead S-strain bacteria to live R-strain bacteria, changed the live bacteria's genetic characteristics (d). This change—known as transformation—resulted in the harmless R strain becoming a virulent, pneumonia-causing strain. Years later, this transforming substance was shown to be DNA.

(a) Live R strain — Injected into mouse — Healthy mouse — Mouse lives

(b) Live S strain — Injected into mouse — Healthy mouse — Mouse dies of pneumonia — Living S strain isolated from dead mouse

(c) Heat-killed S strain — Injected into mouse — Healthy mouse — Mouse lives

Heat-killed S strain / Live R strain (d) — Injected into mouse — Healthy mouse — Mouse dies of pneumonia — Living S strain isolated from dead mouse

During the course of his work, Griffith found that injecting mice with the R strain of the bacteria did not produce pneumonia, but when mice were injected with the S strain, they developed pneumonia and died. Griffith then went through the procedures required to test his results. He reisolated the bacteria from the dead mice and found the S form, as he had expected.

The next test involved killing a sample of the S-strain bacteria with a heat treatment, and injecting a mouse with the mixture. The mouse lived. Then he combined the heat-treated, killed, smooth bacteria with live, R-strain bacteria and again injected mice with the combined material. What do you think happened? Logically, you might expect the mice to live and not contract pneumonia since the virulent S strain (the disease-causing form) had been killed and only the R strain (non-disease-causing form) was alive. That was what Griffith expected to happen.

However, things did not turn out that way. Instead, the unexpected happened — the mice sometimes died of pneumonia! What's more, when the bacteria were reisolated from the blood and other tissues of the dead mice, Griffith found living S forms (Figure 11.2). How did this happen? How were R-strain bacteria transformed into S-capsulated bacteria?

The process of changing one strain of bacteria into another is called **transformation.** But what material was capable of transforming harmless R bacterium into the deadly S form? Whatever the transforming principle (substance) was, it actually had to change the hereditary makeup of the bacteria, because the rough bacteria were now smooth, and they continued to produce only smooth bacteria.

Discoveries Overlooked

■ **Years later, the phenomenon of transformation was explained in terms of DNA, but the importance of the discovery was overlooked because most biologists thought protein was the heredity molecule.**

Three American bacteriologists, Oswald T. Avery, Colin MacLeod, and Maclyn McCarty, read Griffith's study, and it stimulated their interest. They began trying to identify the chemical nature of the substance that resulted in the permanent genetic change in the pneumococcus bacteria. The work lasted 10 years, and finally, in 1944, they determined that the substance that had the ability to change the hereditary patterns of the bacteria was DNA. Their discovery was probably one of the most significant biological discoveries of the century.

Given the significance of the work of Avery and his colleagues, you would have expected the scientific community to be excited about this finding, but for the most part it was not. At this time many scientists believed that the genetic material in the cell was protein, not DNA. Most biologists reasoned that the molecules that directed heredity must be capable of a tremendous amount of variety, given the number of alternatives involved in life processes. Proteins are large complex molecules, made up of various arrangements of 20 components, the amino acids. On the other hand, nucleic acids, the elements of DNA, are also complex, but they were made up of only four components, the nucleotides. Therefore the complexity of the protein molecule led many biologists to believe it was the most likely candidate for the molecule of heredity, and that DNA was too simple to encode the wide variety of traits seen in the living world.

Learning More:
This Time from Virology

■ **Using radioactive isotopes, Hershey and Chase determined that it was the DNA of the bacteriophage that entered the host cell.**

The next key to the identity of the hereditary substance came from an unexpected area. Viruses are smaller than bacteria and are believed by some biologists to be the smallest forms of life. (As some wits have commented, viruses that cause diseases are "a bit of bad news wrapped in protein.") **Viruses** are composed of a protein coat and a nucleic acid core, usually DNA (or, in the case of plant viruses and some others, like the AIDS virus, RNA). Sometimes they have structures through which material is injected into the host cell. In a sense, viruses are infectious agents that can only reproduce when they infect living cells. There are a number of types of viruses; some infect plant cells, others infect animal cells, and still others infect bacterial cells. Members of this last group are called **bacteriophages,** or *phages* for short (see Chapter 18).

In 1930, Max Delbruck, Alfred Hershey, and Salvador Luria were investigating the mechanism by which bacteriophages infected bacterial cells. Eventually they found that the bacteriophage could bind to the bacterial membrane. When this occurred, within 60 seconds the bacterial cell (the host cell) stopped making its normal protein and instead produced an entirely new set of foreign proteins. These proteins were enzymes involved in the synthesis of new bacteriophages within the bacterial cell.

Then, in as short a period of time as 30 minutes, the infected bacterial cell would burst (lyse), releasing newly made bacteriophages. Bacteriophages had reproduced themselves by somehow taking over the metabolic activities of the host cell, dictating that the host should no longer follow its own hereditary instructions but instead should follow the hereditary instructions of the virus. In other words, instead of producing more bacteria, the host would produce more phages.

The next logical question was: What does the bacteriophage inject into the host cell that is capable of providing instructions for the production of bacteriophages? In 1952, Alfred Hershey and Martha Chase provided the answer. It was already known that a bacteriophage consisted of a protein coat and a central core of DNA. Therefore, there were three possibilities: The bacteriophage could inject the protein coat, the DNA, or both substances into the host cell. Biologists already knew that proteins contain sulfur but DNA does not, and that DNA contains phosphorus but proteins generally do not. Hershey and Chase reasoned that if they labeled the proteins with radioactive sulfur (^{35}S) and labeled the DNA with radioactive phosphorus (^{32}P), they could then determine which molecule entered into the cell and was responsible for directing the activity of the host cell.

Hershey and Chase infected one batch of bacteria with bacteriophage whose DNA had been labeled and another batch with bacteriophage whose protein had been labeled. They incubated each batch long enough for the phage to infect the bacteria but not long enough for the phage to begin to reproduce, and then they dumped each batch into a blender. The blender separated the bacteria from any viral material that remained outside the bacteria. They then used centrifugation to separate the intact bacteria from the remaining viral material. As you might have guessed, the liquid suspension containing the viral material contained only radioactive sulfur and the suspension with the bacterial cells contained only radioactive phosphorus.

Therefore, the results demonstrated that only the radioactive phosphorus of the DNA entered into the bacteria cell, and the radioactive sulfur of the protein coat remained outside (Figure 11.3). This proved that DNA alone entered the cell, thus demonstrating that DNA, not protein, was the genetic material capable of encoding the instructions for life. The search for the genetic material was over. The molecule known as DNA was scientifically established as the genetic material.

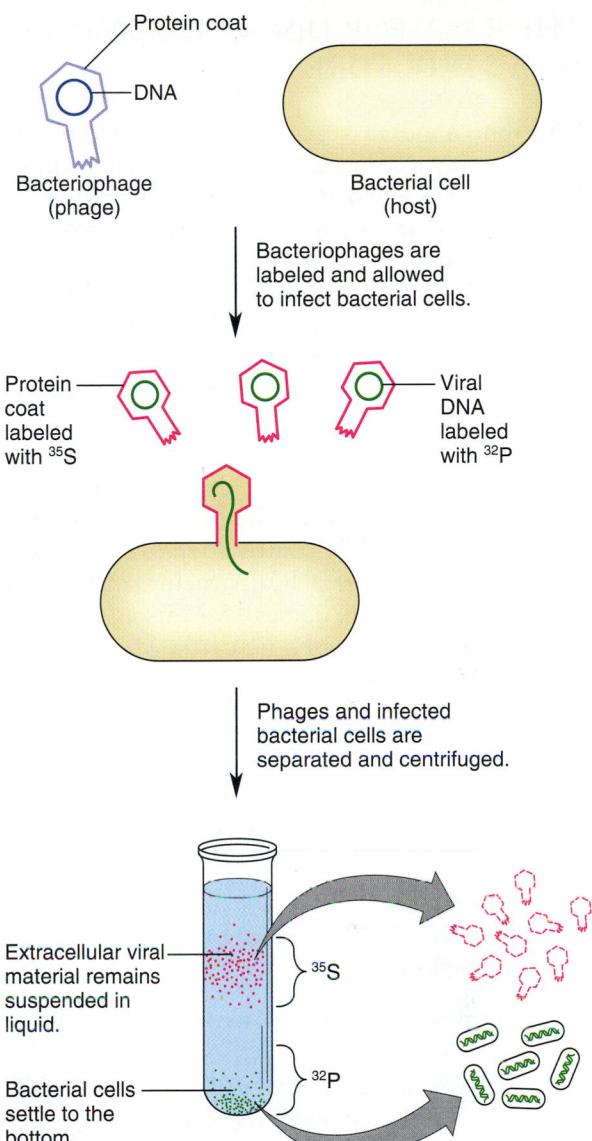

FIGURE 11.3 The Hershey-Chase Experiment
Bacteriophages labeled with radioactive ^{32}P and ^{35}S were allowed to infect bacteria. The protein coats were then mechanically separated from the bacterial cells and each tested for radioactivity. The results showed that most of the ^{32}P had been deposited inside the host cells, whereas most of the ^{35}S remained outside. This demonstrated that it is the DNA, not the protein, that is injected into the bacteria.

Biologists now knew the identity of the molecule of heredity. But what was the structure of DNA, and how would an understanding of DNA structure help us to understand how a DNA molecule replicates itself and controls cellular activities? Those questions were the ones that now needed answering.

THE RACE FOR DISCOVERING DNA STRUCTURE

After one hundred years of work, theorizing, and happy accidents, the search for the genetic material had now developed into a race to determine the exact molecular structure of DNA. The prize was prestige, more research grants, and even more money. (The same type of race is going on now in many areas of science, including genetic research and AIDS research. The prizes are the same, except the monetary reward that will come with patenting — in other words, having ownership of the results of the work — may be truly mind-boggling.) Many people entered the race, but the best known was Linus Pauling (see the Profile in Chapter 3), who had just made a major contribution to science with his description of the helical structure of proteins. Working in England, James D. Watson (from the United States), Francis H. C. Crick, Maurice H. F. Wilkins, and Rosalind Franklin (all from England) entered the race to explain the three-dimensional structure of DNA. How much did they already know?

Basics About Nucleic Acids

■ **Nucleic acids are large molecules consisting of nucleotides, which are formed from a pentose sugar, a phosphate, and a nitrogen base.**

By the 1940s, biologists already knew a fair amount about nucleic acids. As we noted in Chapter 3, nucleic acids are large polymers composed of subunits called nucleotides. A nucleotide is composed of a

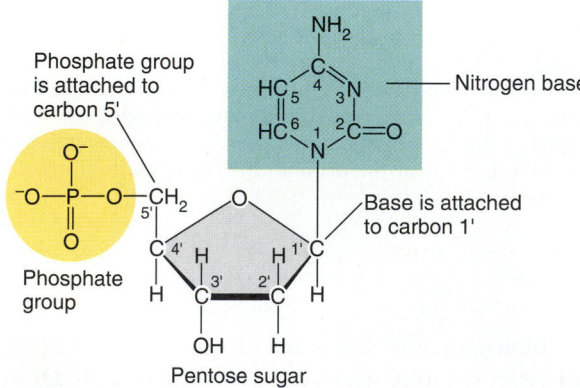

FIGURE 11.4 Basic Nucleotide Structure
A nucleotide—the basic subunit of nucleic acids—is composed of a nitrogen base, a phosphate group, and a 5-carbon (pentose) sugar.

five-carbon sugar called a *pentose sugar* (either **deoxyribose** in DNA or **ribose** in RNA), a *phosphate group*, and a *nitrogen base* (Figure 11.4).

Only four different kinds of nucleotides are found in DNA. Nucleotides differ in the type of nitrogenous base they possess; each nucleotide contains one nitrogenous base. Some nucleotides are composed of a single-ringed nitrogen base called a **pyrimidine.** Cytosine (C) and thymine (T) are pyrimidines. **Purines** are double-ringed nitrogen bases. Adenine (A) and guanine (G) are purines. The three components (sugar, phosphate, and nitrogenous base) combine to form a nucleotide having a structure illustrated in Figure 11.5.

Note in Figure 11.4 that the carbons in the sugar and the base are numbered for convenient identification. The carbon atoms in the base are numbered 1, 2, and so on, while the carbon atoms in the sugar are numbered 1', 2', and so forth (read one prime, two prime, and so on), so that the bonding within the nucleotide can be discussed without too much confusion. Look carefully at the illustrations in Figures 11.4 and 11.6. You will see that the phosphate functional group is attached to the 5' carbon in the sugar, and the base is attached to the 1' carbon in the sugar. In the process of forming a sugar-phosphate backbone, the phosphate can attach to the 3' or 5' carbon.

One Last Hint: Balanced Quantities of Bases

■ **Chargaff's investigations established that (1) each species had a distinctive proportion of four nitrogen bases; (2) the amount of adenine equaled the amount of thymine (A = T); and (3) the amount of guanine equaled the amount of cytosine (C = G).**

By the early 1950s biologists knew that DNA was composed of four different nucleotides. There also were a few other clues. In 1949, Erwin Chargaff investigated DNA in several types of organisms in order to determine the relative amounts of bases in each type (see Table 11.1). His research provided two important observations:

1. The relative amounts of adenine, thymine, guanine, and cytosine varied for different species but were the same for a given species.

2. The amount of adenine was equal to the amount of thymine present (A = T) and the amount of guanine was equal to the amount of cytosine (C = G). (This observation is called **Chargaff's rule.**)

FIGURE 11.5   The Four Nucleotides of DNA
Four different nucleotides comprise the DNA molecule.

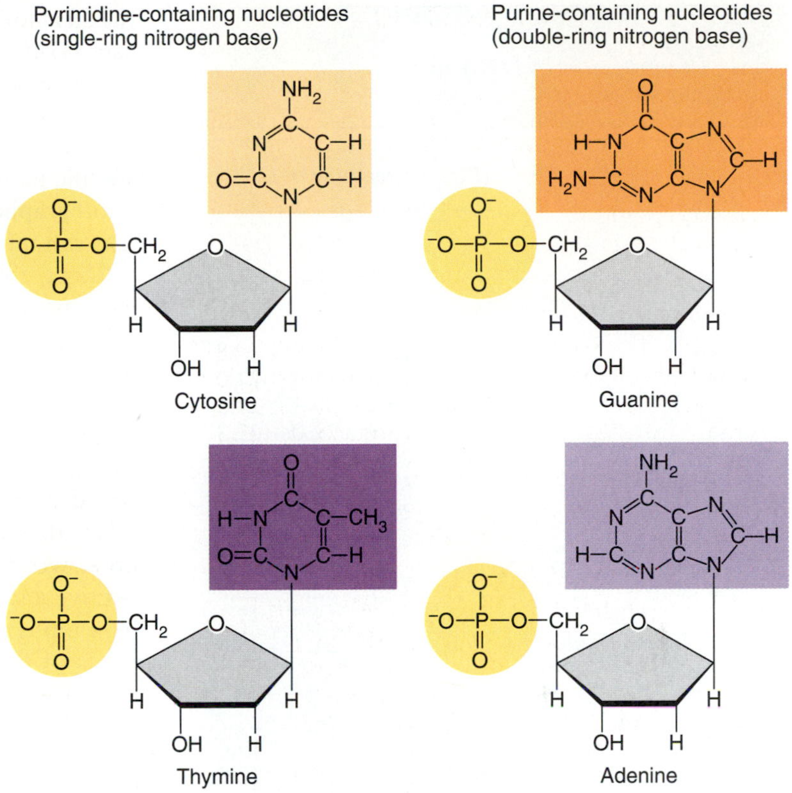

Pyrimidine-containing nucleotides (single-ring nitrogen base)

Purine-containing nucleotides (double-ring nitrogen base)

Cytosine

Guanine

Thymine

Adenine

FIGURE 11.6   Structure of a Single DNA Strand
The backbone of the DNA molecule consists of a strand of alternating sugar and phosphate molecules. Note that the phosphate can attach to either the 5' or 3' carbon atom in the sugar.

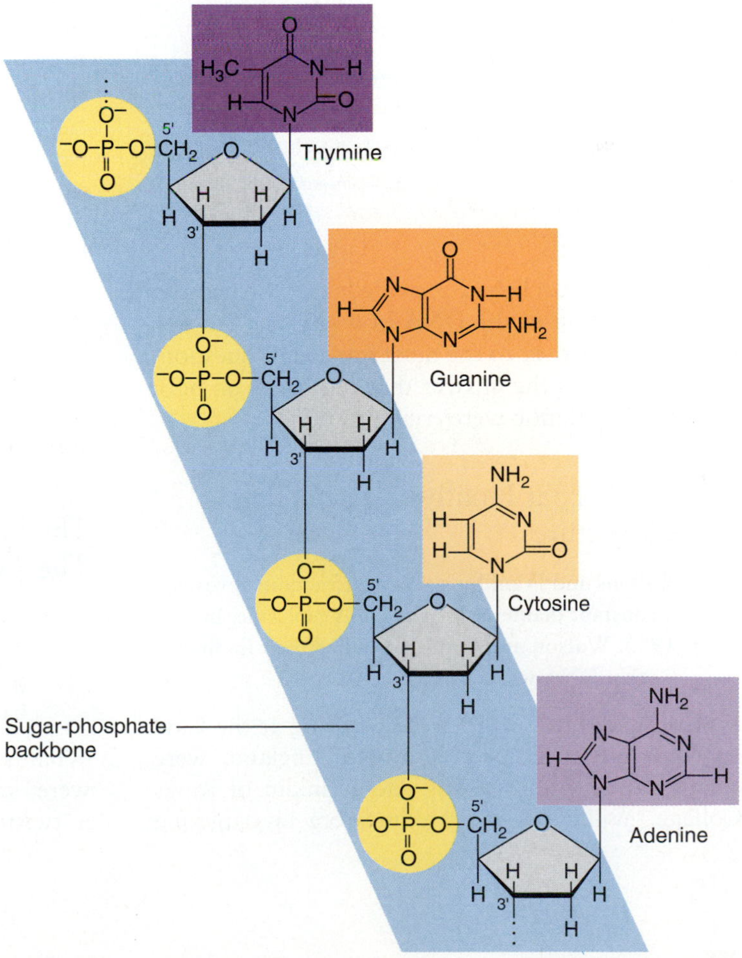

Thymine

Guanine

Cytosine

Adenine

Sugar-phosphate backbone

TABLE 11.1
Base Composition of DNA in Different Organisms

	(Percent per mole of DNA)			
	*A	T	G	C
Animals				
Human	30.9	29.4	19.9	19.8
Sheep	29.3	28.3	21.4	21.0
Hen	28.8	29.2	20.5	21.5
Turtle	29.7	27.9	22.0	21.3
Salmon	29.7	29.1	20.8	20.4
Sea urchin	32.8	32.1	17.7	17.3
Locust	29.3	29.3	20.5	20.7
Plants and Fungi				
Wheat germ	27.3	27.1	22.7	22.8
Yeast	31.3	32.9	18.7	17.1
Aspergillus niger (mold)	25.0	24.9	25.1	25.0
Bacteria				
Escherichia coli	24.7	23.6	26.0	25.7
Staphylococcus aureus	30.8	29.2	21.0	19.0
Clostridium perfringens	36.9	36.3	14.0	12.8
Brucella abortus	21.0	21.1	29.0	28.9
Sarcina lutea	13.4	12.4	37.1	37.1
Bacteriophages				
T7	26.0	26.0	24.0	24.0
λ	21.3	22.9	28.6	27.2

* A = adenine, T = thymine, G = guanine, C = cytosine.

Source: Adapted from Albert L. Lehninger, *Biochemistry,* 2nd ed. (Worth Publishers, Inc.; 1970, 1975).

Chargaff's research had provided a set of clues to the construction of the DNA in organisms but had supplied no answer to the most important question: Why? That was the answer that biologists on both sides of the Atlantic were trying to find.

The Work That Finally Determined DNA Structure

- Wilkins and Franklin's x-ray diffractions revealed a constant diameter for the DNA molecule. In 1953, Watson and Crick published their findings — that DNA was a double helix.

Watson and Crick, who were working at the Cavendish Laboratory in Cambridge, England, were collaborating with Wilkins and Franklin of Kings College. Wilkins and Franklin were investigating the structure of DNA by using a technique known as *x-ray diffraction.* X-ray diffraction is a technique in which a fine beam of x-rays is aimed at a crystal of a substance, behind which is a photographic film. The atoms in the crystal bend the x-ray beam in a specific pattern, and the x-ray diffraction pattern is photographed. By studying the photographs, scientists can infer what the structure of the crystal is.

Using this technique, Rosalind Franklin, while working in Wilkins's laboratory, obtained excellent photographs that provided an important clue. She found that the DNA molecule was long, thin, and had a constant 2-nanometer (remember that 1 nm = one-billionth of a meter) diameter along its entire length. She also found that it was most likely a coiled structure consisting of two structural elements that were repeated every 0.34 and 3.4 nm. In other words, the structure of DNA was highly repetitive and very regular. Biologists now know that 0.34 nm is the distance between each base pair (the rungs of the ladder). The other structure repeated every 3.4 nm, a distance now known to equal one complete twist of the coil of DNA.

Now the quest was almost complete. In fact, in February 1953, Pauling and his colleagues thought they had the answer and published an article presenting a model that featured a triple helix structure. That model was found to be erroneous. Instead, it was Watson and Crick who emerged victorious. On April 25, 1953, they published an article entitled "A Structure for Deoxyribose Nucleic Acid" and won the race for elucidation of the structure of DNA. They had put together a number of pieces of information (from experiments conducted by others) to come to their conclusions. They had remembered the findings of Chargaff and had put that information with what they knew about the hydrogen bonding potential of purines and pyrimidines (and added a tremendous amount of insight). The pieces all came together, and Watson and Crick stated that DNA was a **double helix.**

The Findings of Watson and Crick: The Twisted Ladder

- A double helix looks somewhat like a twisted ladder. The long uprights are made up of the sugar-phosphate backbone. The rungs are made of the base pairs.

What is a double helix structure? Watson and Crick were saying that the DNA molecule resembles a twisted ladder (Figure 11.7). The uprights of

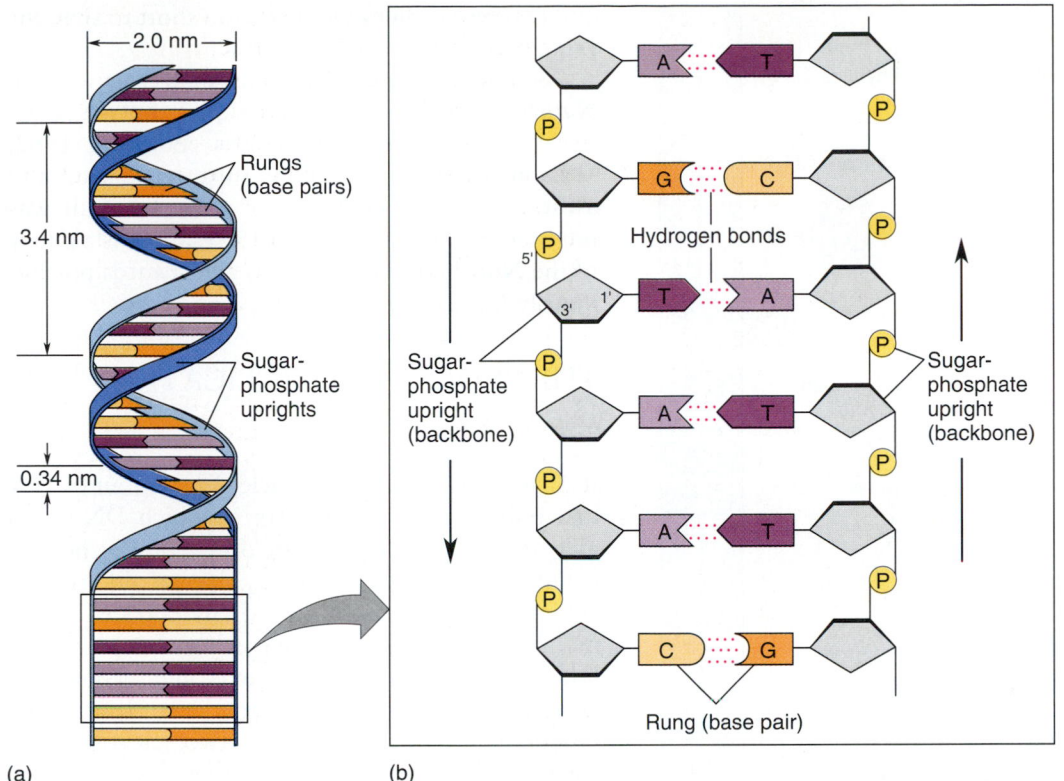

(a)

(b)

Rungs (base pairs)

2.0 nm

3.4 nm

0.34 nm

Sugar-phosphate uprights

Hydrogen bonds

A — T
G — C
T — A
A — T
A — T
C — G

Sugar-phosphate upright (backbone)

Sugar-phosphate upright (backbone)

Rung (base pair)

FIGURE 11.7   The Structure of DNA
(a) The DNA double helix resembles a twisted ladder, with the sugar-phosphate backbone forming the uprights of the ladder and the base pairs forming the rungs. One complete twist of the ladder is 3.4 nm long and the distance between the rungs is 0.34 nm. This pattern repeats itself along the entire length of the molecule. (b) Each pair of bases is connected by hydrogen bonds. Note that adenine always pairs with thymine and cytosine always pairs with guanine. Also note that the sugar-phosphate backbones run in opposite directions (antiparallel), as indicated by the arrows. (c) A scanning-tunneling microscope image of single-stranded DNA magnified 25,000,000 ×.

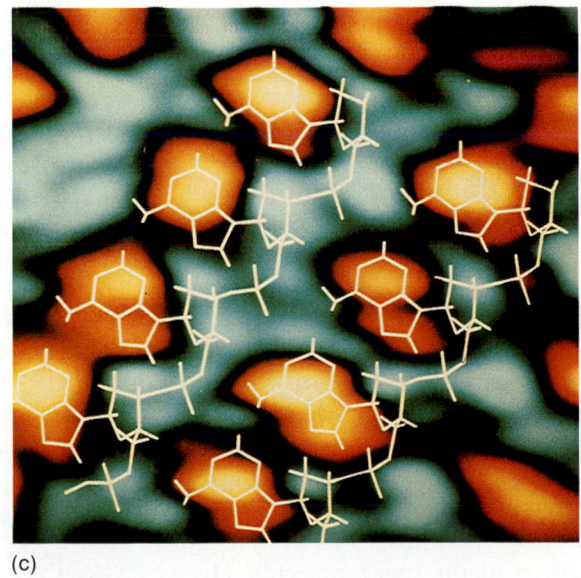

(c)

the ladder are made up of the sugars and phosphates. The rungs or steps of the ladder consist of a purine bonded to a pyrimidine, which explained Chargaff's ratios — specifically, adenine bonded to thymine and guanine bonded to cytosine (Figure 11.8).

How are all these pieces joined together? To understand the bonding that occurs you need to remember that the individual carbons of the sugar molecule are numbered 1', 2', 3', 4', and 5'. When you study the bonds in each upright of the ladder, you find that the position of the bond between the

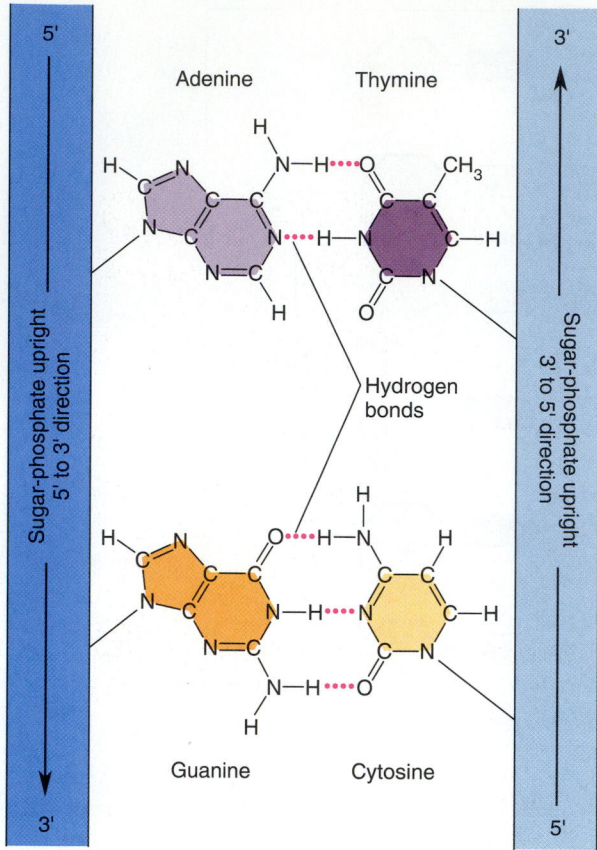

FIGURE 11.8 Base Pairing in DNA
In forming the rungs of the DNA ladder, the nitrogen bases always pair in a specific manner. Because of their chemical structures, adenine always forms two hydrogen bonds with thymine and cytosine always forms three hydrogen bonds with guanine.

phosphates and the sugar occurs alternately between the 3′ and the 5′ carbon atoms of the sugar molecule. The uprights are said to be *antiparallel*, which means they run parallel to each other but in opposite directions. One upright runs from the 5′ to 3′ direction and the other from the 3′ to 5′ direction. (Think of this in terms of a two-lane highway. Both lanes are parallel to each other, but they run in opposite directions.) The nitrogen base attaches to the 1′ carbon atom of the sugar molecule. Hydrogen bonds between the bases form the rungs of the ladder. When Watson and Crick built their scale model of DNA, they found that there was a base pair every 0.34 nm and a complete twist of the coil every 3.4 nm, just as the data of Wilkins and Franklin had suggested. They also found that a purine and a pyrimidine together would make a rung of the ladder 2 nm in width, again as Wilkins and Franklin suggested. Two purines would be too long

and two pyrimidines would be too short to span the 2-nm width of the molecule.

After Watson and Crick published their work in *Nature* on April 25, 1953, the quest for the structure of the DNA molecule was ended. In 1962, Watson, Crick, and Wilkins were recognized and awarded the Nobel Prize. Rosalind Franklin was not honored. She had died in 1958, and it is a policy of the Nobel committee not to give awards posthumously (see Profile).

THE PROCESS OF REPLICATION IN DNA

In that momentous first article, Watson and Crick had only alluded to the way in which DNA was able to replicate itself — in other words, how it could reproduce itself in a way that would produce copies, which would allow for its encoded information to be consistently passed on from one cellular generation to the next and from one generation of organisms to the next. Just a few months later, still in 1953, the two scientists published a second article in which they described how the double helical structure of DNA would allow for replication.

Complementary Strands

■ **Because adenine pairs only with thymine and guanine only with cytosine, the two strands of DNA are complementary. Thus one strand can serve as a template for forming its complementary strand.**

How does the replication of the DNA molecule occur? A partial answer lies in the concept of **base pairing.** To understand the reasoning, keep these elements in mind:

1. Franklin determined that the diameter of DNA was 2 nm. Watson and Crick found in their scale model of DNA that when a purine and a pyrimidine were hydrogen-bonded together to form the rungs of the ladder, the rungs were exactly 2 nm wide.

2. Chargaff discovered that the amount of adenine (A) equals the amount of thymine (T), and the amount of guanine (G) equals the amount of cytosine (C).

On the basis of these pieces of information, Watson and Crick reasoned that A would bond with T (or vice versa) to form a rung across the uprights of the double helix, and that G would bond with C (or

Rosalind Franklin
Biophysicist

In 1962, the Nobel Prize in Medicine and Physiology was shared by Francis Crick, James Watson, and Maurice Wilkins in recognition for their work on the structure of DNA. Rosalind Franklin was not recognized, even though she, along with Crick, Watson, and Wilkins, was responsible for the discovery of DNA's molecular structure. Her controversial omission will probably never be resolved.

One reason that the Nobel committee did not recognize Rosalind Franklin is a fairly straightforward one. Franklin died in 1958, at the early age of thirty-seven, and the Nobel Prize committee does not honor those who have died. Not everyone believes that reason, however — even though the committee's policy is quite clear. Some believe that Franklin was not recognized because she was a woman. She was a victim of the times.

Although criticism of the Nobel committee may be misplaced, the dissenters seem to present a valid point when they discuss the climate of the scientific community in England in the 1940s and 1950s. They present a fair amount of evidence that Franklin faced constant prejudice during her career. A woman in a field dominated by men, she was a social outcast in the scientific community. At professional luncheons, the men would retire to their private clubs to socialize, and she was excluded. She was never allowed to hold a permanent academic position, and always worked under the direction of a male investigator.

It was while Franklin was working in the laboratory of Maurice Wilkins that she used the technique of x-ray diffraction to elucidate the three-dimensional configuration of DNA. She had collected large amounts of data about its structure. Meanwhile in their own laboratory, Watson and Crick were pondering the fact that the amount of adenine equaled thymine and the amount of guanine equaled cytosine. They knew that this information about base pairs had to be explained by their model for the structure of DNA.

Watson and Crick were fortunate enough to have access to mimeographed reports of the precise measurements of DNA made in Wilkins's laboratory before Wilkins and Franklin had published the information in any scientific journal. In addition, Crick, who has an almost photographic memory, would visit Franklin's laboratory and use the opportunity to commit her data tables to memory. With these data and other information, Watson and Crick were able to build a model of DNA structure. On April 25, 1953, they published an article entitled, "A Structure for Deoxyribose Nucleic Acid," and with that they had won the race for elucidation of the structure of DNA.

So who should get credit? Two books have added fuel to the debate. In 1968, J. D. Watson published *The Double Helix,* a book in which he presents his personal account of the human side of the scientific process. There he describes Rosalind Franklin as a secretive, vindictive woman, and he minimizes the importance of her work. Taking the other side of the issue is Anne Sayre,

in her book *Rosalind Franklin and DNA,* published in 1975. There Sayre states that she wrote the book "to set the record straight — to restore to Rosalind Franklin not only her glory as a scientist, but her warmth and fascination as a person." The positions of the two authors are totally opposite.

Both of these books are well worth your time, because together they will help you gain an insight into the human side of scientific research — not only the positive side, in which there is the free exchange of scientific information, but also the negative side, in which there is secrecy, rivalry, and competition. The competitiveness of science is fostered by the funding system that supports research. And the rewards can be many: Grants, prizes, patents, tenure, and scientific prestige generally go to the scientist who wins the competition and becomes famous.

vice versa) to perform the same function. The pairs would always function together — find one of the nitrogen bases, and you will find its counterpart.

With base pairing that was this predictable, one long strand (the upright with its attached bases) would dictate the structure of the other upright, the other long strand. In other words, the two uprights, the two long strands, form **complementary images.** Anytime there is an adenine on one strand, there is a thymine at the corresponding position on the opposite strand, and vice versa. The same is true of cytosine and guanine. Because of their structures, the adenine and thymine would form a pair of hydrogen bonds to hold each other together, and again because of their structures, the guanine and cytosine would form three hydrogen bonds with each other. In so doing the base pairs form the rungs of the ladder.

For a more complete understanding of the meaning of a complementary strand, let's look at it closely for a moment. We know that:

1. We have a molecule that resembles a twisted ladder.

2. Each upright runs antiparallel.

3. Each upright is composed of alternating sugars and phosphates.

4. The half of the rung reaching across consists of one member of a base pair.

5. The bases can occur in any sequence.

This final point is crucial because it means that any base can follow any other base. Thus, the long strand can be of any length. The limitations on DNA structure come only in the process of base pairing — in creating the rung between the two uprights. A bonds with T, forming two hydrogen bonds, and G bonds with C, forming three hydrogen bonds (see Figure 11.8).

For example, in one small portion of the molecule the sequence might be (··· represents a hydrogen bond)

A:::T
T:::A
C:::G

or it might be

T:::A
G:::C
G:::C
C:::G

Therefore, any base pair can follow any other base pair, but A always bonds with T, and G with C. (See Focus 11.1.)

Replicating the DNA Molecule

■ During replication, DNA is split lengthwise, through enzyme action, at a replication fork. This is called semiconservative replication, because one of the original strands is conserved in the new molecule of DNA.

So we have a double-stranded molecule of indeterminant length, whose two strands are complements of each other. How can such a molecule replicate itself and pass this pattern on as cells or organisms reproduce? The answer is that the two strands act as templates, or patterns, to create a new copy of DNA. To understand what this means, consider the Jell-O mold. The inside of this particular mold is hollowed out to create a shape like a cluster of grapes — it is a template for the shape. Pour the Jell-O in and it takes on the shape of the hollows of the mold — a cluster of grapes.

The base that extends out to create half of the rung (because of its shape and the number of hydrogen bonds it can form) can only join with one other base, its partner in the pattern of base pairing. An extended adenine will only pair with a thymine, and vice versa. The same is true of guanine and cytosine. The template nature of DNA forms the basis of its role in passing the hereditary nature of an organism from a parent cell to its daughters.

Prior to cell division, enzymes help break the weak hydrogen bonds holding the base pairs together, forming what is called the replication fork, and the double helix begins to unwind, exposing the inner nitrogen base pairs to the nuclear environment. In other words, the molecule begins to unzip along its length as the replication fork moves along the molecule, resulting in two incomplete uprights. Both strands are now available as templates — they are conserved for use in the two daughter DNA molecules being formed. Within the nuclear environment, nucleotides are available: adenine, cytosine, thymine, and guanine. For example, if the sequence in one of the old strands is A, G, T, then, with the help of the appropriate enzymes, a new deoxyribonucleotide T will bond to A, a new C will bond to G, and a new A will bond with T, to form a new strand built on the requirements of the old strand (Figure 11.9).

Thus, the original molecule has replicated itself, forming two new DNA molecules, each of which includes one upright from the original parent molecule (which has been conserved) and one newly synthesized strand. This type of replication is called **semiconservative replication** because each strand of the original molecule DNA has served as a template for the linear sequence of nucleotides in the new

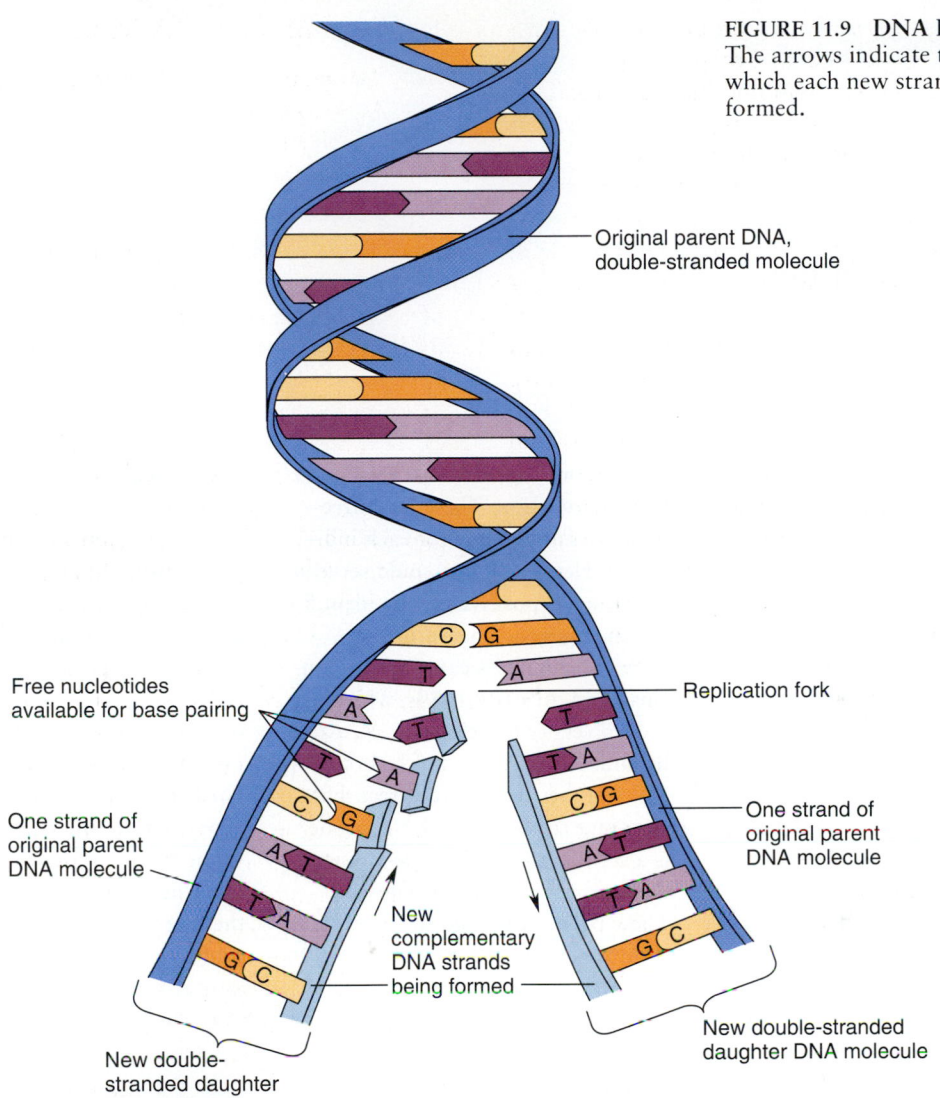

Original parent DNA, double-stranded molecule

Free nucleotides available for base pairing

Replication fork

One strand of original parent DNA molecule

One strand of original parent DNA molecule

New complementary DNA strands being formed

New double-stranded daughter DNA molecule

New double-stranded daughter DNA molecule

strand; thus, each new molecule consists of one strand from the original DNA molecule and one new strand. (If the replication had been conservative, the original DNA molecule would still be intact, and the new DNA molecule would consist of two newly constructed strands.)

Verification: Work with *Escherichia coli* Bacterium

■ **Watson and Crick's theory on the replication process was verified by work done with a nitrogen isotope and *Escherichia coli* bacteria.**

The evidence on hand supported the theory that Watson and Crick had formulated for DNA structure, but some of the theory still needed more testing. In 1958, Matthew Meselson and Frank Stahl

at the California Institute of Technology published their experimental evidence demonstrating that the replication of DNA is semiconservative.

Meselson and Stahl grew the single-celled bacterium known as *Escherichia coli* in a nutrient medium containing a heavy isotope of nitrogen, ^{15}N. The bacterium absorbed the nitrogen for, among other things, replication of its DNA for reproduction. Eventually ^{15}N was incorporated in the DNA of the entire *E. coli* population, which meant that the *E. coli* DNA was heavier than the more common form of DNA containing the nitrogen isotope ^{14}N.

Meselson and Stahl then transferred the ^{15}N *E. coli* and placed them in a medium containing only the nitrogen isotope ^{14}N. The bacteria again absorbed nitrogen, just as before. Analysis of the DNA of the daughter bacteria produced after the

DNA Fingerprinting

We began this chapter by stating that our uniqueness is due to our genetic makeup, more specifically our DNA. The sequence of our complementary base pairs—adenine (A) and thymine (T), cytosine (C) and guanine(G)—provides us with a four-letter alphabet from which our individual genetic signature is produced. It is known that much of the DNA in eukaryotic chromosomes is not expressed—that is, it is not used to code for messenger RNA, which eventually results in protein production. The portion of the DNA that is not expressed includes repetitive sequences or "stutters" that seem to have no function. These repetitive sequences are unique for each individual. Hence, stutters could serve as a method of positive genetic identification, a genetic fingerprint.

Dr. Alec Jeffreys came to this conclusion in the early 1980s, while at the University of Leicester in Great Britain. He discovered a class of DNA probes that could detect these repetitive sequences. A DNA probe is a short segment of single-stranded DNA whose sequence is known. These probes can bind to a specific complementary sequence on an unknown sequence of DNA. Jeffreys reasoned that these probes could be used to develop a technique of DNA "fingerprinting" (an expression he coined). These probes could be used to positively identify individuals involved in paternity disputes or crimes such as rape or murder or other circumstances in which definite identification is needed. In Great Britain, Robert Melias was convicted of rape because of semen stains found on the victim's clothing. The DNA was extracted from the sperm cells, and a DNA fingerprint was used to positively identify

FIGURE A
Creating a DNA fingerprint.

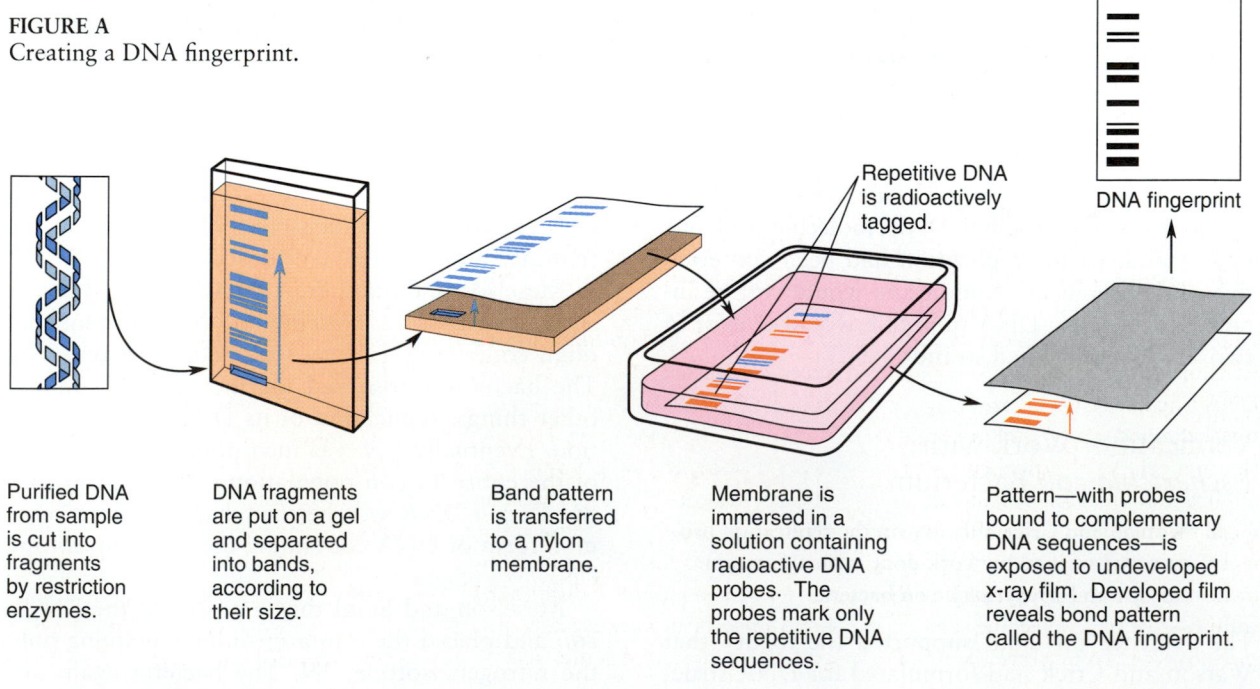

Purified DNA from sample is cut into fragments by restriction enzymes.

DNA fragments are put on a gel and separated into bands, according to their size.

Band pattern is transferred to a nylon membrane.

Repetitive DNA is radioactively tagged.

Membrane is immersed in a solution containing radioactive DNA probes. The probes mark only the repetitive DNA sequences.

DNA fingerprint

Pattern—with probes bound to complementary DNA sequences—is exposed to undeveloped x-ray film. Developed film reveals bond pattern called the DNA fingerprint.

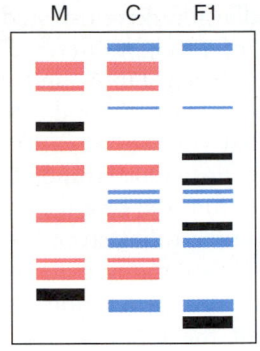

FIGURE B
Once the maternal bands on the child's DNA fingerprint have been identified, the remaining bands can be used to identify the father. All of these remaining bands must be matched on the father's DNA fingerprint to prove paternity.

POSITIVE IDENTIFICATION
Remaining bands in blue show the patterns that are identical in both the child and the father. A complete match for the child has been made,

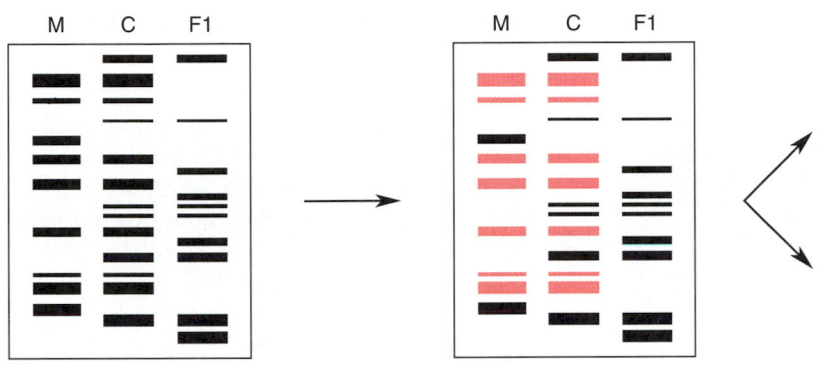

DNA fingerprints of a mother (M), child (C), and the child's father (F1).

Bands in pink show the patterns that are identical in both the mother and child.

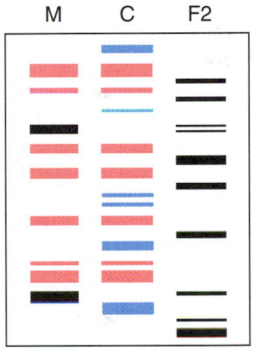

NEGATIVE IDENTIFICATION
Remaining bands in blue do not match up with band pattern in this adult male (F2), showing that this is not the child's father.

Melias as the rapist. Today even smaller amounts of DNA can be detected by means of a technique known as polymerase chain reaction (PCR). With this technique, the amount of DNA available for analysis can be amplified (increased) through a series of temperature-cycling reactions (see Chapter 13).

To produce a DNA fingerprint, a small sample of DNA-bearing cells such as semen, blood, or hair is obtained. The DNA is extracted, amplified, purified, and cut with restriction enzymes (Figure A). These DNA fragments, which are negatively charged, are placed in a small slot of a gel. An electrical current is applied and the DNA fragments move through the gel; smaller fragments move rapidly, larger fragments move more slowly, so that the still-invisible bands become separated by size. To preserve the band

pattern, however, the bands are transferred onto a nylon membrane. The Jeffreys DNA probes are made radioactive and added to the solution in which the nylon membrane is immersed. The radioactive probes attach to the specific invisible complementary bands of repetitive DNA. The probes fit only with the repetitive DNA; the other DNA is ignored. The membrane is placed next to an undeveloped x-ray film, and the probes, which are radioactive, irradiate the film to produce a specific band pattern (similar to the band price code you see on many items), the DNA fingerprint.

Let us look at a simple application of this technique. As we discuss in Chapter 15, ABO blood groups can be used to establish innocence, but not guilt, in a paternity dispute. But DNA fingerprinting can establish guilt in a paternity dispute. The

DNA fingerprints from the mother, child, and alleged father are compared. The DNA bands of the child are matched with the bands of the mother. Those bands which do not match the mother's must match those of the father. If the alleged father is in fact the father, then they will match. If they do not match, he is not the father (see Figure B). The chance that two people (other than identical twins) will have the same pattern is one in 30 billion. As you can see, DNA fingerprinting (the sequence of nucleotides) is the ultimate identification test.

first cell division demonstrated that their DNA was a hybrid, of a weight intermediate between the ^{15}N- and ^{14}N-labeled DNA and therefore containing equal amounts of ^{15}N and ^{14}N (Figure 11.10). This is just what you would expect from semiconservative replication. One strand of each new DNA molecule was from the original molecule — containing ^{15}N — and the other strand was newly synthesized from the ^{14}N-containing medium.

The second generation contained bacteria in which half the DNA molecules were hybrids (of intermediate weight) and the other half were composed of only the nitrogen isotope ^{14}N (light). Again, this is what you would expect. Those molecules resulting from the ^{15}N template would be hybrid, and those resulting from the ^{14}N would be light. Thus, the results supported Watson and Crick's hypothesis. DNA replication is semiconservative.

DNA Replication — A Summary

1. The linear sequence of the nitrogenous bases provides the basis for the genetic code.

2. Prior to cellular reproduction, the DNA is replicated to ensure that the information contained within the DNA molecule is passed on from cell to cell.

3. The replication of DNA is semiconservative. As the molecule "unzips," the two original strands serve as the templates for the synthesis of two new strands. The result is two new molecules, each consisting of one strand from

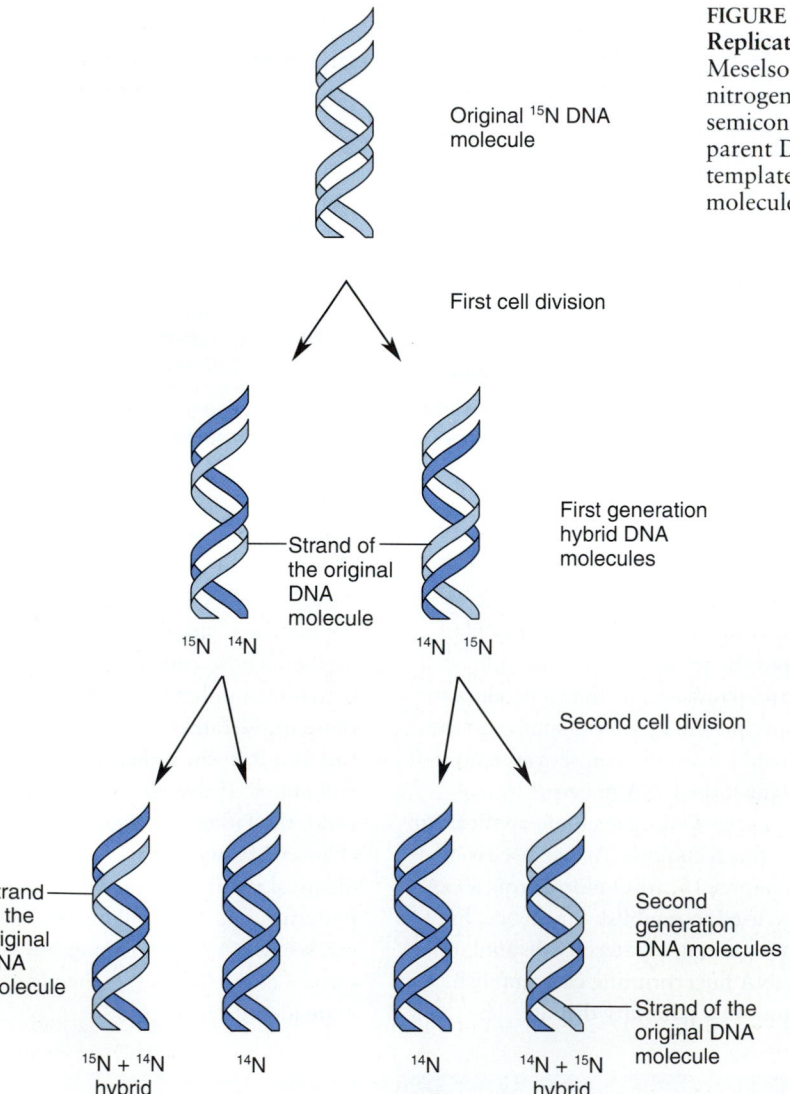

FIGURE 11.10   Semiconservative DNA Replication
Meselson and Stahl used radioisotopes of nitrogen to show that DNA replication is semiconservative. Each strand of the original parent DNA molecule continues to serve as a template for the synthesis of a new DNA molecule.

Original ^{15}N DNA molecule

First cell division

Strand of the original DNA molecule

First generation hybrid DNA molecules

^{15}N ^{14}N ^{14}N ^{15}N

Second cell division

Strand of the original DNA molecule

Second generation DNA molecules

Strand of the original DNA molecule

^{15}N + ^{14}N hybrid ^{14}N ^{14}N ^{14}N + ^{15}N hybrid

the parent molecule and one newly synthe-sized strand. Each daughter molecule contains one half of the original molecule and one half of the newly synthesized molecule.

4. As the molecule unwinds along its length, enzymes regulate the reaction in which the adenine nucleotide (a purine) bonds with thymine (a pyrimidine) and the guanine (purine) nucleotide bonds with the cytosine (pyrimidine).

5. The result of DNA replication is that the genetic code is reproduced and can be passed on from one cell generation to another.

ENZYME ACTIVITY IN DNA REPLICATION

■ **Many kinds of enzymes are involved in the replication of DNA. Some repair damage, others facilitate hydrogen bonds, and others couple and construct new DNA molecules.**

DNA is often referred to as a self-replicating molecule. However, this is only a half-truth, because in reality DNA will not replicate itself without the presence of specific enzymes. In 1956, Arthur Kornberg at Washington University in St. Louis demonstrated that DNA replication could occur in a test tube if the enzyme **DNA polymerase** was present.

Other research established that several enzymes in eukaryotic cells are responsible for unwinding the DNA strand at many different points within the DNA molecule to expose many small regions along the molecule to the nuclear environment. As the strands are forced apart by these enzymes, other enzymes, such as DNA polymerases, link the nucleotides in the complementary sequence to form the new strand. In addition, as you will see in the next chapter, other enzymes are responsible for catalyzing the reactions by which the information in the complementary base pairs is passed on to special forms of RNA called messenger RNA (mRNA) and other RNA forms as well. These forms of RNA are involved in protein synthesis.

Enzymes play other roles in DNA functioning as well. Some enzymes, called **ligase enzymes,** repair broken ends of DNA molecules to form intact strands. Other enzymes clip out and edit portions from the DNA molecule. The editing or "proofreading" functions of some DNA polymerases are responsible for the amazing accuracy of DNA replication. When there is a mismatch of base pairs in the replicating DNA molecule, the enzymes clip out the mismatched base pair and replace them with the correct one. As researchers have learned more and more about the role of enzymes in DNA activities, they have developed the technology of genetic engineering and recombinant DNA. This field is having so much impact on us today that we devote Chapter 13 to its applications and relevance in our world.

DNA is the molecule that serves as a common denominator among members of the living world. It is the source of unity, because all members of a species have a shared genetic code composed of nucleotides. But it is also the source of diversity, because alternative sequences in base pairs spell out unique individual genetic signatures.

This is all very well, you are no doubt thinking, but just how can DNA control life's activities? How can it provide those invisible molecular instructions that dictate life's processes? We know that a fundamental property of the hereditary molecule must be its ability to determine which proteins are going to be synthesized, and that ability in turn controls what a cell is able to do. The challenge now is to determine how the information encoded in DNA's sequence of base pairs is translated into the amino acid sequence in proteins and how these proteins regulate reactions. In fact, scientists working in the late 1950s and 1960s were involved in meeting that challenge. As you might have guessed, our next chapter will try to answer these questions.

SUMMARY

1. Deoxyribonucleic acid (DNA) is the molecule of heredity — the blueprint of life.
2. DNA is composed of nucleotides. A nucleotide consists of a sugar (deoxyribose in DNA) bonded to a phosphate and one of four nitrogen bases: adenine and guanine, which are purines, and thymine and cytosine, which are pyrimidines.
3. DNA is a double helix (a shape rather like a twisted ladder). The two uprights of the ladder consist of antiparallel strands of deoxyribose bonded to phosphates, and the rungs of the ladder consist of a purine bonded to a pyrimidine.
4. Each strand of DNA is complementary to the other. Adenine always bonds with thymine and cytosine with guanine, producing a double helical molecule with the base pairs in a linear sequence.
5. When the DNA molecule replicates, enzymes break the weak hydrogen bonds holding the purines and the pyrimidines together, and the base pairs "unzip." Each half of the molecule functions as a template to which the new complementary nucleotides bond, thereby synthesizing the new strand of DNA from the original strand. Thus, DNA replication is semiconservative.

KEY TERMS

deoxyribonucleic acid (DNA)
transformation
virus
bacteriophage
deoxyribose
ribose
pyrimidine
purine

Chargaff's rule
double helix
base pairing
complementary image
semiconservative replication
DNA polymerase
ligase enzyme

REVIEW QUESTIONS

True/False

1. The molecular basis of heredity is encoded in the linear sequence of nucleotides in the DNA molecules.
   true   false   (page 241)

2. A nucleotide is composed of a pentose sugar, a phosphate group, and a nitrogen base.
   true   false   (page 246)

3. During replication of the DNA molecule the original strand serves as a template for the synthesis of the daughter strand.
   true   false   (page 250)

4. The amount of the four nitrogen bases may vary greatly in a DNA molecule, but the amount of thymine is always equal to the amount of adenine and the amount of guanine is always equal to the amount of cytosine.
   true   false   (page 248)

Fill in the Blank

1. Adenine bonds with _____ and cytosine bonds with _____. *(page 252)*

2. Hershey and Chase determined that the material injected by the bacteriophage was _____, not the protein coat. *(page 245)*

3. The _____ enzymes repair broken ends of DNA. *(page 257)*

4. The structure of the DNA molecule is described as a _____, a "twisted ladder" configuration. *(page 248)*

Multiple Choice

1. In the *pneumococcus* transformation experiment, rough *pneumococcus* bacteria without capsules were transformed into smooth ones with capsules by means of an extract of capsule-covered bacteria. This experiment led to the discovery by others that:
   a. RNA exists.
   b. Genes exist.
   c. DNA has a definite molecular structure.
   d. DNA can be genetic material.
   (pages 245–246)

2. The concept of nucleotide base pairing means that:
   a. A specific purine will always bond with a specific pyrimidine.
   b. Any base can pair with any other base.
   c. Bases are always bonded to a specific sugar.
   d. a and b.
   (page 252)

3. The important conclusion for genetics that arose from the results of the Hershey and Chase experiments with ^{35}S- and ^{32}P-labeled bacteriophages was that:
   a. DNA is the genetic material.
   b. Protein is the genetic material.
   c. RNA is the genetic material.
   d. Individual nucleotides are the genetic material.
   (page 245)

4. In the Watson-Crick model of a double helix, the "rungs" of the "twisted ladder" are composed of
   a. Sugars
   b. A purine and a pyrimidine
   c. Two purines
   d. Two pyrimidines
   (page 252)

Discussion

1. Discuss why it is valid to state that "the living organism is in reality the gene on display."

2. What are the two basic functions of the genetic material DNA?

3. How was the transformation experiment of Griffith important in the search for genetic material?

4. What contributions were provided by Hershey and Chase with respect to the search for the genetic material?

5. Discuss the structure of DNA. Of what significance is the base pair concept?

6. Discuss the semiconservative replication of DNA.

Chapter 11   DNA Replication

DNA and Protein Synthesis

In 1910, a young, black West Indian student went to see his doctor in Chicago, complaining of severe pain. He showed symptoms of clogged blood vessels, pneumonia, inflammation of the soft tissues, physical weakness, and an enlarged spleen. The puzzled physician took a blood sample, prepared a slide by ringing a drop of blood with petroleum jelly, covered the specimen with a coverslip, and excluded any air from the blood. What he discovered was very interesting: The patient's red blood cells, which normally are shaped like spheres, had collapsed into a shape much like a sickle. The physician called the young student's disorder (de-

Above: A photograph of RNA polymerase, captured by using the technique of scanning-tunneling microscopy.

scriptively enough) *sickle-cell disease* or *sickle-cell anemia.*

Further investigations determined that there were actually two general forms of the disease. In one form, known as *sickle-cell trait,* the individual shows mild to moderate degrees of the disturbing symptoms; in the more severe form, **sickle-cell anemia,** the destruction of the body is severe. In fact, an individual who has sickle-cell anemia has only a 50 percent chance of living until the age of ten, and those who live past ten are usually crippled for life.

What was happening in the red blood cells of that West Indian patient to cause them to sickle? By the 1940s, biologists knew that a red blood cell is literally filled with the oxygen-carrying protein called *hemoglobin (Hb)*. In 1949, Linus Pauling, working at the California Institute of Technology, speculated that the disease resulted from an abnormality in the structure of the hemoglobin molecule.

Pauling's hypothesis was demonstrated to be correct when he and other colleagues, such as Vernon Ingram at Massachusetts Institute of Technology, succeeded in demonstrating that the amino acid sequence of the sickle-cell hemoglobin differed from that in normal, nonsickled hemoglobin by only one amino acid.

As we saw in Chapter 3, hemoglobin is a large protein molecule that consists of four polypeptide chains, two called alpha chains and two called beta chains (Figure 12.1). Further investigation of the beta chain determined that in normal hemoglobin (Hb^A) the sequence of the first six amino acids is valine-histidine-leucine-threonine-proline-*glutamic acid*. However, in sickle-cell hemoglobin (Hb^S), the sequence for this series of amino acids in the beta chain is valine-histidine-leucine-threonine-proline-*valine* (Figure 12.1b).

That one substitution has enormous ramifications. With the substitution of valine (which has an uncharged side chain) for glutamic acid (which has a charged side chain), the electrical charges are altered and the hemoglobin tends to form crystals. As a result, the red blood cells buckle and assume the sicklelike shape, which greatly reduces their oxygen-carrying capacity, as well as producing several other serious symptoms. (See Chapter 15.)

Quite early, biologists knew that sickle-cell anemia was an inherited disorder, passed on genetically from parent to offspring. Today, over 2 million African Americans have the sickle-cell trait condition. Fortunately, these patients manifest only mild symptoms of the disease. However, when two sickle-cell trait individuals marry and have children, there is a 25 percent probability that each child will have sickle-cell anemia. In the United States, one child in every 170 born will have sickle-cell anemia. This is a sobering statistic when you realize that one-half of those children usually die before age ten.

The illness called sickle-cell anemia is due to an alteration of the amino acids in the beta chain of the hemoglobin molecule. But why does the alteration occur? In a more general sense, what is the molecular basis of genetically transmitted disorders? What is the molecular code that dictates the linear sequence of amino acids in a protein? In other words, how does the hereditary material, DNA, control cellular activities and how is this information processed and transmitted from generation to generation? In this chapter we discuss the genetic basis of protein synthesis. Then in the following chapters we investigate the ways in which genetic characteristics are passed on from one generation to another.

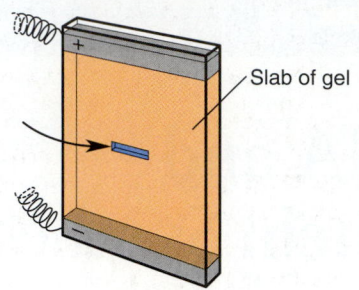

1. A mixture of the organic molecules and detergent is placed in a well in a slab of gel.

Slab of gel

2. An electrode with a positive charge is placed at one end of the gel, and an electrode with a negative charge is placed at the other. The voltage is turned on. Negatively charged molecules are attracted to the positive pole, and positively charged molecules are attracted to the negative pole.

3. Because Hb^S has fewer negatively charged amino acids, it migrates toward the positive pole more slowly than the normal hemoglobin Hb^A.

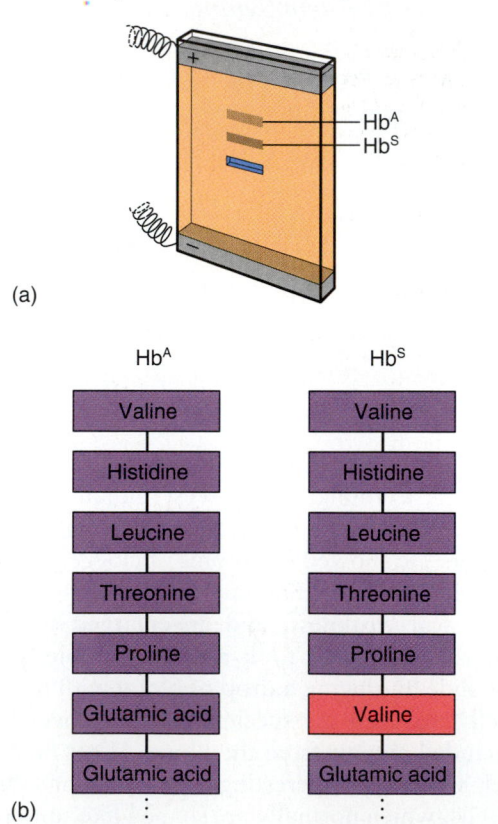

Hb^A
Hb^S

(a)

Hb^A	Hb^S
Valine	Valine
Histidine	Histidine
Leucine	Leucine
Threonine	Threonine
Proline	Proline
Glutamic acid	Valine
Glutamic acid	Glutamic acid

(b)

FIGURE 12.1   Gel Electrophoresis
(a) Gel electrophoresis separates organic materials (proteins) based on their migration through an electrical field. This technique was used to determine the sequence of amino acids in sickle-cell and normal hemoglobin. (b) Sickle-cell hemoglobin: valine is substituted for the glutamic acid found in normal hemoglobin.

THE SEARCH FOR THE MECHANISM OF PROTEIN SYNTHESIS

■ **In the early twentieth century, Archibald Garrod hypothesized that "factors" (his term for genes) were involved in the coding of enzyme structures.**

The idea that some metabolic disorders were the consequences of genetic errors was first proposed by Archibald Garrod in the early part of the twentieth century. Garrod was investigating a heredity disease in humans known as *alkaptonuria* that, among other things, results in the urine turning black on exposure to the air. Garrod proposed that the dark urine was due to a defect or absence of an enzyme normally involved in metabolism of the amino acid phenylalanine. He went on to hypothesize that the defect in the enzyme was somehow caused by a corresponding defect in the hereditary material. Biologists did not, in Garrod's time, know about genes, which are "blocks" of information contained in a chromosome that control a certain function, so Garrod referred to a "factor." However, as sometimes happens, Garrod's proposal was far ahead of its time. His hypothesis was not proved for over 30 years.

The Beginning of Understanding: One-Gene–One-Enzyme Theory

■ **In the 1940s, Beadle and Tatum, studying metabolism in a bread mold, determined that exposing the cells to x-rays (thus causing a gene mutation in the cell) made the cell unable to produce certain enzymes. As a result, the cell's metabolic processes were interrupted. On the basis of their work, they theorized that one gene codes for the synthesis of one enzyme.**

In the 1940s, George W. Beadle and Edward Tatum, at Stanford University, were working with a pink bread mold called *Neurospora crassa*. They were investigating the biochemical pathways that allow this mold to grow and to synthesize the materials it needs for life from only the most minimal source of nutrients, what they called a *minimal medium*.

Beadle and Tatum already knew that x-rays could alter the genes (specific regions of the DNA molecule) of an organism. By exposing the bread mold to x-rays, they produced mutant forms of the mold that were not able to synthesize the nutrients necessary for life from a minimal medium. The sci-

entists reasoned that some of these mutants had lost the ability to synthesize certain molecules necessary in their metabolic pathways. That meant that altering a specific gene in turn altered a specific enzyme. Therefore, Beadle and Tatum concluded that one gene codes for the synthesis of one enzyme.

Refinements of Existing Theories: One-Gene–One-Polypeptide

■ **In the 1950s, Vernon Ingram, who was working with the amino acid structure of hemoglobin, modified Beadle and Tatum's hypothesis to one-gene–one-polypeptide.**

Beadle and Tatum's theories were altered because of the work that Vernon Ingram was conducting in the 1950s on the amino acid sequence of the hemoglobin in sickle-cell individuals. Based on his findings — that a difference existed between normal and sickle-cell individuals in the sixth amino acid in the beta chain — he expanded Beadle and Tatum's theory to include polypeptides.

Ingram's contribution was based on the fact that although hemoglobin is indeed a protein, the beta chain of the hemoglobin is not a complete protein; instead it is a polypeptide chain, a part of the protein. As a result of Ingram's work, the theory was changed to one-gene–one-polypeptide, to indicate that one gene codes for the sequence in one polypeptide chain, any polypeptide, not just those that function as enzymes.

With Ingram's work, biologists knew more about protein synthesis than they ever had. The next question that had to be answered was this: How can information be translated from the gene (DNA) to protein? Read on.

DNA AND PROTEIN SYNTHESIS: THE CENTRAL DOGMA

■ **The central dogma states that the linear sequence of nitrogenous bases in DNA is transcribed onto an RNA molecule. The information encoded on RNA is then translated into a linear sequence of amino acids, which results in a protein.**

For a moment, look back over the knowledge that biologists had accumulated over the years. Recall that Beadle and Tatum provided evidence from their studies with *Neurospora* that a specific gene is responsible for the production of a specific enzyme (protein), and by so doing they had supported the ideas of Garrod, who had postulated the same idea years earlier. In Chapter 3 we discussed other investigators who established that proteins are composed of specific linear sequences of amino acids, and in Chapter 11, we discussed several investigators who demonstrated that DNA transmits genetic information. In that chapter you also learned that the biochemical blueprint for life is the specific linear sequence of nitrogen bases on a strand of DNA. Many other biochemists, molecular biologists, and geneticists worked on similar projects over the years.

From this collection of knowledge came the concept that explains how DNA codes for proteins. Crick called this concept the **central dogma.** The central dogma states that the linear sequence of nitrogenous bases in the DNA molecule is *transcribed* onto an RNA molecule (see the Human Endeavors box). The information encoded on the RNA molecule is then *translated* into the linear sequence of amino acids, which results in a protein. The process of protein synthesis is summarized in Figure 12.2. But in biology, as in any area of study, even dogmas can be altered. Biologists now know that, for the most part, the central dogma is valid, but experiments in 1979 demonstrated that in some viruses containing RNA, RNA can code for DNA and can carry genetic information. Even more recently some investigators have demonstrated that some RNAs themselves can even act as enzymes (see Focus 16.1).

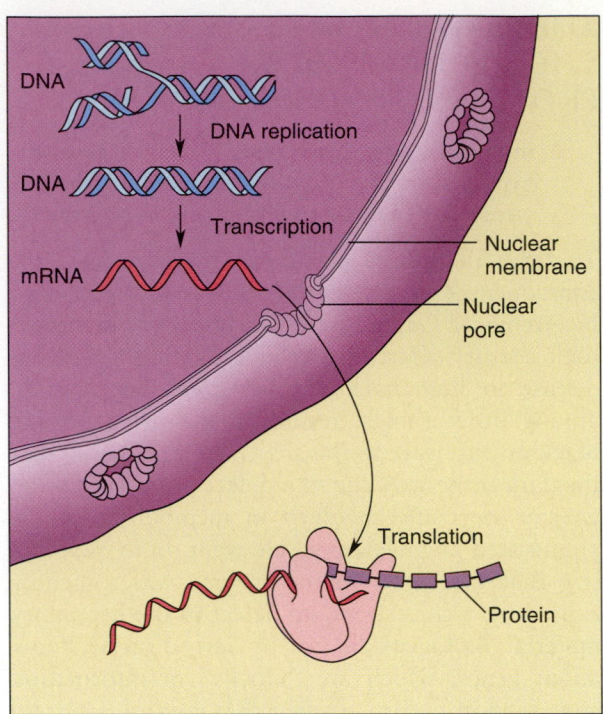

FIGURE 12.2   The Central Dogma
The information encoded in the DNA molecule is transcribed onto RNA. The sequence of bases in RNA are then translated into a specific protein.

The Living Code — A Triplet

■ **George Gamow thought that if the genetic code consisted of three nucleotides, each containing one of four possible bases, there would be enough combinations ($4^3 = 64$) to code for 20 biologically important amino acids. Therefore, he hypothesized that the genetic code was a three-letter code, a triplet code. Each three-letter set is called a codon. Further research determined which codons signaled which amino acids.**

The problem that faced biologists studying the code was much like that which faced the Japanese military in World War II. The U.S. military was using Navajo Indians to transmit messages about ship maneuvers and overall strategy over the radio — in Navajo. The Navajo language is unique — it has no close relatives, and the structure of its sentences is unlike that of any other language. It follows no known rules, so none of the proved methods for

breaking secret codes applied to the Navajo tongue. There had to be a new approach. The Japanese never succeeded in breaking the language of the Navajo "code talkers." Biologists, however, did crack the code of DNA.

There are 20 biologically important amino acids that make up proteins. In DNA there are four nucleotides (those containing adenine, thymine, guanine, and cytosine) that determine the organization of those amino acids. How do four nucleotides send out signals to specify 20 different amino acids? Just how is life "spelled"?

It was the Nobel Prize–winning physicist George Gamow and others who came up with the answer. Gamow reasoned that if only one nucleotide coded for one amino acid, then proteins would consist of only four amino acids. (Adenine would always signal, for instance, valine, and so on.) If two nucleotides were used, the possible combinations would be 4^2, or 16. For example, adenine and thymine together might signal valine, but adenine and guanine together would signal glutamic acid. But a two-nucleotide system still could not supply enough possible combinations to signal for the 20 biologically important amino acids.

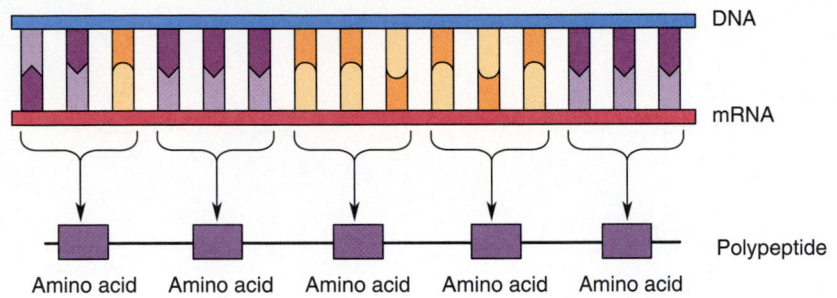

DNA

mRNA

Polypeptide

Amino acid   Amino acid   Amino acid   Amino acid   Amino acid

FIGURE 12.3   The Triplet Code
A series of three nucleotides in the DNA (codon) controls the placement of an amino acid in a polypeptide. Only one strand of the DNA is coded onto the RNA.

Gamow reasoned, if the code consisted of three nucleotides, there would be more than enough combinations ($4^3 = 64$) to code for the 20 amino acids. Therefore, he hypothesized that the code of life was a three-letter code, a triplet code. In other words, three nucleotides in sequence along the DNA or RNA molecule, called a **codon,** must serve as a code for one specific amino acid (Figure 12.3). DNA had its own unique language, based on a four-letter alphabet (the four nitrogen bases) and containing words having only three letters.

Code Breaking

Gamow's hypothesis about the triplet code of life makes sense, but is there evidence for it? What are the specific codons that indicate a specific amino acid will be bonded in place? In 1961, Marshall Nirenberg and J. H. Matthaei at the National Institutes of Health published the first in a series of papers on polypeptide synthesis. They had been working with an artificial form of RNA, a messenger molecule that operates with DNA (we will be discussing RNA and its various forms shortly). Nirenberg and Matthaei had found that when they constructed their artificial RNA with repeated segments of the codon UUU (polyuracil) and placed it in a test tube with a reaction mixture of ribosomes, amino acids, and other essential structures, the result was a polypeptide containing only the amino acid phenylalanine. Hence they reasoned that UUU coded for phenylalanine (Figure 12.4).

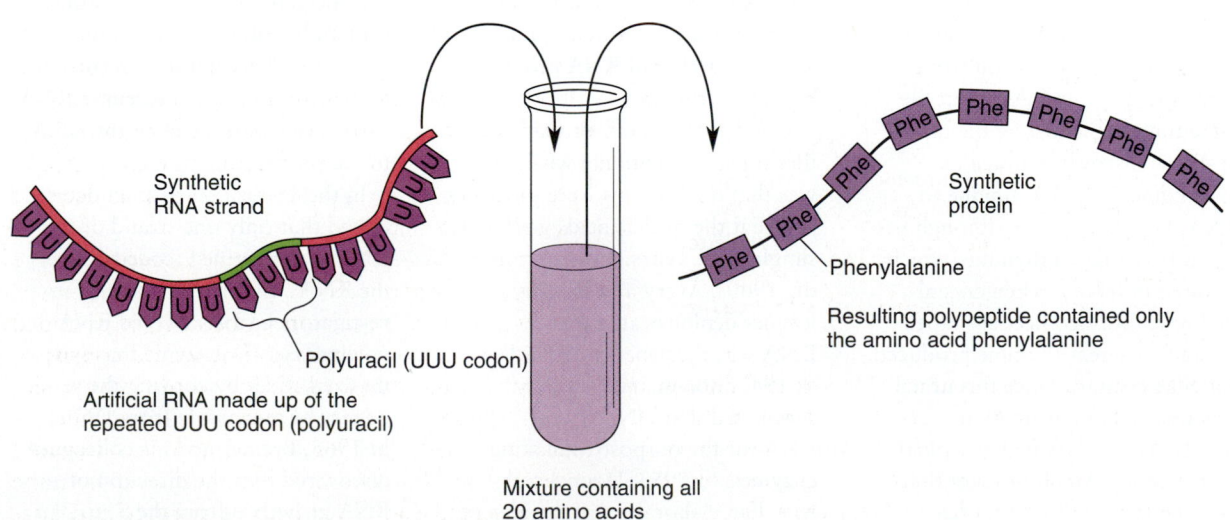

Synthetic RNA strand

Polyuracil (UUU codon)

Artificial RNA made up of the repeated UUU codon (polyuracil)

Mixture containing all 20 amino acids

Synthetic protein

Phenylalanine

Resulting polypeptide contained only the amino acid phenylalanine

FIGURE 12.4   Evidence for the Triplet Code of Life
When placed in a mixture containing all 20 amino acids, RNA consisting only of polyuracil (what we now know to be the codon UUU) synthesized a polypeptide consisting only of phenylalanine.

DNA and Protein Synthesis: The Central Dogma

Transcription

Because of the endeavor of many investigators, we now have a better understanding of transcription — the metabolic pathway from gene to protein.

In the early 1900s, the French scientist Garnier, while observing cells under the microscope, noticed that a nonchromosomal substrate made by the nucleolus migrated to the cytoplasm and combined with the cytoplasmic granules prior to protein synthesis. Garnier didn't know that what he was observing was the RNA and ribosome interaction that is necessary for protein synthesis — because at that time, scientists thought that there was only one type of nucleic acid. At about the same time, a chemist by the name of Ascoli discovered uracil (a pyrimidine), which is unique to RNA. He noticed that although uracil was similar to thymine (which is found in DNA), when crystalized uracil formed needle-shaped crystals, whereas thymine produced platelike crystals. Since this uracil was first isolated from plant cells (yeast — classified as a plant at that time), Ascoli thought that plants had one kind of nucleic acid and animals another. What he didn't know was that every cell has both DNA and RNA.

Between 1909 and 1930, the chemistry of nucleic acids became better understood. Levene in 1909 found ribose sugar in one nucleic acid and in 1930 found deoxyribose sugar in the other. In 1924, Feulgen developed a staining technique that demonstrated that DNA is present in the nucleus of all cells. During that time, ultraviolet absorption techniques were being used by Casperson to determine that both plant and animal cells contained RNA. Casperson and Brachet, working independently, demonstrated that the RNA was most concentrated in cytoplasmic structures that contain protein. These structures were later called *ribosomes*. They also noted that there was a strong correlation between the rate of protein synthesis and the number of RNA protein (ribosome) structures.

One has to remember that at that time, conventional wisdom was that the proteins were the genes and that the nucleic acids were too simple to be genes. However, in the 1940s, Avery and his colleagues demonstrated that DNA was the genetic material. In 1947, Boivin and Vendrely speculated that DNA makes RNA for the purpose of making enzymes. In 1953, Dounce published an elaboration of the idea of DNA → RNA → enzymes that became better accepted.

In 1950, after Watson and Crick arrived at the double helical model for DNA, many investigators developed what was called the *central dogma* — that DNA determined the linear sequence of amino acids in proteins and that the ribosomal RNA determined protein structure. Investigators thought that one-gene—one-ribosome—one-protein was the method of transcription. In other words, each gene (DNA) gave rise to one specialized ribosome.

However, in 1958, while studying the infection of *E. coli* by bacteriophage T2, Astrachen and Volkin discovered a new type of RNA. Jacob and Monod later named this RNA *messenger RNA*. In 1961, Jacob and Monod argued against the one-gene—one-ribosome—one-protein hypothesis. They stated that ribosomes were not the precursors to protein; rather a messenger RNA carried the message from the DNA to the protein.

In the 1960s, Spiegelman determined that only one strand of the DNA is transcribed (coded onto the RNA). In addition, several investigators had discovered RNA polymerase, the enzyme that synthesized RNA by copying the sequence of nucleotide from DNA. In 1965, Bremer and his colleagues discovered that the direction of RNA growth is from the 5′ to 3′ end. Other investigators found that the translation (the conversion of the genetic code on the messenger RNA at the ribosomes to a linear sequence of amino acids in a polypeptide) consists of an initiation, elongation, and termination sequence.

In the 1970s, by using RNA polymerase inhibitors, investigators determined that each type of RNA — ribosomal, messenger, and transfer — had its own type of RNA polymerase. Also in the 1970s, techniques were developed that enabled investigators to actually visualize transcription on lampbrush chromosomes (large, commonly occurring chromosomes that appear puffed when DNA loops out from the chromosome as RNA is being transcribed). In 1977, Sharp and Roberts, working independently, discovered that intervening sequences (introns) are spliced out before the RNA is translated.

From the 1980s to the present, human endeavors to understand transcription have concentrated on the molecular event that regulates transcription and on how initiation, elongation, and termination occur.

Source: Adapted from "Evolutions: Transcription 1900–1990's," *The Journal of NIH Research,* vol. 2, pp. 102–104, June 1990. Adapted from original artwork by Terese Winslow.

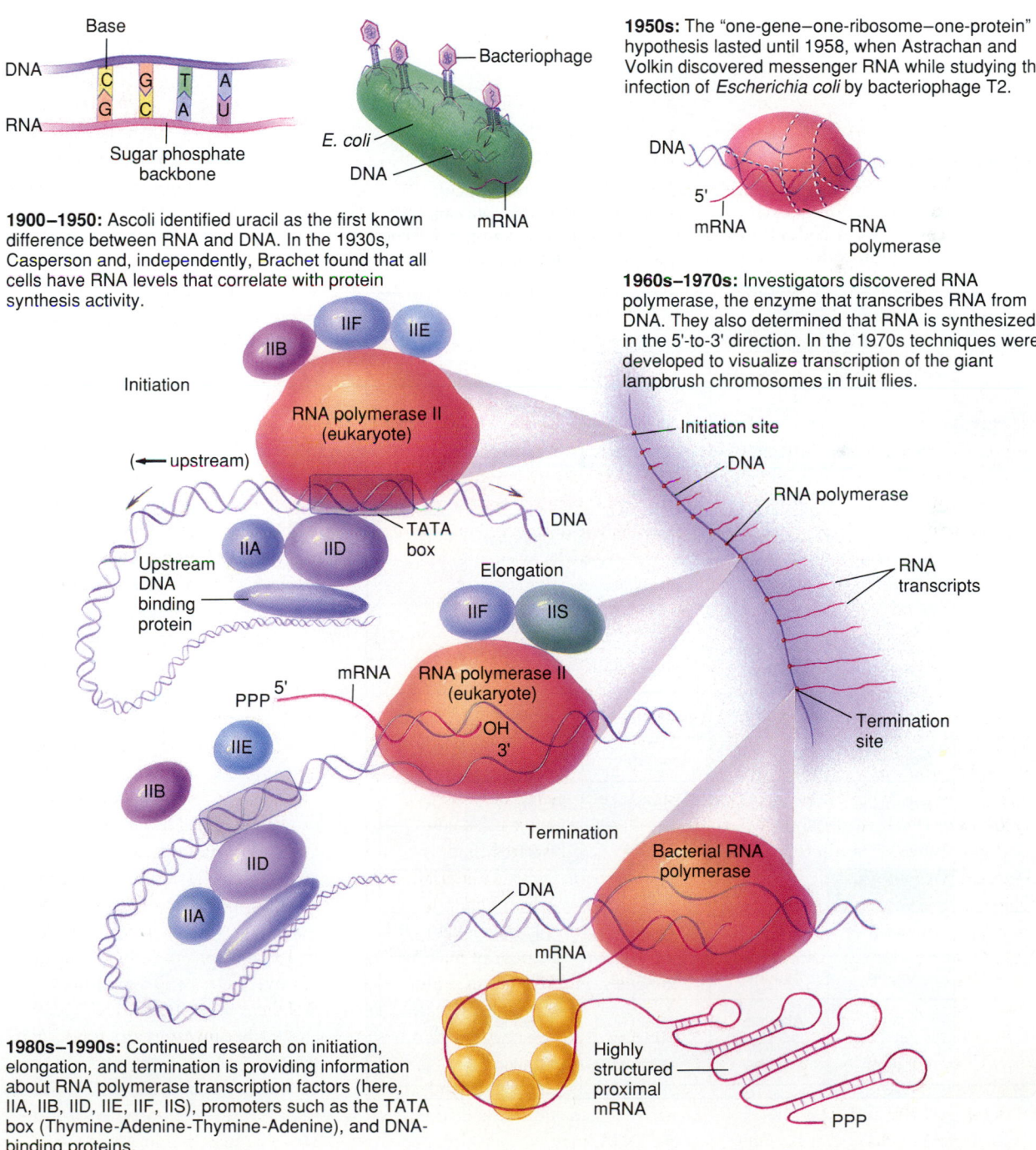

1950s: The "one-gene–one-ribosome–one-protein" hypothesis lasted until 1958, when Astrachan and Volkin discovered messenger RNA while studying the infection of *Escherichia coli* by bacteriophage T2.

1900–1950: Ascoli identified uracil as the first known difference between RNA and DNA. In the 1930s, Casperson and, independently, Brachet found that all cells have RNA levels that correlate with protein synthesis activity.

1960s–1970s: Investigators discovered RNA polymerase, the enzyme that transcribes RNA from DNA. They also determined that RNA is synthesized in the 5'-to-3' direction. In the 1970s techniques were developed to visualize transcription of the giant lampbrush chromosomes in fruit flies.

1980s–1990s: Continued research on initiation, elongation, and termination is providing information about RNA polymerase transcription factors (here, IIA, IIB, IID, IIE, IIF, IIS), promoters such as the TATA box (Thymine-Adenine-Thymine-Adenine), and DNA-binding proteins.

Numerous other investigators working along the same line, such as Severo Ochoa at New York University and Har Gobin Khorana at the University of Wisconsin, participated in deciphering the code of life. They found that in some cases more than one codon codes for a given amino acid. In fact, there are 61 codons that signal for the 20 amino acids. Some codons, UAA, UAG, and UGA, are *terminator codons* (that is, they signal that the message is over) and others are initiation codons. The code of life is shown in Figure 12.5.

RNA AND PROTEIN SYNTHESIS

What is the role of RNA in the process of protein synthesis? There are three types of RNA, each with a particular role to play in protein synthesis: *messenger RNA (mRNA)*, *transfer RNA (tRNA)*, and *ribosomal RNA (rRNA)*. Before we describe the synthesis of RNA, let us quickly review what we already know about this molecule.

First, remember that RNA is ribonucleic acid, a nucleic acid, and that its nucleotides contain the

FIGURE 12.5 The Genetic Code
A specific sequence of three nucleotides codes for a specific amino acid. Following the intersection of each row and column in the table, we can interpret the code: UCA codes for the amino acid serine, for example. Often, more than one codon codes for a specific amino acid. Serine has six codons.

First letter of code	Second letter of code				Third letter of code
	U	C	A	G	
U	Phenylalanine	Serine	Tyrosine	Cysteine	U
	Phenylalanine	Serine	Tyrosine	Cysteine	C
	Leucine	Serine	STOP	STOP	A
	Leucine	Serine	STOP	Tryptophan	G
C	Leucine	Proline	Histidine	Arginine	U
	Leucine	Proline	Histidine	Arginine	C
	Leucine	Proline	Glutamine	Arginine	A
	Leucine	Proline	Glutamine	Arginine	G
A	Isoleucine	Threonine	Asparagine	Serine	U
	Isoleucine	Threonine	Asparagine	Serine	C
	Isoleucine	Threonine	Lysine	Arginine	A
	Methionine (start)	Threonine	Lysine	Arginine	G
G	Valine	Alanine	Aspartate	Glycine	U
	Valine	Alanine	Aspartate	Glycine	C
	Valine	Alanine	Glutamic acid	Glycine	A
	Valine	Alanine	Glutamic acid	Glycine	G

Chapter 12   DNA and Protein Synthesis

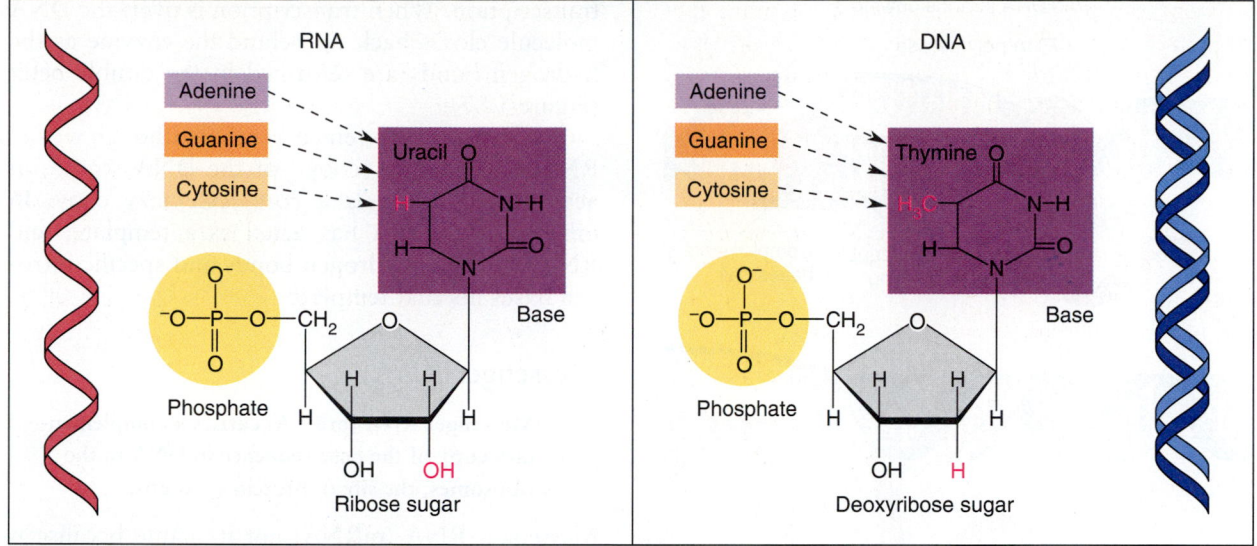

FIGURE 12.6 Comparison of RNA and DNA
The molecules of DNA and RNA are made up of strands of nucleotides, each composed of a phosphate, a sugar, and a base. The double-stranded DNA has the sugar deoxyribose and bases A, G, C, and T; single-stranded RNA has the sugar ribose and bases A, G, C, and U.

pentose sugar ribose (from which it gets its name) rather than deoxyribose (which DNA contains). Second, the nucleotides in RNA contain the bases adenine, guanine, and cytosine, just like DNA, but RNA does not contain the pyrimidine thymine. Instead RNA contains the pyrimidine uracil (U) (Figure 12.6). Another important difference is that RNA is a single-stranded molecule, not a double-stranded molecule. Finally, RNA can be located in either the nucleus or the cytoplasm.

You also need to recall that when the base in a nucleotide is unattached, it can pair with only one other base — in other words, in DNA adenine will pair only with thymine. In the case of RNA, of course, uracil acts like thymine. This specificity is a crucial factor in the process of protein synthesis.

The Basic Process of RNA Formation

■ **RNA polymerase catalyzes the transcription of RNA from a section of one of the DNA strands. As the DNA unzips, an RNA transcript forms through base pairing.**

In 1965, biologists isolated a complex enzyme known as **RNA polymerase,** which catalyzes the reaction in which RNA is copied from DNA. The process, known as **transcription,** begins when the DNA double helix is unwound. RNA polymerase attaches to the DNA strand at the position above the sequence required for a given protein, where a special codon signals that RNA transcription should begin (this type of codon is called an *initiation codon*). The DNA molecule then begins to "unzip" — that is, the hydrogen bonds connecting the two strands are broken. The DNA continues to unzip as the enzyme moves along the molecule, proceeding in a 5′ to 3′ direction (see Figure 10.8, pages 224–225).

Only one of the strands of the DNA molecule actually codes for a protein. Following the same base pair principle involved in DNA replication, this strand of DNA acts as a template for the formation of a single strand of RNA, matching up base pairs just as in DNA replication. (There is one exception: In RNA a uracil nucleotide will bond to the adenine in DNA.) A special codon called a *termination codon* ends the RNA transcription.

As it is formed, the RNA strand peels off from the DNA strand, freeing the DNA for further RNA transcription. Many RNA molecules can be transcribed in a very short period of time from the same portion of DNA, in a process called *simultaneous*

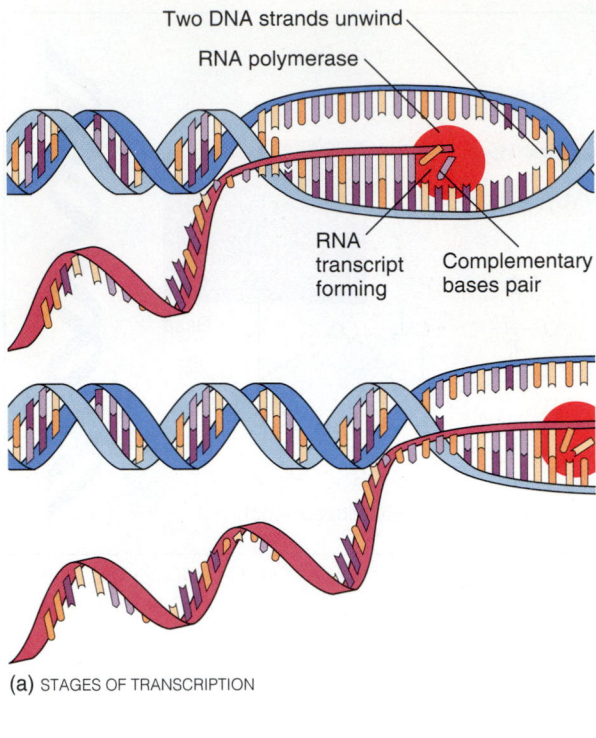

Two DNA strands unwind

RNA polymerase

RNA transcript forming

Complementary bases pair

(a) STAGES OF TRANSCRIPTION

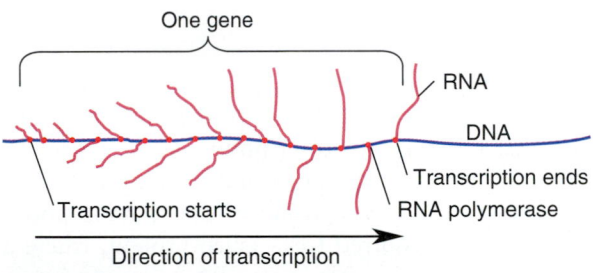

One gene

RNA

DNA

Transcription ends

RNA polymerase

Transcription starts

Direction of transcription

(b) TRANSCRIPTION PROCESS

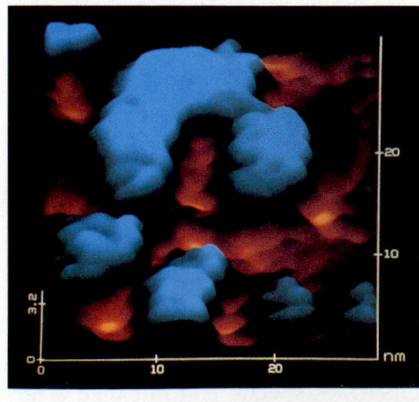

(c)

FIGURE 12.7 mRNA Transcription
(a) and (b) The genetic message of DNA is transcribed onto molecules of mRNA. RNA polymerase is the enzyme responsible for the synthesis of mRNA from one strand of DNA. (For the sake of simplicity the mRNA molecules have been shortened considerably.) (c) A scanning-tunneling microscope image of RNA polymerase from *E. coli*, viewed from the front.

transcription. When transcription is over, the DNA molecule closes back up behind the enzyme as the hydrogen bonds are reformed in the double helix (Figure 12.7).

It is important to remember that although we say RNA has made a "copy" of the DNA strand, in actuality the copy is a *complementary* copy. In other words, DNA has acted as a template, and RNA due to its hydrogen bonds and specific nitrogen bases fits that template.

Messenger RNA

- **Messenger RNA (mRNA) carries a complementary copy of the base sequence in DNA to the ribosomes, the site of protein synthesis.**

Messenger RNA (mRNA) got its name because it takes a complementary copy of the codon sequence on DNA (the message) from the nucleus to the ribosomes, which are the sites of protein synthesis. The molecule is single-stranded and usually quite long — from 300 to 10,000 nucleotides (these will code for from 100 to 3333 amino acids). The message carried in mRNA ultimately will be translated into a protein.

Transfer RNA

- **Transfer RNA carries individual amino acids to the ribosomes for insertion in a protein molecule. tRNA "reads" mRNA by means of an anticodon, a three-nucleotide sequence on the tRNA that pairs with a codon on the mRNA.**

Transfer RNA (tRNA) is a small molecule, consisting of 73 to 95 nucleotides. At one end tRNA carries a three-nucleotide sequence that is a mirror image of one of the specific codons that may occur in a strand of mRNA. At the other end of a tRNA molecule is a structure designed to transport a single specific amino acid. There are over 60 different kinds of tRNAs in a cell, one or more for each specific amino acid.

When viewed two-dimensionally, tRNA appears to have a shape similar to a cloverleaf, which in three dimensions forms an L-shaped molecule (Figure 12.8). An analysis of this structure will help to clarify the molecule's function. On one of the exposed loops, or cloverleaves, is a three-nucleotide sequence called the **anticodon**. Each tRNA has its own specific anticodon, allowing it to bind through base pairing with a specific codon of complementary bases on an mRNA molecule. The loop itself is called the *anticodon segment*, because it matches up with the mRNA codon. At another end of the mol-

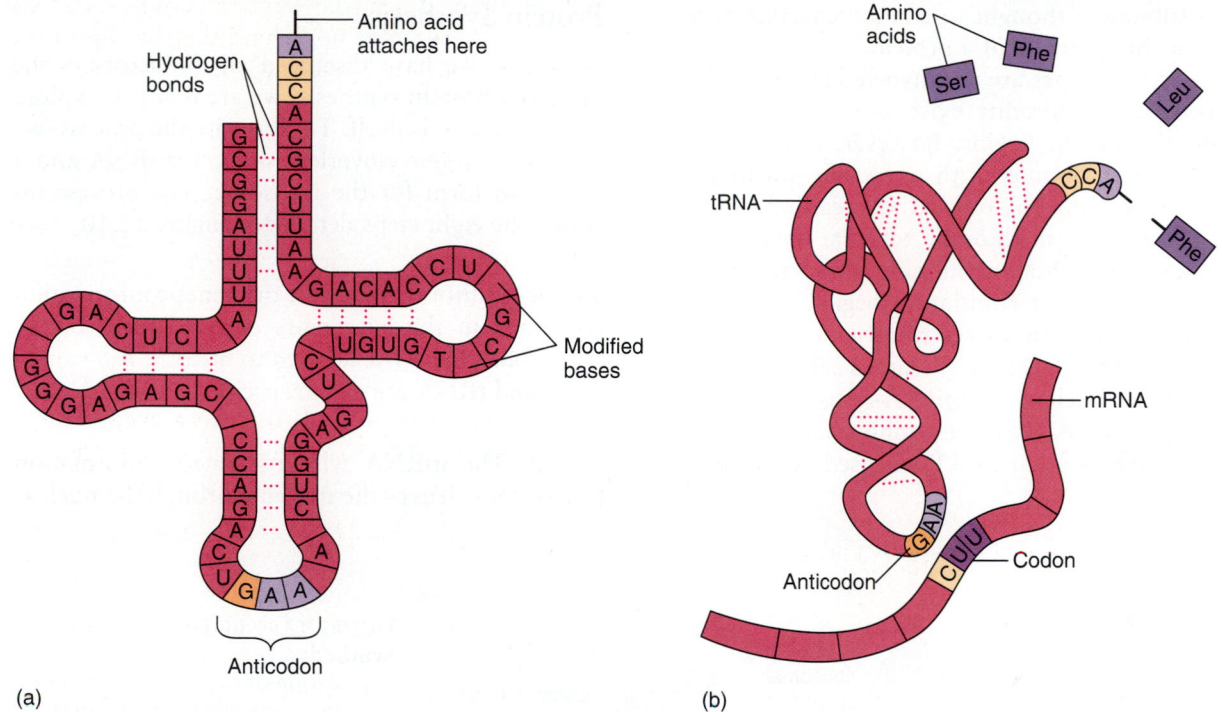

FIGURE 12.8 tRNA
(a) The primary and secondary structure of a tRNA molecule. (b) In its tertiary form, the tRNA molecule forms a characteristic shape that can function in protein synthesis.

ecule is a sequence CCA to which an amino acid attaches in an enzyme-mediated reaction. ATP is the major energy source used to attach an amino acid to a tRNA molecule.

One way to visualize the relationship between mRNA and tRNA is to think of mRNA in terms of a shopping list sent out by DNA. mRNA carries the list to the ribosome, the location where the order will be filled. Each tRNA is responsible for a particular item (that is, a particular amino acid). The tRNA anticodons read the list in the order in which it is presented on mRNA and deliver the necessary items in the order in which they are needed. This process is the fundamental basis of **translation** — the transferring of the genetic code to a linear sequence of amino acids in a protein chain.

Ribosomal RNA

■ **Ribosomes are composed of ribosomal RNA (rRNA) and proteins. There are two ribosome subunits, a large-heavy subunit and a small-light subunit. mRNA attaches to the lighter subunit; the heavier subunit then joins the complex. The heavy ribosome subunit contains the enzymes necessary for protein formation.**

In Chapter 4 we discussed the fact that the ribosome is the site of protein synthesis. The ribosomes of prokaryotic cells are smaller than those of eukaryotic cells. What is more, in eukaryotic cells the ribosomes are often attached to the endoplasmic reticulum or to the cytoskeletal structure, while in prokaryotic cells, ribosomes float free in the cytoplasm.

An individual ribosome can be used to synthesize any protein, an approach that is much more functional than requiring specific ribosomes for specific proteins. Each ribosome consists of two subunits — one that is light and one that is heavy. The terminology arises from the fact that when these subunits are placed in a viscous substance like a glucose solution, the two types settle to the bottom at different rates. Each subunit, in turn, is composed of about two-thirds RNA and one-third protein. Logically enough, **ribosomal RNA (rRNA)** is the type of RNA of which ribosomes are composed. In the light (small) subunit, there is one strand of rRNA and 21 proteins. The heavy (large) subunit contains two rRNA molecules and up to 50 specific proteins. Biologists believe that the smaller ribosome functions almost like a stage, a location where the "play" of translation takes place. The larger ribosome subunit is thought to contain enzymes necessary for the synthesis of a protein.

When ribosomes are not involved in synthesizing a protein, the subunits exist as separate entities. When translation begins, however, a small ribosomal subunit bonds with a specific starting sequence of the mRNA plus tRNA bearing a special first amino acid methionine to form an initiation complex. (The initial methionine amino acid is usually discarded after translation begins.) This complex then joins with a large ribosomal subunit, and protein synthesis begins (Figure 12.9). Only two tRNAs can attach to the mRNA-ribosome complex at the same time. Guanine triphosphate (GTP), an energy carrier similar to ATP, is used as the energy source.

Protein Synthesis

Now that we have discussed all the actors in the drama of protein synthesis, we are ready to explore protein synthesis itself. To illustrate the process, we will use a simple cloverleaf shape for tRNA and a simplified form for the ribosome. The process includes the eight steps detailed in Figure 12.10.

Step 1. Within the nucleus, the genetic information contained in the base pairs of one strand of the DNA molecule is *transcribed* to mRNA. Ribosomal RNA and tRNA are also transcribed from DNA.

Step 2. The mRNA with its genetic information (the codons) leaves the nucleus through the nuclear

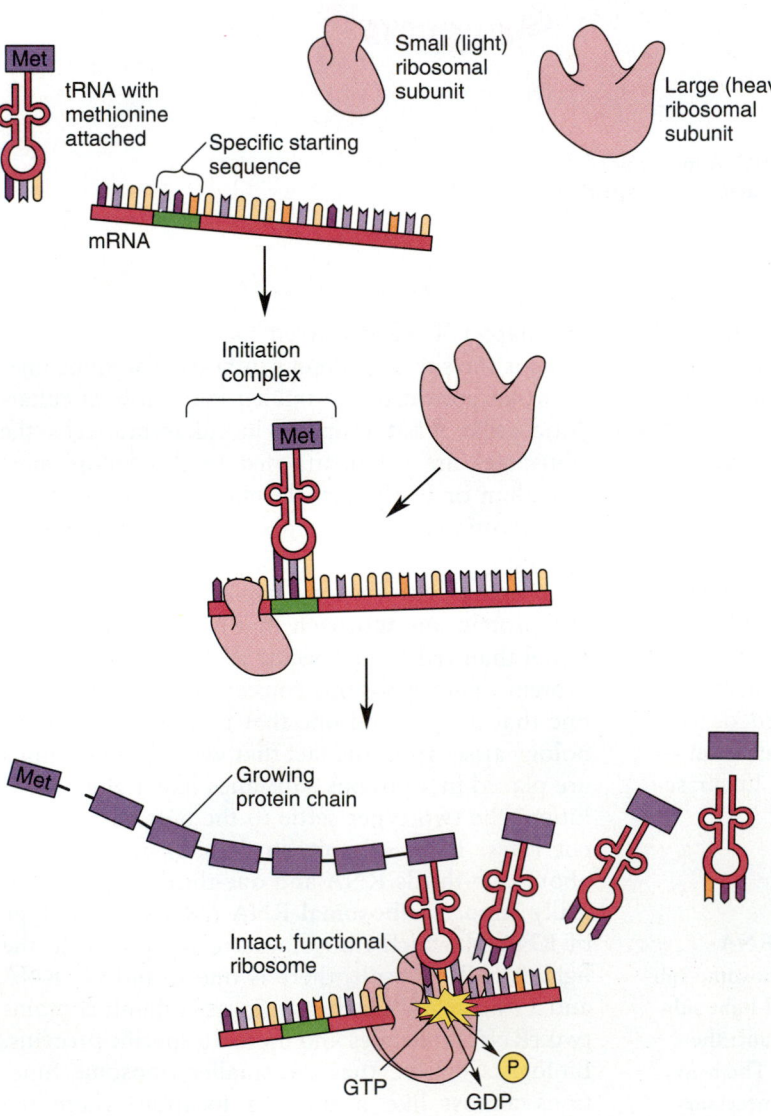

FIGURE 12.9 Initiation of Protein Synthesis
Translation of the genetic code into protein begins when an initiation complex composed of mRNA, tRNA, and a small ribosomal subunit forms and attaches to a large ribosomal subunit.

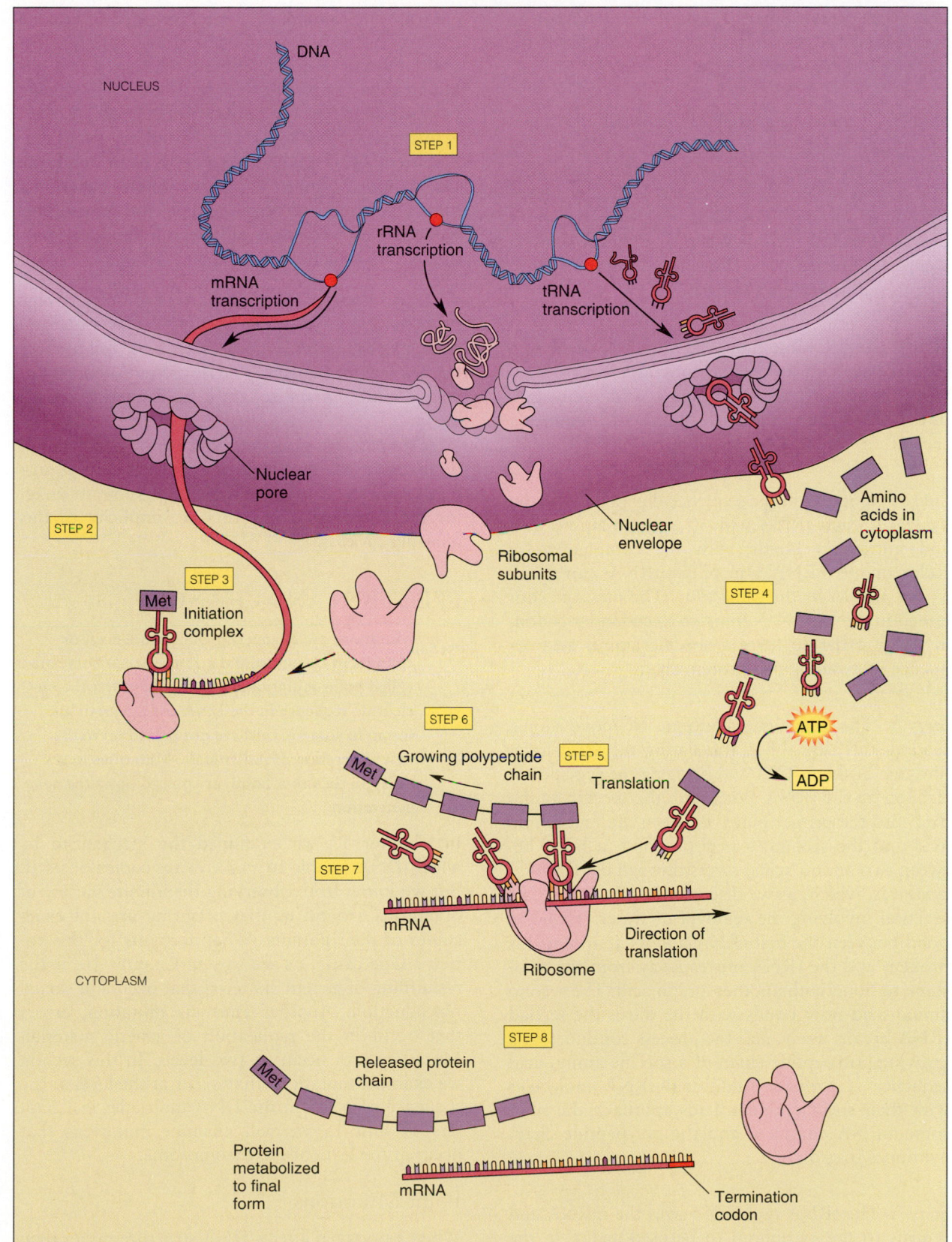

FIGURE 12.10 Summary of Protein Synthesis

pores and is transported through the cytoplasm to a ribosome. rRNA and tRNA can also leave the nucleus and are present in the cytoplasm.

Step 3. Within the cytoplasm, a small ribosomal subunit complexes with mRNA and a tRNA bearing a special form of methionine to form the initiation complex. The complex then is joined by a large ribosomal subunit. The ribosome begins to move from one end of the mRNA to the other. Usually a complex of at least five ribosomes (polyribosomes) is involved in protein synthesis.

Step 4. At the same time that step 3 is occurring, each tRNA with its specific anticodon forms a co-valent bond with a specific amino acid. Activation of the tRNA with its specific amino acid requires a specific enzyme and energy from ATP.

Step 5. Working in the sequence dictated originally by DNA, each tRNA with its specific amino acid and anticodon bonds with the complementary codon on the mRNA. One or two tRNAs can attach to the mRNA at the same time. The message contained in the mRNA is *translated,* codon by codon, in linear order to form a specific amino acid sequence in an elongated protein chain.

Step 6. The ribosomes continue to move along, reading three by three, translating the three-letter message of the mRNA and bringing the specific anticodon on the tRNA along with the specific amino acid into the proper place. Between the first amino acid and the second, a peptide bond (a bond between two amino acids) forms through dehydration synthesis, which, as we discussed in Chapter 3, is a reaction involving the removal of water. Then the bond between the first tRNA and its amino acid is broken, and the tRNA moves away from the ribosome to bond with another amino acid. The second amino acid now bonds with the third, the second tRNA breaks away, and the process continues as a growing polypeptide chain of a specific amino acid sequence is formed. Additional tRNA molecules with their specific amino acids approach the ribosome-mRNA complex, and the polypeptide chain continues to grow.

Step 7. The tRNA molecule leaves the mRNA and returns to the cytoplasm to be used again. In the cytoplasm the tRNA complexes with its specific amino acid and is activated again by enzymes and ATP and can return to the mRNA.

Step 8. One amino acid after another is bonded together as the polypeptide chain grows, until eventually the last codon in the mRNA serves as a terminator to stop the synthesis of the growing polypeptide chain. The polypeptide (protein) chain is then released. It folds into the tertiary configuration dictated by the chemistry of its elements. If the cell requires a more complex protein containing more than one polypeptide chain, the polypeptide will complex with other polypeptides and assume a quaternary structure.

At this point protein synthesis is complete. Some proteins remain in a cell to function as structural proteins or as enzymes; others leave the cell to serve as chemical messengers. The ribosomal subunits uncouple and remain separate until they are used again.

A simple listing of the sequence of steps cannot do justice to the awe-inspiring phenomenon of protein synthesis. It seems almost impossible to imagine the complexity and yet the simplicity of this amazing process.

PRECISION AND CHANGE

■ **Mutations are changes in DNA sequences, or changes in chromosomal arrangements. Some are called point mutations, because they alter the nucleotide sequence in the DNA code. Mutations found in somatic cells are not passed on to the new generation of individuals. Mutations in sex cells, on the other hand, are passed on to the next generation.**

In Chapter 11 we examined the mechanism by which DNA is able to make exact copies of itself. But we know from observing the infinite variety of organisms around us that offspring are not exact copies of their parents. What accounts for this variety? One cause, as we saw in Chapter 10, is the reshuffling of genetic material that occurs in sexual reproduction. Another cause is **mutation,** errors that occur in the replication of genetic material. Mutations can occur at two levels. In this section we examine mutations that occur at the level of the sequence of bases in the DNA molecule. In Chapters 14 and 15 we will examine mutations that occur at the level of the chromosome.

Point Mutations

When a mutagen affects DNA structure rather than the entire chromosome, it is called a **point mutation** or *gene mutation.* An alteration or substitution in the sequence of the DNA base pairs changes

Chapter 12   DNA and Protein Synthesis

the codon. If a base pair (or pairs) is omitted from the sequence, the change is called a *deletion*. If the wrong bases pair or base pairs are inserted, the change is called an *insertion* or a *substitution* (Figure 12.11). In Chapter 10 we mentioned that some forms of cancer may be due to point mutations, or chromosomal aberrations which result in changes in the cell membrane. These changes remove contact inhibition, and the cell divides wildly.

In **frameshift mutations,** one or several nucleotides are omitted from the sequence of the DNA strand, but that omission can have far-reaching ramifications. For example, consider the following message, which consists of a number of three-letter words:

THEMANCANRUNNOW (THE MAN CAN RUN NOW)

As long as you are aware that the message is based on three-letter words, you can decipher what is being said. But what happens if you delete one letter early in the message?

THEANCANRUNNOW (THE ANC ANR UNN OW)

Because of a single, very simple deletion (in this case the letter *M*), the message has lost all its sense — it no longer communicates. In a frameshift mutation, because of the simple deletion or addition of a single base, the three-base groupings within the message shift position, and the message no longer communicates. You can easily see how a misspelling or misreading of a genetic code can be the basis of point mutations.

One cause of point mutations is the action of environmental agents called **mutagens.** Radiation and a wide variety of chemicals can alter the DNA molecule. This is why physicians are using increasing restraints in the number of x-rays they request for their patients, and why close scrutiny is being given to the chemical pollutants in our environment.

Somatic Mutations and Sex-Cell Mutations

Errors in DNA replication that occur in body cells are called **somatic mutations.** (Remember that somatic cells are body cells.) Somatic mutations are passed on only to the daughter cells produced by the mutated parent cell. For example, the Delicious apple is the result of a somatic mutation. Somehow the somatic cells on one branch of a tree mutated, and all the apples on that branch showed a new set

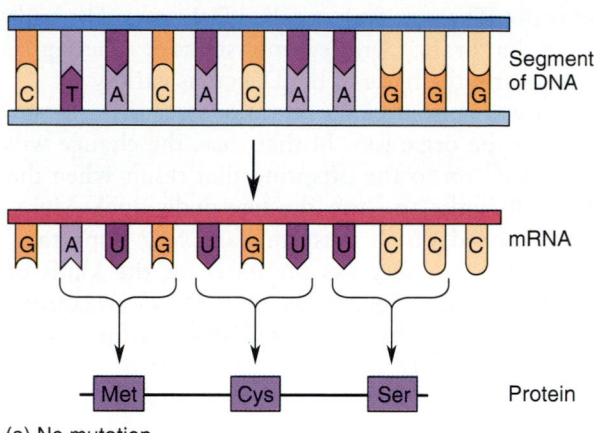

(a) No mutation

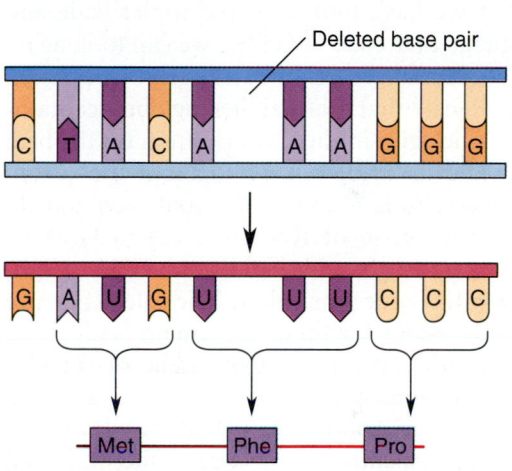

(b) Point mutation—deletion

(c) Point mutation—insertion

FIGURE 12.11 Point Mutation
(a) Transcription of mRNA from a segment of DNA results in translation into protein. In a point mutation, a base pair is either (b) deleted from or (c) inserted into the DNA segment. This change in the sequence of nucleotides alters the sequence of amino acids in the final protein.

of traits. The new fruit is called Delicious. The buds from that branch, grafted onto stems of other apple trees, form the basis of the Delicious variety.

However, sometimes mutations occur in the sex cells of the organism. In that case, the change will be passed on to the offspring that result when the sex cells unite to form the new individual. Mutations passed on to offspring are very important, since they can serve as the basis for the kinds of alterations that occur as groups of organisms evolve. We will discuss the role of mutation in evolution in Chapter 17.

Substitution in Sickle-Cell Hemoglobin

Now that we have looked at the triplet code and the language that code indicates, we can look again at the problem of genetic diseases such as sickle-cell anemia. Recall that normal hemoglobin contains glutamic acid (glu) in the sixth position on the beta chain, while in sickle-cell hemoglobin the amino acid valine (val) is found in the sixth position. In the messenger form of RNA, the codon GAA or GAG codes for glutamic acid, and GUA, GUC, GUG, or GUU codes for valine. The sole difference between the codes for these two amino acids is the presence of the adenine in the middle of the glutamic acid codons and the presence of a uracil in the middle of the codons for valine. Think of it — one simple misspelling in the code of life may spell life or death for an individual (see Figure 12.1b).

THE CONTROL OF PROTEIN SYNTHESIS IN PROKARYOTIC CELLS

Your own unique genetic makeup is spelled in the DNA molecules that reside in the nucleus in every cell of your body. For example, if you have black hair, the cells in your hair follicles are producing pigment that gives your hair its black coloring. However, you also have the genetic information to produce black hair in the DNA molecules located in your liver. (Remember, all the DNA is present in every cell.) Does this mean that your liver has the same black coloring as your hair? Do I have a black, hairy liver? A ridiculous question, you say? Perhaps not so ridiculous after all, for this question leads to one of the most basic questions that biologists are working very hard to answer. What turns DNA molecules on and off? At the moment we have only partial answers to these questions, answers we will now discuss.

Regulation: The Control Mechanism

■ **Regulation is the mechanism by which protein synthesis is increased, decreased, initiated, or stopped.**

The mechanism by which protein synthesis is increased, decreased, initiated, or stopped is called *regulation*. Both prokaryotic and eukaryotic cells depend on regulatory mechanisms to determine which proteins are going to be synthesized. Regulation can occur at several points in the mechanism of protein synthesis. In transcriptional regulation, mRNA synthesis from DNA is inhibited; in translational inhibition, protein synthesis from mRNA is inhibited within the region of the cytoplasm and ribosomes. Current evidence supports the theory that both transcriptional and translational mechanisms of protein regulation occur in both prokaryotic and eukaryotic cells. However, our understanding of gene regulation in prokaryotic cells is more complete than our understanding of gene regulation in eukaryotic cells.

The Operon Hypothesis

One of the best-understood examples of transcriptional gene regulation in prokaryotic cells comes from the work of François Jacob and Jacques Monod. During the early 1950s Jacob and Monod were investigating enzymes involved in the nutritional requirements of *Escherichia coli*, a common bacteria found in the guts of mammals, including humans and animals.

As you might imagine, when a young mammal like a kitten is nursing, the gut contains a lot of milk. Because of its abundance, the milk sugar (lactose) is the principal source of energy for the *E. coli*. In order to metabolize the lactose, a large amount of the appropriate enzymes must be present in the cell. But when the kitten is weaned, the *E. coli* are subjected to an abrupt change in diet, for they no longer have lactose to metabolize. Somehow the metabolic machinery switches over to metabolize other sugars. How do these bacteria switch off the genes that regulate synthesis of lactose-degrading enzymes and switch to genes that can induce the synthesis of other sugar-degrading enzymes? Jacob and Monod's investigation was designed to find the answers to these questions.

First, the two researchers grew *E. coli* on a medium that included no lactose. They then moved the bacteria to a medium in which lactose was the only energy source — and lactose-metabolizing enzymes appeared for the first time. Over a short period of time their number increased dramatically.

Jacob and Monod's experiments revealed that in certain organisms enzymes are inducible. The presence of lactose induces the synthesis of the three enzymes involved in lactose metabolism. It was discovered that all the genes coding for these three enzymes are located next to one another on the DNA molecule of *E. coli*. As a result, the genes are transcribed as a subunit onto the mRNA molecule. These genes are called **structural genes** because they code for the synthesis of a specific protein structure.

Other investigators had determined that the transcription of RNA from DNA is controlled by the enzyme RNA polymerase. As a result of their own research, Jacob and Monod proposed a mechanism that explained control over the activities of RNA polymerase, a mechanism that would result in control of the synthesis of the enzymes involved in lactose metabolism. In their explanation, which is called the *operon hypothesis* (Figure 12.12), Jacob and Monod proposed specific control elements involved in the lactose operon. (An **operon** is a specific set of regulatory genes and structural genes found in bacteria.) Their theory was expanded in 1961, and biologists now recognize three elements of the operon hypothesis.

1. A **regulator:** A regulator is a portion of the DNA molecule that codes for repressor proteins. A *repressor protein* can repress or block mRNA transcription.

2. A **promoter:** A promoter is an *initiation site* on the DNA to which the RNA polymerase enzyme binds in order to initiate transcription.

3. An **operator:** An operator is the site of attachment on the DNA molecule to which the

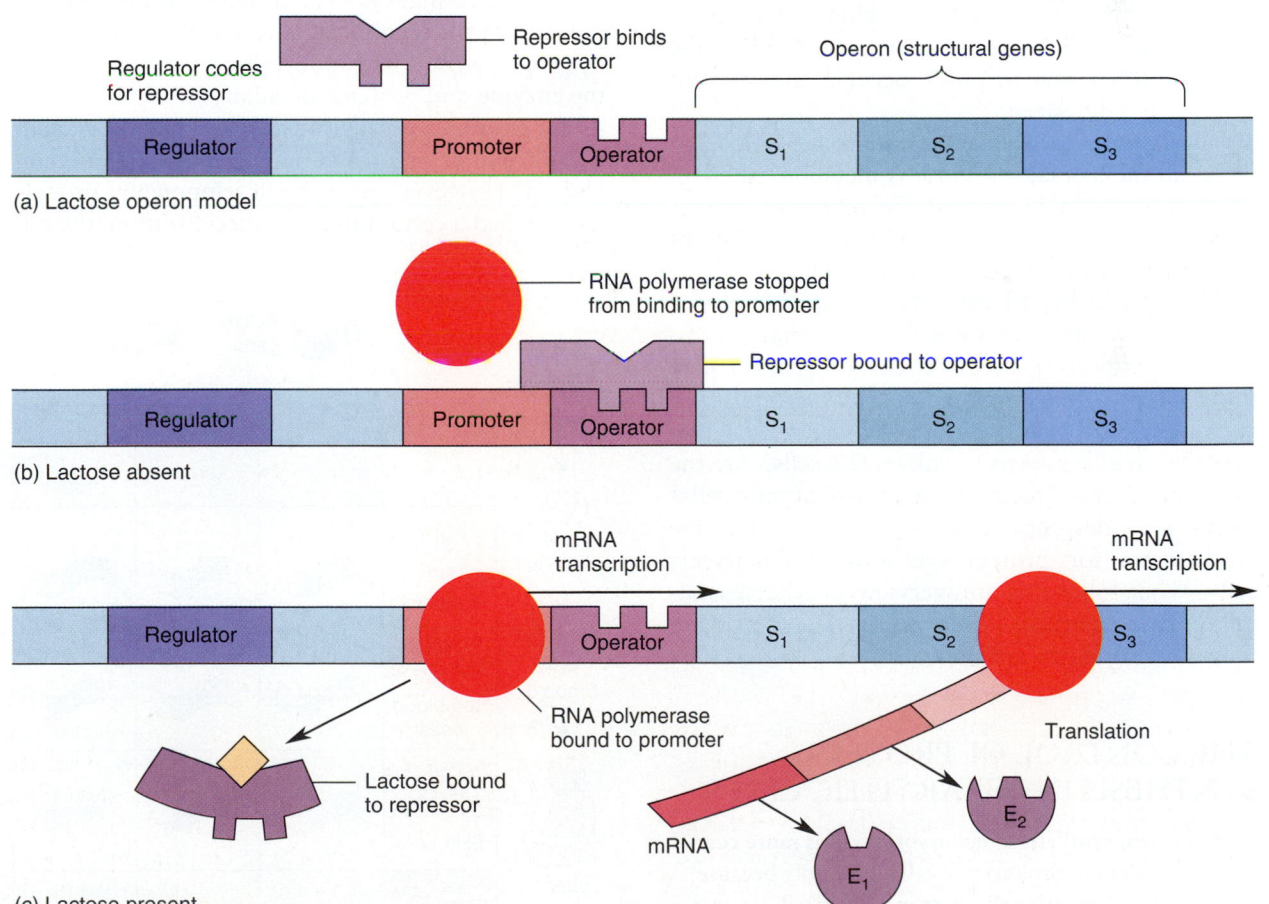

(a) Lactose operon model

(b) Lactose absent

(c) Lactose present

FIGURE 12.12   The Lac Operon
(a) In prokaryotic cells, the lactose operon is responsible for the production of three enzymes needed for lactose metabolism. (b) When lactose is not present in the cell, a repressor molecule inhibits RNA polymerase from binding to the promoter gene, blocking synthesis of these enzymes. (c) When lactose is present, it will bind to the repressor, thereby freeing the promoter and allowing enzyme synthesis to proceed.

repressor can bind. When the repressor is bound to the operator, RNA polymerase cannot bind to the promoter; as a result, transcription is turned *off*. When the repressor is removed from the operator, the RNA polymerase can bind to the promoter; as a result, transcription is turned *on*.

The specific set of genes that induces (turns on) and represses (turns off) the synthesis of enzymes involved in lactose metabolism is called the **lac operon.** How does this operon function? When lactose is not present in the cell, the genes are turned off. This is because a large repressor molecule is bound to the operator on the DNA molecule. As a result, RNA polymerase cannot bind, and transcription is repressed (the gene is turned off). When lactose is present in the cell, the lactose (actually a modified form of it) binds to the repressor and induces a change in the repressor. The repressor can no longer function — it loses its ability to bind to the operator. Therefore, RNA polymerase can bind to the DNA, mRNA transcription occurs, and the enzymes are synthesized.

The operon mechanism again comes into play when all the available lactose is metabolized. With no lactose available to bind the repressor, the repressor can again bind with an operator to prevent transcription, and the gene is repressed again.

The operon hypothesis offered a description of DNA control elements in prokaryotes that coordinate the synthesis of a set of structural genes. Today this concept is known as the *operon mechanism of gene regulation in prokaryotic cells*. But what regulates protein synthesis in eukaryotic cells? Are the mechanisms the same as those in prokaryotic cells? There is widespread interest in discovering the mechanisms for protein synthesis in eukaryotic cells. What follows is an overview of what has already been learned — but there are still many things biologists do not know.

THE CONTROL OF PROTEIN SYNTHESIS IN EUKARYOTIC CELLS

■ **Protein synthesis in eukaryotic cells is more complex than in prokaryotic cells, primarily because most eukaryotic cells exist in a multicellular environment and are specialized.**

At first, biologists hypothesized that the operon mechanism also regulated protein synthesis in eukaryotic cells, but further research indicated that this was not the case. It was soon discovered that the regulator and operator control elements so essential to the function of the operon mechanism are not present in eukaryotic cells.

There are several reasons for the differences in protein synthesis in prokaryotic and eukaryotic organisms. In the first place, structural genes normally are not found in convenient clusters adjacent to one another in eukaryotic cells. In prokaryotic cells, however, they are. A second reason has to do with the task of maintaining homeostasis. As you may remember, all living organisms, both prokaryotes and eukaryotes, must maintain a constant internal environment. Since prokaryotic cells are unicellular, they must adapt constantly to changes in their environment; as a result, control mechanisms must be turned on and off rapidly. On the other hand, most eukaryotic cells are found in multicellular organisms, and, because of their multicellularity and the cellular specialization of individual cells, each individual cell is not usually subject to the drastic changes in cellular environment that are typical of prokaryotes. As a result, the mechanisms of protein synthesis produce long-term changes in the enzyme and proteins in eukaryotic cells.

For example, during the development of a cell in a multicellular organism, many genes are turned on and off. However, once the developmental process has reached a certain point in specialization, the cell

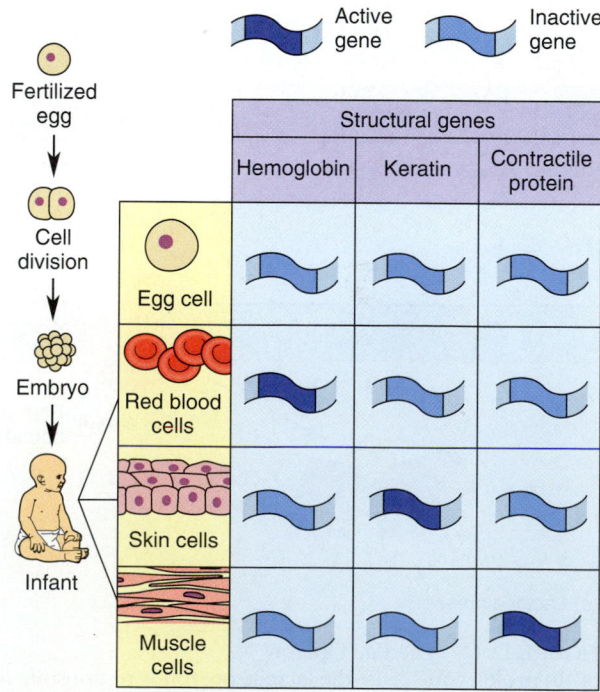

FIGURE 12.13 Gene Regulation During Development
Each cell has the individual's unique blueprint for life, but as cells become specialized during development, some genes are turned on or off depending on the cell.

is committed to that specialty, and it usually cannot return to the unspecialized state. The cell may become committed to the formation necessary to produce, let's say, nerve cells, muscle cells, or leaf cells. This process is called *specialization*. Specialization in a multicellular organism is dictated by the genes found in the DNA, and that genetic control over protein synthesis is long-term (Figure 12.13).

Regulation of Gene Activity in Eukaryotic Cells

As illustrated in Figure 12.14, gene regulation in eukaryotic cells can be thought of as occurring at any of four levels.

1. *Regulation at the Transcriptional Level:* Several mechanisms have evolved to control the amount and rate of mRNA transcribed from a gene.

2. *Regulation at the Posttranscriptional Level:* Some of the mRNA must be processed and modified before it can leave the nucleus.

3. *Regulation at the Level of Translation:* The rate at which mRNA is translated can be regulated by various mechanisms: by blocking access of the mRNA to the ribosomes, or by blocking elongation of the polypeptide chain.

4. *Regulation at the Posttranslational Level and Regulation of the Activity of the Protein.* The polypeptide produced by translation may

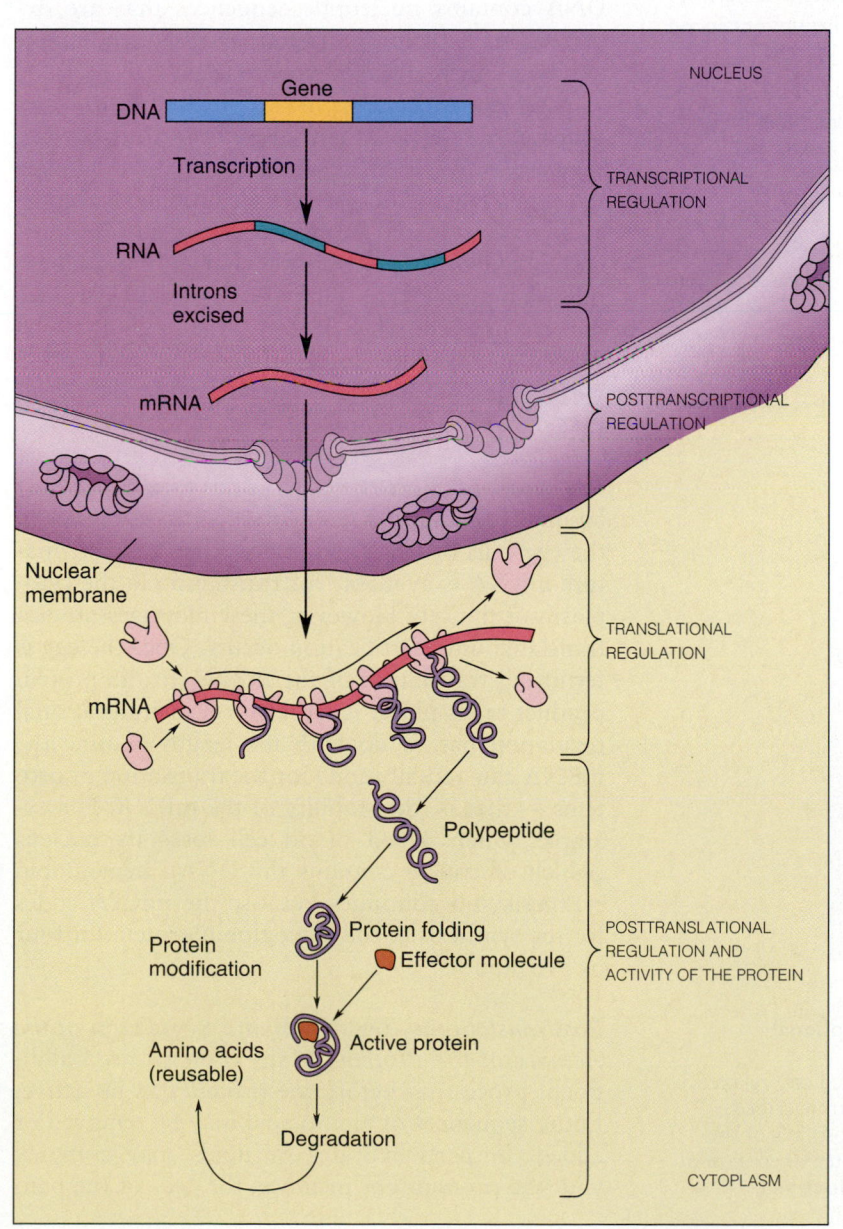

FIGURE 12.14 Gene Regulation in Eukaryotic Cells
A summary of the four types of gene regulation in eukaryotic cells.

need to undergo modification before it can become a functional protein. Environmental conditions, the end product of an enzymatic reaction, and metabolic pathways can reduce an enzyme's activity. Remember the allosteric enzymes we discussed in Chapter 5 in which the end product feeds back on the enzyme and decreases its activity.

Let us now elaborate on the control mechanisms.

Transcriptional Regulation. Regulation of transcription at the molecular level can occur due to regulatory proteins. Some regulatory proteins can activate or repress the synthesis of RNA. Scientists believe that because the majority of genes are turned off in a cell, the regulatory protein functions in turning on genes, and most do (however, some turn off or block transcription). For example, scientists have found regulatory proteins in eukaryotic cells that attach to the DNA before the RNA poly-

merase can attach to the gene to initiate transcription. Therefore, the regulatory protein allows the gene to be turned on — or transcribed. Eukaryotic DNA also contains sequences that enhance the rate of structural gene transcription. Regulation of transcription can occur at the packaging level as well. For example, as the chromatin is condensed to form chromosomes (see Chapter 10), the RNA polymerase enzyme cannot bind to the DNA. It is not until the chromatin unwinds, or "relaxes," that the DNA is available for transcription (Figure 12.15).

Posttranscriptional Regulation. The mRNA must be processed before it leaves the nucleus and is capable of regulating translation. Walter Gilbert at Harvard and others determined that eukaryotic DNA contains nucleotide sequences that are *ex*pressed, called **exons**, and *in*tragene sequences that are not expressed, called **introns**.

After the mRNA is transcribed, this *primary transcript,* as it is called, must be processed. Enzymes remove or splice out the introns and splice the exons together to produce a continuous message to be translated and expressed as protein. This process is called *RNA splicing*; it not only occurs for mRNA but for tRNA and rRNA as well. How does the enzyme complex identify the region to be spliced out? At either end of the intron, most eukaryotic genes have identifying nucleotide sequences that signal for RNA splicing (Figure 12.16).

Translational Regulation. Various types of translational regulation have evolved. For example, in the egg cells of frogs and many other animals, mature mRNA is available for translation in the cytoplasm of the cell. However, these messages are not translated until fertilization occurs. Once the egg is fertilized, translation of the mRNA to the protein product takes place. Another form of translational regulation has to do with the length of time the mRNA can actually code for the translation of proteins — that is, the stability of the mRNA. For example, when a red blood cell loses its nucleus (which of course contains the DNA), hemoglobin synthesis still continues because the mRNA codes for the synthesis of the hemoglobin protein for four or five months.

Posttranslational Regulation and Regulation of the Activity of the Protein. Many proteins need additional processing before the protein can be active. Entire sequences of amino acid may be removed or added, or carbohydrates or lipids may complex with the protein. For instance, the cells of the pan-

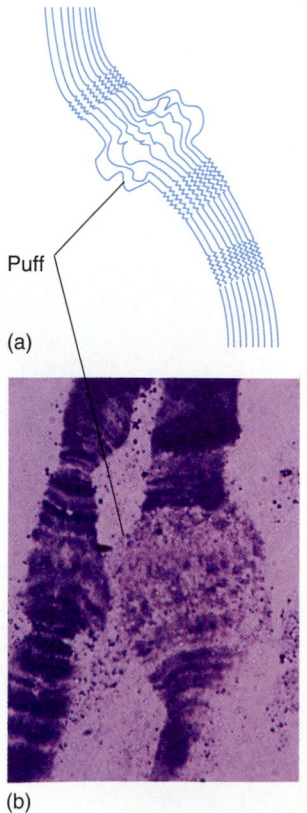

Puff

(a)

(b)

FIGURE 12.15   An Example of Transcriptional Regulation
(a) When the chromatin becomes uncoiled or "relaxed," the DNA within that chromatin can be transcribed.
(b) This photograph of the giant chromosomes of the salivary glands of fruit fly larvae illustrates the region of the "puffs" where the DNA is being transcribed.

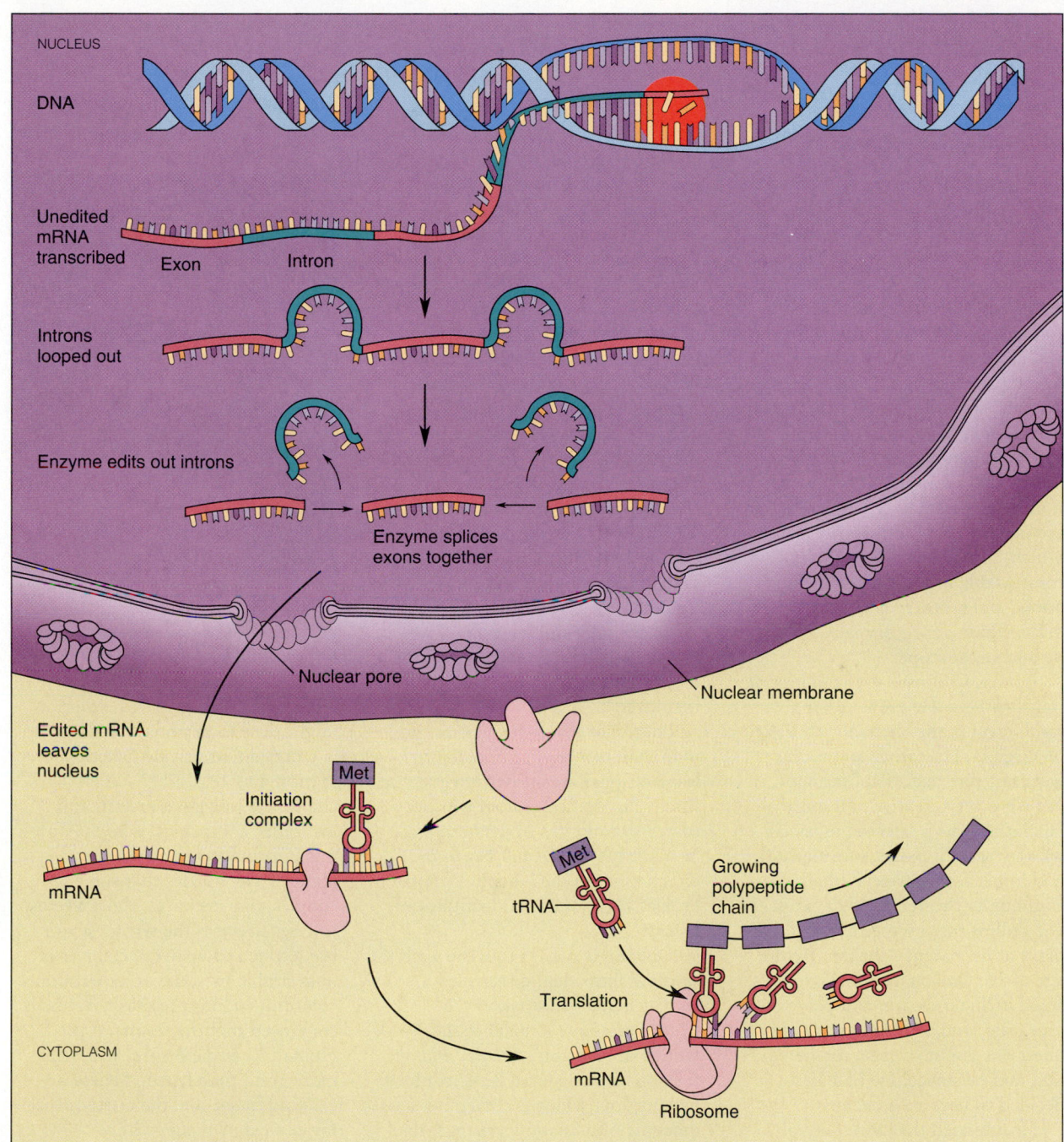

FIGURE 12.16 Editing mRNA

When eukaryotic DNA is transcribed, only certain nucleotide sequences are translated into protein. Specific nucleotide sequences (introns) are first edited out, before the mRNA leaves the nucleus. The remaining sequences (exons) are spliced together, forming the final, edited version of mRNA. It is this edited version that is translated into protein in the cytoplasm of the cell.

The Control of Protein Synthesis in Eukaryotic Cells

Dr. Francis Collins
Molecular Geneticist

"A contradiction in terms" might be the words that best describe Dr. Francis Collins, Director of the Human Genome Project at the National Institutes of Health. The 44-year-old geneticist at times is a compassionate, warm human being and at other times is an aggressive scientific competitor. He is a "prove it to me, show me" kind of scientist, questioning scientific truth. He is also a deeply religious man. Not only is this an apparent contradiction in his personality but also in his appearance and lifestyle. The 6-foot-4-inch-tall Collins is equally comfortable in his jeans, cowboy boots, and motorcycle jacket as he is in his white laboratory jacket with tie and stethoscope.

Francis Collins is the youngest of four brothers. He grew up on a small farm in the Shenandoah Valley of Virginia. His father and mother were not very successful farmers but were very creative and musical. His father was a talented musician who loved folk music. His mother and father also started a small community theater on their farm. Dr. Collins often wrote and directed plays at his parents' theater. By the age of 16, Collins had written plays, acted in Shakespeare, and played bluegrass, and he began to ponder questions about science, mathematics, and existence. He had also decided to become a chemist.

He attended the University of Virginia, from which he graduated with a degree in chemistry. At that time in his life, biology was of no interest. It wasn't until he was a Ph.D. candidate in physical chemistry at Yale University that biology began to intrigue him. This interest developed after he took a course in molecular genetics.

After earning his Ph.D., Dr. Collins enrolled in medical school at the University of North Carolina. After earning his M.D. degree in 1977, he did his residency in internal medicine and then began a teaching and research position at the University of Michigan in Ann Arbor in 1984. His tenure was very productive, and his talents as a scientist and scholar did not go unrecognized by his peers. In 1992, he was asked to direct the multimillion-dollar Human Genome Project, one of the most ambitious scientific projects ever attempted. The object is to find and decipher the entire genetic code of human beings. The decision to leave his position at the University of Michigan Medical School was not easy. After months of deliberation, he made his choice and began the directorship of the Human Genome Project at the National Institutes of Health.

One might ask, "How can such a spiritual individual direct such a secular human endeavor?" Dr. Collins attempts to balance the worlds of medicine, science, and religion. On one level, he is troubled by abortion, which is sometimes an outcome of prenatal screening when genetic defects are discovered. But when he counsels patients, he makes an effort to be sure that his own

personal feelings of right and wrong do not influence the patients' decisions. With respect to the dangers that the genetic revolution might pose to personal freedom, he is worried that private genetic information may be readily available to employers, the government, and insurance companies, who may use this information in an unethical manner. He worries that some couples may decide to abort a fetus just because it is the wrong sex or is predestined to be overweight or is undesirable by some other culturally conditioned interpretation. Yet he is convinced that the results of the Human Genome Project will bring more good than harm. "These are exciting times, and the consequences for clinical medicine will be dramatic," he states.

creas produce a hormone called insulin. As it is translated into a long single-chain polypeptide it is inactive. The insulin hormone undergoes additional modification by enzymes to become the active form. A large center portion of the amino acid chain is enzymatically removed, leaving two shorter chains that are linked together by chemical bonds to produce the active form of the molecule. Many mechanisms have evolved to regulate the activity of the protein. Environmental factors such as temperature can affect the rate of enzymatic reaction. In addition, during metabolism various events can occur that may block a metabolic pathway, hence regulating the activity of the final protein. Some enzymes have allosteric sites to which the end product will bind. This binding of the end product to the allosteric site will change the shape of the enzyme so that it will no longer complex with the substrate at the reaction center.

We have briefly discussed some mechanisms of gene regulation in eukaryotic cells. But, in short, biologists as yet have a limited understanding of the regulatory mechanisms of protein synthesis in eukaryotic cells. However, research in this area is very active. Think of the implications of the understanding of gene regulation. Such knowledge could provide a clue to cell changes that result in cancer as well as many other significant biological events.

Overview

We began this chapter by asking questions about the involvement of DNA in the control of protein synthesis. We explored how the message encoded in DNA could be transcribed into RNA and then translated into the linear sequence of amino acids in a polypeptide chain such as hemoglobin, enzymes, and other proteins. We now have a better understanding of the genetic basis of metabolic disorders. We have also learned about the mechanisms of control of protein synthesis in prokaryotic cells and have discussed some of the evidence available for protein synthesis regulation in eukaryotic cells.

The search for more evidence about gene regulation in eukaryotic cells is expanding (far more than we can discuss in this text) as we learn more about the classes of eukaryotic DNA, multiple genes, gene amplification, jumping genes, genes and their involvement in cancer, and many other aspects of eukaryotic gene regulation. The attempt to understand DNA and protein synthesis is one of the most exciting areas of biology today — simply put, the implications for our everyday life are mind-boggling (see Profile).

SUMMARY

1. Many diseases have their basis in metabolic genetic disorders. One such disorder is sickle-cell anemia, which is due to an abnormality in the linear sequence of amino acids in the hemoglobin molecule, resulting in changes in the electrical charge within the molecule. Those changes within the hemoglobin molecule cause the red blood cells to buckle.

2. Garrod, in the early 1900s, was the first to propose that metabolic disorders resulted from genetic errors.

3. Beadle and Tatum, while working with nutritional mutants of *Neurospora*, formulated the hypothesis that one gene in the DNA molecule coded for the synthesis of one enzyme. This hypothesis was further modified to state that one gene coded for one polypeptide.

4. The central dogma states that the linear sequence of base pairs in the DNA is transcribed onto an RNA molecule. The information encoded on the RNA is translated into the linear sequence of amino acids in a polypeptide chain.

5. The genetic code is a linear sequence of three nucleotides; each triplet is called a codon.

6. RNA, like DNA, is composed of nucleotides, but the sugar unit bonded to a phosphate and a nitrogen base is ribose, not deoxyribose. The bases in RNA are adenine, guanine, cytosine, and uracil instead of the thymine found in DNA. RNA is a single strand rather than a double helix.

7. Three types of RNA are involved in protein synthesis: messenger RNA (mRNA), transfer RNA (tRNA), and ribosomal RNA (rRNA). Ribosomal RNA makes up about 66 percent of each ribosome.

8. During protein synthesis, the linear sequence of nitrogen bases is transcribed from one strand of the DNA onto an mRNA. The mRNA leaves the nucleus and moves to the ribosome.

9. The tRNA with its specific amino acid and anticodon complexes with the ribosome. The message carried on the mRNA is "read" as the anticodon of the tRNA bonds with the complementary codon on the mRNA. The tRNA bonds to the mRNA and the amino acids bond together to form a polypeptide chain. The tRNAs, which are reusable, are released from the ribosome to return to the cytoplasm. There they bond again with their specific amino acids and return to build more chains.

10. Usually the DNA molecule replicates with precision, but sometimes variations or rearrangements occur that produce mutations. Environmental agents that produce mutations in the genetic code are called mutagens. A somatic mutation passes on only to the cell's direct descendants. However, if the mutation occurs in the sex cell, the mutations may be passed on to the descendants of the organism.

11. Protein synthesis can be regulated at either the transcriptional or translational stage. In transcriptional regulation, mRNA synthesis can be inhibited; in

translational regulation, protein synthesis from mRNA can be inhibited.

12. Jacob and Monod proposed the operon concept to explain gene regulation in *E. coli*, a prokaryotic cell. The operon mechanism requires a regulator, a promoter, and an operator.

13. The eukaryotic mechanisms for turning genes on and off are not well understood but are a very active area of investigation. For convenience, gene activity in eukaryotic cells can be thought of as occurring at four levels: transcriptional, posttranscriptional, translational, and posttranslational and at the activity of protein.

KEY TERMS

sickle-cell anemia
central dogma
codon
RNA polymerase
transcription
messenger RNA (mRNA)
transfer RNA (tRNA)
anticodon
translation
ribosomal RNA (rRNA)
mutation
point mutation

frameshift mutation
mutagen
somatic mutation
structural gene
operon
regulator
promoter
operator
lac operon
exon
intron

REVIEW QUESTIONS

True/False

1. The central dogma states that the linear sequence of nucleotides in a DNA molecule provides the information to determine the linear sequence of amino acids in a polypeptide.
   true   *false*   *(page 262)*

2. RNA is found only in the nucleus.
   true   *false*   *(page 267)*

3. Transfer RNA molecules have an anticodon portion and a region where a specific amino acid is attached.
   true   *false*   *(page 268)*

4. Gene regulation in prokaryotic cells and eukaryotic cells is the same.
   true   *false*   *(page 276)*

Fill in the Blank

1. The process by which the nucleotide sequence contained within one strand of the DNA molecule is transferred to an mRNA molecule is called _____. *(page 267)*

2. mRNA contains the nitrogenous base _____ rather than the nitrogenous base thymine that is found in DNA. *(page 267)*

3. The message carried on the codon of the mRNA molecule is _____ at the ribosome to a specific amino acid sequence in a growing polypeptide chain. *(page 272)*

4. As described in the operon hypothesis, a(n) _____ is a molecule that can repress or block mRNA transcription. *(page 275)*

Multiple Choice

1. Sickle-cell anemia may be used as evidence that:
   a. Complete dominance occurs.
   b. Genes control the production of proteins.
   c. There is no relationship between genes and enzymes.
   d. Genes function directly in metabolism.
   (page 274)

2. Transcription of part of a DNA molecule with a nucleotide sequence of AAA CAA CTT results in an mRNA molecule with the complementary sequence of:
   a. GGG AGA ACC
   b. UUU GUU GAA
   c. TTT GAA GCC
   d. CCC ACC TCC
   (page 266)

3. The genetic message of DNA is carried in:
   a. The sequence of the phosphate-sugar components
   b. Its protein
   c. The sequence of the purine-pyrimidine components
   d. The number of the nucleotide components
   e. None of the above
   (page 262)

4. Which of the following would be most consistent with the one-gene–one-enzyme hypothesis?
   a. The alteration of a single gene may result in a metabolic alteration.
   b. Genes are made up of single enzymes.
   c. Genes are made up of several enzymes.
   d. Enzymes are made up of several genes.
   (page 261)

Discussion

1. Discuss the genetic basis of sickle-cell anemia. What are some of the symptoms of sickle-cell disease, and how is this related to the abnormal hemoglobin?

2. Discuss the one-gene–one-polypeptide theory proposed by Ingram and others.

3. What is the central dogma?

4. What is the molecular basis of the living code?

5. How does RNA differ from DNA? Discuss the three basic types of RNA.

6. Briefly describe the events that occur in protein synthesis.

7. What is the operon hypothesis?

8. Why do you suspect that the mechanisms of gene regulation in prokaryotic cells responded more quickly to changes in the environment than the mechanisms of gene regulation in eukaryotic cells?

9. Why is it important to understand gene regulation and protein synthesis?

10. What is the result of the lack of precision of DNA replication that sometimes occurs in the living forms on this earth?

13

Recombinant DNA: Genes by Design

Throughout the history of science, discoveries have occasionally been met with a public outcry. Be it the development of the smallpox vaccine or the splitting of the atom, the public wonders whether or not scientists have finally "gone too far." The technology known as *recombinant DNA* is an example, and the concern expressed, even by some scientists, is not difficult to understand. This time scientists are tinkering with the fundamental genetic blueprints for life. That may be why the public controversy has been especially vociferous. As recently as 1976, newspapers around the country headlined articles about entire cities attempting to ban recombinant DNA experiments.

Above: Using the techniques of recombinant DNA technology, the genes from light-producing bacteria have been incorporated into the cells of this tobacco plant.

What did people fear? They envisioned worldwide, man-made bacterial plagues — an "Andromeda strain," if you will — unresponsive to any known cure and unaffected by any natural enemy. In 1976, the National Institutes of Health (NIH) responded to the widespread anxiety by releasing guidelines that prohibited certain types of recombinant DNA research (see Profile). Now we know that the early fears of Dr. Frankenstein-like experimenters and doomsday bacteria were unfounded. By 1979 there was a general relaxation of the tight NIH restrictions.

But the new technology carries a promise as well as a threat. Imagine for a moment that you could take a few chemicals, some equipment, and some cultured cells and, with those ingredients at hand, could create forms of life that will respond to your command. What would you produce? You could

Dr. Lydia Villa-Komaroff
Molecular Geneticist

Suppose you are sitting in your biology class, waiting to hear a talk given by a molecular biologist with a Ph.D. in molecular genetics from Massachusetts Institute of Technology who is working in recombinant DNA technology. What kind of person comes to mind? Do you have visions of Dr. Frankenstein hovering over a test tube, with lightning all about the room? Such a scientist should be obsessed with projects and should be totally out of touch with colleagues, friends, and family. Fortunately, the truth is sometimes much less dramatic.

Not long ago, students at a college in Massachusetts were awaiting their guest speaker, a person with all the credentials we listed, when in walked Dr. Lydia Villa-Komaroff. Villa-Komaroff, well aware of the contrast between the expectations of the students she lectures and the reality that she represents, is convinced that it is an important part of her role as a scientist and educator to challenge stereotypes. As she puts it, "When I walk into a room and say I'm a molecular biologist, that I am an MIT graduate, then maybe people realize the stereotype of a scientist — male, balding, glasses, like Professor Honeydew of the Muppets — isn't right. I want the students to realize that they can do it too."

Villa-Komaroff, born in Las Vegas, New Mexico, attended school there and in Santa Fe. She has been interested in science since the age of nine. Her mother was interested in biology, and her uncle was a chemist. They both influenced her career choice. In addition, she was encouraged in pursuing her dream of becoming a biologist during the summer of her junior year in high school, when she attended a science training program sponsored by the National Science Foundation. In June 1970, she graduated cum laude from Goucher College in Maryland. By 1975 she had completed the work on a Ph.D. in molecular biology at MIT, where her thesis adviser was the Nobel Prize laureate Dr. David Baltimore.

Receipt of the Ph.D. did not end Villa-Komaroff's education. Instead, she did postdoctoral work at Cold Spring Harbor and was a research fellow at Harvard University with Dr. Fotis Kafatos and Dr. Walter Gilbert (Gilbert is a Nobel Prize laureate — see the Profile on page 280). The time she spent in Dr. Gilbert's lab was "wonderfully exciting."

Villa-Komaroff is an associate professor at the Harvard Medical School. Her days are spent conducting experiments in the morning and reading work-related journals in the library or her laboratory during the afternoon. Her research on insulin and insulin-related proteins found in the brain focuses on gene regulation in brain development. As a result, much of her work involves the techniques of DNA technology, such as gene splicing and cloning.

Listening to Villa-Komaroff discuss her work is an exhilarating experience because her enthusiasm is contagious. Her colleagues consider her to be an exceptional scientist, one who loves her work. However, her work is not her entire professional life. She also counsels students and encourages talented students to enter the sciences. Be-

cause she is of Mexican-American heritage, she feels that she is an important role model for others like herself. As a result, Dr. Villa-Komaroff is actively involved in the Association for Women in Science and the Society for the Advancement of Chicanos and Native-Americans in Science. In her work as a public speaker she stresses the need for equal opportunities for women and minorities in science.

On top of everything else, Villa-Komaroff still finds time for personal interests, such as photography, reading science fiction, and listening to music, both classical and 1960s rock. No, this is no stereotypical scientist. Instead it is a scientist who sees herself as a researcher, teacher, and role model — an active, intelligent, curious scientist who is at home in the classroom and in the laboratory, working within a field that is at the cutting edge of biological research.

"manufacture" a new form of algae that could produce milk protein, thus providing a possible answer to the worldwide food shortage in places like Ethiopia. Perhaps you would form rats that taste like steaks. Or would you use your power to find cures for human genetic disorders? Such possibilities are now within the realm of the probable. Today, or in the very near future, molecular biologists may be able to produce biological forms such as these. In fact, they are able to create life forms that never before existed on earth. Such powers are truly awesome. The field of biotechnology is growing so rapidly it is almost impossible to keep up with all the progress. To provide a historical perspective, the table in the Human Endeavors box provides a recombinant DNA chronology.

To understand the possible benefits and dangers of this new technology, we need to explore this new field of human endeavor. Just what, indeed, is recombinant DNA?

First we need to remember that the "recombining" of DNA is a natural process. In fact, eukaryotic organisms rearrange genetic material naturally through the processes of sexual reproduction, crossing over, and what is called the assortment of chromosomes. You, for example, are the result of the recombination of the genetic information in the sperm of your father and the egg of your mother. Through recombination, the unique genetic signature of life is rewritten in every newborn child.

Simply put, in recombinant DNA technology human beings select a portion of the DNA found in one organism and insert that portion into the DNA of another. In other words, **recombinant DNA technology** means human selection and intervention in the code of life. In this chapter we discuss how the technology works and what benefits and dangers can result from it. Table 13.1 presents a miniglossary of some of the terms used in DNA technology. You may find it helpful to refer to it as the chapter progresses.

TABLE 13.1
Miniglossary for Recombinant DNA Technology

DNA sequencing
Determining the linear sequence of nucleotides in a DNA molecule.

Fertility factor
The plasmid in *E. coli* that codes for the production of the pilus, thus permitting conjugation to occur.

Gene
A unit of heredity; a segment of the DNA molecule coding for a specific function.

Gene cloning
The making of several copies of a gene by inserting the gene into the DNA of a cell and then allowing the cell to multiply.

Gene splicing
Joining segments of DNA from different sources.

Gene therapy
The introduction of a functional gene into an organism in which that gene is defective (nonfunctional).

Ligases
Enzymes that repair breaks in single-stranded DNA.

Plasmid
A small, self-replicating molecule of DNA separate from the main chromosome in bacterial cells, yeast, and some higher organisms.

Recombinant DNA technology
The technology that allows for the cutting apart and splicing of pieces of DNA from different sources.

Resistance factor
Plasmid that contains a gene or genes that impart resistance to one or more antibiotics.

Restriction enzyme
An enzyme that recognizes specific sites (base sequences) in a double-stranded DNA molecule and cuts both strands of the molecule everywhere those particular sites exist.

PROKARYOTIC REPRODUCTION

Much of the work being done right now in recombinant DNA technology utilizes prokaryotic life forms, such as the bacterium *Escherichia coli,* although in recent years more and more investigators have been using eukaryotic cells such as those from yeast and mammals. By imitating many of the reproductive processes of prokaryotes, and by taking advantage of others, biologists have been able to utilize the functioning of prokaryotes in significant ways. As a result, in order to understand recombinant DNA technology, you first must learn about the life cycles of prokaryotic organisms. How do prokaryotic organisms reproduce?

Recombinant DNA Chronology

This chronology lists some of the events of the human endeavor that has resulted in our ever-increasing knowledge of DNA and its technological applications.

Source: From *The DNA Story* by James D. Watson and John Tooze. Copyright © 1981. Reprinted with the permission of W.H. Freeman and Co.

1869   Discovery of DNA in the sperm of trout from the Rhine River.

1944   DNA proved to be a genetic molecule capable of altering the heredity of bacteria.

1953   Postulation of a complementary double-helical structure for DNA.

1956   Genetic experiments support the hypothesis that the sequence of the base pairs in DNA conveys its genetic messages.

1960   Discovery of messenger RNA and demonstration that it carries the information that orders amino acids in proteins.

1961   Use of a synthetic messenger RNA molecule (poly-U) to work out the first letters of the genetic code.

1964   Establishment of the complete genetic code.

1972   Use of the joining enzyme DNA ligase to link fragments of DNA produced by the actions of restriction enzymes. The first recombinant DNA molecules are generated at Stanford University.

1973   Foreign DNA fragments inserted into plasmid DNA to create plasmids that are totally new. Finding that they can be functionally reinserted into the bacterium *E. coli.* Potential now exists for cloning bacteria with any gene.

1973   First public concern that recombinant DNA procedures might generate potentially dangerous, unique microorganisms.

1974   Call for a worldwide moratorium on certain classes of recombinant DNA experiments.

Asexual and Sexual Reproduction in Prokaryotes

■ **The usual means of reproduction in prokaryotes is asexual, a process that is similar to mitosis and cytokinesis. Genetic recombination in *E. coli* takes place by means of conjugation, a process during which F pili form between two bacteria and DNA is exchanged.**

As is the case with most unicellular organisms, bacteria reproduce primarily by asexual means. **Asexual reproduction** (or *binary fission,* as it is sometimes called) is a process similar to mitosis and cytokinesis (Chapter 12) by which new cells are formed. In other words, bacteria reproduce in a way that does not involve any exchange of genetic material between two organisms. A cell simply divides to produce two new cells.

In a prokaryotic organism the chromosome, a self-replicating circular structure, is located in what is called the *nucleoid region.* In *E. coli*, this central chromosome is almost 1.36 mm in length and contains approximately 2.2 million base pairs packaged tightly in the small bacterial cell. Apparently, the chromosome attaches itself to a point along the cellular membrane and replicates. The cell then di-

1976   Release of the first guidelines by the National Institutes of Health; prohibition of many categories of recombinant DNA experimentation. Rising public concern that the guidelines might not be effective. *The New York Times Magazine* article urges prohibiting the awarding of the Nobel Prize for recombinant DNA research.

1975– Formation of the first genetic engineering company (Genentech), specifically founded to use recombinant DNA
77   methods to make medically important products.

1977   Development of procedures for the rapid sequencing of long sections of DNA molecules.

1978   The Nobel Prize in Medicine is awarded for the discovery and use of restriction enzymes.

1978   Production of the first human hormone (somatostatin) through the use of recombinant DNA.

1979   General relaxation of the NIH guidelines allows viral DNAs to be studied by using recombinant DNA procedures.

1980   The Nobel Prize in Chemistry is awarded dually for the cloning of the first recombinant DNA molecules and the development of powerful methods for sequencing DNA.

1980   U.S. Supreme Court rules that a live human-made microbe is patentable.

1981   Offer to the general public of stock in the first recombinant DNA company (Genentech). Valuation by Wall Street in excess of $200 million. This early success indicated the importance of recombinant DNA technology.

1982   Sales of genetically engineered products total about $20 million.

1983   The National Institutes of Health grants permission for the release of genetically engineered microbes into the environment.

1985   The U.S. Board of Patent Appeals and Interferences rules that seeds, plants, and plant tissue cultures can be protected by patents.

1987   Using transgenic techniques, a human protein is synthesized by mouse mammary glands and produced in mouse milk.

1988   First patent issued for a higher life form — to Harvard University for a new breed of genetically altered mice.

1989   Congress gives the National Institutes of Health $28 million to begin the $3 billion Human Genome Project.

1990   Researchers introduce laboratory-produced genes into the cells of human patients. The first step toward actual gene therapy.

1991   The landmark experiment of human gene therapy is conducted. Researchers at the National Institutes of Health insert genes into a four-year-old girl to correct an immune system disorder.

1994   Three years after receiving the landmark human gene therapy, the girl patient appears to be thriving.

The Future?

vides, so that each new daughter cell contains its own DNA molecule. There is no genetic recombination, and the bacteria produced are genetically the same — "a chip off the old block" (Figure 13.1; see also Figure 10.3).

For a long time, biologists thought that bacteria always reproduced asexually. They now know that on some rare occasions bacteria do reproduce by sexual means, and this reproduction does result in genetic recombination. In 1946, the question of whether or not prokaryotic cells (such as bacteria) "have sex" was first addressed by Joshua Lederberg and Edward Tatum. They demonstrated that an ex-

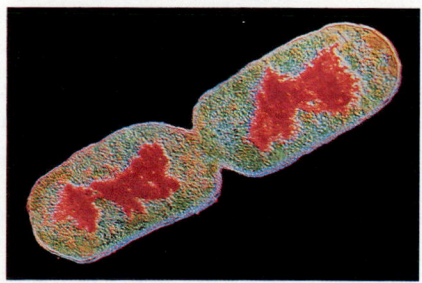

FIGURE 13.1 Dividing Bacterial Cells
Asexual reproduction in prokaryotes, such as the bacterium shown here, results in daughter cells that are genetically identical.

Prokaryotic Reproduction

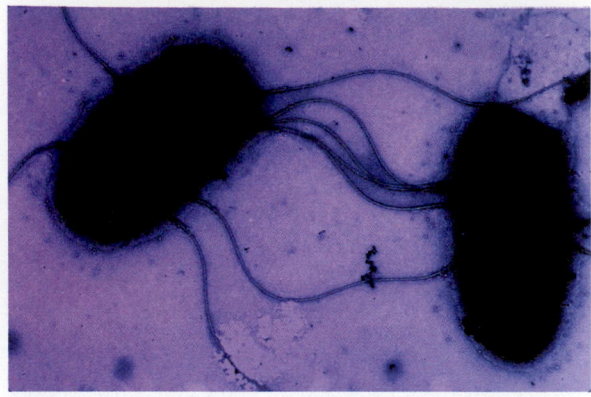

FIGURE 13.2 Conjugation in Bacteria
During sexual reproduction in bacteria, an exchange of genetic material does occur. In this electron micrograph, F pili extend from one *E. coli* cell to another, enabling the transfer of DNA from one bacterium to the other.

English microbiologist, determined that the fertility factor was an extra piece of DNA, called the *fertility plasmid*. The fertility plasmid contains 25 genes, which are involved in the formation of the F pili and the transfer of a replicated F factor from one bacterial cell to another during conjugation.

Resistance in Bacteria — The Result of Recombination. The various methods of genetic exchange in bacteria relate to the genetic basis of some bacteria's resistance to certain antibiotics, such as penicillin or streptomycin. When a group of bacteria is exposed to a given antibiotic, most of the bacteria die, but those that are resistant will survive and produce offspring that are also resistant. As the nonresistant cells die, competition decreases and the number of resistant cells increases. The bacterial

change of genetic material — sexual reproduction — did occur in *E. coli*. Bacteria can undergo *conjugation*, a form of sexual reproduction in which the genetic information of one cell is directly transferred to another cell. During conjugation, *E. coli* cells *congregate* (or pair), and a thin structure called a pilus (or F pilus) is located between them, which facilitates the transfer of DNA between cells (Figure 13.2).

The Role of Plasmids in Bacteria

■ **Plasmids are circular pieces of DNA located away from the nucleoid area in a bacterium. A plasmid called the F factor is the fertility factor. R factors are plasmids that carry resistance. Plasmids can be transmitted through conjugation.**

Later research determined that in addition to the main bacterial chromosome, most bacteria have **plasmids,** small circular forms of DNA located away from the nucleoid region, the area of the cell containing the majority of the cell's DNA. Plasmids are sort of minichromosomes. They are self-replicating and may consist of from 2 to 30 genes (Figure 13.3). Plasmids can also recombine with other plasmids.

Plasmids play an important role in the process of conjugation in bacteria. Much was learned about the role of plasmids when biologists tried to learn the identity of the *fertility factor* (F factor), the factor that induced conjugation. William Hayes, an

(a)

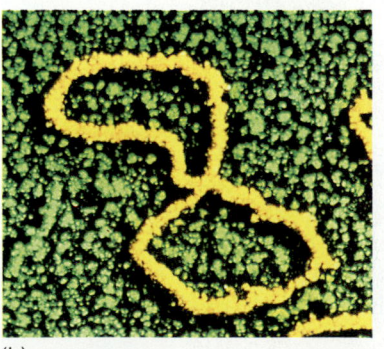

(b)

FIGURE 13.3 Bacterial Plasmids
(a) An electron micrograph of circular plasmids isolated from the bacterium *Neisseria gonorrhoeae* (the bacterium that causes gonorrhea). (b) A single plasmid is shown in the final stages of replication, forming two identical plasmids.

population eventually consists primarily of bacteria resistant to a specific antibiotic.

When antibiotic-sensitive bacteria are mixed with antibiotic-resistant bacteria, the resistance is transferred to the sensitive bacteria within one hour. A group of Japanese biologists discovered that the mechanism for inheriting this resistance involved sexual recombination, which is the result of conjugation. Through conjugation, resistant bacteria transfer plasmids that contain the genetic information, imparting antibiotic resistance to non-resistant strains.

In *E. coli*, a plasmid called R6 (R for resistance), contains genes that impart resistance to six different antibiotics. The R6 plasmid can be transferred from one type of bacteria to another. For example, although *E. coli* normally lives in our intestines and does no harm, it can sometimes pick up plasmids during conjugation and transfer this factor not only to F⁻ *E. coli* but to other types of bacteria such as *Shigella*, which causes dysentery. As a result, the *Shigella* also becomes resistant.

From a medical viewpoint, what we learn from our accumulated knowledge about the abilities of bacteria is caution. We must use antibiotics with restraint, or they will eventually cease to be effective. Otherwise, antibiotic-resistant bacteria will make up the majority of the bacterial population, and our means of controlling disease will be sharply curtailed.

Transduction

Are there ways other than conjugation that genetic recombination can occur in prokaryotic cells? Yes, there are—for example, *transduction*.

Transduction

- **In transduction, bacteriophage DNA infects a bacterium and becomes incorporated into the cell's central chromosome. Upon replication, a portion of the host's DNA may be incorporated within the new phage. When released, the phage may carry a piece of the host cell's DNA. When the phage infects another cell, the DNA of the former host cell can be integrated with the DNA of the present host cell.**

Transduction is the transfer of genetic information through the vehicle of viruses called *bacteriophages* (bacterial viruses, or *phages* for short). The bacteriophage consists of a protein head containing the viral DNA and a protein tail (Figure 13.4). The phage attaches itself to a bacterial cell with its tail. It then injects its DNA into the cell, leaving its protein shell behind. The viral DNA can sometimes take over the bacterial cell's machinery and use it to make copies of itself. The bacterial cell then ruptures and the new phages are released and may infect other bacterial cells.

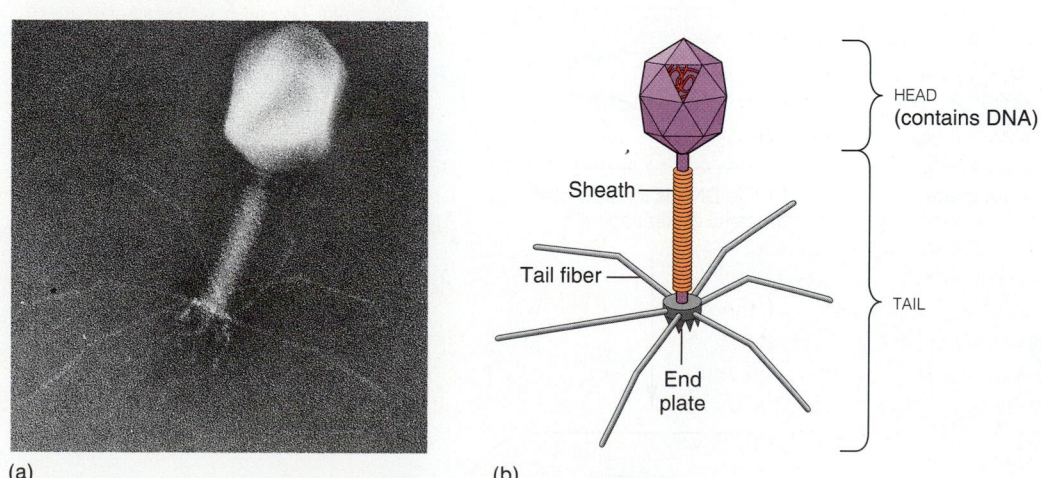

(a) (b)

FIGURE 13.4 Bacteriophage Structure
During infection of a bacterium, the tail fibers and end plate of this bacteriophage known as lambda attach the phage rigidly to the cell. The sheath then contracts, and the DNA in the head is injected into the bacterial cell.

This cycle of infection of bacterial cells by bacteriophages is called a *lytic cycle* (Figure 13.5a). Sometimes, however, the viral DNA remains dormant inside the bacterial cell rather than taking over the cell. The viral DNA becomes incorporated into the bacterial central chromosome (almost like a plasmid in *E. coli*). The result is called a **prophage.** Bacteria that contain this quiet, dormant prophage are called *lysogenic* (Figure 13.5b).

In a prophage the viral DNA is not *doing* much; that is, it is *not* controlling the formation of new bacteriophages. As the bacterium reproduces asexually, the prophages are also replicated. Then, at a certain point, the prophage is induced (recall the inducible gene sequence we discussed in the previous chapter). The viral DNA prophage breaks away from the bacterium's main chromosome and begins to control the synthesis of new phages. The

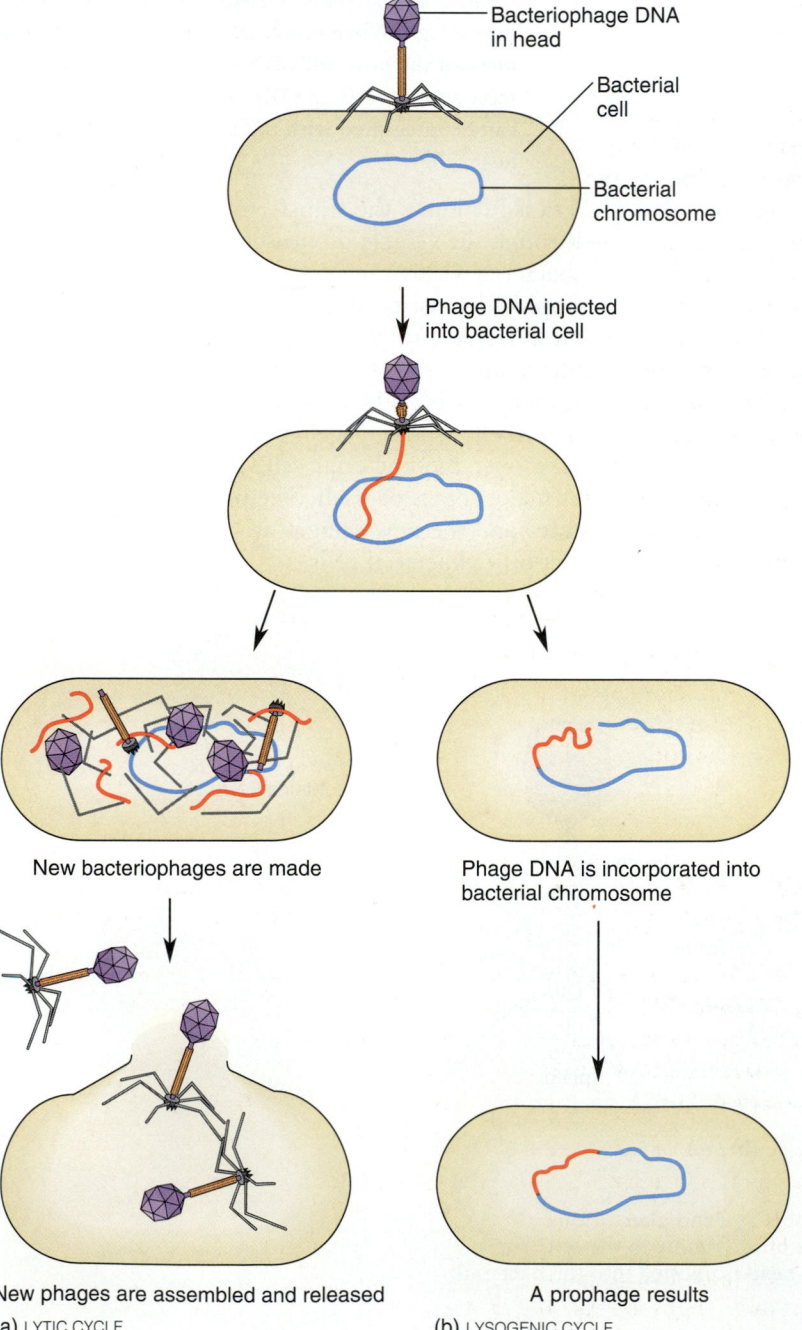

FIGURE 13.5   Bacteriophage Infection Cycles
(a) During a lytic infection, the bacteriophage DNA shuts down normal bacterial genetic activity and directs the production of new phages. The release of these phages results in cell death. (b) Lysogeny occurs when the phage DNA is incorporated into the bacterial DNA and a prophage is formed. The viral DNA is inactive at this point and no new phages are produced. The prophage can become active and enter into a phage-producing lytic cycle.

phages then complete the lytic cycle by causing the bacterial cell to rupture and release the phages to infect other cells.

When the phage exits, it may also take along a piece of the host's DNA. When the phage infects a new cell, it injects not only its own viral DNA but also the genetic information incorporated from the host cell. As a result, there is genetic recombination: The DNA of the former host cell is incorporated with the DNA of the present host cell. The prophage has functioned as the vehicle for recombination. This is the process of *transduction*.

There are two types of transduction: *restricted* and *general* (Figure 13.6). In restricted transduction, the phage carries only those chromosomal genes that are positioned next to the site at which the prophage is attached. Hence, the DNA fragment is not picked up at random but is restricted to certain sites on the DNA molecule.

In general transduction, parts of the bacterial chromosome may be broken into pieces during the lytic phase, and these pieces of bacterial DNA are incorporated into the head of the virus. The virus carries the bacterial DNA to a new host cell. A virus that has undergone general transduction contains little or no viral DNA. Instead, it contains a piece of the DNA from its former host. Such viruses can infect a new host cell but cannot cause it to become lysogenic. Instead, when one of these viruses at-

taches to a new host cell, all it can do is insert the fragmented DNA picked up from the previous host cell.

What does all this mean? Well, there are several means by which bacterial cells undergo genetic recombination. They not only reproduce by asexual means (to become chips off the old block) but can also reproduce by sexual means. Conjugation, transduction, and others we have not mentioned are all mechanisms that result in genetic recombination in prokaryotic cells. Bacteria do "have sex" (although the means are primitive), and the process does, indeed, result in an exchange of genetic material between organisms to produce genetic recombination.

But hasn't this discussion taken us far afield from recombinant DNA technology? How does this knowledge provide scientists with information they need to make these bacterial cells respond to our beckoning? Let us see.

RECOMBINANT DNA TECHNOLOGY

■ **In recombinant DNA technology, a DNA segment, either natural or synthetic, is inserted into a different strand of DNA. The DNA is cleaved through the actions of restriction enzymes and is rejoined through the actions of ligase enzymes.**

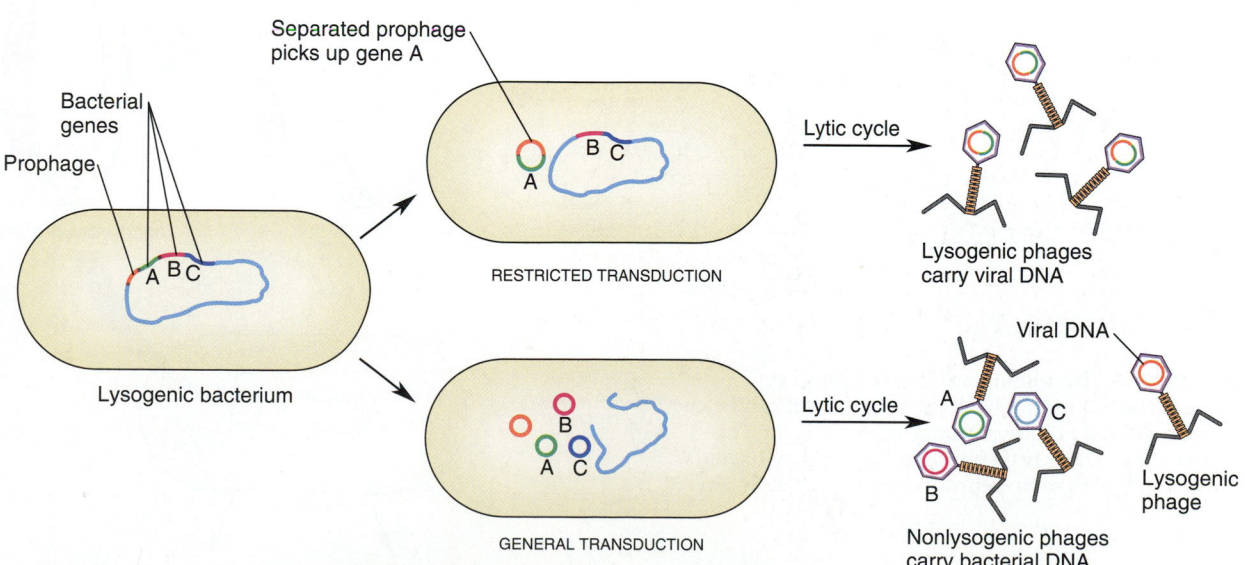

FIGURE 13.6 Genetic Recombination: Transduction
In restricted transduction, the prophage picks up a specific piece of bacterial DNA located next to it on the bacterial chromosome. In general transduction, pieces of bacterial DNA are randomly incorporated into bacteriophages. Most of these phages are not lysogenic because they contain little or no viral DNA.

Recombinant DNA technology is the direct application of our knowledge of molecular genetics and biochemistry in order to produce a particular product (usually a protein). In essence, biologists are able to turn bacterial cells such as *E. coli* into living factories. Biologists use these bacteria because they undergo asexual reproduction so rapidly (they double about once every 20 minutes) that they "manufacture" millions of genes in a 24-hour period.

Figure 13.7 provides a simplified overview of the process of recombinant DNA. In this technology, a piece of DNA segment is taken from a natural or a synthetic source and is inserted into another segment of DNA, in either the same species or a number of other types of organisms. What does that mean, in more specific terms? It means that it is possible to take a piece of human DNA and splice it into the DNA of a prokaryotic plasmid.

Gene Splicing

How do we insert the natural or synthetic foreign gene into a bacterial plasmid? As in all chemical reactions in living organisms, *enzymes* are involved in the gene splicing. Some are used to cleave or cut DNA molecules, and others are used to mend the breaks in the DNA strand.

A group of enzymes called **restriction enzymes** (short for *restriction endonucleases*) can cut DNA molecules at specific sequence sites. The restriction enzyme *Eco*RI was the first restriction enzyme isolated from *E. coli*. *Eco*RI cuts or cleaves the DNA molecule only between guanine and adenine when they appear in a specific sequence. If you were looking at this particular section as it appeared in the double strand of DNA, the sequence would look like this:

```
...G | A A T T  C...
     |   | | | |  |
...C   T T A A | G...
```

When the *Eco*RI cleaves the molecule between G and A, the cuts are not opposite each other. The break that results is asymmetrical:

```
...G                    A A T T C...
   |                            |
...C T T A A                    G...
```

As you can see, because the cut is asymmetrical, one single strand protrudes from either side. The protruding end is called the *sticky end*, because its bases are exposed and available to pair with other bases. One sticky end can readily bind with other sticky ends cut by the same enzyme, through com-

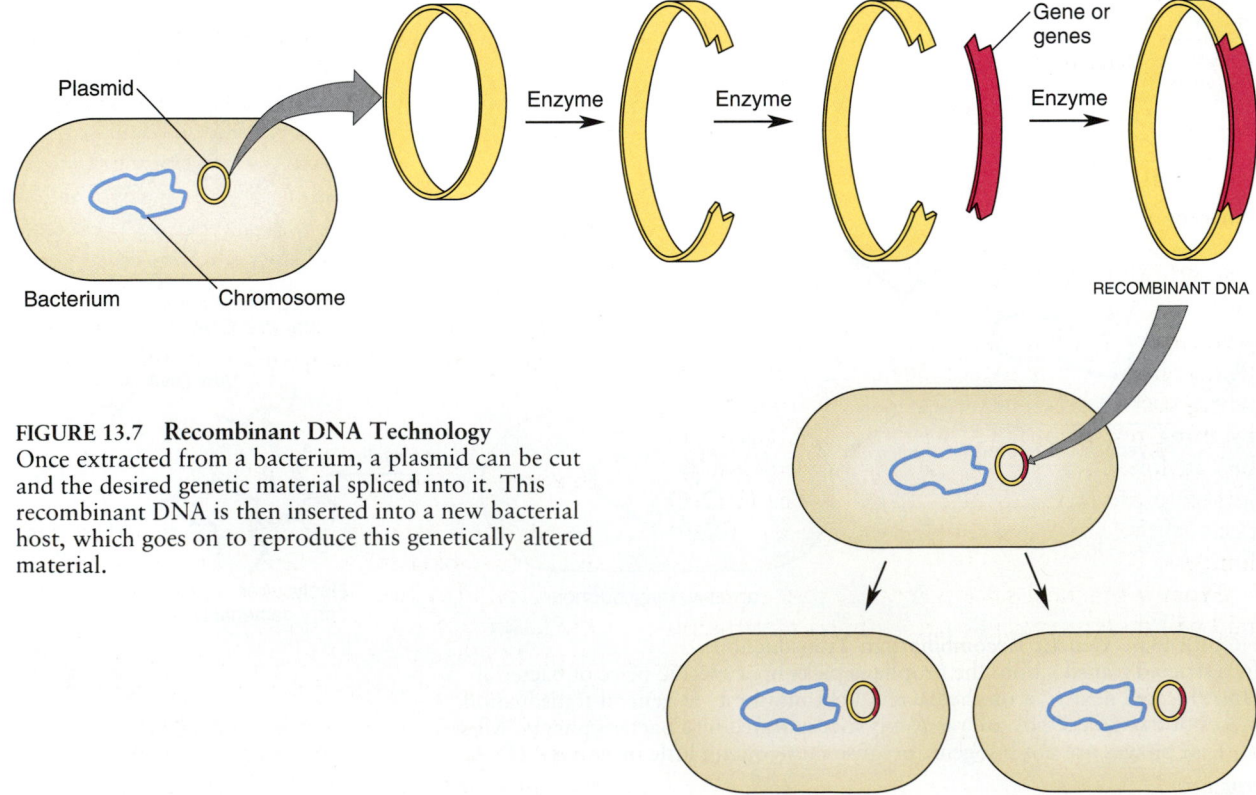

FIGURE 13.7 Recombinant DNA Technology
Once extracted from a bacterium, a plasmid can be cut and the desired genetic material spliced into it. This recombinant DNA is then inserted into a new bacterial host, which goes on to reproduce this genetically altered material.

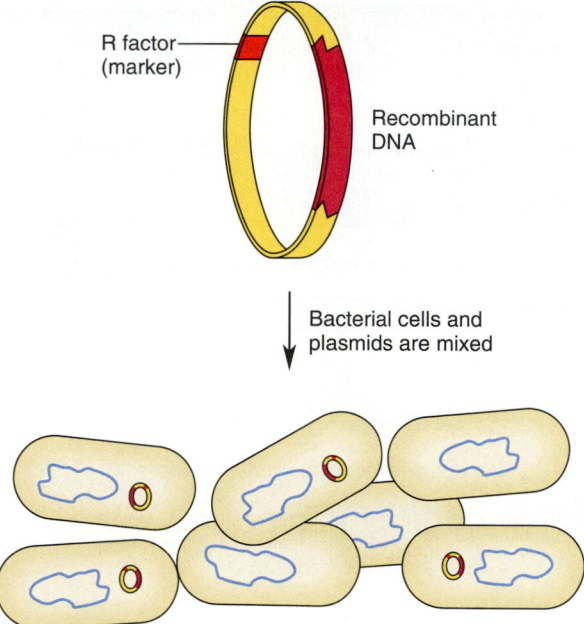

R factor
(marker)

Recombinant
DNA

Bacterial cells and
plasmids are mixed

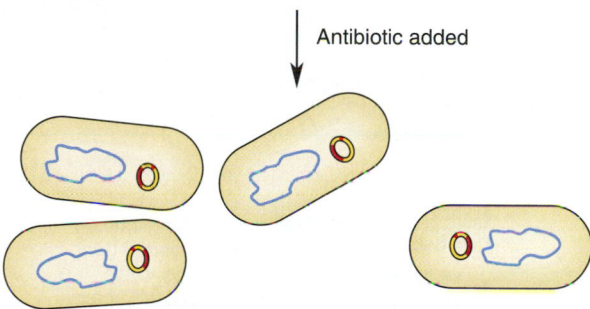

Some bacteria incorporate a plasmid with R factor marker;
others don't

Antibiotic added

Only the bacteria with R factor plasmids are not destroyed by
the antibiotic

FIGURE 13.8   Using Resistance as a Marker
Recombinant DNA is synthesized using plasmids
marked with a resistance factor R and thus known to be
resistant to antibiotic. Once exposed to antibiotic, only
the bacteria containing the recombinant DNA (and the
R factor) will survive.

plementary base pairing. A group of enzymes called
ligase enzymes repairs, or ligates, the breaks left
when sticky ends adhere by base pairing. Thus,
by using restriction and ligase enzymes, DNA can
be cut, foreign genes inserted, and the molecule rejoined to produce recombinant DNA. The applications of the technique of **gene splicing** are almost
limitless.

From a practical standpoint, not just any plasmid will do for gene splicing. To fulfill the requirements of recombinant DNA technology, the
plasmid should replicate quickly, and, in addition,
it should carry genes that make it easily identifiable
(for example, as we will see, the resistance to an
antibiotic).

In fact, prokaryotic bacterial cells are not the
only ones that biologists are using to produce
a desired protein. Because they too reproduce
quickly, yeast cells (specifically, brewer's yeast
cells), some molds, and even mammalian cells,
which are eukaryotic cells, can also be used as living
cell factories, responding to our beck and call. But
that is another story. Back to our workhorse, *E.
coli.*

Using Resistance as a Marker

When you are performing a gene-splicing experiment, how can you tell whether you have succeeded? How can you tell whether the resulting
plasmid contains the portion of DNA you had
hoped to incorporate? One way scientists answer
this question is to include a marker, a section of
DNA whose presence is easy to detect. In bacteria,
resistance to antibiotics is such a marker. If you
expose the bacteria to the antibiotic, only those bacteria that have incorporated the gene for resistance
will survive. Thus if you include the gene for resistance along with the other DNA you want to incorporate, exposing the bacteria to the antibiotic will
quickly tell you whether the plasmid has been incorporated.

Stanley Cohen at Stanford Medical School isolated a small bacterial plasmid that contains only
one instance of the GAATTC sequence in the entire
molecule. This plasmid, called pSC101, carries an
R-factor gene that makes the bacteria resistant to
the antibiotic tetracycline. This plasmid is ideal for
recombinant DNA gene-splicing methods, because
when the bacteria are exposed to the antibiotic tetracycline only those with the R factor are resistant
and can survive. Those that do not have the R factor of course die. Researchers therefore splice genes
into easily identifiable plasmids like R-factor plasmids. Then they can use antibiotic resistance as a
method for identifying the spliced material (Figure
13.8).

Inserting a Gene into a Bacterium

To see how scientists proceed in an experiment
using recombinant DNA technology, let us look at
the work of Arthur Riggs and Keiichi Itakura. In
1977, Itakura and his colleagues at the City of
Hope Medical Center published an article in *Science* in which they described their work with somatostatin, a simple polypeptide consisting of 14
amino acids. Somatostatin is also called a *growth
hormone inhibiting hormone.* Since Itakura and his

colleagues already knew the sequence of the amino acids in somatostatin, they took the necessary chemicals off a shelf and produced an artificial DNA sequence for somatostatin. They then spliced the synthetic DNA sequence into the *E. coli* plasmid. As the bacteria replicated, so too did the hybrid *E. coli* plasmid, including the spliced human DNA. As a result the implanted DNA sequence was passed on to succeeding generations, producing **clones** — multiple copies of the organism, all having the same genetic makeup. And all the while the *E. coli* were carrying on their life functions, they were also doing something no *E. coli* had ever done. They were producing human somatostatin. Somatostatin is being used in the treatment of certain types of diabetes, as well as a condition called acromegaly, in which there is an abnormal enlargement of the face, hands, and feet.

OTHER RECOMBINANT DNA TECHNIQUES

Polymerase Chain Reactions

Imagine yourself as a molecular biologist with an extremely small sample of DNA to sequence. What would be your wish? Probably to have a method to obtain large amounts of that DNA. Today your wish has come true. A technique called the **polymerase chain reaction** (**PCR**) is a technique that enables you to make billions of copies of your targeted sequence; it is a method of selected gene amplification (Figure 13.9). First, the DNA is heated to separate it into two single strands. Then two short sequences are synthesized on both of the sequences (gene) to be copied. These are called *primers*. These primer sequences are chosen because the DNA to be copied lies between them. So the primers mark where the replication is to begin. When the primers are added they attach to the complementary sequence. Then DNA polymerase is added and double-stranded DNA is produced. The cycle is repeated, and with each cycle the amount of DNA is doubled. After 30 cycles, over a billion copies of the desired DNA sequence are available. You might wonder why the enzyme DNA polymerase is not destroyed when it is heated. Think for a moment. How do bacteria that live in hot springs replicate their DNA? Obviously nature selected a variation of DNA polymerase that could remain active at higher temperatures, and it is that DNA polymerase that is used — a variation that is thermostable.

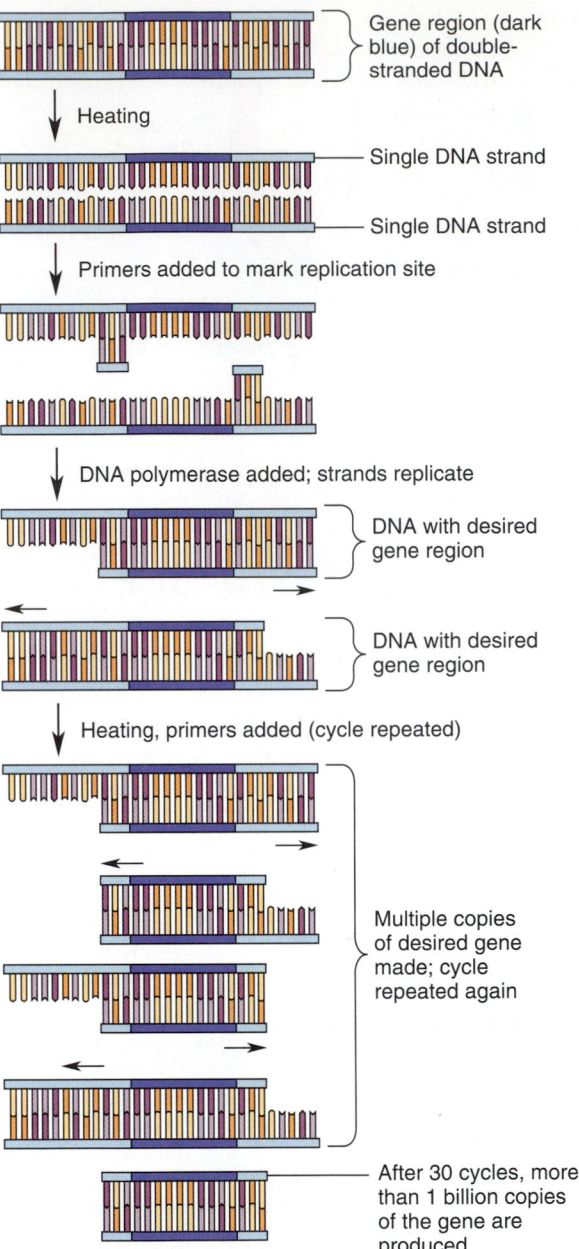

Gene region (dark blue) of double-stranded DNA

Heating

Single DNA strand

Single DNA strand

Primers added to mark replication site

DNA polymerase added; strands replicate

DNA with desired gene region

DNA with desired gene region

Heating, primers added (cycle repeated)

Multiple copies of desired gene made; cycle repeated again

After 30 cycles, more than 1 billion copies of the gene are produced

FIGURE 13.9 Polymerase Chain Reactions (PCR)
As described in the text, biologists can use the technique of PCR for making millions of copies of a segment of DNA. These copies then can be separated from the rest of the DNA by the use of gel electrophoresis.

Today there are many modifications of this technique that are enabling researchers to investigate a wide range of genetic questions and disorders. These include applications of PCR to the detection of genetic disease. For example, the gene for sickle-cell hemoglobin and the alleles for phenylketonuria, hemophilia, cystic fibrosis, and even some cancers have been identified. So PCR can be used as a prenatal (see Chapter 15) technique to know the genetic makeup of the child prior to birth by using the cells of the fetus obtained by amniocentesis or chorionic villus sampling.

Restriction Fragment Length Polymorphism (RFLP)

As we mentioned, sometimes a point mutation, deletion, or insertion occurs within a DNA sequence, resulting in DNA polymorphism ("many forms"). This minor difference in sequence can be detected by using restriction enzymes and gel electrophoresis. For example, if the DNA sequence on a pair of homologous chromosomes is slightly different when a restriction enzyme is used to cut the DNA, the fragments will be of different lengths because restriction enzymes were used to cut these polymorphs and the fragment length varies. The technique is called **restriction fragment length polymorphism** or RFLP (pronounced "riflip"). Figure 13.10 illustrates RFLP analyses. As you can see, in two homologous chromosomes, A and A¹, the DNA sequence in the A chromosome has one more restriction site with the CC sequence than the A¹ chromosome. Because of this difference, when a re-

striction enzyme that cuts between CC is used, the strands differ: A is cut into two fragments of shorter lengths, and A¹ is cut into one large fragment. When gel electrophoresis is used to separate these strands, the patterns are different. How can we use this technique to locate a specific gene? It has been determined that often a particular RFLP is almost always inherited with a specific gene, because that gene and the RFLP are close together; therefore, the RFLP can be used as a marker for that gene. So, if you find that RFLP or a set of RFLPs, you should find the gene. Using this technique, today over 1000 RFLPs have been identified in the human genome. RFLP technology has enabled biologists to test for genes in the fetus or in individuals to see if the gene is present (see Chapter 15). Not only can RFLP analysis be used for the detection of a specific gene, but it can also be used to help map and sequence the entire genome (see Focus 13.1).

As you can see, from learning about reproduction in bacterial cells and DNA came much of the knowledge and techniques that will now enable us to mass-produce a variety of important biological compounds, sequence the human genome, detect genetic disorders, and conduct gene therapy. Let us discuss some of these potential applications.

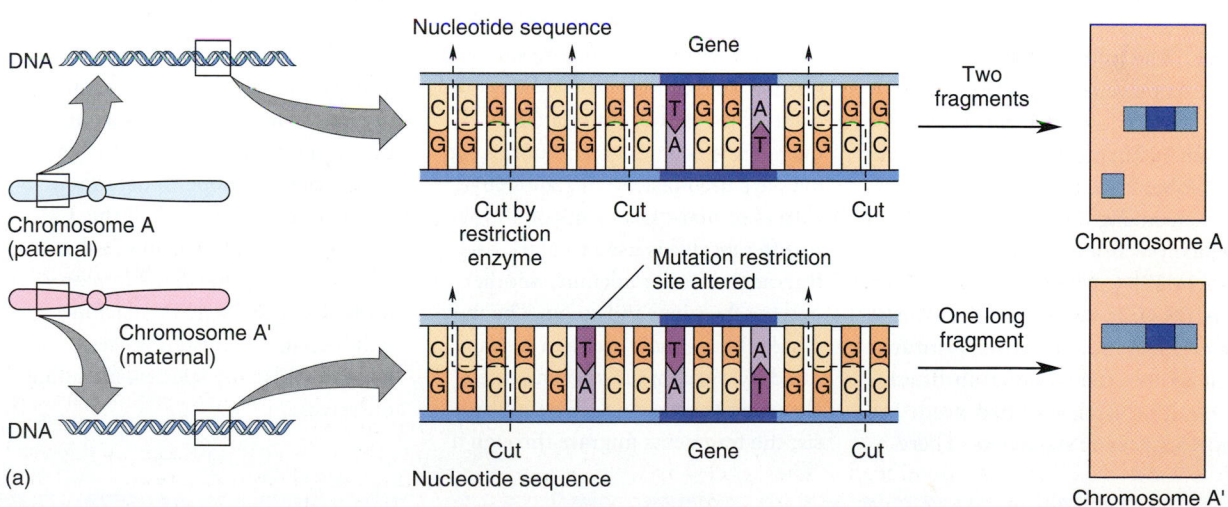

FIGURE 13.10   Restriction Fragment Length Polymorphism (RFLP)
Restriction fragment length polymorphisms are due to point mutations in the DNA sequence. Because of these mutation restrictions, enzymes cut the DNA into different lengths or numbers. This size change or number of fragments can be detected by the use of gel electrophoresis. In this illustration the paternal chromosome (A) has three restriction sites and the maternal only two due to a point mutation. Because of this, the fragment produced by the maternal chromosome (A') is longer than the restriction fragment produced by (A), and two fragments are produced in the DNA of chromosome (A). These differences in fragments can serve to mark (identify) certain genes that are contained within these fragments. Today we have over a thousand RFLPs that can serve as genetic markers.

(b) Gel electrophoresis is used to separate fragments and detect fragments with the gene sequence being investigated.

The Human Genome and Gene-Sequencing Machines

Today, James Watson heads the office at the National Institutes of Health that oversees the human endeavor that will map and sequence the human genome within the next 15 years. Almost every day new techniques are developing to allow us to sequence the genome more effectively and rapidly. Today even scanning-tunneling microscopes and laser cameras are being used to develop new sequencing techniques. And we are using machines to actually determine the order of the nitrogen-containing bases A, T, G, and C. The following is a description of one gene-sequencing method.

When Frederick Sanger (see the Profile in Chapter 12) was awarded the Nobel Prize for his methodology for sequencing nucleic acids, sequencing was a time-consuming process. In 1984, when the first strain of AIDS virus was isolated in America at the National Cancer Institute, a division of the National Institutes of Health (NIH), scientists needed to know its genetic sequences. There-

fore, they sought out the scientists who could decipher its genetic alphabet quickly. Six months later the laborious procedures had been completed and the sequence of the virus's 9000 nucleotides had been determined. How rapidly their crash project was completed! Today, however, six months to sequence 9000 nucleotides would be considered excessive. The work that took six months only a few years ago can now routinely be done in a day.

Using automation and high technology, gene-sequencing machines have compressed the time to sequence genes into hours and days. In some of the newer automated sequencing machines green fluorescent dyes are used instead of radioactive tracers to mark fragments of DNA. A different dye is used to mark each fragment: one for adenine, another one for thymine, and so on. The individual fragments are then separated by a technique known as gel electrophoresis. In gel electrophoresis, the fragments migrate through a

jellylike material exposed to an electrical field. The distance each fragment travels is determined by its electrical charge and size.

A laser beam identifies each of the fluorescent dyes (and hence the nucleotide) as it moves through the gel. A detector differentiates the shades of green and a computer determines the original sequence (see Figure A).

The day has arrived when scientists can do 5000 to 10,000 genetic letters per day. The speed and accuracy of the new machines have allowed scientists to begin a project that would have been unthinkable before the technology was available: the sequencing of the human genome. Scientists hope to determine the precise sequence of all the 3.5 billion A's, T's, G's, and C's that make up the genes in the human body. Gene-sequencing machines will help us establish libraries of the genomes of many species, including ourselves.

THE POTENTIAL USES OF RECOMBINANT DNA TECHNOLOGY

To give you an idea of ways in which biotechnology can affect our lives, we need to explore the directions of research today and to recognize the problems that scientists are trying to address.

Agricultural Applications

For all the attendant fears and misgivings, humans assuredly have much to gain from the new genetic technology. One has only to consider the possible improvements in food production that could be achieved by placing new genes into food crops and livestock. Will genetic recombinant technology solve the hunger problems of the world?

Chapter 13   Recombinant DNA: Genes by Design

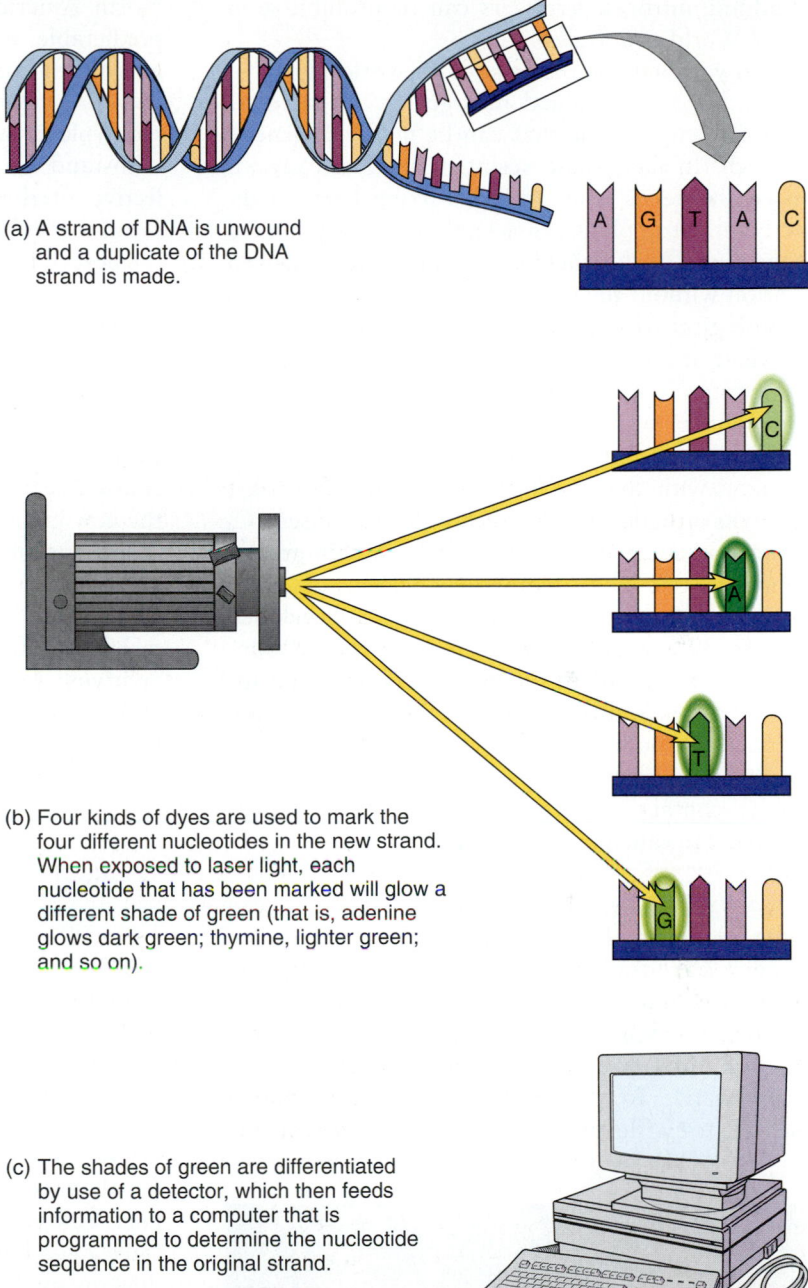

(a) A strand of DNA is unwound and a duplicate of the DNA strand is made.

(b) Four kinds of dyes are used to mark the four different nucleotides in the new strand. When exposed to laser light, each nucleotide that has been marked will glow a different shade of green (that is, adenine glows dark green; thymine, lighter green; and so on).

(c) The shades of green are differentiated by use of a detector, which then feeds information to a computer that is programmed to determine the nucleotide sequence in the original strand.

Research with Plant Crops. Much work is going on in agricultural genetic research. For instance, all plants require a particular form of nitrogen found in the soil in order to develop normally. They cannot use the nitrogen in the air. Some plants, specifically legumes (such plants as soybeans, peas, and others), can convert the nitrogen from the air into the nitrogen usable by plants through the actions of *nitrogen-fixing bacteria* found in nodes or nodules on the plants' roots. In other words, the bacteria provide a natural fertilizer for the plants as the plants grow. Unfortunately, all plants do not have nitrogen-fixing bacteria, so only the nitrogen already present in the soil is available to them. Nitrogen-poor soil, as a result, produces very minimal crops, and the expense, in both money and energy, of adding nitrogen fertilizers can be prohibitive in Third World countries.

Can you imagine the impact on world hunger if all food crops contained nitrogen-fixing bacteria? That challenge is one that can be met with genetic research. In fact, some scientists have already inserted the genes from nitrogen-fixing bacteria directly into crop plants, thus enabling these plants to make some of the chemicals required for nitrogen fixation without the presence of the bacteria.

Biologists are exploring other ways of improving the yield of the world's food crops (Figure 13.11). In Chapter 7, we discussed the advantages that C_4 plants have over C_3 plants in harvesting carbon dioxide for the process of photosynthesis. Today, working with the methods of genetic technology, scientists are developing techniques to insert C_4 photosynthetic genes into C_3 plants, thereby greatly improving the photosynthetic efficiency of many food crops. In addition, plants are being developed that produce what nutritionists refer to as "complete proteins." In this context, complete means that the fruits and vegetables produced will contain adequate amounts of all the amino acids required by the human diet. At present, most fruits and vegetables contain inadequate amounts. Hence their proteins are called incomplete proteins.

Research in Livestock Production. Work is also being done in livestock production. Livestock breeders have traditionally controlled the improvement of commercial livestock production through selective breeding. For instance, in the 1800s ranchers cross-bred two types of cattle. Wild longhorn cattle, native to the western states, were tough enough to withstand the harsh weather of the North American prairies, but they were also unpredictable and irascible and carried a limited amount of muscle. (Perhaps they were too mean to die!) European beef cattle, on the other hand, were more placid and more meaty, but they could not withstand the harsh weather of the open plains. Selective interbreeding gave rise to the beef cattle we see in the United States today — cattle that combine the best of their two ancestral breeds.

Livestock breeding, however, is a long process, sometimes requiring a number of generations before results are achieved. Genetic technology can change all that. Today it is conceivable to insert new or recombined genes into livestock, thus drastically shortening the time required to change the characteristics of an organism. For example, using the new genetic technology, it is possible to greatly reduce the amount of time needed to "design" dairy cattle that produce increased amounts of milk. Further, recent work with farm animals is leading to techniques for cloning (or reproducing) desirable embryos. The combination of recombinant DNA and cloning technologies is considerably speeding up the development of new and better livestock breeds.

Medical Applications: Gene Therapy

The scope of genetic research continues to expand. Right now, scientists have had some success in inserting genes into embryo cells. Working with fertilized mouse eggs, Frank Ruddle and his coworkers used a micropipette to insert the gene for the production of the enzyme thymidine kinase into the genes found in the nucleus of the developing embryos. They then inserted the eggs into the uteri of female mice, where the embryos developed. Of 180 mice embryos in which the genes were inserted, three mice had incorporated the gene for the production of thymidine kinase. This is a step on the way toward performing gene therapy by inserting genes into embryos with missing or defective genes.

In other experiments, Ruddle and his colleagues were trying to "manufacture" interferon, a natural virus fighter normally produced in small amounts by white blood cells. Scientists are looking for ways to increase the supply of this protein. Ruddle succeeded in inserting the gene for human interferon into mouse embryos. When these embryos developed into adult mice and were bred, they were able to pass the gene for human interferon to their offspring. Ruddle's technique offers great potential.

FIGURE 13.11 Genetic Research in Agriculture
Genetically altered strains of plant crops have been developed that make the crops more resistant to disease.

FIGURE 13.12 Biotechnology
Advances in biotechnology allow for the mass production of human protein. Here scientists are involved in the purification of tumor necrosis factor, which is used to kill cancer cells.

The process of transforming bacteria cells into living cell factories has numerous medical applications in the synthesis of protein products for humans, especially in the production of pharmaceuticals. For instance, some diabetics suffer from a low level of the polypeptide insulin. To compensate, many require daily doses of insulin. In the past, they have been given animal insulin, which sometimes causes allergic reactions. Today biotechnology companies are synthesizing human insulin by means of recombinant technology (Figure 13.12). Other important human proteins such as growth hormone and interleukin-2 (a protein normally made by the human immune system) can be produced in large amounts by using recombinant DNA technology (Figure 13.13).

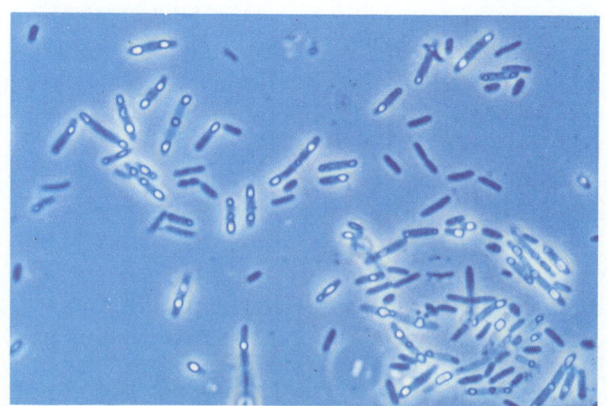

FIGURE 13.13 Interleukin-2
This micrograph shows *E. coli* bacteria containing "bubbles" of interleukin-2.

Working with Eukaryotic Cells

As should be obvious by now, biologists are no longer limited to working with prokaryotic cells like *E. coli*. They are also inserting foreign genes into eukaryotic cells, such as yeast cells, plant cells, and animal cells. For example, Paul Berg at Stanford has inserted genes into animal cell cultures in order to stimulate them to produce the enzyme missing in Lesch-Nyhan syndrome, a childhood metabolic disease that often results in self-mutilation or mental retardation and is usually fatal.

Of course, getting a foreign gene to work in a few cultured cells in a test tube is not the same as inserting a gene into a human being; however, today we are doing human gene therapy. In 1991 Dr. French Anderson and his coworkers used the techniques of modern genetic engineering to introduce a missing gene into the cells of a girl suffering from a genetic immune system disorder (see Focus 15.1, pages 356–357). These experiments raise obvious ethical questions about the use of gene therapy. It is clear that this technique will most likely be used in the foreseeable future to control many genetic disorders; it is also clear that medical ethics require that the benefits of such therapy must outweigh the risks before the therapy is used.

ETHICAL QUESTIONS ABOUT BIOTECHNOLOGY

Because of the potential for good and for ill in the processes and the results of biotechnology, ethical questions are of great concern in the area of genetic research. Many organizations like the National Institutes of Health are involved in setting up guidelines involving procedures, safeguards, and standards that will oversee research in this area.

What is safe when one is discussing recombinant DNA research? Some scientists say that the following conditions must be met:

1. Research in animal experiments must demonstrate that (a) the foreign gene is inserted in the target cell and (b) it remains there.

2. There must be evidence that the gene is properly regulated so that it makes the correct quantity of the substance.

3. It must be demonstrated that the gene will not harm the cell or the organism.

In other words, researchers must proceed in the same careful manner that they have followed before. Life is a delicately coordinated process, and scientists must be certain that their actions do not result in side effects that are ultimately more damaging than the problem they initially set out to combat.

However, gene therapy does hold the potential for changing our own genetic core. Genetic therapy may indeed be able to reduce the suffering and death of individuals afflicted with genetic disease. On the other hand, the question of whether or not humans should tinker with the very essence of human existence remains. Obviously, there is no easy answer.

The Business of Recombinant DNA Technology

With the wide-ranging potential of molecular genetics has come the establishment of several genetic engineering companies, all hoping to use genetic biotechnology to manufacture a marketable, profitable product. What effect does such an industry have on the scientists involved in it? Many university geneticists now are either stockholders in or employees of such companies. Because of the potential profits from the discoveries and techniques developed by any company, advances and discoveries made within genetic engineering companies are considered to be "trade secrets." As a result, molecular scientists and geneticists must keep some information secret.

But that "secret environment" is directly at odds with one of the primary tenets of the scientific method. Traditionally, scientists have been eager to share their knowledge, and much has been learned about the mechanisms of life as a result of the debate, discussion, and research that arose out of the discovery by one or two individuals. In science the interchange of ideas, the public debate and analysis, has functioned, in some ways, like a college "bull session." One person's idea or discovery has generated new ideas and new perspectives in other individuals. The result has often been even greater discoveries. Just consider the chain of events that led to the discovery of the DNA helix.

In the highly competitive market of biotechnology, however, it is financially advantageous to withhold critical information that may simplify, improve, or make a particular procedure more efficient, even though these findings may be of a basic biological nature. To consider the possible impact of such secrecy, imagine how difficult research would have been up to the present if the process of freeze fracturing (Chapter 5) had been kept secret rather than shared with the biological community.

Biotechnology has allowed for the potential to produce commercial amounts of *tissue plasminogen activator* (*TPA*), a protein that can dissolve blood clots, the major cause of heart attacks. Clinical trials were very promising, and researchers were excited, feeling that TPA could save the lives of thousands—heart attacks cause 25 percent (about 400,000) of the deaths in this country every year. However, in May 1987 the Federal Food and Drug Administration (FDA) rejected use of the drug, arguing that not enough information was made available to allow for its approval. An early study had indicated that about 4 percent of patients using the drug suffered brain hemorrhages. This decision caused a furor in the medical and financial community. Why the financial community? TPA is expected to be the first billion-dollar product to emerge from the biotechnology industry. In November of 1987 the FDA approved TPA for immediate use in hospitals. In addition, the National Institutes of Health and a private biotechnology company announced an interesting way of making human proteins like TPA in mouse milk (see Focus 13.2).

Truly the advances are astounding and the dilemmas posed are problematical. Will society lose its best geneticists to corporate interests? Will the university departments of molecular genetics and biochemistry become so strongly linked to commercial ties that academic and intellectual freedom will be lost?

Indeed, some of the questions that arise from recombinant DNA technology are almost unanswerable. Do we want to change human beings, alter our basic genetic core? Who should determine how our new creative power will be used? So far,

FOCUS 13.2
The Land of Milk and Money

Scientists at the National Institutes of Health and Integrated Genetics (a private biotechnology firm) have developed a novel genetic engineering technique to produce a specific human protein in the milk of mice. The human protein, called tissue plasminogen activator (TPA), is an anticlotting agent used in the treatment of certain types of heart attacks and blood clotting disorders.

Scientists first isolated and characterized the specific genetic material that regulated the formation in mice of the milk protein known as acidic protein. This regulatory genetic segment can be fused with the human gene that codes for TPA. The hybrid gene is then injected into newly fertilized mouse eggs. The eggs are returned to a "foster mother" mouse and the embryos brought to term. The resulting female offspring contain this hybrid gene and will produce TPA in their breast milk. The mice are milked and the TPA is extracted from the milk through isolation and purification procedures.

This technique of having genes from one species become a permanent part of the genes of another species by splicing the two together and inserting them into the egg or embryo is called the transgenic technique (Figure A). This technique is significantly more efficient for producing substances like TPA than the older technique of bacterial fermentation, in which *E. coli* were used to synthesize protein products. Trans-

genic mice produce one to ten grams of TPA for every liter of milk whereas only milligrams (thousandths of a gram) are produced by a liter of broth resulting from the bacterial fermentation technique. Obviously, this is a thousandfold increase.

Just think of the implications. Using these techniques, genes can be injected into larger milk-producing animals such as cows. A herd of such transgenic cows could produce large amounts of the desired protein. For example, Factor VIII (a protein-based drug injected by hemophiliacs to prevent bleeding) could be produced in sufficient amounts to treat all the hemophiliacs in the world by a herd of only a hundred cows.

From the commercial perspective, Integrated Genetics administrators believe that this transgenic technique can be applied to produce several protein-based drugs. Using transgenic animals to produce these drugs will become a billion-dollar-per-year business. Truly, the biotechnology firms using this technique will see the near future as "the land of milk and money."

FIGURE A ▶
Here the transgenic technique is used to insert a gene for rabbit beta globulin into a mouse egg. The egg is held by a micropipette while the insertion is made. This egg will be implanted into a recipient mouse whose offspring will have the new gene incorporated in their DNA.

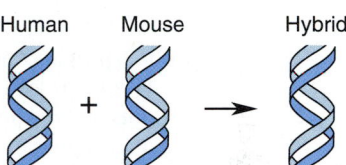

(a) A human gene is fused with a mouse gene that can only be activated in mouse mammary tissue. For example, the gene for tissue plasminogen activator (TPA).

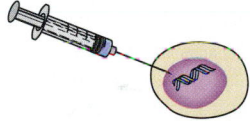

(b) This hybrid gene is injected into a recently fertilized mouse egg, and the hybrid gene becomes a part of the mouse embryo's genetic makeup.

(c) The embryo is implanted into a foster mother, where it develops and is born.

(d) When the female pup with the hybrid gene becomes mature and starts producing milk, her milk contains the human TPA. The TPA in the milk is concentrated and can be used to treat heart attacks.

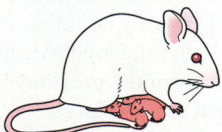

(e) TPA mice are bred to produce generations of TPA-producing mice.

these are questions without definite answers. As the understanding of genetics increases, geneticists are going to thrust on our society an unprecedented power over life itself. We owe it to ourselves as educated individuals to attempt to understand the explosion of genetic knowledge, because its application is surely destined to have an impact on our lives in ways that we have yet to imagine.

SUMMARY

1. In recombinant DNA technology biologists select a portion of the DNA found in one organism and insert that portion into a strand of DNA of another organism.
2. Much of the work being done in recombinant DNA technology utilizes prokaryotic life forms, such as the bacterium *E. coli*. Prokaryotes usually reproduce asexually. Genetic recombination does not occur in asexual reproduction.
3. Genetic recombination in *E. coli* takes place by means of conjugation. During conjugation, *E. coli* cells congregate, and thin tubes called F pili connect the cells and facilitate the transfer of DNA from one cell to another.
4. Plasmids are circular pieces of DNA located away from the nucleoid area in a bacterium, rather like minichromosomes. One of these, called the F factor, is the fertility factor and is transferred from one bacterium to another during conjugation.
5. R factors carry genes responsible for resistance to antibiotics.
6. In transduction, a virus called a bacteriophage infects a cell, then becomes incorporated into the bacterium's central chromosome and remains dormant. The virus is then called a prophage. Bacteria containing prophages are called lysogenic. When induced, the phage replicates, the cell ruptures, and the phages are freed. When released, the phage may carry a piece of host DNA. When it infects another cell, the DNA of the former host cell is intermingled with the DNA of the present host cell.
7. Sometimes the phage DNA immediately initiates synthesis of the new phages. Such a cycle is called an infective lytic cycle.
8. In restricted transduction, a phage carries only those chromosomal genes positioned next to its attachment site.
9. In general transduction, the bacterial chromosome may be broken up during the lytic phase, and these pieces of bacterial DNA are incorporated into the virus. A virus that has undergone general transduction primarily contains a piece of its former host's DNA. In attaching to a new host cell, it inserts the DNA from the previous host cell, but no new phages are produced.
10. In recombinant DNA technology, a DNA molecule is cleaved through the actions of restriction enzymes, a segment of natural or synthetic DNA is inserted, and the molecule is rejoined through the actions of ligase enzymes.
11. Restriction endonucleases, or restriction enzymes, can cut DNA molecules at specific sequence sites. The cut is asymmetrical; a strand, called the sticky end, protrudes from one side. A sticky strand can readily combine with other sticky ends through complementary base pairing. A gene is spliced into the cut DNA through the actions of ligase enzymes to repair sticky ends. The rejoined DNA molecule is called a spliced gene.
12. To fulfill the requirements of recombinant technology, the organism should replicate easily, should carry genes that make it easy to identify, and should have a DNA sequence that allows for the splicing process.
13. To easily identify cells once they have taken up a sequence of foreign DNA, researchers splice the DNA next to a marker. For example, because bacteria containing the R plasmid will continue to grow when exposed to antibiotics, the cells can be detected easily.
14. Polymerase chain reactions can be used to obtain large amounts of a gene. RFLPs can be used to cut polymorphic DNA into fragments that can be used to identify genes.
15. The potential uses of recombinant DNA technology include the following: (1) the insertion of a eukaryotic gene into an *E. coli* cell in order to study it free from the complexity of the eukaryotic cellular environment; (2) applications in plant crops and livestock production; (3) medical applications in gene therapy; and (4) the production of pharmaceuticals such as insulin, interferon, pituitary growth hormones, interleukin-2, and TPA.
16. To determine that genetic research is productive and safe, some biologists have set these requirements: (1) research in animal experiments must demonstrate that the foreign gene is inserted in the target cell and that it remains there; (2) there must be evidence that the gene is properly regulated so that it makes the correct quantity of the substance; (3) it must be demonstrated that the gene will not harm the cell or the organism.
17. The issues of secrecy and profit pose difficult questions for the field of DNA recombinant technology in particular and biology in general.

KEY TERMS

recombinant DNA technology	gene splicing
asexual reproduction	clone
plasmid	polymerase chain reaction (PCR)
transduction	restriction fragment length polymorphism (RFLP)
prophage	
restriction enzyme	

REVIEW QUESTIONS

True/False

1. Bacteria usually reproduce by asexual means; however, some bacteria may reproduce sexually.
   true   false   (page 288)

2. Bacterial conjugation can be considered a type of sexual reproduction.
   true   false   (page 290)

3. Restriction enzymes splice DNA molecules so that one strand protrudes from each end.
   true   false   (page 294)

4. Gene splicing can only occur in prokaryotic cells.
   true   false   (page 300)

Fill in the Blank

1. The "minichromosomes" in bacteria consisting of small circular forms of DNA located away from the central DNA region are called _____.
   (page 287)

2. Some _____ are capable of imparting resistance to antibiotics in the organism that contains them. Historically the first identified factor was called _____ because it imparted resistance to six antibiotics in the recipient bacterium. *(page 291)*

3. During bacterial conjugation, some genetic material can be transferred by the _____.
   (page 290)

4. A group of enzymes known as _____ can cut DNA molecules at specific points. *(page 294)*

Multiple Choice

1. In recombinant DNA research, the plasmids are taken from:
   a. Human cells
   b. Animal cells
   c. Bacterial cells
   d. Any kind of cell
   (page 294)

2. The observation that genes could be incorporated into a plasmid and then introduced into a bacterial cell is the foundation of:
   a. Gene splicing
   b. Organ transplant
   c. Toxicology
   d. Amniocentesis
   (page 294)

3. The transfer of genetic material from one bacterium to another using a bacteriophage as the vehicle for transmission is called:
   a. Transformation
   b. Transduction
   c. Conjugation
   d. Sexduction
   (page 291)

4. The viral DNA that is incorporated into the host's chromosomal makeup is called:
   a. Lysogenic
   b. Prophage
   c. F factor
   d. Hfr
   (page 292)

Discussion

1. Why do you think there was such public controversy regarding recombinant DNA technology?

2. What is the new recombinant DNA technology, and how may it influence our lives?

3. Briefly discuss bacterial conjugation as a method of sexual recombination.

4. What are plasmids and F factors?

5. What is the relationship between F factors and antibiotic resistance in bacteria?

6. Discuss transduction and transformation. Why are these phenomena important to molecular geneticists?

7. Compare and contrast restricted and general transduction.

8. What is gene splicing? What types of enzymes are involved?

9. In your opinion, what are some of the applications of PCR and RFLP techniques?

10. Present your own opinion about the new biotechnology and about human intervention in nature. Do you feel the benefits outweigh the potential moral, social, and ethical risks?

11. Should life forms be patentable?

14

Genetics: Patterns of Inheritance

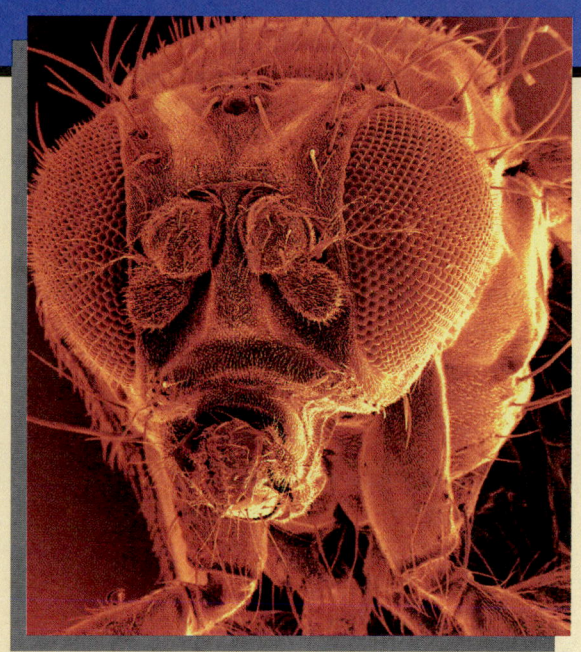

It is fall now, as I sit here writing, and nature is exhibiting its splendor. The leaves are turning the brilliant colors of autumn, the air is crisp, and it is a beautiful Indian summer day. It is time to harvest ears of Indian corn, with their multicolored kernels of red, yellow, and purple, to gather in various-shaped squash and bring them into the house to make decorative table settings in preparation for Thanksgiving. In fall the colors of nature decorate our homes and landscapes. In all this glory, the biologist's fancy naturally turns to — genetics, what else?

Just what do the differences in color and texture of Indian corn kernels have to do with genetics? The fact is that in the fall of 1983 a quiet woman named Barbara McClintock won the Nobel Prize for her genetic studies of the purplish maize, or Indian corn (see Profile). Her endeavors led to the discovery of *mobile genetic elements* (jumping genes), a discovery that some geneticists consider to be second only to the discovery that DNA is the molecular basis of the gene (Chapter 11).

We have already discussed the role of the DNA molecule in the regulation of protein synthesis. In this chapter we discuss the role of DNA in passing information from one generation to the next. Through the process called **heredity,** the basic genetic blueprint of the parents is passed on to their offspring. However, we all know that there are subtle, recognizable differences between parents and offspring. Impressed upon each individual is a unique genetic signature. These differences are called **variations,** and genetics is the study of heredity and variation.

Above: A colorized scanning electron microphotograph of a fruit fly. The fruit fly is very useful in genetic experiments.

Dr. Barbara McClintock
Geneticist and Nobel Prize Laureate

As early as 1951, while conducting research with Indian maize, Dr. Barbara McClintock proposed that genes are not fixed on chromosomes like pearls on a string but rather can move around in an unpredictable pattern. When she presented her findings to the scientific community, most geneticists disregarded her conclusions. In part the reaction was due to the time at which she made her proposal — her most important results were published before the discovery of the double helical structure of DNA, and her ideas contradicted the established genetic concepts of the day.

Dr. McClintock was born in 1902. In 1919, despite her mother's disapproval, she enrolled in the Cornell University College of Agriculture. Although she originally wanted to study plant breeding, she was not permitted to major in that specialty. Women were not allowed in that department. Reluctantly, she enrolled in botany, considered the more socially acceptable field for a woman. Her interest in plant breeding never waned, however, and she became interested in the genetics of Indian corn, or maize. She chose to study Indian corn because the changes in the colors of the corn kernels provided easily identifiable evidence of genetic patterns of inheritance.

After earning her degree, and taking on many short-term jobs, she was offered a position in the Carnegie Institute's laboratories in Cold Spring Harbor on Long Island, where she has remained until today.

During 1944 and 1945, while conducting her experiments in corn genetics, she concluded that genes were mobile and could jump from one site on a chromosome to another. However, in 1951, when she presented her findings on "jumping genes," the scientific community was not prepared to accept her reasoning. Although her concept of mobile genes was considered unacceptable, she steadfastly maintained that the evidence was as plain as the color pattern on the kernels of Indian corn.

But about ten years later, molecular geneticists using modern technology began to confirm the "jumping genes" that McClintock had discovered using the classic Mendelian methods. Today her discovery of what she calls transposable elements has an established place in science. For example, we now know that R factors (Chapter 13) can transmit antibiotic resistance from bacterium to bacterium. Researchers have also proposed that the transposable elements may play a role in transforming normal cells into cancer cells. Other theorists have proposed that transposable elements may speed up the rate of evolution.

In 1981, recognition of McClintock's long endeavors finally came. She was awarded the prestigious $15,000 Albert Lasker Basic Medical Reward, as well as $50,000 from the Wolf Foundation, and received numerous other forms of recognition. In the fall of 1983, the culmination came. The Nobel Prize committee awarded McClintock the

Nobel Prize in Medicine. The prize was one more indication of her uniqueness. She was the first woman in the 82-year history of the Nobel Prize to win the award in medicine for her independent research.

McClintock learned about her Nobel Prize while she was listening to the radio (she has no telephone). On hearing the news, she is said to have murmured, "Oh, dear." Then, dressed as usual in her baggy pants, heavy shirt, and sturdy walking shoes, she went for a walk to pick walnuts in the woods of Cold Spring Harbor Laboratories.

Barbara McClintock passed away in 1992, but her work still lives on, and many people still say that the quiet, solitary, brilliant researcher, Barbara McClintock, was truly a modern Mendel.

MENDEL AND HIS GENETIC CONCEPTS

Is it possible to apply experimental methods to a process as complex as inheritance in order to discover the underlying concepts and principles about heredity and variation? In the late nineteenth and early twentieth century, many investigators felt that the answer was no. They had attempted to unravel the mysteries of inheritance but had failed, primarily because the life forms they chose to study exhibited many traits, some of which were interacting, and their life span was often long, which made it difficult to observe a number of generations. Because of the complexity of inheritance, biologists believed they would never understand how organisms breed their own kind or how the patterns of variation are transmitted. (As a matter of fact, the term *genetics* was not used until the English biologist William Bateson coined it in 1906, and Wilhelm Johannsen introduced the term *gene* only in 1909.)

So up into the early twentieth century, progress in genetics was slow, and the work was mostly descriptive. Almost no one was aware of the work done in the 1860s by an Austrian monk named Gregor Mendel. Mendel believed that inheritance could be studied, and by applying his mathematical background to the study of inheritance, he had taken the first important steps toward an understanding of inheritance. Although he published his results in 1866, the importance of Mendel's work was not recognized until 1900 (this lag time was similar to that experienced by Barbara McClintock). Today, biologists recognize Mendel's work as the foundation of modern genetics.

Gregor Mendel was born in 1822 in what is now Czechoslovakia. At twenty-one, he entered the monastery in Brunn, Austria, where, for a while, he was relieved from his pastoral duties to serve as a temporary high school science teacher. In the summer of 1850 he took an examination to qualify as a permanent teacher; however, he failed the natural science portion of the examination (ironic, that a man who could not pass the standardized science test of his day was later to become the father of genetics).

Subsequently, in what was a turning point in his life, Mendel was sent to the University of Vienna to study the sciences. There he developed his understanding of basic mathematical concepts, as well as concepts of plant and animal breeding. His work at the university completed, Mendel returned to the monastery in what is now Brno, Czechoslovakia. In his garden there, he began experiments to investigate heredity in pea plants.

The Structure of Mendel's Work

■ **Mendel's experiments were carefully designed. He worked with pea plants because they produce large numbers of seeds, have a short life cycle, are normally self-pollinating, can be cross-pollinated easily, and exhibit a number of contrasting traits. He kept careful records and analyzed his results mathematically.**

Mendel conducted his experiments on sweet pea plants. He believed they were the ideal subject because of five major characteristics:

1. They produce large numbers of seeds.
2. They have a short life cycle.
3. They are normally self-pollinating (which would prevent accidental crossbreeding).
4. They can be easily cross-pollinated.
5. They exhibit a number of contrasting traits that are easy to observe.

There were reasons for Mendel's success. He planned his experiments carefully and studied only very obvious contrasting alternatives of a trait over several generations. Most importantly, he *quantified* (counted) and analyzed his results mathematically. Because his experiments were so clearly explained and because of the mathematical analysis of the results, others were then able to replicate his work and verify its accuracy. (There is an interesting footnote to the story of Mendel. Recently, several authors have reported that it is unlikely that Mendel actually achieved the results he reported. In fact, they feel that he may have fabricated some of his data, although not everyone agrees with their conclusion — see the Suggested Readings for this part.)

A Few Horticultural Basics

■ **Mendel grew pure-breeding plants for his experiments. He relied on techniques of self-pollination and cross-pollination to control the inherited traits he was observing.**

Before Mendel was ready to begin his plant-breeding experiments, he made sure that he was working with plants that were pure-breeding. In other words, a tall plant *always* produced offspring that were tall, and a dwarf plant *always* produced offspring that were dwarf. How did he acquire pure-breeding plants?

Self-Pollination. Mendel followed a procedure that horticulturists call **self-pollination.** (This procedure was important throughout his work.) In Mendel's time, it was known that both male and female parents contributed to their offspring and that the gametes, or sex cells, were necessary for sexual reproduction. In flowering plants, the *pollen grains* contain the male gametes (the sperm), and the *ovules* contain the female gametes (the eggs). During pollination, the sperm in the pollen fertilizes the eggs within the ovules; the seeds then form.

Just before the plants were matured and ready to go through the process of pollination, Mendel made sure that the ovules of a given plant would be pollinated only *by its own pollen.* To *self-pollinate* a plant, Mendel removed the *anthers,* which form the pollen, and brushed the pollen over the *stigma,* which is on the top of the egg-containing pistil. He then enclosed the flower in a protective covering, to be sure that the pollen from another plant did not fertilize the developing seeds. To ensure that a plant was pure-breeding, the process must be repeated over a number of generations.

Cross-Pollination. In addition to self-pollination, Mendel also used the process called **cross-pollination** in some phases of his work. To cross-pollinate, he would again remove the anthers from his plants, just before they were mature, thus preventing any possibility of self-pollination. Then, instead of brushing the pollen over the same plant (as in self-pollination), he would brush it over the stigma of a different plant, in order to see what would happen when the two different traits he was investigating were joined together in the seeds.

The Experiments: An Analysis of Families

■ **Mendel crossbred plants with the traits he wanted to follow and recorded changes that appeared in the F_1 (first filial) and F_2 (second filial) generation in order to determine patterns of inheritance.**

After growing pea plants and analyzing them carefully for several years, Mendel selected seven traits to study. For example, the seeds that the plants produced were either round or wrinkled, or the plants produced stems that were either tall or dwarf.

With his beds of pure-breeding pea plants in readiness, Mendel began to experiment to learn about patterns of inheritance. The focus of one of his experiments was to determine what patterns (if any) existed in the inheritance of the pure-breeding

trait he called size (or stem length). Mendel removed the pollen-containing anthers from tall plants (those that produced long stems) and from dwarf plants (those that produced short stems). This was the original parent generation (P) of the generations to come. He then dusted "tall pollen" onto the "dwarf stigmas," and vice versa, and covered the flowers to prevent further pollination from an uncontrolled, outside source (Figure 14.1). After the seeds had developed in the pea pod, he collected them, counted and labeled them for identification, and saved them for planting next season.

F_1 *Generation.* During the winter, as Mendel thought about these seeds, he may have formulated the hypothesis that they would produce plants of a size that was in between tall and dwarf — an intermediate or middle-sized plant. (The idea of the blending of characteristics was commonly accepted in Mendel's time.) However, to his surprise, when he planted the seeds in the spring, he observed that the plants that developed were *all tall.* It didn't make any difference if the seeds came from seeds that combined tall pollen and dwarf ovules or the other way around. What had happened to the dwarf condition in this first generation, known as the **first filial generation,** or F_1? Had the alternative dwarf condition disappeared, or had it been *dominated* by the tall condition so that it was unable to express itself?

F_2 *Generation.* To answer this question, Mendel crossed F_1 plants with other F_1 plants. Again, he collected the seeds for the next year's planting. When he planted these seeds, he found that this generation, the **second filial generation** (or F_2), produced 787 tall pea plants and 277 dwarf pea plants. Dwarfness had reappeared! Mendell calculated the ratio of tall to dwarf plants and found that it was 3:1. (To calculate that mathematical ratio yourself, set up a fraction 787/277, then reduce. Your results will be about 3:1, or 2.84:1 to be exact.)

Mendel's Concepts

■ **Once Mendel had analyzed his data, he formulated four concepts basic to heredity patterns: the concept of dominance, the concept of unit characters, the concept of segregation, and the concept of independent assortment.**

As Mendel analyzed his results, he formulated basic genetic concepts that now underlie much modern work in genetics. We discuss three of Mendel's con-

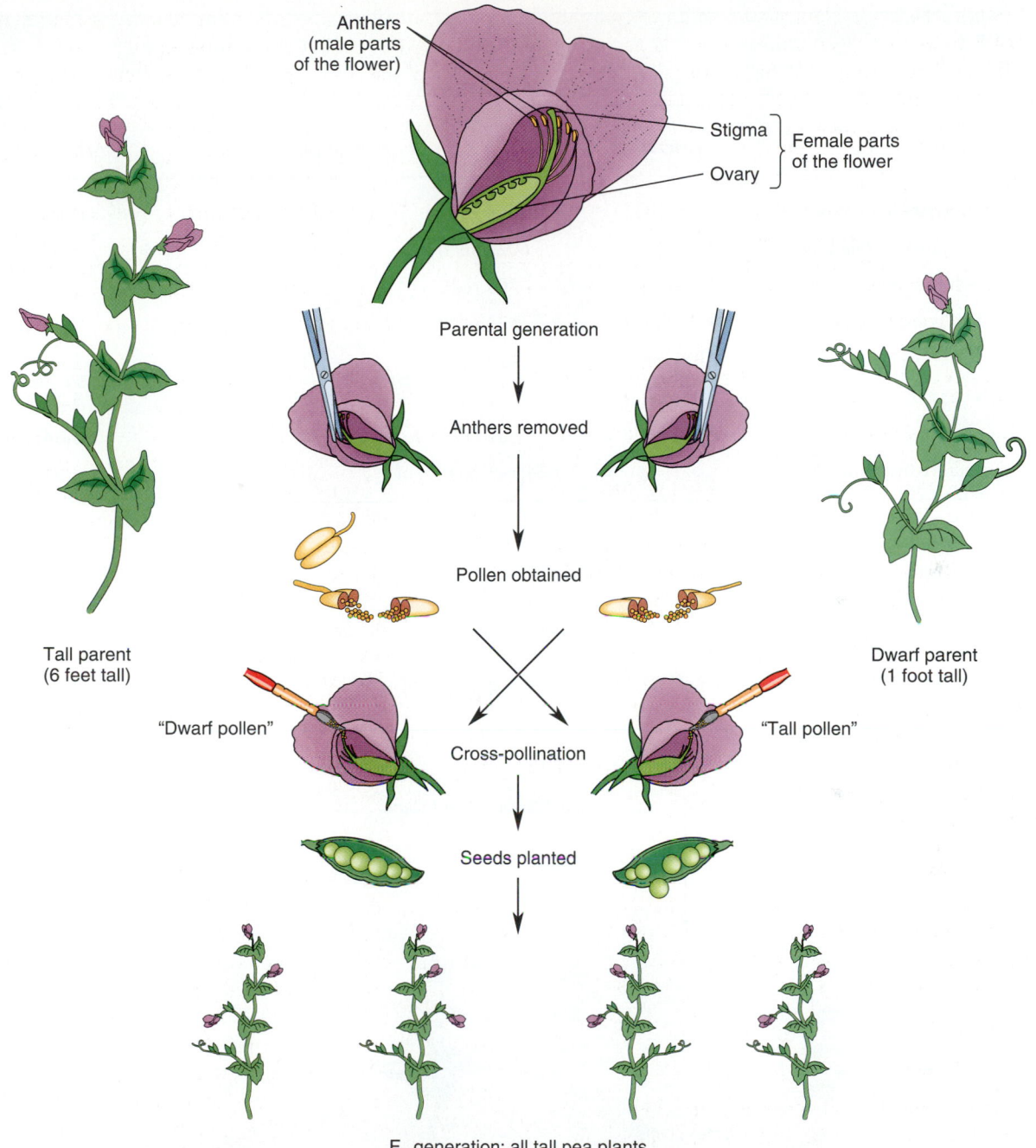

Anthers
(male parts
of the flower)

Stigma

Ovary

Female parts
of the flower

Parental generation

Anthers removed

Pollen obtained

Tall parent
(6 feet tall)

Dwarf parent
(1 foot tall)

"Dwarf pollen"

"Tall pollen"

Cross-pollination

Seeds planted

F_1 generation: all tall pea plants

FIGURE 14.1 Cross-Pollination
Mendel used this process to transfer pollen from the flower of one type of
plant to the flower of another type. As you can see in this example, all the
first-generation offspring were tall—no matter whether the pollen came from
the tall or dwarf parent.

cepts here and will discuss the concept of independent assortment later in this chapter.

The Concept of Dominance. On the basis of his
results, Mendel reasoned that the condition for

dwarfness had not, in fact, disappeared in the F_1
generation, since it had obviously been able to reex-
press itself in the F_2. The trait that was masked, or
prevented from expressing itself, he called **recessive,**
whereas the condition that masked the other he

TABLE 14.1
Mendel's Results

Parental cross	F₁ generation	F₂ generation		Actual ratio
Round × wrinkled seeds	All round	5474 round	1850 wrinkled	2.96:1
Yellow × green cotyledons	All yellow	6022 yellow	2001 green	3.01:1
Red × white flowers	All red	705 red	224 white	3.15:1
Inflated × constricted pods	All inflated	822 inflated	299 constricted	2.75:1
Green × yellow pods	All green	428 green	152 yellow	2.82:1
Axial × terminal flowers	All axial	651 axial	207 terminal	3.14:1
Tall × dwarf stems	All tall	787 tall	277 dwarf	2.84:1
All characteristics combined		14,949 dominant	5010 recessive	2.98:1

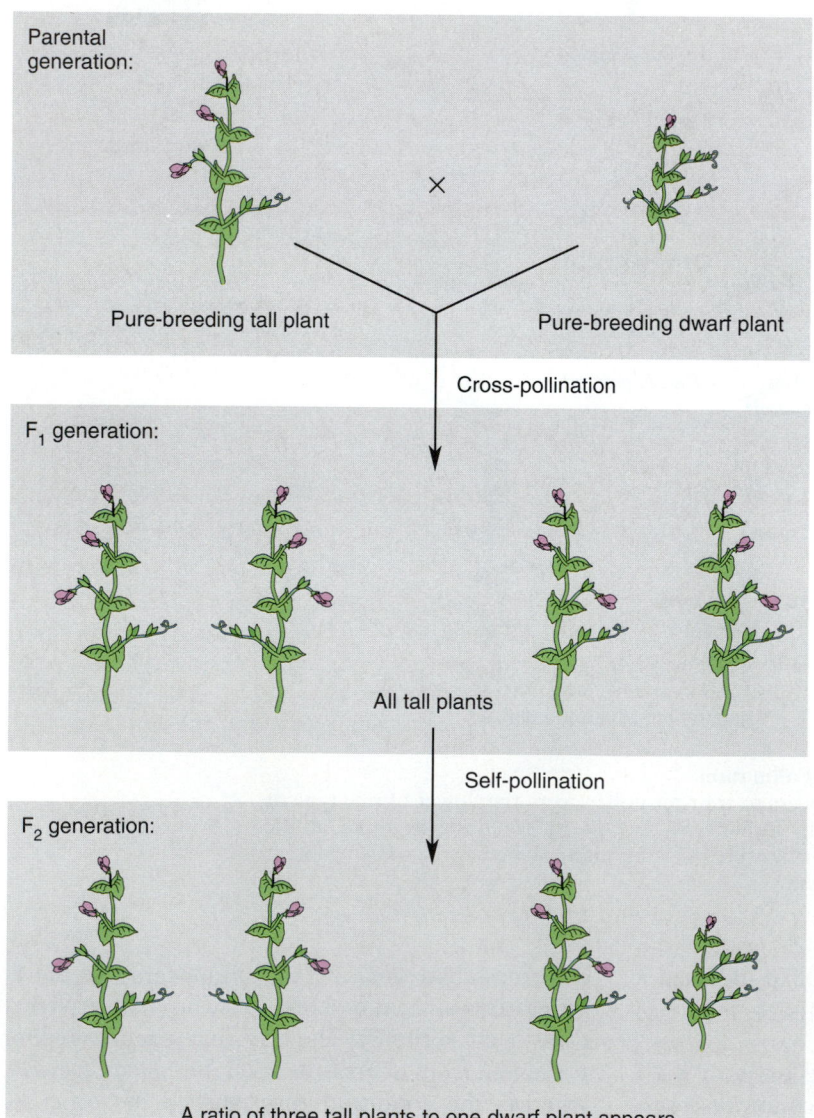

Parental generation:

×

Pure-breeding tall plant   Pure-breeding dwarf plant

Cross-pollination

F₁ generation:

All tall plants

Self-pollination

F₂ generation:

A ratio of three tall plants to one dwarf plant appears.

FIGURE 14.2 The Concept of Dominance
In this example, Mendel observed that one alternative for a specific trait (in this case tallness) dominated and masked the expression of the other alternative (in this case dwarfness). However, this recessive trait reappeared in the F₂ generation, in the ratio of three dominant to one recessive.

Chapter 14 Genetics: Patterns of Inheritance

called **dominant.** Moreover, as Table 14.1 shows, when Mendel analyzed the results from other experiments, he found that the recessive condition always reappeared in the F_2 generation in a ratio of about 3:1.

From the results of this experiment and others just like it, Mendel formulated the **concept of dominance** (Figure 14.2). This concept states that some genetic *factors* (Mendel called them factors because the term *gene* did not exist at that time) can mask the expression of others. In genetic shorthand an *uppercase letter* indicates the dominant trait and a *lowercase letter* indicates the recessive trait. So *T* symbolizes the dominant characteristic of tallness, and *t* indicates the recessive trait of dwarfness.

The Concept of Unit Characters. How could recessive conditions disappear for a generation, only to reappear in a second generation — and always in a constant pattern of 3:1? As he interpreted these data, Mendel came to the conclusion that hereditary characteristics were controlled by discrete, or separate, factors (genes) that occur in *pairs,* one factor from each parent. This conclusion is called the **concept of unit characters.**

The Concept of Segregation. It was the process of segregation, which occurred during the formation of gametes, that gave Mendel's hypothesis about inheritance its predictability. The **concept of segregation** states that contrasting factors for a given trait in an individual segregate (separate) when gametes (sex cells) are formed, and any single gamete receives only one factor for the trait.

MENDEL'S CONCEPTS IN MODERN TERMS

Mendel's conclusions were based on his observations. He did not know the actual mechanisms at work in the processes he was recording, but because of the research that illuminated the role of DNA and its functions, we can discuss the genetic mechanisms behind Mendel's discoveries. (For a convenient listing of the specialized terminology throughout this chapter, refer to the genetics glossary in Table 14.2 and Figure 14.3.)

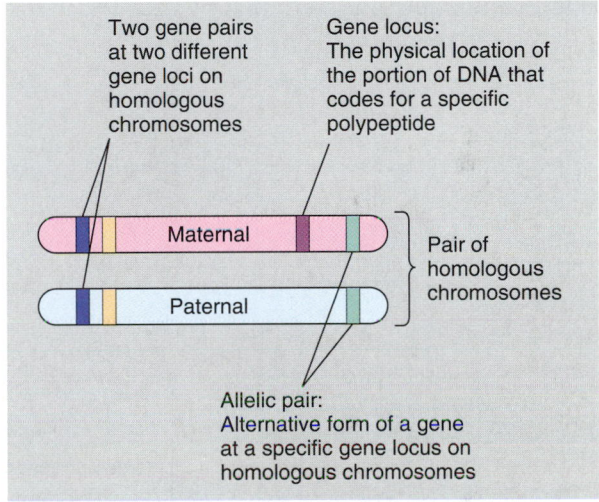

FIGURE 14.3 A Representation of Some Genetic Terms In this pair of homologous chromosomes, gene locus, alleles, and gene pairs are defined.

TABLE 14.2
The Vocabulary of Genetics: A Simple Glossary

Allele	One or more alternative forms of a gene at specific gene locus on homologous chromosomes
Dihybrid cross	A cross between organisms in which the experimenter is interested in only two traits
Dominant allele	An allele that masks the expression of the other allele in a gene pair
Gene	A specific segment of DNA—the physical basis of heredity
Genotype	The genetic makeup of an organism
Heterozygote	An organism in which the paired alleles for a given gene are different
Homozygote	An organism in which the paired alleles are the same
Locus	A particular location of a gene on a chromosome
Monohybrid cross	A cross between individuals in which the experimenter is only interested in one trait
Phenotype	The observable physical appearance of the organism, which is determined by the genetic makeup of the organism and influenced by the environment
Recessive allele	An allele whose expression is masked when paired with a dominant allele

The Activity of Genes in Mendel's Experiments

■ **The contrasting characteristics of a single trait in Mendel's peas are now called alleles of a certain gene. An organism in which both alleles for a trait are identical is called homozygous, while an organism in which the alleles are not identical is called heterozygous. The way in which the gene expresses itself in the organism is called its phenotype; the actual genetic makeup of the organism is called its genotype.**

When Mendel crossed tall and dwarf plants, he was interested in only one trait, that of size. Today, biologists would say that he was undertaking a **monohybrid cross** (involving one trait). To follow Mendel's reasoning, recall the discussion of *diploid* and *haploid* phases in the life cycles of organisms that reproduce sexually (see Chapter 10). In the *diploid* stage, the cell of an organism contains pairs of genes, one copy from each parent. In the *haploid* stage, as during the formation of sex cells (or gametes), the cells have undergone meiosis, so only one gene of each pair is present.

Alternative forms of a gene are called **alleles.** For example, for the trait size in pea plants, one alternative form of the gene expresses tallness (T) and the other form expresses dwarfness (t). Therefore, geneticists talk about an allele for tallness and an allele for dwarfness.

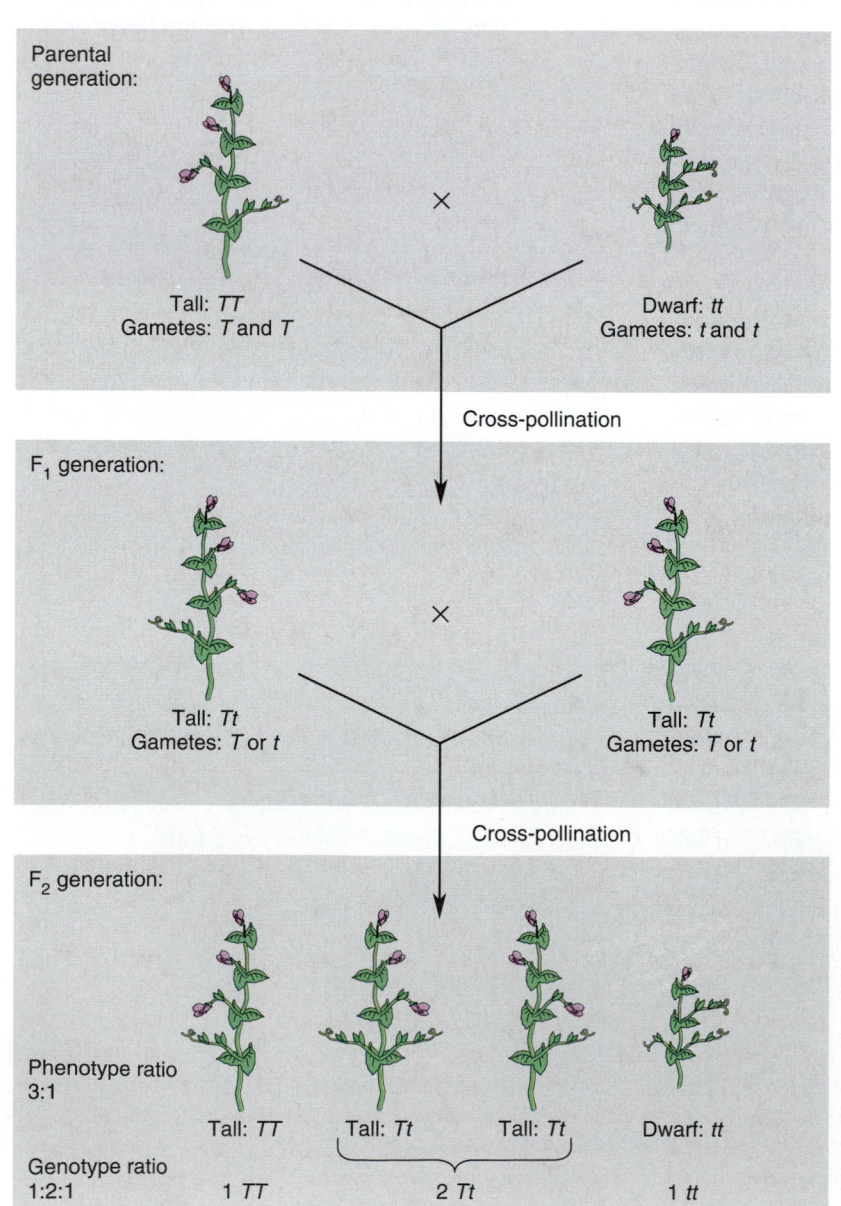

Parental generation:

Tall: TT
Gametes: T and T

Dwarf: tt
Gametes: t and t

Cross-pollination

F_1 generation:

Tall: Tt
Gametes: T or t

Tall: Tt
Gametes: T or t

Cross-pollination

F_2 generation:

Phenotype ratio
3:1

Tall: TT　　　Tall: Tt　　　Tall: Tt　　　Dwarf: tt

Genotype ratio
1:2:1

1 TT　　　　2 Tt　　　　1 tt

FIGURE 14.4 Monohybrid Cross
A modern interpretation of Mendel's monohybrid cross includes the genotype of each parent and the allele(s) each contributes.

Mendel's experiments began with pure-breeding plants. In cases where the alleles of a gene pair are the same, the individual is said to be **homozygous.** In Mendel's experiment, the parents were homozygous tall (TT) and homozygous dwarf (tt). (Follow Figure 14.4 as we discuss the allelic explanation of Mendel's experiments.) When Mendel crossed the homozygous tall (TT) and homozygous dwarf (tt) parents, the gametes joined and formed a zygote — the first cell of an individual. In this case, the zygote received one allele for T and one allele for t. Hence it was **heterozygous**; in other words, the two alleles in the gene pair for size were different (Tt). Each allele regulated for a contrasting condition for the trait size. The plant grown from these seeds would be heterozygous tall (Tt) because the allele for tallness (indicated with an uppercase T) is dominant over the allele for dwarfness (indicated with a lowercase t).

The terms homozygous and heterozygous describe the genetic makeup of an organism, or its **genotype.** However, both the parent plants and the F_1 plants appear tall. The outward expression of a genotype is called the **phenotype** (see Focus 14.1). The phenotype of an organism does not necessarily reveal its genotype. Even though a plant may look tall (its phenotype), it may not be homozygous tall; its genotype may be actually Tt, or heterozygous.

Thus, knowing the phenotype of an organism will not necessarily tell you the genotype. This was true with Mendel's F_1 generation, which produced heterozygous plants. They looked tall only because the allele for tallness dominated the allele for dwarfness. Then he crossed two heterozygous F_1 plants and observed that the phenotypic ratio in the F_2 generation was approximately three tall to one dwarf. But what were the genotypes of those seeds?

Segregation of Alleles. First, remember that the genotype of the F_1 generation is Tt. The sperm within the pollen of each plant will have either a T allele or a t allele to contribute to the next generation. The eggs within the ovules will have either a T or a t to contribute. What combinations are possible in this case? If the sperm within the pollen contains the T allele and the egg also has the T allele, the combination will be TT, and the plant will be tall. If the sperm has the T allele and the egg has the t allele, the combination will be Tt, and again the plant will be tall. If the sperm contains a t allele, then what are the possible combinations? If the egg is a T allele, the combination is Tt, and the plant will be tall. If, however, the egg has a t allele, the combination is tt, and the plant will be dwarf.

That means that there are three possible genotypes in the F_2 generation: TT, Tt, and tt. What phenotypes would these combinations give rise to? There would be three tall plants and one dwarf plant. If you look back to check this answer with Mendel's observations, you will see that the ratio of his F_2 generation was 3:1 — three tall plants to one dwarf plant. Mendel was noting phenotypes. Today, biologists know that the ratio of the genotypes produced in the F_2 generation was one homozygous dominant (TT) to two heterozygous (Tt) to one homozygous recessive (tt), or 1:2:1.

We have covered a lot of vocabulary in a short space; so before we continue, let's work through a simple exercise involving the inheritance of flower color. Mendel's peas were either yellow (Y) or green (y). Were the pure-breeding parent plants *heterozygous* or *homozygous?* They were homozygous — either YY or yy. Were the F_1 plants *heterozygous* or *homozygous?* They were heterozygous — Yy. What was the *genotype* of the parent plants? They were homozygous — either YY or yy. What was the *genotype* of the F_1 plants? They were heterozygous — Yy. What was the *phenotype* of the parent plants? They were either yellow or green. What was the *phenotype* of the F_1 plants? They were yellow.

Biologists also know that any monohybrid cross in which complete dominance is a factor produces the pattern of inheritance that Mendel described. This is true not only in peas but in all diploid organisms. Therefore, Mendel's concepts allow us to predict the probability that a certain genetic combination will occur in the offspring of any diploid parents.

Punnett Square. There is a simple method for keeping track of the possible combinations that can result in the offspring of two parents. The tool is a checkerboard known as a **Punnett square,** named after the geneticist (yes, you guessed it) R. C. Punnett. Typically, the symbols for the alleles of the male gametes are indicated down the side rows of the checkerboard, one by one. The possible female gametes are arranged across the top of the columns of the checkerboard. The genotypes of the next generation fill the squares of the checkerboard.

Let us use the trait of size in Mendel's peas as an example. If we cross two heterozygous parents (Tt) there is a 50 percent chance (or ½) that the sperm carries the allele for tallness (T) and a 50 percent chance (or ½) that it carries the allele for dwarfness (t). The probabilities in the eggs are the same. So along the side of the square, one row is marked T and the other is marked t. The same happens across the top of the box: One column is labeled T and the other t.

FOCUS 14.1
From Gene to Phenotype: An Overview

Sometimes it is difficult for us to "see the forest for the trees." The concept is lost in the details. We have been discussing DNA, mRNA, genes, chromosomes, Mendel, patterns of inheritance, phenotype, and many other terms and concepts. But how can we summarize or provide an overview of the complex pathway from the DNA (gene) to its effect—the phenotype of the organism and eventually the phenotype expressed in the population (a population consists of members of the same species)? The summary illustration in Figure A helps us to understand the series of interactions that result in the phenotype of the organism.

As we can see, the DNA (gene) message is transcribed onto the mRNA, which is translated at the ribosome into a polypeptide (protein). Numerous regulating mechanisms can affect transcription or translation. Some are biochemical; others are environmental, like temperature or light. Hormones and other chemicals, the environment, and gene products can regulate at the transcriptional or translational level. Cells also interact with each other and the tissues. All these interactions result in the phenotype. In addition, the organism itself has an effect on other organisms, and all have an effect on the phenotype. For example, an organism may have the genetic

potential to be large and heavy, but if it is in an overcrowded environment with a limited food supply, its phenotype is affected and it is thin and small rather than heavy and large.

So, as we have discussed, there is a flow of events that lead from the gene to its final expression in what we refer to as the phenotype. This illustration should help you to conceptualize the numerous interactions involved in the flow of information and activities that result in the appearance of the organism, its phenotype.

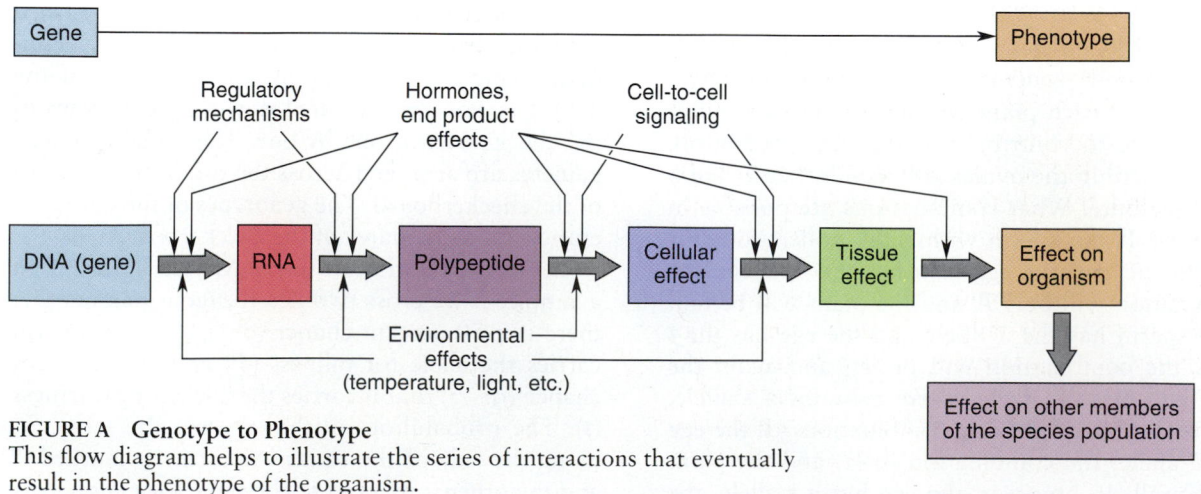

FIGURE A   Genotype to Phenotype
This flow diagram helps to illustrate the series of interactions that eventually result in the phenotype of the organism.

Now fill in each square of the checkerboard with the appropriate allele that occurs in the horizontal rows and vertical columns of the square. For example, when the T sperm fertilizes the T egg, the resulting zygote will be TT. Once the squares of the checkerboard are filled in, you can list the possible combinations. They will correspond to Mendel's results: 1 TT:2 Tt:1 tt.

Eggs

	T	t
T	TT	Tt
t	Tt	tt

Sperm

TT = homozygous tall
Tt = heterozygous tall
tt = homozygous dwarf

Now what would happen if we crossed some of the F_2 plants? A Punnett square will help you visualize the combination. For instance, the square below presents a cross between a heterozygous tall (Tt) and a homozygous dwarf (tt) plant.

Eggs

	t	t
T	Tt	Tt
t	tt	tt

Sperm

Tt = heterozygous tall
tt = homozygous dwarf

As you can see, the probabilities for this cross are 50 percent heterozygous tall (Tt) and 50 percent short (tt).

Probability and Genetics

■ **The mathematical concept called the product rule of probability can be applied to a genetic cross to predict the combination of events that may result.**

When Mendel worked with mathematics in his experiments, he was simply applying the *laws of probability* to genetics. For example, when you flip a coin, what is the chance (or probability) of its landing heads up? What is the probability of its landing tails up? The answer is that there is a 50 percent chance of heads and a 50 percent chance of tails. Keep in mind, though, that if you flip that coin only four times, you might not get exactly two heads and two tails. However, if you flipped that coin 1000 times, you probably would get close to the expected 500 heads and 500 tails. You must use a large enough *number of examples*, or sample size, to get the expected results. If Mendel had planted only four seeds instead of the hundreds that he did, he probably would not have reached his significant conclusions.

The principle illustrated by coin tossing works the same way in a heterozygous individual: The probability that the dominant allele will segregate into a gamete is 50 percent, and the same probability holds true for the recessive allele.

Now, consider what would happen if you flipped two coins simultaneously. What is the probability that they would both land heads? What is the probability that one would be heads and the other tails? What is the probability that the reverse would happen or that both would land tails? There is a mathematical rule called the product rule of probability that can help you figure out the odds for each of these combinations. The rule states that the probability of the combination of two or more independent events occurring is the product of their separate probabilities. (Independent events are events whose outcomes do not affect each other. For example, how the first of two coins lands does not affect how the second coin lands.)

To determine the probability of tossing two heads, look back at the probability of tossing heads with only one coin; it was 50 percent, or ½. Here we have two coins, each with the probability of 50 percent that it will land heads up. So to determine the probability of tossing both coins heads up, you multiply ½ times ½, which gives you a probability of ¼ (Figure 14.5).

Let's look at it this way:

Let H = heads and T = tails.

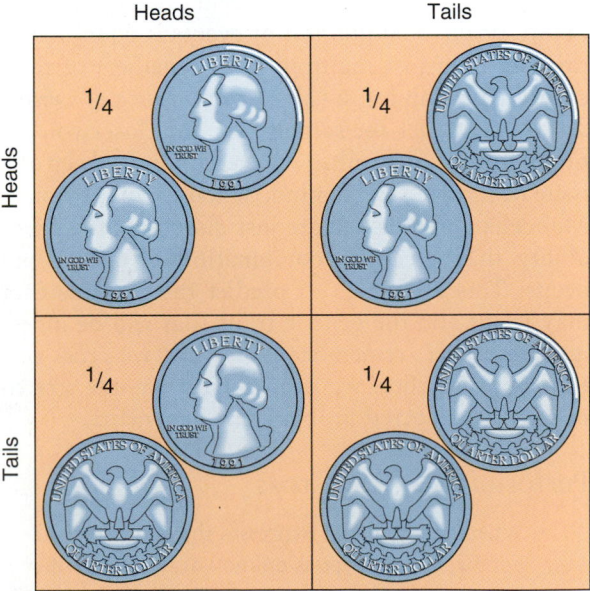

	Heads	Tails
Heads	¼	¼
Tails	¼	¼

FIGURE 14.5 The Product Rule of Probability
Here the product rule of probability is applied to two independent events: the flip of two coins together.

For coin *A:*
½ *H* + ½ *T* = 1 (the total possibilities)

For coin *B:*
½ *H* + ½ *T* = 1 (the total possibilities)

For coins *A* and *B* to be *H:*
½ *AH* × ½ *BH* = ¼ *HH*

For coins *A* and *B* to be *T:*
½ *AT* × ½ *BT* = ¼ *TT*

For coin *A* to be heads and coin *B* to be tails:
½ *AH* × ½ *BT* = ¼ *TH*

For coin *A* to be tails and coin *B* to be heads:
½ *AT* × ½ *BH* = ¼ *HT*

¼ *HH* + ¼ *TT* + ¼ *TH* + ¼ *HT* = 1

Those of you who have had algebra will recognize that the mathematics involved here is the binomial expansion of $(½ H + ½ T)^2 = ¼ HH + ¼ HT + ¼ TH + ¼ TT = 1$.

What would happen if you were tossing three coins? Then the probability of the combination of any particular combination would be ½ × ½ × ½, or ⅛. With four coins, the probability of any particular event is ½ × ½ × ½ × ½, or ¹⁄₁₆, and so on. Although these ratios may not be apparent if you toss the coins only a few times, they will become clear with enough attempts. As long as the sample size is sufficiently large, the ratio will follow the mathematical breakdown.

To summarize, we have learned two important principles of applying mathematics to genetics: (1) the *product rule of probability* — whenever you combine two or more independent events, the probability of the combination of events is equal to the product (you just multiply them together) of their separate probabilities; (2) the *importance of sample size* — the larger the sample size, the closer you get to the expected results. As we continue, keep the issue of sample size in mind. When we make simple Mendelian crosses, we are just discussing the probability that a certain combination of events will occur. This allows us to predict the patterns that may result and the genetic traits that will be inherited. However, remember that when we cross one trait with another, we need a large sample size to obtain the expected probabilities in genetics.

Performing a Test Cross

■ **When an individual expresses the dominant phenotype, its genotype is unknown; it can be either heterozygous or homozygous. To determine the genotype, geneticists perform a test cross. An individual of unknown genotype is crossed with an individual having a recessive phenotype, because to express the recessive trait, the individual must be homozygous. The results of the test cross determine the genotype of the unknown individual.**

An individual that possesses the dominant phenotype could be homozygous dominant, or it could be heterozygous, with the dominant allele overriding the recessive allele. That is what happened in the F_1 generation of Mendel's pea plants. In appearance these heterozygous tall plants did not differ from a tall homozygous plant, because tall was the dominant trait. How can we determine if a plant with a tall phenotype is homozygous or is heterozygous?

To determine the genotype of such an organism, biologists perform what is known as a **test cross** — a cross between an individual expressing the dominant condition and a homozygous recessive individual to determine if the dominant individual is heterozygous or homozygous for a particular trait. Now, we have already learned that a plant exhibiting the dominant trait has an uncertain genotype. However, the genotype of a plant exhibiting a recessive trait (such as dwarfness) is certain. It must be *tt*. Otherwise the recessive trait would be masked by the dominant trait, and the plant would grow tall.

So a test cross requires two elements: (1) an organism of dominant phenotype but unknown genotype and (2) another organism having a recessive phenotype, which is therefore homozygous recessive. When the two organisms are cross-fertilized, there are two possible outcomes (Figure 14.6). If all the offspring exhibit the dominant trait, the genotype of the unknown organism had to be homozygous dominant, since the allele for the dominant trait was the only one available. If the offspring exhibit a ratio of about ½ dominant phenotype and ½ recessive phenotype, the organism in question has to be heterozygous. Only in this way could the recessive trait express itself in the offspring. In other words, an offspring expressing the recessive trait must inherit a recessive allele from both parents.

What if you were given a white ram (a male sheep) and asked to develop a strain of pure-breeding white sheep? You were also told that in sheep, white coat (*W*) is dominant to black coat (*w*). How could you determine whether the ram was homozygous (*WW*) or heterozygous (*Ww*)? Given what you have learned so far about test crosses, you would mate your ram with several female sheep (ewes) who were black, because you know that ewes with that coloring would be homozygous recessive (*ww*). Assume your ewes gave birth to 11 white lambs and 2 black lambs. What would you now know about the genotype of the ram? You

FIGURE 14.6   Test Cross
(a) When an organism of the dominant phenotype but unknown genotype is crossed with the homozygous recessive and all the offspring express the dominant phenotype, the dominant partner must have been homozygous for the condition. (b) If offspring with the recessive phenotype appear, then the unknown parent had to be heterozygous for the trait.

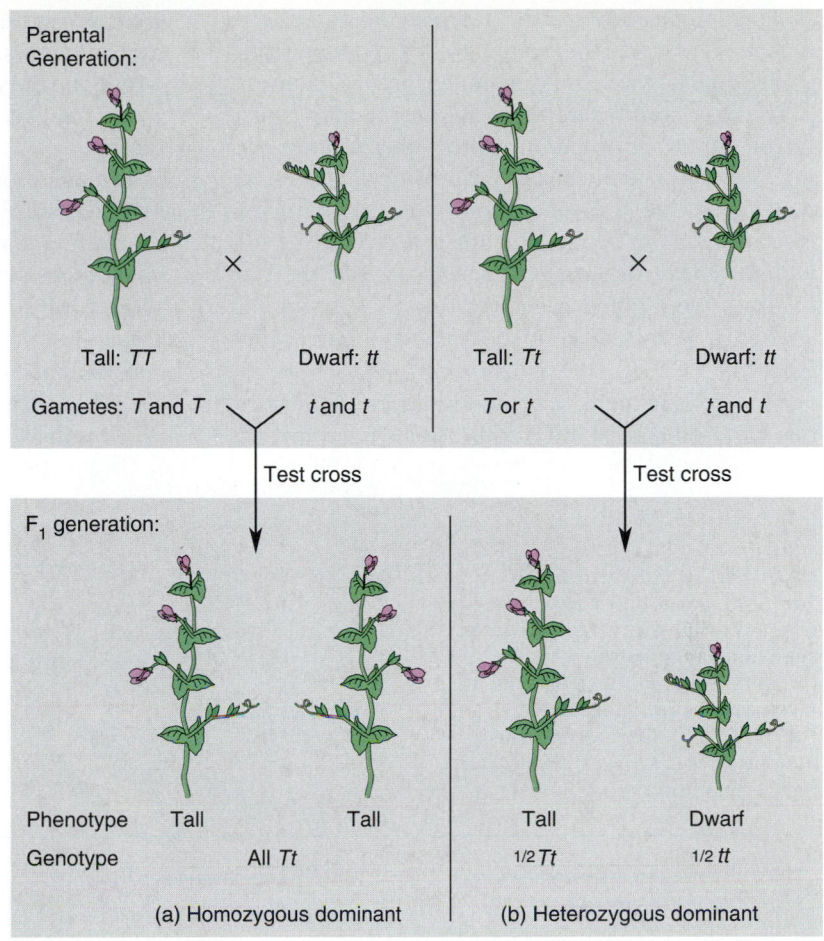

Parental Generation:

Tall: *TT*   Dwarf: *tt*   Tall: *Tt*   Dwarf: *tt*

Gametes: *T* and *T*   *t* and *t*   *T* or *t*   *t* and *t*

Test cross   Test cross

F₁ generation:

Phenotype   Tall   Tall   Tall   Dwarf
Genotype   All *Tt*   ½ *Tt*   ½ *tt*

(a) Homozygous dominant   (b) Heterozygous dominant

---

would know that he is heterozygous, because that's the only way he could sire a black lamb. Otherwise the recessive allele for black coloring would be masked by the dominant allele for white coloring. Query: Will you be able to complete your task? *

Eggs

		w	*w*
Sperm	*W*	*Ww*	*Ww*
	w	*ww*	*ww*

MENDEL'S CONCEPT OF INDEPENDENT ASSORTMENT

■ **Investigating the inheritance pattern of two traits in a single mating is called a dihybrid cross. When the genes for these traits are on two separate chromosomes, the genes will assort independently of each other.**

---

* Not with this ram.

So far, the experiments of Mendel that we have discussed involved only one trait (monohybrid). Mendel also made crosses that involved the inheritance patterns of two traits in a single mating — a **dihybrid cross.** Does the inheritance of one trait affect that of another?

In monohybrid crosses, Mendel had already determined that round seeds (*R*) were dominant over wrinkled seeds (*r*) and that yellow (*Y*) seeds were dominant over green (*y*). What would happen if a plant homozygous for both round seeds *and* yellow seed color (*RRYY*) was crossed with a plant homozygous for both wrinkled seeds and green seed color (*rryy*)? Would the F₁ offspring all have round, yellow seed, or would there be degrees of roundness and yellowness?

Mendel performed the cross and found that all the F₁ plants exhibited round, yellow seeds, indicating that round seeds and yellow coloring were both dominant traits. The pattern followed that of a monohybrid cross, indicating that F₁ generation plants were heterozygous both for seed shape (*Rr*) and for seed color (*Yy*).

Mendel was now ready to cross-pollinate the F₁ plants. What was he trying to learn? Remember that Mendel was thinking in terms of "factors." This cross could determine if there was a single factor for *seeds,* including both their shape and their color (for convenience, we will call this factor Super S). If Super S determined the two characteristics of seed appearance (shape and color), Super S would behave like a single entity — as if the two qualities were glued together. Round seeds would *always* be yellow. Wrinkled seeds would *always* be green.

However, if there were two independent factors, one for seed shape and another for seed color, the seeds produced in the F₂ would display a mixture of characteristics. Seeds that were yellow could also be wrinkled, thus combining a dominant and a recessive trait. Seeds that were round could also be green.

Mendel crossed the heterozygous round and yellow (*RrYy*) plants and harvested the seeds. When he planted the F₂ seeds, the plants exhibited four phenotypes: round yellow (315 plants), round green (108), wrinkled yellow (101), and wrinkled green (32). The phenotypes demonstrated a ratio of approximately 9:3:3:1. The Punnett square in Figure 14.7 illustrates the combinations that occurred as a result of the dihybrid cross, showing the way

FIGURE 14.7   Dihybrid Cross
The results of round yellow pea plants crossed with wrinkled green pea plants demonstrate in the F₁ generation that all the offspring are heterozygous round and yellow. When these F₁ plants are crossed, the phenotypic ration is 9:3:3:1 in the F₂ generation.

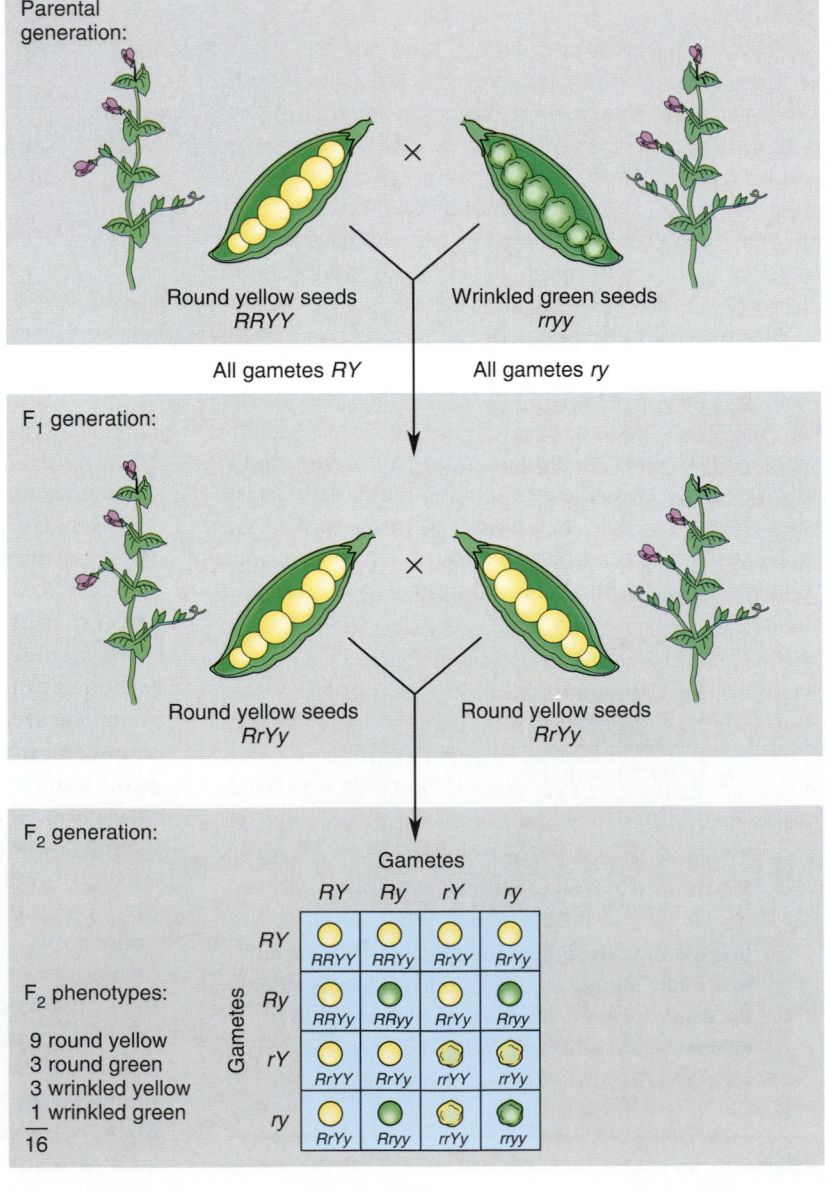

in which Mendel's factors separated when the heterozygous F_1 plants were crossed to form the F_2 seeds.

When you examine the results of the F_2 planting, you can see two phenomena: (1) there are seeds that are homozygous recessive; (2) there are two phenotypic combinations that are not found in any of the earlier generations — wrinkled yellow seeds (*rrYy*) and round green seeds (*Rryy*). What are the implications of Mendel's results?

Mendel's results showed that there was no Super S factor that regulated seed size. Instead, there were two factors — one for shape and one for color — acting independently as unrelated characteristics. Based on the results of his dihybrid experiments, Mendel formulated the concept of independent assortment, which states that factors for one trait segregate independently of the factors for another trait.

Given what we now know about the mechanisms of genetics, what does independent assortment tell us about the chromosomes of Mendel's pea plants? We know that the homologous chromosomes that contain the genes separate during meiosis. As a result, it follows that the genes that regulate seed shape and seed color cannot be located on the same chromosome. Because they separate independently of each other, they must be located on different chromosomes.

The modern interpretation of Mendel's **concept of independent assortment** is as follows: During meiosis, allelic genes from gene pairs on one pair of homologous chromosomes segregate independently from allelic genes on other pairs of chromosomes. In other words, they separate independently.

For practice, draw a Punnett square that will illustrate this cross for traits that assort independently. You have two plants, one that is *RrYy* and one that is *rryy*. What would be the possible offspring, and what would be the ratio? (When you work through the possibilities, the results would reveal these percentages: 25 percent round yellow (*RrYy*); 25 percent round green (*Rryy*); 25 percent wrinkled yellow (*rrYy*); and 25 percent wrinkled green (*rryy*).

In summary, Mendel provided us with the basis for four important concepts in genetics:

The concept of dominance

The concept of unit characters

The concept of segregation

The concept of independent assortment

Perhaps the most impressive feature of Mendel's achievement is that he came to his conclusions without knowing the cytological nature of chromosomes, or the molecular basis of genetics. Mendel's insight and his ability to interpret his results were remarkable.

GENETICS AFTER MENDEL

It is amazing how far the field of genetics has progressed since Mendel's work in the 1860s. As biologists have formulated new concepts, there has been a modification of Mendel's basic tenets. We will discuss some of these modifications in the following section.

Incomplete Dominance

■ **Not all the alternative alleles for a trait are completely dominant or recessive. Instead, the dominance may be incomplete.**

As we mentioned earlier, you would think that when Mendel crossed a tall-stemmed and a dwarf-stemmed pea plant, the results in the F_1 generation would have been medium-stemmed plants — a blending of the traits of both parents. In other words, it would seem that the phenotype of the heterozygous plant should fall somewhere between the two contrasting traits of the parents. Neither trait would dominate the other.

Although it was not the case with Mendel's pea plants, **incomplete dominance,** a partial expression of traits rather than dominance of one over the other, does occur. For example, in the flowers of snapdragons there are three phenotypes: red, pink, and white (Figure 14.8). When you cross a red snapdragon (*RR*) and a white snapdragon (*R'R'*), the heterozygous F_1 generation (*RR'*) has pink flowers, indicating a partial expression of the red and the white. If you cross two F_1 pink snapdragons, the offspring in the F_2 generation has a ratio of 1 red: 2 pink: 1 white. In cases like this, the phenotype is a clear indication of the genotype. The Punnett square below indicates the pairing that takes place.

Eggs

Sperm	R	R'	
R	RR	RR'	RR = red
R'	RR'	R'R'	RR' = pink
			R'R' = white

In the heterozygous individual, the incompletely dominant red allele cannot provide enough genetic instruction for the synthesis of sufficient amounts

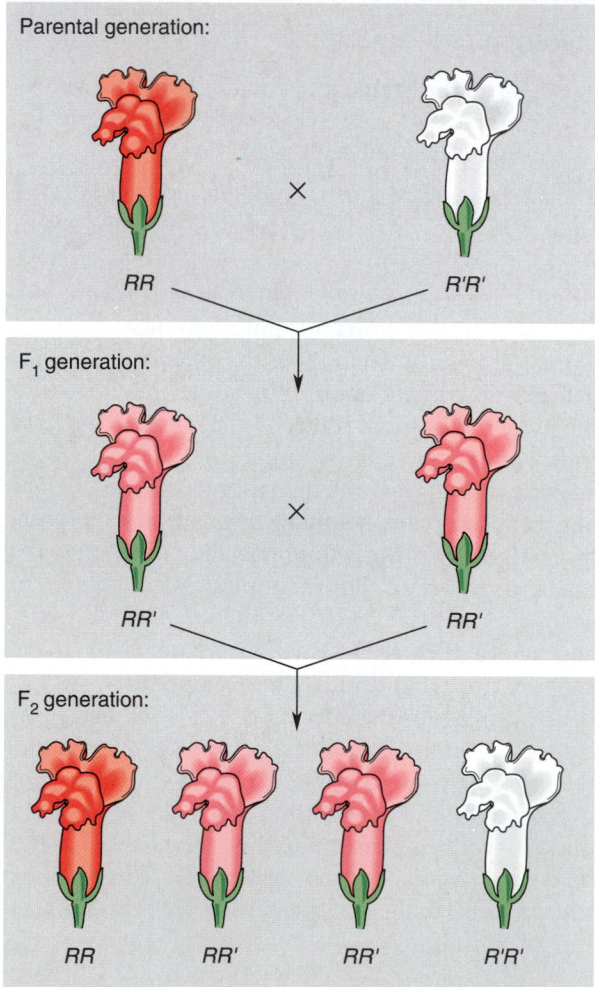

Parental generation:

RR × R'R'

F₁ generation:

RR' × RR'

F₂ generation:

RR   RR'   RR'   R'R'

FIGURE 14.8   Incomplete Dominance
In snapdragons, the heterozygous condition results in a pink phenotype, which is intermediate between the two alternatives of the trait. Neither allele is dominant. *R* represents the allele for red and *R'* represents the allele for white.

of red pigment to result in overall red coloring. As a result, both alleles are expressed, and the pigment color is partial rather than total. An inheritance pattern such as this is called *incomplete* or *partial dominance* or *blended inheritance,* and it often occurs in plants, animals, and humans.

Pleiotropy: The Effect of a Single Gene on Many Traits

■ **Pleiotropy occurs when an individual gene affects more than one trait in an individual.**

Mendel assumed that each gene affected one single trait, but today we know that an organism's phenotype is the result of the metabolism of many substances, as well as several biochemical reactions that result in color production, size, shape, and so on. In other words, the phenotype is an end result of the central dogma (see Chapter 12) — one gene codes for a particular protein. If that protein is an enzyme that controls one step in a metabolic pathway involving more than one system or structure, changes in that protein can affect many traits. When a gene can affect a phenotype in more than one way, the condition is called pleiotropy.

Pleiotropy is the ability of a gene to produce a variety of phenotypic effects. To see the implications of pleiotropy, let us look at our sickle-cell patient again. In Chapter 12 we discussed the difference between the hemoglobin of a sickle-cell anemia patient (Hb^S) and the hemoglobin of a normal adult (Hb^A). That difference is the result of the action of two different alleles for that beta chain. In a non-sickle-cell patient, an allele codes for glutamic acid in the sixth position of the beta chain. In the sickle-cell patient, the allele codes for valine. In sickle-cell anemia the alteration in one gene results in the sickle shape of the red blood cell, which in turn causes a variety of effects, including damage to the heart, kidneys, brain, and lungs. For a relatively complete listing of the effects of this disease, see Figure 14.9.

Polygenic Inheritance: Genes Acting Together

■ **The expression of most traits in living organisms is not a matter of either/or. Instead the trait expressed manifests itself in a continuum of expression from one extreme to the other, with the preponderance of examples clustered between the two extremes. Continuous-variation inheritance patterns are the result of several genes acting together. The pattern is called polygenic inheritance.**

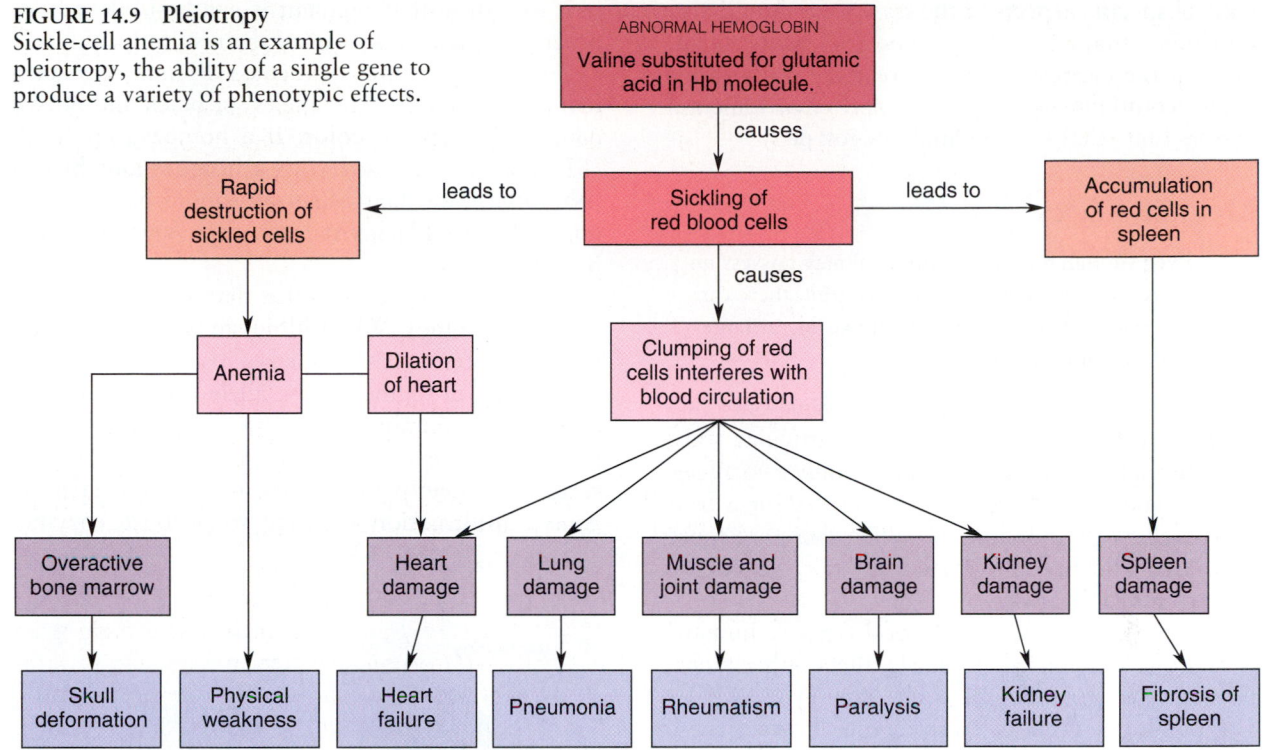

FIGURE 14.9 Pleiotropy
Sickle-cell anemia is an example of pleiotropy, the ability of a single gene to produce a variety of phenotypic effects.

When you look at the various traits exhibited by organisms in the living world, you notice that many traits exhibit what geneticists call **continuous variation**. In other words, the individuals of a group exhibit differences, some showing one extreme, some the other, with the majority of the individual organisms clustered around an average, forming a bell-shaped curve typical of *normal distribution* (Figure 14.10). Inheritance patterns like this are caused by **polygenic inheritance**, in which the interaction of several genes produces a range of variation for a given trait. Many conditions in humans and other organisms (such as eye color, skin color, and fruit size) are due to polygenic inheritance.

The bell-shaped distribution of continuously varying traits is due to the effect of many genes that

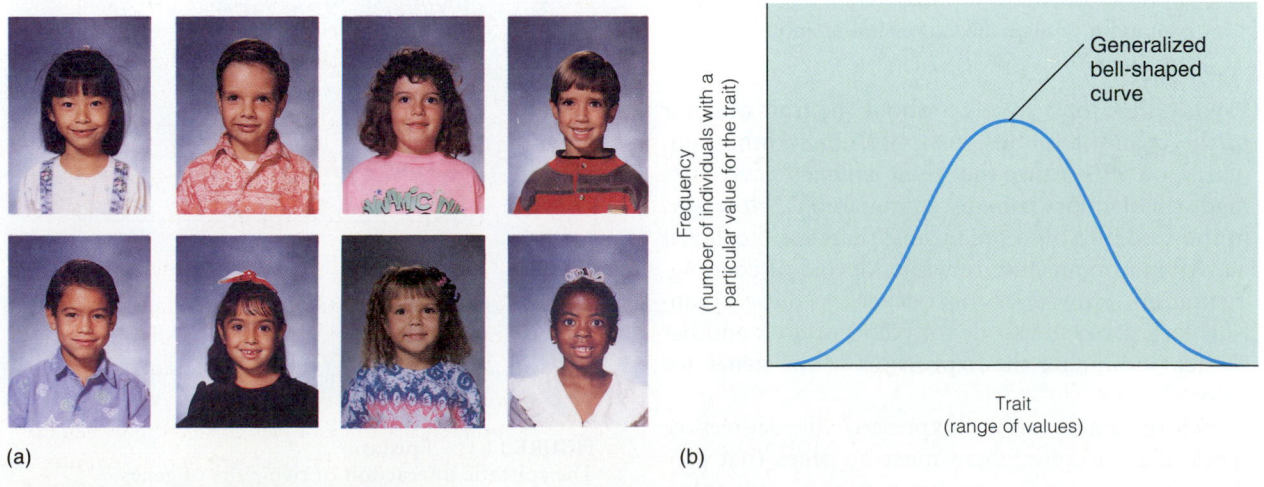

FIGURE 14.10 Polygenic Inheritance
(a) These children exhibit the range of skin, eye, and hair colors typical of polygenic inheritance. (b) A bell-shaped curve typical of the continuously varying nature of a trait that is the result of polygenic inheritance.

Genetics After Mendel

323

control specific aspects of the trait. One should also remember that the environment does have an impact on the expression of a hereditary trait, and it is the combination of genetic and environmental factors that results in the final phenotype.

Multiple Alleles

■ **Even though a diploid individual may possess no more than two alleles for a given trait, there can be multiple alleles for that trait existing at large within a population.**

So far, we have discussed genes that have only two alternative forms, or alleles, for a particular trait. Even though an individual can have only two alleles for a single trait, more than two alleles for a trait may exist in the population. When there are more than two alleles for a particular trait, geneticists refer to **multiple alleles.**

The number of different blood types in humans is a prime example of multiple allelic inheritance. As you may know, humans can have A, B, AB, or O type blood. What is the difference between these types? One gene is responsible for the synthesis of a polysaccharide that occurs on the cell surface membranes of the red blood cells. However, because there are multiple alleles for the gene, the structure of this polysaccharide can vary, and the differences in structure result in the expression of the phenotypes known as blood types A, B, AB, or O. In the next chapter, which focuses on human genetics, we discuss the phenomenon of blood type in more detail.

Epistasis: Interaction between Gene Pairs

■ **Epistasis occurs when the action of one allelic pair can mask or alter the expression of another nonallelic pair.**

If you think for a moment about the trait of color of the coats of rabbits, cats, and many other animals, you will realize that these animals have coats made up of colors in many combinations. However, in the case of albino organisms, there is no color at all. Albino organisms are white, devoid of coat pigmentation. How does this occur? Do these organisms lack genes that code for color, or does another element dominate the expression of the genes for color?

Before genes can be expressed for degrees or types of coat color, there must be genes that produce the *pigment* that in turn results in coat color. In other words, in order for an organism to have color, it requires at least two gene pairs — one gene

pair to ensure that pigment is synthesized and another gene pair to determine the resulting color.

For example, black coat color (*B*) in mice is dominant to brown (*b*). The alleles for black and brown control the type of color. If a homozygous black (*BB*) mouse is crossed with a homozygous brown (*bb*) mouse, all their offspring should be heterozygous (*Bb*), or black. At least, that is what should happen.

However, sometimes mice that should be black are white albinos. Why? Albinism is due to another gene that affects coat color because it regulates the production of pigment. The allele for pigment (*C*), or color, is dominant over the allele for lack of pigment (*c*), or albino. If a mouse is homozygous recessive for no-pigment (*cc*), it does not have the genetic information required to produce enzymes

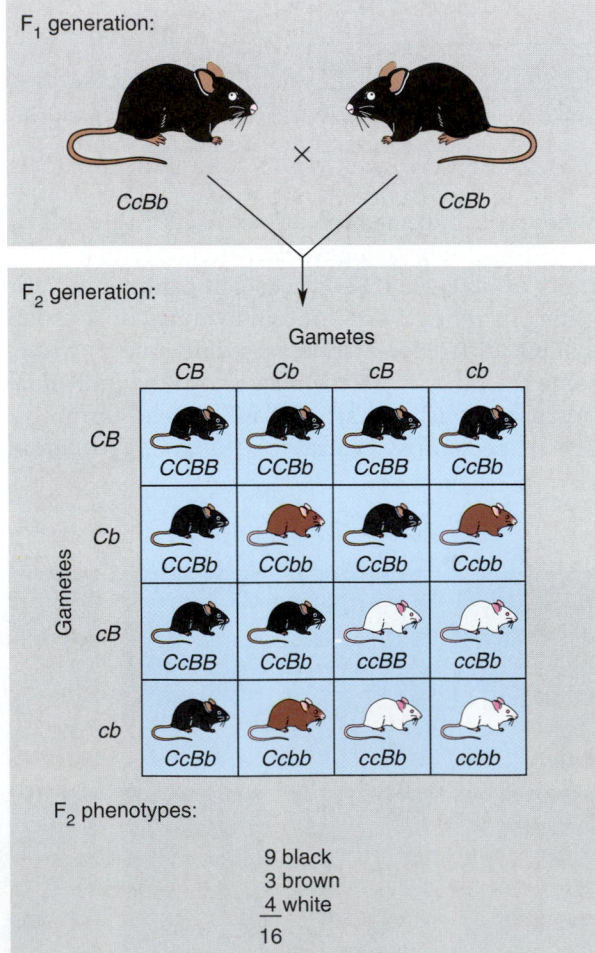

F₁ generation:

CcBb   ×   CcBb

F₂ generation:

Gametes

CB   Cb   cB   cb

CB   CCBB   CCBb   CcBB   CcBb

Cb   CCBb   CCbb   CcBb   Ccbb

cB   CcBB   CcBb   ccBB   ccBb

cb   CcBb   Ccbb   ccBb   ccbb

F₂ phenotypes:

9 black
3 brown
<u>4</u> white
16

FIGURE 14.11   Epistasis
The epistatic interaction of two pairs of genes responsible for coat color influence each other so that the resulting ratio of phenotypes is 9:3:4 (*C* = pigment, *c* = albino, *B* = black, *b* = brown).

necessary for pigment-producing reactions. Even if the mouse has the genetic information for black or brown coat color, it makes no difference, because its system is not supplied with pigment. An albino gene (*cc*) can *mask* the effect of the gene for coat color. The masking of the effect of one gene pair by another gene pair is called **epistasis.**

Figure 14.11 illustrates the epistatic interaction of the two pairs of genes that affect coat color. Note that the ratio that arises from a typical dihybrid cross is not the 9:3:3:1 expected in a typical dihybrid cross, but rather 9 black: 3 brown: 4 white, because of the effects of epistasis.

CHROMOSOMES, GENES, AND HEREDITY

In Chapter 10, we discussed *meiosis*, the nuclear chromosomal events that result in the orderly reduction division of the chromosomes in a diploid cell (2*n*) to form the haploid sex cell (*n*). Today's biologists also know that, for the most part, genes are physically attached to chromosomes. Consequently, the ways in which chromosomes move are going to affect the probability of certain genes being transmitted to offspring. We also know that DNA is the information-carrying molecule of the gene. With this information in hand, let us now begin to further unravel the mysteries of genes, chromosomes, and heredity.

Chromosomes and Sex Determination

■ **Sexual identity in many vertebrates and some nonvertebrates is determined by chromosomes called X and Y.**

In 1900, Mendel's work was rediscovered and reinterpreted on the basis of what was known about cell biology. In 1903, Walter Sutton, a graduate student at Columbia University, was studying meiosis in grasshopper testes (the male sex organ). He observed that the homologous pairs of chromosomes separated *independently* of one another as the gametes formed. These observations paralleled Mendel's concept that factors came in pairs and the concept that these factors must assort independently. Because of his observations, Sutton hypothesized that the factors that Mendel proposed were carried on chromosomes.

As we all know, a hypothesis has to be subjected to experimental testing to determine if it is valid or invalid. Therefore, let us discuss some of the evidence that has led biologists to the conclusion that genes are found on chromosomes.

Several investigators in the early 1900s noticed that often there was a chromosomal difference between males and females. Males and females had similar members for all chromosome pairs except one pair. Pairs that are consistent between males and females are called **autosomes** (nonsex chromosomes). The one pair that is inconsistent is called the **sex chromosomes.**

In many species, including humans, the female has two homologous sex chromosomes known as XX. In the male there is one **X chromosome,** but the other member of the pair is a shortened chromosome called the **Y chromosome.** Hence, in humans and many other animals, females are designated by XX and males as XY. In these species the male is said to be heterogametic, since he can produce one of two types of gametes (X and Y); the female is said to be homogametic, since she can produce only one type of gamete (X).

In humans, males are XY, meaning that there is a 50 percent probability that a Y chromosome

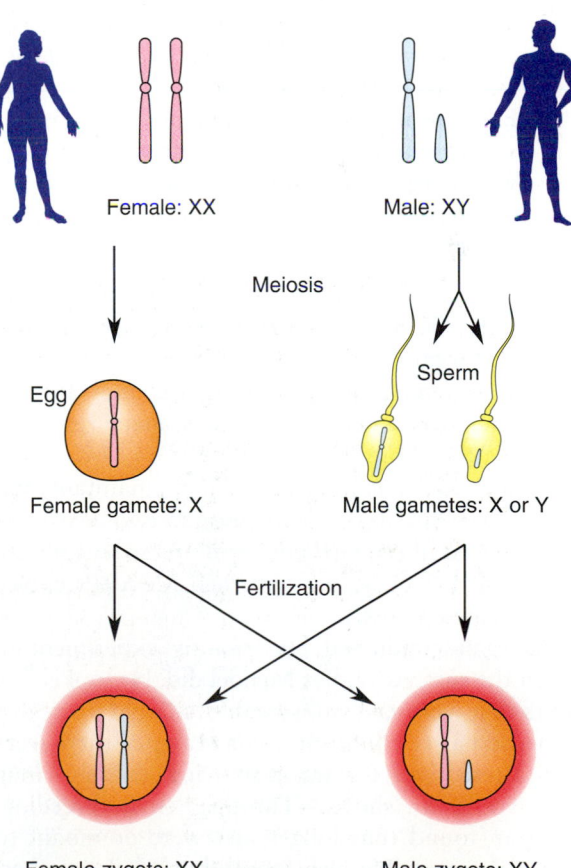

FIGURE 14.12 Sex Determination in Humans
Each individual inherits an X chromosome from the mother and either an X or a Y from the father, resulting in an XX (female) or XY (male) zygote.

might be assorted in a particular sperm cell, and 50 percent that an X chromosome might be assorted to a different sperm cell (Figure 14.12). In the female, which is XX, all the eggs have only X chromosomes. When a sperm fertilizes an egg, the genetic sex of the now-developing individual is determined. An XY combination results in a male, and an XX combination results in a female. Because the probability is 50 percent that the sperm carries an X or Y chromosome, one would expect that there would be equal numbers of males and females conceived for the species. (The Punnett square below illustrates this process.) Sure enough, the expected ratio of 50 percent males to 50 percent females results. These observations provide evidence that genes must be physically attached to chromosomes.

```
              Eggs
           X      X
        ┌──────┬──────┐
     X  │  XX  │  XX  │   XX = female
Sperm   ├──────┼──────┤   XY = male
     Y  │  XY  │  XY  │
        └──────┴──────┘
```

Sex Linkage

- ■ **Experiments conducted with the fruit fly *Drosophila melanogaster* confirmed Mendel's work with independent assortment, and also proved that certain genes were linked to the sex chromosomes of organisms.**

In the early 1900s, another researcher at Columbia University, T. H. Morgan, was studying inheritance. Morgan decided to use the fruit fly *Drosophila melanogaster* as his experimental animal. His preference has been followed by many researchers since that time, for several reasons. *Drosophila* is inexpensive, it is easy to breed and maintain, and it produces hundreds of offspring in two weeks—a short period of time. (If you recall, these reasons are similar to the ones that governed Mendel's choice of sweet pea plants.)

Morgan conducted his fruit fly experiment in much the same way that Mendel did. He looked for traits that were easily observable and that possessed two alternative conditions. For example, there were two alternatives for the trait wing length — long and vestigial (short). Through cross breeding, Morgan found that long wings were dominant to vestigial wings. He also found that when two heterozygous parents were crossed, the offspring showed the typical Mendelian phenotypic ratio of three dominant long-winged flies to one recessive vestigial-winged fly.

Morgan also found a mutant male *Drosophila* that had white eyes, rather than the normal red. He bred the white-eyed male with a homozygous red-eyed female (RR). As he expected, all the F_1 flies were red-eyed, and he concluded that the white-eyed fly had to be just another example of Mendelian dominance — that the fly's genotype had to be Rr and that red eyes are dominant over white eyes. When he crossed the heterozygous red-eyed offspring, however, his results in the F_2 were not quite the ratio of 3 dominant to 1 recessive. Instead the ratio was more like 4:1. Later research demonstrated that white-eyed flies have a higher mortality rate, and this fact was responsible for Morgan's inability to obtain the expected 3:1 ratio.

However, when Morgan looked at the sex and phenotype of the F_2 individuals, he found something that he could not explain. The data are illustrated below.

	Males	*Females*
Red eyes	1011	2459
White eyes	782	0

Why were there white-eyed males but no white-eyed females? Why was this ratio dependent on the sex of the organism? To determine which of *Drosophila*'s four chromosomes carried the gene for eye color, Morgan did a test cross between the original white-eyed male (rr) and a heterozygous F_1 female (Rr). He obtained these results:

	Males	*Females*
Red eyes	132	129
White eyes	86	88

What kind of inheritance pattern did this reveal? Morgan and his coworkers knew that sex in *Drosophila* is determined in a manner similar to that in humans: The males are XY and the females XX. Simple deduction leads to the conclusion that if the gene for eye color were linked to the Y chromosome, only males would have any eye color, because only males have the Y chromosome. Therefore, the gene for eye color must be physically attached to the X chromosome, because both males and females have eye color. **Sex linkage** occurs when a gene is located within the X sex chromosome. Figure 14.13 diagrams the original crosses and then the test cross, linking the allele for red eyes to X chromosomes of the original parent female.

Furthermore, the heterozygous F_1 generation had demonstrated that the allele for white eye color is recessive, because all the F_1 flies had red eyes. Morgan reasoned that there was no corresponding allele on the Y chromosome for eye color. Therefore, the allele for white eyes could express itself more often in the male because it could not be masked by a dominant allele on the Y chromosome.

Morgan and other investigators have confirmed Sutton's hypothesis to be correct. Genes are linked to chromosomes. But, as you know from the profile of Dr. Barbara McClintock included in this chapter, genes can also move. These are the mobile, jumping genes.

Dependent Assortment

We already know that Mendel demonstrated that certain pairs of alleles, such as the alleles for seed shape and color in peas, would assort independently of one another. Because of independent assortment during meiosis, when two organisms heterozygous for two individual traits are mated, the offspring demonstrate a phenotypic ratio of 9:3:3:1.

FIGURE 14.13 Sex Linkage
Crosses between mutant white-eyed male fruit flies and red-eyed females reveal a pattern of sex-linked inheritance. The gene for eye color in *Drosophila* is located within the X chromosome.

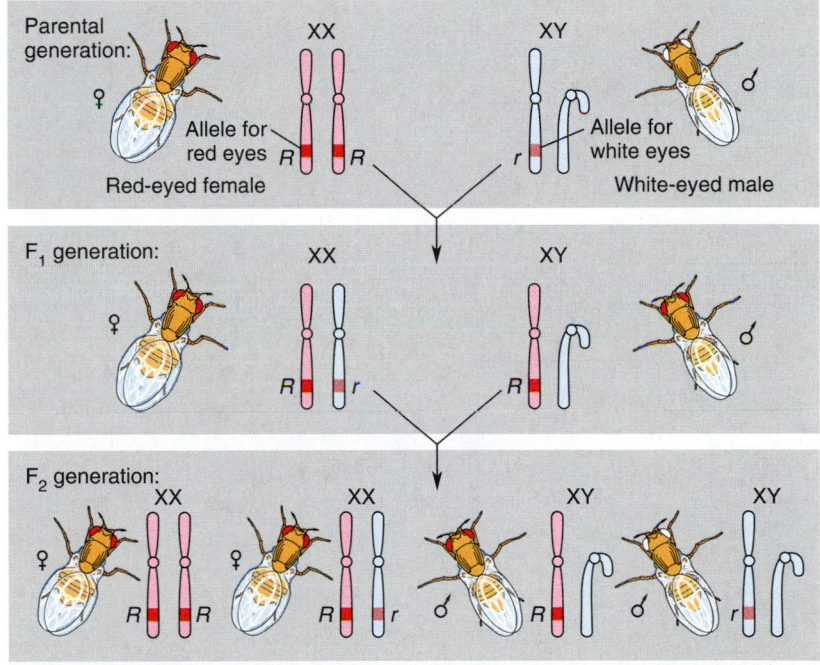

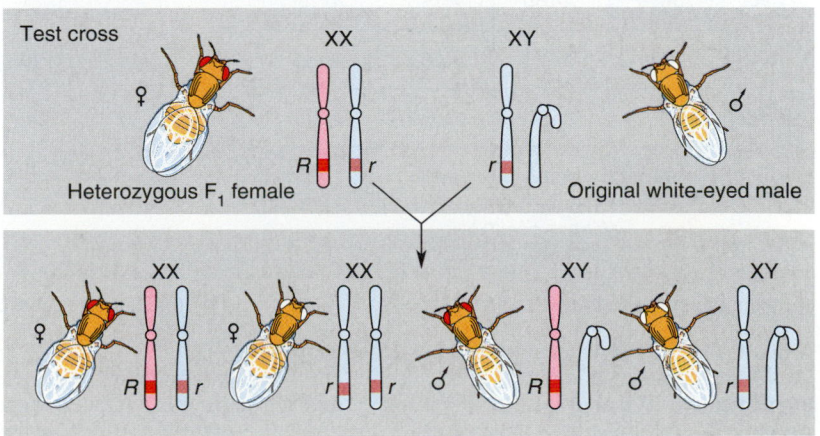

As we mentioned earlier, humans have 23 pairs of chromosomes. Pea plants have seven pairs of chromosomes, and *Drosophila melanogaster* has only four pairs of chromosomes. However, humans have more than 23 traits, peas have more than seven, and *Drosophila* has more than four. In other words, each chromosome must contain literally hundreds of different genes.

Genes that are present in the same chromosome are said to be **linked,** and genes that tend to remain together on a homologous chromosome are said to be members of **linkage groups.** Members of a linkage group cannot assort independently of one another; instead, they remain as part of the chromosome of which they are members. During meiosis, it is the *chromosomes,* not the genes, that segregate to separate gametes.

Assortment can be independent only when genes are on separate chromosomes, so that they can move without regard to genes on another chromosome. For linked genes, however, assortment is **dependent.** If two genes are on the same chromosome, they cannot move independently of one another. (Mendel was trying to determine this when he conducted dihybrid crosses to determine if a superfactor for seed characteristics existed.) Linkage is responsible for traits, such as red hair and freckles in humans. Figure 14.14 illustrates the comparison between independent assortment and dependent assortment due to gene linkage.

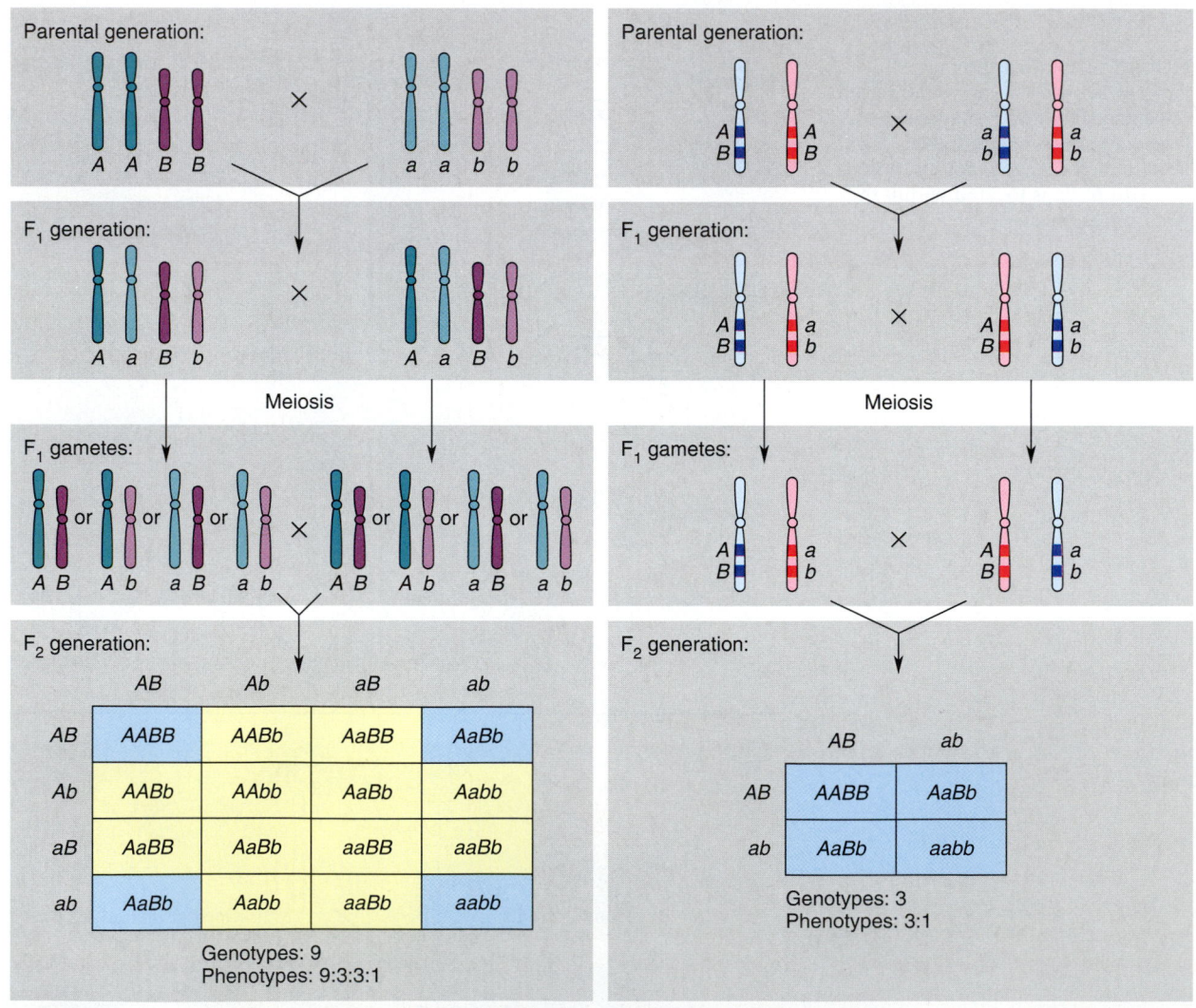

(a) Independent assortment

(b) Dependent assortment

FIGURE 14.14   Independent and Dependent Assortment
(a) When genes are on separate chromosomes, they follow Mendel's law of independent assortment. (b) When genes are linked to the same chromosome, however, the assortment is dependent, and thus the number of genotypes (and phenotypes) is greatly reduced.

Chapter 14   Genetics: Patterns of Inheritance

Crossing Over

Investigations have revealed that genes on linked chromosomes do not always remain linked. Sometimes they cross over. In *Drosophila*, for example, a gray body (*G*) is known to be dominant over a black body (*g*), and long wings (*L*) are dominant to vestigial (short) wings (*l*). When a fruit fly that is homozygous for gray body color (*GG*) and long wings (*LL*) is bred to a fruit fly that is homozygous for a black body (*gg*) and short wings (*ll*), what do you suppose is the result in the F$_1$? Their phenotype is gray and long-winged because gray is dominant to black and long is dominant to short. In terms of genotype, they are all heterozygous (*GgLl*).

As a result of independent assortment, when the F$_1$ offspring are crossed, you would expect the F$_2$ offspring to demonstrate the typical Mendelian ratio of 9:3:3:1. However, that is not the case. Instead, the ratio is closer to ¾ gray long-winged and ¼ black short-winged. Because these results are not those that would result from independent assortment, they show evidence of linkage.

However, as biologists repeated the crosses, unexpected things happened. When researchers conducted a test cross between a heterozygous F$_1$ and a fruit fly that was homozygous for recessive traits (*ggll*), they expected to get a ratio that supported dependent assortment (linkage). Instead, two new combinations appeared. There were 10 percent gray short-winged and 10 percent black long-winged individuals (see Figure 14.15). How could these new combinations appear? The traits seemed to be assorting in ways that defied logic.

To learn why these traits behaved as they did required close observation of the chromosomes during meiosis. Remember that at the time of meiosis, each chromosome has already replicated, and the two chromatids making up the chromosome are joined by the centromere. As the cell begins the process of meiosis, each chromosome pairs up with its sister homologue. (Remember, the members of each pair of chromosomes are called homologues.) During meiosis, a phenomenon called *crossing over* can take place. During this process, genes on one chro-

FIGURE 14.15   Evidence Leading to the Discovery of Crossovers
The genes for body color and wing length are located on the same chromosome—they are linked. However, in the cross shown here two phenotypes result that cannot be explained by simple Mendelian inheritance.

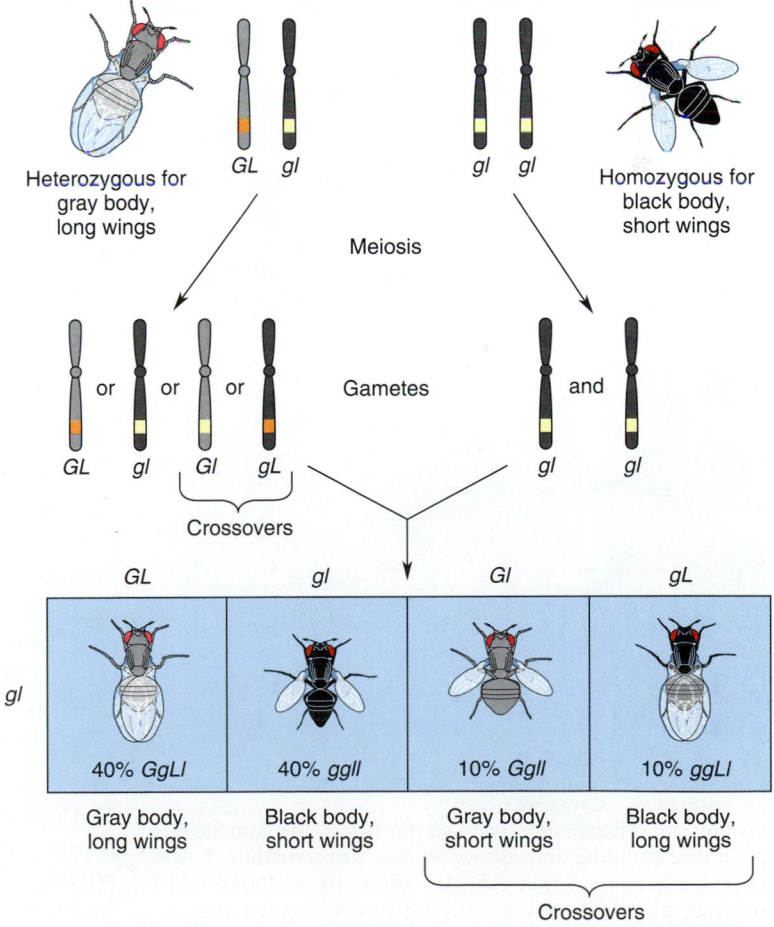

matid may be exchanged for comparable genes on the chromatid on the homologous chromosome.

The homologous chromatids can wrap around each other, and pieces can break off, exchange position with comparable gene sequences, and reattach.

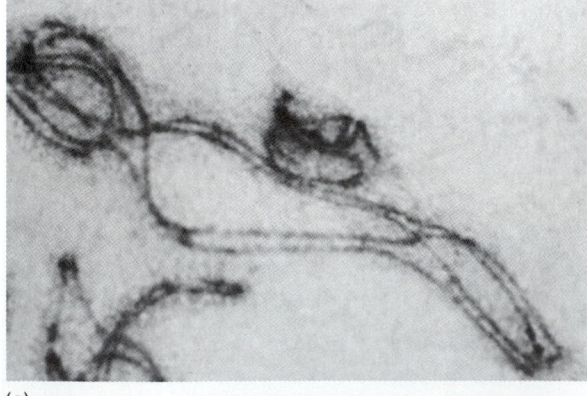

Crossover

Meiosis

Meiosis

Gametes

AB or ab

AB or Ab or aB or ab

(a) No crossover

(b) Crossover

(c)

FIGURE 14.16 Crossing Over
(a) Note that if crossing over did not occur, the gametes produced could be only *AB* or *ab*. (b) With crossing over, the gametes can be *AB*, *Ab*, *aB*, or *ab*. (c) Photomicrograph of chromatids crossing over during meiosis.

Note in Figure 14.16 that if crossing over did not occur, only gametes with either *AB* or *ab* would be produced. With crossover, however, four types of gametes are produced: *AB*, *Ab*, *aB*, and *ab*. Crossing over increases genetic variability. Evidence indicates that crossing over occurs during prophase I of meiosis I.

Chromosome Mapping: Distance Between Genes

■ **Research revealed that crossing over occurs with predictable frequencies. As a result, geneticists began to map the position of genes on the chromosomes of organisms.**

In light of all the chromosomal genetic work, it was becoming clear that not only were genes linked to chromosomes, but they were located at particular positions on the chromosome — loci (the plural form of the term **locus**). In other words, each gene is located at a particular locus along the length of a chromosome. The alleles of any gene must be located at the same locus on each of the homologous chromosomes. If not, crossing over would result in a genetic mixup, not genetic exactness.

While working in Morgan's laboratory, A. H. Sturtevant made two important observations: (1) the crossover combinations always occurred with predictable frequencies, and (2) the percentage of crossing over between two genes was different from the percentage of crossing over for two different genes on the same chromosome. He reasoned that these differences had to do with the physical distance between genes. As a result of his observations, he reached three conclusions:

1. Genes are arranged in a linear sequence on a chromosome like pearls on a string.

2. The percentage of crossing over between two genes has to do with the distance between them. In other words, the closer together the genes are, the less frequently the genes cross over. The farther apart the genes are, the greater the frequency of their crossing over.

3. By using the frequency of their crossing over, one can map the relative positions of genes along a chromosome. In other words, the frequency of crossing over is usually directly related to the relative distance between genes on a chromosome.

In 1913, while still an undergraduate student, Sturtevant constructed chromosome maps for *Drosophila* that indicated the relative position of gene loci along the length of the chromosomes. Today researchers have mapped numerous chromosomes in fruit flies and other organisms. Figure 14.17 illustrates a portion of a genetic map for two chromosomes in *Drosophila*.

Considering the work of Morgan combined with evidence for sex linkage, the discovery of linkage groups crossing over, chromosome mapping, and the study of other chromosomal abnormalities, it is safe to state that Sutton's hypothesis that genes are physically linked to chromosomes was correct. Remember, though, recent evidence from Dr. McClintock has modified this concept slightly, as we know that some genetic elements can move from one chromosome to another as well. Now that we understand some basic principles of genetics, in Chapter 15 we will turn to human genetics.

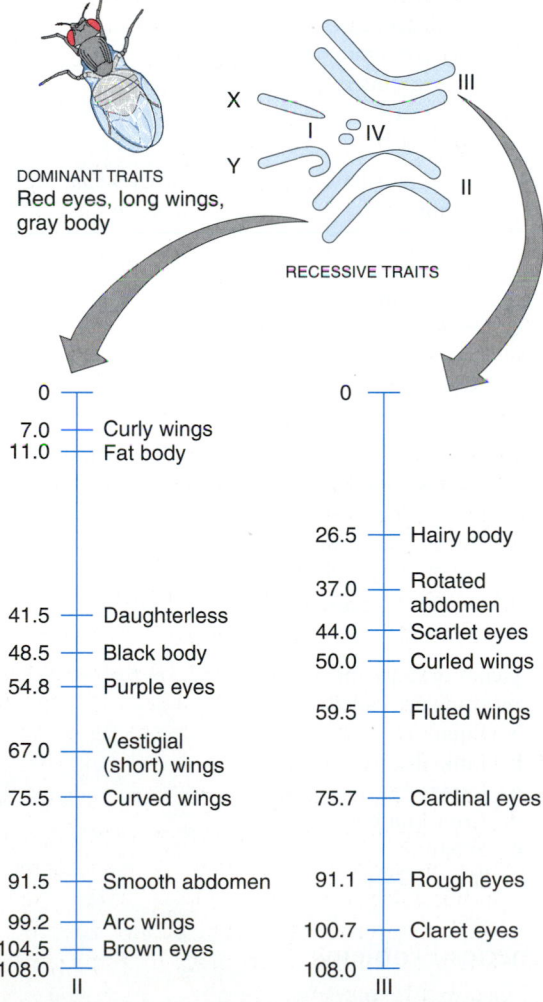

FIGURE 14.17   Genetic Map
A portion of the genetic map of chromosomes 2 and 3 of the fruit fly *Drosophila melanogaster*. Included here are the relative positions of some of its recessive traits unmasked by cross breeding.

SUMMARY

1. Genetics is the study of the mechanisms by which information is passed from parents to offspring.

2. Gregor Mendel, an Austrian monk, is often called the father of genetics. He discovered four major genetic concepts that serve as the basis for modern genetics.

3. The modern interpretation of Mendel's four genetic concepts is:

   a. The concept of dominance: One allele of a trait is dominant over the contrasting allele of that same trait; the masked allele is called recessive.

   b. The concept of unit characters: Hereditary traits are controlled by genes; in diploid organisms they occur in pairs.

   c. The concept of segregation: The paired alleles for a given trait in an individual organism segregate when gametes are formed. Any single gamete receives only one allele for the trait.

   d. The concept of independent assortment: During meiosis, the alleles on one pair of homologous chromosomes segregate independently from alleles on other pairs of homologous chromosomes.

4. Mendel applied the product rule of probability in his genetic crosses with pea plants. This rule states that whenever you combine two or more independent events, the probability of the combination of events is equal to the product of their separate probabilities. The larger the number of trials (or sample size), the more likely it becomes that the researcher will find each possible event occurring at approximately the expected frequency.

5. In incomplete dominance one allele for a trait is not completely dominant over the alternative for the same trait. This results in a blending of the two traits.

6. The ability of one gene to produce a variety of phenotypic effects is called pleiotropy.

7. Genes can work together to produce a single trait, resulting in a continuum of phenotypic expression as represented by the typical bell-shaped curve of normal distribution. Height in humans, for example, is regulated by many pairs of genes. This pattern of inheritance is known as polygenic inheritance.

8. Even though a given individual can have only two alleles for a given trait, more than two alleles may exist in the population for that trait. Traits having more than two alleles are known as multiple alleles. ABO blood type inheritance in humans is an example.

9. Gene pairs may interact with one another. This is the case with epistasis, in which one gene pair masks the effect of another.

10. In many organisms there is a difference in the chromosomal makeup of males and females. In these organisms, the chromosome pair that controls the

sex of the offspring is called the sex chromosomes. (The other chromosomes are called autosomes.) In humans, the sex chromosomes in males are called the XY chromosomes; those in females are called XX.

11. In his work investigating the inheritance of white eyes in male fruit flies, Morgan determined that the gene for eye color was linked to the X chromosome. This discovery provided evidence that the genes are found on chromosomes.

12. Because organisms consist of hundreds of genetic traits and possess relatively few chromosomes, biologists conclude that more than one gene is linked to a particular chromosome. Genes that are found on the same chromosome are said to be members of linkage groups. Because they are located on the same chromosome, their assortment is dependent.

13. When you cross an organism that is heterozygous for two traits in separate characteristics (*AaBb*) with an organism that is homozygous for those characteristics (*aabb*), the genes assort independently, and the probable results are offspring that are ¼ *AaBb*, ¼ *Aabb*, ¼ *aaBb*, and ¼ *aabb*. This ratio is due to the fact that the genes are located on separate chromosomes; therefore, they can assort independently of one another. However, when two pairs of genes are located on the same chromosome, the alleles of the genes assort dependently and the expected results are ½ *AaBb* and ½ *aabb*.

14. The ratios that are expected of a genetic cross are often not obtained because of the phenomenon of crossing over. During the early stages of meiosis I, homologous chromosomes may wrap around one another. Pieces can break off and reattach to the other homologous chromosome, producing unexpected results. Crossing over increases genetic variation.

15. The frequency of crossing over depends on the relative distance between genes on a chromosome. The farther apart the genes are on a chromosome, the greater the probability of crossing over. By using crossing over as an indicator, biologists can construct chromosome maps.

KEY TERMS

heredity
variation
self-pollination
cross-pollination
first filial generation (F₁)
second filial generation
  (F₂)
recessive
dominant
concept of dominance
concept of unit characters
concept of segregation

monohybrid cross
allele
homozygous
heterozygous
genotype
phenotype
Punnett square
test cross
dihybrid cross
concept of independent
  assortment
incomplete dominance

pleiotropy
continuous variation
polygenic inheritance
multiple allele
epistasis
autosome
sex chromosome

X chromosome
Y chromosome
sex linkage
linked gene
linkage group
dependent assortment
locus

REVIEW QUESTIONS

Multiple Choice

1. Genetics is the study of:
   a. Inheritance
   b. Cell structure
   c. Only plants
   d. Only animals
   (page 307)

2. Mendel's concept of segregation implies that the two members of an allelic pair of genes:
   a. Are distributed to separate gametes
   b. May contaminate one another
   c. Are segretated in pairs
   d. Are linked
   (page 313)

3. If two parents have the genotypes $AA \times aa$, the probability of having an *aa* genotype in the F₁ generation is:
   a. 25 percent
   b. 50 percent
   c. 75 percent
   d. 100 percent
   e. None of the above
   (pages 315–316)

4. Eye color in the fruit fly is said to be sex-linked. This simply means that the gene for eye color is:
   a. On the Y chromosome.
   b. On an autosome.
   c. On the X and Y chromosomes.
   d. On the X chromosome.
   (page 326)

5. Genes that are inherited together on the same chromosome are called:
   a. Equal
   b. Linked
   c. Linear
   d. Coordinated
   e. Synaptic
   (page 328)

Genetics Problems

The answers to these questions can be found in the Appendix.

1. In rabbits, short hair is due to a dominant allele (*S*), and long hair to a recessive allele (*s*). A cross between a short-haired female and a long-haired male

Chapter 14   Genetics: Patterns of Inheritance

produces a litter of one long-haired and seven short-haired offspring.

    a. What are the genotypes of the parents?

    b. What phenotypic ratio was expected in the offspring generation?

    c. How many of the eight offspring were expected to be long-haired?

2. In pea plants, yellow flowers (*Y*) are dominant to white (*y*). Predict the ratio of genotypes and offspring produced by a cross between a homozygous yellow-flowered pea plant and a heterozygous yellow-flowered pea plant. What is the phenotypic ratio?

3. One gene has alleles *A* and *a*; another gene on another chromosome has alleles *B* and *b*. For each of the following genotypes, what types(s) of gametes will be produced?

    a. *AA, BB*

    b. *Aa, BB*

    c. *Aa, bb*

    d. *Aa, Bb*

4. Let us say that large, free earlobes are controlled by the dominant allele (*F*) and attached earlobes by the recessive allele (*f*). In addition, the ability to curl the tongue into a U shape is regulated by the dominant allele *T*. The recessive allele (*t*) produces a non-tongue-rolling condition. These genes are on separate chromosomes; hence they assort independently.

A woman who is a free-earlobed tongue roller marries a man who has attached earlobes and is a non-tongue roller. Their only child, a boy, has attached earlobes and cannot roll his tongue. What are the genotypes of the mother, father, and child?

5. Referring to the previous problem, what would be the probability of producing an attached-earlobe non-tongue-roller child, if two individuals mated who were heterozygous for both free earlobes and tongue-rolling?

6. Flower color in snapdragons follows a pattern of incomplete dominance. A pink-flowered snapdragon is crossed with a white-flowered individual. What phenotypes and what ratio of occurrence would you expect in the offspring?

7. Recall that in most fruit flies, the sex chromosomes are XX for females and XY for males.

    a. Does a male fly inherit his X chromosome from his mother or father?

    b. With respect to an X-linked gene, how many different types of gametes can a male produce?

    c. If a female is homozygous for an X-linked gene, how many different types of gametes can be produced for this gene?

    d. If a female is heterozygous for an X-linked gene, how many different types of gametes can be produced for this gene?

15

Human Genetics

Her name is Shirley Liebowitz, and her husband's name is Allen. She is 35 years old, a professional woman who is ready to have a family. He agrees. She begins to look at plans for redoing the spare bedroom (paint life-size branches along this wall, filled with leaves and birds and koala bears). She sketches floor plans while she waits to see her obstetrician. When the doctor tells her she is indeed pregnant, she is overjoyed with the idea of her baby's spending its first years living in a tree room.

Her doctor then begins to talk about what ultimately will become a nightmare for the Liebowitzes. You see, these two people are of Jewish descent. Their grandparents emigrated to the United States from Russia. They belong to a group known as Ashkenazi Jews. And the doctor knows that children born to Ashkenazi Jews, more than any other group of people in the world, suffer from a genetic disease called Tay-Sachs, a disease that results in mental retardation, paralysis, and death in early childhood. If their baby has Tay-Sachs disease, it will develop normally for about seven to eight months, then begin to regress as lipids accumulate in the brain cells because of the absence of an enzyme usually found in the lysosome of the cell (Figure 15.1). It will live, perhaps, three to five years more, becoming increasingly retarded, paralyzed, and blind. Then it will die.

The obstetrician suggests that the couple have their blood tested for genetic analysis. If both parents are *carriers* of Tay-Sachs, their baby, he tells them, will have one chance in four of suffering from the disease (remember the 3:1 ratio of Mendel's experiments with dominant and recessive genes?). If only one of them is a carrier, the child will not

---

Above: This student is studying human inheritance.

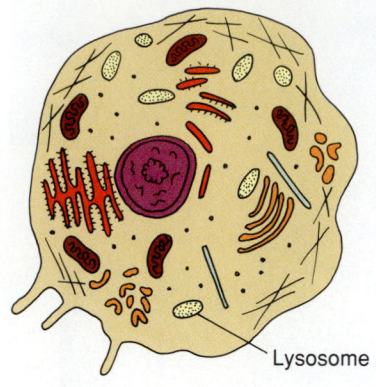
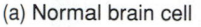

(a) Normal brain cell

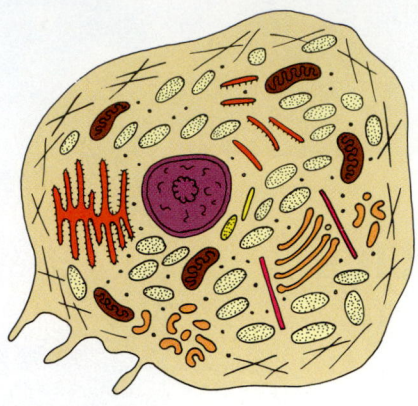

(b) Tay-Sachs brain cell—
excess lysosomes

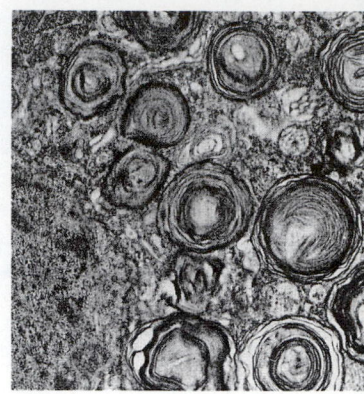

(c)

FIGURE 15.1   Tay-Sachs Disease
(a) A normal cell in which the enzyme is present so that the individual has the normal amount of lipids in the lysosomes. (b) A cell from a Tay-Sachs individual. Note that because of the missing enzyme, lipids accumulate in the lysosome. (c) An electron microphotograph of a brain cell from a child with Tay-Sachs disease. Note the accumulation of lipids in the lysosome.

have the disorder, but the obstetrician still will want to do an amniocentesis. And after all, he tells them, the amniocentesis will also show whether the child suffers from another genetic disease known as Down syndrome, which is much more likely to strike the baby of a woman who is thirty-five or older.

Suddenly biology — genetics and chromosomes and all those other foreign-sounding terms — comes home, sits down in the living room, and becomes the Liebowitzes' constant companion. They learn a new vocabulary, and the words are heavy with lethal meaning because Shirley and Allen, who are both intelligent, educated people, did not know what might happen to them. Like many of us, they thought biology and all its ramifications had little to do with their lives. But more and more issues of genetics are touching our lives. In this chapter we will begin to understand why.

STUDYING HUMAN INHERITANCE

■ **Human inheritance patterns can be studied through the use of pedigree charts and karyotypes. Biologists are learning much about genetic inheritance in humans and the chromosomal and molecular basis of inherited human disorders.**

For a long time, biologists found it difficult to observe the mechanisms of human inheritance (see Profile). Unlike sweet peas or *Drosophila* or *Escherichia coli*, humans have a long life span, and they cannot be bred at will. So what are some of the

tools in the hands of geneticists that enable them to warn the Liebowitzes about the possible problems with their unborn child?

Pedigree

One of the first tools used in human genetics was the pedigree. A **pedigree,** such as that illustrated in Figure 15.2, is a chart or diagram representing the

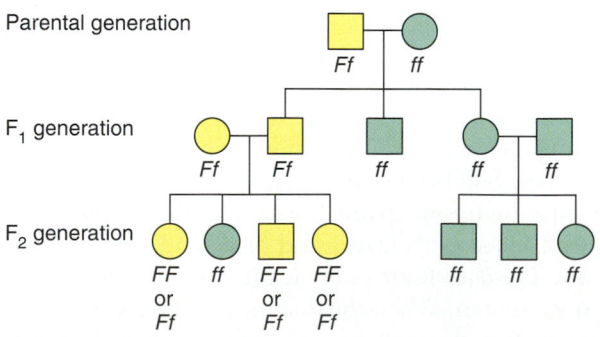

KEY

F   Dominant trait: free earlobes
f   Recessive trait: attached earlobes
▢ Male with free earlobes
◯ Female with free earlobes
▢ Male with attached earlobes
◯ Female with attached earlobes

FIGURE 15.2   A Pedigree
By tracing the progress of a dominant trait through successive generations, geneticists can often determine the genotypes of the family members.

Dr. Nancy Wexler
Psychologist, Geneticist

At the age of 22, Nancy Wexler's life was changed; her mother was diagnosed as having the genetically inherited disease known as Huntington's disease (HD) and later died from it. Huntington's disease is a neurological disorder and is inherited as an autosomal dominant. The disease usually strikes around middle age. The victim gradually loses all control of voluntary muscle, the body and mind degenerate, and the victim slowly becomes worse and dies. The story of Nancy Wexler is one in which an individual has a personal reason to investigate a genetic disorder. Wexler knew that, since the disease was inherited as an autosomal dominant, her chance of having the disease was 50 percent. She made a vow and began a quest to learn as much as she could about HD in order to eventually find a cure. Her work has ranged from the remote fishing villages in Venezuela, where she has used the classical techniques of locating genes through family pedigrees, to the laboratory, where she has used the techniques of molecular biology to discover the actual molecular location of the gene within a chromosome.

Wexler's mother had a master's degree in genetics from Columbia University and her father was a clinical psychologist. Nancy decided to become a clinical psychologist, even though she had a very strong interest in biology, especially genetics. Her decision to study clinical psychology was based on the misconception that biology — genetics was "hard science" — would be too difficult for a woman. Because of that, her formal education in biology was

limited to an introductory biology course she was required to take as an undergraduate at Radcliffe College. Her Ph.D. is in clinical psychology. Although unusual for a clinical psychologist to become a geneticist, Wexler's desire to conquer HD was compelling. She learned genetics from informal discussions, workshops, seminars, and extensive reading. Her exposure to genetics was due largely to her father, Milton Wexler, Ph.D., who founded the Hereditary Disease Foundation in 1968 in Los Angeles. It was during Wexler's participation in the foundation and her association with these geneticists that she received practical training and experience.

In 1972, Wexler found out about a study conducted by Dr. Americo Negrette of Venezuela, in which he documented a very high incidence of HD in small fishing villages that line the shores of Lake Maracaibo. In 1979, Wexler began her studies of the individuals of these villages. She has established a family pedigree for more than 10,000 individuals and has identified at least 650 individuals living or dead with the HD gene. The establishment of the pedigrees and the drawing of blood samples from affected individuals is one of the first steps in locating the gene.

In 1983, Jim Gusella of Massachusetts General Hospital and his colleagues were very lucky and found a marker for HD on chromosome 4. Today, however, there is some controversy as to whether the gene has actually been located on the tip of chromosome 4, whether the gene was located near the middle of the chromosome, or whether

there is more than one gene for this trait. Nevertheless, a marker for HD has been found, even though the location of the gene or genes is still being investigated.

Even though the techniques are available to locate a marker for HD that is 96 percent accurate, and even though Nancy Wexler has a 50 percent chance of having HD, which usually strikes people in their forties — and Nancy Wexler is in her forties — it is a very difficult choice whether to be tested. According to Wexler, some people have said that HD is like a time bomb in a coil of DNA waiting to explode. Wexler has avoided taking this test; rather, she would prefer to channel her energies into her quest for a cure for HD. Today Wexler is an Associate Professor of Neuropsychology at New York City's Columbia Presbyterian Medical Center and president of the Hereditary Disease Foundation in Los Angeles.

history of the occurrence of a particular trait in a given family. During the preparation of a pedigree, females are represented by circles and males by boxes. If an individual has the inherited condition, his or her symbol is darkened. If the individual does not have the condition, the symbol is left blank. A special marker (usually a crosshatched design) is used to indicate individuals who are carriers. **Carriers** are heterozygous for the condition. They have a gene on one chromosome that is the recessive allele for the condition. They do not express the condition themselves, but they are capable of passing it on to their offspring.

Over the last hundred years or so a great deal has been learned about inheritance patterns through the use of pedigrees. The most famous is that of the royal family of Queen Victoria, who was a carrier for the disease called *hemophilia*. We will look at diseases like hemophilia more closely later in this chapter.

Karyotype

Another tool that biologists have at their disposal is the **karyotype.** Until 1956, it was very difficult to observe human chromosomes. Although scientists knew that they should be able to see individual chromosomes once meiosis I was occurring, it was difficult to find cells locked in at that particular stage. In addition, the human nucleus contains so many chromosomes that they often lie on top of each other, making it hard to determine the exact number (and any other characteristics).

All that changed with the development of the techniques of karyotyping, a procedure that can be used to analyze the chromosomes of humans and other organisms. In this technique, human white blood cells are drawn from an individual, and the cells are placed in a medium to stimulate mitosis (Figure 15.3). Because of interactions between the cells and the salt solution, the red blood cells settle out, to the bottom of the test tube. The scientist then adds a chemical called colchicine, which interferes with the formation of the spindle apparatus. This stops the remaining white blood cells in mid-division. These cells are then placed in water, which causes the cells to burst (remember Chapter 5, where we discussed hypertonic and hypotonic solutions and their effect on a cell). When the cells burst, the chromosomes are spread out and then stained.

The next step is to photograph the chromosomes. Once the photograph is enlarged and developed, each chromosome is cut out, and they are pasted on a separate sheet according to size, from the largest to the smallest, with the centromere region aligned along a pencil line. Once the chromosomes are aligned, the pairs that belong together are easily identified — and some abnormalities may be evident. Much that we have learned about the causes of genetic diseases has come from an analysis of the karyotypes of affected individuals.

ABNORMALITIES IN THE AUTOSOMES

■ **Many birth defects are the result of errors occurring during the steps of meiosis. The chromosomal makeup of a sex cell can be affected by at least five types of abnormalities: nondisjunction, translocation, duplication, inversion, and deletion.**

Before we begin to discuss the causes of various birth defects, you should look back to Chapter 10 and review the processes of mitosis and meiosis in diploid organisms. Remember that in humans the diploid number is 46, and that those 46 chromosomes consist of 23 pairs of homologous chromosomes, one from each parent. During the formation of gametes, or sex cells, the homologous chromosomes separate — in other words the diploid number of chromosomes is halved to 23, one member from each pair of the original 46 chromosomes. When abnormalities occur, they usually take place during the formation of haploid gametes, especially during meiosis I.

As you may recall, *meiosis I* involves the separation of homologous chromosomes. During *prophase I,* the homologous chromosomes join and begin to move toward the equator of the cell. Crossing over usually occurs at this time. The homologues then separate from each other and move to the opposite poles. Each newly forming nucleus has the haploid number of chromosomes.

During the four stages of *meiosis II,* the chromatids of each lone homologue separate. The result is four daughter cells, each having the haploid (n) number of chromosomes.

Because meiosis is a complex process, something can go wrong at any of a number of stages. There are five major types of chromosomal abnormalites that might affect the phenotype of an individual: *nondisjunction, translocation, duplication, inversion,* and *deletion.* They can affect both the sex chromosomes, those that determine the sex of the offspring, and the autosomes, any of the other chromosomes. In this section we discuss the mechanisms causing the abnormalities and some examples of autosomal abnormalities. In the next section we discuss abnormalities in the sex chromosomes.

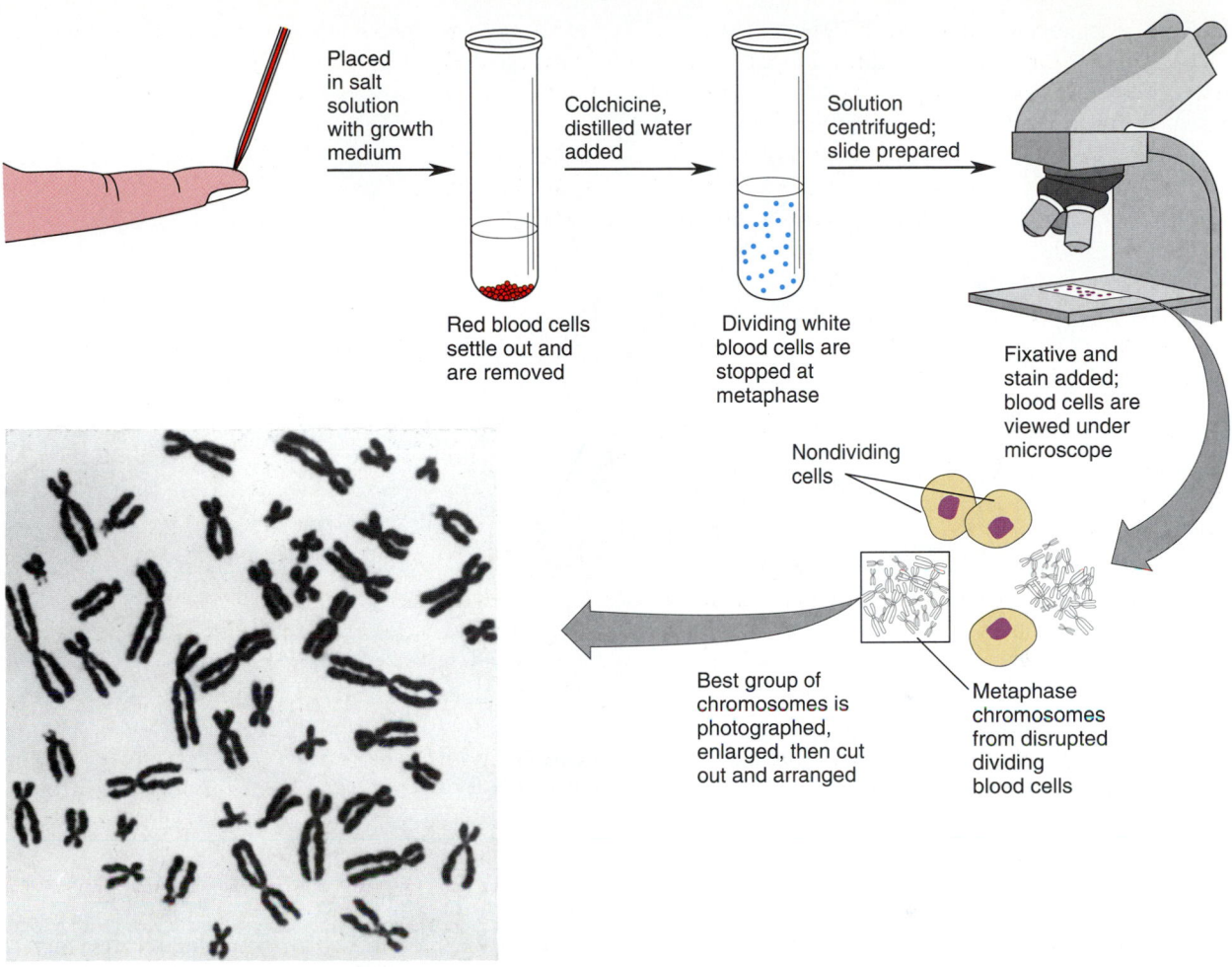

Placed in salt solution with growth medium

Colchicine, distilled water added

Solution centrifuged; slide prepared

Red blood cells settle out and are removed

Dividing white blood cells are stopped at metaphase

Fixative and stain added; blood cells are viewed under microscope

Nondividing cells

Best group of chromosomes is photographed, enlarged, then cut out and arranged

Metaphase chromosomes from disrupted dividing blood cells

1 2 3 4 5
6 7 8 9 10 11 12
13 14 15 16 17 18
19 20 21 22 Sex

Karyotype

FIGURE 15.3 Preparing a Karyotype.
The karyotype shown is that of a normal human male.

Nondisjunction and Translocation

■ **Nondisjunction occurs when homologous chromosomes or sister chromatids fail to separate during meiosis I or II. Translocation occurs when a portion of one chromosome becomes connected to another nonhomologous chromosome.**

From a human genetics perspective, nondisjunction in the formation of sex cells is an important chromosomal abnormality. **Nondisjunction** takes place because of an error during meiosis, when a pair of sister chromatids fails to separate, or *disjoin*. As a result, one sex cell has an extra chromosome and is called $n + 1$; in humans, an $n + 1$ gamete contains 24 chromosomes. The other gamete lacks one of the chromosomes and is called $n - 1$; in humans, an $n - 1$ gamete has 22 chromosomes. Nondisjunction can occur in the formation of either egg cells or sperm cells (Figure 15.4).

When two normal gametes join to form an embryo, the embryo is diploid, or $2n$. An embryo formed from one normal gamete (n) and one $n - 1$ gamete is called a $2n - 1$ embryo. What happens if

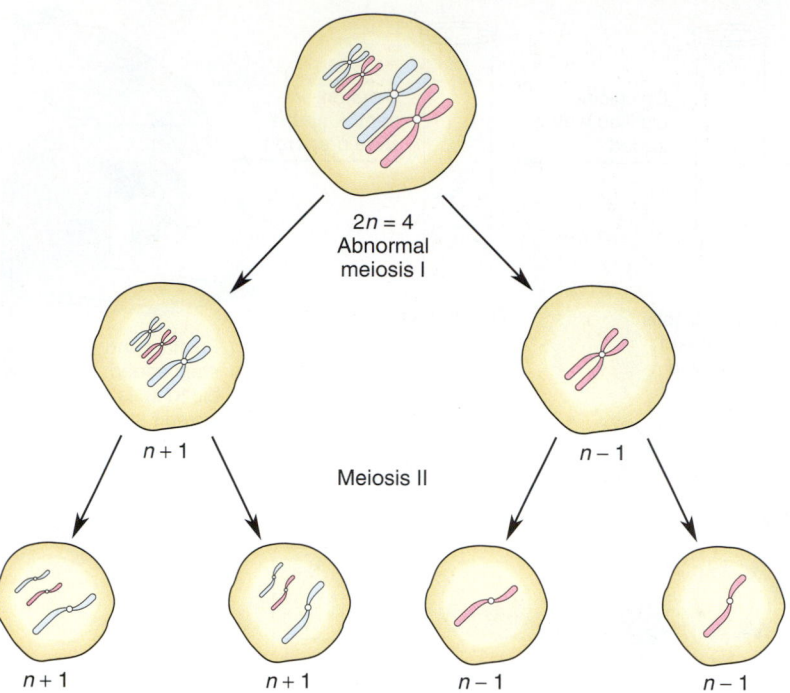

FIGURE 15.4 Nondisjunction During Egg-Cell Formation
Nondisjunction, or the failure of homologous chromosomes to separate
properly during meiosis, occurs during meiosis I and can result in an $n + 1$ or
$n - 1$ gamete.

an embryo is $2n - 1$? An embryo that is missing
one entire chromosome is missing a considerable
amount of genetic information; therefore, these embryos are usually not viable. When a woman mis-
carries or has a spontaneous abortion (in other
words, when she loses the baby naturally) very early
in her pregnancy, the reason often can be traced to
an embryo that is $2n - 1$.

When a normal gamete (n) and an $n + 1$ gamete
join, the resulting zygote is $2n + 1$. This genetic
condition is called **trisomy,** because instead of the
usual two chromosomes for a particular set, the
zygote has three.

Another important type of chromosomal aberration is translocation. In **translocation,** a part of
one chromosome becomes detached and joins another nonhomologous chromosome. In some genetic diseases, the translocations that occur are
fairly predictable.

Down Syndrome

■ **Down syndrome is a genetic condition due to trisomy 21 or chromosomal translocation. Maternal
age may be one factor in the nondisjunction leading to the trisomic condition.**

Most of us have heard of the genetic birth defect
that was once called mongolism and is now known
as **Down syndrome.** (A syndrome is a disorder that
consistently manifests a set of specific symptoms.)
In 1866, J. Langdon Down first described this disorder, the major symptom of which is mental retardation. Although there is much variation in the
expression of the syndrome, the individual with
Down syndrome is usually short in stature and has
short, stubby hands. The eyes are slanted, caused
by a fold of skin between the eyes and brows. This
fold is called an *epicanthic* fold (it is this characteristic that gave the disorder its original misnomer of
mongolism). A person with Down syndrome may
also suffer from heart and digestive system abnormalities.

*Trisomy and Translocation Resulting in Down Syn-
drome.* What causes the spectrum of symptoms
associated with Down syndrome? In 1959,
J. Lejeune determined that a karyotype of a Down
syndrome cell contained the expected 23 pairs of
chromosomes, plus one extra chromosome — an
extra chromosome number 21 (Figure 15.5). This
condition is known as trisomy 21 because there are
three chromosomes of the number 21 designation.

(a)

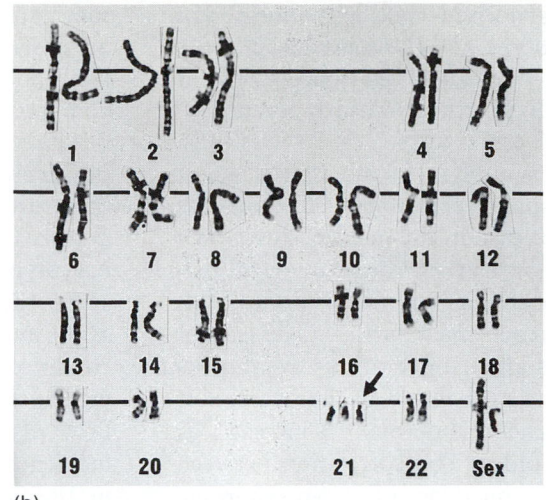
(b)

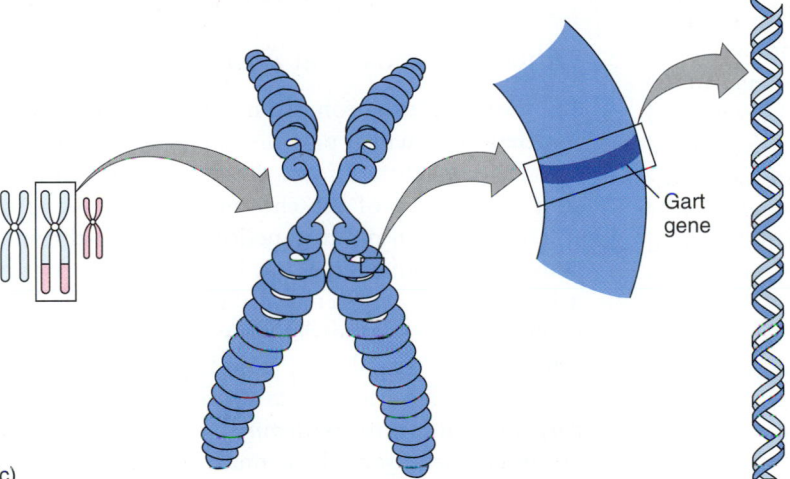

Gart gene

(c)

FIGURE 15.5 Down Syndrome
(a) A child with Down syndrome.
(b) This karotype of a Down syndrome male shows the extra twenty-first chromosome (trisomy 21). (c) Today researchers have identified several genes that are believed to be responsible for the syndrome. For example, the Gart gene leads to high levels of purines, which may be responsible for the mental retardation in Down syndrome individuals.

Approximately 95 percent of all Down syndrome individuals exhibit trisomy 21.

About 5 percent of Down syndrome individuals, however, have the normal pair of chromosome 21. Instead the karyotype of one of these individuals reveals that a piece of chromosome 21 has translocated and attached to another chromosome, usually number 14 but sometimes number 15. Often when the Down syndrome is due to translocation 14/21, a karyotype analysis reveals that the father (this translocation rarely occurs in the mother) has 45 separate chromosomes, and one of those chromosomes is made up of 14 and 21 (Figure 15.6). Geneticists have now determined that the probability

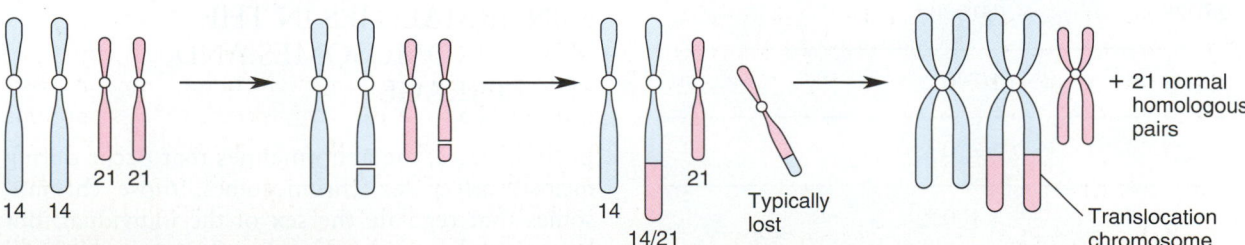

FIGURE 15.6 Translocation 14/21
The translocation of a portion of chromosome 21 results in an elongated chromosome 14/21. This can result in a Down syndrome child with 46 chromosomes.

Abnormalities in the Autosomes

of having a Down syndrome child when one parent has a normal karyotype and the other has a 14/21 karyotype is about one out of six conceptions, or about one of three live births from such parents.

The Issue of Maternal Age. In the United States many medical personnel are concerned about the possibility of an increase in the incidence of Down syndrome babies. For a variety of sociological reasons, some women are electing to postpone childbearing until they reach their thirties. (The case of Shirley Liebowitz is an example of this trend.) It is now known that the probability of the occurrence of some forms of genetic abnormalities increases as the mother grows older. The correlation between maternal age and the incidence of Down syndrome has received special attention. Table 15.1 illustrates how risk varies according to the age of the mother.

Why should the age of a child's mother make any difference? Biologists know that the eggs in the ovaries of the human female remain in prophase I of meiosis from the time the female is a fetus until the individual becomes sexually mature and ovulates. For example, when a woman conceives a child at the age of thirty-three, the egg that is fertilized had been in prophase I in her ovary for more than thirty-three years. Biologists theorize that the longer the elapsed time before fertilization, the greater the probability of damage to the chromosomes in the egg from environmental factors, disease, and cell-process irregularities. However, not all investiga-

tors agree with this explanation. Some think that an older woman who has had children is more likely to carry a Down syndrome child to term, as compared to a younger woman.

Other genetic disorders are also believed to be positively correlated with maternal age. Two of these are the *Patau syndrome* and *Edwards syndrome*. Children who have Patau syndrome have a karyotype exhibiting *trisomy 13* (an extra chromosome 13). This abnormality occurs in about 1 in 5000 live births, and the problems caused by the condition are so severe that most of the individuals die shortly after birth or in early childhood. In the Edwards syndrome, the infant has *trisomy 18*. The incidence of this particular disorder is rare, about 1 in 18,000 births, and the child usually dies within six months after birth.

Other Chromosomal Abnormalities

Another type of chromosomal abnormality occurs in **deletion,** in which part of a chromosome is missing. Deletion differs from nondisjunction: In deletion only a part of the chromosome is missing or deleted, while in nondisjunction an entire chromosome may be missing. What happens when part of a chromosome is deleted? In 1963, J. Lejeune described a genetic disorder in which an affected infant would cry like a cat: Lejeune called the disorder *cri-du-chat* (which is French for *cry of the cat*). In addition to its distinctive cry, a child with cri-du-chat syndrome is mentally and physically retarded. It was determined that the karyotype of a cri-du-chat infant had a deletion in the short arm of chromosome 5.

The two other forms of abnormal chromosomal separation are duplication and inversion. When **duplication** occurs, a portion of a chromosome is duplicated; when **inversion** occurs, the linear sequence of genes on the chromosome becomes rearranged or inverted when a chromosome loops around itself and rejoins to itself in the reverse sequence.

ABNORMALITIES IN THE SEX CHROMOSOMES AND SEX LINKAGE

In some cases, the abnormalities that occur during meiosis affect sex chromosomes, those chromosomes that regulate the sex of the individual. But before we discuss the disorders that may occur, we first need to look at the particular nature of sex chromosomes in all individuals.

TABLE 15.1 Maternal Age and Incidence of Down Syndrome	
Maternal age	*Risk of occurrence per live birth*
Under 25	1/2500
25–29	1/1000
30–34	1/700
35–39	1/200
40	1/100
45	1/25

The Question of the X Chromosome in Males and Females

■ **In females, there is evidence that one of the X chromosomes (called the Barr body) arbitrarily switches off. As a result, the phenotype of a female can be a mosaic, with the X from one parent functioning in some cells and the X from the other parent functioning in the remaining cells.**

As you learned in Chapter 15, in humans one pair of chromosomes regulates the sex of an individual. In males, the sex chromosomes are XY; in females the sex chromosomes are XX. Does it seem logical that the traits regulated by genes on the X chromosome would be expressed twice as strongly in females as in males? Biologists too have asked this question, because, in fact, females do *not* express traits carried on the X chromosome twice as strongly as males. Nor do females produce twice as many of the proteins that are known to be regulated by the X chromosome.

In 1949, while staining nerve cells from cats, Dr. Murray Barr noticed that when the nuclei in female cat cells were stained, there was a darkly stained body along the nuclear membrane. In the nuclei of male cat cells, this body was absent. This particular structure became known as the Barr body (Figure 15.7).

What is the genetic significance of the Barr body? Working with what they already knew about protein regulation in females, scientists reasoned that perhaps one of the two X chromosomes remained inactive and that the **Barr body** represented that inactive X chromosome. However, which X chromosome was inactive, the one donated by the father or the one donated by the mother?

In 1960, Mary F. Lyon proposed a possible answer. Her *Lyon hypothesis* stated that in early embryonic life, purely by chance, either the maternal or the paternal X chromosome becomes inactive in each cell. All the cells would have the same X inactivated as the cell from which they descended. If Lyon is correct, a female is a kind of genetic mosaic because, through chance, in some cells the genes of the X chromosome from the father determine the production of certain proteins, and in other cells, again by chance, the genes on the maternal X chromosome determine the production of these proteins. Lyon's hypothesis offers an answer to the question of X inactivation, but her theory is still questioned by many geneticists because it does not explain the molecular reactions within the cell that would be required to attain random inactivation of an X chromosome.

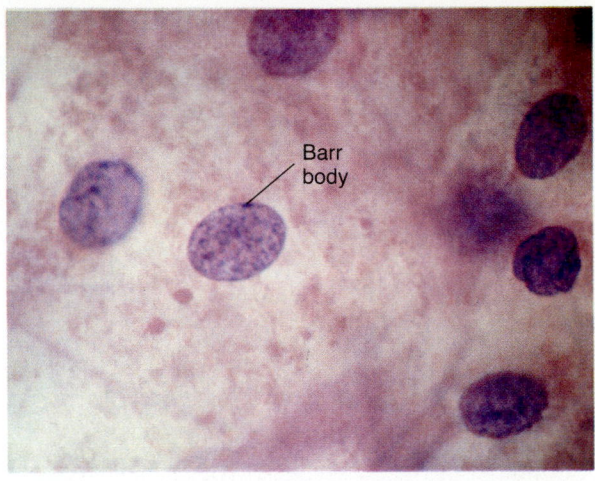

FIGURE 15.7 Barr Body
In this light micrograph of human female epithelial cells, Barr bodies appear as darkly stained bodies along the nuclear membrane.

The classic example of a genetic mosaic in a sex-linked trait is the calico cat (Figure 15.8(a)). The coat pattern of a calico cat is a random arrangement of irregularly shaped black and yellow patches. This random characteristic of the coloring is due to the fact that the cat is heterozygous. One X chromosome carries the allele for black, and the other X chromosome carries the allele for yellow. The random inactivation of X chromosomes in the individual cells of the coat creates the distinctive calico pattern. From this you should be able to deduce that only females can be calico-colored, and for the most part you would be right.

On rare occasions, however, a male cat will have calico coloring. (Because it is so rare, cat enthusiasts will pay dearly for one.) Can you figure out how a male cat could be calico-colored? Remember, there must be two X chromosomes for the calico coloring and a Y chromosome for maleness. That means that nondisjunction must have occurred, and the male has inherited an XXY set of sex chromosomes.

An example of X-chromosome inactivation in human females is a skin disorder called *anhidrotic ectodermal dysplasia* in which there is an absence of sweat glands (Figure 15.8b). In this condition a mosaic pattern of sweat glands results: patches of skin in which sweat glands are normal because of the activation of the X chromosome with the allele for normal sweat glands, and patches of skin in which the X chromosome having the mutant allele for no sweat glands is activated.

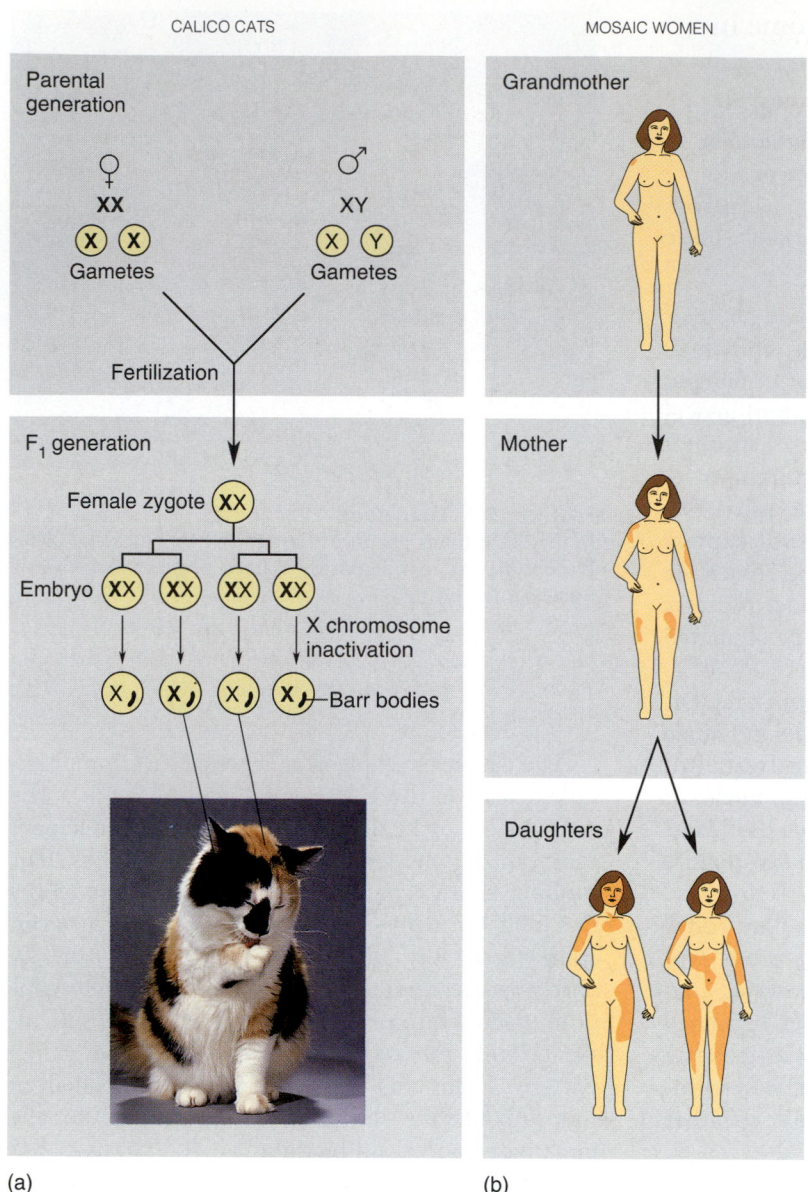

CALICO CATS

Parental generation

♀
XX
(X)(X)
Gametes

♂
XY
(X)(Y)
Gametes

Fertilization

F₁ generation

Female zygote (XX)

Embryo (XX)(XX)(XX)(XX)

X chromosome inactivation

(X♪)(X♪)(X♪)(X♪)—Barr bodies

(a)

MOSAIC WOMEN

Grandmother

Mother

Daughters

(b)

Nondisjunction and Other Sex Chromosome Abnormalities

■ **Nondisjunction can occur in sex chromosomes as well as in autosomes. Two common disorders that result are the Klinefelter syndrome in males and the Turner syndrome in females. In some cases parts of the X chromosome can break off, resulting in fragile-X-chromosome mental retardation.**

Nondisjunction occurs not only in autosomes but also in sex chromosomes. In such cases, the resulting individual will have an extra sex chromosome or, in some cases, be missing one sex chromosome. The more common sex chromosome syndromes are *Klinefelter* and *Turner*.

The Klinefelter Syndrome. In the Klinefelter syndrome, the X sex chromosomes fail to disjoin during gamete formation. The result is a male zygote that is XXY. A baby with the Klinefelter syndrome occurs in about 1 in 500 male births. The individual with the syndrome has the external genitalia of a male, but the testes are small and sterile, the breasts are enlarged, and body and facial hair is sparse. Occasionally the male is also mentally deficient.

The Turner Syndrome. When nondisjunction of the sex chromosomes occurs in *oogenesis* (formation of the egg cell), the result is an abnormal egg — either one *n* + 1 egg, which carries two X chromosomes, or one *n* − 1 egg, which contains no X chromosomes. If an $n - 1$ egg is fertilized by a sperm carrying an X chromosome, the resultant zygote is XO, or $2n - 1$. Because the organism is missing an entire chromosome, much genetic information is lost. As a result, 90 percent of such pregnancies abort naturally. The remaining 10 percent of the XO embryos are born with the *Turner syndrome*. These individuals survive, because one activated X chromosome provides enough genetic information for the individual's system to function.

Phenotypically, a Turner individual is female, but her breasts and reproductive system are undeveloped, and normally she is sterile. In addition, the afflicted individual usually has an IQ lower than that of her siblings (although not necessarily lower than that of the population as a whole) and has poor physical coordination. The Lyon hypothesis concerning the randomly inactivated X chromosome helps to explain why some XO individuals can survive. The incidence of occurrence is about 1 in 3500 live births.

The Klinefelter and Turner syndromes have helped validate Barr's hypothesis concerning Barr bodies. The nuclei of the cells of a male with the Klinefelter syndrome (XXY) do have Barr bodies, and a female with the Turner syndrome (XO) does not have Barr bodies in the nuclei of her cells. Thus, it is accepted that a Barr body is, in fact, a condensed inactive X chromosome.

Another sex chromosome abnormality is called **fragile-X syndrome.** In this condition, the tips of the chromatids on the X chromosome break off at a specific location. Because the genes present on this tip are lacking, this deletion is serious and results in mental retardation. This condition was first discovered in the mid-1980s and is one of the leading causes of genetic mental retardation (Figure 15.9).

Sex-Linked Inheritance in Humans

- Inheritance patterns indicate that some disorders such as hemophilia are carried only on the X chromosome. These disorders are said to be sexlinked. Recessive sex-linked disorders are more commonly seen in males than in females, although females may express the condition. Heterozygous females are considered carriers of such disorders and can pass the condition on to their children, especially to their male offspring.

In Chapter 14, we discussed sex linkage in *Drosophila*. In cases of sex linkage, a gene is located on the X chromosome. Sex-linked inheritance also occurs in humans. In humans, the Y chromosome, which is characteristically small, carries those genes that govern the reactions that result in maleness.

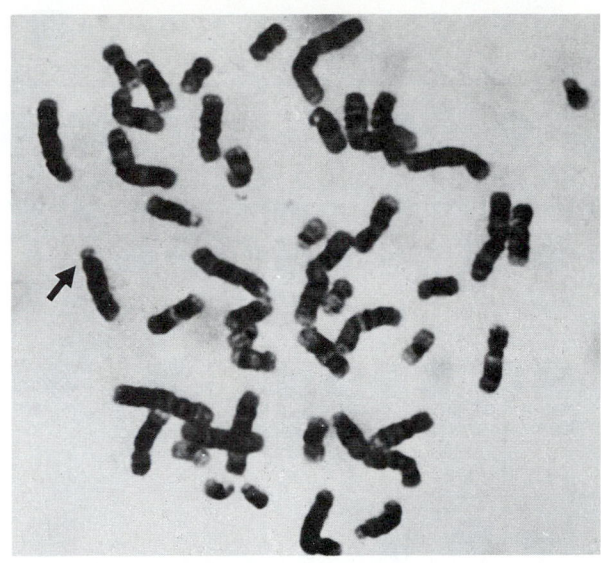

FIGURE 15.9 Fragile-X Syndrome
As you can see, the tip of this X chromosome of a male has broken off. This fragile X is associated with mental retardation.

(The exact mechanism by which the Y chromosome controls maleness is not well understood, but we are learning more about it.) The X chromosome, however, contains genes for many other traits. The work done on some familiar inherited diseases has identified some of the genes that must be linked to the X chromosome, simply because of the disorder's sex-related inheritance pattern.

Hemophilia. The classic example of sex-linked inheritance in which a recessive allele is transmitted on the X chromosome is *hemophilia*. **Hemophilia** (sometimes called bleeder's disease) is the name given to a group of diseases in which the blood fails to clot normally. The hereditary nature of one type of hemophilia was first studied in the royal family of Queen Victoria, the British monarch who reigned during the last half of the nineteenth century (Figure 15.10).

How does a sex-linked inheritance pattern like hemophilia continue through the generations? Queen Victoria's daughters could pass the recessive allele on to their daughters and sons in the same way that the two women had originally received that allele from their mother. Would a hemophiliac father, such as Queen Victoria's son Leopold, pass his X chromosome with the allele for hemophilia both to his daughters and to his sons? Because he contributes only a Y chromosome to his son, he can pass an X chromosome carrying the recessive allele

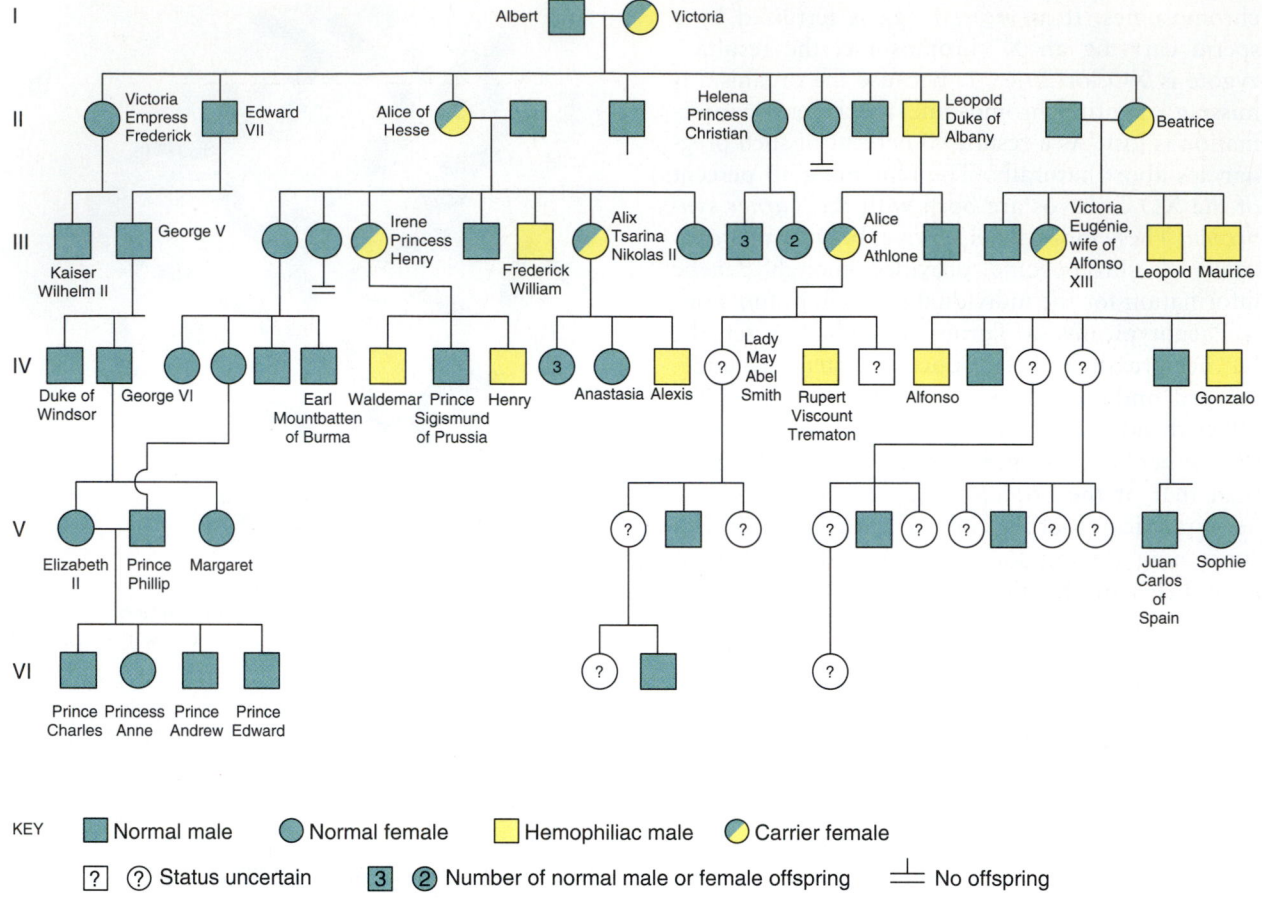

KEY ■ Normal male ● Normal female □ Hemophiliac male ◑ Carrier female

? ⊘ Status uncertain 3 2 Number of normal male or female offspring ⊥ No offspring

FIGURE 15.10 Hemophilia in the Royal Line
A partial pedigree of six generations of the family of Queen Victoria shows
the sex-linked pattern of inheritance of hemophilia.

for hemophilia only to his daughters. Is it possible
for a father with hemophilia to have a hemophiliac
daughter? Yes, if he mates with a woman who is
herself a hemophiliac or is heterozygous for the
condition. Such matings happen very rarely.

Red-Green Color Blindness. There are several
types of color blindness: blue-green, blue-red, and
red-green. The most common type is red-green
color blindness. It is also the only one of these disorders that is sex-linked. Approximately 2 out of
25 white males are red-green color blind, compared
with less than 1 in 200 white females. Do these data
suggest sex linkage? Yes. It has been determined
that the gene for red-green color blindness is recessive and is carried on the X chromosome.

If we let X represent the X chromosome and Y
represent the Y chromosome, and if N represents
the allele for normal color vision and n represents
the allele for red-green color blindness, the following genotypes are possible:

$X^N X^N$ = normal-vision female
$X^N X^n$ = normal-vision female who is a carrier
$X^n X^n$ = color-blind female
$X^N Y$ = normal-vision male
$X^n Y$ = color-blind male

See Figure 15.11 for a pedigree describing the pattern
of inheritance for this sex-linked characteristic.

INBORN ERRORS OF METABOLISM

We began this chapter with the case of a couple
whose doctor was concerned that their unborn
child would suffer from Down syndrome and/or
Tay-Sachs disorder. We have discussed the causes
and symptoms of Down but have yet to look at
Tay-Sachs. To understand this disorder, we need
first to consider what Archibald Garrod early in the
twentieth century called "inborn errors of metabolism."

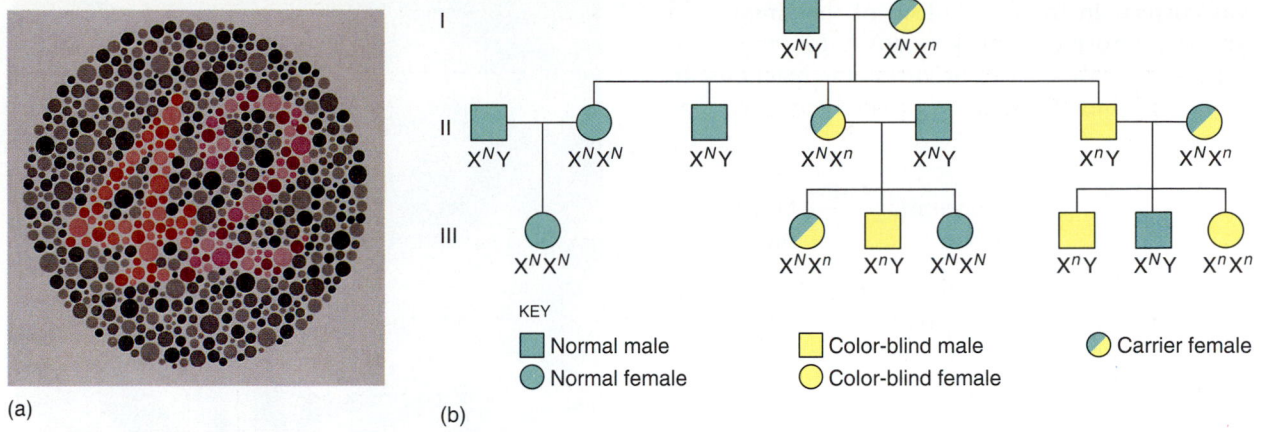

FIGURE 15.11 Color Blindness
(a) Take a color-blindness test. What number do you see? If you see only the number 4, you are green-blind; the number 2, red-blind. If you see the number 42, you have normal color vision. (b) A pedigree showing the sex-linked pattern of inheritance of red-green color blindness. Normal color vision is denoted by N. Note that the allele for color blindness (n) is linked to the X chromosome.

In Chapter 12 we discussed what happens to the metabolic pathways of an organism when a gene fails to function. Because the gene is absent or malfunctioning, a particular protein (usually an enzyme) is missing or nonfunctional. The result is a blocked metabolic pathway, which causes a deficiency of the end product of the affected pathway and an accumulation of certain substances that cannot be processed. (Because these substances often represent one step that *precedes* the final product of a pathway, they are called precursors.) In some cases, the resulting abnormal protein (like the altered hemoglobin structure that causes sickle-cell anemia) can result in an entire series of phenotypic changes.

Many inheritable human diseases are caused by errors at the molecular level of the gene that result in a defective enzyme or protein. Let us now discuss some of these disorders in more detail.

Pleiotropy: Wide-Ranging Effects of Abnormalities in a Single Gene

■ **In inborn errors of metabolism, an individual inherits alleles that cause a malfunction in a metabolic pathway. The single malfunction affects a number of physical systems. Among these disorders are alkaptonuria, albinism, PKU, and sickle-cell anemia. The alleles are recessive; the heterozygous individual either is not affected as severely as are homozygous recessive individuals or is not affected at all.**

In Chapter 14 we discussed the phenomenon of pleiotropy. Now we need to look at this inheritance pattern more closely. The alleles for alkaptonuria, abinism, PKU, and sickle-cell anemia are recessive alleles. As a result, the trait is expressed only in the homozygous recessive condition. In the heterozygous condition, the presence of one normal gene can code for the synthesis of sufficient amounts of the necessary enzyme. Metabolism can then proceed in a normal manner. These genetic disorders are examples of pleiotropy, because one mutant gene or pair of genes can produce a wide variety of phenotypic effects.

Alkaptonuria. When Archibald Garrod was doing his work on metabolic disorders in 1901, he was trying to understand a disease known as *alkaptonuria,* which he realized ran in families and could well be an inherited disorder. Individuals with alkaptonuria develop a number of symptoms. The whites of the eyes and the cartilage just below the skin (such as the cartilage of nose and ears) acquire blue-black discolorations. In addition, the urine turns black after exposure to oxygen. Garrod found that the discoloration of urine and cartilage was due to an increased amount of *homogentisic acid* in the tissues.

On the basis of his observations, Garrod reasoned that alkaptonuria was due to a defect in a gene that resulted in the manufacture of a defective enzyme (homogentisic acid oxidase), which in turn prevented the metabolism of homogentisic acid. He was correct. In the metabolism of the amino acid tyrosine, tyrosine is broken down to other substances through a series of enzymatic reactions. In the normal individual the simplified pathway is as follows:

$$\text{Tyrosine} \xrightarrow{\text{Enzyme}} \underset{\text{acid}}{\text{Homogentisic}} \xrightarrow{\text{Enzyme}} \underset{\text{substances}}{\text{Other}}$$

In an individual with alkaptonuria, the enzyme that breaks down homogentisic acid is lacking and results in a metabolic block. This block can be illustrated as follows:

$$\text{Tyrosine} \xrightarrow{\text{Enzyme}} \underset{\substack{\text{acid} \\ \text{accumulation}}}{\text{Homogentisic}} \xmapsto{\;\;\;} \underset{\substack{\text{substances} \\ \text{unable} \\ \text{to be} \\ \text{synthesized}}}{\text{Other}}$$

The block in this metabolic pathway prevents further metabolism of homogentisic acid, resulting in its accumulation in the tissues, where it produces the characteristic symptoms.

Albinism. Garrod later demonstrated that *albinism* (lack of the brown pigment known as melanin) was also due to a defect in the metabolism of tyrosine. The cause of the most common forms of albinism is a lack of the enzyme that converts the amino acid tyrosine to melanin pigment. This pigment may appear black, brown, yellow, or red.

Albinism in humans sometimes has interesting sociological ramifications. In the United States, the incidence of albinism is higher than normal among the Hopi Indians of the Southwest (Figure 15.12). This phenomenon is due in part to the fact that the breeding population is small. The Hopi Indians consider it good fortune to have an albino in the family. Such an individual is not required to work outside the village, partly because of the danger of illness from exposure to intense sunlight and partly because the individual is honored because of the rare coloring.

Hence, because of a perceived advantage of albino individuals and because albinos were left in the village with the females while the other males worked in the fields, albino males had the opportunity for greater reproductive activity than the other males. Mating was not at random; it was assortative. In other words, the selection of a partner is based on genetic reasons — in this case, the society's preference for a particular phenotype, and the albino males' increased opportunity to reproduce.

FIGURE 15.12 Albinism
An albino Hopi Indian.

Phenylketonuria (PKU). When I was a child in the 1950s, there was a poster that showed two sisters. The older one was about six or seven and was severely retarded; the younger sister was only three or four years old and possessed normal intelligence. Both had the metabolic genetic disorder known as *phenylketonuria,* or PKU, which affects about 1 in 18,000 infants. The poster was notifying prospective parents about the discovery of a simple test that could detect the victims of PKU. It was that test and a new treatment program that made the difference in the lives of the two sisters in the poster.

Under normal circumstances, an enzyme (phenylalanine hydroxylase) converts the amino acid phenylalanine to tyrosine. In 1953, G. A. Jervis determined that PKU resulted from a recessive allele that prevented the synthesis of that enzyme. The two metabolic pathways illustrated in Figure 15.13 demonstrate the differences between the normal process and that in an individual suffering from PKU.

Because of the blocked metabolic pathway 1, phenylalanine is not converted to tyrosine. Instead, the cell produces alternative catabolites of phenylalanine, such as phenylpyruvic acid. It is hypothesized that an accumulation of phenylpyruvic acid (block 2) blocks the synthesis of the brain chemical called *serotonin.* The absence of serotonin produces irreversible brain damage, resulting in mental retardation.

Today, doctors can test the blood to detect the presence of high levels of phenylalanine in the new-

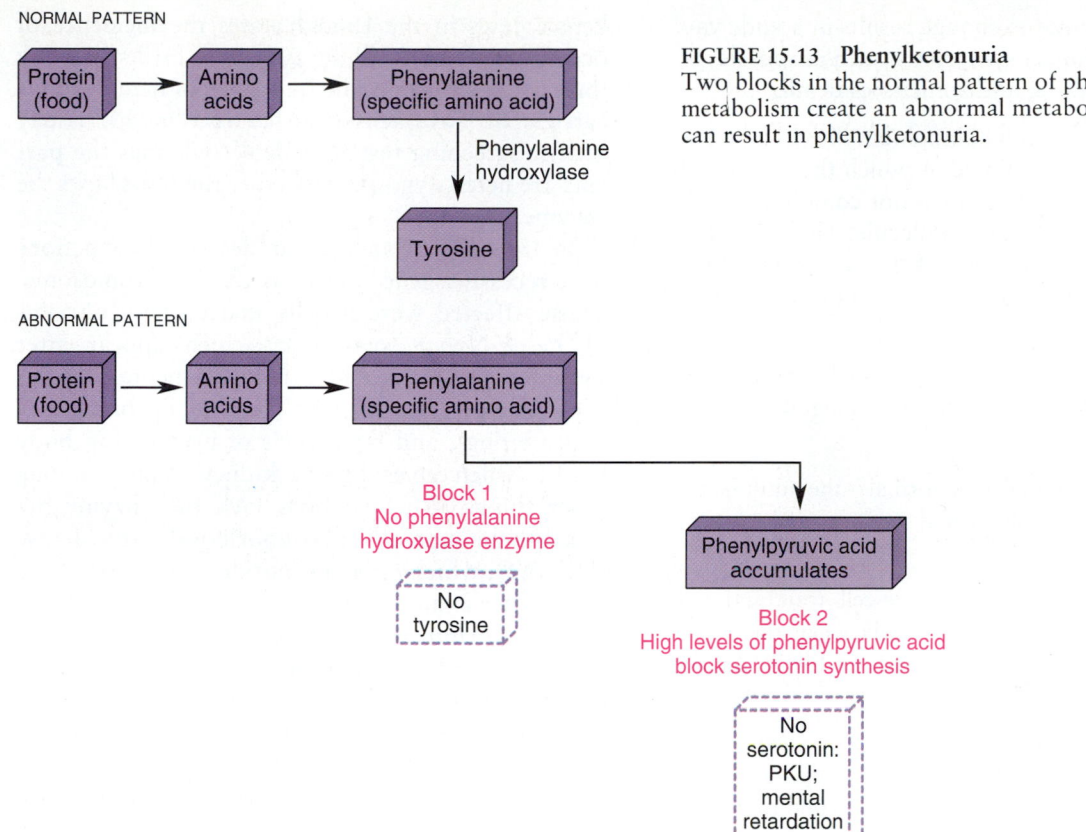

NORMAL PATTERN

Protein (food) → Amino acids → Phenylalanine (specific amino acid)

Phenylalanine hydroxylase

Tyrosine

ABNORMAL PATTERN

Protein (food) → Amino acids → Phenylalanine (specific amino acid)

Block 1
No phenylalanine hydroxylase enzyme

No tyrosine

Phenylpyruvic acid accumulates

Block 2
High levels of phenylpyruvic acid block serotonin synthesis

No serotonin: PKU; mental retardation results

FIGURE 15.13 **Phenylketonuria**
Two blocks in the normal pattern of phenylalanine metabolism create an abnormal metabolic pathway that can result in phenylketonuria.

born infant. An infant found to have PKU can be placed immediately on a diet low in the amino acid phenylalanine to prevent the buildup of phenylpyruvic acid. Once the diet is established, the treatment for PKU is to monitor the blood level of the amino acid phenylalanine and maintain that level in amounts just sufficient for protein synthesis but not so high as to allow phenylpyruvic acid to accumulate. In most cases, the diet of a PKU child needs to be controlled only for the first five or six years, because by that time the growth of the brain is almost complete. Some physicians, however, prefer to continue their PKU patients on the restricted diet for life.

Sickle-Cell Anemia. We have used sickle-cell anemia as an illustration of certain concepts in several chapters, including the concept of one gene, one polypeptide at the beginning of Chapter 12. Sickle-cell anemia is also an example of the pleiotropy of an inborn genetic disorder — an instance in which the actions of a single allele of a gene has several phenotypic effects. You already know that there are two alleles for this gene, Hb^A (normal) and Hb^S

(sickle). The sickle-cell allele (Hb^S) results in a substitution of the amino acid *valine* in place of glutamic acid at the sixth position in the beta chains of the molecule. This one change causes a series of events that can be lethal in their effect. We can look at these events to see the way they accumulate:

1. Because of the difference in this one amino acid, the intermolecular bonding patterns are such that the Hb^S molecules stack up, become rigid, and form long crystals when the oxygen concentration is low.

2. The formation of these fibrous crystals causes the red blood cells to assume their elongated sickle shape and become rigid.

3. Because of their distorted shape and lack of flexibility, the red blood cells cannot easily flow through the microscopic blood vessels (capillaries).

4. Because the capillaries are blocked, gases, nutrients, and waste products cannot be exchanged freely between the cells and the blood, and the cells are affected.

5. This impaired exchange results in a wide variety of clinical symptoms, such as muscle cramps, coma, and sometimes death.

The gene for sickle-cell hemoglobin illustrates a pattern of genetic inheritance in which the normal allele is fully functional but cannot completely mask the expression of the Hb^s molecule. Therefore, heterozygotes can show a few of the symptoms of this disease. The genotypes and resultant phenotypes are as follows:

Hb^A = the allele for normal hemoglobin
Hb^s = the allele for sickle-cell hemoglobin

GENOTYPE

Hb^AHb^A Homozygous normal—the adult hemoglobin is normal

Hb^AHb^s Heterozygous carrier—the individual suffers from sickle-cell trait and will show symptoms only when the hemoglobin is subjected to lower oxygen concentration

Hb^sHb^s Homozygous sickle-cell anemia—the individual produces only sickle-cell hemoglobin; clinical conditions are more severe than those manifested in individuals possessing the sickle-cell trait

An estimated two million African Americans have sickle-cell trait, and only about 1 in 17 shows mild symptoms of the disease. The probability of two sickle-trait (heterozygous) individuals marrying and the probability of their offspring having sickle-cell anemia is about 1 in 1060. A child born with sickle-cell anemia usually has about a 50 percent chance of reaching the age of twenty. This expected life span is increasing every day, however, as doctors work to develop better treatment methods.

Other Metabolic Disorders. Tay-Sachs is a disorder characterized by progressive nerve cell deterioration and is the result of the lack of the enzyme hexosaminidase A. This enzyme would normally break down GM_2 ganglioside, a type of lipid that is important in nerve impulse conduction. Children with a double dose of this gene (those who are homozygous recessive) lack the enzyme necessary to convert GM_2 ganglioside to other substances. As a result, GM_2 ganglioside accumulates in the nerve cells, including those in the brain, and gradually destroys them. Death usually occurs within five years after birth.

The frequency of Tay-Sachs is about 1 in 360,000 in the population at large, but among Ash-kenazi Jews in the United States the incidence of occurrence is about 1 in 3000 live births. Among this particular group, the frequency of heterozygous carriers of Tay-Sachs disorder is 1 in 30. Today genetic screening tests can detect whether the parents are heterozygous or whether the fetus lacks the enzyme.

In 1964, Lesch and Nyhan described the actions of a recessive gene linked to the X chromosome. Those affected were usually male. The symptoms of *Lesch-Nyhan syndrome,* which appear after the first two months of life, are severe mental retardation, severe self-mutilation (such as lip biting and finger biting), and high levels of uric acid in body fluids, which gives rise to kidney stones, among other things. Affected boys lack the enzyme hypoxanthine-guanine phosphoribosyl transferase (HGPRT), which regulates purine metabolism.

Another example of a recessive metabolic disorder is *cystic fibrosis,* a condition in which the sweat glands and other glands produce abnormally high or low amounts of secretion. This disorder occurs in about 1 in 2000 births among the white population of the United States, and at a lower incidence in the nonwhite population. In an affected child, thick, viscous mucus is produced. In the respiratory tract, that mucus makes the child highly susceptible to respiratory disorders. If both the mother and father are heterozygous for cystic fibrosis, the probability of each of their children having cystic fibrosis is 1 in 4. The probability that the child would be a heterozygous carrier is 2 in 4, and the probability that the child would be homozygous normal is only 1 in 4.

The Advantage of Being Heterozygous

How can it be that an inherited disorder such as sickle-cell anemia or Tay-Sachs disorder becomes so concentrated in one particular group of individuals? After all, many of these are lethal conditions — homozygous individuals do not survive. As unlikely as this may sound, genes cannot be characterized as completely good or completely bad. In fact, having one allele for sickle-cell anemia or for Tay-Sachs disorder confers benefit in heterozygous individuals in certain environments.

Sickle-cell-trait individuals are less susceptible to malaria than individuals who do not possess one Hb^s allele. As a result, in groups of persons whose family tree originates in parts of the world where malaria is a common health hazard, it is what we call normal individuals who are at risk for malaria, not sickle-cell-trait individuals. Normal individuals succumb to malaria more often. Heterozygous individuals survive. Over time, the sickle-cell allele can become quite common in a population because the heterozygous individuals who possess it are most likely to survive and reproduce.

The same set of circumstances applies to Tay-Sachs disorder. Heterozygous individuals are apparently less susceptible to tuberculosis, which was one of the primary health dangers in the areas of eastern Europe from which the Ashkenazi Jews came, as well as the most lethal element on the trip across Europe and the Atlantic on the way to this country. Homozygous "normal" individuals, who did not possess an allele for Tay-Sachs disorder, were more at risk for tuberculosis. Heterozygous individuals were less so. Tuberculosis selectively weeded out many more normal individuals than heterozygous individuals. Thus, the number of heterozygous individuals within that particular population increased. That is the situation today.

POLYGENIC INHERITANCE: THE ACTIONS OF MANY GENES ACTING TOGETHER

■ **Many human traits, such as skin color, eye color, and height, are controlled by a number of genes working together in polygenic inheritance.**

Not all inheritance patterns in human genetics are controlled by a single gene, like the ones we have just described. Many traits, such as height and skin coloring, are controlled by more than one gene — polygenic inheritance.

In polygenic inheritance, several genes play a role in the expression of a trait. These genes add their effects to each other, and the result is that the trait they control is expressed along a continuum, or a normal bell-shaped curve. Biologists have studied the actions of polygenic inheritance very carefully. In 1931, Dr. Charles Davenport studied skin color patterns in children of black parents, white parents, and mixed (black and white) parents in Jamaica and Bermuda. He determined that the matings of pure black to pure black or pure white to pure white always produced offspring with the same coloration as the parents. When there was a mixed union, black with white, the color of the offspring was intermediate. However, when two "intermediate" parents had offspring, there was a wide range of skin color. Some were darker than either parent; some were lighter than either parent.

With this evidence he proposed a model for the inheritance of skin color in which two genes, *A* and *B*, were responsible for skin color. Allele *A* produced pigment, allele *a* produced no pigment. Allele

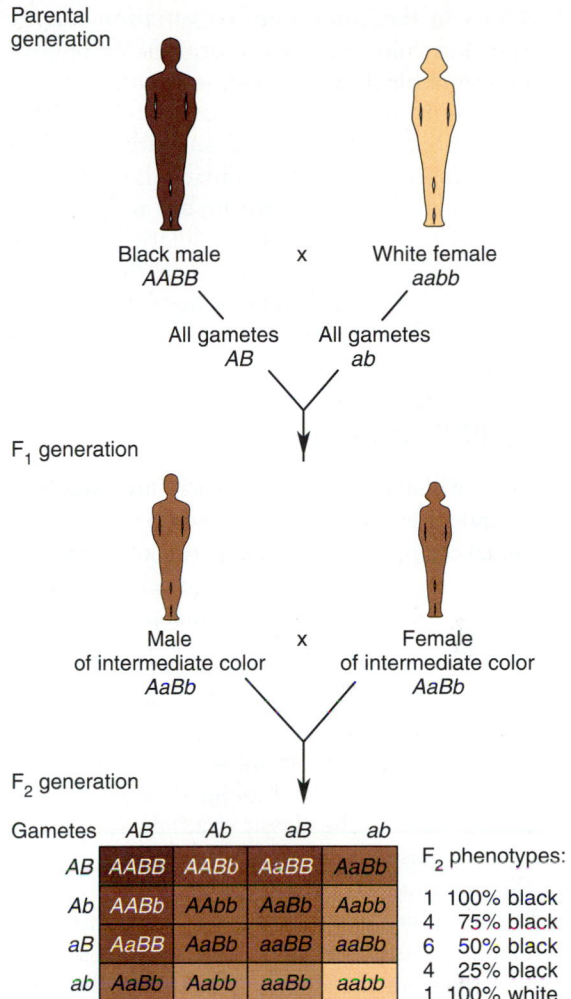

FIGURE 15.14. Polygenic Inheritance of Skin Color Davenport's model of the polygenic inheritance of skin color includes the interaction of two genes, each with two alleles, producing a range of skin color.

B produced pigment and allele *b* produced no pigment. In this model, skin color is determined by the interaction of two genes, each with two alleles as shown in Figure 15.14. Although Davenport's model helped to establish the polygenic pattern of inheritance in human skin color, the actual number of genes involved is still questioned. Various investigators have proposed that anywhere from 2 to 20 genes are involved in the production of various amounts of melanin, which result in the amount of skin pigmentation in humans.

The same interactions are at work in the degree of expression of eye color, height, weight, or any trait that exhibits a gradation of phenotypes. It is the dosage, the cumulative effect of multiple genes, that results in the continuum of variation for the trait. For skin color and eye color, it is the relative amount of the black and brown pigment, melanin. The range is from eyes that appear almost black to eyes that have no melanin and are pink. (In pink eyes, found in albino individuals, the color in the iris is not due to pigment but to the color of the blood circulating in the capillaries.) In skin, the color ranges from very black to the extreme pinkness of the albino, who lacks melanin.

ABO BLOOD GROUPS — MULTIPLE ALLELES

■ **Often within a group of individuals there will be several alleles for a trait, even though a given individual can possess at most only two alleles for a given trait (one on each chromosome). Human blood types are an example of multiple alleles.**

In Chapter 14 we discovered that even though an individual of a specific group of organisms can have only two alleles for a specific trait, within the populations of the entire species there can be more than two alleles. In such a case, biologists say that there are *multiple alleles*. The classic example of multiple alleles in humans is ABO blood types.

In the inheritance of blood type, there is a single blood group *locus*, or location on a chromosome, that regulates the presence or absence of a specific polysaccharide on the surface of the red blood cell membrane. There are several alleles for this gene, and *A*, *B*, and *O* are the most common. Alleles *A* and *B* are **codominant** — both of these alleles will

be expressed in a heterozygous individual. The allele for *O*, however, is recessive. How do these alleles express themselves?

If an individual possesses at least one *A* allele (called I^A), the surface of the red blood cell includes the polysaccharide called A. If the individual possesses at least one *B* allele (I^B), the red blood cells have the polysaccharide B. The allele for *O* (*i*) produces no polysaccharides. Thus an individual with either I^AI^A or I^Ai alleles will produce only polysaccharide A and will have type A blood. An individual with I^BI^B or I^Bi alleles will produce polysaccharide B and have type B blood. An individual with I^AI^B alleles will produce both polysaccharides A and B and will have type AB blood. And an individual with *ii* alleles will produce no polysaccharides and will have type O blood (Table 15.2).

Antigens and Their Role in Blood Type

The polysaccharides on the red blood cell membrane are called **antigens,** a short form of *anti*body-*gen*erating substances. They are part of the cell recognition system that helps the body recognize self and nonself. When an immune system recognizes what it believes to be nonself, it can stimulate an *immune response*, in which molecules called *antibodies* are generated to inactivate the nonself structures. This mechanism is one way in which your body fends off the bacteria that cause disease. What does this mean in terms of blood type?

Let us say you have blood type A and you are given a blood transfusion of type B. Polysaccharide B (an antigen) is foreign to your system. As a result the type B blood will stimulate your body's immune system to generate or mobilize antibodies against it. The antibodies will cause the type B red blood cells to stick together (clump) or *agglutinate*. This clumping can plug up microscopic blood vessels and cause a transfusion reaction that leads to death (Figure 15.15).

Thus, for a blood transfusion to be successful, the donor's blood type must be one that will not cause an agglutination reaction with the recipient's blood. An individual with type A blood can receive blood that is either type A (that is, having the same antigen) or type O (having no antigens). An individual with type B blood can receive either type B or type O. An individual with type AB blood can receive type A, type B, or type O, and hence is called a universal recipient. An individual with type O can receive only type O. Because anyone can receive type O blood, type O individuals are called universal donors.

TABLE 15.2
Genetics of ABO Blood Types

Genotype	Antigen	Blood type	Antibody
I^AI^A or I^Ai	A	A	anti-B
I^BI^B or I^Bi	B	B	anti-A
I^AI^B	AB	AB	—
ii	O	O	anti-A and anti-B

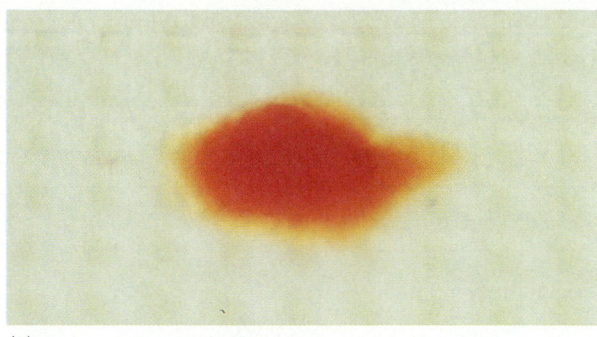

(a)

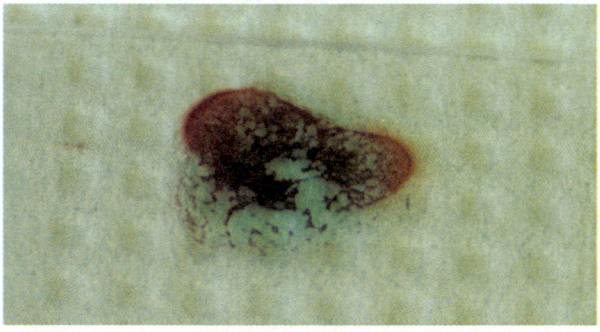

(b)

FIGURE 15.15   Transfusion Reaction
(a) When two blood types are compatible, there is no transfusion reaction. (b) When two blood types are incompatible, clumping occurs that can block the capillaries.

Two discoveries have made blood transfusions practical and safe: blood typing, discovered in the 1900s by Karl Landsteiner, a Viennese physician; and the blood-preservation techniques developed by Dr. Charles R. Drew during the Second World War (see the Profile in Chapter 24). The development of the techniques of blood transfer has been one of the major medical advances of this century.

Some Practice Problems in Blood Typing. To determine how well you understand the possible combinations of blood types, see if you can answer the following questions correctly:

1. If you have blood type A, must one of your parents have contributed the *A* allele to your genotype? (Yes.)

2. Could you have blood type A and one of your parents have blood type O? (Yes, as long as the other parent contributed an *A* allele.)

3. If both parents were AB (I^AI^B), could their child be type O? (No. A child's parents each would have had to contribute an *i* allele.)

4. For a child to have blood type O (*i*), in the previous case, what genotype might the parents have? (I^Ai or I^Bi or *ii*.)

The reasoning required to answer these questions is characteristic of the reasoning used to decide paternity suits. In these lawsuits, blood typing cannot demonstrate guilt; it can only prove innocence, because it can demonstrate with a high degree of probability only that a particular individual could not be the parent of a particular child.

At this point, we have discussed several examples of human genetics (Table 15.3, on the following page, summarizes some of these), emphasizing those that have medical consequences. Can we prevent these disorders? Probably not, but we can sometimes diagnose the condition in the fetus before birth. We can also identify heterozygous carriers in the population and then counsel them on the probability of having an affected child, and we can even insert genes into the cells of the affected individuals — gene therapy (see Focus 15.1). We discuss these techniques in the following section.

PRENATAL AND POSTNATAL GENETIC DETECTION

■ **Genetic counseling can help individuals determine the probability of having children with a particular genetic disorder. Genetic testing, such as amniocentesis, fetoscopy, ultrasound, chorionic villi analysis, and several new molecular diagnostic techniques such as PCR and RFLP, can determine whether a fetus at risk is affected by the disorder in question. Postnatal treatment, especially in the case of metabolic errors, can help alleviate some problems.**

It is estimated that in the United States today 125,000 children are born with genetic defects each year, and that the incidence is rising. In part, this is because parents are electing to postpone the birth of their first child and partly because of an increase in chemical substances in the environment that can cause genetic defects. Ironically, another factor that contributes to the increase in the number of birth defects is improved medical technology. As a result of what we have learned, some individuals with genetic defects can now live to maturity and reproduce. Therefore, they pass their genetic abnormalities on to their offspring.

Given the symptoms of genetic abnormalities, the medical and emotional costs of dealing with birth defects are great. However, by using modern

TABLE 15.3
Common Genetic Disorders

Disorder	Cause	Effect	Occurrence	Genetic basis
Down syndrome	Autosomal chromosome abnormality	Range of mental retardation	1 in 800	Trisomy 21
Klinefelter syndrome	Sex chromosome abnormality	Defect in sexual differentiation	1 in 2000	XXY
Cystic fibrosis	Abnormal ion transport membrane	Complications of excessively thick mucus secretion	1 in 2000 Caucasians	Autosomal recessive
Huntington's disease	Abnormal gene	Regressive mental and neurological degeneration	1 in 2500	Autosomal dominant
Duchenne muscular dystrophy	Probably abnormal Ca^{++} membrane transport proteins in muscle	Muscular degeneration, weakness	1 in 7000	X-linked, recessive
Sickle-cell anemia	Abnormal hemoglobin	Impaired circulation, anemia, pain attacks	1 in 625— mostly blacks	Autosomal recessive
Hemophilia	Defect in blood clotting factors	Uncontrolled bleeding	1 in 10,000	X-linked, recessive
Phenylketonuria	Enzyme deficiency	Mental deficiency	1 in 12,000— mostly Caucasians and Asians	Autosomal recessive
Tay-Sachs disease	Absence of an enzyme	Buildup of fatty deposits in brain, leading to early death	1 in 3000 Ashkenazi Jews	Autosomal recessive
Lesch-Nyhan syndrome	Enzyme deficiency	Mental retardation, self-mutilation	1 in 100,000	X-linked, recessive

medical techniques to diagnose the fetus before birth and by counseling parents about possible birth defects, we can reduce these costs.

Genetic Screening

The best treatment for genetic disorders is preventive — that is, prevention of the union of sperm and egg that might result in an individual with a genetic defect. Many people are aware that their family histories reveal the possibility of inherited disorders. In other cases, a physician may realize that certain individuals are at risk of having offspring with a certain disorder, even if the individuals themselves are unaware of any potential problems. Individuals at risk can take advantage of genetic screening techniques, techniques for determining an individual's genetic makeup. The development of a pedigree for a couple at risk can help determine if a pattern of inheritance does exist.

If there are reasons for concern, the possible next step is to develop a karyotype for one or both prospective parents or to run a series of biochemical tests to screen for the presence of a critical allele. These tests can determine if an individual is heterozygous for sickle-cell trait, Tay-Sachs disorder, cystic fibrosis, or has Down syndrome. If tests determine that one or both parents carry a defective gene or chromosome, they can be counseled about the probability of producing a defective child. They may voluntarily elect not to have their own biological child and to adopt. As a matter of fact, because of genetic screening the incidence of children with Tay-Sachs has been reduced 85% in the United States — a true example of the effectiveness of genetic screening.

Prenatal Diagnosis

Screening is of no help once a child is conceived, as was the case with Shirley and Allen Liebowitz. What resources are available once a woman is pregnant?

The most common test is *amniocentesis* (Figure 15.16a). In 1955, Dr. Povl Riis at the University of Copenhagen was involved in developing a technique to determine the sex of a child while it is still in the womb. As you may know, the developing fetus floats in a watery fluid, which is called the

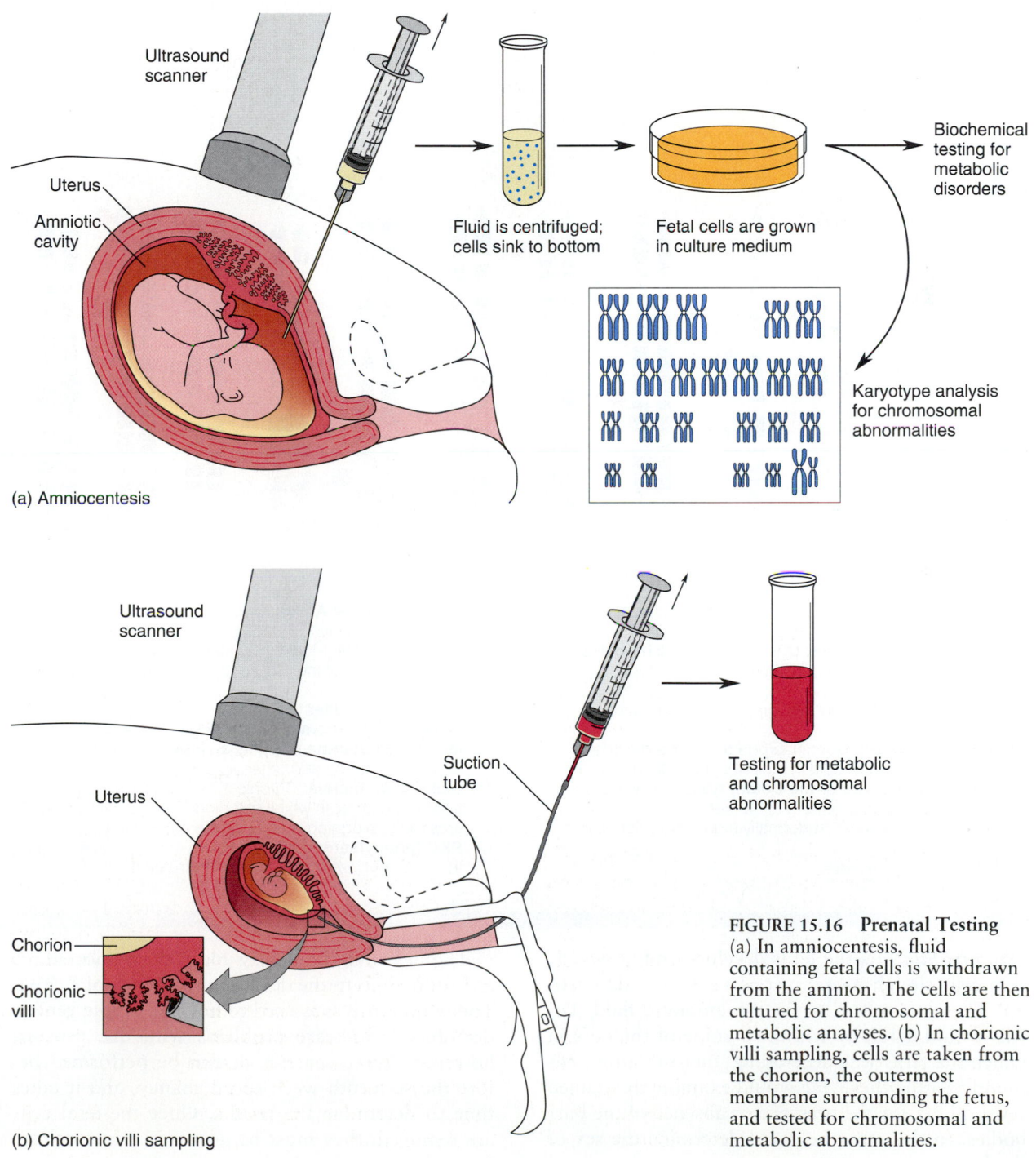

Ultrasound scanner

Uterus

Amniotic cavity

Fluid is centrifuged; cells sink to bottom

Fetal cells are grown in culture medium

Biochemical testing for metabolic disorders

Karyotype analysis for chromosomal abnormalities

(a) Amniocentesis

Ultrasound scanner

Uterus

Suction tube

Testing for metabolic and chromosomal abnormalities

Chorion

Chorionic villi

(b) Chorionic villi sampling

FIGURE 15.16 Prenatal Testing
(a) In amniocentesis, fluid containing fetal cells is withdrawn from the amnion. The cells are then cultured for chromosomal and metabolic analyses. (b) In chorionic villi sampling, cells are taken from the chorion, the outermost membrane surrounding the fetus, and tested for chromosomal and metabolic abnormalities.

The Human Genome

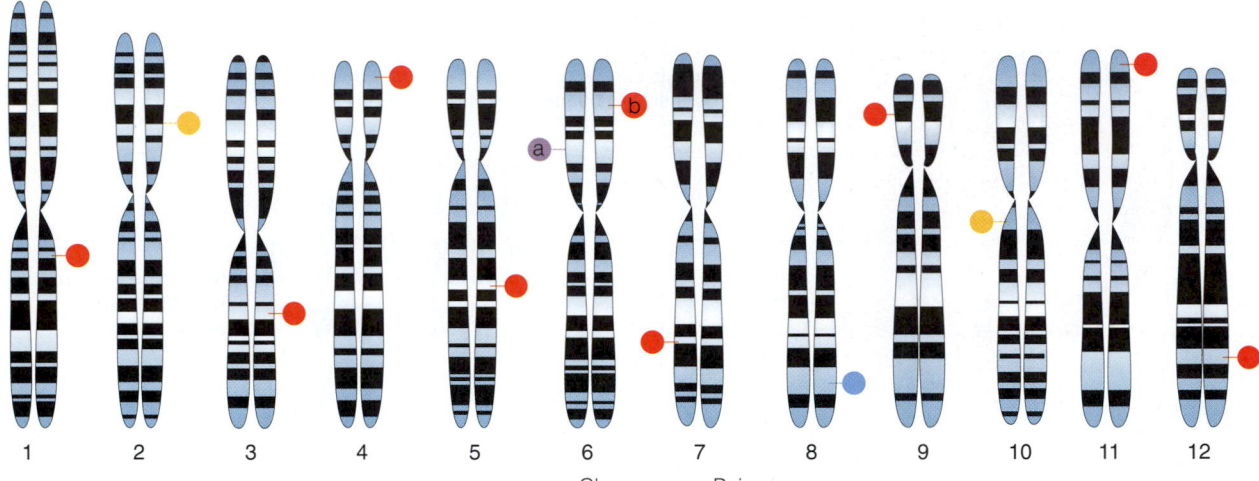

Chromosome Pairs

1. **Gaucher's Disease:** A chronic enzyme deficiency occurring frequently among Ashkenazi Jews
2. **Familial Colon Cancer:** One in 200 people have this gene: of those, 65 percent are likely to develop the disease
3. **Retinitis Pigmentosa*:** Progressive degeneration of the retina
4. **Huntington's Disease:** Neurodegenerative disorder tending to strike people in their 40s and 50s
5. **Familial Polyposis of the Colon:** Abnormal tissue growths frequently leading to cancer
6a. **Hemochromatosis:** Abnormally high absorption of iron from the diet
   *One form of the disease

6b. **Spinocerebellar Ataxia:** Destroys nerves in the brain and spinal cord, resulting in loss of muscle control
7. **Cystic Fibrosis:** Mucus fills up the lungs, interfering with breathing. One of the most prevalent genetic diseases in the U.S.
8. **Multiple Exostoses*:** A disorder of cartilage and bone
9. **Malignant Melanoma:** Tumors originating in the skin
10. **Multiple Endocrine Neoplasia, Type 2:** Tumors in endocrine glands and other tissues
11. **Sickle Cell Anemia:** Chronic inherited anemia, primarily affecting blacks, in which red blood cells sickle, or form crescents, plugging arterioles and capillaries
12. **PKU (phenylketonuria):** An inborn error of metabolism that frequently results in mental retardation

amniotic fluid. As the fetus develops and grows, its skin cells slough away, just as your own skin cells do. The skin cells float in the amniotic fluid. Dr. Riis would carefully withdraw some of the fluid in which the fetus was developing, then stain the cells found floating there. He would examine the stained cells to look for the presence or absence of the Barr bodies. In this way, he could determine the sex of the fetus.

The determinations possible with amniocentesis are not limited to the detection of the sex of a child. Today the process is used to detect possible genetic disorders. There are problems with this process, however. Amniocentesis cannot be performed before the sixteenth week of pregnancy, and it takes time to determine the results. Once the fetal cells are removed, they must be grown in tissue culture for at least two weeks before there are enough to

Using automated DNA sequencers and rough chromosome maps, scientists are finding human genes at the rate of one per day. This knowledge has great implications for the future. Once the genes have been located and sequenced, scientists can use vectors to insert them into human cells for gene therapy, or other techniques may possibly be devised to correct human genetic disorders. The map of the human genome below shows the location of a few of the genes responsible for certain diseases.

Source: Adapted from work by Dr. Victor A McKusick, Johns Hopkins University, and *Time,* January 1994.

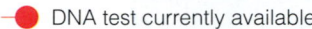

- 🔴 DNA test currently available
- 🔵 Gene mapped but not yet isolated
- 🟣 Diagnosis available through family-linkage study of DNA markers
- 🟡 DNA test under development

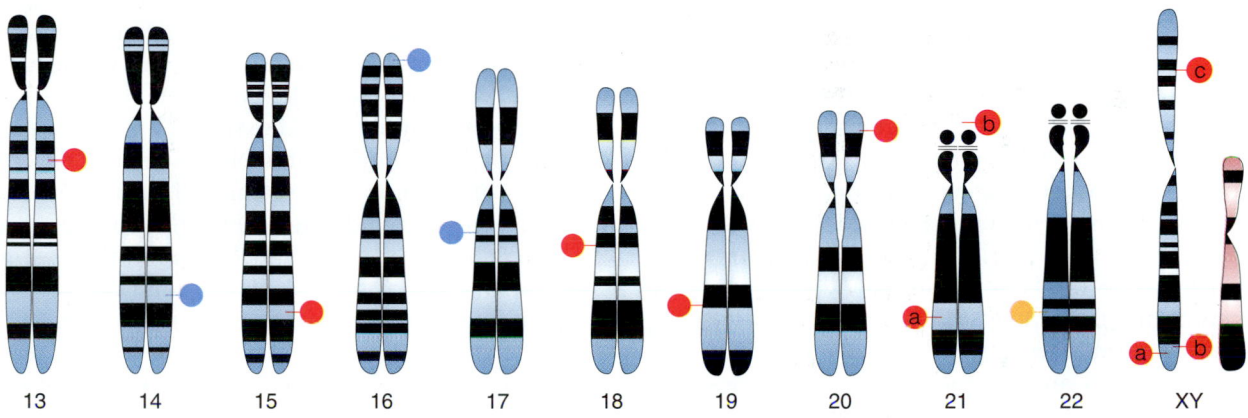

13   14   15   16   17   18   19   20   21   22   XY

13. **Retinoblastoma:** A relatively common tumor of the eye, accountng for 2 percent of childhood malignancies
14. **Alzheimer's Disease*:** Degenerative nerve disease marked by premature senility
15. **Tay-Sachs Disease:** Fatal hereditary disorder involving lipid metabolism, often occuring in Ashkenazi Jews and French Canadians
16. **Polycystic Kidney Disease:** Cysts resulting in enlarged kidneys and renal failure
17. **Breast Cancer*:** 5 percent to 10 percent of cases (Note that the gene was isolated in September 1994)
18. **Amyloidosis:** Accumulation in the tissues of an insoluble fibrillar protein
19a. **Myotonic Dystrophy:** Frequent form of adult muscular dystrophy
19b. **Familial Hypercholesterolemia:** Extremely high cholesterol
20. **ADA Deficiency:** Severe susceptibility to infections. First hereditary condition treated by gene therapy
21a. **Amyotrophic Lateral Sclerosis* (Lou Gehrig's Disease):** Fatal degenerative nerve ailment
21b. **Down Syndrome:** Congenital mental deficiency condition marked by three copies of chromosome 21
22. **Neurofibromatosis, Type 2:** Tumors of the auditory nerves and tissues surrounding the brain
Ya. **Hemophilia:** Blood defect making it difficult to control hemorrhaging
Yb. **Muscular Dystrophy (Duchenne and Becker types):** Progressive deterioration of the muscles
Yc. **ALD (adrenoleukodystrophy):** Nerve disease portrayed in movie *Lorenzo's Oil*

allow karyotyping and an analysis of enzymes. Should the results prove positive and indicate that the developing child is abnormal, a therapeutic abortion is relatively dangerous, simply because the pregnancy is so advanced.

A more recently developed procedure that allows for an earlier genetic analysis is called *chorionic villi analysis* (Figure 15.16b). In this procedure, cells are withdrawn for analysis from the chorion (the outermost membrane surrounding the fetus). This procedure is less risky and can be done as early as the eighth week of pregnancy.

Another prenatal diagnostic tool is *fetoscopy.* In this process, a *fetoscope* (a fibroscopic instrument somewhat like a lighted needle) is inserted into the uterus, and the developing fetus is photographed. The photograph can then be inspected for obvious physical abnormalities. Physicians also use *ultra-*

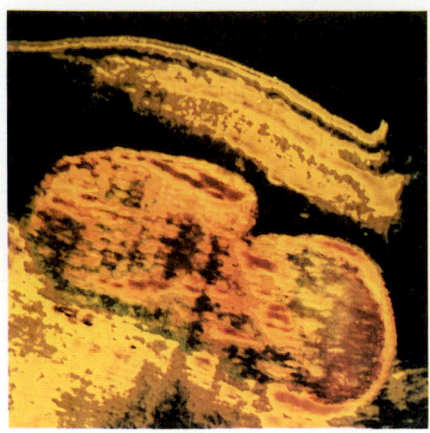

FIGURE 15.17 Ultrasound
A color-enhanced ultrasound image of a
human fetus at 21 weeks.

sound, a method that combines a sophisticated form of sonar (the use of sound waves to determine a form) and computer technology to formulate a picture of the developing fetus on a specially designed television monitor (Figure 15.17). Both amniocentesis and chorionic villi analysis are conducted with the help of ultrasound. The resulting ultrasound image may also be scanned for possible fetal abnormalities.

Geneticists are developing new techniques that allow the direct removal of placental cells for immediate examination of enzyme levels, chromosomal irregularities, and other metabolic conditions. Let us discuss some of these new techniques.

PCR, RFLPs, and Genetic Testing. In Chapter 13 we discussed the polymerase chain reaction (PCR) technique. Using this technique, a molecular biologist can make billions of copies of a targeted DNA sequence (the amount of gene is amplified). As you might suspect, the application of this technique to the detection of genetic disease is ever increasing. The technique can be so specific that it can be used to detect alleles varying in only one base pair, such as the sickle-cell hemoglobin allele or the allele for phenylketonuria. To use this technique, cells of the fetus are obtained by amniocentesis or chorionic villus sampling, the DNA is extracted, and the amount of a specific DNA sequence is amplified using PCR for further study. Today laboratories are routinely using PCR to test for hemophilia and cystic fibrosis. But how do we determine what sequence we want to amplify?

Techniques like restriction fragment length polymorphism (RFLP) can help us to mark (locate) the gene in question.

We also discussed RFLP in Chapter 13, but here we want to emphasize its use in human genetic disease. As we mentioned, restriction enzymes cut DNA between specific base sequences, producing a characteristic fragment. Also, remember that a point mutation—a change in the DNA sequence — can occur. If this happens near the gene in question, the gene becomes part of a different length restriction fragment. This change in length, or polymorphism ("many forms"), can be detected by separating out these fragments using gel electrophoresis, and the gene in question can be located because of its different length. So a specific restriction fragment can sometimes be associated with a specific gene. We then say that the restriction fragment can serve as a marker. For example, one of the first RFLPs used for genetic detection was for the gene that causes Huntington's disease. Huntington's disease is due to a dominant allele that causes degeneration of the nervous system and death when a person reaches middle age. Since the disease does not manifest itself until middle age, in the past a person would not have known if they carried the gene until after they had children. Now, however, RFLP analysis makes it possible to detect the gene for Huntington's disease at an earlier stage. But do those who have the gene really want to know? When people with a high probability of having the gene were asked if they wanted to be tested, over 50 percent decided not to be. Figure 15.18 illustrates how RFLPs could be used to detect the presence of a gene defect in the fetus.

The Question of Action. Doctors can now detect chromosomal abnormalities in the fetus, and even

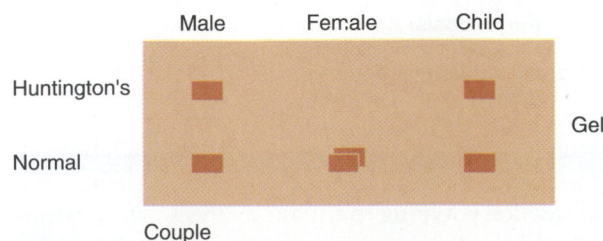

FIGURE 15.18 RFLP and Genetic Testing
In this pattern of gel electrophoresis, the male has the RFLPs linked with Huntington's disease, and so he has the gene for Huntington's; his wife is homozygous normal; and unfortunately their unborn child also has Huntington's. What an ethical dilemma. What would you do?

the absence of necessary enzymes or increased levels of some detrimental substances. Unfortunately, there is no treatment for many of the disorders diagnosed through our new technology. Thus prospective parents are faced with the very emotional question of whether to bear a child with the defect or abort the fetus.

Part of the question is financial. The care of a Tay-Sachs child costs more than $150,000 a year. Few couples can bear such expense over five years. The cost for the care of children suffering from other birth defects varies according to the particular disorder, but the question remains the same. Should society bear the cost of the medical expenses required by these genetically affected individuals? Obviously the social and moral dilemmas are great.

Postnatal Treatment

Numerous screening tests detect genetic defects that can be treated *postnatally* (after birth). Every year in this country, a simple urine or blood test to diagnose PKU can be used to prevent a life of mental retardation in the 200 or 250 infants found with the condition by just modifying their diet to exclude phenylalanine. Similar tests exist for the early detection of cystic fibrosis, and extensive treatment of this condition will enhance the chances for a child to survive the respiratory problems that would kill him or her at an early age.

THE FUTURE

With the genetic information explosion, many new techniques are being developed that expand our ability to detect and alter genes. Today we can even use viruses to insert genes into humans — gene therapy — and every day new ways of inserting genes into cells are being developed (see Focus 15.1). At this very moment, laboratories all over the world are using techniques to sequence the human genome (see Focus 13.1), the hundreds of thousands of genes that make up humans. This project, at a cost of $3 billion over a 15-year period, is one of the most ambitious biological research efforts ever conducted by humans. Where will the future lead in human genetic research? Who knows? But as educated individuals we do know that we have to be aware of the political and ethical nature of applying our genetic knowledge to control our genetic destiny.

SUMMARY

1. Many human genetic disorders are due to abnormal chromosomal makeup or abnormal coding of genetic information by the DNA molecule.

2. Down syndrome is a classic example of abnormal chromosomal nondisjunction in humans. Nondisjunction occurs when homologous chromosomes fail to separate during meiosis. Nondisjunction results in abnormal chromosome numbers; Down syndrome is usually a result of trisomy 21 — an extra chromosome 21.

3. Other forms of chromosomal abnormalities are deletion, in which a portion of the chromosome is missing; duplication, in which a portion of the chromosome is duplicated; translocation, in which a portion of one chromosome breaks off and attaches to another chromosome; and inversion, in which the genes on the chromosome become rearranged.

4. In females, only one X chromosome functions in any cell. The other X chromosome is inactive and is called the Barr body. The presence or absence of the Barr body is the basis for genetic sex determination tests. If the Barr body is present in the cell, the individual is a genetic female. If the Barr body is absent, the individual is a genetic male.

5. According to the Lyon hypothesis, female body cells form a genetic mosaic, depending on whether the paternal or maternal X chromosome is the active chromosome in a particular cell.

6. Nondisjunction of the sex chromosomes can produce the Klinefelter syndrome (XXY; feminized male) or the Turner syndrome (XO; sterile female).

7. Sex-linked inheritance is inheritance governed by genes located on the X chromosome. When a recessive allele is carried on the X chromosome, the recessive trait will be expressed more often in males than in females, because only females have a second X chromosome that can carry a dominant allele to mask the recessive one.

8. Many inheritable human diseases are caused by errors at the molecular level of the gene. Alkaptonuria, albinism, phenylketonuria, and sickle-cell hemoglobin are examples. An error in DNA replication may result in the lack of a particular protein (usually an enzyme), leading to a blocked metabolic pathway, or in an abnormal protein like sickle-cell hemoglobin. Blocking a metabolic pathway or producing an abnormal protein may result in a pleiotropy of effects.

9. Many traits (including human traits) are expressed along a continuum from one extreme to another. For example, skin color ranges from the very light to the very dark. This continuum is the result of several genes coding for skin color, known as polygenic inheritance.

10. ABO blood groups are determined by one gene with three alleles in the population. This is known as the pattern of multiple allelic inheritance. Antigen production (I) is dominant to no antigen production (i), and antigen A (I^A) production is codominant to antigen B (I^B) production. Hence the blood types can be A, B, AB, or O (which indicates no antigen).

11. Many new technological advances have resulted in methods for prenatal and postnatal detection of genetic disorders. This technology has resulted in many moral dilemmas.

KEY TERMS

pedigree
carrier
karyotype
nondisjunction
trisomy
translocation
Down syndrome
deletion

duplication
inversion
Barr body
fragile-X syndrome
hemophilia
codominance
antigen

REVIEW QUESTIONS

Multiple Choice

1. Sickle-cell anemia is an example of a genetically inherited disease resulting from:
   a. Feedback inhibition
   b. The improper sequence in the DNA code for hemoglobin
   c. Competitive inhibition
   d. Improper sequence of sugars in a carbohydrate
   (*page 349*)

2. Mrs. Brown said her baby was exchanged with Mrs. Doe's. Mrs. Brown's baby was blood type O and the Does are both AB. The baby could belong to the Browns if:
   a. Mrs. Brown is homozygous A, and Mr. Brown is homozygous O.
   b. Mrs. Brown is heterozygous B and Mr. Brown is heterozygous A.
   c. Mrs. Brown is AB, and Mr. Brown is O.
   d. Mrs. Brown is O, and Mr. Brown is homozygous B.
   (*page 353*)

3. Down syndrome is caused by a trisomic condition, which means that the cells of the affected persons have:
   a. Three nuclei
   b. Three somatic mutations
   c. Three sets of chromosomes
   d. Three of a specific chromosome, instead of two
   (*page 340*)

4. Assume that dimples are dominant over smooth cheeks and brown eyes dominant over blue eyes. A blue-eyed woman with no dimples marries a man with brown eyes and a dimple (his mother had blue eyes and no dimple). What is the probability that their child will have blue eyes and dimples?
   a. One out of four
   b. Three out of four
   c. Eight out of sixteen
   d. Nine out of sixteen
   (*page 320*)

5. The frequency of the gene for albinism is high in the Hopi Indian population because:
   a. The gene has been selected by cultural conditions of Hopi society.
   b. Isolation without selection caused the frequency of the gene to increase in the Hopi population but not in other populations.
   c. There is a higher mutation frequency to the albino condition in the Hopi populations.
   d. None of the above.
   (*page 348*)

6. Tay-Sachs disorder is a genetic condition associated with a/an:
   a. Autosomal dominant gene
   b. Autosomal recessive gene
   c. Trisomic condition
   d. Monosomic condition
   (*page 354*)

7. If a woman is a carrier for the color-blind gene that is sex-linked and her husband is perfectly normal, among their male offspring, what are their chances of having a color-blind son?
   a. None, since the father is normal
   b. One in two, since the mother is only a carrier
   c. 100 percent because the mother has the gene
   d. One in four because the mother is a hybrid
   (*pages 345-346*)

Genetics Problems

The answers to these questions can be found in the Appendix.

1. In humans, the ability to taste phenylthiocarbamide (PTC) is due to a dominant gene (T) and the inability to taste it is due to its recessive allele (t). A man who can taste PTC, but whose father and sister could not, marries a woman who can also taste PTC, but whose mother could not.
   a. The genotype of the man is:
   b. The genotype of his wife is:
   c. The genotype of the man's father is:

2. In humans, short fingers (brachydactyly) (B) and woolly hair (W) are dominant over the recessive alleles for normal fingers (b) and normal hair (w). A man with brachydactyly and normal hair

marries a woman with normal fingers and woolly hair. Their first child has normal fingers and normal hair.

   a. The genotype of the man is:

   b. The genotype of the woman is:

   c. The genotype of the child is:

   d. The probability of any child of the above matings having normal fingers is:

3. A woman has a rare abnormality of the eyelids called ptosis, which makes it impossible for her to open her eyes completely. The condition has been found to depend on a single dominant gene (P). The woman's father had ptosis but her mother had normal eyelids. Her father's mother had normal eyelids.

   a. The probable genotype of the woman is:

   b. The probable genotype of the father is:

   c. The probable genotype of the woman's mother is:

4. In humans, males have XY sex chromosomes, females have XX, and color blindness is sex-linked. What are the expected sexes and color vision of children from the following crosses?

   a. Normal-color-vision (homozygous) female × color-blind male.

   b. Normal-color-vision (heterozygous) female × color-blind male.

5. From the accompanying pedigree, answer the following questions. Circles indicate females; squares indicate males. Blue figures indicate normal phenotypes; yellow figures indicate hemophiliac individuals.

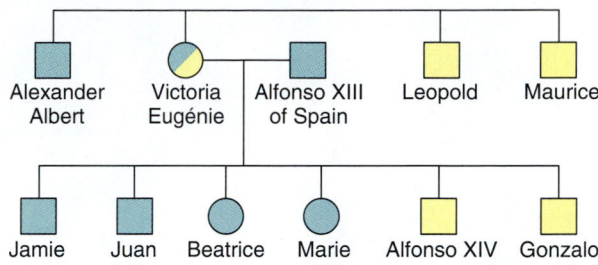

   a. If X^H is the gene causing hemophilia and X is its normal allele, the genotype of Victoria Eugénie is _____.

   b. In the case of the female children, the probability that Marie is a carrier of hemophilia is _____.

6. What are the possible genotypes of the parents of a person with blood type AB?

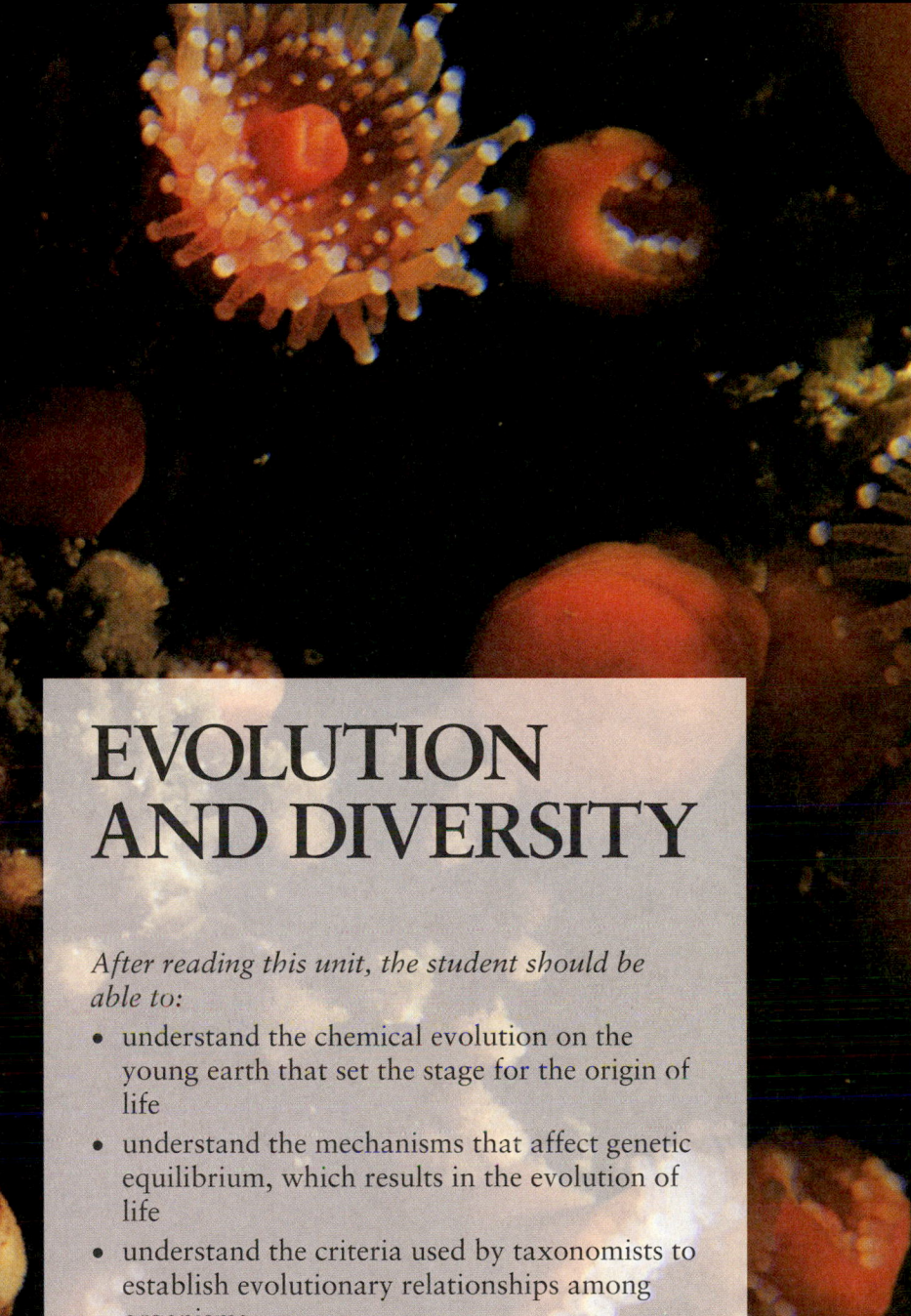

EVOLUTION AND DIVERSITY

After reading this unit, the student should be able to:

- understand the chemical evolution on the young earth that set the stage for the origin of life
- understand the mechanisms that affect genetic equilibrium, which results in the evolution of life
- understand the criteria used by taxonomists to establish evolutionary relationships among organisms
- understand the importance of mutations and genetic recombination, which generate heritable variation, and which are subject to natural selection
- understand the classical and modern explanations of evolutionary change

16

Life: How Did It All Begin?

As we gaze up at the sky on a hot summer night, it is difficult not to ask, "How did it all begin?" How did life originate on planet earth? Humans have pondered this question since the beginning of their time on earth, and they have developed many explanations for life's origins.

SPONTANEOUS GENERATION VERSUS LIFE FROM LIFE

■ **Early scientists believed that life arose from nonlife, a theory called abiogenesis. However, during the mid-nineteeth century, research showed that life as we know it could only arise from life. This theory is called biogenesis.**

Above: Lightning, volcanic activity, thermal energy from the earth's interior, and even radioactivity served as energy sources that resulted in the first life on earth.

Aristotle (384–322 B.C.) and other ancient Greeks believed that life could arise from nonlife, that modern species (flies, mice, and so on) arose fully evolved — a theory now called **spontaneous generation,** or *abiogenesis* (life from nonlife). Spontaneous generation seemed to be confirmed by everyday observations: Flies grew from rotting meat, frogs grew from mud and eels from sand, and cockroaches grew from garbage.

Although it may be difficult to believe today, the idea of abiogenesis was widely believed, and it formed the basis for conventional wisdom even among respected scientists until the mid-nineteenth century. Even today, some people rely on abiogenesis to explain their observations. When I was a child, my grandfather used to tell me that if I pulled a horsehair (including the root) from the mane or tail of a horse and placed it in water, it would turn into a worm. I tried it. For a time it seemed that my room was filled with fruit jars containing water and horsehairs. Needless to say, not one horsehair turned into a worm! My grandfather did not know that there was a worm, which resembles a horsehair, that lives in pools. He did not know that the life we see around us does not come from nonlife.

Biogenesis

Aligned against the biologists who supported abiogenesis were others who supported the idea of **biogenesis** — the theory that life could only arise from life. Still others tried to straddle the fence. They were willing to accept the idea that the macroscopic organisms (those large enough to see with the unaided eye) arose only biogenetically. But what about microorganisms like bacteria? They must arise spontaneously. The controversy lasted until 1862, when the noted microbiologist Louis Pasteur performed what is now called his swan-necked flask experiment.

It had already been observed that if beef broth were boiled and left to stand in the open air, bacteria soon began to grow in it. This was taken as evidence of spontaneous generation. Pasteur performed two experiments that showed that spontaneous generation was not in fact responsible for the growth in the flasks (Figure 16.1). First, he boiled the broth and kept it sealed in a sterile flask. No microbes grew, so Pasteur knew that exposure to air was necessary for the growth to occur. Next, he boiled the broth in flasks with long necks curved into an S shape through which air could enter. No bacteria grew. If he broke off the long necks, bacteria would soon appear.

Pasteur reasoned that although air could enter these flasks, the curved neck trapped small particles like dust and airborne microorganisms. Thus in the open flasks the bacteria grew not from the nonliving contents but from the organisms deposited in the broth by the air. Pasteur's experiment laid the idea of abiogenesis to rest. (In fact, some of his swan-necked flasks are still intact and sterile in the Pasteur Institute in Paris.)

PANSPERMIA HYPOTHESIS

■ **Panspermia is the theory that life came to this planet from elsewhere in the universe, either deliberately, through the intervention of intelligent beings, or accidentally, through the actions of meteorites.**

Biogenesis is the accepted theory today, and all evidence supports the hypothesis that life comes from life. But where did the first life come from? One interesting hypothesis on the origin of life, popular in the nineteenth century and, in some cases, even now, is the idea known as **panspermia,** which postulates that earth was a lifeless planet until life came here from elsewhere in the universe. Some proponents of panspermia feel that the process was deliberate, through the intervention of alien intelligent beings. Others feel that it was merely fortuitous that the basic building blocks for life were "seeded" on earth. For example, meteorites that struck the earth early in its history may have contained living bac-

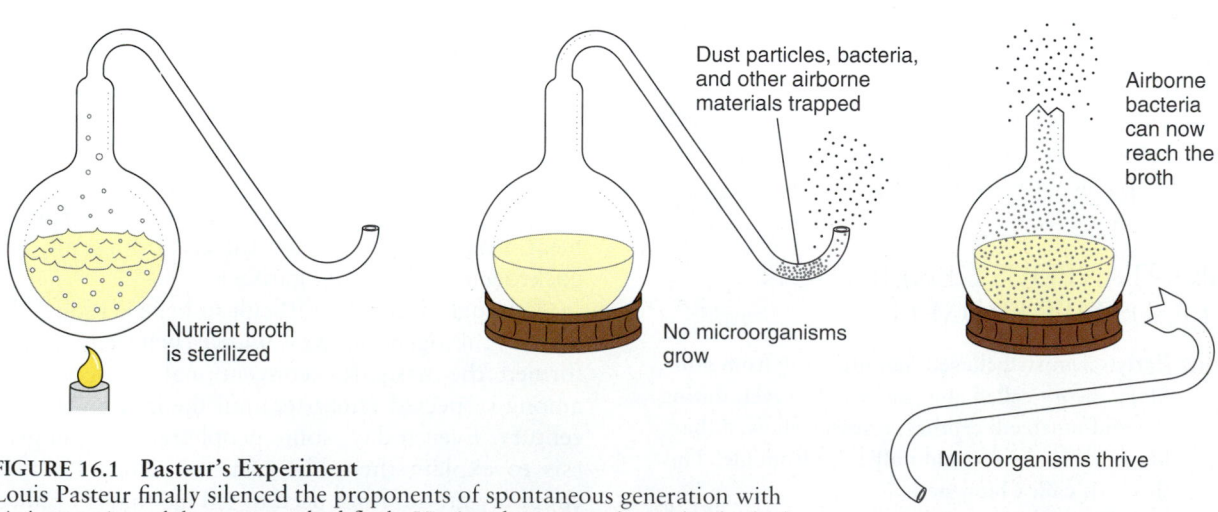

FIGURE 16.1 Pasteur's Experiment
Louis Pasteur finally silenced the proponents of spontaneous generation with his innovation of the swan-necked flask. Having drawn out the neck of a flask into an S shape, he sterilized the nutrient broth it contained. This allowed air to reach the broth but presented a barrier to any airborne material. Nothing grew in this oxygen- and nutrient-rich environment. Only after the neck was cut and the broth exposed to the airborne bacteria did the microorganisms thrive. Pasteur concluded that life arises exclusively from life.

Chapter 16   Life: How Did It All Begin?

terial spores, some of which germinated to become the first living things on this earth.

What support do the adherents of panspermia bring to their hypothesis? They postulate that, if life evolved on the planet earth, the more abundant elements would be required in essential reactions. In reality, however, rare elements such as molybdenum are essential for many enzymatic reactions. Why did nature favor the inclusion of a rare element in an essential reaction? Perhaps the reason is that unusual elements, so rare on the earth, were more prominent on other planets.

This theory has the support of some notable scientists, such as Nobel laureate Francis Crick, best known for his work in describing the molecular structure of the gene. Dr. Cyril Ponnamperuma (see Profile) believes that panspermia was and is possible. Working with meteorite fragments, he has isolated some of the building blocks of life. He states that these findings help to support the idea that life can begin in other parts of the universe and indeed may be plentiful. At the America Chemical Society meeting in the fall of 1986, Dr. Ponnamperuma said, "Life is a natural and inevitable course of chemical evolution. To me, the universe appears to be in the business of making life."

Although it does have some adherents, the theory of panspermia has gained little support in the scientific community. Even if it were widely accepted, it still only postpones the question of the origin of life, since we then must ask how life originated on that planet in the other solar system. Panspermia is not considered an acceptable hypothesis by the majority of scientists today.

CHEMICAL EVOLUTION

Many other hypotheses and a few theories on the origin of life on earth have been suggested. Much modern theory is based on the ideas developed by A. I. Oparin and J. B. S. Haldane in the 1930s. This work describes the formation of the planet and the chemical evolution that led to the first forms of life.

The Primordial Cloud Theory

■ The primordial cloud theory is part of a theory about the origin of the universe. This theory states that the universe began as a great explosion and that the bodies now found in the universe, our sun and planet included, arose as a result of cooling and gravity.

To understand Oparin and Haldane's reasoning, we need to look at the origin of our solar system and

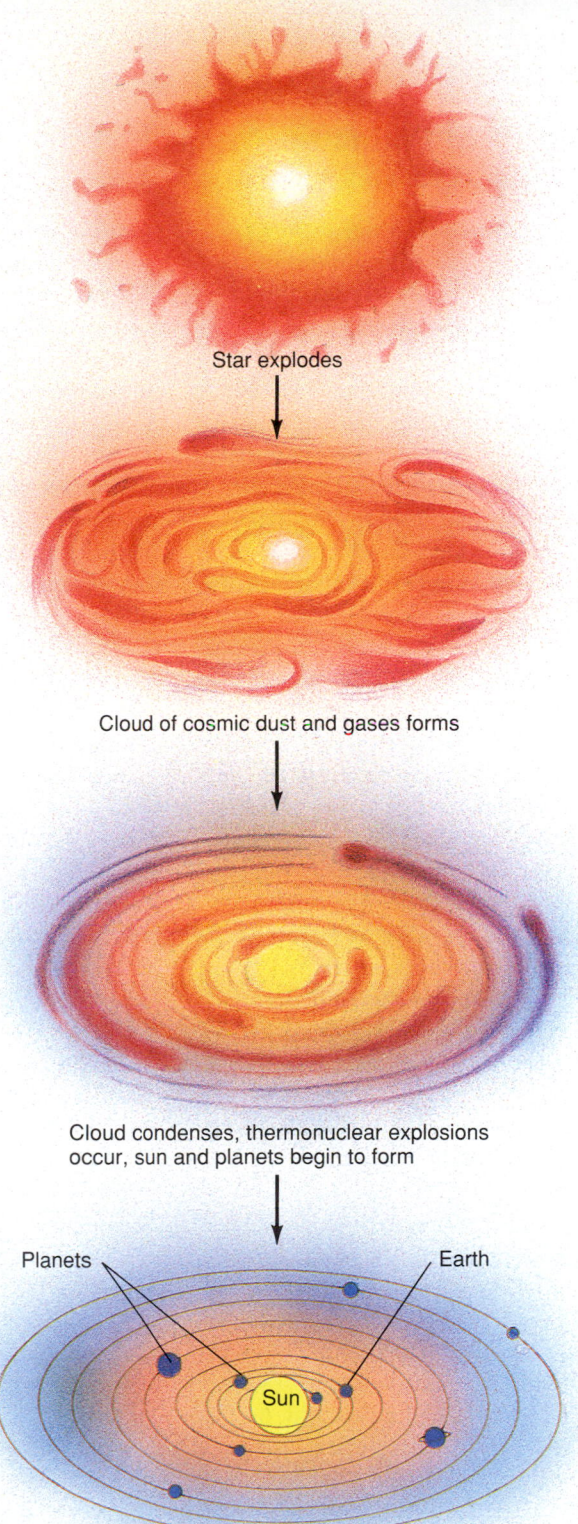

Star explodes

Cloud of cosmic dust and gases forms

Cloud condenses, thermonuclear explosions occur, sun and planets begin to form

Planets / Earth / Sun

Recent solar system

FIGURE 16.2 Primordial Cloud Theory
Our solar system had its beginning billions of years ago when a stellar explosion created a massive cloud of dust and gases. Over time, this cloud collapsed on itself, with its substance being ever more concentrated into the sun and the planets.

Dr. Cyril Ponnamperuma
Chemist, Exobiologist

How, when, and where did life begin? Did life originate on this planet or were life's chemicals carried through space and seeded on earth? Will an understanding of the origin of life on this planet provide us with information about the possibility of life in other parts of the cosmos? These are a few of the questions that concern Dr. Cyril Ponnamperuma, director of the University of Maryland's Laboratory of Chemical Evolution. Dr. Ponnamperuma is an exobiologist, a scientist who studies life outside the confines of earth.

Now in his sixties, Cyril Ponnamperuma was born and raised on the island in the Indian Ocean once called Ceylon and now known as Sri Lanka. As a boy he was inspired primarily by his uncle, who was a scientist. His uncle would perform simple chemical experiments, like dipping a copper coin into mercury and turning its color silver, or cleaning marbles in vinegar. Ponnamperuma now believes, "That sort of thing I suppose influenced me, and it drove home the fun of doing science."

Ponnamperuma first earned a degree in philosophy at the University of Madras in India. His interests then switched to chemistry, so he turned to the University of London, at which he earned a bachelor's degree with honors in that field. In 1962, he earned a doctorate in biochemistry at the University of Cali-

fornia at Berkeley. While he was studying there, he worked with Dr. Melvin Calvin, the Nobel laureate best known for his work in photosynthesis. Calvin was also interested in the origin of life, and his interest fired Ponnamperuma's interest in the chemical origin of life.

Ponnamperuma soon concentrated on one main question: If we study meteorites, the alien dust of other planets, can we find evidence of life's chemical evolution? In 1963, as a measure of the seriousness of his research, he joined a research team at the National Aeronautics and Space Administration, Exobiology Division. He was appointed chief of the branch concerned with chemical evolution. This position gave Ponnamperuma access to many important resources. He was able to look for the signs of life in meteorites, and in 1970 while analyzing a particular meteorite called the Murchison meteorite, he identified amino acids, the building blocks of protein. He had found the essential beginnings for the chemicals of life.

The discovery in 1970 was only a beginning. Recently Ponnamperuma made an even more important discovery. In the powder of a meteorite he found adenine, cytosine, guanine, thymine, and uracil, the nitrogen bases required for the genetic instructions in RNA and DNA. There, in a meteorite from out in space, were major components of the mole-

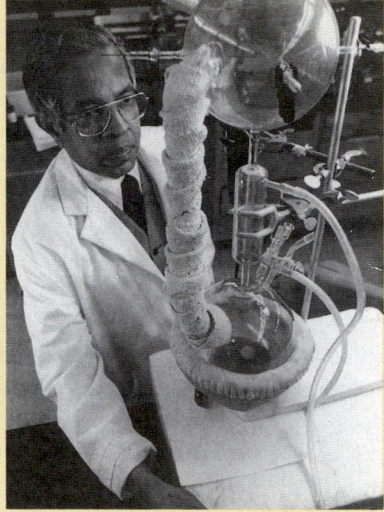

cules that control and regulate life in living things.

Has Ponnamperuma found evidence for panspermia? Does life chemically similar to ours exist elsewhere in the universe? Ponnamperuma would probably say yes, but many other scientists, experts on the prebiotic synthesis of life, would say that the meteorites were contaminated once they reached the earth. In other words, it was earthly matter that accounted for the presence of the components of DNA, not the evidence that these compounds were found naturally in meteorite dust.

So we come back to the same questions. How did life originate on this planet? Is life commonplace in the universe? The quest for these answers goes on in laboratories all over the world. What do you think?

our planet. Most cosmologists (scientists who study the universe) support what is called the **primordial cloud theory** (Figure 16.2). If we could travel back in time to watch the beginning of our solar system, what would we see? We would see clouds of cosmic dust and gases, the result of gigantic stellar explosions. One of these clouds took the form of a flattened disc having a diameter measuring one light-year (the distance light travels in one year's time).

Over time the primordial cloud condensed because of gravitational forces and compression. Then, 5 billion years ago, owing to the exertion of tremendous gravitational pressures, there was a series of thermonuclear reactions, causing the main mass of the cloud to become the luminous body that is our sun. Other centers of concentration formed the planets, including the one we call home.

As earth journeyed through space, it kept contracting, forming a dense, hot core with a temperature of 1000 to 3000°C. Eventually, about 3.7 billion years ago, this thin-crusted planet began to cool. A primitive atmosphere formed from gases emitted from cracks in the crust and from volcanic activity (see Focus 16.1).

Primitive Atmosphere and Primitive Seas

■ There are many theories about the atmosphere of primitive earth. Most theories hold that the primitive atmosphere contained compounds made of hydrogen, carbon, nitrogen, and sulfur, but little if any free oxygen. With cooling, rains came, forming primitive oceans that contained the elements necessary for carbon-based life.

Many scientists speculate that prebiological earth's atmosphere consisted principally of water vapor (H_2O), methane (CH_4), ammonia (NH_3), hydrogen cyanide (HCN), molecular hydrogen (H_2), and smaller amounts of hydrogen sulfide (H_2S), carbon dioxide (CO_2), and carbon monoxide (CO). Many scientists speculate that free oxygen (O_2) was not present in the atmosphere; instead most of it was bonded to nonvolatile silica compounds in the earth's crust.

However, other scientists — like Dr. Joel Levine, an atmospheric scientist who has utilized computer analysis in his research on earth's early atmosphere — feel that the early atmosphere did not contain large amounts of ammonia or methane. Instead Levine says that it consisted primarily of water vapor, carbon dioxide, and nitrogen. Levine and his coworkers also feel that the early atmosphere contained a million times more oxygen than was previously assumed. What is the basis of these statements? It is probable that the early sun produced more ultraviolet light than it does today. The amount of ultraviolet light coming into the atmosphere chemically decomposed water and carbon dioxide to produce oxygen. Therefore, given the calculations Levine has made, the oxygen content of the atmosphere would be larger than the present figures indicate.

Even though there are several theories about the composition of the early atmosphere, most scientists feel that it was a *reducing atmosphere,* which means that it contained much more hydrogen than the atmosphere of today. Free oxygen, the element so essential to our energy utilization, was probably only present in small amounts. However, according to Levine's studies, the concentration of oxygen was probably a million times greater than was first suspected. More and more scientists suspect that Levine may be correct. At that time lower levels of oxygen were an advantage, because free oxygen is a highly corrosive chemical that might have destroyed or prevented reactions leading to the first living precells.

Over long periods of heating and cooling, the water vapor in the atmosphere condensed, rain fell, and larger and larger pools of water accumulated on the earth's surface. These first seas contained dissolved gases and minerals obtained through the washing of the earth's crust. This mixture formed what is called an organic soup, containing the raw materials required for the beginning of carbon-based life forms.

With the formation of the earth's primordial seas, the history of life on earth was ready for the next logical step — the formation of simple organic compounds. As Charles Darwin said in 1871, "In some warm little pond with all sorts of ammonia and phosphoric salts, light, heat and electricity present, a protein compound was chemically formed to begin a process of biological evolution."

Monomer Formation

■ Experimental evidence supports the idea that the input of energy from lightning and from the cooling core of the earth was sufficient to result in the formation of simple compounds from the elements found in the primitive oceans.

How were the raw materials of the early seas transformed into an organic soup composed of simple organic molecules? Many scientists had asked this

FOCUS 16.1
Earth, the Living Planet

Viewed from the distance of the moon, the astonishing thing about the earth, catching the breath, is that it is alive. The photographs show the dry, pounded surface of the moon in the foreground, dead as an old bone (Figure A). Aloft, floating free beneath the moist, gleaming membrane of bright blue sky, is the rising earth, the only exuberant thing in this part of the cosmos. If you could look long enough, you would see the swirling of the great drifts of white cloud, covering and uncovering the half-hidden masses of land. If you had been looking for a very long, geological time, you could have seen the continents themselves in motion, drifting apart on their crystal plate, held afloat by the fire beneath. It has the organized, self-contained look of a live creature, full of information, marvelously skilled in handling the sun.

Source: Lewis Thomas, *The Lives of a Cell: Notes of a Biology Watcher*, Viking Press, New York, 1974.

FIGURE A
The earth as seen from the moon by Apollo 11 astronauts.

question, but in 1953, Stanley Miller, a graduate student working under Dr. Harold Urey at the University of Chicago, decided to put the question to the test. Miller designed an ingenious apparatus that simulated the conditions presumed to exist on primitive earth. Into a sterile, airtight apparatus, he introduced the gases methane (CH_4), ammonia (NH_3), water (H_2O), and hydrogen (H_2). He then needed an energy source to drive chemical reactions. What might this energy be?

Miller knew that the primitive atmosphere contained no free oxygen, nor was there the ozone layer that forms the protective cover in today's atmosphere. That meant that ultraviolet light bombarded primitive earth. There was also thermal energy from the earth's interior, electrical discharges from lightning, and even radioactivity from the decay of atoms in the earth's crust.

Miller chose the electrical energy found in lightning as his energy source. Within his apparatus the gases were circulated, boiled, and exposed to electrical discharge. When he analyzed the resulting mixture, it contained organic monomers important to life, such as amino acids, urea, and acetic acid (vinegar) (Figure 16.3).

Moving from Monomers to Polymers

■ **Formation of complex molecules is enhanced by an enclosed membrane that isolates the contents from the external environment, thus promoting interactions between compounds and preventing dilution of a newly formed molecule.**

The next step in the chemical evolution of life involved moving from simple organic molecules to more complex molecules. This process, which involves binding together simple molecules, or monomers, to form more complex molecules, or polymers, is called *polymerization*. But in order to increase the probability that monomers will com-

Chapter 16 Life: How Did It All Begin?

(a)

FIGURE 16.3 Miller's Experiment
(a) Stanley Miller was able to synthesize simple organic molecules abiotically using an apparatus that simulated the earth's early atmosphere. (b) The lower portion simulated the primitive oceans; the upper part, the primitive atmosphere.

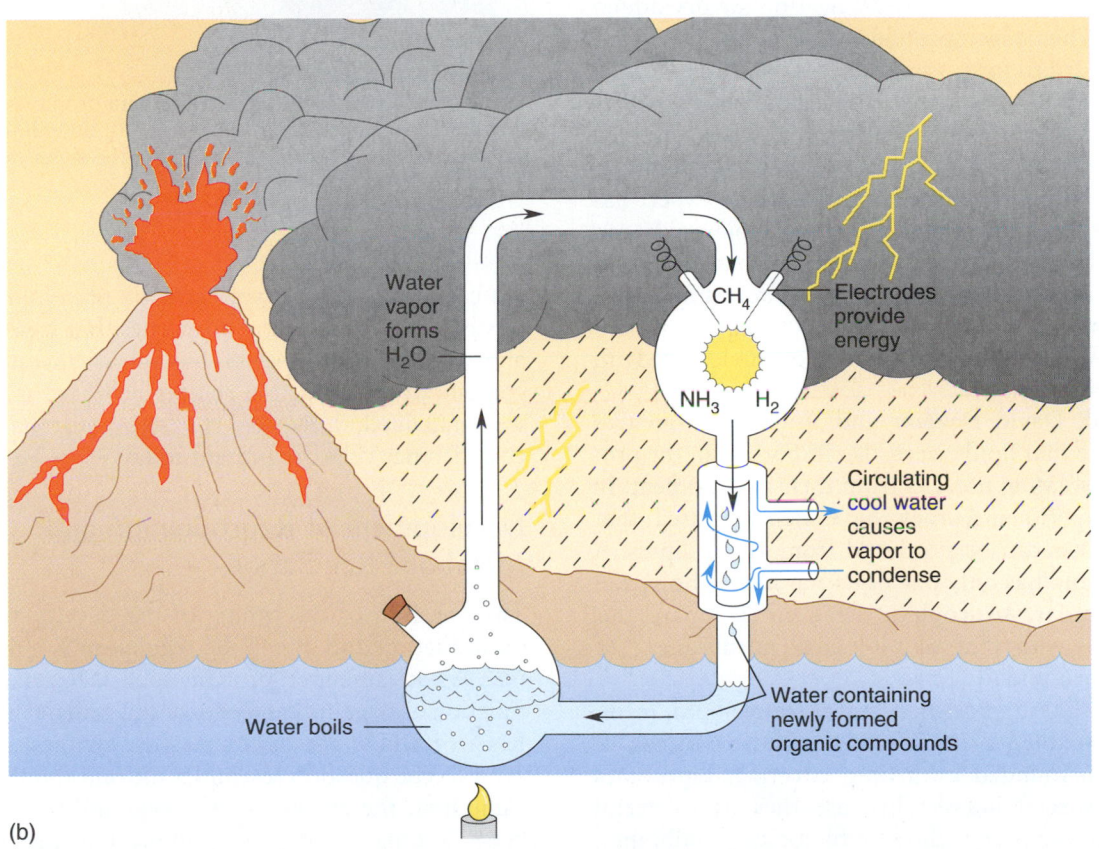

(b)

bine, they must first be isolated from the environment.

Isolation: The Basis of Precell Formation. The formation of precells is an essential step, because it provided the necessary isolation. In other words, each precell had an interior region that was separated from the outer environment. The segregation and concentration of the components of living things into discrete droplets was necessary in order to avoid diluting the newly formed molecules.

How did the **protobionts,** or *first precells,* come into being? Scientists speculate that the first protobionts were microscopic droplets suspended in the organic soup of primitive earth. They must have had simple water-repellent, membranelike lipid or protein coverings that could isolate the contents of the droplet from the watery surroundings.

We have not found fossil examples of the precursors to cells, nor do we expect to, but we have long known of three systems that could be considered similar to protobionts. First, Oparin suggested in

1938 that proteinlike substances aggregated into droplets and formed enveloping membranes. He called these droplets **coacervates.** Coacervates are formed by the suspension of particles in a liquid medium, a colloid. These droplets are similar to the small fat droplets that form in nonhomogenized milk. Scientists have found that if enzymes and other substances are added to coacervate droplets, simple reactions occur. These reactions are similar to reactions in contemporary cells (Figure 16.4).

A second form of sphere was discovered by Dr. Sidney Fox and his colleagues at the University of Miami. In 1959, they conducted a series of experiments elaborating on Miller's work. They reported that they had been able to synthesize small proteins from amino acids simply by heating the dry amino acids. When these proteins were placed in water, they tended to form small spheres, which Fox called proteinoid microspheres. **Proteinoid microspheres** are double-membrane droplets, and when observed under the electron microscope, they are so similar to some primitive bacteria that even an expert has difficulty distinguishing the two. Proteinoids can absorb material from the liquid surrounding them and increase in size. They also produce small bud-like projections that pinch off to form new proteinoids. Are these new droplets an example of growth and reproduction?

Third, David Deamer and William Hargreaves speculate that lipids were also important in the process of isolating precells from their environment. In their experiments, they have demonstrated that membranes can be produced from simple lipids. A simple lipid has only one nonpolar hydrophobic tail and a polar, hydrophilic head (to compare this structure with that of biological membranes, see Chapters 3 and 5).

The sphere produced by a simple-lipid membrane is called a *liposome,* a microscopic bag of water surrounded by a lipid covering. Liposomes are of special interest because they have several properties similar to those of biological membranes. For instance, without the need for outside intervention, liposomes are capable of self-assembly into spheres. They are selectively permeable — in other words, they are permeable to water and impermeable to chemical ions. Finally, because the membrane is fluid and elastic, the spheres are capable of self-repair.

From the research conducted thus far, scientists now speculate that the first precells (protobionts) either were similar to coacervates, proteinoid microspheres, or liposomes or were combinations of the three. Of course, it is possible that the first cells were totally unlike anything proposed thus far, but

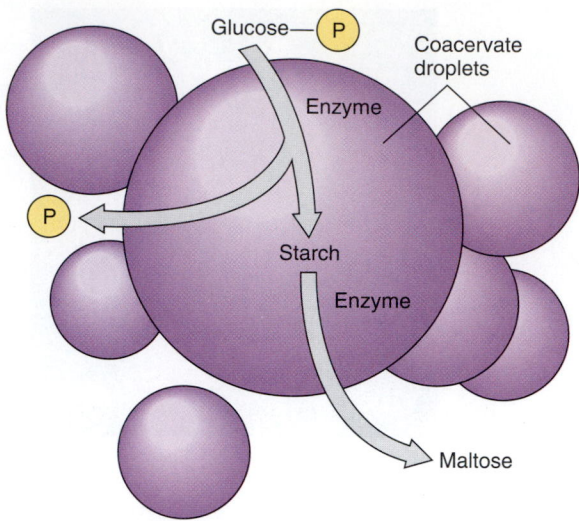

FIGURE 16.4 Coacervates
In the presence of enzymes, a coacervate droplet will polymerize a simple sugar (glucose-phosphate) into starch, and then break that down into the sugar maltose.

we cannot dismiss coacervates or proteinoids because they have so many properties that were probably prevalent in the prebiological ancestors of contemporary cells. No matter how the isolation was managed, however, the step must have occurred, since cells are present today.

Development of Reproduction and Metabolic Regulation

The question of the timing of the development of reproduction and metabolic regulation is a difficult one. What came first, proteins such as enzymes that function as organic catalysts, or the genetic material RNA or DNA, which controls the synthesis of enzymes? The question resembles the one about which came first, the chicken or the egg, and the controversy is unresolved. We examine the question in Focus 16.2; here we will operate on the basis of the most likely explanation: The protobiont contained some genetic material within it.

Self-Replicating RNA. Today many scientists propose a theory about the origin of life based on the findings of Thomas Cech (see Profile, Chapter 1) and Sidney Altman (see Focus 16.2) that RNA has both catalytic (enzyme) and genetic information storage properties (Figure 16.5). The **RNA-based theory** states that the first form of life was based on simple versions of RNA. In other words, the first genetic material to regulate the life activities in the

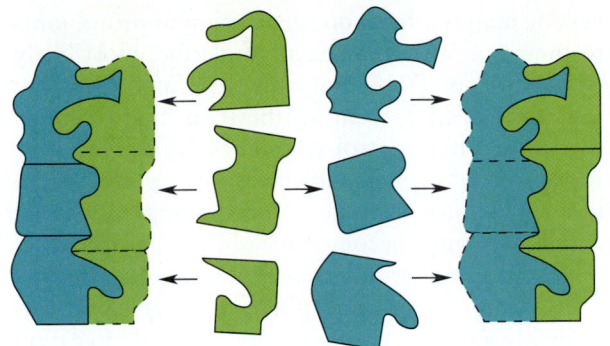

FIGURE 16.5   Self-Replication
In self-replicating molecules, the original molecule or a portion of it serves as the template from which the new molecule is formed. One of the major steps in the origin of life was the evolution of self-replicating molecules.

first precells was not DNA, but RNA. This RNA could not only carry the genetic information required for protein synthesis, but it could also make copies of itself (see Focus 1.1, page 20).

According to this theory, the primordial world consisted of RNA and pre-RNA molecules that dominated the world for 400 million years and served as both the genetic material and the enzymes mediating the reactions that resulted in life. Some biologists, like Nobel laureate Walter Gilbert (see Profile, Chapter 1), have called the primordial world "the RNA world." Many investigators, such as Stanley Miller and Gerald Joyce from the University of California at San Diego, support this theory. They propose that for about 400 million years, RNA dominated life. Then about 3.8 billion years ago DNA probably evolved from RNA and became the molecular basis of the gene, so that today, all forms of life are based on DNA (Figure 16.6). You're probably asking why some scientists think that RNA came before DNA. There are many reasons, but basically it boils down to the simple fact that DNA is just too complicated a molecule to come together by chance in the primordial seas. It is more logical that there was a series of develop-

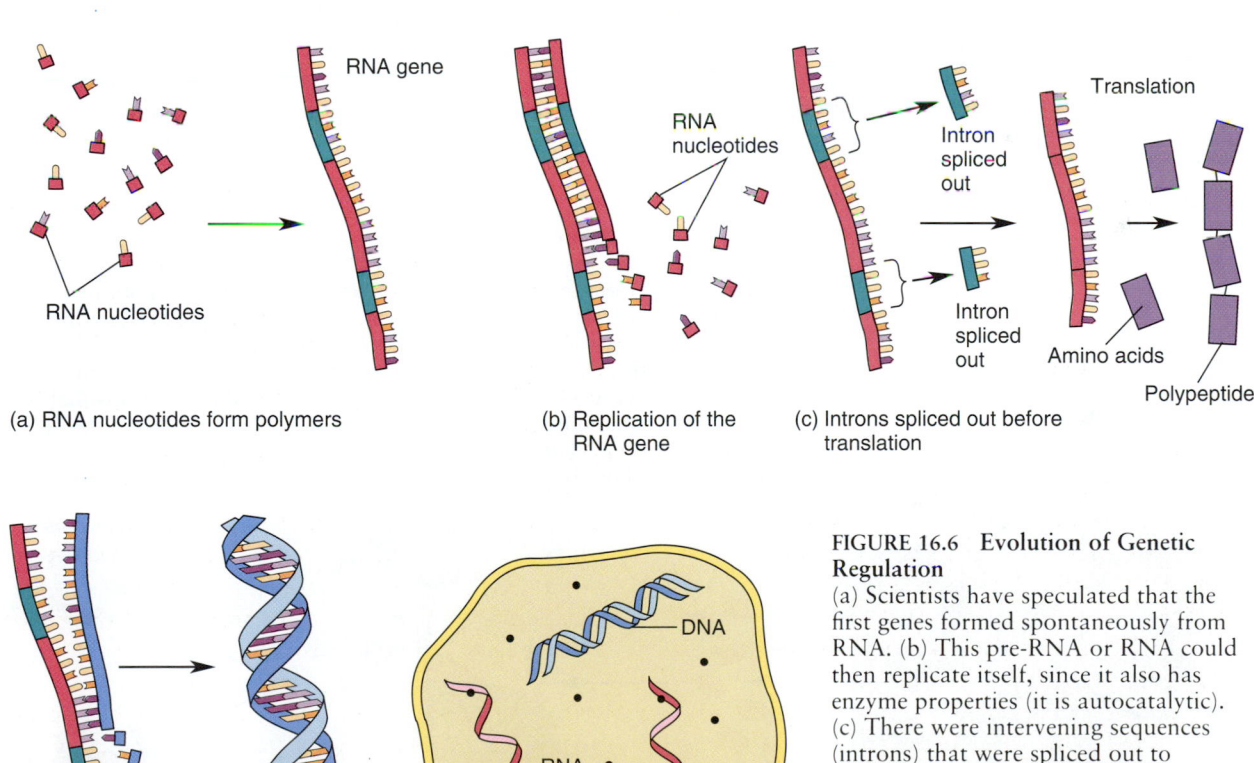

(a) RNA nucleotides form polymers

(b) Replication of the RNA gene

(c) Introns spliced out before translation

(d) Reverse transcriptase enables DNA to form from RNA

(e) The earliest DNA cell

FIGURE 16.6   Evolution of Genetic Regulation
(a) Scientists have speculated that the first genes formed spontaneously from RNA. (b) This pre-RNA or RNA could then replicate itself, since it also has enzyme properties (it is autocatalytic). (c) There were intervening sequences (introns) that were spliced out to produce the active form used to translate the genetic code into a polypeptide. (d) Eventually reverse transcriptase evolved; this is an enzyme capable of synthesizing DNA from RNA so that DNA then evolved. (e) From then on the living world was dominated by DNA.

ments that eventually led to DNA. Also, RNA is simpler and can function as an enzyme. Some even think that RNA is too complex, so that there must have been RNA-like molecules. Those who have argued against the idea of the RNA-based origin of life have asked why, if indeed the first life was RNA-based, scientists can't synthesize RNA-like molecules in the type of apparatus used by Miller (see Figure 16.3) to duplicate the primitive atmosphere. In 1991, however, scientists synthesized **self-replicating RNA.** Dr. Jennifer A. Doudna and her colleagues at Massachusetts General Hospital in Boston reported that they had assembled a self-replicating RNA molecule. They took a piece of RNA from a bacteriophage, modified it, and found that three of the resulting fragments could bind together to form an RNA molecule that could synthesize each of the fragments from sugars, phosphates, and nitrogen bases—the materials that might have been available in the prebiotic world. These experiments suggest that the spontaneous formation of RNA replicases (self-replicating molecules) from the prebiotic "soup" was almost certainly possible and not as complicated as many investigators have believed.

If the activities of the protobiont occurred as stated above, reproduction was not a haphazard event; instead there was a reliable reproduction of genetic material from one generation of protobionts to the next. Since the genetic molecule, most likely RNA and not DNA, not only replicated itself but also controlled protein synthesis, it could control reactions. That control would result in greater organization of organic molecules into definite structures, because energy from energy-liberating reactions would become available for use in reactions within the protobionts that required energy.

Remember, the atmosphere of early earth was probably a reducing atmosphere with little free oxygen (O_2). Therefore, the first cells had to be capable of obtaining energy stored in the organic molecules in their primordial soup without using oxygen — a form of *anaerobic respiration*. With the coming of reproduction, regulation, and respiration, carbon-based primitive cells had now been formed. The earth had come alive (see Human Endeavors).

THE EVOLUTION OF AUTOTROPHS

■ **As early cellular life forms consumed the nutrients available in the early oceans, they depleted the supply. This gave selective advantage to life forms that could form their own nutrients, the autotrophs. Present life forms called methanogens may be closely related to the early autotrophs.**

FIGURE 16.7   Chemosynthetic Bacteria
Evidence that the first forms of life may have used H_2S as their hydrogen source comes from the archaebacteria that live on the ocean floor near the vents that erupt from the inner earth. These autotropic bacteria use chemical H_2S to produce simple organic compounds. They are chemosynthetic. The products of their synthesis — the simple organic molecules — serve as the basis of the food chain for the organisms that live in these communities. For example, the archaebacteria live in the gills of tubeworms and clams, providing food. Then filter feeders or carnivores feed on the tubeworms or clams.

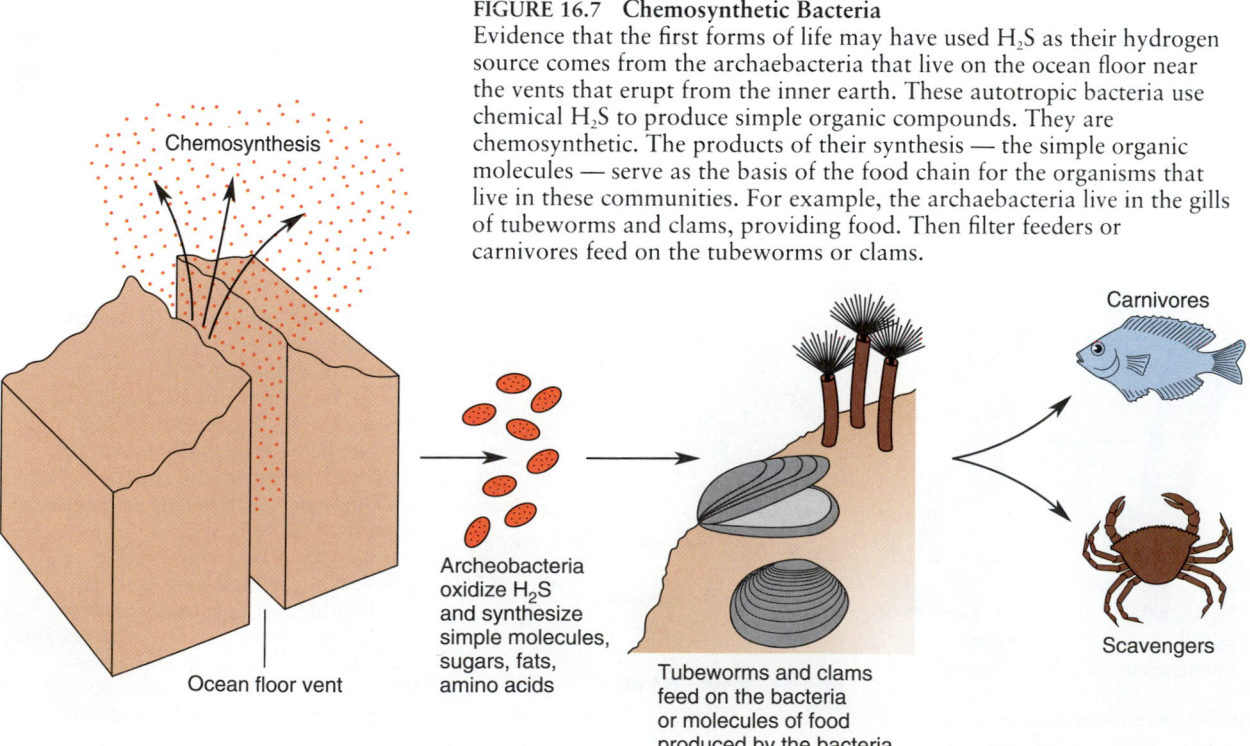

Chemosynthesis

Ocean floor vent

Archeobacteria oxidize H_2S and synthesize simple molecules, sugars, fats, amino acids

Tubeworms and clams feed on the bacteria or molecules of food produced by the bacteria

Carnivores

Scavengers

The lifestyle of these first cells was a heterotrophic one; that is, they relied on an external source of complex organic nutrients, such as proteins, sugars, and fats, for their energy. However, the voraciousness of the cells' appetites began to deplete the supply of nutrients. As the necessary nutrients disappeared, some primitive cells evolved the ability to synthesize some essential complex materials such as proteins, carbohydrates, fats, and nucleic acids from the simpler substances in the soup.

Primitive cells that could synthesize their food from the molecules in the organic soup of the primitive oceans became the self-feeders, or *autotrophs,* and their abilities gave them a selective advantage in that environment.

In 1977, scientists located groups of archaebacteria ("ancient bacteria") that live around the hot vents on the ocean floor (Figure 16.7). These archaebacteria are among the oldest forms of life today. They utilize hot hydrogen sulfide (H_2S) as the hydrogen source in reducing carbons to synthesize complex carbon compounds that can be used as a source of energy. Eventually these first cells were able to use the hydrogen in water rather than in hydrogen sulfide as their hydrogen source to reduce carbon dioxide.

In the 1950s, biologists discovered some primitive bacteria called *methanogens* (Figure 16.8). They were given this name because they use carbon dioxide (CO_2) plus hydrogen (H) to form methane (CH_4) and to release energy. Methanogens live in oxygen-free environments, such as sewage and the intestines of people, sheep, and cows. Some scientists speculate that these cells are the distant relatives of the first autotrophs. Indeed, fossils of the methanogen-like bacteria have been found in rocks in South Africa that date back more than 3.4 billion years, which would make them one of the oldest forms of life known. Recently Dr. William Schopf reported finding fossil cyanobacteria in rocks in western Australia that date from 3.5 to 3.7 billion years ago.

Light energy and carbon dioxide from anaerobic respiration were used to develop another form of autotrophic lifestyle known as *photosynthesis.* This was a critical development. Because photosynthesizing organisms produced oxygen as a waste product, they were polluting the methane atmosphere of primitive earth. Although it may seem strange, the transformation of the atmosphere was detrimental to some organisms, to whom an oxidizing atmosphere was harsh and poisonous. In order to live, these organisms required an oxygen-free environment, such as we could find today in, for example, a mud flat.

To other organisms, however, the change in the makeup of the atmosphere was an advantage because they had developed the ability to use oxygen to break down their foodstuffs. For instance, in the energy-liberating reactions of some organisms, as electrons were transported along the electron transport chain, oxygen was used as the final electron acceptor. When the electrons combined with hydrogen ions to form hydrogen atoms, these hydrogen atoms could combine with oxygen to form water molecules. Aerobic respiration had evolved. As we discussed in Chapters 7 and 9, aerobic respiration is a much more effective way of synthesizing ATP. In other words, more units of energy can be released per molecule of fuel in the presence of oxygen than without it.

There was one other effect of the development of the photosynthetic autotrophic lifestyle: It provided a food source on which the heterotrophs could feed.

THE COMING OF THE EUKARYOTES

■ **With the arrival of primitive cells containing membrane-bound nuclei and other organelles, the first eukaryotes became an active force in early life.**

The first primitive cells were probably like the bacteria and cyanobacteria (blue-green) forms similar to the algae of today, especially in that they lacked a membrane-bound nucleus or other intracellular membrane-bound structures. We call these types of

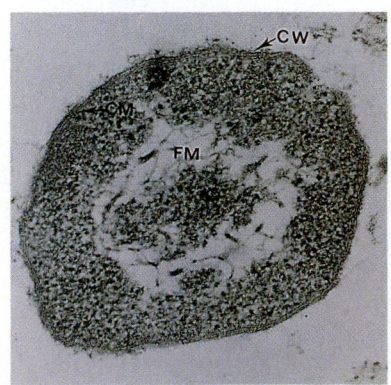

FIGURE 16.8 The First Autotrophs
The first self-feeders were a primitive form of bacteria like this methanogen. Methanogens can be found in oxygen-free environments (like these that were isolated from the sediments of the Great Salt Lake in Utah), where they respire anaerobically.

B.C.: The biblical account of Genesis describes the creation of life by God from the dust of the earth.

3000 B.C.–1650 A.D.: Nearly all cultures embraced some form of the theory of spontaneous generation, in which creatures arose from mud or earth, sometimes with a divine helping hand.

1862: Pasteur's sterilization experiments disproved that microbes spontaneously form in organic broths.

Sterilized broth

Condensation of solar system

Interplanetary dust

Atmosphere

1870–1900: By the end of the 19th century, some scholars embraced the idea that life on earth resulted from seeds transported from other worlds.

Meteorite

Chemical precursors

Ammonia (NH_3)

Methane (CH_4)

Hydrogen (H_2)

Water (H_2O)

1920–1960: Oparin and Haldane in the 1920s proposed that, in the primitive atmosphere, methane, ammonia, and hydrogen gases could react to form simple organic molecules that collected in a primordial soup. In 1953, Miller applied electrical discharges to a mixture of water vapor and gases to produce amino acids. Oró, in 1960, found that adenine could also form under similar conditions.

Tidal pools

Primitive cells

Proteins

Amino acids

Cyanide (CN)

Ocean

Stromatolites

Nitrogenous bases

Sugars

1970s: In 1970, Katachalsky polymerized amino acids into polypeptide chains in the presence of ATP and montmorillonite clay. In 1977, Corliss found the first hydrothermal vents and proposed that simple organic molecules formed in them and deposited on clay, forming more complex molecules.

Double-stranded DNA

RNA

Self-splicing RNA

Spliced RNA

1980s: Cech and Altman discovered catalytic RNA, which cleaves RNA in the absence of protein. Cech found that an intron spliced from *Tetrahymena* RNA, the ribozyme, possesses both polymerase and nuclease activities.

Organic deposits on clay

Ribozyme

Intron

Exon

Exon

Source: Adapted from "Evolutions: Origin of Life," *The Journal of NIH Research*, vol. 5, pp. 95–96, January 1993. Adapted from original artwork by Terese Winslow.

Simple organic molecules formed

Hydrothermal vent

1990s–present: Several theories attempt to explain the origin of life. One proposes that the first organic compounds arrived from interstellar space via interplanetary bodies or dust and collected in the oceans or tidal pools. Another holds that the first organics were created from the primitive gases in the earth's atmosphere and collected in the oceans. Cairns-Smith speculates that certain clays carrying simple organic molecules can crystallize, or "replicate," to promote formation of complex molecules. Many believe that RNA-like polymers formed and somehow evolved into early life forms eventually containing both DNA and protein.

Core of Earth

Clay crystal monomers

Replicating clays

Hydrated metal ion

Possible organic sites

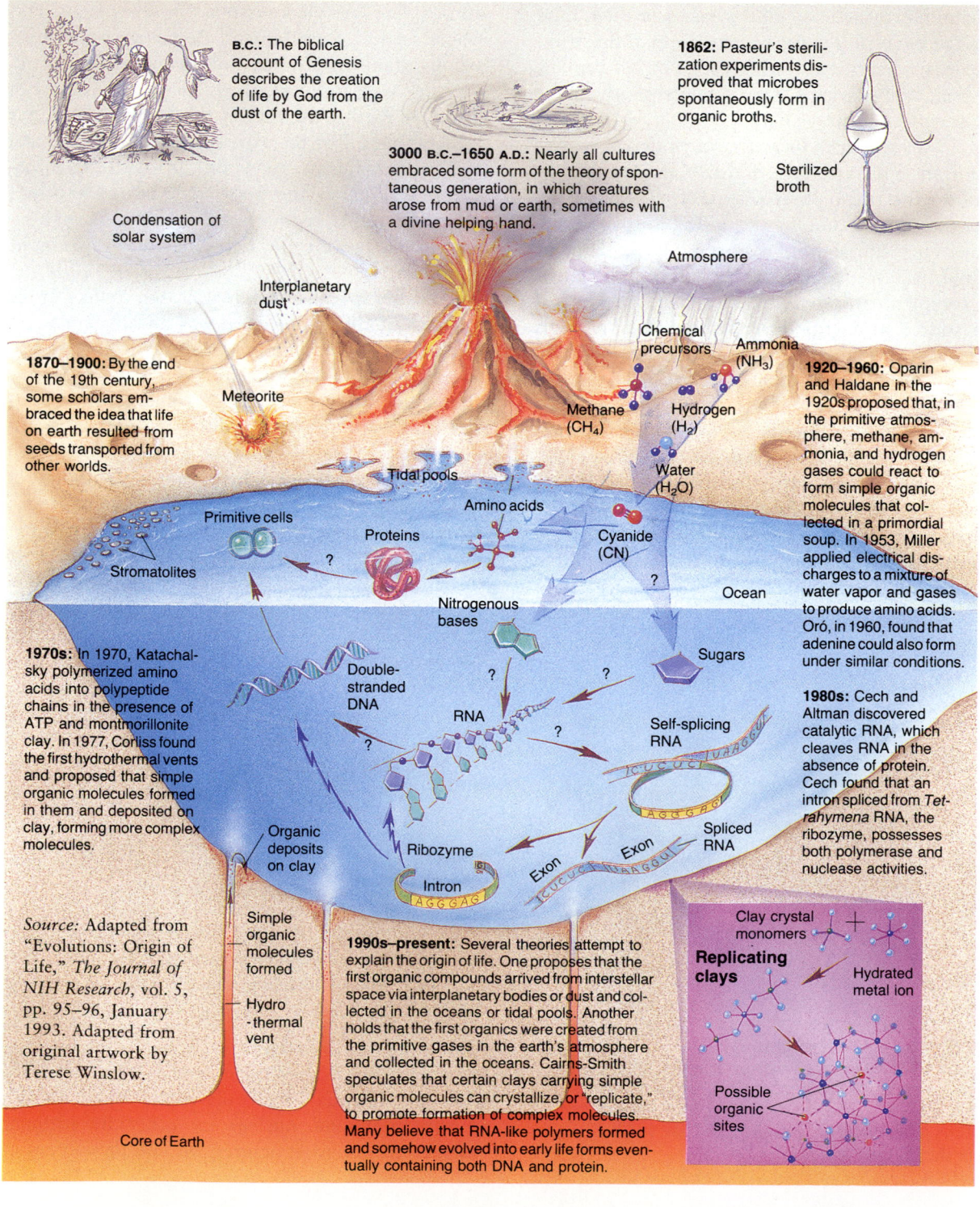

cells *prokaryotic* (primitive cells lacking membrane-bound structures).

Then, about 1.45 billion years ago, another type of cell evolved — the *eukaryotic* cell. Eukaryotes were given their name, which means "true cell," because they had a membrane-bound nucleus and membrane-bound organelles. With the exception of the bacteria and cyanobacteria, eukaryotes make up all the other life forms today. As we learned in Chapter 4, in 1967, Lynn Margulis of Boston University, proposed the *endosymbiotic hypothesis* as an explanation of the rise of eukaryotic life forms. This hypothesis suggests that eukaryotes evolved when small prokaryotes took up permanent residence inside larger prokaryotes. For example, a prokaryote in which a smaller, photosynthesizing prokaryote was trapped could have been the ancestor of modern chloroplast-containing eukaryotic cells.

Table 16.1 summarizes the major events that are hypothesized to have taken place in the chemical evolution of life on earth. Although this table looks fairly authoritative, and many biologists would agree with everything set forth in it, it is important to note that there is still a lot of scientific debate regarding the nature of earth's early atmosphere and the origin of life. Not everyone is in agreement.

We have devoted much of our discussion to the chronological order of the events that resulted in life on earth. Still it is often very difficult to imagine how much time is meant by figures like "1.3 or 1.5 billion years ago." It may help to condense the age of the earth to a 12-hour period, as presented in Figure 16.9.

An example of our changing ideas about the origin of life on the earth is found in an explanation of how complex molecules, cell structure, and reproduction might have evolved. These ideas were developed by A. G. Cairns-Smith and Paul S. Braterman, both of Glasgow University in Scotland.

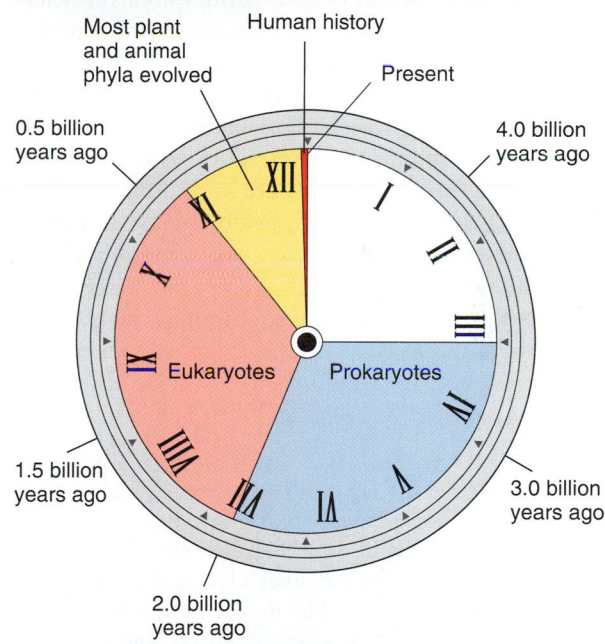

Midnight—Formation of the earth
3:00 A.M.—Life appears: first prokaryotic cells
6:45 A.M.—First eukaryotic cells
10:45 A.M.—Primitive animals
10:54 A.M.—Terrestrial plants
11:00 A.M.—First vertebrates
11:30 A.M.—The dinosaur age
11:50 A.M.—The age of mammals
11:59 A.M.—First humans
12:00 noon—Present

FIGURE 16.9 Sunrise, Sunset
Condensing 4.6 billion years' worth of earth's history into a single 12-hour period, we can see how much of the day is taken up by the formation of primitive life forms. It is only during the final seconds of daylight that all human history is recorded.

TABLE 16.1
Chronology of the Origin and Development of Life on Earth

Events in the origin of life	Years before today
Origin of the universe	15–20 billion
Origin of the sun and our solar system	5–6 billion
Origin of the earth and very primitive life in a pre-RNA world	4.6 billion
RNA dominates the world in very primitive precells	4.0 billion
DNA dominates the world	3.8 billion
First cells (prokaryotic bacteria) appear	3.5 billion
Photosynthesis begins	3.3 billion
First eukaryotes	1.45 billion

The Coming of the Eukaryotes

The two propose that the chemical evolution leading to life may have occurred in self-replicating clays. These clays are capable, under certain circumstances, of replicating their crystalline patterns.

Cairns-Smith and Braterman continue their explanation by stating that these clays often contain cell-like forms that incorporate organic compounds. That means the clays could have made numerous copies of the organic compounds as the clays replicated their crystalline pattern. The crystalline pattern served to catalyze reactions (some scientists believe that because these clay crystals may have acted like enzymes, some enzymes need rare elements such as molybdenum in order to function) and finally assisted in the construction of cells. The **crystal gene hypothesis** may help explain how complex molecules and cells evolved, but as yet there is no laboratory evidence to support their hypothesis.

As you have read, there is considerable controversy over the actual origin of life on earth. Scientists are continuing to gather more information about the origin of life, and we are gaining a better understanding of how life evolved. But no matter which position on the origin of life you subscribe to, one fact remains constant: There *is* life on earth, and the life forms on earth are very different. That fact leads us to our next question: How did those primitive cells in the ancient seas eventually evolve into the diverse life forms we see all around us? That is the subject of the next chapter.

SUMMARY

1. Humans have devised a variety of explanations for the origin of life, such as spontaneous generation, or abiogenesis (a theory that proposes that life comes from nonlife), and biogenesis (a theory that life comes from life). Pasteur's classical swan-necked flask experiments in 1862 demonstrated that microorganisms arise from microorganisms and established the idea of biogenesis.
2. Panspermia is the theory that life was seeded on earth from other planets, either through the deliberate actions of intelligent aliens or by accident. This theory is not widely accepted.
3. Our solar system was probably created by a large stellar explosion, which resulted in a primordial cloud of cosmic dust and gases. The cloud then condensed, forming a sun, about 5 billion years ago, and smaller bodies or planets, such as our earth, about 4.6 billion years ago.
4. As earth cooled, a primitive atmosphere formed. Many scientists speculate that this primitive atmosphere consisted of water, ammonia, methane, hy-

drogen cyanide, hydrogen, carbon dioxide, hydrogen sulfide, and carbon monoxide. This atmosphere probably contained some free oxygen.
5. The chemicals produced by the atmospheric gases collected in the primitive seas, creating an organic soup. The addition of energy from ultraviolet light, radioactivity, and lightning fostered the abiotic synthesis of complex organic monomers and polymers.
6. Today we know that RNA has both genetic information and enzyme properties. Most scientists speculate that RNA or pre-RNA-like molecules were the first regulating systems.
7. The first droplet cell-like forms established a membrane around themselves, isolating inside from outside, probably similar to the coacervates formed by Oparin or the proteinoids produced by Fox or the liposomes produced by Deamer and Hargreaves. These scientists and others have demonstrated that the proteinoid microspheres can mimic simple reactions that occur in modern cells and that proteinoid microspheres and liposomes can bud off and grow.
8. At first respiration was conducted without oxygen, but in time some cells became autotrophs, using light to photosynthesize. The appearance of photosynthetic organisms introduced free oxygen into the atmosphere. The atmosphere then became an oxidizing one, and some organisms evolved the capacity for aerobic respiration.
9. Photosynthesizing organisms had a selective advantage because the nutrients in the primitive seas were being depleted. Photosynthesis provided a new energy source.
10. Primitive cells gradually developed into eukaryotes, cells with membrane-bound nuclei and organelles.

KEY TERMS

spontaneous generation	coacervate
biogenesis	proteinoid microsphere
panspermia	RNA-based theory
primordial cloud theory	self-replicating RNA
protobiont	crystal gene hypothesis

REVIEW QUESTIONS

True/False

1. Today the majority of biologists support the theory of panspermia.
   true   *false*   *(page 366)*
2. The earth is believed to be about 5 billion years old.
   true   *false*   *(page 369)*
3. The primitive atmosphere is speculated to have consisted of water, methane, ammonia, hydrogen cya-

nide, molecular hydrogen, hydrogen sulfide, carbon dioxide, and carbon monoxide.
true   *false*   *(page 369)*

4. The first logical step in the chemical evolution of life was polymer formation.
true   *false*   *(page 370)*

Fill in the Blank

1. Another term for precell is _____. *(page 371)*

2. An important step in the evolution of the first precell was the _____ step in which the membrane evolved. *(page 371)*

3. Organisms that cannot synthesize their own food are called _____. *(page 375)*

4. Primitive cells lacking a membrane-bound nucleus and membrane-bound organelles are called _____. *(page 377)*

Multiple Choice

1. The most logical sequence in the chemical evolution of life would be:
   a. C, H, O, N→monomers→polymers
   b. Monomers→polymers→C, H, O, N
   c. Isolation→C, H, O, N→reproduction
   d. DNA→polymers→C, H, O, N
   (pages 369–373)

2. S. Miller's experiment supported the hypothesis that:
   a. The primitive atmospheric gases could form into monomers.
   b. Electricity could create organic molecules.
   c. Life could be created in a glass flask.
   d. The first life forms required oxygen for respiration.
   (page 370)

3. Which of the following provides the best summary of the evolution of heterotrophs:

   a. Coacervates were the first forms of life.
   b. Organic compounds in the primitive soup could serve as nutrients for the primitive cells.
   c. Primitive cells could make their own food.
   d. Protobionts respire without oxygen.
   (page 375)

4. As the protobionts evolved, genetic reproduction did not occur until the chemical evolution of:
   a. RNA or pre-RNA
   b. Proteins
   c. Carbohydrates
   d. Water
   (pages 372–373)

Discussion

1. Define abiogenesis and spontaneous generation. What is biogenesis? How was the concept of biogenesis eventually accepted by the scientific community?

2. Discuss the hypothesis known as panspermia. Why is it not accepted by most of the scientific community today?

3. What is the primordial cloud theory?

4. Discuss the probable steps in the formation of the first cell.

5. Why do you suppose the first organisms were heterotrophs and not autotrophs?

6. Why do you think there is so much controversy as scientists speculate on origin of life?

7. Do you think there is life on other planets? Why or why not?

8. Why was the development of photosynthesis, or the autotrophic lifestyle, such an important event?

9. Explain how natural selection — "survival of the fittest" — applied to the evolution of the first cell forms.

10. Was the first cell a prokaryote or a eukaryote?

17

Evolution: Living Things Changing Over Time

In his laboratory at Yale University's Peabody Museum of Natural History, Jon Ahlquist has an extract of human spleen in a container. He adds some ethyl alcohol and stirs with a glass rod. A thick, viscous mixture forms, and his colleague, Charles Sibley, quips, "Life on a stick." His comment is correct — the viscous mixture is, in fact, human DNA. The unpretentious blob coating the end of Ahlquist's glass rod contains the genetic blueprint for human life (Figure 17.1).

With procedures such as these, Ahlquist, Sibley, and many other investigators are exploring another line of evidence for evolution—the pattern of molecular similarities and differences among species. By comparing the degree of similarities and differences in the DNA molecules or proteins of various

groups of organisms, they can estimate when two species diverged from a common ancestor. (A **species** is a group of organisms that breed with each other and normally produce fertile offspring.) DNA or protein can be used as a "molecular clock," timing major evolutionary milestones. Thus scientists are writing and rewriting the evolutionary history of many organisms — including humans.

For a long time, an organism was classified only after its structural components had been carefully analyzed. Did it have bones? Gills? Cilia? Did it have blood, and if so, how did the blood circulate? How did it breathe? Only after a lengthy examination could a decision be made about which species the organism belonged to. But comparative analysis of DNA has added a new component to the process of classifying organisms, and some of the findings have caused problems.

For example, in traditional taxonomy humans and chimps are classified in separate families, *Hom-*

Above: All life forms are the result of evolution, including this nudibranch.

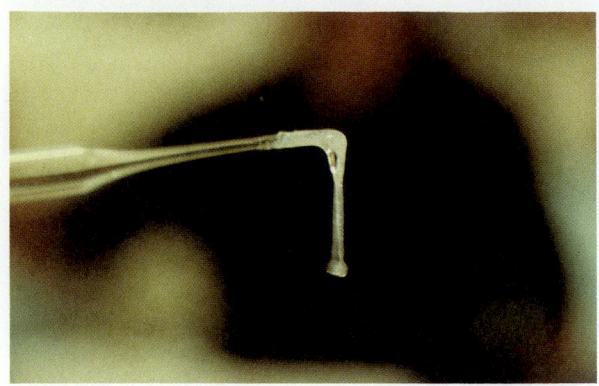

FIGURE 17.1 Human DNA — "Life on a Stick"

inidae (humans) and *Pongidae* (apes). However, scientists doing biochemical studies have demonstrated that the DNA of humans and that of chimpanzees are about 98 percent identical. Since only organisms that are closely related show this degree of similarity, are we, in fact, more closely related to apes than we think?

The accumulating biochemical evidence has heated the long-standing controversy concerning evolution. As Emile Zuckerkandal, editor of the *Journal of Molecular Evolution*, has stated: "Since Copernicus, there has always been resistance to learning the truth of man's place in the universe." This is certainly true now. In fact, not since 1859, when Charles Darwin published his book *On the Origin of Species by Means of Natural Selection, or the Preservation of Favoured Races in the Struggle for Life,* has so much discussion centered on the origin of humans. Today the controversy in scientific circles occurs between the molecular evolutionists and the traditional *paleontologists,* the scientists who study the fossil records to learn about the evolution of life forms. Indeed, some traditional paleontologists regard the findings of molecular biologists as unacceptable — even heretical.

Of course, this is not the first time that the cry of heresy has been heard in discussions of evolution. In Darwin's time, the term *heresy* was also applied to his theory that species were not the product of special creation but rather evolved through natural selection. The controversy arose when Darwin stated that evolution through natural selection also had affected humans, that we are neither more special nor less special than the other living forms with whom we share our planet.

Today, the theory of evolution is accepted by scientists as one of the most important processes in our understanding of life on earth. But what exactly is meant by the term *evolution,* and what are its origins? Why is Darwin's theory regarded as one of the fundamental principles in biology? What is the evidence for evolution and what are its mechanisms? What about our own evolutionary history — where do humans fit in? We have plenty of questions, so let us now explore some answers.

THE ORIGINS OF EVOLUTIONARY THOUGHT

In Chapter 16, we discussed the origins of life and stated that all living things are related by the fact that they contain DNA, RNA, and ATP, that they metabolize, that they are alive, and so on. The cell is the fundamental unit of life; thus cellularity is a common link that connects organisms. Although all living things are alike in certain ways, they differ in other ways. There are trees, whales, birds, humans, worms. The diversity of life forms suggests an evolutionary process from our unicellular past.

Simply speaking, *evolution* means change over time. Almost everything we encounter evolves or changes: fashions, dance styles, aircraft, electronics, computers. Living things also evolve, or change, through time. Ability to evolve is one of the unifying characteristics of living things. However, in this unity there is also diversity. Evolutionary theory simplifies and explains biological relationships among the diverse forms of life on earth (see Profile).

Ancient Evolutionary Thought

■ **Early philosophers such as Aristotle explained the origin of life and its diversity in terms of a ladder-like hierarchy. Later thinkers believed that organisms were related but often held the idea that creation followed biblical chronology.**

Galileo is believed to have said, "If I have seen further than any man, it is because of the numerous shoulders I have stood upon." In other words, no scientist works in a total vacuum: Concepts build on the works of others. Even erroneous ideas can lead ultimately to major contributions as scientists try to disprove the faulty theory. Darwin's theory was no exception. His ideas did not arise in isolation but were grounded in a body of thought that stretched back for centuries.

In Chapters 25 through 30 of the book of Genesis, Moses tells a story about Jacob and the sheep of Laban, which may be the first written analysis of the environment's influence on heredity. The story goes as follows: In return for 20 years of service, Jacob was to receive from his father-in-law all the

Dr. Francisco Jose Ayala
Geneticist, Evolutionist

Francisco Jose Ayala is considered to be one of the foremost authorities on genetics and evolution. He is also a successful businessman, an art collector, a philosopher, and a linguist who speaks five languages fluently (and can converse in five or six more). What is more, he recently turned his energy to viticulture, the art of winemaking.

Francisco Ayala was born into a wealthy family in Madrid, Spain. Like Mendel, he chose the monastic life, entering the monastery at the age of twenty to begin his studies to become a Dominican priest. His decision to enter the church was influenced greatly by the writings of Teilhard de Chardin, a theologian who saw no conflict between science and religion. Teilhard reasoned that science glorified God by allowing humans to understand the wonders of nature and God's creation in more detail.

By the age of twenty-seven, Ayala was having doubts about his religious convictions and was thinking of science as a new pursuit. One of the questions that plagued him was, "How could a God who was all good allow severe birth deformities in the world?" A traditional theological answer is that God allows such defects to exist in order to test the parents' faith. As they wrestle with the problems that arise from their child's deformity, their faith and strength will grow. Ayala could not accept this reasoning.

Reading and independent studies with fruit flies increased Ayala's desire to pursue genetics. He convinced a friend to write to the eminent geneticist Theodosius Dobzhansky, working at Columbia University, to request that Ayala be accepted as a graduate student. Dobzhansky accepted the young man, even though Ayala had not taken any formal biology courses since college.

In 1961, Ayala left Spain and began school in New York. His collaboration with Dobzhansky continued for the next 14 years, until Dobzhansky's death in 1975. At present Ayala is working in genetics and evolution at the University of California at Irvine. Tall and thin, he has an amazing reservoir of energy. He usually sleeps only 4 hours at night, and he travels more than a quarter million miles a year giving lectures and seminars all over the world. His research interests still center around the genetic/evolutionary mechanisms that cause populational variation.

It is difficult to point to the one most significant accomplishment in the life of a man who has accomplished so much. One incident does stand out, however, and that has to do with a theory proposed by the Nobel Prize laureate H. J. Muller. Muller proposed that there was a "best" gene for each trait and that natural selection would select against the variance until only the best remained. Through his work, Ayala has demonstrated the contrary. His hypothesis was that genetic variation within a species is great, that there is no one "best" gene. Working with fruit flies (because they breed rapidly and pro-

duce many offspring), Ayala showed that populations of flies that exhibit more genetic diversity evolve more quickly than those with a more limited genetic makeup.

Ayala has also worked with other organisms to learn that one-third to one-half of the genes exist in two or more forms or variations, even in humans. Because of genetic diversity, it is inevitable that a certain number of offspring will be born as variants or deformed.

Even today, although he understands and is helping increase our awareness of the genetic evolutionary process, he still feels emotionally uncomfortable with birth defects. He can deal with the need for genetic variation at an intellectual level, but not at an emotional level. He states that "science does not give answers to ethical or religious questions, just as ethics and religion do not give answers in science."

brown sheep and the spotted, speckled, and banded goats. So, working ahead of time so that his reward would be increased, Jacob stripped the bark from branches of green poplar trees and hazel or chestnut trees so that they appeared white with brown stripes. He then had the animals mate in the presence of the branches. According to Moses, the mating of the goats in this environment influenced the coloring of the goats, increasing the proportion of spotted, speckled, and banded goats in the herd.

Greek philosophers also attempted to explain the diversity of life on earth. The most notable was Aristotle (384–322 B.C.), who is sometimes called the "father of biology." Aristotle thought that life had always existed on earth. He also believed that all living things could be placed on a "ladder of life," with the simple creatures at the bottom and humans at the top. In this hierarchy, known as the *Scala Naturae*, all organisms had their proper place.

The Swedish scientist Carolus Linnaeus (1707–1778) devised a system of biological nomenclature (or biological names) for the enormous variety of life. He developed a system of two-part names in two of his works, *Systema Naturae* and *Species Plantarum*. (Linnaeus's work is still important today, and we will discuss it in more detail in Chapter 18.) Despite the fact that he supported the idea that certain organisms were related, he still held the philosophy that God had created all life forms on the sixth day of creation and that the number and types of organisms had been fixed since creation.

Linnaeus's position was modified by Georges-Louis Leclerc de Buffon (1707–1788), who proposed that in addition to those creatures produced by divine creation on the sixth day, "lesser creatures were conceived by nature and produced by time." These modifications were further expanded by Charles Darwin's grandfather, Erasmus Darwin (1731–1802), who proposed that species are related to one another and that this relationship is connected in their history. He also suggested that the environmental changes or modifications of an organism could be transmitted to offspring — in other words, offspring would inherit acquired characteristics.

Geologists

■ **Seventeenth- and eighteenth-century geologists were mainly concerned with determining the age of the earth. Catastrophists followed the timing of the Bible. Uniformitarians used analysis and the fossil record to push the earth's age back by millions of years.**

At the same time that the biologists of the seventeenth and eighteenth centuries were refining theories about the diversity of life on earth, *geologists* (scientists who study the earth's structure) were trying to determine the age of the earth by studying the fossil record. For the most part early geologists and naturalists believed that life and the earth itself were the fixed products of divine creation. Two schools of thought emerged, the *catastrophists* and the *uniformitarians*.

The catastrophists supported the idea proposed by Georges Cuvier (1769–1832), who believed that the earth was formed by a series of catastrophes, such as the Great Flood related in the Bible. The uniformitarians, on the other hand, believed geological change had proceeded gradually, according to natural laws, and that the forces of wind and water and volcanic activity that we now observe are the same forces that have produced changes down through the ages (Figure 17.2). The first significant

FIGURE 17.2 Uniformitarianism Uniformitarians believed that the earth was constantly changing over time, the result of natural forces. They argued that the earth was much older than the Bible suggested. Today we have geological evidence to support this thinking. These layers of rock in the Grand Canyon record over 2 million years in geological time.

Chapter 17 Evolution: Living Things Changing Over Time

supporter of uniformitarianism was James Hutton (1726–1797), who in 1788 published a theory proposing that the earth was much older than the 6000 years the biblical calculations suggested. He further suggested that geological processes were constantly operating, producing a world of change, not only through degeneration by means of erosion but through the formation of "new worlds" by means of stratification and upheaval. In other words, *change was normal.*

Hutton's theory was supported by the English surveyor William Smith (1769–1839), whose interpretation of the fossil record supported the idea that the earth was very ancient. Unfortunately, Hutton's theory was not accepted by many of his contemporaries.

In 1830, the geologist Charles Lyell (1797–1875) published the first volume of *Principles of Geology.* Lyell was a uniformitarian, and his arguments convinced most of the geological community that slow, steady, natural change had occurred during the earth's history over a long period of time. Lyell's work struck a final blow to the idea of catastrophism and set the foundation for modern geology.

Darwin's Contemporaries

■ **Lyell influenced Darwin by providing the time frame Darwin felt was required for evolution. Thomas Malthus, with his ideas about population increase and diminishing food supplies, provided Darwin with a mechanism for evolution. Unlike Darwin, Jean-Baptiste de Lamarck believed that changes occurring in an organism during its life span could be passed on to its offspring.**

The work of Charles Lyell had a great influence on Darwin, for Lyell had pushed the age of the earth back by millions of years. Darwin reasoned that if the earth's geological history was one of change over time, then it was possible that change over time was also a significant factor in the history of organisms living on the earth. Lyell's work provided the time frame Darwin believed was essential for the gradual processes involved in evolution.

Another significant influence on Darwin was Thomas Malthus (1766–1834), who in 1798 had written an article entitled *An Essay on the Principle of Population.* Malthus stated that growth in all populations, human as well as otherwise, was geometric — in other words, two organisms would have a total of four offspring, those four would have a total of eight offspring, those eight would have sixteen, and so on. The supply of food available to support those organisms, however, would increase only arithmetically — that is, from 1 to 2 to 3, and so on. The difference in growth rate meant that populations would eventually outgrow their environmental resources, which in turn would cause a tremendous struggle for food and for existence in general. Darwin read Malthus's essay in 1838 and realized that some individuals had inherited variations that gave them an advantage in the "struggle for existence." These individuals would be more successful than others at raising offspring, and so the variations that gave them the advantage would increase, while other variations decreased. This realization became the basis for what Darwin called *natural selection.*

In 1801, Jean-Baptiste de Lamarck (1744–1829), an outstanding naturalist of his time, proposed that acquired traits could affect heredity. In other words, as an organism adapted to environmental changes, it acquired new characteristics that could be passed on to its offspring. A classic example is his explanation of the evolution of the giraffe's long neck. Lamarck explained that the giraffe acquired this characteristic through continuous stretching to reach leaves on the lower branches of trees. The acquired long neck could then be inherited by the offspring.

As Lamarck formulated his evolutionary theory, he hypothesized that the fossil life forms found in rocks provided evidence that more complex forms of life arose progressively from simpler forms. To explain his hypothesis, Lamarck formulated the following ideas for the causes of evolution:

1. A characteristic could be acquired through *use;* conversely, a characteristic could be lost (or regress) through *disuse.*

2. An acquired characteristic could be passed from one generation to the next. The loss of a characteristic could also be passed on from one generation to the next.

3. During the course of evolution, forms of life move toward greater complexity. In other words, each form that arises is more complex than the form from which it evolved.

4. A metaphysical force in nature constantly strives upward in the *Scala Naturae* toward more complexity.

Lamarck was a great scientist and naturalist but, unfortunately, he is usually remembered for his explanation of evolution, which has since been demonstrated to be wrong. Numerous experiments have shown that acquired characteristics are not trans-

FIGURE 17.3 The Giraffe and Lamarck's Theory of Evolution
Lamarck stated that a characteristic acquired during an organism's lifetime could be passed on to its offspring. He reasoned that the giraffe's unusually long neck was a case in point.

mitted to offspring (Figure 17.3). Only changes in the genetic makeup, the DNA, of *sex* cells can be transmitted from parent to offspring. Changes in somatic DNA (that is, changes in the DNA of a body cell) do not affect an organism's offspring. However, Lamarck's endeavor was not in vain, since his ideas helped to set the stage for Charles Darwin.

Darwin's Theory

■ **Darwin's contribution to evolutionary thought was his theory of natural selection.**

The name Charles Darwin (1809–1882) is almost synonymous with the theory of evolution (Figure 17.4). Although he was not the first to propose that organisms evolved over long periods of time, his concept of natural selection provided a mechanism for change and made Darwin the central figure in evolutionary thought.

Charles Darwin was born into a wealthy family of physicians and clergymen. He seemed destined to follow the family tradition. First, he entered medical school. However, since anesthetics had not

been discovered, he needed only to observe one operation (conducted, of course, on an awake and alert patient) to decide that his temperament was not suited to being a physician. He then turned to the clergy for a possible profession, and his studies went well.

But there was another interest in Darwin's life, one that he had been engrossed in since childhood. Darwin was an amateur naturalist, gathering specimens, classifying them, and writing up observations. In 1831 when Captain Robert Fitzroy began looking for individuals to accompany him aboard the scientific research vessel H.M.S. *Beagle,* Darwin was recommended. As a social equal, he went along to be Captain Fitzroy's dinner companion. But they didn't get along, so Darwin had time to pursue his interest in natural history. The voyage entailed charting the coastline of South America, then continuing around the world.

Without his family's wealth the trip would have been impossible. Although the position was comparable in rank to a first mate, Darwin would receive no pay. Instead he would have the chance to gather samples and observe animals and plants that few Europeans had ever seen. He was twenty-two years old when the *Beagle* sailed.

Darwin's journey around the world took nearly five years. At first he observed exotic forms of life along the coast of South America, and as he made his observations he noted similarities between those species and totally different species in Europe. The *Beagle* then traveled to the Galápagos Islands, a

FIGURE 17.4 Darwin
Charles Darwin, shortly after he returned from his trip aboard the *Beagle.*

comparatively young group of islands located 600 miles off the coast of Ecuador.

The Galápagos had been formed by volcanoes, long after the formation of the continents, so the life forms there were considered to be immigrants — organisms that had either been blown there on the trade winds or floated there on ocean currents. Darwin found animal species on the islands, like the marine iguana, that existed nowhere else on earth. The commonplace finches he found there proved later to be one of the cornerstones of his theory of natural selection.

After the *Beagle*'s return to England, Darwin began to analyze the data he had collected. Two years later his finch material was analyzed by a leading ornithologist, who said that the Galápagos finches were closely related to each other and to a species of finch Darwin had found on the western coast of South America. That information, along with his own observations, led Darwin to formulate his hypothesis of descent with modification (which many biologists refer to as natural selection) in 1844. (We discuss Darwin's finches in more detail later in this chapter.)

Darwin's theory of **natural selection** was based on the following assumptions:

1. Population size tends to increase geometrically, which means that organisms produce more offspring than the environment can support.

2. Because of their birthrate, organisms have the potential for tremendous population increase; however, populations remain fairly constant in size.

3. Population size remains fairly constant because all members must compete for available resources — there is a struggle for survival.

4. Variations exist among the individuals of a species. No two individuals are exactly alike.

5. In the struggle for existence, organisms expressing variations best suited to a given environment survive.

6. Organisms that survive pass on their favorable variations to their offspring. In time, these population variations increase, and the individuals become so different from the original that they constitute a new species.

Simply put, those organisms that are best suited to the changing environment survive and produce more offspring. This process is called natural selection.

The wing lengths of fruit flies provide a simple illustration of natural selection (Figure 17.5). Those that have wings as long as their bodies can fly very well. Others, which have very short wings, cannot fly; they just hop around. In the natural environment, long-winged fruit flies are better adapted; that is, they are better equipped to compete for food and mates than are their short-winged, hopping counterparts. In other words, long-winged fruit flies are *selected for*. However, suppose we change the environment by placing the flies in a large container with food on the bottom and flypaper suspended from the top of the container. Which variation will have the selective advantage, long- or short-winged fruit flies? Correct — this environment selects for short-winged flies. Since they cannot fly, they do not get caught on the flypaper and die; instead they survive and reproduce. Biologically speaking, this example demonstrates that genetic variation provides the raw material for evolution and that the environment shapes, through selection, the forms of life that will survive.

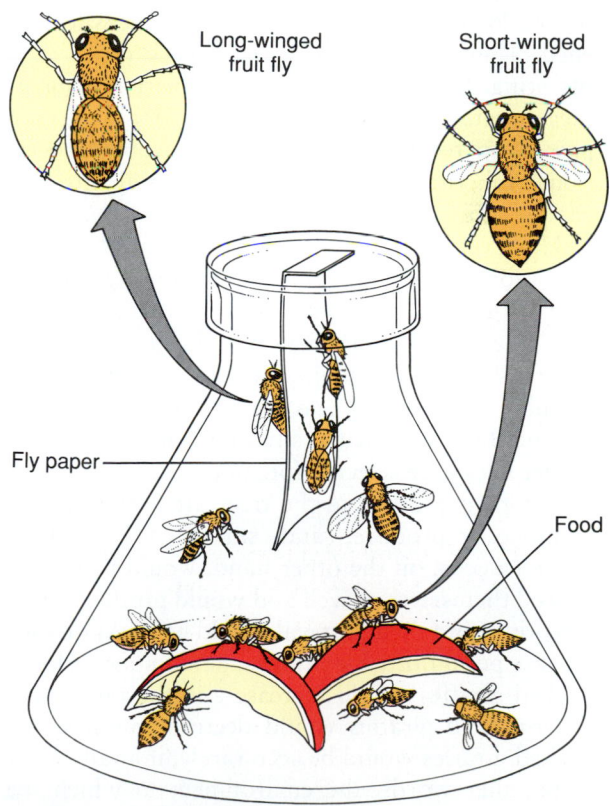

FIGURE 17.5   A Simple Illustration of Natural Selection
In this flask environment, short-winged fruit flies are selected for because the long-winged fruit flies are more likely to get stuck on the flypaper (and are, therefore, selected against).

Alfred Russel Wallace. In 1857, while writing his manuscript, Darwin corresponded with the Royal Society and learned that the English naturalist Alfred Russel Wallace (1823–1913) was ready to present a theory for evolution almost identical to Darwin's. In fact, some scientists refer to the Darwin-Wallace theory. Although Wallace was prepared to withdraw his manuscript once he realized the similarity between his work and Darwin's, Darwin would not allow Wallace to do so. Instead the two men allowed their friends, Charles Lyell and Charles Hooker, a botanist, to read excerpts from their manuscripts at a scientific meeting. The theory of evolution by natural selection was praised overwhelmingly, in great part because of the detailed analysis that Darwin had included. In 1859, Darwin completed his work, and the *Origin of Species* was published.

A Comparison Between Darwin and Lamarck. How did Darwin's theory differ from Lamarck's? To see the differences, let us look at Lamarck's explanation for the giraffe's long neck. He stated that because giraffes stretched their necks to reach leaves on the lower branches of trees, their necks lengthened. They then passed the long neck on to their offspring. We now know this explanation is incorrect because changes in somatic DNA are not passed on to the offspring of succeeding generations. Inheritance of traits by offspring involves changes only in the DNA of sex cells.

However, in Darwin's scheme of things, the explanation would work this way. Within a population of giraffes, random genetic variations occur. As a result, some individuals have longer necks than others. Because of changes in the environment and population pressures, less food is available on the ground. In this particular situation, individuals with longer necks are selected for, since giraffes that can obtain their food from the trees are better able to compete, reproduce, and survive. Those with shorter necks, on the other hand, would be unable to feed themselves as well and would produce fewer offspring, and their survival would be questionable. Over a period of generations, the number of long-necked giraffes would increase and the number of short-necked giraffes would decrease, until short-necked giraffes would be seen rarely, if at all.

In other words, the environment in which the giraffes exist makes all the difference. If you plotted a normal distribution of neck lengths of giraffes for three hypothetical populations A, B, C, as in Figure 17.6, you would see that the distribution or fre-

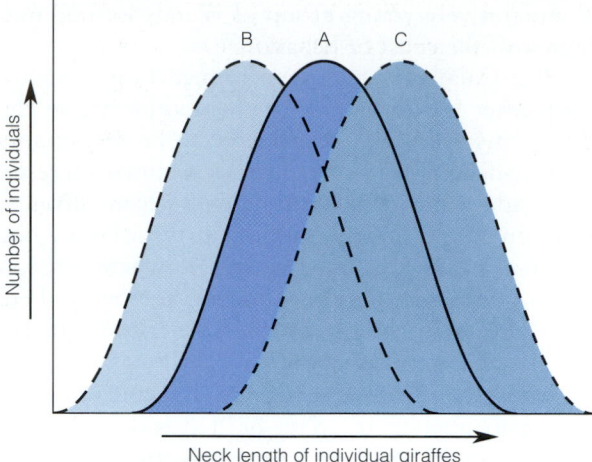

FIGURE 17.6 The Giraffe and Darwin's Theory of Evolution
Curve A represents the normal distribution of neck length in giraffes. If natural selection favored shorter necks, curve B would represent the distribution. If natural selection favored longer necks, curve C would represent the normal curve. Once selection favors a new neck length, a new distribution is established.

quency of phenotypes shifts as the environment changes. In order for evolution to occur, phenotypic changes have to be transmitted genetically to offspring. If reproduction cannot occur, evolution will not proceed.

NATURAL SELECTION — A MODERN INTERPRETATION

■ **In modern terms, the theory of evolution by natural selection follows what we have learned about genetics and inheritance. Variation occurs through recombination and chance mutations in the formation of sex cells and the zygote. Certain variations enhance the survival possibilities of the organism within the environment; that organism is better able to reproduce and pass its genes on to the next generation.**

One cannot help but be amazed by the scientific insight possessed by Darwin and Wallace. To formulate such a theory at a time when genetics and the role of DNA were completely unknown was truly a mark of genius. We are now able to interpret natural selection in light of our present knowledge of modern genetics:

1. Variations within a species are due to changes in the genetic makeup of the individ-

uals through mutation and genetic recombination.

2. Mutation, the basic source of variation, can be due to molecular or chromosomal rearrangements.

3. Individuals with favorable genetic variations have a better chance of survival in a changing environment. As a result, they produce more offspring. That means that over time the number of individuals bearing the favorable variation increases, and the number of individuals not bearing the favorable variation decreases.

4. Since change occurs within a group over a period of time, it is not individuals but populations of the same species that evolve. Individuals retain their own genetic makeup throughout their life. Within a population, however, one can find any number of variations. Over time the environment selects for or against certain variations. Therefore, it is the genetic makeup within the group that is selected for or against; the *population,* not the individual, evolves.

5. Over long periods of time, genetic changes in a population can produce a new species.

In summary, the modern interpretation of evolutionary thought is that evolution is based on random genetic variation and the selection of those variations by the environment. But what are the forces that drive natural selection? It is this question that we now address.

MEASURING CHANGE

■ **The Castle-Hardy-Weinberg equation determines the frequency of the alleles of a given trait within the gene pool. Genetic equilibrium, or constancy in the frequency of alleles from one generation to another, indicates that the population is not changing. Evolution occurs when the allele frequencies in a gene pool change.**

In genetic terms, evolution is a change in the frequency of the alleles in a population. A **population** can be defined as an interbreeding group of organisms of a specific species found within a specific location — all the buffaloes in a herd or all the fish of the same species in a pond. All the alleles of all the genes of the individuals of a population make up its **gene pool,** and the gene pool holds the variations from which the environment selects. The en-

vironment determines the *fitness* of a given genetic combination — simply put, if an organism can reproduce itself in a given environment, it has fulfilled its evolutionary destiny. The success of an organism is measured by the number of survivors.

Castle-Hardy-Weinberg Genetic Equilibrium

How can we calculate the frequency of alleles in a population? And why is it that dominant alleles do not completely dominate a population? If, for example, the allele for brown eyes is dominant over the allele for blue eyes, why don't all members of a population eventually have brown eyes? Why do both dominant and recessive alleles remain in a population?

Mendel provided one answer. Even though recessive alleles are hidden or masked by dominant alleles, they may still remain in the genotype of the individual. In 1908, two scientists, working independently, provided a mathematical answer. G. H. Hardy from England, and W. Weinberg from Germany, are usually given credit for the concept; however, an American, W. E. Castle, had already published a paper in 1903 on the same concept. As a result, this principle is sometimes called the Hardy-Weinberg principle, but at other times the Castle-Hardy-Weinberg principle (which we will use here).

In simplest terms, the **Castle-Hardy-Weinberg principle** states: In a large population of sexually reproducing diploid organisms, the frequency of alleles in the population will remain constant from generation to generation if there are no disturbing factors. More specifically, **genetic equilibrium** will exist if the following conditions are met:

1. The population is large enough so that chance alone does not alter the gene frequencies.

2. No mutations occur, or if they do, they are not common and occur equally in opposite directions so that the net change is zero.

3. No new individuals enter or leave the population.

4. Mating is at random, and all alleles are equally viable.

5. All genotypes have equal selective value and can survive and reproduce.

The Castle-Hardy-Weinberg principle can be expressed as a mathematical equation. By applying

FOCUS 17.1
A Sample Problem

The following example will illustrate the workings of the Castle-Hardy-Weinberg principle. Let us say that you are interested in the frequency of the alleles for the trait "tongue rolling" in humans. This trait controls the ability to shape the tongue to form the letter *U*. From preliminary research, you already know that the ability to roll the tongue (the allele *T*) is dominant over the inability to roll the tongue (the allele *t*). (For our purposes we will treat the tongue-rolling trait as an example of simple Mendelian dominance, even though there is some evidence that this ability may be learned.)

You sample 1000 individuals at random and ask each to demonstrate the ability or inability to roll the tongue. The results are these:

Phenotype	Genotype	Number
Non-tongue rollers	*tt*	160
Tongue rollers	*TT* or *Tt*	840
Total sample		1000

You now have some numbers to plug into the Castle-Hardy-Wein-berg equation: $p^2 + 2pq + q^2 = 100$ percent. In this particular instance, p represents the dominant allele *T* and q represents the recessive allele *t*. Of the 1000 people sampled, the 160 individuals who could not roll the tongue must represent the homozygous recessive condition (*tt*), since only a *tt* individual could express the recessive condition. Therefore, 160 represents q^2 in our equation: $q^2 = 160/1000$, or 0.16, or 16 percent.

If $q^2 = 0.16$, what is q? Since $q^2 = q \times q$, then q must equal the square root of 0.16, or 0.4.

---

the equation, we can calculate the frequencies of alleles in a population. (Before you continue, you might review the discussion of the rules of probability in Chapter 13.)

Determining Frequencies

In the Castle-Hardy-Weinberg formula for a trait with two alleles, the letter p represents the frequency of the dominant allele and the letter q represents the frequency of the recessive allele. For a trait with two alleles in which there is simple Mendelian dominance, the allele could be either p or q. Therefore, $p + q = 100$ percent, or (since 100 percent = 1.00), $p + q = 1$.

Because two diploid individuals are involved in sexually reproducing organisms, when organisms mate we have a cross between two organisms, each with two alleles. If two heterozygous individuals mate, we have $(p + q) \times (p + q.)$ The result of the equation is 1, since $1 \times 1 = 1$. Another way of writing the formula is $(p + q)^2 = 1$. When we perform the multiplication necessary in this equation, we first multiply $(p + q)$ by p; the result is $p^2 + pq$. We then multiply $(p + q)$ by q; the result is $pq + q^2$. Finally we combine the two individual calculations; the final total is

$$p^2 + 2pq + q^2 = 1$$

But what does this set of numbers represent? They tell us this:

p^2 = frequency of homozygous dominant individuals

$2pq$ = frequency of heterozygous individuals

$q^2 = 0.16$

$q = \sqrt{0.16} = 0.4$

According to the first part of the Castle-Hardy-Weinberg principle, $p + q = 1$; so since we now know that $p + 0.4 = 1$, what must the value of p be? That's right, p must equal 0.6.

With the value of p and q now determined, we can plug those numbers into our equation to learn the frequency of genotypes for tongue rolling in the next generation, as long as mating is random:

$p^2 + 2pq + q^2 = 1$

$(0.6)(0.6) + 2(0.6)(0.4) +$
$\qquad\qquad (0.4)(0.4) = 1$

$0.36 + 0.48 + 0.16 = 1$

These calculations tell us that

36 percent of the offspring of the entire group will express the TT genotype (p^2).

48 percent of the offspring of the entire group will express the Tt genotype ($2pq$).

16 percent of the offspring of the entire group will express the tt genotype (q^2).

But we were interested in determining the frequency of alleles in the gene pool, not the frequency of genotypes. We must now look at the equation to determine the frequency of the T allele. In the genotypes for TT, all 0.36 are T alleles. In the genotypes for the heterozygous Tt, one-half of the alleles are for T (0.48 / 2 = 0.24). The total number of T alleles is 0.36 + 0.24 = 0.60, or 0.6. The same process will indicate that the frequency of t alleles is 0.4. The important fact to note here is that the frequency for both T and t has remained the same — 0.6 and 0.4, respectively. The same will be true in any population that satisfies the five requirements of the Castle-Hardy-Weinberg principle: The frequency of the alleles in the gene pools remains the same from generation to generation.

A PUNNETT SQUARE FOR THIS PROBLEM

If you work out a Punnett square to represent the typical mating for your sample, you will find the infor-

mation parallels that provided by the Castle-Hardy-Weinberg formula:

		Eggs	
		0.6T	0.4t
Sperm	0.6T	0.6T ×0.6T 0.36TT	0.6T ×0.4t 0.24Tt
	0.4t	0.6T ×0.4t 0.24Tt	0.4t ×0.4t 0.16tt

Therefore the Castle-Hardy-Weinberg principle provides biologists with a tool for determining gene frequencies in a population. Its formula can be expanded to include traits with more than two alleles. For instance, if we were interested in analyzing the alleles for blood types (multiple alleles), the algebraic expression would become $(p + q + r)^2$. Obviously, in such a case, the mathematics would be more complex.

q^2 = frequency of homozygous recessive individuals

We can use the Castle-Hardy-Weinberg equation to determine frequencies of an allele in a population. Suppose we want to measure the frequency of albino alleles in a population. (Albinism is a recessive trait, inherited according to simple Mendelian dominance.) Knowing the percentage of albino individuals in the population tells us the number of homozygous recessives, or q^2. Taking the square root of that number tells us q. If we plug the value of q into the equation $p + q = 1$, we can calculate the value of p as well: $p = q - 1$. Knowing the frequencies of p and q will enable us to calculate $p^2 + 2pq + q^2$. Thus the equation can tell us how often these alleles will occur in a population that is in equilibrium. See Focus 17.1 for a sample problem.

Using Castle-Hardy-Weinberg in Evolution Studies

What is the evolutionary significance of the Castle-Hardy-Weinberg principle? It gives us a way of measuring or quantifying evolution. Suppose we measure gene frequencies in a population over several generations and find that they are not what the equation predicts — that is, $(p + q)^2 \neq 1$. Then we know that the population is not at equilibrium. Factors such as the ability to compete for mates, level of fertility, and survival rate can all produce a *differential rate of reproduction and survival within populations*. This may result in a change in the frequencies of alleles in a population over time, and that means evolution is taking place. Mechanisms that disturb the Castle-Hardy-Weinberg equilibrium produce the changes in the gene pool, and *the change in frequency of alleles in a population over time is what is meant by evolution.*

MECHANISMS AFFECTING GENETIC EQUILIBRIUM

■ **Genetic equilibrium can be affected by genetic drift, gene flow, mutation, and natural selection.**

Certain genes increase in a population, *not* because of selection, but because of chance. In this case evolution is said to be **nonselective.** There are two viewpoints on the significance of nonselective evolution. The *neutralists* take the position that many of the variations at the molecular level are nonadaptive. In other words, the variations are incidental and do not affect an organism's ability to survive and reproduce. They are simply neutral changes. The allele frequencies drift at random. On the other hand, the *selectionists* take the position that all variations within a population have selective significance.

We have defined evolution as a change in the allele frequencies in a population. Obviously, if genetic equilibrium remained constant there would be no evolution. That is unlikely, since genetic drift, gene flow, mutation, and natural selection all can change the frequency of alleles in a population and result in evolution and the formation of new species (Figure 17.7).

Genetic Drift

We mentioned earlier that the Castle-Hardy-Weinberg equilibrium holds only if the population is large. Since genetics is based on the probability of an event's occurring, the larger the sample size, the more likely you are to get the expected results. For example, if you flip a coin only 10 times, it is unlikely that you will get the expected one-half heads and one-half tails. If, however, you flip that coin 500 times, you will definitely get results that are closer to the expected probability. The same holds true with a large population, since the frequency of an allele for a given condition is randomly distributed within that population.

Genetic drift is a change in gene frequencies in a population as a result of chance. For example, let us say that the allele for green hair (if there were such a thing) has a frequency of 2 percent in a population of 500,000. Therefore, 10,000 alleles for this trait would be in the gene pool. If the population were small, let us say 100, only two green-hair

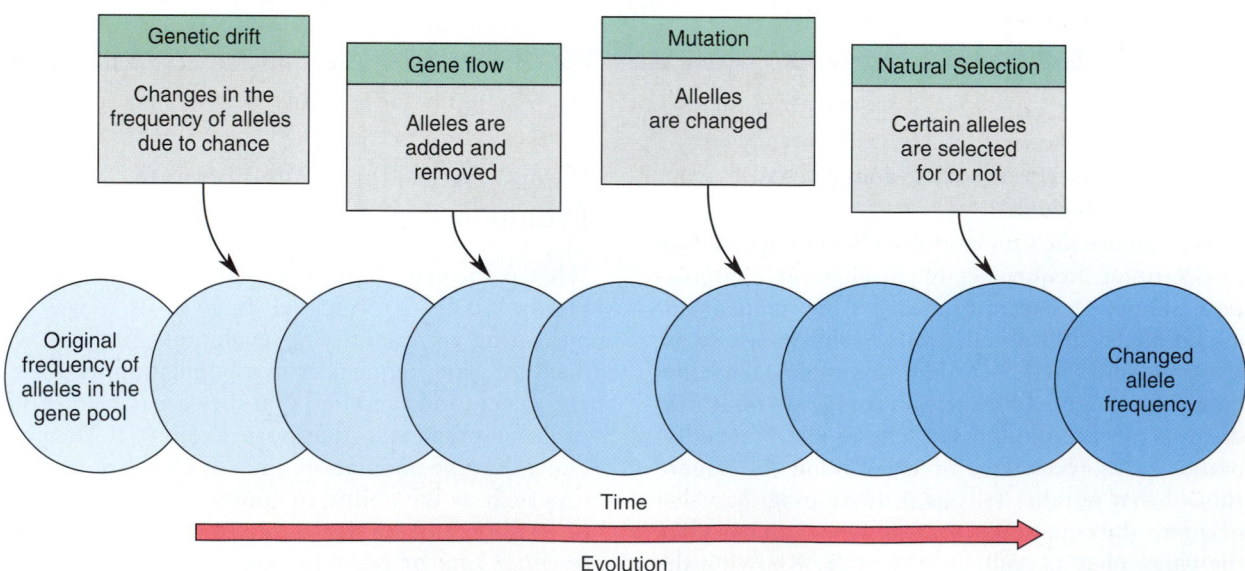

FIGURE 17.7 Mechanisms That Affect Genetic Equilibrium
Genetic drift, gene flow, mutation, and natural selection are mechanisms that can change the frequency of alleles in the gene pool. Over time, as the genetic pool of descendant populations changes, reproductive isolation leads to the evolution of new species.

Chapter 17 Evolution: Living Things Changing Over Time

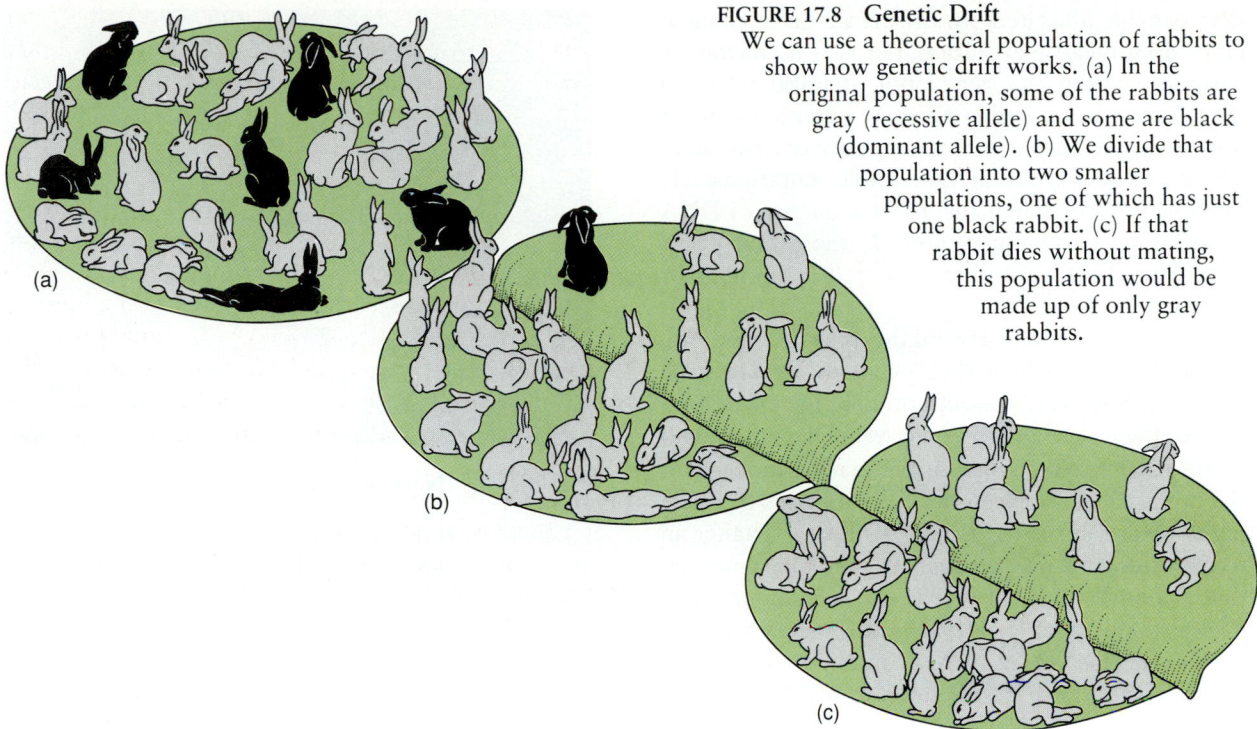

FIGURE 17.8 Genetic Drift
We can use a theoretical population of rabbits to show how genetic drift works. (a) In the original population, some of the rabbits are gray (recessive allele) and some are black (dominant allele). (b) We divide that population into two smaller populations, one of which has just one black rabbit. (c) If that rabbit dies without mating, this population would be made up of only gray rabbits.

alleles would be available. What if, by chance, both of the individuals died, or were of the same gender, or disliked each other intensely, or lived so far apart that they never met? The green-hair allele would be lost from the gene pool. Figure 17.8 shows how genetic drift would affect a rabbit population.

A special type of genetic drift called **founder effect** occurs when a new population is begun by a small group of individuals, one of whom carries a very special gene. An example of founder effect in human populations occurs among the Amish people of Pennsylvania. The Amish population is small, and by social custom, members do not marry individuals outside the religious community. Hence, few, if any, new genes are added to the population.

Once the community reaches a certain size, it is customary for a few Amish founders to leave and organize a new colony. One of these founding fathers was heterozygous for a rare form of dwarfism that is coupled with *polydactyly* (an allele for extra fingers and toes, which is dominant over the allele for five fingers or toes). Today, among the 8000 living members of the colony he helped found, 50 individuals have this condition. In other Amish colonies, the condition is absent. In this instance, the gene that became concentrated in the new colony was relatively harmful. However, in other instances, the trait can be neutral or even beneficial.

Gene Flow

Organisms can move from one population to another, taking their genetic makeup with them. This process is called **gene flow**. Gene flow can alter the frequency of alleles in the gene pool of the population left behind and the new population just joined.

In nature, one group of organisms of a specific species is often isolated from another group of the same species. These isolated groups are called **demes**. For example, a population of sunfish in one pond may be isolated from a population of sunfish in another pond. Should there be a period of heavy flooding, the two ponds may become connected. Then one pond may lose some individuals to the other. This phenomenon can affect the gene pool, changing the frequency of the alleles in both groups. When *immigrants* (those individuals that enter a population) add new genes to the gene pool while *emigrants* (those individuals that leave a population) remove their genes from the gene pool, the process is called gene flow.

Mutation

Another factor that can randomly alter the frequency of alleles in the gene pool of a population is mutation. **Mutations** are changes in the genotype that can be inherited by offspring. Such changes alone do not cause evolution, but they do introduce variation within the gene pool. Natural selection then determines whether the new variation is fit or not fit to survive and reproduce in the environment. There are two types of inheritable mutations: (1) those that occur when the linear sequence of DNA is disrupted or changed, and (2) those that occur when the chromosomes are altered in number or structure.

Mutations are always taking place at random, and the rate at which they occur in a given gene varies. If, however, the mutations of the alleles of a specific gene occur with equal frequency, there will be no observable change in the genetic makeup of the population. For instance, if *A* mutates to *a* at a higher rate than *a* to *A,* there will be a change in frequencies, but if *A* changes to *a* at the same rate that *a* changes to *A* (or vice versa), the frequency of alleles in the population stays the same. You can appreciate the importance of mutations by understanding that they provide the raw material for evolution.

Natural Selection

We are all familiar with *artificial selection,* the process by which humans interbreed plants or animals with given characteristics to produce offspring that combine those characteristics. In artificial selection, the experimenter or breeder is the one who decides which characteristic is desirable and which is not. In natural selection, the environment selects for or against a particular variation.

The peppered moth (*Biston betularia*), a common moth in England, is a good example of natural selection at work (Figure 17.9). At one time, most peppered moths were light colored. But during the middle and late nineteenth century, more and more dark moths were found, particularly near large cities. And over time, the dark moths became far more numerous than the light ones. Why? H. B. D. Kettlewell, an English naturalist, gave us the answer. Peppered moths are usually found on lichen-covered trees and rocks. Before England was industrialized, the light moths were camouflaged by their color. Dark moths, lacking that camouflage, were easy prey for birds, and so their number was small. With the Industrial Revolution came smoke from the increasing number of factories, which darkened the lichen on trees and rocks. Now the dark moths were camouflaged and the light-colored moths were not. In industrialized areas of England, then, the environment has selected for dark moths.

FIGURE 17.9   Natural Selection at Work: The Peppered Moth
Look closely at this photograph. There is a light-colored and a dark-colored peppered moth. In the countryside, the light-colored moth is camouflaged by the light-colored lichen on trees and rocks. Near industrial centers, the dark-colored moth is protected from predators such as birds by blending in with the darker lichen.

To more fully understand the importance of environment in the course of evolution, we need to look at how the changing environment influences the evolution of a new species.

MECHANISMS OF SPECIES FORMATION

■ **The formation of a new species, or speciation, requires that a group of organisms becomes reproductively isolated from another group of organisms.**

The process that results in the formation of a new species from one that already exists is referred to as **speciation.** We have already defined *species* as a particular type of organism that reproduces with members of the same group and produces fertile offspring. Biologists say that the members are reproductively isolated from all other species. Even if members of different species should mate, they cannot produce fertile offspring. For instance, horses and donkeys are not the same species, but they can mate and produce offspring, called mules. However, mules usually are not fertile. Therefore, donkeys and horses are different species because they are reproductively isolated. (I used mules in this example because most of you know about this hybrid. In the rare instances when mules are fertile, they can produce mule offspring.)

TABLE 17.1
Reproductive Isolating Mechanisms That Can Result in Speciation

Prezygotic mechanisms

1. Isolation caused by the formation of geographical barriers
2. Seasonal or temporal differences in reproductive maturity or mating patterns
3. Different courtship patterns or behaviors
4. Structural differences in reproductive organs that prevent or disrupt mating

Postzygotic mechanisms

1. A hybrid that is inviable
2. A hybrid that is infertile
3. A hybrid that weakens and dies

Reproductive Isolation

There are two major types of reproductive isolation: *prezygotic* (before zygote formation) and *postzygotic* (after fertilization). **Prezygotic mechanisms** prevent or inhibit mating and egg fertilization. Some of these are obvious — simple size differences make it impossible for a house cat to mate with an elephant, for instance. Mating rituals differ from one organism to another, which results in a very real barrier to mating. A female seagull is unlikely to choose a male blue-footed booby, no matter how long he performs his particular mating dance, because she does not recognize the signals he is sending out. Table 17.1 lists prezygotic mechanisms.

Table 17.1 also summarizes the **postzygotic mechanisms** of reproductive isolation. These come into effect after mating has occurred, and they affect the development of the hybrid. The infertility of the mule offspring of a donkey and a horse is one postzygotic mechanism. In other instances, the offspring is inviable (that is, it is unable to develop before birth), and in others, the offspring is so weakened that it dies shortly after birth or before it can reproduce.

In essence, speciation is development of a new gene pool that is isolated by the lack of reproductive exchange with other gene pools and that will remain isolated even when mixed in with members of its former species. Evolution describes how a new gene pool (hence, a new species) can be created.

Evolutionary Patterns

■ There are three main patterns of evolutionary change: divergent, convergent, and parallel. In divergent evolution, populations of the same species become increasingly different. In convergent evolution, species look alike and have adapted to similar environmental conditions but are unrelated. In parallel evolution, species in the same environment have evolved in similar ways.

Natural selection has produced a wide variety of organisms on earth. Some look alike but are very different — and vice versa. Basically, investigation has revealed three major patterns of evolution: divergent, convergent, and parallel (Figure 17.10).

Divergent Evolution. **Divergent evolution** occurs when a population becomes isolated from others of its species and eventually follows a different evolutionary path. Over a long period of time, then, isolation leads to the development of different species from a common ancestor.

Divergent evolution can take place over a period of time because of geographical isolation, as was the case with Darwin's finches. Or it can happen very rapidly, especially in plants when meiosis is incomplete. During the formation of the egg cells, the chromosomes in one egg cell will duplicate, but two daughter cells do not form. That means that the egg has four copies of the chromosomes, a condition called *tetraploid* ($4n$). Because the resulting plant has a chromosome number different from that found in its original species, it cannot reproduce with plants that developed from normal seeds. The tetraploid plant is genetically isolated from its former species, and a *new species* has been formed. In this particular case, speciation has been virtually instantaneous.

Convergent Evolution. In **convergent evolution**, two species look very much alike and are adapted to similar environmental conditions, but they are evolutionarily unrelated. For instance, in the deserts of Africa and the American Southwest there are cactuslike plants that are adapted for the same environment, but they are not closely related.

Parallel Evolution. In **parallel evolution** there is similar adaptation between two *closely* related organisms. As a result of natural selection, the adaptations are almost the same — they parallel each other. Geese and ducks are good examples, since they are closely related and have adapted to similar environments.

(a) Divergent evolution

(b) Convergent evolution

(c) Parallel evolution

FIGURE 17.10 Evolutionary Patterns
(a) These guenon monkeys are of different species as a result of divergent evolution. (b) The American sidewinder and South American sidewinder rattlesnakes look alike and are adapted to similar environments; however, they are not closely related. They are examples of convergent evolution. (c) The evolution of the marsupial kangaroo and placental zebra are examples of parallel evolution.

Darwin's Finches

Darwin described one of the classic examples of speciation when he analyzed his observations of the finch species in the Galápagos Islands. All 13 species of finches (and a fourteenth species on Cocos Island) had similar body types and markings, but they had characteristics that reflected their specialized feeding habits. Some species were ground feeders, eating seeds and cacti; others fed on berries and insects in trees (Figure 17.11). In fact, the major differences between species were their beak shapes.

As we noted earlier, it was not until Darwin returned to England that he learned that the finches on the Galápagos were all related and that they were also related to finches found in South America.

Darwin then theorized that the original finches on the Galápagos had come to the islands purely by chance — by being caught in heavy winds or seas during a severe storm. (The incident must have been a rare one, given the fact that the Galápagos Islands are 600 miles from the mainland of South America.) Seeds and spores had also blown to the Galápagos, and they provided a food supply to the new arrivals. With no natural enemies, the finches flourished in their new surroundings. Their population expanded quickly, as Malthus predicted — until the food began to be in short supply.

It is at this point that natural selection — the role of the environment in the selection of traits — began to take place. Chance mutations had occurred from the very beginning of the finches' ar-

FIGURE 17.11 Darwin's Finches
Over time, the finches on the Galápagos Islands, such as the large cactus finch above *(Geospiza sorirostrus)*, have adapted to their food sources. (a) The warbler finch *(Certhidia olivacea)* uses its slender bill to grasp insects on the branches and leaves of trees. (b) The large ground finch *(Geospiza magnirostris)* uses its bill to crush seeds. (c) The vegetarian tree finch *(Platyspiza crassirostris)*, which has a parrotlike bill, is the only Galápagos finch that feeds on berries. (d) The woodpecker finch *(Camarhynchus pallidus)* holds a cactus spine or twig in its beak to probe below the bark of trees for insects.

(a) Warbler finch

(b) Large ground finch

(c) Vegetarian tree finch

(d) Woodpecker finch

rival on the Galápagos, and some provided variations that were advantageous in acquiring food. A beak that is shaped like a pair of pliers is more efficient in cracking seeds open, while a beak that is chisel-shaped can be used to hold a cactus spine or twig in order to probe behind the bark of trees for the insects living there. The chisel-shaped beak belongs to the woodpecker finch, a rare example of a tool-using bird. Thus mutations provided variations that enabled certain finches to become specialized in their feeding behavior. Birds with a ready access to a certain type of food will have a better chance to survive, reproduce, and pass on their mutated characteristics to their offspring.

Other environmental factors besides a limited food supply were operating in the Galápagos Islands. The predominant type of food varies from one island to another. Darwin theorized that at some time in the past some of the finches had flown between the islands. The type of food source most abundant on a particular island exerted an additional selective pressure on the newly arrived finches, with environmental demands skewing their evolution in that direction.

Once the finches were geographically isolated on the separate islands, there was little or no gene flow between groups. Through the development of reproductive isolation mechanisms over time, what once had been one single species became 14 different species. Even when members of different species were combined in the same environment, they could not mate and produce viable offspring.

Darwin's finches illustrate several concepts that are an important part of the theory of evolution and of environmental studies. For instance, the specialization of the various types is a good example of diversification of a single type to fill what is called a **niche,** a specific role played by an organism in an environment. *Camarhynchus crassirostris* is a Galápagos finch that feeds on the tough seeds of the cacti of the islands. It does not compete with insect-eating species, such as *Certhidia olivacea.* Each species plays a particular role in its environment; each fills its own niche.

Another concept illustrated by Darwin's finches is that of **adaptive radiation,** the rapid evolution of several different specialized forms from a single primitive ancestor.

Mechanisms of Species Formation

EVIDENCE OF EVOLUTION

■ **Much of the evidence for evolution comes from the fossil record, taxonomy, and molecular analysis.**

Numerous scientific disciplines, from molecular biology to geology, provide evidence for evolution. Many of the sciences that study morphology (the structure of organisms), such as comparative anatomy, comparative embryology, and comparative plant morphology, provide data on the structural similarities among organisms. Information such as this demonstrates genetic evolutionary relationships (Figure 17.12).

Evidence in the Fossil Record

Fossils are the preserved remains of organisms that lived in the distant past. For the most part they consist only of the bone structure of the animals and the cellulose-based structure of plants, since soft tissues do not last very long after an organism dies. Fossils allow scientists to study the structure of ancient organisms, to compare them with existing life forms, and to hypothesize about the environment in which they existed. The fossil record also makes it possible to construct a time frame for the evolution of various life forms (Table 17.2).

When biologists study fossils, they are assisted by geologists in determining the age of the specimens. Geologists have provided evidence that our present-day continents were once part of a single large land mass called *Pangaea*. Over a period of more than 100 million years, the continents began to drift apart (Figure 17.13). The northern section of Pangaea (North America, Europe, and Asia) was called Laurasia; the southern section (South America, Africa, Australia, and Antarctica), Gondwanaland. Approximately 65 million years ago, these two supercontinents broke up, with North America moving away from Europe and Asia, and South America and Africa separating from each other and Antarctica. India, which started out in the southern region of Pangaea, was now moving north, to its position today. This continental drift established barriers that fostered reproductive isolation and gave rise to new species.

How can geologists prove this? For one thing, they look at layers of earth called *sediment layers* — the layers of rock that you see when you look at

(a) Comparative embryology

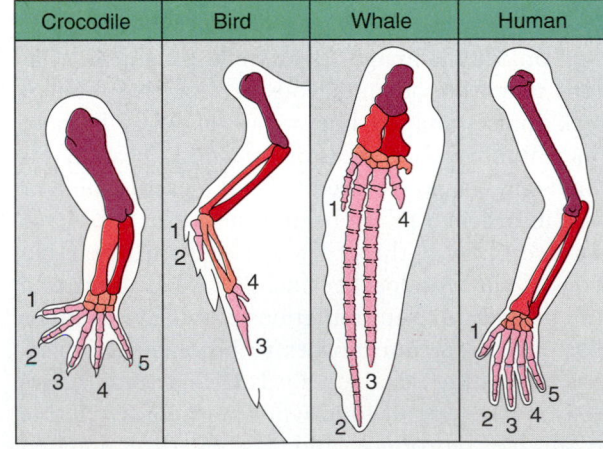

(b) Comparative anatomy

FIGURE 17.12 Morphology: Studying the Structure of Organisms
Scientists are able to demonstrate evolutionary relationships by studying the structure of different organisms. (a) We can see similarities in the embryos of vertebrates in the early stages of development. (b) In the forelimbs of the crocodile, bird, whale, and human, we find evidence of both their common ancestry and the adaptations that have been made to their environments and lifestyles.

Chapter 17   Evolution: Living Things Changing Over Time

TABLE 17.2
The Geological Time Scale

Era	Period	Epoch	Millions of years from start to present	Major events
CENOZOIC	Quaternary	Recent (Holocene)	0.01	Repeated glaciations; extinctions of large mammals; evolution of *Homo*; rise of civilizations.
		Pleistocene	2.0	
	Tertiary	Pliocene	5.1	Radiation of mammals, birds, angiosperms, pollinating insects. Continents nearing modern positions. Drying trend in mid-Tertiary.
		Miocene	24.6	
		Oligocene	38.0	
		Eocene	54.9	
		Paleocene	65.0	
MESOZOIC	Cretaceous		144	Most continents widely separated. Continued radiation of dinosaurs. Angiosperms and mammals begin diversification. Mass extinction at end of period.
	Jurassic		213	Diverse dinosaurs; first birds; archaic mammals; gymnosperms dominant; ammonite radiation. Continents drifting.
	Triassic		248	Early dinosaurs; first mammals; gymnosperms become dominant; diversification of marine invertebrates. Continents begin to drift. Mass extinction near end of period.
PALEOZOIC	Permian		286	Reptiles, including mammal-like forms, radiate; amphibians decline; diverse orders of insects. Continents aggregated into Pangaea; glaciations. Major mass extinction, especially of marine forms, at end of period.
	Carboniferous (Pennsylvanian and Mississippian)		360	Extensive forests of early vascular plants, especially lycopsids, sphenopsids, ferns. Amphibians diverse; first reptiles. Radiation of early insect orders.
	Devonian		408	Origin and diversification of bony and cartilaginous fishes; trilobites diverse; origin of ammonoids, amphibians, insects. Mass extinction late in period.
	Silurian		438	Diversification of agnathans, origin of placoderms; invasion of land by tracheophytes, arthropods.
	Ordovician		505	Diversification of echinoderms, other invertebrate phyla, agnathan vertebrates. Mass extinction at end of period.
	Cambrian		570	Appearance of most animal phyla; diverse algae.
PRE-CAMBRIAN (SINIAN)	Vendian		670	Origin of life in remote past; origin of prokaryotes and later of eukaryotes; several animal phyla near end of era.
	Sturtian		800	

Source: Douglas J. Futvyma, *Evolutionary Biology,* 2nd ed. (Sunderland, MA: Sinauer, 1986), p. 320.

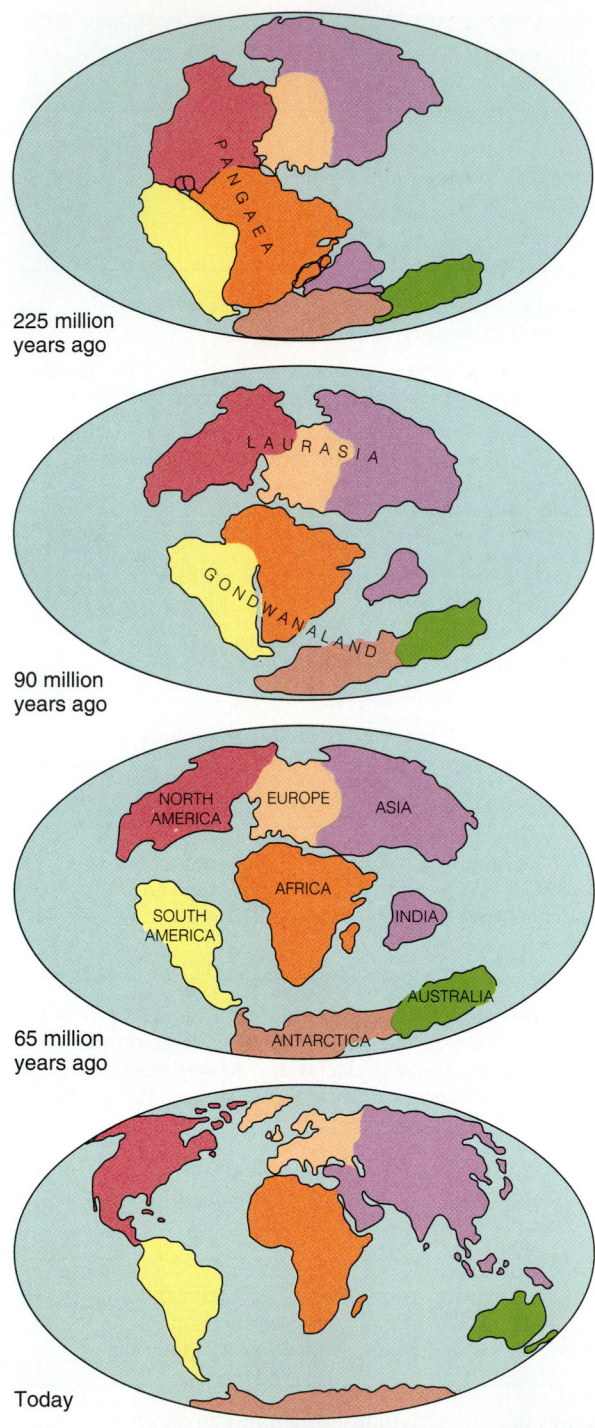

225 million
years ago

90 million
years ago

65 million
years ago

Today

FIGURE 17.13   Continental Drift
In 1912, German scientist Alfred Wegener drew fire
from his colleagues when he suggested that modern
continents had once been joined in a large land mass he
called Pangaea. He struck on the idea after observing the
"jigaw puzzle" fit of South America's east coast and
Africa's west coast. Wegener's theory of continental drift
resurfaced in the 1960s to be refined and adopted by a
new generation of geologists.

the face of a wall in the Grand Canyon (see Figure
17.2) or places on a highway in which road crews
have cut through existing hills or mountains as they
laid a roadbed. For instance, limestone layers were
formed by the bodies of millions upon millions of
tiny creatures living in an ancient sea. When the
animals died, their calcium-containing shells drifted
to the bottom of that sea. The shells accumulated
over a long period of time, were compacted by the
weight of water and of the shells above them, and
ultimately solidified into a layer of limestone. If you
view a piece of limestone under a microscope, you
can often see the outlines of those ancient shells.

So even if geologists had no way to arrive at the
age of a layer of limestone, by examining its parts
they could determine that at one time the land was
covered with water. Given the rate of sedimentation, they can calculate how long the sea existed. By
analyzing the layers of rock above the limestone,
they can find out approximately how long it has
been since the sea disappeared, and they can determine what kinds of environments followed that of
the ancient sea. When biologists find fossils in a
certain layer of rock, then they can turn to geological information to determine its age. By studying
rock formation and fossil evidence, scientists now
know, for instance, that a great swamp once extended from Oklahoma to Pennsylvania, and that it
was the decay of this swamp's plant life that formed
the fossil fuels like coal and oil that are found so
abundantly there.

How does this relate to continental drift? If
matching layers with their special fossil types are
found hundreds of miles apart, that is evidence that
masses of land have moved. Because of continental
drift, there are places on earth (even in the United
States) where you can walk 1000 miles with a single
step. Impossible? Not at all. In California, for example, if you leap westward over the San Andreas
fault, you cross from northern California into what
was once part of Old Mexico. Similarly, if you step
over a fault line in Alaska, you can leave behind the
frozen north for what was once a tropical island.
Thus continental drift helps explain, for example,
why there are fossil tropical species in the Antarctic.

Interestingly enough, there is some dramatic new
evidence that the continents may drift and form a
thin mantle or covering, rather than the thick,
crusty continental masses. Therefore, instead of
thick pieces of crust (about 60 to 70 miles deep),
the regions of some continents may be made up of
much thinner layers of rock spread over a continental plate, something like the frosting on your favorite chocolate cake.

Chapter 17   Evolution: Living Things Changing Over Time

Evidence from Taxonomy

Other procedures such as taxonomy, or classification, also provide evidence for evolution. Since taxonomy proceeds according to similarities between the structures found in various kinds of organisms, analysis often reveals similarities that otherwise would not be anticipated. For instance, even though it seems at first that whales should be related to other marine life forms such as sharks, we know that there is a distant relationship between whales and humans because both breathe by means of lungs. We will return to taxonomic evidence in the next chapter.

Evidence from Molecular Analysis

Today, biologists have another tool to use in the reading of evolutionary history. Emile Zuckerkandl and Linus Pauling discovered that the number of amino acid differences in the hemoglobin molecule between two species seemed proportional to the amount of time since they diverged from a common ancestor. Also, comparisons of the DNA sequences can help establish relationships between species. That means that evidence of evolutionary relationships may exist not only in the fossil record but also in the molecular similarities and differences of the proteins and genes of different species.

THE PACE OF EVOLUTIONARY CHANGE

■ **Gradualism is the theory that evolution takes place slowly and steadily, as an accumulation of small changes. Punctuated equilibrium is the theory that evolution takes place in rapid bursts followed by long periods of no change. Many biologists today think that evolution takes place at different rates, depending on the environmental conditions.**

Since Darwin's time the prevailing belief had been that evolutionary change is constant for all organisms and is a rather slow and gradual process. This theory is called **gradualism.** However, the fossil record seldom supports this position. In the 1970s, Stephen Jay Gould and Niles Eldredge concluded that the pace of evolutionary change varies greatly. Periods of slow change are interspersed, or *punctuated*, with periods of rapid change. This theory is called **punctuated equilibrium.**

The theory of punctuated equilibrium states that many groups of organisms have evolved rapidly during the early part of their history. Once they were successful within their environment, the rate of evolution slowed down — as long as the environment remained the same and food was readily accessible. For example, as the early seas diminished in size and organisms evolved on land, they experienced rapid evolutionary changes because of the demands of the new environment. Once they had undergone the radical alterations necessary, however, evolutionary changes were slower and more modest.

But the case for gradualism — the theory that evolution is sometimes slow and steady — remains strong. For example, recent studies of trilobites (small, shelled animals living in the oceans half a billion years ago) indicate gradual change. Dr. Peter Sheldon, studying trilobites in Wales, found that some characteristics — notably the number of ribs —changed steadily over 3 million years.

Both theories may in part be correct. Evolution may sometimes occur slowly and steadily and sometimes in fits and starts, depending on environmental pressures.

Extinctions. If evidence indicates that some evolution comes in waves, **extinction,** or the end of an organism's lineage, also appears to come in waves. Perhaps the extinction of the great dinosaurs is the most famous case, but the fossil record demonstrates that there have been at least six major waves of mass extinctions (Figure 17.14). Since the extinctions are so major, biologists use them to divide time into geological periods.

The causes of these great extinctions are unknown, but obviously there must have been rapid and drastic changes in the environment to have such a radical effect on so many different types of organisms at the same time. During these mass extinctions the dominant organisms died out and their ecological niche was gradually filled by other groups of organisms that previously had been fairly insignificant.

For example, mammals did exist during the age of the dinosaurs, but they were small and relatively minor organisms in the overall scheme of things. Then something — scientists are not sure exactly what — happened about 65 million years ago. Almost all scientists agree that there was a general cooling of the earth, although there is not universal agreement about the cause of this change in climate. Some think that an asteroid collided with the earth, producing a large dust cloud that encompassed the globe and blocked sunlight. Whatever the reason, weather conditions cooled all over the planet, the

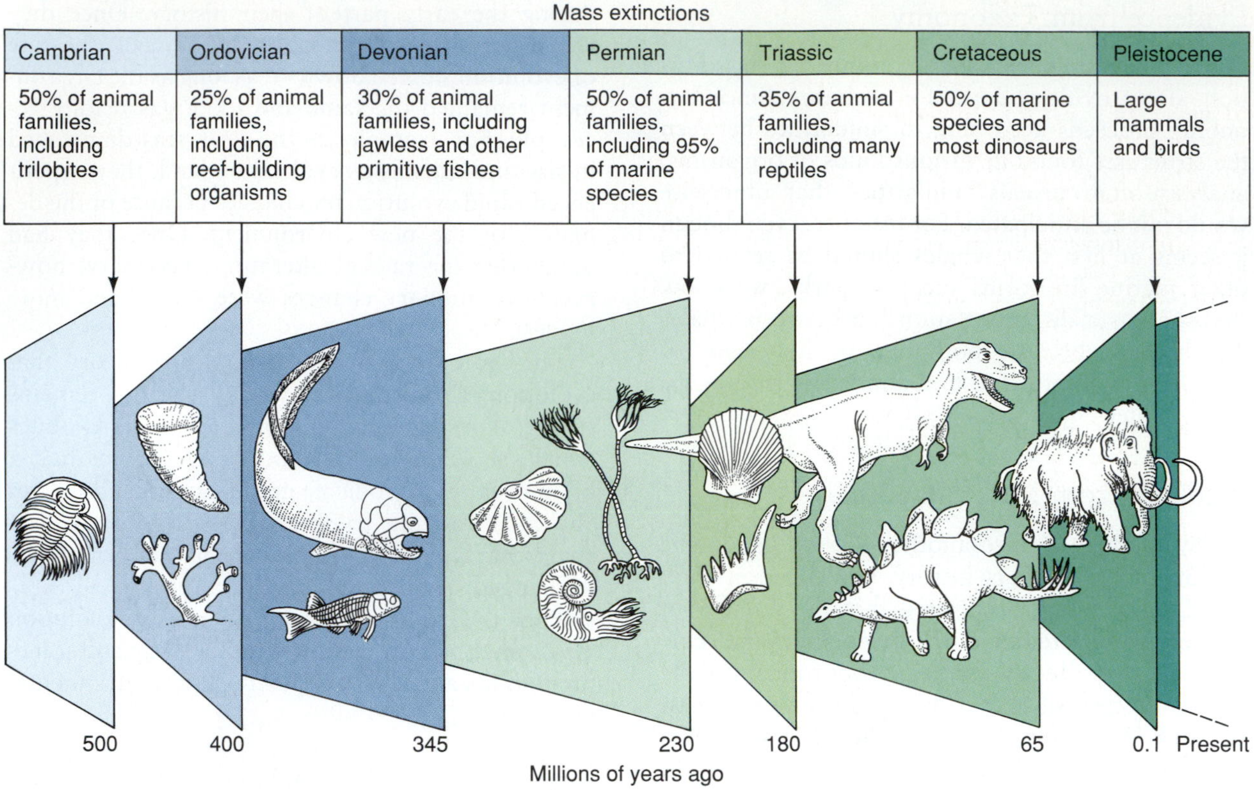

Mass extinctions

Cambrian	Ordovician	Devonian	Permian	Triassic	Cretaceous	Pleistocene
50% of animal families, including trilobites	25% of animal families, including reef-building organisms	30% of animal families, including jawless and other primitive fishes	50% of animal families, including 95% of marine species	35% of anmial families, including many reptiles	50% of marine species and most dinosaurs	Large mammals and birds

500   400   345   230   180   65   0.1 Present

Millions of years ago

FIGURE 17.14   Mass Extinctions
The fossil record shows that there have been at least six major extinctions in the history of the earth.

dominant plant forms died because of the climate change, and the dinosaurs died because of the loss of their food source. Mammals then began to occupy the ecological positions vacated by the dinosaurs, because they could feed on other food sources, such as other mammals.

OUR EVOLUTIONARY PATH

■ **Paleontologists once believed that the human species began to separate from the other primates as long as 20 million years ago, but recent findings suggest that *Homo* separated only approximately 5 million years ago. Before that there was an organism that walked upright but had a very small brain.**

Just in case you are beginning to think that all the information and studies are pat and full of agreement, we now are going to look at one particular evolutionary path: our own. (See Focus 17.2.) The data from molecular genetics are especially instrumental in rewriting the history of our own evolution. In 1987, Dr. Richard Leakey was giving a presentation at the University of Puerto Rico, and

when asked when humans split from the apes, he jokingly remarked, "We haven't yet."

In a sense, however, Leakey was not joking, because molecular scientists have found that human hemoglobin and chimpanzee hemoglobin amino acid sequences are virtually identical. Their hypothesis came from immunological tests of the protein albumin and from other molecular studies that demonstrated that human and chimpanzee DNA is 98 percent identical (Figure 17.15). As a result of this evidence, many biologists have concluded that the split between humans and apes may have occurred about 5 million years ago — or perhaps even more recently than that (Figure 17.16).

Determining Differences

What are the characteristics that will allow us to distinguish between humans and other primates? Most scientists used to believe that small canines (or eyeteeth), large brains, the use of tools, and our upright posture are truly human characteristics and that all these characteristics *evolved together*. However, paleoanthropologists (scientists who study the

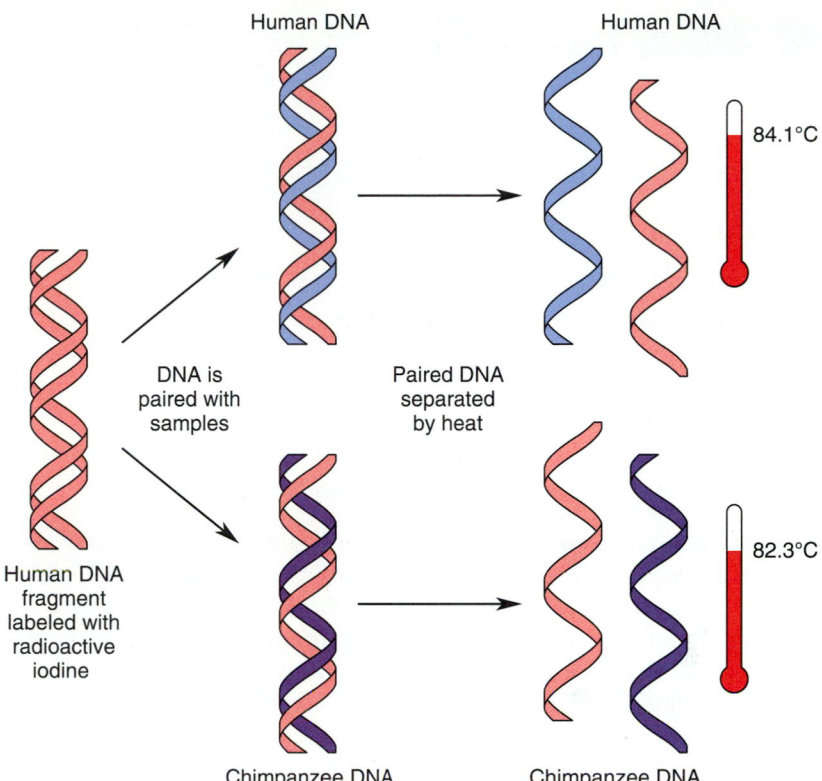

FIGURE 17.15 Using DNA to Show Evolutionary Relationships
By comparing the DNA of two different species, scientists are able to measure the genetic relationship between them. A radioactively labeled sample of human DNA is combined with another sample of human DNA and with chimpanzee DNA. The closer the genetic match between the base pairs of DNA, the tighter the bond. Scientists measure the strength of a bond by heating: Tightly bound DNA must be heated to a higher temperature to separate the strands. In this experiment, the human-human strands separated at 84.1°C; the human-chimpanzee strands, at 82.3°C. These findings imply a close relationship between the two species.

Human DNA

Human DNA

84.1°C

DNA is paired with samples

Paired DNA separated by heat

Human DNA fragment labeled with radioactive iodine

82.3°C

Chimpanzee DNA

Chimpanzee DNA

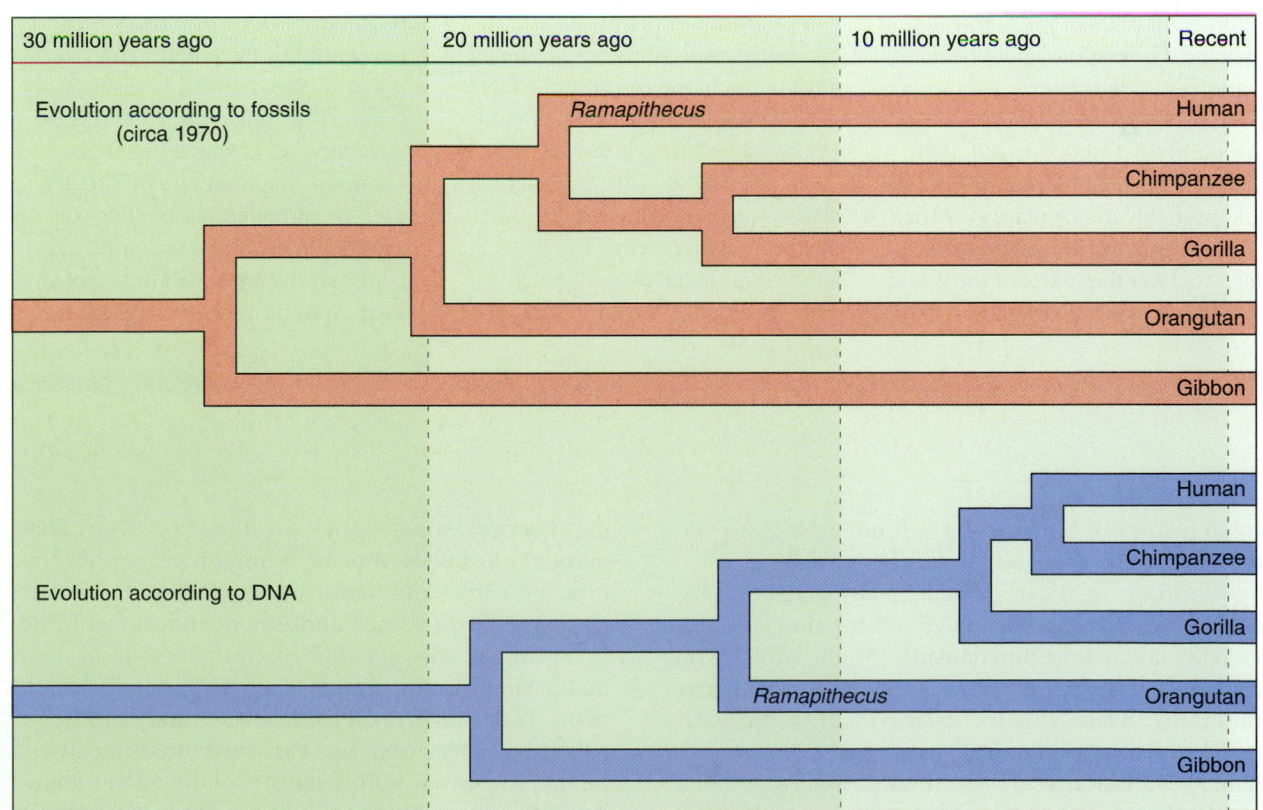

FIGURE 17.16 Human Evolution: The Fossil Record Versus Molecular Analysis
The fossil record and analyses of DNA give us conflicting stories on the evolution of humans. Using the fossil record, geologists date the separation of human and chimpanzee species at 20 million years ago. Molecular biologists, with the evidence of DNA, claim that separation took place 5 million years ago at most.

Human Origins: New Evidence Is Shaking the Roots of Our Human Tree

Scientific controversy continues and nowhere is it more apparent than in trying to answer the fundamental questions about human origins, or our roots. The controversy exists between the traditional anthropologists (sometimes called the stone and bone group) and the molecular anthropologists. The molecular anthropologists propose that humans didn't evolve slowly and gradually in different parts of the world (the view held by traditional anthropologists), but rather that the evolution from archaic to modern *Homo sapiens* occurred 140,000 to 290,000 years ago, probably from one group and probably in one place — Africa. The descendants of this group then migrated to other parts of the world 90,000 to 180,000 years ago.

Anthropologists reached this conclusion based on evidence that indicates that all living humans can trace their mitochondrial DNA back to one woman. As we have mentioned in earlier chapters, mitochondria are self-replicating organelles with their own DNA. At the moment of conception, the sperm injects its nuclear DNA into the egg. The tail of the sperm, which contains mitochondria to fuel its movement, falls off. Thus all the mitochondria in the egg (and the DNA they contain) come from the mother. As stated by Mark Stoneking from the University of California, Berkeley, "Genes in the nucleus are inherited from both parents, but mitochondrial DNA is inherited only maternally, directly from the egg. That means any change in the mitochondrial DNA can be due only to mutation. And we have a pretty good idea that the mutation rate is only 2 to 4 percent per million years."

With this information in mind, Rebecca Cann and her colleagues at the University of Hawaii studied the placentas of 147 women from five geographical regions. The team used restriction enzymes to cleave the mitochondrial DNA into easily identifiable patterns. In comparing these patterns, they found 133 distinct types. How long did it take to produce these 133 types by mutation? Using the mutation rate of 2 to 4 percent per million years (this concept of a regular rate is considered controversial by some but accepted by most anthropologists), the team

fossil history of humans) have found new fossil evidence that has changed all that.

Working in East Africa, Mary Leakey, the widow of the famous paleoanthropologist Louis Leakey, states that bipedalism (upright walking on two feet rather than four) evolved long before larger brain size. Her evidence is provided by the footprints of two hominids fossilized in volcanic ash dated 3.6 million years ago (Figure 17.17a).

Another fossil find, nicknamed Lucy (because just after the discovery the research team listened to the Beatles' song "Lucy in the Sky with Diamonds"), is believed to be 3 million years old (Figures 17.17b). This fossil of a female, which was discovered by a French and American team working in Ethiopia, was capable of upright walking and had a small brain, again suggesting that early humans may have evolved from a chimpanzee-like ancestor. Therefore, we can infer that the early human ancestors with brains no larger than chimpanzees were walking upright long before the larger brain size evolved.

determined that Eve (scientists' name for the mitochondrial mother) must have lived 140,000 to 290,000 years ago.

Just think of the implications: Scientists are proposing that a common ancestor to whom we can all trace our mitochondrial DNA is a single woman, in a single place, who probably lived 140,000 to 290,000 years ago — our 10,000th great-grandmother. As Stephen Jay Gould says, "It makes us realize that all human beings, despite differences in external appearance, are really members of a single entity that's had a very recent origin in one place. There is a kind of biological brotherhood that's much more profound than we ever realized."

It should be noted that the woman described was probably among the first group of modern humans, or their direct ancestors. By chance she was the only woman who had at least one female descendant in every generation, so that her line of mitochondrial DNA continues until today. Where did this woman live (Figure A)?

Cann and her colleagues determined that the placentas from the African pool had the largest amount of diversity in their mitochondrial DNA. This suggests that the African population is probably the oldest. Also, studies of nuclear DNA point

FIGURE A
These African sites have yielded fossilized remains of *Australopithecus*.

to the African origin, and the major fossil finds have been in Africa. Other investigators, such as Douglas Wallace of Emory University and his team, think that Eve may have lived in Asia. They base their conclusion on the study of mitochondrial DNA obtained from the blood of 700 individuals from four continents. Using methods different from those of the Cann group, the Wallace group believes that the human tree goes back to one woman who lived 150,000 to 200,000 years ago. Because the DNA from the Asian sample is most similar to our nearest primate ancestors, they concluded that Asia was the likely location of our roots.

The dispute goes on. Recently, investigators have determined that the Neanderthal people are probably not direct ancestors of modern humans. This conclusion is based on the technique of thermoluminescence (TL). When rock is heated, certain products of radioactive decay are emitted as bursts of light. The amount of light emitted is correlated with the length of time the decay products have been accumulating in the rock. Using flint rocks found beside modern-appearing human skeletons discovered in caves in Israel, researchers have determined that the bones were about 92,000 years old, not the 35,000 to 40,000 years old previously believed. This finding indicates that anatomically modern humans were here before the earliest accepted dating of the Neanderthals, believed to be 35,000 to 40,000 years ago. Modern humans were here about 50,000 years before the Neanderthals. What happened to the latter? That is obviously another story.

So our evolutionary tree is constantly being shaken. The evidence of these studies of mitochondrial DNA suggests that we are a relatively recent species, and our ancestral mother may be one woman who lived about 200,000 years ago in Africa. We are indeed all one great family.

Some scientists do not accept the African evidence as convincing. Instead they push the ancestor of humans back much farther to another fossil find, called *Ramapithecus,* which was an apelike creature dating from between 8 and 14 million years ago. Scientists who support this position believe that *Ramapithecus* used tools and had a partially upright posture. Therefore, as in Figure 17.17, the fossil records show *Ramapithecus* directly in the human line. However, this position has been questioned and will continue to be. In fact, molecular studies imply that *Ramapithecus* evolved long before the divergence of ape and human, about 5 million years ago.

Even though the evolutionary history of humans is still fraught with controversy and is constantly being revised, current evidence supports the idea that the divergence of humans and apes was more recent than we think. It probably occurred closer to 3 to 5 million years ago rather than 14 to 20 million years ago as previously thought. Furthermore, our closest living relative may be the rare pygmy chim-

(a)

(b)

FIGURE 17.17 Fossils and the Path of Human Evolution
(a) Mary Leakey, working in East Africa, discovered the footprints of two hominids dating back 3.6 million years. (b) From fossil remains, paleoanthropologists were able to "reconstruct" Lucy. She was only 3½ feet tall and weighed just 65 pounds. And her brain was only about one-third the size of the human brain today.

panzee, *Pan paniscus*. After all, molecular genetic analyses demonstrate a greater than 98 percent similarity of DNA in humans and pygmy chimpanzees. The ape in our past may be closer than we think, and controversy still reigns in evolution, especially about our own evolutionary path.

SUMMARY

1. Evolution of living things is a basic biological concept that explains how they change through time.
2. The ancient Greeks attempted to find order in the diversity of living things. Aristotle described a *Scala Naturae* in which he delineated a hierarchy of living things with simple creatures on the bottom and humans at the top.
3. Numerous philosophers and scientists have speculated about how life forms and the earth evolved. Carolus Linnaeus formulated a method for classifying all living things into an organized system; Georges Buffon felt that some creatures were created on the sixth day but that the lesser creatures were conceived by nature and produced by time; Erasmus Darwin suggested that species are related to one another through their evolutionary history.
4. Early geologists speculated about the earth's formation and age. Most believed that the earth and its life forms were the fixed and static products of divine creation. However, some supported the idea of uniformitarianism — that change occurred gradually according to natural laws.
5. The arguments and proofs of geologist Charles Lyell convinced the scientific community that only unifor-

mitarianism could adequately explain the age and evolution of the earth. His work put the idea of catastrophism to rest.
6. Thomas Malthus proposed the idea that population growth is geometric while food supply grows arithmetically. The difference creates a shortage of food, hence a struggle for survival.
7. Jean-Baptiste de Lamarck formulated a theory of evolution based on the ideas that acquired characteristics could be inherited, that acquired characteristics could become more pronounced through use or could regress through disuse, and that life forms tended to become more complex.
8. Charles Darwin and Alfred Russel Wallace proposed a mechanism for evolution in the theory of natural selection. They concluded that the organisms best adapted to their environment are best able to compete for the limited necessities of life. The environment selects, or determines, which organisms will survive. Inheritable genetic variations provide the variations expressed by individuals of a population.
9. The modern interpretation of the theory of natural selection is essentially the same as its originators proposed. However, based on current information we know more about the genetic mechanism of evolution. Genetic variation is a result of genetic recombination, fertilization, mutations, and chromosomal aberrations.
10. Populations, not individuals, evolve.
11. The variations for a given gene in a population are contained in its gene pool.
12. Evolution is the result of a change in allele frequencies in a gene pool. This concept can be measured by applying the Castle-Hardy-Weinberg principle.

This mathematical principle states that the frequency of alleles in a population remains the same, unless acted on by an outside force like selection. Changes in allele frequencies result in phenotypic changes in the population — evolution.

13. Factors that affect this genetic equilibrium are genetic drift, gene flow, mutations, and natural selection.

14. When a population becomes reproductively isolated by a geographical or physiological mechanism, its members cannot interbreed with other members of the species to which it once belonged. Such isolation eventually leads to the formation of a new species. Prezygotic and postzygotic isolating mechanisms are types of reproductive isolation mechanisms.

15. There is much evidence for evolution, from the continental drift theory to molecular biology.

16. The rate of evolutionary change is probably not constant. It is punctuated by waves of activity and inactivity. Extinction also comes in waves.

17. There are three major patterns of evolution: divergent, convergent, and parallel.

18. Based on new techniques from molecular biology and the fossil record, our understanding of human evolution is increasing. Current evidence supports the idea that our closest living relative is the pygmy chimpanzee and that the divergence of the human line from the ape line occurred 5 million years ago at most.

KEY TERMS

species	mutation
natural selection	speciation
population	prezygotic mechanism
gene pool	postzygotic mechanism
Castle-Hardy-Weinberg	divergent evolution
principle	convergent evolution
genetic equilibrium	parallel evolution
nonselective evolution	niche
genetic drift	adaptive radiation
founder effect	gradualism
gene flow	punctuated equilibrium
deme	extinction

REVIEW QUESTIONS

True/False

1. In convergent evolution, different species evolve from a common ancestor.
   true   false   (page 395)

2. Lamarck proposed an explanation for the evolution of organisms on the earth that emphasized inheritance of acquired traits.
   true   false   (page 385)

3. The theory of natural selection states that certain genetic variations increase an organism's chance of surviving and passing its genes on to its offspring.
   true   false   (page 387)

4. Punctuated equilibrium is the theory that evolution is slow and steady.
   true   false   (page 401)

Fill in the Blank

1. All the fish of the same species in a pond are an example of a _____. *(page 389)*

2. A change in gene frequencies in a population as a result of chance is called _____ _____. *(page 392)*

3. Speciation is the result of _____ _____. *(page 395)*

4. The formation of different species from a common ancestor is an example of _____ evolution. *(page 395)*

Multiple Choice

Suppose that in a certain community in London there were equal numbers of red-haired and green-haired individuals. (None of them dyed their hair.) This hair-color trait was controlled by a single pair of alleles that followed the patterns of simple Mendelian dominance. The information for answering the following questions is on pages 390 and 391.

1. To calculate the gene frequencies we have to know:
   a. The percentage of individuals expressing both phenotypes
   b. Which characteristic was dominant
   c. a and b
   d. None of the above

2. Two people with red hair sometimes have children with green hair, but two people with green hair never have children with red hair. We can conclude that:
   a. Green hair is dominant to red hair.
   b. Red hair is dominant to green hair.
   c. There is a lack of dominance.
   d. a and b are correct.

3. A count of people in the community revealed that there were 510 people with red hair and 490 people with green hair. What was the approximate frequency of the gene for green hair?
   a. 0.1
   b. 0.3
   c. 0.5
   d. 0.7

4. What is the approximate frequency of the gene for red hair?
   a. 0.1
   b. 0.3
   c. 0.5
   d. 0.7

5. Approximately what percentage of the people had genotypes homozygous for the dominant gene?
   a. 3 percent
   b. 9 percent

   c. 30 percent
   d. 50 percent

6. Approximately what percentage of the people had heterozygous genotypes?
   a. 9 percent
   b. 21 percent
   c. 42 percent
   d. 51 percent

Answers are given in the Appendix.

Chapter 17   Evolution: Living Things Changing Over Time

18

Diversity and Taxonomy

Naming and Classification of Organisms
Whittaker's Five-Kingdom System
Summary

The great diversity of life on earth is remarkable, especially when you realize that whenever we think about it, we usually consider only the organisms that we can see. Add all the invisible microscopic organisms that we usually disregard or that live in the soil or ocean's depths, and the result is a wondrous variety of living things.

NAMING AND CLASSIFICATION OF ORGANISMS

■ **Taxonomy is the branch of biology that is primarily concerned with the classification of organisms. The major divisions of classification are kingdom, phylum, class, order, family, genus, and species.**

Above: This transparent ghost shrimp is but one example of the diversity of living forms on earth.

Although it may be difficult to discern, there is, in fact, order in this diversity, and biologists are able to group organisms in a meaningful way. We can develop classifications according to any number of criteria — color, size, economic worth, or the environment in which an organism lives, to name just a few (Figure 18.1). However, such systems would not necessarily be very useful from a biological perspective, because they would be *artificial*. If, on the other hand, we use a system that is *natural*, based on the organisms' evolutionary history or phylogeny (the pattern of descent of a group of organisms), we have a more useful system of classification. Humans have always been naming organisms and living things — we seem to enjoy it. What is the first thing you ask when you see a newborn infant? Usually you are curious to know its name. We have the same curiosity about all living things — we want to know their names.

FIGURE 18.1   The Diversity of Life

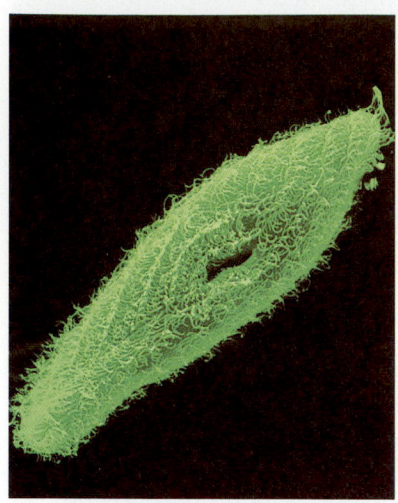

Chapter 18   Diversity and Taxonomy

Binomial Classification

In Chapter 17 we mentioned the system of binomial nomenclature that biologists use for naming organisms. There we noted that every organism belongs to a *species* (a group of organisms that interbreed with one another), and that several species belong to a larger category known as a *genus* (a group of closely related species). The science of classifying organisms is called **taxonomy.** Taxonomists are the ones who are (in part) responsible for the long names that seemed designed to completely frustrate any nonbiologist — *Drosophila melanogaster,* for instance, or *Escherichia coli* (see Profile). Actually, these names are designed to minimize confusion, not cause it, for it is a fact that different languages and cultures have different names for the same organism. The food crop that people in the United States commonly call *corn* is called *maize* in England. To the English *corn* means what we call *grain.*

Taxonomic names help biologists avoid this confusion. They are based on Latin (or sometimes ancient Greek) terms. The meanings of the terms are well defined, and since the languages are no longer spoken by a given population, the meanings do not change over time. The terminology is accepted by everyone in the field. (Though that is not to say there is an absence of disagreement about where a particular organism belongs, or whether different organisms are related or not.)

Systematics is the study of the evolutionary relationships among organisms as the basis for the natural classification system used in biology. We have already looked at some of the processes involved in taxonomy; this field also helps us to create order out of the wide diversity of life forms. In fact, in biology a discipline is devoted to the simple naming of things, called *nomenclature* (*nomen* is Latin for "name"). If we are going to have a universally acceptable system of classifying, we must have agreement on exactly how we are going to do it. Thus, an international congress, which meets every few years, develops the rules for classification. A specific congress develops the rules for classifying animals. These rules are published in the *Bulletin of Zoological Nomenclature.* Another congress and journal do the same for plants, and still another for microorganisms. In biology, the name of an organism must be accurate, descriptive, and universally accepted, so that all scientists understand it.

The two-part name, or **binomial system of nomenclature,** was introduced by Carolus Linnaeus during the eighteenth century, a time when the scientific world desperately needed a way to organize and classify the types of organisms from all over the world that were being observed by European scientists for the first time. Most important to us is the way in which his system indicates the particular *species* of an organism and the *genus* to which that species belongs. Hence the scientific name of an organism consists of two names, one indicating genus and the other species.

Species and Genus. What does that mean — species and genus? A species is a particular type of organism that breeds with other organisms of the same type, and a genus is a closely related group of species. As a matter of convention, the term for genus is a noun and is capitalized, and the term for species is usually an adjective and is not capitalized. Both names are italicized. The scientific name for a human being is *Homo sapiens,* which in Latin means "man the wise" (a name that sometimes seems fairly egotistical!). In other words, humans belong to the genus *Homo* and the species *sapiens.*

To understand how biologists classify organisms, let us look at examples of animals that belong to what we commonly call the cat family: the domesticated house cat and the tiger. We all accept the fact that these two animals are distant relatives. Biologists would say that they all belong to the biological family called *Felidae,* flesh-eating, predacious mammals. A tiger is a member of the genus called *Panthera,* a group of closely related mammals that are large, fierce flesh eaters. A house cat, however, is a member of the *Felis* genus, a group of small, lithe, soft-furred mammals. Would a lion be a member of the genus *Panthera* or *Felis*? It would be *Panthera.*

So the first name of a tiger in taxonomic terminology is *Panthera.* The second term in a binomial system usually is a descriptive name, indicating the species of the organism. In the case of the tiger, the term is *tigris.* The name of this particular species, then, is *Panthera tigris.* If we read about *Panthera leo,* are we still talking about tigers? No, because the second name is different. *Panthera leo* is the taxonomic name for the lion. The taxonomic name for a house cat, on the other hand, is *Felis cattus.* Its name indicates that there are many more differences between it and the members of *Panthera* than there are between *Panthera tigris* and *Panthera leo* (Figure 18.2).

Unfortunately, using only two names to classify an organism does not adequately demonstrate the relationship of one organism to another. How can we establish categories of classification to demonstrate the degree of evolutionary relationships between organisms?

(a)

(b)

(c)

FIGURE 18.2 The Cat Family
The (a) tiger *(Panthera tigris)* and (b) lion *(Panthera leo)* are members of the same genus; (c) the house cat *(Felis cattus)* is a distant relation. All belong to the same biological family, Felidae.

Classification Categories

Linnaeus established his complex system for describing the relationships between various organisms in the eighteenth century. Since then, his system has been adapted, but the basic organizing principles have remained fairly constant. In many ways, classification categories are similar to the categories we use when we address a letter. For example, proceeding from the most specific category to the most inclusive, an address lists the specific name of the individual, the house number, the street, the city, the state, and finally the country.

Similarly, the modern system of classification uses the **kingdom** as the most inclusive category, and the kingdom is then made up of several **phyla** (the singular form of this word is *phylum*). In most instances in the kingdoms of the plants and the fungi, biologists use the term **division** rather than phylum. Phyla are made up of several related **classes,** and classes include several **orders.** Orders are made up of several related **families;** families are composed of several closely related **genera;** and a genus is made up of several related **species.** A way to remember the sequence of classification categories from the broadest to the most specific is *K*ing *P*hilip *C*atches *O*ysters *f*or *G*ood *S*port — *k*ingdom, *p*hylum, *c*lass, *o*rder, *f*amily, *g*enus, *s*pecies.

When necessary, each category can be further divided into subcategories and supercategories. In some cases, depending on the needs of the investigator, these special categories can be subdivided even further. Table 18.1 illustrates the commonly used classification categories, and Table 18. 2 dem-

**TABLE 18.1
Classification Categories**

Animals	Plants
Kingdom	*Kingdom*
Subkingdom	Subkingdom
Phylum	*Division*
Subphylum	Subdivision
Superclass	
Class	*Class*
Subclass	Subclass
Superorder	
Order	*Order*
Suborder	Suborder
Superfamily	
Family	*Family*
Subfamily	Subfamily
Supergenus	
Genus	*Genus*
Subgenus	Subgenus
	Section
	Subsection
Species	*Species*
Subspecies	Subspecies
	Variety
	Form

Chapter 18   Diversity and Taxonomy

TABLE 18.2
Classification for Two Different Organisms

Classification	Human	Domestic Rose (Rosa gallica)
Kingdom	Animalia	Plantae
Phylum (Division)	Chordata	Tracheophyta
Subphylum (Subdivision)	Vertebrata	Pteropsida
Superclass	Tetrapoda	
Class	Mammalia	Dicotyledonae
Order	Primates	Rosales
Family	Hominidae	Rosaceae
Genus	*Homo*	*Rosa*
Species	*Homo sapiens*	*Rosa gallica*

onstrates specific classification of humans and the domestic rose.

It is important to remember that each category indicates a hierarchy of evolutionary relationships. Organisms of the same order show a higher degree of relationship than organisms of the same phylum. For example, humans and fish belong to the same phylum, which indicates that they share a rather distant evolutionary relationship. Monkeys and humans, on the other hand, belong to the same order, that of the primates, and as a result, their classification indicates that they share a closer evolutionary kinship. The most important question for us right now, then, is this: How do biologists determine the categories to which an organism belongs?

Classification Criteria

■ Taxonomists consider many criteria when they classify organisms. These criteria include the presence or absence of a membrane-bound nucleus, the number of cells, body form, homology and analogy, embryological development, the type of body cavity, molecular sequence in DNA and amino acids, behavior, and nutritional patterns.

To classify anything, you must begin with some sort of criteria. When you arrange the things in your room (and I know you all have neatly ordered rooms), you use certain reasons for organizing things as you do. For example, clothes go in the closet, books on the shelves, current notebooks on the desk, old notes in the file (even if the file trans-

lates into "under the bed"). The same holds true for taxonomists, the organizers of biology. They use a specific set of criteria to determine the degree of similarity and differences among organisms. Several criteria can be used to demonstrate evolutionary relationships among organisms. We will discuss only a few here, with emphasis on those used to classify animals.

Membrane-Bound Nucleus. Cells fall into two major categories, those that have a membrane-bound nucleus and those that do not. As we have mentioned before, prokaryotic cells do not have a membrane-bound nucleus. Eukaryotic cells do have a membrane-bound nucleus, as well as other membrane-bound organelles, such as the mitochondria and chloroplasts. A basic taxonomic criterion is the presence or absence of a cellular nucleus.

Number of Cells. Number of cells is also useful as a taxonomic criterion. Some organisms are unicellular (one-celled), and some are multicellular (many-celled). A few fall somewhere in between: They are *colonies*, groups of individual organisms that live together in close association. Some colonies are simple aggregates of independent cells. *Gonium*, for example, is a disc-shaped group of cells whose individual flagella beat and move the colony as a group. Other colonies are more complex. In these, for example the Portuguese man-of-war, individual cells specialize to perform certain functions (Figure 18.3). In multicellular organisms,

FIGURE 18.3   The Portuguese Man-of-War
The individual organisms that make up the Portuguese man-of-war *(Physalia pelagica)* are bound together structurally at the base of a gas-filled sac. Each of these organisms is specialized to carry out a specific function — feeding, reproducing, stinging — that is necessary for this floating colony to survive.

Dr. Stephen Jay Gould
Paleobiologist, Evolutionary Theorist, Taxonomist

The primary motivating force in Stephen Jay Gould's life is intense intellectual curiosity, and his curiosity has led to the publication of more books and articles than most people could write in a lifetime. His productivity is prodigious. He is a field scientist, taxonomist, historian of science, lecturer, and probably the strongest New York Yankee supporter in the city of Boston.

Gould's research in systematics and taxonomy deals with the evolutionary history of and relationships among land snails in the Bahamas. From this work he is learning more about the rate of evolution, systematic relationships among organisms, and other mechanisms of evolution that have led to the diversity of life forms on earth. Perhaps his greatest scientific contribution is the work he did with Niles Eldredge. They developed the theory of punctuated equilibrium, in which they suggested that evolution is not a gradual process but rather one that consists of waves of rapid evolutionary change followed by periods of very little change. The idea of punctuated equilibrium is controversial; in fact, many taxonomists and paleobiologists still do not accept it.

But it is not enough to talk about Gould's research and scientific awards (awards such as the Schuchert Award, presented to outstanding paleobiologists under the age of forty, and the MacArthur Foundation Award). In addition to all that, Gould is an extraordinarily gifted writer, and he has received awards

for that ability as well. In 1980, his columns in *Natural History* magazine won the National Magazine Award for essays and criticism. In 1981, his collection of essays entitled *The Panda's Thumb* received the American Book Award for Science. We encourage you to investigate the world of Stephen Gould. The following excerpts will give you an idea of the work that he does:

From *Emerson's Darwin:*
"Science is not a heartless pursuit of objective information. It is a creative human activity. Its genius is acting more as artists than as information processors."

From *The Panda's Thumb:*
"I am, somehow, less interested in the weight and convolution of Einstein's brain than in the near certainty that people of equal talent have lived and died in cotton fields and sweat shops."

Stephen Gould was born in a lower-middle-class neighborhood in New York City. His father, who was a self-taught intellectual, and his mother encouraged their son's interest in science. He says that "nobody in my whole vast family or circle of acquaintances knew a college professor." At the age of five, his father took him to the American Museum of Natural History in Manhattan; when he saw the dinosaur fossils his passion for paleontology was ignited. Most young children are interested in dinosaurs. With Stephen Gould the difference

was that he did not lose his interest or turn to other pursuits as he grew older. Gould attended Antioch College, then went on to do his graduate work at Columbia University. Today he is Professor of Geology and Biology at Harvard University.

If Gould's theories on evolution are unconventional, so are some of his teaching techniques. His passion for baseball invades his lectures on evolutionary processes, so during the World Series his seminars might include an analysis of the games and the reasons (evolutionarily considered, of course) for the falling batting averages since the game originated. Steven Gould is anything but a conventional scientist.

Sources: The Chronicle of Higher Education, June 6, 1984; *Discover,* January 1982, pp. 56–63.

there is a much greater degree of cellular specialization. Some cells function in reproduction, for example, some in nerve conduction, and so on. The cells are dependent on one another for survival.

Body Form. When we look at the **morphology** of an organism (the form of its body or structure), we find that organisms have a definite shape and structure, which we can categorize according to their *symmetry*. That is, if we draw an imaginary line through the structure, there are some obvious, clearly definable characteristics. Some organisms are **asymmetrical.** That means that there is no relationship between one side of the body of the organism and the other. The amoeba has an asymmetrical structure. Some organisms are shaped like spheres and are said to possess **spherical symmetry** (in other words they have a shape like a basketball). *Volvox* is a type of green alga that exhibits spherical symmetry. Organisms such as a starfish possess **radial symmetry**; its body is constructed like spokes radiating out from the hub of a bicycle wheel. Organisms like ourselves have **bilateral symmetry,** in which the right half is almost a mirror image of the left half (Figure 18.4a).

Homology and Analogy. The term *homo* means the same kind, or alike. **Homologous** structures are

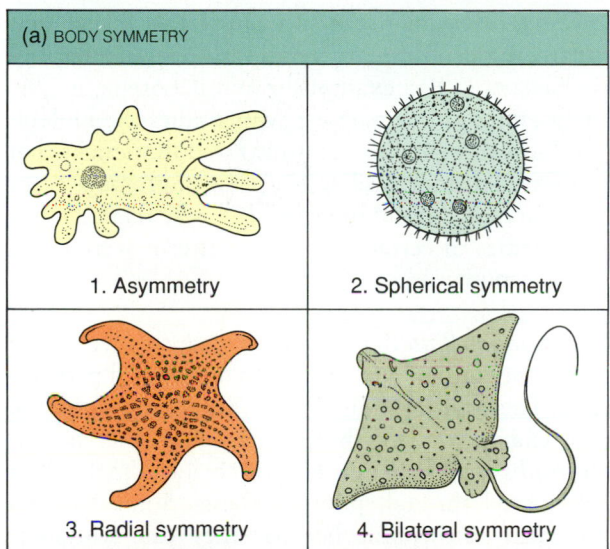

(a) BODY SYMMETRY

1. Asymmetry
2. Spherical symmetry
3. Radial symmetry
4. Bilateral symmetry

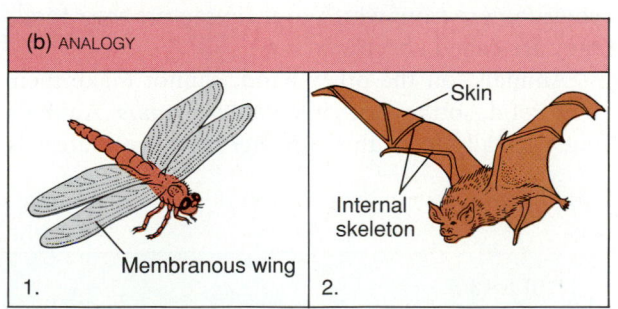

(b) ANALOGY

1. Membranous wing
2. Skin / Internal skeleton

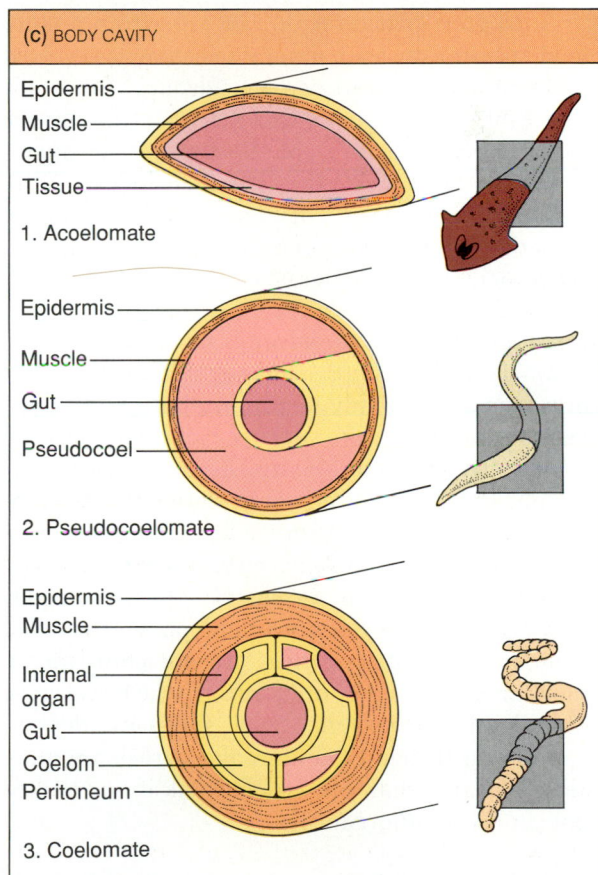

(c) BODY CAVITY

Epidermis
Muscle
Gut
Tissue
1. Acoelomate

Epidermis
Muscle
Gut
Pseudocoel
2. Pseudocoelomate

Epidermis
Muscle
Internal organ
Gut
Coelom
Peritoneum
3. Coelomate

FIGURE 18.4 Taxonomic Critera
(a) **Body Symmetry:** Body symmetry is one criterion biologists use to classify organisms. (1) The amoeba is asymmetrical. (2) *Volvox globator*, a freshwater protozoan, is spherical in shape. (3) The starfish is radially symmetrical, with its arms extending out from a central point. (4) The eagle ray and other fish are bilaterally symmetrical, each side a mirror image of the other.
(b) **Analogy:** The wings of (1) the dragonfly and (2) the bat are analogous structures; they serve the same function but their evolutionary history is different. (c) **Body Cavity:** (1) Acoelomates do not have a body cavity. (2) In pseudocoelomates, the body cavity is not lined with peritoneum. (3) The body cavity in coelomates is lined with peritoneum.

Naming and Classification of Organisms

structures that are similar. In most cases, the closer the structural similarities, the closer the evolutionary relationship. For instance, the bone structures of a monkey's paw and the human hand are very similar, revealing a fairly recent common evolutionary history. Homologous structures do not have to be similar in function. For example, the flipper of a whale and the hand of a human are similar in anatomical structure, but their function is not the same. Flippers are used for swimming and the hand for grasping. This is an example of divergent evolution, that is, evolution in which two organisms with a common origin become increasingly different. The two structures are known as homologous structures because their developmental origin is similar even though they function differently.

Analogous structures do not arise from a common ancestor. Instead, natural selection has evolved different structures to perform identical functions. The wing of a dragonfly and the wing of a bat are very different structures, but both enable the organism to fly (Figure 18.4b). These analogous structures are examples of convergent evolution, evolution in which two unrelated organisms become increasingly similar.

Embryological Development. The structural evolutionary relationships demonstrated by the adult organism are certainly important, but the developmental stages of the embryo can also illustrate evolutionary relationships and become a useful taxonomic criterion. For example, during the development of certain organisms, a *notochord* forms, a structure that in complex vertebrates is replaced by the vertebral column, or spinal column. Animals that pass through the notochord phase are called *chordata.* Although starfish are not chordata, in their larval stages they are similar to the larval stage of primitive *Chordata.* Such similarity demonstrates that the chordates and the starfish are more closely related than one would think if one looked only at the adult organisms.

Body Cavity. In most animals, the type of body cavity, or **coelom**, can be a useful taxonomic criterion (Figure 18.4c). Some animals — like the flatworms — lack a body cavity and are referred to as acoelomates. Others have a body cavity, but it is not lined with the thin membrane called a *peritoneum.* (This is the membrane that becomes infected when a person's appendix ruptures and the patient develops *peritonitis.* Peritonitis is an infection of the peritoneum lining the abdominal cavity.) A coelom lacking a peritoneum is considered a "false coelom." Roundworms and other animals that have such an

internal structure are called pseudocoelomates. A true coelom (one that is lined with a peritoneum) is found in coelomates. Earthworms, fish, and humans are examples of coelomates.

Molecular Sequence. In Chapter 17, we discussed the use of molecular sequencing to determine the relationship between organisms. Biologists are now using DNA sequencing, the sequence of amino acids in proteins, and immune response reactions as they develop taxonomic criteria. The degree of similarity of DNA base sequences or amino acid sequences can help to establish the degree of evolutionary relationship among and between organisms (Table 18.3).

Behavior. Living organisms can further reveal their degree of evolutionary relationship by similarities in behavior. For example, Konrad Lorenz, a German *ethologist* (a scientist who studies the biological basis of behavior) demonstrated that ducks that appeared distantly related because of plumage color were actually found to be more closely related when similarities of certain behavior patterns were taken into account.

Nutritional Patterns. In earlier chapters we talked about nutritional patterns from a chemical perspective. These patterns also serve as taxonomic criteria. For example, generally speaking, all plants are *autotrophs* (self-feeders), because they can make their own food through photosynthesis. Some bacteria are also autotrophic because they can synthesize their own food. Since they use the chemicals in the environment from which to manufacture their food, they are called *chemosynthetic.*

Animals, on the other hand, cannot make their own food, so they are called *heterotrophs.* Animals must usually feed, through *ingestion,* on materials

TABLE 18.3 Degree of Similarity of the DNA Sequence of Some Organisms	
Organisms compared	*Percentage of difference in DNA sequences*
Human–chimpanzee	1.6
Human–rhesus monkey	5.5
Mouse–rat	20.0
Cow–pig	20.0

produced by other organisms. In other words, they eat other organisms. The fungi (mushrooms, bread molds, and others) are heterotrophs that *absorb* their food. What is the difference between these two methods of obtaining food? Usually in **ingestion** the molecules taken in are complex and have to be broken down by the digestive system. In **absorption** the molecules are not as complex and usually can be absorbed directly through the membrane of the fungus.

We now have a set of taxonomic criteria. How can we use these criteria to identify the proper classification of an organism? We will look at this process in the following sections.

Taxonomic Keys and Phylogenic Lines

If you held an unknown organism in your hands, how would you identify it? Simple. You would use a taxonomic key. *Taxonomic keys* are rather like road maps that can guide you through the maze of taxonomic information gathered by biologists. They can assist you in finding the taxonomic position of your unknown organism. Some of you have no doubt used simple taxonomic keys found in various popular guidebooks, perhaps when you were trying to identify the species of birds, trees, or flowers on a nature trip.

When a taxonomist describes a new organism or a group of organisms, or obtains new information about a known organism, the information is published in one of the various scientific journals and is incorporated in the revision of the key for the organism or organisms. This process ensures that biologists will have accurate keys at their disposal.

Taxonomic keys are based on evolutionary relationships among organisms, so their construction reflects these relationships. Table 18.4 provides an example of a simple key. The arrangement is called *dichotomous*. The groups are divided into two groups: those that have a given characteristic and those that do not. Depending on the answer, the user is directed to go on to a next step. By using the various taxonomic techniques available to obtain information about organisms and taxonomic keys, you can begin to understand the evolutionary history of organisms.

Determining evolutionary relationships enables biologists to develop phylogenetic trees, which graphically represent the degree of relationship among organisms. A phylogenetic tree also shows the ancestry of a particular group of organisms. It is rather like the family tree that you can develop to learn about your relatives and ancestors. Both illustrate the relationships and ancestry of a given group

TABLE 18.4
Sample Taxonomic Key for Classes of Chordates

1. Fins or no appendages; marine or aquatic fishes: See no. 2.

1a. Legs, wings, or flippers as appendages, but not fins; mostly terrestrial: See no. 3.

2. Slimy, scaleless skin; no paired appendages; reduced fins; no jaws; snakelike: Cyclostomata.

2a. Leathery skin with small scales; prominent fins; jaws; no bones: Chondrichthyes.

2b. Scaly skin with prominent scales; jaws; bone present: Osteichthyes.

3. No scales, feathers, or hair: Amphibia.

3a. Scales, feathers, or hair: See no. 4.

4. Scales: Reptilia.

4a. Feathers: Aves (birds).*

4b. Hair: Mammalia.

*May be grouped as reptiles by some biologists.

of organisms. But as you might expect, developing phylogenetic trees that accurately represent the evolution of organisms is not an easy task.

Phenetics. Traditionally, systematists have used the school of thought that organisms should be classified on the basis of phenotype or appearance. This approach is called **phenetics** (which comes from the Greek *phainein*, meaning "to appear"). Pheneticists use many taxonomic criteria to determine the degree of evolutionary relationship. Those groups that share the most characteristics are assumed to be the most closely related. From this type of information, pheneticists create evolutionary trees called *phenograms*. For example, using taxonomic criteria, the pheneticists would group the great apes together in the family Pongidae and humans in a separate family, the Hominidae (Figure 18.5a). But as we all know, animals or plants that look alike often do not share similar evolutionary relationships. We learned this when we read about adaptive radiation and convergent evolution in Chapter 17. For instance, many types of cacti from Africa and the southwestern deserts in the United States look alike but have different evolutionary histories. So, as you might expect, there are other schools of taxonomic thought.

Cladistics. **Cladistics** classifies organisms on the basis of shared homologous characteristics and length of time since separation from a common ancestor. In other words, characteristics are separated into two groups: those recently evolved and those that evolved long ago. From this information a phylogenetic tree is constructed. This method relies heavily on the fossil record to determine whether a characteristic is recently evolved or is an old characteristic. In other words, this scheme emphasizes time of descent from a common ancestor. Structural characteristics are less important. For example, using the cladistic method of taxonomy, humans, chimps, and gorillas would be grouped together because they share a recently evolved unique characteristic. Orangutans, on the other hand, evolved and branched off the main branch at an earlier point in time (Figure 18.5c).

So, as you can see, it is not easy to establish evolutionary relationships, and there is still controversy as to where organisms should be grouped. Nevertheless, taxonomic keys and phylogenetic trees are an attempt to understand the evolutionary background of organisms.

WHITTAKER'S FIVE-KINGDOM SYSTEM

When you think about how to group organisms into the most inclusive category — the kingdom — you can see quite quickly that there are at least two kingdoms: the plants and the animals. However, a little more thought will reveal that two kingdoms are not enough because some organisms simply will not fit neatly into either category. Is a mushroom a plant? It does not produce food through the process of photosynthesis. On the other hand, no one would ever say that it is an animal. What about microorganisms? Are they plants, animals, or fungi? It can get confusing.

In 1969, R. H. Whittaker at Cornell University, who was involved in trying to resolve the question of the number of kingdoms, recommended that biologists adopt a system containing five kingdoms (Figure 18.6), based on two major criteria: (1) cel-

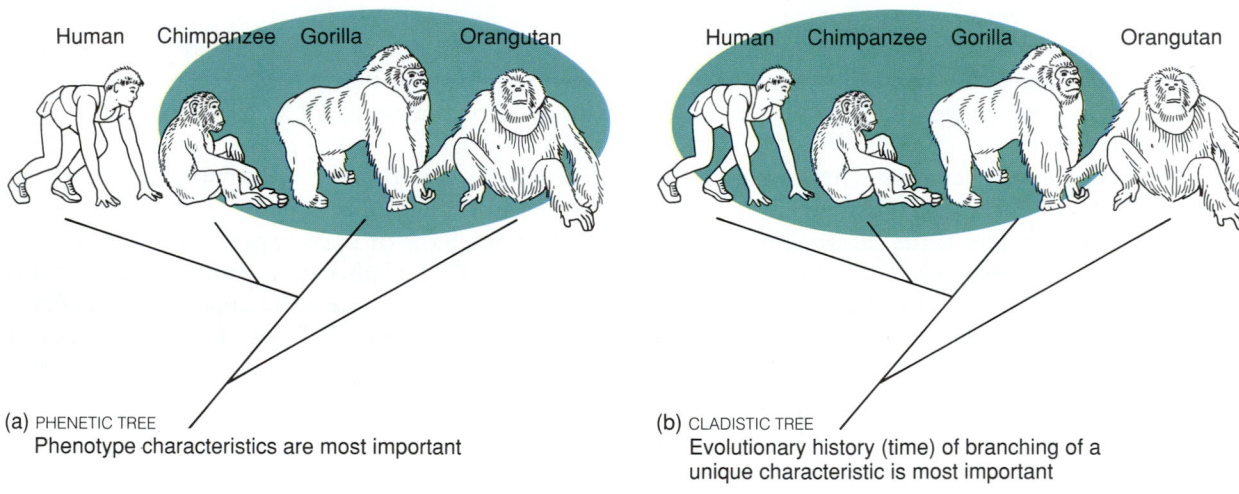

(a) PHENETIC TREE
Phenotype characteristics are most important

(b) CLADISTIC TREE
Evolutionary history (time) of branching of a unique characteristic is most important

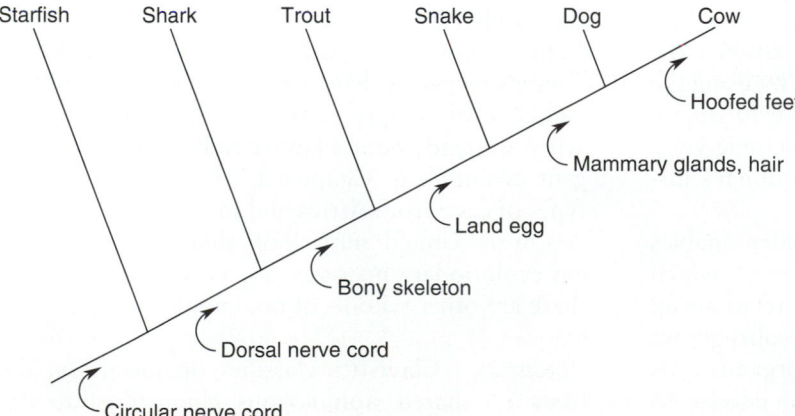

(c) Each branch represents a unique characteristic that makes it different than the one before.

FIGURE 18.5 **Phylogenetic Trees**
(a) According to the phenetic school of thought, chimpanzees, gorillas, and orangutans are grouped into the family Pongidae and humans into a separate family because the Pongidae still look like apes and humans have undergone changes that make them look less like apes. (b) Cladistics classify organisms on the basis of evolutionary time. Therefore, humans, chimpanzees, and gorillas are grouped together because they are more recent than orangutans. (c) Cladestic branch, showing the branching of unique characteristics that set one animal group apart from the others.

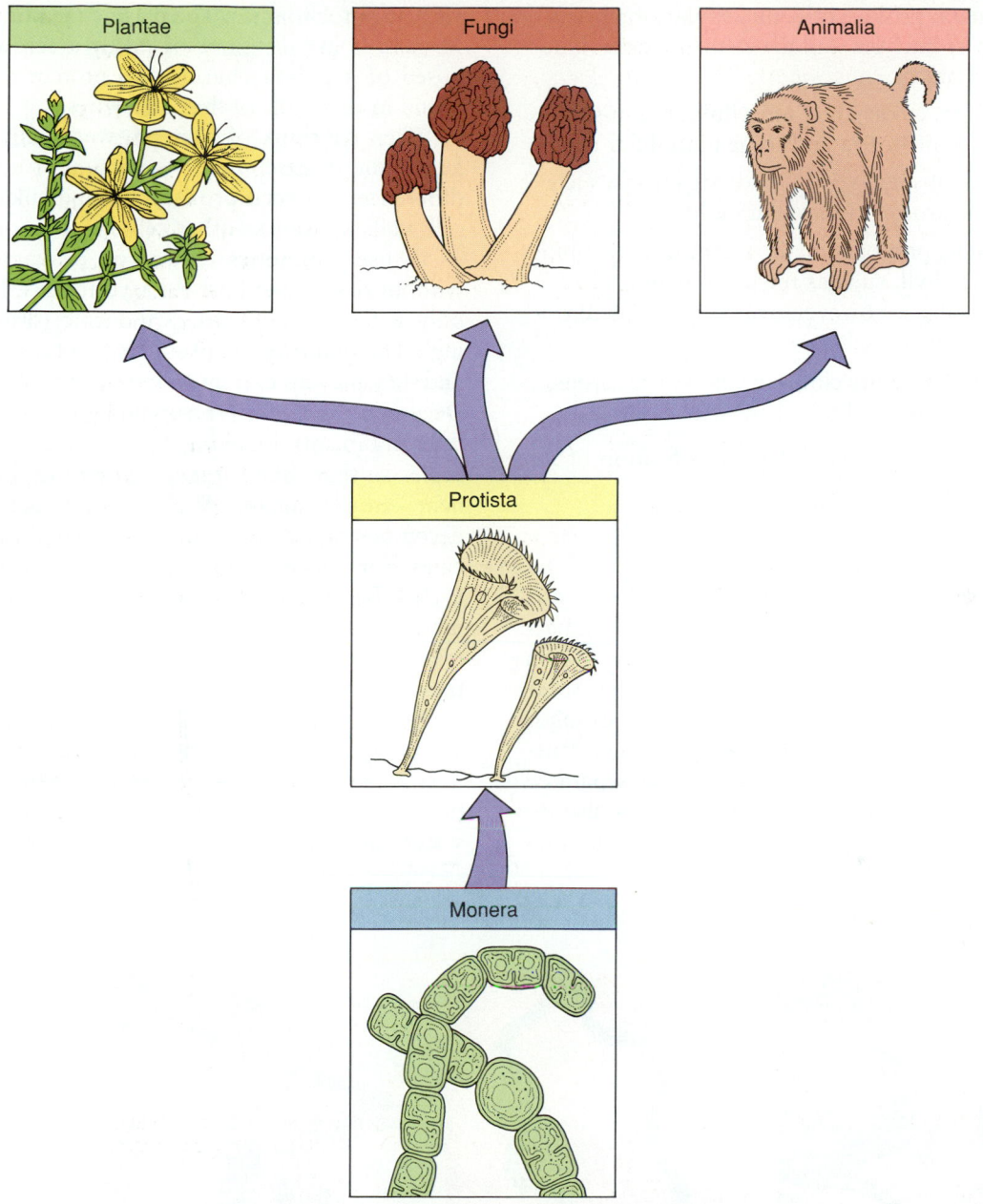

FIGURE 18.6 Phylogenetic Tree: The Five Kingdoms
This simplified phylogenetic tree shows the relationship
of the five kingdoms: Monera, Protista, Fungi, Plantae,
and Animalia.

lularity — whether or not the cell is prokaryotic or
eukaryotic, or is unicellular or multicellular; and (2)
the type of nutrition — photosynthetic, ingestion,
or absorption. Of all the systems that have been
developed to date, Whittaker's is the one that has
been most universally accepted (see Focus 18.1).
These are the five kingdoms he identifies (though
some taxonomists prefer to separate the Monera
into the kingdoms Archaebacteria and Eubacteria
and to call the Protista the Eukaryota):

1. **Monera,** prokaryotic, unicellular organisms such as bacteria and the cyanobacteria (blue-green algae).

2. **Protista,** eukaryotic, unicellular organisms, such as protozoa and some unicellular algae.

3. **Fungi,** absorptive plantlike organisms, such as mushrooms and bread molds.

4. **Plantae,** photosynthetic organisms containing chlorophyll, such as the green plants. (The green algae, however, are placed in the Protista in many schemes.)

5. **Animalia,** multicellular ingestive organisms, such as insects, birds, fish, and mammals.

In the appendix is a complete classification of the five kingdoms.

Monera

The kingdom Monera consists of the prokaryotic cells: bacteria and cyanobacteria (Figure 18.7). Bacteria represent one of the oldest forms of life on earth. The Monera are usually single-celled. Some grow in colonies, but there is no division of labor within each colony. Since they have no membrane-bound nucleus, the nuclear material is not separated from the rest of the cell. The prokaryotes also do not have complex membrane-bound organelles

such as mitochondria. The plasma membrane of the Monera is usually surrounded by a cell wall composed of a polysaccharide and protein that is not found in cell walls of the eukaryotes.

When we think of bacteria, we usually think of germs and disease. That is sometimes true, and the differences between prokaryotes and eukaryotes in cell walls as well as other cellular reactions enable us to use antibiotics to destroy prokaryotic cells without destroying host eukaryotic cells. However, only a few bacteria are pathogenic (disease-causing). The majority are producers and decomposers, and they play an extremely important role on earth. **Decomposers** are heterotrophic bacteria that live on dead organisms, breaking down their complex molecules so that other organisms can use the component atoms again. Without the activities of decomposing bacteria (and many fungi), the molecules from once-living organisms could not be recycled. Because of their activities, there is a sort of molecular reincarnation. The decomposers are a microscopic world essential to the biosphere — they are nature's invisible recyclers.

The bacteria are most metabolically diverse and most abundant and occupy the widest range of habitats of any group of organisms on earth. In plant photosynthesis, the energy of light is used to split water molecules. Some autotrophic bacteria also

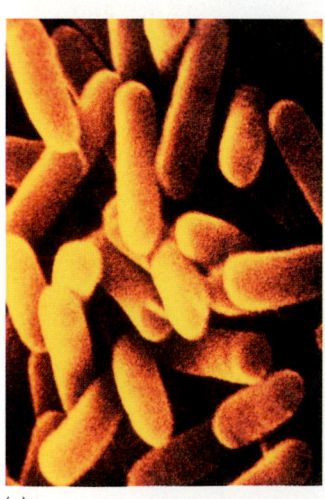
(a)
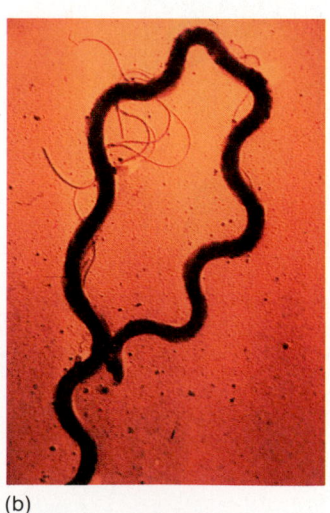
(b)
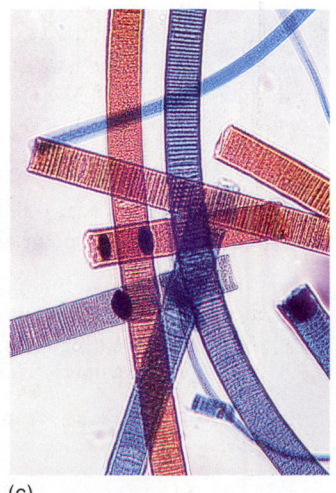
(c)

FIGURE 18.7 Kingdom Monera
Bacteria and cyanobacteria make up the kingdom Monera, the oldest and largest of the five kingdoms. (a) Certain bacteria, among them the Pseudomonas shown here, function as decomposers, recycling organic matter. (b) *Treponema pallidum* is a spirochete, a spiral-shaped bacterium with an undulating movement. It causes syphilis. (c) *Oscillatoria* is a filamentous cyanobacterium that lives in fresh water.

Chapter 18   Diversity and Taxonomy

conduct a form of photosynthesis; however, in this case light energy is used to split hydrogen sulfide (H_2S) rather than water. The pigmentation found in such bacteria differs from that found in plants. Chemosynthetic bacteria can take chemical compounds in the environment and produce the complex molecules required for energy and structural components of the cell through a series of reactions.

The **cyanobacteria** (blue-green algae) are prokaryotic cells that perform a type of photosynthesis similar to that found in higher plants (Chapters 7 and 8). Since these organisms do not have membrane-bound organelles, they have no chloroplasts. Instead, the chlorophylls and other photosynthetic pigments are located within the membrane of the organisms.

Like all other kingdoms, the Monera are divided into several phyla. These divisions are based on cell wall composition, the method of mobility, and the mode of nutrition.

Protista

From an evolutionary perspective, Protista can be considered a transitional kingdom. This position is based on the hypothesis that protista were the first eukaryotic cells and as a result were the direct evolutionary ancestors of all advanced life forms: fungi, plants, and animals. This endosymbiotic hypothesis was discussed in Chapter 4.

Protista are eukaryotic, unicellular or colonial organisms with a membrane-bound nucleus and complex membrane-bound organelles such as mitochondria and plastids. Mitosis in these cells goes through the stages found in other eukaryotic cells. If they possess flagella or cilia, they have the characteristic 9 + 2 filament structure found in other eukaryotic cells, rather than the simpler structure of prokaryotic flagella (see Chapter 4). Protista are a varied group of several phyla from protozoa to slime molds (Figure 18.8 and Table 18.5).

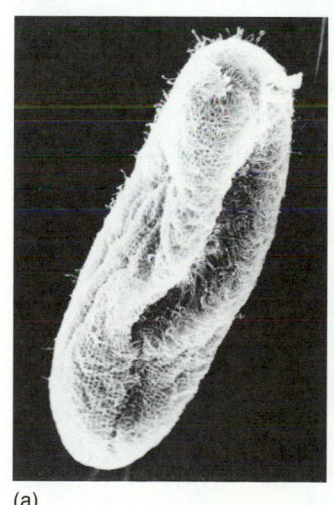
(a)

(b)

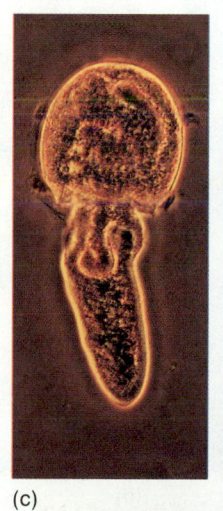
(c)

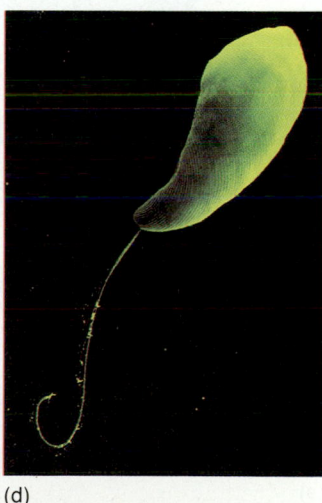
(d)

FIGURE 18.8 Kingdom Protista
Protozoa is a phylum in Protista. Flagellated protozoans propel themselves through the water by means of a flagellum, a whiplike protrusion; other protozoans are amoebalike, using pseudopodia (false feet) to move. (a) *Paramecium caudatum* uses cilia to move. (b) The pseudopodia of *Globigerina bulloides* are long sticky threads that extend from a colorful shell. (c) *Didinium nasutum* is a ciliated protozoan that feeds on the paramecium. (d) *Euglena gracilis* looks like a flagellated protozoan, but many taxonomists classify it as a form of alga because of its chloroplast. Slime molds create the same problem for taxonomists, sharing certain features with protistans and others with fungi. (e) *Stemonitis splendens* is an amoebalike organism that feeds on organic matter. To reproduce, individual organisms mass together, then form the spore-bearing stalks you see here.

(e)

TABLE 18.5
Major Phyla of the Kingdom Protista

Phylum Protozoa	Unicellular heterotrophs, including flagellates, amoebalike organisms, ciliates, and sporozoans
Phylum Euglenophyta	*Euglena* and related algae
Phylum Chrysophyta	Diatoms and related algae
Phylum Pyrrophyta	Dinoflagellates and related algae
Phylum Chlorophyta	Green algae
Phylum Phaeophyta	Brown algae, including the kelps, all multicellular
Phylum Rodophyta	Red algae, all multicellular
Phylum Gymnomycota	Slime molds

Fungi

The kingdom Fungi is a large, diverse kingdom. We are all familiar with many of its members, from athlete's foot to the mold that appears on fruit that we sometimes find in the back of the refrigerator. Other examples in this diverse group are mushrooms, yeast, bread mold, mildew, and bracket fungi (Figure 18.9).

Fungi are usually multicellular, composed of masses of filaments known as *hyphae* (singular *hypha*). A mass of hyphae form the *mycelium,* the fuzzy material you see when you look at bread mold. The common feature of all fungi is that they obtain their organic nutrients through absorption. This means that, like bacteria, fungi are decomposers and are major recyclers of the earth.

Fungi are also economically important for humans: Fungi such as mushrooms are an important

(a)

(b)

(c)

(d)

FIGURE 18.9 Kingdom Fungi
(a) *Morchella esculenta* (morels) are an unusual but delicious fungus.
(b) *Lactiporus sulphareus,* a bracket fungus, grows on decaying wood.
(c) *Penicillium roqueforti* gives Roquefort cheese its sharp flavor. Penicillin is derived from another form of *Penicillium*.
(d) *Phytophthera infestans* is a fungal mildew that attacks potatoes. It was the cause of the potato famine in Ireland in the mid-1800s.

Chapter 18   Diversity and Taxonomy

ingredient in our diets, we use fungi to flavor cheeses, and fungi have a role in medicine. On the one hand, they provide us with a source of many antibiotics; on the other, they are the cause of several animal and plant diseases. Fungi have affected the history of many countries in the world. If you are of Irish heritage, it is likely that your ancestors were members of the more than 1 million immigrants who left Ireland for America during the great potato blight famines between 1845 and 1847. Potato blight is a plant disease caused by a fungal mildew that destroys potatoes, the major food crop in Ireland. The disease was so severe that in one week in the summer of 1845, the entire potato crop of Ireland was wiped out. The famine lasted 2 years and was responsible for more than 1 million deaths due to starvation and related illnesses.

Plantae

The kingdom Plantae consists of eukaryotic, multicellular, autotrophic organisms (Figure 18.10). The

FIGURE 18.10 Kingdom Plantae
The algae in kingdom Plantae are green, red, and brown. (a) *Ulva lobata* (sea lettuce) is a multicellular green alga. In the upper left is *Chondrus crispus*, a red alga found in seawater. (b) *Fucus vesiculosus* is a brown alga that grows on shells and stones. Algae are nonvascular plants. Bryophytes do not have vascular tissue, although they do have rhizoids. (c) Moss, like the one shown here, is the most common form of bryophyte. Ferns, flowering plants, and conifers are all vascular plants. (d) The ancestors of *Nephrolepis cordifolia* (sword fern) first appeared about 400 million years ago. (e) Lily, a flowering plant. (f) *Picea Engelmannii* (Engelmann spruce), a conifer. The female produces its seeds on cones, which are fertilized by the male's pollen.

FOCUS 18.1
Viruses: A Unique Case

Although this may seem difficult to believe, even a system as broad as Whittaker's five kingdoms cannot include all forms of life. For instance, what do we do with the viruses? Indeed, these organisms are a special taxonomic case since they are *acellular* organisms — that is, they are organisms without cells. They also lack the metabolic machinery to conduct life processes — they do not have organelles, for instance. As a result, many scientists are not certain that viruses are even alive. Perhaps they are only simple fragments of cells.

Why are viruses so controversial? The reason concerns their structure. Essentially a virus consists of a nucleic acid core (either DNA or RNA) and a surrounding protein coat (Figure A). Some viruses (especially those that infect animal cells) are more complex, having a membrane-like lipid layer that surrounds the protein coat. In any case, they are the smallest forms of life. Some are even smaller than large protein molecules. Because viruses lack the metabolic machinery necessary for life processes, they are *intracellular obligate parasites*. That means that they are obliged to live within a living cell; otherwise they cannot reproduce.

A particular virus will usually infect only a particular type of cell. The influenza virus infects cells lining the respiratory tracts; the poliomyelitis virus infects nerve cells; the tobacco mosaic virus infects the leaves of tobacco plants. Typically, a virus will attach itself to a host cell and inject its genetic material into the host, leaving its protein coat outside. In some instances, however, the protein coat also enters the host cell. In these viruses the protein coat is broken down by enzymes found in the host cell.

Once inside the cell, the genes coded in the viral genetic material are replicated and transcribed, thus producing more viruses. In other words, to make new viruses, viruses take over and direct the metabolic machinery of the host cell. The new virus or viral particles are released from the host cell as it lyses, or bursts, and new cells are infected. In cells that do not burst as they release the virus particles, the particles are released through exocytosis, a process by which the cell membrane folds inward and around the virus particle. Once the particle is enclosed, the membrane fold bulges outward and releases the particle from the cell.

As you know, the viruses are the cause of many human diseases, from the AIDS virus to herpes zoster (in other words, from A to Z). Viruses have also been implicated as the cause of some human cancers, such as a rare form of leukemia that occurs mainly in the southern islands of Japan. There is also a link between a type of human papilloma

---

autotrophic nutritional pattern is one of photosynthesis. In other words, by using the green chlorophylls, CO_2, and H_2O, plants can trap light energy and use this light energy to form high-energy carbon compounds that function as food.

The evolution of the plants of the earth is a story of slow diversification from the salt oceans to fresh water streams and lakes, and finally to the dry land. As green algae of the ocean became multicellular, they began to exhibit a division of labor that allowed for more efficient nutrient procurement and reproduction. Eventually some of these primitive

virus and cancer of the cervix. As stated by Massachusetts Institute of Technology biologist Nancy Hopkins, "Cancer arises from a number of insults to the DNA. Viruses are one insult. They start the process rolling." Unfortunately, since viruses are so simple and lack metabolic machinery and cell walls, the diseases that they cause are not affected by antibiotics. The body fights viral infection by producing a substance known as interferon. Interferon is a protein molecule created by the infected cell that stimulates noninfected cells to produce an antiviral protein (sort of like a messenger).

It is usually through this system that viruses can be combated.

We do not know the origin of viruses. One school of thought holds that they may have been simple parasitic cells that gradually became so dependent on their hosts that they lost the ability to conduct their own metabolism and hence became totally dependent on life within the hosts. Another theory is that they may have just been gene fragments from other cells that took up an independent lifestyle requiring other living cells in order to reproduce themselves.

We continue to learn about viruses. In 1981, the laboratory of Dr.

Robert Gallo at the National Institutes of Health discovered the human T-lymphotrophic virus HTLV-1. Since that discovery an entire family of viruses has been identified. These HTLV viruses are associated with human leukemias and AIDS. They attack and parasitize the white blood cells called T-lymphocytes that are critical to the body's defense against disease. When infected by the HTLV virus, T-cells can become cancerous (causing various forms of leukemia) or they can fail to defend against disease (leading to AIDS). As of this writing, five HTLV viruses have been discovered.

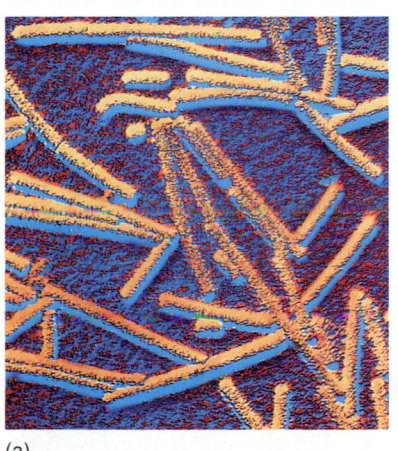
(a)

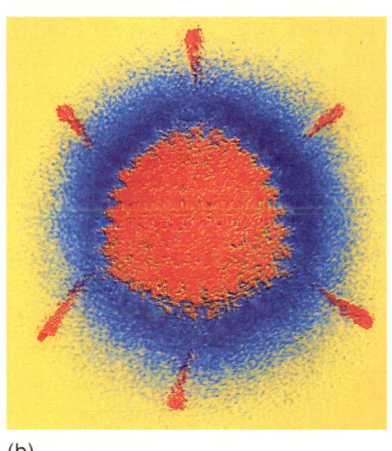
(b)

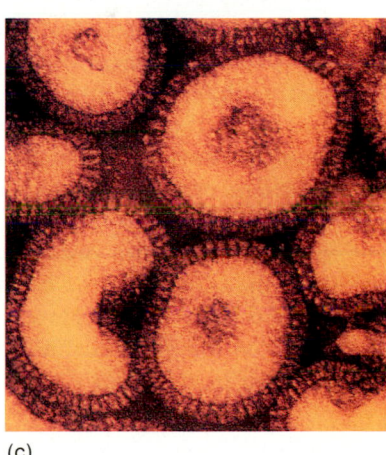
(c)

FIGURE A
The protein coats that enclose the viral nucleic acid core vary considerably in shape. (a) The tobacco mosaic virus has a rod-shaped coat; (b) the adenovirus (one virus responsible for the common cold) is a polyhedron with protein spikes; (c) the influenza virus is contained in a membranous envelope studded with protein spikes.

multicellular plants lived along the coastline, where they were exposed to a terrestrial environment. Because of the differences in the aquatic and terrestrial environments, there are two major evolutionary trends in the kingdom Plantae: **nonvascular plants,** which do not have true roots, stems, leaves, or transporting vascular tissues, and **vascular plants,** those that do possess those structures. Nonvascular plants usually live in or near the water, while vascular plants usually live on the land.

The complex structures that arose in vascular plants serve a number of purposes. Because of the special requirements of land-based existence, natural selection favored the development of these structures:

1. Leaves or leaflike structures as specialized photosynthetic organs

2. Stems or stemlike structures to support plants from the pull of gravity and to expose the leaves to the sun's light

3. Rootlike structures for support and the absorption of water and minerals

4. Vascular tissues to connect the roots, stems, and leaves to ensure the efficient transportation of the required minerals, gases, and water between leaves and roots

The environment selected for plants having the maximum opportunity for survival through increased photosynthetic capability and reproductive success.

Land-based existence is harsher than that of a watery world. Organisms that live on land are subjected to extreme changes in temperature. Without the support that water provides, they are fully exposed to the pull of gravity. Water supply is variable, and organisms must face periods in which their cells have less water than is required for life processes. Finally, without the flow and currents present in a watery world, organisms cannot rely on water for the transportation of gametes. Therefore, those plants that could reproduce without water were selected for and became the ancestors of the predominant land plants.

Animalia

One of the major characteristics shared by the members of the kingdom Animalia is their mode of nutrition. Animals are eukaryotic, multicellular organisms that obtain their nutrients through ingestion. Tissues or other organisms are taken into the body of the animal, where they are digested into simpler molecules that can be used as the raw materials for growth and repair and as sources of energy. In order to maintain the ingestive lifestyle, it is advantageous to be able to move from place to place both to hunt to obtain food and also to escape from becoming food! As a result, a wide variety of mobile lifestyles and methods of locomotion evolved, all tied to the requirements of food procurement and escape from predation.

As always, there are some exceptions to this general observation. Parasites, such as tapeworms, live within their food supply (called the *host*). Other animal groups are *sessile,* which means that they do not move from place to place. Examples are the sponges and sea anemones.

However, most animals move. That means that specialized structures that enabled them to move had a selective advantage and were selected for. These include not only muscular systems and skeletal systems but also nervous systems, sensory systems, and other complex organ systems necessary for procurement and ingestion of nutrients, all to allow the organisms to obtain food and escape predation. What is more, a mobile lifestyle also requires reproductive systems and methods that do not interfere with the means of acquiring food. This demand favored reproductive patterns that (1) provided for the development of the young and (2) did not interfere with the feeding habits of the adults.

The reproductive adaptations that animals have evolved to satisfy these requirements are the primary taxonomic criteria used in classifying animals. Some lay eggs that need no further attention; some lay eggs in trees and fly back and forth between the nest and their food source; some have developed methods of internal development and can therefore carry their developing young inside their bodies as they search out a food supply.

In each of the kingdoms there are specialized taxonomic criteria that can help categorize its members and determine the degree of relationship among them. In the kingdom Animalia, these criteria include such things as differences in the skeletal system, segmentation (repeating body units), types of body cavities, digestive systems, embryological development, and symmetry. Biologists use these distinctions to classify the various organisms into distinct phyla, classes, orders, families, genera, and species. The major evolutionary relationships among the major phyla of kingdom Animalia are illustrated in Figure 18.11. This illustration also demonstrates some of the major trends in animal evolution. Humans are members of the phylum Chordata, subphylum Vertebrata. Within our own phylum, as in other phyla, there are many classes. Figure 18.12 and Table 18.6 illustrate some of the major classes of the subphylum Chordata.

In this chapter you have been introduced to taxonomic and systematic criteria and to organismic diversity. In other parts of this book we investigate the details of the changes that have occurred in plants and animals. Indeed, we can devise systems based on scientific evidence to illustrate the degree of relationship among organisms. We can make order out of what first appears to be chaos among the living things on earth.

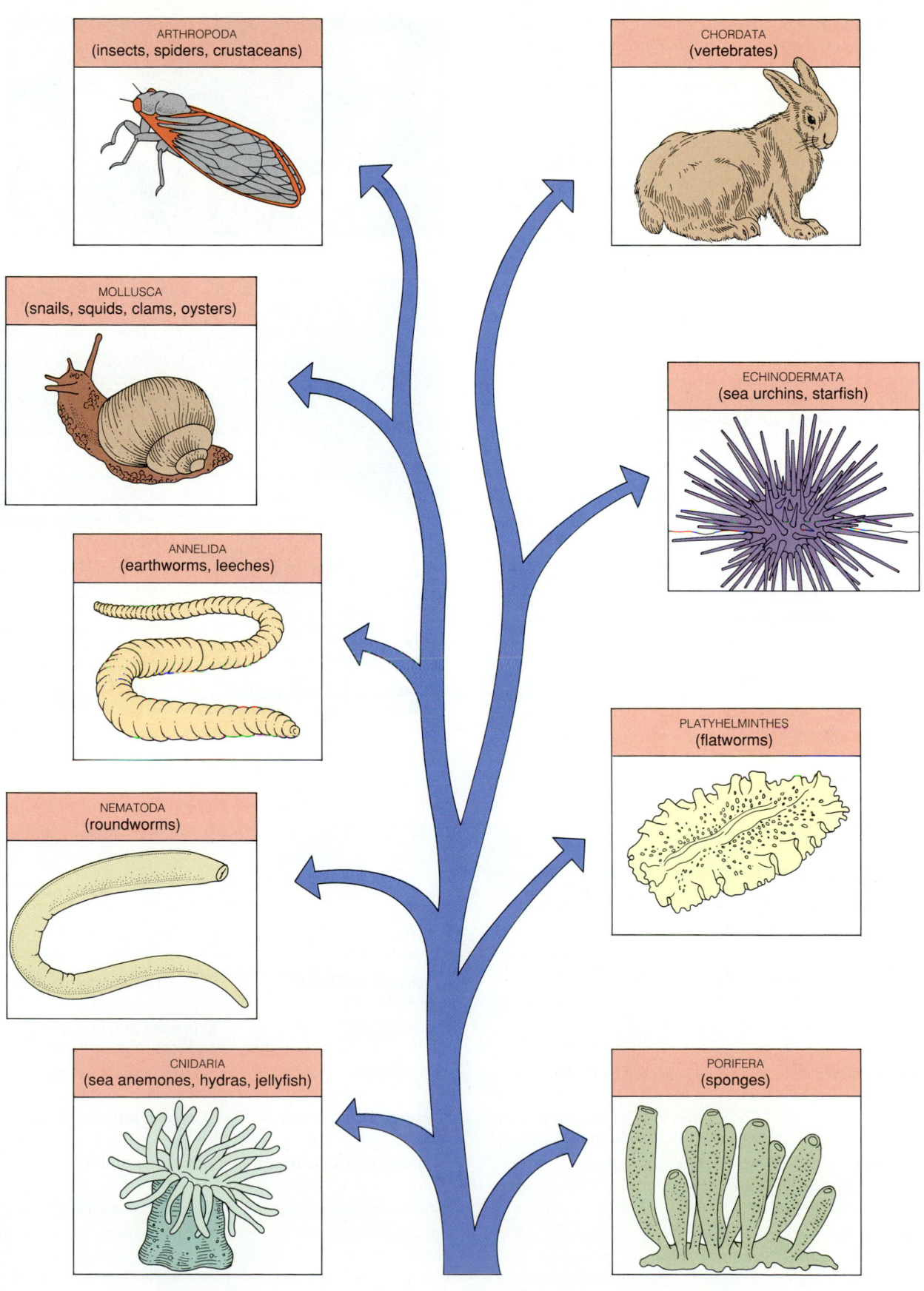

FIGURE 18.11   Phylogenetic Tree: Kingdom Animalia
Biologists theorize that the kingdom Animalia evolved from flagellated protistans. By following the tree upward, you can see the order of evolution and the relationship among the different phyla.

ARTHROPODA
(insects, spiders, crustaceans)

CHORDATA
(vertebrates)

MOLLUSCA
(snails, squids, clams, oysters)

ECHINODERMATA
(sea urchins, starfish)

ANNELIDA
(earthworms, leeches)

PLATYHELMINTHES
(flatworms)

NEMATODA
(roundworms)

CNIDARIA
(sea anemones, hydras, jellyfish)

PORIFERA
(sponges)

FIGURE 18.12 The Chordates
There are seven major classes of animals in the phylum Chordata. (a) The oldest is Cyclostomata — fish without jaws or scales. *Petromyzon marins* is an adult lamprey, a fish shaped much like an eel. *Petromyzon* feeds by clamping on to its prey (another fish), scraping away the scales with its tongue, then swallowing the blood and juices. (b) Chondrichthyes are fish with cartilaginous skeletons. They have primitive jaws and paired fins, and are covered with scales. Sharks — among them the *Ginglymostoma cirratum* — are chondrichthyes, as are rays and chimaeras (ratfishes). (c) Osteichthyes are fish with bony skeletons, like this tropical marine fish. Their brain is more developed than that of the Chondrichthyes. (d) Amphibians are a transitional group, using their two pairs of limbs to move back and forth between aquatic and terrestrial environments. The frog seen here is laying eggs. (e) The earliest class of terrestrial animals was Reptilia. Like amphibians, all reptiles (except snakes) have two pairs of limbs. *Crotaphytus Dickersonae* from Bahía Kino, Sonora, México. This is the first photograph ever published of this species. (f) Aves (birds) evolved from small, slender, long-tailed dinosaurs. Their feathers, now an important factor in their ability to fly, were originally a source of insulation, and they still carry out that function today. Here we see flamingos. (g) Mammals are the highest class of chordates. All have hair for insulation, heightened senses (particularly smell), and developed brains that allow them to adapt to all kinds of environments. The koala bear is a herbivorous animal.

(a)

(b)

TABLE 18.6
Major Vertebrate Classes and Their Characteristics

Class	Respiration	Circulation	External body covering
*Agnatha**	Gills	Two-chambered heart	Skin with mucous glands
Chondrichthyes	Gills	Two-chambered heart	Skin with mucous glands and scales
Osteichthyes	Gills covered by operculum	Two-chambered heart	Skin with mucous glands and scales
Amphibia	Lungs, gills in larvae and some adults; also through skin	Three-chambered heart	Skin with mucous glands
Reptilia	Lungs	Incomplete four-chambered heart	Skin scales and few glands
Aves	Lungs	Four-chambered heart	Skin with feathers
Mammalia	Lungs	Four-chambered heart	Skin with hair

Chapter 18 Diversity and Taxonomy

(c)

(d)

(e)

(f)

(g)

Feeding	Reproduction	Habitat
Mouth sucking or rasping	External fertilization	Marine and freshwater (lampreys); marine (hagfish)
Biting jaws with teeth	Internal fertilization	Marine; very few freshwater
Mouth with teeth	Usually external fertilization	Marine and freshwater
Mouth without prominent teeth, tongue used to catch prey	External or internal fertilization	Freshwater, terrestrial
Mouth with teeth; beak	Internal fertilization	Terrestrial primarily; some freshwater, marine
Beak	Internal fertilization	Terrestrial
Mouth with teeth modified according to diet	Internal fertilization	Terrestrial; some marine, some freshwater

Source: Modified from S. Wolfe, *Biology: The Foundations* (Belmont, CA: Wadsworth Publishing Company, 1982), p. 408
*Not considered a true class by all biologists. Some consider Agnatha a group.

Whittaker's Five-Kingdom System

SUMMARY

1. Taxonomy is the science of classification. Systematics is the study of the evolutionary relationship among organisms and serves as the basis for a natural classification scheme.
2. The scientific name of an organism consists of two names, the genus and species. This system of using two names is called binomial nomenclature.
3. Living things are organized according to a hierarchical system of classifications, ranging from the kingdom, which is the most inclusive, to the species, which is the most specific.
4. Many criteria are used to classify organisms and to determine the degree of evolutionary relationships among them, including: the presence or absence of a membrane-bound nucleus, the number of cells, body form, homology and analogy, embryological development, the type of body cavity, molecular sequence in DNA and amino acid sequences, behavior, and nutritional patterns.
5. Whittaker's five-kingdom system is widely accepted as the basis for the classification of organisms. The kingdoms are: (1) Monera, (2) Protista, (3) Fungi, (4) Plantae, and (5) Animalia.

KEY TERMS

taxonomy	homologous
systematics	analogous
binomial system of	coelom
nomenclature	ingestion
kingdom	absorption
phyla	phenetics
division	cladistics
class	Monera
order	Protista
family	Fungi
genera	Plantae
species	Animalia
morphology	decomposer
asymmetry	cyanobacteria
spherical symmetry	nonvascular plant
radial symmetry	vascular plant
bilateral symmetry	

REVIEW QUESTIONS

True/False

1. A natural classification scheme is based on the evolutionary relationship among organisms.
   true   *false*   *(page 409)*
2. The color and size of an organism are useful taxonomic criteria.
   true   *false*   *(page 409)*

3. Homologous structures are structures sharing an evolutionary origin.
   true   *false*   *(page 416)*
4. The kingdom Animalia includes single-celled and multicellular organisms.
   true   *false*   *(page 426)*

Fill in the Blank

1. In biology we use a system of naming organisms in which two names are used. This system is referred to as _____ nomenclature. *(page 411)*
2. The scientific name of an organism consists of the _____ name and _____ name. *(page 411)*
3. Humans possess a type of symmetry referred to as _____ symmetry. *(page 415)*
4. The kingdom _____ consists of prokaryotic unicellular organisms. *(page 420)*

Multiple Choice

1. Classification places emphasis on:
   a. Similarities
   b. Differences
   c. Similarities and differences
   d. Color
   (page 413)
2. In making biological classifications, which of the following is true:
   a. Anyone can tell a plant from an animal.
   b. Any classification system must be generally accepted to be useful.
   c. Classifications never change.
   d. Classification is unnecessary.
   (page 411)
3. The kingdom Protista is primarily made up of organisms that are:
   a. Eukaryotic and multicellular
   b. Prokaryotic and multicellular
   c. Prokaryotic and single-celled
   d. Eukaryotic and single-celled
   (page 421)
4. Of the following, the most specific or least inclusive taxonomic category is the:
   a. Family
   b. Phylum
   c. Species
   d. Class
   (page 412)

Discussion

1. Why is classification important?

2. Compare an artificial classification scheme with a natural classification scheme.

3. Distinguish among nomenclature, taxonomy, and systematics.

4. List and discuss some of the taxonomic criteria used in the systematic classification of living organisms.

5. Why are nutritional patterns important in the classification of living things?

6. Define each of the principal taxonomic classification categories. Give an example of the complete classification of humans or the rose.

7. List and briefly describe some of the major characteristics of each of the five kingdoms as proposed by Whittaker.

PLANTS

After reading this unit, the student should be able to:

- describe the major groups of plants on earth and their evolutionary relationships
- understand the major trends in the evolution of plants onto the land
- understand alternation of generations in major plant groups
- discuss the major plant structures and functions at the cell, tissue, and organ levels
- understand the relationship of structure and function in transport and regulation in plants

19

Plants: A Survey

Imagine, if you will, a barren landscape without plants — a sort of moonscape. It is beautiful in its own right, yet it is not the beauty with which we are familiar or the beauty that has existed on much of the earth for a long time. The fossil records show that the earth turned green when the first land plants evolved about 425 million years ago. Since then plants have provided much of the loveliness that we equate with our planet. However, we are dependent on plants for far more than aesthetics. In fact, we are dependent on plants for our very existence. They produce our food and the oxygen we require for life. They produce the fibers that we use in our cloth and building materials. Plants are also

marvelous chemists and produce many chemicals from medicines to hallucinogens.

All too often we take the world around us for granted, and that is especially true of the green living things that surround us. What were the major evolutionary schemes that led to the evolution of the diverse groups of plants on earth? What are the characteristics of the major plant groups? How do plants reproduce? This chapter will attempt to answer these questions about the plant world (see Profile).

PLANTS: SOME DEFINITIONS

- Plants are defined as eukaryotic, multicellular, photosynthetic organisms whose cells are surrounded by a cell wall composed primarily of cellulose.

Above: The plants in the rain forest consist of a variety of species.

Dr. Elma Gonzalez
Cell Biologist, Plant Cell Organelle Biogeneticist

For an aspiring Mexican American migrant worker, the journey from the fields of south Texas to the department of biology at the University of California, Los Angeles, to become a researcher and professor can be long and difficult. There are many reasons. Earning a Ph.D. in the sciences is difficult enough; sometimes, because of social pressures, it is more difficult for a woman than for a man. A woman who is a member of a minority group must overcome obstacles in both the mainstream culture and her native culture. Pressures come from the minority culture as well, especially when it holds strict definitions about a woman's role. Nevertheless, Dr. Elma Gonzalez made the journey. Born in Mexico and raised in the United States, the oldest daughter of four in a family of migrant workers broke the cycle of illiteracy and poverty prevalent in most migrant worker families. How? By hard work and by believing that luck "favors the prepared individual."

From the time she was very young, Dr. Gonzalez loved science and math and took every available course to satisfy her curiosity. Thriving on academic challenge, she graduated from a small high school in south Texas at the top of her class. Without outside financial support, she could not have gone any further. But she received the aid she needed and attended the Texas Women's University, where she majored in biology and chemistry. Her goal was to be a high school science teacher, a position that she believed would be the best she could attain. However, after her student teaching experience, she discovered that her strengths and priorities were in research.

Then, through a series of fortunate accidents, Dr. Gonzalez was taken on at Baylor University, to work as a technician in a pharmacology laboratory. It was here that her scientific talents blossomed, and the staff encouraged her to apply to graduate school, a suggestion that far surpassed her aspirations. She entered the Ph.D. program in cell biology at Rutgers University, where she completed her Ph.D. in 1972.

Today, Dr. Gonzalez's research is internationally known. She is currently investigating questions about how organelles develop in the plant cells of germinating plants, especially those organelles in which there is no genome (in other words, those that contain no DNA or RNA). As you know, some cellular organelles like the chloroplasts and mitochondria contain their own DNA, and as the cell develops they can replicate themselves. Some organelles, however, develop without direct genetic control. One of these is the *glyoxysome,* a small membrane-bound organelle consisting of a protein mass surrounded by a membrane. Glyoxysomes are essential in fat metabolism. In the course of her research Dr. Gonzalez has determined that the membranes are synthesized from the endoplasmic reticulum. Her work provides pieces of evidence that help biologists as they try to develop a clear picture of the puzzle of organelle biosynthesis.

Dr. Gonzalez states wryly that science is a "hell of a mistress" but a loving one, and it occupies a great deal of her emotional and academic time. She also sits on a number of scientific committees in her research area. However, all that does not prevent her from gardening and reading science fiction. Since she feels individuals are responsible for their own mental health, she works to maintain a balance between her work and her hobbies. Very involved with politics at the national level, she is working to rectify some of the discrimination directed toward minority women in science.

What are plants? The answer at first seems obvious, but as we mentioned in Chapter 18, it is not always easy to classify the wide variety of life forms on earth into distinct kingdoms. Even today, many botanists (scientists who study plants) disagree about the correct way to define, sort, and classify the organisms that should be included in the plant kingdom. For example, many taxonomists prefer to place the algae in the kingdom Protista, although we will not follow that pattern here. For the most part, and granting that there are some exceptions, they agree that plants are eukaryotic, multicellular, photosynthetic organisms whose cells are surrounded by a cell wall composed primarily of cellulose.

Botanists also agree (with minor exceptions) that the life cycle of plants consists of two stages — a haploid and a diploid phase — that alternate with one another, a phenomenon called *alternation of generations*. Because this process is so typical of plants, our discussion will begin with it. Table 19.1 provides a miniglossary to help you learn the many new terms we will use as our discussion of plants progresses.

TABLE 19.1
A Miniglossary of Plant Reproductive Terms

Alternation of generations
The pattern of reproduction exhibited by most plants in which a diploid spore-producing phase (the *sporophyte* phase) alternates with a haploid gamete-producing phase (the *gametophyte*)

Antheridium (plural form: antheridia)
A sperm-producing cell or organ in the male gametophyte in the nonflowering plants

Archegonium (plural form: archegonia)
An egg-producing cell or organ in the female gametophyte in the nonflowering plants

Asexual reproduction
A process for the production of individuals that does not involve the exchange of genetic material between two organisms

Fertilization
The fusion of a sperm nucleus (n) with an egg nucleus (n), resulting in the zygote — the first cell ($2n$) of the new individual

Gametes
Haploid sex cells that unite, forming a diploid zygote

Gametophyte
See *alternation of generations*

Heterogametes
Gametes that do not appear to be the same, that show morphological distinctions between female and male

Heterospory
A type of spore production in which two different types of spores are produced: a large megaspore, which develops into the female gametophyte, and a smaller microspore, which develops into the male gametophyte

Homospory
A type of spore production found in some plants in which only one type of spore is produced; see *heterospory*

Isogametes
Gametes that appear to be the same, with no morphological distinctions between male and female

Megaspore
See *heterospory*

Microspore
See *heterospory*

Pollination
The transfer of a pollen grain (the male structure) to an egg-containing structure (the female structure)

Sexual reproduction
A process for the production of an individual that involves the exchange of genetic material between organisms

Spore
A reproductive cell that can develop into a gametophyte. Not to be confused with a bacterial spore, which is a dormant protected stage of the bacterium

Spore mother cell
A diploid cell that undergoes meiosis to form haploid spores

Sporophyte
See *alternation of generations*

Zygote
See *fertilization*

Alternation of Generations

■ **Plants exhibit two types of multicellular plant bodies, or generations, following one another in sequence during their life cycles. In the sporophyte generation the plant's cells are diploid (2n); during this phase the plants produce spores (through meiosis), which are haploid cells capable of germinating to form the gametophyte generation. In the gametophyte generation, the plant cells are haploid (n); at the end of this phase the plant produces male and female gametes. Once the female gamete is fertilized, a new 2n sporophyte is initiated and the cycle begins all over again.**

When we discussed mitosis and meiosis in Chapter 10, you saw that in animals, haploid (n) cells were produced (through meiosis) only during the formation of the sex cells, or gametes. The somatic cells of animal bodies are diploid (2n). Plants differ from animals in that during their life cycle there are two kinds of multicellular bodies. One, called the *sporophyte*, is diploid and produces haploid spores through meiosis. The other, called the *gametophyte*, is haploid and produces gametes by mitosis. Alternation of generations, then, simply means that plants pass through two separate and specific phases during their life cycles.

More specifically, the **sporophyte generation** (which means a spore-producing plant) is characterized as a multicellular, diploid (2n) plant body (the sporophyte) that produces haploid (n) spores through meiosis. A **spore** is capable of producing a new plant without undergoing fertilization. The new plant is representative of the other generation of the plant's life cycle, called the **gametophyte generation** (which means gamete-producing plant). The gametophyte generation is characterized by a haploid multicellular plant body that produces haploid gametes (n), which unite through fertilization to form a 2n zygote (first cell of the new generation). The zygote then develops mitotically into the sporophyte, and the cycle begins all over again.

Figure 19.1 illustrates the basic form of alternation of generations in plants. These are the steps in such a cycle:

1. When the diploid sporophyte is mature, specialized cells called *spore mother cells* (or *sporocytes*) produce haploid spore cells by meiosis.

2. A spore germinates, producing the haploid (n) gametophyte plant body. This structure develops and grows.

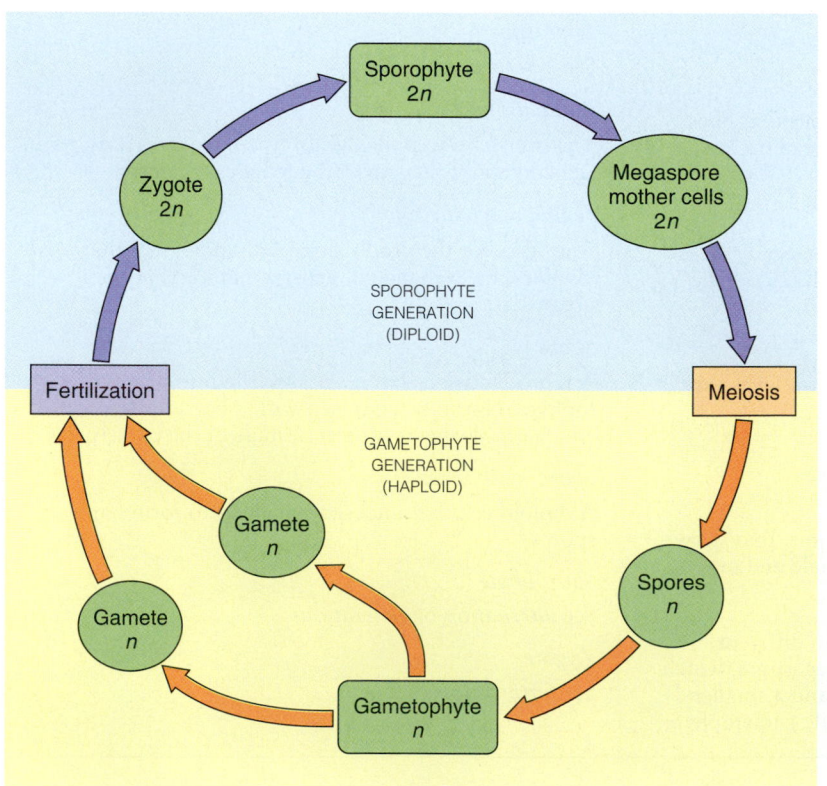

FIGURE 19.1 Alternation of Generations
This generalized diagram of a plant life cycle shows the alternation of the sporophyte generation (the diploid or 2n phase) with the gametophyte generation (the haploid or n phase).

3. Through the process of mitosis, the gameto-phyte plant produces haploid gametes (usually sperm and egg).

4. Through the union of the haploid gametes, a diploid (2*n*) zygote is formed. The zygote divides through mitosis to produce the mature multicellular sporophyte.

The cycle then returns to step 1, the mature sporophyte produces the haploid spores, and the process continues.

The relationship between the sporophyte and gametophyte generations varies among plants; either can be predominant. For example, in some less familiar plants like the mosses, the gametophyte generation is the dominant generation — the one that is the largest, the most obvious. In other words, when you look at moss growing around the base of a tree, you are looking at the gametophyte generation. The sporophyte generation is dependent on the gametophyte generation and is literally attached to it, for it is the gametophyte generation that conducts photosynthesis. On the other hand, when you look at a flowering plant like a rosebush, you are seeing the sporophyte generation. In this case, the gametophyte generation is reduced to an almost microscopic structure hidden within the larger, more obvious reproduction structures of the flower (Figure 19.2).

The fact that the gametophyte generation is dominant in the lower plants and the sporophyte generation in the higher plants is an important trend in the evolutionary movement of plants onto the land. As we survey the major plant divisions (remember, all botanists prefer to use the term *division* rather than *phylum*), we will discuss the increasing predominance of the sporophyte generation as plants become more complex.

The Plant Kingdom: Major Groups

■ **According to most biologists, the plant kingdom can be divided into two major groups: the algae and the land plants. The land plants can be further subdivided into nonvascular plants, which lack specialized tissues to transport material, and vascular plants, which have these tissues.**

The land plants can be divided into two major groups. When botanists discuss these two groups, they are referring to whether or not the organism has *vascular tissues,* which are specialized conductive tissues that transport materials throughout a plant or produce an embryo. Simpler plants are nonvascular; that is, they lack vascular tissue and do not form an embryo. More complex plants, on the other hand, do have vascular tissue and give rise to an embryo.

(a)

(b)

FIGURE 19.2   Gametophytes and Sporophytes
(a) The dominant generation of a moss is the gametophyte — the green, leafy base; the shoots are the sporophytes. (b) In a rosebush, the gametophytes are hidden within the centermost part of the blossom; it is the sporophyte generation that we see.

The majority of the nonvascular plants are small. They live in or near water and absorb water directly through their surfaces. As a result they do not require a complex of vascular tissues to transport fluids. The vascular plants have vascular tissues, which enable them to transport fluids long distances through their bodies. That means that vascular plants are well adapted to live on the land, and most do. Because vascular plants can transport materials great distances, they can be much larger than nonvascular plants. Have you ever looked at a giant oak tree? Obviously, the presence of a vascular system is a major requirement for evolutionary success on dry land, while it is not in an aquatic environment.

From an evolutionary perspective, how are the nonvascular and vascular plants related? What are some of the evolutionary linkages between the major plant divisions?

ALGAE

■ The algae are nonvascular, eukaryotic, non embryo-forming, photosynthetic organisms that require an aquatic environment for support against the pull of gravity and for distribution of gametes. The three most prominent divisions are the Rhodophyta (red algae), Phaeophyta (brown algae), and Chlorophyta (green algae).

The **algae** are *nonvascular, eukaryotic, non-embryo-forming, photosynthetic organisms,* and they represent the oldest plant groups on earth. They have flourished in the waters of the earth for over a billion years. The algae exhibit a large variety of forms and varying degrees of complexity. Some are fairly simple, unicellular organisms; others are colonies of individuals, or aggregations, containing nonspecialized cells; still others are large multicellular organisms consisting of specialized cells that form long filaments; and still others include very large multicellular plants such as the giant seaweeds. These seaweeds have specialized cells that create structures that look and function much like leaves, roots, and stems.

The algae can be classified primarily by using (1) their types of photosynthetic pigments and (2) their modes of reproduction. By these criteria and others (such as cell wall composition), the three major divisions of algae are the Rhodophyta (red algae), Phaeophyta (brown algae), and Chlorophyta (green algae). Figure 19.3 provides photographs of algae for each division.

Rhodophyta

The Rhodophyta or red algae are usually thought of as the delicate seaweeds (as opposed to the brown algae, which we will discuss shortly). About

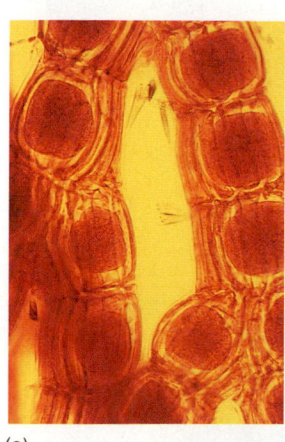
(a)

(b)

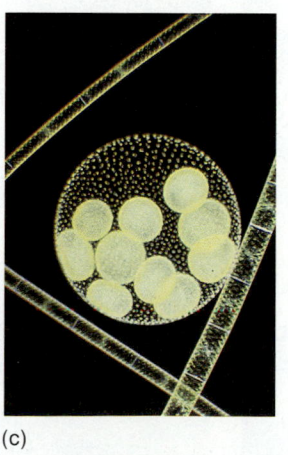
(c)

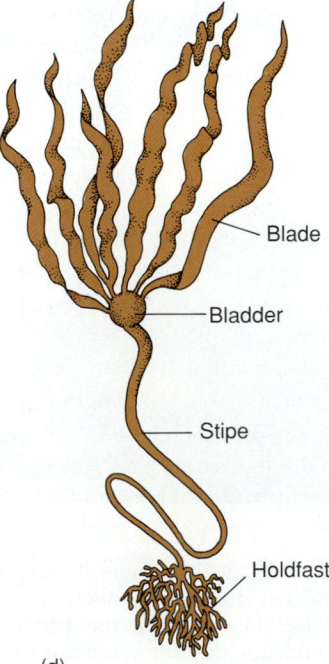
(d)

Blade

Bladder

Stipe

Holdfast

FIGURE 19.3   Red, Brown, and Green Algae
(a) These beadlike cells belong to the red alga *Polysiphonia*. (b) Each blade of the brown alga *Macrocystis pyrifera* (kelp) is attached to a bladder that floats it toward the water's surface, nearer the light. Although the kelp appears green in the photograph, it is actually a dull brown. (c) *Spirogyra* is a filamentous colony of elongated cells. The cells of the *Volvox* form a spherical colony. New daughter cells can be seen within the larger colony. (d) The structure of a brown alga.

4000 species of red algae exist. Although a few are unicellular, most members of this group are multicellular, nonvascular marine plants that live in water along rocky coasts and in deep ocean waters of the temperate zones. Some Rhodophyta, however, live in freshwater lakes, and a few unicellular species live on land in moist regions.

The red algae are distinguished by their red and blue pigments. In some red algae, a special chlorophyll is found called *chlorophyll d*. Because of their red pigment, called *phycoerythrin*, some species can live at ocean depths of 150 meters (500 feet) or more. To understand why, we need to review the nature of light waves as they pass through water. (You may wish to refer to Chapters 7 and 8, in which we discussed light waves.) Red waves of light are longer than the shorter, more energetic green, blue, and violet waves. Water filters out the longer waves of light so that only the shorter waves penetrate to the greater depths. The red pigments of Rhodophyta can absorb green, violet, and blue light, an ability that gives them a selective advantage in the deep-water environment. Even though the red algae are called "red," some appear green and others even black owing to the distribution of the other pigments in their cells.

Red algae share several characteristics. They store their food as a carbohydrate called floridean starch. They lack flagellated cells. The sperm of the red algae are carried passively by water currents, fertilizing the plant's eggs.

From an evolutionary perspective, there is no direct link between the red algae and other algae or the higher plants. There are several reasons for this. First, they are the only algal group that lacks flagellated cells. Second, their chloroplasts more closely resemble the photosynthetic membranes of prokaryotic bacteria (remember that prokaryotic cells have no chloroplasts) than those of other algal groups or higher plants. Finally, molecular analyses of the chlorophyll of the red algae show that the chloroplasts are not related to the *protists*, higher plants, or other algae.

As a result, taxonomists theorize that the red algae evolved separately because of a symbiotic relationship that formed between the cyanobacteria and the ancestral red algae. (In a symbiotic relationship, there is an intimate association between two or more organisms of different species.) In other words, some of the prokaryotic cyanobacteria were incorporated into the eukaryotic cell types that eventually evolved into the organisms that we know as the red algae.

Phaeophyta

When you walk along the rocky beaches of the world's oceans and examine the seaweeds washed onto the shore, most of the specimens you see belong to the division Phaeophyta, or, as they are more commonly called, the brown algae. Brown algae are generally found attached to the ocean floor or to the rocks and reefs along the shoreline. Members of this group dominate the intertidal zone in the earth's temperate regions, and most of the more than 1500 species are found in deeper cold marine waters. (The intertidal zone, also called the littoral, is the area between the high and low tide marks.)

Included in this group are the kelps and other seaweeds. Some of the giant kelps can grow to over 30 meters (about 100 feet) in length. Even though most are found along the seashore, some are found offshore. For example, the seaweed *Sargassum* (from which the Sargasso Sea in the North Atlantic gets its name) is found in large floating masses offshore.

The brown algae are distinguished by their pigments: These include chlorophylls *a* and *c*, as well as the brown pigment called *fucoxanthin*. Brown algae are multicellular rather than unicellular, and they do not come in colonial forms. Many of them, such as the giant kelps, are quite large. Very few live in fresh water.

The structure of a brown alga is distinctive. When you look at the plant body, you see a structure called a *holdfast* that resembles a root. The holdfast helps bind the plant to the ocean floor or to rocks along the shoreline. This structure is very effective in anchoring the plant, even in the face of tides and the pounding of heavy surf. The Phaeophyta also possess structures that look like stems and leaves. These are called *stipes* and *blades*. Attached to the stipes and blades are often hollow, gas-filled floats called *bladders* that hold the algae upright and enable the blades to float on the surface of the waters, close to the sunlight (Figure 19.3d).

Although holdfasts, stipes, and blades seem similar to the roots, stems, and leaves of the higher plants, from an evolutionary perspective they are not closely related. Instead, they are only analogous to the corresponding parts of higher plants, since the structures have similar functions but a different evolutionary origin.

Brown algae have another structure that is analogous to one found in the higher plants. Even though Phaeophyta are technically nonvascular plants, they have evolved a system that is similar in function but not in structure to the vascular system characteristic of higher plants. This system conducts the products of photosynthesis from the blades near or on the surface down into the holdfast anchored to the ocean floor. However, these brown algae tissues do not show the degree and types of differentiation found in the vascular plants.

The brown algae also exhibit examples of *convergent evolution* with vascular plants — that is, the independent evolution of similar structures in distantly related organisms (Chapter 17). Some of these characteristics make the brown algae appear to be fairly specialized. For instance, in many brown algal species the gametophyte develops completely within the sporophyte (as within the rose). The reduction of the dominance of the gametophyte phase is considered an advanced trait. Brown algae may have two types of gametes (sperm and egg), called *heterogametes,** in which (1) both gametes are motile or (2) the eggs may be large and stationary while the sperm are small and flagellated.

Chlorophyta

You may have heard the adage that one way to tell directions in the woods is to look for the "moss" on the trees. That would tell you which way north was. In reality, that green plant material may not be a moss but a green alga. The division Chlorophyta, comprising the green algae, is the largest algal division. To date, botanists have classified about 7000 species of green algae. Their color is due to the large amounts of chlorophylls (*a* and *b*) contained within their chloroplasts. They store their sugars as the complex carbohydrate starch. Some green algae are unicellular; others grow in colonies; still others form filaments. Most are freshwater organisms, but there is a large number of marine species. Some are even terrestrial, living in moist soil or on the moist side of tree trunks. It is the north side of a tree trunk that is not directly exposed to sunlight.

Some of the simpler unicellular green algae look like members of the kingdom Protista (Chapter 18), because they are highly mobile and at first glance look like microscopic one-celled organisms. However, the Chlorophyta form a natural group with the land plants because of their chemical makeup.

Like the higher plants, the simple green algae have the chlorophylls *a* and *b* and carotene pigments, and they store their sugars as starches.

One of the best-known filamentous freshwater forms of green algae is *Spirogyra*. *Spirogyra* cells are elongated and contain one or more spiral-shaped chloroplasts, from which they get their name. The best-known large multicellular green alga is *Ulva,* which is called sea lettuce because its leaves resemble those of the leaf lettuce you may grow in your garden. *Volvox* is one of the best-known colonial forms of green algae. *Volvox* is composed of tiny flagellated cells that are held together in a gelatinous matrix that forms a hollow sphere (see Figure 19.3c).

Chlamydomonas. One of the most intensely studied unicellular green algae is *Chlamydomonas*. *Chlamydomonas* is a small oval organism that contains a cup-shaped chloroplast and a red eyespot that is sensitive to light. Two flagella on the front end of the cell pull it through the water.

The life cycle of *Chlamydomonas* includes diverse and complex reproductive mechanisms. In the haploid phase (the gametophyte generation of the life cycle), a single cell, without fertilization, produces from 2 to 32 daughter cells by asexual means. Asexual reproduction begins when the flagella of the cell either drop off or are absorbed by the cell. The nucleus of *Chlamydomonas* then divides by mitosis, forming two daughter nuclei, and the cytoplasm of the cell divides.

If the parent cell membrane breaks down, the two daughter cells are released and the reproductive process is over. However, mitosis can occur again, up to four times. That means that a single *Chlamydomonas* can produce 2, 4, 8, 16, or 32 daughter cells, depending on how many times it undergoes mitosis. Because there is no change in the number of chromosomes during asexual reproduction (this is a mitotic process), all of the resulting daughter cells are haploid. Under favorable environmental conditions, this type of asexual reproduction is primarily responsible for population increases in *Chlamydomonas*.

When the conditions are unfavorable, great numbers of *Chlamydomonas* come together and carry out sexual reproduction. All of these offspring are virtually identical, because they are the product of asexual reproduction. In other words, one cannot distinguish between "male" and "female" cells. (Botanists theorize that individuals exhibit differences in their flagella, but that theory has not been proved yet.) In any event, since all cells are identical

---

*Some botanists prefer the term *oogamy* to *heterogamy*, as we have noted.

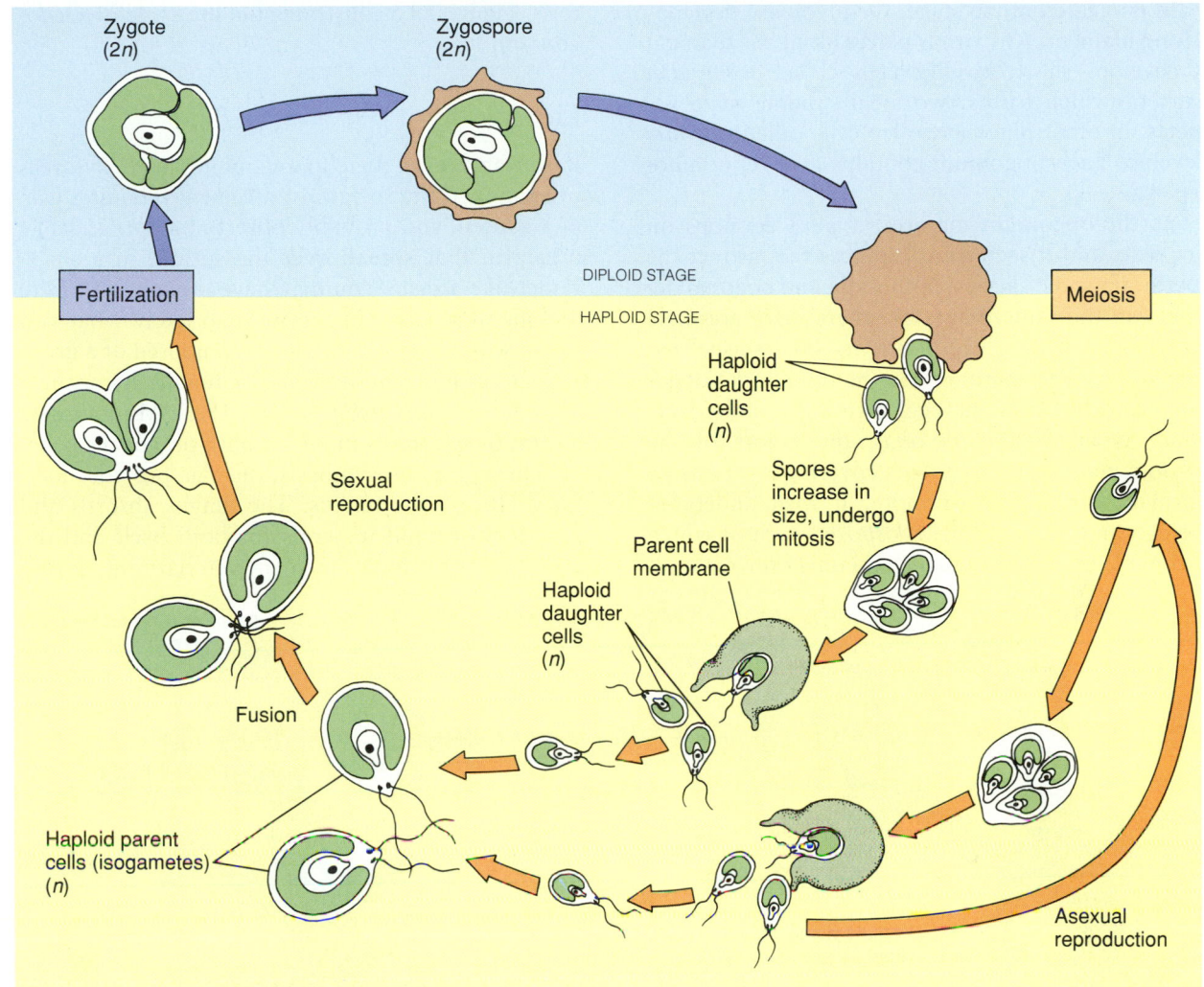

FIGURE 19.4   Life Cycle of *Chlamydomonas*
This green alga provides an example of the production of like gametes (isogametes). Asexual reproduction is most common for this unicellular organism. The haploid daughter cells reproduce sexually by fusing to form a diploid zygote, which then divides through meiosis, forming four haploid spores.

and can function as sex cells, they are said to be **isogametes.** Members of different mating types (let's call them plus and minus) pair off and fuse, forming a diploid zygote. Following fertilization the metabolic activities of the zygote slow down and a tough, thick, bumpy coating forms about the zygote. This structure, called a *zygospore,* remains dormant until environmental conditions are once again more favorable for growth. At that time, meiosis takes place, and the diploid nucleus divides, forming four haploid cells. The zygospore membrane breaks down and the four *Chlamydomonas* cells are released. Figure 19.4 illustrates the life cycle of *Chlamydomonas.*

Oedogonium. Many green algae follow a pattern similar to that of *Chlamydomonas. Oedogonium,* however, does not. This unbranched, filamentous green alga produces heterogametes, which are a pair of gametes that differ in form, size, or behavior. One gamete (the egg) is usually large and nonmotile, and the other (the sperm) is typically small and motile. When the egg is nonmotile, many botanists prefer to call it an *oogamete.* Thus *Oedogonium* is described as *oogamous* because its egg is nonmotile. The formation of two distinctive gametes is considered an advanced characteristic. Because of *Oedogonium*'s heterogamous nature, it is regarded as one of the most advanced green algae.

In its gametophyte stage, *Oedogonium* develops a long filament. At various places along the filament it develops short, boxlike cells called *antheridia*, each of which forms two small, motile male gametes. In other places, swollen cells called *oogonia* develop. Each oogonium contains a large, nonmotile egg.

As the oogonium matures, a pore develops on one side, and it secretes substances that attract the sperm. (In some species, antheridia and oogonia develop on the same plant; in others there are male and female plants, each with its own specialized structures.) The sperm enters the pore in the oogonium and fertilizes the egg, forming a diploid zygote. As in *Chlamydomonas*, the zygote of the *Oedogonium* develops a thick covering and remains dormant for a year or more. It then undergoes meiosis and releases four *zoospores*, each of which is capable of producing a new filament through mi-

tosis. Figure 19.5 illustrates the life cycle of *Oedogonium*.

Lichens

If you have ever done any climbing in rocky areas or examined the surface of a cement retaining wall in a garden, you probably observed a flat, scalelike organism that spread over the surface in a shape rather like a dish. You may have thought you were looking at a moss, but more than likely what you found was a **lichen,** a life form composed of a green alga or cyanobacterium and a fungus living in a *mutualistic relationship* (Figure 19.6). **Mutualism** is a form of symbiosis in which both organisms benefit. Through photosynthesis, the alga provides food for itself and the fungus. The fungus absorbs and retains water and minerals for both itself and the alga, and it protects the alga from harmful, drying

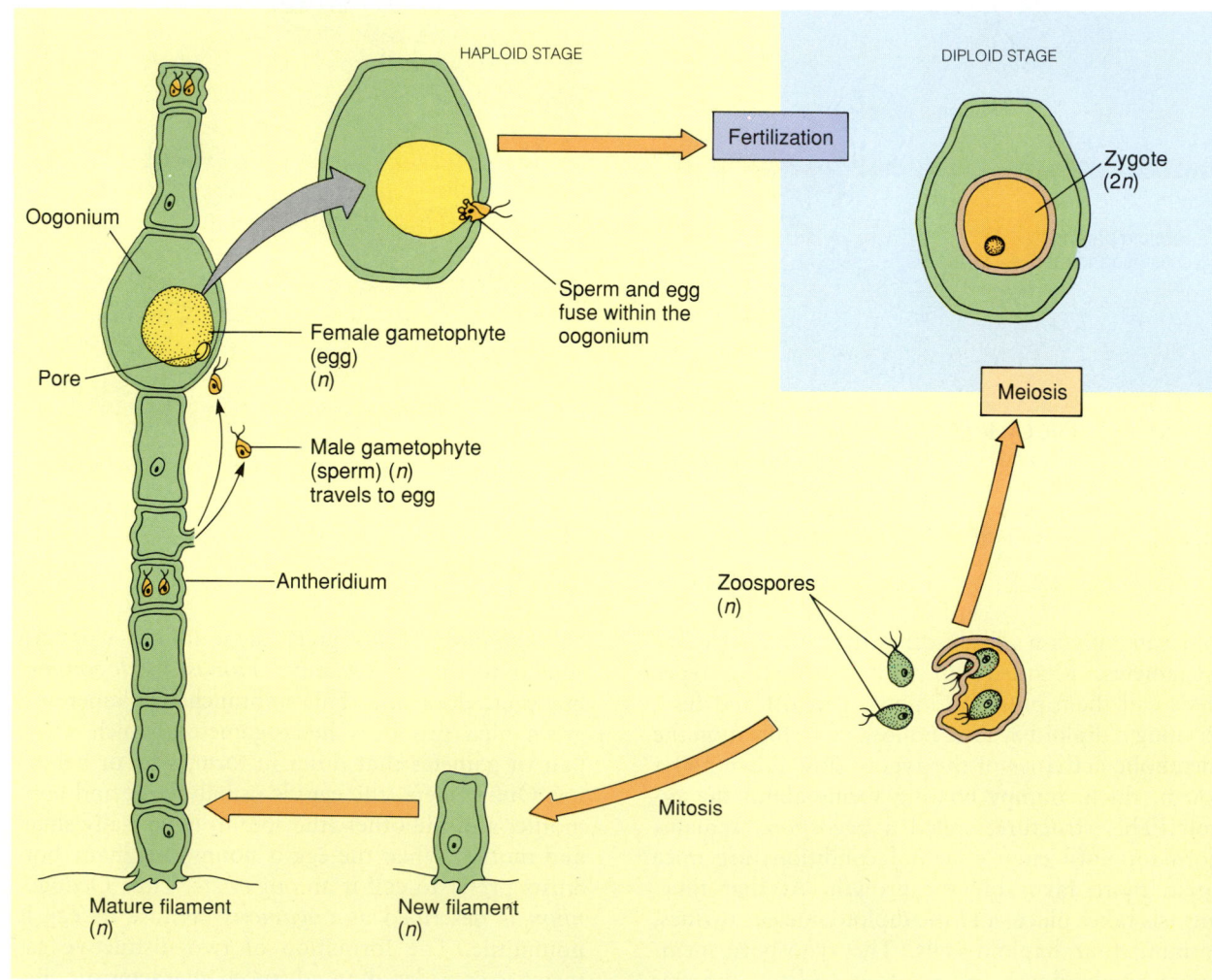

FIGURE 19.5 Life Cycle of *Oedogonium*
Considered a more highly developed organism than *Chlamydomonas*, this green alga produces heterogametes, or distinctly male and female gametes.

(a)

(b)

(c)

FIGURE 19.6 Lichens
Lichens generally take one of three forms. (a) Crusty lichens can grow on bare rock or dead trees. (b) Leafy and (c) shrubby lichens are shown here growing on trees. Most often what you see when you look at a lichen is the fungus. Its filaments surround and draw nutrients from an inner layer of algal or bacterial cells.

light intensities and produces substances that accelerate photosynthesis.

Lichens are amazing life forms. They can live in the desert or in the Arctic, and they live for a very long time — sometimes up to 4500 years. After a volcano erupts and the lava beds cool, the first life forms that appear are lichens, which initiate soil creation by breaking the lava into small particles in which other plants can take root. They are useful in other ways as well, such as by providing a source for antibiotics.

THE MOVE ONTO THE LAND

With an overview of the algae in hand, we are ready to try to answer two questions: How did the algae evolve into the great diversity of land plants that exist on earth, and when? It is easier to answer the question concerning timing. According to the fossil record, it was about 405 to 425 million years ago that the aquatic plants made a successful transition to the land. Table 19.2 provides a geological timetable of the evolution of plants on earth.

How did the transition onto land happen? Most botanists feel that it was the green algae that evolved into the land plants. One hypothesis is that the *bryophytes* (simple, nonvascular plants such as mosses and liverworts) and the *tracheophytes* (vascular land plants) evolved independently from different green algal ancestors. Another hypothesis is that only the tracheophytes arose from the green algae and that the bryophytes are simple descendants of the early tracheophytes.

Let's now take a phylogenetic view as we climb the evolutionary tree. As we examine the changing history of plants, we will stress both the common patterns that unite the plant kingdom and the diversity that exists in the life cycles of the various plant divisions.

Bryophytes: Simple Nonvascular Plants

■ **The first land plants were the bryophytes: nonvascular plants such as mosses that live near water or in moist areas. In these plants, the gametophyte generation is predominant.**

As plants adapted to the terrestrial environment, some transitional land plants such as the mosses, liverworts, and hornworts evolved. The transitional plants comprise a group called the bryophytes. Bryophytes are small nonvascular plants, most of which live near water or in moist land areas. They

TABLE 19.2
Geological Timetable for Plants

Era	Period(s)	Environment	Plants	
Cenozoic	(Millions of years ago)	Glaciation, mountain building, cooling	Extensive grasslands	**Age of Angiosperms**
	—65—		—Asteroid collision—	
Mesozoic	Cretaceous	Rocky mountains form / Extensive lowlands	Flowering plants	
	—135—		Conifers	**Age of Gymnosperms**
	Jurassic	Lowlands, inland seas	Cycads	
	—197—			
	Triassic	Mountains, drying		
	—225—		—Continental drift begins—	
Paleozoic	Permian	Glaciers form / Inland seas dry up	Earliest conifer fossils / Earliest fern fossils	Primitive gymnosperms **Great Paleozoic Forests**
	—280—			
	Carboniferous	Mountains forming		
	—345—		Earliest bryophyte fossils	
	Devonian		Earliest brown algae fossils	
	—405—		Earliest vascular plant fossils (Rhyniophytes)	Plants invade land **Age of Algal Plants and Protists**
	Silurian	Extensive shallow seas, warm, mild climate		
	—425—			
	Ordovician			
	—500—			
	Cambrian		Red and green algae	
	—570—			
Precambrian				

absorb water directly through their surfaces. In the bryophytes the sporophyte generation is attached to and dependent on the gametophyte generation, from which it derives its nutrition.

The Mosses

■ **In mosses, the first structure of the gametophyte generation is the protonema, which develops from a spore. The protonema then develops into the "leafylike" gametophyte, which bears the sex organs. The archegonium is an egg-producing and the antheridium is a sperm-producing structure. Fertilization requires water as a vehicle. The spo-rophyte generation produces a sporangium, which, through meiosis, forms spores. The spores are distributed on the winds.**

To learn the life cycle of a typical bryophyte, we will discuss the life cycle of a moss (Figure 19.7). During the gametophyte phase of a moss's life cycle, the haploid spore germinates into a green threadlike structure called the *protonema*. The protonema forks repeatedly and eventually develops buds that enlarge to become the leafy, aerial new gametophyte *(n)* plant. Since these plants do not possess vascular tissue (which would hold the stems erect), the buds develop close together and the aerial stems provide mutual support through their dense packing.

The mature gametophyte contains the reproductive structures. In some moss species, the gametophyte bears both male and female sex organs in the same plant; in other species, the male and female sex organs are found on separate plants. The multicellular, flasklike, egg-producing structure is called the *archegonium*. Each archegonium produces a single egg. The multicellular, oval-shaped, sperm-producing structure (the *antheridium*) may produce hundreds of flagellated sperm.

The mosses are only one step away from plants that live in an aquatic environment and rely on water for the sperm to reach the egg. Mosses still require water for fertilization, even though they live

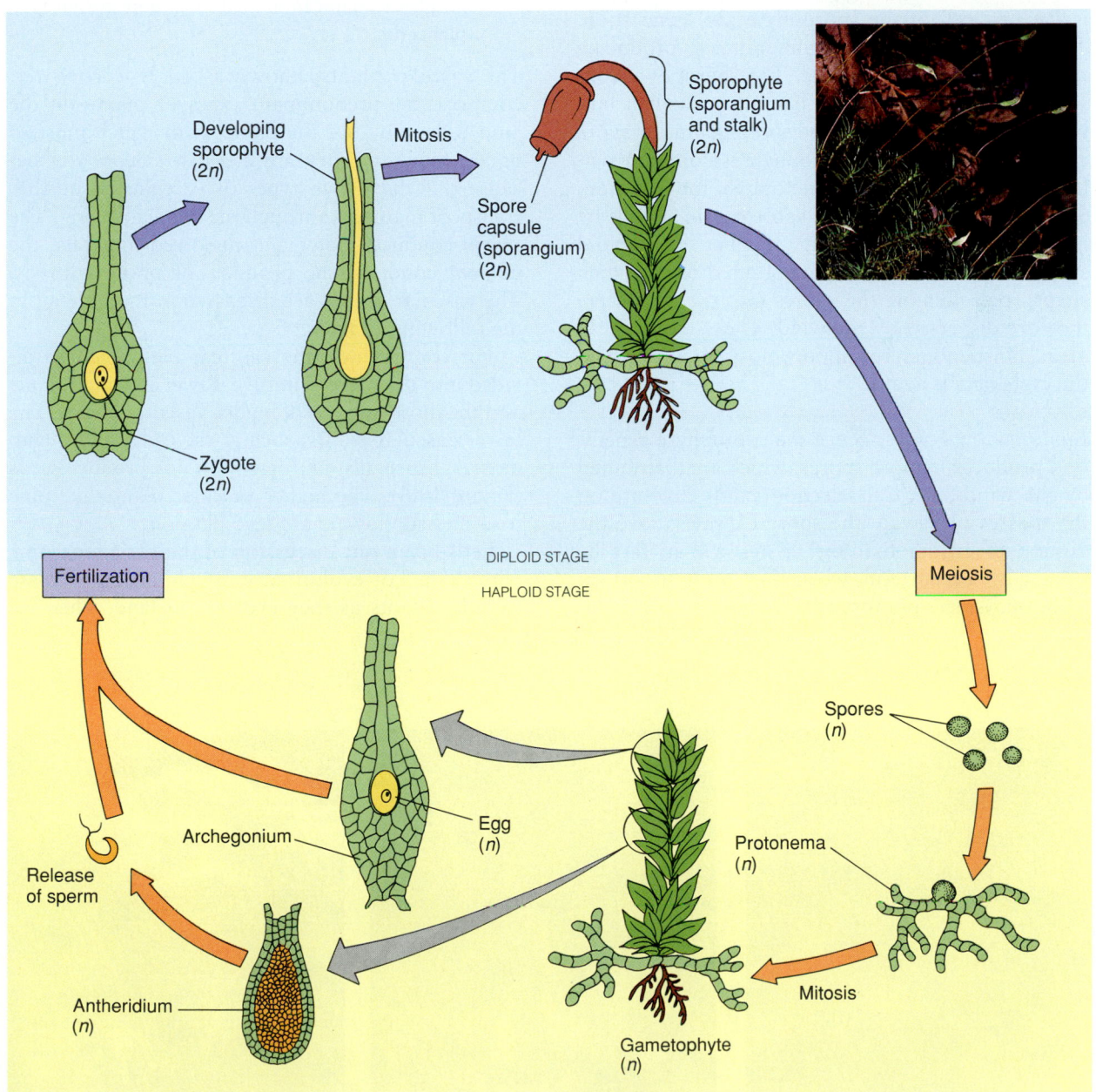

FIGURE 19.7 The Life Cycle of a Moss
Notice how the sporophyte remains attached to and dependent on the gametophyte in the sporophyte generation. Some of the cells within the sporangium divide by meiosis to produce many haploid spores. At maturity, the sporangium dries and then snaps open, flinging the spores into the air, where they are dispersed by the wind. Those that land in areas moist enough to support moss life germinate, and the cycle begins again.

on the land. The flagellated sperm are not released until the antheridium absorbs a sufficient quantity of water. In some species the spattering of falling raindrops disperses the sperm. In other species the sperm swim through the water and are attracted to the egg, possibly because of the higher concentration of sugar near the egg.

Once the sperm reaches the archegonium, it enters and fertilizes the egg, forming a zygote, which is the first cell of the sporophyte (2n) generation. The diploid zygote undergoes mitosis, producing the multicellular sporophyte. The sporophyte consists of an elongated stalk bearing a swollen bulbous structure known as the sporangium, or *spore capsule*, at its tip. The sporophyte is nonphotosynthetic and relies on the gametophyte for nutrition. Some of the cells within the spore capsule divide by meiosis, producing a great number of haploid spores. At maturity, the sporangium dries out, then snaps open, flinging the spores into the air, where they are dispersed by the wind. Those that land in areas moist enough to support moss germinate, and the cycle begins again.

Evolutionary Forces. Since the sporophyte generation produces haploid spores, which are distributed by the winds, natural selection exerts pressure on the mosses to elevate the spore capsule above the ground to ensure that the spores will in fact be caught by air currents when they are released. In addition, since photosynthesis requires sunlight, there is additional pressure on the mosses to grow tall so that sunlight is not blocked by other plants.

THE VASCULAR PLANTS

■ **The tracheophytes, or vascular plants, possess vascular tissues called xylem and phloem that function in the circulation of water, minerals, and the products of photosynthesis. The primitive vascular plants do not form seeds; the more advanced plants do.**

The vascular plants, known as the *tracheophytes*, comprise the predominant types of plants on the land today and are the plants with which most of us are familiar. They are called *tracheophytes* because they have two types of vascular tissue that transport materials throughout the plant body. The *xylem* conducts water and dissolved minerals; the *phloem* conducts the products of photosynthesis. The vascular tissues are discussed in more detail in the following chapter.

For convenience, the vascular plants can be divided into the more primitive, lower vascular plants and the more advanced, higher vascular plants. The lower vascular plants, such as the psilophytes, club mosses, horsetails, and ferns, do not produce seeds (Figure 19.8). The higher vascular plants like pine trees and sunflowers are seed plants.

Let's begin our discussion of the vascular plants with the major evolutionary developments in these vascular plants as they evolved on land. Then we

(a) (b) (c)

FIGURE 19.8 Primitive Vascular Plants
(a) Club mosses, (b) horsetails, and (c) ferns are all examples of primitive vascular plants, which do not produce seeds.

Chapter 19   Plants: A Survey

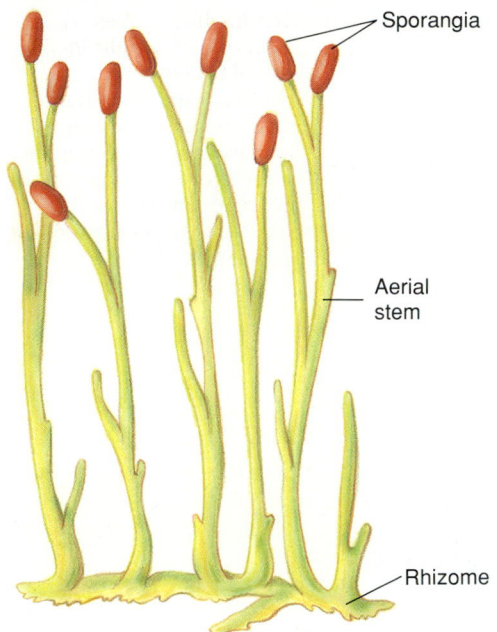

Sporangia

Aerial
stem

Rhizome

FIGURE 19.9   The First Known Vascular Plant
The *Rhynia*, a species that flourished 350 million to 400
million years ago, is the earliest fossil evidence of a true
vascular plant.

will discuss the ferns as examples of lower vascular
plants, followed by the pine tree and a flowering
plant as examples of higher vascular plants.

Major Trends in the Evolution of the Vascular Plants

In order to live on the land, organisms must overcome major problems, such as *desiccation,* or
drying, because the land environment lacks the
water that surrounds the marine organism. Another
problem is the need to support the plant body
against the pull of gravity. In a marine environment,
water buoys up the organism as it floats. The development of a central cylinder of vascular tissue in
the stem was one of the first adaptations of the
earliest land plants, such as the *Rhynia*. (This organism first appeared about 400 million years ago.
It belongs to the now-extinct psilophytes. See Figure
19.9.) Vascular tissue can conduct water and minerals up to the tips of the plant and can conduct the
products of photosynthesis down to the roots. Vascular tissue also aids in supporting the plant against
the pull of gravity.

Another important evolutionary development as
the plants adapted to the land environment was the
development of roots and leaves. Roots are specialized structures for water and mineral absorption,

anchoring the plant to the soil and storing food.
Leaves are highly specialized organs for photosynthesis. With roots, stems, and leaves, as well as efficient conducting systems, the plants had solved
several of the basic problems of life on land, those
of acquiring water and food and delivering these
substances to all the cells of the organism.

In the ferns and some other primitive vascular
plants, like horsetails and club mosses, the gametophyte generation develops independently of the sporophyte. Environmental pressure influenced the
evolution of height to ensure the distribution of
spores. Because of the development of vascular tissues in the sporophyte generation, the diploid phase
became predominant, and the size of the gametophyte was greatly reduced (Figure 19.10). The gametophyte has become virtually microscopic in size
in more complex plants like pine trees and rose
bushes and is dependent on the parent sporophyte
for protection and nutrition.

LIFE CYCLE OF A FERN

■ **In ferns, the gametophyte and sporophyte generations occur as independent plants. The most
prominent of these is the sporophyte. Sori found
on the underside of the fronds (leaves) produce
spores. The prothallus is the young gametophyte
generation. Antheridia and archegonia borne on
the underside of the prothallus produce sperm
and eggs, respectively, which rely on water for
fertilization.**

In the ferns, the sporophyte and gametophyte generations are independent. When you look at a pot-
ted fern growing at home, you are seeing the
sporophyte generation. The large leaves are called
fronds, and if you look carefully at the underside of
one, you might find many small dots, called sori
(singular, *sorus*). (My grandmother used to think
that sori were insect eggs and would carefully remove them, one by one, so that the insects would
not harm her ferns.) Each sorus actually is a group
of *sporangia* (spore-producing structures). Cells
called *spore mother cells* are found in each individual *sporangium*. The spore mother cells divide by
meiosis and produce from 48 to 64 haploid spores
in each sporangium.

When the mature sporangium ruptures, the
spores are carried away by the wind and fall to the
ground. If the spore lands in a moist spot, it germinates and develops into a prothallus, a small, heart-
shaped gametophyte about the size and thickness of
a dime. On the underside of the prothallus are rhi-

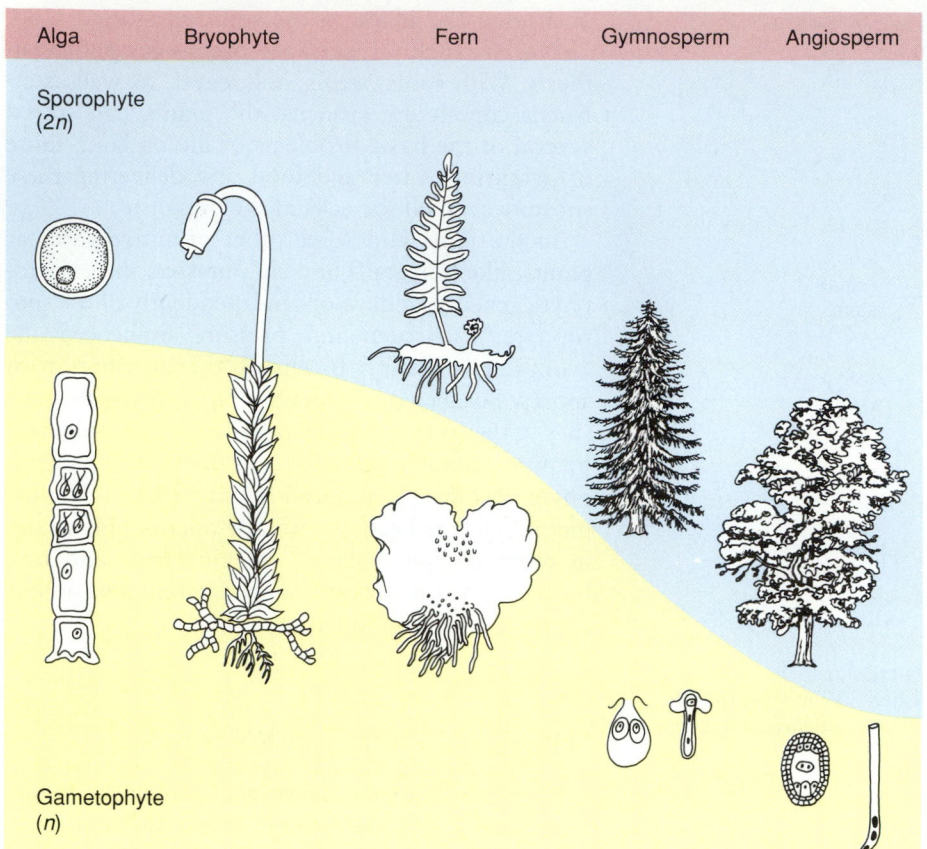

| Alga | Bryophyte | Fern | Gymnosperm | Angiosperm |

Sporophyte (2*n*)

Gametophyte (*n*)

FIGURE 19.10 The Evolution of Height in Vascular Plants
This diagram illustrates the evolutionary trend toward increasing height dominance of the sporophyte generation and the reduction of the gametophyte generation in vascular plants.

zoids (unicellular absorptive structures that attach the prothallus to the ground). The antheridia, which produce sperm, are located among the rhizomes. The archegonia are found near the notch of the prothallus. One egg is formed in each archegonium. Even though these structures are located on the same prothallus, the male and female gametes mature at different times, so that self-fertilization is unlikely. Cross-fertilization ensures a greater probability of genetic recombination from different individuals, as well as greater variability.

No matter how many of the eggs are fertilized by sperm swimming to them through the water after a rain, only one diploid embryo is formed on each prothallus. The prothallus then disintegrates and the embryo develops into the mature sporophyte, a large, conspicuous plant. This completes the life cycle (Figure 19.11).

Even though the ferns are vascular plants, they are still dependent on water for the distribution of sperm. Ferns do not form seeds. To see the development of seeds, we need to look at the gymnosperms and angiosperms, the most advanced plants.

SEED PLANTS, NAKED AND COVERED

■ **The two groups of higher, vascular land plants are the gymnosperms and angiosperms. The sporophyte generation dominates the life cycle in these plants, and the gametophyte generation is borne on the sporophyte generation, primarily within the cones of gymnosperms and within the flowers of angiosperms. These plants produce separate male and female gametophytes, both of which are required for the production of seeds.**

As you munch on peanuts some time, split one of the nuts and take a close look. At one end of the seed, you can see a tiny embryo, the young sporophyte. In a raw peanut, the young sporophyte will develop into a new peanut plant if planted. You are looking at a primary characteristic of the advanced plant families — the seed.

Seed plants are subdivided into two major groups, the *gymnosperms* and *angiosperms*. **Gymnosperms** are so called because their seeds have no

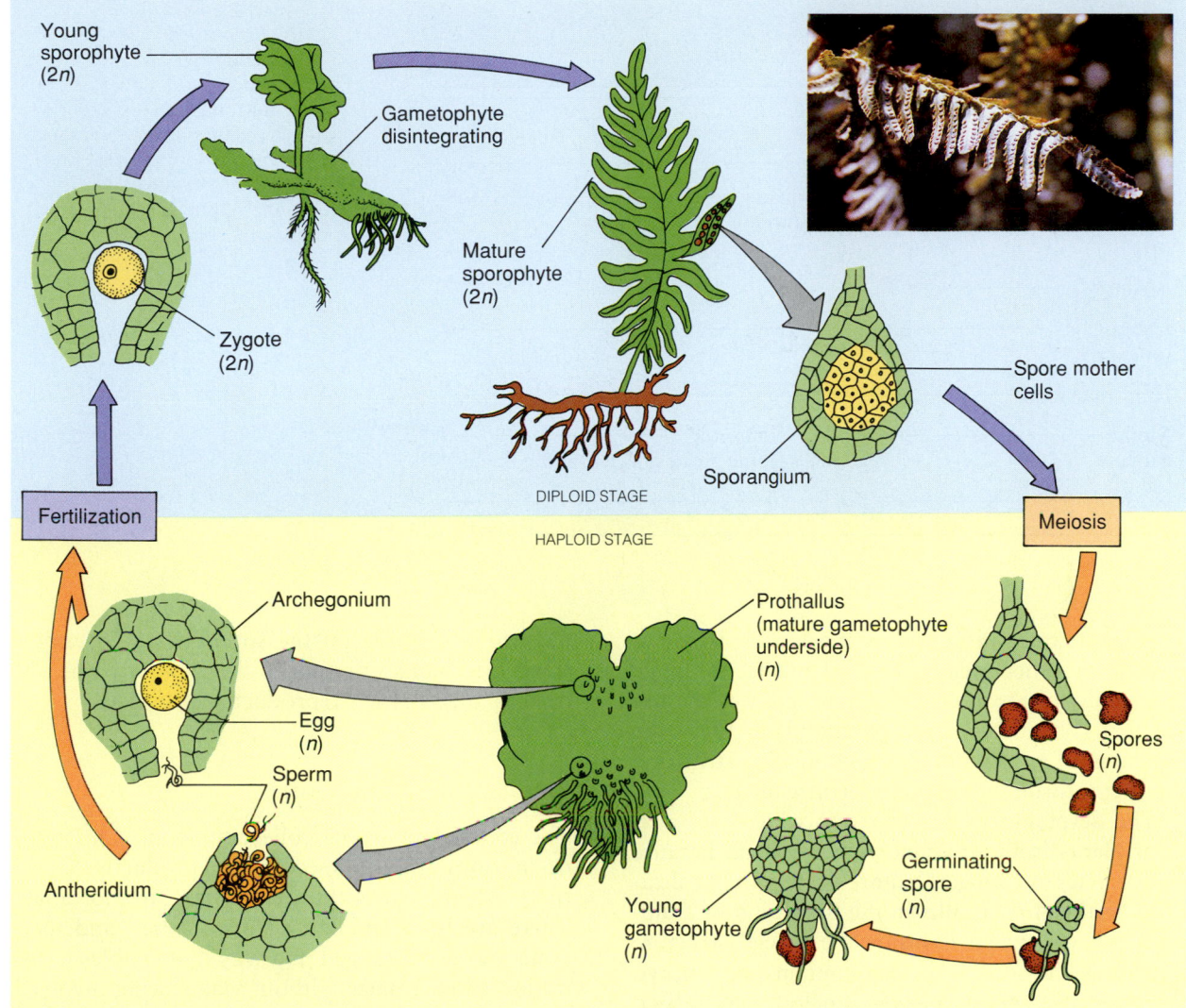

FIGURE 19.11   The Life Cycle of the Fern
When the fern's mature sporangium erupts, spores are swept to the ground by
the wind. Some germinate into the gametophyte, producing eggs (in the
antheridia) and sperm (in the archegonium). Sperm enter the archegonium of
another prothallus, and fertilization occurs. A diploid embryo, which develops
into a mature sporophyte, is the result. The fern reaches maturity, produces
more spores, and the cycle is repeated.

covering that encloses them. (In Greek the word
gymno means naked. What is a gymnasium? Right,
the place where naked exercise takes place — which
is how exercise was conducted in ancient Greece.)
Conifers, which are cone-producing plants such as
the pine trees, are gymnosperms. The other group
of seed plants, the **angiosperms,** have covered seeds.
These seeds develop in a vessel, the fruit, which is
actually one or more ripened ovaries. Angiosperms
are called the "flowering plants," because the gametophyte is located in the flower.

Other characteristics separate the seed plants

from their more primitive relations. Many primitive
vascular plants, such as the ferns, are **homosporous**
— that is, they produce only one type of spore.
Every spore can develop into either a male or female
gametophyte. Some lower vascular plants and all
the seed plants are **heterosporous** — that is, they
produce two types of spores. The larger, called a
megaspore (meaning "large spore"), gives rise to
the female gametophyte; the smaller, called a *microspore* (meaning "small spore"), develops into the
immature male gametophyte, the **pollen grain.** This
structure is then released by the male part of the

TABLE 19.3
Summary of Reproductive Trends in Plants

	Green algae	*Ferns*	*Gymnosperms*	*Angiosperms*
Gameotphyte dominance (*n*)	————→	Sporophyte dominance (2*n*) ——————————————————————————→		
Isogamy	————————→			
Heterogamy (oogamy)	———→			
Homospory	——————→	Heterospory ————————————————————————————→		
Motile gametes	——————→	Usually nonmotile gametes (some groups have motile gametes) ————→		
No seeds	——————→	Seeds ——————————————————————————————————→		

sporophyte to be dispersed, in most cases, by wind, insects, or other means. In seed plants, the male gametophyte, the pollen grain, is not yet mature when it is shed from the parent plant. Once the pollen grain joins with an egg-containing structure, the sperm form, mature, and fertilize the egg.

The development of heterospory freed the vascular plants from an aquatic environment, because the gametes no longer require water for fertilization. Fertilization results in the formation of a diploid embryo, which is inside the special structure known as the **seed.** The seed contains the embryo surrounded by food reserves. Enclosing these two elements is a protective seed coat. Table 19.3 summarizes the key reproduction traits in plant life cycles.

The arrival of heterospory and of the seed was a very important evolutionary step in the plants' colonization of the land. Let's first discuss the life cycle of a gymnosperm. Then we will look at the angiosperms.

GYMNOSPERMS: NAKED SEEDS

■ **Gymnosperms are plants that produce unenclosed seeds in reproductive structures usually called cones.**

The spruce and pine trees that keep forests green through the coldest winters are prime examples of the mature sporophytes of the gymnosperms. Gymnosperms demonstrate a further evolutionary advance in plants in that their gametophytes are reduced and borne on the sporophyte. To understand the changes present in the gymnosperms, we need to analyze their reproductive processes.

Female and Male Cones

The cones that people collect to use in Christmas decorations are very specialized reproductive structures. All the cones on a single tree are not alike. There are both female and male cones, and their structures are arranged in a way that enables fertilization to take place without water being involved as a vehicle.

Female Cones

■ At the base of the scales of female (or ovulate) cones are two ovules containing nutritive tissues, a thick covering, micropyle, and spore mother cells. The spore mother cells produce haploid megaspores. One megaspore develops into a female gametophyte containing two or more archegonia, each with an egg.

The female (or *ovulate*) cones are found at the upper levels of the tree and are much larger and longer-lived than male cones. Those of the sugar pine can reach a length of two feet. The ovulate cones remain on the tree for up to two years. Instead of the papery scales found on male cones, the scales on female cones are firm and woody.

At the base of each scale is a pair of **ovules.** Each ovule contains multicellular nutritive tissue; a thick covering; a **micropyle,** which is a channel leading through the covering to the center of the ovule; and a **spore mother cell,** which will become the source of the eggs that develop. As the female cone matures, the spore mother cell undergoes meiosis, producing a row of relatively large haploid megaspores. The haploid megaspores are a product of the sporophyte generation of pine tree.

Only one of the megaspores will develop further. All the rest degenerate fairly soon after the end of meiosis. It takes several months for the remaining megaspore to develop into the female gametophyte of the pine tree. By the time it nears maturity, the megaspore may consist of several thousand cells and two to six archegonia, which are located at the end of the structure that is near the micropyle. Each archegonium contains a single egg. *The archegonium structure is the female gametophyte of the pine tree.*

Male Reproductive Structures

■ **The male (or staminate) cones, which are found near the base of the pine tree, produce pollen grains within the pollen sacs located on the lower surface of the cone scales. The microspore mother cells located within the pollen sacs divide, producing the pollen grains.**

The male (or *staminate*) cones of a pine tree are found in clusters on the lower branches and are fairly short-lived, living only a few weeks before they fall from the tree. During that time, the cones produce the pollen necessary for the ultimate pollination of the female gametophyte.

Two pollen sacs are located on the lower surface of each cone scale, or *microsporophyll.* The pollen sacs are also called *microsporangia.* Located in the microsporangia are the diploid **microspore mother cells,** each of which divides by meiosis, producing four haploid microspores. Each microspore develops into a pollen grain, consisting of an immature male gametophyte and two air sacs that add buoyancy when the grain is released into the air. The pollen grains produced by the pine tree are the spores of the sporophyte generation.

Once they are mature, literally millions of pollen grains are released by the male cones. (Have you ever parked your car near pine trees in the early spring, only to return to find it covered with a fine yellow coat of pollen?) The grains are released into the wind, and with luck, they land on a female cone, filter down between the scales, and are caught in the sticky secretions produced by the unfertilized ovules.

Pollination

■ **Pollination is the transfer of pollen from the pollen sacs on the male cones to the ovules on the female cones. Once a pollen grain reaches an ovule, it is drawn through the micropyle in the pollen drop, produces the pollen tube, and lives off the nutrients in the ovule until the eggs are ready for fertilization. Then two sperm are formed.**

Pollination, which in conifers results from wind blowing pollen through the air, is the process by which the male pollen comes into contact with the female structures of a plant. (Do not confuse pollination with fertilization — fertilization occurs only when a sperm produced by the pollen grain joins with an egg in the archegonium.) The pollen filters down between the cone scales and becomes trapped in a sticky fluid called the *pollen drop* secreted by the female gametophyte. Up until this point, the scales of the ovulate cone were widely separated. Once pollination has occurred, the scales begin to grow to close these gaps and to protect the ovules.

After pollination, the pollen drop begins to evaporate. As the pollen drop shrinks in size, it retreats through the micropyle, drawing the attached pollen grains with it. When the pollen grains reach the nutritive tissue, they germinate, each producing a structure called the **pollen tube.** The pollen tube extends through the nutritive tissue of the ovule and withdraws nutrients from it. As many as 12 to 15 months may elapse between pollination and fertilization. At the end of this period, the egg is reaching maturity, and the male gametophyte divides, producing two haploid sperm. This structure (pollen grain, pollen tube, and sperm cells) represents the male gametophyte of the pine tree.

Fertilization

■ **Fertilization occurs when sperm from the pollen grain unites with an egg, forming a zygote. Seed development follows.**

About 15 months after pollination of the female scale, the pollen tube reaches an archegonium. One sperm fertilizes the egg, forming a zygote. The other sperm disintegrates. In some cases more than one egg is fertilized and starts to form an embryo. Only one embryo will survive. After fertilization has occurred, the male gametophyte degenerates. The outer protective covering hardens, becoming the *seed coat.* A wing extends from the end of the seed opposite the micropyle. Once the seed is fully developed, the wing will aid in the seed's dispersal.

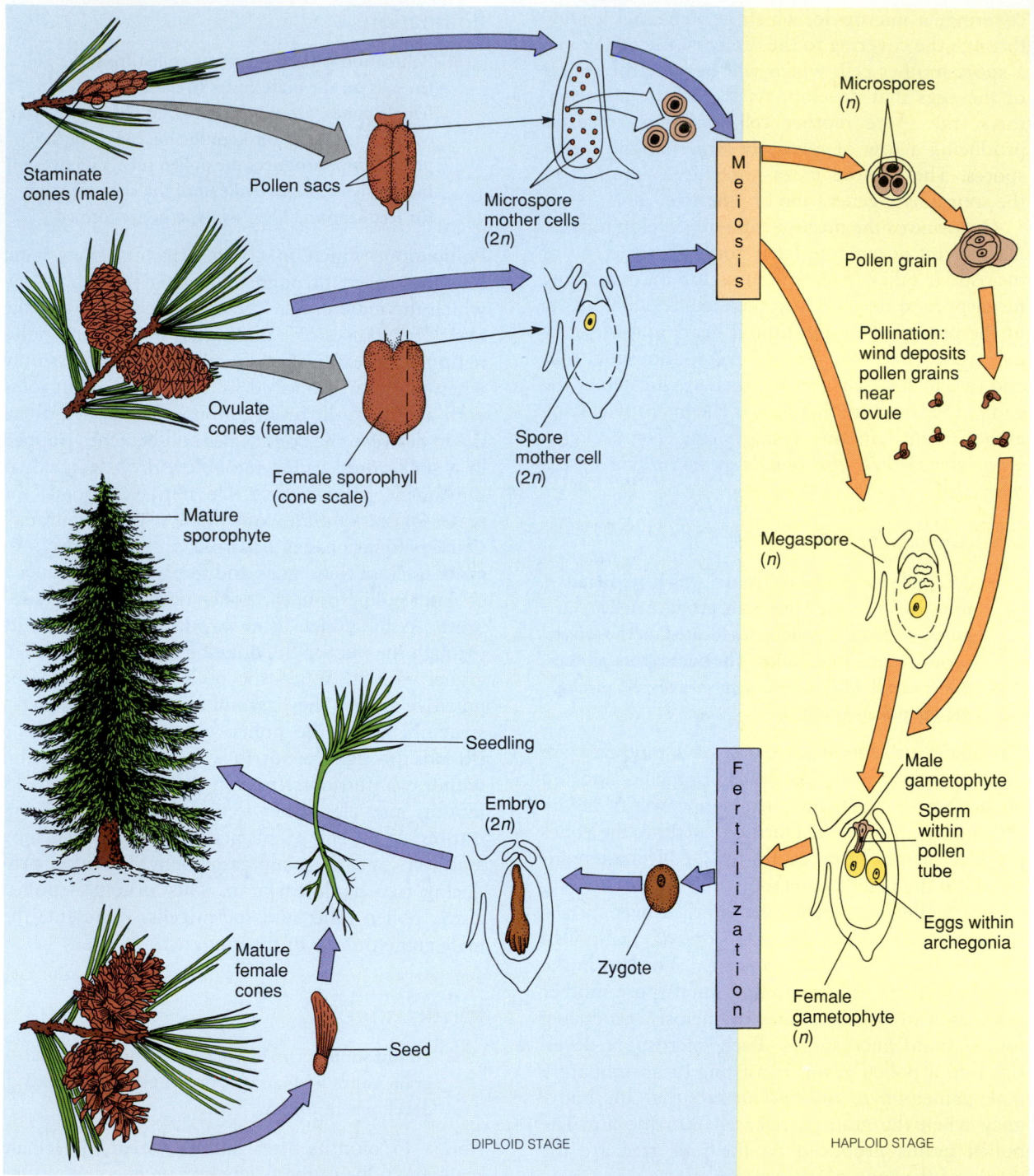

FIGURE 19.12   The Life Cycle of the Pine Tree
Each female cone contains a pair of ovules, in which a megaspore mother cell develops. After meiosis, the mother cell contains a row of large haploid megaspores; one develops into the egg. The male cones produce pollen in sacs, each containing diploid microspore mother cells, which divide by meiosis. The microspores then develop into millions of airborne pollen grains, one of which comes in contact with the female gametophyte and eventually reaches the gametophyte's nutritive tissue, where it germinates and forms the pollen tube. The pollen grain divides to produce two haploid sperm that travel through the pollen tube to the archegonium, fertilizing the egg. The resulting diploid zygote develops a hard seed coat, from which a wing sprouts. The wind carries the seed off; it is now well on its way to becoming a seedling that will complete the life cycle.

Chapter 19   Plants: A Survey

The seed will then give rise to the sporophyte generation of another pine tree. Figure 19.12 presents the life cycle of the pine tree.

ANGIOSPERMS: COVERED SEEDS

■ **In angiosperms the gametophytes are contained within a special structure — the flower borne on the sporophyte, which is dominant.**

When you look at a flowering plant such as a rose bush or an apple tree, you are actually looking at the sporophyte generation of an angiosperm. The beautiful rose or apple blossom is, in reality, the angiosperm's reproductive structure. The evolution of flowers provided a method for pollen distribution that is not dependent on water or even solely on the wind. The evolution of fruits provided a means of seed protection and dispersal that again was not dependent solely on water and wind.

Like all other plants, angiosperms undergo alternation of generations. As with gymnosperms, the gametophyte generation is borne on and protected by the sporophyte generation. To learn about reproduction in gymnosperms, we needed to examine the cones that these plants produced. To learn about reproduction in angiosperms, we need to examine their two most distinctive characteristics: their flowers and their fruits.

Flowers

■ **The flower contains the stamens and/or pistils, which are the male and female reproductive structures, respectively.**

Flowers are not only beautiful but lure unwitting pollinators like bees or birds or even, in some cases, bats into them as they search for food. Their structures are similar, although there is, it seems, endless diversity in the forms those structures can take. In some plants, such as the tulip, the flower is said to be *perfect,* because it contains both male and female reproductive structures. In others, such as the willow, the flower is said to be *imperfect,* because it has only female or male parts. Therefore, some imperfect flowers contain only female reproductive structures (called *pistils*), and other flowers contain only male reproductive structures (called *stamens*). Basically, however, most flowers contain these elements: calyx, corolla, stamens, and pistil(s).

First, envision a closed bud, or unopened flower. The peduncle is the flower stalk. The receptacle is the expanded end of the peduncle to which the flower parts attach. The calyx, which is composed of several sepals, is the leafy covering that protects the bud. Now envision an opened flower. As the calyx opens, the petals forming the corolla are exposed. Inside the petals are the stamens, and at the very heart of the flower are one or more pistils. All these elements may be present, whether you are looking at a lily or a blossom from an apricot tree (Figure 19.13).

The Stamen: The Male Structure

■ **An anther of a stamen produces microspore mother cells that are the source of pollen. Each pollen grain contains two nuclei, and its surface consists of an often-ornamented wall perforated by apertures.**

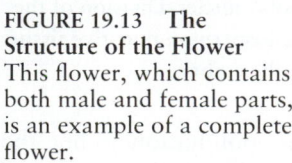

FIGURE 19.13 The Structure of the Flower
This flower, which contains both male and female parts, is an example of a complete flower.

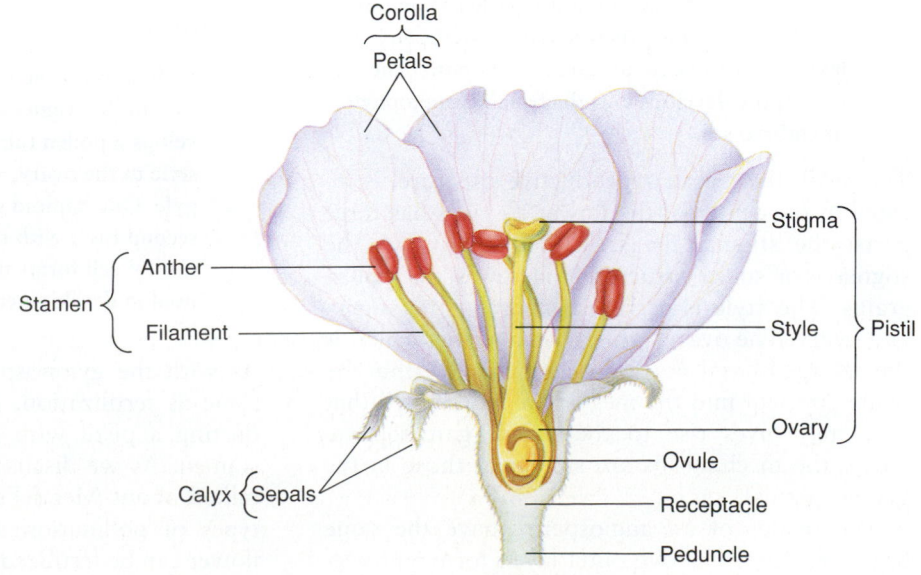

The **stamens** are the male reproductive structures of the flower. An **anther,** which contains numerous microspores, is often a cylindrical structure that is found at the top of a sterile stalk, the *filament.* The anther contains microspore mother cells, which divide, producing pollen grains. The pollen grains develop into the male gametophytes, just as the pollen grains of the pine tree did.

If you examine a developing anther, you may discern four distinctive regions, each of which turns into a chamber lined with a layer of nutritive tissues. Within each chamber are numerous microspore mother cells. By maturity the four chambers have become two, for two of the dividing walls have degenerated.

Figure 19.14 details the development of the pollen of a flower, formation of the egg, and fertilization. While the anther is maturing, the microspore mother cells are also developing. First they each undergo meiosis, forming four haploid microspores, which are the first cells of the gametophyte generation. The microspores may either be clustered together or separate, depending on the plant species. Then each microspore forms. Its nucleus divides by mitosis, and cell walls develop around each nucleus. The result is the mature pollen grain: a microspore containing two nuclei. Along the surface of some pollen grains are indentations, or thin areas, called apertures. Once the pollen grains are mature, they are shed from the anther.

The Pistil: The Female Structure

■ **The ovary of the flower is found at the base of the pistil. Within the ovary are ovules containing megaspore mother cells, which divide, forming haploid megaspores. A mature ovule has an outer covering (which is penetrated by a micropyle), a layer of nutritive tissue, and a megaspore. The megaspore develops into the female gametophyte, or embryo sac.**

The **pistil,** the female reproductive structure, is located in the center of the flower. A pistil has three parts: the stigma, the style, and an ovary. The **stigma** is a sticky structure that traps the pollen grains. The **style** is a stalk connecting the stigma and ovary. The ovary, which contains the ovules, is the enlarged basal portion of the pistil. Within the ovule you will find the megaspore mother cell that ultimately gives rise to the female gametophyte (again the mechanisms are similar to those in the gymnosperms).

The ovules of all angiosperms have the same basic structure. The two outer layers form a protective covering that ultimately will constitute the seed

coat. A micropyle appears at the end of the covering opposite the ultimate location of the developing egg cell. Beneath the outer protective layer is a layer of nutritive tissue. At the center of each ovule is a single megaspore mother cell.

The megaspore mother cell undergoes meiosis to produce four haploid megaspores, three of which degenerate. The one remaining megaspore now undergoes a complex process. The nucleus divides three times, producing one large cell containing eight nuclei. These nuclei are located in groups of four at either end of the megaspore. One nucleus from each grouping migrates to the center of the megaspore to form the *polar nuclei*. These nuclei may either fuse or remain separate until fertilization. In either case, once they fuse, they form a diploid cell. When that cell is fertilized, it becomes a triploid cell called the *primary endosperm nucleus*. Cell walls then form around the other nuclei. The result is an embryo sac consisting of eight (sometimes seven) nuclei in seven cells. See Figure 19.14 for details of the development of the female gametophyte (also called an embryo sac) of a flowering plant.

There are now three groups of nuclei. The two polar nuclei are in the central cell of the megaspore. The three cells nearest the micropyle form the central *egg* and two *synergids* (which are on either side of the egg). Usually synergids either degenerate or are destroyed when the egg is fertilized. The three cells at the opposite end of the megaspore are called *antipodals*. The antipodals ultimately degenerate. The entire structure is the female gametophyte, called an **embryo sac.**

Pollination

■ **Pollination is the transfer of pollen from the anther to the stigma of a plant. The pollen grain develops a pollen tube, which extends down the style to the ovary, where it penetrates the micropyle. One haploid sperm fertilizes the egg and the second fuses with the polar nuclei. Division of the central cell forms the endosperm, a nutritive tissue used in development of the embryo.**

As with the gymnosperms, pollination is not the same as fertilization. Pollination is the process of dusting a pistil with the pollen developed in the stamen. As we discussed in Chapter 14, when we talked about Mendel's experiments, there are two types of pollination: self-pollination, in which a flower can be fertilized by its own pollen, and cross-pollination, in which the flower is fertilized by the

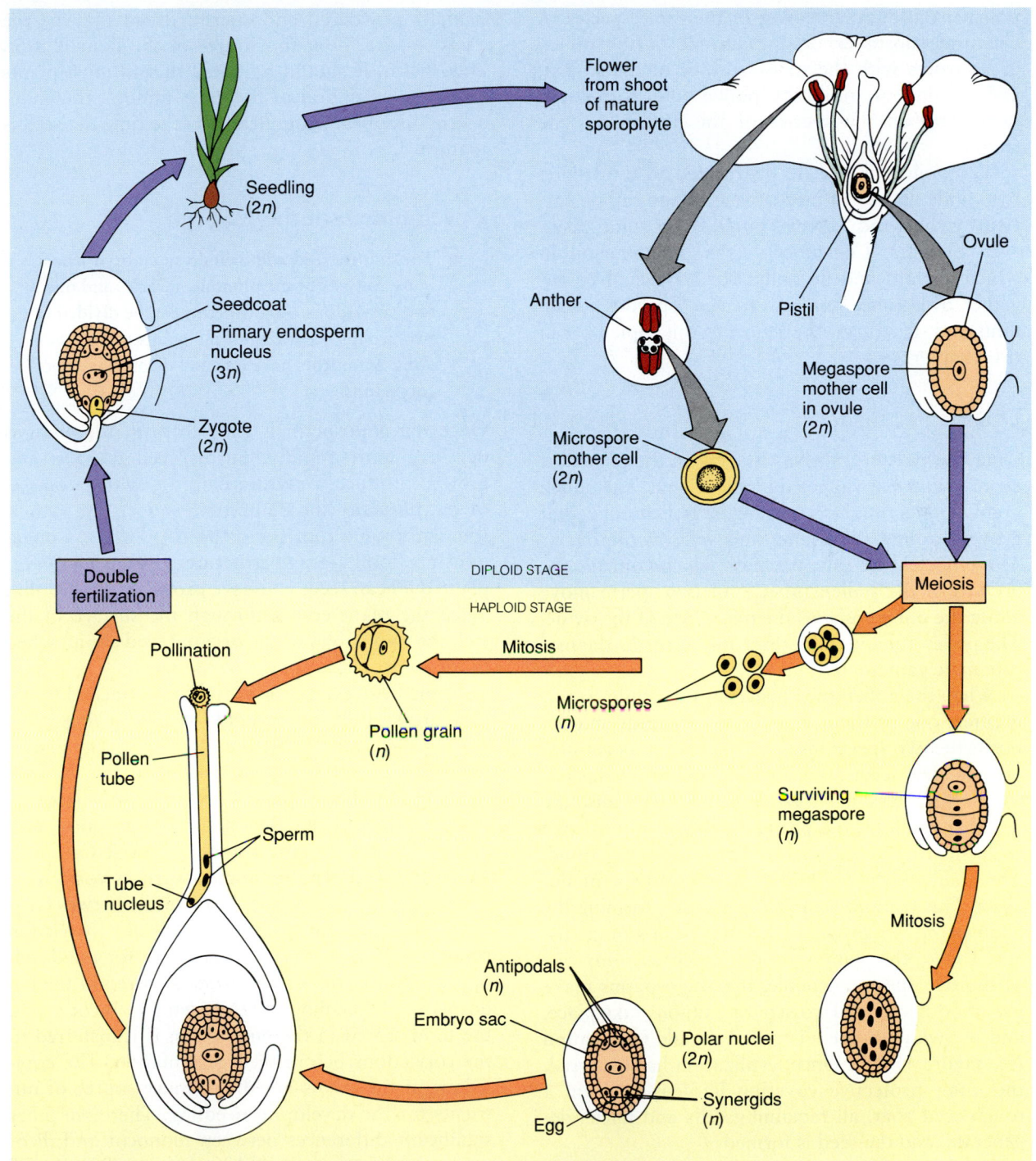

FIGURE 19.14 Development of the Male and Female Gametophytes and Fertilization
The left portion of this diagram details the development of a pollen grain. The microspore mother cell develops inside the maturing anther and undergoes meiosis to form four haploid microspores. Each microspore then divides mitotically. The pollen grain is released from the anther as an immature male gametophyte. Pollination occurs when the grain lands on the sticky stigma of the pistil.

The right side of this diagram illustrates the development of the egg. The diploid megaspore mother cell, housed within the ovule, undergoes meiosis to produce a single megaspore containing eight nuclei. Two of these, the polar nuclei, fuse to form the primary endosperm nucleus; three other nuclei form the antipodal cells and two others the synergids, which appear on either side of the eighth nucleus, the egg. Fertilization begins when the pollen grain germinates, forming two sperm and a pollen tube that eventually enters the ovary. One sperm fertilizes the egg, the other the polar nuclei.

pollen from another flower of the same species. A chemical substance on the surface of the pollen grain reacts with the sticky surface of the stigma, encouraging or inhibiting pollen tube formation, depending on the species of the pollen and the stigma.

The pollen grains are distributed to the stigma by a wide variety of mechanisms. They can be carried by the wind, insects, birds, or in some cases, even by bats. *Coevolution* is a phenomenon in which a plant and its pollinator have evolved together over time, resulting in very specific attractants and structures. We return to this topic later in this chapter.

Double Fertilization

Once the pollen grain has lodged on the stigma, it absorbs nutrients secreted by the stigma. The pollen grain will germinate, forming a pollen tube that grows through the stigma and style to the ovary. Three nuclei, the tube nucleus (which controls the growth of the pollen tube), and two sperm move along the pollen tube to the micropyle of the ovule. The pollen tube with its three nuclei forms the mature male gametophyte.

When the pollen tube penetrates the embryo sac, it enters one of the degenerating synergids and deposits the two sperm inside it. One sperm penetrates the egg. The second sperm joins with the polar nuclei in a process called *triple fusion* — because the cell that results is the product of the fusion of two polar nuclei and one sperm. Three cells have fused. The result is the triploid primary endosperm nucleus. This triploid central cell divides, forming the **endosperm,** which is multicellular food storage tissue. Unlike the gymnosperms, which use only one of the two sperm available, the angiosperms make use of both sperm. Two separate unions take place, and the whole process is called **double fertilization** (see Figure 19.14). Once fertilization has occurred, the outer protective covering hardens to form a tough seed coat, all remaining cells and nuclei degenerate, and the seed is formed

Endosperm. In many plants, at the same time that the seed is being formed, the triploid endosperm is developing. The endosperm surrounds the embryonic plant and provides it with nourishment. Until the endosperm is fully formed, the embryo develops slowly. When you look at a seed of a grain plant such as wheat, you really are looking at a body that is about 80 percent endosperm. When we grind wheat to produce flour to make bread, the result is

actually powdered endosperm. Bread may be the "staff of life," but the source of the flour is a $3n$ endosperm. It should be noted that in most plants (with the exception of the grass family), the endosperm disappears completely by the time the seed is mature.

Development of the Embryo

■ **The mature seed will include an embryo whose body consists of the plumule, radicle, and one or two cotyledons. Angiosperms can be divided into two groups according to the number of cotyledons. Monocots have one and dicots have two cotyledons.**

Once the endosperm is fully formed, the embryo develops more rapidly. Further cell division and growth produce a mature embryo, which consists of the *plumule* (the embryonic shoot), the *radicle* (the embryonic root), and the *cotyledons*. **Cotyledons** are food storage structures. If you have ever planted a bean seed, you have probably noticed that when the plant erupts through the surface of the soil, the two halves of the original seed are attached to the stem. These are the cotyledons.

In the bean plant, the cotyledons attached to the stem grow above the ground for a short time, then wither, die, and fall off the plant. In other plants, however, the cotyledons remain intact, so they are considered to be the plant's first functional leaves.

Botanists divide the angiosperms into two groups according to the number of cotyledons present in the seed. The **monocots,** such as corn, have only one cotyledon, while the **dicots** have two. Generally speaking, in monocot seeds the cotyledon produces enzymes that are necessary for the developing seed to utilize the nutrients stored in the endosperm for seedling development. In dicot seeds, the food stored in the endosperm is transferred to the cotyledons before seed germination. The cotyledons of the dicots are the primary source of nutrients for the developing seedling. There are other significant differences between monocot and dicot plants, but since most are based on the growth and development of the plant after germination, we save that discussion until the next chapter.

Fruits

Would you believe that the U.S. Supreme Court was once called upon to determine whether the tomato was a fruit or a vegetable? The distinction had to be made because at the time there was a 100 percent tariff on vegetables — but none on fruits. In all their wisdom, the Supreme Court Justices recognized that from a botanical perspective, the tomato is a fruit, just as are green beans. On the other hand, the justices felt that according to common usage, fruits are sweet and vegetables are not. In that sense, the tomato is considered a vegetable. Given the fact that the justices were not making a botanical distinction but a legal one, they felt bound by common usage, not by botanical fact. Therefore, tomatoes were not exempt from taxation.

Here we are concerned with biological distinctions, so even though we think of fruits as juicy, sweet, and edible apples, peaches, and plums, from a botanical perspective a chili pepper is also a fruit, as are grains, many types of nuts, and (of course) tomatoes. A **fruit** is the ripened ovary of most flowering plants, and its purposes are to retain and protect the seeds until they are fully matured and to aid in seed dispersal.

In evolutionary terms, fruits evolved for more reasons than the protection of seeds. They also serve as a means of seed dispersal. Animals and birds eat the fruit, and the seeds pass through their digestive system, to be distributed later in the organism's droppings. This method of distribution is responsible for the appearance of so-called *volunteer* tomato plants that you find growing in a yard or field far from any garden. They are called volunteers because they were not deliberately planted by a gardener.

In fact, the seeds of some plants are unable to germinate until they have passed through the digestive tract of a living organism. Such seeds must be *scarified,* a process during which the seed coat is softened. Otherwise they will not germinate, no matter how appropriate the environment may be. Once seeds have undergone the appropriate conditions, they do germinate.

Seed Germination

In many ways, the seed coat acts like a mechanical barrier, preventing the entrance of water and nutrients necessary for germination. In some cases, the seed coat even produces chemical inhibitors that prevent growth. However, once a seed is mature, the seed coat begins to allow water to penetrate. The seed then absorbs large quantities of water and may actually double in size. Ultimately the seed coat splits, and the embryo of the young sporophyte begins germination and development.

A survival mechanism that has allowed seed plants to colonize the land successfully is that a seed can remain dormant for years, germinating only when environmental conditions are correct. In fact, a few seed plants have been successfully germinated after hundreds or even thousands of years of dormancy. Botanists germinated 10,000-year-old seeds of a small flowering plant called *Lupinus arcticus*, found frozen in the ice in the Yukon.

Normally, however, the longevity of well over 99 percent of all seeds is less than 100 years. During their dormant stage the cells of the seed do not demonstrate any evidence of appreciable respiration or other metabolic activity, existing in a virtual state of suspended animation. Then, when conditions are right — in other words, when the appropriate amounts of moisture, temperature, oxygen, and light are present — the seeds resume metabolic activity, begin to grow, and finally germinate.

COEVOLUTION

■ Coevolution is the process by which two different organisms evolve in such a way that each can bring selective pressure on the other. This is the case with flowers and the birds, insects, and mammals that function as their pollinators.

In this chapter we have mentioned on several occasions that the life cycles of plants are highly integrated with those of other organisms, whether bees or bats for pollination or the digestive system of bears for scarification. When two highly different organisms, such as a plant and a particular kind of ant, evolve in such a way that one cannot succeed without the other, biologists say that the organisms have undergone **coevolution.** Through a process of chance, genetic mutation, and environmental pressures, the organisms come to exert heavy selective pressures on each other. Over generations a change in the plant — in its overall structure or its flowers — gives animals (or pollinators) with particular abilities a selective advantage in obtaining the food they need. Changes in the insect or animal pollinator have given certain blossom shapes or certain surface textures selective advantage. The process continues until, in some instances, the interdependence between plant and pollinator is so strong that the plant has a high probability of extinction if the pollinator disappears, and vice versa. Figure 19.15 and Focus 19.1 provide us with some interesting information regarding the relationship of the pollinator and the flower.

As we said at the beginning of this chapter, humans have a tendency to think plants are duller than, say, wolves or whales or tigers. After all, no

Plants, Life's Sexy Creatures

Normally, people don't think of plants as sexy, but they are. Some are asexual, others bisexual, and still others prefer a threesome. Some prefer mating with themselves, rather than mating with others. For example, there is a fern called the "walking fern," which clones little fernlets at the tip of each frond. Whenever the fernlets touch the ground they separate from the frond and become new plants.

Our favorite lawn foe, the dandelion, also prefers doing it alone. Through an asexual process called *vegetative propagation*, fragments of roots are able to develop into entire plants that are clones of the original. Dandelions even carry their asexuality further, by, believe it or not, producing asexual seeds. Without fertilization, the egg develops into an embryo — a process called *apomixis*.

Then there are the plants that prefer to mate with the assistance of a third party — the pollinator. Since mating is difficult when you have your roots firmly attached to the ground, plants have evolved devious schemes to attract the unwary assistant whose responsibility is to ensure the cross-pollination necessary to continue the plant's life cycle.

The giant arum lily, for instance, gives off a very putrid smell, much like the smell of rotting animal flesh. The odor is so strong that some people literally faint from the stench. However, one animal's poison is often another's perfume — in this case, the carrion beetle, which is a carrion eater. When this insect smells carrion, it thinks "food!" and is off to investigate. It searches the blossom, rubs its body against its pollen-carrying anthers, then carries that pollen to another flower. Because of the specialized nature of the attractant in the arum lily, the carrion beetle serves as the lily's only pollinator.

Plants utilize other subtle machinations to gain the cooperation of others in pollination. A species of orchid, for instance, produces such a potent nectar that it intoxicates any bee that drinks it. As the bee staggers about, it falls into the blossom, which is a very slippery surface. There is a way out, however — through a small channel at the base of the flower. The catch is that as the bee squeezes out through the channel it must brush against the pollen-containing structures along the way. When it flies off to another orchid, the process is repeated. When the bee falls into the blossom, it pollinates the stigma. When it escapes through the channel, it carries more pollen with it. Thus the process of cross-fertilization is carried out by a drunken bee (Figure A).

Some plants even go so far as to attract nocturnal rats, monkeys, and other mammals to serve as carriers of their gametes. The pendulous flowers of *Blakea chlorantha,* a New World jungle plant, produce sweet nectar only at night. The nectar attracts nocturnal rats, who serve as pollinating agents as they go from flower to flower.

Some flowers are very picky about the identity of the third party in their love triangle. When a particular plant and a particular pollinator depend on each other for survival, they are said to have undergone *coevolution*. In other words, the pollinator and the plant have been affected by similar selective pressures. This mechanism has produced relationships in which only one type of pollinator can pollinate one type of plant. This exactness prevents "intermarriage," or cross-fertilization, and so each plant protects its genetic continuity.

Coevolution is evidenced in a wide variety of forms. Some plants use deceit, rather than reward, to entice their pollinators. Some orchids sexually arouse male insects by mimicking the smell, touch, color, or appearance of a female insect. The males of the species are attracted to the blossom, and when they try to copulate with it, they are brushed with pollen and leave pollen from an earlier encounter. So instead of perpetuating its own species, the insect actually ensures fertilization of the blossom. This mechanism is called a *pseudocopulation attraction device* (Figure B).

Some plant blossoms utilize a *prey deception device*. In other words, they mimic the prey of an insect pollinator. When the pollinator stings the "prey," it transfers pollen. Other flowers mimic and serve as food. One of the most interesting forms of deception (called *pseudoantagonism*) is exhibited by the orchid blossom of *Oncidium stipitatum*. As the blossom blows in the wind, it looks like a swarm of bees. Real bees perceive this as an infringement on their territory and attack the orchids. In the process, they also conduct pollination.

Sometimes coevolution is so complex that the relationship is a symbiotic one — a relationship that is of mutual benefit to both organisms. This is the case with the yucca plant, which grows in the American southwestern desert, and its pollinator, the yucca moth. The proboscis, or sucking mouth part, of the adult female moth exhibits the special curve that is needed to harvest the pollen of the plant's cream-colored flowers. The moth rolls the pollen into a ball, then flies off to another flower. There the moth goes to the bottom of the flower and inserts its eggs into the base of the pistil, in several of the ovules. The moth then climbs to the top of the stigma and inserts the pollen ball into the structure to ensure that the flower's seeds will be fertilized and develop. The moth repeats the process until she must be exhausted.

The ovules go through their process of development, and those that contain moth eggs grow particularly large. When the eggs hatch, the caterpillars eat through the seeds. Not all seeds contain moth eggs, however, and so they develop normally, thus ensuring that the plant species will continue. Without the yucca, the moth would lose the source of nutrition for its caterpillars. Without

FIGURE A
This rare bucket orchid produces a potent nectar that intoxicates bees that drink it. As the bee falls into the flower, the only way out is through a channel in the center of the flower. The struggling bee backs up against the anthers, which contain pollen; the drunken bee is now laden with pollen, ready to pollinate another orchid.

FIGURE B
The slipper orchid emits a scent that mimics the odor of the female ichneumon wasp. The male wasp is attracted to the flower and tries to mate with it, picking up pollen which it then carries to other orchids.

the moth, the yucca would lose its pollinator. The extinction of one would result in the extinction of the other.

Nowhere in the living world has coevolution operated to such a great degree as between pollinator and flower. Each participant serves as a strong selective force on the other. So consider the alternatives in the plant world. Plants can be sexually interdependent, or they can be dependent on their pollinator. Or they can be completely nonsexual and simply break into pieces and produce replicas of themselves. Given their variety, plants are truly nature's sexy creatures.

Source: Adapted from *Science Digest*, October 1983, pp. 58–61.

(a)

(b)

(c)

(d)

(e)

FIGURE 19.15   Coevolution in the Plant World
Flowering plants depend on the assistance of a variety of pollinators in completing their life cycles. (a) The blossoms of the saguaro cactus of the American southwest get a boost from gila woodpeckers, which are attracted by their whiteness. (b) The color of many flowers also attracts bees. (c) The coevolutionary relationship of the sphinx moth and the tobacco blossom is evident in this photo. What other insect could have a tongue long enough to reach the blossom's nectar? (d) The hummingbird's beak plays a similar role in obtaining nectar from this flower. (e) Normally the substance produced by milkweeds keeps predators at bay — all but monarch butterflies, which lay their eggs on milkweeds and pollinate the flowers.

one has ever been attacked by a marauding violet. However, when you consider the fact that most plants would get along quite well without us and that we would most surely become extinct without them, they deserve a little more respect than we realize. We need to know the variety of ways in which they propagate to ensure their continuation. Knowing the numbers of adaptations that have brought plants from the sea to the land gives us an appreciation for the seemingly infinite number of ways that life has persisted on our planet. Humans need seeds and fruits. We need algae and even lichens. We also need mature plants, and in the next chapter we will examine the ways in which plants grow and develop to form the grasses, the food crops, and the trees we find all around us.

Once a seed germinates, the plant begins its process of growth and development. We end Chapter 19 at this point, since we will discuss the structure and growth of vascular seed plants in the next chapter.

SUMMARY

1. Plants are eukaryotic, photosynthetic, multicellular organisms. Plant cells are surrounded by a cell wall composed primarily of cellulose. With minor exceptions, plants have a life cycle in which the sporophyte and gametophyte generations alternate.

2. In the gametophyte generation, plant cells are haploid and, through mitosis, produce male and female gametes. In the sporophyte generation plant cells are diploid and produce spores through meiosis.

3. For the most part, aquatic plants rely on the water around them to support them against gravity and to provide a medium for fertilization.

4. The algae represent the oldest plant groups on earth. These plants are virtually nonvascular, have no true roots, stems, or leaves, and do not produce an embryo. They exhibit a variety of forms, from unicellular organisms to large sheetlike plants, and in this book are grouped into three major divisions, depending on pigmentation and reproductive method. The divisions are the Rhodophyta (red algae), Phaeophyta (brown algae), and Chlorophyta (green algae).

5. In the algae the gametophyte and sporophyte generations may appear alike. In addition, some species produce identical gametes, called isogametes. Other plants produce gametes that are not identical, called heterogametes.

6. Bryophytes, such as the mosses, were among the first plants to invade the land. Fertilization takes place through the medium of water, and the gametophyte generation occurs in a small structure close to the ground. The stemlike sporophyte generation enables the spores to be distributed by the wind.

7. Vascular plants have true roots, stems, and conductive tissue. Nonvascular plants do not.

8. In nonvascular plants like bryophytes, specialized multicellular structures develop in the gametophyte generation. The antheridium produces sperm and the archegonium produces the egg. In most of these plants the embryonic sporophyte generation is nourished by the gametophyte generation as it develops from the fertilized egg in the archegonium. The gametophyte is the dominant generation in the mosses and liverworts.

9. In primitive vascular plants like the ferns, the sporophyte generation is dominant and the gametophyte generation is reduced to a multicellular structure known as the prothallus.

10. Homosporous plants produce only one kind of spore, which develops into the body of the gametophyte generation. Most vascular plants are heterosporous; they produce two kinds of spores. The megaspore develops into the female gametophyte and the microspore develops into the male gametophyte. Heterospory is typical of seed-producing plants.

11. There are two major types of seed plants. Gymnosperms produce unprotected seeds located on the scales of cones and angiosperms produce seeds enclosed within a fruit.

12. The gymnosperms, such as pine trees, are nonflowering seed plants. The sporophyte generation bears both female (or ovulate) and male (or staminate) cones. The female cones produce the ovules that contain the eggs, and the male cones produce pollen. The gametophyte is highly reduced and borne inside the ovule.

13. The angiosperms are the most advanced land plants, and as with gymnosperms, the sporophyte generation is dominant. The female and male reproductive structures are found in the flower.

14. The female reproductive structure, the pistil, consists of a stigma, a style, and an ovary containing an ovule. Within the ovule the spore mother cells produce four haploid megaspores through meiosis.

15. One of the four megaspores will develop into the female gametophyte, which consists of the haploid egg, the central cell (containing the two polar nuclei), and five other cells (two synergids and three antipodals).

16. The male reproductive structure is the stamen, which consists of an anther and a filament. Within the anther, the microspore mother cells undergo meiosis, forming haploid microspores that develop into pollen grains.

17. After pollination, the pollen grain develops a pollen tube that grows to the ovary. Two sperm are found in the pollen tube. This structure is the male gametophyte.

18. During fertilization, one sperm merges with the egg and the other sperm joins with the polar nuclei of

the central cell, producing a triploid endosperm. The endosperm grows and provides food for the embryonic plant.

19. Seeds form in the ripening ovary or fruit in flowering plants. The seed consists of the seed coat, the triploid endosperm (in monocots), and the diploid sporophyte embryo with its seed leaves (cotyledons). Some plants like corn have only one seed leaf and are called monocots; others have two seed leaves and are called dicots.

20. Seeds can remain dormant for many years. Then through a combination of environmental and chemical processes they can germinate. During germination the seed coat splits and the young plant emerges.

KEY TERMS

sporophyte generation	microspore mother cell
spore	pollination
gametophyte generation	pollen tube
alga	stamen
isogamete	anther
lichen	pistil
mutualism	stigma
gymnosperm	style
angiosperm	embryo sac
homosporous	endosperm
heterosporous	double fertilization
pollen grain	cotyledon
seed	monocot
ovule	dicot
micropyle	fruit
spore mother cell	coevolution

REVIEW QUESTIONS

True/False

1. The life cycle of most plants involves two sequential generations that alternate one with another.
   true   false   (page 437)

2. The algae are vascular plants.
   true   false   (page 440)

3. The bryophytes are simple nonvascular plants that can be considered transitional land plants (sort of the amphibians of the plant world).
   true   false   (page 445)

4. The antheridium is the plant reproductive structure in which the sperm are produced.
   true   false   (page 447)

Fill in the Blank

1. The green alga Oedogonium produces _____ gametes. *(page 443)*

2. In alternation of generations, meiosis occurs in the formation of the _____. *(page 438)*

3. The development of the vascular system occurred in the _____ generation as the plants evolved onto the land. *(page 449)*

4. The flowering plants have covered seeds and are called _____. *(page 455)*

Multiple Choice

1. During the life cycle of a pine tree meiosis takes place in the
   a. Formation of the seed
   b. Formation of spores
   c. Formation of the gametes
   d. Germination of the gamete
   (page 454)

2. Compared with the mosses, the gymnosperms are better adapted to life on land because:
   a. They are homosporous.
   b. They have vascular tissue present in the sporophyte generation.
   c. They have vascular tissue present in the gametophyte generation.
   d. They have external fertilization.
   (page 450)

3. In rosebuds the gametophyte generation
   a. Is haploid
   b. Is not the dominant generation
   c. Is limited to structures within the flower
   d. All of these
   (page 457)

4. In the seed plants the gametophyte containing the sperm is found in
   a. The embryo
   b. The germinated pollen grain
   c. The megaspore mother cells
   d. The cotyledon
   (page 457)

5. One of the major evolutionary developments that has allowed for the survival of land plants is
   a. The seed
   b. Photosynthesis
   c. Diffusion
   d. Isogamy
   (page 459)

Discussion

1. What are plants?

2. Discuss the major evolutionary trends that occurred as the plants moved onto the land.

3. What are the major characteristics of the green algae?

4. In a generalized sense, discuss alternation of generations in the plant life cycle.

5. Review the life cycle of a moss and fern. Be sure to include the events that occur in the gametophyte and sporophyte generations.

6. From an evolutionary perspective, why are the gymnosperms better adapted for life on land than the ferns?

7. Summarize the life cycle of a pine tree and discuss the difference between a pine tree life cycle and a moss and fern life cycle.

8. Draw and label a typical flower. Be sure to indicate which structures produce the male gametophytes and which produce the female gametophytes.

9. Discuss pollination in flowering plants. Be sure to include the events that result in fertilization of the egg and polar nuclei.

10. Draw and label a typical seed, providing both an internal and an external view. From an evolutionary perspective, why are seeds considered an advanced plant characteristic?

20

Plant Structure and Growth

All of us have planted seeds and marveled at the way they germinate and develop into mature plants. Most of us have driven through rich farming areas and seen our agricultural abundance. Almost everyone has been awed by the beauty of a rose. Many of us have purchased new seedlings at a store or a garden center, removed them from their containers, looked at their roots, cut back their stems and leaves, soaked them in water, and planted them in our home or garden. And as you have gone through all these activities, you may have wondered what the major structures that make up the typical seed plant are, and you may have been curious about how plants grow.

The answers to these questions will be our main concern in this chapter. We will emphasize the flowering plants (the angiosperms) as we examine the plant body and study its growth. Remember that there are two major types of flowering plants — those that have one seed leaf (the *monocotyledons*) and those that have two seed leaves (the *dicotyledons*). The differences between these two major types of flowering plants are summarized in Figure 20.1.

THE TISSUES MAKING UP THE TYPICAL PLANT BODY

■ The typical flowering plant consists of two types of tissues: meristematic tissues, which consist of undifferentiated cells, and permanent tissues, made up of differentiated cells.

Above: Plant structures such as this rose add beauty to the world.

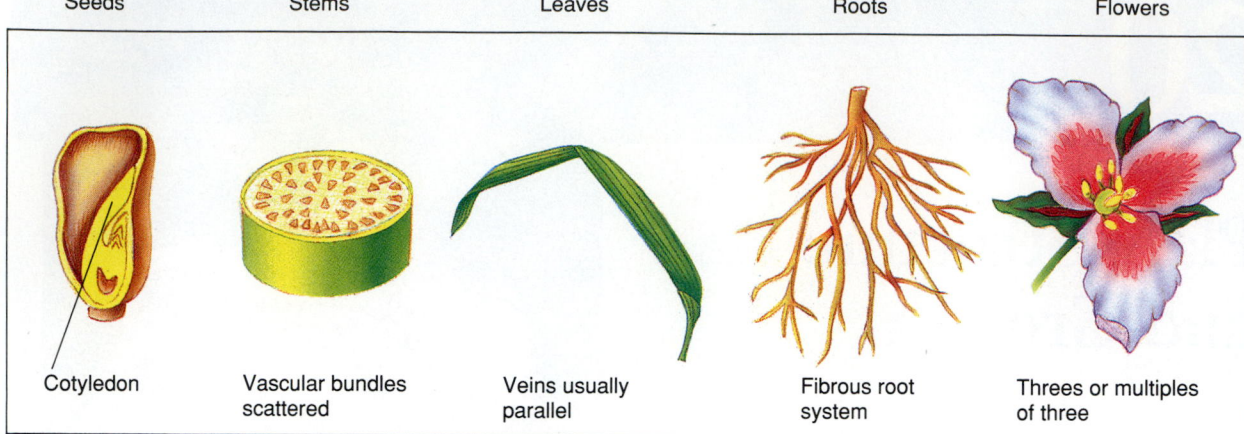

Seeds	Stems	Leaves	Roots	Flowers
Cotyledon	Vascular bundles scattered	Veins usually parallel	Fibrous root system	Threes or multiples of three

Monocots

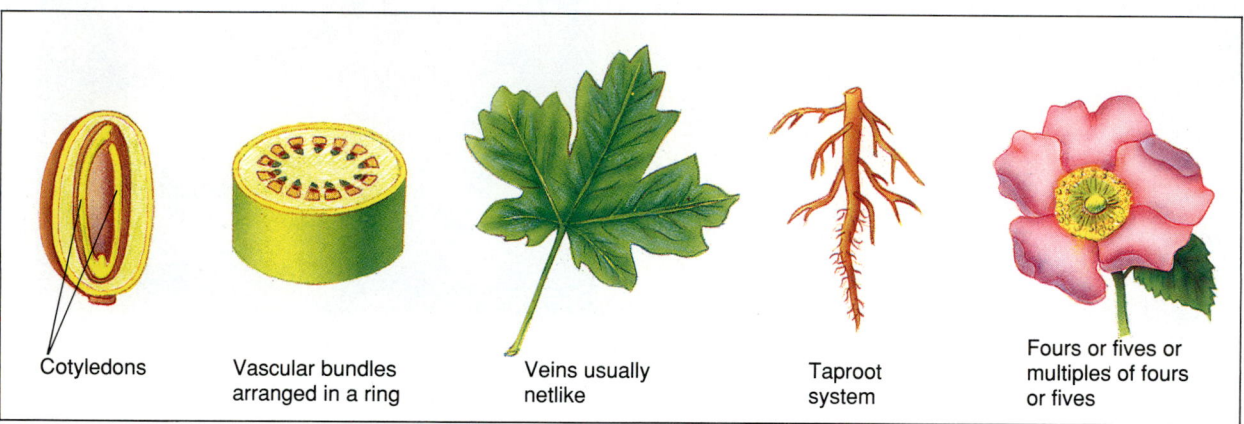

Cotyledons	Vascular bundles arranged in a ring	Veins usually netlike	Taproot system	Fours or fives or multiples of fours or fives

Dicots

FIGURE 20.1 One Seed Leaf or Two
As seeds, monocots have one seed leaf, or cotyledon; dicots have two. In monocots, the vascular bundles — the plant's supporting and conducting tissues — are scattered throughout the stem, while in dicots, they are arranged in a concentric ring beneath the outer layer (the epidermis). The monocot's leaf veins are parallel; the dicot's veins are netlike. The roots of monocots, which are fibrous, also differ from the taprooted dicots. The monocot's flower parts are in threes or multiples of three; the dicot's are in fours or fives, or multiples of four or five.

The typical mature flowering plant body consists of the following plant organs: roots, stems, leaves, and flowers (Figure 20.2). These organs are composed of **tissues,** or groups of cells that perform similar functions. Your body, for instance, has many specialized tissues, such as those that line the stomach or those that make up the liver. In plants there are two basic kinds of tissues: *meristematic tissues* and *permanent tissues.*

Meristematic tissues are made up of undifferentiated cells that give rise to new cells by means of mitosis — in other words, they are the tissues that provide all the new cells that are added to the plant body as it grows. Meristematic tissues are composed of *unspecialized cells,* cells that have not been incorporated into a particular organ and structured for a particular function.

Permanent tissues, on the other hand, are the result of differentiation of the cells produced by the dividing meristematic tissues. They are specialized, so that they perform a particular function. Within each of these general groupings are subcategories, which we need to know before we can discuss plant growth and development.

Flower

Fruit

Node

Internode

Node

Blade

Petiole

Stem

Leaf

Shoot system

Root system

FIGURE 20.2   A Generalized Drawing of a Flowering Plant Body
In this plant, the above-ground shoot system is made up of the stem, leaves, flowers, and fruit. Leaves appear at the stem's nodes; the area of the stem between nodes is called the internode. A leaf is composed of the blade and the petiole, which connects the blade to the stem. The below-ground root system supplies water and nutrients to the shoot system.

Meristematic Tissues

■ **There are three basic types of meristematic tissues. Apical meristems, found at the tips of roots and stems, are responsible for the lengthwise growth of these structures. Lateral meristems include the vascular cambium and cork cambium and are responsible for growth of the plant's diameter. Intercalary meristems are responsible for growth at the nodes in the grasses.**

There are three basic types of meristems: *apical meristems, lateral meristems* (including the *vascular cambium* and *cork cambium*), and *intercalary meristems.* **Apical meristems** are found at the apex, or tip, of a developing root or shoot. If you have ever grown a plant from a seed, you saw that plants grow outward from the tip. What you were observing was the result of the activities of the apical meristem, which is responsible for lengthening the root or stem (Figure 20.3).

Did you ever carve your initials on the bark of a tree? If you returned after several years, you would find that even though the tree had grown taller, the initials were still at the same height at which you had carved them. Obviously the tree trunk didn't grow lengthwise all over; it grew only from

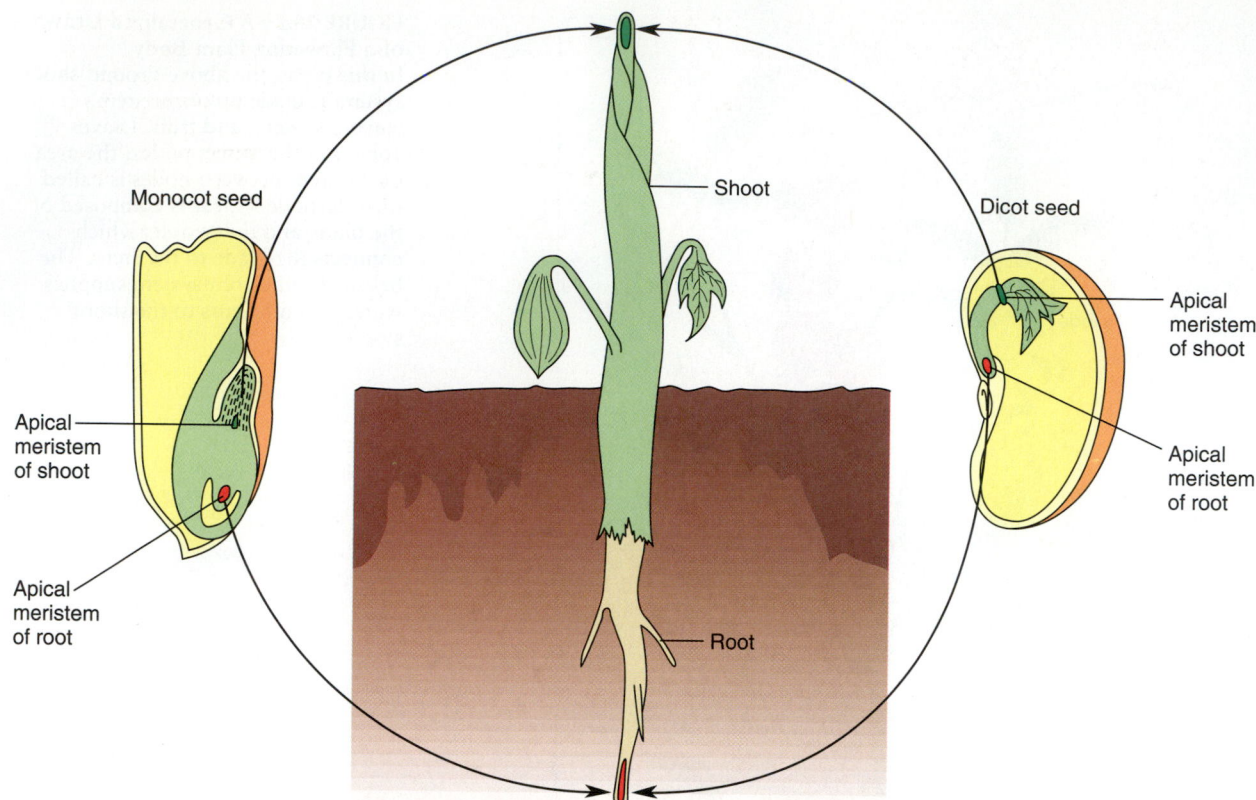

Monocot seed

Apical meristem of shoot

Apical meristem of root

Shoot

Root

Dicot seed

Apical meristem of shoot

Apical meristem of root

FIGURE 20.3 The Apical Meristem
As seeds germinate, the apical meristem cells divide, causing the shoot and the root to elongate. In this illustration, you can see the young leaves forming from the shoot and the roots forming below ground.

its tips, through the division of apical meristems. The term *primary growth* describes lengthwise growth resulting from the activities of apical meristems.

The second type of meristem is called the **lateral meristem.** There are two lateral meristems, **vascular cambium** and **cork cambium.** Vascular cambium and cork cambium are shaped much like long tubes running from the tips of the developing shoots to the tips of the developing roots. Cork cambium forms a casing for vascular cambium, like a straw within a straw. They are responsible for the development of the girth, or diameter, of a plant. Botanists refer to this type of growth as *secondary growth*.

Intercalary meristems are found in grasses and are responsible for the growth of their leaves or blades, which takes place at the bases of these structures. This growth pattern explains why you have to mow your lawn more than once. Grass leaves grow from their base, providing for the continuous elongation of their leaves.

Permanent Tissues

As we said earlier, permanent tissues have become specialized and perform a specific function. As with meristems, there are several types of permanent tissues: *surface tissues; storage, support, and secretory tissues;* and *conducting tissues* (Figure 20.4). Each of these categories can also be subdivided.

Surface Tissues

■ **Epidermal cells, which are highly variable structures, protect the outer surface of plant structures. They secrete fatty substances called cutin and wax, which also provide protection. Epidermal cells are the source of root hairs found on growing roots, and guard cells on either side of stomata on leaves. In plants having secondary growth, epidermal cells are replaced by the periderm.**

Surface tissues, called *epidermal tissues,* found at the outer surface of the plant interact with the environment. In most cases, the epidermis is only one cell thick, although there are exceptions, like the

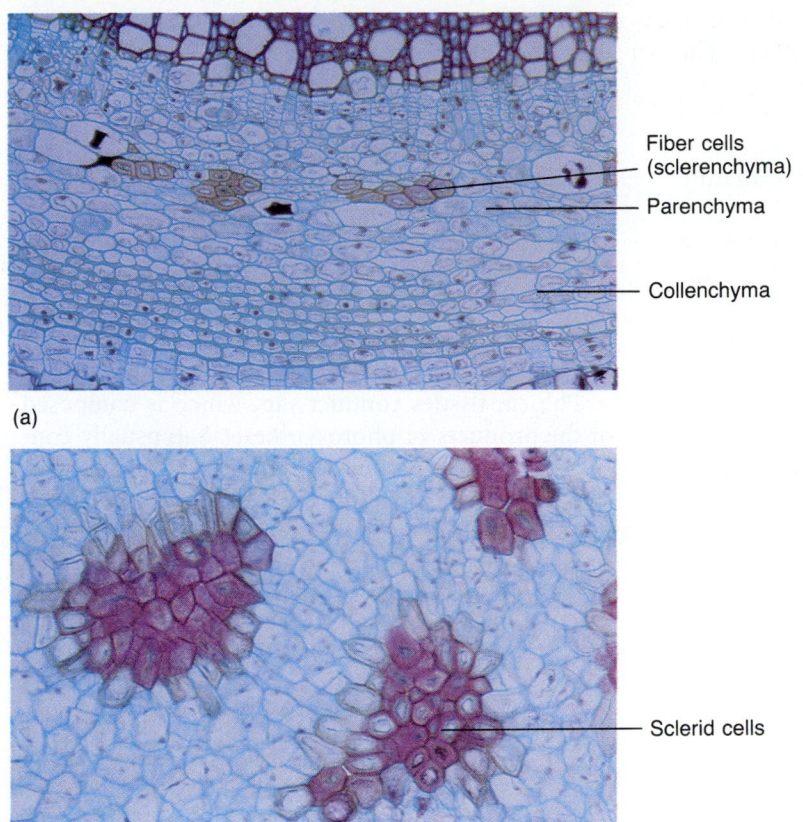

Fiber cells
(sclerenchyma)

Parenchyma

Collenchyma

(a)

Sclerid cells

(b)

FIGURE 20.4 **Types of Plant Cells**
(a) In this cross-section of a stem, three types of plant cells can be seen. The parenchyma cells, which are unspecialized with flexible cell walls, can assume a variety of shapes. The collenchyma cells, which are less flexible (note the thicker cell walls), provide support for the development of the young growing plant. Sclerenchyma cells are less flexible still, and are of two types; fiber cells, which provide support and strength and are made rigid because their cell walls are impregnated with lignin, and sclerid cells. (b) In this slide the thick cell walls of the sclerid cells can be seen. Sclerid cells are often found in seeds, fruits, and stems. The sclerenchyma cells provide support and protection to the plant.

leaves of some houseplants such as the rubber plant. Epidermal cells are the source of the guard cells on either side of a leaf pore called a *stoma* and are also the source of *root hairs*, both of which we will discuss shortly.

Periderm *tissues* are found in woody plants. Once the vascular cambium begins to develop, increasing the diameter of the plant, the epidermal cells slough off and are replaced by periderm cells produced by the cork cambium. Most periderm cells are cork cells, and they form the protective layer of bark that covers woody plants. They produce a fatty substance that makes them water resistant. Certain periderm cells form pores called *lenticels* that allow air and plant gases into and out of the interior areas of the plant.

Storage, Support, and Secretory Tissues

■ Parenchyma cells, the most abundant cells in plants, function in several ways. Some carry out photosynthesis; others store food. Collenchyma cells have thicker walls than parenchyma cells and function in plant support during primary growth. Sclerenchyma cells have very thick cell walls and are important in plant support once primary growth has ceased. Secretory cells produce plant hormones and other substances.

Parenchyma cells are the most abundant cell type in plants. Many contain chlorophyll and carry out photosynthesis; these are called *chlorenchyma.* Other parenchyma cells provide storage areas for water and food. The walls of parenchyma cells are very flexible and can assume a variety of shapes. Unlike cork cells of the periderm, which are dead at maturity (that is, they lack cytoplasm and cellular contents), parenchyma cells live for a long time after specialization. Because they retain the ability to divide, parenchyma cells are important in the healing of wounds and the initiation of new roots in stem cuttings.

Collenchyma cells are not as flexible as parenchyma cells and are less rigid than the cork cells of the periderm. In addition, collenchyma cells are often longer and have thicker walls (which are rich in pectins) than parenchyma cells. They often occur beneath the epidermal cells of the plant and are important in providing support for the developing primary plant body.

Sclerenchyma cells are tough, thick-walled cells. There are two basic types: *sclereids* and *fibers*. The cell walls of both are very thick and usually contain a hardening substance called *lignin*. Usually dead when they are mature, fibers and sclereids provide mechanical support and protection in the mature plant body.

Secretory cells secrete substances that perform a variety of functions in a plant. Some release *plant hormones,* regulatory chemicals that affect the plant's growth and development as well as many of its responses to its environment. (Plant hormones are one of the subjects of Chapter 21.) Other cells secrete waste products. Still others are responsible for producing plant products that we all use, such as rubber and oils. Secretory cells are found in a variety of locations in a plant, depending on the species.

Conducting Tissues

- **Xylem and phloem are vascular, or conducting, tissues. The branched tubular tissue called xylem consists of fibers, parenchyma cells, ray cells, tracheids, and vessel elements. It is responsible for conducting water and dissolved minerals from the roots to all other live cells of the plant. The branched tubular tissue called phloem consists of sieve tube elements, companion cells, and sometimes fibers. It is responsible for conducting the products of photosynthesis to nonphotosynthetic structures.**

There are two types of conducting **vascular tissues:** *xylem* and *phloem*. Xylem conducts water and dissolved minerals from the roots to the leaves. Because the cells of xylem have thick cell walls, they also provide support for the plant. Phloem, on the other hand, conducts the products of photosynthesis throughout the plant, usually from the photosynthetic organs to the roots. During the spring, when the sap rises in the plant, the phloem conducts the stored products of photosynthesis from their storage place (usually in the roots) upward through the plant. In some plants, such as the sugar maple, the xylem also stores large amounts of carbohydrate. In the spring the stored carbohydrate is transported by the xylem as well as the phloem.

Xylem tissue consists of fibers, parenchyma cells, ray cells, tracheids, and vessel elements. At maturity, both vessel elements and tracheids are dead. The cell walls remain, however, and function as tubes for the continuous upward transport of water and minerals. Ray cells, composed primarily of parenchyma cells, are the plant's primary means for transporting minerals from side to side.

Tracheids are long, thick-walled cells having tapered ends that overlap one another. The overlapping areas contain thin, porous spots called *pits,* which lack secondary cell walls. Water and dissolved minerals pass from one cell to another through the pits of the tracheids.

Generally speaking, vessel elements are shorter and wider than tracheid cells. At maturity, the ends of a vessel element break down completely. As a result, vessel elements are stacked end to end like oil drums, forming a continuous tube for conducting water and dissolved minerals (Figure 20.5).

Phloem tissues conduct sap, which is composed of the products of photosynthesis. Sap usually consists of sucrose and minor amounts of other organic substances. Sap moves through the phloem from the plant's photosynthetic regions (usually the leaves) to the nonphotosynthetic regions. Two major cell types are found in the phloem of angiosperms; sap-conducting cells called *sieve tube elements,* and *companion cells* (Figure 20.6).

The end walls of each sieve tube element (called *sieve plates*) have openings through which the products of photosynthesis flow from one element to the next. Sieve tube elements, stacked end to end, form a long column called the *sieve tube*. At maturity, a sieve tube element loses its nucleus. This is an advantage, because it leaves few cellular structures to block the sap's movment. Despite the absence of a nucleus, sieve tube elements are not dead. The action of their cytoplasm is a contributing factor in sap movement. So the question is, how does a sieve tube element continue the metabolic activities necessary to remain alive without its nucleus?

Botanists hypothesize that the answer lies in the **companion cells,** specialized parenchyma cells adjacent to the sieve tube elements. Since a companion cell has a nucleus and contains all the plant cell organelles, botanists believe that the nucleus of the companion cell functions for the sieve cell as well as for itself, providing the energy that sieve tube elements require for life. The way in which companion cells communicate with sieve tube elements, however, is not completely understood.

GROWTH IN SEED PLANTS

- **Some plant cells, such as parenchyma cells, are totipotent and can regenerate. Plants are capable of open growth. Plants exhibit two types of growth: primary and secondary.**

If you have ever had to do any yardwork, you have probably been amazed at how fast grass (and weeds) grow in a lawn. Without constant yard-

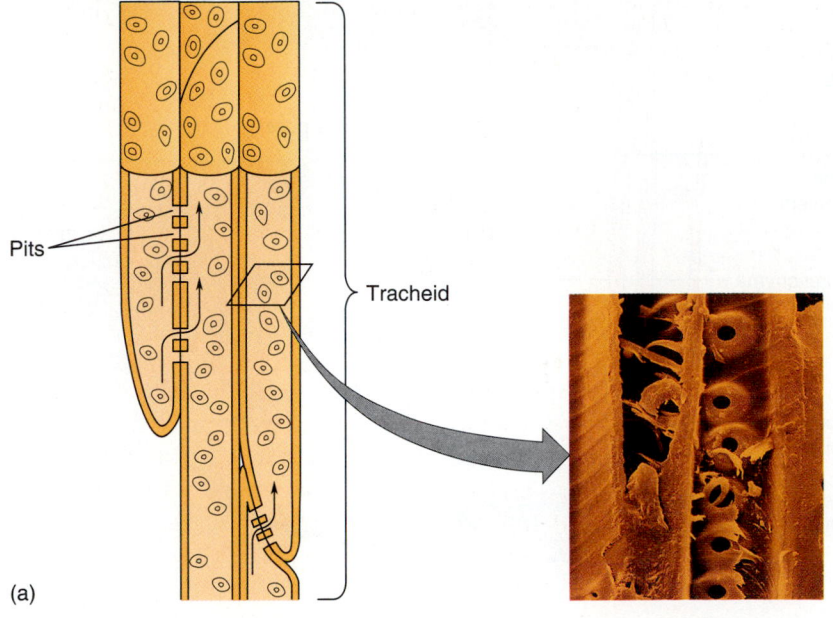

Pits

Tracheid

(a)

Perforated end plate

Vessel element

(b)

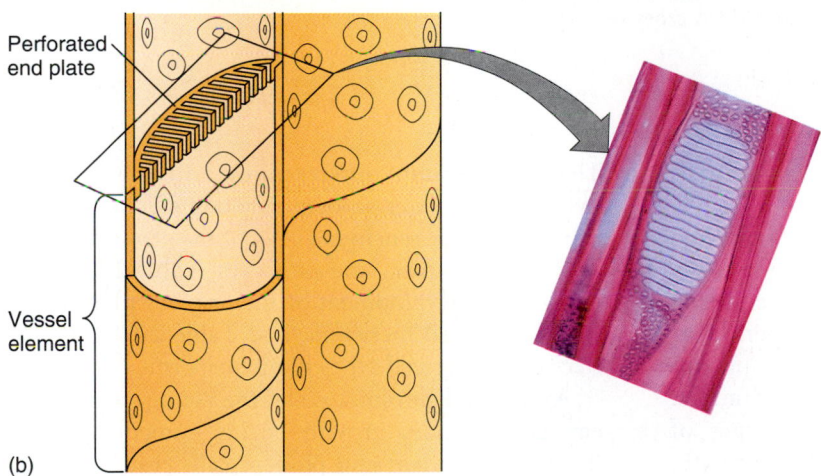

FIGURE 20.5 Xylem Tissue
In flowering plants, tracheids and vessel members are both dead at maturity. However, the cell walls remain to pass water from the roots throughout the shoot system. (a) Water moving from one tracheid to the next passes through pits, areas that have no secondary cell wall. (b) Water moves differently through vessel members, which have perforated primary walls where joined to other vessel members. If the walls break down entirely as the cells reach maturity, a single tube opening is formed.

work, you cannot keep plant growth in check. Obviously plants grow as animals do, but the pattern of plant growth is different. Early in the embryonic life of most animals the destiny of the cell is predetermined. That means that some cells become specialized nerve cells, others become muscle cells, and so on. Once an animal cell is committed to becoming a particular cell type, it usually cannot change its "direction" and develop into a different cell type. In addition, when the mature animal organism reaches a genetically predetermined size, the organism for all practical purposes stops growing.

In plants, however, the pattern of growth is very different. Some plant cells, such as the parenchyma, have the ability to regenerate (a capability also seen in some more primitive animals such as flatworms and sponges). In fact, studies have determined that many plant cells are *totipotent*, which means that one plant cell has the potential to regenerate an entire new plant. If you (or someone you know) ever started a new plant from a *cutting*, which is a single stem or leaf, you have observed the capacity of plants to form new roots, stems, and leaves from a different plant part, without the need for a seed.

Not only do plants have the capacity to regenerate, they also have the potential for continuous, or

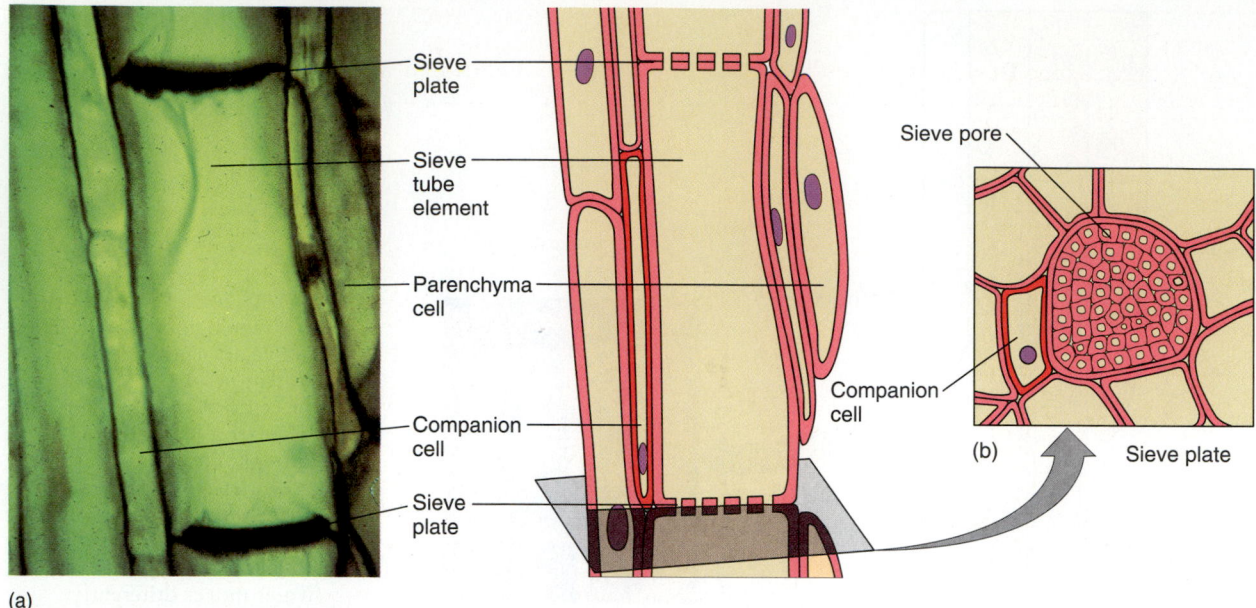

FIGURE 20.6   Phloem Tissue, Sieve Tube Elements, and Companion Cells
(a) A longitudinal view of a mature sieve tube element; its companion cell appears to the left. Note the thickened end walls. (b) A cross-section of a sieve plate.

open, growth. As long as the apical meristem is alive, most plants never stop growing completely for the duration of their life span, even at maturity. Do not think, however, that all plants can live forever. Annual plants die after one complete season of growth. Perennials can live over a period of years, but their meristematic tissues are inactive during the winter in temperate or cooler climates. As we saw earlier, plants exhibit two types of growth patterns: primary and secondary growth. We will look at those patterns next.

Primary Growth

■ **Primary growth results from cell division in the apical meristem. The cells produced by these divisions form three primary meristematic tissues: The protoderm differentiates into the epidermis; the procambium differentiates into the vascular tissue; and the ground meristem differentiates into the cortex, pith, and pith rays.**

During **primary growth,** the cells of the apical meristem divide, producing the three primary meristematic tissues: the protoderm, ground meristem, and procambium. The primary meristematic tissues will differentiate and form the mature primary tissues. The **protoderm** differentiates into the epidermis.

The **procambium** differentiates into the primary xylem and primary phloem. **Ground meristem** is the source of the parenchyma cells of the ground tissues — the cortex, pith, and pith rays. Table 20.1 summarizes the development of the primary plant body from the apical meristem, and Table 20.2 summarizes the major cell types in angiosperms.

As the seed germinates and the young sporophyte emerges, primary growth begins immediately. As the primary xylem and phloem cells elongate, they begin to differentiate into the sieve tube elements of the phloem and the vessel elements of the xylem. Two extra primary tissues are found in the root. An *endodermis*, or "inner skin," forms from the innermost cell layer of the cortex, enclosing the vascular tissue. Inside the endodermis is the *pericycle*, consisting of one or more layers of parenchyma.

Secondary Growth

■ **Secondary growth results from cell divisions in the vascular and cork cambium, which contribute to growth in the girth (or diameter) of the plant. The vascular cambium produces secondary xylem and secondary phloem. Cork cambium produces the layers of cork.**

Secondary growth starts with cell divisions in the undifferentiated meristematic cells between the

Chapter 20   Plant Structure and Growth

TABLE 20.1
A Summary of the Development of the Primary Plant Body from the Apical Meristem

Embryonic source	Develops into	Develops into	Functions
Apical meristem	Protoderm	Epidermis	Protection of internal tissues; in roots, water and mineral uptake; in leaves and stems, carbon dioxide through the stomata and oxygen exchange
	Ground meristem	Cortex	Stores food, supports plant parts
		Pith	Food storage
		Leaf mesophyll	Major photosynthetic region of leaf
	Procambium	Vascular tissue:	
		Primary phloem	Conducts products of photosynthesis
		Vascular cambium	Produces secondary phloem and xylem
		Primary xylem	Conducts water and dissolved minerals; supports plant parts

TABLE 20.2
Major Cell Types in Angiosperms

Cell type	Structure	Function	Location
Apical meristem	Small, thin-walled, six-sided	Growth in length	Stem and root tips
Lateral meristem	Long, six-sided tubes	Growth in diameter	Tips of developing shoots to tips of developing roots
Parenchyma	Many-sided, with thin, flexible walls	Photosynthesis, respiration, storage, healing, secretion, protection	Throughout the plant
Collenchyma	Long, thick-walled, slightly rigid	Support the developing plant	Between epidermal cells
Sclereids	Variable, tough, thick-walled	Mechanical support and protection	Throughout the plant
Fibers	Long, narrow, thick-walled; usually dead at maturity	Mechanical support and protection	Primary and secondary xylem and phloem
Tracheids	Long, thick-walled, with tapered ends and pitted walls; dead at maturity	Transportation of water and dissolved minerals	Primary and secondary xylem
Vessel element	Wide, with pitted and perforated walls; dead at maturity	Transportation of water and dissolved minerals	Primary and secondary xylem
Sieve tube element	Long, stacked in columns, with porous end walls and no nucleus at maturity	Transportation of products of photosynthesis	Primary and secondary phloem
Companion cell	Variable, long, connected to sieve tube element	Believed to execute metabolic activities for sieve tube element	Adjacent to sieve tube element in the phloem

phloem and xylem. Divisions in the vascular cambium help increase the girth of the plant, since they give rise to secondary xylem (wood) and secondary phloem (inner bark). The cells produced on the inside of the vascular cambium differentiate into the secondary xylem, and the cells produced on the outside of the vascular cambium differentiate into the secondary phloem (Figure 20.7). The secondary xylem and phloem function just like the primary xylem and phloem, moving water and minerals and food molecules throughout the plant.

As new vascular tissue is formed, the root, stem, or branch increases in diameter. The tissues located toward the outside of the vascular cambium compose the bark of a tree. As a tree grows older, the primary and secondary xylem that are toward the center of the trunk are increasingly nonfunctional and become the heartwood. Secondary phloem, on the other hand, becomes the innermost layer of the bark. If you have ever cleared a field of trees, you may have girdled a tree (cut away a strip of bark and some of the wood below it). In effect, what you

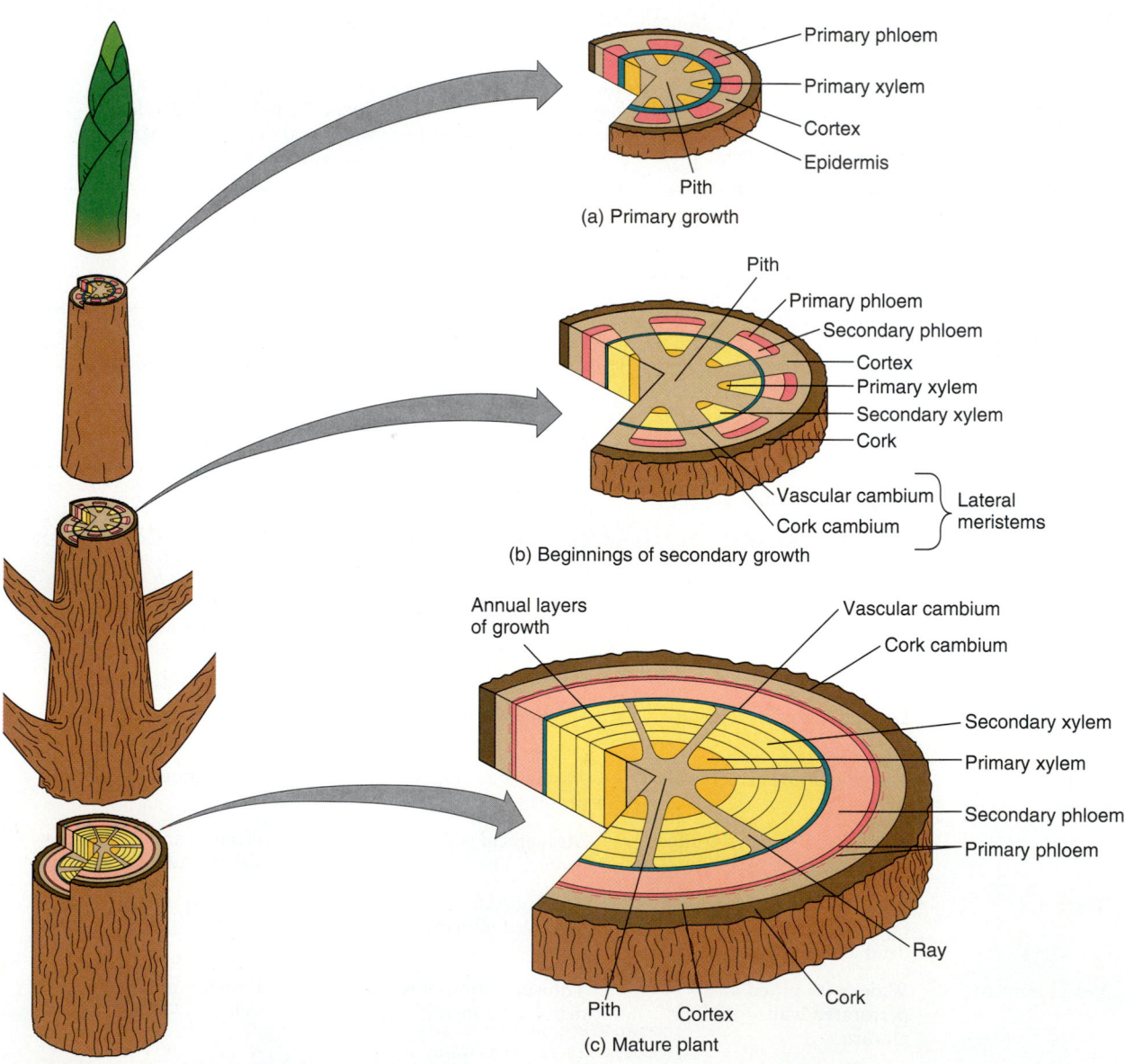

(a) Primary growth

(b) Beginnings of secondary growth

(c) Mature plant

FIGURE 20.7 **Growth in a Young Dicot**
(a) Primary growth features maturation of the shoot's vascular system into distinct vascular bundles. (b) At the beginning of secondary growth, the vascular cambium begins to produce secondary xylem and phloem. The epidermis gradually is replaced by cork from the cork cambium. (c) At maturity, growth layers are evident. Rays of living cells transport water and minerals laterally across the plant's stem.

Chapter 20   Plant Structure and Growth

were doing was cutting away the living phloem cells and the vascular cambium cells beneath it. In so doing, you cut off the transportation of food materials in the tree and killed the vascular cambium cells that could have replaced these phloem cells, thereby starving the roots until the entire tree died.

As the stems and roots increase in diameter because of secondary growth, the epidermis stretches and cracks. If the epidermis were not replaced, the young root or stem would die from lack of protection and dehydration. It is the function of the cork cambium to produce **cork,** a new protective outer covering. Cork cambium is formed by the pericycle of the root and in the cortex or other cells in the stem.

If you have ever cut down a tree or cut through a branch, you have seen the concentric circles that are often called *annual growth rings.* One ring, which is formed by new secondary growth, is produced during each growing season, when the plant grows in diameter. In fact, you can determine the age of a tree by counting the rings (see Focus 20.1).

Growth rings tell more than a tree's age. Each growing season leaves evidence of that year's climatic conditions etched in its annual ring. The width of each ring varies with environmental conditions. For example, in a year when there is just the right combination of sun, rain, and nutrition to ensure optimal growth, the width of the growth layer will be much greater than that of a ring developed in a year in which the systems of the plant could barely maintain life.

Growth rings can even be used to help date archaeological finds. Archaeologists sometimes discover pieces of wood of an unknown age at sites that they are investigating. By comparing the rings in the wood pieces with rings in samples with a known age from the area, they can determine the age of the site itself. The larger cells of the ring (the lighter region) are produced in the spring. The smaller cells (the darker region) are produced in the late summer.

Often it is difficult to visualize how all the tissues of a woody dicot plant relate to one another. Figure 20.8 illustrates the relationships between the secondary tissues in a tree trunk. Now that we know something about plant tissues and growth patterns, we are ready to look at the main plant structures, the roots, stems, and leaves. We discussed the flower, the fourth main structure, in Chapter 19.

FIGURE 20.8   The Relationships of Various Layers of a Tree Trunk

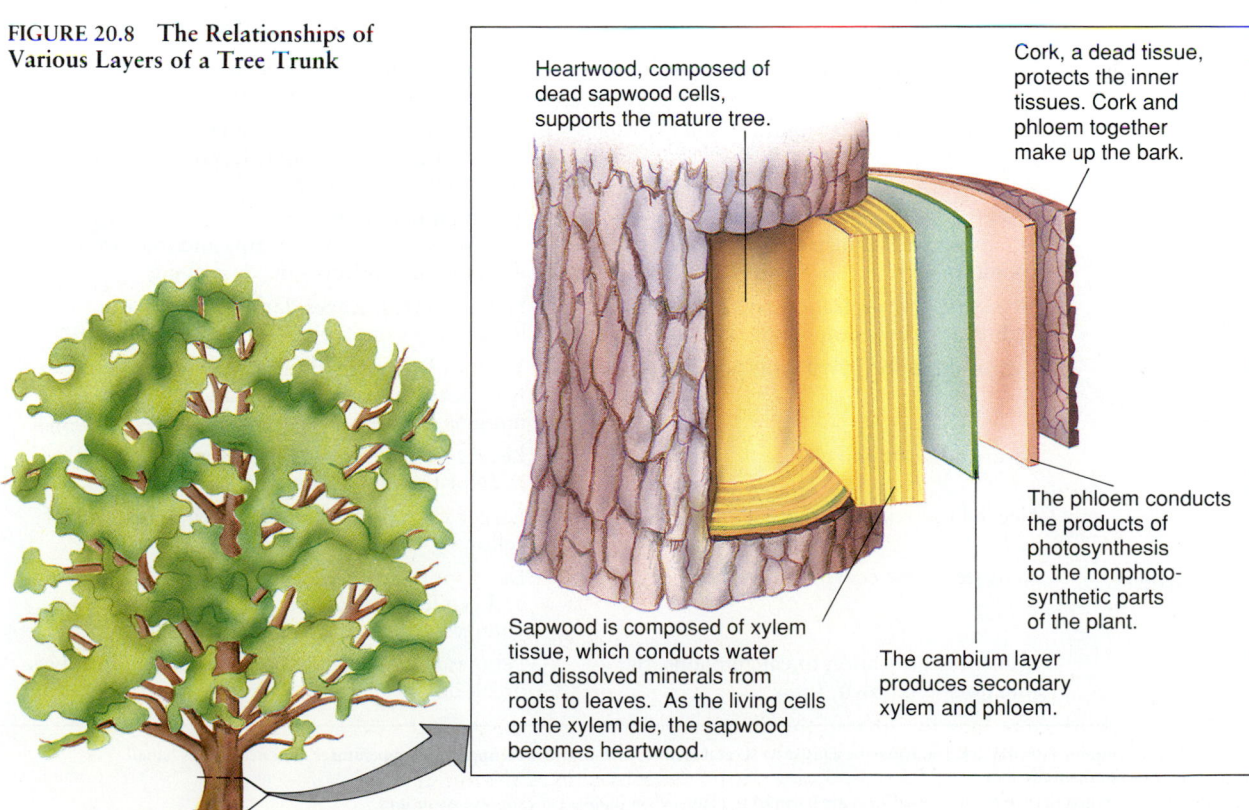

Heartwood, composed of dead sapwood cells, supports the mature tree.

Cork, a dead tissue, protects the inner tissues. Cork and phloem together make up the bark.

The phloem conducts the products of photosynthesis to the nonphotosynthetic parts of the plant.

The cambium layer produces secondary xylem and phloem.

Sapwood is composed of xylem tissue, which conducts water and dissolved minerals from roots to leaves. As the living cells of the xylem die, the sapwood becomes heartwood.

ROOTS

■ The primary functions of roots are to anchor the plant and to absorb water and dissolved minerals from the soil. Specialized root structures also function in food storage and gaseous exchange.

If you have ever pulled weeds, you know that one of the functions of roots is to anchor the plant in the soil. As you pull weeds or harvest garden vegetables, you may have noticed that their root systems are not all alike. Some plants, like carrots, dandelions, or sugar beets, have a *taproot,* while others, like the grasses, have a *fibrous root system* (Figure 20.9).

Taproots consist of a dominant primary central root with comparatively few lateral roots, all of which are much smaller than the dominant one. Taproots function in food storage as well as in the absorption of water and dissolved minerals.

In the fibrous root system, there is no central

Grass   Dandelion

(a) Fibrous root system   (b) Taproot system

FIGURE 20.9   Root Systems
(a) Fibrous root system, characteristic of monocots like grass. (b) Taproots in dicots, such as a dandelion.

TABLE 20.3
Uses of Essential Elements in Plants

Element	Some functions	Deficiency symptoms
Nitrogen	Part of proteins, nucleic acids, chlorophyll	Relatively uniform loss of color in leaves, occurring first on the oldest ones
Potassium	Activates enzymes; concentrates in meristems	Yellowing of leaves, beginning at margins and continuing toward center; lower leaves mottled and often brown at tip
Calcium	Essential part of middle lamella; involved in movement of substances through cell membranes	Terminal bud often dead; young leaves often appearing hooked at tip; tips and margins of leaves withered; roots dead or dying
Phosphorus	Necessary for respiration and cell division; high-energy cell compounds	Plants stunted; leaves darker green than normal; lower leaves often purplish between veins
Magnesium	Part of the chlorophyll molecule; activates enzymes	Veins of leaves green but yellow between them with dead spots appearing suddenly; leaf margins curling
Sulfur	Part of some amino acids	Leaves pale green with dead spots; veins lighter in color than the rest of the leaf area
Iron	Needed to make chlorophyll and in respiration	Larger veins remaining green while rest of leaf yellows*
Manganese	Activates some enzymes	Dead spots scattered over leaf surface; all veins and veinlets remain green; effects confined to youngest leaves
Boron	Influences utilization of calcium ions, but functions unknown	Petioles and stems brittle; bases of young leaves break down

* *Note:* the symptoms of iron deficiency may be caused by several factors, such as overwatering, cold temperatures, and nematodes (small roundworms) in the roots.
   It should be noted that all the micronutrients are harmful to plants when supplied in excessive quantities.

Source: Stern, *Introductory Plant Biology, 3rd ed.* (Dubuque, Iowa: Brown, 1985). All rights reserved. Reprinted by permission.

---

taproot. Instead the root system is a mass of small roots of similar size arising from the base of the stem. A fibrous mat of roots is highly efficient in harvesting water and dissolved minerals from the soil and is often instrumental in anchoring not only the plant itself but also the soil in which it grows.

There are other types of root systems. Corn plants, for instance, develop prop roots at the surface of the soil. These roots help support the plant in high winds. Aerial roots develop from mangrove trees found growing in southern swamplands. Since mangrove trees grow in standing water, aerial roots function in the absorption of oxygen from the air. Some plants in African subdesert areas develop taproots that function in water storage rather than food storage. Apple trees, and many others, develop what are called suckers, branches that arise from roots below the soil. If you cut away the sucker with its root system, you can grow a new plant.

No matter what system is employed, the primary purpose of roots is to absorb water and dissolved minerals. The root system of a plant provides a tremendous surface area for the efficient absorption of water and dissolved minerals. Table 20.3 lists some of the important minerals absorbed by roots.

Growth Patterns in Roots

■ **The root is the first organ to emerge from a germinating seed. The protective epidermis covers the surface and forms root hairs, increasing the root's absorptive capabilities. Ground tissue develops into cortex, which functions in storage. The endodermis regulates water movement into the vascular cylinder. The pericycle gives rise to lateral roots and the cork cambium.**

The first plant organ to emerge from the germinating seed is usually the root. When the root is cut into thin lengthwise sections (called longitudinal sections) for observation under a microscope, you can see four major regions in the young primary root (Figure 20.10). The *root cap* covers and protects the second region, the apical meristem, as the root grows and pushes through the soil. Through division, the apical meristem provides all the new cells of the growing root. Located just behind the apical meristem is the *zone of elongation*, where the cells originating in the apical meristem increase in length up to 20 times their original size. In the *zone of differentiation*, which lies farthest away from the

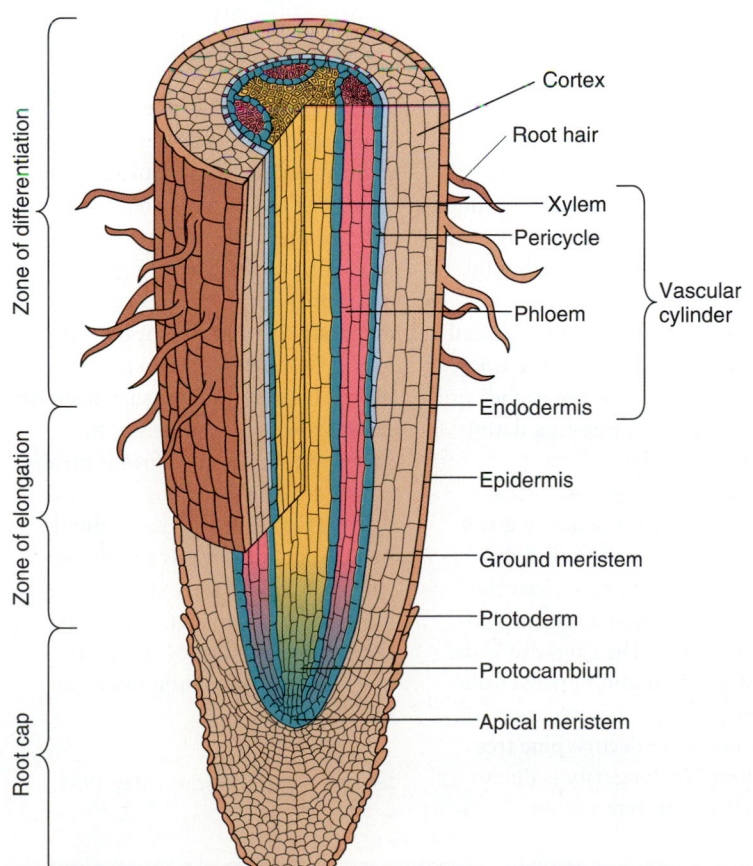

FIGURE 20.10 A Longitudinal View of the Young Primary Root of a Dicot

FOCUS 20.1
Aging Ancients

Four thousand nine hundred years old (going on 5000) — it is hard to believe that anything can live that long. In the 1960s, however, near Wheeler Peak in eastern Nevada, Donald Currey discovered a misshapen bristlecone pine tree that is over 4900 years old. Bristlecone pines, which are generally gnarled and hardly look alive, resemble living driftwood. They are found in the subalpine coniferous forests that grow between 9500 and 11,800 feet above sea level along the western Rocky Mountains and the White Mountain range in California, near the Nevada state line.

How did Currey determine the age of that bristlecone pine tree — or any other tree, for that matter? We touched on the answer to this question in this chapter. During each growing season a tree adds a layer of secondary xylem to its trunk and branches, which form concentric annual growth rings. In the spring, when the rains guarantee an abundance of water, new secondary xylem cells arise from the living vascular cambial cells that separate the xylem and phloem. As the summer progresses, the secondary xylem cells become smaller and smaller. In the fall, growth ceases, and new secondary xylem cells are not added until the next year. The same cycle repeats itself each year.

When you examine wood cut for a fireplace or a woodstove, you can see the layers of secondary xylem cells as a series of rings called annual growth rings. The difference in ring color is partly due to the variations in water supply during the course of the year. In the spring, when there is ample water, the cells produced by the cambium are large, but as the water supply diminishes during the summer and fall months, the cells become smaller. A square inch of the portion of the ring developed in the spring therefore has fewer cells than a square inch produced later on in the season. Fewer cells means fewer cell walls, so the area is lighter in color than an area developed by the production of many cells.

The contrast between the last layer of small cells, produced in the fall of one year, and the first layer of the new large cells, produced in the spring of the next year, is what causes the color changes that we call growth rings. If you count the rings, you can determine the age of the tree. The science of tree-ring dating is called *dendrochronology*.

Dendrochronologists can ascertain more than just the age of a tree. By examining the width of each annual growth ring, they chart the climatic variations in a specific geographical region. The rings are wider during a wet year and narrower during a dry year.

How can a bristlecone pine tree live so long? Its longevity is due to its specialized growth pattern. Botanists now know that trees that are more than 1500 years old have only a narrow ring of living tissue within the trunk, along with a few living twigs and needles on a single branch. Most of the tree has already died over the years. This pattern is typical of a bristlecone pine. Each tree grows very slowly, its diameter increasing only about 1 inch over a 100-year period. For example, one bristlecone pine 3 feet tall with a 3-inch trunk diameter was dated to be over 700 years old.

The environment itself is another factor that contributes to the age of the trees. Individual trees are spread over a wide area, eliminating competition for available moisture and light. The oldest trees are found in a type of limestone called dolomite. Very little vegetation can grow in dolomitic, which further reduces competition from other plants for available nutrients and water. In addition, because there is such a sparse undergrowth of vegetation, the probability of ground fires is greatly reduced.

It makes sense that there should be a variety of reasons for any organism to attain the extreme age of a bristlecone pine. These trees are indeed living antiques. They are among the oldest living things on earth.

Source: Natural History, May 1985, p. 39.

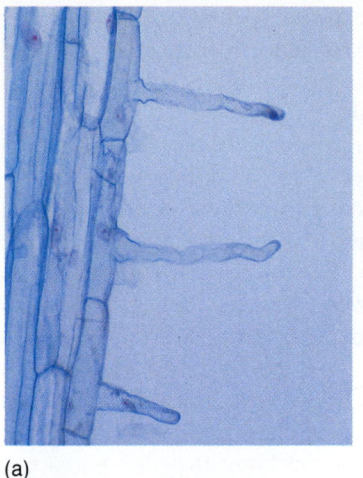
(a)

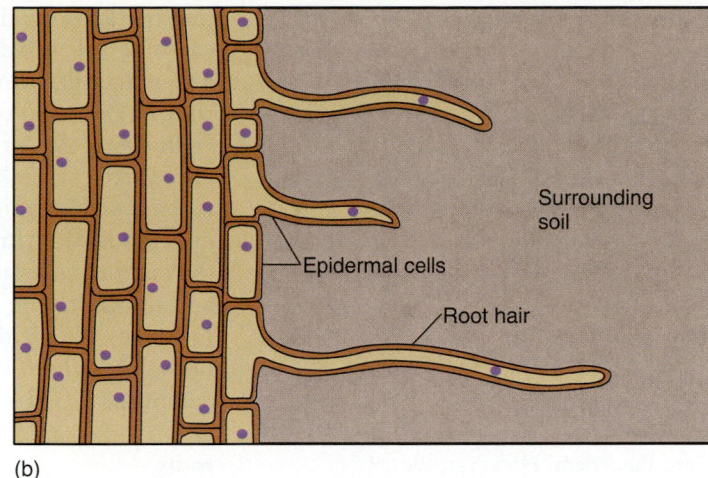
(b)

FIGURE 20.11 Root Hairs
(a) The growing root in this photograph has developed numerous root hairs.
(b) The long, thin root hairs absorb dissolved nutrients and water from the surrounding soil.

root cap and just beyond the zone of elongation, the elongated cells differentiate into the epidermis, cortex, and vascular cylinder (Figure 20.10).

Epidermis. The protoderm differentiates into the epidermis, which covers and protects the soft tissue of the root. In the zone of differentiation, the epidermal cells develop extensions called **root hairs,** which dramatically increase the surface area of the root over which absorption can occur. Root hairs are long, thin extensions of the epidermal cells (Figure 20.11). In fact, the nucleus of the epidermal cell is often found in the root hair. Since a plant can have millions of root hairs, these tiny structures increase the root's surface area tremendously. In one study, botanists estimated that the root hairs of just one rye plant, if linked end to end, would extend more than 10 kilometers, or 6 miles. (If you have ever run a 10-kilometer marathon, you have some idea of how long that is.)

Cortex. Enclosed within the epidermis are the cortex and the vascular cylinder (Figure 20.12). The cells of the ground meristem differentiate into the **cortex,** which consists of loosely packed, specialized parenchyma cells that make up the largest area within a young root. The parenchyma cells of the cortex function in the storage of starch and other organic substances. Because its cells are loosely packed, the cortex includes many air spaces that take in and trap oxygen from the soil to be used in plant cell respiration. (Remember that all living plant cells must respire to release the chemical en-

ergy present in high-energy compounds and to form ATP.)

Endodermis. The *endodermis,* the innermost layer of the cortex, is instrumental in controlling the movement of water and dissolved minerals into the xylem. Unlike those in the rest of the cortex, the cells of the endodermis are packed tightly together — there are no air spaces between them. Embedded in each endodermal cell wall is a waxy substance called *suberin.* The continuous strip of suberin sur-

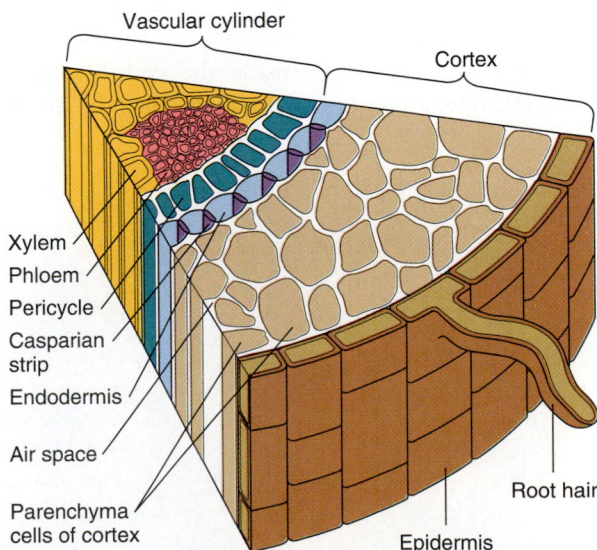

FIGURE 20.12 Cross-Section of a Root
The cortex of a root lies between the epidermis and the vascular cylinder and functions in the storage of starch and other substances.

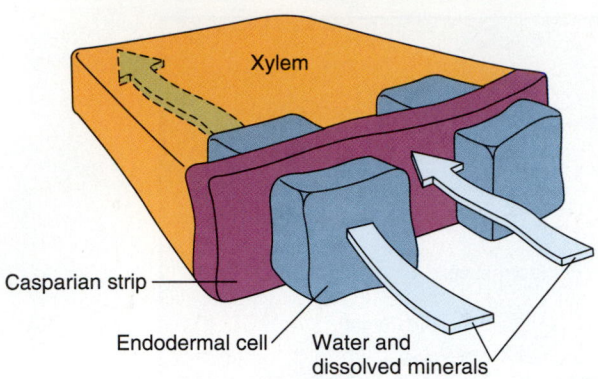

Xylem

Casparian strip

Endodermal cell Water and
dissolved minerals

FIGURE 20.13 The Casparian Strip
Water and dissolved minerals cannot pass through the
Casparian strip into the xylem. However, the
endodermal cells are permeable to water and its
dissolved minerals and allow their passage.

rounding each cell of the epidermis is known as a
Casparian strip.

Suberin is impermeable to water. Consequently,
water and the minerals dissolved in it cannot enter
the xylem by passing between adjacent endodermal
cells. Instead, water and dissolved minerals must
pass through the cells' cytoplasm (Figure 20.13).
Passage into the cytoplasm is controlled by the
plasma membrane. As you remember from Chapter
5, a plasma membrane is differentially permeable
and can regulate the passage of nutrients into and
out of the cell through active and passive transport.

In plant roots, the plasma membrane of the endodermal cells regulates the passage of water and
dissolved minerals into the vascular tissues of the
root.

Vascular Cylinder. Just inside the endodermis is
the **pericycle,** a layer of tissue that is usually about

one cell layer thick. Branch roots and the vascular
cambium of woody roots form from the pericycle.
The remainder of the procambium gives rise to the
root's xylem and phloem.

In roots, the arrangement of the primary phloem
and xylem resembles a pie that has been cut into pieces. The phloem forms the pieces of the pie,
and the xylem forms the broad cuts that divide
the pie into sections. The number of sections varies
according to the plant species being examined.
In some plants, the inner circle of the root can
be solid xylem (Figure 20.14). In others, it is
composed of parenchyma cells collectively known
as *pith.* Secondary growth occurs in some plant
roots.

STEMS

■ **The functions of stems include raising the photosynthetic structures of the plant into the light;
transporting food, water, and minerals; and ele-
vating reproductive structures into the wind or
into sight to ensure pollination.**

As plants evolved from an aquatic habitat to a terrestrial one, those best able to harvest light, disperse
seeds, and transport materials were the most successful. The stem is the major plant structure that
performs these functions. Stems support and
display the photosynthetic structures to light,
transport materials to and from the roots and photosynthetic structures, and elevate the reproductive
structures into the wind for pollination and seed
dispersal. Most stems are above ground, but in
some instances, such as the white potato plant, they
function as specialized storage and/or reproductive
structures located underground.

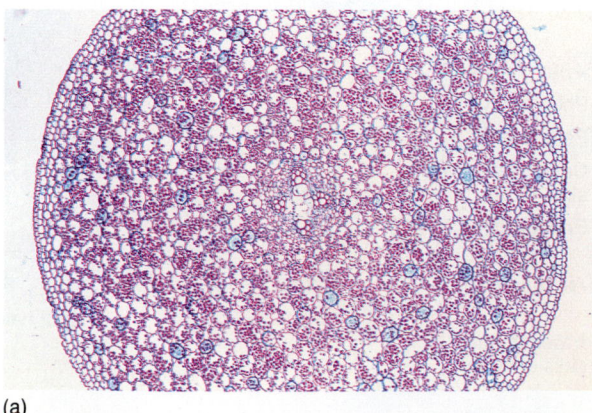

(a)

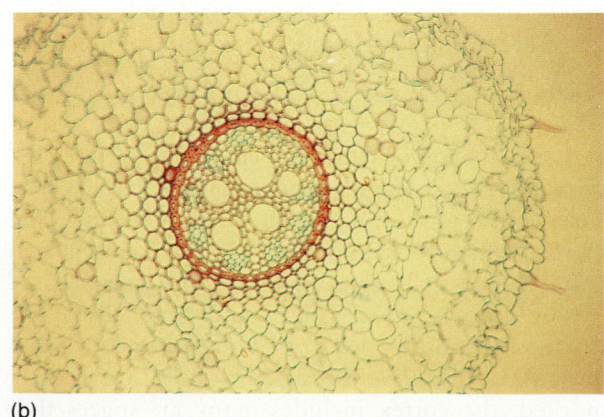

(b)

FIGURE 20.14 Comparative Cross-Sections of Roots
(a) A cross-section of a vascular cylinder from a root lacking pith. (b) A cross-
section of a corn root. Note that the vascular cylinder surrounds the pith.

Growth Patterns in Stems

■ The stem is enclosed and protected by the epidermis. The cortex, composed of parenchyma cells (and sometimes collenchyma cells), provides support and functions in food storage. Vascular tissues form either a solid cylinder, a ring of vascular bundles, or a scattered mass of vascular bundles. Vascular cambium will be present in plants having secondary growth. The pith, located at the center of the stem, consists of storage parenchyma cells. Leaf and bud primordia produced by the apical meristem result in the development of leaves and branches.

Primary growth in the stem follows the same basic pattern as in roots. First, cell division takes place within the apical meristem, then the cells elongate, and finally they differentiate. However, because stems give rise to new leaves, branches, and flowers, the development pattern of stems is more complicated than that of roots.

The outer protective layer of a young stem is the *epidermis.* The epidermal cells secrete the cuticle. Both the epidermis and the cuticle protect the plant's stem from injury, reduce water loss, and prevent invasion by parasites. The epidermis itself is usually one cell layer thick. It may contain stomata. If the stem is green, the outer parenchyma cells of the cortex contain chloroplasts and thus are photosynthetic. This is the case with cactus.

Beneath the epidermis is the *cortex,* which consists primarily of parenchyma cells that function in food storage. The parenchyma cells provide support for the young plant by means of turgor pressure. Collenchyma cells may also aid in support of the stem. Some stem parenchyma cells can contain chloroplasts and carry on photosynthesis. The cells, with chloroplasts, form the chlorenchyma.

The *vascular tissues* of monocot and dicot plants are not arranged in the same way. In monocots, the vascular tissues are scattered throughout the stem in what are called *vascular bundles* (Figure 20.15a). In the young dicot stem, vascular bundles form a hollow ring beneath the cortex (Figure 20.15b). Within the vascular bundles are the primary phloem and xylem.

At the center of the dicot stem is the *pith,* the stem's core, which is composed of parenchyma storage cells. In monocots, the scattered vascular bundles are surrounded by the parenchyma cells that form the ground tissue. The monocot stem lacks pith.

The *vascular cambium* forms from the procambial cells between the xylem and phloem in stems of dicots and gymnosperms. Cell divisions in the

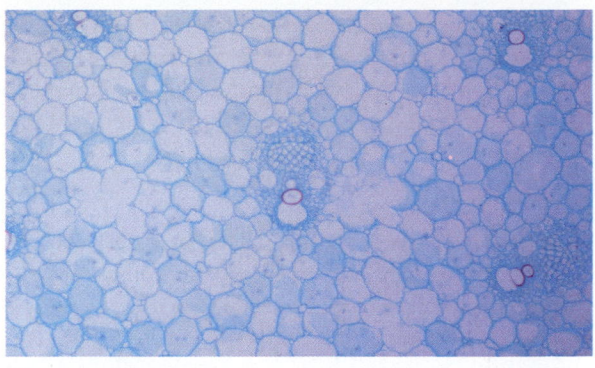

(a)

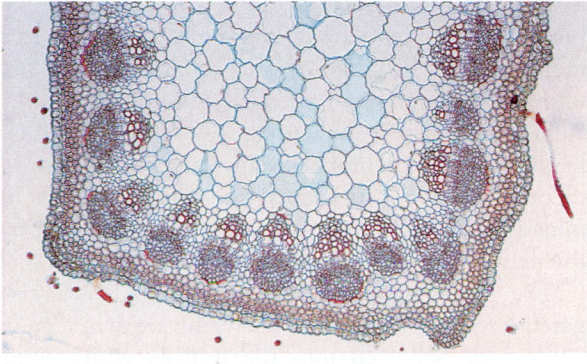

(b)

FIGURE 20.15 Cross-Sections of Young Stems
(a) In young monocot stems, the vascular bundles are scattered throughout the ground tissue. (b) In young dicot stems, the vascular tissue forms a continuous circular cylinder around the pith.

vascular cambium give rise to the secondary xylem and phloem. As in roots, cellular division of vascular cambium results in an increase in the stem's diameter. Monocots do not have secondary growth. Cork cambium forms in plants having secondary growth and produces the periderm.

Leaf Buds

As the primary tissues of a stem develop, leaf primordia and bud primordia arise along the length of the stem. These structures will develop, respectively, into leaves and branches or flowers. Strands of xylem and phloem branch off the main vascular cylinder and lead into the developing leaf and bud primordia. The strands are called leaf and bud traces.

LEAVES

The leaf is the major photosynthetic organ of the plant. It is the original "green machine," converting solar energy to chemical energy that is stored in the high-energy carbon compounds we call food. As we discussed in Chapters 7 and 8, the process of converting light energy, CO_2, and H_2O into chemical energy stored in food is called photosynthesis.

Leaf Structure

■ In most cases, leaves have flat, thin blades that maximize light harvesting and control gaseous exchange. Leaf blades are usually attached to the stem by a petiole.

The typical leaf has evolved into a thin, flat blade, a structure modified to provide the maximum surface area with which to trap light energy and to facilitate the exchange of photosynthetic gases. In monocots the base of the leaf may surround the stem, as you can see when you pull a blade of grass loose from the plant's body in order to chew it. In dicots and some monocots, the flattened blade is joined to the stem and oriented toward the sun by the **petiole,** or stalk.

The vascular tissue located in the blade is aggregated into veins. In dicots, the veins are arranged in a network and feed into the midrib, which joins with the petiole. In monocots, like that blade of grass you chewed on, the veins run parallel to each other. Figure 20.16 illustrates the differences between monocot and dicot leaves. In both types of leaves, the veins serve three functions: They supply the blade with water and minerals, they take away the products of photosynthesis, and they strengthen the leaf, allowing it to retain its form.

The Internal Organization of the Leaf

■ A leaf typically has an upper and lower epidermis, which secrete cutin and waxes and contain stomata and their guard cells. The mesophyll forms the bulk of the tissue within the leaf. The upper layer of the mesophyll consists of the photosynthetic palisade parenchyma cells. Beneath this layer are the photosynthetic spongy parenchyma cells, which are loosely packed and function to provide areas for gas and water exchange.

The typical leaf blade has two surfaces covered by the *upper epidermis* and the *lower epidermis,* respectively. These protect the inner tissue of the leaves. The *cuticle,* secreted by the epidermal cells, covers the leaf and protects it from mechanical injury and from drying out.

Stomata. The epidermal layers contain *stomata,* small pores through which gases are exchanged between the mesophyll and the external environment. On each side of a stoma (the singular form of *stomata*) is a bean-shaped cell known as a **guard cell.** Stomata may occur in both the upper and lower epidermis, although more are usually found in the lower leaf surface than in the upper. In some aquatic flowering plants, like the water lily, the situation is reversed. Stomata are found in the upper epidermis because the lower epidermis is under water.

Guard cells expand or contract based on the amount of water they contain. By changing shape,

(a)

(b)

FIGURE 20.16   Monocot and Dicot Leaves
(a) In this monocot leaf, the veins run parallel. (b) In this dicot leaf, the veins form a netted pattern.

FIGURE 20.17 A Look Inside a Leaf

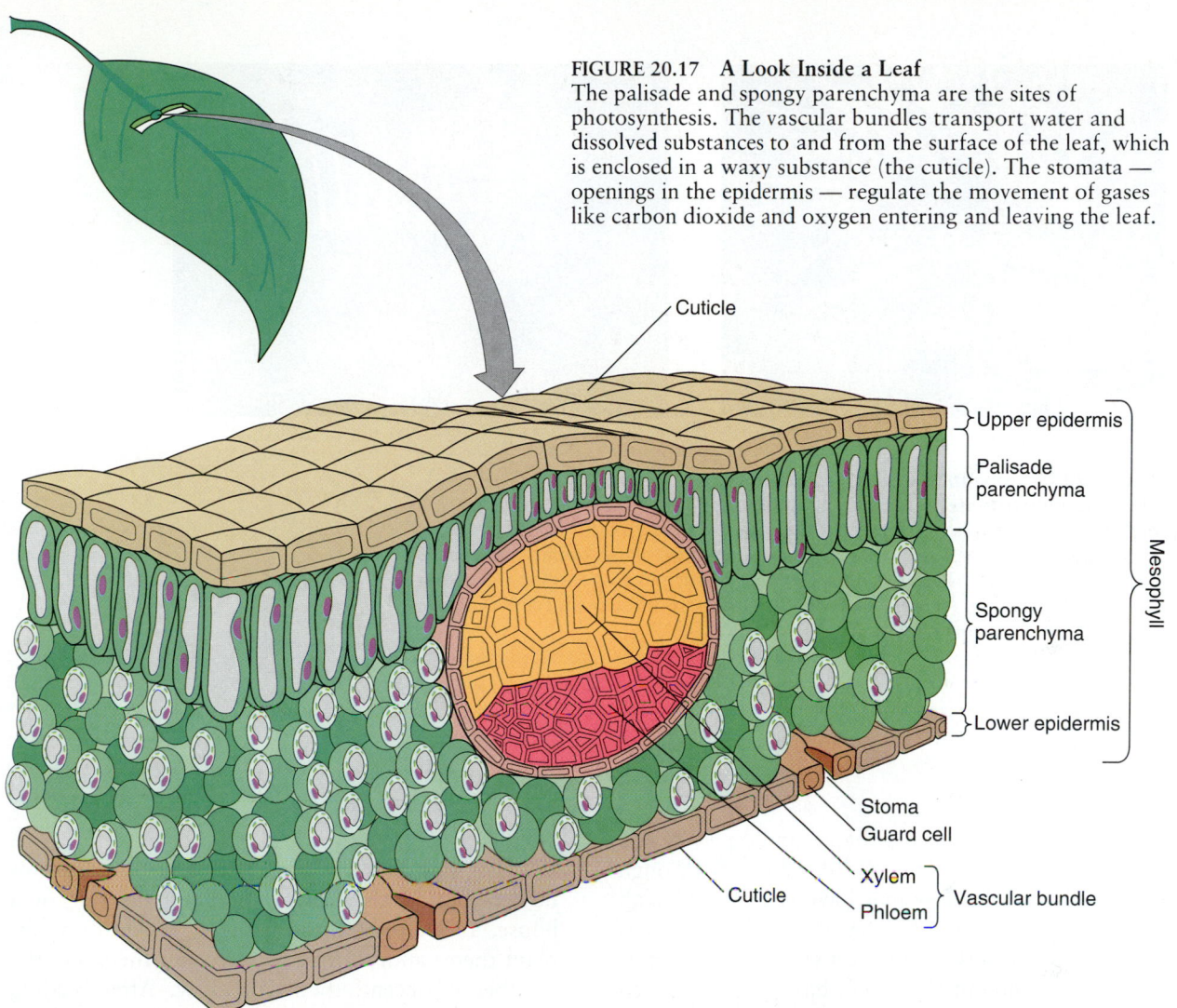

FIGURE 20.17 A Look Inside a Leaf
The palisade and spongy parenchyma are the sites of photosynthesis. The vascular bundles transport water and dissolved substances to and from the surface of the leaf, which is enclosed in a waxy substance (the cuticle). The stomata — openings in the epidermis — regulate the movement of gases like carbon dioxide and oxygen entering and leaving the leaf.

Cuticle

Upper epidermis

Palisade parenchyma

Spongy parenchyma

Mesophyll

Lower epidermis

Stoma

Guard cell

Cuticle

Xylem

Phloem

Vascular bundle

the guard cells control the opening and closing of the stoma and regulate the passage of gases into and out of the leaf. In the next chapter, we discuss the mechanisms of stomata functions.

Trichomes. The surface of the leaf may also be covered by *trichomes,* which arise from the epidermis. The trichomes can assume many forms. They may be unicellular or multicellular, as well as branched or unbranched. Glandular trichomes are common. If you have ever wandered through a field and suffered from the pain and itching of stinging nettle, you have experienced one instance of the actions of glandular trichomes on the leaves of a plant.

Mesophyll. The internal region of the leaf, sandwiched between the two epidermal layers, is known as the **mesophyll,** or middle leaf. The mesophyll is made up of the two types of chloroplast-containing parenchyma cells, which are arranged in layers: *palisade parenchyma* and *spongy parenchyma* (Figure 20.17).

The palisade and spongy cells each exhibit a particular orientation and function within the leaf. Those that are elongated and arranged in an upright manner beneath the upper epidermis of the leaf are called **palisade parenchyma cells,** since the cells are arranged like a palisade (the wall built of upright logs that surrounded army forts in the old west). Most plant photosynthesis takes place within the palisade parenchyma cells. Their shape and arrangement maximize the number of chloroplasts that are exposed to sunlight, an important consideration in photosynthesis.

Between the palisade parenchyma cells and the lower epidermis are the cells known as **spongy parenchyma cells,** which are irregular in shape, loosely packed, and surrounded by large intercellular spaces. Because they are packed together very loosely, about 50 percent of the space in which they are found is air space. This loose organization enhances the exchange of carbon dioxide, oxygen, and water vapor between the air and the leaf tissues during photosynthesis and respiration.

(a)   (b)

FIGURE 20.18　Examples of Leaf Adaptations
Like other things in nature, leaves adapt to their environments. (a) The spines (highly adapted leaves) on a cactus make a good substitute for large flat leaves in the desert that lose a lot of water. (b) Carnivorous plants obtain nitrogen from the protein of insects or small vertebrates, such as frogs — with the help of their leaves. In the Venus flytrap, a leaf closes around any object moving on its surface, such as this frog.

Leaf Adaptations

Leaves come in numerous shapes and types, as you have probably noticed every time you walk through the woods. All those shapes have emerged through natural selection. In the dry desert, the leaf has rolled up to form the spine you see on every cactus. In this environment the leaves have been sacrificed in favor of water conservation. In desert plants, much of the photosynthesis is carried out by the chlorophyll-containing cells of the stem. In contrast, the leaves of tropical rain forest shrubs are large and broad, especially at the bottom of the forest canopy, where light is scarce and water plentiful. You can see this same characteristic in many of the large-leaved houseplants you may have growing in your house or apartment.

Some plants, such as cabbage, have acquired compact leafy heads that serve as specialized structures for food storage. The tendrils of pea plants are modified leaflets. In general, however, the leaf consists of layers of chlorophyll-containing cells sandwiched between two epidermal layers. Figure 20.18 illustrates examples of leaf adaptations.

THE ROLE OF ROOTS, STEMS, AND LEAVES IN ASEXUAL REPRODUCTION

■ **The three main plant organs also function in asexual reproduction by such means as runners, rhizomes, and cuttings, depending on the species involved.**

We all know that if we want to grow a border of flowers beside a driveway or along the side of a house, all we need to do is buy a packet of seeds, plant them, and, as long as we water them and the weather is decent, they will grow. After reading Chapter 19, we know how those seeds come about in the first place. But there are other ways to produce new plants, ways that require no pollination, no fertilization, and no seeds. When a plant can be reproduced without fertilization, the process is called asexual reproduction. The three plant organs we have studied in this chapter — roots, stems, and leaves — are involved in asexual reproduction in a variety of plants. All techniques do not apply to all plants.

Specialized root structures provide for asexual reproduction. Horseradish plants, for instance, send out *adventitious roots* from various points along their stems. New plants will form at intervals along the adventitious roots. Root cuttings from these plants will produce the same result. That is why you cannot get rid of horseradish by rototilling or chopping off the top area of the root. New plants arise from the remaining root pieces.

Stems can produce new plants in some species. Strawberries, for instance, produce stems called runners (or *stolons*). Runners are stems produced from the base of a mature strawberry plant. The

Dr. Booker T. Whatley
Plant Scientist, Horticulturist

In these days, when most plant scientists use recombinant DNA technology and gene splicing to produce superplants and increase plant production, it is sometimes a relief to find a more practical, down-to-earth, back-to-basics approach. This is the method of Booker T. Whatley, retired professor of horticulture at Tuskegee Institute in Tuskegee, Alabama, an institution that serves a predominantly African-American student body.

Booker T. Whatley, who was named after Booker T. Washington, the founder of the institute, has established several plant-breeding programs, all designed to produce a better plant product. His approach is a traditional one. He worked for five years developing the Carver sweet potato (considered one of the highest-quality sweet potatoes) following methods that plant breeders have used for centuries—simple artificial selection. He planted only the best seeds. He then analyzed the plants that grew and the potatoes they produced, and collected seeds only from those plants that were the most prolific and produced the tastiest, firmest sweet potatoes. He followed the same procedure for subsequent generations and ultimately produced the Carver sweet potato. In a similar breeding program, he developed the "Foxy Lottie" grape, named after his wife.

This fast-talking gentleman is known not only for his successful plant-breeding experiments but also

for his contributions to the theory of the layout of the small farm. His plans for small farms provide information on achieving the most productivity from a limited space. Whatley has written a book on small farming in which his farm models have allowed numerous small farmers to achieve success, and he is often credited with saving the small farmer in these days of huge agribusiness.

Whatley downplays his ability, saying he just uses common sense when it comes to farming. As a result, he has been called the "guru of common sense," a title that makes him shake his head and chuckle, for he knows that his commonsense solutions are based on modern agricultural technology and homespun knowledge. For example, he uses an integrated, balanced approach to pest management, in which he relies on both insecticides and natural predators, like ladybugs and praying mantises. He also has suggested that farmers incorporate the guinea fowl into their complement of livestock, because these birds are gluttonous when it comes to eating insects and can help cut down on the amount of pesticide needed. To keep deer and rabbits away from his plots, he has found that all that is necessary is to string bags of human hair around the plants. The smell of the hair is enough to do the job, he says.

Whatley is a tall, thin man in his 70s, with an infectious smile and laugh. His fascination with agricul-

ture developed while he was living on his parents' farm. In the early 1940s he studied at Alabama A&M, and after numerous years of plant science and horticultural experience, he entered Rutgers University, where he completed his Ph.D. in 1957. He taught for 12 years at Southern University in Baton Rouge, Louisiana, and then moved to Tuskegee Institute, where he remained until retirement. Booker T. Whatley is truly an uncommon man of common sense, and his understanding of plant science has sometimes led him to be called the "Plant Producer from Tuskegee."

Source: B. H. Seeber, "The Producer," *Science 84,* 5:40–47, July/August, 1984.

TABLE 20.4
Asexual Reproductive Mechanisms in Plants

Type of mechanism	Plant example	Characteristics
Adventitious roots	Horseradish	New plants arise from points along the roots
Runners	Strawberry	New plants arise from horizontal stems above the ground
Rhizomes	Iris	New plants arise from horizontal stems underground
Cuttings	Philodendron	New plants arise from cut leaves that grow roots
Bulb	Tulip	New bulb arises from underground stems
Tuber	Potato	New plants arise from buds or modified rhizomes
Parthenogenesis	Rose	New plants arise from unfertilized eggs

runner grows along the surface of the ground, and every so often it produces the normal stems and leaves of a new plant. Once the plant has developed leaves, the runner can be severed and you have a new strawberry plant that is genetically identical to the parent plant (see Profile).

Rhizomes are horizontal stems that grow underground, and they are found in plants such as Bermuda grass and irises. Like runners, rhizomes grow out from the plant body for a short distance, then sprout the leaves and roots of new plants. Tubers and bulbs are also modified stems that can give rise to new plants.

Finally, leaves can produce new plants through *cuttings*. This method is one often used to propagate houseplants, such as philodendrons and African violets. Simply remove a leaf, place it in a cup of water, and in a relatively short time, roots will form at the base of the petiole, and a new plant is ready for potting.

There are many ways that plants reproduce asexually. Table 20.4 lists several methods of asexual reproduction in seed plants.

The growth patterns we have just covered are similar whether you are studying the root tips of a tomato seedling or of a 100-year-old oak tree in your yard, the stems of asparagus or a banyan tree, or the leaves of a violet or a rubber plant. The processes and the tissue types we have discussed can be found in all representatives of the seed plants. Consider the number of ways in which the basic unity of these plants is expressed — the buttercup in a Minnesota meadow, a prickly pear cactus in an Arizona desert, a palm tree in Puerto Rico, a red-

wood in northern California. These are only four examples out of hundreds. How do all these plants survive in such a variety of environments? In the next chapter we look at the ways in which growth, development, and physiological processes take place in plants, as well as the role of the stomata and plant hormones.

SUMMARY

1. The typical seed plant body is composed of two major tissue types. Meristematic tissues are made up of undifferentiated cells. They divide to produce permanent tissues, which are made up of specialized cells.
2. There are three kinds of meristems. Apical meristems are found at the tips of developing roots and stems. Lateral meristems include the vascular cambium and cork cambium. Cell divisions in the vascular cambium result in growth in the diameter of the plant. Cell divisions within the cork cambium give rise to the periderm. Intercalary meristems allow the grasses to grow at their nodes.
3. There are three general groups of permanent tissues: surface tissues; support, storage, and secretory tissues; and vascular tissues.
4. Surface tissues include the epidermis and periderm. The epidermis covers the stems, roots, and leaves. The periderm is produced by the cork cambium and forms the protective layer that covers woody plants.
5. Parenchyma cells are the principal vehicles of storage and secretion in plants. Parenchyma cells containing chlorophyll carry out photosynthesis. Collenchyma cells help support many plant structures. Sclerenchyma cells are fibrous and are im-

portant in plant support and vascular strength. Secretory cells release hormones and other plant products.

6. Xylem is a vascular tissue responsible for transporting water and dissolved minerals to the photosynthetic cells. The xylem is composed of fibers, parenchyma cells, ray cells, tracheids, and vessel elements. Tracheids are long, thin cells with tapered ends that overlap with one another, forming a column. The overlapping ends of the tracheids contain pits through which water and dissolved minerals pass. Vessel elements are short and wide with perforated end walls. They are stacked to form a continuous column called a vessel.

7. Phloem is a tissue that transports the products of photosynthesis throughout the plant. The phloem is composed of sieve tube elements and companion cells. Sieve tube elements are long cells with porous end walls called sieve plates. At maturity the sieve tube element loses its nucleus. The companion cell (which has a nucleus) adjacent to the sieve tube member regulates its own life processes as well as those of the sieve tube member.

8. Most plant cells are totipotent, which means that under certain conditions they can produce a complete new plant. Growth in plants is continuous. A mature plant always has regions of active growth.

9. Cell divisions in the apical meristem cause roots and stems to grow in length (primary growth). The cells produced by the apical meristem form the three primary meristematic tissues. Protoderm gives rise to the epidermis. Procambium forms the primary xylem and phloem, as well as the vascular cambium. Ground meristem is the source of the parenchyma cells that form the cortex and pith. In the roots of some plants, an endodermis develops around the vascular tissue. The pericycle is enclosed by the endodermis.

10. The vascular cambium gives rise to the secondary xylem and phloem. The cork cambium produces the periderm. Cells produced by the vascular cambium and cork cambium increase the diameter of the plant (secondary growth). The production of secondary xylem results in the formation of wood and of annual growth rings.

11. There are two major types of root systems — fibrous and taproot systems. The roots absorb water and dissolved minerals from the soil and anchor the plant in the soil.

12. A newly forming root consists of a root cap, a meristematic region, a zone of elongation, and a zone of differentiation.

13. As the root matures, it develops an epidermis, which covers and protects the internal structures and produces root hairs. Inside the epidermis is the cortex, which consists of specialized storage and/or photosynthetic parenchyma cells. The endodermis is the innermost layer of the cortex and regulates the flow of water and dissolved minerals into the vascular tissues. The endodermis encircles the vascular cylinder. The outer layers of the vascular cylinder make up the pericycle, which gives rise to vascular cambium and lateral roots. The rest of the vascular cylinder produces the xylem and phloem.

14. Water and dissolved minerals move easily through the epidermis and cortex of a root but not through the endodermis. Each endodermal cell is surrounded by a Casparian strip that prevents water from passing between cells. Water and dissolved minerals must pass through the plasma membrane of the endodermal cells to enter the vascular cylinder.

15. Stems display and support the photosynthetic and reproductive structures and transport water, minerals, and food between the roots and these structures.

16. The stem is covered by a protective epidermis, which secretes the cuticle and may include stomata. Below the epidermis is the cortex, which consists of specialized storage, support, and photosynthetic parenchyma cells. The vascular tissues of the monocots are scattered in vascular bundles throughout the stem. In the dicots, the vascular bundles form a hollow ring beneath the cortex. The pith, located in the center of the dicot stem, consists of storage parenchyma cells.

17. Leaf and bud primordia are formed on the developing stem. They will become new leaves and branches. The vascular tissue that runs into a leaf or branch is called a trace.

18. Leaves are the major photosynthetic organs of the plant. In dicots the leaf consists of a flat blade, a network of veins, and a midrib, all attached to the stem by a petiole. The leaf is made up of two epidermal layers, the upper and lower epidermis, with an inner section called the mesophyll sandwiched in between.

19. The mesophyll layer of the leaf is composed of an upper layer of palisade parenchyma cells, in which most of the plant's photosynthesis occurs, and a lower layer of spongy parenchyma that contains loosely packed cells of irregular shape.

20. In most leaves the lower epidermis contains stomata enclosed by two guard cells. The gases and water vapor produced by or required for respiration and photosynthesis pass into and out of the leaf through the stomata.

21. Plant organs can be involved in asexual reproduction.

KEY TERMS

tissue	vascular cambium
meristematic tissue	cork cambium
permanent tissue	intercalary meristem
apical meristem	periderm
lateral meristem	parenchyma cell

collenchyma cell
sclerenchyma cell
vascular tissue
xylem
phloem
companion cell
primary growth
protoderm
procambium
ground meristem
secondary growth

cork
root hair
cortex
Casparian strip
pericycle
petiole
guard cell
mesophyll
palisade parenchyma cell
spongy parenchyma cell

REVIEW QUESTIONS

True/False

1. It is through the epidermal tissues of a plant that materials are transported.
   true   false   (page 470)

2. The palisade layer of the leaf is composed of photosynthetic parenchyma cells.
   true   false   (page 485)

3. The cells that make up the meristematic region of a plant are specialized and differentiated.
   true   false   (page 469)

4. Xylem and phloem are specialized conductive tissues that transport materials through the plant.
   true   false   (page 472)

Fill in the Blank

1. The _____ tissues are a category of tissues that function in support, storage, and secretion in a plant. *(page 471)*

2. Within the phloem, _____ cells are found in association with sieve cells. *(page 472)*

3. The _____ meristem is located at the tips of the root stem or branch. *(page 469)*

4. Secondary growth in plants is due to the activities of the _____ _____. *(page 476)*

Multiple Choice

1. The mesophyll of a leaf consists of:
   a. Spongy parenchyma cells
   b. Palisade parenchyma cells
   c. Both spongy and palisade parenchyma cells
   d. Pith cells
   (page 485)

2. A plant shoot's growth in length is due to cell division in the:
   a. Vascular cambium
   b. Apical meristem
   c. Cortex
   d. Cork cambium
   (page 469)

3. The vascular tissues of the plant function in:
   a. Support
   b. Support and transport of materials
   c. Secretion of plant hormones
   d. All of the above
   (page 472)

4. When comparing sieve tube elements with companion cells, which of the following statements is true:
   a. Xylem cells are alive at maturity.
   b. Companion cells lack cytoplasmic material and a nucleus at maturity.
   c. Companion cells contain a nucleus and cytoplasm at maturity.
   d. Sieve tube elements are found in xylem.
   (page 472)

Discussion

1. Compare and contrast the monocotyledonous and the dicotyledonous seed plants.

2. Discuss three major types of tissues found in seed plants.

3. Draw and label a leaf. Be sure to label the principal cell types and their functions.

4. From an evolutionary perspective, speculate as to why the palisade parenchyma cells are arranged in an upright position.

5. Suppose that when you were 15 years old you carved your initials in a mature tree trunk, 1.5 m above the ground. At the age of 21 you returned to the same tree and your initials were still 1.5 m above the ground. Explain why, relative to the nature of plant growth.

6. List the principal meristematic tissues, and explain how they are responsible for the growth of plants.

7. Draw and label a cross-section of a young dicot stem.

8. Discuss the structure and function of root hairs.

9. Discuss primary and secondary growth in plants.

10. We require plants not only as our source of oxygen but also for food. What plant organs are you eating as you consume each of the following: carrots, beets, lettuce, tomatoes, chili peppers, beans, celery, and spinach?

21

Transport and Regulation

If you have ever driven through the giant redwood tree country of northern California or lain in the grass looking upward through the trees to the sky, you have probably been amazed by the enormous size of some trees. In fact, the needles and leaves of many of them are located hundreds of feet above the ground. How do water and minerals absorbed by the roots reach those lofty leaves? How do the products of photosynthesis travel from the leaves or needles to the cells of the roots deep underground?

There are other questions about trees and other plants that intrigue us. Why do the houseplants in your home or your dorm room curve toward the light? As you walk through barren winter fields, you may find yourself anxiously awaiting spring or

Above: This redwood tree, one of nature's largest living things, contains complex transport systems that distribute water, minerals, and other materials throughout, despite its size.

summer, when the fields will be in full flower. But why do plants flower only during certain times of the year? Why do flowers open and close during certain times of the day? Why do roots grow downward and stems (usually) upward? In other words, what mechanisms have evolved in plants to regulate their activities? In this chapter we try to answer some of the questions concerning transport in plants and the mechanisms that regulate plant activities.

In Chapter 20, we talked about roots, stems, and leaves. Now we investigate the ways in which they interact to transport materials throughout the plant. It will help you understand the material to come if you review the section on water in Chapter 2 and the discussion of osmosis and passive and active transport in Chapter 5. Before we begin to discuss the mechanisms at work in transport in plants, we first need to discuss in detail the functioning of stomata in leaves.

THE FUNCTION OF
THE STOMATA

As we discussed in Chapter 20, the stoma is the opening through which gaseous exchange (usually carbon dioxide, oxygen, and water vapor) occurs. Over 90 percent of transpiration — that is, water vapor loss by plants — occurs through the stomata. On either side of each opening is a bean-shaped guard cell. The structure of the walls of the guard cells is significant in their function. The wall on the side away from the stoma is usually thinner than the wall on the side adjacent to the stoma, and this difference in wall thickness plays an important role, since the guard cell regulates the diameter of the opening of the stoma.

Turgor Pressure

■ **When water passes into the guard cells, their turgor pressure increases, the guard cells bend, and the stoma opens. When water passes out of the guard cells, turgor pressure decreases, the guard cells relax, and the stoma closes.**

When water moves into the guard cells, the turgor pressure inside each cell increases, and the guard cells expand. (**Turgor pressure** is the force with which the water presses against the cell wall.) If you have ever played with water balloons, you have experienced different degrees of turgor pressure. The more water you add when you fill the balloon, the greater the turgor pressure, and the more rigid the balloon becomes. The more water you remove, the lower the turgor pressure and the more flexible the balloon.

Turgor pressure operates in guard cells in a similar way. (As you read through this explanation, keep in mind that solutes are substances that are dissolved in a solvent; an illustration is the solute salt when it is dissolved in the solvent water.) The movement of water into and out of a guard cell depends on whether its cytoplasm is *hypertonic* or *hypotonic* in comparison with the cytoplasm of the leaf cells adjacent to it. (Remember that osmosis operates along a concentration gradient.) When the cytoplasm of the guard cell is hypertonic (that is, when it contains more solutes than the adjacent cells), water moves by osmosis from the adjacent cells into the guard cell. The vacuoles of the guard cells fill with water, turgor pressure increases, and the cells expand. As the guard cells expand, they buckle so that they appear somewhat curved, taking on a shape resembling a kidney bean. When the guard cells swell and buckle, the stoma opens, allowing carbon dioxide and oxygen to pass into and out of the leaf and water vapor to pass out of it.

On the other hand, when the cytoplasm of the guard cell is hypotonic in comparison with the leaf cells around it (that is, when there is a lower concentration of solutes in the guard cells), water moves out of the guard cells into the adjacent leaf cells. The turgor pressure decreases in the guard cells, the cells collapse and lose their turgor pressure, and the stoma closes (Figure 21.1).

Mechanisms That Regulate How Stomata Function

■ **Other factors, such as potassium ion concentration, carbon dioxide concentration, sunlight, temperature, and plant hormones, affect the movement of water into and out of the guard cells.**

Several factors influence how guard cells function in the opening and closing of the stoma, from the ion concentration, the amount of light, and the temperature to plant hormones. Let's discuss some of these factors.

The Role of Potassium Ions (K^+) in the Functioning of the Stomata. Turgor pressure in guard cells is regulated to a large degree by the level of *potassium ions (K^+)* in them. As the guard cells take up potassium ions from adjacent cells, water moves into the guard cells and the stomata open, because the increase in potassium ions causes the cytoplasm of the guard cells to become hypertonic in comparison with the cytoplasm of the adjacent cells. When potassium ions leave the guard cells, their cytoplasm becomes hypotonic. Water moves out of the guard cells and into the adjacent cells, the guard cells lose their turgor pressure, and the stomata close.

But why do the potassium ions move into and out of the guard cells? Although the exact reason is not known, botanists believe it is related to the concentration of carbon dioxide, one of the important raw materials in photosynthesis. Guard cells contain chloroplasts and carry out photosynthesis. During the day, as carbon dioxide is used up by the photosynthetic process, the concentration of carbon dioxide in the guard cells decreases. When that happens, potassium ions are actively transported into these cells. As the concentration of potassium ions increases in the guard cells, the cytoplasm becomes hypertonic, establishing a concentration gradient. Water moves from the leaf cells into the guard cells. The guard cells swell, and the stoma opens.

FIGURE 21.1   Turgor Pressure and Guard Cells
Guard cells regulate the passage of gases through the leaf stoma. (a) As the turgor pressure within the guard cells increases, the guard cells swell and their thick inner walls buckle, opening the stomata (as viewed in this scanning electron microphotograph of a soybean leaf). A single stoma and several waxy particles can be seen in this photograph. The waxy particles minimize water loss through the leaf. (b) As turgor pressure decreases, the guard cells collapse and the stoma closes (as viewed in this light microphotograph).

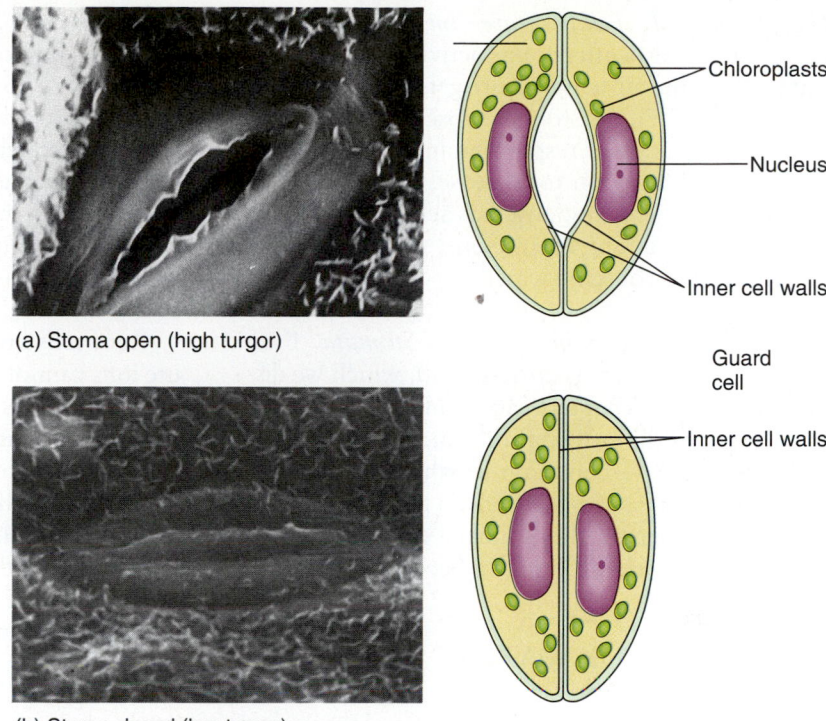

(a) Stoma open (high turgor)

(b) Stoma closed (low turgor)

The interrelationship between carbon dioxide concentration, potassium ion concentration, and water movement requires that the guard cells remain open during daylight hours when photosynthesis takes place. The process of turning light energy into chemical energy keeps the concentration of carbon dioxide low. A low carbon dioxide concentration causes the guard cells to absorb potassium ions. The increase in potassium ion concentration in the guard cells causes water to move into these cells, opening the stoma. With the stoma open, the plant can continue to take up carbon dioxide and release water vapor.

At night, when the light source is removed, carbon dioxide is no longer used for photosynthesis. Because respiration in the guard cells is producing carbon dioxide, the concentration of the gas in the guard cells rises. Under these conditions potassium ions move out of the guard cells. The guard cells become hypotonic, water moves out of these cells, the cells collapse, and the stoma closes.

The Effect of Light on the Stomata. As is clear from the discussion of the role of carbon dioxide and potassium ion concentration in guard cell functioning, environmental factors can stimulate the opening and closing of the stomata. High concentrations of CO_2 inside the leaf cause the stomata to close. Low concentrations of CO_2 cause them to open. Because light initiates photosynthesis, light also causes the stomata to open, by stimulating carbon dioxide utilization.

However, in some plants (the cactus species of the southwestern deserts, for instance) the response to light is reversed. Photosynthesis in these plants is similar to the Hatch-Slack C_4 photosynthesis discussed in Chapter 8. The stomata open only at night and CO_2 diffuses into them, where it can be converted to one of two four-carbon acids, either isocitric acid or malic acid. (In other words, although the plant is taking in carbon dioxide, the gas does not build up within the cell because the gas is integrated into other compounds.) The acids are stored in the chloroplasts until the daylight hours, when the stomata close. The CO_2 is then released from these acids and used in photosynthesis.

This photosynthetic process is known as Crassulacean acid metabolism (CAM), and the succulent desert plants that use it are called CAM plants. Why would this occur in plants in arid deserts? In evolutionary terms, CAM has selective value. Because the stomata are closed during the day, the plants lose less water, and water conservation is a primary concern in desert-dwelling organisms.

The Function of the Stomata

The Effect of Temperature on the Stomata. Temperature also affects the activities of the stomata. When the air temperature gets higher than 30 to 35°C (or 95°F), the stomata close. As the temperature rises, the rate of respiration increases, causing the CO_2 concentration to increase. As a result, the stomata close. This response has selective value for plants that live in hot, dry regions, since closing the stomata would conserve water.

The Effect of Hormones on the Stomata. Even plant hormones such as *abscisic acid,* which we discuss later in this chapter, can regulate the opening and closing of the stomata. In some plants, moderate water stress (a condition in which there is less available water) causes leaf cells to produce abscisic acid, which signals the stomata to close before water stress or dehydration can become severe.

TRANSPORT OF WATER AND MINERALS

■ **Transpiration is the process by which water vapor is lost from the plant body. The leaves are the principal organ and the spongy mesophyll the principal tissue of transpiration.**

If you have ever tried to grow anything outdoors in a dry desert region where irrigation is required, you know that plants require tremendous amounts of water in order to live. For example, an acre of corn requires about 4.5 million liters of water per growing season (a liter is about the same as a quart; so that means one acre of corn requires about a million gallons of water). Why so much? When animals take in water, most of it remains in their bodies. On the other hand, while plants do absorb a great deal of water through their root system, they retain only about 10 percent of it. The other 90 percent is lost to the air as water vapor through a process called *transpiration.* If you have ever walked through a cornfield on a hot, sultry summer day, you probably noticed the high humidity in the field. Most of the water vapor in the air was due to transpiration from the corn.

Root Pressure

■ **The roots absorb water from the soil. The xylem then transports the water to the remainder of the plant. The difference in solute concentration between the root cells and the surrounding soil causes water to move into the root, establishing a force in the xylem called root pressure.**

The water required by a plant enters through its roots. As the roots remove water from the soil immediately surrounding them, it is replaced by water moving through the soil by diffusion and capillary action. Root cells contain a higher concentration of solutes than exists in the soil; that is, root cells have more solutes (hypertonic) in comparison with the soil around them, and, conversely, water concentration in root cells is less than that in the surrounding soil.

Root cells are always hypertonic because the solute ions cannot escape through the Casparian strips and their passage out of the vascular cylinder through the endodermal cells is greatly reduced. Therefore, water moves into the root through *osmosis,* establishing a positive pressure called *root pressure,* which forces both water and dissolved minerals into the xylem. Root pressure alone can account for the movement of water up the stems of some small plants such as grasses.

The Cohesion-Tension Theory of Water Movement

■ **The cohesion-tension theory of water movement states that water moves through a plant because water molecules cohere to each other and adhere to xylem cell walls. As water molecules evaporate through the stomata of the leaves, a tension develops on the water column in the xylem as it is pulled upward to replace the evaporated molecules.**

Is it atmospheric pressure that pushes water up a tree? Although it may seem likely, atmospheric pressure alone cannot be responsible for the upward movement of water in a tree because at sea level atmospheric pressure can only push a column of water in a pipe or tube to the height of 11 meters (about 34 feet). Remember that while atmospheric pressure is pushing water up, the force of gravity is pulling it down. After the column of water is about 34 feet high at sea level, the pull of gravity on the column is equal to the atmospheric pressure drawing up, and the water will not rise any farther. Obviously, many trees are taller than 11 meters, so another explanation is required. To understand what causes the transportation of water and minerals in plants, we need to look at an explanation proposed by an Irish botanist, Henry Dixon. Dixon's explanation for the ascent of water in plants is called the *cohesion-tension theory of water transport.*

The Characteristics of Water. Henry Dixon observed that water appeared to be pulled up the stem from the roots to the leaves in a continuous manner. Understanding the effects of hydrogen bonding in water, Dixon proposed that under certain conditions water molecules would cohere so tightly that they would become as strong as a piece of wire. (Remember that *cohere* refers to the mutual attraction of like molecules.) To see this principle of water at work, watch water bugs skim over the surface of a pond. *Cohesion* is the force that holds the water molecules together against the weight of the water bug.

Another property of water is *tensile strength*, the resistance of a material against being pulled apart. The tensile strength of a thin column of water has been measured at 140 kilograms per square centimeter. That means it would take a force of 140 kilograms (or about 308.5 pounds) per square centimeter (about ⅜ square inch) to pull the column of water apart. Today we know that under certain conditions a column of water molecules has a tensile strength that exceeds that of a thin steel wire. These are the required conditions:

1. The water must be contained in a continuous unbroken column, one that is free of air bubbles.
2. The water must be in a tube of almost microscopic diameter.
3. The tube must be made of material to which the water molecules can *adhere* (*adhesion* means the mutual attraction of unlike molecules).

The conditions found within the xylem of plants meet these requirements. The microscopic-diameter tracheids and vessels carry a continuous column of water and provide an inner surface to which water molecules can adhere. Biologists who have measured the tensile strength of water in xylem have found that it can exceed 140 kilograms per square centimeter, more than enough to allow for the pull of water through the plant.

The Transpiration Process. We now have a continuous column of water molecules that exhibits cohesion, adhesion, and tensile strength. But what happens to move that column upward in a vascular plant? First, you must remember that the column has been developing since the seed first germinated. Therefore, the plant has a continuous column of water in place when it breaks through the surface of the soil. The molecules are linked in a chainlike

fashion through the xylem. Second, the stomata release water vapor from the spongy mesophyll cells of the plant's leaves into the air.

Because of the drying effect of sunlight and wind, water evaporates, molecule by molecule, from the plant's stem and the stomata of its leaves. This process is called **transpiration.** The evaporation produces a water deficit in the mesophyll cells of the leaf. To replenish the deficit, water is drawn upward through the xylem into these water-deficient cells.

In other words, water ascends in a continuous column through the vessels of the xylem from the roots to the leaf. Because the xylem cells have small diameters, and because the water molecules are held in a continuous column, the water molecules cohere, and they are literally pulled through the plant by the act of transpiration. Figure 21.2 illustrates the cohesion-tension theory of water transport in a tree.

The Influence of Environmental Factors. A number of environmental factors influence a plant's rate of water loss. Temperature and humidity are extremely important. As the air temperature increases, the rate of transpiration by the plant increases. As the relative humidity (water concentration of the air) decreases, water loss from the plant increases. In addition, the amount of wind blowing across the surface of the leaf affects the rate of transpiration. The more wind, the more rapidly the water vapor is blown away from the leaves, increasing the transpiration rate.

Mineral Uptake

■ Dissolved minerals are very important in plant nutrition. These minerals are absorbed by active transport, sometimes with the assistance of mycorrhizae, a symbiotic fungus-root association.

The minerals plants require are inorganic substances that are found in the soil. Soil consists of tiny rock particles, decayed organic material, silica particles, and other substances. The minerals occurring in soil dissolve in water. Plants cannot absorb them unless they are in solution, and in order to survive, growing plants require these minerals. In the technique known as hydroponics, plants are grown only in water that has been supplemented with the proper minerals.

Because plants take the minerals that they need from the soil, the soil minerals often become depleted. That is why we periodically fertilize the soil

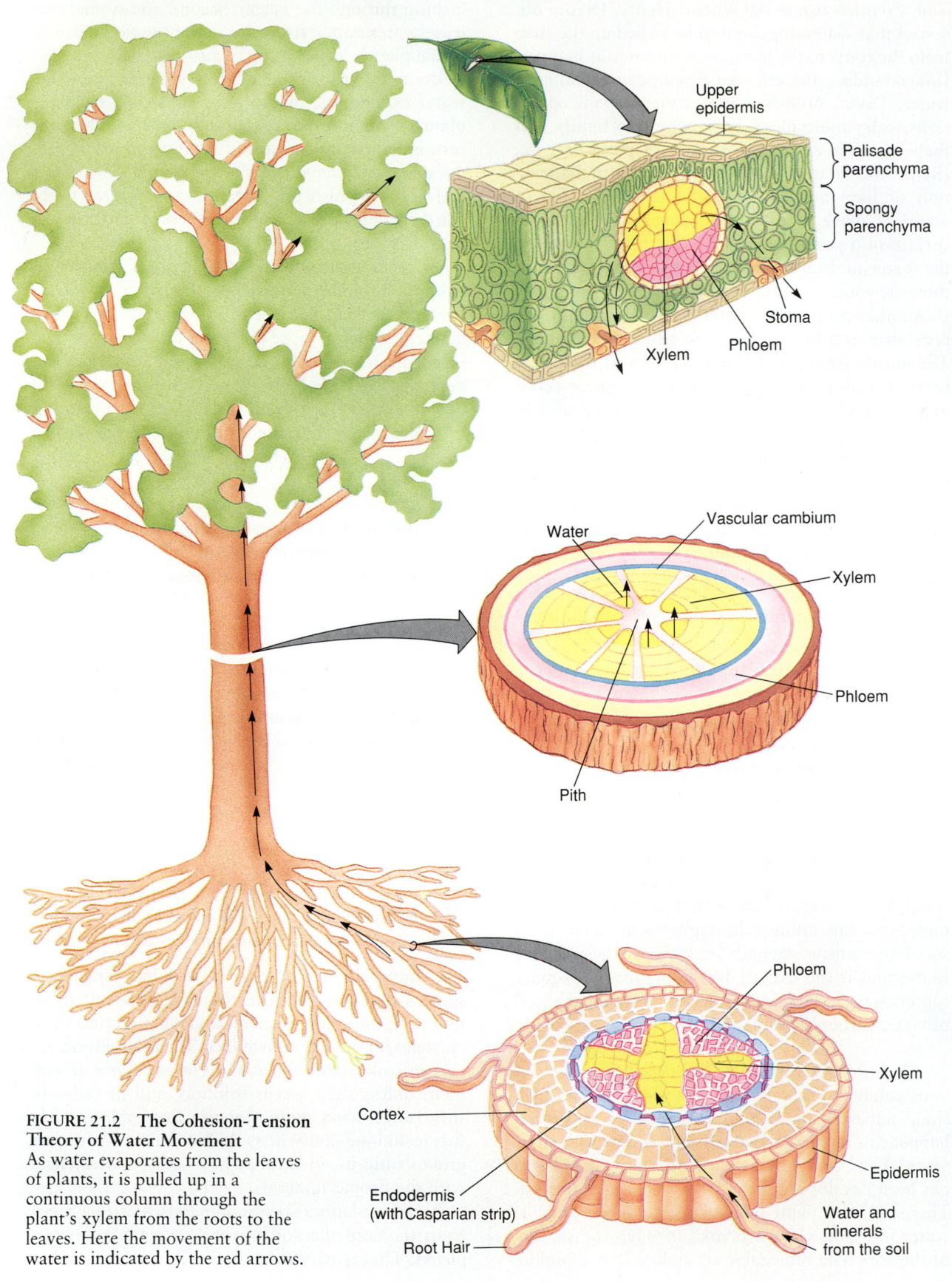

Upper
epidermis

Palisade
parenchyma

Spongy
parenchyma

Stoma

Xylem

Phloem

Water

Vascular cambium

Xylem

Phloem

Pith

Phloem

Xylem

Cortex

Epidermis

Endodermis
(with Casparian strip)

Water and
minerals
from the soil

Root Hair

**FIGURE 21.2 The Cohesion-Tension
Theory of Water Movement**
As water evaporates from the leaves
of plants, it is pulled up in a
continuous column through the
plant's xylem from the roots to the
leaves. Here the movement of the
water is indicated by the red arrows.

with additional minerals. Gardeners often provide a *mulch*, or covering of organic material, over the soil surface to retard weed growth, reduce moisture loss, and eventually to provide organic materials to the soil. Organic materials increase the soil's ability to absorb water. Table 20.1 lists many of the minerals plants require and indicates the function of each.

Active Transport. Plant roots absorb water by means of osmosis, but that is not the case with the absorption of minerals. Instead, minerals are usually absorbed by *active transport*. Once minerals dissolve in water, they move from a place of lower concentration (in the solution found in the soil) to a place of higher concentration (the solution found in the tissues of the plant). As you may remember, active transport (Chapter 5) requires an expenditure of energy. Energy from ATP activates the membrane pumps that move molecules from the soil into the xylem cells.

Mycorrhizae. Fungi often play a part in mineral uptake in the roots of certain plants. Research has shown that even when grown in solutions that contain the correct mix and proportions of nutrients, the seedlings of many trees, such as pine trees, will often die of malnutrition. However, if forest soil is added to the solution, even in small amounts, the seedlings grow well. Through experimentation and observation, botanists have determined that forest soil contains fungi that are critical to seedling growth. These fungi establish a symbiotic association with the plant's roots called a *mycorrhizal association*.

FIGURE 21.3 The Effect of Mycorrhizae on Plant Growth
Studies show that mycorrhizae (fungus roots) are essential to the growth of certain plants. Pictured here are pine seedlings. The ones on the right developed without the fungus. The one on the left developed with the mycorrhizae.

"Fungus roots" like this are called **mycorrhizae** (singular, **mycorrhiza**). The fungi can either surround the root or actually penetrate it. In either case, the fungi extend into the soil farther than the root hairs and increase the surface area for the absorption of water and dissolved minerals. In some instances the mycorrhizae break down complex molecules into substances the roots can absorb. Therefore, association with the fungus benefits the root. The root, in return, is a source of sugars and nitrogen for the fungus. Figure 21.3 illustrates the effect of mycorrhizae on the growth of a pine tree.

TRANSLOCATION: THE MOVEMENT OF THE PRODUCTS OF PHOTOSYNTHESIS

■ **The process of transporting sucrose — the ultimate product of photosynthesis — throughout the plant in the phloem is called translocation.**

We now know about the movement of water and dissolved minerals through the vascular plant, but how are the products of photosynthesis moved to nonphotosynthetic cells? How do the nonphotosynthetic cells obtain the high-energy carbon compounds required to produce ATP? The answer is by **translocation,** which is the transport of materials like *sucrose* (a sugar), the major high-energy carbon compound found dissolved in the sap, from one plant tissue to another.

Sap: Composition and Movement

■ **Phloem sap travels through the sieve tubes of the phloem. Botanists have used radioactive carbon dioxide and the activities of aphids (small sucking insects) to study the nature of sap and its speed of movement.**

In Chapter 20 we noted that **sap,** which is a thick fluid composed of the products of photosynthesis and associated substances dissolved in water, is transported in the phloem. Specifically, translocation of photosynthetic products occurs in sieve tubes. Sieve tube elements, unlike the vessel elements and tracheids of the xylem, are alive at maturity. The sieve tube elements are connected end to end to form the long tubes called sieve tubes, which are used to conduct the sap. Because the sieve tube elements are alive, it is very difficult to study translocation in plants directly. If a researcher disturbs the sieve tubes through experimentation, the pores

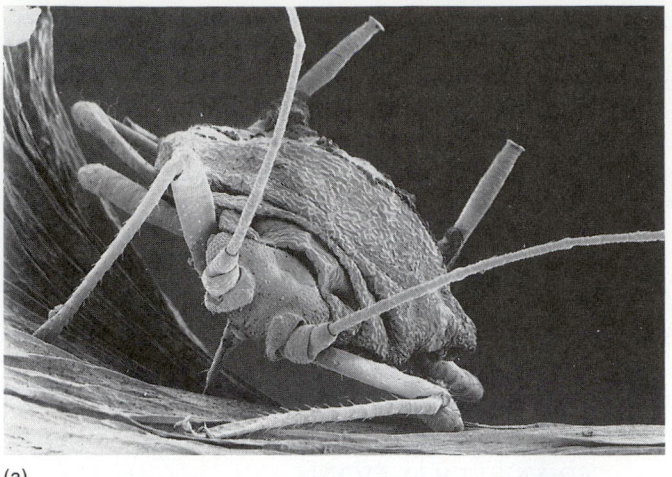
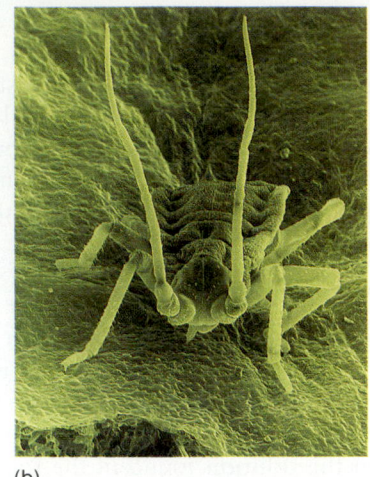

(a)

(b)

FIGURE 21.4 Aphids and Sap
Scientists use aphids to study sap and the translocation process in plants.
(a) Aphid feeding on the leaf of a tree. (b) Another view of an aphid inserting its stylet into the leaf to feed on the sugar in the phloem tissues.

in the sieve plates become plugged, and translocation stops.

However, by using CO_2 synthesized with radioactive isotopes of ^{14}C, experimenters can follow the course of the sugars made from radioactive CO_2 through the plant, learning much about the speed and direction of translocation. Researchers have also studied translocation by using aphids, which are small sucking insects that feed on the sap of plants. (You may have seen aphids on rosebushes, which they seem to be particularly attracted to.) When an aphid feeds, it inserts its mouth parts into a sieve tube in the phloem. The plant sap is under such high pressure that it flows out of the tube and into the mouth parts and then on into the aphid's body. In fact, the pressure is so great that some of the sap even drips out of the insect's posterior (Figure 21.4). Sap consists primarily of water, with a percentage of solutes that runs between 10 and 25. Sucrose accounts for 90 percent of the solutes; amino acids and other nitrogen compounds are found in low concentrations.

From research like this, botanists have determined the composition of sap, as well as the pressure and rate of translocation. They know that sap moves at about 1 meter (about 39 inches) per hour. This movement is much too rapid to be accounted for by diffusion, and so how can it be explained?

Pressure-Flow Hypothesis

■ **The pressure-flow hypothesis states that phloem sap moves from a source (the photosynthetic cells) to a sink (the location at which the sucrose**

is used or stored). Pressure exists at the source because sucrose is actively pumped into the phloem and because of the osmosis of water into the phloem. A low concentration of sap at the sink causes a chemical pull on the column of sap.

The **pressure-flow hypothesis** explains the phenomenon of translocation. In essence, the movement of sugars and other organic solutes during translocation follows a pattern of fluid flow called the *source-to-sink pattern*. The photosynthesizing cells or cells that store the products of photosynthesis serve as the *source* of sugars and other organic solutes. The plant cells that are using or storing the sugars act as the *sink*. Figure 21.5 illustrates the movement of photosynthetic products from the leaves to the roots.

Conversion of Glucose to Sucrose. The pressure-flow hypothesis follows these lines. (It will help you to think of the cells of the leaf in three categories: photosynthetic cells at the surface of the leaf, intermediate cells between the photosynthetic cells and the phloem, and sieve tubes in the phloem.) The process begins with the production of glucose in the photosynthetic parenchyma cells of the leaf. Glucose then diffuses into the intermediate cells, where it is converted to sucrose. After that, the intermediate cells actively pump the sucrose from a place of low concentration (themselves) to a place of high concentration (the sieve tube, which already contains sucrose formed elsewhere). Active transport of this sort requires energy from the cells.

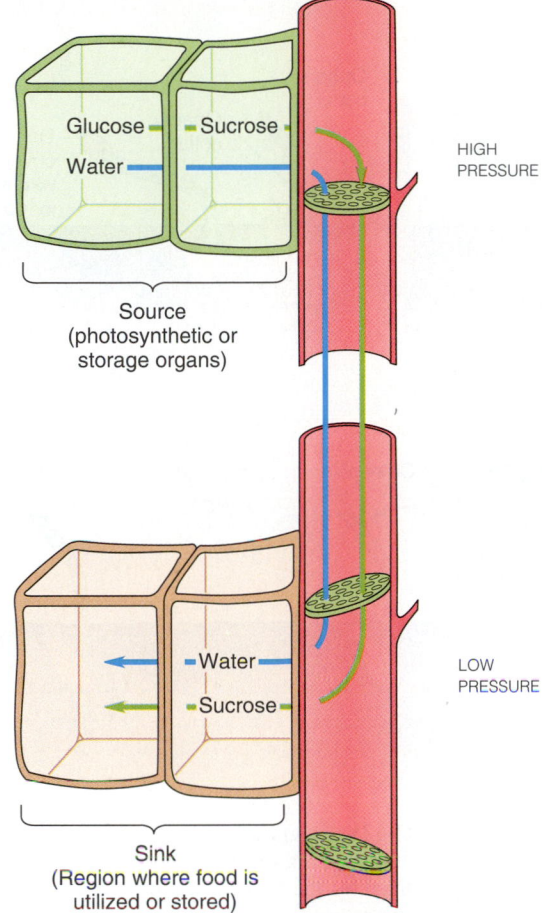

HIGH PRESSURE

Glucose — Sucrose
Water

Source
(photosynthetic or
storage organs)

LOW PRESSURE

Water
Sucrose

Sink
(Region where food is
utilized or stored)

FIGURE 21.5   The Pressure-Flow Hypothesis
Sucrose is actively transported into the sieve tube at the
source and removed at the sink. Water moves into the
sieve tube by osmosis, raising the pressure in the tube at
the source. As water moves out of the sieve tube at the
sink, the pressure in the tube drops. The difference in
pressure along the sieve tube moves the sucrose from the
source to the sink.

Role of Water. Because of the high concentration
of sucrose (and a correspondingly low concentration of water) in the sieve tube, water will move
into it by osmosis. As a result, the water pressure in
the sieve tube increases — in other words, there is
an increase in turgor pressure in the sieve tube, the
source.

Activity at the Sink. At the same time, other cells
at the *sink* are removing the sucrose from the sieve
tube. Water follows sucrose out of the sieve tube by
osmosis, decreasing its water pressure. The result is
a difference in pressure along the sieve tube, which
causes its contents to flow from the source to the
sink.

You can visualize movement due to differences
in pressure flow if you think about what happens
when you turn on a faucet connected to a garden
hose. When you turn the faucet on, water flows
from the high pressure in the water pipe (the source)
to the low pressure in the empty hose (the sink).

The interrelationship between transpiration, active solute absorption, water transport, and translocation accounts for the transportation of the
materials required for plant growth and maintenance. However, what regulates plant growth?
How are plant activities coordinated? Why do certain plants flower only during the day? How do
plant cells communicate with one another? Read
on.

PLANT HORMONES

■ **Plant hormones are molecules that regulate plant
growth and development. Hormones induce physiological changes in the cells they enter.**

The growth and development of plants, like animal
growth and development, are dependent on the
interplay of internal and external environmental
factors. Plant hormones (information-carrying molecules, or "chemical messengers," if you will) are
the major internal regulators of growth and development. They are usually produced by one cell or
tissue and carried to other cells or tissues, where
they exert their effects. The cells affected by a hormone are called *target cells*. Hormones influence the
target cells by inhibiting or stimulating their activities. Hormones exert their effects in very low concentrations.

Plant hormones are usually less specific than animal hormones in terms of the cells that they affect
and the physiological changes that they induce. The
responses initiated by plant hormones also do not
occur as rapidly as those of animal hormones. For
example, when an animal is frightened, the adrenal
glands produce hormones such as epinephrine that
reach target cells within seconds and effect an almost instantaneous response. Even though some
plant responses occur within minutes, most responses take much longer. This is because transport
(usually translocation) in plants is simply not as
rapid as circulation in vertebrate animals.

Plant hormones exert some of their control
through stimulating or inhibiting plant growth. The
hormones are manufactured by meristematic tissues
(Chapter 20) and are transported by the phloem to
other parts of the plant. Plant physiologists divide
plant hormones into five groups: auxins, gibberellins, cytokinins, abscisic acid, and ethylene.

Auxins

■ **Auxins influence cell elongation. They also influence phototropism, abscission, and apical dominance. Synthetic auxins are used as weed killers.**

Phototropism. The first plant hormone to be discovered and isolated was **auxin.** The influence of auxin was first described by Charles Darwin and his son Francis in 1881, although the hormone was not identified and named until 1926. The Darwins were investigating *phototropism* (the way a plant bends toward the light) in grass seedlings. They observed that the grass shoot bent toward the light at a point just below the tip of the coleoptile (the hollow sheath that surrounds and protects the shoot tip of a monocot seedling). The Darwins tested their observation by covering the tip of the coleoptile with an opaque material. When light was directed toward one side of the shoot, the shoot did not bend. However, if the opaque covering was placed below the tip, bending did occur. Finally, when they punched a hole in the covering before placing it over the tip, bending still occurred. Darwin and his son concluded that when plants are exposed to lateral light (light from one side), the tip produces a substance that is transported to the lower part of the stem and causes the lower part to bend (Figure 21.6a).

It was Dutch plant physiologist Frits W. Went (see Profile) who isolated the chemical product that produced the bending. Went cut the coleoptile tips off many oat seedlings and placed the tips on blocks of agar (a gelatinlike substance derived from algae) for an hour. He then discarded the tips and placed a section of the agar block on one side of each of the coleoptile stumps. The seedlings bent away from the side on which the agar block was placed, even if the seedlings were kept in the dark. An untreated block of agar did not produce this bending, nor did agar blocks on which cut portions of the lower shoot had been placed. Went concluded that the coleoptile tip produced a chemical substance (a hormone) that induced the shoot to bend toward the light (Figure 21.6b).

Went called this substance *auxin,* from the Greek *auxein,* meaning "to increase." The auxin was stimulating cell elongation. On the side having the agar block, excess auxin caused increased cell elongation, and the stem bent away from the agar block. Light causes auxin to diffuse to the dark side of the stem. Therefore, the light side of the plant has less auxin than the dark side. The cells on the dark side elongate more and the plant bends toward the light.

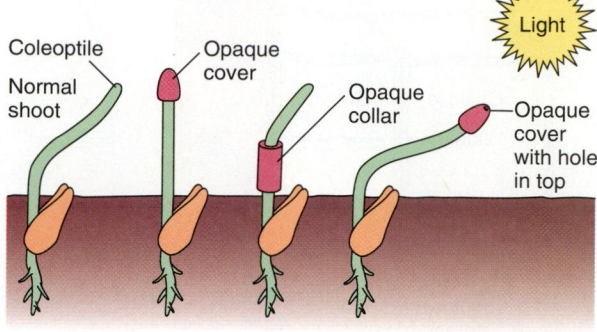

(a) The Darwins

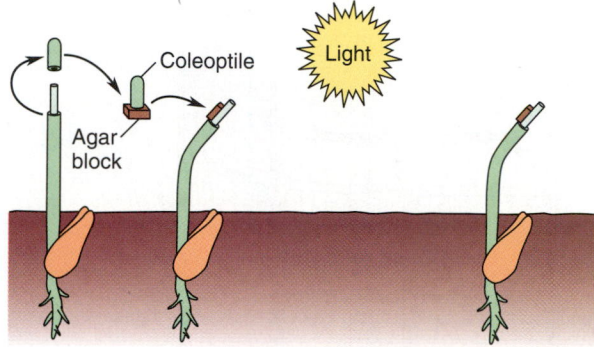

(b) Went

FIGURE 21.6 The Discovery of Auxin
(a) Charles Darwin and his son Francis proposed that a substance produced by the tip of a seedling (the coleoptile), when exposed to light, stimulates the seedling to bend toward the light source. (b) In experiments some forty years later, Frits W. Went concluded that the coleoptile produces a chemical substance that induces a shoot to bend toward a light source. He called this substance *auxin.*

Botanists now know that there are several types of auxin, all of which are growth-stimulating hormones responsible for cell elongation. The best-known auxin is *indoleacetic acid* (IAA). Another auxin is *phenylacetic acid.* Auxin causes the cell wall of the target cell to soften, and the turgor pressure on the inside of the cell wall causes the cell to elongate. Auxins can have numerous effects on plant cells, depending on how the specific cells respond to that hormone.

Leaf and Fruit Drop. If you have ever had to rake the lawn, you can put the ultimate blame on a shortage of auxin in the trees around your house. When leaves fall away from a plant, the process is called **leaf abscission.** Auxin prevents or delays leaf abscission. At the base of the petiole, the point where the leaf joins the stem, a decrease in auxin

Dr. Frits Warmolt Went

Botanist

The late Dr. Frits Warmolt Went, one of the codiscoverers of the plant hormone auxin, had a long and varied carreer. Born and educated in the Netherlands, where he received a Ph.D. in botany at the University of Utrecht, his career led him to the deserts of Nevada, with a number of stops in between. After teaching at Utrecht from 1923 to 1927, Went took a position at the Botanical Gardens in Java, working on the revegetation of Krakatoa (the volcano that exploded in 1883). Then, in 1933, Went emigrated to the United States and joined the faculty of the California Institute of Technology. During his career he also worked as director of the Missouri Botanical Gardens in St. Louis and at the University of Nevada at Reno and its Desert Research Group Institute.

During the first 10 years of his research career, Went concentrated on the mechanism of internal regulation of plant growth, developing several techniques for the study of plant hormones. During the rest of his career, he primarily studied the external factors that control plant growth, especially the effects of climate and light.

Went is probably best known for his research on phototropism, the response of plants to unidirectional light (light that shines on only one side of the shoot). First, he succeeded in diffusing the hormones produced by the coleoptiles (the sheath surrounding the shoot) of plants into agar blocks. He found that with the plant hormones contained in the agar blocks he could then manipulate plant development. He called these plant hormones auxins.

Using oat seedlings as the experimental model, Went was able to demonstrate a number of properties of auxins. He showed that the hormones are stable even when exposed to light and heat. He was also able to explain why a plant stem curves toward the light. He did so by demonstrating that unilateral light resulted in a lower concentration of auxin in the cells of the plant shoot on the side facing the light, and a higher concentration of auxin in the cells on the side away from the light. Because of the increased concentration of auxin, the cells on the side away from the light elongated faster than the cells on the side facing the light. The differences in the rate of growth resulted in the curvature of the stem toward the light.

By means of his experiments, Went succeeded in explaining light curvature in plants. Because N. Cholodny, a Russian botanist, reached the same conclusion through theoretical work at the same time, the theory explaining plant curvature toward light is called the Cholodny-Went theory. Botanists later determined that light induces auxin to move from the light side to the dark side of the plant. Hence, the side facing the light grows more slowly. Unequal growth causes the plant to grow toward the light — a reaction called positive phototropism.

At the California Institute of

Technology, Went worked on a number of plant hormone problems. He demonstrated the role of hormones in the production of roots on plant stems. He also investigated the general effects of climate on plants, by breaking down climate into its individual components: temperature, light, wind, and so on. Through his work he demonstrated that plants are thermoperiodic — that is, they grow at their maximum rate when the daytime temperature is higher than the nighttime temperature.

Went also investigated the volatile chemicals relased by plants into the atmosphere. These chemicals produce the blue heat or summer haze over many desert regions of sagebrush. Frits Went is only one example of a scientist whose desire to know and to understand led him from one discovery to others as he explored the world around him.

concentration, simply stated, results in the weakening of the glue holding the cells at the base of the petiole together, forming the *abscission zone*. Leaves drop off trees in the fall because the abscission zone breaks the connection between the stem and petiole. Treatment with auxin can delay the development of the abscission zone.

In a similar way, auxins also delay fruit drop. Under normal conditions, an abscission zone forms at the point where the fruit stalk meets the plant stem. Auxins prevent or delay the formation of the abscission zone. If fruit trees are sprayed with synthetic auxinlike compounds, fruit can be induced to cling to the trees for several days. In the spring, greater auxin synthesis stimulates the vascular cambium to divide, increasing the diameter of the plant.

Weed Killers. Plant chemists have developed a variety of synthetic auxins called herbicides that are used to kill unwanted dicots, usually described as broad-leafed weeds. For instance, grass is a monocot, and dandelions are dicots. The synthetic auxin called 2,4-D (2,4-dichlorophenoxyacetic acid) will kill dicots (since it is toxic in high concentrations) but not monocots when sprayed on a field in the proper concentrations. When a broad-leafed dicot absorbs 2,4-D, the substance stimulates wild, rapid, and unregulated growth of some plant parts and inhibits the growth of others. Biologists do not understand the exact mechanism yet. Nonetheless, the substance is a good weed killer. (Weeds, by the way, are plants growing where you do not want them — for example, a rosebush in a flower bed beside a home would not be a weed, but if it is growing in a cornfield, that is a different story.)

As far as we know, 2,4-D does not have any harmful effects on humans. However, we have all heard about the harmful effects of another synthetic plant growth regulator called Agent Orange, which was used as a defoliant to clear the jungles during the Vietnam War. Agent Orange is a mixture of equal parts of 2,4-D and 2,4,5-T (2,4,5-trichlorophenoxyacetic acid). The problems with this defoliant lie not with 2,4,-D but with 2,4,5-T. The harmful effects of Agent Orange are believed to result from the fact that during the synthesis of 2,4,5-T small amounts of a substance called TCDD (2,3,7,8-tetrachlorodixbenzoparadioxin) is formed as a contaminant. It is TCDD that can have damaging effects. Experiments with laboratory animals have demonstrated that 2,4,5-T can cause leukemia, miscarriages, birth defects, and lung and liver diseases. The substance has also been implicated in other human cancers.

Gibberellins

■ **Gibberellins are hormones that promote stem elongation and seed germination.**

Found in plants and fungi, **gibberellins** are a group of hormones that promote stem elongation. Unlike auxin, which promotes only cell elongation, the gibberellins promote both cell division and cell elongation. They were first isolated from a fungus called *Gibberella fujikuroi,* which causes the "foolish seedling disease" in rice. Plants affected with the disease are spindly and tend to fall over. In addition, they do not produce normal flowers. Between 1926 and 1935, Japanese scientists studied and isolated the chemical substance causing the foolish seedling disease and named the substance gibberellin, since they were working with *G. fujikuroi.* (English-speaking scientists did not learn of the Japanese work until the 1950s, first because of World War II and later because of the language barrier, since the research was reported in Japanese journals.) Since that time botanists have isolated and identified more than 65 specific gibberellins, each of which has a different chemical structure. Gibberellic acid, the form produced by *G. fujikuroi,* is the best known.

Dwarf Varieties. If dwarf plants, such as the corn mutant shown in Figure 21.7, are supplied with gibberellin, they will grow to a normal height. Because the plants do have the potential to grow tall when gibberellin is added, the results of the experiment depicted in Figure 21.7 suggest that this plant's small size may be due to its inability to synthesize the hormone. In corn, four different dwarf mutants have been identified. None of these mutants can synthesize gibberellin.

FIGURE 21.7 Gibberellin: A Growth Hormone
When a plant is treated with gibberellin (right), its stem elongates. The plant on the left is the normal plant (the control).

Seed Germination. Gibberellins are found in very high concentrations in seeds and play a role in seed germination. In some seeds, the hormone stimulates the production of an enzyme called *alpha amylase,* which breaks down starch to glucose. The germinating seed then uses the glucose for growth and development until the embryo can produce its own glucose through photosynthesis.

Cytokinins

■ **Cytokinins are hormones that stimulate cell division. In conjunction with auxin, cytokinins help determine the differentiation and specialization of the primary meristematic tissue.**

Cytokinesis means cell division. So, as you might guess, the **cytokinins** are a group of plant hormones that stimulate cell division. The molecular structure of cytokinins is similar to the nitrogen-base adenine. The first biologically active cytokinin, *kinetin,* was isolated from a DNA preparation that was "aged" by heating it in an acid solution. Working with corn kernels (*Zea mays*), researchers isolated the first naturally occurring cytokinin, which they named *zeatin.* By varying the ratio of auxins and cytokinins, and providing tissue cultures of undifferentiated cells with different concentrations of auxins and cytokinins, botanists have been able to determine that cytokinins working in conjunction with auxin are responsible for the development of the different plant organs.

For example, through tissue culture techniques botanists have determined that an undifferentiated meristematic cell can either elongate or enlarge and divide. A cell that divides repeatedly remains undifferentiated, but a cell that elongates usually becomes differentiated or specialized. By varying the ratio of auxin to cytokinin, plant physiologists have determined the effects of these two hormones working together. When the relative concentration of auxin is higher than that of cytokinin, the tissue develops into roots. If the concentration of cytokinin is higher than that of auxin, the tissue develops into buds. When the concentrations of auxin and cytokinin are equal, the cells of the tissue culture remain meristematic (Figure 21.8). Hence, the relative concentration of auxin and cytokinin is responsible for cell differentiation and development in seed plants.

Growth and development are not quite that simple, however. Recent evidence demonstrates that the presence of calcium ions modifies the action of auxin and cytokinin. If a tissue culture contains a high concentration of calcium ions, cell division is favored over elongation. The cell divides, even if the auxin-cytokinin concentration would normally stimulate cell enlargement. Hence, the role of hormones in plants, as in animals, involves the interaction of several substances. Environmental influences also can affect metabolic and developmental processes within the cells that receive the hormones.

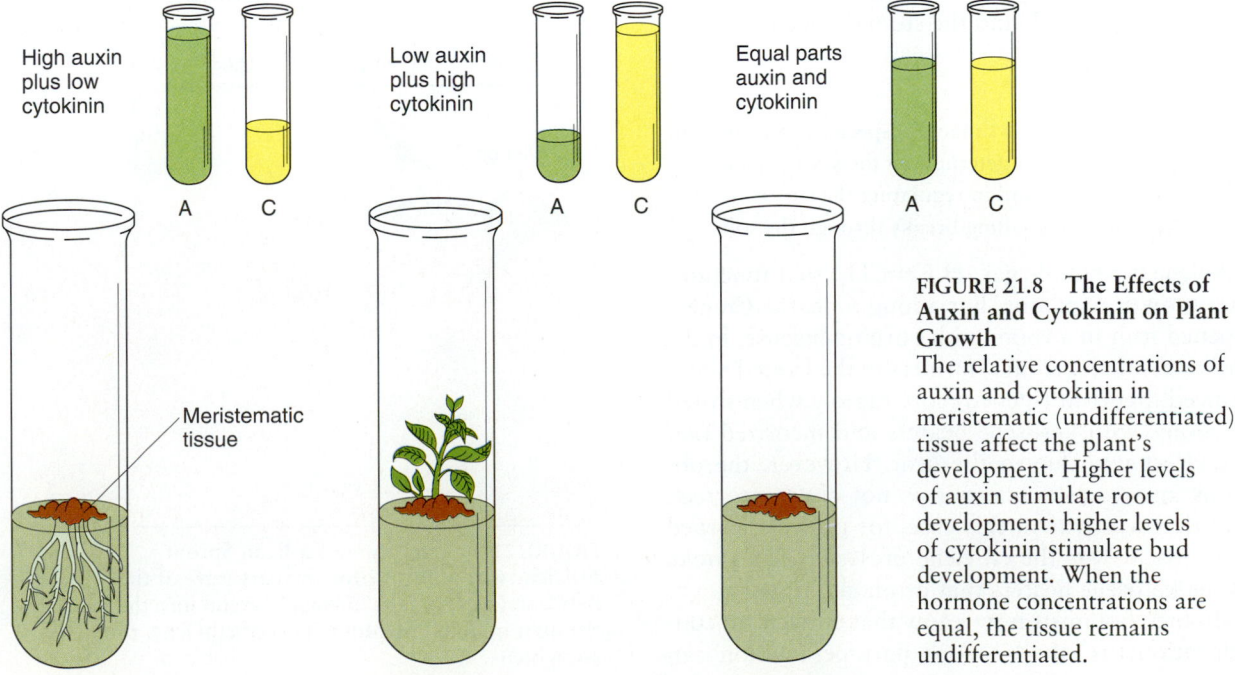

High auxin plus low cytokinin

Low auxin plus high cytokinin

Equal parts auxin and cytokinin

A C

Meristematic tissue

FIGURE 21.8 The Effects of Auxin and Cytokinin on Plant Growth
The relative concentrations of auxin and cytokinin in meristematic (undifferentiated) tissue affect the plant's development. Higher levels of auxin stimulate root development; higher levels of cytokinin stimulate bud development. When the hormone concentrations are equal, the tissue remains undifferentiated.

Plant Hormones

Abscisic Acid

■ **Abscisic acid is a hormone that promotes dormancy in plants. It prevents buds from producing leaves and is instrumental in closing stomata during very dry conditions.**

Abscisic acid (ABA) is a plant hormone that promotes abscission by weakening the abscission zone. Today, however, botanists believe that ABA's primary role is to promote winter dormancy. High levels of abscisic acid will stimulate the formation of winter buds. If abscisic acid is applied to a developing bud, the parts of the bud that would normally develop into leaves will be converted to bud scales. In this state, they are said to be in a dormant stage, functioning as if environmental conditions were not conducive to growth. Nurseries take advantage of this to protect saplings and seedlings that could be damaged during transit. They spray the plant with abscisic acid, send it to its destination, then spray again, this time with gibberellin, which counteracts the dormancy brought on by abscisic acid.

Abscisic acid is also responsible for the closing of stomata. Wilted leaves produce much more abscisic acid than leaves not experiencing water stress. The increase in abscisic acid concentration interferes with the transport and retention of potassium ions by the guard cells. Because the guard cells have a lower concentration of potassium ions than the adjacent epidermal cells, water diffuses out of the guard cells and the stomata close. As more water becomes available, plant cells can metabolize more efficiently, breaking down abscisic acid in the process. Water then diffuses back into the guard cells, they become turgid, and the stomata open.

Ethylene

■ **Ethylene is a growth factor especially influential in fruit ripening, in determining the sex of single-sexed flowers, and in regulating the curvature of the stem as a seedling breaks through the soil.**

Ethylene is a simple gas ($H_2C{=}CH_2$) that functions as a growth regulator. For a long time, the Chinese ripened fruit in a room with burning incense. In the early twentieth century, farmers in the United States noticed that fruit ripened more rapidly when stored in rooms with kerosene heaters and theorized that the effect was due to the heat. However, the obvious answer to a question is not always correct, and that was true in this case, for growers learned later that it was the ethylene present in the smoke of the kerosene heaters that promoted ripening.

In fact, botanists now know that ethylene in concentrations as small as one part per million can stimulate fruit ripening. Consider how small a ratio that is! It is as if you had a pile of one million pebbles, one of which is an uncut diamond.

Ethylene is given off by fruit as it ripens. Have you ever placed fruit in a brown paper bag to ripen? Because the fruit is enclosed, the concentration of the ethylene given off by the fruit is increased, which in turn increases the rate of ripening. Spoiling fruit also gives off ethylene. (Now you should know the answer to the old question: Can one rotten apple spoil the entire barrel?)

Ethylene has many other functions in addition to fruit ripening. For example, in plants that have separate male and female flowers, such as pumpkins, an application of ethylene before the flower bud is formed will produce an increased number of female flowers (such flowers are called *pistillate*). On the other hand, if you treat the buds with gibberellins, they become *staminate* (male).

If you have ever watched a bean sprout break through the soil, you may have noticed that the first thing that pushes through the soil is a curved stem located just below the cotyledons (Figure 21.9). This portion of the plant is called the *hypocotyl*, and ethylene is believed to be responsible for its

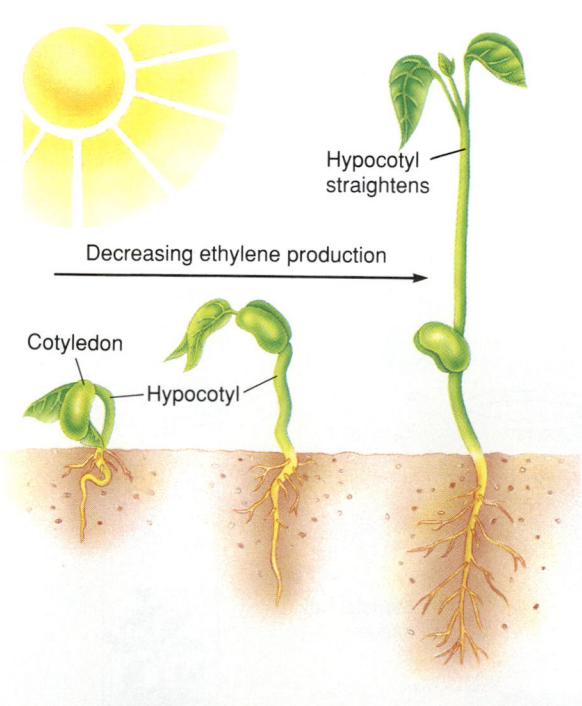

FIGURE 21.9   Curvature of a Bean Sprout
Ethylene is responsible for the curvature of the hypocotyl as it breaks through the soil into the light. As the light inhibits the production of ethylene, the shoot straightens.

curvature. Once the hypocotyl is exposed to light, which inhibits the production of ethylene, the shoot begins to straighten.

As you can see, hormones are responsible for the coordination of many metabolic and developmental activities in plants. Table 21.1 and Figure 21.10 summarize some plant hormones and their functions.

FIGURE 21.10 Hormones and Plants
An illustration of some of the hormonal activities in a plant.

TABLE 21.1
Major Plant Hormone Groups and Their Functions

Auxins	Growth-stimulating, responsible for cell elongation
Gibberellins	Stimulate stem growth
Cytokinins	Stimulate cell division
Abscisic acid	Stimulates fruit and leaf drop, triggers dormancy
Ethylene	Involved in fruit ripening and sex determination of some flowers

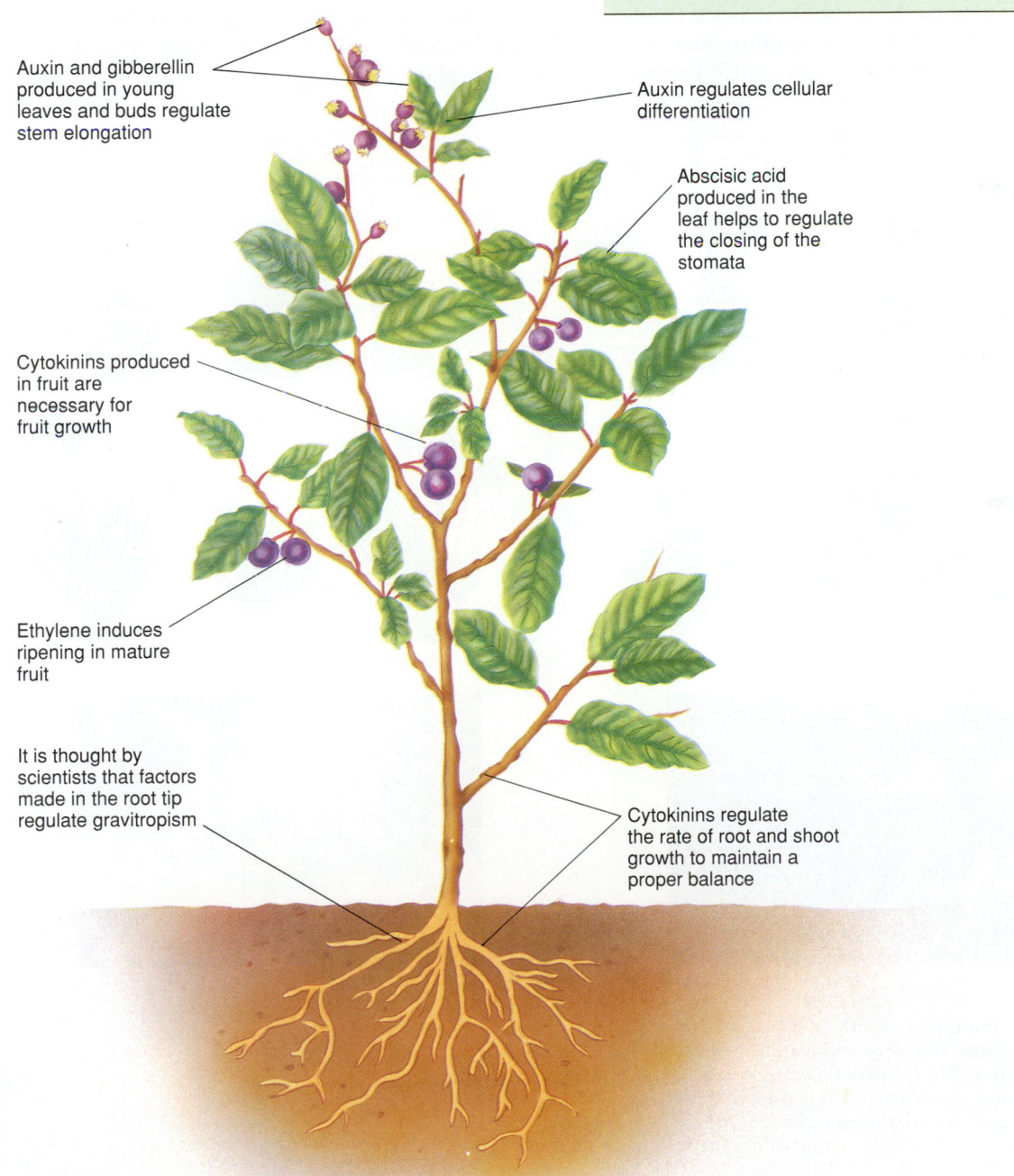

Auxin and gibberellin produced in young leaves and buds regulate stem elongation

Auxin regulates cellular differentiation

Abscisic acid produced in the leaf helps to regulate the closing of the stomata

Cytokinins produced in fruit are necessary for fruit growth

Ethylene induces ripening in mature fruit

It is thought by scientists that factors made in the root tip regulate gravitropism

Cytokinins regulate the rate of root and shoot growth to maintain a proper balance

PHOTOPERIODISM: PLANT RESPONSES TO THE DURATION OF LIGHT

■ The way plants respond to the changing ratio between light and darkness is called photoperiodism. Based on their response to light, plants fall into one of three general types: day-neutral, short-day, and long-day plants.

Why do morning glory flowers open during the day and close at night? How do plants anticipate the changing of the seasons or sense the relative periods of light and darkness? How are florists able to have available 10,000 to 30,000 chrysanthemums at the peak of their bloom for homecoming celebrations, regardless of the fact that they are not in season?

The way in which plants respond to the duration of periods of light and darkness is known as **photoperiodism** (which means "period of light"). In terms of photoperiod, there are three kinds of flowering plants: *day-neutral*, *short-day*, and *long-day*. Plants that flower regardless of the day length are **day-neutral**. Garden beans, dandelions, and carnations are examples of day-neutral plants. **Short-day plants** flower in the early spring or in the fall, when the hours of daylight fall short of a critical period. Strawberries, cocklebur, and ragweed plants are all short-day plants. **Long-day plants** flower in summer, when there is more than a critical period of light. Clover and spinach are long-day plants.

(a)   (b)

FIGURE 21.11 Photoperiodism
(a) Short-day plants, like chrysanthemums, require an extended period of darkness to flower. They do not flower if the period of darkness is less than eight hours or if it is interrupted by a brief flash of light. (b) Long-day plants, like the iris, require less darkness to flower. Although normally they do not flower if exposed to the longer period of darkness required by short-day plants, they do flower if that period of darkness is disrupted by a flash of light.

Affecting Blooming by Manipulating Light and Darkness

■ Botanists have discovered that flowering is dependent on an uninterrupted period of darkness. Interrupting the period of darkness with light prevents flowering. Interrupting the period of light with darkness has no effect.

Unfortunately, the terms short-day plant and long-day plant are misleading because in reality it is not day length but the length of the period of darkness that causes a response — a fact determined almost by accident while researchers were working with experiments that involved growing plants in total darkness. Using the cocklebur (a short-day plant), Karl C. Hammer and James Bonner were investigating photoperiodism. They determined that short-day length (16½ hours of light or less) induced flowering. However, they learned two other things: (1) if they interrupted the dark periods with as little as 1 minute of light, flowering did not occur; (2) if they interrupted the periods of light with darkness, flowering still occurred (Figure 21.11).

Therefore, it is the length of uninterrupted darkness, not the length of uninterrupted light, that regulates flowering. Somehow plants possess an internal "biological clock" that enables them to measure length of time in darkness. Biological clocks (which have a role in the life cycles of many kinds of organisms, including plants and animals) are believed to be information-controlling mechanisms that establish *circadian rhythms* (daily cycles). The controlling mechanisms for these daily rhythms are still poorly understood.

How do plants measure darkness? How does the amount of light influence growth and flowering? We'll turn to these questions next.

Phytochrome

■ Phytochrome is a pale blue pigment that botanists believe is involved in the biological clocks of plants. This pigment has two forms: P_r, which absorbs red light, and P_{fr}, which absorbs far-red light. These forms are interconvertible — that is, they can change back and forth from one to the other.

Early in this century a group of plant physiologists in Beltsville, Maryland, began to study the phenomenon of photoperiodism in plants. They isolated a light receptor molecule known as **phytochrome**, a pale-blue pigment that they believed was involved in plant photoperiodism. There are two forms of phytochrome. One is most active when absorbing

wavelengths in the red range (660 nanometers); hence it is called *phytochrome-red,* or P_r for short. The other form is most active when absorbing light in the far-red wavelength range of 730 nanometers; hence it is called *phytochrome-far-red,* or P_{fr}. Their wavelengths are at the far limits of visible light. When the P_r form of the pigment absorbs red light, it is transformed into P_{fr}; when P_{fr} absorbs far-red light, it is transformed into P_r. Figure 21.12 illustrates this conversion.

In sunlight, an equilibrium is achieved between the amount of P_r and P_{fr} in the plant. In darkness, the P_{fr} is slowly converted back to P_r. For short-day plants to flower, the P_{fr} must all convert back to P_r during the period of darkness. A burst of light will prevent the conversion and will also prevent flowering. For long-day plants to flower, some P_{fr} must remain after the period of darkness. A burst of light will prevent all the P_{fr} from converting to P_r, and flowering will occur. Research has shown that other

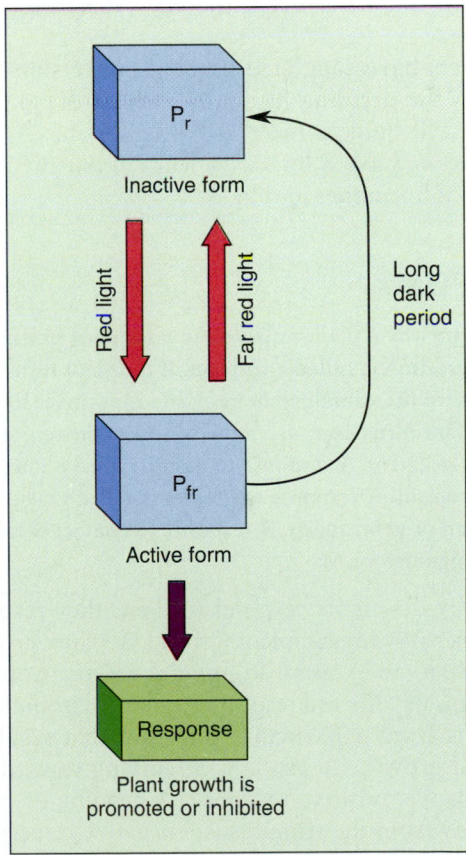

FIGURE 21.12 Phytochrome Conversion
Phytochrome plays a part in promoting and inhibiting plant growth. The absorption of red light, which predominates in daylight, converts phytochrome to its active form (P_{fr}); at night, when far-red light predominates, phytochrome converts back to its inactive form (P_r).

factors must work with phytochrome to control flowering. This is another area where exciting botanical research is being carried out.

Hormonal Work of Phytochrome. Phytochrome does not work alone, however. It works in conjunction with the plant hormones just described to regulate many aspects of plant physiology and development. For example, in some plants phytochrome also stimulates germination, and in others it stimulates leaf growth and branch length. Because of its variety of functions, botanists suggest that phytochrome is the mechanism that turns hormones on or off.

If they are right, how does phytochrome do this? The mechanism of phytochrome action is not completely understood. However, several hypotheses have been proposed. Because phytochrome is found primarily in the plasma membrane, some scientists believe that the substance alters membrane permeability. This determines the materials that will pass into and out of the cell, thereby regulating plant activities.

Others have suggested that the conversion of P_r to P_{fr} is the deciding factor in whether or not hormones will bind to the plasma membrane. As you can see, we have a lot yet to learn about the interaction of hormones and light.

TROPISMS

■ **The way a plant responds to a stimulus from one direction is called a tropism. If the plant turns toward the stimulus, the response is positive. If the plant turns away from the stimulus, the response is negative. A response to light is called phototropism. A response to gravity is called gravitropism or geotropism. A response to contact is called thigmotropism.**

Not only do plants respond to light, they respond to numerous environmental cues. A plant growth response to an external stimulus is called a **tropism**. Tropisms usually are responses to an environmental stimulus from a particular direction that result in unequal growth. A bending or curving toward the stimulus is a positive response. A bending or curving away from the stimulus is a negative response.

Phototropism

Let's look again at the example of that plant in your room that grows and bends toward a lighted window. This plant behavior is due to the fact that the auxin produced in the shoot tip (primarily IAA)

diffuses to the unlighted side, where it stimulates cell elongation. The cells on the lighted side receive less of the hormone, and so they elongate less than the cells on the dark side. Because the cells on the side away from the light elongate faster than those facing the light, the result is an unequal growth rate on opposite sides of the stem. The ultimate result is the curving of the plant toward the light (Figure 21.13).

The response of a plant to the light is called **phototropism** (light response). In most cases, a stem will grow toward the light, and so the response is said to be *positive* phototropism. Roots, however, grow away from the light, so that their response is an example of *negative* phototropism.

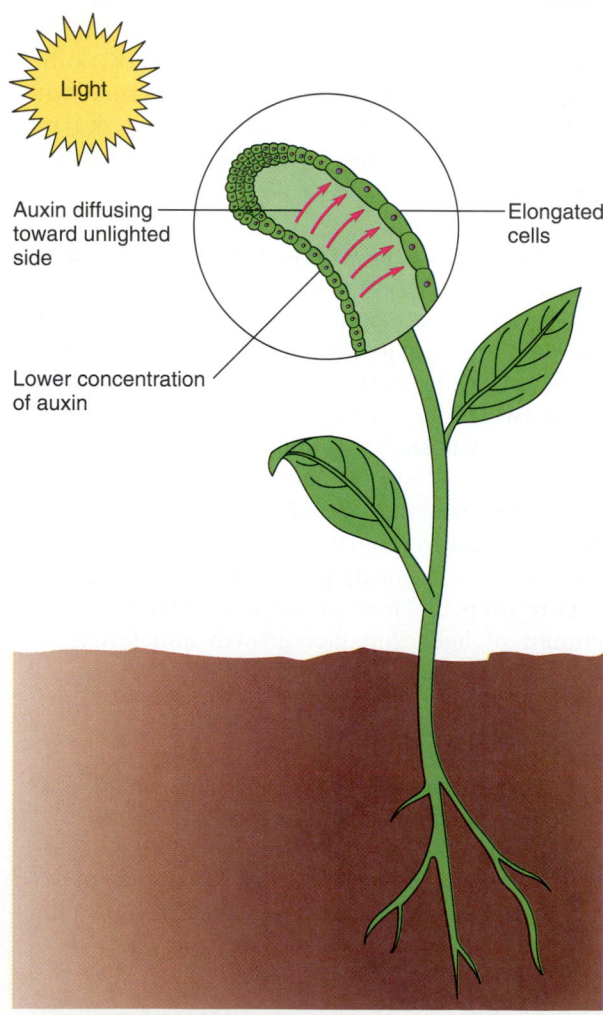

Light

Auxin diffusing toward unlighted side

Elongated cells

Lower concentration of auxin

FIGURE 21.13 **Phototropism**
The concentration of auxin in the shoot tip explains why plants grow toward light. The hormone diffuses from the lighted side of the tip to the unlighted side. The higher concentration of auxin on the unlighted side causes the cells there to elongate faster than the cells on the lighted side, bending the plant toward the light.

Gravitropism or Geotropism

Onions are grown in most gardens from what are called "onion sets," very small onion bulbs shaped like a teardrop. The pointed top of the set is the shoot end, and the rounded bottom is the root end. When I was a child, my father used to tell me to make sure I placed the onion set in the ground with the shoot facing up and the root facing down. Being mischievous at times, I used to set some with the roots facing up. Regardless, the sets would send up their green spiky leaves and when I dug them up to eat, their roots were growing down. What happened?

The effect of gravity on the structures of a plant is called **gravitropism** or *geotropism*. Roots show positive geotropism, because they grow toward gravity. Stems, on the other hand, are negatively geotropic. In roots, the root cap seems to be able to detect gravity, produce hormones that cause unequal cell elongation, and make the root turn down.

The importance of gravity was clearly demonstrated by experiments carried out as part of the space program. In experiments conducted in space capsules, plants grown on earth were placed in the spacecraft. They already had an established up-down axis that was determined by gravity prior to being launched into space. Changing their orientation in space did not change the direction of their growth, as it would have on earth. Therefore, gravity must influence normal growth.

Other Tropisms

Have you ever touched the leaf of the sensitive plant *Mimosa pudica* (Figure 21.14) or noticed how the leaves of the insect-eating Venus flytrap close around an insect when it lands on the leaf? These movements are due to changes in turgor pressure; they are called *turgor movements*. The movements of tendrils and twining stems are called *thigmotropisms*. *Nastic* movements are those that occur in flat organs such as leaves or petals more or less at random, such as movement of a leaf as it buds. They have no overall effect on the position of the leaf or the basic shape of the plant. *Hydrotropism* is movement in response to the presence of water. When your household water pipes become clogged with roots, you are seeing the effects of hydrotropism, since the roots have grown toward the water in the pipes. *Chemotropism* is movement in response to a chemical stimulus. An excellent example of chemotropism is the development of the pollen tube, which follows along a concentration gradient for sugar that leads to the ovule.

As the last three chapters have shown, plants are not the simple, uncomplicated organisms that we humans often think they are. In fact, their life cycles are quite complex, and they demonstrate amazing adaptations. Their beauty and the intricacy of their functions are especially awe-inspiring.

SUMMARY

1. The stomata of the leaf are opened and closed by changes in the turgor pressure within the guard cells. These changes are caused by the pumping of potassium ions and other solutes into and out of the guard cells. When the concentration of solutes is low, water moves out of the guard cells to the adjacent cells, and the stomata close. When the concen-

FIGURE 21.14   The Sensitive Plant (a) *Mimosa pudica*, the sensitive plant, responds to touch by folding its leaves (b). This movement is a thigmotropism that is caused by changes in turgor pressure.

(a)   (b)

tration of solutes is high, water moves into the guard cells from the adjacent cells, increasing the turgor pressure and causing the stomata to open.

2. Photosynthesis depletes carbon dioxide. A low carbon dioxide concentration stimulates potassium ions to move into the guard cells. Water follows, stimulating the stomata to open and absorb carbon dioxide.

3. Water and dissolved minerals enter the roots through the root hairs. Water enters by osmosis and the dissolved minerals usually enter through active transport.

4. Water and dissolved minerals move through the xylem from the roots to the leaves. This movement is best explained by the cohesion-tension theory, which states that loss of water from the plant body (transpiration) starts the movement of water up through the xylem.

5. Translocation in the phloem is the movement of the products of photosynthesis through the plant tissue, between regions of production, storage, and use. The movement of sap is best explained by the pressure-flow hypothesis, which states that the sap in the sieve tubes moves from regions of higher turgor pressure to regions of lower turgor pressure.

6. Plant growth and development are regulated by the interaction of the environment and plant hormones. There are five basic plant hormones: auxins, gibberellins, cytokinins, abscisic acid, and ethylene.

7. Auxins constitute a group of plant hormones produced in the meristematic tissues. These hormones stimulate cell elongation, prevent leaf and fruit abscission, and promote apical dominance.

8. Gibberellins promote stem elongation and seed germination.

9. Cytokinins stimulate cell division.

10. Abscisic acid promotes winter dormancy and the closing of the stomata.

11. Ethylene stimulates fruit ripening.

12. The way plants respond to changes in light and darkness is called photoperiodism. A pigment called phytochrome regulates photoperiodism.

13. Plant responses to unidirectional stimuli are called tropisms. The response of a plant to light is called phototropism. Its response to the pull of gravity is called gravitropism or geotropism.

KEY TERMS

turgor pressure
transpiration
mycorrhiza
translocation
sap
pressure-flow hypothesis
auxin
leaf abscission
gibberellin
cytokinin

abscisic acid
ethylene
photoperiodism
day-neutral plant
short-day plant
long-day plant
phytochrome
tropism
phototropism
gravitropism

REVIEW QUESTIONS

True/False

1. The loss of water vapor by a plant is called transpiration.
   true   false   (page 494)

2. Under certain conditions, water can have a tensile strength that exceeds the tensile strength of thin steel wire.
   true   false   (page 495)

3. Translocation is the transportation of the products of photosynthesis.
   true   false   (page 497)

4. Cytokinins assist in the ripening of fruit.
   true   false   (page 503)

Fill in the Blank

1. Ninety percent of the water lost by the plants during transpiration is through the _____ of the leaf. *(page 492)*

2. The opening and closing of the stomata are regulated primarily by the concentration of _____ ions in the guard cells. *(page 492)*

3. Gibberellins were first isolated from _____. *(page 502)*

4. A plant pigment known as _____ is involved in the phenomenon of photoperiodism. *(page 507)*

Multiple Choice

1. The movement of water and dissolved minerals from the roots to the leaves is best explained by:
   a. Cohesion-tension theory
   b. Translocation
   c. Tensile strength
   d. Pressure-flow hypothesis
   (page 494)

2. When you place fruit in a paper bag to accelerate fruit ripening, the confinement of the bag increases the concentration of:
   a. Ethylene
   b. Auxin
   c. Cytokinins
   d. Abscisic acid
   (page 504)

3. If you plant onion sets with the roots pointing up, the roots of the mature plants still grow downward. This plant response is an example of:
   a. Geomagnetism
   b. Gravitropism
   c. Photoperiod
   d. Magnetics
   (page 509)

Chapter 21   Transport and Regulation

4. In short-day plants like the cocklebur, flowering can be inhibited if dark periods are interrupted by light, but if you interrupt light periods with darkness, flowering will occur. What does this prove?
   a. Light is more important in flowering.
   b. Darkness is more important in flowering.
   c. The duration of light has no effect on flowering.
   d. No meaningful conclusions can be drawn from the available information.
   (page 506)

Discussion

1. Speculate on how plants measure darkness.

2. The herbicide Agent Orange was used as a defoliant (caused plants to drop their leaves) during the Vietnam War. What do you think about the use of defoliants such as Agent Orange?

3. If you have ever seen the giant redwood trees, you have marveled at their majesty. What is the mechanism of transport of water and dissolved minerals through the stems of these magnificent trees?

4. Discuss the mechanism of action regarding plant responses to the duration of darkness.

5. Discuss translocation and the pressure-flow hypothesis.

6. Discuss some of the earlier historical experimental evidence that suggested that auxin is involved in phototropism.

7. Discuss in a general sense how plant hormones affect plant growth and other cellular activities.

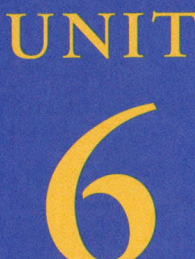

ANIMALS

After reading this unit, the student should be able to:

- understand the relationship between structure and function in various animal organ systems
- understand the comparative relationships between various animal groups with respect to: protection, support, and locomotion; digestion; transportation; immune response; temperature regulation, osmoregulation, and gas exchange; reproduction; and development
- understand the various regulatory mechanisms that help animals to maintain homeostasis
- understand how scientists study the evolutionary relationships between major animal groups
- discuss the adaptive features that have contributed to the success of various animals on land

22

An Introduction to Animal Form and Function: Protection, Support, and Motion

Animal Cells and Tissues
Protection
Support
Motion
Summary

During the evolution of multicellular organisms from their single-cell ancestors, specialization of tissues developed. In multicellular organisms, there may be specialized cells for support, reproduction, neural coordination, photosynthesis, and so on. An organism's efficiency in functioning in its environment is increased through specialization, but, like all things, specialization has its price. As cells become more specialized, they become more dependent on each other for the maintenance of life functions.

A nerve cell, for example, normally cannot live when divorced from the organism from which it is obtained (unless grown in tissue culture). On another level, a cat's stomach cannot function unless the animal's brain is functioning, and vice versa.

Above: The muscles of this bodybuilder illustrate the interaction of muscles and the skeleton, which results in movement.

The same interdependence exists at a number of levels in the biosphere. Tigers cannot exist without grass, in part because plants release oxygen, which animals breathe, and in part because the tiger's primary food source is animals that feed on grass.

We are now going to study specialized systems in animals. Before we look at the special structures involved in animal protection, support, and locomotion, we will investigate the specialized cells and tissues that form these systems.

ANIMAL CELLS AND TISSUES

Tissues are cells that are woven together and have a similar function. (In fact, the word *tissue* comes from the Latin word meaning to weave.) *Histologists,* the scientists who study cells and tissues, usually group human tissue into four groups: *epithelial, connective, muscle,* and *nerve.* Complex multicellular animals like humans have approximately 200 kinds of cells, nearly all of which fit into these four tissue groups.

Epithelial Tissue

■ Epithelial tissue is the tissue that covers organisms or parts of organisms. There are three basic types of epithelial cells classified according to shape: squamous, cuboidal, and columnar. The primary functions of epithelial tissue are protection, sensory reception, and glandular secretion.

The cells that cover an organism and line its cavities are organized into **epithelial tissue.** These aggregated cells form sheets of cells that protect the organism. Layers of epithelial cells not only form on the external surface of the organism but also line the internal organs — the gut, lungs, and other internal structures. (*Gut* is the delicate term that biologists use for the digestive tract.) Because epithelial cells function in covering surfaces and offer protection from abrasion, they are often thin and flat to facilitate the job of covering large areas. (Remember that form follows function.)

The Structures of Epithelial Tissues. Have you ever looked at a human cheek cell under a microscope? Such cells have an irregular shape and are thin and flat. This shape is appropriate because their function is to cover and line the inside of the mouth, almost like floor tiles. Biologists classify epithelial cells according to their shape. Cells that are flat and irregular, like those inside your cheek, are called *squamous cells. Cuboidal cells,* which line the tubules of the kidney, are cube-shaped. *Columnar cells,* which make up the lining of the gut and respiratory tracts, are column-shaped. Some of the columnar epithelial cells that line our respiratory tract also bear cilia, which beat and sweep a cleansing film of mucus upward to the back of your mouth in order to keep your respiratory passages clear. Smoking can paralyze these cilia, leading to many respiratory disorders, including lung cancer. Figure 22.1 illustrates the three types of epithelial cells and their functions.

Epithelial cells form tight junctions where the individual cell membranes come together. The more tightly woven the junctions are between cells in a tissue, the stronger the barrier the junctions provide. The cells fit together like the pieces of a jigsaw puzzle, and the aggregated structure allows the epithelial layer to expand and stretch without breaking the barrier.

The complexity of the epithelial layer varies. In some instances, epithelial tissue is one layer thick, or *simple;* in others, such as human skin, the epithelium is layered, or *stratified.* Simple epithelium is usually found in regions of the body where materials must cross through the tissues easily, like the air sacs of the lungs or the lining of blood vessels. Stratified epithelial tissue is located in regions where protection is the major function, like the skin. In either case, the epithelium is attached to the underlying connective tissue by a layer called the *basement membrane.*

The Function of Epithelial Tissues. Biologists separate epithelial tissues into three functional types. (1) **Protective tissues** function as covers, such as the lining of the gut and lungs. (2) **Sensory tissues** contain highly specialized sensory receptor cells to detect appropriate signals. The taste buds of your tongue or the heat and pain sensors in your skin are examples of sensory tissues. (3) **Glandular tissues** contain cells that specialize in secreting substances like milk or perspiration. (Interestingly, mammary glands are nothing more than modified sweat glands.) Keep in mind that these cell types are not unique to human beings. They are also found in worms, caterpillars, elephants, and most other multicellular animals.

Connective Tissue

■ Connective tissue binds structures together. Loose connective tissue supports internal organs and holds them in place. The cells of adipose tissue function in energy storage, insulation, and cushioning. Dense connective tissue connects bones and other structures; the two main forms are tendons and ligaments. Supportive connective tissue forms cartilage and bone; bone tissue comes in two types, spongy and compact. Blood is a connective tissue that functions in the distribution of materials and in immunity.

The function of **connective tissue** is to bind and support other tissues. Unlike the cells in epithelial tissue, the cells making up connective tissue are not tightly packed. Instead they are widely separated from one another by large amounts of extracellular substances, or, in biological terms, connective tissues are scattered throughout a complex extracellular matrix. (The matrix is a gel-like or fibrous substance composed of polysaccharides and specialized proteins.) There are four major types of connective tissue: *connective tissue proper, cartilage, bone,* and *blood.* The types differ in the shapes of

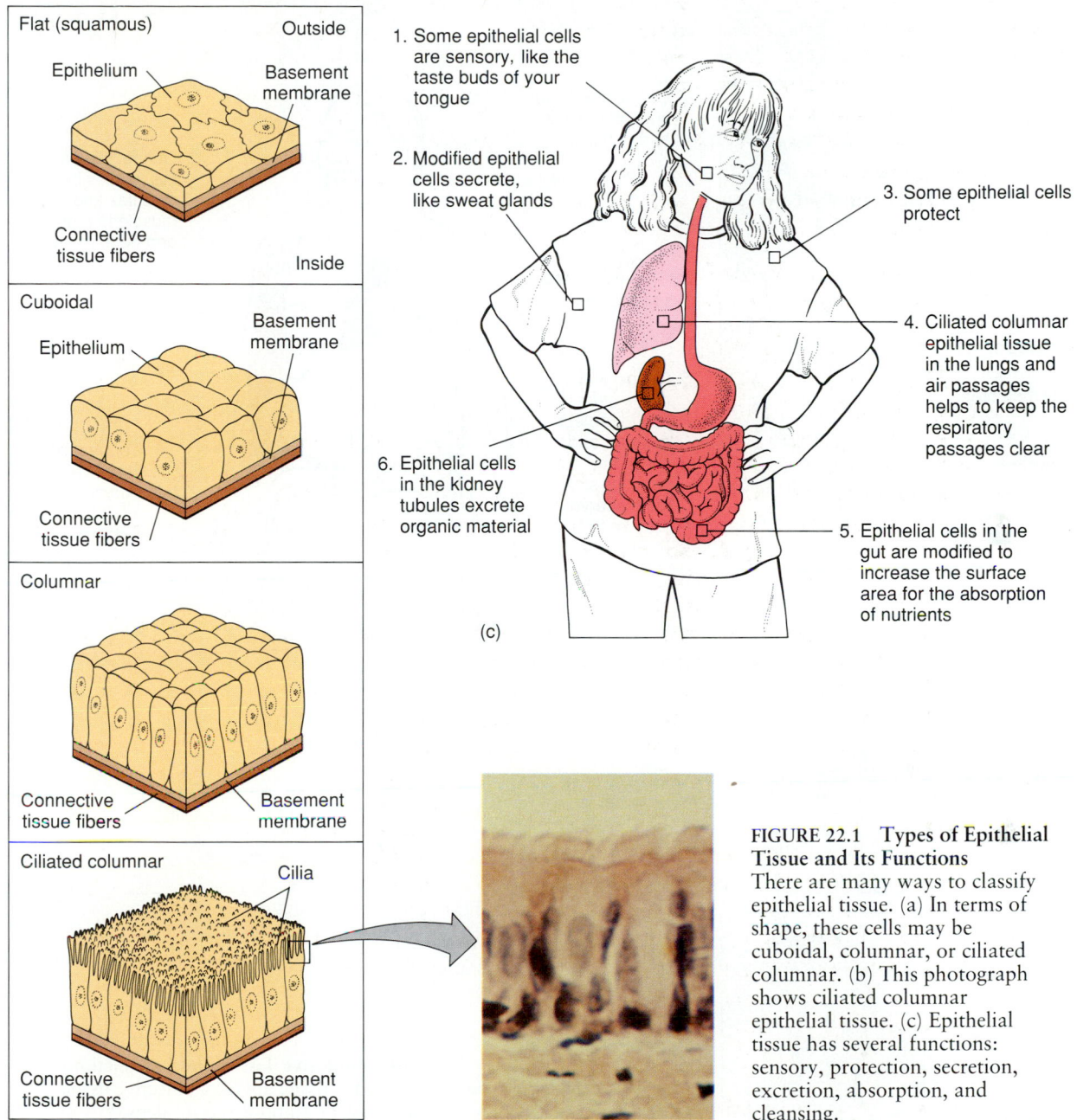

Flat (squamous)

Epithelium

Basement membrane

Outside

Connective tissue fibers

Inside

Cuboidal

Epithelium

Basement membrane

Connective tissue fibers

Columnar

Connective tissue fibers

Basement membrane

Ciliated columnar

Cilia

Connective tissue fibers

Basement membrane

(a)

(b)

(c)

1. Some epithelial cells are sensory, like the taste buds of your tongue

2. Modified epithelial cells secrete, like sweat glands

3. Some epithelial cells protect

4. Ciliated columnar epithelial tissue in the lungs and air passages helps to keep the respiratory passages clear

5. Epithelial cells in the gut are modified to increase the surface area for the absorption of nutrients

6. Epithelial cells in the kidney tubules excrete organic material

FIGURE 22.1 Types of Epithelial Tissue and Its Functions
There are many ways to classify epithelial tissue. (a) In terms of shape, these cells may be cuboidal, columnar, or ciliated columnar. (b) This photograph shows ciliated columnar epithelial tissue. (c) Epithelial tissue has several functions: sensory, protection, secretion, excretion, absorption, and cleansing.

their cells and the type and structure of matrix that they secrete (Figure 22.2).

Connective Tissue Proper. There are three types of connective tissue proper: loose connective tissue, adipose (or fatty) tissue, and dense connective tissue. The function of **loose connective tissue** is to support and maintain the position of internal organs. It also binds muscle cells together and binds skin to the underlying tissue. The cells of loose connective tissue are generally called **fibroblasts.** They

secrete flexible protein fibers called *collagen* and *elastin.* These fibers are embedded in the gel-like matrix between cells. Collagen is the most common of all human and animal proteins. Humans become aware of one aspect of collagen when they observe one of the effects of aging — wrinkled, sagging skin. The fibroblasts are producing less collagen, and the collagen becomes less flexible as the intermolecular bridges break. As a result, the binding that holds skin to muscle — the loose connective tissue — weakens.

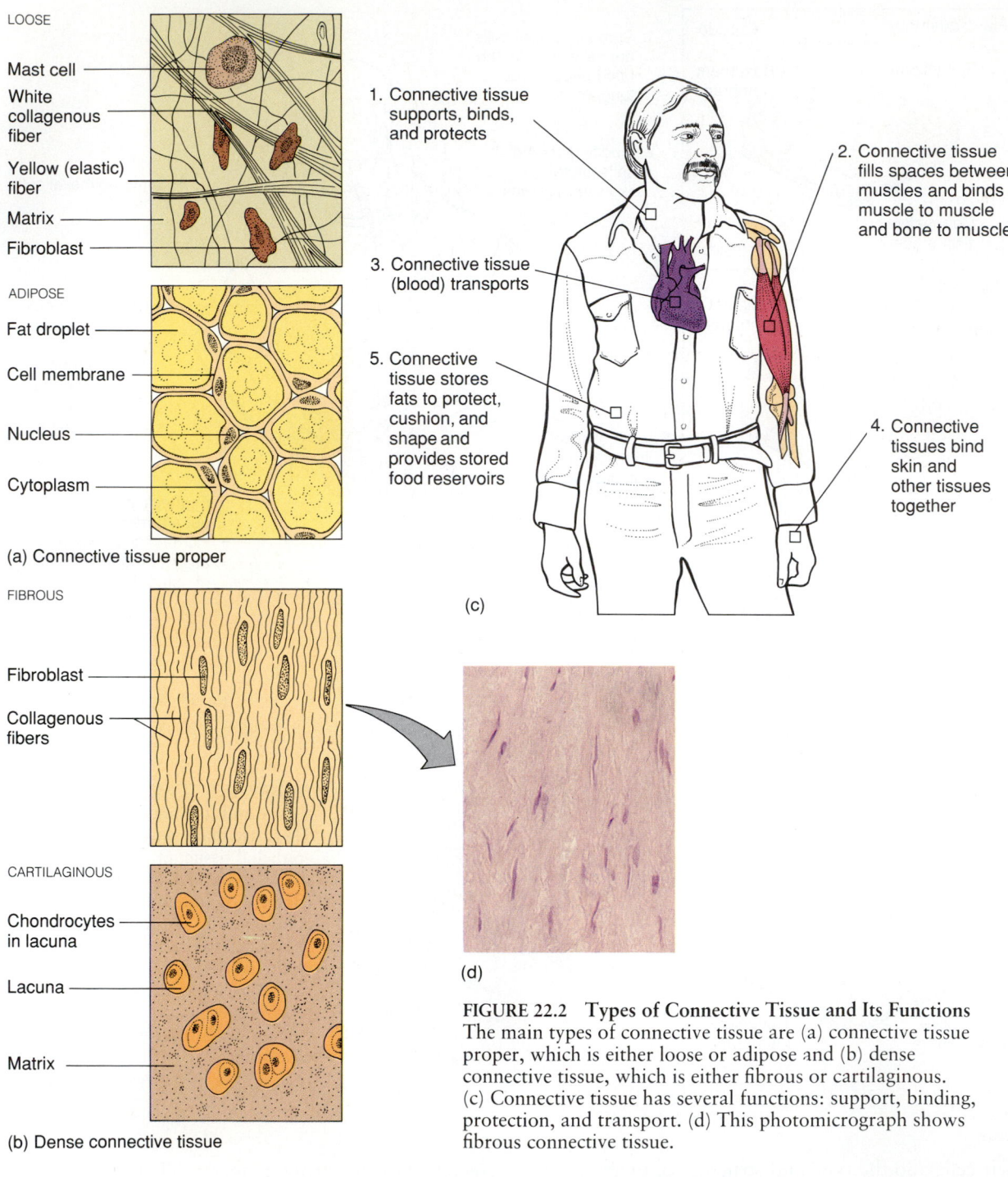

LOOSE

Mast cell

White collagenous fiber

Yellow (elastic) fiber

Matrix

Fibroblast

ADIPOSE

Fat droplet

Cell membrane

Nucleus

Cytoplasm

(a) Connective tissue proper

FIBROUS

Fibroblast

Collagenous fibers

CARTILAGINOUS

Chondrocytes in lacuna

Lacuna

Matrix

(b) Dense connective tissue

1. Connective tissue supports, binds, and protects

2. Connective tissue fills spaces between muscles and binds muscle to muscle and bone to muscle

3. Connective tissue (blood) transports

5. Connective tissue stores fats to protect, cushion, and shape and provides stored food reservoirs

4. Connective tissues bind skin and other tissues together

(c)

(d)

FIGURE 22.2 **Types of Connective Tissue and Its Functions** The main types of connective tissue are (a) connective tissue proper, which is either loose or adipose and (b) dense connective tissue, which is either fibrous or cartilaginous. (c) Connective tissue has several functions: support, binding, protection, and transport. (d) This photomicrograph shows fibrous connective tissue.

The cells of **adipose tissue,** or fatty tissue, have become specialized for fat storage. The nucleus and cytoplasm of these cells form a ring around the fat droplets stored within them. Adipose tissue functions in energy storage, insulation, and shock absorption. Penguins that sit on their eggs for weeks without feeding are living off the fat reserves in their adipose tissue. Walruses are protected from the cold Arctic waters by their heavy layers of adipose tissue. The adipose tissue of the buttocks and the palms of your hands and bottoms of your feet both serves as padding and acts as a shock absorber.

Adipose tissue also gives form to the body. Fat cells are usually found under the skin in the buttocks, breasts, and abdomen. In humans as well as other animals, the fat deposit location varies with the gender of the individual. (I don't have to remind men that the abdomen has a fat pad or women that the buttocks, thighs, and breasts consist of large amounts of adipose tissue.) Since the number of fat cells in a body varies from individual to individual, some investigators feel that obesity may be partly related to the number of fat cells, not just to excessive eating. If they are right, obesity may be the result of an individual's genetic tendency to develop more fat cells or may be linked to an oversupply of fat in the diet at early developmental stages.

Dense connective tissue is usually composed of parallel collagen fibers that are tightly packed, almost like ropes and cables. This analogy provides a hint of their function — dense connective tissues *connect*. Basically there are two types of dense connective tissues. Those that connect muscle to bone are called **tendons**, and those that connect bone to bone are called **ligaments**. Dense connective tissue arranged in sheets (fascia) also holds muscles in place in animals.

Cartilage and Bone. **Supportive connective tissue** forms the skeletal system that supports the bodies of vertebrate animals. There are two types: **cartilage** and **bone.** Cartilage gives shape to your ear and the tip of your nose. It also forms a cushion at the ends of bones. The cells of cartilage are called *chondrocytes* (Figure 22.3). They secrete an extracellular matrix that has a rubbery texture. The cells themselves are nestled in pockets in the extracellular matrix (see Profile).

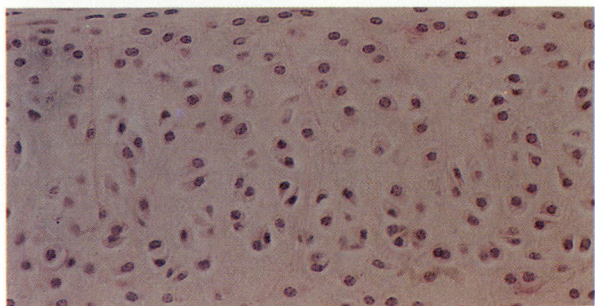

FIGURE 22.3   Hyaline Cartilage
The most common cartilage in the human body, hyaline cartilage helps shape the nose and allows movement at joints.

Bone is also a supportive tissue and, believe it or not, it is not solid, nor is it "as dry as a bone." If it were, your broken leg or arm would never heal. The living cells that form the hard outer casing of bones are called **osseous tissue** (*os* is Latin for bone), and the process of bone formation is called *osteogenesis*. Bone cells, or *osteocytes*, originate as *osteoblasts*. They secrete a dense collagen matrix that is impregnated with calcium and other mineral salts until the cell becomes rigidly encased with its own secretions.

In compact bone, the matrix is laid down in circles called *concentric lamellae* (Figure 22.4). The

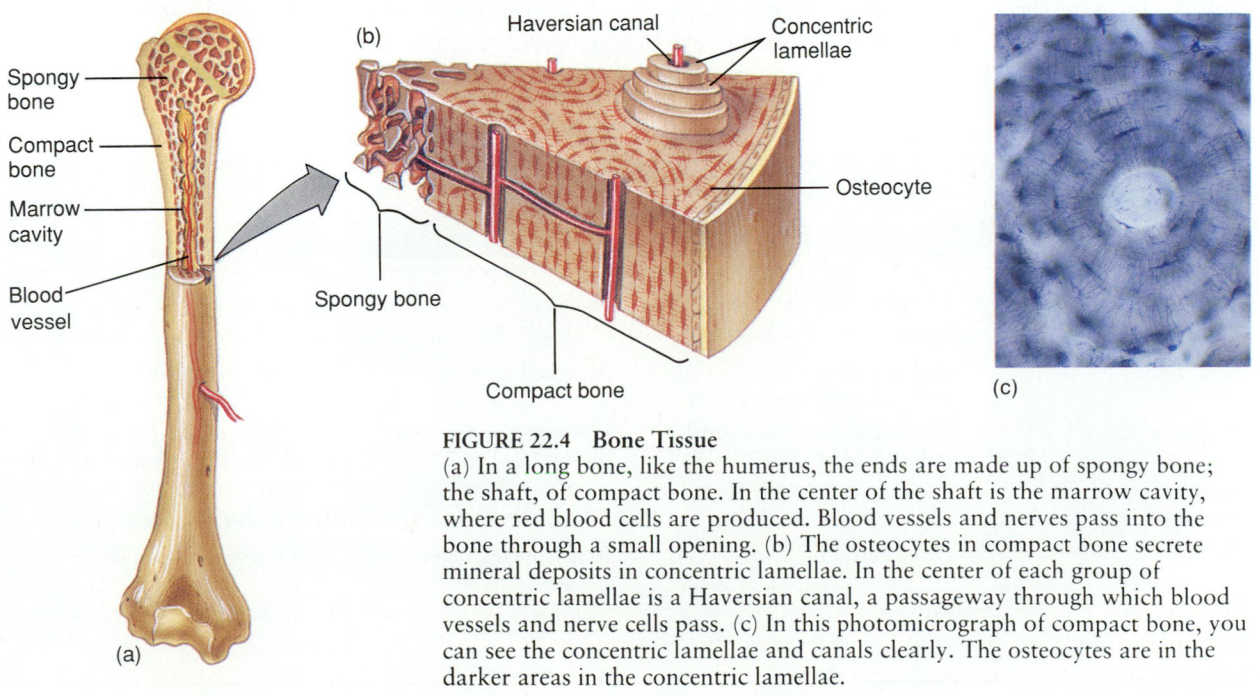

FIGURE 22.4   Bone Tissue
(a) In a long bone, like the humerus, the ends are made up of spongy bone; the shaft, of compact bone. In the center of the shaft is the marrow cavity, where red blood cells are produced. Blood vessels and nerves pass into the bone through a small opening. (b) The osteocytes in compact bone secrete mineral deposits in concentric lamellae. In the center of each group of concentric lamellae is a Haversian canal, a passageway through which blood vessels and nerve cells pass. (c) In this photomicrograph of compact bone, you can see the concentric lamellae and canals clearly. The osteocytes are in the darker areas in the concentric lamellae.

mineral deposits impart strength and rigidity to the bone. Once the osteoblast becomes rigidly encased (ossified), it can no longer produce calcium. However, it is still a living cell. Running almost parallel through the compact bone tissue are channels surrounded by the concentric lamellae. These channels are called *Haversian canals,* and within them are found blood vessels and nerves. Haversian canals supply the substances required by osteocytes and remove wastes.

Unlike the dense compact bone, spongy bone tissue is porous and moist. It looks somewhat like a kitchen sponge. In the ends of long bones the spaces in **spongy bone tissue** are filled with red marrow, which produces red blood cells. Inside the shaft of long bones is yellow marrow, which supplements red marrow in blood production in cases of emergency, as after excessive bleeding.

As people age, the density of their bones is gradually reduced, possibly because of increased calcium (Ca^{++}) reabsorption, which leads to bone reabsorption by the body. This is especially true in women past forty-five. As bone density decreases, the bones themselves become porous and weak and have a tendency to fracture. This condition is called *osteoporosis* (Figure 22.5). In serious cases, bones in the spinal column may break simply as a result of a person's rolling over in bed.

Recently, osteoporosis has been receiving a lot of attention, and much has been made of the supposed role of dietary calcium in preventing this disorder. Many calcium supplements now can be found advertised on television and stocking drugstore shelves. To date, however, there is no conclusive evidence — though there is a lot of supporting evidence — that increased calcium intake can prevent

FIGURE 22.5 Osteoporosis
The disease causes the progressive degeneration of bones, particularly in the spine, hips, legs, and feet.

or correct this disorder. Investigators do find a link between the level of the female hormone estrogen and bone density. (Such a connection would explain why the incidence of osteoporosis increases after the onset of menopause.) As a result, treatment plans for osteoporosis often include estrogen supplements.

Sometimes the process of osteogenesis is imperfect, resulting in a disorder called *osteogenesis imperfecta,* a rare calcium deficiency that produces extremely brittle bones.

TABLE 22.1
Types of Muscle Tissues

	Smooth	Skeletal	Cardiac
Location	Walls of blood vessels, stomach, intestines	Attached to bone	Walls of heart
Shape of the cell	Elongated, spindle-shaped, pointed ends	Elongated, cylindrical, blunt ends	Elongated, cylindrical fibers that branch and fuse
Striations	Absent	Present	Present
Nuclei	One central nucleus	Many nuclei, close to cell membrane	One central nucleus
Control	Involuntary	Voluntary	Involuntary
Speed of contraction	Slow	Fast	Moderate

Blood Tissue. Blood is a fluid connective tissue. In its liquid matrix, called *plasma,* blood cells are suspended. Plasma transports these cells and other substances throughout the body. In Chapter 24 we look at blood more closely.

Muscle Tissue

■ **Muscle tissue functions in movement. Smooth muscle, which moves involuntarily, is found in internal organs. Skeletal muscle moves voluntarily and is primarily responsible for locomotion. Cardiac muscle, made up of large cells fused together, is responsible for the heart's pumping movement.**

Muscle cells are responsible primarily for motion in animals but also provide protection for the internal organs. In general there are three major types of muscles: *smooth, skeletal,* and *cardiac* (Table 22.1 and Figure 22.6).

Smooth Muscle. **Smooth muscle** consists of spindle-shaped cells, each with a single nucleus. Smooth muscle tissue comprises part of the walls of blood vessels and hollow organs such as the intestines and the uterus. Smooth muscles are **involuntary muscles,** because they are not usually under conscious or voluntary control. (Has your stomach or intestine ever moved or growled at the most inopportune time? This is an example of involuntary muscular action.) The contractions of smooth muscle cells are slower and longer in duration than those of skeletal muscles.

Skeletal Muscle. Skeletal muscle tissue is attached to bone. It appears **striated,** or striped, when viewed under the microscope. Skeletal muscle cells are long and cylindrical. Each cell or fiber has many nuclei, which lie close to the cell membrane. Skeletal muscle tissue is made up of parallel fibers. Skeletal muscles are **voluntary muscles;** in other words, we can usually contract them at will.

Cardiac Muscle. As you might expect, **cardiac muscle** forms the wall of the heart of vertebrate animals. Somewhat like skeletal muscle, cardiac muscle is striated. But cardiac muscle cells are shorter and branched, and their membranes are fused end to end at the intercalated discs, allowing direct communication between the cells. This means that the cells do not function independently but contract together. When individual embryo heart cells are isolated in a tissue culture, free of any nervous control, they begin to contract independently. But then they aggregate and contract in unison. In other words, their contraction is intrinsic; they need no outside

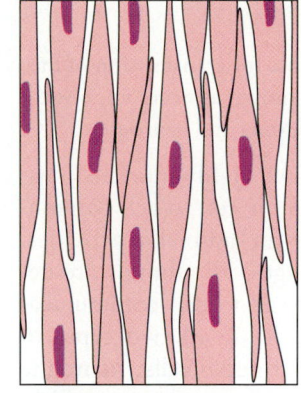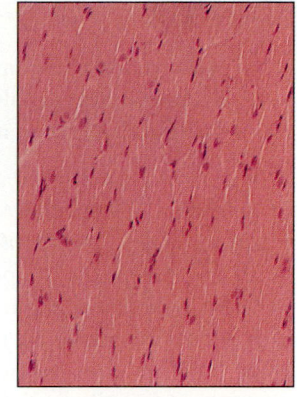

(a) Smooth muscle

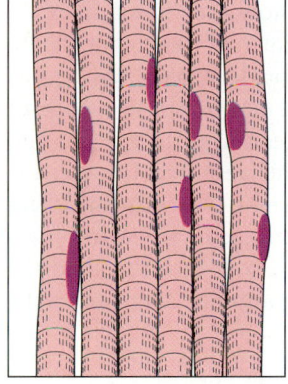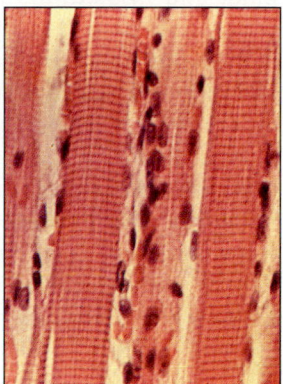

(b) Skeletal muscle

Intercalated discs

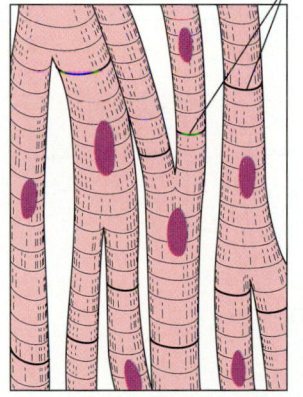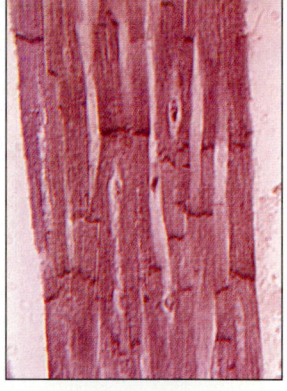

(c) Cardiac muscle

FIGURE 22.6 Muscle Tissue
The three principal types of vertebrate muscle tissue are (a) smooth, (b) skeletal, and (c) cardiac.

signal like a neural or hormonal stimulus. On the other hand, smooth and skeletal muscles will contract only in response to an outside stimulus.

Although cardiac muscle is striated, it is not usually under voluntary control. Instead it is a type of involuntary muscle, although subject to stimulation and inhibition by nerves. Biofeedback techniques can foster some degree of control over heart action.

Nerve Tissue

■ Nerve cells are responsible for transmission of the electrochemical signals known as nerve impulses. There are two types, neurons and neuroglia.

The cells of **nerve tissue** are a highly modified type of epithelial cell specialized for cell-to-cell communication. They coordinate an animal's activities. With the exception of sponges, all animals are dependent on nerve cells for cell-to-cell communication. Nervous tissue is made up of two types of cells. Neurons are individual nerve cells; **neuroglia** — sometimes called *glial cells* — support, nourish, bind, and insulate the neuron. In Chapter 30 we concentrate on nerves and how they function.

Neurons are highly branched cells that create and transmit electrochemical signals known as nerve impulses. Typically, a neuron consists of three parts: the *cell body,* which contains the nucleus and other important components required for life; the *dendrites,* which receive stimuli and transmit them toward the cell body (although in some instances the cell body can receive stimuli directly); and the *axon,* which transmits the impulse away from the cell body (Figure 22.7).

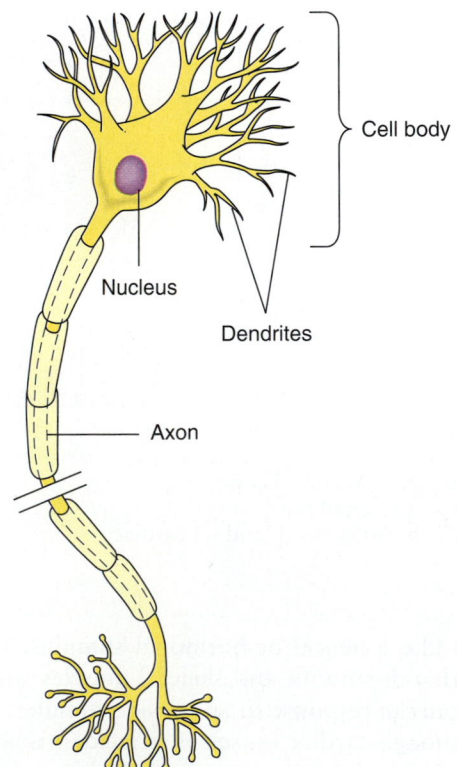

FIGURE 22.7   Neuron
A nervous tissue cell whose form is well suited to its function.

An Overview of Organs and Organ Systems

■ Tissues functioning together form organs. Organs functioning together form organ systems.

Although biologists may talk about tissues as if they exist in isolation, they do not. As we noted earlier, as cells become more specialized, they also become more interdependent. Tissues combine in various configurations to form **organs,** which are composed of more than one kind of tissue. People often think that the largest organ of the human body is the liver, but in fact it is the skin, which consists of epithelial tissues, connective tissues, and sensory receptors (neurons that carry impulses from the skin). All these tissues are functionally related as they protect and cover the organism.

Organs, too, are not independent. In higher animals, they are combined with other organs to form **organ systems.** For example, your brain is a part of the nervous system, and your heart is a part of the circulatory system. A quick look at the chapter titles in this part of the book will show that there are numerous organ systems. All groups of animals possess these systems in one form or another. Table 22.2 lists the names and functions of the major organ systems that occur in the animal kingdom. With this overview in mind, let us now discuss how different tissues and organ systems combine to perform the functions of protection, support, and locomotion in animals.

TABLE 22.2
Major Organ Systems in Animals

Organ system	Primary function
Body covering	Protection from the external environment
Skeletal	Protection, support, movement
Muscular	Movement
Digestive	Ingestion, digestion, and absorption of nutrients
Excretory	Removal of some metabolic wastes; regulation of fluid balance
Circulatory	Internal transport and immune response
Respiratory	Gas exchange
Reproductive	Procreation
Endocrine	Chemical communication, integration of body activities
Nervous	Communication, integrating body activities

FOCUS 22.1
Skin from the Test Tube

Severe burns are one of the most disfiguring and painful injuries that can beset humans. At one time or another we have all read about horrible incidents in which someone has been severely burned. We have followed the person's plight as he or she teeters on the brink of death. Why are severe burns so dangerous? An individual with third-degree burns has lost a great deal of skin and as a result is subject to infection and fluid loss. Unless the skin is replaced, many of the 15,000 people per year in the United States who are severely burned will die. The usual solution to the problem is a skin graft, a process in which new layers of skin are "fastened" over the burn area.

But where can the skin for grafting be obtained? Many burn patients do not have enough healthy skin of their own to replace the burned skin. Scientists have developed synthetic skin, made from a combination of plastic, shark skin, and cowhide, but at present this does not offer a permanent solution. Physicians also have used skin from cadavers, but this technique too is only a temporary solution.

So the problem remains. Where can enough skin be obtained to treat burn victims (Figure A)? Now Dr. Howard Green of Harvard Medical School has found a possible solution. He and his coworkers have developed a cell culture technique that can grow new skin cells. The value of this technique became very clear

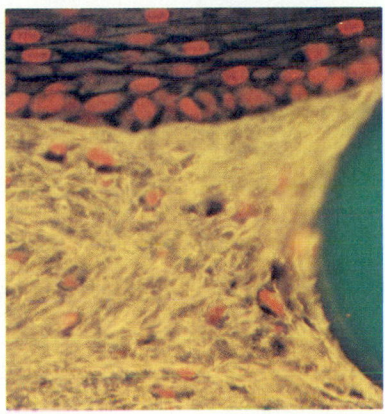

Cross-section of cultured human fibroblasts and keratinocytes grown at Marrow-Tech Corporation.

when it was used in the treatment of two young boys who were suffering from third-degree burns over 97 percent of their body. Neither was expected to live.

To obtain the cells necessary, plastic surgeon G. Gregory Gallico removed healthy pieces of each boy's skin. These were then minced, mixed in enzymes to separate the individual epithelial cells, and grown in cell culture, using Dr. Green's technique. Within 20 days, the pieces of skin had grown to the size of a playing card. They were then grafted onto the burned areas.

Dr. Green's tissue culture technique has several advantages. In the first place, since the skin used for the graft is the individual's own, the patient's body does not "reject" the skin graft. (In other words, the newly grown cells have all the characteristics of the individual, and the

body recognizes the new skin rather than perceiving it as coming from an outside source and attacking it as an invader.) What's more, a relatively large amount of skin can be grown in a relatively short period of time. In this particular case, within 12 weeks over a half square yard of skin was grown.

The success of Dr. Green's technique is indicated by the fact that neither boy died. In fact, 9 months later one of the boys went back to school and the other was recovering nicely. But even though researchers and physicians are optimistic, they will continue to be cautious, since so few patients have received cultured skin. It will take more research and analysis before the long-term effectiveness of cultured skin grafts have been established.

Recently, investigators at the New York Hospital–Cornell Medical Center developed a technique to use skin cloned from one person's cells to transplant on another individual. The problems of transplant rejection are minimized because the investigators have developed a technique that eliminates cell types (remember that the skin is a mixture of many cell types) containing the antigens that induce the acute immune response from the clonal cultured skin. Because researchers can now produce skin that does not trigger an acute immune response rejection in the recipient, the clinical potential for cultured skin cell grafts will improve recovery of burn victims.

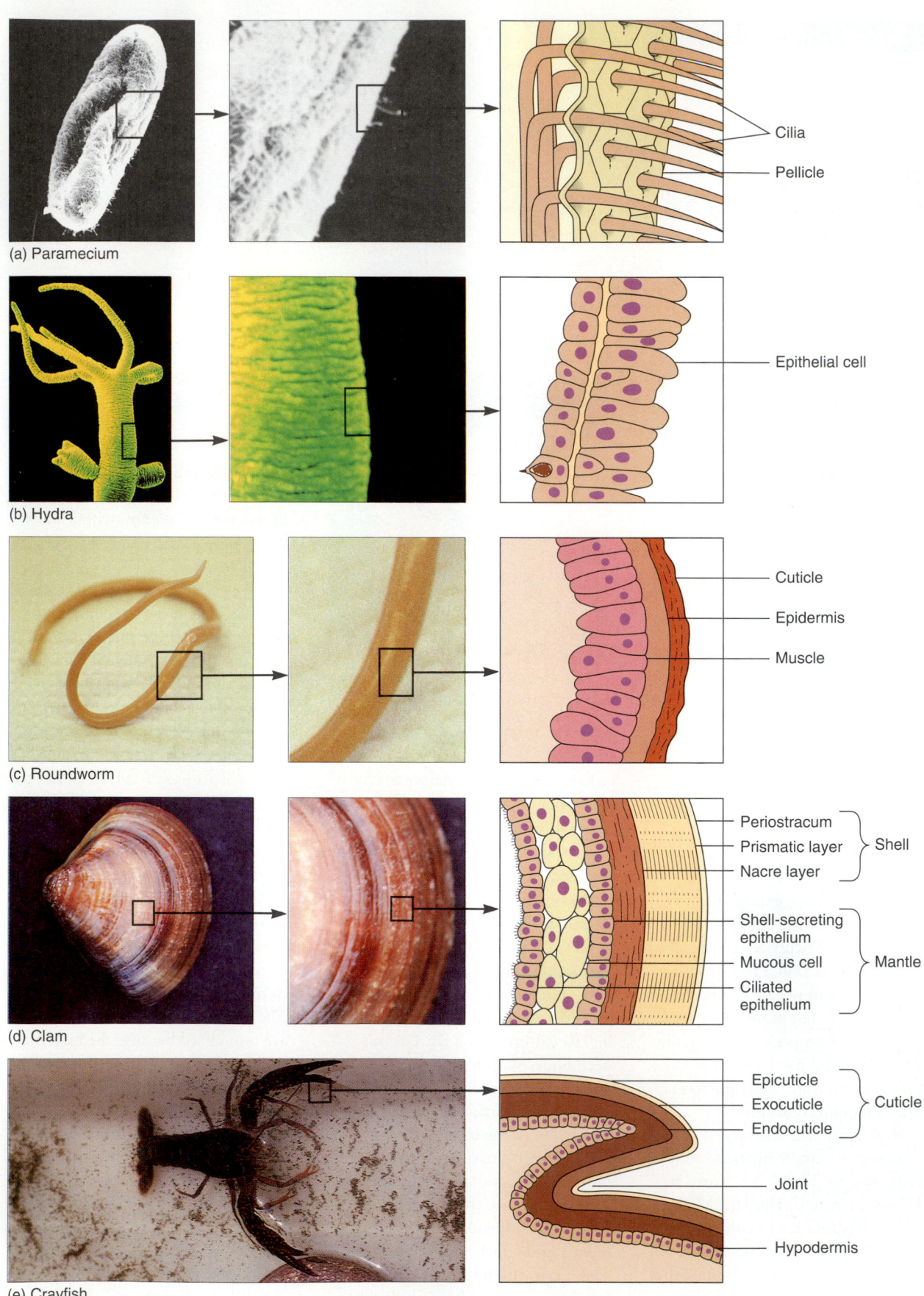

(a) Paramecium

Cilia
Pellicle

(b) Hydra

Epithelial cell

(c) Roundworm

Cuticle
Epidermis
Muscle

(d) Clam

Periostracum ⎫
Prismatic layer ⎬ Shell
Nacre layer ⎭

Shell-secreting epithelium ⎫
Mucous cell ⎬ Mantle
Ciliated epithelium ⎭

(e) Crayfish

Epicuticle ⎫
Exocuticle ⎬ Cuticle
Endocuticle ⎭

Joint

Hypodermis

FIGURE 22.8 Covering Layers of Several Invertebrates
(a) The pellicle of the paramecium gives the animal its shape and protects it from the environment. (b) In the hydra, the epidermis is just one cell thick and is anchored to a gelatinous, elastic layer. (c) In the roundworm, the cuticle is actually several layers. The epidermis is a thin layer of protoplasm with nuclei but no cell membranes. It is attached to a layer of fibrous muscle. Roundworms shed their cuticles several times as they grow. (d) In the clam, the outer layer of the mantle secretes the shell. (e) Epidermal cells in the crayfish secrete a chitinous exoskeleton that protects the animal.

PROTECTION

■ A variety of covering systems have evolved in animals. Some unicellular types form a pellicle, a thickened outer covering. Some epidermal coverings secrete substances that provide added protection, such as a cuticle, a mantle, or a chitinous shell. Feathers and scales are specialized epidermal coverings.

In most unicellular animals the covering of the organism is very simple — it is the cell membrane. However, even these tiny organisms exhibit a number of specializations that have evolved to provide extra protection. Some *protozoa* (single-celled organisms), for instance, secrete a translucent elastic covering called a *pellicle* (Figure 22.8). The pellicle provides a shape for the organism and separates it from the environment. Other protozoa, such as the *radiolarians*, secrete a shell composed of silica for protection and form.

Multicellular animals, from the simplest to the most complex, have the epidermis as their outer covering. In simple aquatic animals such as the hydra, jellyfish, and free-living flatworms, the epidermis is just one cell. In complex invertebrates such as the parasitic flatworms, roundworms, and segmented worms, the epidermal cells secrete a protective covering called a *cuticle.*

Mollusks are animals such as clams and scallops. They are covered by a sheet of tissue called a *mantle.* The mantle's outer layer of epithelial cells secrete the calcium carbonate shell we so often associate with this group of organisms. Some mollusks, like the squid, do not possess an external shell. In these cases, the mantle is thick and muscular and functions as the only external covering.

In *arthropods,* which are jointed-legged animals like the crayfish, shrimp, and spiders and other insects, the epidermis secretes a three-layered cuticle that contains a substance called *chitin.* Chitin, a polysaccharide made up of nitrogen-containing sugars, forms an external skeleton (or **exoskeleton**) to which the muscles are attached. If you have eaten a lobster you probably had trouble penetrating the tough, hard chitin of its exoskeleton. An exoskeleton is almost like a suit of confining armor; so as an arthropod grows it must shed its exoskeleton from time to time, a process called *molting.* It then exposes the newly developed larger exoskeleton.

The *vertebrates* are animals with backbones, such as fish, birds, and of course humans. In this group the skin, or integument, is the outer weatherproofing, self-repairing covering that provides form, temperature regulation, and protection from drying out (or desiccation), trauma, bacterial invasion, and ultraviolet light. The skin is also a very important sensory structure, containing touch, pain, and temperature receptors. In some vertebrates like the frogs, the skin also functions in gas exchange. Some frog species produce special waxes that protect them from drying out, and in some frogs the skin secretes poisons. As vertebrates have diversified, numerous modifications of the outer surface layers of the skin have occurred, resulting in everything from scales to feathers to hair.

Skin

Skin is composed of two layers: the outer layer, the *epidermis,* and the layer below the epidermis, the *dermis* (Figure 22.9). These two layers themselves can also be broken down into layers. The outermost layer of the **epidermis** consists of dead, dried out squamous epithelial cells filled with large amounts of the fibrous protein keratin. Keratin is the substance that imparts the toughness that we associate with the protective function of the skin. The dead cells are continuously being sloughed away. Below the surface layer are layers containing actively dividing epidermal cells that replace the surface cells (see Focus 22.1). In the skin disorder called *psoriasis,* the new cells reach the epidermal surface every three or four days, which is about seven or eight times faster than normal. As a result, the affected areas of skin are constantly flaking and may crack and bleed.

Below the epidermis is the **dermis,** which consists mainly of dense connective tissue. Within the dermis are sebaceous (oil) glands, the ducts of sweat glands, blood vessels, nerves, and muscle. Follicles are areas of surface skin that are folded inward (structures like this are called *invaginations*). Hair, as well as the feathers of birds, grows from these follicles.

Beneath the dermis is a thick layer of tissue called the **subcutaneous layer.** The subcutaneous layer,

FIGURE 22.9 **Human Skin**
Notice the multilayered
organization of the organ.

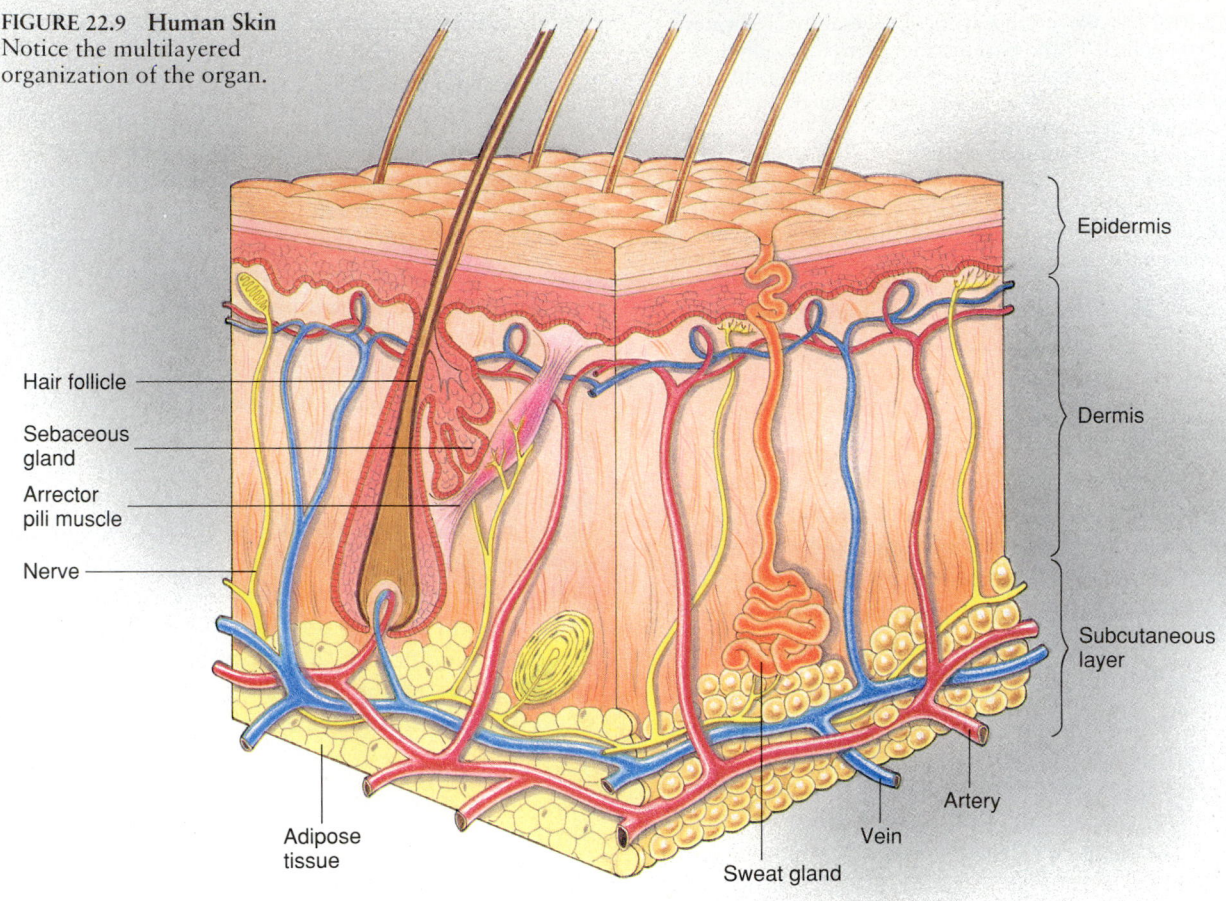

Epidermis

Dermis

Hair follicle

Sebaceous
gland

Arrector
pili muscle

Nerve

Subcutaneous
layer

Adipose
tissue

Sweat gland

Vein

Artery

made up of loose connective tissue and adipose tissue, is not truly a part of the skin. Instead, it is the layer of tissues to which the skin is attached (Figure 22.9).

SUPPORT

■ **The skeletal system supports the animal's body. An exoskeleton is outside the body; an endoskeleton is inside the body. Bones are joined together at joints, which are connected by ligaments. Muscles are joined to skeletal elements by tendons.**

Almost all organisms have a type of framework, or skeleton, that gives their body support and protection and helps them get around. Some protozoans secrete mineral deposits that function something like a skeleton. Primitive invertebrates like the *porifera* (pore-bearing animals such as the sponges) secrete microscopic internal supportive rods called *spicules*. Clams and other mollusks usually have shells for support, and as we just mentioned, arthropods have chitinous exoskeletons.

A Vertebrate Skeleton: The Human

Vertebrates like ourselves possess an internal skeleton made up of bone and cartilage called an **endoskeleton.** We discussed the microscopic composition of bone tissue earlier (see Figure 22.4). The skeleton is responsible for the following functions: (1) support, (2) protection, (3) movement (because it serves as the point of attachment for muscles), and (4) blood cell formation in the bone marrow.

The role of the skeleton in support is fairly obvious — what would happen if all your bones literally turned to jelly? Standing would be impossible. The bones of your spine not only help you stand erect, they also enclose the spinal cord and protect it from damage. Another example of skeletal protection is found in the shape of the rib cage. Here your heart and lungs are protected. We will look at the skeleton's relationship to movement in a moment, and the process of blood formation is a special topic that we pursue in another chapter.

For convenience, the human skeleton can be said to have two main parts: the **axial** and **appendicular**

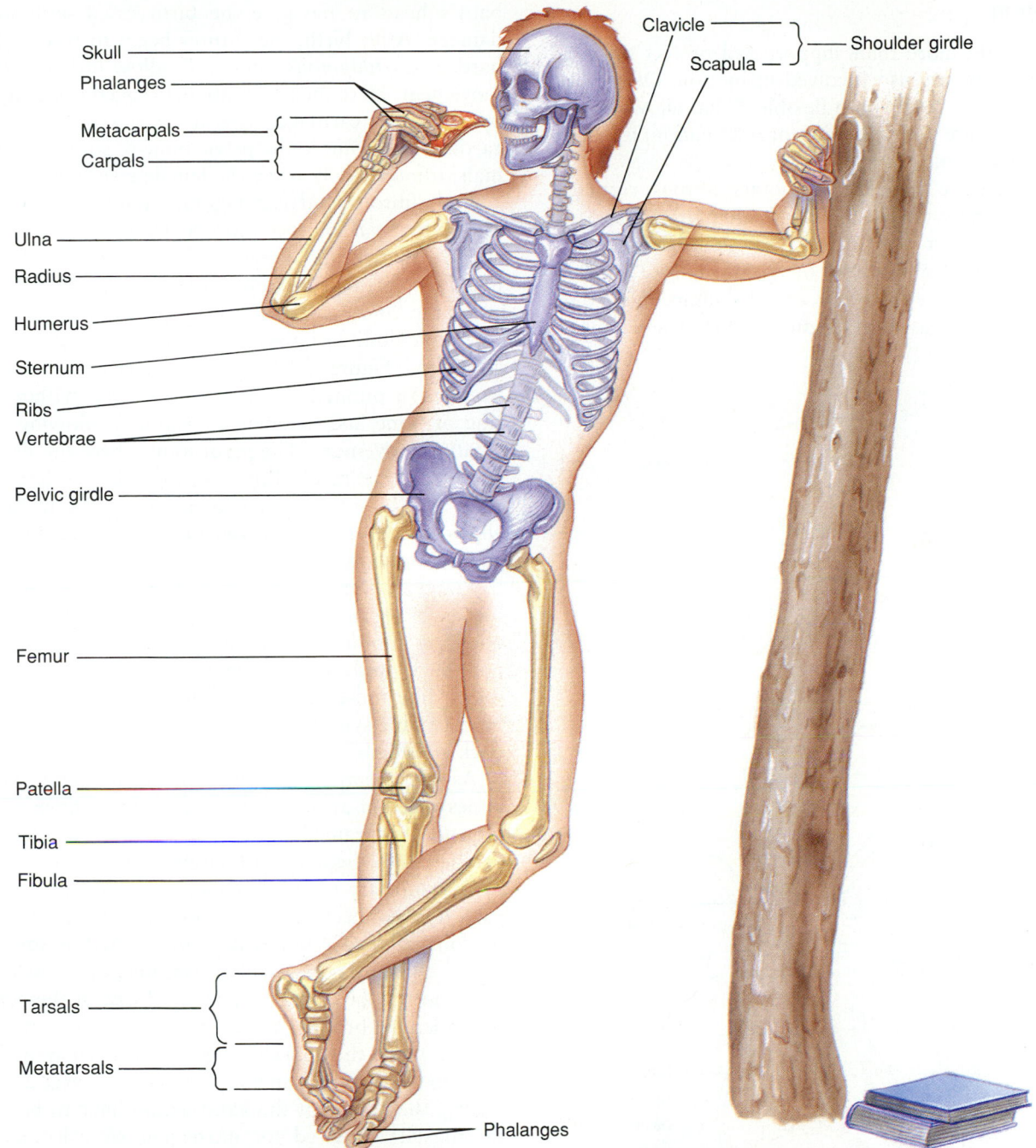

Skull

Phalanges

Metacarpals

Carpals

Ulna

Radius

Humerus

Sternum

Ribs

Vertebrae

Pelvic girdle

Femur

Patella

Tibia

Fibula

Tarsals

Metatarsals

Phalanges

Clavicle

Scapula

Shoulder girdle

FIGURE 22.10   Human Skeleton
The appendicular skeleton is shown here in its natural color. The different
parts of the axial skeleton are shown brightly colored.

skeletons. The axial system is made up of the bones that form the central axis of the vertebrate body: the skull, vertebral column, breastbone, and ribs. The appendicular system consists of the appendages attached to the axial skeleton, such as the shoulders, arms, hips, and legs. Figure 22.10 illustrates the two main divisions of the human skeleton and identifies some of the major bones.

Joints

Bones do more than support and protect the animal; they are also involved in motion. Of course, bone tissue itself is not flexible. What allows bones to move are joints. A **joint** (or *articulation*) is where two bones meet.

There are three kinds of joints: fibrous, cartilaginous, and synovial. In a *fibrous* joint, the bones are held together tightly by dense connective tissue. These joints allow little or no movement. *Sutures,* the area where the bones of the skull meet, are one kind of fibrous joint (Figure 22.11). They allow a

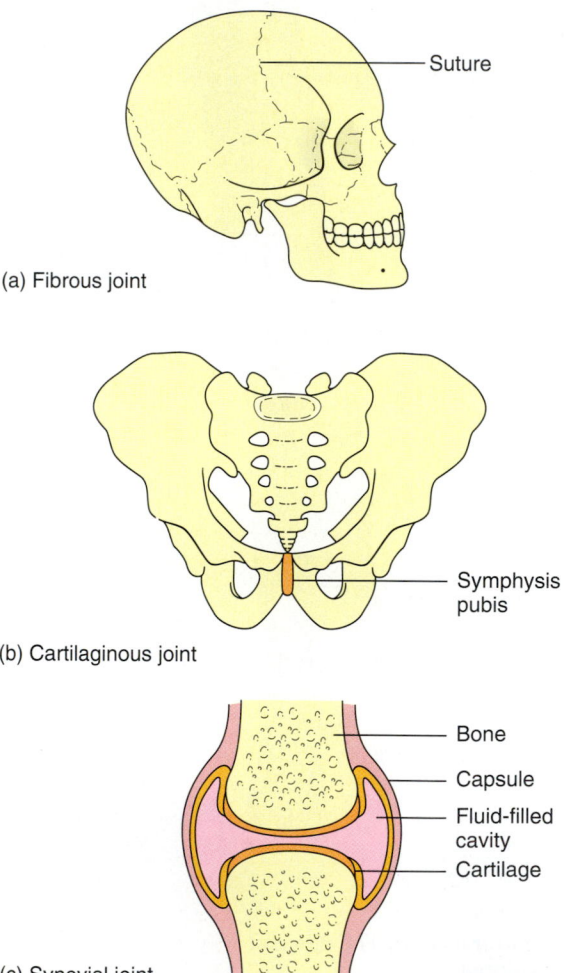

(a) Fibrous joint

— Suture

(b) Cartilaginous joint

— Symphysis pubis

(c) Synovial joint

— Bone
— Capsule
— Fluid-filled cavity
— Cartilage

FIGURE 22.11 Joints
(a) The sutures in the skull, fibrous joints, become immovable after birth. (b) The symphysis pubis, a cartilaginous joint, allows the bones of the female pelvis to move during childbirth. (c) The bones at a synovial joint are held together by a capsule. Notice that the cartilage that covers the ends of the articulating bones does not actually meet.

baby's head to navigate the birth canal without damage. After birth, the sutures begin to fuse and harden. *Cartilaginous* joints also allow little or no movement. Here the bones are held together tightly by a disk of cartilage. The symphysis pubis is a cartilaginous joint in the pelvic bone of some mammals. It moves only when the female gives birth.

Both fibrous and cartilaginous joints are tight joints; *synovial* joints are looser. Here a fluid-filled cavity separates the bones. Holding them together is a capsule of dense connective tissue, which secretes the fluid that lubricates the joint.

The flexibility of a synovial joint is related to its structure (Figure 22.12). In a *pivot* joint, the rounded or pointed end of one bone turns within a ring of bone and ligament in another, allowing a rotating movement. The pivot joint where the first two vertebrae meet allows us to turn our heads from side to side. *Hinged* joints, like those in the elbow and knee, limit movement to back and forth and up and down. The joints in the carpal bones of the hand and tarsal bones of the feet are called *gliding* joints because the small bones at the joint glide over one another. Gliding joints allow side to side and back and forth movement. *Ball-and-socket* joints, like those at the hip and shoulder, are the most flexible synovial joints. They allow movement back and forth, up and down, and around.

As we mentioned earlier, ligaments hold the bones together at the joints. Many of you have experienced a sprained ankle. A sprain is the result of damage to the ligaments of a joint. Tendons, on the other hand, are tough, flexible, often cablelike organs that connect muscle to bone. In some cases, tendons can move bones that are located at some distance from the muscle. For example, the muscles that control the fingers are in the forearm. Finger movement is brought about by long tendons that extend from the muscles through the wrist and hand to the fingers. Can you imagine how large and clumsy the fingers of the hand would have to be if the muscles required for movement were located within the fingers?

The interaction of bones, muscles, tendons, ligaments, and joints is what makes possible the flexibility and movement of vertebrates. Disorders that affect any of the factors will affect an organism's ability to function fully. Arthritis (or rheumatism, as it is sometimes called) generally is the chronic inflammation of one or more joints. Tendinitis, on the other hand, is an inflammation of the ligaments holding bones of a particular joint together or of the tendons that attach muscle to bone.

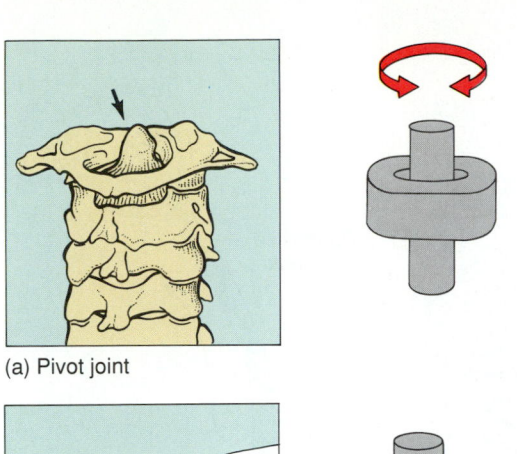

(a) Pivot joint

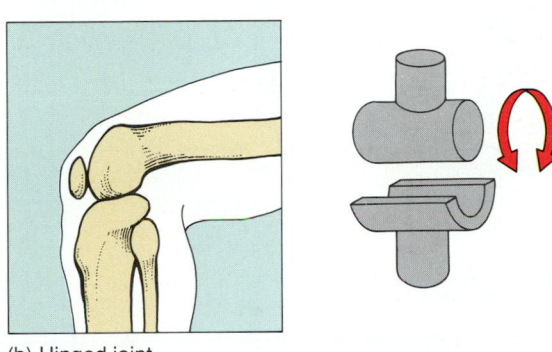

(b) Hinged joint

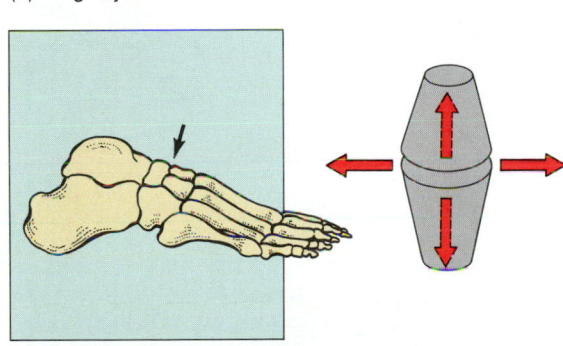

(c) Gliding joint

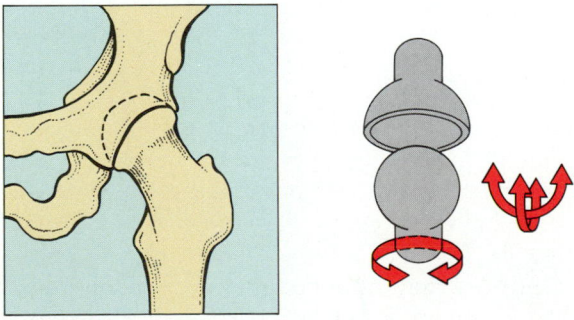

(d) Ball-and-socket joint

FIGURE 22.12   Movement at Synovial Joints
(a) A projection on the axis (the second cervical vertebra) fits through an opening in the atlas (the first cervical vertebra). This pivot joint allows the head to move from side to side. (b) A hinged joint at the elbow. (c) A gliding joint allows the small bones of the foot to move. (d) At the ball-and-socket joint of the hip, the head of the femur fits into a hollow in the pelvic bone.

MOTION

■ Animal organisms have specialized in a number of different methods of locomotion. Muscular movement takes place in skeletal muscular tissue. Atrophy occurs when muscle size diminishes because of disuse. Hypertrophy, an increase in muscle size, occurs when the number of myofibrils that make up the muscle increases.

Motion is one of the fundamental characteristics of life. Whether motion occurs within one cell, such as the contraction of the mitotic apparatus, or within a group of cells, such as the contraction of muscles, the mechanisms are similar. With the use of ATP as an energy source, filamentlike proteins slide past one another, shortening the fiber.

In unicellular organisms like the amoeba, the contraction of the protein filaments in the cell's outer membrane pushes or pulls at the cell's cytoplasm. As the cytoplasm flows within the cell, the organism changes shape and moves (Figure 22.13a). In some protozoa, flagella or cilia contain contractile elements that propel the organisms through their aquatic environment. The protozoan *Vorticella* has a contractile stalk that is attached to an object in the environment, like a piece of decaying plant. In the core of the stalk is a bundle of contractile filaments (Figure 22.13b). When these filaments contract, the stalk shortens. When they relax, the stalk elongates.

In the cnidarians, specialized contractile units have evolved. For example, in the hydra, contractile cells on the body surface control movement. These T-shaped *epitheliomuscular cells* contain contractile fibers at their base. As the fibers in adjacent cells contract along one side of the body, the organism bends and "somersaults," in a kind of "walking" (Figure 22.13c). By coordinating the pattern of contraction and relaxation from side to side, the hydra achieves directed, not just random, movement.

True muscle cells first developed in the flatworm, and with them came the development of opposing layers of muscle cells. These layers allow a range of movement — driven by the contraction/relaxation pattern. In the flatworm, there are three sets of opposing muscles: circular muscles, which line the body; dorsoventral muscles, which extend from the circular muscles at the top and bottom of the body; and longitudinal muscles, which run the length of the body. Although the flatworm usually glides along on cilia, the interaction of these different muscles allows the animal to crawl (Figure 22.13d).

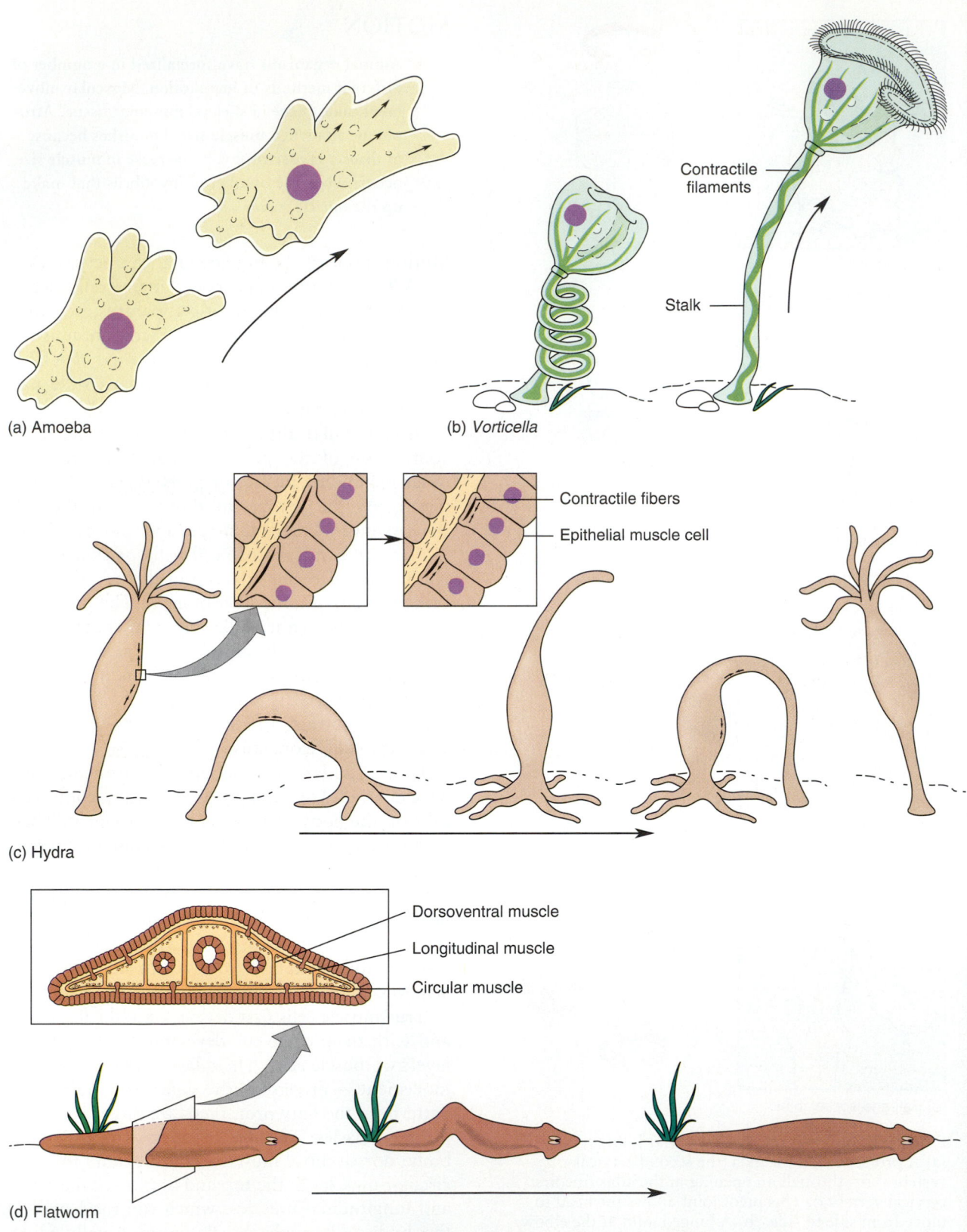

(a) Amoeba

(b) *Vorticella*

Contractile filaments

Stalk

Contractile fibers

Epithelial muscle cell

(c) Hydra

Dorsoventral muscle

Longitudinal muscle

Circular muscle

(d) Flatworm

FIGURE 22.13 Locomotion in Several Invertebrates
(a) In the amoeba, changes in the consistency of the protoplasm change the shape of and move the organism.
(b) As contractile fibers in the stalk of the *Vorticella* contract, so does the stalk. (c) When contractile
fibers in the epitheliomuscular cells of the hydra contract, they shorten the body and the tentacles,
allowing it to "walk" in a series of tumbling movements. (d) When its muscles contract, the body
of the flatworm flattens and elongates, the front part adheres to the surface the worm is crawling on,
and the back part is pulled up. When the front of the worm releases, the body moves forward.

Muscle Groups and Movement

■ Muscles can only contract. Opposing muscle pairs are called antagonistic muscles. The origin is the end of the muscle that is attached to a stationary bone; the belly is the bulk of the muscle; and the insertion is the end of the muscle that attaches to movable bone.

Earlier in this chapter, we discussed the structure, location, and function of the three major types of muscles — smooth, skeletal, and cardiac. But just how do muscles work? What are their characteristics?

Skeletal muscles, those that are attached to bone, pull on the tendons, which in turn pull on and move the bones. Most skeletal muscles are attached to two different bones, across a joint. When these muscles contract, they pull one bone toward the other. That is, one bone remains stationary and the other is pulled toward it. The point at which the muscle is attached to the stationary bone is called the **origin,** and where it is attached to the movable bone, the **insertion.** The mass of muscle between the origin and the insertion is called the **belly** (Figure 22.14).

Although the bone's structure can play a part in keeping a bone stationary, usually another muscle is at work. Skeletal muscles are *antagonistic;* they are arranged in opposing pairs, each pulling in a different direction. Luckily, they are not pulling at the same time. As one contracts, the other relaxes, producing smooth, coordinated movements.

For example, there are two major muscles in the upper arm: the *biceps brachii* at the front of the

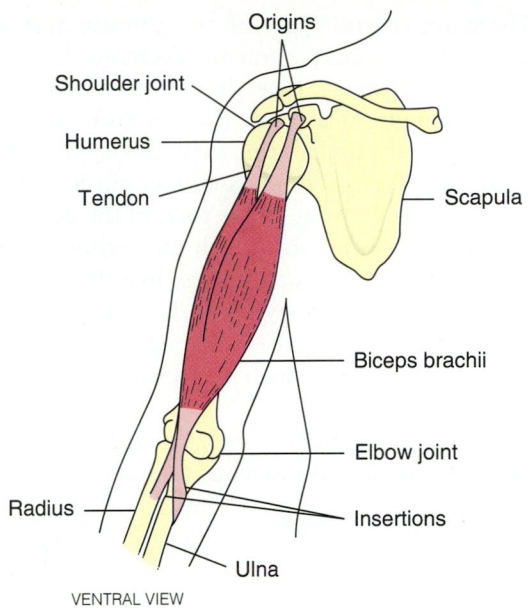

FIGURE 22.14 **Points of Origin and Insertion of a Skeletal Muscle**
The biceps brachii is a skeletal muscle in the upper arm (anterior view).

arm and the *triceps brachii* at the back. Both are attached by tendons to the scapula (shoulder blade) and humerus (upper arm bone) and, below the elbow joint, to the lower arm bones (the radius and the ulna). When you bend your arm, the biceps brachii contracts, pulling the radius up, and the triceps brachii relaxes. When you straighten your arm, the triceps brachii contracts, pulling the ulna down, and the biceps brachii relaxes (Figure 22.15).

FIGURE 22.15
Antagonistic Muscles at Work
(a) When the biceps brachii contracts, the triceps brachii relaxes and the arm bends. (b) When the triceps brachii contracts, the biceps brachii relaxes and the arm straightens. Notice how the belly of the muscle bulges during contraction.

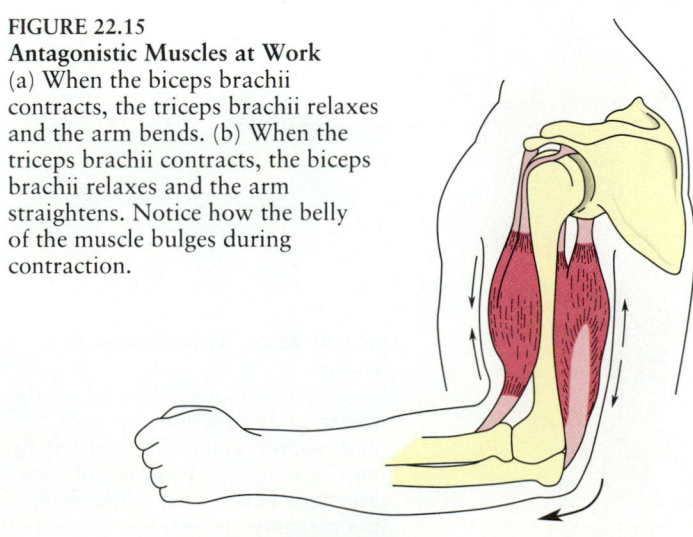

(a) Biceps brachii contracting

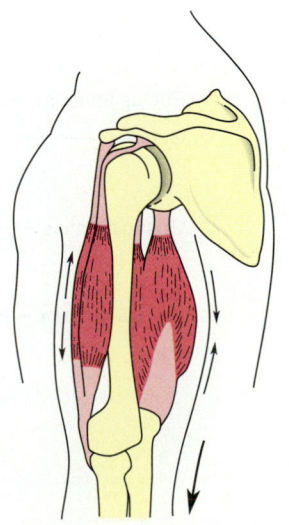

(b) Triceps brachii contracting

There are several types of antagonistic muscles. *Flexors,* like the biceps brachii, decrease the angle of a joint; *extensors,* like the triceps brachii, increase the angle of a joint. Flexors and extensors are located in the upper arm, wrist, ankle, and knee. Antagonistic muscles in the shoulder girdle include *adductors,* which bring things toward the center of the body, and *abductors,* which move things away from it. *Levators* and *depressors,* like those in the jaw, raise and lower body parts. *Pronators* and *supinators,* like those at the ankle and wrist, rotate downward and backward and upward and forward.

Not all muscles move bones. For example, there are muscles that move just your face and help com-municate the unique range of human facial expressions. Combinations of skeletal and smooth muscles form rings around body passages and blood vessels. Circular ringed muscles, called *sphincters,* can close openings like the anus, gut, blood vessel, or mouth; others, called *dilators,* open these passages. The muscles of the abdomen are formed like flattened sheets and are equipped with broad thin tendons that allow you to hold in your stomach. Figure 22.16 illustrates some of the major muscles of the human body.

You may have heard the old saying, "Use it or lose it." When the speaker is referring to muscles, this saying is very appropriate. Muscles atrophy, or shrink in size, with disuse and increase in size with

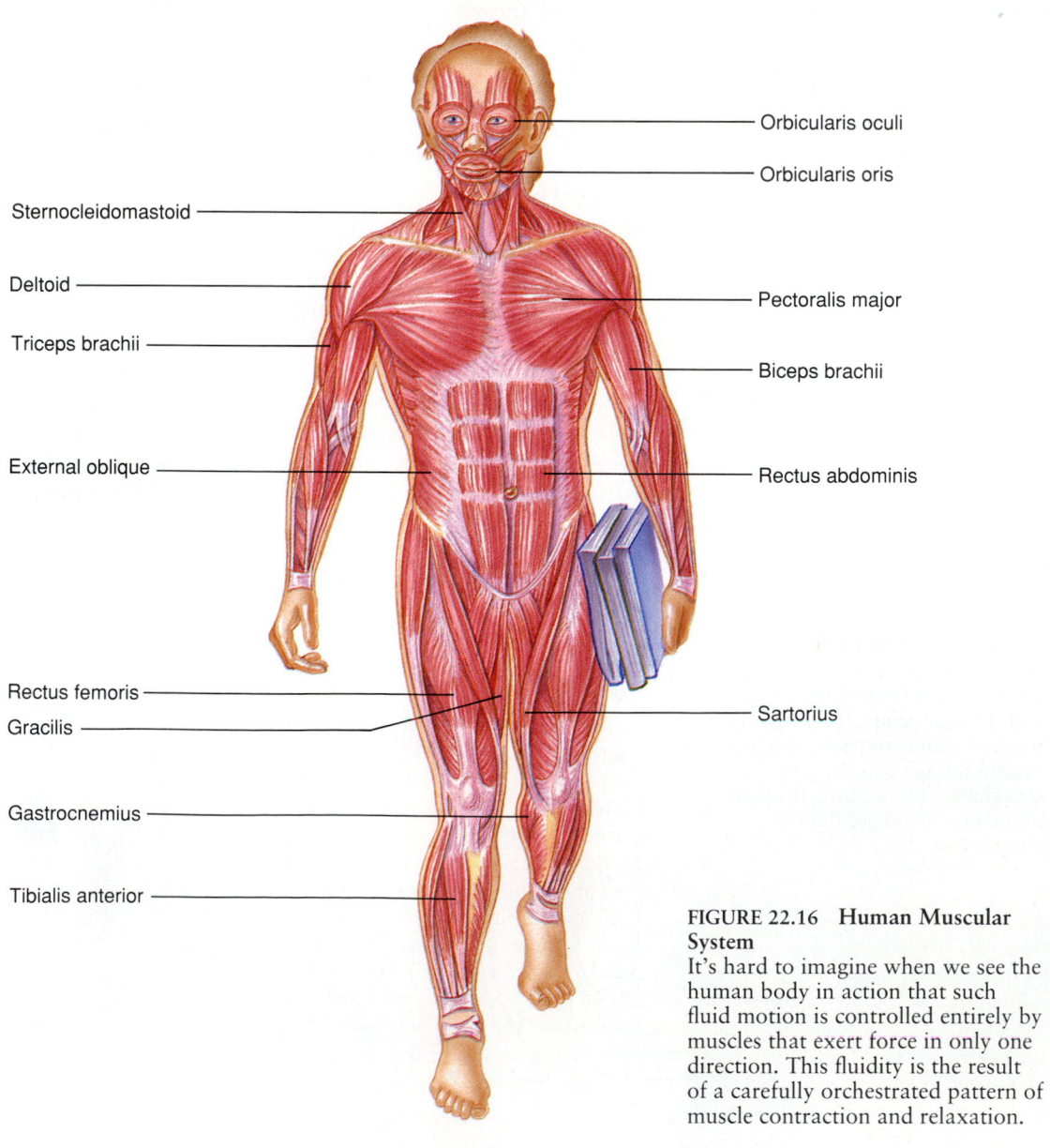

Sternocleidomastoid

Deltoid

Triceps brachii

External oblique

Rectus femoris

Gracilis

Gastrocnemius

Tibialis anterior

Orbicularis oculi

Orbicularis oris

Pectoralis major

Biceps brachii

Rectus abdominis

Sartorius

FIGURE 22.16 Human Muscular System
It's hard to imagine when we see the human body in action that such fluid motion is controlled entirely by muscles that exert force in only one direction. This fluidity is the result of a carefully orchestrated pattern of muscle contraction and relaxation.

Dr. Estralita M. E. Martin

Embryologist — Bone Development and Endocrinology

Dr. Estralita M. E. Martin was born in Hannibal, Missouri, and she has the strength of character and determination often ascribed to individuals from the "Show Me" state. She attributes her drive and persistence to the encouragement, love, and support of her parents. In order to feed and clothe their eight children, Dr. Martin's father worked two jobs and her mother cleaned houses on weekends. Although neither had obtained a formal college degree, both valued education and recognized that only through higher education would their children succeed. Therefore, they were very active in their children's education. In fact, Dr. Martin states, "We were not allowed to have jobs outside of the normal childhood jobs (paper routes, mowing lawns, and the like). My parents felt that our 'job' was to do well in school and their job was to provide us with the best environment they could for that to happen. They were like a tag team. Whatever school event one couldn't attend, the other was there. Whatever homework assignment one couldn't help us with, the other would. Both our parents were there, however, for academic and disciplinary discussions with our principals and teachers. With this much encouragement and support it was only natural that all the children would seek to fulfill their parents' dream — all attended college.

Dr. Estralita Martin has come a long way from her home in Missouri to become a scientific researcher in California. When asked why she entered science, she says, "In part for two reasons; one, because I thought I wanted to go into medicine after one of my sisters was injured in a car accident, and second, my teachers said I shouldn't because it was too difficult." However, her interest in medicine soon changed to biology after taking her first course in embryology at Oberlin College. Her professor noted her abilities and interest in embryology and encouraged her to pursue a graduate degree in the field. While attending Oberlin, Dr. Martin received a MARC scholarship to do research for the summer with Dr. John Browne at Atlanta University. That was all it took to persuade her that research was her niche in life. After graduation, Dr. Martin returned to Atlanta University to earn her masters degree. At that time, she continued under the tutelage of Dr. John Browne, investigating the cellular metabolism of brain development in chick embryos. She was able to demonstrate that administering the drug DMSO (dimethylsulfoxide) to pregnant women with a history of miscarriages caused an accumulation of amino acids and cerebral spinal fluid in the midregion of the developing brain, resulting in the formation of symptoms similar to that seen in hydroencephalitis. Her work did not go unrecognized. She received the E. E. Just Award (see the profile of Just on page 692) for outstanding research and the Samuel M. Nabrit Award for Meritorious Achievement in Developmental Biology.

Dr. Martin's interests in developmental biology eventually centered around bone development, so she wrote to Dr. Fred Wilt, an embryologist at the University of California, Berkeley, to inquire about a possible collaboration. She was accepted into their Ph.D. program and investigated the composition of the skeletal system of sea urchin embryos and the chemical similarities between the skeletal system of sea urchins and humans, although the assembly of amino acids was different than previously believed. She completed her Ph.D. degree in 1984.

After a brief detour investigating the effects of nerve growth factor on

neuronal differentiation, she was accepted as a post-doctoral fellow in the School of Medicine at the University of California, San Diego, where she resumed her studies in bone development with Dr. Leonard Deftos. Today, Dr. Martin continues her research as a research endocrinologist and physiologist at the University of California at San Diego/Veterans Administration Medical Center, where she is investigating the biochemical aspects of human bone growth metabolism. By modifying techniques originally developed for nerve cells, she is able to study individual bone cells and measure their contributions to bone development and breakdown. Dr. Martin hopes to use these techniques to study the effect of strenuous exercise on bone development in children.

Not only does Dr. Martin recognize the importance of her contributions as a researcher, but she also recognizes the importance of teaching and is presently giving courses at San Diego State University. Her commitment and passion for research is very apparent, coupled with her commitment for improving the status of minority women in research. She also enjoys relaxing with her husband.

use. If you have ever broken an arm or leg, you will remember that when the cast was removed, the limb was weak and probably smaller in size than before the accident.

The opposite of atrophy is hypertrophy, or increased muscle size. Muscles increase in size by increasing the mass of the muscle cell, not the number of cells. **Myofibrils** are the fundamental contractile portions of a muscle cell. When you exercise, you increase the number of myofibrils per cell, which results in an increase in the size of the cell and thus an increase in muscle size.

Muscle Contraction

■ **A muscle is a bundle of individual muscle fibers (cells), which in turn are composed of individual myofibrils. Each muscle fiber is surrounded by a membrane called the sarcolemma. The sarcoplasmic reticulum stores calcium ions utilized during muscle contraction. Sarcomeres, consisting of actin and myosin filaments, are the unit of contraction.**

Muscles convert chemical energy to the mechanical energy of movement (see Focus 22.2). What are the molecular and cellular events that lead to a muscle contraction?

Skeletal Muscle Structure. Most of us have taken advantage of vertebrate skeletal muscle in a Thanksgiving turkey drumstick. If you looked at a cross-section of the muscle in that drumstick, you would see groups of muscle bundles surrounded by layers of connective tissue (Figure 22.17). Under a microscope, you would see that each muscle bundle consists of parallel fibers, each a muscle cell. The cells are very large — 100 to 150 micrometers (µm) in diameter and as much as several centimeters long. In addition, each cell has more than one nucleus. Within the cytoplasm of each muscle fiber are up to 200,000 *myofibrils.*

Each muscle fiber is surrounded by a special cell membrane known as the **sarcolemma** (Figure 22.17c), which plays a key role in muscle contraction. Surrounding the muscle fibers encased by the sarcolemma are numerous mitochondria — which, as you know, produce ATP — and a modified endoplasmic reticulum called the **sarcoplasmic reticulum,** which stores calcium ions that are involved in muscle contraction. *Transverse tubules* (T tubules), which extend from the sarcolemma and run through and across the muscle fiber, are associated with the sarcoplasmic reticulum.

Each myofibril in a muscle fiber is made up of an aggregation of parallel filamentous proteins called **myofilaments.** A myofilament may appear thin and be composed of the protein called **actin** or it may appear thick and be made up of many **myosin** protein molecules.

The myofilaments in the myofibril are broken up lengthwise into sections called **sarcomeres** (Figure 21.17d). The sarcomere is the unit of contraction in skeletal muscle. The arrangement of actin and myosin in the sarcomere produces the pattern of striation so typical of this type of muscle. Look at Figure 22.17c. The sarcomere is defined at either end by a narrow dark area called the *Z line.* Two sets of the thin myofilaments — actin — are anchored on either side of the Z line. This creates a light area called the *I band.* Toward the center of the sarcomere, where the actin and myosin myofilaments overlap, is a much darker area, the *A band.* In the center of the A band (also the center of the sarcomere), where the actin myofilaments end, leaving just the myosin, is another light area. This is called the *H zone.*

The pattern we've been talking about describes the sarcomere when it is relaxed. When the muscle contracts, things change dramatically (Figure 22.18). The Z lines and I bands move in toward the center of the sarcomere, increasing the density of the A band, and the H zone disappears from view.

In summary, going from the larger structure to the smaller, muscles are arranged in this way: muscles → muscle bundles → muscle fibers (cells) → myofibrils → sarcomere → actin and myosin. The actual contractile unit is the sarcomere, and movement is the result of the properties of actin and myosin.

Now that we know something about the skeletal muscle structure, let's discuss its function: contraction.

Sliding Filament Model

■ **According to the sliding filament model, muscle contraction results from the actin myofilament sliding past the myosin myofilament. Nerve stimulation activates the release of calcium ions, which bind with molecules along the actin fiber, freeing up sites of attachment. Myosin heads form bridges along the actin. ATP supplies energy to the myosin head, causing a power stroke that slides the actin past the myosin. The process repeats itself, and the sarcomere shortens, resulting in muscle contraction.**

FIGURE 22.17 Skeletal Muscle in a Turkey Leg
(a) The cross-section shows the muscle bundles. Connective tissue separates the groups of bundles and the bundles themselves. In the center, you can see the thigh bone and a tendon. (b) Each bundle is made up of muscle fibers, large multinucleate cells. (c) Muscle fiber itself is made up of many parallel myofibrils. Interspersed between the myofibrils are mitochondria and a network of transverse tubules and sarcoplasmic reticulum — all enclosed by the sarcolemma. (d) The myofibrils are made up of parallel actin and myosin myofilaments, arranged in repeating units called sarcomeres. The striations visible in each myofibril are produced by the arrangement of myofilaments.

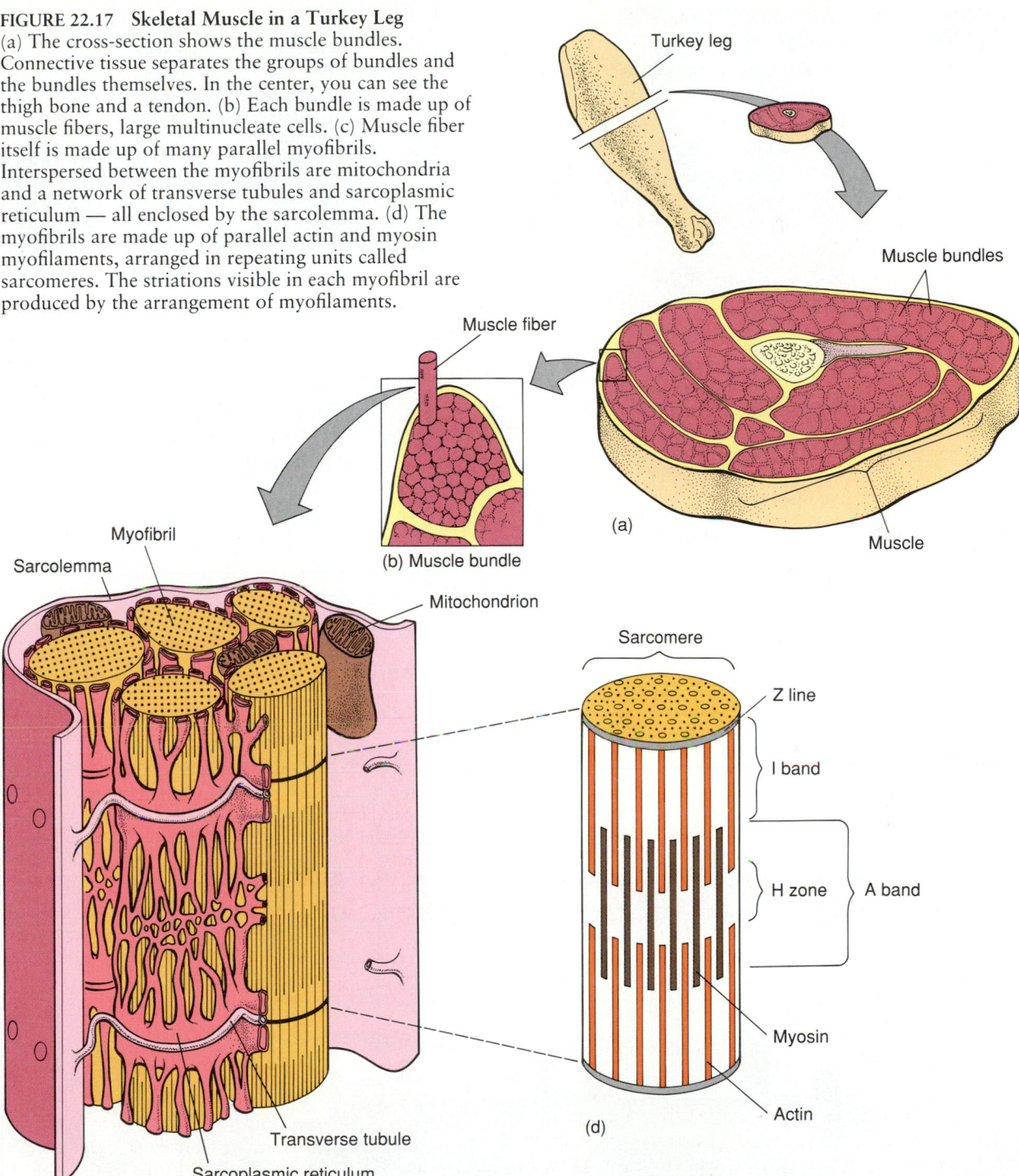

Turkey leg

Muscle bundles

Muscle

Muscle fiber

(b) Muscle bundle

(a)

Myofibril

Sarcolemma

Mitochondrion

Sarcomere

Z line

I band

H zone

A band

Myosin

Actin

(d)

Transverse tubule

Sarcoplasmic reticulum

(c) Muscle fiber

A current explanation of muscle contraction is known as the **sliding filament model.** The actions that occur take place simultaneously in all the sarcomeres of the muscle cell. First proposed in the 1950s by H. E. Huxley and J. Hanson, the sliding filament model holds that when a muscle contracts, the actin myofilaments at either side of the sarcomere move toward one another by sliding over the stationary myosin myofilaments. How does this movement take place?

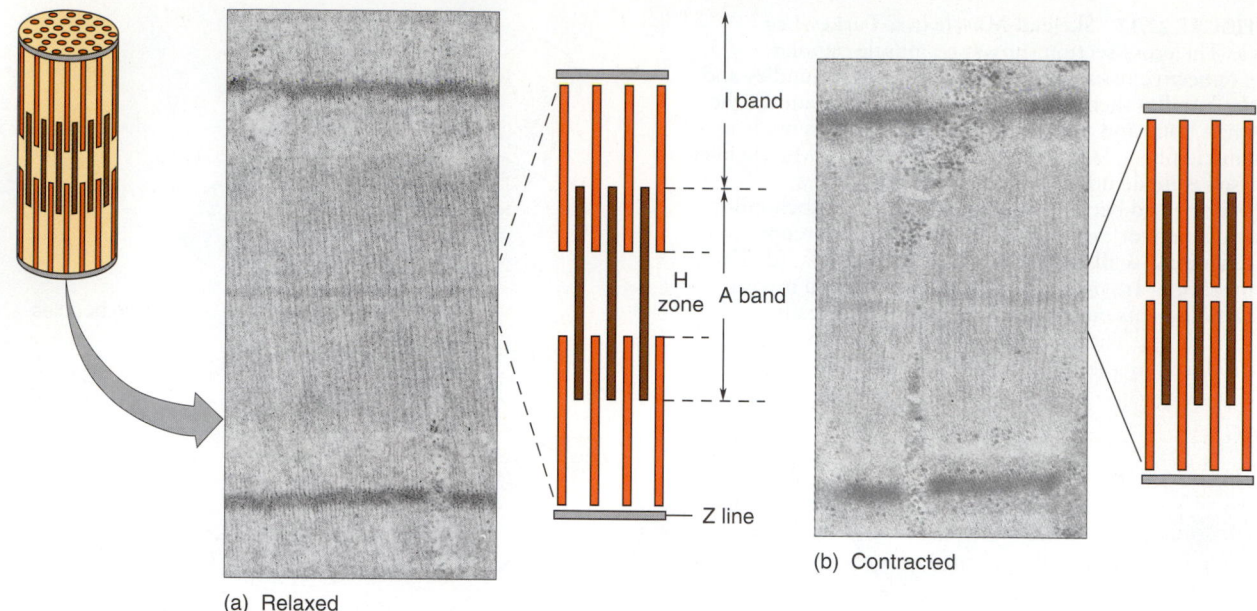

(a) Relaxed

(b) Contracted

FIGURE 22.18   Changes in the Sarcomere During Contraction
(a) The muscle relaxed. (b) The muscle contracted. The myosin remains
stationary. The actin myofilaments do not shorten, but their movement
inward shortens (contracts) the sarcomere and changes the pattern of
striation.

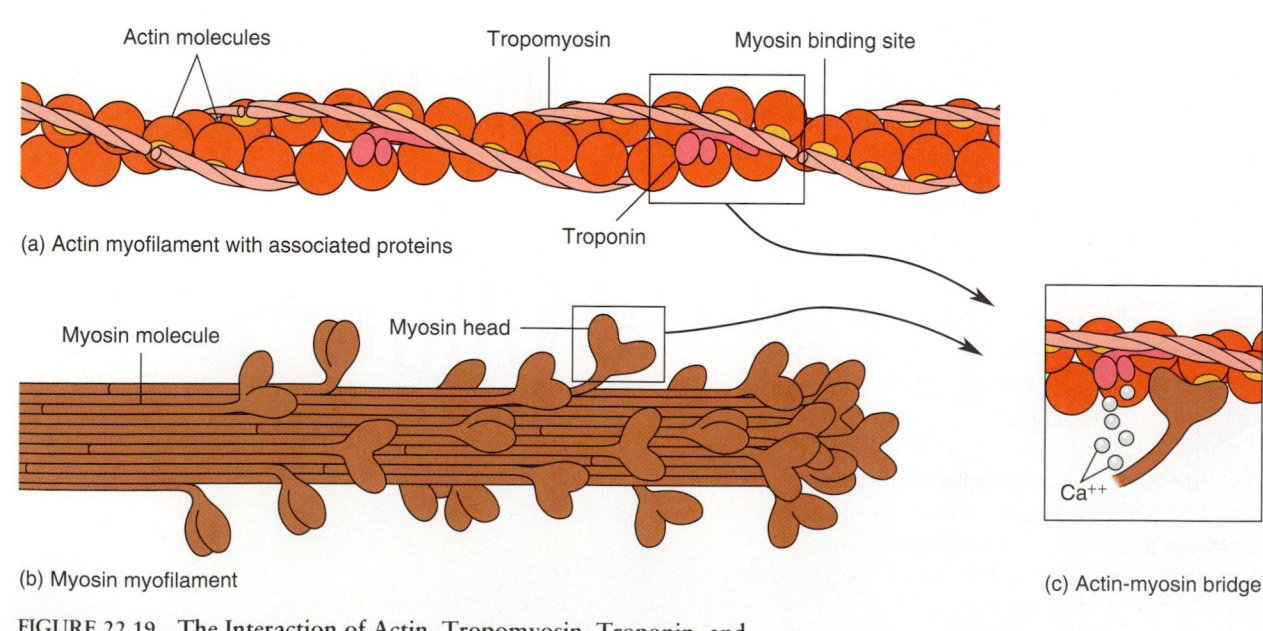

(a) Actin myofilament with associated proteins

(b) Myosin myofilament

(c) Actin-myosin bridge

**FIGURE 22.19   The Interaction of Actin, Tropomyosin, Troponin, and
Myosin in Muscle Contraction**
(a) Actin, tropomyosin, troponin, and (b) myosin in a muscle at rest.
Tropomyosin blocks the attachment sites for the myosin heads. (c) Calcium
ions (Ca^{++}) bind to the troponin, shifting the position of the tropomyosin and
exposing the binding site to the myosin head. A bridge between the two
myofilaments forms.

FIGURE 22.20   The Sliding Filament Model ▶
A nerve impulse sets off a muscle contraction by triggering the release of calcium ions from the sarcoplasmic reticulum. (a) The introduction of calcium ions exposes the binding site on the actin molecule, allowing the myosin head to attach to the actin. (b) The breakdown of ATP to ADP in the myosin head moves the head backward (toward the H zone), and pulls the actin along (c). (d) The myosin head picks up a new molecule of ATP and is ready to form a bridge with another actin molecule.

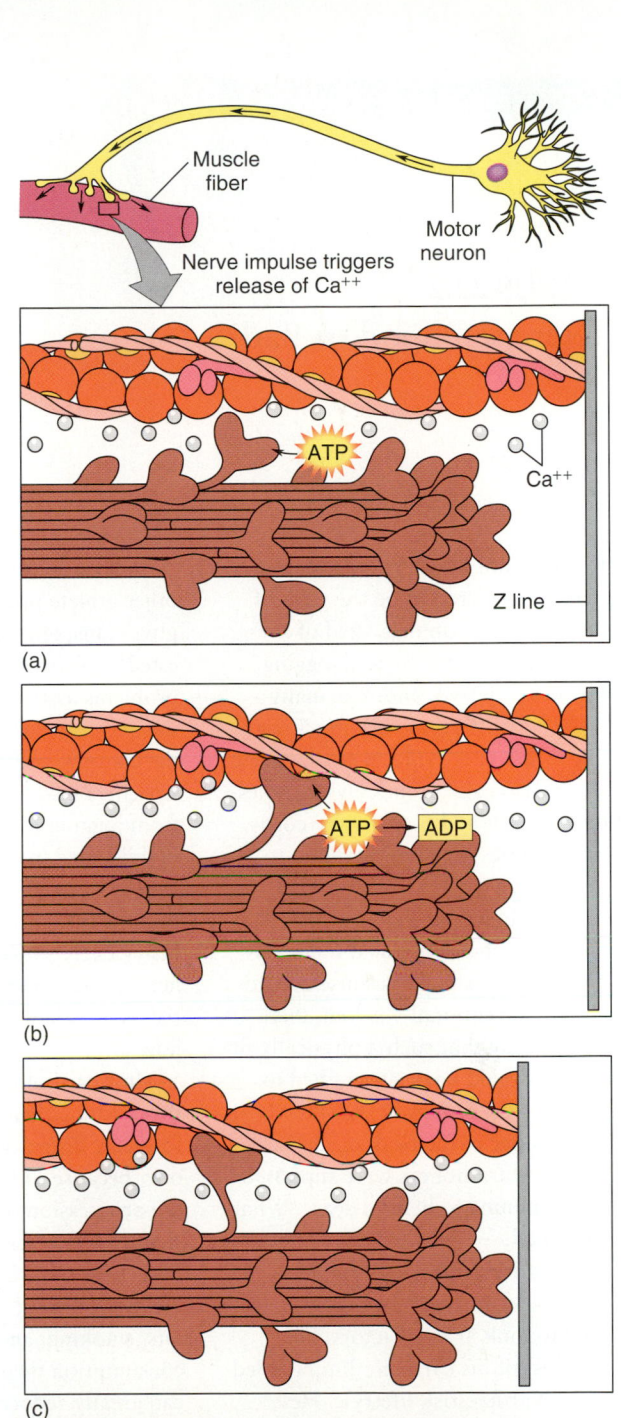

Since the actin myofilaments are connected to the Z lines, their sliding action brings the Z lines closer together. As the distance between the Z lines shortens, the sarcomere shortens, and so does the muscle. The sliding filaments themselves do not shorten.

To understand the contraction mechanism, we must first look at the proteins that make up the myofilaments. The thin actin myofilament is a twisted double strand of actin molecules. Attached to the strand at regular intervals are a rod-shaped protein, **tropomyosin,** and a globular protein, **troponin** (Figure 22.19). The thick myosin myofilament is made up of flexible myosin molecules lying parallel to one another. Each molecule ends in a double rounded head that corresponds to attachment sites located along the actin strands. When the muscle is at rest, these binding sites on the actin are blocked by tropomyosin. In the presence of calcium ions, however, troponin changes its conformation and forces tropomyosin to shift, exposing the myosin binding sites on the actin. The myosin heads can now attach to the actin, forming a bridge, and slide it along. To understand exactly how this happens, we have to look at a process of contraction from its beginning — a nerve impulse.

Muscle contraction is stimulated by a nerve impulse from a motor neuron. The impulse reaches the muscle fiber by means of a chemical messenger (acetylcholine). The impulse travels from the sarcolemma, through the transverse tubules, to the sarcoplasmic reticulum, where it signals the release of stored calcium ions (Ca^{++}) into the sarcomere. The calcium ions are free to bind to troponin and "unlock" the myosin binding sites on the actin.

Although the thick myofilament does not move during a contraction, the action of the myosin heads, coupled with the energy of ATP, is the source of the actin's movement (Figure 22.20). ATP molecules bridge the gap between the actin and myosin. As the ATP breaks down (into ADP), it releases energy, setting off a *power stroke* that propels the myosin head. The action of that head causes the actin to slide toward the center of the sarcomere.

Exercise: Good or Bad?

In 1985, Jim Fuller Fixx, a man many considered to be the guru of long-distance running, died of a heart attack while he was jogging. He was fifty-two, and, like many heart attack victims, his attack was due to the fatty buildup of plaque, a substance made up mainly of cholesterol, which had blocked the coronary arteries that supply blood to the heart. So there was nothing strange about the way in which he died. But when Fixx died there was disbelief and shock, because according to conventional wisdom, there was no way that such a physically fit person should have succumbed to heart disease. After all, he was a marathoner in the peak of physical health. Marathoners were supposed to be immune to heart disease. What happened?

To understand why a runner would succumb to heart disease, we need to look at two factors. First, for most of his early life Jim Fixx led a high-cardiac-risk lifestyle. He smoked, drank, got little exercise, and ate a high-cholesterol diet. If the circulatory system has already been damaged, exercise alone is not enough to reverse that damage. The other factor is that Fixx believed in "working through the pain." In other words, if he suffered chest pains, he did not stop running. He continued, in the tradition of "no pain, no gain." In his case (and with others too, unfortunately), the practice proved to be fatal.

The death of a famous runner or other athlete prompts biologists, physicians, and individuals interested in physical fitness and health problems associated with it to examine exercise more carefully. Amid the concern, there is good news. Researchers have found that death from jogging is very rare. Studies of middle-aged men having no family history of heart disease have shown that the odds are 1 in 5 million that heavy exercise will produce a fatal heart attack in males. In women the risk is even less, about 1 in 17 million.

Most cardiologists believe that those with either an inherited risk for heart disorders or an acquired one (because of lifestyle) can slow the progression of the disease only slightly by exercise. However, they do believe that regular exercise and the avoidance of high-cholesterol fats, smoking, and excessive alcohol consumption throughout one's life can greatly reduce the risk of heart trouble for individuals who do not have the genetic predisposition for heart trouble.

Physicians are not just concerned about problems with heart disease and exercise, however. Another major concern, especially on the part of orthopedists (physicians who specialize in bone and muscle injury), is the increased number of sports injuries to the muscular-skeletal systems and associated ligaments and tendons. In fact, there seems to be an epidemic of strains and sprains, mostly due to jogging, although aerobic exercise is also considered a culprit. What is happening in these cases?

Most of the sports injuries suffered by adults happen to the part-time athlete. This person leads a sedentary life all week, then, without proper conditioning, engages in strenuous exercise on the weekend. Sports physicians recommend proper stretching and training in order to reduce the probability and the frequency of damage. For adults, they recommend 10 minutes of stretching exercise, 10 minutes of moderate exercise, and then 15 minutes cooling down and stretching after exercise. The more common muscular-skeletal injuries are illustrated in Figure A. Many sports physicians now actively question the concept of "no pain, no gain" and recommend that no one ignore severe pain, either in the chest or anywhere else.

Most muscular-skeletal sports injuries in adults can be corrected. When they are incurred by children, however, excessive exercise can produce long-term damage. During the growing years, children's bones lengthen through an accumulation of cartilage near the ends of bones. Damage to the cartilage can stunt bone growth.

So, is exercise good or bad for you? In general, the answer is that a regular, commonsense exercise regime is good for you. In fact, every-

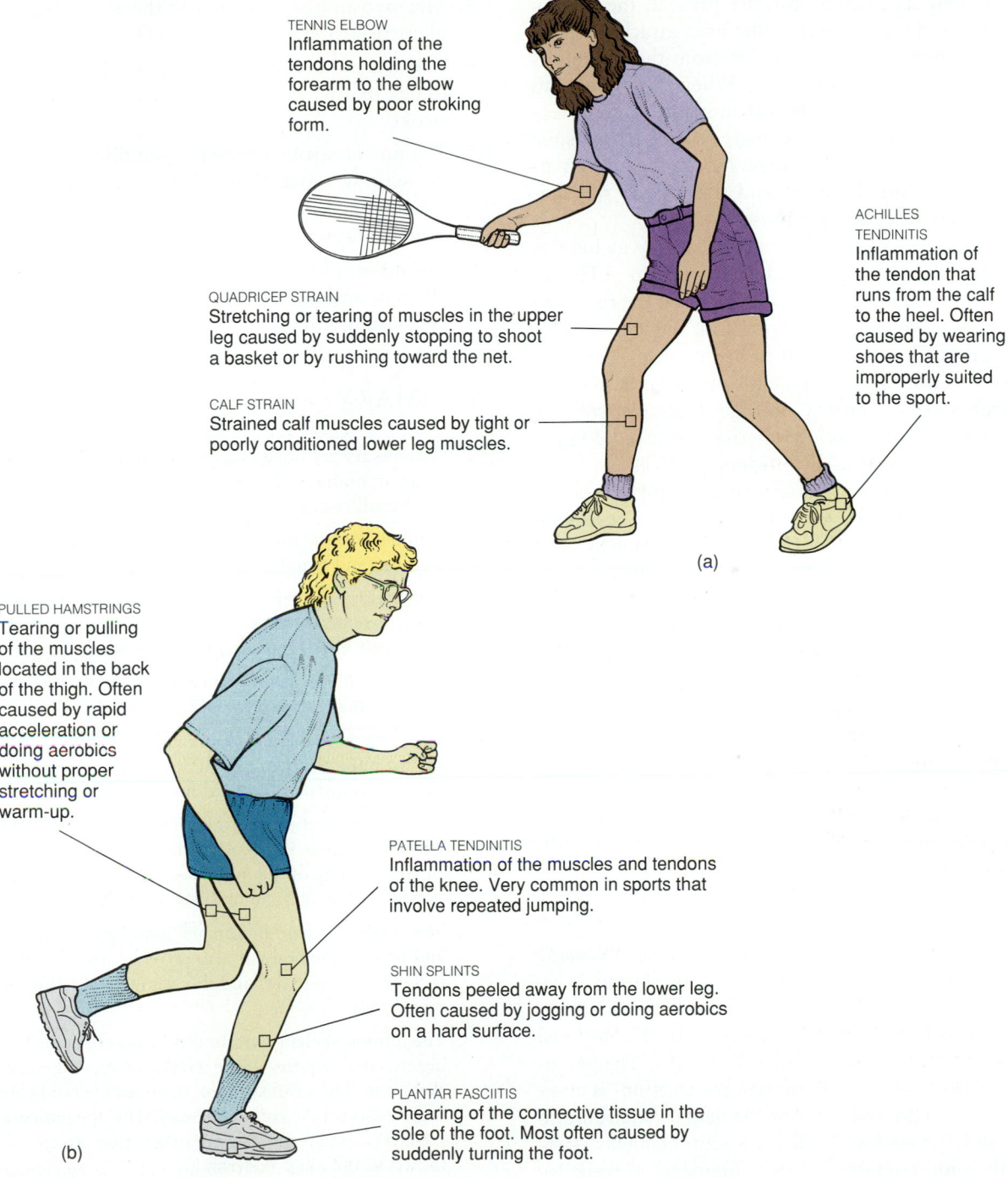

TENNIS ELBOW
Inflammation of the tendons holding the forearm to the elbow caused by poor stroking form.

QUADRICEP STRAIN
Stretching or tearing of muscles in the upper leg caused by suddenly stopping to shoot a basket or by rushing toward the net.

CALF STRAIN
Strained calf muscles caused by tight or poorly conditioned lower leg muscles.

ACHILLES TENDINITIS
Inflammation of the tendon that runs from the calf to the heel. Often caused by wearing shoes that are improperly suited to the sport.

(a)

PULLED HAMSTRINGS
Tearing or pulling of the muscles located in the back of the thigh. Often caused by rapid acceleration or doing aerobics without proper stretching or warm-up.

PATELLA TENDINITIS
Inflammation of the muscles and tendons of the knee. Very common in sports that involve repeated jumping.

SHIN SPLINTS
Tendons peeled away from the lower leg. Often caused by jogging or doing aerobics on a hard surface.

PLANTAR FASCIITIS
Shearing of the connective tissue in the sole of the foot. Most often caused by suddenly turning the foot.

(b)

FIGURE A   Sites of the more common sports injuries.

one should exercise, because it is one element of good health. However, the key to physical fitness is proper training, moderation, and constancy of exercise. Proper train-ing and care will greatly reduce the risk of injury to the muscular-skeletal system. Do not try to work through severe pain — that is one good way to cause serious injury. And finally, in general, keep in mind that exhaustion is not the aim of recreational exercise. You should feel refreshed.

As long as calcium ions are present, the myosin heads continue to bind to the next attachment site and, with the energy available from the hydrolysis of ATP, slide the actin along. When the stimulus to contract is removed, the calcium ions are actively transported from the sarcomere to the sarcoplasmic reticulum; tropomyosin again blocks the myosin attachment sites; the actin and the myosin myofilaments separate; and the muscle relaxes.

A muscle can remain contracted only as long as enough calcium ions and energy from ATP are available. ATP availability depends on the recycling processes of cellular respiration, which we discussed in Chapters 7 and 9. In this case, ATP then serves as the energy source for muscular activity. In vertebrates the energy storage molecule *creatine phosphate* serves as an alternate source of high-energy phosphate used to regenerate ATP.

Under normal circumstances, the energy for the phosphorylation of creatine phosphate comes from aerobic respiration, which functions as long as there is a supply of oxygen. If not enough oxygen is available within the muscle system, lactic acid can serve as a limited source of energy for ATP synthesis. However, this process can only provide ATP for a short period of time, because lactic acid metabolism can lead to lactic acid buildup, which causes fatigue and leads to cramped muscles.

When an animal dies, the actin and myosin are permanently bound together, because no more ATP is produced to allow for the separation of the actin and myosin filaments. The result is a gradual stiffening of the muscles known as *rigor mortis*.

The Steps of Muscle Contraction. The following summarizes the steps in muscle contraction: Muscles are composed of muscle fibers made up of thin strands called myofibrils. A myofibril is composed of several segments called sarcomeres. The sarcomere (the basic unit of muscle contraction) is composed of actin and myosin filaments. According to the sliding filament model, the actin filaments physically slide over the myosin filament, moving toward the center of the sarcomere, thus shortening it. The process follows these steps:

1. Motor neurons stimulate the release of calcium ions from the sarcoplasmic reticulum.
2. Calcium ions diffuse the short distance to the actin filament, where they bind to troponin. Troponin changes its shape and in so doing displaces tropomyosin. That clears binding sites on the actin myofilament for the head of the myosin filament.

3. The myosin head attaches to the actin myofilament, using energy from ATP.
4. The attachment of the myosin head causes it to flex in a wavelike motion called a power stroke.
5. The power stroke of the myosin fiber induces the actin to move along, sliding over the myosin fiber. This shortens the sarcomere to produce muscle contraction.
6. The myosin head releases from the actin myofilament and reattaches at a site further along the actin myofilament.

SUMMARY

1. Various tissues have evolved in multicellular animals. In humans, the usual tissue classifications are epithelial, connective, muscle, and nerve.

2. Epithelial tissue functions in protection, sensory reception, and glandular secretions.

3. The major types of connective tissue are connective tissue proper, cartilage, bone, and blood. Connective tissue binds and supports the animal.

4. There are three general types of muscle tissue: smooth, skeletal, and cardiac.

5. Nerve tissue is composed of neurons that conduct nerve impulses and neuroglia or glial cells, which support, nourish, bind, and insulate the neuron.

6. Various protective covering systems have evolved in animals, such as the exoskeleton for arthropods, shells for some of the mollusks, feathers and scales for birds, and skin, hair, or fur for mammals.

7. The skeletal system in humans functions to support and protect the body. It also provides a site of attachment for muscles and the site of production of blood cells.

8. The human skeletal system can be described as having two major parts — the axial and appendicular skeletons. The axial skeleton includes the bones that form the central axis of the body. The appendicular skeleton consists of the appendages that are attached to the axial skeleton.

9. The points where two bones meet or join are called joints or articulations. Ligaments hold the bones of a joint together, and tendons bind the muscles to bones.

10. Living things move. In some cases, such as single-celled organisms, only the contents of the cell move, or else specialized structures move the cell. In multicellular organisms, specialized contractile cells are responsible for movement.

11. In humans a skeletal muscle usually consists of the site of origin, belly, and point of insertion. Muscles

are found in opposing pairs and are called antagonistic because they move in opposite directions.

12. Vertebrate skeletal muscle cells are surrounded by a membrane called the sarcolemma. The sarcoplasmic reticulum stores calcium ions used during muscle contraction.

13. Skeletal muscle is composed of filamentous muscle proteins known as myofilaments. The proteins in the myofilament are primarily actin and myosin. The protein molecules tropomyosin and troponin are associated with actin.

14. Actin strands are fastened to what are called the Z lines, which delineate the ends of the sarcomere. During contraction, the actin strands slide over the myosin strands, bringing the Z lines closer together.

15. Voluntary muscle contraction is the result of stimulation by a motor neuron. The sarcoplasmic reticulum releases calcium ions, which bind with troponin molecules. Troponin changes shape and displaces tropomyosin from the binding sites of actin, leaving them open to bind with the head of the myosin.

16. ATP energy is released in the myosin head and the myosin fiber moves in a wavelike fashion, sliding the actin strand along in the process. Each head detaches, reattaches further along the actin strand, and slides it further along. The result is the contraction of the sarcomere.

KEY TERMS

epithelial tissue	organ
protective tissue	organ system
sensory tissue	exoskeleton
glandular tissue	skin
connective tissue	epidermis
loose connective tissue	dermis
fibroblast	subcutaneous layer
adipose tissue	endoskeleton
dense connective tissue	axial skeleton
tendon	appendicular skeleton
ligament	joint
supportive connective	origin
tissue	insertion
cartilage	belly
bone	myofibril
osseous tissue	sarcolemma
spongy bone tissue	sarcoplasmic reticulum
smooth muscle	myofilament
involuntary muscle	actin
striated muscle	myosin
voluntary muscle	sarcomere
cardiac muscle	sliding filament model
nerve tissue	tropomyosin
neuroglia	troponin

REVIEW QUESTIONS

True/False

1. Histology is the study of cells and tissues.
   true   false   (page 515)

2. Connective tissue functions in binding and supporting animal tissues.
   true   false   (page 516)

3. Adipose tissue is a type of connective tissue.
   true   false   (page 518)

4. Blood is a type of connective tissue.
   true   false   (page 521)

Fill in the Blank

1. Some animals, such as the arthropods, have a skeleton on the outside of the body. This skeleton is called an _____. *(page 525)*

2. The dense connective tissues that connect muscle to bone are called _____. *(page 519)*

3. Opposing muscle pairs are called _____ because they move in opposite directions. *(page 531)*

4. The current explanation of muscle contraction is known as the _____ _____ theory. *(pages 534–535)*

Multiple Choice

1. The major function or functions of the skeletal system is/are:
   a. Support, protection
   b. Hormone secretion
   c. Digestion
   d. Circulation
   (page 526)

2. The currently accepted model for the explanation of striated muscle contraction is called the
   a. Sliding filament hypothesis or theory
   b. Z-band shortening hypothesis
   c. Fluid mosaic model
   d. Hydrophobic model
   e. Actin-lengthening hypothesis
   (pages 534–535)

3. In comparison with other cells, nerve cells show a higher degree of:
   a. Metabolism
   b. Growth
   c. Contractibility
   d. Irritability
   e. Elasticity
   (page 522)

Discussion

1. Discuss the advantages and disadvantages of multicellularity in animals.

2. List and briefly describe the four types of human tissue.

3. From a phylogenetic perspective, discuss the "covering system" from the protozoa to the vertebrates.

4. How does the skeleton function in protection, support, and motion?

5. Since muscles move the fingers in your hand, why are your fingers not larger due to the large amounts of muscle required to move them?

6. Discuss the biological reasons for the increase in muscle mass observed in bodybuilders.

7. Describe the microscopic structure of muscles.

8. Briefly summarize the sequential events that take place during muscle contraction.

23

Digestion

Digestion — A Phylogenetic Overview
A Closer Look at the Digestive Process
Digestion, Nutrition, and Diet
Summary

Like all other animals, we humans are heterotrophs. Heterotrophs cannot make their own food; they are dependent on other living forms for nutrients. In the last chapter we considered the role of the muscles and skeleton in locomotion, and obviously locomotion is one activity that is involved in the heterotrophic lifestyle. But there are other activities. Heterotrophs must be able to digest the food they eat. Is it true that we are what we eat? This chapter will answer that question and help you understand digestion and the role of proper nutrition in your life. It will also help you to understand the evolution of the digestive system in various animal groups. The structure and function of the human digestive system will also be discussed.

Above: As a heterotroph, this brown bear must feed to stay alive.

DIGESTION — A PHYLOGENETIC OVERVIEW

■ **Digestion is the process of breaking down complex molecules into simpler molecules that can be used by the body for energy, growth, and repair.**

Heterotrophs obtain their food in a variety of ways. Some animals, like certain of the sponges, corals, and sea anemones, are sedentary. Because they are immobile, they must sweep the waters to collect microscopic nutrients. Other organisms are active browsers and predators; they hunt for their food. Examples range from the microscopic *Didinium* to lions, tigers, and people. No matter what method is involved, a heterotroph must obtain its nutrients from organic molecules in either live organisms or newly dead or decaying organisms, be they plant, fungi, or animal.

The next step in the process is **ingestion** — moving food from the environment into the organism's body. The amoeba engulfs food particles; the sponge filters them out of water. Higher-level organisms ingest food through their mouths.

Once a nutrient source is procured and ingested, it has to be broken down into a form the organism can use. This occurs through the process of **digestion**. Digestion can be both chemical and mechanical. **Chemical digestion** is a simple hydrolytic process in which, through the action of enzymes, water is added to the food's complex polymer molecules, breaking them down into simple soluble parts (monomers). Carbohydrates are broken down into monosaccharides, proteins into amino acids, fats into fatty acids and glycerol, and nucleic acids

into nucleotides. Numerous minerals and vitamins are also made available by digestion. These molecules are degraded and sent to the Krebs cycle or another energy-liberating process to provide the energy required to synthesize ATP. In some cases, the materials are used as building blocks to form structural components or are stored as a future energy reserve.

In higher-level organisms — where the breakdown of food substances begins in a digestive cavity or tract — **mechanical digestion** plays an important part in the process. The contractions of the muscles that line the cavity or tract churn and grind the food, helping to liquefy it.

When you eat that hamburger or bean sprout sandwich, your body, through the process of diges-

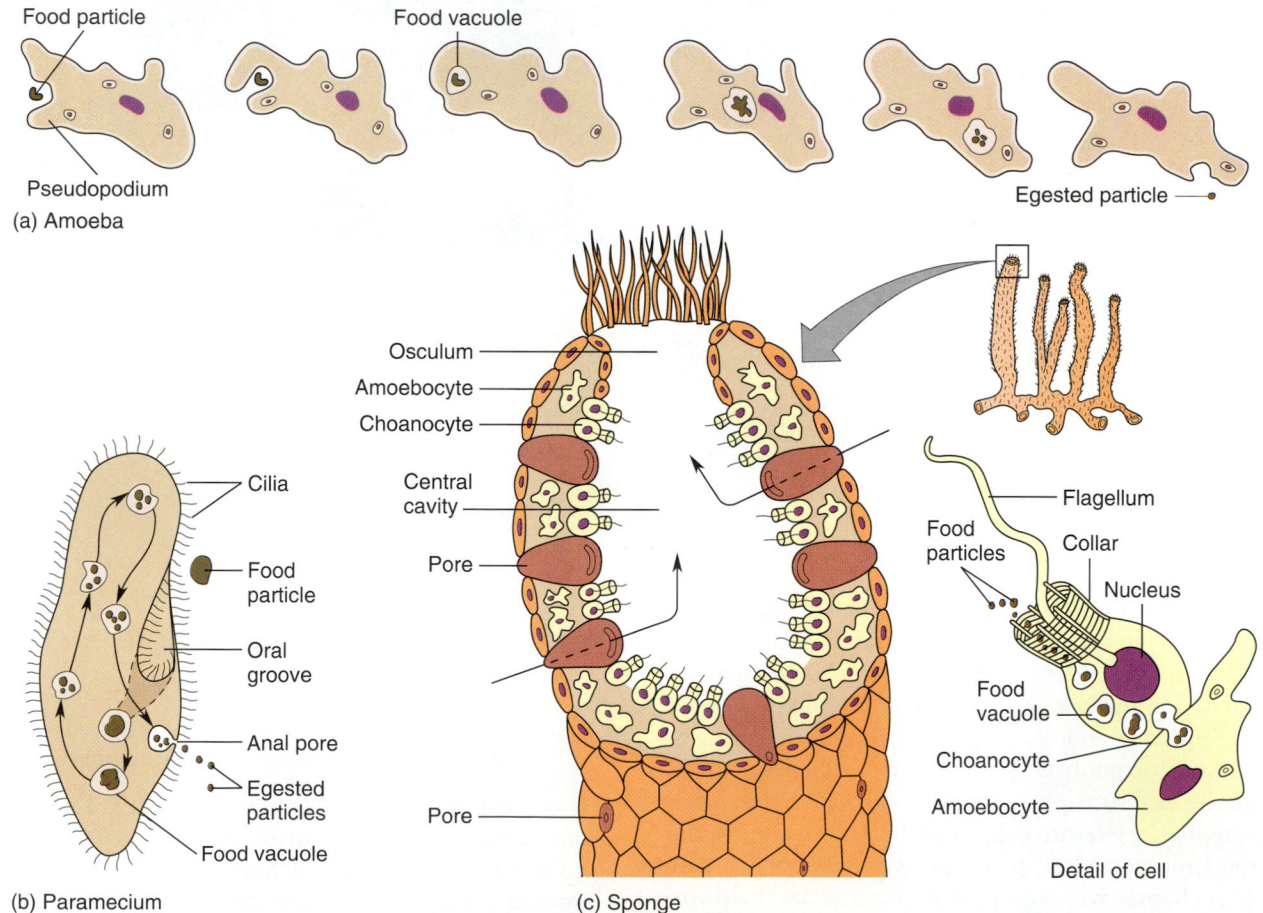

(a) Amoeba

(b) Paramecium

(c) Sponge

Detail of cell

FIGURE 23.1   Intracellular Digestion
(a) The amoeba can ingest and egest food anywhere on its surface. (b) The paramecium has specialized structures for ingestion and egestion. The arrows indicate a path of the food vacuole. (c) In a simple sponge, water and nutrients move into the central cavity through small pores on its surface. Choanocytes trap food particles in their meshlike collars, then ingest them. Some nutrients are broken down and absorbed by the collar cells themselves; others are passed to amoebocytes, which digest them, then distribute them throughout the organism. The arrows indicate the flow of water and nutrients into and out of the animal.

tion and assimilation, turns it into a part of you. **Assimilation** is the process by which the cell uses the digested nutrients for growth, repair, and energy.

Digestion: From Simple to Complex

■ In single-celled organisms, the digestive process occurs within the cell. In more complex organisms, a digestive system evolved — first an incomplete system, then a complete one. With this change, the digestion of food was no longer just an intracellular operation; it began outside the cells, in a digestive cavity or tract.

In unicellular organisms, digestion is *intracellular;* that is, the digestive process begins and ends within the individual cell. The amoeba, a phagocytic organism, can take food in anywhere along its surface (Figure 23.1a). It extends *pseudopodia* ("false feet") that surround and engulf the food particle and a small amount of water, creating a sac called a *food vacuole.* Once the vacuole is inside the cell, it is pinched off the membrane's surface. Lysosomes fuse with the vacuole, releasing enzymes that break down the materials inside it. The movement of the organism's cyctoplasm distributes the digested materials throughout the cell, where they are used for metabolic activities. Undigested materials diffuse or are egested (excreted) through the cell membrane.

Cilia sweep food into the paramecium's oral groove. From there it passes into a gulletlike structure, where a food vacuole forms. The food vacuole moves through the organism, becoming smaller as the food is broken down and assimilated into the cell's cytoplasm (Figure 23.1b). What is not used passes through the organism's anal pore.

Although sponges are multicellular organisms, they have no digestive tract. But here we do find the first division of labor in the digestive process: two different types of cells playing two different roles.

Water and nutrients enter the central cavity (spongocoel) of the organism through small pores on its surface (Figure 23.1c). Flagellated collar cells (choancytes) lining the cavity procure the food. Their collars, a mesh of fibers extending from the body of the cell, act like a net, trapping food particles. When the particles reach the base of the collar, the cell ingests them, forming a food vacuole. The vacuole moves through the cell body and is transferred to an amoebocyte — an amoebalike cell — which continues to break down the nutrients, then distributes them throughout the organism.

Although the collar cells do break down and assimilate some nutrients before passing the vacuole on to the amoebocytes, their primary function is to obtain food for the sponge. Chemical digestion and distribution are the responsibility of the amoebocytes.

Extracellular Digestion: The Evolution of the Incomplete Digestive Tract. In the hydra and jellyfish, we find the beginning of *extracellular digestion* — the digestive process begins outside the cell. These organisms have a primitive digestive cavity called the *gastrovascular cavity,* which has only one opening. In a system like this, nutrients and waste products must enter and exit through the same opening. Because the digestive tract has only one opening, it is considered to be an **incomplete digestive tract.** The hydra (a cnidarian) stings its prey and uses its tentacles to carry food to its mouth. From there, the food moves into the gastrovascular cavity, where specialized cells of the inner lining, or *gastrodermis,* secrete enzymes that begin to break down the nutrients (Figure 23.2a). Usually this digestion is limited to large molecules like proteins. Other nutrients (and some proteins as well) will have to be taken in and digested within the cell. Any wastes are discharged through the mouth.

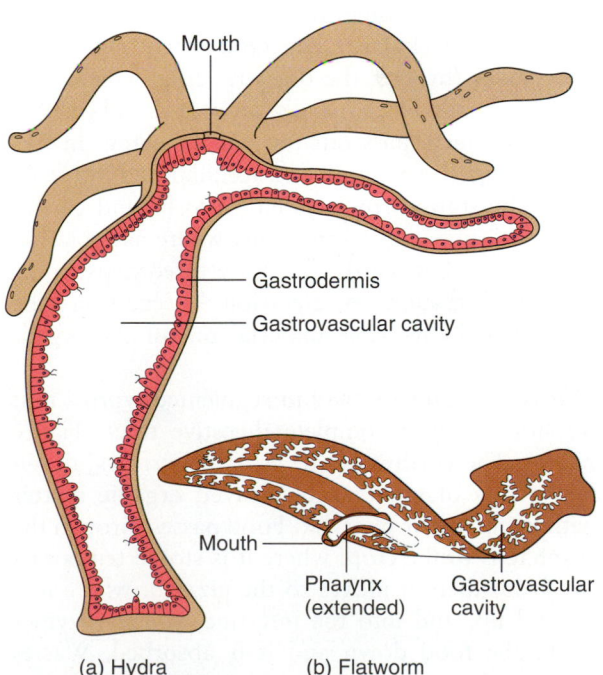

Mouth

Gastrodermis

Gastrovascular cavity

Mouth

Pharynx (extended)

Gastrovascular cavity

(a) Hydra   (b) Flatworm

FIGURE 23.2   Extracellular Digestion: The Incomplete Digestive Tract
In both (a) the hydra and (b) the flatworm, the digestive cavity has just one opening. Although chemical digestion does take place in the cavity, most food substances are broken down within the individual cells of the organism.

In free-living flatworms like the *Planaria,* the gastrovascular cavity is more complex. It is highly branched but still has only one opening (Figure 23.2b). Some extracellular digestion occurs (in other words, digestion takes place in the digestive system, not in the cell), but a high percentage of digestion still happens within the cell. The flatworm captures its prey in a sticky mucous secretion. Its muscular pharynx — a tube that extends from and retracts into the body — functions in the breakdown of food (depositing enzymes on it) as well as the swallowing of food.

Interestingly, some parasitic forms like tapeworms have completely lost their digestive system. These organisms live within a host, absorbing the predigested nutrients from the host organism's intestines or tissue fluids. But not all parasites function in this manner — many of the members of the group called the flukes have their own digestive systems, though they are often rudimentary and nonfunctional.

Extracellular Digestion: The Evolution of the Complete Digestive Tract. A **complete digestive tract** begins with an oral opening and ends with an anal opening and is surrounded by a body cavity. (Think of the digestive tract as a tube within the tube of your body.) The digestive process here moves in one direction, step by step, rather like an assembly line that works backward. The food enters the mouth and passes through the digestive tube, where it is chemically and mechanically processed; indigestible waste products pass out the anal opening. In this system, digestion is both extracellular and intracellular. The transporting system absorbs and carries processed nutrients to the cells, where assimilation occurs. In other words, there are three steps to the process: ingestion → digestion → egestion (the passing of indigestible material out of the organism).

It is in roundworms and segmented worms that we find the first complete digestive tract (Figure 23.3a). The earthworm feeds as it burrows, drawing in bits of plants and decayed organic matter with its muscular pharynx. Food passes through the esophagus to the crop, where it is stored temporarily. From there it moves to the gizzard, where it is ground up, and into the intestine, where enzymes break the food down and it is absorbed. Wastes pass out the animal's anus.

As it eats its way through its environment, the earthworm loosens and turns the soil, moving potassium and phosphorus closer to the surface. These minerals and the worm's own nitrogenous wastes are important nutrients for plant growth.

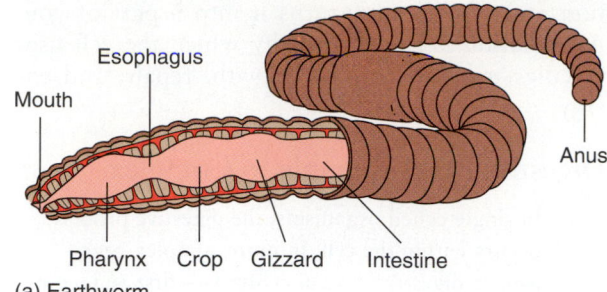

(a) Earthworm

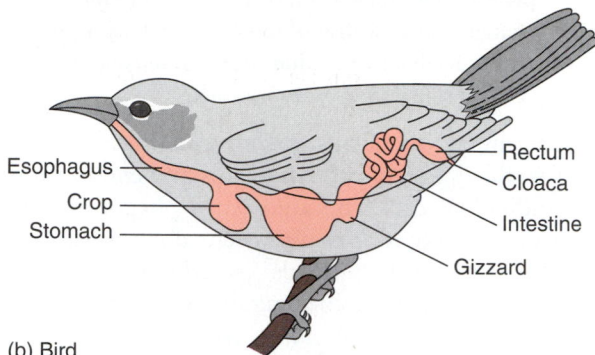

(b) Bird

FIGURE 23.3 Extracellular Digestion: The Complete Digestive Tract
The complete digestive tract has evolved from the primitive system in (a) the earthworm into a system with specialized food-processing regions, like that in (b) the bird. In most organisms with a complete digestive tract, much of the process of digestion takes place in the gut, in a sequence that moves from mouth to anus.

In vertebrates, the complete digestive tract has specialized food-processing regions. In the bird, for example, the stomach separates the crop and gizzard (Figure 23.3b). Here gastric juices help break down the food. That breakdown continues both mechanically and chemically in the gizzard and in the intestines, where nutrients are absorbed and then carried to other cells in the body for assimilation. Wastes pass into the rectum and are discharged through the cloaca.

The complete digestive tract is found in the more complex invertebrates and in all vertebrates. Depending on the lifestyle of the species, digestive systems do exhibit some modifications. In animals without teeth, for example, food is ground up in the gizzard — a sac with thick, muscular walls. The primary food source of *ruminants,* (animals that chew their cud) is bulky foliage or grass or hay. The cellulose content of grass is very high, and it is hard to digest. A ruminant like a cow has a four-part stomach that is highly adapted for the digestion of cellulose, its major source of carbohydrates. The cow's enzymes cannot hydrolyze cellulose, but each stomach chamber contains large numbers of microscopic organisms that can. In humans, the small fingerlike structure called the vermiform appendix is the remnant of a blind sac that might possibly have functioned in cellulose digestion in our ancestors.

Now that we have a brief phylogenetic overview, we concentrate our attention on the human digestive process.

An Overview of Digestion in Humans

■ **In cross-section from the inside to the outside, the walls of the gut are constructed of four tissue layers: the mucosa, the submucosa, smooth muscle, and the serosa. The hollow inner chamber of the** gut is called the lumen. Food moves through the gut by means of wavelike muscular contractions called peristalsis. A series of sphincter muscles keep food moving in the correct direction through the gut.

The human digestive tract is a continuous tube from mouth to anus, consisting of the mouth, esophagus, stomach, small intestine, large intestine, rectum, and anus (Figure 23.4). This tract can also be referred to as the **gut,** *gastrointestinal tract,* or *alimentary canal.* If you removed the gut and stretched it out, it would measure as long as 30 feet. However, inside a living individual it is about 12 to 15 feet long, because its muscles are in a state of partial contraction and are distended with contents.

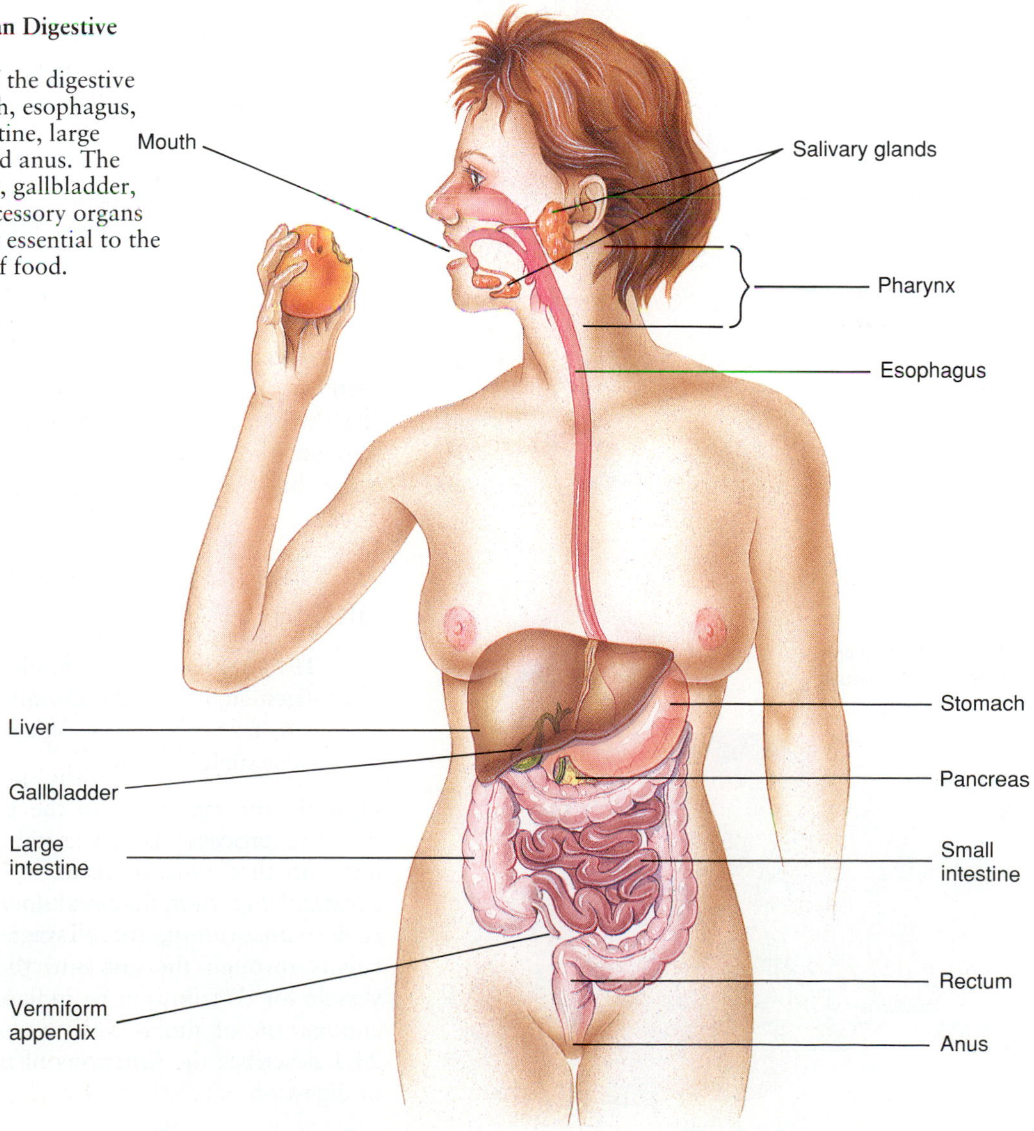

FIGURE 23.4 Human Digestive System
The major organs of the digestive system are the mouth, esophagus, stomach, small intestine, large intestine, rectum, and anus. The salivary glands, liver, gallbladder, and pancreas are accessory organs whose secretions are essential to the chemical digestion of food.

Mouth · Salivary glands · Pharynx · Esophagus · Stomach · Pancreas · Small intestine · Rectum · Anus · Liver · Gallbladder · Large intestine · Vermiform appendix

In the center of the gut, running from the esophagus to the anus, is the *lumen* (the space that makes up the central passageway). The walls of the gut are lined with four layers of tissue (Figure 23.5):

1. The **mucosa** is the innermost layer. It consists of epithelial cells that line the lumen, produce secretions necessary for digestion, and absorb water and nutrients.

2. The **submucosa,** the next layer outward, is made up of loose connective tissue. It contains nerves and blood vessels.

3. **Smooth muscle,** which is on the outside of and surrounding the submucosa, is itself ar-

ranged into two layers. The muscles of the innermost layer are arranged in a circular pattern around the tube; the muscles of the outermost layer are arranged longitudinally, along the length of the tube. These muscles help mix food and move it through the gut.

4. The **serosa,** the thin outer layer of connective tissue, covers the digestive tract and is continuous with the lining of the body cavity.

The process causing movement of food through the digestive tract from the esophagus to the anus is called **peristalsis.** During peristalsis, the circular muscles of the wall of the gut contract behind the bolus (or ball of food), pushing it forward. The process is like squeezing a marble through a rubber tube. As you squeeze the tube behind the marble, you push the marble forward.

Peristaltic waves churn the food, mix it as it is broken down and absorbed, and move it. Food usually passes downward, a direction controlled by valves with sphincter muscles, which are ring-shaped smooth or skeletal muscles located along the digestive tract. The *cardiac sphincter* is a ring of muscle located between the esophagus and stomach, near the heart. Normally the cardiac sphincter is closed, effectively keeping food from moving back up into the esophagus. When it is open, food passes into the stomach.

There are times, however, as you may know from personal experience, when the *vomit reflex* takes over. First, the cardiac sphincter opens; then there are intense contractions of abdominal muscles, which press the stomach against the diaphragm (the large muscle used in breathing). Food passes back up into the esophagus to the mouth.

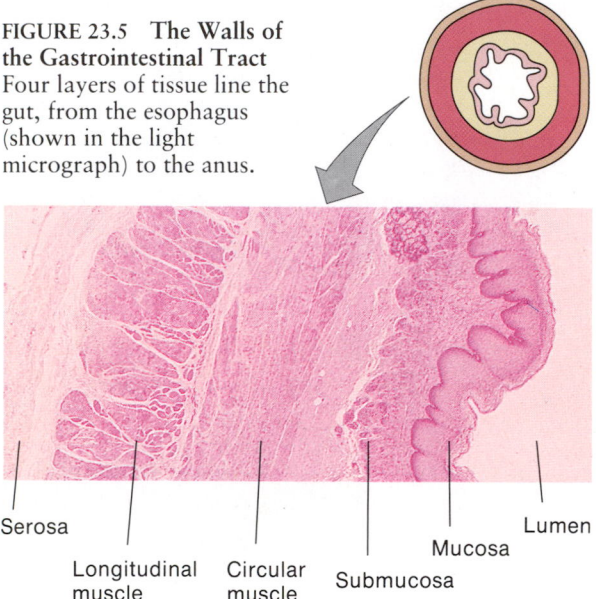

FIGURE 23.5 The Walls of the Gastrointestinal Tract Four layers of tissue line the gut, from the esophagus (shown in the light micrograph) to the anus.

Serosa

Longitudinal muscle

Circular muscle

Submucosa

Mucosa

Lumen

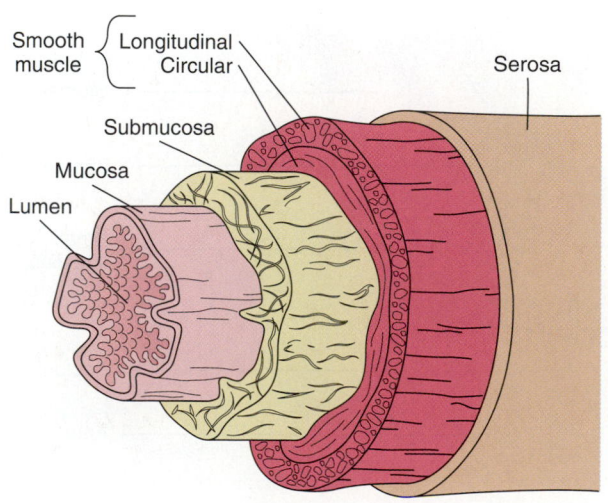

Smooth muscle { Longitudinal Circular

Submucosa

Mucosa

Lumen

Serosa

A CLOSER LOOK AT THE DIGESTIVE PROCESS

■ **The three functions of the digestive system are the digestion of food, the absorption of nutrient molecules for the body's use, and the elimination of indigestible nutrients.**

How do the structures in the gut function in the digestive process? To understand this, we need to look at three major processes: *mechanical* and *chemical digestion,* the breakdown of complex molecules; *absorption,* the passage of processed nutrients through the gut into the blood or lymph vessels for distribution to individual cells; and the *elimination* of indigestible waste products. Table 23.1 describes the functions of the primary organs of digestion.

The Mouth and the Esophagus

■ Teeth break the food into smaller pieces in a mechanical process called mastication. The salivary glands secrete enzymes through ducts that initiate the chemical digestion of carbohydrates. The tongue functions as the major taste organ; it also positions the food for chewing and swallowing. The swallowing reflex moves the food into the esophagus, which connects the mouth and the stomach.

TABLE 23.1
The Major Organs of Digestion

Organs	Functions
Mouth (teeth, salivary glands, tongue)	Mechanical breakdown of food
	Lubrication
	Chemical digestion of carbohydrates (starches)
	Taste
	Movement of food
	Swallowing
Esophagus	Peristaltic movement of food into the stomach
Stomach	Mixing food into chyme
	Initial chemical digestion of proteins
	HCl production
	Peristaltic movement of food into the small intestine
Small intestine	Chemical digestion and absorption of most nutrients
	Peristaltic movement of food into the large intestine
Large intestine	Absorption of water and salts
Rectum	Concentration of fecal mass
Anus	Elimination of undigested residue
Accessory organs	
Liver	Production of chemicals that neutralize (buffer) pH of the small intestine
	Secretion of bile
Gallbladder	Bile storage and concentration
Pancreas	Secretion of enzymes that enter the small intestine to complete chemical digestion
	Production of chemicals that help to neutralize the pH of the small intestine

The first major digestive organs are those that are involved with the initial food intake: the mouth (or oral cavity) and the esophagus. The primary function of the mouth is to break the food into small pieces that the rest of the digestive system can handle. The esophagus delivers the food to the stomach.

In the mouth are the *teeth,* which are involved in *mastication,* or the tearing and chewing of food; the *salivary glands,* which begin the process of chemical digestion; and the *tongue,* which not only moves the food to the esophagus but also contains the taste buds that enable us to perceive the flavor of the things we eat.

The Teeth. Most vertebrates (except birds and some reptiles) have teeth, but the extensive mechanical breakdown (chewing) of food occurs only in mammals. The arrangement and type of teeth in separate animal species are so distinctive that it is possible to identify animals by their dentition alone.

Teeth are shaped to perform certain functions. Some, like the chisel-shaped *incisors* in the front of the mouth, shear and cut (Figure 23.6). Behind the incisors are the pointed *canines,* which are adapted

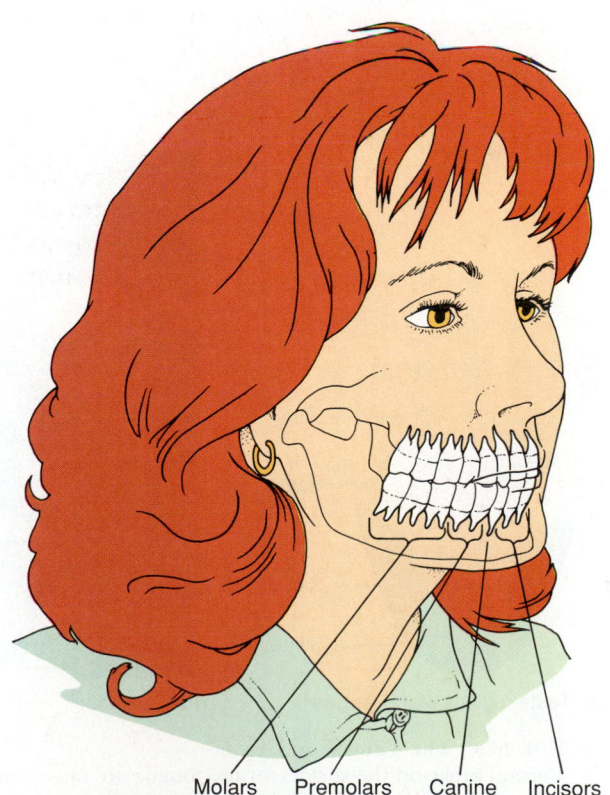

Molars   Premolars   Canine   Incisors

FIGURE 23.6 Teeth
The process of mechanical digestion — the mechanical breakdown of food substances — begins in the mouth with the teeth.

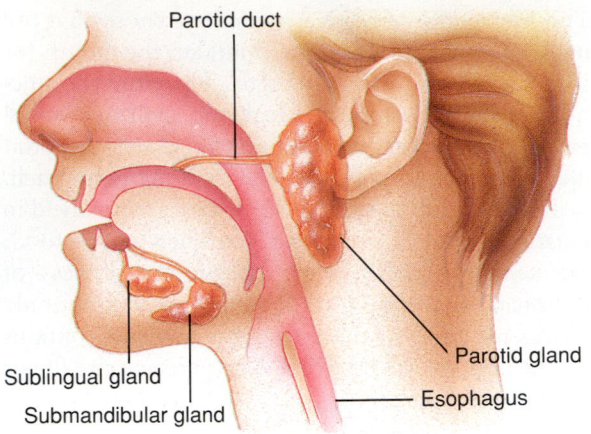

Parotid duct

Sublingual gland

Submandibular gland

Parotid gland

Esophagus

FIGURE 23.7 The Salivary Glands
Although small amounts of saliva are secreted by the glands in the mucous membranes of the mouth, most of the saliva is produced by the salivary glands. The mumps virus infects the parotid glands.

for piercing, tearing, slashing, and ripping. Behind the canines are the *premolars* and *molars,* which grind and crush. Not all species exhibit all four types, and individual species exhibit varying proportions of each type.

The Salivary Glands. Found in the jaw region and the floor of the mouth are three pairs of **salivary glands:** parotids, sublinguals, and submandibulars (Figure 23.7). These glands secrete saliva, which moves through ducts into the mouth and mixes with food. Saliva contains a lubricating substance known as *mucus,* and buffers that help to maintain the acid-base balance. In addition, saliva also contains *salivary amylase,* an enzyme that initiates the hydrolysis of carbohydrates like starch and glycogen.

The Tongue. The tongue moves food around, helping us chew and swallow. The tongue is also a taste organ. On its surface are hundreds of taste buds, structures that detect the salty, sour, sweet, and bitter flavors in food (Figure 23.8). When saliva reaches the taste cells in a taste bud, it triggers a nerve impulse that tells us the taste of what we are eating. (Another major factor involved in taste is smell. Have you ever noticed how food seems almost tasteless when you have a cold and your nose is stuffed up?) The tongue also moves food around in the mouth, serves as a chewing aid, and pushes the food into the back of the mouth as swallowing occurs.

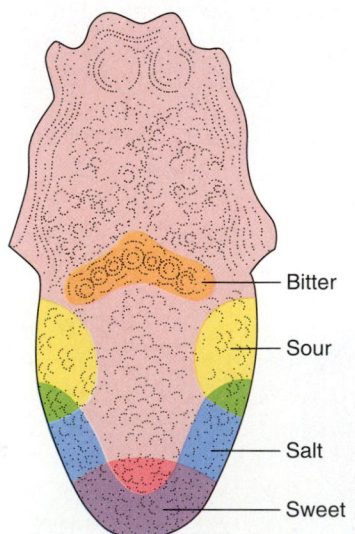

Bitter

Sour

Salt

Sweet

(a) Taste regions of the tongue

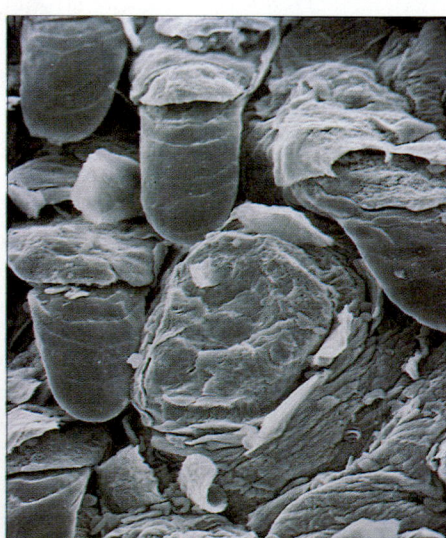

(b) Taste bud

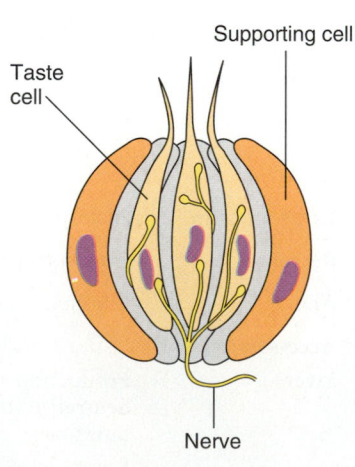

Taste cell

Supporting cell

Nerve

FIGURE 23.8 The Tongue and Taste
(a) Certain areas on the surface of the tongue are more sensitive than others to specific tastes. (b) In a taste bud, supporting cells form a capsule around the taste cells. When food mixes with saliva, the "taste" moves through the taste cells, triggering a nerve impulse that is carried to the brain, which tells us what we're tasting.

The Esophagus. Swallowing is a combination of voluntary and involuntary muscle contractions. In the voluntary phase, the tongue forces the food mass to the back of the mouth (the *pharynx*). Here the swallowing reflex — the involuntary phase — begins, pushing the food into the **esophagus** (the tube that connects the mouth to the stomach), then into the stomach, through peristaltic action. (The esophagus does not act directly in chemical digestion.) The esophagus is located behind the *larynx* and *trachea* (the network that passes air into the lungs). To prevent food from entering the trachea as we swallow, the larynx pushes up and a flap of tissue called the *epiglottis* presses against it, directing the food into the esophagus (Figure 23.9a).

Sometimes a piece of food goes down the trachea instead of the esophagus and interferes with breathing. If the food is not dislodged within four to five minutes, the person will die. You can remove the food from the trachea of a choking person by using the Heimlich maneuver, which is detailed in Figure 23.9b. Never perform the Heimlich maneuver on a person who can speak or is coughing because they are probably not choking. A choking person will not be able to make any noise at all, because he or she cannot get the air supply required to speak or to cough. Make sure the person sees a doctor immediately after receiving the Heimlich maneuver.

Stomach

- **With some exceptions, food molecules are not absorbed in the stomach. The stomach stores and mixes the food. Digestion also occurs in the stomach by means of gastric juices consisting of hydrochloric acid, pepsinogen, and a few other enzymes. These substances break the food down into its constituent molecules. The semiliquid mixture that results is called chyme.**

The esophagus passes through the diaphragm and joins the stomach. The **stomach** is a muscular, expandable, J-shaped sac (Figure 23.10a). It is closed at both ends by sphincter muscles. We have already mentioned the *cardiac sphincter*, located at the junction between the esophagus and the stomach. The lower sphincter, called the *pyloric sphincter*, is found at the junction of the stomach and the small intestine.

The walls of the stomach are lined with the same tissues that line the rest of the gastrointestinal tract. Here, however, there are three layers of smooth muscle, which help the stomach churn and grind food. And the mucosa in the stomach walls is arranged in large folds (Figure 23.10b). On the surface of these folds are mucous cells, which secrete mucus, a substance that protects the stomach walls (Figure 23.10c). In the folds are the glands that secrete the gastric juices and enzymes that break down and liquefy food substances (Figure 23.10d).

The stomach serves two major functions: the mechanical and chemical digestion of food. The three layers of smooth muscle in its walls help the stomach mix food. As the food is being mixed, the glands in the mucosa secrete gastric juices — hydrochloric acid (HCl), mucus, a protein-digesting enzyme called pepsinogen, and other enzymes — and a hormone called gastrin, which helps coordinate digestive activities, primarily HCl production.

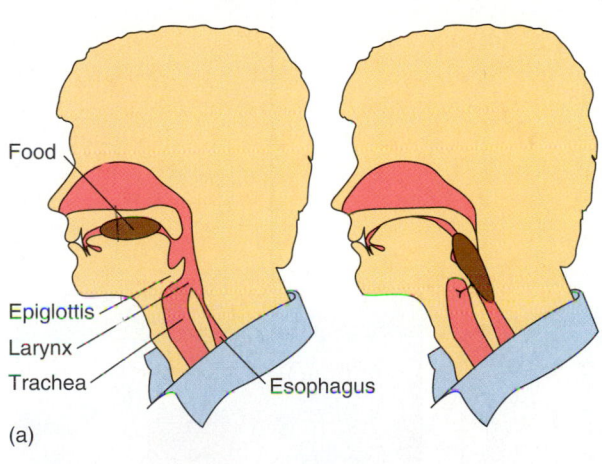

Food
Epiglottis
Larynx
Trachea
Esophagus

(a)

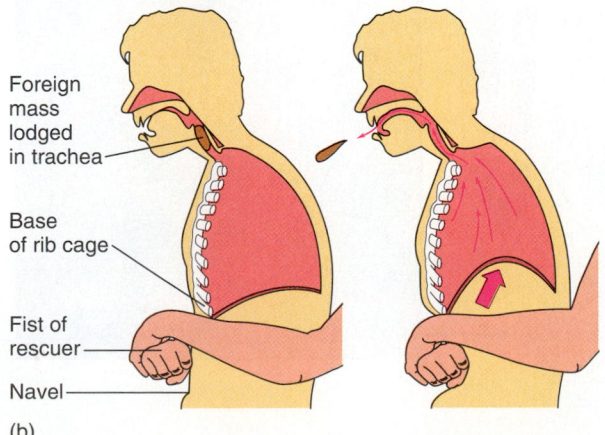

Foreign mass lodged in trachea

Base of rib cage

Fist of rescuer

Navel

(b)

FIGURE 23.9 Swallowing and the Heimlich Maneuver (a) As you swallow food, the larynx moves up and the epiglottis presses against it, blocking the entrance to the trachea and directing food into the esophagus. (b) If food or other foreign matter gets lodged in the trachea, the Heimlich maneuver can be used to dislodge it.

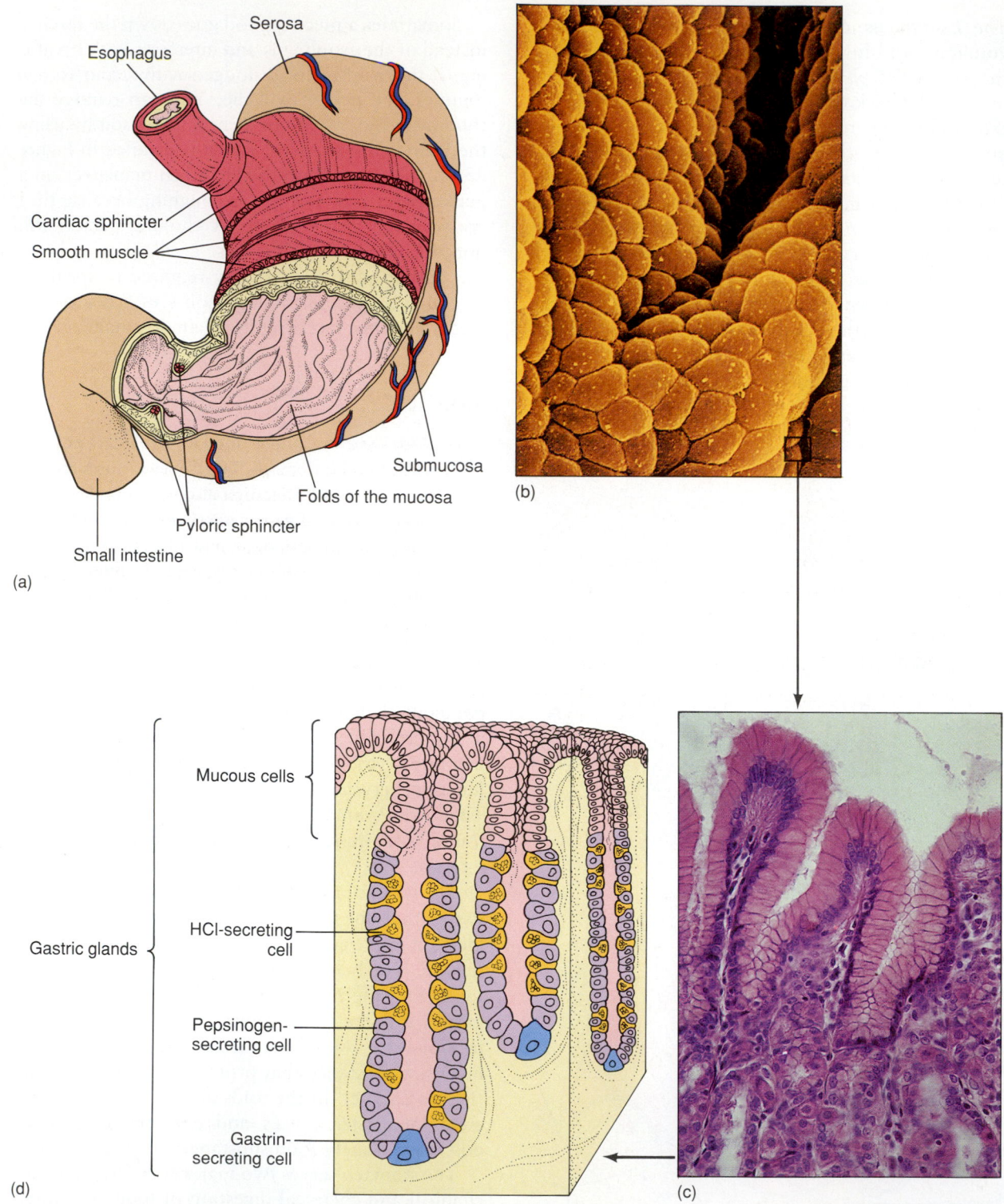

FIGURE 23.10 The Stomach
(a) Sphincter muscles control the flow of food into and out of the stomach. Contractions of the smooth muscle layer of the stomach churn and mix food, breaking it down mechanically. (b) When the stomach is empty, the mucosa, the inside lining, is arranged in folds. (c) The surface of the mucosa contains many mucus-secreting cells. (d) The pits in the mucosa lead to the gastric glands. Gastrin-secreting cells are found only in the glands near the pyloric sphincter.

Through the combination of mechanical and chemical forces in the stomach, the food substance becomes a semiliquid mixture called **chyme.**

The Chemistry of the Stomach. The breakdown of proteins to polypeptides begins in the stomach. When a food containing protein enters the stomach, it signals the release of *gastrin* from the gastric glands. (In fact, sometimes just thinking about food can stimulate stomach movement and the production of gastric juices.) Gastrin is carried locally by the bloodstream in the stomach, inducing certain cells in the gastric glands to increase their production of HCl. HCl serves several purposes. It destroys any bacteria and other organisms in the stomach, and it erodes tough fibers in the substances being digested. It also converts the inactive form of pepsinogen into one of its active forms, known as *pepsin,* an enzyme that breaks protein down into peptides. Although most fats are digested in the small intestine, the breakdown of fats to fatty acids begins in the stomach. At work here is gastric lipase, an enzyme that acts on fat.

The mucosal cells secrete about 2 liters of gastric juices per day. These juices have a pH of between 1.5 and 2.5, which makes them the most acidic of all body fluids. If you have ever had heartburn, the burning sensation in your esophagus was due to the acidic gastric juices passing upward into the esophagus. (Gastric juices occasionally enter the esophagus when the stomach is full and the cardiac sphincter relaxes.)

But if the stomach acids are so concentrated, why are the walls of the stomach unaffected by them? Remember that we noted that the stomach also secretes mucus. Most of the time, mucus forms a thick viscous lining over the stomach wall and protects it. If the mucus lining breaks down, the stomach tissue is damaged, and the sore formed is called an ulcer. (Sometimes the ulcer is in the small intestine or the esophagus.) A stomach ulcer is called a *peptic* or *gastric ulcer.*

The mechanisms that produce high levels of gastric juices are hormonal and neural and as yet are poorly understood. But we do know that several drugs can either inhibit acid secretion or neutralize the acid once it is present. We also know that certain drugs, including aspirin or nonaspirin anti-inflammatory drugs given to arthritis sufferers, can induce lesions that result in ulcers. That is one reason why the instructions for some prescriptions include a note that the drug should be taken with meals — to protect the stomach walls from full exposure to the substance.

Though some substances are absorbed into the bloodstream from the stomach (alcohol, B vitamins, and aspirin, for example), most are not. Instead, food absorption occurs in the small intestine. Movement into the small intestine is regulated by hormonal, chemical, and neural signals that indicate when the pyloric sphincter muscle should relax. Chyme then leaves the stomach by means of stomach muscle contractions and enters the small intestine.

Small Intestine

■ **The small intestine consists of three regions: the duodenum, the jejunum, and the ileum. The duodenum functions primarily in the continued breakdown of chyme and the initial absorption of food molecules. Bile, produced in the liver and stored in the gallbladder, enters the duodenum and breaks up fat molecules. The pancreas produces enzymes and sodium bicarbonate, which helps neutralize the chyme's acidic level. Villi lining the small intestine and microvilli on the cells of the villi assimilate nutrients.**

The **small intestine** is a coiled tube that is about 2 to 3 centimeters (about 1 inch) in diameter and about 6 meters in length (about 20 feet). For convenience, the small intestine is divided into three regions, according to structure and function: the *duodenum,* the *jejunum,* and the *ileum.*

Duodenum, Jejunum, and Ileum. The region nearest the stomach — the first 20 to 30 cm (8 to 10 inches) — is called the **duodenum.** This is the most important part of the small intestine because most of the chemical breakdown and some of the absorption of nutrients occur here. It is also where secretions from the accessory organs of digestion — the liver, gallbladder, and pancreas — enter the gastrointestinal tract through the common bile duct and the pancreatic duct.

Liver, Gallbladder, and Pancreas. The **pancreas** secretes enzymes that act on carbohydrates, proteins, fats, and nucleic acids, and sodium bicarbonate, which helps to neutralize the acid produced by the stomach. The **liver** produces *bile,* which is stored in the **gallbladder** and is then passed into the duodenum. Bile contains salts that break large fat droplets into smaller ones — a process called emulsification — and also help in their absorption. Emulsification increases the surface area of fat, making it more accessible to the intestinal enzymes.

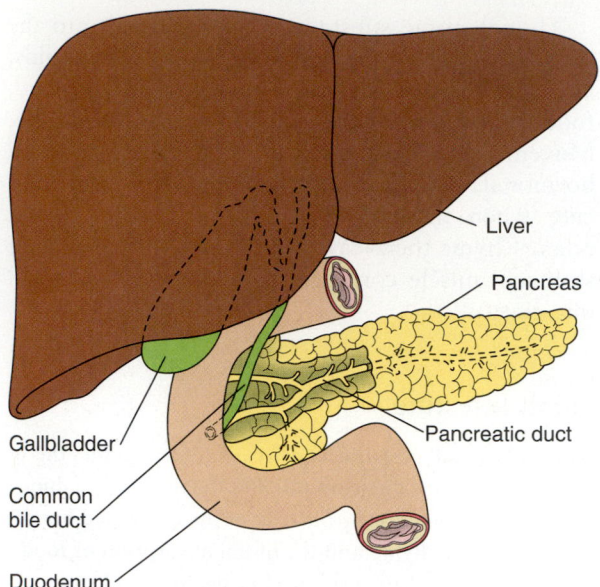

Liver

Pancreas

Pancreatic duct

Gallbladder

Common
bile duct

Duodenum

**FIGURE 23.11   The Relationship of the Liver,
Gallbladder, and Pancreas to the Small Intestine**
Bile from the liver is stored in the gallbladder, then
released into the common bile duct. The duct merges
with the pancreatic duct and empties into the duodenum
of the small intestine.

Following the duodenum, as the small intestine
turns downward, is the **jejunum,** a section that is
about 2.5 meters (8 feet) in length. The jejunum is
followed by the highly coiled **ileum,** which is about
3.6 meters (12 feet) in length and is connected to
the large intestine (Figure 23.11).

Villi and Microvilli. Once the chyme is almost broken down in the duodenum, the digested food and
some water are absorbed in the remainder of the
small intestine. The inner lining of the small intestine contains an immense number of fingerlike projections called **villi** (singular, *villus*), each of which
is covered with mucosal cells (Figure 23.12a). The
plasma membranes of these cells are composed of
numerous microscopic foldings called **microvilli**
(Figure 23.12b). Villi and microvilli greatly increase
the surface area of the small intestine, thus allowing
for a greater absorption of the digested food. You
may have experienced this if you have ever used a
worn terrycloth towel to dry dishes. That towel was
not as absorbent as a new one because it lacked the
many fine projections that would increase its absorptive area. Biologists estimate that the coiling of
the small intestine, the villi, and the microvilli increase the surface area to about 2000 square feet.
That is the size of the floor space in a three- or four-
bedroom home — about 100 times that of your

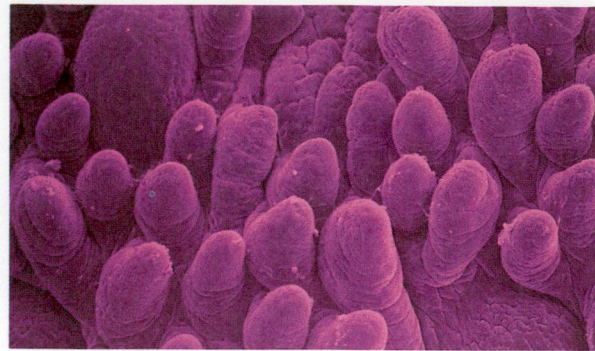

(a) Inner surface of the small intestine

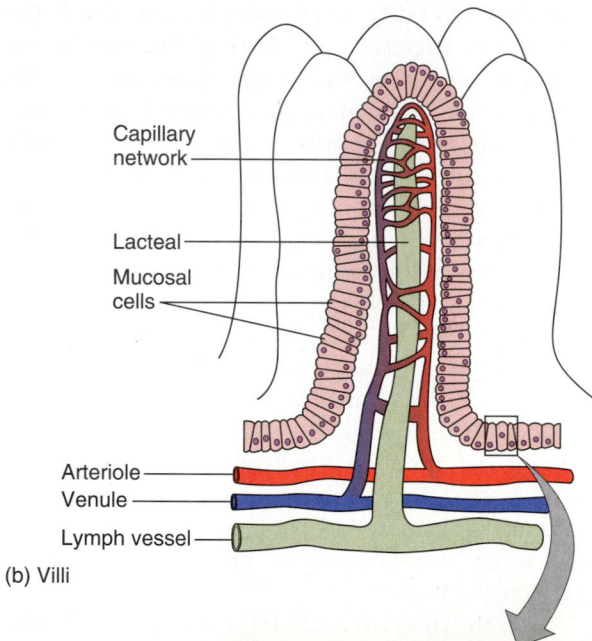

Capillary
network

Lacteal

Mucosal
cells

Arteriole

Venule

Lymph vessel

(b) Villi

Microvilli

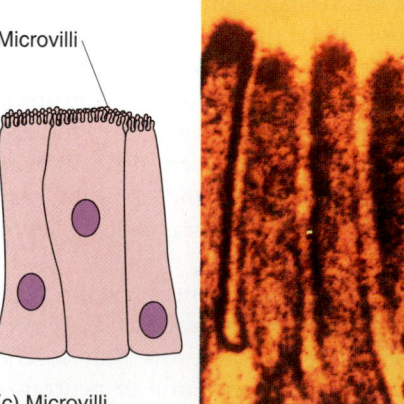

(c) Microvilli

FIGURE 23.12   Villi and Microvilli in the Small Intestine
(a) The small intestine is lined with thousands of villi,
small fingerlike projections. (b) Inside each villus is a
lacteal surrounded by a capillary network. (c) The
plasma membrane of the mucosal cells that line each
villus is made up of microscopic folds called microvilli.
These folds, and villi, and the folds of the intestinal wall
all increase the surface of the small intestine, facilitating
the absorption of nutrients.

skin — truly an example of form following function.

The villi are specialized for the absorption of food molecules, and they consist of a *capillary network,* a network of microscopic blood vessels. (The capillaries are the point of exchange between the bloodstream and the body's cells.) Arterioles feed oxygenated blood into the capillary network; venules carry deoxygenated blood from the area. (We discuss the circulatory system in the next chapter.) Amino acids, sugars, and other nutrients (with the exception of fats) are absorbed into the capillary network of the villus. The network surrounds a lymph vessel called a *lacteal,* which absorbs the products of fat digestion (Figure 23.12b).

At one time biologists believed that the absorption of food molecules into the capillary network and lacteal was a fairly passive process. Now they know that, for the most part, active transport is involved. Molecules are moved from regions of lower concentration in the lumen of the small intestine to regions of higher concentration — namely, the blood. This selective active transport requires energy from ATP. The villi also require energy, for

they are constantly beating and waving to renew their contact with and absorption of the dissolved food substances.

Several substances are involved in the digestive processes of the small intestine. The mucus, for example, is produced by the epithelial cells lining the villi. Mucus serves to lubricate the food, buffer the pH, and protect the lining of the digestive tract. The epithelial cells are also the source of enzymes that act on proteins, carbohydrates, and nucleic acids. Table 23.2 is a list of the major digestive enzymes in the small intestine and other parts of the gut.

Epithelial cells are constantly being sloughed off by the friction produced by the chyme as it passes through the intestines and is mixed there. Although it seems difficult to believe, biologists estimate that a half pound of epithelial cells are lost through this abrasive action every day. As a result, the epithelia have to be constantly replaced through mitosis.

Previously, it was thought that the epithelial cells secreted their digestive enzymes directly into the lumen, where they would mix freely with the chyme. However, recent evidence shows that most digestive enzymes are actually bound to the plasma

TABLE 23.2
Major Human Digestive Enzymes

Enzyme	Source	Material digested	Action
Pepsins	Stomach mucosa	Protein	Break proteins down into peptide fragments
Trypsin, chymotrypsin	Produced in pancreas; act in small intestine	Protein	Break proteins and polypeptides down into peptide fragments
Carboxypeptidase	Produced in pancreas; acts in small intestine	Protein	Breaks peptide fragments down into amino acids
Aminopeptidase	Produced in intestinal mucosa; acts in small intestine	Protein	Breaks peptide fragments down into amino acids
Pancreatic nucleases	Produced in pancreas; act in small intestine	Nucleic acids	Break DNA and RNA down into nucleotides
Intestinal nucleases	Produced in intestinal mucosa; act in small intestine	Nucleic acids	Break nucleotides down into nucleotide bases and monosaccharides
Salivary amylase	Salivary glands	Carbohydrates	Breaks polysaccharides down into disaccharides
Pancreatic amylase	Produced in pancreas; acts in small intestine	Carbohydrates	Breaks polysaccharides down into disaccharides
Disaccharidases	Produced in pancreas; act in small intestine	Carbohydrates	Break disaccharides down into monosaccharides
Gastric lipase	Stomach mucosa	Fat	Breaks triglycerides down into fatty acids and glycerol
Lipase	Produced in pancreas; acts in small intestine	Fat	Breaks triglycerides down into fatty acids and monoglycerides

A Closer Look at the Digestive Process

membrane of the epithelial cells and, along with the other cell surface molecules, form a covering over the plasma membrane. Biologists speculate that these enzymes are located in precise positions to aid with the digestive process. Their specific positioning also prevents the lining of the small intestine from being digested.

Digestion in the small intestine is also a physical process. Localized contractions of the smooth muscle in the walls of the gut mix the chyme and enzymes, and peristalisis continues to move the chyme along.

Large Intestine, Rectum, and Anus

■ **The large intestine absorbs water and minerals from the materials that leave the small intestine. From the large intestine, the materials move into the rectum and pass out of the body through the anus.**

In our exploration of the digestive process, we come next to the **large intestine,** or colon. The colon is about 1.5 to 2 meters (5 to 6 feet) long and 6 to 7 centimeters (about 2½ inches) in diameter. The large intestine is more than twice as wide as the small intestine. It connects with the small intestine on the right side of the body, at the ileum, then moves up, across, down, and toward the center of the body, ending at the rectum. The rectum and anus are the terminal organs of the gut.

Like the walls of the stomach and small intestine, the walls of the large intestine are arranged in folds (Figure 23.13). These folds increase the area of absorption, and the cells along their surface secrete mucus that lubricates the passageway.

The primary function of the colon is to reabsorb the large volume of water secreted into the gut in the form of enzymes, acids, and bile. This keeps us from becoming dehydrated. The colon is also a storage place for large numbers of intestinal bacteria. Some physiologists hypothesize that the enzymes produced by the bacteria here continue the breakdown and absorption process.

The length of time that digesting material spends in the colon determines the amount of water that is removed. Constipation results from the removal of too much water; the insufficient removal of water results in diarrhea. These conditions can be brought on by many factors, including disease and parasites. Sometimes the lining of the colon, which secretes mucus, is irritated by parasites like the amoeba *Entamoeba histolytica*. This organism causes amoebic dysentery. Other irritating chemicals produced by bacteria or the decay products of spoiled food can cause diarrhea.

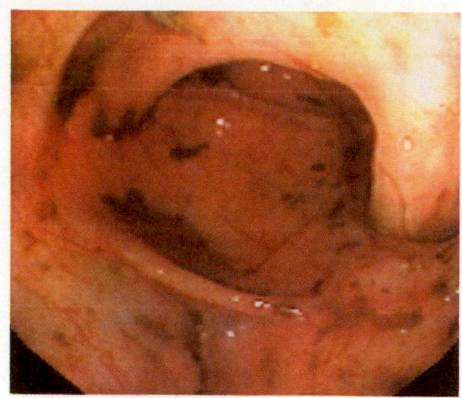

FIGURE 23.13
The Walls of the Large Intestine
You can see the mucus glistening in the folds that line the walls.

Dysentery and related diseases result in severe dehydration because the irritated large intestine cannot reabsorb water and dissolved minerals. One of the leading causes of infant death in many Third World countries is the dehydration that results from diarrhea.

Intestinal Bacteria. Not all bacteria are harmful. Because the billions of bacteria living in the large intestine produce enzymes that humans cannot produce, they aid in the digestion of food not digested in the small intestine. Intestinal bacteria can also synthesize amino acids and vitamins that then can be absorbed by our circulatory system. For example, bacteria produce vitamin K and some of the B vitamins that are useful to us. This relationship is *mutualistic*, since both organisms obtain mutual benefit from this relationship. The bacteria require the environment of the intestine to survive, and we require their by-products to fully utilize the nutritive elements in our food.

Defecation. The large intestine continues to process the semiliquid material, reabsorbing water and minerals until it becomes a semisolid mass known as **feces.** As strange as it may seem, the primary element in feces is living and dead bacteria, along with a little cellulose.

There are two anal sphincter muscles — inner and outer. The outer anal sphincter is skeletal muscle that is under voluntary control, giving us conscious control over the process of defecation, or the expulsion of the feces. Control of the external anal sphincter takes time to develop, as all of you who have toilet-trained small children know. The fecal mass is stored in the rectum until the stretching of its walls signals nerve reflexes that pass the fecal mass through the two sphincter muscles out of the body through the anus.

In the rectum and around the anus are a number of blood vessels, which sometimes become enlarged. The enlarged blood vessels, known as *hemorrhoids,* are a form of varicose veins (see Chapter 24). The actual cause of this condition is not known, but constipation, obesity, abdominal pressure, and genetic predisposition may be the "seat" of the problem. Hemorrhoids can cause pain, itching, and, in some instances, bleeding. In severe cases, they may have to be surgically removed.

Appendicitis. One portion of the colon forms a blind pouch called the *cecum,* and extending from it is a small projection called the *appendix* (Figure 23.14). The appendix is involved in combating infection — in a sense, it is really a part of the immune system. When the appendix becomes inflamed or infected, the condition is known as *appendicitis.* If the attack is mild, it can be treated with antibiotics, but severe cases require surgical removal. If treatment is not begun in time, the appendix may rupture, spilling the bacteria into the body cavity (coelom) and causing an infection of the *peritoneum,* the lining of the coelom. This condition, known as *peritonitis,* often is fatal.

Negative Feedback — Coordinating the Digestive Process

As food moves through the various parts of the gut, three kinds of factors—mechanical, neural, and hormonal—control the secretion of enzymes and

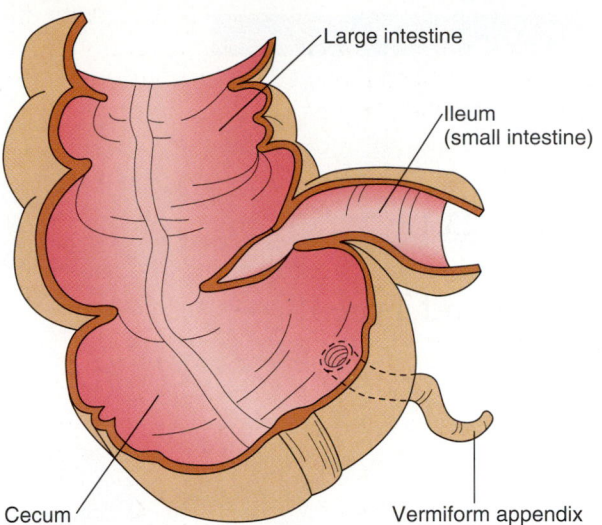

FIGURE 23.14   The Appendix
In our primitive ancestors, the appendix may have functioned in the digestion of cellulose — a substance our bodies cannot digest today.

other substances by the cells of the digestive system. These mechanisms all work together. You may have noticed this integration when at one time or another you found that just thinking of food or even excessive stress can stimulate contractions of the stomach and the secretion of digestive substances. Table 23.3 lists the important digestive hormones involved in the regulation of the digestive process.

In Chapter 1 and in other chapters as well, we discussed *homeostasis,* the ways in which the organism maintains a relatively stable internal envi-

TABLE 23.3
Digestive Hormones

Hormone	Source	Stimulation	Function
Secretin	Epithelial cells of the duodenum	Intestinal acidity; peptides in intestine	Stimulates release of sodium bicarbonate
Enterocrinin	Intestinal mucosa	Stomach materials in intestine	Stimulates intestinal secretions
Enterogastrone	Intestinal mucosa	Intestinal acidity; fats in intestine	Inhibits release of gastric juices; slows stomach contractions
Cholecystokinin (pancreozymin)	Upper intestinal mucosa	Food in small intestine	Stimulates secretions of pancreatic enzymes; stimulates gallbladder to contract
Gastrin	Gastric glands in stomach	Protein in stomach; nerve impulses	Stimulates release of gastric juices

A Closer Look at the Digestive Process

FOCUS 23.1
Diabetes:
Type I and Type II

Most of us have known at least one person who suffered from the disorder known as diabetes. What you may not know is that there is more than one kind of diabetes. In general, there are two major kinds, both characterized by the production of abnormal amounts of urine. *Diabetes insipidus,* which is relatively rare, results in large volumes of urine output (sometimes as much as 10 gallons per day). The disorder is caused by the fact that the pituitary gland does not release antidiuretic hormone. The more common kind of diabetes is called *diabetes mellitus,* or sugar diabetes. There are two forms of sugar diabetes: type I and type II. Both result in the inability of the body's cells to regulate proper glucose utilization. As a result, high levels of glucose accumulate in the blood. From a biochemical perspective, however, type I and type II differ greatly.

In type I diabetes, the pancreas does not produce enough insulin,

the hormone that facilitates the transport of glucose through cell membranes and into the cells. Because of the low levels of insulin, body cells have too little glucose, which is their primary source of cellular energy. The problem lies in the insulin production by the pancreas, not in the structure of the target cell membrane. The target cell membrane has enough chemical receptors to accept insulin if it is available. This distinction is important in diagnosing the disorder and formulating treatment for it. Type I diabetes is *insulin-dependent,* which means that the condition can be regulated by daily insulin injections (Figure A).

In type II diabetes, the pancreas produces enough insulin, but the cell membranes do not have enough receptors to receive the hormone. That means that the cells cannot respond to the insulin. It is hypothesized that obesity is the major cause of type II diabetes. As fat cells grow larger and larger, the number of chemical re-

ceptors for insulin per unit volume decreases. Because there are not enough insulin receptors on the cell membrane, glucose cannot enter the

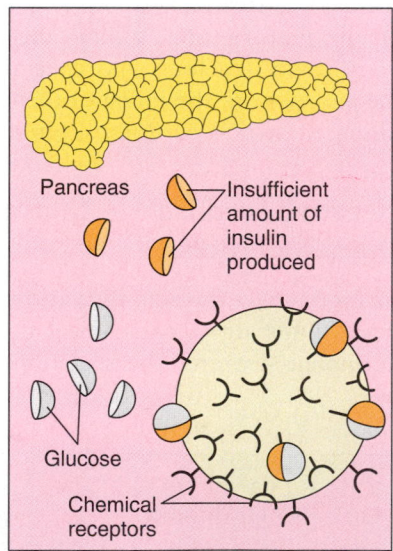

FIGURE A
Type I (insulin-dependent) diabetes

ronment. We also discussed *negative feedback,* which is central to the functioning of the digestive system. One example of negative feedback and its role in maintaining homeostasis is the regulation of blood glucose. As one of the cells' major energy sources, glucose must be maintained at a relatively constant level throughout the day. Since we are not

constantly eating, how do we maintain this regular level of glucose?

Glucose (and other simple sugars as well) is the product of carbohydrate digestion and is absorbed through the capillaries of the villi into the bloodstream. The blood, rich in glucose, goes directly to the liver through a vein called the *hepatic portal*

cell in adequate amounts, and so glucose metabolism goes awry. The level of glucose in the blood can become dangerously high, regardless of type—a condition that can lead to coma and even death (Figure B).

Type II diabetes used to be called *adult onset diabetes*, because it was usually found in adults. Today, however, we know that even children can have this form of the disorder. There is a serious concern about the increase of type II diabetes. Six hundred thousand new cases are diagnosed every year. What is more, the American Diabetic Association estimates that 5 million cases of type II diabetes have been identified, and another 5 million individuals may have the disorder and not even

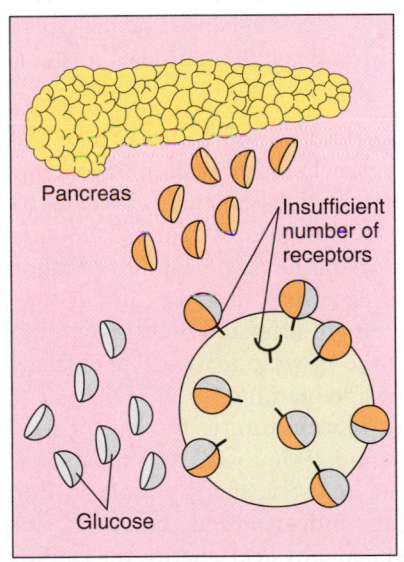

FIGURE B
Type II diabetes

know it, because it can take several years to manifest itself. Some individuals have had it for from 5 to 10 years without realizing what was wrong. Among its symptoms are drowsiness, itchy skin, numbness in the feet, blurred vision, and skin infections that heal slowly. Because the symptoms are mild, they are often overlooked. The only way to detect type II diabetes is through a blood test. The disorder is suspected if the blood exhibits high glucose levels while at the same time the individual has normal levels of insulin.

Any individual can help fend off the onset of type II diabetes. Physicians know that obesity is the number one risk factor for the disorder. In fact, this disorder is known as the disease of overeating, lack of exercise, and stress. Eighty percent of type II diabetics are overweight, and the chances of getting the disease doubles for every 20 pounds of fat. Interestingly enough, obese women who store their fat above the waist are more likely to get type II diabetes than women who store their fat below the waist. This is because fat cells above the waist are larger than those below the waist. Hence, in the first group there are fewer insulin receptors per volume of cell. When fat cells accumulate below the waist, there are more of them but they retain their normal size. As a result, there are usually enough receptor sites for insulin on the cell membrane.

Social pressures can complicate attempts to control type II diabetes

because of prejudice against obesity, which can result in psychological problems in type II diabetics — they may lack self-confidence or may feel guilty, rejected, and lonely. To compensate for these feelings, they may turn to food, which only aggravates the condition. So the vicious cycle continues.

Genetics may also play a role in the incidence of type II diabetes. If an individual has a family history of diabetes, his or her chances are one-third greater that type II diabetes will develop. Ethnicity also seems to be a factor. African Americans, Hispanics, and Native Americans are twice as likely to develop type II diabetes as white Americans. For example, among the Pima Indians of Arizona, half of the population over the age of thirty-five have type II diabetes. The mechanisms involved are still unknown.

Because most of us have heard of diabetes, and because we know that there are treatments for it, we sometimes think the disorder is not serious. That assumption can be dangerous. When nondiabetics are compared with individuals having uncontrolled type I and II diabetes, the diabetic individuals are twenty-five times more likely to suffer blindness and deterioration of the retina. They are also seventeen times more likely to develop kidney trouble, five times more likely to develop gangrene, and two times more likely to have heart disease and/or stroke. In severe cases, diabetes can kill. It is not a disorder to take lightly.

vein. (A *portal vessel* is one that is between two capillary beds. In most circulatory patterns, blood flows through only one capillary bed before returning to the heart.) The liver cells then convert some of the glucose into animal starch, *glycogen*. Some of the glucose is sometimes converted to fat. Fat is stored in several parts of the body, in adipose tissue.

Glycogen not stored in adipose tissue may be retained in skeletal muscle cells. The glycogen stored in skeletal muscle is an even more significant energy reserve than that stored in fat.

Although this may seem like a circular argument, maintenance of the blood's glucose level is determined by the blood's glucose level. In other words, the system functions in the same way as the thermostat in your home. When the level of heat falls, the thermostat controlling your furnace signals the furnace to turn on. Once the temperature reaches a certain level, it signals the furnace to turn off.

This is a *negative feedback system*. Negative feedback is also involved in maintaining the glucose level in your body (Figure 23.15). When the glucose level falls below a certain point, a hormone called *glucagon* (produced by the pancreas) signals the cells of the liver and other glycogen-containing cells to convert glycogen to glucose. When the level of glucose in the blood rises above a certain point, other hormones (such as *insulin*, which is also produced by the pancreas) are secreted into the bloodstream. These hormones are responsible for the facilitated transport of glucose through cell membranes, and so the glucose level in the blood falls (see Focus 23.1).

The pancreatic hormones are involved in other functions in addition to maintaining glucose levels. For example, some pancreatic hormones regulate the release of bile by the gallbladder. Three mechanisms stimulate the release of digestive enzymes in a complex interrelationship: the mechanical stimuli of the presence of food; neural stimuli (the idea of food, for instance); and hormonal mechanisms. Receptors in the digestive tract detect the chemical presence of food. Neural messages reach the stomach through the vagus nerve. The neural stimulus causes some cells to release the hormone gastrin into the blood. When this hormone reaches certain cells of the gut, it stimulates them to release specific enzymes. In all cases, these hormones (as well as others) work in concert with nerves to maintain internal regulation.

DIGESTION, NUTRITION, AND DIET

■ **Changes in our lifestyle have resulted in a growing health problem: obesity. They have also given rise to a new profession: the diet-nutrition expert. The best way to lose weight is to understand the evolution of our body form and the nutritional needs of the human system.**

Early humans, living in caves and hunting and gathering seeds and nuts for their daily ration of protein, expended a great deal of physical energy in order to survive. From the perspective of nature, slimness and beauty were not the ideals. A certain amount of fat storage was necessary, because our history was fraught with times of feast and times of famine. Those able to store the extra food available in times of feast could survive the famine.

With the advent of agriculture, the rise of efficient food-preservation techniques, and mass production in the food industry, many people in this country and other industrialized countries no longer need to expend physical energy to acquire food and they eat more. The amount we eat, however, has not changed, nor have the bodily functions that evolved over eons. As a result, our countries' biggest nutritional problem today is obesity. A quick look at best-seller lists provides a survey of the latest diets. Popular magazines are loaded with seductive promises to lose 10 pounds in 10 days without taking any pills or increasing your exercise.

The relative ease of our lifestyle is now coupled with the American public's obsession with youth, sexiness, and the ideal body shape. In America we have come to the problem of the average American's perpetual search for the ideal form presented

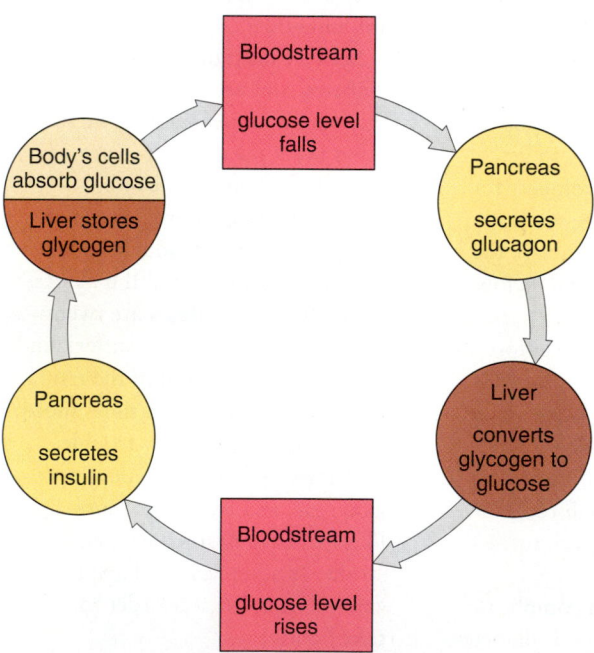

FIGURE 23.15 A Negative Feedback System: Maintaining Blood Glucose Levels
The level of glucose in the bloodstream is self-regulating. As the glucose level falls, it triggers the secretion of glucagon by the pancreas. The glucagon signals the liver to convert stored glycogen to glucose, raising the level of glucose in the bloodstream. As the glucose level in the bloodstream rises, it triggers the secretion of insulin in the pancreas. Insulin increases the rate at which glycogen is stored in the liver and the rate at which glucose is absorbed by the body's cells, lowering the glucose level in the bloodstream.

Chapter 23 Digestion

every hour of the day in the mass media. Consider the amount of money we spend on diet books, diet aids, exercise machines and programs, and you will have some idea of how much we have been influenced in our search for physical perfection.

These days it is trendy — and, perhaps more importantly, lucrative — to be an "expert" on nutrition, and so most of the diet plans available are based on "nutritional wonders" like nutrients that eat fat cells. Generally speaking, the long-term effectiveness of any of the popular diet plans is limited, which is unfortunate for those who have invested their time and money in one or another.

However, scientific research has revealed much about the nutrients we need and the diseases we can avoid through changes in our diets. So, instead of looking for a panacea, perhaps it is more appropriate for us to consider three questions: What is proper nutrition? How can an individual lose or gain weight to achieve and maintain a healthy body? What are some of the successful treatment programs for weight problems — whether one weighs too much or too little? Let's find out.

Nutrients

■ **The human body requires a balance of carbohydrates, lipids, proteins, minerals, vitamins, and water.**

Nutrients are the chemical substances that we usually refer to as food. For convenience, we usually divide nutrients into six major groups: carbohydrates, lipids, proteins, minerals, vitamins, and, of course, water. A well-balanced diet for humans should consist of approximately 50 percent carbohydrates, 30 percent fats, and 20 percent proteins, plus vitamins, minerals, trace elements, and water. These requirements vary slightly with activity, age, and general health conditions.

Carbohydrates and Lipids. Carbohydrates are the body's primary energy source and, to a lesser extent, are also used as structural components, like portions of cell membranes. Lipids (fats) are stored forms of energy and contain about twice the number of Calories (units of heat energy) that carbohydrates do. The digestive process breaks down fats to essential fatty acids, glycerol, and some vitamins. These substances are also used for structural components, such as cell membranes, and to maintain life processes. Vitamins can be converted to coenzymes, which are important components of some enzymes.

Proteins. Proteins are essential as the structural components of the body needed for growth and repair. Sometimes an amino group is removed from an amino acid and the remaining portion of the amino acid can be metabolized into sources of energy or converted to fats or carbohydrates (see Profile). Of the 20 amino acids that our bodies require, 12 can be synthesized by the body. The remaining 8 are called *essential amino acids* because they *must* be provided by diet (Figure 23.16).

Excess amino acids are not stored by the body; therefore, high-protein diets do not necessarily increase the amount of amino acids in the cells. Instead, the liver converts excess amino acids to glucose by the process of *deamination*, which we discussed in Chapter 9. In deamination, the amino group is removed from amino acids and is eventually metabolized to urea, which passes out of the body in the urine. The hydrocarbon skeleton that remains is metabolically converted to glucose and then to glycogen. This process explains how a person can gain weight on a high-protein diet.

Plant proteins are usually described as incomplete because some, especially cereal sources, are

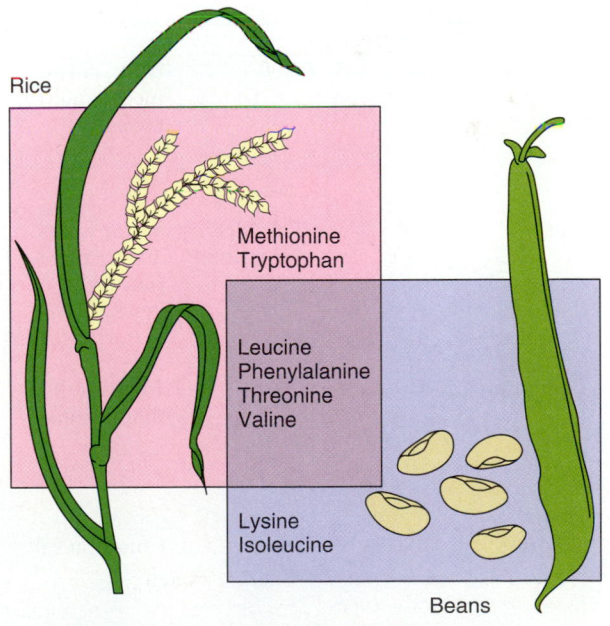

Rice

Methionine
Tryptophan

Leucine
Phenylalanine
Threonine
Valine

Lysine
Isoleucine

Beans

FIGURE 23.16 Essential Amino Acids
Of the 20 amino acids required for good nutrition, 12 can be synthesized by our bodies and 8 must be provided by the diet. This illustration lists the 8 essential amino acids. Rice can supply two of the essential amino acids—methionine and tryptophan—and beans can provide only 6. Therefore, it is important to those on a vegetarian diet to eat both rice and beans to ensure that all 8 amino acids are ingested.

relatively low in lysine and tryptophan, two of the essential amino acids. Because most vegetarian diets rely on a single cereal crop, they are nutritionally deficient, a problem that is quite serious in Third World countries that use plants, not animals, as their major protein source. Corn, for instance, is extremely low in lysine. To meet this problem, researchers have developed a new corn variety that is high in lysine. This new plant will definitely benefit countries that rely heavily on corn and corn products. Also, high-lysine wheat and rice have recently been developed.

All this is not to say that an individual should not adopt a vegetarian diet. However, it is vital for vegetarians to supplement their diet with the essential amino acids, through the use of amino acid supplements or by extending the vegetarian diet to include some combination of legumes, yeast products, nuts, chicken, eggs, and fish.

Minerals. Minerals, which serve as constituents for many cellular structures, are inorganic substances required for proper metabolism. These inorganic materials may be bound to organic molecules. Cal-

TABLE 23.4
Selected Minerals and Their Functions in the Body

Mineral	Amount in adult body (grams)	RDA* (milligrams)	Dietary sources	Major body functions	Possible outcomes of deficiency	Possible outcomes of excess
Calcium	1500	800	Milk, cheese, dark-green vegetables, dried legumes	Bone and tooth formation Blood clotting Nerve transmission	Stunted growth Rickets, osteoporosis Convulsions	Not reported for humans
Phosphorus	860	800	Milk, cheese, meat, poultry, grains	Bone and tooth formation Acid-base balance, ATP formation, etc.	Weakness, demineralization of bone, loss of calcium	Erosion of jaw (fossy jaw)
Sulfur	300	(Provided by sulfur amino acids)	Sulfur amino acids (methionine and cystine) in dietary proteins	Constituent of active tissue compounds, cartilage, and tendons	Related to intake and deficiency of sulfur amino acids	Excess sulfur amino acid intake leads to poor growth
Potassium	180	2500	Meats, milk, many fruits	Acid-base balance Body water balance Nerve function	Muscular weakness Paralysis	Muscular weakness Death
Chlorine	74	2000	Common salt	Formation of gastric juice Acid-base balance	Muscle cramps Mental apathy Reduced appetite	Vomiting
Sodium	64	2500	Common salt	Acid-base balance Body water balance Nerve function	Muscle cramps Mental apathy Reduced appetite	High blood pressure

* Recommended daily allowance, for an adult male in good health.

cium and phosphorus are essential for proper bone formation. Calcium ions also function in several cellular processes such as muscle contraction, and phosphate functions in synthesis of ATP as well as nucleic acids. Minerals such as magnesium and manganese are consitituents of enzymes (Chapter 6).

Sodium and phosphorus help to maintain the pH acid-base balance of the tissue fluid in the body. In addition, sodium, potassium, and other ions are essential in proper nerve cell coordination. They also play a role in maintaining the proper osmotic concentrations within the body. Hemoglobin contains ionic iron, which is required for oxygen transport. Iron is also a component of the cytochrome molecules that transport electrons in the electron transport chain (Chapters 7, 8, and 9). Because some minerals are required only in extremely small amounts (less than 0.01 gram each day), they are called *trace elements*. Trace elements include iodine, copper, fluoride, zinc, cobalt, and silicon. Table 23.4 lists selected minerals and their functions in the body. As you can see, minerals and trace elements are important to our diet.

Mineral	Amount in adult body (grams)	RDA* (milligrams)	Dietary sources	Major body functions	Possible outcomes of deficiency	Possible outcomes of excess
Magnesium	25	350	Whole grains, green leafy vegetables	Activates enzymes involved in protein synthesis	Growth failure Behavioral disturbances Weakness, spasms	Diarrhea
Iron	4.5	10	Eggs, lean meats, legumes, whole grains, green leafy vegetables	Constituent of hemoglobin and enzymes involved in energy metabolism	Iron-deficiency anemia (weakness, reduced resistance to infection)	Siderosis Cirrhosis of liver
Fluorine	2.6	2	Drinking water, tea, seafood	May be important in maintenance of bone structure	Higher frequency of tooth decay	Mottling of teeth Increased bone density Neurological disturbances
Zinc	2	15	Widely distributed in foods	Constituent of enzymes involved in digestion	Growth failure Small sex glands	Fever, nausea, vomiting, diarrhea
Copper	0.1	2	Meats, drinking water	Constituent of enzymes associated with iron metabolism	Anemia, bone changes (rare in humans)	
Iodine	0.011	0.14	Marine fish and shellfish, dairy products	Constituent of thyroid hormones	Goiter (enlarged thyroid)	Very high intakes depress thyroid activity
Cobalt	0.0015	(Required as vitamin B$_{12}$)	Organ and muscle meats, milk	Constituent of vitamin B$_{12}$	None reported for humans	Industrial exposure; dermatitis and diseases of red blood cells

Vitamins. Organic molecules required in small amounts for proper growth, maintenance, and metabolism, vitamins are also essential for proper nutrition. As discussed in Chapter 6, many vitamins function as coenzymes, so that they are essential for efficient acceleration of enzymatic reactions (Table 23.5). When vitamins are missing from the diet, metabolic reactions cannot be completed, which

TABLE 23.5
Selected Vitamins and Their Function in the Body

Vitamin	RDA* (milligrams)	Dietary sources	Major body functions	Possible outcomes of deficiency	Possible outcomes of excess
Water-Soluble					
Vitamin B_1 (thiamine)	1.5	Pork, organ meats, whole grains, legumes	Coenzyme (thiamine pyrophosphate) in the removal of carbon dioxide	Beriberi (peripheral nerve changes, edema, heart failure)	None reported
Vitamin B_2 (riboflavin)	1.8	Widely distributed in foods	Constituent of two flavin nucleotide coenzymes involved in energy metabolism (FAD and FMN)	Reddened lips, cracks at corner of mouth (cheilosis), lesions of eye	None reported
Niacin	20	Liver, lean meats, grains, legumes (can be formed from tryptophan)	Constituent of two coenzymes involved in oxidation-reduction reactions (NAD$^+$ and NADP$^+$)	Pellagra (skin and gastrointestinal lesions, nervous and mental disorders)	Flushing, burning and tingling around neck, face, and hands
Vitamin B_6 (pyridoxine)	2	Meats, vegetables, whole grain cereals	Coenzyme (pyridoxal phosphate) involved in amino acid metabolism	Irritability, convulsions, muscular twitching, kidney stones	None reported
Pantothenic acid	5–10	Widely distributed in foods	Constituent of coenzyme A, which plays a central role in energy metabolism	Fatigue, sleep disturbances, impaired coordination, nausea (rare in humans)	None reported
Folacin (folic acid)	0.4	Legumes, green vegetables, whole wheat products	Coenzyme (reduced form) in carbon transfer in nucleic acid and amino acid metabolism	Anemia, gastrointestinal disturbances, diarrhea, red tongue	None reported
Vitamin B_{12}	0.003	Muscle meats, eggs, dairy products	Coenzyme in carbon transfer in nucleic acid metabolism	Pernicious anemia, neurological disorders	None reported

* Recommended daily allowance, for an adult male in good health.

can result in a series of symptoms called *diet deficiency disorders*. For example, a deficiency in vitamin B$_1$, known as thiamin, can affect the proper metabolism of nervous and muscular tissue, which can, in turn, affect the cardiovascular system. This can produce mental confusion, muscle weakness, an enlarged heart, and in some cases death due to heart failure.

Vitamin	RDA* (milligrams)	Dietary sources	Major body functions	Possible outcomes of deficiency	Possible outcomes of excess
Biotin	Not established. Usual diet provides 0.15–0.3	Legumes, vegetables, meats	Coenzyme in fat synthesis, amino acid metabolism, glycogen formation	Fatigue, depression, nausea, dermatitis, muscular pains	None reported
Choline	Not established. Usual diet provides 500–900	All foods containing phospholipids (egg yolk, liver, grains, legumes)	Constituent of phospholipids Precursor of putative neurotransmitter acetylcholine	None reported for humans	None reported
Vitamin C (ascorbic acid)	45	Citrus fruits, tomatoes, green peppers, salad greens	Maintains intercellular matrix of cartilage, bone, and dentin Important in collagen synthesis	Scurvy (degeneration of skin, teeth, blood vessels, epithelial hemorrhages)	Relatively nontoxic Possibility of kidney stones
Fat-Soluble					
Vitamin A (retinol)	1	Provitamin A in green vegetables Retinol in milk, butter, cheese, margarine	Constituent of rhodopsin (visual pigment) Maintenance of epithelial tissues	Xerophthalmia (keratinization of ocular tissue), night blindness, permanent blindness	Headache, vomiting, peeling of skin, anorexia, swelling of long bones
Vitamin D	0.01	Cod liver oil, eggs, dairy products, margarine	Promotes bone growth, mineralization Increases calcium absorption	Rickets (bone deformities) in children Osteomalacia in adults	Vomiting, diarrhea, weight loss, kidney damage
Vitamin E (tocopherol)	15	Seeds, green leafy vegetables, margarines	Functions as an antioxidant to prevent cell membrane damage	Possibly anemia; never observed in humans	Relatively nontoxic
Vitamin K (phylloquinone)	0.03	Green leafy vegetables Small amount in cereals, fruits, and meats	Important in blood clotting (involved in formation of active prothrombin)	Deficiencies associated with severe bleeding, internal hemorrhages	Synthetic forms at high doses may cause jaundice

So there is no doubt that vitamins are essential to maintain proper enzymatic function. The recommended daily allowances of these vitamins are spelled out fairly clearly in government publications. However, in the United States many unfounded commercial claims are made about the importance of vitamins. Most revolve around the idea that a specific vitamin will cure a given symptom of a disease or that megadoses of certain vitamins (doses far above the daily requirements) will prevent such things as aging, colds, and a host of other problems.

To date, no scientific evidence supports the premise that megadoses of vitamins and/or minerals are beneficial. The body can store only certain amounts of minerals and vitamins. Any excesses are metabolized away and excreted. If you have ever taken a megadose of vitamins, you may have noted that your breath or urine smelled like vitamins. That was because your body was getting rid of the excess vitamins.

Megadoses of some vitamins — especially vitamins A, D, E, and K — can injure your health because of the way that they participate in body functions and, in some cases, the ways in which they are stored. Some vitamins, like A and D, are fat-soluble, while others like C and B, are water-soluble. It is "easier" for the body to rid itself of water-soluble vitamins than fat-soluble vitamins. Therefore, physicians usually find that excessive amounts of fat-soluble vitamins can lead to more dietary problems than water-soluble vitamins.

For example, excessive amounts of vitamin A (10 times the recommended intake) can result in bone destruction, fatigue, nausea, loss of appetite or hair loss, liver enlargement, and a yellowish color to the skin. Excessive amounts of vitamin D (5 times the recommended amount) can result in increased calcium loss in bones, kidney stones and/or damage, calcification of soft tissues such as the lungs, kidneys, and blood vessels, and even death.

The Centers for Disease Control in Atlanta, Georgia, stated in its *Morbidity and Mortality Weekly* that "with the general increase in use of vitamins and mineral supplements in this country, the public and medical community should be aware of the potential for toxicity." In other words, vitamins and minerals are required in specific amounts. Too much or too little can upset that delicate balance that maintains good physiological health. Contrary to the popular belief that "if a little is good, more is better," the human body needs vitamins, minerals, and nutrients in proper amounts. Both too much and too little may be unhealthy.

Proper Nutrition

■ **Our diet consists of large amounts of fats, cholesterol, and carbohydrates, which have resulted in many diet-related health problems. Practicing proper nutrition can lessen the danger of diseases such as heart disease and cancer.**

If you look at the human animal from an evolutionary perspective, you see that humans developed primarily as fruit and vegetable gatherers who ate meat only as an occasional supplement. You also see that our bodies are not designed for a diet low in fiber and high in fat, animal protein, processed foods, and additives. According to the Department of Agriculture, even with the current focus on diet and nutrition, the average urban American's diet is not as balanced as was that of his or her grandparents. Today, fat makes up 40 percent of our total caloric intake, compared with the levels in 1910, which were 32 percent. Our carbohydrate intake today is down to 25 from 37 percent. In addition, we also consume 70 percent more beef, 180 percent more chicken, 820 percent more turkey, 56 percent more animal and vegetable oil, and 33 percent less fruit than in 1910.

Many of our modern health problems are due to the changes in our diet. Some Americans are literally eating themselves to death. For example, a 10-year study conducted by the National Institutes of Health concluded that there is a definite correlation between heart attacks and high fat intake (especially the fat called cholesterol) in the diet. Lower the cholesterol levels and you reduce the risk of heart attack and of death from heart attack. In fact, a 1 percent decrease in cholesterol can result in a 2 percent decrease in the probability of a heart attack.

Some of the methods for reducing cholesterol are really quite simple. As much as two teaspoonful of cholesterol are found in bile, which represents about one-half of the total circulating levels of cholesterol in the blood. Biologists have long known that vegetable fiber — especially pectin, a fiber found in apples, for instance — traps cholesterol. It follows, then, that trapping cholesterol with dietary fibers such as pectin or oat bran can lower cholesterol by as much as 15 percent. In addition, dietary fiber speeds the passage of chyme through the colon, thus reducing absorption. Fiber also retards the absorption of possible cancer-causing agents in the diet. Maybe the old saying "an apple a day keeps the doctor away" has some merit. Dietary fiber is certainly essential to proper digestion.

Dr. R. Bruce Merrifield
Chemist, Nobel Prize Laureate

Dr. R. Bruce Merrifield is a modest, soft-spoken father of six and a former Boy Scout leader. His favorite pastime is tinkering around his New Jersey home — in fact, he once said that he could spend "the rest of his life raking leaves." His working hours are spent in his laboratory at Rockefeller University in New York City. And he does his work well. On October 17, 1984, he stepped off the elevator on the fourth floor of Flexner Hall (as he has done since 1949) and was informed by a lab assistant that he had been awarded the Nobel Prize in Chemistry. This elevator was the same one in which, over 25 years before, he was discussing a new method of synthesizing proteins with a colleague. The technique he proposed, one for synthesizing proteins on a solid matrix, brought him that Nobel Prize.

Before Dr. Merrifield developed his technique, synthesizing proteins from amino acids was a painstaking, almost impossible technique that could take months, sometimes even years, to produce only a small polypeptide chain. Researchers struggling with this method realized the need for a more efficient and productive system. What Merrifield was looking for was "a rapid, quantitative, automatic method for synthesis of long-chain peptides." He filled

that need by developing an automatic technique in which the first amino acid in a peptide chain was bound to a microscopic polystyrene bead. The bead acted as a foundation from which the polypeptide chain could grow through the addition of subsequent amino acids. His first protein-synthesizing device was built in a home basement and looked like something out of a Rube Goldberg cartoon. But today his computer-controlled, automated protein synthesizers are sophisticated instruments used in laboratories around the world.

Merrifield's protein synthesizers have had a tremendous impact on science, opening new vistas in biochemistry, pharmacology, and medicine. For example, hundreds of thousands of diabetics are dependent on daily injections of insulin to sustain life. Until Merrifield's techniques were perfected, most of the insulin these diabetics used was isolated from the pancreases of pigs, and the substance sometimes caused side effects. Synthetic insulin could have alleviated the problem, but the prevailing methods required years to synthesize the substance. Today, using Merrifield's technique, the synthesis of human insulin is a practical reality. Although recombinant DNA techniques are in place to cre-

ate human insulin, the process cannot produce the hormone in large commercial amounts. The same is true of the production of many important proteins, including other hormones, neural peptides, toxins, and enzymes. The Merrifield method allows them to be synthesized in amounts that can be useful to humans.

When someone asked Merrifield what he would do with the Nobel Prize money, he said, "I could use a new car." The response was consistent with the character of the man — someone who wants to rake the leaves and who built his revolutionary machines in his garage. The search is the reason he is a scientist.

Calories and Their Role in Weight Management

■ **Weight problems are partly due to an individual's basic metabolic rate, which dictates the amount of energy required to maintain life processes. Dietary plans must be adjusted according to the individual, and long-term procedures are the most effective, whether the individual weighs too much or too little.**

Even though many elements are required in an adequate diet, most dieters are not concerned about anything besides the calories in their meals. How many should they consume? How can they "get rid" of them? What is amazing is that, despite all the emphasis on weight management, most people still do not know what to do in order to lose weight — or gain it, if that is what is desired.

To understand the caloric needs of the average human, we need to discuss what is known as the basal metabolic rate (BMR), or the amount of energy required to maintain minimal life processes. At rest, the "average body" burns about 1500 to 1800 Calories a day. Of course, the BMR varies with gender, age, size, and other factors. For example, per pound of body weight, an infant requires 50 Calories and an adult at rest 18 Calories. An active adult requires 35 Calories per pound of body weight.

We also must acknowledge the influence of genetic factors on the body. Normally, our bodily processes balance energy use, food intake, and fat storage quite precisely. This set point varies with each individual. As we all have observed, some people are biologically set almost never to gain weight. They may have a high BMR or perhaps other genetic factors that seem to prevent weight gain. Others just "look" at food and gain weight. These individuals seem to be biologically set to produce fat deposits. For them, only continual restraint of food intake can cause weight loss. Otherwise, weight will readily accumulate.

Dealing with Obesity. Generally speaking, when caloric intake exceeds the caloric needs of the body, the excess nutrients are converted to fats and glycogen to be stored in the cells as energy reserves. In other words, the individual gains weight. To lose weight, a person must ensure that caloric intake is less than caloric needs. Only in that way will the stored fat be metabolized.

Proper treatment of obesity involves a realistic plan and a nutritionally adequate diet. Often the dieter should be under a doctor's supervision to ensure good health during the diet period. Serving sizes should be small, and meals must be regular — in other words, fasting is not the most efficient way to achieve long-term weight control. Weight loss should be limited to a pound a week. Despite the claims of most best-selling authors, the plan must be tailored to the individual's basic likes and dislikes. Do not expect miracles. No one gains all those pounds overnight; no one can lose them overnight either.

Many people can lose weight, but less than one-third manage to maintain the level they have achieved once they go off the diet. The behavior and attitudes that have led to the overeating and a sedentary way of life have to be changed. To treat obesity successfully (in other words, to take it off and keep it off), a regime of fewer calories and more exercise must be maintained. A workable treatment of obesity involves proper diet, increased activity, and behavioral modification.

In some special cases, physicians have resorted to surgical or hormonal treatments or even liposuction, the procedure in which a suction device is inserted through an incision in the skin and fat is "vacuumed" away. The use of such treatment plans is questionable because they may harm an individual's health over a long period. For example, some patients undergo surgery that involves relocating the duct that carries bile from the liver to the duodenum so that the duct empties into the large intestine instead. This operation is based on the idea that bile is needed to digest fats properly and fats are absorbed in the small intestine. If bile is not present, fats cannot be digested and absorbed. Hence the calories contained in the fats are lost. The operation has also been used to lower cholesterol levels in the blood, since bile is a major source of cholesterol.

However, some investigators have found that the digestive detour increases the concentration of bile salts in the large intestine, which is not constructed to tolerate the direct dripping of bile salts, and irritation may result. Clinical studies in Germany and England and at the Mayo Clinic here in the United States suggest that the dripping of bile in the large intestine causes the rate of colon cancer to double. This indicates the need for long-term clinical studies of individuals who have had digestive detour surgery for obesity treatment, cholesterol reduction, or gallbladder trouble. The main point for individuals to remember is that they should be reasonable and seek competent medical advice if they are considering drastic surgical or hormonal procedures to control obesity.

Dealing with the Problem of Being Underweight. Individuals who are underweight also must change their caloric intake, activity levels, and behavioral patterns. Obviously, they have to increase their caloric intake, reduce activities to a logical level, and change behavior so that caloric intake exceeds requirement. Along with diet, a good exercise program is usually needed. Because weight loss can be an early indicator of a health problem, people should see a physician before beginning any changes in lifestyle. In severe cases, psychological counseling may be required to achieve the behavioral changes necessary for success.

Whether you want to gain weight or lose it, keep your goals realistic, and remember that genetics is a major determinant of body shape. Very high protein diets are of questionable value. Then too, weight problems of any kind can be due to psychological problems as well as physical ones. Sometimes stress reduction alone can correct the problem.

Our obsession with the "body beautiful" is producing some of our major nutritionally related health problems. So how does one lose or gain weight while following a good nutritional plan? In general, follow these recommendations: (1) avoid excess use of vitamins and minerals, (2) use moderation and common sense, (3) avoid fad diets, and (4) exercise regularly. Stored energy in the form of fat is essential to our survival, as are exercise and the proper intake of fruits and vegetables. Also keep in mind that pursuit of the "body beautiful" sometimes flies in the face of the evolutionary history of the human body form, which evolved in an environment of alternating cycles of feast and famine.

SUMMARY

1. Animals are heterotrophs. Several methods of food procurement have evolved to guarantee them a source of nutrients. Mechanical digestion is the breakdown of food as a result of chewing and the churning of smooth muscle in the walls of the gut.

2. Chemical digestion is a hydrolytic enzymatic reaction in which complex polymers are broken down to monomers. These are further metabolized to be used for energy, growth, and repair.

3. In unicellular animals, digestion occurs within the cell. In more complex animals, digestion is both extracellular and intracellular. Certain parts of the process go on in the digestive tract, after which the nutrients enter the cell where they are used.

4. The lower invertebrates such as the cnidarians and flatworms possess an incomplete digestive tract. Food enters through the same opening that is used for excretion of undigested elements. Animals that evolved after the roundworms have a complete digestive tract — one that has both oral and anal openings.

5. The human gut is a continuous tube from oral to anal openings, consisting of the mouth, esophagus, stomach, small intestine, large intestine, rectum, and anus.

6. In cross-section, the gut consists of four layers: the mucosa, submucosa, smooth muscle, and serosa. The contraction of the muscle layers produces peristalsis, a wavelike movement that propels food through the gut.

7. In the gut, food is broken down mechanically and chemically, then absorbed. The food residue is then eliminated.

8. The digestive process is coordinated by mechanical, neural, and hormonal stimuli.

9. In the mouth, the teeth mechanically break down food. The salivary glands lubricate the food and initiate carbohydrate digestion. The tongue moves food toward the esophagus; the swallowing reflex moves the food into the esophagus, which connects the mouth and the stomach.

10. In the stomach the food is mixed with hydrochloric acid and enzymes to become a semiliquid mixture called chyme. The hydrolytic breakdown of protein to polypeptides and fats to fatty acids begins in the stomach.

11. The small intestine consists of four regions: the duodenum, the jejunum, and the ileum. Most protein, carbohydrate, and fat digestion occurs in the duodenum. The folded walls of the small intestine, the villi, and the microvilli increase the surface area, facilitating absorption.

12. Within each villus is a capillary network and a lacteal. The products of protein and carbohydrate digestion are absorbed into the bloodstream by the capillaries, and the products of fat digestion are absorbed into the lacteals.

13. The large intestine functions primarily in the absorption of water and dissolved minerals. It is also the storage place for large numbers of intestinal bacteria. Some of these bacteria produce enzymes that aid in the digestive process.

14. As the water and dissolved minerals are absorbed from the large intestine, the chyme becomes a semisolid mass known as feces. Feces are eliminated through the anus.

15. The maintenance of proper glucose levels in the blood is regulated by negative feedback involving the pancreatic hormones insulin and glucagon. Negative feedback regulates other chemical levels as well.

16. Obesity is one of the major dietary health problems in the United States.

17. Proper weight management requires commonsense attitudes about caloric intake, exercise, and behavior modification.

18. For proper nutrition, it is important to eat a well-balanced diet consisting of about 50 percent carbohydrates, 28 to 30 percent fats, and 20 percent proteins, as well as vitamins, minerals, trace elements, and water.

KEY TERMS

ingestion	stomach
digestion	chyme
chemical digestion	small intestine
mechanical digestion	duodenum
assimilation	pancreas
incomplete digestive tract	liver
complete digestive tract	gallbladder
gut	jejunum
mucosa	ileum
submucosa	villi
smooth muscle	microvilli
serosa	large intestine (colon)
peristalsis	feces
salivary gland	nutrient
esophagus	

REVIEW QUESTIONS

True/False

1. In a general sense, digestion is simply hydrolysis of complex polymers to monomers.
   true   false   (page 544)

2. A complete digestive tract consists of an oral and an anal opening.
   true   false   (page 546)

3. The contraction of the smooth muscles within the digestive tract known as peristalsis moves the food through the gut.
   true   false   (page 548)

4. In humans, protein digestion is completed in the mouth.
   true   false   (page 549)

Fill in the Blank

1. The upper sphincter muscle of the stomach is called the _____ sphincter. *(page 551)*

2. The semiliquid mixture of partially digested food found in the stomach is called _____. *(page 551)*

3. The _____ and _____ that line the small intestine greatly increase the surface area available for the absorption of nutrients. *(page 554)*

4. The hormones _____ and _____ are the major hormones involved in the regulation of blood glucose levels. *(page 560)*

Multiple Choice

1. Chemical digestion can be described as a:
   a. Synthesis reaction
   b. Dehydration reaction
   c. Hydrolysis reaction
   d. Rearrangement reaction
   (page 544)

2. The major function of the small intestine is:
   a. Absorption of minerals
   b. Digestion and absorption of nutrients
   c. Absorption of water
   d. Production of gastrin
   (page 553)

3. The major hormones involved in the maintenance of blood glucose levels are produced by the:
   a. Liver
   b. Pancreas
   c. Spleen
   d. Gallbladder
   (page 560)

4. Today the average American diet is excessively high in:
   a. Fats and proteins
   b. Carbohydrates and fats
   c. Vitamins and proteins
   d. Fiber
   (page 566)

Discussion

1. Compare and contrast ingestion, digestion, and assimilation.

2. Discuss a phylogenetic overview of the digestion in the animal kingdom.

3. Speculate on the evolutionary significance of a complete digestive tract.

4. Would you expect the fishes to have salivary glands? Why or why not?

5. What are the major functions of the mouth, esophagus, stomach, small intestine, and large intestine?

6. Describe what happens to a meal of steak and potatoes when it passes through the gut. Be sure to include the enzymes involved, location of digestion, and location of absorption.

7. Discuss at least three of the hormones involved in digestion.

8. Give the names and discuss the functions of five digestive enzymes.

9. If you were asked by a friend for advice on how to lose weight, from a biological perspective what would your answer be?

24

Circulation

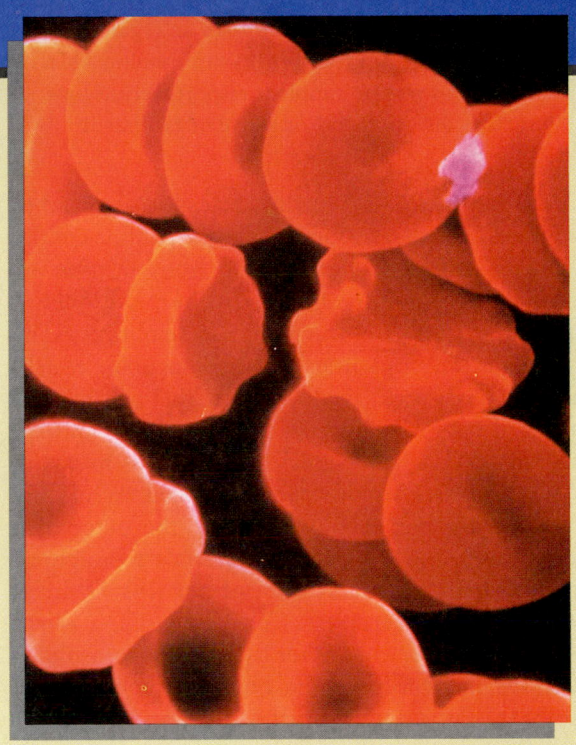

You go to the doctor for a checkup and the nurse takes a blood sample for testing. You read articles about the fact that aerobic exercise strengthens your heart muscles. A professorial, white-haired actor in a television commercial tells you to use a certain cooking oil because it is low in cholesterol and therefore cannot damage your arteries. An aunt suffers a stroke and loses the use of her right hand. What do all these things have in common? They all relate to the circulatory system.

Circulation of molecules is essential for life, whether the organism has only one cell like a paramecium or is a multicellular, multisystem human being. In all organisms, in order for life to endure,

materials must be transported through the organism. This task is the responsibility of the *circulatory system.*

The circulatory system also plays a crucial role in the body's *immune response,* which is the way the body protects itself from bacterial and other foreign substances. The workings of the immune system are involved with vaccination programs that have, until recently, virtually eradicated such illnesses as smallpox and measles from our society. The reason that AIDS is so frightening is that it destroys the immune system.

In this chapter we discuss transport mechanisms in lower animals and then investigate the vertebrate circulatory system as exemplified by humans. Because there is an intimate interrelationship between the blood and the immune response, we also introduce the immune system in this chapter (we discuss it in more detail in Chapter 25).

Above: A scanning electron microphotograph of red blood cells reveals the cell's three-dimensional structure.

TRANSPORT MECHANISMS IN SIMPLE ANIMALS

■ **Materials are circulated through an organism by various means of transport. Simple multicellular organisms rely on simple transport mechanisms such as diffusion and/or active or facilitated transport.**

In the single-celled protistans, transport is a rather simple process that occurs within the confines of the cell. The cell engulfs materials from outside its boundaries through *phagocytosis.* In this process a depression forms in the plasma membrane. The depression develops into a saclike vacuole, which, once inside the cell, is pinched off the membrane's surface. Lysosomes fuse with the vacuole and release their hydrolytic enzymes into it, thus breaking down the materials that were engulfed. While the materials are being broken down, the cytoplasm itself moves and distributes materials throughout the cell for metabolic activities by means of diffusion and active transport. Conversely, CO_2, nitrogen wastes, water, and other by-products of cellular processes move out of the body through the cell membrane by means of diffusion, active or facilitated transport, or *exocytosis,* which, in a general sense, is the reverse of endocytosis (see Figure 23.1).

Because of their structure and small size, simple multicellular animals such as the sponges, cnidarians (hydras and jellyfish relatives), and planarians and other flatworms also have a relatively simple process of internal transport. Although they are multicellular, their structures are rudimentary, their cells are close to their environment, and they have lower metabolic requirements. As a result, simple transport mechanisms such as diffusion and/or active or facilitated transport can effectively carry nutrients and waste products into and out of these cells. An elaborate transport system is not necessary (see Figures 23.2 to 23.4).

Open Circulatory Systems

■ **More complex organisms have an open circulatory system, which contains a heart, dorsal artery, and sinus cavities. In this system, the blood is contained within vessels for only part of the circulatory process.**

As animals became more complex, their circulatory systems became more intricate. Some of the invertebrates, such as the arthropods (animals with jointed legs, such as insects and crayfish), have an

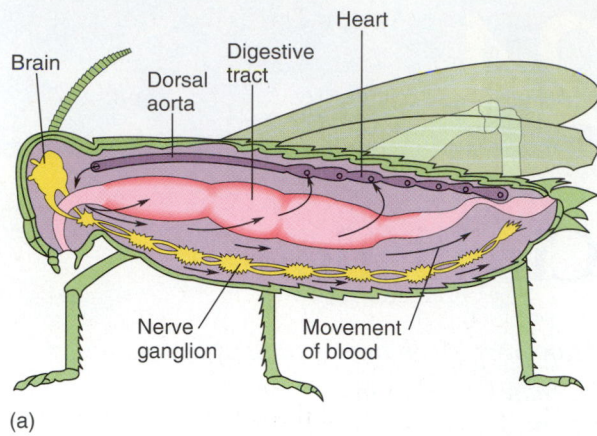

(a)

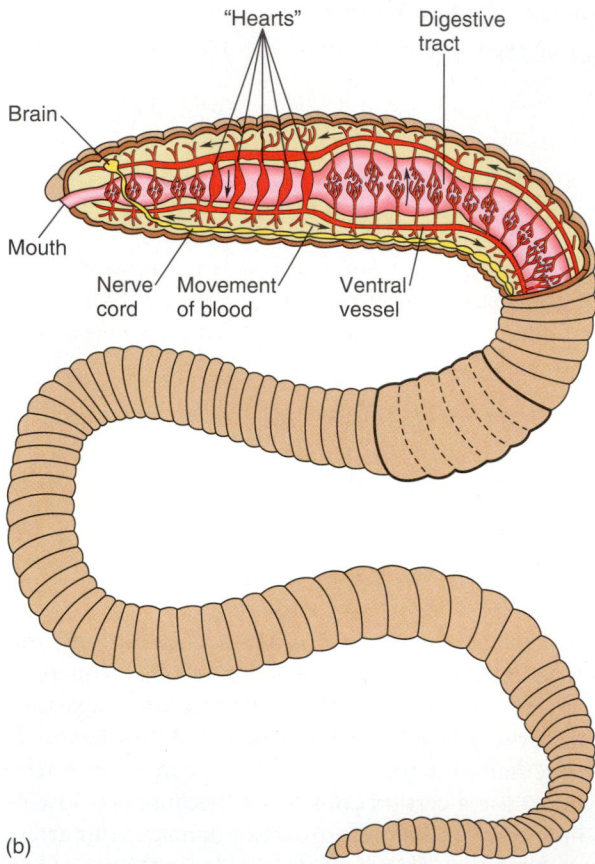

(b)

FIGURE 24.1   Open and Closed Circulatory Systems
(a) In an open circulatory system like that of the grasshopper, the heart pumps blood into the dorsal aorta, from which it enters the body cavity and then diffuses into the tissues. (b) In a closed circulatory system like the earthworm's, the five "hearts" and a continuous system of blood vessels direct blood the full length of the body and back again.

open circulatory system. This system features three main elements:

1. A heart, a pump that is a series of pulsating muscular vessels

2. A main artery running along the back of the body, called a *dorsal artery* (an *artery* is a vessel that carries blood *away from* the heart; the term *dorsal* refers to the backside of an organism, in contrast to the ventral side, which refers to the belly of an organism)

3. Various cavities, called *sinuses*

The blood is pumped through the vessels by the heart, then spills out into the blood sinuses and bathes the tissues. As the blood flows over the tissues, nutrients are distributed and wastes collected. The blood then seeps from the tissues into perforations in the heart, where it is recirculated. The grasshopper in Figure 24.1a is a good example of an organism possessing an open circulatory system.

Closed Circulatory Systems

■ **Closed circulatory systems are more complex and consist of a heart, arteries (which carry blood from the heart), and veins, which return the blood to the heart.**

The circulatory system found in some invertebrates, like the segmented worms, and in the vertebrates is even more complex than the open system. In these organisms the blood is contained within a system of vessels or tubes; these systems are known as **closed circulatory systems.** The system that occurs in the earthworm is an example of a closed circulatory system (Figure 24.1b). We will concentrate on the closed circulatory system in this chapter.

Whether open or closed, the circulatory system aids the organism in maintaining homeostasis. It transports nutrients and dissolved ions, controls fluid volume, and regulates pH and temperature. In addition, it transports hormones, gases, nitrogenous metabolic wastes, and cells like red blood cells and those involved in immunity. It should be noted, however, that terrestrial insects have tracheae that allow for gas transport without the involvement of the circulatory system. Aquatic insects, on the other hand, utilize a closed system containing a respiratory pigment called hemocyanin (this blue segment makes the aquatic insects the true "blue bloods" of the living world). All these mechanisms serve to maintain homeostasis within the organism. A closed system in invertebrates carries on an addi-

tional function: It transports antibodies to aid in the body's defense against foreign organisms or their products.

HUMAN CIRCULATION

Before we look at the functioning of the heart, we need to provide an overview of the circulatory process. The transport system found in humans (and all other vertebrates) is called the **cardiovascular system,** taken from the words *cardio,* which refers to the heart, and *vascular,* an adjective that means "vessel." This system consists of the heart, which functions as a pump, and blood vessels, which are tubes containing the blood (Figure 24.2). The heart pumps blood that is rich in oxygen and nutrients throughout the body through the major **arteries.**

From the arteries the blood travels into smaller blood vessels called *arterioles,* and then finally into the smallest blood vessels, which are called **capillaries.** It is at the capillary level that the exchange of gases, waste products, nutrients, and other substances actually occurs between the blood and cells. From the capillaries, the blood collects into small veins called *venules,* which then converge into larger vessels called **veins,** which in turn return the blood to the heart.

Blood Vessels

■ **Blood vessels are tubes for the transport of blood. Arteries carry oxygenated blood to the cells of the body. Capillaries, spread out in beds, are the sites of exchange. Surrounding the beds is interstitial fluid, through which transport takes place between the capillaries and the cells. Veins carry the blood back to the heart.**

Basically, *blood vessels* are tubes that transport blood with its materials to the cells and return it to the heart for recirculation. These vessels vary greatly in size. For example, the major artery in the body — that having the largest diameter — is the **aorta,** whose diameter is 2.5 centimeters (about 1 inch). The aorta is attached to the heart and transmits blood from the heart to the body. The largest veins in the body are the superior and inferior *vena cava,* which are about 3 cm in diameter. These blood vessels return blood to the heart.

On the other hand, the smallest blood vessels, the *capillaries,* are 6 micrometers (μm) in diameter. (To picture what this means, consider the edge of a dime

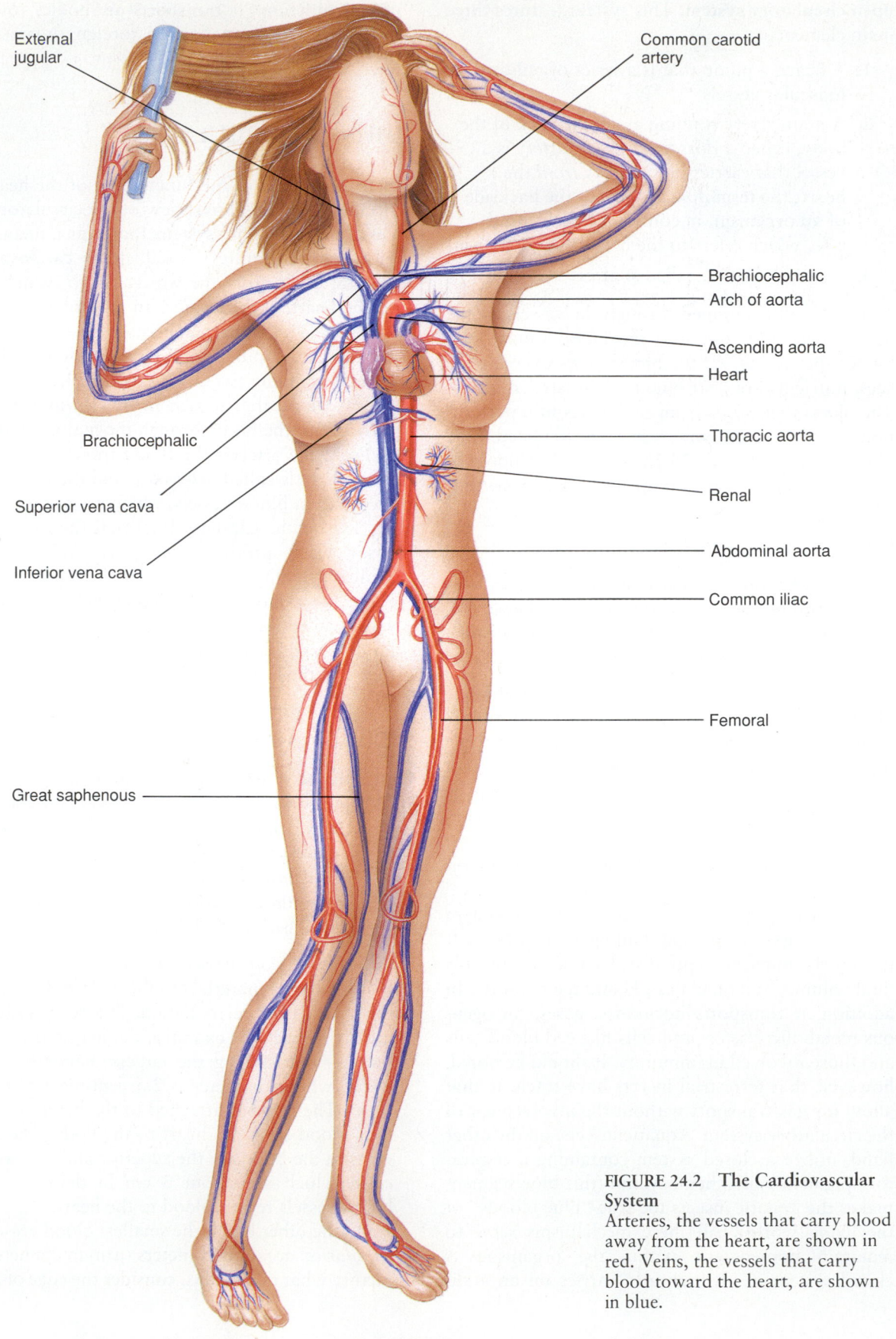

External jugular

Common carotid artery

Brachiocephalic

Arch of aorta

Ascending aorta

Heart

Thoracic aorta

Renal

Abdominal aorta

Common iliac

Femoral

Brachiocephalic

Superior vena cava

Inferior vena cava

Great saphenous

FIGURE 24.2 The Cardiovascular System
Arteries, the vessels that carry blood away from the heart, are shown in red. Veins, the vessels that carry blood toward the heart, are shown in blue.

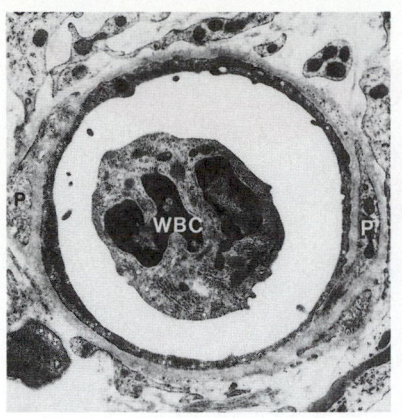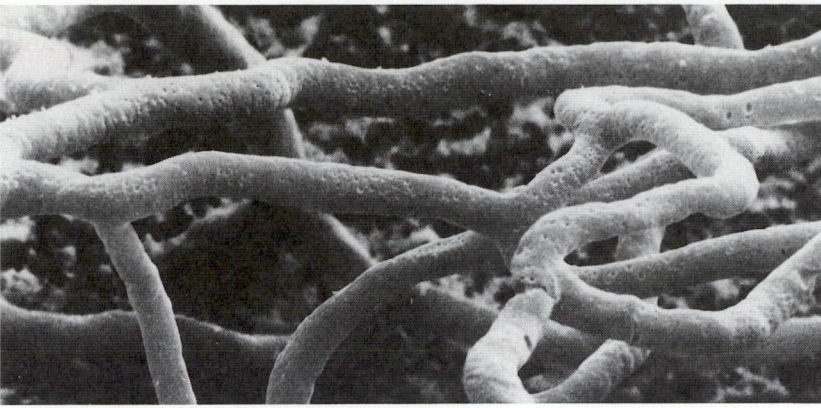

FIGURE 24.3 Blood Cells in a Capillary and a Capillary Bed
(a) Capillaries are so small that blood cells, such as this white blood cell (WBC), must move through them in single file, as shown in this cross-section of a capillary. (b) Using a technique known as casting, a cast of a capillary network can be produced as illustrated.

sliced into a thousand equal slices, then take six of those slices.) Some capillaries are so small that red and white blood cells must pass through them in single file (Figure 24.3). Because of the size differential, the pressure of the blood within different vessels varies, and so the structure of their walls is related to the resistance and pressure of the blood passing through them.

Arteries. *Arteries* carry blood away from the heart. As a result, the blood they receive is under the highest pressure. The walls of these vessels must be able to withstand this pressure, and they do. The vessel walls of arteries, composed of connective tissue, smooth muscle, and an inner lining (the endothelium), are thick, impermeable, and elastic (Figure 24.4a). The smaller *arterioles*, which are micro-

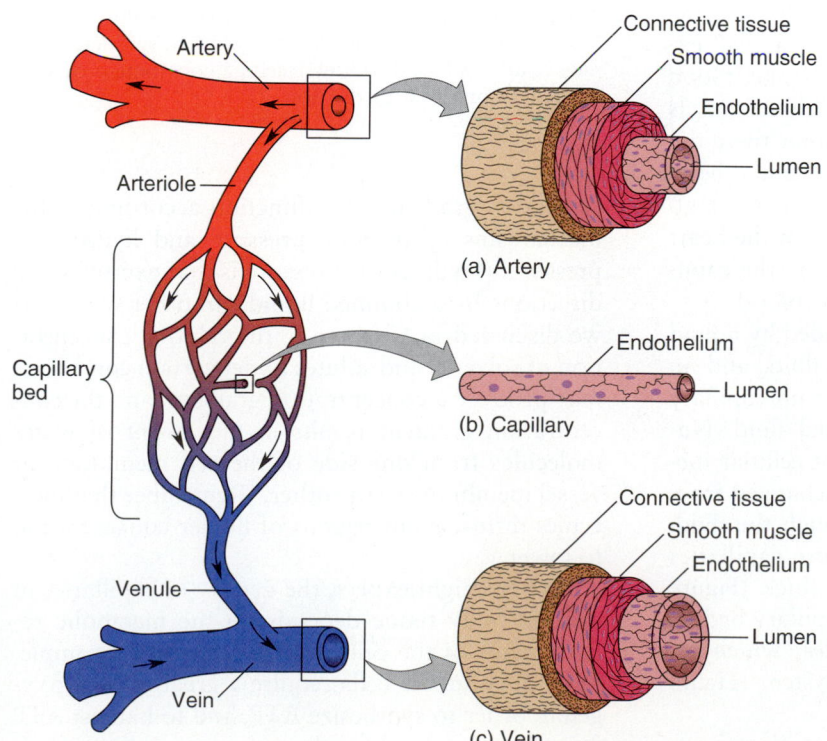

FIGURE 24.4 The Structure of Blood Vessels
Form follows function in the structure of blood vessels. (a) The walls of the arteries are thick and can withstand the pressure of blood pumped directly from the heart. (b) Capillary walls are just one cell thick, which facilitates the exchange of gases, nutrients, and other substances between the blood and the cells. (c) The walls of veins are thinner than they are in arteries and do not have to withstand the same amount of pressure. Notice how much larger the lumen (passageway) is in the vein. The arrows show the movement of the blood.

Human Circulation

scopic subdivisions of arteries, possess a similar organization, but on a smaller scale.

Contractions or relaxations of the smooth muscle increase or decrease the diameter of arterial vessels, thus controlling the flow of blood through them. Contractions result in a decrease in the diameter of the vessel, or *vasoconstriction*. Relaxations result in an increase in the diameter of the vessel, or *vasodilation*. The autonomic nervous system, which is that part of the nervous system generally not under conscious control, regulates actions of the smooth muscles, thus coordinating and regulating the distribution of blood to the brain, digestive system, muscles, and so on.

How does this affect your daily life? Consider the fact that after a large meal you get a little sleepy. Why? What is happening within your body to promote this reaction? The response is due in part to the fact that the digestive system demands more blood; consequently, some blood is redirected from the brain and muscles to the digestive system. As a result, you feel tired. In a sense, your digestive system is asserting a higher priority than your muscles. This explanation also is the basis for the time worn rule about not going swimming right after a large meal. When blood flow to the muscles is decreased, the muscle cells are deprived of oxygen, and lactic acid builds up, resulting in muscle cramps.

The Capillaries. Since the large vessels of the body are muscular tubes that do not allow for the exchange of materials, where does the actual exchange of materials between the cells and blood occur? It occurs in the microscopic blood vessels known as *capillaries*. It is estimated that there are over 50,000 miles of capillaries in the human body — an amazing distribution system that reaches almost every cell. All cells, even the cells of the heart and blood vessels themselves, depend on the capillary system for their exchange with the blood.

The cells of your body are surrounded by a fluid known as **interstitial fluid,** or tissue fluid, and an extensive network of capillaries make up capillary beds extending through the interstitial fluid. Nutrients, oxygen, the waste products of cellular metabolism, and other substances are exchanged from cell to capillary (and vice versa) through the fluid. The process is possible in part because capillaries have thin walls, only one cell layer thick (Figure 24.4b). The flow of blood into the capillary beds is controlled by small sphincter muscles, which respond to signals from the nervous system (Figure 24.5).

In most cases, exchange takes place along con-

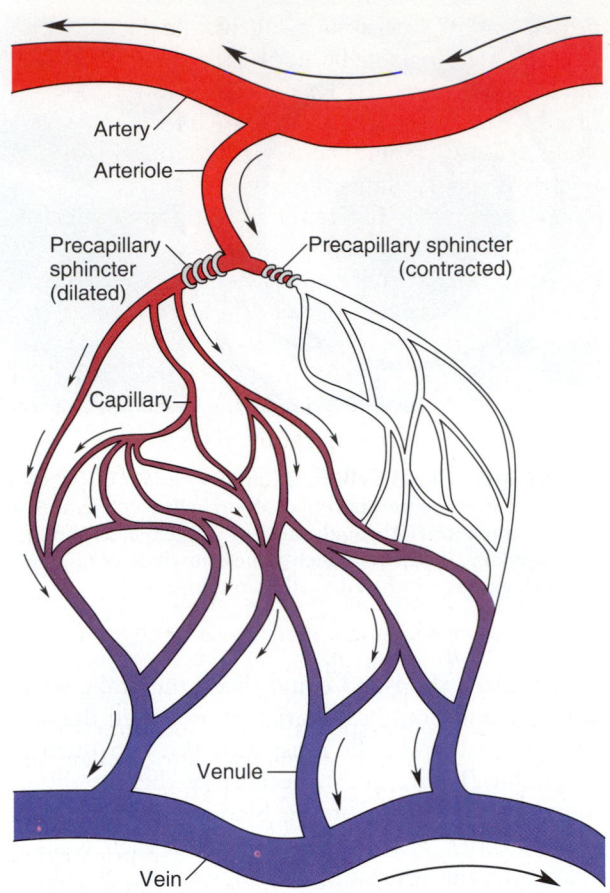

FIGURE 24.5 Controlling the Flow of Blood into a Capillary Bed
The action of sphincter muscles controls the flow of blood into capillary beds. A dilated (open) sphincter muscle allows blood to flow freely; a constricted (closed) sphincter muscle stops the flow.

centration gradients that function according to the mechanisms of osmotic pressure and hydrostatic pressure. (Hydrostatic pressure is that exerted in all directions by a confined liquid.) In other words, as we discussed in Chapter 5, the relative concentration of solvents and solutes under certain conditions may produce a concentration gradient, and the concentration gradient results in movement of water molecules from one side of the cell membrane or vessel membrane to the other. (Remember that molecules diffuse from regions of higher concentration to lower.)

As you might expect, the density of capillaries in a given body tissue depends on the metabolic requirements of the cells in that tissue. For example, the kidney tubule cells require a great deal of oxygen in order to synthesize ATP, and to harvest ATP for energy required for the active transport of molecules against diffusion gradients. In fact, kidney tubules have higher energy requirements than cardiac muscle cells. To be able to function, kidney tubules must have extensive capillary beds.

To a large extent, the concentration of proteins in the blood regulates the proper balance of fluid. Because blood in the capillaries contains a higher concentration of proteins (or relatively less water concentration) than is present in the interstitial fluid (which has relatively higher water concentration), fluid tends to flow into the bloodstream. This process regulates the concentration of the interstitial tissue fluid. (Remember that water is one of the by-products of cellular respiration.)

When the solvent-solute balance is disturbed, tissue fluid is not reabsorbed. Instead, it collects in the interstitial fluid between the cells, water is retained, and swelling occurs. This is called edema. Shock, a condition that sometimes occurs after severe injury, is characterized by an increased permeability of capillary walls. This allows blood proteins to enter into the tissue fluid. Because the proteins have drained into the tissues, the concentration within the blood is reduced and the osmotic gradient between the blood and tissue fluid lessens. As a result, the volume of blood is reduced, which in turn affects the return of blood to the heart. If not treated —usually with blood transfusions or glucose solution—shock can cause death, because ultimately there is an inadequate supply of blood to the brain, heart, and other vital organs.

Veins. After the blood leaves the capillary bed, it collects in small microscopic *venules.* The venules empty their contents into the *veins,* which return the blood to the heart. Veins are very elastic and are composed of less smooth muscle than arteries (Figure 24.4c). Because of our upright posture, humans, like some other animals, have *one-way valves* in the veins to prevent gravity from pulling blood back down as the heart relaxes (see Figure 24.6).

We now know something about the vessels that transport and distribute materials, so let us move on to the "heart" of the matter and discuss this wondrous pump.

THE EVOLUTION OF THE HEART

■ **In fishes, the heart is a single-circuit pump, consisting of two vessels, the atrium (which receives the blood from the vessels) and the ventricle (which pumps the blood through the gills and on through the body).**

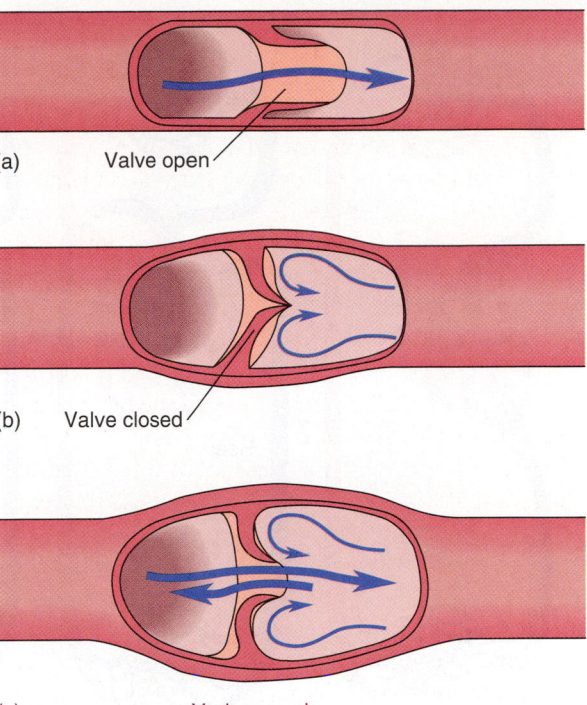

(a) Valve open

(b) Valve closed

(c)   Varicose vein

FIGURE 24.6   Valves of the Veins
(a) When the one-way valves in the veins open, blood can move toward the heart. (b) When these valves close, the backflow of blood is prevented. (c) When the valves in the veins weaken, blood flows backward, pools, and forms distended veins known as varicose veins.

The heart is a pump, and so the evolution of the heart was one basically of modifying a pumping, pulsating vessel. In the earthworm, a series of five pairs of pulsating vessels called aortic arches serve to circulate the blood through the body. In vertebrates, the heart has undergone a metamorphosis from a *single-circuit heart,* which is found in fish, for instance, to a *double-circuit heart,* found in humans. The changes that have occurred are associated with another evolutionary change required for life on land — namely, the change in respiratory structures from gills to lungs.

Single-Circuit Heart

In a *single-circuit heart,* the heart consists of the **atrium,** the receiving chamber for blood, and the **ventricle,** which is the chamber that pumps the blood via the vessels throughout the body. In this system, the blood circuit is a single pathway from the heart, to the gills (where the exchange of gases occurs between the water and the blood), to the

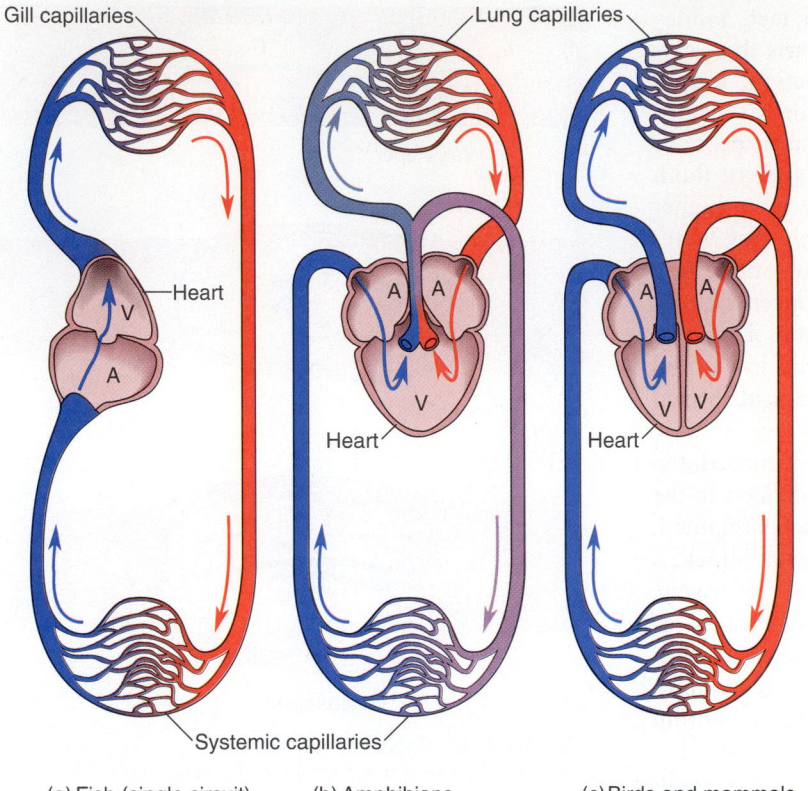

Gill capillaries

Lung capillaries

Lung capillaries

Heart

Heart

Heart

Systemic capillaries

(a) Fish (single circuit)

(b) Amphibians
(double circuit)

(c) Birds and mammals
(double circuit)

**FIGURE 24.7 The Evolution of the
Circulatory System**
Over time, the circulatory system
evolved from a single-circuit to a
double-circuit system. The chambers
of the heart are represented by V
(ventricle) and A (atrium). Red
indicates the oxygenated blood;
blue, the deoxygenated blood.

body, then back to the heart. This is the origin of
the term *single circuit* (Figure 24.7a).

Double-Circuit Heart

■ **A double-circuit circulatory system has a four-
chambered heart. In the pulmonary circuit, carbon
dioxide is removed from the blood and oxygen is
added. In the systemic circuit, the oxygenated
blood is cycled through the body, ultimately returning to the heart to begin the cycle again.**

In amphibians and reptiles, the circuit of blood vessels is more complex than in fishes. Amphibians
possess a three-chambered heart, which has two
atria (the plural of *atrium*) but only one ventricle
(Figure 24.7b). Because the three-chambered heart
allows oxygenated blood from the lungs or gills to
mix slightly with deoxygenated blood from the
body, this system is not as efficient as the four-
chambered heart found in birds and mammals.
However, if you think a moment, you will realize
that amphibians have a lower metabolic rate, so
they demand less oxygen. Their skin also functions
in CO_2-O_2 exchange.

Reptiles may represent a transitional life form,
because some have three-chambered hearts, some
(specifically alligators and crocodiles) have four,
and still others have a "three and one-half" chambered heart, in which the ventricle is almost divided.
This system is more efficient in preventing oxygenated and deoxygenated blood from mixing.

It is in birds and mammals that we see a double-
circuit circulatory system consisting of two pathways (Figure 24.7c). One extends from the heart to
the lungs and back to the heart. The second pathway extends to the body cells and then back to the
heart. Not only is the heart circuit double, but the
heart itself is a *double pump*, consisting of two
halves, one on the right and the other on the left.
The right half is responsible for **pulmonary circulation,** or circulation to the lungs. The left half is
responsible for **systemic circulation,** distributing
blood through the remainder of the body (except
the lungs).

Though the process may seem complicated, it
follows a basically simple pattern. Blood that has a
low concentration of oxygen and a high concentration of carbon dioxide enters the atrium (the thin-
walled upper chamber) on the right side of the heart

through the vena cava. It passes to the right ventricle (the thick-walled lower chamber), which pumps it into the lungs, where the blood releases CO_2 and picks up O_2. The oxygenated blood then returns to the atrium on the left side of the heart. From there it is passed to the left ventricle, which then pumps the oxygenated blood to the body. A schematic of the circuit would look like this: body → heart → lung → heart = first circuit (pulmonary); heart → body → heart = second circuit (systemic).

Figure 24.7 compares the single circuit in the fishes with a double circuit in the birds and mammals. But, as you can see, not until the evolution of the four-chambered heart in some reptiles, the birds, and mammals was there an efficient structure for keeping blood from the lungs and body separate. It should be noted that most reptiles have four-chambered hearts, although the ventricles are incompletely separated by an opening between the right and left ventricle that allows the mixing of oxygenated and unoxygenated blood.

THE HUMAN HEART

Contrary to popular myth, the human heart is not the seat of the soul or the organ of love. Rather it is a muscular, hollow, saclike structure about the size of a fist (Figure 24.8). Covering the entire organ is a tough protective outer membrane known as the *pericardium* ("around the heart"). The major portion of the heart consists of the *myocardium* ("mus-

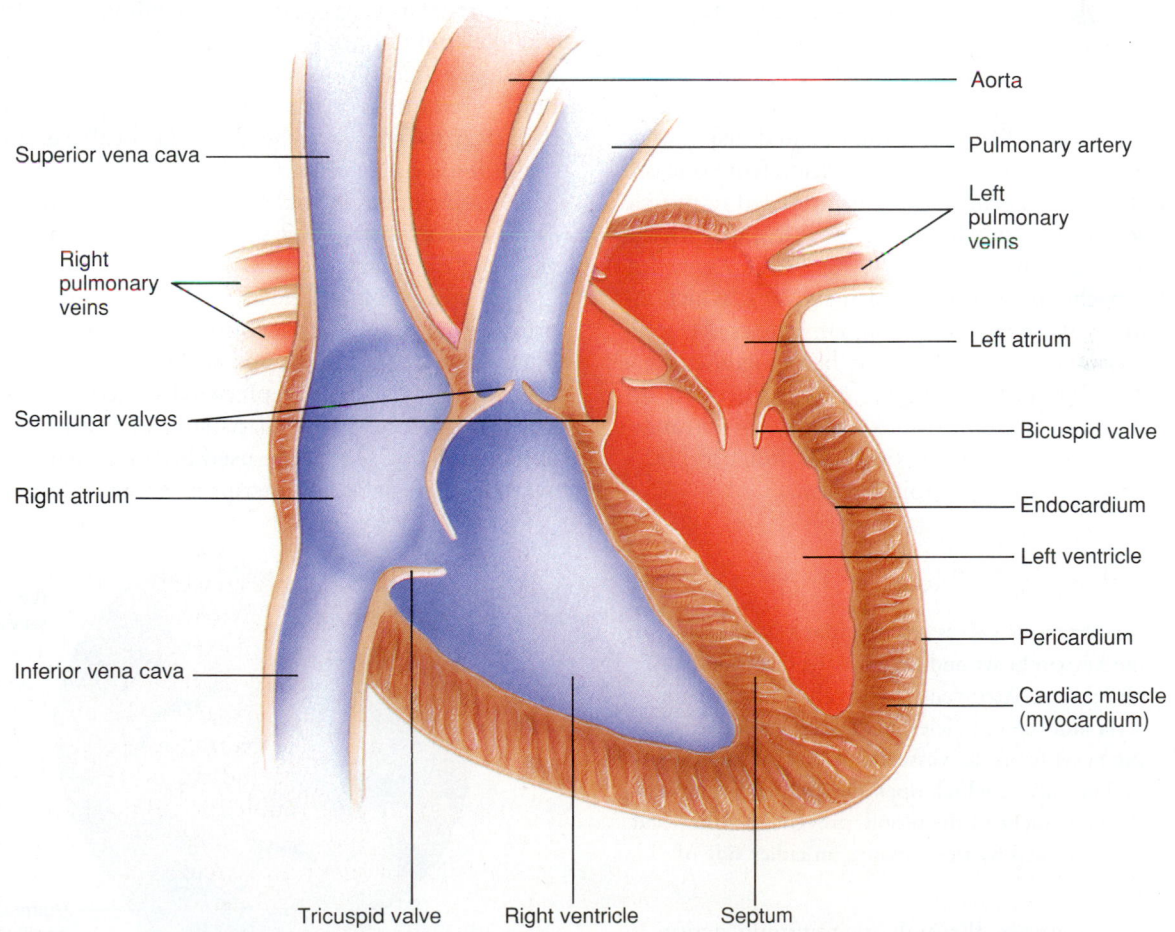

FIGURE 24.8   The Human Heart
Inside the thick muscular walls of the heart are the two atria and two ventricles. The pulmonary artery, leading from the right ventricle to the lungs, is the only artery in the body that carries deoxygenated blood. The pulmonary veins, leading from the lungs to the left ventricle, are the only veins in the body that carry oxygenated blood.

cle heart"). The myocardium is composed of cardiac muscle, which provides the contractile force to pump the blood out of the heart. The inner chambers of the myocardium are lined with the *endocardium* ("inside heart"). The endocardium is made up of connective tissue and endothelial tissue, the specialized tissue that forms the inner lining of the heart and of blood and lymph vessels. The *septum* separates the right and left halves of the heart, which are themselves further divided into upper and lower chambers. The atria are the thin-walled upper chambers that receive blood from veins; the ventricles are the thick-walled lower chambers that pump blood into the arteries going to the lungs or the body.

One of the amazing characteristics of heart muscle cells is their intrinsic contractile properties. Even in tissue culture, independent embryonic cardiac muscle cells contract regularly, and when one or more muscle cells come in contact, they beat in synchrony. The joined cells work together because cardiac muscle cells branch and join with one another end to end in a tight junction called an *intercalated disk* (See Figure 22.6). The intercalated disk allows for the communication of information from one cell to another cell, because the ions involved in muscle contraction are transferred directly from one muscle fiber to another.

Direct chemical transfer of these ions, which results in an electrochemical signal, causes the individual cardiac muscle cells to beat together as a unit. The fibrils of the cardiac muscle fibers are arranged in a spiral, whorl-like pattern. This arrangement causes the contractions to function in a squeezing, rotating, pumping action.

The Valves of the Heart

■ **Membranous valves separate the four chambers of the human heart and regulate the direction of blood flow. Atrioventricular valves separate the atria and the ventricles. Semilunar valves separate the heart from the vessels that lead into and out of it. The valves, which open only in the direction of the movement of the blood, prevent backflow and are regulated by the pressure on either side of them.**

As the heart goes through its pumping action, it contracts and relaxes, contracts and relaxes. Within the heart's structure are a series of four valves, all of which are involved in preventing the blood from flowing backward during the brief moments when the pumping action relaxes. Each valve is made up of flaps of tough, thin-walled membranes, which

open when the heart contracts and close when the heart relaxes. Two *atrioventricular valves* separate the upper and lower chambers on the right and left side of the heart; two *semilunar valves* separate the heart itself and the blood vessels to which the chambers are attached.

The Atrioventricular Valves. The two atrioventricular valves each have specific structures. The *tricuspid valve*, comprised of three "cusps" (or parts), separates the right atrium and right ventricle. The *bicuspid valve* (or mitral valve), comprised of two cusps, separates the left atrium and ventricle. As the atria contract, blood is forced from the atria, through the tricuspid and bicuspid valve, and into the ventricles.

The Semilunar Valves. Semilunar valves, so called because they are shaped like half moons, separate the heart from the blood vessels leading away from it. The **pulmonary artery,** which is the artery through which blood leaves the heart to go to the lungs, contains a one-way valve at the point where the artery merges with the heart. This valve is called the *pulmonary semilunar valve.* The valve separating the aorta and the heart is called the *aortic semilunar valve.* Figure 24.9 shows the structure of the valves of the heart.

The valves of the heart respond solely to the pressure of the heart's contractions. Because no musculature is involved, nor any response to neural signals, it is possible to replace a diseased valve with an identical one made of plastic, or with a valve taken from a pig. Pigs are used because their valves are similar in size and function to those of humans.

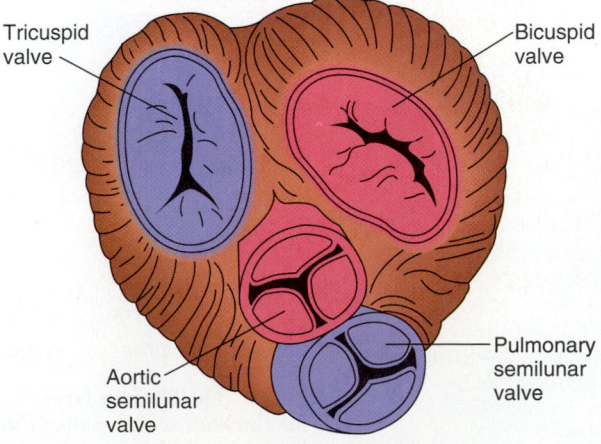

FIGURE 24.9 The Valves of the Heart
The valves of the heart ensure that blood flows in only one direction. This illustration of a heart cross-section shows the valves.

Cardiac Cycle

■ The cardiac cycle consists of two phases. In the first phase, the atria are relaxed, and blood enters them under venous pressure; the atria then contract weakly, forcing the blood into the relaxed ventricles. In the second phase, the ventricles contract, forcing the blood out of the heart and into arteries. Systole refers to the contractions of the heart muscles; diastole refers to the relaxation of the heart muscles. Blood pressure is a measurement of the amount of pressure at work during the systole and diastole.

During the average 70-year lifetime, the heart will beat about 2.5 billion times. Yes, that is *billions,* and given the fact that the heart only relaxes momentarily between beats, you can see that this muscle is a remarkably efficient pump. However, not all fibers contract in each beat. They take turns, relaxing for two to three beats. When there are heavy demands, such as during exercise or in high blood pressure patients, more fibers contract. In a normal, healthy adult at rest, the heart beats about 70 times per minute. That rhythm increases by more than double during strenuous exercise. (These figures vary according to the fitness of the individual, and according to other factors as well, both medical and genetic.) The coordinated muscle contractions and relaxations that make up the heartbeat are known as the *cardiac cycle.*

The cardiac cycle falls into two distinct phases. In the first step, the heart is relaxed and blood is drawn into the atrial chambers from the body and lungs. The atria then contract, forcing the blood through the atrioventricular valves into the ventricles. In the second part of the heart cycle, the ventricle walls contract, forcing the blood through the semilunar valves (which prevent backflow into the heart when the atria contract) of the pulmonary artery or the aorta and away from the heart.

The Pattern of Systole and Diastole. To understand the relationship between the various heart contractions, we need to look at the two steps of the cardiac cycle in more detail. To do so we use the terms systole and diastole, which you may have heard used in connection with someone's blood pressure. **Systole** refers to a contraction of the heart muscle; **diastole** refers to a relaxation of the heart muscle. An atrial systole is followed by an atrial diastole, and a ventricular systole is followed by a ventricular diastole. The contraction-relaxation cycle of the atria alternates with the cycle of the ventricles. In other words, when the atria are in systole (when they contract), the ventricles are in diastole (they are relaxed), and when the atria are in diastole, the ventricles are in systole (Figure 24.10).

When the atria undergo systole, the atrioventricular valves are forced open, and they do not close until the ventricles undergo systole. Once the atria relax, backflow pressure from the ventricular systole forces the atrioventricular valves closed. At the same time, the contraction of the ventricles forces the semilunar valves open, and they remain so until the ventricles relax. When they do, the backflow pressure in the aorta or the pulmonary artery pushes the semilunar valves closed. At the same

FIGURE 24.10 **Systole and Diastole**
(a) As the atria contract, they force blood into the relaxed ventricles. (b) Contractions of the ventricles move blood into the aorta and pulmonary artery. While the ventricles are contracted, the atria relax and fill with blood.

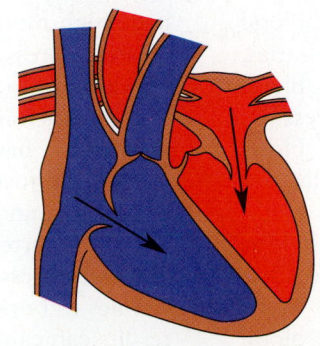

(a) Atrial contraction (systole),
ventricular relaxation (diastole)

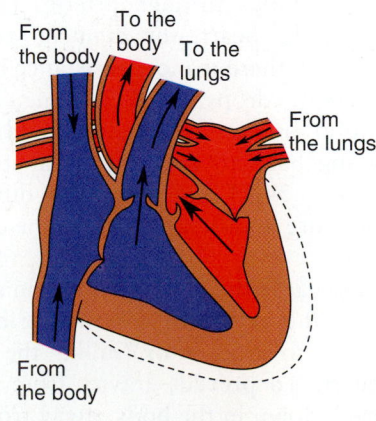

(b) Ventricular contraction (systole),
atrial relaxation (diastole)

time, blood flows into the atria, and the cycle begins again.

As you know, you can hear your heartbeat, which is generally called a "lub-dup" rhythm. But you are not hearing the contractions and relaxations of the muscle. What you do hear is the closing of the heart's valves. The lub sound, which is deep and slow, is made by the closing atrioventricular valves; the dup sound, which is higher and sharper, is made by the closing semilunar valves.

Blood Pressure. At one time or another, you probably have had your blood pressure measured during the course of a physical examination; so you already know some of the elements involved, especially the device with the tongue-twisting name of *sphygmomanometer.* What does this device—with its column of mercury and the heavy band that is placed around your arm—do and what does it measure?

Blood pressure is the result of the contraction of the ventricles of the heart. This force is required to keep blood moving through the body. Blood pressure is force per unit area, measured against a specific standard: the height to which the pressure can push a column of mercury. The band of the sphygmomanometer measures pressure in the *brachial artery* in the arm. The blood pressure of a normal young adult can push a column of mercury 120 mm when the heart contracts and 80 mm when the heart relaxes. *Systolic pressure* is registered by the contraction of the heart, and *diastolic pressure* is registered by the heart's relaxation, and the ratio is expressed as 120/80.

As you would expect, blood pressure varies with distance from the heart. The blood pressure in the pulmonary arteries and aorta is relatively high, since they are close to the heart and receive the full force of the contraction of the ventricles. As the blood flows through arteries farther and farther from the heart, the diameter of the arteries decreases, there is more resistance to flow, and blood pressure drops. Farther along, in the capillaries, blood pressure is lower still, and in the veins blood going back to the heart is at its lowest pressure. Figure 24.11 illustrates the differences in blood pressure in different parts of the cardiovascular system.

Because the blood pressure in the veins is so low, it is the action of contracting body muscles that serves to squeeze, or "milk," the blood back to the heart. To prevent gravity from pulling the blood back down in the body, away from the heart, there

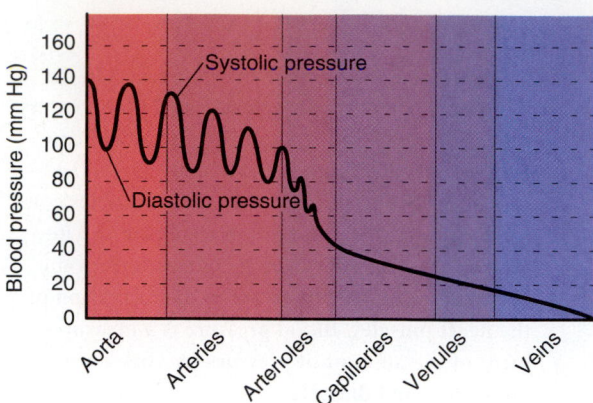

FIGURE 24.11 Blood Pressure in the Circulatory System Blood pressure drops as the blood moves from the arteries toward the capillaries.

are one-way valves in the veins (see Figure 24.6). If these valves weaken, blood can pool in the veins, resulting in a condition known as varicose veins. In order to reduce the pull of gravity on the valves in varicose veins, the patient is usually told to elevate the legs with a pillow when lying down. This method allows gravity to assist in returning the blood to the heart. It also decreases the pressure in the chest cavity during breathing, thus producing a pressure gradient that helps return blood to the heart. In essence, blood flows from higher pressure to lower, from the high pressure in the ventricle to pressures of almost 0 in the right atrium of the heart as the circuit is completed.

As you may know, high arterial blood pressure, or *hypertension,* is the most prevalent of all cardiovascular disorders. Hypertension means that the person's blood pressure at rest is usually 140/90 or greater. This reading must hold true for three consecutive readings before the individual is said to have high blood pressure. Of major concern is the diastolic pressure (the second of the two figures) because this indicates the pressure while the heart is not contracting. That means that the walls of all vessels are under a great deal of pressure at all times, not just when the heart contracts. Hence pressure cannot be zero as blood flows into the heart. Hypertension can cause serious heart damage. The pressure can cause a wall of one of the venules to burst. The bleeding, or hemorrhage, that results can cause a stroke if it occurs in the brain or can cause other cardiovascular problems.

It is estimated that about 23 million Americans are hypertensive. Most are not being treated, and about 180,000 will die each year. The actual causes of hypertension are unknown, but lack of exercise and improper diet increase the probability of the disease. Hypertension is found in higher frequencies among the African American population as compared with the white, and the cause for this difference in incidence is unknown. Researchers suspect that a combination of inherited tendencies, dietary choices, and stress is responsible for this high incidence.

Regulation of the Heartbeat

■ **The sinoatrial node (SA node) within the atrial wall of the heart controls the rhythm of contraction and relaxation that typifies heart functioning. External factors such as neural signals and hormonal actions control the rate and strength of the heart rhythm.**

Regulating cardiac muscle contraction involves two independently functioning factors, *internal* (those originating within the heart) and *external* (those coming from outside the heart). External factors, like hormones and the autonomic nerves, regulate only the *rate* of contraction. Contractions are the result of internal factors. The ancient Aztecs were very aware of this phenomenon, for in their sacrifices they would remove the heart from an individual's body and offer it, still beating, to their gods. If you have ever removed a frog's heart during a laboratory experiment, you should have noticed the same phenomenon. What initiates cardiac muscle contraction? To answer this question, we need to study the functioning of certain internal structures of the heart.

Internal Regulators. Within the atrial wall is a special area of cells known as the **sinoatrial node** (SA node), or, as it is more commonly known, the *pacemaker* (Figure 24.12). The cells of the SA node are capable of independent spontaneous depolarization. That means that they are capable of generating an electrical impulse. This results in an action potential, similar to a nerve impulse (we discuss nerve functioning in Chapter 30). The action potential causes the muscle to contract. The action potential is expressed as a wave of depolarization that spreads from the right atrium into the left, initiating contraction of the atria.

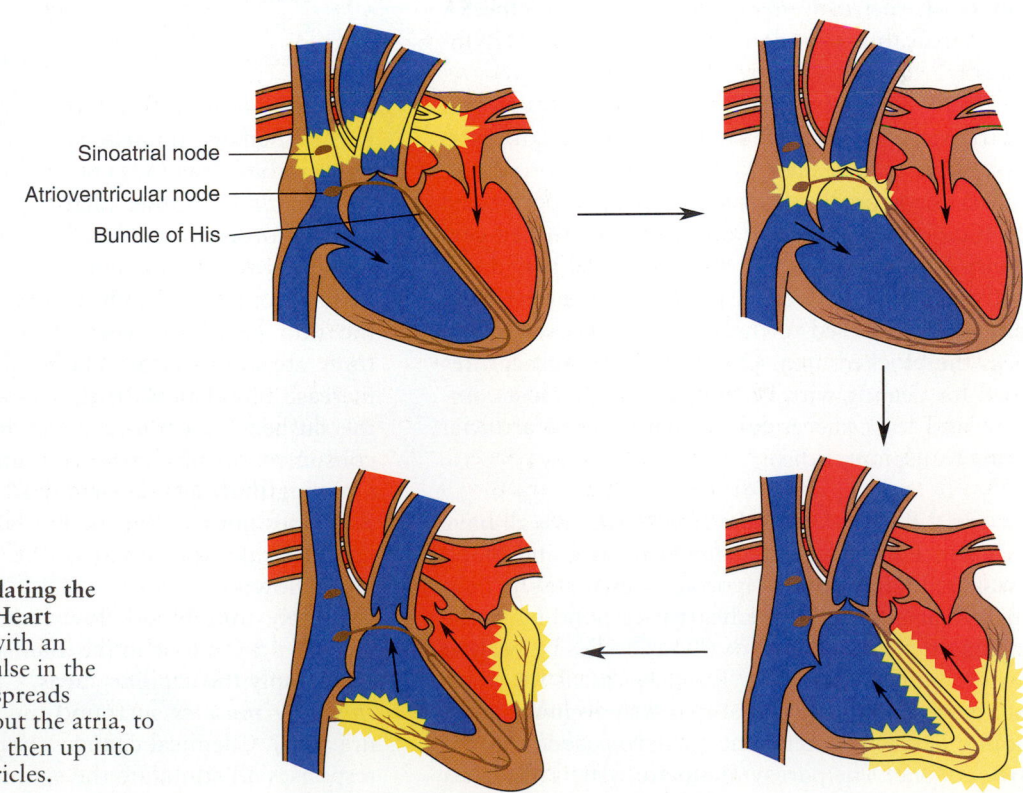

Sinoatrial node
Atrioventricular node
Bundle of His

FIGURE 24.12 Regulating the Contractions of the Heart Contraction begins with an electrochemical impulse in the sinoatrial node that spreads downward, throughout the atria, to the base of the heart, then up into the walls of the ventricles.

The action potential also stimulates another node, called the *atrioventricular node* (AV node), located in the lower portion of the right atrium. Special muscle fibers known as the *bundles of His* transmit the impulses from the AV node, stimulating the ventricles to contract simultaneously. Because the SA node impulses travel faster through the atria than to the ventricles, there is a delay between the contraction of the atria and that of the ventricles. The result is the contraction-relaxation rhythm associated with the heartbeat.

One of the most serious forms of heart attack occurs when the heart's internal regulators fail to work properly. As a result, the heart does not contract efficiently but only quivers or flutters. This condition is known as *fibrillation*. You may have seen rescue workers or emergency room personnel in television programs use a defibrillator to normalize the heartbeat of a patient. In this procedure, they stop the heart's abnormal fluttering by administering a very short-term electrical shock. The shock interrupts the abnormal pattern, giving the regulators responsible for normal nerve impulses an opportunity to return to the regular pattern.

When the natural pacemaker system goes awry, physicians can insert an artificial pacemaker into the patient's chest cavity. The pacemaker's small battery powers an electrical device that sends small bursts of electrical energy to the tissues of the SA node area, thus stimulating the heart to beat rhythmically.

There have been numerous other advances in heart surgery, from the simple procedure of insertion of a pacemaker to the complex procedure called coronary bypass surgery. This is performed on individuals who have one or more blocked coronary blood vessels. The blocked vessel is tied off, and a replacement vessel, usually from a vein from the leg, is attached to the vessel and is used to bypass the blocked area. Of all the surgical measures used for dealing with heart diseases, the most complex and least successful are natural and artificial heart transplantations.

External Regulation of the Heartbeat. We all have experienced the fact that our heart rate and blood pressure increase after vigorous exercise. Why does this occur? Actually, the heart is responding to the increased cellular demands of our bodies by beating faster and more strongly. This response is the result of the stimulation of one of two systems in the autonomic nervous system, the *parasympathetic* or the *sympathetic*. The parasympathetic system releases a neurotransmitter *acetylcholine* into the heart, causing the heartbeat to slow down. The *sympathetic* nervous system, on the other hand, releases a chemical called *norepinephrine* (formerly called noradrenalin), which causes the heartbeat to speed up. We discuss the workings of the nervous system fully in Chapter 30.

In addition to neural control, hormones such as epinephrine (what we commonly call adrenalin) can cause the heartbeat to increase. (Adrenalin is actually a trade name for the hormone epinephrine, much like the name Xerox, a trade name for a photocopying process.) You have all heard of the adrenalin surge that has allowed people to perform amazing feats of strength during emergency situations. What is the role of the heart in these phenomena? Adrenalin accelerates the heart rate, shunts blood from other organs to skeletal muscles, and causes an instant release of glucose into the blood from the liver. The result is a sudden increase of energy.

Regulation of Blood Flow

■ **The rate of blood flow through the human system is determined in part by the body's response to changes in the environment. As body temperatures increase, the arterioles that supply the capillary beds dilate, which increases blood flow to the skin. When temperature decreases, the arterioles constrict. Precapillary sphincter muscles are the mechanisms involved in this action.**

The flow of blood is dependent on the body's activities. When we discussed cellular respiration, you learned that strenuous exercise increases blood flow to the muscles. Other factors can also affect blood flow. In order to regulate body temperature, blood flow increases to the skin when the temperature of the environment is high. Conversely, blood flow to the skin decreases when environmental temperatures are low. Certain chemicals, such as alcohol, increase blood flow to the skin — which explains the flushed appearance of an individual who has consumed a substantial amount of alcohol. But knowing that outside elements can influence blood flow does not explain the mechanisms that control the physical responses involved. Just what regulates blood flow?

Simply put, blood flow is regulated by an increase or a decrease in the diameter of the arterioles that supply the capillary beds. Circular *precapillary sphincter muscles* surround the arterioles (see Figure 24.5). Chemical changes, hormones, and neural responses all stimulate the sphincter muscles to expand (dilate) or contract. When they contract,

Chapter 24   Circulation

blood flow to that capillary bed ceases; when they dilate, blood flow to the capillary bed increases.

BLOOD, THE SEA WITHIN

When I was in the tenth grade, I remember seeing a movie called *Hemo the Magnificent*. In this particular film, a cartoon character called Hemo (meaning "blood") asked Mr. Scientist a question that had to be answered before Hemo would divulge the secrets of circulation. Mr. Scientist had to describe blood, and his response was "Seawater." The other cartoon animals laughed, because they assumed Mr. Scientist to be wrong, but Hemo was impressed and, according to their agreement, began to divulge the secrets of circulation within the human body.

Mr. Scientist was right. The makeup of the blood in the circulatory system of all of us does have a consistency vaguely similar to that of seawater. In a sense, the blood in our systems bathes each of our cells in a wash of highly modified seawater. To understand what this means, we need to explore the components and chemicals that combine to make up blood.

Blood Components

■ **Blood consists of a liquid substance called plasma and three types of formed elements: red blood cells, platelets, and white blood cells. Plasma consists primarily of water and dissolved materials. Red blood cells are primarily responsible for oxygen transport. Platelets are primarily responsible for blood clotting. White blood cells are primarily responsible for protection against disease.**

A 145-pound male contains about 5 liters of blood (slightly more than 5 quarts), or about the same as the amount of oil required for an automobile oil change. Blood has two main components: formed elements, or cells (that is, red blood cells, platelets, and white blood cells), and *plasma,* which is the liquid in which the formed elements are found. Plasma comprises about 55 percent of the total volume of whole blood, and the formed elements make up the other 45 percent. Red blood cells, platelets, and white blood cells comprise the formed elements in blood. Each group carries out specific functions within the body as they make their daily rounds through the circulatory system. Red blood cells primarily function in oxygen transport, platelets function in clot formation, and white blood cells fight infections.

Plasma. If you place a flask of whole blood in a centrifuge and spin it down, the formed elements will settle to the bottom of the flask, leaving a straw-colored liquid at the top (Figure 24.13). This liquid is called **plasma,** which makes up about 55 percent of whole blood. Plasma consists of about 90 percent water and 7 percent blood proteins such as albumin, globulin, and fibrinogen. The remaining 3 percent is dissolved gases, ions, hormones, and various nutrients.

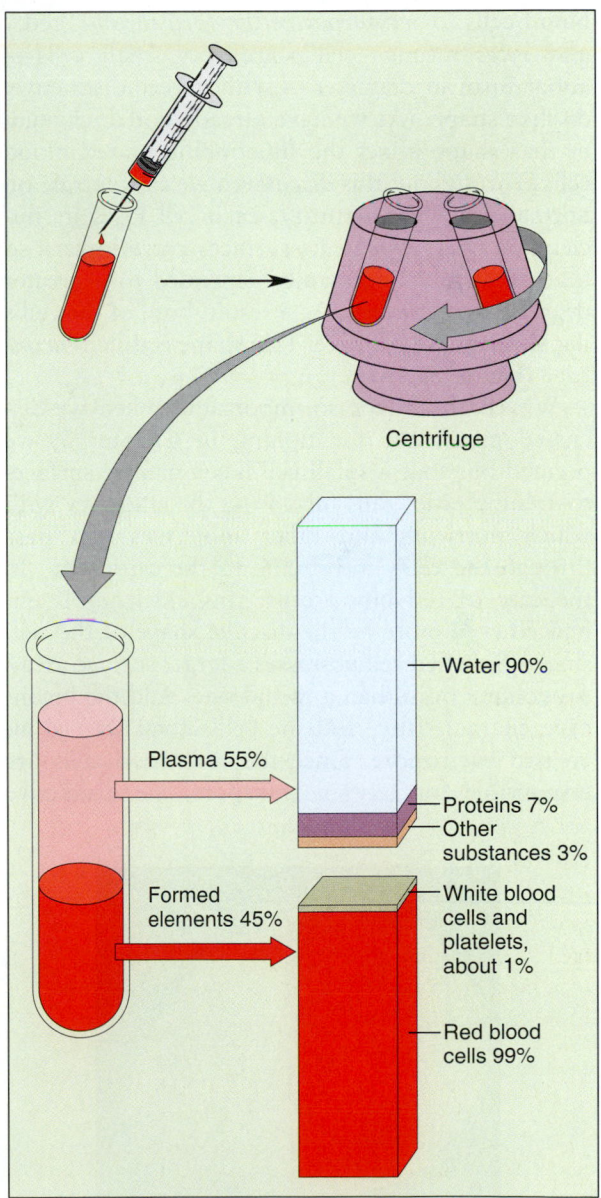

Centrifuge

Water 90%

Plasma 55%

Proteins 7%
Other substances 3%

Formed elements 45%

White blood cells and platelets, about 1%

Red blood cells 99%

FIGURE 24.13   The Composition of Blood
When whole blood is centrifuged, the formed elements settle to the bottom. The liquid portion, called plasma, makes up about 55 percent of whole blood. Plasma consists of water (90 percent), proteins (7 percent), and other substances (3 percent).

The plasma protein, *albumin*, regulates the osmotic condition of the blood. *Globulin* transports lipids like cholesterol and fat-soluble vitamins; it also functions in the immune response. *Fibrinogen*, another globular protein, is essential for blood clotting.

Red Blood Cells. Imagine a cube the thickness of three dimes (3 mm³). If that cube were filled with blood, it would contain between 4.5 and 5 million red blood cells floating in the plasma fluid. **Red blood cells,** or *erythrocytes* (*erythro* means "red," and *cytes* means "cell") are very small cells — about 8 μm in diameter — with a very distinctive, disclike shape. (As we have already noted, changes in that shape affect the functioning of red blood cells. Throughout this discussion we concentrate on normal cells.) At maturity, each cell loses its nucleus. In other words, its surfaces curve inward so that the center is generally depressed to a greater degree than the edges. As a result, both of the cell's flat surfaces are concave. This shape is called *biconcave* (Figure 24.14).

Why is this shape so important? When we discussed membrane functioning in Chapter 5, we pointed out that a small cell has a greater surface-to-volume ratio, thus increasing the efficiency with which nutrients and other molecules can pass through the cell's membrane to the cytoplasm. In the case of red blood cells, this efficiency is enhanced even more by the disclike shape of the cell, since a flattened cell possesses a larger surface area–to–volume ratio than a round one. Add the biconcave characteristic, and the cell's total area is increased even more. Since the red blood cells are responsible for oxygen transport, the distinctive

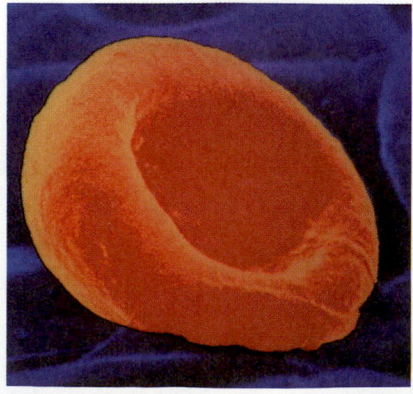

FIGURE 24.14 Red Blood Cells
A central depression on each surface of the red blood cell (where the nucleus used to be) gives it a biconcave shape.

shape is a useful characteristic, for it increases the efficiency with which chemicals can be moved from one side of the membrane to the other.

The means by which red blood cells transport oxygen is a chemical called hemoglobin. In fact, the red blood cell is like a hemoglobin sac. **Hemoglobin** is a complex protein molecule composed of four polypeptide chains. Included in each chain is a heme group, which contains iron atoms. The hemoglobin molecule can readily combine with oxygen to form *oxyhemoglobin*. Oxyhemoglobin gives the blood its bright red color. When the blood has less oxygen, it appears darker, almost bluish, and the individual will take on a bluish tinge.

Have you ever heard of a "blue baby"? The term is accurate, because such an infant appears blue, or even dark purple, owing to a lack of sufficient oxygen in the blood. (The medical term for this condition is *cyanosis*.) The usual reason for this condition in a newborn is the failure of its heart to change from a fetal circulatory pattern, which does not include the lungs (to any large extent), to the pattern necessary after birth, which does.

Most of the CO_2 produced as a waste product of cellular respiration is carried in the blood plasma, dissolved in the blood or as a bicarbonate ion (CO_3), but hemoglobin can also play a role in the process. About 11 percent of the CO_2 forms a complex with the amino acid portion of the hemoglobin molecule to create a substance known as *carbaminohemoglobin*.

Earlier we mentioned that mature red blood cells have no nucleus. That means that they cannot carry out protein synthesis, nor can they repair themselves if they are damaged. As a result, the cells are fairly short-lived. Without protein synthesis, of course, they cannot "feed themselves." Then too, the coursing of the plasma through the blood vessels causes a fair amount of turbulence, which inevitably damages the red blood cells, and since they cannot carry out repairs, they die quite quickly. As a result, the average red blood cell lives about 120 days. After that, it is destroyed by the liver and spleen and must be replaced by new red blood cells.

In order to maintain a constant level of red blood cells, feedback mechanisms either increase or decrease red blood cell production, which is carried out by *stem* cells in the bone marrow. The major mechanism for initiating red blood cell synthesis is a hormone known as *erythropoietin*. When the kidneys sense a drop in the oxygen level in the blood (therefore a drop in red blood cells), they produce a substance that acts on a protein found in the plasma. The result is erythropoietin, which stimu-

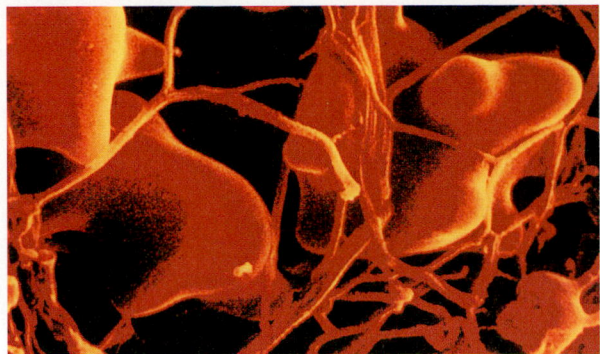

FIGURE 24.15 Platelets and Fibrin
Platelets are trapped in a network of fibrin as a blood clot forms.

lates an increase in red blood cell production by the bone marrow. When the kidneys detect high levels of oxygen in the blood, erythropoietin levels fall, and the red blood cell production slows down.

Platelets. **Platelets** are also called *thrombocytes* (*thrombo*, "clot"; *cytes*, "cell"), because they are involved in the blood clotting process. Platelets are small, colorless, disc-shaped fragments of larger white blood cells known as *megakaryocytes*. Megakaryocytes are produced in the bone marrow. Platelets are responsible for blood clotting, which is necessary to stop bleeding (Figure 24.15).

The clotting process consists of several steps. In essence these are the events that take place when a blood vessel is cut or damaged and a clot forms.

1. The smooth muscle walls of the damaged vessel constrict, which reduces or contains blood flow.

2. Platelets attach and clump together at the wound site, temporarily forming a plug, which helps to slow or stop blood flow.

3. The platelets rupture and release serotonin, a chemical that causes the blood vessel to constrict even more and gives rise to enzymes (called thromboplastins) involved in clot formation.

4. The presence of thromboplastins and calcium ions brings about a series of reactions that result in the formation of a sticky protein called fibrin.

5. Fibrin forms a clot and the clot contracts, pulling the wound together, thus preventing further bleeding and stimulating healing.

In essence, the platelets release substances that initiate the clotting process, eventually resulting in the formation of the protein *fibrin* that forms a woven mesh in which blood cells are trapped, and the clot forms.

White Blood Cells. **White blood cells,** also known as *leukocytes* (*leuko*, "white"; *cytes*, "cell"), function in defending the body from infection (see Chapter 25). There are five major types of white blood cells: *neutrophils, eosinophils, basophils, lymphocytes,* and *monocytes* (Figure 24.16). The different types are distinguished from one another

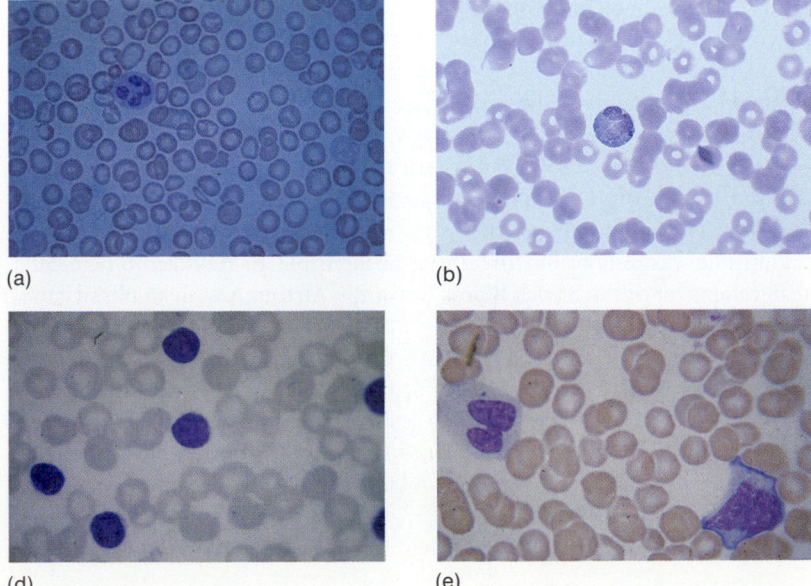

(a)

(b)

(c)

(d)

(e)

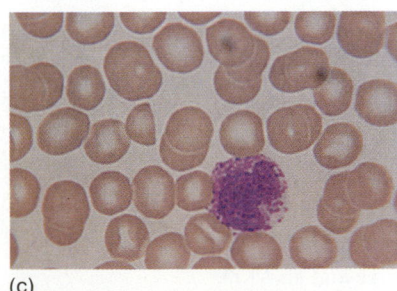

FIGURE 24.16 White Blood Cells
(a) Neutrophils, the most common white blood cells, are phagocytes. Both (b) eosinophils and (c) basophils play a part in the inflammatory response. (d) Lymphocytes are a critical component of the immune system. (e) Monocytes, like neutrophils, are phagocytes.

Dr. Charles R. Drew
Surgeon, Teacher, Blood Preservation Pioneer

When you think of the major medical breakthroughs in surgery over the last forty years, you most likely think of antibiotics and sophisticated technical advances. However, many physicians and scientists feel that the major technical contribution to surgery was the blood plasma preservation techniques developed by the physician-surgeon Dr. Charles R. Drew. On September 9, 1976, Dr. Donald S. Fredrickson, director of the National Institutes of Health (NIH), in Bethesda, Maryland, hung the portrait of Charles R. Drew at NIH in recognition of his work on blood preservation. Drew was the first African American to be so honored.

The oldest son of a building tradesman, Charles Drew was born June 3, 1904, in an impoverished section of Washington, D.C. Despite his surroundings, he dreamed of going beyond. His athletic talents were discovered early at Washington's Dunbar High School, and he earned an athletic scholarship to Amherst College. His interest in science developed there, and he decided to become a physician, an ambitious dream, given the racial climate in the United States in the 1930s. Drew was accepted at McGill Medical School in Montreal, Canada. In 1933, he graduated second in his class with a master's degree and an M.D.

While in Canada, Charlie (as his friends knew him) worked with Dr. John Beattie, who was interested in blood. At this time there was quite an interest in hematology (the branch of biology that deals with blood). Karl Landsteiner had just won the Nobel Prize for his discovery of human blood types. Nevertheless, people were still dying of severe hemorrhage following surgery or accidents because there was no way to store a sufficient supply of blood. Whole blood simply broke down in storage. Dr. Drew, along with other medical researchers, was working on solutions to this problem.

Dr. John Scudder and Dr. Drew developed an experimental blood bank at Presbyterian Hospital in New York, but after a week the red blood cells always broke down. Then it occurred to Drew that every blood storage problem involved the breakdown of red blood cells. What if they were removed? What if only the liquid plasma portion of blood was used in transfusions? Could plasma be stored indefinitely? More important, would plasma be sufficient in transfusions? Drew and Scudder transfused plasma into hemorrhaging individuals, and it worked. In some cases, especially in violent injury and in cases of severe shock, it worked even better than whole blood. Moreover, since plasma contained no red blood cells (remember that the blood-type antigens are located on the red blood cell membrane), the blood did not need to be typed.

During World War II Drew became the director of the Plasma for Britain Project, and in a span of only 5 months he was responsible for the development of the British Blood Bank, supervising the collection of over 14,500 pints of plasma. This act saved the lives of several thousand injured individuals. Eventually, Drew developed a system for collecting and storing millions of pints of plasma, and because of his efforts,

adequate plasma was available for American soldiers when Pearl Harbor was attacked.

Ironically, though the technique developed by an African American scientist saved the lives of thousands, the armed forces ordered the Red Cross to segregate blood supplies and collect only Caucasian blood. The blood of African Americans would not be acceptable for members of the armed forces. Because of this racist attitude, Drew resigned his post at the Red Cross and became professor of surgery at the predominantly African American Howard University Medical School in Washington, D.C. His fame as a teacher grew, and he was personally responsible for training 50 percent of the African American physicians between 1941 and 1950. Dr. Charles R. Drew would tell his students to "dream high" and work hard, following his own philosophy. Tragically, Drew was killed in an automobile accident in 1950, at the age of forty-five.

on the basis of size and staining properties. All are produced by the same stem cells in the bone marrow that produce red blood cells. Unlike red blood cells, they retain their nuclei at maturity.

Under normal (that is, healthy) circumstances, the ratio of white blood cells to red is about one white cell for every 500 red cells. Deviations from that ratio indicate that something is wrong with the body. If the white cell count falls below 5000 per mm³, physicians know that the bone marrow cells that produce white blood cells have been damaged. On the other hand, an increase in the number of white cells indicates that the body is suffering from an infection. Counts of 10,000 per mm³ signal a serious infection, and in patients suffering from leukemia the white blood cell count can be over 100,000 per mm³.

Some white blood cells are amoebalike and can migrate through the tissues, engulfing foreign substances like bacteria. As white blood cells are destroyed, they are replaced by new cells, usually in the bone marrow, spleen, and certain other body organs.

The most numerous white blood cell type are the *neutrophils*, which function by engulfing invading microorganisms through phagocytosis. Neutrophils aggregate at infection sites. *Monocytes* are activated at the site of infection. They then change to macrophages, which function like the neutrophils in phagocytizing foreign cells and cellular debris. *Basophils* are involved in the release of histamines and serotonin at infection sites. This serves to initiate the inflammatory response, which we discuss in Chapter 25. *Eosinophils* are also involved in the inflammatory response, especially in inflammation resulting from parasitic infections. The *lymphocytes* are the major "workers" of the immune system. Found in lymphatic tissue, they perform various functions that result in the defense of the body from invading infections.

Blood Transfusion

- Blood transfusions are done on the basis of blood type, be it A, B, AB, or O. The Rh factor, indicated by a + or a −, must also be identified. Otherwise the transfused blood will agglutinate. Individuals with O⁻ blood are known as universal donors. Individuals with AB⁺ blood are known as universal recipients.

Sometimes because of accidents, disease, or surgery, there is excessive blood hemorrhaging, and the blood has to be replaced. This is accomplished through blood transfusion. As we mentioned in

Chapter 15, there are four major human blood types (A, B, AB, or O), determined by the type of complex protein (an antigen, which is an *anti*body *gen*erating substance that triggers the formation of a specific antibody) located on the red blood cell membrane. Blood also contains another factor that is distinctive: a series of antigens known as the *Rh factor*. Eighty-five percent of the white population, and varying degrees in other ethnic groups, possess the Rh factor and are said to be Rh⁺. The other 15 percent lack these antigens and are called Rh⁻. (Figure 24.17).

When an individual is transfused with a blood type different from his or her own, the transfused blood will react with products of the recipient's immune system and will agglutinate (or clump together). If that happens, the capillaries become clogged, depriving tissues of nutrients and oxygen. Ultimately, the patient may die. Therefore, the same type of blood must be transferred from donor to recipient, which means that before a transfusion can take place, blood is carefully typed for A, B, AB, or O, and for Rh⁺ or Rh⁻ factors.

In emergency situations, however, physicians can give a transfusion of type O⁻. Individuals with type

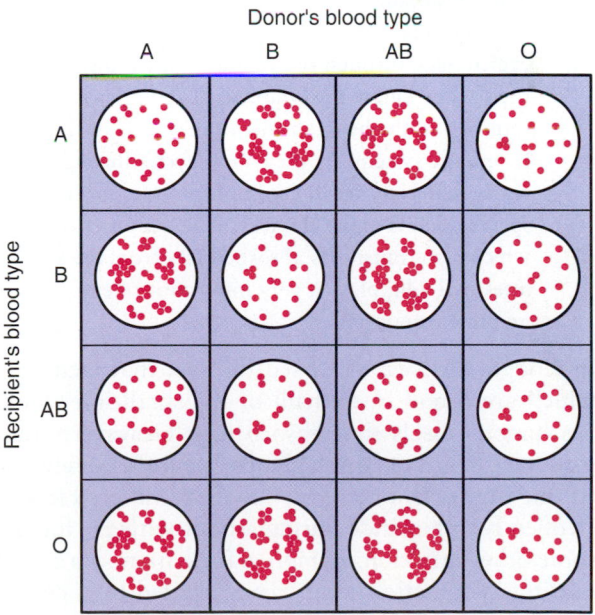

FIGURE 24.17 Blood-Type Compatibility
Since people with blood type O have no blood-type antigens on the surface of their red blood cells, they will form antibodies to other blood types and the blood will agglutinate. Blood type A can receive O or A and will agglutinate with B or AB. Blood type B can receive B or O. Individuals with blood type AB are called universal recipients because they can receive A, B, AB, or O.

O⁻ blood are called universal donors, because their blood lacks the antigens for A, B, or AB, as well as those for the Rh factor. A universal recipient is an individual who has type AB⁺ blood, for this person's blood possesses all the antigens possible, hence will not produce antibodies to the received blood's antigens. Therefore, universal recipients can be transfused with any type of blood, since their own blood does not contain antibodies to react against the new blood.

Scientists during the Second World War knew of the various blood types, but they could not conduct blood transfusions efficiently in large numbers of people because they had no means of storing quantities of whole blood. However, Dr. Charles R. Drew pioneered blood preservation techniques using blood plasma rather than whole blood. Because blood plasma does not contain the red blood cells, blood typing is not necessary. His techniques led to the formation of blood banks, which today save the lives of countless hundreds of thousands of individuals. His techniques for preserving blood plasma are thought by some to be one of the major surgical advances of this century (see Profile).

Lymphatic Vascular System

■ **The lymphatic system is a secondary circular system in humans. Lymph is blood plasma that is not cycled back through the venous system. Instead the plasma enters the lymphatic capillaries and is returned to the heart through the vessels of the lymphatic system. As the lymph cycles through, waste materials are filtered out in nodes located at specific points along the way. The lymph system, which also transports absorbed fats from the small intestinal villi, plays a very important role in the immune response.**

In Chapter 23, we discussed the fact that lacteals, which are found in the villi of the small intestine, are lymph vessels that absorb fats. However, you may have heard of lymph glands in another context. At one time or another you may have suffered from swollen lymph glands during a bout with the flu or some other infection. Lymph glands are nodes built into the lymphatic system. They are fed by a complex system of vessels and lymphatic capillaries. The lymphatic system works in concert with the blood circulatory system. The interaction occurs in the capillary beds of the body.

We have already discussed the fact that the blood supply flows through the interstitial fluids of the capillary beds, returns to the venous capillaries, and is ultimately returned to the heart to be reoxygen-

ated and repeat the cycle. However, not all the blood plasma that leaves the capillaries returns to them. Some of the fluid remains in the interstitial tissue. The colorless fluid, which is filtered out of the capillaries and into the interstitial tissues, is called **lymph.** Lymph is collected by the lymphatic capillaries, which are also embedded in the interstitial tissue fluid. It is the function of the lymphatic system to return excess tissue fluid to the bloodstream and to protect the body from disease.

The Structure of the Lymphatic System. Lymphatic capillaries are about the same size as those of the blood supply system (Figure 24.18a and b). The ends of the capillaries that extend into the interstitial tissue are called "blind" — in other words, the ends are closed, not open. At the other end, the capillary feeds into the vessels of the lymphatic circuit. These vessels ultimately empty into the collecting veins of the *thoracic duct.* The thoracic duct, in turn, empties into the *left subclavian vein,* and that vein empties into the *superior vena cava,* thus returning the lymph to the heart.

Lymph nodes are masses of spongy tissue located at various intervals within the lymphatic system. Your tonsils are lymphatic nodes, as are the spleen and thymus (Figure 24.18c). The function of these nodes is to filter cellular debris from the lymph. Lymph nodes also produce a high number of lymphocytes. The lymph, lymphocytes, and other white blood cells located in the lymph nodes (such as the neutrophils) clear the lymph nodes of foreign substances that have been filtered out by the nodes. When you have an infection, your lymph nodes often swell as the lymphocytes and other white blood cells combat the infectious agent.

Lymph vessels are similar to veins, and, as with veins, it is the squeezing of muscles against adjacent lymph vessels and pressure difference between the lymph vessels and the rib cage that returns the lymph to the collecting veins. Recently, researchers have demonstrated that lymph vessels can contract, and those contractions also help move the lymph through the system.

In summary, the lymphatic system has four distinct functions:

1. It transports absorbed fats from the lacteals in the small intestinal villi.

2. It returns excess tissue fluid back into the circulatory system.

3. It disposes of cellular particles and other substances through the lymph nodes.

4. It plays a very important role in the immune response.

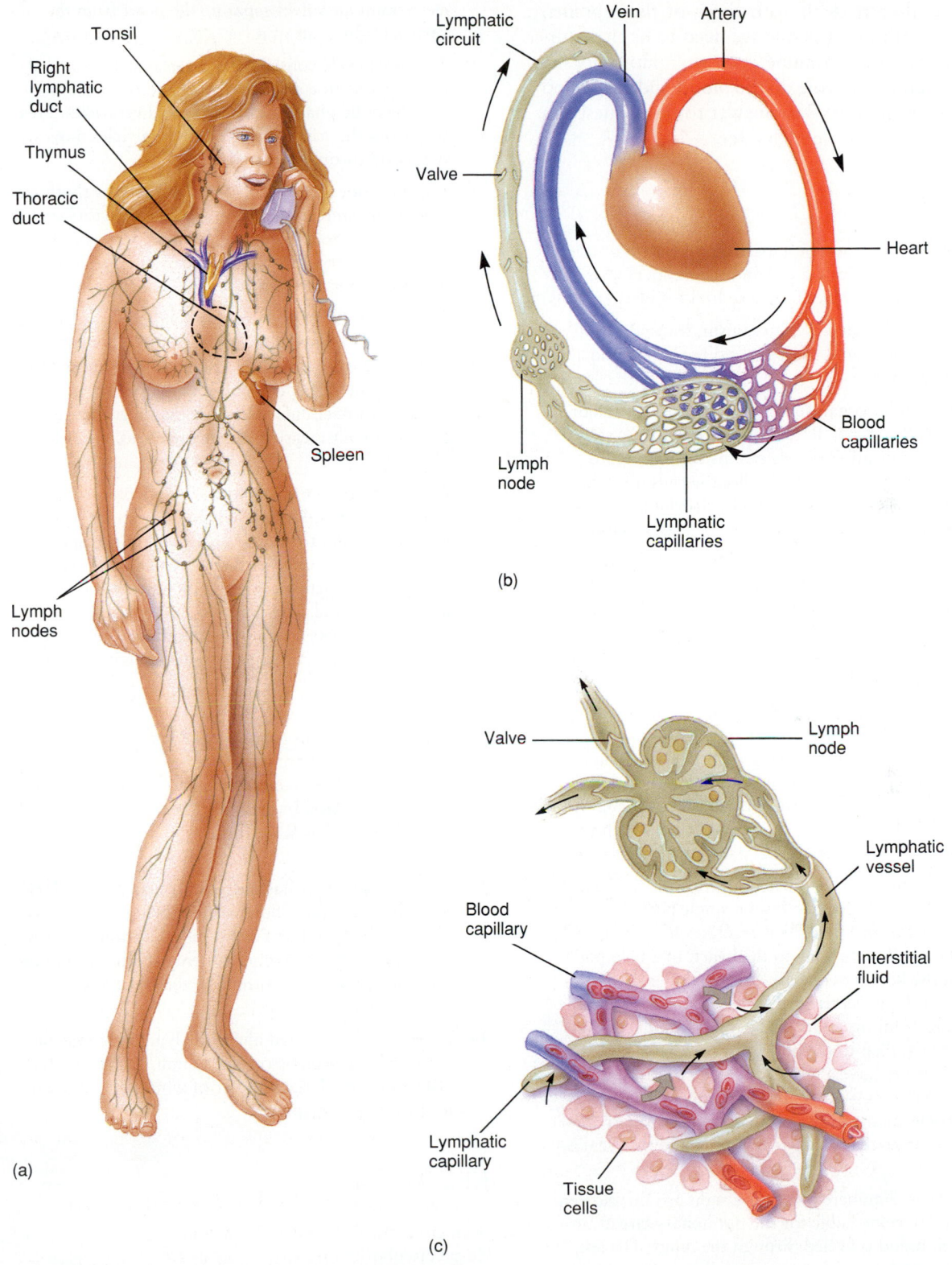

Tonsil

Right lymphatic duct

Thymus

Thoracic duct

Lymph nodes

(a)

Spleen

Lymphatic circuit

Vein

Artery

Valve

Heart

Lymph node

Blood capillaries

Lymphatic capillaries

(b)

Valve

Lymph node

Blood capillary

Lymphatic vessel

Interstitial fluid

Lymphatic capillary

Tissue cells

(c)

FIGURE 24.18   The Lymphatic System
(a) The lymphatic system is as extensive as the cardiovascular system. (b) A schematic of the lymphatic system shows the "blind" end of the lymphatic capillaries. The valves along the lymphatics prevent backflow. (c) The lymph nodes filter cellular debris from the lymph fluid. They also have other functions; for example, they play a role in the immune system.

We have already dealt with three of these points, but the fourth point is one we need to explore further. What is the immune response? How does it function, and what role do the blood and the lymph play in its processes? The answer to these questions are explored in the next chapter.

SUMMARY

1. The transportation of nutrients and essential substances to or within a cell and the removal of waste products from the cell are essential for life.

2. Within the single-celled organism, transportation relies on diffusion, osmosis, and facilitated and active transport.

3. Multicellular organisms exhibit one of two major transport systems. In the open circulatory system, a series of pulsating vessels pumps the blood into body cavities; the blood bathes the cells, then returns to the heart. In the closed circulatory system, the blood is enclosed in a series of vessels. Humans have a closed circulatory system.

4. Circulatory systems perform several functions: They maintain homeostasis, transport essential substances and waste products, aid in temperature regulation and pH balance, and defend against foreign organisms (in the higher organisms).

5. The human circulatory system (the cardiovascular system) consists of the heart and major types of blood vessels: arteries, which carry blood away from the heart to the body; capillaries, which are microscopic blood vessels that exchange materials between the cells and blood; and veins, which return the blood to the heart.

6. In fishes the heart consists of one atrium and one ventricle. The blood follows a single pathway from the heart to the gills, then circulates through the body, finally returning to the heart. In the amphibians, the heart consists of three chambers — two atria and one ventricle. The blood follows a double circuit, from the heart to the lungs or gills, back to the heart, then to the body and back to the heart. The reptiles have a heart consisting of three and a half chambers — two atria and an almost divided ventricle. In the birds and mammals the heart consists of four chambers — two atria and two ventricles.

7. In a four-chambered heart the right atrium and ventricle are responsible for the pulmonary circuit, in which blood is cycled through the lungs. The left atrium and ventricle are responsible for the systemic circuit, in which blood is circulated through the body.

8. The human heart has four valves that prevent the backflow of blood into the heart chambers. Atrioventricular valves separate the atria and the ventri-

cle. Semilunar valves separate the heart from the vessels leading out of it.

9. The heart cycle consists of a series of synchronized contractions (the systolic phase) and relaxations (the diastolic phase) of the atria and ventricles. Pressure from the contractions of the ventricles moves the blood through the body.

10. Various nodes within the heart regulate its rhythmic relaxations and contractions. External factors such as the CO_2 levels, hormones, and the nerves of the autonomic nervous system control the rate established by internal factors.

11. The flow of the blood through the body is regulated by the body's activities. At rest, blood flow is decreased. During strenuous exercise, blood flow is increased. The physical regulator is the diameter of the blood vessel. Increasing the diameter increases blood flow. Decreasing the diameter decreases blood flow.

12. About 55 percent of the total volume of whole blood is plasma. Plasma, which is about 90 percent water, provides the liquid medium that transports material. The formed elements, which comprise the remaining 45 percent, are red blood cells, platelets, and white blood cells. Red blood cells transport oxygen; platelets aid in blood clotting; and white blood cells protect the body from foreign organisms or their products.

13. Blood is classified as A, B, AB, or O, according to the antigen present on the red blood cell membrane. Another series of antigens, known as the Rh factor, is present in 85 percent of the white population of the United States. Individuals having these antigens are categorized as Rh^+; those who do not are said to be Rh^-.

14. Blood plasma that has passed through the capillary walls into the tissue fluid is called lymph. Lymph is taken up by lymphatic capillaries that drain into the lymph system. The lymph system eventually empties into the veins to be returned to the heart and general circulation.

15. Lymph nodes situated along the lymphatic system filter cellular debris from the lymph. Lymph nodes also produce a large number of white cells that aid in the immune response.

KEY TERMS

open circulatory system	atrium
closed circulatory system	ventricle
cardiovascular system	pulmonary circulation
artery	systemic circulation
capillary	pulmonary artery
vein	systole
aorta	diastole
interstitial fluid	sinoatrial node

plasma
red blood cell
hemoglobin
platelet

white blood cell
lymph
lymph node

REVIEW QUESTIONS

True/False

1. Only the multicellular organisms require transporting mechanisms.
   true   *false*   *(page 572)*

2. Humans have an open circulatory system.
   true   *false*   *(page 573)*

3. The exchange of nutrients and waste products between the blood and cells occurs within the arteries.
   true   *false*   *(page 576)*

4. The liquid portion of the blood is called plasma.
   true   *false*   *(page 583)*

Fill in the Blank

1. The major function of the _____ blood cells is to transport oxygen. *(page 586)*

2. The pressure of the heart as it contracts is referred to as the _____ pressure. *(page 581)*

3. Arteries carry blood _____ from the heart. *(page 575)*

4. The cells of our body are surrounded by a fluid known as _____ fluid. *(page 576)*

Multiple Choice

1. The major site of biological action of the human circulatory system occurs in:
   a. The arteries
   b. The veins
   c. The capillary bed
   d. The heart
   (page 573)

2. The major function of the animal transport system is:
   a. Movement of nutrients
   b. Movement of wastes
   c. Neither A nor B
   d. Both A and B
   (page 571)

3. In comparison with a single-circuit heart, a double-circuit heart is:
   a. Always larger
   b. More efficient, since it keeps oxygenated and unoxygenated blood separate
   c. Found only in birds
   d. Found in frogs
   (pages 577–578)

4. The control of heartbeat pattern is due to:
   a. Both internal and external factors
   b. Only internal factors
   c. Only external factors
   d. Neither internal nor external factors
   (page 583)

Discussion

1. Distinguish between open and closed circulatory systems.

2. What are the functions of any circulatory system?

3. Compare the structure and function of the arteries, capillaries, and veins.

4. Discuss the exchange of nutrients and waste products between the blood in the capillaries and the cells of the tissues.

5. Distinguish between single circulatory systems and double circulatory systems, pulmonary circulation and systemic circulation, and diastolic pressure and systolic pressure.

6. Discuss the evolution of the heart from a pulsating blood vessel to a four-chambered organ.

7. Discuss the pathway of blood through the human heart.

8. Discuss the regulation of the heartbeat. Be sure to include internal and external factors.

9. Why do our veins have valves?

10. What is hypertension?

11. Describe the components of blood.

25

The Immune System

Introduction
Nonspecific Immune Responses
Specific Immune Responses
Summary

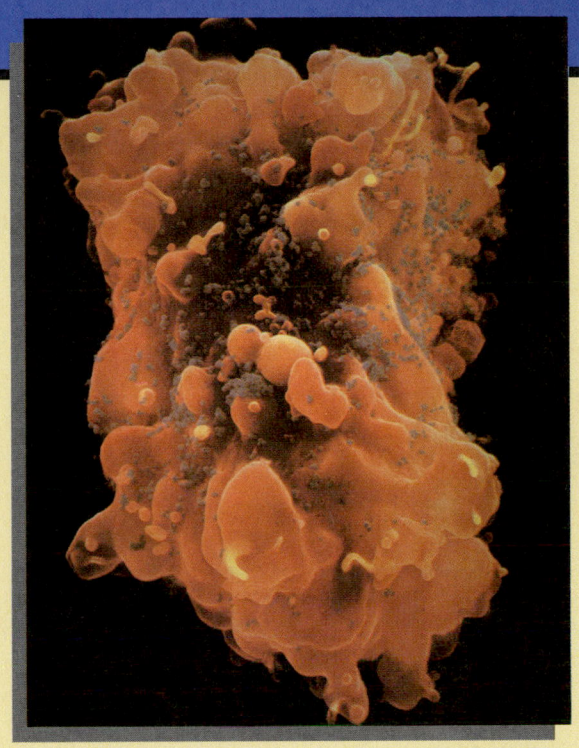

We all know that various agents — such as viruses, bacteria, fungi, and parasites — can cause infectious diseases. Some of us have had allergies. We know that if donor A gives his or her kidney to recipient B, there is a chance of tissue rejection. AIDS is something we hear about on a regular basis. How do we and other animals combat infection? What is the immune system and how does it work?

INTRODUCTION

■ The immune system enables us to defend against disease and recognize self. There are two major categories of immune responses: the nonspecific, which are immediate and very general, and the specific, which are longer-term and very specific.

Above: A colorized scanning electron microphotograph of the AIDS virus infecting a human white blood cell.

In general, the immune system consists of those components of the body that enable us to defend against disease and recognize self from nonself. We can divide immune responses into two major types: the **nonspecific responses** (Figure 25.1), which are general and immediate in nature, and the **specific responses,** which defend against a specific invader.

We'll begin our discussion of the immune system by discussing nonspecific responses, then we'll turn to specific responses. Some useful vocabulary is included in Table 25.1.

NONSPECIFIC IMMUNE RESPONSES

■ Nonspecific responses involve the skin, secretions of chemicals such as mucus and lysozyme, phagocytosis, and inflammation.

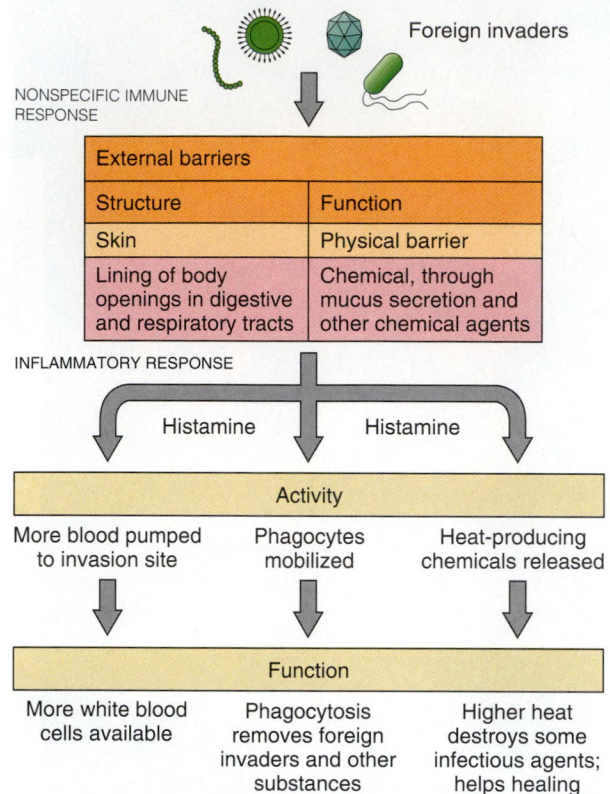

Foreign invaders

NONSPECIFIC IMMUNE RESPONSE

External barriers	
Structure	Function
Skin	Physical barrier
Lining of body openings in digestive and respiratory tracts	Chemical, through mucus secretion and other chemical agents

INFLAMMATORY RESPONSE

Histamine   Histamine

Activity
More blood pumped to invasion site	Phagocytes mobilized	Heat-producing chemicals released

Function
More white blood cells available	Phagocytosis removes foreign invaders and other substances	Higher heat destroys some infectious agents; helps healing

FIGURE 25.1 Nonspecific Responses
Nonspecific immune responses are general and immediate. They can involve physical barriers like the skin or chemical ones like the production of mucus. During the inflammatory response, histamine is released; this makes the capillaries more permeable, so that more blood is sent to the injury site and the temperature at the site increases. Phagocytic white blood cells engulf the invaders.

Skin and Enzymes

Nonspecific responses are barriers to invasion and can be thought of as physical or chemical. For example, the skin for the most part protects against invasion by infectious agents; it serves as a physical barrier. There are many types of nonspecific chemical defenses. The lining of the gut, respiratory tract, and other body cavities contains cells that produce mucus. These mucous membranes trap the invaders and prevent infection, and so in a sense they function as both physical and chemical barriers. The mucous membranes also produce an enzyme called **lysozyme** (from *lysis*, which as you know means to split or break apart). Lysozyme is an enzyme that breaks down the cell boundaries of the invading cells. Tear ducts also secrete lysozyme. In addition, some of the sebaceous glands (oil glands of the skin) produce chemicals that inhibit the growth of bacteria and fungi.

Phagocytosis

Phagocytic cells, which are formed from the stem cells of the bone marrow, also engulf and destroy the invading foreign elements (see Human Endeavors box). Small white blood cells called *monocytes* are attracted to the invaders. The monocytes are then transformed into larger white blood cells called **macrophages** ("big eaters"), which devour the invaders (Figure 25.2). Other phagocytic cells also carry out this function. These include neutrophils and eosinophils (see Chapter 24).

Inflammation

■ **Inflammation is a nonspecific response to the intrusion of foreign elements. The intrusion stimulates the production of histamine. Histamine in turn dilates capillaries to increase the blood supply and increases the permeability of capillary membranes. With the expanded blood supply, the affected area becomes hot, preventing bacteria from multiplying. White cells and complement proteins pass through the capillary membrane to ward off the intruding bacteria.**

Another type of nonspecific response is called *inflammation*. The nonspecific inflammatory response often occurs in conjunction with the specific response. For example, when you have a red swollen scab, not only is there general inflammation at work, but the specific response as well. Let's discuss inflammation in more detail.

Perhaps one of the most commonly spoken orders from a parent to a child is the one that goes "Don't pick at that scab — it'll get infected!" The order is disobeyed as often as not, so all of us know what an infection looks like. The area becomes reddened and swollen to some degree. Below the scab that forms over the injury there is an accumulation of a whitish-yellow mucuslike material. When you touch the area, it feels hot. All these characteristics are an indication that your **inflammatory response** is at work. What are the mechanisms involved in inflammation?

If the skin and/or mucous membrane is invaded, the injured cells are stimulated to secrete several substances, the most important of which is **histamine**. Histamine has two effects on the circulatory system: It dilates the blood vessels in the injured area, thus increasing the supply of blood, and it causes the capillaries to become more permeable.

Increasing the blood supply to the injured area serves several purposes. First, the temperature around the site of the injury rises, which makes it difficult for the invading microorganisms to repro-

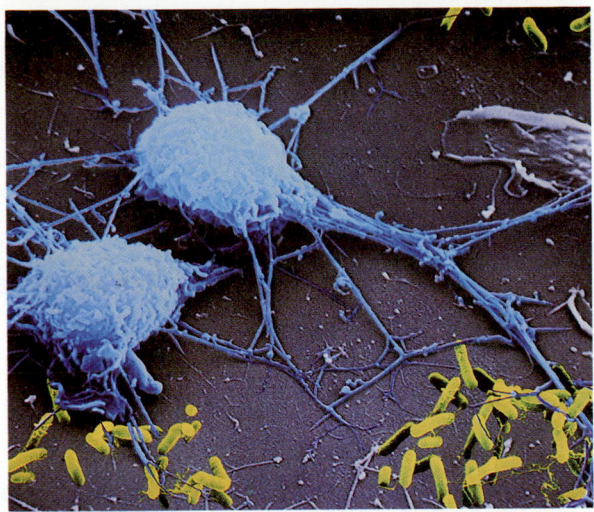

FIGURE 25.2 Phagocytic Macrophage
A scanning electron microphotograph of macrophages engulfing bacteria.

duce. And second, there is an increase in the supply of phagocytic white blood cells and other plasma components — usually proteins called **complement.** These are able to attack the invading microorganisms because the capillaries have become more permeable.

About 30 different types of complement proteins have been identified. These special proteins amplify the inflammatory response. Some increase capillary permeability and blood vessel dilation. Others produce chemicals that attract white blood cells, and still others coat the foreign substances, making them more "desirable" to the phagocytotic cells. Some complements also lyse, or break apart, the cell (Figure 25.3).

Complement acts through two pathways to lyse cells, the **alternative** and the **classical.** The alternative pathway does not require antibodies and is nonspecific, while the classical pathway does require antibodies and acts specifically against an invader.

Next, the large white blood cells called *macrophages* move into the interstitial tissues and engulf the foreign substances through the process of phagocytosis. Once the bacteria are ingested, the cellular processes of the macrophage break the bacteria into molecules.

Although you may not realize it, you take advantage of our understanding of the mechanisms of the inflammatory process whenever you try to ward off the effects of a bad cold or hay fever. Either condition brings on a stuffy nose. It is histamine that causes the swelling of the tissues lining the nasal

FIGURE 25.3 Functions of
Complement Proteins

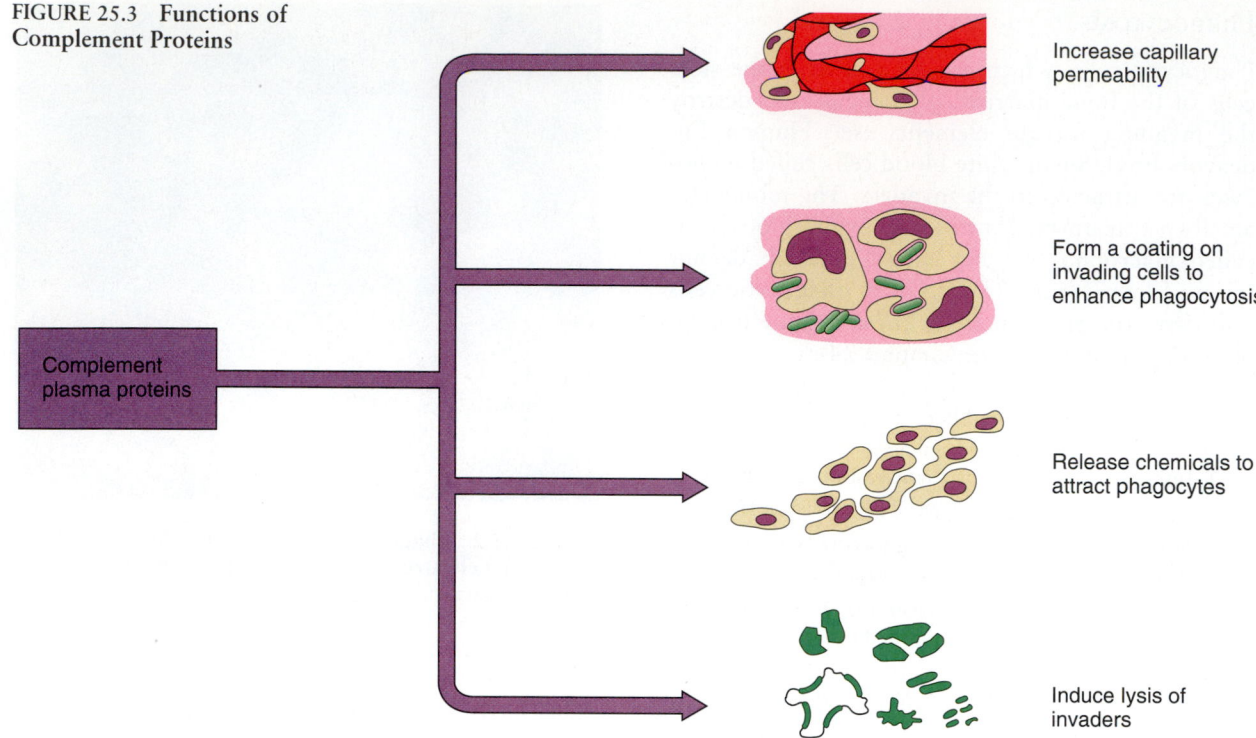

Complement plasma proteins

Increase capillary permeability

Form a coating on invading cells to enhance phagocytosis

Release chemicals to attract phagocytes

Induce lysis of invaders

passages. And the antihistamine you take is a chemical designed to reduce the effects of histamine and enable you to breathe more easily. Antihistamines suppress histamine production, hence reducing tissue swelling.

It should be noted that antihistamines are only a temporary measure. Once they are absent, the condition returns. Many people overuse antihistamines to such a degree that they virtually become "addicted" to them. Overuse can also mask more serious medical problems. This over-the-counter medicine must be used with the same care as any prescription medicine.

As we mentioned earlier, in inflammation the nonspecific and specific responses can occur in conjunction with each other.

SPECIFIC IMMUNE RESPONSES

■ Specific immune responses can be subdivided into two responses — a cell-mediated response, in which specialized white blood cells attack a specific target, and a "humoral" or antibody response, in which the white blood cells "learn" the identity of an invader by identifying chemical markers on the invader's membrane surface and promptly form antibodies against it.

Every year, schoolchildren around the United States troop off to their doctors' offices to receive their vaccinations against any number of diseases such as measles, polio, tetanus, and whooping cough. Vaccination programs work because they utilize the mechanisms of the body's specific line of defense, called the specific immune response, a defense that — as the term suggests — is a very *specific* process.

When you receive a vaccination, you are artificially infected with a weak or dead form of an infectious agent. The dosage is low enough that in most instances, you will notice no ill effects from the injection. However, the mere presence of the infectious agent sets your immune system into action. Lymphocytes (types of white blood cells) not only fight off the infection, but they also "remember" the identity of the invader; this is called *immunologic memory*. Should your body ever be invaded by the same agent (or in some cases, a similar agent), the lymphocytes would identify it immediately and destroy it before it could cause any harm, thus preventing the disease.

The same process occurs if you contract a disease like mumps or chicken pox in the normal course of your life. Your immune system identifies the invader and then fights it off. If you are ever exposed to the disease again, your system recognizes the invader and rejects it immediately. Biologists know that the system works. How it works is another question. As you might expect, the mechanics of the immune system is the subject of a rapidly growing field of research. Many pieces of the puzzle are missing, but some are known. This defense system is very *specific*. The process involves mechanisms for recognizing a particular foreign substance (called an *antigen*) and mounting an attack to destroy it.

■ The specific immune response arises in conjunction with inflammation or when an infection or disease cannot be handled adequately by inflammation alone. Two specialized lymphocytes are involved in the immune response: T-cells and B-cells. Killer T-cells fight infection directly. Helper T-cells work with B-cells to transform B-cells into plasma cells. B-cells move through the blood system and produce antibodies that fight infection.

Suppose you stepped on a rusty nail laden with bacteria (Figure 25.4). How would your body fight this infection? First, the inflammatory response would be activated. Histamine and other chemicals would increase circulation to the wound, and complement and macrophages would work at destroying the bacteria. However, the process is not totally reliable. Some of the foreign bacteria may escape the phagocytic white blood cells, and molecules from those that are engulfed can escape and become attached to the white blood cell's surface.

It is at this point that an important mechanism of the immune response is activated — *the "learning" phase*. Other white blood cells, similar to lymphocytes, literally "taste," or "sample," the molecular tidbits and become *activated*. Those that do have been genetically preprogrammed to become sensitized.

Lymphocytes

Many types of white blood cells are involved in the process that we call immunity. Among these are the lymphocytes. Lymphocytes are small, colorless white blood cells. There are two categories of lymphocytes: T-cells and B-cells.

T-Cells. Some preprogrammed lymphocytes are called **T-cells.** These white cells arise from stem cells in the bone marrow (as do all blood cells, both red and white). They then move into the thymus, from which they take their name: T-cells, or thymus cells. While in the thymus, the lymphocytes acquire specific receptor sites on their membranes. These sites will fit with a particular foreign molecule, rather

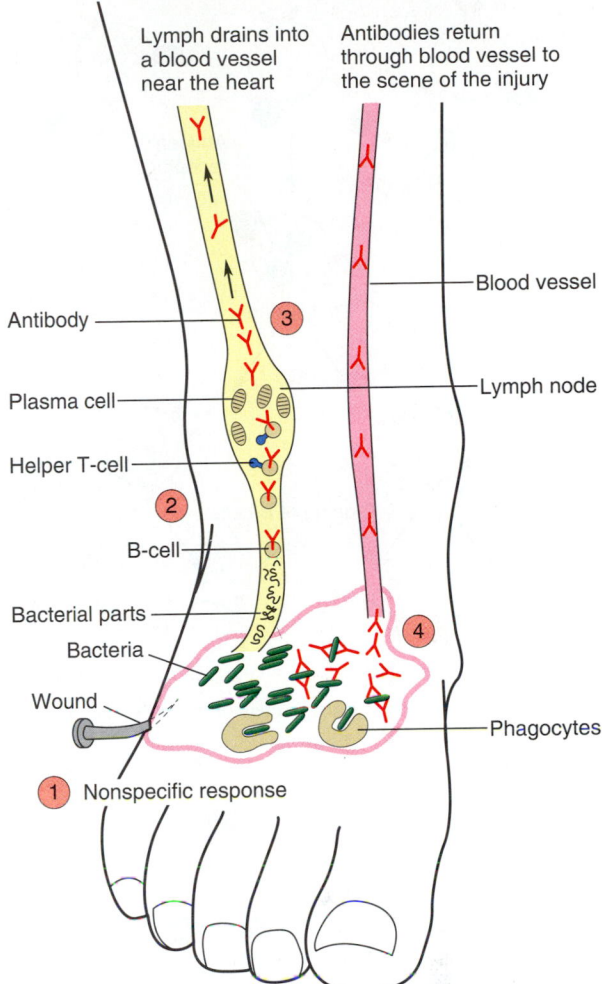

FIGURE 25.4 Specific Immune Responses
If nonspecific immune responses are not successful, specific immune responses are activated. (Specific responses may sometimes occur in conjunction with nonspecific ones.) Step 1: Inflammation and other types of nonspecific responses. Step 2: B-cells and T-cells begin the cellular and humoral responses. T-cells identify antigens on the bacterial invaders and kill them. Step 3: The B-cells are converted into plasma cells that produce the specific antibodies to destroy the invaders. Step 4: The antibodies and macrophages eventually eliminate the invading elements.

like the corresponding shapes of two puzzle pieces. There are two types of T-cells. Some become *killer T-cells*, which directly attack and kill foreign organisms. Other T-cells are called *helper* cells; their function is to help B-cells, another group of sensitized cells, mature into what are known as *plasma cells*.

B-Cells. **B-cells** are lymphocytes that are so named because they were first discovered in the *bursa of*

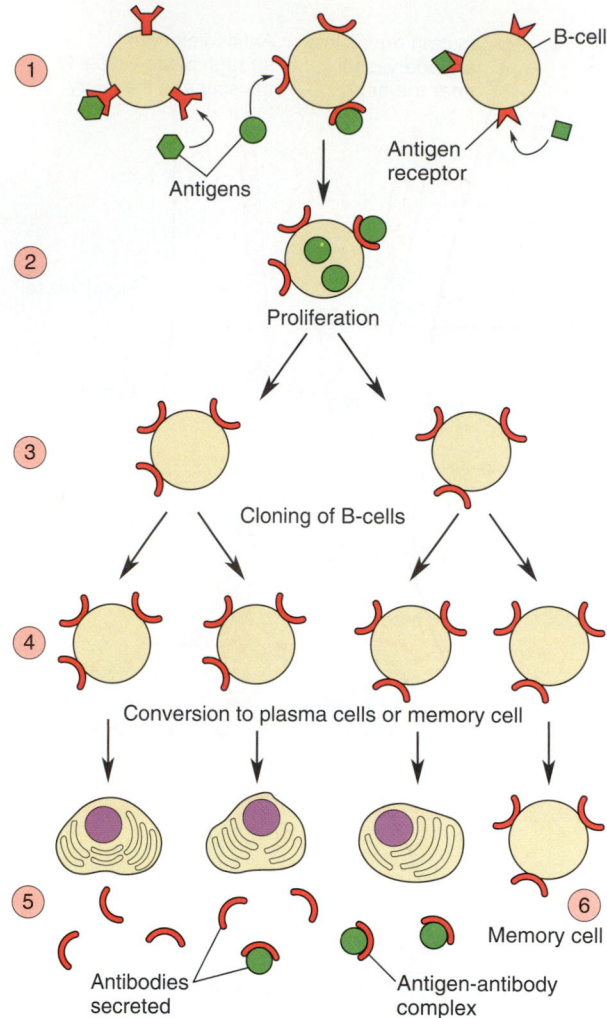

FIGURE 25.5 The Sequence of Events in B-Cells That Result in Antibody Production and the Production of Memory Cells

Steps 1–5 lead to the production of antibodies by the plasma cells. First, of the millions of different B-cells with their individual recognition sites, a specific antigen binds with the B lymphocyte and that lymphocyte is selected to be activated. The cells of that type proliferate, resulting in clones of identical B-cells. These B-cells differentiate into plasma cells that have a well-developed endoplasmic reticulum, which synthesizes the specific antibody — thousands and thousands per minute. Step 6, the memory cell retains an immunologic memory and can respond almost immediately to a second exposure of the antigen months or even years later.

Fabricius, a structure found in birds. In mammals, B-cells are formed in the bone marrow, after which they migrate to adjacent lymphoid tissue: lymph glands, spleen, or gut-associated lymph tissue. Like T-cells, B-cells are precommitted to do a specific

job, but so far scientists do not completely understand how or where they acquire their precommitment to do battle with a specific foreign substance (though there is a lot of evidence that this occurs in the bone marrow).

When B-cells are stimulated by the presence of an antigen (remember that an antigen is an *anti*body-*gen*erating substance), they differentiate into plasma cells. Plasma cells are capable of secreting **antibodies,** Y-shaped protein molecules that circulate in the plasma. Like enzymes and T-cells, antibodies are highly specific; they will complex only with a specific substance. Antibodies are produced after the body has detected the presence of a specific foreign antigen. When an antigen is detected, plasma cells become antibody factories, able to produce about 2000 molecules of a specific antibody per second per cell.

The antibodies then circulate within the blood plasma, reacting with a specific antigen (Figure 25.5). Once an antibody binds with an antigen, it is easier for a macrophage to engulf the antigen or for complement to aid in the destruction of the antigen. There are five major classes of antibodies, each having two or more combining sites that can bind to an antigen. The best understood is *immunoglobulin G (IgG)*, the structure of which is illustrated in Figure 25.6a. When very young infants come down with measles, a common procedure is to give them an injection of gamma globulin, which increases the supply of antibodies in their system and helps fight off the disease.

Scientists have determined that the Y-shaped antibodies have two identical light protein chains and two identical heavy protein chains joined together by disulfide bonds. Both chains form parts of the antibody that are known as the *constant* and *variable* regions. For example, IgG contains two sites that combine with the specific antigens. This combining site is formed by the variable regions of the light and heavy chain located near the tip of the Y-shaped antibody. This antigen recognition and combination is very specific. Antibody combining sites can recognize molecules that differ in only the arrangement of a few atoms — the antibodies are that specific (Figure 25.6b).

One major question about antibodies has been how the immune system can make such a wide array of antibodies. In the late 1970s, Susumu Tonegawa (see Profile) determined that bunches of genes on chromosome six coded for the variable regions, and that these genes shuffle around and rearrange themselves to form different genetic sequences in the B-cells. This eventually leads to the wide diversity of antibodies (Figure 25.6c).

FIGURE 25.6 **Antibody Structure and Synthesis** ▶
(a) Computer-generated image of an IgG antibody molecule and (b) an illustration of the antibody molecule. As you can see, the Y-shaped antibody consists of two heavy and two light chains joined by disulfide bonds. It is at the tip of each that the variable regions exist. The assortment and rearrangement of the genes expressed determine the structure of the antigen binding sites and the particular antigen that will bind with the antigen binding site. (c) The rearrangement of gene sequences can result in millions of variable site combinations, producing millions of different kinds of antibodies from only a few genes.

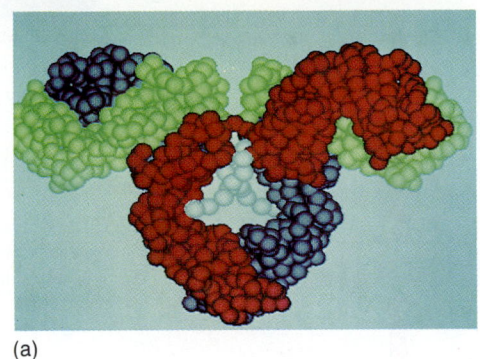

(a)

Two Phases of the Immune Response

The immune response can be described as having two stages.

The Primary Response of the Immune System. The **primary response** encompasses the system's shift from the nonspecific defense of inflammation to T- and B-cell formation. It takes about 10 hours for B-cells to develop into plasma cells, and the process continues for about 4 or 5 days. After this time, usually enough antibodies are present within the system to destroy the infectious agent or organism.

The Secondary Response of the Immune System. The **secondary response** comprises the actions of long-term immunity that result from vaccination programs or accidental exposure to an antigen. The B-cells that performed during the primary response phase of the immune system do not all die off once their work is finished. Some remain within the system for an indefinite time. In a sense, their particular cell surface antibodies are a form of molecular "memory" of infectious agents.

These cells are called, appropriately enough, *memory cells,* and are responsible for the long-term immunity that makes us resistant to subsequent infections for almost a lifetime. If we are exposed to mumps, for instance, a year or so after we have the disease, even though we are exposed to the same type of infectious agents, our B *memory cells* become activated (see Figure 25.5). They bind with the mumps antigen, activating the rapid cloning of plasma cells. The plasma cells produce antibodies to destroy the infectious agents. The second exposure also stimulates the memory lymphocytes to clone killer T-cells required to destroy the invader. When stimulated by a specific antigen, sensitized T- and B-cells divide rapidly, producing offspring that are clones, all having the same membrane surface receptors.

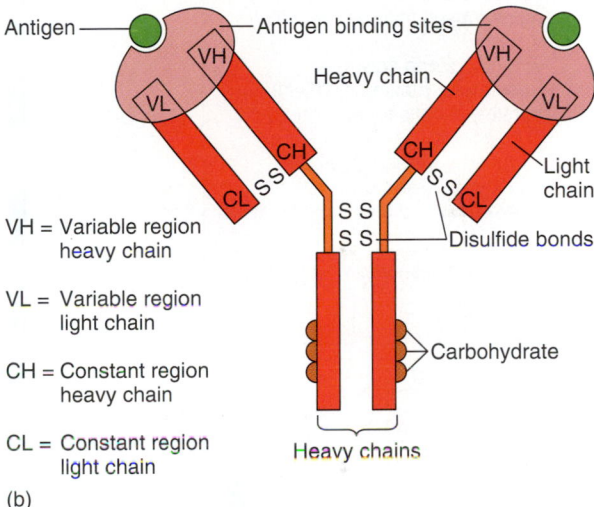

VH = Variable region heavy chain

VL = Variable region light chain

CH = Constant region heavy chain

CL = Constant region light chain

(b)

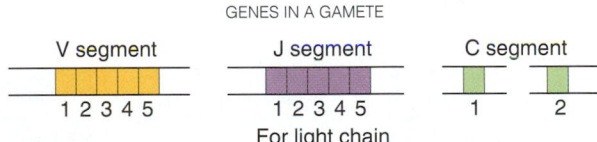

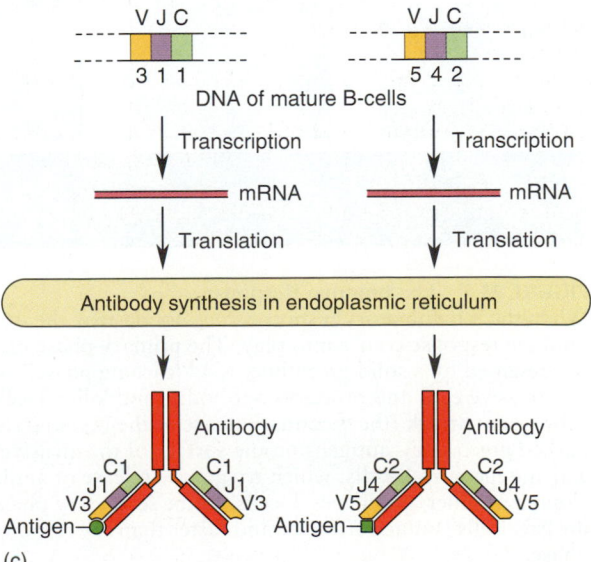

(c)

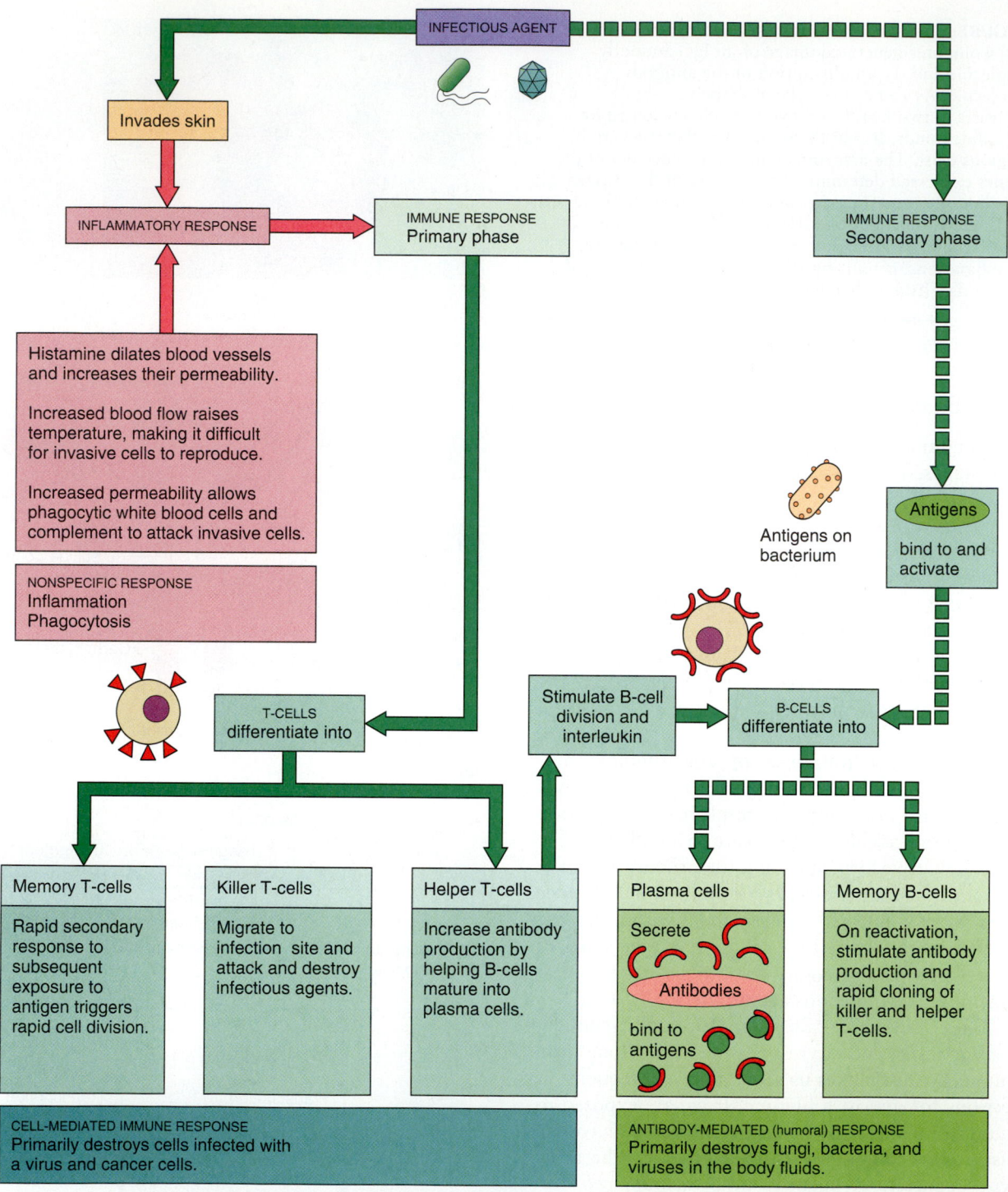

FIGURE 25.7   The Immune Response
When the inflammatory response does not destroy the infectious agent, the
immune response comes into play. The primary phase of the response
(represented by a solid green line) is a "learning phase" — the body identifies
the invasive cells and produces antibodies and killer T-cells to destroy it. In a
subsequent attack (the secondary phase of the immune response shown by a
dashed green line), antigens on the surface of the invasive cells bind to and
activate memory B-cells, which trigger the release of antibodies and the
cloning of killer and helper T-cells. In the secondary phase, the reaction to the
invasive cells is much stronger and faster than the response in the initial
phase.

An important element in the secondary response of the immune system is that it amplifies the immune response. The body produces more antibodies because sensitization already occurred during the primary response. In other words, the body already "knows" the identity of the antigen, and antibodies are ready and waiting to destroy the specific infectious invader. Figure 25.7 summarizes the immune response.

One way to differentiate between T- and B-cells is to remember that T-cells are responsible for the *cell-mediated response* and B-cells are associated with what is called the *humoral response* or *antibody response*. The humoral response involves the production of free IgG antibodies, which are found in blood plasma. (The term *humoral* comes from the medieval concept of the four humors that were believed to rule human nature.)

The Immune Response and Transplant Operations

■ **When a patient undergoes a transplant operation, the immune system recognizes the new organ as a foreign element in the body and attacks the organ. This response is called rejection. Drug therapy is designed to turn off the immune response in such cases.**

T-cells can directly attack, destroy, and kill foreign cells. Organ transplants are foreign cells, for the organs have come from the body of an individual other than the patient. As a result, T-cells can also destroy the cells of the transplant. This phenomenon is called *rejection,* and a tremendous research effort is underway to find medical means that will paralyze and/or destroy T-cells that prevent successful organ transplants. Cyclosporin is one of the drugs that helps suppress tissue rejection, as do x-rays, which destroy killer T-cells.

Not only do drugs like cyclosporin suppress the immune response, but the body must also have its own mechanism for suppressing T-cells. Various subpopulations of T-cells regulate turning the immune system on and off. Helper T-cells are in part responsible for turning on the immune response, and for a while many investigators thought that there were *suppressor T-cells* to help turn off the lymphocytic reactions that initiate the immune response. However, suppresssor T-cells have never been found, and so now scientists speculate that it is the destruction of helper T-cells that turns off the immune response. It is known that in part the virus responsible for acquired immune deficiency syndrome, or AIDS, is somehow involved in inhibiting the immune response (see Focus 25.1 and Figure 25.8).

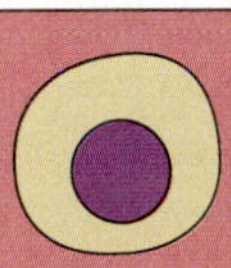

Functions of helper T-cells
NORMAL HELPER T-CELL
1. Stimulation of B-cells to differentiate into plasma cells and memory cells.
2. Increase the rate of production of macrophages from stem cells.
3. Stimulate production of killer T-cells.
4. Clone memory T-cells.

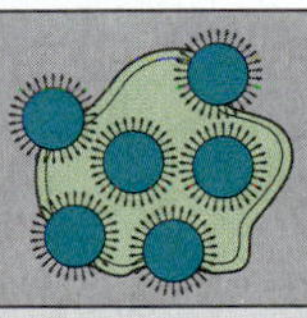

Actions of the virus	Symptoms of AIDS
AIDS VIRUSES DESTROY HELPER T-CELLS	
1. B-cells are not converted into plasma and memory cells.	1. Decreased number of plasma cells produces less antibodies to combat infections.
2. Decreased production of lymphocytes and macrophages.	2. Lessened immune response.
3. The number of killer T-cells reduced greatly.	3. Cancer and rare type of pneumonia develops.
4. No memory T-cell clones produced.	4. Lowered immune resistance to secondary infections.

FIGURE 25.8   The Mechanism of Immune Deficiency in AIDS
Because the AIDS virus destroys the helper T-cells, the normal functions of helper T-cells are inhibited. This results in the symptoms of the disease we call AIDS.

FOCUS 25.1
AIDS

It's called acquired immune deficiency syndrome (AIDS). As the term indicates, the deficiency in the immune system is acquired, not inherited. Because the deficiency exists, a group of rare diseases can take advantage of the infected individual's weakened defenses. The diseases that occur are common enough to constitute what is called a syndrome (a group of symptoms and disorders commonly accompanying a condition). Patients do not die of AIDS; instead they die of one of the illnesses that are part of the syndrome. People with AIDS die because their bodies cannot fight off disease.

AIDS has become probably the most feared disease in this country, primarily because no cure has been found. Although it is impossible to overrate the danger that threatens an individual who has contracted AIDS, scientists continue to stress that the disease is difficult to acquire unless one comes into sexual contact or has contact with the blood of someone from a high-risk group. According to the Centers for Disease Control in Atlanta, Georgia, of all those exposed to the disease, only 25 to 35 percent will develop AIDS.

The virus that causes AIDS is not transmitted through the air or through the phlegm of a cough or a sneeze. Rather, it is transmitted through bodily fluids, mainly blood, semen, and vaginal secretions. What is more, the virus must come in contact with the blood before an individual can be infected, because the receptors on the virus's cell membrane are only compatible with those on certain blood cells. That means that simply breathing in the virus does not endanger any individual. Finally, the individual's immune system must be weakened enough by other health factors to be susceptible to the disease for it to take hold.

At present, the major populations at risk in the United States are male homosexuals, intravenous drug abusers and their sexual partners, and hemophiliacs. However, the number of heterosexual individuals diagnosed as having the disease is on the rise, and there is evidence that the trend will continue. In central Africa, almost 50 percent of the victims are heterosexual females. In the central African countries of Zaire and Zambia in particular, AIDS is rampant among middle-class heterosexuals. As reported by the Centers for Disease Control, from 1981 to February 1988, 53,814 AIDS cases were diagnosed and 30,158 victims died of this incurable disease in the United States. Through October of 1988 there were 29,000 cases of AIDS reported and 13,000 deaths for that year, a definite increase.

Patients with the disease have normal amounts of antibodies and antibody-producing cells. However, the white blood cells known as helper T-cells, which assist the antibody-producing B-cells, are low in number. This low number of T-cells turns off, or *inhibits*, the immune system by immobilizing killer T-cells, resulting in an overall inability of the immune system to function — an immune deficiency.

A blood test can reveal if a person has antibodies to the HIV virus responsible for AIDS. People who test positive fall into two groups: those who have a full-blown case of AIDS and those who have ARC, or AIDS-related complex. ARC patients have symptoms like oral thrush (throat infections) and mild viral problems that cause diarrhea and other infections. Since they do have the HIV virus, their T-cell ratios are abnormal, and so they are more prone to all infections. Although ARC patients do have a difficult set of symptoms, they will not necessarily develop a full-blown case of AIDS, as the Atlanta Centers for Disease Control statistics show.

How does AIDS progress? At first, an individual with AIDS feels as if he or she has a severe hangover or the flu. However, the feeling lasts for from 6 to 18 months, during which time the immune system is progressively weakened. As a result, the individual becomes vulnerable to a variety of diseases, such as a rare form of cancer called Kaposi's sarcoma. Kaposi's begins with small purplish spots on the feet and legs; eventually the spots spread over the skin, and the cancer spreads to the internal organs. Some develop an unusual form of pneumonia caused by a protozoan. Patients also can suffer from one or more infectious diseases caused by viruses, bacteria, protozoa, or fungi.

In 1984, two groups of researchers, one in France headed by Luc Montagnier, and the other at the National Institutes of Health in the United States, headed by Frank Gallo reported that they had isolated viruses responsible for AIDS. One, called HTLV III, for human T-cell leukemia/lymphoma virus number three, was isolated by Dr. Frank Gallo and his research team. Antibodies to HTLV-III virus were found in 90 percent of all AIDS victims and in 80 percent of ARC patients (those with the symptoms that precede the disease). A research team in France isolated a similar virus from the lymph nodes of an ARC patient. The French called this virus the LAV virus, for lymphadenopathy-associated virus. Although the viruses have different names, they are identical. Today both viruses are called the HIV virus, and Gallo has stated that the French team has priority in finding the HIV virus (Figure A).

The identification of the virus responsible for AIDS provided a glimmer of hope that a vaccine may be found against the disease. That hope became more intense when researchers identified the protein receptors located on the helper T-cells to which the AIDS virus attaches. When antibodies are synthesized against the receptor proteins, the antibodies complex with the receptor protein on the helper T-cells. With the antibodies blocking the attachment site, the AIDS virus cannot take hold of the cell, and hence infection is prevented. Recently, a research group began testing a new drug that appears to inhibit the progress of the disease by preventing opportunistic infections from developing. (An opportunistic infectious organism is one that takes advantage of the depressed immune system's inability to fend off disease that under normal circumstances would pose no danger.) The testing of the drug is being expanded, but

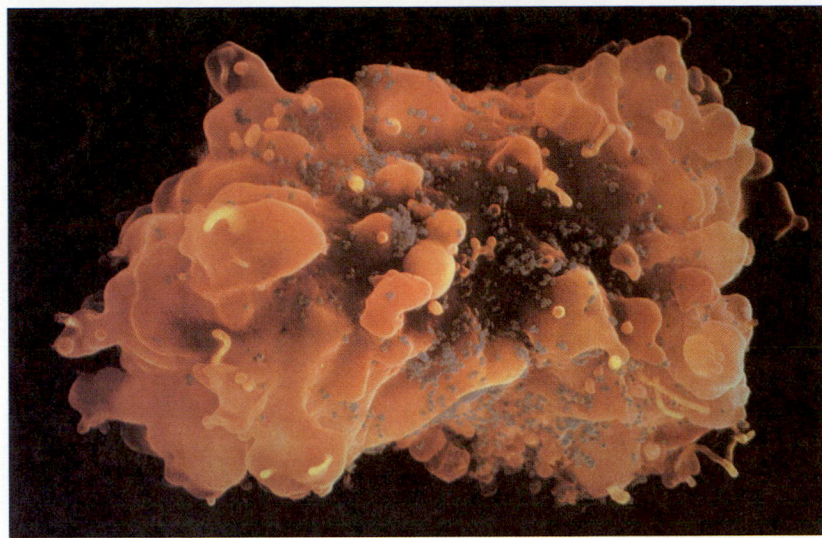

FIGURE A
A scanning electron microphotograph of a T-cell infected with the HIV virus.

its developers stress that it is still experimental and that it does not prevent the transmission of the virus from one individual to another.

Since 1986, the experimental antiviral drug azidothymide (AZT) has been made more widely available for AIDS patients as a result of successful experiments in which AIDS patients received either AZT or a placebo. (It is a known fact that when a group of patients are given medication, even if that medication is actually a "sugar pill" that contains no medicine, the condition of some of those patients will improve. Therefore, in order to measure the effectiveness of any drug being tested, a control group of patients receive a *placebo*, or sugar pill, in order to determine the true effectiveness of the medication under investigation.) In the course of the AZT tests, researchers found that those patients receiving the AZT fared much better clinically than those receiving the placebo. The results were encouraging enough that the United States Senate appropriated $47 million to enable 10,000 AIDS patients to participate in drug trials during fiscal year 1987.

Is AZT the miracle drug that will "cure" AIDS? Not exactly. Biolo-

gists know that AZT stops the growth of the HIV virus by interrupting the metabolic pathway that the virus needs to replicate itself. HIV relies on the enzyme *reverse transcriptase* to produce its genetic material. AZT appears to inhibit this enzyme, thus inhibiting the virus's ability to reproduce itself. The important point to be made about this treatment is that it is just that — a treatment. AZT does not destroy the AIDS virus already existing within the patient's bloodstream. Only continued research will produce a way of destroying the virus entirely.

Scientists know much more about AIDS now than they did 10 years ago and vaccines are being tested. However, even though there is no cure, it is known that monogamous individuals in long-term relationships (assuming they are not intravenous drug users) are at low risk. Also, the use of condoms reduces the risk.

But there is still much more to be learned before AIDS is conquered. For this disease, the answers are not ones that scientists will come by easily.

Susumu Tonegawa

Immunologist/Molecular Biologist, Nobel Prize Laureate

It was 6:30 A.M. on Monday, October 12, 1987. Susumu Tonegawa was asleep at his Newton, Massachusetts, home when the telephone rang. He got up and answered it. A Japanese journalist was calling to ask him about winning the Nobel Prize in Physiology and Medicine. Tonegawa was in a state of disbelief. He says, "I thought it must have been a mistake. Journalists are known to make mistakes." The phone rang a second time. This time it was his father calling from Japan to congratulate him. Tonegawa knew then that indeed he had won the Nobel Prize, and soon he was giving a press conference from his research laboratory at the Massachusetts Institute of Technology (MIT).

Winning the Nobel Prize may have been a surprise for Tonegawa, but it came as no surprise to his colleagues. For he is known as a brilliant, hard-working scientist truly dedicated to the pursuit of exploring the complex workings of the immune system. Tonegawa won the Nobel Prize for discovering how only a limited number of genes can code for such a wide variety of antibodies.

He began working on the problem of how genes regulate antibody synthesis in the mid-1970s at the Basel Institute for Immunology in Switzerland. At that time, he demonstrated that much of the antibody diversity was caused by assembling antibody coding genes from three or four separate segments of DNA — sort of like rearranging the letters of the alphabet to form new and different words.

As we mentioned in our discussion of antibodies in this chapter, antibodies consist of two pairs of light and heavy protein chains that form a Y-shaped molecule. The regions of the heavy and light chains at the tips of the "Y" are the variable regions. They form the recognition sites that bind to the specific antigens. These tips or outer arms show significant differences in their amino acid sequences. For antibody molecules, the remainder of the molecule is called the constant region; it is made up of only heavy chains.

The gene in the B-cells must dictate how these cells produce antibodies. Many researchers thought that a separate gene in the B-cells caused each protein chain to create an antibody, but this would be impossible, because the number of genes would be too great; the genome would be too large. Between 1976 and 1978, Tonegawa published research papers that demonstrated that bunches of genes code for the variable arms; the recognition sites shuffle around and rearrange themselves to form different genetic sequences in the B-cells, eventually leading to the wide diversity of antibodies.

Tonegawa's human endeavor has resulted in our understanding of how approximately 1000 genes recombine to form the 10 million to 1

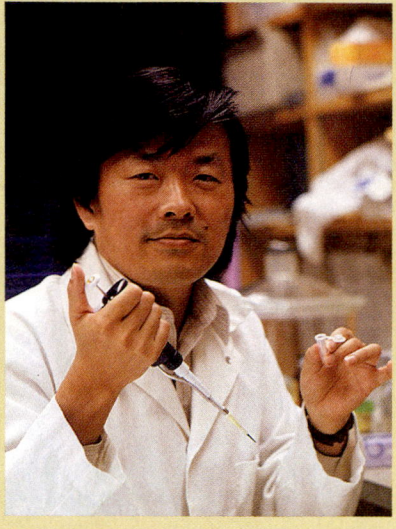

billion antibody genes for the millions of antigens our bodies may be exposed to. These findings have shed important light on how the immune system works and have opened the doors to numerous possibilities for regulating antibody production. This will help increase the efficacy of vaccines, decrease or inhibit the rejection of transplanted organs, and mitigate the severity of autoimmune disorders.

Tonegawa spent 10 years in Basel, then moved to MIT in 1981, where he is a professor of biology. His work has not only been recognized by the Nobel committee. In 1987, for example, he won the prestigious Lasker Award for basic medical research, which he shared with several other biologists.

Is it possible that biologists could find even more efficient means of preventing organ rejection if they could determine how T-cells learn to distinguish "self" from "nonself"? The answer is not clear. Investigators do know, however, that the DNA of nucleated mammalian cells contains regions of genes known as the *major histocompatibility complex (MHC)*. MHCs control the synthesis of molecules, known as *surface recognition factors,* on the surfaces of cells. In a sense, MHCs tag or label a cell as "self." The T-cells "learn" how to read this molecular "name" on the cell surface during their development in the thymus. If they recognize a cell as self, they leave it alone. But if a cell does not have the correct "molecular password," it is recognized as nonself and is destroyed.

Monoclonal Antibodies

■ **Monoclonal antibodies are produced by a specifically programmed white cell to respond to a specific antigen. These antibodies can now be produced by recombinant DNA techniques and may play a particular role in fighting diseases.**

The antibodies produced by B-cells are Y-shaped structures that precisely fit with a specific antigen. However, in response to the presence of a given antigen, the body may produce several different types of antibodies, only a few of which will bind to a specific region of the antigen. In essence, the immune system's response is almost a matter of overkill. Today, however, researchers have developed techniques to produce larger amounts of very specific antibodies, called **monoclonal antibodies,** that bind only to a specific portion of any particular antigen.

Monoclonal antibodies are produced by a very specific white blood cell, and all are clones of a single cell. Because they target in only on a specific antigen, they can be thought of as "magic bullets." Researchers have been growing monoclonal antibodies to fight cancerous tumors and have high hopes for their effectiveness. How does one grow antibodies?

To produce monoclonal antibodies, researchers fuse two cellular components from mice: (1) cancerous lymphocytes, and (2) an antibody producing B-cells from the mouse's spleen, which has been injected and thus "immunized" with the specific antigen of interest. The hybrid cells, called *hybridomas,* inherit abilities from both parents. The B-cells pass on the capacity to make antibodies, and the cancerous cells pass on the capacity to carry on almost endless cell division. The individual hybrid-

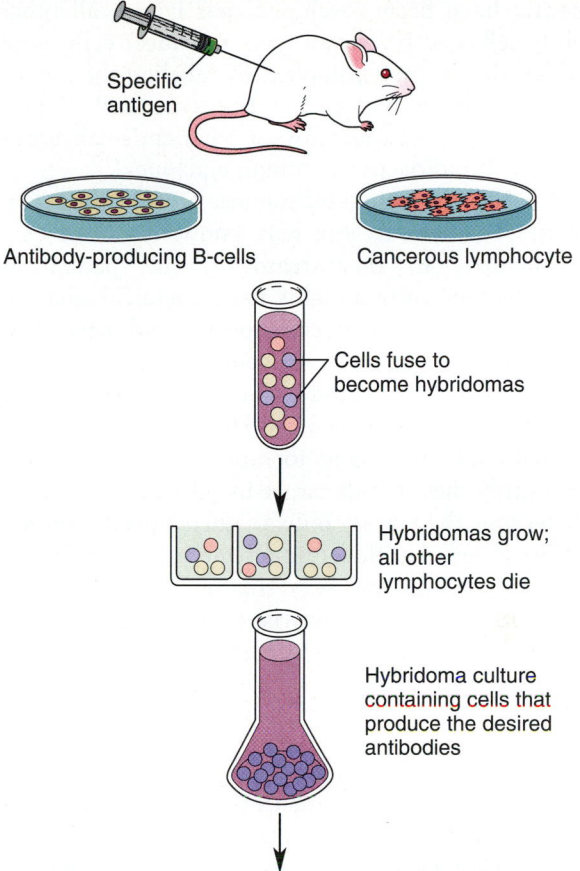

Specific antigen

Antibody-producing B-cells   Cancerous lymphocyte

Cells fuse to become hybridomas

Hybridomas grow; all other lymphocytes die

Hybridoma culture containing cells that produce the desired antibodies

Antibody-producing clones can be grown in unlimited quantity either in culture or in mouse tumor

FIGURE 25.9 Production of Monoclonal Antibodies
In the production of monoclonal antibodies, cancerous lymphocytes and antibody-producing B-cells, which have been injected with a specific antigen, are fused. These fused cells, called hybridomas, provide an almost endless source of the desired antibodies.

omas are grown and then tested for the desired antibody. Once researchers have isolated those hybridomas that produce the desired antibody, they culture the cells and ultimately freeze them, or they inject the hybridomas into a mouse. There they produce large amounts of antibodies in the body fluids (Figure 25.9).

With the monoclonal antibodies scientists can concentrate the immune system's attack on one specific target, instead of depending on the random, all-inclusive method that the system ordinarily employs. In fact, using monoclonal antibodies is almost like using a sharpshooter target pistol, rather than a shotgun, to hit a target.

The more specific the antibody, the greater use it would have in *immune therapy,* a method used to stimulate a patient's own immune response to combat disease. For example, today monoclonal antibodies have been developed that ignore all other body cells except certain types of cancer cells, with which they form a complex. By labeling the monoclonal antibody with radioactive tracers, physicians can find the exact location of a patient's cancerous cells with radioactive scanning machines.

One of the major techniques used in the treatment of cancer, as you may know, is cancer chemotherapy. In this treatment, the patient is administered certain highly toxic chemicals that do not discriminate between normal and cancerous cells. As a result, there are some severe side effects to chemotherapy, including hair loss, severe nausea, and general loss of energy. However, if chemical toxins could be bound to monoclonal antibodies, the antibodies would target in on cancerous cells, delivering the toxins only to their specific target. Normal cells would be unaffected. Targeted drug therapy is only one way that medical science is using the special characteristics of monoclonal antibodies. The ways in which they can be used continue to grow.

Interferon and Interleukin-2

■ **Interferon is the substance produced by the body to fight off viruses by interfering with the virus's replication process. Interleukin-2 is a protein that helps to actuate the immune system. Interferon and interleukin-2 can now be produced through recombinant DNA techniques, and their medical usefulness is now being investigated.**

Before we leave the subject of the immune system, we need to touch on two of the immune system's major mechanisms for defending the body against viruses.

Interferon. When a vertebrate cell like a human cell is attacked by a virus, the immune system responds by producing a chemical substance called **interferon.** This chemical interferes with the infection phenomena in such a way that viruses cannot infect the cell. Biologists know that interferon inhibits viral DNA-RNA transcription, thus counteracting the virus's ability to infect and reproduce within the host cell.

Interferon can also be carried to other cells by means of the circulatory system, making them resistant to viral infections, too. Interferon does not target in on a specific type of virus. Instead it inhibits the attack of any virus. Its use in combating viral infection, on the other hand, is still being researched. It is particular to a specific species. Human interferon can be used only in humans, mouse interferon in mice, and so on.

There are three types of human interferon: alpha, beta, and gamma. There are subcategories of each type; for example, there are 12 types of alpha interferon alone. By using the recombinant DNA technology discussed in Chapter 13, all types of interferon can be synthesized in large amounts for use in combating certain diseases. For example, alpha interferon helps fight certain cancers and viral infections. Gamma interferon can be used to treat cancer and the autoimmune disease rheumatoid arthritis.

The uses of interferon in fighting viral infections are still being researched. For a while, scientists thought that interferon would also provide an effective treatment for cancer, because their research showed that interferon might reduce the size of some cancerous tumors, but it was not a panacea or a cure-all, since the amount of tumor reduction was questionable. However, recent research demonstrates that some of the earlier disappointments with interferon as a cancer-fighting drug may have been premature. For example, alpha interferon has been demonstrated to be very effective against a particular type of leukemia, called hairy cell leukemia (because of the physical appearance of the cells). In addition, it has proved to be effective in treating certain types of bone marrow cancer, kidney cancer, and Kaposi's sarcoma, often found in people with AIDS.

Interleukin-2. Even though some investigators have been disappointed with the effectiveness of interferon in combating cancer, there is still room for cautious optimism. **Interleukin-2,** a naturally occurring protein that functions as an immune-system activator, and another naturally occurring protein

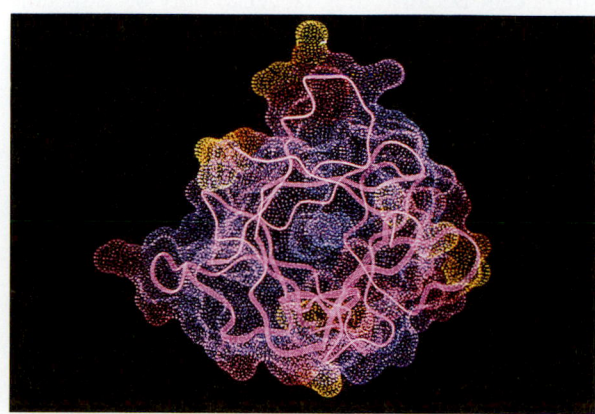

FIGURE 25.10 Interleukin-2
A computer-generated image of Interleukin-2.

Chapter 25   The Immune System

Phagocytosis

Since the late 1800s, scientists have tried to understand how certain cells involved in the immune response identify, engulf, and eliminate foreign invaders. The role of phagocytosis in the cellular immune response is the subject of this Human Endeavors box.

Source: Adapted from "Evolutions: Phagocytosis," *The Journal of NIH Research*, vol. 2, pp. 93–94, October 1990. Adapted by Elizabeth Morales-Denney from original artwork by Terese Winslow.

1880s–1920s: Metchnikoff studied phagocytosis in transparent organisms such as *Daphnia*. He also classified human phagocytic cells as either "macrophage" or "microphage." He thought the irregular nuclei of microphages helped them squeeze through capillary walls.

1920s–1950s: Immunologists focused on antibodies and made little progress in understanding cellular defenses against infection. For example, in 1933, Baldridge and Gerard discovered that phagocytizing cells increase their oxygen consumption two to four times, but it was not until the 1960s that this "respiratory burst" was related to oxidative killing of bacteria.

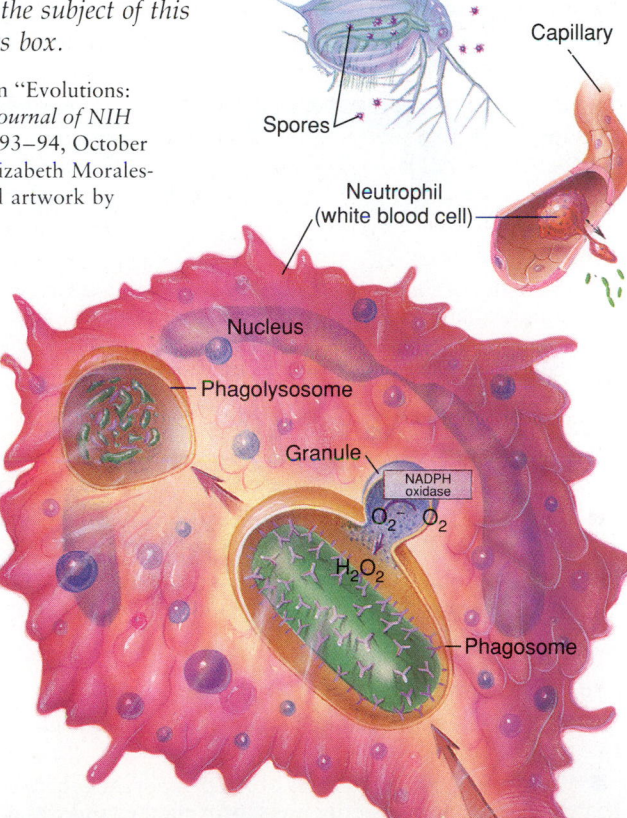

Daphnia

Spores

Capillary

Neutrophil (white blood cell)

Nucleus

Phagolysosome

Granule

NADPH oxidase

O_2 O_2

H_2O_2

Phagosome

Lymphocyte

Antibody

Receptor

Phagocytic cell plasma membrane

Phagocytic vesicle

1960s: With a revived interest in cellular processes, scientists rapidly discovered more about phagocytosis. Hirsch and Cohn isolated granules from neutrophils and found various hydrolytic enzymes and antibacterial substances in them. Berken and Benacerraf found antibody receptors on macrophage—leading to the understanding that macrophage recognize bacteria that have been coated with antibody (opsonized). Senda and colleagues isolated actin and myosin from neutrophils, which provided an understanding of how phagocyte pseudopodia surround their targets.

1970s: Griffin and colleagues showed that, for uptake to occur, receptors on phagocytes must bind sequentially to ligands—such as complement or antibody—over the entire surface of the target cell. In what they call a "zippering" mechanism, the plasma membrane continues to extend around the target cell only as long as receptor-ligand complexes continue to form.

1980s: Scientists have now isolated and cloned several forms of receptors for the constant fragment (Fc) of antibody molecules and are studying the signaling mechanisms that stimulate phagocytic action. Other scientists are studying the oxidative and nonoxidative mechanisms that phagocytes use to kill bacteria. Several diseases that result from deficient phagocytic processes have been identified.

Antibodies

Fc receptor in neutrophil cell membrane

Opsonized bacterium

called *tumor necrosis factor* (*TNF*) show promise as possible cancer treatments. These substances have one very special advantage. They are both proteins, which means that they can be manufactured in large amounts through the genetic engineering techniques we discussed in Chapter 13. Because biologists have access to a large supply of the proteins, they can use them in cancer research with human subjects (Figure 25.10).

As you can see, an understanding of the immune system may have long-range applications in fighting cancer and other problems. For that reason, research on its workings continues. The research will tell us more about the involvement of the immune response with a variety of very serious disorders.

When the Immune System Goes Wrong

■ Autoimmune diseases are those in which, for reasons not yet known, the immune system attacks its own body rather than an intruder. Acquired immune deficiency syndrome (AIDS) is a virally caused disorder that prevents the immune system from functioning.

As we have seen, when the body's immune cells are developing, they learn to distinguish self from nonself. In a sense, they are educated to tolerate and not react against self. But this process of education is not infallible, and for reasons that are not yet known, the immune system sometimes turns against the self, giving rise to a group of diseases called

---

TABLE 25.2
A Decade of AIDS Milestones

1981
■ Five cases of *Pneumocystis carinii* pneumonia (PCP) among young homosexual men in Los Angeles and 26 cases of Kaposi's sarcoma among young homosexual men in New York and California are reported to the Centers for Disease Control.

1982
■ Collection of clinical conditions recognized as new disease known as acquired immune deficiency syndrome (AIDS).
■ Case definition of AIDS for national reporting is first published in the CDC's *Morbidity and Mortality Weekly Report.*
■ First cases of AIDS reported among patients with hemophilia who had been treated with clotting factor concentrates and among people who had received blood transfusions.

1983
■ U.S. Public Health Service and other groups issue guidelines for the prevention of AIDS.
■ First well-documented cases of AIDS reported in heterosexual partners of male intravenous drug users, proving the virus can be transmitted from men to women as well as from men to men.

1984
■ A retrovirus, at first called HTLV-III or LAV, is firmly identified as the cause of AIDS; it is later renamed the human immunodeficiency virus, or HIV.

1985
■ Laboratory tests to detect presence of antibodies to the AIDS virus widely available.
■ First patients given AZT (zidovudine) in clinical trials.

Source: American Medical News, Jan. 5, 1990, p. 24. Reprinted with permission of the American Medical Association.

autoimmune diseases. Most of the autoimmune diseases are well known. Among them are arthritis, myasthenia gravis, and systemic lupus erythematosus (SLE). In an autoimmune disease, an individual's immune system no longer tolerates self. Instead, the immune response is aimed inward and attacks the individual's own tissues as if they were foreign. In certain types of arthritis, the immune system attacks joints; in other autoimmune diseases, the immune system attacks mucous membranes, muscles, and other tissues. Treatment of autoimmune disorders covers a wide range of methods, ranging from alleviating pain to suppressing the activities of the immune system with powerful drugs. As yet, biologists do not know why the immune system begins to attack the cells of its own body.

One other immune system disorder has taken on what many people feel are quite frightening proportions: acquired immune deficiency syndrome, or AIDS. AIDS is caused by a specific virus, which interferes with those mechanisms employed by the immune system to signal that the body is under attack. The AIDS virus infects the helper T-cells that are important in generating an immune response. As a result, an individual with AIDS cannot fend off infections that in a normal person would rarely cause any illness at all. Focus 25.1 discusses AIDS in detail, and Table 25.2 lists a decade of AIDS milestones.

1986
- Clinical studies at 12 medical centers demonstrate that AZT can improve survival and quality of life for people with severe HIV infection.

1987
- Food and Drug Administration approves AZT as a prescription drug for severe HIV infection.
- President Reagan orders all immigrants and federal prisoners to be tested for HIV.
- A federal advisory panel recommends that wider testing for HIV be done on a voluntary, not mandatory, basis.
- Clinical trials of a possible vaccine against HIV begin in United States.

1988
- Community-based clinical trials in San Francisco prove that aerosol pentamidine can prevent the onset of PCP.

1989
- FDA approves licensure of aerosol pentamidine for prophylaxis against PCP.
- In Phase I clinical trials, people taking the antiviral drug DDI show decrease in p24 antigen and rise in helper T-cells.
- Bristol-Myers Co. begins wide, free distribution of DDI in new "parallel track" distribution program as the government launches phase II clinical trials.
- Government announces that AZT appears to be effective in delaying onset of AIDS in adults with asymptomatic HIV infections, slows the progression of HIV disease in adults with early AIDS-related complex (ARC), and prolongs survival and relieves symptoms in children with AIDS.
- More than 100,000 people diagnosed with AIDS in the United States.

1991
- Some 270,000 U.S. citizens are predicted to have been diagnosed with AIDS.

SUMMARY

1. The immune system consists of those components of the body that protect us from disease and allow us to recognize self.

2. In general, the immune response can be thought of as nonspecific or specific. Nonspecific responses involve physical barriers such as the skin, as well as processes like phagocytosis and inflammation. These are very general. As the term suggests, specific responses are specific to particular types of invaders. Specific responses can be either cellular or can involve the production of antibodies.

3. The skin and mucous membranes provide a physical barrier to bacteria and dirt, and they constitute the first line of defense against infection and disease. Another type of defense involves the inflammatory response, a localized response to tissue damage. The initial attack prompts a release of histamine, which in turn increases blood supply to the affected area and increases the permeability of the capillaries.

4. Another line of defense against infection and disease is the specific immune response, in which white blood cells engulf invading bacteria or produce antibodies to help destroy the invaders or their products. In the primary response, the white cells learn the identity of the invading antigen and destroy it. The secondary response is activated by subsequent exposure to the same antigen.

5. The specificity of the immune response is the result of specialized white cells called T-cells and B-cells. Killer T-cells destroy the invading antigen. Helper T-cells help to convert B-cells into plasma cells. B-cells produce antibodies that aid in the destruction of the antigen.

6. Antibodies are Y-shaped proteins consisting of two light protein chains and two heavy ones. At the tips of the "arms" of the antibody are variable regions that allow for the millions of possible antigen binding sites. The stem portion of the molecule consists of the constant region, which maintains a constant structure.

7. The reshuffling or recombination of a few genes during the development of the B-cells produces the thousands and thousands of genetic combinations that result in the specific antigen binding site of an antibody. The B-cell then converts into an antibody-secreting plasma cell.

8. Some B-cells develop into plasma cells that possess a "molecular memory" of the antigen. These memory cells are activated on subsequent exposure to the antigen. The rapidity of the process accelerates and enhances the immune response to a specific antigen.

9. Because the body will reject most foreign substances, including foreign organs, a great deal of research is devoted to understanding the immune response. Such an understanding will enable physicians to control the immune response so that transplanted organs will not be rejected.

KEY TERMS

nonspecific immune
  response
specific immune response
lysozyme
macrophage
inflammatory response
histamine
complement
alternative pathway
classical pathway

T-cell
B-cell
antibody
primary response
secondary response
monoclonal antibody
interferon
interleukin-2
autoimmune disease

REVIEW QUESTIONS

True/False

1. The immune system is involved in protecting the body from disease and the recognition of self.
   true   *false*   *(page 595)*

2. B-cells are produced in the thymus.
   true   *false*   *(pages 599–600)*

3. When they are exposed to a specific antigen, B-cells are converted into plasma cells that secrete antibodies.
   true   *false*   *(page 600)*

4. The symptoms of AIDS are the result of the HIV virus destroying helper T-cells.
   true   *false*   *(page 603)*

Fill in the Blank

1. During the inflammatory response a substance called _____ increases blood flow to the site of injury or infection. *(page 597)*

2. The rearrangement of the genes that code for the _____ region of the antibody is what makes such a large number of antibodies possible. *(page 600)*

3. The substance that stimulates the white blood cell to produce antibodies is called a(n) _____. *(page 596)*

4. When stimulated by an antigen, B-cells can be converted into _____ cells that produce antibodies. *(page 600)*

Multiple Choice

1. In this chapter, the term *complement* means:
   a. Proteins that can enhance the immune response

b. A form of flattery
c. Produced by red blood cells
d. Produced by the thymus
   (page 597)

2. Monoclonal antibodies are:
   a. General in their response to an antigen
   b. Precise or very specific in their response to an antigen
   c. Produced only by T-cells
   d. Antibodies
      (page 607)

3. When a person is infected by the HIV virus that causes AIDS, the number of helper T-cells decreases. This results in:
   a. The failure of B-cells to convert to plasma cells
   b. A reduction in the number of killer T-cells
   c. Little or no memory T-cells produced
   d. All of the above
      (page 603)

4. A person who had mumps as a child is again exposed to the mumps virus but does not get the disease. His or her immunity is due to:
   a. The inflammatory response
   b. Phagocytosis alone
   c. The activation of memory cells, which destroy the mumps virus
   d. Helper T-cells, which destroy the virus directly
      (page 601)

Discussion

1. Distinguish between nonspecific and specific immune responses.

2. Define antigen, antibody, monoclonal antibody, T-cell, B-cell, and plasma cells.

3. Suppose you stepped on a nail. Trace the immune response events that would be activated to eliminate the infectious agents.

4. Discuss the structure of an antibody. How is the structure related to its function?

5. Explain how only a limited number of genes can code for the large number of different protein sequences in the variable regions of an antibody.

6. Compare the cell-mediated response with the humoral or antibody immune response.

7. How do cells recognize self?

8. Explain how memory cells prevent reinfection.

9. Many people are failing to have their children inoculated against polio. What do you suspect will happen to the incidence of polio? Why?

10. The HIV virus infects helper T-cells. How is the destruction of helper T-cells related to the symptoms of AIDS?

11. What are interleukin-2 and interferon?

26

Temperature Regulation, Osmoregulation, and Gas Exchange

You eat in the morning, and again at noon and in the evening. You sleep (or you should) every night. You breathe in air containing oxygen and exhale air containing carbon dioxide about 12 times every minute. Your body eliminates the waste products that its system cannot digest. It is constantly supplying you with the energy you need in order to breathe, to have a constant heartbeat. In fact, your body must supply materials to your cells for them to be able to maintain their membrane enclosure and conduct repairs, to name just two cellular functions. Should any of the functions just mentioned cease, you would become ill and might even die. By carrying out these processes over and over again, your body establishes a steady-state

Above: This lizard is basking in the sun to aid in the regulation of its body temperature.

condition — not too little or too much oxygen or carbon dioxide, an ample supply of ATP molecules, and an elimination of wastes. That steady state is called homeostasis.

All living organisms constantly change in order to remain the same. In some ways, you can think of this in terms of someone who is trying to remain upright in the face of a 40-mile-per-hour wind. To maintain a steady state (the upright position) the person must face into the wind, muscles constantly tensed and pushing against the force of the wind. If he or she does not exert the correct amount of energy, the individual will be forced backward and lose control. To move forward, the person must exert additional energy. The same is true in the living processes of any organism, be it a college student or a paramecium swimming in pond water. To sustain life, all organisms must maintain constant internal conditions.

In Chapter 1 we said that Claude Bernard, working in 1870, was one of the first individuals to develop the concept of homeostasis. We have also touched on the concept in several other chapters. An organism is homeostatic when (1) its internal environment has the appropriate physiological levels of gases, nutrients, and fluids, and (2) its internal environment is held at an optimal temperature range and pH. If the homeostatic condition is upset, illness or even death can result.

Any number of things can alter an organism's homeostasis, many of which you have probably experienced. For example, what happens to the cells of your body when you take a dive into a cold, icy stream, or when you heat up during exercise? How are your fluid levels changed after you consume too much beer or run a marathon? What is the result if your body cells are not provided with adequate amounts of oxygen? In this chapter we answer these questions as we discuss the physiological mechanisms that help maintain a constant internal environment. We deal with temperature regulation, fluid balance, and the proper gas concentrations required for the maintenance of life. But before we look at the workings of specific systems in the body, we need to examine feedback mechanisms — the mechanisms that regulate these homeostatic activities.

FEEDBACK MECHANISMS

■ **Usually negative feedback mechanisms regulate homeostatic processes. A receptor detects a stimulus (usually the absence of a product). The stimulus prompts the appropriate response (usually production of a supply of the missing product). When the stimulus ceases (when enough of the product is present), the response also ceases (no more of the product is supplied).**

What mechanisms are involved in maintaining this homeostatic condition? The general answer is as follows: The coordination of cellular activities is due to chemical communication among cells (either neural or hormonal) and is regulated through feedback mechanisms. Feedback mechanisms can be classified as either negative or positive. In a **negative feedback loop,** the initial stimulus that causes a change is reversed. It begins with an imbalance (too much or too little of a certain substance) and ends with the reversal of that imbalance (a decrease or increase in the production of that substance). In a **positive feedback loop,** the stimulus is amplified and made stronger. If the imbalance is not corrected, it is intensified.

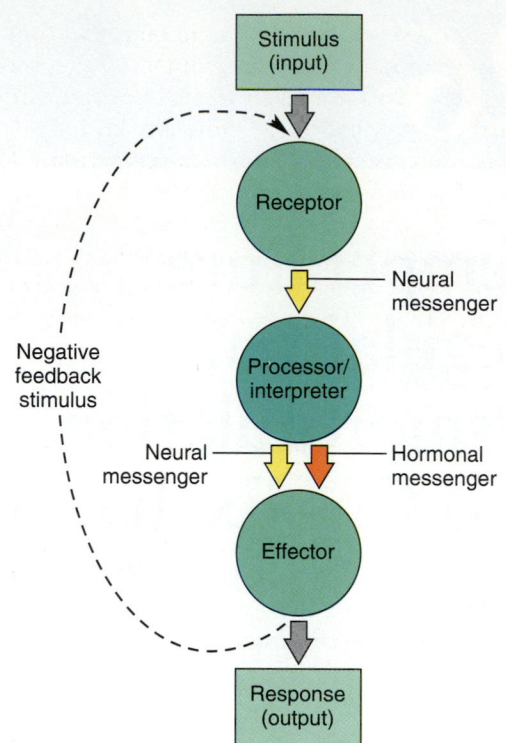

FIGURE 26.1 A Feedback Loop
Components of a feedback system at the organ level. The feedback is considered negative if the change (response) is great enough to turn off, or cancel, the original stimulus. The feedback is considered positive if the original stimulus is amplified by the response.

Feedback is a circular process, a loop (Figure 26.1). It begins with a stimulus and ends with a response that either cancels or intensifies the stimulus. There are several components involved:

1. A *stimulus* (a homeostatic imbalance)
2. A *receptor,* which detects the stimulus (and later, the response)
3. A *processor/interpreter,* which processes information about the stimulus and compares the information with an internal standard
4. An *effector,* which responds to the processed information and feeds back information to the processor/interpreter
5. *Messengers,* the nerves and hormones that carry information from the receptor to the processor and from the processor to the effector
6. A *response* (usually involving correction of the imbalance) to maintain the steady state

A quick example will illustrate how a negative feedback loop functions. When your body temperature gets too hot, the excess heat (the stimulus) is detected by receptors in the skin and other organs. The information is then transmitted to the processing center (in this case, a specific portion of the brain), which calls for the necessary hormonal and neural adjustments to keep your temperature from rising. The nerve impulses and hormones are carried to the appropriate organs, sweat glands, skeletal muscles, and the circulatory system, all of which respond in the ways necessary to reduce body temperature. Once the body temperature is back to normal, the receptors detect this fact, and the adjustments are turned off.

This is an example of negative feedback. We begin with an imbalance (too much heat) and end with the correction of that imbalance (less heat). Most homeostatic coordination in living organisms is controlled by negative feedback loops; however, there are some positive feedback mechanisms, most of which result in abnormal conditions. For example, some behavioral disorders are the result of the activities of positive feedback loops. Depression is one example. When some people are depressed, they become worried about the fact that they are depressed. The worry itself causes more depression, which in turns causes more worry, so that they are caught in a vicious, positive feedback cycle.

Our example centered on temperature control mainly because it is one of the major homeostatic mechanisms in many organisms. Among the others are osmoregulation (fluid regulation) and gas exchange. We will spend the rest of this chapter concentrating on these three homeostatic mechanisms.

TEMPERATURE REGULATION

Anyone who has suffered the extreme cold of the midwest or the extreme heat of the desert areas of the southwest knows all too well that life exists within rather broad temperature ranges. Some fish and invertebrates can live in Arctic waters having a temperature of −1.8°C (29°F), which is cold enough to freeze body fluids. Other organisms can live in hot springs having temperatures that are over 82.9°C (181°F). Some organisms live in environments (Chicago, for instance) in which the temperature fluctuates greatly from below-freezing Arctic temperatures to humid, almost jungle-level heat of over 100°F. All these life forms survive, in spite of the fact that living cells can function only within the temperature ranges of 0°C (32°F) to 40°C (104°F). This means that organisms must have mechanisms that can maintain acceptable internal body temperatures.

Ectotherms and Endotherms

■ Temperature regulation is less critical in organisms that live in water than in those living on land, because the temperature of water remains fairly stable, while the temperature of air is much more variable. The internal temperature of ectotherms (cold-blooded organisms) fluctuates because it is influenced by the temperature of the environment, while the internal temperature of endotherms (warm-blooded animals) is constant, independent of the environment.
Their internal temperature is maintained in large part by the energy released during cellular respiration.

It takes a large input of energy to change the temperature of water. As a result, the temperature of the aquatic environment rarely fluctuates, remaining, in most circumstances, within the range conducive to life. Thus there was little environmental pressure on aquatic organisms that could have encouraged the evolution of temperature-regulating mechanisms. However, as animals evolved onto the land, they encountered much greater fluctuations in temperature, both during the day-night cycle and during the rotation of the seasons. The ability to regulate body temperature became advantageous.

Some animals — like the fishes and their terrestrial descendants, the amphibians and reptiles — have body temperatures that fluctuate with their surroundings. These creatures are commonly called **ectothermic** (*ecto-* means "outside" and *therm* means "heat") or "cold-blooded" organisms. Ectotherms have virtually no mechanisms to conserve or regulate heat. They depend on the external environment to raise or lower their body temperature (Figure 26.2a).

The fact that an organism is cold-blooded affects the way it conducts its daily life. If you have ever watched a lizard on a cool morning, you may have noticed how sluggishly it moves. It will bask in the sun for a time, and as its body temperature increases, it becomes more active. The lizard's activity level depends on the outside environment. If the animal becomes too warm, it will move out of the sunlight; if it becomes too cool, it moves back into the open. By varying the amount of heat that it absorbs, it can maintain a relatively stable temperature. The organism depends on its environment for the rate of its metabolic processes.

Birds and mammals are commonly called "warm-blooded," which means that they have mechanisms that maintain relatively constant internal temperatures. As a result, the body temperature

(a)

(b)

FIGURE 26.2   Ectotherms and Endotherms
(a) Ectotherms, like this chuckawalla, have no internal mechanisms to control their body temperature. They depend on the sun—moving into and out of its heat—to regulate that temperature and the rate of their activity.
(b) Endotherms are able to maintain a constant body temperature, which helps them adapt to wide shifts in external temperature. In the cold, for example, a bird's feathers fluff up, trapping air and holding it close to the body to warm it. Birds also pull in their wings to conserve heat.

of these organisms is not influenced to any great extent by their surroundings. Biologists often use the term **endothermic** (which simply means internal regulation of body temperature) to describe warm-blooded organisms (Figure 26.2b).

Previously, scientists believed that only the higher animals (birds and mammals) were capable of complete control of their internal temperature, while lower animals did not exhibit this distinction. Today, however, they know that the distinction between ectothermic and endothermic organisms is not related to phylogenetic level alone. For example, some predatory fishes, like the bluefin tuna, are able to maintain high body temperatures even in cold water.

As you might expect, animals that can maintain a constant internal temperature have great advantages in environments that exhibit significant temperature extremes or fluctuations. As we saw in Chapter 6, the rate of enzymatic reactions is a function of temperature range. When the temperatures become too high, enzymes are destroyed (by being denatured); when temperatures are too low, the rate of enzyme reactions slows down. Maintaining a relatively constant temperature provides optimal conditions for efficient enzymatic metabolism.

During the process of cellular respiration, as molecules are oxidized in the cells, some of the released energy is in the form of heat, which serves as

the source of the heat energy in endotherms. To conserve and transport this heat, the endotherms have a four-chambered heart that provides more efficient circulation of the warmed blood. In this way, the organisms maintain a steady body temperature. Endotherms have also developed various insulating devices to retain heat, such as fur, feathers, and fat. Conversely, mechanisms have evolved in endotherms that provide means of ridding the body of too much heat, such as sweating.

Humans are endotherms, and, as everyone knows, 37°C or 98.6°F (with normal fluctuations during the day and during exercise or rest) is considered normal body temperature. How do we maintain this constant temperature? The following discussion will center on the phenomenon of body temperature regulation.

Regulation by Metabolic Activity and Through the Skin

■ **Thermoreceptors in the skin and glands detect the body's temperature, and neural and hormonal responses are activated to correct any imbalance. To prevent heat loss, blood supply to the skin's surface is limited. To promote heat loss, the blood supply to the skin surface is increased. Sweat is excreted by sweat glands. Evaporation of the sweat results in further cooling.**

Two major mechanisms in mammals (including humans) are involved in maintaining body temperature. The rate of metabolic activity controls the amount of heat that is produced. The amount of heat retained is regulated through its loss or retention by the body surface. For these mechanisms to work in a coordinated fashion, two systems — the nervous system and the endocrine system (whose hormonal secretions play an important role in controlling body functioning) — must interact.

We will discuss the operation of the endocrine and nervous systems in detail in Chapters 29 and 30. However, the interrelationship of these two systems can be illustrated by a simple example: the way your body responds when you jump into very cold water. What you consciously experience is that your skin color becomes visibly paler and you begin to shiver. What are the physical reactions that bring about these responses?

Two structures in the brain — the hypothalamus and the pituitary gland — play a major role in temperature regulation. The hypothalamus has several functions. One of these is that of body-temperature regulator; it is, in effect, the body's thermostat. *Thermoreceptors* (temperature detectors) in the skin and in the hypothalamus detect the drop in body temperature. The hypothalamus then takes corrective action to increase body temperature.

Nerve impulses are sent through the autonomic nervous system to decrease blood flow to the skin, which causes important body heat to be retained in the deeper organs. The blood vessels near the skin's surface are called *arterioles*. The sphincter muscles in the arterioles respond by contracting, thus reducing blood flow to the skin. (This is what causes you to look pale.) The *medulla* (center) of the adrenal gland produces the hormone epinephrine (remember that in Chapter 24 we said that epinephrine is the chemical name for the more common term adrenalin). The epinephrine causes a sudden increase in metabolic activity, which in turn results in more heat production. Epinephrine and the autonomic nervous system (that part of the nervous system not under conscious control) signal the breakdown of fat for more energy. Other hormones play a role in these processes. The outer region of the adrenal gland (the cortex) produces a hormone called *cortisol,* which aids in temperature regulation. Cortisol is involved in the conversion of fatty acids and amino acids into glucose, and as the glucose is metabolized, more heat is produced.

Finally, the hypothalamus secretes a substance that causes the pituitary gland to produce thyroid-stimulating hormone (TSH). TSH makes the thyroid secrete the hormone *thyroxine,* which also increases cellular metabolism and heat production on a longer-term basis than the hormones secreted by the adrenal gland do. Thyroxine directly induces the mitochondria to produce more ATP and heat energy (Figure 26.3).

When you shiver, your muscles contract and relax rapidly. The contractions require the utilization of the energy stored in ATP molecules, but because the system is not completely efficient (remember the laws of thermodynamics), some of the

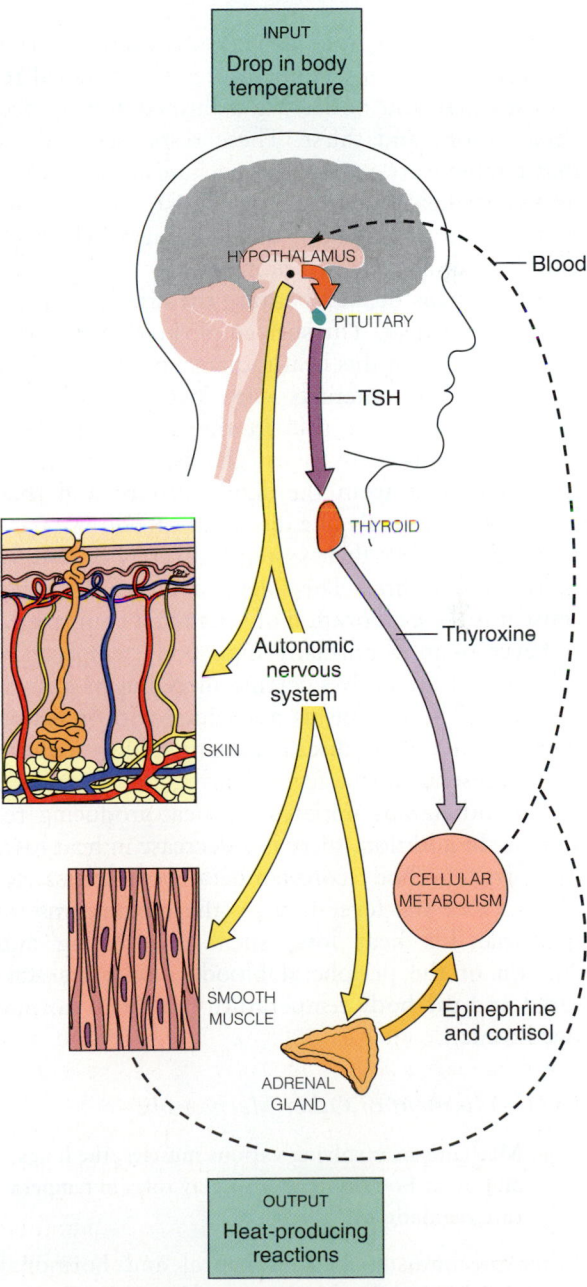

FIGURE 26.3 Heat Production and Negative Feedback
The coordinated neural and hormonal mechanisms involved in temperature regulation.

energy is lost as body heat. Previously it was thought that shivering was controlled only by the hypothalamus. However, recently it was demonstrated that the spinal cord also plays a role.

Heat-producing reactions work together to raise your body temperature. Without the actions of negative feedback, the process could continue until you would be too warm. However, as the body responds and temperature increases, the thermal receptors in the skin and hypothalamus detect the increase. The corrective measures then cease or are turned off.

Some activities, such as vigorous exercise, bring about a rise in body temperature. The physical responses that you notice are a flushed appearance, perspiration, and thirst. These responses indicate that temperature-regulation mechanisms are acting to reduce heat conservation. Thirst also involves fluid-regulation reactions of the kidneys, which we will discuss in detail in a moment.

You appear flushed because the arterioles dilate and give off heat. The surface of the skin becomes hot, and some of this heat radiates into the air. The remaining heat interacts with sweat, which also plays an important role in the cooling process. Sweat is produced by sweat glands. The sweat absorbs the heat from the skin's surface and then evaporates. Animals like dogs and cats do not have sweat glands, but their systems also rely on evaporation for cooling. They depend on panting, because it is the evaporation of saliva that cools them.

Fever — an abnormally high body temperature — is one of the body's defense mechanisms against infection. The organisms associated with infection seem to have their disease-causing ability reduced in the presence of the fever. During a fever episode the hypothalamus initiates the heat-producing reactions. In addition, there is a decrease in heat loss; therefore, the body core temperature becomes elevated. Once the fever is over, the mechanisms responsible for heat loss, such as sweating and dilation of the peripheral blood vessels, are activated and the body temperature returns to normal (see Profile).

The Involvement of Other Mechanisms

- **Mechanisms involving various muscles, the lungs, and other body systems also play roles in temperature regulation.**

Other mechanisms besides neural and hormonal ones are also involved in heat exchange. For example, most birds and mammals have *piloerector* muscles that cause the hair or feathers to stand on end

or to fluff up. This phenomenon traps air and conserves body heat. Since humans do not have well-developed body hair, we get "goosebumps" when our autonomic nervous system signals the piloerector muscles of the hair follicles to contract (see Figure 22.9).

The lungs, with their large surface area (about 100 square meters), are also instrumental in heat retention and loss. Even though they are inside the body, they contain air from the environment. This means that, in a sense, they are exposed to the environment. The lungs' large surface area serves to release heat to the environment or retain heat within the body. When the outside air is cold, the blood circulating in nasal passages, as well as the warm air exhaled through the nasal passages and mouth, can warm the cold inhaled air almost to body temperature before it ever reaches the lungs.

As we have seen, body-temperature regulation involves a coordinated interaction of neural and hormonal signals that can either increase heat production and storage or increase heat loss. This is what enables warm-blooded animals to survive in environments with widely divergent temperatures.

OSMOREGULATION

- **Osmoregulation is the regulation of the concentration of ions and water in the body. In simple organisms, diffusion provides an adequate means for regulating fluids. In more complex animals, special tubes and receptacles are involved in osmoregulation.**

As we mentioned in the previous chapter, our cells are bathed by a sea within (the extracellular fluid), and our systems can function only within narrow chemical limits, called the *physiochemical balance*. To maintain the necessary chemical constancy, various homeostatic mechanisms have evolved to provide for **osmoregulation,** or the regulation of ions and water in the cells or in extracellular spaces.

Phylogenetic Overview of Osmoregulatory Mechanisms

In the single-celled animals, such as the paramecium and amoeba, gases and dissolved nitrogen wastes are eliminated by simple diffusion through the cell membrane. In addition, several protozoa, like the paramecium, possess a fluid-controlling structure called a contractile vacuole, which has the ability to expand and contract (Figure 26.4a). Any excess water that diffuses into the organism, and to a lesser

Chapter 26 Temperature Regulation, Osmoregulation, and Gas Exchange

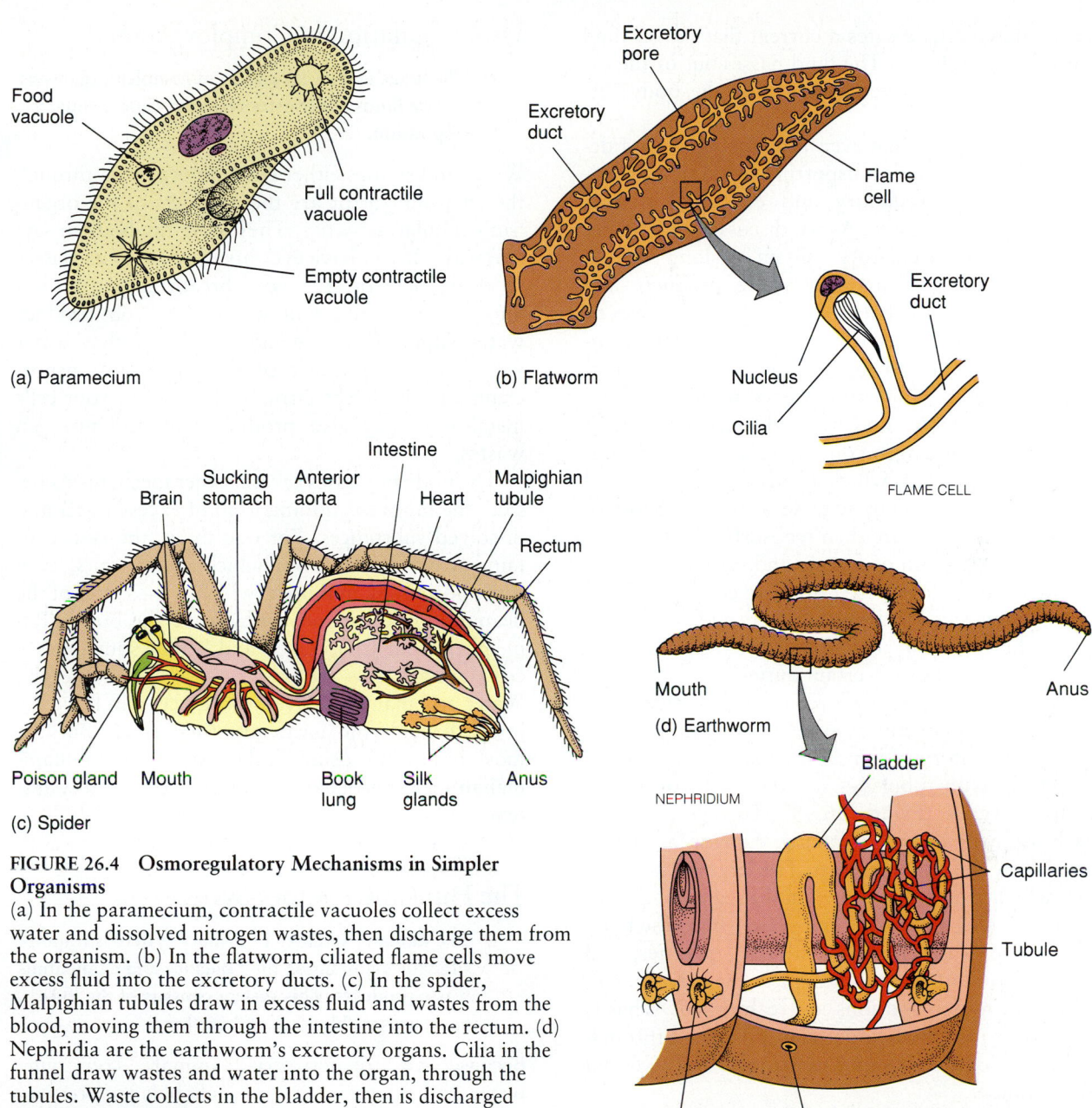

(a) Paramecium

Food vacuole

Full contractile vacuole

Empty contractile vacuole

(b) Flatworm

Excretory pore

Excretory duct

Flame cell

Excretory duct

Nucleus

Cilia

FLAME CELL

(c) Spider

Brain

Sucking stomach

Anterior aorta

Intestine

Heart

Malpighian tubule

Rectum

Poison gland

Mouth

Book lung

Silk glands

Anus

(d) Earthworm

Mouth

Anus

Bladder

NEPHRIDIUM

Capillaries

Tubule

Funnel

Nephridiopore

FIGURE 26.4 Osmoregulatory Mechanisms in Simpler Organisms

(a) In the paramecium, contractile vacuoles collect excess water and dissolved nitrogen wastes, then discharge them from the organism. (b) In the flatworm, ciliated flame cells move excess fluid into the excretory ducts. (c) In the spider, Malpighian tubules draw in excess fluid and wastes from the blood, moving them through the intestine into the rectum. (d) Nephridia are the earthworm's excretory organs. Cilia in the funnel draw wastes and water into the organ, through the tubules. Waste collects in the bladder, then is discharged through a nephridiopore.

extent dissolved nitrogen wastes, is collected in the vacuole. The contractile vacuole increases in size as it fills, and then it contracts, ejecting its contents to the outside. This structure is a very primitive, yet adequate, osmoregulatory pump, and it exhibits the two functions of an excretory system: It maintains water balance and it rids the organism of nitrogenous wastes.

In the simple aquatic invertebrates, like the sponges and cnidaria, simple diffusion accounts for the intake of water and other substances into the organism's cells and the movement of waste products out of them. As animals became more complex, so did their systems, and a primitive excretory system, including cells specialized for maintaining fluid constancy, developed. For example, in the free-living flatworms two networks of branching tubules function as excretory ducts, collecting and disposing of fluid (Figure 26.4b). At the end of these tubules are ciliated cells that are called *flame cells*, because the movements of the cilia make the cells appear to flicker like flames. The beating of the cilia in the flame cells creates a current that moves fluid through the tubules. The fluid passes out of the organism through excretory pores on the body surface.

As animals became even more complex and developed systems for transporting blood, the association of the circulatory and excretory systems became more intimate. As we discussed in Chapter 24, an open circulatory system — along with a mechanism for exchanging waste products and maintaining fluid balance — evolved in insects. Two or more slender tubes, called *Malpighian tubules,* are connected to the insects' digestive tract (Figure 26.4c). Fluid and wastes from the blood enter these tubules and move through the gut. In the process, nitrogen-containing wastes are converted to uric acid, which forms into crystals and passes out of the anus with fecal matter. Most of the water and ions are then reabsorbed by the gut.

The exoskeleton of insects and certain other arthropods also plays a part in the excretory process. Nitrogen wastes are deposited there and "excreted" when the organism molts. And uric acid has been found in the wings of certain butterflies.

We find the same close relationship between the circulatory and excretory systems in organisms with closed circulatory systems. *Nephridia* — funnel-like structures with tubules — are the earthworm's excretory organs (Figure 26.4d). There are two in each segment of the organism, running alongside and below the digestive tract. The large ciliated funnel (which is actually located in the segment in front of the tubules) draws in fluid wastes from the body cavity and water from the blood capillaries, and passes them into the tubules. Here they are filtered, and some water and ions are reabsorbed. What is left moves into the bladder and is excreted through a nephridiopore, a small opening to the external environment.

In vertebrates like ourselves, the principal excretory organs are the paired kidneys and the collecting ducts (ureters) extending from them. Each kidney collects fluid wastes from the blood and filters them, reabsorbing water and essential ions. The waste — in the form of urine — moves into the bladder and is excreted through a single tube, the urethra. In birds and reptiles, the waste appears like a white paste; it passes into the cloaca and is excreted through the anus with the feces.

In humans and other vertebrates, the management of osmoregulation and of the excretion of nitrogen wastes is even more complex. Let's now examine the osmoregulatory mechanisms in the complex animals, with a focus on humans.

Osmoregulation in Complex Animals

- **The lungs and sweat glands of complex organisms release fluids and wastes resulting from cellular respiration.**

Water and solutes either enter an organism through the environment or are the products of the organism's cellular activities. They leave the body in several ways. If you have ever breathed on your glasses to clean them or seen your breath as you took a long winter walk, you know that you produce water vapor as you exhale. Some of that water vapor is a by-product of the electron transport chain and the Krebs citric acid cycle. As your cells metabolize they also produce ions and nitrogen wastes.

The body possesses several other mechanisms besides the lungs for ridding itself of excess water and dissolved substances. For one thing, the skin contains sweat glands from which water, salts, even some hormones, and urea (which is the result of the deamination of amino acids) leave the body. (Remember from Chapters 7 and 9 that sometimes a cell's amino acids are used as a source of energy.) Water and many dissolved substances can leave the body through the gastrointestinal tract. But the most important osmoregulatory system in mammalian vertebrates like humans is the excretory system.

The Human Excretory System

- **The human excretory or urinary system is responsible for maintaining fluid balance and eliminating waste products. This system consists of the kidneys, ureters, bladder, and urethra.**

Because we are terrestrial animals, conserving water and maintaining ionic salt balance are very critical to our well-being. Witness the fact that kidney failure leads to disease or death. The human excretory system, or the **urinary system** as it is sometimes called, is responsible for maintaining the necessary balance. This system consists of *two kidneys, two ureters,* the *bladder,* and the *urethra* (Figure 26.5).

The **kidneys** are paired structures that excrete nitrogen wastes and help maintain water and ion balance in the blood. Located in the lower back and red-brown in color, each is about the size of a fist and is shaped like a kidney bean (or is it the kidney bean that is shaped like the kidney?). Three tubes are attached to the kidney: the *renal artery* (*renal* is from the Latin term for kidney), which carries

Dr. Reynaldo S. Elizondo
Physiologist/Temperature Regulation

In 1957, Reynaldo S. Elizondo was doing what he liked best, listening to Johnny Cash on a country and western radio station and playing his guitar. His practice was not only recreational, he also had his own band.

Elizondo is the oldest of three children from a lower-middle-class Mexican-American family. He attended Roy Miller High School in Corpus Christi, Texas, never thinking that biology would become his passion. At that stage of his life, country and western music and football were his highest priorities. However, through planned circumstances and by happenstance, Elizondo earned his doctorate in physiology and became a recognized authority in the physiology of temperature regulation and fluid balance.

Elizondo became interested in science while taking a course in the natural history of animals at Del Mar Junior College. He was excited by both the lecture material and the field trips to learn about various animal life cycles. He discovered that biology could be fun and enjoyable. As his interest in biology grew, Elizondo enrolled at Texas A&M University and majored in biology with the intent of becoming a laboratory technician. He earned his B.S. in 1963. His talent in biology was recognized by the chairman of the biology department, who suggested that Elizondo apply to medical school. Elizondo had no desire to enter medicine, nor did he think his family could afford the expense of medical school. The chairman encouraged him to apply anyway, assuring him that most medical schools had financial assistance for students. Elizondo was accepted by Tulane School of Medicine, but he still wanted to become a physiologist, not a physician. To resolve this conflict, he entered the Tulane Ph.D program in physiology.

After earning his doctorate, Elizondo wanted additional training in physiology. He worked with Dr. Robert Bullard, a leading authority on temperature regulation and fluid balance at the Indiana University School of Medicine in Bloomington. After completing his postdoctoral research, Elizondo was offered a position in the physiology department at the same institution. He remained there for 20 years doing research and eventually becoming a full professor and department chair.

Although he was very happy at Indiana University, Elizondo thought of one day returning home to Texas. The opportunity to work there occurred when the University of Texas at El Paso asked him to serve as visiting professor. Two years later, he was offered the position of dean of the College of Sciences at El Paso. Today Elizondo is Dean of the College of Sciences at the University of Texas at San Antonio.
nity service.

Elizondo feels that it is very important to encourage students to be academically successful, especially students who have not had the opportunity or exposure to careers such as science. He donates much of his time to the national biomedical research effort, and he serves on the advisory council of the Institute of General Medical Sciences at the National Institutes of Health. He is dedicated to encouraging minorities to pursue the opportunities in science, reminding them that we are all a product of circumstance, but we must be prepared to take advantage of the opportunities, through hard work and education.

After experiencing a busy day as dean, researcher, or when he returns from a business trip to Washington, Elizondo still likes to relax by listening to country and western music and playing his guitar. He often reflects, "who would have thought that a former high school quarterback like me, with a love of country and western music, would find himself a scientist?"

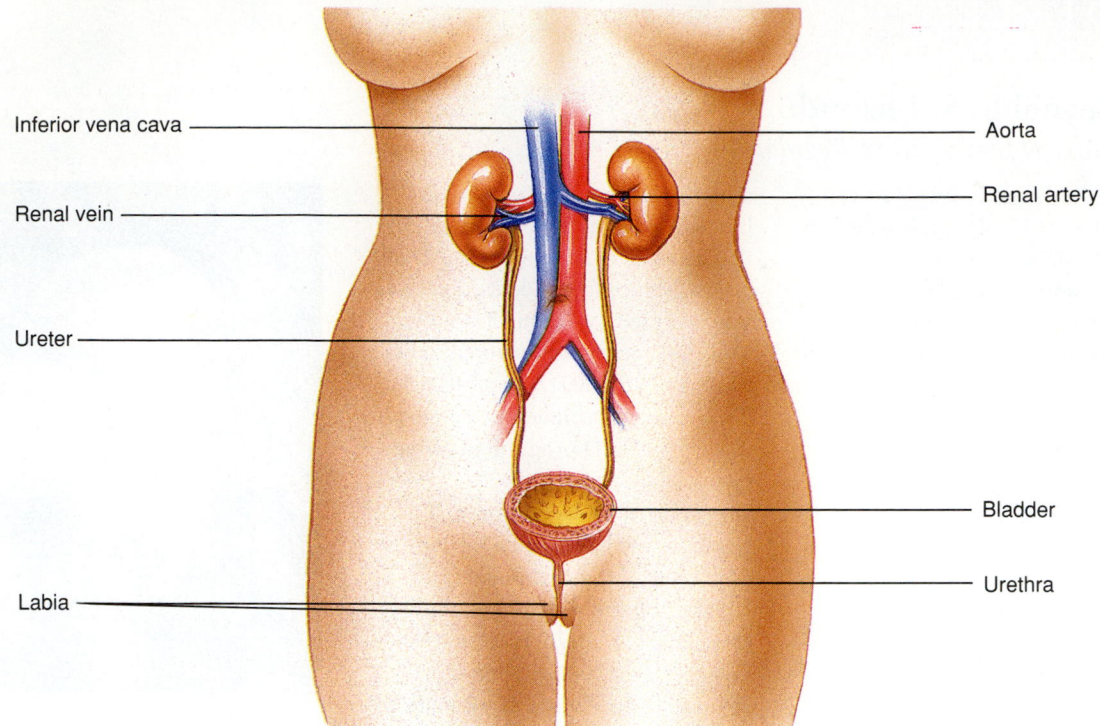

FIGURE 26.5 The Human Urinary System
Urine—the by-product of the kidney's functioning—passes through the
ureters to the bladder and is discharged through the urethra. In the female
(shown), the urethra empties directly to the exterior through the folds of skin
(labia) surrounding the vagina; in the male, the urethra passes through the penis.

blood from the aorta to the kidney; the *renal vein,*
which transports blood from the kidney to the inferior vena cava (which then leads to the heart); and
the *ureter,* the tube that carries urine to the bladder.
The **bladder** is a hollow sac made of smooth muscle.
It expands as it fills with urine and contracts as it
empties. Urine passes from the bladder to the **urethra** — a tube that connects to the exterior of the
body. In the female, the urethra opens directly to
the outside, in front of the vaginal opening; in the
male, it passes through the penis.

Each kidney consists of three major regions: an
outer layer called the **cortex,** an inner layer called
the **medulla,** and a hollow central region called the
renal pelvis (Figure 26.6). It is in the cortex and
medulla that the kidneys carry out their critical
functions of removing wastes from the blood, maintaining the fluid balance in our bodies, and regulating the composition of the blood — processes we
call *filtration, reabsorption,* and *tubular secretion.*
The renal pelvis is where the urine — the by-product of these processes — collects before it moves
into the ureter.

Filtration, Reabsorption, and Tubular Secretion

Thousands and thousands of *nephrons,* each of
which is a microscopic filter, are the workhorses of
the system. It is in these tiny structures that filtration, reabsorption, and tubular secretion occur (see
Focus 26.1).

Nephrons, the Primary Filtering Structure

■ **Within the kidneys, nephrons process the blood in
three steps: filtration, reabsorption, and tubular
secretion. Filtration, which takes place in the
glomerulus of the nephron, removes products
from blood plasma. Reabsorption, which occurs
in the tubule of the nephron, returns the filtrate —
along with selected ions — to the circulatory system. Tubular secretion, also occurring in the tubule, involves the active transport of substances
such as hydrogen ions from the blood to the fluid
within the nephron. Water is reabsorbed and
urine collects in the bladder and is excreted
through the urethra.**

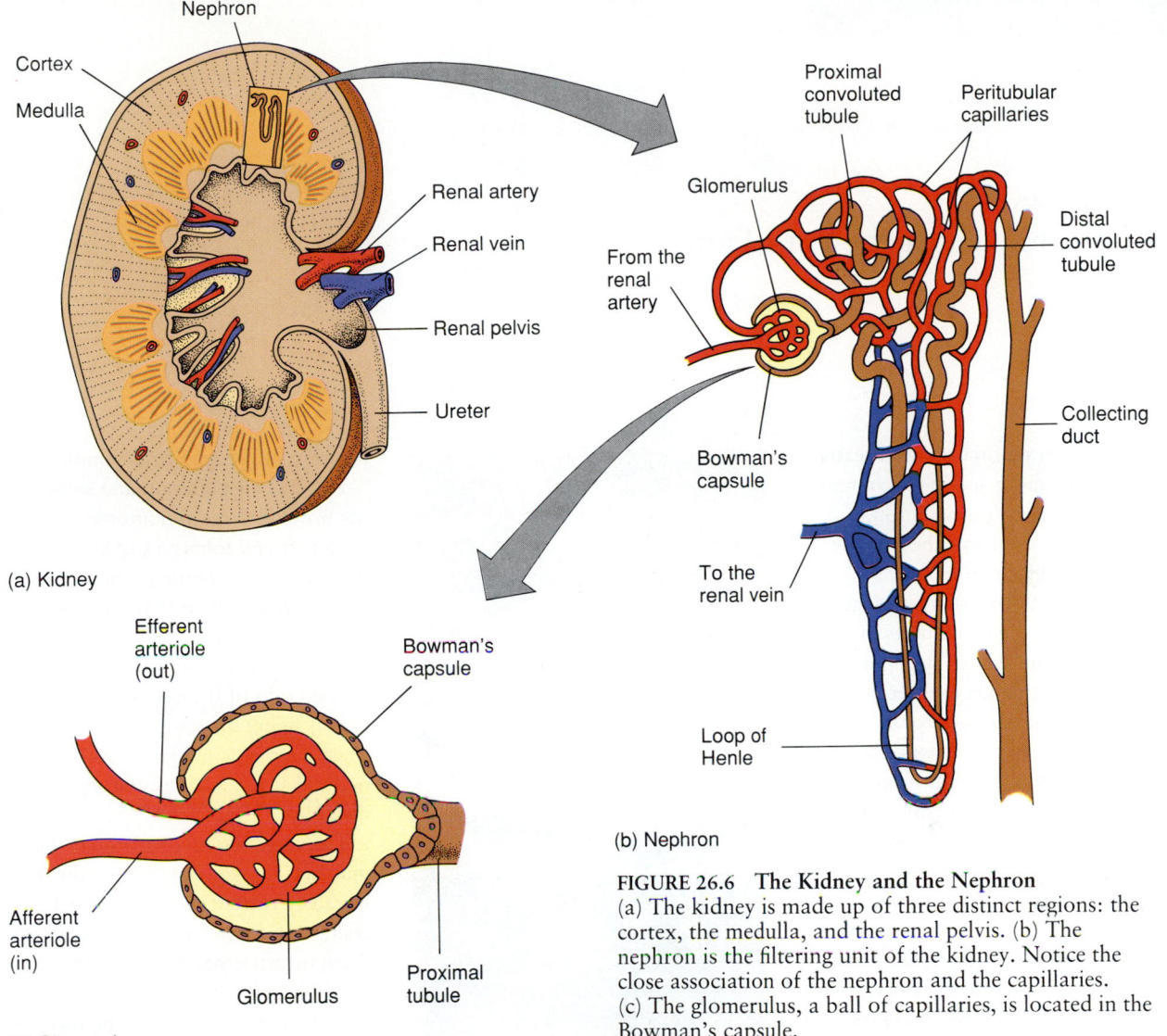

(a) Kidney

Nephron
Cortex
Medulla
Renal artery
Renal vein
Renal pelvis
Ureter

(b) Nephron

Proximal convoluted tubule
Peritubular capillaries
Glomerulus
From the renal artery
Distal convoluted tubule
Bowman's capsule
Collecting duct
To the renal vein
Loop of Henle

(c) Glomerulus

Efferent arteriole (out)
Bowman's capsule
Afferent arteriole (in)
Glomerulus
Proximal tubule

FIGURE 26.6   The Kidney and the Nephron
(a) The kidney is made up of three distinct regions: the cortex, the medulla, and the renal pelvis. (b) The nephron is the filtering unit of the kidney. Notice the close association of the nephron and the capillaries. (c) The glomerulus, a ball of capillaries, is located in the Bowman's capsule.

Every day, the kidneys can process over 180 liters of blood, remove the nitrogen wastes, and maintain the proper osmotic fluid balance in the body. How can they do this? The answer, in part, has to do with the 1,000,000 microscopic units called nephrons, which are found in the cortex of the kidney (Figure 26.6b). The **nephrons** are microscopic filtering units that are essential to kidney functioning. If the nephrons of both kidneys were linked end to end, they would produce a tube over 50 miles long.

A nephron is a hollow tube that is open at one end and closed at the other. The closed end of the nephron tubule forms a double-walled cup called **Bowman's capsule**, the inner portion of which is filled with fluid. The tube-shaped portion of the nephron is divided into three regions. The part nearest Bowman's capsule is called the *proximal*

convoluted (twisted) *tubule* or region. Next is a long loop called the *loop of Henle* (named after its discoverer). Finally, the region farthest from Bowman's capsule is called the *distal convoluted tubule*. This section twists and meanders until it joins a collecting duct that eventually connects with the renal pelvis. The renal pelvis connects with the main ureter of the kidney.

In close association with each nephron is a cluster of blood vessels. The renal artery, which carries blood to the kidneys from the aorta, branches off into smaller arteries, the arterioles. Each afferent arteriole ends in a ball of capillaries called a **glomerulus** in a Bowman's capsule, (Figure 26.6c). Unlike most capillary beds, a glomerulus does not empty into a venule; instead, it merges with a second arteriole, the efferent (outgoing) arteriole. This

When Smoke Gets in Your Lungs

It is no secret that the air we breathe is becoming more and more contaminated with chemical pollution. In the course of a lifetime the average person will inhale more than 50 pounds of airborne particles (Figure A). Every day the inhabitants of New York City take in the equivalent of the pollutants in three-fourths of a pack of cigarettes, even if they do not smoke. Mexico City, Tokyo, London, and Los Angeles are only a few of the urban centers in which the air is unfit to breathe most days of the year (Figure B).

Not surprisingly, respiratory diseases — such as bronchitis, sinusitis, laryngitis, asthma, emphysema, and lung cancer — are on the rise in all industrialized countries. Diseases like bronchitis and emphysema were rare 25 or 30 years ago. Today, however, more than 11 million people suffer from these particular disorders. It is estimated that, worldwide, 40 to 50 million people a year experience respiratory problems of one kind or another that are the direct result either of air pollution or of smoking (including the in-

FIGURE A   The lung tissue of a non-smoker (top) as compared with that of a smoker (bottom) reveals the extent to which smoking can dirty the lungs.

halation of "secondhand" smoke).

How do air pollution and smoking produce these conditions? Dirty urban air and tobacco smoke consist of numerous airborne particles, aerosol particles (those that are less than $\frac{1}{2}$ μm in size), and toxic gases. These substances can paralyze the ciliated cells of the mucous membranes lining the respiratory tract. Then a series of interrelated events occur that result in respiratory damage.

The cilia's inactivity prevents the mucous blanket from leaving the respiratory tract. The mucus, which contains bacteria and other particles that under normal circumstances would be moved out of the respiratory tract, accumulates in the trachea and bronchi and forms mucus plugs that block off the smaller tubes. This initiates a vicious cycle. The ciliated mucous cells are exposed to more chemicals and particular pollutants, which causes coughing. The coughing irritates the bronchial walls, thus aggravating the condition. The bronchi become infected and inflamed. The respira-

---

arteriole divides into a second set of capillaries, called *peritubular capillaries,* which wind over and around the tubular regions of the nephron. The peritubular capillaries eventually empty into a venule, which merges with the renal vein. The renal vein

carries the blood to the inferior vena cava, which leads to the heart.

The relationship of nephron and blood vessels is critical to the filtration, reabsorption, and tubular secretion processes. *Filtration* (the movement of sol-

FIGURE B A smoggy day in Los Angeles.

tory tract responds by producing more mucus-secreting cells. The number of cilia decreases, and more tissue damage of the respiratory tract occurs. The result is *bronchitis,* which is an inflammation of the bronchioles.

If the individual continues to be exposed to the pollution, which, as noted, can include the secondhand smoke of others (or if he or she continues to chain-smoke), fibrous scar tissue forms, and the bronchitis worsens. We have all heard the typical hacking, mucus-laden cough that accompanies heavy smoking. That cough is the body's response to the violation of its respiratory tract.

If the exposure to air pollution or tobacco smoke continues, one possible result is *emphysema,* which is caused by mucus plugging the bronchioles. Air becomes trapped in the alveoli; the alveoli balloon out and form blisters. Some alveoli even burst from the pressure of the trapped air. The alveoli and the lungs lose their elasticity. The lungs cannot inflate as efficiently as before, and the size of the respiratory surface is decreased. The lungs try to compensate for the loss of functional surface by increasing in size.

Because the lungs have stretched, the emphysema patient develops a barrel-shaped chest. Breathing is difficult, and oxygen deprivation produces damage to the brain and other vital organs. The heart becomes enlarged and weakened. Physical activity is impaired (even walking becomes difficult). At its extreme, emphysema can cause death. Imagine the discomfort and agony — yet some emphysema patients continue to smoke. We do not know the mor-

tality rate for emphysema or bronchitis due to polluted air, but we do know that the cigarette smoker has a 700 percent higher death rate from these diseases than the nonsmoker.

Tobacco smoke and polluted air also contain chemical compounds that are known to cause lung cancer. For example, biologists know that methylcholanthrene, found in coal tar and tobacco smoke, is transformed through normal body chemistry into cancer-causing substances called *carcinogens*. Carcinogens alter the lung cells' DNA in such a way that the result is cancer. There is no doubt that the major cause of lung cancer is smoking — it is the cause of over 80 percent of all lung cancer deaths. At present, nine out of ten lung cancer patients will die.

Cigarette smoking is responsible for more than 320,000 deaths in the United States each year, almost two times more deaths than those due to alcohol (125,000), 80 times more than those due to the opiates including heroin (4,000), and 160 times more than those due to cocaine (2,000). In 1988, the U.S. Surgeon General C. Everett Koop pointed out that 171 separate studies demonstrated that nicotine is addictive. If he had his way, he maintained, he would add to the labels of cigarette packages that "cigarettes are just as addictive as heroin and cocaine."

Given all that science can tell us about the dangers of cigarette smoking, why do people still smoke? Why is smoking increasing among women? Why do we continue the excessive use of fossil fuels as energy sources? Maybe all this smoke is clouding our thinking.

vents and other substances across a membrane in response to pressure) takes place in the Bowman's capsule (Figure 26.7). It involves interaction between the glomerulus and the capsule. *Reabsorption* (the transfer of useful substances from the nephron to the bloodstream) and *tubular secretion* (the transfer of additional wastes from the bloodstream to the nephron) take place in the tubular portion of the nephron. These processes involve the peritubular capillaries.

Osmoregulation

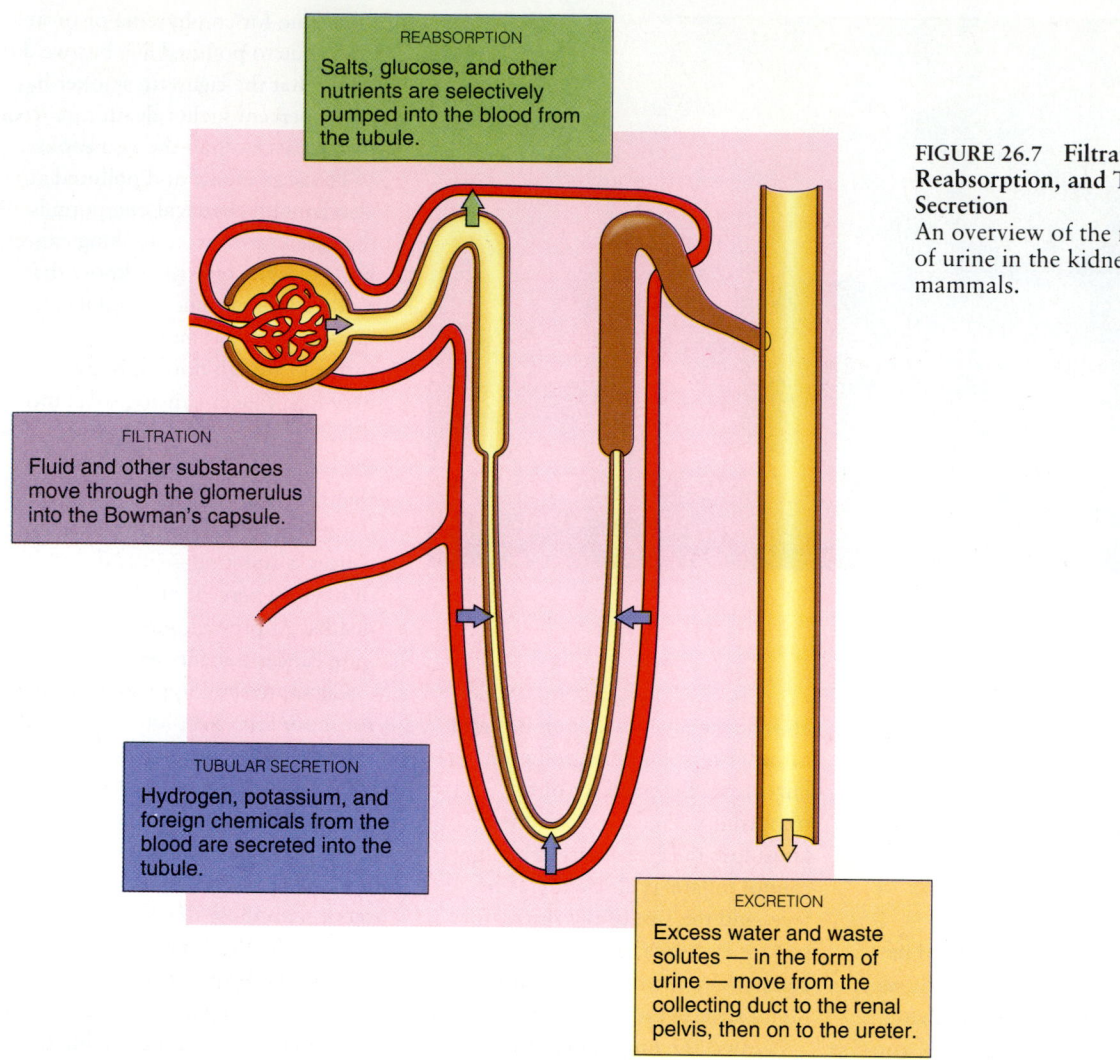

REABSORPTION
Salts, glucose, and other nutrients are selectively pumped into the blood from the tubule.

FILTRATION
Fluid and other substances move through the glomerulus into the Bowman's capsule.

TUBULAR SECRETION
Hydrogen, potassium, and foreign chemicals from the blood are secreted into the tubule.

EXCRETION
Excess water and waste solutes — in the form of urine — move from the collecting duct to the renal pelvis, then on to the ureter.

FIGURE 26.7 Filtration, Reabsorption, and Tubular Secretion
An overview of the formation of urine in the kidneys of mammals.

Filtration. Osmoregulation in the urinary system begins in the Bowman's capsule with the process of **filtration** — the movement of solvents and dissolved substances across a selectively permeable membrane from an area of higher pressure to an area of lower pressure. The blood pressure in the glomerulus is greater than that in the Bowman's capsule. The difference in pressure moves the blood plasma and dissolved substances in the blood (sodium, glucose, urea) out of the capillary ball and into the capsule. The structure of the glomerulus facilitates the process. Its walls are very porous, making them especially permeable to water and small solutes. Elements that are too large to move through the pores — blood cells and proteins — remain in the blood. Since the process of filtration utilizes no ATP, molecules move by means of passive transport, not active.

The fluid that is now in the nephron is a crude filtrate *(glomerular filtrate)* that includes both waste molecules and useful substances (water and glucose, for example). Table 26.1 compares the composition of blood plasma, glomerular filtrate, and urine.

Reabsorption. Filtration is not selective: Both waste products and useful materials move from the blood into the Bowman's capsule. **Reabsorption** is the process of returning useful materials to the blood. It begins as the filtrate moves into the proximal convoluted tubule, with the active transport of sodium (Na^+) ions, chloride (Cl^-) ions, glucose, and some small amino acids from the tubule to the peritubular capillaries. The movement of sodium and chloride changes the osmotic gradient in the tubule. Water, following that gradient, moves out of the tubule and diffuses into the peritubular cap-

TABLE 26.1
Comparison of the Composition of Blood Plasma, Glomerular Filtrate, and Urine

Component	Blood plasma (g/liter)	Glomerular filtrate (g/liter)	Urine (g/liter)
Glucose	1.0	1.0	0
Sodium	3.0	3.0	3.0
Chloride	3.5	3.5	6.0
Potassium	0.15	0.15	1.5
Uric acid	0.003	0.03	0.3
Urea	0.25	0.25	20.0
Sulfate	1.0	1.0	1.0
Creatinine	0.15	0.15	0.7
Phosphate	0.03	0.03	1.0
Amino acids	0.3	0.3	0
Protein	70.0	0.2	0

illaries. (About 80 percent of the water in the glomerular filtrate is reabsorbed in the proximal convoluted tubule; the rest is reabsorbed in the loop of Henle.)

Reabsorption is the process that regulates the composition of the blood. Unlike filtration, it is a selective process. Only certain elements are returned to the bloodstream. The active transport mechanisms involved in reabsorption require a great deal of energy. The source of that energy is the ATP produced by the large number of mitochondia in the cells of the tubule.

Tubular Secretion. Reabsorption carries substances out of the filtrate back into the blood. **Tubular secretion** — the third process occurring in the nephron — carries substances from the blood into the filtrate. These substances include hydrogen and potassium ions, penicillin and other drugs, ammonia, and even some pesticides.

Tubular secretion serves two functions. First, it removes additional wastes from the bloodstream. Second, it regulates the pH level of the blood. As we discussed in Chapter 2, the pH level of a substance depends on its concentration of hydrogen ions. By transporting hydrogen ions from the blood into the filtrate, the tubular secretion process helps maintain the correct pH level in the blood.

The by-product of the filtration, reabsorption, and tubular secretion processes is urine. Urine is

carried by the collecting ducts to the renal pelvis, then through the ureters to the bladder. Here it is stored, then excreted through the urethra. In a sense, the kidney cleans the blood almost as you clean your desk drawers. You begin by emptying almost everything out of the drawers, then selectively put everything but waste back into them.

Mechanisms of Nephron Regulation

■ The amount of urine secreted is a function of environmental and physical factors that are independent of nephron processes. The amount of fluid present in the body is determined by osmoregulators, which respond to the concentration level of salts and solutes in the blood. Also, antidiuretic hormone secreted by the adrenal gland regulates the permeability of the kidney tubules, controlling the movement of fluid out of the tubules.

We all know from personal experience that the amount of urine we produce varies with activity, fluid intake, and disease. After vigorous activity that brings on a lot of perspiration or a day's hike in the desert, we notice that our urine output is reduced; its color is very dark yellow. What causes the intensified color? The water in your body is being reabsorbed after undergoing filtration and is used for temperature regulation. As a result, the concentration of the products of metabolic activity is high. After we consume a lot of fluids, we note that our urine production has increased; its color is clear. The color is pale because the metabolic wastes are not concentrated.

The amount of fluid (an indication of the volume of water) in the body also determines the relative concentration of salts. The more water that is present, the more dilute and less concentrated the salts. On the other hand, the less water that is present, the greater the concentration of salts. Hence, water volume affects solute (salts and other dissolved substances) concentration, and vice versa.

Let's deal with the situation in which you have worked outside in the summer sun for hours without drinking any water. In this case, the total volume of body fluid is low, so the body must conserve water to maintain the proper fluid level and solute concentration. How does your body respond, in order to preserve the water content already existing in your body? The absorption of water (or water loss) through the nephrons is regulated by two major factors, hormonal and neural. The hypothalamus at the base of the brain contains cells called osmoreceptors, which are sensitive to the levels of salts and other solutes. When the body fluids are low and salt concentrations high, osmoreceptors are stimulated. Nerve fibers carry the stimulus to the posterior (back region) of the pituitary gland, signaling the release of *antidiuretic hormone (ADH),* a hormone that regulates against water loss from the blood. ADH is carried by the blood to the kidney tubules, where it acts on the cells of the convoluted distal tubule and the nephrons' collecting ducts. ADH makes the cells more permeable to water, so that more water is reabsorbed into the blood. Urine volume decreases and salt concentration increases (Figure 26.8).

The body's response to a normal or excessive amount of water is like a mirror of its response to a lack of water. When ADH levels are low, the cells of the kidney tubules become less permeable, the absorption of water decreases, and urine volume increases.

The Phenomenon of Thirst

■ **The thirst center is located in the hypothalamus. It is stimulated by the same solute levels as are the osmo regulatory sensors.**

There are other factors that can affect osmoregulation. Alcohol inhibits the production of ADH. Therefore, when you drink alcohol-containing beverages, the levels of ADH decrease, reabsorption of water decreases, and you urinate in great quantities. (You probably are very thirsty the next morning, too.) A form of diabetes known as *diabetes insipidus* is characterized by a deficiency in the ADH supply. People with diabetes insipidus have a daily output of urine of almost 10 gallons. Obviously they are always very thirsty and consequently drink tremendous amounts of water.

These situations are not what most people consider normal. On the other hand, all of us have experienced thirst. What actually regulates thirst? For most individuals, water intake is voluntary — we drink when we want to. The signal to drink, however, is an unconscious one produced by our *thirst center,* which is located within the hypothalamus. When the osmoregulatory cells are stimulated by high solute levels in the blood, the thirst-center cells are also stimulated. That stimulus generates neural signals that we interpret as feelings of thirst.

Sodium Reabsorption. We have discussed water reabsorption, but what mechanisms regulate sodium reabsorption? To understand the processes, we need a quick review of osmosis. First, the amount of dissolved solutes (like salts) that water contains determines the concentration of a solution.

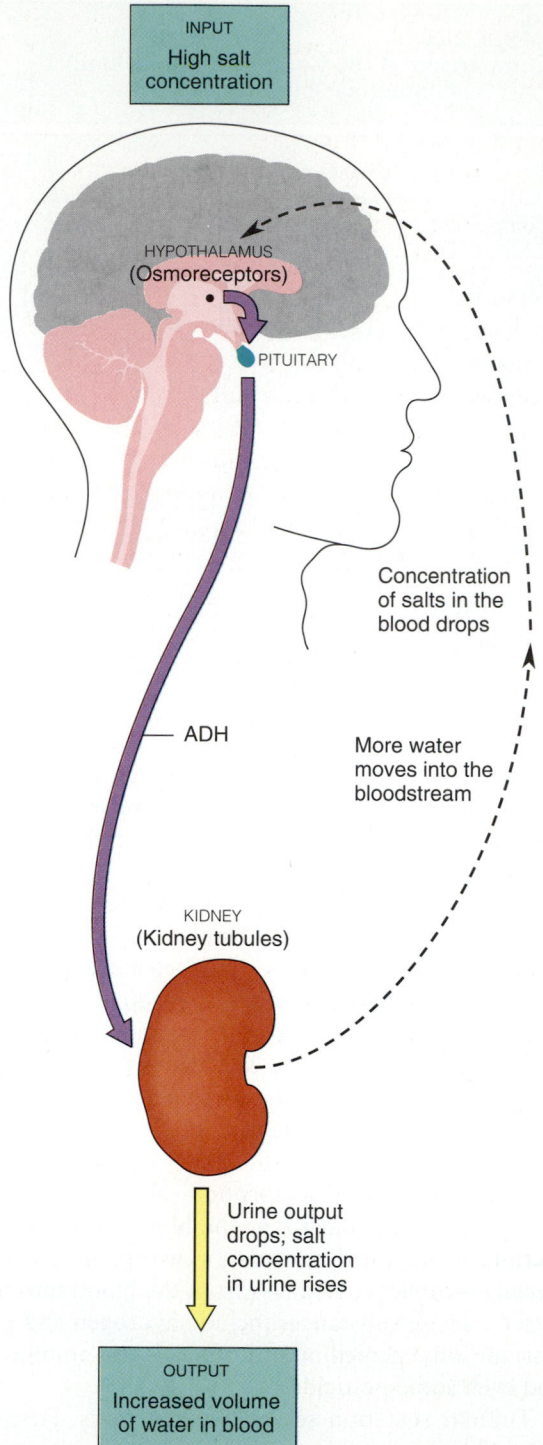

FIGURE 26.8 ADH and the Volume of Water in Urine
When the concentration of water in the body is low, the concentration of salts is high. Osmoreceptors in the hypothalamus detect the high concentration of salts and stimulate the secretion of ADH by the pituitary gland. The rate at which water is reabsorbed into the bloodstream increases. This reduces the volume of water in the urine, increasing the concentration of salts there; at the same time, it increases the volume of water in the blood, reducing the concentration of salts in the body.

In other words, if the concentration of salts is high, the concentration of water is low, and vice versa. Second, remember that *concentration* is the major determinant of the movement of water (osmosis) from one side of a membrane to the other. Water moves from a place of higher concentration to one of lower concentration. The major mechanism of sodium reabsorption is triggered by a hormone produced by the adrenal glands (two small glands, each of which is located above a kidney). This hormone, secreted by the cortex (outer layer) of the adrenal gland, is called *aldosterone*. Aldosterone is produced when special cells lining the arterioles — those located just before the arterioles end in the glomerulus — release an enzyme called *renin*. Renin acts on a protein that is normally circulating in the blood plasma, converting it to a hormone known as *angiotensin*. Angiotensin then stimulates aldosterone secretion, in a process similar to positive feedback.

Something must start the cycle, of course, and so the logical question is "What stimulates renin secretion?" Unfortunately, biologists do not have the complete answer, although we do know that a drop in sodium levels makes blood pressure fall. Baroreceptors (blood pressure receptors), which line the arteries, detect the lowered pressure and stimulate the nervous system to activate renin-producing cells.

Scientists do not completely understand the renin-angiotensin-aldosterone mechanism, but we do know that aldosterone acts on the cells of the nephrons' distal tubules and collecting ducts to promote the active transport of sodium back into the blood. Thus, aldosterone promotes reabsorption of sodium.

Biologists are aware, too, that angiotensin can also cause the constriction of arterial walls, which results in higher blood pressure. A tremendous amount of research is devoted to understanding the relationship of the renin-angiotensin mechanism to high blood pressure, since high blood pressure is a "silent killer" that is responsible for the deaths of thousands of individuals each year.

GAS EXCHANGE

- In aerobic organisms, cellular respiration involves the processes within the cell by which energy is released, oxygen is utilized, and carbon dioxide is given off. Internal respiration is the mechanism by which these gases move between cells and the blood. External respiration, or breathing, is the process by which oxygen and carbon dioxide are exchanged with the air.

Gas exchange is the process whereby oxygen and carbon dioxide are exchanged in cells. Those fortunate enough to have given birth or to have witnessed the birth of an infant remember waiting for the newborn baby's first breath (which results in gas exchange) after the umbilical cord — its lifeline to the mother — was cut. What happens after the umbilical cord is cut that causes that first breath? When the infant is separated from the placenta or afterbirth, it loses its source of oxygen, which results in an increase in the carbon dioxide (CO_2) level in the blood. An increase in CO_2 signals the respiratory centers of the body to send nerve impulses to the muscles of the rib cage and the **diaphragm,** the large muscle separating the chest cavity from the abdominal cavity. (You know where your diaphragm is located if you have ever had the wind knocked out of you.) In response, the muscles work to increase and decrease the size of the chest cavity, which brings about the baby's first breath. Its bluish color changes to a lively, healthy pink as blood rich in oxygen circulates through its body. It is obvious that we need to breathe, but why?

We can live without food for weeks and without water for days, but without oxygen we can sustain life only for precious minutes. In that respect, the cells of our bodies are no different from those of all other organisms that utilize aerobic respiration for the release of chemical energy in the high-energy carbon compounds. We all require oxygen for this metabolic activity (Chapters 7 and 9). High-energy carbon compounds (food) enter our cells, and through a series of reactions, most of them requiring oxygen, chemical bonds are broken and the energy released is used to synthesize ATP. The by-products of the process are water and carbon dioxide, a poisonous gas that must be removed from the cells and the body continually. Conversely, oxygen cannot be stored by cells and must constantly be supplied.

Various mechanisms have developed in animals to ensure a constant supply of oxygen and the removal of carbon dioxide. The result is an exchange of gases between the organism and the environment in a physical process called **respiration.** Respiration can occur at three levels — cellular, internal, and external. We discussed respiration at the *cellular level* in Chapters 7 and 9; *internal* respiration is the exchange of carbon dioxide and oxygen between the circulatory system and the tissues of the body; *external respiration* (breathing) occurs when carbon dioxide and oxygen are exchanged between the blood and the environment.

In all organisms and in all forms of respiration, the exchange of oxygen and carbon dioxide is a product of *diffusion*, the random movement of molecules from an area of higher concentration to one of lower concentration. There are limitations on the diffusion process, however. First, diffusion is only efficient over a short distance: 1 millimeter (about the thickness of a dime), to be exact. Second, the respiratory surface — the membrane through which the gases diffuse — must be thin and moist and must provide a large surface area. Third, the process requires that oxygen and carbon dioxide be in solution. This means that some form of transporting fluid (water or blood) must be available.

Respiratory Mechanisms: A Phylogenetic Perspective

■ **In simple organisms, gases diffuse readily across the cell membrane or body surface. As animals became more complex, and as the environment in which they live changed, the mechanisms of respiration changed too, adapting in different ways to the limitations of the diffusion process.**

All organisms rely on diffusion for the exchange of gases between the environment and their cells. But remember that diffusion is only efficient over a short distance. This does not pose a problem for microscopic organisms. In the amoeba and the paramecium, for example, gases move easily across the cell membrane. Nor is it a problem for the sponge, the hydra, the jellyfish, and other multicellular organisms whose cells lie close to the body surface and whose level of activity (and metabolic needs) is low.

In certain animals, body form facilitates the exchange of gases between environment and cells. It won't surprise you to learn that flatworms are flat. What you may not know is that their body form increases the surface-to-volume ratio, cutting down the distance that gases and other substances have to travel between the environment and the cells (Figure 26.9a).

In more complex animals, some mechanism became necessary to allow diffusion to operate. In the earthworm, for example, the skin is a respiratory surface; that is, it is thin and moist, allowing the ready diffusion of gases across it. But the cells of the organism are some distance from the skin. Here we find blood vessels acting as a sort of bridge. Oxygen moves through the skin into the bloodstream, then into the body cells; carbon dioxide moves out of the cells into the bloodstream, then diffuses out of the body through the skin (Figure 26.9b).

Insects breathe through a network of chitin-lined tubules called *tracheae*. These tubules are actually infoldings of the exoskeleton. They carry air into the organism's body, closer to the interior cells. At their closed ends are fluid-filled cells through which oxygen diffuses into the body's cells and carbon dioxide diffuses out (Figure 26.9c).

We hinted earlier that the environment in which the organism lives also plays a part in shaping its respiratory mechanisms. An aquatic environment

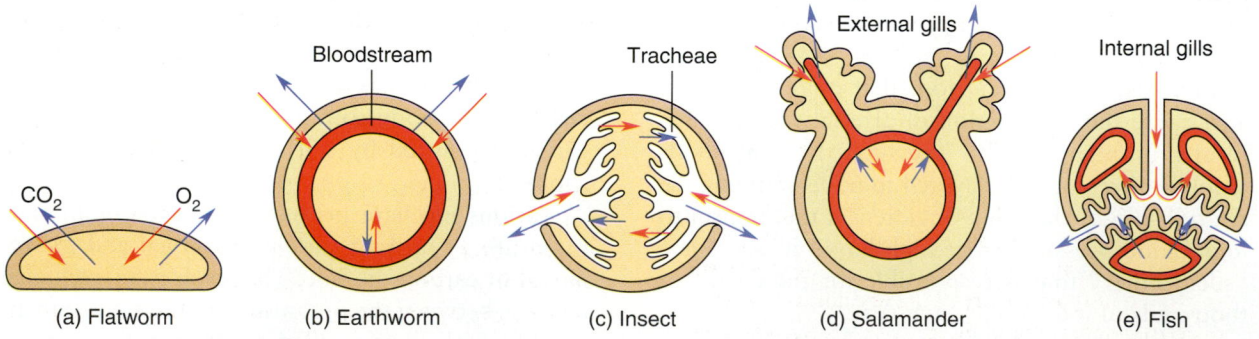

FIGURE 26.9 Respiratory Systems
The complexity of the organism, its body structure, and the external environment all play a part in the form of its respiratory system. (a) The flatworm's shape increases the surface-to-volume ratio of its body, facilitating the diffusion of gases into and out of the body. (b) In the earthworm, the bloodstream bridges the distance between the organism's respiratory membrane (its skin) and tissue cells. (c) In insects, the gas exchange takes place in tracheae, fine tubules that branch through the body. (d) In some salamanders, external gills increase the area of the respiratory membrane. (e) In the fishes, gills are internal. (f) All terrestrial vertebrates have lungs that are closely related to a network of capillaries. It is through these capillaries that gases are exchanged with the body's tissues.

certainly facilitates diffusion in some ways. The respiratory membrane is always moist, and there is a constant source of circulatory fluid at hand. But water is a poor source of oxygen, since it is less than 1 percent oxygen by volume. In stagnant water the percentage is even lower. Air, on the other hand, is about 21 percent oxygen by volume. In addition, oxygen's rate of diffusion in water is about 300,000 times slower than it is in air.

In some aquatic animals, *gills* — filamentous structures filled with blood capillaries — function as the moist respiratory surface. External gills, like those in certain salamanders and mollusks, extend outward from the organism's body (Figure 26.9d). Oxygen diffuses in from the water to the capillaries, and moves through the bloodstream to the body's tissues. Carbon dioxide follows a similar path in reverse as it moves out of the body. The gills in most fish are inside the body, in chambers near the pharynx. Here water moves from the mouth to the gills, carrying oxygen that diffuses into the blood capillaries (Figure 26.9e).

All terrestrial vertebrates — reptiles, birds, and mammals — have *lungs,* internal chambers lined with moist epithelium and closely associated with a network of blood capillaries (Figure 26.9f). Oxygen and carbon dioxide diffuse into and out of the lungs and the bloodstream by way of these capillaries. The structure of the lungs and the number of capillaries associated with them change with the metabolic activity of the animal. The lungs are more complex and the capillaries more numerous in birds and mammals, organisms with high levels of metabolic activity.

Gas-Exchange Structures: Gills and Lungs

■ **The respiratory interface is the site where gas exchange occurs. Gills, filamentous infoldings of surface cells containing numerous blood vessels, function in watery environments. Water washes over the filaments, producing a steady oxygen supply. Lungs are an internal respiratory system found in most air-breathing organisms. Their internal nature maintains moist surfaces, and their surface area is increased by alveoli, which are microscopic air sacs enclosed in capillary sheaths.**

To achieve more efficient exchange, several methods have evolved that increase the area of the respiratory surface where it meets the environment — the *respiratory interface.* The two major structures are *gills* and *lungs.* **Gills** are modifications of the body wall, and are thin and rich with blood vessels. Their structure provides a tremendous increase in the surface area exposed to the environment, maximizing gas exchange. The gills of invertebrates are rather simple extensions of the body surface; in vertebrates like the fishes, gills are thin filaments well supplied with blood (Figure 26.10). The flow of blood through the capillaries runs in the opposite direction from the flow of water across the filaments. The countercurrent created by the opposing flows increases the efficiency of the oxygen exchange.

The surface area devoted to gills varies with the amount of oxygen available in the environment. For example, salamanders living in swamps or stagnant water (containing little oxygen) have larger gills than salamanders that live in running water (which is richer in oxygen). Gill size also varies with the metabolic needs of the animal. A fast-moving mackerel has a much larger gill area (in proportion to its body weight) than does the goosefish, a fish that is not very active.

Fishes can be grouped into two major divisions, those that have bony skeletons and those with skeletons composed of cartilage. (A tuna, for instance, is a bony fish, while a shark is a cartilaginous one.) The gills of bony fishes are covered with a bony flap known as the *operculum* (Figure 26.10d). If you have ever gone fishing, you probably noticed this flap as you strung the line of your stringer through the mouth and gills of the fish you caught. The movement of the operculum (somewhat like the movement of a bellows), the rhythmic opening and closing of the mouth, and the slow fanlike movement within the gill chambers combine to produce a one-way current of water across the gills. Because the cartilaginous fishes lack this bony operculum, they usually must keep moving through the water to resupply their gills with oxygen. However, it should be noted that contrary to popular belief, sharks (a type of carilaginous fish) can rest and do not need to be constantly moving in order to respire (Figure 26.10e).

It is interesting to note that even though gills are very efficient respiratory surfaces, fishes use 20 percent of their metabolic energy to obtain their oxygen, because of the low concentration of oxygen in the water, its viscosity, and its mass (water is hard to move). In comparison, air-breathing animals with lungs, like ourselves, use only about 2 percent of their energy to obtain oxygen. It is this environmental pressure that underlies the fact that air-breathing animals that returned to the water (such as seals, whales, and sea snakes) retained air-breathing respiration. Air is a much richer source of oxygen.

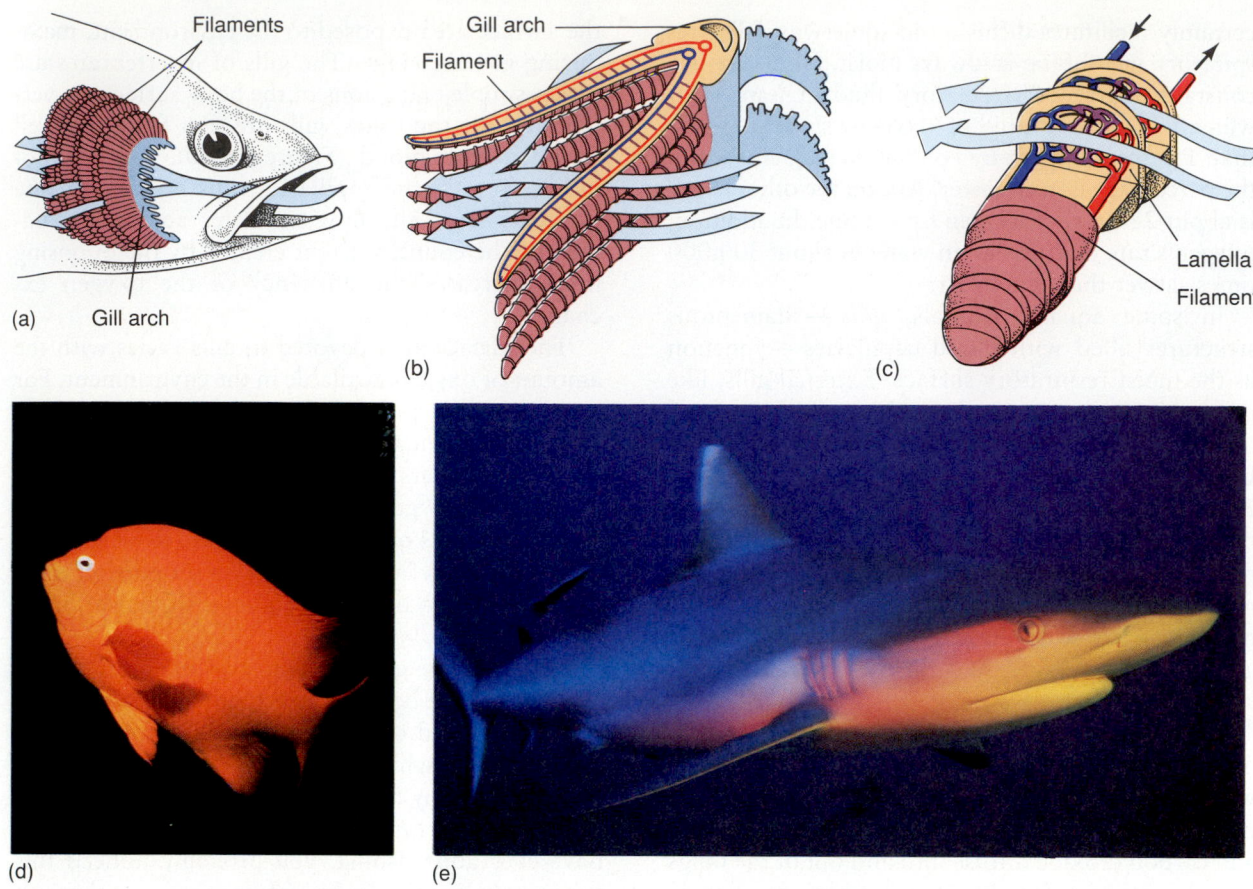

FIGURE 26.10 Internal Gills and Gill Structure in Bony and Cartilaginous Fishes
(a) There are four gills in the pouch on each side of the body. (b) Each gill is a double row of slender filaments attached to a cartilaginous arch. In the cutaway, you can see the system of blood vessels running through the filaments. (c) In the lamellae are networks of capillaries. It is in these capillaries that the gas exchange takes place. The flow of blood through the capillaries (small arrows) runs directly opposite the flow of water across the gill (large arrows). (d) In this bony fish, the opercula cover the gill arches and help in respiration. (e) In the shark, a cartilaginous fish, the opercula are absent.

Living on land does have advantages in terms of respiration, since air contains a greater concentration of oxygen. However, everything has its price, and land-based life is no exception. The major difficulty of life on land is keeping the thin membranes moist, for, as we pointed out earlier, oxygen must be in solution to move across the respiratory surface. The simple solution to this problem was the evolution of an interior respiratory surface, where the membranes could be kept moist. A variety of internal structures developed to meet this requirement.

Insects, which were the first land animals, developed a series of infoldings of the body surface that formed internal tubes and chambers to serve as an internal respiratory surface. Chitin-lined air tubes called *tracheae* extend from the body surface to inside the body cavity, where they branch into smaller tubules that lead to individual cells responsible for the direct exchange of gases (Figure 26.11). The external openings (spiracles) of the tracheae open or close according to the need for oxygen.

In land vertebrates, large internal cavities known as **lungs** are found, each of which consists of a large, moist surface area available for gas exchange. Lungs are chambers that receive air through a series of tubes, and their surface area is increased by a series of tiny microscopic air pockets known as **alveoli** (the plural form of the term *alveolus*). Each alveolus consists of a sac formed of epithelial tissue and a surrounding layer of capillaries that cover the air sac, as a glove covers your hand.

It is across the alveolar membrane and capillaries that the actual exchange of gases occurs. If laid end to end, the capillary network of the alveoli within lungs would extend hundreds of miles, a remarkable fact when you consider that in each pound of fatty tissue there is about 6 to 15 miles of blood vessels. The lungs, with their hundreds of miles of capillaries, weigh only about 2 pounds.

The lungs provide the moist respiratory surface for gas exchange on land. The human respiratory system is a good example of air breathing in vertebrates, and it is to this subject that we now turn.

HUMAN RESPIRATION

■ **Like other mammals, humans rely on lungs for gas exchange. The respiratory system includes nasal passages, pharynx, epiglottis, larynx, trachea, bronchi, bronchioles, alveolar ducts, and alveoli. The pleura encloses the lungs.**

As we breathe (undergo the mechanical act of moving gases into and out of the lungs in a process

sometimes called ventilation), our systems go through several steps:

1. The bulk flow of air moves into and out of the lungs.

2. There is an exchange of gases between the air in the alveoli and the blood in the capillaries of the alveoli.

3. Gases are transported by the blood to and from the cells.

4. There is an exchange of gases between the blood and tissue fluid.

5. There is an exchange of gases between the tissue fluid and the cells.

Breathing, or ventilation, and the exchange of gases between the alveoli of the lungs and blood occur in the respiratory system. The events of gas transport on the one hand, and the actual exchange of gases between the blood and cells on the other hand, are the responsibility of the circulatory system.

Respiratory Structures

The human respiratory system (typical of most mammals) consists of several interrelated structures, illustrated in Figure 26.12.

Structures within the nasal passages warm, filter, clean, and moisten the air. Following the nasal passages is the *pharynx,* which connects with two tubes at the back of the mouth, the *esophagus,* which goes to the stomach, and the *larynx,* which leads to the lungs. The larynx is covered by the **epiglottis,** a flaplike structure that covers the opening of the larynx during swallowing, so that food goes into the stomach and not the lungs. Contained within the larynx are two vocal cords that vibrate from air forced over them to produce sounds.

After air passes through the larynx, it enters the *trachea,* or windpipe (not to be confused with the insect tracheae). The walls of the trachea are reinforced by cartilage rings that prevent it from collapsing. The trachea branches into two *bronchi,* one that enters the right lung and one that enters the left. As the bronchi enter the lungs, they branch into smaller *bronchioles,* which eventually branch to form the *alveolar ducts.* These ducts lead into the grapelike clusters, or air sacs, of the alveoli, of which there are over 300 million. The alveolar epithelium, if spread out, would measure 100 square meters (which happens also to be the size of the average college biology laboratory).

Lungs are located within a thin-layered sac called the *pleura.* The pleura is a membrane composed of

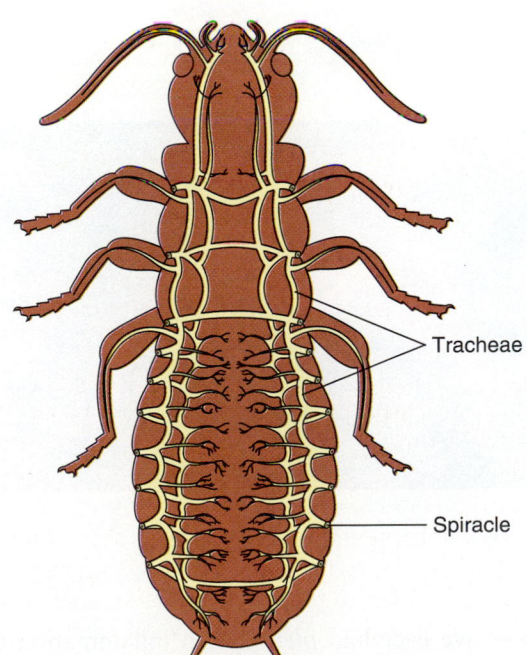

FIGURE 26.11   The Tracheal System in Insects
Air moves into the organism through paired openings (spiracles) in each segment of the body. The spiracles open or close according to the needs of the insect. Tracheae—chitinous tubules—carry the air to the internal tissues. Oxygen leaves the tracheae and carbon dioxide enters them through specialized cells in their terminal branches.

Tracheae

Spiracle

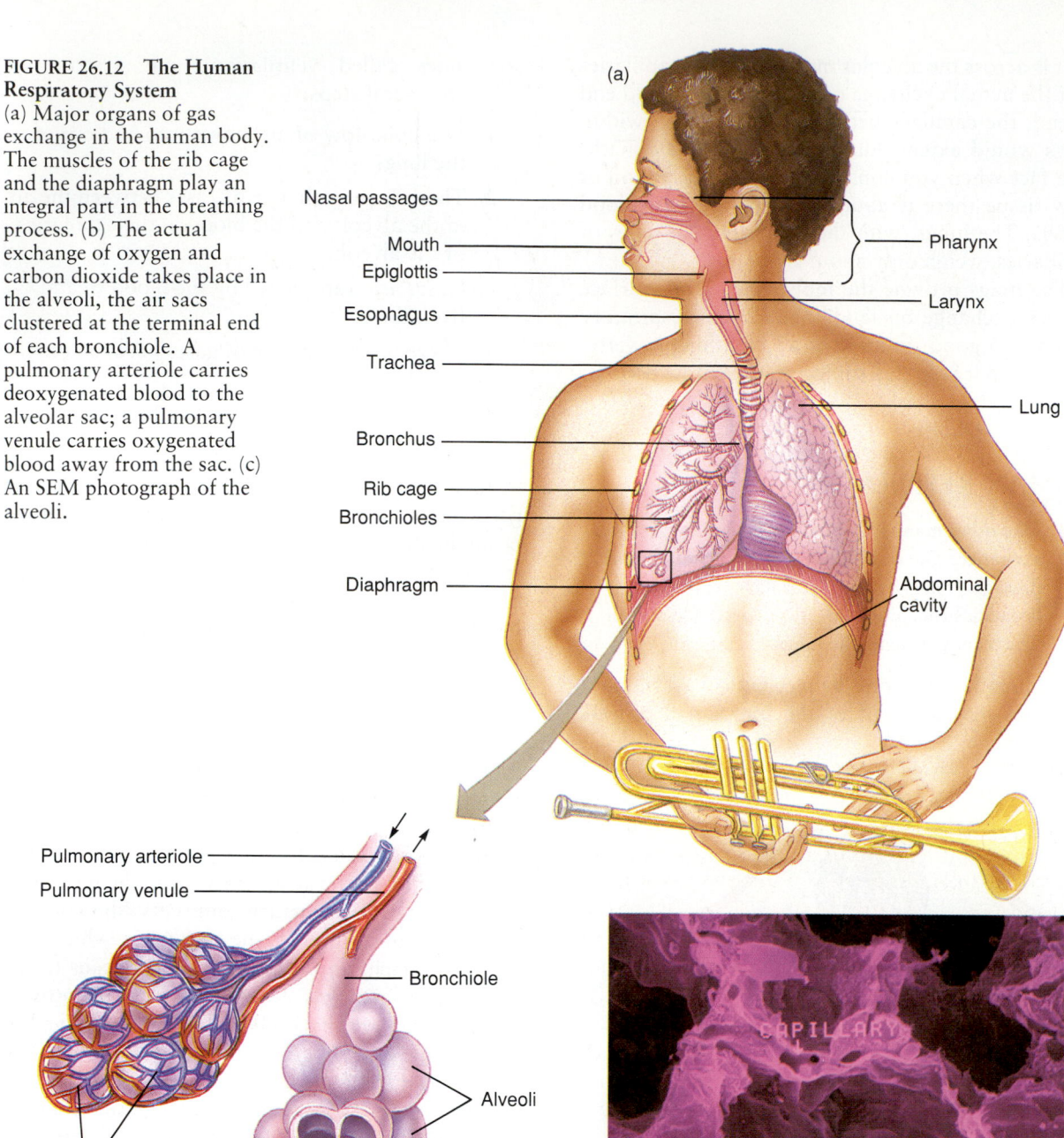

FIGURE 26.12 The Human Respiratory System
(a) Major organs of gas exchange in the human body. The muscles of the rib cage and the diaphragm play an integral part in the breathing process. (b) The actual exchange of oxygen and carbon dioxide takes place in the alveoli, the air sacs clustered at the terminal end of each bronchiole. A pulmonary arteriole carries deoxygenated blood to the alveolar sac; a pulmonary venule carries oxygenated blood away from the sac. (c) An SEM photograph of the alveoli.

epithelial and loose connective tissue cells forming a thin sac that surrounds each lung. The lungs inflate and deflate within this membrane. The cells of the pleural membrane secrete a lubricating fluid into the space betwen the layers that reduces friction as the lungs move up and down within the chest cavity. Normally this fluid is absorbed by the lymph system as rapidly as it is produced. However,

if you have ever had *pleurisy* (an inflammation of the pleura that can result in an accumulation of fluids in the pleural sac), you know that when the pleura is irritated breathing can be painful as the fluid accumulates and as the lungs rub up against the inflamed pleura. An excessive accumulation of the lubricating fluid can result in an increased pressure in the pleural sac. This increased pressure may

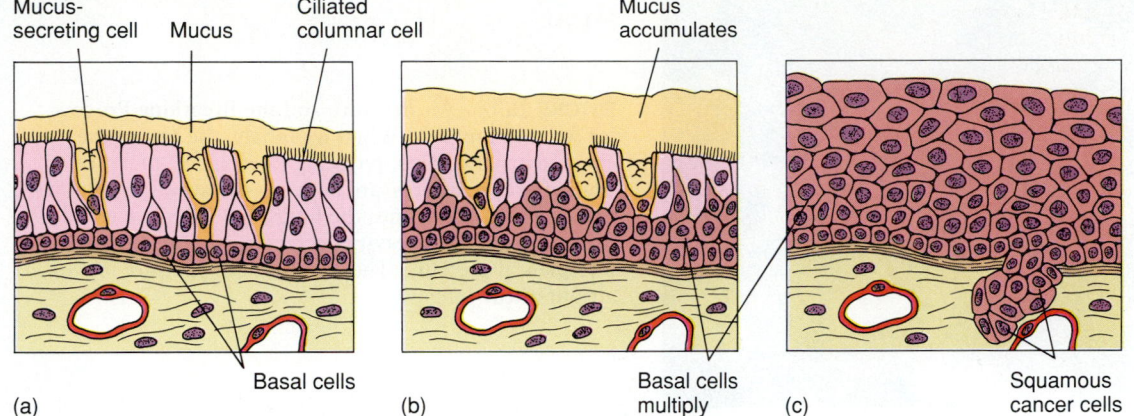

Mucus-secreting cell Mucus Ciliated columnar cell

Mucus accumulates

Basal cells

(a)

Basal cells multiply

(b)

Squamous cancer cells

(c)

FIGURE 26.13   Smoking and the Respiratory Passages
(a) Under normal conditions, specialized cells in the epithelium lining the respiratory passages produce mucus that traps dust and other particles. The ciliated columnar cells move the mucus up and out of the passageway. The basal cells divide continually, replacing the columnar cells as they wear out. (b) Smoke paralyzes the cilia, allowing mucus to accumulate. If cancer is induced, the basal cells begin to multiply quickly, displacing the columnar cells. (c) As the disease progresses, the basal cells continue to divide, breaking through the basement membrane, and cancerous squamous cells replace the columnar cells completely.

be enough to exceed the air pressure within the lung, causing the lung to collapse.

The nasal passages, trachea, and bronchi are lined with epithelial cells that secrete mucus and are covered by cilia. The sticky mucus traps foreign particles, and the beating action of the cilia in the trachea and bronchi sweeps the particles upward and out of the respiratory passage. Smoke and other chemical pollutants in the air can damage and paralyze the cilia, resulting in an accumulation of foreign substances that can bring on a whole array of respiratory disorders, including lung cancer (Figure 26.13). Focus 26.2 discusses lung cancer and other lung disorders.

Having surveyed the anatomy of the lungs, we can now discuss the mechanisms of breathing.

Mechanisms of Breathing

■ **Breathing is a mechanical process that is based on the air-pressure gradient inside and outside the chest cavity. When the air pressure outside is greater than that inside the lungs, air flows into the lungs — we inhale. When the air pressure inside the chest cavity is greater than outside, air flows out of the lungs — we exhale. The pressure within the lungs is governed by the diaphragm and the external intercostal muscles, which control the size of the chest cavity.**

When I was a child, a friend of mine was stricken with polio and placed in an iron lung, an enclosing cylinder that enabled him to breathe. I remember looking at him and wondering why he couldn't breathe and how an iron lung could breathe for him. I also wondered how he could live his entire life in that machine.

I now know that the iron lung works by simply changing the air pressure within the chamber relative to the air pressure outside it (Figure 26.14a). As the pressure increased, there was more pressure inside than outside the chamber, and so air would flow out of my friend's lungs. As the pressure in the iron lung decreased, there was more air pressure outside the chamber, so that air moved into his lungs.

The principles that govern the mechanics of an iron lung are similar to those of normally functioning lungs. When the air pressure in the alveoli is greater than atmospheric air pressure, air moves out of the lungs; when the pressure within the alveoli is less than that in the atmosphere, air moves into the lungs. What would happen if the alveolar and atmospheric pressure were equal? Correct, there would be no air flow.

The pressure within an iron lung is maintained by an apparatus similar to the bellows used by an old-fashioned blacksmith. How do we increase or decrease the air pressure in our lungs? The mecha-

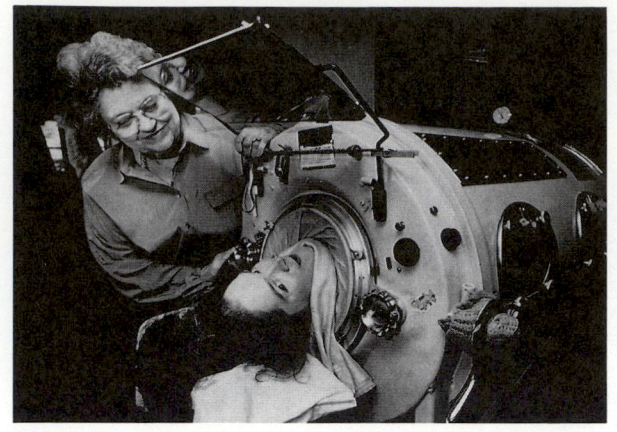

(a)

FIGURE 26.14 Air Pressure and the Breathing Process
(a) An iron lung. (b) When we inhale, the size of the chest cavity increases, reducing interpleural pressure. The difference in pressure between the atmosphere and the lungs moves air into the lungs. (c) When we exhale, the size of the chest cavity decreases, increasing interpleural pressure. The difference in pressure moves air out of the lungs.

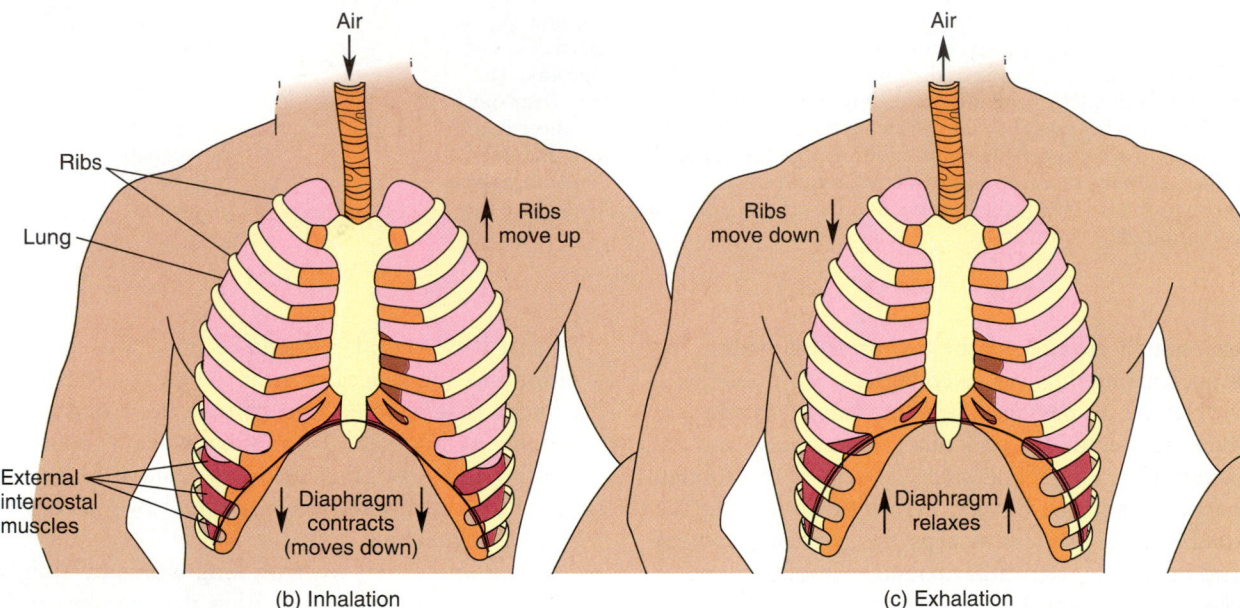

(b) Inhalation

(c) Exhalation

nism we employ is based on changing the volume of the chest cavity through the contractions and relaxations of the muscles of the diaphragm and the *external intercostal muscles* (*inter*, between; *costia*, ribs) (Figure 26.14b and c).

When we *inhale*, or inspire, the diaphragm contracts, flattens out, and moves downward. At the same time, the external intercostal muscles move the ribs upward. The sum of the movement is to increase the volume of the chest cavity, and interpleural pressure drops. Because the pressure inside the chest cavity is less than that of the surrounding environment, air flows in and the lungs expand. When we *exhale*, or expire, the external intercostal muscles move the ribs downward and the diaphragm relaxes. The size of the chest cavity decreases, increasing the air pressure in the lungs, and air moves out. Thus, the act of breathing is a mechanical process involving changes in air pressure of the chest cavity.

We don't completely deflate our lungs when we exhale. In fact, only about 10 percent of the total volume of air in the lungs is exchanged in each normal breath. However, deliberate deep breathing or exercise can increase the volume of air inhaled and exhaled. It can also alter the rhythmic rate of breathing. So the next question to ask is, how is the rate of respiration controlled?

Regulation of Respiration

■ **Respiration is regulated by both voluntary and involuntary mechanisms. Chemoreceptors in the arterial walls and the pneumotaxic center in the spinal cord detect the high levels of CO_2 and low pH. This stimulus is relayed to the medulla of the brain, which then sends neural signals regulating the depth and rate of breathing to the diaphragm and external intercostal muscles.**

As may be obvious by now, both voluntary and involuntary mechanisms are involved in controlling respiratory rate. Through voluntary control, you can consciously regulate the rate and depth of respiration by consciously sending nerve impulses to the muscles that increase and decrease the size of the chest cavity. My friend with polio was in an iron lung because he could not control his breathing. The polio virus had damaged either his spinal cord, his *phrenic nerves* (which lead from the spinal cord to the diaphragm), or his intercostal nerves (which lead from the spinal cord to the intercostal muscles), or all of them. Because of the paralysis, he could not contract and relax his diaphragm or external intercostal muscles in order to produce that rhythmic pattern of contractions and relaxations that we call breathing.

The rate and depth of breathing are regulated by that portion of the brainstem called the *medulla*. through the activities of the phrenic and intercostal nerves. Chemoreceptors in the arterial walls detect changes in carbon dioxide and pH level in the blood and transmit that information to the medulla. The medulla also receives information from the *pneumotaxic center,* which is a set of chemoreceptors that monitor pH levels and carbon dioxide concentration in the spinal fluid. (A secondary control monitors the oxygen level in the carotid arteries. This mechanism is seldom activated but could be important at high altitudes.) The information is processed by the medulla, nerve signals are sent to the motor nerves of the muscles that control respiration, and the rate and depth of breathing are regulated.

In simplest terms, when the concentration of carbon dioxide in the blood or spinal fluid increases, nerve signals increase the rate of respiration. That is why we yawn after a period of inactivity and shallow breathing. The carbon dioxide concentration is high, so the nervous system signals an action (yawning) designed to decrease the amount of carbon dioxide in the blood by facilitating gas exchange. As you can see, a delicate homeostatic mechanism carefully monitors and regulates gas exchange.

The Role of Blood in Gas Transport

■ **The blood moves gases through the body. Oxygen enters the blood in the lungs and combines with hemoglobin in the red blood cells to form oxyhemoglobin. It is then carried to the tissues, where it is released. Carbon dioxide diffuses from the cells to the capillaries, where it reacts with the water in plasma to form bicarbonate and hydrogen ions. The blood carries carbon dioxide to the lungs, where it leaves the body. Because of the Bohr effect, gas exchange is concentrated in areas in which the demand for energy is highest.**

Oxygen Transport. The movement of oxygen from the environment through the body begins in the lungs. The pulmonary arteries carry deoxygenated blood from the heart to the lungs. The oxygen concentration in the alveoli is higher than that in the capillaries surrounding them. This difference creates a diffusion gradient (Figure 26.15). Oxygen diffuses out of the alveoli into the capillaries, returning to the heart through the pulmonary veins. The heart pumps the oxygenated blood through the body, where the diffusion gradient works in reverse — oxygen diffuses out of the blood into the tissue fluid, then into the cells.

We know that blood picks up and releases oxygen. But how? Oxygen is relatively insoluble in water, and plasma, the fluid portion of the blood, consists primarily of water. Hence, plasma is not an

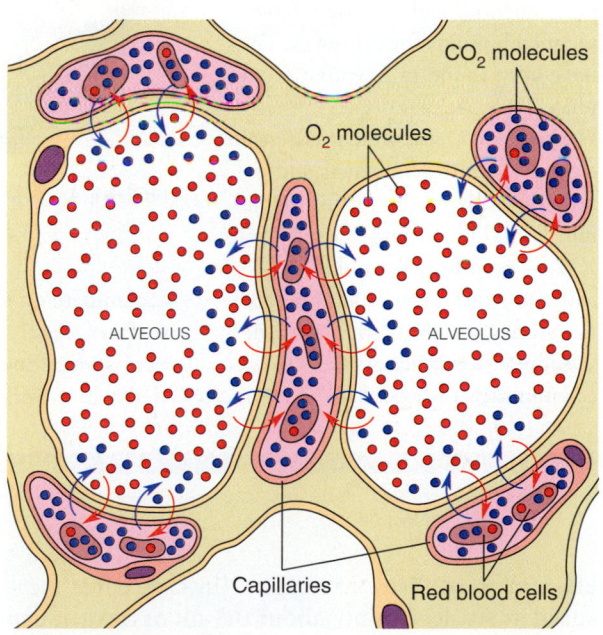

FIGURE 26.15 Diffusion Gradients and the Exchange of Gases Between the Blood and Alveolus
Blood rich in CO_2 enters the capillary network of the alveolus. Because the CO_2 concentration is higher in the blood of the capillary, CO_2 diffuses into the alveolus. The alveolus has a higher concentration of O_2 as compared to the blood in the capillary. Therefore, O_2 diffuses into the capillary, where it combines with the hemoglobin in the red blood cell to be transported throughout the body. The blue arrows represent the movement of CO_2 molecules and the red arrows represent the movement of O_2 molecules.

FOCUS 26.2
Kidney Stones: A Real Pain in the Back

Kidney stones cause some of the most excruciating pain known to humans. Spasms of pain rack the bladder and lower back. The sufferer doubles over in pain — often in indescribable agony. One patient said that on a scale of 0 to 10, kidney pain is an 11. This painful affliction is not only a modern phenomenon. Kidney stones have plagued humans for centuries. They have been found in Egyptian mummies. They are mentioned in the Hippocratic Oath, which reflects ancient Greek medical beliefs and is now taken by all physicians. Benjamin Franklin, Peter the Great, Louis XIV — even Burt Reynolds — all suffered from them. What are kidney stones, and how to they arise?

Kidney stones are crystals of calcium salts or other salts that

FIGURE A   A kidney stone.

accumulate on the kidneys' inner surfaces (Figure A). This accumulation is often due to a citrate deficiency in the urine. (Citrate

normally prevents the salts from crystallizing.) Presently the U.S. Food and Drug Administration is testing a pill composed of potassium citrate that is designed to prevent calcium salts from crystallizing in the urinary tract. If the testing is successful, this medication will help those patients who suffer from chronic kidney stone attacks.

Physicians can also administer other drugs that dissolve the kidney stones. The grainlike particles can then pass out of the body — painfully — in the urine. However, for about 100,000 sufferers of kidney stones whose stones are too large to pass through the urinary tract, surgery is the only answer. If they are not removed, they will block the passage of urine, which in turn will cause kidney infection.

efficient transporter of oxygen. (In fact, when measured at sea level, only about 0.3 ml of oxygen can dissolve in 100 ml of blood plasma.) However, hemoglobin greatly increases the blood's oxygen-carrying efficiency because hemoglobin consists of four polypeptide chains and four heme groups. Within each heme group is an iron atom that can bind with one molecule of oxygen (Figure 26.16).

As we discussed in Chapter 24, red blood cells are little sacs of hemoglobin (each cell contains about 250 million hemoglobin molecules), and hemoglobin functions in oxygen transport, carrying over 60 times more oxygen per unit volume than

plasma alone. When the blood picks up oxygen, the hemoglobin molecules complex with the oxygen gas to form *oxyhemoglobin*. When the reverse happens, oxyhemoglobin loses its oxygen to become hemoglobin. We can summarize this reaction in this way:

$$\text{Hb} \quad + \quad 4(\text{O}_2) \rightleftharpoons \text{Hb.4(O}_2)$$

Hemoglobin   Oxygen   Oxyhemoglobin

[The period in Hb.4(O_2) indicates that the oxygen is loosely bound to the hemoglobin.]

The actual amount of oxygen carried by a hemoglobin molecule is dependent on the partial pressure

In the past, kidney stone surgery was major, involving a large, deep incision in the lower back and surgical removal of the stones. In some instances, the urologist could insert a hollow tubular instrument into the bladder and snare the stone with a basketlike device. Today, however, technology has come up with a new procedure with the impressive name of extracorporeal shockwave lithotripter (out-of-body stone crusher, or ESWL).

Developed in West Germany, the ESWL is being used to literally blast kidney stones to pieces with shockwaves. The patient is mildly anesthetized and is lowered into a large steel tank full of water. The precise location of the stone is found with x-rays and is viewed on a television monitor. A brass reflector is aimed at the stone, the surgeon pushes a button, and a sparkplug at the bottom of the tank fires, vaporizing the water near it and sending a pressure shockwave to the reflector, which bounces it at the stone. The shockwave (sort of a miniature sonic boom) passes harmlessly through the patient's soft tissue and shatters the stone (Figure B).

The stone is not destroyed immediately. Depending on its size, more than 1000 shockwaves may be required to smash it into pieces small enough to wash out in the patient's

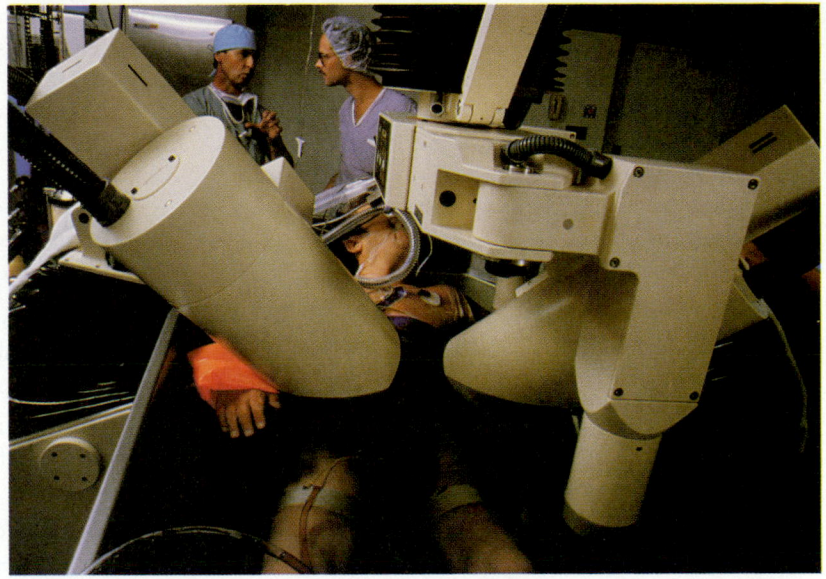

FIGURE B   Patient undergoing ESWL treatment for kidney stones.

urine without causing the pain usually associated with passing a kidney stone. Physicians estimate that ESWL can be successful in 70 percent of the kidney stone cases that would normally require surgery.

Another new surgical method also does not require major surgery. An ultrasonic probe is inserted into the body through a small incision in the patient's back. X-rays are used to locate kidney stones, and once they are pinpointed, the surgeon blasts the stones with high-frequency sound waves. In cases involving larger stones, both ESWL and ultra-

sound can be used to disintegrate the stones.

The advantages of these new methods should not be underestimated. The average kidney stone operation involves the formation of a 10-inch scar, 4 or 5 days in the hospital, and 4 to 8 weeks of recuperation. The costs average out at $21,000, and that does not include the loss of productivity of the patient during recuperation. The cost of ESWL is $8000, one day in the hospital, and a return to productivity in a week or less. Which would you prefer?

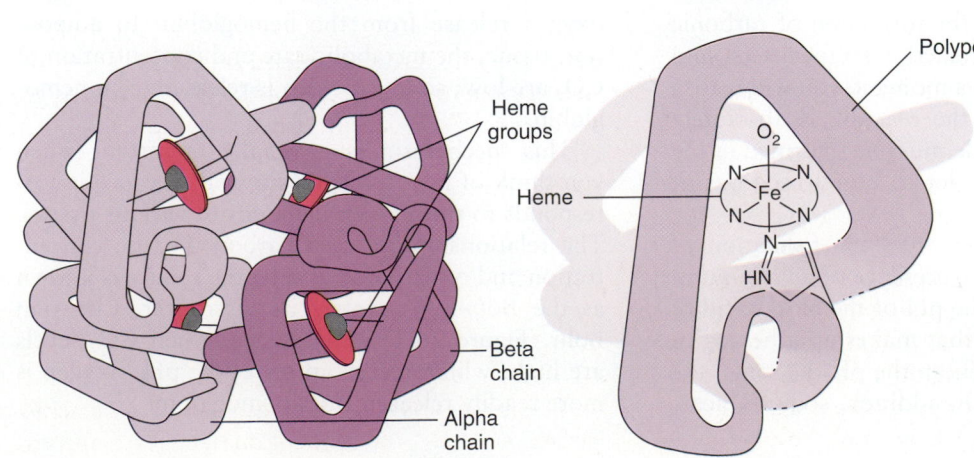

FIGURE 26.16   The Molecular Structure of Hemoglobin Hemoglobin functions well in transporting oxygen. Iron atoms within its four heme groups easily bind to oxygen.

of the gas in which the oxygen is contained. For example, at sea level, where the partial pressure of the atmosphere is high, the binding sites of the hemoglobin's iron atoms can pick up oxygen until the molecule is saturated. At higher levels, like the mountains of Boulder, Colorado, the partial pressure of the atmosphere is low; so fewer of the molecule's oxygen-binding sites pick up oxygen and the hemoglobin molecule is less saturated.

Carbon Dioxide Transport. In the tissues of actively metabolizing cells, carbon dioxide concentration is higher than that in the capillaries. This difference establishes a concentration gradient, and the carbon dioxide diffuses from the cells into the blood of the capillaries for transport to the alveoli of the lungs, where it is exhaled. In order to get from the tissues to the lungs, the carbon dioxide is carried in the blood in one of three ways:

1. 7 percent is dissolved in the blood plasma.
2. 23 percent can combine with hemoglobin of the red blood cells to form *carbaminohemoglobin* ($HbCO_2$), a large protein.
3. The remaining 70 percent is transported in the form of the bicarbonate ions (HCO_3^-) contained in blood plasma.

Let's discuss the third method in more detail. When the carbon dioxide combines with water, the result is carbonic acid, which dissociates into a bicarbonate ion and hydrogen ion as follows:

$$CO_2 + H_2O \rightleftharpoons H_2CO_3 \rightleftharpoons HCO_3^- + H^+$$

| Carbon dioxide | Water | Carbonic acid | Bicarbonate ion | Hydrogen ion |

This reaction can occur in the plasma, although the process is very slow. However, red blood cells contain an enzyme, *carbonic anhydrase,* which speeds up the reaction 250 times. Carbonic anhydrase not only catalyzes the formation of carbonic acid but can reverse the reaction to form CO_2 and H_2O. (If you consider for a moment, you will realize that the reversibility of the reaction is absolutely necessary. Carbon dioxide must be reformed in the alveoli of the lungs in order to be released out of the body.)

Part of the reaction involves the formation of hydrogen ions. Would the presence of a large number of these ions make the pH of the blood acidic? Fortunately, the protein that makes up a hemoglobin molecule can *buffer* (keep the pH the same) the levels of hydrogen ions. In addition, sodium bicarbonate is an important acid-base-buffering system. In other words, it helps keep the pH of the blood constant, at a level that is almost neutral. The bicarbonate ions move into the plasma, where they combine with sodium ions to form sodium bicarbonate ($NaHCO_3$ — like baking soda).

Once the blood, rich in carbon dioxide and low in oxygen, gets back to the alveoli of the lungs, the following occurs:

1. The cabon dioxide in the blood diffuses into the alveoli.
2. The carbaminohemoglobin releases its carbon dioxide, since the hemoglobin has a greater affinity for oxygen than carbon dioxide.
3. The carbonic acid dissociates back to CO_2 and water.
4. The carbon dioxide diffuses out of the blood capillaries in the alveoli of the lungs and is exhaled, oxygen diffuses from the alveoli into the blood capillaries where it is combined with hemoglobin to form oxyhemoglobin, and the circuit continues.

The result is the maintenance of a steady-state, homeostatic condition.

It is interesting to note that in larger organisms, in areas where carbon dioxide concentration is high in the body, the release of oxygen from hemoglobin occurs more easily. If oxygenated blood is circulating through tissues in which the cells are metabolizing slowly, there is a lower level of oxygen released from hemoglobin. When cells are metabolically active, more carbon dioxide is produced, hence the acidity (H^+ concentration) increases and this readily triggers the oxygen release. For example, the metabolism rate is high in the kidney tubules, requiring high levels of ATP for active transport. As a result, CO_2 concentration is also high. This results in an increased acidic state, which readily triggers oxygen release from the hemoglobin. In adipose (fat) tissue, the metabolic rate and concentration of CO_2 are low; so less oxygen is released from hemoglobin.

This mechanism is especially important when you think of it in terms of survival, because it corresponds to the oxygen demands of various tissues. The relationship between carbon dioxide concentration and oxygen demand by the tissues is known as the *Bohr effect,* after its discoverer, Christian Bohr (Figure 26.17). In essence, when CO_2 levels are high, which results in an acidic pH, oxygen is more readily released from hemoglobin.

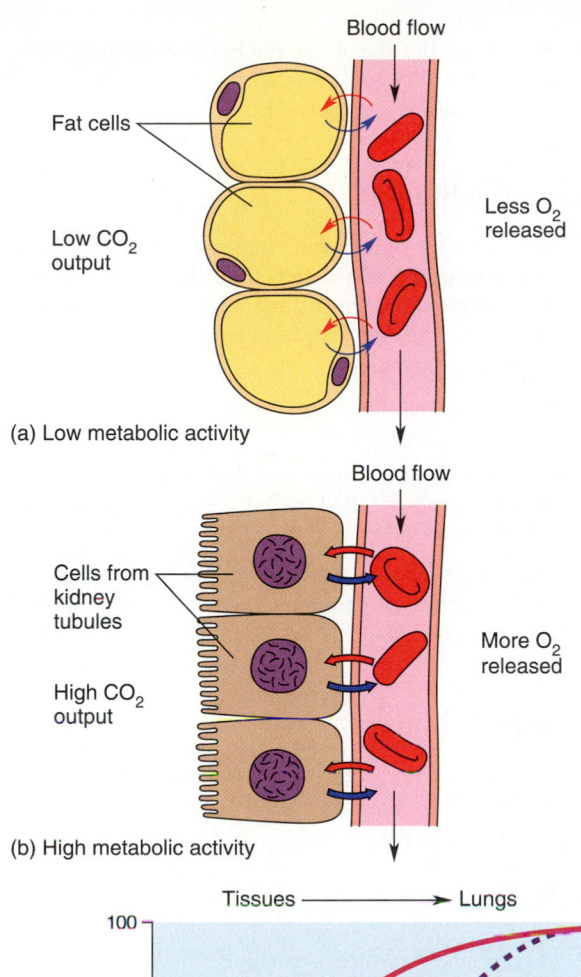

Fat cells

Low CO$_2$ output

Blood flow

Less O$_2$ released

(a) Low metabolic activity

Cells from kidney tubules

High CO$_2$ output

Blood flow

More O$_2$ released

(b) High metabolic activity

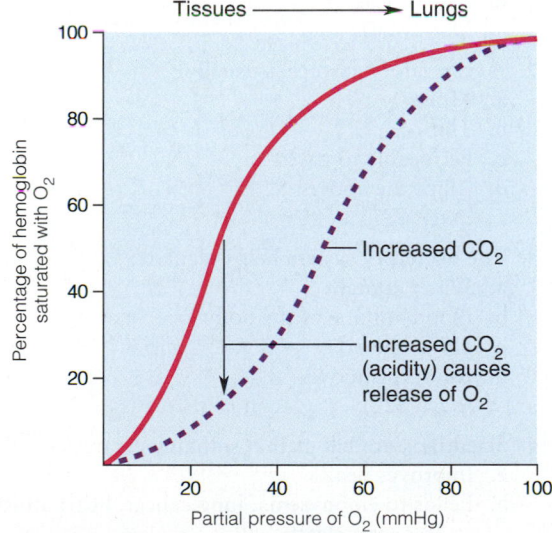

Tissues ⟶ Lungs

Increased CO$_2$

Increased CO$_2$ (acidity) causes release of O$_2$

Partial pressure of O$_2$ (mmHg)

Percentage of hemoglobin saturated with O$_2$

(c) The Bohr effect

FIGURE 26.17   The Bohr Effect
According to the Bohr effect, hemoglobin surrenders its oxygen more easily in areas where the carbon dioxide concentration is high. This means that (a) metabolically inactive cells receive less oxygen because they produce less carbon dioxide, and (b) metabolically active cells receive more oxygen because they produce more carbon dioxide. (c) The graph shows that, in general, the higher the concentration of carbon dioxide, the greater the release of oxygen from hemoglobin.

SUMMARY

1. Temperature regulation, osmoregulation, and gas exchange are some of the major mechanisms necessary to maintain homeostasis.

2. In negative feedback, a stimulus is detected and produces a response that corrects the imbalance. When the stimulus stops, the response also stops. In this way, cells communicate to maintain a steady state.

3. Body temperatures of cold-blooded animals (ectotherms) fluctuate with the environment. Warm-blooded animals (endotherms) like ourselves maintain a constant temperature despite environmental fluctuations. The ability to maintain a constant temperature increases the types of environments in which an organism can exist.

4. In warm-blooded animals, two major mechanisms maintain a constant body temperature. The rate of metabolic activity controls the amount of heat production, and the body surface regulates the loss or retention of that heat.

5. The neural and endocrine systems work together to regulate body temperature.

6. Regulation of the chemical balance of body fluids is called osmoregulation. In primitive organisms, various osmoregulatory mechanisms evolved, from simple contractile vacuoles and diffusion in some protozoa to branching collecting tubules in the flatworms. Eventually primitive kidneylike structures emerged that collect and dispose of fluids.

7. In more complex animals, the kidneys are the principal organs of excretion. The kidneys contain microscopic units called nephrons.

8. The processing of the blood in the nephron can be thought of as occurring in three phrases: filtration, reabsorption, and tubular secretion.

9. Neural and hormonal mechanisms regulate the absorption and loss of water, sodium, ions, and other substances through the nephrons. These are examples of negative feedback mechanisms.

10. Respiration occurs at three levels: cellular (the chemical processes involving oxygen use, energy release, and carbon dioxide output), internal (between the circulatory system and the tissues), and external (between the blood and the environment). External respiration, or breathing, involves the exchange of gases between the organism and the environment. In single-celled organisms, simple diffusion serves as a mechanism of gas exchange. In multicellular organisms, higher metabolic requirements and a greater body mass necessitate a specialized respiratory surface.

11. A respiratory surface, whether it is the skin of an earthworm, the gills of a fish, or the lungs of a human, consists of a large, thin, moist, membranous respiratory interface with a large surface area exposed to the environment, and intimate association with the transporting fluid.

12. For most land vertebrates, lungs provide the major respiratory surface.
13. The human respiratory system is typical of most mammals and consists primarily of a series of air passages: the nasal passages, pharynx, epiglottis, larynx, trachea, bronchi, bronchioles, alveolar ducts, and alveoli.
14. When we inhale, the size of the chest cavity increases, which results in less pressure within the cavity, and air flows in. When we exhale, the size of the chest cavity decreases, which results in higher air pressure inside the cavity, and air flows out.
15. Oxygen is carried in the hemoglobin of the red blood cells, and carbon dioxide is carried primarily in the plasma portion of the blood in the form of bicarbonate ions.
16. Gas exchange occurs between the cells, tissue fluid, and capillaries. Capillaries carry oxygen to the cells and receive carbon dioxide, which is carried back to the lungs. In the alveoli of the lungs the gas exchange between the capillaries of the alveoli and the environment occurs. In the areas of the body where carbon dioxide concentration is high the release of oxygen takes place more easily.
17. Nerves and hormones interact to regulate respiratory rate and volume.

KEY TERMS

negative feedback loop	nephron
positive feedback loop	Bowman's capsule
ectotherm	glomerulus
endotherm	filtration
osmoregulation	reabsorption
urinary system	tubular secretion
kidney	diaphragm
bladder	respiration
urethra	gill
cortex	lung
medulla	alveolus
renal pelvis	epiglottis

REVIEW QUESTIONS

True/False

1. Living organisms must maintain a constant internal environment.
   true   *false*   *(page 615)*
2. The fishes, amphibians, and reptiles are vertebrate animals that are usually considered ectothermic.
   true   *false*   *(page 617)*
3. In humans and other endothermic mammals, the hypothalamus is important in temperature regulation.
   true   *false*   *(page 619)*

4. In humans the alveoli are the functioning units of external respiration.
   true   *false*   *(page 634)*

Fill in the Blank

1. As your body temperature increases, the excess heat is sent to processing centers in the brain, proper temperature adjustments are made, and your temperature decreases. This is an example of _____ feedback. *(page 616)*
2. The functional unit of the mammalian kidney is the _____. *(page 624)*
3. The portion of the nephron tubule nearest the Bowman's capsule is called the _____ convoluted tubule. *(page 625)*
4. Red blood cells are literally little sacs of _____ molecules that function in oxygen and carbon dioxide transport. *(page 640)*

Multiple Choice

1. In humans the kidney's excretory function involves:
   a. Filtration
   b. Reabsorption
   c. Secretion
   d. All of these
   (page 624)
2. A respiratory surface is usually:
   a. Moist
   b. Thin
   c. Permeable to gases
   d. All of the above
   (page 631)
3. The excretory system helps regulate:
   a. Water content
   b. Blood volume of the body
   c. pH of the body
   d. All of the above
   (page 622)
4. Scientific data tell us that smoking cigarettes:
   a. Improves health
   b. Leads to emphysema, lung cancer, heart attack, and/or early death
   c. Is harmful only in lower animals
   d. Is harmful only to the individual involved
   (page 637)

Discussion

1. What is meant by negative feedback? What are the four components of a negative feedback system?
2. Why are feedback mechanisms necessary to homeostasis?

3. From an evolutionary perspective, what are the advantages to an organism of being warm-blooded?

4. Discuss the physiological events that occur as you try to regulate your body temperature. Be sure to include the organs or systems involved in temperature regulation.

5. What are goosebumps, and why do they form when you are cold?

6. From a phylogenetic perspective, briefly describe the evolution of osmoregulatory devices from the protozoa to the mammals.

7. List and describe the structure and function of the major organs of the human excretory system.

8. Relative to kidney function, describe filtration, reabsorption, and tubular secretion.

9. What is high blood pressure and how is it related to the renin-angiotensin mechanism?

10. Define respiration, cellular respiration, internal respiration, and external respiration.

11. What are the major components of any respiratory surface?

12. List and describe the structure and function of the major components of the human respiratory system.

13. What happens to you, from a respiratory perspective, when you get the "wind knocked out" of you?

14. How are oxygen and carbon dioxide transported in the blood?

27

Human and Animal Reproduction

Reproduction: An Overview
Human Reproduction
Male Reproductive System
Female Reproductive System
Human Sexual Response
Reproductive Technology:
Contraception and Infertility
Sexually Transmitted Diseases
Summary

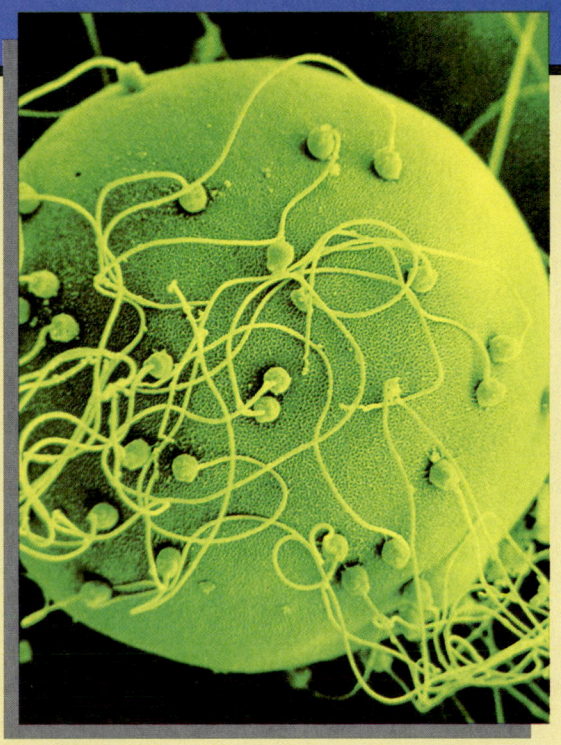

You need only look around you to realize that a spring meadow is literally bursting with life. Think of a lush green meadow. Birds sing; flowers bloom; frogs croak. The trees are covered with the fresh green of new leaves.

What is less apparent in a spring meadow is the fact that most of the life forms you see are moving toward death, even those that have just been born. Indeed, most will die before you do, but they will be replaced. In a nest cradled in a maple tree you can see a new generation of robins. You look into a pond and see tadpoles — the new generation of frogs. Individual life forms are being replaced; life is being renewed.

Above: As these sperm attach themselves to this egg, the timeless biological ritual of reproduction of the species continues.

How does this constant renewal happen? The continuity of life is the result of the timeless biological ritual of reproduction, in which egg and sperm unite (*sexual reproduction*), or in which parent cells divide and disappear into their own progeny (*asexual reproduction*).

Humans have always been fascinated with the phenomenon of reproduction. This fascination is not simply a matter of idle curiosity. Obviously, each species must be able to reproduce itself if it is to survive. Plants produce plants, dogs produce dogs, and amoebas produce amoebas. In earlier chapters we discussed asexual and sexual reproduction in plants, and the process of fission in single-celled organisms. This chapter will concentrate on reproduction in animals. First we describe the types of reproduction exhibited by various animal groups, and then we turn to a discussion of human reproduction.

REPRODUCTION: AN OVERVIEW

There are two major types of reproduction: *asexual*, in which there is no exchange of genetic material between organisms, and *sexual*, in which this exchange does occur. Asexual reproduction provides very little opportunity for variation. Since genetic material is not exchanged between organisms, the offspring are literally "chips off the old block." In sexual reproduction, there is genetic recombination and, as a result, genetic variation, because two individuals with their own unique genetic makeup contribute one-half of their genes to the new offspring. Generally speaking, the more variation you have within a species, the greater the likelihood of survival in a changing environment. This is the major advantage of sexual reproduction. However, when organisms are well adapted to a constant environment, asexual reproduction is advantageous. In some instances, it is desirable to have sameness.

Asexual Reproduction in Animals

■ **Asexual reproduction involves no exchange of genetic material between organisms and can occur in several ways, three of which are fission, budding, and fragmentation. In fission, the organism divides both nuclear material and cytoplasm to produce two daughter organisms. In budding, the organism forms buds, which then break away from the parent organism. In fragmentation, a piece of an organism has the capability of forming an entire new organism.**

In general, asexual reproduction in animals is limited to the invertebrate forms and occurs in several ways. Three of these are fission, budding, and fragmentation. Many unicellular organisms reproduce by **fission,** a type of cell division in which the one-celled organism splits to produce two individuals. Paramecia, amoebas, and many single-celled organisms reproduce through fission (Figure 27.1).

The cnidarians, such as the freshwater hydra, sometimes reproduce through a process called **budding.** Hydra are simple multicellular animals. During their life span some of their undifferentiated cells will form into an outgrowth. The structure, which is in essence a miniature hydra, is called a *bud* (Figure 27.1b). The bud breaks off the parent and develops into a new individual.

Fragmentation is also a variety of asexual reproduction. A piece of the organism breaks off; then, through mitosis (nuclear division that maintains the number of chromosomes from one generation to the next), it develops into an independent adult. Sea anemones and flatworms reproduce in this manner. An interesting modification of fragmentation is *regeneration,* the process by which some animals regenerate lost parts. If a blue-tailed lizard loses its tail, the animal will eventually regrow, or regenerate, a new one (Figure 27.1c). Many organisms possess this very useful trait. In the starfish, the process works in the extreme. Clam fishers dislike starfish because starfish eat clams, thus decreasing the size of the catch. For a time, the fishers would capture starfish and cut each one in two or three pieces. They soon learned that this process did not kill the starfish. In fact, it only made the problem worse because each of the arms could regenerate a completely new organism.

However, we all know that if a bird loses a wing or a dog loses its ear, the structures do not regenerate. The ability to reproduce through fragmentation or regeneration depends on the complexity of the organism. The more complex and specialized it is the less its ability to reproduce through fragmentation. Instead, more complex animals reproduce through sexual reproduction.

Sexual Reproduction

■ **Sexual reproduction, which does involve an exchange of genetic material, can occur in four major ways. In conjugation, two individuals fuse in order to exchange materials. In hermaphroditism, a single organism produces both male and female gametes. Various mechanisms ensure that hermaphroditic organisms do not undergo self-fertilization but undergo cross-fertilization between individuals. In parthenogenesis, even though both male and female individuals are present, the egg does not require fertilization to develop into a new individual. In biparentalism, male and female individuals produce haploid gametes, which must fuse in order to give rise to a new individual.**

Sexual reproduction involves an exchange of genetic material between two organisms. There are a variety of methods by which the exchange can take place. The four major methods found in the animal world are *conjugation, hermaphroditism, parthenogenesis,* and *biparentalism.*

Conjugation. **Conjugation** occurs when two individuals fuse and exchange genetic material. Many protozoans, simple eukaryotic one-celled animals, can reproduce asexually through fission, or sexually through conjugation. For example, *paramecia* are single-celled, slipper-shaped organisms having two

FIGURE 27.1 Asexual Reproduction Among Animals
(a) Fission in the paramecium, (b) budding in the hydra, and (c) regeneration in a lizard.

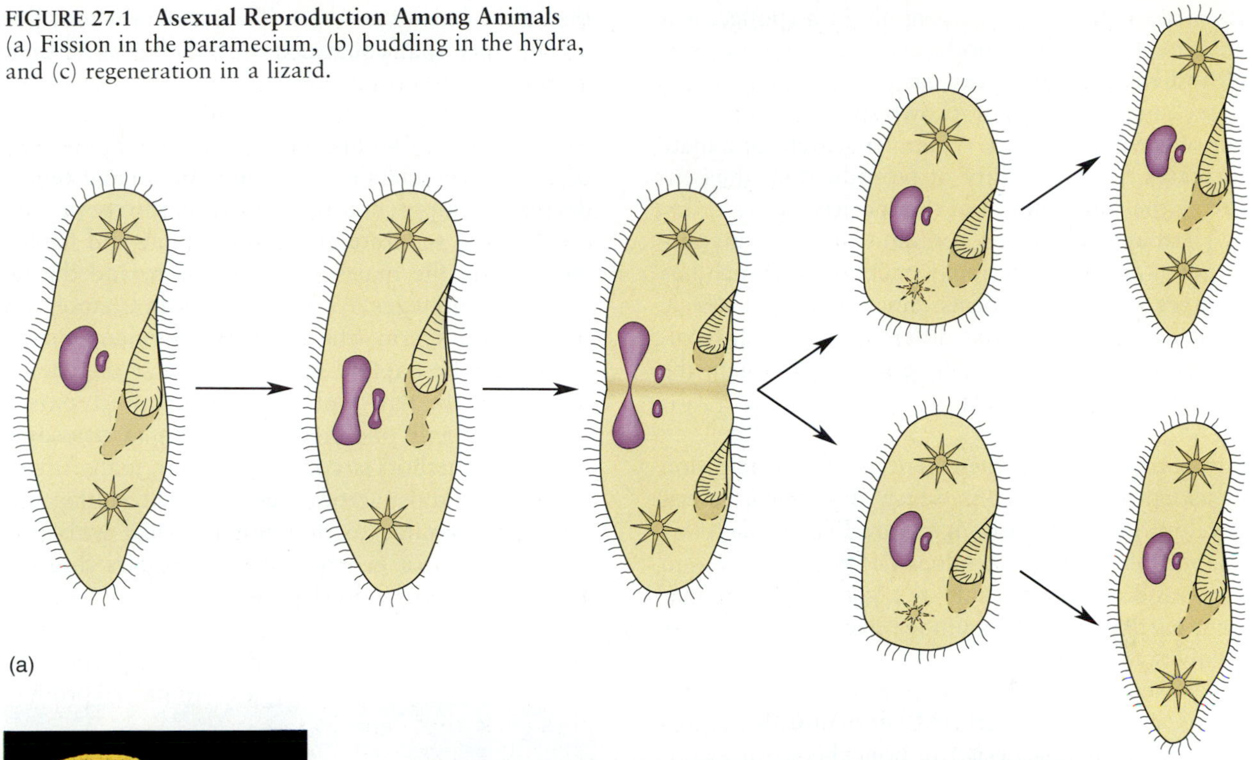

(a)

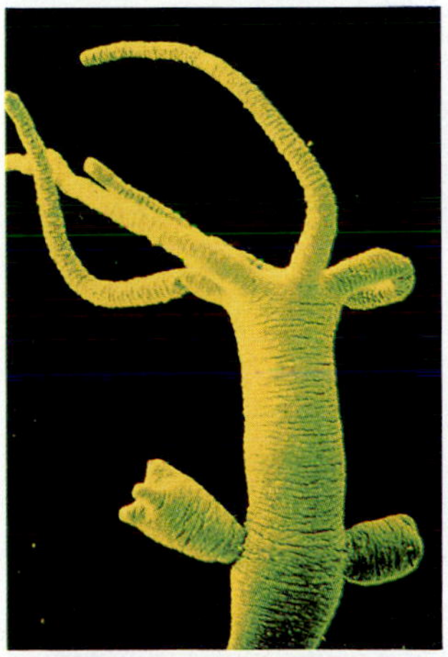

(b)

(c)

nuclear components, one large and one small. Although individual paramecia look the same *morphologically* (in body form), they do differ genetically. In fact, there are eight mating types of paramecia. Any two types can mate (Figure 27.2).

Hermaphroditism. In most cases, animal groups express two separate sexes — there are both males and females. In some instances, however, both sexes are found in one individual. These organisms are said to be **hermaphroditic,** or *monoecious.* There are evolutionary pressures that selected for hermaphroditism. When the sexes are separate, an animal must move about in order to seek out the other sex. On the other hand, when the animal is *sessile*

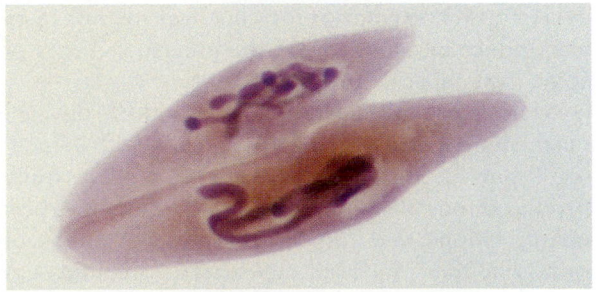

FIGURE 27.2 Conjugation
Conjugation is a process through which one paramecium exchanges genetic material with another. Once the exchange takes place, the conjugants separate, and each produces four daughters through asexual reproduction (fission).

(nonmoving), like a sea anemone or a sponge, it is advantageous to have both sexes expressed in a single individual. In that way, an individual only needs to seek out a member of its own species, not a member of the opposite sex. A simpler search for a mate increases the probability of reproduction, thus ensuring the continuation of the species.

Hermaphroditism is common in many invertebrates (animals without backbones), such as sponges and many worms. Since male and female gametes are produced at different times, self-fertilization does not occur. These species still gain the benefits of cross-fertilization.

Parthenogenesis. **Parthenogenesis** is a modification in sexual reproduction in which an *unfertilized* egg develops by itself, without sperm. Even though this method of reproduction does not require a second party and does not result in a genetic recombination, parthenogenesis is unique because the process involves an egg.

Parthenogenesis is very common among bees, wasps, and other insects. (Many plants also reproduce by parthenogenesis.) In honeybees, for example, the queen is inseminated only once during her life. She stores the sperm in a sac within her reproductive tract that is closed off by a small muscular valve. As she lays her eggs, she can either open the valve to have her eggs fertilized, or she can keep the valve closed. If the eggs are fertilized, they become females (the queens and workers). If the valve remains closed and the eggs are unfertilized, they develop into males (drones) by parthenogenesis.

The mechanism that determines whether or not fertilization occurs is not well understood. It is interesting to note that in the laboratory environment, human intervention can bring about artificial parthenogenesis in the eggs of various species, such as frogs and bees. In some instances, development can be stimulated by such simple mechanical means as a pinprick. Radiation or chemical methods can also induce development of the unfertilized eggs of some lower species like sea urchins.

Mammals normally do not reproduce through parthenogenesis, but Peter Hoppe and Karl Illmensee reported in 1982 that they had successfully produced four parthenogenetic mice. In their experiment, Hoppe and Illmensee used two strains of mice, one that produced the expected haploid eggs and one that produced diploid eggs. (Occasionally, the diploid eggs can develop *without* fertilization, although the process does not proceed very far.) Hoppe and Illmensee removed the nuclei of mouse eggs fertilized in the usual way, substituted nuclei from the diploid eggs, and then placed the eggs in the uteri of foster mothers. The result was four live female mice that contained only the genes of their mother.

Biparentalism. We are most familiar with the type of sexual reproduction known as **biparental reproduction.** In biparental reproduction, the species are divided into separate and distinct male and female individuals; the males produce sperm, and the females produce eggs. As we discussed in Chapter 10, the haploid sperm (the result of spermatogenesis) fuses with the haploid egg (the result of oogenesis) to produce the new diploid individual. Each parent contributes only one member of each chromosome pair. This method is common among many invertebrates and most vertebrates, and the resulting offspring are similar to (but not identical with) the parents, because genetic variation occurs through genetic recombination (Figure 27.3).

(a)

(b)

FIGURE 27.3 Biparentalism and Genetic Diversity
The genetic recombination that is basic to biparental reproduction results in offspring who are like, but not identical, to their parents.

Chapter 27 Human and Animal Reproduction

Sexual Versus Asexual Reproduction

From nature's perspective, there are many advantages to sexual reproduction, the most important of which is the fact that genetically dissimilar gametes are produced through segregation of the genes, crossing over, recombination, and independent assortment during meiosis (Chapter 10). The union of genetically dissimilar gametes during fertilization forms a new organism, thus greatly increasing genetic variability. Genetic variability, in turn, provides a "pool," or reservoir, of genetic types that increases the chance of a species' survival in a changing environment.

The major disadvantage of sexual reproduction is that two individual gametes must unite. To ensure this union, elaborate mechanisms of courtship and reproductive timing have evolved in all species. In addition, especially among the higher animals such as the birds and mammals, parental cooperation in the care of the young of these species has enhanced the survival of their offspring. So, although sexual reproduction is generally more complex than asexual reproduction, the evolutionary advantages in most environments clearly outweigh the disadvantages.

HUMAN REPRODUCTION

More than 5 billion people are living on earth right now, and about 128 million more are born every year. All of us begin life with the union of the haploid sperm of our father and the haploid egg of our mother. These sex cells arise in the gamete-producing organs (the *gonads*) through the meiotic process of either *spermatogenesis* in the male or *oogenesis* in the female (Chapter 10). Let's consider the male and female reproductive systems in more detail.

MALE REPRODUCTIVE SYSTEM

The male reproductive system is less complex than that of the female, primarily because it has just one task — producing male gametes, or *sperm*. Spermatogenesis takes place in the testes, two small organs enclosed in the scrotum. From the testes, the sperm moves into the epididymis, where it matures, then travels through the vas deferens and ejaculatory duct to the urethra. The urethra runs the length of the penis, emptying outside the body (Figure 27.4).

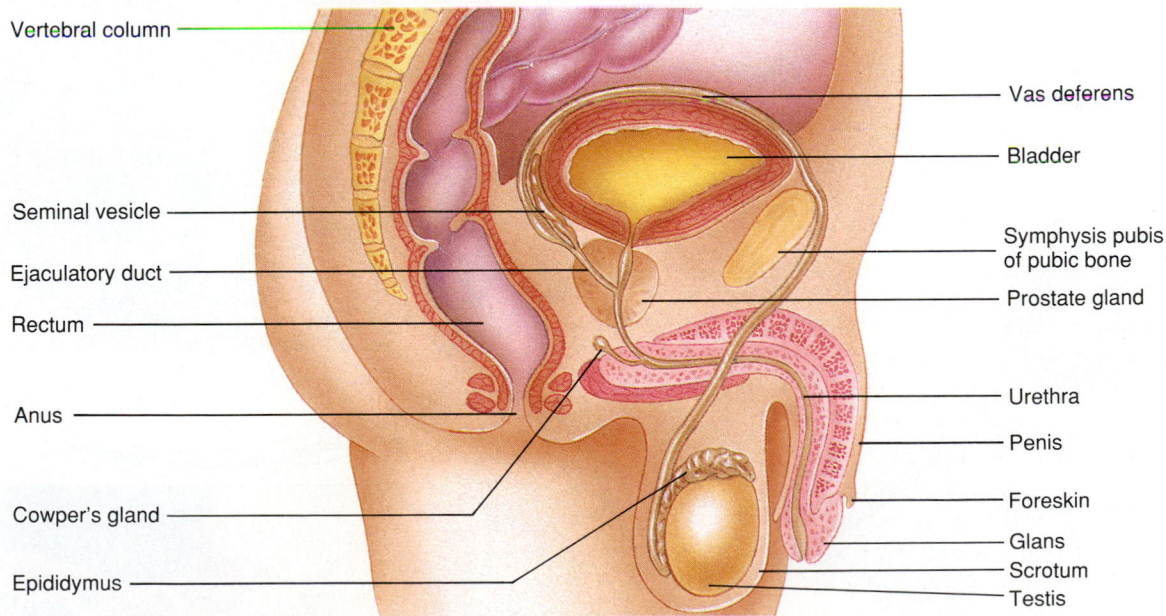

FIGURE 27.4 The Male Reproductive System
Sperm are formed in the testes, two small organs in the scrotum. From the testes, they move into the epididymis, where they mature. The sperm then travel through a series of ducts (the vasa deferentia and the ejaculatory duct) to the urethra, which carries the sperm through the penis, out of the body. The glands of the reproductive system — the seminal vesicles, the prostate gland, and the bulbourethral (Cowper's) glands — secrete the components of seminal fluid, the liquid in which the sperm are contained. At birth, a protective foreskin hangs over and covers the glans penis. This sheath of skin may be surgically removed by circumcision.

The Testes, Site of Sperm Formation

■ The testes (male gonads) are small, almond-shaped organs contained within the scrotum. Enclosed in each testis are 15 to 20 seminiferous tubules, in which spermatogonia develop into sperm cells. Each sperm consists of a cap containing receptors that function in fertilization, a head containing DNA, a middle section containing mitochondria, and a flagellum, which propels the sperm.

The **testes** (*testis*, in singular form) are the male gonads, producing about 500 million sperm every 24 hours from the time of adolescence through old age. The paired testes develop within the fetus and, shortly before birth, descend into an external sac between the thighs called the *scrotum*. The location of the testes suspended in the scrotum outside of the body provides the optimal temperature for sperm production, because sperm production requires a slightly lower temperature than that existing inside the body.

The relationship between temperature and sperm production is the basis for physical responses of the testes to temperature change. When the temperature of the testes falls, the muscles of the scrotal sac

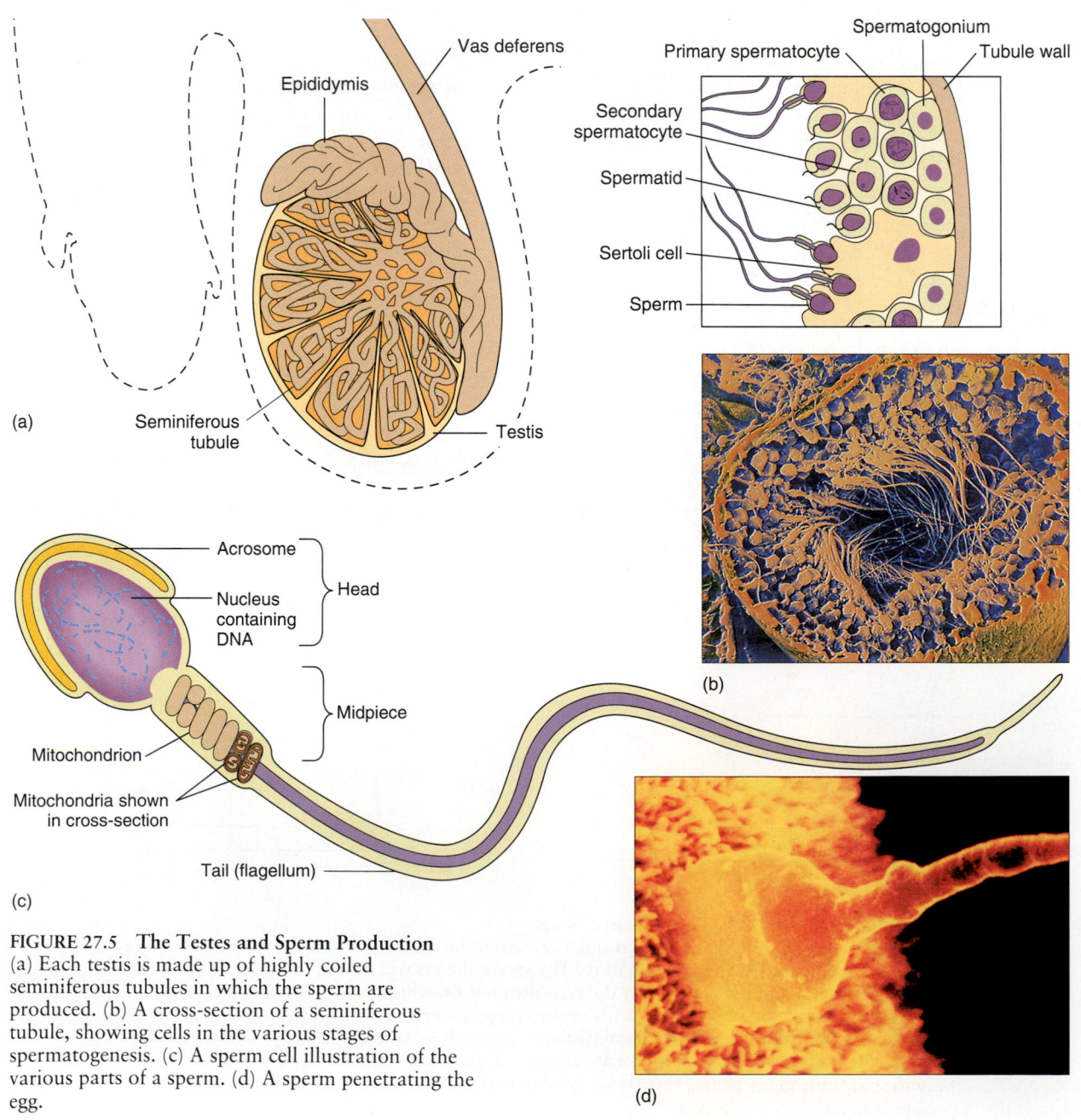

FIGURE 27.5   The Testes and Sperm Production
(a) Each testis is made up of highly coiled seminiferous tubules in which the sperm are produced. (b) A cross-section of a seminiferous tubule, showing cells in the various stages of spermatogenesis. (c) A sperm cell illustration of the various parts of a sperm. (d) A sperm penetrating the egg.

Chapter 27   Human and Animal Reproduction

tighten, thus hugging the testes close to the body in order to conserve heat. When the temperature of the testes rises, as in the case of a hot bath or a fever, the muscles relax, allowing them to hang farther from the body to bring about a reduction of temperature. These reactions are involuntary, for too low or high a temperature may result in lowered levels of sperm production, which in turn can cause infertility. (In fact, some physicians think that the skin-tight style of some jeans, especially in the crotch area, may hug the testes too close to the body, resulting in high temperatures that can interfere with sperm production.)

The coiled **seminiferous tubules** found within each testis actually produce the sperm (Figure 27.5a). Within these tubules are undifferentiated cells called *spermatogonia*. As these cells undergo meiosis, they grow larger and develop into *spermatocytes*. The spermatocytes continue to experience meiosis, each producing four haploid *spermatids*. The spermatids undergo further development and maturation to become sperm (Figure 27.5b).

The sperm is a specialized cell with a design that reflects its sole function: its union with an egg to initiate reproduction. An *acrosome* (or cap) at the front of the head of the sperm contains enzymes that help the sperm penetrate the egg. The cell's nucleus (also in the head) carries tightly packed DNA. In the *midpiece*, which connects the head to the tail, are the mitochondria that produce the energy needed to move the tail — a single flagellum that propels the swimming sperm (Figure 27.5c). A single sperm can swim about 600 times its body length in only 8 minutes — that means that, rela-

tively speaking, a sperm can swim faster than a marathon runner can run.

The Journey from Formation of Sperm to Ejaculation

■ Once the sperm are formed, they travel through the epididymis to the vas deferens, the major sperm duct, which, in turn, empties into the ejaculatory duct. During this time the sperm are activated and join with secretions from the seminal vesicles and the prostate gland to form seminal fluid. The seminal fluid includes phospholipids to activate the sperm cells, energy molecules to provide energy to the sperm, and buffers to neutralize the acid content of the penis and the vagina. The ejaculatory duct then fuses with the urethra, which extends through the penis.

The pathway of sperm from origin to ejaculation is as follows. Formed in the seminiferous tubules, the sperm enter the *epididymis*, a coiled tube about 6 meters in length that functions in sperm storage. The majority of the epididymis's length lies above the testis, with one end extending down one side of the testis. When sperm enter the epididymis, they are immobile, and it is there that they mature and become mobile, possibly because of chemical secretions of the epididymis or simply because of an internal maturation process. The epididymis then connects into the major sperm duct called the *vas deferens* (plural, *vasa deferentia*). During the method of birth control known as a *vasectomy*, (Figure 27.6), the vas deferens is surgically cut and tied.

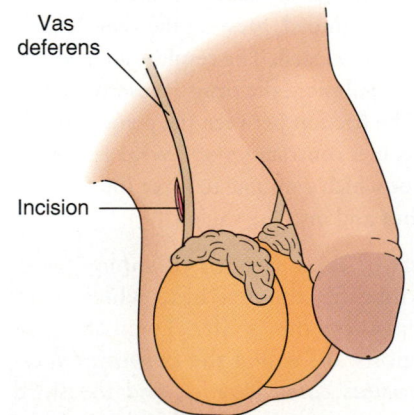
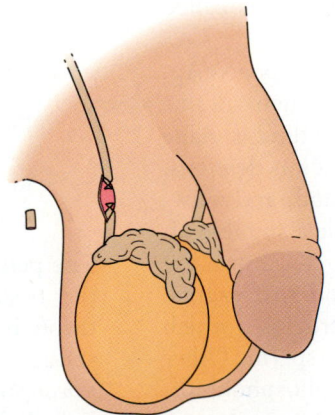
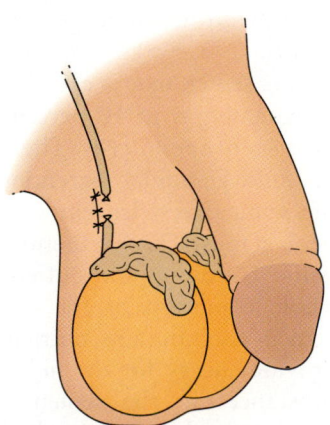

Vas deferens

Incision

FIGURE 27.6 Vasectomy
A vasectomy is a relatively safe and painless method of sterilization in which the vas deferens of each testis is cut and tied. This prevents the sperm from entering the seminal fluid. Because the testes are left intact, the individual continues to produce normal levels of male hormones, so sexual performance and secondary sex characteristics remain normal too. A vasectomy should be thought of as a permanent procedure, although there are cases where the procedure has successfully been reversed.

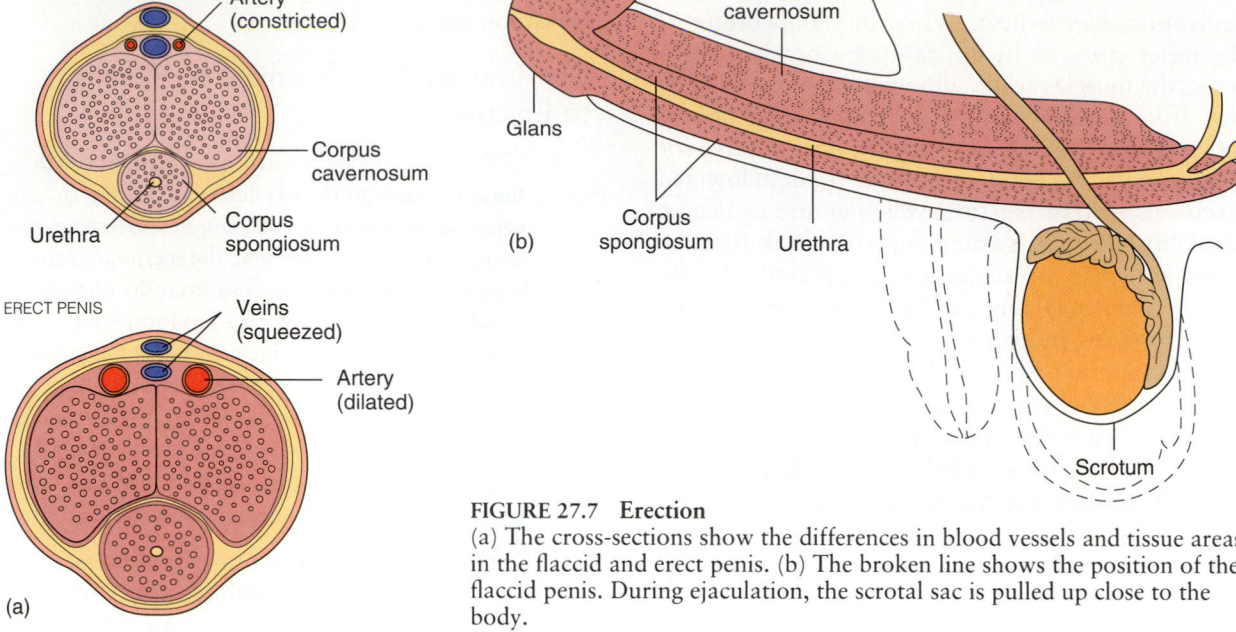

FIGURE 27.7 Erection
(a) The cross-sections show the differences in blood vessels and tissue areas in the flaccid and erect penis. (b) The broken line shows the position of the flaccid penis. During ejaculation, the scrotal sac is pulled up close to the body.

From the lower abdominal cavity, the vas deferens leaves the testis and passes around the bladder, where it receives secretions through ducts from glands called the *seminal vesicles*. Then each vas deferens fuses with an **ejaculatory duct**, which passes through the *prostate gland,* where more secretions are added to form **seminal fluid.** Located below the prostate are two small glands called *Cowper's glands* or the *bulbourethral glands*. On sexual arousal, these glands secrete a mucus fluid into the urethra. Once the secretions of the Cowper's gland are added to seminal fluid, it is called **semen.**

It is semen that is ejaculated from the penis. Approximately 5 to 10 percent of the semen volume consists of sperm cells. The remaining fluid is made up of the glandular secretions from the vas deferens, the seminal vesicles, the prostate gland, and Cowper's glands. These secretions provide a suitable alkaline condition and buffers to help neutralize the acidic condition of the urethra and the vagina. Semen also contains nutrients like fructose, which provide energy for the highly motile sperm. Other substances in the semen include phospholipids, which help the sperm to become motile, and prostaglandins, which may cause muscle contractions in the female, thus speeding the transport of sperm during sexual intercourse.

After leaving the prostate gland, the ejaculatory duct fuses with the *urethra* from the bladder. Moving through the urethra, the sperm, now suspended in the semen, make their journey through the penis. The urethra also functions in the excretion of urine, just as the penis functions both as the organ for internal fertilization and as the structure through which urine passes.

Penis Structure

- **The penis consists of the urethra and three surrounding layers of erectile tissues. At the tip of the penis is the glans. On sexual stimulation, the blood vessels in the penis dilate and the erectile tissues become engorged, pressing the veins leading from the penis closed. The result is an erection. During copulation, the penis is inserted into the vagina. The friction between the penis and the vagina results in a contraction of several sets of muscles, all of which combine to force the semen out the urethra and into the vagina.**

The penis consists of three regions of spongy erectile tissues surrounding the urethra, which runs through the length of the shaft. Two of these spongy tissue regions are called the *corpora cavernosa* (singular, *corpus cavernosum*), and the third is called the *corpus spongiosium*. The tip of the penis forms the glans. This structure is covered by a foreskin in an uncircumcised male. The glans contains many nerve endings and is important in sexual arousal. Incidentally, contrary to popular belief, there is no correlation between penis size and sexual performance or fertility.

During sexual excitement, nervous stimulation causes the arteries of the penis to dilate, providing an increased flow of blood to the spongy erectile tissues that surround the urethra. The turgor pressure of the blood affects the penis in two ways. First, it causes the penis to enlarge. Second, it places pressure on the veins of the penis, which inhibits blood flow back into the body. The two together cause the spongy erectile tissues of the penis to become filled with blood, which in turn causes the penis to become enlarged and erect. These reactions constitute an erection (Figure 27.7).

During **copulation,** or sexual intercourse, the erect penis is inserted into the vagina of the female. The friction of the movement of the penis in the vagina triggers muscles in the testes and vasa deferentia to contract, forcing the sperm into the urethra. At the same time, contractions of the seminal vesicles and prostate expel fluid from the glands into the seminal fluid. A third set of muscles forces the semen from the urethra, a process called **ejaculation.** About 3 to 5 ml (about a teaspoonful) of semen is ejaculated.

The ejaculation of sperm is usually accomplished with pleasurable sensations called the climax, or **orgasm.** After orgasm there is usually a sense of relaxation, sometimes to the point of sleep. The veins of the penis relax, and the penis becomes flaccid, or soft. (The human sexual response will be discussed in more detail later in this chapter.)

Hormonal Regulation

- **Three hormones are primarily responsible for the regulation of sperm production and sexual response in males. Testosterone, produced by the interstitial cells in the testes, is the major male sex hormone, regulating sperm production and development of secondary sex characteristics. The hypothalamus produces gonadotropin-releasing hormone (GnRH), and the pituitary gland produces interstitial cell–stimulating hormone (ICSH), which stimulates testosterone production, and follicle-stimulating hormone (FSH), which regulates the development and maturation of sperm. All are influential at the onset of puberty, after which their level remains fairly constant in the adult male.**

Sex hormones are chemical messengers that play an important role in sperm production and reproduction in many invertebrate and all vertebrate animals, including humans. In the human male, small masses of *interstitial cells* secrete male sex hor-

(a)

(b)

(c)

(d)

FIGURE 27.8 Secondary Sex Characteristics in Vertebrate Animals
Sex hormones are what often make males and females look different. Among most species, the male is larger than the female. Other differences have to do with color, plumage, and hair.

mones in the testes that are collectively called **androgens.** The most important of these is **testosterone,** which is essential for (1) sperm production, (2) the development and maturation of sex organs during adolescence, and (3) the production of the secondary sex characteristics of the male, such as facial hair, skeletal proportions, and distribution of fat and muscle (Figure 27.8).

In most vertebrate animals, testosterone is also essential for maintaining the sex drive and initiating courtship patterns. The role of testosterone in

human sexual behavior is not well understood, but some investigators feel that high levels of the hormone may be responsible for some abnormalities in sexual behavior.

Testosterone production is regulated by structures located at the base of the brain, the *hypothalamus* and the *pituitary* gland, which is attached to the hypothalamus. These glands are also involved in producing two hormones involved in sperm production. One is called **interstitial cell–stimulating hormone** (ICSH), and the other is **follicle-stimulating hormone** (FSH). (ICSH is also found in females, although there it is called luteinizing hormone. The two names reflect the targets that the chemical acts on.) Testosterone, ICSH, and FSH are all involved in sperm production.

The pituitary consists of two parts, an anterior (front) section and a posterior (back) section. The gland produces several hormones, two of which are called **gonadotropins** because they affect the gonads. ICSH is one of the gonadotropins. This hormone acts on the interstitial cells of the testes, stimulating them to produce testosterone. The concentration level is controlled by a negative feedback mechanism. As the concentration level of testosterone in the blood rises, the higher concentration inhibits the pituitary's release of ICSH (Figure 27.9).

The other gonadotropin is FSH, which acts on the seminiferous tubules, stimulating proper development and maturation of sperm. The hypothalamus signals and regulates the pituitary's secretion of FSH and ICSH by releasing a hormone known as the **gonadotropin-releasing hormone** (GnRH) (see Chapter 29).

Puberty in males is regulated primarily by ICSH, FSH, and testosterone. Around the ages of 13 to 16, the male undergoes a period of sexual maturation called **puberty.** These changes are initiated by several hormonal functions that stimulate the pituitary to release high levels of FSH and ICSH, resulting in higher levels of testosterone and other aspects of sexual maturity.

In our society, biologists believe that the behavioral changes associated with what is referred to as "male menopause" (middle-aged males seeking varied and/or more youthful sex partners) are due more to social factors than to physiological changes. While males do experience a decline in testosterone levels after the age of forty, they do not experience the hormonal, physiological, and (sometimes) psychological changes that we associate with female menopause (we will discuss female menopause in a moment). Males can remain fertile (they continue to produce sperm) well into old age, while in menopausal females, egg production ceases.

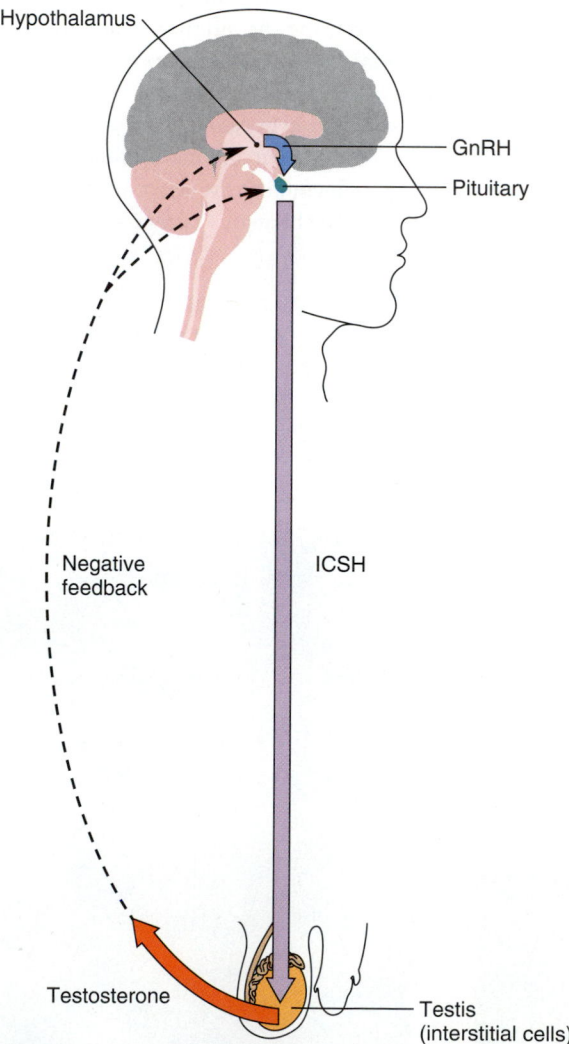

FIGURE 27.9 Negative Feedback and Testosterone Production
The production of testosterone in the testes is controlled by a negative feedback mechanism. When the testosterone level falls, the hypothalamus releases GnRH, which triggers the secretion of ICSH in the pituitary. ICSH, a gonadotropin, acts on the interstitial cells of the testes, increasing their production of testosterone. As the testosterone level in the bloodstream increases, the production of GnRH and ICSH decreases.

FEMALE REPRODUCTIVE SYSTEM

The human female reproductive system is more complex than that of the male. Not only does the female produce the female gamete (the *ovum*, or egg); she also retains and protects the embryo within the *uterus* (womb) during its development.

The female reproductive system consists of the ovaries, which produce the eggs and sex hormones; the fallopian tubes, which transport the eggs to the uterus; the uterus itself; and the vagina, the tube that functions as both a receptacle for the penis and as the birth canal (Figure 27.10). The mammary glands are also a part of the system.

FIGURE 27.10 **The Female Reproductive System**
(a) From the front. (b) A side view.

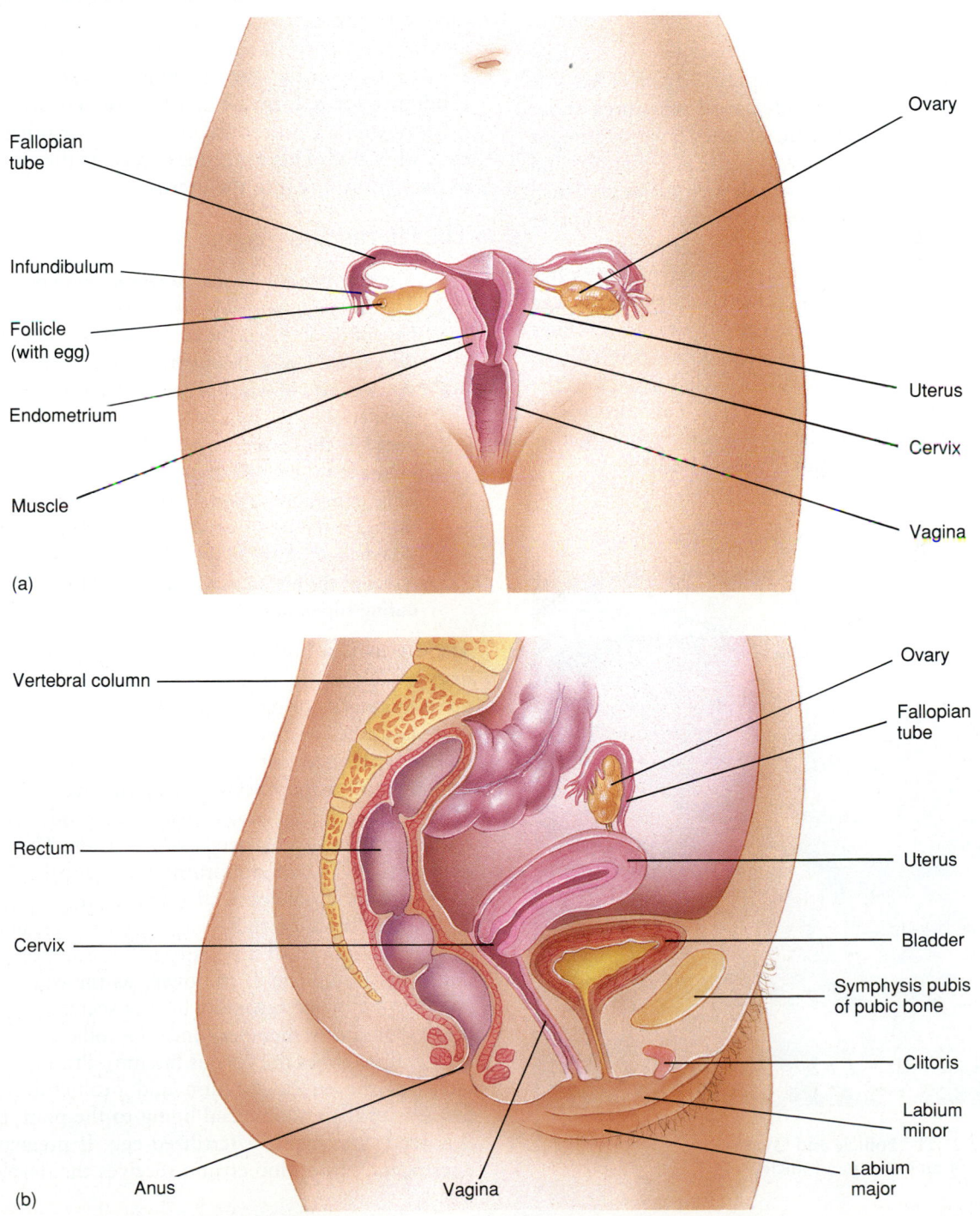

The Ovaries

■ The ovaries (the female gonads) are two small organs that produce the eggs. They also produce the female sex hormones, estrogen and progesterone. The menstrual cycle takes approximately 28 days. One egg develops within a follicle, and midway during the cycle the follicle bursts, releasing the egg — a process called ovulation. The egg then moves through the infundibulum to the fallopian tube, which leads to the uterus. The follicle, after the egg is released, becomes a corpus luteum.

The female gonads are called **ovaries**. The ovaries, located in the lower abdominal cavity on either side of the uterus, are almond-shaped structures about 5 cm (2 inches) in length. Ovaries have two major functions: One is to produce the egg (ovum), and the other is to produce female sex hormones, such as those collectively called **estrogens** and another called **progesterone** ("prolonging gestation").

The ovaries release **ova** (eggs) in a monthly cycle known as the **menstrual cycle**. (Usually only one egg is released per cycle.) The egg matures in a **follicle** (a blisterlike structure located within the ovary). Once the egg is mature, the follicle bursts, and the egg is released (Figure 27.11). This process is known as **ovulation**. Once the follicle bursts, it is called the **corpus luteum** ("yellow body").

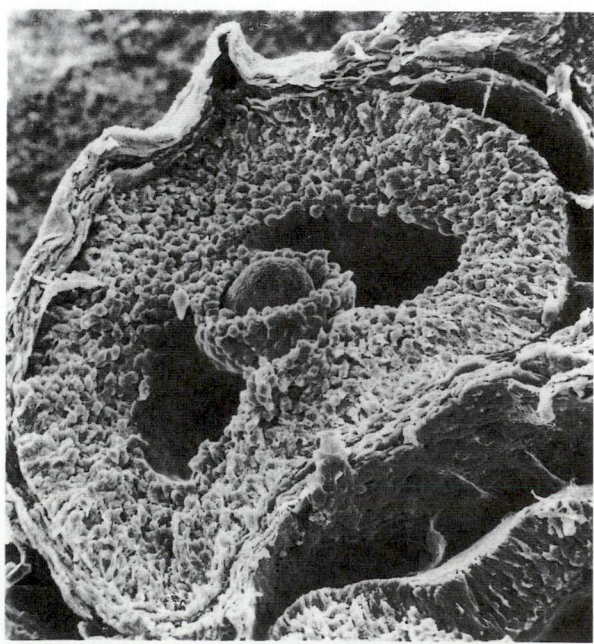

FIGURE 27.11 Follicle and Ovum
A SEM of an ovum in a follicle.

The mature egg leaves the ovary and moves into a **fallopian tube** (oviduct), one of the two ducts that connect with the uterus. The end of these ducts, near the ovaries, is funnel-shaped. This area of the tubes is called the **infundibulum.** There is a small gap between each ovary and infundibulum. As the egg leaves the ovary, the fringed ends of the infundibulum help sweep the egg across the gap into the fallopian tube. Ciliated columnar cells and gentle contractions of the muscles in the wall of the oviduct continue to move the egg through the fallopian tube to the uterus. Very rarely, the egg is lost in the abdominal cavity, where it usually disintegrates. Sometimes women who no longer desire to bear children elect a form of sterilization known as a *tubal ligation,* a procedure in which their tubes are cut and/or tied. This procedure prevents the passage of the mature egg into the uterus.

The Uterus and Vagina

■ The uterus is a pear-shaped reproductive organ. The fallopian tubes join the uterus at the top. The base of the uterus is the cervix, which unites with the vagina. Lining the uterus is the endometrium, the thickness of which is regulated by estrogen (produced by the follicle) and progesterone (produced by the corpus luteum). A fertilized egg lodges in the endometrium, which provides nourishment and protection for the developing fetus. If the egg is not fertilized, the endometrial lining is sloughed off as the menstrual flow. The vagina serves as the birth canal and receives the penis during copulation.

The **uterus** is a hollow, muscular organ about 7.5 by 5 cm (about the size of a pear) that can increase up to 6 times its original size during pregnancy. The fallopian tubes enter on either side of the top of the uterus. The region of the uterus where it joins the vagina is called the *cervix.* The uterine wall consists of three layers. The innermost layer produces the spongy, highly vascularized *endometrium.* The middle layer consists of smooth muscle, and the outer layer is made up of connective tissue.

The thickness of the endometrium is regulated by the secretions of the estrogen hormones produced by the follicle of the ovary as the egg develops, and by progesterone, which is secreted by the corpus luteum (remember, once the follicle releases its egg, it becomes the corpus luteum). Progesterone increases the vascularization and proliferation of the cells of the endometrial lining to the point that it is ready to receive a fertilized egg. If pregnancy does occur, the endometrium receives the developing embryo and nourishes and protects it; if pregnancy does not occur, the endometrium is sloughed off, and the menstrual flow begins. The cycle then repeats itself. Because the menstrual cycle is continuous, the human female is fertile during a certain period every month throughout her reproductive years. This is not the case with the females among most other higher animals, whose fertility comes at only certain times during the year.

At the lower end of the uterus is the cervix, which is the opening of the uterus through which sperm pass on their way to the fallopian tubes to fertilize an egg, or through which the fetus passes during the birth process. Physicians recommend that women have a pap smear taken every year in order to test for cervical cancer. In this procedure, cells are scraped from the outer cervix for examination under a microscope. The test is very effective in early detection of a form of cancer that, if unobserved, can be life-threatening.

The uterus leads into the **vagina,** which is a muscular tube about 7.5 cm (3 to 4 inches) long connecting the uterus to the outside of the body. The vagina has two major functions: (1) it is the organ that receives the penis during copulation for internal fertilization, and (2) it is also the canal through which the fetus passes during the birth process. The

structure is lined with glands that secrete an acid mucus, producing a slightly acidic environment that works to prevent infection (remember, most organisms thrive in an environment with a neutral pH). Some infections still occur — yeast infections, for example, are rather common.

External Genitalia

■ **The vulva, the external female genital structure, includes the labia majora and minora and the clitoris. All function in sexual arousal. The labia majora are the outer folds of flesh that surround the vagina. The labia minora protect the vaginal opening and produce secretions that function as lubricants during copulation. The clitoris is a highly sensitive structure that is associated with female orgasm.**

Collectively, the external genital organs are called the *vulva* (Figure 27.12). The *labia* (lips) are folds of skin that surround the vagina: The *labia majora* are the fleshy folds of skin lined with fatty tissue that cushion the vagina and in the mature female are covered with pubic hair; the *labia minora* are smaller skin folds that surround the vaginal opening. During sexual excitement, the labia minora and

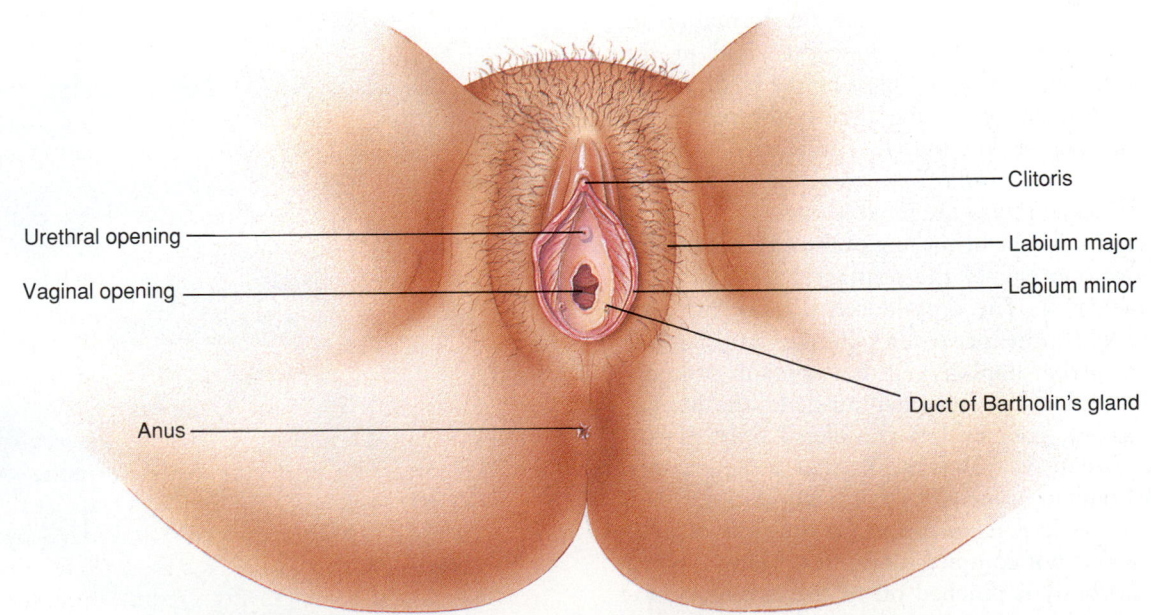

FIGURE 27.12 The Vulva
The external genital organs of the human female. Notice that the urethral opening is distinct from the vaginal opening. In the female, the urethra is not a part of the reproductive system.

mucus-secreting *Bartholin's glands,* located just beneath the labia minora, provide mucus secretions that lubricate the vagina.

Located within the anterior folds of the vagina just behind the junction of the labia minora is the *clitoris.* This small (about 2 cm long) sensitive structure is comparable with the glans of the penis in the male. (In fact, in the developing embryo, the glans of the penis and the clitoris of the female are identical.) During sexual arousal, the sensitive clitoris becomes engorged with blood and, together with the vagina, provides most of the neural stimulation that results in the female orgasm. Contrary to popular literature, there is no evidence for the role of the "G spot" (an area about 1 to 1½ inches inside the vagina, on the upper surface toward the navel) in higher sexual arousal or orgasmic response.

Egg Formation

■ **The eggs within the ovaries begin to form during fetal development, and by birth they have already undergone the meiotic steps through prophase I. The remaining steps do not occur until the female has reached puberty, and then usually only one egg matures at a time. During oogenesis, cytoplasmic development is unequal, resulting in a larger viable egg and a smaller nonfertile polar body.**

The meiotic process resulting in the formation of the haploid egg is called *oogenesis,* and in the human female, oogenesis begins before birth. At the time of birth, the ova have already completed the prophase step of meiosis I. The ovaries contain about 2 million *primary oocytes* (the diploid cell that will develop into the haploid egg).

It is not until the beginning of the menstrual cycle that hormones signal the primary oocyte to complete meiosis I. The diploid cell divides into two haploid cells: the *secondary oocyte,* and a *polar body* (a smaller haploid cell that contains little cytoplasm and usually disintegrates). In the human female, then, meiosis I is completed only shortly before ovulation, which means that some primary oocytes remain in meiosis I for 30 to 50 years. The second meiotic division begins at the time of ovulation and is not completed until fertilization, when the polar body is pinched off of the secondary oocyte after the sperm penetrates the egg (Figure 27.13). Usually only one egg is released during the course of the 28-day menstrual cycle. Therefore, only about 350 to 400 of the 2 million primary oocytes ever reach meiosis II.

Look at Figure 27.14. It shows the progression of events in the ovary throughout a menstrual cycle. At point 1, a follicle is beginning to form around a primary oocyte. As the follicle develops (point 2), the diploid primary oocyte completes meiosis I, dividing into the haploid secondary oocyte and polar body. Meiosis II begins with ovulation (point 3), when the blisterlike follicle bursts, releasing the secondary oocyte; meiosis is completed only if the egg is fertilized. After ovulation, the follicle cells form the corpus luteum (point 4), a glandular body that secretes progesterone, the hormone that prepares the lining of the uterus to receive the fertilized egg and maintains hormone levels during pregnancy. If the egg is not fertilized, the corpus luteum degenerates (point 5), and the cycle begins again. Oogenesis differs from spermatogenesis in that the primary oocyte produces only one egg, or functional gamete. In spermatogenesis, a primary spermatocyte produces four sperm, or functional gametes. The nuclear division in oogenesis is equal (that is, each new cell contains 23 chromosomes), but the cytoplasmic division is unequal. The result is two cells of unequal size. The larger one becomes the egg, con-

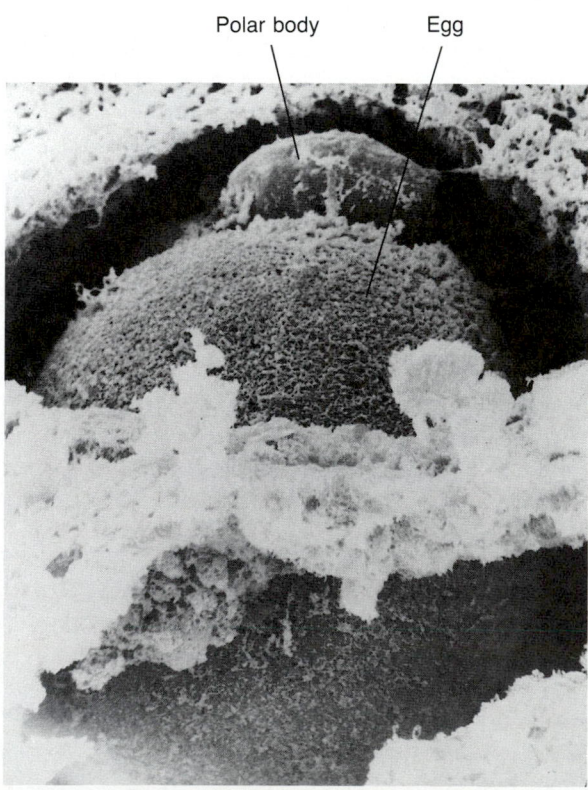

Polar body · Egg

FIGURE 27.13 Egg with Polar Body
A SEM of an unfertilized rat egg with the primary polar body still attached.

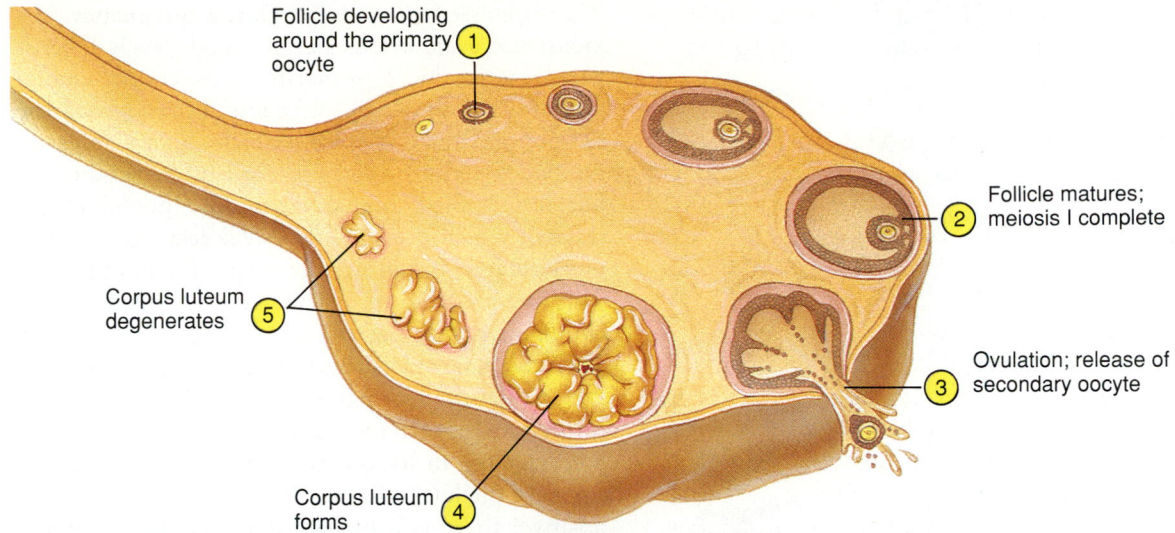

FIGURE 27.14   The Ovarian Cycle
As the cycle starts, a follicle begins to form around a primary oocyte. As the oocyte completes meiosis I, the follicle matures. At ovulation, the follicle bursts, releasing the secondary oocyte and triggering meiosis II. The cells of the follicle form the corpus luteum. If pregnancy does not occur, the corpus luteum disintegrates. Although the diagram shows the follicle moving around the ovary, in fact it stays in one position throughout the cycle.

taining the nutrient-rich cytoplasm. The smaller becomes a polar body, which never is fertilized. Sometimes the first polar body may undergo meiosis to form secondary polar bodies. These polar bodies usually remain in contact with the egg.

Hormonal Regulation of the Menstrual Cycle and Pregnancy

■ **The timing of the 28-day menstrual cycle is determined by gonadotropin-releasing hormone (GnRH), which stimulates the production of follicle-stimulating hormone (FSH) and luteinizing hormone (LH). The follicle produces the egg and estrogens, which promote development of secondary sex characteristics and the thickening of the endometrium. Progesterone, produced by the corpus luteum, completes and maintains the endometrium. In pregnancy, the developing fetus produces human chorionic gonadotropin (HCG), which maintains the corpus luteum during the first two months of pregnancy (then the placenta produces the necessary hormones). Pregnancy also stimulates the production of two hormones that regulate milk production in the mammary glands and initiate labor and childbirth.**

Hormones and the Menstrual Cycle. All the events that occur during the menstrual cycle — the maturation and release of the egg and the increased thickness of the endometrium — are intricately timed by a complex interaction of hormones, some of which we have already mentioned in the section concerning male sexual processes: gonadotropin-releasing hormone (GnRH), follicle-stimulating hormone (FSH), and **luteinizing hormone** (LH), which is identical to the interstitial cell–stimulating hormone (ICSH) active in males.

GnRH, produced by the hypothalamus, regulates the release of the gonadotropins — FSH and LH — from the anterior pituitary (Figure 27.15). FSH and LH are involved in the regulation of the events in the ovary that bring on maturation of the egg in the follicle and the release of the egg from the ovary.

The human female is fertile only during the few days of her cycle around the time of ovulation. This fact is the basis of the *rhythm method* of birth control. If there is no sexual intercourse during the three or four days before and after ovulation, the probability of pregnancy is minimized. As many of you may know, the rhythm method is not a very effective method of birth control because it is sometimes difficult to pinpoint the exact time of ovula-

Female Reproductive System

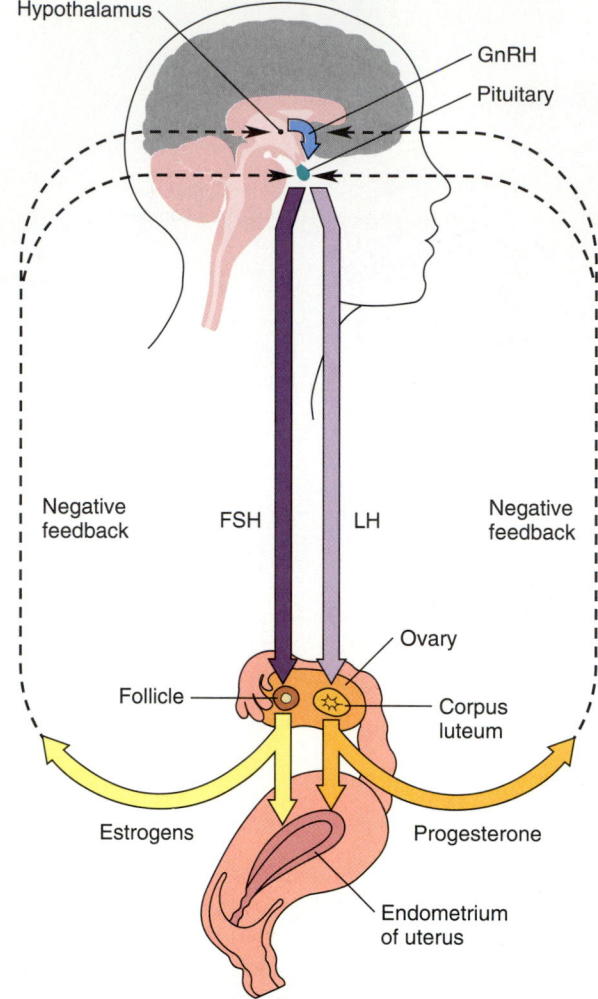

FIGURE 27.15   Interaction of the Hormones Involved in Menstruation
In the female, the levels of estrogens and progesterone in the blood regulate the secretion of GnRH by the hypothalamus and FSH and LH in the pituitary. Negative feedback is also involved.

The Effects of Pregnancy. During pregnancy, the menstrual cycle ceases because high levels of progesterone produced primarily by the corpus luteum inhibit FSH and LH secretion. In addition, the developing embryo itself, which is implanted in the endometrium and nourished by it, produces a hormone called **human chorionic gonadotropin (HCG)**, which prevents the breakdown of the corpus luteum. This allows the corpus luteum to go on secreting progesterone, which in turn maintains the endometrium for the first two months of pregnancy. Then the placenta produces progesterone and estrogen, which maintain the endometrium until childbirth. Early pregnancy diagnostic test kits, which are available in local drugstores today, chemically measure the level of HCG in the urine. If the test is positive, there is a high probability that a human embryo is producing HCG, which means that the woman is probably pregnant.

Numerous other hormonal changes occur during pregnancy. The anterior pituitary gland increases its production of the hormone *prolactin*, and the posterior pituitary increases the production of the hormone *oxytocin*. These hormones are involved in the production and release of milk by the mammary glands after childbirth. Oxytocin also initiates the uterine contractions that result in the birth of the baby. Prostaglandin hormones produced by the fetus itself also function in initiating labor. Table 27.1 lists the major hormones involved in human reproduction.

The Hormonal Cycle. In summary, these are the rhythmic hormonal events regulating the menstrual cycle. (Day 1 of the cycle is the first day of menstrual flow. These steps are indicated in Figure 27.16.)

1. The hypothalamus releases GnRH, stimulating the anterior pituitary to release FSH. FSH stimulates one of the immature follicles in the ovary to mature. Maturation takes about the first 14 days of the cycle.

2. The follicle produces estrogens, which stimulate an increase in the number of cells in the endometrium, building it up and enriching it. High estrogen levels also inhibit the secretion of FSH, so that no more follicles will be stimulated to develop.

3. Around the twelfth to fourteenth day, a surge of LH causes the follicle to rupture, releasing the egg (ovulation). The female is fertile at this time. The follicle is converted to the cor-

tion. Also, human sperm can live in the female reproductive tract for two to four days.

The ovarian follicle also produces hormones, primarily the estrogens, which maintain the human female secondary sex characteristics and also initiate the thickening of the endometrium early in the cycle. In the process of releasing the egg, the follicle bursts. The corpus luteum, formed by the ruptured follicle, secretes progesterone, which completes and maintains the endometrial lining.

TABLE 27.1
Major Human Reproductive Hormones

Hormone	Source	Function
Male		
Testosterone	Interstitial cells of the testes; regulated by levels of ICSH	Increases sperm production; stimulates development of male secondary sex characteristics
Interstitial cell–stimulating hormone (ICSH) (see LH in the female)	Pituitary; secretion regulated by GnRH	Stimulates the secretion of testosterone by interstitial cells of the testes
Follicle-stimulating hormone (FSH)	Pituitary; secretion regulated by GnRH	Stimulates the development and maturation of the sperm in the semniferous tubules
Female		
Follicle-stimulating hormone (FSH)	Pituitary; release regulated by GnRH	Maturation of the egg and follicle; development of the endometrium after menstrual flow increases secretion of estrogens
Luteinizing hormone (LH)	Pituitary; release regulated by GnRH	Continued development of follicle; corpus luteum formation; signals ovulation; increases secretion of progesterone
Progesterone	Corpus luteum	Thickens and maintains the endometrium
Estrogens	Ovarian follicle	Initiate thickening of endometrium; stimulate development of female secondary sex characteristics
Prolactin	Pituitary; regulated by hypothalamic hormones	Milk production by mammary glands
Oxytocin	Pituitary; regulated by hypothalamus	Stimulates uterine contractions during labor and milk release
Embryo		
Human chorionic gonadotropin (HCG)	Chorion of the human embryo	Helps to maintain the corpus luteum; stimulates progesterone secretion in the corpus luteum and some estrogen secretion
Placental estrogen and progesterone	Human placenta: from the third month after fertilization	Helps to maintain pregnancy

pus luteum, and it begins to secrete progesterone. The endometrium, in the presence of both estrogens and progesterone, continues to thicken. Progesterone completes the endometrium's increased thickness and development and maintains it for about another 10 to 14 days.

4. If the egg is not fertilized, the corpus luteum degenerates, and the levels of progesterone and estrogen fall.

5. As the level of progesterone drops, the endometrium, with its assorted blood cells, passes out of the uterus through the cervix and vagina. This process, called menstruation (or the menstrual period), usually lasts about four to five days.

6. Decreased levels of progesterone and estrogens signal the pituitary to increase FSH secretion, initiating the development of another follicle, and the cycle repeats itself.

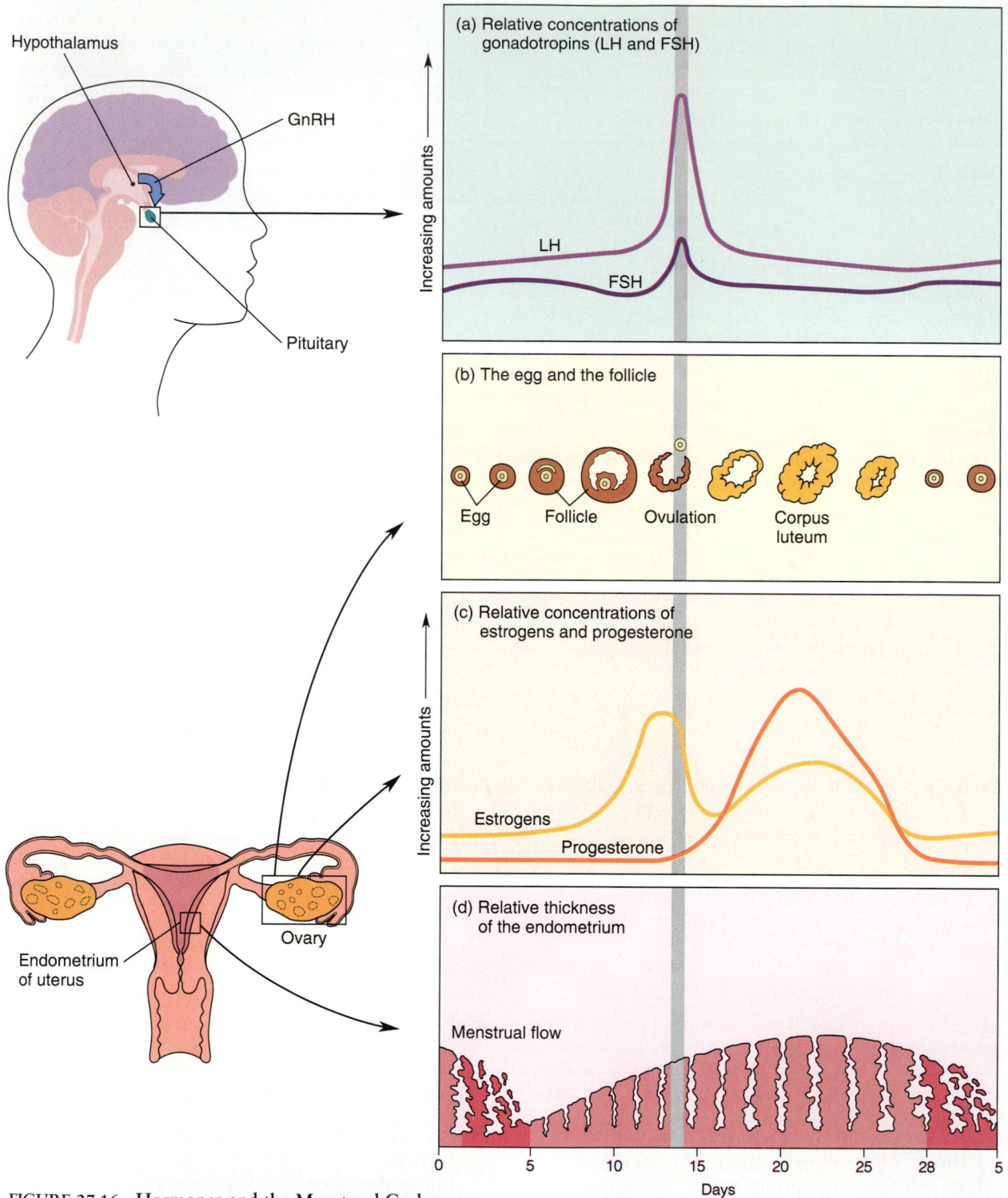

Hypothalamus

GnRH

Pituitary

(a) Relative concentrations of gonadotropins (LH and FSH)

Increasing amounts →

LH

FSH

(b) The egg and the follicle

Egg

Follicle

Ovulation

Corpus luteum

(c) Relative concentrations of estrogens and progesterone

Increasing amounts →

Estrogens

Progesterone

(d) Relative thickness of the endometrium

Menstrual flow

0   5   10   15   20   25   28   5

Days

Endometrium of uterus

Ovary

FIGURE 27.16   Hormones and the Menstrual Cycle
There is a very definite pattern to hormonal activity during the menstrual cycle.
The graphs show the relative concentration in the blood plasma of (a) the
gonadotropins and (b) the events in the follicle. In the first half of the cycle, the
follicle develops, releasing a mature egg at the point of ovulation (day 14). If the
egg is not fertilized, the follicle (now the corpus luteum that secretes progesterone)
disintegrates, and a new one develops. The relative concentration of estrogens and
progesterone is shown in (c). (d) The relative thickness of the endometrium. High
levels of progesterone after ovulation maintain the thickness of the endometrium.
Notice that the cycle begins (day 1) with the first day of menstruation.

7. If the egg is fertilized, it becomes embedded in the endometrium. Both the corpus luteum and the placenta (the sac enclosing the developing fetus) secrete high levels of progesterone, which prevent the secretion of FSH and LH and maintain the endometrium.

The fact that high levels of progesterone prevent FSH and LH secretion underlies the mechanisms of many birth control pills. By increasing the progesterone level in the blood, these medications inhibit FSH secretion and the LH surge, thus inhibiting ovulation. Since no egg is produced, the woman cannot become pregnant. Some birth control pills include estrogens, thus stimulating the increased thickness of the endometrium so that a menstrual flow will occur.

Menarche and Menopause

■ **In the human female, puberty usually begins between the ages of 10 and 14. At this time, menstruation starts (menarche) and the secondary sex characteristics begin to develop. Menarche is brought about by gonadotropins and estrogens. Menopause, which occurs between forty-five and fifty-five years of age, is the end of the menstrual cycle and is accompanied by changes in hormonal levels, as well as physiological and sometimes psychological changes.**

Puberty in the human female usually begins between the ages of ten and fourteen. During that time, pituitary gonadotropins and ovarian estrogens initiate the onset of sexual maturity. The pelvis widens, fat deposits begin to produce the characteristic female form, the breasts enlarge, and the first menstrual flow, **menarche**, is activated. In some instances, although the cycle has begun, actual ovulation does not occur until a year after menarche.

The menstrual cycle naturally ceases between the ages of forty and fifty, signaling an end to the reproductive cycle in the female. This event is called **menopause**. The timing of menopause seems to be tied to the timing of menarche, for studies have shown that the earlier menarche occurs, the later menopause occurs, and vice versa. Menopause is a gradual process. At first, the most prominent symptom is the onset of irregular periods. Hormonal levels drop and physiological changes occur, and both these phenomena can have psychological and physical consequences as well. Women may experience emotional changes or hot flashes, caused by dilation of blood vessels of the skin. Although physicians sometimes prescribe estrogen (or estrogen and progesterone) to relieve the symptoms of menopause,

there is no universal agreement in medical circles about their use because of the slight possibility that they may bring on cancer. Some physicians feel that since estrogen slows the process of osteoporosis (the loss of calcium in bones), the benefits outweigh the dangers. As a result, most cases are decided on an individual basis.

Once a woman has passed through menopause, her reproductive cycle has ceased, and she is no longer fertile. However, her sex drive does not necessarily decrease. In fact, some women report an increase in sexual desire, perhaps because they are released from the possibility of pregnancy.

HUMAN SEXUAL RESPONSE

■ **There are four steps in human sexual response: excitement, plateau, orgasm, and resolution.**

Our understanding of the human sexual response cycle is due primarily to the research conducted by William H. Masters and Virginia E. Johnson. From their study of 382 women and 312 men, they concluded that the human sex cycle in males and females consists generally of four phases: *excitement, plateau, orgasm,* and *resolution*. The response pattern is basically the same, no matter what type of sexual activity leads to the sexual response. However, the intensity of the response can vary with different types of stimulation, at different times, and with different individuals.

During the excitement phase, erotic thoughts or feelings and physical stimuli cause physiological changes that prepare the body for sexual activity. In both male and female, the heart rate, breathing rate, and blood pressure increase. In the male, the flaccid penis becomes erect, and the scrotal sac thickens and lifts the testes within the scrotum. Cowper's glands secrete a fluid to lubricate the glans. In the female, the vagina moistens, the labia majora become reddish in color, and the outer two-thirds of the vagina expand. The clitoris becomes erect, and its sensitivity increases.

The plateau phase is characterized by an increased stage of excitement. In the male, the penis becomes firmer and enlarged. In the female, the clitoris withdraws. The skin may redden, and a rash-like appearance will occur, beginning with the lower chest. The uterus may elevate, pulling the cervix up and producing a tenting, or enlargement, of the vagina. Bartholin's glands secrete mucus.

During the orgasmic phase, the abrupt release of sexual excitement leads to climax, or orgasm, a series of voluntary and involuntary muscle contractions that produce feelings of intense pleasure and

Retroviruses

Modes of Retroviral Transmission

Infected mouse — Horizontal →

Vertical ↓

Offspring

1950s: Gross isolated a mouse leukemia virus and showed horizontal infection and vertical transmission to offspring. Manaker and Groupé discovered transformation of cultured chicken-embryo cells.

1960s: Crawford and Crawford showed that Rous sarcoma virus (RSV) is an RNA virus. Hanafusa, Hanafusa, and Rubin discovered defective sarcoma viruses that cannot replicate unless "helper" leukosis viruses are present (see diagram). Temin suggested his "provirus" hypothesis that viral genes incorporate into the host-cell genome, but no one believed him.

Cell Transformation with Defective Retrovirus

Helper virus

Transformed virus-producing cell

Defective transforming virus

Uninfected cell

Transformed virus-producing cell

Pre-1950: Rous, in 1910, discovered a virus that transmits sarcomas in chickens (RSV), but researchers soon found that genetic and environmental factors can also contribute to cancer.

Infecting HIV-1 virion

Envelope proteins:

gp120

gp41

p17

Viral core

p24

Viral RNA

Source: Adapted from "Evolutions: Retroviruses," *The Journal of NIH Research*, vol. 4, pp. 159–160, July 1992. Adapted from original artwork by Terese Winslow.

1970s: Temin and Mizutani found reverse transcriptase in RSV, and Baltimore found it in mouse leukemia virus, upsetting the central dogma of molecular biology that genetic information could not flow from RNA to DNA. Researchers soon determined the major steps of the retroviral life cycle and discovered more animal retroviruses.

Virion Attaches to Cell Surface

HIV-1 Life Cycle

CD4 receptor

Fusion

Double-stranded viral DNA (provirus) in nucleoprotein complex

Reverse Transcription

Long terminal repeat (LTR)

Viral DNA

Viral RNA

Single strand viral RNA

Reverse transcriptase

Uncoating

Integration

Integrated provirus

Host DNA

Nucleus

Proviral DNA

RNA polymerase

Translation and protein processing

Viral mRNA

Viral mRNA

Protein

Ribosome

Viral mRNA

Virion assembly

T4 cell

Viral genome RNA

Viral proteins

Viral envelope glycoprotein

Budding of Immature Virion

RNase H degrades RNA

5'

Viral DNA

3' OH

3' OH

DNA elongation

Viral RNA

Reverse Transcriptase

1980s–1990s: Poeisz, Gallo, and colleagues discovered the first human retrovirus, human T cell lymphotrophic/leukemia virus–type 1 (HTLV-1). Several human retroviruses are now known, including the AIDS virus, human immunodeficiency virus–type 1 (HIV-1). Researchers are devising treatments for AIDS and other retroviral diseases on the basis of knowledge of the viruses and their components.

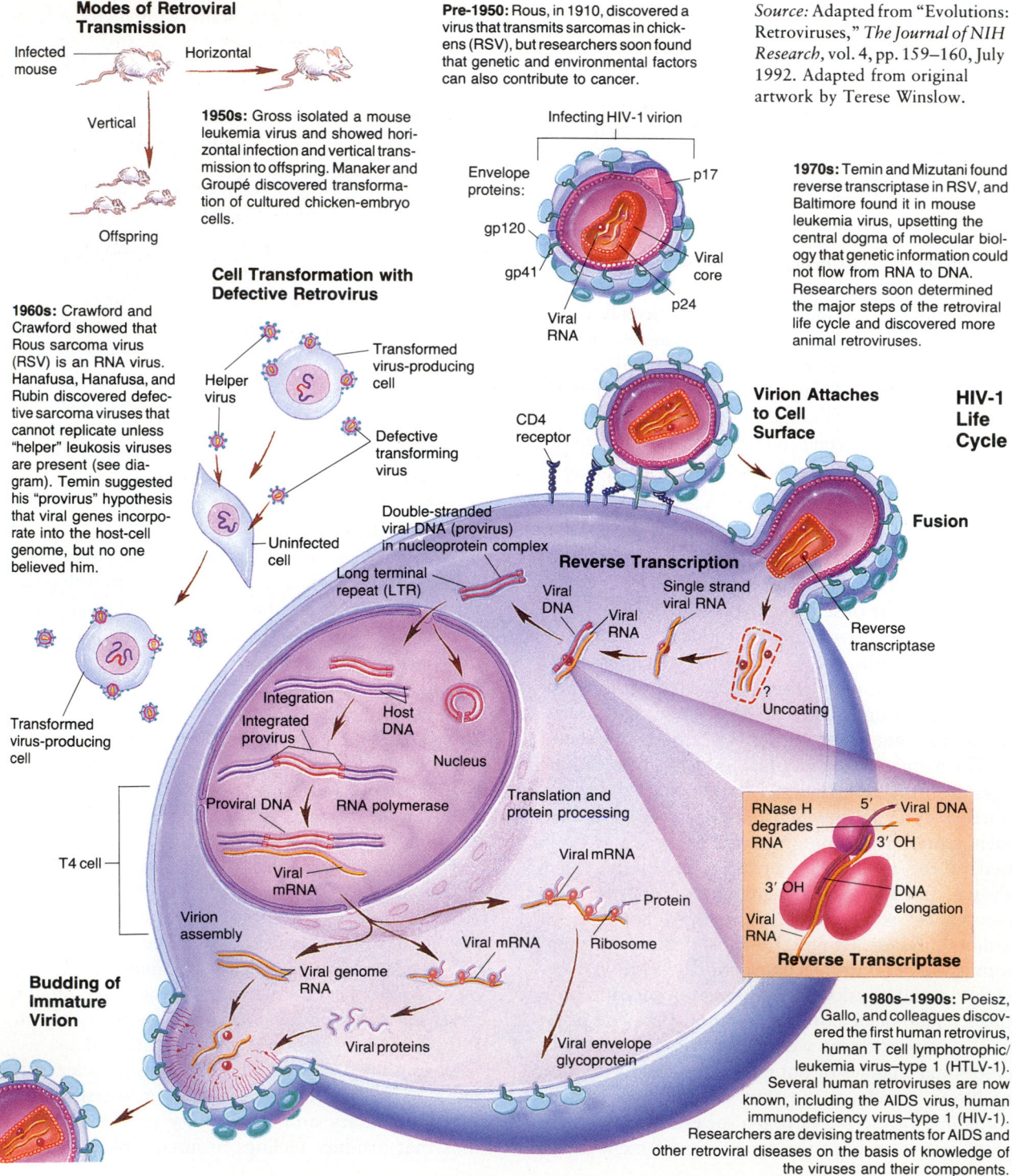

Dr. Flossie Wong-Staal

Immunologist, Molecular Biologist

Her first love was poetry; reading Shelley's beautiful lyricism was her passion. Who would have ever thought that the Catholic schoolgirl in Hong Kong, Yee Ching Wong, would become a world-renowned researcher on the sexually transmitted disease AIDS (Acquired Immune Deficiency Syndrome).

As a child, Dr. Wong-Staal did not enjoy science, but the nuns recognized her scientific abilities. At that time in China the pursuit of science was considered an honor, an honor that the young girl believed was beyond her grasp or interest. However, when asked directly by the nuns to pursue science, she couldn't say no, and her career in science began.

In 1965, she and her family immigrated to Los Angeles. She later became a student at the University of California, Los Angeles, where she became a research assistant in bacteriology. From UCLA she went to the National Institutes of Health (NIH), where she worked closely with Dr. Robert Gallo, the codiscoverer of the human immunodeficiency virus (HIV). Her work was outstanding, and she was responsible for much of the progress relevant to an understanding of the virus that causes AIDS. Together Dr. Wong-Staal and Dr. Gallo built the HIV AIDS laboratory at NIH. Prior to going to NIH, Dr. Wong-Staal did postdoctoral work at the University of

California, San Diego. Because of her desire to go back to academia and head her own research program, Dr. Wong-Staal decided to return to the West Coast to continue her research. Her outstanding work brought her several offers, but she decided that she wanted to return to the University of California, San Diego.

Today at U.C.S.D., she and her colleagues are working on developing a vaccine for the AIDS virus and a therapy to interfere with the replication of the virus in infected individuals. She and her colleagues at Northern Illinois University have developed a kind of molecular scissors called a *ribozyme*. A ribozyme is a type of ribonucleic acid that has enzyme properties. The particular ribozyme that they have isolated is called a "hairpin" ribozyme, which has been demonstrated to cut up the genetic instruction of the HIV virus in the test tube. When the ribozyme is injected into cultured HIV-infected white blood cells, more than 99 percent of the viral activity is stopped.

Dr. Wong-Staal and her laboratory staff are also investigating the nine genes that make the virus function, especially the tat and rev genes. They are trying to find compounds that inhibit the activity of these genes, hence interfering with the ability of the HIV virus to remain infectious.

She is also looking for a vaccine against AIDS by using many different strategies, including the use of a part of the AIDS virus outer coat known as a protein or projection of amino acid called the V-3 loop, which invokes a very strong immune response in infected laboratory animals. She is also using weakened forms of the AIDS virus that will elicit a strong immune response from the host but will not cause disease.

As you can see, Dr. Wong-Staal is very motivated and is directed by scientific curiosity. Her quest to understand AIDS was initially due to direct scientific curiosity, not the emotional need to cure AIDS. However, as she learns more about the epidemic and its victims, this curiosity has become a passion to win the battle against AIDS.

satisfaction. In the male, orgasm is linked with the ejaculation of semen. The accessory glands contract, releasing their secretions into the urethra, and the urethra contracts, ejaculating the semen. In the female, there is no ejaculation. Rather, the uterus contracts, as do the pelvic floor and vagina. One of the major differences between the sexual responses of males and females is that females may experience multiple orgasms, whereas males usually have only one orgasm per cycle (Figure 27.17).

The resolution phase follows orgasm. In the male, the erection is lost, and breathing and blood pressure return to normal. There is also a *refractory* period, during which the male cannot be aroused again. The length of this period varies tremendously with individuals and circumstances. In females, the resolution phase is not as abrupt as in males; instead women experience a gradual return to pre-excitement stages.

As you might suspect, the above describes a general pattern of human sexual response. In real life there is individual variation in the intensity and duration of each phase, as well as variation for each cycle.

REPRODUCTIVE TECHNOLOGY: CONTRACEPTION AND INFERTILITY

■ **As biologists learn more about reproductive processes, they can devise more effective means of contraception (measures designed to prevent conception). They are also devising means of dealing with infertility (the inability to conceive).**

Contraception ("against conception") can be defined as anything that interferes with, or prevents, the fertilization or implanting of an egg. *Abortion* refers to anything that prevents the birth of a fetus once conception occurs. The decision to use contraceptive techniques is a personal matter.

Contraception, or birth control, is widely used to prevent pregnancy. About 33.4 million American women use some method of birth control; only 3 million do not. A number of contraceptive technologies are available to limit or prevent pregnancy. They are described in Table 27.2.

About 10 percent of couples are *infertile* — that is, they cannot have children for a variety of physiological reasons. New reproductive technology has provided partial solutions to some of the causes of infertility (see Focus 27.1). Hormonal imbalances can sometimes be regulated to allow for conception. Surgical methods can correct various obstructions of the fallopian tubes or other physical defects of

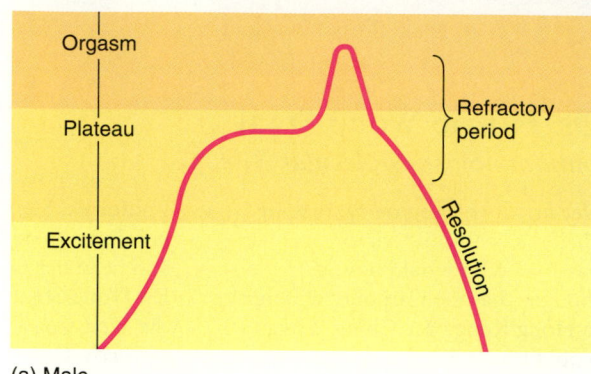

(a) Male

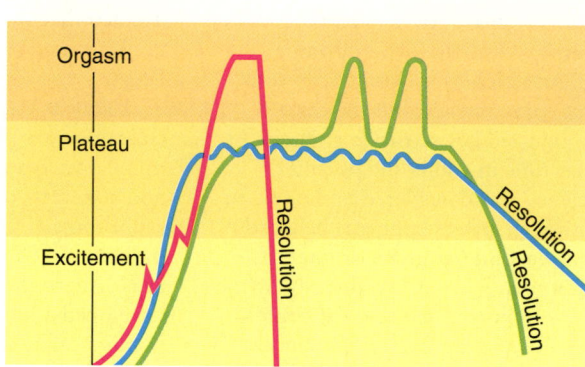

(b) Female

FIGURE 27.17   The Sexual Response Cycle
(a) The male usually has one orgasm per cycle, followed by a period of resolution. (b) Although the female may have just one orgasm during a sexual response cycle (line A), multiple orgasms can be common. These orgasms can occur as several minor episodes (line B) or as fewer, more intense ones (line C).

the male or female reproductive systems. Medical techniques have been developed to increase the number and quality of sperm, which increases the probability of fertilization.

The most advanced method of helping infertile individuals is the process known as *in vitro* ("in glass" or "outside of the body") fertilization. After the woman has undergone hormone treatment to stimulate the release of more than one egg, eggs are removed from the follicles of her ovary and placed in a solution in a glass dish. Next, sperm is added to the solution, and the eggs are fertilized and begin dividing. After two to four days, fertilized eggs develop into a clump of cells (called a blastocyst). The blastocyst is inserted into the woman's uterus and implants in the endometrium. On July 25, 1978, the first "test-tube" baby, Louise Brown, was born in Oldham, England — the first human to have been conceived in glass. Since that time thousands of children have been conceived by this method.

TABLE 27.2
Methods of Contraception

Method	Mode of action	Advantages	Disadvantages
Vasectomy	Prevents release of sperm	Very reliable; permanent	Relatively irreversible; possible complications; expensive
Tubal ligation	Prevents egg cell from entering uterus	Very reliable; permanent	Relatively irreversible; possible complications; expensive
Birth control pill (21-day administration)	Prevents follicle maturation and ovulation	Very reliable; some health benefits	Side effects; continual cost; must be taken every day
Minipill	Inhibits follicle development	Thought to have fewer side effects than other birth control pills	Breakthrough bleeding; daily use; continual cost
Hormonal implant (Norplant)	Prevents follicle development and ovulation	Long-term release of hormone; can be implanted for up to five years	Expensive ($600–$700); requires minor surgery to remove the implant
Morning-after pill	Arrests pregnancy by preventing implantation (50 times normal dose of estrogen)	Possible treatment for emergency situations (like rape)	Nausea; vomiting; possible cause of vaginal cancer in female offspring
RU 486	Interferes with progesterone synthesis	Can be taken once and induces miscarriage	Not approved for use in the United States at this time; experimental; causes spontaneous abortion
Intrauterine device (coil, loop)	Prevents implantation	Very reliable; no motivation or memory necessary	Bleeding; cramping; expulsion; risk of pelvic inflammatory disease
Cervical cap	Prevents sperm from entering uterus	Can wear for 48 hours at a time	May irritate cervix; may be difficult to insert
Diaphragm and spermicidal jelly or foam	Prevents sperm from entering uterus; jelly kills sperm	No major health risks; inexpensive	Can interfere with spontaneity; some aesthetic objections
Sponge	Prevents sperm from entering uterus; contains spermicide to kill sperm	No major health risks; 24-hour protection; inexpensive; no prescription required	Can interfere with spontaneity; some aesthetic objections
Condom	Prevents sperm from entering vagina	Highly reliable; no major health risks; helps protect against sexually transmitted diseases; inexpensive; no prescription required	Required interruption of sexual activity; some aesthetic objections
Spermicide	Kills sperm	No major health risks; inexpensive; no prescription required	Unreliable; some aesthetic objections
Douching	Washes out sperm	Inexpensive	Extremely unreliable
Rhythm	Determines ovulation time by measuring body temperature and indicates abstinence during that time	No health risks; no cost; acceptable to Catholic church	Unreliable; requires motivation and periods of abstinence
Withdrawal	Remove penis from vagina before ejaculation	No health risks; no cost	Unreliable; reduces sexual pleasure

FOCUS 27.1
The Causes and Treatments of Infertility:
The Difficult Questions

It often happens that a scientific discovery or a new scientific technique raises legal and ethical questions that researchers never considered in their wildest imaginings. In most cases, biologists have said that those questions must be resolved by the institutions in our country in which legal, ethical, and moral questions are best discussed — schools, churches, courtrooms. That has been the case with the new methods of treating fertility problems. The techniques have taken scientific research from the laboratory to the courtroom.

Consider test-tube babies. Is it ethical to experiment with the production of human babies outside a woman's body, in vitro (which originally meant "in glass" but now means "outside the normal biological environment")? What are the legal rights of the frozen embryos of deceased parents — are they entitled to their deceased parents' estate? A woman on fertility pills gives birth to sextuplets — can she sue her doctor for doing too good a job? If a woman signs a contract to be a surrogate mother, is the child hers or does it belong to the couple with whom she made the agreement? All these questions have arisen out of the new, sometimes bold, treatments for human infertility.

In the United States, from 3 to 4 million American couples are *infertile*. The problem may lie with the man, the woman, or both. In men, common causes of infertility are low sperm count, abnormal sperm, failure to achieve an erection, diabetes, and varicoceles (varicose veins of the testicles, which cause the temperature of the scrotum to increase, interfering with proper sperm production). In women, the common fertility problems are a failure to ovulate (produce eggs), blocked fallopian tubes (which prevent the sperm from reaching the egg), allergic reactions to sperm, and failure of the endometrium to receive and maintain the developing embryo (Figure A).

Once the specific problem is diagnosed, physicians can choose from a number of treatments to increase the likelihood of pregnancy. For instance, blocked fallopian tubes and sperm ducts can now be opened with microsurgery and laser technology. Microsurgery can be used to treat varicoceles, which some physicians believe are the cause of 40 percent of male infertility.

If a woman is having problems with egg production, the physician can prescribe synthetic hormones such as clomiphene and Pergonal, which stimulate the release of gonadotropins by the pituitary, resulting in egg production. But these drugs are not without side effects, one of which is that 10 percent of the women treated with clomiphene will release more than one egg. In other words, there may be a multiple birth. Obviously, if too many embryos result, they cannot be carried to term. In one instance, drug treatment produced 15 fetuses, all of which were miscarried. The other 90 percent of women treated, however, will deliver only one child.

If a woman's menstrual cycle and ovulation are erratic, the physician may implant a hormone-releasing micropump to serve as an artificial gland, releasing regulated amounts of hormones that can restore normal functioning. Incorporating male hormones, the same type of pump can increase sperm production in males.

A number of causes can bring on organic impotence (the inability to achieve an erection). These include circulatory abnormalities that obstruct blood flow to the penis, neurological disorders, hormonal disorders, diabetes, prostate problems, or heavy drug and/or alcohol use. Some physicians estimate that 80 to 85 percent of the causes of impotence are psychological, others feel that the percentage is closer to 40 percent, but all agree that there are both physical and psychological causes. One way to correct organic impotence involves a bold technique: the inflatable penile implant, which allows the impotent male to achieve an erection. As of this writing, about 85,000 men have received penile implants.

Artificial insemination, a technique in which sperm is collected and artificially placed in the woman's vagina, next to the cervix, was once considered rare but today is a rather commonplace treatment. It is used to increase the sperm count of a donor whose count is low, or it can be used if the female has a vaginal condition that is hostile to sperm. The sperm may be that of the husband or may be acquired from a sperm bank. One sperm bank in this country actively collects donations from Nobel Prize winners and other men who have excelled in their fields, following through on the theory that the donor's traits may be inherited by their offspring.

Another technique is "test-tube babies" (or, more correctly, in vitro fertilization), a process that does not take place in a test tube at all. First, the eggs are removed from the mother and placed in a laboratory dish. They are then fertilized with the father's sperm. After a certain length of time (usually five to eight days), the developing embryo is placed in the mother's uterus. If the embryo then implants in the uterine walls, the pregnancy can continue as a normal one. Variations on this basic technique include GIFT (or gamete intrafallopian transfer), in which eggs and sperm are placed on the mother's fallopian tubes, and ZIFT (zygote intrafallopian transfer), in which a fertilized egg is placed in the tubes.

In recent years, artificial insemination and in vitro fertilization have been used in situations involving a "surrogate mother," who is working for the natural parents by bearing a child for them. Either the surrogate is artificially inseminated with the father's sperm, or the mother's eggs are fertilized in vitro with the father's sperm and the eggs are transferred to the surrogate's uterus. The legal and ethical issues raised by surrogacy are complex, as was clearly demonstrated in the widely publicized "Baby M" case.

Seminal vesicle contributes nutrients and fluid for sperm transport. Too much or too little fluid can be detrimental to fertility.

Vas deferens carries sperm out of the testis. If it is blocked by scars from infection or surgery, the sperm are trapped.

Bladder

Prostate gland can harbor infections that affect sperm motility.

Urethra may harbor infections that can spread and damage the sperm duct.

Rectum

Penis may be unable to deliver sperm into the vagina if the erection is incomplete.

Epididymis can be infected. Infections here can prevent sperm from reaching the sperm duct.

Scrotum may contain a varicose vein, or varicocele, that allows for backward flow of warm blood. This heats up the testis, interfering with sperm production.

Testis may produce sperm that are too weak or too few in number to reach the egg in the fallopian tube. Hormonal signals from hypothalamus and pituitary may be weak.

(a)

Uterus may not be hospitable to the egg. Most often, hormone imbalances affect the endometrium and prevent the egg from implanting.

Ovary can fail to produce eggs because it receives no hormone signal from the hypothalamus and pituitary.

Fallopian tube may be blocked by scars from infection. If so, the egg cannot move or sperm cannot reach the egg.

Bladder

Cervix may be hostile to sperm. Some women develop a plug of mucus within the passageway that blocks the sperm.

Vagina may harbor infections that destroy sperm or prevent the cervix from accepting sperm. (Some vaginal infections also may spread to the fallopian tubes.)

(b)

FIGURE A
Causes and treatments of infertility in the male and female, respectively.

Scientists will continue to develop new types of reproductive technology, no matter how many thorny questions accompany such research, for one simple reason: They can bring hope to those infertile couples who will pursue any and all means of having their own child. It is those hopes, too, that will force us to consider again and again what it means to be a parent, what it means to be a child, and how much some of us will pay in order to take part in the phenomenon of childbirth.

FOCUS 27.2
Sexually Transmitted Diseases

In 1990 the rate of gonorrhea was about 520 cases per 100,000. In 1990 the rate for syphilis was 20 cases per 100,000. It is estimated that genital herpes, ranked as the third most communicable disease (right behind the common cold and gonorrhea), is twice as common as syphilis. Chlamydia affects as many as 20 percent of college-age women. As is plain by these rankings, sexually transmitted diseases are highly contagious. In this Focus we consider four of the most important sexually transmitted diseases. (Focus 25.1 discusses AIDS, which can also be transmitted sexually.)

Chlamydia
Each year there are 3–10 million new cases of the STD urinary infection known as chlamydia. The bacterium *Chlamydia trachomatis* causes this disease. This bacterium infects the urinary tract and can produce irreversible damage to the female reproductive system. This can include scarring of the oviducts, which may result in sterility. Chla-

mydia is difficult to detect, since the symptoms are often subtle and infected individuals — especially women — can unknowingly carry the disease for years. For this reason, and because chlamydia is easily transmitted during sexual activity, it has become very common among sexually active individuals. It is estimated that as many as 20 percent of college-age females have the infection, and the incidence of the disease is increasing.

Herpes Simplex Type II: Genital Herpes
Most cases of genital herpes are caused by the herpes simplex type II virus, although 10–20 percent are caused by the herpes simplex type I virus, which is also responsible for cold sores and blisters that occur above the waist. Infection results through sexual activity with an infected person during the contagious stage, when the small, fluid-filled blisters are present. In the male, the blisters usually appear around the penis, and in the female, the blisters

can appear externally, on the genitalia, or internally, in the cervix or uterus. Because the blisters can be internal, it is often difficult for a woman to know she is infected.

The blisters usually appear two to ten days after infection, and they usually clear up in about ten days. Along with the blisters, there is an itching, burning, and tingling around the blister area. Sometimes an individual has only one incident of genital herpes, but others may suffer from outbreaks again and again, through their entire life.

An infected pregnant woman can give the herpes virus to her child as it passes through the birth canal. In about one of four births to infected mothers, a child may die or be seriously injured by the virus. If a woman is suspected of having herpes, the physician usually delivers the child by cesarean section to avoid infecting the child. At present there is no cure for genital herpes, but some drugs—like acyclovir, taken orally—can reduce the severity and length of the eruptions.

Gonorrhea

Gonorrrhea is an infectious venereal disease caused by the bacterium called *Neisseria gonorrhoea*. This bacterium infects various mucus membranous tissues, normally those of the genitalia, throat, or rectum. Which mucus tissue becomes infected depends on the sexual activity leading to infectious contact. The disease is spread during sexual contact with an infected person, because the bacterium is very delicate and cannot live without the human host. In other words, it cannot live out in the open environment — that means you cannot catch gonorrhea from toilet seats.

When a man contracts gonorrhea, the symptoms are easily diagnosed. Three to five days after contact, there is a greenish-yellow discharge from the urethra of the penis. However, when a woman contracts gonorrhea, she may not realize the problem, because many times women have no symptoms in the earlier stages of the disease. She may not be aware that she has the disease until she has severe abdominal pains, arising from the fact that the bacteria usually settle in the uterus or oviducts.

It is relatively simple to diagnose gonorrhea. The physician takes a sample of the discharge from the penis or vagina and examines it for the presence of the bacterium. If it is present, the individual has the disease. Prompt treatment with penicillin, tetracycline, or erythromycin will clear it up in 90 percent of those infected. However, physicians are now finding a penicillin-resistant strain of gonorrhea called "super clap." This strain is causing gonorrhea to become a worldwide health problem.

In about 1 percent of the cases, a painful arthritic condition known as gonorrheal arthritis can result from infection. If the condition is not treated, it can cause sterility in the patient. If a child is born to a woman who has the gonorrhea bacterium in her vagina, the newborn can contract the bacterium as it passes through the birth canal. One result is a form of blindness called *ophthalmia neonatum*. To prevent this disease, it is compulsory to add silver nitrate eyedrops or penicillin ointment to the newborn's eyes.

Syphilis

In 1905, Fritz Richard Schaudinn, of Germany, identified *Treponema pallidum,* the spiral-shaped bacterium that causes syphilis. Like gonorrhea, syphilis is contracted through sexual activity with an infected person, and like gonorrhea, it is not contracted from a toilet seat, a wet towel, or other similar items. In the early stages, a skin lesion called a chancre (a hard, round ulcer — usually painless) appears at the site of contact, usually on the genitals, anus, mouth, or breast. Even if the chancre is untreated, it will go away in a few days.

Just because the chancre is gone, however, the disease is not. If untreated, syphilis can progress to the secondary stages, or even to the tertiary stages. Several weeks or even months can pass before the symptoms of the secondary stage appear. Then the patient will develop a skin rash, headache, fever, and pain in the muscles and joints. These symptoms will disappear after a period of time, signaling the beginning of a latent period.

During the latent period, the patient will notice no symptoms. The bacteria, however, are not inactive. They are burrowing into the blood vessels, bones, and central nervous system. The tertiary stage can occur years after initial infection, and it sometimes lasts until the patient dies. Its severity depends on how much organ damage has occurred. A patient in this phase can even die of the disease. Even in the late stages, treatment with penicillin is possible, but whatever organ damage has taken place is irreversible. Certain types of senility are the result of untreated syphilis infection.

An infected mother can transmit syphilis through the placenta (the structure that surrounds the fetus, formed by the inner lining of the mother's uterus and the fetal membranes) to the developing fetus. This condition is called congenital syphilis. In nine out of ten cases, a fetus infected in this way is miscarried. If it survives until birth, it may come into the world with deformities in its vision, hearing, bones, or neural system. Early treatment with penicillin can cure or reduce many of the symptoms of congenital syphilis.

Reduction of the incidence of STDs can be achieved only by educating people about (1) the nature of the disease, (2) the mechanisms of spread, (3) prevention, (4) symptoms, and (5) cures. Of no use to anyone is ignorance, fear, or social prejudice against infected individuals.

SEXUALLY TRANSMITTED DISEASES

■ **Sexually transmitted diseases are those that are transmitted by means of sexual activities. The most common are chlamydia, herpes simplex type II, gonorrhea, syphilis, and AIDS.**

Diseases that are spread through sexual intercourse and semen are known as *sexually transmitted diseases* (STDs). Contrary to popular myth, you cannot catch a sexually transmitted disease from toilet seats or casual contact. The means must almost always be through direct sexual contact. The most common diseases are *chlamydia, herpes simplex type II, gonorrhea, syphilis,* and *AIDS* (see Profile and Human Endeavors), which we discussed in Chapter 25. (See Focus 27.2 for a survey of the more common sexually transmitted diseases.)

SUMMARY

1. Asexual reproduction involves no exchange of genetic material between organisms. In fission, the organism divides to form two organisms. In budding, the organism produces a bud, which breaks away from the parent to become a new organism. In fragmentation, a piece of an organism forms a new organism.

2. In sexual reproduction, organisms exchange genetic material. Conjugation involves the fusion of two individuals. Hermaphroditism involves organisms that produce both male and female gametes in the same individual (although self-fertilization is extremely rare). In parthenogenesis, an unfertilized egg develops into a new organism. In biparental reproduction, males and females both produce haploid gametes, and reproduction requires a fusion of those gametes. Sexual reproduction provides for increased genetic variability.

3. Humans reproduce through biparental sexual reproduction. Both male and female produce sex cells (gametes); the male produces the highly mobile *sperm*; and the female, the *egg* or ovum.

4. In the human male, sperm and sex hormones are produced in the testes, the male gonads. The anatomy of the male reproductive system is designed to deliver sperm to the female through internal fertilization; this system consists of the testes, various glands and ducts, and the penis.

5. Sperm are produced in the seminiferous tubules of the testes. Each sperm consists of a head, a midpiece, and a tail. The head includes both acrosome and nucleus. The acrosome (cap) contains enzymes that help the sperm penetrate the egg. The nucleus encloses the DNA. The midpiece contains mito-

chondria that produce the energy to move the flagellum (tail), which propels the sperm.

6. Once formed, sperm pass to the epididymis, where they mature and become motile. They then move through the vas deferens into the ejaculatory duct. The seminal vesicles and the prostate gland secrete fluids into the duct as the sperm pass through, forming the seminal fluid. The ejaculatory duct fuses with the urethra at the penis.

7. Semen leaves the penis through the urethra. Three layers of spongy erectile tissues surround the urethra. During sexual excitement, blood vessels within the spongy tissues fill with blood, causing the penis to enlarge and become erect and closing the veins that would move the accumulated blood back to the body. Erection allows the penis to be inserted into the vagina for internal fertilization. Muscular contractions of the urethra during orgasm propel the sperm into the vagina.

8. Hormones regulate sperm production and reproduction. Testosterone, the major male sex hormone, is produced in the interstitial cells of the testes. Pituitary hormones involved in the control of the secretions of the hormones of the gonads are called gonadotropins. One of these, interstitial cell–stimulating hormone (ICSH), stimulates testosterone production. Another, follicle-stimulating hormone (FSH), regulates sperm development and maturation. The hypothalamus produces gonadotropin-releasing hormone (GnRH), which regulates the pituitary gland.

9. In the human female, the reproductive system is more complex. In addition to producing the sex cell (the egg) and the female hormones, women also must retain, protect, and nourish the embryo.

10. The ovaries usually produce one egg approximately every 28 days. The ovaries also produce estrogens and progesterone, female sex hormones.

11. The haploid egg, produced through meiosis oogenesis, matures in the follicle. When the egg is mature, usually midway through the monthly menstrual cycle, it is released (ovulation). The follicle is then called the corpus luteum. The egg passes into the infundibulum, through the fallopian tube, a structure leading to the uterus.

12. The uterus is designed to enclose a developing fetus. The fallopian tubes enter at the top of the uterus, and the cervix is found at the bottom. When an egg is fertilized, it lodges in the endometrium, the uterine lining that protects and nourishes the embryo. An unfertilized egg passes out through the vagina, the tubelike structure in which sperm are deposited during sexual intercourse. This structure also serves as the birth canal.

13. The female's external genitalia (vulva) consist of labia, fleshy folds protecting the vaginal opening, and the clitoris (an erectile structure functioning in sexual stimulation).

14. The menstrual cycle is timed by the interaction of hormones. GnRH regulates the secretion of follicle-stimulating hormone (FSH) and luteinizing hormone (LH), both of which regulate ovulation. Estrogens produced by the follicle govern the thickness of the endometrium. The corpus luteum produces progesterone, which thickens and maintains the lining.

15. If fertilization occurs, the embryo produces a hormone called human chorionic gonadotropin (HCG), which helps to prevent the breakdown of the corpus luteum. HCG and progesterone maintain the uterine lining through embryonic development. Other hormones initiate labor and childbirth and regulate milk production.

16. Menarche is the name given to the first menstrual flow, usually occurring between the ages of ten and fourteen. Between forty-five and fifty-five years of age, the menstrual cycle stops. This is called menopause.

17. The human sexual response consists of four general phases: excitement, plateau, orgasm, and resolution.

KEY TERMS

fission	gonadotropin
budding	gonadotropin-releasing
fragmentation	hormone
conjugation	puberty
hermaphrodite	ovary
parthenogenesis	estrogen
biparental reproduction	progesterone
testis	ovum
seminiferous tubule	menstrual cycle
ejaculatory duct	follicle
seminal fluid	ovulation
semen	corpus luteum
copulation	fallopian tube
ejaculation	infundibulum
orgasm	uterus
androgen	vagina
testosterone	luteinizing hormone
interstitial cell–	human chorionic
stimulating hormone	gonadotropin
follicle-stimulating	menarche
hormone	menopause

REVIEW QUESTIONS

True/False

1. One advantage of sexual reproduction is that it allows for genetic sameness.
   true   false   (page 648)

2. In humans, sex cells are produced in the gonads.
   true   false   (page 651)

3. The only function of the testes is to produce sperm.
   true   false   (page 652)

4. If pregnancy occurs, the endometrium is shed in the menstrual flow.
   true   false   (page 662)

Fill in the Blank

1. In many invertebrate organisms, both sexes are found in the same individual. This is called _____. *(page 649)*

2. The _____ cells of the testes produce androgens. *(page 656)*

3. During a pap smear, the physician removes cells from the _____ to examine for possible early stages of cancer. *(page 659)*

4. In the human sexual response cycle, the _____ phase or stage usually follows orgasm. *(page 668)*

Multiple Choice

1. The birth control pill prevents ovulation by inhibiting the production of:
   a. FSH and LH
   b. Testosterone
   c. Estrogen and progesterone
   d. FSH
   (page 665)

2. Menarche is:
   a. The gradual decrease of menstruation
   b. The first menstrual flow
   c. The first stage of labor
   d. A type of reproductive surgery
   (page 665)

3. Wearing tight jeans may interfere with sperm production because of:
   a. Negative feedback on the pituitary
   b. Increased testicular temperature
   c. Production of varicose veins of the testes
   d. Lowering the levels of ICSH
   (page 653)

4. In the generalized human sexual response cycle, the usual sequence of events is as follows:
   a. Plateau, excitement, orgasm, resolution
   b. Excitement, plateau, orgasm, resolution
   c. Plateau, resolution, orgasm, excitement
   d. Orgasm, resolution, excitement, plateau
   (page 665)

Discussion

1. If you wrote a paragraph that provided an overview of the major concepts of reproduction, what would it include?

2. Discuss and give examples of asexual reproduction in animals.

3. Discuss the journey of a sperm from formation to ejaculation. Through what structures does it pass?

4. Can you think of an evolutionary reason for the development of the penis and the vagina?

5. Discuss the hormonal regulation of the human menstrual cycle.

6. Why does the menstrual cycle cease during pregnancy?

7. Discuss contraception.

Animal Development

Developmental Processes
Principles of Early Embryonic Development in Animals
Human Development
Summary

The theme of this chapter might well be "From Fertilization to Fertilizers." Ridiculous, you say? Think a moment. All asexually and sexually reproducing organisms undergo a series of processes that result in formation of one or more new individuals. In all sexually reproducing organisms — plants, animals, and yes, even humans — development is an ongoing process that continues from fertilization to death. Birth is only an event in a continuum of development — a punctuation mark in the cycle of life. Human development occurs from the womb to the tomb — a continuous process including embryonic development, birth, childhood, adolescence, maturity, old age, and death.

In more primitive organisms, such as bacteria and one-celled eukaryotic organisms, development

Above: Within the pouch located on the back of this marsupial frog, the frog embryos develop.

is not particularly obvious since the offspring almost immediately resembles its parent. In this chapter, however, we center our discussion of development on the sexually reproducing animals, concentrating our attention on humans. To begin, we examine the developmental processes common to most organisms.

DEVELOPMENTAL PROCESSES

■ **In general, organisms develop by three processes: cell division, morphogenesis, and differentiation. Mitosis results in cell division. During morphogenesis, the embryo assumes the shape of its species. In differentiation, particular cells become specialized to perform certain functions.**

In general, three major processes occur as a fertilized egg develops into a mature individual: *cell division*, *morphogenesis*, and *differentiation*. The processes do not take place in sequence. Instead they are interrelated, and so any one process may dominate the others, depending on the developmental stage.

Cell Division

Since developmental processes involve cell division, we need to refresh our memories about the details of that procedure. In Chapter 10 we discussed *mitosis,* during which genetic material is equally distributed to daughter nuclei and (in most cases) cells, and *meiosis,* during which the chromosome number is reduced from $2n$ to n in two sequential nuclear and cell divisions. Mitosis results in an increase in the number of somatic cells; meiosis results in the production of gametes, or sex cells. Refer to Chapter 10 for a review of the basic events of those processes.

Cell division plays an important role throughout any organism's development. For example, after fertilization, the zygote (the fertilized egg) begins to divide, resulting in growth of the individual. (In fact, the majority of cells continue to divide throughout the entire life span of the organism, but never as rapidly as they did in the early embryonic stages.) Not only does the number of cells in the organism increase, but the cells themselves grow to a genetically predetermined size. Since living things are not just masses of cells, development involves the emergence of groups of organized structures with recognizable forms and cellular specialization.

Morphogenesis

The word **morphogenesis** means the rise (*genesis*) of a specific form and shape (*morpho*). Through morphogenesis, the complex structure and form that are unique to a particular species develop. This is why dogs look like dogs, apple trees look like apple trees, and so on. Morphogenesis gives rise to primary germ layers, from which specialized cells develop into the specialized tissues we discussed in Chapter 22.

Morphogenesis is a prime example of gene regulation at work (Chapter 12). At certain times during development, certain genes are activated, resulting in biochemical reactions leading to the formation of specific structures. Once the structure is completed, some genes may deactivate. Others do not.

Differentiation

We all begin life as a single zygote the size of the head of a pin. From that one cell, through a mar-velous series of events, undifferentiated cells become **differentiated,** or specialized, turning into muscle cells, nerve cells, blood cells, and so on. The more highly evolved the organism, the more difficult it is for a specialized cell to revert to its undifferentiated form. Once morphogenesis and cellular differentiation have occurred in animals, the "clock" cannot revert (or be reset) to an earlier stage. In other words, in highly evolved organisms like mammals, for the most part development is irreversible.

For instance, if a human being loses a finger, an arm, a leg, or any other part of the body in an accident, that structure will not regenerate (redevelop). Some animals, however, like lizards and salamanders, can regenerate lost parts. For example, in some species of lizards, if an individual loses a tail, it is capable of growing a new one (see Figure 27.1).

Interestingly enough, one of the major theories held by cancer researchers today is that cancer is a developmental problem — an *oncodevelopmental* question. Scientists think that, for some reason, genes that were active in cell division, rapid growth, and development before birth are reactivated. As a result, a normal specialized cell reverts to a condition much like that of an undifferentiated embryonic cell, where rapid growth is the norm. Therefore, an understanding of gene regulation in embryonic cells is leading to a better understanding of the biological basis of cancer.

PRINCIPLES OF EARLY EMBRYONIC DEVELOPMENT IN ANIMALS

The process of embryonic development incorporates the processes of cellular division, morphogenesis, and differentiation. As reproduction progresses through the stages of fertilization, cleavage, and gastrulation, the various mechanisms predominate. Each is instrumental in the development of vertebrate animal embryos.

Fertilization

■ **During fertilization, the egg and sperm fuse, producing a $2n$ zygote enclosed in a fertilization membrane. In higher vertebrates, like humans, once the sperm's head penetrates the egg, the tail drops off. The egg completes meiosis, producing a haploid egg that includes most of the cytoplasm and a small polar body. The zygote usually consists of yolk (nutrient supply), cytoplasm, and a diploid nucleus.**

Chapter 28   Animal Development

As we discussed in Chapter 10, fertilization is the union of gametes to form a zygote. The female gamete is the ovum (egg); the male gamete is the sperm. The egg is much larger than the sperm and it contains the *yolk*, the nutrients that support the developing embryo until the fertilized egg implants in the wall of the uterus. Both gametes contain genetic material in their nuclei.

In humans, a membrane called the *vitelline membrane* surrounds the entire egg cell. A gelantinous coating (the zona pellucida) separates membrane from an outer layer of follicle cells (Figure 28.1a).

When a sperm cell penetrates the vitelline membrane, microvilli—small projections on the surface of the membrane—surround and trap the head of the sperm, pulling it into the egg. This triggers a chemical reaction in the egg: Vesicles near the plasma membrane of the cell (called the plasmalemma in mammals) release a substance that changes the vitelline membrane into a *fertilization membrane* (Figure 28.1b). This membrane, which rises up from the surface of the egg, prevents other sperm cells from entering the fertilized egg (see Human Endeavor box).

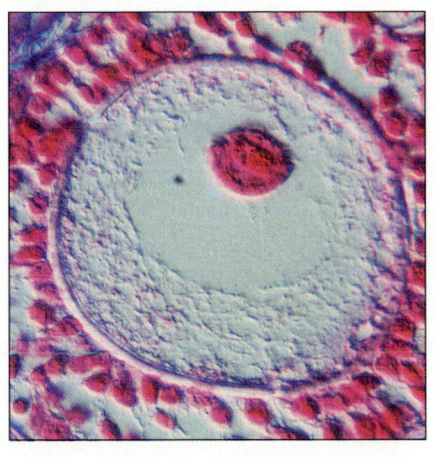

(a) An unfertilized human egg

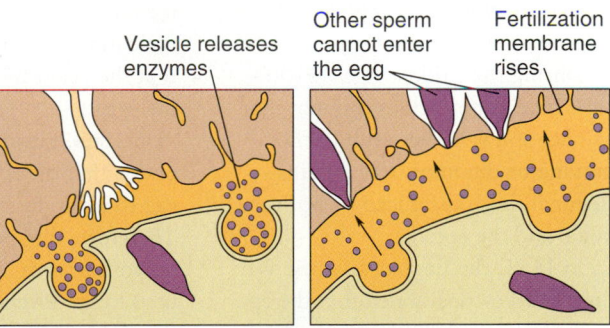

- Follicle cells
- Zona pellucida
- Vitelline membrane
- Plasma membrane
- Polar body
- Nucleus
- Cytoplasm

Sperm — Vitelline membrane — Plasma membrane / Egg cytoplasm / Vesicle — Microvilli trap the sperm — Vesicle releases enzymes — Other sperm cannot enter the egg — Fertilization membrane rises

(b) Penetration and the fertilization membrane

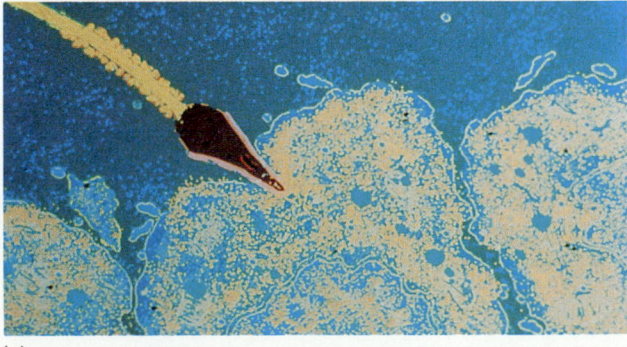

(c)

FIGURE 28.1 Fertilization
(a) An unfertilized human egg. (b) Once a sperm cell penetrates the vitelline membrane, enzymes from the vesicles in the egg transform that membrane into the fertilization membrane. (c) A sperm penetrating the egg.

Principles of Early Embryonic Development in Animals

Fertilization

Down through history, philosophers and scientists have speculated on the nature of the reproductive process. Let's explore the evolution of our understanding of fertilization, that moment when sperm unites with the egg.

Aristotle—the Greek philosopher and scientist who is sometimes called the "Father of Biology"—studied chick embryos in the fourth century B.C. He concluded that organs developed in sequence (epigenesis) and that the complete organism was not preformed.

During the Middle Ages (1100–1500 A.D.), physicians thought that conception was the result of the mixing or union of sexual fluids—seminal fluids produced by both the male and the female.

In 1651, the English scientist William Harvey postulated that the egg is the product of both uterine desire and sexual desire in the brain that results in the generation of the egg by the uterus.

Also in the seventeenth century, the Italian anatomist Malpighi claimed that he observed the complete embryo preformed in a chicken egg. This statement accelerated a debate that lasted for centuries as to whether embryos were preformed or developed through epigenesis, an orderly sequence of events. In the late 1800s and early 1900s, preformation was proved to be incorrect (or shall we say was proved to be a *misconception*).

With the advent of the microscope, the Dutch naturalist Leeuwenhoek was able to observe and describe sperm in human semen. But being a proper gentleman, he wrote a letter to Britain's Royal Society stating that his observations of sperm might be considered "nauseous or likely to bring scandal to learned men."

Hartsoecker said that he observed the complete preformed male or female in the sperm. Others said no, that the preformed embryo was in the egg.

In 1824, Prevost and Dumas suggested that the sperm penetrated the egg to form the nervous system. Therefore, the sperm contributed the nervous system and the egg contributed to the formation of the rest of the embryo. Several other scientists reported that the sperm penetrated the egg, so at least some scientists agreed on this point.

Between 1876 and 1879, Fol correctly described the events of the maturation, division, and origin of the egg nucleus; the "attraction cone" on the surface of the egg; and the formation of the fertilization membrane that blocks other sperm from entering the egg.

After a biological observation is made, the chemical mechanisms are often questioned. In 1910, Lillie proposed that the egg produced a substance called *fertilizin* that served as a receptor to link with the *antifertilizin* of the sperm. In 1939, Tyler and Fox demonstrated that fertilizin was present in the jelly coat of most aquatic eggs, and in the 1950s, it was determined to be a glycoprotein. At the same time, Tyler and Frank isolated an acid protein from the head of the sperm antifertilizin that precipitates fertilizin.

And the quest goes on. Our understanding of fertilization from 1980 to the present is summarized in the accompanying diagram. In brief, fertilization can be thought of as occurring in six steps. These extend from attachment to penetration of the egg by the sperm and the release of the sperm's pronuclei genetic material into the egg to initiate cell division and the restoration of the $2n$ genetic makeup.

As you can see, our understanding of conception has replaced a lot of earlier misconceptions. And as we have said before, it is important to question answers, not just answer questions.

Source: Adapted from "Evolutions: Fertilization," *Journal of NIH Research,* vol. 2, pp. 94–96, Sept. 1990. Illustration adapted by Elizabeth Morales-Denney from original artwork by Sally Bensusen.

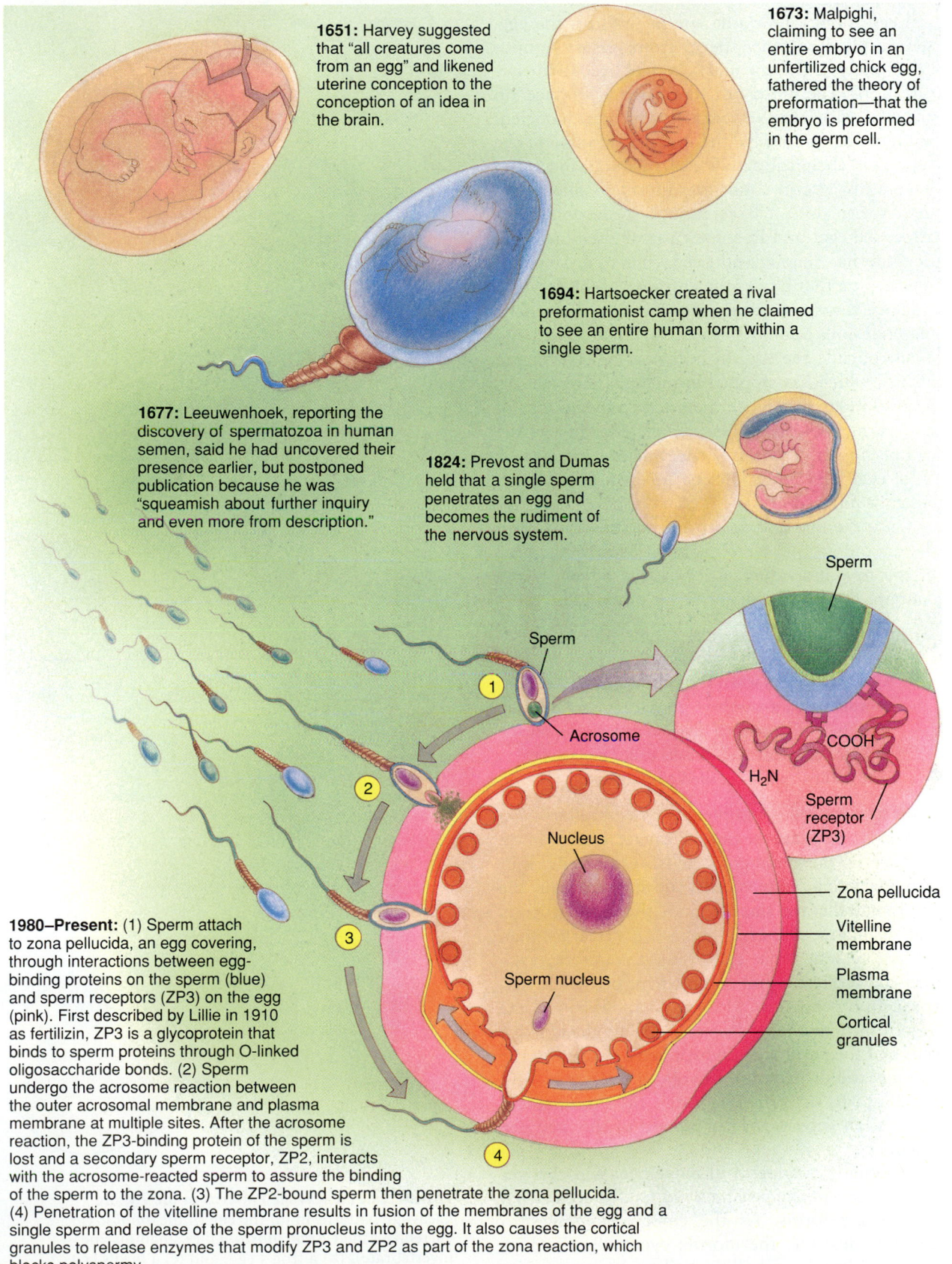

1651: Harvey suggested that "all creatures come from an egg" and likened uterine conception to the conception of an idea in the brain.

1673: Malpighi, claiming to see an entire embryo in an unfertilized chick egg, fathered the theory of preformation—that the embryo is preformed in the germ cell.

1694: Hartsoecker created a rival preformationist camp when he claimed to see an entire human form within a single sperm.

1677: Leeuwenhoek, reporting the discovery of spermatozoa in human semen, said he had uncovered their presence earlier, but postponed publication because he was "squeamish about further inquiry and even more from description."

1824: Prevost and Dumas held that a single sperm penetrates an egg and becomes the rudiment of the nervous system.

Sperm

Sperm

Acrosome

COOH

H_2N

Sperm receptor (ZP3)

Nucleus

Zona pellucida

Vitelline membrane

Plasma membrane

Sperm nucleus

Cortical granules

1980–Present: (1) Sperm attach to zona pellucida, an egg covering, through interactions between egg-binding proteins on the sperm (blue) and sperm receptors (ZP3) on the egg (pink). First described by Lillie in 1910 as fertilizin, ZP3 is a glycoprotein that binds to sperm proteins through O-linked oligosaccharide bonds. (2) Sperm undergo the acrosome reaction between the outer acrosomal membrane and plasma membrane at multiple sites. After the acrosome reaction, the ZP3-binding protein of the sperm is lost and a secondary sperm receptor, ZP2, interacts with the acrosome-reacted sperm to assure the binding of the sperm to the zona. (3) The ZP2-bound sperm then penetrate the zona pellucida. (4) Penetration of the vitelline membrane results in fusion of the membranes of the egg and a single sperm and release of the sperm pronucleus into the egg. It also causes the cortical granules to release enzymes that modify ZP3 and ZP2 as part of the zona reaction, which blocks polyspermy.

Principles of Early Embryonic Development in Animals

Even though only one sperm fertilizes the egg, several hundred other sperm are necessary in initiating many of the chemical changes needed for fertilization. The heads of the sperm produce an enzyme called *hyaluronidase,* which in sufficient amounts helps dissolve the outer membrane of the egg. This chemical action explains why men who have a low sperm count are infertile. Collectively, their sperm do not have enough hyaluronidase to affect the egg, and in some cases the seminal fluid does not have enough other enzymes to dissolve the mucus plug that blocks the entrance of the cervix.

Once the sperm penetrates the egg, the genetic material spills out of its head. The tail drops off and is left outside the egg. In humans, the egg (Figure 28.1c), which has stopped in metaphase of meiosis II, is stimulated to finish meiosis by the entrance of the sperm. Meiosis is then completed. Most of the cytoplasm remains with one nucleus, forming a large egg. The other nucleus forms a polar body, which is pinched off.

Once the egg and sperm nuclei fuse, they form the diploid zygote, which contains the genetic plan of the new individual. The nucleus of the zygote contains the genetic blueprint for a new life.

Cleavage

■ **Cleavage occurs when the zygote divides. At the morula stage, the embryo appears as a solid ball of cells. Continued cell division and development produce a hollow ball of cells called a blastula. The hollow center is called a blastocoel.**

Once fertilization is complete, a type of cell division known as **cleavage** begins. Cleavage results in an increase in the number of cells — two cells, four cells, eight cells, sixteen cells, and so on. Usually the resulting cells are smaller than the initial zygote, and so initially the early embryo is no larger than the zygote was. Since cell division is mitotic, each new cell receives a full complement of genetic material. In other words, cleavage is a division of the zygote into smaller cells. The size of the divided cells and the pattern of division have to do with the amount and distribution of yolk in the zygote (Figure 28.2). In some invertebrates, the site of sperm penetration determines the pattern of cleavage (see Profile).

Eventually, through cell movement and changes in the shape of the cells, a solid ball of cells called the *morula* forms. As the embryo continues to undergo cleavage, the morula eventually develops into the **blastula,** a hollow ball enclosing a cavity called the *blastocoel* (Figure 28.3a). In mammals, the blastula is called a *blastocyst.*

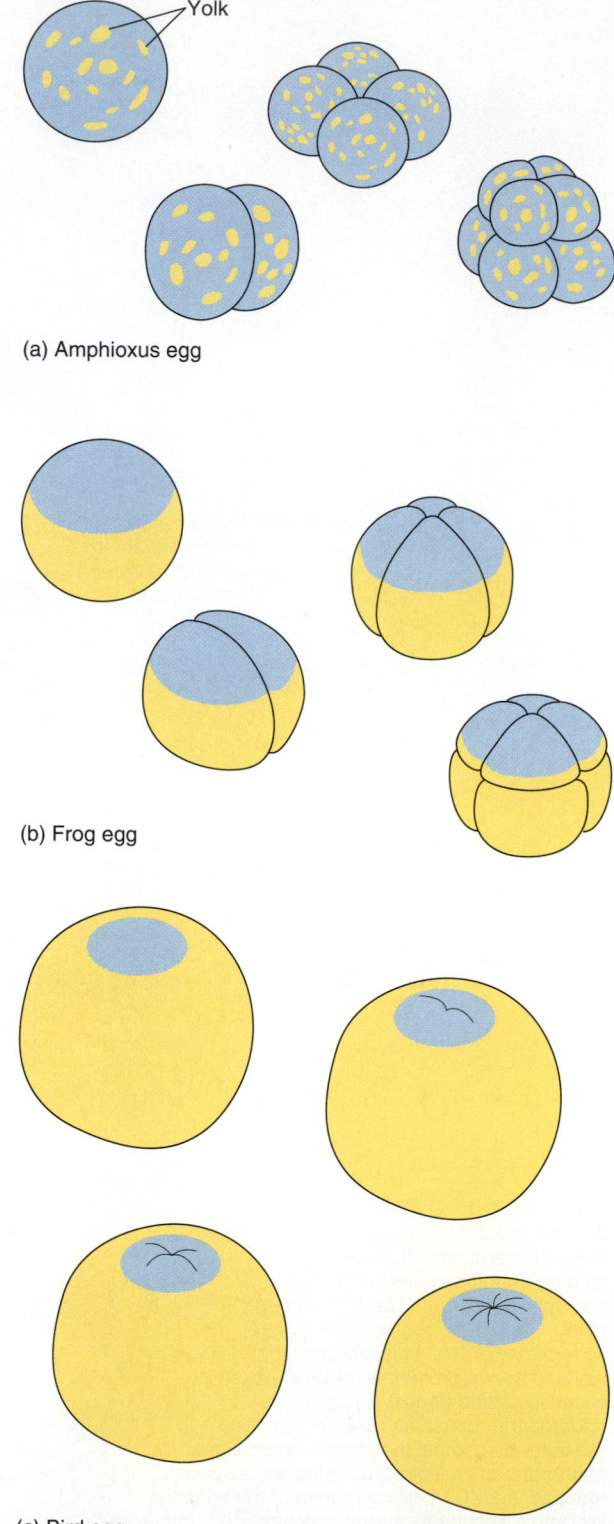

(a) Amphioxus egg

(b) Frog egg

(c) Bird egg

FIGURE 28.2 Cleavage Patterns in the Zygote
Cleavage patterns in (a) the egg of the *amphioxus,* currently called *Branchiostoma* (a small marine invertebrate), (b) a frog's egg, and (c) a bird's egg. A large amount of yolk (shown here in yellow) can impede cell division. Because the bird's egg is almost entirely yolk, only a small section at the top of the egg divides.

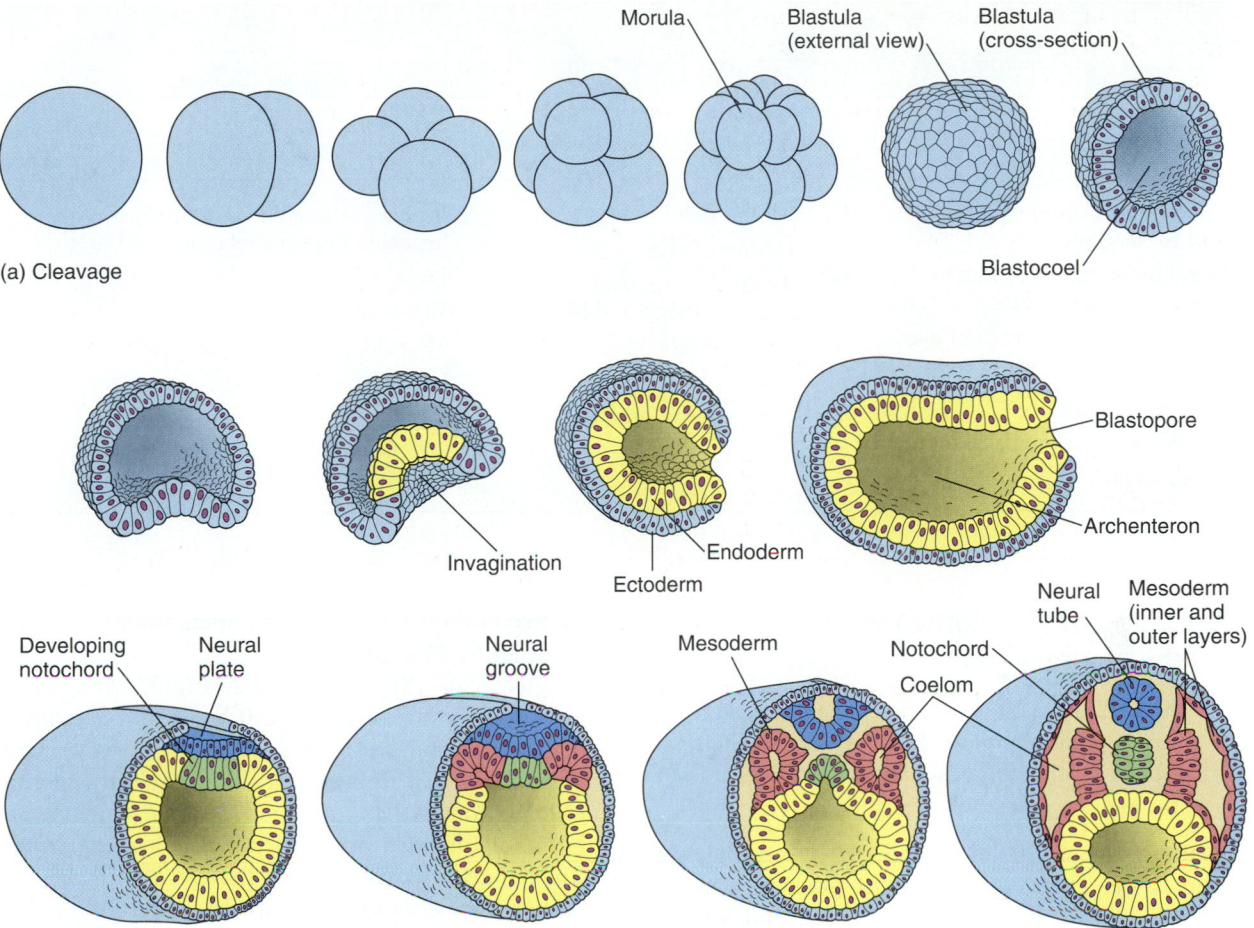

(a) Cleavage

Morula

Blastula (external view)

Blastula (cross-section)

Blastocoel

Invagination

Ectoderm

Endoderm

Blastopore

Archenteron

Developing notochord

Neural plate

Neural groove

Mesoderm

Notochord

Coelom

Neural tube

Mesoderm (inner and outer layers)

(b) Gastrulation and the development of the neural tube

FIGURE 28.3 Stages in the Development of *amphioxus (Branchiostoma)*
(a) During cleavage, the egg does not appreciably change in size; it just divides into smaller and smaller cells. (b) During gastrulation, the blastula folds in (invaginates), forming the ectoderm and endoderm. The mesoderm and notocord develop from the endoderm. The neural tube develops from the ectoderm.

Gastrulation

■ **During gastrulation, a primitive gut is formed. The cells continue to divide and rearrange to form primary germ layers called the ectoderm (outer layer), the endoderm (inner layer), and the mesoderm (middle layer).**

As the cells of the developing embryo continue to divide, the blastula enters into a stage known as gastrulation. Gastrulation is the process of coordinated cell migration that results in a dramatic rearrangement of the cells in the embryo. The actual process is different in different types of organisms, but in all, the inward migration of cells produces a multilayered embryo with an ectodermal layer surrounding the outside and the endodermal and me-

sodermal layers on the inside. (Figure 28.3b shows the process as it occurs in *amphioxus (Branchiostoma)*, a small invertebrate.)

The blastocoel becomes a flattened cavity between the ectoderm and the endoderm, and a primitive gut, the *archenteron*, forms. The archenteron, which will become the digestive tract of the animal, is lined with endoderm and opens to the outside through the *blastopore*. Eventually, the mesoderm forms two layers. The space between the two layers of mesoderm is called the *coelom*.

Initially, the three germ layers are undifferentiated; they have no distinct or special characteristics. Over time, however, they will develop into specific organs and tissues. Table 28.1 lists the vertebrate structures that derive from the primary germ layers.

TABLE 28.1
Organs That Develop in Vertebrates from the Three Primary Germ Layers

Ectoderm	Mesoderm	Endoderm
Epidermis of skin including hair, nails, and sweat glands	All muscles	Lining of digestive tract, trachea, bronchi, lungs, gallbladder, and urethra
Nervous system including brain, spinal cord, ganglia, nerves	Dermis of skin	Liver
Retina, lens, and cornea of eye	All connective tissue including bone, cartilage, and blood	Pancreas
Inner ear	Blood vessels	Thyroid, parathyroid, and thymus glands
Pituitary gland	Kidneys	Urinary bladder
Lining of nose, mouth, and anus	Reproductive organs	
Teeth enamel		

Formation of the Notochord and the Neural Tube

■ **In the chordates the notochord is a primitive structure that supports the embryo. It develops from a section of the mesoderm that lines the gut.**

At about the time the mesoderm is forming, we begin to see the development of other structures. The *notochord*, firm supportive tissue that will be replaced by the vertebral column (in the vertebrates), begins to form on the roof of the archenteron (see Figure 28.3b). The area of the ectoderm above it (the neural plate) starts to fold inward, creating a groove on the dorsal (back) side of the embryo. As the folds come together, the hollow *neural tube* forms. This structure eventually differentiates into the brain and spinal cord.

Embryonic Induction

■ **In induction, one group of cells in the developing embryo can influence another group of cells to develop into certain structures. As cells migrate to certain positions, their cellular interactions produce distinct structures; the process is called morphogenesis.**

What determines the direction of development and the positioning of body parts? Why are arms where they are supposed to be? What about the backbone? The lens of the eye? The question of how developing parts coordinate to form the whole was first answered in part during the early part of this century

by two embryologists—Hans Spemann and his student, Hilde Mangold.

Specifically Spemann and Mangold were interested in investigating how cell layers in the embryo became organized into specific structures. They were studying formation of the notochord. During gastrulation in the amphibian, a portion of the embryo rolls inward in a process called *invagination*. (Invagination is the process that results in the formation of the endoderm and the underlying mesoderm.) The place at which invagination begins is called the *dorsal lip of the blastopore*. During the course of normal development, the notochord begins to form next to the dorsal lip of the blastopore.

In an early experiment, Spemann and Mangold found that an amphibian embryo could be split in half and would develop into two normal embryos — as long as the division was made through the dorsal lip of the blastopore. This led them to study the dorsal lip more closely. Working with two developing embryos, they removed the dorsal lip of the blastopore from embryo A and transplanted it into the belly region of embryo B. When they did this, what do you suppose happened? Embryo B developed two neural plates and two notochords, one set in normal placement adjacent to its own dorsal lip and a second set at the site where the lip of embryo A was transplanted (Figure 28.4). The transplanted lip cells had acted as *organizers* and had *induced* the ectoderm cells in the transplanted region to form a neural plate and notochord.

With further investigation, Spemann and Mangold learned that even if they transplanted only the mesodermal cells from the dorsal lip of the blastopore into another embryo, the same thing hap-

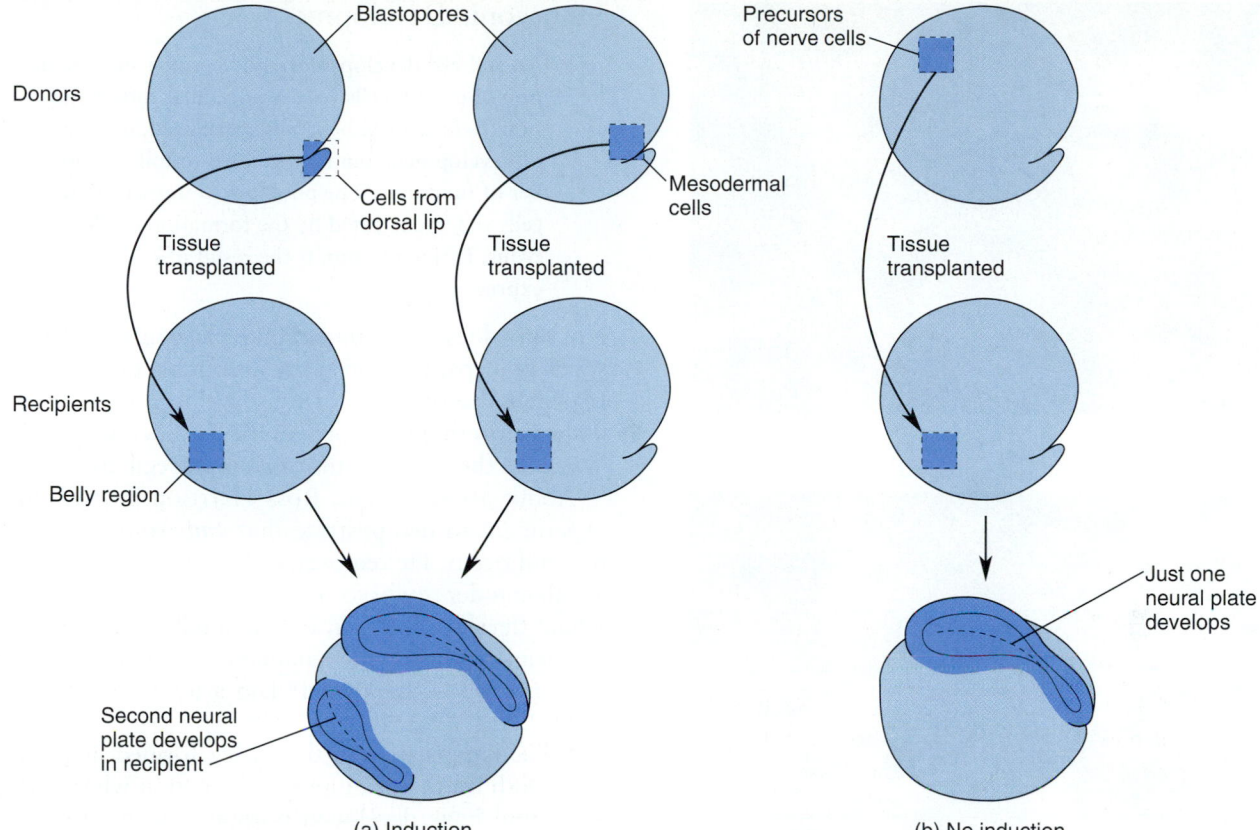

FIGURE 28.4 Induction and the Role of Organizers
(a) Spemann and Mangold identified cells of the dorsal lip and nearby
mesodermal cells as organizers. They transplanted these cells from donor
embryos to areas on recipient embryos that would not normally develop a
neural plate or notochord. The recipient embryos each developed two neural
plates and notochords: one set in the dorsal lip area; the other where the cells
had been transplanted. (b) In contrast, when cells that normally develop into
nerve cells were transplanted to the belly area of a recipient embryo, they
developed into belly tissue, not nerve tissue.

pened. The embryo developed two neural plates
and notochords. Their experiments and those of
many others led to the conclusion that in many
cases the direction of the development of a group of
tissues can be induced by another group of tissues.
This phenomenon is called *induction*.

In all animal embryos, inductive interactions
occur between cells, determining the direction of
orderly development. One tissue induces the formation of a second tissue, and the second influences
the development of yet another. As a result, morphogenesis (body structuring) takes place in sequential, orderly development.

Homeo Box Genes. But what induces this morphogenesis at the genetic level? Are there master genes

that control for the expression of hundreds of other
genes? In other words, is there a genetic hierarchy
that results in the formation of the leg or some other
body part? A leg is composed of many types of cells
and tissues; are there master genes that regulate the
activity of the hundreds of genes that must work in
concert to form this structure? As early as 1915,
researchers were proposing the existence of "master
control genes." The evidence to support this idea
came from rather extreme mutations in fruit flies.
In some cases there were bizarre mutants whose
antennas or mouth parts became legs (Figure 28.5).
Since one body part replaced another, these mutations were called *homeotic mutations* (*homeotic*
means replacing like parts). But what was happening at the genetic level? Researchers discovered that

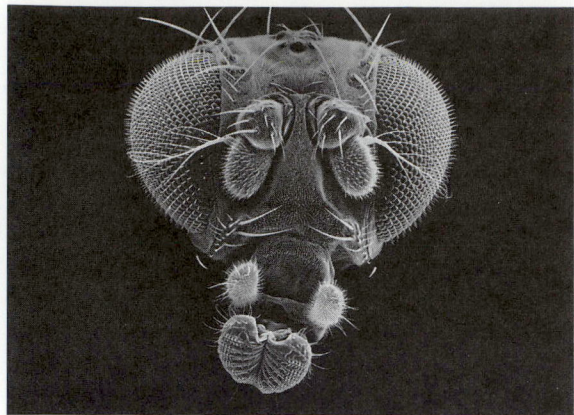

(a)

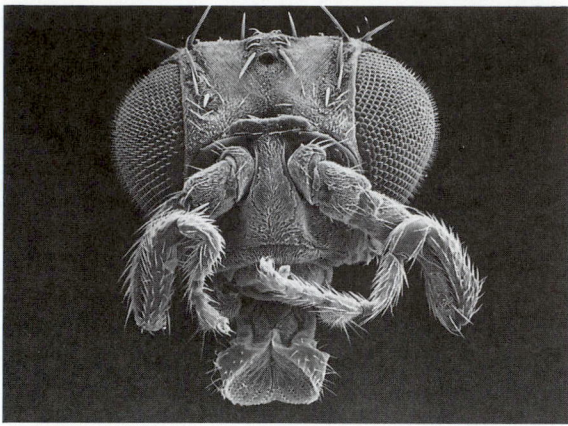

(b)

FIGURE 28.5 Homeotic Mutants
(a) A SEM of a normal fruit fly. (b) A homeotic mutant in which the antennas developed into legs.

many of these mutated genes shared 180 base pairs, sort of like a box filled with base pairs. A scientist by the name of Walter Gehring called it a *homeo box*. He and others discovered that homeo box genes coded for about six amino acids that could possibly serve as a switch regulating development. More important, this homeo box sequence was found in many different homeotic mutants. Therefore, the homeo box genes serve as master genes that may regulate transcription of other genes and thereby regulate or control development.

Scientists have found similar homeo box sequences in many species of animals, including ourselves. The universality of the sequence of DNA in the homeo box in many species suggests that we do share a genetic relationship and do share several genetic control mechanisms with many of our evolutionary ancestors.

Embryonic Development

■ **An embryo develops normally because the interaction of certain cells results in cellular differentiation. Before a cell becomes specialized, its course of development can be regulated to follow a number of functions. The potential of unspecialized cells is demonstrated by the formation of identical twins. Differentiation is the result of selective gene expression.**

Not only does development take place in an orderly series of steps, time or age is also important. What regulates the schedule of the development of cells and ensures that they form in their proper location? What are the mechanisms of embryo regulation?

About 50 years ago, Ross Harrison devised an experiment to demonstrate that *embryonic regulation* did occur. He removed cells from the region in a salamander embryo from which the forelimb would develop. He then transplanted these cells to the head region of the salamander embryo. What do you suppose happened? Did a leg form on the head?

What happened was that a leg formed at both sites, both on the head and in the region where the leg would have developed normally. Somehow, in the area of normal leg formation, the embryonic cells compensated for the missing cells, and a leg formed where it was supposed to. Harrison reasoned that embryonic cells have the capacity to compensate for missing or extra parts.

A number of experiments support this idea. If some embryos are split at the proper time, two complete, identical individuals (identical twins) will develop. Likewise, in humans the inner mass of cells sometimes splits, developing two embryos rather than one. The result is identical twins, sometimes called *monozygotic* twins, because they have arisen from one zygote. In such a case, both individuals arise genetically from one fertilized egg, and both inherit the same chromosomal makeup. Fraternal twins are sometimes called *dizygotic* twins because they arise from two zygotes. Two different eggs are fertilized by two different sperm. The twins are not identical, because each has its own unique genetic makeup (Figure 28.6). And, because fraternal twins develop from two different zygotes, they can be different sexes.

Selective Gene Expression. As the cells of the embryo differentiate into the hundreds of different types of specialized cells, selective gene expression is required at specific times. Some cells, such as nerve cells, differentiate before birth and are rarely replaced. Such cells are not usually damaged, and so they do not require replacement. Other cells, such as those that make up the skin and blood and those lining the inside of the intestines, differentiate before birth and continue to do so during the organism's entire life.

Cells that are constantly differentiating, even in the adult, are formed from stem cells. Stem cells still have the capacity to replicate and differentiate. Through mitosis, stem cells replace cells that are lost during the organism's life span. Other cells (like the developing egg cells in the ovaries of the female embryo) differentiate early in the embryo but do not mature until years later.

The bottom line of the developmental process is that differentiation is dependent on regulation through a complex interaction of the nucleus, cytoplasm, hormones, and other chemical substances. This interaction determines when, how, and whether a cell will differentiate. As you might expect, a great deal of scientific research is devoted to understanding the molecular events involved in embryonic development. How does a fertilized egg develop into the individual of its particular species? This is a fascinating phenomenon that has amazed and puzzled humans since the beginning of time, especially when we explore our own development.

HUMAN DEVELOPMENT

In the preceding chapter, we discussed human reproduction, stating that during sexual intercourse semen, which contains sperm, is released into the vagina. The sperm swim through the uterus into the fallopian tube, where fertilization occurs. Now starts a remarkable process . . . the development of a human being.

Early Embryonic Development

■ **Approximately 36 hours after fertilization, the first cleavage is complete. The zygote continues to divide as it moves through the fallopian tube on its way to the uterus. By the time it reaches the uterus, an inner mass of cells consisting of two layers (the ectoderm and endoderm) has formed, from which the embryo will develop. Approximately six days after fertilization, the blastocyst implants on the uterine wall, burrowing into the endometrium. In the next week, the inner cell mass differentiates into the germinal disc and amnion. Within the germinal disc, cells migrate to a location between the ectoderm and endoderm to form the mesoderm. The amnion will become a membranous sac that protects the embryo.**

FIGURE 28.6 Twins
(a) Monozygotic (identical) twins develop from one fertilized egg that separates in two during an early stage of development. (b) Dizygotic (fraternal) twins develop from two separate eggs, each fertilized by a different sperm cell. Notice that identical twins share a single placenta (the structure that nourishes the developing embryo) while fraternal twins have individual placentas.

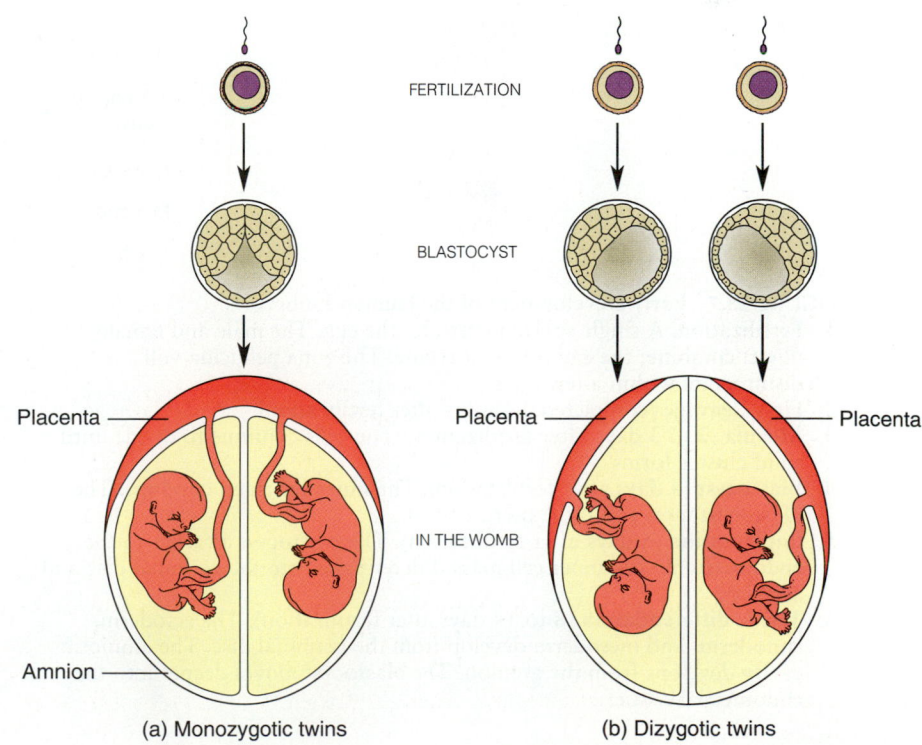

FERTILIZATION

BLASTOCYST

Placenta —

IN THE WOMB

Placenta —   — Placenta

Amnion —

(a) Monozygotic twins

(b) Dizygotic twins

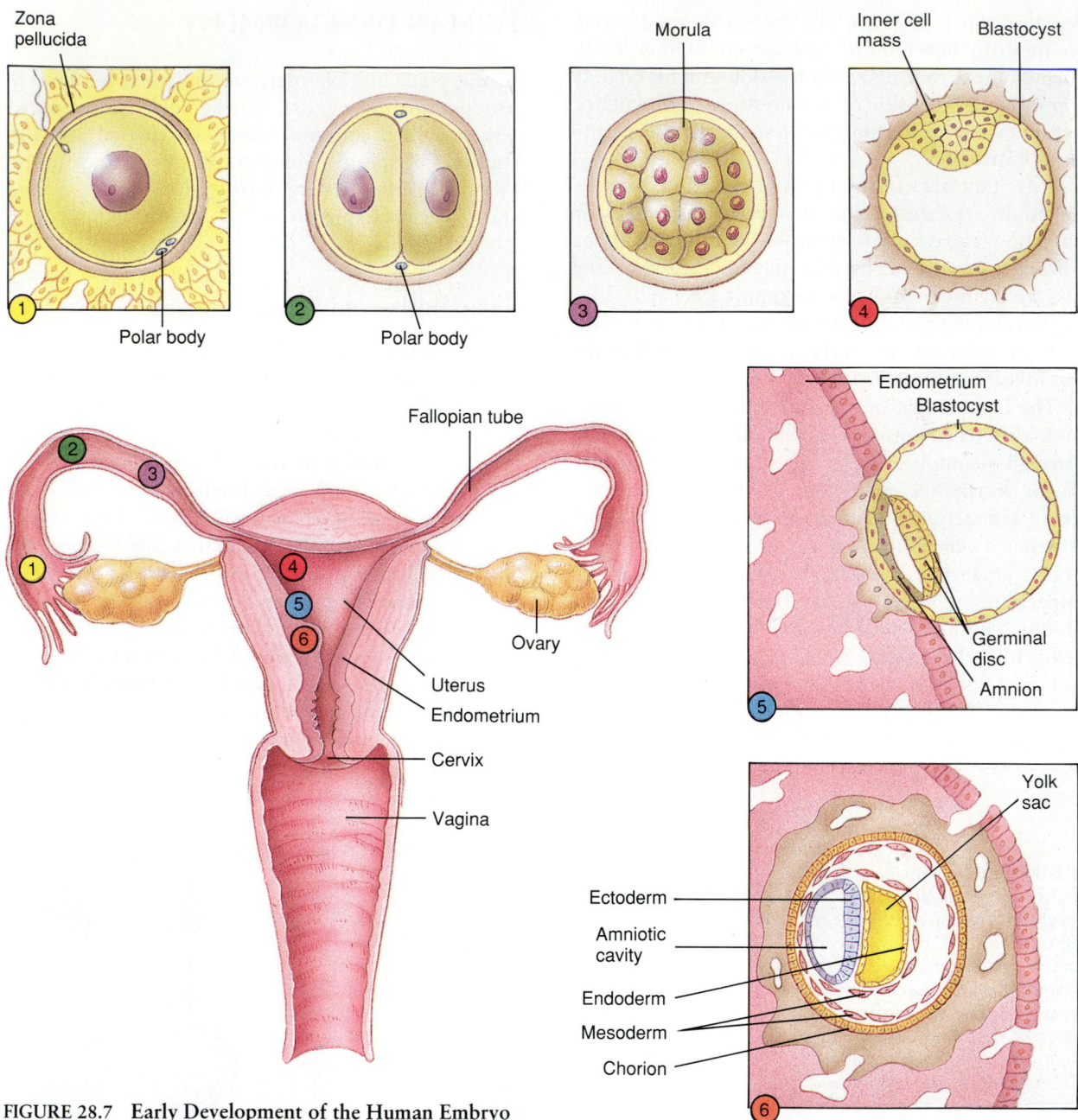

FIGURE 28.7 Early Development of the Human Embryo

1. **Fertilization.** A single sperm penetrates the egg. The male and female nuclei combine; the egg is now a zygote. The zona pellucida will disintegrate within a few days.
2. **First cleavage** (completed 36 hours after fertilization).
3. **Morula** (2 to 3 days after fertilization). The cells continue to divide until a solid cluster forms.
4. **Blastocyst** (4 days after fertilization). The inner cell mass develops. The blastocyst moves into the uterus.
5. **Implantation** (6 days after fertilization). The blastocyst attaches to the endometrium. The inner cell mass differentiates into the germinal disc and amnion.
6. **Tissue differentiation** (8 to 14 days after fertilization). The ectoderm, endoderm, and mesoderm develop from the germinal disc. The amniotic cavity develops from the amnion. The blastocyst moves deeper into the endometrial tissue.

Fertilization is the point at which the sperm penetrates the egg. It sets off a fascinating series of events. Here we look at the development of the embryo through the first two weeks following fertilization (Figure 28.7).

Fertilization triggers the completion of meiosis in the egg and the fusion of genetic material from the nuclei of the sperm and egg. Fusion—the union of the *n* (haploid) sperm and the *n* (haploid) egg—restores the 2*n* (diploid) chromosome number. The egg is now called a *zygote*.

Cell division begins immediately after the zygote is formed. In about 36 hours, the first cleavage is complete. The process continues for two to three days until a morula forms. All the time, the zygote is moving through the fallopian tube on its way to the uterus.

By the time the zygote reaches the uterus, it is a blastocyst, the human equivalent of a blastula. Unlike the blastula, however, the blastocyst has already formed an inner mass of cells. From these cells, the embryo will develop. The outer cells help the blastocyst attach to the uterine wall; they become the *chorion*, a membrane that nourishes the embryo in the early stages of development and that eventually becomes part of the placenta.

The blastocyst remains in the uterus for two to four days before attaching itself to the uterine wall. The endometrium (the inner lining of the uterus) is ready to receive the egg, growing thicker and becoming engorged with blood. As we pointed out in Chapter 27, the development of the endometrium is regulated by progesterone and estrogen, which are produced by the ovaries. Through a complex series of chemical interactions between the blastocyst's cells and those of the endometrium, the blastocyst attaches itself to the lining and begins to penetrate the nourishing endometrial tissue until it is completely surrounded. The *implantation* process is technically called pregnancy.

In the next week, the blastocyst changes dramatically. The inner cell mass develops into the germinal disc and amnion. The *germinal disc*, which at first consists of a single layer of ectoderm and a single layer of endoderm, will become the embryo. The *amnion* is the membrane that will enclose and form the fluid-filled sac that protects the embryo from physical shock. By the fourteenth day after fertilization, the embryo has developed three distinct layers of tissue: ectoderm, endoderm, and mesoderm. These three primary germ layers give rise to structures typical of the adult vertebrate (see Table 28.1).

Extraembryonic Membranes and the Placenta

■ **Extraembryonic membranes enable land animals to reproduce outside of the water. In reptiles, birds, and mammals, a chorion encloses the embryo. The human chorion, the embryonic portion of the placenta, forms chorionic villi, which extend into the endometrium. In placental mammals, blood vessels run through the chorion into the villi and connect with the embryo through the umbilical cord. The endometrium is the maternal portion of the placenta. It connects to the mother's circulatory system. Therefore, the placenta consists of the embryonic chorionic tissue and the mother's endometrial tissue. Materials pass between mother and child by means of diffusion and other mechanisms of molecular transport. The amniotic sac surrounds the organism. The placenta also produces hormones. During the first eight weeks of development, the embryo is especially sensitive to teratogens, substances that can produce malformations.**

Extraembryonic membranes are tissues that are not part of the developing organism before birth. In order to understand the structures and the functions of human extraembryonic membranes, we should discuss their evolution in other vertebrate groups. As the land vertebrate animals evolved, mechanisms that enabled them to reproduce without the presence of a watery environment were selected for. The change did not take place immediately, of course. It arose slowly, over a long period of time, and its development can be traced in the life cycles of various species.

Amphibians. In many ways, amphibians are a transitional group of organisms, for they spend part of their lives in water and the other part on land. The duality of their lives is reflected in the structures that they exhibit during their life spans. Frog tadpoles, for instance, possess gills, which are lost when they develop into mature adults.

But what truly demonstrates the duality of amphibians is the fact that they must return to a watery environment in order to *spawn* (the process by which they lay their eggs and fertilize them). Amphibians exhibit external fertilization, not internal fertilization. In other words, eggs and sperm do not fuse until they are released from the bodies of the mating pair. The female releases her eggs, bound together in a jellied mass, into the water (Figure 28.8a). The thin transparent membrane surround-

(a)

(b)

FIGURE 28.8 Lower Vertebrates and Their Eggs
(a) The frog and other amphibians release their eggs in water, in a jellied
mass. Fertilization here is external; the sperm never enter the female's body.
(b) A female alligator with her eggs.

ing each egg is highly permeable, and so if the eggs were removed from the water, their water molecules would diffuse through the membranes and the eggs would dry out completely. As the eggs are laid, the male releases his sperm over them. Water currents are the medium of fertilization. When the fertilized eggs hatch, the young develop in their watery environment until they are mature adults.

Reptiles and Birds. With the rise of reptiles (like snakes and lizards) and birds, we see the development of two processes that freed animals from an aquatic environment. Internal fertilization is one, and the other is the land egg. Because internal fertilization takes place within the female's body, organisms are no longer dependent on water currents to ensure the fusion of sperm and egg. The land egg is distinctive because its system of membranes that encloses the developing egg holds moisture inside the egg, forming an artificial "aquatic environment." The system is so efficient that even if a reptile like a sea turtle or an alligator spends the largest part of its life in water, it will still return to land to lay its eggs (Figure 28.8b).

The outer shell of reptile eggs has a leathery consistency. Bird eggs, however, exhibit another modification of egg development. The shell of a bird egg is a hard, porous cover that protects the developing embryo and provides for gas exchange.

Lying directly below the hard shell is the **chorion** (Figure 28.9). This structure contains a large num-

ber of blood vessels and functions in gas exchange between the embryo and the environment. The yolk sac is a membrane that surrounds the **yolk,** the substance that provides nutrients for the developing embryo. The **allantois** is a membrane sac that receives and collects the nitrogen wastes produced as the embryo deaminates proteins.

As the allantois grows, it develops from a little pouch to a continuous membrane. Eventually the allantois fuses with the chorion to form the chorioallantois membrane. The extraembryonic membrane called the **amnion** produces amniotic fluid, providing an aquatic environment that surrounds the developing embryo. Amniotic fluid also prevents temperature extremes and acts as a shock absorber.

Placental Mammals. Extraembryonic membranes are also found in *placental mammals,* such as humans. The placenta is the structure through which nutrients, gases, and wastes move between embryo (later, the fetus) and mother (Figure 28.10). In humans, the **placenta** is formed by membranes arising from both the developing embryo and the endometrium of the mother. Once the embryo embeds in the uterine wall, it produces a chorion similar to that surrounding the chick embryo. The chorion then develops numerous fingerlike projections known as *chorionic villi,* which become embedded in the endometrium. Blood vessels also form in the chorion. Capillaries extend from these vessels into

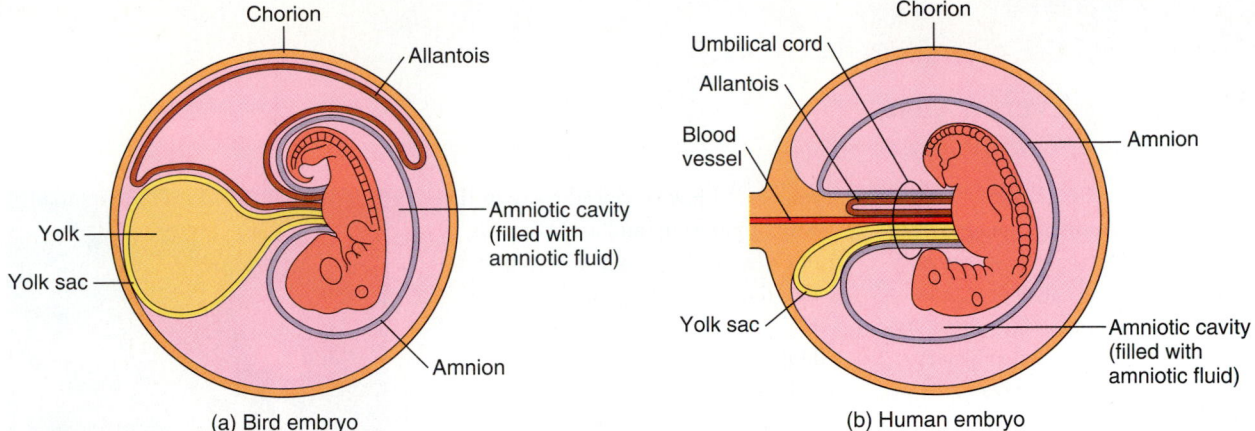

FIGURE 28.9 A Comparison of Extraembryonic Membranes: Bird and Human Embryos
Although the same membranes appear in both (a) the bird embryo and (b) the human embryo, their form and function are somewhat different. For example, the yolk sac in the bird embryo is much larger than it is in the human embryo because the bird embryo receives all of its nutrients from the yolk. In the human embryo, the extraembryonic membranes function in the exchange of nutrients, gases, and wastes.

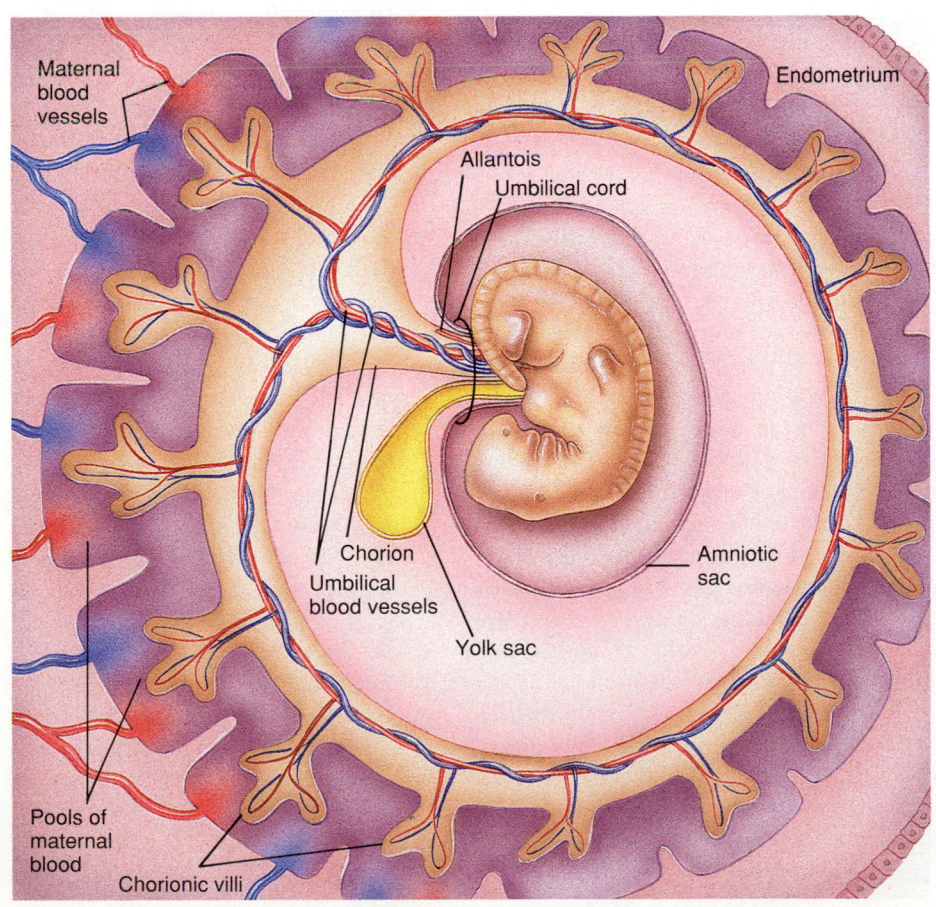

FIGURE 28.10
The Placenta
The placenta is formed from the chorion of the embryo and the endometrial tissue of the mother. In this drawing of an embryo at about four weeks after fertilization, you can see the chorionic villi extend into pools of maternal blood in the endometrium. The exchange of gases, nutrients, and wastes between mother and embryo occurs through the membranes of the chorionic villi. The fetal and maternal blood supplies do not merge — they are separate.

Dr. Ernest Everett Just
Embryologist, Cell Biologist

At 23 years of age, Ernest Everett Just graduated magna cum laude from Dartmouth — the only magna cum laude that year. His major was biology. With these credentials and his brilliance, energy, and demonstrated research talents, Just was ready to succeed in the professional world of science. However, Just was African American. In 1907, there was very little career mobility for African Americans except in the professions of teaching or the ministry — in the African American community. Just followed in that tradition, taking a teaching position as professor of English and zoology at Howard University.

But Just was not happy at Howard because of its limited research opportunities. He sought sources of financial support that allowed him to spend summers at the Marine Biological Laboratory in Woods Hole, Massachusetts. While working there, his interest in experimental embryology grew. Returning for postgraduate study, he earned his Ph.D. from the University of Chicago in 1916. His dissertation was on the mechanisms of fertilization.

Just gained a reputation as a scientist, but he was still thought of primarily as the gifted young assistant of Frank R. Lillie, the man who had been his major professor and was the second director at Woods Hole. Just's growing frustration with the social climate led him to leave the United States. After working a year at the Kaiser Wilhelm Institute for Biology in Europe, he returned. His bitterness with his home country became clear during a birthday celebration for Frank Lillie at Woods Hole. After presenting his scientific paper, he told his colleagues, "I have received more in the way of paternity and assistance in one year at the Kaiser Wilhelm Institute than in all my years at Woods Hole put together." He left Woods Hole, never to return.

During the period after World War I, many people became disenchanted with life in the United States and traveled to Europe to live. Just exiled himself to Europe. For a brief while he found the scientific acceptance he craved, but in the 1930s, Europe was a political nightmare. Both his personal and scientific life became embroiled in the turmoil. He narrowly escaped the Nazis as they gained power in Germany.

In 1940 Just returned to the United States to teach at Howard University. A year later, he died of pancreatic cancer at the age of 58. During his career he published over 80 scientific research papers and two books on embryology and cell biology. His best-known work, published in 1939, was titled *The Biology of the Cell Surface*. He left a legacy in the classic study of embryology. Before Just's work, biologists thought that the cell membrane was simply a passive barrier separating the inside of a cell from the outside. Through his work on fertilization, he recognized that, in actuality, the cell membrane was an active regulator and determinant of cellular activities. It was Just who demonstrated that the first cleavage plane in the zygote of many marine invertebrates is determined by the sperm's point of entry — no matter where the sperm penetrates. Before Just, biologists believed that the embryo was preformed in miniature in the egg or the sperm. He established that instead the embryo developed

epigenetically — in other words it developed through a series of sequential cell divisions.

In his biography *Black Apollo of Science: The Life of Ernest Everett Just*, the Massachusetts Institute of Technology historian Kenneth R. Manning quotes Lillie's obituary of Just: "An element of tragedy ran through all Just's scientific career due to the limitations imposed by being a Negro in America, to which he could make no lasting psychological adjustment in spite of earnest efforts on his part. . . . In Europe he was received with universal kindness, and made to feel at home in every way. . . . Hence, in part at least, his prolonged self-imposed exile on many occasions. That a man of his ability, scientific devotion, and of such strong personal loyalty as he gave and received, should have been warped in the land of his birth must remain a matter for regret."

the chorionic villi, and a vein and two arteries running through the **umbilical cord** connect the network of embryonic blood vessels to the circulatory system of the developing embryo.

Maternal tissue from the endometrium also surrounds the embryo. The endometrium forms the maternal portion of the placenta. Small pools of maternal blood surround the chorionic villi. These pools are fed by maternal blood vessels, which connect to the circulatory system of the mother. The chemical processes that are conducted by the cells of the developing embryo involve gas exchange, nutrient acquisition, hormone stimulation, and waste removal, just as with any independent organism. Notice that the circulatory systems of the embryo and mother do not merge; they are totally separate. The exchange of nutrients, gases, and wastes between embryo and mother takes place through the placenta.

Molecules of oxygen and food (such as amino acids and simple sugars) diffuse from the mother's capillaries into the capillaries of the developing embryo. Waste products such as CO_2 and nitrogen wastes diffuse from the embryo's capillaries into the pools of maternal blood, then are carried by the mother's bloodstream to her lungs (which eliminate CO_2) and her kidneys (which eliminate nitrogen). Diffusion and other mechanisms of molecular transport ensure that materials are exchanged between the mother and the developing embryo.

There are several important facts to remember about early pregnancy. First, the embryo has its own circulatory system, complete with a heart that started beating only 24 days after conception — perhaps even before the woman is certain she is pregnant. Second, the developing embryo is receiving its nutrients in a preprocessed form. In other words, it is receiving amino acids rather than protein, simple sugars rather than complex carbohydrates, and so on. This means that the embryo does not have to digest its food and that it does not contain solid waste products from the direct digestion of food. In other words, there is no defecation. The wastes that are produced are nitrogen wastes, carbon dioxide, and others that are the result of cellular metabolism.

As the embryo develops, the yolk sac and allantois form into the umbilical cord, which contains one vein and two arteries. (Remember, a vein takes blood to the heart and an artery takes blood away from the heart.) The cord serves as a lifeline through which nutrients and waste products pass between the placenta and embryo. The umbilical vein also carries blood from the placenta via the embryo's liver to the embryo's heart. As the embryo

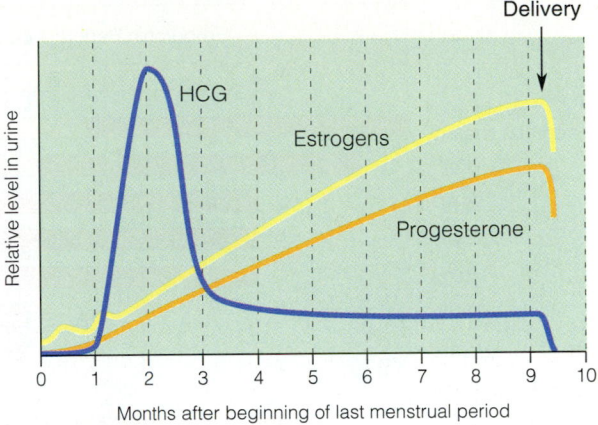

FIGURE 28.11 Levels of Hormones in the Urine During Pregnancy

grows, it is surrounded by the *amniotic sac* (bag of waters) in which the embryo floats. *Amniotic sac* is the term used to describe the amnion, which surrounds the embryo. The cells of the amnion secrete the amniotic fluid in which the embryo floats, protected from drying out and physical shock.

Hormone Production. As we mentioned in the last chapter, the chorion produces human chorionic gonadotropin (HCG), which in turn stimulates the corpus luteum to secrete progesterone and estrogens for the first two months of pregnancy. Progesterone maintains the thickness of the endometrium and hence prevents menstruation. Not only does the placenta protect and nourish the embryo, but, as mentioned, it also secretes progesterone and estrogens to maintain the pregnancy (Figure 28.11).

Teratogens. The first two months of pregnancy are the most sensitive time of the embryo's development, and it is also the time when the embryo is most threatened by outside factors (Figure 28.12). Unfortunately, in those first two months a woman may not know she is pregnant. Agents that can produce malformations in the developing embryo are called *teratogens*.

A number of substances can upset the course of embryonic development. Some teratogens are *direct*, such as many drugs and x-rays. Other teratogens are *indirect*, such as a deficiency in the hormone production of the mother's thyroid gland. Biologists and physicians know that if a mother contracts certain infections, such as *rubella* (German measles), between the fourth and twelfth weeks of pregnancy, the result can be heart, eye, ear, or brain malformation in the developing embryo.

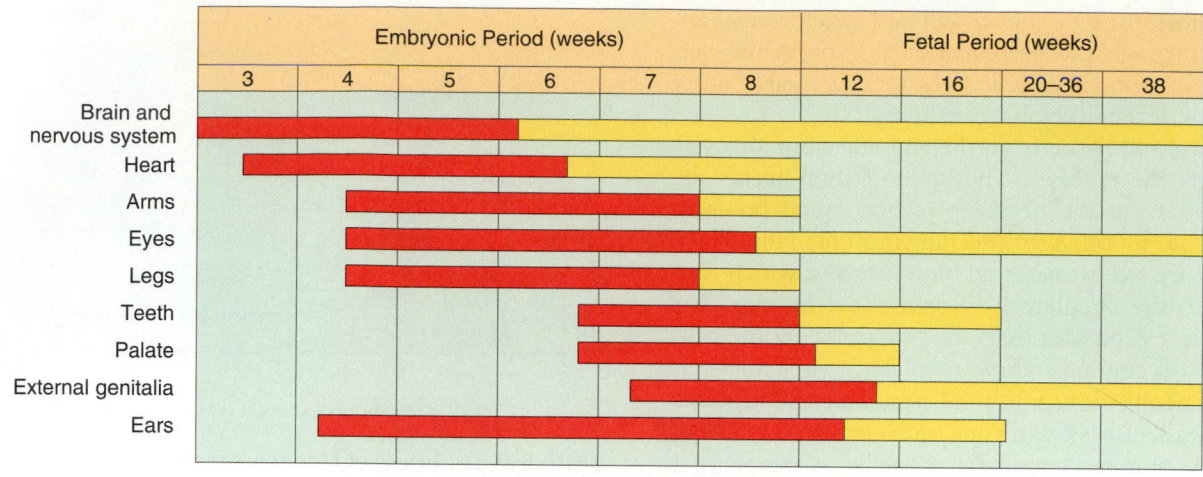

FIGURE 28.12 Teratogens and the Critical Periods of Development
From the third through the twelfth week of development, the major organs of
the body are forming in the embryo (then fetus). Exposure to teratogens
during this time can lead to severe abnormalities (red bars). Later exposure
can also result in defects, although the effects here are less severe (yellow bars).

There are other hazards as well. Alcohol consumption, even in small quantities, can bring on a condition called fetal alcohol syndrome (discussed in Chapter 9), which is characterized by both mental and physical abnormalities. The actual amount of alcohol consumption required to produce this condition is not yet known. However, the potential health risk is great enough that the Surgeon General has posted a warning on alcoholic beverages. Drug consumption, be it tobacco, marijuana, cocaine, heroin, or any number of prescription drugs, can also harm a developing embryo. There are a wide range of problems. They can include low birth weight, drug addiction of the newborn child, physical malformations, neurological defects, or a combination of all four. A pregnant woman should be cautious about the possible risks to which she may be exposing her developing child.

Later Human Fetal Development

■ **After eight weeks, the embryo is recognizable as a human being and is thereafter called a fetus. Further development involves growth and maturation of the fetus's immature organs.**

In the early embryo there is evidence of the physiological relationship of humans to other chordates. For example, the early embryo possesses a notochord, a tail, and gill slits, providing evidence of our link to other chordates. By the end of the first eight weeks, the developing embryo has increased in mass by about 500 times, weighs about as much as a peanut, and is about an inch long. The embryo is now recognizable as a human, and it is referred to as a **fetus.**

Once the fetal stage is reached, all the organs are in place. Now the process of human development is primarily one of growth and maturation of those organs. For the next seven months the developing fetus continues to grow, its body becomes more defined, and its organ systems mature.

After the first three months, the fetus definitely looks human. By the fourth month, the mother can sense its movements. The color portfolio on human development (Photo Essay 28.1) chronicles development up to the time of birth.

Childbirth

■ **After nine months, the fetus is ready for birth. Hormonal reactions initiate labor, resulting in birth. The embryo produces prostaglandins, and the mother produces oxytocin, both of which cause the uterus to contract. Birth occurs in three stages. Dilation is a period of uterine contractions that expand the diameter of the cervix. The second stage is expulsion, during which the child passes through the birth canal. During the placental stage, the placenta is expelled through the birth canal. Before birth, the infant received oxygen and expelled carbon dioxide through the placenta. Once the umbilical cord is cut, changes occur in the infant's heart and lungs, and independent breathing begins.**

Chapter 28   Animal Development

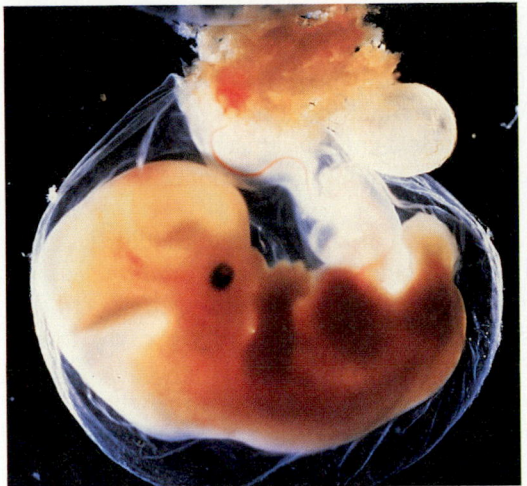

At five weeks

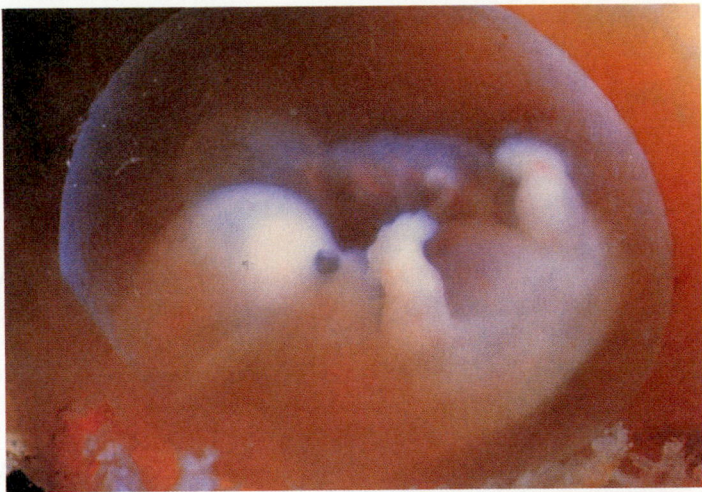

At seven weeks

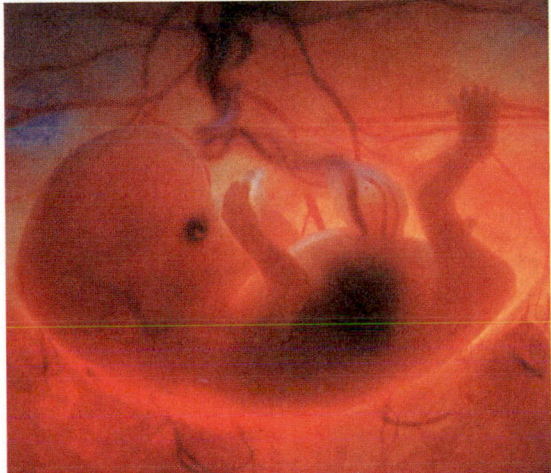

At twelve weeks

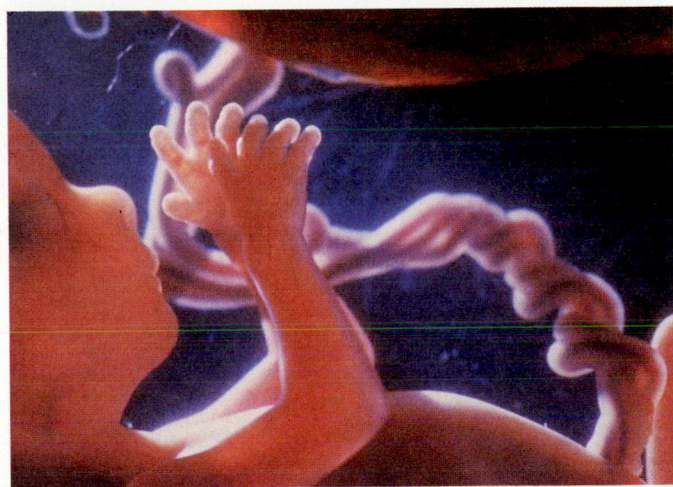

At sixteen weeks

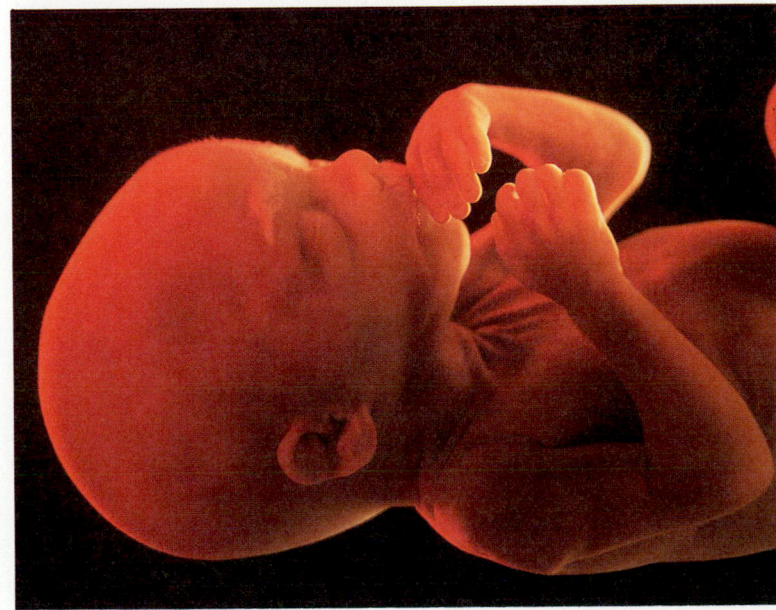

At seven months

PHOTO ESSAY 28.1   Human Development

Fifth week (⅓ inch). Spinal nerves form. Stomach begins to form. Premuscle masses are visible in the head, trunk, and limbs.

Seventh week (¾ inch). Back straightens. Muscles differentiate. Eyelids form. External genitalia start to form.

Twelfth week/third month (3 inches). Teeth buds and sockets form. Nails begin to form. Kidneys are functioning. External ears are visible. The fetus is able to move in the amniotic sac.

Sixteenth week/fourth month (6½–7 inches). Skin is pink and transparent, covered with fine hair. Skeleton is visible; body is catching up in size with head. Muscles are active.

Fifth month (10–12 inches). Internal organs are maturing. Skin is bright red.

Seventh month (11–14 inches). Eyelids separate; eyelashes form. Skin is wrinkled. Internal organs are in normal position.

Nine months (or 280 days) after the beginning of the last menstrual period (or 266 days after conception) the developing fetus is said to be *at term* — in other words it is ready for birth. When this point is reached, a complex interaction of hormonal events between the fetus and mother signals that childbirth is ready to begin. What are these events?

Experimental evidence suggests that the adrenal cortex (portions of glands located on top of the kidneys) of the fetus is involved in the initiation of labor. One of the hormones, called cortisone, stimulates the placenta to produce other hormones called *prostaglandins*. Although prostaglandins alone can cause the mother's uterus to contract, they also are believed to signal the mother's posterior pituitary to secrete a hormone called *oxy-*

tocin, which causes the uterus to contract. Biologists believe that both prostaglandins produced by the fetus and oxytocin are involved during childbirth. When a woman does not go into labor normally, oxytocin can be used to begin the contractions (or, technically, to induce labor).

Stages of Labor. For convenience, childbirth is described as involving three stages: *dilation, expulsion,* and *placental.* The dilation stage refers to the process of the expansion of the cervix, the lower section of the uterus that leads to the vagina (Figure 28.13a and b). Dilation begins with rhythmic contractions of the uterus. At first the contractions are weak and short-lived, and they occur at 15- to 20-minute intervals. In some instances, there will be a

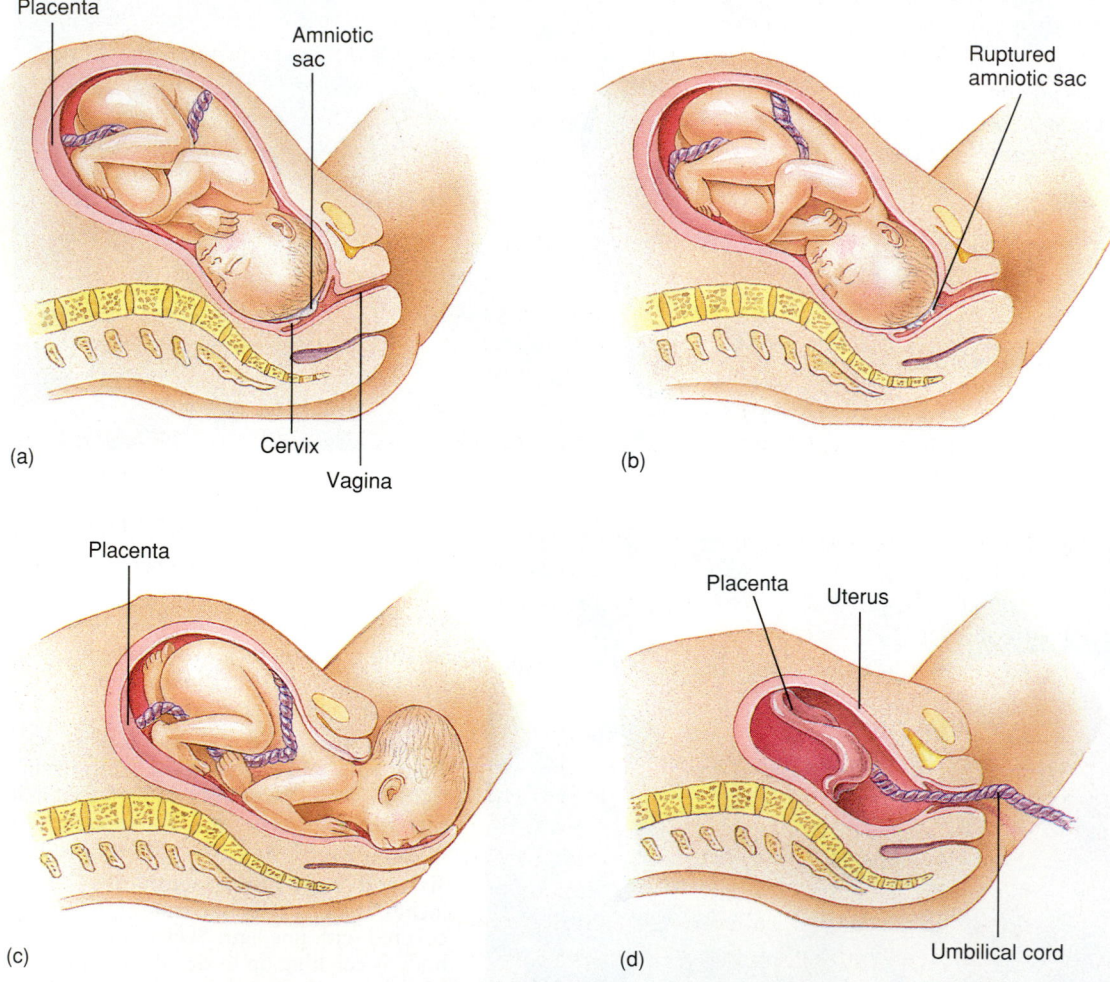

FIGURE 28.13 Childbirth
(a) In early dilation, the amniotic sac pushes through the cervix. (b) When dilation is complete, the sac usually ruptures. (c) In the expulsion stage, uterine contractions move the baby through the dilated cervix and vagina. (d) In the placental stage, uterine contractions expel the umbilical cord and placenta.

Chapter 28 Animal Development

slight bloody discharge from the vagina, sometimes called *show*. As dilation progresses, the contractions increase in strength and duration. By the time dilation is complete, the contractions are quite strong and occur at 1- to 2-minute intervals.

By this time, just prior to expulsion of the newborn, the cervix is fully dilated to about 10 cm. In a first pregnancy especially, it is common at this point for the amniotic sac to burst, releasing the amniotic fluid. (In everyday terms, the woman's water has burst.) The length of time it takes for a woman to go through dilation, and the point at which the amniotic membrane breaks, vary greatly from woman to woman, and the number of pregnancies she has experienced is also a factor. Generally speaking, first pregnancies take longer to deliver than later ones.

The expulsion stage itself, which involves the actual delivery of the child, is rather short, lasting from 2 to 45 minutes. The birth canal, made up of both the cervix and the vagina, is fully dilated, and uterine contractions push the baby, usually head first, out of the vagina (Figure 28.13c). Thus the newborn (biologically speaking, the *neonate*) enters the world to celebrate his or her birthday. (As you know, we have only one birthday; others are anniversaries of our birthday.)

The placental stage occurs immediately after birth, when the newborn is still attached to the placenta by the umbilical cord. The cord is cut (because there are no nerves in the cord, the procedure does not hurt the infant or mother), and the infant begins its separate and independent life. The uterus continues to contract, expelling the placenta (sometimes called the *afterbirth*) and other fluids through the vagina (Figure 28.13d).

For a period of time after birth is concluded, the uterus continues to contract. (The length of time involved varies from woman to woman.) These contractions help control bleeding at the site from which the placenta detached. Eventually the uterus returns to its original size. With birth, the wonders of human conception and prenatal development are complete.

The Newborn — Circulation and Respiration. During the birth process, the infant must shift from the fetal method of breathing (without the use of lungs) to postnatal (lung-dependent) breathing. From then on, throughout its life, blood high in carbon dioxide enters the right chambers of the heart and is pumped to the lungs, where the exchange of gases occurs. Then the blood, now rich in oxygen, returns to the left side of the heart to be pumped throughout the body. Before birth, however, since the fetus floats in its amniotic ocean, it does not use its lungs to breathe in the uterus. Rather, the placenta serves as the organ of respiration for the exchange of gases. Through it, the fetus obtains oxygen and eliminates carbon dioxide. What mechanisms are involved in this process?

Since the fetus developing in the uterus does not use its lungs, circulatory shortcuts that for the most part bypass the lungs and emphasize the placenta as the breathing organ have evolved. In the fetal heart, there is an oval opening, or *foramen ovale*, between the upper chambers (atria) that allows the right atrial blood to pass to the left atrium without going through the right ventricle or the lungs. In other words, deoxygenated and oxygenated blood can mix. Covering the foramen ovale is a flap of skin that serves as a valve. Another shortcut, called the *arterial duct* or *ductus arteriosus,* is a small blood vessel that carries blood from the fetus's pulmonary artery to the aorta (the major artery of the body). As a result of this system, the lungs for the most part are bypassed.

There is also a liver bypass, called the *ductus venosus,* which allows some of the blood from the umbilical vein to bypass the fetus's liver and enter its heart. After birth, the bypass closes and blood is routed directly to the liver. The locations of the fetal bypasses and the changes in the circulatory pattern in the newborn are shown in Figure 28.14.

How is lung breathing initiated at birth? The neonate can be helped to take its first breath by holding it upside down (in order to drain any fluid out of the breathing passages) and either spanking it very gently or exposing it to a small whiff of carbon dioxide. An increase in the level of carbon dioxide in the blood lowers the blood's pH (makes the blood more acidic). The pneumotaxic centers of the brain detect the acidity and signal breathing to occur involuntarily.

Involuntary lung breathing means that the fetal system is no longer necessary. At birth, the foramen ovale and ductus arteriosus close, allowing blood to travel directly to the lungs. If the foramen ovale fails to close, oxygenated and deoxygenated blood can mix, resulting in a low concentration of oxygen in the blood. This deprives the respiring cells of oxygen, which in turn affects the energy-liberating respiratory reactions. The result is cellular suffocation, which may cause the baby's skin to appear bluish (hence the name blue baby). Circulatory defects of this type can usually be repaired surgically a few days after birth, initiating normal adultlike circulation in the infant. However, surgery is dan-

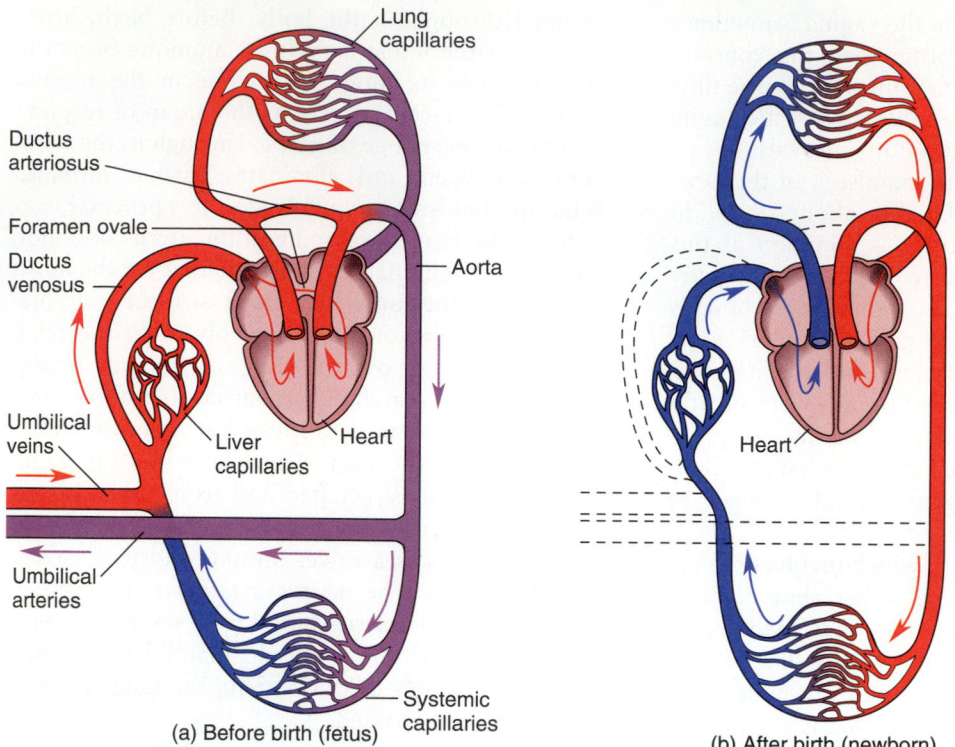

FIGURE 28.14 Fetal Circulation
(a) In the foramen ovale and the ductus arteriosus, deoxygenated blood bypasses the fetal lungs; the ductus venosus bypasses the liver. (Actually, a small amount of blood passes through both the lungs and the liver, to keep the tissue alive.) In the umbilical vein, oxygenated blood is moved into the fetal circulatory system; the umbilical arteries carry blood out to the placenta, where it is oxygenated. In the schematic, oxygenated blood is shown in red; deoxygenated blood, in blue; and mixed blood, in purple. (b) After birth, the foramen ovale and the bypass ducts close. Blood circulates in the postnatal pattern, and the neonate breathes through the lungs.

gerous for premature infants, who are as threatened by the procedure as by the condition.

Failure of the ductus arteriosus to close produces a similar respiratory syndrome that results in inadequate amounts of oxygen reaching the cells. Recently, researchers have determined that the drug indomethacin (an inhibitor of prostaglandin synthesis) can be used in such cases to close the ductus arteriosus.

Lactation — Milk Production

- **During pregnancy, prolactin and estrogens stimulate the mother's milk-producing glands, and placental hormones suppress milk release. After birth, the placenta no longer produces hormones, so prolactin can stimulate milk secretion. Oxytocin stimulates the milk release, and it also stimulates uterine contractions that return the uterus to its original size.**

During pregnancy, in order to develop and prepare her body for **lactation** (milk production and secretion), the mother's mammary glands are stimulated by an interaction of two hormones, prolactin (produced by the pituitary) and estrogens. At the same time, hormones produced by the placenta and protein-inhibiting hormones produced by the mother inhibit the secretion of the milk. Once a child is born and the placenta is sloughed away, protein-inhibiting hormone levels fall and the breasts can produce their milk. Prolactin is involved in milk secretion, and another pituitary hormone, oxytocin, is involved in the release of milk (Figure 28.15). Oxytocin secretion is stimulated by the newborn's sucking action. This hormone also acts on the uterine walls, stimulating contractions that aid the uterus in returning to its normal size and shape.

On the average, the nursing mother will produce 1.5 to 2 liters (a liter equals about a quart) of milk each day. If the mother does not nurse, milk pro-

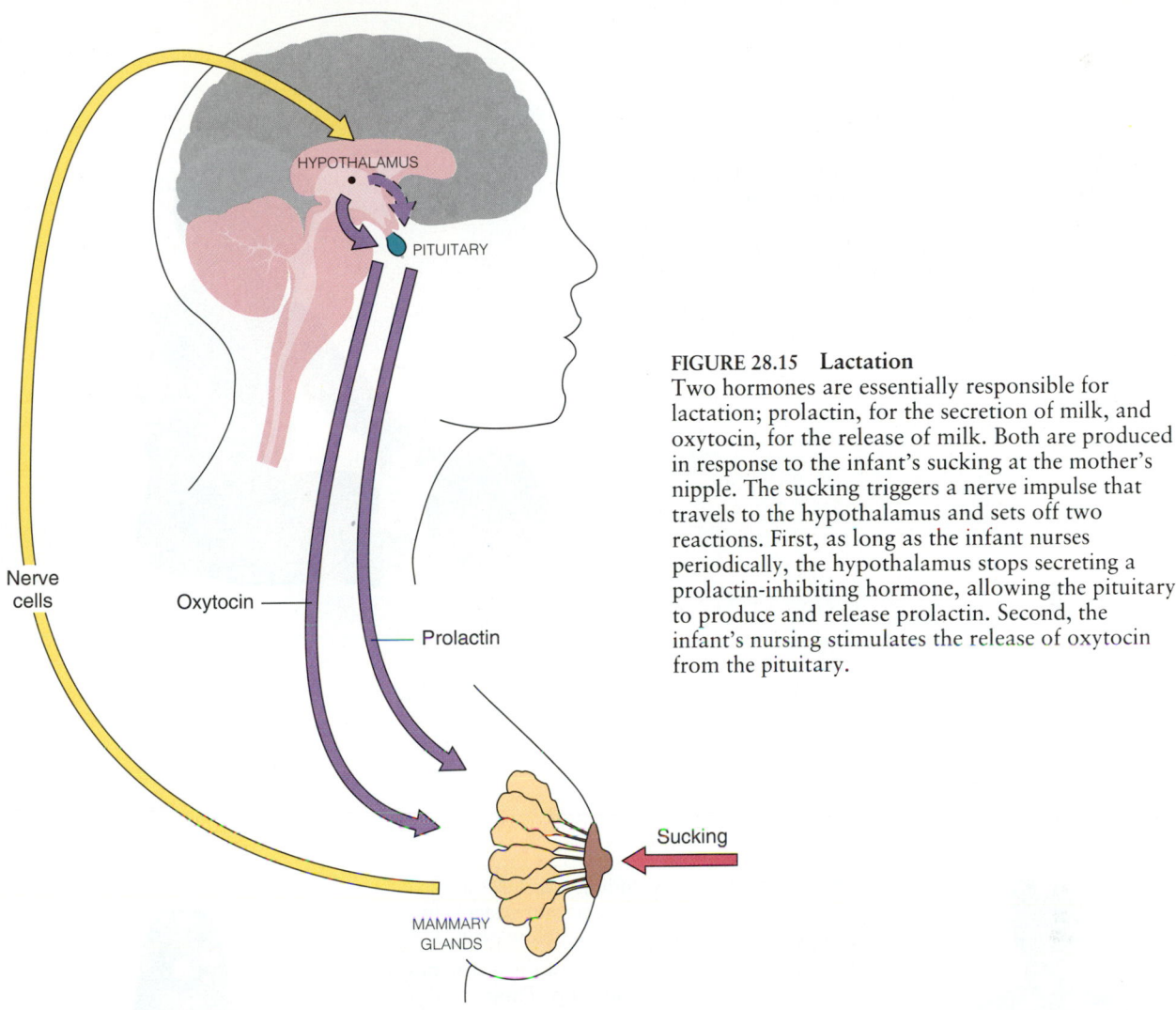

FIGURE 28.15   Lactation
Two hormones are essentially responsible for lactation; prolactin, for the secretion of milk, and oxytocin, for the release of milk. Both are produced in response to the infant's sucking at the mother's nipple. The sucking triggers a nerve impulse that travels to the hypothalamus and sets off two reactions. First, as long as the infant nurses periodically, the hypothalamus stops secreting a prolactin-inhibiting hormone, allowing the pituitary to produce and release prolactin. Second, the infant's nursing stimulates the release of oxytocin from the pituitary.

duction will cease, and milk stored in the glands will be reabsorbed. It is not true that a woman cannot become pregnant as long as she is nursing. Even though, because of hormonal changes in her body during lactation, she will not have a menstrual period, she still ovulates, and as is often the case with a fertile, sexually active woman, she can become pregnant.

As one might suspect, human milk is better for the human infant than cow's milk. Human milk contains all the required vitamins, and compared with cow's milk human milk is free of bacteria. For this reason, nursing babies are less likely to develop anemia, vitamin deficiencies, and diseases. For the first month of nursing, the mother's antibodies are present in her milk, and so in this way she transfers her immunity to her child. Cow's milk and other infant milk formulas, however, can be an adequate substitute, when required.

Development After Birth

■ **The neonatal stage of development happens during the first four weeks after birth. Organ systems become fully functional. Infancy extends from the fourth week to the first year; childhood extends from the first year to the end of puberty. Adolescence extends between puberty and adulthood. The final stage of development is senescence, during which the aging process is predominant, leading ultimately to death.**

We now move on to the neonatal stage, the period of development from birth to four weeks. During this period, organ systems such as the respiratory system, excretory system, and digestive system are becoming fully functional in order to provide an independent existence for the neonate. From the end of the first four weeks to the first year, a child is called an infant. During this time, growth is tri-

PHOTO ESSAY 28.2   The Process of Human Aging
Human development is an ongoing process from conception
to death. During this time, we are continuously aging. This
photo essay describes several of the major stages of aging
that occur during the human lifespan.

Source: Adapted from "Life Spans," *Science Digest*, February
1984.

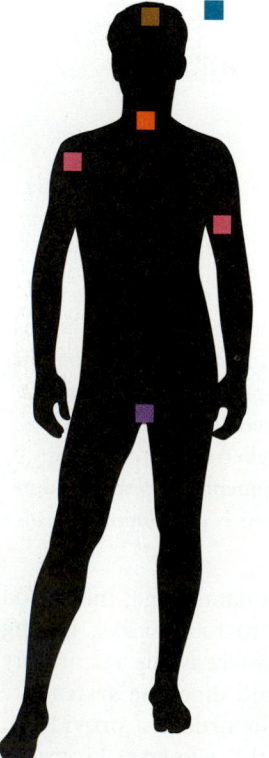

◀ **Birth to age 29: the ascending years.** Although
old age is many years away, certain systems in
the body are already operating below peak.

■ In late childhood, the level of the hormones
in the thymus begins to drop. These
hormones help regulate the immune system.

■ Girls reach puberty between the ages of 10
and 14; boys, between the ages of 13 and
16. By age 18, men are secreting more
testosterone than at any other time during
their lives.

■ Hair is thick, and the hairline is intact.

■ At age 25, muscular strength is at its peak.

■ At age 25, height may already begin
decreasing.

Age 30 to 39: reaching a plateau and passing the ▶
physical peak. After age 30, the functional
capacity of the body declines by about 0.8 percent
each year.

■ Heart muscle begins to thicken.

■ Hearing begins to decline (it peaked at age
10).

■ Skin loses its elasticity. Laugh lines and frown
lines begin to appear.

□ At age 30, the musculature is still intact, but
the vertebral disks begin to deteriorate,
moving the bones closer together.

■ Women reach their sexual peak.

Birth to 29

Age 30–39

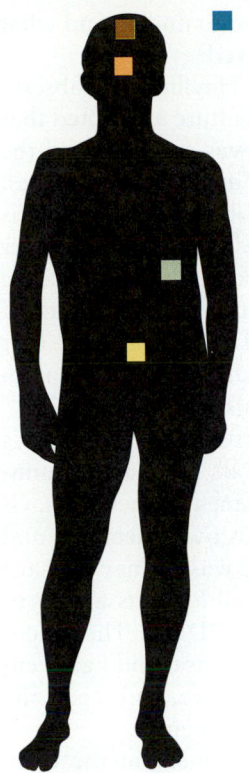

Age 40–49

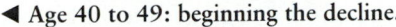

◄ **Age 40 to 49: beginning the decline.**

- ☐ A 40-year-old man is probably carrying 10 to 20 pounds more weight than he did at age 20.
- ☐ A 40-year-old man stands about one-eighth inch shorter than he did at 20.
- ☐ The body's natural defenses begin to weaken. Lymphocytes show a marked decrease in their ability to fight cancer cells. Other infection-fighting cells are less effective.
- ☐ Hair is graying and thinner; men may start to bald.
- ☐ Most men in their late forties are farsighted.

Age 50 to 54: the clock speeds up. ►

- ☐ The skin begins to loosen and sag.
- ☐ Eyes begin to fail at close range.
- ☐ Most women pass through menopause.
- ☐ The pancreas produces less insulin, so diabetes is more of a threat.
- ☐ Fingernails grow slowly.
- ☐ The sense of taste becomes less acute.

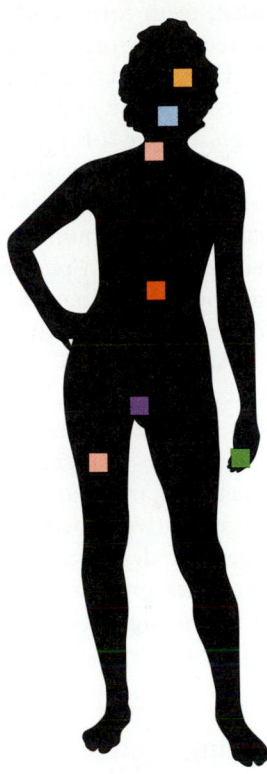

Age 50–54

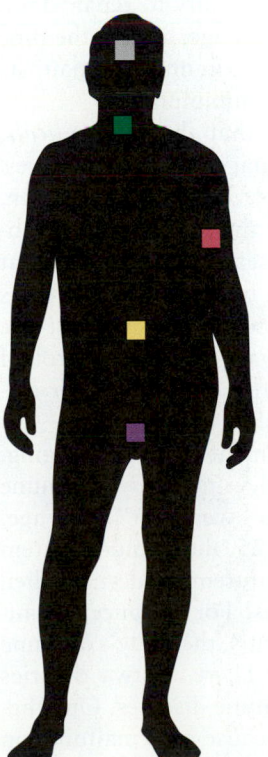

Age 55–59

◄ **Age 55 to 59: rapid changes.**

- ☐ Muscles and other tissues begin to deteriorate.
- ☐ The body begins to decrease in weight, but the slowdown of metabolism may mean a greater accumulation of fat.
- ☐ In men, sperm levels and semen volume fall.
- ☐ Men's voices may rise.
- ☐ Billions of nerve cells in the brain become inactive (though the effect in the healthy adult is usually just a slight memory loss).

Age 60 to 79: the clock slows down. On average, ► white males live to age 71.4; white females, to age 78.7. Past age 75, African Americans have a greater life expectancy than whites.

- ☐ At age 60, muscle strength is about half of what it was at age 25.
- ☐ By age 70, an inch of height is lost.
- ☐ The capacity of the lungs is decreased by half.
- ☐ Only 36 percent of taste buds remain active.

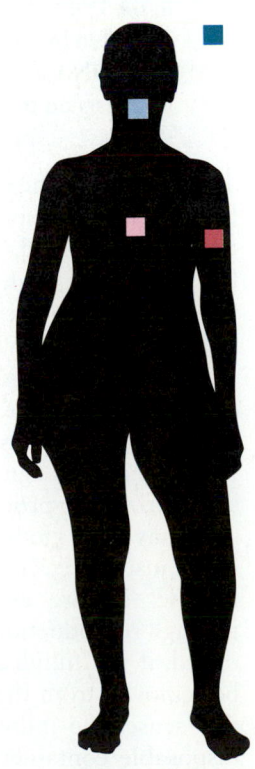

Age 60–79

pled, teeth form, and many neuromuscular coordinated behaviors and learning develop.

Childhood begins at the end of the first year and lasts until the end of puberty. During childhood, a high degree of neuromuscular control develops, along with many social developmental stages. Adolescence is the period of sexual maturity that occurs between puberty and adulthood (the teenage years). This is a period of rapid growth and of numerous physiological and emotional changes resulting in the adult body form and behavior.

Adolescence is followed by adulthood, which lasts until old age, or senescence, which is the term for the changes associated with aging, usually degenerative changes. During adulthood, developmental changes are gradual, eventually resulting in degenerative processes that cause the individual to age. The degenerative changes that take place during senescence are more apparent, and the body has a lessened ability to maintain life, a fact that eventually results in death.

Aging

■ **Aging is a complex process that is not fully understood. Many cells seem to be capable of dividing only about 50 times before they die, a characteristic that may be due to an accumulation of errors in the DNA. Another factor may be free radicals, produced by oxidative respiration, which may cause DNA damage. Another possible explanation rests on the mechanisms of the immune system.**

The last part of the life cycle is not one that humans accept easily — especially in developed nations of the Western world. We recognize that the cycle of life beginning at fertilization is completed with death. But we spend an inordinate amount of money and time trying to stave off the aging process. Indeed, medical developments (as well as changes in lifestyle, diet, and working environments) have increased the length of human life. Must we age? Is aging inevitable? Photo Essay 28.2 summarizes the process of human aging.

It may seem gloomy to say that "aging is nature's death insurance," but, unfortunately, the statement is true. Nature ensures survival of the species through reproduction. Once an organism has reproduced, it has fulfilled its biological destiny and can be removed from the population through death. In this sense, the individual can be thought of as a disposable container of DNA. One way to ensure death is through aging. Before we can find out if

aging is inevitable, we first need to understand what it is. What processes are involved?

In the early 1960s Leonard Hayflick was observing lung cells grown in tissue culture and noted that growth ceased, the cell cycle was arrested, and the cells died after they had divided about 50 times. From these observations, he concluded that aging is genetically programmed within the cell. In other words, each cell has a predetermined, finite life span. The question was, what caused the death of the cell? What was it about DNA replication that resulted in its failure to replicate after a certain period of time? What mechanisms were involved?

Currently, researchers investigating these issues have proposed that aging results from an accumulation of DNA errors. Environmental chemicals, viruses, radiation, and the body's own chemistry and its accumulation of metabolic wastes may damage the DNA. Progressively, the cell loses its ability to repair the damage, especially to DNA. They theorize that, for some reason, restriction and ligase enzymes cannot edit (cut out), remove, and repair the damaged portion as efficiently as before.

This theory was supported when, in the early 1970s, Dr. Ron Hard discovered that rates of DNA repair were correlated with the maximum life span for the individual of the species — for example, 45 years for chimpanzees and 110+ years for humans. In other words, the ability to repair damage to the DNA decreases with age, and by the time an individual reaches the maximum life span, its ability to repair DNA is at a minimum.

Biochemists also suspect that the *free radicals*, highly charged ions, can damage cellular structures. Free radicals are produced by oxidative respiration. Normally, enzymes convert them to harmless substances; however, some researchers propose that enzymes' ability to convert free radicals decreases with age. They have located the position of the genes involved in DNA repair and the control of free radicals. These genes are called the *major histocompatibility complex* (MHC).

In addition to the involvement of MHC in aging, researchers think aging involves the body's immune system, which also seems to "wear out" with age. As we discussed in Chapter 25, the immune system directs specialized cells and proteins to destroy alien bacteria, viruses, and tissues. For instance, in autoimmune diseases like arthritis, the body's immune system attacks its own cells. There are two theories about the causes of autoimmune diseases. One theory states that they arise because of a malfunction of the immune system. The other states that autoimmune diseases are caused by an immunological involvement in aging. This theory maintains that as cells age, an accumulation of genetic damage produces abnormal molecules on the surface of the cell membrane. As a result, they appear foreign, like an antigen. Hence the body attacks its own cells.

The previously discussed theories of aging have emphasized the cellular and biochemical levels. However, some *gerontologists* (scientists who study the aging process) feel that cultured cells should not be the only unit of aging that is investigated. Rather, researchers should explore the aging of the total organism. For example, many gerontologists feel that there may be a biological clock in the hypothalamus that controls and regulates hormone secretion of the pituitary gland and other endocrine glands.

For example, the Pacific salmon is full of life and vigor when it spawns in the freshwater stream it calls home. However, within two weeks of spawning, the salmon ages and dies. Researchers believe that the fish's adrenal glands produce an aging hormone. As a matter of fact, some investigators, such as the Australian scientist A. V. Everett and the American scientist Donner Denckla, have proposed that the pituitary, working through other glands, produces a death hormone. At this point, however, such a hormone (or hormones) has not been found.

In other words, aging is, indeed, inevitable, at least for the present. Over time, every organism undergoes progressive deterioration, leading inevitably to death. (Consider how crowded the world would be without it.) Although several theories of aging have been proposed, as of now aging is a process that biologists do not understand completely. With a better understanding of the developmental process called aging, humans may make major breakthroughs that will result in life span extension. Who knows how long we may eventually live.

SUMMARY

1. Development is a continuous process from conception through old age.
2. The development of an organism from a fertilized egg into a mature individual is made up of three major processes — cell division and growth, morphogenesis, and differentiation. Cell division refers to normal cellular mitosis. Morphogenesis refers to the development of the body shape of a particular species. Differentiation is the process by which certain genes in a cell are turned off so that the cell can perform specialized functions.
3. Many investigators think cancer is an oncodevelopmental question: that something has happened to cancer cells to return them to the type of cellular division typical of undifferentiated cells.
4. Fertilization is the fusion of the nuclei of a haploid sperm and egg to form the $2n$ zygote. Once the sperm penetrates the egg's outer membrane, its tail drops off. In most vertebrates, when the egg is fertilized, meiosis is complete. The zygote consists of yolk (which contains nutrient materials), cytoplasm, and a diploid nucleus.
5. The division of the zygote is called cleavage. First a solid ball, called a morula, forms. Then a hollow ball, called a blastula, is created. The center is called a blastocoel.
6. During gastrulation, a primitive gut forms. Then three germ layers develop — the ectoderm (outer layer), endoderm (inner layer), and mesoderm (middle layer).
7. The positioning of body parts in their proper places is determined by embryonic induction, regulation, and cellular differentiation. Some cells act as inducers, producing chemicals that influence neighboring cells to develop into certain structures. Genes control regulation and cellular differentiation, and the elements of time, age, and neighboring cells produce the correct morphogenesis. Homeo box genes function almost as master genes, regulating the activities of the many genes that must work in concert to control development.
8. In humans, the germinal disc, which will become the embryo, consists of the ectoderm and endoderm. Cells migrate between the endoderm and ectoderm to form the mesoderm.
9. Around the sixth day after fertilization the developing embryo implants itself in the endometrium. This process is called pregnancy.
10. The development of the extraembryonic membranes was an important evolutionary step in the evolution of the land animals. The land egg of reptiles and birds contains membranes that enclose a watery environment in which the embryo can develop.
11. The extraembryonic membranes in placental mammals are used to protect the developing fetus and to ensure that nutrients can be provided and waste materials can be eliminated. The placenta derives from the endometrium and the chorion, a structure made up of the embryo's outer membrane. The chorion implants in the endometrium, and it functions in gas exchange, nutrient and hormone supply, and waste exchange between the fetus and mother by diffusion and other methods of molecular transport. The blood of the mother and that of the fetus do not connect directly. The amnion is a membrane surrounding the fetus. It secretes amniotic fluid, in which the fetus floats.
12. After about the eighth week of development the embryo appears humanlike and is called the fetus. The yolk sac and allantois eventually develop into and

form the umbilical cord, which contains two arteries and one vein — the lifeline between the fetus and the placenta through which materials are transported.

13. The placenta also functions as an endocrine gland, secreting hormones such as human chorionic gonadotropin, which helps maintain high levels of progesterone and estrogens in the mother's body.

14. The first two months of pregnancy are a sensitive developmental period. Numerous substances, called teratogens, can affect the embryo's development, causing physical and/or mental defects.

15. Later fetal development consists of growth and maturation of the developing organ system.

16. Childbirth usually occurs 280 days after the beginning of the last menstrual cycle (or 266 days after fertilization). Hormones secreted by the fetus and the mother signal the beginning of labor. Childbirth can be divided into three stages: dilation, in which rhythmic contractions of the uterus open the cervix; expulsion, which is the passage of the child through the birth canal; and the placental stage, in which the placenta is expelled.

17. After birth, numerous physiological changes have to occur to enable the newborn to breathe in air. High levels of carbon dioxide in the blood, along with other events, stimulate the foramen ovale and the ductus arteriosus to close. With these changes the postnatal circulatory pattern begins.

18. During pregnancy, the hormones estrogen and prolactin stimulate the mother's mammary glands to prepare for milk production and secretion. Placental hormones inhibit secretion. With the birth of the child, the inhibiting hormones are no longer present. The anterior pituitary begins to secrete prolactin and the posterior pituitary begins to secrete oxytocin, which stimulates lactation and triggers the release of milk.

19. Development continues through the newborn or neonatal stage (birth to four weeks), the infant stage (four weeks to one year), childhood (one year through puberty, adolescence (teenage years), adulthood, and senescence (old age).

20. Numerous advances have occurred in the study of aging. Many theories are centered at the molecular and cellular level. Many investigators feel that large accumulations of highly charged ions damage the DNA, and that accumulation of this damage leads to aging. Another theory holds a breakdown of the immune system responsible. Still other investigators think that aging is a glandular event and that there may even be death hormones.

KEY TERMS

morphogenesis	blastula
differentiation	chorion
cleavage	yolk
allantois	umbilical cord
amnion	fetus
placenta	lactation

REVIEW QUESTIONS

True/False

1. Animal development is limited to the period prior to birth or hatching.
   true   *false*   *(page 677)*

2. Some scientists feel that cancer is a developmental question.
   true   *false*   *(page 678)*

3. The amnion develops into the umbilical cord in placental mammals.
   true   *false*   *(page 690)*

4. The placenta consists of a maternal portion, the endometrium, and an embryonic portion, the chorion.
   true   *false*   *(page 690)*

Fill in the Blank

1. The development of the _____ egg and _____ fertilization freed the animals from the aquatic environment for reproduction and development. *(page 690)*

2. Substances that can produce malformations in the developing embryo are called _____. *(page 693)*

3. During the birth process, the pituitary hormone _____ signals the uterus to contract. *(page 696)*

4. The hormones _____ and _____ are involved in the secretion and release of milk by the nursing mother's breasts. *(page 698)*

Multiple Choice

1. An example of embryonic induction would be:
   a. The transplantation of the dorsal lip of the blastopore to a different site on a salamander embryo, which results in the induction of notochord formation at that site
   b. The mesoderm inducing the formation of ectoderm
   c. The induction of the morula
   d. The closure of gill slits in a mature frog
   (page 684)

2. The hollow ball stage of the embryo is called the:
   a. Blastula
   b. Gastrula
   c. Archenteron
   d. Morula
   (page 682)

3. The early diagnostic pregnancy test kits measure the levels of _____ in the urine.
   a. Human chorionic gonadotropin
   b. Testosterone
   c. Estrogens
   d. Progesterone
   (page 693)

4. It is known that salmon always age and die within two weeks after spawning. This observation has caused some scientists to hypothesize that this rapid aging and death is due to:
   a. Death hormones
   b. The production of too many free radicals
   c. DNA breakdown
   d. Polluted water
   (page 703)

Discussion

1. Provide some examples of experiments that demonstrate embryonic induction, regulation, and differentiation.

2. To understand the evolution of the land egg, why is it important to understand the need for extraembryonic membranes in the placental mammals such as humans?

3. What is the placenta? Is it composed only of maternal tissue?

4. Why can the placenta be called an endocrine gland?

5. What are teratogens? Do you feel that alcohol and cigarette smoke are teratogens?

6. List and describe the three stages of childbirth.

7. What physiological events occur in the adaptation of the newborn for breathing in air?

8. What is lactation? How is it hormonally regulated?

9. Do you think that aging is inevitable? Is it nature's death insurance?

10. Discuss some current theories of aging.

29

Cell-to-Cell Communication: Hormonal Signals

Evolution of Coordinating Systems
The Human Endocrine System:
A Mammalian Example
Summary

As we watch a basketball player or see a jockey, we cannot help but appreciate the extremes of height. When we watch the care a mother provides her newborn, or how she responds so strongly to the cry of a child long after her own children are grown, it is easy to see how the behavior is brought about in part because of biological mechanisms residing in the glands that produce the chemical messengers known as hormones.

The same biological mechanisms are at work when we see a dog chase a cat up a tree. It is amazing how rapidly the nervous system communicates information to the muscles resulting in the chase (exhibited by the dog) and the escape (exhibited by the cat). All these activities, such as growth, parental care, behavior, development, neuromuscular co-

ordination, intracellular coordination, and sexual maturity, are influenced by chemical secretions. Even nerve impulses are transmitted chemically between nerve cells.

Two systems are involved in coordinating bodily activities. One is the nervous system, that web of nerve cells that extends from the brain to every part of your body. (We discuss the nervous system in the next chapter.) The other is the **endocrine system,** an interrelated complex of glands and cells and the hormones that they produce. Both these systems interact with each other, and so as we proceed through this chapter and the next, it is important to remember that we are discussing them separately only for our own convenience.

In Chapter 28 we discussed the way hormones regulate activities throughout an organism's development, from the time of conception until death. For instance, you learned that during the develop-

---

Above: The size of this giraffe is due in part to complex hormonal interactions.

ment of the human embryo, sexual differentiation occurs at about the seventh week after conception. At that time, the gonads begin to change in response to steroid hormones. Androgens stimulate the changes that ultimately produce a male by entering specific cells and inducing specific genes to produce particular proteins that result in the development of the testes and other male sex characteristics. If no androgens are present, the fetus develops into a female (Figure 29.1). Hormones are instrumental in all stages of human development.

Hormones may regulate development and other body activities, but what regulates hormones? Most hormones are regulated by the mechanism of negative feedback, which we discussed in Chapter 23. Negative feedback involves four steps: (1) detection of a stimulus, (2) interpretation of the stimulus, (3) response to the stimulus, and (4) adjustments or modifications once the stimulus has been responded to. (In other words, once the thermostat has detected a change in temperature and signaled the air conditioner to run, the thermostat must adjust to

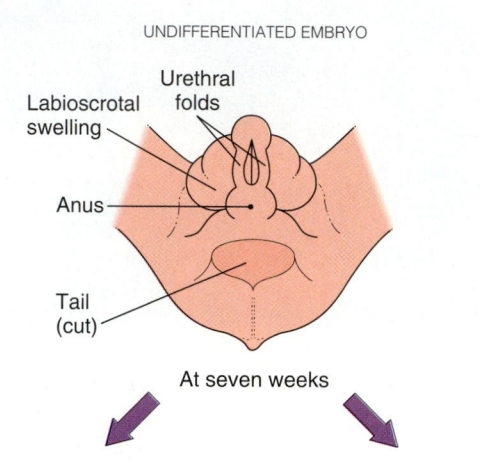

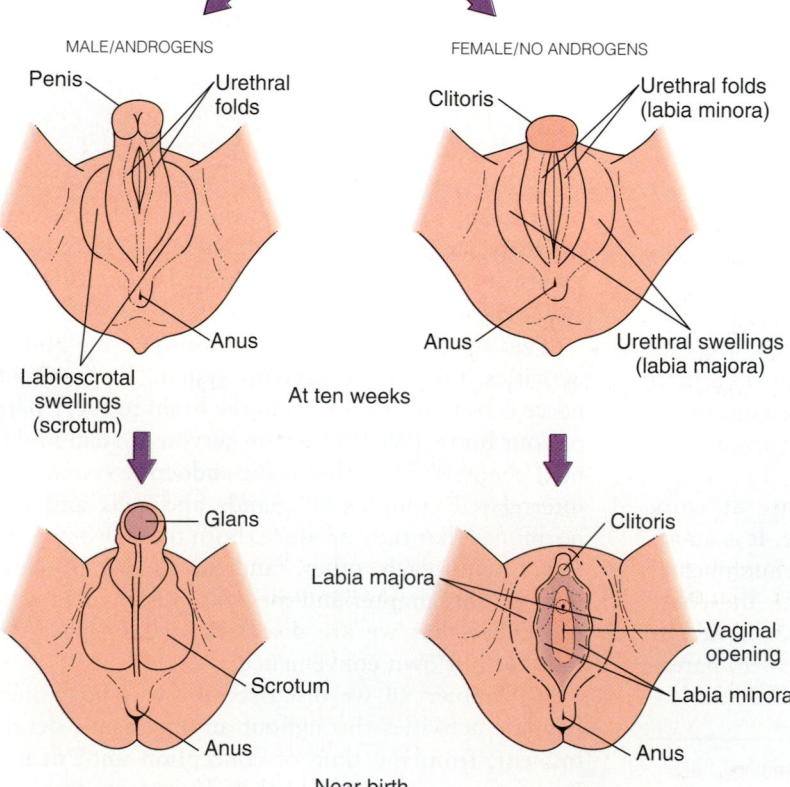

FIGURE 29.1 Hormones and Sexual Development
The sexual organs of all human embryos, male or female, develop from the same structures — the urethral folds and the labioscrotal swellings. It is the presence or absence of the male sex hormones — androgens — that determines the gender of the embryo. If androgens are present, a penis forms and the labioscrotal swellings become the scrotum. If androgens are not present, a clitoris forms, the urethral folds develop into the labia minora, and the labioscrotal swellings become the labia majora.

the new conditions and signal the air conditioner to turn off once the air has been cooled to the desired level.)

In order for these steps to occur, there must be a way to transmit and receive information. In both multicellular and unicellular organisms, the cell must "talk" to its parts, and, in multicellular organisms, cells must "talk" to each other. What is the means of this communication? (See Human Endeavors box.) All cell-to-cell communication is chemical. In multicellular organisms, in some instances, specialized cells (or *sources*) produce chemical signals that are transported to **target cells** — other cells at a distance from the source that have receptor sites for a particular chemical message. Once the target cells receive the message, they respond. Even more specialized cellular communications systems, such as the nervous system, have evolved in animals, although not in plants. (Plants do have hormonal systems, however.) These systems function in the integration and control of activities on a broad scale (see Profile).

No matter how complex the system may be and regardless of the actual message — whether a hormone secreted by a specialized group of cells or glands, or nerve impulses moving from one nerve cell to another — the means is *chemical*. In this chapter we examine the coordinating, regulating mechanisms responsible for homeostasis, survival, reproduction, response to physical threat, behavior, and even thought — the marvels of chemical coordination.

EVOLUTION OF COORDINATING SYSTEMS

■ In early unicellular organisms, communication occurred between organisms through chemicals called pheromones, which were released into the environment. In multicellular organisms, the communicating chemicals function within the organism rather than without and are called hormones. Hormones are produced by cellular sources or ductless glands called endocrine glands. Secretions are regulated by the nervous system and by specialized endocrine glands or endocrine sources.

We have already discussed the evolution of sexual reproduction. As one result, it became necessary for organisms to be in close proximity to one another in order to allow the exchange of genetic material. Over time there evolved a group of chemical mechanisms called *pheromones* (Figure 29.2). Pheromones are chemicals that the organism discharges

FIGURE 29.2 Pheromones
Pheromones are chemicals produced by one organism that can affect the behavior of others. Many pheromones are sex attractants. For example, the female gypsy moth emits a very powerful chemical called bombykol that attracts the male moth. The receptors for the pheromone are found on the male moth's large antennas (shown above). They can detect minute quantities of bombykol (1 molecule per 100,000,000,000,000,000 molecules of air) from as much as half a mile away. Other pheromones signal danger or mark territorial boundaries.

into the environment (usually water or air). The chemicals then make contact with other organisms. This system for sending and receiving messages between organisms of the same species is still in operation today.

With the rise of multicellularity, the process of communication became much more complex. Not only did cells have to communicate with their parts, they also had to communicate with other groups of cells within the same organism. Also necessary was the integration and control of the activities of cells within the organism. Finally, there was a need for a mechanism that could send the messages. Through the process of evolution, more and more complex systems for accomplishing cellular communication evolved.

Primitive multicellular organisms, such as the sponges, jellyfish, and hydra, for instance, literally took a portion of the sea within their saclike body cavities to function as the mechanism for the transfer of chemical messages. In these organisms, regulatory chemicals were no longer discharged into the environment outside the organism but were discharged within the body sac. To signal the difference between chemicals secreted into the environment and those secreted within the organism, biologists refer to the latter as hormones. They define hormones as internal regulatory chemicals pro-

Signal Transduction

How do cells communicate with one another?

Source: Adapted from "Evolutions: Signal Transduction," *Journal of NIH Research,* vol. 2, p. 104, Jan./Feb. 1990. Adapted from original artwork by Sally Bensusen.

A generic signal-transducing protein has four different functions: (1) the detector interacts with the primary signal, initiating the response, (2) the generator makes the response, (3) a timer determines how long the generator will keep responding, and (4) a modulator may detect secondary signals and modify the response. [Adapted from H.R. Bourne, "Summary: Signals Past, Present, and Future," *Cold Spring Harb. Symp. Quant. Biol.* 53, 1019 (1988).]

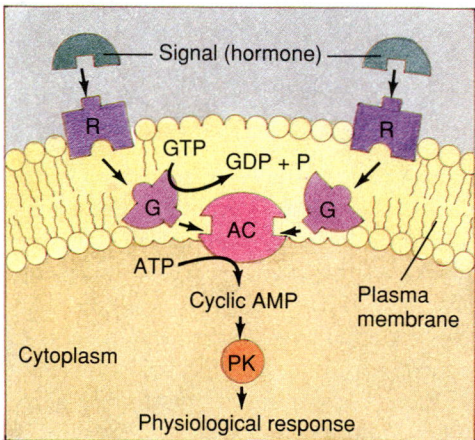

1900–1940: Hormones, as blood-borne chemical messengers, were identified and isolated. In 1905 Oliver and Shafer found that extracts from the posterior pituitary at the base of the brain strongly constricted arteries and capillaries.

1940–1970: Most elements of the cyclic AMP (cAMP) second messenger system, except for some of the binding proteins, were discovered. (R, receptor; G, proteins and protein subunits; GTP, guanosine triphosphate; PK, protein kinase.)

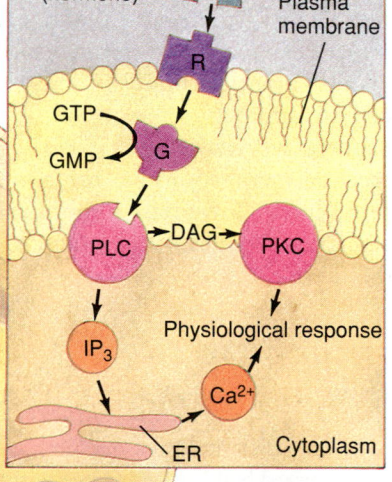

1970–1980s: Researchers discovered GTP-binding proteins and several new second messengers, such as calcium ions and many others (abbreviated in this illustration as IP$_3$, PKC, and so on).

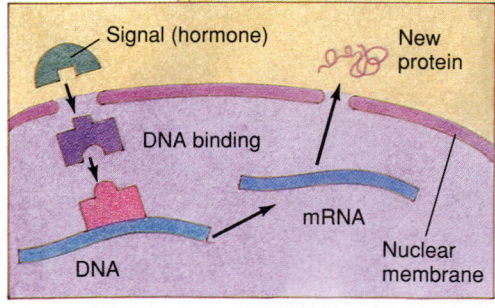

1980s: The molecular structures of both membranous (such as that for epithelial growth factor, or EGF) and soluble receptors (such as those for glucocorticoids) are revealing how signals are transmitted to the cell's interior. Several membranous receptors apparently phosphorylate cytoplasmic proteins via a tyrosine kinase activity, while some signals are transferred directly to the nucleus by cytoplasmic receptors. Proto-oncogene products may regulate gene expression in response to signal systems.

duced by specialized cells or groups of cells. Hormones are produced by specialized cells (in most cases cells in *endocrine glands*) and are carried to other cells by the fluids within the body cavity.

The control and integration of chemical communication are achieved by a number of systems. In some instances, the function of the endocrine elements is directly regulated by an element of the nervous system; in other instances, secretions from central regions of the nervous system regulate the endocrine elements. For example, in both the invertebrates and the vertebrates there are special nerve cells called *neurosecretory cells* that secrete hormonelike substances. In mammals, including humans, interactions of an area of the brain called the *hypothalamus* and a gland called the *pituitary* provide one of the major means of hormone control and regulation. Much of this interaction is due to neurosecretory cells. The interaction of these two structures also provides a major piece of evidence supporting the theory that regulatory structures evolved over the centuries. In addition, the link provides evidence that both the nervous and endocrine systems are intimately associated.

How did the linkage between the hormonal and neural systems evolve? As animals became more complex, biologists speculate that the environment selected for variations that integrated the chemical activities within organisms. Some biologists think that predation and avoidance of being prey (that is, eating and avoiding being eaten) became such a strong selection factor that those animals with the most rapid methods of communication, both internal and external, survived.

In other words, those that could coordinate their own activities and exchange information among members of their own species could catch prey and escape predators much more successfully. Thus the complex coordinating systems we see in the higher organisms evolved.

Comparisons and Contrasts Between the Nervous System and the Endocrine System

- Both nerve cells and cells that produce hormones communicate chemically. Nerve cells form a communication system called the nervous system, which functions very rapidly. Specialized nerve cells called neurosecretory cells produce hormone-like substances that are carried through the interstitial fluid or the bloodstream. Hormones themselves are also carried through the bloodstream and interstitial fluid, which results in a slower rate of communication.

Before we concentrate on the hormonal system, we need to look at the relationships between it and the nervous system a little more closely. Nerve cells and hormone-secreting cells are strongly related, since they both communicate chemically. The first nerve cells were cells that, when electrochemically stimulated, secreted substances much like hormones. The hormonelike chemicals then stimulated a certain response.

In higher species such as humans, nerves are, in actuality, message transmission lines. The cells that form nerves are adjacent to one another (although they do not actually touch), forming a pathway for rapid internal chemical coordination. This pathway is called the *nervous system*. The space between nerve cells, called the *synapse*, is a microscopically short distance. A chemical called a *neurotransmitter* is produced by the tip of a nerve cell and is transmitted across the synapse to the membranes on the tip of adjacent nerve cells. When the chemical is received, the information is transmitted to the next nerve cell.

In general, nerves produce messages that are rapid and short-lived. The neurotransmitter functions only for a brief moment, crossing the synapse to adjacent nerve cells and delivering the appropriate message. The chemical message is then degraded through the actions of enzymes or reabsorbed by the delivering nerve cell. Some specialized nerve cells, however, serve as neurosecretory cells, which function much like the neurosecretory cells found in lower organisms. These cells produce substances called *neurosecretory hormones,* which, unlike neurotransmitters, are carried by tissue fluid or the blood to nonneural cells far removed from the source.

Hormones, on the other hand, are chemicals produced (in most cases) by structures called **glands.** The chemicals are usually secreted directly into the bloodstream or interstitial fluid. Because hormones travel through the interstitial fluid or through the bloodstream, their movement is slower than that of nerve impulses.

Hormonal Chemical Coordination

- Hormones are usually produced by endocrine glands and are secreted into the bloodstream. There are two basic types of hormones, steroids and proteinlike substances. Hormones are a fairly slow means of regulating the metabolic activities of cells.

Hormones are chemical regulatory substances produced by specialized endocrine cells. In some cases,

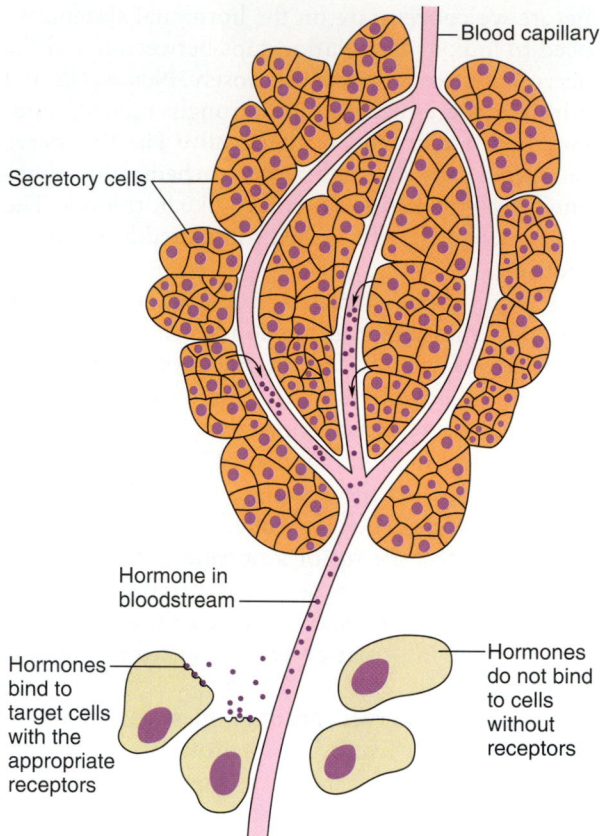

Blood capillary

Secretory cells

Hormone in bloodstream

Hormones bind to target cells with the appropriate receptors

Hormones do not bind to cells without receptors

(a) Endocrine gland

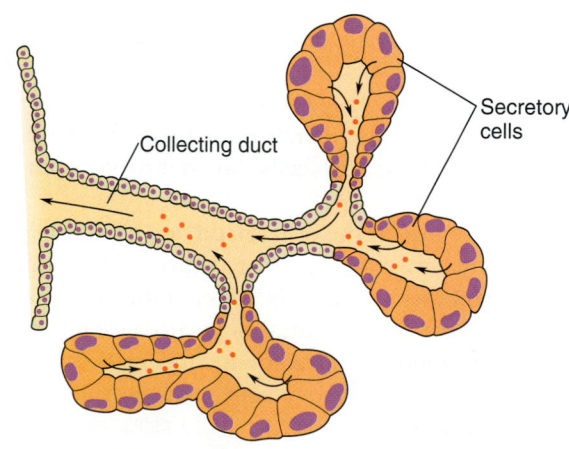

Collecting duct

Secretory cells

(b) Exocrine gland

FIGURE 29.3 Endocrine and Exocrine Glands: A Comparison
(a) Endocrine glands are clusters of cells whose products are secreted directly into the bloodstream or the interstitial fluid. (b) In exocrine glands, the secretory cells release their products into ducts, where they are stored until they are secreted. The arrows indicate the movement of secretions.

the endocrine cells work alone; in others, they are organized into tissues that form ductless glands called **endocrine glands.** (Glands having ducts, such as the salivary glands, are called **exocrine glands**— *ex* means "out of"—because their cellular products are released through a tube into areas of the body or onto its surface.) Since endocrine glands lack ducts, the hormones they produce are secreted directly into the bloodstream. The differences between the endocrine and exocrine glands are illustrated in Figure 29.3.

What are the properties of hormones? In general, hormones have these characteristics and produce these effects:

1. Hormones are secreted by the endocrine cells or sources directly into the bloodstream or the interstitial fluid.

2. Potent in very small amounts, hormones are usually found in two forms: steroids — which have chemical properties much like those of lipids — and modified amino acids, small polypeptides, or proteins.

3. Hormones are usually transported some distance through the body to target cells, which have specific receptor molecules that receive and recognize a specific hormone. Despite the fact that hormones come into contact with most cells, only target cells respond to the hormones.

4. Once within the target cell, hormones regulate the cell's metabolic activities, resulting in a change in physiology or even in behavior. The target cell's response to the hormone's stimulus may be short-lived, as is the case with epinephrine (commonly known as adrenalin), or long-lived, as in the case of hormones that are influential in growth and development.

5. Compared with the speed of nerve signals, hormonal regulation is slow.

Now that we know a little about the properties of hormones, let us see how they affect the target cell.

Hormone Action

■ **There are two chemical categories of hormones: steroid and proteinlike substances. Endocrine glands that produce lipid-soluble steroids arise from the mesoderm. Steroids can pass through the cellular membrane. They then bind with a receptor in the target cell's cytoplasm, forming a hor-**

mone-receptor complex that is carried to the nucleus. Once there, the complex affects transcription and gene expression. Proteinlike hormones, called first messengers, cannot penetrate the cell's membrane, because they are not lipid-soluble. Instead, they bind with receptor molecules on the surface of the cell, prompting formation of second messengers such as cAMP, which alter cellular metabolism.

What happens at the molecular level, once the hormone reaches the target cell? Many endocrinologists (individuals who study hormones) have tried to answer this question, but as is often the case, they ultimately learned that there was not just one mechanism. Much of this knowledge was due to the work of Dr. E. W. Sutherland, of Vanderbilt University, who in 1971 was awarded the Nobel Prize for his work on hormone mechanisms. Before we look at his findings, we need to look at the basis for them.

In Chapter 28 we discussed the stages of development of the fertilized egg, which at one point differentiates into layers called the endoderm, the ectoderm, and the mesoderm. From each layer certain specific structures arise. You might think that endocrine glands, since they all have similar functions, would arise from the same layer of the embryo. However, by studying the embryological development of hormone-producing tissues, endocrinologists learned that certain glands originate from specific layers. Those originating from the mesoderm, such as the gonads and the adrenal cortex, produce chemical regulators that are steroids (lipidlike molecules). Glands derived from the ectoderm or endoderm, like the pituitary, pancreas, and thyroid, produce regulatory substances that are protein or protein derivatives, such as modified amino acids or polypeptides.

The chemical properties of hormones, whether they are steroid or protein, affect how they penetrate the plasma membranes of the target cells. To understand how this works, we need to refer to Chapter 5, in which we discussed the fluid mosaic model for the cell membrane. The fluid mosaic model consists primarily of proteins, phospholipids, and carbohydrates. The most notable characteristic of this model is that the proteins are embedded and float in the phospholipid bilayer comprising the continuous portion of the membrane. The fact that the bilayer is made up of phospholipids determines, to a large extent, what type of molecule can penetrate through the membrane.

Because steroid hormones are lipid-soluble, they are able to penetrate through the phospholipid-rich cell membrane quite easily, entering the cytoplasm of the cell. Most proteinlike hormones, on the other hand, are unable to penetrate the cell's plasma membrane because proteins are not readily soluble in lipids.

Since proteins do not pass through the phospholipid bilayer, hormones must act through at least two mechanisms: one in which the hormone enters the cell and the other in which the hormone remains outside of the cell. Endocrinologists began to investigate two mechanisms of hormone action, one for steroid hormones and the other for proteinlike hormones.

Mechanism of Steroid Hormones. Because steroid hormones are lipids, they can easily penetrate the plasma membrane of the target cell in order to regulate its activities (Figure 29.4a). Once inside the cell, the steroid hormone binds with a cytoplasmic receptor molecule to form a hormone-receptor complex, which is then carried to the cell nucleus.

Once inside the nucleus, the hormone-receptor complex binds to specific sites on the DNA to initiate the transcription of specific genes. As a result, specific mRNA molecules are transcribed. These enter the cytoplasm, and at the ribosomes their message is translated to produce a specific protein. The protein then may modify the metabolism of the target cell or the production and secretion of the chemicals that are the end product of the entire process.

Mechanism of Proteinlike Hormones. Sutherland was the first individual to explain the mechanism of action of proteinlike hormones. Since proteinlike hormones cannot penetrate the lipid layers of the plasma membrane, they must rely on a second set of messenger molecules to produce their effect in the target cell. How does this mechanism work?

When biologists studied the functioning of these molecules, they discovered that polypeptide hormones would bind to receptors on the membrane of the target cell. Since these hormones carried the message to the cell but did not actually enter the cell, they are called *first messengers*. The hormones then rely on a secondary set of molecules, called the *secondary messengers*, to affect the functioning of the target cell. Because two messengers are involved, the process is called the *second messenger mechanism* of hormone action.

Typically binding of the hormone to the receptor on the outer portion of the cell membrane activates

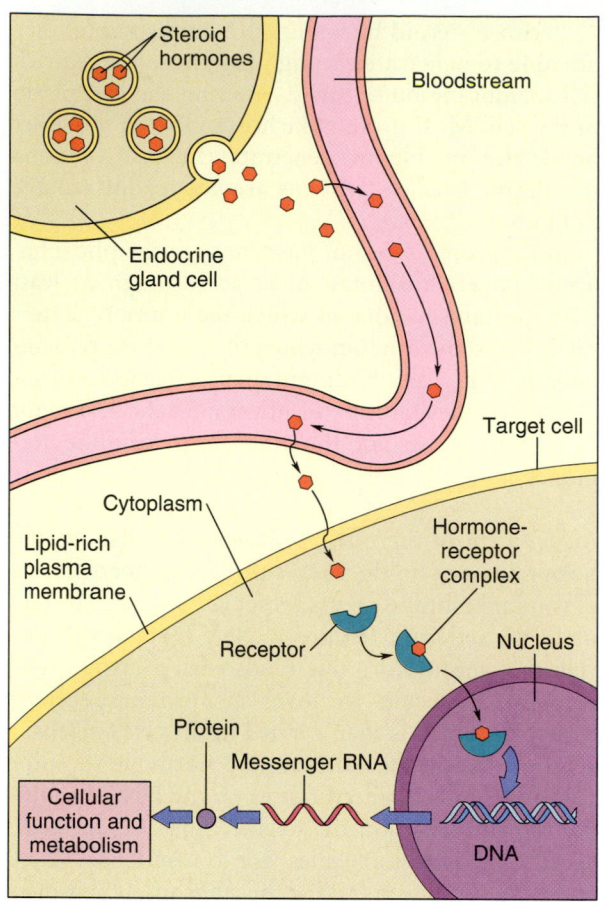

(a) Steroid hormones

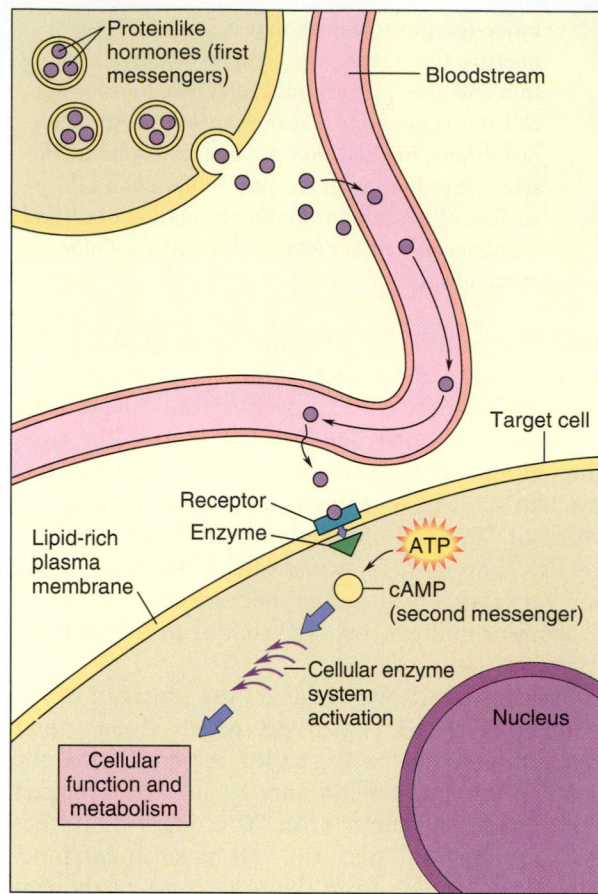

(b) Proteinlike hormones

FIGURE 29.4   Hormones and Their Target Cells
(a) A steroid hormone crosses the lipid-rich plasma membrane directly; it then
binds with a receptor in the cytoplasm, which moves it into the nucleus. Here,
the hormone acts on DNA, which produces messenger RNA. The "message"
is translated into proteins that modify the cell's metabolism, resulting in the
effect. (b) Proteinlike hormones cannot cross the target cell's plasma
membrane; they must bind to receptors on the surface of the membrane. This
activates an enzyme (usually adenyl cyclase) on the inner surface of the
membrane to convert ATP to cyclic AMP (cAMP) — the second messenger
(the hormone is the first messenger). cAMP triggers a chain of reactions in the
target cell that results in changes in the cell's function or metabolic activities.

an enzyme located on the inner surface of the cell
membrane known as *adenyl cyclase* (Figure 29.4b).
Adenyl cyclase transforms the cell's ATP to a substance called **cyclic adenosine monophosphate,** or
cyclic AMP for short (often written cAMP). The
chemical cAMP is the second messenger involved in
the mechanism of proteinlike hormones. It should
be noted that there are other second messengers but
cAMP is the best understood.

The second messenger is the chemical that affects
cell functions. cAMP initiates a series of reactions
involving cellular enzymes that alter the cell's metabolic activities.

For example, when the proteinlike hormone ep-

inephrine complexes with the receptor site of the
target cell, adenyl cyclase is activated, triggering the
formation of cAMP. cAMP then initiates the conversion of inactive protein kinase to the active form.
This conversion initiates a series of events in which
phosphorylase enzymes are activated. Phosphorylase enzymes catalyze the reactions in which glycogen (animal starch) is broken down to glucose. This
glucose can then be transported through the blood
to other cells for emergency sources of energy. That
is why epinephrine is called the "fight, fright, or
flight hormone" — it initiates molecular responses
that provide immediate emergency energy sources
for the cells.

Hormone Activities

■ **Hormones are first synthesized, after which they are secreted into the bloodstream. Once they are transported to their specific target cells, they exert their influence on them, after which they are degraded by enzymes.**

What processes are involved in hormone activities? In general, there are five steps:

1. The hormone must be *synthesized* from a chemical precursor of the hormone.

2. It is *secreted* into the bloodstream.

3. It is *transported* and may be *activated* on its way to the target cells.

4. It exerts its particular effect.

5. It is *degraded* by enzymes.

Should there be insufficient amounts of a raw material required to synthesize a given hormone, the effects can be quite widespread. For example, iodine is necessary for the formation of the thyroid hormone known as thyroxine. If there is not adequate iodine in the diet, the thyroid produces insufficient thyroid hormone. An insufficient level of thyroxine stimulates the thyroid to produce more, so the gland overcompensates by growing large. The result is simple thyroid goiter, pictured in Figure 29.5. This condition is especially prevalent in parts of the world such as the European Alps and the Great Lakes area of the United States, where insufficient iodine occurs naturally in the water and the food supply. To avoid the condition, individuals in these areas can include iodized salt in their diet to supplement their iodine supply, for an intake of only 4 grams a year is sufficient. Nonetheless, 200 million people still have this condition.

Once the hormone is synthesized by the endocrine cell, it is secreted to the outside. As you might remember, ribosomes are involved in the production of proteins and the smooth endoplasmic reticulum is involved in the synthesis of lipids and hence steroid hormones. Rough endoplasmic reticulum is involved in the synthesis of proteinlike hormones. The Golgi body is involved in packaging and storage of these secretory products. Once the hormone is synthesized, it is packaged by the Golgi bodies, vesicles form, and the hormone is secreted, usually by the formation of an exocytotic vesicle. Sometimes the active form of the hormone is secreted, and in other cases, inactive forms, or precursors, are secreted.

Hormones are not always transported in their active form. Instead they are often bound to carrier

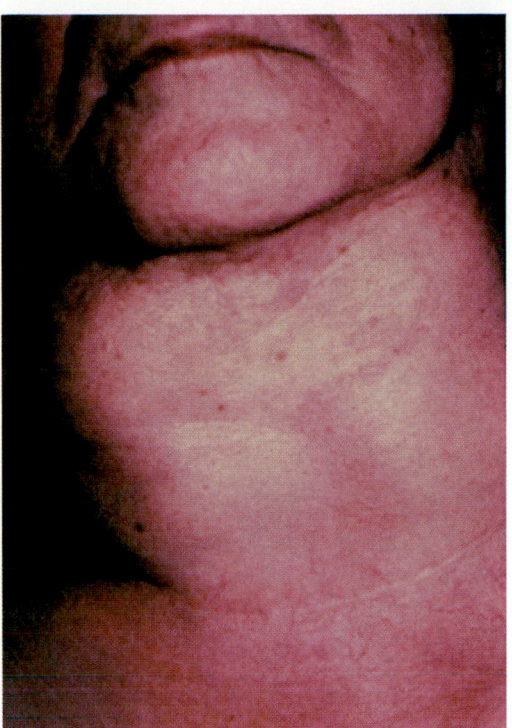

FIGURE 29.5 Simple Thyroid Goiter
An enlarged thryroid gland is the result of low levels of iodine in the diet.

molecules. In such cases, they have to be converted by enzymes to their active form, either within the blood plasma or interstitial tissue or within the target cell. For example, the male steroid hormone testosterone is converted to its active form, *dihydrotestosterone,* once it is inside the target cell and has formed its hormone-receptor complex.

The final step of hormone regulation is degradation. The cells of the liver are primarily responsible for the degradation of hormones. Once degraded, they pass into the blood and are carried to the kidney, where they are excreted. As a result, hormones do not accumulate in the body.

THE HUMAN ENDOCRINE SYSTEM: A MAMMALIAN EXAMPLE

All vertebrate endocrine systems, including that of humans, consist of endocrine tissues that secrete hormones. The major components of the human endocrine system are illustrated in Figure 29.6. Because of the breadth of the activities of various hormones, we cannot discuss all the glands and their secretions in this book. As a matter of fact, as

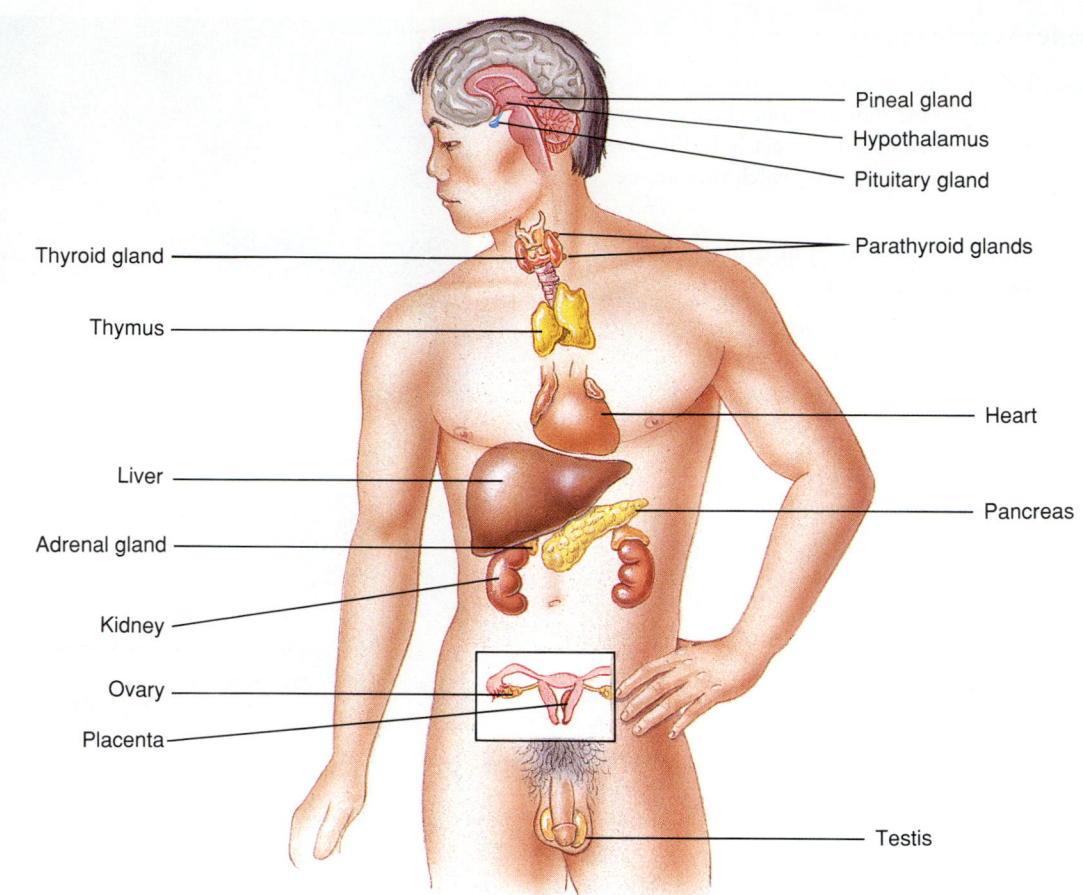

FIGURE 29.6   The Human Endocrine System
The pituitary, pineal, thyroid, parathyroid, and adrenal glands are clusters of secreting cells — true endocrine glands. The pancreas functions as both an endocrine (ductless) and exocrine (duct) gland. Although the testes and ovaries have glandular tissue in them, they are primarily reproductive organs. Other organs in the body — including the thymus gland, liver, heart, and kidneys — contain some hormone-secreting cells.

recently as 1970 only 20 human hormones had been identified. Today, however, most researchers feel that there are over 200 hormones. In addition, the definition of hormones has expanded. Not only are they defined as chemical substances produced by specific glands to perform specific functions, they are also defined as any chemical substance produced by one cell that is carried to another cell and alters or changes the activity of the cell that receives it.

In Table 29.1 we illustrate the role and the source of the principal human hormones. The remainder of this chapter provides an overview of the major components of the human endocrine system. Remember, although we concentrate on the hormones found in humans for the rest of this chapter, hormones are also present in invertebrates, plants, and even microorganisms (see Focus 29.1).

The Pituitary and the Hypothalamus: The Link Between Body and Mind

■ **A major link between the body and the mind is established by the pituitary gland and the region of the brain called the hypothalamus. The pituitary has three lobes: the anterior, the intermediate, and the posterior. The anterior lobe produces at least six hormones: those regulating growth and reproduction, and tropic hormones, which regulate other endocrine glands. The posterior lobe stores and secretes two neurosecretory hormones produced by the hypothalamus. The hypothalamus produces at least nine neurosecretory hormones. Hypothalamic-releasing hormones stimulate the anterior pituitary to release hormones. Hypothalamic-inhibiting hormones inhibit the release of hormones in the anterior pituitary.**

TABLE 29.1
Important Human Hormones

Gland or organ	Hormone	Effect
Anterior pituitary	Luteinizing hormone (LH), or interstitial cell–stimulating hormone (ICSH)	Regulates secretion of sex hormones by the gonads; in the female, stimulates growth of the corpus luteum; signals ovulation; increases secretion of progesterone
	Follicle-stimulating hormone (FSH)	Stimulates follicle growth in females and spermatogenesis in males
	Adrenocorticotropic hormone (ACTH)	Regulates hormone secretion by the adrenal cortex
	Thyroid-stimulating hormone (TSH), or thyrotropin	Regulates hormone secretion by the thyroid
	Prolactin	Stimulates mammary glands to produce milk
	Growth hormone	Stimulates growth
Hypothalamus	Gonadotropin-releasing hormone (GnRH)	Stimulates anterior pituitary to release LH and FSH
	ACTH-releasing hormone	Stimulates anterior pituitary to release ACTH
	Thyroid stimulating–releasing hormone (TSRH)	Stimulates anterior pituitary to release TSH
	Prolactin-inhibiting hormone (PIH)	Regulates secretion of prolactin by anterior pituitary
	Growth hormone–releasing hormone	Regulates secretion of growth hormone by anterior pituitary
	Growth hormone–inhibiting hormone	Inhibits release of growth hormone by anterior pituitary
Released by the posterior pituitary; synthesized in the hypothalamus	Oxytocin	Stimulates uterus to contract during childbirth; stimulates mammary glands to release milk
	Antidiuretic hormone (ADH)	Regulates water reabsorption by kidney tubules
Thyroid	Thyroxine	Controls the basal metabolic rate of cells
	Calcitonin	Inhibits the release of calcium from bone storage sites
Parathyroids	Parathyroid hormone (PTH)	Stimulates bone storage sites to release calcium
Pancreas	Insulin	Stimulates tissues to take in glucose
	Glucagon	Activates breakdown of glycogen to glucose
Gastrointestinal tract	Gastrin	Stimulates release of gastric juices
Adrenal cortex	Glucocorticoids	Regulate carbohydrate, lipid, and protein metabolism; suppress the immune response
	Mineralocorticoids	Regulate mineral levels in the body
	Sex hormones	Influence development of secondary sex characteristics
Adrenal medulla	Epinephrine	Increases muscle strength and sensory alertness
	Norepinephrine	Stimulates constriction of blood vessels to maintain blood pressure
Testes	Testosterone	Increases sperm production; stimulates development of male secondary sex characteristics
Ovaries	Estrogens	Initiate preparation of uterus for pregnancy; stimulate development of female secondary sex characteristics
	Progesterone	Retains and supports embryo during pregnancy
Thymus	Thymosins	Regulate development of lymphatic tissue and functioning of certain lymphocytes
Heart	Atrial natriuretic factor (ANF)	Reduces blood volume and increases water and salt excretion

FOCUS 29.1
Insect Hormones

The only organisms more interested in insect reproduction and development than the insects are humans. The reasons for our concern are fairly simple. For one thing, insects are our chief competitors for food — in large numbers they are capable of untold damage to our food crops. Some, such as mosquitoes, are carriers of serious diseases that affect whole populations. Others, like honeybees, are essential because they are the primary fertilizers of valuable plants. Still others provide products that humans use in making goods. It is only logical that in order to develop mechanisms to control the growth of insect populations, we have to understand how they reproduce and develop. Endocrinology has provided our understanding of the regulatory hormones responsible for the development of the fertilized egg into the various larval and pupal stages and the metamorphosis that eventually brings about the mature adult insect.

The silkworm moth, which provides the raw silk for silk cloth, is one of the best-understood examples of the role of hormones in insect metamorphosis (Figure A). The early embryo develops into a larva (or caterpillar) encased in an exoskeleton of chitin. The caterpillar eats and increases in size until it reaches the limits of its exoskeleton. Once the exoskeleton becomes too small

the larva sheds it in a process called molting. (You can think of the shedding of the exoskeleton as similar to outgrowing your favorite pair of jeans. The exoskeleton does not increase in size, so it allows only a certain degree of growth.) It then grows rapidly, develops another exoskeleton, and the process is repeated.

After the fourth molt, the larva forms a pupa (the encased developmental stage in some insects between the larva and adult). The developmental events that result in the shaping of the adult form occur within the pupal stage. Eventually the pupal exoskeleton deteriorates and is shed, and the adult silkworm moth emerges.

Three hormones are involved in regulating the symphony of events that result in metamorphosis and molting in the silkworm (Figure B). Specialized neurosecretory cells in the brain secrete brain hormones. These are tropic hormones that stimulate the prothoracic glands in the thorax to produce a mixture of two steroid hormones. Collectively, the two are called *ecdysone* (from the Latin meaning "to strip"). Ecdysone causes molting, as well as stimulating the acceleration of growth that the larva undergoes immediately after molting.

Located behind the moth's brain is another endocrine gland called the *corpus allatum*. Tropic brain hor-

mones stimulate the corpora allatum to produce a hormone called *juvenile hormone*. High levels of juvenile hormone cause the larva to remain in its larval form and to develop an exoskeleton after the first three molts.

After the fourth molt, however, no juvenile hormone is produced. Therefore, instead of a larval exoskeleton, a stiffer case forms, woven from silk. This structure, called the pupal case, is more environmentally resistant than the exoskeleton. When ecdysone is released again, the pupal case disintegrates and the adult moth emerges. As you can see, insect metamorphosis and development are coordinated by the interaction of these three hormones.

Some of what biologists have learned about insect hormones has enabled us to produce substances that act as natural insecticides by inhibiting maturation. In the 1950s a group of European entomologists brought some European insects to America for study. However, the European insects would not mature. Instead they remained in their larval state. Eventually, the investigators discovered that the paper used to line the insect cages contained a chemical substance very similar to insect juvenile hormone. When the insects ate this "paper factor," it maintained their juvenile state. Why had this happened in America but not in Europe? American paper is often manufactured from hemlock, balsam, and fir trees. Europeans used different trees, in which the chemical is absent.

Once biologists realized that plants can produce chemical substances that serve as natural insecticides, they could begin to explore new methods of controlling populations of insect pests, methods that would not rely on highly toxic insecticides that endangered other organisms, including ourselves. With the discovery of the pheromone secreted by the female gypsy moth to attract a mate, biologists acquired a strong weapon for protecting the coniferous forests of New England. They could set out traps that gave off the sex attractant. The male moths would fly to the trap, believing it was a receptive female. Instead the males were trapped, the females went unfertilized, and the population of the moths diminished. Perhaps some day, owing to continued work on the insect hormone system, we will not have to worry about insect pests.

FIGURE A
A newly hatched silkworm.

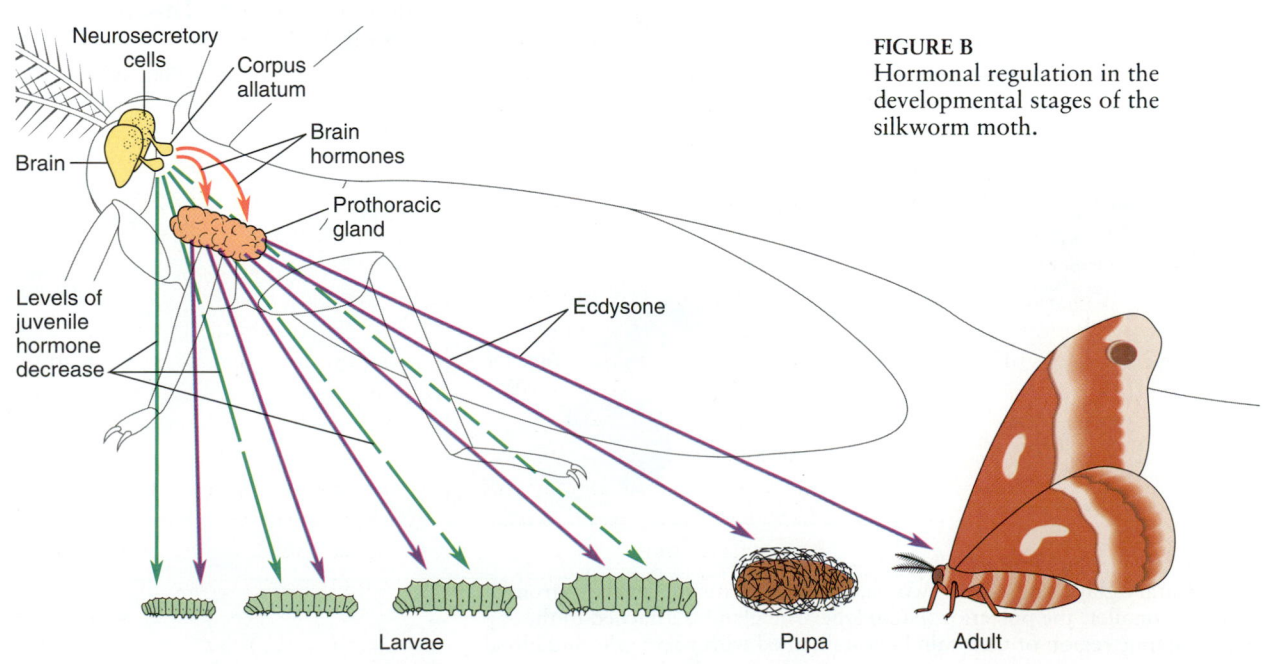

FIGURE B
Hormonal regulation in the developmental stages of the silkworm moth.

Regulation in humans is the result of both the nervous system and the endocrine system. Nerve receptors detect a stimulus, both external and internal, and carry the message to the brain. The region of the brain known as the hypothalamus and/or the pituitary gland detects hormone levels already present in the body. For example, when a high level of gonadal hormones feeds back to the pituitary/hypothalamus, the pituitary/hypothalamus signals the glands that produce the hormone to lower the production level or to stop production completely. If, on the other hand, the level of gonadal hormones is low, the pituitary/hypothalamus signals the glands that produce the hormones to increase production.

The process is, of course, more involved than this overview. A closer examination of the pituitary/hypothalamus complex will demonstrate the steps involved, and the way that chemical information passes from one part of the body to another.

The Pituitary. Since we have referred to the pituitary gland in earlier chapters, you should have some sense of its importance. The **pituitary,** which is about the size of a small pinto bean, is located at the base of the brain, just above the nasal passages. Because of its anatomical position, the Flemish anatomist Vesalius (1514–1564) believed that the pituitary's function was to produce mucus nasal secretions. Today, however, we know that the pituitary serves as the "conductor" of the endocrine orchestra, the master gland.

The pituitary is actually two glands in one, the *anterior lobe* and the *posterior lobe* (Figure 29.7). There is also an intermediate portion of the pituitary that plays an important role in many verte-

brates. In lower vertebrates, the intermediate pituitary produces hormones involved in changes in skin color, such as the changes you can see in the skin of frogs and salamanders. In humans the function of the intermediate lobe is unclear.

In humans (as well as other vertebrates) the functions of the anterior and posterior pituitary are quite different. The **anterior pituitary** (also called the *frontal lobe*) consists primarily of glandular tissue and is responsible for the production of at least six pituitary hormones. This portion of the pituitary is essential for life. It produces not only hormones that affect growth and reproduction but also a group of stimulatory hormones called *tropic hormones,* which regulate the secretory activities of the other endocrine glands.

The **posterior pituitary** (also called the *rear lobe*) is comprised largely of nervous tissue and serves to store and release two neurosecretory hormones that actually are produced by the hypothalamus. One of the hormones stored in the posterior pituitary is *oxytocin,* which stimulates the smooth muscles of the uterus to increase contractions during labor and childbirth. In most cases, a woman's body produces enough oxytocin to carry out the birthing process. However, when uterine contractions are weak or their rhythm is irregular, the attending physician may administer synthesized oxytocin to initiate or to strengthen uterine contractions to facilitate labor. Oxytocin is also involved in breast feeding. It causes the smooth muscles of the milk ducts of the breast to contract when the infant begins to nurse, and the milk is released. This process is sometimes called "milk letdown."

The other hormone stored and released by the

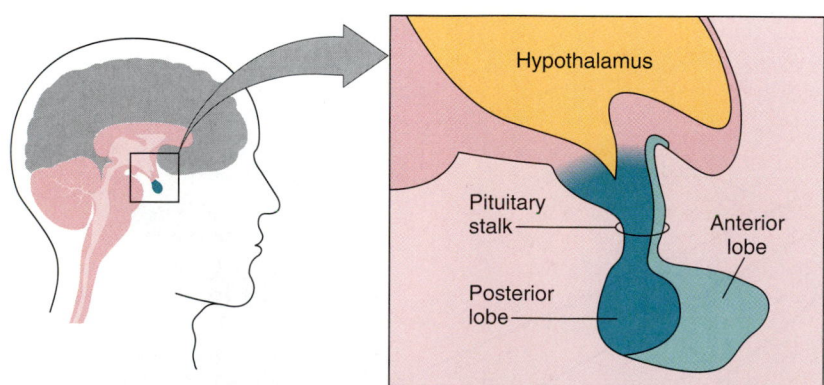

FIGURE 29.7   The Pituitary Gland
The pituitary gland consists of two parts: The larger is the anterior (front) lobe; the smaller, the posterior (back) lobe. The gland is attached to the hypothalamic region of the brain by a stalk filled with nerve cells and blood vessels.

Dr Elba Serrano
Cell Biologist, Biophysicist

When Dr. Elba Serrano is at a social gathering and asked what she does, she answers that she is a professor of biology at New Mexico State University with an interest in how plants and animals respond and adapt to their environment. To the person with little or no interest in biology, these seem like completely separate and unrelated topics, but at the cellular and molecular level, both plants and animals respond to their environment with changes in gene expression and membrane-mediated signal transduction events that require activity of transport proteins.

Dr. Serrano cannot remember a time when, as a child, she was not interested in math, physics, and chemistry, although, ironically, biology held absolutely no appeal as a subject area. It was "too descriptive. I enjoy quantitative science and the study of systems and processes, and biology was not presented as challenging in that way." Although she enjoyed physics very much, particularly quantum mechanics, she felt that the atmosphere for women and minorities was not encouraging. "When I looked around and saw the other students and professor, I could not see myself as a peer in any sense." She took a graduate course in biochemistry in her senior year in college, and this sparked an interest in biophysics that matured into a vocation in graduate school.

When asked how she became a biologist, her answer is somewhat different from that of other biologists, because she was a physicist first. In 1973, Elba Serrano graduated from the University of Rochester in physics with distinction. As her interests in science grew, she became more and more interested in biophysics, eventually earning her Ph.D. from Stanford

University with a specialization in neuroscience and membrane physics. For her doctoral research, Serrano used two microelectrode voltage clamps to study potassium channel diversity in identified molluscan neurons. Her postgraduate work was with the late Susumu Hagiwara at UCLA, where she developed an interest in sensory cell biology: the auditory and vestibular cells of the ear and the guard cells of plants. What was initially an interest in membrane biophysics gradually evolved into larger research questions. Today, Dr. Serrano's lab uses integrated approaches to study how plants from the tropics and arid regions sense and adapt to their environment, and also how sensory hair cells of the ear function, develop, and proliferate.

Dr. Serrano was born into a Puerto Rican family and grew up as a "Mili-Rican," as her father was in the U.S. Army. Because of her father's military career, Dr. Serrano lived in many different countries. She grew up in Asia, lived in Central America, and graduated from Nuremburg American High School. Because of all of this moving and the fact that her mother emphasized the importance of higher education, Elba and her two sisters completed college. All are first-generation college graduates. This emphasis on education in the Serrano family did not go unheeded by the parents. While in their 40s both of Dr. Serrano's parents returned to school and earned their bachelor's degrees while working days and going to school at night.

Dr. Serrano feels at home in her laboratory conducting reasearch, training students, and teaching. But she also has many other interests. She enjoys dancing and gymnastics,

and as often as not she may be seen in the white jacket of a practioner of martial arts as the white laboratory coat of a scientist.

Dr. Serrano is very happy with her research and especially happy with the opportunity to mentor young budding scientists. Recently, one of her students, Paula Mora, a Native American of the Navajo tribe, was accepted to Stanford Medical School. Paula says she especially enjoyed working in Dr. Serrano's laboratory beacause she treats students as equals and encourages them to develop critical thinking skills.

In sum, Dr. Serrano says, "I have felt discouraged about pursuing this career many times, as a graduate student, as a postdoc, and even as a faculty member. But it always comes back to the same thing. Although the work environment may sometimes be oppressive, the research never is! This is a constant exploration, where the boundaries are established by my curiosity, imagination, and initiative — but the best part is, I am never bored!"

posterior pituitary is the *antidiuretic hormone* known as ADH. ADH regulates water reabsorption by the kidney tubules. Alcohol inhibits the release of ADH — that is why you become diuretic and urinate often if you have drunk alcohol. Water reabsorption in turn regulates salt concentration in the cells and blood. Antidiuretic hormone in vertebrates other than humans is sometimes called *vasopressin* because it increases blood pressure. In humans, ADH has this effect in cases of severe blood loss due to hemorrhage.

The Hypothalamus. The pituitary is attached to the hypothalamus, which is a region at the base of the brain. The **hypothalamus** directly influences the activities of the pituitary and also mediates neural signals from other brain centers. So, while the pituitary gland may be the master gland, interacting with the other endocrine glands, its activities are directed by a critical balance of feedback controls mediated by the hypothalamic region of the brain. The hypothalamus signals the pituitary to release or not to release certain hormones (see Profile).

The fact that the pituitary is attached by a stalk to the hypothalamus at the base of the brain and that nerve tracts pass from the hypothalamus into the pituitary tells us that there is an anatomical link between the nervous system and the endocrine system. Forming a system for the interchange of information (similar to the off-ramps and on-ramps between major highways), a series of capillaries are physically attached to both the brain and the hypothalamus.

At present, researchers have isolated nine neurosecretory hormones secreted by the hypothalamus. These hormones are very small peptides, some no longer than three amino acids in length. Also called brain hormones, neurosecretory hormones stimulate or inhibit the release of the hormones by the anterior pituitary. Those that stimulate the re-

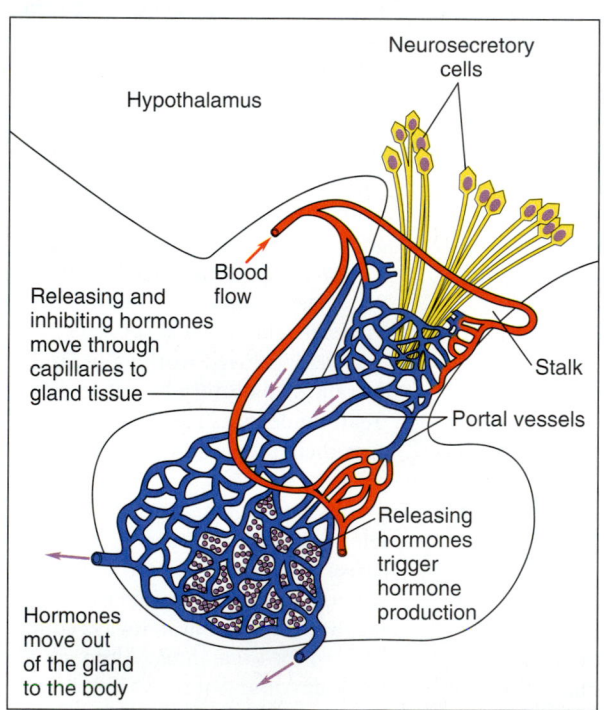

(a) Anterior pituitary

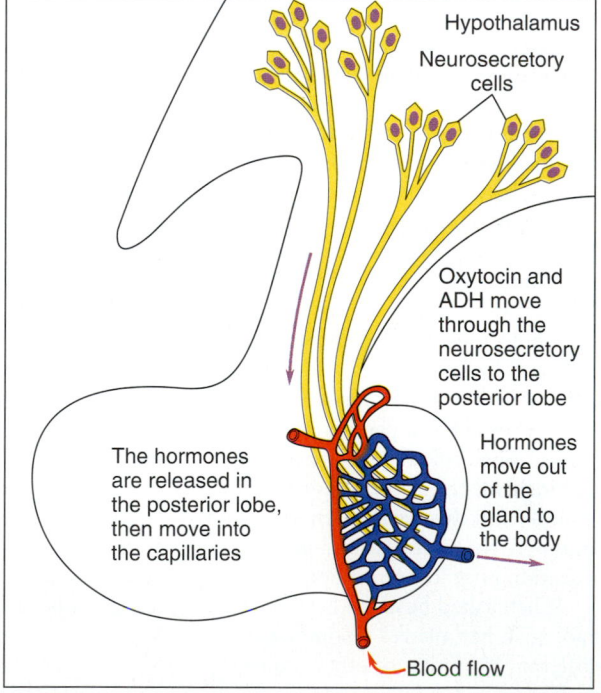

(b) Posterior pituitary

FIGURE 29.8 The Pituitary and Hypothalamus: The Physiological Links
The anterior and posterior lobes of the pituitary gland are very different, as is their relationship with the hypothalamus. (a) Hypothalamic-releasing and hypothalamic-inhibiting hormones reach the anterior pituitary through a network of blood capillaries and portal vessels. The releasing hormones trigger the secretion of pituitary hormones, which are transported through the bloodstream to their target cells. (b) The posterior lobe stores and releases oxytocin and ADH, two substances that are synthesized in the hypothalamus and transported to the pituitary by the axons of neurosecretory cells. When the hormones are released into a network of capillaries, they are transported out of the posterior lobe to their target cells.

Chapter 29   Cell-to-Cell Communication: Hormonal Signals

lease of pituitary hormones are called *hypotha-lamic-releasing hormones;* those that inhibit the release of pituitary hormones are called *hypotha-lamic-inhibiting hormones.*

Neurosecretory hormones produced by the hypothalamus discussed in the chapter on reproduction included gonadotropin-releasing hormone (GnRH), which regulates the release of the gonadotropins.

Other hypothalamic hormones inhibit the release of hormones produced by the anterior pituitary. Prolactin-inhibiting hormone (PIH), for instance, inhibits the release of *prolactin,* the hormone that stimulates milk production.

Today, researchers using recombinant DNA techniques can synthesize several brain (hypothalamic) hormones. In Chapter 13 we discussed the use of recombinant DNA techniques to synthesize the hormone somatostatin, which is a hypothalamic-inhibiting hormone that prevents the release of the growth hormone by the pituitary. Somatostatin can be used to regulate some forms of giantism.

The Pituitary and the Hypothalamus: Physiological Links. The activities of both the anterior pituitary and the posterior pituitary are controlled by the hypothalamus. The physiological connections between the lobes and the hypothalamus reflect the differences in the functions of the lobes.

The anterior pituitary is made up of glandular tissue. The secretion of hormones here is regulated by the releasing and inhibiting hormones produced in the neurosecretory cells of the hypothalamus. The neurosecretory hormones move into a network of capillaries surrounding the hypothalamus and stalk, then travel through blood vessels (called portal vessels) to the capillaries of the anterior pituitary (Figure 29.8a). Once in the anterior pituitary, the hormones leave the capillaries to act on the gland's secretory cells. Hypothalamic-releasing hormones signal those secretory cells to produce and release hormones, which move out of the gland through the bloodstream. Hypothalamic-inhibiting hormones stop the secretory cells from releasing hormones.

The physiological link between the posterior lobe and the hypothalamus is different from that between the anterior lobe and the hypothalamus. Remember, the posterior lobe does not actually produce hormones; it simply stores and releases oxytocin and ADH, two substances produced by the hypothalamus. Here, the axons of the hypothalamic neurosensory cells extend through the stalk into the posterior lobe, where they release oxytocin and

ADH (Figure 29.8b). When the hormones are needed, they move into the network of capillaries in the posterior lobe and are carried by the blood to their target cells.

The intimate functional relationship between the pituitary and hypothalamus clearly demonstrates that the nervous system and endocrine system are not two separate entities. Instead they are part of an integrated system of chemical communication that coordinates and maintains an organism's homeostatic condition. Figure 29.9 summarizes this feedback relationship among the hypothalamus, the pituitary, and other endocrine glands.

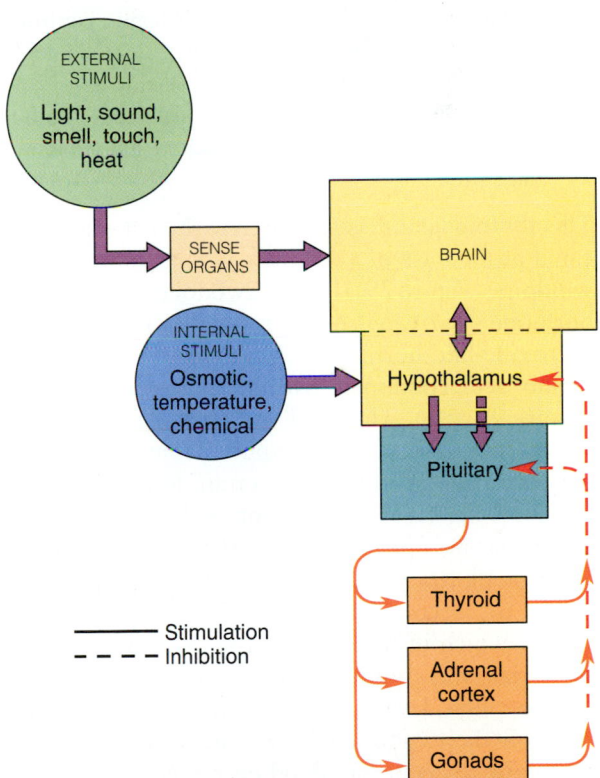

FIGURE 29.9 The Interaction of the Hypothalamus, the Pituitary, and Other Endocrine Glands
The body's neural and hormonal systems work together to maintain homeostasis. Receptors detect a stimulus (either external or internal) and transmit the message to the brain. The hypothalamus, through its hormones, regulates the secretions of the anterior pituitary, whose secretions, in turn, control those of other endocrine glands in the system. The hormones produced by these other glands enter the bloodstream and are detected by the hypothalamus and/or pituitary, which adjusts the production of the hormones accordingly. High levels of hormones in the bloodstream decrease the production of hypothalamic-releasing hormones and tropic hormones; low levels increase the production of both types of hormones.

Other Endocrine Glands and Elements

If the pituitary is the master gland, what are the other glands involved in the endocrine system? There are a number of important endocrine glands and tissues, including the thyroid and parathyroid glands, endocrine tissues in the pancreas and the gastrointestinal tract, the adrenal glands, the gonads (or sex glands), the endocrine tissues in the kidneys, the pineal gland, and the thymus gland. Also important are a series of hormones collectively known as the prostaglandins and the endorphins and even hormones produced by the heart.

The Thyroid and Parathyroid Glands

■ **The thyroid produces thyroxine. This hormone, which is monitored by the hypothalamus, controls the basal metabolism rate of cells. Calcitonin, another thyroid hormone, and parathyroid hormone, produced by the parathyroid glands, inhibit and stimulate the storage of calcium in bone storage sites.**

The **thyroid gland** is a butterfly-shaped gland located at the base of your neck just in front of your windpipe (Figure 29.10a). Its two lobes produce the hormone *thyroxine* (sometimes called T$_4$) and other thyroxine-like hormones. **Thyroxine** is a modified amino acid combined with four atoms of iodine (Figure 29.10b). The level of thyroxine in the system is monitored by the hypothalamus. When the concentration falls below a certain level, the hypothalamus increases the amount of thyroid stimulating–releasing hormone (TSRH) being sent to the anterior pituitary. The anterior pituitary is stimulated to secrete thyroid-stimulating hormone (TSH), which in turn signals the thyroid to secrete thyroxine and other hormones similar to it. Conversely, high levels of thyroxine negatively feed back on the hypothalamus/pituitary complex, and thyroxine levels in the blood decrease.

Thyroxine controls the basal metabolic rate of cells. An individual who produces too much thyroxine is said to have *hyperthyroidism* (the prefix *hyper* means "over"), resulting in high rates of metabolism. When thyroxine concentration is low, the individual is said to have *hypothyroidism* (the prefix *hypo* means "under"), and the metabolic rate is correspondingly low. People who have a hyperthyroid condition are usually nervous and hyperactive, tend to have insomnia, and usually are thin. Those with a hypothyroid condition have a generally slow metabolism, resulting in symptoms of obesity, have a bloated appearance due to fluid retention, and have slow responses, both physically and mentally.

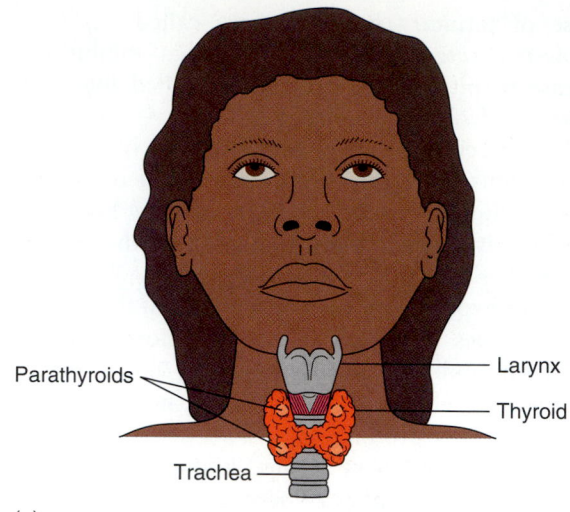

(a)

(b)

FIGURE 29.10   The Thyroid and Parathyroid Glands
(a) The thyroid gland is located at the base of the larynx and consists of two lobes connected by a band of tissue. Two small parathyroid glands are embedded in the back of each side of the thyroid. (b) The chemical structure of thyroxine.

Special cells within the thyroid produce a hormone called **calcitonin,** which inhibits the release of calcium from bone storage sites. Calcium is vital for many cellular reactions, especially muscle and nerve membrane functions.

The function of calcitonin is balanced by **parathyroid hormone** (PTH), which is produced by the **parathyroid glands** — four pea-sized glands embedded in the back side of the thyroid (see Figure 29.10a). Parathyroid hormone is a small polypeptide that functions to counteract the action of calcitonin. Proper levels of calcium in the blood, therefore, are due to the interactions of calcitonin and PTH.

Endocrine Tissues of the Pancreas and the Gastrointestinal Tract

■ **Cells within the pancreas and the gastrointestinal tract produce hormones. The islets of Langerhans, clumps of tissues in the pancreas, produce three major hormones. Of the three, two — insulin and glucagon — regulate glucose levels in the bloodstream. Hormones of the gastrointestinal tract function in digestion.**

In Chapter 23 we discussed the fact that the **pancreas** is an *exocrine gland* that secretes digestive enzymes through the common bile duct into the small intestine. Those enzymes play an important role in the breakdown of food. However, enzyme production is not the only function of the pancreas. This gland also includes elements called the **islets of Langerhans.** The islets of Langerhans are over 2 million isolated clusters or clumps of endocrine tissue cells. In a slide of pancreatic tissue they resemble small islands (Figure 29.11).

Separate cells in each clump of islets of Langerhans produces three hormones. Biologists know the function of two, *insulin* and *glucagon,* which are released in response to circulating plasma levels of glucose and amino acids. The two work together to maintain glucose levels in the system. The third, somatostatin, is also produced by the hypothalamus. Its role in digestion is not completely clear. However, many investigators feel that somatostatin regulates the release of both insulin and glucagon and helps to control gastrointestinal function. The receptors that monitor, stimulate, and inhibit glucose concentrations are located in the pancreas itself.

Glucagon is a polypeptide hormone of 29 amino acids. It is primarily responsible for the breakdown of glycogen (animal starch) into glucose and is released when the level of glucose concentration in the blood plasma is low. **Insulin** is a polypeptide hormone consisting of 51 amino acids. Its major function is to move glucose through the cell membrane, but it also aids in fatty acid metabolism,

glycogen synthesis, and proper capillary function. When blood glucose levels are high, the islets of Langerhans release insulin, and the rate at which cells take in glucose is increased.

How do these two substances, glucagon and insulin, interact? When the concentration level of glucose in the blood is low, the pancreas secretes glucagon. The hormone travels to the liver, where glucose is stored. Once in the liver, glucagon activates the breakdown of glycogen to glucose, raising the concentration level of glucose in the bloodstream (see Figure 23.15).

When the levels of blood glucose are high, on the other hand, the pancreas secretes insulin, which increases the amount of glucose taken in by various tissues. In the liver and the muscles, glucose is converted to glycogen. The disorder known as diabetes results when the feedback mechanisms involved in the insulin-glucagon process function incorrectly (see Focus 23.1 in Chapter 23).

Some of the cells in the glandular epithelium of the stomach and small intestine produce hormones that are involved in timing the release of digestive enzymes and the bile stored in the gallbladder. The major hormones produced by the gastrointestinal tract are *gastrin, cholecystokinin* (CCK), and *secretin,* all of which were discussed in Chapter 23. CCK seems to also act as an appetite suppressant, signaling the appetite centers in the brain that the stomach is full. Some investigators suggest that CCK may one day be used to treat genetically obese individuals whose condition may be due to a deficiency of CCK.

Adrenal Glands

■ **The two adrenal glands each consist of two portions, the adrenal cortex and the adrenal medulla. The adrenal cortex produces three steroid hormones: glucocorticoids, which regulate metabolism; mineralocorticoids, which regulate mineral levels; and sex hormones (in small amounts). The adrenal medulla produces epinephrine and norepinephrine, which are modified amino acids.**

In humans the **adrenal glands** (*ad* means "above" and *renal* means "kidney") are perched above each kidney. Each gland actually represents two glands in one: the *adrenal cortex,* which is the outer layer of tissue, and the *adrenal medulla,* which is the inner layer (Figure 29.12). Because the **adrenal cortex** is embryologically derived from the mesoderm, it produces steroid hormones. The adrenal medulla, on the other hand, develops embryologically from the ectoderm; it produces modified amino acid hormones.

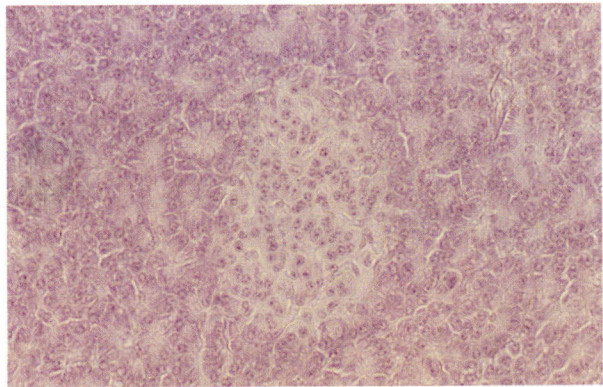

FIGURE 29.11   The Islets of Langerhans
The pancreas is both an endocrine gland and an exocrine gland. In this cross-section of pancreatic tissue, islets of Langerhans — the endocrine tissue — are the smaller, lighter cells in the center. The exocrine tissue, which secretes digestive enzymes, is the larger, darker tissue at the sides.

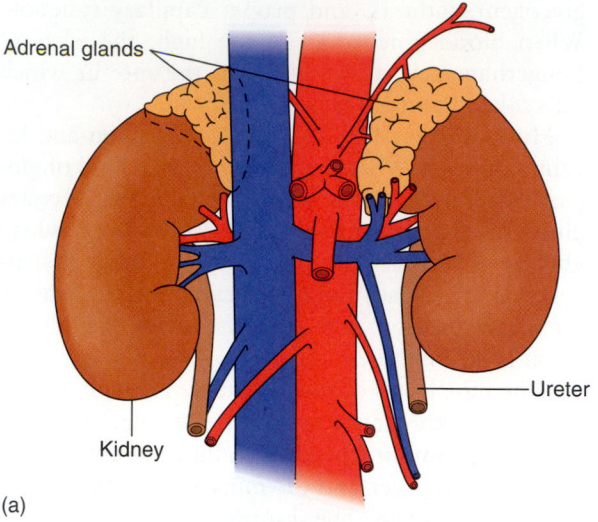

(a)

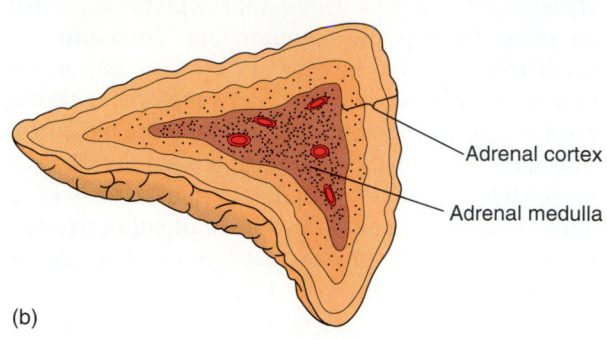

(b)

FIGURE 29.12   The Adrenal Glands
(a) From the front. (b) In cross-section.

The anterior pituitary regulates the secretion of adrenal cortex hormones through the release of a tropic hormone called *adrenocorticotropic hormone* (ACTH). Figure 29.13 illustrates the negative feedback mechanism regulating the secretion of ACTH and adrenal cortex hormones. The adrenal cortex hormones fall into three categories, according to their function: the glucocorticoids, the mineralocorticoids, and sex hormones. The glucocorticoids, one of which is *cortisol,* are involved in the regulation of carbohydrate, lipid, and protein metabolism and suppression of the immune response. Some of you may have come into contact with hydrocortisone at one time or another, either as an injection or as a cream. Hydrocortisone is one of the glucocorticoids, and it functions as an anti-inflammatory agent — that is, it suppresses the inflammatory response.

The second category of adrenal cortex hormones, called mineralocorticoids, influence the levels of minerals in the body. They directly affect the salt and water balance in the body. One of the best-known mineralocorticoids is *aldosterone,* which (as we mentioned in Chapter 26) promotes the reabsorption of sodium ions and the release of potassium ions by the kidney tubules. Some mineralocorticoids also work in concert with the glucocorticoids to inhibit the inflammatory response.

The third category of hormones produced by the adrenal cortex are the sex hormones: androgens and estrogens. Both male and female sex hormones are produced by males and females. The difference lies in the amount of each type that the two sexes produce. In general, males produce larger amounts

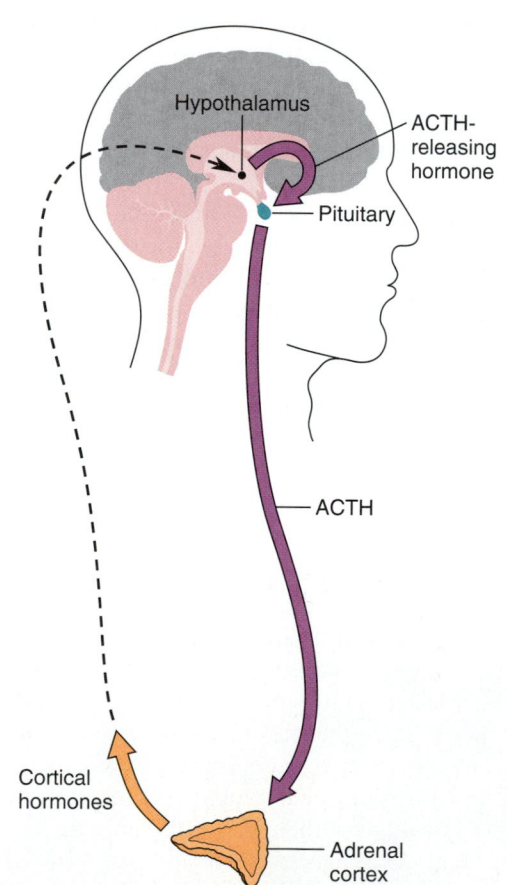

FIGURE 29.13   ACTH and the Adrenal Cortex
ACTH, a tropic hormone produced in the anterior pituitary, regulates the secretion of hormones in the adrenal cortex. The production of ACTH is controlled by a hormone secreted in the hypothalamus. Again, a negative feedback mechanism is at work, with the level of the adrenal cortex hormones in the blood increasing or decreasing the rate of their production.

of androgens than estrogens, whereas females produce more estrogens than androgens. In any case, the adrenal cortex produces a rather small amount of sex hormone — most of it is produced by the ovaries or the testes.

The **adrenal medulla** produces modified amino acid hormones known as **epinephrine** and **norepinephrine** (both substances are better known by their trade names adrenalin and noradrenalin). As we noted earlier, the adrenal medulla arises from the ectodermal layer of the embryo, and that fact provides a clue to its nature. In the last chapter, we noted that the ectodermal layer also develops into the nervous system. In a sense, the cells of the adrenal cortex are modified nerve cells. (In fact, the two hormones that it produces are also produced by some nerve cells as neurotransmitters.) That makes the adrenal medulla another structure (like the hypothalamus) that supports the close association between the nervous and endocrine systems.

Epinephrine and norepinephrine function in the same way, varying only slightly in structure. Both are known as the fight, fright, and flight hormones, since they adjust the body performance to these activities. When they are present in the system, they reduce blood flow to the skin, shunting it instead to the muscle and the brain. In this way they increase muscle strength and sensory alertness. (This function results in those stories of amazing feats of strength during emergency situations.) The hormones also increase the rate of the flow of blood, blood pressure, and carbohydrate metabolism.

What regulates the release of hormones of the adrenal medulla? Brain centers, including the hypothalamus, are directly connected to the adrenal medulla by sympathetic nerves (those over which, in most instances, we have no conscious control). In this way there is direct neural connection to an endocrine gland, and direct neural control regulating the gland's secretions. Since neural control is much quicker than control through the bloodstream, hormone release occurs much more rapidly, a fact that gives a tremendous survival advantage. Can you imagine what would happen to animals that had to wait for their circulatory system to carry a chemical messenger to their adrenal medulla before they responded? By the time they were ready to act, it would be too late.

Gonads: The Sex Glands

■ **Hormones produced by the gonads influence the formation of secondary sex characteristics and gametes, and maintain pregnancy in the female.**

In Chapter 27 we discussed the ways in which hormones regulate male and female reproductive cycles. As you remember, the gonads — the testes (male) and ovaries (female) — not only produce the sperm and egg but also function as endocrine glands. The ovaries are the major source of the female hormones categorically known as estrogens and progesterone, which are involved in the maintenance of the secondary sex characteristics such as breast development, preparation of the uterus for the implantation of the embryo, and retention and support of the embryo during pregnancy.

The testes produce the male hormones known as androgens, the major one being testosterone. Not only is testosterone a necessary hormone in proper sperm formation, it is also responsible for the maintenance of the male's secondary sex characteristics, like body hair and muscle development (see Focus 29.2).

In Chapter 27, you recall, we discussed the structure and function of the gonads in some detail, noting that they developed from the mesoderm of the embryo. Because of this, the hormones that they secrete are steroids. In addition to affecting the development of gametes and secondary sex characteristics, sex hormones also affect behavior such as sexual awakening. At the present time, in the midst of a fair amount of controversy, researchers are experimenting with the use of Depo-Provera (a synthetic estrogen and cytroperone acetate that blocks testosterone receptor sites on target cells) to see if it can serve as a chemical method for modifying the behavior of sex offenders. (Psychiatric care alone has not proved to be as effective as hoped.)

Some women suffering from premenstrual syndrome (PMS) have successfully argued in courts of law that their violent behavior was the result of the balance and interaction of sex hormones that occur just before menstruation. Some evidence suggests that PMS patients showed lower levels of thyroxine. When some patients were given thyroxine, some reported complete relief, although researchers are not sure why this response takes place. PMS sufferers also may have unusually high levels of endorphins (morphinelike substances secreted by the brain that relieve pain) during most of their monthly cycle. When these levels fall at the time of menstruation, the patients react as if they were going through drug withdrawal.

Are all of us prisoners of our hormones when it comes to our behavior? I think not, but our learning experiences and hormones are surely a very important part of human behavior. Behavior does have a biological basis.

Steroids in Sports: Fool's Gold?

We have all heard of athletes who were barred from competition or stripped of their medals because of the illegal use of steroids. In a much-publicized case, the Canadian sprinter Ben Johnson was stripped of his Olympic gold medal in 1988 for using steroids (Figure A). Not only do well-known athletes abuse steroids; it is estimated that 2 to 3 million Americans do this (this figure includes as many as 6.6 percent of high school–age boys). Why do people use steroids? Are they right when they say that steroids give the competitor a keen edge and improve athletic performance?

When athletes say they take steroids, they are not referring to steroid hormones like estrogens or the mineralocorticoids. Actually they are talking about testosterone, the male hormone responsible (in part) for the differences in muscle development between males and females. Since testosterone produces greater muscle mass in males than in females, athletes reason that taking the steroid will give them greater muscle mass. A study published in *Science* in 1969 provided even more support for steroid use. The study compared two groups of athletes for six weeks. One of the groups (the

control group) participated only in a weight training program. The other group also participated in the training, but they were given steroids as well. At the end of the study, the researchers found that the group that received steroids demonstrated increased muscle strength compared with the control group.

Why does testosterone make such a difference? There is no question that testosterone can increase muscle size. But how? Some investigators feel that the increase in the size of muscles is due to the fact that the muscle cells retain water. If they are right, any report of increased

---

Kidney, Pineal Gland, Thymus Gland, and Heart

■ **In addition to its main function, the kidney secretes hormones involved in regulating sodium ion levels in the blood and red blood cell production. The pineal gland is involved in timing reproduction in lower animals; its function in humans is not known. The thymus produces hormones that biologists hypothesize regulate development of lymphatic tissue and lymphocyte functioning. The upper chambers of the heart produce hormones that relax the smooth muscles of blood vessels and reduce blood volume.**

In the past, biologists did not think of the kidneys, pineal gland, thymus gland, and heart as endocrine glands, but they do have endocrine functions. For example, as we learned in Chapter 26, the arterioles

of the kidneys produce renin, an enzyme that acts on a plasma protein produced by the liver to form a hormone called *angiotensin*. Not only is angiotensin a strong vasoconstrictor, it also acts on the adrenal cortex, signaling the release of aldosterone, which regulates sodium ion levels in the blood.

The kidneys produce a glycoprotein substance that aids in the conversion of a plasma protein to the hormone *erythropoietin*. This hormone stimulates the bone marrow cells to increase the production of erythrocytes (red blood cells).

The pineal gland, a tiny structure near the brain, plays a very important role in timing the reproduction of lower vertebrates such as amphibians and reptiles, but its function in humans is not known. Biologists do know that the pineal gland produces hormones in response to the photoperiod (the ratio

strength is questionable. That would mean, as many investigators suspect, that the reported increase in strength is psychological. In other words, because athletes think that the steroids are making them stronger, subconsciously they work out harder and perform harder. This is an example of the placebo effect.

There are many reasons why steroid use is banned from amateur sports, some of which are fairly philosophical. But here we need to look at the physiological reasons why biologists frown on the use of testosterone by athletes, or by anyone else for that matter. Remember how a steroid like testosterone works. It passes through the cell membrane and combines with a receptor complex. The complex moves into the nucleus where it binds with DNA, signaling RNA synthesis and ultimately causing the synthesis of new proteins. Any chemical that affects the basic chemical productions of a cell is going to have a number of effects, not just one. That is the case with testosterone.

When a male athlete takes in high levels of testosterone, he can suffer from numerous side effects, which

FIGURE A
For illegally using steroids, Ben Johnson was stripped of his Olympic gold medal.

may include increased aggression (the so-called "roid rage") and decreased sperm production. Because there are high levels of testosterone in the blood, natural functioning of the testes is reduced and the testes shrink. Since the liver changes excess testosterone into estradiol, which is

a female hormone, breasts may develop. Excess testosterone in the system can result in liver damage because it is the function of the liver to rid the body of the excess steroids. Many of these side effects are irreversible.

Female athletes can also suffer from side effects if they use steroids. Aggressiveness can increase. They can grow a beard, go bald, develop severe acne, and become temporarily infertile. Female athletes who take testosterone are usually easy to spot. They take on a body shape that looks more masculine than feminine. They also may develop a five o'clock shadow and leathery skin, shortened life span, adrenal problems, and other effects.

At present no conclusive evidence supports the idea that steroids give an athlete a competitive edge. However, as long as athletes hear other athletes rave about the "benefits" of steroid use, they will continue to use them in increasing amounts. No physician or endocrinologist who says otherwise will be able to convince them. Steroids will continue to be "fool's gold."

of daylight hours to those of darkness). The photoperiod changes as the seasons change, and those changes determine the release of hormones associated with reproduction and the amount of hormones in the bloodstream. Although the actual hormone mechanism is not known, some investigators feel that the pineal secretion plays a role in the timing of reproduction.

The **thymus** is a two-lobed structure located behind the breastbone. In the newborn, the thymus is much larger (in relation to the other organs) than it is in an adult. As we discussed in the chapter on the immune system (Chapter 25), the thymus produces T-cells, the lymphocytes that aid in the fight against infection and/or the detection and elimination of damaged cells. The thymus also secretes a group of hormones called *thymosins*. Although the

research being done is not conclusive so far, investigators believe at this time that thymosins affect the development of lymphatic tissue and the functioning of certain lymphocytes.

Even the heart may serve as an endocrine gland. The cells of the atria — the upper chambers of the heart — secrete peptides called (as you might suspect) *atrial peptides*. (More specifically, they are known as *atrial natriuretic factor* or *ANF*.) These molecules act to reduce blood volume, increase salt excretion, and relax the smooth muscles that line the blood vessels. The last effect dilates the vessels, thus contributing to a marked reduction in blood pressure. Already scientists have produced antihigh-blood-pressure drugs, such as auriculin, that are based on atrial peptides. Preliminary studies have shown that auriculum may be effective in treating some forms of heart failure and high blood pressure.

Prostaglandins

■ **Prostaglandins are produced by many different tissues throughout the body. Biologists hypothesize that prostaglandins mediate between membrane receptors and cAMP.**

In 1981, the Nobel Prize was awarded to a group of investigators for their work with regulatory chemicals known as **prostaglandins.** As you might suspect, prostaglandins were first discovered in the prostate fluid, from which they got their name. Today, however, we know that prostaglandins are produced by almost all types of tissues.

Prostaglandins function much like the protein-like hormones, serving as mediators between membrane receptors and cAMP. It is suggested that prostaglandin production is an intermediate step between the external binding of the hormone to the receptor cell membrane and cAMP formation within the cell. In other words, prostaglandins may almost act as an additional messenger, ensuring that the protein hormone's message can produce the appropriate change in the target cell.

Some endocrinologists feel that prostaglandins are not true hormones because their effects are more local and they are not synthesized at the specific site. Instead, these specialized lipid-regulating molecules are produced in all parts of the body and have a wide variety of physiological effects. Some prostaglandins affect areas only in the specific region where they were produced, others affect the cell that produced them, and still others operate at sites distant from their source.

There are many different types of prostaglandins. Some cause contractions of the uterus, as well as contractions of other smooth muscle. (They are probably one of the major causes of menstrual cramps.) As a result, prostaglandins are being studied for use as contraceptives, since some investigators hypothesize that by causing the uterus to contract they would prevent implantation of the embryo.

Some prostaglandins are involved in regulating hormone secretions, and others are involved in the body's reaction to inflammation or tissue injury. Some prostaglandins, such as *thromboxane,* cause platelets to stick together to form clots. Others, such as *prostacyclin,* prevent blood clotting and arterial closure.

Researchers have found that aspirin suppresses the formation of certain prostaglandins, which in turn reduces the pain associated with inflammation.

Prostacyclin secretion is very sensitive to blood levels of aspirin and is produced in response to low levels of the painkiller. As a result, people who take aspirin have high levels of prostacyclin, and so they may experience excessive bleeding and the associated failure of the arteries to constrict. Since it has been found that people who use a lot of aspirin have a lower rate of arterial disease and heart attack, large, long-term studies by 2000 physician volunteers to see if "one aspirin a day will keep the heart attack away" have been completed. It appears that aspirin can reduce the incidence of heart attack. (Of course, you realize that you should not take any medication without a doctor's advice.)

Endorphins

■ **Endorphins are morphinelike substances produced throughout the body in response to pain. In some cases, they operate like hormones; in others, they function like neurotransmitters.**

Endorphins are often referred to as the body's "natural opiates." Some individuals have proposed that they are the major source of what is called "runner's high," that state at which long-distance runners no longer feel the pain and stress of prolonged activity and sometimes even feel as if they are undergoing a mystical experience. Researchers who believe endorphins are responsible also report that when individuals cease their long-distance running, their bodies undergo a series of changes similar to withdrawal from an addictive drug.

What are these magic molecules? Endorphins are morphinelike substances that are produced by the human brain in response to pain. In fact, their name means *the morphine within the body.* Chemically, they seem to function as both hormones and neurotransmitters, depending on the distance they travel.

Most of what has been learned about endorphins has come from working backward, so to speak. Researchers wondered why brain cells would have receptor sites suitable for the chemicals produced by poppies and other addictive plants if the body did not produce similar chemicals. (As you may know, the poppy flower serves as the source for many of the opiates.) The answer was found when researchers discovered endorphins, which bind to the same receptor sites.

All endorphins are small peptides, and all share a common five-amino-acid sequence. Further, the endorphins include six related peptides, four specifically called endorphins and two others that are called *enkephalins.* It is difficult to classify endorphins because they function as both neurotransmitters and/or hormones, depending on how the terms are defined. They may well be intermediates between the two.

At first biologists thought that endorphins could be used as natural painkillers that were only mildly addictive, since they are natural body substances. Unfortunately, endorphins are not as effective as the other natural opiates, and they are even more addictive.

Endocrinology is rapidly merging with the neural sciences. For example, in the mid-1970s, researchers determined that the brain itself secreted endorphins, its own hormones. Today, as stated by Dr. Sidney Ingbar of Harvard Medical School, "Our idea of what a hormone is is changing . . . the lines between hormone and other body chemicals are blurring."

As we mentioned in the beginning of this chapter, separating the nervous system and the endocrine system is simply for our convenience. The endorphins again demonstrate that this is true, since they illustrate the close intimacy between hormonal secretion and neurosecretion. These substances function both as hormones and, in some cases, as neurotransmitters, depending on the distance they travel. The nervous system and endocrine system in all probability had a common evolutionary origin, that of chemical regulation and control.

SUMMARY

1. Hormones regulate cellular activities from conception to death, not only in embryological development but also in cell-to-cell communication in the adult.
2. Homeostasis is maintained within an organism through chemical communication. This chemical coordination can be hormonal or neural. Because both employ similar means, biologists speculate that the nervous and endocrine systems had the same evolutionary origin.
3. Endocrine cells or glands produce hormones, which are regulatory chemicals. Neural impulses, which travel across the space between one nerve cell and the next, are also chemical in nature.
4. Hormones are released directly into the bloodstream or interstitial fluid.
5. Glands that are mesodermal in origin, like the gonads, produce hormones that are steroids. Glands or cells that develop from the ectoderm or endoderm, such as the pituitary, produce hormones that are modified amino acids or proteins.
6. Steroid hormones pass through the cellular membrane (since steroids are lipid-soluble). Once inside the cell they are received by cytoplasmic receptor molecules. This hormone-receptor complex is then carried to the nucleus, where the hormone initiates the DNA transcription of specific genes. The result is a series of chemical reactions that make up the hormonal response.
7. The mechanism of action for proteinlike hormones is called the second messenger mechanism. Protein hormones bind with a receptor molecule on the surface of the cell membrane (proteins are not lipid-soluble). This process activates a second messenger, called cAMP. cAMP then initiates a series of chemical reactions, or the hormonal response.
8. Hormone levels are regulated by the synthesis, secretion, transport, and activation of the hormone, and by the hormone's rate of degradation by metabolic processes.
9. The pituitary consists of an anterior and posterior portion. The anterior portion produces at least six hormones: some that affect growth and reproduction and some (called tropic hormones) that regulate hormone secretion by other endocrine glands. The posterior pituitary stores and releases two neurosecretory hormones produced by the hypothalamus.
10. Within complex vertebrates such as humans, the link between the pituitary and the hypothalamus is critical. The hypothalamus secretes brain hormones, which signal that a certain hormone should be or should not be secreted by the pituitary.
11. The thyroid gland produces two major hormones, thyroxine and calcitonin. Thyroxine regulates the rate of cellular metabolism, and calcitonin inhibits the release of calcium from bone.
12. Embedded within the thyroid gland are the parathyroid glands. These glands produce parathyroid hormone, which stimulates the release of calcium from bone storage sites. Calcium release and storage, therefore, are regulated by the calcitonin produced by the thyroid and parathyroid hormone, produced by the parathyroids.
13. The islets of Langerhans of the pancreas contain cells that produce insulin and glucagon, which help to maintain the levels of glucose concentration in the blood. Some of the cells of the small intestine produce hormones like cholecystokinin and secretin, which are involved in the timing and release of digestive enzymes.
14. The adrenal glands are actually two glands in one organ. The adrenal cortex (the outer layer) produces three major categories of steroid hormones: glucocorticoids, mineralocorticoids, and sex hormones. The adrenal medulla (inner portion) produces epinephrine and norepinephrine.
15. The gonads produce sex hormones involved in secondary sex characteristics and reproduction.
16. The kidney, pineal gland, thymus gland, and heart can also be considered endocrine glands, because they release hormones directly into the blood. Kidney hormones affect water reabsorption and sodium ion concentration levels. Pineal hormones affect the timing of reproductive cycles in lower vertebrate an-

imals. The thymus produces hormones called thymosins, whose functions are not yet known. The heart produces atrial peptides (also known as atrial natriuretic factor or ANF) that reduce blood volume and relax the smooth muscles that make up blood vessels.

17. Prostaglandins, a group of regulatory chemicals first identified in prostate fluid, are produced by many different types of cells. There are many kinds of prostaglandins, all of which have different regulatory effects.

18. Brain cells also produce the endorphins, a group of regulatory chemicals released in response to pain.

KEY TERMS

endocrine system	parathyroid hormone
target cell	parathyroid gland
gland	pancreas
hormone	islet of Langerhans
endocrine gland	glucagon
exocrine gland	insulin
cyclic adenosine	adrenal gland
monophosphate	adrenal cortex
pituitary	adrenal medulla
anterior pituitary	epinephrine
posterior pituitary	norepinephrine
hypothalamus	thymus
thyroid gland	prostaglandin
thyroxine	endorphin
calcitonin	

REVIEW QUESTIONS

True/False

1. Cell-to-cell communication, whether neutral or hormonal, is fundamentally chemical.
   true   false   *(page 709)*

2. Hormones are chemical regulating substances produced by specialized cells; the definition is rapidly changing.
   true   false   *(page 712)*

3. Proteinlike hormones can penetrate the plasma membrane.
   true   false   *(page 713)*

4. The hypothalamus produces hormones that regulate the secretion of hormones by the pituitary.
   true   false   *(page 720)*

Fill in the Blank

1. Endocrine cells that originate from _____ embryonic tissues secrete steroid hormones. *(page 713)*

2. The mechanism of action for proteinlike hormones involves a second messenger known as _____. *(page 713)*

3. During early embryonic development the human fetus is sexually bipotent; however, during the seventh week the gonads begin to differentiate. In the fetus, _____ regulate the development of male genitalia. *(page 708)*

4. The _____ are often referred to as the body's natural opiates. *(page 730)*

Multiple Choice

1. Which of the following represents the mechanism of cellular coordination found in all organisms?
   a. Muscles
   b. Nerve impulses
   c. Chemical agents
   d. Sense organs
   (page 707)

2. Pheromones are:
   a. Molecules involved in chemical communication between members of a species
   b. Plant hormones
   c. A group of sponges
   d. Present in all forms of life
   (page 709)

3. Most hormones are transported:
   a. Along nerves
   b. In the bloodstream
   c. Within muscles
   d. Along bones
   (page 711)

4. The major difference between a nerve cell and a neurosecretory cell is:
   a. The neurosecretory cell produces hormones.
   b. The nerve cell produces hormones.
   c. The neurosecretory cell produces neurotransmitter substances only at the synapse.
   d. There is no difference.
   (page 711)

Discussion

1. Speculate on the origin of the endocrine system.

2. Why can it be said that all cell-to-cell communication is chemical?

3. List and discuss the properties of seven hormones and their effects.

4. Based on the chemical categories of hormones, why was it hypothesized that there should be at least two mechanisms of hormone action?

5. Discuss the mechanisms of action at the cellular level for proteinlike hormones and steroid hormones.

6. Discuss the hypothalamic pituitary complex.

7. What are brain hormones? What are their functions?

8. Why is goiter prevalent among people who have deficiencies of iodine in their diet?

9. What is the evolutionary advantage of having the adrenal medulla directly connected by nerves to the brain centers?

10. Speculate on the hormonal mechanisms that result in "runner's high."

30

Cell-to-Cell Communication: Neural Signals

The Evolution of the Nervous System

The Neuron — An Electrochemical Messenger Cell

Message Transmission, Propagation, and Conduction

Nerve Cell Communication

The Result of the Transmission of Neural Messages

The Human Nervous System

Sense Organs

Drugs and Their Effects

Summary

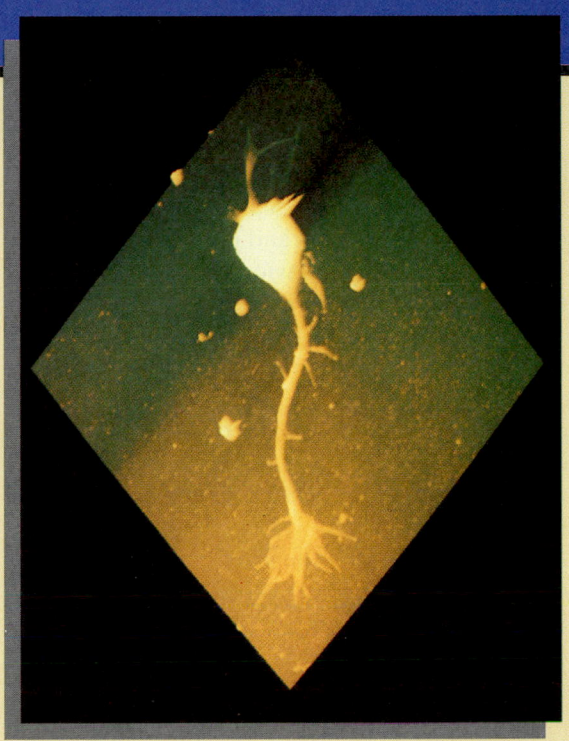

Human motion — such things as the beauty of dance or the first steps of an infant — is a marvel to behold. Those motions are the result of chemical bursts that result in cell-to-cell communication coordinated and synchronized by billions of nerves and muscles. In the same way, when we have an idea, that act of thought is the result of a trip through the brain involving an electrochemical maze stored in billions of neural fibers.

When you fully envision the processes involved within the brain, thought is an awesome phenomenon, for in a poetic sense the brain is the center of the soul and of being. Indeed, all thought and all neural coordination are due to the specialized electrochemical secretory nerve cells organized into what is known as the nervous system. Regardless of

the degree of an organism's complexity, all nervous systems have four basic functions:

1. Stimulus detection, during which information is collected

2. Stimulus conduction, during which messages are transmitted

3. Interpretation and analysis of the information, during which the message is processed and integrated

4. Response, which involves acting on the information received and coordinating the activities of the effectors, such as muscles and glands

In this chapter, we investigate the marvelous chemistry involved in message detection, conduction, and interpretation as carried out by the human nervous system, including the brain and sense organs.

Above: Using an atomic force microscope, scientists are able to image a neuron from the brain.

THE EVOLUTION OF
THE NERVOUS SYSTEM

■ One-celled organisms have no nerve cells. Early multicellular organisms exhibit a primitive nervous system called a nerve net.

In Chapter 29 we discussed the fact that the endocrine system and its regulatory chemicals (hormones) control, coordinate, and integrate many aspects of the metabolism and physiology as well as the development and behavior of an organism. But the communication occurring in the endocrine system involves a lag time of seconds or even hours before a hormone affects its target cells, and hormones may circulate for days before they are degraded. That means that the endocrine system provides slow, long-term cell-to-cell communication that leads to integrated behavior and homeostasis.

Coordination via the endocrine system does not operate alone. It is complemented in animal organisms by the nervous system, which is primarily responsible for the rapid transmission of information. The nervous system has evolved over time, and it has become more and more complex as organisms themselves have become increasingly complex.

One-Celled Organisms

The *neuron* is an individual nerve cell, and since the protistans are made up of only single-celled individuals living either in isolation or in colonies, they cannot contain nerve cells as such. However, some protozoans do contain certain receptor organelles such as sensory bristles, stigmata (light receptors), contractile fibers, and specialized effectors such as cilia and flagella (Figure 30.1a).

The porifera (the sponges) appear to be multicellular animals, but actually they are not. Instead, each is a colony of many unicellular organisms with a limited amount of cellular specialization. Responses to stimuli are very limited, centering around the individual cell. Therefore, each cell functions almost independently, and an elaborate communication system has not developed.

Rise of a Nerve Net

The cnidarians — hydras and jellyfish — are multicellular organisms, and it is in them that we see the simplest form of a nervous system — a **nerve net.** In the hydra (Figure 30.1b), this network of primitive nerve cells runs throughout the body, connecting receptors in the epidermis to contractile fibers

at the base of the epitheliomuscular cells (see Figure 22.13c). These fibers are effectors; when stimulated by the nerve cells, they contract, shortening the body and tentacles.

In the jellyfish, we find two specialized receptor cells: *ocelli* (singular, *ocellus*) and *statocysts*. Ocelli are eyelike structures that are sensitive to light; statocysts help the organism keep its balance (Figure 30.1c).

The jellyfish, like the hydra, exhibits a nerve net. In part, this is a function of its body shape. The *radial symmetry* of the organism means that there is no head in which major sense organs can cluster or a brain can form. Its nervous system, then, is not centralized.

Moving to a Central Nervous System

■ Bilateral symmetry led to the development of complex nervous systems because it allowed for cephalization and a complex nervous system that had paired parts in the right and left side of the organism.

From an evolutionary perspective, an elaborate nervous system has not developed in radially symmetrical animals because there is no **cephalization** (the concentration of sensory structures in a head end). Organisms exhibiting *bilateral symmetry* have a head end in which there is a concentration of sensory structures. Therefore bilaterally symmetrical animals can have paired nerve cells, muscles, sensory structures (eyes, ears), and regions of the brain. This body shape allows an increase in the nervous system's complexity. Bilateral symmetry permits a more efficient nervous system. This means the organism can respond quickly to stimuli in order to capture prey or to escape from predators.

The move from a nerve net to a nervous system involved a series of evolutionary trends that can be summarized as follows:

1. The development of a nerve cord — movement from a random nerve net to a centralized bundle of nerve cells

2. Specialization of the nerve cord for message transmission

3. Development of and increase in the number of connecting neurons, called association neurons or interneurons

4. The evolution of the brain, resulting from the increase in the concentration of interneurons and the coordinating function in the anterior end (the end that goes first) of the organism

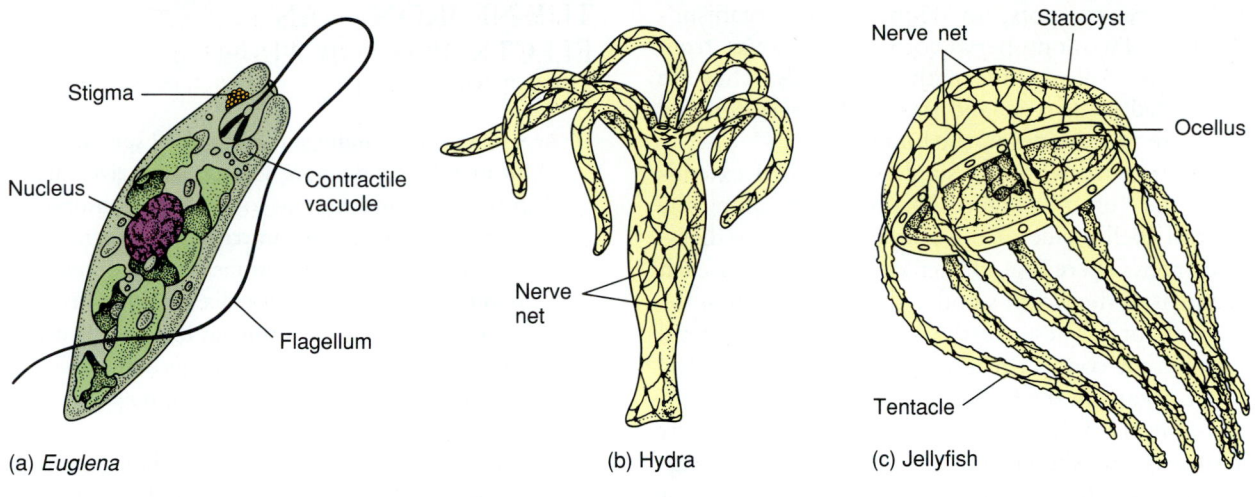

(a) *Euglena*

Stigma

Nucleus

Contractile vacuole

Flagellum

(b) Hydra

Nerve net

Nerve net

(c) Jellyfish

Nerve net

Statocyst

Ocellus

Tentacle

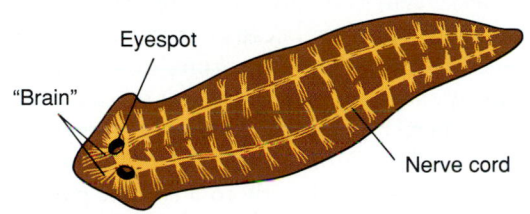

Eyespot

"Brain"

Nerve cord

(d) Flatworm

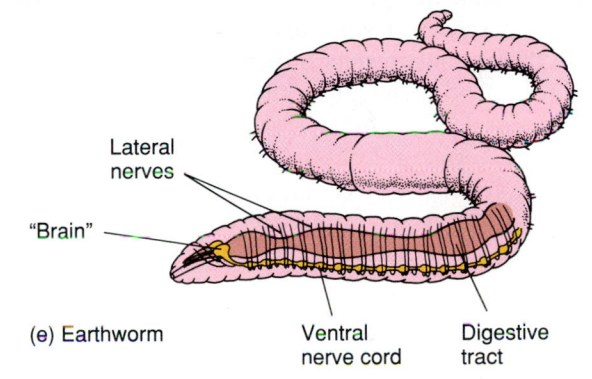

Lateral nerves

"Brain"

(e) Earthworm

Ventral nerve cord

Digestive tract

FIGURE 30.1 Sensory Receptors in a Single-Celled Organism and in Invertebrate Nervous Systems
(a) The *Euglena* is a flagellated protozoan. The red stigma functions as a light receptor. Its long flagellum functions as an effector, and it responds to the stimulus. (b) In the hydra, primitive nerve cells form a network—called a nerve net—throughout the body. (c) The jellyfish has both a nerve net and specialized sensory structures that respond to light and gravity. (d) In the flatworm, the nervous system is both centralized and linear. Two nerve cords extend the length of the body, from the organism's "brain." In both (e) the earthworm and (f) the crayfish, a single nerve cord runs along the ventral length of the body, below the digestive tract. Clusters of nerve cells along the cord give off small nerves that connect to all parts of the body.

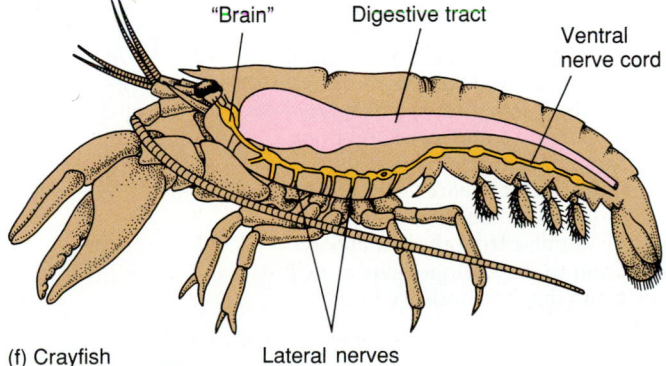

"Brain"

Digestive tract

Ventral nerve cord

(f) Crayfish

Lateral nerves

5. The placement of the centralized nervous system deeper into the body, for more protection

6. The evolution of sensory, association, and motor neuromessage pathways

7. The evolution of specialized sensory receptors

In invertebrates, the first centralized nervous system is found in the free-living flatworm (Figure 30.1d). Two clusters of nerve cells in the head, below the eyespots, function as the organism's "brain." Two long nerve cords extend away from the "brain," along the length of the body. They are connected by nerves, and there are nerves running off them to the sides of the body.

In the earthworm, a single nerve cord extends from the "brain" (Figure 30.1e). It runs along the ventral (belly) side of the animal, below the digestive tract. There is a cluster of nerve cells in each segment of the worm's body, along the length of the ventral cord. These clusters give off pairs of lateral nerves that connect to the epidermis, muscles, and other organs in the body.

In the crayfish and other arthropods, the nervous system is very similar to that of the earthworm (Figure 30.1f). The "brain" is actually two clusters of nerve cells, joined together. From the "brain," nerves extend forward to the eyes and antennas. A ventral nerve cord runs toward the back of the body. Here, too, clusters of nerve cells in each segment of the body send nerves to the animal's appendages, muscles, and other organs.

In the vertebrates like ourselves, the complex nervous system evolved in the dorsal side (back side) of the body and became increasingly protected by the skull and backbone.

THE NEURON — AN ELECTROCHEMICAL MESSENGER CELL

■ **A neuron is a modified epithelial cell specialized for impulse conduction. These cells receive, integrate, relay, and respond to messages. Sensory neurons receive stimuli, interneurons integrate information, and motor neurons carry information to muscles or glands. Neuroglia are supportive nervous tissue. Each neuron has three parts: dendrites, cell body, and axon. Groups of neurons surrounded by connective tissue form nerves.**

In Chapter 22, we discussed the fact that a **neuron**, or nerve cell, is a modified epithelial cell that is specialized for impulse conduction. The neuron (individual nerve cell) is highly excitable and capable of transmitting impulses because of its specialized plasma membrane. Adjacent neurons are organized in such a way that they butt up against one another, forming a specific message pathway. Therefore, messages providing information to the body about change are transmitted directly to a specific part or parts involved. Neurons receive, integrate, relay, and even respond to messages.

FIGURE 30.2 Neurons
There are three kinds of neurons. (a) Sensory neurons act as receptors or are activated by receptors in the skin and other sense organs, transmitting a stimulus to the spinal cord or brain. (b) Interneurons, which are located in the spinal cord and brain, integrate information about the stimulus and influence other nerve cells. (c) Motor neurons carry the nerve impulse from the spinal cord or brain to the effector—the muscle or gland that responds to the stimulus.

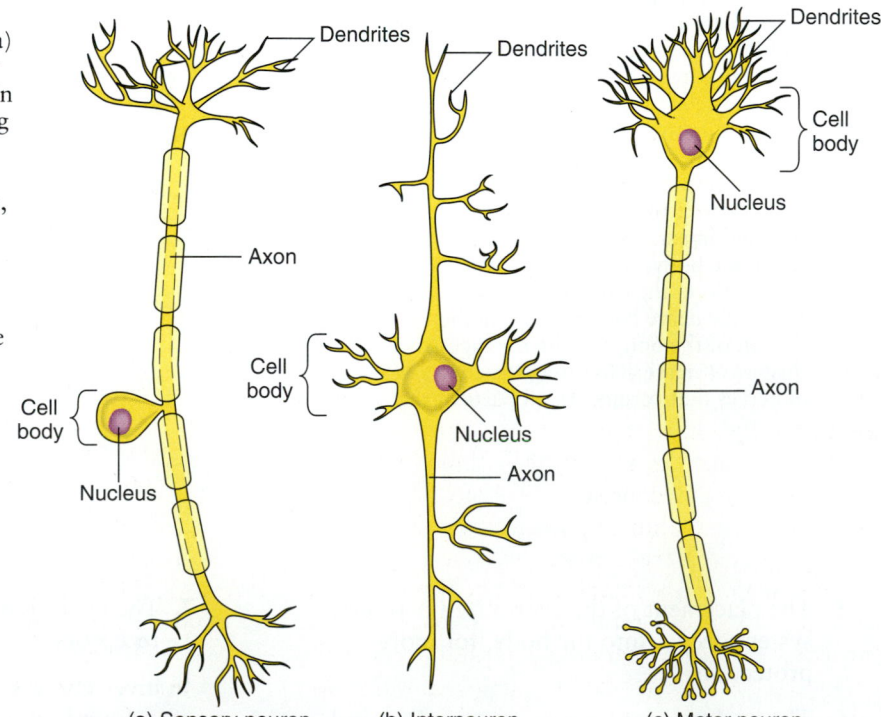

(a) Sensory neuron   (b) Interneuron   (c) Motor neuron

Types of Nerve Cells and Their Structure

There are two major classes of cells in the nervous system: *neurons* and *neuroglia* (sometimes called glial cells).

There are three types of neurons. **Sensory neurons** receive stimuli, **interneurons** (sometimes called *association neurons*) integrate information and influence other neurons (the majority of nerve cells are interneurons), and **motor neurons** carry information to an effector muscle or gland (Figure 30.2).

Although all three types of neurons have specialized characteristics, they contain three similar parts: the *dendrites, cell body,* and *axon.* The **dendrites** are cytoplasmic processes that receive input, which is then passed on to the cell body. The **cell body** contains the nucleus and other cytoplasmic organelles. Information is integrated in the cell body, then travels to the **axon,** a long cylindrical process that transmits the information to the dendrite of an adjacent nerve cell or to a muscle or gland.

Neuroglia are cells closely associated with neurons and are considered to be *supporting cells* that are important in the maintenance and function of the neuron. Because neurons carry out only a minimal amount of nonimpulse cellular activity, they exist on the metabolites produced by the neuroglia. This means that the neuroglia support the existence of the highly specialized neurons. Neuroglia comprise at least one-half of the total volume of nervous tissue, and 90 percent of the nervous tissue in the brain.

Nerves

Individual neurons are grouped into nerves. **Nerves** are bundles of nerve cells (dendrites, cell bodies, and axons), surrounded by connective tissue, which serves to protect them (Figure 30.3). If you have ever seen a nerve in a piece of drumstick or in a dissected frog, you know that they are light, thin, tough threads. Nerves function like pathways over which the axons and dendrites of several nerve cells travel together throughout the body. Some nerves are composed only of motor or sensory neurons, conducting impulses in only one direction. Other nerves are mixed, consisting of both sensory and motor neurons within the same nerve fiber. Moreover, in mixed nerves, there can be a two-way traffic system, with some neurons delivering messages to and others returning information from a particular site. The nerve cell bodies are usually grouped together to form masses called **ganglia** (singular, *ganglion).*

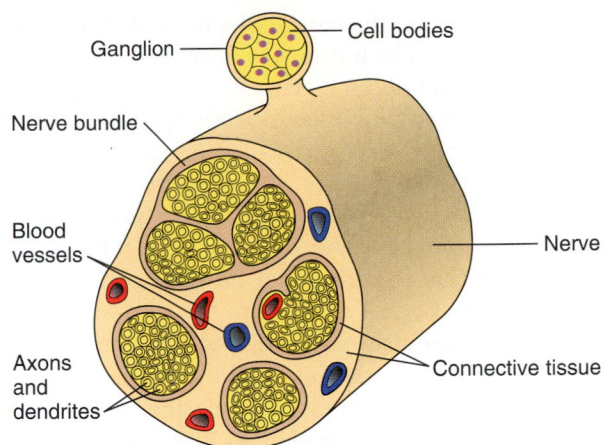

FIGURE 30.3   A Nerve
A nerve is a collection of bundles of sensory and motor neurons. Each bundle is surrounded by a layer of connective tissue, and the bundles are held together by another layer of connective tissue. The dendrites and axons of the neurons run the length of the bundles; the cell bodies are clustered in ganglia.

The structure of a typical neuron is instrumental in transmitting information. How does this occur? What happens to the plasma membrane of a nerve cell that allows for the transmission of an electrochemical impulse?

MESSAGE TRANSMISSION, PROPAGATION, AND CONDUCTION

- Nerve impulses travel through a neuron by means of a change in the electrical charge across the cell membrane. In a resting nerve, there is a difference in electrical charge across the membrane, the result of the differences in the amounts of sodium and potassium ions inside and outside the cell. Hence the cell is said to be polarized. When the cell receives a nerve impulse, it becomes depolarized. After the impulse has been transmitted, it is said to repolarize, or return to its original electrical potential.

Information moves through the nerve cell in an impulse. The impulse reaches the neuron through its dendrites, then travels the length of the axon and is passed along to yet another neuron. Although the process is very fast, it is not instantaneous. Instead, the impulse spreads along the nerve cell in a wave.

To understand the nature of a nerve impulse and the process by which it moves through a nerve cell, we have to talk about electrical charges and potentials. There are two kinds of electrical charges: positive and negative. Positive charges and negative charges attract each other. Thus, negatively charged ions tend to move toward areas where positive ions are concentrated and vice versa. *Electrical potential* is the difference in electrical charge between two areas.

A nerve impulse moves through the neuron as the electrical potential inside and outside the neuron changes. As an impulse spreads, the distribution of positive and negative charges inside and outside the neuron is altered. The neuron's plasma membrane plays a critical role in this change.

When a neuron is resting (no impulse), the cell is polarized: There is a difference in electrical charge between the cytoplasm and the external environment. The cytoplasm is negatively charged, and the external environment, positively charged. A nerve impulse depolarizes the cell: Both cell and environment are positively charged. Once the impulse moves on, the cell is repolarized: The original potential difference is restored.

Resting Potential: Polarization

When a nerve cell is at rest, it is polarized; there is a difference in the electrical charge inside and outside the cell. This difference is called the **resting potential.** In a neuron, the resting potential is roughly -70 millivolts (mV). This means that the cytoplasm inside the plasma membrane is more negative than the material outside the plasma membrane (Figure 30.4a).

The electrical potential here has to do with the distribution of sodium and potassium ions inside and outside the plasma membrane. And that distribution has to do with the selective permeability of the membrane to sodium and potassium ions and negatively charged proteins inside the cell that are too large to diffuse through the membrane.

When the neuron is resting, the outside of the cell has a higher concentration of sodium ions than the inside of the cell, and the inside of the cell has a higher concentration of potassium ions than the outside of the cell. The key to this distribution is in the plasma membrane of the neuron.

There are two mechanisms in the plasma membrane that move ions into and out of the cell. The ATP-driven **sodium-potassium pump** actively transports sodium (Na^+) ions out of the cell and potassium (K^+) ions into the cell (Figure 30.4b).

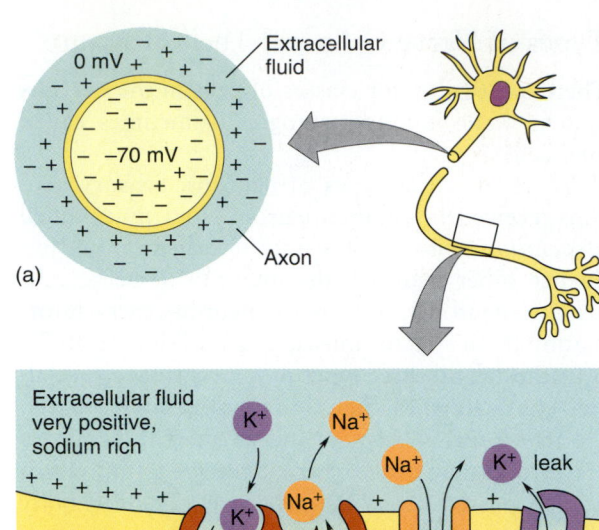

(a)

(b)

FIGURE 30.4   Resting Potential and the Sodium-Potassium Pump
(a) The nonconducting neuron is polarized—that is, there are more sodium ions in the extracellular fluid, hence the outside of the axon is more positive than the inside. This results in a resting potential of -70 mV inside the axon. (b) This resting potential is maintained by the sodium-potassium pump (1) in which, by active transport, Na^+ ions are transported out of the cell and K^+ ions are transported into it. There are also gated sodium channels (2) and gated potassium channels (3), which allow sodium to enter the cell and potassium to leave when the nerve cell is stimulated. At rest, however, the sodium channel is closed as well, but some K^+ ions leak out. The sodium-potassium pump also pumps proteins into the membrane, helping to maintain the negative electrical potential inside the cell.

Research suggests that with each operation of the pump, three Na^+ ions move out of the cell and two K^+ ions move in. There are also gated sodium and potassium channels in the plasma membrane that allow sodium to enter the cell and potassium to leave it. These channels are protein-lined pores that play an important role in nerve impulse transmission and restoring the resting potential.

The action of the sodium-potassium pump explains the concentration of ions in the neuron. But

FIGURE 30.5   The Nerve Impulse
(a) At rest, the action of the sodium-potassium pump and the permeability of the plasma membrane keep the nerve cell polarized—negative inside and positive outside. The concentration of sodium ions is higher outside the cell; the concentration of potassium ions is higher inside the cell. (b) A threshold stimulus triggers depolarization, which initiates the action potential. The gates of the sodium channels open, and sodium ions rush into the cell, creating a positive charge there. (c) Repolarization is almost instantaneous. The potassium channels open, and the sodium channels close. The sodium-potassium pump continues to move sodium ions out of the cell (and potassium ions in). The resting potential is restored.

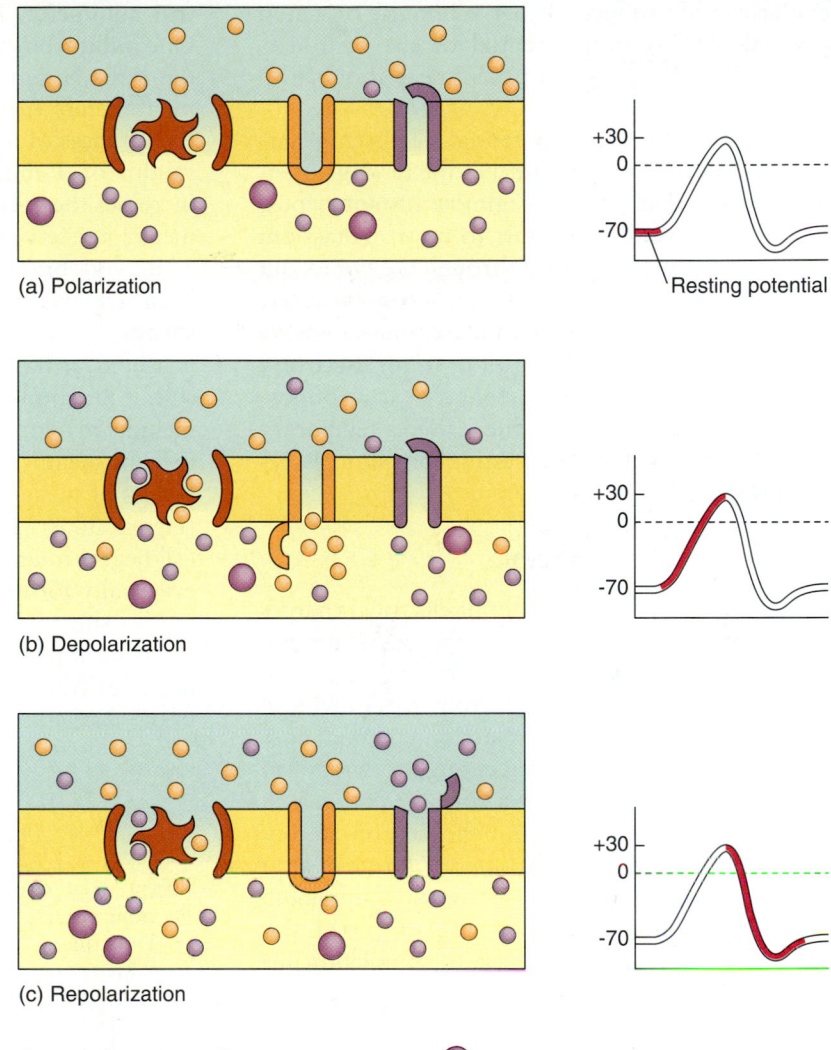

(a) Polarization

(b) Depolarization

(c) Repolarization

Resting potential

○ Sodium ions   ● Potassium ions   ● Charged protein ions

what about the electrical balance? Both Na^+ and K^+ ions are positively charged. Why, then, is the cytoplasm negatively charged? As the sodium-potassium pump moves ions into and out of the neuron, it creates two concentration gradients: one for sodium and one for potassium. As a result, sodium ions tend to diffuse back into the cell, and potassium ions tend to diffuse out. The ions diffuse through the plasma membrane, but at very different rates. The membrane is more permeable to the K^+ ions, which leak out of the cell much faster than the Na^+ ions leak in. What happens? There is a net loss of positive charge in the cytoplasm, as the K^+ leak out. (Remember, the negatively charged proteins cannot diffuse out of the cell.)

In sum, when a nerve cell is resting, there is a separation of charges across the cell membrane. Be-

cause the nerve cell membrane is more permeable to potassium ions than to sodium ions at rest, the diffusion of potassium ions across the plasma membrane to the outside leaves the inside of the cell negatively charged (Figure 30.5a).

Action Potential—Depolarization and Repolarization

A stimulus changes the permeability of the plasma membrane, making it more permeable to sodium ions by opening the gates of the Na^+ channels. As Na^+ ions move into the neuron (through the sodium channels), the electrical balance there changes, becoming positive. This process is called *depolarization* (Figure 30.5b). The change in ion balance in the nerve cell and the early stages of depolarization produce the nerve impulse or **action potential.** The resting potential of the neuron is about −70 mV; its action potential is about +30 mV.

As soon as the action potential has been generated, *repolarization* — a return to the resting potential — begins. The process is almost instantaneous, just 0.5 millisecond from start to finish. Potassium ions move out of the neuron through the potassium channels. The gates to the sodium channels close, and the sodium-potassium pump continues to move Na^+ ions out of the cell (and K^+ ions in). Recovery is so quick that there is a momentary excess of positive ions outside the membrane, called a temporary overshoot, just before the resting potential is restored (Figure 30.5c).

The All-or-None Principle

Although we've talked a lot about electrical charges here, it's important to remember that an action potential (nerve impulse) is different from an electrical impulse. As it moves farther from its source, an electrical impulse weakens; an action potential does not. Once a stimulus is strong enough to initiate an action potential (we call this a *threshold stimulus*), the impulse moves throughout the neuron at constant strength. This is the *all-or-none principle* at work. If the stimulus does not reach a minimum threshold, the nerve impulse is never triggered. But once it does, the impulse goes on to completion.

Changing the Velocity of a Nerve Impulse

■ **There are two major mechanisms for increasing the velocity of a nerve impulse. Increasing the diameter of the axon is found in the mollusks. In vertebrates, some axons are enclosed in an insulating layer of myelin. Interruptions in this sheathing, called nodes of Ranvier, allow the nerve impulse to skip from node to node, reducing the amount of time required for the movement of the impulse.**

As organisms evolved, certain structural modifications of the neuron have occurred that have increased the speed (velocity) by which a nerve impulse is propagated. If the diameter of an axon is increased, there is less resistance to the movement of ions, so the impulse moves more quickly. If the diameter of the axon is decreased, there is more resistance to the movement of ions, so the impulse moves more slowly. For instance, as invertebrates such as the squid and octopus evolved, their systems included axon fibers whose diameter had increased

to 1 millimeter. These are known as "giant axons." One millimeter may not seem large, but the diameter of the largest mammalian axon is only about 20 micrometers (μm). Remember that a millimeter is the thinness of a dime, and a micrometer is a thousandth of a millimeter. The increase in diameter increases the speed of nerve conduction to about 30 meters per second.

In vertebrates, quicker neural responses have been achieved by wrapping the axon in a **myelin sheath,** a concentric layer of lipid. Because myelin is a lipid, it has great electrical resistance. As a result, it functions by insulating the axon. Axons that include myelin are called myelinated axons. The myelin sheath is formed from the plasma membranes of specialized neuroglia called *Schwann cells* (Figure 30.6). During development, the Schwann cell begins to envelop and surround the nerve axon, eventually forming a membrane encircling the lipid myelin sheath. The myelin sheath is interrupted at frequent intervals, called **nodes of Ranvier.** The nodes of Ranvier are small areas of exposed membrane that enable electrochemical signals to "jump"

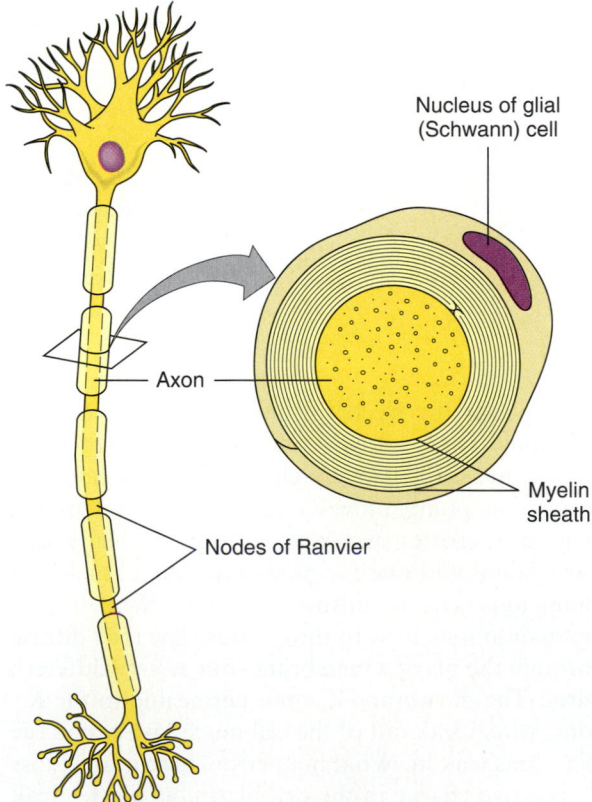

Nucleus of glial (Schwann) cell

Axon

Nodes of Ranvier

Myelin sheath

FIGURE 30.6 Myelin Sheath
The myelin sheath increases the speed at which a nerve impulse can travel. Instead of moving down the cell membrane in a continuous wave, the impulse jumps or skips from one node of Ranvier to the next.

Chapter 30 Cell-to-Cell Communication: Neural Signals

from one node to the next, skipping over large segments of the nerve cell membrane. This skipping action is called *saltatory* (from the Latin for "jump") *conduction*. The action acts as a "booster" because it increases the velocity of nerve conduction. Because of saltatory conduction, the speed of nerve conduction in myelinated nerve axons can reach as high as 120 meters per second (about 270 miles per hour), much more rapid than the 30 meters per second in nonmyelinated invertebrate (octopus) axons (see Focus 30.1).

NERVE CELL COMMUNICATION

■ **The space between two adjacent nerve cells is called the synapse, or the synaptic cleft. Neurotransmitter chemicals are housed in synaptic knobs found at the end of the axon. When the impulse reaches the end of the axon, the membrane becomes permeable to calcium ions in the synaptic cleft and releases the neurotransmitters, which transmit the impulse to the dendrites of the next neuron.**

As a nerve impulse travels from the axon of one neuron to the dendrite or cell body of another, it must cross a gap, or junction, called the **synapse** (something like a spark plug gap). Ordinarily nerve cells do not touch each other. However, in a few exceptions, adjacent neurons do make cytoplasmic contact with one another. The sending neuron is called the *presynaptic neuron* and the receiving neuron is called the *postsynaptic neuron*. The microscopic extracellular space between the presynaptic and postsynaptic neurons is known as the *synaptic cleft*. Because of the existence of the synapse, the nerve impulse cannot be directly propagated and transmitted from one nerve cell to another. Instead, the neuron releases chemicals called **neurotransmitters,** which serve to propagate the nerve impulse (see Profile). *Acetylcholine, norepinephrine, epinephrine,* and *dopamine* are common examples of neurotransmitters. How do neurotransmitters function?

At the end of the axon are microscopic knobs, known as *synaptic knobs*. When biologists observe these knobs under the microscope, they can see that

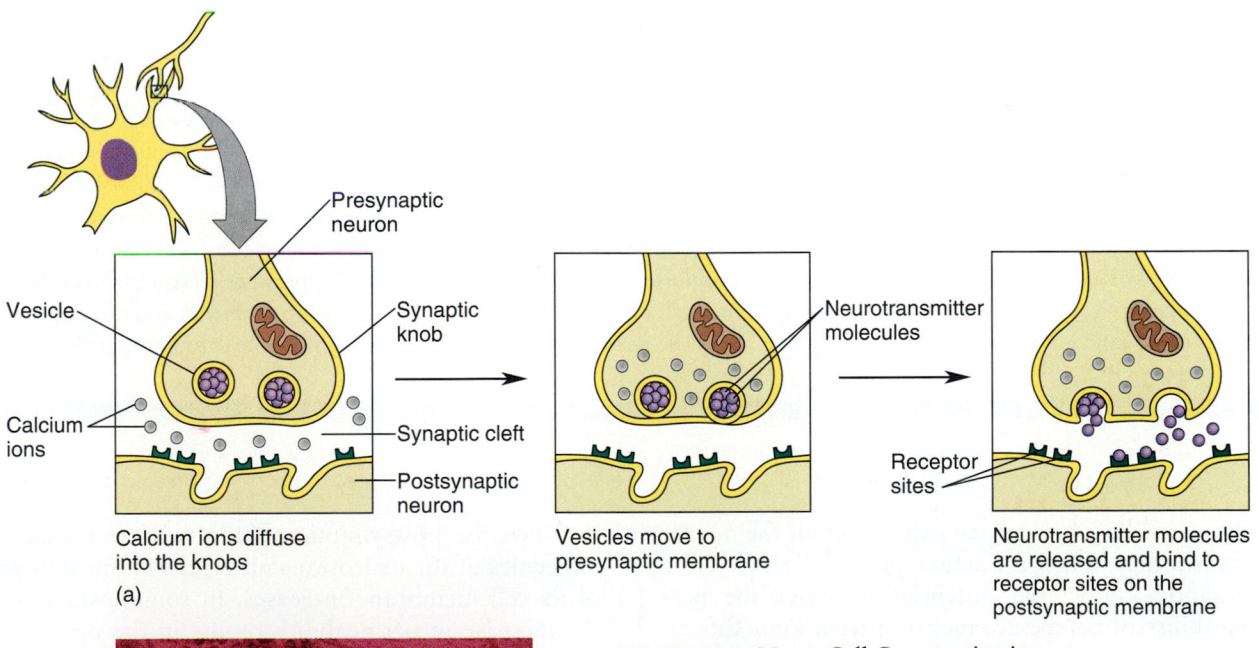

Vesicle
Calcium ions

Presynaptic neuron
Synaptic knob
Synaptic cleft
Postsynaptic neuron

Calcium ions diffuse into the knobs

(a)

Vesicles move to presynaptic membrane

Neurotransmitter molecules
Receptor sites

Neurotransmitter molecules are released and bind to receptor sites on the postsynaptic membrane

(b)

FIGURE 30.7 Nerve Cell Communication
Nerves communicate with one another (or with effectors) across a synapse. (a) As the action potential reaches the synaptic knob, it changes the permeability of the presynaptic membrane, allowing calcium ions to diffuse into the knob. These ions act on vesicles in the knob, which in turn move to the presynaptic membrane and release molecules of neurotransmitter. The molecules move across the synaptic cleft and bind to receptor sites on a dendrite of the postsynaptic neuron. (b) A transmission electron photograph of a neuronal synapse.

FOCUS 30.1
The Calamari Connection

If people in the United States have a fascination with squid (*Loligo*), it is probably based on the fear we felt as we watched or read *20,000 Leagues Under the Sea*. Or it may have arisen from our culinary delight as we consumed calamari (squid) at our favorite Italian restaurant. However, many neurologists base their fascination on the squid's giant axons (1 mm in diameter, or about the same diameter as a pencil lead). The squid also has a well-developed brain and sensory and motor neurons. Because of its large axons and its well-developed nervous system, neurobiologists can study the animal's nervous system quite easily.

If you have watched squid, you notice that it flees its predators by using an aquatic jet-propulsion system (Figure A). The squid "puts itself in reverse" by drawing seawater into a chamber between its long cylindrical body and its outer envelope, known as the mantle. It then forcibly contracts the mantle, ejecting the water out through an adjustable nozzle, to propel itself backward. The neural activities of the jet-propulsion system resulting in the mantle's forcible contractions interest neurobiologists most.

The common Atlantic squid (*Loligo pealei*) is about 1 to 1.5 feet long. In this squid, messages from the brain are transmitted to a pair of ganglia known as the stellate ganglia (Figure B). From the stellate ganglia the nerve impulse travels through two axons, each of which is about 5 inches long. The nerve impulses speed through these axons to the muscles of the mantle. There they secrete neurotransmitters across the neuromuscular junctions, causing the muscles of the mantle to contract.

By implanting electrodes in these large axons, researchers have learned about the biochemical properties of a nerve impulse. Much of our understanding of the movement of sodium and potassium ions through membrane channels and about other mechanisms of the neurochemical nature of the nerve cell impulse is the direct result of the experiments with squid nerve axons.

At present, much of the research into the nerve cell axon involves the small neurofilaments and microtubules inside the cytoplasm of the

they are filled with vesicles that contain the neurotransmitter. When an action potential reaches the synaptic knobs, the potential increases the permeability of the membrane to calcium ions (Ca^{++}). The calcium ions then enter the cytoplasm of the synaptic knobs from the extracellular fluid. The ions cause contractions of the cytoskeleton, which cause the vesicles to move to the plasma membrane and fuse with it. Once fusion is achieved, the vesicles release their contents into the synaptic cleft through exocytosis. The neurotransmitter molecules diffuse across the synaptic cleft and are received by specialized receptor sites on the dendrites of the postsynaptic neuron (Figure 30.7).

When the postsynaptic neuron receives enough molecules of the neurotransmitter, the permeability of its cell membrane increases. In some instances, the increase in permeability results in the opening of the sodium channels. Once the neurotransmitter is received, it is immediately destroyed by enzymes to prevent the continuous formation of action potentials in the neuron.

Since this mechanism results in the depolarization of the postsynaptic neuron, the synapse can be thought of as an *excitatory synapse*. That is, the receiving neuron was excited, and the nerve impulse was propagated. However, sometimes the changes at the synapse result in the inhibition of a nerve

axon called the axoplasm. Within the axoplasm are the secretory structures that synthesize and release the neurotransmitter substances across the synaptic gaps from one nerve cell to another.

Because of the size of the squid axon, scientists have an abundant supply of large axons for the collection of information about nerve structure and function. Substances such as analgesics and hypnotics can be pumped into these axons to show how the drugs function. For example, Toshio Narahashi of Northwestern University Medical School has demonstrated that xylocaine, a local anesthetic, works within the nerve fiber, blocking nerve conduction by closing sodium and potassium channels.

Using the squid enables us to understand molecular mechanisms of nerve physiology that would be impossible to study in humans. Because the molecular mechanism of nerve conduction is similar in squid and humans, the squid provides us with a "calamari connection" for our understanding of the fascinating cellular communicators — the nerve cells. This understanding will further our ability to treat many degenerative nerve diseases that can result in paralysis, brain function impairment, and even death.

FIGURE A
A group of squid *(Loligo pealei)*.

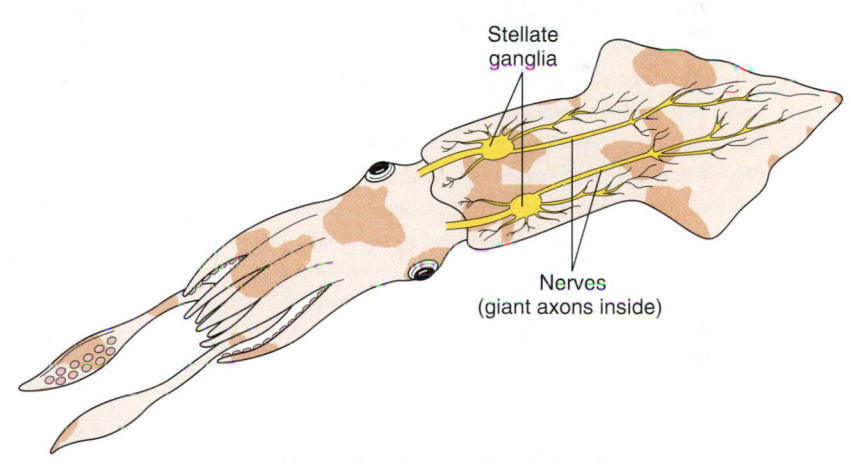

FIGURE B
The giant nerves (axons) and stellate ganglia of the squid.

impulse. In such a case the mechanism is called an *inhibitory synapse*. When a nerve impulse is inhibited, changes occur in the membrane of the receiving neuron that allow the negatively charged chloride ions to move into the neuron. This results in an increase in negative charges inside the neuron. In other words, chemical ionic changes occur so that a stimulation greater than normal is required to trigger the action potential threshold. As a result, the impulse is inhibited.

Because of the amazing complexity of nerve action, then, in order to have coordinated communication and regulation of cell-to-cell communication, your brain is constantly sending out and receiving thousands of bits of chemical information. Just think of all the action along the way. The message is excitatory at some neuronal synaptic junctions and inhibiting at others. Nerve cell impulses are being turned on and off in an integrated way, resulting in the coordination of body functions.

THE RESULT OF THE TRANSMISSION OF NEURAL MESSAGES

Neurons do not communicate only with one another. They also communicate with muscles and glands.

The Neuromuscular Synapse

In Chapter 22, we talked about muscle contraction — the effect of a nerve impulse on a muscle fiber. Motor neurons carry impulses to muscle cells (Figure 30.8). The process by which they pass along impulses to muscles is much the same as that used to pass along impulses to other neurons. As the action potential reaches the synaptic knob of the motor neuron, calcium ions move into the nerve cell and act on the vesicles there. The vesicles move into the presynaptic membrane, then empty their load of neurotransmitter molecules. At a neuromuscular synapse, the neurotransmitter is acetylcholine (ACh). The ACh molecules move across the synaptic cleft, then bind with receptor sites on the membrane of the muscle cell. The process triggers a muscle contraction (see Human Endeavors box).

The Reflex Arc: A Simple Pathway

■ **The reflex arc is a simple nerve pathway in which there is a specific response to a specific stimulus. Because conscious thought is not involved, the response is called involuntary. In most cases, a sensory nerve transmits information about the stimulus to an interneuron in the spinal cord. The interneuron transmits the information to a motor neuron, which stimulates an effector such as a muscle, which responds.**

You have no doubt heard the statement, "She is a good athlete, because she has good reflexes." In fact, this statement is not quite true. Actually, the individual possesses good eye-hand coordination or other types of complex coordination as well as a fast response time that has been learned and perfected through years of practice.

A reflex, on the other hand, is a simple unlearned automatic response to a particular stimulus. The response takes place through a simple nerve pathway called the reflex arc. Your body responds before you are even conscious of what is happening. Reflexes are at work when you close your eyes when a bug flies near or when your leg jerks when your doctor taps just below your kneecap. All these actions are called involuntary responses, because no conscious thought is required for them to take place.

A **reflex arc** is a simple nerve pathway (which can consist of as few as two neurons) in which stimulus detection, integration, and response occur and which may not directly involve the brain. Some of you may have observed this response in your high school biology class, where you learned that a frog with its brain destroyed would still respond when you put a drop of dilute acid on its leg. Obviously the response did not require a functioning brain.

In the knee-jerk reflex, only two types of neurons are involved. The stretch receptors located in the muscles above the knee are stretched as the tendon

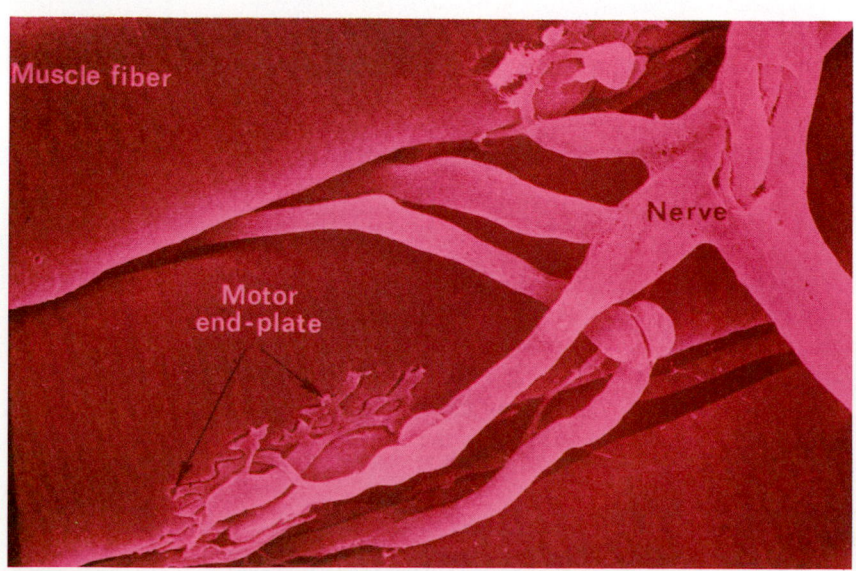

FIGURE 30.8   The Neuromuscular Junction
The photograph shows several axon branches laying across skeletal muscle fibers at the motor end plate.

Acetylcholine Receptor

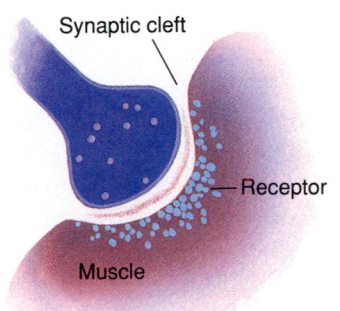

These illustrations summarize the conceptual evolution of the human endeavor resulting in our understanding of the acetylcholine receptor.

Source: Adapted from "Evolutions: Acetylcholine Receptor," *The Journal of NIH Research,* vol. 1, p. 152, Sept./Oct. 1989. Adapted from original artwork by Sally Bensusen.

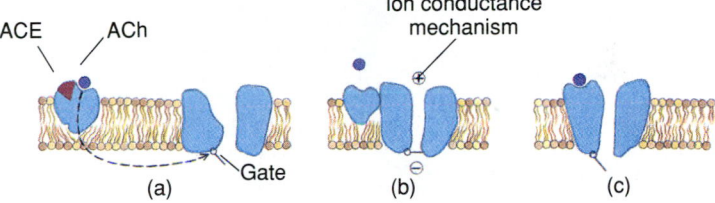

1906–1950s: Early work established that acetylcholine (ACh) mediates chemical transmission of nerve impulses at the neuromuscular junction.

1950–1970s: Electrophysiologists distinguished binding, ion conductance, and acetylcholinesterase (ACE) activities but could not tell if one receptor mediated all three functions.

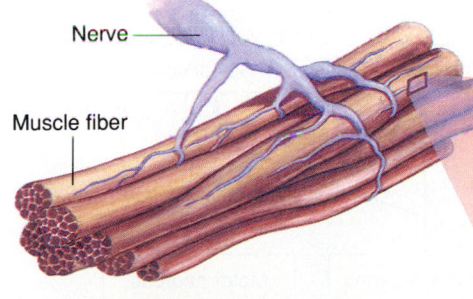

During the same period, between 1950 and 1970, anatomists used electron microscopy to characterize the neuromuscular junction in greater detail.

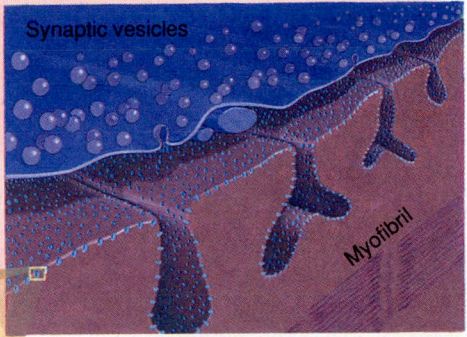

1970s: The release of ACh from synaptic vesicles in the nerve ending triggers the influx of about 10^6 sodium ions into each of 15,000 to 20,000 channels/μm^2 on muscle membrane.

1989: The ion influx occurs as ACh binds to the α subunits of the receptor and opens the channel; 200,000 single channels open simultaneously.

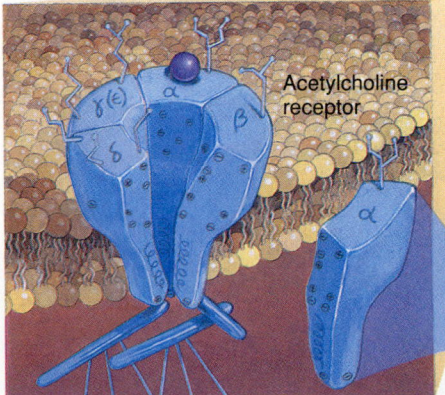

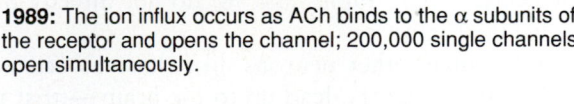

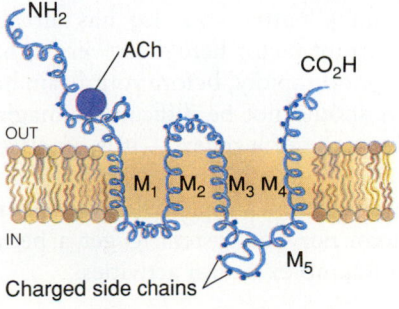

1980s: Recent evidence suggests that each subunit has four domains that cross the membrane and that the channel is uncharged.

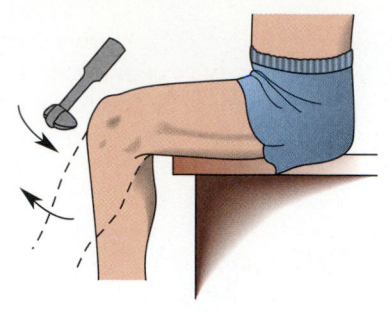

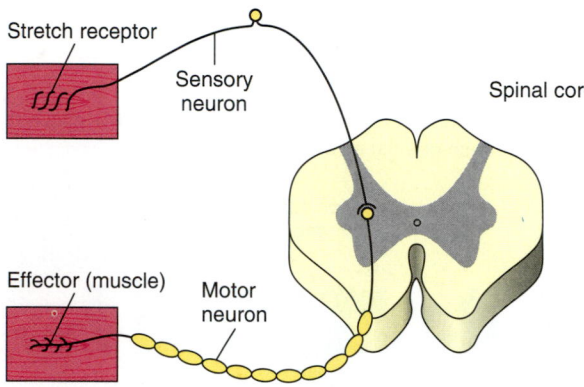

FIGURE 30.9 A Reflex Arc
The knee jerk is a simple reflex arc, involving just two neurons—a sensory neuron and a motor neuron. Like all reflexes, the response occurs without intervention from the brain.

is hit with a hammer. A *sensory neuron* receives the stimulus and sends it to the spinal cord. There, it connects with a *motor neuron*. The motor neuron, without any communication with the brain, then takes a message out to the muscle, and the muscle contracts (Figure 30.9). Reflexes are simple neural responses that (1) are usually involuntary, (2) always perform the same way, (3) help the organism adapt to the environment, (4) involve only a few neurons, (5) are rapid, and (6) do not directly involve the brain.

Of course, other neurons, like those associated with pain receptors, lead up to the brain — that is why you say "ouch" after your leg has already jerked. The important factor here, however, is that your reflexes respond rapidly, before your brain has time to think. It should not be difficult to imagine the survival advantage of a reflex — it is, in effect, nature's insurance policy against danger. Now that we know something about simple reflex arcs, let us discuss the human nervous system to get a better understanding of complex neural activities.

THE HUMAN NERVOUS SYSTEM

As illustrated in Figure 30.10, the typical vertebrate nervous system is divided into the *central nervous system* (brain and spinal cord) and the *peripheral nervous system* (neurons or any of their parts located outside the central nervous system). The peripheral nervous system can be further divided into the *somatic* and the *autonomic nervous systems*. We begin our discussion with the central nervous system and then turn to a discussion of the peripheral nervous system and its subdivisions.

Central Nervous System

■ The central nervous system includes the brain (encased in the skull) and the spinal cord, encased in bony vertebrae. The brain can be divided into the hindbrain, midbrain, and forebrain, or cerebrum. The cerebrum is divided into the right and left hemispheres. The spinal cord extends from the base of the brain to the second lumbar vertebra. Both structures contain gray- and white-colored

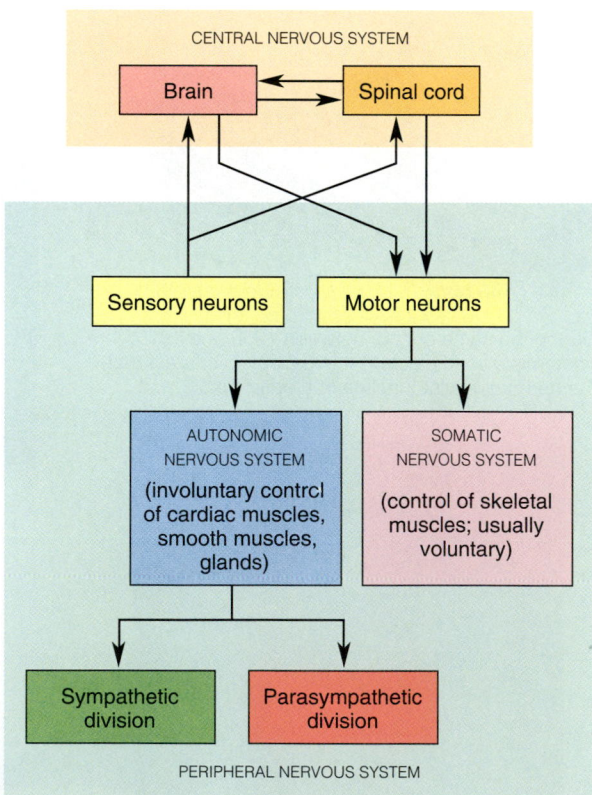

FIGURE 30.10 The Vertebrate Nervous System
A representation of the major divisions of the vertebrate nervous system.

regions. The white region consists of nerve cell axons; the gray region consists of nerve cell bodies and synapses.

We pointed out earlier that two of the major trends in the evolution of the nervous system were cephalization (the collection and concentration of neural matter and sense receptors in the head end of organisms) and the internalization (increased protection) of the nervous system. The result of these trends is clearly demonstrated in the human **central nervous system** (Figure 30.11).

In humans, the central processing unit is the brain, a structure that is encased in the skull. Cerebrospinal fluid within the skull cushions the brain. The **spinal cord,** which in a sense is an extension of the brain, is encased within the central canal of the individual vertebrae that make up the vertebral column. The central nervous system functions in integrating, interpreting, analyzing, and directing all messages within the body. Those messages are sent through nerves to the spinal cord and brain, where the information is analyzed and sent out to the body.

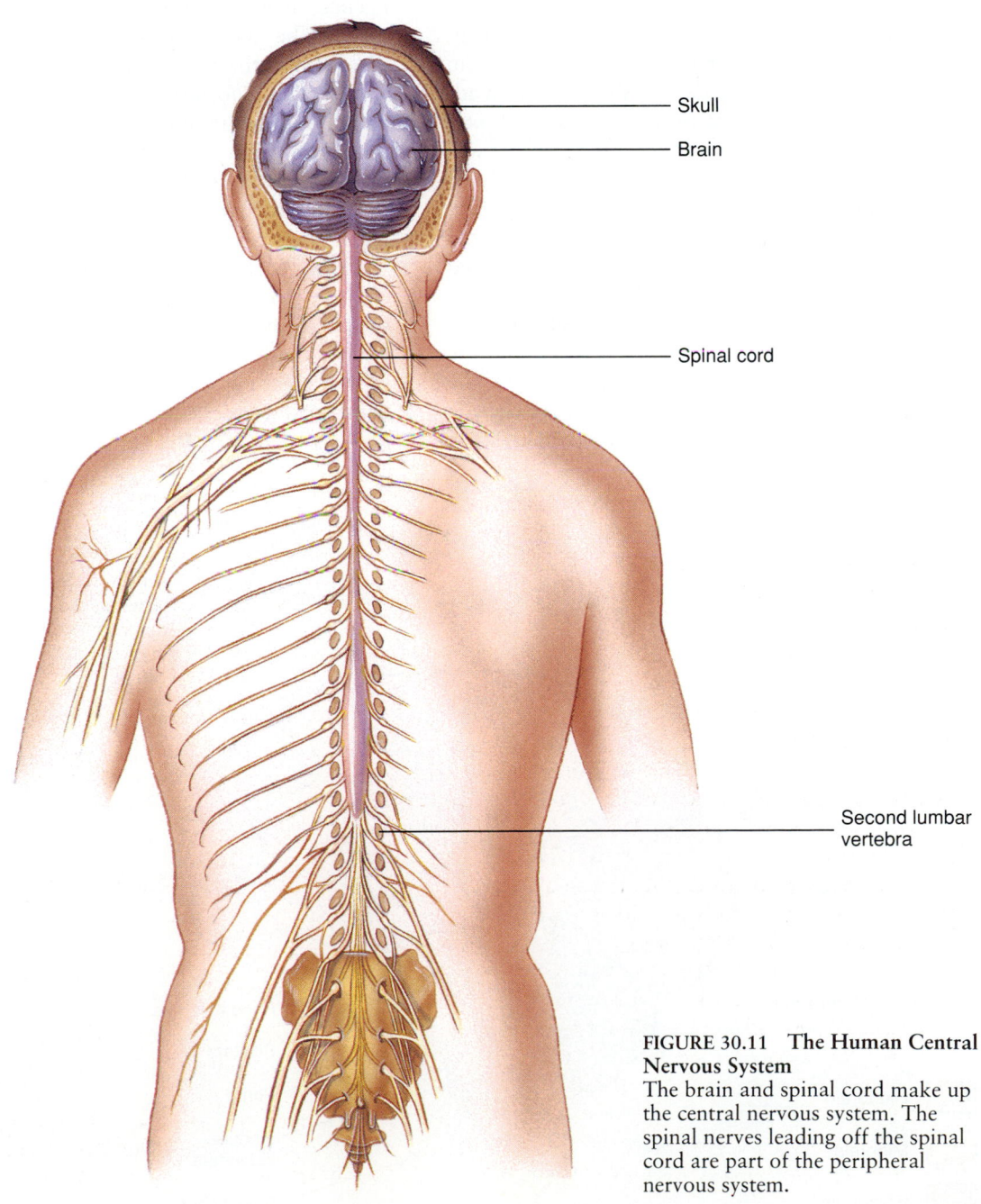

Skull

Brain

Spinal cord

Second lumbar vertebra

FIGURE 30.11   The Human Central Nervous System
The brain and spinal cord make up the central nervous system. The spinal nerves leading off the spinal cord are part of the peripheral nervous system.

The Brain. If you have ever seen a human brain or the brain of another animal, you know that it is not very impressive to look at. The adult human brain weighs about 1.4 kg (3 pounds). Its outer region (the exterior) is gray and its inner region (the interior) is white. It is a jellylike mass, mostly made up of water and fat and having the consistency of a thick custard or mush. Despite its unimpressive appearance, the brain is, in a poetic sense, the "seat of the soul," the mysterious computer wizard, the director of human emotions that allows us to experience joy and sorrow. It is an extraordinarily complex organ (see Profile).

The brain utilizes 25 percent of the oxygen available in the body and has a circulatory system equipped with regulatory devices that ensure an adequate supply of blood, with its nutrients and oxygen. These regulatory devices are arterioles that will not constrict, although they can be dilated easily by various drugs. In times of extreme physical, emotional, or environmental stress to the body, the brain takes precedence over all other systems, and blood is rerouted from muscles to the brain.

For convenience, the brain is usually divided into three regions: the hindbrain, the midbrain, and the forebrain. The hindbrain, sometimes called the brain stem, consists of the *medulla oblongata,* the *pons,* and the *cerebellum.* The cerebellum is a paired structure that resembles a walnut and is involved in muscle coordination, balance, and equilibrium. The pons contains the ascending and descending tract between the spinal cord and the brain, and the medulla oblongata is an extension of the spinal cord that controls essential functions like heart rate, breathing rate, and blood pressure.

The midbrain is the middle region, which connects the hindbrain to the forebrain. The forebrain consists of the *cerebrum,* the *thalamus,* and the *reticular system.* The cerebral cortex, the thin layer of nerves that covers the cerebrum, is the seat of intelligence, memory, personality, speech, and judgment. The area beneath the cerebral cortex is divided into the right and left cerebral hemispheres. The *corpus callosum* connects the two hemispheres and allows communication between them.

At the base of the forebrain is the thalamus, which is responsible for communication between the different parts of the brain. The thalamus contains the reticular system, an interconnected network of neurons that monitors brain activities. The *reticular activating system* is the part of the reticular system responsible for activating the appropriate part of the brain when a stimulus is received. This same system allows us to ignore stimuli that do not demand responses, like the hum of the refrigerator (as I write this chapter) or a conversation going on

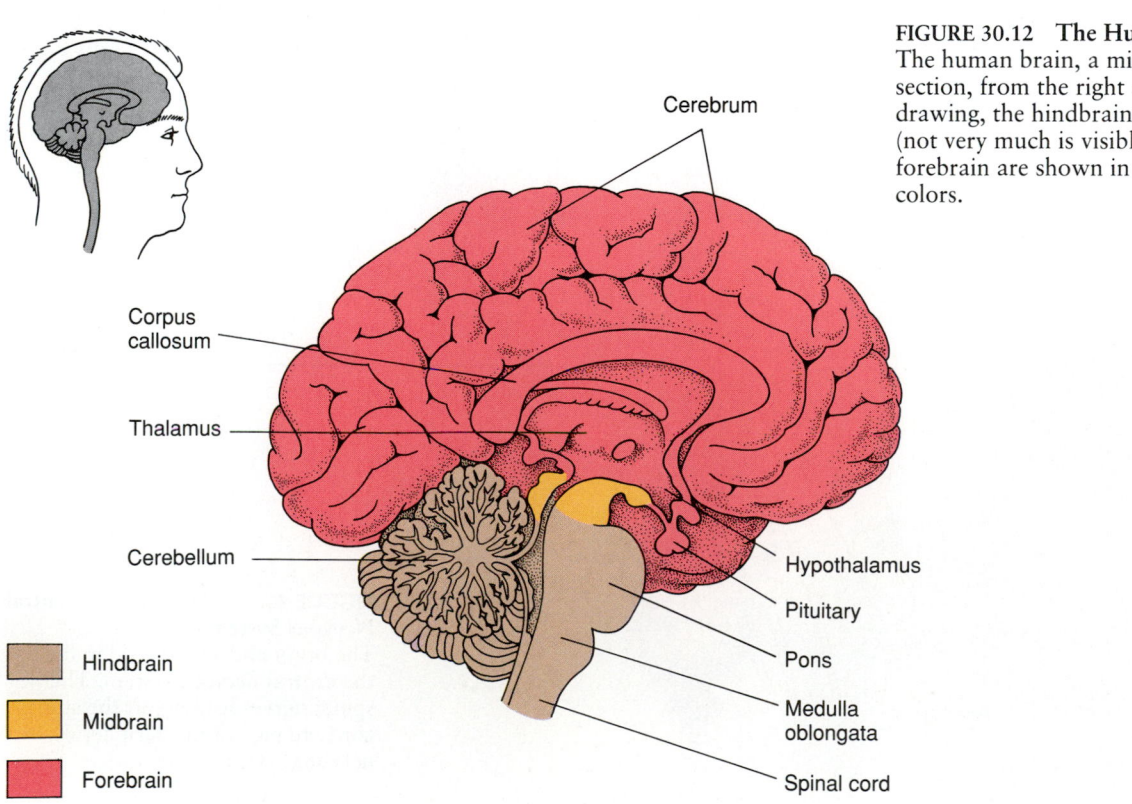

FIGURE 30.12 The Human Brain
The human brain, a midsagittal section, from the right side. In this drawing, the hindbrain, midbrain (not very much is visible), and the forebrain are shown in different colors.

Cerebrum

Corpus callosum

Thalamus

Cerebellum

Hindbrain

Midbrain

Forebrain

Hypothalamus

Pituitary

Pons

Medulla oblongata

Spinal cord

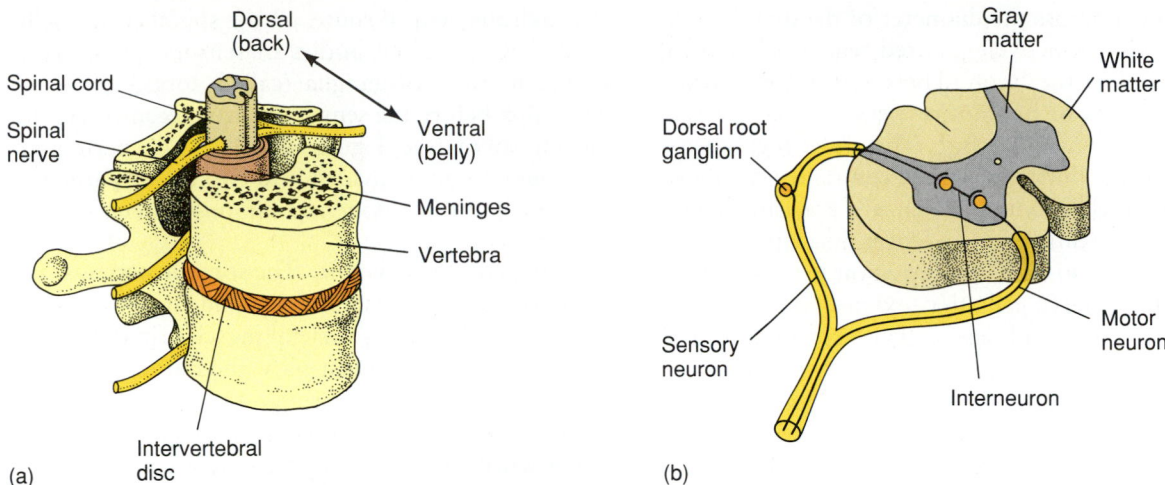

FIGURE 30.13   The Human Spinal Cord
(a) The vertebrae form a column that surrounds and protects the spinal cord.
(b) The white matter in the spinal cord is made up of nerve cell axons; the
gray matter, of nerve cell bodies and synapses. Sensory neurons carry impulses
to the cord; interneurons in the gray matter transmit those impulses to motor
neurons; and the motor neurons carry them to effectors (muscles or glands).

in another room. The forebrain also contains the *limbic system,* a system of pathways that influences some aspects of emotion and behavior. Figure 30.12 and Table 30.1 illustrate and describe the functions of the major regions of the brain.

Spinal Cord. The spinal cord extends from the base of the brain to the second lumbar vertebra (the lum-

bar vertebrae are in your lower back). The cord is about 18 inches long and is about as thick as your thumb. A delicate structure, the spinal cord is surrounded by cerebral spinal fluid that acts as a cushion. The bony vertebrae protect the spinal cord (Figure 30.13a). Both the brain and the spinal cord are covered by a series of three protective membranes called the *meninges.*

TABLE 30.1
Major Structures of the Human Brain

Structure	Function	Structure	Function
Cerebrum	Represents the center of sensory-motor coordination and the seat of memory, intelligence, personality, conscious thought, speech, and judgment	Medulla oblongata	Helps control heart rate, respiration, and gastrointestinal function
Corpus callosum	Connects the hemispheres of the cerebral cortex	Pons	Relays information from the medulla to higher brain centers and links the parts of the brain to each other
Thalamus	Helps regulate sleep; relays information related to sight, hearing, touch, and taste to the cerebral cortex	Pituitary	Produces hormones that activate other glands and responds to negative feedback mechanisms
Cerebellum	Helps regulate motor coordination and equilibrium	Hypothalamus	Regulates blood pressure, body temperature, heart rate; controls basic human drives like eating, drinking, sleeping, and sex

The Human Nervous System

If you cut across the diameter of the spinal cord, you see two distinctively colored regions, similar to those found in the brain. The outer white-colored region consists of bundles of nerve axons covered with myelin, which is white. This outer region surrounds a gray, butterfly-shaped central region made up of nerve cell bodies, which are grayish (Figure 30.13b). The difference in color indicates differences of function. The white region is made up of the axons of nerve cells that extend up and down the spinal cord and are called *ascending tracts* (paths of nerve impulses toward the brain) and *descending tracts* (paths of nerve impulses away from the brain). The butterfly-shaped gray region consists primarily of nerve cell bodies and the synapses.

The spinal cord controls most of the involuntary reflexes of the body and integrates neural activities from other parts of the body. The spinal cord also serves as the route through which nerve impulses travel to and from the brain. Thirty-one pairs of spinal nerves pass through openings between the vertebrae. They then branch and innervate the body. These spinal nerves and their branches make up the peripheral nervous system — so-called because they are outside the central nervous system.

Peripheral Nervous System

■ **The peripheral nervous system includes sensory and motor neurons and encompasses two systems: the somatic and the autonomic. The somatic nervous system governs voluntary actions. The autonomic nervous system governs responses that are involuntary. The autonomic nervous system is divided into two divisions, the sympathetic and parasympathetic, and the effects of these two divisions are usually antagonistic, or opposite.**

The **peripheral nervous system** of vertebrates consists of those nerves outside the central nervous system. The nerves of this system occur in pairs, one extending to the right side of the spinal cord and the other to the left (see Figure 30.11). (This arrangement is a basic element in bilateral symmetry.) The nerves of this system conduct impulses from the body to the central nervous system and from the central nervous system to the body.

The nerves of the peripheral nervous system are mixed nerves, consisting of the sensory and motor neurons within the same structure. Just before it enters the spinal cord, the nerve divides into two portions. The axons of the sensory neurons enter the dorsal route, on the back side of the spinal cord. The motor neuron axons leave the spinal cord through the ventral route, on the spinal cord's belly side. The nerve cell bodies of sensory neurons are found in rows of ganglia (called dorsal root ganglia), just before the sensory neuron axon enters the dorsal route (see Figure 30.13b). The peripheral nervous system is divided into two major parts, the *somatic* and the *autonomic nervous system*.

Somatic Nervous System. The **somatic nervous system** consists of those sensory and motor neurons that move the body's skeletal muscles. In the higher vertebrates, for the most part, the motor responses travel from the brain, through the spinal cord, and to the effectors. These responses are *voluntary* — in other words, they are under conscious control. You perform voluntary actions when you contract your skeletal muscles to throw a baseball or to stir a pot of stew. The rate of your heartbeat, on the other hand, is an *involuntary* action. (The terms *voluntary* and *involuntary* are not cut and dried, as you may know — a yogi can voluntarily control the heart rate.)

The cell body of each motor neuron of the somatic system is found within the central nervous system, and the axon of each runs uninterrupted to its effector. For example, when you wiggle your big toe, those motor neurons that are found in the nerve to your big toe have axons that go unbroken from the base of your spinal cord down your leg to the toe — a very long axon.

Autonomic Nervous System. As you would suspect, neurons that make up the **autonomic nervous system** (self-managed) are responsible primarily for the involuntary automatic responses such as controlling the activities of the viscera (the internal organs) — the heart rate, blood vessel constriction, smooth muscle contraction, and glandular secretions. Neuromessages sent from the brain and spinal cord through the autonomic nervous system are not generally under conscious control. They are automatic, or involuntary.

The autonomic nervous system consists only of motor neurons, is involuntary, and is divided into two divisions: the **sympathetic division** and the **parasympathetic division.** These two divisions are structurally and functionally different and arise in different parts of the spinal cord. The effects of the two systems are generally opposite, or *antagonistic.* If you examine most organs of the body, you note that they are innervated by both parasympathetic and sympathetic neurons. This dual innervation might lead you to guess that the activities of each neuron would be different. It might seem wasteful

Dr. Floyd Bloom
Neuroscientist

Dr. Floyd Bloom, the son of a pharmacist, was born in Minneapolis and raised near Dallas. His interest in medicine began at an early age. Even though he felt lukewarm about being a practicing physician, he entered medical school, with a specialization in internal medicine. However, during his residency, Bloom changed his mind when the opportunity arose to do research on a drug called reserpine. B. B. Brodie at the National Institutes of Health discovered that reserpine worked on the brain and on blood vessels to reduce blood pressure. Although Bloom continued with his studies and eventually became a licensed physician, he found he was addicted to research, and it was there that he concentrated his efforts.

Floyd Bloom is currently one of the world's foremost leaders in the area of brain chemistry — especially the neurotransmitters of the brain. He is best known for his early research, in which he delineated the function of the part of the brain known as the *locus cerulus* (a pinpoint portion of the brain that communicates with the rest of the brain by secreting norepinephrine). Today, using techniques from molecular genetics, he continues to search for the neurotransmitters of the brain. He worked at the Salk Institute in La Jolla, California, for a period of time before moving on to the equally famous Scripps Clinic and Research Center, located only a mile north. There Bloom has teamed with Dr. Richard Lerner, a molecular geneticist, and others to develop cloning techniques that will enable biologists to investigate the brain's biochemistry. Bloom believes that cloning and other recombinant DNA techniques will lay a foundation for new knowledge of the functioning of the brain and the nervous system.

Bloom is a very energetic individual, as evidenced by his prodigious research output. He has published more than 350 research papers. An excellent speaker, he is actively involved in everything from the establishment of federal research policies to expertise on California wines. It does not bother him to be involved with more than one project at the same time. He has a life outside the lab as well — relaxing by reading Aristotle and Plato and listening to jazz. Still, it is the laboratory and his research that are his real loves.

to have a dual innervation. If the function was the same for both, why would there be two?

The autonomic nervous system has a dual innervation system because the messages transmitted by each pathway are different. Depending on the organ, the signal can be either stimulatory or inhibitory. In other words, one of the two systems may serve to increase the organ's activity; the other, to reduce it. For example, parasympathetic stimulation through the *vagus nerve* slows, or decreases, the heart rate. On the other hand, sympathetic stimulation through the *cardioaccelerator nerve* produces the opposite effect — it increases the heart rate. Similarly, parasympathetic stimulation increases the movement of the smooth muscles of the gastrointestinal tract; sympathetic stimulation decreases that movement (Figure 30.14).

The reason that these two types of neurons have opposite effects is the type of neurotransmitter that they produce. Sympathetic neurons generally secrete *norepinephrine*. As we pointed out in Chapter 29, this chemical tends to speed up activities involved in fight, fright, or flight. It increases pulse rate, blood pressure, respiration, and dilation of sphincter muscles, such as those in the iris of the eye. It also can affect the sphincter muscles of the rectum and bladder — that is why extreme excitement or fear can result in defecation or urination. (Obviously, defecation would have a survival advantage — it would discourage or distract a predator, since the predator may smell or eat the excrement.)

The sympathetic neurons are involved in excitation and emotional stress. But the neurons of the parasympathetic system are involved in relaxation and the maintenance of body activities — sort of the rest and rumination neurons. The parasympathetic neurons (except in a few instances) secrete a neurotransmitter called *acetylcholine,* which stimulates functions that slow down and suppress activities.

The neurons that make up the sympathetic and parasympathetic systems also differ in their anatomic locations: The parasympathetic nerve fibers originate in the brainstem and the lower portions of the spinal cord (located at the small of the back). The sympathetic nerves originate in gray matter, the middle portion of the spinal cord. The ganglia in the parasympathetic nervous system are near or in the effector organ. In the sympathetic nervous system, the ganglia are near and run parallel to the spinal cord on the ventral (belly) side.

As you can see, the parasympathetic and sympathetic nervous systems work in a coordinated manner to maintain the conditions necessary for life.

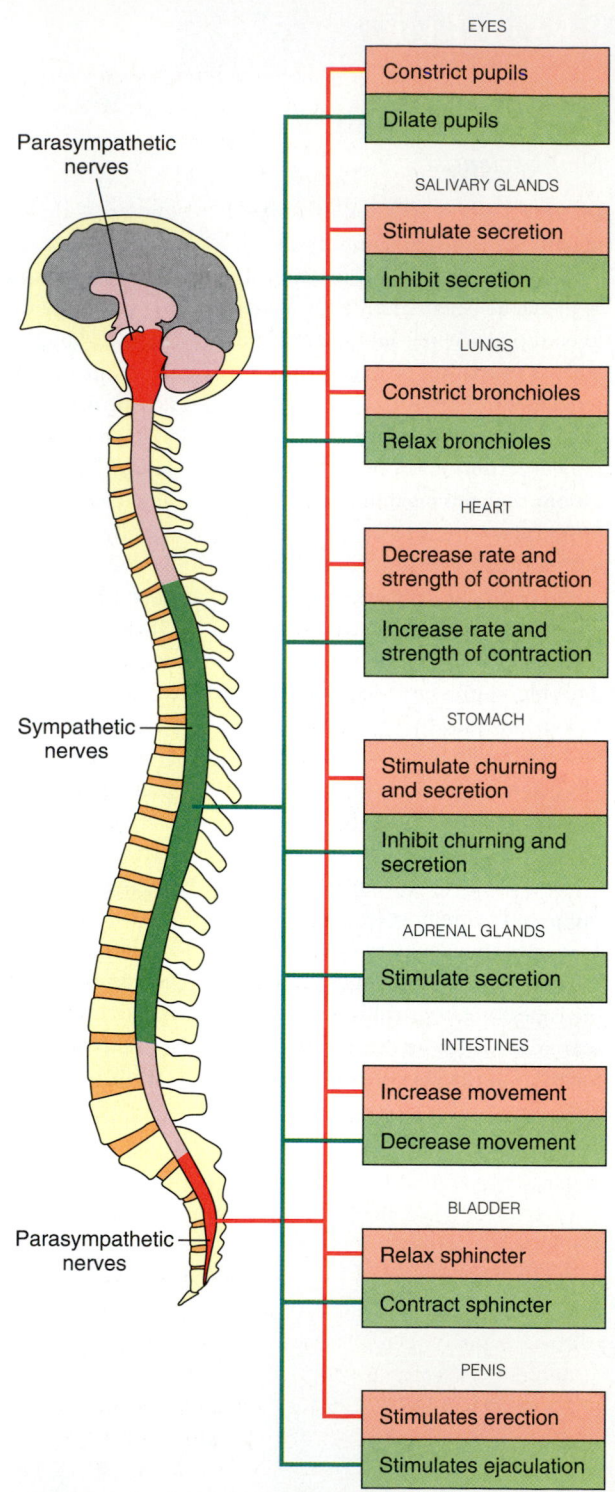

FIGURE 30.14 The Autonomic Nervous System
The autonomic nervous system is made up of two parts: sympathetic nerves (in green) and parasympathetic nerves (in red). In most cases, both sets of nerves innervate the same organs. Their actions, however, are usually antagonistic. For example, sympathetic nerves dilate the bronchioles of the lungs; parasympathetic nerves constrict them.

Chapter 30 Cell-to-Cell Communication: Neural Signals

SENSE ORGANS

■ Sensory structures receive stimuli and then transmit the information to the brain. Receptors can also focus and amplify weak stimuli. Generally speaking, humans have five senses, which can function alone or together: touch, hearing, sight, taste, and smell.

Can you imagine what it would be like to be born both deaf and blind? Your sensory perceptions of the world would be very limited (although individuals with these conditions report that their other senses are heightened to compensate for the lost senses). However, when we grasp the extent of our sensory structures, we may find that our perception of the world is also very limited, because we can only experience what our sensory structures will allow.

In essence, sensory structures are *stimuli receptors* (they receive information about a change in energy form or a variation in energy form) and *energy transducers* (structures that selectively convert energy from one form to another). Once sensory structures receive the information and convert it to another form, they transmit that energy along neuropathways. For example, when you look at a flower, photoreceptors in your eyes detect the radiant energy from the visible portion of the electromagnetic spectrum and transduce it into chemical energy. The chemical energy then becomes neurochemical energy, and the nerve impulse is carried to the brain, where it is interpreted and analyzed. In everyday terms, you see the flower, but we humans don't see what an insect sees when it looks at a flower because, unlike humans, insects can perceive short wavelengths. In other words, we can only perceive what our senses will allow.

Receptors not only detect stimuli, they can also focus and amplify weak stimuli. For example, our night-accommodated vision can detect the light of a single candle that is over a mile away, which we think is fairly good. But our vision is much less acute than a bird of prey's. If that bird were sitting on top of the Empire State Building, it could detect the date of a dime lying on the sidewalk one hundred plus stories below it. We notice the smell of a pot boiling over in the kitchen, but we cannot detect the scent that a hunting dog follows through the woods or the residual body heat that some snakes follow in search of prey. The ability of an organism to detect, focus, and amplify weak stimuli and to determine the direction of the origin of the stimuli varies among organisms, and enables each to survive in its particular environment. Refer to

TABLE 30.2
Major Receptors in Vertebrates

Chemoreceptors	Detect the chemical energy in ions or molecules. Taste receptors, odor receptors, and receptors that detect levels of chemicals in the blood and tissue fluids are examples of chemoreceptors.
Mechanoreceptors	Detect mechanical energy such as changes in pressure, position with respect to gravity, or acceleration. Touch, stretch, equilibrium, and sound are detected by mechanoreceptors.
Thermoreceptors	Detect energy associated with heat—the infrared wavelengths of light. Infrared receptors are examples.
Photoreceptors	Detect the energy from the visible spectrum. The photoreceptor cells of the eye are examples.
Electroreceptors	Detect electrical energy. Some electroreceptors detect disturbances in self-produced electrical fields (for example, those in the electric eel). Other electroreceptors detect changes in electromagnetic fields (for example, those that aid birds in migration).
Nociceptors	Detect pain. Free nerve endings in the skin are nociceptors.

Table 30.2 for a listing of the major receptors in vertebrates and their functions. Humans are said to have five "senses": touch, hearing, sight, taste, and smell. These senses often blend to provide even more complex information than we usually associate with them. In the following discussion we discuss the five senses, with a special concentration on the mechanisms of hearing and sight. However, we should keep in mind that we can detect information in more ways than with our five senses. For example, there are pH, CO_2, O_2, and hot-cold receptors.

Skin Receptors

■ The skin contains numerous sensory structures, such as free nerve endings that detect pain, mechanoreceptors that detect pressure, and thermoreceptors that detect heat and cold.

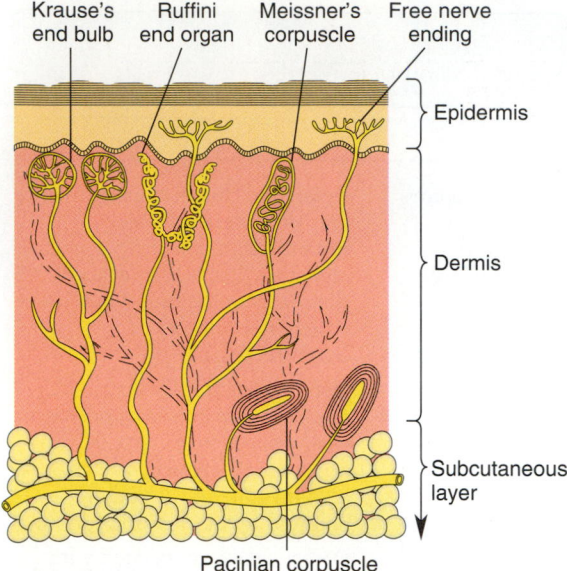

Krause's end bulb Ruffini end organ Meissner's corpuscle Free nerve ending

Epidermis

Dermis

Subcutaneous layer

Pacinian corpuscle

FIGURE 30.15 The Skin: A Sensory Organ
Structures in the skin are sensitive to pain, pressure, and temperature. These receptors transmit stimuli to the brain.

In Chapter 22, we described the skin as an organ that protects us (see Figure 22.9). In Chapter 26, we talked about its role in maintaining homeostasis. Here we look at the skin as a sensory organ.

Structures in the skin allow us to feel pain, pressure, and heat and cold (Figure 30.15). **Free nerve endings,** which extend into the epidermis, are nociceptors; they detect pain. *Pacinian corpuscles* and *Meissner's corpuscles* are mechanoreceptors. Pacinian corpuscles, which are actually below the skin, in the subcutaneous layer, are sensitive to heavy pressure. Meissner's corpuscles, which are found at the very top of the dermis, are sensitive to light touch. Thermoreceptors are sensitive to temperature. *Krause's end bulbs* respond to cold, and *Ruffini end organs,* to heat.

The Ear

■ **The external ear gathers and focuses sound through the pinna, the auditory canal, and the tympanic membrane. The middle ear transmits sound vibrations through structures known as the hammer, anvil, and stirrup to the inner ear. The inner ear consists of the cochlea and the vestibular apparatus. The cochlea contains sensory neurons that transmit the information to the brain's hearing centers. The vestibular apparatus determines body position by detecting the position of its fluids and otoliths in relation to the pull of gravity.**

Human beings are compulsive communicators. The production of sound (speech) and the reception of sound (hearing) have facilitated our organized societal structure and advanced evolutionary status. Any sound, such as clapping or talking, causes the molecules in the air to change their vibration pattern, which, in turn, produces invisible waves in the air, like the ripples that radiate outward on a pond when you throw a pebble in the water. The changes in vibration are energy changes that we perceive with our ears and call *sound.*

Hearing. The human sense receptors that detect changes in vibration are found mainly in the ear. This structure consists of three regions (Figure 30.16). The external ear is the portion that is visible. It is comprised of the pinna, which collects and directs sound waves into the ear, the auditory canal (sometimes called the ear canal), and the tympanic membrane, usually called the eardrum. The tympanic membrane vibrates and responds to sound waves. These vibrations cause corresponding vibrations in three small bones in the middle ear.

The three bones of the middle ear are the hammer, anvil, and stirrup (technically called the malleus, incus, and stapes). These structures act in concert as a lever system, increasing the force of the vibrations (sound waves) to a thin membrane in the cochlea of the inner ear. At one end of the middle ear is the Eustachian tube, which leads to and enters the back of the mouth.

The *inner ear* consists of the cochlea and vestibular apparatus. The *cochlea* is the organ of hearing, and the *vestibular apparatus,* the organ of balance. The *cochlea* is a coiled bony tube. A thin layer of bone inside the tube divides it into three channels. The two larger channels are filled with fluid. Actually, these channels form one continuous tube that doubles back on itself. The tube connects at one end with the oval window and at the other, with the round window.

As sound strikes the eardrum, the vibrations are transferred to the small bones of the middle ear and from the stapes to the oval window of the cochlea of the inner ear. These vibrations create a wave of motion in the fluid of the tube, causing the round window at the other end to bulge. The round window's change in shape dissipates the sound energy. As the fluid moves through the tube, it pushes against the membranes that surround the third channel in the cochlea, the cochlear duct.

To understand what happens next, we have to look at the cochlea in cross-section (Figure 30.17). Resting on the basilar membrane of the cochlear

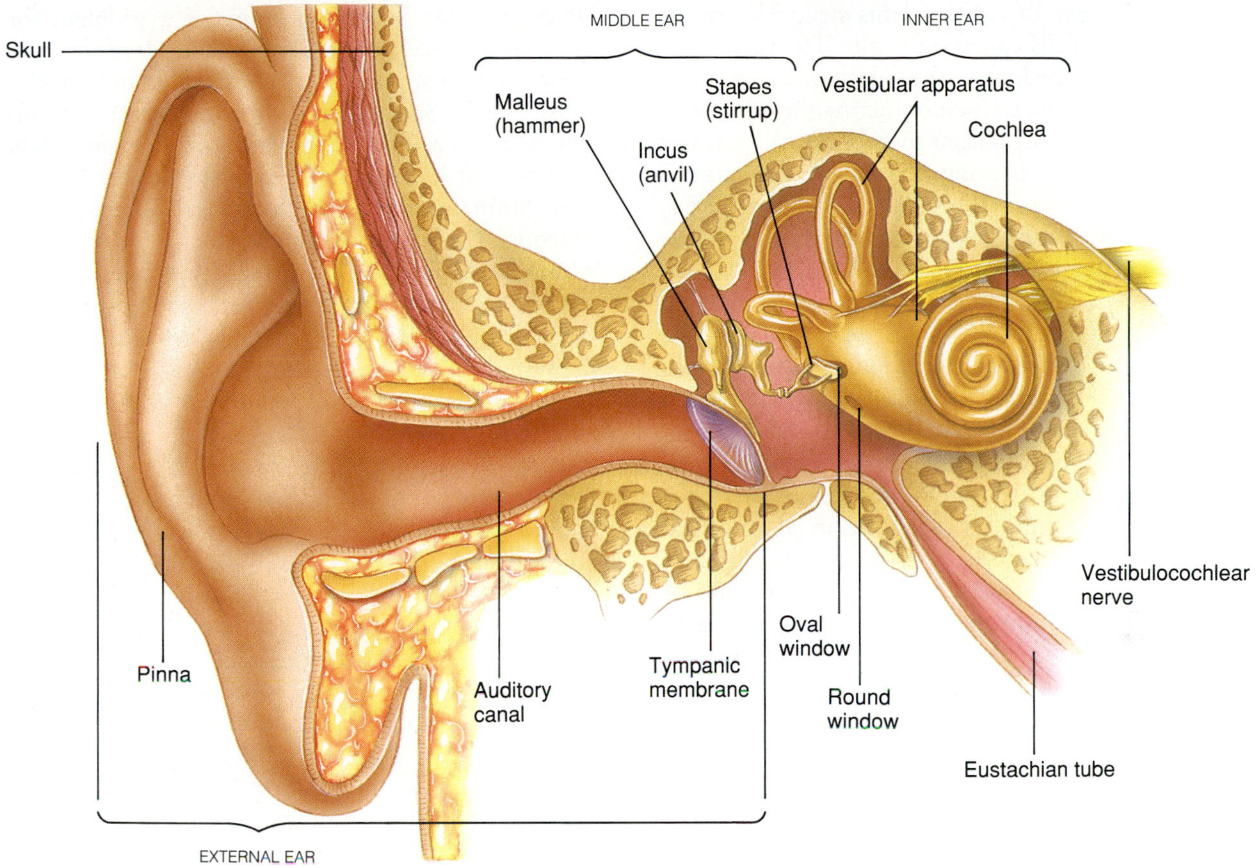

Skull

Malleus (hammer)

Incus (anvil)

Stapes (stirrup)

Vestibular apparatus

Cochlea

Vestibulocochlear nerve

Pinna

Auditory canal

Tympanic membrane

Oval window

Round window

Eustachian tube

EXTERNAL EAR

FIGURE 30.16 The Human Ear
There are three regions of the ear: external, middle, and inner. The organs of the inner ear play a role in both hearing and balance.

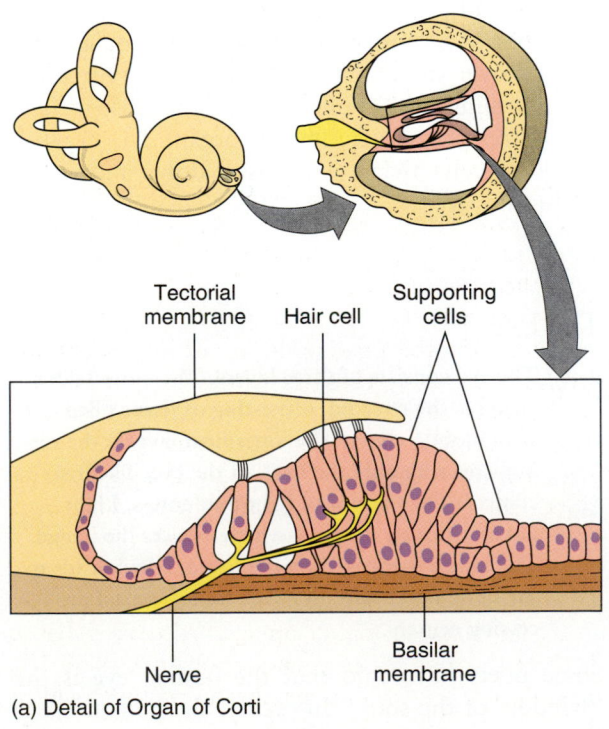

Tectorial membrane

Hair cell

Supporting cells

Nerve

Basilar membrane

(a) Detail of Organ of Corti

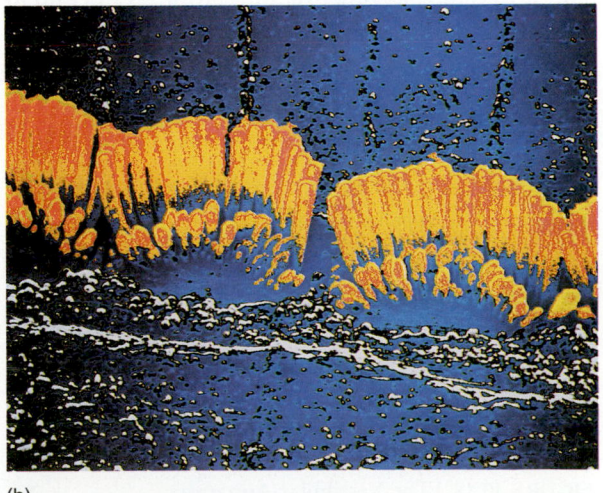

(b)

FIGURE 30.17 The Organ of Corti
(a) In a cross-section of the cochlea, the organ of Corti is visible. The hair cells are auditory receptors. They move in response to the movement of fluid through the cochlea, generating nerve impulses that carry "sound" to the brain. (b) A scanning electron micrograph of the organ of Corti, showing the pattern that the hairs form.

Sense Organs

757

duct is the *organ of Corti*. In this structure are ciliated receptor cells called *hair cells*. The tips of the hair cells are embedded in the gelatinous membrane known as the *tectorial membrane*. As the vibrations in the fluid move the basilar membrane, the sensory hair cells bend, activating neurons leading to the cochlear branch of the vestibulocochlear nerve, which in turn leads to the hearing centers in the cerebral cortex of the brain.

Body Position. The **vestibular apparatus** of the inner ear functions in determining body position. This structure contains the fluid-filled *saccule, utricle,* and *semicircular canals* (Figure 30.18). The saccule and utricle contain small ear stones, or *otoliths,* which are made of calcium carbonate. These stones shift within the fluid as the position of the head moves in relation to the pull of gravity. Sensory hairs within these structures detect the position of the fluid and otoliths, sending the information to the cerebellum, where it is integrated. The cerebellum then sends out stimuli to muscles that help maintain the position and equilibrium of the body with respect to the force of gravity.

The semicircular canals are arrayed in three spatial dimensions. They provide information to the brain about motion, as when a child spins to get dizzy. The fluid in the semicircular canals sloshes around, stimulating hair cells within the canals. The information about motion is transmitted through nerves to the brain, and motion is detected.

Our motion-detecting system evolved on earth and is designed to function in relation to the pull of gravity. Much of the thrill of riding a roller coaster is the result of a disruption of this system, for you are, for a moment, in a weightless environment. The effect of weightlessness is more of a problem for astronauts, since they are in a gravity-free environment far longer than the few seconds on a roller coaster ride. Space sickness, a condition that includes nausea and disorientation, is apparently caused by the fact that the otoliths do not sense a gravitational pull. The earth-evolved system cannot function. There is conflicting sensory input from the vestibular apparatus, eyes, and proprioceptors (receptors that detect changes in body position) in skeletal muscles. The result demonstrates how all the systems in the human body interact.

Stretch receptors provide animals with information about body position and skeletal muscle contraction. These receptors, for instance, provide us with the information that enables us to tie our shoes in the dark. The eyes also play a role in determining body position. (Have you ever tried to stand on one leg with your eyes closed?) All play a role in helping us maintain and coordinate our body position, balance, and movement with respect to gravity.

As you would suspect, invertebrates also have organs of equilibrium. For example, the crayfish, or "crawdad," places a small grain of sand in its organ of equilibrium during *molting* (the shedding of the exoskeleton). The pull of gravity on the sand grain then stimulates the sensory hairs in the organ of equilibrium, providing the crayfish with information about its body position. When I was a child, I would catch a crayfish that was molting, place it in a bucket without sand, and then add some small pieces of iron. The crayfish would proceed to place the pieces of iron in its organ of equilibrium. Then, more out of curiosity than mischief, I would put a magnet over the crayfish. As the magnet attracted the iron filings, they pulled against the sensory hairs at the top of the equilibrium organ. The crayfish would perceive that it was upside down and would try to turn over.

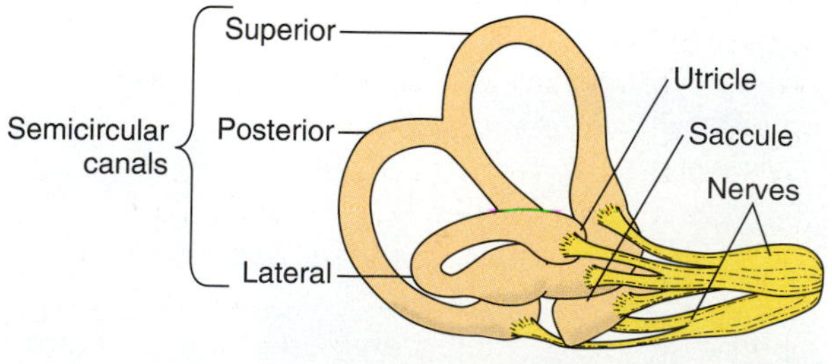

FIGURE 30.18   The Vestibular Apparatus
The vestibular apparatus is the organ of balance. Within the saccule and utricle are layers of calcium crystals called otoliths. These crystals move as the head moves. Hair cells detect their movement and generate a nerve impulse that travels to the brain, and from there to the muscles, which act to restore balance. Hair cells in the semicircular canals also play a role in maintaining equilibrium.

The Eye

■ **The eye consists of three layers. The outer sclera encloses the choroid, consisting of tissue filled with blood vessels. The innermost layer is the retina, covering only the back of the eye. The iris controls the amount of light that enters. Light passes through the lens, which focuses the image. The image is focused on the fovea, which leads to the optic nerve and on to the brain's visual processing centers.**

Some poets have said that the human eye is the "window of the soul." In reality, the human eye is

Chapter 30   Cell-to-Cell Communication: Neural Signals

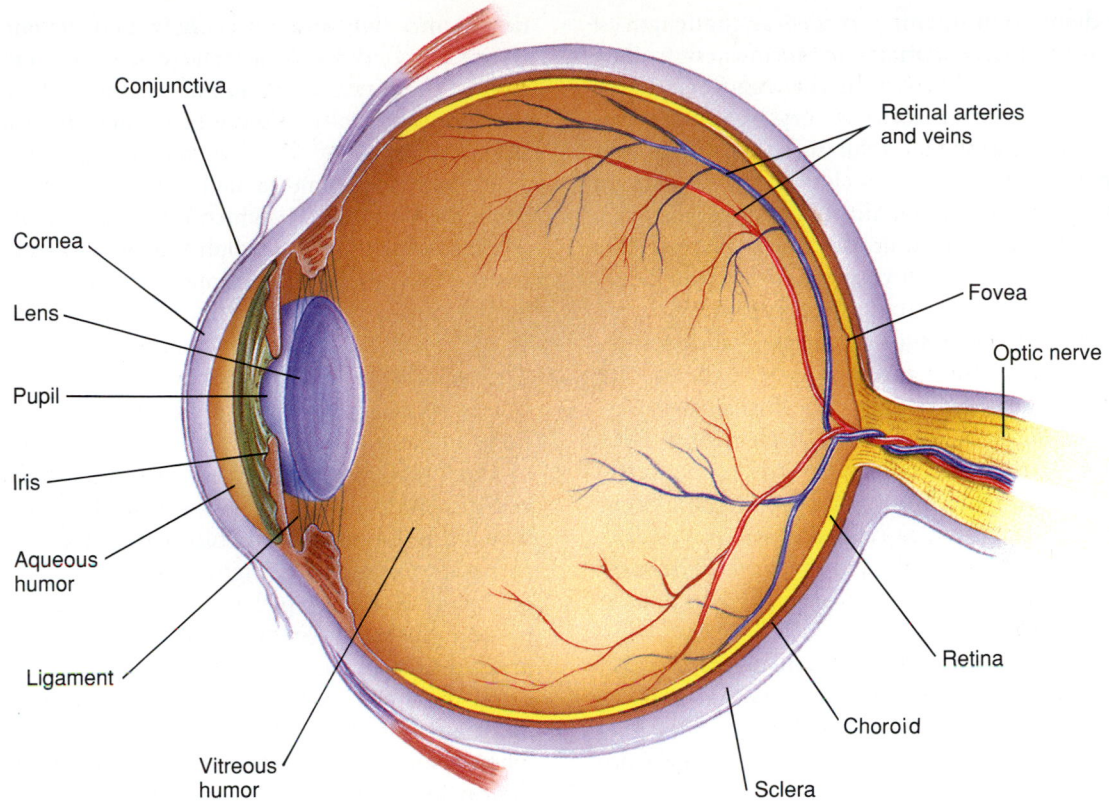

FIGURE 30.19   The Human Eye
The major structures of the human eye.

just a bulge in the brain that protrudes out of the skull. (Nice thought, isn't it? Believe it or not, I prefer the poets' description.) Most animals have *photoreceptor cells,* which detect light. In some animals, like the planarians, these cells are concentrated and organized into structures called eyes. In these organisms, the eye can detect motion but not shape or images. However, in the mollusks, like the octopus, and in almost all vertebrates, the eyes are not only light detectors. They also can perceive images.

The human eye consists of the eyeball and the accessory structures that surround it. The adult human eye is about 1 inch in diameter. Only about one-sixth of it is exposed. The rest is contained within the eye socket of the skull. The accessory structures that surround the eye are the eyelids, eyelashes, eyebrows, lacrimal glands (tear glands), muscles that rotate the eye, and *conjunctiva,* which is the clear membrane covering the front of the eyeball. (If you have ever had "pinkeye," you actually had conjunctivitis, which is an infection of the conjunctiva.)

Figure 30.19 shows the structure of the human eye. Below the conjunctiva is the *cornea,* a fibrous layer that covers the *iris,* the pigmented (colored) portion of the eye. The black hole in the middle of the iris is the *pupil.* It is through the pupil that light enters the eye. Behind the pupil is the *lens,* a transparent layer of protein fiber. The lens is held in place by ligaments. The fluid-filled chamber in front of the lens is called the *aqueous humor;* that behind the lens is called the *vitreous humor.*

Surrounding the vitreous humor are three layers of tissue. The white fibrous outer layer is called the *sclera.* (The cornea is actually an extension of the sclera.) The middle layer, the *choroid,* consists of dark pigmented tissue filled with blood vessels. It connects at the front of the eye with the iris. The *retina,* the inner layer, covers only the back of the eye chamber and is made up of nerves and pigmented tissue. The *optic nerve* passes from the retina, through the choroid and sclera out the back of the eye, to the brain.

Many have described the vertebrate eye as a camera, and, in a sense, the comparison is valid. In a camera, reflected light passes through the lens onto photographic film, which reacts chemically with the light energy to reproduce the scene. In "nonautomatic" cameras you can control the size of the diaphragm opening as well as the length of time the diaphragm remains open, thus controlling the amount of light passing through the lens to strike the film. If the scene is very bright, the diaphragm opening is very small and opens for a very short period. If the scene is cloudy, the diaphragm is opened farther and remains open longer.

The process involved in the camera is much like that which takes place in your eye. When you "see," light is reflected from the object you are viewing. That reflected light passes through the cornea, pupil, and lens, then is focused on the retina, which contains photosensitive cells. The focusing process involves three actions: Small muscles control the curvature of the lens, the iris constricts to control the amount of light entering the eye (much like the diaphragm of a camera), and the eye moves. The image formed on the retina is inverted. (The same is true of the image projected onto the film in your camera.)

The iris, the pigmented portion of the eye surrounding the pupil, contains muscles that regulate the diameter of the pupil opening, thus controlling the amount of light that enters. Have you ever noticed how large one's pupils become when entering a darkened room or how they decrease in diameter in bright sunlight? This adjustment is called dark-light accommodation.

The retina contains two types of photoreceptor cells, called **rods** and **cones** (Figure 30.20). Rods respond to low-intensity light and are responsible for night vision. They also contain a light-sensitive pigment known as *rhodopsin* (visual purple). When rhodopsin absorbs light, it temporarily breaks down into two subunits, *opsin* and *retinal*. The chemical breakdown somehow triggers transmission of the visual nerve impulse. Opsin and retinal are then chemically reformed to restore the rhodopsin in the cells.

Retinal (sometimes called visual yellow) is a derivative of vitamin A, which explains why vitamin A is necessary for good night vision — it is needed for the resynthesis of rhodopsin. (That fact should not encourage you to take megadoses of vitamin A — too much vitamin A can cause vitamin A poisoning, which destroys lysosomal membranes and causes spontaneous bone fractures and lesions in bone and cartilage.)

Have you ever noticed that in dim light you cannot discern color? That is because the *cone* cells, which are responsible for color vision, respond only to higher intensities of light. There are three types of cone cells. Each type has photopigments sensitive to certain wavelengths. There are red-sensitive, green-sensitive, and blue-sensitive cone cells. Because these cells have overlapping sensitivities, humans are able to discern many colors. For example, when both red and blue cone cells are stimulated, their responses overlap, so we interpret the color as purple.

Cone cells also provide a "sharper" image. This mechanism results in a hawk's sharp vision — the retinas of its eyes have a million cones per square millimeter. As a result, it has over eight times the visual acuity of humans. The retina also determines visual acuity through the density of cones. Cones are most highly concentrated in the region of the retina known as the fovea. This is the region of

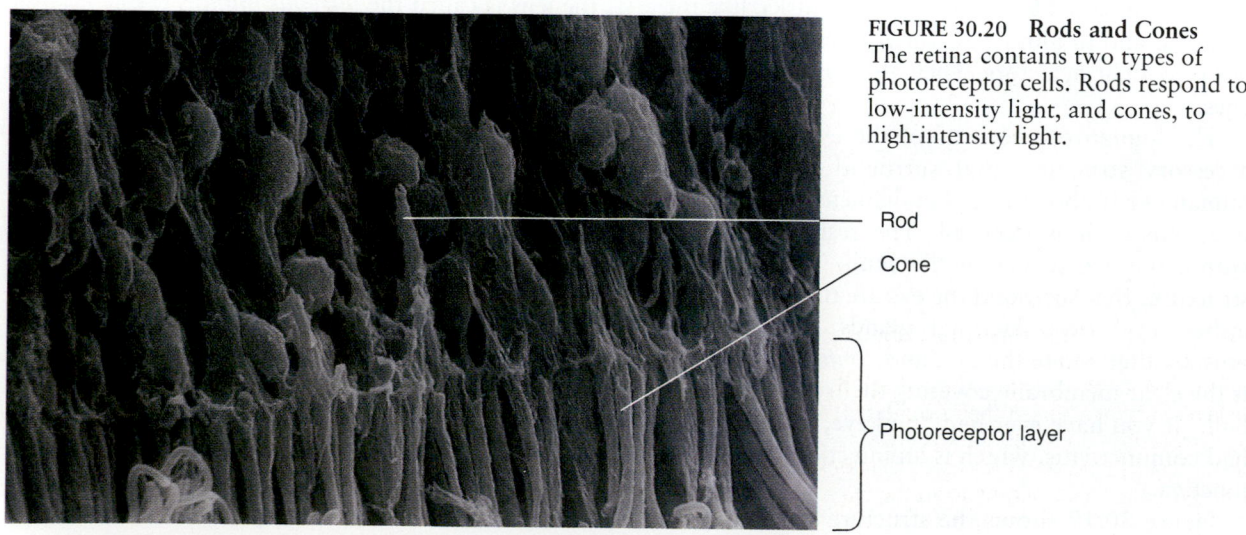

FIGURE 30.20 Rods and Cones
The retina contains two types of photoreceptor cells. Rods respond to low-intensity light, and cones, to high-intensity light.

Rod

Cone

Photoreceptor layer

Chapter 30   Cell-to-Cell Communication: Neural Signals

greatest visual acuity because it consists almost exclusively of cone cells.

In a sense, the arrangement of the layers of the retina is backward (in birds, however, the receptors form the first layers). You would expect the rods and cones to be nearest the incoming light, but they are not. Instead, they are at the very rear of the retina, resting on a layer of pigmented epithelial cells. As light enters, it first strikes a transparent layer of ganglion cells. It then strikes a transparent layer of bipolar cells, and then the rods and cones.

When rods and cone cells are stimulated, they transmit their impulses through synapses with the bipolar cells that lie in front of them. The bipolar cells transmit the visual impulse to ganglion nerve cells. The axons of these nerve cells form the optic nerve, which enters the thalamus and then other visual processing centers of the brain. Figure 30.21 shows the process schematically.

The mechanisms of rods and cones grant special selective advantages, and their prevalence has responded to selective pressures. For instance, would you expect nocturnal animals to have more rods than cones? You are correct; they have more rods, which enables them to have keener night vision.

Taste and Smell — Chemoreception

■ **Olfactory receptors detect chemical stimuli, and taste receptors are modified olfactory receptors.**

As mentioned in Chapter 29, chemical coordination probably began with pheromones, chemicals secreted into the environment. When the receptors of another organism detect these chemicals, a response is elicited (see Figure 29.2). Because pheromones were the basis of the first chemical communication between organisms, their ability to receive, detect, or smell them evolved early. Hence olfaction — the sense of smell — is one of the most primitive senses.

Olfactory (smell) receptors are modified epithelial cells that respond to molecular stimulation. Taste receptors are modified olfactory receptors. Both types function as chemoreceptors and are responsible for the sensations that we describe as taste and smell. Olfactory receptors in the nasal passages send their information through the olfactory nerves into the olfactory bulb of the brain, where the information is processed (Figure 30.22). Taste buds in the mouth synapse with sensory nerve fibers that transmit signals to the taste centers of the brain, where the information is interpreted.

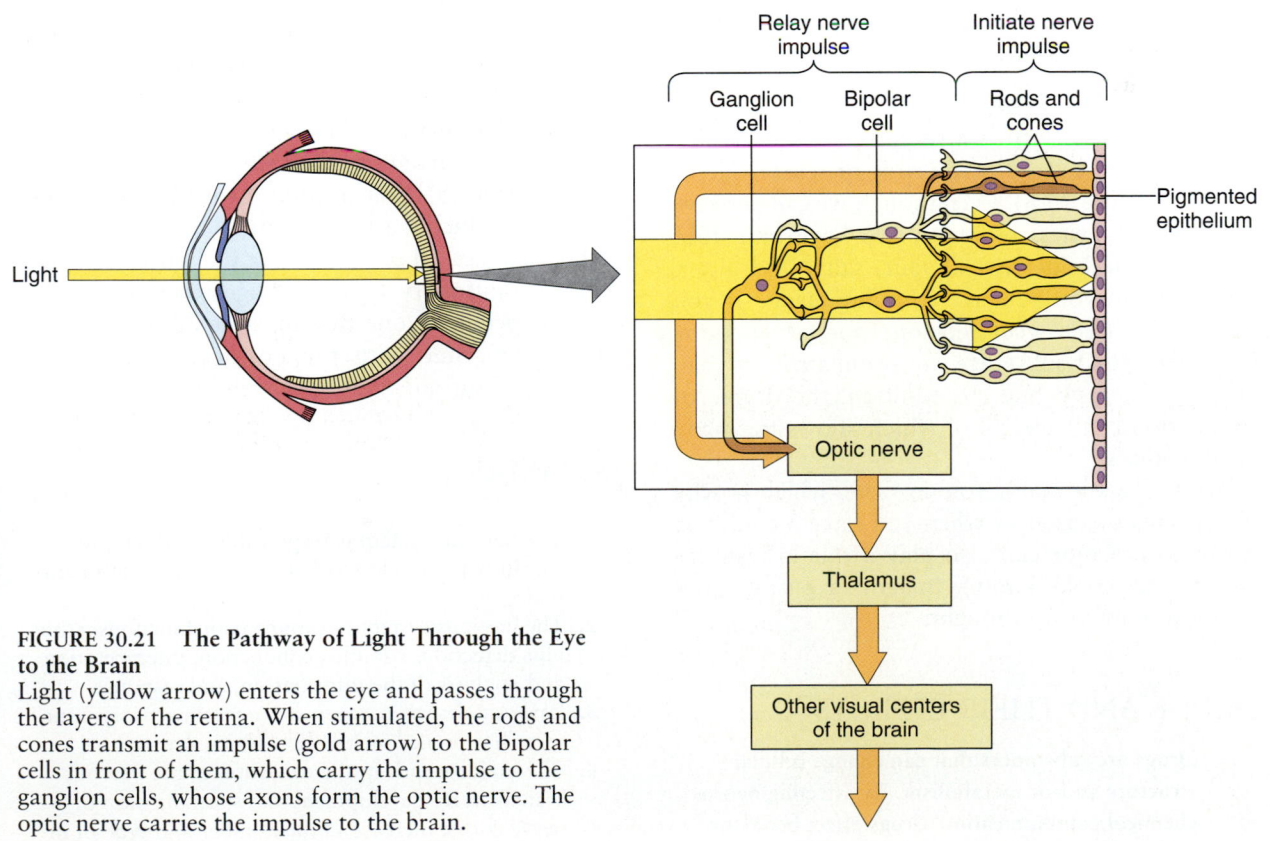

FIGURE 30.21   The Pathway of Light Through the Eye to the Brain
Light (yellow arrow) enters the eye and passes through the layers of the retina. When stimulated, the rods and cones transmit an impulse (gold arrow) to the bipolar cells in front of them, which carry the impulse to the ganglion cells, whose axons form the optic nerve. The optic nerve carries the impulse to the brain.

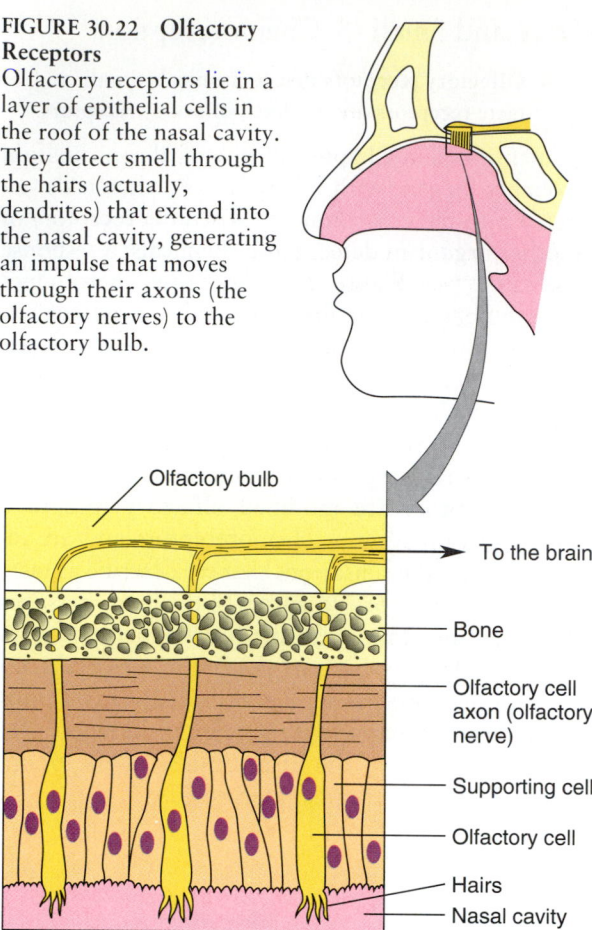

FIGURE 30.22 Olfactory Receptors
Olfactory receptors lie in a layer of epithelial cells in the roof of the nasal cavity. They detect smell through the hairs (actually, dendrites) that extend into the nasal cavity, generating an impulse that moves through their axons (the olfactory nerves) to the olfactory bulb.

Olfactory bulb

To the brain

Bone

Olfactory cell axon (olfactory nerve)

Supporting cell

Olfactory cell

Hairs

Nasal cavity

Taste buds consist of a receptor portion (which is not a nerve cell) plus the ends of sensory nerve cells (see Figure 23.8). Even though we can perceive hundreds of tastes, in reality we have only four types of taste buds: sweet, sour, salty, and bitter. But how can you perceive so many tastes? If you have ever eaten sweet-and-sour pork, you know that a particular substance can stimulate more than one type of taste bud. In addition, the degree of stimulation can vary, all of which allows for many combinations.

Smell plays a major role in taste, which is why food seems so tasteless when you have a cold or a stuffy nose. Color can also play a role in taste (as all gourmet cooks know). Imagine a purple steak — not too inviting a thought.

DRUGS AND THEIR EFFECTS

■ **Drugs are substances that can change cellular structure and/or metabolism. By affecting neurochemical communication, drugs affect behavior and bodily functioning.**

We are constantly bombarded by news articles about the drug abuse epidemic. Star athletes die of cardiac problems associated with cocaine abuse. Teenagers die of crack abuse. Movies are made about the drug and alcohol addiction of the wife of a former president. So let us briefly discuss drugs. A *drug* is a chemical agent that is not a nutrient but that can effect changes in cellular structure and/or metabolism. Some drugs have a marked effect in neurochemical communication and can mask, edit, speed up, slow down, and mix up neurochemical communication.

For example, many drugs that affect the nervous system act on the synapse. Amphetamines (uppers, or methamphetamines) increase the synthesis and/or release of excitatory neurotransmitters, such as norepinephrine, across the synapse. That is the basis for the state of intense excitability of an individual abusing amphetamines. Other drugs inhibit, modify, or enhance the release of chemical messengers in the brain. The majority of tranquilizers (downers) block receptor sites for brain chemicals such as dopamine and target cells from taking up norepinephrine and serotonin. Table 30.3 lists some of the more commonly abused drugs and summarizes their effects on the brain and nervous system.

We have discussed chemical coordination at the neural level, investigated the marvelous chemistry of a neural impulse, described the human nervous system, and discussed the senses. The nervous system is indeed a complex entity, a wonder of evolution. Today, with our ability to distort reality and shear the fabric of our social structure through drug abuse, we may be at a point in our history where we are producing a large population of individuals who have ceased to care about the world or about their evolutionary legacy. Drug abuse is indeed a serious problem, one that must be addressed if we are to continue with the positive legacy of our human evolution.

SUMMARY

1. The nervous system is responsible for the rapid transmission of electrochemical information in animals.
2. The nervous system has four major functions: stimulus detection, stimulus conduction, interpretation and analysis of the information, and response.
3. Early single-celled organisms contain sensory structures but do not include actual neurons.
4. The major trends in the evolution of the nervous system were: development and specialization of nerve cords, increase in the number and type of interneurons, cephalization and evolution of the

TABLE 30.3
Mood-Altering Drugs

Drug	Mechanism of action	Immediate effects	Long-term effects
Caffeine	Facilitates synaptic transmission	Stimulant; promotes alertness; increases motor activity; produces insomnia	Increases chance of heart disease and heart failure; contributes to hypertension; can cause ulcers and peripheral vascular disorders; has been linked to cancer of the pancreas
Nicotine	Facilitates synaptic transmission	Stimulant	Increases chance of heart disease, heart failure, stroke, peripheral vascular disorders, emphysema, and ulcers; contributes to hypertension and cancer
Alcohol	Depresses central nervous system	Sedative, after short-lived stimulant effects; decreases motor coordination and alertness; produces euphoria	Contributes to atherosclerosis, hypertension, and clinical depression; increases chance of heart disease and heart failure; aggravates diabetes; can cause chronic organic brain syndrome, liver cirrhosis, and ulcers
Marijuana, hashish	Unknown	Enhances sensory perception; produces mild euphoria; promotes drowsiness and poor motor coordination	Produces psychological dependence
Cocaine	Inhibits uptake of norepinephrine	Promotes local anesthesia, euphoria, and alertness	Causes addiction, breakdown of nasal membrane, vitamin deficiency, anemia, severe weight loss; overdose can cause cardiac arrest
Amphetamines	Increase synthesis or uptake of norepinephrine and dopamine; increase activity of biogenic amines in brain	Stimulants; lessen fatigue and depression; suppress appetite	Cause malnutrition, exhaustion, poor judgment, hallucinations, strong addiction, and psychosis; death can result from overdose
Barbiturates	Inhibit synthesis and action of norepinephrine and dopamine	Sedatives; promote drowsiness, euphoria, and irritability	Cause strong addiction; withdrawal causes convulsions; overdose can cause death
Psilocybin, mescaline	May enhance effects of biogenic amines	Increase sensory awareness and cause hallucinations	Inspire suicidal tendencies, unpredictable behavior, and psychosis; may have neurological effects
LSD	Inhibits brain serotonin	Promotes sensory awareness and rich visual hallucinations	Causes psychosis, unpredictable behavior, suicidal tendencies; may have neurological effects
Narcotics (opium, morphine, heroin, methadone, Demerol, codeine)	Depress central nervous system	Relieve pain; induce euphoria, drowsiness, lethargy; produce constipation and loss of appetite	Cause strong addiction, liver dysfunction, and systemic infection; can kill by depressing respiratory center

brain, greater protection for nerves, one-way nerve impulse conduction, and development of specialized sensory receptors.

5. There are three major types of neurons. Sensory neurons detect stimuli; interneurons interpret information; and motor neurons carry information to an effector.

6. Neurons receive, integrate, relay, and respond to messages.

7. There are two classes of cells in nervous tissue: neurons and neuroglia.

8. Neurons have three basic parts: the dendrites, which receive a nerve impulse; the cell body; and the axon, which transmits the information to the dendrites of an adjacent cell or to an effector. Myelin often surrounds the axon and acts as an insulator.

9. Nerves are composed of bundles of axons and dendrites surrounded by a protective covering of connective tissue. These structures form neural pathways over which nerve impulses travel.

10. The transmission of a nerve impulse is due to changes in the permeability of the nerve cell plasma membrane.

11. In a typical resting neuron, the inside of the nerve cell has a lower electrical charge than the outside. This produces a voltage difference, called the resting potential, across the plasma membrane.

12. The sodium-potassium pump actively transports sodium and potassium through the membrane of the neuron.

13. As the nerve impulse is being propagated, the nerve cell membrane depolarizes in sequence, allowing for the impulse to move along the entire length of the axon. Once the impulse has passed, the neuron is repolarized.

14. The speed of the nerve impulse is directly proportional to the diameter of the axon. In vertebrates, interruptions in the myelin sheath called the nodes of Ranvier increase the speed of nerve impulses by allowing impulses to skip from one node to another.

15. As the nerve impulse travels from the axon of one neuron to the dendrites of the next, in most cases it has to cross a gap called a synapse. To enable it to do so, the axon releases neurotransmitter molecules that are received by the adjacent neuron.

16. One of the most primitive nerve pathways is the simple reflex arc. Reflexes are involuntary. They usually perform the same way, involve only a few sensory, association, and motor neurons, and do not involve the brain.

17. The human nervous system consists of the central nervous system (brain and spinal cord) and the peripheral nervous system (those nerve paths lying outside the central nervous system).

18. The central nervous system functions in the interpretation and analysis of neural information.

19. The spinal cord extends from the brain through the vertebral column.

20. The peripheral nervous system is divided into two major parts. The somatic nervous system consists of sensory and motor neurons that transmit information to and from the central nervous system. Most of the motor responses are voluntary. The autonomic nervous system is responsible for the activities that we normally consider involuntary.

21. The autonomic nervous system can be divided into the sympathetic and parasympathetic divisions. In general, these systems work in opposite ways. The sympathetic neurons secrete the neurotransmitter norepinephrine, which usually speeds up activities, and the neurons of the parasympathetic system secrete acetylcholine, which generally slows down activities.

22. The sense organs are specialized for the reception of stimuli.

23. The skin contains pain, touch, and heat and cold receptors.

24. The ear is a specialized sensory receptor to detect vibratory stimuli (sound waves).

25. The inner ear of mammals contains the vestibular apparatus, which detects body position in relation to the pull of gravity and motion. Stretch receptors in the muscles provide information about body position and skeletal muscle contraction.

26. The eye is specialized for the transduction of wavelengths of light to nerve impulses that our brain interprets as vision.

27. The nose and tongue function in the detection of olfactory stimuli and taste.

28. Drugs can effect changes in the brain and nervous system.

KEY TERMS

nerve net	synapse
cephalization	neurotransmitter
neuron	reflex arc
sensory neuron	central nervous system
interneuron	spinal cord
motor neuron	peripheral nervous system
dendrite	somatic nervous system
cell body	autonomic nervous
axon	system
nerve	sympathetic division
ganglion	parasympathetic division
resting potential	free nerve ending
sodium-potassium pump	vestibular apparatus
action potential	rod
myelin sheath	cone
node of Ranvier	

Chapter 30 Cell-to-Cell Communication: Neural Signals

REVIEW QUESTIONS

True/False

1. From a functional perspective, the nervous system provides slow, long-term coordination.
   true   false   (page 736)

2. Only the vertebrates have a nervous system.
   true   false   (pages 736–737)

3. The propagation of a nerve impulse is due to changes in the permeability of the nerve cell membrane that allow for a voltage difference across the membrane.
   true   false   (pages 740–742)

4. The central nervous system consists of the brain and spinal cord.
   true   false   (page 749)

Fill in the Blank

1. Neurons that carry information to an effector are called _____ neurons. *(page 739)*

2. The initial depolarization of the nerve cell membrane involves the movement of _____ ions into the cell. *(page 741)*

3. The _____ of the neuron secretes the neurotransmitter substance. *(pages 743–744)*

4. For convenience, the brain is divided into regions. *(page 750)*

Multiple Choice

1. Coordination via the nervous system tends to differ from that produced by the endocrine system because the nervous system:
   a. Is quick, precise, and localized
   b. Is slower and more pervasive
   c. Does not require conscious activity
   d. Has long-lasting effects
   (page 736)

2. In comparison with other cells, nerve cells show a higher degree of:
   a. Metabolism
   b. Growth
   c. Contractility
   d. Irritability
   e. Elasticity
   (page 736)

3. When you tie your shoes in the dark, you are probably depending primarily on your:
   a. Proprioceptors
   b. Vestibular apparatus
   c. Rods and cones
   d. a and b
   (page 758)

4. The photoreceptor cells of the eye are located in the:
   a. Sclera
   b. Iris
   c. Retina
   d. Optic nerve
   (page 759)

5. The peripheral nervous system consists of:
   a. Nerves within the spinal cord
   b. Only sensory nerves
   c. Only motor nerves
   d. Both motor and sensory nerves
   (page 752)

6. Which of the following receptors is *incorrectly* paired with what it senses?
   a. Chemoreceptors — chemicals
   b. Photoreceptors — pain
   c. Thermoreceptors — heat
   d. Nociceptors — pain
   (page 755)

7. Gray matter is composed of:
   a. Nerve cell bodies
   b. Nerve axons
   c. Pons
   d. Cerebellum
   (page 751)

8. A substance that increases the synthesis and/or release of excitatory neurotransmitters is called:
   a. Amphetamine
   b. Alcohol
   c. Barbiturate
   d. Heroin
   (page 763)

9. Which of the following brain structures is *incorrectly* matched with its function?
   a. Cerebrum — thought
   b. Thalamus — sleep
   c. Cerebellum — memory
   d. Pituitary — hormones
   (page 751)

10. Primitive nerve nets are characteristic of what animal group?
    a. Sponges
    b. Cnidarians
    c. Vertebrates
    d. Flatworms
    (page 736)

Discussion

1. Discuss the evolution of the nervous system.

2. What are the basic functions of all nervous systems?

3. Speculate on the evolutionary significance of cephalization.

4. What is the structure of a typical neuron? Describe three types of neurons.

5. Briefly summarize the molecular events involved in the transmission and propagation of a nerve impulse.

6. Relate the nature of the nerve cell membrane to the nerve impulse.

7. How do nerve cells communicate with each other? In other words, what activities occur at the synapse?

8. Describe and discuss the reflex arc.

9. Discuss the major divisions of the human nervous system and their functions.

10. Speculate on the evolutionary significance of the sympathetic and parasympathetic nervous systems.

11. What are sensory structures?

12. What senses are involved when you find your keys in the dark and open the door?

13. Discuss some of the effects of four or five drugs on the central nervous system.

31

Animals: An Evolutionary Perspective — Invertebrates

What Is an Animal?
The Origin of Animals
Important Animal Characteristics
Major Phyla
Echinodermata: Sea Stars and Their Relatives
Summary

We have devoted a lot of our discussion of biology to other organisms, without discussing our own kingdom — the kingdom of animals (Figure 31.1). What are animals? How did animals originate? How can we develop a system to help explain the evolutionary relationships among the animal phyla? What are the major animal phyla, and what are some of their characteristics? These are the types of questions we will answer in this chapter.

WHAT IS AN ANIMAL?

- **Animals are multicellular heterotrophs. Biologists have grouped animals into more than 30 phyla.**

Above: These sea anemones at first glance appear to be flowers, but in fact are invertebrate animals.

As mentioned in Chapter 18, animals are called animals because they are "animated"— they move. But why do they move? They move for two key reasons: to avoid being eaten or to find something to eat. Yes, the major characteristics of all animals, whether microscopic or macroscopic, is that they are *heterotrophs* — they cannot make their own food. Rather, they must take food into their bodies to live. The majority of animals ingest their food in the form of other organisms or their decomposing remains. In addition, animals are multicellular organisms. Because most move, nearly all (except for sponges, jellyfish, and their relatives) have nervous tissue, which is responsible for the rapid conduction of information, and muscle tissue, which contracts and moves the parts of their bodies. Biologists have grouped animals into more than 30 phyla. However, in this chapter and the next, we will limit our discussion to the animal phyla listed in Table 31.1. Now that we

FIGURE 31.1 Animals
A fish swimming within a coral reef illustrates the two major groups of animals: the animals without backbones — the invertebrates — represented by the coral, and the animals with backbones — the vertebrates — represented by the fish.

have an idea of what animals are, let us discuss how animals originated and how they became different from each other.

THE ORIGIN OF ANIMALS

■ **Animals are hypothesized to have originated from heterotrophic colonial flagellates. The first protoanimal was probably a flattened multicellular, multilayered animal similar to the planuloid larval forms found in some animal groups today.**

There are several schools of thought as to how animals originated. Some biologists speculate that they evolved from a multinucleate protistan with cilia. As you might remember from your high school biology

TABLE 31.1
Animal Phyla Discussed

Phylum	Examples	Number of known species
Porifera (sponges)	Tubular, cuplike, vaselike sponges	9,000
Cnidaria (cnidarians)	Jellyfishes, corals, sea anemones, hydra	11,000
Ctenophora (comb jellies)	Comb jellies	100
Platyhelminthes (flatworms)	Planaria, flukes, tapeworms	20,000
Nemertea (ribbon worms)	Ribbon-shaped worms with a proboscis	800
Nematoda (roundworms)	Pinworms, hookworms	80,000
Rotifera (rotifers)	Small animals with cilia around the mouth	1,800
Mollusca (mollusks)	Chitons, snails, slugs, clams, squids, octopuses	50,000
Annelida (annelids)	Earthworms, leeches, polychaetes	15,000
Arthropoda (arthropods)	Crabs, crayfish, spiders, insects	1,000,000+
Echinodermata (echinoderms)	Sea stars, sea urchins, sea cucumbers	6,000
Chordata (chordates)	Invertebrate chordates: Tunicates, lancelets	2,100
	Vertebrates: Fishes	21,000
	Amphibians	3,900
	Reptiles and birds	15,750
	Mammals	4,500

course or from Chapter 18, some single-celled protistans have cilia and a large and small nucleus like the *paramecium*. It is hypothesized that this multinucleate protistan became subdivided by membranes into small cells and became the ancestor of the first multicellular animals. But as we study the animal kingdom further, you will see that this hypothesis only helps to explain why the animals are multicellular; it does not explain the development of radial and bilateral body forms. In other words, this **syncitial hypothesis** that animals arose from a multinucleate protistan is not well accepted by most biologists today. Another hypothesis, called the **colonial hypothesis,** is based on the idea that the ancestral animals were heterotrophic colonial flagellates (Figure 31.2a). In essence, the colonial flagellates were free-swimming hollow balls of eukaryotic cells. This configuration would also promote a front (anterior) and rear (posterior) orientation. These colonies had distinct body cells and reproductive cells, indicating the beginning of **cellular specialization.** The colonial flagellate probably resulted in *choanoflagellate* or **collar cells** found in the choanocyte cells of modern sponges (page 776). Most supporters of the colonial hypothesis agree that these hollow colonies gave rise to all animals, including the radial symmetrical jellyfish and sea anemones and members of the bilateral phyla. But how did this happen? It is speculated that as the cells divided in these hollow colonies, an inner layer of cells developed in the center, producing the first protoanimal with an outer layer of flagellated cells that could be used for locomotion and an inner layer of cells that could be used for digestion and reproduction. This hypothetical animal is called a **planuloid** and is very similar to the larval form of some animal groups (Figure 31.2b).

If the animal ancestor were planuloid-like, what major evolutionary events took place to produce the varied forms of animals that exist today? How do we as biologists decide which characteristics are important in categorizing animals?

IMPORTANT ANIMAL CHARACTERISTICS

■ **From an evolutionary perspective, biologists consider body symmetry and cephalization, number of germ layers, type of body cavity, and embryonic development important in the classification of animals.**

As mentioned in Chapter 18, in a natural taxonomy scheme, biologists decide on characteristics that demonstrate evolutionary relationships. With re-

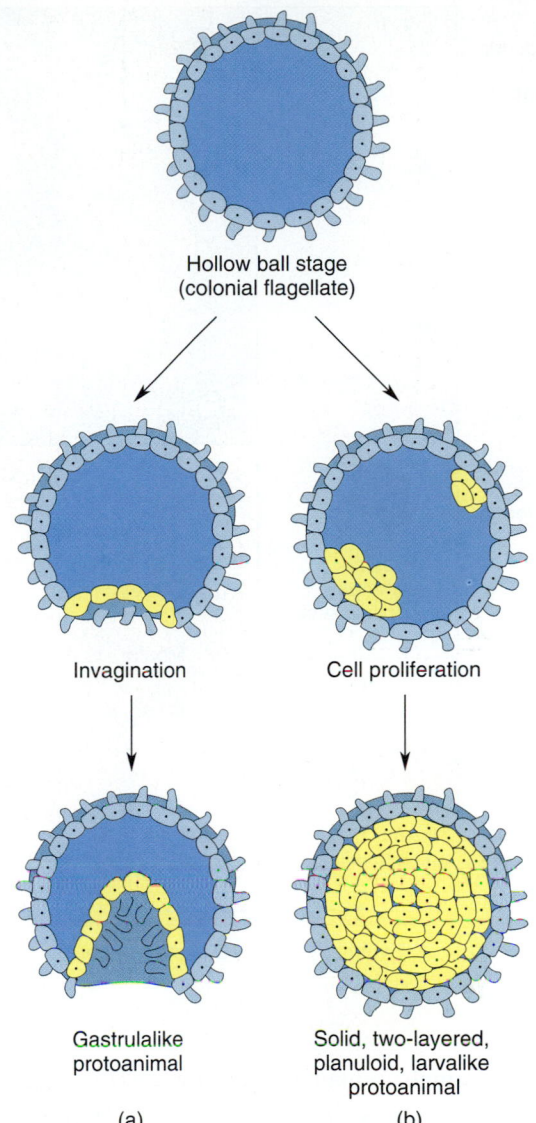

FIGURE 31.2   Hypothetical Models of Protoanimals
(a) In one model, the colonial flagellates produced a hollow-ball, two-layered protoanimal through invagination, resulting in a gastrulalike protoanimal.
(b) A more accepted model is that the protoanimal evolved from cell proliferation, resulting in a solid, two-layered, planuloid, larvalike form.

spect to the increasing diversity of animal groups, the following characteristics are considered important: body symmetry and cephalization, number of germ layers, type of body cavity, and embryonic development (Figure 31.3). In Chapter 18, on pages 415 and 416, we introduced you to some of these classification criteria.

Body Symmetry and Cephalization

You might wonder why body symmetry is an important characteristic, but think about it (Figure

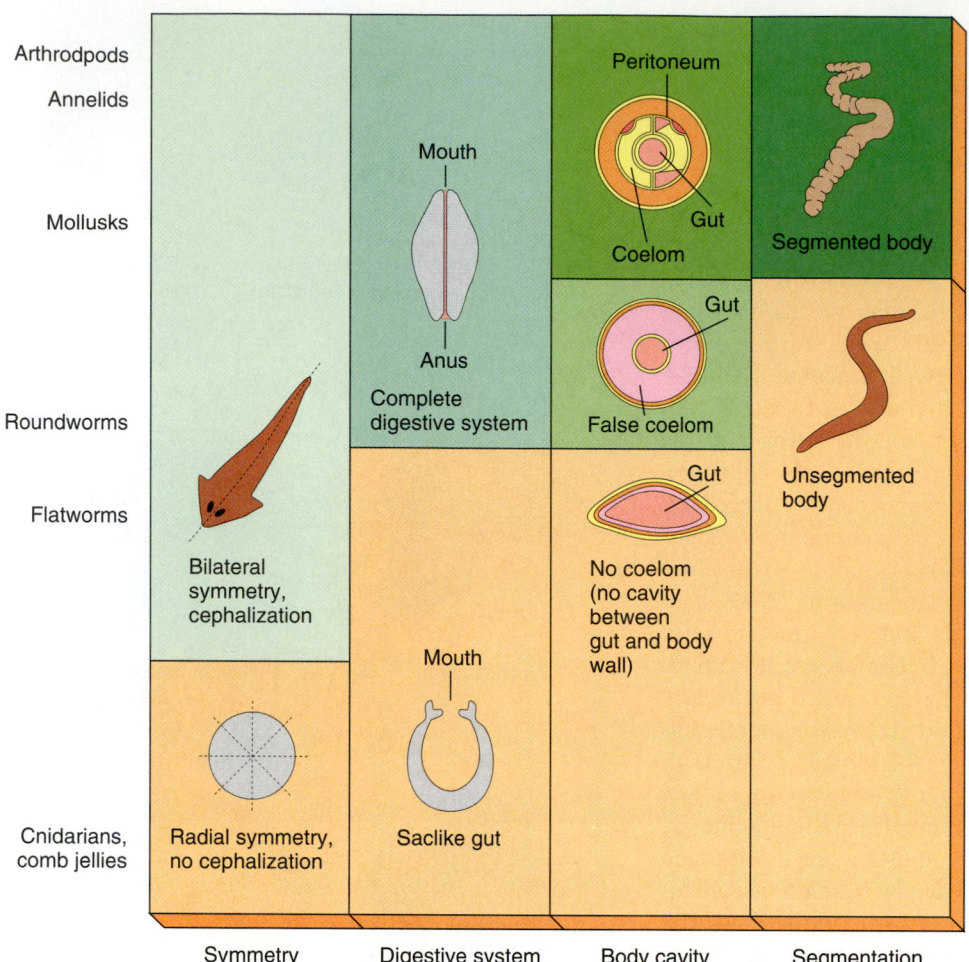

Arthrodpods
Annelids

Mollusks

Roundworms

Flatworms

Cnidarians, comb jellies

Peritoneum
Mouth
Gut
Coelom

Segmented body

Anus
Complete digestive system

Gut
False coelom

Unsegmented body

Gut
No coelom (no cavity between gut and body wall)

Bilateral symmetry, cephalization

Mouth

Radial symmetry, no cephalization

Saclike gut

Symmetry   Digestive system   Body cavity   Segmentation

31.4). If animals remained radially symmetrical, they would have never had a front or back end. *Radial symmetry* refers to the observation that the body parts radiate out from a central axis, like slices from a round pizza. In animals such as a sea anemone, all parts radiate out from the center. Animals like us exhibit *bilateral symmetry*. This is because we have a right and left half that are mirror images of each other — and more important, because we have a head end or an end that goes first. This evolution of a head end, within which most of the sensory organs are found, allows an animal, as it crawls or moves around, to become selective in approaching its environment. It can sense whether food is present, in which case it should continue to move forward, unless danger is present, when it should withdraw. This concentration of the senses in the anterior end is called **cephalization.** Not only does bilateral symmetry allow for cephalization, but it also allows for right and left half pairs of structures, such as paired nerves, muscles and brain parts.

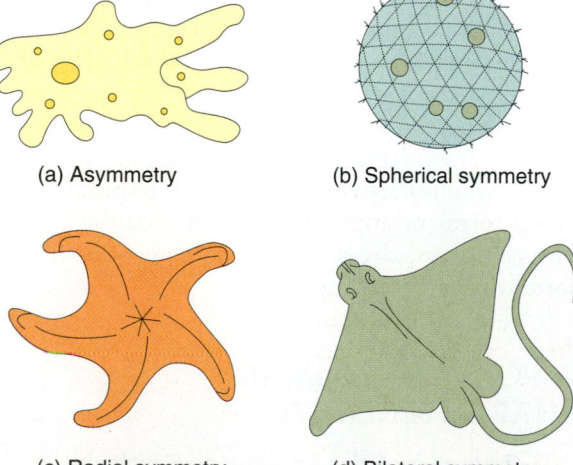

(a) Asymmetry

(b) Spherical symmetry

(c) Radial symmetry

(d) Bilateral symmetry

FIGURE 31.4 **Symmetry**
Body symmetry is an important characteristic for understanding evolutionary relationships among animals. (a) An amoeba is without symmetry. (b) *Volvox globator* — a freshwater organism — is spherical. (c) The starfish is radially symmetrical, with parts extending from a central axis. (d) This eagle ray is bilaterally symmetrical, each side a mirror image of the other.

Number of Germ Layers

Germ layers are the embryonic layers of cells in an animal. As an animal embryo develops, it forms a *gastrula* (except sponges) which gives rise to layers of cells termed *germ layers*. The outermost layer is called the *ectoderm,* which gives rise to the outer covering of an animal, and the *endoderm,* which develops into the inner structures — the gut and its outpockets, such as the lungs, gills, or liver. The *cnidarians* (hydra, coral, sea anemones) have only two embryonic germ layers, so they are called *diploblastic.* Remember that the cnidarians are also radially symmetrical and do not move much. Animals groups other than the cnidarians and the sponges have three germ layers. Their embryos are called *triploblastic* and have a third germ layer that develops between the ectoderm and endoderm. This third layer is called the *mesoderm* or middle layer; it is an important evolutionary development. It is within this layer that muscles and a tube-within-a-tube body plan develops.

Type of Body Cavity

All cells must obtain nutrients and get rid of wastes. Diffusion has limits; the cells in the innermost part of an organism's body have difficulty in ridding themselves of wastes or obtaining food. With a flattened body form, as in the flatworms, the various cells are close to the environment and waste and food exchange can occur more easily and efficiently. Flatworms have no body cavity (coelom), so they are called **acoelomates** and belong to the phylum Platyhelminthes (flatworms are triploblastic animals that exhibit this body form). The other triploblastic animals have a tube-within-a-tube body structure. A body cavity lies between the gut and the outer body. Some organisms, like the roundworms (nematodes), have a coelom that is not lined with a sheet of epithelial cells called the *peritoneum.* This type of coelom is called a false coelom or **pseudocoelom.** Other animals, such as humans, have a true body cavity or coelom because it is lined with a peritoneum (Figure 31.5). If you have ever known someone who had a ruptured appendix, they may have had an infection of their body cavity called *peritonitis.* Or if you ever stuffed a turkey, you placed the stuffing in the coelom of the turkey. Body cavities have important functions, such as allowing the internal organs like your gut or heart to move or beat independently of your outer wall. Therefore, as you can see, the type of body cavity is an important taxonomic criterion.

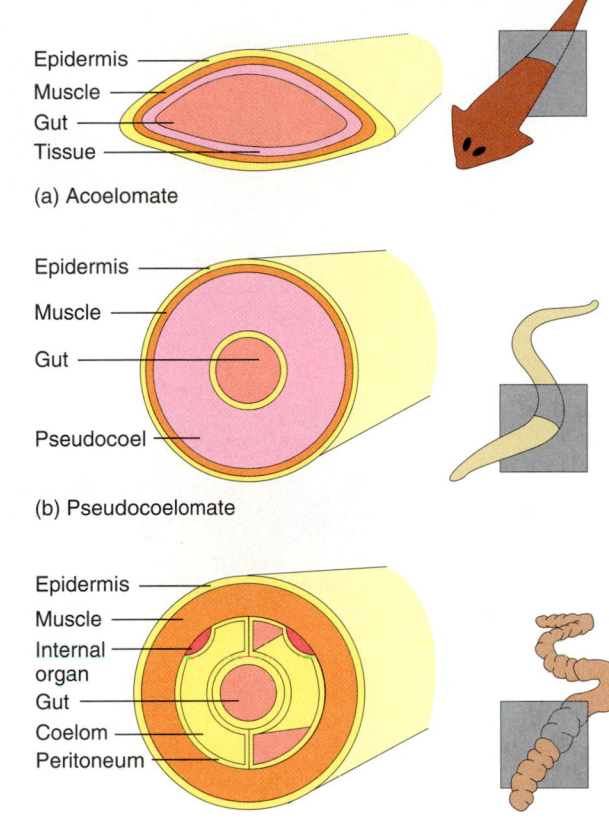

(a) Acoelomate
Epidermis
Muscle
Gut
Tissue

(b) Pseudocoelomate
Epidermis
Muscle
Gut
Pseudocoel

(c) Coelomate
Epidermis
Muscle
Internal organ
Gut
Coelom
Peritoneum

FIGURE 31.5 Body Cavity
(a) Acoelomates lack a body cavity; the flatworm is an example. (b) In the pseudocoelomates, the body cavity is not lined with a peritoneum, as in this roundworm. (c) The coelomates have a body cavity lined with a peritoneum. The earthworm, mollusks, echinoderms, and chordates are examples.

Embryonic Development

The coelomate animals can be subdivided into two major evolutionary lines, the **protostomes** — which include the mollusks, annelids, and arthropods — and the **deuterostomes,** which include the echinoderms and chordates. These two distinct evolutionary lines are determined by the nature of embryonic development, more specifically the way the mouth and anus form, cleavage patterns, and coelom formation (Figure 31.6). As you may remember, we discussed the formation of the blastula and the indentation known as the blastopore in Chapter 28. In the protostomes ("first mouth"), the blastopore develops into the mouth. In the deuterostomes ("second mouth"), the blastopore develops into the anus, then another opening develops into the mouth. Cleavage patterns also differ between these groups.

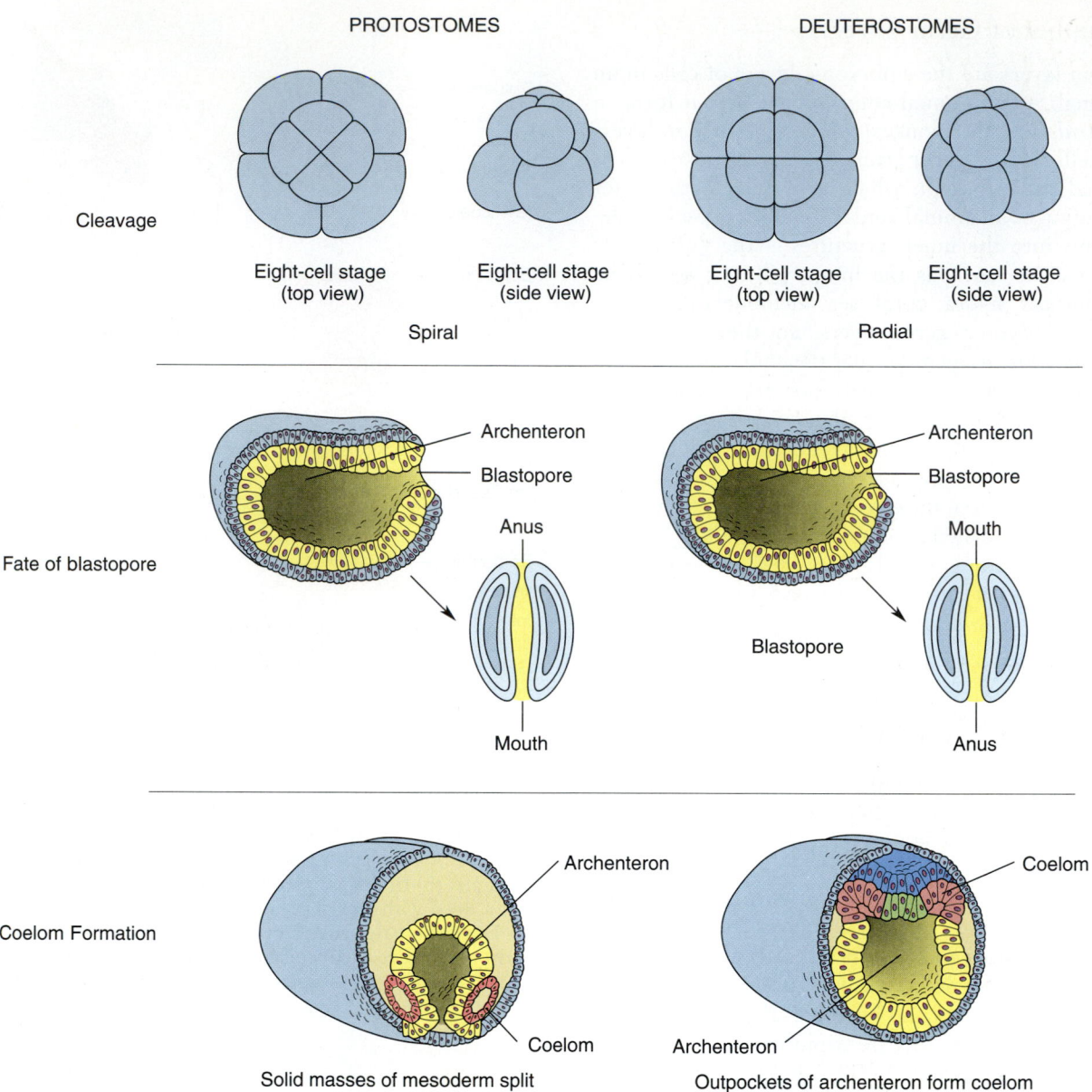

FIGURE 31.6 A Comparison of Protostomes and Deuterostomes
Protostomes and deuterostomes differ in terms of development, cleavage, fate of the blasto-pore, and coelom formation (ectoderm = blue, mesoderm = red, endoderm = yellow).

In the protostomes, cleavage patterns are *spiral,* so that the plane of cell division is diagonal to the vertical axis. In the deuterostomes, cleavage patterns are *radial,* with cleavage planes parallel or perpen-dicular to the vertical axis. This is like peeling an or-ange and cutting it perpendicular to the vertical axis (in other words, cutting it in half in the middle).

In the protostomes, the coelom forms as solid masses of mesoderm from the primitive gut — the *archenteron* — split to form the coelom. In the deuterostomes, outpockets of the primitive gut form

the coelom. The key trends in embryonic develop-ment and evolutionary relationships are noted in Figure 31.7.

MAJOR PHYLA

Now that we have discussed the major evolutionary trends in the animal kingdom and proposed origins of the animals, let us discuss examples from the major phyla.

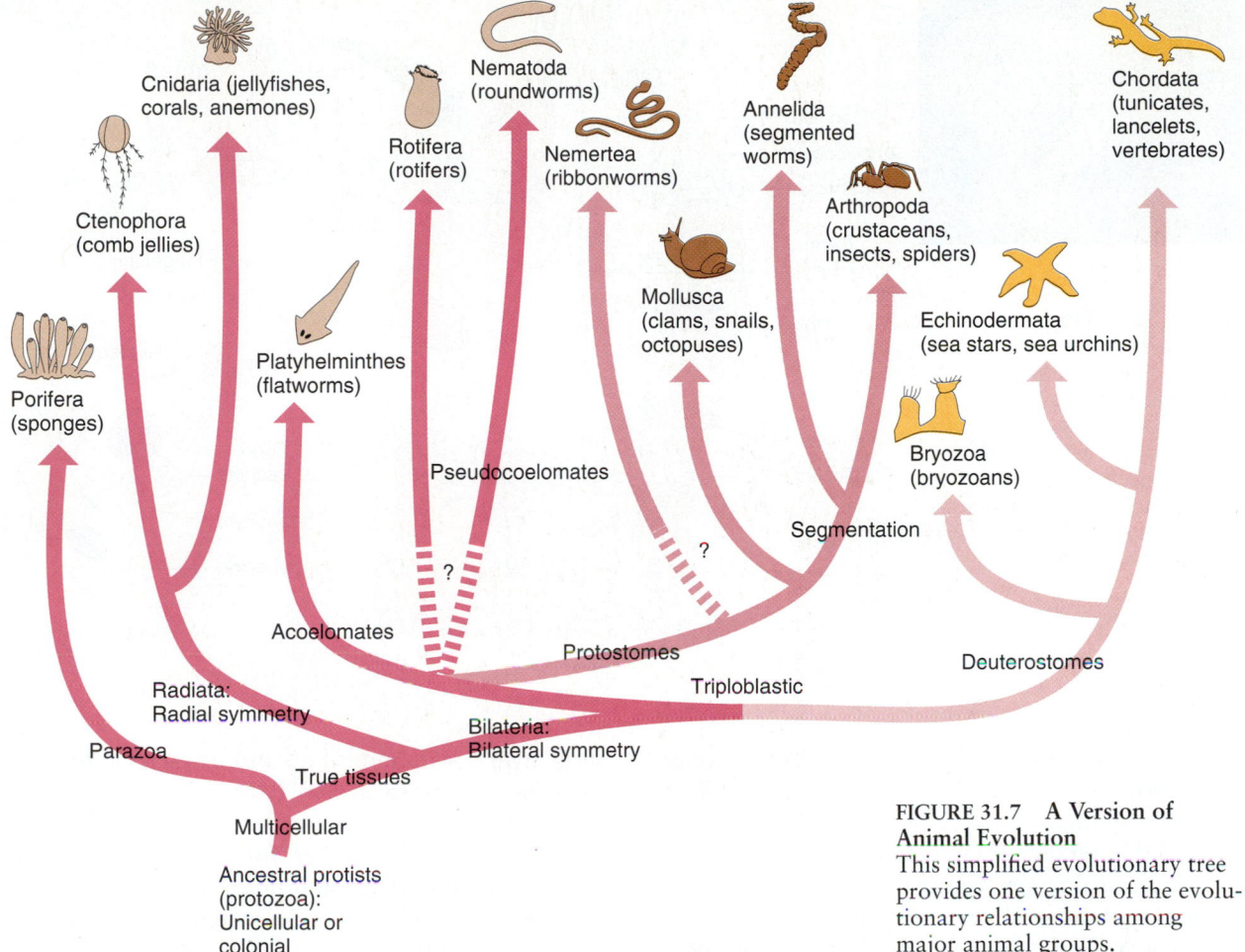

FIGURE 31.7 **A Version of Animal Evolution**
This simplified evolutionary tree provides one version of the evolutionary relationships among major animal groups.

Porifera: An Evolutionary Aside

■ **Sponges — members of the phylum Porifera — are pore-bearing animals and are abundant in marine waters. Because of their simple body plan, they are not even multicellular. Most biologists think that they are not on the main evolutionary line of the animals.**

Sponges, which belong to the phylum Porifera, are abundant marine animals (there are about 9,000 species); a few freshwater types exist as well. As their name suggests, they are "pore-bearing" animals. It is hard to think of sponges as animals, but remember that even though they are *sessile* (nonmoving), they are animals primarily because they are heterotrophs. Most biologists do not think they are multicellular; their cells function independently, and they appear more like a colony of independent cells than a multicellular organism with cellular specialization. As a matter of fact, because they are so different, most biologists refer to them as *parazoans* ("beside the animals"). The sponges are believed to

be a separate evolutionary lineage that did not evolve into other animal groups. Let us now discuss a typical sponge body plan.

Sponges usually appear like a sac with numerous holes or pores. Water enters the pores into the central cavity — the *spongocoel* — and then exits through the large opening known as the *osculum* (Figure 31.8). Sponges are filter feeders and feed on organic material that enters their body. The inner layer of the spongocoel is lined with flagellated collar cells called **choanocytes**. The flagella whip and create a current that traps the food particles, which the collar cells then trap and ingest. In the middle of the soft, jellylike region are amoebalike cells called *amoebocytes*. These cells take the food from the choanocytes and carry the nutrients to the other cells. Some amoebocytes secrete the skeletal fibers or *spicules* that support the sponge and give the different classes of sponges their characteristics.

With respect to reproduction, the majority of sponges are *hermaphrodites* — organisms that produce both male and female sex cells. The amoebo-

(a)

FIGURE 31.8 Porifera
(a) A photograph of a typical sponge. Notice the pores. (b) The typical sponge body plan consists of a saclike body with many pores. There is limited cellular specialization; the sponges are more like colonial unicellular organisms than true multicellular animals.

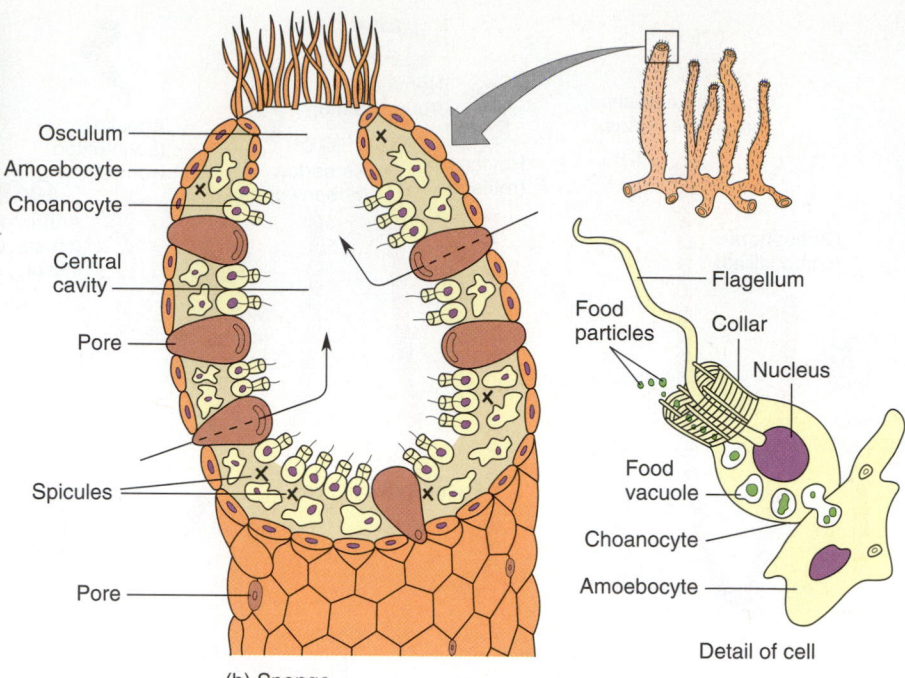

(b) Sponge

cytes or the choanocytes develop into the sex cells. Some sponges reproduce by *regeneration,* in which a piece of a sponge cell develops into an adult sponge.

Even though the sponges are simple animals, they are a successful group of animals and have been in existence for about a billion years. The ancestors of the sponge were probably colonial collared protistans.

Cnidaria: Radial Symmetry and the Beginning of Multicellularity

■ **Cnidarians exhibit the beginning of multicellularity, with their specialized stinging cells, nerve net, and two germ layers.**

If you have ever been stung by a jellyfish, you know our next animal group intimately. Cnidarians are called the "stinging-cell" animals because they have cells specialized for stinging called **cnidocysts.** The phylum Cnidaria consists of three major classes: the *scyphozoans* (jellyfish), *anthozoans* (sea anemones and corals), and the *hydrozoans* (hydra) (see Figure 31.9a). All are aquatic, with most being marine. They all exhibit radial symmetry and a saclike gut, tentacles, a diffuse nerve net, and two germ layers — ectoderm and endoderm. Because they lack a mesoderm, they are limited to a simple body plan.

Most cnidarians exhibit two body forms during their life cycle. The **polyp** (Figure 31.9b) looks like a little hydra, with mouth up and tentacles extended, attached to the substrate. Polyps do not float but rather wait and sweep prey into their mouths with their tentacles. The other form is known as the **medusa** (Figure 31.9c) and floats in the water like a little jellyfish, with tentacles extended and mouth down to capture prey.

Their body plan consists of a saclike gut called the *gastrovascular cavity,* with one opening that functions both for the entrance and exit of nutrients. The gastrovascular cavity is lined with endodermal cells that make up the *gastrodermis.* The epidermal cells cover the body and contain specialized cells that sting (especially in the tentacles) called *cnidocysts* (Figure 31.9d). The cnidocysts function in capturing prey. Between the epidermis and endoderm is a jellylike substance called the *mesoglea.* The mesoglea gives the jellyfish its jellylike consistency. Cnidarians exhibit the beginning of multicellularity and true cellular specialization. Within the body is a diffuse nerve net consisting of specialized nerve cells and contractile fibers that work together for movement. Some specialized sensory cells also exist in some species.

Some cnidarians have simple reproductive organs called *gonads* located within the epidermis or gastrodermis, which break open and release the gametes into the water where fertilization takes place. The zygote usually develops into a flat larval form with cilia that creeps along the bottom, attaches, and forms a

(a)

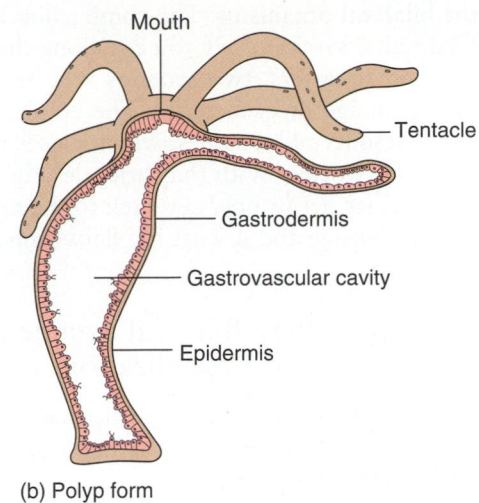

(b) Polyp form

Mouth

Tentacle

Gastrodermis

Gastrovascular cavity

Epidermis

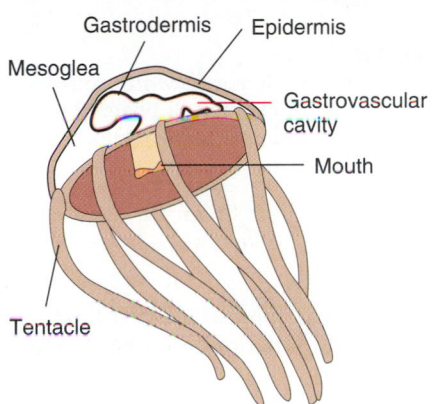

Gastrodermis   Epidermis

Mesoglea

Gastrovascular cavity

Mouth

Tentacle

(c) Medusa form

Prey

Thread
Cnidocyte
Nematocyst

Thread

Before discharge   After discharge

(d) Cnidocyte

FIGURE 31.9   Cnidaria
(a) This sea anemone is a common example of a cnidarian. Most cnidarians exhibit two body forms during their life cycle: the polyp form that looks like a hydra (b) or the jellyfish-like form, the medusa (c). The tentacles have specialized stinging cells — the cnidocysts (d).

polyp. This larval form is called the **planula.** The planula then forms the polyp or medusa, and the cycle begins again.

Ctenophora: Modified Radial Symmetry Without Stinging Cells

■ **The comb jellies have a modified radial symmetry and mesodermlike tissue, which places them close to the flatworms.**

The Ctenophora or comb jellies are a small group of animals (consisting of only about 100 species); they are all marine and resemble small medusae. They have rows of comblike structures of fused cilia that beat to move the animal forward mouth first to feed (Figure 31.10). These comblike structures do not have cnidocysts (stinging cells). From an evolutionary perspective, the comb jellies fall somewhere between the cnidarians that have radial symmetry and

FIGURE 31.10   Comb Jelly
Members of this small group resemble jellyfish. However, on closer observation their comblike structures lack cnidocysts. Therefore they are not cnidarians.

the bilateral organisms. The comb jellies have modified radial symmetry. If you cut along their vertical axis into quarters, two sections will be nearly the mirror image of each other. The comb jellies also have mesodermlike tissues, which places them close to the flatworms. With that brief description of the comb jellies, let us now get back to the main evolutionary lineage and discuss the flatworms.

Platyhelminthes: Bilateral Symmetry, Mesoderm, and Cephalization

■ **The flatworms exhibit cephalization and have three germ layers. Many are free living, but a large number are parasites.**

The flatworms (*Platyhelminthes*) are bilateral triploblastic acoelomate organisms with a head end, an evolutionary feature that makes them very different from the cnidarians. Because they have a mesoderm, which develops into more complex muscles and organ systems, they are capable of complex movement. Since they have bilateral symmetry and cephalization, they are capable of movement in one direction. Of those flatworms that have a gut, the gut is still cnidarianlike and consists of a saclike gastrovascular cavity with one opening.

There are about 20,000 species of flatworms; most are parasitic (they live on or within a living host). The free-living forms are marine, freshwater, or live in a damp environment. Almost all flatworms have a body form that is flattened along the belly-back axis.

The flatworms are divided into four classes: free-living, nonparasitic *turbellarians*; flukes, including *trematodes* and *monogeneans*; tapeworms, or *cestodes*. The flatworms that most students who have taken a high school biology course know are the turbellarian worms called *planarians* (Figure 31.11). Planarians are free-living carnivores that feed on smaller animals or their dead remains. They have a bilateral arrangement of nerve cells with clusters of nerve cells in the head end. Because their bodies are flattened, they can exist without specialized structures for gas exchange and circulation. In addition, because their gastrovascular cavity has branched cells, they are not far away from their source of nutrients. Nitrogenous wastes and CO_2 can diffuse directly from the cells to the surrounding environment. Flatworms have a relatively simple method of regulating osmotic balance through the use of *flame cells*. These flame cells are found in *protonephridia,* which are small branched tubes with pores that extend from the body surface to inner tissue. Flame

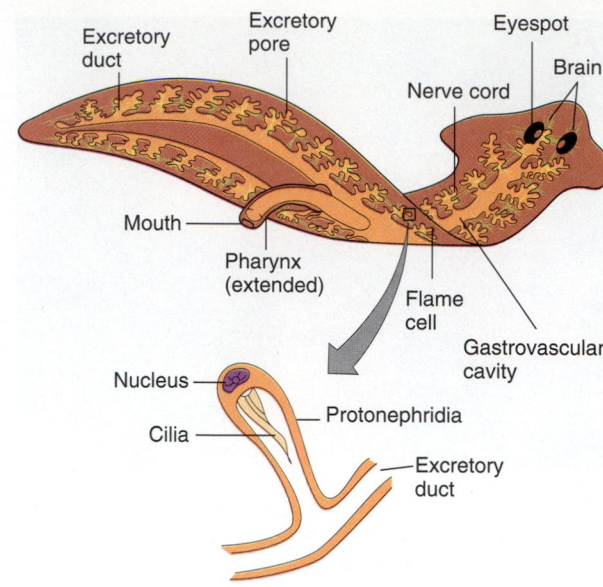

FIGURE 31.11 **Platyhelminthes**
The typical body plan of a free-living flatworm consists of simple organ systems — for example, a bilateral nerve cord, gastrovascular cavity, and excretory system.

cells consisting of tufts of cilia are located within these protonephridia and function in moving excess water to the outside.

Planarians move by using the cilia present on their ventral surface and glide on a film of mucus that they secrete. Reproduction in planarians can be asexual through regeneration, in which adults constrict in the middle and each half regenerates the missing end. You may remember doing a regeneration experiment with planarians in high school. Planarians also reproduce sexually. They are hermaphroditic, each producing egg and sperm. When they mate, they cross fertilize, which results in increased genetic diversity.

Trematoda and Monogenea: Flukes

Adult flukes are parasites that infect various organs of their animal hosts. They attach themselves to the organs by suckers and literally suck nutrients from the host. Because they are absorbing nutrients that have already been digested, their digestive systems are simple. The reproductive organs are well developed. Flukes generally have complex life cycles, which include a larval stage that infects an intermediate host (usually a snail). The larval stage becomes encysted in the intermediate host, where it develops

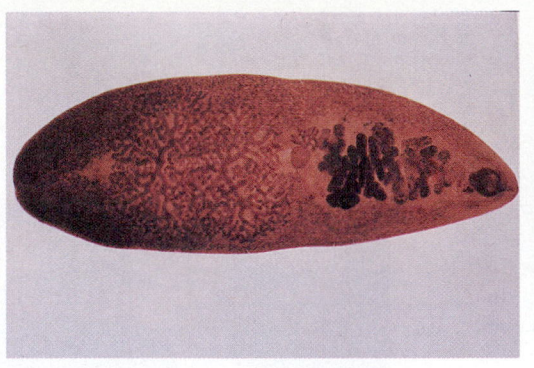
(a)

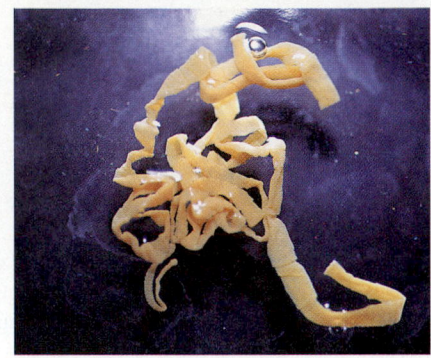
(b)

FIGURE 31.12 Flukes and Tapeworms
(a) This liver fluke is an example of a flatworm belonging to the class Trematoda. (b) This ribbon-like tapeworm is an example of the class Cestoda.

into a form that can infect the primary host. While in the primary host, it becomes an adult (Figure 31.12a). Blood flukes cause schistosomiasis, a disease infecting over 200 million people worldwide (see Profile).

Cestoda: Tapeworms

Tapeworms are parasitic flatworms that as adults live in their host, which is usually a vertebrate (Figure 31.12b). As you would expect, humans are also hosts to tapeworms. Tapeworms evolved toward a simple body plan due to the selection pressures of their parasitic lifestyle — for example, they lack a digestive system. The body plan of an adult consists of a head or a **scolex,** which is lined with suckers or suckers and hooks, enabling them to attach to the gut of their host.

Behind the tapeworm's scolex are individual units called **proglottids,** which are literally reproductive sacs. These proglottids trail off behind the head and form the typical ribbonlike tapeworms. Proglottids pass through the animal and leave the body through the feces. These eggs or embryos can then infect other hosts as the cycle continues. The larval form encysts in the muscle or other organs, and humans can acquire the larval form by eating raw or undercooked meat. Some human tapeworms can be over 30 feet in length and can cause severe medical problems, because of intestinal blockage and malnutrition due to absorption of the host's nutrients.

The flatworms, as you can see, are a group that demonstrate several important evolutionary developments. Because some free-living forms and larval stages of the flukes and tapeworms resemble the planulas of the cnidarian life cycle, this helps to support the speculation that the bilateral animals evolved from planula-like animals.

Nemerta: Ribbonworms — A Complete Digestive Tract

■ Because ribbonworms have a small fluid-filled coelomlike sac, they are placed on the protostome line of evolution. They also have a complete digestive tract.

Ribbonworms or proboscis worms, as they are sometimes called, present an interesting evolutionary question. They are believed to be related to the protostomes. Ribbonworms have a body plan like an acoelomate; however, the body has a small fluid-filled sac that resembles a coelom. This fluid-filled sac allows them to extend or withdraw their proboscis. Primarily for this reason, ribbonworms are placed in the coelomate line of evolution (Figure 31.13). In addition, ribbonworms have a complete digestive tract, with an oral and anal opening.

FIGURE 31.13 Ribbonworms
Ribbonworms are one of the lesser-known worms. They have a fluid-filled sac that resembles a coelom.

Dr. George V. Hillyer
Parasitologist

In 1982, Dr. George Hillyer was helping at the Guajataka Boy Scout Camp in Puerto Rico. When he returned home, his wife Josefina told him that he had just received a letter from the Secretary of the American Society of Parasitologists informing him that he had been selected to receive the prestigious Henry Baldwin Ward Medal. The medal is the highest research award given by the society and is given in recognition of meritorious contribution to parasitology.

In 1987, Dr. Hillyer received the Bailey K. Ashford Award from the American Society of Tropical Medicine and Hygiene. This award is made to a young scientist who has done outstanding work in the field of tropical medicine. The medal is named in honor of Bailey Ashford, whose work with hookworm infections in children helped to greatly improve the quality of life for children in Puerto Rico.

Dr. Hillyer has followed in the steps of the great parasitologists whose awards he has received. He is a pioneer in investigative parasitology — more specifically, the immunology of worms leading to a possible vaccine for human schistosomiasis (an infectious parasitic disease caused by flatworms).

Dr. Hillyer and his research group are investigating the genetic and molecular basis of antigen production of the blood flukes and the liver flukes. Because of immunological similarities, he has found that using antigens produced by the liver fluke can produce immunity against the blood fluke that causes schistosomiasis. This finding opens up the potential of developing a vaccine against the disease.

Dr. Hillyer was born in San Juan, Puerto Rico, in 1943. He says that several events influenced his choice of science as a profession. His neighbor and family friend José Oliver-Gonzales was a parasitologist. When Hillyer attended St. John's Preparatory High School, his science teacher also strengthened his interest in parasitology and helped him get one of his first summer jobs — with the U.S. Army Tropical Research Medical Laboratory in parasitology. He completed his bachelor's degree in biology at the University of Puerto Rico. In 1967, he began working toward his Ph.D. at the University of Chicago with Dr. Lewert, investigating schistosomal DNA. In 1972, he completed his Ph.D., and the next day he was back at the University of Puerto Rico establishing his research laboratory in an old classroom. With the assistance of his dean, Ismael Almodovar, and support from an NIH Minority Biomedical Support Program, directed by Dr. Ciriaco Gonzalez, he was able to develop one of the most successful research programs in Puerto Rico. Today, Dr. Hillyer is recognized as one of the world leaders in immunodiagnosis and immunology of schistosomiasis of humans and fascioliasis of animals

and people. When he accepted the Ward Medal in Canada, which is a bilingual country, he felt it was only fitting, because Puerto Rico is a bilingual island and Spanish is Dr. Hillyer's first language, to quote the words of Spanish Nobel laureate Salvador Ramón y Cajal, who said on being inducted into the Spanish Royal Academy of Science in 1897 that a scientist must possess the following qualities:

1. *Independencia de juicio* (independence of judgment)
2. *Perseverancia en el trabajo* (perseverance in work)
3. *Pasión por la gloria* (passion for glory)
4. *Patriotismo* (patriotism)

And Dr. Hillyer's work truly reflects these qualities.

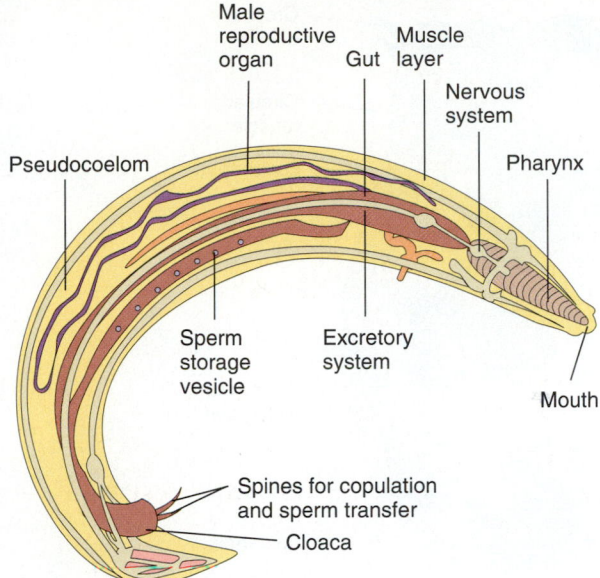

Male reproductive organ

Gut

Muscle layer

Nervous system

Pharynx

Pseudocoelom

Sperm storage vesicle

Excretory system

Mouth

Spines for copulation and sperm transfer

Cloaca

FIGURE 31.14   Nematoda
The body plan of a roundworm is well adapted for feeding and reproduction.

Nematoda: Roundworms — Development of a Body Cavity

■ **Roundworms are among the most numerous worms. They have a complete digestive tract and a pseudocoelom.**

Roundworms are extremely numerous. There are approximately 80,000 species (Figure 31.14). You have probably never seen one — unless you have had a child with pinworms — since most live underground in moist soil. They are usually very small, even though some can be a meter in length. The important evolutionary development in roundworms is that they have a complete digestive tract and a pseudocoelom. They are bilaterally symmetrical, with a round body covered by a tough *cuticle*. Reproduction is usually sexual; in most species, the sexes are separate and fertilization is internal. Some nematodes are important parasites even of humans; we are host to over 50 species, like the common pinworm or the more serious *Trichinella spiralis,* which causes *trichinosis*. Trichinosis is contracted by eating raw or undercooked pork infected with the encysted *Trichinella* larva. Some roundworms, like the tomato root worm, are serious agricultural pests. Others, like the soil worm, *Caenorhabditis elegans,* serve as research models for animal development.

Rotifera: "Wheel-Bearers"

■ **Although microscopic, rotifers possess complex organ systems; they also have cilia around their mouths.**

Rotifers are very small animals that usually live in fresh water, although some are found in the oceans or damp soil (Figure 31.15). They are remarkable in that even though most are microscopic, they are true multicellular animals. They have a complete digestive tract, pseudocoelom, and other specialized organ systems.

They are called rotifers because they have cilia surrounding their mouths that beat and produce a whirlpool current, drawing food into their mouth. Reproduction in some rotifers is **parthenogenetic** (the unfertilized egg develops). In some species, the unfertilized egg develops into a female. In other species, two types of unfertilized eggs are produced — one that develops into a female and one that develops into a simple degenerate male that is basically a male gonad. It cannot even feed itself.

Rotifers probably did not evolve along the main animal line and are likely to be a fork in the road, even though they are complex and have a pseudocoelom.

Let us now move to the major evolutionary path, consisting of the protostomes and deuterostomes. We will begin with the first major protostome phylum, Mollusca.

Mollusca: Soft-Bodied Animals — The Development of the True Coelom

■ **Mollusks are soft-bodied animals that exhibit adaptations of a similar body plan. The foot, visceral mass, mantle, and radula are modified. The classes of mollusks are based on these body-plan modifications.**

When looking at a chiton, slug, clam, or octopus, you probably do not recognize any similarities and wonder why biologists group them into the same phylum. Mollusks (soft-bodied animals) all share a similar body plan (Figure 31.16). As we mentioned, mollusks have a coelom. Other important features include the muscular *foot*, which is used for locomotion, the *visceral mass*, which includes most internal organs, the *mantle*, which is a tissue that covers the visceral mass. Some species, like the clam, secrete a calcium carbonate shell and have a *radula* — a rasplike structure that surrounds the mouth and

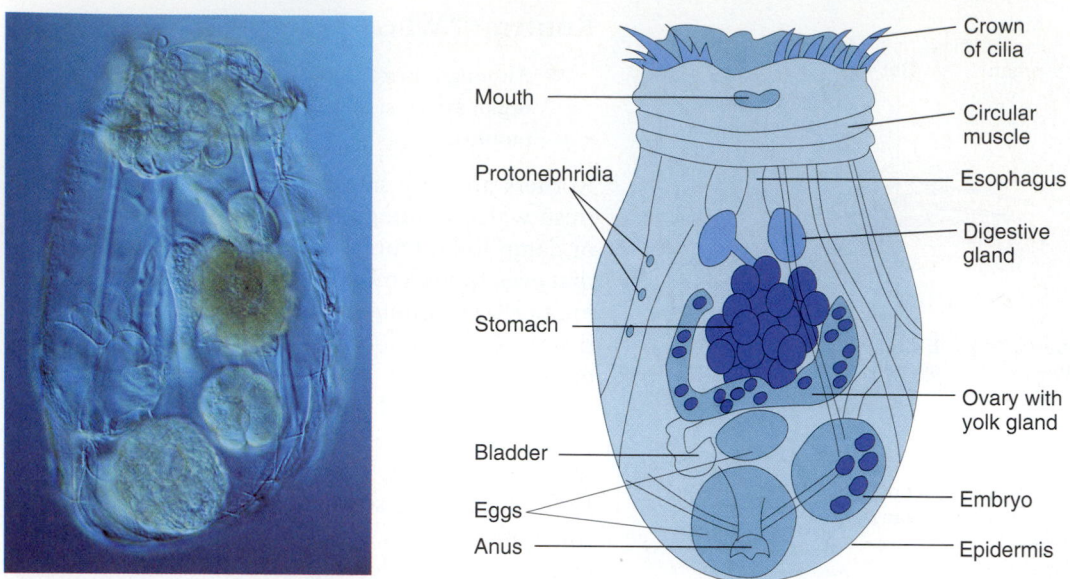

FIGURE 31.15 Rotifera
Even though the rotifers are almost microscopic, they have highly specialized
organ systems and a crown of cilia surrounding their mouths.

is used to scrape up food. There are over 50,000 species of mollusks. Most are aquatic, but some, like snails and slugs, live on land. Biologists recognize eight classes of mollusks, but we will discuss four classes as examples of how their major body plan has evolved. This approach allows us to group the mollusks into four major classes (Figure 31.17).

Polyplacophora: Chitons. The chitons are a major example of the class Polyplacophora. They are marine animals with elongated bodies that have eight plates of shells on their back (dorsal) side. Because of their separate plates, most think that the chitons are segmented, but the body is actually unsegmented. (Note: The first group we will discuss that ex-

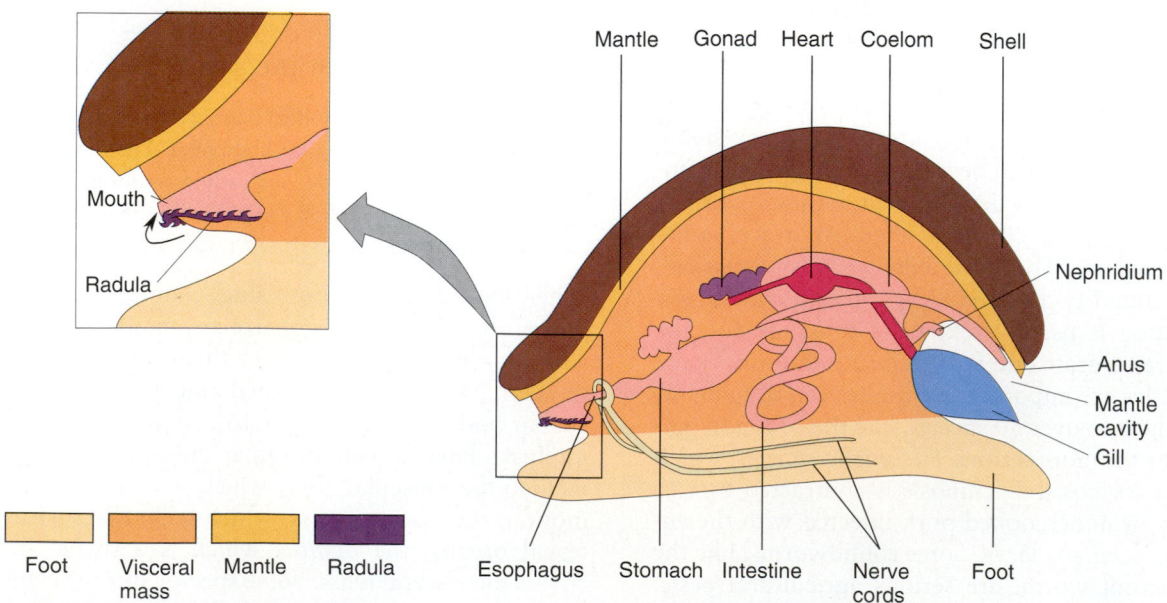

FIGURE 31.16 Mollusca Body Plan
This generalized body plan of the mollusks highlights the four major characteristics of the phylum: foot, visceral mass, mantle, and radula.

Chapter 31   Animals: An Evolutionary Perspective — Invertebrates

(a)

(b)

(c)

(d)

FIGURE 31.17   Representation of Four Classes of Mollusks
(a) The chiton is a representative of the class Polyplacophora. (b) The nudibranch, often
called a sea slug, is a type of snail without a shell and is an example of the class Gastropoda.
(c) This clam is covered with sea anemone. Clams are members of the class Bivalvia.
(d) Squids are examples of mollusks that are grouped into the class Cephalopoda.

hibits true segmentation will be the annelids or segmented worms.) Chitons cling to rocks and reefs as they scrape their mouths and radula over the surface to remove algae and other food. If you have ever tried to remove a chiton from a rock, you can appreciate how tightly they are attached.

Gastropods: Snails.   The snails and slugs are called the "stomach-footed" animals, since they glide along on a mucus trail secreted by the muscular foot. Another characteristic of most snails is that their body and shell are twisted and spiral-like. This spiral body plan is the result of a developmental process called *torsion,* in which body parts grow at different rates, resulting in the mantle along with the visceral mass being twisted above the head. This results in the form we associate with snails. However, as you probably know, some slugs and their marine equivalent, the *nudibranchs,* do not have this twisted body plan.

Bivalvia: Clams, Scallops, and Oysters.   The bivalves, as you would suspect, are mollusks with a shell divided into two halves and a hatchetlike foot (some biologists prefer to use the name *Pelecypoda* or hatchetfoot as the class name for this group). When you eat clams, you are eating the muscular foot. The halves of the shell are hinged, and powerful adductor muscles hold the clam closed — which explains the expression "clam up." You probably eat these powerful adductor muscles in your clam chowder. The bivalves also have gills located in their bodies that they use for respiration and feeding (Figure 31.18). Bivalves are the only group of mollusks with no distinct head end, even though they are bilaterally symmetrical. The next group have a very distinct head and so are called head-footed: the cephalopods.

Cephalopoda: Octopuses, Squids, Nautiluses, and Cuttlefish.   The members of this class are all fast-swimming predators with the mantle modified into

FIGURE 31.18
Anatomy of a Clam
In this illustration, the
shell has been removed
so that the internal
organs can be seen.

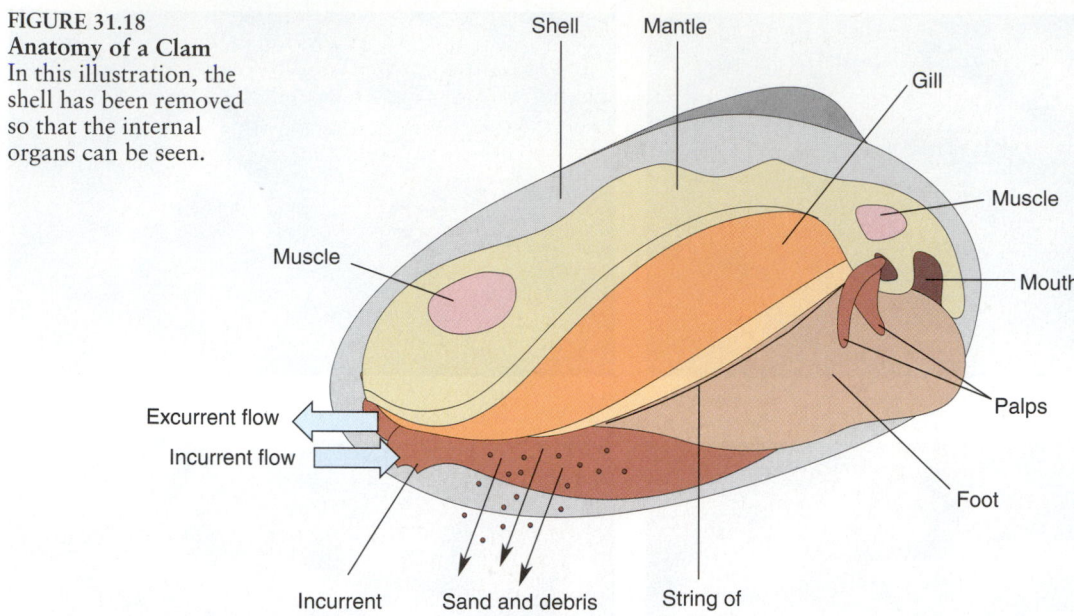

Shell Mantle Gill Muscle Mouth Palps Foot

Muscle

Excurrent flow
Incurrent flow

Incurrent
siphon

Sand and debris
being ejected

String of
mucus

---

a head with tentacles. Their body form is excellent for predation. For example, the squid can move rapidly by a type of jet propulsion in which the water that has been drawn into the mantle cavity is forced out through a siphon.

The cephalopods have a well-developed nervous system, which aids in predation. They also are the only mollusks whose blood is contained within vessels — a *closed circulation system*. Because of their complex nervous system, they can even learn to avoid certain body shapes when presented with electrical shock. All in all, the cephalopods are among the most complex animals without backbones.

Annelida: Segmented Worms

■ **The annelids are segmented worms. Segmentation leads to specialized body sections.**

When we think of the segmented worms, the familiar earthworm comes to mind (Figure 31.19a). The earthworm and its relatives, the marine worms and leeches, exhibit some key evolutionary developments, the most important being **segmentation.** These repeating units or segments are a significant evolutionary development because they can lead to specialized body sections, as we will see when we discuss our next phylum, the Arthropoda. Annelids

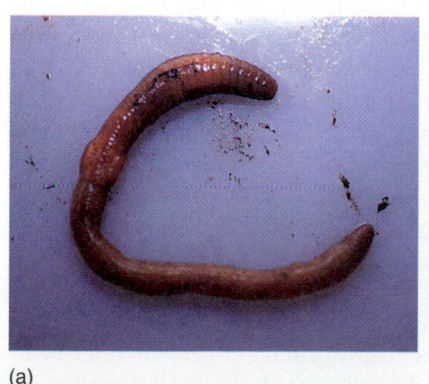

(a)

(b)

(c)

FIGURE 31.19   Annelida
These worms are representative examples of the major classes of Annelida. (a) The earthworm is a member of the class Oligochaeta, (b) the marine worm belongs to the class Polychaeta, and (c) the leech belongs to the class Hirudinea.

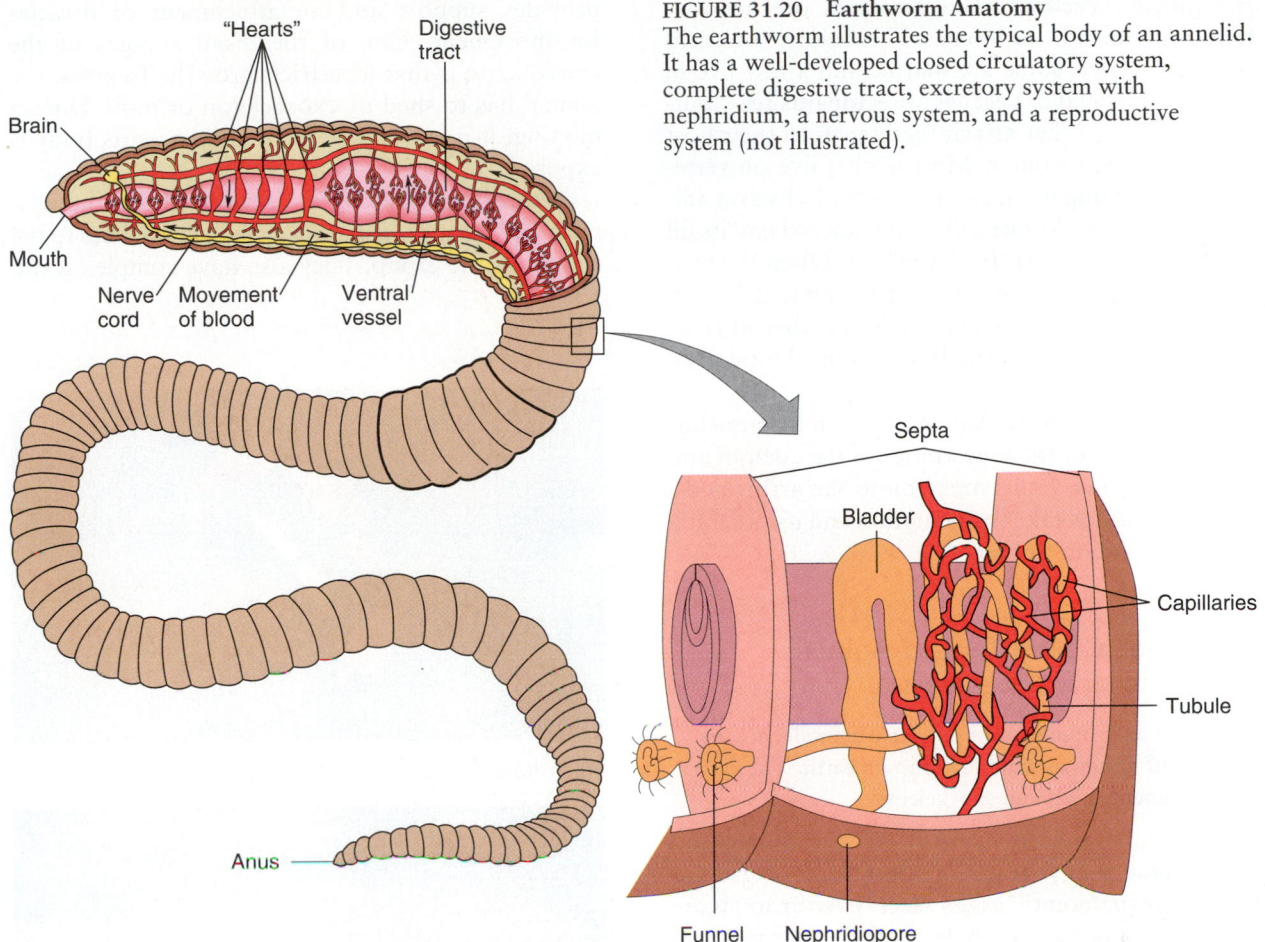

FIGURE 31.20 **Earthworm Anatomy**
The earthworm illustrates the typical body of an annelid.
It has a well-developed closed circulatory system,
complete digestive tract, excretory system with
nephridium, a nervous system, and a reproductive
system (not illustrated).

also have a closed circulatory system, which is more
efficient for nutrient, waste, and oxygen transport
(Figure 31.20). This more efficient circulatory system allows for a larger, more complex body plan.
Annelids also have a ventral (bellyside) nerve cord
with clusters of nerve cell bodies or **ganglia**. The
coelom of the annelids is partitioned by *septa* (cross-
walls), but the digestive tract, nervous system, and
circulatory system penetrate these septa. The excretory system is also well developed and consists of excretory tubes called *nephridia,* which regulate the
volume and composition of body fluids. The coelom
allows for the longitudinal muscles and circular
muscles to contract independently of the digestive
tract, allowing for the characteristic crawling associated with these segmented worms. The coelom
also is filled with a fluid that presses against the body
wall, resulting in *hydrostatic* pressure (fluid pressure) that provides for movement and support.

The outer body is covered with a thin protective
covering called a *cuticle*. The skin below the cuticle
is well supplied with blood vessels and the thin outermost layer of skin serves as an organ for gas exchange. Along the skin of most annelids are fused
bristles called *setae*. These are used for locomotion,
crawling, and swimming. If you have ever run your
fingers over an earthworm as you baited a hook, you
have felt the setae. The classes of annelids are based
in part on the number of setae a worm has per body
segment.

Oligochaeta: Few Setae. This class includes the
common earthworm as well as many marine worms.
They belong to the class Oligochaeta because they
have a few setae.

Polychaeta: Many Setae. The many-setae worms
have their setae modified into a pair of paddlelike
structures per segment that function in swimming
and also to increase the surface area of the skin for
gas exchange. Almost all polychaetes are marine and
are free swimming or live in the sand. The majority
of sand dwellers form a tube around them by mixing mucus and sand.

Hirudinea: Leeches. The members of the class Hirudinea include the leeches. The majority of leeches live in water; some are marine and a few live in moist areas on land. Leeches are ectoparasites ("outside parasites") that attach themselves to their host and suck blood from it. Most leeches live on vertebrates, including humans. A few species live on animals without backbones. Once a leech has had its fill of blood, it frees itself from the host. Often the host does not even know that a leech has attached itself, because it produces an anesthetic. Leeches also secrete *hirudin* — a chemical that prevents blood from clotting.

Now that we know something about the annelids and are aware of the importance of the coelom and of segmentation, let us move on to the arthropods. These organisms take segmentation and its potential for specialization further.

Arthropoda: Joint-Legged Animals with Specialized Segments

■ **The arthropods, or joint-legged animals, are the most abundant animal group on earth. They are segmented with an exoskeleton.**

If you have ever seen a bug or eaten a lobster, you are familiar with the arthropods or joint-legged animals (*arthro,* "joint"; *poda,* "feet"). Arthropods are the most numerous animals on the earth; they are also diverse and well adapted for survival. Even though at first a butterfly, lobster, and dragonfly seem to have nothing in common, they do (Figure 31.21). Let us discuss their common features.

The most basic feature of all arthropods is that they have jointed appendages, which exhibit diverse forms and functions. For example, the head of an arthropod contains specialized appendages used for sensory functions (antennae) and others specialized for feeding. Other appendages on the body are used for locomotion, defense, and reproduction. The important evolutionary step is that these specialized appendages are built on the important characteristic of segmentation.

As the arthropods evolved, segmentation led to specialization. This specialization, as in the development of the head, thorax, and abdomen, provided arthropods with more variability, allowing them to be more capable of adapting to a variety of environments.

Another important characteristic of arthropods is that they have a protective outer skeleton, an exoskeleton made of chitin. The exoskeleton provides protection from enemies and dessication; it also provides support and an attachment of muscles for movement. One of the disadvantages of the exoskeleton is that it restricts growth. To grow, the animal has to shed its exoskeleton or molt. During molting, it is vulnerable to predation as its body is exposed.

Arthropods have highly developed nervous systems and sensory structures, which contribute to the success of the group. They also have complex respi-

(a)

(b)

(c)

FIGURE 31.21   Representative Arthropods
There are many classes within the phylum Arthropoda. (a) The butterfly represents the class Insecta, (b) the lobster belongs to the class Crustacea, and (c) the dragonfly is also a member of the class Insecta.

ratory structures — gills, lungs, and trachea. The trachea are pores on the body surface that open to the outside and allow for the exchange of CO_2 and O_2 (see page 632). The efficient respiratory system of the arthropods allows for high metabolic rates, which can increase evolutionary success.

The life cycles of arthropods are varied and demonstrate a division of labor. For example, in some insects there is the *larval stage,* which is specialized for eating and growing, and the *pupal stage,* which is the resting stage involving a lot of embryonic development to produce the final adult. This division of labor in the life cycle has contributed to the success of the arthropods by allowing them to adapt to seasonal changes.

Now that we have touched on the common features of the arthropods, let us discuss the major classes.

Crustacea: Shrimp, Lobster, Crayfish, and Crabs

As you might suspect, the crustaceans are major contributors to the food web, from the large lobsters we eat to the small crabs that comprise the diet of several marine organisms.

Crustaceans all appear crusty — that is, they have a well-defined hard exoskeleton. (Have you ever sat in a seafood restaurant and used a hammer to break open a soft-shelled crab for dinner?)

Crustaceans have two pairs of antennae and three or more highly modified appendages used as mouthparts. A group of crustaceans often overlooked are the *copepods,* which are among the most numerous aquatic animals and play an important part in the food web. Most are only about 2mm long and have the characteristic single eye in their head — hence the name *copepod.*

Barnacles are also crustaceans, but they are unusual in that their exoskeleton is highly modified and contains calcium, which almost makes it shell-like.

Arachnida: Spiders, Mites, Ticks, and Scorpions

The arachnids have a fused head and thorax called a *cephalothorax* with six pairs of appendages; they also have *chelicerae,* which inject poison, *pedipalps* used in feeding, and four pairs of walking legs (Figure 31.22). (Remember what you learned in grade school — insects have six legs and spiders, eight.) Spiders also have unusual lungs called *book lungs,* which consist of extensive folded walls that look like pages in a book. These folds increase the surface area available for gas exchange (Figure 31.22b). There are about 60,000 species of arachnids, with more being discovered every day. As a matter of a fact, most zoologists think that the actual number may be eight to ten times greater, since the majority of arachnids are the almost microscopic mites.

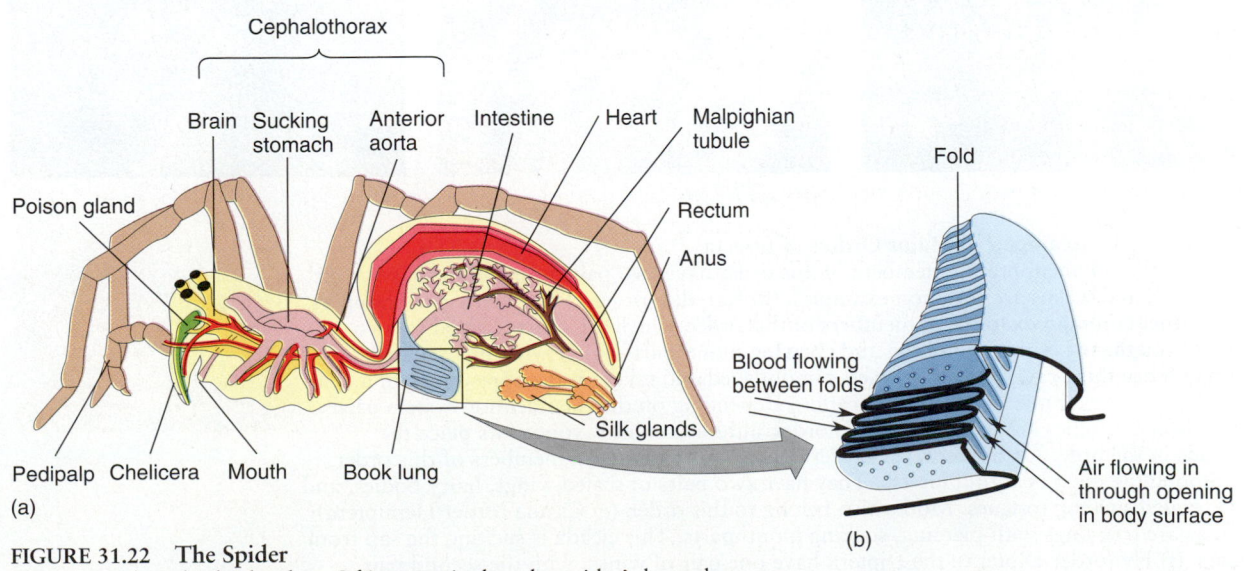

FIGURE 31.22 The Spider
(a) A typical spider body plan. Of interest is that the spider's lungs have a folded wall structure that look like pages in a book, the book lungs (b).

Insecta: Insects

The insects are without a doubt the most successful (if we measure success by sheer numbers) animals on the earth. They live in almost every habitat imaginable, from the oceans to the sky. Their success is due primarily to their diversity and great reproduction ability. Almost 1 million species of insects have been identified. Figure 31.23 shows some photographs of the major orders of insects. Figure 31.24 provides an illustration of a generalized insect body plan, using the grasshopper as a representative. In this typical plan, we see the three body regions: the head, thorax, and segmented abdomen.

Myriapoda: Centipedes and Millipedes

Almost as familiar to us as the insects are the millipedes and centipedes. Millipedes and centipedes have a flattened wormlike body with legs. The centipedes have one pair of appendages per body segment and the millipedes have two.

Of interest to biologists is the observation that there is a separate phylum of worms called the

(a)

(b)

(c)

(d)

(e)

(f)

FIGURE 31.23   Examples of Major Orders of Insecta
(a) Bee (order Hymenoptera): members of this order have two pairs of membranous wings. Ants, bees, and wasps are common examples. (b) Katydid (order Orthoptera): grasshoppers are another common example of members of this order, which are characterized by two pairs of toughened wings and biting and chewing mouthparts. (c) Praying mantis (order Dictyoptera): this praying mantis is well camouflaged and takes on the appearance of a dead leaf. This is an example of cryptic coloration. Depending on the classification system used, mantids and roaches are members of this order, although some taxonomists place the mantids in the order Orthoptera. (d) Butterfly (order Lepidoptera): members of this order are among the most beautiful insects. They have two pairs of scaled wings, hairy bodies, and long, coiled sucking tongues. Moths also belong to this order. (e) Cicada (order Hemiptera): cicadas are true bugs with piercing, sucking mouthparts. This cicada is sucking the sap from plants. (f) Fly (order Diptera): the Diptera have one pair of wings, with the second pair modified into short, winglike structures called "halteres," and sucking or piercing mouthparts. If you have ever been bitten by another member of this order, the mosquito, you know that they have piercing mouthparts.

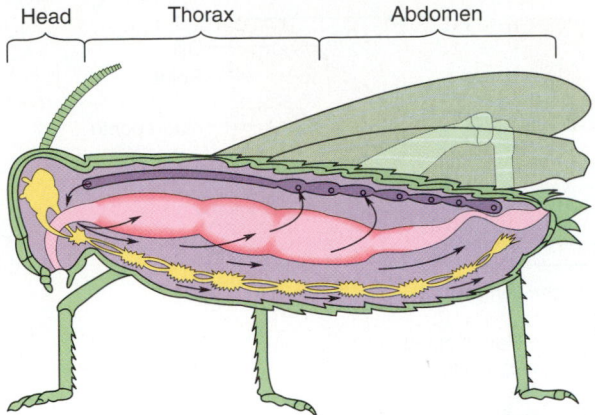

FIGURE 31.24   Grasshopper
The grasshopper illustrates the typical body plan of the insects, consisting of head, thorax, and abdomen.

FIGURE 31.25   Onychophora
The members of the small phylum Onychophora have characteristics resembling both annelids and arthropods.

Onychophora (Figure 31.25), which are segmented like the annelids but have multiple appendages that are unjointed. However, the Cambrian fossils of joint-legged animals resemble segmented worms. This observation provides evidence that the arthropods probably evolved from the annelids.

ECHINODERMATA: SEA STARS AND THEIR RELATIVES

■ **The echinoderms are marine animals that have a spiny skin. Of importance is the observation that the echinoderm larvae are deuterostomes and directly related to the phylum Chordata.**

The most obvious features of the sea stars and their relatives is their spiny skin, from which their scientific name (*echinas,* "spiny"; and *derm,* "skin") originates. Echinoderms are all marine and include the sea star, sea urchins, brittlestars, and sea cucumbers (Figure 31.26). Since they all are marine, they have a unique *water vascular system,* which functions in movement and in the capture of prey (Figure 31.27). When water is withdrawn from the small tube feet

(a)

(b)

FIGURE 31.26   Echinoderms
(a) Sea stars, (b) brittle stars, (c) sea urchins, and (d) sea cucumbers are all members of the various classes of the spiny-skinned animals, the Echinodermata.

(c)

(d)

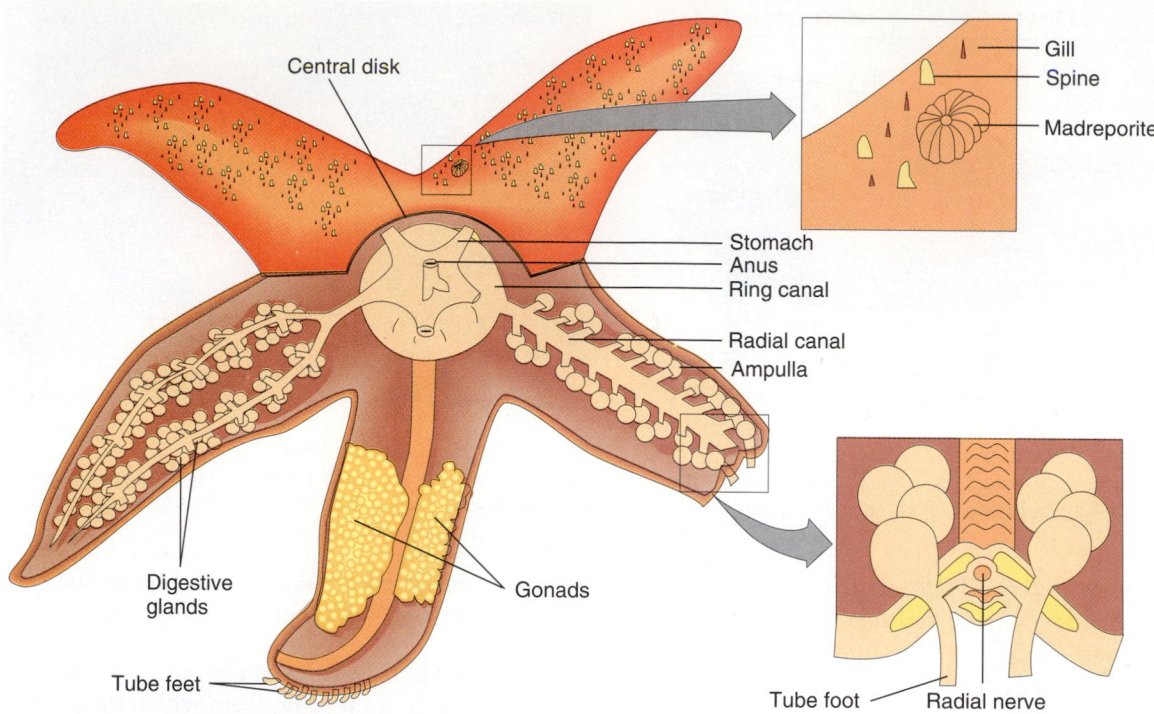

FIGURE 31.27 Body Plan of a Sea Star
Unique to the phylum Echinodermata are the spiny skin and the water vascular system that
functions in locomotion and the capture of prey. The other systems are also illustrated.

into the main canal, a partial vacuum is formed
(somewhat like squeezing the bulb of an eyedrop-
per), resulting in a suction. This suction allows the
echinoderm to cling to substrate or to open the muscles of an oyster. Once the prey (a clam in this example) is open, the sea star pushes its stomach out
of its mouth and digests the soft body of the clam before ingesting its partially digested remains.

Of evolutionary interest to biologists are not the
radial adults but rather the *bilateral* nature of the
larval form of echinoderms. Also, the echinoderm
larvae are deuterostomes, as are the members of the
phylum Chordata, to which humans belong. There
are even more similarities between the echinoderms
and the chordates — for example, as we have just
mentioned, the deuterostome (anus first, mouth second) embryological development and the development of the coelom. We will conclude the chapter
at this point and discuss the chordates in the next
chapter.

SUMMARY

1. Animals are multicellular eukaryotic heterotrophs,
   and most move.
2. There are two major hypotheses as to the origin of
   the animals. The syncitial hypothesis speculates that

animals arose from a multinucleate protozoan that
became subdivided, hence multicellular; the colonial
hypothesis is based on the idea that the ancestors of
animals were heterotrophic colonial flagellates.

3. Important evolutionary features of animals include
   bilateral symmetry, the development of a head
   (cephalization), number of embryonic layers,
   formation of the gut, type of body cavity, and
   embryonic development of the mouth, anus, and
   cleavage patterns.
4. Sponges are simple animals that consist of a porous
   saclike body; they have only two germ layers and
   little cellular specialization. Most zoologists place
   the sponges on a side branch of the main evolutionary line of the animals.
5. The jellyfish and their relatives (Cnidaria) exhibit
   radial symmetry and the beginning of true multicellularity with distinct cellular specialization. A
   unique feature of cnidarians are their stinging cells,
   the cnidocysts.
6. The comb jellies (Ctenophora) are of evolutionary
   interest since they have a radial symmetry that is
   almost bilateral.
7. The flatworms (Platyhelminthes) are bilateral, have
   a third germ layer called the mesoderm, have a true
   head end, and lack a coelom. Some are free living,
   but most are parasites, like the flukes and
   tapeworms.
8. The ribbonworms (Nemerta) have a complete
   digestive tract with a mouth and anus and the begin-

nings of a coelom, even though technically they are considered acoelomates.

9. Roundworms (Nematoda) are bilateral, triploblastic, and have a complete digestive tract with a pseudocoelom.

10. The rotifers are microscopic multicellular animals, with a complete digestive tract, other complex organ systems, and a pseudocoelom.

11. The Mollusca (soft-bodied animals) exhibit a true coelom lined with peritoneum. All members share a body plan that includes a foot, visceral mass, mantle, and radula. These features are very diverse in the four major classes discussed.

12. Segmented worms (Annelida) are characterized by segmented bodies and a true coelom. Body segmentation is the basis for more specialization of the body parts. The classification of the segmented worms is based on the number of setae or their modification: few setae (Oligochaeta), many setae (Polychaeta), and the leeches (Hirudinea).

13. The joint-legged animals (Arthropoda) are the most abundant animals on the earth. They exhibit true specialization of body segments like the head, thorax, and abdomen. The also have complex organ systems and are at the apex of the protostome line of evolution.

14. Spiny-skinned animals (Echinodermata) are radially symmetrical as adults but pass through a bilateral larval stage. This bilateral larval stage and other embryonic features are very similar to those of the primitive chordates.

KEY TERMS

syncitial hypothesis	medusa
colonial hypothesis	polyp
cellular specialization	planula
collar cell	Ctenophora
planuloid	Platyhelmin
cephalization	scolex
acoelomate	proglottids
pseudocoelom	Rotifera
protostome	parthenogenetic
deuterostome	Mollusca
Porifera	Annelida
choanocyte	segmentation
Cnidaria	ganglia
cnidocyst	Arthropoda
Echinodermata	Onychophora

REVIEW QUESTIONS

True/False

1. All animals are heterotrophs.
   true   *false*   *(page 767)*

2. Most biologists accept the colonial hypothesis as an explanation for the origins of the animals.
   true   *false*   *(page 769)*

3. Sponges, because of their lack of specialized cells, are in the direct line of animal evolution.
   true   *false*   *(page 773)*

4. An important evolutionary development in the phylum Platyhelminthes was bilateral symmetry.
   true   *false*   *(page 776)*

Fill in the Blank

1. The ribbonworms have a complete digestive tract with a _____ opening and opening. *(page 777)*

2. The _____ allows for independent movement of the gut and body wall. *(page 771)*

3. The members of the phylum Mollusca have a body cavity lined with a _____. *(page 780)*

4. Segmentation is an important evolutionary development, since it leads to _____ body parts. *(page 782)*

Multiple Choice

1. Which of the following is not an important evolutionary characteristic when grouping animals?
   a. Symmetry
   b. Number of germ layers
   c. Embryonic development
   d. Type of mitochondria
   (page 769)

2. While collecting animals in a tidepool, you find an animal that looks like a cnidarian medusa; however, it lacks stinging cells. In all probability, the animal belongs to the phylum or is a(n) . . .
   a. Ctenophora
   b. Planula larvae
   c. Porifera
   d. Cnidaria
   (page 775)

3. The two major evolutionary lines of animals, the protostomes and deuterostomes, are classified on the basis of . . .
   a. Nature of embryonic development
   b. Formation of the mouth and anus
   c. Cleavage patterns
   d. All of the above
   (pages 771–772)

4. The arthropods are the most successful animal group on earth (if we base success on sheer numbers), and they share all the following characteristics except . . .
   a. Endoskeleton
   b. Exoskeleton
   c. Jointed appendages
   d. Complex nervous system
   (page 784)

Discussion

1. List the major characteristics of the animals.

2. What are the important characteristics of the animals that biologists use to try to develop evolutionary relationships?

3. Why is embryonic development an important taxonomic criterion?

4. Why can it be said that the coelomate animals can be divided into two major evolutionary lines?

5. As compared to the animals that have a saclike gut, what are the advantages of a one-way digestive tract?

6. Pick any two invertebrate phyla and list their major characteristics.

7. Why is the coelom considered an important evolutionary development?

8. Why is segmentation important?

9. What are the basic features of the body plan of a mollusk? How are these features modified in various classes of mollusks?

10. What is the evolutionary basis for the following statement: "The echinoderms are the direct ancestors of the phylum Chordata"?

32

Animals: An Evolutionary Perspective — The Chordates

We have all wondered about our family trees — our grandparents and their grandparents, but have you ever thought about the evolutionary tree that led from animals to humans? We discussed human evolution in Chapter 17 (pages 402 to 406). In the last chapter we discussed animal phylogeny from the Porifera to the Echinodermata. Now we will discuss the evolutionary history of our phylum, the phylum Chordata. We will begin with a discussion of the origin of the chordates, then present chordate characteristics and evolutionary trends and end with a survey of the major classes.

Above: This Pacific White-Sided Dolphin is an example of a vertebrate animal.

THE CHORDATES: THEIR ORIGIN AND CHARACTERISTICS

■ **The chordates probably evolved from echinoderm-like larval forms that retained their larval forms as adults.**

The **chordates** are bilateral deuterostomes that exhibit great variations in appearance. Some chordates lack a vertebral column (backbone) and are categorized as *invertebrates*. The majority of chordates have a vertebral column, and are categorized as *vertebrates*. Because of several embryonic similarities between the invertebrate chordates and the echinoderms, most zoologists postulate that the larval forms of some echinoderm-like animals became sexually mature and yet retained their larval form as adults. Eventually, sometime in the Cambrian period, these sexually mature larval forms developed segmented muscles and stronger skeletons and became the first chordates. But what characteristics do animals as diverse as the saclike sea squirts and humans have in common? Let us see.

MAJOR CHORDATE FEATURES

■ **The major chordate features are the notochord, dorsal hollow nerve cord, gill slits, and postanal tail.**

All chordates have four major features at some time in their lives (Figure 32.1):

1. A **notochord,** a flexible rodlike structure that runs along the length of the back. In the invertebrate chordates, the notochord supports the animal. In vertebrates, the notochord only exists in the embryonic stages and is replaced by the boney vertebral column.

2. A dorsal hollow nerve cord that enlarges in the head region to form the brain.

3. Pharyngeal slits (gill slits), which form from the digestive tube just behind the mouth, and which appear as several slits opening to the outside. In fish, these develop into gill slits used in respiration. In humans and other species, these gill slits are only present in the embryo. (Our ear opening is a gill slit.)

4. A tail that extends beyond the anal region.

A few of the other common features are a true coelom, three embryonic germ layers, and a distinct head.

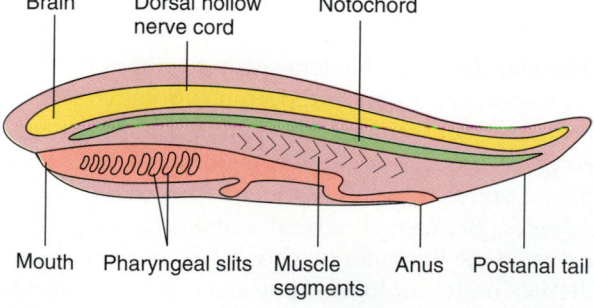

Brain   Dorsal hollow   Notochord
         nerve cord

Mouth   Pharyngeal slits   Muscle   Anus   Postanal tail
                           segments

FIGURE 32.1   Chordate Characteristics
The chordates all have the following four major characteristics at some time during their lifetime: a notochord; a dorsal, hollow nerve cord and brain; pharyngeal gill slits, and a postanal tail.

THE INVERTEBRATE CHORDATES: CHORDATES WITHOUT A VERTEBRAL COLUMN

■ **Some chordates lack a vertebral column and are placed in two subphyla: the *urochordates*, the sea squirts, and the *cephlochordates*, the lancelets.**

The chordates without vertebrae include two subphyla, the **urochordates,** the sea squirts, or tunicates, and the **cephlochordates,** the lancelets (these small marine animals are called lancelets because they look like a small blade — a lance). The sea squirts are small, sessile marine animals that live in shallow water. When touched, they squirt water from their excurrent (outgoing) siphon; hence, they are called sea squirts. When looking at an adult sea squirt, you may find it difficult to determine why they are even called chordates. However, the larval form clearly embodies the chordate characteristics (Figure 32.2). The urochordates were probably a branch of the direct evolutionary line leading to the vertebrates. The other invertebrate subphylum, the cephlochordates, consists of only a few living species, the most common being the lancelets or *Branchiostoma* (formerly *Amphioxus*). It is postulated that these small marine animals evolved from larval forms of sea squirt-like animals that failed to complete metamorphosis and remained fishlike as adults (Figure 32.3). It is not known what animals are the direct ancestors of vertebrates, but one can speculate that the lancelets (because of their similarities to primitive vertebrates, the jawless fish) are probably close to the ancestral vertebrates.

THE VERTEBRATES

■ **The subphylum Vertebrata consists of animals that have a vertebral column.**

The vertebrates are what we usually think of when we think of animals — fish, frogs, snakes, birds, and cows. Although varied, all share the vertebrate common body plan. Vertebrates have a hollow dorsal nerve cord protected and surrounded by skeletal subunits called **vertebrae** (singular, vertebra), which form the vertebral column or spine. This vertebral column is a prime example of segmentation. Vertebrates are cephalized with a well-developed brain and sensory organs. The skeletal system is internal and consists of a central column (*axial skeleton*) and the appendages skeleton (*appendicular skeleton*). The skeleton is made up of cartilage or of bone and cartilage. Vertebrates have closed circulatory sys-

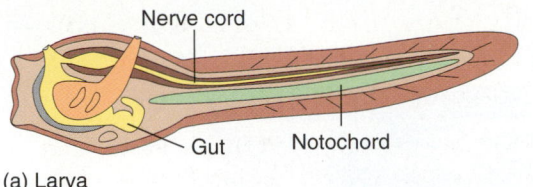

(a) Larva

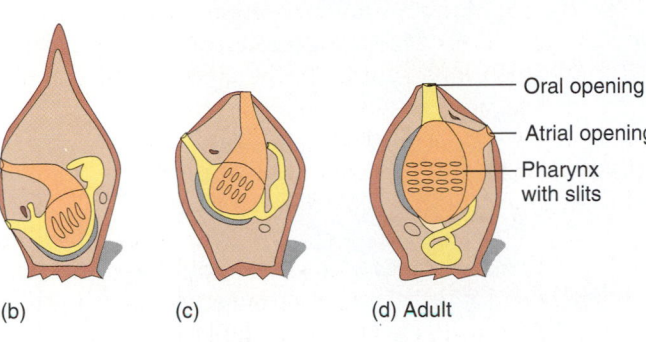

(b) (c) (d) Adult

Oral opening
Atrial opening
Pharynx with slits

FIGURE 32.2   **Tunicates**
(a–c) At first glance, it is difficult to see why an adult tunicate, or sea squirt, would be classified as a chordate. However, during embryonic development the notochord and other chordate features are very noticeable in the tadpole-like larva. (d–e) As development continues, the tail, the notochord, and most of the nervous tissue are reabsorbed, and the organs rotate into their final adult position. (e) A group of adult sea squirts.

(e)

tems with a ventral heart that pumps hemoglobin-containing blood. The skin is very complex, and its outermost covering ranges from scales to hair. The sexes are usually separate, with external and internal embryonic development. The other organ systems are also well developed.

There are several classes in the subphylum Vertebrata. Zoologists disagree as to the actual number. Some place the primitive jawless fish in classes of their own. In addition, the exact evolutionary relationship and placement of the birds are also in question. Therefore, we will emphasize the following: agnatha, chondrichthyes, osteichthyes, amphibia, reptiles, and mammals. Figure 32.4 illustrates one version of vertebrate evolution. Table 32.1 also provides information about the major classes of vertebrates and their characteristics. Remember, as with

all taxonomy and systematics, these groupings are subject to change as we gather more information about evolutionary relationships.

The Jawless Fish: The Agnatha

- **Some fish lack a jaw and are grouped as the agnatha.**

The oldest vertebrate fossils are of jawless fish called the **agnatha.** As the name agnatha implies, the members of this group lack a jaw. The eel-shaped lamprey and hag fishes are living examples of this group (Figure 32.5). They have rounded mouths with circular or slitlike openings that are jawless. The lamprey mouths are suckerlike, and they usually feed by attaching themselves to fish and literally sucking the

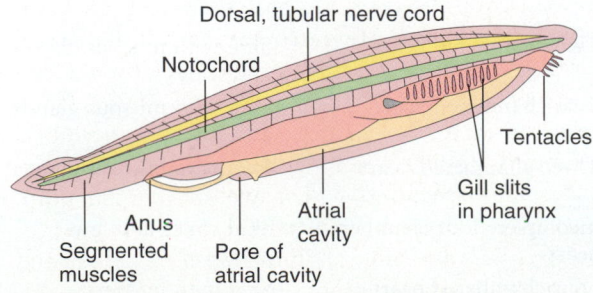

Dorsal, tubular nerve cord
Notochord
Tentacles
Gill slits in pharynx
Atrial cavity
Pore of atrial cavity
Anus
Segmented muscles

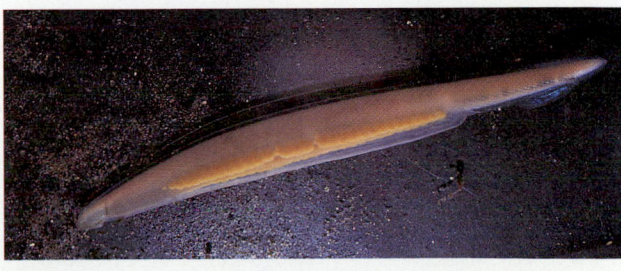

FIGURE 32.3   **Lancelet**
Photo and illustration of a lancelet showing its chordate characteristics.

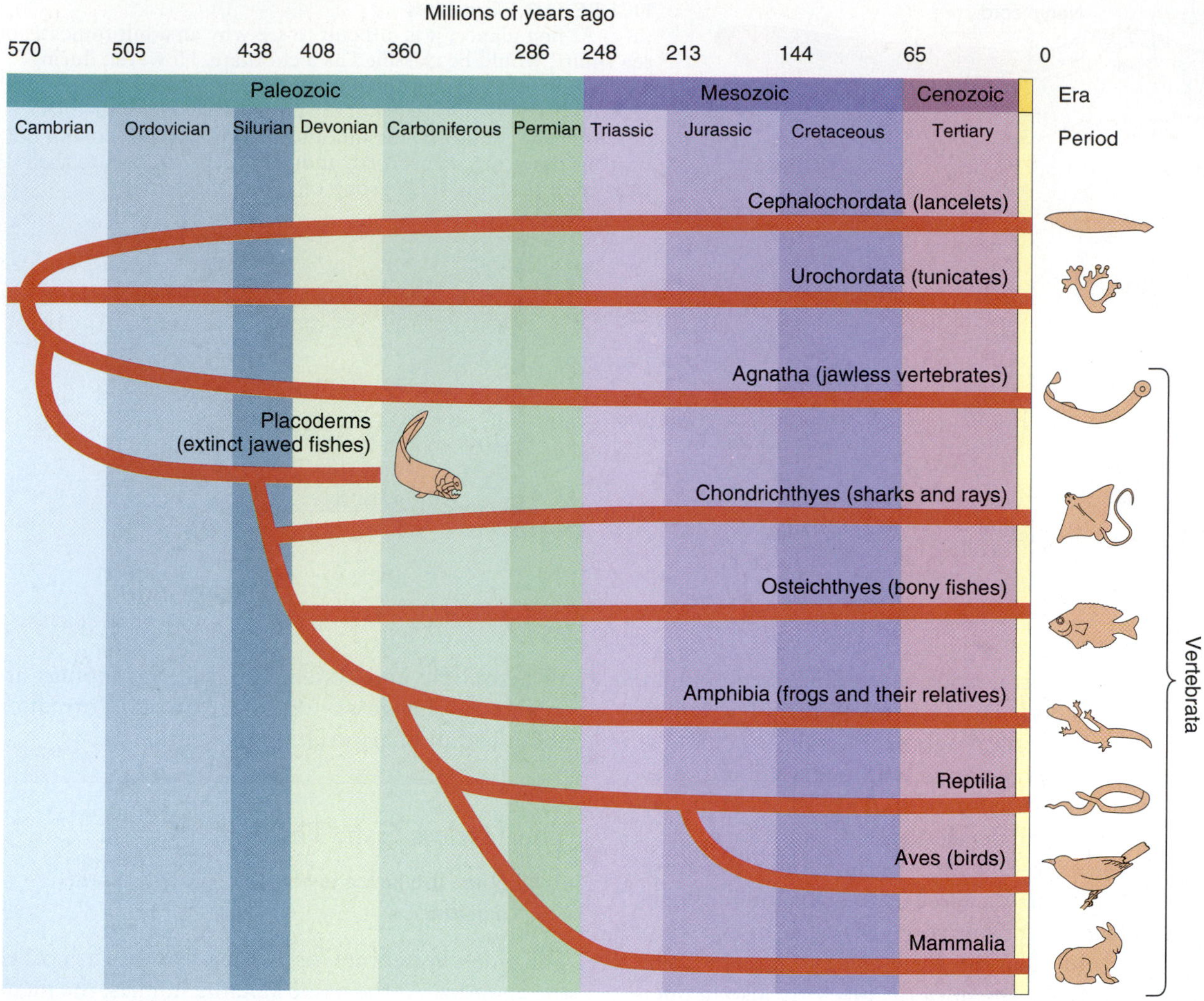

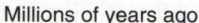

Millions of years ago

| 570 | 505 | 438 | 408 | 360 | 286 | 248 | 213 | 144 | 65 | 0 |

| Paleozoic | | | | | | Mesozoic | | | Cenozoic | Era |
| Cambrian | Ordovician | Silurian | Devonian | Carboniferous | Permian | Triassic | Jurassic | Cretaceous | Tertiary | Period |

Cephalochordata (lancelets)

Urochordata (tunicates)

Agnatha (jawless vertebrates)

Placoderms
(extinct jawed fishes)

Chondrichthyes (sharks and rays)

Osteichthyes (bony fishes)

Amphibia (frogs and their relatives)

Reptilia

Aves (birds)

Mammalia

Vertebrata

FIGURE 32.4   Chordate Evolution
One version of the evolutionary relationship among major chordate groups.

TABLE 32.1
Major Vertebrate Classes and Their Characteristics

Class	Respiration	Circulation	External body covering
*Agnatha**	Gills	Two-chambered heart	Skin with mucous glands
Chondrichthyes	Gills	Two-chambered heart	Skin with mucous glands and scales
Osteichthyes	Gills covered	Two-chambered heart	Skin with mucous glands and scales
Amphibia	Lungs, gills in larvae and some adults; also through skin	Three-chambered heart	Skin with mucous glands
Reptilia	Lungs	Incomplete four-chambered heart	Skin scales and few glands
Aves	Lungs	Four-chambered heart	Skin with feathers
Mammalia	Lungs	Four-chambered heart	Skin with hair

Source: Modified from S. Wolfe, *Biology: The Foundations* (Belmont, CA: Wadsworth Publishing Company, 1982), p. 408.
*Not considered a true class by all biologists. Some consider Agnatha a group.

FIGURE 32.5 Lamprey
The lamprey is an example of a jawless fish. Note the rounded, suckerlike mouth.

juices out of them. The hag fishes are opportunistic feeders and scavengers rather than bloodsuckers. The earliest jawless fish were the **ostracoderms** (armor skin). These small fish had a body covered with external body plates made up of bone. The ostracoderms were the most prevalent during the Silurian and Devonian periods.

The First Jawed Fish: The Placoderms

■ **In extinct primitive fish, the gill-supporting structures developed into jaws.**

The earliest jawed fish were the **placoderms.** The placoderms were bottom-feeders that scavenged and also preyed on the ostracoderms. Most placoderms were less than 1 meter long. One of their major features was a strengthened dermal skeleton with enlarged gill-supporting structures. These gill-supporting structures became hinged and modified with

teethlike structures and eventually evolved into the first jaws (Figure 32.6). Not only did these fish have jaws, they also had paired fins that stabilized their bodies and increased their swimming abilities. Because they had jaws, their feeding habits could expand from simple filtering, sucking, and rasping to biting large pieces of prey. Ingesting larger prey was presumably more effective in their environment.

Toward the end of the Devonian period, the placoderms radiated and evolved into other fish types, such as sharks, and eventually became extinct before the end of the Carboniferous period.

Cartilaginous Fish: The Chondrichthyes

■ **The chondrichthyes are the cartilaginous fish; sharks, skates, and rays are examples.**

The most common examples of **chondrichthyes** or cartilaginous fish are sharks and rays (Figure 32.7). They are called cartilaginous because their endoskeletons consist of flexible cartilage instead of bone. There are over 850 species of cartilaginous fish. They have streamlined bodies, five to seven gill slits, and paired fins. Their large caudal fin or tail helps in propulsion.

If you have ever touched a sharkskin wallet, you know that sharks have small scales that feel like sandpaper covering their body. Sharks have irregular cone-shaped scales, whereas the skates and rays have only a few rows of scales on their backs. If you have ever been stung by a ray or stepped on one, you know that sometimes these scales on the dorsal fin can be modified into stinging spines on the tail.

Feeding	Reproduction	Habitat
Mouth sucking or rasping	External fertilization	Marine and freshwater (lampreys); marine (hag fish)
Biting jaws with teeth	Internal fertilization	Marine; very few freshwater
Mouth with teeth	Usually external fertilization	Marine and freshwater
Mouth without prominent teeth, tongue used to catch prey	External or internal fertilization	Freshwater, terrestrial
Mouth with teeth; beak	Internal fertilization	Terrestrial primarily; some freshwater, marine
Beak	Internal fertilization	Terrestrial
Mouth with teeth modified according to diet	Internal fertilization	Terrestrial; some marine, some freshwater

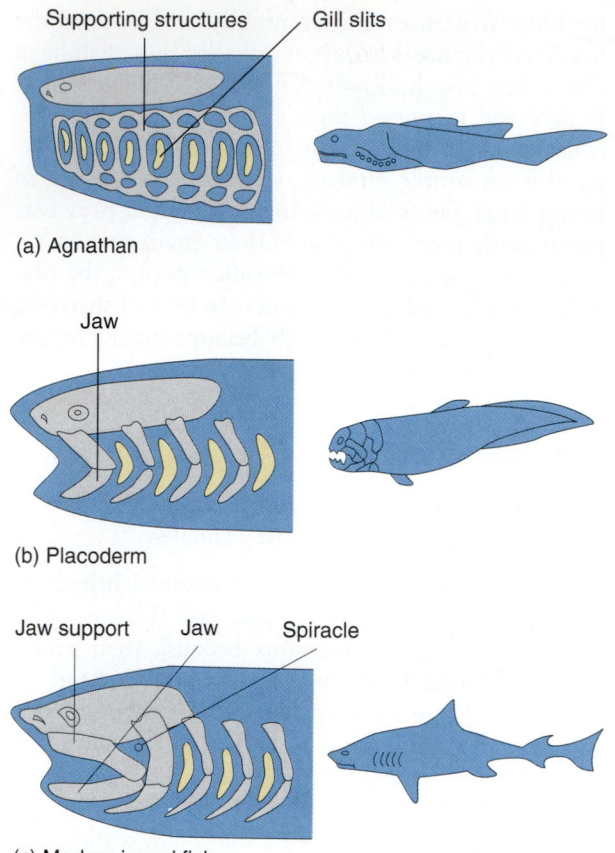

(a) Agnathan

(b) Placoderm

(c) Modern jawed fish

FIGURE 32.6   Gill-Supporting Structures
Gill-supporting structures as found in (a) the jawless fish, (b) the first primitive jawed fish, the placoderms, and (c) the modern jawed fish. Notice that in the modern jawed fish, the gill slit functions as a spiracle through which water is drawn.

(a)

(b)

FIGURE 32.7   Sharks and Rays
(a) The sharks and (b) the rays are examples of the cartilaginous fish, the class Chondrichthyes.

The Great White shark, over 15 meters in length, is one of the largest living vertebrates. Most sharks are excellent predators with strong jaws, large sharp teeth, and acute senses. Rays are almost all bottom dwellers. They have flat bodies as well as teeth and jaws well adapted for crushing crustaceans present in the sand.

Bony Fish: The Osteichthyes

■ **The osteichthyes are the bony fish. The bony fish are divided into two major groups characterized by fin types, ray-finned or lobe-finned. The lobe-finned fish evolved into the amphibians.**

When we say that we are going fishing, we are thinking of the fish most familiar to us. The **osteichthyes** are the bony fish, and the name means their endoskeleton is composed of bone that replaced (in part) the cartilage. Bony fish are the most numerous

of all vertebrates. There are more than 30,000 species, and they inhabit almost every aquatic environment. Zoologists state that bony fish go back as far as the Silurian period. The bony fish diverged into two major lineages characterized by fin type. The **ray-finned fish,** the fish we know as bass, trout, and so on have paired fins that are supported by long flexible rays. These paired fins are highly flexible and allow for complex, rapid movement. Figure 32.8 illustrates the anatomy of a typical ray-finned fish. It should be noted that the ray-finned fish moved from their freshwater environment to sea water and then back to fresh water again during their evolution. Several osmotic regulating adaptations have occurred (see Chapter 26, which discusses the solution to the problem of living in salt water).

The other major evolutionary line is the **lobe-finned fish** (Figure 32.9). Lobe-finned fish species are all extinct except for the coelacanth (*Latimeria*). Most lobe-finned fish lived in fresh water and prob-

FIGURE 32.8 Anatomy of a Bony Fish
(a) A group of bony fishes. (b) An illustration of the internal anatomy of a typical bony fish.

(a)

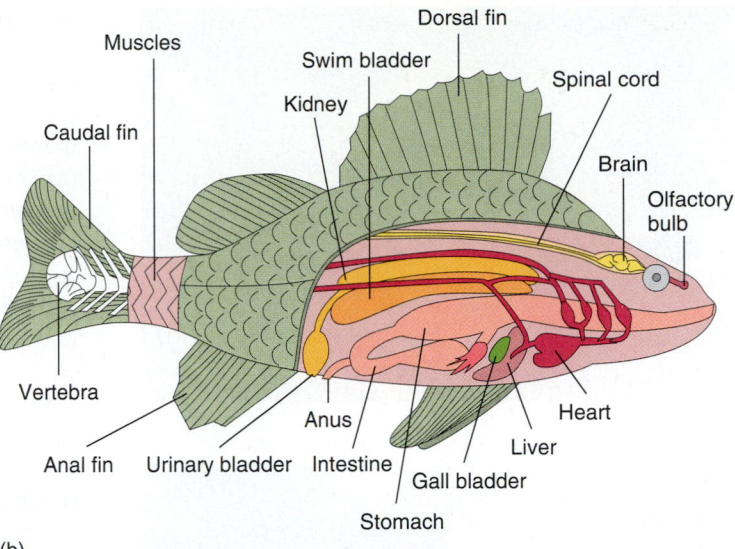

(b)

ably had lungs. The living lobe-finned fish, the coelacanth, belongs to a saltwater, gill-breathing lineage. Lobe-finned fish have paired fins with fleshy and muscular bases. The lobe-finned fish of the Devonian period probably had to move across land from pond to pond as their environment alternately flooded and dried. The ability to use these fins to support their body allowed them to move, "walk," or waddle. In addition, these fish were well adapted to use their moist skin and modified swim-bladder "lungs" as respiratory structures. Today we also find lung fishes that can live in stagnant water that is low in oxygen. Lung fish swim to the surface and gulp air into their lungs, which are connected to the pharynx of the digestive tract. The lobe-finned fish are of special evolutionary importance because they gave rise to our next group, the **amphibians**.

FIGURE 32.9 *A Preserved Latimeria*
The *Latimeria* is the only known living representative of the extinct lobe-finned fishes. *Latimeria* live in the oceans off the coast of Madagascar.

The First Land Vertebrates: The Amphibians

■ **Amphibians were the first land vertebrates. They are well adapted for life both in water and on land. Because of external fertilization and the type of eggs they produce, the amphibians require water for fertilization and embryonic development.**

During the Devonian period, there was a cycle of rain and drought, and ponds would fill and then dry. The animals that lived in these ponds were subject to the pressures of survival in water and land. As we mentioned, the lobe-finned fish could use their fins as paired appendages to wiggle across land from drying pond to drying pond. These ancestors were probably the first amphibians.

Amphibians usually have two life cycles (*amphibian* literally means "two lives"): the larval form, like the tadpole that lives in water, and the adult form that lives on land (Figure 32.10). This schizophrenic lifestyle has advantages and disadvantages. The amphibians were the first land vertebrates, and they enjoyed a distinct advantage because they were first: In their new environment they found new sources of food in the available terrestrial plants and insects, but there was no competition from other vertebrates for the food. In addition, because of the higher oxygen content of air as compared to water, their respiration was more efficient on land, resulting in more energy and mobility and increased ability to capture prey. Life on land also presented new challenges, such as the pull of gravity (water provides buoyancy), respiration in air versus water, temperature extremes, and drying out.

(a)

(b)

(c)

FIGURE 32.10 Life Cycle of a Frog
(a) The male frog grasps the female, and this behavior stimulates her to lay her eggs. As the eggs are laid, the male releases his sperm, and fertilization occurs in the water. (b) The eggs develop into tadpoles, which are fishlike herbivores that respire with gills. (c) As the tadpole continues to develop, hind legs form, then forelegs, and eventually the tail is absorbed and the tadpole becomes a young frog.

The amphibian body plan, as it evolved, reflected strategies that were best suited for this new half-water and half-land lifestyle. To stand and move on land, amphibians developed a stronger skeletal system and two pairs of appendages (four legs). For gas

exchange, they retained gills for water and grew lungs for land. Also, the moist skin could be used for respiration. Amphibians evolved behavioral patterns that kept them living in moist humid environments near water to prevent drying out as well as other behavioral patterns that prevented death from temperature extremes. Another adaptation was an efficient closed circulatory system.

Reproduction is well adapted to this dual lifestyle. Fertilization is external, with sperm and egg uniting in the water. The amphibian egg has no shell, so embryonic development occurs in the water or in a moist environment. It is not well adapted for land. As another solution to the requirement for moisture for embryonic development, some species of frogs retain the eggs within or on the body of the adult as the tadpoles develop. Such frogs are called *live-bearing,* meaning that the tadpoles develop in the body of the adult (Figure 32.11). Most often, survival of the species is assured by safety in numbers. The female lays hundreds of eggs, but only a few reach adulthood. In other species, only a few eggs are laid, and various types of parental care have evolved to increase the survival rate. Amphibians also exhibit complex behavioral patterns associated with courtship, territoriality, reproduction, and homing (returning to their home).

There are three major orders of amphibians living today (Figure 32.12). These orders are distinguished by their methods of locomotion and their anatomy: the *urodeles* have a tail and are typified by salamanders; *anurans* don't have a tail, as exemplified by frogs and toads, and *apodans* (remember when we discussed the arthropods, that *a* means "without" and *poda* means "leg") are legless amphibians, the *caecilians.*

There are about 400 species of salamanders, which belong to the order *urodeles.* When on land they exhibit the characteristic side-to-side bending that probably resembles the movement of early amphibians, lungfish, and coelocanths. Of interest is that some species of salamanders do not complete metamorphosis, rather, the sexually mature adult resembles the aquatic larval form. For example, the Mexican *axolotl* never develops the adult body form. It retains its tail and external gills.

The frogs and toads lack an external tail as adults and are members of the order Anurans. There are over 3500 species. The frogs and toads are especially suited for life on land, because they have well-developed limbs, a tongue for capturing insect prey, adaptation for protection from predators like poisonous mucus glands in the skin, and moist skin and lungs for gas exchange.

FIGURE 32.11   Livebearing Amphibians
Pipa pipa is a livebearing frog. As the eggs are laid, the male pushes the fertilized eggs into the soft flesh of the female's back. The eggs are embedded and develop within pockets on the female's back. The tadpoles complete their metamorphosis within the pockets and emerge as small frogs.

A group of amphibians that most of us have never seen is the legless amphibians, the Apodans. There are 150 species of apodans called caecilians. These amphibians resemble earthworms, because they are legless and have wormlike bodies. Upon closer examination, these burrowing amphibians exhibit an endoskeleton with a backbone and other vertebrate characteristics typical of amphibians. Some caecilians even have small scales, which suggests that they are similar to the reptiles.

As you can see, the amphibians are the descendants of the lobe-finned fish that ventured out of the water onto the land sometime during the Devonian period. During the late Devonian and early Carboniferous periods, amphibians were the only terrestrial vertebrates. During the late Carboniferous period, amphibians began to decline and new group, the Reptiles, began to emerge.

Reptiles: The First True Land Vertebrates

■ **Reptiles were the first true land vertebrates due to adaptations that suited them for land such as scales, internal fertilization, and the development of the land egg.**

The **reptiles** consist of a diverse group of animals. About 7000 species exist today (Figure 32.13). However, if you include birds within the phylum Reptilia — and most zoologists would — there would be more than 15,750 species. Of course, all of us know of some of the extinct species of the late Triassic period, the dinosaurs and their relatives.

(a)

(b)

(c)

FIGURE 32.12   Amphibian Orders
(a) This salamander is an example of the order Urodeles. (b) This colorful Poison Arrow frog is a member of the order Anurans. (c) This legless amphibian, the caecilian, belongs to the order Apodans.

Reptiles probably evolved from amphibians, as evidence from the fossil record of about 300 million years ago shows. Reptiles were the first vertebrate animals to complete the transition to land; hence their bodies, when compared to amphibians, have several adaptations for living on land.

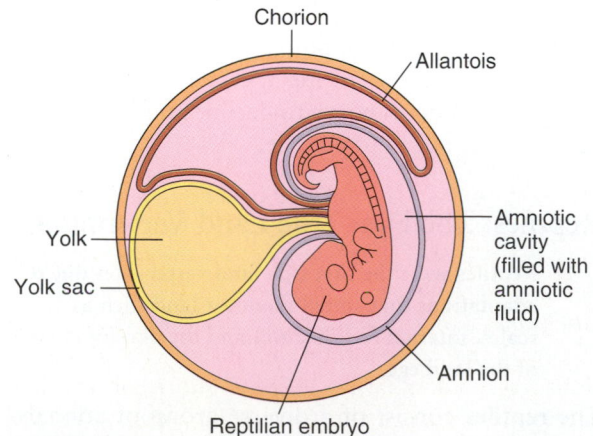

(a)

(b)

(c)

(d)

FIGURE 32.13   Reptiles
(a) This slider turtle belongs to the order Chelonia. Turtles are an ancient group of reptiles that have changed very little since the Triassic period. (b) This whiptail lizard is a common example of the order Squamata to which the snakes such as this cape water snake (c) also belong. (d) This young albino crocodile is a member of the order Crocodilian.

The skin of reptiles is covered by scales consisting of keratin that helps prevent water loss. They have well-developed lungs. They have internal fertilization (so they don't require water for the sperm to reach eggs), and the eggs of reptiles are covered by a tough water-resistant shell that allows for exchange of gases but prevents drying out.

The embryo develops within the fluid secreted by a membrane called the *amnion;* hence the **amniote egg** provides for a self-contained aquatic environment complete with food (Figure 32.14). We discussed the importance of this egg on page 690. With this shelled egg, reptiles could complete their life cycles on land. Reptiles are called cold-blooded or *ectothermic* (page 617), but several behavioral mechanisms have evolved for regulating body temperature. (Have you ever seen a lizard basking in the sun or laying in the shade of a tree?) Now that we know something about the reptilian adaptations for life on land, let us discuss three orders of living reptiles.

FIGURE 32.14   Amniote Egg
The amniote egg of the reptiles and birds provides a self-contained environment, which frees them from returning to the water for embryo development. The membrane provides protection, and the amnion secretes the fluid that surrounds the embryo.

Turtles (Chelonia)

Turtles evolved from the early reptiles during the Triassic period and have remained rather unchanged since that time. The most predominate features of the turtle are the shell and the beaklike jaw, which help protect it from predators. Some turtles have returned to the water to live in, but even so, they return to land to lay their eggs on beaches, where the eggs are incubated by the heat of the sun.

Lizards and Snakes (Squamata)

When we think of reptiles, we frequently think of lizards and snakes. They are the most familiar and most common reptiles. Over 6000 species of lizards and snakes exist. Lizards are usually small and very common in the deserts and tropics (see Profile). It is of interest that some lizards as well as some snakes give birth to offspring that are fully formed rather than laying eggs. Some lizards burrow and live in the soil. In these species, like the "skink" (a lizard with a long body and tail), the legs are shorter and the body has become elongated. Even though snakes are legless, some species still have remnants of a pelvic girdle and limb bones, which provide evidence that the snakes probably evolved from burrowing lizards. Snakes are carnivorous predators and are well suited for hunting and capturing prey. Some species have well-adapted heat-detecting organs. All have an acute sense of smell and loose jaws that enable them to open their mouths wide enough to swallow prey larger in diameter than their bodies. Some species are poisonous and inject their venom through hollow fangs. Others coil themselves around their prey and squeeze them to death.

Alligators and Crocodiles (Crocodilians)

The alligators and crocodiles are large reptiles that live in many swampy regions of the earth. The crocodilians look almost like living dinosaurs. Physiologically, the crocodilians are of interest because rather than a three-chambered heart typical of the amphibians and most reptiles the crocodilians have a heart that is almost four-chambered. Their ventricle is almost completely divided into right and left halves. Blood from the right ventricle goes to the lungs, where the exchange of gases occurs, and then returns to the left side of the heart. The blood from the left ventricle goes to the body. This dual system (page 578) allows for more efficient circulation of oxygenated blood and is more typical of our next groups, birds and mammals.

The Birds (Aves) — Flying Dinosaurs

■ **Because of current data, most biologists group the birds with reptiles rather than as a separate class. Birds all share a common body plan well-suited for flight. Birds are also warm-blooded, with a four-chambered heart.**

Zoologists who understand evolution would consider it appropriate to call birds flying dinosaurs. Their ancestors were reptiles that ran on two legs, were probably *endothermic* (warm-blooded), and presumably had complex behaviors associated with courtship, parental care, and social groups. The fossil remains of *Archaeopteryx lithographia,* a reptile, dating from 150 million years ago in the Jurassic period, provide evidence for the identification of birds with reptiles. *Archaeopteryx* had feathers, clawed forelimbs, teeth, wings, and a long vertebrate tail. Most recently, in 1986, a fossil found in Texas called *Protoavis* ("first bird"), which is a hybrid of dinosaur and avian features, has supported the idea of the birds as flying dinosaurs. *Protoavis* lived about 225 million years ago, long before *Archaeopteryx.* As with most evolutionary data, the actual origin of the birds is still subject to controversy, but nearly all zoologists recognize birds as a branch of the reptilian lineage.

There are over 8750 species of **birds** alive today, and all share a similar body plan (Figure 32.15). The most obvious feature is that all birds have feathers. As one of my professors said when I was a college student, "Feathers are glorified scales," and they are. Feathers are made up of the protein keratin, which also forms scales, and even our hair. Feathers function in flight and insulation. Birds in flight capture air with their feathered wings. Have you ever watched a bird on a cold day fluff up its feathers to capture air to keep warm? Birds are light weight, with a number of physiological modifications for that purpose. For example, the bones have air cavities, which make them light. Other adaptions to create a lightweight body include a small skull and toothless head (birds have a gizzard to break down food mechanically; see Chapter 23), and the reduction or absence of some organs. Female birds only have one large ovary. The wing is a marvelous adaption for flight, being long in respect to body size, and light.

As the old joke states, "I just flew in from New York and boy are my arms tired." Flying requires a great expenditure of energy. Birds have a four-chambered heart, which separates oxygenated blood from unoxygenated blood. This circulatory pattern provides more oxygen for more efficient respiration

L. Lee Grismer
Herpetologist

The clouds looked threatening as Lee and his 10-year-old son Jesse looked across the Sea of Cortés from the small deserted island of Coloradito. They were there to study a new species of lizard found on the island. They were dropped off earlier that day by a Mexican fisherman who promised to return two days later to take them back to the Baja California peninsula. As the evening approached, so did a small *chubasco* (hurricane), and the inevitable thunder, lightning, and rain began crashing all around, forcing them to seek safety and shelter. The sea lions were roaring almost as thunderously as the storm, and Lee and Jesse's only shelter was a low shallow cave on the beach at the edge of a sea lion colony. The mouth of the cave was not a welcoming sight, however, as several deer mice and biting arthropods stood their ground. Lee and Jesse crawled into the cave on their stomachs. They spent a sleepless night, safe from the storm but constantly fighting off the mice and insects.

Many believe the stereotype in the media of biologists as introverted, bumbling nerds with no life other than a Dr. Frankenstein-like obsession with work. Obviously, as with most stereotypes, that is not true. Most biologists have adventures that far exceed the imagination of movie script writers. The example of Dr. Grismer and his son Jesse sharing adventures in their search for an understanding of amphibians and reptiles would make Indiana Jones envious.

Dr. Grismer loves the isolation and excitement of going where no man has gone before. On his many collecting trips — which range from traveling through the Amazon Basin in a dugout canoe to being lowered by helicopter onto desolate islands in the Ryukyu Archipelago off the coast of China — he has collected specimens that have never been formally described to science.

Dr. Grismer began his love for the study of lizards and snakes when he was 5 years old. His family owned a raceway for off-road vehicles in Southern California. During the long hot summers, as he helped his family with the raceway, he would explore the chaparral-covered hillsides looking for lizards and snakes. This passion consumed him, and he always knew that he would wind up studying natural history — especially reptiles. The passion for biology kept growing, and after he graduated from high school, he began his formal education in the science. He was the first in his family to earn a college degree.

His interest in Baja California also began at an early age. His father was very adventurous and one of the original Baja bush pilots. As a young boy, Dr. Grismer spent a great deal of time flying with his father into the wilderness of Baja California. On one of these trips when Lee was 13 years old, he dreamed of becoming one of the leading experts in the natural history of Baja California, and today his dream has come true.

Dr. Grismer studies patterns and processes of herpetological evolution and uses the Baja California peninsula and its associated islands as a natural laboratory, which has furthered our knowledge of island biogeography. He has written several research papers and has two books in progress on the herpetology of Baja California and the Sea of Cortés. He is considered an expert on the reptiles and amphibians of this region.

His love of herpetology is contagious and infectious. Not only does he infect his students with his passion but also his family. His wife, his son Jesse, and 9-year-old daughter Lacy have accompanied him on many of his research trips.

Jesse has already become a respected herpetologist in his own right. When Jesse was in the fourth grade, he won an award from the state of California for an essay he wrote on a "special species." He wrote about an endangered species, the barefoot banded gecko. Of interest is that ten years prior, on the same date as Jesse's award, his father published his very first paper on the same species. Jesse loves to travel with his father, and as a matter of fact, this year when Dr. Grismer travels to study the reptiles of an isolated island in the Galápagos, Jesse will be his field assistant. Jesse published his first solo research paper this year at the age of 10. His paper was on the feeding habits of a poorly known snake in the mountains of Baja California and appeared in the journal *Herpetological Natural History.*

Lee and Jesse Grismer disapprove of the misconception that in order to be a good scientist you must sacrifice your personal and family life. When Dr. Grismer is asked about his son, or when he hears his son talk with other herpetologists with a level of knowledge far beyond his age, Dr. Grismer's chest swells with pride.

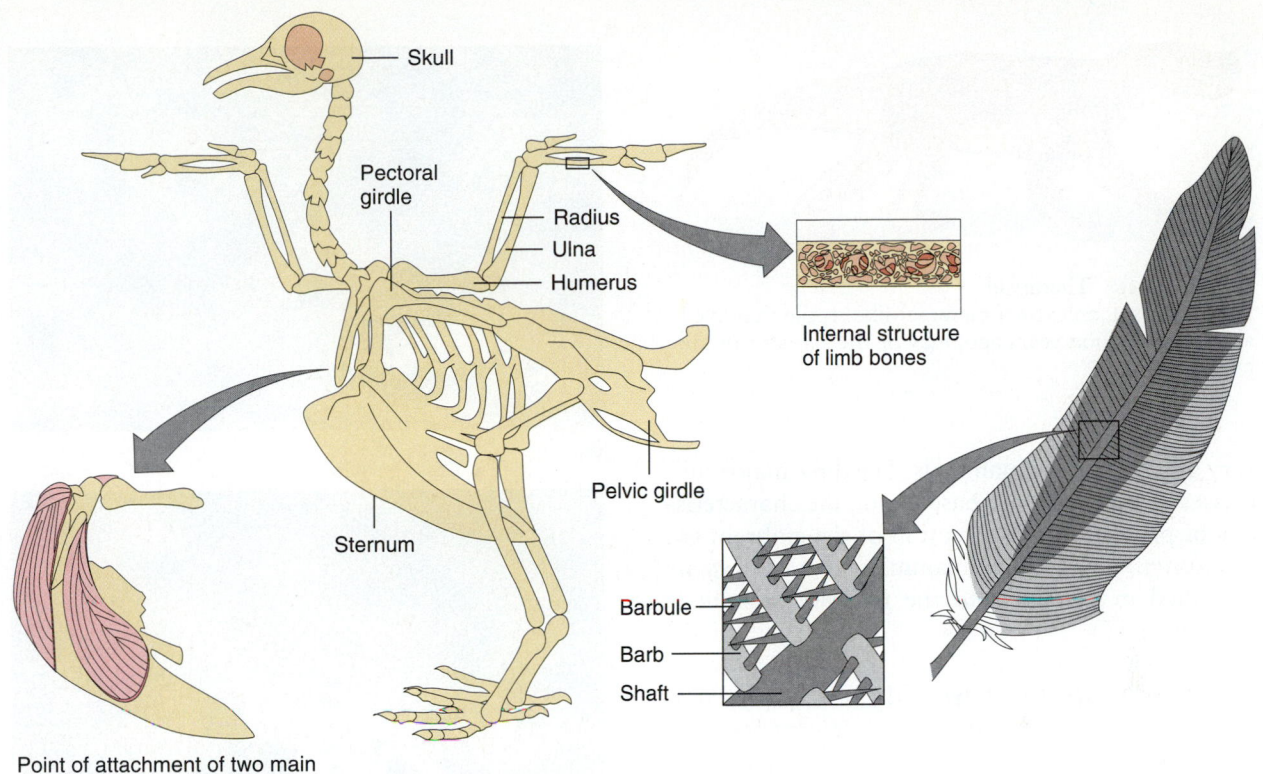

Skull

Pectoral
girdle

Radius
Ulna
Humerus

Internal structure
of limb bones

Pelvic girdle

Sternum

Barbule

Barb

Shaft

Point of attachment of two main
flight muscles to keel of sternum

FIGURE 32.15   Birds
The body plan of the bird is very well adapted for flight. The large flight
muscles of the bird attach to the keel of the sternum, which provides for the
movement of the wings, resulting in flight. The feathers are light and strong.
The interlocking of the barbules forms a lattice-like arrangement giving
strength to the feathers. The bones are strong and light.

to generate the high energy required for flying. These
high-energy requirements are most evident in birds'
large pectoral (breast) muscles, which provide the
powerful downsweep of the wing that results in
flight. These muscles are attached to a large sternum
(breastbone). Other reptilian characteristics shared
by birds are the amniote egg and scales on the legs.

The Mammals: Mammary Glands and Hair

■ **Mammals are warm-blooded animals with hair
and mammary glands, which secrete milk. The
orders of the class Mammalia are characterized by
the increase in embryo protection and the way the
embryo is retained by the mother.**

Humans are mammals, and most of the animals we
commonly think of are: our trusty dog, our cat, and
the cute monkeys. The major characteristic of **mammals** is their modified sweat glands that secrete milk,
the **mammary glands.** A secondary characteristic is
that mammals have hair. Hair is made up of the pro-

tein keratin, which also makes up the scales of reptiles. In addition, mammals are characterized by an
efficient four-chambered heart, which provides for
efficient circulation of nutrients, waste products,
and gases and helps to maintain a constant body
temperature. If you have ever hugged a warm puppy,
you know that mammals are endothermic, or warm-
blooded. Mammals have a diaphragm that functions
in efficient breathing, which helps to ventilate the
lungs. Mammals have a very well-developed brain.
Their teeth are specialized — from the shearing
teeth of the meat eaters to the large, grinding, crushing molars of grass-eaters — and they can be used
for a variety of diets.

As to the origin of the mammals, evidence suggests that they evolved from reptilian ancestors even
before birds evolved. The ancestors of the mammals
were mammal-like reptilians called *therapids,* which
lived about 220 million years ago. These small in-
sectivores were probably shrewlike (Figure 32.16).

One of the major trends in reproduction, the increased protection and retention of the embryo, is

FIGURE 32.16 Therapsid
The therapsid, an extinct mammal-like reptile that lived about 220 million years ago, may be the ancestor of the mammals.

(a)

(b)

(c)

very apparent in the mammals. The three major subclasses of mammals are based upon the characteristics of protection and retention of the embryo: the *monotremes* (egg-laying mammals), the *marsupials* (pouched mammals), and the *placental* mammals (Figure 32.17).

The Monotremes: Egg-Laying Mammals. The monotremes are the only living mammals that lay eggs. Platypuses and spiny anteaters are examples. The monotremes lay eggs (which are reptilelike) and incubate them. Once the eggs hatch, the young platypuses lick milk from the fur on the underside of the mother. She just "sweats" milk onto the fur because her mammary glands have no nipples. The monotremes descended from a very early branch of the mammalian line and follow a very elementary method of protecting the embryo — in the egg. They don't even retain the embryo. Let us look at another group of mammals, the marsupials, which retain the embryo a bit differently — in a pouch.

The Marsupials: Pouched Animals. Kangaroos, opossums, and koalas are all examples of the marsupials. The embryo of the marsupial develops within the mother's reproductive tract, then at a very early age (for example, in the kangaroo at about 33 days after fertilization) the small embryo begins the most important journey of its life. It crawls from its mother's vagina along her fur to the pouch where it attaches to a teat, and sucks milk as it develops. As you can see, in the marsupial we have the beginning of retention of the embryo within the body of the mother during development.

Of interest is that the native mammals of Australia are either marsupials or monotremes. This is explained by the fact that as Pangaea (page 398) broke up Australia and South America became island continents dominated by marsupials. Australia has been isolated for 65 million years. The early Australian mammals were marsupials that evolved and radiated to fill the habitats usually filled by placen-

FIGURE 32.17 Three Subclasses of Mammals
(a) The duck-billed platypus is an example of an egg-laying mammal, which is a reptile-like characteristic.
(b) Kangaroos are examples of marsupials. The young of a marsupial develop within the pouch of the mother.
(c) This zebra and her young are examples of placental mammals. The embryo develops within the placenta formed by the tissues of the mother's uterus and the membranes of the developing embryo.

tal mammals in other parts of the world — a convergent evolution (page 395). You might ask how the opossums got to the United States. South America was also an island continent until about 12 million years ago, when the land bridge (Central

America) formed, connecting South and North America. This land bridge served as a conduit for the fauna to distribute themselves between the two land masses.

The apex of protection and retention of the embryo is exhibited by the placental mammals. In the placental mammals, such as ourselves, development of the embryo occurs within the uterus of the female. Within the uterus, the endometrium of the mother and the outer membrane of the embryo join to form the placenta. The development of placental mammals is discussed in Chapter 28. The placenta protects and nourishes the embryo, rids the embryo of waste products, and functions to exchange gases between the embryo and the mother. The divergence of the marsupial line and the placental line probably occurred 90 million years ago.

The placental mammals are very diverse and incled 4500 species. Some are very exotic, like the manatees, and some do not appear like mammals at first glance, like the whales. But most are very familiar: mice, rats, dogs, cows, and humans.

10. The amphibians were the first land vertebrates and probably evolved from the lobe-finned fish, whose fins could support the body against the pull of gravity. Frogs, salamanders, and legless caecilians are examples of living amphibians.

11. The reptiles consist of turtles, snakes, lizards, alligators, crocodiles, and birds. Reptiles are well adapted for life on land. The evolution of the reptilian egg and internal fertilization freed reptiles from the water for reproduction and development.

12. Most biologists believe that birds are really warm-blooded reptiles because of the numerous characteristics that they share: the egg and leg scales, for example. Birds have a body plan well-adapted for flight.

13. Mammals also evolved from the reptilian line, in all probability from a warm-bloodied reptilian such as the extinct *Therapsid*. Mammals all have mammary glands and hair.

14. Mammals are grouped into orders based on the degree of protection and retention of the embryo. The monotremes (egg-laying mammals) have the least; the marsupials show an intermediate degree of protection and retention of the embryo; and the placental mammals reflect the apex of maximum protection and retention of the embryo.

SUMMARY

1. Chordates are bilateral deuterostomes that consist of two subphyla, those that lack a vertebral column, the invertebrates, and those that do, the vertebrates.

2. Because of the embryonic similarities between the invertebrate chordates and echinoderm larvae, zoologists surmise that the echinoderms gave rise to the chordates.

3. The major characteristics of the phylum Chordata are (1) the presence of a notochord, (2) gill slits at some time during development, (3) a dorsal hollow nerve cord with a brain, and (4) a tail.

4. The invertebrate chordates are the urochordates (sea squirts) and the cephlochordates (lancelets).

5. In the vertebrates, the notochord is replaced or degenerates into the vertebral column (the backbone).

6. Agnathans, the jawless fish, are the earliest vertebrates, and the extinct ostracoderms were the earliest forms. Modern forms include the lamprey and the hag fish.

7. The earliest jawed fish were the armored placoderms, in which the gill-supporting bodies became modified into jaws and teeth.

8. The chondrichthyes, or cartilaginous fish, have a skeleton consisting of cartilage rather than bone. Sharks, rays, and skates are modern examples.

9. The osteichthyes, or bony fish, are the most numerous of all vertebrates. There are two major types of bony fish, the lobe-finned and the ray-finned fish.

KEY TERMS

chordates	osteichthyes
notochord	ray-finned fish
urochordates	lobe-finned fish
cephlochordates	amphibians
vertebrae	reptiles
agnathans	amniote egg
ostracoderms	birds
placoderms	mammals
chondrichthyes	mammary glands

REVIEW QUESTIONS

True/False

1. All members of the phylum Chordata have vertebrae.
   true   *false*   *(page 791)*

2. Chordates are bilateral deuterostomes that exhibit a variety of body plans.
   true   *false*   *(page 791)*

3. Because of embryonic similarities, most zoologists speculate that the chordates originated from the echinoderms.
   true   *false*   *(page 791)*

4. The notochord is a feature common to all chordates at some time in their life cycle.
   true   *false*   *(page 792)*

Fill in the Blank

1. Members of the subphylum _____ all have a vertebral column. *(page 792)*

2. Members of the group _____ are called jawless fish. *(page 793)*

3. The jaw of the jawed fish developed from modified _____ _____ _____ that contained teeth and were hinged. *(page 795)*

4. The lobe-finned fish are of evolutionary significance because they gave rise to members of the class_____. *(page 797)*

Multiple Choice

1. A significant evolutionary development that freed the reptiles from the water for reproduction was
   a. Scales
   b. The amniote egg
   c. External fertilization
   d. The amniote egg and internal fertilization
   (page 800)

2. Reptiles and birds share all of the following characteristics except:
   a. Scales
   b. Amniote egg
   c. Warm-bloodedness
   d. Lungs
   (page 801)

3. If you found a new species of vertebrate animal that looked liked a fish but had hair, you would place it in the class:
   a. Chondrichthyes
   b. Osterichthyes
   c. Aves
   d. Mammalia
   (page 803)

4. When grouping the mammals into orders, the following characteristic is most often emphasized:
   a. Number of appendages
   b. Method of embryonic development
   c. Method of locomotion
   d. Type of teeth
   (page 804)

Discussion Questions

1. What are the major characteristics of the phylum Chordata?

2. Even though at first observation one would not think of the sea squirts and lancelets as chordates, they are. Why?

3. Compare and contrast the four major groups of fish, agnathans, placoderms, chondrichthyes, and osteichthyes.

4. From an evolutionary perspective, why are the lobe-finned fish important?

5. List and discuss several adaptations among the amphibians that suit them for living on land.

6. Why is the amniote egg important?

7. As a zoologist, where would you place the birds? With the reptiles, or in a separate class? Why?

8. What characteristics are used for placing the mammals into specific orders?

UNIT

7

BEHAVIOR AND ECOLOGY

After reading this unit, the student should be able to:

- understand the biological basis of behavior
- understand the basic concepts involved in ecosystem dynamics
- understand how biotic and abiotic factors affect community structure and ecosystem function
- understand the cycling of elements and water through the biosphere and how organisms affect this cycling of elements
- discuss the global and local issue of humans affecting the delicate balance of the biosphere

33

Behavior

The Study of Behavior
Innate and Learned Behaviors
Memory: The Brain and Behavior
Behavior and Communication
Sociobiology
Summary

So far we have described, dissected, weighed, chemically probed, and analyzed the brain, nervous system, endocrine systems, and their parts. However, we are still left with a question: How can the brain (three pints of grayish matter, 100 billion or more nerve cells embedded in a matrix of supporting cells) and the secretions from these cells, tissues, and glands regulate our motion, emotion, thought, ideas, and behavior?

That question leads to a number of others. Everyone knows what behavior is, but on what is it based? How much control do we have over it? How can we define it? How can it be studied? What are some of the biological bases of behavior?

Above: The behavioral response of this lion-tail monkey to a threat is very obvious.

Researchers suggest that there may be a biological basis for even complex human behaviors such as sexual preference. In 1991 investigators at the Salk Institute determined that a region of the hypothalamus (known to be involved in sexual behavior) is smaller in the brains of homosexual men than heterosexual men. However, the study is not conclusive and, as with all aspects of science, one should always question answers, not only answer questions. We cannot say that human sexual preference is due solely to a biological basis. But this study does demonstrate the possibility of a biological basis for human sexual preference.

So as you may suspect, there are no easy answers to any of these questions. Nevertheless, in this chapter we will search for clues that can untangle some of the mysteries of behavior by discussing instinct, memory, and communication.

THE STUDY OF BEHAVIOR

■ **There are three main avenues to studying behavior: the psychological approach, which emphasizes philosophical concepts or experimental techniques and uses nonhuman animals to understand human behavior; the ethological approach, which emphasizes the study of an organism's behavior in its natural environment; and the biological approach — sometimes called the biophysiological approach — which emphasizes the biological and physiological underpinnings of behavior.**

Behavior is a series of actions that depend on chemical and physical coordination systems. If you study animals, you will observe that they respond to their external and internal environment in a coordinated fashion. Those responses, internal or external, are **behavior,** and they are the result of the organism's biological nature. How can we study behavior? There are several approaches: the *psychological, ethological,* and *biological* (or *biophysiological*).

Many psychologists have developed explanations of human behavior that are rooted more firmly in philosophy than in experimental evidence. The psychoanalytical approaches of Freud and Jung were based on insight and philosophical conclusions, not on experimental fact. That does not mean that their conclusions were not valid — only that they did not rely on the scientific method.

Other psychologists base their explanations of behavioral phenomena on the results of experiments. But although the work may be designed according to the scientific method, with observation, analysis, and control of variables, the subjects (usually rats) are placed in an unnatural environment and subjected to strange and novel stimuli. In addition, although such experiments study animal behavior, the results are interpreted in terms of human behavior, and this final step may or may not be valid.

There is another approach to the study of behavior, one that was developed in Europe in the mid-1930s. The method is called **ethology,** from the Greek word *ethos,* which means "habit," and *logy,* for "study of." Ethologists study animal behavior from a biological perspective. That is, whether working in an animal's natural environment or in the laboratory, they systematically observe and test animal behavior to learn how that behavior relates to the animal's role in its natural habitat. In 1973, three of the original European ethologists, Konrad Lorenz, Niko Tinbergen, and Karl von Frisch, shared the Nobel Prize for developing the ethological approach to studying animal behavior.

That these three men were not recognized for their work until 1973 is a good indication that biologists were slow to recognize ethology as a legitimate scientific discipline. In part, the reason for the skepticism is the absence of hardware. Ethologists rarely use sophisticated equipment in their studies. Instead, they rely on their powers of observation. Today, these observations can be augmented by sound and visual recording devices that allow the researcher to study and restudy a certain behavioral pattern. Then too, today's computers allow for a more thorough statistical analysis of behavioral pathways.

Often it is not enough to simply philosophize or observe and describe behavior. Other questions need to be resolved. What are the biological, chemical, and physical bases of the behavior? How can a form of behavior be passed on from generation to generation? How can a series of neural impulses or hormonal activities produce a coordinated response? How are thoughts stored?

To answer these types of questions, a new approach, based on biophysiology, has been taken up by neurologists, neurophysiologists, and behavioral endocrinologists. In fact, even molecular biologists and geneticists are studying the biological basis of behavior. Their results, when integrated with the findings of psychologists and other social scientists and ethologists, ultimately will give us a better understanding of behavior.

INNATE AND LEARNED BEHAVIORS

■ **Some behaviors are innate; that is, they are inherited and can be performed by the organism with little analysis or interpretation of stimuli. Other behaviors are learned.**

No matter what approach researchers use in the study of behavior, be it psychological, ethological, or biological, all agree that in the simplest terms, behavior is a set of actions — a response to a received signal, called a **stimulus.** In some organisms, or for some behaviors, the response is almost like a reflex. The organism performs very little analysis or interpretation of the stimulus. Other behaviors are the result of learning.

Innate Behavior

■ **Innate behaviors are stereotyped and are species-specific.**

Some behaviors are called instinctive or innate behaviors. **Innate behaviors,** which are shaped by evolution, are inherited by offspring. Such activities are said to be *highly stereotyped,* meaning that they are always the same and not usually modified by learning. They are *species-specific.* In other words, the instinctive response of a monarch butterfly to food is not the same as the response of a white-tailed deer or a starling.

Taxis. The simplest pattern of innate behavior is a taxis (plural form, *taxes*). A **taxis** is the directed movement of an organism toward or away from a stimulus. (We studied several types of taxes in Chapter 21, when we looked at plant behaviors.) Consider the single-celled organism *Euglena.* This animal has a photoreceptor structure that perceives light and a flagellum that enables it to move. In most cases, the animal moves toward light. But if the light is too intense, the *Euglena* will move away. The differentiation is of important selective value to the organism, because too bright a light could kill it.

Taxis in *Euglena* is well understood. An opaque structure called the *stigma* casts a shadow onto the photoreceptor region (the paraflagellar swelling) that is below the stigma. When the *Euglena* is lined up with the light, the shading of the swelling is at a maximum. This swelling in turn controls the flagellum and thus steers the *Euglena.*

Some taxes are dependent on an organism's age. For example, the larval stages of some insects exhibit negative phototaxis, which means their movements are directed away from light. Going toward dark and leaving the light helps to ensure their survival because it enables them, in essence, to hide from predators, although no one is suggesting that a conscious thought process is involved. As the insects mature into adults, however, they may exhibit positive phototaxis, moving toward light for food gathering and mating. Therefore, the developmental stage of the organism can play a role in determining how it responds to a stimulus.

These examples tell biologists that behaviors, including taxes, depend on the strength and quality of a stimulus and the age or condition of the organism (Figure 33.1). In other words, the process is not simply stimulus → response, but it is stimulus → processing center → response. An organism's processing center also includes a *filtering element.* An organism does not respond to all the stimuli to which it is exposed. Some stimuli are filtered out. For example, as you read this text you are aware of the pressure of your body against the chair only if you consciously think about it. Under normal cir-

FIGURE 33.1 Light Taxis in *Euglena*
Euglena has the ability to perceive light, recognize differences in the intensity of light, and direct its movements accordingly.

cumstances, you filter out this stimulus in your brain.

Fixed-Action Patterns

■ **Fixed-action patterns consist of several behavioral elements performed in a rigid, fixed sequence.**

Each type of organism inherits a set of specific **fixed-action patterns,** behavioral patterns consisting of several elements performed in a coordinated, rather rigid sequence. Some duck species are distinguished not by their appearance (which is almost identical) but by the fixed-action patterns of their courtship rituals, which differ among species. The differences are great enough that biologists (not to mention ducks) can determine the species by the species-specific behavior.

As an example of a fixed-action pattern, there is a sea slug (an ocean snail without a shell) called *Tritonia* that manages to escape by swimming away in a fixed-action pattern after it has come into contact with a starfish or other sea star. *Tritonia* moves by a series of up-and-down coordinated contractions of its body. A neurophysiologist discovered that single neurons in the brain of *Tritonia* fire with a frequency pattern and intensity that correlate with the swimming movements. Moreover, even if the brain is removed from the animal and the neurons are then electrically stimulated, they respond with the same series of coordinated impulses. The neural centers, in a sense, can be called innate releasing mechanisms (IRMs), since they release, or stimulate, the sequence of the innate fixed-action pattern. In other words, a specific pattern of behavior, an innate response, has its origin in the preprogrammed neural circuits of the brain.

Innate releasing mechanisms permit the fixed-action patterns of innate responses to unfold. Innate responses are inherited. But how are these responses or behavioral patterns preprogrammed? And how are they passed on from generation to generation? Since animal offspring inherit DNA from their parents, DNA or genes must influence innate behavior.

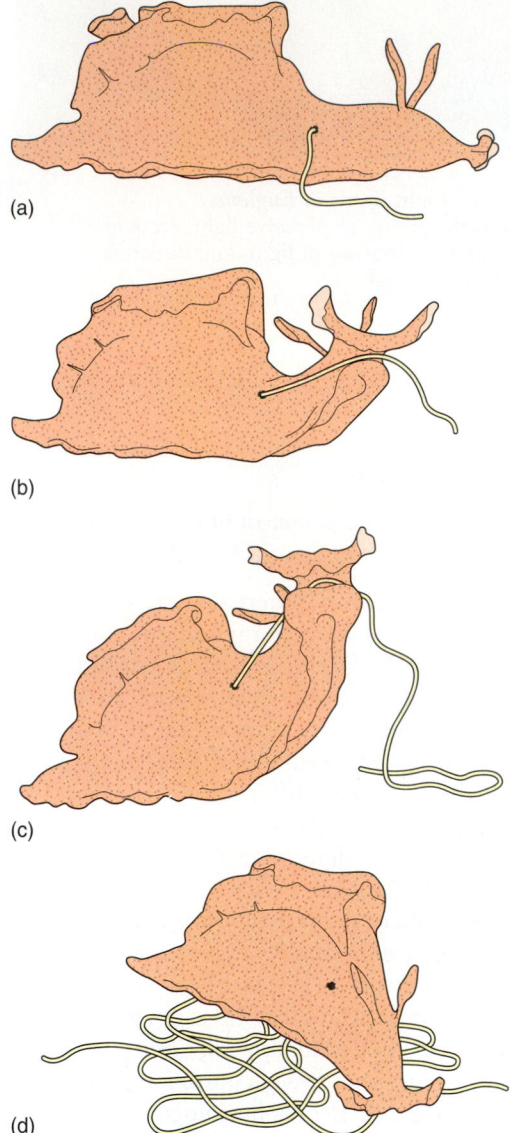

(a)

(b)

(c)

(d)

FIGURE 33.2 Fixed-Action Behavior: *Aplysia*
Aplysia, a marine slug, (a) lays its eggs in a long string, (b) uses its mouth, (c) shakes its head to pull the string out of its reproductive duct, (d) and then affixes the eggs to a solid surface. Researchers have demonstrated that this egg-laying behavior is a fixed-action behavior, an innate response that follows a rigid sequence. This behavior is caused by a protein called egg-laying hormone, whose synthesis is controlled by genes.

To test this hypothesis, scientists again turned to a species of sea slug, one called *Aplysia*. *Aplysia* exhibits an innate fixed-action pattern of egg-laying behavior (Figure 33.2). Through experimental study, researchers determined that specialized neurons in an abdominal ganglion (clump of nerve cells) called *bag cells* elicited the response. Continued research revealed that if just the peptides pro-

duced by bag cells were injected into another slug, the egg-laying response was stimulated. Because the peptide was acting as a hormone, it was called (logically enough) *egg-laying hormone*.

Investigations have identified the genes that influence the secretion of this peptide. Just think of it. We now have evidence that a group of genes can code for a set of related peptides (in this case neuropeptides) that coordinate, release, or control a fixed-action pattern. This means that we know some genes regulate innate behavior, and we have a mechanism demonstrating that some behavior is indeed inherited.

However, even though genes can program a body to behave, not all behavior is preprogrammed. Animals are not simply robots performing to ancestral genetic demands. They can adapt to their environment. Adaptation or modification of behavior due to environmental experiences is called *learning*.

Learning

■ **Learning is a complex adaptive behavior that is the result of a combination of genetic inheritance and the environment.**

Learning is an adaptation or modification of behavior as a result of environmental experience. Although this may seem like a strange statement, learning is based on genetics and it occurs within the limits of the animal's genes. In order for higher animals to learn, certain regions of the brain and species-specific "wiring" are necessary, which is genetically controlled. Thus, two species from the same genus can have similar brain regions but different learning patterns. For example, in order to play the piano, you must have fingers and the various coordinating centers in the brain to move the muscles of the fingers in a way that results in playing music. All these requirements are genetically determined. But more is involved. I have all the mechanisms required to play the piano, but I can't. Why not? Because the ability to play the piano comes about through learning, which I have not undergone.

Thus we can say that *genes* and the *environment* work together in the development of learning. Here we discuss some different types of learning: habituation (imprinting), conditioned or associative responses, trial and error, and insight.

Habituation

■ **Habituation is the decline of a response to stimuli that are frequently repeated.**

| Inexperienced chick | Inexperienced chick | Experienced chick |

(a)

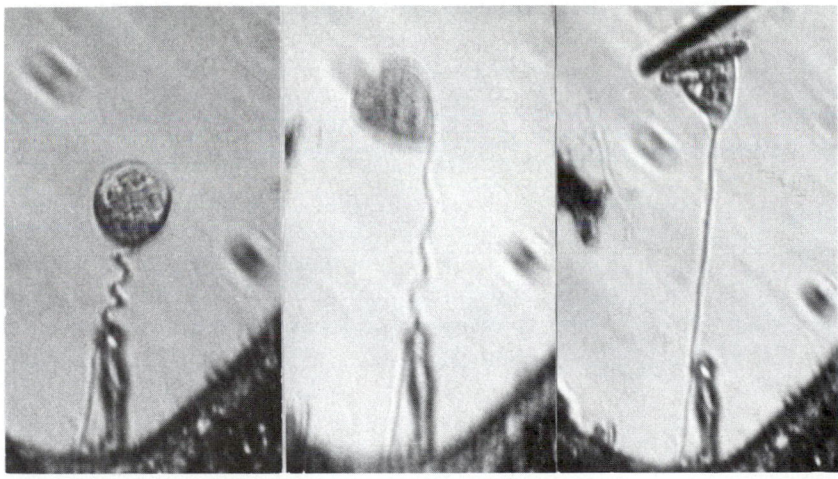

(b)

FIGURE 33.3 Habituation
Habituation is the process of learning not to respond to a repeated stimulus. (a) Lorenz and Tinbergen found evidence of habituation in the response of gull chicks to overhead silhouettes. The chicks learned to differentiate between hawklike short-necked shadows and the long-necked silhouettes of nonthreatening birds. When this model was flown to the right (so that it looked like a hawk), the chicks would exhibit fear and crouch and run, but when the same model was flown to the left, they did not respond. (b) Other experiments have uncovered habituation in *Vorticella*, microscopic organisms that usually contract when they feel vibrations.

Habituation is the waning of a response to a repeated stimulus, which can be instantly reversed by an irrelevant stimulus. Usually this stimulus is nonhazardous or monotonous. In a sense, habituation is an unlearning process — learning to ignore a stimulus. For example, young chickens and turkeys normally crouch and flee from any overhead flying object. Scientists infer that this action is a species-specific response to flying predators like chicken hawks. However, it is to the animals' advantage not to respond to every overhead stimulus. They learn to *discriminate*, to respond only to appropriate stimuli. There-

fore, as the young birds mature, they learn to ignore some stimuli, like falling leaves or the silhouette of a duck flying overhead. A hawk silhouette, however, still elicits the instinctive response (Figure 33.3a)

The unlearning that takes place gives an animal a distinct survival advantage. It does not waste its time and energy responding to familiar, nonthreatening stimuli. Unlearning is not total. The response will reoccur when the stimulus is encountered after a long interval or if the stimulus is presented at a different intensity or in a different pattern.

Even primitive organisms like the stalked proto-

(a)

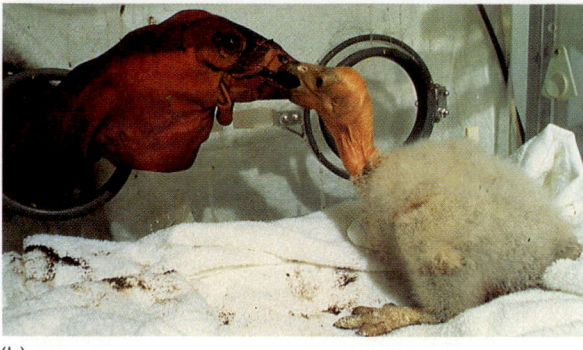
(b)

FIGURE 33.4 Imprinting
(a) The late Konrad Lorenz discovered that newly hatched goslings and chicks form a strong attachment to the first moving object they are exposed to — usually their mother. By seeing to it that he was the first moving object the goslings encountered, Lorenz became their "mother" and they followed him everywhere. (b) Keepers at the San Diego Wild Animal Park are using imprinting to help condors born at the zoo survive in the wild. These baby condors are being fed with condor puppets.

zoan *Vorticella* can habituate to a vibrating stimulus, which usually causes them to retreat (Figure 33.3b). Initially, any vibration causes the single-celled animal to retract its contractile stalk. Continued exposure to this stimulus results in an elimination of the response.

Imprinting

■ **Imprinting is the tendency of a newborn to become attached to the first object it is exposed to.**

One of the most famous photographs of the late Konrad Lorenz shows him being followed by a flock of young geese (Figure 33.4a). The photograph illustrates the learning phenomenon known as **imprinting,** the tendency of newborns to follow the first moving object they sense, usually a parent. Lorenz had discovered that newly hatched chickens or goslings would trail after the first moving object they encountered, forming a very strong attachment. If Lorenz was this object, the goslings would follow him and accept him as "mother." Lorenz also observed that if he waited a day and then tried to get them to imprint on him, they would not do so.

Imprinting also works in reverse. In these instances, it enables the parents to identify offspring. Reverse imprinting enables families to be maintained in social situations such as herds.

Lorenz's work on imprinting revealed several important facts. The nervous system seems to be ready to go through imprinting only during certain times in its development. That special period is called the

critical, or sensitive, period. The critical period varies according to species. For example, in mallard ducks, imprinting is strongest between 13 and 16 hours after hatching, but in chickens, the critical period can last for several days. In nature, imprinting obviously enhances the survival of the newborn organism, because its parent(s) can protect it and serve as role models for other species-specific behavior.

If imprinting occurs accidentally, it can be detrimental to the offspring. For example, attempts in Georgia to release eagles raised by humans failed because the birds imprinted on their human "parents" and did not learn normal bird feeding behaviors. The Georgia problem taught a lesson to those in the San Diego Zoo Wild Animal Park who are attempting to hatch condor eggs and raise the birds for release into the wild. In this endeavor, the keepers feed the hatchlings with a puppet that looks like a parent condor, to ensure that they will imprint on a normal stimulus (Figure 33.4b). It is hoped that this process will benefit the condors when they are released into the wild.

Mammals also can imprint, and today many psychologists are speculating about the importance of imprinting in human infants. In recent years there has been a growing trend toward hospital birthing rooms. Instead of the old practice, in which the mother was anesthetized and the father remained in the waiting room during delivery, birthing is now often an experience shared by mother, father, and newborn. The change is due in part to our increased understanding of the importance of imprinting and the effect of a child's very early experiences on its subsequent behavior and personality.

Conditioned or Associative Responses

■ In a conditioned response, an animal learns to make a reflexive response to a stimulus that would not normally elicit that response, or a familiar stimulus gives rise to a new response. Operant conditioning teaches an animal to make new responses to a particular stimulus. The teaching is reinforced through a system of rewards and punishments during the training period. If reinforcement ceases, the new behavior becomes extinct and the old patterns reoccur.

A **conditioned response** is a reaction that an animal will have for no other reason than that it has been trained to do so. For example, in the 1920s, Russian physiologist Ivan Pavlov observed that dogs salivated when meat extract was placed on their tongue. The meat was a *common stimulus*, and the response (salivation) was the *unconditioned response,* one that any untrained dog would automatically have.

Pavlov then began a training period. Every time he gave the dogs the extract, he rang a bell. After a period of time, he found that if he rang the bell just before giving the meat extract, the dogs would salivate to the bell alone. The bell became a *conditioned stimulus*. In other words, the dogs were conditioned (trained) to associate the ringing of the bell with the meat extract. Because of the association, they responded by salivating to the bell alone. Pavlov called this phenomenon a conditioned response. Creating an association between a stimulus that causes a reaction and one that does not cause it is called **classical conditioning.**

Another type of associative or conditioned response is called **operant conditioning.** In this process, the organism takes an action (operates) to gain a reward or to avoid punishment. If the response is the one desired by the trainer, the subject is rewarded — a process called positive reinforcement. If the response is incorrect or not desired by the trainer, the subject is not rewarded — a process called negative reinforcement.

The most famous work on operant conditioning was carried out by B. F. Skinner. His first work was done in the 1930s. at Harvard University. Skinner noted that if an animal motivated by a very basic need like hunger (let's say a hungry pigeon) was placed in a box, it would wander about, actively seeking to satisfy its need. The box was so constructed that if the pigeon accidentally tripped a lever or pecked a certain object, food was released. The food was a positive reinforcement for the behavior. In a very short period, the animal learned

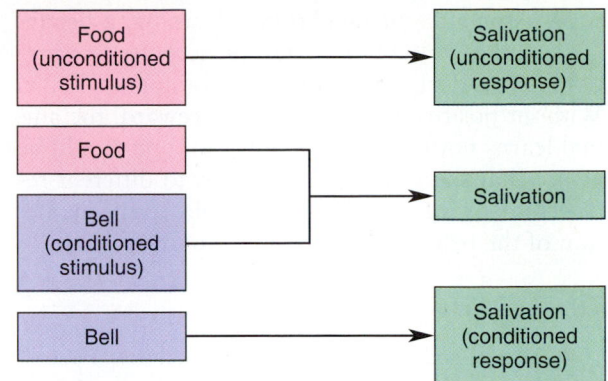

(a) CLASSICAL CONDITIONING

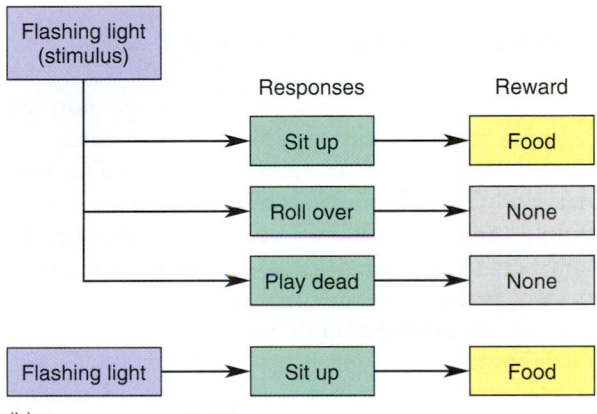

(b) OPERANT CONDITIONING

FIGURE 33.5   Conditioned Responses
(a) Classical conditioning creates an association between a stimulus that naturally causes a reaction (unconditioned stimulus) and one that does not (conditioned stimulus). The conditioned stimulus must precede the unconditioned stimulus. By associating the conditioned stimulus (bell) with the unconditioned stimulus (food), Pavlov was able to elicit a conditioned response (salivation). (b) Operant conditioning uses specific response to an unconditioned stimulus. Using food as a positive reinforcement, for example, a dog can be taught to sit up in response to a flashing light.

(or was conditioned) to associate tripping the lever or pecking the object with acquiring food.

Reinforcement like food or water is a positive reward, which teaches the animal to approach the stimulus. Negative reinforcement (like an electrical shock) conditions the animal to avoid the stimulus. The basic difference between classical and operant conditioning is that in classical conditioning, the subject is not required to respond in order to receive the reward, while in operant conditioning, the animal is taught to behave so as to produce the desired effect on its environment (Figure 33.5).

In associative or conditioned learning, a behavior becomes *extinct* (it wanes, or becomes less predominant) if the reinforcement is not available. Without positive reinforcement or reward, the animal learns not to respond. However, one should be aware that the response of animals to different reinforcement regimes is quite variable, so the extinction of the behavior is also quite variable.

Trial and Error

■ **In trial-and-error learning, an animal tries a number of behaviors before discovering the one that is successful. This type of learning is dependent on memory.**

In **trial-and-error learning,** an animal learns, after several unsuccessful attempts (trials), that a certain behavior satisfies a need and others do not. By learning from its mistakes the animal figures out how to respond successfully. You may think that trial-and-error learning is similar to operant conditioning, and you would be right. (Some scientists define them as the same.) The main difference between the two is that trial-and-error learning occurs in the natural environment and operant conditioning in the laboratory.

For example, if a frog sees an insect, its feeding behavior is activated. The frog sticks out its tongue to capture the insect. But if the insect is a wasp, the frog may be stung. It learns from this experience, modifying its behavior to exclude wasps or even insects that resemble them. This is an example of trial-and-error learning — the frog has learned from its mistake.

Insight

■ **Insight learning takes place when an animal tests possible responses through mental processes rather than actual physical activities.**

Insight learning is the most complex form of learning. It involves the ability to solve problems without going through the mechanical process of trial and error. Instead the testing is mental. Past experiences are retained as an accumulation of stored responses. By drawing on this reserve, the animal determines how to solve a new problem or respond in a new situation. The process of drawing on old experiences and cognitive (instead of mechanical) processes is called insight learning. The animal reasons that a particular response might or might not work. It then reacts accordingly. For example, if a chimpanzee is placed in a cage with boxes and a banana hanging from the roof of the cage, out of its reach,

the chimpanzee will use insight learning to solve the problem by placing the boxes one upon the other to reach the banana.

It is important to remember that the ability to learn a behavior or set of actions is dependent on the developmental stage of the organism. The organism's developmental stage involves the interaction of genetic makeup and acquired environmental experiences — the learning that has already occurred. Thus learning has two components, one genetic and the other acquired.

MEMORY: THE BRAIN AND BEHAVIOR

In a sense, **instinct** involves genetic memory — the existence of genetic information that can activate and coordinate a species-specific fixed-action pattern that passes from one generation to another. Simply put, instinct is something like a memory passed from parent to offspring. We also said an organism alters its behavior; it learns in part through building on previous experience, which can be stored or remembered. In fact, memory — the ability to recall stored information — is one of the fundamental principles of learning. If learning is to be retained, experience must be remembered. What is memory? Where are memories stored? Let's discuss some of the recent findings about memory in humans.

Memory Storage

■ **Memory is the storage of information for later recall. Most research concentrates on the areas of the brain and the chemical pathways involved in memory.**

In the 1920s, Karl Lashley, one of the early pioneers of brain research, tried to find the "Engram," by which he meant specific memory traces, physical changes, or regions of the brain that store memory. His search seemed logical, because it was already known that if a specific region of the brain was electrically stimulated during surgery, the patient would recall past events very clearly. The individual would remember, for instance, the smell of bacon cooking. Lashley continued his work for 30 years before concluding in his report, *The Search for the Engram,* that he had been unsuccessful.

Then in the 1960s, as knowledge of the information-storing molecules such as DNA grew, brain researchers hypothesized that there might be "memory molecules." This hypothesis was supported by a series of experiments with planarians (free-living flatworms). First, a group of planarians were trained to travel through a T-maze. The length of time it took for them to learn the maze was carefully recorded. The trained planarians were then ground up and fed to untrained planarians, which were in turn placed in the T-maze, and their learning time was recorded. Researchers found that the second group required less time to learn the maze than the first group. Perhaps the second group had ingested "memory molecules" that provided them with learning. Nonetheless, the excitement over the search for memory molecules died, because other scientists had difficulty in replicating the results of the experiments with the planarians.

Today biologists study memory by studying sea slugs as well as amnesiacs and other memory-damaged patients (see Profile). What have biologists learned from these diverse sources? Researchers have found chemicals that seem to play a role in memory retention and have detected areas of the brain that appear to be involved in particular memory skills. The molecular machinery for memory seems to be located within neurotransmitters, the chemical messages produced by nerve cells in the brain. These chemicals are the molecular basis of memory.

For example, the neurotransmitter serotonin is essential to learning in some marine snails. In mammals such as rats, epinephrine plays an important role in the regulation of learning. When rats receive injections of epinephrine, they learn faster and retain information longer. This applies to old individuals as well as to younger ones. In addition, scientists have established the role of receptors on brain cells in the memory process and have even isolated the genes that code for these receptors. Who knows whether there are "memory genes." But the research is advancing.

Acetylcholine, another neurotransmitter, is also thought to be involved in memory. Patients suffering from the disorder known as Alzheimer's disease (a type of memory loss usually associated with senility) have lower levels of acetylcholine in the brain because of the gradual death of cells that produce the neurotransmitter. Peter Davis at Albert Einstein Medical School determined that Alzheimer's patients also either lack the enzyme *choline transferase* (which is involved in acetylcholine synthesis) or have it only in low levels.

It is perhaps ironic that as biologists study the diminishing memory of patients with Alzheimer's disease, their work increases our knowledge about the chemical basis of memory. Such research also may soon enable physicians to modify the effects of Alzheimer's and develop medication to improve the memory loss and personality disorders of those who suffer from it. Over 2.5 million Americans are estimated to have this disease, which kills more than 150,000 every year (see Focus 33.1).

Skill Memory and Fact Memory

■ **Skill memory involves remembering how to perform a mechanical process learned through practice. Intellectualization is not involved in skill memory and in fact can slow the learning process. Fact memory involves remembering specific facts. This knowledge is acquired and forgotten quickly. When a memory becomes consolidated, it is retained.**

Amnesiacs are not all affected in the same way. Some amnesiacs retain their pretrauma memories but cannot form new ones. For example, they cannot learn a new address or telephone number, but they do remember their old ones. However, if they are given particular kinds of skill puzzles, they can both solve them and remember how to do them again, even though they cannot tell you they remember how to do them or even remember doing them at all. The observation that some amnesiacs can perform a skill (puzzle solving) but yet cannot remember that they did the puzzle has led scientists at the National Institute of Mental Health and at various universities to conclude that there are two categories of memory — skill memory and fact memory.

Skill memory deals with learning skills such as riding a bicycle, swimming, or skiing, or solving a skill puzzle — behavioral patterns acquired through practice. You do not consciously remember how to ride a bicycle or tie your shoes. Once you learn how to do these things, you perform without consciously thinking about the processes involved. For example, when small children learn how to ski, they learn rather quickly by imitation and practice. Adults, on the other hand, learn how to ski much more consciously. They intellectualize the process and think about putting pressure on this edge or that ski, and so on. The result is that it takes longer for adults to learn the skill. Many skill memories are acted out when the individual does not think about them on a conscious level.

Fact memory is the ability to remember explicit facts such as names, places, and dates. Fact memory is easy to acquire and is often short-lived, which means that the information is easy to forget. The characteristics of fact memory explain why you

Dr. Richard Jed Wyatt
Neuroscientist

Dr. Richard Wyatt is a pioneer in the area of brain tissue transplants. Recently he and his colleagues took brain tissues from fetal rats and grafted them into the brains of rats having a deficiency of dopamine—a neurotransmitter—similar to that found in patients with Parkinson's disease. (The procedure is less difficult than other transplants, because the brain does not produce antibodies against transplanted brain tissue from the same species.) The transplanted substantia nigra (the area of the brain that makes dopamine) from the fetal rats corrected the abnormal behavior the other rats had developed because of the dopamine deficiency.

Although it sounds like a Frankenstein movie, there are valid reasons for Wyatt's experiments. Low dopamine levels prevent brain cells from communicating with one another, resulting in paralysis and other neurological disorders. Wyatt's research and that of others suggest that transplanted human fetal brain tissue might help treat deficiencies that result in Parkinson's disease, as well as other diseases such as Alzheimer's and Huntington's disease, although the ethical issues of transplanting human fetal tissue are complex.

Wyatt's interest in the brain began early in life, after he visited the brain research laboratories at the University of Chicago, where he saw researchers doing fascinating experiments with goldfish. He spent hour upon hour at the Museum of Science and Industry in Chicago, which happened to be across the street from his elementary school. Because of his interests, he noted the physiological and behavioral changes taking place in himself and his friends as they went through puberty; he found himself wondering how the brain was involved.

One reason Wyatt might have been interested in the role of the brain in life processes and disorders is that he suffers from dyslexia. Dyslexics tend to invert letters and numbers when they read and write. Wyatt wanted to enter medical school, but he knew that in order to do so, he had to do well during his undergraduate premedical studies. Knowing he would do poorly on written essay examinations but could do very well on multiple-choice examinations, he elected to do his premedical work at the University of Michigan in Ann Arbor — because for the first two years, examinations there were primarily multiple choice. After his sophomore year he was accepted into a special premedical-medical program at Johns Hopkins Medical School. Wyatt jokes about the fact that if he had remained at the University of Michigan and been required to take essay examinations, he would never have been accepted to medical school.

He obviously learned to adapt to his handicap, and it has not stymied his scientific creativity or brilliance. He has consistently published major research articles and has been involved in the writing of several books. At present Wyatt is the chief of the adult psychiatry branch at the National Institute of Mental Health, and he has served as a consultant to various Head Start programs. During the 1970s, when the disadvantaged protested in Washington, Wyatt volunteered to provide medical and psychiatric care for the thousands of demonstrators who lived in the tent city known as Resurrection City. His work in his laboratory is yet another way of serving people.

often do not retain information you learn when you cram for an examination. It includes such things as remembering a telephone number only long enough to dial it.

Biologists now theorize that, from an evolutionary perspective, the more sophisticated circuitry of fact memory was built on the more primitive skill memory, because skill memory involves a set of activities that require little intellectualization. Skills involve physical actions rather than mental activities like analysis. Fact memory, even though short-term, does involve intellectualization.

Biologists also theorize that fact memory develops later than skill memory in childhood. This is a break with some common psychological assumptions. Freud thought that childhood memories were vague because early memories were repressed; other psychologists have theorized that the cause of this vagueness is a lack of language. Today neurophysiologists believe that childhood memories are hazy simply because the neuromechanisms involved in fact memory take time to mature.

Short- and Long-Term Memory. Another way of categorizing memory is as **short-term memory** and **long-term memory**. Learning appears to occur in two phases. Short-term memory occurs first and can be followed by long-term memory. Think of the way you remember an address. The first time you look up an address, you probably remember it only long enough to write it down. This is short-term memory. If you write the address several times or repeat it several times, you can remember it for a long time — long-term memory.

Why are there differences in the length of time you remember an address? The differences are partially explained by the hypothesis that short-term memory involves the repeating of a specific electrochemical circuit within the brain. The memory lasts as long as the circuit is active. If this hypothesis is correct, short-term memory is delicate and can easily be interrupted or lost through the intrusion of other thoughts, other stimuli, or physical trauma.

Long-term memory, on the other hand, is thought to be structural in nature. This means that new, permanent synaptic junctions are formed between nerve cells. Once these new links between neurons are formed, the memory lasts almost indefinitely. But how is short-term memory converted into long-term memory? Scientists use the mechanism of **consolidation** to explain this process.

It is hypothesized that in the process of consolidation, the cerebral cortex (the outer surface of the brain — the most highly evolved portion) stores preexisting concepts. When an event takes place, a neuroconnection is made between the cerebral cortex and the hippocampus (region of the brain for processing and storing new information). New information is consolidated with previous memories in the hippocampus and becomes long-term memory. As you probably already know from experience (like studying for an exam), consolidating memory is subject to all sorts of disturbances, and in some cases the information can be lost.

Researchers at the National Institute of Child Health and Human Development have determined that the number of synapses between brain cells is greatest at birth. After birth, a pruning process begins and specific brain cells die. Through research, scientists are learning more about how the body "weeds out" excess nerve cells, leaving a tightly knit series of mental pathways resulting in a healthy, functioning brain. Indeed, recent research seems to indicate that many of the problems suffered by individuals with Down syndrome are the result of failure of the weeding process.

BEHAVIOR AND COMMUNICATION

Many animals live in groups rather than alone. In some species grouping is seasonal; in others it is lifelong. The degree of association varies from simple aggregations, like schools of fish, to very structured caste systems, like those found among bees and other social insects.

Group living does have advantages. Individuals have better protection from predators, better mate selection and availability, and increased reproductive success. The young have adults to protect them, and the acquisition of food is enhanced. Since natural selection favors behaviors that provide increased survival advantage, social interaction or group living is widespread (Figure 33.6).

When we study groups of animals, we must remember that animals do not perceive the world as we do. Each animal has its own perception of the world, or *Umwelt* (the German word for "environment"). If you think about your friendly family dog, you will realize that its *Umwelt* is one of smell and sound, not the visual world of color that you and I experience. An animal's *Umwelt* is based on its evolutionary history, during which natural selection favored sensory adaptations that provided its species with the greatest chance of survival. To understand how individuals relate to others in the same species, one must understand the special adaptations that have evolved for that species.

(a)

(b)

(c)

FIGURE 33.6   Social Groups
Social groups help ensure the survival of the individual.
(a) Scientists believe that the nestlings' gaping mouths trigger a feeding response in the parent bird.
(b) Baboons live in cooperative groups, forming individual families within these groups. Male baboons often band together to ward off predators, protecting the females and young. (c) Musk oxen form a defensive circle around their young when threatened.

The Role of Communication and Cooperation

■ **Communication among groups of animals is based on signals and releasers. Physical signals involve coloring, posture, and gesture. Pheromone signals involve scent. Communication is necessary in matters of aggression, food location, mating, and all other group interactions.**

Living in groups requires communication — an exchange of messages between individuals. There are many types of signals that serve to communicate information within and between species. Some are visual; others are chemical, auditory, or tactile. The limitation of the communication signal depends on the animal's ability to perceive it. Blind cavefish do not rely on visual communication; in this species nonvisual signals would be more important.

A wide diversity of communication signs and signals have evolved (Figure 33.7). *Signs, signals,* or *releasers* are stimuli that influence the behavior of other individuals. When you hear a growling or barking dog and see that it is crouched and baring its teeth, you know that these auditory and visual signals indicate the dog's aggressive state.

Visual Signals. Much work on communication has been done with three-spined sticklebacks, which are small minnowlike fish. During most of the life cycle of adult male sticklebacks, their coloring is silver. During mating season, however, the underbelly of a male becomes tinged with red. In this species, the males are nest builders. Thus each selects a territory and begins to construct his nest, a tube-shaped construction, and each will defend his territory from any intruding male. Researchers have learned that the signal that brings on a male's aggressiveness is the red underside. The color serves as a signal that *releases* the fixed-action pattern of aggression in the resident territorial male. Even a model can elicit the response.

Chemical Signals. **Pheromones** are chemicals that are produced by one animal and that alter the behavior of another animal of the same species. These chemicals serve as communication devices. Animals detect them by either taste or smell. Have you ever watched ants trail through your kitchen? They are following a pheromonal highway (a scent trail) produced by fellow ants. Mammals also communicate through scent — which is one reason that neighborhood dogs urinate on every corner. Even solitary animals such as mountain lions mark out their territories by releasing scents at their boundaries.

All of us already know the importance of visual signs and signals in human sexual and courtship behavior. It may surprise you to learn that some biologists believe that pheromones also play a role in human sexual and aggressive behavior, although the evidence is not conclusive. Certain steroids produced by the apocrine glands (a type of sweat gland) located in the hairy area of the genital region and/or armpit of males and females may elicit sexual arousal in some individuals. (In one of Napoleon's love letters to Josephine, he wrote, "Home in three days — don't wash.") Based on tentative ex-

FIGURE 33.7   Communication
(a) The male three-spined stickleback relies on visual signs such as the red belly of another male to protect its territory. (b) Termites use both visual and chemical signs to communicate within a colony. These insects live in tunnels. When a tunnel is threatened, worker termites bang their heads against the walls of the tunnel, creating vibrations that alert soldier termites. When they fight a predator, soldier termites release a chemical that attracts other soldier termites. (c) Dolphins use their voices to "talk" — to warn others of danger or to attract a mate. (d) The honeybee's dance is a form of tactile communication, describing the distance, direction, and quality of a particular food source to other bees in the hive.

perimental results, some perfume companies have synthetically produced pheromones that seem to duplicate the effect of human musk. The perfume "Muscone pH5," as it is called, was the first perfume on the market to use synthetic human pheromones, rather than animal pheromones.

Messages

■ **The primary messages communicated between animals involve species recognition and individual recognition. They maintain hierarchies, provide a means of handling aggression between members of a species, and convey a willingness to mate.**

Individual and Species Recognition. Have you ever wondered how fish recognize other members of their species? They do it through **species recognition** — recognizing specific signals produced by their own species. Some of these signals are visual, others are auditory, and still others are chemical. Species recognition has obvious survival advantages, since individuals can produce viable offspring only when they mate with the opposite sex of their own species. The signs and signals that we have just mentioned serve as behavioral isolating mechanisms that ensure specific-species recognition. In Chapter 17 we discussed some other physiological and geographical reproductive isolating mechanisms.

Individual recognition is very important in social species since it facilitates mate selection, offspring and parental recognition, and the maintenance of social hierarchies. Look at a group of penguins, for example. They all look alike to us, but to a penguin, it is quite another matter. Mates recognize each other, and parents recognize their own offspring.

Hierarchies and Aggression. Among some animals, the establishment of **hierarchies** is important to maintaining the group. In such groups, each member establishes a rank or position within the social structure. Individuals in the group recognize both

Senile Dementia — Alzheimer's Disease

One set of disorders that plague the human race is a group known as senile dementia disorders, which produce brain damage or impairment resulting in mental confusion. Less than half of these cases are due to causes such as thyroid disease, stroke, drug reactions, anemia, and alcoholism. Over half the patients suffering from senile dementia have Alzheimer's disease, a disease that afflicts 2.5 million Americans and claims over 150,000 lives per year. It is the fourth leading cause of death among the elderly, right behind heart disease, cancer, and stroke.

Alzheimer's disease is an incurable degenerative brain disorder that gradually results in loss of memory and brain function. Although individuals may experience some memory loss due solely to aging,

someone with Alzheimer's disease is affected much more severely. The memory loss begins to affect a person's work and social life. They can't remember names, places, dates. Their personalities change. As the disease progresses, patients lose the ability to make judgments. They lose control of their bodily functions, such as bladder control. They lose their ability to speak and eventually cannot walk.

Several memory tests have been developed to diagnose Alzheimer's disease. The cause of the disease is unknown, but researchers are getting closer and closer. They now know that Alzheimer's victims have low levels of the enzyme responsible for the synthesis of the brain chemical transmitter known as acetylcholine. When volunteers are given a drug called scopolamine, which

blocks acetylcholine action, they show memory lapses. Alzheimer's victims exhibit a significant loss of brain neurons in an area known as the basal nucleus, where most acetylcholine is produced. Other brain neurohormones may be included.

Autopsies have shown that the brains of Alzheimer's victims are clogged with plaques and twisted nerves called *neurofibrillary tangles*. The plaques are composed of a protein called *beta amyloid protein*. In 1987, researchers determined that production of this protein is governed by a gene located on chromosome 21, the chromosome associated with Down syndrome. The same protein deposits found in the brains of Alzheimer's victims have also been found in the brains of Down syndrome individuals over the age of 35. The discovery of this

the individual and its rank within the social hierarchy. A structure such as this reduces the amount of fighting over food, mates, and other resources, because the subordinate will usually yield to a more dominant member.

Establishing social hierarchies often involves aggressive behavior, usually between individuals of the same species and of the same sex. After all, members of the same species want and need the same things — mates, food, space. What does a tur-

key have that you are willing to fight for? But obviously a turkey does share needs with other turkeys, and it is willing to fight to satisfy those needs. In turkeys (as in chickens), social hierarchies are expressed in pecking orders. Dominant individuals will peck those of lower status. The highest member of the flock is not pecked at all. Lower-ranking individuals are pecked by certain individuals, and the lowest ranking chicken is pecked by all. Because an individual knows its rank in a peck-

genetic marker is important because it proves that at least one kind of Alzheimer's disease is hereditary. Scientists estimate that at least 10 to 30 percent of Alzheimer's cases fit into this category, and some suspect that other types can be inherited, too. Researchers have recently reported that beta amyloid protein can kill neurons.

Scientists have not found a genetic marker for A68, the protein found in the neurofibrillary tangles, but they're looking. This protein has also been detected in the brains of human infants, where it works to eliminate excess neurons from the brain. When the process is complete, A68 disappears. In Alzheimer's victims, this protein reappears.

How does the A68 protein contribute to Alzheimer's disease? There are two theories. Either it stimulates growth of the neurofibrillary tangles in which it has been found, or it kills specific neurons in the brain, as it does in infants. Scientists are now trying to discover what causes A68 to reappear in Alzheimer's victims, and whether A68 production, like beta amyloid production, is governed by a particular gene.

When the gene or genes responsible for Alzheimer's disease have been identified, scientists should be able to use genetic engineering technology to alter the gene or its effects. Identification of this gene could also help people with the disease in their families to determine their chances of coming down with it, if they want to know. There are also less radical measures in sight: If scientists find that A68, beta amyloid, or both are responsible for the disease, they could probably develop a drug to keep production of these proteins down.

In the meantime, researchers are trying to develop treatment that will improve the memory and prevent continued memory loss in Alzheimer's patients. Physostigmine, a drug that increases levels of acetylcholine by blocking the enzyme that normally breaks it down, has produced some improvement in these patients. Researchers are testing newer methods of administering acetylcholine, like injecting drugs directly into the ventricles of the brain through a system of pumps and catheters. Patients using these infusion pumps have significant improvement. In 1984, a drug called bethanechol chloride was also fed into the brain through pumps and drew national attention as four patients seemed to show marked improvement. Unfortunately, further studies demonstrated no significant improvement.

In 1987, a new drug called tetrahydroaminoacridine (THA) was used on an Alzheimer's patient. THA can modify the symptoms of

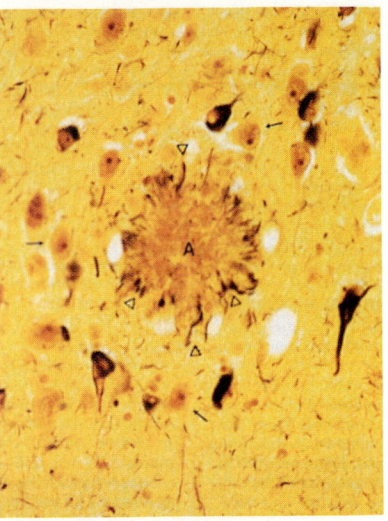

FIGURE A
In this photograph of stained brain tissue from an Alzheimer's individual, a senile plaque (A) with compacted amyloid is apparent. Also note the dark stained neurofibrillary tangles present.

the disease by preventing the fast breakdown of acetylcholine. The results of the chemical trials are not known, but if favorable the data will be presented to the FDA for approval. Even limited successes lead researchers to believe that one day they will have a greater understanding of the mechanisms of Alzheimer's disease and will be able to develop effective treatment for it.

ing order, aggressive behavior is decreased and more time can be used to find food, mates, or the other necessities of life.

Although members of the same sex fight over access to mates, the fighting is usually **ritualized behavior,** characterized by exaggerated actions that usually follow a stereotyped, fixed-action pattern (Figure 33.8). The rules of the game have evolved in such a way that battles usually result in little tissue damage to either combatant. Ritualized behavior involves aggressive postures or signals that initiate the fight and **appeasement behaviors** or submissive signals that end the fight. Have you ever watched a dog turn and expose its belly or neck to its opponent in a dogfight? Such behavior signals appeasement and turns off the fighting.

In some species, especially when the dispute is between members of different social groups, fighting can be to the death. For example, if a strange rat is placed in a cage containing an established

(a)

(b)

FIGURE 33.8 Aggressive Behaviors
Most of the aggressive behaviors animals show within their own species are ritualized and harmless. The winner usually wins by tiring out the loser, not killing him. (a) Male giraffes compete for social dominance by wrapping and unwrapping their necks around each other. They do not use their hooves. (b) Although a clash of bighorn sheep is frightening, the animals are well protected by their skull and horns.

group of resident rats, the residents will kill the intruder. Of course, you have to remember that in its normal environment, the intruder rat would probably be able to escape, whereas in the artificial confinement situation of the cage, the intruder was trapped.

Courtship and Mating. Recognizing individuals within a social structure is a very important component in courtship and mating. In many species, males produce a variety of signals that serve as stimuli to attract females (Figure 33.9). We are all familiar with various types of courtship behavior,

whether it is the singing of robins, the croaking of frogs, the head bobbings of lizards, or various postures and sniffing of dogs and cats. No matter what the specifics of the courtship pattern, the result is that the organisms orient themselves into positions that will allow for the exchange of gametes and the continuity of the species.

As we said earlier, behavior (except in the protozoans and sponges) is the result of the cooperation of the multicellular neural and hormonal coordinating systems. The interweaving of physiology and behavior is especially clear in reproduction. Reproduction requires physiological readiness, and hormones are responsible for this readiness. Biologists now know that behavior can affect hormones and hormones can affect behavior. The courtship dance of some male birds raises the levels of repro-

(a)

(b)

FIGURE 33.9 Courtship Behaviors
Courtship behaviors attract mates and ready the individuals for mating. (a) Albatross use a sequence of formal movements to show their readiness for mating. Other ritualistic behaviors help them communicate when they nest and care for their young. (b) The courtship pattern of the Arctic hare includes a highly stylized "boxing match."

FIGURE 33.10 Territoriality
Birds establish territories to protect both their food sources and their nests. Male birds are particularly defensive of their territories during the mating season. Gulls have a very small territory, the size of their nests. Notice that each bird on this cliff is just far enough away from its neighbor not to be pecked at.

ductive hormones and induces ovulation in females observing the pattern. Similarly, the mere injection of sex hormones into some vertebrates, including lower mammals, can induce sexual behavior.

Territoriality. When resources are limited in an environment, there are intense selective pressures on the organisms within that environment. One behavior that has arisen as a result is **territoriality**, in which an animal claims a certain area, a **territory**, as its own and defends it from others of its own species. (Rarely are territories established in social groups that have unlimited resources.) Territories reduce competition between individuals and can serve as mechanisms of population control, since only those with territories have access to food and to mates for breeding. Those without territories do not usually have these advantages and in all probability will die (Figure 33.10).

Territorial organisms can become accustomed to the presence of some neighbors and will not behave aggressively toward them. If a new individual intrudes upon their territory, however, their aggressive behavior will arise again.

SOCIOBIOLOGY

■ **Sociobiology is a discipline concerned with the social behavior of animal groups; it places special emphasis on the genetic basis and evolutionary history of this behavior.**

We cannot leave a discussion of social grouping without a discussion of sociobiology. **Sociobiology**

is just what you would think it is: the biology of social behavior. Sociobiologists study the effects of natural selection on the social grouping in animals, including humans. As you would suspect, the phrase that causes the problem is "including humans."

Although the concept of natural selection's effect on behavior had been around for a long time, it was not a biological issue until the early 1960s, when V. C. Wynne-Edwards and W. D. Hamilton discussed the effect of natural selection on group as well as on individual behavior. In the late 1970s, E. O. Wilson, a Harvard scientist best known for his work on insect societies, wrote a milestone publication entitled *Sociobiology.*

Of the topics Wilson presented in his book, the issue of **altruism,** or acting in a totally unselfish manner, turned out to be one of the most controversial (Figure 33.11). There are those who consider reproduction and caring for offspring to be an example of altruism. Altruism is also at work when a mother endangers or even forfeits her life by running into a burning building to save her children. Some animals die protecting their offspring, and some insects and spiders die after mating.

FIGURE 33.11 Altruism
Although controversy continues regarding altruism in animals, there is no question that animals can and do sacrifice themselves for the good of their "societies." The male praying mantis gives his life for his offspring. In fact, the female often begins to eat his head before they've finished mating.

Are these acts truly unselfish? On the surface we would say yes, but you must remember two things: (1) biologists believe that much altruism is a matter of genetics, not consciously reasoned sacrifice; and (2) from nature's perspective, living things are "disposable containers of DNA." When you look at altruistic behavior from this perspective, what seems to be an unselfish act is, in reality, a selfish one. By ensuring that its own offspring survive, the animal ensures the greater survival of its genes.

The basic biological selfishness of an apparent altruistic act can even be extended to include close genetic relatives in what is called *inclusive fitness*. Not only do our children share our genes, but so do our genetic relatives. By sharing our life or resources with these individuals, we are ensuring the survival of our own genes. Therefore, altruism, even if it involves reproductive self-sacrifice, is selectively advantageous to the organism's genes since without altruism its offspring might not survive. As you would suspect, the interpretation of altruism as a selfish genetic act is not shared by everyone.

In one sense the battle about sociobiology is over the issues of free will and determinism. Many individuals feel that the sociobiological explanation of human behavior revives the notion that social behavior is genetically determined and, therefore, unchangeable. The implication of this point of view is that we cannot change undesirable social behavior such as racism, sexism, war, and aggression, because they are genetically or biologically determined.

Sociobiologists argue that understanding the biological determination of social behavior will provide us with a better understanding of all social organizations, including those of humans. How can we gain an understanding of the social behavior that some of us find morally objectionable, like prejudice, if we do not attempt to see if it has a biological basis?

Part of the controversy involves the extension of findings about animal social groups to human social groups. An example of how this works is the sociobiological interpretation of the phenomenon of *infanticide*, or child murder. There is a growing body of information about infanticide in a number of animal species, including primate groups like baboons. In most instances, there is one dominant male, and he fathers many of the offspring of the group. If the male is deposed and another male becomes dominant, he often kills the infants in the group. Why? Doesn't such action endanger the preservation of the group?

There are a number of explanations of infanticide — and even more questions — but the one we are concerned with here follows this logic: The existing offspring carry the genes of the original dominant male. Females with infants, especially in primate groups, will not be ready to mate for a long time. If the infants are killed, the females become ready to mate. The new dominant male then can impregnate them, thus ensuring that his genetic heritage will be passed on.

This explanation is not widely accepted, for much more study needs to be done before any accurate conclusions can be drawn. However, some use the explanation of infanticide in primate groups to account for infanticide and child abuse in human groups. A certain percentage of child abuse involves a stepfather or stepmother. Is such action a result of genetic patterns evolved over millions of years? Are abusive parents following genetic dictates? If so, can our government prosecute individuals who are responding to genetic controls?

Issues such as these are hotly debated, not only by sociobiologists but in society at large. Research in sociobiology is growing rapidly. New findings are being analyzed and synthesized, and new results are emerging all the time.

SUMMARY

1. The coordinated responses of an organism to its environment, whether internal or external, are behavioral responses.
2. The three major approaches to the study of behavior are the psychological approach, which often emphasizes the study of animal behavior to learn more about human behavior; the ethological approach, which emphasizes the study of animal behavior in relation to its natural environment; and the biological or biophysiological approach, which emphasizes the physiological underpinnings of behavior.
3. A behavioral response depends on the strength of the stimuli and the age and condition of the organism.
4. Instinctive or innate behaviors are shaped by evolution and inherited by offspring. They are highly stereotyped, are not usually modified by learning, and are species-specific.
5. The innate releasing mechanism releases fixed-action patterns, ritualistic behaviors that are species-specific.
6. Evidence that genes can control some behavior has been found. Genes regulate the synthesis of specific peptides (hormones) that in turn elicit a fixed-action pattern. In some cases (like the egg-laying behavior of the sea slug *Aplysia*), these behaviors can be directly inherited.
7. Learning is behavior modified by experience. The genetic component of learning involves develop-

ment of mental and physical structures. The experiential component is training and practice.

8. The major types of learned responses can be categorized as habituation, imprinting, conditioned or associative responses, trial and error, and insight.

9. Today, many researchers believe that there are two major types of memory: skill memory and fact memory. Skill memory deals with mechanical skills that improve through practice, like learning to ski or riding a bicycle. Fact memory is the ability to remember explicit information such as dates, places, and material for an examination. Another way to categorize memory is as short-term and long-term memory.

10. Memory is believed to be due to chemical communication between brain cells.

11. Behavior is subject to natural selection.

12. Group living requires communication between group members. Various signs, signals, and releasers have evolved to serve as communicative devices.

13. Animals communicate for species recognition and individual recognition.

14. Dominance hierarchies help maintain order within certain animal groups. The hierarchy establishes the individual's position within the group and helps to eliminate fighting and energy expenditure over available resources.

15. Often, because resources are limited within an environment, animals establish territories so that successful individuals will have access to limited resources and mates. This ensures that the genes of the most competitive individuals will be passed on to offspring, increasing the chances of the species' survival.

16. Various courtship patterns have evolved that release sexual behavior in organisms, thus ensuring reproduction and the genetic continuity of the species.

17. Hormones aid in giving rise to these reproductive courtship behaviors, and the behaviors themselves can also stimulate or inhibit the secretion of hormones.

18. Sociobiology is the study of the biology of social behavior. It places special emphasis on the genetic basis and evolutionary history of this behavior.

19. Altruism can be considered a kind of selfishness, because the individual willing to sacrifice its life for its close genetic relatives is ensuring the survival of its own genes or those of close relatives within the gene pool.

KEY TERMS

behavior
ethology
stimulus
innate behavior
taxis
fixed-action pattern

learning
habituation
imprinting
conditioned response
classical conditioning
operant conditioning

trial-and-error learning
insight learning
instinct
skill memory
fact memory
short-term memory
long-term memory
consolidation
pheromone

species recognition
individual recognition
hierarchy
ritualized behavior
appeasement behavior
territoriality
territory
sociobiology
altruism

REVIEW QUESTIONS

True/False

1. Ethologists study the biological basis of animal behavior.
   true   *false*   *(page 812)*

2. Innate behaviors are usually learned.
   true   *false*   *(page 813)*

3. Today we can state that the innate releasing mechanism has been located in some marine invertebrates.
   true   *false*   *(page 814)*

4. Habituation, in a sense, is learning to unlearn.
   true   *false*   *(page 815)*

Fill in the Blank

1. A _____ is a directed movement of an organism toward or away from a stimulus. *(page 813)*

2. During operant or instrumental learning the reinforcement _____ the response stimuli. *(page 817)*

3. The ability to remember how to ride a bicycle is an example of _____ memory. *(page 819)*

4. Stimuli that serve to release a response are called _____. *(page 822)*

Multiple Choice

1. Behavior that does not usually change with regard to environmental influences is most likely:
   a. Learned
   b. Innate
   c. Stereotyped
   d. Conditioned
   (page 813)

2. A releaser, sign, or cue:
   a. Makes animals less inhibited
   b. Allows and stimulates only courtship behavior among animals
   c. Provokes a fixed-action pattern of behavior
   d. Is required of all behaviors
   (page 822)

3. The sensory limitations of an organism determine the organism's perception of the world. This concept is known as the:
   a. *Umwelt*
   b. Engram
   c. Fixed-action pattern
   d. taxis
   (page 821)

4. Most neurochemists believe that the chemical basis of memory has to do with:
   a. Neurotransmitters
   b. Fats
   c. Phospholipids
   d. Adenosine triphosphate
   (page 819)

Discussion

1. Define behavior. What are the major categories of behavior?

2. Why can it be stated that behavior has a biological basis? Provide examples.

3. Compare the psychological and ethological approaches to the study of animal behavior.

4. Why can it be stated from experiments with *Tritonia* that direct evidence for the existence of the innate releasing mechanism has been provided?

5. Is behavior inherited? Provide evidence for your answer.

6. Although genes can program a body to behave, not all behavior is preprogrammed. Discuss the types of behavior that are adaptive and that are the result of modification due to experience and learning. Limit your discussion to learning in the general sense.

7. "Behavior is subject to natural selection." Provide examples to support this statement.

8. Discuss skill memory and fact memory.

9. Speculate on the use of brain grafts of fetal tissue to correct neurological brain disorders.

10. What are the advantages to an organism of living in groups? What are the disadvantages?

11. When designing an experiment to understand an animal's behavior, why is it important to consider the animal's *Umwelt*?

12. Do you agree that altruistic behavior is a selfish genetic act? Why or why not?

13. Speculate as to why the topic of sociobiology is considered controversial by some.

34

Introduction to Ecology: Biomes

We hear about the environment every day and how we are destroying the very planet that gives us life. We read about the politics of an environmental impact statement. And difficult questions arise: Are we depleting the ozone layer with manufactured atmospheric chemicals, or is the ozone hole really a natural cyclic phenomenon? Should we jeopardize the economy of a small lumber town to save the spotted owl? How does one determine if insect diversity directly impacts the distribution of plants?

Obviously, there are no easy answers to these questions, but we can use **ecology** — the scientific study of the interaction of the organism and the en-

vironment — to help arrive at informed decisions. Many times, important ecological questions are complex and multidisciplinary, in which case ecologists draw on other areas of biology — biochemistry, physiology, genetics, evolution, and behavior — to find answers.

SOME BASIC DEFINITIONS

Let's begin with a few definitions. In terms of the interaction of the organism and the environment, we all know what *organisms* are: living entities such as humans, plants, and microorganisms. *Interaction* means that the activities of organisms are reciprocal with the environment and with other organisms. For example, if we humans spew sulfur and hydrocarbons into the environment from smokestacks, it

---

Above: Joshua trees are examples of plants well adapted to living in the desert biome.

831

affects us when we breathe polluted air or see forests and lakes destroyed by acid rain. The **environment** is the surroundings of an organism. These surroundings include living things, or **biotic components,** and nonliving things, or **abiotic components** such as light, water, temperature, space, and nutrients. Life can be organized in a hierarchy beginning with the atomic level and increasing in complexity through the molecular level, cellular level, tissue level, organ and system levels, and eventually to the level of organisms — in fact, we have followed that organization in this text. Now we are going to discuss how individual organisms interact at the population level to form the various species and communities of the living planet.

In our chapter on evolution and genetics, we discussed populations from a genetic/evolutionary perspective. There we defined **populations** as groups of interbreeding individuals, all of whom belong to the same species and occupy a specific, geographically defined area. All the bluegills in one pond belong to one population; those in another pond belong to a different population.

In nature, populations sometimes live in isolation, or they live with other populations of other species in interacting communities. **Communities** consist of all the populations of different species living in a specific, defined area. For example, all the populations within the physical confines of our pond — water lilies, aquatic insects, worms, microorganisms, and other species of fish — are members of the pond community.

The abiotic elements of a community (its nonliving components) are such things as the soil, temperature, rainfall, available organic matter, and available inorganic matter. The members of specific populations within a community interact not only with these abiotic elements but also with biotic elements — members of their own and other species.

The place where an organism lives is called its **habitat.** There are a number of different types of habitats, each with its own characteristics. For example, the habitat of an intestinal parasite is the intestine of its host. Sometimes the term *habitat* refers to a particular place in an area, such as a tropical forest canopy. We will discuss habitats in more detail in Chapter 36.

An **ecosystem** is an ecological unit of geography consisting of all the included communities and abiotic components. The sum of the ecosystems — all the living organisms on the earth and their abiotic components — is often viewed as one single ecosystem called the **biosphere** (Figure 34.1).

In a sense, the biosphere is the stage upon which all the acts of the living are played. The physical boundaries of the biosphere are rather limited, including only a few kilometers above and below sea level on our planet. The sun supplies energy to sustain this system. Other than this input of energy, the biosphere is self-contained and recycles life's critical components within its own system — water, organic and inorganic nutrients, and gases.

How do we study ecology? Ecologists use observation and experimentation to learn more about

FIGURE 34.1 The Biosphere
The earth as viewed from the moon. Life exists from the ocean depths to several kilometers above the earth's surface. If the earth were the size of a basketball, this thin layer of space and its inhabitants — the biosphere — would be approximately as thick as a single layer of paint.

Chapter 34   Introduction to Ecology: Biomes

the interaction of organisms and the environment. Ecology, as is true with most science, began as an **observational science.** Humans have always been interested in observing their environment. What foods are edible? How can an understanding of the habits of the animals lead to a more successful hunt? Also, how does the weather affect the availability of plants and animals? How do animals affect one another?

As you can see, observation alone is not enough to answer some of these questions. Ecology is also an **experimental science,** in which ecologists design experiments to find answers to their questions. And today, with the advantage of computer technology and satellites, ecologists are able to develop mathematical models to predict changes over large distances or long periods of time.

ECOLOGY AND EVOLUTION

◼ **The environment of an organism is a factor in natural selection, which changes the frequency of alleles in a population, resulting in evolution.**

Earlier in this chapter we defined populations from an evolutionary perspective. Obviously, the environment serves as a selection force to change the frequency of alleles in the population, which results in evolutionary change. For example, as a result of genetic variation some mice in the high desert of New Mexico are white and some are brown. On the brown desert soil, brown mice blend into the desert background and are not as easily preyed upon by hawks, whereas the white mice are easily seen and preyed upon by hawks. However, on the white sands near Alamogordo, New Mexico, white mice have an advantage against predators because they blend into the white background. So the frequency of alleles for coat color has changed because of selection pressures from the environment. Not only are predator-prey environmental conditions responsible for evolution, but so are geological factors. For example, the distribution of animals can be explained in part by the breakup of Pangaea (page 398), which isolated Australia as an island continent and separated India and Africa into discrete land masses.

But how can geological factors and evolution lead to a variety of communities and populations on earth? The biosphere is composed of all the terrestrial and aquatic ecosystems, and within this biosphere are large geographical regions with common characteristics. These collections of communities are called **biomes.** What are the factors responsible for these biomes? Why deserts, why grasslands, why oceans? As you would suspect, most of the factors are abiotic — water, sunlight, nutrients — in a word, *climate.*

The distribution of biomes on the earth results from the actual geographical features, such as the type of soil, the location and extent of oceans, the presence of mountains or valleys, and so on. But there are two key physical features in addition to the geographical factors: (1) The amount of heat in a particular area and (2) the global circulation of the atmosphere (wind patterns) and the resulting ocean currents. These are the major factors resulting in climate (see Focus 34.1, pages 842 to 843), which directly affects individual world biomes.

TERRESTRIAL BIOMES

◼ **The earth's land masses can be divided into several realms according to geographic features. Within those realms, there are a number of biomes, or major communities. The typical terrestrial biomes are deserts, scrublands, grasslands, forests, and tundras. Biomes are characterized by similar conditions of climate and soil and by similar flora and fauna.**

To help them study ecosystems, biologists divide the natural world into **biomes,** distinct communities based on the dominant type of vegetation adapted to live in that specific environment. The major terrestrial biomes are *deserts, scrubland, grasslands, forests,* and *tundras* (Figure 34.2). It is important to keep in mind that the concept of a biome is an abstraction. It does not define a particular geographic region, but rather describes regions that have similar climates and soil conditions and are characterized by the same biota. A biome is a category or a class, not a place.

Because climatic and soil factors are similar in specific biomes, organisms living within a given community have similar adaptations, even though they may not be closely related evolutionarily or have similar genetic backgrounds. This is an example of *convergent evolution* in which similar life forms evolve in different regions, even though there is no close evolutionary link between the species.

For example, the South American sidewinding pitviper, and the North American sidewinder rattlesnake are adapted to similar environments although they are not directly related. Now that we know something about the major biomes of the planet, let us devote our discussion to a more thorough description of each biome.

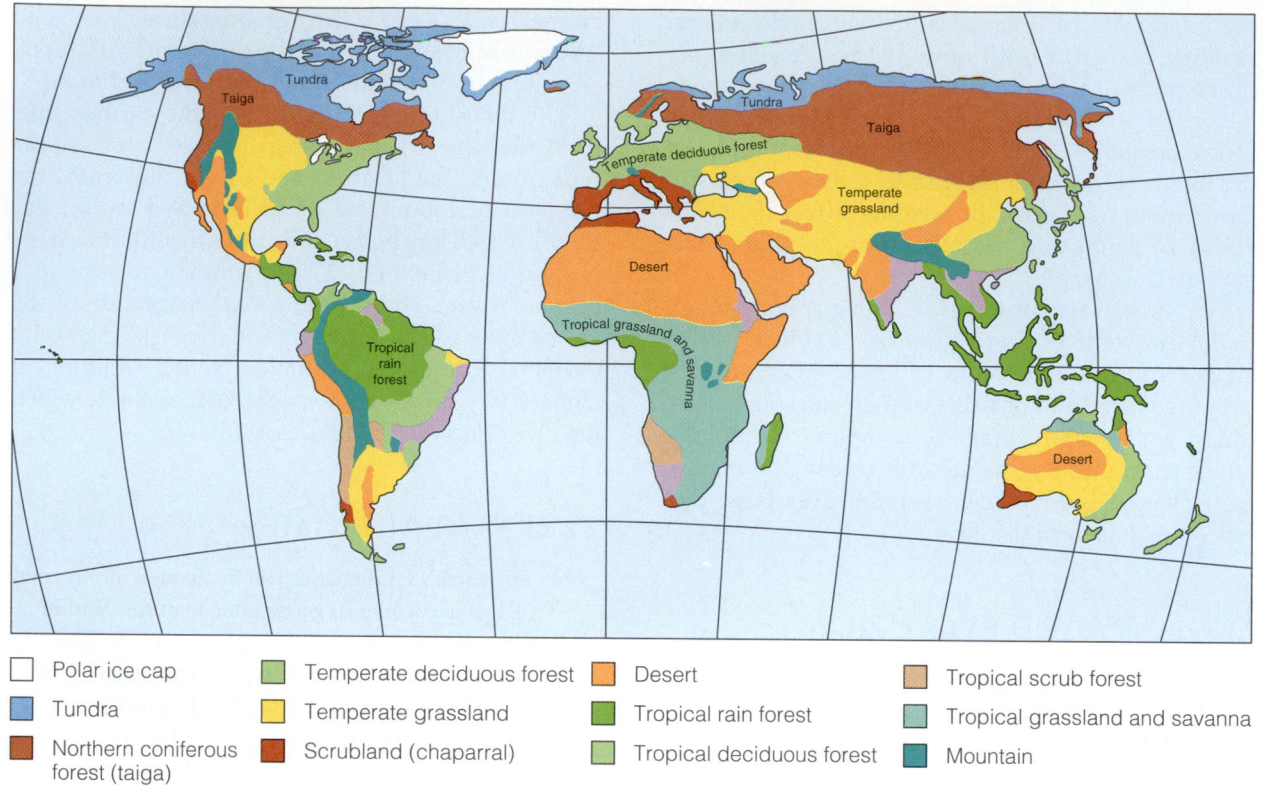

☐ Polar ice cap	🟩 Temperate deciduous forest	🟧 Desert	🟫 Tropical scrub forest
🟦 Tundra	🟨 Temperate grassland	🟩 Tropical rain forest	🟩 Tropical grassland and savanna
🟥 Northern coniferous forest (taiga)	🟥 Scrubland (chaparral)	🟩 Tropical deciduous forest	🟦 Mountain

FIGURE 34.2   Biomes of Our World
Biomes are characterized by the dominant types of plants that occupy the area.
This illustration depicts some of the world's major biomes.

Deserts

> ■ **Desert biomes are characterized by little rainfall or water sources, high daytime temperatures, and cold nighttime temperatures. Vegetation is sparse, so topsoil is not retained. Plants and animals have adapted to minimize water loss and protect themselves from temperature extremes.**

Deserts are regions of descending warm air, often in rain shadows of mountain ranges. If you have ever been to the Sierra Nevada Mountains in California, you have seen how the mountain range serves as a great wall. On the west side, rain clouds and rain are prevalent, but by the time clouds are blown over the mountains and reach Death Valley, most of the moisture has precipitated out, and a desert biome is the result.

Desert climatic conditions are most prevalent at a latitude of about 30 degrees north of the equator and 30 degrees south of the equator. At this latitude in North America are the Great Basin, Mojave, Sonoran, and Chihuahuan Deserts (Figure 34.3).

Desert areas receive less than 10 inches (25.4 cm) of rain per year. Often when rain comes, it comes in the form of violent cloudbursts and windstorms, both of which wash and blow away unprotected soils. Daytime temperatures are usually very hot, and at night, because of low humidity and a lack of clouds, the heated soil quickly radiates its heat to the atmosphere. The result is much colder nights. The desert then is a region of daily temperature extremes. The range between the hottest hour and the coldest may be as much as 30°C (the record is 126°F to 26°F within 24 hours in the Sahara Desert).

Vegetation in these conditions is sparse, and deserts have the lowest net productivity rate of all the biomes. Plants are adapted to this environment in a variety of ways. Many, like the barrel cactus, are *succulents,* plants whose structures are designed to hold water. The cacti are plants adapted to conserve water and whose leaves have been converted to spines. Photosynthesis is conducted in cells of the trunk or "stem." Other deciduous (leaf-dropping) plants drop their leaves at times of drought —

Great Basin
Desert

Mojave
Desert

Death
Valley

Sonoran
Desert

Chihuahuan
Desert

(a)

FIGURE 34.3   Deserts
Unlike most other biomes, where
temperature and day length
determine cycles of growth and
reproduction, rainfall is the prime
factor for desert organisms. (a)
Mojave Desert. (b) Sonoran Desert.

(b)

again, a mechanism for the conservation of water. And as you may remember from Chapter 8, many desert plants have C_4 carbon-fixation processes, which assist in the conservation of water. Desert plants must be capable of germinating, maturing (flowering), and producing seeds during the brief periods when there is enough available water and there are favorable growing conditions. Perhaps you have traveled the southwestern deserts in March and April and marveled at their carpets of flowers? If so, you have seen the wonder of a desert in bloom.

The animals of the desert are also specifically adapted to living in an extremely hot, dry climate. The external coverings of reptiles and insects protect them from drying out. Some of the small mammals, like the kangaroo rat, eat plants and never drink, surviving on the water produced by their own metabolism and contained in the plants they eat. Most desert animals are nocturnal, so they are out in the cool temperature of evening and nighttime. They spend most of the hot daylight hours in a protected location either in burrows in the ground or in caves (Figure 34.4).

(a)   (b)

FIGURE 34.4   Nocturnal Desert Dwellers
Temperatures at night in the desert are far less severe than during the day.
The majority of desert animals are nocturnal. They avoid being out during
the heat of the day. (a) Rattlesnake. (b) Kangaroo rat.

Scrublands

■ **Scrubland biomes are communities that are swept periodically by fire. This biome exists in coastal areas in which there are cool, wet winters and dry, arid summers.**

You have all read about the fires that occur regularly in southern California, fires that threaten human lives and homes. Periodic brush fires like these are typical of the **scrubland** (or chaparral) biome (Figure 34.5). The majority of plants are well adapted to the condition, and recovery after a fire is rapid. In fact,

(a)

FIGURE 34.5   Scrublands
(a) Wet winters and dry summers pose the major problem for plants living in this environment. Water is most often available when temperatures are least favorable for photosynthesis. Shrubs with tough drought-resistant leaves represent an adaptive response to these conditions. (b) Many animals, such as these toads, have become adapted to living in this environment.

(b)

(a)

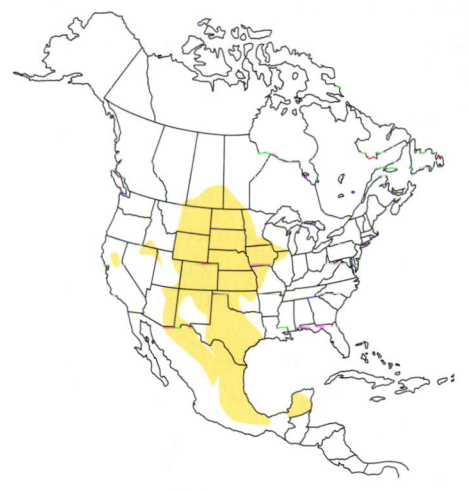

(b)

FIGURE 34.6 Savannahs and Other Types of Grasslands
(a) Herbaceous plants of the savannahs are adapted to the pressures of heavy browsing. Savannahs support some of the largest herbivores in the world, like these zebras. (b) Many flowering plants live among the grasses, and in late summer the prairie can be awash with color.

some seeds must be exposed to fire before they will germinate.

Scrublands typically occur in regions with wet, cool winters followed by long, hot, dry summers. The rains that fall in the winter give rise to large amounts of plant growth at the end of the season. There is little rain during the dry season, so the plants and animals of the region are almost like those of the desert — well adapted to periods of very little water.

Most scrublands are found along coastal areas and occur along the Pacific coast of the United States and in Chile, the Mediterranean, South Africa, and southern Australia. These biomes are dominated by small trees and spiny, thick-leaved evergreen shrubs. The animals of the scrublands include lizards, snakes, birds, and some mammals like rabbits, mice, mule deer, and coyotes.

Grasslands

■ Temperate grasslands and tropical savannas are characterized by rolling flat terrains and alternating seasons of either hot and cold weather or wet and dry weather. The rainfall patterns in both types of grassland support sod-forming grasses and legumes.

Grasslands cover large areas of the world, in both temperate and tropical areas. Temperate grasslands occur in the central plains and parts of California in the United States and in northern Asia (Russia). The grasslands of southeast Asia and Africa, on the other hand, are tropical grasslands and are often referred to as *savannas* (Figure 34.6a).

Temperate grasslands are characterized by rolling flat terrain and alternating hot and cold seasons. The savannas have the same physical characteristics but have alternating wet and dry seasons. For a number of reasons, grasslands cannot support forest growth (though some trees survive): The total rainfall is too low, the seasons are irregular, or the wet and dry rainfall pattern cannot support tree growth. Grasslands receive from 10 to 30 inches (25.4 to 76.2 cm) of rain per year, depending on their location.

The vegetation in a grassland biome consists primarily of bunch-forming grasses and legumes such as clover. Although a grassland may seem to be fairly uniform, there are around 10,000 species of grasses in the world, which makes this environment one that is highly diverse. The types of grass in a grassland depend upon the area's annual rainfall. In North America, the western, short-grass prairie grasslands are dryer than the moist, tall-grass/corn belt region of the United States (Figure 34.6b).

Grasses form a dense mass of roots, stems, and leaves, which support communities of ants, beetles, grasshoppers, and other primary consumers. These insects are in turn preyed upon by insect-eating predators such as the giant anteater of South America, which consumes as many as 30,000 termites a day. Other mammal species are herbivores like bison or wildebeests, which travel in herds, feeding upon the grasses. Herbivores are assured of a continued supply of food because grasses grow from the base of the stem, not the tip (which is the normal plant growth pattern), so that grazing does not kill the plant. When grazing or weather conditions limit the food supply, herbivores migrate to new areas.

The large number of herbivores can support carnivores, and these too are found in grasslands. In the savannas of Africa there are predators such as lions and cheetahs. In South America there are hawks, foxes, and the maned wolf. In the United States there are foxes, coyotes, and (once upon a time) wolves.

Forests

■ **There are three major types of forest biomes: deciduous, coniferous, and tropical. Deciduous forests are populated by trees that shed their leaves**

and occur in temperate regions having warm, mild springs and cold, often freezing winters. Coniferous forests are found in areas with long, harsh winters and short, warm seasons. Tropical rain forests are found in areas characterized by hot, humid, wet conditions all through the year.

Forest biomes arise in areas that receive enough light and water during the year to support tree growth. The types of trees in a forest biome depend upon the amount of light and water they receive, as well as the length of the growing season and the temperatures of both winter and summer seasons. There are three major types of forest biomes: deciduous, coniferous, and tropical rain forests.

Deciduous Forests. **Deciduous forests** are dominated by trees that shed their leaves. The growing season of warm, mild springs and summers, followed by cool, often cold, freezing winters that are less suitable for growing, at which time leaves are shed. Rainfall is moderate, from 30 to 60 inches (76.2 to 152.4 cm) per year.

Plants that lose their leaves have a selective advantage in such a climate. When there are fewer leaves, water is conserved. Leaf drop also is a way of cycling nutrients efficiently, because accumulated litter on the ground provides organic matter for the decomposers (if you have ever raked leaves in the fall you know what a large organic mass it is). The action of the decomposers on the organic matter produces a very rich organic source of nutrients within the topsoil.

The tree species that dominate deciduous forest biomes depend upon the biome's location. In the southeastern United States, deciduous forests are dominated primarily by species of oak and hickory. However, in wet, cool ravines of the southeast, as well as farther west in Ohio, beech and maples predominate (Figure 34.7). Deciduous forests contain a variety and abundance of animal life, from small and large mammalian herbivores to large predators. To survive the winter, many bird species migrate to warmer areas. Many mammals either migrate or hibernate.

The rich topsoil of deciduous forest land is one reason why forest biomes are shrinking in size. Humans have cleared these forests and converted them to agricultural areas. In fact, because of this human intervention, deciduous forests are rapidly disappearing. The southern Appalachian mountains and Great Smokies of Tennessee are two of the best examples of a once abundant, now rapidly disappearing biome.

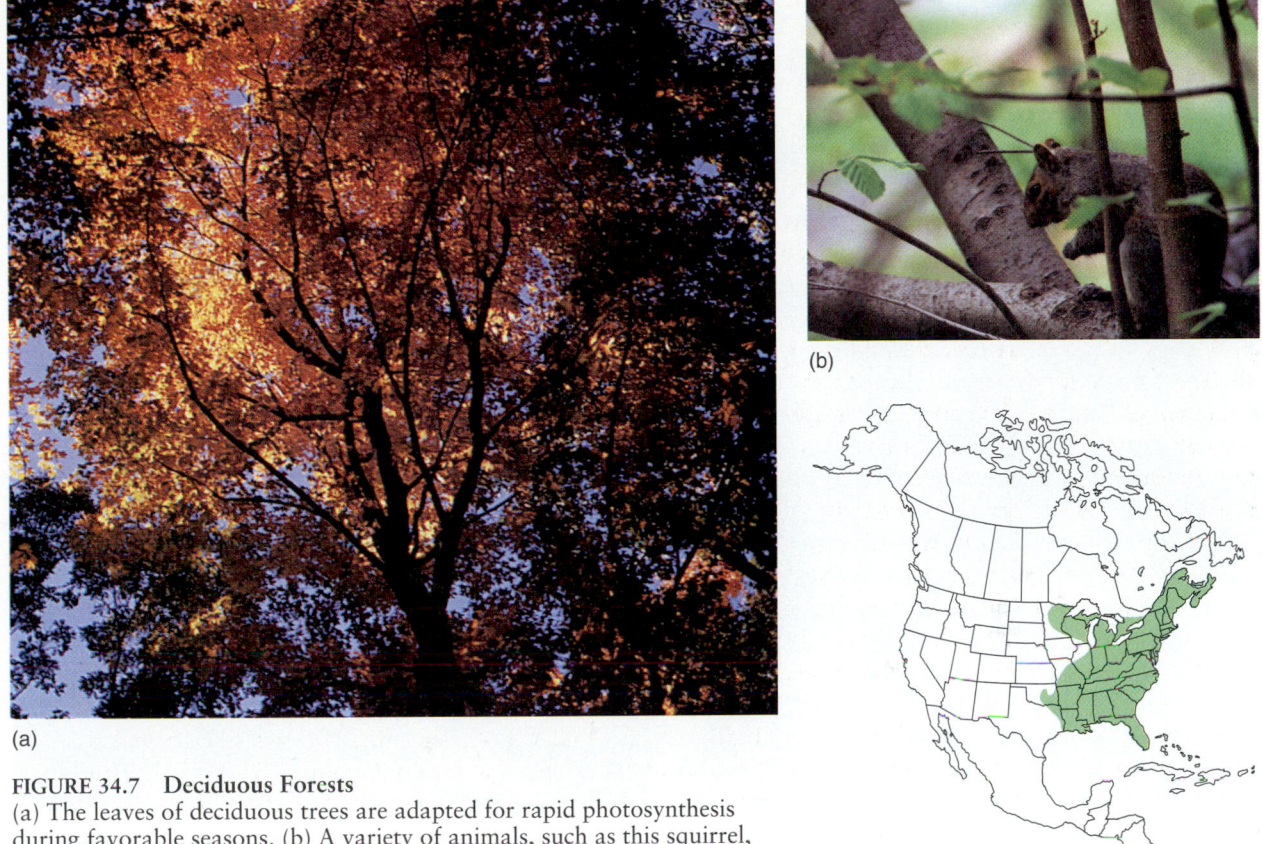

FIGURE 34.7 Deciduous Forests
(a) The leaves of deciduous trees are adapted for rapid photosynthesis during favorable seasons. (b) A variety of animals, such as this squirrel, live in deciduous forests.

Coniferous Forests. In the northern regions of North America and Asia, there are large coniferous forests called the **taiga.** Once you have seen the coniferous forests of the United States you will never forget their majestic splendor (Figure 34.8). The taiga is almost totally limited to the northern hemisphere. Taiga organisms have evolved in areas with long, severe winters and constant snow cover.

Conifers are evergreens having small, compact, needle-like leaves that are an adaptation preventing water loss. Losing leaves is a more efficient adaptation to prevent water loss than are needle-shaped leaves; so when deciduous trees and conifers occupy the same niche, conifers are usually competitively excluded. However, deciduous trees require a period of warmth and a steady source of water in order for leaves to regrow, and in areas characterized by coniferous forests, growing seasons are often too short for leaf reformation. In fact, because winters are very long and very cold, the only water is frozen and not available to the trees. Coniferous forests must often withstand long periods of drought as severe as those in a desert biome.

The short summers of coniferous forests allow a rapid burst of vegetative growth. The soil is not as rich as that of deciduous forests because less leaf drop and the low temperatures of their extreme winters do not support an array of the abundant and efficient decomposers. Also, evergreen needles are acid-rich and difficult to decompose in comparison with deciduous leaves. Some fungi form a symbiotic relationship with the conifer root systems by breaking down the nutrients. The soil acidity is an advantage for conifers, since they can absorb the acidic nutrients, while other plants cannot.

During the winter months, there is a limited amount of animal activity in coniferous forests because of the extreme cold. During the spring and summer months, however, activity is abundant. These species are well adapted to the rigorous winters by migrating. Those that remain hibernate, or they possess heavy coats of fur.

Terrestrial Biomes

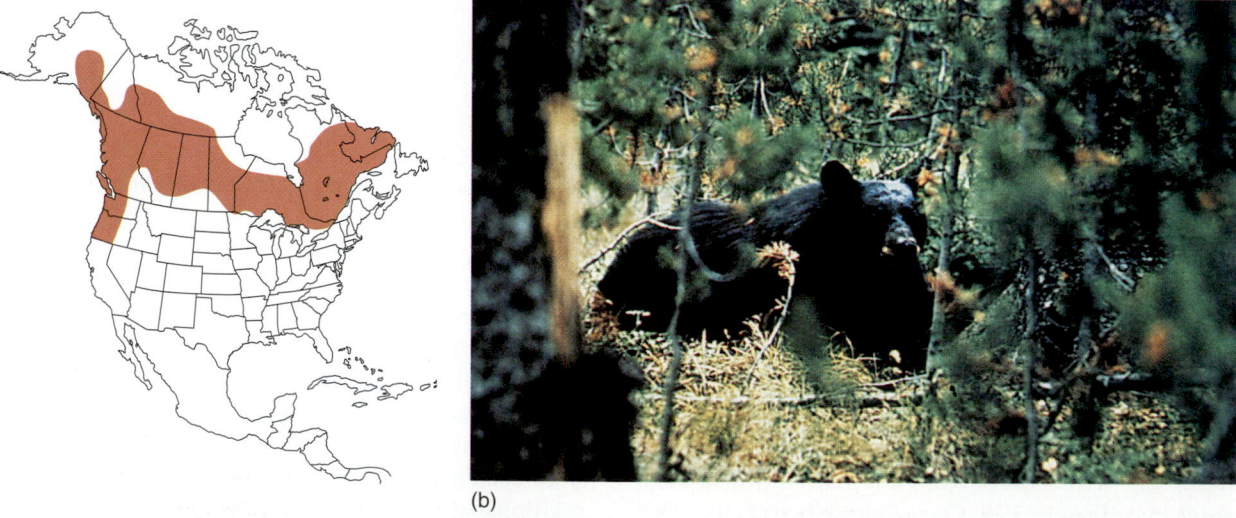

FIGURE 34.8   Coniferous Forests
(a) Although expansive, coniferous forests consist of a small number of different species, including spruce trees. (b) Many types of animals, such as this bear, inhabit coniferous forests.

Tropical Rain Forests.   **Tropical rain forests** are located near the equator (Figure 34.9). These communities are considered to have existed for tens of millions of years, which is one reason for the density and great abundance of life they contain. Unfortunately, because of the interaction of humans —

namely, lumbering and clearing for farming — the rain forests are shrinking at an alarming rate.

Tropical rain forests are characterized by warm, humid conditions. Abundant rainfall and sunlight support a great diversity of plants and animals. Over 16,000 rain showers a year feed the lush vegetation

FIGURE 34.9 Tropical Rain Forests
(a) A greater variety of plant and animal species exists in tropical rain forests than in any other biome. (b) A typical inhabitant of the rain forest.

in the tropical rain forests of El Yunque in Puerto Rico. In a temperate forest there may be three to ten dominant tree species, but in a tropical rain forest there may be over 100. One of those trees can be the home of more than 950 species of beetles alone. (The number of species living in one tree can be greater than the total number of species found living in an entire coniferous forest!) There also are over 1600 species of birds in El Yunque (see Profile).

The dominant trees are tall, forming a dense canopy that produces constant shade on the forest floor. The canopy produces a vertically stratified environment. In the top regions there is abundant light and less moisture. Many animals in tropical rain forests spend their entire life at one level of the canopy, never touching the forest floor, and some live only in the top of the canopy. On the forest floor, there is shade and constant high humidity.

The soil of a tropical rain forest is poor because organic matter decomposes rapidly and the growth of the biomass outstrips the limited amount of available nutrients. All available nutrients are consumed and very few are stored in the soil. Given the heavy amount of rainfall in these biomes, nutrients in the soil would be washed away into rivers and lost if they were not taken in quickly by the plants.

Biologists do not know about all the life forms that live in tropical rain forests, simply because there are so many that all have not been discovered yet. The spreading deforestation in countries of South

FOCUS 34.1
Climate

Common sense tells us that climate affects the distribution of organisms within an environment. But what is climate? **Climate** refers to those abiotic factors related to prevailing weather conditions. But what is weather? **Weather** consists of various factors: the temperature, which is due to the intensity of sunlight striking a given region; the amount of moisture in the area, known as the *humidity*; the direction of the wind and the speed of the wind; the amount of cloud cover; and the amount of rainfall. As we just mentioned, climate is determined by many factors, but for the purpose of this Focus box we will discuss and illustrate the following:

1. The amount and intensity of sunlight (Figure A)
2. The distribution of land masses and oceans (Figure B)
3. Elevation of land masses (Figure C)
4. The earth's daily rotation and seasonal rotation around the sun (Figure D)

These factors are responsible for the prevailing winds and the ocean currents, which in turn determine the worldwide patterns of climate that influence rainfall, nutrients available in the soil, temperature, and other factors that determine the environment in a given region where certain plants, animals, and microorganisms have adapted to live. Plants and animals from similar environments show many features in common even

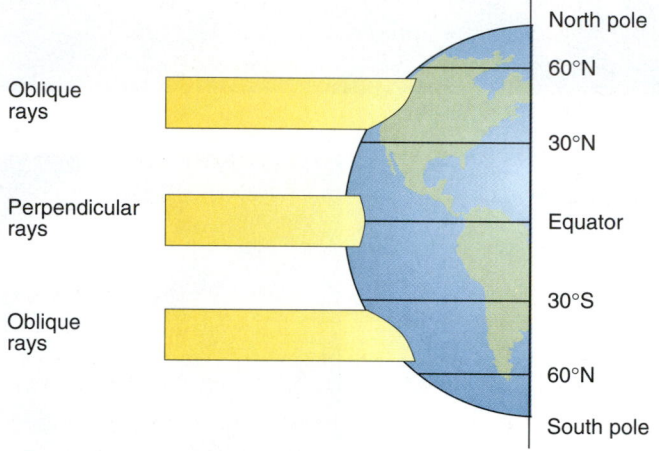

FIGURE A   The Amount and Intensity of Sunlight Affect Climate

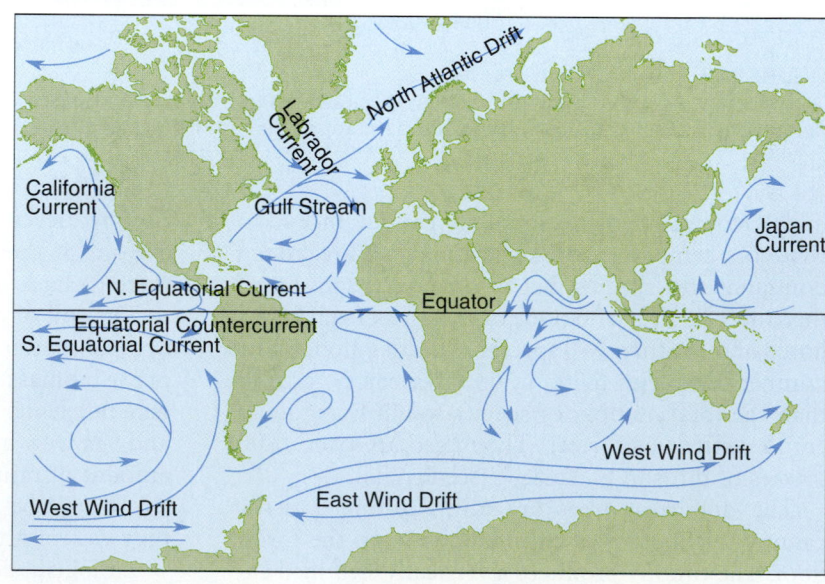

FIGURE B   The Major Prevailing Currents of the World

though they are distantly related. This is called *convergent evolution*, as discussed on page 395.

Not only are there global weather patterns, but there are also seasonal and local climatic effects. I live about 25 miles from the Pacific Ocean, and the amount of humidity is rather low. Just 10 miles toward the ocean the humidity becomes much higher. Factors such as proximity to bodies of water, mountain ranges, and even smaller geological features such as ponds or canyons impact the humidity level of environments. The seasonal effects we are familiar with (winter, spring, summer, and fall) in the temperate regions are all related to the amount of solar radiation striking a specific region of the earth. These rhythmic changes in the amount of solar radiation striking a region, called seasonal influences, weather conditions, and environmental conditions, determine the temperature ranges of specific biomes By studying these conditions, we can better understand the biomes of the world.

As you can see, the distribution of the earth's biomes in part depends on climate. But one should remember that as a part of climate there are distinct patterns of weather. For example, when reading scientific reports, a report might give the mean or average rainfall for a given area for a given year. You may also notice that the mean amount of rainfall for another area is the same. One should be cautious to conclude that the two biomes are the same because not only is the amount of rainfall important, but the distribution of rainfall over time. For example, if all the rain comes in two large cloudbursts, the biome may be a desert; while on the other hand if there are continuous light showers during a four-month period during the year, and then little or no rain for the remainder of the period, the biome may be scrubland (chaparral). Therefore, when interpreting weather or climatic data one has to consider patterns of temperature, precipitation, and other abiotic factors. Climate is therefore a complex interaction of several factors that help to influence the distribution of organisms, resulting in biomes.

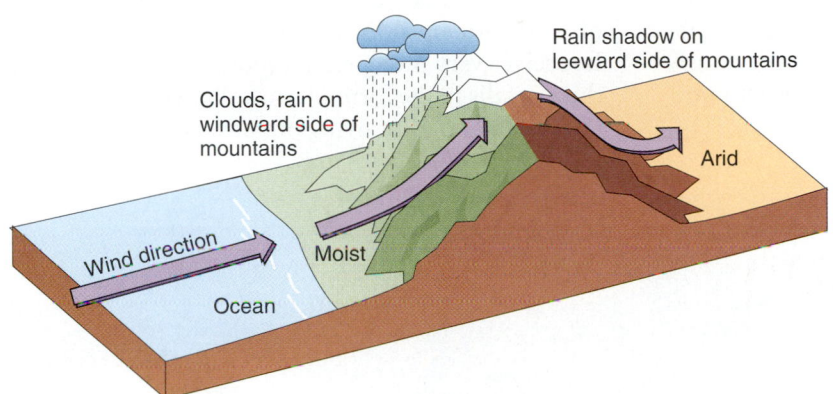

FIGURE C   Rain Shadows Caused by Elevated Land Masses

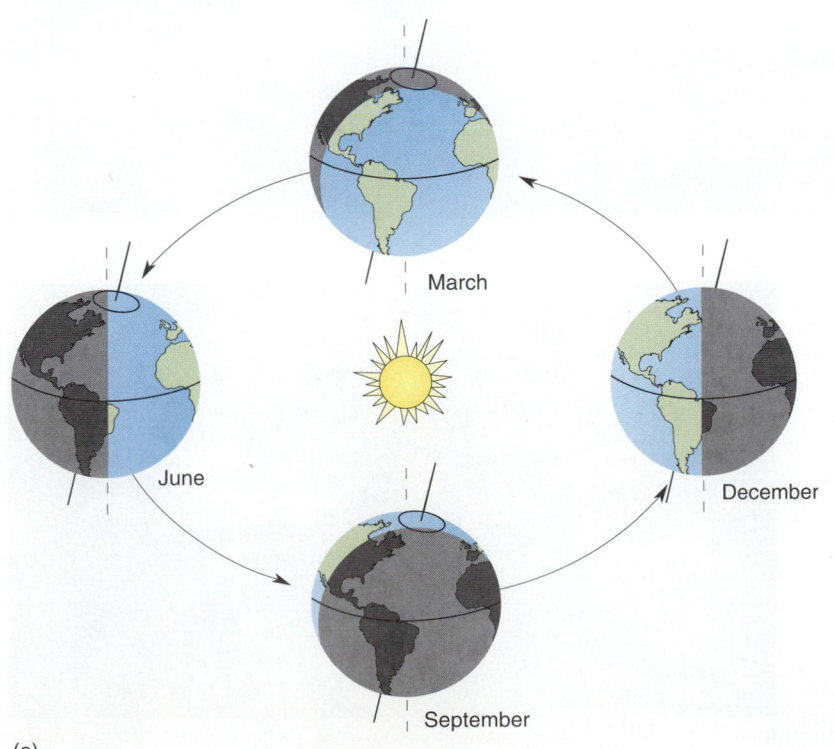

(a)

(b)

FIGURE D   Seasonal Effects and Temperature
(a) The tilt of the earth as it rotates around the sun affects the amount of exposure of the different hemispheres to solar radiation. (b) World temperature zones.

America and other areas threatens to destroy species before we even know of their existence. Millions of square kilometers of land are being deforested as governments and individuals clear these areas by a method called *slash and burn*. Slash and burn works just like it sounds; all the trees and vegetation are cut down and hauled away, and any remaining growth is burned off. The result is a bare soil, totally exposed to the intense tropical sun and rains. Since the nutrients contained within the biomass of the forest have been removed, nutrients of the biome are not recycled. Since the soil of this biome is nutrient-poor, it cannot support the agricultural growth that replaces the former community. Farms fail and are abandoned, and secondary succession (the reestablishment of organisms in a site that was previously occupied), which is a lengthy process in such nutrient-poor soil, must begin (see page 874).

Tundra

■ **Tundra biomes are harsh regions of rolling flatlands, bogs, and marshes in which the subsoil is permanently frozen. This subsoil is called permafrost. Most precipitation is frozen, resulting in low amounts of available water during most of the year.**

Tundra, a Russian term meaning north of the timberline, is used to describe those regions of gently rolling flatlands, bogs, and marshes where the subsoil is permanently frozen. This kind of subsoil is called *permafrost.*

The tundra is located in the far northern regions of North America, Europe, and Asia (Figure 34.10). It receives very little precipitation, and most of what there is falls in the form of snow. Water is frozen and unavailable to most living organisms for most of the

(a)

(b)

FIGURE 34.10 Tundra
(a) Most plant life in the undra is scrubby and low or matlike. The growing season is short and, for a time, colorful. (b) The tundra supports some animal life.

Dr. Alexander Cruz
Tropical Ecologist, Ornithologist

It may seem strange that Dr. Alexander Cruz, an internationally respected tropical ecologist and ornithologist, developed his interest in natural history while he was growing up in the Puerto Rican neighborhoods of New York City. Nonetheless, even as a child Cruz was fascinated by the invertebrate animals he observed in the city's vacant lots. He literally filled his room with his collection of the animals he found. He kept his live turtle collection in his mother's washtub. When he visited his grandparents in Puerto Rico, his interest in wildlife was kindled even more. There he observed tropical wildlife in a natural setting. Even today, tropical ecology and especially tropical birds are his major research interest.

Cruz's formal biological training began at the City University of New York, where he earned a B.S. degree in 1964. Like many people, he did not go to graduate school immediately. Instead, he had to get a job to support himself and his family. Cruz worked as a microbiologist with the New York City Health Department from 1964 to 1968, studying viruses and their effect on the nervous system.

Though the work fascinated Cruz, he spent his spare time studying nature, especially birds. He visited museums, observed organisms in the park, and read as much as he could about tropical and island ecosystems. Island ecology was his fascination. His future, he believes, was affected profoundly by the book by R. H. MacArthur and

E. O. Wilson, *The Theory of Island Biogeography.* By 1968, his interest in island ecology had become so strong that he resigned his position with the health department and was accepted as a Ph.D. student at the University of Florida. There he concentrated his studies on a comparison of a mainland and island woodpecker species.

Cruz's work set out important concepts about resource exploitation, behavioral patterns, and population parameters of the birds he studied. He spent time working in the field in the tropical forest, and came to love the solitude and intimacy with nature that the tropics provided someone who had lived his life in a temperate zone. Cruz was "hooked" on island ecology.

He received his Ph.D. in 1973. The same year, he took a position at the University of Colorado, and with the university as a base he continues his research on tropical and island biology. He has spent time studying and doing fieldwork not only in Colorado but on the islands of Trinidad, Hispaniola, and Puerto Rico. His interests have expanded to include the destruction of tropical rain forests by lumbering and agricultural expansion, and work on these problems has taken him to the mainland tropics of Venezuela and Mexico.

Cruz believes, as do many others, that the world's tropical forests are disappearing at an alarming rate. As an institute scientist, he has studied the impact of deforestation at the Luquillo Experimental Forest

Biosphere Reserve in Puerto Rico. "It is extremely important that we understand more about the dynamics of nutrient cycling and productivity in the natural tropical forest before we can endorse the development of monoculture (single-species plantations) for economic purposes," he says.

When Cruz is not in the field in Colorado or in the tropics, or teaching at the University of Colorado, he can often be found working as a consultant for the National Science Foundation in Washington, D.C. For example, he was appointed to the Committee on Equal Opportunities in Science and Technology for the National Science Foundation in 1981. He has come a long way from the vacant city lots in New York City. More important, Cruz is truly doing what he enjoys.

year. In fact, the shortage of available water makes the tundra much like a desert.

However, during the short growing season (usually less than two months), when the temperature rises and the snow melts, the tundra becomes marshy and several varieties of perennial flowering plants bloom. Typically, the tundra has sparse vegetation, mainly mosses, lichens, and grasses. There are very few trees, and those trees that do exist are dwarf willows that grow in regions along streams, ponds, or lakes.

The tundra biome supports herbivorous reindeer, caribou, and hares, and there are also several mammalian predators, such as foxes and Arctic wolves. The tundra is also the summer home of flock upon flock of migratory waterfowl. And if you have ever visited the tundra regions of the Great Rocky Mountains in Colorado, like those in Rocky Mountain National Park or the Never Summer Range, you know that during the summer this environment supports an abundance of insects and flies. (As you campers of the tundra know, mosquitoes — on which tundra birds feed — are particularly abundant.) Usually biomes refer to geological regions that fall in rough bands around the globe (**latitudinal biomes**). But the fact that there are tundra biomes in states like Colorado points out that the biomes change with altitude. A mature terrestrial community that varies with altitude is called an **altitudinal biome.** Therefore, one can say that as you move upward from the equator to the pole, or upward from sea level, you would find the same sequence of biomes, mainly because of similar temperatures and precipitation.

The region around Flagstaff, Arizona, is an example. At the bottom of the Grand Canyon, you find a desert biome. Climb upward to the canyon's inner rim, and you are in a scrubland. Continue to the top, and you have moved into a transitional forest. After that, travel to the top of the San Francisco peaks near Flagstaff, and you are in a coniferous forest, which is followed by an alpine, tundra-like biome at the very top of the mountains. Figure 34.11 illustrates the correspondence between latitudinal and altitudinal biomes.

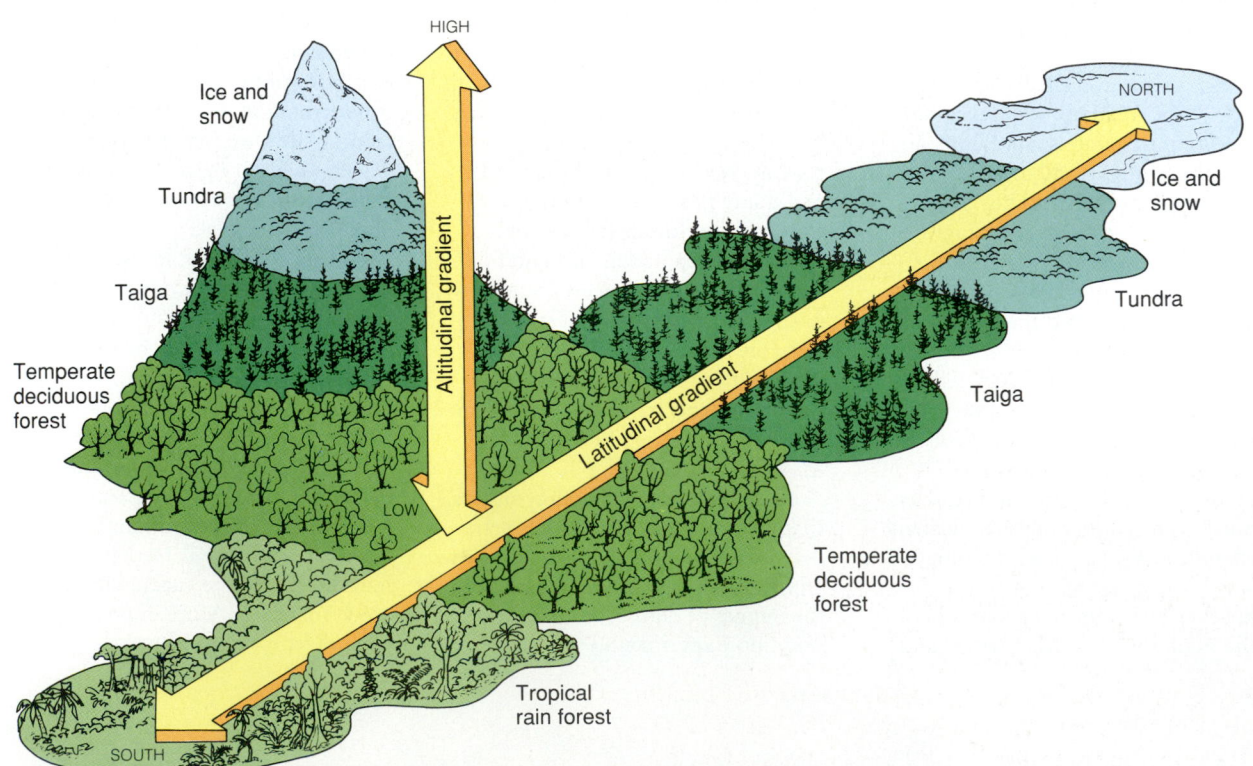

FIGURE 34.11 Altitudinal Biomes
Because temperature and precipitation change with altitude as well as with latitude, biomes that exist at different altitudes generally reflect the same features as biomes governed by their geographical position on the globe. Higher altitudes are colder and support biomes similar to those that exist at cold latitudes.

Chapter 34   Introduction to Ecology: Biomes

AQUATIC BIOMES

■ There are two major types of aquatic biomes: freshwater and saltwater.

There is a vast variety of conditions in the earth's waters. Aquatic biomes fall into two broad categories: freshwater and saltwater. Each of these can be divided into a number of divisions.

Freshwater Biomes

■ There are two major divisions of freshwater biomes: still-water biomes and moving-water biomes.

Limnologists are people who study the earth's freshwater biomes. They have divided the **freshwater biomes** into two major divisions; **still-water biomes,** or stationary water biomes (lakes and ponds), and **moving-water biomes** (streams and rivers).

Still-Water Biomes

■ Still-water biomes are comprised of three zones: the littoral (around the edges of the lake), the limnetic (in open water and penetrated by light), and the profundal (depths light does not penetrate). Oligotrophic lakes are nutrient-poor and eutrophic lakes are nutrient-rich.

Although ponds are obviously smaller than lakes and so include less variation in temperature and environmental conditions, many of the elements found in lakes apply to both habitats. The major one is that still-water biomes are comprised of three zones (Figure 34.12). The *littoral zone* lies at the water's edge and is the area of the lake where sunlight can reach the bottom. This zone supports an abundance of plant life — such things as cattails, willows, and rushes — along the edge and water lilies and duckweed in the water. The latter serve as habitats for water insects, fish, and amphibians.

Out from the water's edge is the *limnetic zone,* the zone of open water as far down as light penetrates. Floating algae are found here, along with the free-swimming fishes. The *profundal zone,* located below the limnetic zone, is an area that light does not penetrate. Here, in the lake's deepest and darkest region, there is no plant life. The life forms here consist of lake scavengers (called *detritivores*) and the decomposers that feed on the organic debris constantly falling from the waters above. That debris is often described as a rain of food.

Just as altitude can determine terrestrial biomes, depth can determine biomes in lakes because as water becomes deeper, it becomes colder. Surface

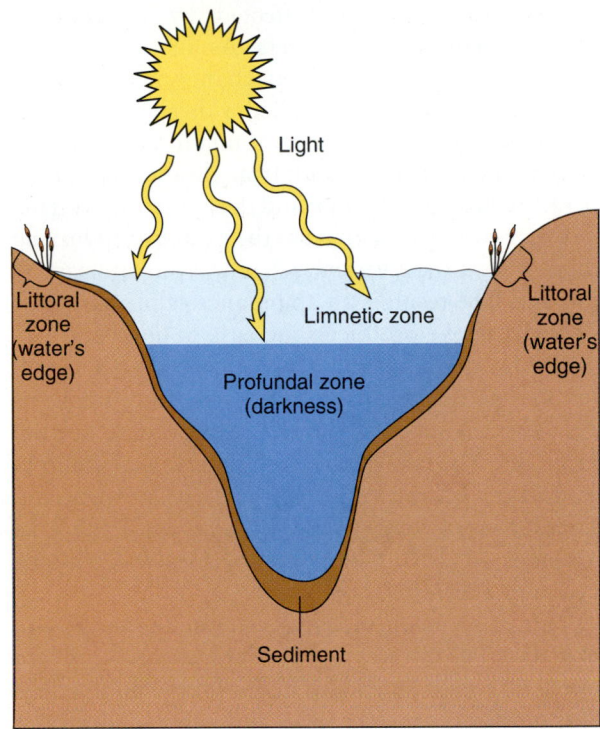

(a)

(b)

FIGURE 34.12   Still-Water Biomes
(a) A springtime surge of productivity occurs in most temperate lakes, because water turnover coincides with longer days and warmer temperatures. (b) Ponds are also still-water biomes.

waters also are more affected by temperature changes than are deep waters. As air temperatures and the intensity of sunlight change with the seasons, they cause what is called a thermal overturn, which occurs in either spring or fall. In thermal overturn, vertical currents result from differences in temperature between shallow and deep waters, bringing nutrient-rich waters up from the bottom and mixing them with surface waters rich in oxygen and carbon dioxide. The result is an abundance of primary productivity as the producers increase in numbers.

Freshwater biomes undergo succession (see Chapter 36). As a lake or pond becomes more nutrient-rich, the amount of sediment falling to the bottom increases. The lake becomes shallower and shallower, and the littoral zone enlarges. The process continues over a long period of time until the lake becomes a marsh. Ultimately the freshwater biome becomes a terrestrial one.

Humans, of course, can accelerate the succession of a lake. The nitrates and phosphates we dump into our sewage system end up in our lakes and function like fertilizer, causing algae and other underwater plants to bloom. The bloom then increases the plant population. When the plants die, their remains drift to the bottom, speeding up the rate at which sediment accumulates. The normal steps of succession have been accelerated by human waste procedures.

The process by which water becomes more and more rich with nutrients is called **eutrophication.** Lakes can be classified on the basis of available nutrients. *Eutrophic lakes* are nutrient-rich, and *oligotrophic lakes* are nutrient-poor. Human activity can make lakes oligotrophic as well as eutrophic. (For example, the use of fertilizers may change oligotrophic lakes into eutrophic ones. This is called *cultural eutrophication.*) Acid rain increases acid levels in lakes, interfering with fish reproduction cycles and slowing the rate at which organic matter decays, probably because the decomposer population falls. Ultimately, the waters become brilliant blue, perfectly clear, and totally lifeless (Figure 34.13).

Moving-Water Biomes

■ **In moving-water biomes, the abundance of life is determined by the strength of the current.**

Streams and rivers, which consist of continuously running water, are moving-water biomes. Communities in moving-water biomes are affected a great deal by the strength of the current. Very few plants and animals can establish themselves in swiftly moving water; so one finds few large photosynthetic algae, few plants, and few fish or other animals. Slower

FIGURE 34.13   Eutrophication
As more nutrients become available, more organic material is produced. This material accumulates, eventually filling the lake.

streams and rivers or protected pools contain more animals and plants (as all fishing enthusiasts know).

Marine Biomes

■ **The major types of saltwater biomes are the estuaries, the neritic biomes, and the oceanic biomes.**

Seventy-five percent of the earth's surface is covered by water, and 70 percent of that consists of salt- or marine-water realms. Just like freshwater realms, marine environments are divided into biomes. Where rivers empty into oceans, there are *estuaries,* the areas where fresh and salt water mix. Farther out are shallow waters along the coastline called *neritic waters,* and *oceanic waters,* which are open-ocean, deeper waters beyond the continental shelf.

Estuaries.   Salt and fresh waters mix in **estuaries,** the mudflat and tidal marshes found where rivers empty into saltwater bodies. An estuary is a con-

FIGURE 34.14   A Clean Estuary
Unfortunately, many of the beautiful and crucial breeding grounds are becoming victims of environmental pollution.

stantly changing environment — flushed with fresh water full of mud and nutrients, subject to tides that result in an increase and decrease in the amount of salt in the water. Organisms living in an estuary must have adapted to a wide range of environmental conditions in order to survive.

Despite their changeable nature, estuaries are rich with nutrients and support a wide variety of life forms. In fact, estuaries serve as the breeding grounds and nurseries for many of the ocean's organisms, who lay their eggs near estuary waters and whose young develop in estuary tidal marshes (Figure 34.14).

Neritic Biomes.   **Neritic biomes** can be divided into two main types, depending on the type of ocean and shoreline — sandy or rocky. Animals and plants living along a rocky coast are adapted to cling fast to the rocks so that they will not be washed away by the pounding surf. Some of these organisms are trapped in small tidepool communities that form within the rocks as the tide recedes. On sandy beaches,

FIGURE 34.15   Coral Reefs
Coral reefs are home to a multitude of species.

Aquatic Biomes

communities are not subject to the pounding surf. Most animal species are adapted to use burrowing as a protection from the changing tides. Both types of shorelines are inhabited at some time of the year by large flocks of seabirds who feed on fish and other organisms that live on or near the coast.

In tropical oceans, the calcium and the mineral secretions of millions of coral animals produce coral reef biomes. The largest may be the Great Barrier Reef, which extends for a full 1200 miles along the east coast of Australia (Figure 34.15). Coral reefs may be as productive as tropical rain forests because their structures provide havens in the open ocean for a myriad of life forms, both plant and animal. Reefs receive ample light and warmth. The nooks and crannies formed by the skeletons of coral provide a foothold for plants, thus assuring an ample supply of oxygen in the waters. The plants attract primary consumers, providing the base for an extensive food web (Chapter 37).

The Open Ocean. **Open-ocean biomes** are much lower in productivity than the shoreline. As with lakes, the oceans are divided into zones, according to how far light penetrates the waters (Figure 34.16). The *euphotic zone* is the zone in which light penetrates. The *aphotic zone* is the zone that receives no light. The deepest parts of the ocean, called the *abyssal region,* include areas like the Marianas Trench, which lies at a depth of over six miles. As you would expect, these waters are extremely cold and the water pressure is high because of the weight of the water above.

You would expect few life forms to exist here because of the tremendous depth. However, life forms do exist. These bottom-dwelling creatures, called benthic organisms, consist of various scavengers and predators. The remains of organisms living in waters above and the nutrients they contain rain down continually to the abyssal region. The bottom-dwelling organisms feed on these nutrients.

In the region of the ocean that overlies the continental shelf (the neritic region), a process called *upwelling* provides the requirements of life for a variety of phytoplankton and fish. Deep waters are rich with nutrients that drift down from the waters above. In upwelling, the deeper, colder waters with their nutrient-rich sediments are moved up toward the surface

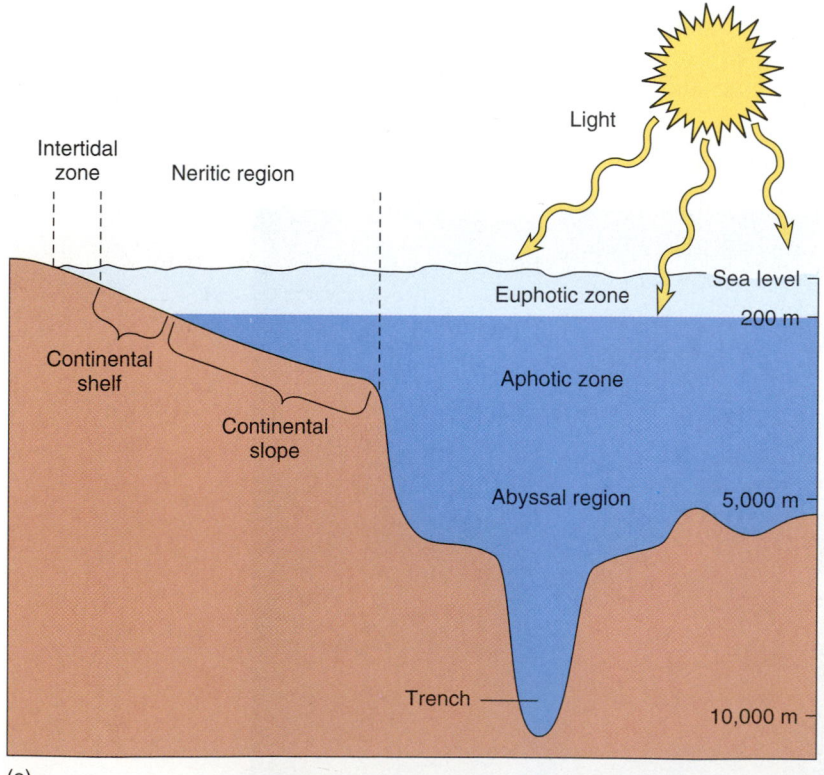

(a)

(b)

FIGURE 34.16 Open-Ocean Biomes
(a) The open ocean is divided into euphotic and aphotic zones. (b) This anglerfish lives in the aphotic zone. Bacteria that produce light are contained in the bulblike structure at the end of the stalk projecting from its forehead. The light attracts prey.

by ocean currents and wind patterns. Upwelling is usually most prevalent along the west side of continents, owing to the patterns of ocean currents. In part, that is why there are kelp forests along the coasts of California and why the western coasts are the nursery for a number of fish and ocean mammals. The upward movement of mineral nutrients from the sediments on the ocean bottom provides food for the organisms that serve as the primary producers in the ocean's food chain, the phytoplankton and seaweed.

The phytoplankton are one of the major sources of photosynthetic activity on the earth. It has been estimated that 80 to 90 percent of the earth's photosynthetic activity, including oxygen production, is carried out by these microscopic organisms. They are truly an important component of the oceanic biomes and of the biosphere itself.

SUMMARY

1. Ecology is the scientific study of the interaction of organisms with their environment.
2. The environment consists of nonliving (abiotic) components and living (biotic) components.
3. Ecology is multidisciplinary and uses the methods of observation and experimentation to investigate the interaction of organisms and their environment.
4. The environment is one of the selective forces that affects the evolution of organisms.
5. Climate and weather influence the distribution and types of organisms within a particular community.
6. Biomes are regions of the world that incorporate similar groups of communities. A biome is an abstract concept that defines regions with similar organisms, similar climates, and similar soil factors.
7. The major terrestrial biomes on the earth are deserts, scrublands, grasslands, forests, and tundra.
8. The aquatic biomes on earth are the freshwater biomes, both moving and still, and the marine (saltwater) biomes: estuaries, neritic, and oceanic.
9. Energy flows through ecosystems, communities change over time, and there are large collections of communities known as biomes that occupy the terrestrial and aquatic environment of the earth's biosphere.

KEY TERMS

ecology
environment
biotic components
abiotic components
populations
communities
habitat
ecosystem

biosphere
observational science
experimental science
biomes
climate
desert
scrubland
grasslands

deciduous forest
taiga
tropical rain forest
tundra
latitudinal biome
altitudinal biome
freshwater biome

still-water biome
moving-water biome
eutrophication
saltwater biome
estuary
neritic biome
open-ocean biome

REVIEW QUESTIONS

True/False

1. Ecology is the scientific study of the interaction of organisms with each other and the environment.
   true   *false*   *(page 831)*

2. The abiotic components of the environment are the living factors.
   true   *false*   *(page 832)*

3. The place where an organism lives is its habitat.
   true   *false*   *(page 832)*

4. Organisms have no impact on their environment.
   true   *false*   *(page 832)*

Fill in the Blank

1. Climate refers to the prevailing _____ conditions. *(page 833)*

2. The worldwide pattern of climate is due largely to _____ winds and _____ currents. *(page 842–843)*

3. _____ biomes are characterized by harsh regions of subsoil that is permanently frozen. *(page 844)*

4. Two major types of aquatic biomes are the _____ biome and the _____ biome. *(page 847)*

Multiple Choice

1. As a biologist, if you become very interested in the study of the interaction of organisms with each other and the environment your subspecialty would be:
   a. Zoology
   b. Ecology
   c. Botany
   d. Herpetology
   (page 831)

2. As you hike down the Grand Canyon of Arizona, you notice definite life zones. These life zones illustrate the concept of:
   a. Latitudinal life zones
   b. Altitudinal life zones
   c. Aquatic life zones
   d. Neritic life zones
   (page 846)

3. As we continue to use detergents and fertilizers, our freshwater lakes are becoming nutrient rich. The process resulting in the formation of nutrient-rich lakes is called:
   a. Eutrophication
   b. Oligotrophic
   c. Succession
   d. Climate
      (page 848)

4. All of the following are examples of terrestrial biomes except:
   a. Deserts
   b. Grasslands
   c. Estuaries
   d. Rain forests
      (page 833)

Discussion

1. Why is the study of ecology important?

2. Someone says, "it is the study of recycling." How would you respond to that definition of ecology?

3. How does the environment affect evolution?

4. Discuss and describe the environmental and geographic conditions that produce a desert biome.

5. Discuss and describe the three major types of forests.

6. Discuss the two major types of aquatic biomes.

7. Why are open-ocean biomes less productive than shoreline regions?

35

Populations

Population Density and Distribution
Population Dynamics and Growth
Summary

We previously mentioned populations when we discussed genetics and evolution (Chapter 17). There we defined a population as an interbreeding group of individuals of the same species in a given place at a given time. But what are the factors that affect a population's density, distribution, and dynamics? In addition, what about the human population growth problem? Let us see.

POPULATION DENSITY AND DISTRIBUTION

■ **Population density is the number of individuals per unit of area. Distribution is the spacing of the individuals within that habitat. Distribution patterns can be uniform, clumped, or random.**

Above: These dolphins are all members of the same species, hence they are a population.

A population has characteristics that do not apply to individuals. Individuals have fixed dates of birth, age, and death. Populations have additional dynamics: *population size, density, growth,* and *biotic potential,* to mention a few. These population dynamics are influenced by environmental relationships or interactions.

The size of a population in itself tells us nothing about the *fitness* of individual organisms in the environment. A particular environment may house a population of millions of individuals. However, just knowing the numbers of individuals in a population tells us very little about the interactions in that population.

From an ecological perspective, **population density** is a more meaningful concept than population size because density tells us about the relationship of a population to the environment. For example, 25 oak trees per acre or 50 humans per square mile is

Uniform distribution	Clumped distribution	Random distribution

FIGURE 35.1   Distribution of Organisms
Although clear-cut patterns are not always apparent in nature, organisms tend to be found in uniform, clumped, or random distribution.

useful to ecologists because it can be compared with other population densities.

Knowing the density of individuals in a given area is only a beginning point. The data are useful, but information about how organisms are spaced or clustered within the area and the size of the clusters tells ecologists even more about the relationship between the organism and its environment. That type of information is called *ecological distribution.*

Not all areas within a defined region afford suitable conditions for life. For example, if you look across a prairie, you may note that the grass grows in clumps. Those clumps may indicate places where cow manure has added natural fertilizer to the soil. Manure contains nitrogenous wastes, which in-

crease the available nitrogen in a particular spot. The result is the clumped nature of grass distribution. The distribution of nutrients varies according to area or time of year. As a result, the distribution of individual organisms also varies. Knowing the ecological distribution of organisms in a particular environment also gives ecologists a better idea of an organism's pattern of resource distribution and social interaction in its environment. Ecological distribution can be *uniform, clumped,* or *random,* depending on the availability of resources (abiotic factors) and competition (biotic factors) (Figure 35.1).

Uniform Distribution

■ **Uniform distribution occurs when organisms of a population are evenly distributed within a region.**

Uniform distribution, in which organisms of a population are evenly distributed throughout a region, is rare in nature because this arrangement is dependent upon uniform distribution of resources in the habitat. In addition, uniform distribution often reflects the result of intense competition among individuals in the population. For example, individual plants of some species (like avocado trees or desert creosote bushes) produce toxic substances that prevent other plants from germinating within a specific distance (Figure 35.2).

Cliff-dwelling gulls serve as a good example of uniform distribution. The conditions within the habitat do not vary much from place to place, and competition for nesting sites is intense. To reduce competition for available space, the birds establish and defend territories. Once territories are established,

FIGURE 35.2   Uniform Distributions
The uniform distribution of these creosote bushes is due to the limited supply of water in the desert environment. Each creosote bush competes for water and minerals, resulting in a uniform distribution.

aggression is reduced, allowing each bird access to the resources — in this case nesting space. Because each territory is a particular size, the resulting distribution is uniform. Uniform distribution arises as each individual establishes a territory, maximizing its access to resources.

Clumped Distribution

■ **The clumped distribution pattern is the most common distribution pattern in nature. This clumped pattern is due to the unequal distribution of resources in the environment.**

Clumped distribution is a pattern characterized by numbers of individuals clustered in particular locations within a region (Figure 35.3). This pattern is more common than uniform distribution in plants and animals for three basic reasons: *environmental conditions, reproductive patterns,* and *predator defense.*

It is very rare that environmental conditions are uniform throughout an area. Some locations provide better living conditions than others because the resources themselves are clumped. In the world's oceans, for instance, most life forms either require sunlight for photosynthesis, feed upon the photosynthesizers, or feed upon each other. As a result, most aquatic organisms are found near the surface regions of the oceans. There are far fewer organisms living in the cold depths of the oceans, which receive little if any light at all.

Reproductive patterns in plants and animals often require individuals of the same species to live in close proximity. Individuals are clumped to facilitate breeding and also to disperse pollen or gametes. We saw an example of this living pattern in Chapter 19 when we discussed how sperm swam through rainwater to reach the egg in the reproductive cycle of mosses. On the other hand, there are so few members of some endangered species, such as the Galápagos tortoise, that males and females do not meet often enough to breed. The pattern of clumped distribution has broken down.

Social groupings, such as herds, flocks, and schools, whether temporary or permanent, provide the group with a survival advantage. Individuals have greater protection from predators, and offspring often receive better paternal and maternal care. If you have ever seen films about the interactions within one of the large herds of wildebeests in Africa, you have seen the value of a large group. Although wildebeests do not care for one another's offspring, they do work together to fend off predators, and the sheer mass of the herd itself is a protection.

Random Distribution

■ **Random distribution occurs when individuals within a population are spaced at random.**

Random distribution, in which individuals occur in no discernible pattern throughout a region, is common when environmental conditions are the same in all parts of the habitat and organisms do not interact socially. As you would suspect, random distributions are relatively rare in nature. The random distribution of ticks on the forest floor is an example. Another example is the distribution of many organisms that inhabit the ocean floor (Figure 35.4).

FIGURE 35.3 Clumped Distributions
The agave cacti in the photograph exhibit a clumped distribution. This distribution is due to the minor variations in soil conditions that provide better conditions in one area as compared to the other.

FIGURE 35.4 Random Distribution
The distribution of some intertidal-dwelling organisms in the ocean is random.

Dr. George M. Woodwell
Ecologist

"Leisure is hard to deal with when what interests me is labor," states George Woodwell. To some people, Woodwell seems to lead a job rather than a life. His job (if you want to call it that) is crusading for the earth, and his crusade is his life.

Woodwell is the director and president of the Woods Hole Research Center, an organization he founded after founding and directing for ten years the Ecosystems Center at the Marine Biological Laboratory, also in Woods Hole. He is also the ecologist who has led several historic battles to pressure industry and government agencies to be aware of the cumulative effects of pollution on the environment. In the late 1960s and early 1970s, he played a critical role in uniting scientists to convince the Environmental Protection Agency to ban the use of DDT in 1972. He is one of the founders of the Environmental Defense Fund. He is also a founding trustee of the Natural Resources Defense Council and the World Resources Institute, a fellow of the American Academy of Arts and Sciences, a member of the National Academy of Sciences, and the Director of the Center for Marine Conservation. He has led several crusades to inform the public of the dangers of the "greenhouse effect," the global warming of the earth that may result from the buildup of CO_2 in the atmosphere. He and other scientists also produced a major report warning our government and the public of the nuclear winter that may result from the aftermath of a nuclear war.

Woodwell feels that scientists have an obligation to speak out and provide the public with the scientific data and solutions to environmental problems. However, he is not just an environmental advocate; he is a brilliant and thorough basic researcher. Woodwell has devised many experiments to explain the delicate relationships existing within an ecosystem. From these experiments has come an understanding of what happens when these systems are disrupted. When our activities interrupt these ecosystems, we threaten our own sources of biological energy and life.

One of Woodwell's landmark experiments began in 1961. The work was a study of the effects of radiation on forest ecosystems. He exposed 14 acres of forest to radiation. The result was a systematic disturbance of the ecosystem. First the pine trees died, then the oaks, then the tall, leafy shrubs. That was followed by some grasses. Ultimately all that was left were some lichens, mosses, and bacteria. Woodwell's study helped emphasize the concept of ecological succession and demonstrated that breakdown followed the same sequence as succession. Radiation-sensitive species were destroyed first, followed by those that were less sensitive, until only those species resistant to the stress survived.

Woodwell was raised in Cambridge, Massachusetts, but he spent his summers on a farm in York, Maine, where he developed his interest in natural ecological cycles. In 1946 he entered Dartmouth and majored in biology. Following graduation he spent three years in the Navy. He then entered Duke University to earn a Ph.D. in ecology. After three years at the

University of Maine, he took a position as visiting scientist at the Brookhaven National Laboratories on Long Island, New York. In 1961, he became a full-time ecologist there.

Eventually Woodwell was asked to serve as a founding member and to direct a research institute solely for ecological research at Woods Hole. The ecosystem center there opened in 1975. Woodwell continues his research and his crusade against the cumulative effects of pollutants on our fragile ecosystem. He recognizes the high cost of safeguarding the ecosystem from exposure to pollutants. He also recognizes that well-meaning governments are often controlled by the very economic interests that governments should control and regulate.

It is to the benefit of the human race that we have politically active, respected scientists like George Woodwell crusading for the protection of our earth. Perhaps we should listen to them very carefully.

Distribution over Time

■ **The pattern of distribution varies over time with changes in environmental conditions.**

Population distribution also varies with time. For example, the distribution of organisms varies with changes in temperature from season to season. In winter, many migrating birds fly to warmer climates. Some animals move locally between habitats to accommodate seasonal changes. During periods of drought, many animals clump along streambeds or other places that hold water. With the coming of the rains, the distribution becomes more dispersed once again.

POPULATION DYNAMICS AND GROWTH

■ **Populations change over time because of births, deaths, immigration, and emigration.**

Zero population growth has been put forth as the goal humans should strive for in order to avoid the problem of overpopulation. Simply stated, zero population growth means that factors that increase the density of a population are balanced by factors that decrease the density of the population. The result is no net change in the population. *Births* are balanced by deaths; *immigration* (moving into a population) is balanced by *emigration* (moving out of a population). If any change in any one of these factors is not balanced by a change in another, the population will either increase or decrease.

It is logical that if more individuals die or leave a population than are born or move in, the population decreases. If more individuals are born than die, or more individuals enter a population than leave it, the population will increase. The difference between the number of individuals entering a population and those leaving over a particular time determines the *rate of population growth.*

Growth Rate Curves

■ **Growth rate curves reflect changes in the number of individuals in a population. There are two major types of growth rate curves: the exponential J-shaped curve and the logistic S-shaped curve. The carrying capacity of the environment is the population density that a particular habitat or environment can support.**

Changes in population density are usually indicated in growth rate curves. Growth rate curves can help predict future population size. Let us discuss these curves in more detail.

Exponential Growth Curve: The J-Shaped Curve. If you study the dynamics of natural populations and plot their increases in size over time under ideal conditions, a characteristic growth pattern emerges. Ideal conditions assume that the members of a population reproduce to their full **biotic potential,** unhindered by shortages of food or raw materials and free of disease. When researchers project the biotic potential of any population, they are theorizing about that population's maximum growth rate. Assume that a man and a woman produce the maximum number of offspring possible during the woman's reproductive years, which fall between the ages of 13 and 45. Then suppose that all the offspring live, and all go on to produce their own maximum number of offspring. This generation also lives and continues the pattern. If this pattern continues for 100 years, scientists have determined that (at least theoretically) one man and one woman could be responsible for the births of over 200,000 individuals!

You must admit that such a figure is rather astounding. Growth like this is called **exponential growth.** In exponential growth, the number of individuals increases by doubling, going from 2 to 4, to 8, to 16, to 32, to 64, and so on. Exponential growth increases slowly at first, then quickly gathers momentum, and finally bursts forth like an explosion. Theoretically, exponential growth continues to infinity. When plotted against time, exponential growth produces a **J-shaped curve.**

Scientists can study growth rates more easily in single-celled organisms than in human beings, simply because single-celled organisms reproduce in a very short period of time. Under specific conditions, a population of *Escherichia coli* doubles in size every 30 minutes, compared with the human growth rate, which doubles the population every 39 years. If you place one *E. coli* bacterium in a culture flask containing unlimited resources and space, within 30 minutes you have two bacteria. Within 1 hour you have four *E. coli,* and within 1½ hours there will be eight. After 5 hours, there will have been 10 doublings — or 1024 individuals. In 10 hours there would be over a million (Figure 35.5). Of course, this assumes that none of the bacteria die or leave the population.

No matter what the growth rate — a doubling over 30 minutes or over 39 years — the final shape of the curve for an exponentially growing population is the same: It looks like a J. In reality, though,

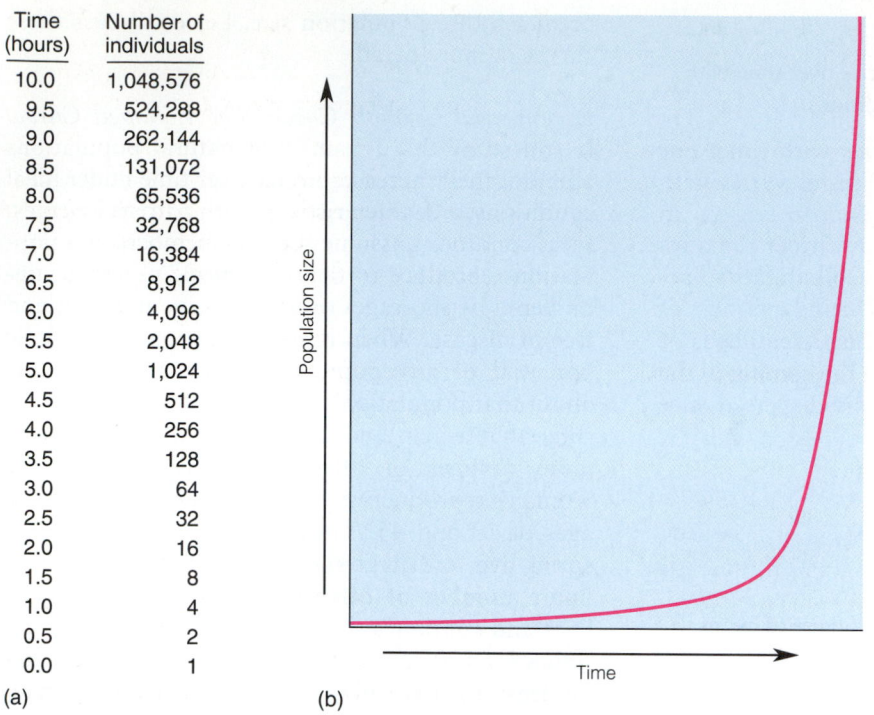

Time (hours)	Number of individuals
10.0	1,048,576
9.5	524,288
9.0	262,144
8.5	131,072
8.0	65,536
7.5	32,768
7.0	16,384
6.5	8,912
6.0	4,096
5.5	2,048
5.0	1,024
4.5	512
4.0	256
3.5	128
3.0	64
2.5	32
2.0	16
1.5	8
1.0	4
0.5	2
0.0	1

(a)

(b)

FIGURE 35.5 Exponential Growth of Populations
(a) The figures show how the number of individuals would increase in a population of *E. coli* if unchecked. (b) When plotted on a graph, the curve representing the number of individuals in the population over time is shaped like a J. The per capita growth rate remains the same, the population doubles each half hour, but the number of individuals added at each doubling increases dramatically.

a J-shaped curve does not happen. A population can reach infinity only in theory. What limits natural population growth? Let us take a closer look at a real growth curve in nature.

Logistic Growth Curve: The S-Shaped Curve. If unlimited growth occurred, we would be buried in cows, flies, or horses, or even humans. There are

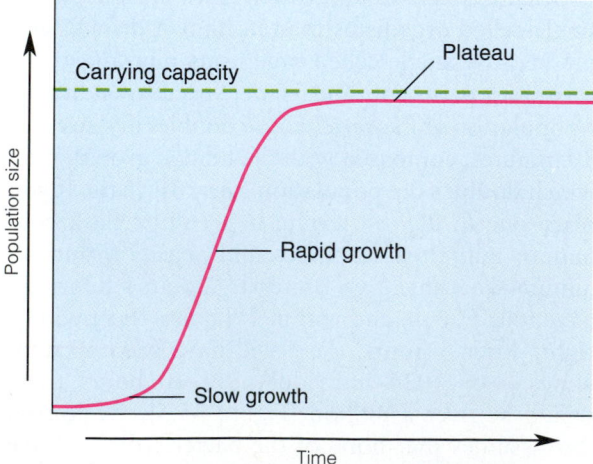

FIGURE 35.6 Logistic Growth of Populations
Until a population becomes large enough to strain the capacity of the environment that supports it, the population will grow exponentially. As limiting factors begin to come into play, the population's growth slows, eventually reaching a plateau.

checks and balances in nature that prevent the continued acceleration of the exponential J-shaped growth curve. Because of a system of checks and balances, population growth in nature oscillates, producing what is called a **logistic growth** rate; it produces an **S-shaped curve** (Figure 35.6). What does an S-shaped curve reflect?

In nature, a population will grow slowly at first (because there are few breeding individuals) and then pick up speed. As exponential growth continues, population size grows more rapidly. At some point, however, the population becomes large enough to strain the resources of its environment. Limiting factors such as decreased nutrients, accumulated toxins, and decreased living space begin to slow the growth rate. The curve of the growth rate becomes less and less steep, finally reaching a plateau. The population density at which the curve flattens out or where the growth rate is constant represents the population at growth equilibrium — there is a balance between births and deaths.

A population at equilibrium has reached a particular density; the environment is supporting as many organisms as it can. **Carrying capacity** is determined by both biotic and abiotic components of the environment. If the environment cannot meet the needs of a certain population density, the population begins to decrease. If the carrying capacity of the environment increases, population size also increases.

Though a logistic curve is called S-shaped, in reality it does not level off smoothly. Instead it

858

oscillates, producing a series of peaks and valleys as population growth and population density rise above and fall below carrying capacity.

Density-Dependent Mechanisms

■ **Density-dependent mechanisms are selective pressures that vary in intensity as a result of the density of individuals in a population.**

All mechanisms that regulate population growth are **density-dependent mechanisms.** For example, as population density increases, so does competition for food. As the amount of food available to each individual declines, the death rate rises because of starvation, until the death rate exceeds the birth rate. The population decrease is reflected in a downward curve in the growth rate. A population decrease in turn means that more food is available to the survivors. With a greater amount of food, the rate of reproduction increases, and the population increases again. Figure 35.7 illustrates the fluctuating curve resulting from this interaction.

An example of density-dependent growth is a crowded row of radishes. As the individual plants compete for sunlight, water, and available nutrients, some survive and others die. As some die, more resources are available to the survivors, and the curve rises again.

Density-Independent Mechanisms

■ **Density-independent factors affecting population growth are factors not related to population density.**

Sometimes the population crashes — there is a sudden, drastic drop in a population's size before it reaches a new equilibrium with the environment (Figure 35.8a). For many short-lived animals (the housefly, for one) a sudden cold spell in early fall will cause a population crash even though the population is below the environment's carrying capacity. In this case, abiotic factors not dependent on density bring about the high mortality. Obviously, a cold spell is not dependent on the density of the population. Elements of this type are called **density-independent mechanisms.** Density-independent factors are usually catastrophic and include such things as floods, extremes in the weather, and fire (Figure 35.8b).

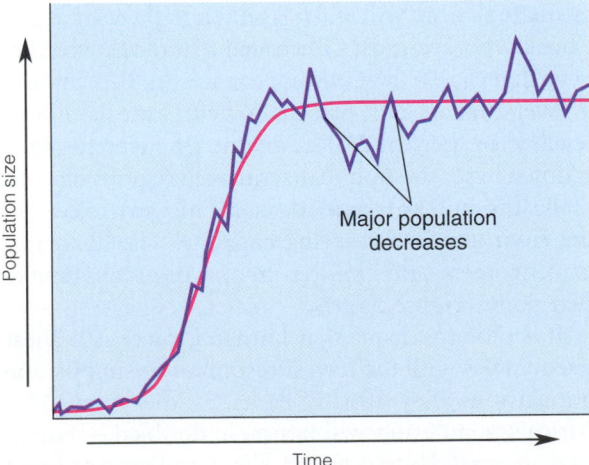

FIGURE 35.7 Density-Dependent Mechanisms
When all other environmental conditions remain relatively constant, the size of a population near carrying capacity can change because of conditions brought on by the density of the population's individuals in the habitat. For example, as populations become denser, competition for food becomes more intense. Notice that the curve of the graph generally follows the logistic growth curve and then begins to fluctuate around the plateau.

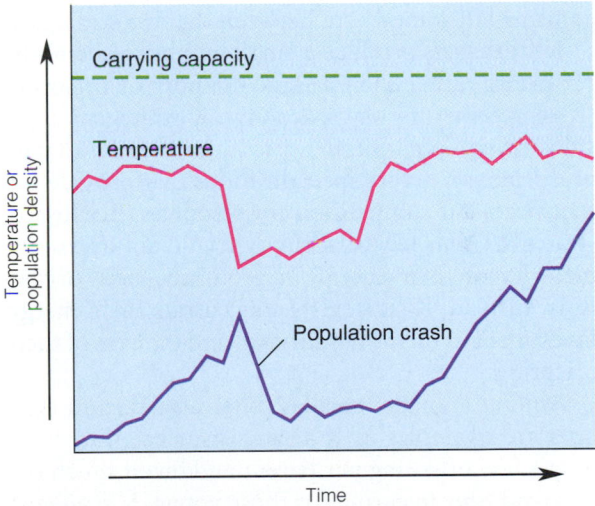

(a)

(b)

FIGURE 35.8 Density-Independent Mechanisms
(a) Changes in the environment can affect populations in ways unrelated to its density. (b) A prolonged freeze in an orange grove might kill a particular portion of trees whatever the density of the population.

Theoretically, it is possible for a population to reach an unlimited size, expressed by a J-shaped exponential growth curve. In nature, however, unlimited population growth is impossible, because resources are limited. Growth rates are, in fact, limited by both density-dependent and density-independent factors in the environment. Graphically, actual population growth rates result in an S-shaped logistic curve.

K and r Strategists

■ **K strategists produce a small number of better-fit offspring; r strategists produce a large number of offspring.**

Through the course of evolution, different adaptations among organisms have brought about "strategies" that help to ensure the survival of individuals within their environment. There are two extreme types of reproductive strategies, the *K strategy* and the *r strategy*. Both refer to the amount of energy parents invest in their offspring. In nature, most organisms fall somewhere between the two strategies.

K strategists produce a small number of better-fit offspring, rather than a large quantity of offspring (K represents carrying capacity). K strategists typically have population densities at or near K, so natural selection favors specializations that provide for the successful competition for resources. Because K strategists bear few offspring, they do not invest the majority of their energy in producing new organisms. Instead, K strategists concentrate their energy investment in their own survival and the care of their offspring.

Among animals, gorillas, blue whales, and condors are examples of K strategists, since they have only a few offspring per parent and invest much energy and time in caring for those young. K strategist plants are characterized by the evolution of seed distribution methods and the maintenance of energy reserves that greatly increase their survivability and their ability to germinate. Trees have "K-selected" characteristics in constant woodland habitats. They live a long time, do not reproduce until mature (maturity requires a long number of years in many trees), have large seeds, grow to a larger size, and reproduce several times during their life span. In small animals and herbs, we see more of the "r-selected" features, especially in those plants that grow in disturbed, unpredictable habitats.

R strategists exhibit other reproductive adaptations — they have a high rate (r) of reproduction (here r represents population growth rate). R strate-

gists ensure their survival by investing more energy in reproduction than in parental care. Because r strategists usually are small, have short life spans, and generally do not protect their young, these populations are more easily influenced by large-scale changes in density-independent factors such as frost, fire, drought, or other catastrophic events. For example, a fruit fly (unlike, say, a wolf) does not cuddle its young to protect them from a sudden temperature drop. Any density-independent factor can reduce a population of r strategists in one fell swoop. Thus, they are usually at densities well below K. Their growth rate curves tend to follow a "boom-and-bust" pattern.

Insects are typical r strategists. These organisms are rarely at densities near K, typically have short life spans, and produce large numbers of small offspring. In a sense, one can say that K strategists maximize the *quality* of offspring, while r strategists maximize the competitive *quantity* of offspring.

Human Population Growth

When people hear that the human population growth rate has started to drop from approximately 2.1 percent in the late 1960s to 1.7 percent today, most believe that the human population bomb has been defused. Unfortunately, these figures hide the fact that human population continues to grow very rapidly. In 1993, our population was estimated to be 5.6 billion. Estimates in 1989 stated that by the year 2023 the population will reach 8 billion. Today, we estimate that we will reach 8 billion in the year 2020. Those who are most concerned are those who believe that the earth's carrying capacity is 10.2 billion, a level that at our present growth rate could be reached by the year 2100. Anyone who watches television newscasts soon realizes that in regions like Somalia the human population has already exceeded the environment's carrying capacity. The deaths of men, women, and children by the thousands make that point (Figure 35.9).

It is perhaps ironic that human growth is highest in countries with the fewest resources to support the populations they already have. By the year 2025, Africa's population will not have doubled — it will have increased by 3.5 times! The population of Latin America will have doubled. In the United States, on the other hand, the population will grow only about a third by 2025. Solutions to the human population bomb are available. Because of the political forces in China that limit population growth to one child per family, growth is being restrained. But in Africa, population growth is increasing at a runaway speed

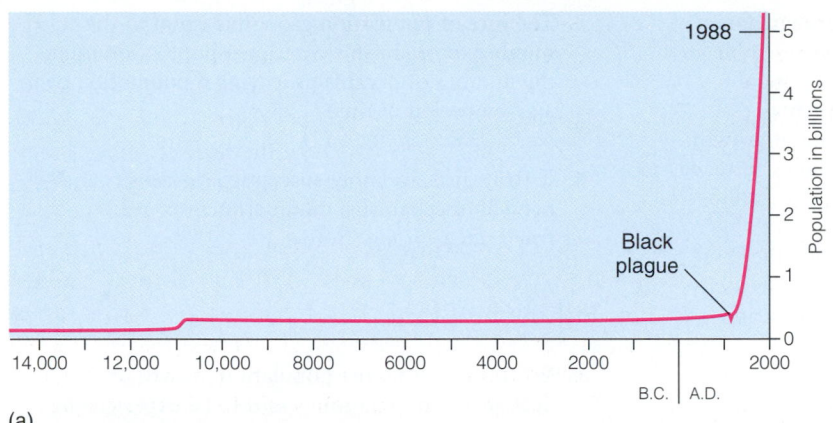

(a)

FIGURE 35.9 Human Population
(a) Since the Middle Ages, the size of the human population has increased exteremely rapidly. The graph reflects a steep logarithmic growth curve. As we have seen, in nature these curves approach a carrying capacity and reach a plateau. The environmental, social, and political consequences of reaching that plateau for our species are impossible to predict. (b) This crowded street scene in Beijing, China, illustrates overpopulation.

(b)

and is proportional to environmental degradation. Why do cultures fail to take advantage of possible solutions to the population problem?

Although it is true that birth control technology is available to regulate population size, in many cultures the social, cultural, and religious atmosphere is not conducive to accepting that technology. For instance, if a culture values life so much that it bars eating eggs because they may have been fertilized, it is not going to adopt practices that might limit life-producing activities in its people. Agriculture-based populations and cultures may continue the "tradition" of having many children to provide a labor force.

What we all must realize is that if we do not regulate our own growth, nature will. Once we reach or overshoot the environment's carrying capacity, the inevitable result will be a population crash that will make present problems in certain countries look like

a picnic. The human population bomb is still ticking, whether or not we want to listen to the sound.

SUMMARY

1. A population has characteristics not found in its individuals: size, growth, density, and biotic potential.
2. Population density is the number of individuals per unit area.
3. Individuals within a population are grouped in several ways. Some are grouped uniformly within the environment, others are clumped, and still others are grouped at random. The grouping of organisms depends on the availability of resources within the environment.
4. The J-shaped curve is an exponential growth curve that assumes that the population will reach its full biotic potential under ideal conditions.

5. Because of environmental factors that regulate population size, natural growth curves resemble an S-shaped curve, or logistic curve. Exponential growth occurs until limiting factors such as decreased resources slow down the rate of growth and the population growth curve becomes flattened. This plateau represents the population at equilibrium within the environment.

6. A population at equilibrium has reached the environment's carrying capacity. The carrying capacity of the environment is dependent on abiotic and biotic components.

7. Density-dependent mechanisms are regulated by population density. For example, as population density increases, available nutrients decrease, causing the population to decrease. Density-independent mechanisms are not regulated by population density. For example, a flood or drought can affect population density.

8. The organisms whose population curve is S-shaped produce fewer offspring that are better adapted to the environment. These organisms are called K strategists. Organisms like insects, whose growth curves follow a boom-and-bust pattern, are r strategists. They ensure their survival by producing large numbers of small offspring.

9. Human population growth is very rapid, and it is estimated that the human population will reach 8 billion by the year 2020. We as humans have to be concerned about our population growth relevant to the carrying capacity of the earth.

KEY TERMS

population density	S-shaped curve
uniform distribution	carrying capacity
clumped distribution	density-dependent
random distribution	mechanisms
biotic potential	density-independent
exponential growth	mechanisms
J-shaped curve	K strategist
logistic growth	r strategist

REVIEW QUESTIONS

True/False

1. Population is defined as all the communities living in a particular area.
   true false (page 853)

2. Both uniform and random distribution occur when resources are spread evenly throughout an environment.
   true false (page 854)

3. The rate of population growth is equal to the number of organisms entering a population minus the number of organisms leaving a population over a given period of time.
   true false (page 857)

4. K strategists are more susceptible to density-independent mechanisms than r strategists are.
   true false (page 860)

Fill in the Blank

1. When there is no net population growth or decline, the population is said to be experiencing _____ population growth. *(page 857)*

2. In nature, organisms usually follow the population growth pattern known as the _____-shaped curve. *(page 858)*

3. K reproductive strategists typically produce _____ numbers of offspring than r strategists. *(page 860)*

4. Density-independent factors affecting population growth are not related to the _____ of the population. *(page 859)*

Multiple Choice

1. Population density is of more value to ecologists than population size because:
   a. It tells them more about the population and its interaction with the environment
   b. It tells them how many organisms are being born
   c. It can be used to predict the amount of immigration
   d. It can be used to determine the age of the population
   (page 854)

2. In a stable ecosystem, which of the following factors are important in maintaining population size?
   a. Competition
   b. Predation
   c. Limited resources
   d. All of the above
   (page 859)

3. Social groupings provide:
   a. Protection from predation
   b. Increased parental care
   c. Both a and b
   d. Neither a nor b
   (page 855)

4. Human population growth:
   a. Is growing very rapidly
   b. Will strain the resources of the environment
   c. Both a and b
   d. Neither a nor b
   (page 861)

Discussion

1. Why is the study of populations important to an understanding of ecology?

2. Discuss populations, communities, ecosystems, and the biosphere.

3. Discuss some of the factors that limit population growth. Be sure to include both density-dependent and density-independent factors.

4. At our present growth rate, how long will it be before the human population doubles? Discuss some solutions to the problem of increasing human population.

5. Compare and contrast J- and S-shaped population growth curves.

6. What is the selective advantage to organisms that are K strategists?

7. From an ecological perspective, why can it be said that predator-prey relationships are reciprocal?

36

Communities

Competitive Interaction
Ecological Succession
Summary

A **community** is a collection of populations living in a specifically defined area. For example, all the organisms in a specific pond — the fishes, algae, microorganisms, plants, amphibians, and so on — form a community (Figure 36.1). As you would suspect, these populations all interact with each other. There are direct and indirect interactions within the populations making up the community. Each population within the environment is adapted for living within that community as long as the given conditions within that environment remain within certain limits.

The place where an organism lives is its **habitat.** As we mentioned earlier (Chapter 34), there are many different types of habitats, each with its own abiotic (nonliving) and biotic (living) characteristics. When studying habitats and communities, several

Above: The organisms living within this coral reef provide an example of a collection of populations in a community.

questions come to mind. What determines how many species share a habitat? How are species distributed within the habitat? What determines how many individuals of a given species will live in a given habitat? Let us try to answer some of the questions by listing factors that affect the habitats within communities. Habitats are influenced by

1. Climatic conditions such as rainfall, soil conditions, temperature, and other abiotic characteristics of the environment.

2. The availability of nutrients throughout the year.

3. The ability of organisms within the habitat to adapt to changing abiotic and biotic conditions.

4. Interactions between the species within a habitat such as competition, predator-prey relationships, and the sharing of resources, just to mention a few.

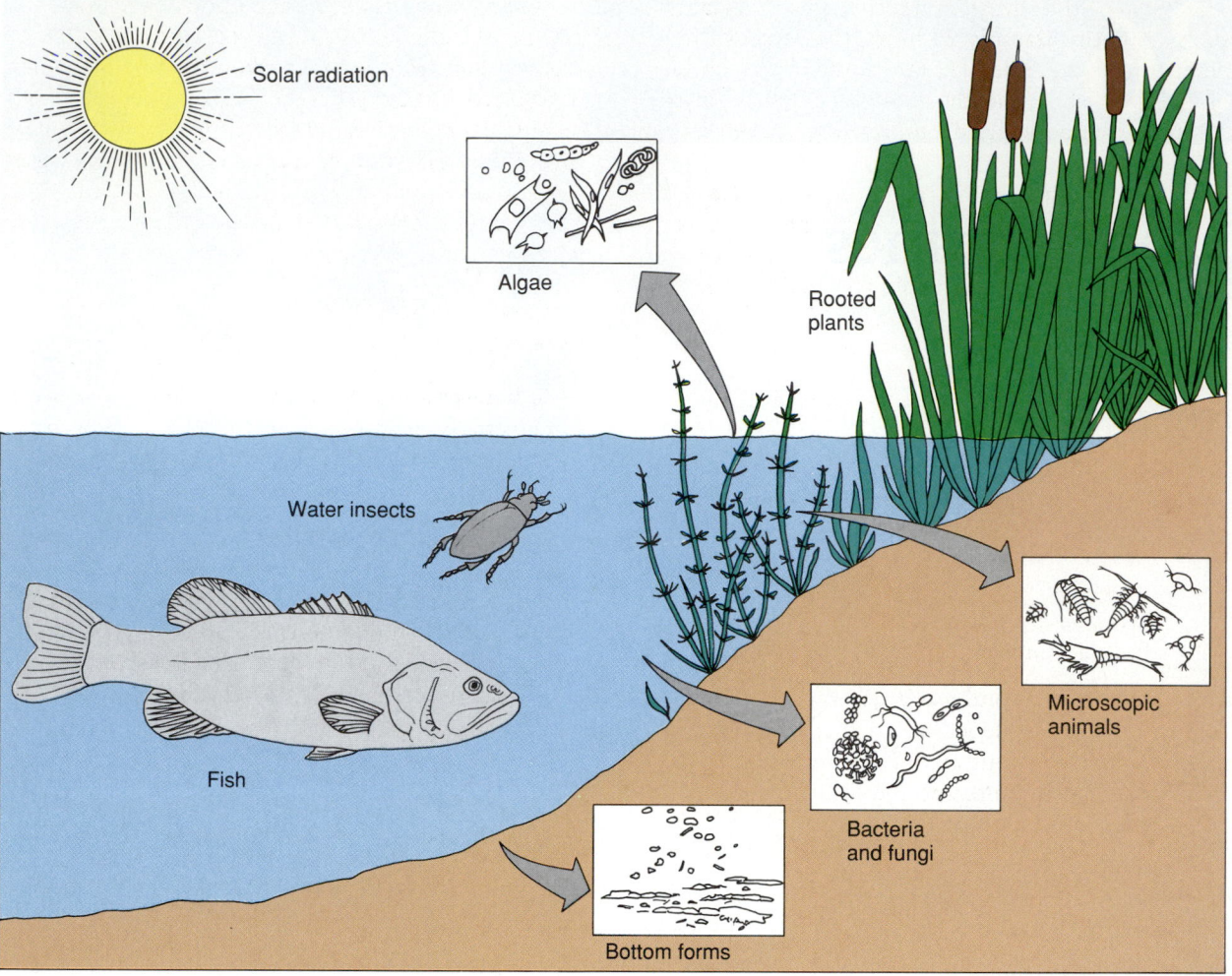

FIGURE 36.1 A Pond Ecosystem
The organisms living in a pond may vary greatly in size, but their lives are
all interrelated.

5. Long-term and short-term patterns of disturbance, such as immigration and extinction
patterns.

Let us examine these factors and their contributions in more detail so that we can understand the
properties of communities. For example, these factors can determine the "richness" of a community,
that is, the number of species present. When I lived
in Puerto Rico, the environmental conditions of high
humidity, warm temperature, and other factors of
the tropical environment led to a rich variety of
plants and other organisms (Figure 36.2a). When I
lived in the high mountain regions of Colorado in almost Arctic conditions, I noticed that the community was less rich (Figure 36.2b). In addition to species

richness, the diversity of species in the community
and the number of individuals in each species are determined by the combination of these five factors.

But just knowing where an organism lives does
not tell us much. We need to know how an organism fits into its community; in other words, what
does it do? These questions are answered in part by
the concept of the **niche.** You may have heard niche
defined as the job or role an organism has in its community, but that definition is only partly true. A
niche is the sum total of the interactions of the organism and the environment — the full range of
abiotic and biotic conditions under which it can live
and reproduce. For example, I live in the chaparral
biome of southern California, and as I write this
chapter I am looking at a lizard basking on a rock.

The niche of that lizard consists of many variables, such as the availability of food, the range of temperatures it can tolerate, the number and species of insects it eats, and the number of other lizards of the same species and other living species in its environment.

But how do species interact with each other? Obviously, in many ways. They can *compete* with each other for available resources, the interaction can be *neutral* (neither species directly affects the other), or the interaction can be *beneficial*.

COMPETITIVE INTERACTION

■ There are two forms of competition: intraspecific competition (between individuals of the same species) and interspecific competition (between individuals of different species). Intraspecific competition is over resources and mates, and it increases as population density increases. Interspecific competition arises from niche overlap.

Biotic factors that influence communities usually function as density-dependent regulators. There are many such factors, but most fall into the categories of intraspecific and interspecific competition.

When we discussed the evolution of organisms by natural selection, we said an organism's ability to compete successfully for the requirements of life and reproduction assures its survival. When organisms of the same species compete among themselves for mates and available resources, this is known as **intraspecific competition**. Competition of individuals of different species is called **interspecific competition**.

Intraspecific Competition

■ Intraspecific competition occurs between members of the same species. This type of competition is intense, because members of the same species have similar needs.

Since individuals of the same species have more in common than individuals of different species, their needs are more similar and the competition between them is more intense. The competition between robins and sparrows for nesting space, for instance, is not as great as between two birds of the same species. As population density increases, the intraspecific competition for any limited resource also increases. Because of genetic variation, some organisms

(a)   (b)

FIGURE 36.2 Richness of Communities
The number of species present within a community depends upon several environmental factors. (a) Because the environmental conditions of the rain forest are favorable to many organisms, species diversity is high. (b) Within the high mountain regions of Colorado, species diversity is less.

cannot compete as successfully as others of their species, and this can limit population growth.

Competition for limited resources between plants helps to limit their size, density, and distribution. For example, if tomato seeds are planted too close to one another, there is intense competition for a limited amount of resources. Because of their genetic makeup, some of the plants will not be as competitive and may die. Those whose genetic makeup enables them to live in the dense environment will survive, although they will be small and spindly and will produce very small fruits. Their survival chances will be enhanced by the fact that the others died; the decrease in population density will have left more resources for the survivors.

Territoriality

- **The ability of an individual to take and hold a territory acts as a selective pressure, assuring that the genes of the most fit are passed on within the population.**

A **territory** is an area defended by an individual in order to reserve resources for itself. Establishing a territory is usually an example of intraspecific competition. As we said in Chapter 33, a territory is the particular area that an animal occupies and defends against intruders. When an animal establishes a territory, it reserves the resources within that territory for its own use. In locations with limited resources, a territory helps assure that the individual will survive, mate, and produce offspring. In a sense, an individual's ability to establish a territory signals other animals that the individual is a "genetic winner" — that is, a healthy and competitively successful organism.

In the course of the evolution of some species, the genetic attractiveness of an individual who has established a territory exerts selective pressure on mating. In some bird species, like the Bower bird, the male's splendid coloring and courtship rituals are not enough to attract a female. He also must have an attractive territory. In this particular species, once the male has built his nest, he goes through a process much like "exterior decorating." He spends hours gathering feathers, pebbles, stones, and other objects and arranging them around his nest. Perhaps his ability to move freely within his territory and gather his ornaments is an indication of his competitive ability. The female responds to the nesting area as well as to the nest itself and to courtship rituals.

Dominance Hierarchies

- **In some animals there is a selective pressure to form a dominance hierarchy because the dominant, most-fit individuals have first access to resources and mates. Less dominant individuals may not survive, but the most dominant do, thus maintaining a strong gene pool.**

A **dominance hierarchy** is another form of intraspecific competition. A hierarchy provides a mechanism by which a society can function. It is a mechanism of population regulation. Once a hierarchy is established, the dominant individual has first access to both resources and mates. Subordinate individuals, usually those that are less well adapted to their environment (genetically weaker) or sexually immature, must wait (Figure 36.3). If resources become scarce, subordinates are the first to die. Their genes are removed from the gene pool, leaving a concentration of genes from the strongest and most competitive individuals.

Interspecific Competition

- **Organisms of different species also compete for the environment's limited resources. The intensity of interspecific competition is greatest among those organisms that share a similar niche.**

In a pond habitat, there are different types of organisms, such as algae, bluegill sunfish, frogs, insects, water lilies, bacteria, and worms. When many species share the same habitat, each has its own

FIGURE 36.3 Dominance Hierarchies
Wolf packs are organized around dominance hierarchies. In the photograph, the wolf lying on the ground occupies a lower status than the wolf standing above it. Lowered ears, submissive posture, and exposure of vital areas are commonly observed when a subordinate wolf interacts with a wolf of higher status.

niche — its own strategy for obtaining its life requirements. Sometimes these niches overlap. When this happens, the species involved will compete for the available resources. For instance, both bluegill sunfish and small-mouth bass may compete for the same kind of insects. The more that niches overlap, the greater the interspecific competition.

A Russian biologist, G. F. Gause, was the first to describe the concept of increased competition that arises with niche overlap. Gause performed a simple but ingenious experiment. First, he grew two different species of paramecia, *Paramecium aurelia* and *Paramecium caudatum*. He chose these two because they competed for the same resources in the same habitat. Each species did well in its isolated environment. Then Gause mixed them together, and the *P. caudatum* rapidly died out. He concluded that no two species can occupy the same niche for an extended time. One species will be competitively excluded. This is called Gause's **competitive exclusion principle** (Figure 36.4).

In reality, the principle of competitive exclusion describes a condition that will never exist, because no two species could completely occupy the same niche at the same time. If they did, the species would have to be identical in every way and would therefore be one species, not two. In nature, however, when two or more species occupy a similar niche, several adaptations subdivide the niche so that the species can coexist.

An example of how a niche is subdivided was studied by the American ecologist R. H. MacArthur. MacArthur studied five species of North American warblers, all of which eat insects and, in many instances, in the same fir tree. Why didn't the different species competitively exclude one another? MacArthur determined that even though the warblers are specialized for a particular resource, they have different feeding zones on the tree. That means they exploit slightly different resources. The subdivision of the niche enables the five species to coexist (Figure 36.5).

Predator-Prey Interaction

- **Predators survive by eating other organisms, their prey. But the relationship between a predator and its prey is reciprocal, resulting in the S-shaped logistic curve.**

A **predator-prey interaction** is a form of interspecific competition. A predator is an organism that eats other living organisms; so it is obvious that predators have a definite effect on the numbers of their prey. But the relationship between a predator and its prey is reciprocal. If the number of prey diminishes greatly, predators cannot survive. In fact, predation

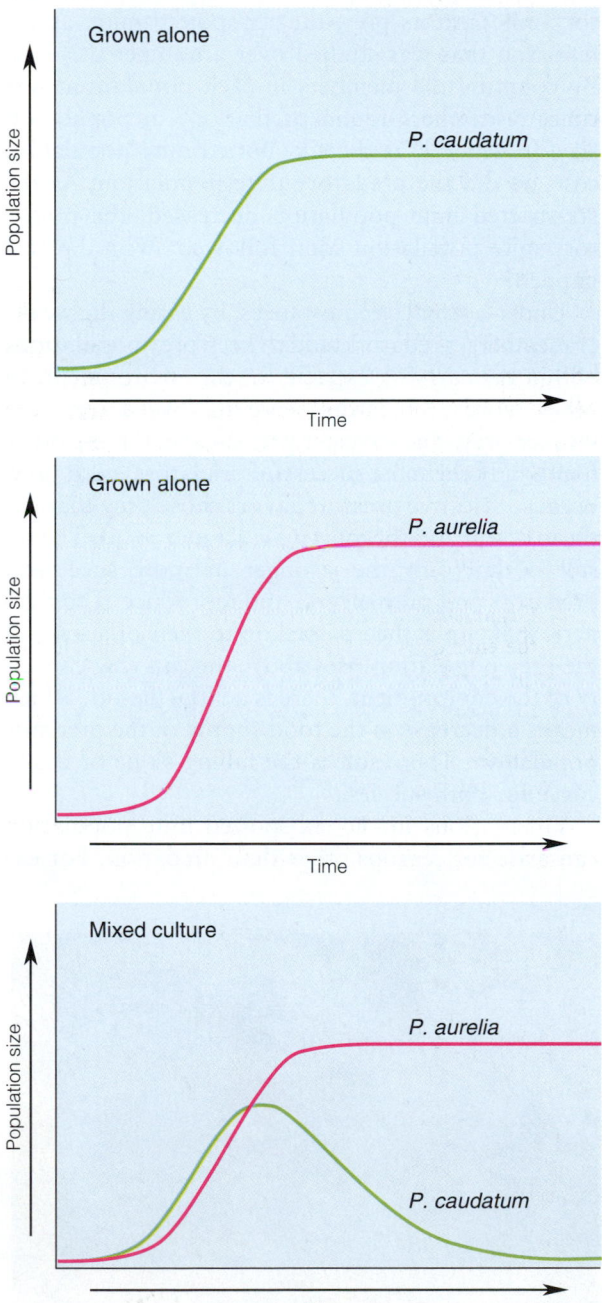

FIGURE 36.4   Competitive Exclusion
In Gause's experiments, each species of paramecium followed a typical logistic growth curve when grown alone. But when the two species were forced to exist together, *P. aurelia* drove the population of *P. caudatum* into extinction.

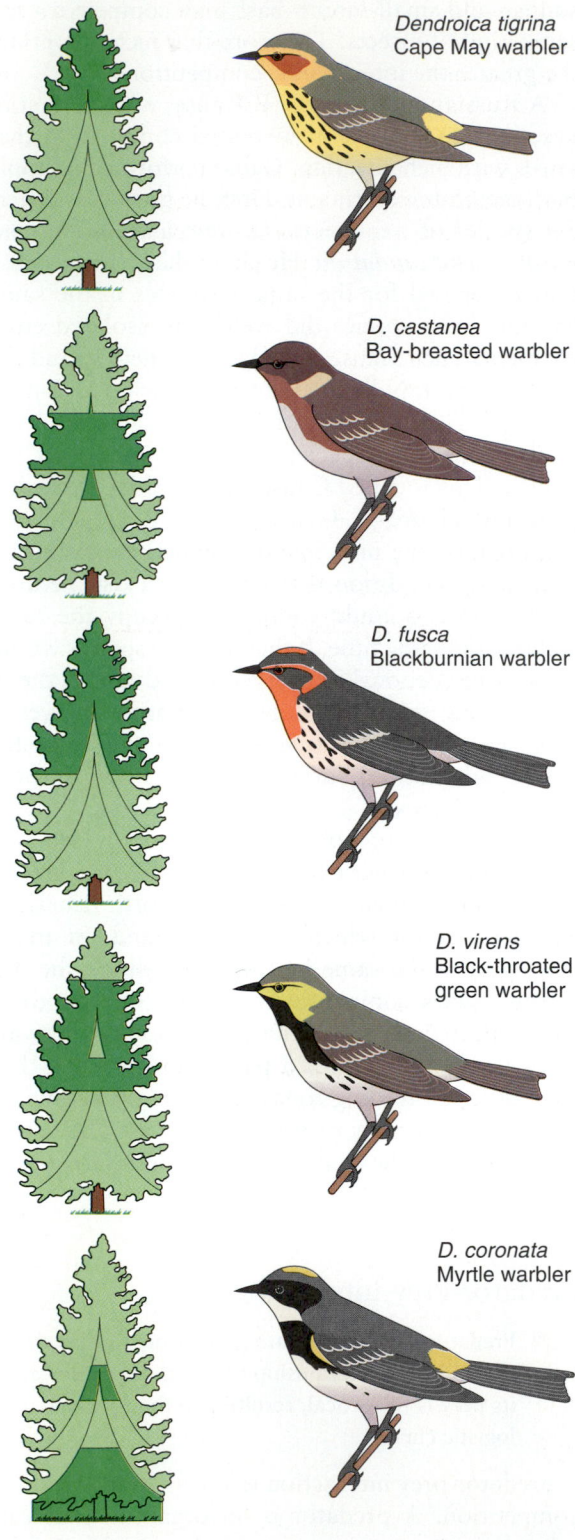

FIGURE 36.5 **Separation of Niches**
While the food resources of these five warbler species seem to be the same, each species spends more than half its time feeding within a different zone on spruce trees. Each tree is, in effect, divided into five different niches.

Dendroica tigrina
Cape May warbler

D. castanea
Bay-breasted warbler

D. fusca
Blackburnian warbler

D. virens
Black-throated
green warbler

D. coronata
Myrtle warbler

is an important density-dependent factor in most species. (Of course, a predator cannot eliminate all of its prey. If it did, it too would ultimately be eliminated.)

A classic example of a predator-prey relationship are the fluctuations in the populations of the predatory mite and its prey, the six-spotted mite, an interaction that was studied over a number of years. By counting the members in each population over time, researchers found oscillations in population size. In general, as the six-spotted mite population rose, so did the predatory mite population. As the six-spotted mite population decreased, the predatory mite population soon followed. Why did this happen?

Under normal circumstances, by eating the weaker members, predators tend to keep prey populations within the carrying capacity of the environment. In other words, predators serve to "weed out" the weaker prey. The stronger prey escape. On the other hand, only the most successful predators catch prey, because selective pressure favors those prey that are the quickest and the most agile (Figure 36.6). The result is that only the stronger or more successful predators and prey survive and reproduce. Their genetic makeup is then passed on to their offspring. If the prey population rises above the carrying capacity of the environment, there is a large die-off, which means a decrease in the food supply of the predator population. The result is the falling slope of the S-curve for both species.

Fluctuations in the six-spotted mite population can arise for reasons other than predation. For ex-

FIGURE 36.6 **Predator and Prey**
Predators keep the poulations of many insect pests in check. The ladybird beetles are among the most voracious predators on earth, consuming countless numbers of aphids and thus helping to keep these serious pests in check. The photo shows a ladybird beetle preying on aphids.

ample, a decrease in the amount of food or other requirements can cause the six-spotted mite population to decrease. That event is, in turn, reflected in a
decrease in the predatory mite population. In all
probability, predator-prey interaction is the major
cause of population oscillations, but other factors
can influence such trends. Other biotic characteristics that affect population size are parasitism, disease, and physiological factors.

Parasitism and Other Physiological Factors

■ **Parasitism and disease can serve as population-
limiting factors. Coevolution and its mechanisms
tend to balance predator-prey interactions.**

Parasitism is a form of **symbiosis,** a relationship between organisms that live in close association. Some
ecologists regard **parasitism** as a special form of predation. Unlike more visible predators, such as a lion,
a parasite usually lives either within its prey or on its
surface. As with more conventional predator-prey
relationships, it is not advantageous for the parasite
to destroy its host, since by doing so it would also
destroy the source of the requirements for maintaining its life. Natural selection functions on parasite
species in complex ways. Mechanisms select for parasites that either are not fatal to their hosts or that
can easily move to other hosts if the original host
dies. Coevolution selects for hosts that are resistant
to the parasite.

Parasites regulate population density because, in
some instances, a parasitic infection can lead to a
weakened condition or to a disease in the host. This
in turn makes the host more susceptible to predation. During periods of environmental stress, disease and parasitism work together to eliminate the
less strong. When a person has a tapeworm, he or
she rarely dies directly from the parasitic infection.
Even filarial roundworm parasites, which cause the
limb disfigurement associated with elephantiasis,
do not, in most cases, destroy their human host.
However, these parasites do weaken their hosts,
leaving them more susceptible to death by other
diseases.

Parasitism and disease are density-dependent
population-regulating mechanisms. Increased population density leads to a closer proximity of individuals within the population. Proximity aids in the
distribution of diseases and/or parasites. Have you
ever ridden in a crowded subway car in winter? You
are well aware that this close proximity aids in the
spread of diseases like the common cold.

Physiological characteristics of organisms can
also regulate population density. Although biologists do not yet understand the mechanisms involved, it seems that increased population density
also increases an organism's susceptibility to disease.
For some reason, the immune system changes. Animal studies have also shown that the stress of overcrowding gives rise to hormonal fluctuations that
modify behavior and reproduction.

Predator-Prey Adaptations

■ **The interactions of predators with prey lead to
diverse adaptations such as cryptic coloration and
mimicry.**

Obviously, not all predators are successful because
continual selection pressures have resulted in devices
to protect prey against attack, while predators have
evolved in ways that enhance capturing and consuming of prey. This coevolution is continuous. As
we said earlier, coevolution is the evolution of two
or more species that share a close ecological relationship. Several mechanisms of prey defense or
predator enhancement such as cryptic coloration,
mimicry, and mechanical and chemical devices have
evolved.

Cryptic coloration refers to camouflage, which
many animals use as a passive defense to avoid being
consumed by their predator (Figure 36.7). Not only
is color similarity useful in this type of defense, but
in some cases, the organism takes on the shape of
elements in the background, such as the praying
mantis or walking stick that resembles a twig, or
even bats that look like thorns (see page 10).

FIGURE 36.7 Cryptic Coloration
The grasshopper on this leaf blends very well with its
background, a prime example of cryptic coloration.

Dr. Ariel Lugo
Tropical Ecologist

Like many undergraduate students with an interest in biology, Ariel Lugo was a pre-med major and hoped to become a physician. He had studied hard at the University of Puerto Rico, San Juan and had done well. But the summer before he was to apply for medical school, he took a job with Howard T. Odum, an ecologist of national reputation. Ariel's job was to help with a study of the effect of atomic irradiation on a tropical ecosystem. He spent the entire summer working on that project and became increasingly aware that his interest was really in the areas of tropical ecosystems and not medicine. As he prepared for his medical school interviews, Ariel Lugo began to have doubts about his real dedication to medicine, and instead decided to pursue a Ph.D. in ecology at the University of North Carolina. After earning his degree, Dr. Lugo decided to take a position at the University of Florida as a biology professor.

Like many Puerto Ricans who move to the mainland, his heart remained in Puerto Rico, and Ariel looked for an opportunity to return home. Because of his research and growing expertise in tropical ecosystems, he was offered a position with the United States Department of Agriculture (USDA) in Puerto Rico. His lab was located in the only tropical forest within the United States, the Caribbean National Forest. Dr. Lugo's job was to become the Director of the Luquillo experimental forest, El Yunque. El Yunque is located in the northeastern quadrant of the island and is a rich piece of tropical rain forest. El Yunque is not only a special place with respect to the study of tropical rain forest ecosystems but also has a special place in the hearts of the people of Puerto Rico. El Yunque is popularly esteemed as a sacred place and a rich natural treasure.

Dr. Lugo is considered by many as an ecologist whose dedication to the preservation of the rain forest is tempered with reason. Many individuals who have concerns about saving our natural resources fail to take into account the economic and social consequences of strict limitations on uses of rain forest resources. For example, millions of people live within tropical rain forests; and they of course have their own perspective on what uses should be made of the resources in their habitat. Their very lives depend on the rain forest ecosystem. Often ecologists and outsiders want the rain forest to be left alone because we have scientific interests or strong beliefs about the preservation of the rain forest. The natives, on the other hand, do not want to destroy the rain forest, but they do want to improve the quality of their lives.

Contrary to popular belief, the major reason for deforestation is not because of increasingly ambitious logging companies but, rather, the need to expand agriculture. More than 50 percent of the deforestation is because nations need more land to grow more food. There are other reasons for deforestation, such as the need for lumber, fuel, and room for expansion in an increasingly crowded country to name a few. But

as Dr. Lugo states, "Even though there is a real concern about the 17 million hectares of tropical rain forest that we are losing every day, we must use reason as we try to address the needs of the diverse groups interested in the tropical rain forest." In several of his articles, he declares that we cannot study ecosystems in a vacuum. We must be aware of the social consequences of our actions, and science has to be done in context. We as scientists have to explain to the people of the rain forests and the countries that own them how our investigations will improve their lives directly and, indirectly, the overall health of our global biosphere. We cannot just march into the rain forest as ecologists and force our priorities on the people who live there without concern for the social consequences of our activities on these people. Communication is the key.

(a)

(b)

FIGURE 36.8 Batesian Mimicry
(a) The red, white, and black markings of this poisonous coral snake are often mimicked by harmless snakes such as (b) the shovel-nosed snake.

Mimicry refers to mimicking or imitation. A predator may mimic another species that doesn't frighten the prey, or a prey species may mimic another species that doesn't stimulate the predator to attack it. In one type of mimicry, called **Batesian mimicry,** a palatable prey mimics an unpalatable model. For example, many harmless snakes mimic the color patterns of poisonous snakes such as the red, white, and black markings of the coral snake (Figure 36.8). In **Müllerian mimicry,** two or more unpalatable species advertise the model, and they mimic each other. For example, the cuckoo bee and the yellow jacket both produce a toxic sting, and they both gain an adaptive advantage because predators will learn to avoid anything that looks like either one of them.

Neutral Interaction

■ Not all interactions between organisms are predator-prey interactions. Some are neutral, and neither species affects the other.

The majority of species' interactions are neutral; neither species directly affects the other. For example,

hawks and pine trees have no direct effect on each other. Their effects are only indirect, such as the hawk producing CO_2 and the tree producing O_2.

Beneficial Interaction

There are two major types of beneficial interactions among species. In the first, **commensalism,** one species directly benefits from the other, but the other species is not harmed or helped. In the second, **mutualism,** both species benefit from the interaction. Let us first discuss mutualism.

In Focus 19.1, "Plants, Life's Sexy Creatures" (pages 460–461), we examined the coevolution of a specific plant with a specific pollinator, as exemplified by several excellent examples of mutualism. In the case of the yucca plant and the yucca moth, the sucking mouth parts of the female moth are curved so as to fit into the floral parts of the pollen-producing flowers. The moth then takes the balls of pollen to the flower of another yucca, where she lays her eggs in the base of the flower. The eggs are laid into that part of the flower that produces seeds. The seeds and the eggs develop together, and when the eggs hatch, the larvae eat through the seeds. Without the moth, the yucca would not be pollinated, and without the yucca, the moth young would not have a source of nutrition. The extinction of one would result in the extinction of the other. Another example of mutualism is found within our own intestines — the *E. coli* bacterium — which helps us produce vitamin K, which is essential for blood clotting. We, on the other hand, provide the bacterium with a warm environment and the nutrients required for its survival.

Commensalism is a type of beneficial interaction between two species that is sort of "one-sided." In other words, only one species benefits, but the other species is not helped or harmed. For example, as cattle graze, some birds become closely associated with them and follow them, because the grazing disturbs insects upon which the birds feed. The birds benefit from the grazing, but it is hard to see how the cattle benefit. Also, when birds nest in trees, the birds benefit, but the tree is usually neither harmed nor helped. So as you can see from these examples, commensalism benefits one but not the other. But if you think about it, it is hard to imagine that the tree does not benefit in some way. Maybe the birds eat parasites off the tree, or maybe their droppings fertilize the tree. Therefore, it is not very probable that absolute communalism exists, just as it is difficult to imagine how one species cannot be affected by its interactions with another.

ECOLOGICAL SUCCESSION

■ **Communities change over time in a pattern called ecological succession.**

The composition of communities varies over time in response to changes in energy flow, nutrients, temperature, soil condition, water resources, and so on. For example, a pond may gradually dry up, thereby reducing the number of aquatic plants or producers, which in turn reduces the number and types of consumers. The composition of the communities changes, and eventually, the pond becomes a grassland community. When a community changes and some organisms die off, the organisms that replace them do so in a manner called **ecological succession.** Although the specific communities involved and the rate of succession may differ, all maturing ecosystems have the following in common:

1. Their total biomass increases.

2. As the system matures, there is a decrease in the rate of production of biomass — in other words, fast-growing plants are replaced by plants that grow more slowly.

3. The more mature the system, the greater its capacity to hold nutrients.

4. In the first stages of succession, there is a greater diversity of species than is found during later stages of maturity. At later stages, there are fewer species, and those that are there are more stable.

There are two types of succession: *primary succession* and *secondary succession*. Both have orderly characteristics that are fairly well known.

Primary Succession

■ **Primary succession takes place in an area of virgin soil or a virgin aquatic environment. The first organisms to appear are called pioneer organisms, and they break up rocky ground and ultimately form soil in which other plants can grow.**

When islands are formed in the ocean or in lakes, or when there are newly formed lava flows, glaciers, or ponds, they eventually become occupied by living organisms. The orderly process involved is known as **primary succession;** that is, establishing communities in virgin soil or in aquatic environments where none existed before (Figure 36.9).

Even though succession occurs in both terrestrial and aquatic environments, we will limit our discussion to the terrestrial environment. Before primary succession can begin, the bare, rocky surface is subjected to expansion and contraction, resulting from daily changes in temperature. The cracking that results forms broken areas in which hardy species, capable of living in harsh environments, can catch hold. These species are aptly called **pioneer organisms.** Pioneers are characterized by wide tolerance of environmental stress and by high vagility (i.e., seed dispersal). In most cases, the first life forms to become established are lichens (organisms comprised of algae and fungi; see Chapter 19). These organisms produce acid secretions that break up the rocks upon which they live. As lichens die, their organic matter mixes with the broken rock to begin soil formation.

As organic matter continues to accumulate, lichens are followed by mosses, and seeds, blown in by winds or present in the droppings of birds, lodge in the developing soil. Those that survive have little competition, and their seeds establish more growth. The first true plants are usually those that send a single taproot deep into the rocks, breaking up the rocks' structure and utilizing its nutrients for growth. When such plants die, the nutrients they spread while decomposing contribute to a growing layer of soil.

The soil can soon accommodate plants with massed root structures that do not penetrate very deep. Larger plants or producers become established; they can support primary consumers, which, in turn, support secondary consumers. As a community matures, some species are better able to compete than others and become dominant. They remain dominant as long as environmental conditions remain constant.

Secondary Succession

■ **Secondary succession takes place when land or water that has been the site of previous growth is disturbed and repopulated by new organisms. In this type of succession, the environment returns to its natural ecological composition.**

Another type of ecological succession is **secondary succession,** in which a site previously occupied is disturbed and then reestablished. Secondary succession occurs, for example, when a farm is abandoned for a period of time. The original community was cleared away to make way for agriculture, but when the farm is abandoned, the original community will eventually reestablish itself. It takes much less time to reach a community in secondary succession than in primary succession, since in most cases the soil is already present. The length of time required depends upon the severity of the disturbance.

My yard in California is an excellent example of secondary succession. If I do not work on it con-

(a)

(b)

FIGURE 36.9 Primary Succession
(a) The ash and mud that issued from the massive eruption of Mount St. Helens created vast areas of virgin soil. (b) After a time, pioneer life forms began to take hold on the barren landscape. The pioneers in soil of this type are usually different from the lichens and mosses that first occupy barren rocks.

stantly, those plants native to southern California will begin to reappear practically overnight. If you have ever planted a backyard garden, you have had to deal with secondary succession every time you went out to weed the rows.

Ecologists once thought that succession was an orderly process that ultimately reached an end point or climax. Ecologists called this end point a climax community. However, if you think a moment, you will realize that communities continue to change and that a permanent final stage in most cases does not really exist. Many communities are not stable; they continue to change. Many communities are disturbed by fire, drought, and other environmental factors. For these reasons, many ecologists view the concept of the climax community as too simplistic to explain succession. In these examples of succession, we have discussed the results of the organisms in a community reacting to both abiotic and biotic factors in their environment. Therefore, communities are disturbed by outside factors during succession.

Human disturbance has a tremendous impact on community succession. In Chapter 38, we will discuss this in more detail. Logging, deforestation of tropical rain forests, and strip mining are but a few examples of obvious community disturbance. We also introduce species into different regions, either on purpose or by accident. Sometimes these introductions are useful, and other times they can result in disastrous consequences. For example, aggressive honeybees brought from Africa into South America were cross-bred with less aggressive honeybees. It was thought that this cross-breeding would result in a variety that would be a more aggressive pollinator; hence, more crops. However, the descendants were aggressive and have been migrating north into the United States. Not only are these bees aggressive about pollinating, but they are also aggressive about swarming and stinging and have killed many animals, including humans. So this introduction of a species from one environment into another has resulted in the "killer bees" that we read about in the news.

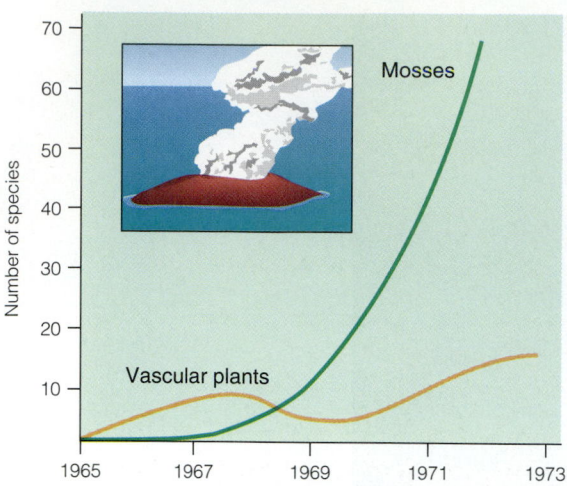

FIGURE 36.10   Species Diversity
The number of species recorded for the new volcanic island of Surtsey from 1965 to 1973.

Species Diversity

■ **Species diversity is affected by the rate of immigration and colonization of the new area as well as the rate of extinction of the species.**

What affects the number and kinds of species within a community? Island communities provide an excellent opportunity to answer this question. We have all heard about or seen volcanic eruptions in Hawaii. Similar eruptions around the earth can result in the formation of new islands, which act as natural laboratories for the study of species diversity. One of the best-known island studies is the study of Surtsey. Surtsey is a new island formed by a volcanic eruption in 1965 southwest of Iceland (Figure 36.10).

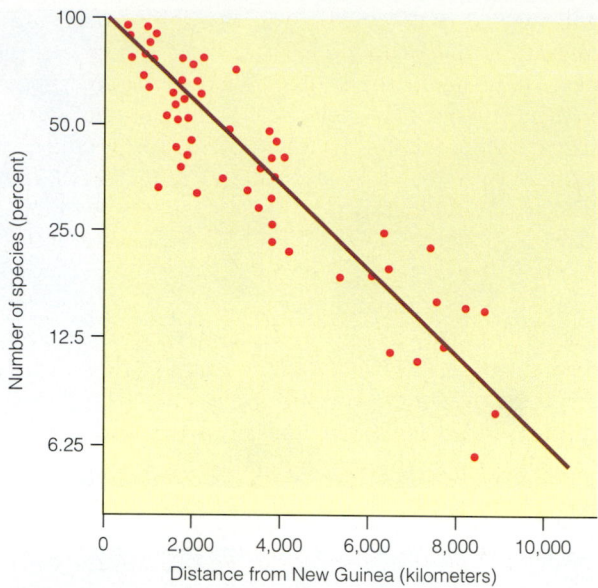

FIGURE 36.11   Distance Effect
The number of different species colonizing a new island decreases with the distance of the new island from a colonizing source; in this case, New Guinea.

What factors determine the number of species that will inhabit the island? There are two important factors:

1. The rate at which species immigrate and successfully colonize the new island.
2. The rate of extinction of the species.

Therefore, the number of species existing on an island community is a balance between immigration rates and extinction rates. If you think further, you

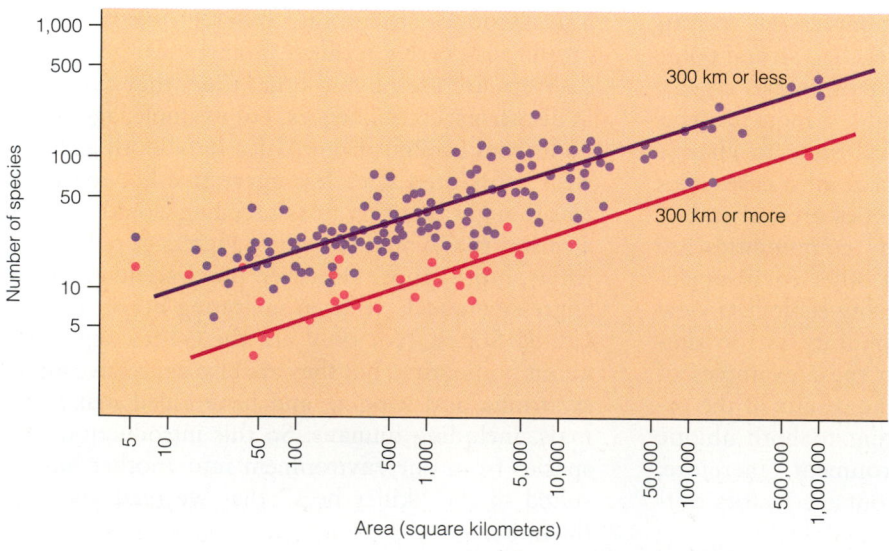

FIGURE 36.12   Area Effect
As can be seen from this graph, larger islands support larger numbers of species. The upper line represents islands 300 km or less from the colonizing source, and the lower line 300 km or more from the colonizing source.

will surmise that two important variables affect immigration and extinction. First, the distance of the new island from the source of potential colonists; that is, how far it is from the mainland. Second, the size of the island. On the island of Surtsey, within six months of its emergence, bacteria, fungi, seeds, insects, and sea birds began to colonize it. All species came from neighboring Iceland. This shows that the distance from the colonizing source affects the species composition of a new island. For example, islands far away from sources of organisms are colonized at a slower rate and by species adapted for dispersal over large distances (Figure 36.11). Such species would be things like palm (coconut) seeds and birds. Also, the size or area of the island will affect species diversity. Smaller islands generally have slower immigration rates simply because they are harder to find (i.e., they make a smaller target) because they consist of a smaller area. Additionally, because they are smaller, they have fewer resources. Because many species can only live in certain habitats, the chances of finding a favorable habitat are greater on larger islands. In addition, the larger the island, the greater the chance a species will land on the island.

Extinction also plays a role in the number and kinds of species on an island. For example, because smaller islands far from the source of species have a lower rate of immigration, the effects of extinction are high. Extinctions keep species diversity lower on smaller islands more distant from the source of colonists (Figure 36.12).

SUMMARY

1. A community is a collection of populations living in a specifically defined area.
2. A habitats is the place where an organism lives. Each habitat has its own abiotic and biotic characteristics.
3. Habitats are influenced by climate, availability of nutrients, the adaptability of the organisms within the habitat, the organisms' interactions, and short- and long-term patterns of disturbances.
4. Competitive interaction can result between individuals of the same species or individuals of different species. These biotic factors regulate population growth.
5. Examples of interspecific and intraspecific competition include territoriality and dominance hierarchies. Predator-prey interactions and parasitism are also forms of interspecific competition.
6. Several mechanisms of prey defense or predator enhancement have evolved, such as cryptic coloration, mimicry, and chemical devices.

7. Interaction among species can be neutral, beneficial, or competitive.
8. As an ecosystem changes, the composition of the community changes in an orderly manner, called succession. In primary succession, a bare environment is colonized by living organisms. In secondary succession, a previously occupied site is disturbed and repopulated.
9. In theory, a climax community is one in which succession is no longer a factor; that is, the ecosystem is in equilibrium.
10. Island communities provide excellent opportunities to study species diversity within a community. The two major factors that determine the number of species that will inhabit an island are the rate at which species immigrate and the rate at which they go extinct.

KEY TERMS

community	parasitism
habitat	cryptic coloration
niche	Batesian mimicry
intraspecific competition	Müllerian mimicry
interspecific competition	commensalism
territory	mutualism
dominance hierarchy	ecological succession
competitive exclusion	primary succession
principle	pioneer organisms
predator-prey interaction	secondary succession

REVIEW QUESTIONS

True/False

1. A community is made up of only one type of population.
   true   false   (page 865)

2. Parasitism is a type of symbiosis.
   true   false   (page 871)

3. The distance an island is from the source of colonization has no effect on the rate of immigration.
   true   false   (page 877)

4. Communities within an ecosystem are fixed and never change.
   true   false   (page 874)

Fill in the Blank

1. When two or more organisms of the same species compete for similar resources, this is an example of _____ competition. *(page 867)*

2. During _____ succession, organisms are established where none existed before. *(page 874)*

3. Once community succession ceases or slows down, the mature community is called a _____ community. *(page 875)*

4. Islands far away from a source of colonization are colonized _____ than islands closer to a colonizing source. *(page 877)*

Multiple Choice

1. Five different birds of similar species can all live in the same forest because they:
   a. Are predator and prey
   b. Are parasites
   c. Have niches
   d. Belong to the same populations
   (page 869)

2. In a stable ecosystem, which of the following factors are important in maintaining population size?
   a. Competition
   b. Predation
   c. Limited resources
   d. All of these
   (page 871)

3. When two closely related species of paramecia are grown together in a culture flask, one species dies out. This is an example of what ecological principle?
   a. Abiotic population regulation
   b. Competitive exclusion
   c. Competitive selection
   d. Non–density-dependent abiotic factors
   (page 869)

4. Sometimes a palatable prey may mimic an unpalatable model. This is an example of:
   a. Müllerian mimicry
   b. Protective coloration
   c. Batesian mimicry
   d. Cryptic coloration
   (page 873)

Discussion

1. Compare and contrast the terms *habitat* and *niche* of a species.

2. What are some of the environmental features that influence habitats?

3. An ecologist observes that five similar species occupy a similar niche. How can they coexist within a similar niche?

4. What is interspecific competition? Does competition increase with niche overlap? Give an example to support your answer.

5. What are some biotic factors that affect population density?

6. Compare and contrast cryptic coloration and mimicry.

7. Do you agree that the concept *climax community* is part of succession? Why or why not?

8. Why are islands useful models when studying species diversity and the colonization of a new area?

37

Ecosystems

Energy in Ecosystems
Abiotic Factors that Regulate Population
Summary

Populations do not exist alone in nature. Rather, they are found in communities made up of many populations coexisting within a defined space, as for example, the communities that reside in a pond. Together with the physical environment, these communities constitute an *ecosystem*, which forms a functional unit in nature.

Ecosystems vary in size. Some are very small, like the small water ecosystem that collects in a bromeliad high up in a tree in Puerto Rico's tropical rain forests. Many microorganisms — even tree frogs — can live within such an ecosystem (Figure 37.1). The life forms that coexist in a rotten log also are part of the ecosystem of the log. Other ecosystems are very large; for example, a large forest can be considered an ecosystem. An astronaut looking down on the

Above: This desktop ecosystem depicts the interactions of the organism with the environment.

earth could view the entire earth as an ecosystem — the *biosphere*.

How can we study ecosystems? There are various ways, and we will discuss some of them in this chapter. Did you know that a construction company must have an independent team of biologists draw up an environmental impact study before it receives permits to build a new shopping center, a highway, or a university? Why is such a study necessary? You will understand the reasons for this requirement after you finish this chapter.

ENERGY IN ECOSYSTEMS

- Solar energy drives the water cycle, weather systems, and ocean currents; it is absorbed by the soil and reradiated back as heat. It is also utilized by photosynthesizers to produce food.

FIGURE 37.1 Ecosystems
Some ecosystems are limited to very small defined regions such as the leaves of the tree in Puerto Rico, where this female tree frog (*Coqui*) lives.

The ultimate source of energy on our planet is the sun. About 1.94 Calories/cm²/minute, or about 102 Calories/cm²/year, strike the upper atmosphere of our planet. This figure is called the *solar constant* and represents a tremendous amount of energy. Only about 50 percent of this total energy actually reaches the surface of the earth. The rest is reflected into space (30 percent) or absorbed by the atmosphere (20 percent).

Solar Energy

■ Solar energy is largely responsible for global weather conditions, due to the heat it generates. The sun is also the source of energy used in photosynthesis.

Some of the solar energy that reaches the earth's surface drives the water cycle — the never-ending pattern of evaporation, cooling, precipitation, and evaporation again. The part of the solar energy known as heat is largely responsible for weather by forming high- and low-pressure systems, winds, ocean currents, and seasons.

Some solar energy is absorbed by the soil and then reradiated back into the atmosphere in the form of heat. If you have ever been in a house made of adobe (bricks made of sun-dried mud), you realize how much solar energy the soil can absorb and reradiate back into the atmosphere. As the adobe reradiates the heat it has absorbed during the day, it warms the

house in the cool of the night. It is the reradiation of heat that you feel in the summer when you step onto the hot black asphalt of a shopping center parking lot. If you have ever watched birds or hang gliders soaring on air currents, you have seen the effects of rising air thermals, rising currents of air heated by the sun that are the result of this reradiation of heat.

Of all the solar energy striking the earth, only about one-tenth of 1 percent is used for photosynthesis and actually drives the wheels of life. Nonetheless, the photosynthesizers on this planet produce (with water and carbon dioxide as raw materials and light as a source of energy) about 170 billion tons of organic material per year.

Another source of energy just discovered in 1977 is energy from *hydrothermal vents* under the ocean. These vents are located where plates of the earth's crust split and the heat inside the earth spews forth superheated water. The primary producers in these communities cannot use sunlight because it is unable to reach these depths. Instead, they are chemosynthetic bacteria that obtain energy through reactions using hydrogen sulfide (see page 169).

Gross and Net Primary Production. The total amount of organic material per unit time produced in an ecosystem is called its **gross primary productivity.** Of course, plants must use some of the food they produce for their own needs. It is estimated that plants use, by their own respiration, 15 to 20 percent of the food they produce. The remaining energy, known as the **net primary productivity,** serves as the energy source for earth's heterotrophic organisms.

As you would imagine, the gross and net primary productivity of an ecosystem vary. Ecosystems with an abundance of plants generally produce more per unit volume than those with fewer plants. Several other factors also play a role in the productivity of an ecosystem. Light, carbon dioxide concentration, temperature, water, and inorganic nutrients all affect the rate of photosynthesis and thereby the primary productivity of an ecosystem (Table 37.1).

The flow of energy through an ecosystem is one of the most important factors in determining the structure and organization of its communities. Let us now discuss the various levels involved as energy passes from one organism to another.

Trophic Levels

■ Trophic levels are hierarchies within a food chain; that is, the feeding pattern in a community. Organisms fall into three general categories: producers, consumers, and decomposers.

TABLE 37.1
Estimated Gross Primary Production (Annual Basis) of the Biosphere and Its Distribution Among Major Ecosystems

Ecosystem	Area $(10^6\ km^2)$	Gross primary productivity $(kcal/m^2/year)$	Total gross production $(10^{16}\ kcal/year)$
Marine			
Open ocean	326.0	1,000	32.6
Coastal zones	34.0	2,000	6.8
Upwelling zones	0.4	6,000	0.2
Estuaries and reefs	2.0	20,000	4.0
Subtotal	362.4	—	43.6
Terrestrial			
Deserts and tundra	40.0	200	0.8
Grasslands and pastures	42.0	2,500	10.5
Dry forests	9.4	2,500	2.4
Boreal coniferous forests	10.0	3,000	3.0
Cultivated lands with little or no energy subsidy	10.0	3,000	3.0
Moist temperate forests	4.9	8,000	3.9
Fuel-subsidized (mechanized) agriculture	4.0	12,000	4.8
Wet tropical and subtropical (broadleaved evergreen) forests	14.7	20,000	29.0
Subtotal	135.0	—	57.4
Total and average			
(Column 2) for biosphere (not including ice caps) (round figures)	500.0	2,000	100.0

Source: Odum, *Basic Ecology* (Holt, Rinehart and Winston, 1983). Reprinted by permission of the publisher.

We have already used the suffix -*trophic,* which means "feeding" (remember *heterotrophic,* which means "different feeding"). Here we discuss **trophic levels,** which are hierarchical categories of energy transference within a food chain. A **food chain** is a feeding pattern within a community concerned with who eats what or whom. Usually represented as a straight line: A produces, B eats A, C eats B, and so on. However, in reality, food chains are more cross-connected, each trophic level being connected with another to form **food webs,** which are patterns of energy exchange throughout an ecosystem. The trophic position that an organism occupies within a food chain is related to its ecological niche. In general, there are three positions: *producer, consumer* (which can involve more than one level), and *decomposer* (Figure 37.2).

The first trophic level is occupied by **producers,** the autotrophic organisms in an ecosystem. Most producers are photosynthetic organisms, but as we just mentioned, there are some rare food chains, like those found deep in the Pacific Ocean, where producers are chemosynthetic bacteria. That is, they use chemical energy rather than solar energy. Whether they are photosynthetic or chemosynthetic, primary producers convert energy to food.

Consumers are the heterotrophic organisms in an ecosystem. They can occupy different trophic levels. For example, **primary consumers,** the herbivores (or plant-eating animals), consume producers. **Secondary consumers** are those that consume the primary consumers. The lions (carnivores) that eat wildebeests (which are herbivores), the parasites that line the wildebeests' stomachs, and scavengers such as

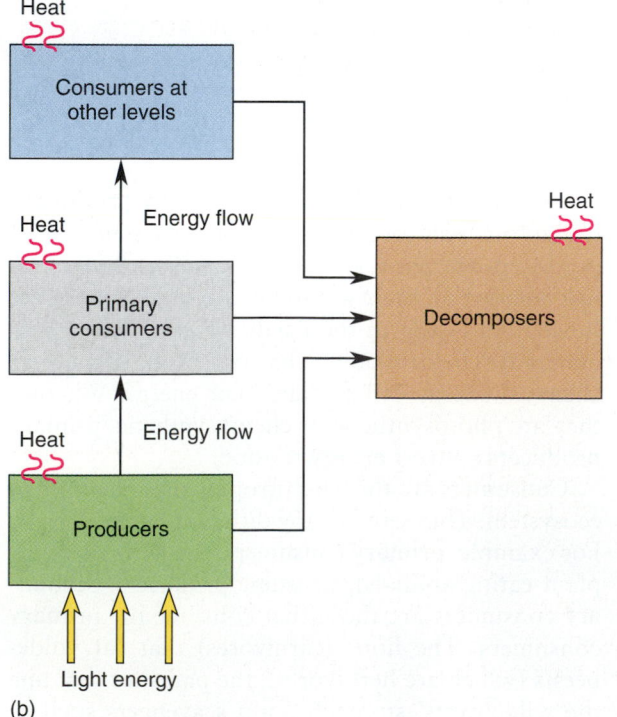

(a)

Consumers at other levels

Primary consumers

Producers

(b)

Heat

Consumers at other levels

Heat

Energy flow

Primary consumers

Heat

Energy flow

Producers

Light energy

Heat

Decomposers

FIGURE 37.2 Trophic Levels and Food Webs
(a) Light energy is constantly converted to chemical energy by producers. In this chemical form the energy can be used by producers and consumers alike. Decomposers obtain their energy by breaking down dead organisms. (b) The overall pattern of energy flow proceeds from primary producers to consumers to decomposers. Along the way some energy is lost to the system as heat, reducing the amount of energy available to organisms at the next trophic level.

vultures that pick at carcasses of dead wildebeests are all examples of secondary consumers.

Finally, there are the **decomposers,** a specialized kind of consumer able to break down many of the chemicals (like protein) in dead organisms and return nutrients to the ecosystem. The bacteria and fungi growing on dead logs in a forest are decomposers. The nutrients released by decomposers are taken up by producers to complete the cycle.

Of course, there are other trophic levels, such as third-level consumers. You may have seen a cartoon of a tiny fish about to bite a fisherman's hook — and behind it is a larger fish and behind that an even larger fish, and so on. Each organism in the cartoon represents a trophic level. The organism at the top of the food chain is the "final consumer" — and in the cartoon the question is whether the final consumer will be the man or the largest fish at the end of the sequence.

An organism's trophic level varies with its feeding habits. Humans are omnivores, organisms that eat both plants and animals. An omnivore can be either a primary or secondary consumer, depending upon whether it is eating a plant or an animal.

So each food chain, no matter how complex, begins with autotrophic producers, usually photosynthetic plants, and continues to include consumers at various levels. Upon the death of any organism in the food chain, the decomposers — bacteria and fungi — chemically change its molecules into forms that can be recycled through the ecosystem and used again.

But what happens to energy as we move from one trophic level to another? Is it all consumed, or is there energy loss? To find the answer, we must return to the discussion of thermodynamics, which we began in Chapter 6.

Ecosystem Thermodynamics

■ **In general, only about 10 percent of the energy available in one trophic level is transferred to the next. As a result, any amount of biomass in one trophic level can support only a substantially smaller amount of biomass at the next level above it. The result is an ecological pyramid.**

In Chapter 6 we discussed the second law of thermodynamics, which states that during energy transformation some energy is dissipated in an unusable form, usually as heat. That principle operates in every energy transformation, including food chains, so usable energy is lost from one trophic level to another. In many food chains, the loss is about 90 percent per

level. That means that only about 10 percent of the energy available at one trophic level is passed on to the next.

Because there is a loss of available energy at each level, less **biomass** (the mass of all living organisms in a defined region) can be supported at each level. Thus, the collective biomass of carnivores is usually less than the biomass of producers. For instance, the average brown bat in a Mississippi River town is a small animal — with its wings folded it would fit easily into the palm of your hand. Every night that bat eats almost two-thirds of its body weight in insects. That means that over the summer it takes a substantial number of insects to support the life of only one bat. If you could gather all the insects and all the bats together and compare their biomasses, you would find that the biomass of the insects far exceeds that of the bats.

The energy loss from one trophic level to the next is the reason that most food webs have only third- or fourth-level consumers. The more levels that exist between a trophic level and the level of producers, the less energy and biomass are available to that consumer. Bats eat insects, as do birds and fish. Cats eat bats, as do hawks. What eats cats or hawks? In their food chains they are at the top trophic level.

When the trophic levels of an ecosystem are diagrammed, the loss of energy and biomass results in a pyramid shape. So (logically enough), such diagrams are called ecological pyramids. Some pyramids are used to illustrate the decreasing numbers of individuals at each trophic level (Figure 37.3a), others are used to illustrate the quantity of organic material in an ecosystem (Figure 37.3b), and still others are used to illustrate the transfer of energy (Calories) from one level to another. The last are called pyramids of energy (Figure 37.3c).

Ecological pyramids explain more than energy flow in ecosystems. For one thing, they explain why pesticides sprayed on fields and forests have resulted in endangered species of birds such as the bald eagle (we will discuss this in more detail in the next chapter). Pesticides like DDT were widely used in the 1940s and '50s. Rains washed those pesticides into lakes, streams, and rivers, and river plants absorbed the substances, as did the organisms that fed on those plants. Perhaps only a tiny amount was in each plant, but many plants were required to support one primary consumer, like a snail. Thus, the primary consumer ended up with a considerably higher concentration of the pesticide than the plants on which it fed. Small fish ate primary consumers and bigger fish ate the small fish. Ultimately, fish that had a concentrated amount of DDT were eaten by

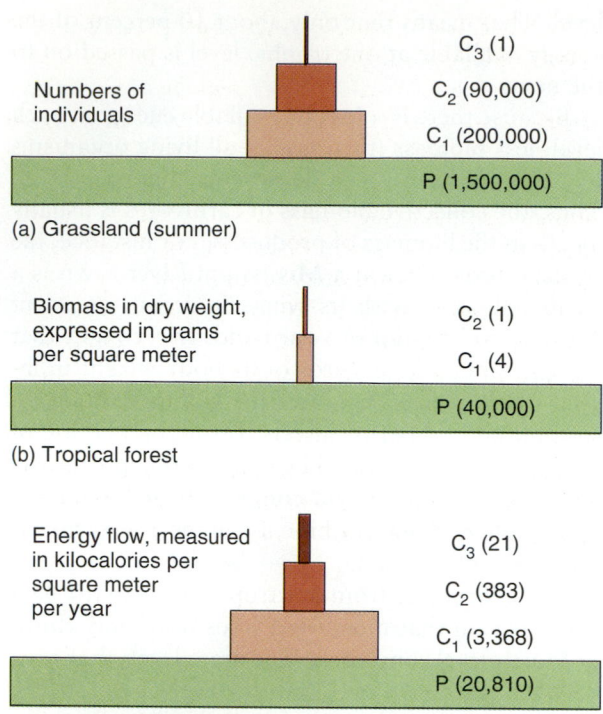

Numbers of individuals

C₃ (1)
C₂ (90,000)
C₁ (200,000)
P (1,500,000)

(a) Grassland (summer)

Biomass in dry weight, expressed in grams per square meter

C₂ (1)
C₁ (4)
P (40,000)

(b) Tropical forest

Energy flow, measured in kilocalories per square meter per year

C₃ (21)
C₂ (383)
C₁ (3,368)
P (20,810)

(c) Freshwater community

Key: P represents producers
 C represents consumers

FIGURE 37.3   Ecological Pyramids
(a) This diagram represents a pyramid of numbers of individuals in a grassland community. In some situations the number of consumers can be greater the the number of producers. This would occur if, for example, many caterpillars were feeding on the same tree. (b) This illustration portrays the quantity of organic matter (in grams per square meter) in a tropical forest. (c) This diagram illustrates the energy flow (in Calories per square meter per year) in a freshwater aquatic community.

an eagle. It takes many fish to support one eagle. DDT accumulated in the eagle's system and weakened the calcium bonds in the shells of its eggs, so that they broke when the mother sat on them before they hatched. As you move up the food chain, the concentration of toxins increases. This concentration of toxins is responsible for many of the ecological problems we face today.

ABIOTIC FACTORS THAT REGULATE POPULATION

■ **Abiotic elements in the environment are the nonliving factors that influence the size and density of populations.**

As you remember, living organisms are, in essence, homeostatic chemical factories that are constantly changing to remain the same. Homeostasis requires an input of energy and of essential organic and inorganic substances. The availability of these nutrients affects population size and distribution, as well as their density. Most of the elements required for plant and animal life are found either in the soil of terrestrial habitats or dissolved in the water of aquatic habitats. Some arise from general conditions on earth — the amount and duration of sunlight, the temperature of a given region, and the composition of the soil.

Sunlight

■ **Sunlight is the ultimate source of energy on earth, and it is also responsible for the length of daylight. These variations in amount of daylight help to regulate breeding cycles and flowering cycles.**

Sunlight is the ultimate source of energy on this planet, and so the availability of light is a major factor affecting the distribution and abundance of organisms. The amount of sunlight an area receives is in part determined by day length. At the equator, light and dark periods are divided into nearly equal 12-hour periods. As one moves toward either pole, that division changes so that at the North or South Pole, there is almost continuous night or continuous day at given times of the year.

Would you expect a variation in day length to affect breeding cycles? Well, it does. As you learned earlier, there are long-day and short-day plants (Chapter 21). Therefore, the amount of daylight selects for certain reproductive patterns, restricting plants to environments of either long or short days. You do not find cocklebur (short-day plants) in the tropics because short-day plants would not be able to reproduce and would be selected against.

Water

■ **Water is essential for life. The distribution of water directly affects the type of organism that will live in a particular environment.**

The 25 percent of the earth's surface that is not covered with water — terrestrial habitats — contains a diversity of habitats and supports a variety of ecosystems. Although these are terrestrial, the abundance and distribution of organisms they include are in large part dependent on the availability of water. Tropical rain forests, where there is considerable rain, have an abundance of plant and animal life (Figure 37.4a). Little rainfall, on the other hand, produces deserts like those of the Southwest, where there is a limited supply of life forms relative to the tropics (Figure 37.4b).

The availability of water, in turn, is determined by other components in the environment. The rocky soil of some mountain areas contains very little organic material. As a result, it does not hold water well. In meadowlands, the soil is less rocky, contains much organic matter, and holds water much more efficiently. So, even though the two areas may receive the same amount of rainfall, the composition of the soil (which is dependent on the size of the particles that make it up) determines its ability to hold water. The amount of water available for plants affects the amount of vegetation in an area, and that affects the

distribution of animals, which directly or indirectly require plants to survive.

Temperature also interacts with water to affect distribution. In terrestrial environments like tropical rain forests, high temperatures produce humid conditions, which affect the distribution of organisms. Some organisms, like the tree frogs that live in the mists of waterfalls, are found in more humid regions. Others, like monkeys, live in less humid regions of the forest. In the tundra, above timberlines, or near the poles of the earth, temperatures are low, so water is found mainly in the form of ice. Frozen water is not a usable form for plants, so vegetation is sparse.

Temperature also affects the distribution of organisms in aquatic environments. If you have ever had a bowl of goldfish, you know that they survive

FIGURE 37.4 **Water**
Without water, life could not exist. The amount of usable water in the environment plays a major role in determining the forms of life that will survive and flourish in that environment, whether (a) rain forest or (b) desert.

(a)

(b)

Dr. Jane Lubchenco and Dr. Bruce Menge

Ecologists

Jane Lubchenco and Bruce Menge are the quintessential dual-career couple. Both are university professors, well-known marine ecologists, and devoted parents. How do they manage to "have it all"? The two of them decided long ago that having a family was as important as their careers and that they would find a way to do both well. Their innovative solution was for each of them to work part-time while their children were young.

Jane grew up in Denver. Her parents were both physicians, and her mother provided a positive role model for Jane, working part-time while raising six daughters. Jane, the oldest, was full of energy and curiosity, traits that served her well in her chosen field. She attended Colorado College, majoring in biology. A summer fellowship to the Marine Biological Laboratory in Woods Hole, Massachusetts, first introduced her to the world of marine biology. She became fascinated with the interactions of marine organisms with each other and with their environment, and that fascination still drives her today.

Bruce's childhood in Minnesota was more traditional. His father was a busy agronomist, and his mother devoted herself mostly to family and home. He became interested in marine biology when his family took a trip to Florida. He remembers looking down into the blue waters of the Florida Keys and seeing a manta ray swim by. That memory was still with him when he graduated from the University of Minnesota with a degree in zoology. He decided that his desire to spend time outdoors and also avoid mosquitoes could be met by graduate study in marine biology.

Jane and Bruce met at the University of Washington, where Bruce earned a Ph.D., and Jane an M.S. degree. After they married, they moved to the East Coast, where Jane continued her graduate studies at Harvard and Bruce began teaching and doing research at the University of Massachusetts at Boston. After completing her Ph.D. degree, Jane accepted a full-time position at Harvard, but neither she nor Bruce were content. They wanted to have children, but their jobs left no time for a family. They dreamed of the perfect solution: two part-time but tenure-track positions. Fortunately, in 1976, the Department of Zoology at Oregon State University agreed to split a single faculty position into two separate half-time positions. Eighteen years and two children later, the experiment has proved to be a resounding success. Lubchenco and Menge are both now full-time professors with outstanding records in teaching and research. Their two sons are teenagers with active, energetic lives.

The dedicated teamwork exhibited by Bruce and Jane's family life is characteristic of their partnership at work. In 1979 Jane and Bruce were jointly awarded the Mercer Award, given by the Ecological Society of America for an outstanding scientific paper, for collaborative research they did on New England shores.

Jane and Bruce have also been successful individually. Several years ago, Jane led the Sustainable Biosphere Initiative, an effort to rank ecological research priorities and to provide guidance to government funding agencies. As her reputation for sound scientific advice has grown, national and international organizations have sought her out for expert scientific and organizational information. She has been tapped to advise members of Congress, the Cabinet, and White House staff and has served on numerous committees of the National Academy of Sciences and the National Research Council. In

1993, Dr. Lubchenco received the prestigious MacArthur Fellowship, often called the "genius grant." In addition to this coveted award, she has been named Pew Scholar in Conservation and Environment, Fellow in the American Academy of Arts and Sciences, and Oregon Scientist of the Year in 1994. She was elected President of the Ecological Society of America for 1992–1993.

Bruce is a widely respected scientist and is particularly noted for his work in predator-prey interactions and in food-web dynamics. He is especially skilled at advising graduate students on their own research projects. In 1994 Bruce was awarded a prestigious Guggenheim Fellowship in support of a sabbatical year conducting marine research on the coast of New Zealand.

Both Jane and Bruce are active members of professional societies and other organizations, requiring frequent travel. Each invitation to a meeting is checked against the family calendar and both office calendars. Scheduling conflicts over baseball games, dentist appointments, and music lessons are negotiated and accommodated creatively.

Jane Lubchenco and Bruce Menge are living proof that a dual-career couple can be successful in both work and family. All it takes is dedication, commitment, and the cooperation of faculty and administrators.

well even in water that is at room temperature. Tropical fish, on the other hand, require water that is warmer than room temperature, and their tolerance for temperature fluctuations is very limited.

Approximately 75 percent of the earth's surface is covered by water. Even though this seems like a more than adequate supply, the amount of water is limited because, in a sense, no new water is being made.

Our water supply constantly recycles in the environment, making it available for all living organisms.

In the **water cycle,** water evaporates from various sources because of solar radiation. The water condenses in the atmosphere, falls back to earth in some form of precipitation, and the cycle begins again. Figure 37.5 illustrates the water cycle.

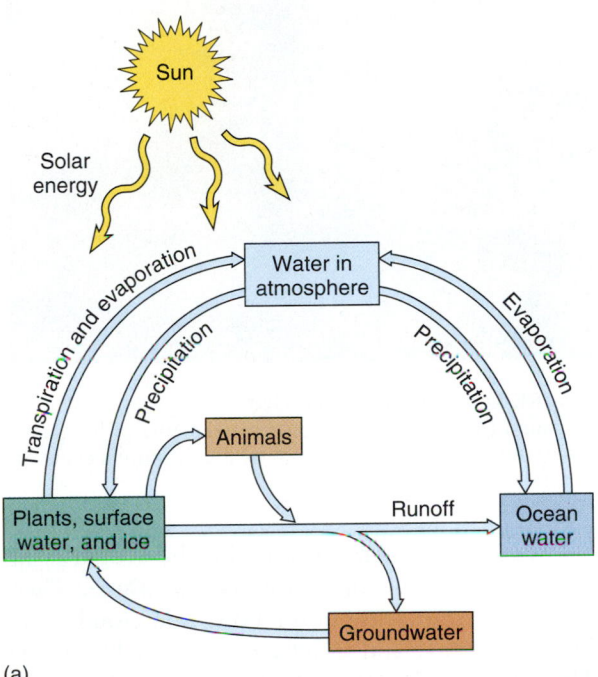

(a)

(b)

(c)

FIGURE 37.5   The Water Cycle
(a) Diagram of the water cycle. (b) At any time, only about 14,000 cubic kilometers are in the atmosphere. (c) Of the nearly 1.5 billion cubic kilometers of water on our planet, nearly 1.4 billion cubic kilometers are in the oceans.

Abiotic Factors that Regulate Population

Mineral Cycles

■ **The minerals that form many of the important chemicals within living systems also recycle through the environment. The most important environmental cycles are the carbon cycle, the oxygen cycle, the phosphorus cycle, and the nitrogen cycle. Trace minerals also recycle.**

Water is not the only substance on earth that recycles through the system. In a sense, all biologically useful elements are recycled. They pass from the soil, air, or water through living organisms and return to the soil, air, or water. If biological elements did not recycle, they would be locked up in a "biotic compartment" — the organism. Death and decomposition allow for the "unlocking" and recycling of molecules through the biosphere (Figure 37.6). Just imagine — there may be molecules in your body that once belonged to George Washington, who exhaled carbon dioxide and water vapor, and who produced nitrogenous wastes that were recycled back into the environment — a molecular reincarnation.

There are also several important **biogeochemical cycles**, cycles of elements found both in the environment and in living organisms. Among them are the carbon cycle, the oxygen cycle, the phosphorus cycle, and the nitrogen cycle. Trace minerals such as iron, manganese, and copper also cycle through the environment, the living organism, and back to the environment.

The Carbon Cycle. In the **carbon cycle**, carbon dioxide in the air and water is cycled through the plants, where the CO_2 is fixed into organic molecules (food) during photosynthesis. As they respire, consumers (animals), decomposers (bacteria and fungi), and plants release CO_2 back into the environment. Burning fossil fuel and the weathering of rocks also return carbon to the environment (Figure 37.7).

Since 1860, there has been a 30 percent increase in the carbon dioxide concentration in the atmosphere. It is estimated that within the next 70 years the concentration of CO_2 in the atmosphere will double its present level. Increased concentration of CO_2 in the atmosphere interferes with the radiation of heat back into space, the absorption of energy, and even changes in the wavelength of light. As a result, the temperature of the earth rises in a process called the *greenhouse effect.* It is predicted that by the year 2055 the earth's temperature will rise from 3.7 to 8.1 degrees Fahrenheit, owing to the greenhouse effect.

FIGURE 37.6 Elements Recycled
Through the recycling process, the elements of life pass from animal to plant to animal — a continuous cycle of replenishment.

An increase in temperature like this would result in a dramatic imbalance of the biosphere. There would be an increase in humidity and cloud cover. Ocean currents and wind currents would change. The amount of permanent ice at the poles would decrease or increase. Ocean levels would rise, bringing about widespread flooding of low-lying coastal areas, where many of the world's large cities are located, or would decrease and disrupt shipping. The distribution of precipitation would change. Some moist regions would become dry and vice versa. Most scientists have predicted that ocean levels would rise as the earth's temperature rose and the permanent ice at the poles melted. However, recent evidence suggests that increased atmospheric temperature would cause wind and ocean currents to change course, moving more water to the poles, where it would turn to ice, resulting in lower ocean levels.

Critics of the greenhouse theory state that scientists have measured only a slight warming of the atmosphere, not a significant change. Although scientists may not agree about the significance of the small amount of temperature increase, they do caution that the oceans take a long time to warm, and not until they have done so will we detect significant changes in atmospheric temperatures.

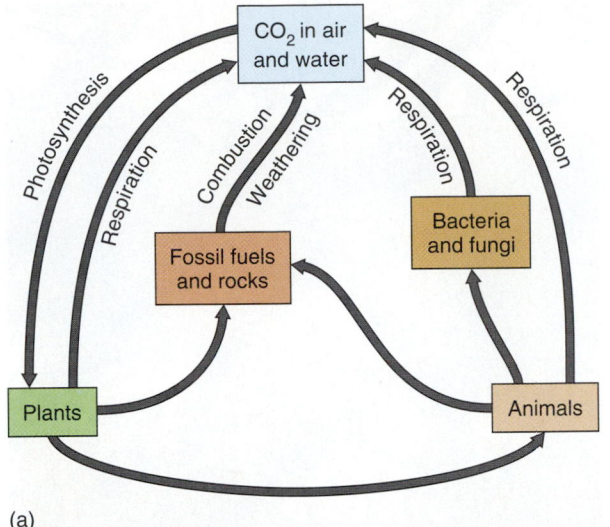

(a)

(b)

(c)

FIGURE 37.7   The Carbon Cycle

(a) Diagram of the carbon cycle. (b) The reduced carbon was originally produced by plants such as the ferns through photosynthesis millions of years ago and is now being oxidized rapidly as (c) industry burns fossil fuels to power modern civilizations. The result has been a dramatic increase in the level of CO_2 (oxidized carbon) in our atmosphere.

The Oxygen Cycle.   The **oxygen cycle,** which follows much the same pattern as carbon dioxide, is also linked to photosynthesis and respiration. Photosynthetic organisms use CO_2 and give off oxygen. All aerobic organisms utilize this oxygen through cellular respiration and give off CO_2. These cycles are, then, complementary (Figure 37.8).

The Phosphorus Cycle.   Phosphorus is another element essential to living things, and the **phosphorus** cycle keeps this element in the environment. The phosphorus cycle is a sedimentary cycle — phosphorus is dissolved in water while on land and is then deposited in the sea. Then, after eons of time, these deposits are exposed to the land again by various geological events (uplifting, for example) and are weathered and eroded. They again dissolve in water and return to the sea. Usable forms of phosphorus are found dissolved in the soil and aquatic systems and in the bones and teeth of animals.

(a)

(b)

(c)

FIGURE 37.8   The Oxygen Cycle
(a) Diagram of the oxygen cycle. Plants produce oxygen as a by-product of photosynthesis. All aerobic cells, including plant cells, require oxygen for cellular respiration. Without oxygen, aerobic organisms like this fish (b) or giraffe (c) cannot derive energy from food and cannot survive.

Plants, animals, and microorganisms usually obtain phosphates in a dissolved form and incorporate the mineral into their phosphate-requiring molecules (Figure 37.9).

The Nitrogen Cycle.   Nitrogen is an essential mineral in life cycles because it is present in amino acids, nucleic acids, and chlorophyll. However, even though 78 percent of the atmosphere is composed of nitrogen gas, only a few microorganisms, like some cyanobacteria (blue-green algae) and nitrogen-fixing bacteria, can use the gas in this form. The **nitrogen cycle** involves these particular organisms (Figure 37.10).

Nitrogen-fixing bacteria found in the nodules on the roots of legumes function in the nitrogen cycle by fixing (or binding) atmospheric nitrogen (N_2) to hydrogen to form ammonia (NH_3) or by fixing it to oxygen to form nitrogen oxides like *nitrite* (NO_2) or *nitrate* (NO_3) ions. When the plant dies, the nitrogen is released into the soil through the action of decomposers. Animal nitrogenous wastes and other sources also add to the nitrogen store through decomposition, to be cycled through other living organisms.

If the nitrogen in the soil of a certain area is depleted, plants can no longer photosynthesize or synthesize the amino acids and proteins they require.

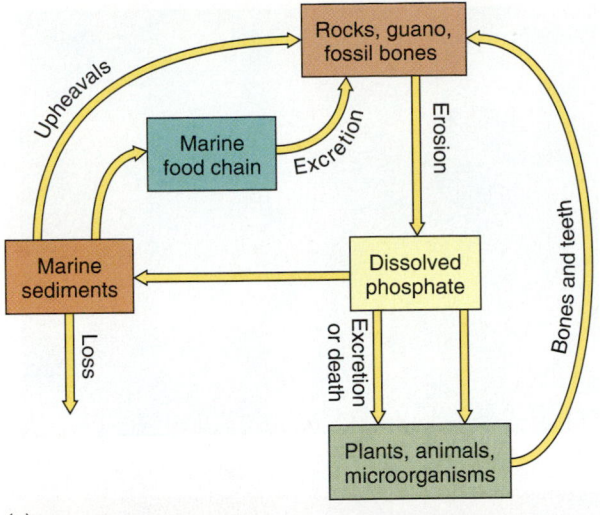

(a)

(b)

(c)

FIGURE 37.9   The Phosphorus Cycle
(a) Diagram of the phosphorus cycle. Phosphorus is one of the elements vital
to life forms. (b) Phosphorus is found in bones and (c) deposited in soil,
dissolved by water, and returned to the ocean.

Hence their number is reduced or they die. As the population of photosynthetic organisms decreases, the consumers dependent on these plants also decrease, either by emigrating to another area or by dying.

Farmers deal with the problem of nitrogen depletion all the time. To maintain a sufficient supply of nitrogen in their fields, they use nitrogenous fertilizers or plant legume crops such as clover or soybeans, either for a short time during the growing season or for an entire year. The bacteria in the root nodules of legumes fix nitrogen in the soil, thus providing it for later crops.

Many other abiotic environmental factors affect population growth and distribution — soil pH, toxic substances, and salt spray, for example.

In the next chapter, we will discuss a major biotic factor that is affecting the ecosystem, humans.

Abiotic Factors that Regulate Population

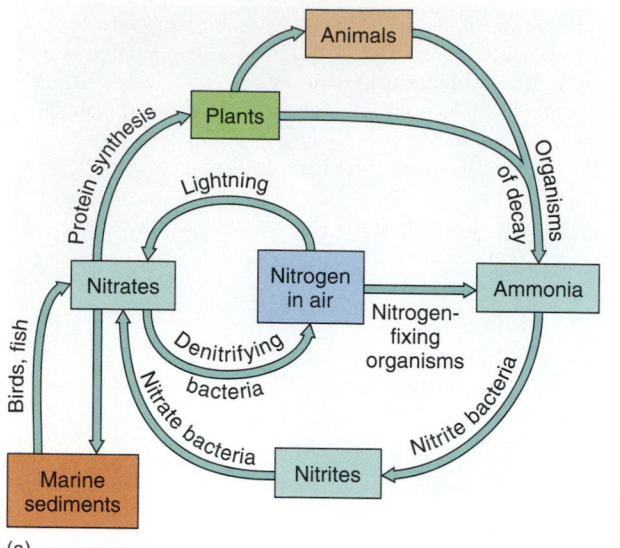

(a)

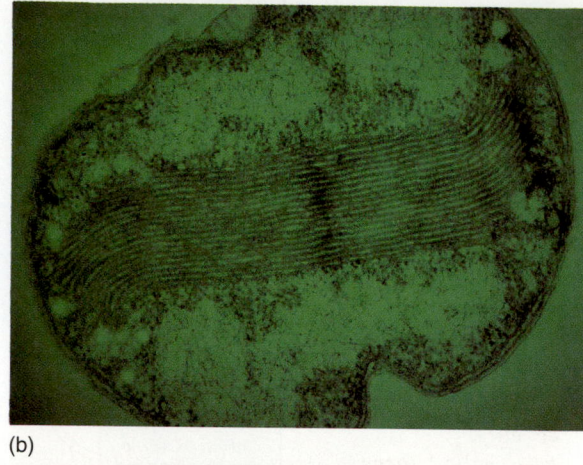

(b)

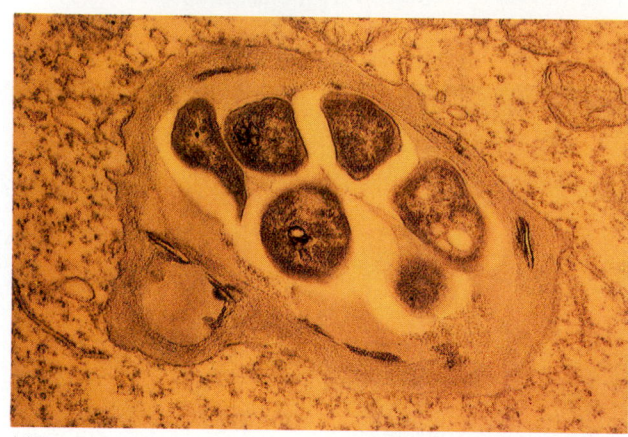

(c)

FIGURE 37.10   The Nitrogen Cycle
(a) Diagram of the nitrogen cycle. Without
prokaryotes, which change the atmospheric nitrogen
into nitrogen compounds that plants can use, the
only sources of usable nitrogen would be lightning
and the industrial production of fertilizers — a tiny
fraction of the fixed nitrogen produced on our
planet. (b) An electron micrograph of bacteria of
the genus *Nitrococcus*. These oxidize ammonia to
nitrite. (c) An electron micrograph of cells of
Rhizobium leguminosarum, a nitrogen-fixing
bacterium found in the vacuoles of pea plant root
cells.

SUMMARY

1. The sun is the earth's ultimate source of energy.
   Photosynthetic producers convert solar energy into
   chemical energy to be stored as food. The amount of
   organic matter or food produced per unit of time is
   the gross primary productivity of an ecosystem.

2. As energy is passed from one organism to another
   some is lost. In a food web, there are three major
   trophic levels: producers, consumers, and decomposers.

3. As energy is transferred from one trophic level to
   another, about 90 percent of the available energy is
   lost.

4. Abiotic factors in the environment such as sunlight,
   water, and temperature influence the size and density
   of a population.

5. Minerals are important chemicals within living
   systems that cycle through the environment.

6. Other important abiotic factors that regulate population growth include the availability of water and
   light and the cycling of carbon, oxygen, phosphorous, and nitrogen.

KEY TERMS

gross primary productivity
net primary productivity
trophic level
food chain
food web
producer
consumer
primary consumer
secondary consumer

decomposer
biomass
water cycle
biogeochemical cycle
carbon cycle
oxygen cycle
phosphorus cycle
nitrogen cycle

REVIEW QUESTIONS

True/False

1. An ecosystem is made up of one type of community.
   true   false   (page 879)

2. In general, food webs consist of producers, consumers, and decomposers.
   true   false   (page 881)

3. The amount of usable energy remains constant as it is passed from one trophic level to another.
   true   false   (page 883)

4. The energy within an ecosystem is fixed and never changes.
   true   false   (page 883)

Fill in the Blank

1. The total amount of _____ _____ per unit time produced in an ecosystem is called the gross primary productivity. *(page 880)*

2. The hierarchies within a food web are called _____ levels. *(page 881)*

3. Without the _____ in a food web many chemicals would not be recycled. *(page 883)*

4. Because there is a loss of energy at each trophic level _____ biomass can be supported at each successive level. *(page 883)*

Multiple Choice

1. As energy is passed from one trophic level to another, the amount of usable energy:
   a. Increases
   b. Decreases
   c. Remains the same
   d. Energy is not passed from one trophic level to another
   (page 883)

2. An example of a producer in the aquatic food web would be:
   a. Duckweed
   b. Ducks
   c. Fish
   d. Insects
   (page 881)

3. When you were a child, you may have heard the old poem, "Fishy, fishy in the brook, daddy catch it with a hook, mama fries it in the pan, baby eats it like a man." This poem illustrates the biological concept of:
   a. Trophic levels in a food web
   b. Primary productivity
   c. The decomposition of organic matter
   d. A poem that would be considered politically incorrect today
   (page 882)

4. In the biosphere, which of the following is the ultimate source of energy?
   a. Carbon
   b. Water
   c. Sunlight
   d. Nitrogen
   (page 880)

Discussion

1. Discuss the three major trophic levels. Be sure to include the energy source for each level.

2. Define *gross primary productivity, net primary productivity,* and *biomass.*

3. We mentioned that living organisms are energy transducers and are constantly trying to overcome entropy. Does this statement hold true for ecosystems? Why or why not?

4. Why would you expect variation in day length to affect breeding cycles?

5. What are some biotic factors that affect population density?

6. Draw and discuss any three biogeochemical cycles.

7. Do you agree that a climax community is part of succession? Why or why not?

8. Why are islands useful models for studying species diversity and colonization of a new area?

38

Human Interaction with the Biosphere

Atmospheric Change
The Impact of Agriculture
Waste Disposal
Deforestation
Extinction
An Environmental Perspective
Summary

We can't help but be concerned about our impact on the biosphere. We see evidence of our destructive actions every day. One must question, "Does the pillage of our planet go unabated?" In Southern California, as we drive on our freeways we see, taste, and smell the impact of air pollution. In the Midwest, flood after flood remind us of what happens when we convert grassland to agriculture without much forethought about the flood plains and river flows. We see pictures on television of starving children in Africa as well as pictures of acid rain, deforestation, overpopulation, ozone depletion, and other images. All are concerns of ecologists and environmentalists because they are the negative effects of humans on the biosphere. In this chapter, we will discuss the impact of humans on the delicate interaction of ecosystem dynamics.

Above: An aerial view of New York City illustrates the negative impact of air pollution and overcrowding in our environment.

ATMOSPHERIC CHANGE

■ **In part because of the human population explosion and expanding technology, we have had a tremendous impact on the biosphere. Carbon dioxide and other atmospheric pollutants are damaging air quality, contributing to a global warming, and destroying the earth's ozone layer.**

Almost everyone is aware of our impact on the quality of our atmosphere. We are damaging this delicate, finite veil of protection that is only miles thick. In the United States alone, we produce more than 700,000 metric tons of **atmospheric pollutants** every day. That's over 14 million pounds per day. There are 6 major types of air pollutants that are harmful to our health and that impact our ecosystem (Table 38.1). These pollutants present major atmospheric intrusions. First, because of increased manufacturing and because of our dependence on fossil fuels (coal and oil) and wood, we are burning a tremendous amount

TABLE 38.1
Six Major Classes of Air Pollutants

Examples	Class of Air Pollutants
Nitric oxide (NO)	Nitrogen oxides
Nitrogen dioxide (NO_2)	
Nitrous oxide (N_2O)	
Sulfur dioxide (SO_2)	Sulfur oxides
Sulfur trioxide (SO_3)	
Carbon monoxide (CO)	Carbon oxides
Carbon dioxide (CO_2)	
Solid particles (asbestos, soot, dust, lead)	Suspended particles
Liquid droplets (dioxins, oils, sulfuric acid, pesticides)	
Methane (CH_4)	Volatile organic compounds
Benzene (C_6H_6)	
Chlorofluorocarbons (CFCs)	
Ozone (O_3)	Photochemical oxidants
Peroxyacyl nitrates (PANs)	
Hydrogen peroxide (H_2O_2)	

18°C. It is predicted that an increase of CO_2 would increase the atmospheric temperature as well. Some models predict that the doubling of CO_2 concentration would increase atmospheric temperature by 3°C to 4°C. Obviously, such an increase would have a significant effect on weather patterns, which would change the distribution of precipitation and hence the distribution of organisms. Some models predict that because of the warming conditions, the solar ice caps would melt and flood coastal areas as the ocean's levels rise by approximately 100 m. Other models show that because of the increased temperature, weather patterns would move the water towards the poles as it evaporates. Therefore, ice caps would increase in size and ocean water levels would decrease. No matter what the scenario, increasing CO_2 concentrations in the atmosphere would have major effects on the ecosystems of our planet. But CO_2 is not the only atmospheric pollutant that should concern us.

Chlorofluorocarbons (CFCs) are used as propellants in aerosol spray cans, reactants in industrial solvents, ingredients in plastic foams, and coolant in air conditioners and refrigerators. Other chemical CFCs are responsible for the reduction of the **ozone** layer and the increase in the size of the ozone hole over our poles (see Focus 38.1). The lower stratosphere (about 17 to 25 km above the earth) contains a protective layer of ozone (O_3). Ozone absorbs harmful ultraviolet light, limiting the amount reaching the surface of the earth. Some of the consequences of an increase in ultraviolet light are an increase in skin cancer, cataracts, and a weaker immune system in humans. Also, increased ultraviolet light harms phytoplankton, which are the largest group of photosynthesizers and oxygen producers on our planet.

of fuel, which results in an increase of CO_2 (carbon dioxide) in the atmosphere. Burning forests as a means of clearing land (deforestation) has produced huge amounts of excess CO_2 in the atmosphere. In addition, by burning fossil fuels to generate electricity and other forms of energy, our internal combustion engines abnormally increase atmospheric CO_2.

Since 1850, when the estimated concentration of CO_2 in the atmosphere was about 274 parts per million (ppm), the concentration has been increasing. In 1958, a monitoring station on Hawaii measured the CO_2 concentration as 316 ppm, and today the measurement is 351 ppm (Figure 38.1). If this increase of more than 11 percent in the last 30 years continues, the amount of CO_2 in the air will have more than doubled since 1850 to 702 ppm by the year 2075. What are the consequences of an increased concentration of atmospheric CO_2? One possible consequence is the **greenhouse effect** (Figure 38.2). As mentioned previously, the majority of solar radiation (infrared heat) is reflected back into space, but water and CO_2 capture and absorb heat and reflect it back to the earth's surface. It is this phenomenon that helps maintain the earth's temperature. Without this process, the average temperature would be only

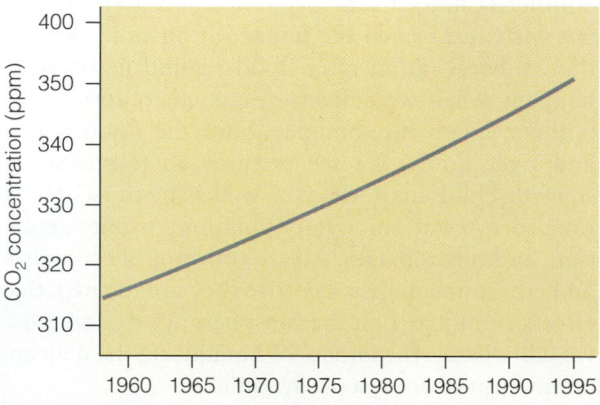

FIGURE 38.1   Increasing CO_2 in the atmosphere since 1958.

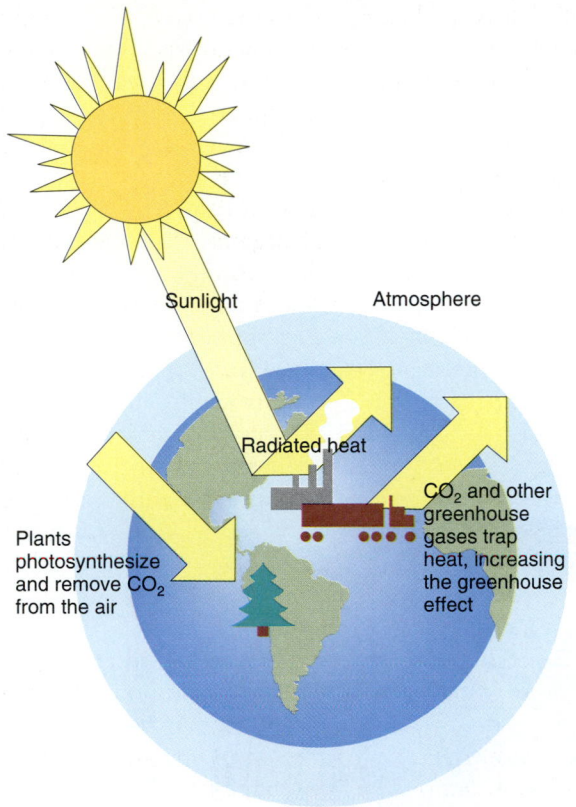

The solution to this problem is obvious: Stop using CFCs. But, as you might imagine, from a political perspective the solution is more difficult. In 1992, at the Earth Summit Conference in Brazil, leaders of industrialized nations signed a treaty to phase out the use of CFCs. Hopefully the nations will adhere to this treaty. But what about the CFCs already in the atmosphere? How long will they continue to deplete the level of ozone? Most scientists predict it will take at least 100 years before the CFCs break down.

The burning of fossil fuels not only adds to the increase of CO_2 but also produce oxides of sulfur and nitrogen. Power plants, especially those that burn high-sulfur coal, are the chief sources of sulfur dioxide. Last year, over 24.1 million metric tons of sulfur dioxide were emitted into the atmosphere in the United States alone. Once in the atmosphere, these gases combine with water and oxygen to form carbonic, sulfuric, and nitric acid. These acids may condense and fall back to the earth as acid precipitation in the form of rain, snow, hail, mist, or fog — what is commonly termed **acid rain.** Automobiles and nitrogen fertilizer are the chief source of oxides of nitrogen.

The pH level of "normal" rain is slightly acidic, usually falling between 4.0 and 5.6. Today the precipitation in the United States is 40 times more acidic than normal rain (Figure 38.3). We know how damaging and corrosive acid rain can be to buildings and other structures, but the damage to the ecosystem is even more severe.

FIGURE 38.2   The Greenhouse Effect
The burning of fossil fuels adds CO_2 and other gases to the atmosphere. Because the amount of CO_2 is more than the plants can utilize, the amount of CO_2 in the atmosphere is increasing. The increased CO_2 acts like the glass window in a greenhouse, trapping solar energy and heating the air. This global warming alters the climate.

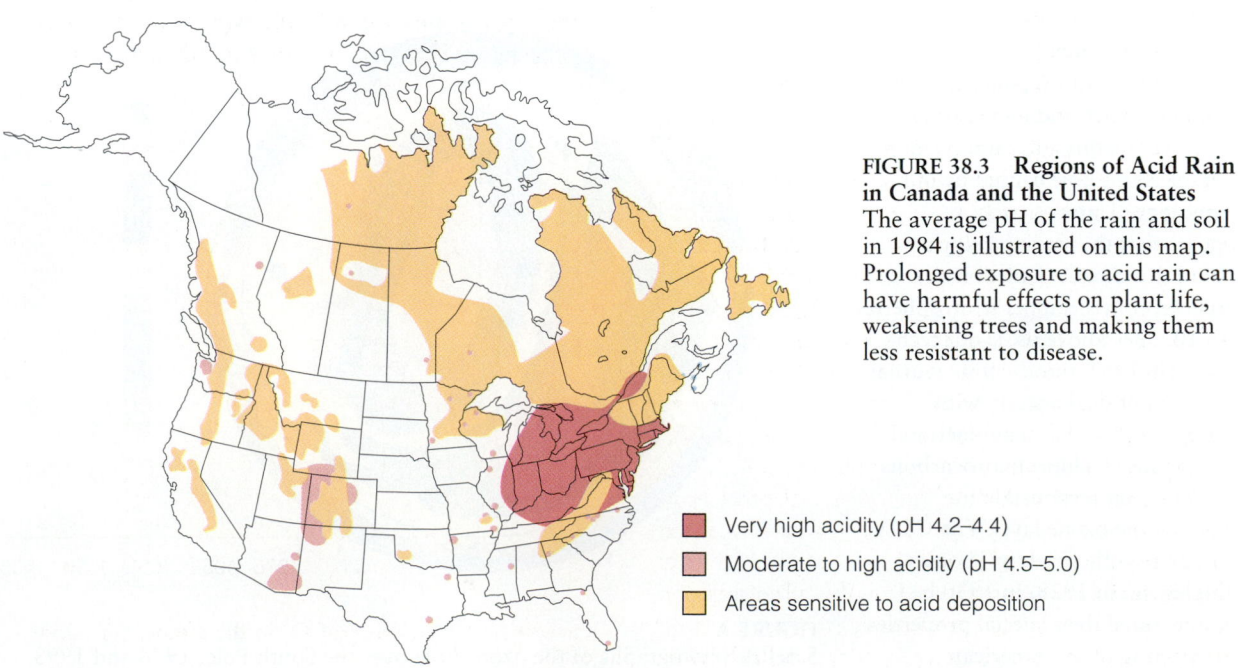

FIGURE 38.3   Regions of Acid Rain in Canada and the United States
The average pH of the rain and soil in 1984 is illustrated on this map. Prolonged exposure to acid rain can have harmful effects on plant life, weakening trees and making them less resistant to disease.

Very high acidity (pH 4.2–4.4)

Moderate to high acidity (pH 4.5–5.0)

Areas sensitive to acid deposition

FOCUS 38.1
The Hole in the Bottom of the Earth

The ozone layer over the Antarctic has been decreasing each spring for more than 17 years (see Figure A). A decrease was first measured by special, high-flying aircraft in 1976. Since that time, the ozone layer over Antarctica has continued to thin. In 1985, a team of British Antarctic survey scientists reported that the ozone layer in the atmosphere over Halley Bay, Antarctica, had decreased by almost 50 percent over the preceding ten years. Their findings have been confirmed in the years since their survey. In 1992, the area of the ozone hole was more than three times the area of the continental United States. In 1993, the actual level of the ozone was the lowest ever recorded.

The ozone around the South Pole is at its thinnest in September, stabilizes in October, and increases in November. Is this a natural phenomenon or is this hole caused by our technology? Unfortunately, it appears that the leading culprit behind this drastic depletion of the ozone layer in the upper stratosphere, 6 to 10 miles above us, is our technology. In 1987, the scientific journal *Nature* published a study with strong evidence that manufactured compounds — chlorofluorocarbons (CFCs) — are responsible for depleting the ozone layer. The CFCs were first synthesized by Thomas Midgley, Jr., in 1928. In 1930 he demonstrated their special properties to a meeting of the American

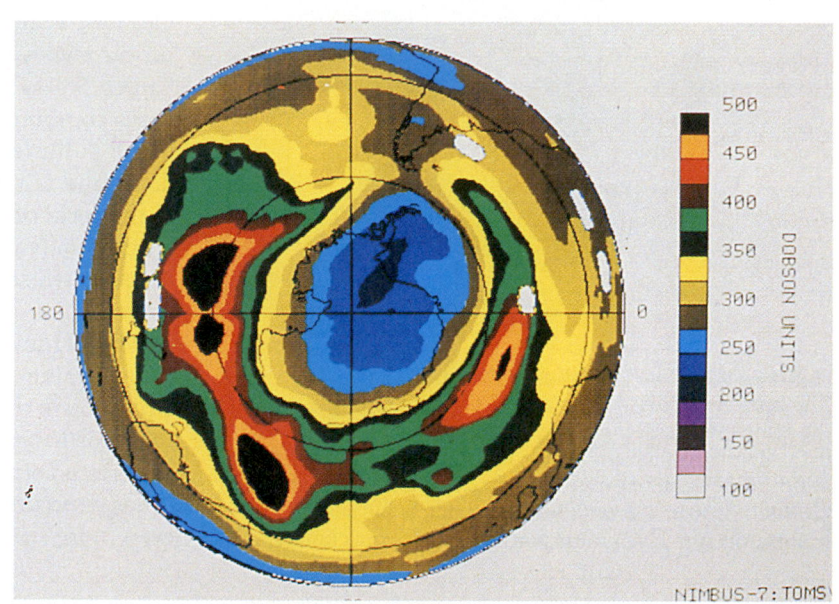

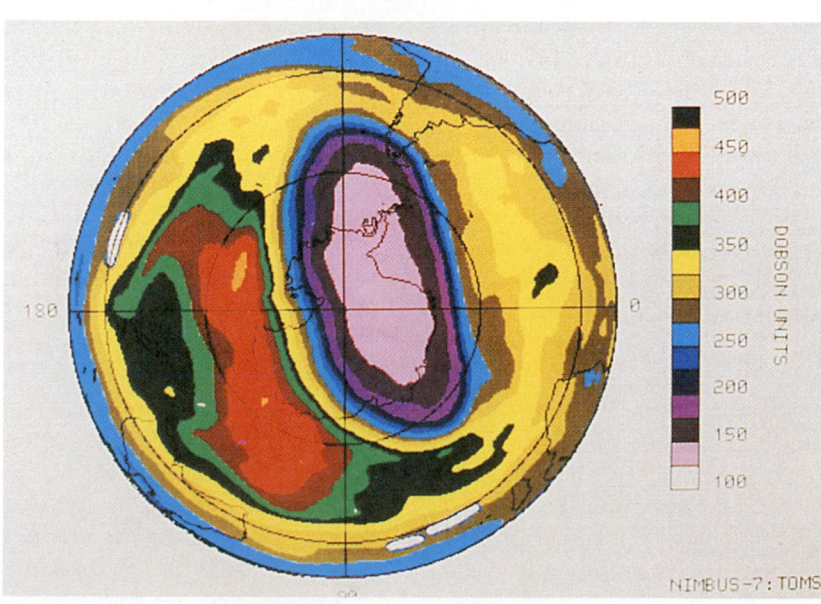

FIGURE A
Satellite photographs of the ozone hole over the South Pole, 1976 and 1993.

Chemical Society. Today CFCs serve as coolants for refrigerators, "blowing agents" for plastic foams, and solvents for cleaning computer microchips. They are essential to our way of life, but they are also destroying the ozone layer, which is essential to all life on this planet.

The ozone layer is the protective veil that shields us from dangerous solar radiation — ultraviolet (UV) light. With less ozone, more UV light will strike the surface of the earth. UV radiation has sufficient energy to mutate DNA, and it can increase the incidence of human skin cancer, cataracts, and immune deficiencies. But the global harm to our biosphere is more serious. UV light can harm photosensitive phytoplankton (microscopic plants) on the ocean surface, which are at the base of many food chains. Disturbing the phytoplankton could have grave consequences.

Investigators in West Germany suggested in 1988 that a 5 percent increase in UV light could cut the life span of some phytoplankton in half, with effects that would reverberate throughout the biosphere's food chains and even cause a rise in the temperature of the earth. Phytoplankton consume large quantities of CO_2 as they photosynthesize. Destroying phytoplankton would raise CO_2 levels, which would contribute to the greenhouse effect and raise the earth's temperature.

How was it determined that CFCs were at fault? In part, the evidence came in 1986 from converted U-2 spy planes, which scientists flew directly into the ozone hole to gather data. The most significant finding was that the high levels of chlorine- and fluorine-containing substances are correlated with the decrease in ozone. Chlorine might be increased by natural phenomena like volcanic eruptions, but the only source for fluorine in the upper atmosphere had to be the CFCs.

How are CFCs involved in destroying the ozone? Ozone is

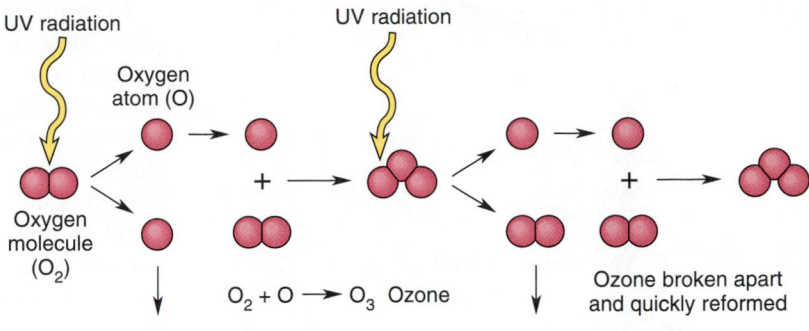

$$O_2 + O \longrightarrow O_3 \quad \text{Ozone}$$

Ozone broken apart and quickly reformed

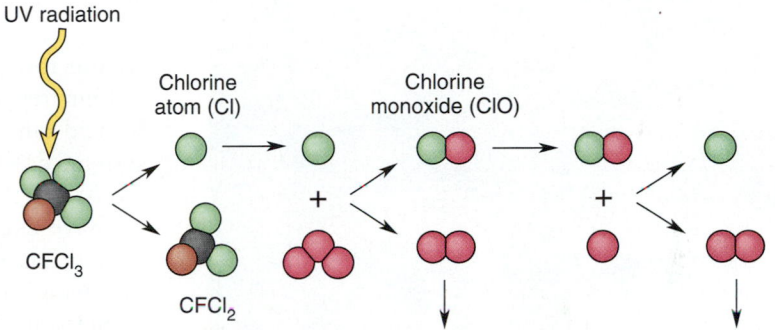

The presence of chlorofluorocarbons (CFCs) speeds up the conversion of ozone to oxygen molecules, thus destroying the ozone

FIGURE B
Ozone formation and the destruction of the ozone by the presence of CFCs.

formed in the atmosphere when UV or visible light strikes an oxygen molecule (O_2), freeing its oxygen atoms (O) to combine with other oxygen molecules (O_2) to form ozone (O_3). The ozone thus formed is again broken apart by the UV- or visible-light photons, again absorbing UV light and forming and reforming ozone many times (see Figure B). On the other hand, chlorine from the CFCs speeds up conversion of O_3 to O_2, and one chlorine atom can destroy about 100,000 ozone molecules. When a chlorine atom (Cl) collides with an O_3 molecule, the third oxygen atom combines with the Cl to form ClO (chlorine monoxide) and an oxygen molecule (O_2). Because this radical ClO has an odd number of electrons, it is very reactive, and when the ClO collides with a free oxygen atom, the oxygen atom combines with the oxygen from ClO to form O_2. The free chlorine is now available to

repeat the process continuously, destroying the ozone.

Not only is ozone thinning apparent at the South Pole, but new evidence prepared in 1988 by the National Aeronautics and Space Administration confirmed that ozone is being destroyed over the Northern Hemisphere more rapidly than predicted by computer models.

As you can see, the only solution to this problem is to reduce the quantity of CFCs that we use and release into the atmosphere. The industrialized countries, led by the United States, today use 30 percent of the 2.1 billion tons of CFCs produced each year. Science and industry have begun curbing manufacture of CFCs, but environmentalists feel they are not doing so aggressively enough. Faster action is needed, because the CFCs linger in the atmosphere more than 100 years, all that time destroying the ozone.

FIGURE 38.4 Acid Rain
Acid rain is not only destroying large tracts of forest;
it is also changing the pH of lakes, streams, and ponds
enough to destroy or threaten the organisms living there.

In some soils, the hydrogen ions in acid rain displace some essential elements such as calcium, magnesium, and potassium, which are required for tree growth. The process is called *leaching*. As a result, trees are dying. Biologists have calculated that more than 7 million hectares of trees have been lost in the United States and Europe due to acid rain. In some parts of Europe, such as Germany, more than 50 percent of the forest has been affected. More disturbing is the fact that in the same time, the biomass per acre for spruce has fallen from 30,000 to just about 8,000 — a decline of 73 percent. Just driving through the mountains around Los Angeles, one can see our forests dying (Figure 38.4).

Our freshwater resources are also being affected by acid rain. Under most circumstances, freshwater systems are highly resilient to disturbances in acid levels, but the problem is such that in many places we have exceeded the thresholds, and the natural system has become unbalanced. Living organisms are dying, and lakes are dying. Even our coastal habitats are beginning to be affected by acid water flowing into them from rivers. Our crops are being damaged, as is our health.

Critics of this view contest the conclusion that acid rain is destroying our ecosystem, insisting that such a finding is premature and unfounded and that more

FIGURE 38.5 Sources of Acid Pollution
Even as concerns grow and evidence accumulates, we continue to inject
massive amounts of sulfur dioxide and other pollutants into our atmosphere.

Chapter 38 Human Interaction with the Biosphere

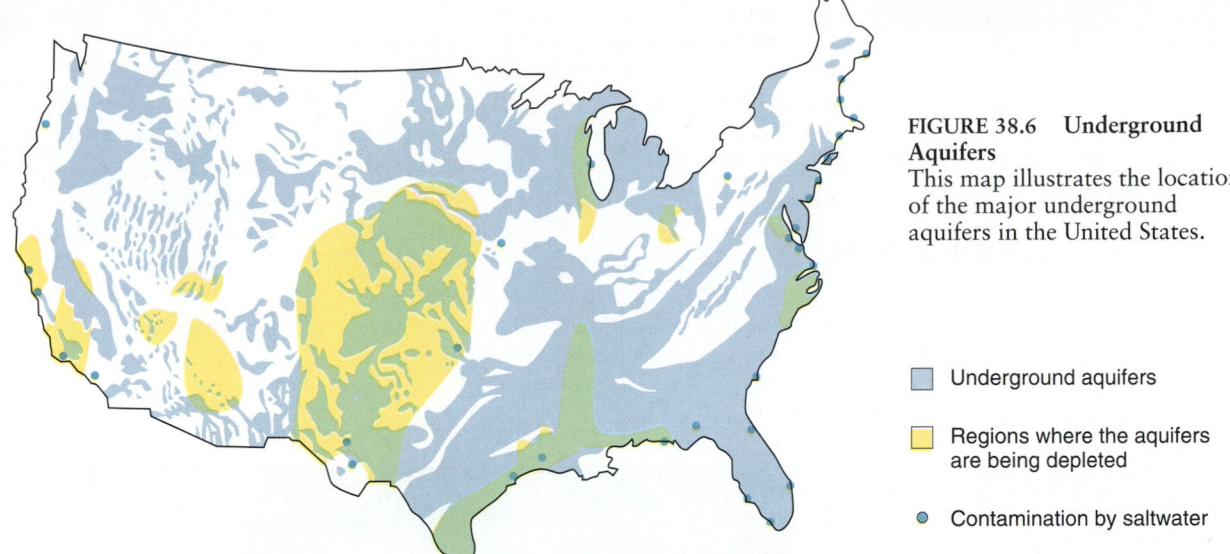

FIGURE 38.6 **Underground Aquifers**
This map illustrates the location of the major underground aquifers in the United States.

- ▪ Underground aquifers
- ▪ Regions where the aquifers are being depleted
- ● Contamination by saltwater

research needs to be done. It is true that more research should be conducted, but that facts should not be used as an excuse not to act now to reduce the emissions responsible for acid rain (Figure 38.5).

THE IMPACT OF AGRICULTURE

■ **The consequences of agricultural development have resulted in activities that are changing our biosphere and water usage. Additionally, pesticides and fertilizers are upsetting the delicate ecosystem.**

At first blush, the expansion of our agricultural lands and increased food production seem like good ideas. But as we think about the consequences, we have to ask ourselves, "How long can we continue to use technology to convert nonagricultural land to agricultural land to feed the increasing human population? What is the impact of this intrusion on water supply, soil conditions, and nutrient cycles? In short, what is the impact of agriculture on the biosphere?"

Let us now discuss water quality and usage. Today over one-third of agricultural land is irrigated. We pipe water from underground wells or we drain lakes to provide water for our crops. Irrigation not only depletes the underground water supply (**aquifers**) but the irrigation water contains salts that increase the salinity of the soil. As the water evaporates, the soil becomes increasingly salty. This increased salinity can decrease plant growth and eventually kill crop plants.

In the San Joaquin Valley and the Imperial Valley of California, the depletion of underground aquifers has caused the level of the soil to drop as the water is pumped from the aquifers. In Texas, Oklahoma, and Kansas (Figure 38.6), farmers are depleting the Ogallala aquifer at an annual rate that exceeds the annual flow of the Colorado River. Where will the water come from when we pump the aquifers dry?

We also use a tremendous amount of water for recreational uses. Golf courses consume many more times the amount of water per acre than agricultural land does (Figure 38.7). If you have ever been to Palm Springs, California, you can see the extensive waste of water and fertilizer needed to make the desert blossom. Not only does cultivation impact water usage, what about the fertilizer and pesticides we use to increase crop productivity?

FIGURE 38.7 **Golf Course**
The water required to maintain this golf course helps to illustrate the excessive amount of water being depleted from underground aquifers for recreational use.

FOCUS 38.2
The Vanishing Rain Forest

Because of our overzealous use of the plants of the rain forest we are upsetting the delicate ecosystem of the rain forest. The ecosystem of the rain forest is very sensitive to change and cannot recover very rapidly if at all from extensive deforestation. If any one of the components of the ecosystem is destroyed the rain forest is changed forever.

The rain forest ecosystem consists of three major levels: the canopy, the understory, and the forest floor (Figure A). The canopy is 100–300 feet high and consists of the closely spaced tree tops that form the crown of the forest.

The understory, from 50–80 feet, consists of the trunks of the canopy trees, young trees, shrubs, small trees, and other plants. The lowest level is the forest floor, where vegetation is very sparse. The dead organic material, leaf litter, and animal remains are broken down by insects, microorganisms, and fungi to organic material. Because the organic material decomposes so rapidly, the soil is poor and the roots are shallow. In addition, because the soil is so poor the destruction of the forest ecosystem can upset the mineral cycle. In the mineral cycle of an undisturbed jungle, the vital mineral nutrients circulate in an almost closed cycle. All the minerals

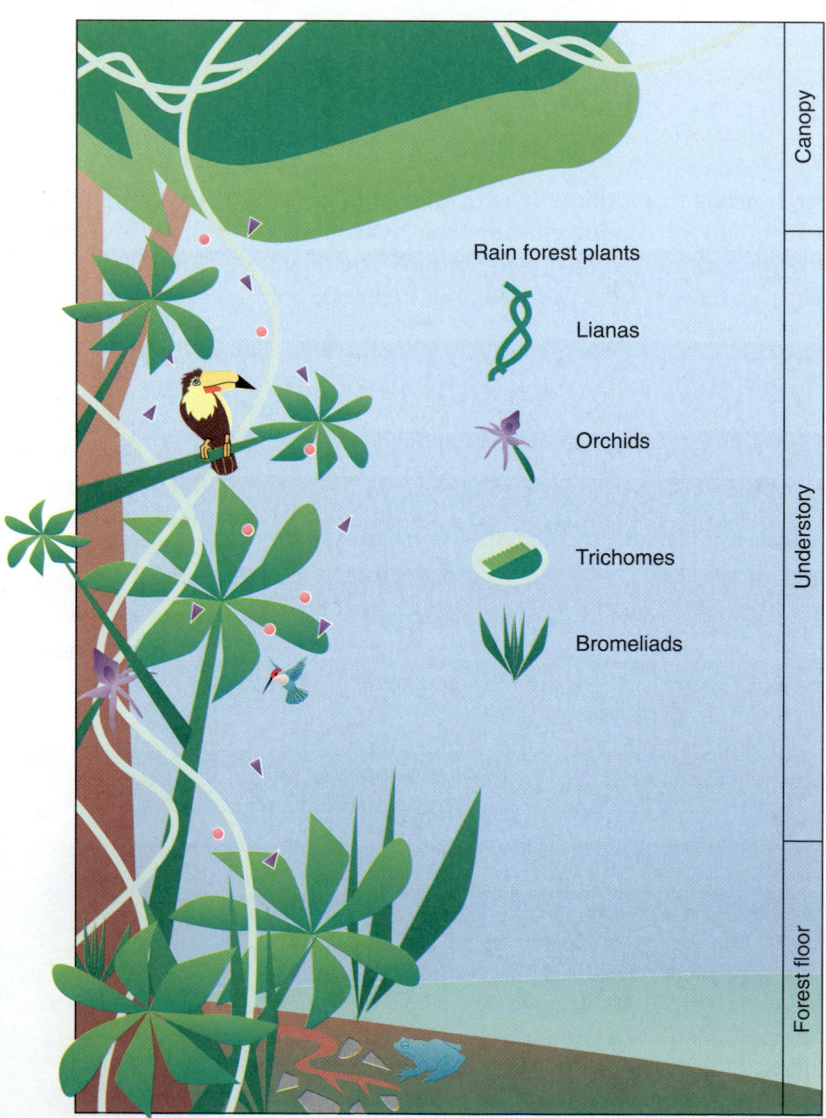

Rain forest plants

Lianas

Orchids

Trichomes

Bromeliads

Canopy

Understory

Forest floor

FIGURE A

from dead plants and animal waste are washed by the rain down to the soil, where they can be reused. In a healthy rain forest ecosystem, the small amount of minerals lost in the drainage of water is balanced by the continuing release of minerals dissolved from rock particles. Plants absorb these minerals and the cycle starts again.

The rain forest consists of thousands of species of plants of various types and habitats; among them are many plants that attach their roots to the host plants (usually trees). If the trees are cut down or burned, the plants that are attached to the host plants die. Among the various groups of plants or adaptations of plant organs found in the rain forest are:

Lianas: 90 percent of all vine species in the world are in tropical rain forests. Some grow out of the floor; others wind around or climb young trees.

Orchids: These are epiphytes with two kinds of roots. One kind anchors the orchid to the host; the other hangs free, to absorb water and nutrients from the air and plants around them.

Trichomes: Trichomes are specialized hairs on plants that collect water and extract nutrients from decayed material in the rain washing through the canopy.

Bromeliads: They form a holding place to store water for nourishment. Many small creatures find refuge inside the leaves. Mosquitoes and frogs lay their eggs in the water. Mammals and birds eat the insects.

As human populations increase, the rain forests are increasingly cleared for agriculture, timber, some cattle ranching, and mining (Figure B). Hardwood trees such as teak, mahogany, rosewood, balsa, and sandalwood are found in the jungle and have great commercial value. Farmers burn and clear, in the process destroying ecosystems. Because rain forest soil is not suited to agriculture and is depleted within a few years, fields are eventually abandoned. This destruction of the world's rain forests is accelerating, with growing planetary implications.

Once the trees are gone, monkeys, tree frogs, flying squirrels, and other canopy dwellers lose their livelihoods. Because there are no longer trees or shade, the moist, litter-strewn forest floor is replaced by expanses of bare soil that dries out. Almost all the nitrogen and carbon disappears into the atmosphere. Eventually the clearing is covered by a maze of herbs, grasses, ferns, bushes, and young trees, few of the species that grow in the interior of the jungle. Lacing it all together are tangles of vines studded with fierce thorns. The rain forest is lost (Figure C).

Deforestation affects us as follows:

- Seventy percent of the 2,000 plants identified as containing cancer-healing properties are found only in the rain forest.

- Deforestation could alter the climate, affecting rainfall patterns and air temperature worldwide.

- Clearing and burning the rain forest releases large amounts of carbon dioxide, methane, and nitrous oxide, contributing to global warming (Figure D).

Source: Adapted from the *San Diego Union Tribune,* February 2, 1994

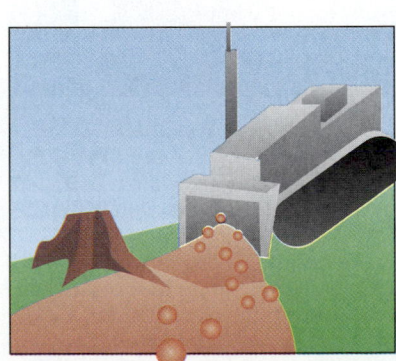

FIGURE B

FIGURE C

FIGURE D

WASTE DISPOSAL

■ The waste produced by manufacturing and other technologies has resulted in a disposal problem. In addition, our use of fertilizers and pesticides is upsetting the ecosystem. It is important for us to develop agricultural and waste disposal methods that have the least impact on the environment or to change our technology so that we do not produce so much ecosystem-damaging waste.

Once land has been cleared for agriculture, the land can only provide enough nutrients for crop production for a short time, because the increased number of plants will quickly deplete the available nutrients. Once the nutrients are depleted, fertilizer must be added. The fertilizers are quickly leeched from the soil and appear in creeks, streams, and rivers, and eventually lakes. The increase of nutrients in aquatic ecosystems rapidly changes the nutrient composition of lakes. The enrichment of the lake ecosystem causes *eutrophication* (page 848). The increased supply of nutrients increases the density of photosynthetic organisms, which increase oxygen production during the day. However, as the photosynthetic organisms respire at night, oxygen is depleted. Therefore, algae and other organisms begin to die. This type of pollution also kills commercially valuable fish.

Not only are fertilizers upsetting the ecosystem, but the extensive use of pesticides and other poisons is too. The primary impact is that their concentration in the food web increases. For example, as the pesticide DDT (banned in the U.S.) passes from one trophic level to another, it becomes more concentrated. This process is called **biological magnification** (Figure 38.8a and b).

Not only do pesticides impact on the ecosystem, so do the by-products and chemical wastes produced by manufacturing. Lead, mercury, and other poisons enter food webs and become concentrated as they pass through different trophic levels. In other cases, the poisons have a direct effect on many organisms, including humans. The issue of disposal of toxic wastes is a very difficult one. The products produced — medicine, fertilizers, chemicals, plastics, and so on — all benefit the human condition, but on the other hand, the waste by-products are also destroying the quality of life in some areas (Figure 38.9).

Not only do we produce waste from agriculture and manufacturing, we also produce nuclear wastes. Our increasing population has increased demands for electricity. In the 1950s, the solution to the demand for more energy was to build nuclear power plants. Many industrialized nations, including the United States, did so. Unfortunately, as we know, these nuclear power plants are not only more ex-

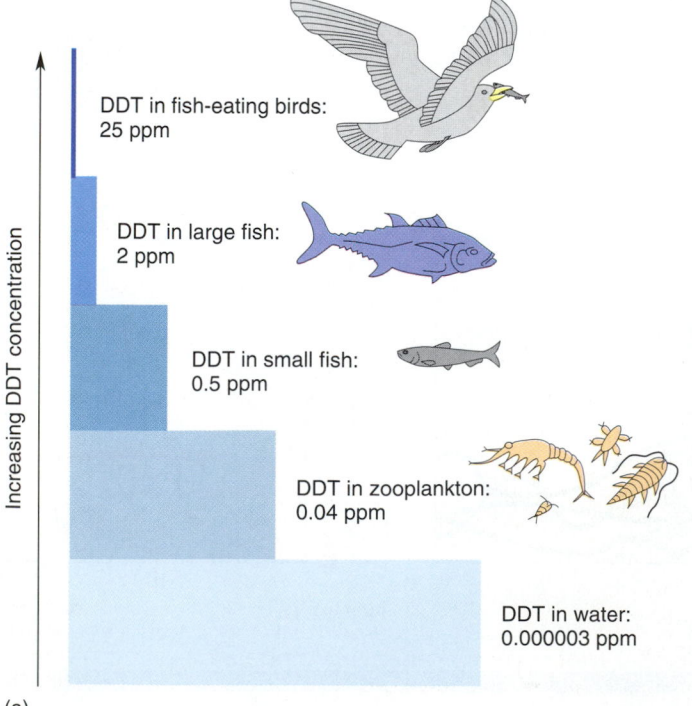

DDT in fish-eating birds: 25 ppm

DDT in large fish: 2 ppm

DDT in small fish: 0.5 ppm

DDT in zooplankton: 0.04 ppm

DDT in water: 0.000003 ppm

Increasing DDT concentration

(a)

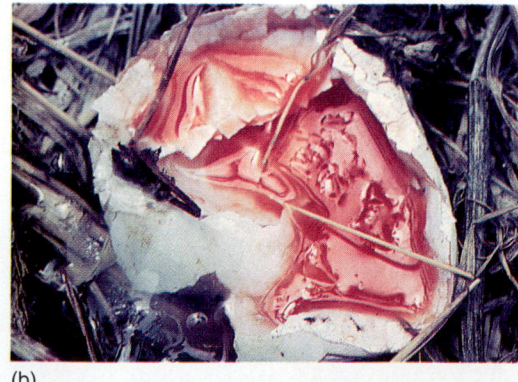

(b)

FIGURE 38.8   Biological Magnification of DDT (a) In one study of a Long Island, New York, food web, the amount of DDT increased from 0.000003 parts per million (ppm) in the seawater to 25 ppm in the fish-eating birds. Thus, as DDT passes through the food web it becomes more concentrated. (b) When DDT enters a food chain, it is retained in the bodies of the consumers and can result in thin-shelled eggs. At higher levels in the pyramid the biomass of consumers decreases and the concentration of DDT increases, as was the case of this pelican's egg.

FIGURE 38.9
These barrels are leaking toxic wastes into the environment, resulting in a contamination of the soil, water, and air.

pensive than coal-burning power plants, but the potential for a nuclear accident is always present.

The most fearful nuclear accident is a **meltdown.** In order to generate electricity, nuclear reactions generate heat, which is used to turn water to steam. The steam drives the electrical generators that produce the electricity. The danger is that an overheated reactor core could literally *melt down* and release radioactive material into the environment. In April 1986, in the town of Chernobyl in the Ukraine, explosions rocked the nuclear power plant there and decimated the 1,000-metric-ton, concrete nuclear reactor. Radiation was released into the environment. More than 31 people died immediately, and who knows how many more people died or will continue to die as a result of their exposure to radioactivity. The danger is not only to the immediate area, because winds blow radioactive waste hundreds of miles (Figure 38.10). It has been estimated that over 400 million people have been exposed to radiation due to this accident. These people are now at risk for cancer and other radiation sicknesses. Not only did the accident take a toll on humans, it extensively damaged plants and animals in the area.

Nuclear waste from physics and radioisotope research, manufacturing, and atomic weapons poses serious disposal problems. Unlike coal, nuclear waste is not burned to ash. **Radioactive wastes,** however, take a long time to decay. For example, the radioactive isotope plutonium (^{239}Pu) takes 25 million years to decay. Some scientists have proposed that radioactive wastes be placed in ceramic vessels, sealed

in steel containers, and buried deep underground in solid rock formations, but as of yet this has not been done, nor do we know the consequences. The radioactive waste problem is a serious one, and no one agrees on the safest method of disposal.

Even though we have discussed nuclear energy in a negative light, we should at least mention that nuclear power plants, when compared to coal-burning power plants, have benefits. They produce electricity at lower cost, and they release less carbon dioxide and sulfur oxides into the air. As you can see, however, neither coal-burning nor nuclear power plants are the solution to our increasing need for electrical energy.

Hopefully, by the year 2000 we will see more and more solar energy power plants with natural gas backups producing our electricity. These plants will be less damaging to our environment.

Solid waste disposal is also having an impact on our environment. We in developed countries have a "throw away" mentality. We hardly recycle anything. In less developed countries, very few things are ever thrown away. City dwellers do not recycle, but primitive tribes do. What are we to do with the millions of metric tons of garbage that we produce every year (Figure 38.11)? We must think more

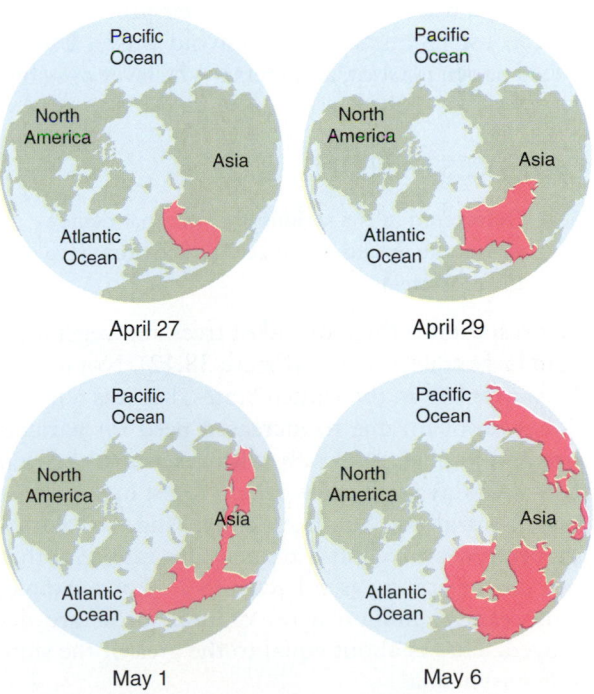

FIGURE 38.10 Worldwide Distribution of Radioactivity
After the explosions and fire at the Chernobyl nuclear power plant on April 26, 1986, the radioactive fallout was measurable hundreds of miles away.

FIGURE 38.11 Solid Waste Pollution
This unsightly garbage dump illustrates how our self-centered consumption and careless disposal of solid waste affect our environment.

about the consequences of our actions. Obviously, the only solution to the problem is to recycle whenever possible and to use less disposable items. We as consumers can put pressure on companies that encourage waste by not buying their products, and we can use common sense to avoid waste. Use recycled paper, plastics, and metals whenever possible.

DEFORESTATION

■ **Deforestation is a serious problem that demonstrates the upsetting impact our activities have on the ecosystem.**

Deforestation is the removal of trees and vegetation from large tracts of land (Figure 38.12). Not only is this a problem in the United States, but it is a major global problem, due to increased need for agriculture and lumber, especially with respect to the rain forests. Today large areas of rain forest in Southeast Asia, Central and South America, Asia, and Africa are being destroyed at a tremendous rate, and that rate is accelerating (see Focus 38.2). For example, each year 25 million acres of rain forest are destroyed. That is about equal to the area of the state of Pennsylvania!

But why should we be concerned about the global destruction of the rain forest? Because photosynthetic plants absorb CO_2 and produce O_2, and because animals absorb O_2 and produce CO_2, forests help to maintain an atmospheric gas balance. In ad-

dition, the burning of forests adds more CO_2 to the environment, which adds to the greenhouse effect, which in turn affects the climate. Water cycles are directly affected by rain forest plants, because they produce large amounts of water vapor through transpiration and evaporation. They also use large amounts during photosynthesis. So as you can see, the tropical rain forests use much of the water provided on this earth. When the rain forests are destroyed, water that would have been absorbed by the rain forest is carried to other areas, resulting in floods and damage to agriculture. In my opinion, however, even more important than the direct ecological impact of deforestation is the destruction of so many irreplaceable plant and animal species. The rain forest is a virtual storehouse for new drugs, foods, and other useful products (see Profile, page 908). For example, some drugs that are useful against cancer came from plants of the rain forest.

Because of our increased awareness of the negative impact of our lifestyles on the environment, in July 1992, an Earth Summit was convened in Brazil that was designed to bring attention to and find a solution for the destruction of our planet. Many countries participated and agreed to establish rules and regulations that will decrease the rapid destruction of our environment. At least the summit brought attention to the problems, but unfortunately the destruction goes on. The stakes are high; deforestation has the potential of changing global weather patterns and causing the extinction of plants that could cure disease. Also, because of the political and economic impact, peace can be threatened. Today, in the jungles of Central America, some of the armed forces are trying to enforce laws against illegal logging. Thus the army is a partial solution in some countries. In the Philippines and Thailand, however, the military turns a blind eye to illegal logging and actually is responsible for smuggling teakwood. In some cases, the military provides protection for the illegal logging and even does the cutting.

If the deforestation of rain forests continues at the present rate, most will be gone by the year 2035.

EXTINCTION

■ **The extinction of many species of organisms is an indication that our biosphere is in trouble.**

Not only are we facing a crisis over deforestation, which is responsible for the elimination and **extinction** of many plant species, but the animals dependent on these habitats are also becoming extinct.

Chapter 38   Human Interaction with the Biosphere

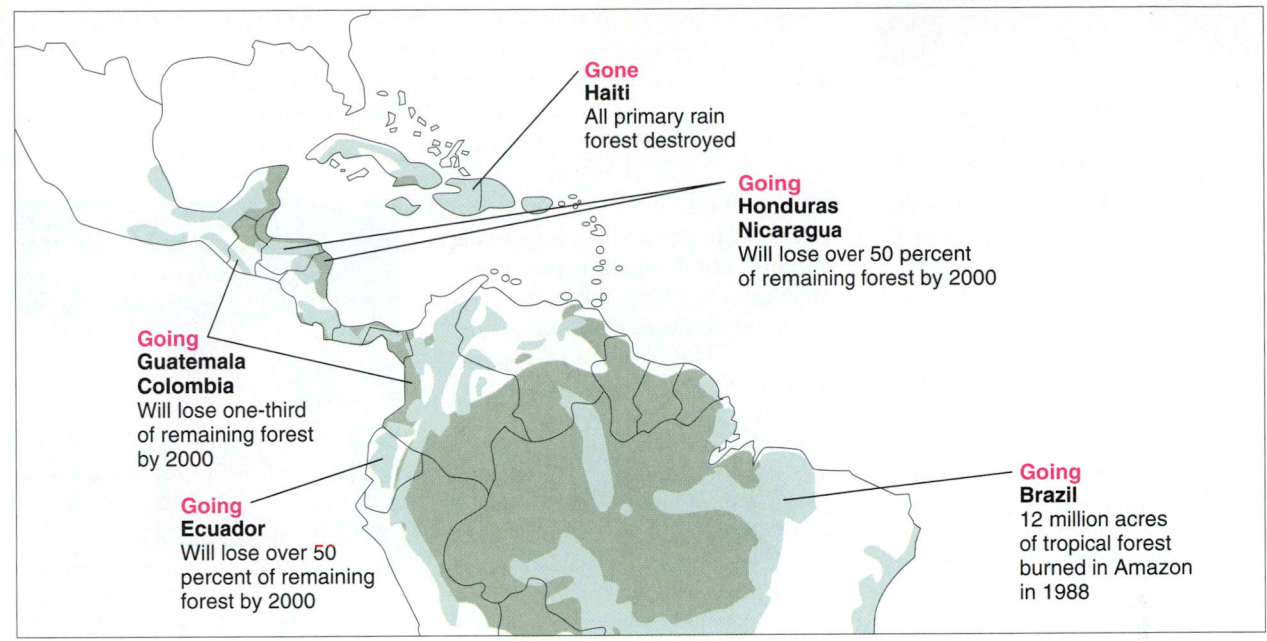

(a)

Gone
Haiti
All primary rain
forest destroyed

Going
Honduras
Nicaragua
Will lose over 50 percent
of remaining forest by 2000

Going
Guatemala
Colombia
Will lose one-third
of remaining forest
by 2000

Going
Ecuador
Will lose over 50
percent of remaining
forest by 2000

Going
Brazil
12 million acres
of tropical forest
burned in Amazon
in 1988

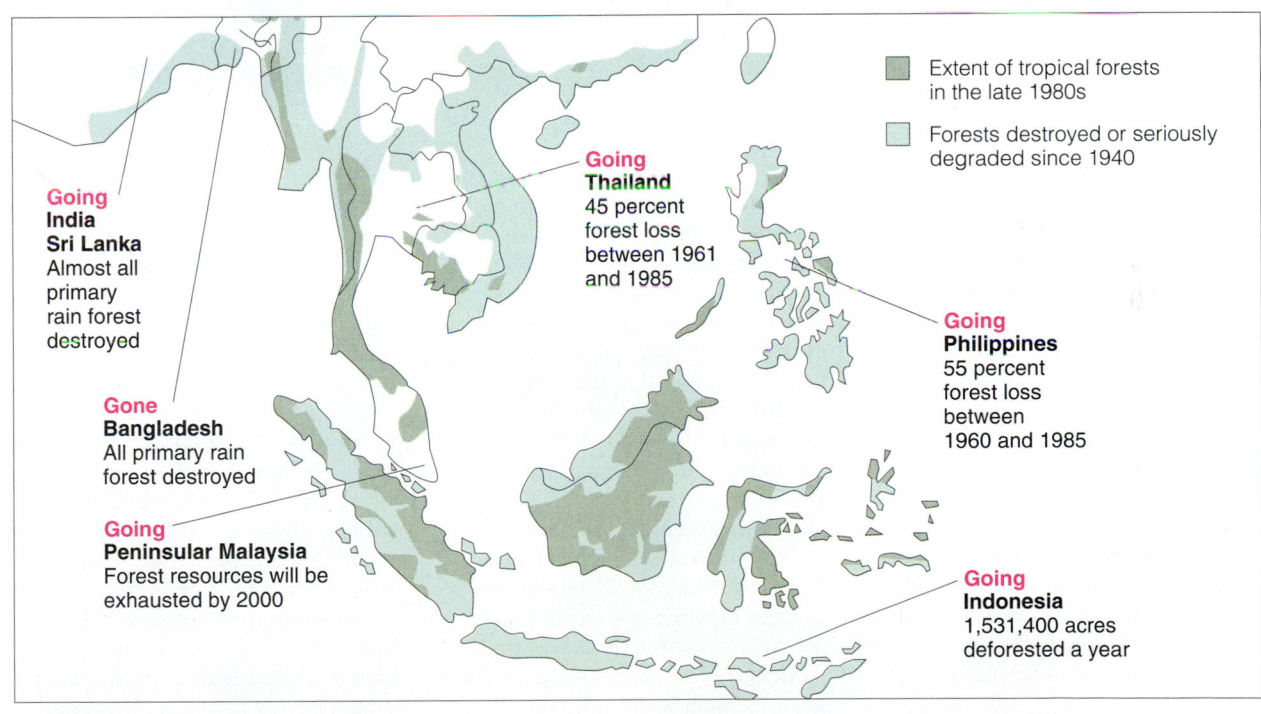

(b)

Going
India
Sri Lanka
Almost all
primary
rain forest
destroyed

Gone
Bangladesh
All primary rain
forest destroyed

Going
Peninsular Malaysia
Forest resources will be
exhausted by 2000

Going
Thailand
45 percent
forest loss
between 1961
and 1985

Going
Philippines
55 percent
forest loss
between
1960 and 1985

Going
Indonesia
1,531,400 acres
deforested a year

Extent of tropical forests
in the late 1980s

Forests destroyed or seriously
degraded since 1940

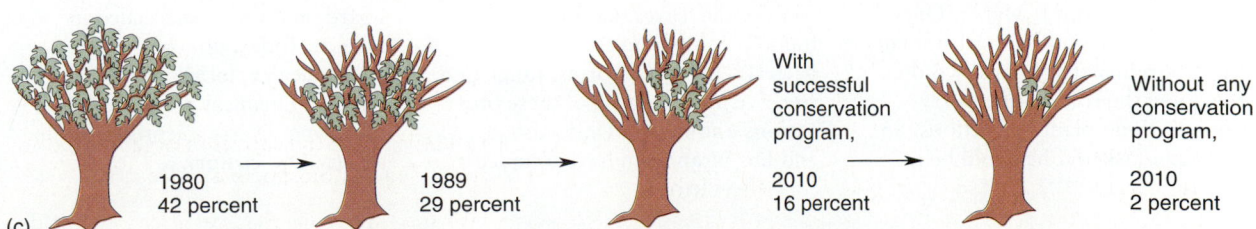

(c)

1980
42 percent

1989
29 percent

With
successful
conservation
program,

2010
16 percent

Without any
conservation
program,

2010
2 percent

FIGURE 38.12   Deforestation
Every year, the wholesale destruction of our rain forests is accelerating at an
alarming rate. (a) Destroyed forests of the Americas. (b) Destroyed forests of the
East Indies. (c) The disappearing forest in Guatemala, 1980 to 2010 (projected).
Sources: World Wildlife Fund Atlas of the Environment and the U.S. Agency for
International Development, Congressional Office of Technology Assessment.

Dr. Eloy Rodriguez
Ecologist-Phytochemist

As far back as he can remember, Dr. Eloy Rodriguez has been interested in biology. He was born and raised in Hildalgo County, Texas, the county with the lowest average income per person in the United States. His origins were humble from an economic perspective, but from the perspective of family, love of nature, and education his life was rich. His family was very close — 67 cousins lived within a five-block radius. Even though Dr. Rodriguez's father dropped out of school in the first grade and his mother in the seventh, the family emphasized the importance of education, and 64 of the 67 cousins earned degrees. Four eventually earned a Ph.D.

Dr. Rodriguez has not forgotten the stereotype that he had to face, which cast Hispanics and other minorities as people with limited futures in science and education. He remembers (even though he graduated in the top 5 percent of an accelerated program in high school) his high school counselor encouraging him not to go to college but rather to attend a good trade school more suited for students like him. He says, "Thank God I had enough wisdom to ignore such nonsense." Instead, he was accepted at the University of Texas at Austin. He worked summers in an automobile factory in Michigan, swept floors, washed dishes, and worked several other odd jobs to pay for his education. One of those odd jobs changed his life. He was hired to do general cleanup in the laboratory of plant chemist Tom J. Mabry. One day, one of the postdoctoral students told Rodriguez that he really didn't like doing the routine chemistry required to do plant extractions. So Rodriguez asked if he could help.

The postdoctoral student taught Eloy how to extract the compounds, and he loved doing the experiments. He also says, "My work in that laboratory changed my life."

Today Dr. Rodriguez is recognized as one of the most innovative scientists in the country. His work in phytochemistry and cell biology is internationally known. He had been a professor of biology at the University of California, Irvine, but recently moved to Cornell University as the James Perkins Endowed Chair of Environmental Biology. He and his numerous students are investigating chemicals derived from plants and their potential uses in medicine and in chemistry as alternative sources of energy and other useful products. His research has taken him all over the world.

In the deserts, he has studied plants for natural pesticides, oils, rubber, and food. In the rain forests of South America and Africa he has learned a lot about the medical uses of plants from the native medicine men. He truly is a real-life "medicine man," as depicted in the movie of the same name. Not only has Dr. Rodriguez learned about medical plants from indigenous peoples but also indirectly from wild animals like chimpanzees and gorillas.

Harvard anthropologist Richard Wrangham was observing sick chimpanzees in Africa and noted that they would choose the leaves from a particular plant very carefully and swallow them whole. He contacted Dr. Rodriguez and asked him to analyze the leaves. Dr. Rodriguez found a compound that kills bacteria, parasitic worms, fungi, and some viruses. Because of these observations and others, Dr. Rodriguez and Dr. Wrangham have founded a

new discipline they call *zoopharmacognosy* — the chemistry of plants used medically by animas.

Not only is Dr. Rodriguez a respected scientist, he is also committed to academic excellence in science and has been funded for numerous programs to rectify the underrepresentation of minorities in science. He is well known for his programs in science education from elementary schools to postdoctoral studies. He says, "As diversity continues to grow in our population base, so does the importance of tapping into these sources." He knows this from his own experience. He remembers while in elementary school having to write, "I will not speak Spanish in class" on the blackboard, and yet he succeeded in science. Today, not only is Dr. Rodriguez exploring diverse plants for their chemical properties, he is nurturing diverse scientists for the twenty-first century.

Today there is a serious biodiversity crisis. Animals and plant species are becoming extinct at an alarming rate due to the impact of humans. It is estimated that 3.5 million species of organisms are present on earth, that 2.5 million of them are yet unfound, and that 95 percent of them live in rain forests. Because of deforestation, some species will become extinct before we ever discover them. Because of habitat destruction, several species are in danger of extinction and several others have already become extinct.

Habitat destruction is not the only human cause of extinction and the endangerment of many species. Overhunting threatens many species (Figure 38.13). In Africa, elephants and rhinoceroses are almost extinct due to the black market in ivory and rhinoceros horns.

AN ENVIRONMENTAL PERSPECTIVE

- **As educated people, we must understand the natural world and respect its ecological principles in order to make responsible ecological decisions.**

In these last five chapters, we have discussed the interaction of organisms with their environment and with each other, and we have learned that these interactions are complex and delicate. We should be very cautious about the ways in which we intervene in these interactions. Otherwise, we create imbalances that result in destruction or change within the system. We saw in our discussion of lakes that we can destroy a lake not only by decreasing its nutrients (making it oligotrophic) but also by increasing its nutrients (making it eutrophic). Acts that seem unimportant can ultimately have far-reaching consequences.

For instance, clawed frogs were first brought to the United States for people who stock private aquariums and for laboratories to conduct pregnancy tests. However, some were released into the environment by accident, or owners just wanted to get rid of them. With no natural predators and a genetic aggressiveness, the frogs now pose a hazard to the native species in the rivers and lakes of Florida and other states.

Humans are living creatures, no greater or lesser than any other living creature in nature. However, as you all know, we are blessed (or damned) with the power of forethought and therefore with an unprecedented power to influence the natural world. Still seeing ourselves as rulers of the earth, we consume energy and resources at breakneck speed. Unless we

(a)

(b)

(c)

(d)

FIGURE 38.13   Endangered Species
The effects of the destructive impact of humans on the environment is effecting the extinction of species at an alarming rate. Over 25,000 species of organisms are considered endangered. Among them are (a) rhinoceroses, (b) tigers, (c) elephants, and (d) sea turtles.

reverse ourselves, our rape of the earth can only lead to our end — the death of our population. Biological principles do not change. A population can only exist within the carrying capacity of its environment, and that principle applies to humans as well as rabbits, mites, and all other species. Unchecked population growth and the attitude that our nonrenewable resources are unlimited have placed us in the position of rapidly reaching our carrying capacity.

Biological principles generally do not change, but fortunately our values can. It is a biological imperative that we channel our reasoning powers and endeavors to preserve or improve the quality of life on this earth for all organisms and that we realize the implications of our interventions in the delicate balance of life. Our long-term goals should be to understand life's principles and to live in harmony with those principles. Then and only then will we as a species be assured that our contributions to the earth and all the living creatures on it have been the result of a positive human endeavor.

SUMMARY

1. Because of our increased dependency on technology, we are producing huge amounts of pollutants that affect the quality of the atmosphere.
2. Some of these pollutants — CO_2, for example — contribute to an increasing atmospheric warming pattern called the "greenhouse effect."
3. Chlorofluorocarbons, which are used as propellants and coolants, are the major chemicals contributing to the breakdown of the ozone layer of the atmosphere. This breaking down of the ozone layer allows far more ultraviolet light to strike the earth. Increased amounts of ultraviolet light can increase skin cancer in humans and impact the reproductive success of some animals.
4. The oxides of nitrogen and sulfur that result from burning fossil fuels are released into the atmosphere. When combined with atmospheric water, they produce acid rain, which is damaging to plants and other organisms within the ecosystem.
5. Agriculture is also having an impact on the ecosystem, because we use huge quantities of water, fertilizers, and pesticides, which upset its delicate balance.
6. As pesticides pass from one trophic level to another, they become concentrated.
7. Solid wastes, radioactive wastes, and chemical wastes are difficult to dispose of without damaging the environment. The irresponsible disposal of waste has produced a serious ecological problem.

8. One of the major environmental problems today is deforestation, especially large areas of rain forest. This global destruction of forests is upsetting the delicate balance of our ecosystem.
9. We as humans must make every endeavor to resolve the political and ecological problems that our population is producing. We must modify our lifestyles to include a concern for the increasing damaging effect we have on our environment.

KEY TERMS

atmospheric pollutants
greenhouse effect
chlorofluorocarbons (CFCs)
ozone
acid rain
aquifers

biological magnification
meltdown
radioactive wastes
deforestation
extinction

REVIEW QUESTIONS

True/False

1. Human population and technology are having a destructive impact on the biosphere.
   true   false   (page 895)

2. The oxides of nitrogen and sulfur, when mixed with water in the air, are the chief components of acid rain.
   true   false   (page 897)

3. If we are going to live in harmony with our ecosystem, we must learn to respect the environment.
   true   false   (page 910)

4. Radioactive wastes produced by nuclear power plants (i.e., ^{239}Pu, plutonium) are easy to dispose of.
   true   false   (page 905)

Fill in the Blank

1. In the United States alone, we produce over _____ pounds of atmospheric pollutants per day. *(page 895)*

2. Each year we destroy over _____ million acres of rain forest. *(page 907)*

3. The destruction of the rain forest is also destroying many irreplaceable _____ species that may be used for medical purposes. *(page 907)*

4. Our throw-away philosophy is resulting in large amounts of _____ waste. *(page 905)*

Multiple Choice

1. Mass extinction of species has been a trend in the evolutionary history of the earth. However, if we have a mass extinction of many species in the next century, this would be different, because most likely:
   a. Human technology would be the cause
   b. A solar discharge would be the cause
   c. The climate would change rapidly within a year or two
   d. Extinction would be due to a giant earthquake
   (page 909)

2. Of the following, which consumes more energy per capita?
   a. South America
   b. Asia
   c. United States
   d. Africa
   (page 895)

3. In Puerto Rico, there was a real concern about mercury poisoning and public health. This concern illustrates what principle?
   a. Biological magnification
   b. Eutrophication
   c. Biomass
   d. Population distribution
   (page 904)

4. One solution to our increasing energy needs is solar energy because:
   a. Solar energy is not renewable
   b. It can be used directly as a fuel
   c. It depends on ^{239}Pu
   d. It is constantly available
   (page 905)

Discussion

1. As members of a technical society, we are used to creature comforts such as automobiles, electricity, and so on. What is your opinion of the environmental impact of these creature comforts?

2. We have emphasized the scope of the human destructive impact on our biosphere. Pick one environmental problem due to human interaction and provide a solution. For example, what solution would you propose to reduce the increasing concentration of CO_2 in the atmosphere?

3. If you were asked to give a presentation on the consequences of human activity for the biosphere, which topic would you discuss, and what solutions would you provide?

4. In 1992 at the Earth Summit Conference in Brazil, leaders of industrialized nations signed a treaty to phase out the use of CFCs. What do you think will become of that treaty?

5. Describe some potential consequences of acid rain, the greenhouse effect, ozone layer reduction, and deforestation.

6. After reading this chapter and other articles on the effects of human activity on the biosphere, how will you change your lifestyle to have a less destructive impact on the environment?

7. Native Americans generally embrace the philosophy that the earth is our mother and brings forth all life, that humans should respect and live in harmony with nature, and that we don't inherit the earth from our ancestors but borrow it from our children. Do you agree or disagree with this philosophy? Why or why not?.

Appendix
Classification of Organisms

KINGDOM MONERA

Prokaryotic, unicellular organisms. Nutrition mostly by absorption, with some photosynthetic and chemosynthetic organisms. Reproduction mostly asexual (fission, cell division, budding), with some sexual reproduction (conjugation).

Division Archaebacteria: methanogens, halophiles, thermophiles

Division Eubacteria: cyanobacteria

KINGDOM PROTISTA

Eukaryotic, unicellular organisms. Nutrition by photosynthesis, absorption, and ingestion. Reproduction asexual (cell division) and sexual (conjugation).

Phylum Sarcodina: amoebas

Phylum Gymnomycota: slime molds

Phylum Euglenophyta: euglenoids

Phylum Chrysophyta: diatoms, golden algae, yellow-green algae

Phylum Pyrrophyta: dinoflagellates

Phylum Charophyta: stoneworts

KINGDOM FUNGI

Eukaryotic, mostly multicellular organisms. Nutrition mostly by absorption. Reproduction both sexual and asexual (binary fission and budding).

Phylum Chytridiomycota: aquatic fungi

Phylum Oomycota: egg-forming fungi; water molds

Phylum Zygomycota: zygospore-forming fungi; bread mold

Phylum Ascomycota: mildews, truffles, yeasts

Phylum Basidiomycota: mushrooms, toadstools, club fungi

Phylum Deuteromycota: fungi imperfecti; *Penicillium*

KINGDOM PLANTAE

Eukaryotic, multicellular organisms. Nutrition mostly by photosynthesis. Reproduction primarily sexual, with some asexual reproduction (modified stems, parthenogenesis, vegetative propagation).

Division Chlorophyta: green algae

Division Phaeophyta: brown algae

Division Rhodophyta: red algae

Division Bryophyta: liverworts, hornworts, mosses

Division Psilophyta: whisk ferns

Division Lycophyta: club mosses

Division Sphenophyta: horsetails

Division Pterophyta: ferns

Division Coniferophyta: evergreens

Division Cycadophyta: cycads

Division Ginkgophyta: ginkgo

Division Anthophyta: flowering plants

Class Monocotyledones: monocots; grasses, lilies, corn

Class Dicotyledones: dicots; beans, daisies, daffodils

KINGDOM ANIMALIA

Eukaryotic, multicellular organisms. Nutrition primarily ingestive. Reproduction primarily sexual, with some asexual reproduction (budding, fission).

Phylum Porifera: sponges

Phylum Cnidaria: jellyfish, sea anemones, corals

Phylum Ctenophora: comb jellies and sea walnuts

Phylum Platyhelminthes: free-living flatworms, parasitic flukes, parasitic tapeworms

Phylum Nematoda: roundworms

Phylum Nematomorpha: horsehair worms

Phylum Rotifera: rotifers

Phylum Mollusca: snails, mussels, squids, octopuses

Phylum Annelida: marine worms, earthworms, leeches

Phylum Tardigrada: water bears

Phylum Anthropoda: joint-legged animals with exoskeletons

Class Crustacea: lobsters, shrimps, barnacles

Class Arachnida: spiders, scorpions, ticks

Class Insecta: dragonflies, cockroaches, crickets

Phylum Echinodermata: sea stars, sea urchins, sand dollars, sea lilies

Phylum Chordata: animals who at some time have a notochord, a dorsal nerve cord, and pharyngeal gill slits

Subphylum Vertebrata: vertebrates

Class Chondrichthyes: cartilaginous fishes; sharks, rays, skates

Class Osteichthyes: bony fishes

Class Amphibia: frogs, toads, salamanders

Class Reptilia: snakes, lizards, turtles, crocodiles

Class Aves: birds

Class Mammalia: bats, cats, elephants, humans

Glossary

abiogenesis (ăb′ē-ō-jĕn′ə-sĭs). See *spontaneous generation*.

abiotic components (ā′bī-ŏt′ik). The nonliving factors in the environment.

abscisic acid (ăb-sĭs′ĭk). A hormone that promotes dormancy and leaf drop in plants. Abbreviated *ABA*.

absorption. A method of taking in food in which the food molecules move directly through a membrane and into the organism. Common to the fungi.

absorption spectrum. A graphic representation of light-wave absorption patterns of a particular molecule.

acid. A substance that releases hydrogen ions in water; a proton donor, which lowers the pH of a solution.

acoelomates (ā-sēl′ō-māts). Animals that lack a body cavity between the gut and body wall.

actin (ăk′tən). A cytoplasmic contractile protein that is found in the thin myofilaments of the cytoskeleton or in the sarcomere of a muscle.

action potential. The quick depolarization and repolarization of a neuron that results in the transmission and propagation of a nerve impulse.

activation energy. See *minimum energy of activation*.

active site. The specific area of an enzyme that interacts with the molecule being catalyzed (the substrate).

active transport. The use of energy by carrier proteins to move substances against a concentration gradient from an area of lower concentration to an area of higher concentration.

adaptation. A change in an existing trait of an individual organism to accommodate new environmental conditions that promotes the survival and reproduction of the individual.

adaptive radiation. The evolution of several different specialized forms from a single primitive ancestor.

adenosine diphosphate (ə-dĕn′ə-sēn′ dī′fŏs′fāt). See *ADP*.

adenosine monophosphate (ə-dĕn′ə-sēn′ mŏn′ə-fŏs′fāt). See *AMP*.

adenosine triphosphate (ə-dĕn′ə-sēn′ trī′fŏs′fāt). See *ATP*.

adipose tissue. The type of connective tissue proper that specializes in fat storage.

ADP. *Adenosine diphosphate*, the molecule composed of adenine, ribose, and two phosphate groups that remain when one of ATP's energy-rich bonds is broken; the remaining phosphate bond can be broken to release energy.

adrenal cortex (ə-drē′nəl kôr′tĕks). The outer layer of the adrenal glands, which produces glucocorticoids, mineralocorticoids, and other steroids, as well as sex hormones.

adrenal glands (ə-drē′nəl). Endocrine glands located above the kidneys; composed of the adrenal cortex and the adrenal medulla.

adrenal medulla (ə-drē′nəl mə-dŭl′ə). The inner layer of the adrenal glands, which produces epinephrine and norepinephrine.

aerobic respiration (âr-ō′bĭk). A type of energy exchange that utilizes oxygen as the final electron acceptor.

Agnatha (ăg-năth′ə). The jawless fish, one of the oldest groups of fish on the earth; the eel-shaped lamprey and the hagfish are living examples of this group.

AIDS. *Acquired immune deficiency syndrome*, a sexually transmitted viral disease that renders its victims unable to fend off infections.

alga (ăl′gə). A nonvascular, eukaryotic, photosynthetic organism; may be single-celled, colonial, or multicellular.

allantois (ə-′lant-ə-wəs). A membrane sac that receives and collects nitrogenous wastes produced by a developing bird or reptile embryo.

allele (ə-lēl′). An individual member of a gene pair; one or more alternative forms of a gene at a specific gene locus.

allosteric enzyme (ăl′ə-stĕr′ĭk). An enzyme that has more than one binding site, which allows it to bind with a molecule other than the enzyme's substrate.

allosteric site (ăl′ə-stĕr′ĭk). The place on an allosteric enzyme where the nonsubstrate molecule binds to regulate enzyme activity.

alternation of generations. The pattern of reproduction exhibited by most plants, in which a diploid spore-producing phase alternates with a haploid gamete-producing phase.

altruism (ăl′trōō-ĭz′əm). Unselfish behavior that is beneficial to others at some risk to the altruistic individual.

alveoli (ăl-vē′ə-lī′). Microscopic air pockets that increase the surface area of the lungs and in which the exchange of gases occurs.

amino acid (ə-mē′nō). One of 20 monomers that combine to make proteins; each amino acid consists of a central carbon atom, an amine group, a carboxyl group, a hydrogen molecule, and an R-group.

amnion (ăm′nē-ən). In reptiles, birds, and mammals, an extraembryonic membrane that encloses the developing embryo and produces the fluid that surrounds the embryo.

amniote egg (ăm′nē-ŏt). The land egg common in birds and reptiles, consisting of a membrane called the amnion, which secretes a fluid and provides for a self-contained aquatic environment.

AMP. *Adenosine monophosphate,* the molecule that remains when a phosphate group is removed from ADP, consisting of adenosine, ribose, and one phosphate group.

amphibians. The first land vertebrates well-adapted for life both in water and on land; frogs and salamanders are common examples.

anaerobic respiration (ăn′ə-rō′bĭk). A type of energy exchange that does not use oxygen as the final electron acceptor.

analogous (ə-năl′ə-gəs). In morphology, describing structures that are similar in form but different in evolutionary origin.

anaphase (ăn′ə-fāz). The third stage of mitosis or meiosis II, in which the centromere separates and the chromosomes move to the opposite poles of the cell.

androgens (ăn′drə-jənz). A collective term used to describe the male sex hormones.

angiosperm (ăn′jē-ə-spûrm′). A flowering plant with covered seeds.

Animalia. One of the five kingdoms of life, consisting of multicellular ingestive organisms.

Annelida (ă′nə-lĭ′də). The phylum of the segmented worms; the earthworm is an example.

anterior pituitary (pĭ-tōō′ə-tĕr′ē). The portion of the pituitary gland that produces hormones that are important in growth and reproduction and that regulate the secretory activities of other endocrine glands.

anther (ăn′thər). The male part of the flower, which contains the microspore cells that develop into pollen; the pollen-producing part of the flower.

antibody (ăn′tĭ-bŏd′ē). A Y-shaped protein molecule produced by the immune system that binds with an antigen.

anticodon (ăn′tĭ-kō′dŏn). A three-nucleotide sequence on a tRNA molecule that corresponds to a specific codon on an mRNA molecule.

antigen (ăn′tə-jĕn). An antibody-generating substance; a polysaccharide or protein that stimulates the production of a specific antibody by the immune system.

aorta (ā-ôr′tə). The main artery that carries blood from the heart to the body.

apical meristem (ā′pə-kəl mĕr′ə-stĕm). In most plants, the structure at the tip of a developing root or shoot that is responsible for the primary growth of the root or stem.

appeasement behavior. Animal behavior characterized by submissive signals.

appendicular skeleton (ă-pĕn-dĭk′yə-lər). In vertebrates, the skeletal system formed by the appendages, such as the shoulders, hips, and legs.

aquifer. Underground water supply.

artery (är′tər-ē). A blood vessel that carries blood away from the heart to the arterioles and capillaries.

Arthropoda. The most diverse and most abundant phylum, consisting of animals with jointed limbs; it includes the insects, crabs, spiders, and millipedes.

asexual reproduction. A type of reproduction that does not involve the exchange of genetic material between two organisms.

associative neuron. See *interneuron.*

asters (ăs′tərz). Microtubules and microfilaments arranged in starburst-like structures that radiate from the centrioles of a cell during cell division.

asymmetry. A morphological category describing an organism with no resemblance between one side of its body and the other.

atmospheric pollutants. Gases and particles produced by human technology that alter or pollute the atmosphere.

atom. The smallest indivisible unit of an element that can enter a chemical reaction.

atomic mass. The number of protons in an atom plus the number of neutrons.

atomic number. The number of protons in an atom.

ATP. *Adenosine triphosphate,* a nucleotide consisting of ribose, adenine, and three phosphate groups that is involved in many biological energy-exchange reactions.

atrium (ā′trē-əm). A chamber of the heart that receives blood from the body.

autoimmune disease (ô′tō-ĭ-myōōn′). A disorder that causes the immune system to attack and destroy its own cells.

autonomic nervous system (ô′tə-nŏm′ĭk). In vertebrates, the network of nerve cells and their ganglia arising from the central nervous system, responsible for involuntary responses.

autosomes (ô′tə-sōmz). Nonsex chromosomes.

autotroph (ô′tə-trŏf). An organism capable of synthesizing its own food from simple inorganic substances (self-feeder).

auxin (ôk′sĭn). A class of natural or synthetic plant hormones that can influence cell elongation, phototropism, abscission, and apical dominance.

axial skeleton. In vertebrates, the system formed by the bones along the central axis of the body; the skull and spinal column.

axon (ăk′sŏn). The part of a neuron that transmits information away from the nerve cell body.

bacteriophage (băk-tîr′ē-ə-fāj′). A virus that infects bacterial cells. Also called a *phage.*

Barr body. A structure in the nuclei of female animal cells representing a condensed inactive X chromosome.

basal body. The organelle responsible for the formation of a cilium or flagellum, consisting of nine triplets of microtubules arranged in a circle.

base. A substance that releases hydroxide ions when dissolved in water.

base pairing. The combination pattern of the nitrogenous bases that form DNA and RNA; adenine always bonds with thymine in DNA and uracil in RNA, and guanine always bonds with cytosine.

Batesian mimicry. A palatable prey mimics an unpalatable species.

B-cell. A lymphocyte that produces antibodies in the presence of an antigen.

behavior. An animal's coordinated responses to its external and internal environment.

bilateral symmetry. A morphological classification describing an organism whose halves mirror each other.

binomial nomenclature (bī-nō′mē-əl nō′mən-clā′chər). The taxonomic convention of assigning two names to each organism, a genus name and a species name.

biogenesis (bī′ō-jĕn′ə-sĭs). The theory that life can arise only from life.

biogeochemical cycle. A pathway of elements through the environment.

biological magnification. The increasing concentration of chemicals in an organism as the chemical passes from one trophic level to another.

biomass. The mass (total dry weight) of all living organisms in a defined region.

biome (bī′ōm). A type of mature community characterized by a specific form of vegetation, as determined by environmental conditions.

biosphere. The sum of all living organisms on earth and their interactions with the physical and chemical environments.

biotic components (bī-ŏt′ĭk). The living factors in the environment.

biotic potential (bī-ŏt′ĭk). A population's maximum growth rate, unhindered by disease or shortages of raw materials.

biparental reproduction. The fusion of a haploid male sperm with a haploid female egg to produce a new diploid individual.

birds. A group of animals that all share a common body plan well-suited for flight; birds are warm-blooded, and some biologists group them with the reptiles.

blastula (blăs′chə-lə). In many animals, a stage in embryonic development consisting of a fluid-filled ball one cell layer thick.

blood complements. Plasma proteins that interact with antigens and antibodies to destroy foreign cells.

Bowman's capsule. A double-walled cup filled with fluid at the beginning of a nephron in the kidney.

budding. A type of asexual reproduction in which some of the undifferentiated cells of one individual form an outgrowth that breaks off and develops into a new individual.

bulk flow. A transport method that moves fluid through organisms along a pressure gradient from areas of greater water pressure to areas of less pressure.

C$_3$ cycle. See *Calvin-Benson cycle.*

C$_4$ carbon-fixing process. A light-independent scheme used by many desert plants in which carbon dioxide is fixed into four-carbon compounds.

calcitonin (kăl′sə-tō′nĭn). A hormone secreted by the thyroid gland that inhibits the release of calcium from bone storage sites.

Calorie. A kilocalorie; the amount of heat required to raise the temperature of 1000 grams of water 1 degree Celsius.

Calvin-Benson cycle. In photosynthesis, the light-independent reaction in which CO_2 from the air is reduced to form food with the aid of NADPH and ATP. Also called the C$_3$ *cycle.*

cAMP. See *cyclic adenosine monophosphate.*

capillaries. The smallest blood vessels in a closed circulatory system, where the exchange of gases, waste products, and other substances between the blood and cells occurs.

carbohydrate. Any organic molecule with the empirical formula $[C(H_2O)]_n$; an important energy source for and structural component in plants and animals.

carbon cycle. The biogeochemical cycle for carbon, in which plants take in carbon dioxide from the air, animals eat plants, and animals, plants, and fungi release carbon dioxide back into the environment.

carcinogen (kär-sĭn′ə-jən). A cancer-causing agent.

cardiac muscle. Muscle that forms the walls of the heart.

cardiovascular system. The circulatory system of vertebrates.

carrier. In genetics, an individual who is heterozygous for a particular condition; the individual does not express the condition but can pass it on to future generations.

carrying capacity. The maximum number of organisms that an ecosystem can support.

cartilage. The type of supportive connective tissue that serves as a precursor to bone, gives shape to flexible body parts like the ear, and cushions the ends of bones.

Casparian strip (kăs-pĕr′ē-ən). In plant roots, a layer of waxy impermeable material on the walls of the cells surrounding the vascular cylinder.

Castle-Hardy-Weinberg principle. A genetics principle stating that the frequency of alleles will remain constant from generation to generation in a large population of sexually reproducing diploid organisms if there are no disturbing factors.

catalyst. An agent that lowers the amount of energy necessary for a reaction to occur without becoming involved in the reaction.

cell. The fundamental unit of life; every cell is enclosed by a membrane and contains cytoplasm, which may contain a variety of organelles.

cell body. The part of a nerve cell that contains the nucleus and other cytoplasmic organelles.

cell cycle. The cell's pattern of growth, maintenance, and reproduction; the three phases of the cell cycle are interphase, mitosis, and cytokinesis.

cell theory. The statement that all living things come from preexisting cells and are composed of cells.

cellular respiration. The process in which cells release the energy stored in molecules.

cellular specialization. In true multicellular organisms, cells become specialized for special functions, for example, nerve cells, muscle cells, and so on.

cell wall. In the cells of plants, fungi, and bacteria, a structure that surrounds the plasma membrane and provides strength and rigidity to the cell.

central dogma. An explanation of how DNA codes for proteins, suggesting that the linear sequence of nitrogenous bases in the DNA molecule is transcribed onto an RNA molecule. The RNA is then translated into the linear sequence of amino acids, which results in protein.

central nervous system. In vertebrates, the information-processing centers in the brain and spinal cord.

centriole (sĕn′trē-ōl′). Paired microtubular organelles found near the nucleus in the cells of animals, protistans, fungi, and some plants. The centriole establishes the plane of cell division.

centromere (sĕn′trə-mîr′). A region on the chromosome that connects the two strands of a chromosome, and to which spindle fibers are attached.

cephalization (sĕf′ə-lə-zā′shən). The concentration of sensory structures in the head end of an organism.

cephlochordates (sĕf′lō-kôrd′āts). A subphylum of the phylum Chordata consisting of invertebrate animals; for example, the lancelets.

CF₁ complex. In plants, a structure on the outside of the thylakoid that connects the lumen to the stroma and contains the enzymes involved in phosphorylating ADP.

chaparral (shăp′ə-răl). See *scrubland*.

Chargaff's rule. In DNA structure, a rule holding that the amount of adenine always equals the amount of thymine and the amount of guanine always equals the amount of cytosine.

chemical bond. An energy relationship that holds the atoms in a molecule together.

chemical reaction. The breaking and reforming of chemical bonds.

chemiosmotic theory (kĕm′ē-ŏz-mŏt′ĭk). An explanation of the phosphorylation of ADP suggesting that the necessary energy comes from an energy gradient established across the inner membranes of the mitochondrion and the thylakoid membranes of the chloroplast of a cell.

chemosynthetic (kĕm′ō-sĭn-thĕt′ĭk). Using energy from inorganic reactions to synthesize food (many bacteria are chemosynthetic).

chiasma (kī-ăz′mə). During crossing over in meiosis, the point where the chromosomes break and exchange sections of genetic material (plural, *chiasmata*).

chlorophyll (klôr′ə-fĭl). The major photosynthetic green pigment in plants.

chloroplasts (klôr′ə-plăsts′). Plant cell organelles that trap light energy to be used in photosynthesis.

chondrichthyes (kŏn-drĭk′thīz). The cartilaginous fishes, such as the sharks, skates, and rays; these fish are called cartilaginous because their endoskeletons consist of cartilage rather than bone.

chordates. Animals that are bilateral deuterostomes characterized by a notochord, gill slits, dorsal hollow nerve cord, and a postanal tail; it consists of two subphyla: those that lack a vertebral column, the invertebrates, and those that have one, the vertebrates.

chorion (kôr′ē-ŏn). The outermost membrane around an embryo in reptiles, birds, and mammals that functions in gas exchange between the embryo and its environment. The chorion is part of the placenta in placental mammals.

chromatid (krō′mə-tĭd). One side, or arm, of a chromosome; one of the two identical replicas of a chromosome held together by a centromere.

chromatin (krō′mə-tĭn). Masses of DNA and protein in a cell's nucleoplasm that condense to form chromosomes.

chromosome (krō′mə-sōm). In eukaryotic cells, genetic material and protein in its condensed form during nuclear division; the vehicle for transporting genetic information from one generation to the next.

chyme (kīm). A light mixture of food and gastric juices produced in the stomach; the product of chemical and mechanical digestion.

cilia (sĭl′lē-ə). Thin, hairlike projections that function in cell movement; similar to flagella but smaller and more numerous.

citric acid cycle. See *Krebs cycle*.

cladistics (klə-dĭs′tĭks). A taxonomic approach based on the chronological order in which branches arise on a phylogenetic tree, not based on morphology.

class. The third most inclusive taxonomic category, below *phylum* or *division* and above *order*.

classical conditioning. Creating an association between a stimulus that causes a reaction and one that does not cause the reaction.

cleavage. (1) The division of the zygote into smaller cells; (2) the breaking down of larger molecules into simpler subunits.

climate The general pattern of weather conditions over time in a specific region.

climax community. A mature community, in which the basic composition of the populations does not change unless the climate changes.

clone. An identical genetic copy of an organism produced from a single cell.

closed circulatory system. A complex system for circulating blood through the body, in which the blood remains in the blood vessels.

clumping. A pattern of distribution in which members of a species cluster in particular locations within a region.

Cnidaria (nī-dâr′ĭ-ə). The "stinging-cell" animals, such as the jellyfish, are multicellular and have two germ layers.

cnidocysts (nī′dō-sĭsts). Specialized stinging cells found in members of the phylum Cnidaria.

coacervate (kō-ăs′ər-vāt′). A droplet formed by the suspension of particles in a liquid medium; important to some theories of the evolution of life.

codominance. A genetic condition in which both alleles are expressed in a heterozygous individual; neither allele is dominant over the other.

codon. Three nucleotides in a sequence along the DNA molecule that serve as a code from which the mRNA is transcribed; or the three sequential nucleotides in mRNA that code for a specific amino acid.

coelom (sē′lōm). A body cavity lined with peritoneum; the presence of coelom is an important taxonomic criterion.

coenzyme. A small organic molecule essential for some enzymatic reactions.

coevolution. The process by which two different organisms evolve so that one is influenced by the other.

cofactor. An ion that is necessary for an enzyme to function.

collar cell. A flagellated cell lining the spongocoel in sponges, sometimes called a choanocyte.

collenchyma cells (kə-lĕng′kə-mə). Slightly rigid cells that provide support for a plant.

colon. See *large intestine.*

colonial hypothesis. A well-accepted hypothesis that animals arose from heterotropic colonial flagellates.

commensalism. A beneficial interaction between two species in which one species benefits more than the other, and the other is not helped or harmed.

community. In ecology, all the populations living in a specified area.

compact bone. Dense bone cells with mineral deposits that impart strength and rigidity to the bone.

companion cell. In plants, a specialized parenchyma cell with a nucleus that appears to carry out the metabolic activities of the sieve tube elements, which lack their own nuclei.

competitive exclusion principle. A theoretical principle that no two species occupy the same niche at the same time.

complement. See *blood complements.*

complete digestive tract. A digestive system that begins with an oral opening, ends with an anal opening, and is surrounded by a body cavity.

compound. In chemistry, a substance composed of two or more elements bonded in a fixed ratio.

concentration gradient. The difference in concentration between the place where most of a substance is and the place toward which the substance is moving.

conditioned response. A behavioral response that becomes associated with a new stimulus; a form of learning.

cones. (1) Photoreceptor cells in the eye that are responsible for color vision and visual acuity; (2) the reproductive structure of conifers.

coniferous forest. See *taiga.*

conjugation. A form of sexual reproduction in which the genetic information of one cell is directly transferred to another cell.

connective tissue. In animals, loosely packed cells that bind and support other tissues.

consolidation. The transfer of information from short- to long-term memory.

consumer. A heterotroph; an animal that eats plants, fungi, or other animals.

continuous variation. A type of inheritance in which a trait manifests itself on a continuum of expression from one extreme to the other.

convergent evolution. The formation of different structures that look alike and are adapted to similar environmental conditions but do not share a common ancestor.

copulation. Sexual intercourse.

cork. The protective outer covering of woody trees.

cork cambium. In plants, a type of lateral meristem that encases the vascular cambium and gives rise to the periderm.

corpus luteum (kôr′pəs lōō′tē-əm). A glandular structure that results after ovulation has occurred in the follicle. The corpus luteum secretes the hormones estrogen and progesterone, which help to maintain the thickness of the endometrium.

cortex (kôr′tĕks). (1) In plants, a group of loosely packed, specialized parenchyma cells that make up the largest area of a young root and take in oxygen from the soil; (2) in animals, the outer layer of an organ. In general, cortex means "rind."

cotyledon (kŏt′ə-lēd′n). The food storage structure in the seeds of flowering plants; the "seed leaf."

covalent bond (kō-vā′lənt). A chemical bond between two or more atoms in which a pair or pairs of valence electrons are shared.

crossing over. The exchange of genetic material when adjacent chromatids of a homologous pair break during meiosis and exchange material; this can result in genetic recombination.

cross-pollination. The process of removing the anthers from plants, then brushing the anthers of one plant over the stigma of another.

cryptic coloration. A type of protective camouflage in which the animal appears similar to elements in its biome, like a walking stick that looks like a twig.

Ctenophora (těn-ō-fôr′ə). The group of animals known as the comb jellies, which resemble the Cnidaria.

cyanobacteria (sī′ə-nō-băk-tĭr′ē-ə). Prokaryotic single-celled organisms that perform a type of photosynthesis. Formerly called the *blue-green algae*.

cyclic adenosine monophosphate (sĭk′lĭk ə-děn′ə-sĭn mŏn′ə-fŏs′fāt). A nucleotide that functions as a chemical messenger involved in the mechanism of action of proteinlike hormones. Abbreviated *cyclic AMP* or *cAMP*.

cyclic photophosphorylation (sĭk′lĭk fō′tō-fŏs′fôr-ə-lā′shən). An electron transfer pathway that moves electrons excited by light energy from photosystem I reaction centers down an electron transport chain and back to the same type of chlorophyll molecule from which they came.

cytochromes (sī′tə-krōmz). Proteins and enzyme molecules containing iron that accept electrons during electron transfer.

cytokinesis (sī′tō-kĭ-nē′sĭs). The cytoplasmic division that usually follows nuclear division.

cytokinins (sī′tō-kīn′ĭnz). Plant hormones that stimulate cell division.

cytoplasmic region (sī′tə-plăz′mĭk). The area of a cell enclosed by the plasma membrane but not including the nucleus.

cytoskeleton. A fibrous network consisting of microtubules, microfilaments, and other structures in a cell's cytoplasm that gives shape to the cell, provides locations for the attachment of organelles, and arranges structures within the cell.

daughter cells. The cells that result from cell division.

day-neutral plant. A plant that flowers in spring, summer, or fall, regardless of day length.

deciduous forest (dĭ-sĭj′oo-əs). A terrestrial biome dominated by trees that shed their leaves.

decomposers. Bacteria or fungi that live on dead organisms and break down complex molecules so that other organisms can use the component atoms again.

deforestation. The removal of trees and vegetation from large tracts of land.

dehydration synthesis. A reaction that results in the synthesis of larger molecules from smaller ones by removing two hydrogen atoms and one oxygen atom, which form water molecules.

deletion. A chromosomal abnormality in which part of a chromosome is missing.

deme. A group of organisms of a species that is isolated from another group of the same species.

dendrites (děn′drīts). Structures on nerve cells that receive information and pass it on to the cell body.

dense connective tissue. The type of connective tissue that connects muscle to bone (tendons) or bone to bone (ligaments).

density-dependent mechanism. A selective pressure affecting population size, resulting from the number of individuals living in a habitat.

density-independent mechanism. A factor (for example, fire) unrelated to the size of a population that decreases the number of individuals in the population.

deoxyribonucleic acid (dē-ŏk′sī-rī′bō-noo-klā′ĭk). See *DNA*.

deoxyribose (dē-ŏk′sī-rī′bōs). The five-carbon sugar in DNA nucleotides; similar to ribose except that it has one less oxygen.

dependent assortment. The assortment pattern of genes that are on the same chromosome and usually cannot assort independently of one another.

dermis (dŭr′məs). In animals, the layer of skin below the epidermis, containing oil glands, the ducts of sweat glands, blood vessels, nerves, and muscle.

desert. A region of falling warm air that gets less than 10 inches of rain a year; one of the major terrestrial biomes.

deuterostomes (doo′tĕ-rō-stōmz). An evolutionary branch of coelomate animals based on embryonic development that includes the echinoderms and the chordates.

diaphragm (dī′ə-frăm). In mammals, the large muscle separating the chest cavity from the abdominal cavity that is involved in breathing.

diastole (dī-ăs′tə-lē′). The stage of the cardiac cycle when the heart muscle relaxes.

dicot (dī′kŏt). Short for *dicotyledon*; a flowering plant with two cotyledons present in its seed.

differentiation. The process by which a cell becomes specialized for one particular role.

diffusion (dĭ-fyoo′zhən). The random movement of molecules from an area of higher concentration to an area of lower concentration, resulting in a uniform distribution of the molecules.

digestion. (1) *Chemical digestion*, the enzymatic process in which water is added to complex food molecules, breaking them down into smaller molecules; (2) *mechanical digestion*, the process in which the teeth and the muscles that line the digestive cavity or tract churn and grind the food, helping to liquefy it.

dihybrid cross (dī-hī′brĭd). In classical genetics, a cross involving only two traits.

diplohaplontic life cycle. The life cycle of plants that combines a haploid stage and a diploid stage.

diploid number. In organisms with homologous pairs of chromosomes, the number of chromosome parts. Also called the *2n number*.

diplontic life cycle. The life cycle that occurs in most multicellular animals, in which the adult is diploid.

disaccharide (dī-săk′ə-rīd). A carbohydrate formed when two monosaccharides (two sugar units) are joined together.

divergent evolution. The formation of different species from a common ancestor.

division. The second most inclusive taxonomic category in the kingdoms Plantae and Fungi; analogous to *phylum* in the other kingdoms.

DNA. *Deoxyribonucleic acid*, the double-stranded,

helically coiled nucleic acid that contains an organism's genetic material.

DNA polymerase (pŏl′ə-mə-rās′). An enzyme necessary for DNA replication.

dominance, concept of. One of Mendel's laws, which states that some genetic factors can mask the expression of others.

dominance hierarchy. A form of intraspecific competition that allows many societies to function.

dominant allele. In classical genetics, a member of a gene pair that masks the expression of the other member.

double fertilization. The process of seed formation in flowering plants; one sperm penetrates the egg, and another sperm joins with the polar nuclei to create the endosperm.

double helix. The structure of a DNA molecule; a twisted ladder configuration with the uprights made of a sugar-phosphate combination and the rungs made of base pairs.

Down syndrome. A genetic birth defect in which there are three chromosomes at location number 21, instead of two. This syndrome can also be due to translocation.

duodenum (doo′ə-dē′nəm). The section of the small intestine nearest the stomach, where most of the chemical breakdown and some absorption of food occurs.

duplication. A chromosomal abnormality in which a portion of a chromosome is accidentally reproduced.

Echinodermata (ĭ-kī′nə-dûrm-ä′tə). The animal phylum consisting of sea stars and their relatives, the "spiny-skin" animals.

ecological succession. The transitions of a species in a biological community, usually after a biological disturbance.

ecology. The study of the interactions between organisms and the environment.

ecosystem. All the communities and raw materials within a particular region and the interactions that occur between the organisms and the environment.

ectoderm. The outermost layer of an animal embryo, which eventually develops into the skin and nervous system.

ectotherm. An animal whose body temperature varies according to the temperature of its surroundings.

egg. The female sex cell.

electromagnetic spectrum. The continuous range of low-frequency to high-frequency waves of radiation from the sun.

electron. A negatively charged particle of negligible mass that revolves around an atom's nucleus and takes part in chemical reactions.

electron carrier. A molecule commonly found in electron transport systems that transfers electrons from one reaction site to another.

element. In chemistry, a substance made up of only one kind of atom.

embryo sac. (1) The female reproductive structure of a flowering plant, containing the egg and the polar nuclei; (2) a general term for the sac in which an animal embryo develops.

endergonic (ĕn-dər-gŏn′ĭk). Describing a reaction that requires energy.

endocrine gland (ĕn′dō-krĭn). A structure without ducts that secretes hormones directly into the bloodstream.

endocrine system (ĕn′dō-krĭn). In animals, an interrelated complex of glands, tissues, cells, and the hormones they produce.

endocytosis (ĕn′dō-sī-tō′sĭs). A form of active transport in which the cell membrane surrounds a substance and brings it into the cell.

endoderm. The inner layer of the gastrula in an animal embryo, which develops into the gut and other internal organs.

endoplasmic reticulum (ĕn′dō-plăs′mĭk rĭ-tĭk′yə-ləm). A double-membraned system of tubes that functions as the cell's interconnecting transport system.

endorphins (ĕn-dôr′fənz). Morphinelike substances produced by the brain in response to pain, causing a state of euphoria.

endoskeleton. In vertebrates, the internal skeleton made up of bone and cartilage.

endosperm. In the seeds of flowering plants, a tissue that surrounds the embryonic plant and provides it with nourishment.

endotherm. An animal whose body maintains a relatively constant internal temperature.

energy. The capacity or potential to do work.

energy level. A path traced by electrons as they travel around an atom's nucleus; the location in which the electrons of an atom are usually found. Also called an *energy shell.*

energy-rich bond. A phosphate bond on the ATP or ADP molecule that can be broken to release a large amount of energy.

entropy (ĕn′trə-pē). The tendency to become disordered and unavailable for useful work; a measure of the tendency toward disorder in a system.

environment. The surroundings of an organism.

enzyme. An organic catalyst, a special protein molecule in living things that lowers the amount of energy necessary for a chemical reaction to occur.

epidermis (ĕp′ə-dər′mĭs). (1) In animals, the outer layer of the skin; (2) in plants, the outer layer of cells in the leaves, root, and stems.

epiglottis (ĕp′ə-glôt′əs). A flap of tissue that covers the opening of the larynx during swallowing.

epinephrine (ĕp′ə-nĕf′rən). A hormone produced by the adrenal medulla and nerve synapses of the nervous system that has many functions: in general, it increases muscle strength and sensory alertness.

epistasis (ĭ-pĭs′tə-sĭs). The masking of the effect of one gene pair by another gene pair.

epithelial tissues (ĕp′ə-thē′lē-əl). Sheets of cells that cover an organism and line its cavities.

erythrocyte (ĭ-rĭth′rō-sīt). See *red blood cells*.

estrogen. A collective term for a group of hormones produced by the ovaries that are usually involved in the development of female secondary sex characteristics.

estuary (ĕs′chōō-ĕr′ē). A region, like mudflats or tidal marshes, where salt and fresh water mix.

ethology. The study of animal behavior, usually in the natural environment.

ethylene (ĕth′ə-lēn). A simple gas influential in fruit ripening, sex determination in single-sexed flowers, and regulation of stem curvature.

eukaryote (yōō-kăr′ē-ōt). A type of cell with a membrane-bound nucleus; all life forms except for viruses, bacteria, and cyanobacteria have eukaryotic cells.

eutrophication (yōō-trō′fə-kā′shən). The process by which water becomes more and more rich with nutrients.

eutrophic lake (yōō-trō′fĭk). A nutrient-rich lake.

evolution. Gradual change; in biology, the changes in the frequency of the alleles of a population over time.

excretion. The passing of water and solutes out of an organism's body, as through the kidneys in vertebrates.

exergonic (ĕk′sər-gŏn′ĭk). Describing a reaction that releases energy.

exocrine gland (ĕk′sə-krĭn). A structure with ducts that releases secretions into areas of an animal's body or onto its surface.

exocytosis (ĕk′sō-sī-tō′sĭs). The process in which a vesicle fuses with a plasma membrane and releases its contents to the outside of the cell.

exon (ĕk′sŏn). A nucleotide sequence that is expressed as part of a cell's genome.

exoskeleton. An external skeleton composed of chitin, such as in the arthropods.

experimental science. A method of science based upon experimentation.

exponential growth. A pattern of population growth in which the number of individuals increases at a continuously accelerating rate (2, 4, 16, for instance).

extinction. The end of the lineage of a species.

F_1 generation. See *first filial generation*.

F_1 complex with ATP synthetase. A structure that connects the mitochondrion's outer and inner compartments; the site of ATP synthetase enzymes involved in phosphorylation.

F_2 generation. See *second filial generation*.

facilitated transport. Passive transport through a cell membrane in which carrier proteins move molecules along a concentration gradient.

fact memory. The ability to retain explicit pieces of information like names, places, and dates.

family. In taxonomy, the category below *order* and above *genus*.

fermentation. A form of anaerobic respiration; for example, in yeast fermentation, the process that converts pyruvate to carbon dioxide and ethyl alcohol.

fertilization. The fusion of a sperm nucleus with an egg nucleus, resulting in the zygote.

fetus. The human embryo eight weeks after conception.

fibroblasts. Loose connective tissue cells that secrete flexible protein substances such as collagen.

first filial generation. In classical genetics, the first set of offspring resulting from a cross.

first law of thermodynamics. A fundamental law of physics that states that energy cannot be created or destroyed, but only changed in form.

fission. A type of asexual reproduction in which a one-celled organism splits to produce two or more individuals.

fixed-action pattern. A behavior pattern consisting of several elements performed in a coordinated, rigid sequence.

flagella (flə-jĕl′ə). Thin, hairlike projections that function in cell movement; similar to cilia but longer and fewer in number per cell.

fluid mosaic model. The currently accepted description of cell membranes, which states that all membranes consist of a fluid lipid bilayer.

follicle (fŏl′ĭ-kəl). In mammals, the structure in the ovary that houses the ovum.

follicle-stimulating hormone. The hormone produced in the anterior pituitary that aids in sperm production in males and maturation of the follicle in females. Abbreviated *FSH*.

food chain. The feeding pattern of the organisms in a community; who eats whom.

food web. A pattern of energy exchange throughout an ecosystem, formed by the interaction of many food chains.

founder effect. A type of genetic drift that occurs when a new population is begun by a group of individuals (or founders).

fragile-X syndrome. A syndrome involving mental retardation usually in males, associated with the breaking off of the tip of an X chromosome.

fragmentation. A type of asexual reproduction in which a piece of the organism breaks off and develops into an independent adult through cell division.

frameshift mutation. A change in genetic constitution that occurs when a single nucleotide is omitted from or inserted into the sequence of a DNA strand, changing the three-letter base groupings.

freshwater biome. Any mature aquatic community existing around still or moving fresh water.

fruit. One or more ripened ovaries of a flowering plant; retains and protects the seed and aids in seed dispersal.

functional group. A group of atoms that provides a distinctive chemical property to any molecule of which the group is a part.

Fungi (fŭn′jī). One of the five kingdoms of life, consisting of heterotrophic, absorptive organisms, such as molds, mushrooms, and mildew.

G₁ phase. The first stage of interphase in the cell cycle, when the cell maintains homeostasis and prepares for DNA replication.

G₂ phase. The last phase of interphase in the cell cycle, when the cell prepares for organelle replication after DNA replication has occurred.

gallbladder. A saclike structure in the digestive system where bile is stored.

gametes (găm′ētz). Haploid sex cells that unite to form a diploid zygote.

gametogenesis (gə-mēt′ə-jĕn′ə-sĭs). In adult diploid animals, the process in which specialized cells in the gonads undergo meiosis to become haploid gametes.

gametophyte generation (gə-mē′tō-fīt). In plants that demonstrate alternation of generations, a haploid multicellular plant body that produces gametes which unite through fertilization to form a diploid zygote.

ganglia (găng′glē-ə). Masses of nerve cell bodies grouped together.

gastrula (găs′trōō-lə). In many animals, the stage of embryonic development when a primitive gut and three germ layers form.

gene. A sequence of nucleotides that functions as a discrete unit; the mechanism of transmission of a particular quality from parents to offspring.

gene flow. A change in allele frequencies created when organisms move out of or into a population.

gene locus. See *locus.*

gene pool. All the alleles of all the genes of the individuals of a population.

gene splicing. The use of specific restriction and ligase enzymes to cut DNA, insert foreign DNA, and rejoin DNA molecules.

genetic continuity. A biological principle stating that the perpetuation of each species leads to a continuation of the genetics of that species.

genetic drift. Change in gene frequencies in a population as a result of chance rather than natural selection.

genetic equilibrium. The maintenance of a constant frequency of alleles in a large population of sexually reproducing diploid organisms from generation to generation.

genetic recombination. The rearrangement of genetic information to produce genetically unique offspring.

genome (jē′nōm). All of a cell's or an organism's genetic characteristics.

genotype (jĕn′ō-tīp). The genetic makeup of an organism.

genus (jēn′ŭs). The second most specific taxonomic category, and the first term in a species' binomial name.

geotropism (jē-ŏt′rə-pĭz′əm). See *gravitropism.*

gibberellins (jĭb′ə-rĕl′ənz). A family of hormones found in plants that promote stem elongation, seed germination, differentiation, and other effects.

gills. Modifications of the body wall that maximize gas exchange; a respiratory organ.

gland. In animals, a structure that is specialized to secrete substances.

glandular tissue. The type of tissue whose cells specialize in secretion.

glial cells (glē′əl). See *neuroglia.*

glomerulus (glə-mĕr′ə-ləs). A ball of capillaries within Bowman's capsule in the kidney, where blood filtration begins.

glucagon (glōō′kə-gŏn). A hormone produced in the pancreas that activates the breakdown of glycogen to glucose.

glycolysis (glī-kŏl′ə-sĭs). A series of respiratory reactions in which glucose is broken down into two molecules of pyruvate, releasing energy to make ATP from ADP; glycolysis can occur in either the presence or absence of oxygen.

Golgi apparatus (gōl′jē). A membranous organelle that receives molecules produced in the endoplasmic reticulum and processes them into complex macromolecules.

gonadotropin-releasing hormone (gō-năd′ə-trō′pĭn). A hormone produced in the hypothalamus that signals the pituitary to release follicle-stimulating hormone and interstitial cell–stimulating hormone. Abbreviated *GnRH.*

gonadotropins (gō-năd′ə-trō′pĭnz). Interstitial cell–stimulating hormone and follicle-stimulating hormone, hormones produced in the pituitary that affect the gonads and their secretions.

gonads (gō′nădz). The site of sex cell production in adult diploid animals.

gradualism. The belief that evolutionary change is constant for all organisms and is a slow and gradual process.

granum. A stack of thylakoids in the chloroplast.

grasslands. Temperate or tropical regions of rolling flat terrain; one of the major terrestrial biomes.

gravitropism (grăv′ĭ-trō′pĭz′əm). The effect of gravity on the structures of a plant, resulting in directional growth to or away from gravity.

greenhouse effect. The rise in the temperature of the earth caused by increased carbon dioxide in the atmosphere that traps heat radiating from the earth.

gross primary productivity. The total amount of organic material produced in an ecosystem per unit time.

ground meristem. In plants, a meristematic tissue that differentiates into the ground tissues.

guard cells. In plants, cells on either side of the stomata that change shape and control the opening and closing of the stomata, which regulates the passage of gases into and out of the leaf.

gymnosperm (jĭm′nə-spûrm). A nonflowering seed plant whose seeds have no protective covering.

habitat. The place where an organism lives.

habituation. The process of learning not to respond to a stimulus that is frequently repeated.

haploid number. The number of chromosomes that a parent contributes to the offspring; since there are two parents, the haploid number represents half the chromosomes the offspring will ultimately possess. Also called the *n number.*

haplontic life cycle. The life cycle of the lower unicellular plants, protistans, and many fungi, in which the adult is haploid.

Hardy-Weinberg principle. See *Castle-Hardy-Weinberg principle*.

hemoglobin. A complex globular protein molecule that transports oxygen through the body; usually found in red blood cells.

hemophilia. A group of sex-linked diseases in which the blood fails to clot normally.

heredity. The inheritance of genetic information by offspring from parents.

hermaphrodite (hər-măf′rə-dīt). An organism that has both male and female reproductive organs.

heterospory (hĕt′-ə-rə-spôr′-ē). In plants, a type of spore production in which two different types of spores are produced: a large megaspore, which develops into the female gametophyte, and a smaller microspore, which develops into the male gametophyte.

heterotroph (hĕt′ər-ə-trŏf). An organism that cannot manufacture its own food and so gets its energy by eating other organisms.

heterozygote (hĕt′ər-ə-zī′gōt). An organism in which each member of a pair of alleles is different for a particular trait, usually dominant and recessive.

hierarchy. With respect to behavior, an organization within groupings in which there are dominant and subordinate individuals.

histamine (hĭs′tə-mēn). A protein secreted by cells that dilates the blood vessels and causes the capillaries to become more permeable.

histone. A positively charged protein molecule associated with eukaryotic DNA and possibly involved in inhibiting transcription in eukaryotic cells.

homeo box. Master genes that regulate the activation of other genes involved in development; the sequence of DNA in these genes is very similar in many species.

homeostasis (hō′mē-ə-stā′sĭs). In higher animals, the maintenance of a constant internal environment in the face of external changes.

homologous (hŏ-mŏl′ə-gəs). Describing structures that are similar and have the same evolutionary origin.

homologue. One member of a chromosome pair.

homospory (hŏm′ə-spôr′ē). A type of spore production found in some plants in which only one kind of spore is produced.

homozygote (hō′mə-zī-gōt). An organism whose paired alleles are the same for a particular trait.

hormone. A chemical produced by specialized glands, tissues, or cells in animals or by special areas in plants that regulates an organism's internal activities.

human chorionic gonadotropin. A hormone released by the chorionic membrane of the developing embryo that prevents the breakdown of the corpus luteum. Abbreviated *HCG*.

hydrogen bond. A weak electrostatic attraction between two or more polar molecules in which the negatively charged section of one molecule is attracted to the positively charged section of another.

hydrolysis (hī-drŏl′ĭ-sĭs). The enzymatic addition of water to macromolecules, causing the macromolecules to split into their component subunits.

hydrophilic. "Water-loving," describing polar molecules that form hydrogen bonds with water and thus dissolve.

hydrophobic. "Water-fearing," describing molecules that are not polar and therefore do not dissolve in water.

hypertonic. Describing a solution that has a higher concentration of solutes than another solution.

hypothalamus (hī′pə-thăl′ə-məs). A region at the base of the brain that influences the activities of the pituitary and mediates neural signals from other brain centers, resulting in the regulation of many visceral activities.

hypothesis (hī-pŏth′ə-sĭs). A possible explanation for a scientific problem that is based on familiarity with the problem; a scientific guess.

hypotonic. Describing a solution that has a lower concentration of solutes than another solution.

immune response. The body's ability to recognize an antigen and destroy or inactivate it in a specific way.

imprinting. In behavior, the tendency of newborn animals to follow the objects they sense early in their development.

incomplete digestive tract. A digestive system with only one opening, present in organisms like hydra and jellyfish.

incomplete dominance. In genetics, a partial expression of traits, rather than dominance of one allele over the other.

independent assortment, concept of. Mendel's second law, which states that during meiosis, alleles from one pair of homologous chromosomes separate independently from alleles on other pairs of chromosomes.

induced-fit theory. An explanation of enzyme function suggesting that an imperfect fit between an enzyme's active site and its substrate produces physical stress that weakens the substrate's bonds and allows the reaction to occur.

inflammatory response. The body's nonspecific response to the intrusion of foreign substances, resulting in the mobilization of phagocytic cells and a series of reactions that restore damaged tissues or cells.

ingestion. Moving food from the environment into an organism's body; eating.

innate behavior. Highly stereotyped, species-specific instinctive behavior that is shaped by evolution and inherited by offspring.

insight learning. The process of drawing on old experiences and cognitive activities to solve a problem.

instinct. Unlearned behavior with a genetic basis that can activate and coordinate a species-specific fixed-action pattern and that passes from one generation to another.

insulin. A hormone produced by the islets of Langerhans cells of the pancreas that facilitates the transport of glucose through cell membranes.

intercalary meristem. In grasses, a structure responsible for the growth of leaves or blades from the base.

interferon (ĭn'tər-fēr'ŏn). A chemical substance produced by the immune system that increases the resistance of cells to viral infections.

interleukin-2 (ĭn'tər-lōō'kən). A protein that functions as an immune-system activator.

intermediate fibers. Long, thin structures composed of fibrous proteins that are found in cells that require or provide mechanical strength.

interneuron. A nerve cell that integrates information and influences other nerve cells. Also called an *associative neuron*.

interphase. The nondividing phase of the cell cycle, devoted to growth and homeostasis; divided into the G_1 phase, the S phase, and the G_2 phase.

interspecific competition. Competition between individuals of different species.

interstitial cell–stimulating hormone (ĭn'tər-stĭsh'əl). The hormone (identical to the female hormone LH) produced in the anterior pituitary that stimulates the interstitial cells of the testes to produce testosterone. Abbreviated *ICSH*.

interstitial fluid. A fluid in tissues that surrounds the cells.

intraspecific competition. Competition between individuals of the same species.

intron. An unexpressed intragene sequence contained within the nucleotide sequences of a cell's DNA; an intervening sequence.

inversion. A chromosomal abnormality in which the linear sequence of genes on a chromosome becomes rearranged.

involuntary muscle. Muscle that is not under conscious control.

ion. An atom or molecule that has lost or gained one or more electrons and therefore carries a charge.

ionic bond. An electrostatic attraction between ions with opposite charges.

islets of Langerhans. Cells in the pancreas that produce the hormones insulin, glucagon, and somatostatin.

isogametes (ī'sō-găm'ēts). Sex cells with no morphological differences between male and female.

isotonic. Describing a solution with the same concentration of solutes as another solution.

isotopes. Atoms of the same element with different numbers of neutrons, and therefore different atomic masses.

J-shaped curve. In exponential growth, the graph produced by plotting the number of individuals against time.

karyotype (kăr'ē-ō-tīp). A technique for studying chromosomes by pairing micrographs of the chromosomes and arranging the pairs according to size.

kidneys. In vertebrates, organs that excrete nitrogenous wastes and help maintain water and ion balance in the blood.

kinetic energy. Energy of motion.

kingdom. The most inclusive taxonomic category.

Krebs cycle. In cellular respiration, the final common pathway for the oxidation of pyruvate, fatty acids, and the carbon-hydrogen chains of amino acids. Also called the *citric acid cycle*.

K strategist. An organism that produces a small number of quality offspring, usually adapted to a constant environment.

lac operon. In the bacterium *E. coli*, the set of genes that induces and represses the synthesis of enzymes involved in lactose metabolism.

lactation. Milk production and secretion.

large intestine. A coiled tube responsible for removing water and minerals from the materials that leave the small intestine and passing the materials on to the rectum. Also called the *colon*.

lateral meristem. Either vascular cambium or cork cambium, responsible for increasing the diameter in most plants.

latitudinal biome. The distribution of biomes based on latitude.

learning. An adaptation or modification of behavior as a result of environmental experience.

leukocytes (lōō'kə-sītz). See *white blood cells*.

lichen (lī'kən). A life form composed of a green alga or cyanobacterium and a fungus living in a mutualistic relationship.

ligament. The type of dense connective tissue that connects bone to bone.

ligase enzyme (lĭg'ās). An organic catalyst that repairs broken ends of DNA molecules.

light-dependent reaction. The first stage of photosynthesis, when light energy is converted to chemical energy, which is used to synthesize ATP only (cyclic pathway), or to split water, reduce $NADP^+$, and synthesize ATP (noncyclic pathway).

light-harvesting antennas. In plants, clustered pigment molecules in the thylakoid membrane that absorb light.

light-independent reaction. The stage of photosynthesis when carbon dioxide is reduced to form food, using the ATP and NADPH produced by the light-dependent reactions.

light microscope. A magnifying instrument that works by passing visible light through a specimen and then through lenses that magnify and focus the image.

linkage group. A group of genes that are linked to the same homologous chromosome.

linked genes. Genes that are present on the same chromosome.

lipid. Any nonpolar organic molecule that functions in energy storage, structural support, or protection; lipids consist of triglycerides, phospholipids, steroids, terpenes, and glycolipids.

liver. An accessory organ for the digestive system that has many functions, including bile production, food storage, and blood protein production.

locus. A particular location of a gene on a chromosome. Also called the *gene locus*.

logistic growth. The oscillating pattern of population growth in nature, produced by checks and balances in the environment.

long-day plant. A plant that flowers in summer, when the days are relatively long and daylight exceeds darkness.

long-term memory. Relatively permanent memory that is stored differently from the way short-term memory is.

loose connective tissue. The type of connective tissue proper that supports and maintains the position of internal organs.

lumen. Any cavity in a hollow structure.

lungs. Internal cavities in land vertebrates that provide a large, moist surface area for gas exchange.

luteinizing hormone. A hormone released by the anterior pituitary that regulates the maturation of the egg in the follicle, the release of the egg from the ovary, and hormone production by the corpus luteum. Abbreviated *LH*.

lymph. A component of blood plasma that is filtered out of the capillaries and added to the interstitial tissue fluid.

lymph nodes. Masses of spongy tissue that produce lymphocytes and filter cellular debris from the lymph.

lysis. The rupture of a cell due to the destruction of the cell membrane.

lysosome (lī′sə-sōm). A single-membraned cytoplasmic organelle that breaks open to release hydrolytic enzymes that digest the cell or cellular components.

lysozyme (lī′sə-zīm). An enzyme that lyses bacteria by breaking down the cell wall, a type of immune defense produced by some cells and contained in tears.

macromolecule. Any large polymer, like a protein, carbohydrate, or lipid.

macrophages (măk′rō-fāj-əz). Large white blood cells that engulf foreign substances through phagocytosis.

major histocompatibility complex (MHC). A complex of glycoproteins coating the surface of the cell membrane that identifies individuals as unique.

mammary glands. Modified sweat glands that secrete milk, localized into the breasts of mammalian females.

mantle. A tissue present in mollusks and, in some mollusks, a tissue that secretes the shell.

matrix. (1) A thick fluid that fills the inner compartment of the mitochondrion; (2) the intercellular substances of tissue.

mechanist. One who believes that life processes are subject to natural laws.

medusa. The jellyfish-like stage in the lifecycle of some Cnidaria, with the mouth and tentacles facing down.

meiosis (mī-ō′sĭs). The process of chromosome reduction during cellular reproduction, in which a diploid cell contributes one member of a chromosome pair to each haploid cell.

menarche (mən′är′kē). The onset of menstruation and the development of female secondary sex characteristics in puberty.

Mendel's laws. See *dominance, concept of; unit characters, concept of; segregation, concept of;* and *independent assortment, concept of.*

menopause (mĕn′ə-pôz′). The end of a woman's reproductive cycle, usually occurring around age 50.

menstrual cycle. The monthly female reproductive cycle that is regulated by hormonal events, in which a follicle matures, the endometrium thickens, and the mature egg is released from the follicle; if the egg is not fertilized it passes out of the uterus and the endometrium degenerates and is shed.

meristematic tissue. A type of plant tissue made up of undifferentiated cells that give rise to new cells by mitosis; meristem tissues provide all the new cells that are added to the plant body as it grows.

mesoderm. The middle germ layer of the three primary germ layers of the gastrula.

mesophyll. The region of a leaf between the two epidermal layers.

messenger RNA. A type of RNA transcribed from modified DNA that transports a copy of a DNA codon sequence from the nucleus to the ribosomes. Abbreviated *mRNA.*

metabolism (mə-tăb′ə-lĭz′əm). All the chemical and physical reactions that normally occur in a living organism.

metaphase. The second and shortest phase of cell division, in which the chromosomes are brought to the center of the cell by the centrioles and spindle fibers.

metastasis (mə-tăs′tə-sĭs). The spreading of cancer cells broken off from an original mass to other parts of the body through the circulatory system.

microfilaments. Very thin protein threads that are composed of actin and function in cell contraction and cell movement.

micropyle. A channel in seed plants through which sperm enters the ovule.

microspore mother cells. Cells within the pollen sacs of the male that divide by meiosis to produce four haploid microspores, which form a pollen grain.

microtubules. Long, thin, hollow structures that are composed of tubulin and are partly responsible for cell movement.

minimum energy of activation. The amount of energy required to initiate a chemical reaction.

mitochondrion (mī′tō-kŏn′drē-ŏn). A double-membraned self-replicating eukaryotic organelle that functions in oxidative respiration; the site of ATP synthesis.

mitosis (mī-tō′sĭs). The second phase of the cell cycle, resulting in nuclear division in which the number of chromosomes is maintained from cell generation to cell generation.

molecule. Two or more atoms joined by a covalent bond.

Mollusca (mŏ-lŭs′kə). The phylum of the soft-bodied animals that have a common body plan consisting of the foot, visceral mass, mantle, and radula; clams, snails, and squids are examples.

Monera. One of the five kingdoms of life, consisting of prokaryotic, unicellular organisms; bacteria and cyanobacteria.

monocot. Short for *monocotyledon;* a flowering plant with one cotyledon in its seed.

monohybrid cross. In classical genetics, a cross involving only one trait.

monosaccharide. The simplest carbohydrate unit; a simple sugar (for example, glucose).

morphogenesis (môr′fō-jĕn′ə-sĭs). The rise of a specific form or shape in the embryo of an organism.

morphology. The form of an organism's body or structure; an important taxonomic criterion.

motor neuron. A nerve cell that carries information to a muscle or gland.

moving-water biome. A freshwater aquatic biome consisting of streams and rivers.

mRNA. See *messenger RNA.*

mucosa. The innermost layer of the human digestive tract.

Müllerian mimicry. Two or more unpalatable species advertise the model and they mimic each other.

multiple alleles. In genetics, the condition existing when there are more than two alleles for a single trait in the population.

mutagen. An environmental agent that causes changes in an organism's genetic composition.

mutation. (1) A change in genetic makeup that can be passed on to new cells or offspring; (2) in DNA replication, an error that occurs in the replication of genetic material that is passed on to the new cells or offspring.

mutualism. A form of symbiosis in which all organisms benefit.

mycorrhizae (mī′kə-rī′zē). Fungi that establish a mutualistic relationship with a plant's roots by attaching themselves to the root and increasing its surface area.

myelin sheath. In vertebrates, a layer of lipids that surrounds the axons of some nerves.

myofibrils. The tubular subunits of a muscle fiber; the fundamental contractile portion of a muscle cell.

myofilaments. Parallel filamentous proteins (actin and myosin) that make up the myofibrils of a muscle.

myosin. A protein involved in cell movement.

NAD⁺. *Nicotinamide adenine dinucleotide* (oxidized form), an electron carrier that accepts hydrogen ions and electrons from molecules as they are metabolized.

NADP⁺. *Nicotinamide adenine dinucleotide phosphate* (oxidized form), an electron and hydrogen carrier that functions in photosynthesis by transferring electrons and hydrogen ions from sites where the energy from sunlight is harvested to sites where that energy will be converted to the chemical energy found in molecules.

natural selection. The theory that organisms expressing variations best suited to their environment survive and reproduce, and those expressing variations poorly suited to their environment eventually die out. Sometimes referred to as *survival of the fittest.*

negative feedback. A process in which the product of a reaction inhibits or regulates the reaction that produced it.

nephrons (nĕf′rŏnz). Microscopic units in the cortex of the kidney; the functional unit of the kidney.

neritic biome (nĭ-rĭt′ĭk). Any mature sandy or rocky community located along the shoreline.

nerve. A group of nerve cells surrounded by a protective sheath that forms a pathway for axons and dendrites to transmit their messages.

nerve net. A web of interrelated specialized cells that conduct stimuli in the cnidarians; the simplest form of nervous system.

nerve tissue. In animals, highly modified epithelial cells specialized for cell-to-cell communication.

net primary productivity. In ecology, the amount of energy available to the earth's heterotrophic organisms; equal to the gross primary productivity minus the amount of energy plants use to conduct photosynthesis.

neuroglia (nûr′rŏg-lē′-ə). In animals, cells that support, nourish, bind, and insulate nerve cells.

neuron. A modified epithelial cell in most animals that is specialized for impulse conduction; a nerve cell.

neurotransmitter. Any chemical in the body that propagates the nerve impulse.

neutron. In chemistry, a particle within an atom's nucleus that has no electrical charge.

niche. The specific role an organism plays in its environment.

nitrogen cycle. The pathway of nitrogen through the environment, in which nitrogen-fixing bacteria on the roots of certain plants take in nitrogen, animals eat plants, and animals return their excreted nitrogenous wastes to the soil.

nodes of Ranvier. In vertebrates, small exposed areas in myelinated neurons that enable the electrochemical signals to jump from one node to the next.

noncyclic photophosphorylation (fō′tō-fŏs′fôr-ə-lā′shən). In plants, a combination of the mechanisms of photosystem I and photosystem II that forms a proton gradient to provide energy for ATP synthesis, splits hydrogens and water to reduce NADP⁺ to NADPH, and passes NADPH to the light-independent reactions to serve as the hydrogen source to reduce carbon dioxide.

nondisjunction. A chromosomal abnormality that occurs when sister chromatids fail to separate during meiosis, resulting in daughter cells with an $n+1$ or $n-1$ chromosome number.

nonselective evolution. The change of the frequencies of alleles in a population not because of natural selection, but because of chance.

nonspecific immune response. A series of chemical and physical responses that protect the body from infection or injury, not involving the specific activity of B- and T-cells.

nonvascular plant. A member of the kingdom Plantae without true roots, stems, leaves, or transporting vascular structures.

norepinephrine (nôr-ĕp′ə-nĕf′rən). A synaptic neurotransmitter produced in the adrenal medulla that stimulates constriction of blood vessels to maintain blood pressure.

notochord (nōt′ə-kôrd). A flexible, rodlike structure that runs along the length of the back in chordates. In vertebrates the notochord is only present in embryonic stages.

nuclear envelope. A double membrane that encloses the nuclear region in a eukaryotic cell.

nucleic acid (noo-klā′ĭk). Any polymer composed of nucleotides; DNA or RNA.

nucleolus (noo-klē′ō-ləs). A large, dark region in the nucleoplasm that is composed of RNA and protein and functions in the synthesis of ribosomes.

nucleosome (noo′klē-ō-sōm). Globular proteins in the eukaryotic chromosome in which DNA is wound around.

nucleotide (noo′klē-ō-tīd). A component of a nucleic acid that consists of a nitrogen base, a sugar, and a phosphate and can function as an energy carrier, a chemical messenger, or an electron transporter.

nucleus (noo′klē-əs). (1) In chemistry, the core of an atom, containing protons and neurons; (2) in cell biology, the eukaryotic cellular component that is surrounded by a double membrane and contains the cell's genetic material.

observational science. A method of science based upon observation.

oligotrophic lake. A nutrient-poor lake.

Onychophora (ŏn′ĭ-kŏf′ər-ə). A phylum of wormlike animals that are segmented; in fossil members of this group, the worms had jointed legs similar to the arthropods.

oncogene. A term describing a gene that can induce the uncontrolled cell division that we call cancer.

oogenesis (ō′ə-jĕn′ə-sĭs). The process of female sex cell formation.

open circulatory system. A transporting system in which the blood passes from the vessels into the sinus cavities and then to the heart.

open-ocean biome. Any mature ocean community.

operant conditioning. A form of learning in which desired behaviors are rewarded and undesired behaviors are punished.

operator. The location on the DNA molecule where the repressor binds in order to stop transcription.

operon. In prokaryotic cells, a sequence of structural genes and regulating genes.

orbital. Within each energy level, a specific hypothetical region where an electron is most likely to be found.

order. The taxonomic category below *class* and above *family*.

organ. A body structure made up of more than one kind of tissue that performs a specific function or group of functions.

organelle. Any distinct structure in a cell's cytoplasm that performs a specific cellular function.

organic. Describing molecules that contain carbon and are related to life.

orgasm. In humans, a series of pleasurable sensations that can accompany sexual stimulation.

osmoregulation. The maintenance of chemical constancy within an organism through regulation of ions and water in the cells.

osmosis (ŏz-mō′sĭs). Diffusion of a solvent (usually water) through a membrane from a region of higher concentration to a region of lower concentration.

osteichthyes (ŏs-tē-ĭk′thîz). The bony fishes.

osteous tissue. The living cells that form the hard outer casing of bones.

ostracoderms (ŏs-tră′kō-dûrmz). An extinct group of fishes that were prevalent during the Silurian and Devonian periods; these small fish had a body covered with external bony plates.

ovaries. (1) In animals, the female gonads, which produce the ova as well as female sex hormones; (2) in plants, the structure in which the ovule matures and megaspore formation occurs.

ovulation. The stage of the menstrual cycle when the follicle in the ovary bursts and releases one or more eggs.

ovule. A structure in the ovary of female seed plants that contains nutritive tissue, a skinlike covering, a micropyle, and megaspore mother cells.

ovum. The egg in animals.

oxidation. The loss of electrons or hydrogen atoms by an atom or molecule.

oxygen cycle. The pathway of oxygen through the environment, in which animals take in oxygen from the air and release carbon dioxide, and plants take in carbon dioxide and release oxygen.

P680. In photosystem II, the chlorophyll *a* molecule that absorbs wavelengths of light up to 680 nanometers.

P700. In photosystem I, the chlorophyll *a* molecule that absorbs wavelengths of light up to 700 nanometers.

palisade parenchyma cells (pə-rĕng′kə-mə). Elongated, upright, chloroplast-containing cells in the upper epidermis of a leaf that carry out most of a plant's photosynthesis.

pancreas. A large digestive and endocrine gland that produces the hormones insulin and glucagon and secretes digestive enzymes into the small intestine.

panspermia. The theory that life came to earth from elsewhere in the universe.

parallel evolution. The development of similar adaptations by organisms that are closely related.

parasite. An organism that either lives within another organism or on its surface and depends on that organism for its energy supply.

parasympathetic division. The cells of the autonomic nervous system that promote relaxation and maintenance of body activities.

parathyroid glands. Four glands embedded in the back side of the thyroid that produce parathyroid hormone.

parathyroid hormone. The hormone that counteracts calcitonin by stimulating bone storage sites to release calcium. Abbreviated *PTH*.

parenchyma cells (pə-rĕng′kə-mə). Plant cells involved in photosynthesis and food storage.

parent cell. In cell reproduction, the original cell that divides.

parthenogenesis. A type of reproduction in which an unfertilized egg can develop into a new individual.

passive transport. A method of moving molecules through a membrane without any expenditure of cellular energy.

pedigree. In genetics, a chart or diagram representing the history of the occurrence of a particular trait in a given family.

peptide linkage. A dehydration synthesis reaction that covalently joins the carboxyl group of one amino acid to the amino group of another.

pericycle. A layer of plant tissue that surrounds the stele of the root and gives rise to branch roots.

periderm. The protective layer that covers woody plants.

peripheral nervous system. In vertebrates, those nerves outside the central nervous system, including the somatic and autonomic nervous systems.

peristalsis (pĕr′ə-stăl′sĭs). (1) The muscle action that moves food through the digestive tract from the esophagus to the anus; (2) successive contractions of any hollow muscle tubes that move the contents.

permanent tissue. Plant tissues that are specialized to perform a particular function.

peroxisome. A small organelle containing enzymes that break down peroxides.

petiole. In dicots and some monocots, the structure that attaches the flattened blade of the leaf to the stem.

pH. A logarithmic scale that runs from 0 to 14 and indicates the relationship between hydrogen ions and hydroxide ions in solution.

phage. See *bacteriophage.*

phagocytosis (făg′ə-sī-tō′sĭs). A special kind of endocytosis in which a cell engulfs a large particle or another cell.

phenetics (fĭ-nĕt′ĭks). Taxonomic schemes based on similarities and differences in phenotype; homology, analogy, and phylogeny are not considered in these schemes.

phenotype (fē′nə-tīp). The observable appearance of an organism, which is determined by the genetic makeup of the organism and influenced by the environment.

pheromone (fĕr′ə-mōn). A chemical produced by animals and released to the environment that alters the behavior of other animals of the same species.

phloem (flō′əm). Tissues of higher plants that conduct sap from photosynthetic regions to nonphotosynthetic regions.

phospholipid. A lipid that contains a glycerol, two fatty acids, and a phosphate group; the major type of lipid found in brain cells, nerve cell coverings, and plasma membranes.

phosphorous cycle. The pathway of phosphorus through the environment, in which phosphorus is dissolved in water while on land, deposited in the sea, and eventually exposed to the land again.

phosphorylation (fŏs′fŏr-ə-lā′shən). (1) The process of adding a phosphate to ADP to form ATP; (2) the addition of phosphate to any molecule.

photolysis (fō-tŏl′ə-sĭs). A process in which light energy is indirectly used to split a water molecule into one oxygen atom, two hydrogen atoms, and two electrons.

photon. In the particle theory of light, a particle or unit of light energy.

photoperiodism. An organism's pattern of response to periods of light and darkness.

photosynthesis. A reduction reaction in which chlorophyll-containing organisms convert light energy to chemical energy.

photosystem I. In photosynthesis, the light-receiving system containing the molecule P700, including the reduction of NADP to NADPH.

photosystem II. In photosynthesis, the light-receiving system containing the molecule P680, involving the photolysis of water.

phototropism. The directional response of a plant to light.

phylum. The second most inclusive taxonomic category in the kingdoms Animalia, Monera, and Protista; analogous to *division* in the other kingdoms.

phytochrome. A light-sensitive pigment believed to trigger hormonal activities that function in the regulation of many plant activities.

pinocytosis (pĭn′ō-sī-tō′sĭs). A special type of endocytosis in which a cell engulfs a small, liquid substance; "cell drinking."

pioneer organism. During ecological succession, the first life form to establish itself in a new community; usually a hardy species able to tolerate severe environmental stress.

pistil. The female reproductive structure of a flowering plant, composed of one or more ovaries, style, and stigma.

pituitary (pĭ-tōō′ə-tĕr′ē). "The master gland"; a gland at the base of the brain that produces many important hormones, consisting of an anterior and posterior portion.

placenta. In some mammals, the structure formed by the extraembryonic membrane of the developing embryo and the endometrium of the mother.

placoderms (plă′kō-dûrmz). Members of this extinct group of fishes were the first fish to have jaws.

Plantae. One of the five kingdoms of life, consisting of eukaryotic, multicellular, photosynthetic organisms whose cells are surrounded by a cell wall.

plasma. A liquid that makes up about 55 percent of whole blood, containing water, blood proteins, dissolved gases, ions, hormones, and various nutrients.

plasma membrane. The boundary between a cell's internal environment and its external environment.

plasmids. In most bacteria, small circular forms of DNA located away from the nucleoid region.

plastid. In plant cells, a double-membraned self-replicating organelle involved in pigmentation, food production, and storage.

platelets. Fragments of white blood cells involved in blood clotting. Also called *thrombocytes.*

Platyhelminthes (plăt′ĭ-hěl-mĭnth′ēz). The phylum of the bilateral, triploblastic flatworms that exhibit cephalization.

pleiotropy (plī-ŏt′rə-pē). The ability of one gene to produce a variety of observable effects.

point mutation. A change in the nucleotide sequence in DNA.

polar. Describing a molecule with slight electrostatic differences between its parts.

polar bodies. The small daughter cells produced through the unequal cleavage during oogenesis that do not function as eggs; they nourish the egg cell and eventually disintegrate.

polar covalent bond. A type of covalent bond in which different electrostatic charges draw the shared electrons to one section of the molecule more than to the other, resulting in a polar molecule.

pollen grain. The immature male gametophyte in seed plants.

pollen tube. The structure of internal fertilization in seed plants, produced by the pollen grains, that extends into the ovule and releases sperm nuclei.

pollination. In plants, the transfer of a pollen grain to an egg-containing structure.

polydactyly (pŏl′ē-dăk′tə-lē). The genetic condition of having extra fingers or toes.

polygenic inheritance. An inheritance pattern in which the interaction of several genes produces a range of variations for a given trait.

polymer (pŏl′ə-mər). See *macromolecule.*

polymerase chain reaction (PCR) (pŏl′ə-mə-rās′). A method of amplifying (copying) a gene sequence a millionfold. A very useful technique of molecular genetics.

polyp. The stationary body form of some members of the phylum Cnidaria; mouth and tentacles face upward.

polypeptide. A medium-sized macromolecule composed of amino acids; bigger than a dipeptide but smaller than a protein.

polysaccharide. A carbohydrate containing more than two monosaccharides.

population. In ecology, an interbreeding group of organisms occurring in a specific area.

population density. The number of individuals of a species per unit area.

Porifera (pô-rĭ′fĕr-ə). The "pore-bearing" animals, the sponges.

positive feedback loop. Physiological regulation mechanism in which a change in the variable causes mechanisms to amplify the response.

posterior pituitary (pĭ-tōō′ə-tĕr′ē). The portion of the pituitary gland that stores and releases two hormones that are produced by the hypothalamus.

postzygotic mechanism. A type of reproductive isolation that comes into effect after fertilization has occurred.

potential energy. Stored energy; energy available to do work.

predator. An organism that eats other living organisms, its prey.

pressure-flow hypothesis. An explanation for the movement of sap through a plant suggesting that differences in pressure cause sap to move from the photosynthetic cells (the source) to the place where the sucrose is stored (the sink).

prezygotic mechanism. A type of reproductive isolation that prevents mating and egg fertilization.

primary consumer. An animal that feeds on the producers.

primary growth. The initial growth in length of roots and stems in plants.

primary response. The first stage of the immune response, when the system shifts from the nonspecific defense of inflammation to T- and B-cell formation.

primary structure. The linear sequence of the amino acids in a protein.

primary succession. The sequential vegetational events that result in the establishment of a climax community in virgin soil.

primordial cloud theory. An evolutionary theory supported by most scientists, suggesting that the universe began as a great explosion and took shape as a result of cooling and gravity.

procambium (prō-kăm′bē-əm). In plants, a division of the meristematic tissue that differentiates into the primary xylem and the primary phloem.

producer. An organism that makes its own food from the simple organic molecules in the environment.

progesterone (prō-jĕs′tə-rōn). A hormone produced in the corpus luteum of the ovaries responsible for maintaining the thickness of the endometrium during pregnancy.

proglottid (prō-glăd′əd). The individual unit of a tapeworm.

prokaryote (prō-kăr′ē-ōt). A cell that lacks a membrane-bound nucleus or other membrane-bound organelles.

promoter. An initiation site on the DNA molecule to which the RNA polymerase binds in order to initiate transcription.

prophage. Viral DNA that remains dormant inside a bacterial cell and becomes incorporated into the bacterial central chromosome.

prophase. (1) The first stage of mitosis, when the chromatin begins to form distinct chromosomes; (2) in meiosis I, the stage when crossing over is most likely to occur.

prostaglandins. A group of regulatory chemicals produced by almost all types of body tissues derived from long-chain fatty acids.

protective tissue. The types of epithelial tissue that lines body cavities.

protein. A molecule composed of one or more chains of amino acids; proteins function as enzymes, carrier molecules, hormones, structural components, and receptors.

proteinoid microsphere. A double-membraned droplet formed spontaneously by heating amino acids; the possible precursors of the first cells.

Protista. One of the five kingdoms of life, consisting of eukaryotic, unicellular organisms.

protobiont. A hypothetical simple precell; a microscopic droplet capable of life.

protoderm. In plants, a division of the apical meristem in primary growth that differentiates into the epidermis.

proton. A positively charged particle within an atom's nucleus.

proto-oncogene. A normal gene with the potential to be converted to an oncogene (cancer gene) by environmental insults such as viruses, chemicals, and radiation.

protostomes (prō′tō-stōmz). An evolutionary branch based on embryonic development that includes the mollusks, the annelids, and the arthropods.

pseudocoelomates (soo′dō-sēl′ō-māts). Animals that have a false body cavity between the gut and body wall; this cavity is false because it is not lined with a peritoneum.

puberty. The time of sexual maturation, usually between the ages 14 and 16 in human males and 10 and 14 in females.

pulmonary artery. The vessel that transports blood from the heart to the lungs.

pulmonary circulation. The circulatory pathway that pumps blood from the right side of the heart to the lungs, and then back to the left side of the heart.

punctuated equilibrium. The theory that periods of slow evolutionary change are interspersed with periods of rapid evolutionary change.

Punnett square. In genetics, a grid used to keep track of the possible combinations that can result in the offspring of two parents.

purine. A double-ringed nitrogenous base, such as adenine and guanine.

pyrimidine (pĭ-rĭm′ĭ-dēn). A single-ringed nitrogenous base, such as cytosine, thymine, and uracil.

pyruvate (pī′roo-vāt). A three-carbon compound that contains stored chemical energy, the ionized form of pyruvic acid.

quaternary structure. The configuration of a complex protein molecule formed by two or more polypeptide globular chains.

radial symmetry. The type of symmetry in an organism with similar parts arranged like the radii of a circle; a starfish and a bicycle wheel exhibit radial symmetry.

radioisotope. An unstable isotope that breaks down and emits ionizing radiation.

random distribution. A distribution pattern in which individuals of the same species occur in no discernible pattern throughout a region.

ray-finned fish. A group of bony fishes, now extinct, except for the coelacanth. Lobe-finned fishes have paired fins with fleshy and muscular bases.

receptor site. A molecule (usually a protein) on the plasma membrane that receives chemical messages from other molecules.

recessive allele. In classical genetics, an allele whose expression is masked when paired with a dominant allele.

recombinant DNA technology. A general term used to describe human selection, intervention, and manipulation of the DNA molecule.

red blood cells. Small biconcave blood cells responsible for oxygen transport. Also called *erythrocytes*.

reduction. In chemical reactions, the addition of one or more electrons to an atom or molecule.

reflex arc. A simple nerve pathway in which stimulus detection, integration, and response occur and that may not involve the brain directly.

regulator. A portion of the DNA molecule that codes for proteins that block mRNA transcription.

reproduction. The natural ability of living organisms to produce offspring.

reptiles. A diverse group of animals characterized by adaptations that suited them for life on land, such as scales, internal fertilization, and the land egg.

resolving power. The property of light that results in the ability to distinguish between two points as separate entities; the limiting factor in developing light microscopes with high magnification.

respiration. (1) *Cellular respiration*, an oxidation reaction in which organisms release the energy in food molecules; (2) *internal respiration*, the exchange of carbon dioxide and oxygen between the circulatory system and the tissues of the body; (3) *external respiration*, the exchange of carbon dioxide and oxygen between the blood and the environment.

resting potential. The steady voltage difference across the plasma membrane of a nerve cell that is not transmitting an impulse.

restriction enzyme. An enzyme that cuts DNA molecules at a specific sequence site.

restriction fragment length polymorphism (RFLP). A technique useful in establishing markers for genetic mapping, based on differences in DNA sequence length on homologous chromosomes due to treatment with restriction enzymes.

R-group. A specific atom or group of atoms that varies from one type of amino acid to another.

ribonucleic acid (rī′bō-nōō-klā′ĭk). See *RNA.*

ribose. The five-carbon sugar in RNA nucleotides.

ribosomal RNA. The type of RNA of which ribosomes are composed. Abbreviated *rRNA.*

ribosome. An organelle that provides the site of protein synthesis in eukaryotic and prokaryotic cells.

ritualized behavior. Animal behavior characterized by exaggerated actions that usually follow a specific pattern.

RNA. Any single-stranded nucleic acid whose nucleotides contain the sugar *ribose.*

RNA-based theory. With respect to the origin of life, many investigators propose that RNA rather than DNA encoded the genes for the first life forms, since RNA is a smaller polymer, capable of self-replication, and has enzymelike properties.

RNA polymerase. The enzyme or enzyme complex that catalyzes the reaction in which RNA is copied from DNA.

rods. Photoreceptor cells in the eye that respond to low light intensity and are responsible for night vision.

root hairs. In plants, extensions of a root's epidermal cells that increase the surface area of a root over which absorption can occur.

Rotifera (rō-tĭf′ər-ə). Microscopic multicellular animals, known as the "wheel-bearers."

rRNA. See *ribosomal RNA.*

r strategist. An organism that produces a large number of offspring and does not invest much energy in their care.

SA node. See *sinoatrial node.*

saltwater biome. An aquatic biome consisting of saltwater.

sap. In plants, a thick fluid composed of the products of photosynthesis and associated substances.

sarcolemma (sär′kō-lĕm′ə). The membrane surrounding a muscle fiber.

sarcomere (sär′kō-mēr). The contractile unit of skeletal muscle.

sarcoplasmic reticulum (sär′kō-plăz′mĭk rə-tĭk′yōō-ləm). A modified endoplasmic reticulum that surrounds the myofibrils of muscle fibers and accumulates substances involved in muscle contraction.

saturated fat. A fatty acid with hydrogen atoms covalently bonded to all possible carbon atoms on its carbon skeleton.

scanning electron microscope. A magnifying instrument that passes a thin beam of electrons over a specimen, presenting the illusion of a three-dimensional image.

scientific method. A logically ordered procedure for solving problems and answering questions that involves stating a problem, developing a hypothesis, testing the hypothesis, and drawing conclusions.

sclerenchyma cells (sklə-rĕng′kə-mə). Thick-walled cells that provide mechanical support and protection in a mature plant body.

scolex (skō′lĕks). The head of a tapeworm, which is lined with suckers or suckers and hooks, and which enables the tapeworm to attach to the gut of the host.

secondary consumer. An animal that eats primary consumers.

secondary growth. Growth in the diameter of vascular plants.

secondary response. The second phase of the immune response, activated when the B-cells recognize an invader that the body has been exposed to before.

secondary structure. The helical pattern of a protein molecule's configuration.

secondary succession. The repopulation of land that was once occupied by life forms and was then disturbed.

scrubland. A coastal region that is periodically swept by fire; one of the major terrestrial biomes.

second filial generation. In classical genetics, the set of offspring resulting from the reproduction of the first filial generation.

second law of thermodynamics. A fundamental law of physics which states that part of the energy released in any reaction is dissipated into the surroundings and is no longer available to do work.

seed. The ripened ovule, which consists of the embryo, one or two cotyledons, and the embryo's food reserves enclosed by a protective seed coat.

segregation, concept of. Mendel's first law, which states that contrasting factors for a given trait in an individual separate when sex cells are formed, and any single sex cell receives only one factor for the trait.

selectively permeable. Describing a membrane that can biochemically or biophysically determine which molecules can bind, enter, or leave the cell.

self-pollination. The act of pollination in some seed plants in which the pollen from a plant is used to fertilize the eggs in the same plant.

semen. In mammals, the sperm-containing liquid formed when the seminal fluid is joined by the secretions of Cowper's glands; the white viscous liquid that the male ejaculates during copulation.

semiconservative replication. The method of DNA reproduction; each new molecule consists of one strand from the original molecule and one new strand.

seminal fluid. The liquid formed when sperm is joined by the secretions of the seminal vesicles and the prostate gland.

seminiferous tubule (sĕm′ə-nĭf′ə-rəs tōō′byōōl). The tubule within the testes where spermatogenesis (sperm production) occurs.

sensory neuron. A nerve cell that receives sensory stimuli from a receptor and sends the stimuli to the spinal cord or brain.

sensory tissue. The type of epithelial tissue containing specialized sensory receptor cells.

serosa. The layer of connective tissue that covers the digestive tract and is continuous with the lining of the body cavity.

sex chromosomes. The pair of chromosomes that determines whether an embryo develops into a male or a female.

sex linkage. The attachment of a gene to the X sex chromosome, which has no allelic partner on the Y chromosome.

sexual reproduction. A process for the production of an individual that involves the exchange of genetic material between organisms.

short-day plant. A plant that flowers in early spring or fall, when the days are relatively short.

short-term memory. The kind of memory at work the first time an organism encounters a piece of information; if the information is not converted into long-term memory, it will be forgotten.

sickle-cell anemia. A genetic disease resulting in an abnormality in the hemoglobin that causes the red blood cells to buckle, reducing their oxygen-carrying capacity.

sinoatrial node. A group of cells in the atrial wall of the heart that controls the rhythm of contraction and relaxation. Abbreviated *SA node.*

skeletal muscle. Striated multinucleated muscle tissue that is under conscious control.

skill memory. The ability to retain behavioral patterns acquired through practice.

skin. In vertebrates, the outer covering that provides form, temperature regulation, and protection from dessication, trauma, bacterial infection, and ultraviolet light.

sliding filament model. An explanation of muscle contraction suggesting that when a muscle contracts, the actin myofilaments at either side of the sarcomere move toward one another by sliding over the stationary myosin myofilaments.

small intestine. The region of the digestive tract between the stomach and large intestine in which most of the food digestion and absorption occur.

smooth muscle. (1) Uninucleate, unstriated muscle cells in the walls of blood vessels and hollow organs; (2) the layer of the digestive tract below the serosa.

sodium-potassium pump. The active transport mechanism for moving sodium and potassium molecules through the cell membrane.

solute. A substance dissolved in another substance.

solution. A mixture of a solvent and a solute.

solvent. The component of a solution in which the other component (the solute) is dissolved.

somatic mutation (sō-măt′ĭk). An error in DNA replication that occurs in body cells but is not transmitted to offspring.

somatic nervous system (sō-măt′ĭk). In humans, the sensory and motor neurons that move the skeletal muscles.

speciation. The evolutionary process that results in the formation of a new species from an already-existing species.

species. (1) In sexually reproducing organisms, a group of organisms that breed with each other and normally produce fertile offspring; (2) in taxonomy, the most specific taxonomic category.

specific heat. The amount of heat required to raise the temperature of a substance by a specific amount; a method of determining the amount of heat that a substance can hold.

specific immune response. An immune response to a specific antigen mediated both by cells and by antibodies, involving the T- and B-cells.

sperm. The male sex cell.

spermatogenesis (spûr-măt′ə-jĕn′ə-sĭs). The forming of male sex cells.

S phase. In the cell cycle, a phase of interphase when the DNA is synthesized.

spherical symmetry. The type of symmetry exhibited by an organism that is shaped like a ball and appears similar at all points equidistant from the center of that ball.

spinal cord. The part of the central nervous system that runs through the spinal column in vertebrates and controls most of the body's reflexes, integrates neural activities from other parts of the body, and transmits neural impulses to and from the brain.

spindle. A structure formed of microtubules that extends between the centrioles of a cell and the chromosomes and is involved in mitosis and meiosis.

spongy bone. Porous, moist bone tissue filled with red or yellow bone marrow.

spongy parenchyma cells (pə-rĕng′kə-mə). In plants, loosely packed cells of the mesophyll that enhance the exchange of carbon dioxide, oxygen, and water vapor between the air and leaf tissues.

spontaneous generation. The belief that life can arise from nonliving things.

spore. (1) In plants, a reproductive cell or group of haploid cells that can develop into a new plant without undergoing fertilization; (2) in bacteria, a dormant protected stage of the bacterium; (3) in fungi, a reproductive cell.

spore mother cell. In female seed plants, a diploid cell that undergoes meiosis to form haploid megaspores.

sporophyte generation. In plants exhibiting alternation of generations, a multicellular, diploid plant body that produces haploid spores through meiosis.

S-shaped curve. When a logistic population growth is plotted, it appears as an S.

stamen. The male reproductive structure of a flowering plant.

steroid. A group of lipid-soluble proteins consisting of 17 carbon atoms arranged in four rings.

stigma. The top part of the pistil of a flower, with a sticky covering to which pollen grains stick.

still-water biome. A freshwater aquatic biome consisting of lakes or ponds.

stimulus. In behavior, a received signal.

stomach. In animals, a muscular, expandable sack that stores, mixes, and digests food.

stomata. Pores in plant leaves that allow for the exchange of gases and water vapor.

striated muscle. Skeletal muscle, which appears striped when viewed under the microscope. See *skeletal muscle.*

stroma. A dense enzyme-containing solution in the chloroplasts of green plants that surrounds the thylakoids.

structural gene. A gene that codes for the synthesis of a specific protein structure.

style. The stalk of the pistil covering the stigma and ovary in the female reproductive structure of flowering plants.

submucosa. The layer of the digestive tract beneath the mucosa that contains nerves and blood vessels.

substrate. The specific molecule or group of molecules catalyzed by an enzyme.

succulent. A desert plant capable of holding water.

supportive connective tissue. In vertebrates, the type of connective tissue that forms the skeletal system.

survival of the fittest. The idea, based on Darwin's theory of natural selection, that organisms well suited to their environment survive and reproduce and organisms that are not well suited cannot compete for limited resources or reproduce and do not survive.

symbiosis (sĭm′bī-ō′sĭs). A biological relationship between organisms that live in close association.

sympathetic division. A division of the autonomic nervous system that promotes excitation or emotional stress.

sympatric (sĭm-păt′rĭk). Describing species that live in the same geographical region and cannot interbreed.

synapse (sĭn′ăps). The space between two adjacent nerve cells.

synapsis (sĭn-ăp′sĭs). The first stage of meiosis I, in which each homologous chromosome pairs up with its partner.

syncitial hypothesis (sĭn-sĭsh′ē-əl). A hypothesis that the multicellular animals arose from a multinucleate protozoan that became subdivided by membranes into small cells. This hypothesis for the origin of animals is not widely accepted by biologists.

systematics. The study of the evolutionary relationships among organisms.

systemic circulation. The circulation pathway that pumps blood from the left side of the heart to all body parts except the lungs, and back to the right side of the heart.

systole (sĭs′tə-lē). The stage of the cardiac cycle when the heart muscle contracts.

taiga (tī′gə). A terrestrial biome, usually in the northern hemisphere, dominated by evergreens with small, needle-like leaves.

target cell. In cell-to-cell communication, a cell at a distance from the source that has receptor sites for a particular chemical message.

taxis. The directed movement of an organism toward or away from a stimulus.

taxonomy. The science of classifying organisms. The taxonomic categories, from most general to most specific, are kingdom, phylum or division, class, order, family, genus, and species.

T-cell. A lymphocyte that influences cell-mediated responses.

telophase. The final phase of mitosis, in which a new nuclear membrane forms around each set of chromosomes, the nucleolus reappears, and the chromosomes uncoil into their chromatin form.

tendon. The type of dense connective tissue that connects muscle to bone.

territoriality. The behavior of an animal defending its territory.

territory. An area that an animal claims as its own and defends from others of its own species.

tertiary structure. The three-dimensional pattern of folding of a polypeptide, resulting in a specific configuration.

test cross. In genetics, a cross between an individual expressing the dominant condition and a homozygous recessive individual to determine if the dominant individual is heterozygous or homozygous for a particular trait.

testes. The male gonads, the site of sperm production.

testosterone. A category of male hormones essential for sperm production, maturation of the male sex organs during adolescence, and development of male secondary sex characteristics.

tetraploid. A term used to describe a cell that has four copies of chromosomes.

thrombocytes. See *platelets.*

thylakoid. In plants, the important photosynthetic membranous structure in the stroma that contains chlorophylls, other pigments, electron acceptors, enzymes, and CF_1 particles.

thymus. A gland located behind the sternum that produces thymosins, which promote T-cell lymphocyte production.

thyroid gland. A butterfly-shaped endocrine gland located at the base of the neck that secretes thyroxine and calcitonin.

thyroxine. A hormone secreted by the thyroid gland that controls many functions, including the basal metabolic rate of cells.

tissue. A group of cells that perform similar functions.

tonicity. The relationship between the level of dissolved substances (solutes) in a cell and the level of dissolved substances (solutes) outside the cell.

transcription. The reaction that copies RNA from DNA.

transduction. The transfer of genetic material through bacteriophages.

transfer RNA. A type of RNA whose individual amino acids and anticodons complex with the codons on the mRNA when proteins are synthesized at the ribosome. Abbreviated *tRNA.*

transformation. The direct incorporation of DNA from the bacterial medium into the bacterium.

translation. The transferral of the genetic code to a linear sequence of amino acids in a protein chain.

translocation. (1) In genetics, a chromosomal aberration occurring when a part of one chromosome becomes detached and joins another nonhomologous chromosome; (2) in plants, the process of transporting sucrose in a soluble form from one plant tissue to another.

transmission electron microscope. A magnifying instrument that works by directing electrons through a specimen.

transpiration. Water loss through the stomatal openings of a leaf.

trial-and-error learning. The practice of trying a number of different behaviors before discovering the one that is successful.

trisomy (trī′sō-mē). An abnormal genetic condition occurring when a normal gamete and an $n + 1$ gamete join, producing a set of three chromosomes instead of two, for a specific set of chromosomes $(2n + 1)$.

tRNA. See *transfer RNA*.

trophic level (trō′fĭk). An organism's position in a food chain.

tropical rain forest. A terrestrial biome located near the equator with abundant rainfall and sunlight and a great diversity of plant and animal life.

tropism. A plant growth response either toward or away from an external stimulus.

tropomyosin. (trăp′ə-mī′ə-sən). A small filamentous protein that with actin comprises the actin microfilaments.

troponin. (trōp′ə-nĭn). A protein molecule involved in muscle contraction; calcium ions mediate the movement of the troponin molecules and expose the actin site so that the muscle filaments can slide over one another, resulting in a muscle contraction.

tundra. A region of flatlands, bogs, and marshes where the subsoil is permanently frozen; one of the major terrestrial biomes.

turgor pressure. In plant cells, the force with which water presses against the cell wall; the hydrostatic pressure.

umbilical cord. In placental mammals, a vascular cord that connects the fetus with the placenta.

uniform distribution. A distribution pattern in which organisms of the same species occur at regular intervals throughout a region.

unit characters, concept of. One of Mendel's laws, which states that hereditary characteristics are controlled by separate factors that occur in pairs, one factor from each parent.

unsaturated fat. A fatty acid in which some carbon atoms form double bonds with other carbon atoms.

urinary system. The system responsible for conserving water and maintaining the body's ionic salt balance.

urochordates (yoor′ə-kôrd′āts). A subphylum of the phylum Chordata consisting of invertebrate animals; for example, the sea squirts.

uterus. In female mammals, the structure that protects and nurtures the embryo during pregnancy.

vacuole (văk′yoo-ōl). A large space within a cell, usually filled with fluid that contains dissolved substances.

vagina. (1) A muscular tube leading from the uterus to the outside of a female mammal's body, serving as the female's copulatory organ and the birth canal; (2) in other animals, a canal of similar structure and function.

valence electron. An electron in an atom's outermost orbital.

variation. A genetic difference between parents and offspring.

vascular cambium. In plants, a type of lateral meristem that gives rise to secondary growth, producing xylem on its inner side and phloem on its outer side, thereby increasing the diameter of the plant.

vascular plant. A member of the kingdom Plantae with true roots, stems, leaves, and transporting vascular tissues.

vascular tissue. Groups of cells responsible for conducting water, dissolved minerals, and the products of photosynthesis through a plant.

vein. A blood vessel that carries blood back to the heart.

ventricle. A large muscular chamber of the heart that receives blood from an atrium and pumps the blood to the rest of the body.

vertebra (vûrt′ə-brə). One of a series of subunits of bone surrounding the nerve cord, the individual bones that make up the backbone.

vesicle. A small droplet in the cytoplasm bound by a single membrane.

villi (vĭl′ī′). Projections in the inner lining of the small intestine that increase its surface area and allow for greater absorption of the digested food.

virus. An intracellular obligate parasite, composed of a protein coat and a nucleic acid core.

visible light. The portion of the electromagnetic spectrum that can be seen with the unaided human eye.

vitalist. One who believes that living things possess a vital force that leaves the body at death.

vitamin. A molecule that is usually a part of a coenzyme.

voluntary muscle. Muscle that can usually be contracted at will; skeletal muscle.

water cycle. The cycle of water through the environment, in which water from the earth evaporates, condenses in the atmosphere, and falls back to earth in the form of precipitation.

white blood cells. Blood cells responsible for defending the body from infection.

X chromosome. A sex chromosome; human females have two X chromosomes and human males have one.

xylem tissue (zī′ləm).   A type of vascular tissue that conducts water and dissolved minerals from the roots of a plant to the photosynthesizing cells.

Y chromosome.   A sex chromosome; the homologous chromosome to the X chromosome in human males.

yolk.   The structure that provides nutrients for a developing embryo.

zygote (zī′gōt).   The diploid cell of the new generation that forms when sperm and egg fuse.

Suggested Readings

UNIT 1
THE CONCEPTS OF LIFE

The Journal of NIH Research, "Fifth Anniversary Issue," May 1994. A must read for the student interested in the key findings in biomedical research.

Levenson, T. "The Next Step: 25 Discoveries That Could Change Our Lives," *Science 85,* vol. 6, no. 9, November 1985. Almost the entire issue is devoted to 25 important scientific discoveries that will affect our future. In addition, the first article, "Paradoxes of Prediction," discusses predictions about the science of the future.

Moore, J. A. *Science as a Way of Knowing: The Foundation of Modern Biology,* Cambridge, MA: Harvard University Press, 1993. A famous developmental biologist provides an interesting perspective of biology.

Morse, S. S. "Crossing Over: The Interspecies Traffic of Emerging Infections," *The Journal of NIH Research,* October, 1994. A good article about new viral infections.

Mulligan, R. C. "The Basic Science of Gene Therapy," Science, May 14, 1993.

Stephens, T., and D. M. Barnes. "Leaks Plague OSI Misconduct Inquires," *Journal of NIH Research,* vol. 3, no. 5, pp. 24–25, May 1991. An insight into how the press sometimes distorts scientific fraud findings, resulting in trail by press.

UNIT 2
UNITY: A MOLECULAR/CELLULAR PERSPECTIVE

Alberts, B., and others. *Molecular Biology of the Cell,* Garland, New York, 1989. Detailed but yet well-written textbook on cell biology.

Allen, R. D. "The Microtubule as an Intracellular Engine," *Scientific American,* February 1987. This article presents useful material about the microtubule.

Baker, J. J., and G. E. Allen. *Matter, Energy, and Life: An Introduction to Chemical Concepts,* 4th ed., Addison-Wesley, Reading, Mass., 1981. An excellent elementary book for students with little or no chemistry or physics background. Available in paperback.

Becker, W. M., and D. W. Deamer. *The World of the Cell,* 2nd ed. Redwood City, CA: Benjamin/Cummings, 1991. An excellent text for undergraduates.

de Duve, C. *A Guided Tour of the Living Cell,* Scientific American Library, Freeman, New York,1984. Christian de Duve, one of the founders of modern cell biology, takes the reader on a journey through the cell in a beautifully illustrated two-volume set.

Ezzel, C. "Sticky Situations," *Science News,* June 13, 1992. How cells are held together.

Hanawalt, P. C., ed. *Molecule to Living Cells: Readings from Scientific American,* Freeman, San Francisco, 1980. A collection of interesting and informative articles from *Scientific American.* Highly recommended.

Hendry, G. M. "Making, Breaking, and Remaking Chlorophyll," *Natural History,* May 1990. A good current review of photosynthesis.

Hinkle, P. C., and R. E. McCarty. "How Cells Make ATP," *Scientific American* 1978. A very good summary and explanation of Mitchell's chemiosmotic theory of ATP formation. The illustrations are helpful in explaining chemiosmosis.

Lehninger, A. L. *Principles of Biochemistry,* Worth, New York, 1982. A clear, well-written text of biochemistry for advanced students. Emphasizes the living cell.

Lewis, R. "Mitochondria — Eclectic Organelles," *Biology Digest,* October 1989. A discussion of respiration within the mitochondria.

Margulis, L. *Symbiosis in Cell Evolution,* Freeman, San Francisco, 1981. A summation of the endosymbiotic hypothesis as an explanation for the origin of eukaryotic cellular organelles.

Miller, K. R. "The Photosynthetic Membrane," *Scientific American,* October 1979. A good description of the thylakoid membrane and how its structure is related to its function of converting light energy to chemical energy.

Porter, K. R., and M. A. Bonneville, *An Introduction to the Fine Structures of Cells and Tissues,* 4th ed., Lea & Febiger, Philadelphia, 1973. A classic atlas of electron microphotographs.

Scientific American, "The Molecules of Life," entire October 1985 issue. A must for the student intersted in cell physiology.

Singer, S., and G. Nicolson. "The Fluid Mosaic Model of the Structure of Cell Membranes," *Science*, vol. 175, pp. 720–731, 1972. The original publication of the fluid mosaic model of the cell membrane.

Stossel, T. P. "The Machinery of Cell Crawling," *Scientific American*, September 1994. A discussion of cell locomotion.

Vogel, S. "The Shape of Proteins to Come," Discover, October 1988. Current directions of protein engineering.

Youvan, D. C., and B. L. Mars. "Molecular Mechanisms of Photosynthesis," *Scientific American*, June 1987. An excellent article on photosynthesis.

UNIT 3
GENETIC CONTINUITY

Avery, O. T., et al. "Studies on the Chemical Nature of the Substance Inducing Transformation of Pneumococcal Types," *The Journal of NIH Research*, March 1994. A reprint and discussion of the landmark paper leading to an understanding of DNA as the genetic material.

Beardsley, T. "Diagnosis by DNA," *Scientific American*, October 1992.

Ezzell, C. "Breast Cancer Genes," *The Journal of NIH Research*, October 1994. A current article about breast cancer genes.

Gilbert, W. "Toward a Paradigm Shift in Biology," *Nature*, January 10, 1991. An interesting article about the impact of genome dissection on biological thought.

Gilbert, W., and L. Villa-Komaroff. "Useful Proteins from Recombinant Bacteria," *Scientific American*, April 1980. A description of recombinant DNA techniques and their applications, written by two of the pioneers in the field.

Glover, D. M., C. Gonzales, and J. W. Raff. "The Centrosome," *Scientific American*, June 1993. A good article about the role of the centrosome in forming the mitotic spindle and cytoskeleton.

Grunstein, M. "Histones as Regulators of Genes," *Scientific American*, October 1992. A well-written article about gene regulation.

Hall, S. S. "Biologists Zero in on Life's Very Essence," *Smithsonian*, February 1990. A discussion of the human endeavor known as sequencing the human genome.

Hammer, S., A. Dorfman, and A. Wilber. "To Conquer Cancer," *Science Digest*, vol. 93, p. 31ff., August 1985. A series of articles in which scientists discuss their approach to conquering cancer.

Hartl, D. L. *Genetics*, 3rd ed. Jones and Bartlett Publishers, Boston, 1994. An excellent introductory genetics text.

Hunter, T. "The Proteins of Oncogenes," *Scientific American*, August 1984. An excellent article about oncogenes for the serious student.

Johnson, J. "Gene Therapy: After the Dream World, the Serious Science," *Journal of NIH Research*, vol. 3, no. 6, p. 26, June 1991. A realistic appraisal of gene therapy.

Journal of NIH Research, January 1991. A special section on cancer beginning on p. 41. An excellent summary of our war on cancer that discusses the genetic, molecular, and cellular processes that dominate cancer research.

Judson, H. E. *The Eighth Day of Creation*, Simon and Schuster, New York, 1979. An enjoyable and interesting presentation of the history of DNA.

Lewis, R. "DNA Fingerprints: Witness for the Prosecution," Discover, June 1988. Light reading about the application of DNA principles to identification.

Morell, V. "Huntington's Gene Finally Found," *Science*, April 2, 1993. A good article on the search and location of the gene for Huntington's.

National Institutes of Health, National Cancer Institute, Office of Cancer Communication, Bethesda, MD, 1-800-4-CANCER. Students interested in knowing more about cancer can call this toll-free number for answers to their questions.

Pelecih, S. "When Cells Divide," *The Sciences*, July-August 1990. A good overview of cell division.

Sayre, A. *Rosalind Franklin and DNA*, Norton, New York, 1975. Ann Sayre wrote this book "to set the record straight." The race for the structure of DNA is examined from the perspective of restoring credit to Rosalind Franklin for her role in the process. This is not the same viewpoint espoused by Watson in his book *The Double Helix*.

Stern, C., and E. R. Sherwood. *The Origin of Genetics: A Mendel Source Book*, Freeman, San Francisco, 1966. A translation of Mendel's original major paper presented to the Braun Society and a collection of some minor works. Included is a discussion of whether or not Mendel fabricated some of his work.

Tetrault, S. M. "The Student with Sickle-Cell Anemia," *Today's Education*, pp. 54–57, April-May 1981. The everyday adjustment required for patients with sickle-cell anemia is discussed.

Watson, J. D. *The Double Helix,* Atheneum, New York, 1968. Interesting reading about how science is really conducted. A personal discussion by Watson on the events that led to the discovery of the double helical structure of DNA.

Watson, J. D., and F. H. C. Crick. "Molecular Structure of Nucleic Acids: A Structure of Deoxyribose Nucleic Acid," *Nature*, vol. 171, p. 737, 1953. This is the article that set the foundation for all of molecular genetics. Probably the single most important page in the history of biology.

Watson, J. D., J. Tooze, and D. T. Kurtz. *Recombinant DNA: A Short Course*, Freeman, San Francisco, 1983. An introduction to the concepts of genetic engineering. For the serious student.

UNIT 4
EVOLUTION AND DIVERSITY

Attenborough, D. *Life on Earth*. Little, Brown, Boston, 1979. An easy-to-read, well-illustrated account of the evolution of life. Based on popular *Life on Earth* PBS television series.

Cairn-Smith, A. G. "The First Organisms," *Scientific American*, June 1985. An interesting article defending the premise that self-replicating clays, not the primordial soup, were the source of materials that allowed for the origin of life.

Desmond, A., and J. Moore. *Darwin*, Warner, New York, 1992. One of the more recent biographies.

Gould, S. J. *Ever Since Darwin*, Norton, New York, 1977.

Gould, S. J. *The Panda's Thumb*, Norton, New York, 1980. This and the previous work are collections of imaginative and interesting essays, primarily about evolution.

Gould, S. J. "Darwinism Defined: Sifting Fact from Theory," *Discover*, January 1987. An essay about why some nonscientists egregiously misunderstood evolution.

Gould, S. J. "We Are All Monkey's Uncles," *Natural History*, June 1992. An interesting discussion of cladistics and human classification.

Johanson, D., and M. Edey. *Lucy: The Beginnings of Humankind*, Simon and Schuster, New York, 1981. The significance of the fossil Lucy to an understanding of early human evolution.

Margulis, L., and K. U. Schwartz. *Five Kingdoms: An Illustrated Guide to the Phyla of Life on Earth*, 2nd. ed., Freeman, San Francisco, 1988. An excellent account of the diversity of organisms on the earth.

Poole, R. J. "Closing the Gap Between Proteins," *Science*, June 29, 1990. Scientists discuss how our understanding of self-replication RNA is helping us to understand the origin of life.

Rebek, J. R. "Synthetic Self-Replicating Molecules," *Scientific American*, July 1994. Helps to answer the question as to how life began.

Rennie, J. "Are Species Specious?" *Scientific American*, November 1991. Some of the problems associated with the concept of species.

Shipman, P. "Baffling Limb on the Family Tree," *Discover*, September 1986. Light, interesting reading with respect to the recent finds that are changing our understanding of the human lineage.

Touchette, N. "Sonic Hedgehog Signals the Shape of Things to Come." *The Journal of NIH Research*, April 1994. A brief review of the genes involved in vertebrate morphogenesis.

Vogel, S. "Face-to-Face with a Living Fossil," *Discover*, March 1988. An account of the discovery of the primitive fish, the coelacanth.

Weisburd, S. "Australia's Animals of the Past," *BioScience*, September 1988. An account of some of Australia's ancient creatures.

Whittaker, R. H. "New Concepts of Kingdoms of Organisms," *Science*, vol. 163, 1969. The article that led to the five-kingdom system for classifying animals.

UNIT 5
PLANTS

Bazzazz, F. A., and E. D. Fajer. "Plant Life in a CO_2-Rich World," *Scientific American*, January 1992. Will global warming affect C_3 and C_4 plants?

Dale, J. "How Do Leaves Grow?" *BioScience*, June 1992. A molecular approach to the understanding of plant growth.

Jordan, W. "The Bee Complex," *Science*, May 1984. An article that discusses the evolutionary insect-flower relationship.

Lehner, R., and J. Lehner. *Folklore and Odysseys of Food and Medicinal Plants*, Tudor, New York, 1962. A folklore approach to some interesting stories about the ways humans use plants.

Mooney, H. A., and others. "Plant Physiological Ecology Today," *BioScience*, vol. 35, pp. 18–67, 1988. An excellent review of plant physiology.

Moses, P. B. and C. Nam-Hai. "Light Switches for Plant Genes," *Scientific American*, April 1988. An interesting explanation of new research about how genes are turned on and off in response to light.

Raven, P. H., R. F. Evert, and S. E. Eichhorn. *Biology of Plants*, 5th ed., Worth, New York, 1992. A good introductory text.

Rost, T., and others. *Botany: A Brief Introduction to Plant Biology*, 2nd. ed., Wiley, New York, 1984. A good introductory botany text.

Sisler, E. C., and S. F. Yang. "Ethylene, the Gaseous Plant Hormone," *BioScience*, vol. 33, pp. 233–238, 1984. An excellent review of this important plant hormone.

Stern, K. *Introductory Plant Biology*, Brown, Dubuque, Iowa, 1985. A succinct plant biology text, with helpful appendixes on biological controls, useful and poisonous plants, and houseplant care.

Stewart, D. "Green Giants," *Discover*, pp. 61–64, April 1990. An interesting explanation of the workings of the giant sequoias.

Taubes, G. "Unmasking Agent Orange," *Discover*, April 1988. A discussion of the effect of Agent Orange on health. Some students may want to challenge the conclusions of this article.

Went, F. W. *The Plants*, Life Nature Library/Time, New York, 1963. An old, yet beautifully illustrated and readable approach to botany.

Whatley, F. R., and J. M. Whatley. *Light and Plant Life*, Arnold, London, 1980. A brief text that summarizes the effects of light on germination.

Zimmerman, M. H. *Xylem Structure and the Ascent of Sap,* Springer-Verlag, New York, 1983. An excellent discussion of the functional, structural relationship between xylem and sap ascent.

Zimmerman, M. H. "How Sap Moves in Trees," *Scientific American,* May 1984. An interesting description of how aphids can be used as research tools to investigate sap movement.

UNIT 6
ANIMALS

Caldwell, M. "How Does a Single Cell Become a Whole Body?" *Discover,* November 1992. An excellent discussion of development.

Crooks, R., and K. Baur. *Our Sexuality,* 5th ed., Benjamin/Cummings, Redwood City, CA, 1993. A good textbook about human sexuality.

DeRoberts, E. M., and others. "Homeobox Genes and the Vertebrate Body Plan," *Scientific American,* July 1990. An excellent discussion of the family of genes known as homeobox genes and their role in development.

Diamond, J. "Survival of the Sexiest," *Discover,* May 1988. Light reading with respect to sexual selection.

Gottlieb, D. I. "GABAergic Neurons," *Scientific American,* February 1988. This article will provide the reader with a better understanding of the inhibitory aspects of neural transmission.

Henig, R. M. *How a Woman Ages and How a Man Ages,* Ballantine Books, New York, 1985. Insights into the human aging process.

Houston, C. S. "Mountain Sickness," *Scientific American,* October 1992. High altitude and how it interferes with homeostasis.

Jaret, P. "Our Immune System, the Wars Within," *National Geographic,* June 1986. A well-illustrated, easy-to-understand article about the human immune system.

Karetz, J. F., and G. H. Handelman. "How the Human Eye Focuses," *Scientific American,* July 1988. The effect of aging on the eye's ability to focus.

Lawrence, J. "The Immune System and AIDS," *Scientific American,* December 1985. A must, for those students interested in learning more about the biological basis of AIDS.

Nilsson, L., and J. Lindberg. *Behold Man,* Little, Brown, Boston, 1974. Even though old, this text is a wonderful collection of photographs of human development and the human body.

Old, L. J. "Tumor Necrosis Factor," *Scientific American,* May 1988. The anticancer potential of TNF is discussed.

Science, "AIDS: The Unanswered Questions," May 28, 1993. This entire issue is devoted to a discussion of HIV and AIDS.

Shipman, P. "Human Origins: New *Australopithecus* Finds Revive Debate," *The Journal of NIH Research,* February 1994.

Smith, H. W. *From Fish to Philosopher,* Doubleday, Garden City, N. Y., 1961. A novel approach to the evolution of the vertebrate kidney.

Snyder, S., and D. Bredt. "Biological Roles of Nitric Oxide." *Scientific American,* May 1992. A discussion of the recent research that demonstrates that NO is an important chemical signal.

Tonewaga, S. "The Molecules of the Immune System," *Scientific American,* October 1985.

Vander, A., J. Sherman, and D. Luciano. *Human Physiology: The Mechanisms of Body Function,* 4th ed., McGraw-Hill, New York, 1985. An excellent basic physiology text.

Wolkomir, R. "Chilling Out for Science," *Discover,* February 1988. This article provides some insight into human thermoregulation.

Zimmer, C. "Ruffled Feathers," *Discover,* May 1992. A very interesting discussion about the controversy associated with the origin of birds.

UNIT 7
BEHAVIOR AND ECOLOGY

Alcock, J. *Animal Behavior: An Evolutionary Approach,* 3rd ed., Sinauer Associates, Inc., Sunderland, Mass., 1984. An excellent introductory animal behavior text.

Attenborough, D. *The Living Planet: A Portrait of the Earth,* William Collins Sons and British Broadcasting Corporation, London, 1984. A beautiful coffee table book of our planet and its biomes.

Begon, M. J., L. Harper, and C. R. Townsend. *Ecology: Individuals, Populations, and Communities,* Sinauer Associates, Inc., Sunderland, Mass., 1986. A good general ecology text.

Broeker, W. "Global Warming on Trial," *Natural History,* April 1992. Controversy related to the evidence for global warming.

Burman, A. "Saving Brazil's Savannas," *New Scientist,* March 2, 1991. The politics of deforestation.

Daily, G. C., and P. R. Ehrlich. "Population, Sustainability, and Earth's Carrying Capacity." *BioScience,* November 1992.

Harwell, M. A. Nuclear *Winter: The Human and Environmental Consequences of Nuclear War,* Springer-Verlag, New York, 1984. A discussion of the impact of nuclear war and the resulting nuclear winter on our environment.

Houghton, R. A., and G. M. Woodwell. "Global Climate Change," *Scientific American,* April 1989. A discussion of the current research relevant to global warming.

Lent, C. M., and M. H. Dickinson. "The Neurobiology of Feeding in Leeches," *Scientific American,* June 1988. An interesting discussion of how a single neurotransmitter orchestrates feeding behavior in the leech.

MacArthur, R. H., and and E. O. Wilson. *The Theory of Island Biogeography,* Princeton University Press, Princeton, N. J., 1967. A book that helped to develop the disci-

Suggested Readings

pline of island biogeography and was so influential in the career choice of Dr. Cruz, who is profiled in our text.

O'Shea, T. J. "Manatees." *Scientific American*, July 1994. Discusses how humans are responsible for the endangerment of these mammals.

Repetto, R. "Deforestation in the Tropics," *National Geographic*, pp. 36–42, April 1990. The ecological consequences of massive deforestation in the tropics is discussed.

Shashoua, V. "The Role of Extracellular Proteins in Learning and Memory,. *American Scientist,* vol. 73, 1985. Speculation about the role of proteins in synapse formation during learning.

Stolarski, R. S. "The Antarctic Ozone Hole," *Scientific American*, January 1988. An interesting article that presents the facts of an important topic.

Verrell, P. "When Males are Choosy," *New Scientist,* pp. 46–50, January 1990. Interesting discussion of the behavior of males.

Wilson, E. O. *Sociobiology: The New Synthesis,* Belknap Press, Cambridge, Mass., 1973. One of the most important and controversial books on behavior.

Wilson, E. O. *The Diversity of Life*, Harvard University Press, Cambridge, MA, 1992. The author explores the theme as to why we should be concerned about the extinction of species.

Answers to Review Questions

Chapter 1
True/False: 1. true   2. true   3. true
Fill in the Blank: 1. enzymes   2. DNA
   3. metabolism
Multiple Choice: 1. a   2. a   3. c   4. b

Chapter 2
True/False: 1. true   2. false   3. false   4. false
Fill in the Blank: 1. ion   2. polar   3. isomers
   4. polar
Multiple Choice: 1. c   2. d   3. b   4. c

Chapter 3
True/False: 1. true   2. true   3. false   4. true
Fill in the Blank: 1. synthesis   2. monosaccharide
   3. minimum energy   4. primary
Multiple Choice: 1. c   2. d   3. c   4. b

Chapter 4
True/False: 1. false   2. false   3. true   4. true
Fill in the Blank: 1. resolving power   2. rough
   endoplasmic reticulum   3. Golgi apparatus
   4. cytoskeleton
Multiple Choice: 1. a   2. b   3. c   4. c

Chapter 5
True/False: 1. true   2. true   3. false   4. false
Fill in the Blank: 1. lower, higher   2. carrier, energy
   3. phagocytosis   4. proteins, carbohydrates
Multiple Choice: 1. a   2. d   3. d   4. d

Chapter 6
True/False: 1. true   2. true   3. true   4. false
Fill in the Blank: 1. entropy   2. ATP   3. reduction
   4. active site
Multiple Choice: 1. a   2. c   3. b   4. d

Chapter 7
True/False: 1. true   2. true   3. true   4. false
Fill in the Blank: 1. light-dependent, light-independent
   2. carbon   3. chemiosmotic   4. absence
Multiple Choice: 1. d   2. c   3. d   4. b

Chapter 8
True/False: 1. true   2. true   3. false   4. false
5. true
Fill in the Blank: 1. NADPH, ATP   2. photolysis
   3. protons (or H^+)   4. CO_2
Multiple Choice: 1. b   2. d   3. a   4. c

Chapter 9
True/False: 1. false   2. false   3. false   4. false
Fill in the Blank: 1. lactic acid   2. cristae   3. CO_2
   4. protons
Multiple Choice: 1. e   2. d   3. d   4. a

Chapter 10
True/False: 1. true   2. true   3. true   4. true
Fill in the Blank: 1. genetic   2. karyokinesis,
   cytokinesis   3. haploid   4. Cancer
Multiple Choice: 1. b   2. d   3. a   4. a

Chapter 11
True/False: 1. true   2. true   3. true   4. true
Fill in the Blank: 1. thymine, guanine   2. DNA
   3. ligase   4. double helix
Multiple Choice: 1. d   2. a   3. a   4. b

Chapter 12
True/False: 1. true   2. false   3. true   4. false
Fill in the Blank: 1. transcription   2. uracil
   3. translated   4. repressor protein
Multiple Choice: 1. b   2. b   3. c   4. a

Chapter 13
True/False: 1. true   2. true   3. true   4. false
Fill in the Blank: 1. plasmids   2. plasmids, R6
   3. pili (or F pili)   4. restriction enzymes
Multiple Choice: 1. c   2. a   3. b   4. b

Chapter 14
Multiple Choice: 1. a   2. a   3. e   4. d   5. b

Genetics Problems:
1a. female = *Ss*, male = *ss*   1b. ½ long-haired, ½ short-
   haired   1c. four
2.   genotypic ratio; ½ *YY*, ½ *Yy*; phenotype; all yellow
3a. *AB*   3b. *AB, aB*   3c. *Ab, ab*   3d. *AB, aB, Ab, ab*
4.   Female: *FfTt*   male: *fftt*   child: *fftt*
5.   ¹⁄₁₆
6.   ½ pink (RR') and ½ white (R'R')
7. a. Males inherit their X chromosome from their
      female parent.
   b. A male can produce two types of gametes for an
      X-linked gene. One type will lack this gene and
      possess a Y chromosome. The other will have an
      X chromosome and the linked gene.
   c. A homozygous female will produce just one type of
      gamete containing an X chromosome with the gene.

d. A heterozygous female for an X-linked gene will produce two types of gametes. One will contain an X chromosome with the dominant allele and the other gamete will contain the X chromosome with the recessive allele.

Chapter 15
Multiple Choice: 1. b   2. b   3. d   4. a   5. a
  6. b   7. b

Genetics Problems:

1a. *Tt*   1b. *Tt*   1c. *tt*

2a. *Bbww*   2b. *bbWw*   2c. *bbww*   2d. ½

3a. *Pp*   3b. *Pp*   3c. *pp*

4a. all normal-color-visioned, ½ male and ½ female

4b. ¼ normal-color-visioned female, ¼ color-blind female, ¼ normal-visioned male, ¼ color-blind male

5a. XX^H 5b. 50 percent

6. The blood types of the parents would have had to be either $I^A i$ or $I^B i$ or $I^A I^A$ or $I^B I^B$.

Chapter 16
True/False: 1. false   2. true   3. true   4. false
Fill in the Blank: 1. protobiont   2. isolation
  3. heterotrophs   4. prokaryotic
Multiple Choice: 1. a   2. a   3. b   4. a

Chapter 17
True/False: 1. false   2. true   3. true   4. false
Fill in the Blank: 1. population   2. genetic drift or gene flow   3. reproductive isolation   4. divergent
Multiple Choice: 1. c   2. b   3. d   4. b   5. b   6. c

Chapter 18
True/False: 1. true   2. false   3. true   4. false
Fill in the Blank: 1. binomial   2. genus, species
  3. bilateral   4. monera
Multiple Choice: 1. c   2. b   3. d   4. c

Chapter 19
True/False: 1. true   2. false   3. true   4. true
Fill in the Blank: 1. heterogamous   2. spores
  3. sporophyte   4. angiosperms
Multiple Choice: 1. b   2. b   3. d   4. b   5. a

Chapter 20
True/False: 1. false   2. true   3. false   4. true
Fill in the Blank: 1. storage, support, and secretory
  2. companion   3. apical   4. lateral meristem
Multiple Choice: 1. c   2. b   3. b   4. c

Chapter 21
True/False: 1. true   2. true   3. true   4. false
Fill in the Blank: 1. stomata   2. potassium   3. fungi
  4. phytochrome
Multiple Choice: 1. a   2. a   3. b   4. b

Chapter 22
True/False: 1. true   2. true   3. true   4. true
Fill in the Blank: 1. exoskeleton   2. tendons

  3. antagonistic   4. sliding filament
Multiple Choice: 1. a   2. a   3. d

Chapter 23
True/False: 1. true   2. true   3. true   4. false
Fill in the Blank: 1. cardiac   2. chyme
  3. villi, microvilli   4. insulin, glucagon
Multiple Choice: 1. c   2. b   3. b   4. b

Chapter 24
True/False: 1. false   2. false   3. false   4. true
Fill in the Blank: 1. red   2. systolic   3. away
  4. interstitial
Multiple Choice: 1. c   2. d   3. b   4. a

Chapter 25
True/False: 1. true   2. false   3. true   4. true
Fill in the Blank: 1. histamine   2. variable   3. antigen
  4. plasma
Multiple Choice: 1. a   2. b   3. d   4. c

Chapter 26
True/False: 1. true   2. true   3. true   4. true
Fill in the Blank: 1. negative   2. nephron
  3. proximal   4. hemoglobin
Multiple Choice: 1. d   2. d   3. d   4. b

Chapter 27
True/False: 1. false   2. true   3. false   4. false
Fill in the Blank: 1. hermaphroditism   2. interstitial
  3. cervix   4. resolution
Multiple Choice: 1. a   2. b   3. b   4. b

Chapter 28
True/False: 1. false   2. true   3. false   4. true
Fill in the Blank: 1. land, internal   2. teratogens
  3. oxytocin   4. prolactin, oxytocin
Multiple Choice: 1. a   2. a   3. a   4. a

Chapter 29
True/False: 1. true   2. true   3. false   4. true
Fill in the Blank: 1. mesodermal   2. cAMP
  3. androgens   4. endorphins
Multiple Choice: 1. c   2. a   3. b   4. a

Chapter 30
True/False: 1. false   2. false   3. true   4. true
Fill in the Blank: 1. motor   2. sodium   3. axon
  4. forebrain
Multiple Choice: 1. a   2. d   3. d   4. c   5. d
  6. b   7. a   8. a   9. c   10. b

Chapter 31
True/False: 1. true   2. true   3. false   4. true
Fill in the Blank: 1. oral   2. body cavity
  3. peritoneum   4. specialized
Multiple Choice: 1. d   2. a   3. d   4. a

Chapter 32
True/False: 1. false   2. true   3. true   4. true
Fill in the Blank: 1. Vertebrata   2. Agnatha
  3. gill-supporting structures   4. Amphibia
Multiple Choice: 1. d   2. c   3. d   4. b

Chapter 33
True/False: 1. true   2. false   3. true   4. true
Fill in the Blank: 1. taxis   2. follows   3. skill
   4. releasers
Multiple Choice: 1. b   2. c   3. a   4. a

Chapter 34
True/False: 1. true   2. false   3. true   4. false
Fill in the Blank: 1. weather   2. prevailing, ocean
   3. Tundra   4. freshwater, saltwater
Multiple Choice: 1. b   2. b   3. a   4. c

Chapter 35
True/False: 1. false   2. false   3. true   4. false
Fill in the Blank: 1. zero   2. s   3. smaller   4. density
Multiple Choice: 1. a   2. d   3. c   4. c

Chapter 36
True/False: 1. false   2. true   3. false   4. false
Fill in the Blank: 1. intraspecific   2. primary
   3. climax   4. slower
Multiple Choice: 1. c   2. d   3. b   4. c

Chapter 37
True/False: 1. false   2. true   3. false   4. false
Fill in the Blank: 1. organic material   2. trophic
   3. decomposers   4. less
Multiple Choice: 1. b   2. a   3. a   4. c

Chapter 38
True/False: 1. true   2. true   3. true   4. false
Fill in the Blank: 1. 14 million   2. 25   3. plant
   4. solid
Multiple Choice: 1. a   2. c   3. a   4. d

Credits and Acknowledgments

Page xxv: San Diego State University Foundation.

Unit 1 Tom Van Sant/The GeoSphere Project.

Chapter 1 Page 3: Cetus Corporation. Fig. 1.1: (a) Don Martin/San Diego State University; (b) Robert Frye. Fig. 1.2: (a) Bill Smith/Stanford Visual Arts, Stanford University; (b) NASA; (c) Ann Kelly/Photo Nats; (d) G. France/Visuals Unlimited. Fig. 1.3: (a) A. Lesk/Photo Researchers, Inc.; (b) Secchi-Lecague & Roussel/Photo Researchers, Inc.; (c) Don Martin/San Diego State University. Fig. 1.4: Ray Clark & Mervyn Goff/Photo Researchers, Inc. Fig. 1.5: Peter Parks/Animals Animals. Fig. 1.6: (a) Alfred Pasieka/Photo Researchers, Inc.; (b) Don Martin/San Diego State University. Fig. 1.7: (a) Robert Pearcy/Animals Animals; (b) Patricia Avila/San Diego State University. Fig. 1.8: (a) Nicole Duplaix/Peter Arnold, Inc.; (b) Don Martin/San Diego State University. Fig. 1.9: (a–e) Don Martin/San Diego State University. Fig. 1.10: Photo Researchers, Inc. Fig. 1.11: Tom McHugh/Photo Researchers, Inc. Profile: (left) Gerry Gropp; (middle) Wide World Photos, Inc.; (right) Public Relations, University of Colorado, Denver. Fig. 1.12: Stephen Dalton/Photo Researchers, Inc. Fig. 1.13: (a,b) Don Hinrischen/Institute for Local Self-Reliance; Fig. 1.14: Don Martin/San Diego State University. Fig.1.15: (a, b) Joy Zelder/San Diego State University. Fig. 1.16: Elaine Corets/San Diego State University. Focus 1.2: (A) Joe McDonald/Visuals Unlimited; (B) Alfred Paisieka/Photo Researchers, Inc.

Unit 2 J. Camissa, GEMM Program/NCI National Institutes of Health and Focus Graphics.

Chapter 2 Page 29: Computer Graphics Laboratory/University of California, San Francisco. Human Endeavors: (a) The Bettmann Archive; (b) Darin Norfleet/San Diego State University; (c) Don Fawcett/Visuals Unlimited; (d) Photo Researchers, Inc.; (e) University of Santa Barbara. Focus 2.1: (A) J. James Donady/Wesleyan University; (B) Simon Fraser/Photo Researchers, Inc. Profile: Wide World Photos, Inc. Fig. 2.10: (b) Phil Borden/Photo Edit. Fig. 2.16: (a) Hermann Eisenbeiss/Photo Researchers, Inc.; (b) Vernon Avila/San Diego State University.

Chapter 3 Page 51 and Focus 3.1: (A) Don Eigler/IBM Alamaden Research Center. Fig. 3.6: (a) Don Martin/San Diego State University; (b) Carson Baldwin Jr./Animals Animals. Fig. 3.9: John Crum/San Diego State Unversity. Fig. 3.11: (b) American Heart Association. Profile: Wide World Photos, Inc. Fig. 3.16: (b) David Dunlap & Carlos Bustamante/University of Oregon.

Chapter 4 Page 75: G. Musil/Visuals Unlimited. Fig. 4.2: (a) Darin Norfleet/San Diego State University; (b, c) D. W. Fawcett/Visuals Unlimited. Fig. 4.3: Photo Researchers, Inc. Focus 4.1: (A–C) Darin Norfleet/San Diego State University. Fig. 4.5: Photo Researchers, Inc. Fig. 4.6: (b, c) Max Planck Institüt. Profile: J. Guyden/City College of New York. Fig. 4.7: Aldolfo Plazaola/University of Puerto Rico. Fig. 4.8: Richard Weiss Bizzoco/San Diego State University. Fig. 4.10: K. Gorgas & S. K. Krisans/San Diego State University. Fig. 4.11: (a) P. Bagavandoss/The Upjohn Co.; (b) Richard Weiss Bizzoco/San Diego State University. Fig. 4.12: Photo Researchers, Inc. Fig. 4.13: (a) Darin Norfleet/San Diego State University; (b) Jeremy Burgiss/Photo Researchers, Inc. Fig. 4.14: (a, c) Don W. Fawcett/Visuals Unlimited. Fig. 4.15: (a, b) Darin Norfleet/San Diego State University. Fig. 4.16: (b, c) Biological Photo Service; (d) Richard Weiss Bizzoco/San Diego State University. Focus 4.2: P. Bagavandoss/The Upjohn Co.

Chapter 5 Page 101: Don Fawcett/Visuals Unlimited. Fig. 5.1: D. W. Fawcett/Photo Researchers, Inc. Fig. 5.2: Roger Sabbadini/San Diego State University. Fig. 5.11: Joseph Hedrick/San Diego State University. Fig. 5.12: Jerry Goldstein. Focus 5.1: (A) Computer Graphics Laboratory/University of California, San Francisco. Profile: Wilfred Denetclaw/University of California, Berkeley.

Chapter 6 Page 123: Hank Levine/Visuals Unlimited. Fig. 6.1: J. N. Reichel/Photo Researchers, Inc. Fig. 6.2: Don Martin/San Diego State University. Fig. 6.3: Angelica Brandt/University of Oldenburg. Fig. 6.8: Kenneth Lucas/Biological Photo Service. Fig. 6.11: (a, b) Robert Stodola/Fox Chase Cancer Center and Donald Voet/University of Pennsylvania. Profile: Gordon Swan and Jay Van Rensselaer/The John-Hopkins University. Focus 6.1: (A) Barry Finzel/The Upjohn Co.

Chapter 7 Page 145: L. Lee Grismer/La Sierra University. Fig. 7.3: (b, c) James Neel/San Diego State University. Profile: Wide World Photos, Inc. Fig. 7.10: (a) Electron Microscope Facility/San Diego State University.

Chapter 8 Page 163: Andrew Borsar/Images LTD. Fig. 8.1: (a) J. Robert Paterek & Paul H. Smith/University of Florida, Gainesville; (b) Bob Gladden/San Diego State University. Fig. 8.3: (all) James Neel/San Diego State University. Fig. 8.6: (b) Christine Case/Visuals Unlimited. Profile: University of California, Berkeley, News Bureau. Fig. 8.15: (a) Carolina Biological Supply Company.

Chapter 9 Page 187: Carolina Biological Supply Company Fig. 9.1 : Stephen J. Krassman/Photo Researchers, Inc. Fig. 9.2: (a) John Thaw/San Diego State University; (b) Bob Gladden/San Diego State University; (c) D. M. Phillips/Visuals Unlimited; (d) A. M. Siegelman/Visuals Unlimited. Fig. 9.8: (a) Electron Microscope Facility/San Diego State University. Fig. 9.13: Bruno J. Zehnder/Peter Arnold, Inc. Profile: Roger Sabbadini/San Diego State University. Focus 9.1: (left) F. Hossler/Visuals Unlimited; (right) John D. Cunningham/Visuals Unlimited.

Unit 3 James Crouch/San Diego State University.

Chapter 10 Page 213: Dr. Joe Gray/University of California, San Francisco. Fig. 10.1: (a, b) C. H. Fox/Photo Researchers, Inc. Fig. 10.3: (a) Photo Researchers, Inc. Profile: Wide World Photos, Inc. Fig. 10.4: James T. Mascarello/Genetics Laboratory, Children's Hospital, San Diego. Fig. 10.5: (a) Visuals Unlimited. Fig. 10.7: (a–e) Darin Norfleet/San Diego State University. Fig. 10.11: (a) Don Martin/San Diego State University. Fig. 10.12: (a) Richard Weiss Bizzoco/San Diego State University. Fig. 10.13: (a) Don Martin/San Diego State University.

Chapter 11 Page 241: J. Cammisa, GEMM Program/NCI, National Institutes of Health and Focus Graphics. Fig. 11.1: (a–c) Don Martin/San Diego State University. Fig. 11.7: (c) David Dunlap & Carlos Bustamante/University of Oregon. Profile: From: *Biology: Discovering Life*, p. 414, Levine/Miller, D. C. Heath & Co.

Chapter 12 Page 259: Rebecca Keller/University of New Mexico, Albuquerque. Fig. 12.7: (c) Rebecca Keller/University of New Mexico, Albuquerque. Fig. 12.15: (b) Jack Bostrack/Visuals Unlimited. Profile: Francis Collins/National Institutes of Health.

Chapter 13 Page 285: Keith W. Wood/University of California, San Diego. Profile: Lydia Villa-Kamaroff/Harvard Medical School. Fig. 13.1: Photo Researchers, Inc. Fig. 13.2: Fred Hossler/Visuals Unlimited. Fig. 13.3: (a, b) K. G. Murti/Visuals Unlimited. Fig. 13.4: Robley Williams. Fig. 13.11: Cetus Corporation. Fig. 13.12: Cetus Corporation. Fig. 13.13: J. C. Hunter-Cevera/Cetus Corporation.

Chapter 14 Page 307: Steve Barlow/San Diego State University. Profile: Wide World Photos, Inc. Fig. 14.8: (a) Vernon Avila/San Diego State University. Fig. 14.10: (a) Courtesy of Vernon Avila. Fig. 14.16: (c) J. Kezer/University of Oregon, Eugene.

Chapter 15 Page 335: Bill Smith/Stanford Visual Arts, Stanford University. Fig. 15.1: (c) John S. O'Brian. Profile: Nancy Wexler. Fig. 15.3: (top) Science Source/Photo Researchers, Inc.; (bottom) James T. Mascarello/Genetics Laboratory, Children's Hospital, San Diego. Fig. 15.5: (a) Rose Frye/Primrose School, Fontana, CA; (b) James T. Mascarello/Genetics Laboratory, Children's Hospital, San Diego. Fig. 15.7: Lester Bergman. Fig. 15.8: (a) Mike & Moppet Reed/Animals Animals. Fig. 15.9: James T. Mascarello/Genetics Laboratory, Children's Hospital, San Diego. Fig. 15.11: (a) American Psychological Association. Fig. 15.12: Charles Carpenter/Field Museum of Natural History, Chicago. Fig. 15.15: (a, b) Don Martin/San Diego State University. Fig. 15.17: Photo Researchers, Inc.

Unit 4 Bob Gladden/San Diego State University

Chapter 16 Page 365: Sigurgeir Jónasson. Profile: Al Daneggar/University of Maryland. Focus 16.1: (A) NASA. Fig. 16.3: (a) Roger Ressmeyer/Starlight. Fig. 16.8: Robert Paterek & Paul H. Smith/University of Florida, Gainesville. Fig. 16.9: Don Martin/San Diego State University.

Chapter 17 Page 381: Bob Gladden/San Diego State University. Fig. 17.1: Don Martin/San Diego State University. Profile: Courtesy of Francisco Ayala/University of California, Irvine. Fig. 17.2: Robert Frye, Fontana, CA. Fig. 17.3: Mark Lung/Southwestern College. Fig. 17.4: Photo Researchers, Inc. Fig. 17.9: John D. Cunningham/Visuals Unlimited. Fig. 17.10: (a) (both) Tom McHugh/Photo Researchers, Inc.; (b) (both) L. Lee Grismer/San Diego State University; (c) (left) Breck P. Kent/ Animals Animals; (c) (right) James Crouch/San Diego State University. Fig. 17.11: George Cox/San Diego State University. Focus 17.2: (A) James Crouch/San Diego State University. Fig. 17.17: (a) John D. Cunningham/Visuals Unlimited.

Chapter 18 Page 409: Bob Gladden/San Diego State University. Fig. 18.1: (a) Karl Aufderheide/Visuals Unlimited; (b, d, e) Don Martin/San Diego State University; (c) James Crouch/San Diego State University. Fig. 18.2: (a, c) Don Martin/San Diego State University; (b) James Crouch/San Diego State University. Fig. 18.3: John D. Cunningham. Profile: Harvard University Press Office. Fig. 18.7: (a) D. M. Phillips/Visuals Unlimited; (b) Visuals Unlimited; (c) Michael Simpson/San Diego State University. Fig. 18.8: (a) P. Bagavandoss/The Upjohn Co.; (b) John D. Cunningham/Visuals Unlimited; (c, e) Visuals Unlimited; (d) D. M. Phillips/Visuals Unlimited. Fig. 18.9: (a) John D. Cunningham/Visuals Unlimited; (b) Richard Thom/Visuals Unlimited; (c) Barbara Hemmingsen/San Diego State University; (d) W. E. Fry/Visuals Unlimited. Fig. 18.10: (a) D. P. Wilson, Eric & David Hasking/Photo Researchers, Inc.; (b) Wm. S. Ormerod, Jr./Visuals Unlimited; (c) Michael Simpson/San Diego State University; (d, e) Don Martin/San Diego State University; (f) George Cox/San Diego State University. Focus 18.1: (A, B) Photo Researchers, Inc.; (C) K. G. Marti/Visuals Unlimited. Fig. 18.12: (a) Breck P. Kent/Animals Animals; (b) Fred Bavendam/ Peter Arnold, Inc.; (c, f, g) Don Martin/San Diego State University; (d, e) L. Lee Grismer/San Diego State University.

Unit 5 Andrew Borsar: Images LTD/Rockport, MA.

Chapter 19 Page 433: Robert Ross/Cayey University College. Profile: Dr. Emma Gonzalez, University of California, Los Angeles. Fig. 19.2: (a) Michael Simpson/San Diego State University; (b) Don Martin/San Diego State University. Fig. 19.3: (a, c) Visuals Unlimited; (b) Bob Gladden/San Diego State University. Fig. 19.6: (a) Don Martin/San Diego State University; (b) S. D. Sharnoff/Visuals Unlimited; (c) Michael Simpson/San Diego State University. Fig. 19.7: Michael Simpson/San Diego State University. Fig. 19.8: (a) Michael Simpson/San Diego State University; (c) Don Martin/San Diego State University. Fig. 19.11: Don Martin/San Diego State University. Focus 19.1: (A) David Thompson/OSF; (B) J. Alcock/Visuals Unlimited. Fig. 19.15: (a, c, d, e) Peter Arnold, Inc.; (b) Robert Frye/Fontana, CA

Chapter 20 Page 467: Don Martin/San Diego State University. Fig 20.4: (a–d) Don Martin/San Diego State University. Fig. 20.5: (a) Jerry Burgess/Photo Researchers, Inc.; (b) J. Litway/ Visuals Unlimited. Fig. 20.6: Randy Moore/Visuals Unlimited. Profile: Courtesy of Booker T. Whatley, Montgomery, AL. Fig. 20.11: Darin Norfleet/San Diego State University. Fig. 20.14: (a, b) Michael Simpson/San Diego State University. Fig. 20.15: (a) Darin Norfleet/San Diego State University; (b) Michael Simpson/San Diego State University. Fig. 20.16: (a, b) Don Martin/San Diego State University. Fig. 20.18: (a) Don Martin/ San Diego State University; (b) Peter Parks/Animals Animals.

Chapter 21 Page 491: Robert Frye, Fontana, CA. Fig. 21.1: (a) John Crum/San Diego State University; (b) Darin Norfleet/ San Diego State University. Fig. 21.3: Visuals Unlimited. Fig. 21.4: (a, b) Photo Researchers, Inc. Profile: Wide World Photos, Inc. Fig. 21.7: Visuals Unlimited. Fig. 21.11: (a) Vernon Avila/San Diego State University; (b) Don Martin/San Diego State University. Fig 21.14: (a, b) Robert Ross/Cayey University College.

Unit 6 Robert Ross/Cayey University College.

Chapter 22 Page 515: Carlo Luciano. Fig. 22.1: (b) Gerald Collier/San Diego State University; (d) Don Martin/San Diego State University. Fig. 22.3: Don Martin/San Diego State University. Fig. 22.4: (c) Harry Plymale/San Diego State University. Fig. 22.5: George Gardner/The Image Works. Fig. 22.6: (a, b) Don Martin/San Diego State University; (c) Shelly Metten. Focus 22.1: (A) Marrow-Tech Inc./La Jolla, CA. Fig. 22.8: (a) P. Bagavandoss/The Upjohn Co.; (b) Stanley Flegler/Visuals Unlimited; (c) Don Martin/San Diego State University; (d) Li-Jang Chen/Tung Kang Marine Laboratory, Taiwan; (e) Mark Lung/Southwestern College. Profile: Estralita Martin/San Diego State University. Fig. 22.18: (a, b) Robert S. Hikida/University of Ohio.

Chapter 23 Page 543: Carolina Biological Supply Company. Fig. 23.5: (a) Harry Plymale/San Diego State University. Fig. 23.8: Peter Arnold, Inc. Fig. 23.10: (b) Michael Webb/Visuals Unlimited; (c) Ed Reschke/Peter Arnold, Inc. Fig. 23.12: (a) Michael Webb/Visuals Unlimited; (c) W. R. Btnagar/Visuals Unlimited. Fig. 23.13: Mark Johnson/Scripps Institute. Profile: Wide World Photos, Inc.

Chapter 24 Page 571: Photo Researchers, Inc. Fig. 24.3: (a, b) Steven Myers/Wayne State University. Fig. 24.14: Bill Longcore/ Photo Researchers, Inc. Fig. 24.15: David M. Phillips/Visuals Unlimited. Fig. 24.16: (a, c, d, e) Harry Plymale/San Diego State University; (b) Darin Norfleet/San Diego State University. Profile: With permission from artist Al Laoangi/National Institutes of Health, Bethesda, MD.

Chapter 25 Page 595: Boehringer Ingelheim International/ GMBH. Fig. 25.2: Manfred Kage/Peter Arnold, Inc. Fig. 25.6: (a) R. Feldman/Visuals Unlimited. Focus 25.1: (A) Boehringer Ingelheim International/GMBH. Profile: Allen Green/Photo Researchers, Inc. Fig. 25.10: Barry Finzel/The Upjohn Co.

Chapter 26 Page 615: L. Lee Grismer/La Sierra University. Fig. 26.2: (a) L. Lee Grismer/San Diego State University; (b) Don Martin/San Diego State University. Profile: Office of News & Publications, University of Texas, El Paso. Focus 26.1: (A) M. Auerback/Visuals Unlimited; (B) Frank Hanna/Visuals Unlimited. Fig. 26.10: (d) Bob Gladden/San Diego State University; (e) Peter Arnold, Inc. Fig. 26.14: Bettmann Newsphotos. Focus 26.2: (A) Bruce Iverson/Visuals Unlimited; (B) Peter Arnold, Inc.

Chapter 27 Page 647: D. M. Phillips/Visuals Unlimited. Fig. 27.1: (b) Stanley Flegler/Visuals Unlimited; (c) L. Lee Grismer/ San Diego State University. Fig. 27.2: Darin Norfleet/San Diego State University. Fig. 27.3: (a) Angelisa Brandt/University of Oldenburg; (b) James Crouch/San Diego State University. Fig. 27.5: (b) Secchi, Lecague & Roussel/Photo Researchers, Inc.; (d) D. M. Phillips/Visuals Unlimited. Fig 27.8: (a) L. Lee Grismer/ San Diego State University; (b) Gerald Collier/San Diego State University; (c) D. W. Fawcett/Visuals Unlimited; (d) Richard Tenaza/University of the Pacific. Fig. 27.11: P. Bagavandoss/ Cancer Research, Upjohn Labs, Kalamazoo. Fig. 27.13: P. Bagavandoss/Cancer Research, Upjohn Labs, Kalamazoo. Profile: Flossie Wong-Staal, University of California, San Diego.

Chapter 28 Page 677: John H. Tashjian/Oklahoma City Zoo. Fig. 28.1: (a) Alfred Pasieka/Peter Arnold, Inc.; (c) C. Edelmann, La Villete/Photo Researchers, Inc. Fig. 28.5: (a, b) F. R. Turner/ University of Indiana. Fig. 28.8: (a) John Serrao/Visuals Unlimited; (b) Frank Godwin/Gatorland, Kissimmee, FL. Profile: Scurlock/ Moorland/Springarn Research Center, Howard University. Photo Essay 28.1: (all top and middle) Photo Researchers, Inc.; (bottom left) J. Stevenson/Photo Researchers, Inc.

Chapter 29 Page 707: Mark Lung/College of the Virgin Islands. Fig. 29.2: Thomas W. Martin/Photo Researchers, Inc. Fig. 29.5: Ken Greer/Visuals Unlimited. Profile: Elba Serrano/ New Mexico State University, Las Cruces. Focus 29.1: (A) Don Martin/San Diego State University. Fig. 29.11: Don Martin/San Diego State University. Focus 29.2: (A) Focus On Sports.

Chapter 30 Page 735: Valdimir Parpura, Eric Henderson, and Philip Hayden/Iowa State University. Fig. 30.7: (b) J. Heuser, D. W. Fawcett/Visuals Unlimited. Focus 30.1: (A) Bob Gladden/ San Diego State University. Fig. 30.8: J. Desaki, Y. Uehara, D. W. Fawcett/Visuals Unlimited. Profile: Courtesy of Floyd Bloom/ Scripps Clinic & Research Foundation. Fig. 30.17: (b) Goran Bredberg/Photo Researchers, Inc. Fig. 30.20: R. C. Eaque/Photo Researchers, Inc.

Chapter 31 Page 767: Bob Gladden/San Diego State University. Fig. 31.1: Bob Gladden/San Diego State University. Fig.31.8 (a) and Fig. 31.9 (a): Bob Gladden/San Diego State University. Fig. 31.10: James R. McCullagh/Visuals Unlimited. Fig. 31.12: (a) Ronald Monroe/San Diego State University; (b) Bob Gladden/San Diego State University. Fig. 31.13: Kjell B. Sandred/Visuals Unlimited. Profile: George Hillyer/University of Puerto Rico Medical Center. Fig. 31.15: John Gilbert/Dartmouth College. Fig. 31.17: (a, b, c, d) Bob Gladden/San Diego State University. Fig. 31.19: (a, b) Bob Gladden/San Diego State University; (c) C. P. Hickman/Visuals Unlimited. Fig. 31.21: (a) Kathy Williams/San Diego State University; (b) Bob Gladden/ San Diego State University; (c) Don Martin/San Diego State University. Fig. 31.23: (a, b, c, d, e, f) Kathy Williams/San Diego State University. Fig. 31.25: R. F. Ashley/Visuals Unlimited. Fig. 31.26: (a, b, c, d) Bob Gladden/San Diego State University.

Chapter 32 Page 791: R. H. Defran/San Diego State University. Fig. 32.2: (e) Bob Gladden/San Diego State University. Fig. 32.3: Bob Gladden/San Diego State University. Fig. 32.5: Breck P. Kent/Animals Animals. Fig. 32.7: (a, b) Bob Gladden/San Diego State University. Fig. 32.8: Bob Gladden/San Diego State University. Fig. 32.9: California Academy of Sciences. Fig. 32.10: (a) L. Lee Grismer/La Sierra University; (b, c) Carolina Biological Supply Company. Fig. 32.11: John H. Tashjian/Fort Worth Zoo. Fig. 32.12: (a, b) L. Lee Grismer/La Sierra University; (c) Richard L. Carlton/Visuals Unlimited. Fig. 32.13: (a, b, c) L. Lee Grismer/ La Sierra University; (d) Valentine Lance/ Zoological Society of San Diego. Profile: L. Lee Grismer/La Sierra University. Fig. 32.17: (a) C. Williams/Visuals Unlimited; (b) Breck P. Kent/ Animals Animals; (c) James Crouch/San Diego State University.

Unit 7 Andrew Borsar/Images LTD.

Chapter 33 Page 811: Helena Fitch-Snyder/Zoological Society of San Diego, Center for Reproduction of Endangered Species. Fig. 33.1: T. E. Adams/Visuals Unlimited. Fig. 33.3: (b) Shannon Taylor, courtesy of David Chiszar/University of Colorado, Boulder. Fig. 33.4: (a) Thomas McElvoy/Life Picture Service; (b) Ron Garrison/Zoological Society of San Diego. Profile: Richard Wyatt/NIMH Neurosciences Center at St. Elizabeth's, Washington, D.C. Fig. 33.6: (a) Jeff Wells/San Diego State University; (b) G. I. Bernard/Animals Animals; (c) Jeff Lapore/Photo Researchers, Inc. Fig. 33.7: (a) Animals Animals; (b) R. A. Mendez/Animals Animals; (c) R. H. Defran/San Diego State University; (d) Scott Camazine/Photo Researchers, Inc. Focus 33.1: (A) From *Neuron*, Vol. 6: 487–498, April, 1991 ©1991 Cell Press "The Molecular Pathology of Alzheimer's Disease," Dennis J. Selkoe. Fig. 33.8: (a) Gregory G. Dimijian/ Photo Researchers, Inc.; (b) Pat & Tom Laeson/Photo

Researchers, Inc. Fig. 33.9: (a) Frans Lanting/Photo Researchers, Inc.; (b) Stefan Meyers/Animals Animals. Fig.33.10: Glenn Oliver/Visuals Unlimited. Fig. 33.11: Syd Greenberg/Photo Researchers, Inc.

Chapter 34   Page 831: Mike Schum/California Environmental Protection Agency. Fig. 34.1: NASA. Fig. 34.3: (a) Don Martin/San Diego State University; (b) Paul Zedler/San Diego State University. Fig. 34.4: (a) L. Lee Grismer/La Sierra University; (b) Lee McClenaghan/San Diego State University. Fig. 34.5: (a) Don Martin/San Diego State University; (b) L. Lee Grismer/La Sierra University. Fig. 34.6: (a) James Crouch/San Diego State University; (b) Paul Zedler/San Diego State University. Fig. 34.7: (a and b) Don Martin/San Diego State University. Fig. 34.8: (a) Paul Zedler/San Diego State University; (b) John Thaw, courtesy of Paul Zedler/San Diego State University. Fig. 34.9: (a) Robert Ross/Cayey University College; (b) Patricia Pedrozo, courtesy of Don Lundberg/Zoological Society of San Diego, Center for Reproduction of Endangered Species. Fig. 34.10: (a) Paul Zedler/San Diego State University; (b) John Thaw, courtesy of Paul Zedler/San Diego State University. Profile: Alexander Cruz/University of Colorado, Boulder. Fig. 34.12: (a) Robert Frye; (b) Paul Zedler/San Diego State University. Fig. 34.13: Doug Sekell/Visuals Unlimited. Fig. 34.14: Joy Zedler/San Diego State University. Fig. 34.15: Bob Gladden/San Diego State University. Fig. 34.16: (b) W. Gregory Brown/Animals Animals.

Chapter 35   Page 853: R. H. Defran/San Diego State University. Fig. 35.2: George Cox/San Diego State University. Fig. 35.3: Paul Zedler/San Diego State University. Fig. 35.4: Don Martin/San Diego State University. Profile: Woods Hole Research Center. Fig. 35.8: (b) Willard Thompson/*California Grower*. Fig. 35.9: (b) Lo Chai Chen/San Diego State University.

Chapter 36   Page 865: Bob Gladden/San Diego State University. Fig. 36.2: (a) Robert Ross/Cayey University College; (b) George Cox/San Diego State University. Fig. 36.3: Zig Leszczynski/Animals Animals. Fig. 36.6: Michael Atkins/San Diego State University. Fig. 36.7: Kathy Williams/San Diego State University. Profile: Ariel Lugo/USDA Puerto Rico. Fig. 36.8: (a, b) L. Lee Grismer/ La Sierra University. Fig. 36.9: (a) Tom and Pat Laeson/Photo Researchers, Inc.; (b) Janet Jaboda Rodgers/Photo Researchers, Inc.

Chapter 37   Page 879: SEBRA Engineering and Research Association. Fig. 37.1: Robert Ross/Cayey University College. Fig. 37.4: (a) Eloy Rodriguez/Cornell University; (b) Paul Zedler/San Diego State University. Profile: Shan Gordon/Positive Image Photo, Inc. Fig. 37.5: (b) Don Martin/San Diego State University; (c) Angelica Brandt/University of Oldenburg. Fig. 37.6: L Eleanor Radoslovich. Fig. 37.7: (b) Don Martin/San Diego State University; (c) The Image Works. Fig. 37.8: (b) Don Martin/San Diego State University; (c) Mark Lung/College of the Virgin Islands. Fig. 37.9: (b) Don Martin/San Diego State University; (c) Robert Frye. Fig. 37.10: (b) R. G. E. Murray/Visuals Unlimited; (c) Photo Researchers, Inc.

Chapter 38   Page 894: Don Martin/San Diego State University. Focus 38.1: (A) Richard Stolarski/NASA Goddard. Fig. 38.4: Peter Miller/Photo Researchers. Fig. 38.5: Jeffrey W. Meyers/Stock Boston. Fig. 38.7: Barbara Kus/San Diego State University. Fig. 38.8: (b) L. Kitt/Visuals Unlimited. Fig. 38.9: Mike Schum/California Environmental Protection Agency. Fig. 38.11: Don Hinrischen/Institute for Local Self-Reliance. Profile: Eloy Rodriguez/Cornell University. Fig. 38.13: (a, b) Don Martin/San Diego State University; (c) James Crouch/San Diego State University; (d) Bob Gladden/San Diego State University.

Index

Italicized page numbers indicate tables and illustrations.

Growth rings, 477, 480
GTP, 270
Guanine, 217, 246
    base pairing, 250–252, *250*
    DNA, 246, *247, 248*
    meteorite, 368
    oncogene, 234
    RNA, 267
Guanine triphosphate (GTP), 270
Guard cells, 485–486
    turgor pressure, 492
    wall thickness, 492
Guinea fowl, 487
Gulls, 395, 854–855
Gusella, Jim, 337
Gut, 547–548, 683
Guyden, Jerry, 86
Gymnosperms, 451, 452–455
    characteristics, 450–452
    fertilization, 453, 455
    life cycle, *454*
    pollination, 453
    reproductive trends, *452*
    structures, 452–453
Gypsy moths
    pheromones, 709, 719
    research, 22–23
    stimuli responses, 8

Habitats, 832, 853, 865–866
    influences upon, 865–866
    interspecific competition,
        868–869
Habituation, 814–816
Hair, human, *525*
Hair cells, 758
Haldane, J. B. S., 367
Hamilton, W. D., 827
Hammer, Karl C., 507
Hanson, J., 535
Haploid (*n*) number, 225, 226, 314
    plant, 438–439
Haplontic life cycle, 230–231, *231*
Hard, Ron, 702
Hardy, G. H., 389–392
Hargreaves, William, 372
Harrison, Ross, 686
Hashish, *763*
Hatch, M. D., 180
Hatch-Slack scheme. *See* C$_4$
    carbon–fixing cycle
Haversian canals, 520
Hayes, William, 290
Hay fever, 597–598
Hayflick, Leonard, 702
HCG, 662, *663*, 693
HCl, *47, 551*, 553
Hearing, 756–758
Heart, *573*, 577–584
    beat regulation, 583–584
    blood flow regulation, 584–585
    cardiac cycle, 581–583
    double-circuit, 578–579
    as endocrine gland, 729–730
    evolution, *577*
    hormones, *717*

human, 579–584, *579*
    muscles, 580
    oxygen transport, 639
    single-circuit, 577–578
    valves, 580, *580*
Heart disease, 584
    aspirin, 730
    diet, 566
    exercise, 538
Heart surgery, 584
Heart transplantations, 584
Heartwood, 476
Heat, for chemical reactions, 132
Heavy metals, distorting enzymes,
    138
Heimlich maneuver, 551, *551*
HeLa cells, 213
Helium
    energy levels, 39
    reactivity, 40
Helix
    alpha, 69
    double, 71, *219, 221*, 248–250,
        249
Helper T-cells, *596, 599*, 603
    protein receptors, 605
Hemocyanin, 573
Hemoglobin, 586, 640–642
    displacement reaction, 53
    human and chimpanzee, 402,
        403
    oxygen transport, 66, 563
    Pauling's work, 68, 401
    quaternary structure, 69, *69*
    sickle-cell, 259–260, 322, *323*
    species differences, 401
    structure, 260, 322, *641*
    synthesis, 278
Hemophilia, 303, *354*
    carriers, 338
    sex-linked disorder, 345–346, *346*
    testing for, 358
Hemorrhoids, 557
Hepatic portal vein, 558–559
Hepatitis
    alcohol, 205
    B virus vaccine, 141
Herbicides, 502
Hereditary disorders. *See* Genetic
    disorders
Heredity, 307
Hermaphroditism, 649–650
Heroin, *763*
Herpes simplex type I, 672
Herpes simplex type II, 672
Herpetology, 802
Hershey, Alfred, 244–245
Heterogametes, *437*, 442
Heterospory, *437, 451*, 452
Heterotrophs, 187, 207
    animals as, 416–417
    trophic level, 880–883
Heterozygote, *313*, 315
    carriers, 338
Hexokinase, 193
Hexosaminidase A, 350

HGPRT, 350
Hierarchies, 823–826
    dominance, 868
High altitudes, 639, 642
Hillyer, George V., 778
Hindbrain, 750
Hinged joints, 528
Hippocampus, 821
Hispanics, and diabetes type II, 559
Histamine, 589, 597–598, *597*
Histochemistry, 80
Histologists, 515
Histone proteins, 218, *219*
    types, 218
Hitting the wall, 200, 203
HIV virus, 424, 425, 667
    AZT's action on, 605
    discovery, 605
    genetic sequencing, 298
    HTLV, 425, 605
    testing positive, 604
Holism, 14
Homeo box genes, 685–686
Homeostasis, 6, 7, *7*, 615–616
    blood glucose regulation,
        558–560
    cell, 108
    circulatory system, 573
    conditions, 616
    interphase, 220
    maintaining, 616–644
Homeostasis and regulation concept,
    12–13
Hominidae, 382, 417
Homogentisic acid, 347–348
Homologues, 217, 218
Homology (homologous), structures,
    415–416
Homo sapiens. See Human beings
Homospory, *437*, 451
Homozygote, *313*, 315
Hooke, Robert, 76
Hopi Indians, 348
Hoppe, Peter, 650
Hormones, 707–732. *See also*
    Insulin; Plant hormones
    action, 712–714
    activities, 715
    characteristics, 711–712
    digestive, *557*
    effects in cells, 118
    embryo, 693
    evolution, 709, 711
    female, 658
    historical perspective, 710
    implanted, 669
    male sexual, 655–656
    measuring, 38
    menstrual cycle, 661–665
    pancreatic, 560
    principal, *717*
    proteins as, 66
    regulating, 13
    stomach, 551
    travel medium, 711
Horseradish, 486

LAV virus (lymphadenopathy-
associated virus), 605
Laws of probability, 317–318, *317*
Laws of thermodynamics, 126–127
   ecosystem, 883–884
   electron transport chain, 139–140
   first, 126
   second, 6, 126–127
LDL receptors, 115
LDLs, 115
Leaching, 900
Lead poisoning, 138
Leaf. *See* Leaves
Leaf abscission, 500, 502
Leaf buds, 483
Leaf mesophyll, *475*
Leaf primordia, 483
Leakey, Mary, 404–406
Leakey, Richard, 402–404
Learned behavior, 814–818
Leaves, 449, 483–486
   adaptations, 486, 838–839
   asexual reproduction, 488
   dropping, 838
   stimulating growth, 508
   structure, 484–485
Lederberg, Joshua, 289–290
Leeuwenhoek, A. van, 32, 75
Left cerebral hemisphere, 750
Left subclavian vein, 590
Legumes, 837, 890–891
Lejeune, J., 340, 342
Lens, *759*
Lenticels, 471
Lesch-Nyhan syndrome, 301, 350,
   354
Leucippus, 30
Leucoplasts, 92
Leukemia, 424, 425, 608
Leukocytes. *See* White blood cells
Levators, 532
Lichens, 444–445, 874
Life
   biological principles, 11–18
   characteristics, 4–10
   mechanism versus vitalism, 10–11
   origin of, 376
   philosophy, 14
   studying, 18–22
Life cycle
   diplohaplontic, 231–232, *232*
   diplontic, 230, *230*
   haplontic, 230–231, *231*
Life span
   Edwards syndrome, 342
   Patau syndrome, 342
   sickle-cell anemia, 350
   Tay-Sachs disease, 350
Ligaments, 519, 528
Ligase enzymes, 20, 257, 287, 295,
   702
Light
   characteristics, 146–147, *147*
   in photosynthesis, 146–147,
     164–165

population ecology, 884
   wavelengths, 32–33, 77–78, 146,
   147, 164–165
Light-dependent reactions, 146–150,
   170–178, *180*
Light energy. *See* Radiant energy
Light-harvesting antennas, 170, *171*
Light-harvesting reactions, *180*
Light-independent reactions, 146,
   150–153, 178–181, *180*
Light microscope, 77–78
Lignin, 472
Lily
   arum, 460
   water, 484
Limbic system, 751
Limnetic zone, 847
Limnologists, 847
Linkage group, 328
Linked genes, 328
Linnaeus, Carolus, 384, 411
Lipase, *555*
Lipids, 52, 60–66
   breaking down, 553, 561
   characteristics, 60
   dietary, 561
   metabolizing, 726
   plasma membrane, 102, 107
   synthesizing, 87
   transporting in plasma, 586
Lipoproteins, 64
Liposomes, 372
Liquids, 30
Littoral zone, 847–848
Liver, 553, *554*
   converting amino acids, 561
   function, 204–205, *549*
   maintaining blood glucose,
     558–560
   synthesizing cholesterol, 65
Liverworts, 445
Lives of a Cell (Thomas), 98
Livestock production, 300
Lobster, 525
Locomotion, 529–540
Locus, *313*
   of genes on chromosomes, 330,
     356–357
Locus cerulus, 753
Logistic growth curve. *See* S-shaped
   curve
Logistic growth rate, 858–859
Loligo, 525, 744–745
Long-day plants, 506, 884
   flowering, 507
Long-term memory, 821
Loop of Henle, 625
Loose connective tissue, 517, *518,*
   526
Lorenz, Konrad, 416, 812, 816
Low-density lipoproteins (LDLs),
   115
Lowe, Douglas, 234
LSD, *763*
Lubchenco, Jane, 886

Lucy, 404, 513
Lugo, Ariel, 872
Lumen, 548
Lung cancer, 627, 637
Lungs, 634–635
   anatomy, 635–637
   damaged, 627
   temperature regulation, 620
Lupinus arcticus, 459
Luria, Salvador, 244
Luteinizing hormone (LH), 661,
   665. *See also* Interstitial cell-
   stimulating hormone (ICSH)
   during pregnancy, 662
   function, *663, 717*
   source, *663, 717*
Lyell, Charles, 385, 388
Lymphadenopathy-associated virus
   (LAV), 605
Lymphatic capillaries, 590
Lymphatic vascular system, 590,
   590–592, *591*
Lymph glands, 590
Lymph nodes, 590
Lymphocytes, 587, *587,* 589, 590,
   596, 599–600
   vaccinations, 598
Lyon hypothesis, 343
   XO individuals, 345
Lyon, Mary F., 343
Lysine, 562
Lysogenic bacteria, 292
Lysosomes, 79, 89–90
   in digestion, 545
   in phagocytosis, 572
Lysozyme, 596
Lytic cycle, 292, *292*

MacArthur, R. H., 869
MacLeod, Colin, 244
Macromolecules, 53, 54, 59,
   370–372
Macrophages, 589, 596, 597, *597,*
   600
   from monocytes, 589
Magnesium
   dietary, 563, *563*
   leaching, 900
   in living things, *35*
   plant element, *478*
Maintenance, 6–7. *See also*
   Metabolism
Major histocompatibility complex
   (MHC), *596, 607, 702*
Make-ready rearrangement
   reactions, 206
Malaria, 350–351
Male (human)
   causes of infertility, 670
   hormones, *663*
   sex chromosomes, 325–326
   sexual response cycle, 665, 668,
    668
Male menopause, 656
Malpighian tubules, 622